JANEWAY'S

8TH EDITION

IMMUNOBIOLOGY

Kenneth Murphy
Washington University School of Medicine, St. Louis

With acknowledgment to:

Charles A. Janeway Jr.

Paul Travers
MRC Centre for Regenerative Medicine, Edinburgh

Mark Walport
The Wellcome Trust, London

With contributions by:

Allan Mowat
University of Glasgow

Casey T. Weaver
University of Alabama at Birmingham

LONDON AND NEW YORK

Vice President: Denise Schanck
Assistant Editor: Janete Scobie
Text Editor: Eleanor Lawrence
Production Editor: Ioana Moldovan
Typesetter and Senior Production Editor: Georgina Lucas
Copy Editor: Bruce Goatly
Proofreader: Sally Huish
Illustrations and Design; Animations Programming: Matthew McClements, Blink Studio, Ltd.
Permissions Coordinator: Becky Hainz-Baxter
Indexer: Medical Indexing Ltd.
Director of Digital Publishing: Michael Morales
Associate Editor: Monica Toledo

ISBN 978-0-8153-4243-4

Library of Congress Cataloging-in-Publication Data

Murphy, Kenneth P.
Janeway's immunobiology / Kenneth Murphy with acknowledgment to Paul Travers, Mark Walport ; with contributions by Allan Mowat, Casey T. Weaver.
p. cm.
ISBN 978-0-8153-4243-4 (alk. paper)
1. Immunology. I. Travers, Paul, 1956- II. Walport, Mark. III. Janeway, Charles. IV. Title.
QR181.J37 2011
616.07'9--dc23

2011023486

Published by Garland Science, Taylor & Francis Group, LLC, an informa business,
711 Third Avenue, 8th floor, New York, NY 10017, USA,
and 2 Park Square, Milton Park, Abingdon, OX14 4RN, UK.

Printed in the United States of America

15 14 13 12 11 10 9 8 7 6 5 4 3 2 1

Visit our website at http://www.garlandscience.com

Preface

Janeway's *Immunobiology* is intended for undergraduate and graduate courses in immunology, as well as for medical students. The book can be used as an introduction to immunology but its scope is sufficiently comprehensive and deep to be useful for more advanced students and working immunologists. *Immunobiology* presents immunology from the consistent viewpoint of the host's interaction with an environment full of microbes and pathogens, and illustrates that the loss of any component of this system increases host susceptibility to some particular infection. The companion book, *Case Studies in Immunology*, provides an additional, integrated discussion of clinical topics (diseases covered in *Case Studies* are indicated by a symbol in the margin of *Immunobiology*).

This eighth edition retains the overall organization of the previous edition, and chapters in which the field has made important and rapid developments have been extensively revised. The discussion of innate immunity has been substantially expanded and its mechanisms are now treated in two separate chapters, presented in the order in which a pathogen would encounter innate defenses as it attempts to establish an infection. The immediate and soluble defenses are treated in Chapter 2. The complement system is introduced in the context of innate immunity, with the lectin pathway presented before the classical pathway of activation. The induced defenses of innate immunity—including a completely updated treatment of innate sensing—follows in Chapter 3, where various innate cell subsets and their receptors are also described. Signaling pathways are now presented as they are encountered, and not confined to a single chapter. Signaling pathways of the Toll-like receptors and other innate sensors are described in Chapter 3, while antigen receptor signaling pathways and cytokine and apoptotic pathways are retained in Chapter 7. Chapter 10 has been revised to place more emphasis on the trafficking of B cells in peripheral lymphoid organs and the locations at which they encounter antigen. Mucosal immunology (Chapter 12) has been expanded to include more discussion of responses to the commensal microbiota and the role of specialized dendritic cells and regulatory T cells in maintaining tolerance to food antigens and commensal bacteria. The last four chapters—the clinical chapters (Chapters 13–16)—reinforce the basic concepts discussed earlier with our latest understanding of the causes of disease, whether by inherited or acquired immunodeficiencies or by failures of immunological mechanisms. Chapter 16 describes how the immune response can be manipulated in attempts to combat infectious diseases, transplant rejection, and cancer. This chapter includes a complete update of the immunotherapeutics and vaccine sections. Aspects of evolution, which were confined to the last chapter of previous editions, are now discussed throughout the book as the relevant topics are encountered.

The eighth edition has benefited again from the contributions of Allan Mowat, who extensively revised and updated Chapter 12. I welcome Casey Weaver's new contributions to Chapters 13 and 15, and Robert Schreiber's and Joost Oppenheim's revisions to the appendices on cytokines and chemokines. I thank Barry Kay for his suggestions in revising Chapter 14. Most importantly, I acknowledge Charles A. Janeway Jr., Paul Travers, and Mark Walport for their pioneering work on the previous editions.

The editors, illustrators, and publishers have contributed in many ways. Eleanor Lawrence's editorial skills give the book its consistent style and ensure the orderly and didactically sound presentation of concepts. Matt McClements has transformed the author's clumsy sketches into the informative yet artistic diagrams that define Janeway's text. Janete Scobie, Bruce Goatly, Sally Huish, Georgina Lucas, and Ioana Moldovan have brought skill and dedication to the editing, proofreading, and typesetting of this edition. Monica Toledo and Michael Morales were key in updating and generating new animations. I thank Adam Sendroff and Lucy Brodie, who are instrumental in communicating information about this book to immunologists around the world, and most of all I thank the publisher Denise Schanck for her incredible patience and support.

I would like to thank all those people who read parts or all of the chapters of the seventh edition and advised on the revision plan for this edition. I would also like to thank the many instructors and students who have taken the time to write to me with their suggestions on how to improve the book. I hope I have done those suggestions justice in this edition. Every effort has been made to write a book that is error-free. Nonetheless, you may find them, and I would greatly appreciate it if you let me know.

Kenneth Murphy

Instructor and Student Resources Websites

Accessible from www.garlandscience.com, these websites provide learning and teaching tools created for Janeway's *Immunobiology, Eighth Edition*. Below is an overview of the resources available for this book. On the websites, the resources may be browsed by individual chapters and there is a search engine. You can also access the resources available for other Garland Science titles.

Instructor Resources: The following resources are available on the Instructor Site:

The Art of Janeway's Immunobiology, Eighth Edition
The images from the book are available in two convenient formats: PowerPoint® and JPEG. They have been optimized for display on a computer. Figures are searchable by figure number, figure name, or by keywords used in the figure legend from the book.

Animations and Videos
The 40 animations and videos that are available to students are also available on the Instructor's Site in two formats. The WMV-formatted movies are created for instructors who wish to use the movies in PowerPoint presentations on Windows® computers; the QuickTime-formatted movies are for use in PowerPoint for Apple computers or Keynote® presentations. The movies can easily be downloaded to your PC using the 'download' button on the movie preview page.

Figure Integrated Lecture Outlines
The section headings, concept headings, and figures from the text have been integrated into PowerPoint presentations. These will be useful for instructors who would like a head start in creating lectures for their course. Like all of Garland Science's PowerPoint presentations, the lecture outlines can be customized. For example, the content of these presentations can be combined with videos on the website to create unique lectures that facilitate interactive learning in the classroom.

Student Resources: The following resources are available on the Student Site at www.garlandscience.com/students/immunobiology:

Animations and Videos
The 40 animations and videos dynamically illustrate important concepts from the book, and make many of the more difficult topics accessible. Icons located throughout the text indicate the relevant media.

Flashcards
Each chapter contains a set of flashcards, built into the website, that allow students to review key terms from the text.

Glossary
The complete glossary from the book is available on the website and can be searched and browsed as a whole or sorted by chapter.

Acknowledgments

We would like to thank the following experts who read parts or the whole of the seventh edition chapters and provided us with invaluable advice in developing this new edition.

Chapter 1: Hans Acha-Orbea, Université de Lausanne; Elizabeth Godrick, Boston University; Michael Gold, University of British Columbia; Derek McKay, University of Calgary.

Chapter 2: Shizuo Akira, Osaka University; Lewis Lanier, University of California, San Francisco; Gabriel Nunez, University of Michigan Medical School; Philip Rosenstiel, University of Kiel, Germany; Hung Bing Shu, Wuhan University, China; Caetano Reis e Sousa, Cancer Research UK; Tada Taniguchi, University of Tokyo; Andrea Tenner, University of California, Irvine; Eric Vivier, Université de la Méditerranée Campus de Luminy.

Chapter 3: Bernard Malissen, Centre d' Immunologie Marseille-Luminy; Ellis Reinherz, Harvard Medical School; Robyn Stanfield, The Scripps Research Institute; Ian Wilson, The Scripps Research Institute.

Chapter 4: Michael Lieber, University of Southern California; Michael Neuberger, University of Cambridge; David Schatz, Yale University School of Medicine; Barry Sleckman, Washington University School of Medicine, St. Louis; Philip Tucker, University of Texas, Austin.

Chapter 5: Siamak Bahram, Centre de Recherche d'Immunologie et d'Hematologie; Peter Cresswell, Yale University School of Medicine; Mitchell Kronenberg, La Jolla Institute for Allergy & Immunology; Philippa Marrack, Howard Hughes Medical Institute; Hans-Georg Rammensee, University of Tubingen, Germany.

Chapter 6: Oreste Acuto, University of Oxford; Leslie Berg, University of Massachusetts Medical Center; Doreen Cantrell, University of Dundee, UK; Andy Chan, Genentech, Inc.; Vigo Heissmeyer, Helmholtz Center Munich; Steve Jameson, University of Minnesota; Gabriel Nunez, University of Michigan Medical School; Takashi Saito, RIKEN; Larry Samelson, National Cancer Institute, NIH; Pamela Schwartzberg, National Human Genome Research Institute, NIH; Art Weiss, University of California, San Francisco.

Chapter 7: Michael Cancro, University of Pennsylvania School of Medicine; Robert Carter, University of Alabama; Richard Hardy, Fox Chase Cancer Center; Kris Hogquist, University of Minnesota; John Monroe, Genentech, Inc.; Nancy Ruddle, Yale University School of Medicine; Marc Veldhoen, National Institute for Medical Research, London.

Chapter 8: Michael Bevan, University of Washington; Frank Carbone, University of Melbourne, Victoria; Gillian Griffiths, University of Oxford; Bill Heath, University of Melbourne, Victoria; Anne O'Garra, The National Institute for Medical Research, London; Steve Reiner, University of Pennsylvania School of Medicine; Brigitta Stockinger, National Institute for Medical Research, London.

Chapter 9: Katherine Calame, Columbia University; Michael Cancro, University of Pennsylvania School of Medicine; Robert H. Carter, The University of Alabama, Birmingham; Jason Cyster, University of California, San Francisco; John Kearney, The University of Alabama, Birmingham; Garnett Kelsoe, Duke University; Michael Neuberger, University of Cambridge.

Chapter 10: Michael Bevan, University of Washington; Marc K. Jenkins, University of Minnesota; Robert Modlin, University of California, Los Angeles; Michael Oldstone, The Scripps Research Institute; Michael Russell, University at Buffalo; Federica Sallusto, Institute for Research in Biomedicine, Switzerland.

Chapter 11: Chuck Elson, University of Alabama; Michael Lamm, Case Western Reserve University; Thomas MacDonald, Barts and The London School of Medicine and Dentistry; Kevin Maloy, University of Oxford; Maria Rescigno, University of Milan; Michael Russell, University at Buffalo.

Chapter 12: Jean-Laurent Cassanova, Groupe Hospitalier Necker-Enfants-Malades, Paris; Mary Collins, University College London; Alain Fischer, Groupe Hospitalier Necker-Enfants-Malades, Paris; Raif Geha, Harvard Medical School; Paul Klenerman, Oxford University; Luigi Notarangelo, Harvard Medical School; Sarah Rowland-Jones, Oxford University; Adrian Thrasher, London Institute of Child Health.

Chapter 13: Cezmi A. Akdis, Swiss Institute of Allergy and Asthma Research; Barry Kay, National Heart and Lung Institute; Raif Geha, Harvard Medical School; Gabriel Nunez, University of Michigan Medical School; Albert Sheffer, Harvard Medical School.

Chapter 14: Anne Davidson, Albert Einstein College of Medicine; Robert Fairchild, Cleveland Clinic; Fadi Lakkis, University of Pittsburgh; Wayne Hancock, University of Pennsylvania School of Medicine; Rikard Holmdahl, Lund University; Laurence A. Turka, University of Pennsylvania School of Medicine.

Chapter 15: Benny Chain, University College London; James Crowe, Vanderbilt University; Glen Dranoff, Dana Farber Cancer Institute; Giuseppe Pantaleo, Université de Lausanne; Richard O. Williams, Imperial College of London.

Chapter 16: Jim Kaufman, University of Cambridge; Gary W. Litman, University of South Florida; Martin Flajnik, University of Maryland, Baltimore; Robert Schreiber, Washington University School of Medicine, St. Louis; Casey Weaver, University of Alabama at Birmingham.

Contents

Detailed Contents

Part I AN INTRODUCTION TO IMMUNOBIOLOGY AND INNATE IMMUNITY

Chapter 3 The Induced Responses of Innate Immunity 75

Part II THE RECOGNITION OF ANTIGEN

Chapter 4 Antigen Recognition by B-cell and T-cell Receptors 127

Part III THE DEVELOPMENT OF MATURE LYMPHOCYTE RECEPTOR REPERTOIRES

PART I

AN INTRODUCTION TO IMMUNOBIOLOGY AND INNATE IMMUNITY

Basic Concepts in Immunology

1

Immunology is the study of the body's defense against infection. We live surrounded by microorganisms, many of which cause disease. Yet despite this continual exposure we become ill only rarely. How does the body defend itself? When infection does occur, how does the body eliminate the invader and cure itself? And why do we develop long-lasting immunity to many infectious diseases encountered once and overcome? These are the questions addressed by immunology, which we study to understand our body's defenses against infection at the cellular and molecular levels.

Immunology is a relatively new science. Its origin is usually attributed to **Edward Jenner** (Fig. 1.1), who observed in the late 18th century that the relatively mild disease of cowpox, or vaccinia, seemed to confer protection against the often fatal disease of smallpox. In 1796, Jenner demonstrated that inoculation with cowpox could protect against smallpox. He called the procedure **vaccination**, and this term is still used to describe the inoculation of healthy individuals with weakened or attenuated strains of disease-causing agents to provide protection from disease. Although Jenner's bold experiment was successful, it took almost two centuries for smallpox vaccination to become universal, an advance that enabled the World Health Organization to announce in 1979 that smallpox had been eradicated (Fig. 1.2), arguably the greatest triumph of modern medicine.

When Jenner introduced vaccination he knew nothing of the infectious agents that cause disease: it was not until late in the 19th century that **Robert Koch** proved that infectious diseases are caused by microorganisms, each one responsible for a particular disease. We now recognize four broad categories of disease-causing microorganisms, or **pathogens**: viruses, bacteria, fungi, and the unicellular and multicellular eukaryotic organisms collectively termed parasites.

Fig. 1.1 Edward Jenner. Portrait by John Raphael Smith. Reproduced courtesy of Yale University, Harvey Cushing/John Hay Whitney Medical Library.

The discoveries of Koch and other great 19th-century microbiologists extended Jenner's strategy of vaccination to other diseases. In the 1880s, **Louis Pasteur** devised a vaccine against cholera in chickens, and developed a rabies vaccine that proved a spectacular success upon its first trial in a boy bitten by a rabid dog. These practical triumphs led to a search for the mechanism of protection and to the development of the science of immunology. In the

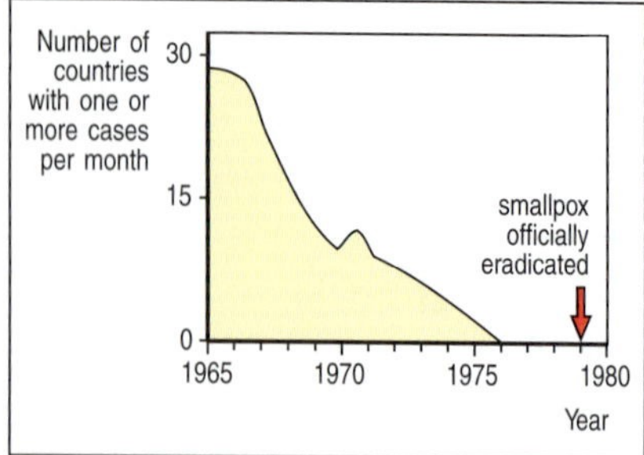

Fig. 1.2 The eradication of smallpox by vaccination. After a period of 3 years in which no cases of smallpox were recorded, the World Health Organization was able to announce in 1979 that smallpox had been eradicated, and vaccination stopped (upper panel). A few laboratory stocks have been retained, however, and some fear that these are a source from which the virus might reemerge. Ali Maow Maalin (lower panel) contracted and survived the last case of smallpox in Somalia in 1978. Photograph courtesy of Dr. Jason Weisfeld.

early 1890s, **Emil von Behring** and **Shibasaburo Kitasato** discovered that the serum of animals immune to diphtheria or tetanus contained a specific 'anti-toxic activity' that could confer short-lived protection against the effects of diphtheria or tetanus toxins in people. This activity was due to the proteins we now call **antibodies**, which bound specifically to the toxins and neutralized their activity.

The responses we make against infection by potential pathogens are known as **immune responses**. A specific immune response, such as the production of antibodies against a particular pathogen or its products, is known as an **adaptive immune response** because it is developed during the lifetime of an individual as an adaptation to infection with that pathogen. In many cases, an adaptive immune response also results in the phenomenon known as immunological memory, which confers lifelong **protective immunity** to reinfection with the same pathogen. This is just one of the features that distinguish an adaptive immune response from the **innate immune response**, or innate immunity, which is always immediately available to combat a wide range of pathogens but does not lead to lasting immunity and is not specific for any individual pathogen. At the time that von Behring was developing serum therapy for diphtheria, innate immunity was known chiefly through the work of the great Russian immunologist **Elie Metchnikoff**, who discovered that many microorganisms could be engulfed and digested by phagocytic cells, which he called 'macrophages.' These cells are always present and ready to act, and are a front-line component of innate immune responses. In contrast, an adaptive immune response takes time to develop and is highly specific; antibodies against the influenza virus, for example, will not protect against poliovirus.

It quickly became clear that antibodies could be induced against a vast range of substances. Such substances were called **antigens** because they could stimulate *anti*body *gen*eration. Much later, it was discovered that antibody production is not the only function of adaptive immune responses, and the term antigen is now used to describe any substance that can be recognized and responded to by the adaptive immune system. The proteins, glycoproteins, and polysaccharides of pathogens are the antigens normally responded to by the immune system, but it can recognize and make a response to a much wider range of chemical structures—hence its ability to produce allergic immune responses to metals such as nickel, drugs such as penicillin, and organic chemicals in the leaves of poison ivy. The innate and adaptive immune responses together provide a remarkably effective defense system. Many infections are handled successfully by innate immunity and cause no disease; in those that cannot be resolved, the activities of the innate immune system trigger an adaptive immune response. If the disease is overcome, the adaptive immune response is often followed by lasting immunological memory, which prevents disease if reinfection occurs.

This book describes the mechanisms of both innate and adaptive immunity and illustrates how they are integrated into an effective overall system of defense against invasion by pathogens. Although the white blood cells known as **lymphocytes** possess the most powerful ability to recognize and target pathogenic microorganisms, they need the participation of the innate immune system to initiate and to mount their offensive. Indeed, the adaptive immune response and innate immunity use many of the same destructive mechanisms to finally destroy invading microorganisms.

In this chapter we first introduce the principles of innate and adaptive immunity, the cells of the immune system, the tissues in which they develop, and the tissues through which they circulate. We then outline the specialized functions of the different types of cells and the mechanisms by which they eliminate infection.

Principles of innate and adaptive immunity.

The body is protected from infectious agents and the damage they cause, and from other harmful substances such as insect toxins, by a variety of effector cells and molecules that together make up the **immune system**. In this part of the chapter we discuss the main principles underlying immune responses and introduce the cells and tissues of the immune system on which an immune response depends.

1-1 The immune system recognizes infection and induces protective responses.

To protect the individual effectively against disease, the immune system must fulfill four main tasks. The first is **immunological recognition**: the presence of an infection must be detected. This task is carried out both by the white blood cells of the innate immune system, which provide an immediate response, and by the lymphocytes of the adaptive immune system. The second task is to contain the infection and if possible eliminate it completely, which brings into play **immune effector functions** such as the complement system of blood proteins, the antibodies produced by some lymphocytes, and the destructive capacities of lymphocytes and the other white blood cells. At the same time the immune response must be kept under control so that it does not itself do damage to the body. **Immune regulation**, or the ability of the immune system to self-regulate, is thus an important feature of immune responses, and failure of such regulation contributes to conditions such as allergy and autoimmune disease. The fourth task is to protect the individual against recurring disease due to the same pathogen. A unique feature of the adaptive immune system is that it is capable of generating **immunological memory**, so that having been exposed once to an infectious agent, a person will make an immediate and stronger response against any subsequent exposure to it; that is, they will have protective immunity against it. Finding ways of generating long-lasting immunity to pathogens that do not naturally provoke it is one of the greatest challenges facing immunologists today.

When an individual first encounters an infectious agent, the initial defenses against infection are physical and chemical barriers, such as antimicrobial proteins secreted at mucosal surfaces, that prevent microbes from entering the body. If these barriers are overcome or evaded, other components of the immune system come into play. The complement system can immediately recognize and destroy foreign organisms, and phagocytic white blood cells, such as macrophages and neutrophils of the innate immune system, can ingest and kill microbes by producing toxic chemicals and powerful degradative enzymes. Innate immunity is of ancient origin—some form of innate defense against disease is found in all animals and plants. The macrophages of humans and other vertebrates, for example, are presumed to be the direct evolutionary descendants of the phagocytic cells present in simpler animals, such as those that Metchnikoff observed in the invertebrate sea stars.

Innate immune responses occur rapidly on exposure to an infectious organism. In contrast, responses by the adaptive immune system take days rather than hours to develop (summarized in Fig. 1.34). However, the adaptive immune system is capable of eliminating infections more efficiently because of the exquisitely specific recognition functions of lymphocytes. These cells can recognize and respond to individual antigens by means of highly specialized **antigen receptors** on the lymphocyte surface. The billions of lymphocytes present in the body collectively possess a vast repertoire of antigen receptors, which enables the immune system to recognize and respond to virtually any antigen a person is likely to be exposed to. In this way, adaptive immunity can more effectively focus its resources to overcome pathogens that have evaded or overwhelmed innate immunity. Antibodies and activated lymphocytes produced by an adaptive immune response can also persist after the original infection has been eliminated. They help to prevent immediate reinfection and also provide for long-lasting immunity, allowing a faster and more intense response to a second exposure even when it occurs many years later.

1-2 The cells of the immune system derive from precursors in the bone marrow.

Both innate and adaptive immune responses depend upon the activities of white blood cells or **leukocytes**. These cells all originate in the **bone marrow**, and many of them also develop and mature there. Once mature, they migrate to guard the peripheral tissues: some of them reside within tissues, while others circulate in the bloodstream and in a specialized system of vessels called the **lymphatic system**, which drains extracellular fluid and free cells from tissues, transports them through the body as **lymph**, and eventually empties back into the blood system.

All the cellular elements of blood, including the red blood cells that transport oxygen, the platelets that trigger blood clotting in damaged tissues, and the white blood cells of the immune system, derive from the **hematopoietic stem cells** of the bone marrow. Because these stem cells can give rise to all the different types of blood cells, they are often known as pluripotent hematopoietic stem cells. They give rise to cells of more limited developmental potential, which are the immediate progenitors of red blood cells, platelets, and the two main categories of white blood cells, the **lymphoid** and **myeloid** lineages. The different types of blood cells and their lineage relationships are summarized in Fig. 1.3.

1-3 The myeloid lineage comprises most of the cells of the innate immune system.

The **common myeloid progenitor** is the precursor of the macrophages, granulocytes, mast cells, and dendritic cells of the innate immune system, and also of megakaryocytes and red blood cells, which we will not be concerned with here. The cells of the myeloid lineage are shown in Fig. 1.4.

Fig. 1.3 All the cellular elements of the blood, including the cells of the immune system, arise from pluripotent hematopoietic stem cells in the bone marrow. These pluripotent cells divide to produce two types of stem cells. A common lymphoid progenitor gives rise to the lymphoid lineage (blue background) of white blood cells or leukocytes—the natural killer (NK) cells and the T and B lymphocytes. A common myeloid progenitor gives rise to the myeloid lineage (pink and yellow backgrounds), which comprises the rest of the leukocytes, the erythrocytes (red blood cells), and the megakaryocytes that produce platelets important in blood clotting. T and B lymphocytes are distinguished from the other leukocytes by the possession of antigen receptors, and from each other by their sites of differentiation—the thymus and bone marrow, respectively. After encounter with antigen, B cells differentiate into antibody-secreting plasma cells, while T cells differentiate into effector T cells with a variety of functions. Unlike T and B cells, NK cells lack antigen specificity. The remaining leukocytes are the monocytes, the dendritic cells, and the neutrophils, eosinophils, and basophils. The last three of these circulate in the blood and are termed granulocytes, because of the cytoplasmic granules whose staining gives these cells a distinctive appearance in blood smears, or polymorphonuclear leukocytes, because of their irregularly shaped nuclei. Immature dendritic cells (yellow background) are phagocytic cells that enter the tissues; they mature after they have encountered a potential pathogen. The common lymphoid progenitor also gives rise to a minor subpopulation of dendritic cells, but for simplicity this developmental pathway has not been illustrated. However, as there are more common myeloid progenitor cells than there are common lymphoid progenitors, the majority of the dendritic cells in the body develop from common myeloid progenitors. Monocytes enter tissues, where they differentiate into phagocytic macrophages. The precursor cell that gives rise to mast cells is still unknown. Mast cells also enter tissues and complete their maturation there.

Macrophages are resident in almost all tissues and are the mature form of **monocytes**, which circulate in the blood and continually migrate into tissues, where they differentiate. Together, monocytes and macrophages make up one of the three types of phagocytes in the immune system: the others are the granulocytes (the collective term for the white blood cells called neutrophils, eosinophils, and basophils) and the dendritic cells. Macrophages are relatively long-lived cells and perform several different functions throughout

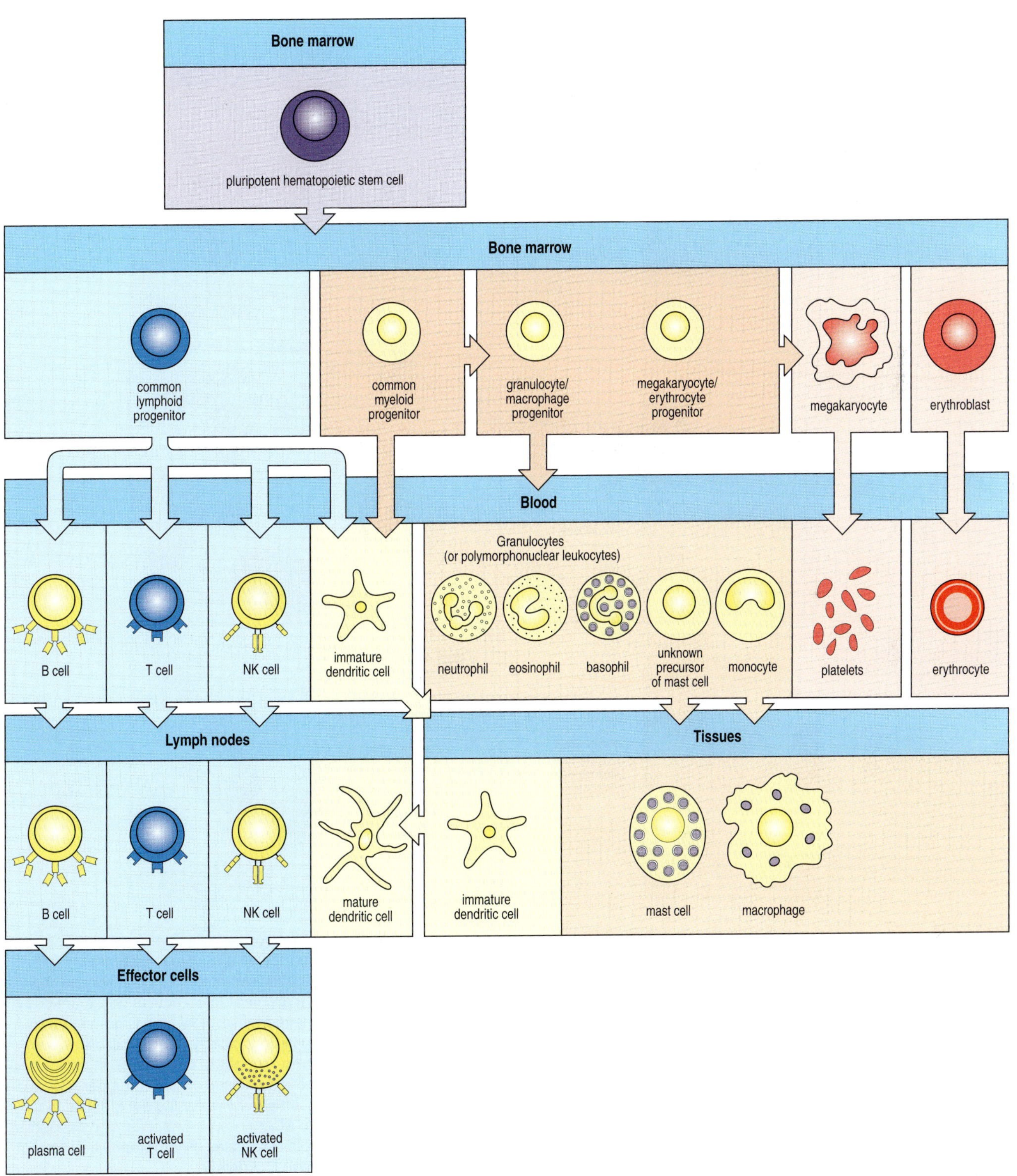

the innate immune response and the subsequent adaptive immune response. One is to engulf and kill invading microorganisms. In this phagocytic role they are an important first defense in innate immunity and also dispose of pathogens and infected cells targeted by an adaptive immune response. Both monocytes and macrophages are phagocytic, but most infections occur in the tissues, and so it is primarily macrophages that perform this important protective function. An additional and crucial role of macrophages is to

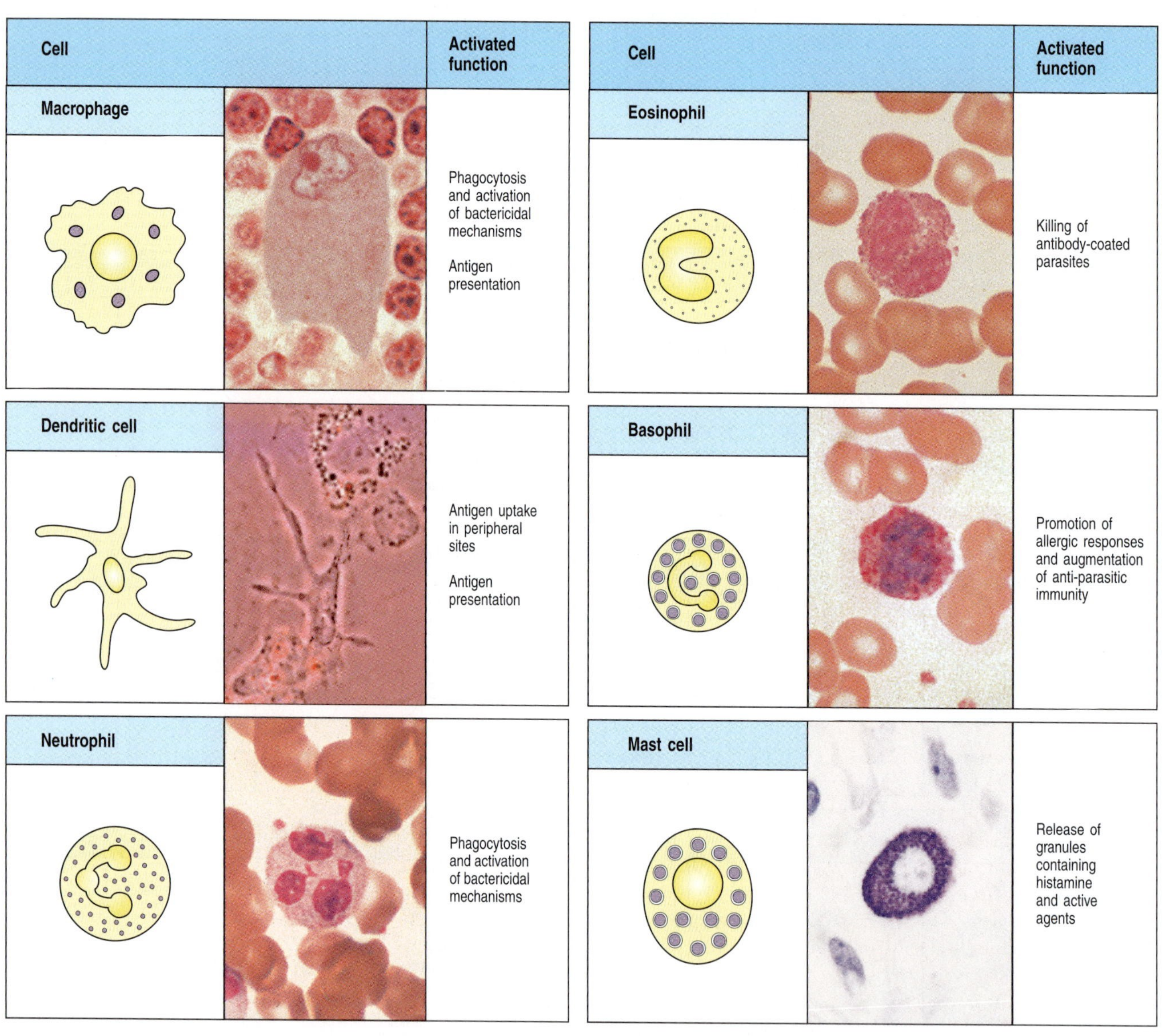

Fig. 1.4 Myeloid cells in innate and adaptive immunity. Cells of the myeloid lineage perform various important functions in the immune response. In the rest of the book, these cells will be represented in the schematic form shown on the left. A photomicrograph of each cell type is shown in the center panels. Macrophages and neutrophils are primarily phagocytic cells that engulf pathogens and destroy them in intracellular vesicles, a function they perform in both innate and adaptive immune responses. Dendritic cells are phagocytic when they are immature and can take up pathogens; after maturing, they function as specialized cells that present pathogen antigens to T lymphocytes in a form they can recognize, thus activating T lymphocytes and initiating adaptive immune responses. Macrophages can also present antigens to T lymphocytes and can activate them. The other myeloid cells are primarily secretory cells that release the contents of their prominent granules upon activation via antibody during an adaptive immune response. Eosinophils are thought to be involved in attacking large antibody-coated parasites such as worms; basophils are also thought to be involved in anti-parasite immunity. Mast cells are tissue cells that trigger a local inflammatory response to antigen by releasing substances that act on local blood vessels. Mast cells, eosinophils, and basophils are also important in allergic responses. Photographs courtesy of N. Rooney, R. Steinman, and D. Friend.

orchestrate immune responses: they help induce inflammation, which, as we shall see, is a prerequisite to a successful immune response, and they secrete signaling proteins that activate other immune-system cells and recruit them into an immune response. In addition to their specialized role in the immune system, macrophages act as general scavenger cells in the body, clearing dead cells and cell debris.

The **granulocytes** are so called because they have densely staining granules in their cytoplasm; they are also called **polymorphonuclear leukocytes** because of their oddly shaped nuclei. There are three types of granulocytes—neutrophils, eosinophils, and basophils—which are distinguished by the different staining properties of the granules. In comparison with macrophages they are all relatively short-lived, surviving for only a few days, and are produced in increased numbers during immune responses, when they leave the blood to migrate to sites of infection or inflammation. The phagocytic **neutrophils** are the most numerous and most important cells in innate immune responses: they take up a variety of microorganisms by phagocytosis and efficiently destroy them in intracellular vesicles using degradative enzymes and other antimicrobial substances stored in their cytoplasmic granules. Their role is discussed in more detail in Chapter 3. Hereditary deficiencies in neutrophil function lead to overwhelming bacterial infection, which is fatal if untreated.

Eosinophils and **basophils** are less abundant than neutrophils, but like neutrophils they have granules containing a variety of enzymes and toxic proteins, which are released when the cells are activated. Eosinophils and basophils are thought to be important chiefly in defense against parasites, which are too large to be ingested by macrophages or neutrophils. They can also contribute to allergic inflammatory reactions, in which their effects are damaging rather than protective. We discuss the functions of these cells in Chapter 10 and their role in allergic inflammation in Chapter 14.

Mast cells, whose blood-borne precursors are not well defined, differentiate in the tissues. Although best known for their role in orchestrating allergic responses, which is discussed in Chapter 14, they are believed to play a part in protecting the internal surfaces of the body against pathogens, and are involved in the response to parasitic worms. They have large granules in their cytoplasm that are released when the mast cell is activated; these help induce inflammation.

There are several kinds of **dendritic cells**, which form the third class of phagocytic cell of the immune system. Most dendritic cells have long finger-like processes, like the dendrites of nerve cells, which give them their name. Immature dendritic cells migrate through the bloodstream from the bone marrow to enter tissues. They take up particulate matter by phagocytosis and also continually ingest large amounts of the extracellular fluid and its contents by a process known as **macropinocytosis**. Like macrophages and neutrophils, they degrade the pathogens they take up, but their main role in the immune system is not the clearance of microorganisms. Instead, the encounter with a pathogen stimulates dendritic cells to mature into cells that can activate a particular class of lymphocytes—the T lymphocytes—which are described in Section 1-4. Mature dendritic cells activate T lymphocytes by displaying antigens derived from the pathogen on their surface in a way that activates the antigen receptor of a T lymphocyte. They also provide other signals that are necessary to activate T lymphocytes that are encountering their specific antigen for the first time, and for this reason dendritic cells are also called **antigen-presenting cells** (**APCs**). As such, dendritic cells form a crucial link between the innate immune response and the adaptive immune response (Fig. 1.5). In certain situations, macrophages can also act as antigen-presenting cells, but dendritic cells are the cells that are specialized in initiating adaptive immune responses. We describe the types and functions of dendritic cells in Chapters 6, 9, 11, and 12.

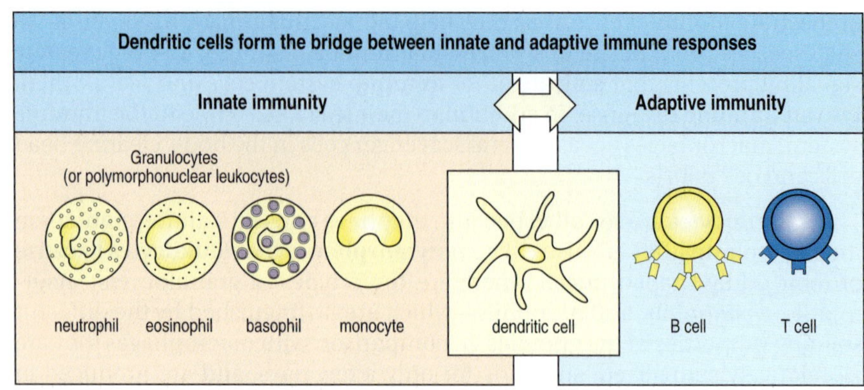

Fig. 1.5 Dendritic cells form a key link between the innate immune system and the adaptive immune system. Like the other cells of innate immunity, dendritic cells recognize pathogens via invariant cell-surface receptors for pathogen molecules, and are activated by these stimuli early in an infection. Dendritic cells in the tissues are phagocytic and are specialized to ingest a wide range of pathogens and to display their antigens at the dendritic cell surface in a form that can be recognized by T cells. As described later in the chapter, activated dendritic cells also produce molecules that enable T cells to be activated by antigen.

1-4 The lymphoid lineage comprises the lymphocytes of the adaptive immune system and the natural killer cells of innate immunity.

The **common lymphoid progenitor** in the bone marrow gives rise to the antigen-specific lymphocytes of the adaptive immune system and also to a type of lymphocyte that responds to the presence of infection but is not specific for antigen, and is thus considered to be part of the innate immune system. This latter is a large cell with a distinctive granular cytoplasm and is called a **natural killer cell** (**NK cell**) (Fig. 1.6). These cells can recognize and kill some abnormal cells, for example some tumor cells and cells infected with herpesviruses, and are thought to be important in holding viral infections in check before the adaptive immune response kicks in. Their functions in innate immunity are described in Chapter 3.

We come finally to the key components of adaptive immunity, the antigen-specific lymphocytes. Unless indicated otherwise, we shall use the term lymphocyte from now on to refer to the antigen-specific lymphocytes only. The immune system must be able to mount an immune response against any of the wide variety of different pathogens a person is likely to encounter during their lifetime. Lymphocytes collectively make this possible through the highly variable antigen receptors on their surface, by which they recognize and bind antigens. Each lymphocyte matures bearing a unique variant of a prototype antigen receptor, so that the population of lymphocytes expresses a huge repertoire of receptors that are highly diverse in their antigen-binding sites. Among the billion or so lymphocytes circulating in the body at any one time there will always be some that can recognize a given foreign antigen.

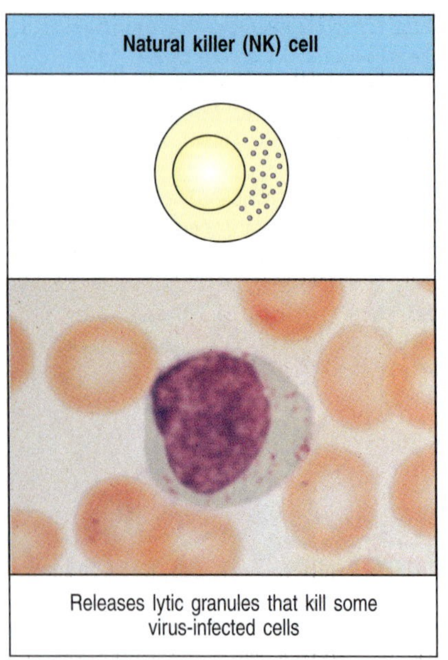

Fig. 1.6 Natural killer (NK) cells. These are large granular lymphoid-like cells with important functions in innate immunity, especially against intracellular infections, being able to kill other cells. Unlike lymphocytes, they lack antigen-specific receptors. Photograph courtesy of B. Smith.

In the absence of an infection, most lymphocytes circulating in the body are small, featureless cells with few cytoplasmic organelles and much of the nuclear chromatin inactive, as shown by its condensed state (Fig. 1.7). This appearance is typical of inactive cells. It is hardly surprising that until the 1960s, textbooks described these cells, now the central focus of immunology, as having no known function. Indeed, these small lymphocytes have no functional activity until they encounter their specific antigen. Lymphocytes that have not yet been activated by antigen are known as naive lymphocytes; those that have met their antigen, become activated, and have differentiated further into fully functional lymphocytes are known as **effector lymphocytes**.

There are two types of lymphocytes—**B lymphocytes** (**B cells**) and **T lymphocytes** (**T cells**)—each with quite different roles in the immune system and distinct types of antigen receptors. After antigen binds to a **B-cell antigen receptor**, or **B-cell receptor** (**BCR**), on the B-cell surface, the lymphocyte will proliferate and differentiate into **plasma cells**. These are the effector form of B lymphocytes and they produce antibodies, which are a secreted form of the B-cell receptor and have an identical antigen specificity. Thus the antigen that activates a given B cell becomes the target of the antibodies produced by that cell's progeny. Antibody molecules as a class are known as **immunoglobulins** (**Ig**), and so the antigen receptor of B lymphocytes is also known as **membrane immunoglobulin** (**mIg**) or **surface immunoglobulin** (**sIg**).

The **T-cell antigen receptor**, or **T-cell receptor** (**TCR**), is related to immunoglobulin but is quite distinct in its structure and recognition properties. After a T cell has been activated by its first encounter with antigen, it proliferates and differentiates into one of several different functional types of **effector T lymphocytes**. Effector T-cell functions fall into three broad classes: killing, activation, and regulation. **Cytotoxic T cells** kill cells that are infected with viruses or other intracellular pathogens. **Helper T cells** provide essential additional signals that influence the behavior and activity of other cells. Helper T cells provide signals to antigen-stimulated B cells that influence their production of antibody, and to macrophages that allow them to become more efficient at killing engulfed pathogens. We return to the functions of cytotoxic and helper T cells later in this chapter, and their actions are described in detail in Chapters 9 and 11. **Regulatory T cells** suppress the activity of other lymphocytes and help to control immune responses; they are discussed in Chapters 9, 11, 12, and 15.

During the course of an immune response, some of the B cells and T cells activated by antigen differentiate into **memory cells**, the lymphocytes that are responsible for the long-lasting immunity that can follow exposure to disease or vaccination. Memory cells will readily differentiate into effector cells on a second exposure to their specific antigen. Immunological memory is described in Chapter 11.

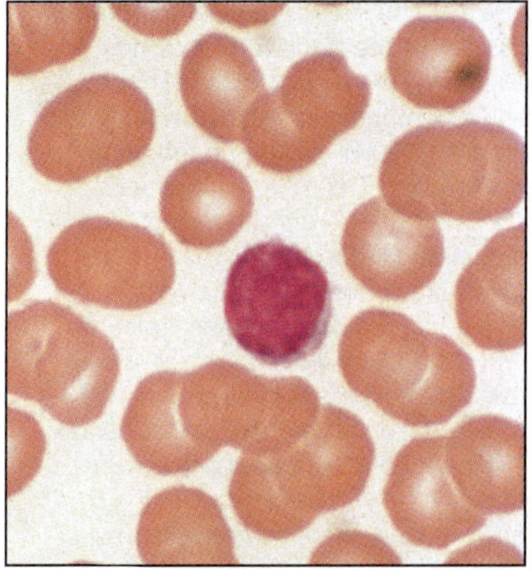

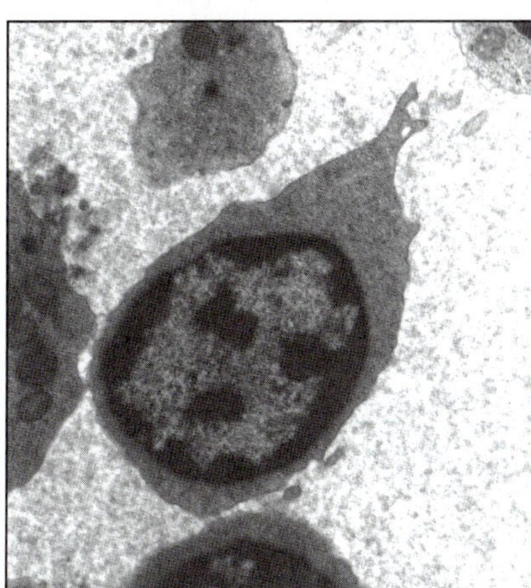

Fig. 1.7 Lymphocytes are mostly small and inactive cells. The upper panel shows a light micrograph of a small lymphocyte in which the nucleus has been stained purple by the hematoxylin and eosin dye, surrounded by red blood cells (which have no nuclei). Note the darker purple patches of condensed chromatin of the lymphocyte nucleus, indicating little transcriptional activity, the relative absence of cytoplasm, and the small size. The lower panel shows a transmission electron micrograph of a small lymphocyte. Again, note the evidence of functional inactivity: the condensed chromatin, the scanty cytoplasm, and the absence of rough endoplasmic reticulum. Photographs courtesy of N. Rooney.

1-5 Lymphocytes mature in the bone marrow or the thymus and then congregate in lymphoid tissues throughout the body.

Lymphocytes circulate in the blood and the lymph and are also found in large numbers in **lymphoid tissues** or **lymphoid organs**, which are organized aggregates of lymphocytes in a framework of nonlymphoid cells. Lymphoid organs can be divided broadly into the **central** or **primary lymphoid organs**, where lymphocytes are generated, and the **peripheral** or **secondary lymphoid organs**, where mature naive lymphocytes are maintained and adaptive immune responses are initiated. The central lymphoid organs are the bone marrow and the **thymus**, an organ in the upper chest. The peripheral lymphoid organs comprise the **lymph nodes**, the **spleen**, and the **mucosal lymphoid tissues** of the gut, the nasal and respiratory tract, the urogenital tract, and other mucosa. The location of the main lymphoid tissues is shown schematically in Fig. 1.8, and we describe the individual peripheral lymphoid organs in more detail later in the chapter. Lymph nodes are interconnected by a system of lymphatic vessels, which drain extracellular fluid from tissues, through the lymph nodes, and back into the blood.

Both B and T lymphocytes originate in the bone marrow, but only the B lymphocytes mature there. The precursor T lymphocytes migrate to the thymus, from which they get their name, and mature there. The 'B' in B lymphocytes originally stood for the **bursa of Fabricius**, a lymphoid organ in young chicks in which lymphocytes mature; fortunately, it can stand equally well for bone marrow derived. Once they have completed maturation, both types of lymphocytes enter the bloodstream as mature naive lymphocytes. They circulate through the peripheral lymphoid tissues, in which an adaptive immune response is initiated if a lymphocyte meets its corresponding antigen. Before this, however, an innate immune response to the infection has usually occurred, and we now look at how this alerts the rest of the immune system to the presence of a pathogen.

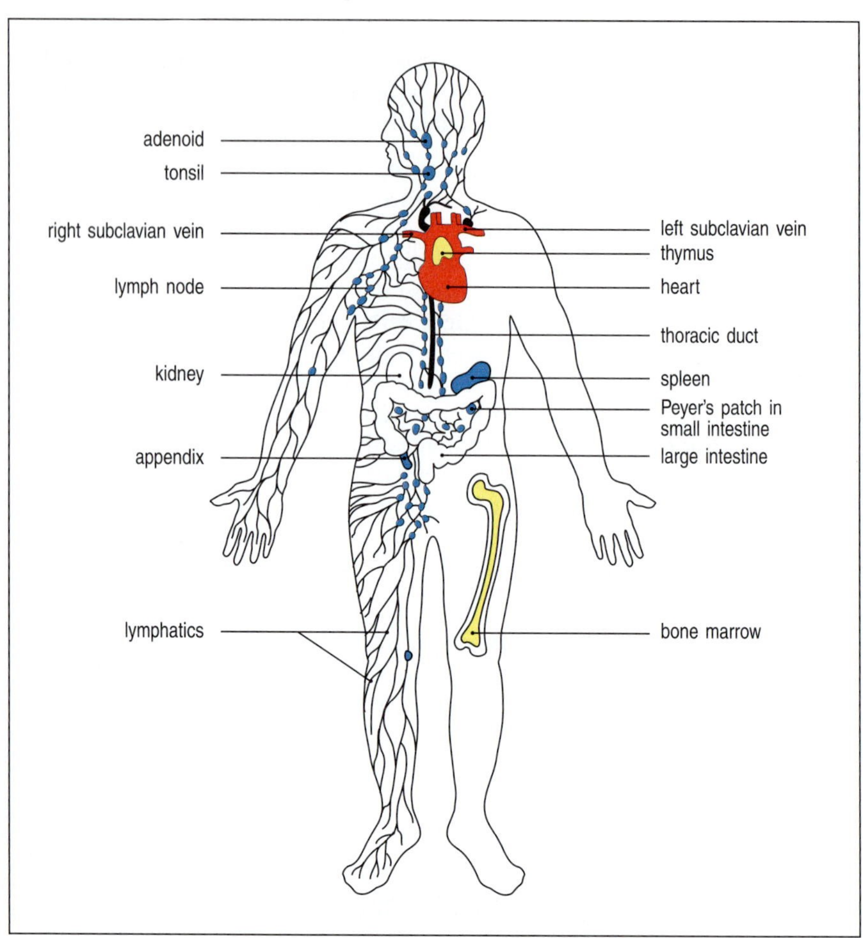

Fig. 1.8 The distribution of lymphoid tissues in the body. Lymphocytes arise from stem cells in bone marrow and differentiate in the central lymphoid organs (yellow)—B cells in the bone marrow and T cells in the thymus. They migrate from these tissues and are carried in the bloodstream to the peripheral lymphoid organs (blue). These include lymph nodes, spleen, and lymphoid tissues associated with mucosa, such as the gut-associated tonsils, Peyer's patches, and appendix. The peripheral lymphoid organs are the sites of lymphocyte activation by antigen, and lymphocytes recirculate between the blood and these organs until they encounter their specific antigen. Lymphatics drain extracellular fluid from the peripheral tissues, through the lymph nodes and into the thoracic duct, which empties into the left subclavian vein. This fluid, known as lymph, carries antigen taken up by dendritic cells and macrophages to the lymph nodes, as well as recirculating lymphocytes from the lymph nodes back into the blood. Lymphoid tissue is also associated with other mucosa such as the bronchial linings (not shown).

1-6 Most infectious agents activate the innate immune system and induce an inflammatory response.

The skin and the mucosal epithelia lining the airways and gut are the first defense against invading pathogens, forming a physical and chemical barrier against infection. Microorganisms that breach these defenses are met by cells and molecules that mount an immediate innate immune response. Macrophages resident in the tissues, for example, can recognize bacteria by means of receptors that bind common constituents of many bacterial surfaces. Engagement of these receptors triggers the macrophage both to engulf the bacterium and degrade it internally, and to secrete proteins called cytokines and chemokines that convey important signals to other immune cells. Similar responses occur to viruses, fungi, and parasites. **Cytokine** is a general name for any protein that is secreted by cells and affects the behavior of nearby cells bearing appropriate receptors. **Chemokines** are secreted proteins that act as chemoattractants (hence the name 'chemokine'), attracting cells bearing chemokine receptors, such as neutrophils and monocytes, out of the bloodstream and into infected tissue (Fig. 1.9). The cytokines and chemokines released by activated macrophages initiate the process known as **inflammation**. Inflammation is beneficial to combating infection by recruiting proteins and cells from the blood into infected tissues that help to directly destroy the pathogen. In addition, inflammation increases the flow of lymph carrying microbes and antigen-presenting cells from the infected tissue to nearby lymphoid tissues, where they activate lymphocytes and initiate the adaptive immune response. Once adaptive immunity has been triggered, inflammation also recruits the effector components of the adaptive immune system—antibody molecules and effector T cells—to the site of infection.

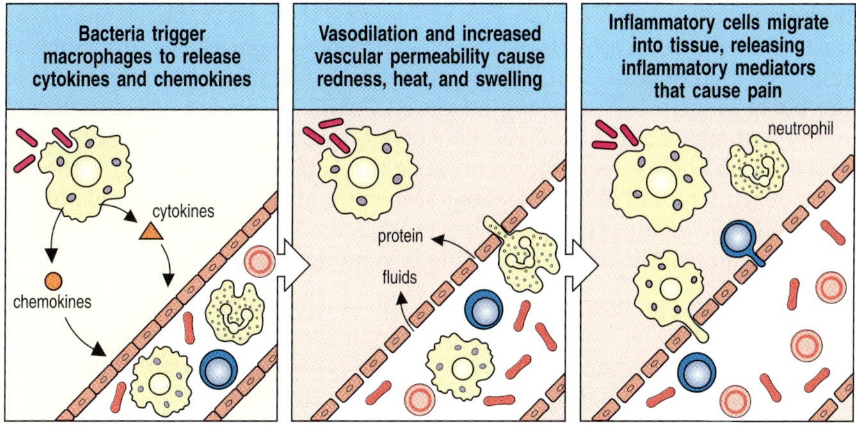

Bacteria trigger macrophages to release cytokines and chemokines
cytokines
chemokines
Vasodilation and increased vascular permeability cause redness, heat, and swelling
protein
fluids
Inflammatory cells migrate into tissue, releasing inflammatory mediators that cause pain
neutrophil

Fig. 1.9 Infection triggers an inflammatory response. Macrophages encountering bacteria or other types of microorganisms in tissues are triggered to release cytokines that increase the permeability of blood vessels, allowing fluid and proteins to pass into the tissues. They also produce chemokines, which direct the migration of neutrophils to the site of infection. The stickiness of the endothelial cells of the blood vessel wall is also changed, so that cells adhere to the wall and are able to crawl through it; first neutrophils and then monocytes are shown entering the tissue from a blood vessel. The accumulation of fluid and cells at the site of infection causes the redness, swelling, heat, and pain known collectively as inflammation. Neutrophils and macrophages are the principal inflammatory cells. Later in an immune response, activated lymphocytes can also contribute to inflammation.

Local inflammation and the phagocytosis of invading bacteria can also be triggered as a result of the activation of a group of plasma proteins known collectively as **complement**. Activation of the complement system by bacterial surfaces leads to a cascade of proteolytic reactions that coats microbes, but not the body's own cells, with complement fragments. Complement-coated microbes are recognized and bound by specific **complement receptors** on macrophages, taken up by phagocytosis, and destroyed.

Inflammation is described clinically by the Latin words *calor, dolor, rubor,* and *tumor*, meaning heat, pain, redness, and swelling. Each of these features reflects an effect of cytokines or other inflammatory mediators on the local blood vessels. Heat, redness, and swelling result from the dilation and increased permeability of blood vessels during inflammation, leading to increased local blood flow and leakage of fluid and blood proteins into the tissues. Cytokines and complement fragments have important effects on the **endothelium** that lines blood vessels; the **endothelial cells** themselves also produce cytokines in response to infection. The pro-inflammatory cytokines produce changes in the adhesive properties of the endothelial cells, in turn causing circulating leukocytes to stick to the endothelial cells and migrate between them into the site of infection, to which they are attracted by chemokines. The migration of cells into the tissue and their local actions account for the pain.

The main cell types seen in the initial phase of an inflammatory response are macrophages and neutrophils, the latter being recruited into the inflamed, infected tissue in large numbers. Macrophages and neutrophils are thus also known as **inflammatory cells**. Like macrophages, neutrophils have surface receptors for common bacterial constituents and for complement, and they are the principal cells that engulf and destroy the invading microorganisms. The influx of neutrophils is followed a short time later by the increased entry of monocytes, which rapidly differentiate into macrophages, thus reinforcing and sustaining the innate immune response. Later, if the inflammation continues, eosinophils also migrate into inflamed tissues and contribute to the destruction of the invading microorganisms.

1-7 Pattern recognition receptors of the innate immune system provide an initial discrimination between self and nonself.

The defense systems of innate immunity are effective in combating many pathogens, but rely on a limited number of invariant receptors that recognize microorganisms. The pathogen-recognition receptors of macrophages, neutrophils, and dendritic cells recognize simple molecules and regular patterns of molecular structure known as **pathogen-associated molecular patterns** (**PAMPs**) that are present on many microorganisms but not on the

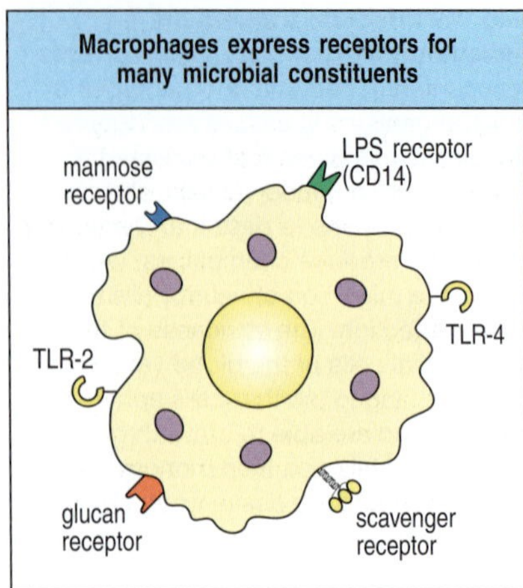

Fig. 1.10 Macrophages express a number of receptors that allow them to recognize different pathogens. Macrophages express a variety of receptors, each of which is able to recognize specific components of microbes. Some, like the mannose and glucan receptors and the scavenger receptor, bind cell-wall carbohydrates of bacteria, yeast, and fungi. The Toll-like receptors (TLRs) are an important family of pattern recognition receptors present on macrophages and other immune cells, and they are able to bind different microbial components; for example, TLR-2 binds cell-wall components of Gram-negative bacteria, whereas TLR-4 binds cell-wall components of Gram-positive bacteria. LPS, lipopolysaccharide.

body's own cells. The receptors that recognize PAMPs are known generally as **pattern recognition receptors** (**PRRs**), and they recognize structures such as mannose-rich oligosaccharides, peptidoglycans, and lipopolysaccharides in the bacterial cell wall, and unmethylated CpG DNA, which are common to many pathogens and have been conserved during evolution, making them excellent targets for recognition because they do not change (Fig. 1.10). We consider this system of innate recognition in detail in Chapter 3. Much of our knowledge of innate recognition has emerged only within the past 10 years, and this is one of the most exciting areas in modern immunology.

The pattern recognition receptors allow the innate immune system to distinguish self (the body) from nonself (pathogen). Detection of nonself activates innate cells and initiates adaptive immunity: macrophages are triggered to engulf microbes; immature dendritic cells are triggered to activate naive T lymphocytes. The molecules recognized by pattern recognition receptors are quite distinct from the individual pathogen-specific antigens recognized by lymphocytes. The fact that microbial constituents were needed to stimulate immune responses against purified proteins highlights the requirement that an innate response must precede the initiation of an adaptive response (see Appendix I, Sections A-1–A-4). This requirement was recognized long before the discovery of dendritic cells and their mode of activation. It was known that purified antigens such as proteins often did not evoke an immune response in an experimental immunization—that is, they were not **immunogenic**. To obtain adaptive immune responses to purified antigens, killed bacteria or bacterial extracts had to be mixed with the antigen. This additional material was termed an **adjuvant**, because it helped the response to the immunizing antigen (*adjuvare* is Latin for 'to help'). We know now that adjuvants are needed, at least in part, to activate dendritic cells to full antigen-presenting status in the absence of an infection. Finding suitable adjuvants is still an important part of vaccine preparation, as we discuss in Chapter 16.

1-8 Adaptive immune responses are initiated by antigen and antigen-presenting cells in secondary lymphoid tissues.

Adaptive immune responses are initiated when antigens or antigen-presenting cells, particularly dendritic cells bearing antigens picked up at sites of infection, reach the secondary lymphoid organs. Like the neutrophils and macrophages described earlier, dendritic cells have pattern recognition receptors that recognize molecular patterns common to microorganisms, such as bacterial lipopolysaccharide. Microbial components binding to these receptors stimulate the immature dendritic cell to engulf the pathogen and degrade it intracellularly. Immature dendritic cells also take up extracellular material, including virus particles and bacteria, by receptor-independent macropinocytosis, and thus internalize and degrade pathogens that their cell-surface receptors do not detect. In addition to the display of antigens that activates the antigen-receptors of lymphocytes, mature dendritic cells also express cell-surface proteins called **co-stimulatory molecules**, which provide signals that act together with antigen to stimulate the T lymphocyte to proliferate and differentiate into its final fully functional form (Fig. 1.11). Free antigens can also stimulate the antigen receptors of B cells, but most B cells require 'help' from activated helper T cells for optimal antibody responses. The activation of naive T lymphocytes is therefore an essential first stage in virtually all adaptive immune responses.

Movie 1.1 

1-9 Lymphocytes activated by antigen give rise to clones of antigen-specific effector cells that mediate adaptive immunity.

Cells of the innate immune system express many different pattern recognition receptors, each recognizing a different feature shared by many pathogens.

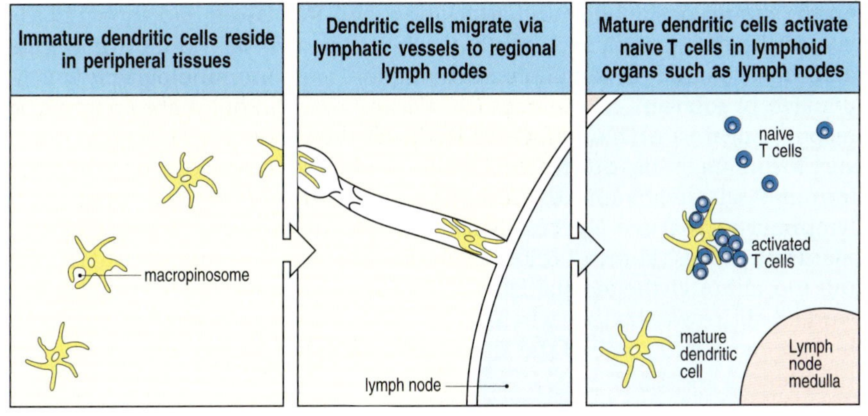
Immature dendritic cells reside in peripheral tissues
macropinosome
Dendritic cells migrate via lymphatic vessels to regional lymph nodes
lymph node
Mature dendritic cells activate naive T cells in lymphoid organs such as lymph nodes
naive T cells
activated T cells
mature dendritic cell
Lymph node medulla

Fig. 1.11 Dendritic cells initiate adaptive immune responses. Immature dendritic cells resident in a tissue take up pathogens and their antigens by macropinocytosis and by receptor-mediated endocytosis. They are stimulated by recognition of the presence of pathogens to migrate through the lymphatics to regional lymph nodes, where they arrive as fully mature non-phagocytic dendritic cells that express both antigen and the co-stimulatory molecules necessary to activate a naive T cell that recognizes the antigen, stimulating lymphocyte proliferation and differentiation.

In contrast, lymphocyte antigen receptor expression is 'clonal'—in other words, each mature lymphocyte emerging from the central lymphoid organs differs from the others in the specificity of its antigen receptor. When that lymphocyte proliferates it forms a **clone** of identical cells bearing identical antigen receptors. The diversity in antigen receptors is generated by a unique genetic mechanism that operates during lymphocyte development in the bone marrow and the thymus to generate millions of different variants of the genes encoding the receptor molecules. This ensures that the lymphocytes in the body collectively carry millions of different antigen receptor specificities—the **lymphocyte receptor repertoire** of the individual.

Lymphocytes are continually undergoing a process akin to natural selection; only those lymphocytes that encounter an antigen to which their receptor binds will be activated to proliferate and differentiate into effector cells.

This selective mechanism was first proposed in the 1950s by **Macfarlane Burnet** to explain why a person produces antibodies against only those antigens to which he or she has been exposed. Burnet postulated the preexistence in the body of many different potential antibody-producing cells, each having the ability to make antibody of a different specificity and displaying on its surface a membrane-bound version of the antibody: this serves as a receptor for the antigen. On binding antigen, the cell is activated to divide and to produce many identical progeny, a process known as **clonal expansion**; this clone of identical cells can now secrete **clonotypic** antibodies with a specificity identical to that of the surface receptor that first triggered activation and clonal expansion (Fig. 1.12). Burnet called this the **clonal selection theory** of antibody production.

1-10 Clonal selection of lymphocytes is the central principle of adaptive immunity.

Remarkably, at the time that Burnet formulated his theory, nothing was known of the antigen receptors of lymphocytes; indeed, the function of lymphocytes themselves was still obscure. Lymphocytes did not take center stage until the early 1960s, when **James Gowans** discovered that removal of the small lymphocytes from rats resulted in the loss of all known adaptive immune responses. These immune responses were restored when the small lymphocytes were replaced. This led to the realization that lymphocytes must be the units of clonal selection, and their biology became the focus of the new field of **cellular immunology**.

Clonal selection of lymphocytes with diverse receptors elegantly explained adaptive immunity, but it raised one significant conceptual problem. If the antigen receptors of lymphocytes are generated randomly during the lifetime of an individual, how are lymphocytes prevented from recognizing antigens

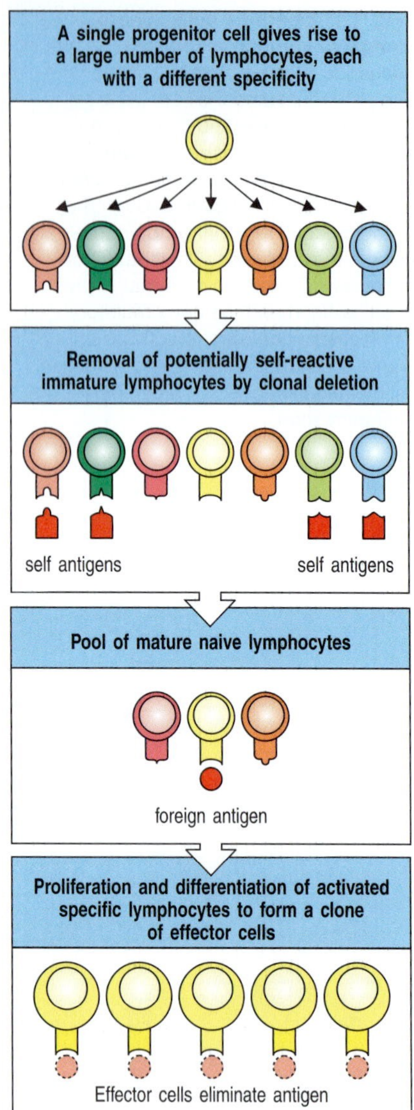
A single progenitor cell gives rise to a large number of lymphocytes, each with a different specificity
Removal of potentially self-reactive immature lymphocytes by clonal deletion
self antigens
self antigens
Pool of mature naive lymphocytes
foreign antigen
Proliferation and differentiation of activated specific lymphocytes to form a clone of effector cells
Effector cells eliminate antigen

Fig. 1.12 Clonal selection. Each lymphoid progenitor gives rise to a large number of lymphocytes, each bearing a distinct antigen receptor. Lymphocytes with receptors that bind ubiquitous self antigens are eliminated before they become fully mature, ensuring tolerance to such self antigens. When a foreign antigen interacts with the receptor on a mature naive lymphocyte, that cell is activated and starts to divide. It gives rise to a clone of identical progeny, all of whose receptors bind the same antigen. Antigen specificity is thus maintained as the progeny proliferate and differentiate into effector cells. Once antigen has been eliminated by these effector cells, the immune response ceases, although some lymphocytes are retained to mediate immunological memory.

on the tissues of the body and attacking them? **Ray Owen** had shown in the late 1940s that genetically different twin calves with a common placenta, and thus a shared placental blood circulation, were immunologically unresponsive, or **tolerant**, to one another's tissues: they did not make an immune response against each other. **Peter Medawar** then showed in 1953 that exposure to foreign tissues during embryonic development caused mice to become immunologically tolerant to these tissues. Burnet proposed that developing lymphocytes that are potentially self-reactive are removed before they can mature, a process known as **clonal deletion**. He has since been proved right in this too, although the mechanisms of **immunological tolerance** are still being worked out, as we shall see when we discuss the development of lymphocytes in Chapter 8, and some of the situations in which tolerance breaks down in Chapters 14 and 15.

Clonal selection of lymphocytes is the single most important principle in adaptive immunity. Its four basic postulates are listed in Fig. 1.13. The last of the problems posed by the clonal selection theory—how the diversity of lymphocyte antigen receptors is generated—was solved in the 1970s, when advances in molecular biology made it possible to clone the genes encoding antibody molecules.

1-11 The structure of the antibody molecule illustrates the central puzzle of adaptive immunity.

As discussed above, antibodies are the secreted form of the B cell's antigen receptor. Because they are produced in very large quantities in response to antigen, antibodies can be studied by traditional biochemical techniques; indeed, their structure was understood long before recombinant DNA technology made it possible to study the membrane-bound antigen receptors of B cells. The startling feature that emerged from the biochemical studies was that antibody molecules are composed of two distinct regions. One is a **constant region** that takes one of only four or five biochemically distinguishable forms; the other is a **variable region** that can be composed of a seemingly infinite variety of different amino acid sequences, forming subtly different structures that allow antibodies to bind specifically to an equally vast variety of antigens. This division is illustrated in Fig. 1.14, where the antibody is depicted as a Y-shaped molecule. The variable region determines the antigen-binding specificity of the antibody. There are two identical variable regions in an antibody molecule, and it thus has two identical **antigen-binding sites**. The constant region determines the effector function of the antibody: that is, how the antibody will interact with various immune cells to dispose of antigen once it is bound.

Each antibody molecule has a two-fold axis of symmetry and is composed of two identical **heavy chains** and two identical **light chains** (see Fig. 1.14). Heavy and light chains each have variable and constant regions; the variable regions of a heavy chain and a light chain combine to form an antigen-binding site, so that both chains contribute to the antigen-binding specificity of the antibody molecule. The functional properties of antibodies conferred by their constant regions are considered in Chapters 5 and 10.

The T-cell receptor for antigen shows many similarities to the B-cell antigen receptor, and the two molecules are clearly related to each other evolutionarily. There are, however, important differences between the two that, as we shall see, relate to their different roles within the immune system. The T-cell receptor, as shown in Fig. 1.14, is composed of two chains of roughly equal size, called the T-cell receptor α and β chains, each of which spans the T-cell membrane. Each chain has a variable region and a constant region, and the combination of the α- and β-chain variable regions creates a single site for binding antigen. The structures of both antibodies and T-cell receptors are

Postulates of the clonal selection hypothesis
Each lymphocyte bears a single type of receptor with a unique specificity
Interaction between a foreign molecule and a lymphocyte receptor capable of binding that molecule with high affinity leads to lymphocyte activation
The differentiated effector cells derived from an activated lymphocyte will bear receptors of identical specificity to those of the parental cell from which that lymphocyte was derived
Lymphocytes bearing receptors specific for ubiquitous self molecules are deleted at an early stage in lymphoid cell development and are therefore absent from the repertoire of mature lymphocytes

Fig. 1.13 The four basic principles of clonal selection.

described in detail in Chapter 4, and how the diversity of the antigen-receptor repertoire is generated is detailed in Chapter 5.

There is, however, a crucial difference in the way in which the B-cell and T-cell receptors bind antigens: the T-cell receptor does not bind antigen molecules directly but instead recognizes fragments of antigens bound on the surface of other cells. The exact nature of the antigen recognized by T cells, and how the antigens are fragmented and carried to cell surfaces, is the subject of Chapter 6. A further difference from the antibody molecule is that there is no secreted form of the T-cell receptor; the function of the receptor is solely to signal to the T cell that it has bound its antigen, and the subsequent immunological effects depend on the actions of the T cells themselves, as we describe in Chapter 9.

1-12 Each developing lymphocyte generates a unique antigen receptor by rearranging its receptor gene segments.

How are antigen receptors with an almost infinite range of specificities encoded by a finite number of genes? This question was answered in 1976, when **Susumu Tonegawa** discovered that the genes for immunoglobulin variable regions are inherited as sets of **gene segments**, each encoding a part of the variable region of one of the immunoglobulin polypeptide chains. During B-cell development in the bone marrow, these gene segments are irreversibly joined by DNA recombination to form a stretch of DNA encoding a complete variable region. How the complete antigen receptors are assembled from incomplete gene segments is the topic of Chapter 5.

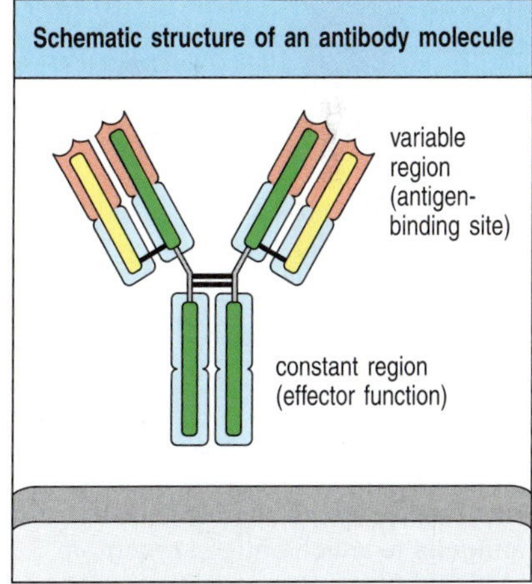

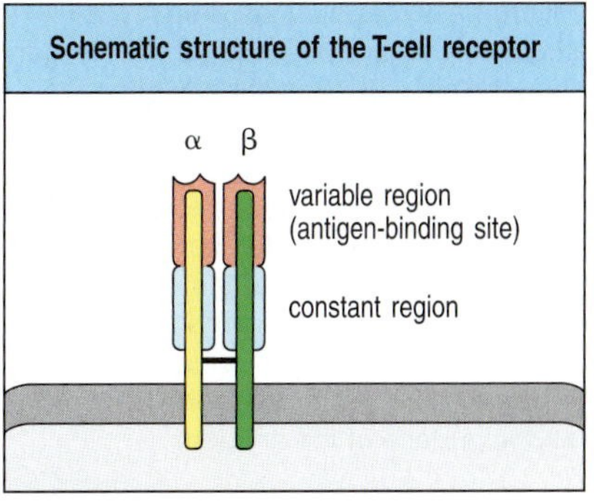

Fig. 1.14 Schematic structure of antigen receptors. Upper panel: an antibody molecule, which is secreted by activated B cells as an antigen-binding effector molecule. A membrane-bound version of this molecule acts as the B-cell antigen receptor (not shown). An antibody is composed of two identical heavy chains (green) and two identical light chains (yellow). Each chain has a constant part (shaded blue) and a variable part (shaded red). Each arm of the antibody molecule is formed by a light chain and a heavy chain such that the variable parts of the two chains come together, creating a variable region that contains the antigen-binding site. The stem is formed from the constant parts of the heavy chains and takes a limited number of forms. This constant region is involved in the elimination of the bound antigen. Lower panel: a T-cell antigen receptor. This is also composed of two chains, an α chain (yellow) and a β chain (green), each of which has a variable and a constant part. As with the antibody molecule, the variable parts of the two chains create a variable region, which forms the antigen-binding site. The T-cell receptor is not produced in a secreted form.

Here, we need only mention two features of these mechanisms. First, it is the combinatorial assembly of a large number of different gene segments that makes possible the enormous size of the antigen receptor repertoire. This means that a finite number of gene segments can generate a vast number of different proteins. Second, the assembly process is regulated in a manner that ensures that each lymphocyte expresses only one receptor specificity. Third, because gene segment rearrangement involves an irreversible change in a cell's DNA, all the progeny of that cell will inherit genes encoding the same receptor specificity. This general scheme was later also confirmed for the genes encoding the antigen receptor on T cells.

The potential diversity of lymphocyte receptors generated in this way is enormous. Just a few hundred different gene segments can combine in different ways to generate thousands of different receptor chains. The diversity of lymphocyte receptors is further amplified by junctional diversity, created by adding or subtracting nucleotides in the process of joining the gene segments, and by the fact that each receptor is made by pairing two different variable chains, each encoded by distinct sets of gene segments. A thousand different chains of each type could thus generate 10^6 distinct antigen receptors through this combinatorial diversity. In this way, a small amount of genetic material can encode a truly staggering diversity of receptors. Only a subset of these randomly generated receptor specificities survive the selective processes that shape the peripheral lymphocyte repertoire; nevertheless, there are lymphocytes of at least 10^8 different specificities in an individual human at any one time. These provide the raw material on which clonal selection acts.

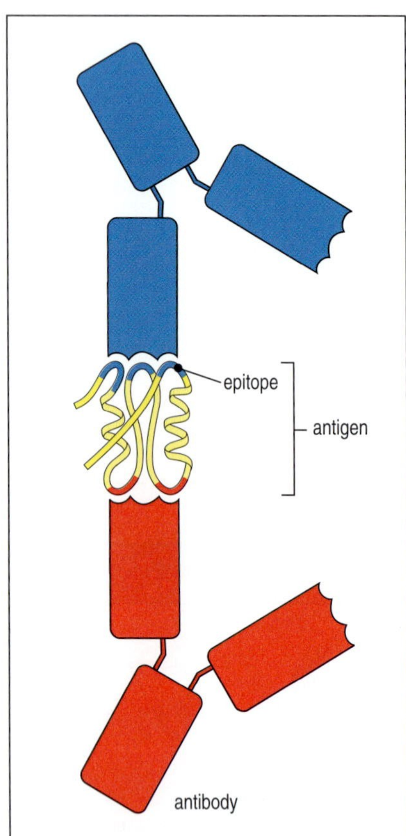

Fig. 1.15 Antigens are the molecules recognized by the immune response, while epitopes are sites within antigens to which antigen receptors bind. Antigens can be complex macromolecules such as proteins, as shown in yellow. Most antigens are larger than the sites on the antibody or antigen receptor to which they bind, and the actual portion of the antigen that is bound is known as the antigenic determinant, or epitope, for that receptor. Large antigens such as proteins can contain more than one epitope (indicated in red and blue), and thus may bind different antibodies. Antibodies generally recognize epitopes on the surface of the antigen.

1-13 Immunoglobulins bind a wide variety of chemical structures, whereas the T-cell receptor is specialized to recognize foreign antigens as peptide fragments bound to proteins of the major histocompatibility complex.

In principle, almost any chemical structure can be recognized by the adaptive immune system as an antigen, but the usual antigens encountered in an infection are the proteins, glycoproteins, and polysaccharides of pathogens. An individual antigen receptor or antibody recognizes a small part of the molecular structure of an antigenic molecule, which is known as an **antigenic determinant** or **epitope** (Fig. 1.15). Macromolecular antigens such as proteins and glycoproteins usually have many different epitopes that can be recognized by different antigen receptors.

The antigen receptors of B cells and T cells recognize antigen in fundamentally different ways. B cells directly recognize the native antigen that either has been secreted by a pathogen or is expressed on its surface (see Fig. 1.15). B cells eventually differentiate into effector plasma cells that secrete antibodies that will bind to and neutralize these antigens and pathogens. In contrast, T-cell receptors do not directly recognize native antigens. Rather, they recognize antigens that have been processed, partly degraded, and displayed as peptides bound to proteins on the surface of antigen-presenting cells (Fig. 1.16). A main source of the antigens recognized by T cells is cells infected with a pathogen, commonly a virus. In this case, the antigen that is recognized by the effector T cells is derived from within the infected cell. Importantly, T-cell receptors will only recognize antigen-derived peptides when these are bound to particular cell-surface glycoproteins called **MHC molecules**, which are encoded in a cluster of genes called the **major histocompatibility complex** (**MHC**). The antigen recognized by T-cell receptors is thus a complex of a foreign peptide antigen and an MHC molecule (see Fig. 1.16). We shall see how these compound antigens are recognized by T-cell receptors and how they are generated in Chapters 4 and 6, respectively.

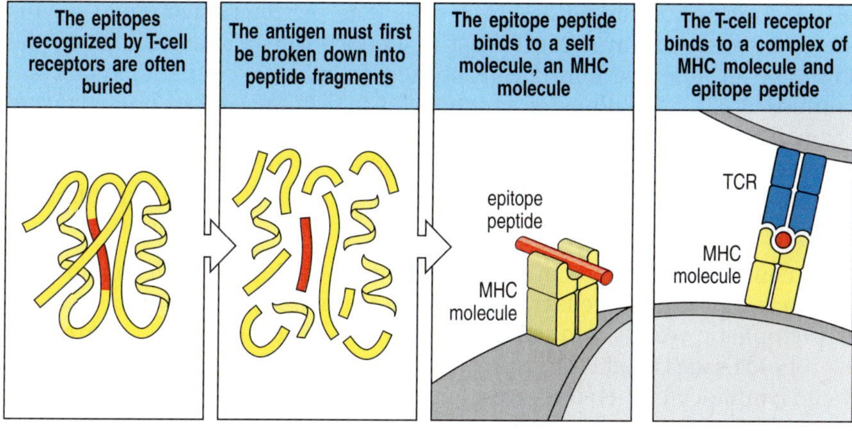

The epitopes recognized by T-cell receptors are often buried
The antigen must first be broken down into peptide fragments
The epitope peptide binds to a self molecule, an MHC molecule
The T-cell receptor binds to a complex of MHC molecule and epitope peptide
epitope peptide
MHC molecule
TCR
MHC molecule

Fig. 1.16 T-cell receptors bind a complex of an antigen fragment and a self molecule. Unlike most antibodies, T-cell receptors can recognize epitopes that are buried within antigens (first panel). These antigens must first be degraded by proteinases (second panel), and the peptide epitope delivered to a self molecule, called an MHC molecule (third panel). It is in this form, as a complex of peptide and MHC molecule, that antigens are recognized by T-cell receptors (fourth panel).

1-14 The development and survival of lymphocytes is determined by signals received through their antigen receptors.

The continuous generation of lymphocytes throughout life creates a problem of keeping total numbers of peripheral lymphocytes relatively constant. In addition, with so many different antigen receptors being generated during lymphocyte development, it is inevitable that potentially dangerous receptors that can react against an individual's own **self antigens** will be produced. Both these problems seem to be solved by making the survival of a lymphocyte dependent on signals received through its antigen receptor. Lymphocytes that react strongly to self antigens during development are removed by clonal deletion, as predicted by Burnet's clonal selection theory, before they mature to a stage at which they could do damage. The complete absence of signals from the antigen receptor during development can also lead to cell death. Lymphocytes that receive either too much or too little signal during development are eliminated by a form of cell suicide called **apoptosis** or **programmed cell death**.

Apoptosis, derived from a Greek word meaning the falling of leaves from the trees, is a general means of regulating the number of cells in the body. It is responsible, for example, for the death and shedding of old skin and intestinal epithelial cells, and the turnover of liver cells. Every day the bone marrow produces millions of new neutrophils, monocytes, red blood cells, and lymphocytes, and this production must be balanced by an equal loss. Most white blood cells are relatively short lived and die by apoptosis. The dying cells are phagocytosed and degraded by specialized macrophages in the liver and spleen.

Lastly, if a lymphocyte's receptor is not used within a relatively short time of its entering the repertoire in the periphery, the cell bearing it dies, making way for new lymphocytes with different receptors. In this way, self-reactive receptors are eliminated, and receptors are tested to ensure that they are potentially functional. The mechanisms that shape and maintain the lymphocyte receptor repertoire are examined in Chapter 8.

1-15 Lymphocytes encounter and respond to antigen in the peripheral lymphoid organs.

Antigen and lymphocytes eventually encounter each other in the peripheral lymphoid organs—the lymph nodes, spleen, and the mucosal lymphoid tissues (see Fig. 1.8). Mature naive lymphocytes are continually recirculating through these tissues, to which pathogen antigens are carried from sites of infection, primarily by dendritic cells. The peripheral lymphoid organs are specialized to trap antigen-bearing dendritic cells and to facilitate the initiation of adaptive immune responses. Peripheral lymphoid tissues are composed of aggregations of lymphocytes in a framework of nonleukocyte stromal cells, which provide the basic structural organization of the tissue and provide survival signals to help sustain the life of the lymphocytes. Besides lymphocytes, peripheral lymphoid organs also contain resident macrophages and dendritic cells.

When an infection occurs in a tissue such as the skin, free antigen and antigen-bearing dendritic cells travel from the site of infection through the afferent lymphatic vessels into the **draining lymph nodes** (Fig. 1.17), peripheral lymphoid tissues where they activate antigen-specific lymphocytes. The activated lymphocytes then undergo a period of proliferation and differentiation, after which most leave the lymph nodes as effector cells via the efferent lymphatic vessel. This eventually returns them to the bloodstream (see Fig. 1.8), which then carries them to the tissues where they will act. This whole process takes about 4–6 days from the time that the antigen is recognized, which means that an adaptive immune response to an antigen that has not been encountered before does not become effective until about a week after infection (see Fig. 1.34). Naive lymphocytes that do not recognize their antigen also leave through the efferent lymphatic vessel and are returned to the blood, from which they continue to recirculate through lymphoid tissues until they recognize antigen or die.

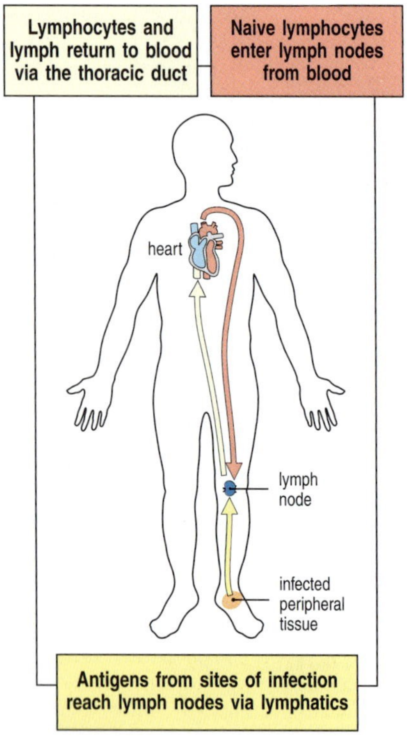

Lymphocytes and lymph return to blood via the thoracic duct

Naive lymphocytes enter lymph nodes from blood

heart

lymph node

infected peripheral tissue

Antigens from sites of infection reach lymph nodes via lymphatics

Fig. 1.17 Circulating lymphocytes encounter antigen in peripheral lymphoid organs. Naive lymphocytes recirculate constantly through peripheral lymphoid tissue, here illustrated as a popliteal lymph node—a lymph node situated behind the knee. In the case of an infection in the foot, this will be the draining lymph node, where lymphocytes may encounter their specific antigens and become activated. Both activated and nonactivated lymphocytes are returned to the bloodstream via the lymphatic system.

The **lymph nodes** are highly organized lymphoid organs located at the points of convergence of vessels of the lymphatic system, which is the extensive system that collects extracellular fluid from the tissues and returns it to the blood (see Fig. 1.8). This extracellular fluid is produced continuously by filtration from the blood and is called lymph. Lymph flows away from the peripheral tissues under the pressure exerted by its continual production, and is carried by lymphatic vessels, or **lymphatics**. One-way valves in the lymphatic vessels prevent a reverse flow, and the movements of one part of the body in relation to another are important in driving the lymph along.

Afferent lymphatic vessels drain fluid from the tissues and carry pathogens and antigen-bearing cells from infected tissues to the lymph nodes (Fig. 1.18). Free antigens simply diffuse through the extracellular fluid to the lymph node, while the dendritic cells actively migrate into the lymph node, attracted by chemokines. The same chemokines also attract lymphocytes from the blood, and these enter lymph nodes by squeezing through the walls of specialized blood vessels called **high endothelial venules** (**HEV**). In the lymph nodes, B lymphocytes are localized in **follicles**, which make up the outer **cortex** of the lymph node, with T cells more diffusely distributed in the surrounding **paracortical areas**, also referred to as the deep cortex or **T-cell zones** (see Fig. 1.18). Lymphocytes migrating from the blood into lymph nodes enter the paracortical areas first and, because they are attracted by the same chemokines, antigen-presenting dendritic cells and macrophages also become localized there. Free antigen diffusing through the lymph node can become trapped on these dendritic cells and macrophages. This juxtaposition of antigen, antigen-presenting cells, and naive T cells in the T-cell zone creates an ideal environment in which naive T cells can bind their specific antigen and thus become activated.

As noted earlier, activation of B cells usually requires not only antigen, which binds to the B-cell receptor, but also the cooperation of activated helper T cells, a type of effector T cell (see Section 1-4). The location of B cells and T cells within the lymph node is dynamically regulated by their state of activation. When they become activated, T cells and B cells both move to the border of the follicle and T-cell zone, where T cells can first provide their helper function to B cells. Some of the B-cell follicles include **germinal centers**, where

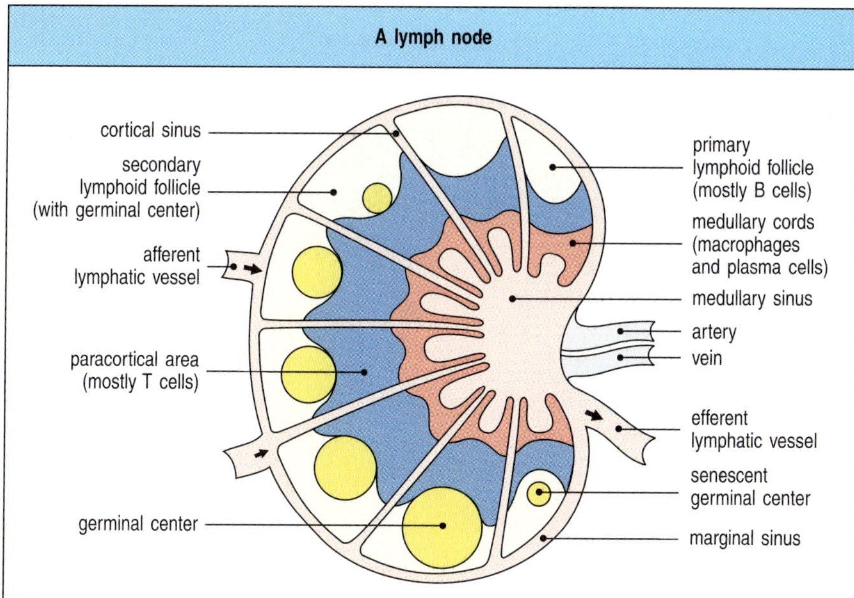

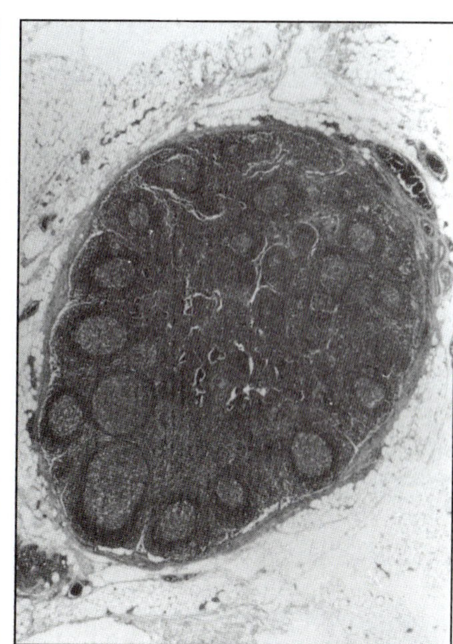

Fig. 1.18 Organization of a lymph node. As shown in the diagram on the left, which shows a lymph node in longitudinal section, a lymph node consists of an outermost cortex and an inner medulla. The cortex is composed of an outer cortex of B cells organized into lymphoid follicles, and deep, or paracortical, areas made up mainly of T cells and dendritic cells. When an immune response is under way, some of the follicles contain central areas of intense B-cell proliferation called germinal centers and are known as secondary lymphoid follicles. These reactions are very dramatic, but eventually die out as senescent germinal centers. Lymph draining from the extracellular spaces of the body carries antigens in phagocytic dendritic cells and phagocytic macrophages from the tissues to the lymph node via the afferent lymphatics. These migrate directly from the sinuses into the cellular parts of the node. Lymph leaves by the efferent lymphatics in the medulla. The medulla consists of strings of macrophages and antibody-secreting plasma cells known as the medullary cords. Naive lymphocytes enter the node from the bloodstream through specialized postcapillary venules (not shown) and leave with the lymph through the efferent lymphatic. The light micrograph shows a transverse section through a lymph node, with prominent follicles containing germinal centers. Magnification ×7. Photograph courtesy of N. Rooney.

activated B cells are undergoing intense proliferation and differentiation into plasma cells. These mechanisms are described in detail in Chapter 10.

In humans, the **spleen** is a fist-sized organ situated just behind the stomach (see Fig. 1.8). It has no direct connection with the lymphatic system; instead, it collects antigen from the blood and is involved in immune responses to blood-borne pathogens. Lymphocytes enter and leave the spleen via blood vessels. The spleen also collects and disposes of senescent red blood cells. Its organization is shown schematically in Fig. 1.19. The bulk of the spleen is composed of **red pulp**, which is the site of red blood cell disposal. The lymphocytes surround the arterioles running through the spleen, forming isolated areas of **white pulp**. The sheath of lymphocytes around an arteriole is called the **periarteriolar lymphoid sheath** (**PALS**) and contains mainly T cells. Lymphoid follicles occur at intervals along it, and these contain mainly B cells. A so-called marginal zone surrounds the follicle; it has few T cells, is rich in macrophages, and has a resident, noncirculating population of B cells known as **marginal zone B cells**, about which little is known; they are discussed in Chapter 8. Blood-borne microbes, soluble antigens, and antigen:antibody complexes are filtered from the blood by macrophages and immature dendritic cells within the marginal zone. Like the migration of immature dendritic cells from peripheral tissues to the T-cell areas of lymph nodes, dendritic cells in the marginal zones in the spleen migrate to the T-cell areas after taking up antigen and becoming activated; here they are able to present the antigens they carry to T cells.

Most pathogens enter the body through mucosal surfaces, and these are also exposed to a vast load of other potential antigens from the air, food, and the natural microbial flora of the body. Mucosal surfaces are protected by an extensive system of lymphoid tissues known generally as the **mucosal immune system** or **mucosa-associated tissues** (**MALT**). Collectively, the mucosal immune system is estimated to contain as many lymphocytes as all the rest of the body, and they form a specialized set of cells obeying somewhat different rules of recirculation from those in the other peripheral lymphoid organs. The **gut-associated lymphoid tissues** (**GALT**) include the **tonsils**, **adenoids**, **appendix**, and specialized structures in the small intestine called **Peyer's patches**, and they collect antigen from the epithelial surfaces of the gastrointestinal tract. In Peyer's patches, which are the most important and highly organized of these tissues, the antigen is collected by specialized epithelial cells called

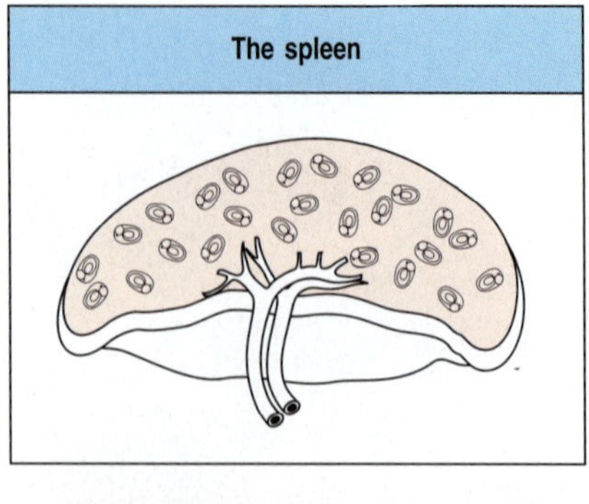

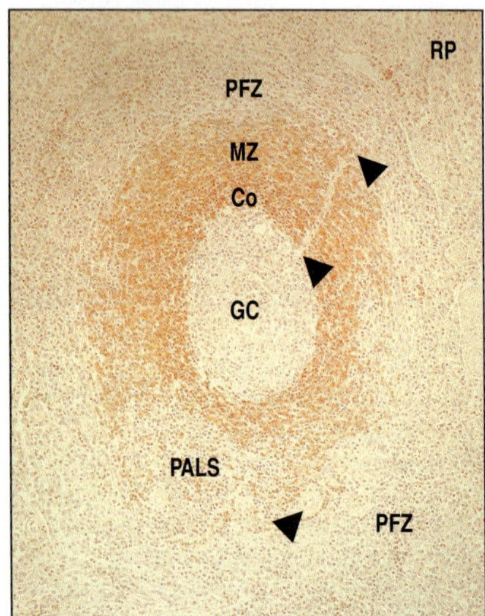

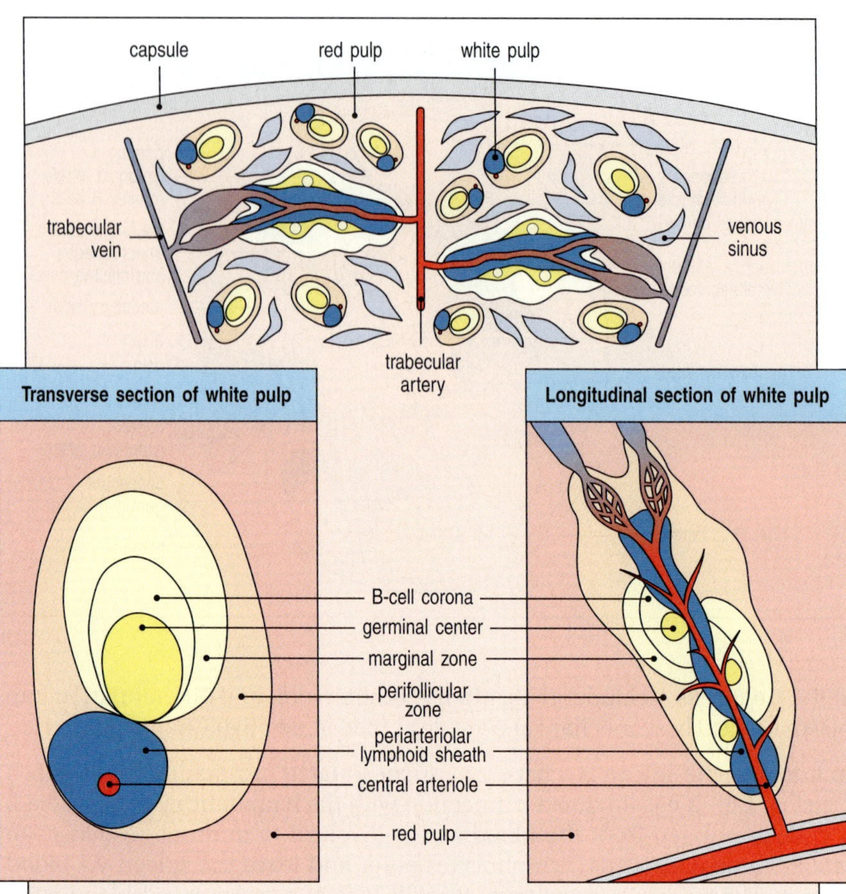

Fig. 1.19 Organization of the lymphoid tissues of the spleen. The schematic at top left shows that the spleen consists of red pulp (pink areas), which is a site of red blood cell destruction, interspersed with the lymphoid white pulp. An enlargement of a small section of a human spleen (top right) shows the arrangement of discrete areas of white pulp (yellow and blue) around central arterioles. Most of the white pulp is shown in transverse section, with two portions in longitudinal section. The bottom two schematics show enlargements of a transverse section (lower center) and longitudinal section (lower right) of white pulp. Surrounding the central arteriole is the periarteriolar lymphoid sheath (PALS), made up of T cells. Lymphocytes and antigen-loaded dendritic cells come together here. The follicles consist mainly of B cells; in secondary follicles a germinal center is surrounded by a B-cell corona. The follicles are surrounded by a so-called marginal zone of lymphocytes. In each area of white pulp, blood carrying both lymphocytes and antigen flows from a trabecular artery into a central arteriole. From this arteriole smaller blood vessels fan out, eventually terminating in a specialized zone in the human spleen called the perifollicular zone (PFZ), which surrounds each marginal zone. Cells and antigen then pass into the white pulp through open blood-filled spaces in the perifollicular zone. The light micrograph at bottom left shows a transverse section of white pulp of human spleen immunostained for mature B cells. Both follicle and PALS are surrounded by the perifollicular zone. The follicular arteriole emerges in the PALS (arrowhead at bottom), traverses the follicle, goes through the marginal zone and opens into the perifollicular zone (upper arrowheads). Co, follicular B-cell corona; GC, germinal center; MZ, marginal zone; RP, red pulp; arrowheads, central arteriole. Photograph courtesy of N.M. Milicevic.

microfold or **M cells** (Fig. 1.20). The lymphocytes form a follicle consisting of a large central dome of B lymphocytes surrounded by smaller numbers of T lymphocytes. Dendritic cells resident within the Peyer's patch present the antigen to T lymphocytes. Lymphocytes enter Peyer's patches from the blood and leave through efferent lymphatics. Effector lymphocytes generated in Peyer's patches travel through the lymphatic system and into the bloodstream, from where they are disseminated back into mucosal tissues to carry out their effector actions.

Similar but more diffuse aggregates of lymphocytes are present in the respiratory tract and other mucosa: **nasal-associated lymphoid tissue** (**NALT**) and **bronchus-associated lymphoid tissue** (**BALT**) are present in the respiratory

Peyer's patches are covered by an epithelial layer containing specialized cells called M cells, which have characteristic membrane ruffles

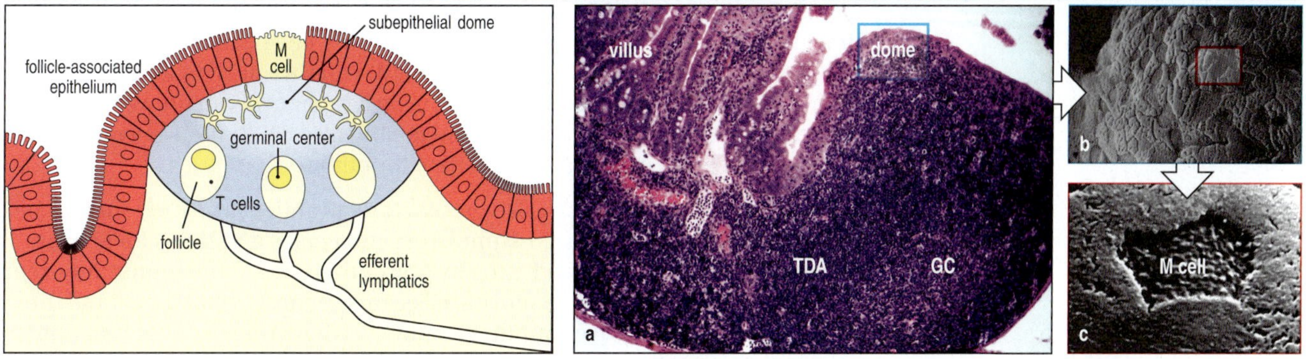

tract. Like the Peyer's patches, these mucosal lymphoid tissues are also overlaid by M cells, through which inhaled microbes and antigens that become trapped in the mucous covering of the respiratory tract can pass. The mucosal immune system is discussed in Chapter 12.

Although very different in appearance, the lymph nodes, spleen, and mucosa-associated lymphoid tissues all share the same basic architecture. They all operate on the same principle, trapping antigens and antigen-presenting cells from sites of infection and enabling them to present antigen to migratory small lymphocytes, thus inducing adaptive immune responses. The peripheral lymphoid tissues also provide sustaining signals to lymphocytes that do not encounter their specific antigen immediately, so that they survive and continue to recirculate.

Because they are involved in initiating adaptive immune responses, the peripheral lymphoid tissues are not static structures but vary quite markedly, depending on whether or not infection is present. The diffuse mucosal lymphoid tissues may appear in response to infection and then disappear, whereas the architecture of the organized tissues changes in a more defined way during an infection. For example, the B-cell follicles of the lymph nodes expand as B lymphocytes proliferate to form germinal centers (see Fig. 1.18), and the entire lymph node enlarges, a phenomenon familiarly known as swollen glands.

Finally, specialized populations of lymphocytes can be found distributed throughout particular sites in the body rather than being found in organized lymphoid tissues. Such sites include the liver and the lamina propria of the gut, as well as the base of the epithelial lining of the gut, reproductive epithelia, and, in mice but not in humans, the epidermis. These lymphocyte populations seem to have an important role in protecting these tissues from infection, and are described further in Chapters 8 and 12.

Fig. 1.20 Organization of a Peyer's patch in the gut mucosa. As the diagram on the left shows, a Peyer's patch contains numerous B-cell follicles with germinal centers. T cells occupy the areas between follicles, the T-cell dependent areas. The layer between the surface epithelium and the follicles is known as the subepithelial dome, and is rich in dendritic cells, T cells, and B cells. Peyer's patches have no afferent lymphatics and the antigen enters directly from the gut across a specialized epithelium made up of so-called microfold (M) cells. Although this tissue looks very different from other lymphoid organs, the basic divisions are maintained. As in the lymph nodes, lymphocytes enter Peyer's patches from the blood across the walls of high endothelial venules (not shown), and leave via the efferent lymphatic. The light micrograph in panel a shows a section through a Peyer's patch in the gut wall of the mouse. The Peyer's patch can be seen lying beneath the epithelial tissues. GC, germinal center; TDA, T-cell dependent area. Panel b is a scanning electron micrograph of the follicle-associated epithelium boxed in panel a, showing the M cells, which lack the microvilli and the mucus layer present on normal epithelial cells. Each M cell appears as a sunken area on the epithelial surface. Panel c is a higher-magnification view of the boxed area in panel b, showing the characteristic ruffled surface of an M cell. M cells are the portal of entry for many pathogens and other particles. Panel a, hematoxylin and eosin stain; magnification ×100; panel b, ×5000; panel c, ×23,000. Source: Mowat, A., Viney, J.: *Immunol. Rev.* 1997, **156**:145–166.

1-16 Lymphocyte activation requires additional signals beyond those relayed from the antigen receptor when antigen binds.

Peripheral lymphoid tissues promote the interaction between antigen-bearing APCs and lymphocytes, but antigen alone is not sufficient to initiate an adaptive immune response. Lymphocytes require other signals to become activated and to acquire effector functions. These signals are delivered to lymphocytes by another cell through cell-surface molecules known generally as co-stimulatory molecules (see Section 1-8). For naive T cells, an activated dendritic cell usually delivers these signals, but for naive B cells, the second signal is delivered by an activated helper T cell (Fig. 1.21). We discuss the nature of these signals in detail in Chapter 7.

The induction of co-stimulatory molecules is important in initiating an adaptive immune response because contact with antigen without accompanying co-stimulatory molecules inactivates naive lymphocytes rather than activating them, leading either to clonal deletion or an inactive state known as **anergy**. We return to this topic in Chapter 8. Thus, we need to add a final postulate to the clonal selection theory. A naive lymphocyte can only be activated by cells that bear not only specific antigen but also co-stimulatory molecules, whose expression is regulated by innate immunity.

Macrophages and B cells can also present foreign antigens on their surface and can be induced to express co-stimulatory molecules and thus can activate T cells. These three specialized antigen-presenting cells of the immune system are illustrated in Fig. 1.22. Dendritic cells are the most important of the three in initiating the adaptive immune response, whereas the others function as antigen-presenting cells at later stages, when T cells have acquired particular effector activities. These circumstances are discussed in Chapters 9 and 10.

1-17 Lymphocytes activated by antigen proliferate in the peripheral lymphoid organs, generating effector cells and immunological memory.

The great diversity of lymphocyte receptors means that there will usually be at least a few that can bind to a given foreign antigen. However, this number will be very small, certainly not enough to mount a response against a pathogen. To generate sufficient antigen-specific effector lymphocytes to fight an infection, a lymphocyte with an appropriate receptor specificity is activated first to proliferate. Only when a large clone of identical cells has been produced do these finally differentiate into effector cells. On recognizing its specific antigen on an activated antigen-presenting cell, a naive lymphocyte stops migrating, the volume of the nucleus and cytoplasm increases, and new mRNAs and new proteins are synthesized. Within a few hours, the cell looks completely different and is known as a **lymphoblast**.

Dividing lymphoblasts are able to duplicate themselves two to four times every 24 hours for 3–5 days, so that a single naive lymphocyte can produce a clone of around 1000 daughter cells of identical specificity. These then differentiate into effector cells. In the case of B cells, the differentiated effector cells are the **plasma cells**, which secrete antibody; in the case of T cells, the effector cells are **cytotoxic T cells** able to destroy infected cells, or **helper T cells** that activate other cells of the immune system. Effector lymphocytes do not recirculate like naive lymphocytes. Some effector T cells detect sites of

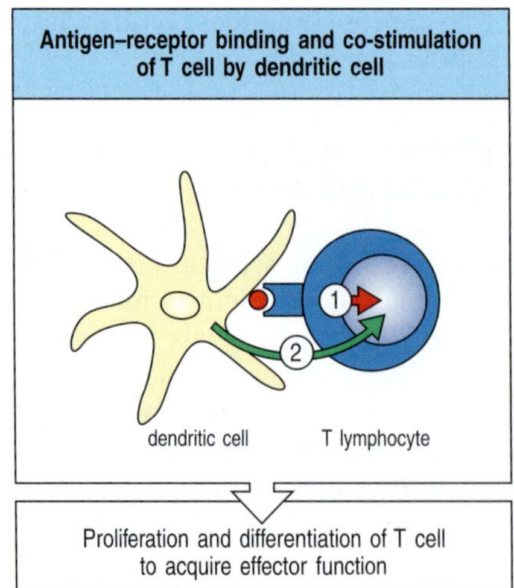
Antigen–receptor binding and co-stimulation of T cell by dendritic cell
1
2
dendritic cell
T lymphocyte
Proliferation and differentiation of T cell to acquire effector function

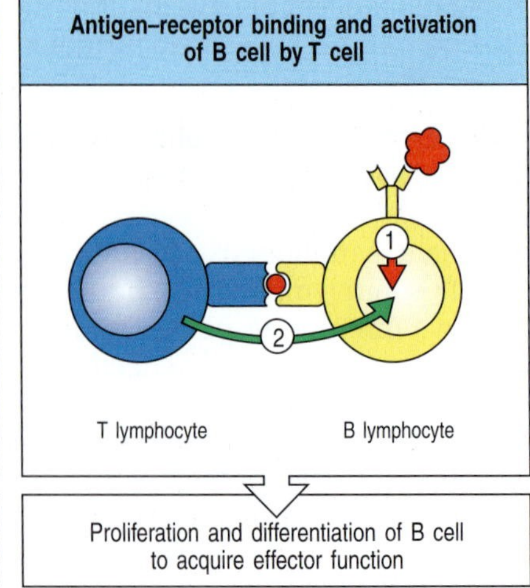
Antigen–receptor binding and activation of B cell by T cell
1
2
T lymphocyte
B lymphocyte
Proliferation and differentiation of B cell to acquire effector function

Fig. 1.21 Two signals are required for lymphocyte activation. In addition to receiving a signal through their antigen receptor (signal 1), mature naive lymphocytes must also receive a second signal (signal 2) to become activated. For T cells (left panel), this second signal is delivered by an antigen-presenting cell such as the dendritic cell shown here. For B cells (right panel), the second signal is usually delivered by an activated T cell, which recognizes antigenic peptides taken up, processed, and presented by the B cell on its surface.

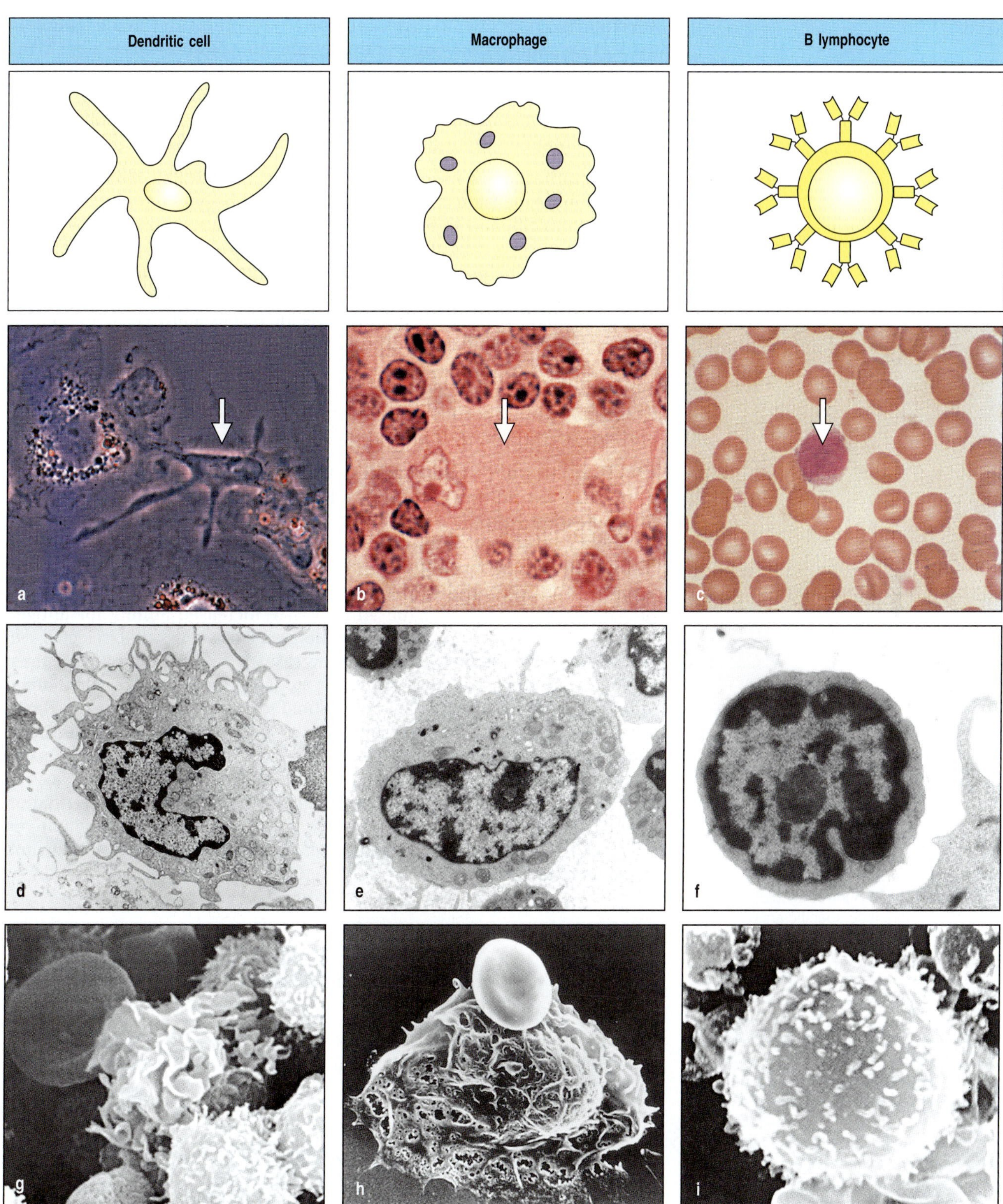

Fig. 1.22 The antigen-presenting cells. The three types of antigen-presenting cells are shown in the form in which they are depicted throughout this book (top row), as they appear in the light microscope (second row; the relevant cell is indicated by an arrow), by transmission electron microscopy (third row) and by scanning electron microscopy (bottom row). Mature dendritic cells are found in lymphoid tissues and are derived from immature tissue dendritic cells that interact with many distinct types of pathogens. Macrophages are specialized to internalize extracellular pathogens, especially after they have been coated with antibody, and to present their antigens. B cells have antigen-specific receptors that enable them to internalize large amounts of specific antigen, process it, and present it. Photographs courtesy of R.M. Steinman (a), N. Rooney (b, c, e, f), S. Knight (d, g), and P.F. Heap (h, i).

infection and migrate into them from the blood; others stay in the lymphoid tissues to activate B cells. Some antibody-secreting plasma cells remain in the peripheral lymphoid organs, but most plasma cells generated in the lymph nodes and spleen migrate to the bone marrow and take up residence there, pouring out antibodies into the blood system. Effector cells generated in the mucosal immune system generally stay within the mucosal tissues.

As noted earlier, after a naive lymphocyte has been activated, it takes 4–5 days before clonal expansion is complete and the lymphocytes have differentiated into effector cells, and so the first adaptive immune response to a pathogen only occurs several days after the infection begins and has been detected by the innate immune system. Most of the lymphocytes generated by the clonal expansion in any given immune response will eventually die. However, a significant number of activated antigen-specific B cells and T cells persist after antigen has been eliminated. These cells are known as **memory cells** and form the basis of immunological memory. They can be reactivated much more quickly than naive lymphocytes, which ensures a more rapid and effective response on a second encounter with a pathogen and thereby usually provides lasting protective immunity.

The characteristics of immunological memory are readily observed by comparing the antibody response of an individual to a first or **primary immunization** with the same response elicited in the same individual by a **secondary** or **booster immunization** with the same antigen. As shown in Fig. 1.23, the secondary antibody response occurs after a shorter lag phase, achieves a markedly higher level, and produces antibodies of higher affinity, or strength of binding, for the antigen. The increased affinity for antigen is called **affinity maturation** and is the result of events that select B-cell receptors, and thus antibodies, for progressively higher affinity for antigen during an immune response. Importantly, T-cell receptors do not undergo affinity maturation, and the lower threshold for activation of memory T cells compared with naive T cells results from a change in the responsiveness of the cell, not from a change in the receptor. We describe the mechanisms of these remarkable changes in Chapters 5 and 10. The cellular basis of immunological memory is the clonal expansion and clonal differentiation of cells specific for the eliciting antigen, and it is therefore entirely antigen specific. It is immunological memory that enables successful vaccination and prevents reinfection with pathogens that have been repelled successfully by

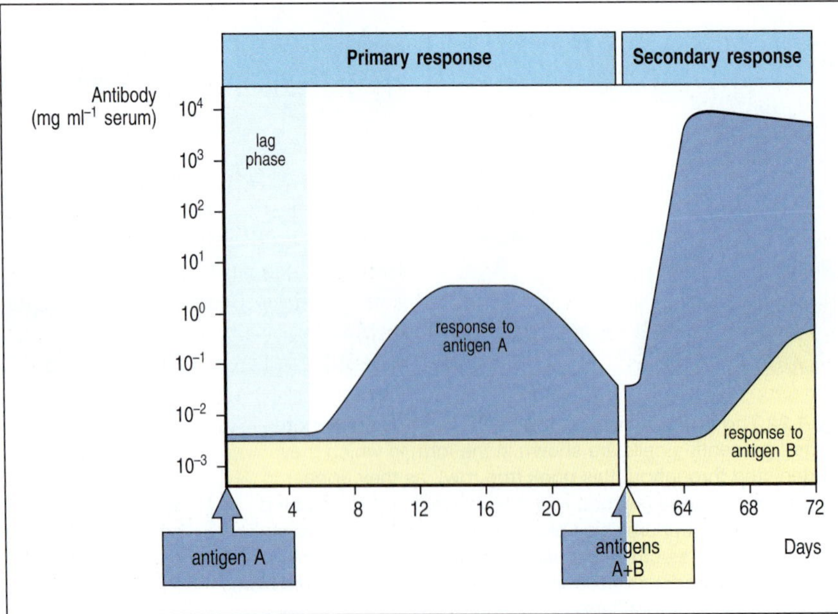

Fig. 1.23 The course of a typical antibody response. The first encounter with an antigen produces a primary response. Antigen A introduced at time zero encounters little specific antibody in the serum. After a lag phase (light blue), antibody against antigen A (dark blue) appears; its concentration rises to a plateau, and then gradually declines. This is typical of a primary response. When the serum is tested for antibody against another antigen, B (yellow), there is little present, demonstrating the specificity of the antibody response. When the animal is later challenged with a mixture of antigens A and B, a very rapid and intense secondary response to A occurs. This illustrates immunological memory, the ability of the immune system to make a second response to the same antigen more efficiently and effectively, providing the host with a specific defense against infection. This is the main reason for giving booster injections after an initial vaccination. Note that the response to B resembles the initial or primary response to A, as this is the first encounter of the host with antigen B.

an adaptive immune response. Immunological memory is the most important biological consequence of adaptive immunity, and its cellular and molecular bases are some of the most active areas of research. The properties of memory cells are described in more detail in Chapter 11.

Summary.

The early innate systems of defense depend on invariant pattern recognition receptors that detect common features of pathogens. Innate defenses are crucially important, but they can be overcome by many pathogens and they do not lead to immunological memory. Recognizing a particular pathogen and providing enhanced protection against reinfection is unique to adaptive immunity. An adaptive immune response involves the selection and amplification of clones of lymphocytes bearing receptors that recognize the foreign antigen. This clonal selection provides the theoretical framework for understanding all the key features of an adaptive immune response. There are two major types of lymphocytes: B lymphocytes, which mature in the bone marrow and are the source of circulating antibodies, and T lymphocytes, which mature in the thymus and recognize peptides from pathogens presented by MHC molecules on infected cells or antigen-presenting cells. Each lymphocyte carries cell-surface receptors of a single antigen specificity. These receptors are generated by the random recombination of variable receptor gene segments and the pairing of distinct variable protein chains: heavy and light chains in immunoglobulins, or the two chains of T-cell receptors. The large antigen receptor repertoire of lymphocytes can recognize virtually any antigen.

Adaptive immunity is initiated when an innate immune response fails to eliminate a new infection, and activated antigen-presenting cells bearing pathogen antigens are delivered to the draining lymphoid tissues. When a recirculating lymphocyte encounters its corresponding antigen in peripheral lymphoid tissues, it is induced to proliferate, and its clonal progeny differentiate into effector T and B lymphocytes that can eliminate the infectious agent. A subset of these proliferating lymphocytes differentiates into memory cells, ready to respond rapidly to the same pathogen if it is encountered again. The details of these processes of recognition, development, and differentiation form the main material of the central three parts of this book.

The effector mechanisms of adaptive immunity.

We have seen in the first part of this chapter how naive lymphocytes are selected by antigen to differentiate into clones of activated effector lymphocytes. We now expand on the mechanisms by which activated effector lymphocytes target different pathogens for destruction in a successful adaptive immune response. The distinct lifestyles of different pathogens (Fig. 1.24) require different responses for both their recognition and their destruction. B-cells recognize native antigens from the extracellular environment and they differentiate into effector plasma cells that secrete antibody back into that environment. T-cells are specialized to detect peptides that have been generated inside the body's cells, whether from extracellular antigens that have been ingested or from proteins produced in situ, and this is reflected in the effector actions of T cells. Some effector T cells directly kill cells infected with intracellular pathogens such as viruses, while others participate in responses against extracellular pathogens by interacting with B cells to help them make antibody.

Fig. 1.24 The major types of pathogens confronting the immune system, and some of the diseases they cause.

The immune system protects against four classes of pathogens		
Type of pathogen	**Examples**	**Diseases**
Extracellular bacteria, parasites, fungi	*Streptococcus pneumoniae* *Clostridium tetani* *Trypanosoma brucei* *Pneumocystis jirovecii*	Pneumonia Tetanus Sleeping sickness *Pneumocystis* pneumonia
Intracellular bacteria, parasites	*Mycobacterium leprae* *Leishmania donovani* *Plasmodium falciparum*	Leprosy Leishmaniasis Malaria
Viruses (intracellular)	Variola Influenza Varicella	Smallpox Flu Chickenpox
Parasitic worms (extracellular)	*Ascaris* *Schistosoma*	Ascariasis Schistosomiasis

Most of the other effector mechanisms used by an adaptive immune response to dispose of pathogens are essentially identical to those of innate immunity and involve cells such as macrophages and neutrophils, and proteins such as complement. Indeed, it seems likely that the vertebrate adaptive immune response evolved by the late addition of specific recognition by clonally distributed receptors to innate defense mechanisms already existing in invertebrates. We begin by outlining the effector actions of antibodies, which depend almost entirely on recruiting cells and molecules of the innate immune system.

1-18 Antibodies protect against extracellular pathogens and their toxic products.

Antibodies are found in the fluid component of blood, or plasma, and in extracellular fluids. Because body fluids were once known as humors, immunity mediated by antibodies is known as **humoral immunity**.

As we saw in Fig. 1.14, antibodies are Y-shaped molecules whose arms form two identical antigen-binding sites. These are highly variable from one molecule to another, providing the diversity required for specific antigen recognition. The stem of the Y is far less variable. There are only five major forms of this constant region of an antibody, and these are known as the antibody **classes** or **isotypes**. The constant region determines an antibody's functional properties—how it will engage with the effector mechanisms that dispose of antigen once it is recognized—and each class carries out its particular function by engaging a distinct set of effector mechanisms. We describe the antibody classes and their actions in Chapters 5 and 10.

The first and most direct way in which antibodies can protect against pathogens or their products is by binding to them and thereby blocking their access to cells that they might infect or destroy (Fig. 1.25, left panels). This is known as **neutralization** and is important for protection against viruses, which are prevented from entering cells and replicating, and against bacterial toxins.

For bacteria, however, binding by antibodies is not sufficient to stop their replication. In this case, the function of the antibody is to enable a phagocytic cell such as a macrophage or a neutrophil to ingest and destroy the bacterium. Many bacteria evade the innate immune system because they have an outer coat that is not recognized by the pattern recognition receptors of phagocytes. However, antigens in the coat can be recognized by antibodies, and phagocytes have receptors that bind the stems of the antibodies coating the

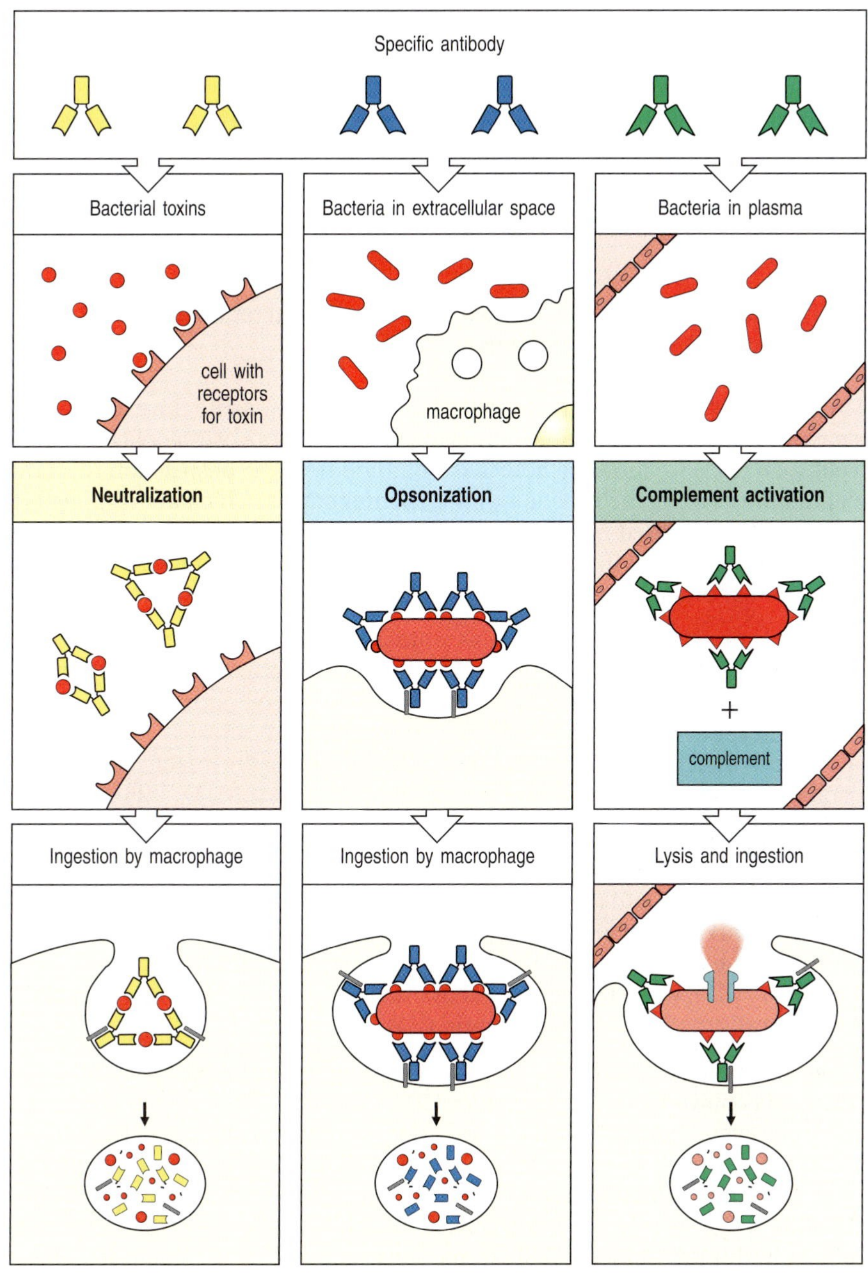

Fig. 1.25 Antibodies can participate in host defense in three main ways. The left panels show antibodies binding to and neutralizing a bacterial toxin, thus preventing it from interacting with host cells and causing pathology. Unbound toxin can react with receptors on the host cell, whereas the toxin:antibody complex cannot. Antibodies also neutralize complete virus particles and bacterial cells by binding and inactivating them. The antigen:antibody complex is eventually scavenged and degraded by macrophages. Antibodies coating an antigen render it recognizable as foreign by phagocytes (macrophages and neutrophils), which then ingest and destroy it; this is called opsonization. The center panels show opsonization and phagocytosis of a bacterial cell. The right panels show activation of the complement system by antibodies coating a bacterial cell. Bound antibodies form a receptor for the first protein of the complement system, which eventually forms a protein complex on the surface of the bacterium that, in some cases, can kill the bacterium directly. More generally, complement coating favors the taking up and destroying of the bacterium by phagocytes. Thus, antibodies target pathogens and their toxic products for disposal by phagocytes.

bacterium, leading to phagocytosis (see Fig. 1.25, center panels). The coating of pathogens and foreign particles in this way is known as **opsonization**.

The third function of antibodies is **complement activation**. Complement, which we discuss in detail in Chapter 2, is first activated in innate immunity by microbial surfaces, without the help of antibodies. But when an antibody binds to a bacterial surface, its constant regions provide a platform to activate the first protein of the complement system. Thus, once antibodies are produced, complement activation against a pathogen can be substantially increased. Complement components that are deposited on a bacterial surface can directly destroy certain bacteria, and this is important in a few bacterial infections (see Fig. 1.25, right panels). The main function of complement, however, is like that of antibodies: it coats the pathogen surface and enables phagocytes to engulf and destroy bacteria that they would not otherwise recognize. Complement also enhances the bactericidal actions of phagocytes; indeed, it is so called because it 'complements' the activities of antibodies.

Antibodies of different classes are found in different compartments of the body and differ in the effector mechanisms they recruit, but all pathogens and free molecules bound by antibody are eventually delivered to phagocytes for ingestion, degradation, and removal from the body (see Fig. 1.25, bottom panels). The complement system and the phagocytes that antibodies recruit are not themselves antigen-specific; they depend upon antibody molecules to mark the particles as foreign. Producing antibodies is the sole effector function of B cells. T cells, by contrast, have a variety of effector actions.

1-19 T cells orchestrate cell-mediated immunity and regulate B-cell responses to most antigens.

Antibodies are accessible to pathogens only in the blood and the extracellular spaces. However, some bacteria and parasites, and all viruses, replicate inside cells, where they cannot be detected by antibodies. The destruction of intracellular invaders is the function of the T lymphocytes, which are responsible for the **cell-mediated immune responses** of adaptive immunity. But T lymphocytes participate in responses to a wide variety of pathogens, including extracellular organisms, and so must exert a wide variety of effector activities.

The most direct action of T cells is cytotoxicity. Cytotoxic T cells are effector T cells that act against cells infected with viruses. Antigens derived from the virus multiplying inside the infected cell are displayed on the cell's surface, where they are recognized by the antigen receptors of cytotoxic T cells. These T cells can then control the infection by directly killing the infected cell before viral replication is complete and new viruses are released (Fig. 1.26).

From the end of their development in the thymus, T lymphocytes are composed of two main classes, one of which carries the cell-surface protein called **CD8** on its surface and the other bears a protein called **CD4**. These are not just random markers, but are important for a T cell's function, because they help to determine the interactions between the T cell and other cells. Cytotoxic T cells carry CD8, while the helper T cells involved in activating, rather than killing, the cells that they recognize carry CD4.

CD8 T cells are destined to become cytotoxic T cells by the time they leave the thymus as naive lymphocytes. Naive CD4 T cells, in contrast, can differentiate into several types of effector T cells after their initial activation by antigen. At least four main types of CD4 effector T cells are distinguished, called $\mathbf{T_H1}$, $\mathbf{T_H2}$, $\mathbf{T_H17}$, and **follicular helper T cells,** or $\mathbf{T_{FH}}$ **cells**, which are all described in Chapter 9. These effector T-cell subsets promote distinct types of responses aimed at different types of infections. For example, T_H1 cells help to control certain bacteria that take up residence in membrane-enclosed vesicles inside macrophages; the T_H1 cells activate macrophages to increase their intracellular killing power and destroy these bacteria. Important infections that are controlled, at least to some extent, by this function of T_H1 cells are tuberculosis and leprosy, which are caused by the bacteria *Mycobacterium tuberculosis* and *M. leprae*, respectively. Mycobacteria survive intracellularly because they prevent the vesicles they occupy from fusing with lysosomes, which contain a variety of degradative enzymes and antimicrobial substances (Fig. 1.27). However, the infected macrophage presents mycobacteria-derived antigens on its surface that can be recognized by activated antigen-specific T_H1 cells, which in turn secrete particular cytokines that induce the macrophage to overcome the block on vesicle fusion (see Fig. 1.27).

T_H2 cells are specialized in promoting responses at mucosal surfaces and particularly in response to parasitic infections, and again this is largely a result of the distinct set of cytokines that T_H2 cells produce. The effector mechanisms involved are shared with what are commonly known as 'allergic' responses, and are characterized by increased production of a protective barrier of

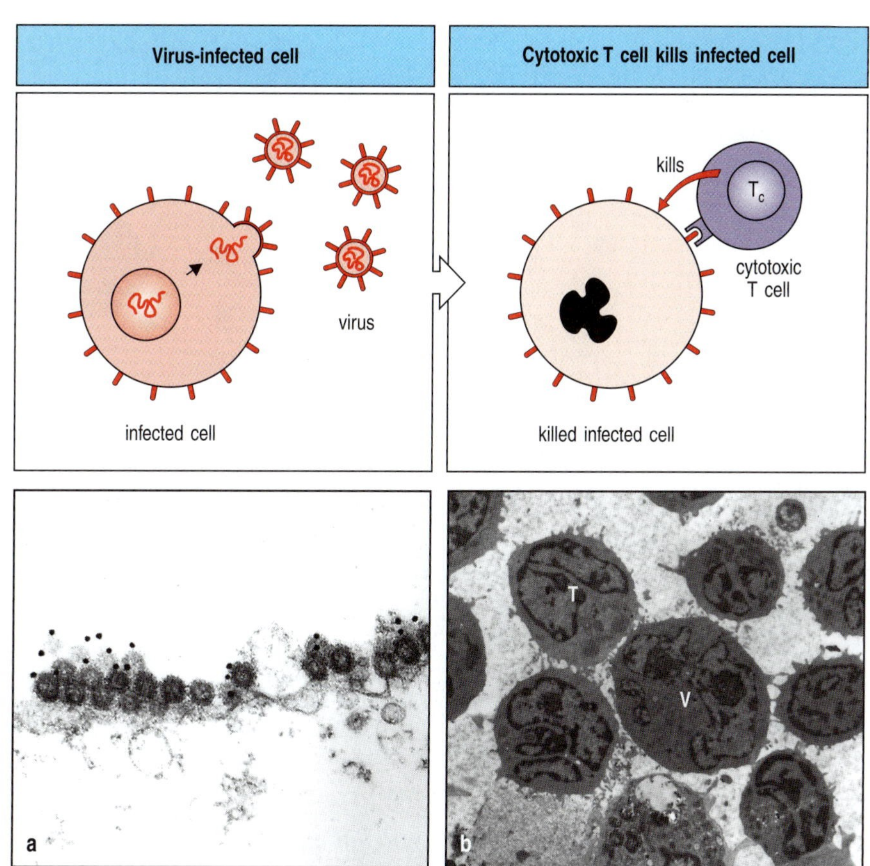

Fig. 1.26 Mechanism of host defense against intracellular infection by viruses. Cells infected by viruses are recognized by specialized T cells called cytotoxic T cells, which kill the infected cells directly. The killing mechanism involves the activation of enzymes known as caspases, which contain cysteine in their active site and cleave after aspartic acid. These in turn activate a cytosolic nuclease in the infected cell, which cleaves host and viral DNA. Panel a is a transmission electron micrograph showing the plasma membrane of a cultured CHO cell (the Chinese hamster ovary cell line) infected with influenza virus. Many virus particles can be seen budding from the cell surface. Some of these have been labeled with a monoclonal antibody that is specific for a viral protein and is coupled to gold particles, which appear as the solid black dots in the micrograph. Panel b is a transmission electron micrograph of a virus-infected cell (V) surrounded by cytotoxic T lymphocytes. Note the close apposition of the membranes of the virus-infected cell and the T cell (T) in the upper left corner of the micrograph, and the clustering of the cytoplasmic organelles in the T cell between its nucleus and the point of contact with the infected cell. Panel a courtesy of M. Bui and A. Helenius; panel b courtesy of N. Rooney.

mucus at mucosal surfaces, the recruitment of eosinophils, and the production of a particular class of antibody, IgE. Although allergies are usually considered an adverse response to an antigen, this type of immune response has an important protective role against many types of pathogens. Finally, T_H17 cells produce distinct cytokines that help to promote responses rich in the recruitment of neutrophils and which are effective in dealing with extracellular bacteria and fungi.

The term 'helper T cell' was coined well before the various CD4 T-cell subsets were known, to specifically describe T cells that 'help' B cells produce antibody. It is now used more generally in connection with functions described above, rather than being restricted to T cells that help B cells. As we shall see in Chapters 11 and 12, a recently discovered subset of CD4 T cells called T_{FH} cells, which is distinct from the T_H1, T_H2, and T_H17 cells, seems to provide much of the help to B cells; it resides in lymphoid follicles and gives unique signals to B lymphocytes that are required for many aspects of antibody production (discussed in Chapters 9 and 10).

1-20 CD4 and CD8 T cells recognize peptides bound to two different classes of MHC molecules.

The different types of effector T cells must be directed to act against the appropriate target cells. Antigen recognition is obviously crucial, but correct target recognition is also ensured by additional interactions between the CD8 and CD4 molecules on the T cells and the MHC molecules on the target cell.

As we saw in Section 1-13, T cells detect peptides derived from foreign antigens after antigens are degraded within cells, their peptide fragments are captured by MHC molecules, and this complex is displayed at the cell surface (see Fig. 1.16). There are two main types of MHC molecules, called **MHC class I**

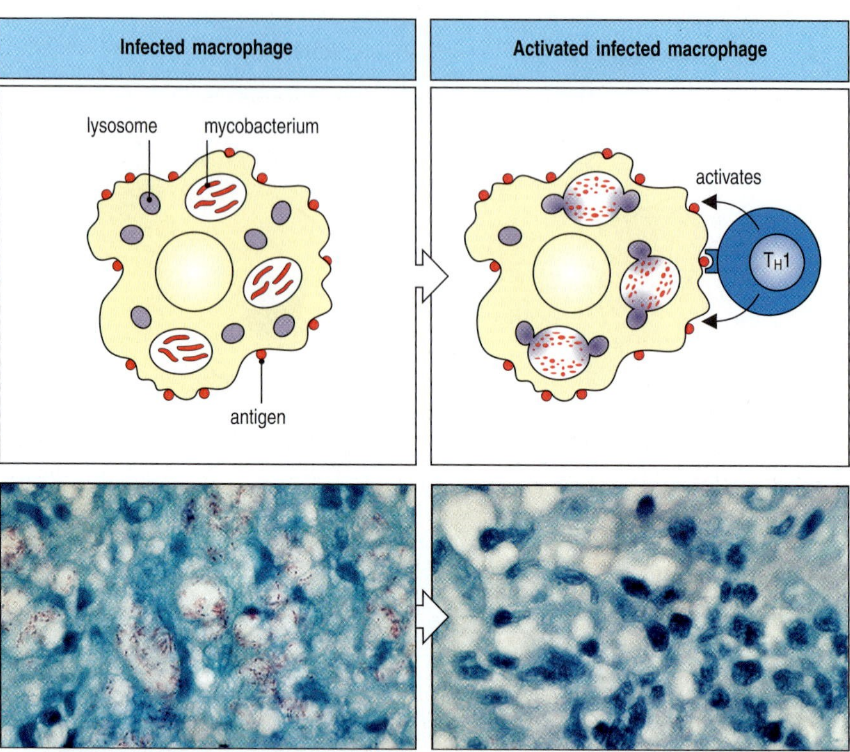

Fig. 1.27 Mechanism of host defense against intracellular infection by mycobacteria. Mycobacteria are engulfed by macrophages but resist being destroyed by preventing the intracellular vesicles in which they reside from fusing with lysosomes containing bactericidal agents. Thus the bacteria are protected from being killed. In resting macrophages, mycobacteria persist and replicate in these vesicles. When the phagocyte is recognized and activated by a T_H1 cell, however, the phagocytic vesicles fuse with lysosomes, and the bacteria can be killed. Macrophage activation is controlled by T_H1 cells, both to avoid tissue damage and to save energy. The light micrographs (bottom row) show resting (left) and activated (right) macrophages infected with mycobacteria. The cells have been stained with an acid-fast red dye to reveal mycobacteria. These are prominent as red-staining rods in the resting macrophages but have been eliminated from the activated macrophages. Photographs courtesy of G. Kaplan.

and **MHC class II**. They have slightly different structures but both have an elongated cleft in the extracellular surface of the molecule, in which a single peptide is trapped during the synthesis and assembly of the MHC molecule inside the cell. The MHC molecule bearing its cargo of peptide is transported to the cell surface, where it displays the peptide to T cells (Fig. 1.28).

There are some important functional differences between the two classes of MHC molecules. First, as noted earlier, the CD8 and CD4 molecules that distinguish different T-cell subsets are not just random markers. These two proteins have the ability to recognize parts of the MHC class I and MHC class II molecules, respectively. CD8 T cells therefore selectively recognize peptides that are bound to MHC class I molecules, and CD4 T cells recognize peptides presented by MHC class II. CD4 and CD8 are known as **co-receptors**, because they are inextricably involved in signaling to the T cell that the receptor has bound the correct antigen. There are also differences in the source of peptides that are processed and eventually bound to the two types of MHC proteins. In most cells, **MHC class I molecules** collect peptides derived from proteins synthesized in the cytosol and are thus able to display fragments of viral proteins on the cell surface (Fig. 1.29). Because MHC class I molecules are expressed on most cells of the body, they serve as an important mechanisms to defend against viral infections. MHC class I molecules bearing viral peptides are recognized by CD8-bearing cytotoxic T cells, which then kill the infected cell (Fig. 1.30). In contrast, **MHC class II molecules** are expressed predominantly by

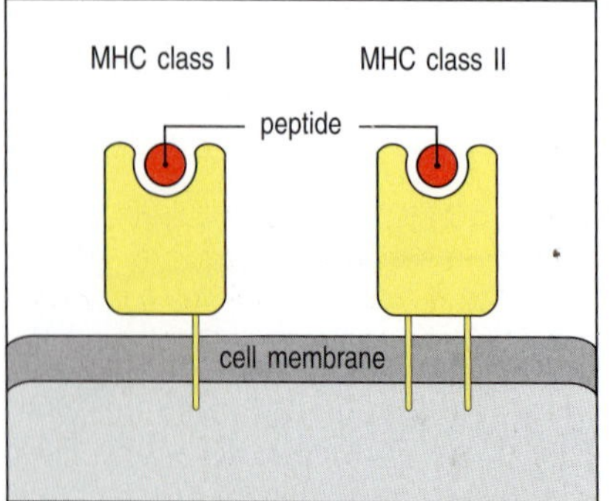

Fig. 1.28 MHC molecules on the cell surface display peptide fragments of antigens. MHC molecules are membrane proteins whose outer extracellular domains form a cleft in which a peptide fragment is bound. These fragments are derived from proteins degraded inside the cell, including both self and foreign protein antigens. The peptides are bound by the newly synthesized MHC molecule before it reaches the cell surface. There are two kinds of MHC molecules—MHC class I and MHC class II—with related but distinct structures and functions. Although not shown here for simplicity, both MHC class I and MHC class II molecules are trimers of two protein chains and the bound self or nonself peptide.

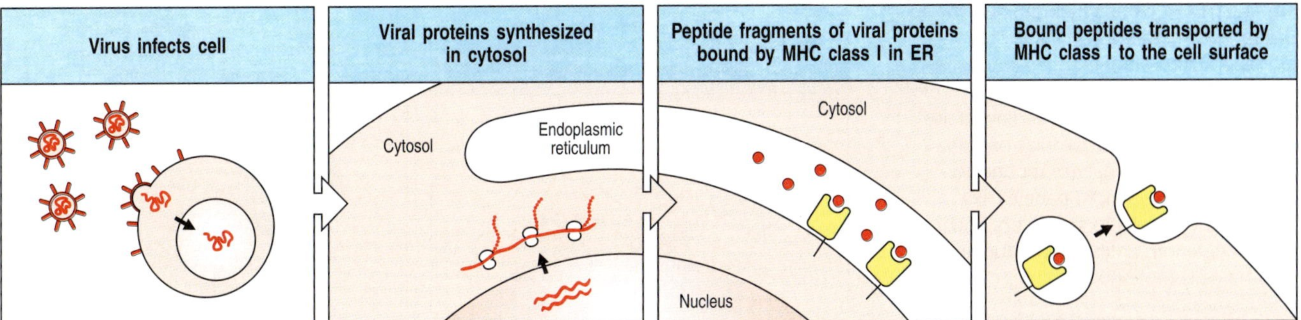
Virus infects cell
Viral proteins synthesized in cytosol
Cytosol
Endoplasmic reticulum
Peptide fragments of viral proteins bound by MHC class I in ER
Cytosol
Nucleus
Bound peptides transported by MHC class I to the cell surface

Fig. 1.29 MHC class I molecules present antigen derived from proteins in the cytosol. In cells infected with viruses, viral proteins are synthesized in the cytosol. Peptide fragments of viral proteins are transported into the endoplasmic reticulum (ER), where they are bound by MHC class I molecules, which then deliver the peptides to the cell surface.

antigen-presenting cells (dendritic cells, macrophages, and B cells) and bind peptides derived largely from proteins in intracellular vesicles. These encompass proteins taken up by phagocytosis, proteins derived from pathogens living within macrophage vesicles, or proteins internalized by B cells by endocytosis. This group of cells must either activate, or be activated by, CD4 T cells (Fig. 1.31). We shall see in Chapter 6 exactly how peptides from these different sources are made available to the two types of MHC molecules.

Because the T-cell receptor is specific for a combination of peptide and MHC molecule, any given T-cell receptor will recognize either an MHC class I molecule or an MHC class II molecule. To be useful, T lymphocytes bearing antigen receptors that recognize MHC class I must also express CD8 co-receptors, whereas T lymphocytes bearing receptors specific for MHC class II must express CD4. The matching of a T-cell receptor with a co-receptor of the appropriate type occurs during of process of selection during lymphocyte development, and naive T cells emerge from the central lymphocyte organs bearing the correct combination of receptors and co-receptors. Exactly how this selective process works and how it maximizes the usefulness of the T-cell repertoire are described in Chapter 8.

As we noted earlier, when the various types of effector T cells are stimulated by antigen, they release different sets of effector molecules that affect the target cells or recruit other effector cells in ways we discuss in Chapter 9. These effector molecules include many cytokines, which can influence clonal expansion of lymphocytes, innate immune responses, and effector actions of most immune cells. Because cytokines are central to understanding immune responses, this book has taken the approach of discussing each cytokine individually as their activities are encountered in each aspect of innate and adaptive immunity, rather than treating them arbitrarily together in a single chapter. For convenience, we have summarized their actions in Appendix III.

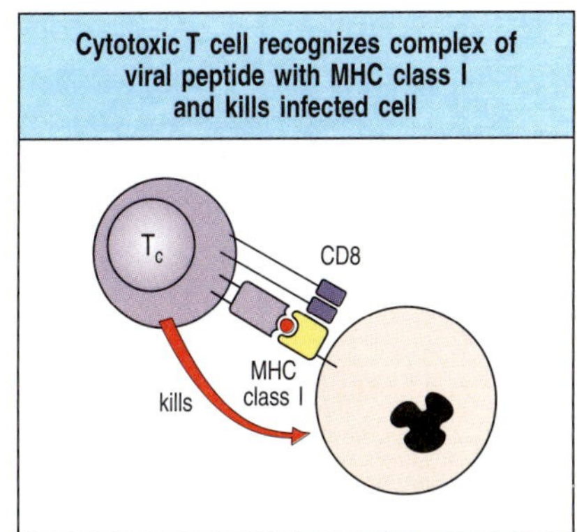
Cytotoxic T cell recognizes complex of viral peptide with MHC class I and kills infected cell
T_c
CD8
MHC class I
kills

Fig. 1.30 Cytotoxic CD8 T cells recognize antigen presented by MHC class I molecules and kill the cell. The peptide:MHC class I complex on virus-infected cells is detected by antigen-specific cytotoxic T cells. Cytotoxic T cells are preprogrammed to kill the cells they recognize.

1-21 Inherited and acquired defects in the immune system result in increased susceptibility to infection.

We tend to take for granted the ability of our immune systems to free our bodies of infection and prevent its recurrence. In some people, however, parts of the immune system fail. In the most severe of these **immunodeficiency diseases**, adaptive immunity is completely absent, and death occurs in infancy from overwhelming infection unless heroic measures are taken. Other less catastrophic failures lead to recurrent infections with particular types of pathogens, depending on the particular deficiency. Much has been learned about the functions of the different components of the human immune system through the study of these immunodeficiencies, many of which are caused by inherited genetic defects. Because understanding the features of immunodeficiencies requires a detailed knowledge of normal immune mechanisms, we have postponed discussion of most of these diseases until Chapter 13, where they can be considered together.

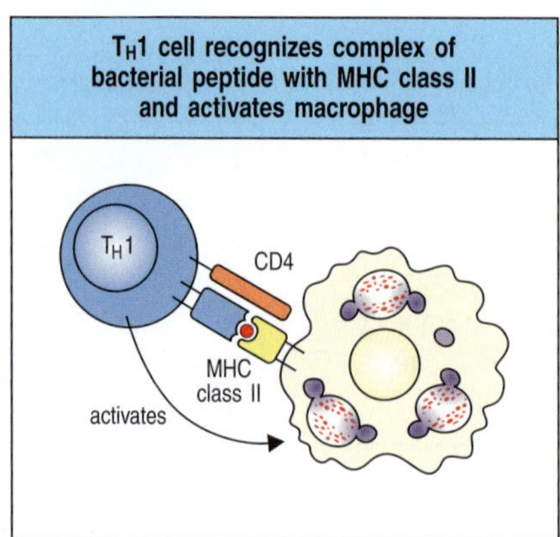
T_H1 cell recognizes complex of bacterial peptide with MHC class II and activates macrophage
T_H1
CD4
MHC class II
activates

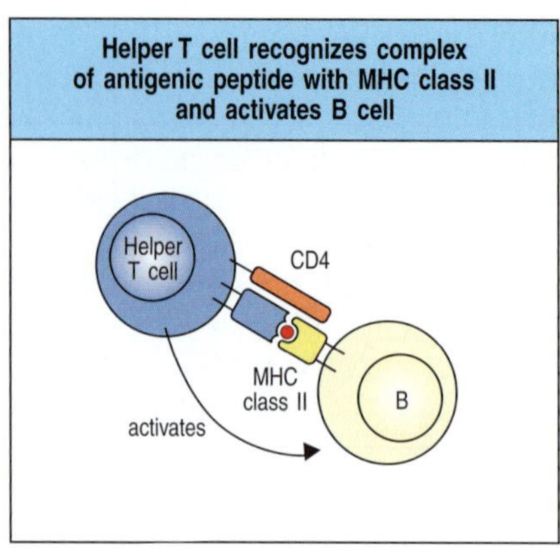
Helper T cell recognizes complex of antigenic peptide with MHC class II and activates B cell
Helper T cell
CD4
MHC class II
activates
B

Fig. 1.31 CD4 T cells recognize antigen presented by MHC class II molecules. On recognition of their specific antigen on infected macrophages, T_H1 cells activate the macrophage, leading to the destruction of the intracellular bacteria (left panel). When T_{FH} cells recognize antigen on B cells (right panel), they activate these cells to proliferate and differentiate into antibody-producing plasma cells (not shown).

More than 30 years ago, a devastating form of immunodeficiency appeared, the **acquired immune deficiency syndrome**, or **AIDS**, which is caused by an infectious agent, the human immunodeficiency viruses HIV-1 and HIV-2. This disease destroys T cells, dendritic cells, and macrophages bearing CD4, leading to infections caused by intracellular bacteria and other pathogens normally controlled by such cells. These infections are the major cause of death from this increasingly prevalent immunodeficiency disease, which is discussed fully in Chapter 13 together with the inherited immunodeficiencies.

1-22 Understanding adaptive immune responses is important for the control of allergies, autoimmune disease, and the rejection of transplanted organs.

The main function of our immune system is to protect the human host from infectious agents. However, many medically important diseases are associated with a normal immune response directed against an inappropriate antigen, often in the absence of infectious disease. Immune responses directed at noninfectious antigens occur in **allergy**, in which the antigen is an innocuous foreign substance; in **autoimmune disease**, in which the response is to a self antigen; and in **graft rejection**, in which the antigen is borne by a transplanted foreign cell (both discussed in Chapter 15). The major antigens provoking graft rejection are, in fact, the MHC molecules, as each of these is present in many different versions in the human population—that is, they are highly **polymorphic**—and most unrelated people differ in the set of MHC molecules they express, a property commonly known as their 'tissue type.' The MHC was originally recognized in mice as a gene locus, the **H2 locus**, that controlled the acceptance or rejection of transplanted tissues, whereas the human MHC molecules were first discovered after attempts to use skin grafts from donors to repair badly burned pilots and bomb victims during the Second World War. The patients rejected the grafts, which were recognized by their immune systems as being 'foreign.' What we call a successful immune response or a failure, and whether the response is considered harmful or beneficial to the host, depends not on the response itself but rather on the nature of the antigen and the circumstances in which the response occurs (Fig. 1.32).

Allergic diseases, which include asthma, are an increasingly common cause of disability in the developed world. Autoimmunity is also now recognized as the cause of many important diseases. An autoimmune response directed against pancreatic β cells is the leading cause of diabetes in the young. In allergies and autoimmune diseases, the powerful protective mechanisms of the adaptive immune response cause serious damage to the patient.

Immune responses to harmless antigens, to body tissues, or to organ grafts are, like all other immune responses, highly specific. At present, the usual way to treat these responses is with immunosuppressive drugs, which inhibit all immune responses, desirable and undesirable alike. If it were possible to suppress only those lymphocyte clones responsible for the unwanted response, the disease could be cured or the grafted organ protected without impeding protective immune responses. At present, antigen-specific immunoregulation is outside the reach of clinical treatment. But as we shall see in Chapter 16, many new drugs, particularly monoclonal antibody therapies, have been developed recently that offer more selective immune suppression to control autoimmune and other unwanted immune responses. We shall discuss the present state of understanding of allergies, autoimmune disease, graft rejection, and immunosuppressive drugs in Chapters 14–16, and we shall see in Chapter 15 how the mechanisms of immune regulation are beginning to emerge from a better understanding of the functional subsets of lymphocytes and the cytokines that control them.

1-23 Vaccination is the most effective means of controlling infectious diseases.

The deliberate stimulation of an immune response by immunization, or vaccination, has achieved many successes in the two centuries since Jenner's pioneering experiment. Mass immunization programs have led to the virtual eradication of several diseases that used to be associated with significant morbidity (illness) and mortality (Fig. 1.33). Immunization is considered so safe and so important that most states in the United States require children to be immunized against up to seven common childhood diseases. Impressive as these accomplishments are, there are still many diseases for which we lack effective vaccines. And even where vaccines for diseases such as measles can be used effectively in developed countries, technical and economic problems can prevent their widespread use in developing countries, where mortality from these diseases is still high.

The tools of modern immunology and molecular biology are being applied to develop new vaccines and improve old ones, and we discuss these advances in Chapter 16. The prospect of controlling these important diseases is tremendously exciting. The guarantee of good health is a critical step toward population control and economic development. At a cost of pennies per person, great hardship and suffering can be alleviated.

Many serious pathogens have resisted efforts to develop vaccines against them, often because they can evade or subvert the protective mechanisms of an adaptive immune response. We examine some of the evasive strategies

Antigen	Effect of response to antigen	
	Normal response	Deficient response
Infectious agent	Protective immunity	Recurrent infection
Innocuous substance	Allergy	No response
Grafted organ	Rejection	Acceptance
Self organ	Autoimmunity	Self tolerance
Tumor	Tumor immunity	Cancer

Fig. 1.32 Immune responses can be beneficial or harmful, depending on the nature of the antigen. Beneficial responses are shown in white, harmful responses in red shaded boxes. Where the response is beneficial, its absence is harmful.

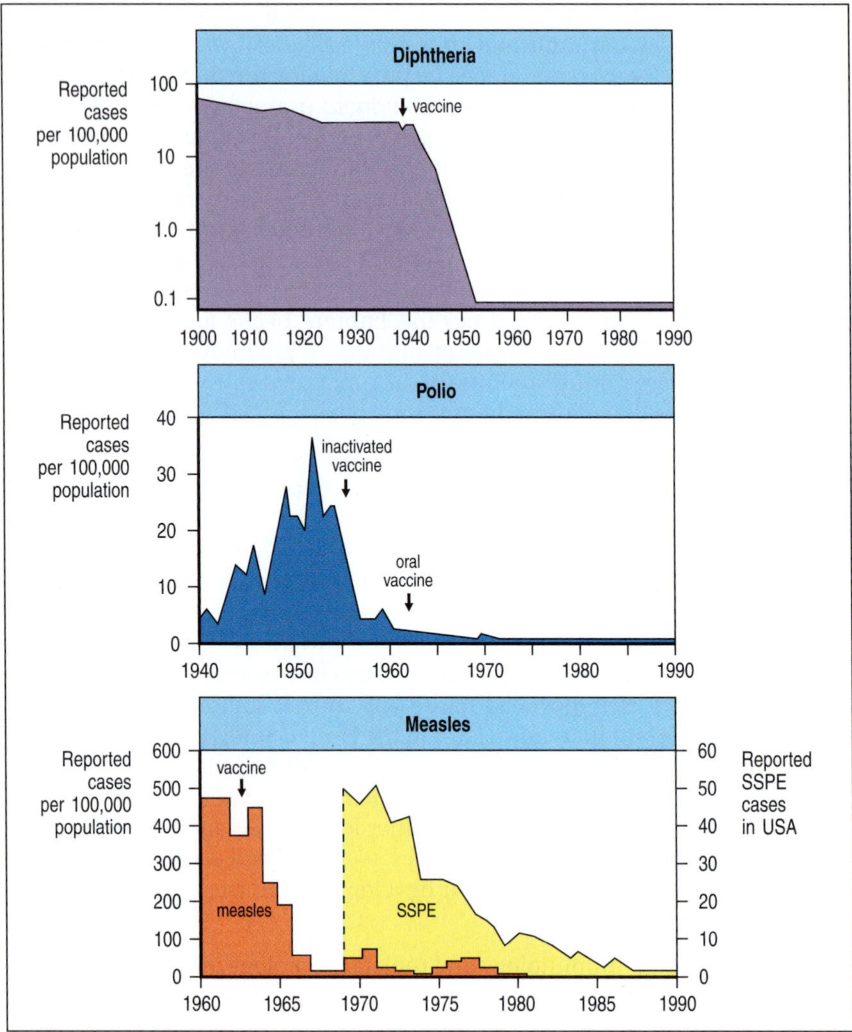

Fig. 1.33 Successful vaccination campaigns. Diphtheria, polio, and measles and their consequences have been virtually eliminated in the United States, as shown in these three graphs. SSPE stands for subacute sclerosing panencephalitis, a brain disease that is a late consequence of measles infection in a few patients. When measles was prevented, SSPE disappeared 15–20 years later. However, because these diseases have not been eradicated worldwide, immunization must be maintained in a very high percentage of the population to prevent their reappearance.

used by successful pathogens in Chapter 13. The conquest of many of the world's leading diseases, including malaria and diarrheal diseases (the leading killers of children) as well as the more recent threat from AIDS, depends on a better understanding of the pathogens that cause them and their interactions with the cells of the immune system.

Summary.

Lymphocytes have two distinct recognition systems specialized for the detection of extracellular and intracellular pathogens. B cells have cell-surface immunoglobulin molecules as receptors for antigen and, upon activation, secrete the immunoglobulin as soluble antibody that provides a defense against pathogens in the extracellular spaces of the body. T cells have receptors that recognize peptide fragments of intracellular pathogens transported to the cell surface by the glycoproteins of the MHC. Two classes of MHC molecules transport peptides from different intracellular compartments to the cell surface to present them to distinct types of effector T cells: cytotoxic CD8 T cells that kill infected target cells, and various types of differentiated CD4 T cells that can activate macrophages, promote allergic responses, mobilize neutrophils, or offer help to B cell for antibody production. Thus, T cells are crucially important for both the humoral and cell-mediated responses of adaptive immunity. The adaptive immune response seems to have grafted specific antigen recognition by highly diversified receptors onto innate defense systems, which have a central role in the effector actions of both B and T lymphocytes. The vital role of adaptive immunity in fighting infection is illustrated by the immunodeficiency diseases and the problems caused by pathogens that succeed in evading or subverting an adaptive immune response. The antigen-specific suppression of adaptive immune responses is the goal of treatment for important human diseases involving inappropriate activation of lymphocytes, whereas the specific stimulation of an adaptive immune response is the basis of successful vaccination.

Summary to Chapter 1.

The immune system defends the host against infection. Innate immunity serves as a first line of defense but lacks the ability to recognize certain pathogens and to provide the specific protective immunity that prevents reinfection. Adaptive immunity is based on clonal selection from a repertoire of lymphocytes bearing highly diverse antigen-specific receptors that enable the immune system to recognize any foreign antigen. In the adaptive immune response, antigen-specific lymphocytes proliferate and differentiate into clones of effector lymphocytes that eliminate the pathogen. Figure 1.34 summarizes the phases of the immune response and their approximate timings. Host defense requires different recognition systems and a wide variety of effector mechanisms to seek out and destroy the wide variety of pathogens in their various habitats within the body and at its external and internal surfaces.

Phases of the immune response

Response		Typical time after infection to start of response	Duration of response
Innate immune response	Inflammation, complement activation, phagocytosis and destruction of pathogen	Minutes	Days
Adaptive immune response	Interaction between antigen-presenting dendritic cells and antigen-specific T cells: recognition of antigen, adhesion, co-stimulation, T-cell proliferation and differentiation	Hours	Days
	Activation of antigen-specific B cells	Hours	Days
	Formation of effector and memory T cells	Days	Weeks
	Interaction of T cells with B cells, formation of germinal centers. Formation of effector B cells (plasma cells) and memory B cells. Production of antibody	Days	Weeks
	Emigration of effector lymphocytes from peripheral lymphoid organs	A few days	Weeks
	Effector cells and antibodies eliminate the pathogen	A few days	Weeks
Immunological memory	Maintenance of memory B cells and T cells and high serum or mucosal antibody levels. Protection against reinfection	Days to weeks	Can be lifelong

Fig. 1.34 Phases of the immune response.

Not only can the adaptive immune response eliminate a pathogen but, in the process, it also generates increased numbers of differentiated memory lymphocytes through clonal selection, and this allows a more rapid and effective response upon reinfection. The regulation of immune responses, whether to suppress them when unwanted or to stimulate them in the prevention of infectious disease, is the major medical goal of research in immunology.

General references.

Historical background

Burnet, F.M.: *The Clonal Selection Theory of Acquired Immunity.* London, Cambridge University Press, 1959.

Gowans, J.L.: **The lymphocyte—a disgraceful gap in medical knowledge.** *Immunol. Today* 1996, **17**:288–291.

Landsteiner, K.: *The Specificity of Serological Reactions*, 3rd ed. Boston, Harvard University Press, 1964.

Metchnikoff, E.: *Immunity in the Infectious Diseases*, 1st ed. New York, Macmillan Press, 1905.

Silverstein, A.M.: *History of Immunology*, 1st ed. London, Academic Press, 1989.

Biological background

Alberts, B., Johnson, A., Lewis, J., Raff, M., Roberts, K. and Walter, P.: *Molecular Biology of the Cell*, 5th ed. New York, Garland Publishing, 2007.

Berg, J.M., Stryer, L. and Tymoczko, J.L.: *Biochemistry*, 5th ed. New York, W.H. Freeman, 2002.

Geha, R.S. and Notarangelo, L.D.: *Case Studies in Immunology: A Clinical Companion*, 6th ed. New York, Garland Science, 2012.

Harper, D.R.: *Viruses: Biology, Applications, Control.* New York, Garland Science, 2012.

Kaufmann, S.E., Sher, A. and Ahmed, R. (Eds): *Immunology of Infectious Diseases.* Washington, DC: ASM Press, 2001.

Lodish, H., Berk, A., Kaiser, C.A., Krieger, M., Scott, M.P., Bretscher, A., Ploegh, H., Matsudaira, P.: *Molecular Cell Biology*, 6th ed. New York, W.H. Freeman, 2008.

Lydyard, P., Cole, M., Holton, J., Irving, W., Porakishvili, N., Venkatesan, P., and Ward, K.: *Case Studies in Infectious Disease.* New York, Garland Science, 2009.

Mims, C., Nash, A. and Stephen, J.: *Mims' Pathogenesis of Infectious Disease*, 5th ed. London, Academic Press, 2001.

Ryan, K.J. (ed): *Medical Microbiology*, 3rd ed. East Norwalk, CT, Appleton-Lange, 1994.

Primary journals devoted solely or primarily to immunology

Autoimmunity
Clinical and Experimental Immunology
Comparative and Developmental Immunology
European Journal of Immunology
Immunity
Immunogenetics
Immunology
Infection and Immunity
International Immunology
International Journal of Immunogenetics
Journal of Autoimmunity
Journal of Experimental Medicine
Journal of Immunology
Nature Immunology
Regional Immunology
Thymus

Primary journals with frequent papers in immunology

Cell
Current Biology
EMBO Journal
Journal of Biological Chemistry
Journal of Cell Biology
Journal of Clinical Investigation
Molecular Cell Biology
Nature
Nature Cell Biology
Nature Medicine
Proceedings of the National Academy of Sciences, USA
Science

Review journals in immunology

Advances in Immunology
Annual Reviews in Immunology
Contemporary Topics in Microbiology and Immunology
Current Opinion in Immunology
Immunogenetics Reviews
Immunological Reviews
The Immunologist
Immunology Today
Nature Reviews Immunology
Research in Immunology
Seminars in Immunology

Advanced textbooks in immunology, compendia, etc.

Lachmann, P.J., Peters, D.K., Rosen, F.S., and Walport, M.J. (eds): *Clinical Aspects of Immunology*, 5th ed. Oxford, Blackwell Scientific Publications, 1993.

Mak, T.W., and Saunders, M.E.: *The Immune Response. Basic and Clinical Principles.* Elsevier/Academic Press, 2006.

Mak, T.W., and Simard, J.J.L.: *Handbook of Immune Response Genes.* New York, Plenum Press, 1998.

Paul, W.E. (ed): *Fundamental Immunology*, 6th ed. New York, Lippincott Williams & Wilkins, 2008.

Roitt, I.M., and Delves, P.J. (eds): *Encyclopedia of Immunology*, 2nd ed. (4 vols). London/San Diego, Academic Press, 1998.

Innate Immunity: The First Lines of Defense

2

Microorganisms that are encountered daily in the life of a healthy individual cause disease only occasionally. Most are detected and destroyed within minutes or hours by defense mechanisms that do not rely on the clonal expansion of antigen-specific lymphocytes (see Section 1-9). These are the mechanisms of **innate immunity**. To recognize pathogens, both the innate and adaptive immune systems can distinguish between self and nonself, but they differ in how they do this. Innate immunity relies on a limited number of receptors and secreted proteins that are encoded in the germline and that recognize features common to many pathogens (see Section 1-8). In contrast, adaptive immunity uses a process of somatic cell gene rearrangement to generate an enormous repertoire of antigen receptors that are capable of fine distinctions between closely related molecules (Section 1-12). Nonetheless, the innate immune system discriminates very effectively between host cells and pathogens, providing initial defenses and also contributing to the induction of adaptive immune responses. The importance of innate immunity is shown by the fact that defects in its components, which are very rare, can lead to increased susceptibility to infection, even in the presence of an intact adaptive immune system.

The response to an encounter with a new pathogen occurs in three phases, summarized in Fig. 2.1. When a pathogen succeeds in breaching one of the host's anatomic barriers, some innate immune mechanisms start acting immediately. These first defenses include several classes of preformed soluble molecules present in blood, extracellular fluid, and epithelial secretions that can either kill the pathogen or weaken its effect. **Antimicrobial enzymes** such as lysozyme begin to digest bacterial cell walls; **antimicrobial peptides** such as the defensins lyse bacterial cell membranes directly; and a system of plasma proteins known as the **complement system** targets pathogens both for lysis and for phagocytosis by cells of the innate immune system such as macrophages. In the second phase of the response, these innate immune cells sense the presence of a pathogen by recognizing molecules typical of a microbe and not shared by host cells—pathogen-associated molecular patterns (PAMPs)—and become activated, setting in train several different effector mechanisms to eliminate the infection. By themselves, neither the soluble nor the cellular components of innate immunity generate long-term protective immunological memory. Only if an infectious organism breaches these first two lines of defense will mechanisms be engaged to induce an adaptive immune response—the third phase of the response to a pathogen. This leads

Fig. 2.1 The response to an initial infection occurs in three phases. These are the innate phase, the early induced innate response, and the adaptive immune response. The first two phases rely on the recognition of pathogens by germline-encoded receptors of the innate immune system, whereas adaptive immunity uses variable antigen-specific receptors that are produced as a result of gene segment rearrangements. Adaptive immunity occurs late, because the rare B cells and T cells specific for the invading pathogen must first undergo clonal expansion before they differentiate into effector cells that migrate to the site of infection and clear the infection. The effector mechanisms that remove the infectious agent are similar or identical in each phase.

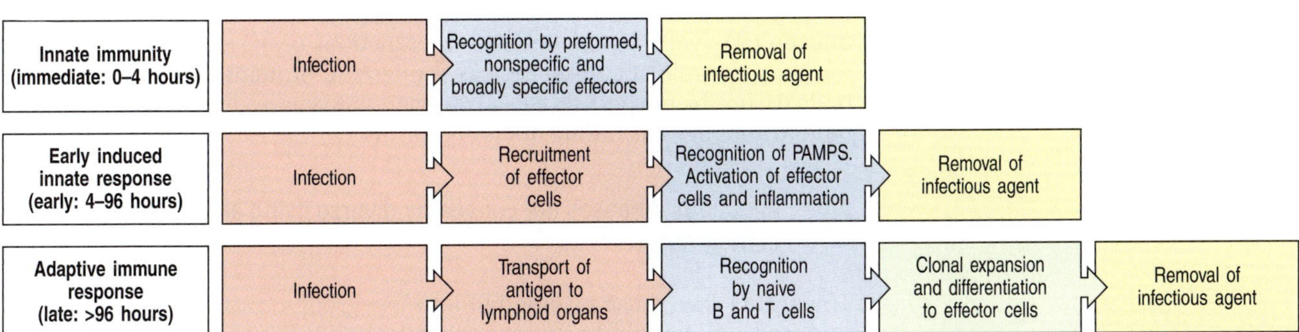

to the expansion of antigen-specific lymphocytes that target the pathogen specifically and to the formation of memory cells that provide long-lasting specific immunity.

In this chapter we consider the first phase of an immune response—the **innate immune response**. In the first part of the chapter we describe the anatomic barriers that protect the host against infection and examine the immediate innate defenses provided by various secreted soluble proteins. The anatomic barriers are fixed defenses against infection comprising the epithelia that line the internal and external surfaces of the body and the phagocytes that lie beneath all epithelial surfaces and that engulf and digest invading microorganisms. Epithelia are also protected by many kinds of chemical defenses, including antimicrobial enzymes and peptides. The phagocytes residing beneath the epithelia act both in the direct killing of microorganisms and in the next phase of the innate immune response: the induction of an inflammatory response that recruits new phagocytic cells and circulating effector molecules to the site of infection. The last part of the chapter is devoted to the complement system. This important element of innate immunity interacts with microorganisms to promote their removal by phagocytic cells. The complement system, and other soluble circulating defensive proteins, are sometimes referred to as **humoral** innate immunity, from the old word 'humor' for body fluids. In Chapter 3 we will examine the cellular components of innate immunity and the system of pattern recognition receptors that regulate cellular responses when they are triggered by pathogens.

The first lines of defense.

Microorganisms that cause disease in humans and animals enter the body at different sites and produce disease symptoms by a variety of mechanisms. Many different infectious agents can cause disease and damage to tissues, or pathology, and are referred to as **pathogenic microorganisms** or **pathogens**. In vertebrates, microbial invasion is initially countered by innate defenses that preexist in all individuals and begin to act within minutes of encounter with the infectious agent. Only when the innate host defenses have been bypassed, evaded, or overwhelmed is an adaptive immune response required. Innate immunity is sufficient to prevent the body from being routinely overwhelmed by the vast number of microorganisms that live on and in it. Pathogens are microorganisms that have evolved ways of overcoming the body's innate defenses more effectively than other microorganisms. Once they have gained a hold, they require the concerted efforts of both innate and adaptive immune responses to clear them from the body. Even in these cases, the innate immune system performs a valuable delaying function, keeping pathogen numbers in check while the adaptive immune system gears up for action. In the first part of this chapter we will describe briefly the different types of pathogens and their invasive strategies, and then examine the innate defenses that, in most cases, prevent microorganisms from establishing an infection. We will examine the defense functions of the epithelial surfaces of the body, the role of antimicrobial peptides and proteins, and the defense of body tissues by phagocytic cells—the macrophages and neutrophils—which bind to and ingest invading microorganisms.

2-1 Infectious diseases are caused by diverse living agents that replicate in their hosts.

The agents that cause disease fall into five groups: viruses, bacteria, fungi, protozoa, and helminths (worms). Protozoa and worms are usually grouped together as parasites, and are the subject of the discipline of parasitology, whereas viruses, bacteria, and fungi are the subject of microbiology. In Fig.2.2, some representatives of the different classes of microorganisms and parasites, and the diseases they cause, are listed. The characteristic features

Fig. 2.2 A variety of microorganisms can cause disease. Pathogenic organisms are of five main types: viruses, bacteria, fungi, protozoa, and worms. Some well-known pathogens are listed.

Routes of infection for pathogens				
Route of entry	**Mode of transmission**	**Pathogen**	**Disease**	**Type of pathogen**
Mucosal surfaces				
Mouth and respiratory tract	Inhalation or ingestion of infective material (e.g. saliva droplets)	Measles virus	Measles	Paramyxovirus
		Influenza virus	Influenza	Orthomyxovirus
		Varicella-zoster	Chickenpox	Herpesvirus
		Epstein-Barr virus	Mononucleosis	Herpesvirus
		Streptococcus pyogenes	Tonsilitis	Gram-positive bacterium
		Haemophilus influenzae	Pneumonia, meningitis	Gram-negative bacterium
	Spores	*Neisseria meningitidis*	Meningococcal meningitis	Gram-negative bacterium
		Bacillus anthracis	Inhalation anthrax	Gram-positive bacterium
Gastrointestinal tract	Contaminated water or food	Rotavirus	Diarrhea	Rotavirus
		Hepatitis A	Jaundice	Picornavirus
		Salmonella enteritidis, S. typhimurium	Food poisoning	Gram-negative bacterium
		Vibrio cholerae	Cholera	Gram-negative bacterium
		Salmonella typhi	Typhoid fever	Gram-negative bacterium
Reproductive tract and other routes	Sexual transmission/ infected blood	Hepatitis B virus	Hepatitis B	Hepadnavirus
		Human immunodeficiency virus (HIV)	Acquired immunodeficiency syndrome (AIDS)	Retrovirus
	Sexual transmission	*Neisseria gonorrhoeae*	Gonorrhea	Gram-negative bacterium
		Treponema pallidum	Syphilis	Bacterium (spirochete)
Opportunistic pathogens				
	Resident microbiota	*Candida albicans*	Candidiasis, thrush	Fungus
	Resident lung microbiota	*Pneumocystis jirovecii*	Pneumonia	Fungus
External epithelia				
External surface	Physical contact	*Trichophyton*	Athlete's foot	Fungus
Wounds and abrasions	Minor skin abrasions	*Bacillus anthracis*	Cutaneous anthrax	Gram-positive bacterium
	Puncture wounds	*Clostridium tetani*	Tetanus	Gram-positive bacterium
	Handling infected animals	*Francisella tularensis*	Tularemia	Gram-negative bacterium
Insect bites	Mosquito bites (*Aedes aegypti*)	Flavivirus	Yellow fever	Virus
	Deer tick bites	*Borrelia burgdorferi*	Lyme disease	Bacterium (spirochete)
	Mosquito bites (*Anopheles*)	*Plasmodium* spp.	Malaria	Protozoan

of each pathogen are its mode of transmission, its mechanism of replication, its mechanism of **pathogenesis**—the means by which it causes disease—and the response it elicits from the host. The distinct habitats and life cycles of different pathogens mean that a range of different innate and adaptive immune mechanisms have to be deployed for their destruction.

Infectious agents can grow in all body compartments, as shown schematically in Fig. 2.3. We saw in Chapter 1 that two major compartments can be defined—extracellular and intracellular. Both innate and adaptive immune responses have different ways of dealing with pathogens found in these two compartments. Many bacterial pathogens live and replicate in extracellular spaces, either within tissues or on the surface of the epithelia that line body cavities. Extracellular bacteria are usually susceptible to killing by phagocytes, an important arm of the innate immune system, but some pathogens, such as *Staphylococcus* and *Streptococcus* species, are protected by a polysaccharide capsule that resists engulfment. This can be overcome to some extent by the help of another component of innate immunity—complement—which renders the bacteria more susceptible to phagocytosis. In the adaptive immune response, bacteria are rendered more susceptible to phagocytosis by a combination of antibodies and complement.

Infectious diseases differ in their symptoms and outcome depending on where the causal pathogen replicates within the body and what damage it does to the tissues (Fig. 2.4). Pathogens can live in either the intracellular or the extracellular compartment. Pathogens that live intracellularly frequently cause disease by damaging or killing the cells they infect. Obligate intracellular pathogens, such as viruses, must invade host cells to replicate. Facultative intracellular pathogens, such as mycobacteria, can replicate either intracellularly or outside the cell. Two strategies of innate immunity defend against intracellular pathogens. One is to destroy pathogens before they infect cells. To this end, innate immunity includes soluble defenses such as antimicrobial peptides, as well as phagocytic cells that can engulf and destroy pathogens before they become intracellular. Alternatively, the innate immune system can recognize and kill cells infected by some pathogens. This is the role of the natural killer cells (NK cells), which are instrumental in keeping certain viral infections in check until the cytotoxic T cells of the adaptive immune

	Extracellular		Intracellular	
	Interstitial spaces, blood, lymph	Epithelial surfaces	Cytoplasmic	Vesicular
Site of infection	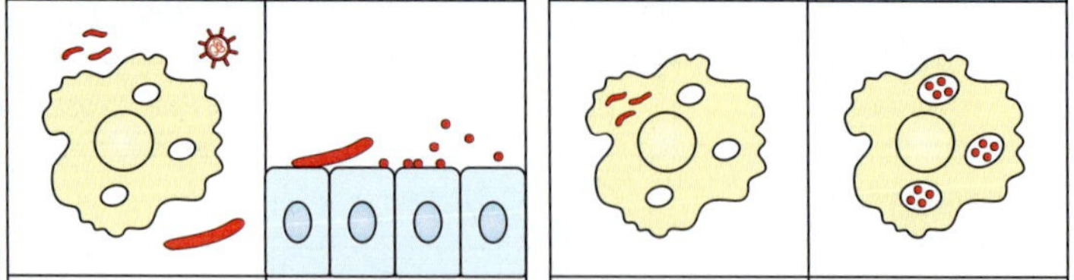			
Organisms	Viruses Bacteria Protozoa Fungi Worms	*Neisseria gonorrhoeae* *Streptococcus pneumoniae* *Vibrio cholerae* *Helicobacter pylori* *Candida albicans* Worms	Viruses *Chlamydia* spp. *Rickettsia* spp. Protozoa	*Mycobacterium* spp. *Yersinia pestis* *Legionella pneumophila* *Cryptococcus neoformans* *Leishmania* spp.
Protective immunity	Complement Phagocytosis Antibodies	Antimicrobial peptides Antibodies, especially IgA	NK cells Cytotoxic T cells	T-cell and NK-cell dependent macrophage activation

Fig. 2.3 Pathogens can be found in various compartments of the body, where they must be combated by different host defense mechanisms. Virtually all pathogens have an extracellular phase in which they are vulnerable to the circulating molecules and cells of innate immunity and to the antibodies of the adaptive immune response. All these clear the microorganism mainly by promoting its uptake and destruction by the phagocytes of the immune system. Intracellular phases of pathogens such as viruses are not accessible to these mechanisms; instead, the infected cell is attacked by the NK cells of innate immunity or by the cytotoxic T cells of adaptive immunity. Activation of macrophages as a result of NK-cell or T-cell activity can induce the macrophage to kill pathogens that are living inside macrophage vesicles.

	Direct mechanisms of tissue damage by pathogens			Indirect mechanisms of tissue damage by pathogens		
	Exotoxin production	Endotoxin	Direct cytopathic effect	Immune complexes	Anti-host antibody	Cell-mediated immunity
Pathogenic mechanism	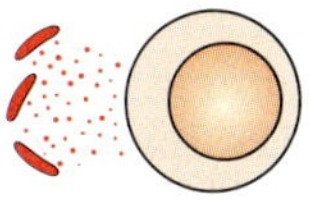	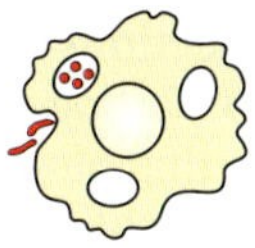	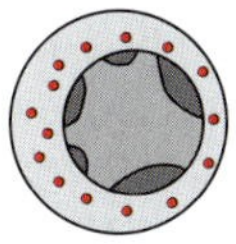	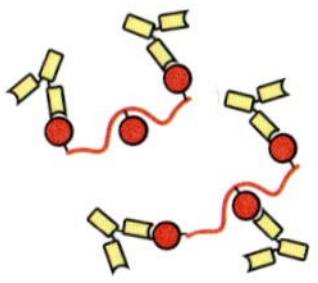	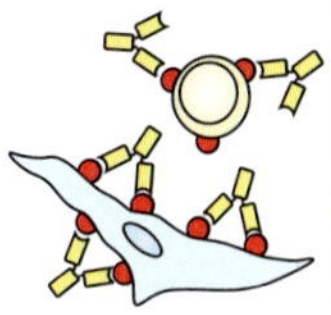	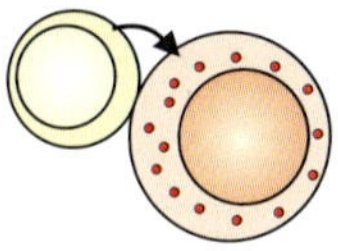
Infectious agent	*Streptococcus pyogenes* *Staphylococcus aureus* *Corynebacterium diphtheriae* *Clostridium tetani* *Vibrio cholerae*	*Escherichia coli* *Haemophilus influenzae* *Salmonella typhi* *Shigella* *Pseudomonas aeruginosa* *Yersinia pestis*	Variola Varicella-zoster Hepatitis B virus Polio virus Measles virus Influenza virus Herpes simplex virus Human herpes virus 8 (HHV8)	Hepatitis B virus Malaria *Streptococcus pyogenes* *Treponema pallidum* Most acute infections	*Streptococcus pyogenes* *Mycoplasma pneumoniae*	Lymphocytic choriomeningitis virus Herpes simplex virus *Mycobacterium tuberculosis* *Mycobacterium leprae* *Borrelia burgdorferi* *Schistosoma mansoni*
Disease	Tonsilitis, scarlet fever Boils, toxic shock syndrome, food poisoning Diphtheria Tetanus Cholera	Gram-negative sepsis Meningitis, pneumonia Typhoid fever Bacillary dysentery Wound infection Plague	Smallpox Chickenpox, shingles Hepatitis Poliomyelitis Measles, subacute sclerosing panencephalitis Influenza Cold sores Kaposi's sarcoma	Kidney disease Vascular deposits Glomerulonephritis Kidney damage in secondary syphilis Transient renal deposits	Rheumatic fever Hemolytic anemia	Aseptic meningitis Herpes stromal keratitis Tuberculosis Tuberculoid leprosy Lyme arthritis Schistosomiasis

Fig. 2.4 Pathogens can damage tissues in a variety of different ways. The mechanisms of damage, representative infectious agents, and the common names of the diseases associated with each are shown. Exotoxins are released by microorganisms and act at the surface of host cells, for example by binding to receptors. Endotoxins, which are intrinsic components of microbial structure, trigger phagocytes to release cytokines that produce local or systemic symptoms. Many pathogens are cytopathic, directly damaging the cells they infect. Finally, an adaptive immune response to the pathogen can generate antigen:antibody complexes that activate neutrophils and macrophages, antibodies that can cross-react with host tissues, or T cells that kill infected cells. All of these have some potential to damage the host's tissues. In addition, neutrophils, the most abundant cells early in infection, release many proteins and small-molecule inflammatory mediators that both control infection and cause tissue damage.

system can take over. Intracellular pathogens can be subdivided further into those that replicate free in the cell, such as viruses and certain bacteria (for example, *Chlamydia*, *Rickettsia*, and *Listeria*), and those that replicate inside intracellular vesicles, such as mycobacteria. Pathogens that live inside macrophage vesicles may become more susceptible to being killed after activation of the macrophage as a result of NK-cell or T-cell actions (see Fig. 2.3).

Many of the most dangerous extracellular bacterial pathogens cause disease by releasing protein toxins—these secreted toxins are called **exotoxins.** The innate immune system has little defense against such toxins, and highly specific antibodies produced by the adaptive immune system are required to neutralize their action (see Fig. 1.26). The damage caused by a particular infectious agent also depends on the site in which it grows; *Streptococcus pneumoniae* in the lung causes pneumonia, for example, whereas in the blood it causes a potentially fatal systemic illness, pneumococcal sepsis. In contrast, non-secreted constituents of bacterial structure that trigger phagocytes to release cytokines with local and systemic efects are called **endotoxins**. An endotoxin of major medical importance is the lipopolysaccharide of the outer cell membrane of Gram-negative bacteria.

Most pathogenic microorganisms can overcome innate immune responses and continue to grow, making us ill. An adaptive immune response is required to eliminate them and to prevent subsequent reinfection. Certain pathogens are never entirely eliminated by the immune system, and persist in the body for years. But most pathogens are not universally lethal. Those that have lived for thousands of years in the human population are highly evolved to exploit their human hosts; they cannot alter their pathogenicity without upsetting the compromise they have achieved with the human immune system. Rapidly killing every host it infects is no better for the long-term survival of a pathogen than being wiped out by the immune response before the microbe has had time to infect someone else. In short, we have adapted to live with our enemies, and they with us. Nevertheless, the recent concern about highly pathogenic strains of avian influenza, and the episode in 2002–2003 of SARS (severe acute respiratory syndrome), caused by a coronavirus from bats that caused severe pneumonia in humans, remind us that new and deadly infections can transfer to humans from animal reservoirs. These are known as **zoonotic** infections—and we must be on the alert at all times for the emergence of new pathogens and new threats to health. The human immunodeficiency virus that causes AIDS (discussed in Chapter 13) serves as a warning that we remain constantly vulnerable.

2-2 Infectious agents must overcome innate host defenses to establish a focus of infection.

Our bodies are constantly exposed to microorganisms present in our environment, including infectious agents that have been shed from other infected individuals. Contact with these microorganisms may occur through external or internal epithelial surfaces. For a microorganism to invade the body, it must first bind to or cross an epithelium. The epithelium lining the respiratory tract provides a route of entry into tissues for airborne microorganisms, and the lining of the gastrointestinal tract does the same for microorganisms ingested in food and water. The intestinal pathogens *Salmonella typhi*, which causes typhoid fever, and *Vibrio cholerae*, which causes cholera, are spread through fecally contaminated food and water, respectively. Insect bites and wounds allow microorganisms to penetrate the skin, and direct contact between individuals offers opportunities for infection through the skin, the gut, and the reproductive tract (see Fig. 2.2). The microorganisms normally inhabiting the healthy intestine—the **commensal microbiota**—provide a first defense against microorganisms invading through the gut by competing with pathogens for nutrients and epithelial attachment sites. Commensal microorganisms also provoke responses that help strengthen epithelial barrier function. Immune responses that occur in the specialized mucosal immune system, such as that of the gut, will be described in greater detail in Chapter 12.

In spite of this exposure, infectious disease is fortunately quite infrequent. The epithelial surfaces of the body serve as an effective barrier against most microorganisms, and they are rapidly repaired if wounded. Furthermore, most of the microorganisms that do succeed in crossing an epithelial surface are efficiently removed by innate immune mechanisms that function in the underlying tissues. Thus, in most cases these defenses prevent infection from becoming established. It is difficult to know how many infections are repelled in this way, because they cause no symptoms and pass undetected. It is clear, however, that the microorganisms that a normal human being inhales or ingests, or that enter through minor wounds, are mostly held at bay or eliminated, because they seldom cause clinical disease.

Disease occurs when a microorganism succeeds in evading or overwhelming innate host defenses to establish a local site of infection, and then replicates there to allow its further transmission within our bodies. In general, pathogenic microorganisms are distinguished from the mass of microorganisms in the environment by having special adaptations that help them do this (some of the ways in which pathogens evade the immune system are described in Chapter 13). In some cases, such as the fungal disease athlete's foot, the initial infection remains local and does not cause significant pathology. In other cases, the infectious agent causes significant damage and serious illness as it spreads through the lymphatics or the bloodstream, invades and destroys tissues, or disrupts the body's workings with its toxins, as in the case of the tetanus bacterium (*Clostridium tetani*), which secretes a powerful neurotoxin.

In the second, induced, phase of the innate immune response, the spread of a pathogen is often initially countered by an **inflammatory response** that

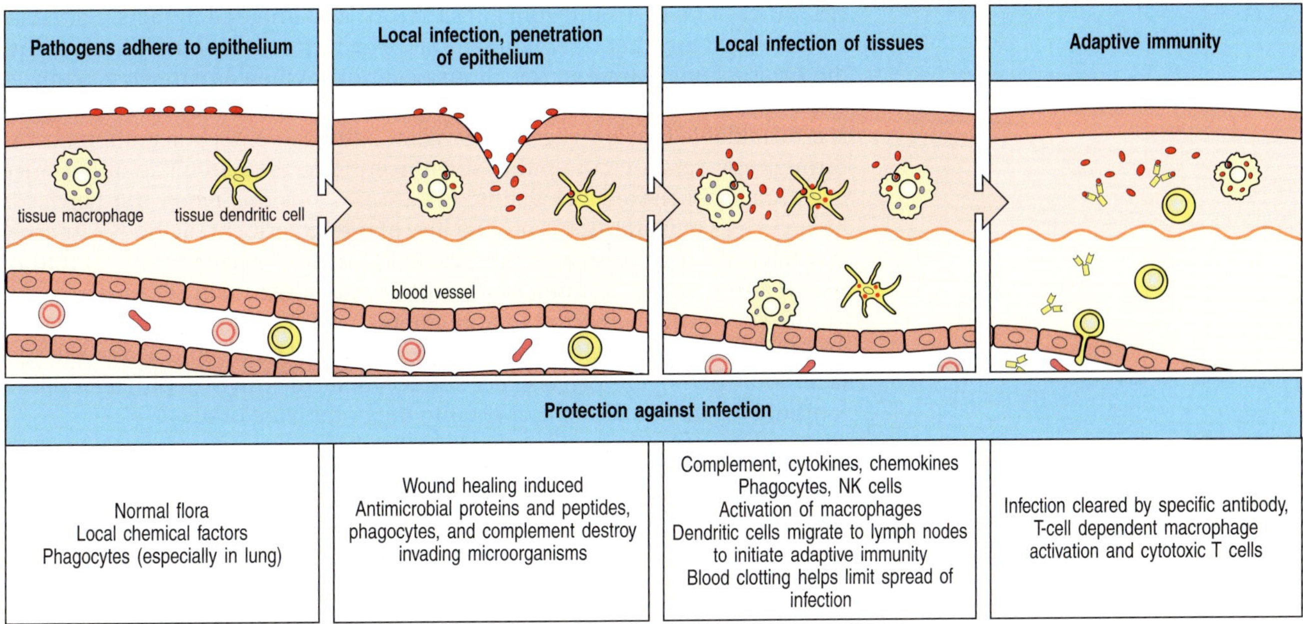

Pathogens adhere to epithelium
Local infection, penetration of epithelium
Local infection of tissues
Adaptive immunity
tissue macrophage
tissue dendritic cell
blood vessel
Protection against infection
Normal flora
Local chemical factors
Phagocytes (especially in lung)
Wound healing induced
Antimicrobial proteins and peptides, phagocytes, and complement destroy invading microorganisms
Complement, cytokines, chemokines
Phagocytes, NK cells
Activation of macrophages
Dendritic cells migrate to lymph nodes to initiate adaptive immunity
Blood clotting helps limit spread of infection
Infection cleared by specific antibody, T-cell dependent macrophage activation and cytotoxic T cells

Fig. 2.5 An infection and the response to it can be divided into a series of stages. These are illustrated here for an infectious microorganism entering through a wound in the skin. The infectious agent must first adhere to the epithelial cells and then cross the epithelium. A local immune response may prevent the infection from becoming established. If not, it helps to contain the infection and also delivers the infectious agent, carried in lymph and inside dendritic cells, to local lymph nodes. This initiates the adaptive immune response and eventual clearance of the infection.

recruits more effector cells and molecules of the innate immune system out of the blood and into the tissues (Fig. 2.5), while inducing clotting in small blood vessels further downstream so that the microbe cannot spread through the circulation. The induced responses of innate immunity act over several days, during which time an adaptive immune response is also getting under way in response to pathogen antigens delivered to local lymphoid tissue by dendritic cells (see Section 1-15). An adaptive immune response differs from innate immunity in its ability to target structures that are specific to particular strains and variants of pathogens. This response will usually clear the infection, and protect the host against reinfection with the same pathogen, by producing effector cells and antibodies against the pathogen and by generating immunological memory of that pathogen.

2-3 Epithelial surfaces of the body provide the first line of defense against infection.

Our body surfaces are defended by epithelia, which impose a physical barrier between the internal milieu and the external world that contains pathogens (Fig. 2.6). Epithelia comprise the skin and the linings of the body's tubular

	Skin	Gut	Lungs	Eyes/nose/oral cavity
Mechanical	Epithelial cells joined by tight junctions			
	Longitudinal flow of air or fluid		Movement of mucus by cilia	Tears Nasal cilia
Chemical	Fatty acids	Low pH Enzymes (pepsin)	Pulmonary surfactant	Enzymes in tears and saliva (lysozyme)
	β-defensins Lamellar bodies Cathelicidin	α-defensins (cryptdins) RegIII (lecticidins) Cathelicidin	α-defensins Cathelicidin	Histatins β-defensins
Microbiological	Normal microbiota			

Fig. 2.6 Many barriers prevent pathogens from crossing epithelia and colonizing tissues. Surface epithelia provide mechanical, chemical, and microbiological barriers to infection.

structures—the gastrointestinal, respiratory, and urogenital tracts. Epithelial cells are held together by tight junctions, which effectively form a seal against the external environment. Infections occur only when a pathogen colonizes or crosses these barriers, and because the dry tough outer layer of the skin is a formidable barrier when not broken, pathogen entry most often occurs through the vast area of epithelial surface inside the body. The importance of epithelia in protection against infection is obvious when the barrier is breached, as in wounds, burns, and loss of the integrity of the body's internal epithelia, in which cases infection is a major cause of mortality and morbidity. In the absence of wounding or disruption, pathogens can set up an infection by specifically adhering to and colonizing epithelial surfaces, using the attachment to avoid being dislodged by the flow of air or fluid across the surface. Some pathogens can also use surface molecules on the epithelial cells as footholds to invade the cells or get into the underlying tissues.

The internal epithelia are known as **mucosal epithelia** because they secrete a viscous fluid called **mucus**, which contains many glycoproteins called **mucins**. Mucus has a number of protective functions. Microorganisms coated in mucus may be prevented from adhering to the epithelium, and in the respiratory tract, microorganisms can be expelled in the outward flow of mucus driven by the beating of cilia on the mucosal epithelium. The efficacy of mucus flow in clearing infection is illustrated by people with abnormally thick mucus secretion or inhibition of ciliary movement, as occurs in the inherited disease cystic fibrosis. Such individuals frequently develop lung infections caused by bacteria that colonize the epithelial surface but do not cross it. In the gut, peristalsis is an important mechanism for keeping both food and infectious agents moving through the body. Failure of peristalsis is typically accompanied by the overgrowth of pathogenic bacteria within the lumen of the gut.

Most healthy epithelial surfaces are also associated with a with a large population of normally nonpathogenic bacteria, known as commensal bacteria or the microbiota, that compete with pathogenic microorganisms for nutrients and for attachment sites on epithelial cells. This microbiota can also produce antimicrobial substances, such as the lactic acid produced by vaginal lactobacilli, some strains of which also produce antimicrobial peptides (bacteriocins). Commensals also help to strengthen the barrier functions of epithelia by stimulating the epithelial cells to produce antimicrobial peptides. When commensal microorganisms are killed by antibiotic treatment, pathogens frequently replace them and cause disease (see Fig. 12.23). Under some circumstances commensal microbes themselves can cause disease if their growth is not kept in check or if the immune system is compromised. The survival of commensal microorganisms on our body surfaces is regulated by a balance between bacterial growth and their elimination by the mechanisms of innate and adaptive immunity (discussed in Chapter 12); failures in this regulation, such as those caused by inherited deficiencies of proteins of innate immunity that we discuss in Chapter 15, can allow normally nonpathogenic bacteria to grow excessively and cause disease.

2-4 Epithelial cells and phagocytes produce several kinds of antimicrobial proteins.

Our surface epithelia are more than mere physical barriers to infection; they also produce a wide variety of chemical substances that are microbicidal or that inhibit microbial growth. For example, the acid pH of the stomach and the digestive enzymes, bile salts, fatty acids, and lysolipids present in the upper gastrointestinal tract create a substantial chemical barrier to infection. One important group of antimicrobial proteins comprises enzymes that attack chemical features specific to bacterial cell walls. Such antibacterial enzymes include **lysozyme** and **secretory phospholipase A_2**, which are secreted in tears and saliva and by phagocytes. Lysozyme is a glycosidase that breaks a specific chemical bond present in the **peptidoglycan** component of the bacterial cell wall. Peptidoglycan is an alternating polymer of *N*-acetylglucosamine (GlcNAc) and *N*-acetylmuramic acids (MurNAc) strengthened by cross-linking peptide bridges (Fig. 2.7). Lysozyme selectively cleaves the β(1,4) linkage between these two sugars and is more effective against Gram-positive bacteria, in which the peptidoglycan cell wall is exposed, than against Gram-negative bacteria, which have an outer layer of lipopolysaccharide (LPS) covering the peptidoglycan (see Fig. 2.7). Lysozyme is also produced by **Paneth cells**, specialized epithelial cells in the base of the crypts in the small intestine that secrete many antimicrobial proteins into the gut. Paneth cells also produce secretory phospholipase A_2, a highly basic enzyme that can enter the bacterial cell wall to access and hydrolyze phospholipids in the cell membrane, killing the bacteria.

The second group of antimicrobial agents secreted by epithelial cells and phagocytes is the **antimicrobial peptides**. These represent one of the most ancient forms of defense against infection. Epithelial cells secrete these peptides into the fluids bathing the mucosal surface, whereas phagocytes secrete them in tissues. Three important classes of antimicrobial peptides in mammals are **defensins**, **cathelicidins**, and **histatins**.

Defensins are an ancient, evolutionarily conserved class of antimicrobial peptides made by many eukaryotic organisms, including mammals, insects, and plants. They are short cationic peptides of around 30–40 amino acids that usually have three disulfide bonds stabilizing a common **amphipathic** structure—a positively charged region separated from a hydrophobic region. Defensins act within minutes to disrupt the cell membranes of bacteria, fungi, and the membrane envelopes of some viruses. The mechanism is thought to involve insertion of the hydrophobic region into the membrane bilayer and

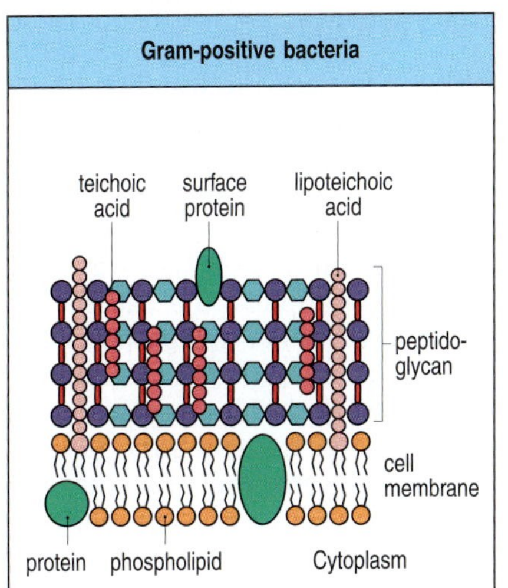

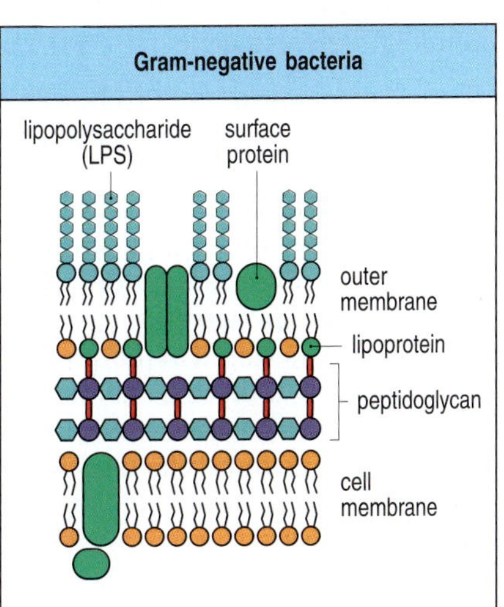

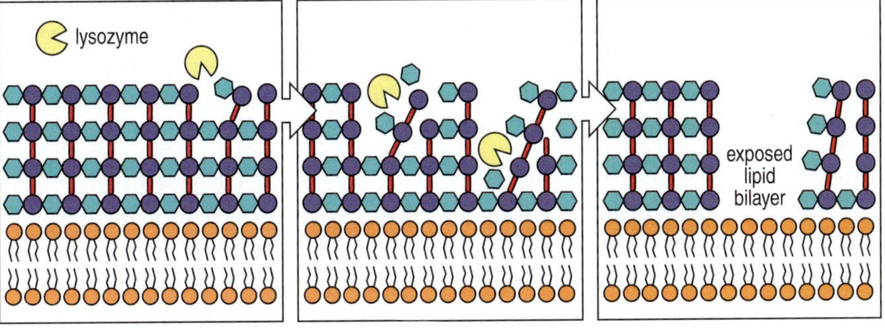

Fig. 2.7 Lysozyme digests the cell walls of Gram-positive and Gram-negative bacteria. Upper panels: the peptidoglycan of bacterial cell walls is a polymer of alternating residues of β-(1,4)-linked *N*-acetylglucosamine (GlcNAc) (large turquoise hexagons) and *N*-acetylmuramic acid (MurNAc) (purple circles) that are cross-linked by peptide bridges (red bars) into a dense three-dimensional network. Peptidoglycan forms the outer layer of Gram-positive bacteria (upper left panel), in which other molecules are embedded such as teichoic acid and the lipoteichoic acids that link the peptidoglycan layer to the bacterial cell membrane itself. In Gram-negative bacteria (upper right panel), a thin inner wall of peptidoglycan is covered by an outer lipid membrane that contains proteins and lipopolysaccharide (LPS). Lipopolysaccharide is composed of a lipid, lipid A (turquoise circles), to which is attached a polysaccharide core (small turquoise hexagons). Lysozyme (lower panels) cleaves β-(1,4) linkages between GlcNAc and MurNAc, creating a defect in the peptidoglycan layer and exposing the underlying cell membrane to other antimicrobial agents. Lysozyme is more effective against Gram-positive bacteria because of the relatively greater accessibility of the peptidoglycan.

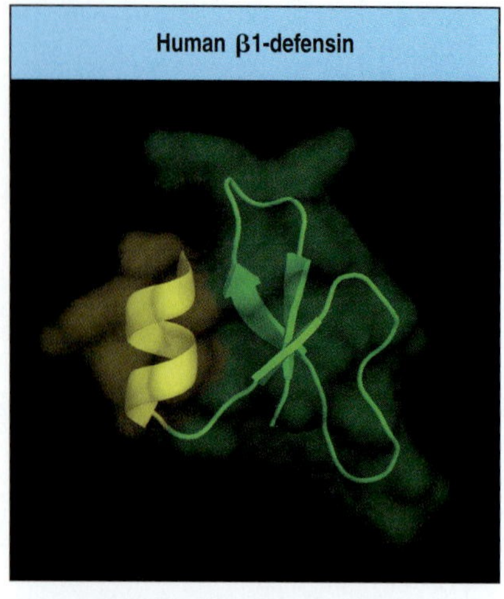

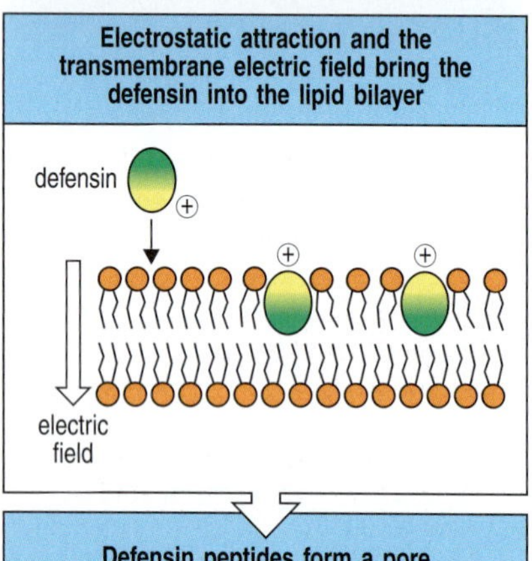

Electrostatic attraction and the transmembrane electric field bring the defensin into the lipid bilayer
defensin
electric field

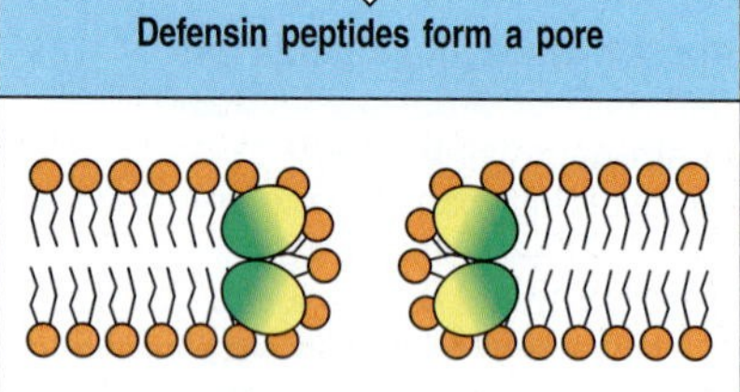

Fig.2.8 Defensins are amphipathic peptides that disrupt the cell membranes of microbes. The structure of human β_1-defensin is shown in the top panel. It is composed of a short segment of α helix (yellow) resting against three strands of antiparallel β sheet (green), generating an amphipathic peptide with separate regions having charged or hydrophobic residues. This general feature is shared by defensins from plants and insects and allows them to interact with the charged surface of the cell membrane and become inserted in the lipid bilayer (center panel). Although the details are still unclear, a transition in the arrangement of the defensins in the membrane leads to the formation of pores and a loss of membrane integrity (bottom panel).

the formation of a pore that makes the membrane leaky (Fig. 2.8). Most multicellular organisms make many different defensins—the plant *Arabidopsis thaliana* produces 13 and the fruitfly *Drosophila melanogaster* at least 15. Human Paneth cells make as many as 21 different defensins, many of which are encoded by a cluster of genes on chromosome 8.

Three subfamilies of defensins—α-, β-, and θ-defensins—are distinguished on the basis of amino-acid sequence, and each family has members with distinct activities, some being active against Gram-positive bacteria and some against Gram-negative bacteria, while others are specific for fungal pathogens. All the antimicrobial peptides, including the defensins, are generated by proteolytic processing from inactive **propeptides** (Fig. 2.9). In humans, developing neutrophils produce **α-defensins** by the processing of an initial propeptide of about 90 amino acids by cellular proteases to remove an anionic propiece, generating a mature cationic defensin that is stored in the so-called **primary granules**. The primary granules of neutrophils are specialized membrane-enclosed vesicles, rather similar to lysosomes, that contain a number of other antimicrobial agents as well as defensins. As we shall see in Chapter 3, when we discuss neutrophil function in more detail, these granules fuse with phagocytic vesicles (phagosomes) after the neutrophil has engulfed a pathogen, helping to kill the microbe. The Paneth cells of the gut constitutively produce α-defensins, called **cryptdins**, which are processed by proteases such as the metalloprotease matrilysin in mice, or trypsin in humans, before being secreted into the gut lumen. The **β-defensins** lack the long propiece of α-defensins and are generally produced specifically in response to the presence of microbial products. β-Defensins (and some α-defensins) are made by epithelia outside the gut, primarily in the respiratory and urogenital tracts, skin, and tongue. β-Defensins made by keratinocytes in the epidermis and by type II pneumocytes in the lungs are packaged into **lamellar bodies**, lipid-rich secretory organelles that release their contents into the extracellular space to form a watertight lipid sheet in the epidermis and the pulmonary surfactant in the lung. The **θ-defensins** arose in the primates, but the single human θ-defensin gene has been inactivated by a mutation.

The cathelicidin family of antimicrobial peptides lack the disulfide bonds that stabilize the defensins. Humans and mice have one cathelicidin gene, but some other mammals have several. Cathelicidins are made constitutively by neutrophils, by macrophages, and by keratinocytes in the skin and epithelial cells in the lungs and intestine in response to infection. They are made as inactive propeptides composed of two linked domains and are processed before secretion (see Fig. 2.9). In neutrophils, the cathelicidin propeptides are stored in the **secondary granules**, another type of specialized cytoplasmic granule that contains antimicrobial agents, and are activated by proteolytic cleavage when these granules fuse with the phagosome and encounter **neutrophil elastase** that has been released from primary granules. This mechanism ensures that cathelicidins are activated only when needed. Cleavage by elastase separates the two domains, and the cleavage products either remain in the phagosome or are released from the neutrophil by exocytosis. The carboxy-terminal peptide is a cationic amphipathic peptide that disrupts membranes and is toxic to a wide range of microorganisms. In keratinocytes, cathelicidins, like β-defensins, are stored and processed in the lamellar bodies. The amino-terminal peptide is similar in structure to a protein called cathelin, an inhibitor of cathepsin L, but it is still unclear what role, if any, it has in immune defense.

Another class of antimicrobial peptides is the histatins, which are constitutively produced by the parotid, sublingual, and submandibular glands in the oral cavity. These short histidine-rich cationic peptides are active against pathogenic fungi such as *Cryptococcus neoformans* and *Candida albicans*.

Another layer of innate immunity is provided by bactericidal proteins that are carbohydrate-binding proteins, or **lectins. C-type lectins** require calcium for their binding activity and have a characteristic carbohydrate-recognition domain (CRD), stabilized by two disulfide bonds, that provides a highly variable interface for binding carbohydrate structures. In mice, the C-type lectin **RegIIIγ** is produced by Paneth cells and secreted into the gut, where it binds to peptidoglycans in bacterial cell walls and exerts direct bactericidal activity. RegIIIγ preferentially kills Gram-positive bacteria, in which the peptidoglycan is exposed on the outer surface and is therefore more accessible (see Fig. 2.7). The precise mechanism of killing is still uncertain but involves direct damage to bacterial cell membranes in a similar manner to the actions of defensins. RegIIIγ is produced in inactive form but is cleaved by the protease trypsin, which removes a short amino-terminal fragment to activate the bactericidal potential of RegIIIγ within the intestinal lumen (see Fig. 2.9). The related human protein REG3 (also called **HIH/PAP** for hepatocarcinoma-intestine-pancreas/pancreatitis-associated protein) and several other members of the Reg family of secreted C-type lectins are also expressed in the intestine, suggesting that there may be a 'lecticidin' family of antimicrobial proteins. It is worth noting that it is a common strategy for the antimicrobial proteins to be produced as propeptides that require cleavage by a protease to complete their activation. This form of activation is also a central feature of the next component of innate immunity that we shall consider, the complement system. Fig. 2.10 summarizes the strategies of innate immunity that involve epithelial barriers and preformed antimicrobial enzymes and peptides.

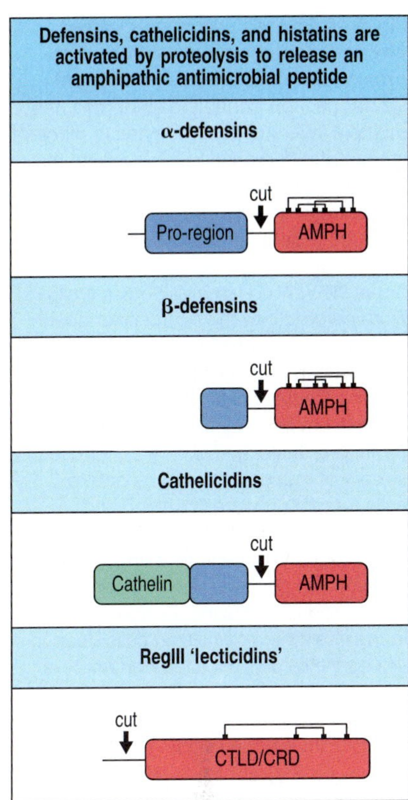

Fig. 2.9 Defensins, cathelicidins and RegIII proteins are activated by proteolysis. When α- and β-defensins are first synthesized, they contain a signal peptide (not shown), a pro-region (blue), which is shorter in the β-defensins, and an amphipathic domain (AMPH, red); the pro-region represses the membrane-inserting properties of the amphipathic domain. After defensins are released from the cell, or into phagosomes, they undergo cleavage by proteases, which releases the amphipathic domain in active form. Newly synthesized cathelicidins contain a signal peptide, a cathelin domain, a short pro-region, and an amphipathic domain; they, too, are activated by proteolytic cleavage. RegIII contains a C-type lectin domain (CTLD), also known as a carbohydrate recognition domain (CRD). After release of the signal peptide, further proteolytic cleavage of RegIII also regulates its antimicrobial activity.

Summary.

For a pathogen to cause disease, it must first make contact with the host and then establish a focus of infection. The mammalian body is susceptible to infection by many pathogens, which reside and replicate in different anatomic locations. Pathogens differ greatly in their lifestyles, the structures of their surfaces, and their mechanisms of pathogenesis, so an equally diverse set of defensive responses from the host's immune system is required. The mammalian immune response to invading organisms proceeds in three phases, from immediate innate defenses, to induced innate defenses, and finally to adaptive immunity. The first phase of host defense consists of those mechanisms that are present and ready to resist an invader at any time. Epithelial surfaces provide a physical barrier against pathogen entry, but they also have specialized strategies. Mucosal surfaces have a protective barrier of mucus. Specialized epithelia protect against colonization and against viruses and bacteria through specialized cell-surface interactions. Defense mechanisms of epithelia include the prevention of pathogen adherence and the secretion of antimicrobial enzymes and peptides. Antimicrobial peptides are usually made as an inactive proprotein that requires a proteolytic step to complete its activation, often by generating a cationic peptide that takes on an amphipathic structure capable of disrupting the cell membrane of a microbe. Antimicrobial lectins, which recognize glycans specific to many pathogens and disrupt microbial cell walls, also require proteolytic cleavage for activity. The actions of antimicrobial enzymes and peptides described in this section often involve binding to the unique glycan/carbohydrate structures on the microbe. Thus, these soluble molecular defenses are both pattern recognition receptors and effector molecules at the same time, representing the simplest form of innate immunity.

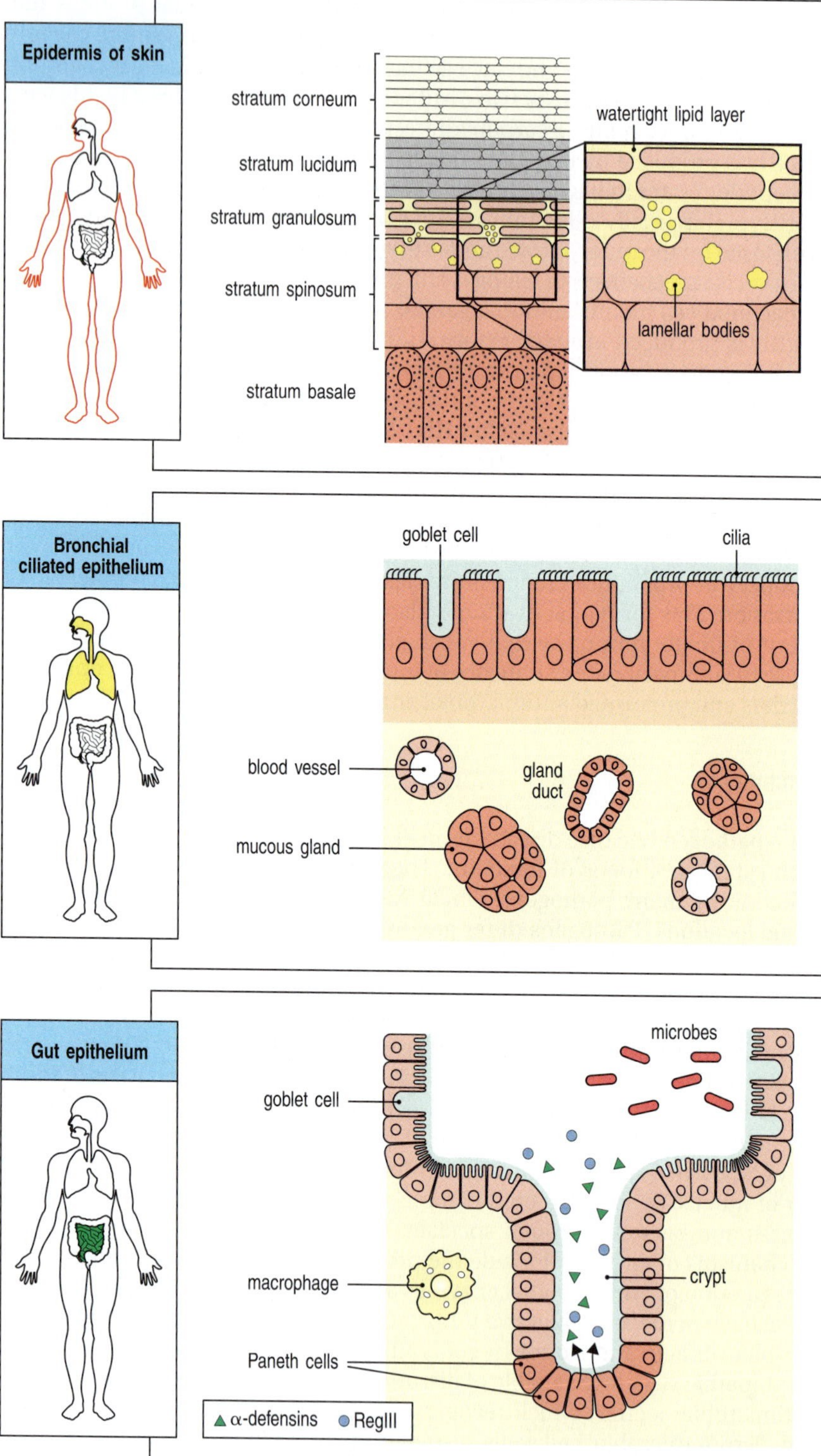

Fig. 2.10 Epithelia form specialized physical and chemical barriers that provide innate defenses in different locations. Top panel: the epidermis has multiple layers of keratinocytes in different stages of differentiation arising from the basal layer of stem cells. Differentiated keratinocytes in the stratum spinosum produce β-defensins and cathelicidins, which are incorporated into secretory organelles called lamellar bodies (yellow) and secreted into the intercellular space to form a waterproof lipid layer (the stratum corneum) containing antimicrobial activity. Center panel: in the lung, the airways are lined by ciliated epithelium. Beating of the cilia moves a continuous stream of mucus secreted by goblet cells (green) outward, trapping and ejecting potential pathogens. Type II pneumocytes in the lung alveoli (not shown) also produce and secrete antimicrobial defensins. Bottom panel: in the intestine, specialized cells deep in the epithelial crypts called Paneth cells produce several kinds of antimicrobial proteins: α-defensins (cryptdins) and the antimicrobial lectin RegIII.

The complement system and innate immunity.

When a pathogen breaches the host's epithelial barriers and initial antimicrobial defenses, it next encounters a major component of innate immunity known as the complement system, or **complement**. Complement is a collection of soluble proteins present in blood and other body fluids.

It was discovered in the 1890s by **Jules Bordet** as a heat-labile component of normal plasma that augmented the opsonization and killing of bacteria by antibodies, and so this activity was said to 'complement' the actions of antibodies. **Opsonization** refers to the coating of a pathogen by antibodies and/or complement proteins so that it is more readily taken up and destroyed by phagocytic cells. Although complement was first discovered as an effector arm of the antibody response, we now understand that it originally evolved as part of the innate immune system and that it still provides protection early in infection, in the absence of antibodies, through more ancient pathways of complement activation.

The complement system is composed of more than 30 different plasma proteins, which are produced mainly by the liver. In the absence of infection, these proteins circulate in an inactive form. In the presence of pathogens or of antibody bound to pathogens, the complement system becomes 'activated.' Particular **complement proteins** interact with each other to form several different pathways of complement activation, all of which have the final outcome of killing the pathogen, either directly or by facilitating its phagocytosis, and inducing inflammatory responses that help to fight infection. There are three pathways of **complement activation**. As the antibody-triggered pathway of complement activation was discovered first, this became known as the classical pathway of complement activation. The next to be discovered was called the alternative pathway, which can be activated by the presence of the pathogen alone, and the most recently discovered is the lectin pathway, which is activated by lectin-type proteins that recognize and bind to carbohydrates on pathogen surfaces.

We learned in previous sections that proteolysis can be used as a means of activating antimicrobial proteins. In the complement system, activation by proteolysis is used to an even greater degree, because many of the complement proteins are proteases that successively cleave and activate one another. The proteases of the complement system are synthesized as inactive pro-enzymes, or **zymogens**, which only become enzymatically active after proteolytic cleavage, usually by another complement protein. The complement pathways are triggered by proteins that act as pattern recognition receptors to detect the presence of pathogens. Detection of pathogen activates an initial zymogen, triggering a cascade of proteolysis in which complement zymogens are activated sequentially, each becoming an active protease that cleaves and activates many molecules of the next zymogen in the pathway. These proteolytic cascades finally generate the effector complement components that aid the removal of the pathogen. In this way, the detection of even a small number of pathogens produces a rapid response that is greatly amplified at each step.

Nomenclature can be an obstacle for students learning the complement proteins, so we will start by explaining their names. The first proteins discovered belong to the classical pathway, and are designated by the letter C followed by a number. The native complement proteins—such as the inactive zymogens—have a simple number designation, for example C1 and C2. Unfortunately, they were named in the order of their discovery rather than the sequence of reactions. The reaction sequence in the classical pathway, for example, is C1, C4, C2, C3, C5, C6, C7, C8, and C9 (note that not all of these are proteases). Products of cleavage reactions are designated by adding a lowercase letter as a suffix. For example, cleavage of C3 produces a small protein fragment called C3a and a bigger fragment, C3b. The larger fragment is always designated by the suffix b, with one exception. For C2, the larger fragment was originally named **C2a**, because it was the enzymatically active fragment, and this name has survived. Another exception to the general rule is the naming of C1q, C1r, and C1s: these are not cleavage products of C1 but are distinct proteins that together comprise C1. The proteins of the alternative pathway were

Functional protein classes in the complement system	
Binding to antigen:antibody complexes and pathogen surfaces	C1q
Binding to carbohydrate structures such as mannose or GlcNAc on microbial surfaces	MBL Ficolins C1q Properdin (factor P)
Activating enzymes	C1r C1s C2a Bb D MASP-2
Membrane-binding proteins and opsonins	C4b C3b
Peptide mediators of inflammation	C5a C3a C4a
Membrane-attack proteins	C5b C6 C7 C8 C9
Complement receptors	CR1 CR2 CR3 CR4 CRIg
Complement-regulatory proteins	C1INH C4BP CR1 MCP DAF H I P CD59

Fig. 2.11 Functional protein classes in the complement system.

discovered later and are designated by different capital letters, for example factor B and factor D. Their cleavage products are also designated by the addition of lower-case a and b: thus, the large fragment of B is called Bb and the small fragment Ba. Activated complement components are sometimes designated by a horizontal line, for example, $\overline{\text{C2a}}$; however, we will not use this convention. All the components of the complement system are listed in Fig. 2.11.

Besides acting in innate immunity, complement also influences adaptive immunity. Opsonization of pathogens by complement facilitates their uptake by phagocytic antigen-presenting cells that express complement receptors; this enhances the presentation of pathogen antigens to T cells, which we discuss in more detail in Chapter 6. B cells express receptors for complement proteins that enhance their responses to complement-coated antigens, as we describe later in Chapter 10. In addition, several of the complement fragments can act to influence cytokine production by antigen-presenting cells, thereby influencing the direction and extent of the subsequent adaptive immune response, as we describe in Chapter 11.

2-5 The complement system recognizes features of microbial surfaces and marks them for destruction by the deposition of C3b.

Figure 2.12 gives a highly simplified preview of the initiation mechanisms and outcomes of complement activation. The three pathways of complement activation are initiated in different ways. The **lectin pathway** is initiated by soluble carbohydrate-binding proteins—mannose-binding lectin and the ficolins—that bind to particular carbohydrate structures on microbial surfaces. Proteases associated with these recognition proteins then trigger the cleavage of complement proteins and activation of the pathway. The **classical pathway** is initiated when the complement component C1, which comprises a recognition protein (C1q) associated with proteases (C1r and C1s), either recognizes a microbial surface directly or binds to antibodies already bound to a pathogen. Finally, the **alternative pathway** can be initiated by spontaneous hydrolysis and activation of the complement component C3, which can then bind directly to microbial surfaces.

These three pathways converge at the central and most important step in complement activation. When any of the pathways interacts with a pathogen surface, an enzymatic activity called a **C3 convertase** is generated. There are various types of C3 convertase, depending on the complement pathway activated, but each is a multisubunit protein with protease activity that cleaves **complement component 3** (**C3**). The C3 convertase is bound covalently to the pathogen surface, where it cleaves C3 to generate large amounts of **C3b**, the main effector molecule of the complement system, and **C3a**, a peptide that helps induce inflammation. Cleavage of C3 is the critical step in complement activation and leads directly or indirectly to all the effector activities of the complement system (see Fig. 2.12). C3b binds covalently to the microbial surface and acts as an opsonin, enabling phagocytes that carry receptors for complement to take up and destroy the C3b-coated microbe. We describe the different complement receptors involved in this function of the complement system later in the chapter. C3b can also bind to the C3 convertases formed by the classical and lectin pathways, forming another multisubunit enzyme, a **C5 convertase** (not shown in Fig. 2.12). This cleaves C5, liberating the highly inflammatory peptide **C5a** and generating **C5b**. C5b initiates the 'late' events of complement activation, in which a further set of complement proteins interact with C5b to form a **membrane-attack complex** on the pathogen surface, creating a pore in the cell membrane that leads to cell lysis.

The key feature of C3b is its ability to form a covalent bond with microbial surfaces, which allows the innate recognition of microbes to be translated into effector responses. Covalent bond formation is due to a highly reactive

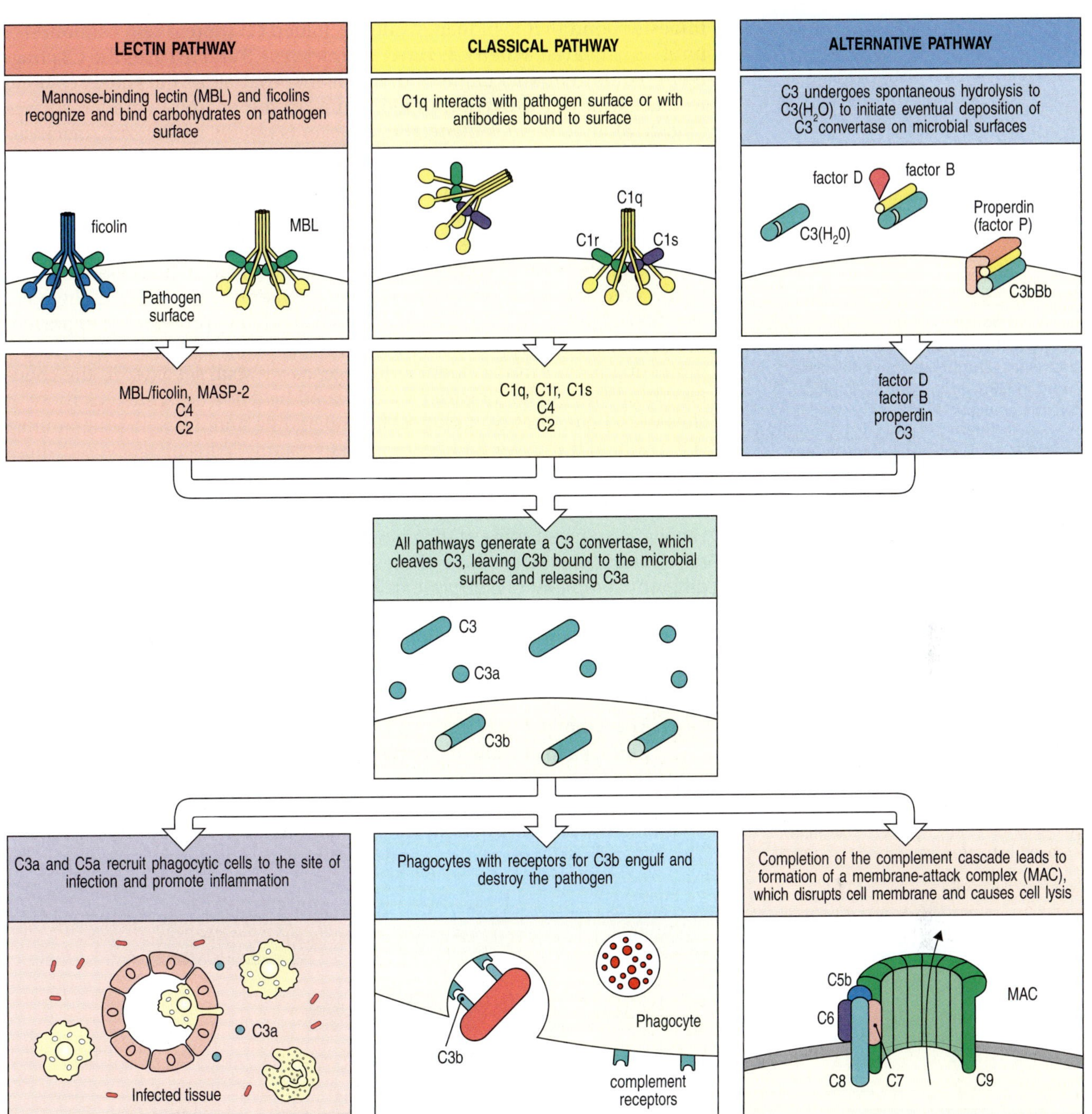

Fig. 2.12 Complement is a system of soluble pattern-recognition receptors and effector molecules that detect and destroy microorganisms. The pathogen-recognition mechanisms of the three complement-activation pathways are shown in the top row, along with the complement components used in the proteolytic cascades leading to formation of a C3 convertase. This enzyme activity cleaves complement component C3 into the small soluble protein C3a and the larger component C3b, which becomes covalently bound to the pathogen surface. The components are listed by biochemical function in Fig. 2.11 and are described in detail in later figures. The lectin pathway of complement activation (top left) is triggered by the binding of mannose-binding lectin (MBL) or ficolins to carbohydrate residues in bacterial cell walls and capsules. The classical pathway (top center) is triggered either by binding of C1q to the pathogen surface or to antibody bound to the pathogen. In the alternative pathway (top right), soluble C3 undergoes spontaneous hydrolysis in the fluid phase, generating C3(H_2O), which is augmented further by the action of factors B, D, and P (properdin). All pathways thus converge on the formation of C3b bound to a pathogen and lead to all of the effector activities of complement, which are shown in the bottom row. C3b bound to a pathogen acts as an opsonin, enabling phagocytes that express receptors for C3b to ingest the complement-coated bacteria more easily (bottom left). C3b can also bind to C3 convertases to produce another activity, a C5 convertase (detail not shown here), which cleaves C5 to C5a and C5b. C5b triggers the late events of the complement pathway in which the terminal components of complement—C6 to C9—assemble into a membrane-attack complex that can damage the membrane of certain pathogens (bottom center). C3a and C5a act as chemoattractants that recruit immune-system cells to the site of infection and cause inflammation (bottom right).

thioester bond that is hidden inside the folded C3 protein and cannot react until C3 is cleaved. When C3 convertase cleaves C3 and releases the C3a fragment, conformational changes occur in C3b that allow the thioester bond to react with a hydroxyl or amino group on the nearby microbial surface (Fig. 2.13). If no bond is made, the thioester is rapidly hydrolyzed, inactivating C3b.

Pathways leading to such potent inflammatory and destructive effects—and which have a series of built-in amplification steps—are potentially dangerous and must be tightly regulated. One important safeguard is that the key activated complement components are rapidly inactivated unless they bind to the pathogen surface on which their activation was initiated. There are also several points in the pathway at which regulatory proteins act to prevent the activation of complement on the surfaces of healthy host cells, thereby protecting them from accidental damage, as we shall see later in the chapter. Complement can, however, be activated by dying cells, such as those at sites of ischemic injury, and by cells undergoing apoptosis or programmed cell death. In these cases, the complement coating helps phagocytes dispose of the dead and dying cells neatly, thus limiting the risk of cell contents being released and triggering an autoimmune response (discussed in Chapter 15).

Having introduced some of the main complement components, we are ready for a more detailed account of the three pathways. To help distinguish the different complement components according to their function, we will use a color code in the figures in this part of the chapter. This code was introduced in Fig. 2.11, where all the components of the complement system are grouped by function.

2-6 The lectin pathway uses soluble receptors that recognize microbial surfaces to activate the complement cascade.

Microorganisms typically bear repeating patterns of molecular structures on their surface, known generally as pathogen-associated molecular patterns. The cell walls of Gram-positive and Gram-negative bacteria, for example, are

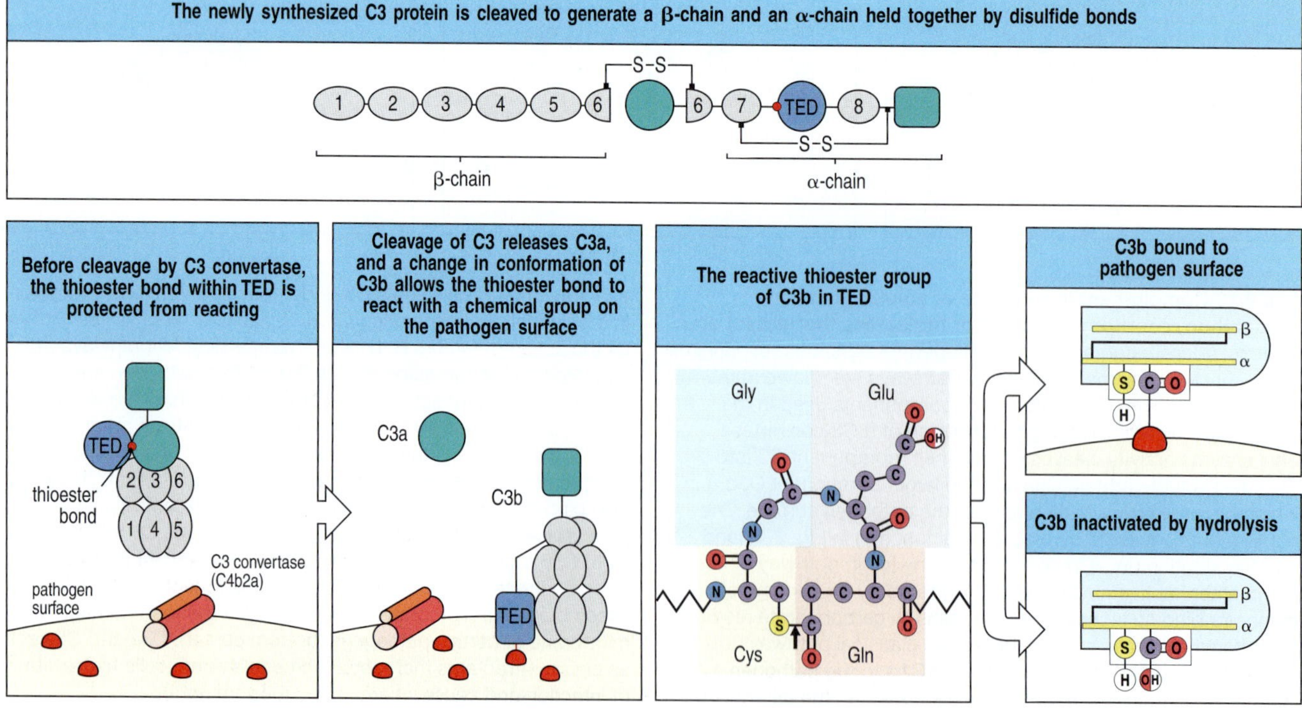

Fig. 2.13 C3 convertase activates C3 for covalent bonding to microbial surfaces by cleaving it into C3a and C3b and exposing a highly reactive thioester bond in C3b. Top panel: C3 in blood plasma consists of an α-chain and a β-chain (formed by proteolytic processing from the native C3 polypeptide) held together by a disulfide bond. The thioester-containing domain (TED) of the α-chain contains a potentially highly reactive thioester bond (red spot). Bottom left panels: cleavage by C3 convertase (the lectin pathway convertase C4b2a is shown here) and release of C3a from the amino terminus of the α-chain causes a conformational change in C3b that exposes the thioester bond. This can now react with hydroxyl or amino groups on molecules on microbial surfaces, covalently bonding C3b to the surface. Bottom right panels: schematic view of the thioester reaction. If a bond is not made with a microbial surface, the thioester is rapidly hydrolyzed (that is, cleaved by water), rendering C3b inactive.

composed of a matrix of proteins, carbohydrates, and lipids in a repetitive array (see Fig. 2.7). The lipoteichoic acids of Gram-positive bacterial cell walls and the lipopolysaccharide of the outer membrane of Gram-negative bacteria are not present on animal cells and are important in the recognition of bacteria by the innate immune system. Similarly, the glycans of yeast surface proteins commonly terminate in mannose residues rather than the sialic acid residues (*N*-acetylneuraminic acid) that terminate the glycans of vertebrate cells (Fig. 2.14). The lectin pathway uses these features of microbial surfaces to detect and respond to pathogens.

The lectin pathway can be triggered by any of four different pattern recognition receptors circulating in blood and extracellular fluids that recognize carbohydrate on microbial surfaces. The first to be discovered was **mannose-binding lectin** (**MBL**), which is shown in Fig. 2.15, and this is synthesized in the liver. MBL is an oligomeric protein built up from a monomer that contains an amino-terminal collagen-like domain and a carboxy-terminal C-type lectin domain. Proteins of this type are called **collectins**. MBL monomers assemble into trimers through the formation of a triple helix by their collagen-like domains. Trimers then assemble into oligomers by disulfide bonding between the cystine-rich collagen domains. The MBL present in the blood is composed of two to six trimers. A single carbohydrate-recognition domain of MBL has a low affinity for mannose, fucose, and *N*-acetylglucosamine (GlcNAc) residues, which are common on microbial glycans, but does not bind sialic acid residues, which terminate vertebrate glycans. Thus, multimeric MBL has high total binding strength, or **avidity**, for repetitive carbohydrate structures on a wide variety of microbial surfaces, including Gram-positive and Gram-negative bacteria, mycobacteria, yeasts, and some viruses and parasites, while not interacting with host cells. MBL is produced by the liver and is present at low concentrations in the plasma of most individuals, but in the presence of infection its production is increased during the acute-phase response. This is part of the induced phase of the innate immune response and is discussed in Chapter 3.

The other three pathogen-recognition molecules used by the lectin pathway are known as **ficolins**. Although related in overall shape and function to MBL, they have a fibrinogen-like domain, rather than a lectin domain, attached to

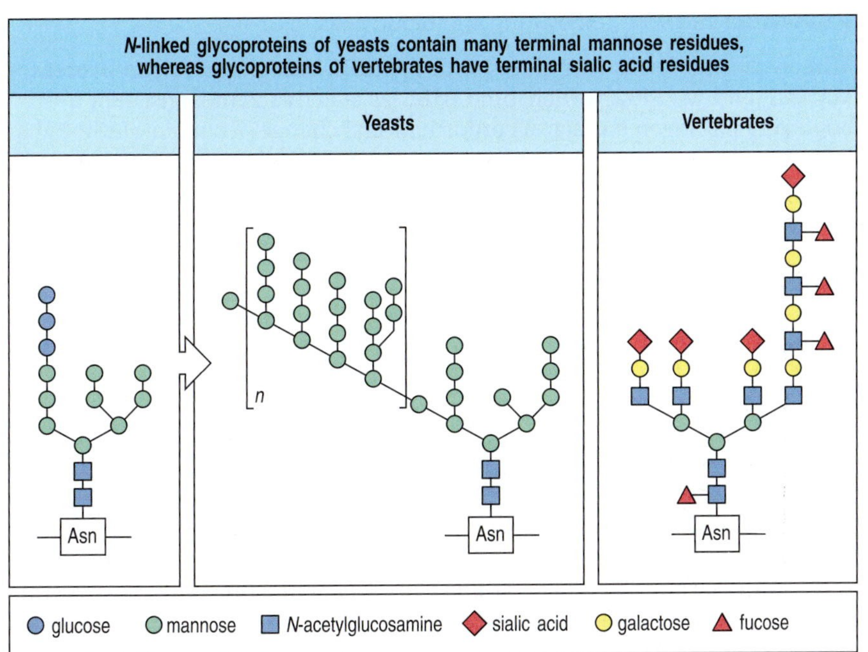

Fig. 2.14 The carbohydrate side chains on yeast and vertebrate glycoproteins are terminated with different patterns of sugars. *N*-linked glycosylation in fungi and animals is initiated by the addition of the same precursor oligosaccharide, Glc_3-Man_9-$GlcNAc_2$ (left panel) to an asparagine residue. In many yeasts this is processed to high-mannose glycans (middle panel). In contrast, in vertebrates, the initial glycan is trimmed and processed, and the *N*-linked glycoproteins of vertebrates have terminal sialic acid residues (right panel).

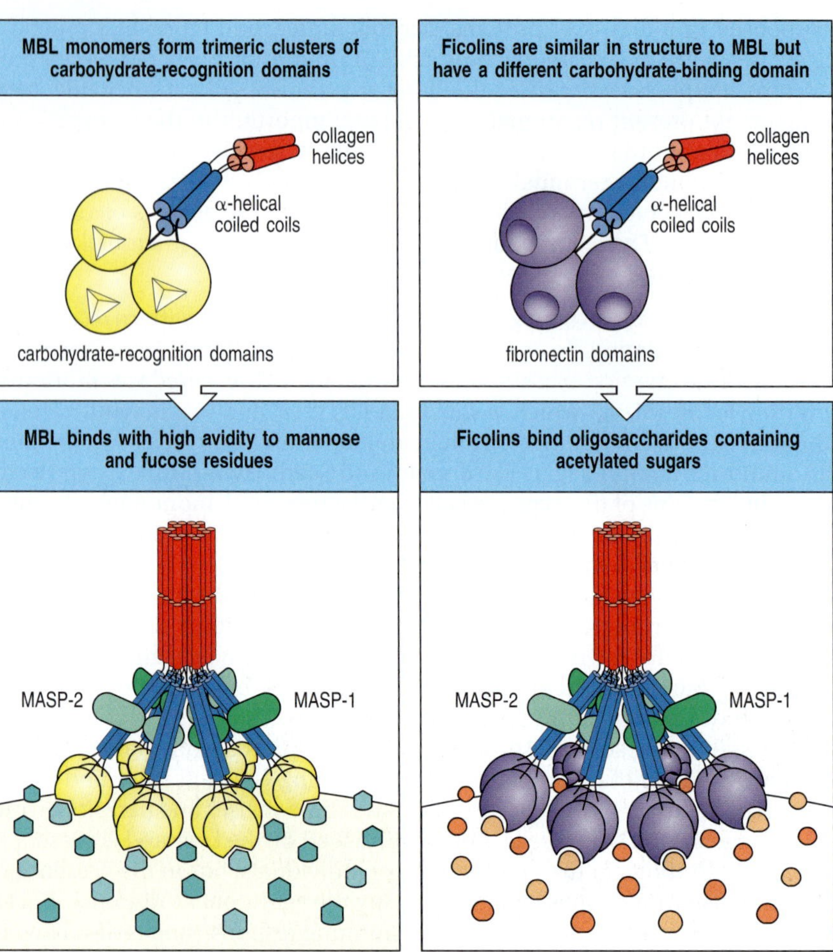

Fig. 2.15 Mannose-binding lectin and ficolins form complexes with serine proteases and recognize particular carbohydrates on microbial surfaces. Mannose-binding lectin (MBL) (left panels) is an oligomeric protein in which two to six clusters of carbohydrate-binding heads arise from a central stalk formed from the collagen-like tails of the MBL monomers. An MBL monomer is composed of a collagen region (red), an α-helical neck region (blue) and a carbohydrate-recognition domain (yellow). Three MBL monomers associate to form a trimer, and between two and six trimers assemble to form a mature MBL molecule (bottom left panel). Associated with the MBL molecule are two serine proteases, MBL-associated serine protease 1 (MASP-1) and 2 (MASP-2). MBL binds to bacterial surfaces that display a particular spatial arrangement of mannose or fucose residues. The ficolins (right panels) resemble MBL in their overall structure, are associated with MASP-1 and MASP-2, and can activate C4 and C2 after binding to carbohydrate molecules present on microbial surfaces. The carbohydrate-binding domain of ficolins is a fibrinogen-like domain, rather than the lectin domain present in MBL.

the collagen-like stalk (see Fig. 2.15). The fibrinogen-like domain gives ficolins a general specificity for oligosaccharides containing acetylated sugars, but it does not bind mannose-containing carbohydrates. Humans have three ficolins: L-ficolin (ficolin-2), M-ficolin (ficolin-1), and H-ficolin (ficolin-3). L- and H-ficolin are synthesized by the liver and circulate in the blood; M-ficolin is synthesized and secreted by lung and blood cells.

MBL in plasma forms complexes with the **MBL-associated serine proteases MASP-1** and **MASP-2**, which bind MBL as inactive zymogens. When MBL binds to a pathogen surface, a conformational change occurs in MASP-2 that enables it to cleave and activate a second MASP-2 molecule in the same MBL complex. Activated MASP-2 can then cleave complement components C4 and C2 (Fig. 2.16). Like MBL, ficolins form oligomers that make a complex with MASP-1 and MASP-2, which similarly activate complement upon recognition of a microbial surface by the ficolin. C4, like C3, contains a buried thioester bond. When MASP-2 cleaves C4 it releases C4a, allowing a conformational change in C4b that exposes the reactive thioester as described for C3b (see Fig. 2.13). C4b bonds covalently via this thioester to the microbial surface nearby, where it then binds one molecule of C2 (see Fig. 2.16). C2 is cleaved by MASP-2, producing C2a, an active serine protease, which remains bound to C4b, forming **C4b2a**, which is the C3 convertase of the lectin pathway. C4b2a now cleaves many molecules of C3 into C3a and C3b. The C3b fragments bond covalently to the nearby pathogen surface, and the released C3a initiates a local inflammatory response. The role of MASP-1 is still somewhat unclear, but some evidence indicates that it is able to cleave C3 directly, although less efficiently than C4b2a. The complement-activation pathway initiated by ficolins proceeds like the MBL lectin pathway (see Fig. 2.16).

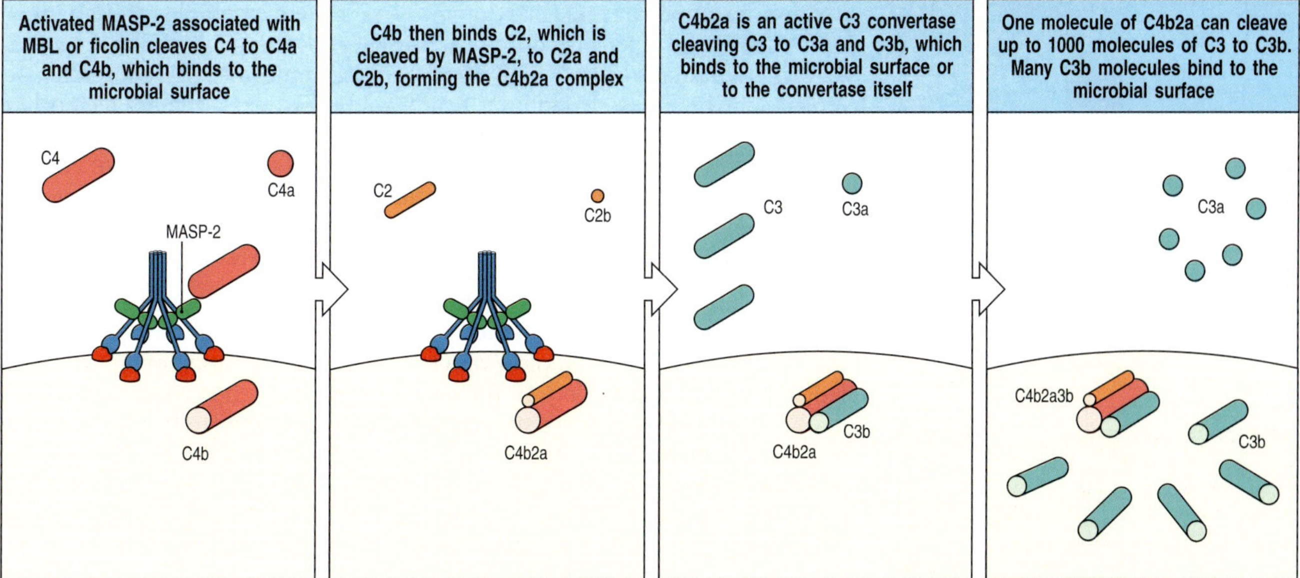

Fig. 2.16 The actions of the C3 convertase result in the binding of large numbers of C3b molecules to the pathogen surface. Binding of mannose-binding lectin or ficolins to their carbohydrate ligands on microbial surfaces activates the serine protease MASP-2. This leads to the cleavage of C4 by MASP-2, which exposes a reactive group on C4b that allows it to bind covalently to the pathogen surface. C4b then binds C2, making C2 susceptible to cleavage by MASP-2. The larger C2a fragment is the active protease component of the C3 convertase. It cleaves many molecules of C3 to produce C3b, which binds to the pathogen surface, and C3a, an inflammatory mediator. The covalent attachment of C3b and C4b to the pathogen surface is important in confining subsequent complement activity to pathogen surfaces.

Individuals deficient in MBL or MASP-2 experience substantially more respiratory infections by common extracellular bacteria during early childhood, indicating the importance of the lectin pathway for host defense. This susceptibility illustrates the particular importance of innate defense mechanisms in early childhood, when adaptive immune responses are not yet fully developed but the maternal antibodies transferred across the placenta and present in the mother's milk are gone. Other members of the collectin family are the **surfactant proteins A** and **D** (**SP-A** and **SP-D**), which are present in the fluid that bathes the epithelial surfaces of the lung. There they coat the surfaces of pathogens, making them more susceptible to phagocytosis by macrophages that have left the subepithelial tissues to enter the alveoli. Because SP-A and SP-D do not associate with MASPs, they do not activate complement.

We have used MBL here as our prototype activator of the lectin pathway, but the ficolins are more abundant than MBL in plasma and so may be more important in practice. L-ficolin recognizes acetylated sugars such as GlcNAc and *N*-acetylgalactosamine (GalNAc), and particularly recognizes lipoteichoic acid, a component of the cell walls of Gram-positive bacteria that contains GalNac. It can also activate complement after binding to a variety of capsulated bacteria. M-ficolin also recognizes acetylated sugar residues; H-ficolin shows a more restricted binding specificity, for D-fucose and galactose, and has only been linked to activity against the Gram-positive bacterium *Aerococcus viridans*, a cause of bacterial endocarditis.

2-7 The classical pathway is initiated by activation of the C1 complex and is homologous to the lectin pathway.

The overall scheme of the classical pathway is similar to the lectin pathway, except that it uses a pathogen sensor known as the **C1 complex**, or **C1**. Because C1 interacts directly with some pathogens but can also interact with antibodies, C1 allows the classical pathway to function both in innate immunity, which we describe now, and in adaptive immunity, which we examine in more detail in Chapter 10.

Like the MBL–MASP complex, the C1 complex is composed of a large subunit (**C1q**), which acts as the pathogen sensor, and two serine proteases (**C1r** and **C1s**), initially in their inactive form. C1q is a hexamer of trimers, composed of monomers that each contain an amino-terminal globular domain and a

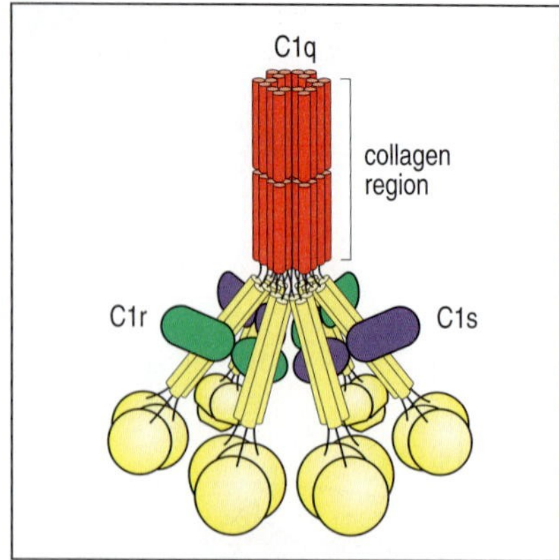

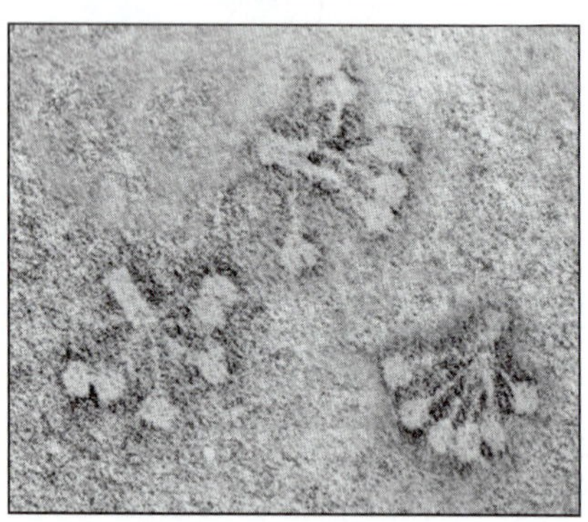

Fig. 2.17 The first protein in the classical pathway of complement activation is C1, which is a complex of C1q, C1r, and C1s. As shown in the micrograph, C1q is composed of six identical subunits with globular heads and long collagen-like tails, and it has been described as looking like "a bunch of tulips." The tails combine to bind to two molecules each of C1r and C1s, forming the C1 complex C1q:C1r$_2$:C1s$_2$. The heads can bind to the constant regions of immunoglobulin molecules or directly to the pathogen surface, causing a conformational change in C1r, which then cleaves and activates the C1s zymogen. The C1 complex is similar in overall structure to the MBL–MASP complex, and has an identical function, cleaving C4 and C2 to form the C3 convertase C4b2a (see Fig. 2.16). Photograph (×500,000) courtesy of K.B.M. Reid.

carboxy-terminal collagen-like domain. The trimers assemble through interactions of the collagen-like domains, bringing the globular domains together to form a globular head. Six of these trimers assemble to form a complete C1q molecule, which has six globular heads held together by their collagen-like tails (Fig. 2.17). C1r and C1s are closely related to MASP-2, whereas MASP-1 is somewhat more distantly related; all four enzymes are likely to have evolved from the duplication of a gene for a common precursor. C1r and C1s interact noncovalently, forming C1r:C1s pairs. Two or more pairs fold into the arms of C1q, with at least part of the C1r:C1s complex being external to C1q.

The recognition function of C1 resides in the six globular heads of C1q. When two or more of these heads interact with a ligand, this causes a conformational change in the C1r:C1s complex, which leads to the activation of an autocatalytic enzymatic activity in C1r; the active form of C1r then cleaves its associated C1s to generate an active serine protease. The activated C1s acts on the next two components of the classical pathway, C4 and C2. C1s cleaves C4 to produce C4b, which binds covalently to the pathogen surface as described earlier for the lectin pathway (see Fig. 2.16). C4b then binds one molecule of C2, which is cleaved by C1s to produce the serine protease C2a. This produces the active C3 convertase C4b2a, which is the C3 convertase of both the lectin and the classical pathways. Because it was first discovered as part of the classical pathway it is often known as the **classical C3 convertase**. The proteins involved in the classical pathway, and their active forms, are listed in Fig. 2.18.

C1q can attach itself to the surface of pathogens in several different ways. One is by binding directly to surface components on some bacteria, including certain proteins of bacterial cell walls and to polyanionic structures such as the lipoteichoic acid on Gram-positive bacteria. A second is through binding to C-reactive protein, an acute-phase protein in human plasma that binds to phosphocholine residues in bacterial surface molecules such as pneumococcal C polysaccharide—hence the name C-reactive protein. We discuss the acute-phase proteins in detail in Chapter 3. However, the main function of C1q in an immune response is to bind to the constant regions of antibodies (the Fc regions) that have bound pathogens via their antigen-binding sites (see Section 1-18). C1q thus links the effector functions of complement to recognition provided by adaptive immunity. This might seem to limit the usefulness of C1q in fighting the first stages of an infection, before the adaptive immune response has generated pathogen-specific antibodies. However, **natural antibodies** will be present, which are antibodies produced by the immune system in the apparent absence of any infection. These antibodies have a low affinity for many microbial pathogens and are highly cross-reactive, recognizing common membrane constituents such as phosphocholine, and even recognizing some self antigens—components of the body's own cells. It is not known whether natural antibodies are produced in response to the commensal microbiota or in response to self antigens, but they do not seem to be the consequence of an adaptive immune response to infection. Most natural antibody is of the class known as IgM, and represents a considerable amount of the total IgM circulating in humans. IgM is the class of antibody most efficient at binding C1q, making natural antibodies an effective means of activating complement on microbial surfaces immediately after infection, leading to the clearance of bacteria such as *Streptococcus pneumoniae* (the pneumococcus) before they become dangerous.

2-8 Complement activation is largely confined to the surface on which it is initiated.

We have seen that both the lectin and the classical pathways of complement activation are initiated by proteins that bind to pathogen surfaces. During the triggered-enzyme cascade that follows, it is important that activating events are confined to this same site, so that C3 activation also occurs on the

Proteins of the classical pathway of complement activation		
Native component	**Active form**	**Function of the active form**
C1 (C1q: $C1r_2$:$C1s_2$)	C1q	Binds directly to pathogen surfaces or indirectly to antibody bound to pathogens, thus allowing autoactivation of C1r
	C1r	Cleaves C1s to active protease
	C1s	Cleaves C4 and C2
C4	C4b	Covalently binds to pathogen and opsonizes it. Binds C2 for cleavage by C1s
	C4a	Peptide mediator of inflammation (weak activity)
C2	C2a	Active enzyme of classical pathway C3/C5 convertase: cleaves C3 and C5
	C2b	Precursor of vasoactive C2 kinin
C3	C3b	Many molecules of C3b bind to pathogen surface and act as opsonins. Initiates amplification via the alternative pathway. Binds C5 for cleavage by C2a
	C3a	Peptide mediator of inflammation (intermediate activity)

Fig. 2.18 The proteins of the classical pathway of complement activation.

surface of the pathogen and not in the plasma or on host-cell surfaces. This is achieved principally by the covalent binding of C4b to the pathogen surface. In innate immunity, C4 cleavage is catalyzed by a C1 or MBL complex bound to the pathogen surface, and so C4b can bind adjacent proteins or carbohydrates on the pathogen surface. If C4b does not rapidly form this bond, the thioester bond is cleaved by reaction with water and C4b is irreversibly inactivated. This helps to prevent C4b from diffusing from its site of activation on the microbial surface and becoming attached to healthy host cells.

C2 becomes susceptible to cleavage by C1s only when it is bound by C4b, and the active C2a serine protease is thereby also confined to the pathogen surface, where it remains associated with C4b, forming the C3 convertase C4b2a. Cleavage of C3 to C3a and C3b is thus also confined to the surface of the pathogen. Like C4b, C3b is inactivated by hydrolysis unless its exposed thioester rapidly makes a covalent bond (see Fig. 2.13), and it therefore opsonizes only the surface on which complement activation has taken place. Opsonization by C3b is more effective when antibodies are also bound to the pathogen surface, as phagocytes have receptors for both complement and antibody (the latter will be described in Chapter 10). Because the reactive forms of C3b and C4b are able to form a covalent bond with any adjacent protein or carbohydrate, when complement is activated by bound antibody a proportion of the reactive C3b or C4b will become linked to the antibody molecules themselves. This combination of antibody chemically cross-linked to complement is likely to be the most efficient trigger for phagocytosis.

2-9 The alternative pathway is an amplification loop for C3b formation that is accelerated by recognition of pathogens by properdin.

Although probably the most ancient of the complement pathways, the alternative pathway is so named because it was discovered as a second, or

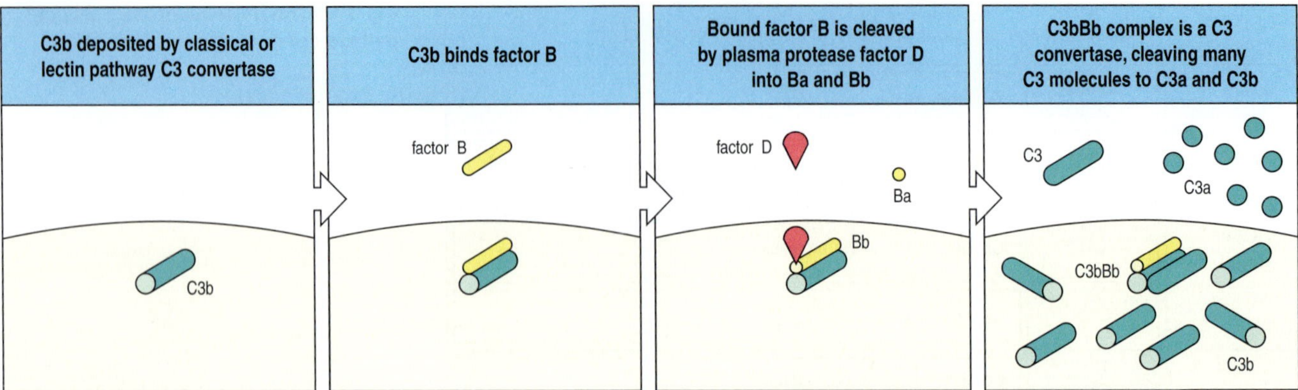

Fig. 2.19 The alternative pathway of complement activation can amplify the classical or the lectin pathway by forming an alternative C3 convertase and depositing more C3b molecules on the pathogen. C3b deposited by the classical or lectin pathway can bind factor B, making it susceptible to cleavage by factor D. The C3bBb complex is the C3 convertase of the alternative pathway of complement activation and its action, like that of C4b2a, results in the deposition of many molecules of C3b on the pathogen surface.

'alternative,' pathway for complement activation after the classical pathway had been defined. Its key features are its ability to be spontaneously activated and its unique C3 convertase, the **alternative pathway C3 convertase**. This is not the C4b2a convertase of the lectin or classical pathways, but is composed of C3b itself bound to Bb, which is a cleavage fragment of the plasma protein **factor B**. This C3 convertase, designated **C3bBb**, has a special place in complement activation because, by producing C3b, it can generate more of itself. This means that once some C3b has been formed, by whichever pathway, the alternative pathway can act as an amplification loop to increase C3b production rapidly.

The alternative pathway can be activated in two different ways. The first is by the action of the lectin or classical pathway. C3b generated by either of these pathways and covalently linked to a microbial surface can bind factor B (Fig. 2.19). This alters the conformation of factor B, enabling a plasma protease called **factor D** to cleave it into Ba and Bb. Bb remains stably associated with C3b, forming the C3bBb C3 convertase. The second way of activating the alternative pathway involves the spontaneous hydrolysis (known as '**tickover**') of the thioester bond in C3 to form C3(H_2O), as shown in Fig. 2.20. C3 is abundant in plasma, and tickover causes a steady low-level production of C3(H_2O). This C3(H_2O) can bind factor B, which is then cleaved by factor D, producing a short-lived **fluid-phase C3 convertase, C3(H_2O)Bb**. Although formed in only small amounts by C3 tickover, fluid-phase C3(H_2O)Bb can cleave many molecules of C3 to C3a and C3b. Much of this C3b is inactivated by hydrolysis, but some attaches covalently via its thioester bond to the surfaces of any microbes present. C3b formed in this way is no different from C3b produced by the lectin or classical pathways and binds factor B, leading to the formation of C3 convertase and stepping-up of C3b production (see Fig. 2.19).

On their own, the alternative pathway C3 convertases are very short-lived. They are, however, stabilized by binding the plasma protein **properdin**

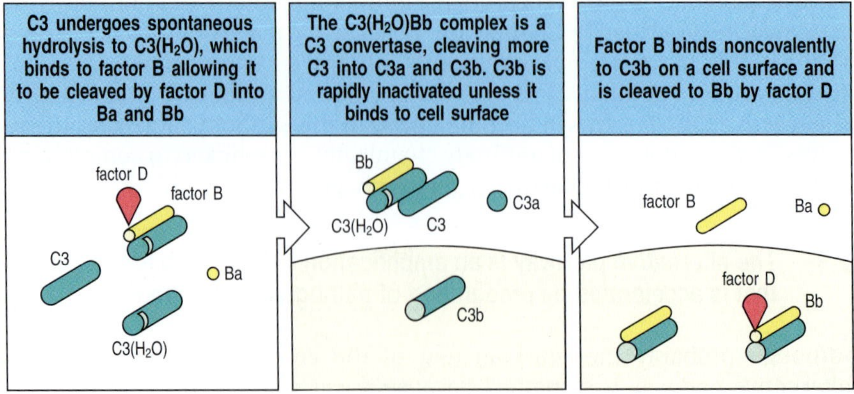

C3 undergoes spontaneous hydrolysis to C3(H2O), which binds to factor B allowing it to be cleaved by factor D into Ba and Bb
The C3(H2O)Bb complex is a C3 convertase, cleaving more C3 into C3a and C3b. C3b is rapidly inactivated unless it binds to cell surface
Factor B binds noncovalently to C3b on a cell surface and is cleaved to Bb by factor D
factor D
factor B
C3
Ba
C3(H2O)
Bb
C3a
C3b

Fig. 2.20 The alternative pathway can be activated by spontaneous activation of C3. Complement component C3 is cleaved spontaneously in plasma to give C3(H_2O), which binds factor B and enables the bound factor B to be cleaved by factor D (first panel). The resulting 'soluble C3 convertase' cleaves C3 to give C3a and C3b, which can attach to host cells or pathogen surfaces (second panel). Covalently bound C3b binds factor B; in turn, factor B is rapidly cleaved by factor D to Bb, which remains bound to C3b to form a C3 convertase (C3bBb), and Ba, which is released (third panel). This convertase is the functional equivalent of C4b2a of the lectin and classical pathways (see Fig. 2.16).

(**factor P**), which binds to C3b or C3(H_2O) (Fig. 2.21). Properdin is made by neutrophils, stored in high concentrations in their secondary granules and released when the neutrophils are activated by the presence of pathogens. As well as its stabilizing function in the alternative pathway, properdin may have some properties of a pattern recognition receptor, promoting binding to microbial surfaces. However, its precise mode of recognition is not known. Properdin-deficient patients are particularly susceptible to infections with *Neisseria meningitidis*, the main agent of bacterial meningitis. It is also known to bind to *N. gonorrhoeae*. By its ability to bind to both bacterial surfaces and to C3b or C3(H_2O), properdin could therefore direct the activity of the alternative complement pathway to pathogen surfaces. Such interactions would bring the alternative pathway into line with the other two complement pathways, which depend on the initial binding of a 'recognition protein' to a pathogen's surface. Properdin can also bind to mammalian cells that are undergoing apoptosis or have been damaged or modified by ischemia, viral infection, or antibody binding, leading to the deposition of C3b on these cells and facilitating their removal by phagocytosis. The distinctive components of the alternative pathway are listed in Fig. 2.22.

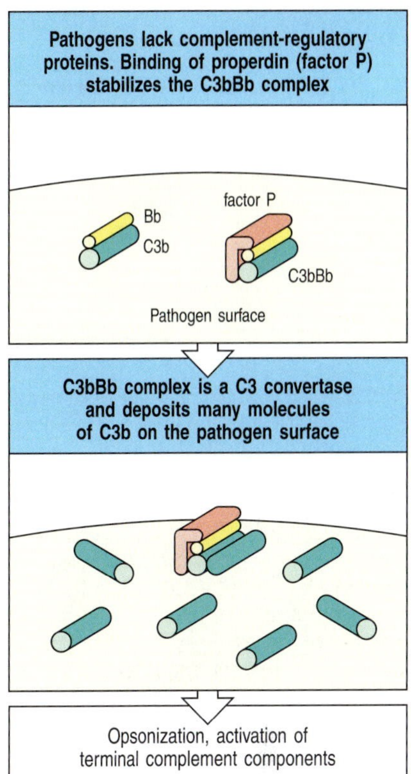
Pathogens lack complement-regulatory proteins. Binding of properdin (factor P) stabilizes the C3bBb complex
Bb
C3b
factor P
C3bBb
Pathogen surface
C3bBb complex is a C3 convertase and deposits many molecules of C3b on the pathogen surface
Opsonization, activation of terminal complement components

Fig. 2.21 Properdin stabilizes the alternative pathway C3 convertase on pathogen surfaces. Bacterial surfaces do not express complement-regulatory proteins and favor the binding of properdin (factor P), which stabilizes the C3bBb convertase. This convertase activity is the equivalent of C4b2a of the classical pathway. C3bBb then cleaves many more molecules of C3, coating the pathogen surface with bound C3b.

2-10 Membrane and plasma proteins that regulate the formation and stability of C3 convertases determine the extent of complement activation under different circumstances.

Several mechanisms ensure that complement activation will proceed only on the surface of a pathogen or on damaged host cells, and not on normal host cells and tissues. After initial complement activation by any pathway, the extent of amplification via the alternative pathway is critically dependent on the stability of the C3 convertase C3bBb. This stability is controlled by both positive and negative regulatory proteins. We have already described how properdin acts as a positive regulatory protein on foreign surfaces, such as those of bacteria or damaged host cells, by stabilizing C3bBb.

Several negative regulatory proteins, present in plasma and in host-cell membranes, protect healthy host cells from the injurious effects of inappropriate complement activation on their surfaces. These **complement regulatory proteins** interact with C3b and either prevent the convertase from forming or promote its rapid dissociation (Fig. 2.23). For example, a membrane-attached protein known as **decay-accelerating factor** (**DAF** or **CD55**) competes with factor B for binding to C3b on the cell surface, and can displace Bb from a

Proteins of the alternative pathway of complement activation

Native component	Active fragments	Function
C3	C3b	Binds to pathogen surface, binds B for cleavage by D, C3bBb is C3 convertase and $C3b_2Bb$ is C5 convertase
Factor B (B)	Ba	Small fragment of B, unknown function
	Bb	Bb is active enzyme of the C3 convertase C3bBb and C5 convertase $C3b_2Bb$
Factor D (D)	D	Plasma serine protease, cleaves B when it is bound to C3b to Ba and Bb
Properdin (P)	P	Plasma protein that binds to bacterial surfaces and stabilizes the C3bBb convertase

Fig. 2.22 The proteins of the alternative pathway of complement activation.

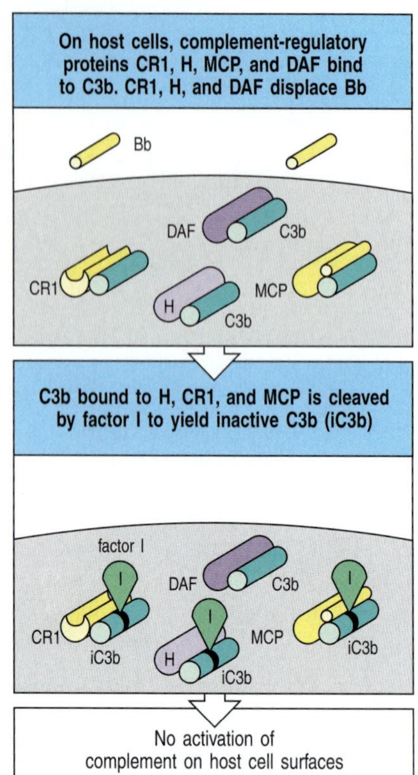
On host cells, complement-regulatory proteins CR1, H, MCP, and DAF bind to C3b. CR1, H, and DAF displace Bb
Bb
DAF
C3b
CR1
MCP
H
C3b
C3b bound to H, CR1, and MCP is cleaved by factor I to yield inactive C3b (iC3b)
factor I
DAF
C3b
CR1
iC3b
MCP
H
iC3b
iC3b
No activation of complement on host cell surfaces

Fig. 2.23 Complement activation spares host cells, which are protected by complement-regulatory proteins. If C3bBb forms on the surface of host cells, it is rapidly inactivated by complement-regulatory proteins expressed by the host cell: complement receptor 1 (CR1), decay-accelerating factor (DAF), and membrane cofactor of proteolysis (MCP). Host cell surfaces also favor the binding of factor H from plasma. CR1, DAF, and factor H displace Bb from C3b, and CR1, MCP, and factor H catalyze the cleavage of bound C3b by the plasma protease factor I to produce inactive C3b (known as iC3b).

convertase that has already formed. Convertase formation can also be prevented by cleaving C3b to an inactive derivative **iC3b**. This is achieved by a plasma protease, **factor I**, in conjunction with C3b-binding proteins that act as cofactors, such as **membrane cofactor of proteolysis** (**MCP** or **CD46**), another host-cell membrane protein (see Fig. 2.23). Cell-surface complement receptor type 1 (**CR1**, also known as CD35) has similar activities to DAF and MCP in the inhibition of C3 convertase formation and the promotion of the catabolism of C3b to inactive products, but has a more limited tissue distribution. **Factor H** is another complement-regulatory protein in plasma that binds C3b and, like CR1, it is able to compete with factor B to displace Bb from the convertase in addition to acting as a cofactor for factor I. Factor H binds preferentially to C3b bound to vertebrate cells because it has an affinity for the sialic acid residues present on the cell surface (see Fig. 2.14). Thus, the amplification loop of the alternative pathway is allowed to proceed on the surface of a pathogen or on damaged host cells, but not on normal host cells or on tissues that express these negative regulatory proteins.

The C3 convertase of the classical and lectin pathways (C4b2a) is molecularly distinct from that of the alternative pathway. However, understanding of the complement system is simplified somewhat by recognition of the close evolutionary relationships between the different complement proteins (Fig. 2.24). Thus the complement zymogens, factor B and C2, are closely related proteins encoded by homologous genes located in tandem within the major histocompatibility complex (MHC) on human chromosome 6. Furthermore, their respective binding partners, C3 and C4, both contain thioester bonds that provide the means of covalently attaching the C3 convertases to a pathogen

Step in pathway	Protein serving function in pathway			Relationship
	Alternative	Lectin	Classical	
Initiating serine protease	D	MASP	C1s	Homologous (C1s and MASP)
Covalent binding to cell surface	C3b	C4b		Homologous
C3/C5 convertase	Bb	C2a		Homologous
Control of activation	CR1 H	CR1 C4BP		Identical Homologous
Opsonization	C3b			Identical
Initiation of effector pathway	C5b			Identical
Local inflammation	C5a, C3a			Identical
Stabilization	P	None		Unique

Fig. 2.24 There is a close evolutionary relationship between the factors of the alternative, lectin, and classical pathways of complement activation. Most of the factors are either identical or the homologous products of genes that have duplicated and then diverged in sequence. The proteins C4 and C3 are homologous and contain the unstable thioester bond by which their large fragments, C4b and C3b, bind covalently to membranes. The genes encoding proteins C2 and factor B are adjacent in the MHC region of the genome and arose by gene duplication. The regulatory proteins factor H, CR1, and C4BP share a repeat sequence common to many complement-regulatory proteins. The greatest divergence between the pathways is in their initiation: in the classical pathway the C1 complex binds either to certain pathogens or to bound antibody, and in the latter circumstance it serves to convert antibody binding into enzyme activity on a specific surface; in the lectin pathway, mannose-binding lectin (MBL) associates with a serine protease, activating MBL-associated serine protease (MASP), to serve the same function as C1r:C1s; in the alternative pathway this enzyme activity is provided by factor D.

surface. Only one component of the alternative pathway seems entirely unrelated to its functional equivalents in the classical and lectin pathways: the initiating serine protease, factor D. Factor D can also be singled out as the only activating protease of the complement system to circulate as an active enzyme rather than a zymogen. This is both necessary for the initiation of the alternative pathway (through the cleavage of factor B bound to spontaneously activated C3) and safe for the host, because factor D has no other substrate than factor B bound to C3b. This means that factor D finds its substrate only at pathogen surfaces and at a very low level in plasma, where the alternative pathway of complement activation can be allowed to proceed.

2-11 Complement developed early in the evolution of multicellular organisms.

The complement system was originally known only from vertebrates, but homologs of C3 and factor B and the existence of a prototypical 'alternative pathway' have been discovered in nonchordate invertebrates. C3, which is cleaved and activated by serine proteases, is evolutionarily related to the serine protease inhibitor α_2-macroglobulin. The amplification loop of the vertebrate alternative pathway, based on a C3 convertase formed by C3 and factor B, is present in echinoderms (sea urchins and sea stars). The echinoderm versions of C3 and factor B are expressed by amoeboid coelomocytes, phagocytic cells in the coelomic fluid, and the expression of C3 is increased when bacteria are present. This simple system seems to function to opsonize bacterial cells and other foreign particles and facilitate their uptake by coelomocytes. C3 homologs in invertebrates are clearly related to each other and all contain the distinctive thioester linkage characteristic of this family of proteins (the thioester proteins, or TEPs). In the mosquito *Anopheles*, the protein TEP1 is induced in response to infection. There is also direct evidence of binding of *Anopheles* TEP1 to bacterial surfaces, and of its involvement in the phagocytosis of Gram-negative bacteria. Some form of C3 function may even pre-date the evolution of the Bilateria—animals with bilateral symmetry—because genomic evidence of C3, factor B, and some later-acting complement components has been found in the Anthozoa (corals and sea anemones).

After its initial appearance, the complement system seems to have evolved by the acquisition of new activation pathways, allowing microbial surfaces to be specifically targeted. The first of these new complement-activating systems to appear is likely to have been the ficolin pathway, which is present both in vertebrates and in some closely related invertebrates, such as the urochordates. Evolutionarily, the ficolins may pre-date the collectins, which are also first seen in the urochordates. Homologs of both MBL and the classical pathway complement component C1q, another collectin, have been identified in the genome of the ascidian urochordate *Ciona*. This suggests that in the evolution of the antibody-mediated classical pathway of complement activation, the ancestral immunoglobulin molecule, which did not appear until much later in evolution, took advantage of an already diversified family of collectins, rather than driving the diversification of C1q from an MBL-like ancestor.

Two distinct invertebrate homologs of mammalian MASPs have been identified in the same ascidian species from which the ficolins were identified. The specificity of the invertebrate MASPs has not been determined, but it seems likely that they are able to cleave and activate C3. This invertebrate ficolin complement system is functionally identical to the ficolin- and MBL-mediated pathways found in mammals. Thus, the minimal complement system of the echinoderms has been supplemented in the urochordates by the recruitment of a specific activation system that can target the deposition of C3 onto microbial surfaces. After the evolution of the specific antigen-recognition molecules of the adaptive immune system, the complement activation system evolved further by diversification of a C1q-like collectin and its associated MASPs to become the initiating components of the classical complement pathway, namely C1q, C1r, and C1s.

2-12 Surface-bound C3 convertase deposits large numbers of C3b fragments on pathogen surfaces and generates C5 convertase activity.

We now return to the present-day complement system. The formation of C3 convertases is the point at which the three pathways of complement activation converge. The convertase of the lectin and classical pathways, C4b2a, and the convertase of the alternative pathway, C3bBb, initiate the same subsequent events—they cleave C3 to C3b and C3a. C3b binds covalently through its thioester bond to adjacent molecules on the pathogen surface; otherwise it is inactivated by hydrolysis. C3 is the most abundant complement protein in plasma, occurring at a concentration of 1.2 mg/ml, and up to 1000 molecules of C3b can bind in the vicinity of a single active C3 convertase (see Fig. 2.19). Thus, the main effect of complement activation is to deposit large quantities of C3b on the surface of the infecting pathogen, where it forms a covalently bonded coat that can signal the ultimate destruction of the pathogen by phagocytes.

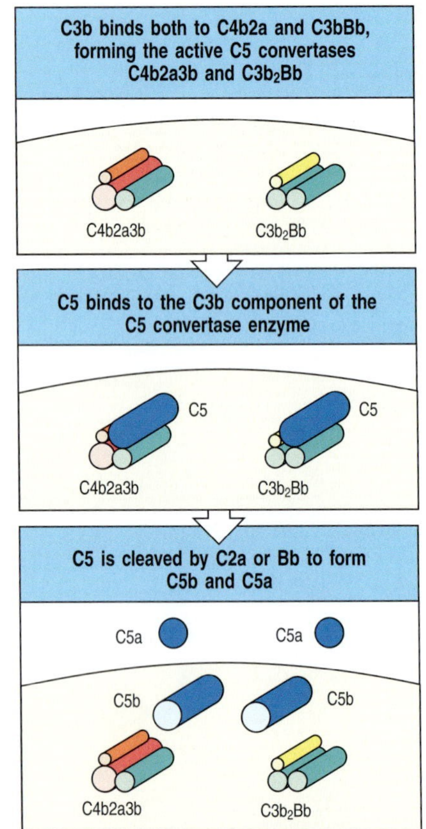
C3b binds both to C4b2a and C3bBb, forming the active C5 convertases C4b2a3b and C3b$_2$Bb
C4b2a3b
C3b$_2$Bb
C5 binds to the C3b component of the C5 convertase enzyme
C5
C5
C4b2a3b
C3b$_2$Bb
C5 is cleaved by C2a or Bb to form C5b and C5a
C5a
C5a
C5b
C5b
C4b2a3b
C3b$_2$Bb

Fig. 2.25 Complement component C5 is cleaved when captured by a C3b molecule that is part of a C5 convertase complex. As shown in the top panel, C5 convertases are formed when C3b binds either the classical or lectin pathway C3 convertase C4b2a to form C4b2a3b, or the alternative pathway C3 convertase C3bBb to form C3b$_2$Bb. C5 binds to C3b in these complexes (center panel). The bottom panel shows how C5 is cleaved by the active enzyme C2a or Bb to form C5b and the inflammatory mediator C5a. Unlike C3b and C4b, C5b is not covalently bound to the cell surface. The production of C5b initiates the assembly of the terminal complement components.

The next step in the complement cascade is the generation of the C5 convertases. C5 is a member of the same family of proteins as C3, C4, α_2-macroglobulin, and the thioester-containing proteins (TEPs) of invertebrates. C5 does not form an active thioester bond during its synthesis but, like C3 and C4, it is cleaved by a specific protease into C5a and C5b fragments, each of which exerts specific downstream actions that are important in propagating the complement cascade. In the classical and the lectin pathways, a C5 convertase is formed by the binding of C3b to C4b2a to yield C4b2a3b. The C5 convertase of the alternative pathway is formed by the binding of C3b to the C3bBb convertase to form C3b$_2$Bb. A C5 is captured by these C5 convertase complexes through binding to an acceptor site on C3b, and is thus rendered susceptible to cleavage by the serine protease activity of C2a or Bb. This reaction, which generates C5b and C5a, is much more limited than cleavage of C3, because C5 can be cleaved only when it binds to C3b that is in turn bound to C4b2a or C3bBb to form the active C5 convertase complex. Thus, complement activated by all three pathways leads to the binding of large numbers of C3b molecules on the surface of the pathogen, the generation of a more limited number of C5b molecules, and the release of C3a and a smaller amount of C5a (Fig. 2.25).

2-13 Ingestion of complement-tagged pathogens by phagocytes is mediated by receptors for the bound complement proteins.

The most important action of complement is to facilitate the uptake and destruction of pathogens by phagocytic cells. This occurs by the specific recognition of bound complement components by **complement receptors (CRs)** on phagocytes. These complement receptors bind pathogens opsonized with complement components: opsonization of pathogens is a major function of C3b and its proteolytic derivatives. C4b also acts as an opsonin but has a relatively minor role, largely because so much more C3b than C4b is generated.

The seven known types of receptors for bound complement components are listed, with their functions and distributions, in Fig. 2.26. The best characterized is the C3b receptor CR1, which has already been described in its role as a negative regulator of complement activation (see Fig. 2.23). CR1 is expressed on many types of immune cells including macrophages and neutrophils. Binding of C3b to CR1 cannot by itself stimulate phagocytosis, but it can lead

Receptor	Specificity	Functions	Cell types
CR1 (CD35)	C3b, C4bi	Promotes C3b and C4b decay Stimulates phagocytosis (requires C5a) Erythrocyte transport of immune complexes	Erythrocytes, macrophages, monocytes, polymorphonuclear leukocytes, B cells, FDC
CR2 (CD21)	C3d, iC3b, C3dg, Epstein–Barr virus	Part of B-cell co-receptor Epstein–Barr virus receptor	B cells, FDC
CR3 (Mac-1) (CD11b/CD18)	iC3b	Stimulates phagocytosis	Macrophages, monocytes, polymorphonuclear leukocytes, FDC
CR4 (gp150, 95) (CD11c/CD18)	iC3b	Stimulates phagocytosis	Macrophages, monocytes, polymorphonuclear leukocytes, dendritic cells
CRIg	C3b, iC3b	Phagocytosis of circulating pathogens	Tissue-resident macrophages Hepatic sinusoid macrophages
C5a receptor	C5a	Binding of C5a activates G protein	Endothelial cells, mast cells, phagocytes
C3a receptor	C3a	Binding of C3a activates G protein	Endothelial cells, mast cells, phagocytes

Fig. 2.26 Distribution and function of cell-surface receptors for complement proteins. A variety of complement receptors are specific for bound C3b and its further cleavage products (iC3b and C3dg). CR1 and CR3 are especially important in inducing the phagocytosis of bacteria with complement components bound to their surface. CR2 is found mainly on B cells, where it is part of the B-cell co-receptor complex and the receptor by which the Epstein–Barr virus selectively infects B cells, causing infectious mononucleosis. CR1 and CR2 share structural features with the complement-regulatory proteins that bind C3b and C4b. CR3 and CR4 are integrins; CR3 is also important for leukocyte adhesion and migration, as we shall see in Chapter 3, whereas CR4 is only known to function in phagocytosis. The receptors for C5a and C3a are seven-span G-protein-coupled receptors. FDC, follicular dendritic cells; these are not involved in innate immunity and are discussed in later chapters.

to phagocytosis in the presence of other immune mediators that activate macrophages. For example, the small complement fragment C5a can activate macrophages to ingest bacteria bound to their CR1 receptors (Fig. 2.27). C5a binds to another receptor expressed by macrophages, the **C5a receptor**, which has seven membrane-spanning domains. Receptors of this type transduce their signals via intracellular guanine-nucleotide-binding proteins called G proteins and are known generally as G-protein-coupled receptors (GPCRs) (see Section 3-2). Proteins associated with the extracellular matrix, such as fibronectin, can also contribute to phagocyte activation; these are encountered when phagocytes are recruited to connective tissue and activated there.

Four other complement receptors—**CR2** (also known as **CD21**), **CR3** (**CD11b:CD18**), **CR4** (**CD11c:CD18**), and **CRIg** (complement receptor of the immunoglobulin family)—bind to inactivated forms of C3b that remain

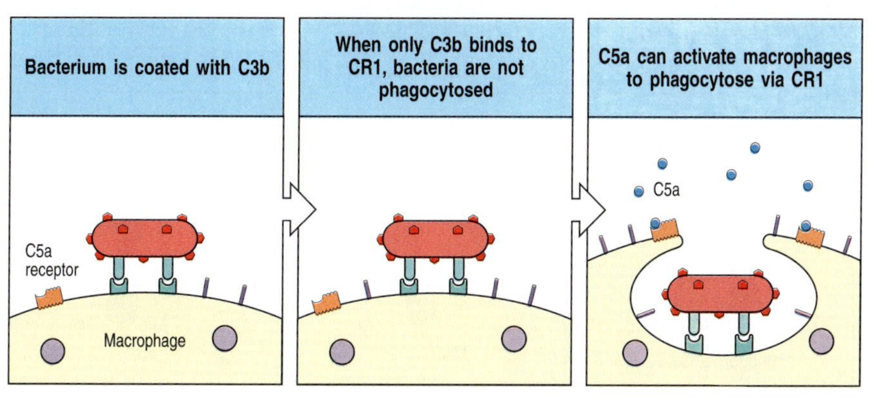

Fig. 2.27 The anaphylatoxin C5a can enhance the phagocytosis of microorganisms opsonized in an innate immune response. Activation of complement leads to the deposition of C3b on the surface of microorganisms (left panel). C3b can be bound by the complement receptor CR1 on the surface of phagocytes, but this on its own is insufficient to induce phagocytosis (center panel). Phagocytes also express receptors for the anaphylatoxin C5a, and binding of C5a will now activate the cell to phagocytose microorganisms bound through CR1 (right panel).

attached to the pathogen surface. Like several other key components of complement, C3b is subject to regulatory mechanisms that cleave it into derivatives, such as iC3b, that cannot form an active convertase. C3b bound to the microbial surface can be cleaved by factor I and MCP to remove C3f, leaving the inactive iC3b form bound to the surface (Fig. 2.28). iC3b is recognized by several complement receptors—CR2, CR3, CR4, and CRIg. Unlike the binding of iC3b to CR1, the binding of iC3b to the receptor CR3 is sufficient on its own to stimulate phagocytosis. Factor I and CR1 cleave iC3b to release C3c, leaving C3dg bound. C3dg is recognized only by CR2. CR2 is found on B cells as part of a co-receptor complex that can augment the signal received through the antigen-specific immunoglobulin receptor. Thus, a B cell whose antigen receptor is specific for a given pathogen will receive a strong signal on binding this pathogen if it is also coated with C3dg. The activation of complement can therefore contribute to producing a strong antibody response, providing an example of the interplay between the innate and adaptive immune responses, as discussed in more detail in Chapter 10.

The importance of opsonization by C3b and its inactive fragments in destroying extracellular pathogens can be seen in the effects of various complement deficiencies. For example, individuals deficient in C3 or in molecules that catalyze C3b deposition show an increased susceptibility to infection by a wide range of extracellular bacteria. We describe the effects of various defects in complement in Chapter 12.

2-14 The small fragments of some complement proteins initiate a local inflammatory response.

The small complement fragments C3a and C5a act on specific receptors on endothelial cells and mast cells (see Fig. 2.26) to produce local inflammatory responses. Like C5a, C3a also signals through a G-protein-coupled receptor (discussed in more detail in Chapter 3). When produced in large amounts or injected systemically, they induce a generalized circulatory collapse, producing a shock-like syndrome similar to that seen in a systemic allergic reaction involving antibodies of the IgE class (which is discussed in Chapter 14). Such a reaction is termed **anaphylactic shock**, and these small fragments of complement are therefore often referred to as **anaphylatoxins**. Of the three, C5a has the highest specific biological activity. All three induce the contraction of smooth muscle and increase vascular permeability, but C5a and C3a also act on the endothelial cells lining blood vessels to induce the synthesis of adhesion molecules. In addition, C3a and C5a can activate the mast cells that populate submucosal tissues to release inflammatory molecules such as histamine and the cytokine tumor necrosis factor (TNF)-α, which cause similar effects. The changes induced by C5a and C3a recruit antibody, complement, and phagocytic cells to the site of an infection (Fig. 2.29), and

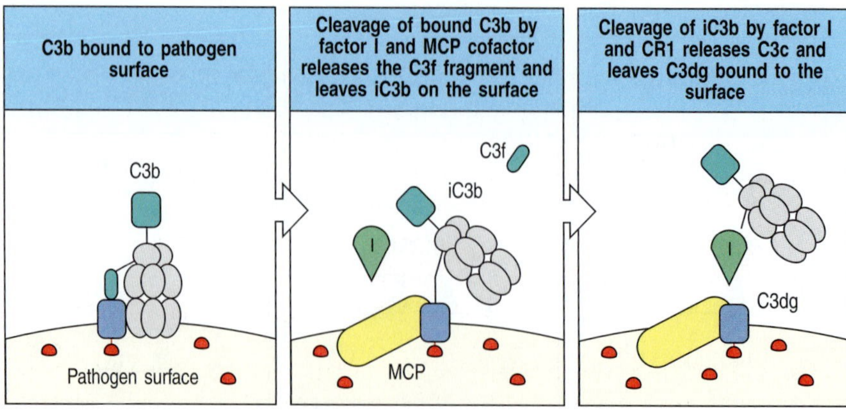

C3b bound to pathogen surface
Cleavage of bound C3b by factor I and MCP cofactor releases the C3f fragment and leaves iC3b on the surface
Cleavage of iC3b by factor I and CR1 releases C3c and leaves C3dg bound to the surface
C3b
Pathogen surface
C3f
iC3b
I
MCP
C3dg

Fig. 2.28 The cleavage products of C3b are recognized by different complement receptors. After C3b is deposited on the surface of pathogens, it can undergo several conformational changes that alter its interaction with complement receptors. Factor I and MCP can cleave the C3f fragment from C3b, producing iC3b, which is a ligand for the complement receptors CR2, CR3 and CR4, but not CR1. Factor I and CR1 cleave iC3b to release C3c, leaving C3dg bound, which is recognized by CR2.

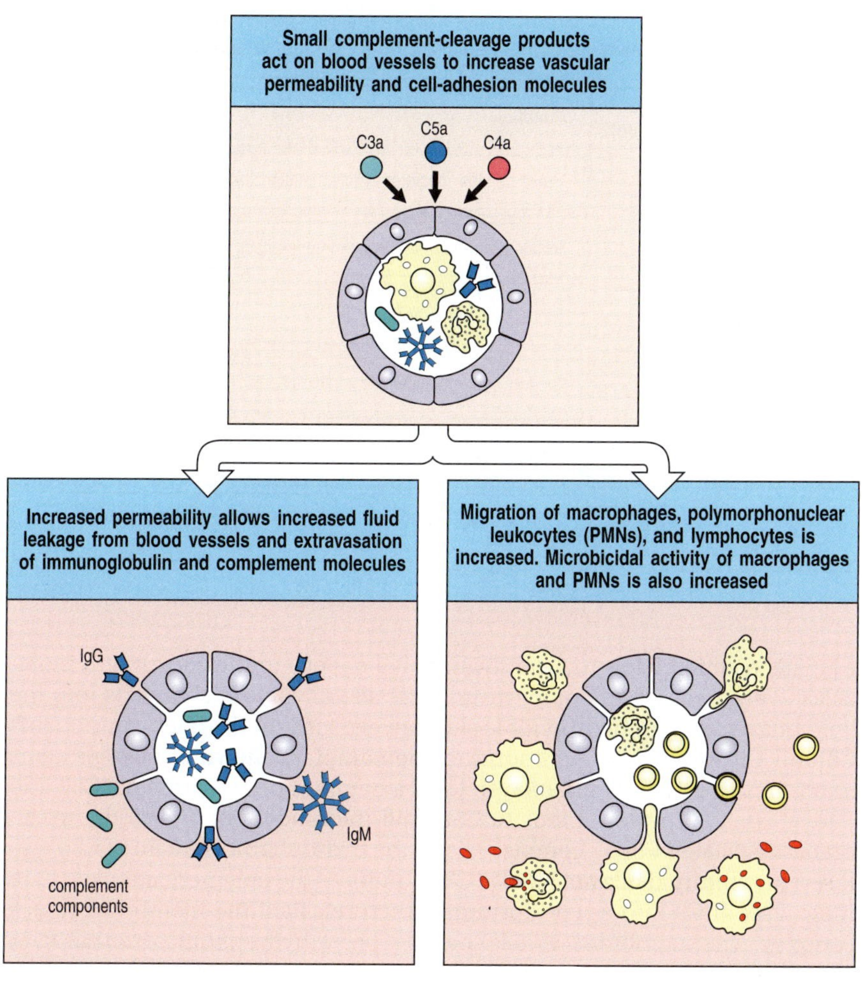

Small complement-cleavage products act on blood vessels to increase vascular permeability and cell-adhesion molecules
C3a
C5a
C4a
Increased permeability allows increased fluid leakage from blood vessels and extravasation of immunoglobulin and complement molecules
IgG
IgM
complement components
Migration of macrophages, polymorphonuclear leukocytes (PMNs), and lymphocytes is increased. Microbicidal activity of macrophages and PMNs is also increased

Fig. 2.29 Local inflammatory responses can be induced by small complement fragments, especially C5a. The small complement fragments are differentially active: C5a is more active than C3a, which is more active than C4a. They cause local inflammatory responses by acting directly on local blood vessels, stimulating an increase in blood flow, increased vascular permeability, and increased binding of phagocytes to endothelial cells. C5a also activates mast cells (not shown) to release mediators, such as histamine and TNF-α, that contribute to the inflammatory response. The increase in vessel diameter and permeability leads to the accumulation of fluid and protein. Fluid accumulation increases lymphatic drainage, bringing pathogens and their antigenic components to nearby lymph nodes. The antibodies, complement, and cells thus recruited participate in pathogen clearance by enhancing phagocytosis. The small complement fragments can also directly increase the activity of the phagocytes.

the increased fluid in the tissues hastens the movement of pathogen-bearing antigen-presenting cells to the local lymph nodes, contributing to the prompt initiation of the adaptive immune response.

C5a also acts directly on neutrophils and monocytes to increase their adherence to vessel walls, their migration toward sites of antigen deposition, and their ability to ingest particles. C5a also increases the expression of CR1 and CR3 on the surfaces of these cells. In this way, C5a, and to a smaller extent C3a and C4a, act in concert with other complement components to hasten the destruction of pathogens by phagocytes.

2-15 The terminal complement proteins polymerize to form pores in membranes that can kill certain pathogens.

One of the important effects of complement activation is the assembly of the terminal components of complement (Fig. 2.30) to form a membrane-attack complex. The reactions leading to the formation of this complex are shown schematically in Fig. 2.31. The end result is a pore in the lipid bilayer membrane that destroys membrane integrity. This is thought to kill the pathogen by destroying the proton gradient across the pathogen's cell membrane.

The first step in the formation of the membrane-attack complex is the cleavage of C5 by a C5 convertase to release C5b (see Fig. 2.25). In the next stages, shown in Fig. 2.31, C5b initiates the assembly of the later complement components and their insertion into the cell membrane. First, one molecule of C5b binds one molecule of **C6**, and the C5b6 complex then binds one molecule of

Fig. 2.30 The terminal complement components.

The terminal complement components that form the membrane-attack complex		
Native protein	**Active component**	**Function**
C5	C5a	Small peptide mediator of inflammation (high activity)
	C5b	Initiates assembly of the membrane-attack system
C6	C6	Binds C5b; forms acceptor for C7
C7	C7	Binds C5b6; amphiphilic complex inserts into lipid bilayer
C8	C8	Binds C5b67; initiates C9 polymerization
C9	$C9_n$	Polymerizes to C5b678 to form a membrane-spanning channel, lysing cell

C7. This reaction leads to a conformational change in the constituent molecules, with the exposure of a hydrophobic site on C7, which inserts into the lipid bilayer. Similar hydrophobic sites are exposed on the later components **C8** and **C9** when they are bound to the complex, allowing these proteins also to insert into the lipid bilayer. C8 is a complex of two proteins, C8β and C8α-γ. The C8β protein binds to C5b, and the binding of C8β to the membrane-associated C5b67 complex allows the hydrophobic domain of C8α-γ to insert into the lipid bilayer. Finally, C8α-γ induces the polymerization of 10–16 molecules of C9 into a pore-forming structure called the membrane-attack complex. The membrane-attack complex has a hydrophobic external face, allowing it to associate with the lipid bilayer, but a hydrophilic internal channel. The diameter of this channel is about 100 Å, allowing the free passage of solutes and water across the lipid bilayer. The disruption of the lipid bilayer leads to the loss of cellular homeostasis, the disruption of the proton gradient across the membrane, the penetration of enzymes such as lysozyme into the cell, and the eventual destruction of the pathogen.

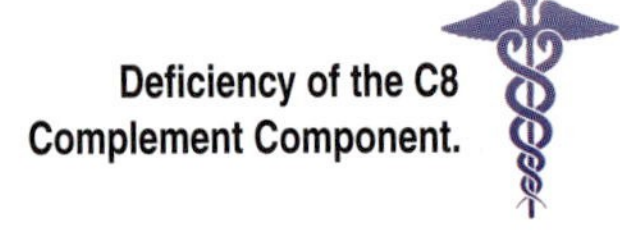

Although the effect of the membrane-attack complex is very dramatic, particularly in experimental demonstrations in which antibodies against red blood cell membranes are used to trigger the complement cascade, the significance of these components in host defense seems to be quite limited. So far, deficiencies in complement components C5–C9 have been associated with susceptibility only to *Neisseria* species, the bacteria that cause the sexually transmitted disease gonorrhea and a common form of bacterial meningitis. Thus, the opsonizing and inflammatory actions of the earlier components of the complement cascade are clearly more important for host defense against infection. Formation of the membrane-attack complex seems to be important only for the killing of a few pathogens, although, as we will see in Chapter 15, it might well have a major role in immunopathology.

2-16 Complement control proteins regulate all three pathways of complement activation and protect the host from their destructive effects.

Given the destructive effects of complement, and the way in which its activation is rapidly amplified through a triggered-enzyme cascade, it is not surprising that there are several mechanisms to prevent its uncontrolled activation. As we have seen, the effector molecules of complement are generated through

C5b binds C6 and C7
C6
C7
C5b
lipid bilayer
Pathogen
C5b67 complexes bind to membrane via C7
C5b67 complex
C8 binds to the complex and inserts into the cell membrane
C8
C9 molecules bind to the complex and polymerize
C9
10–16 molecules of C9 bind to form a pore in the membrane
Membrane lesions—end on (rings)
Membrane lesions—side on (tubes)
Schematic representation of the membrane-attack complex pore
15 nm
3 nm
10 nm

the sequential activation of zymogens, which are present in plasma in an inactive form. The activation of these zymogens usually occurs on a pathogen surface, and the activated complement fragments produced in the ensuing cascade of reactions usually bind nearby or are rapidly inactivated by hydrolysis. These two features of complement activation act as safeguards against uncontrolled activation. Even so, all complement components are activated spontaneously at a low rate in plasma, and activated complement components will sometimes bind proteins on host cells. The potentially damaging consequences are prevented by a series of complement control proteins, summarized in Fig. 2.32, which regulate the complement cascade at different points. As we saw in discussing the alternative pathway of complement activation (see Section 2-9), many of these control proteins specifically protect normal host cells while allowing complement activation to proceed on pathogen surfaces. The complement control proteins therefore allow complement to distinguish self from nonself.

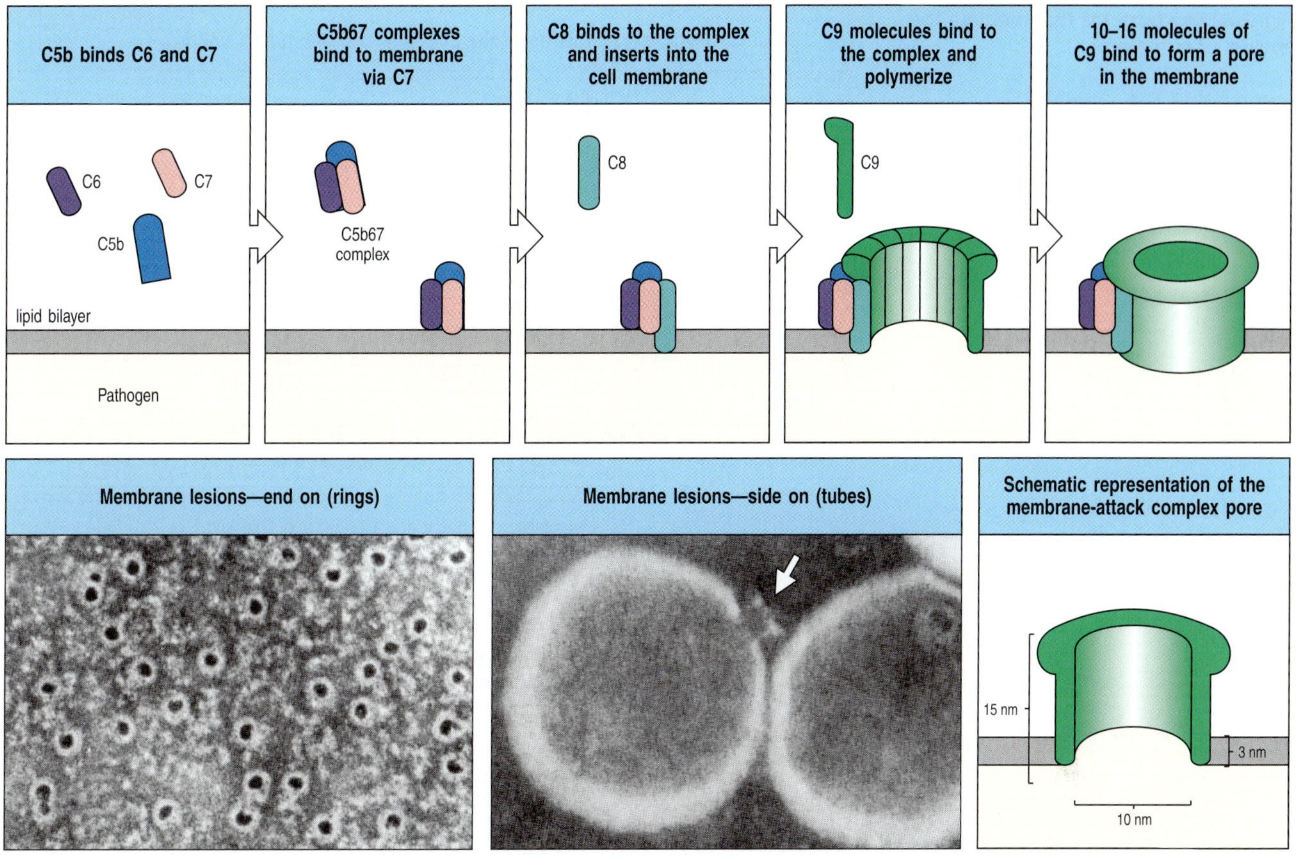

The reactions that regulate the complement cascade are shown in Fig. 2.33. The top two panels show how the activation of C1 is controlled by the C1 inhibitor (**C1INH**), which is a plasma serine protease inhibitor or serpin. C1INH binds the active enzymes C1r:C1s and causes them to dissociate from C1q, which remains bound to the pathogen. In this way, C1INH limits the time during which active C1s is able to cleave C4 and C2. By the same means, C1INH limits the spontaneous activation of C1 in plasma. Its importance can be seen in the C1INH deficiency disease **hereditary angioedema** (**HAE**), in which chronic spontaneous complement activation leads to the production of excess cleaved fragments of C4 and C2. The small fragment of C2, C2b, is further cleaved into a peptide, the C2 kinin, which causes extensive swelling—the most dangerous being local swelling in the larynx, which can lead to suffocation. Bradykinin, which has similar actions to C2 kinin, is also produced in an uncontrolled fashion in this disease, as a result of the lack of

Fig. 2.31 Assembly of the membrane-attack complex generates a pore in the lipid bilayer membrane. The sequence of steps and their approximate appearance are shown here in schematic form. C5b triggers the assembly of a complex of one molecule each of C6, C7, and C8, in that order. C7 and C8 undergo conformational changes, exposing hydrophobic domains that insert into the membrane. This complex causes moderate membrane damage in its own right, and also serves to induce the polymerization of C9, again with the exposure of a hydrophobic site. Up to 16 molecules of C9 are then added to the assembly to generate a channel 100 Å in diameter in the membrane. This channel disrupts the bacterial cell membrane, killing the bacterium. The electron micrographs show erythrocyte membranes with membrane-attack complexes in two orientations, end on and side on. Photographs courtesy of S. Bhakdi and J. Tranum-Jensen.

Hereditary Angioedema

Fig. 2.32 The proteins that regulate the activity of complement.

Regulatory proteins of the classical and alternative pathways	
Name (symbol)	**Role in the regulation of complement activation**
C1 inhibitor (C1INH)	Binds to activated C1r, C1s, removing them from C1q, and to activated MASP-2, removing it from MBL
C4-binding protein (C4BP)	Binds C4b, displacing C2a; cofactor for C4b cleavage by I
Complement receptor 1 (CR1)	Binds C4b, displacing C2a, or C3b displacing Bb; cofactor for I
Factor H (H)	Binds C3b, displacing Bb; cofactor for I
Factor I (I)	Serine protease that cleaves C3b and C4b; aided by H, MCP, C4BP, or CR1
Decay-accelerating factor (DAF)	Membrane protein that displaces Bb from C3b and C2a from C4b
Membrane cofactor protein (MCP)	Membrane protein that promotes C3b and C4b inactivation by I
CD59 (protectin)	Prevents formation of membrane-attack complex on autologous or allogeneic cells. Widely expressed on membranes

inhibition of another plasma protease, kallikrein, a component of the kinin system that we discuss in Section 3-3, which is activated by tissue damage and is also regulated by C1INH. This disease is fully corrected by replacing C1INH. The large activated fragments of C4 and C2, which normally combine to form the C3 convertase, do not damage host cells in such patients because C4b is rapidly inactivated by hydrolysis in plasma, and the convertase does not form. Furthermore, any convertase that accidentally forms on a host cell is inactivated by the mechanisms described below.

The thioester bond of activated C3 and C4 is extremely reactive and has no mechanism for distinguishing an acceptor hydroxyl or amine group on a host cell from a similar group on the surface of a pathogen. A series of protective mechanisms, mediated by other proteins, has evolved to ensure that the binding of a small number of C3 or C4 molecules to host cell membranes results in minimal formation of C3 convertase and little amplification of complement activation. We have already encountered most of these mechanisms in the control of the alternative pathway (see Fig. 2.23), but we will consider them again here because they are important regulators of the classical pathway convertase as well (see Fig. 2.33, second and third rows). The mechanisms can be divided into three categories. Proteins in the first group catalyze the cleavage of any C3b or C4b that does bind to host cells into inactive products. The complement-regulatory enzyme responsible is the plasma serine protease factor I; it circulates in active form but can cleave C3b and C4b only when they are bound to a membrane cofactor protein. In these circumstances, factor I cleaves C3b, first into iC3b and then further to C3dg, thus permanently inactivating it. C4b is similarly inactivated by cleavage into C4c and C4d. Two cell-membrane proteins bind C3b and C4b and possess cofactor activity for factor I; these are MCP and CR1 (see Section 2-10). Microbial cell walls lack these protective proteins and cannot promote the breakdown of C3b and C4b. Instead, these proteins act as binding sites for factor B and C2, promoting complement activation. The importance of factor I can be seen in people with genetically determined **Factor I Deficiency**. Because of uncontrolled complement activation, complement proteins rapidly become depleted and such people suffer repeated bacterial infections, especially with ubiquitous pyogenic bacteria.

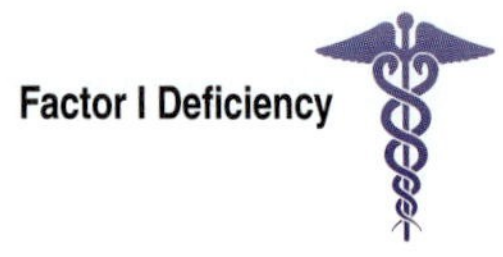

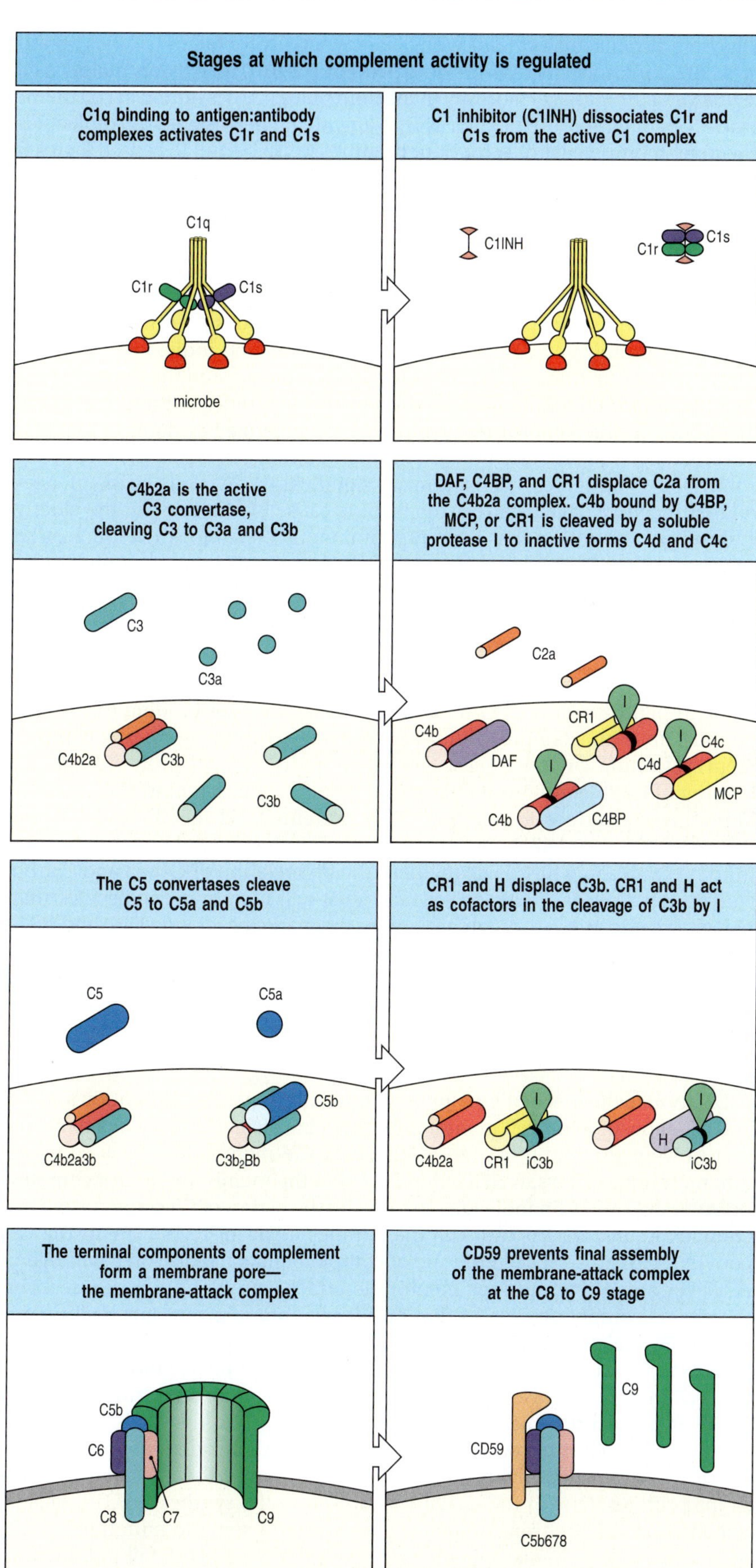
Stages at which complement activity is regulated
C1q binding to antigen:antibody complexes activates C1r and C1s
C1q
C1r
C1s
microbe
C1 inhibitor (C1INH) dissociates C1r and C1s from the active C1 complex
C1INH
C1r
C1s
C4b2a is the active C3 convertase, cleaving C3 to C3a and C3b
C3
C3a
C4b2a
C3b
C3b
DAF, C4BP, and CR1 displace C2a from the C4b2a complex. C4b bound by C4BP, MCP, or CR1 is cleaved by a soluble protease I to inactive forms C4d and C4c
C2a
CR1
I
C4b
DAF
C4c
C4d
MCP
C4b
C4BP
The C5 convertases cleave C5 to C5a and C5b
C5
C5a
C5b
C4b2a3b
C3b2Bb
CR1 and H displace C3b. CR1 and H act as cofactors in the cleavage of C3b by I
C4b2a
CR1
iC3b
H
iC3b
The terminal components of complement form a membrane pore— the membrane-attack complex
C5b
C6
C8
C7
C9
CD59 prevents final assembly of the membrane-attack complex at the C8 to C9 stage
C9
CD59
C5b678

Fig. 2.33 Complement activation is regulated by a series of proteins that serve to protect host cells from accidental damage. These act on different stages of the complement cascade, dissociating complexes or catalyzing the enzymatic degradation of covalently bound complement proteins. Stages in the complement cascade are shown schematically down the left side of the figure, with the regulatory reactions on the right. The alternative pathway C3 convertase is similarly regulated by DAF, CR1, MCP, and factor H.

There are also plasma proteins with cofactor activity for factor I. C4b is bound by a cofactor known as **C4b-binding protein** (**C4BP**), which acts mainly as a regulator of the classical pathway in the fluid phase. C3b is bound at cell membranes by cofactor proteins such as DAF and MCP. These regulatory molecules effectively compete with factor B for binding to C3b bound to cells. If factor B 'wins,' as typically happens on a pathogen surface, then more C3bBb C3 convertase forms and complement activation is amplified. If DAF and MCP win, as occurs on cells of the host, then the bound C3b is catabolized by factor I to iC3b and C3dg and complement activation is inhibited.

The critical balance between the inhibition and the activation of complement on cell surfaces is illustrated in individuals heterozygous for mutations in the regulatory proteins MCP, factor I, or factor H. In such individuals the concentration of functional regulatory proteins is reduced, and the tipping of the balance toward complement activation leads to a predisposition to **atypical hemolytic uremic syndrome**, a condition characterized by damage to platelets and red blood cells and by kidney inflammation, as a result of ineffectively controlled complement activation. A significantly increased risk of **age-related macular degeneration**, the leading cause of blindness in the elderly in developed countries, has been predominantly linked to single-nucleotide polymorphisms in the factor H genes. Polymorphisms in other complement genes have also been found to be either detrimental or protective for this disease. Thus, even small alterations in the efficiency of either the activation or the regulation of this powerful effector system can contribute to the progression of degenerative or inflammatory disorders.

The competition between DAF or MCP and factor B for binding to surface-bound C3b is an example of the second mechanism for inhibiting complement activation on host cells. Several proteins competitively inhibit the binding of C2 to cell-bound C4b and of factor B to cell-bound C3b, thereby inhibiting convertase formation. These proteins bind to C3b and C4b on the cell surface, and also mediate protection against complement through a third mechanism, which is to augment the dissociation of C4b2a and C3bBb convertases that have already formed. Host-cell membrane molecules that regulate complement through both these mechanisms include DAF and CR1, which promote the dissociation of convertase in addition to their cofactor activity. All the proteins that bind the homologous C4b and C3b molecules share one or more copies of a structural element called the short consensus repeat (SCR), complement control protein (CCP) repeat, or (especially in Japan) the sushi domain.

In addition to the mechanisms for preventing C3 convertase formation and C4 and C3 deposition on cell membranes, there are also inhibitory mechanisms that prevent the inappropriate insertion of the membrane-attack complex into membranes. We saw in Section 2-15 that the membrane-attack complex polymerizes onto C5b molecules created by the action of C5 convertase. This complex mainly inserts into cell membranes adjacent to the site of the C5 convertase; that is, close to the site of complement activation on a pathogen. However, some newly formed membrane-attack complexes may diffuse from the site of complement activation and insert into adjacent host-cell membranes. Several plasma proteins including, notably, vitronectin, also known as S-protein, bind to the C5b67 complex and thereby inhibit its random insertion into cell membranes. Host-cell membranes also contain an intrinsic protein, **CD59** or **protectin**, which inhibits the binding of C9 to the C5b678 complex (see Fig. 2.33, bottom row). CD59 and DAF are both linked to the cell surface by a glycosylphosphatidylinositol (GPI) tail, like many other peripheral membrane proteins. One of the enzymes involved in the synthesis of GPI tails is encoded on the X chromosome. In people with a somatic mutation in this gene in a clone of hematopoietic cells, both CD59 and DAF fail to function.

This causes the disease **paroxysmal nocturnal hemoglobinuria**, which is characterized by episodes of intravascular red blood cell lysis by complement. Red blood cells that lack only CD59 are also susceptible to destruction as a result of spontaneous activation of the complement cascade.

Summary.

The complement system is one of the major mechanisms by which pathogen recognition is converted into an effective host defense against initial infection. Complement is a system of plasma proteins that can be activated directly by pathogens or indirectly by pathogen-bound antibody, leading to a cascade of reactions that occurs on the surface of pathogens and generates active components with various effector functions. There are three pathways of complement activation: the lectin pathway, triggered by the pattern recognition receptors MBL and the ficolins; the classical pathway, triggered directly by pathogen or indirectly by antibody binding to the pathogen surface; and the alternative pathway, which provides an amplification loop for the other two pathways, and is augmented by properdin, which also has innate recognition activity. All three pathways can be initiated independently of antibody as part of innate immunity. The early events in all pathways consist of a sequence of cleavage reactions in which the larger cleavage product binds covalently to the pathogen surface and contributes to the activation of the next component. The pathways converge with the formation of a C3 convertase enzyme, which cleaves C3 to produce the active complement component C3b. The binding of large numbers of C3b molecules to the pathogen is the central event in complement activation. Bound complement components, especially bound C3b and its inactive fragments, are recognized by specific complement receptors on phagocytic cells, which engulf pathogens opsonized by C3b and its inactive fragments. The small cleavage fragments of C3, C4, and especially C5 recruit phagocytes to sites of infection and activate them by binding to specific trimeric G-protein-coupled receptors. Together, these activities promote the uptake and destruction of pathogens by phagocytes. The molecules of C3b that bind the C3 convertase itself initiate the late events, binding C5 to make it susceptible to cleavage by C2a or Bb. The larger C5b fragment triggers the assembly of a membrane-attack complex, which can result in the lysis of certain pathogens. The activity of complement components is modulated by a system of regulatory proteins that prevent tissue damage as a result of inadvertent binding of activated complement components to host cells or of spontaneous activation of complement components in plasma.

Questions.

2.1 *What specialized features of the skin, lungs, and intestines are useful in keeping microbes at bay?*

2.2 *What structural feature of antimicrobial peptides prevents them from being inappropriately activated within the host cells that produce them? How do neutrophils and Paneth cells control the activation of these antimicrobial peptides?*

2.3 *Complement proteins are present in the blood plasma in healthy people. Why does complement only become activated in the presence of infection?*

2.4 *Recognition of bacteria by mannose-binding lectin can lead to their destruction by the same mechanisms as recognition by antibody. Explain why this is so, and describe additional common features between these pathways of pathogen recognition.*

2.5 *Spontaneous hydrolysis of the thioester bond in C3 occurs at a continuous low rate even in the absence of infection. Describe the reactions likely to occur in a healthy person and those that occur during an infection.*

2.6 *What is the key feature of the alternative pathway amplification loop? What are the various mechanisms that prevent this amplification process from targeting host tissues? What sort of defects can occur in these regulatory mechanisms? Describe some of the consequences.*

2.7 *People with a deficiency in factor I are susceptible to recurrent bacterial infections. Why is this?*

Section references.

2-1 Infectious diseases are caused by diverse living agents that replicate in their hosts.

Kauffmann, S.H.E., Sher, A., and Ahmed, R.: *Immunology of Infectious Diseases.* Washington, DC, ASM Press, 2002.

Mandell, G.L., Bennett, J.E., and Dolin, R. (eds): *Principles and Practice of Infectious Diseases*, 4th ed. New York, Churchill Livingstone, 1995.

Salyers, A.A., and Whitt, D.D.: *Bacterial Pathogenesis: A Molecular Approach.* Washington, DC, ASM Press, 1994.

2-2 Infectious agents must overcome innate host defenses to establish a focus of infection.

Gorbach, S.L., Bartlett, J.G., and Blacklow, N.R. (eds): *Infectious Diseases*, 3rd ed. Philadelphia, Lippincott Williams & Wilkins, 2003.

Hornef, M.W., Wick, M.J., Rhen, M., and Normark, S.: **Bacterial strategies for overcoming host innate and adaptive immune responses.** *Nat. Immunol.* 2002, **3**:1033–1040.

2-3 Epithelial surfaces of the body provide the first line of defense against infection.

Aderem, A., and Underhill, D.M.: **Mechanisms of phagocytosis in macrophages.** *Annu. Rev. Immunol.* 1999, **17**:593–623.

2-4 Epithelial cells and phagocytes produce several kinds of antimicrobial proteins.

Cash, H.L., Whitham, C.V., Behrendt, C.L., and Hooper, L.H.: **Symbiotic bacteria direct expression of an intestinal bactericidal lectin.** *Science* 2006, **313**:1126–1130.

De Smet, K., and Contreras, R.: **Human antimicrobial peptides: defensins, cathelicidins and histatins.** *Biotechnol. Lett.* 2005, **27**:1337–1347.

Ganz, T.: **Defensins: antimicrobial peptides of innate immunity.** *Nat. Rev. Immunol.* 2003, **3**:710–720.

Zanetti, M.: **The role of cathelicidins in the innate host defense of mammals.** *Curr. Issues Mol. Biol.* 2005, **7**:179–196.

2-5 The complement system recognizes features of microbial surfaces and marks them for destruction by the deposition of C3b.

Gros, P., Milder, F.J., and Janssen, B.J.: **Complement driven by conformational changes.** *Nat. Rev. Immunol.* 2008, **8**:48–58.

Janssen, B.J., Huizinga, E.G., Raaijmakers, H.C., Roos, A., Daha, M.R., Nilsson-Ekdahl, K., Nilsson, B., and Gros, P.: **Structures of complement component C3 provide insights into the function and evolution of immunity.** *Nature* 2005, **437**:505–511.

Janssen, B.J., Christodoulidou, A., McCarthy, A., Lambris, J.D., and Gros, P.: **Structure of C3b reveals conformational changes that underlie complement activity.** *Nature* 2006, **444**:213–216.

2-6 The lectin pathway uses soluble receptors that recognize microbial surfaces to activate the complement cascade.

Bohlson, S.S., Fraser, D.A., and Tenner, A.J.: **Complement proteins C1q and MBL are pattern recognition molecules that signal immediate and long-term protective immune functions.** *Mol. Immunol.* 2007, **44**:33–43.

Gál, P., Harmat, V., Kocsis, A., Bián, T., Barna, L., Ambrus, G., Végh, B., Balczer, J., Sim, R.B., Náray-Szabó, G., *et al*: **A true autoactivating enzyme. Structural insight into mannose-binding lectin-associated serine protease-2 activations.** *J. Biol. Chem.* 2005, **280**:33435–33444.

Wright, J.R.: **Immunoregulatory functions of surfactant proteins.** *Nat. Rev. Immunol.* 2005, **5**:58–68.

2-7 The classical pathway is initiated by activation of the C1 complex and is homologous to the lectin pathway.

McGrath, F.D., Brouwer, M.C., Arlaud, G.J., Daha, M.R., Hack, C.E., and Roos, A.: **Evidence that complement protein C1q interacts with C-reactive protein through its globular head region.** *J. Immunol.* 2006, **176**:2950–2957.

2-8 Complement activation is largely confined to the surface on which it is initiated.

Cicardi, M., Bergamaschini, L., Cugno, M., Beretta, A., Zingale, L.C., Colombo, M., and Agostoni, A.: **Pathogenetic and clinical aspects of C1 inhibitor deficiency.** *Immunobiology* 1998, **199**:366–376.

2-9 The alternative pathway is an amplification loop for C3b formation that is accelerated by recognition of pathogens by properdin.

Fijen, C.A., van den Bogaard, R., Schipper, M., Mannens, M., Schlesinger, M., Nordin, F.G., Dankert, J., Daha, M.R., Sjoholm, A.G., Truedsson, L., *et al.*: **Properdin deficiency: molecular basis and disease association.** *Mol. Immunol.* 1999, **36**:863–867.

Kemper, C., and Hourcade, D.E.: **Properdin: new roles in pattern recognition and target clearance.** *Mol. Immunol.* 2008, **45**:4048–4056.

Spitzer, D., Mitchell, L.M., Atkinson, J.P., and Hourcade, D.E.: **Properdin can initiate complement activation by binding specific target surfaces and providing a platform for de novo convertase assembly.** *J. Immunol.* 2007, **179**:2600–2608.

Xu, Y., Narayana, S.V., and Volanakis, J.E.: **Structural biology of the alternative pathway convertase.** *Immunol. Rev.* 2001, **180**:123–135.

2-10 Membrane and plasma proteins that regulate the formation and stability of C3 convertases determine the extent of complement activation under different circumstances.

Golay, J., Zaffaroni, L., Vaccari, T., Lazzari, M., Borleri, G.M., Bernasconi, S., Tedesco, F., Rambaldi, A., and Introna, M.: **Biologic response of B lymphoma cells to anti-CD20 monoclonal antibody rituximab *in vitro*: CD55 and CD59 regulate complement-mediated cell lysis.** *Blood* 2000, **95**:3900–3908.

Spiller, O.B., Criado-Garcia, O., Rodriguez De Cordoba, S., and Morgan, B.P.: **Cytokine-mediated up-regulation of CD55 and CD59 protects human hepatoma cells from complement attack.** *Clin. Exp. Immunol.* 2000, **121**:234–241.

Varsano, S., Frolkis, I., Rashkovsky, L., Ophir, D., and Fishelson, Z.: **Protection of human nasal respiratory epithelium from complement-mediated lysis by cell-membrane regulators of complement activation.** *Am. J. Respir. Cell Mol. Biol.* 1996, **15**:731–737.

2-11 Complement developed early in the evolution of multicellular organisms.

Fujita, T.: **Evolution of the lectin-complement pathway and its role in innate immunity.** *Nat. Rev. Immunol.* 2002, **2**:346–353.

Zhang, H., Song, L., Li, C., Zhao, J., Wang, H., Gao, Q., and Xu, W.: **Molecular cloning and characterization of a thioester-containing protein from Zhikong scallop *Chlamys farreri*.** *Mol. Immunol.* 2007, **44**:3492–3500.

2-12 Surface-bound C3 convertase deposits large numbers of C3b fragments on pathogen surfaces and generates C5 convertase activity.

Rawal, N., and Pangburn, M.K.: **Structure/function of C5 convertases of complement.** *Int. Immunopharmacol.* 2001, **1**:415–422.

2-13 Ingestion of complement-tagged pathogens by phagocytes is mediated by receptors for the bound complement proteins.

Gasque, P.: **Complement: a unique innate immune sensor for danger signals.** *Mol. Immunol.* 2004, **41**:1089–1098.

Helmy, K.Y., Katschke, K.J., Jr, Gorgani, N.N., Kljavin, N.M., Elliott, J.M., Diehl, L., Scales, S.J., Ghilardi, N., and van Lookeren Campagne M.: **CRIg: a macrophage complement receptor required for phagocytosis of circulating pathogens.** *Cell* 2006, **124**:915–927.

2-14 The small fragments of some complement proteins initiate a local inflammatory response.

Kohl, J.: **Anaphylatoxins and infectious and noninfectious inflammatory diseases.** *Mol. Immunol.* 2001, **38**:175–187.

Schraufstatter, I.U., Trieu, K., Sikora, L., Sriramarao, P., and DiScipio, R.: **Complement C3a and C5a induce different signal transduction cascades in endothelial cells.** *J. Immunol.* 2002, **169**:2102–2110.

2-15 The terminal complement proteins polymerize to form pores in membranes that can kill certain pathogens.

Hadders, M.A., Beringer, D.X., and Gros, P.: **Structure of C8α-MACPF reveals mechanism of membrane attack in complement immune defense.** *Science* 2007, **317**:1552–1554.

Parker, C.L., and Sodetz, J.M.: **Role of the human C8 subunits in complement-mediated bacterial killing: evidence that C8 γ is not essential.** *Mol. Immunol.* 2002, **39**:453–458.

Scibek, J.J., Plumb, M.E., and Sodetz, J.M.: **Binding of human complement C8 to C9: role of the N-terminal modules in the C8α subunit.** *Biochemistry* 2002, **41**:14546–14551.

2-16 Complement control proteins regulate all three pathways of complement activation and protect the host from their destructive effects.

Atkinson, J.P., and Goodship, T.H.: **Complement factor H and the hemolytic uremic syndrome.** *J. Exp. Med.* 2007, **204**:1245–1248.

Blom, A.M., Rytkonen, A., Vasquez, P., Lindahl, G., Dahlback, B., and Jonsson, A.B.: **A novel interaction between type IV pili of *Neisseria gonorrhoeae* and the human complement regulator C4B-binding protein.** *J. Immunol.* 2001, **166**:6764–6770.

Hageman, G.S., *et al.*: **A common haplotype in the complement regulatory gene factor H (HF1/CFH) predisposes individuals to age-related macular degeneration.** *Proc. Natl Acad. Sci. USA* 2005, **102**:7227–7232.

Jiang, H., Wagner, E., Zhang, H., and Frank, M.M.: **Complement 1 inhibitor is a regulator of the alternative complement pathway.** *J. Exp. Med.* 2001, **194**:1609–1616.

Miwa, T., Zhou, L., Hilliard, B., Molina, H., and Song, W.C.: **Crry, but not CD59 and DAF, is indispensable for murine erythrocyte protection in vivo from spontaneous complement attack.** *Blood* 2002, **99**:3707–3716.

Singhrao, S.K., Neal, J.W., Rushmere, N.K., Morgan, B.P., and Gasque, P.: **Spontaneous classical pathway activation and deficiency of membrane regulators render human neurons susceptible to complement lysis.** *Am. J. Pathol.* 2000, **157**:905–918.

Smith, G.P., and Smith, R.A.: **Membrane-targeted complement inhibitors.** *Mol. Immunol.* 2001, **38**:249–255.

Spencer, K.L., Hauser, M.A., Olson, L.M., Schmidt, S., Scott, W.K., Gallins, P., Agarwal, A., Postel, E.A., Pericak-Vance, M.A., and Haines, J.L.: **Protective effect of complement factor B and complement component 2 variants in age-related macular degeneration.** *Hum. Mol. Genet.* 2007, **16**:1986–1992.

Spencer, K.L., Olson, L.M., Anderson, B.M., Schnetz-Boutaud, N., Scott, W.K., Gallins, P., Agarwal, A., Postel, E.A., Pericak-Vance, M.A., and Haines, J.L.: **C3 R102G polymorphism increases risk of age-related macular degeneration.** *Hum. Mol. Genet.* 2008, **17**:1821–1824.

The Induced Responses of Innate Immunity

3

In Chapter 2 we considered the innate defenses—such as epithelial barriers, secreted antimicrobial proteins and the complement system—that act immediately upon encounter with microbes to protect the body from infection. We also introduced the phagocytic cells that lie beneath the epithelial barriers, ready to engulf and digest invading microorganisms that have been flagged for destruction by complement. As well as killing microorganisms directly, these phagocytes also initiate the next phase of the innate immune response, inducing an inflammatory response that recruits new phagocytic cells and circulating effector molecules to the site of infection. In this chapter we take a closer look at the ancient system of pattern recognition receptors used by the phagocytic cells of the innate immune system to identify pathogens and distinguish them from self antigens. We shall see how, as well as directing the immediate destruction of pathogens, stimulation of some of these receptors on macrophages and dendritic cells leads to their becoming cells that can effectively present antigen to T lymphocytes, thus initiating an adaptive immune response. In the last part of the chapter we describe how the cytokines and chemokines produced by activated phagocytes and dendritic cells induce the later stages of the innate immune response, such as the acute-phase response. We shall also meet another cell of the innate immune system, the natural killer cell (NK cell), which contributes to innate host defenses against viruses and other intracellular pathogens. In the later stages of the innate immune response, the first steps toward initiating an adaptive immune response take place, so that if the infection is not cleared by innate immunity, a full immune response will ensue.

Pattern recognition by cells of the innate immune system.

Although the innate immune system lacks the fine specificity of adaptive immunity that is necessary to produce immunological memory, it can distinguish self from nonself. We have already seen an example of this in the recognition of microbial surfaces by complement (see Chapter 2). In this part of the chapter we look more closely at the cellular receptors that recognize pathogens and signal for a cellular innate immune response. Regular patterns of molecular structure are present on many microorganisms but do not occur on the body's own cells. Proteins that recognize these features occur as receptors on macrophages, neutrophils, and dendritic cells, and as secreted molecules, such as the mannose-binding lectin (MBL) described in Chapter 2. The general characteristics of these **pattern recognition receptors** are contrasted with the antigen-specific receptors of adaptive immunity in Fig. 3.1.

Pattern recognition receptors can be classified into four main groups on the basis of their cellular localization and their function: free receptors in the serum (such as MBL), which have been discussed in Chapter 2; membrane-bound

Receptor characteristic	Innate immunity	Adaptive immunity
Specificity inherited in the genome	Yes	No
Expressed by all cells of a particular type (e.g. macrophages)	Yes	No
Triggers immediate response	Yes	No
Recognizes broad classes of pathogens	Yes	No
Interacts with a range of molecular structures of a given type	Yes	No
Encoded in multiple gene segments	No	Yes
Requires gene rearrangement	No	Yes
Clonal distribution	No	Yes
Able to discriminate between even closely related molecular structures	No	Yes

Fig. 3.1 Comparison of the characteristics of recognition molecules of the innate and adaptive immune systems. The innate immune system uses receptors that are encoded by complete genes inherited through the germline. In contrast, the adaptive immune system uses antigen receptors encoded in gene segments that are assembled into complete T-cell and B-cell receptor genes during lymphocyte development, a process that leads to each cell expressing a receptor of unique specificity. Receptors of the innate immune system are in general deployed nonclonally (that is, by all the cells of a given cell type), whereas the antigen receptors of the adaptive immune system are clonally distributed on individual lymphocytes and their progeny.

phagocytic receptors; membrane-bound signaling receptors; and cytoplasmic signaling receptors. Phagocytic receptors primarily stimulate ingestion of the pathogens they recognize. Signaling receptors are a diverse group that includes chemotactic receptors, which guide cells to sites of infection, and receptors that induce the production of effector molecules that contribute to the later, induced responses of innate immunity. In this part of the chapter we first look at the recognition properties of phagocytic receptors and of signaling receptors that activate phagocytic microbial killing mechanisms. Next we describe an evolutionarily ancient pathogen recognition and signaling system, involving receptors called Toll-like receptors (TLRs), which has a key role in defense against infection in vertebrates and many invertebrates. Finally, we look at a more recently discovered class of cytoplasmic signaling receptors, some of which have very similar effects to the Toll-like receptors while others are concerned with antiviral defenses.

3-1 After entering tissues, many pathogens are recognized, ingested, and killed by phagocytes.

If a microorganism crosses an epithelial barrier and begins to replicate in the tissues of the host, in most cases it is immediately recognized by resident phagocytic cells. There are three major classes of phagocytic cells in the innate immune system: macrophages and monocytes, granulocytes, and dendritic cells. **Macrophages** mature continuously from **monocytes** that leave the circulation to migrate into tissues throughout the body, and they are the major phagocyte population resident in normal tissues. Macrophages in different tissues were historically given different names, for example microglial cells in neural tissue and **Kupffer cells** in the liver; generically these cells are referred to as **mononuclear phagocytes**. Macrophages are found in especially large numbers in connective tissue: for example, in the submucosal layer of the gastrointestinal tract; in the submucosal layer of the bronchi, and in the lung interstitium—the tissue and intercellular spaces around the air sacs (alveoli)—and in the alveoli themselves; along some blood vessels in the liver; and throughout the spleen, where they remove senescent blood cells.

The second major family of phagocytes comprises the granulocytes, which include **neutrophils**, **eosinophils**, and **basophils**. Of these, neutrophils have the greatest phagocytic activity and are the most immediately involved in innate immunity against infectious agents. Also called polymorphonuclear neutrophilic leukocytes (PMNs or polys), they are short-lived cells that are abundant in the blood but are not present in healthy tissues. Macrophages and granulocytes have an important role in innate immunity because they can recognize, ingest, and destroy many pathogens without the aid of an adaptive immune response. Phagocytic cells that scavenge incoming pathogens represent an ancient mechanism of innate immunity, as they are found in both invertebrates and vertebrates.

The third class of phagocytes in the immune system is the immature **dendritic cells** resident in tissues. Dendritic cells arise from both myeloid and lymphoid progenitors within the bone marrow, and they migrate via the blood to tissues throughout the body and to peripheral lymphoid organs. Dendritic cells ingest and break down microbes but, unlike macrophages and neutrophils, their primary role in immune defense is not the front-line large-scale direct killing of microbes. There are two main functional types of dendritic cells: **conventional dendritic cells** (**cDCs**) and **plasmacytoid dendritic cells** (**pDCs**). The main role of conventional dendritic cells is to process ingested microbes to generate peptide antigens that can activate T cells and induce an adaptive immune response, and to produce cytokines in response to microbial recognition. Conventional dendritic cells are thus considered to act as a bridge between innate and adaptive immune responses. Plasmacytoid dendritic cells are major producers of the antiviral interferons and are considered to be part of innate immunity; they are discussed in detail later in the chapter.

Because most microorganisms enter the body through the mucosa of the gut and respiratory system, macrophages in the submucosal tissues are the first cells to encounter most pathogens, but they are soon reinforced by the recruitment of large numbers of neutrophils to sites of infection. Macrophages and neutrophils recognize pathogens by means of cell-surface receptors that can discriminate between the surface molecules of pathogens and those of the host. Although they are both phagocytic, macrophages and neutrophils have distinct properties and functions in innate immunity.

All phagocytic cells internalize pathogens by the same process of **phagocytosis**, which is initiated when certain receptors on the surface of the phagocyte bind to components of a microbial surface. The bound pathogen is first surrounded by the phagocyte plasma membrane and then internalized in a large membrane-enclosed endocytic vesicle known as a **phagosome**. The phagosome then becomes acidified, which kills most pathogens. The phagosome fuses with one or more lysosomes to generate a **phagolysosome**, in which the lysosomal contents are released to destroy the pathogen (Fig. 3.2). Neutrophils, which are highly specialized for the intracellular killing of microbes, also contain cytoplasmic granules called primary and secondary granules, which fuse with the phagosome and contain additional enzymes and antimicrobial peptides that attack the microbe (see Section 2-4). Another pathway by which extracellular material, including microbial material, can be taken up into the endosomal compartment of cells and degraded is **receptor-mediated endocytosis**, which is not restricted to phagocytes. Dendritic cells and the other phagocytes can also take up pathogens by a nonspecific process called **macropinocytosis**, in which large amounts of extracellular fluid and its contents are ingested.

Macrophages and neutrophils constitutively express a number of cell-surface receptors that stimulate the phagocytosis and intracellular killing of microbes bound to them, although some also signal through other pathways to trigger responses such as cytokine production. These phagocytic receptors include several members of the C-type lectin-like family (see Fig. 3.2). **Dectin-1** is

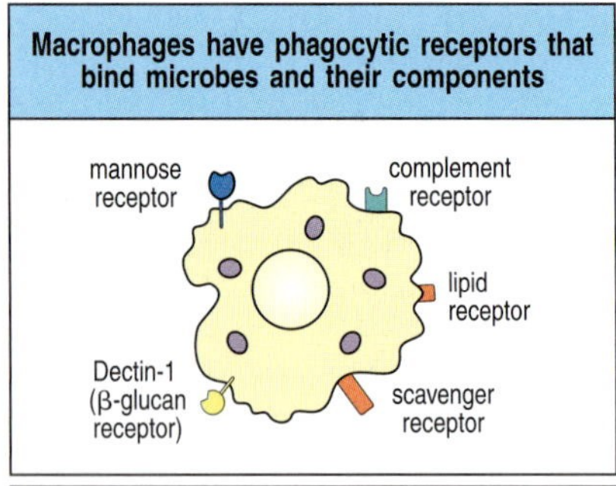

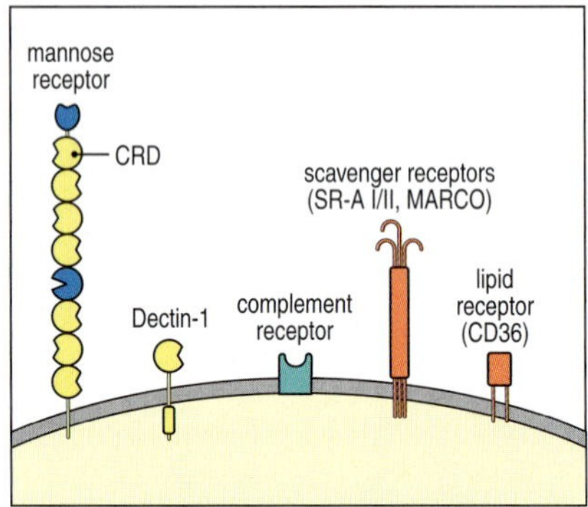

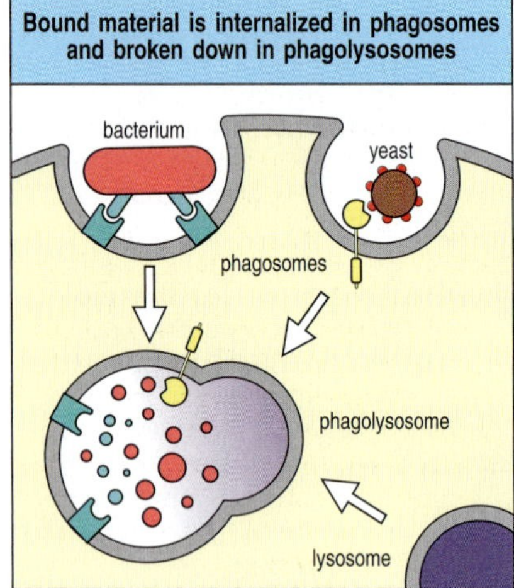

Fig. 3.2 Macrophages express receptors that enable them to take up microbes by phagocytosis. First panel: macrophages residing in tissues throughout the body are among the first cells to encounter and respond to pathogens. They carry cell-surface receptors that bind to pathogens and their components and induce phagocytosis of the bound material. Second panel: macrophages express several kinds of receptors that interact directly with microbial components, in particular carbohydrates and lipids. Dectin-1 is a C-type lectin built around a single carbohydrate-recognition domain (CRD). The macrophage mannose receptor contains many CRDs, with a fibronectin-like domain and cysteine-rich region at its amino terminus. Class A scavenger receptors are built from collagen-like domains and form trimers. The receptor protein CD36 is a class B scavenger receptor that recognizes and internalizes lipids. Macrophages also express complement receptors, which internalize complement-coated bacteria. Third panel: binding of microbes or microbial components to any of these receptors stimulates phagocytosis and uptake into intracellular phagosomes. Phagosomes fuse with lysosomes, forming an acidified phagolysosome in which the ingested material is broken down by lysosomal hydrolases.

strongly expressed by macrophages and neutrophils and recognizes β-1,3-linked glucans (polymers of glucose), which are common components of fungal cell walls in particular. Dendritic cells also express Dectin-1, as well as several other C-type lectin-like phagocytic receptors, which will be discussed in relation to pathogen uptake for antigen processing and presentation in Chapter 9. Another C-type lectin, the **mannose receptor** (**MR**) expressed by macrophages and dendritic cells, recognizes various mannosylated ligands, including some present on fungi, bacteria, and viruses; it was once suspected to have an important role in resistance to microbes. However, experiments with mice that lack this receptor do not support this idea. The macrophage mannose receptor is now thought to function mainly as a clearance receptor for host glycoproteins such as β-glucuronidase and lysosomal hydrolases, which have mannose-containing carbohydrate side chains and whose extracellular concentrations are raised during inflammation.

A second set of phagocytic receptors on macrophages, called **scavenger receptors**, recognize various anionic polymers and acetylated low-density lipoproteins. These receptors are structurally heterogeneous, consisting of at least six different molecular families. Class A scavenger receptors are membrane proteins composed of trimers of collagen domains (see Fig. 3.2). They include **SR-A I**, **SR-A II**, and **MARCO** (macrophage receptor with a collagenous structure), which all bind various bacterial cell-wall components and help to internalize bacteria, although the basis of their specificity is poorly understood. Class B scavenger receptors bind high-density lipoproteins, and they internalize lipids. One of these receptors is CD36, which binds many ligands, including long-chain fatty acids.

A third set of receptors of crucial importance in macrophage and neutrophil phagocytosis are the complement receptors discussed in Chapter 2, which bind complement-coated microbes. The complement receptor CR3 also directly recognizes and phagocytoses microbes bearing β-glucans. All these receptors act together in innate immunity to facilitate the phagocytosis of a wide range of pathogenic microorganisms.

3-2 G-protein-coupled receptors on phagocytes link microbe recognition with increased efficiency of intracellular killing.

Phagocytosis of microbes by macrophages and neutrophils is generally followed by the death of the microbe inside the phagocyte. As well as the phagocytic receptors, macrophages and neutrophils have other receptors that signal to stimulate antimicrobial killing. These receptors belong to the evolutionarily ancient family of **G-protein-coupled receptors** (**GPCRs**), which are characterized by seven membrane-spanning segments. Members of this family are crucial to immune system function because they also direct responses to anaphylatoxins such as the complement fragment C5a (see Section 2-14) and to many chemokines (chemoattractant peptides and proteins), recruiting phagocytes to sites of infection and promoting inflammation.

The **fMet-Leu-Phe (fMLP) receptor** is a G-protein-coupled receptor that senses the presence of bacteria by recognizing a unique feature of bacterial polypeptides. Protein synthesis in bacteria is typically initiated with an *N*-formylmethionine (fMet) residue, an amino acid present in prokaryotes but not in eukaryotes. The fMLP receptor is named after a tripeptide for which it has a high affinity, although it also binds other peptide motifs. Bacterial polypeptides binding to this receptor activate intracellular signaling pathways that direct the cell to move toward the most concentrated source of the ligand. Signaling through the fMLP receptor also induces the production of microbicidal reactive oxygen species in the phagolysosome. The C5a receptor recognizes the small fragment of C5 generated when the classical or lectin pathways of complement are activated, usually by the presence of microbes (see Section 2-14) and triggers similar responses to those of the fMLP receptor. Thus, stimulation of these receptors both guides monocytes and neutrophils toward a site of infection and triggers increased antimicrobial activity, and these cell responses can be activated by directly sensing unique bacterial products or by messengers such as C5a that indicate previous recognition of a microbe.

The G-protein-coupled receptors are so named because ligand binding activates a member of a class of intracellular GTP-binding proteins called **G proteins**, sometimes called **heterotrimeric G proteins** to distinguish them from the family of 'small' GTPases typified by Ras. Heterotrimeric G proteins are composed of three subunits: Gα, Gβ, and Gγ, of which the α subunit is similar to the small GTPases (Fig. 3.3). In the resting state, the G protein is inactive, not associated with the receptor, and a molecule of GDP is bound to the α subunit. Ligand binding induces conformational changes in the receptor that allow it to bind the G protein, which results in the displacement of the GDP from the G protein and its replacement with GTP. The active G protein dissociates into two components, the α subunit and a complex of β and γ subunits; each of these components can interact with other intracellular signaling molecules to transmit and amplify the signal. G proteins can activate a wide variety of downstream enzymatic targets, such as adenylate cyclase (which produces the second messenger cyclic AMP), phospholipase C (whose activation gives

Fig. 3.3 G-protein-coupled receptors signal by coupling with intracellular heterotrimeric G proteins. G-protein-coupled receptors (GPCRs) such as the fMet-Leu-Phe (fMLP) and chemokine receptors signal through GTP-binding proteins known as heterotrimeric G proteins. In the inactive state, the α subunit of the G protein binds GDP and is associated with the β and γ subunits (first panel). The binding of a ligand to the receptor induces a conformational change that allows the receptor to interact with the G protein, which results in the displacement of GDP and binding of GTP by the α subunit (second panel). GTP binding triggers the dissociation of the G protein into the α subunit and the βγ subunit, both of which can activate other proteins at the inner face of the cell membrane (third panel). In the case of fMLP signaling in macrophages and neutrophils, the α subunit of the activated G protein indirectly activates the GTPases Rac and Rho, whereas the βγ subunit indirectly activates the GTPase Cdc42. The actions of these proteins result in the assembly of the NADPH oxidase. Chemokine signaling acts by a similar pathway and activates chemotaxis. The activated response ceases when the intrinsic GTPase activity of the α subunit hydrolyzes GTP to GDP, and the α and βγ subunits reassociate (fourth panel). The intrinsic rate of GTP hydrolysis by α subunits is relatively slow, and signaling is regulated by additional GTPase-activating proteins (not shown), which accelerate the rate of GTP hydrolysis.

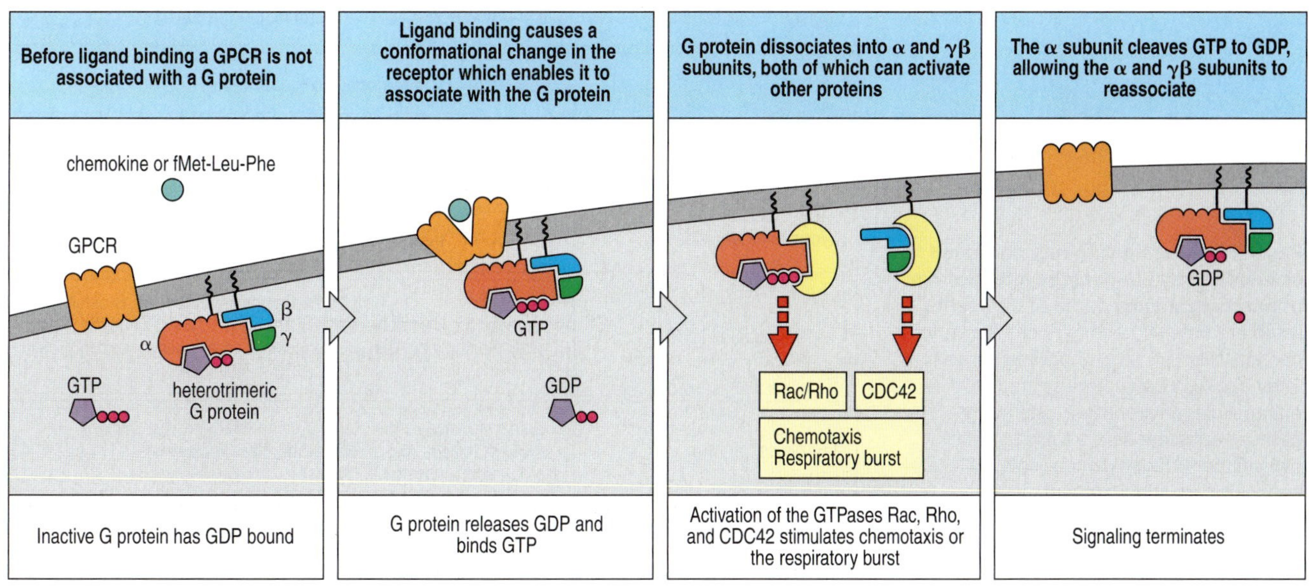
Before ligand binding a GPCR is not associated with a G protein
chemokine or fMet-Leu-Phe
GPCR
β
γ
α
GTP
heterotrimeric G protein
Inactive G protein has GDP bound
Ligand binding causes a conformational change in the receptor which enables it to associate with the G protein
GTP
GDP
G protein releases GDP and binds GTP
G protein dissociates into α and γβ subunits, both of which can activate other proteins
Rac/Rho
CDC42
Chemotaxis
Respiratory burst
Activation of the GTPases Rac, Rho, and CDC42 stimulates chemotaxis or the respiratory burst
The α subunit cleaves GTP to GDP, allowing the α and γβ subunits to reassociate
GDP
Signaling terminates

rise to the second messenger inositol 1,3,5-trisphosphate (IP_3) and the release of free Ca^{2+}), and regulators of Ras-family GTPases, which in turn can affect cell metabolism, motility, gene expression, and cell division.

In the case of fMLP and C5a signaling, the α subunit of the activated G protein indirectly activates the small GTPases Rac and Rho, while the $\beta\gamma$ subunit indirectly activates the small GTPase Cdc42. The actions of these proteins help to increase the microbicidal capacity of macrophages and neutrophils that have ingested pathogens. Upon phagocytosis, macrophages and neutrophils produce a variety of toxic products that help to kill the engulfed microorganism (Fig. 3.4). The most important of these are the antimicrobial peptides described in Section 2-4, reactive nitrogen species such as nitric oxide (NO), and **reactive oxygen species** (**ROS**) such as the superoxide anion (O_2^-) and hydrogen peroxide (H_2O_2). The nitric oxide is produced by a high-output form of nitric oxide synthase, inducible NOS2 (iNOS2), whose expression is induced by a variety of stimuli, including fMLP. Activation of the fMLP and C5a receptors is also directly involved in the generation of reactive oxygen species.

Superoxide is generated by a multicomponent, membrane-associated **NADPH oxidase**, also called **phagocyte oxidase**. In unstimulated phagocytes this enzyme is inactive because it is not fully assembled. One set of subunits, the cytochrome b_{558} complex, is localized in the membranes of neutrophil secondary granules and macrophage lysosomes; the other components are in the cytosol. Activation of phagocytes induces the addition of the cytosolic subunits to the membrane-associated cytochrome b_{558} to form a complete functional NADPH oxidase in the phagolysosome membrane (Fig. 3.5). The fMLP and C5a receptors participate in the process by activating Ras-family GTPases, such as Rac2, that are necessary for the assembly of the active NADPH oxidase.

The NADP oxidase reaction results in a transient increase in oxygen consumption by the cell, which is known as the **respiratory burst**. It generates superoxide anion within the lumen of the phagolysosome, and this is converted by the enzyme superoxide dismutase into H_2O_2. Further chemical and enzymatic reactions produce a range of toxic chemicals from H_2O_2, including

Antimicrobial mechanisms of phagocytes		
Class of mechanism	**Macrophage products**	**Neutrophil products**
Acidification	pH=~3.5–4.0, bacteriostatic or bactericidal	
Toxic oxygen-derived products	Superoxide O_2^-, hydrogen peroxide H_2O_2, singlet oxygen ${}^1O_2^{\bullet}$, hydroxyl radical $^{\bullet}OH$, hypohalite OCl^-	
Toxic nitrogen oxides	Nitric oxide NO	
Antimicrobial peptides	Cathelicidin, macrophage elastase-derived peptide	α-Defensins (HNP1–4), β-defensin HBD4, cathelicidin, azurocidin, bacterial permeability inducing protein (BPI), lactoferricin
Enzymes	Lysozyme: digests cell walls of some Gram-positive bacteria Acid hydrolases (e.g. elastase and other proteases): break down ingested microbes	
Competitors		Lactoferrin (sequesters Fe^{2+}), vitamin B_{12}-binding protein

Fig. 3.4 Bactericidal agents produced or released by phagocytes after uptake of microorganisms. Most of the agents listed are directly toxic to microbes and can act directly in the phagolysosome. They can also be secreted into the extracellular environment, and many of these substances are toxic to host cells. Other phagocyte products sequester essential nutrients in the extracellular environment, rendering them inaccessible to microbes and hindering microbial growth.

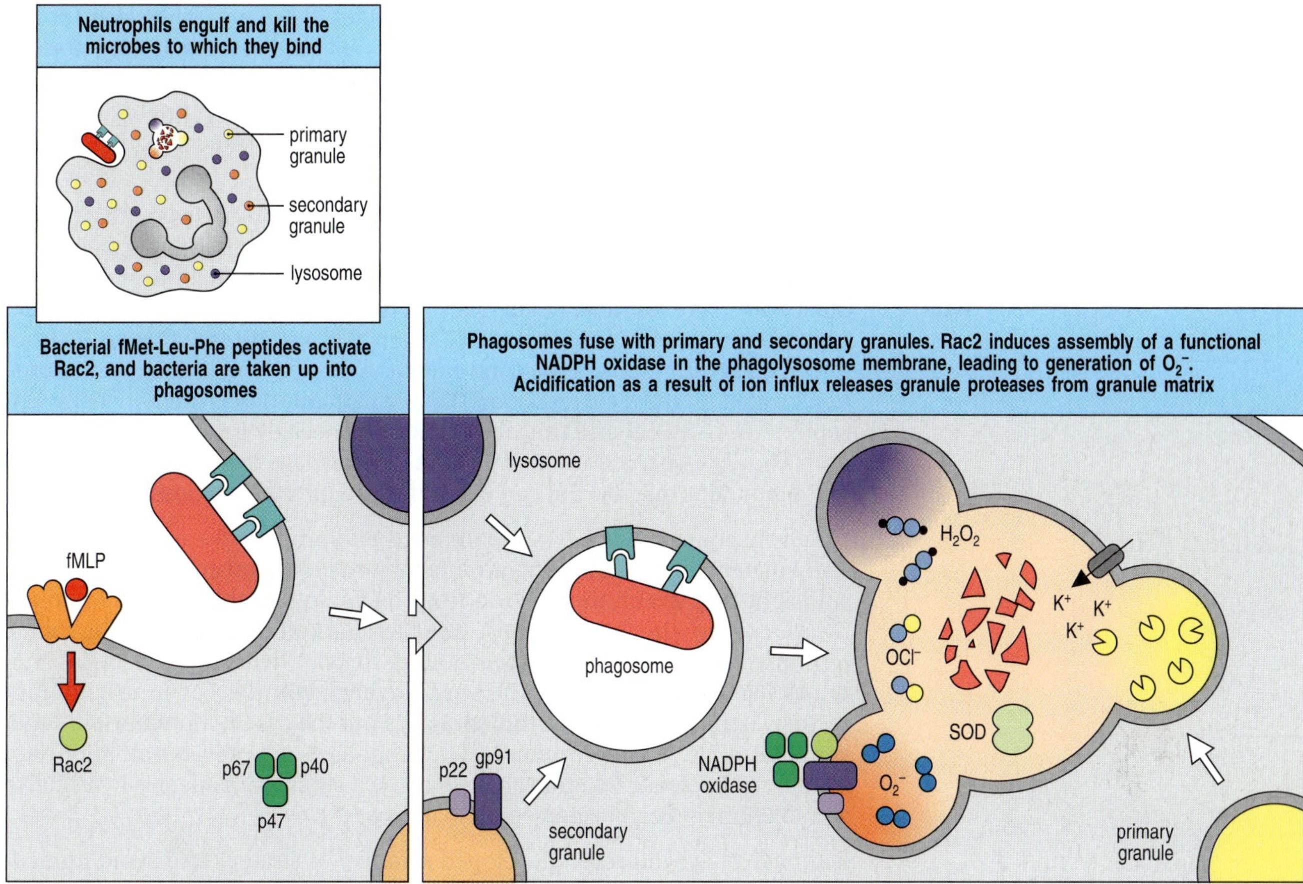

Fig. 3.5 The microbicidal respiratory burst in phagocytes is induced by activation-induced assembly of the phagocyte NADPH oxidase. Neutrophils are highly specialized for the uptake and killing of pathogens, and contain several different kinds of cytoplasmic granules, such as the primary and secondary granules shown in the first panel. These granules contain antimicrobial peptides and enzymes. In resting neutrophils, the cytochrome b_{558} subunits (gp91 and gp22) of the NADPH oxidase are localized in the membranes of secondary granules; the other oxidase components (p40, p47, and p67) are located in the cytosol (second panel). In concert with the actions of phagocytic receptors, signaling by fMLP or C5a receptors activates Rac2. This induces the assembly of the cytosolic subunits with cytochrome b_{558} to form active NAPDH oxidase in the membrane of the phagolysosome, which has been formed by the fusion of the phagosome with lysosomes and primary and secondary granules (third panel). Active NADPH oxidase transfers an electron from its FAD cofactor to molecular oxygen forming the superoxide ion O_2^- (blue) and other oxygen free radicals in the lumen of the phagolysosome (third panel). Potassium and hydrogen ions are then drawn into the phagolysosome to neutralize the charged superoxide ion, increasing acidification of the vesicle. Acidification dissociates granule enzymes such as cathepsin G and elastase (yellow) from their proteoglycan matrix, leading to their cleavage and activation by lysosomal proteases. O_2^- is converted by superoxide dismutase (SOD) to hydrogen peroxide (H_2O_2), which can kill microorganisms, and is also converted by other enzymes and by chemical reactions with ferrous (Fe^{2+}) ions to microbicidal hypochlorite (OCl^-) and the hydroxyl (•OH) radical.

the hydroxyl radical (•OH), hypochlorite (OCl^-), and hypobromite (OBr^-). In this way, the direct recognition of bacterially derived polypeptides or previous pathogen recognition by the complement system activates a potent killing mechanism within macrophages and neutrophils that have ingested microbes via their phagocytic receptors. Because hydrolytic enzymes, membrane-disrupting peptides, and reactive oxygen species can be released into the extracellular environment and are toxic to host cells, phagocyte activation can cause extensive tissue damage.

Neutrophils are not tissue-resident cells and they need to be recruited to a site of infection from the bloodstream. Their sole function is to ingest and kill microorganisms. Although neutrophils are eventually present in much larger numbers than macrophages in some types of acute infection, they are short-lived, dying soon after they have accomplished a round of phagocytosis and used up their primary and secondary granules. Dead and dying neutrophils are a major component of the **pus** that forms in abscesses and in wounds infected by certain extracellular capsulated bacteria such as streptococci and staphylococci, which are thus known as **pus-forming** or **pyogenic bacteria**. Macrophages, in contrast, are long-lived cells and continue to generate new lysosomes.

Patients with a disease called **chronic granulomatous disease** (**CGD**) have a genetic deficiency of NADPH oxidase, which means that their phagocytes do not produce the toxic oxygen derivatives characteristic of the respiratory burst and so are less able to kill ingested microorganisms and clear an infection. The most common form of CGD is an X-linked disease that arises from inactivating mutations in the gene encoding the gp91 subunit of cytochrome b_{558}, located on the X chromosome. People with this defect are unusually susceptible to bacterial and fungal infections, especially in infancy. Autosomal mutations in other subunits of the NADPH oxidase can also cause chronic granulomatous disease, but can be milder and have a later onset.

Macrophages can phagocytose pathogens and produce the respiratory burst immediately on encountering an infecting microorganism, and this can be sufficient to prevent an infection from becoming established. In the nineteenth century the immunologist **Elie Metchnikoff** believed that the innate response of macrophages encompassed all host defense; indeed, invertebrates such as the sea star that he was studying rely entirely on innate immunity to overcome infection. Although this is not the case in humans and other vertebrates, the innate response of macrophages still provides an important front line of defense that must be overcome if a microorganism is to establish an infection that can be passed on to a new host.

Pathogens have, however, developed a variety of strategies to avoid immediate destruction by macrophages and neutrophils. Many extracellular pathogenic bacteria coat themselves with a thick polysaccharide capsule that is not recognized by any phagocytic receptor. In such cases, however, the complement system can recognize microbial surfaces, coat them with C3b and so flag them for phagocytosis via complement receptors, as described in Chapter 2. Other pathogens, for example mycobacteria, have evolved ways to grow inside macrophage phagosomes by inhibiting their acidification and fusion with lysosomes. Without such devices, a microorganism must enter the body in sufficient numbers to simply overwhelm the immediate innate host defenses and to establish a focus of infection.

3-3 Pathogen recognition and tissue damage initiate an inflammatory response.

An important effect of the interaction between pathogens and tissue macrophages is the activation of macrophages and other immune cells to release small proteins called **cytokines** and **chemokines** (chemoattractant cytokines), and other chemical mediators that set up a state of **inflammation** in the tissue, attract monocytes and neutrophils to the infection, and allow plasma proteins to enter the tissue from the blood. An inflammatory response is usually initiated within hours of infection or wounding. Macrophages are stimulated to secrete **pro-inflammatory** cytokines and chemokines by interactions between microbes and microbial products and specific receptors expressed by the macrophage. Before examining such interactions in detail, we will describe some general aspects of inflammation and how it contributes to host defense.

Inflammation has three essential roles in combating infection. The first is to deliver additional effector molecules and cells from the blood into sites of infection, and so increase the destruction of invading microorganisms. The second is to induce local blood clotting, which provides a physical barrier to the spread of the infection in the bloodstream. The third is to promote the repair of injured tissue.

Inflammatory responses are operationally characterized by pain, redness, heat, and swelling at the site of an infection, reflecting four types of change in the local blood vessels, as shown in Fig. 3.6. The first is an increase in vascular diameter, leading to increased local blood flow—hence the heat and redness—and a reduction in the velocity of blood flow, especially along the inner walls of small blood vessels. The second change is that the endothelial cells lining the blood vessel are activated to express **cell-adhesion molecules** that promote the binding of circulating leukocytes. The combination of slowed blood flow and adhesion molecules allows leukocytes to attach to the endothelium and migrate into the tissues, a process known as **extravasation**. All these changes are initiated by the pro-inflammatory cytokines and chemokines produced by activated macrophages.

Once inflammation has begun, the first white blood cells attracted to the site are neutrophils. These are followed by monocytes, which differentiate into tissue macrophages (Fig. 3.7). Monocytes are also able to give rise to dendritic cells in the tissues, depending on the precise signals that they receive from their environment: for example, the cytokine granulocyte–macrophage colony-stimulating factor (GM-CSF), together with interleukin 4 (IL-4), will induce monocytes to differentiate into dendritic cells, whereas macrophage colony-stimulating factor (M-CSF) induces their differentiation into macrophages. In the later stages of inflammation, other leukocytes such as eosinophils and lymphocytes (see Section 1-3) also enter the infected site.

The third major change in local blood vessels is an increase in vascular permeability. Thus, instead of being tightly joined together, the endothelial cells lining the blood vessel walls become separated, leading to an exit of fluid and proteins from the blood and their local accumulation in the tissue. This accounts for the swelling, or **edema**, and pain—as well as the accumulation in tissues of plasma proteins such as complement and MBL that aid in host defense. The changes that occur in endothelium as a result of inflammation are known generally as **endothelial activation**. The fourth change, clotting in microvessels in the site of infection, prevents the spread of the pathogen via the blood.

These changes are induced by a variety of inflammatory mediators released as a consequence of the recognition of pathogens by macrophages, and later by neutrophils and other white blood cells. Both macrophages and neutrophils secrete lipid mediators of inflammation—**prostaglandins**, **leukotrienes**, and **platelet-activating factor** (**PAF**)—which are rapidly produced by enzymatic

Fig. 3.6 Infection stimulates macrophages to release cytokines and chemokines that initiate an inflammatory response. Cytokines produced by tissue macrophages at the site of infection cause the dilation of local small blood vessels and changes in the endothelial cells of their walls. These changes lead to the movement of leukocytes, such as neutrophils and monocytes, out of the blood vessel (extravasation) and into the infected tissue, guided by chemokines produced by the activated macrophages. The blood vessels also become more permeable, allowing plasma proteins and fluid to leak into the tissues. Together, these changes cause the characteristic inflammatory signs of heat, pain, redness, and swelling at the site of infection.

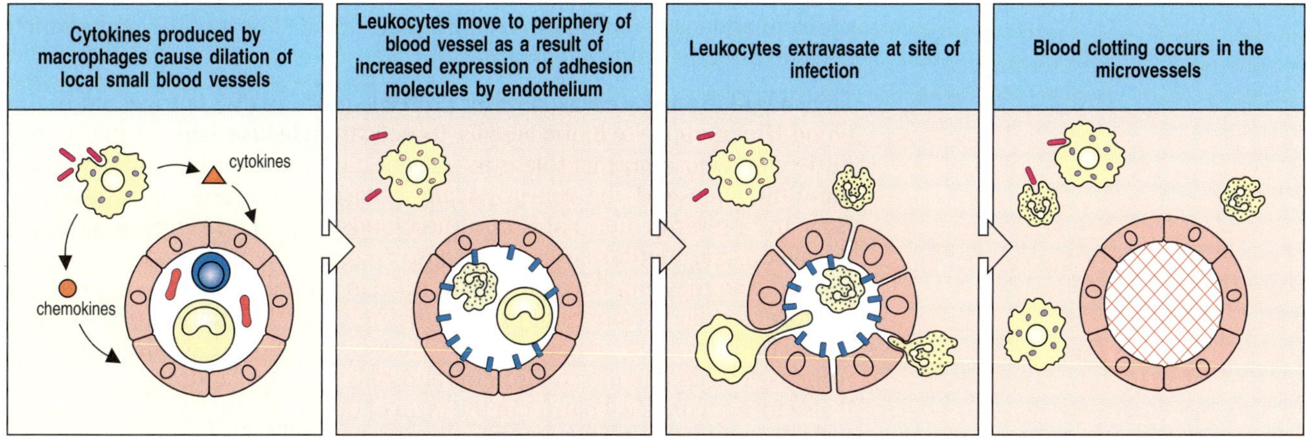
Cytokines produced by macrophages cause dilation of local small blood vessels
cytokines
chemokines
Leukocytes move to periphery of blood vessel as a result of increased expression of adhesion molecules by endothelium
Leukocytes extravasate at site of infection
Blood clotting occurs in the microvessels

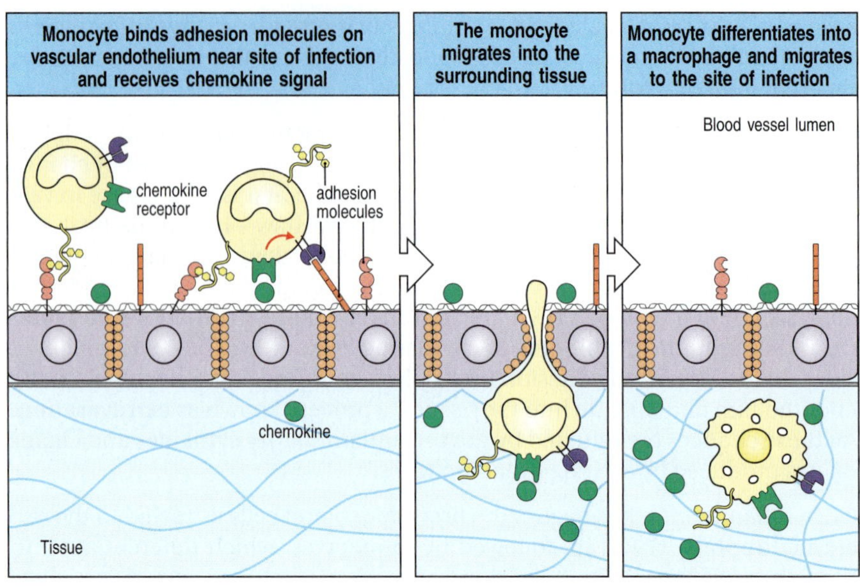
Monocyte binds adhesion molecules on vascular endothelium near site of infection and receives chemokine signal
The monocyte migrates into the surrounding tissue
Monocyte differentiates into a macrophage and migrates to the site of infection
Blood vessel lumen
chemokine receptor
adhesion molecules
chemokine
Tissue

Fig. 3.7 Monocytes circulating in the blood migrate into infected and inflamed tissues. Adhesion molecules on the endothelial cells of the blood vessel wall first capture the monocyte and cause it to adhere to the vascular endothelium. Chemokines bound to the vascular endothelium then signal the monocyte to migrate across the endothelium into the underlying tissue. The monocyte, now differentiating into a macrophage, continues to migrate, under the influence of chemokines released during inflammatory responses, toward the site of infection. Monocytes leaving the blood are also able to differentiate into dendritic cells (not shown), depending on the signals that they receive from their environment.

pathways that degrade membrane phospholipids. Their actions are followed by those of the chemokines and cytokines that are synthesized and secreted by macrophages in response to pathogens. The cytokine **tumor necrosis factor-α** (**TNF-α**, also known simply as **TNF**), for example, is a potent activator of endothelial cells. We describe TNF-α and related cytokines in more detail in Section 3-17.

Besides stimulating the respiratory burst in phagocytes and acting as a chemoattractant for neutrophils and monocytes (see Section 3-2), C5a also promotes inflammation by increasing vascular permeability and inducing the expression of certain adhesion molecules on endothelium. C5a also activates local **mast cells** (see Section 1-3), which are stimulated to release their granules containing the small inflammatory molecule histamine and TNF-α as well as cathelicidins.

If wounding has occurred, the injury to blood vessels immediately triggers two protective enzyme cascades. One is the **kinin system** of plasma proteases that is triggered by tissue damage, and is a protease cascade similar to the complement system, in which the enzymes are initially in an inactive or zymogen form. Tissue injury triggers an enzymatic cascade in which one active protease cleaves and activates the next protease, resulting in the production of several inflammatory mediators including the vasoactive peptide **bradykinin**. Bradykinin causes an increase in vascular permeability that promotes the influx of plasma proteins to the site of tissue injury. It also causes pain, which, although unpleasant to the victim, draws attention to the problem and leads to immobilization of the affected part of the body, which helps to limit the spread of the infection.

The **coagulation system** is another protease cascade that is triggered in the blood after damage to blood vessels. Its activation leads to the formation of a fibrin clot, whose normal role is to prevent blood loss. With regard to innate immunity, however, the clot physically encases the infectious microorganisms and prevents their entry into the bloodstream. The kinin cascade and the blood coagulation cascade are also triggered by activated endothelial cells, and so they can have important roles in the inflammatory response to pathogens even if wounding or gross tissue injury has not occurred. Thus, within minutes of the penetration of tissues by a pathogen, the inflammatory response causes an influx of proteins and cells that may control the infection. It also forms a physical barrier in the form of blood clots to limit the spread of infection and makes the host fully aware of the local infection.

3-4 Toll-like receptors represent an ancient pathogen-recognition system.

Cytokine and chemokine production by macrophages is the result of stimulation of signaling receptors on these cells by a wide variety of pathogen components. Of these receptors, the **Toll-like receptors** (**TLRs**) represent an evolutionarily ancient host defense system. The receptor protein **Toll** was first identified as a gene controlling the correct dorso-ventral patterning embryo of the fruitfly *Drosophila melanogaster*. But in 1996 it was discovered that in the adult insect, Toll signaling induces the expression of several host-defense mechanisms, including antimicrobial peptides such as drosomycin (see Fig. 2.9) and is critical for defense against Gram-positive bacteria and fungal pathogens. Antimicrobial peptides seem to be the earliest form of defense against infection to evolve (see Section 2-4), and so receptors that recognize pathogens and send signals for the production of antimicrobial peptides have a good claim on being the earliest receptors dedicated to defense against infection in multicellular organisms.

It was found that mutations in *Drosophila* Toll or in signaling proteins activated by Toll decreased the production of antimicrobial peptides and led to susceptibility of the adult fly to fungal infections (Fig. 3.8). Subsequently, homologs of Toll, called Toll-like receptors, were found in other animals, including mammals, in which they are associated with resistance to viral, bacterial, and fungal infection. In plants, proteins with domains resembling the ligand-binding regions of TLR proteins are involved in the production of antimicrobial peptides, indicating the ancient association of these domains with this means of host defense.

3-5 Mammalian Toll-like receptors are activated by many different pathogen-associated molecular patterns.

There are 10 expressed *TLR* genes in humans (13 in mice), and each is devoted to recognizing a distinct set of molecular patterns that are not found in healthy vertebrate cells. These patterns are characteristic of components of pathogenic microorganisms at one or other stage of infection and are often called **pathogen-associated molecular patterns** (**PAMPs**). Between them, the mammalian TLRs recognize molecular patterns characteristic of Gram-negative and Gram-positive bacteria, fungi, and viruses. Bacterial cell walls and membranes are composed of repetitive arrays of proteins, carbohydrates, and lipids, many of which are not found in animal cells. Among these, the **lipoteichoic acids** of Gram-positive bacterial cell walls and the **lipopolysaccharide** (**LPS**) of the outer membrane of Gram-negative bacteria (see Fig. 2.7) are particularly important in the recognition of bacteria by the innate immune system, and are recognized by TLRs. Other microbial components also have a repetitive structure. Bacterial flagella are made of a repeated protein subunit, and bacterial DNA has abundant unmethylated repeats of the dinucleotide CpG (which is often methylated in mammalian DNA). Viruses almost invariably produce double-stranded RNA as part of their life cycles, a type of RNA not characteristic of healthy mammalian cells. All these are recognized by TLRs.

Movie 3.3

The mammalian TLRs and the known microbial ligands they recognize are listed in Fig. 3.9. Because there are only a relatively few *TLR* genes, the TLRs have limited specificity compared with the antigen receptors of the adaptive immune system. However, they can recognize elements of most pathogenic microbes and are expressed by many types of cells, including the phagocytic macrophages and dendritic cells, B cells, and certain epithelial cells, enabling the initiation of antimicrobial responses in many tissues.

Fig. 3.8 Toll is required for antifungal responses in *Drosophila melanogaster*. Flies that are deficient in the Toll receptor are dramatically more susceptible than wild-type flies to fungal infection. This is illustrated here by the uncontrolled hyphal growth of the normally weak pathogen *Aspergillus fumigatus* in a Toll-deficient fly. Photo courtesy of J.A. Hoffmann.

TLRs are sensors for microbes present in extracellular spaces. Some mammalian TLRs are cell-surface receptors similar to *Drosophila* Toll, but others are located intracellularly in the membranes of endosomes, where

Innate immune recognition by mammalian Toll-like receptors		
Toll-like receptor	**Ligand**	**Cellular distribution**
TLR-1:TLR-2 heterodimer TLR-2:TLR-6 heterodimer	Lipomannans (mycobacteria) Lipoproteins (diacyl lipopeptides; triacyl lipopeptides) Lipoteichoic acids (Gram-positive bacteria) Cell-wall β-glucans (bacteria and fungi) Zymosan (fungi)	Monocytes, dendritic cells, mast cells, eosinophils, basophils
TLR-3	Double-stranded RNA (viruses)	NK cells
TLR-4 (plus MD-2 and CD14)	LPS (Gram-negative bacteria) Lipoteichoic acids (Gram-positive bacteria)	Macrophages, dendritic cells, mast cells, eosinophils
TLR-5	Flagellin (bacteria)	Intestinal epithelium
TLR-7	Single-stranded RNA (viruses)	Plasmacytoid dendritic cells, NK cells, eosinophils, B cells
TLR-8	Single-stranded RNA (viruses)	NK cells
TLR-9	DNA with unmethylated CpG (bacteria and herpesviruses)	Plasmacytoid dendritic cells, eosinophils, B cells, basophils
TLR-10	Unknown	Plasmacytoid dendritic cells, eosinophils, B cells, basophils
TLR-11 (mouse only)	Profilin and profilin-like proteins (*Toxoplasma gondii*, uropathogenic bacteria)	Macrophages, dendritic cells, liver, kidney, and bladder epithelial cells

Fig. 3.9 Innate immune recognition by Toll-like receptors. Each of the TLRs whose specificity is known recognizes one or more microbial molecular patterns, generally by direct interaction with molecules on the pathogen surface. Some Toll-like receptor proteins form heterodimers (e.g. TLR-1:TLR-2 and TLR-6:TLR-2). LPS, lipopolysaccharide.

they detect pathogens or their components that have been taken into cells by phagocytosis, receptor-mediated endocytosis, or macropinocytosis (Fig. 3.10). They are single-pass transmembrane proteins with an extracellular region composed of 18–25 copies of a **leucine-rich repeat** (**LRR**). These multiple LRRs create a horseshoe-shaped protein scaffold that is adaptable for ligand binding and recognition on both the outer (convex) and inner (concave) surfaces. Mammalian TLRs are activated when binding of a ligand induces them to form dimers or oligomers. All mammalian TLR proteins have a **TIR** (for **Toll–IL-1 receptor**) **domain** in their cytoplasmic tail, which interacts with other TIR-type domains, usually in other signaling molecules. This name comes from the fact that the receptor for the cytokine **interleukin 1β** (**IL-1β**) has a TIR domain in its cytoplasmic tail and signals by the same pathway as that activated by some TLRs, although the extracellular regions of the IL-1 receptors are composed of immunoglobulin-like domains and not LRR repeats. For years after the discovery of the mammalian TLRs it was not known whether they made direct contact with microbial products or whether they sensed the presence of microbes by some indirect means. *Drosophila* Toll, for example, is not a classical pattern recognition receptor. It does not recognize pathogen products directly; instead it is activated when it binds a cleaved version of a self protein, Spätzle. *Drosophila* has other direct pathogen-recognition molecules and these trigger the proteolytic cascade that ends in the cleavage of Spätzle. However, recent X-ray crystal structures of three mammalian dimeric TLRs bound to their ligands show that some at least of the mammalian TLRs make direct contact with microbial ligands.

Mammalian **TLR-1**, **TLR-2**, and **TLR-6** are cell-surface receptors that are activated by various ligands including lipoteichoic acid and the **diacyl** and **triacyl lipoproteins** of Gram-negative bacteria. They are found on macrophages, dendritic cells, eosinophils, basophils, and mast cells. Ligand binding induces the formation of heterodimers of TLR-2 and TLR-1, or of TLR-2 and TLR-6.

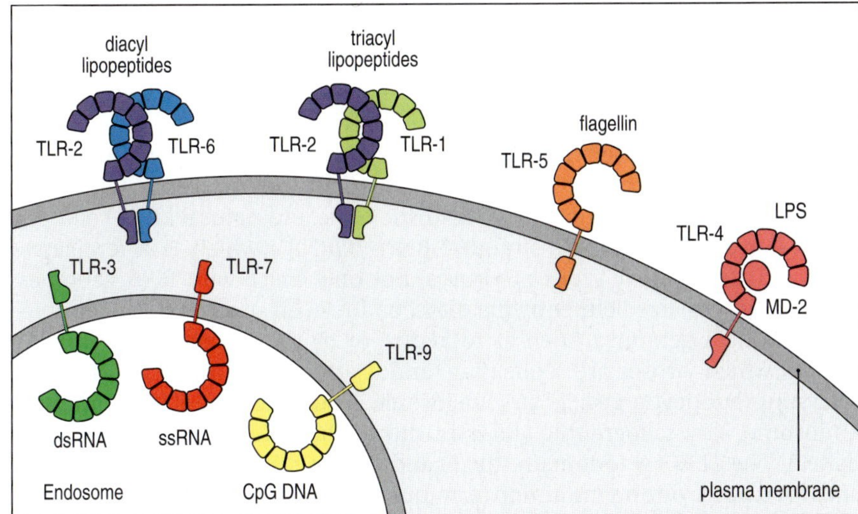

Fig. 3.10 The cellular locations of the mammalian Toll-like receptors. Some TLRs are located on the cell surface of dendritic cells, macrophages, and other cells, where they are able to detect extracellular pathogen molecules. TLRs are thought to act as dimers; only those that form heterodimers are shown in dimeric form here. The rest act as homodimers. TLRs located intracellularly, in the walls of endosomes, can recognize microbial components, such as DNA, that are accessible only after the microbe has been broken down. The diacyl and triacyl lipopeptides recognized by the heterodimeric receptors TLR-6:TLR-2 and TLR-1:TLR-2, respectively, are derived from the lipoteichoic acid of Gram-positive bacterial cell walls and the lipoproteins of Gram-negative bacterial surfaces.

The X-ray crystal structure of a synthetic triacyl lipopeptide ligand bound to TLR-1 and TLR-2 shows exactly how it induces dimerization (Fig. 3.11). Two of the three lipid chains bind to the convex surface of TLR-2, while the third binds to the convex surface of TLR-1. Dimerization brings the cytoplasmic TIR domains of the TLR chains into close proximity with each other to initiate signaling. Similar interactions are presumed to occur with the diacyl lipopeptide ligands that induce the dimerization of TLR-2 and TLR-6. Recognition of some ligands by the TLR-2:TLR-6 heterodimer, such as long-chain fatty acids and cell-wall β-glucans, require an associated co-receptor. The scavenger receptor CD36, which binds long-chain fatty acids, and Dectin-1, which binds β-glucans (see Section 3-1), both cooperate with TLR-2 in ligand recognition.

TLR-5 is expressed on the cell surface of macrophages, dendritic cells, and intestinal epithelial cells; it recognizes flagellin, the protein subunit of bacterial flagella. TLR-5 recognizes a highly conserved site on flagellin that is buried and inaccessible in the assembled flagellar filament. This means that the receptor is only activated by monomeric flagellin, which is produced by the enzymatic breakdown of flagellated bacteria in the extracellular space. Mice, but not humans, express **TLR-11**, which shares with TLR-5 the ability to recognize an intact protein. TLR-11 is expressed by macrophages and dendritic cells, and also by liver, kidney, and bladder epithelial cells. TLR-11-deficient mice develop urinary infections caused by uropathogenic strains of *Escherichia coli*, although a bacterial ligand for TLR-11 has not yet been identified. TLR-11 can be activated by the mammalian actin-binding protein profilin. The protozoan parasite *Toxoplasma gondii* expresses a protein

Fig. 3.11 Direct recognition of pathogen-associated molecular patterns by TLR-1 and TLR-2. TLR-1 and TLR-2 are located on cell surfaces (left panel), where they can directly recognize bacterial triacyl lipoproteins (middle panel). The convex surfaces of their extracellular domains have binding sites for the lipid side chains of triacyl lipopeptides. In the crystal structure (right panel), the ligand is a synthetic lipid that can activate TLR1:TLR2 dimers; it has three fatty-acid chains bound to a polypeptide backbone. Two fatty-acid chains bind to a pocket on the convex surface of the TLR-2 ectodomain, and the third chain associates with a hydrophobic channel in the convex binding surface of TLR-1, inducing dimerization of the two TLR subunits and bringing their cytoplasmic TIR domains together to initiate signaling. Structure courtesy of Jie-Oh Lee.

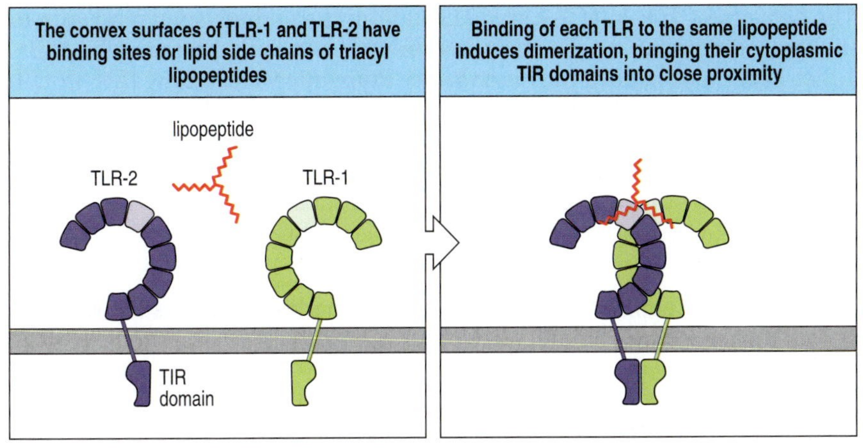

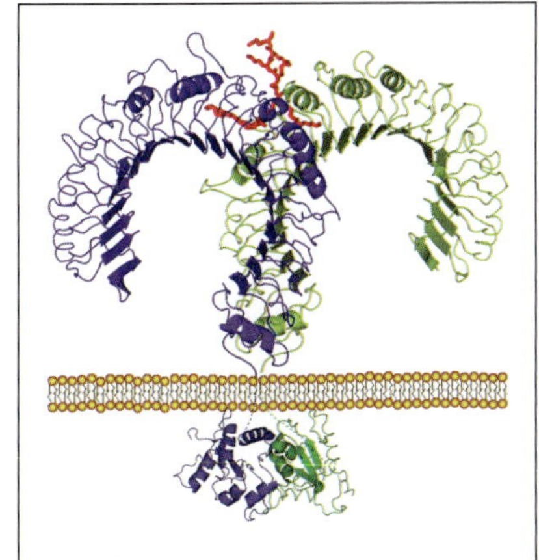

similar to profilin, and mice lacking TLR-11 develop more severe tissue injury on infection with *Toxoplasma*, suggesting that the *Toxoplasma* protein might be a natural ligand for TLR-11.

Not all mammalian TLRs are cell-surface receptors. The TLRs that recognize nucleic acids are located in the membranes of endosomes, to which they are transported via the endoplasmic reticulum. **TLR-3** is expressed by macrophages, intestinal epithelial cells, dendritic cells, and natural killer cells (NK cells); it recognizes double-stranded RNA (dsRNA), which is a replicative intermediate of many types of viruses, not only those with RNA genomes. dsRNA is internalized either by the direct endocytosis of viruses with double-stranded RNA genomes, such as rotavirus, or by the phagocytosis of dying cells in which viruses are replicating, and it encounters the TLRs when the incoming endocytic vesicle or phagosome fuses with the TLR-containing endosome. Crystallographic analysis shows that TLR-3 binds directly to dsRNA. The TLR-3 ectodomain (the ligand-binding domain) has two contact sites for dsRNA: one on the amino terminus and a second near the membrane-proximal carboxy terminus. The twofold symmetry of dsRNA allows it to bind simultaneously to two TLR-3 ectodomains, inducing dimerization that brings its TIR domains together and activates intracellular signaling. Mutations in the ectodomain of human TLR-3, which produce a dominantly acting loss-of-function mutant receptor, have been associated with encephalitis caused by a failure to control the herpes simplex virus.

TLR-7 and **TLR-9**, like TLR-3, are endosomal nucleotide sensors involved in the recognition of viruses. They are found in plasmacytoid dendritic cells, NK cells, B cells, and eosinophils. TLR-7 is activated by single-stranded RNA (ssRNA). ssRNA is a component of healthy mammalian cells, but it is normally confined to the nucleus and cytoplasm and is not present in endosomes. Many virus genomes, for example those of flaviviruses (such as West Nile virus) and rabies, are ssRNA. When extracellular particles of these viruses are endocytosed by macrophages or dendritic cells, they are uncoated in the acidic environment of endosomes and lysosomes, exposing the ssRNA genome for recognition by TLR-7. In abnormal settings, TLR-7 may be activated by self-derived ssRNA. Normally, extracellular RNases degrade the ssRNA released from apoptotic cells during tissue injury. But in a mouse model of lupus nephritis, an inflammatory condition of the kidney, TLR-7 recognition of self ssRNA was observed to contribute to disease. However, it is not yet known whether similar autoimmunity in humans is caused by this pathway.

TLR-9 recognizes unmethylated CpG dinucleotides. In mammalian genomes, CpG dinucleotides in genomic DNA are heavily methylated on the cytosine by DNA methyltransferases. But in the genomes of bacteria and many viruses, CpG dinucleotides remain unmethylated and represent another pathogen-associated molecular pattern.

The delivery of TLR-3, TLR-7, and TLR-9 from the endoplasmic reticulum to the endosome relies on their interaction with a specific protein, UNC93B1, which is composed of 12 transmembrane domains. Mice lacking this protein have defects in signaling by these endosomal TLRs. Rare human mutations in UNC93B1 have been identified as causing susceptibility to herpes simplex encephalitis, similarly to TLR-3 deficiency, but do not impair immunity to many other viral pathogens, presumably because of the existence of other viral sensors, which are discussed later in this chapter.

3-6 TLR-4 recognizes bacterial lipopolysaccharide in association with the host accessory proteins MD-2 and CD14.

Not all mammalian TLRs bind their ligands so directly. **TLR-4** is expressed by several types of immune-system cells including dendritic cells and macrophages; it is important in sensing and responding to numerous bacterial infections. TLR-4 recognizes bacterial lipopolysaccharide (LPS) by a mechanism that is partly direct and partly indirect. LPS is a cell-wall component of Gram-negative bacteria such as *Salmonella* that has long been known to induce a reaction in the infected host. The systemic injection of LPS causes a collapse of the circulatory and respiratory systems, a condition known as **shock**. These dramatic effects of LPS are seen in humans as **septic shock**, which results from an uncontrolled systemic bacterial infection, or **sepsis**. In this case, LPS induces an overwhelming secretion of cytokines, particularly TNF-α (see Section 3-16). Mutant mice that lack TLR-4 function are resistant to LPS-induced septic shock but are highly sensitive to LPS-bearing pathogens such as *Salmonella typhimurium*, a natural pathogen of mice. In fact, TLR-4 was identified as the receptor for LPS by positional cloning of its gene from the LPS-resistant C3H/HeJ mouse strain, which harbors a naturally occurring mutation in the cytoplasmic tail of TLR-4 that interferes with the receptor's ability to signal. When we discuss septic shock more fully later in the chapter, we shall see that it is an undesirable consequence of the same effector actions of TNF-α that are important in containing local infections.

LPS varies in composition between different bacteria but essentially consists of a polysaccharide core attached to an amphipathic lipid, lipid A, with a variable number of fatty-acid chains per molecule (see Fig. 2.7). To recognize LPS, the ectodomain of TLR-4 uses an accessory protein, **MD-2**. MD-2 initially binds to TLR-4 within the cell and is necessary both for the correct trafficking of TLR-4 to the cell surface and for the recognition of LPS. MD-2 associates with the central section of the curved ectodomain of TLR-4, binding off to one side as shown in Fig. 3.12. When the TLR4–MD-2 complex encounters LPS, five lipid chains of LPS bind to a deep hydrophobic pocket of MD-2, but not directly to TLR-4, while a sixth remains exposed on the surface of MD-2. This lipid chain and parts of the LPS polysaccharide backbone bind directly to a region on the convex surface of the ectodomain of a second TLR-4 molecule. Thus, the TLR-4 dimerization required to activate intracellular signaling depends on both indirect and direct interactions with LPS.

TLR-4 activation by LPS involves two other accessory proteins besides MD-2. LPS is an integral component of the outer membrane of Gram-negative bacteria, but during an infection it can become detached from the membrane and is picked up by the **LPS-binding protein** present in the blood and in extracellular fluid in tissues. LPS is transferred from LPS-binding protein to a second protein, **CD14**, which is present on the surface of macrophages, neutrophils, and dendritic cells. On its own CD14 can act as a phagocytic receptor, but on macrophages and dendritic cells it also acts as an accessory protein for TLR-4.

3-7 TLRs activate the transcription factors NFκB, AP-1, and IRF to induce the expression of inflammatory cytokines and type I interferons.

Signaling by mammalian TLRs in various cell types induces a diverse range of intracellular responses that together result in the production of inflammatory cytokines, chemotactic factors, antimicrobial peptides, and the antiviral cytokines **interferon** (**IFN**)-**α** and **IFN-β**, the **type I interferons**. TLR signaling achieves this by being able to activate several different signaling pathways that each activate different transcription factors. One pathway activates the **NFκB** transcription factors (Fig. 3.13), which are related to DIF, the factor activated by *Drosophila* Toll. Mammalian TLRs also activate several members of the **interferon regulatory factor** (**IRF**) transcription factor family through a second pathway, and activate members of the activator protein 1 (**AP-1**) family, such as c-Jun, through yet another signaling pathway involving **mitogen-activated protein kinases** (**MAPKs**). NFκB and AP-1 act primarily to induce the expression of pro-inflammatory cytokines and chemotactic

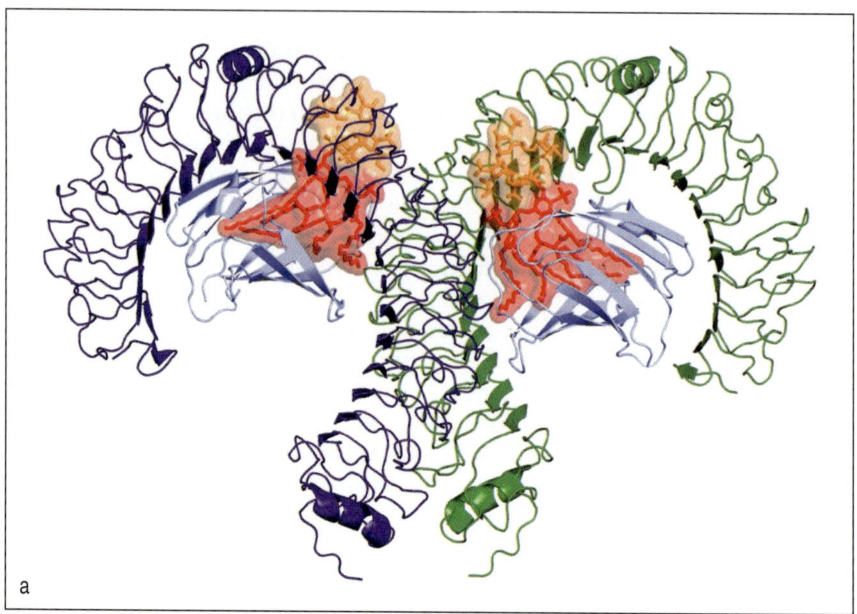

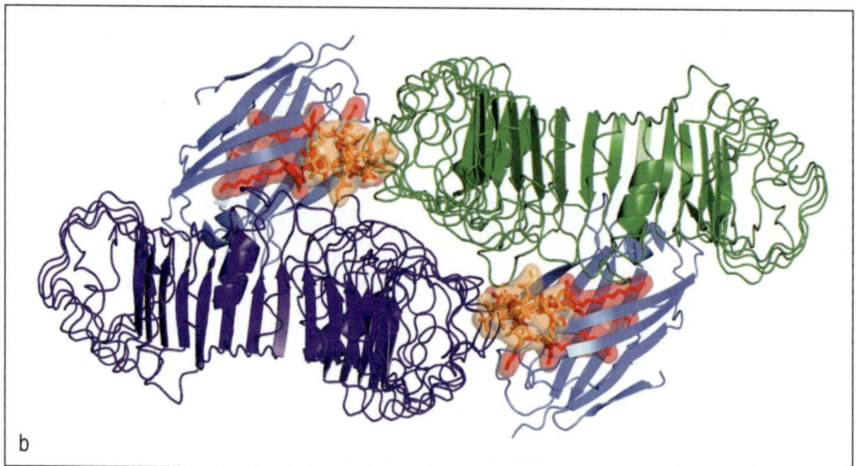

Fig. 3.12 TLR-4 recognizes LPS in association with the accessory protein MD-2. Panel a: a side view of the symmetrical complex of TLR-4, MD-2 and LPS. TLR-4 polypeptide backbones are shown in green and dark blue. The structure contains the entire extracellular region of TLR4, composed of the LRR region (shown in green and dark blue), but lacks the intracellular signaling domain. The MD-2 protein is shown in light blue. Five of the LPS acyl chains (shown in red) are inserted into a hydrophobic pocket within MD-2. The remainder of the LPS glycan and one lipid chain (orange) make contact with the convex surface of a TLR-4. Panel b: the top view of the structure shows that an LPS molecule makes contact with one TLR-4 subunit on its convex (outer) surface, while binding to an MD-2 molecule that is attached to the other TLR-4 subunit. The MD-2 protein binds off to one side of the TLR-4 LRR region. Structure courtesy of Jie-Oh Lee.

factors, whereas IRF factors are particularly important for inducing antiviral type I interferons. We describe here how TLR signaling induces cytokine and interferon production; how these proteins exert their effects is discussed later in the chapter.

TLR signaling is activated by the ligand-induced dimerization of two TLR ectodomains, which brings their cytoplasmic TIR domains close together, allowing them to interact with the TIR domains of cytoplasmic adaptor molecules that initiate intracellular signaling. There are four such adaptors used by mammalian TLRs: MyD88 (myeloid differentiation factor 88), MAL (MyD88 adaptor-like), TRIF (TIR domain-containing adaptor-inducing IFN-β), and TRAM (TRIF-related adaptor molecule). It is significant that different TLRs interact with different combinations of these adaptors. TLR-5, TLR-7, and TLR-9 interact only with MyD88, which is required for their signaling. TLR-3 interacts only with TRIF, which is also required for signaling. Other TLRs either use MyD88 paired with MAL, or use TRIF paired with TRAM. Signaling by the TLR-2 heterodimers (TLR-2/1 and TLR-2/6) requires MyD88/MAL. TLR-4 signaling uses both of these adaptor pairs, MyD88/MAL and TRIF/TRAM. Importantly, the choice of adaptor influences which signals will be activated by the TLR. We shall consider first the signaling pathway triggered by TLRs that use MyD88, and then look at the signaling pathway stimulated by viral nucleic acids that leads to interferon production.

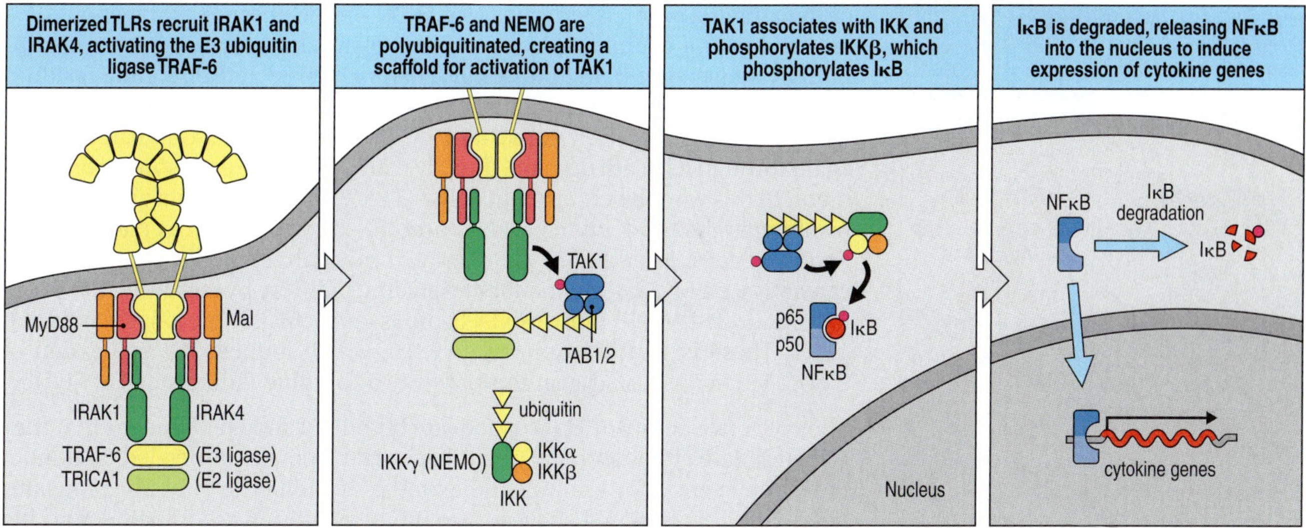
Dimerized TLRs recruit IRAK1 and IRAK4, activating the E3 ubiquitin ligase TRAF-6
MyD88
Mal
IRAK1
IRAK4
TRAF-6
(E3 ligase)
TRICA1
(E2 ligase)
TRAF-6 and NEMO are polyubiquitinated, creating a scaffold for activation of TAK1
TAK1
TAB1/2
ubiquitin
IKKγ (NEMO)
IKKα
IKKβ
IKK
TAK1 associates with IKK and phosphorylates IKKβ, which phosphorylates IκB
p65
p50
IκB
NFκB
Nucleus
IκB is degraded, releasing NFκB into the nucleus to induce expression of cytokine genes
NFκB
IκB degradation
IκB
cytokine genes

Fig. 3.13 TLR signaling can activate the transcription factor NFκB, which induces the expression of pro-inflammatory cytokines. First panel: TLRs signal via their cytoplasmic TIR domains, which are brought into proximity by the dimerization of their ectodomains by ligand. Some TLRs that use the adaptor protein MyD88 use the MyD88/MAL pair. The MyD88 death domain recruits the serine–threonine kinases IRAK1 and IRAK4, in association with the ubiquitin E3 ligase TRAF-6. IRAK undergoes autoactivation and phosphorylates TRAF-6, activating its E3 ligase activity. Second panel: TRAF-6 cooperates with an E2 ligase, TRICA1, to ubiquitinate TRAF-6 itself and NEMO, a component of the IKK complex, creating polyubiquitin scaffolds on these proteins. These polyubiquitin scaffolds are built by attachment of ubiquitin through its lysine 63 (K63) and do not signal for protein degradation in the proteasome (see Section 7-5). Adaptor proteins TAB1 and TAB2, which form a complex with TAK1, bind to the ubiquitin scaffold on TRAF6, allowing TAK1 to be phosphorylated by IRAK. Third panel: TAK1 can then associate with the ubiquitin scaffold on NEMO, allowing TAK1 to phosphorylate IKKβ, which is activated and phosphorylates IκB. Fourth panel: phosphorylated IκB is targeted by ubiquitination (not shown) to undergo degradation, releasing NFκB, which is composed of two subunits, p50 and p65. NFκB enters the nucleus and activates the transcription of many genes, including those encoding inflammatory cytokines. TAK1 also stimulates activation of the mitogen-activated protein kinases (MAPKs) JNK and p38, which phosphorylate and activate AP-1 transcription factors (not shown).

Two protein domains of MyD88 are responsible for its function as an adaptor: a TIR domain at the carboxy terminus and a **death domain** at the amino terminus. Death domains form heterodimers with similar death domains on other intracellular signaling proteins, and we shall encounter them later in other signaling pathways. They are so named because they were first identified in signaling proteins involved in apoptosis or programmed cell death. Both the TIR and the death domain are crucial for MyD88 function, as shown by rare mutations in MyD88 in humans. Mutations in either domain are associated with immunodeficiency characterized by recurrent bacterial infections. The MyD88 TIR domain interacts with the TIR domain of the TLR, while the death domain recruits and activates two serine–threonine protein kinases—**IRAK4** (IL1-receptor associated kinase 4) and **IRAK1**—via their death domains. The IRAK complex recruits TRAF-6, an E3 ubiquitin ligase. The E2 ligase TRICA1, acting in cooperation with TRAF-6, then generates a scaffold of polyubiquitin chains on TRAF-6 itself and on the protein NEMO that recruits and activates the serine–threonine kinase TAK1.

TAK1 has two important actions. It activates certain MAPKs, such as c-Jun terminal kinase (JNK) and MAPK14 (p38 MAPK), which activate AP-1-family transcription factors. TAK1 also phosphorylates and activates the **IκB kinase (IKK)** complex, which is composed of three proteins: IKKα, IKKβ, and IKKγ (also known as **NEMO**, for NFκB essential modifier). Activated IKK phosphorylates IκB, which then releases NFκB. This then moves into the nucleus, where it induces the transcription of genes for pro-inflammatory cytokines such as TNF-α, IL-1β, and **IL-6**. The actions of these cytokines in the innate immune response are described in the second half of this chapter. The outcome of TLR activation also depends on the cell type. Activation of TLR4 via MyD88 in specialized epithelial cells such as the Paneth cells of the intestine (see Section 2-4) results in the production of antimicrobial peptides, a mammalian example of the ancient function of Toll-like proteins.

The ability of TLRs to activate NFκB is crucial to their role of alerting the immune system to the presence of bacterial pathogens. Rare instances of inactivating mutations in IRAK4 in humans cause an immunodeficiency, **IRAK4 deficiency**, which like MyD88 deficiency is characterized by recurrent bacterial infections. Mutations in human NEMO produce a syndrome known as **X-linked hypohidrotic ectodermal dysplasia and immunodeficiency** or **NEMO deficiency**, which is characterized by both immunodeficiency and developmental defects.

The nucleic-acid sensing TLRs—TLR-7, TLR-8, and TLR-9—signal uniquely through MyD88 to activate IRF transcription factors that induce the production

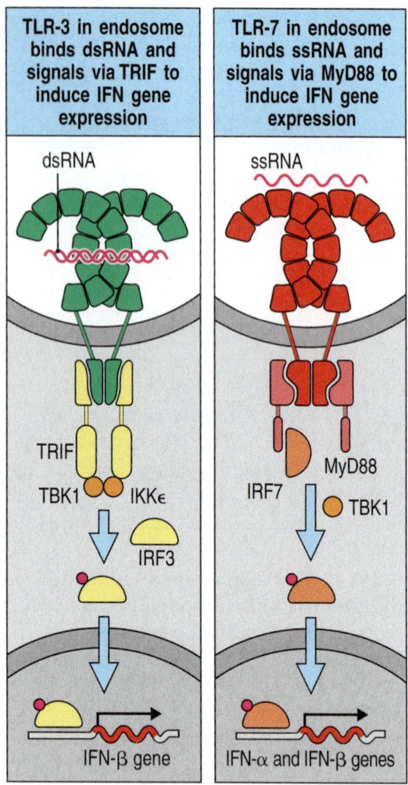

Fig. 3.14 Expression of antiviral interferons in response to viral nucleic acids can be stimulated by two different pathways from different TLRs. Left panel: TLR-3 senses double-stranded viral RNA (dsRNA) and signals through a MyD88-independent pathway that uses the structurally similar adaptor protein TRIF. Via its death domain, TRIF activates the serine–threonine kinases IκKε (IKKε) and TBK1; through the involvement of TRAF 3 and TRAF6 (not shown) this leads to phosphorylation of the transcription factor IRF3, which enters the nucleus and induces expression of the IFN-β gene. Right panel: TLR-7 detects single-stranded RNA (ssRNA) and signals through MyD88 to phosphorylate and activate the transcription factor IRF7. This enters the nucleus and switches on expression of the IFN-β and IFN-α genes.

of the antiviral type I interferons (Fig. 3.14). In the absence of infection, IRFs are held in the cytoplasm in an inactive form and only become transcriptionally active after being phosphorylated on serine and threonine residues located at their carboxy termini. Of the nine members of the IRF family, IRF3 and IRF7 are particularly important in TLR signaling. MyD88 can interact, again through its death domain, directly with IRF7 in the endosomal compartment. In this way, the endosomal TLRs—TLR-7 and TLR-9—can activate IRF7, resulting in the production of IFN-α and IFN-β by cells such as plasmacytoid dendritic cells. In addition, TLR-3, which recognizes double-stranded RNA, can activate a MyD88-independent signaling pathway by recruiting the adaptor protein TRIF. TRIF, unlike MyD88, binds and activates the kinases IκKε and TBK1. These in turn phosphorylate IRF3, which induces the expression of IFN-β. TRL-4 can also trigger this pathway by binding TRIF instead of MyD88.

The collective ability of TLRs to activate both IRFs and NFκB means that they can stimulate either antiviral or antibacterial responses as needed. In human IRAK4 deficiency, for example, no extra susceptibility to viral infections has been noted. This would suggest that IRF activation is not impaired and the production of antiviral interferons is not affected. TLRs are expressed by different types of cells involved in innate immunity and by some epithelial cells, and the responses generated will differ in some respects depending on what type of cell is being activated.

3-8 The NOD-like receptors act as intracellular sensors of bacterial infection.

The TLRs are all located on cell surfaces or in the membranes of intracellular vesicles, and sense the presence of extracellular pathogens. Another large family of receptors that use LRR scaffold domains to detect pathogen products are located in the cytoplasm. These are the **NOD-like receptors** (**NLRs**), which contain a centrally located nucleotide-binding oligomerization domain (NOD) and an LRR domain near their carboxy terminus. NLRs are intracellular sensors for microbial products and activate NFκB to initiate the same inflammatory responses as the TLRs. The NLRs are considered a very ancient family of innate immunity receptors because the resistance (R) proteins that are part of plant defenses against pathogens are NLR homologs.

Subfamilies of NLRs can be distinguished on the basis of the protein domains found near the amino terminus. The NOD subfamily has an amino-terminal **caspase recruitment domain** (**CARD**) (Fig. 3.15), initially recognized from its role in mediating interactions with a family of proteases called **caspases** (cysteine-aspartic acid proteases), which are important in many intracellular pathways, including those leading to cell death by apoptosis. CARD is related to the death domain in MyD88 and can dimerize with CARD domains on other proteins. NOD proteins recognize fragments of bacterial cell-wall peptidoglycans, although it is not known whether this occurs through direct binding or via accessory proteins. **NOD1** senses **γ-glutamyl diaminopimelic acid** (iE-DAP), a breakdown product of peptidoglycans of Gram-negative bacteria such as *Salmonella* and *Listeria*, which enter the cytoplasm of cells, whereas **NOD2** recognizes **muramyl dipeptide**, which is present in the peptidoglycans of most bacteria.

When NOD1 or NOD2 recognizes its ligand, it recruits the CARD-containing serine–threonine kinase **RIPK2** (also known as RICK and RIP2). RIPK2 activates the kinase TAK1, which activates NFκB via the activation of IKK, as shown in Fig. 3.13. NFκB then induces the expression of genes for inflammatory cytokines. In keeping with their role as sensors of bacterial components, NOD proteins are expressed in cells that are routinely exposed to bacteria. These include epithelial cells forming the barrier that bacteria must cross to establish an infection in the body, and the macrophages and dendritic cells

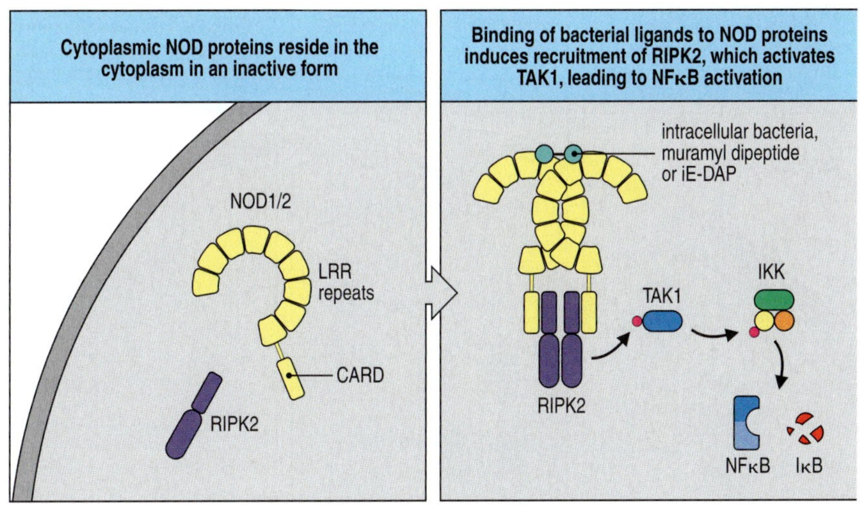

Fig. 3.15 Intracellular NOD proteins sense the presence of bacteria, recognizing bacterial peptidoglycans and activating NFκB to induce the expression of pro-inflammatory genes. Degradation of bacterial cell-wall peptidoglycans produces muramyl dipeptide, which is recognized by NOD2, which therefore acts as a general intracellular sensor of a bacterial presence. NOD1 recognizes γ-glutamyl diaminopimelic acid (iE-DAP), a breakdown product of Gram-negative bacterial cell walls. The presence of these ligands induces the recruitment and activation of the serine–threonine kinase RIPK2, which phosphorylates the kinase TAK1. This leads to activation of NFκB by the pathway shown in Fig. 3.13.

that ingest bacteria that have succeeded in entering the body. Macrophages and dendritic cells express TLRs as well as NOD1 and NOD2, and are activated by both pathways. In epithelial cells, however, TLRs are expressed weakly or not at all, and NOD1 is an important activator of the innate immune response in these cells. NOD2 seems to have a more specialized role, being strongly expressed in the Paneth cells of the gut, where it regulates the expression of potent antimicrobial peptides such as the α- and β-defensins (see Chapter 2).

Consistent with this, human loss-of-function mutations in NOD2 are associated with the inflammatory bowel condition known as **Crohn's disease** (discussed in Chapter 15). Some patients with this condition carry mutations in the LRR domain of NOD2 that impair its ability to sense muramyl dipeptide and activate NFκB. This is thought to diminish the production of defensins and other antimicrobial peptides, thereby weakening the natural barrier function of the intestinal epithelium and leading to the inflammation characteristic of this disease. Gain-of-function mutations in human NOD2 produce an inflammatory disorder called **Blau syndrome**, which is characterized by spontaneous inflammation in the joints, eyes, and skin. Activating mutations in the NOD domain seem to promote the signaling cascade in the absence of ligand, leading to an inappropriate inflammatory response in the absence of pathogens.

Another large subfamily of NLR proteins have a **pyrin** domain at their amino termini and are known as the NLRP family (Fig. 3.16). Pyrin domains are structurally related to CARD and death domains, and interact with other pyrin domains. Humans have 14 NLR proteins containing pyrin domains. The best characterized is **NALP3** (also called **NLRP3** or **cryopyrin**), which is an important sensor of cellular damage or stress. In stressed cells, such as those exposed to infection, it assembles with an adaptor protein and the protease caspase 1 to form a complex called an **inflammasome**. Caspase 1 is required for the proteolytic processing of some pro-inflammatory cytokines, which is needed before they can be secreted. In healthy cells, the LRR domain of NALP3 is bound to accessory proteins that keep NALP3 in a monomeric inactive state. A wide variety of cellular stimuli activate NALP3 and it is thought that the common trigger is the efflux of cytoplasmic K^+ ions that occurs in stressed cells. The consequent low intracellular K^+ concentration causes dissociation of the accessory proteins from NALP3. The inflammasome provides a structural framework in which caspase 1 becomes enzymatically active after ATP-dependent autocatalytic cleavage of the pro-caspase. The related NALP1 and NALP2 proteins similarly form inflammasomes.

Another pyrin domain family member, **AIM2** (absent in melanoma 2), has replaced the LRR domain of NALP3 with an **HIN** (H inversion) domain, named

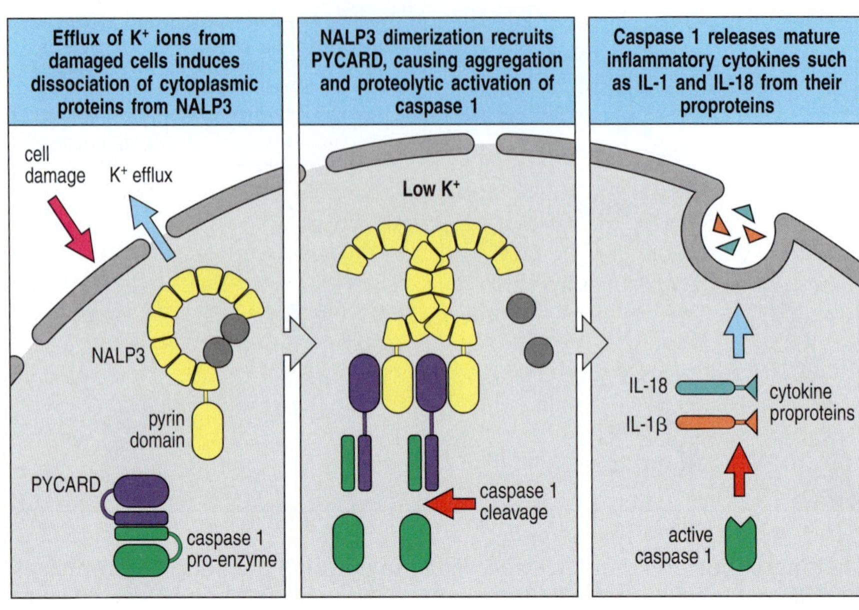

Fig. 3.16 NALP proteins sense cellular damage and activate the processing of pro-inflammatory cytokines. Under normal physiologic conditions the LRR domain of NALP3 associates with cytoplasmic proteins, which are presumed to prevent the dimerization of NALP3 via the LRR region. When cells are injured or put under stress, the typical efflux of K^+ ions is thought to trigger the dissociation of these proteins from NALP3, allowing its dimerization (second panel). The pyrin domains of NALP3 recruit complexes of the adaptor protein PYCARD associated with the caspase 1 pro-enzyme through their CARD (C) domains. Aggregation of the pro-enzymes causes them to undergo autoactivation via proteolytic cleavage to form active caspases. Active caspase 1 cleaves the proprotein forms of pro-inflammatory cytokines to release the mature cytokines, which can then be secreted.

for the HIN DNA recombinase of *Salmonella* that mediates DNA inversion between flagellar H antigens. In AIM2, the HIN domain recognizes double-stranded DNA genomes and triggers caspase 1 activation through the AIM2 pyrin domain. AIM2 is important for cellular responses *in vitro* to vaccinia virus, and its *in vivo* role has been demonstrated by the increased susceptibility of AIM2-deficient mice to infection by *Francisella tularensis*, the causative agent of tularemia.

It is not yet clear whether NALP3 and other NLRPs function as receptors for specific microbial products. The inflammasome has, however, been implicated in the actions of several inflammation-inducing chemicals and in some inflammatory diseases. For example, the stimulatory effects of the immunological adjuvant (see Section 3-10) aluminum hydroxide (alum) rely on NALP3 and the inflammasome pathway. The disease **gout** has been known for many years to be caused by monosodium urate crystals deposited in the cartilaginous tissues of joints, causing inflammation. But how the urate crystals caused inflammation was a mystery. Although the precise mechanism is still unclear, we now know that urate crystals activate the NALP3 inflammasome, and that this induces the inflammatory cytokines that cause the symptoms of gout. Mutations in the NOD domain of NALP2 and NALP3 can activate inflammasomes inappropriately, and they are the cause of some inherited **autoinflammatory diseases**, in which inflammation occurs in the absence of infection. Mutations in NALP3 in humans are associated with the hereditary periodic fever syndomes **familial cold inflammatory syndrome** and **Muckle–Wells syndrome** (discussed in more detail in Chapter 13). Macrophages from patients with these conditions show spontaneous production of inflammatory cytokines such as IL-1β as a result of activation of the NFκB pathway.

3-9 The RIG-I-like helicases detect cytoplasmic viral RNAs and stimulate interferon production.

TLR-3, TLR-7, and TLR-9 can detect viral RNAs and DNAs, but they interact primarily with extracellular material entering the endocytic pathway rather than with nucleic acids present in the cytoplasm of a virus-infected cell as a result of viral replication. Cytoplasmic viral RNAs are sensed by proteins called the **RIG-I-like helicases** (**RLHs**), which have an RNA helicase-like domain that binds to viral RNAs, and two amino-terminal CARD domains that interact with adaptor proteins and activate signaling when viral RNAs are bound. The first of these sensors to be discovered was **RIG-I** (for **retinoic acid-inducible gene I**). RIG-I is widely expressed across tissues and cell types and serves as an intracellular sensor for several kinds of infections. Mice deficient in RIG-I are highly susceptible to infection by several kinds of ssRNA viruses, including paramyxoviruses, orthomyxoviruses, and flaviviruses, but not picornaviruses.

Initially, RIG-I was thought to recognize long dsRNA, but later studies showed that it recognizes specific differences between eukaryotic and viral ssRNA transcripts, although it also seems able to sense the presence of short blunt-ended dsRNAs. When eukaryotic RNA is first transcribed in the nucleus, it contains a 5′-triphosphate group on its initial nucleotide that undergoes enzymatic modification. For example, mRNAs undergo capping by the addition of a 7-methylguanosine to the 5′ triphosphate. Most RNA viruses, however, do not replicate in the nucleus and their transcripts do not undergo this modification. Biochemical studies have determined that RIG-I senses the unmodified 5′-triphosphate end of ssRNA. Flavivirus RNA transcripts have the unmodified 5′ triphosphate, as do the transcripts of many other ssRNA viruses, which explains why they are detected by RIG-I. In contrast, the picornaviruses, which include poliovirus and hepatitis A, replicate by a mechanism that involves the covalent attachment of a viral protein to the 5′ end of the viral RNA, so that the 5′ triphosphate is absent, which explains why RIG-I is not involved in sensing them.

MDA-5 (melanoma differentiation-associated 5), also called helicard, is similar in structure to RIG-I, but it senses dsRNA. In contrast to RIG-I-deficient mice, mice deficient in MDA-5 are susceptible to picornaviruses, indicating that these two sensors of viral RNAs have crucial but distinct roles in host defense. Inactivating mutations in alleles of human RIG-I or MDA-5 have been reported, but these mutations were not associated with immunodeficiency. Genome-wide association studies have, however, linked mutations predicted to inactivate human MDA-5 with a decreased risk of developing type 1 diabetes, a disease caused by immune destruction of the insulin-producing β cells of the pancreas. The reason for this association is still unclear.

Sensing of viral RNAs by RIG-I and MDA-5 induces the production of type I interferons, appropriate for defense against viral infection. When RIG-I or MDA-5 detects a viral RNA ligand in the cytosol, its CARD domain interacts with a CARD-containing adaptor protein called MAVS that is attached to the outer mitochondrial membrane (Fig. 3.17). The association of RIG-I or MDA-5 with MAVS activates a signaling pathway that involves TRAF6 and which

Fig. 3.17 RIG-I and MDA-5 are cytoplasmic sensors of viral RNA. First panel: before detecting viral RNA, RIG-I and MDA-5 are cytoplasmic, and the adaptor protein MAVS is attached to the mitochondrial outer membrane. Second panel: detection of uncapped 5′-triphosphate RNA by RIG-I or viral dsRNA by MDA-5 causes their CARD domains to interact with the amino-terminal CARD domain of MAVS, which induces the dimerization of MAVS. Third panel: a proline-rich region of MAVS is involved in interactions with TRAFs, while a more carboxy-terminal region of MAVS interacts with a complex of TRADD and FADD (death-domain-containing adaptor proteins). The TRAF6 pathway activates IRF3 and IRF7 (not shown) inducing the transcription of a number of genes, principally those encoding IFN-α and IFN-β. FADD leads to the activation of IKK and NFκB as shown in Fig. 3.13, leading to the expression of pro-inflammatory cytokines.

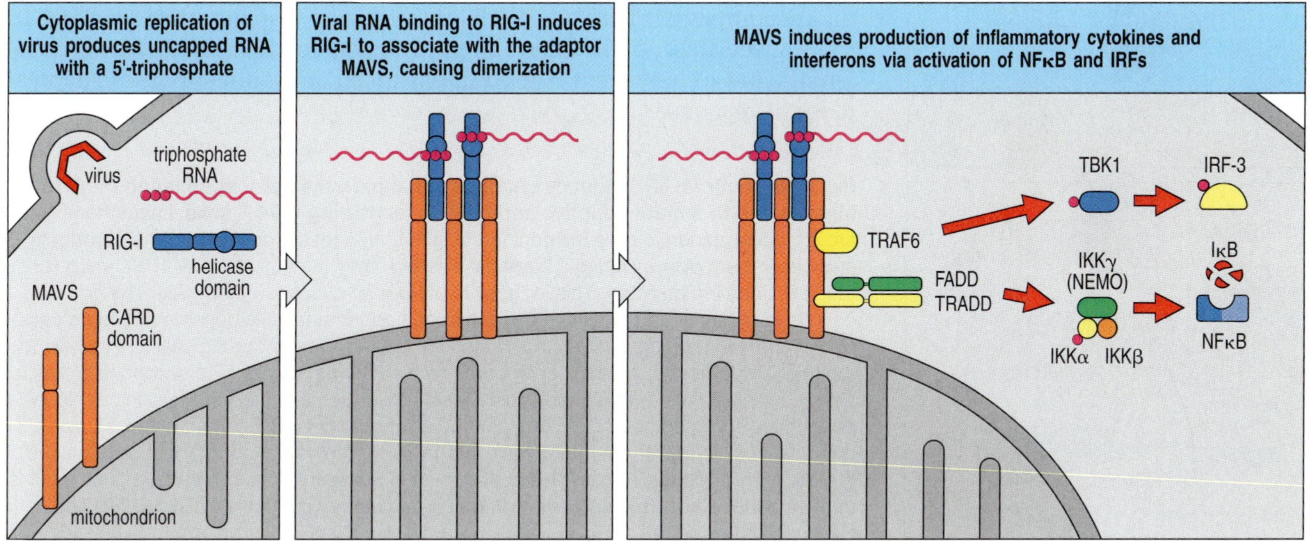

activates TBK1 (see Fig. 3.14). This leads to the activation of IRF3 and induces the production of IFN-β and IFN-α. Activation of RIG-I and MDA-5 can also induce cytokine production through a signaling pathway that involves the adaptor protein FADD, which also contains a death domain. In this case, a complex of FADD and other proteins activates caspases 8 and 10, which leads to the activation of IKK and NFκB, inducing the production of pro-inflammatory cytokines (see Fig. 3.17).

3-10 Activation of TLRs and NLRs triggers changes in gene expression in macrophages and dendritic cells that have far-reaching effects on the immune response.

The cytokines and chemokines produced by macrophages and dendritic cells as a result of activation of NFκB by the TLR and NOD pathways include not only important mediators of innate immunity but some, such as **IL-12**, that influence the subsequent adaptive immune response, as we shall see later in the chapter. Another outcome of the NFκB pathway important for adaptive immunity is the appearance of **co-stimulatory molecules** on tissue dendritic cells and macrophages. These cell-surface proteins, called **B7.1** (**CD80**) and **B7.2** (**CD86**), are expressed by both macrophages and tissue dendritic cells in response to activation by pathogen sensors such as TLRs (Fig. 3.18) and are essential for the induction of adaptive immune responses. The co-stimulatory molecules, together with antigenic microbial peptides displayed by MHC molecules on the surface of dendritic cells, activate naive CD4 T cells, which help to initiate most adaptive immune responses (see Section 1-19). The cytokine TNF-α, also produced as a result of TLR-4 signaling, has numerous functions in innate immunity, but in addition it stimulates the antigen-presenting dendritic cells to migrate into the lymphatic system and enter nearby lymph nodes, through which circulating naive CD4 T cells pass. Thus, the activation of adaptive immunity depends on molecules induced as a consequence of the innate immune recognition of pathogens.

Substances such as LPS that induce co-stimulatory activity have been used for years in mixtures that are co-injected with protein antigens to enhance their immunogenicity. These substances are known as **adjuvants** (see Appendix I, Section A-4), and it was found empirically that the best adjuvants contained microbial components. A range of microbial components (see Fig. 3.9) can induce macrophages and tissue dendritic cells to express co-stimulatory molecules and cytokines. The exact profile of cytokines produced by the macrophage or dendritic cell varies according to the receptors stimulated and, as we shall see in Chapters 9 and 11, the cytokines secreted will in turn influence the functional character of the adaptive immune response that develops. In this way the ability of the innate immune system to discriminate between different types of pathogens is used to ensure an appropriate type of adaptive immune response.

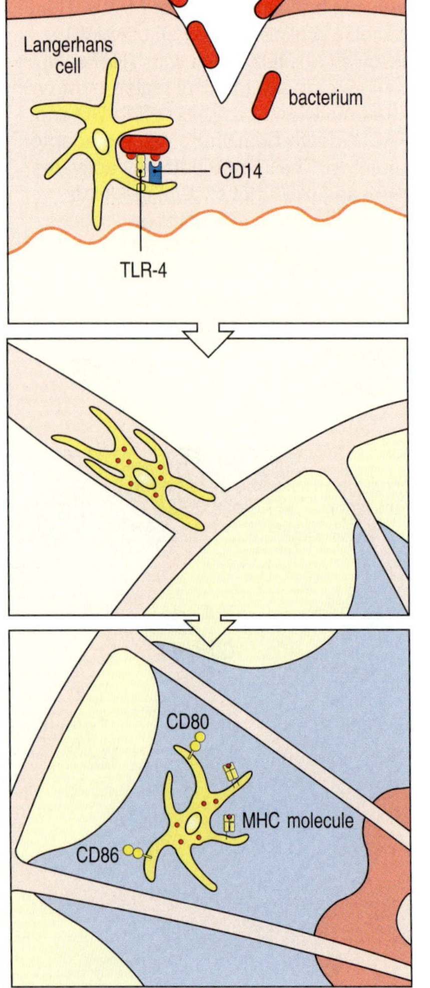

Fig. 3.18 Bacterial LPS induces changes in Langerhans cells, stimulating them to migrate and to initiate adaptive immunity by activating CD4 T cells. Langerhans cells are immature dendritic cells resident in the skin that ingest microbes and their products and break them down. During a bacterial infection they are activated by LPS through the TLR signaling pathway. This induces two types of changes in the cells. The first is a change in behavior and location. The activated Langerhans cells become migrating cells and enter the lymphatic system that drains the tissues. The migrating cells are carried to regional lymph nodes, where they become mature dendritic cells. The second is a drastic alteration in their cell-surface molecules. Resting Langerhans cells in the skin are highly phagocytic and macropinocytic but lack the ability to activate T lymphocytes. Mature dendritic cells in the lymph nodes have lost the ability to ingest antigen but have gained the ability to stimulate T cells. This is due to an increase in the number of MHC molecules on their surface and the expression of the co-stimulatory molecules CD80 (B7.1) and CD86 (B7.2).

3-11 TLR signaling shares many components with Toll signaling in *Drosophila*.

Before we leave the topic of pathogen recognition by the innate immune system, we shall look briefly at how Toll, TLRs, and NODs are used in invertebrate innate immunity. Although it is central to defense against both bacterial and fungal pathogens in *Drosophila*, Toll itself is not a pattern recognition receptor. Instead, it relies for activation on several families of pathogen detector proteins. One comprises the **peptidoglycan-recognition proteins** (**PGRPs**). *Drosophila* has some 13 PGRP genes, encoding proteins that bind the peptidoglycan components of bacterial cell walls (see Fig. 2.7). Another family of pathogen sensors comprises the **Gram-negative binding proteins** (**GNBPs**), which recognize β-1–3-linked glucans and are involved in the recognition of fungi and, rather unexpectedly, Gram-positive bacteria. The family members GNBP1 and PGRP-SA cooperate in the recognition of peptidoglycan from Gram-positive bacteria. They interact with a serine protease called Grass, which initiates a proteolytic cascade that terminates in the cleavage of the protein Spätzle. One of the cleaved fragments forms a homodimer, which induces the dimerization and activation of Toll to induce the antimicrobial response (Fig. 3.19). A fungus-specific recognition protein, GNBP3, also activates the proteolytic cascade, causing cleavage of Spätzle and activation of Toll.

In *Drosophila*, fat-body cells and hemocytes are phagocytic cells that act as part of the fly's immune system. When Toll on these cells binds the Spätzle dimer, the cells respond by synthesizing and secreting antimicrobial peptides. The signaling pathway from Toll is very similar to the NFκB pathway in vertebrates (see Fig. 3.14). It results in the activation of a transcription factor called DIF, which is related to mammalian NFκB. DIF enters the nucleus and induces the transcription of genes for antimicrobial peptides such as drosomycin.

DIF and mammalian NFκB are members of the Rel family of transcription factors, named after the *Drosophila* transcription factor Relish. Relish itself induces the production of antimicrobial peptides in response to a different signaling pathway, the **Imd** (**immunodeficiency**) **pathway**. This pathway is used by *Drosophila* in the recognition of Gram-negative bacteria, and is triggered by particular PGRPs. Relish induces expression of the antimicrobial peptides diptericin, attacin, and cecropin, which are distinct from the peptides induced by Toll signaling. Thus, the Toll and Imd pathways activate effector mechanisms to eliminate infection by different kinds of pathogens. Four mammalian PGRP homologs have been identified, but they do not serve the same function as in *Drosophila*, although they do have a function in innate immunity. One, PGLYRP-2, is secreted and functions as an amidase to hydrolyze bacterial peptidoglycans. The others are present in neutrophil granules and exert a bacteriostatic action through interactions with bacterial cell-wall peptidoglycan.

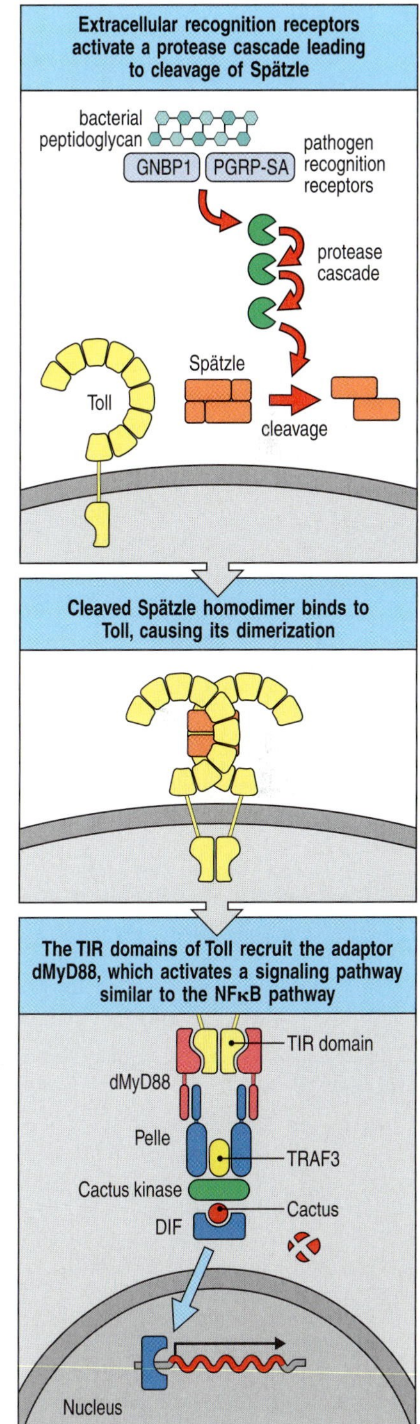

Fig. 3.19 *Drosophila* Toll is activated as a result of a proteolytic cascade initiated by pathogen recognition. The peptidoglycan-recognition protein PGRP-SA and the Gram-negative binding protein GNBP1 cooperate in the recognition of bacterial pathogens, activating the first protease in a protease cascade that leads to cleavage of the *Drosophila* protein Spätzle (first panel). Cleavage alters the conformation of Spätzle, enabling it to bind Toll and induce Toll dimerization (second panel). Toll's cytoplasmic TIR domains recruit the adaptor protein dMyD88 (third panel), which initiates a signaling pathway very similar to that leading to the release of NFκB from its cytoplasmic inhibitor in mammals. The *Drosophila* version of NFκB is the transcription factor DIF, which then enters the nucleus and activates the transcription of genes encoding antimicrobial peptides. Fungal recognition also leads to cleavage of Spätzle and the production of antimicrobial peptides by this pathway, although the recognition proteins for fungi are as yet unidentified.

3-12 TLR and NOD genes have undergone extensive diversification in both invertebrates and some primitive chordates.

Mammalian TLRs are a family of about a dozen receptors that interact directly with pathogen-derived ligands. Some organisms, however, have diversified their repertoire of innate recognition receptors, especially those containing LRR domains, to a much greater degree. The sea urchin *Strongylocentrotus purpuratus* has an unprecedented 222 different *TLR* genes, more than 200 NOD-like receptor genes, and more than 200 scavenger receptor genes in its genome. The sea urchin also has an increased number of proteins that are likely to be involved in signaling from these receptors, with four genes similar to mammalian *MyD88*. Despite this diversification of innate receptors, there is no apparent increase in the number of downstream targets, such as the family of NFκB transcription factors, suggesting that the ultimate outcome of TLR signaling in the sea urchin will be very similar to that in other organisms. Sea urchin TLR genes fall into two broad categories: a small set of 11 divergent genes and a large family of 211 genes, which show high degree of sequence variation within particular LRR regions. Together with the large number of pseudogenes in this family, this implies rapid evolutionary turnover, suggesting rapidly changing receptor specificities, which contrasts with the few stable mammalian TLRs. Although the pathogen specificity of sea urchin TLRs is unknown, the hypervariability in the LRR domains could be used to generate a highly diversified pathogen-recognition system based on Toll-like receptors.

A similar expansion of innate receptors has occurred in some chordates, the phylum to which vertebrates belong. Amphioxus (the lancelet) is a nonvertebrate chordate lacking an adaptive immune system. The amphioxus genome contains 71 TLRs, more than 100 NOD-like receptors, and more than 200 scavenger receptors. And as we will see later in the book, a primitive vertebrate lineage—the jawless fish, which lack immunoglobulin- and T-cell-based adaptive immunity—uses somatic gene rearrangement of LRR-containing proteins to provide a version of adaptive immunity (see Section 5-21).

Summary.

Cells of the innate immune system express several receptor systems for recognizing microbes, which induce both rapid defenses and more delayed cellular responses. Neutrophils, macrophages, and dendritic cells can rapidly eliminate microbes through phagocytosis by using several scavenger and lectin-like receptors. Signaling receptors such as the G-protein-coupled receptors for C5a (which engages the complement system's pathogen-recognition ability) and for the bacterial peptide fMLP synergize with phagocytic receptors, activating the NADPH oxidase in phagosomes to generate antimicrobial reactive oxygen intermediates. Other receptors that detect pathogens induce signaling pathways that turn on a general inflammatory response, inducing local inflammation, recruiting new effector cells, containing local infection, and finally triggering an adaptive immune response. The Toll-like receptors (TLRs) have been highly conserved across evolutionary time and activate host defenses through several signaling pathways. The NFκB pathway from these receptors operates in most multicellular organisms and induces the expression of pro-inflammatory cytokines, including TNF-α, IL-1β, and IL-6. An alternative signaling pathway from some Toll-like receptors activates the IRF transcription factors that induce the expression of antiviral cytokines. TLRs on the cell surface and in the membranes of endosomes sense pathogens present outside the cell, but many cells also possess intracellular receptors that detect pathogens in the cytosol. NOD proteins detect bacterial products within the cytosol, activating NFκB and the production of pro-inflammatory cytokines. The NALP proteins are part of the inflammasome, which is concerned with detecting general cellular stress and damage; the inflammasome activates caspases, enabling the processing and secretion of pro-inflammatory cytokines. The cytosolic proteins RIG-I and MDA-5 detect viral infection by sensing the presence of viral RNAs, leading to induction of the antiviral type I interferons. The signaling pathways activated by these primary sensors of pathogens induce a variety of genes, including those for the cytokines, chemokines, and co-stimulatory molecules, that have essential roles in immediate defense and in directing the course of the adaptive immune response later in infection.

Induced innate responses to infection.

In this part of the chapter we will look at the responses of macrophages and dendritic cells to stimulation via pathogen-sensing receptors such as the TLRs, and the consequences for the innate immune response. One immediate and important outcome of such stimulation is the production and secretion of cytokines and chemokines by macrophages and dendritic cells that help induce and maintain inflammation. The chemokines are a large family of chemoattractant proteins with a central role in leukocyte migration, and the chemokines secreted by activated macrophages attract neutrophils and other immune-system cells to the site of infection. Adhesion molecules on white blood cells and the endothelial cells of blood vessels have an equally important role in the movement of cells out of the blood and into infected tissue, and we will briefly consider the different types of adhesion molecules involved. We will then consider in more detail how macrophage-derived chemokines and cytokines promote the continued destruction of infecting microbes. This is achieved both by stimulating the production and recruitment of fresh phagocytes and by inducing another phase of the innate immune response—the acute-phase response—in which the liver produces proteins that act as opsonizing molecules, helping to augment the actions of complement. We will also look at the mechanism of action of the antiviral interferons, and at a class of lymphoid cells known as NK cells that are activated by interferon to contribute to innate immmune defense against viruses and other intracellular pathogens. We will also consider the innate-like lymphocytes (ILLs), which contribute to the rapid response to infection by acting early but use a limited set of antigen-receptor gene segments (see Section 1-12) to make immunoglobulins and T-cell receptors of limited diversity.

The induced innate response can either succeed in clearing the infection or will contain it while an adaptive response develops. If the infection is not cleared, the adaptive response will harness many of the same effector mechanisms used by the innate immune system, such as complement-mediated phagocytosis, but will target them with much greater precision. Antigen-specific T cells activate the microbicidal and cytokine-secreting properties of macrophages, while antibodies activate complement, act as direct opsonins for phagocytes, and stimulate NK cells to kill infected cells. The effector mechanisms described here therefore serve as a primer for the focus on adaptive immunity in the rest of the book.

3-13 Macrophages and dendritic cells activated by pathogens secrete a range of cytokines that have a variety of local and distant effects.

Cytokines (see Appendix III) are small proteins (about 25 kDa) that are released by various cells in the body, usually in response to an activating stimulus, and they induce responses through binding to specific receptors. They can act in an **autocrine** manner, affecting the behavior of the cell that releases the cytokine, in a **paracrine** manner, affecting the behavior of adjacent cells, and some cytokines are stable enough to act in an **endocrine** manner, affecting the behavior of distant cells, although this depends on their ability to enter the circulation and on their half-life in the blood. Activation of the pathogen-sensing receptors expressed by macrophages and dendritic cells induces the production of specific cytokines that activate the cellular arm of the innate immune system.

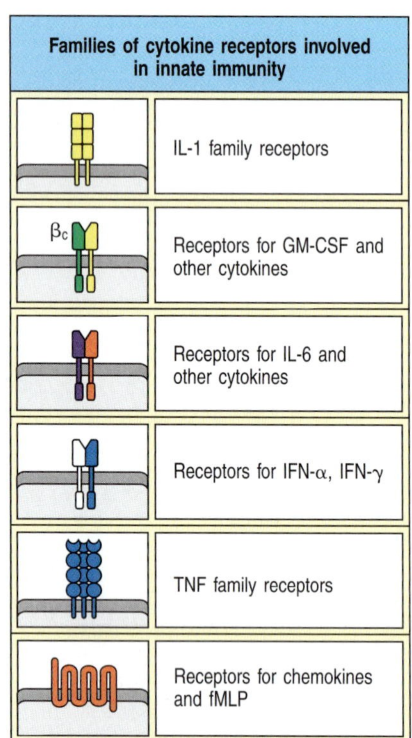

Fig. 3.20 Cytokine receptors important in innate immunity. Members of the IL-1 receptor family have extracellular regions composed of immunoglobulin-like domains and they signal as dimers through TIR domains in their cytoplasmic tails. The hematopoietin receptor superfamily is subdivided into various families on the basis of protein sequence and subunit structure, and not all families are shown here. In the receptors for GM-CSF and IL-6, one chain defines the ligand specificity of the receptor (e.g. the common β chain, β_c, in the GM-CSF receptor), whereas the other confers the intracellular signaling function. These receptors signal through the JAK–STAT pathway (described in Section 7-20). Receptors for interferons and interferon-like cytokines are also heterodimeric receptors that signal via the JAK–STAT pathway but do not have a common chain. TNF family members act as trimers and are mostly associated with cell membranes, although some can be secreted. Ligand binding induces trimerization of TNF receptors, which contain death domains in their cytoplasmic tails and signal via death-domain-containing adaptors such as TRADD and FADD.

At least two broad classes of dendritic cells are recognized, as noted earlier in the chapter: conventional dendritic cells, which seem to participate most directly in antigen presentation and activation of naive T cells; and plasmacytoid dendritic cells. The latter are a distinct cell lineage that produces large amounts of interferon, particularly in response to viral infections, but may not be as important for activating naive T cells. Immature dendritic cells in tissues are phagocytic but can be distinguished from macrophages by their expression of different surface molecules.

The cytokines secreted by dendritic cells and macrophages in response to activation of pattern recognition receptors include a structurally diverse group of proteins that include IL-1β, IL-6, IL-12, TNF-α, and the chemokine CXCL8 (formerly known as IL-8). The name **interleukin (IL)** followed by a number (for example IL-1 or IL-2) was coined in an attempt to develop a standardized nomenclature for molecules secreted by, and acting on, leukocytes. However, this became confusing when an ever-increasing number of cytokines with diverse origins, structures, and effects were discovered, and although the IL designation is still used, it is hoped that eventually a nomenclature based on cytokine structure will be developed. The cytokines are listed alphabetically, together with their receptors, in Appendix III.

Cytokines can be grouped by structure into different families, and their receptors can likewise be grouped (Fig. 3.20). We will focus here on the IL-1 family, the hematopoietins, the TNF family, and the type I interferons, because they include the cytokines active in innate immunity. The IL-1 family contains 11 members, notably IL-1α, IL-1β, and IL-18. Most members of this family are produced as inactive proproteins that are cleaved (removing an amino-terminal peptide) to produce the mature cytokine. IL-1β and IL-18, which are produced by macrophages in response to TLR signaling, are cleaved by caspase 1 (a component of the inflammasome; see Fig. 3.16). The IL-1-family receptors have TIR domains in their cytoplasmic tails and signal by the NKκB pathway described earlier for TLRs (see Fig. 3.13).

The very large **hematopoietin superfamily** includes non-immune-system growth and differentiation factors such as erythropoietin (which stimulates red blood cell development) and growth hormone, as well as interleukins with roles in innate and adaptive immunity. IL-6 is a member of this superfamily, as is the cytokine GM-CSF, which stimulates the production of new monocytes and granulocytes in the bone marrow. The hematopoietin-family cytokines act through dimeric receptors that fall into several structural subfamilies characterized by functional similarities and genetic linkage. We shall discuss the other hematopoietin-family cytokines and their receptors later in the book, in relation to their production by T cells and their effects on the adaptive immune response. The **interferon receptors** comprise a small family of heterodimeric receptors that recognize the type I interferons as well as some cytokines, such as IL-4, that are produced by T cells. The hematopoietin and interferon receptors all signal through the JAK–STAT pathway (described in Fig. 7.29) and activate different combinations of STATs with different effects.

The **TNF family**, of which TNF-α is the prototype, contains more than 17 cytokines with important functions in adaptive and innate immunity. Unlike most of the other immunologically important cytokines, many members of the TNF family are transmembrane proteins, which gives them distinct properties and limits their range of action, although some can also be released from the membrane in some circumstances. They are usually found as homotrimers of a membrane-bound subunit, although some heterotrimers between different subunits occur. TNF-α (sometimes called simply TNF) is initially expressed as a trimeric membrane-bound cytokine but can be released from the membrane. The effects of TNF-α are mediated by either of two **TNF receptors**. TNF receptor I (**TNFR-I**) is expressed on a wide range of cells, including endothelial cells and macrophages, whereas **TNFR-II** is expressed largely by lymphocytes. The signaling pathways stimulated by immunologically important members of the TNF family are described in Chapter 7.

Overall, the structural, functional, and genetic relations between the cytokines and their receptors suggest that they may have diversified in parallel during the evolution of increasingly specialized effector functions. These specific functional effects depend on intracellular signaling events that are triggered by the cytokines binding to their specific receptors. All the cytokines produced by macrophages in innate immune responses have important local and systemic effects that contribute to both innate and adaptive immunity, and these are summarized in Fig. 3.21. The recognition of different classes of pathogens by phagocytes and dendritic cells may involve signaling through different receptors, such as the various TLRs, and can result in some variation in the cytokines expressed by stimulated macrophages and dendritic cells. This is one way in which appropriate immune responses can be selectively activated, as the released cytokines orchestrate the next phase of host defense.

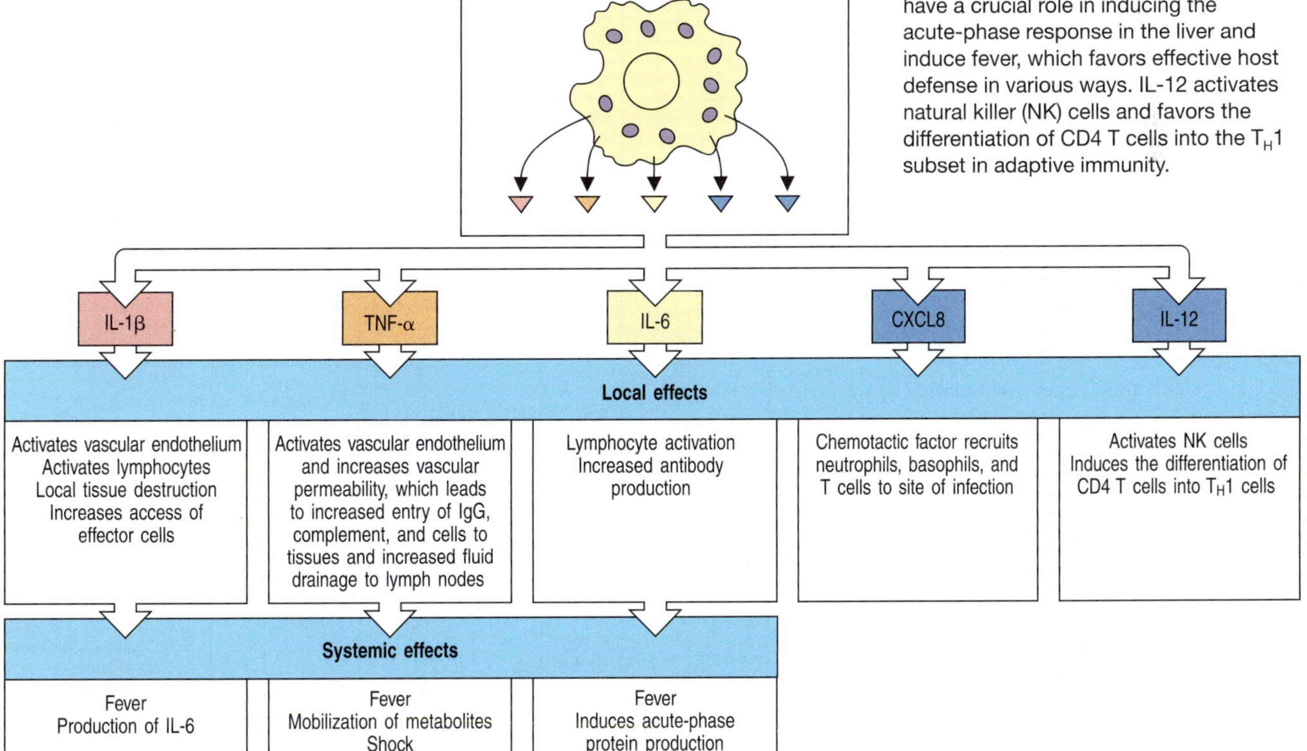

Fig. 3.21 Important cytokines and chemokines secreted by dendritic cells and macrophages in response to bacterial products include IL-1β, IL-6, CXCL8, IL-12, and TNF-α. TNF-α is an inducer of a local inflammatory response that helps to contain infections. It also has systemic effects, many of which are harmful (discussed in Section 3-17). The chemokine CXCL8 is also involved in the local inflammatory response, helping to attract neutrophils to the site of infection. IL-1β, IL-6, and TNF-α have a crucial role in inducing the acute-phase response in the liver and induce fever, which favors effective host defense in various ways. IL-12 activates natural killer (NK) cells and favors the differentiation of CD4 T cells into the T_H1 subset in adaptive immunity.

3-14 Chemokines released by macrophages and dendritic cells recruit effector cells to sites of infection.

Movie 3.5

Among the cytokines released by tissues in the earliest phases of infection are members of a family of chemoattractant cytokines known as chemokines. These small proteins induce directed **chemotaxis** in nearby responsive cells, resulting in the movement of the cells toward the source of the chemokine. Because chemokines were first detected in functional assays, they were initially given a variety of names, which are listed along with their standardized nomenclature in Appendix IV. All the chemokines are related in amino acid sequence, and their receptors are G-protein-coupled receptors (see Section 3-2). The signaling pathway stimulated by chemokines causes changes in cell adhesiveness and changes in the cell's cytoskeleton that lead to directed migration. Chemokines can be produced and released by many different types of cells, not only those of the immune system. In the immune system they function mainly as chemoattractants for leukocytes, recruiting monocytes, neutrophils, and other effector cells of innate immunity from the blood into sites of infection. They also guide lymphocytes in adaptive immunity, as we will learn in Chapters 9–11. Some chemokines also function in lymphocyte development and migration and in angiogenesis (the growth of new blood vessels). There are more than 50 known chemokines, and this striking multiplicity may reflect their importance in delivering cells to their correct locations, which seems to be their main function in the case of lymphocytes. Some of the chemokines that are produced by or that affect innate immune cells are listed in Fig. 3.22 along with their properties.

Fig. 3.22 Properties of selected chemokines. Chemokines fall mainly into two related but distinct groups: the CC chemokines, which have two adjacent cysteine residues near the amino terminus, and the CXC chemokines, in which the equivalent cysteine residues are separated by a single amino acid. In humans, the genes for CC chemokines are mostly clustered in one region of chromosome 4. Genes for CXC chemokine genes are found mainly in a cluster on chromosome 17. The two groups of chemokines act on different sets of receptors, all of which are G-protein-coupled receptors. CC chemokines bind to receptors designated CCR1–9. CXC chemokines bind to receptors designated CXCR1–6. Different receptors are expressed on different cell types, and so a particular chemokine can be used to attract a particular cell type. In general, CXC chemokines with a Glu-Leu-Arg tripeptide motif immediately before the first cysteine promote the migration of neutrophils. CXCL8 is an example of this type. Most of the other CXC chemokines, including those that interact with receptors CXCR3, 4, and 5, lack this motif. Fractalkine is unusual in several respects: it has three amino acid residues between the two cysteines, and it exists in two forms, one tethered to the membrane of the endothelial and epithelial cells that express it, where it serves as an adhesion protein, and a soluble form released from the cell surface, which acts as a chemoattractant for a wide range of cell types. A more comprehensive list of chemokines and their receptors is given in Appendix IV.

Class	Chemokine	Produced by	Receptors	Cells attracted	Major effects
CXC	CXCL8 (IL-8)	Monocytes Macrophages Fibroblasts Epithelial cells Endothelial cells	CXCR1 CXCR2	Neutrophils Naive T cells	Mobilizes, activates and degranulates neutrophils Angiogenesis
	CXCL7 (PBP, β-TG, NAP-2)	Platelets	CXCR2	Neutrophils	Activates neutrophils Clot resorption Angiogenesis
	CXCL1 (GROα) CXCL2 (GROβ) CXCL3 (GROγ)	Monocytes Fibroblasts Endothelium	CXCR2	Neutrophils Naive T cells Fibroblasts	Activates neutrophils Fibroplasia Angiogenesis
CC	CCL3 (MIP-1α)	Monocytes T cells Mast cells Fibroblasts	CCR1, 3, 5	Monocytes NK and T cells Basophils Dendritic cells	Competes with HIV-1 Antiviral defense Promotes T_H1 immunity
	CCL4 (MIP-1β)	Monocytes Macrophages Neutrophils Endothelium	CCR1, 3, 5	Monocytes NK and T cells Dendritic cells	Competes with HIV-1
	CCL2 (MCP-1)	Monocytes Macrophages Fibroblasts Keratinocytes	CCR2B	Monocytes NK and T cells Basophils Dendritic cells	Activates macrophages Basophil histamine release Promotes T_H2 immunity
	CCL5 (RANTES)	T cells Endothelium Platelets	CCR1, 3, 5	Monocytes NK and T cells Basophils Eosinophils Dendritic cells	Degranulates basophils Activates T cells Chronic inflammation
CXXXC (CX_3C)	CX3CL1 (Fractalkine)	Monocytes Endothelium Microglial cells	CX_3CR1	Monocytes T cells	Leukocyte–endothelial adhesion Brain inflammation

Chemokines fall mainly into two related but distinct groups. **CC chemokines** have two adjacent cysteine residues near the amino terminus, whereas in **CXC chemokines** the corresponding two cysteine residues are separated by a single amino acid. The CC chemokines promote the migration of monocytes, lymphocytes, and other cell types. One example relevant to innate immunity is **CCL2**, which attracts monocytes, inducing their migration from the bloodstream to become tissue macrophages. In contrast, neutrophil migration is promoted by CXC chemokines. **CXCL8** induces neutrophils to leave the blood and migrate into the surrounding tissues. CXCL8 and CCL2 therefore have similar but complementary functions in the innate immune response, attracting neutrophils and monocytes respectively.

Movie 3.6

The role of chemokines in cell recruitment is twofold. First, they act on the leukocyte as it rolls along endothelial cells at sites of inflammation, converting this rolling into stable binding by triggering a change of conformation in the adhesion molecules known as leukocyte integrins, which enables them to bind strongly to their ligands on the endothelial cells. This in turn allows the leukocyte to cross the blood vessel wall by squeezing between the endothelial cells, as we will see in Section 3-16 when we describe the process of extravasation. Second, the chemokine directs the migration of the leukocyte along a gradient of chemokine molecules bound to the extracellular matrix and the surfaces of endothelial cells. This gradient increases in concentration toward the site of infection.

Chemokines are produced by a wide variety of cell types in response to bacterial products, viruses, and agents that cause physical damage, such as silica, alum, or the urate crystals that occur in gout. Complement fragments such as C3a and C5a, and fMLP bacterial peptides also act as chemoattractants for neutrophils. Thus, infection or physical damage to tissues induces the production of chemokine gradients that can direct phagocytes to the sites where they are needed. Neutrophils are the first cells to arrive in large numbers at a site of infection, with monocytes and immature dendritic cells being recruited later. The complement fragment C5a and the chemokines CXCL8 and CCL2 activate their respective target cells, so that not only are neutrophils and monocytes brought to potential sites of infection but, in the process, they are armed to deal with pathogens they encounter there. In particular, neutrophils exposed to C5a or CXCL8 are activated to produce the respiratory burst that generates oxygen radicals and nitric oxide, and to release their stored antimicrobial granule contents (see Section 3-2).

Chemokines do not act alone in cell recruitment. They require the action of vasoactive mediators that bring leukocytes close to the blood vessel wall (see Section 3-3) and cytokines such as TNF-α to induce the necessary adhesion molecules on endothelial cells. We will return to the chemokines in later chapters, when they will be discussed in the context of the adaptive immune response. We will now turn to the molecules that enable leukocytes to adhere to the endothelium, and we shall then describe step by step the extravasation process by which monocytes and neutrophils enter infected sites.

3-15 Cell-adhesion molecules control interactions between leukocytes and endothelial cells during an inflammatory response.

Movie 3.7

The recruitment of activated phagocytes to sites of infection is one of the most important functions of innate immunity. Recruitment occurs as part of the inflammatory response and is mediated by cell-adhesion molecules that are induced on the surface of the endothelial cells of local blood vessels.

As with the complement components, a significant barrier to understanding the functions of cell-adhesion molecules is their nomenclature. Most adhesion molecules, especially those on leukocytes, which are relatively easy to analyze functionally, were originally named after the effects of specific monoclonal antibodies directed against them. Their names therefore bear no relation to their structural class. For instance, the **leukocyte functional antigens** LFA-1, LFA-2, and LFA-3 are actually members of two different protein families. In Fig. 3.23, the adhesion molecules relevant to innate immunity are grouped according to their molecular structure, which is shown in schematic form, alongside their different names, sites of expression, and ligands. Three structural families of adhesion molecules are important for leukocyte recruitment. The **selectins** are membrane glycoproteins with a distal lectin-like domain that binds specific carbohydrate groups. Members of this family are induced on activated endothelium and initiate endothelium–leukocyte interactions by binding to fucosylated oligosaccharide ligands on passing leukocytes (see Fig. 3.23).

The next step in leukocyte recruitment depends on tighter adhesion, which is due to the binding of **intercellular adhesion molecules** (**ICAMs**) on the endothelium to heterodimeric proteins of the **integrin** family on leukocytes. ICAMs are single-pass membrane proteins that belong to the large superfamily of **immunoglobulin-like proteins**, which contain protein domains similar to those of immunoglobulins. The extracellular regions of ICAMs are composed of several immunoglobulin-like domains. An integrin molecule is composed of two transmembrane protein chains, α and β, of which there are numerous different types. Subsets of integrins have a common β chain partnered with different α chains. The leukocyte integrins important for extravasation are **LFA-1** (α_L:β_2, also known as **CD11a:CD18**) and **CR3** (α_M:β_2, complement receptor type 3, also known as **CD11b:CD18** or Mac-1; we described CR3 in Section 2-13 as a receptor for iC3b, but that is just one of the ligands for this integrin).

Both LFA-1 and CR3 bind to **ICAM-1** and to **ICAM-2** (Fig. 3.24). Even in the absence of infection, circulating monocytes are continuously leaving the

Family	Example	Name	Tissue distribution	Ligand
Selectins Bind carbohydrates. Initiate leukocyte–endothelial interaction	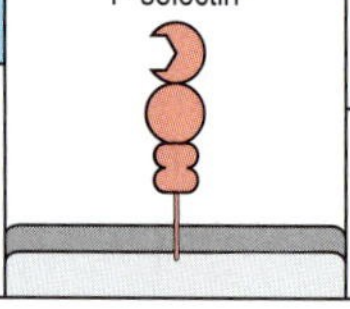 P-selectin	P-selectin (PADGEM, CD62P)	Activated endothelium and platelets	PSGL-1, sialyl-Lewis^x
		E-selectin (ELAM-1, CD62E)	Activated endothelium	Sialyl-Lewis^x
Integrins Bind to cell-adhesion molecules and extracellular matrix. Strong adhesion	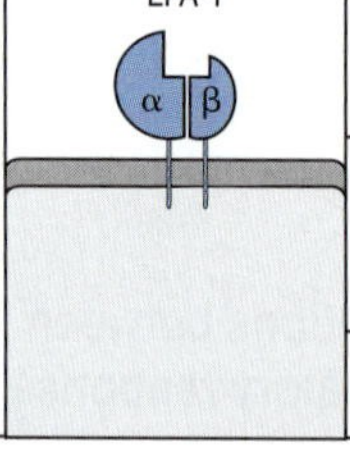 LFA-1 (α, β)	α_L:β_2 (LFA-1, CD11a:CD18)	Monocytes, T cells, macrophages, neutrophils, dendritic cells, NK cells	ICAMs
		α_M:β_2 (CR3, Mac-1, CD11b:CD18)	Neutrophils, monocytes, macrophages, NK cells	ICAM-1, iC3b, fibrinogen
		α_X:β_2 (CR4, p150.95, CD11c:CD18)	Dendritic cells, macrophages, neutrophils, NK cells	iC3b
		α_5:β_1 (VLA-5, CD49d:CD29)	Monocytes, macrophages	Fibronectin
Immunoglobulin superfamily Various roles in cell adhesion. Ligand for integrins	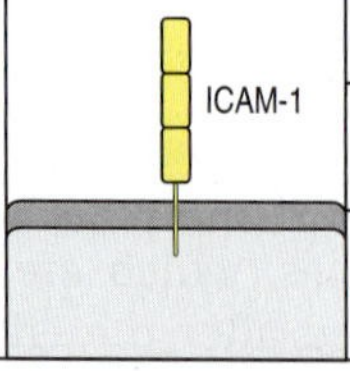 ICAM-1	ICAM-1 (CD54)	Activated endothelium, activated leukocytes	LFA-1, Mac1
		ICAM-2 (CD102)	Resting endothelium, dendritic cells	LFA-1
		VCAM-1 (CD106)	Activated endothelium	VLA-4
		PECAM (CD31)	Activated leukocytes, endothelial cell–cell junctions	CD31

Fig. 3.23 Adhesion molecules involved in leukocyte interactions. Several structural families of adhesion molecules have a role in leukocyte migration, homing, and cell–cell interactions: the selectins, the integrins, and proteins of the immunoglobulin superfamily. The figure shows schematic representations of an example from each family, a list of other family members that participate in leukocyte interactions, their cellular distribution, and their ligand in adhesive interactions. The family members shown here are limited to those that participate in inflammation and other innate immune mechanisms. The same molecules and others participate in adaptive immunity and will be considered in Chapters 9 and 11. The nomenclature of the different molecules in these families is confusing because it often reflects the way in which the molecules were first identified rather than their related structural characteristics. Alternative names for each of the adhesion molecules are given in parentheses. Sulfated sialyl-Lewis^x, which is recognized by P- and E-selectin, is an oligosaccharide present on the cell-surface glycoproteins of circulating leukocytes.

blood and entering tissues, where they become resident macrophages. To navigate out of the blood vessel, they may adhere to ICAM-2, which is expressed at low levels by unactivated endothelium.

Strong adhesion between leukocytes and endothelial cells is promoted by the induction of ICAM-1 on inflamed endothelium together with a conformational change in LFA-1 and CR3 that occurs on the leukocyte. Integrins can switch between an 'active' state, in which they bind strongly to their ligands, and an 'inactive' state, in which binding is easily broken. This enables cells to make and break integrin-mediated adhesions in response to signals received by the cell either through the integrin itself or by other receptors. In the activated state, an integrin molecule is linked via the intracellular protein talin to the actin cytoskeleton. In the case of migrating leukocytes, chemokines binding to their receptors on the leukocyte generate intracellular signals that cause talin to bind to the cytoplasmic tails of the β chains of LFA-1 and CD3, forcing the integrin extracellular regions to assume an active binding conformation. The importance of the leukocyte integrin function in inflammatory cell recruitment is illustrated by the **leukocyte adhesion deficiencies**, which can be caused by defects in the integrins themselves or in the proteins required for modulating adhesion. People with these diseases suffer from recurrent bacterial infections and impaired healing of wounds.

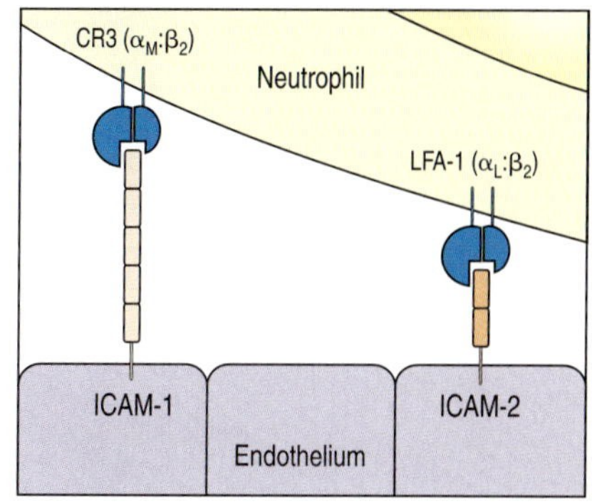

Fig. 3.24 Phagocyte adhesion to vascular endothelium is mediated by integrins. When vascular endothelium is activated by inflammatory mediators it expresses two adhesion molecules, namely ICAM-1 and ICAM-2. These are ligands for integrins expressed by phagocytes—α_M:β_2 (also called CR3, Mac-1, or CD11b:CD18) and α_L:β_2 (also called LFA-1 or CD11a:CD18).

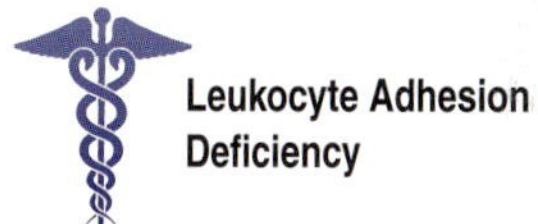

Endothelial activation is driven by macrophage-produced cytokines, particularly TNF-α, which induce the rapid externalization of granules called **Weibel–Palade bodies** in the endothelial cells. These granules contain preformed **P-selectin**, which appears on the surfaces of local endothelial cells just minutes after macrophages have responded to the presence of microbes by producing TNF-α. Shortly after P-selectin gets to the cell surface, mRNA encoding **E-selectin** is synthesized, and within 2 hours the endothelial cells are expressing mainly E-selectin. Both P-selectin and E-selectin interact with the sulfated sialyl-Lewis^x that is present on the surface of neutrophils.

Integrins are also convenient cell-surface markers for distinguishing different cell types. Dendritic cells, macrophages, and monocytes express different integrin α chains and thus display distinct β_2 integrins on their surface. The predominant leukocyte integrin on conventional dendritic cells is α_X:β_2, also known as **CD11c:CD18** or complement receptor 4 (CR4) (see Fig. 3.23). This integrin is a receptor for the complement C3 cleavage product iC3b, fibrinogen, and ICAM-1. In contrast to conventional dendritic cells, monocytes and macrophages express low levels of CD11c, and predominantly express the integrin α_M:β_2 (CR3). Plasmacytoid dendritic cells express lower levels of CD11c and have been identified by specific markers, such as blood dendritic cell antigen 2 (BDCA-2, a C-type lectin) in humans, or the sialic-acid-binding immunoglobulin-like lectin H (Siglec-H) in mice, both of which may function in pathogen recognition. They also express MHC class II molecules.

Cell-adhesion molecules have many roles in the body other than in the immune system, directing many aspects of tissue and organ development. Here we have considered only those functions that participate in the recruitment of inflammatory cells in the hours to days after the establishment of an infection.

3-16 Neutrophils make up the first wave of cells that cross the blood vessel wall to enter an inflamed tissue.

The physical changes that accompany the initiation of the inflammatory response have been described in Section 3-3; here we describe the steps by which effector cells are recruited into the infected tissue. Under normal conditions, leukocytes travel in the center of small blood vessels, where blood flow is fastest. Within sites of inflammation, the vessels are dilated and the consequent slower blood flow allows leukocytes to interact in large numbers with the vascular endothelium. During an inflammatory response, the induction of adhesion molecules on the endothelial cells of blood vessels within the infected tissue, as well as induced changes in the adhesion molecules expressed on leukocytes, recruits large numbers of circulating leukocytes, initially neutrophils and later monocytes, into the site. The migration of leukocytes out of blood vessels, the process known as extravasation, is thought to occur in four steps. We will describe this process in regard to monocytes and neutrophils (Fig. 3.25).

Movie 3.8 

The first step involves selectins. P-selectin appears on endothelial cell surfaces within a few minutes of exposure to leukotriene B4, C5a, or histamine, which is released from mast cells in response to C5a. The appearance of P-selectin can also be induced by exposure of the endothelium to TNF-α or LPS, and both of these have the additional effect of inducing the synthesis of a second selectin, E-selectin, which appears on the endothelial cell surface a few hours later. These selectins recognize the sulfated sialyl-Lewis^x moiety of certain leukocyte glycoproteins that are exposed on the tips of leukocyte microvilli. The interaction of P-selectin and E-selectin with these glycoproteins allows monocytes and neutrophils to adhere reversibly to the vessel wall, so that circulating leukocytes can be seen to 'roll' along endothelium that has been treated with inflammatory cytokines (see Fig. 3.25, top panel). This adhesive interaction permits the stronger interactions of the next step in leukocyte migration.

This second step depends on interactions between the leukocyte integrins LFA-1 and CR3 with adhesion molecules on endothelium such as ICAM-1 (which can be induced on endothelial cells by TNF-α) and ICAM-2 (see Fig. 3.25, bottom panel). LFA-1 and CR3 normally bind their ligands only weakly, but CXCL8 or other chemokines, bound to proteoglycans on the surface of endothelial cells, bind to specific chemokine receptors on the leukocyte and signal the cell to trigger a conformational change in LFA-1 and CR3 on the rolling leukocyte, which greatly increases the adhesive properties of the leukocyte, as discussed in Section 3-15. The cell attaches firmly to the endothelium and its rolling is arrested.

In the third step the leukocyte extravasates, or crosses the endothelial wall. This step also involves LFA-1 and CR3, as well as a further adhesive interaction involving an immunoglobulin-related molecule called **PECAM** or **CD31**, which is expressed both on the leukocyte and at the intercellular junctions of endothelial cells. These interactions enable the phagocyte to squeeze between the endothelial cells. It then penetrates the basement membrane with the aid of enzymes that break down the extracellular matrix proteins of the basement membrane. The movement through the basement membrane is known as **diapedesis**, and it enables phagocytes to enter the subendothelial tissues.

Movie 3.9 

The fourth and final step in extravasation is the migration of leukocytes through the tissues under the influence of chemokines. Chemokines such as CXCL8 and CCL2 (see Section 3-14) are produced at the site of infection and bind to proteoglycans in the extracellular matrix and on endothelial cell surfaces. In this way, a matrix-associated concentration gradient of chemokine is formed on a solid surface along which the leukocyte can migrate to the focus of infection (see Fig. 3.25). CXCL8 is released by the macrophages that first encounter pathogens; it recruits neutrophils, which enter the infected tissue in large numbers in the early part of the induced response. Their influx usually peaks within the first 6 hours of an inflammatory response, whereas monocytes can be recruited later, through the action of CCL2. Once in the inflamed tissue, neutrophils are able to eliminate many pathogens by phagocytosis. In an innate immune response, neutrophils use their complement receptors and the direct pattern recognition receptors discussed earlier in this chapter (see Section 3-2) to recognize and phagocytose pathogens or pathogen components directly or after opsonization with complement (see Section 2-13).

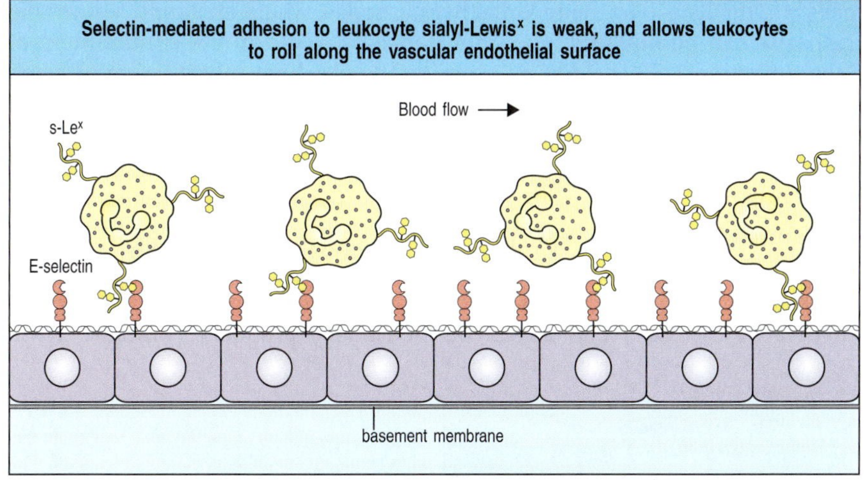

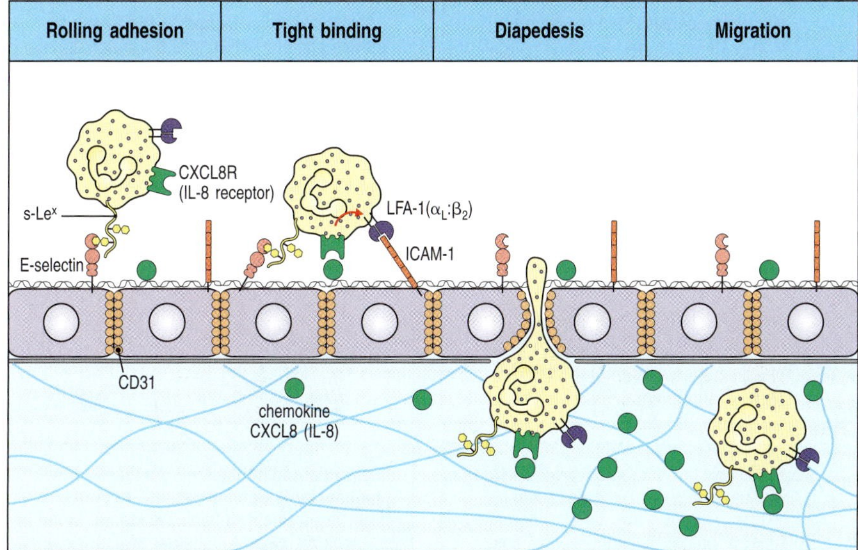

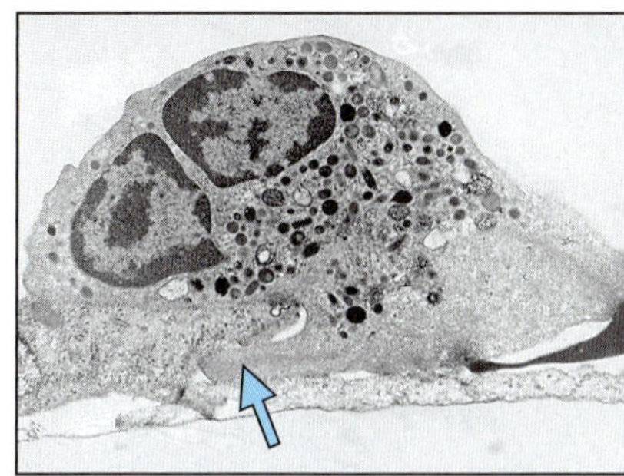

Fig. 3.25 Neutrophils leave the blood and migrate to sites of infection in a multi-step process involving adhesive interactions that are regulated by macrophage-derived cytokines and chemokines. The first step (top panel) involves the reversible binding of a neutrophil to vascular endothelium through interactions between selectins induced on the endothelium and their carbohydrate ligands on the neutrophil, shown here for E-selectin and its ligand the sialyl-Lewis^x moiety (s-Le^x). This interaction cannot anchor the cells against the shearing force of the flow of blood, and they roll along the endothelium, continually making and breaking contact. The binding does, however, allow stronger interactions, which only result when binding of a chemokine such as CXCL8 to its specific receptor on the neutrophil triggers the activation of the integrins LFA-1 and CR3 (Mac-1) (not shown). Inflammatory cytokines such as TNF-α are also necessary to induce the expression of adhesion molecules such as ICAM-1 and ICAM-2, the ligands for these integrins, on the vascular endothelium. Tight binding between ICAM-1 and the integrins arrests the rolling and allows the neutrophil to squeeze between the endothelial cells forming the wall of the blood vessel (i.e., to extravasate). The leukocyte integrins LFA-1 and CR3 are required for extravasation and for migration toward chemoattractants. Adhesion between molecules of CD31, expressed on both the neutrophil and the junction of the endothelial cells, is also thought to contribute to extravasation. The neutrophil also needs to traverse the basement membrane; it penetrates this with the aid of a matrix metalloproteinase enzyme that it expresses at the cell surface. Finally, the neutrophil migrates along a concentration gradient of chemokines (shown here as CXCL8) secreted by cells at the site of infection. The electron micrograph shows a neutrophil extravasating between endothelial cells. The blue arrow indicates the pseudopod that the neutrophil is inserting between the endothelial cells. Photograph (×5500) courtesy of I. Bird and J. Spragg.

In addition, as we will see in Chapter 10, neutrophils act as phagocytic effectors in humoral adaptive immunity, taking up antibody-coated microbes by means of specific receptors.

The importance of neutrophils in immune defense is dramatically illustrated by diseases or medical treatments that severely reduce neutrophil numbers. Such patients are said to have **neutropenia**, and they are highly susceptible to deadly infection with a wide range of pathogens and commensal organisms. Restoring neutrophil levels in such patients by transfusion of neutrophil-rich blood fractions or by stimulating their production with specific growth factors largely corrects this susceptibility.

3-17 TNF-α is an important cytokine that triggers local containment of infection but induces shock when released systemically.

TNF-α acting on endothelial cells stimulates the expression of adhesion molecules and aids the extravasation of cells such as monocytes and neutrophils. Another important action of TNF-α is to stimulate endothelial cells to express proteins that trigger blood clotting in the local small vessels, occluding them and cutting off blood flow. This can be important in preventing the pathogen from entering the bloodstream and spreading through the blood to organs all over the body. Instead, the fluid that has leaked into the tissue in the early phases of an infection carries the pathogen, usually enclosed in dendritic cells, via the lymph to the regional lymph nodes, where an adaptive immune response can be initiated. The importance of TNF-α in the containment of local infection is illustrated by experiments in which rabbits were infected locally with a bacterium. Normally, the infection would be contained at the site of the inoculation; if, however, an injection of anti-TNF-α antibody was also given to block the action of TNF-α, the infection spread via the blood to other organs.

Once an infection has spread to the bloodstream, however, the same mechanisms by which TNF-α so effectively contains local infection instead become catastrophic (Fig. 3.26). Although produced as a membrane-associated cytokine, TNF-α can be cleaved by a specific protease TACE

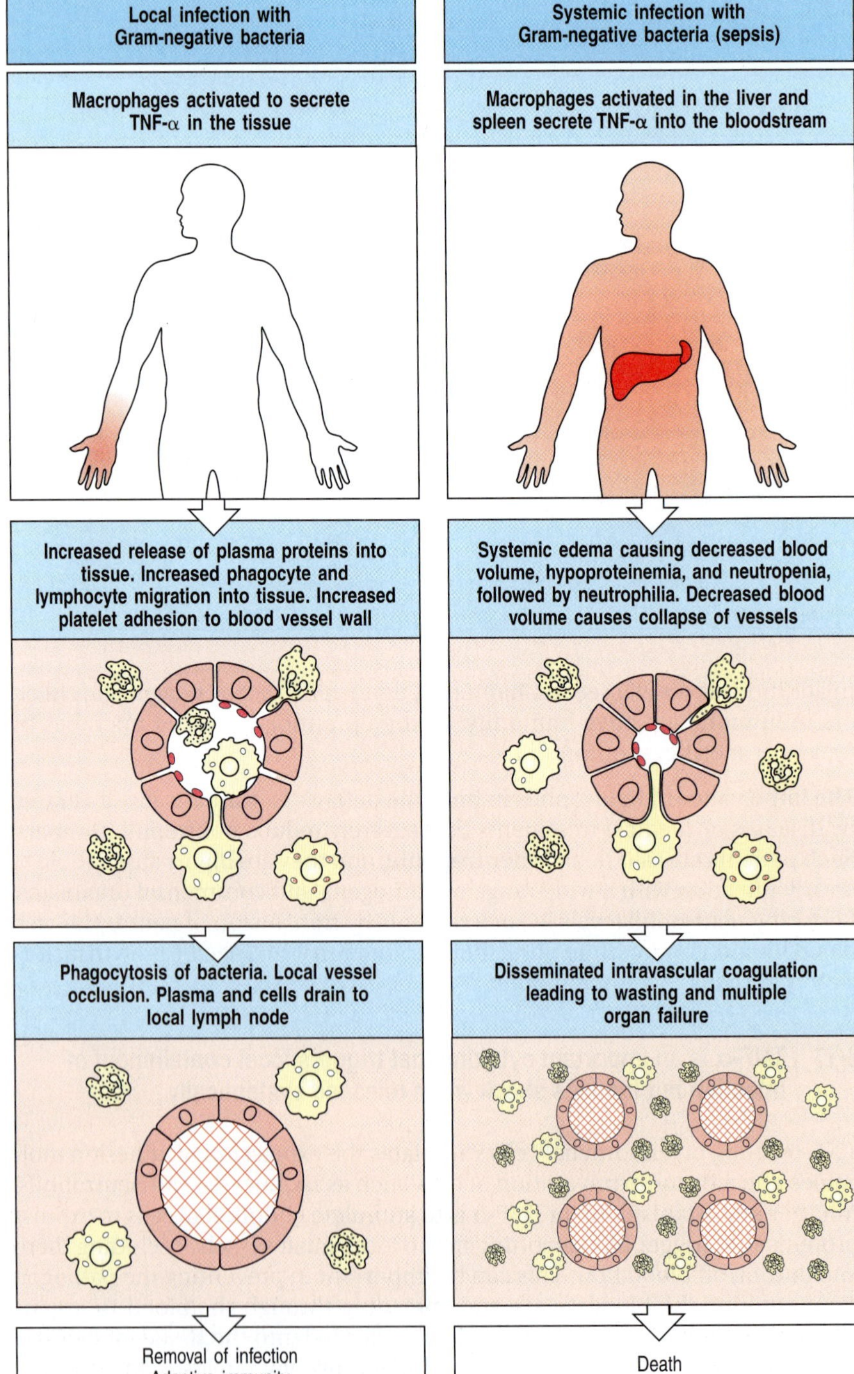

Fig. 3.26 The release of TNF-α by macrophages induces local protective effects, but TNF-α can be damaging when released systemically. The panels on the left show the causes and consequences of local release of TNF-α, and the panels on the right show the causes and consequences of systemic release. In both cases TNF-α acts on blood vessels, especially venules, to increase blood flow and vascular permeability to fluid, proteins, and cells, and to increase endothelial adhesiveness for leukocytes and platelets (center panels). Local release thus allows an influx of fluid, cells, and proteins into the infected tissue, where they participate in host defense. Later, blood clots form in the small vessels (bottom left panel), preventing spread of infection via the blood, and the accumulated fluid and cells drain to regional lymph nodes, where an adaptive immune response is initiated. When there is a systemic infection, or sepsis, with bacteria that elicit TNF-α production, TNF-α is released into the blood by macrophages in the liver and spleen and acts in a similar way on all small blood vessels (bottom right panel). The result is shock, disseminated intravascular coagulation with depletion of clotting factors, and consequent bleeding, multiple organ failure, and frequently death.

(TNF-α-converting enzyme, formerly called ADAM17) and released from the membrane as a soluble cytokine. The presence of infection in the bloodstream, or **sepsis**, is accompanied by a massive release of TNF-α from macrophages in the liver, spleen, and other sites throughout the body. The systemic release of TNF-α into the bloodstream causes vasodilation, which leads to a loss of blood pressure and increased vascular permeability, leading to a loss of plasma volume and eventually to shock, known in this case as **septic shock** because the underlying cause is a bacterial infection. The TNF-α released in septic shock also triggers blood clotting in small vessels throughout the body—known as **disseminated intravascular coagulation**—which leads to the massive consumption of clotting proteins, so that the patient's blood cannot clot appropriately. Disseminated intravascular coagulation frequently leads to the failure of vital organs such as the kidneys, liver, heart, and lungs, which are quickly compromised by the failure of normal blood perfusion; consequently, septic shock has a very high mortality rate.

Mice with defective or no TNF-α receptors are resistant to septic shock but are also unable to control local infection. Mice in which TACE has been selectively inactivated in myeloid cells are also resistant to septic shock, confirming that the release of soluble TNF-α into the circulation both depends on TACE and is the main factor responsible for septic shock. Blocking the action of TNF-α, either with specific antibodies or with soluble proteins that mimic the receptor, is a successful treatment for several inflammatory disorders, including rheumatoid arthritis. However, these treatments have been found to reactivate tuberculosis in some apparently well patients with evidence of previous infection (as demonstrated by skin test), which is a direct demonstration of the importance of TNF-α in keeping infection local and in check.

3-18 Cytokines released by macrophages and dendritic cells activate the acute-phase response.

As well as their important local effects, the cytokines produced by macrophages have long-range effects that contribute to host defense. One of these is the elevation of body temperature, which is caused mainly by TNF-α, IL-1β, and IL-6. These cytokines are termed **endogenous pyrogens** because they cause fever and derive from an endogenous source rather than from bacterial components such as LPS, which also induces fever and is an **exogenous pyrogen**. Endogenous pyrogens cause fever by inducing the synthesis of prostaglandin E2 by the enzyme cyclooxygenase-2, the expression of which is induced by these cytokines. Prostaglandin E2 then acts on the hypothalamus, resulting in an increase in heat production by brown fat and increased vasoconstriction, decreasing the loss of excess heat through the skin. Exogenous pyrogens are able to induce fever both by inducing the production of the endogenous pyrogens and also by directly inducing cyclooxygenase-2 as a consequence of signaling through TLR-4, leading to the production of prostaglandin E2. Fever is generally beneficial to host defense; most pathogens grow better at lower temperatures, whereas adaptive immune responses are more intense at elevated temperatures. Host cells are also protected from the deleterious effects of TNF-α at raised temperatures.

The effects of TNF-α, IL-1β, and IL-6 are summarized in Fig. 3.27. One of the most important of these is the initiation of a response known as the **acute-phase response** (Fig. 3.28). The cytokines act on liver hepatocytes, which respond by changing the profile of proteins that they synthesize and secrete into the blood. In the acute-phase response, blood levels of some proteins go down, whereas levels of others increase markedly. The proteins induced by TNF-α, IL-1β, and IL-6 are called the **acute-phase proteins**. Several of these are of particular interest because they mimic the action of antibodies, but unlike antibodies they have broad specificity for pathogen-associated molecular patterns and depend only on the presence of cytokines for their production.

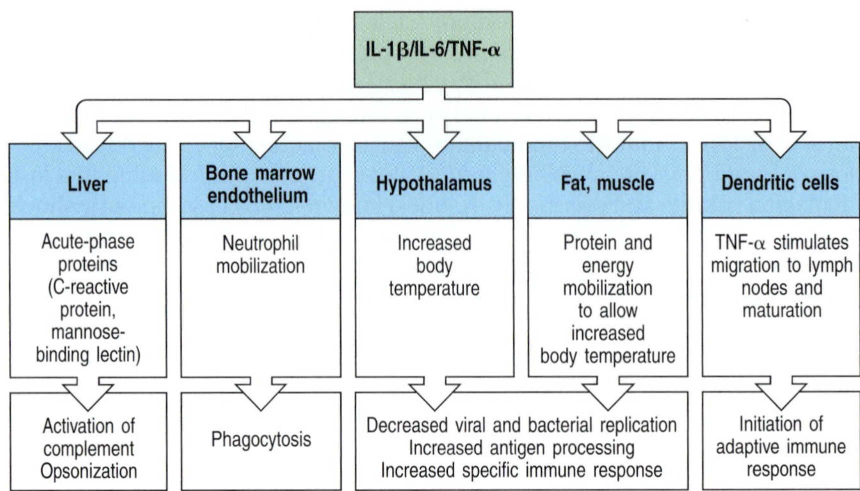

Fig. 3.27 The cytokines TNF-α, IL-1β, and IL-6 have a wide spectrum of biological activities that help to coordinate the body's responses to infection. IL-1β, IL-6, and TNF-α activate hepatocytes to synthesize acute-phase proteins, and bone marrow endothelium to release neutrophils. The acute-phase proteins act as opsonins, whereas the disposal of opsonized pathogens is augmented by the enhanced recruitment of neutrophils from the bone marrow. IL-1β, IL-6, and TNF-α are also endogenous pyrogens, raising body temperature, which is believed to help in eliminating infections. A major effect of these cytokines is to act on the hypothalamus, altering the body's temperature regulation, and on muscle and fat cells, altering energy mobilization to increase the body temperature. At higher temperatures, bacterial and viral replication is less efficient, whereas the adaptive immune response operates more efficiently.

One acute-phase protein, the **C-reactive protein**, is a member of the **pentraxin** protein family, so called because they are formed from five identical subunits. C-reactive protein is another example of a multipronged pathogen-recognition molecule, and it binds to the phosphocholine portion of certain bacterial and fungal cell-wall lipopolysaccharides. Phosphocholine is also found in mammalian cell membrane phospholipids, but it cannot be bound by C-reactive protein. When C-reactive protein binds to a bacterium, it is not only able to opsonize it but can also activate the complement cascade by binding to C1q, the first component of the classical pathway of complement activation, as we saw in Section 2-7. The interaction with C1q involves the collagen-like parts of C1q rather than the globular heads that make contact with pathogen surfaces, but the same cascade of reactions is initiated.

The second acute-phase protein of interest is mannose-binding lectin (MBL), which we have already introduced as a pathogen-binding molecule (see Fig. 2.17) and as a trigger for the complement cascade (see Section 2-6). MBL is present in low levels in the blood of healthy individuals, but it is produced in increased amounts during the acute-phase response. By recognizing mannose residues on microbial surfaces it can act as an opsonin that is recognized by monocytes, which do not express the macrophage mannose receptor. Two other proteins with opsonizing properties that are also produced in increased amounts during an acute-phase response are the surfactant proteins SP-A and SP-D. SP-A and SP-D are produced by the liver and a variety of epithelia. They are, for example, found along with macrophages in the alveolar fluid of the lung, where they are secreted by pneumocytes, and are important in promoting the phagocytosis of opportunistic respiratory pathogens such as *Pneumocystis jirovecii* (formerly known as *P. carinii*), one of the main causes of pneumonia in patients with AIDS.

Thus, within a day or two, the acute-phase response provides the host with several proteins with the functional properties of antibodies but able to bind a broad range of pathogens. However, unlike antibodies, which we describe in Chapters 4 and 10, acute-phase proteins have no structural diversity and are made in response to any stimulus that triggers the release of TNF-α, IL-1, and IL-6. Therefore, unlike antibodies, their synthesis is not specifically induced and targeted.

A final distant effect of the cytokines produced by macrophages is to induce a **leukocytosis**, an increase in the numbers of circulating neutrophils. The neutrophils come from two sources: the bone marrow, from which mature leukocytes are released in increased numbers; and sites in blood vessels where they are attached loosely to endothelial cells. Thus, the effects of these cytokines contribute to the control of infection while the adaptive immune response is

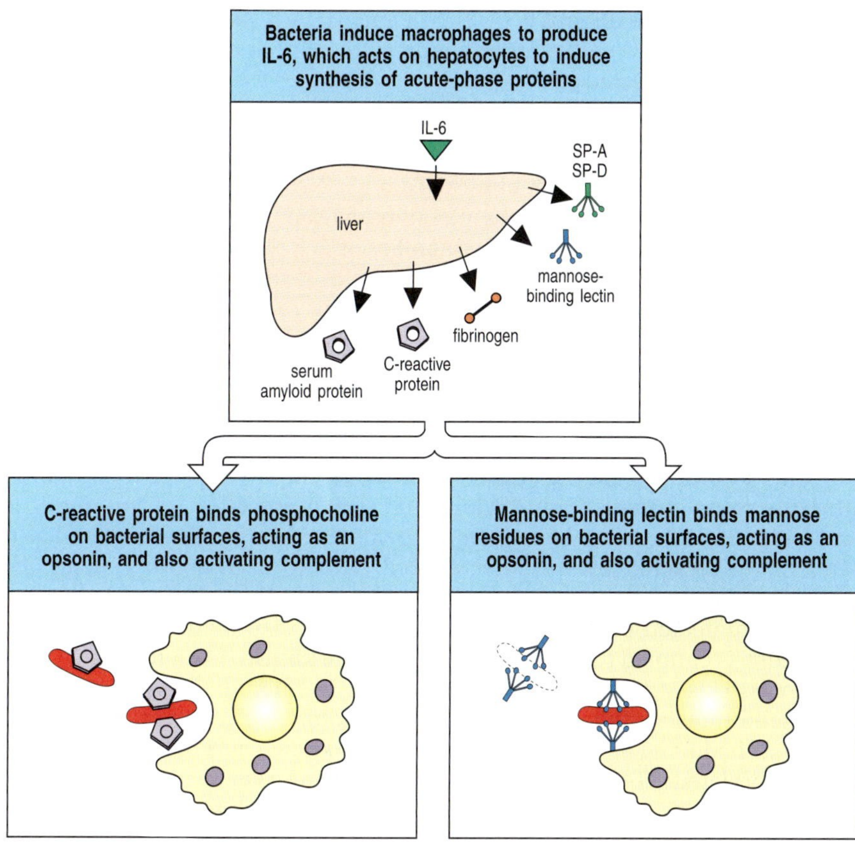

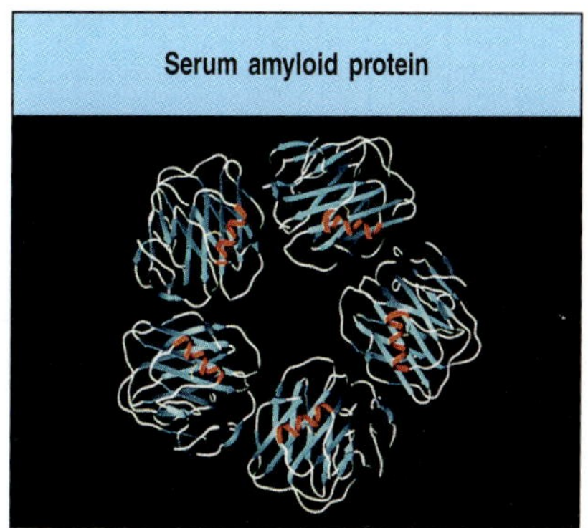

Fig. 3.28 The acute-phase response produces molecules that bind pathogens but not host cells. Acute-phase proteins are produced by liver cells in response to cytokines released by macrophages in the presence of bacteria. They include serum amyloid protein (SAP) (in mice but not humans), C-reactive protein (CRP), fibrinogen, and mannose-binding lectin (MBL). SAP and CRP are homologous in structure; both are pentraxins, forming five-membered discs, as shown for SAP (right panel). CRP binds phosphocholine on certain bacterial and fungal surfaces but does not recognize it in the form in which it is found in host cell membranes. It both acts as an opsonin in its own right and activates the classical complement pathway by binding C1q to augment opsonization. MBL is a member of the collectin family, which also includes the pulmonary surfactant proteins SP-A and SP-D. Like CRP, MBL can act as an opsonin in its own right, as can SP-A and SP-D. Model structure courtesy of J. Emsley.

being developed. As shown in Fig. 3.27, TNF-α also has a role in stimulating the migration of dendritic cells from their sites in peripheral tissues to the lymph node and in their maturation into nonphagocytic but highly co-stimulatory antigen-presenting cells.

3-19 Interferons induced by viral infection make several contributions to host defense.

Virus infection induces the production of interferons, which were originally called this because of their ability to interfere with viral replication in previously uninfected tissue culture cells. They are believed to have a similar role *in vivo*, blocking the spread of viruses to uninfected cells. There are two classes of antiviral, or type I, interferons, IFN-α, which is a family of several closely related proteins, and IFN-β, the product of a single gene. The type I interferons are synthesized by many cell types after infection by diverse viruses.

Almost all types of cells can produce IFN-α and IFN-β, but some cells seem to be specialized for the task. In Section 3-1 we introduced the plasmacytoid dendritic cell (pDC). Also called **interferon-producing cells** (**IPCs**) or **natural interferon-producing cells**, human plasmacytoid dendritic cells were initially recognized as rare peripheral blood cells that accumulate in peripheral lymphoid tissues during a viral infection and make abundant type I interferons (IFN-α and IFN-β)—up to 1000 times more than other cell types. Plasmacytoid dendritic cells express CXCR3, a receptor for the chemokines CXCL9, CXCL10, and CXCL11, which are produced by T cells. This means that pDCs can migrate from the blood into lymph nodes in which there is an ongoing inflammatory response to a pathogen. Their abundant production of type I interferon may not be due to a greater capacity for production as such, but because they efficiently couple viral recognition by TLRs to the production of interferon (see Section 3-7). Type I interferons are also induced by the

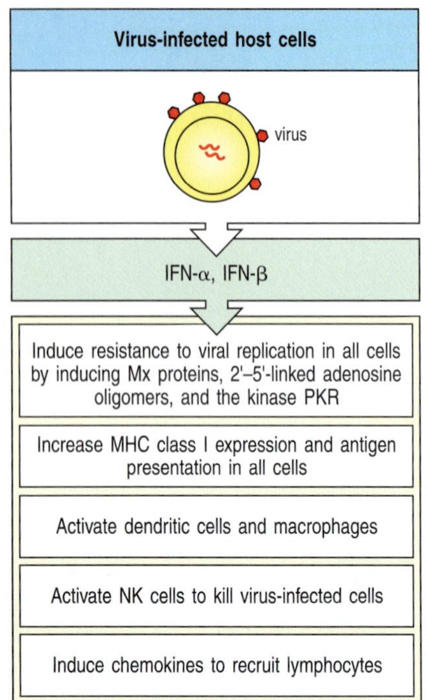

Fig. 3.29 Interferons are antiviral proteins produced by cells in response to viral infection. The interferons IFN-α and IFN-β have three major functions. First, they induce resistance to viral replication in uninfected cells by activating genes that cause the destruction of mRNA and inhibit the translation of viral proteins and some host proteins. Such genes include the Mx proteins, oligoadenylate synthetase, and PKR. Second, they can induce MHC class I expression in most cell types in the body, thus enhancing their resistance to NK cells; they may also induce increased synthesis of MHC class I molecules in cells that are newly infected by virus, thus making them more susceptible to being killed by CD8 cytotoxic T cells (see Chapter 9). Third, they activate NK cells, which then selectively kill virus-infected cells.

cytoplasmic sensors of viral RNA, RIG-I, and MDA-5, and may be generated either by an infected cell or by an uninfected cell that detects viral material produced by an infected cell.

Plasmacytoid dendritic cells express a subset of TLRs that includes TLR-7 and TLR-9, which are endosomal sensors of viral RNA and of the nonmethylated CpG residues present in the genomes of many DNA viruses (see Fig. 3.10). The requirement for TLR-9 in sensing infections caused by DNA viruses has been demonstrated, for example, by the inability of TLR-9-deficient plasmacytoid dendritic cells to generate type I interferons in response to herpes simplex virus. Some cell-surface proteins characteristic of plasmacytoid dendritic cells, such as Siglec-H, may help capture viruses, internalize them and deliver them to the endosomes.

Interferons make several contributions to defense against viral infection (Fig. 3.29). Production of IFN-β is particularly important because it can act on cells to induce them to make IFN-α, thus amplifying and maintaining the interferon response. An obvious and important effect of interferon is the induction of a state of resistance to viral replication in all cells. IFN-α and IFN-β secreted by the infected cell bind to a common cell-surface receptor, known as the interferon receptor, on both the infected cell and nearby uninfected cells. The interferon receptor, like many other cytokine receptors, is coupled to a **Janus-family tyrosine kinase** (**JAK**), through which it signals. This signaling pathway, described in Chapter 7, rapidly induces new gene transcription as the Janus-family kinases directly phosphorylate signal-transducing activators of transcription known as **STAT proteins**. The phosphorylated STAT proteins enter the nucleus, where they activate the transcription of several different genes, including those encoding proteins that help to inhibit viral replication.

One such protein is the enzyme **oligoadenylate synthetase**, which polymerizes ATP into 2′–5′-linked oligomers (whereas nucleotides in nucleic acids are normally linked 3′–5′). These 2′–5′-linked oligomers activate an endoribonuclease that then degrades viral RNA. A second protein induced by IFN-α and IFN-β is a dsRNA-dependent protein kinase called **PKR**. This serine–threonine kinase phosphorylates the eukaryotic protein synthesis initiation factor eIF-2, inhibiting translation and thus further contributing to the inhibition of viral replication. **Mx proteins** are also induced by type I interferons. Wild mice and humans have two highly similar proteins, **Mx1** and **Mx2**, which are GTPases belonging to the dynamin protein family. They are present in the nucleus in multimeric complexes and are thought to interfere with viral replication in some way, but how they do this is not understood. Mice that lack Mx gene are highly susceptible to infection with the influenza virus, whereas mice that can make Mx are not. Oddly, most common laboratory strains of mice have inactivated both Mx genes, and in these mice, IFN-β cannot act to protect against influenza infection. Another way in which interferons act in innate immunity is to activate NK cells, which can kill virus-infected cells, as described in the next section.

Lastly, interferons have more general effects in immune responses. Some TLRs, notably TLR-4, can induce IFN-β in response to the recognition of bacterial cell-wall components. This is still a useful response, because IFN-β promotes the differentiation and maturation of conventional dendritic cells from blood monocytes and induces the expression of co-stimulatory molecules on macrophages and conventional dendritic cells, which enables them to act as effective antigen-presenting cells to T cells.

Interferons also stimulate production of the chemokines CXCL9, CXCL10, and CXCL11, which recruit lymphocytes to sites of infection and increase the expression of MHC class I molecules on all types of cells. The cytotoxic T lymphocytes of the adaptive immune system recognize virus-infected cells by the display of viral peptides complexed to MHC class I molecules on the infected cell surface (see Fig. 1.30). Cytotoxic T cells are also induced to proliferate by type I interferons. By all these effects, interferons indirectly help promote the killing of virus-infected cells by CD8 cytotoxic T cells.

3-20 NK cells are activated by interferon and macrophage-derived cytokines to serve as an early defense against certain intracellular infections.

Natural killer cells (**NK cells**) develop in the bone marrow from the same progenitor cells as T lymphocytes and B lymphocytes and circulate in the blood. They are larger than T and B cells, have distinctive cytoplasmic granules containing cytotoxic proteins, and are functionally identified by their ability to kill certain tumor cell lines *in vitro* without the need for specific immunization. NK cells kill cells by releasing their cytotoxic granules, which are similar to those of cytotoxic T cells and have the same effects (discussed in Chapter 9). In brief, the cytotoxic granules, containing the cytotoxic granzymes and the pore-forming protein perforin, are released onto the surface of the target cell, which the NK cell has recognized and bound via cell-surface receptors, and their contents penetrate the cell membrane and induce programmed cell death. However, unlike T cells, killing by NK cells is triggered by germline-encoded receptors that recognize molecules on the surface of infected or malignantly transformed cells. NK cells are classified as part of the innate immune system because of their invariant receptors.

NK cells can be activated in response to interferons or certain macrophage-derived cytokines. Although NK cells that can kill sensitive targets can be isolated from uninfected individuals, killing activity is increased 20–100-fold when NK cells are exposed to IFN-α and IFN-β, or to IL-12, which is one of the cytokines produced early in many infections by dendritic cells and macrophages. Activated NK cells serve to contain virus infections while the adaptive immune response is generating antigen-specific cytotoxic T cells and neutralizing antibodies that can clear the infection (Fig. 3.30). A clue to the physiological function of NK cells in humans comes from rare patients deficient in these cells, who are frequently susceptible to the early phases of herpes virus infection. A similar finding has recently been made in mice infected with mouse cytomegalovirus, a herpes virus.

IL-12, acting in synergy with the cytokine IL-18 produced by activated macrophages, can also stimulate NK cells to secrete large amounts of the cytokine **interferon** (**IFN**)-γ, and this is crucial in controlling some infections before the IFN-γ produced by activated CD8 cytotoxic T cells becomes available. IFN-γ is quite distinct functionally from the antiviral type I interferons IFN-α and IFN-β and it is not directly induced by viral infection. It is an important and characteristic cytokine produced by some types of T cells; its effects are discussed in detail in Chapter 9. The production of IFN-γ by NK cells early in an immune response may influence the CD4 T-cell response to infectious agents, inducing activated CD4 T cells to differentiate into pro-inflammatory T_H1 cells, which are able to activate macrophages.

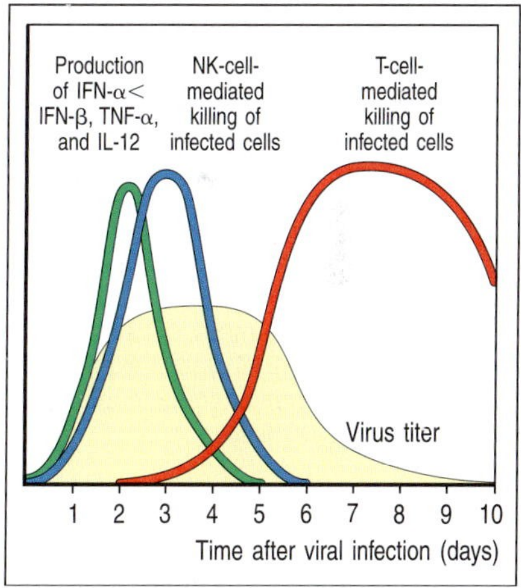

Fig. 3.30 Natural killer cells (NK cells) are an early component of the host response to virus infection. Experiments in mice have shown that IFN-α, IFN-β, and the cytokines TNF-α and IL-12 appear first, followed by a wave of NK cells, which together control virus replication but do not eliminate the virus. Virus elimination is accomplished when virus-specific CD8 T cells and neutralizing antibodies are produced. Without NK cells, the levels of some viruses are much higher in the early days of the infection, and can be lethal unless treated vigorously with antiviral compounds.

3-21 NK cells possess receptors for self molecules that prevent their activation by uninfected cells.

If NK cells are to defend the body against infection with viruses and other pathogens, they must have some mechanism for distinguishing infected cells from uninfected healthy cells. Exactly how this is achieved has not yet been worked out in every case, but it is thought that an NK cell is activated by a combination of direct recognition of changes in cell-surface glycoprotein composition, which is induced by metabolic stresses such as malignant transformation or viral or bacterial infection, together with a recognition of '**altered self**,' which involves changes in the expression of **MHC molecules.** The MHC molecules are a set of glycoproteins expressed on almost every cell of the body that constitute the 'tissue type' of an individual (see Section 1-20). They are discussed in detail in Chapter 5 in connection with their role in antigen recognition by T cells. In brief, there are two main classes of MHC molecules. MHC class I molecules are expressed on most of the cells of the body (except, notably, red blood cells), whereas MHC class II molecules have a much more restricted distribution. Altered expression of MHC class I molecules may be a common feature of cells infected by intracellular pathogens, because many of these pathogens have developed strategies to interfere with the ability of MHC class I molecules to capture pathogen peptides and display them to T cells (discussed in Chapter 6). One mechanism by which NK cells distinguish infected from uninfected cells is by recognizing alterations in MHC class I expression (Fig. 3.31).

NK cells are able to sense changes in the expression of MHC class I molecules by integrating the signals from two types of surface receptors, which together control the NK cell's cytotoxic activity and cytokine production. **Activating receptors** trigger the NK cell to kill its target. Several classes of activating receptors are expressed by NK cells, including members of the immunoglobulin-like and the C-type lectin families. Stimulation of activating receptors causes the release of cytokines such as IFN-γ and the directed killing of the stimulating cell through the release of cytotoxic granules by the NK cell. NK cells also carry receptors for the constant region of immunoglobulins (Fc receptors), and binding of antibodies to these receptors activates NK cells to

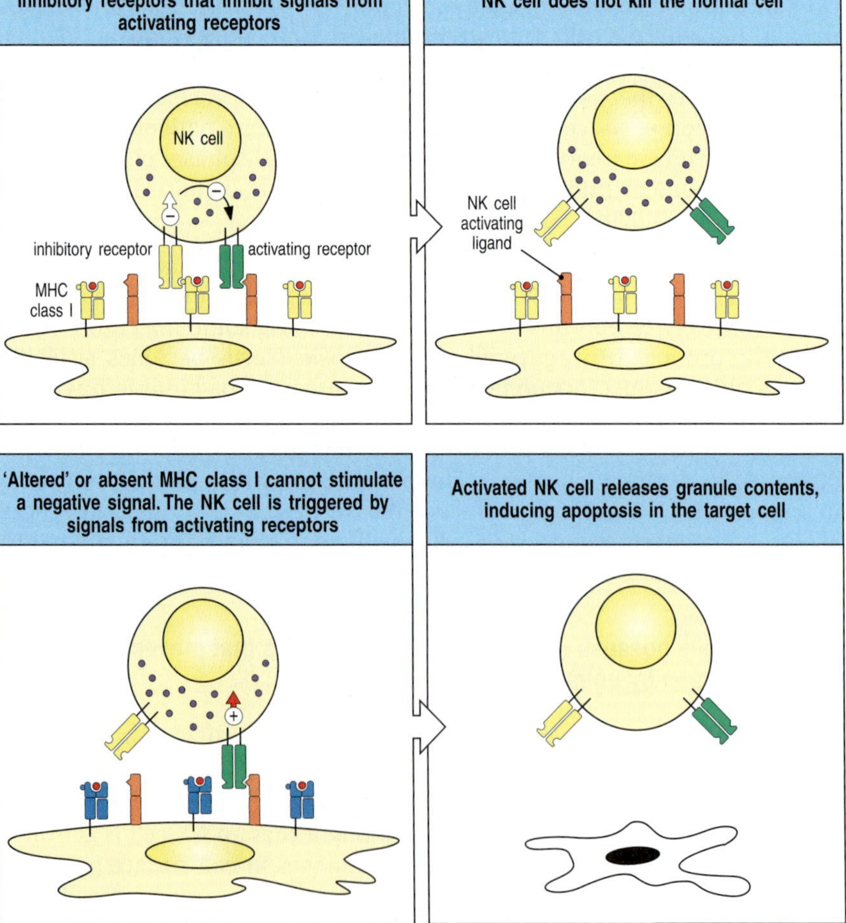

Fig. 3.31 Killing by NK cells depends on the balance between activating and inhibitory signals. NK cells have several different activating receptors that signal the NK to kill the bound cell. However, NK cells are prevented from a wholesale attack by another set of receptors that recognize MHC class I molecules (which are present on almost all cell types) and inhibit killing by overruling the actions of the activating receptors. This inhibitory signal is lost when the target cells do not express MHC class I, such as in cells infected with viruses, many of which specifically inhibit MHC class I expression or alter its conformation so as to avoid recognition by CD8 T cells.

release their cytotoxic granules; this is known as antibody-dependent cellular cytotoxicity, or ADCC, and is described in Chapter 10.

A second set of receptors called **inhibitory receptors** act to prevent NK cells from killing normal host cells. Some of these inhibitory receptors are specific for various MHC class I molecules, which helps to explain why NK cells selectively kill cells bearing low levels of MHC class I molecules but are usually prevented from killing cells with normal levels. The greater the number of MHC class I molecules on a cell surface, the better protected the cell is against attack by NK cells. This is why interferons, which induce the expression of MHC class I molecules, protect uninfected host cells from NK cells. At the same time, the interferon activates the NK cell to kill virus-infected cells.

The receptors that regulate the activity of NK cells fall into two large families that contain a number of other cell-surface receptors in addition to NK receptors (Fig. 3.32). One family of receptors is characterized by immunoglobulin-like domains—hence their name **killer cell immunoglobulin-like receptors** (**KIRs**). Different KIR genes encode proteins with differing numbers of immunoglobulin domains; some, called KIR-2D, have two immunoglobulin domains, whereas others, called KIR-3D, have three. The KIR genes form part of a larger cluster of immunoglobulin-like receptor genes known as the **leukocyte receptor complex** (**LRC**). The other large family of NK-cell receptors consists of the C-type lectin-like proteins; these are called **killer lectin-like receptors** (**KLRs**). The genes for KLRs are found within a gene cluster called the **NK receptor complex** (**NKC**). Mouse NK cells predominantly express several Ly49 receptors, which are members of the KLR family. In contrast, humans lack functional Ly49 genes. Both the NKC and LRC clusters are present in mice and in humans, but mice lack KIR genes and therefore rely on the Ly49 C-type lectin-like receptors of the NKC to control their NK-cell activity.

One complicating factor in understanding NK-cell activation is that both activating and inhibitory receptors are present within the same structural family, for example the KIR receptor family. Whether a KIR protein is activating or inhibitory depends on the presence or absence of particular signaling motifs in its cytoplasmic domain. Inhibitory KIRs have long cytoplasmic tails that contain an **immunoreceptor tyrosine-based inhibition motif** (**ITIM**, with consensus sequence V/I/LxYxxL/V); for example, the cytoplasmic tails of the inhibitory receptors KIR-2DL and KIR-3DL each contain one ITIM (Fig. 3.33). When ligands associate with an inhibitory KIR, the tyrosine in its ITIM becomes phosphorylated, and it can then bind intracellular phosphatases, which become localized near the cell membrane. These phosphatases inhibit signaling by other receptors by removing phosphates from tyrosine residues on other intracellular signaling molecules.

Activating KIR receptors have short cytoplasmic tails, designated for example as KIR-2DS and KIR-3DS (see Fig. 3.33). These receptors lack an ITIM and instead have a charged residue in their transmembrane regions that associates

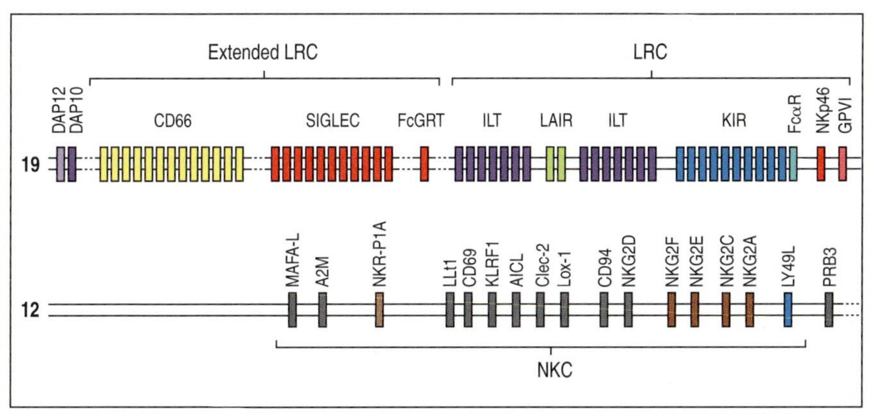

Fig. 3.32 The genes that encode NK receptors fall into two large families. The first, the leukocyte receptor complex (LRC), comprises a large cluster of genes encoding a family of proteins composed of immunoglobulin-like domains. These include the killer immunoglobulin-like receptors (KIRs) expressed by NK cells, the ILT (immunoglobulin-like transcript) class, and the leukocyte-associated immunoglobulin-like receptor (LAIR) gene families. The sialic-acid-binding Ig-like lectins (SIGLECs) and members of the CD66 family are located nearby. In humans, this complex is located on chromosome 19. The second gene cluster is called the NK receptor complex (NKC) and encodes killer lectin-like receptors, a receptor family that includes the NKG2 proteins and CD94, with which an NKG2 molecule pairs to form a functional receptor. This complex is located on human chromosome 12. Some NK receptor genes are found outside these two major gene clusters; for example, the genes for the natural cytotoxicity receptors NKp30 and NKp44 are located within the major histocompatibility complex on chromosome 6. Figure based on data courtesy of J. Trowsdale, University of Cambridge.

Activating and inhibitory receptors of NK cells can belong to the same structural family

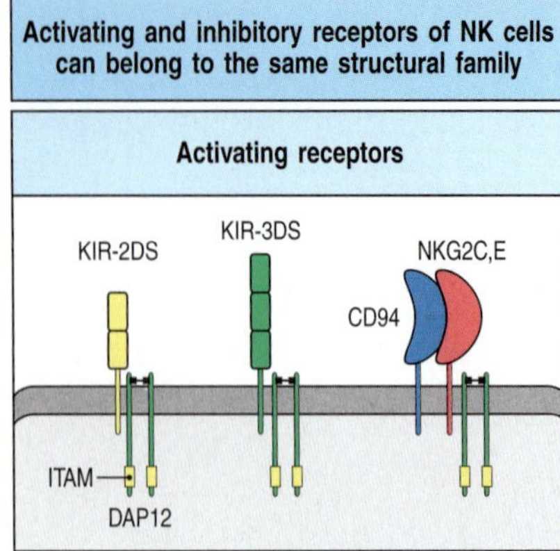

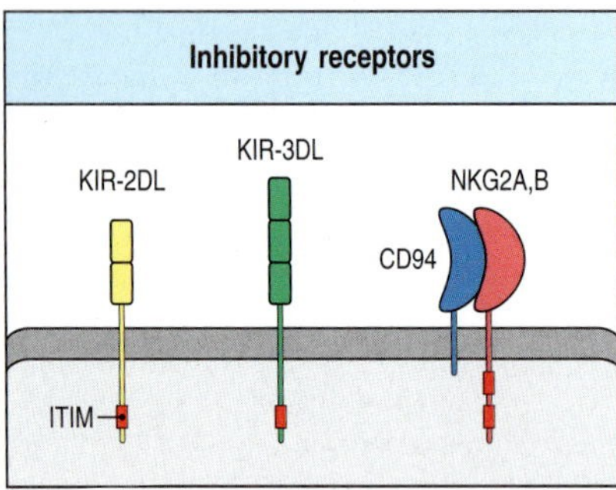

Fig. 3.33 The structural families of NK receptors encode both activating and inhibitory receptors. The families of killer immunoglobulin-like receptors (KIR) and killer lectin-like receptors (KLR) have members that send activating signals to the NK cell (upper panel) and those that send inhibitory signals (lower panel). KIR family members are designated according to the number of immunoglobulin-like domains they possess and by the length of their cytoplasmic tails. Activating KIR receptors have short cytoplasmic tails and bear the designation 'S.' These associate with a signaling protein DAP12 via a charged amino acid residue in the transmembrane region. The cytoplasmic tails of DAP12 contain amino acid motifs called ITAMs, which are involved in signaling. The activating NKG2 receptors (members of the KLR family) are heterodimers with another C-type lectin family member, CD94. The inhibitory KIR receptors have longer cytoplasmic tails and are designated 'L'; these do not associate constitutively with adaptor proteins but contain a signaling motif called an ITIM, which when phosphorylated is recognized by inhibitory phosphatases. Like the activating KLRs, the inhibitory KLRs, NKG2A and NKG2B (a splice variant of NKG2A) form heterodimers with CD94.

with an accessory signaling protein called **DAP12**. DAP12 is transmembrane protein that contains an **immunoreceptor tyrosine-based activation motif** (**ITAM**, with consensus sequence YXX[L/I]X_{6-9}YXX[L/I]) in its cytoplasmic tail and forms a disulfide-linked homodimer in the membrane. When a ligand binds to an activating KIR, the tyrosine residues in the ITAM become phosphorylated, turning on intracellular signaling pathways that activate the NK cell and lead to release of the cytotoxic granules. The phosphorylated ITAMs bind and activate intracellular tyrosine kinases such as Syk or ZAP-70, leading to further signaling events in the cell (discussed in Sections 7-9 and 7-16).

The KLR family also has both activating and inhibitory members. In humans and mice, NK cells express a heterodimer of two C-type lectins, called **CD94** and **NKG2**, which interacts with nonpolymorphic MHC class I-like molecules, including HLA-E in humans and Qa-1 in mice, that bind leader peptide fragments from other MHC class I molecules. This makes CD94:NKG2 able to sense the presence of several different MHC class I variants. In humans there are five NKG2 family proteins, NKG2A, C, D, E, and F. Of these, for example, NKG2A contains an ITIM and is inhibitory, whereas NKG2C has a charged transmembrane residue, associates with DAP12, and is activating (see Fig. 3.33). NKG2D is also activating, but it binds to a distinct class of ligands and will be discussed separately in the next section. Other inhibitory NK receptors specific for the products of the MHC class I loci are rapidly being defined, and all are members of either the immunoglobulin-like KIR family or the Ly49-like C-type lectins. An important feature of the NK-cell population is that any given NK cell expresses only a subset of the receptors in its potential repertoire, and so not all NK cells in the individual are identical. Clearly, the regulation of NK-cell activity is complex, and whether any individual NK cell is activated or inhibited by a target cell will depend on the overall balance of activating and inhibitory receptors that the NK cell is expressing.

The overall response of NK cells to differences in MHC expression is further complicated by the extensive polymorphism of KIR genes; for example, for one of the KIR genes there are two alleles, one of which codes for an activating receptor and the other for an inhibitory receptor. Moreover, the KIR gene cluster seems to be a very dynamic part of the human genome, because different numbers of activating and inhibitory KIR genes are found in different people. What advantage this diversity might have is not yet clear. Some genetic epidemiologic studies indicate an association between certain alleles of KIR genes and earlier onset (although not absolute frequency) of rheumatoid arthritis. As noted earlier, the KIR gene cluster is not present in mice, which use only the Ly49 KLR proteins to regulate NK-cell activity. So whatever the driving force is for the evolution of the KIR genes and their diversity, it may have arisen relatively recently in evolutionary terms.

Signaling by the inhibitory NK receptors suppresses the killing activity and cytokine production of NK cells. This means that NK cells will not kill healthy, genetically identical cells with normal expression of MHC class I molecules, such as the other cells of the body. Virus-infected cells, however, can become susceptible to being killed by NK cells by a variety of mechanisms. First, some viruses inhibit all protein synthesis in their host cells, so that synthesis of MHC class I proteins would be blocked in infected cells, even while their production in uninfected cells is being stimulated by the actions of interferon. The reduced level of MHC class I expression in infected cells would make them correspondingly less able to inhibit NK cells through their MHC-specific receptors, and they would therefore be more susceptible to being killed. Second, some viruses can selectively prevent the export of MHC class I molecules to the cell surface. This might allow the infected cell to evade recognition by cytotoxic T cells but would make them susceptible to being killed by NK cells.

Clearly, much remains to be learned about this innate mechanism of cytotoxic attack and its physiological relevance. The role of MHC class I molecules in allowing NK cells to detect intracellular infections is of particular interest because these same proteins govern the response of T cells to intracellular pathogens. It is possible that NK cells, which use a diverse set of nonclonal receptors to detect altered MHC, represent the modern remnants of the evolutionary forebears of T cells. Those hypothetical T-cell ancestors could have gone on to evolve rearranging genes that encode a vast repertoire of antigen-specific T-cell receptors geared to recognizing MHC molecules 'altered' by binding peptide antigens.

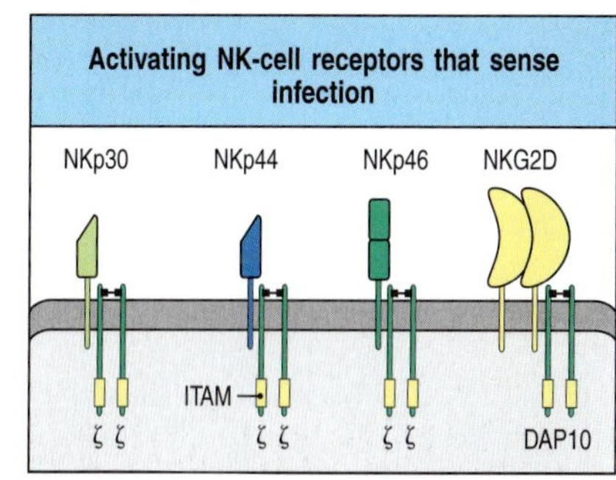

Fig. 3.34 Activating receptors of NK cells are the natural cytotoxicity receptors and NKG2D. The natural cytotoxicity receptors are immunoglobulin-like proteins. NKp30 and NKp44 have an extracellular domain that resembles a single variable domain of an immunoglobulin molecule. NKp30 and NKp44 activate the NK cell through their association with homodimers of the CD3ζ chain or the Fc receptor γ chain (these are signaling proteins that also associate with other types of receptors and are described in more detail in Chapter 7). NKp46 resembles the KIR-2D molecules in having two domains that resemble the constant domains of an immunoglobulin molecule. NKG2D is a member of the C-type lectin family and forms a homodimer, and it associates with DAP10.

3-22 NK cells bear receptors that activate their effector function in response to ligands expressed on infected cells or tumor cells.

In addition to the KIR and KLR receptors, which have a role in sensing the level of MHC class I proteins present on other cells, NK cells also express receptors that more directly sense the presence of infection or other perturbations in a cell. Activating receptors for the recognition of infected cells are the **natural cytotoxicity receptors** (**NCRs**) NKp30, NKp44, and NKp46, which are immunoglobulin-like receptors, and the C-type lectin family member **NKG2D** (Fig. 3.34). The ligands recognized by the natural cytotoxicity receptors are not well defined.

NKG2D seems to have a specialized role in activating NK cells. Other NKG2 family members (NKG2A, C, and E) form heterodimers with CD94 and bind the MHC class I molecule HLA-E. NKG2D does neither; instead, the ligands for the NKG2D receptor are families of proteins that are distantly related to the MHC class I molecules but have a completely different function, being produced in response to stress. The ligands in humans for NKG2D, as shown in Fig. 3.35, are the MHC class I-like MIC molecules, MIC-A and MIC-B, and the RAET1 protein family, which are similar to the α_1 and α_2 domains of MHC class I molecules (which we shall describe when we discuss MHC molecule structure in Chapter 4 and also in Section 6-18). The RAET1 family has 10 members, 3 of which were initially characterized as ligands for the cytomegalovirus UL16 protein and are also called UL16-binding proteins (ULBPs). Mice do not have equivalents of the MIC molecules; the ligands for mouse NKG2D have a very similar structure to that of the RAET1 proteins, and are probably orthologs of them. In fact, these ligands were first identified in mice as the Rae1 (retinoic acid early inducible 1) protein family, and also include related proteins H60 and Mult-1 (see Fig. 6.23).

The ligands for NKG2D are expressed in response to cellular or metabolic stress, and so are upregulated on cells infected with intracellular bacteria or with some viruses, such as cytomegalovirus, as well as on incipient tumor cells that have become malignantly transformed. Thus, recognition by NKG2D acts as a generalized 'danger' signal to the immune system. NKG2D is expressed on NK cells, γ:δ T cells, and activated CD8 cytotoxic T cells, and recognition of NKG2D ligands by these cells provides a potent co-stimulatory signal that enhances their effector functions.

3-23 The NKG2D receptor activates a different signaling pathway from that of the other activating NK receptors.

As well as the ligands it recognizes, NKG2D also differs from other activating receptors on NK cells in the signaling pathway it engages within the cell. The

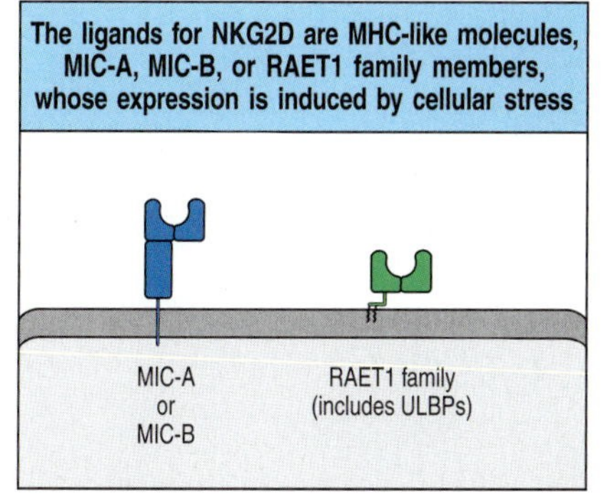

Fig. 3.35 The ligands for the activating NK receptor NKG2D are proteins that are expressed in conditions of cellular stress. The MIC proteins MIC-A and MIC-B are MHC-like molecules induced on epithelial and other cells by stresses, such as heat shock, metabolic stress, or infection. RAET1 family members, including the subset designated as UL16-binding proteins (ULBPs), also resemble a portion of an MHC class I molecule, the α_1 and α_2 domains, and most (but not all) are attached to the cell via a glycophosphatidylinositol linkage.

other activating receptors are associated intracellularly with signaling proteins such as the CD3ζ chain, the Fc receptor γ chain, and DAP12, which all contain ITAMs. In contrast, NKG2D binds a different adaptor protein, **DAP10**, which does not contain an ITAM sequence and instead activates the intracellular lipid kinase phosphatidylinositol-3-kinase (PI 3-kinase), initiating a different series of intracellular signaling events in the NK cell (see Figure 7-5, panel 3). Generally, PI 3-kinase is considered to enhance the survival of cells in which it is activated, thereby augmenting the cell's overall effector activity. In mice, the workings of NKG2D are even more complicated, because mouse NKG2D is produced in two alternatively spliced forms, one of which binds DAP12 and DAP10, whereas the other binds DAP10. Mouse NKG2D can thus activate both signaling pathways, whereas human NKG2D seems to signal only through DAP10 to activate the PI 3-kinase pathway.

3-24 Several lymphocyte subpopulations behave as innate-like lymphocytes.

Receptor gene rearrangements are a defining characteristic of the lymphocytes of the adaptive immune system, and they allow the generation of an infinite variety of antigen receptors, each expressed by a different individual T cell or B cell (see Section 1-11). There are, however, several minor lymphocyte subsets that produce antigen receptors of this type but with only very limited diversity, encoded by a few common gene rearrangements. Because their receptors are relatively invariant and because they occur only in specific locations within the body, these lymphocytes do not need to undergo proliferation (clonal expansion) before responding effectively to the antigens they recognize; they are therefore known as **innate-like lymphocytes** (**ILLs**) (Fig. 3.36). To produce antigen receptors in these cells requires the recombinases RAG-1 and RAG-2; these proteins and their role in gene rearrangement in lymphocytes are described in Chapter 4. Because they express RAG-1 and RAG-2 and undergo the process of antigen receptor gene rearrangement, ILLs are, by definition, cells of the adaptive immune system. They behave, however, more like a part of the innate immune system, and so we discuss them here.

One type of ILL is the subset of **γ:δ T cells** that resides within epithelia, such as the skin. γ:δ T cells are themselves a minor subset of the T cells introduced in Chapter 1. Their antigen receptors are composed of a γ chain and a δ chain, rather than the α and β chains that make up the antigen receptors on the majority subset of T cells involved in adaptive immunity. γ:δ T cells were discovered purely as a consequence of their having immunoglobulin-related receptors encoded by rearranged genes, and their function is still being clarified.

Fig. 3.36 The three main classes of innate-like lymphocytes and their properties.

Innate-like lymphocytes		
B-1 cells	Epithelial γ:δ cells	iNKT cells
Make natural antibody, protect against infection with *Streptococcus pneumoniae*	Produce cytokines rapidly	Produce cytokines rapidly
Ligands not MHC associated	Ligands are MHC class IB associated	Ligands are lipids bound to CD1d
Cannot be boosted	Cannot be boosted	Cannot be boosted

One of the most striking features of γ:δ T cells is their division into two highly distinct subsets. One is found in the lymphoid tissues of all vertebrates and, like the α:β T cells, has highly diversified T-cell receptors. In contrast, intraepithelial γ:δ T cells occur variably in different vertebrates, and commonly display receptors of very limited diversity, particularly in the skin and the female reproductive tract of mice, where the γ:δ T cells are essentially identical in any one site. On the basis of this limited diversity and lack of recirculation, it has been proposed that intraepithelial γ:δ T cells may recognize ligands that are derived from the epithelium in which they reside but which are expressed only when a cell has become infected. Candidate ligands are heat-shock proteins, MHC class Ib molecules (described in Chapter 6), and unorthodox nucleotides and phospholipids; there is evidence of recognition of all these ligands by γ:δ T cells.

Unlike α:β T cells, γ:δ T cells do not generally recognize antigen as peptides presented by MHC molecules; instead, they seem to recognize their target antigens directly, and thus could recognize and respond rapidly to molecules expressed by many different cell types. Recognition of molecules expressed as a consequence of infection, rather than recognition of pathogen-specific antigens themselves, would distinguish intraepithelial γ:δ T cells from other lymphocytes and would place them in the innate-like class.

Another subset of lymphocytes with antigen receptors of limited diversity is the **B-1** subset of B cells which have properties quite distinct from those of the conventional B cells that mediate adaptive humoral immunity. B-1 cells are in many ways analogous to intraepithelial γ:δ T cells: they arise early in embryonic development, they use a distinctive and limited set of gene rearrangements to make their receptors, they are self-renewing in tissues outside the central lymphoid organs, and they are the predominant lymphocyte in a distinctive microenvironment—the peritoneal and pleural cavities. B-1 cells seem to make antibody responses mainly to polysaccharide antigens and can produce antibodies of the IgM class without 'help' from T cells (Fig. 3.37). Although these responses can be enhanced by T-cell cooperation, they first appear within 48 hours of exposure to antigen, when T cells cannot be involved. Thus, B-1 cells are not part of the antigen-specific adaptive immune response. The lack of an antigen-specific interaction with helper T cells might explain why immunological memory is not generated as a result of B-1 cell responses: repeated exposures to the same antigen elicit similar, or decreased, responses with each exposure. These responses, although generated by lymphocytes with rearranging receptors, therefore resemble innate rather than adaptive immune responses.

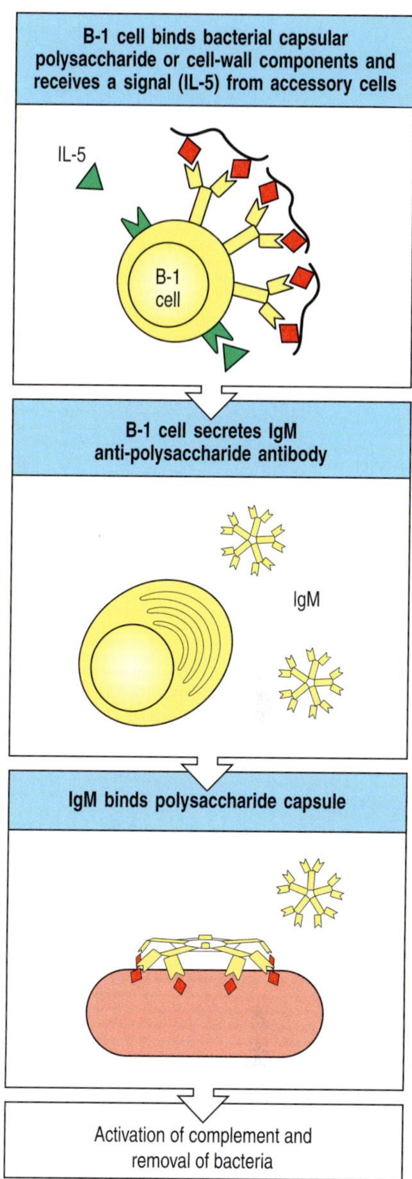

Fig. 3.37 B-1 cells might be important in the response to carbohydrate antigens such as bacterial polysaccharides. These responses occur rapidly, with antibody appearing within 48 hours after infection, presumably because there is a high frequency of precursors of the responding lymphocytes so that little clonal expansion is required. Unlike the responses to many other antigens, this response does not need the 'help' of T cells. In the absence of such help, only IgM is made (for reasons that will be explained in Chapter 9) and, in mice, these responses therefore clear bacteria mainly through the activation of complement, which is most efficient when the antibody is of the IgM isotype.

As with γ:δ T cells, the precise role of B-1 cells in immune defense is uncertain. Mice deficient in B-1 cells are more susceptible to infection with *Streptococcus pneumoniae* because they fail to produce an anti-phosphocholine antibody that provides protection against this bacterium. A significant fraction of the B-1 cells can make antibodies of this specificity, and because no antigen-specific T-cell help is required, a potent response can be produced early in infection with this pathogen. Whether human B-1 cells have the same role is not certain.

A third subset of ILLs, known as **invariant NKT cells (iNKT cells)**, exists both in the thymus and in peripheral lymphoid organs, including the mucosal immune system. These cells express an invariant T-cell receptor α chain, paired with one of three different β chains, and are able to recognize glycolipid antigens presented to them by the MHC-like molecule CD1 (discussed in Section 6-20). The main response of NKT cells to antigen stimulation seems to be the rapid secretion of cytokines, including IL-4, IL-10, and IFN-γ, and it is thought that these cells may have a primarily regulatory function. We shall encounter these unusual cells again in Chapter 9.

From an evolutionary perspective, it is interesting to note that γ:δ T cells seem to defend the body surfaces, whereas B-1 cells defend the body cavity. Both cell types are relatively limited in their range of specificities and in the efficiency of their responses. It is possible that these two cell types represent a transitional phase in the evolution of the adaptive immune response, guarding the two main compartments of primitive organisms—the epithelial surfaces and the body cavity. It is not yet clear whether they are still critical to host defense or whether they represent an evolutionary relic. Nevertheless, because each cell type is prominent in certain sites in the body and contributes to responses against certain pathogens, they must be incorporated into our thinking about host defense.

Summary.

Innate immunity uses a variety of induced effector mechanisms to clear an infection or, failing that, to hold it in check until the pathogen can be recognized by the adaptive immune system. These effector mechanisms are all regulated by germline-encoded receptor systems that are able to discriminate between normal self molecules on uninfected cells and infectious nonself ligands. Thus, the phagocyte's ability to discriminate between self and pathogen controls its release of pro-inflammatory chemokines and cytokines that act together to recruit more phagocytic cells to the site of infection. Especially prominent is the early recruitment of neutrophils that can recognize pathogens directly. Furthermore, cytokines released by tissue phagocytic cells induce fever and the production of acute-phase response proteins, including mannose-binding lectin, C-reactive protein, fibrinogen, and pulmonary surfactant proteins. These cytokines also mobilize antigen-presenting cells that induce the adaptive immune response. Viral pathogens are recognized by the cells in which they replicate, leading to the production of interferons that serve to inhibit viral replication and to activate NK cells. These in turn can distinguish healthy cells from those that are infected by virus or that are transformed or stressed in some way, based on the expression of class I MHC molecules and MHC-related molecules that are ligands for some NK receptors. In addition, several lymphocyte populations have innate-like behavior, forming a kind of 'transitional immunity.' As we will see later in the book, cytokines, chemokines, phagocytic cells, and NK cells are all effector mechanisms that are also employed in the adaptive immune response, which uses variable receptors to target specific pathogen antigens.

Summary to Chapter 3.

The induced responses of the innate system that defend humans and other animals against infection are based on several distinct components. After the initial barriers—the body's epithelia and the soluble antimicrobial molecules described in Chapter 2—have been breached, the most important innate defenses rely on tissue macrophages, which provide a double service. They mediate rapid cellular defense at the borders, through phagocytosis and rapid activation of antimicrobial killing mechanisms that are tuned by soluble messengers such as complement. In addition, these cells have an array of cell-surface and cytoplasmic sensors for microbes and their products—membrane-bound Toll-like receptors, and intracellular NOD-like receptors and RIG-like helicases—which detect pathogens and activate signaling pathways leading to the production of pro-inflammatory and antiviral cytokines. These in turn stimulate innate effector responses while also helping to initiate an adaptive immune response. The uncovering of the pathogen-sensing mechanisms described in this chapter has been one of the most active areas in immunology in the past decade and is still extremely active. It is providing important new insights into human autoinflammatory conditions such as

Crohn's disease and gout. Indeed, the induction of powerful effector mechanisms by innate immune recognition based on germline-encoded receptors clearly has some dangers. It is a double-edged sword, as embodied by the effects of the cytokine TNF-α—beneficial when released locally, but disastrous when produced systemically. This illustrates the evolutionary knife-edge along which all innate mechanisms of host defense travel. The innate immune system can be viewed as a defense system that mainly frustrates the establishment of a focus of infection; however, even when it is inadequate to this function, it has already set in motion—by recruiting and activating dendritic cells—the initiation of the adaptive immune response, which forms an essential part of humans' defenses against infection. Having introduced immunology with a consideration of innate immune function, we now turn our attention to the adaptive immune response, beginning with an explanation of the structure and function of the antigen receptors expressed by lymphocytes.

Questions.

3.1 *The innate immune system uses two different strategies to identify pathogens: recognition of nonself and recognition of self. (a) Give examples of each and discuss how each example contributes to the ability of the organism to protect itself from infection. (b) What are the disadvantages of these different strategies?*

3.2 *"The Toll receptors represent the most ancient pathways of host defense." Is this statement justified? Explain your answer.*

3.3 *Proteolysis and proteolytic cascades are found in several systems of innate defense. Name the systems described in Chapters 2 and 3 where proteolytic cascades are used to amplify the signal downstream of a initiating event. What is the advantage of such a system? What are the disadvantages? Give an example where the tuning of such a system can give rise to an autoinflammatory disease state.*

3.4 *Signaling cascades involving multiple adaptors and numerous intermediate steps seem unnecessarily complex. Why might they have evolved in this way?*

3.5 *Many signaling cascades used in innate immunity involve the interactions of proteins via specific protein domains. Name the kinds of interaction domains used in TLR signaling. How does recognition of pathogen-derived material regulate signaling by TLRs? How are the domains of the intermediates used to propagate these signals? Repeat this question for NOD-like receptors and RLH proteins.*

3.6 *Elie Metchnikoff discovered the protective role of macrophages by observing what happened in a starfish injured by a sea urchin spine. Describe the sequence of events that would follow were you to be speared by a sea urchin spine.*

3.7 *Erythrocytes lack MHC class I; why don't NK cells kill autologous erythrocytes?*

General references.

Beutler, B.: **Microbe sensing, positive feedback loops, and the pathogenesis of inflammatory diseases.** *Immunol. Rev.* 2009, **227**:248–263.

Ezekowitz, R.A.B., and Hoffman, J.: **Innate immunity.** *Curr. Opin. Immunol.* 1998, **10**:9–53.

Gallin, J.I., Goldstein, I.M., and Snyderman, R. (eds): *Inflammation—Basic Principles and Clinical Correlates*, 3rd ed. New York, Raven Press, 1999.

Janeway, C.A., Jr, and Medzhitov, R.: **Innate immune recognition.** *Annu. Rev. Immunol.* 2002, **20**:197–216.

Kawai, T., and Akira, S.: **The roles of TLRs, RLRs and NLRs in pathogen recognition.** *Int. Immunol.* 2009, **21**:317–337.

Section references.

3-1 After entering tissues, many pathogens are recognized, ingested, and killed by phagocytes.

Aderem, A., and Underhill, D.M.: **Mechanisms of phagocytosis in macrophages.** *Annu. Rev. Immunol.* 1999, **17**:593–623.

Goodridge, H.S., Wolf, A.J., and Underhill, D.M.: **Beta-glucan recognition by the innate immune system.** *Immunol. Rev.* 2009, **230**:38–50.

Greaves, D.R., and Gordon, S.: **The macrophage scavenger receptor at 30 years of age: current knowledge and future challenges.** *J. Lipid Res.* 2009, **50**:S282–S286.

Harrison, R.E., and Grinstein, S.: **Phagocytosis and the microtubule cytoskeleton.** *Biochem. Cell Biol.* 2002, **80**:509–515.

Lee, S.J., Evers, S., Roeder, D., Parlow, A.F., Risteli, J., Risteli, L., Lee, Y.C., Feizi, T., Langen, H., and Nussenzweig, M.C.: **Mannose receptor-mediated regulation of serum glycoprotein homeostasis.** *Science* 2002, **295**:1898–1901.

Linehan, S.A., Martinez-Pomares, L., and Gordon, S.: **Macrophage lectins in host defence.** *Microbes Infect.* 2000, **2**:279–288.

McGreal, E.P., Miller, J.L., and Gordon, S.: **Ligand recognition by antigen-presenting cell C-type lectin receptors.** *Curr. Opin. Immunol.* 2005, **17**:18–24.

Peiser, L., De Winther, M.P., Makepeace, K., Hollinshead, M., Coull, P., Plested, J., Kodama, T., Moxon, E.R., and Gordon, S.: **The class A macrophage scavenger receptor is a major pattern recognition receptor for *Neisseria meningitidis* which is independent of lipopolysaccharide and not required for secretory responses.** *Infect. Immun.* 2002, **70**:5346–5354.

Podrez, E.A., Poliakov, E., Shen, Z., Zhang, R., Deng, Y., Sun, M., Finton, P.J., Shan, L., Gugiu, B., Fox, P.L., *et al.*: **Identification of a novel family of oxidized phospholipids that serve as ligands for the macrophage scavenger receptor CD36.** *J. Biol. Chem.* 2002, **277**:38503–38516.

3-2 G-protein-coupled receptors on phagocytes link microbe recognition with increased efficiency of intracellular killing.

Bogdan, C., Rollinghoff, M., and Diefenbach, A.: **Reactive oxygen and reactive nitrogen intermediates in innate and specific immunity.** *Curr. Opin. Immunol.* 2000, **12**:64–76.

Dahlgren, C., and Karlsson, A.: **Respiratory burst in human neutrophils.** *J. Immunol. Methods* 1999, **232**:3–14.

Gerber, B.O., Meng, E.C., Dotsch, V., Baranski, T.J., and Bourne, H.R.: **An activation switch in the ligand binding pocket of the C5a receptor.** *J Biol. Chem.* 2001, **276**:3394–3400.

Reeves, E.P., Lu, H., Jacobs, H.L., Messina, C.G., Bolsover, S., Gabella, G., Potma, E.O., Warley, A., Roes, J., and Segal, A.W.: **Killing activity of neutrophils is mediated through activation of proteases by K^+ flux.** *Nature* 2002, **416**:291–297.

Ward, P.A.: **The dark side of C5a in sepsis.** *Nat. Rev. Immunol.* 2004, **4**:133–142.

3-3 Pathogen recognition and tissue damage initiate an inflammatory response.

Chertov, O., Yang, D., Howard, O.M., and Oppenheim, J.J.: **Leukocyte granule proteins mobilize innate host defenses and adaptive immune responses.** *Immunol. Rev.* 2000, **177**:68–78.

Kohl, J.: **Anaphylatoxins and infectious and noninfectious inflammatory diseases.** *Mol. Immunol.* 2001, **38**:175–187.

Mekori, Y.A., and Metcalfe, D.D.: **Mast cells in innate immunity.** *Immunol. Rev.* 2000, **173**:131–140.

Svanborg, C., Godaly, G., and Hedlund, M.: **Cytokine responses during mucosal infections: role in disease pathogenesis and host defence.** *Curr. Opin. Microbiol.* 1999, **2**:99–105.

Van der Poll, T.: **Coagulation and inflammation.** *J. Endotoxin Res.* 2001, **7**:301–304.

3-4 Toll-like receptors represent an ancient pathogen-recognition system.

Lemaitre, B., Nicolas, E., Michaut, L., Reichhart, J.M., and Hoffmann, J.A.: **The dorsoventral regulatory gene cassette spätzle/Toll/cactus controls the potent antifungal response in Drosophila adults.** *Cell* 1996, **86**:973–983.

Lemaitre, B., Reichhart, J.M., and Hoffmann, J.A.: ***Drosophila* host defense: differential induction of antimicrobial peptide genes after infection by various classes of microorganisms.** *Proc. Natl Acad. Sci. USA* 1997, **94**:14614–14619.

3-5 Mammalian Toll-like receptors are activated by many different pathogen-associated molecular patterns.

Beutler, B., and Rietschel, E.T.: **Innate immune sensing and its roots: the story of endotoxin.** *Nat. Rev. Immunol.* 2003, **3**:169–176.

Hoebe, K., Georgel, P., Rutschmann, S., Du, X., Mudd, S., Crozat, K., Sovath, S., Shamel, L., Hartung, T., Zähringer, U., *et al.*: **CD36 is a sensor of diacylglycerides.** *Nature* 2005, **433**:523–527.

Jin, M.S., Kim, S.E., Heo, J.Y., Lee, M.E., Kim, H.M., Paik, S.G., Lee, H., and Lee, J.O.: **Crystal structure of the TLR1-TLR2 heterodimer induced by binding of a tri-acylated lipopeptide.** *Cell* 2007, **130**:1071–1082.

Liu, L., Botos, I., Wang, Y., Leonard, J.N., Shiloach, J., Segal, D.M., and Davies, D.R.: **Structural basis of toll-like receptor 3 signaling with double-stranded RNA.** *Science* 2008, **320**:379–381.

Lund, J.M., Alexopoulou, L., Sato, A., Karow, M., Adams, N.C., Gale, N.W., Iwasaki, A., and Flavell, R.A.: **Recognition of single-stranded RNA viruses by Toll-like receptor 7.** *Proc. Natl Acad. Sci. USA* 2004, **101**:5598–5603.

Lund, J., Sato, A., Akira, S., Medzhitov, R., and Iwasaki, A.: **Toll-like receptor 9-mediated recognition of herpes simplex virus-2 by plasmacytoid dendritic cells.** *J. Exp. Med.* 2003, **198**:513–520.

Salio, M., and Cerundolo, V.: **Viral immunity: cross-priming with the help of TLR3.** *Curr. Biol.* 2005, **15**:R336–R339.

Yarovinsky, F., Zhang, D., Andersen, J.F., Bannenberg, G.L., Serhan, C.N., Hayden, M.S., Hieny, S., Sutterwala, F.S., Flavell, R.A., Ghosh, S., *et al.*: **TLR11 activation of dendritic cells by a protozoan profilin-like protein.** *Science* 2005, **308**:1626–1629.

3-6 TLR-4 recognizes bacterial lipopolysaccharide in association with the host accessory proteins MD-2 and CD14.

Beutler, B.: **Endotoxin, Toll-like receptor 4, and the afferent limb of innate immunity.** *Curr. Opin. Microbiol.* 2000, **3**:23–28.

Beutler, B., and Rietschel, E.T.: **Innate immune sensing and its roots: the story of endotoxin.** *Nat. Rev. Immunol.* 2003, **3**:169–176.

Kim, H.M., Park, B.S., Kim, J.I., Kim, S.E., Lee, J., Oh, S.C., Enkhbayar, P., Matsushima, N., Lee, H., Yoo, O.J., *et al.*: **Crystal structure of the TLR4–MD-2 complex with bound endotoxin antagonist Eritoran.** *Cell* 2007, **130**:906–917.

Park, B.S., Song, D.H., Kim, H.M., Choi, B.S., Lee, H., and Lee, J.O.: **The structural basis of lipopolysaccharide recognition by the TLR4–MD-2 complex.** *Nature* 2009, **458**:1191–1195.

3-7 TLRs activate the transcription factors NFκB, AP-1, and IRF to induce the expression of inflammatory cytokines and type I interferons.

Hiscott, J., Nguyen, T.L., Arguello, M., Nakhaei, P., and Paz, S.: **Manipulation of the nuclear factor-kappaB pathway and the innate immune response by viruses.** *Oncogene* 2006, **25**:6844–6867.

Honda, K., and Taniguchi, T.: **IRFs: master regulators of signalling by Toll-like receptors and cytosolic pattern-recognition receptors.** *Nat. Rev. Immunol.* 2006, **6**:644–658.

Puel, A., Yang, K., Ku, C.L., von Bernuth, H., Bustamante, J., Santos, O.F., Lawrence, T., Chang, H.H., Al-Mousa, H., Picard, C., *et al.*: **Heritable defects of the human TLR signalling pathways.** *J. Endotoxin Res.* 2005, **11**:220–224.

Von Bernuth, H., Picard, C., Jin, Z., Pankla, R., Xiao, H., Ku, C.L., Chrabieh, M., Mustapha, I.B., Ghandil, P., Camcioglu, Y., *et al.*: **Pyogenic bacterial infections in humans with MyD88 deficiency.** *Science* 2008, **321**:691–696.

Werts, C., Girardinm, S.E., and Philpott, D.J.: **TIR, CARD and PYRIN: three domains for an antimicrobial triad.** *Cell Death Differ.* 2006, **13**:798–815.

3-8 The NOD-like receptors act as intracellular sensors of bacterial infection.

Eisenbarth, S.C., Colegio, O.R., O'Connor, W., Sutterwala, F.S., and Flavell, R.A.: **Crucial role for the Nalp3 inflammasome in the immunostimulatory properties of aluminium adjuvants.** *Nature* 2008, **453**:1122–1126.

Fernandes-Alnemri, T., Yu, J.W., Juliana, C., Solorzano, L., Kang, S., Wu, J., Datta, P., McCormick, M., Huang, L., McDermott, E., *et al.*: **The AIM2 inflammasome is critical for innate immunity to *Francisella tularensis*.** *Nat. Immunol.* 2010, **11**:385–393.

Hornung, V., Ablasser, A., Charrel-Dennis, M., Bauernfeind, F., Horvath, G., Caffrey, D.R., Latz, E., and Fitzgerald, K.A.: **AIM2 recognizes cytosolic dsDNA and forms a caspase-1-activating inflammasome with ASC.** *Nature* 2009, **458**:514–518.

Inohara, N., Chamaillard, M., McDonald, C., and Nunez, G.: **NOD-LRR proteins: role in host–microbial interactions and inflammatory disease.** *Annu. Rev. Biochem.* 2005, **74**:355–383.

Martinon, F., Pétrilli, V., Mayor, A., Tardivel, A., and Tschopp, J.: **Gout-associated uric acid crystals activate the NALP3 inflammasome.** *Nature* 2006, **440**:237–241.

Shaw, M.H., Reimer, T., Kim, Y.G., and Nuñez, G.: **NOD-like receptors (NLRs): bona fide intracellular microbial sensors.** *Curr. Opin. Immunol.* 2008, **20**:377–382.

Strober, W., Murray, P.J., Kitani, A., and Watanabe, T.: **Signalling pathways and molecular interactions of NOD1 and NOD2.** *Nat. Rev. Immunol.* 2006, **6**:9–20.

Ting, J.P., Kastner, D.L., Hoffman, H.M.: **CATERPILLERs, pyrin and hereditary immunological disorders.** *Nat. Rev. Immunol.* 2006, **6**:183–195.

3-9 The RIG-I-like helicases detect cytoplasmic viral RNAs and stimulate interferon production.

Hornung, V., Ellegast, J., Kim, S., Brzózka, K., Jung, A., Kato, H., Poeck, H., Akira, S., Conzelmann, K.K., Schlee, M., *et al.*: **5′-Triphosphate RNA is the ligand for RIG-I.** *Science* 2006, **314**:994–997.

Kato, H., Takeuchi, O., Sato, S., Yoneyama, M., Yamamoto, M., Matsui, K., Uematsu, S., Jung, A., Kawai, T., Ishii, K.J., *et al*: **Differential roles of MDA5 and RIG-I helicases in the recognition of RNA viruses.** *Nature* 2006, **441**:101–105.

Konno, H., Yamamoto, T., Yamazaki, K., Gohda, J., Akiyama, T., Semba, K., Goto, H., Kato, A., Yujiri, T., Imai, T., *et al.*: **TRAF6 establishes innate immune responses by activating NF-kappaB and IRF7 upon sensing cytosolic viral RNA and DNA.** *PLoS ONE* 2009, 4:e5674.

Meylan, E., Curran, J., Hofmann, K., Moradpour, D., Binder, M., Bartenschlager, R., and Tschopp, J.: **Cardif is an adaptor protein in the RIG-I antiviral pathway and is targeted by hepatitis C virus.** *Nature* 2005, **437**:1167–1172.

Pichlmair, A., Schulz, O., Tan, C.P., Näslund, T.I., Liljeström, P., Weber, F., and Reis e Sousa, C.: **RIG-I-mediated antiviral responses to single-stranded RNA bearing 5′-phosphates.** *Science* 2006, **314**:935–936.

3-10 Activation of TLRs and NLRs triggers changes in gene expression in macrophages and dendritic cells that have far-reaching effects on the immune response.

Brightbill, H.D., Libraty, D.H., Krutzik, S.R., Yang, R.B., Belisle, J.T., Bleharski, J.R., Maitland, M., Norgard, M.V., Plevy, S.E., Smale, S.T., *et al.*: **Host defense mechanisms triggered by microbial lipoproteins through Toll-like receptors.** *Science* 1999, **285**:732–736.

Martinon, F., Mayor. A., and Tschopp, J.: **The inflammasomes: guardians of the body.** *Annu. Rev. Immunol.* 2009, **27**:229–265.

Takeda, K., Kaisho, T., and Akira, S.: **Toll-like receptors.** *Annu. Rev. Immunol.* 2003, **21**:335–376.

3-11 TLR signaling shares many components with Toll signaling in *Drosophila*.

Dziarski, R., and Gupta, D.: **Mammalian PGRPs: novel antibacterial proteins.** *Cell Microbiol.* 2006, **8**:1059–1069.

Ferrandon, D., Imler, J.L., Hetru, C., and Hoffmann, J.A.: **The *Drosophila* systemic immune response: sensing and signalling during bacterial and fungal infections.** *Nat. Rev. Immunol.* 2007, **7**:862–874.

Gottar, M., Gobert, V., Matskevich, A.A., Reichhart, J.M., Wang, C., Butt, T.M., Belvin, M., Hoffmann, J.A., and Ferrandon, D.: **Dual detection of fungal infections in *Drosophila* via recognition of glucans and sensing of virulence factors.** *Cell* 2006, **127**:1425–1437.

Kambris, Z., Brun, S., Jang, I.H., Nam, H.J., Romeo, Y., Takahashi, K., Lee, W.J., Ueda, R., and Lemaitre, B.: ***Drosophila* immunity: a large-scale *in vivo* RNAi screen identifies five serine proteases required for Toll activation.** *Curr. Biol.* 2006, **16**:808–813.

Pili-Floury, S., Leulier, F., Takahashi, K., Saigo, K., Samain, E., Ueda, R., and Lemaitre, B.: ***In vivo* RNA interference analysis reveals an unexpected role for GNBP1 in the defense against Gram-positive bacterial infection in *Drosophila* adults.** *J. Biol. Chem.* 2004, **279**:12848–12853.

Royet, J., and Dziarski, R.: **Peptidoglycan recognition proteins: pleiotropic sensors and effectors of antimicrobial defences.** *Nat. Rev. Microbiol.* 2007, **5**:264–277.

3-12 TLR and NOD genes have undergone extensive diversification in both invertebrates and some primitive chordates.

Rast, J.P., Smith, L.C., Loza-Coll, M., Hibino, T., and Litman, G.W.: **Genomic insights into the immune system of the sea urchin.** *Science* 2006, **314**:952–956.

Samanta, M.P., Tongprasit, W., Istrail, S., Cameron, R.A., Tu, Q., Davidson, E.H., and Stolc, V.: **The transcriptome of the sea urchin embryo.** *Science* 2006, **314**:960–962.

3-13 Macrophages and dendritic cells activated by pathogens secrete a range of cytokines that have a variety of local and distant effects.

Larsson, B.M., Larsson, K., Malmberg, P., and Palmberg, L.: **Gram-positive bacteria induce IL-6 and IL-8 production in human alveolar macrophages and epithelial cells.** *Inflammation* 1999, **23**:217–230.

Shortman, K., and Liu, Y.J.: **Mouse and human dendritic cell subtypes.** *Nat. Rev. Immunol.* 2002, **2**:151–161.

Svanborg, C., Godaly, G., and Hedlund, M.: **Cytokine responses during mucosal infections: role in disease pathogenesis and host defence.** *Curr. Opin. Microbiol.* 1999, **2**:99–105.

3-14 Chemokines released by macrophages and dendritic cells recruit effector cells to sites of infection.

Luster, A.D.: **The role of chemokines in linking innate and adaptive immunity.** *Curr. Opin. Immunol.* 2002, **14**:129–135.

Matsukawa, A., Hogaboam, C.M., Lukacs, N.W., and Kunkel, S.L.: **Chemokines and innate immunity.** *Rev. Immunogenet.* 2000, **2**:339–358.

Scapini, P., Lapinet-Vera, J.A., Gasperini, S., Calzetti, F., Bazzoni, F., and Cassatella, M.A.: **The neutrophil as a cellular source of chemokines.** *Immunol. Rev.* 2000, **177**:195–203.

Yoshie, O.: **Role of chemokines in trafficking of lymphocytes and dendritic cells.** *Int. J. Hematol.* 2000, **72**:399–407.

3-15 Cell-adhesion molecules control interactions between leukocytes and endothelial cells during an inflammatory response.

Alon, R., and Feigelson, S.: **From rolling to arrest on blood vessels: leukocyte tap dancing on endothelial integrin ligands and chemokines at sub-second contacts.** *Semin. Immunol.* 2002, **14**:93–104.

Bunting, M., Harris, E.S., McIntyre, T.M., Prescott, S.M., and Zimmerman, G.A.: **Leukocyte adhesion deficiency syndromes: adhesion and tethering defects involving β 2 integrins and selectin ligands.** *Curr. Opin. Hematol.* 2002, **9**:30–35.

D'Ambrosio, D., Albanesi, C., Lang, R., Girolomoni, G., Sinigaglia, F., and Laudanna, C.: **Quantitative differences in chemokine receptor engagement generate diversity in integrin-dependent lymphocyte adhesion.** *J. Immunol.* 2002, **169**:2303–2312.

Johnston, B., and Butcher, E.C.: **Chemokines in rapid leukocyte adhesion triggering and migration.** *Semin. Immunol.* 2002, **14**:83–92.

Ley, K.: **Integration of inflammatory signals by rolling neutrophils.** *Immunol. Rev.* 2002, **186**:8–18.

Vestweber, D.: **Lymphocyte trafficking through blood and lymphatic vessels: more than just selectins, chemokines and integrins.** *Eur. J. Immunol.* 2003, **33**:1361–1364.

3-16 Neutrophils make up the first wave of cells that cross the blood vessel wall to enter an inflamed tissue.

Bochenska-Marciniak, M., Kupczyk, M., Gorski, P., and Kuna, P.: **The effect of recombinant interleukin-8 on eosinophils' and neutrophils' migration *in vivo* and *in vitro*.** *Allergy* 2003, **58**:795–801.

Godaly, G., Bergsten, G., Hang, L., Fischer, H., Frendeus, B., Lundstedt, A.C., Samuelsson, M., Samuelsson, P., and Svanborg, C.: **Neutrophil recruitment, chemokine receptors, and resistance to mucosal infection.** *J. Leukoc. Biol.* 2001, **69**:899–906.

Gompertz, S., and Stockley, R.A.: **Inflammation—role of the neutrophil and the eosinophil.** *Semin. Respir. Infect.* 2000, **15**:14–23.

Lee, S.C., Brummet, M.E., Shahabuddin, S., Woodworth, T.G., Georas, S.N., Leiferman, K.M., Gilman, S.C., Stellato, C., Gladue, R.P., Schleimer, R.P., *et al.*: **Cutaneous injection of human subjects with macrophage inflammatory protein-1 α induces significant recruitment of neutrophils and monocytes.** *J. Immunol.* 2000, **164**:3392–3401.

Worthylake, R.A., and Burridge, K.: **Leukocyte transendothelial migration: orchestrating the underlying molecular machinery.** *Curr. Opin. Cell Biol.* 2001, **13**:569–577.

3-17 TNF-α is an important cytokine that triggers local containment of infection but induces shock when released systemically.

Croft, M.: **The role of TNF superfamily members in T-cell function and diseases.** *Nat. Rev. Immunol.* 2009, **9**:271–285.

Dellinger, R.P.: **Inflammation and coagulation: implications for the septic patient.** *Clin. Infect. Dis.* 2003, **36**:1259–1265.

Georgel, P., Naitza, S., Kappler, C., Ferrandon, D., Zachary, D., Swimmer, C., Kopczynski, C., Duyk, G., Reichhart, J.M., and Hoffmann, J.A.: ***Drosophila* immune deficiency (IMD) is a death domain protein that activates antibacterial defense and can promote apoptosis.** *Dev. Cell* 2001, **1**:503–514.

Pfeffer, K.: **Biological functions of tumor necrosis factor cytokines and their receptors.** *Cytokine Growth Factor Rev.* 2003, **14**:185–191.

Rutschmann, S., Jung, A.C., Zhou, R., Silverman, N., Hoffmann, J.A., and Ferrandon, D.: **Role of *Drosophila* IKKγ in a *toll*-independent antibacterial immune response.** *Nat. Immunol.* 2000, **1**:342–347.

3-18 Cytokines released by macrophages and dendritic cells activate the acute-phase response.

Bopst, M., Haas, C., Car, B., and Eugster, H.P.: **The combined inactivation of tumor necrosis factor and interleukin-6 prevents induction of the major acute phase proteins by endotoxin.** *Eur. J. Immunol.* 1998, **28**:4130–4137.

Ceciliani, F., Giordano, A., and Spagnolo, V.: **The systemic reaction during inflammation: the acute-phase proteins.** *Protein Pept. Lett.* 2002, **9**:211–223.

He, R., Sang, H., and Ye, R.D.: **Serum amyloid A induces IL-8 secretion through a G protein-coupled receptor, FPRL1/LXA4R.** *Blood* 2003, **101**:1572–1581.

Horn, F., Henze, C., and Heidrich, K.: **Interleukin-6 signal transduction and lymphocyte function.** *Immunobiology* 2000, **202**:151–167.

Manfredi, A.A., Rovere-Querini, P., Bottazzi, B., Garlanda, C., and Mantovani, A.: **Pentraxins, humoral innate immunity and tissue injury.** *Curr. Opin. Immunol.* 2008, **20**:538–544.

Mold, C., Rodriguez, W., Rodic-Polic, B., and Du Clos, T.W.: **C-reactive protein mediates protection from lipopolysaccharide through interactions with FcγR.** *J. Immunol.* 2002, **169**:7019–7025.

3-19 Interferons induced by viral infection make several contributions to host defense.

Honda, K., Takaoka, A., and Taniguchi, T.: **Type I interferon gene induction by the interferon regulatory factor family of transcription factors.** *Immunity* 2006, **25**:349–360.

Kawai, T., and Akira, S.: **Innate immune recognition of viral infection.** *Nat. Immunol.* 2006, **7**:131–137.

Liu, Y.J.: **IPC: professional type 1 interferon-producing cells and plasmacytoid dendritic cell precursors.** *Annu. Rev. Immunol.* 2005, **23**:275–306.

Meylan, E., and Tschopp, J.: **Toll-like receptors and RNA helicases: two parallel ways to trigger antiviral responses.** *Mol. Cell* 2006, **22**:561–569.

Pietras, E.M., Saha, S.K., and Cheng, G.: **The interferon response to bacterial and viral infections.** *J. Endotoxin Res.* 2006, **12**:246–250.

3-20 NK cells are activated by interferons and macrophage-derived cytokines to serve as an early defense against certain intracellular infections.

Barral, D.C., and Brenner, M.B.: **CD1 antigen presentation: how it works.** *Nat. Rev. Immunol.* 2007, **7**:929–941.

Godshall, C.J., Scott, M.J., Burch, P.T., Peyton, J.C., and Cheadle, W.G.: **Natural killer cells participate in bacterial clearance during septic peritonitis through interactions with macrophages.** *Shock* 2003, **19**:144–149.

Lanier, L.L.: **Evolutionary struggles between NK cells and viruses.** *Nat. Rev. Immunol.* 2008, **8**:259–268.

Salazar-Mather, T.P., Hamilton, T.A., and Biron, C.A.: **A chemokine-to-cytokine-to-chemokine cascade critical in antiviral defense.** *J. Clin. Invest.* 2000, **105**:985–993.

Seki, S., Habu, Y., Kawamura, T., Takeda, K., Dobashi, H., Ohkawa, T., and Hiraide, H.: **The liver as a crucial organ in the first line of host defense: the roles of Kupffer cells, natural killer (NK) cells and NK1.1 Ag⁺ T cells in T helper 1 immune responses.** *Immunol. Rev.* 2000, **174**:35–46.

Yokoyama, W.M., and Plougastel, B.F.: **Immune functions encoded by the natural killer gene complex.** *Nat. Rev. Immunol.* 2003, **3**:304–316.

3-21 NK cells possess receptors for self molecules that prevent their activation by uninfected cells.

Borrego, F., Kabat, J., Kim, D.K., Lieto, L., Maasho, K., Pena, J., Solana, R., and Coligan, J.E.: **Structure and function of major histocompatibility complex (MHC) class I specific receptors expressed on human natural killer (NK) cells.** *Mol. Immunol.* 2002, **38**:637–660.

Boyington, J.C., and Sun, P.D.: **A structural perspective on MHC class I recognition by killer cell immunoglobulin-like receptors.** *Mol. Immunol.* 2002, **38**:1007–1021.

Brown, M.G., Dokun, A.O., Heusel, J.W., Smith, H.R., Beckman, D.L., Blattenberger, E.A., Dubbelde, C.E., Stone, L.R., Scalzo, A.A., and Yokoyama, W.M.: **Vital involvement of a natural killer cell activation receptor in resistance to viral infection.** *Science* 2001, **292**:934–937.

Long, E.O.: **Negative signalling by inhibitory receptors: the NK cell paradigm.** *Immunol. Rev.* 2008, **224**:70–84.

Robbins, S.H., and Brossay, L.: **NK cell receptors: emerging roles in host defense against infectious agents.** *Microbes Infect.* 2002, **4**:1523–1530.

Trowsdale, J.: **Genetic and functional relationships between MHC and NK receptor genes.** *Immunity* 2001, **15**:363–374.

Vilches, C., and Parham, P.: **KIR: diverse, rapidly evolving receptors of innate and adaptive immunity.** *Annu. Rev. Immunol.* 2002, **20**:217–251.

3-22 NK cells bear receptors that activate their effector function in response to ligands expressed on infected cells or tumor cells.

Gasser, S., Orsulic, S., Brown, E.J., and Raulet, D.H.: **The DNA damage pathway regulates innate immune system ligands of the NKG2D receptor.** *Nature* 2005, **436**:1186–1190.

Moretta, L., Bottino, C., Pende, D., Castriconi, R., Mingari, M.C., and Moretta, A.: **Surface NK receptors and their ligands on tumor cells.** *Semin. Immunol.* 2006, **18**:151–158.

Lanier, L.L.: **Up on the tightrope: natural killer cell activation and inhibition.** *Nat. Immunol.* 2008, **9**:495–502.

Parham, P.: **MHC class I molecules and KIRs in human history, health and survival.** *Nat. Rev. Immunol.* 2005, **5**:201–214.

3-23 The NKG2D receptor activates a different signaling pathway from that of the other activating NK receptors.

Gonzalez, S., Groh, V., and Spies, T.: **Immunobiology of human NKG2D and its ligands.** *Curr. Top. Microbiol. Immunol.* 2006, **298**:121–138.

Upshaw, J.L., and Leibson, P.J.: **NKG2D-mediated activation of cytotoxic lymphocytes: unique signaling pathways and distinct functional outcomes.** *Semin. Immunol.* 2006, **18**:167–175.

Vivier, E., Nunes, J.A., and Vely, F.: **Natural killer cell signaling pathways.** *Science* 2004, **306**:1517–1519.

3-24 Several lymphocyte subpopulations behave as innate-like lymphocytes.

Bos, N.A., Cebra, J.J., and Kroese, F.G.: **B-1 cells and the intestinal microflora.** *Curr. Top. Microbiol. Immunol.* 2000, **252**:211–220.

Chan, W.L., Pejnovic, N., Liew, T.V., Lee, C.A., Groves, R., and Hamilton, H.: **NKT cell subsets in infection and inflammation.** *Immunol. Lett.* 2003, **85**:159–163.

Chatenoud, L.: **Do NKT cells control autoimmunity?** *J. Clin. Invest.* 2002, **110**:747–748.

Feinberg, H., Uitdehaag, J.C., Davies, J.M., Wallis, R., Drickamer, K., and Weis, W.I.: **Crystal structure of the CUB1-EGF-CUB2 region of mannose-binding protein associated serine protease-2.** *EMBO J.* 2003, **22**:2348–2359.

Fraser, D.A., Arora, M., Bohlson, S.S., Lozano, E., and Tenner, A.J.: **Generation of inhibitory NF-κ B complexes and phosphorylated cAMP response element-binding protein correlates with the activity of complement protein C1q in human monocytes.** *J. Biol. Chem.* 2007, **282**:7360–7367.

Fraser, D.A., Bohlson, S.S., Jasinskiene, N., Rawal, N., Palmarini, G., Ruiz, S., Rochford, R., and Tenner, A.J.: **C1q and MBL, components of the innate immune system, influence monocyte cytokine expression.** *J. Leukoc. Biol.* 2006, **80**:107–116.

Gaboriaud, C., Juanhuix, J., Gruez, A., Lacroix, M., Darnault, C., Pignol, D., Verger, D., Fontecilla-Camps, J.C., and Arlaud, G.J.: **The crystal structure of the globular head of complement protein C1q provides a basis for its versatile recognition properties.** *J. Biol. Chem.* 2003, **278**:46974–46982.

Galli, G., Nuti, S., Tavarini, S., Galli-Stampino, L., De Lalla, C., Casorati, G., Dellabona, P., and Abrignani, S.: **CD1d-restricted help to B cells by human invariant natural killer T lymphocytes.** *J. Exp. Med.* 2003, **197**:1051–1057.

Hawlisch, H., and Kohl, J.: **Complement and Toll-like receptors: key regulators of adaptive immune responses.** *Mol. Immunol.* 2006, **43**:13–21.

Hawlisch, H., Belkaid, Y., Baelder, R., Hildeman, D., Gerard, C., and Kohl, J.C.: **5a negatively regulates toll-like receptor 4-induced immune responses.** *Immunity* 2005, **22**:415–426.

Kronenberg, M., and Gapin, L.: **The unconventional lifestyle of NKT cells.** *Nat. Rev. Immunol.* 2002, **2**:557–568.

Poon, P.H., Schumaker, V.N., Phillips, M.L., and Strang, C.J.: **Conformation and restricted segmental flexibility of C1, the first component of human complement.** *J. Mol. Biol.* 1983, **168**:563–577.

Reid, R.R., Woodcock, S., Prodeus, A.P., Austen, J., Kobzik, L., Hechtman, H., Moore, F.D., Jr, and Carroll, M.C.: **The role of complement receptors CD21/CD35 in positive selection of B-1 cells.** *Curr. Top. Microbiol. Immunol.* 2000, **252**:57–65.

Roos, A., Xu, W., Castellano, G., Nauta, A.J., Garred, P., Daha, M.R., and van Kooten, C.: **Mini-review: a pivotal role for innate immunity in the clearance of apoptotic cells.** *Eur. J. Immunol.* 2004, **34**:921–929.

Schumaker, V.N., Hanson, D.C., Kilchherr, E., Phillips, M.L., and Poon, P.H.: **A molecular mechanism for the activation of the first component of complement by immune complexes.** *Mol. Immunol.* 1986, **23**:557–565.

Sharif, S., Arreaza, G.A., Zucker, P., Mi, Q.S., and Delovitch, T.L.: **Regulation of autoimmune disease by natural killer T cells.** *J. Mol. Med.* 2002, **80**:290–300.

Stober, D., Jomantaite, I., Schirmbeck, R., and Reimann, J.: **NKT cells provide help for dendritic cell-dependent priming of MHC class I-restricted CD8⁺ T cells *in vivo*.** *J. Immunol.* 2003, **170**:2540–2548.

Van den Berg, R.H., Faber-Krol, M.C., Sim, R.B., and Daha, M.R.: **The first subcomponent of complement, C1q, triggers the production of IL-8, IL-6, and monocyte chemoattractant peptide-1 by human umbilical vein endothelial cells.** *J. Immunol.* 1998, **161**:6924–6930.

Yamada, M., Oritani, K., Kaisho, T., Ishikawa, J., Yoshida, H., Takahashi, I., Kawamoto, S., Ishida, N., Ujiie, H., Masaie, H., *et al.*: **Complement C1q regulates LPS-induced cytokine production in bone marrow-derived dendritic cells.** *Eur. J. Immunol.* 2004, **34**:221–230.

Zhang, X., Kimura, Y., Fang, C., Zhou, L., Sfyroera, G., Lambris, J.D., Wetsel, R.A., Miwa, T., and Song, W.C.: **Regulation of Toll-like receptor-mediated inflammatory response by complement in vivo.** *Blood* 2007, **110**:228–236.

Zinkernagel, R.M.: **A primitive T cell-independent mechanism of intestinal mucosal IgA responses to commensal bacteria.** *Science* 2000, **288**:2222–2226.

PART II

THE RECOGNITION OF ANTIGEN

Antigen Recognition by B-cell and T-cell Receptors

4

Innate immune responses initially defend the body against infection, but these only work to control pathogens that have certain molecular patterns or that induce interferons and other nonspecific defenses. To effectively fight the wide range of pathogens an individual will encounter, the lymphocytes of the adaptive immune system have evolved to recognize a great variety of different **antigens** from bacteria, viruses, and other disease-causing organisms. An antigen is any molecule or part of a molecule that is specifically recognized by the highly specialized recognition proteins of lymphocytes. On B cells these are the **immunoglobulins** (**Ig**), which are produced by these cells in a vast range of antigen specificities, each B cell producing immunoglobulin of a single specificity (see Sections 1-11 and 1-12). Membrane-bound immunoglobulin on the B-cell surface serves as the cell's receptor for antigen, and is known as the **B-cell receptor** (**BCR**). Immunoglobulin of the same antigen specificity is secreted as **antibody** by terminally differentiated B cells—the plasma cells. The secretion of antibodies, which bind pathogens or their toxic products in the extracellular spaces of the body (see Fig. 1.26), is the main effector function of B cells in adaptive immunity.

Antibodies were the first proteins involved in specific immune recognition to be characterized, and they are still the best understood. The antibody molecule has two separate functions: one is to bind specifically to the pathogen or its products that elicited the immune response; the other is to recruit other cells and molecules to destroy the pathogen once antibody has bound. For example, binding by antibodies neutralizes viruses and marks pathogens for destruction by phagocytes and complement, as described in Section 1-18. Recognition and effector functions are structurally separated in the antibody molecule, one part of which specifically binds to the antigen whereas the other engages the elimination mechanisms. The antigen-binding region varies extensively between antibody molecules and is known as the **variable region** or **V region**. The variability of antibody molecules allows each antibody to bind a different specific antigen, and the total repertoire of antibodies made by a single individual is large enough to ensure that virtually any structure can be recognized. The region of the antibody molecule that engages the effector functions of the immune system does not vary in the same way and is known as the **constant region** or **C region**. It comes in five main forms, each of which is specialized for activating different effector mechanisms. The membrane-bound B-cell receptor does not have these effector functions, because the C region remains inserted in the membrane of the B cell. The function of the B-cell receptor is to recognize and bind antigen via the V regions exposed on the surface of the cell, thus transmitting a signal that activates the B cell, leading to clonal expansion and antibody production. To this end, the B-cell receptor is associated with a set of intracellular signaling proteins, which will be described in Chapter 7.

The antigen-recognition molecules of T cells are made solely as membrane-bound proteins, which are associated with an intracellular signaling complex and function only to signal T cells for activation. These **T-cell receptors** (**TCRs**) are related to immunoglobulins both in their protein structure—having both V and C regions—and in the genetic mechanism that produces their great variability, which will be discussed in Chapter 5. The T-cell receptor differs from the B-cell receptor in an important way, however: it does not recognize and bind antigen by itself, but instead recognizes short peptide fragments of protein antigens, which are presented by proteins known as **MHC molecules** on the surfaces of host cells.

The MHC molecules are transmembrane glycoproteins encoded in the large cluster of genes known as the **major histocompatibility complex** (**MHC**). Their most striking structural feature is a cleft in the extracellular face of the molecule, in which peptides can be bound. MHC molecules are highly **polymorphic**—each type of MHC molecule occurs in many different versions within the population. Most people are therefore heterozygous for the MHC molecules: that is, they express two different forms of each type of MHC molecule, which increases the range of pathogen-derived peptides that can be bound. T-cell receptors recognize features of both the peptide antigen and the MHC molecule to which it is bound. This introduces an extra dimension to antigen recognition by T cells, known as **MHC restriction**, because any given T-cell receptor is specific for a unique combination of a particular peptide and a particular MHC molecule. We shall discuss MHC polymorphism and its consequences for T-cell antigen recognition and T-cell development in Chapters 6 and 8, respectively.

In this chapter we focus on the structure and antigen-binding properties of immunoglobulins and T-cell receptors. Although B cells and T cells recog nize foreign molecules in distinct fashions, the receptor molecules they use for this task are very similar in structure. We will see how this basic structure can accommodate great variability in antigen specificity, and how it enables immunoglobulins and T-cell receptors to perform their functions as the antigen-recognition molecules of the adaptive immune response.

The structure of a typical antibody molecule.

Antibodies are the secreted form of the B-cell receptor. Because they are soluble and secreted into the blood in large quantities, antibodies are easily obtainable and easily studied. For this reason, most of what we know about the B-cell receptor comes from the study of antibodies.

Antibody molecules are roughly Y-shaped molecules consisting of three equal-sized portions connected by a flexible tether. Three schematic representations of antibody structure are shown in Fig. 4.1. In this part of the chapter we will explain how this structure is formed and how it allows antibody molecules to perform their dual tasks—binding on the one hand to a wide variety of antigens, and on the other to a limited number of effector molecules and cells that destroy the antigen. Each of these tasks is performed by different parts of the molecule. The ends of the two arms of the Y—the V regions—vary in their detailed structure between different antibody molecules. These are involved in antigen binding. The stem of the Y—the C region—is far less variable and is the part that interacts with effector cells and molecules.

All antibodies are constructed in the same way from paired heavy and light polypeptide chains, and the generic term immunoglobulin is used for all such proteins. Five different **classes** of immunoglobulins—IgM, IgD, IgG, IgA, and

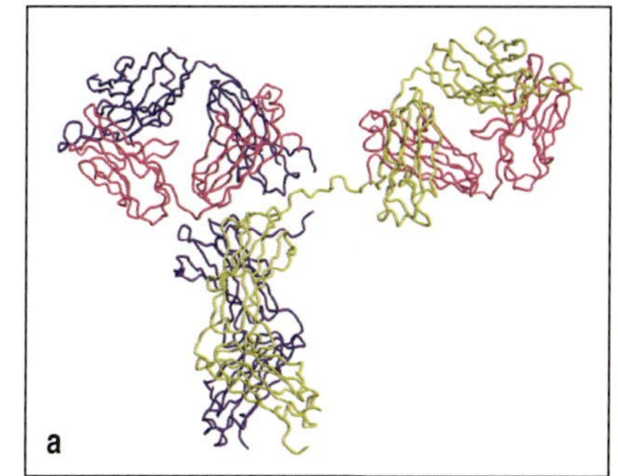

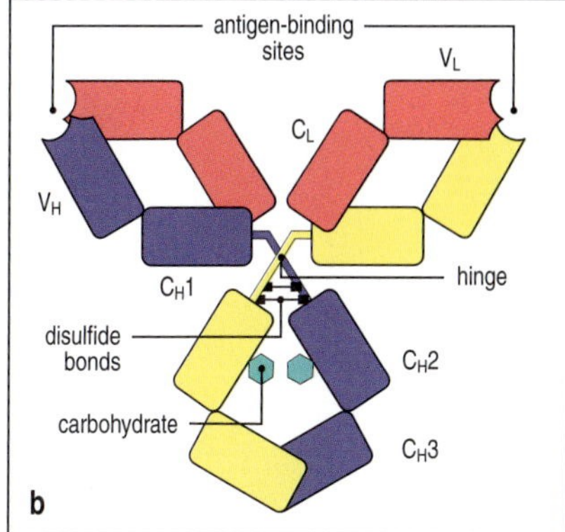

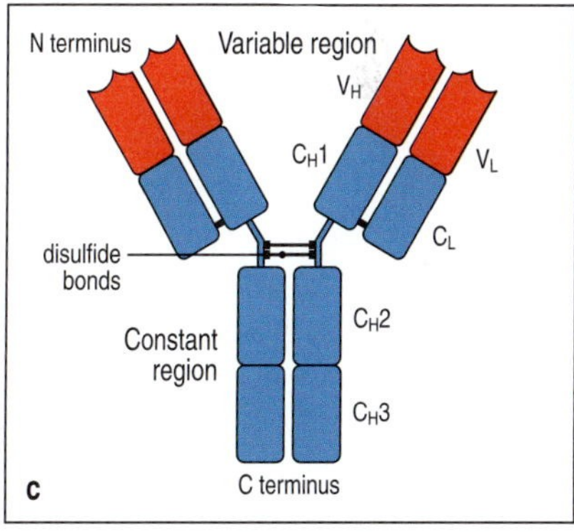

Fig. 4.1 Structure of an antibody molecule. Panel a illustrates a ribbon diagram based on the X-ray crystallographic structure of an IgG antibody, showing the course of the backbones of the polypeptide chains. Three globular regions form a Y shape. The two antigen-binding sites are at the tips of the arms, which are tethered to the trunk of the Y by a flexible hinge region. A schematic representation of the structure in panel a is given in panel b, illustrating the four-chain composition and the separate domains comprising each chain. Panel c shows a simplified schematic representation of an antibody molecule that will be used throughout this book. C terminus, carboxy terminus; N terminus, amino terminus. Structure courtesy of R.L. Stanfield and I.A. Wilson.

IgE—can be distinguished by their C regions. More subtle differences confined to the V region account for the specificity of antigen binding. We will use the IgG antibody molecule as an example to describe the general structural features of immunoglobulins.

4-1 IgG antibodies consist of four polypeptide chains.

IgG antibodies are large molecules with a molecular weight of approximately 150 kDa and are composed of two different kinds of polypeptide chains. One, of approximately 50 kDa, is called the **heavy** or **H chain**, and the other, of 25 kDa, is the **light** or **L chain** (Fig. 4.2). Each IgG molecule consists of two heavy chains and two light chains. The two heavy chains are linked to each other by disulfide bonds, and each heavy chain is linked to a light chain by a disulfide bond. In any given immunoglobulin molecule, the two heavy chains and the two light chains are identical, giving an antibody molecule two identical antigen-binding sites (see Fig. 4.1). This gives it the ability to bind simultaneously to two identical antigens on a surface and hence increase the total strength of the interaction, which is called its **avidity**. The strength of the interaction between a single antigen-binding site and its antigen is called its **affinity**.

Two types of light chains, **lambda (λ)** and **kappa (κ)**, are found in antibodies. A given immunoglobulin has either κ chains or λ chains, never one of each. No functional difference has been found between antibodies having λ or κ light chains, and either type of light chain can be found in antibodies of any of the five major classes. The ratio of the two types of light chains varies from species to species. In mice, the average κ to λ ratio is 20:1, whereas in humans it is 2:1 and in cattle it is 1:20. The reason for this variation is unknown. Distortions of this ratio can sometimes be used to detect the abnormal proliferation of a clone of B cells. These will all express the identical light chain, and thus an excess of λ light chains in a person might indicate the presence of a B-cell tumor producing λ chains.

The class, and thus the effector function, of an antibody is defined by the structure of its heavy chain. There are five main heavy-chain classes or **isotypes**, some of which have several subtypes, and these determine the functional

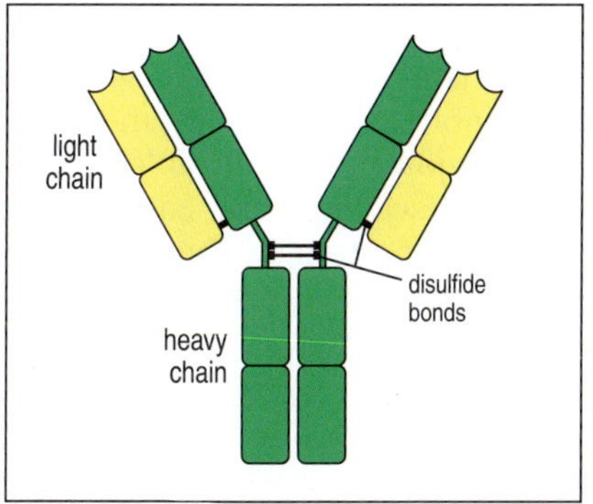

Fig. 4.2 Immunoglobulin molecules are composed of two types of protein chains: heavy chains and light chains. Each immunoglobulin molecule is made up of two heavy chains (green) and two light chains (yellow) joined by disulfide bonds so that each heavy chain is linked to a light chain and the two heavy chains are linked together.

activity of an antibody molecule. The five major classes of immunoglobulin are **immunoglobulin M (IgM)**, **immunoglobulin D (IgD)**, **immunoglobulin G (IgG)**, **immunoglobulin A (IgA)**, and **immunoglobulin E (IgE)**. Their heavy chains are denoted by the corresponding lower-case Greek letter (μ, δ, γ, α, and ε, respectively). IgG is by far the most abundant immunoglobulin and has several subclasses (IgG1, 2, 3, and 4 in humans). The distinctive functional properties of the different classes and subclasses of antibodies are conferred by the carboxy-terminal part of the heavy chain, where it is not associated with the light chain. We will describe the structure and functions of the different heavy-chain isotypes in Chapter 5. The general structural features of all the isotypes are similar, and we will consider IgG, the most abundant isotype in plasma, as a typical antibody molecule.

The structure of a B-cell receptor is identical to that of its corresponding antibody except for a small portion of the carboxy terminus of the heavy-chain C region. In the B-cell receptor, the carboxy terminus is a hydrophobic amino acid sequence that anchors the molecule in the membrane, and in the antibody it is a hydrophilic sequence that allows secretion.

4-2 Immunoglobulin heavy and light chains are composed of constant and variable regions.

The amino acid sequences of many immunoglobulin heavy and light chains have been determined and reveal two important features of antibody molecules. First, each chain consists of a series of similar, although not identical, sequences, each about 110 amino acids long. Each of these repeats corresponds to a discrete, compactly folded region of protein structure known as a protein domain. The light chain is made up of two such **immunoglobulin domains**, whereas the heavy chain of the IgG antibody contains four (see Fig. 4.1a). This suggests that the immunoglobulin chains have evolved by repeated duplication of an ancestral gene corresponding to a single domain.

The second important feature is that the amino-terminal amino acid sequences of the heavy and light chains vary greatly between different antibodies. The variability is limited to approximately the first 110 amino acids, corresponding to the first domain, whereas the remaining domains are constant between immunoglobulin chains of the same isotype. The amino-terminal **variable domains (V domains)** of the heavy and light chains (V_H and V_L, respectively) together make up the V region of the antibody and confer on it the ability to bind specific antigen, whereas the **constant domains (C domains)** of the heavy and light chains (C_H and C_L, respectively) make up the C region (see Fig. 4.1b and c). The multiple heavy-chain C domains are numbered from the amino-terminal end to the carboxy terminus, for example C_H1, C_H2, and so on.

4-3 The antibody molecule can readily be cleaved into functionally distinct fragments.

The protein domains described above associate to form larger globular domains. Thus, when fully folded and assembled, an antibody molecule comprises three equal-sized globular portions joined by a flexible stretch of polypeptide chain known as the **hinge region** (see Fig. 4.1b). Each arm of this Y-shaped structure is formed by the association of a light chain with the amino-terminal half of a heavy chain, whereas the trunk of the Y is formed by the pairing of the carboxy-terminal halves of the two heavy chains. The association of the heavy and light chains is such that the V_H and V_L domains are paired, as are the C_H1 and C_L domains. The C_H3 domains pair with each other but the C_H2 domains do not interact; carbohydrate side chains attached to the C_H2 domains lie between the two heavy chains. The two antigen-binding sites are formed by the paired V_H and V_L domains at the ends of the two arms of the Y (see Fig. 4.1b).

Proteolytic enzymes (proteases) that cleave polypeptide sequences have been used to dissect the structure of antibody molecules and to determine which parts of the molecule are responsible for its various functions. Limited digestion with the protease papain cleaves antibody molecules into three fragments (Fig. 4.3). Papain cuts the antibody molecule on the amino-terminal side of the disulfide bonds that link the two heavy chains, releasing the two arms of the antibody molecule as two identical fragments that contain the antigen-binding activity. These are called the **Fab fragments**, for **F**ragment **a**ntigen **b**inding. The other fragment contains no antigen-binding activity but was originally observed to crystallize readily, and for this reason it was named the **Fc fragment**, for **F**ragment **c**rystallizable. It corresponds to the paired C_H2 and C_H3 domains and is the part of the antibody molecule that interacts with effector molecules and cells. The functional differences between heavy-chain isotypes lie mainly in the Fc fragment.

Another protease, pepsin, cuts on the carboxy-terminal side of the disulfide bonds (see Fig. 4.3). This produces a fragment, the **F(ab′)$_2$ fragment**, in which the two antigen-binding arms of the antibody molecule remain linked. Pepsin cuts the remaining part of the heavy chain into several small fragments. The F(ab′)$_2$ fragment has exactly the same antigen-binding characteristics as the original antibody but is unable to interact with any effector molecule. It thus has potential for therapeutic applications.

Many antibody-related molecules can be constructed with the use of genetic engineering techniques. One is a truncated Fab comprising the V domain of a heavy chain linked by a stretch of synthetic peptide to a V domain of a light chain. This is called **single-chain Fv**, named from **F**ragment **v**ariable. Fv molecules could become valuable therapeutic agents because of their small size, which allows them to penetrate tissues easily. For example, Fv molecules specific for tumor antigens and coupled to protein toxins have potential applications in tumor therapy, as discussed in Chapter 16.

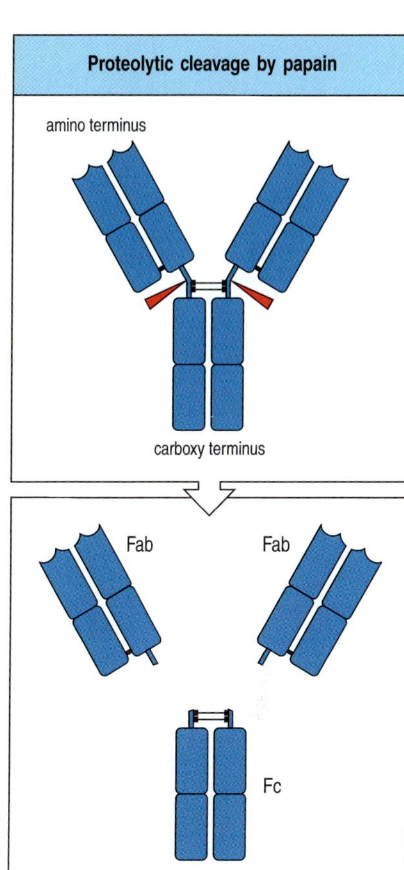

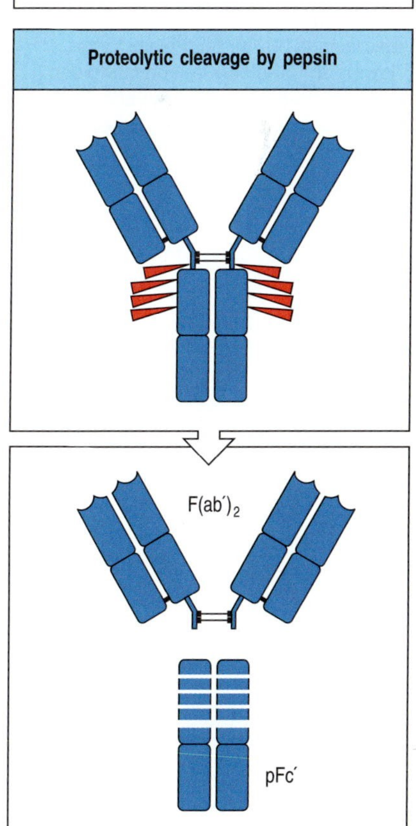

Fig. 4.3 The Y-shaped immunoglobulin molecule can be dissected by partial digestion with proteases. Upper panels: papain cleaves the immunoglobulin molecule into three pieces, two Fab fragments and one Fc fragment. The Fab fragment contains the V regions and binds antigen. The Fc fragment is crystallizable and contains C regions. Lower panels: pepsin cleaves immunoglobulin to yield one F(ab′)$_2$ fragment and many small pieces of the Fc fragment, the largest of which is called the pFc′ fragment. F(ab′)$_2$ is written with a prime because it contains a few more amino acids than Fab, including the cysteines that form the disulfide bonds.

4-4 The immunoglobulin molecule is flexible, especially at the hinge region.

The hinge region that links the Fc and Fab portions of the IgG molecule is in reality a flexible tether, allowing independent movement of the two Fab arms. This flexibility is revealed by studies of antibodies bound to small antigens known as **haptens**. These are molecules of various types that are typically about the size of a tyrosine side chain. Although haptens are specifically recognized by antibody, they can only stimulate the production of anti-hapten antibodies when linked to a protein (see Appendix I, Section A-1). Two identical hapten molecules joined by a short flexible region can link two or more anti-hapten antibodies, forming dimers, trimers, tetramers, and so on, which can be seen by electron microscopy (Fig. 4.4). The shapes formed by these complexes show that antibody molecules are flexible at the hinge region. Some flexibility is also found at the junction between the V and C domains, allowing bending

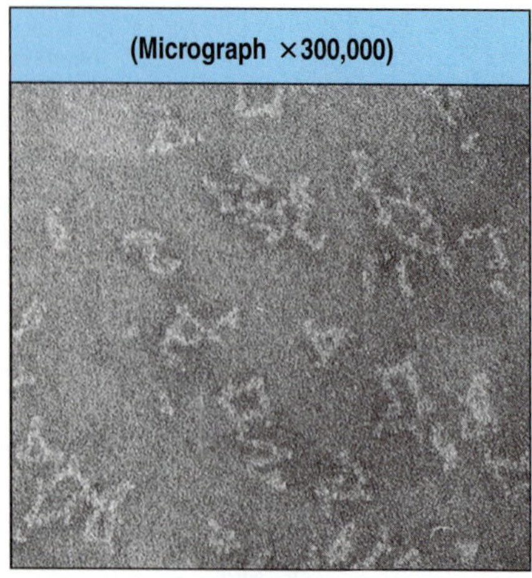

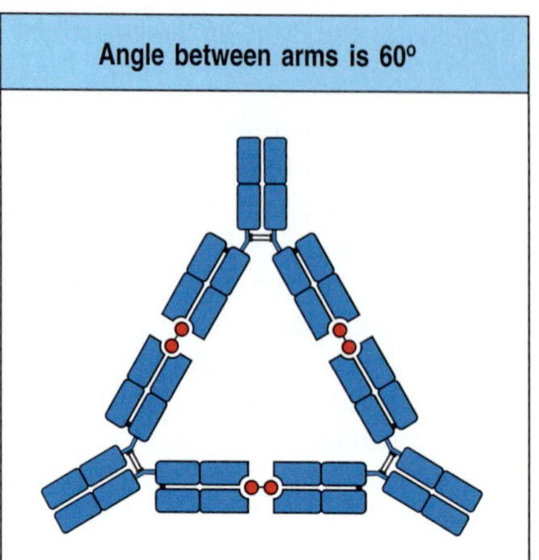

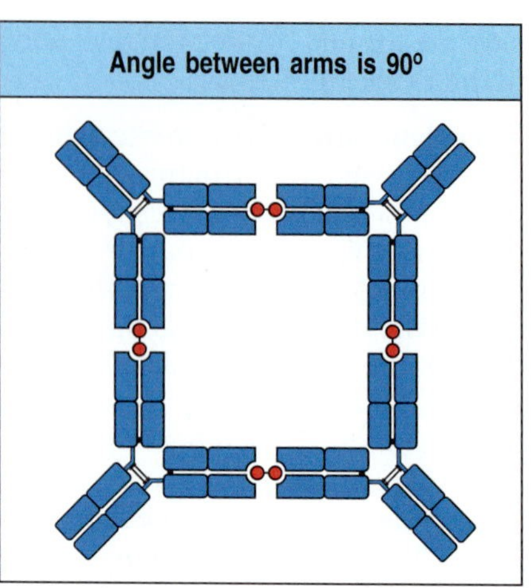

Fig. 4.4 Antibody arms are joined by a flexible hinge. An antigen consisting of two hapten molecules (red balls in diagrams) that can cross-link two antigen-binding sites is used to create antigen:antibody complexes, which can be seen in the electron micrograph. Linear, triangular, and square forms are seen, with short projections or spikes. Limited pepsin digestion removes these spikes (not shown in the figure), which therefore correspond to the Fc portion of the antibody; the $F(ab')_2$ pieces remain cross-linked by antigen. The interpretation of some of the complexes is shown in the diagrams. The angle between the arms of the antibody molecules varies. In the triangular forms, this angle is 60°, whereas it is 90° in the square forms, showing that the connections between the arms are flexible. Photograph (×300,000) courtesy of N.M. Green.

and rotation of the V domain relative to the C domain. For example, in the antibody molecule shown in Fig. 4.1a, not only are the two hinge regions clearly bent differently, but the angle between the V and C domains in each of the two Fab arms is also different. This range of motion has led to the junction between the V and C domains being referred to as a 'molecular ball-and-socket joint.' Flexibility at both the hinge and the V–C junction enables the two arms of an antibody molecule to bind to sites some distance apart, such as the repeating sites on bacterial cell-wall polysaccharides. Flexibility at the hinge also enables antibodies to interact with the antibody-binding proteins that mediate immune effector mechanisms.

4-5 The domains of an immunoglobulin molecule have similar structures.

As we saw in Section 4-2, immunoglobulin heavy and light chains are composed of a series of discrete protein domains, all of which have a similar folded structure. Within this basic structure there are distinct differences between V and C domains. The structural similarities and differences can be seen in the diagram of a light chain in Fig. 4.5. Each domain is constructed from two **β sheets**, which are elements of protein structure made up of strands of the polypeptide chain (β strands) packed together; the sheets are linked by a disulfide bridge and together form a roughly barrel-shaped structure, known as a **β barrel**. The distinctive folded structure of the immunoglobulin protein domain is known as the **immunoglobulin fold**.

Both the essential similarity of V and C domains and the critical differences between them are most clearly seen in the bottom panels of Fig. 4.5, in which the cylindrical domains are opened out to reveal how the polypeptide chain folds to create each of the β sheets and how it forms flexible loops as it changes direction. The main difference between the V and C domains is that the V domain is larger, with an extra loop. The flexible loops of the V domains form the antigen-binding site of the immunoglobulin molecule.

Many of the amino acids that are common to the C and V domains lie in the core of the immunoglobulin fold and are essential for its stability. Other proteins with sequences similar to those of immunoglobulins have been shown to form domains of similar structure, which are called **immunoglobulin-like domains** (**Ig-like domains**). These domains are present in many proteins of the immune system, and in proteins involved in cell–cell recognition and adhesion in the nervous system and other tissues. Together with the immunoglobulins and the T-cell receptors, they make up the extensive **immunoglobulin superfamily**.

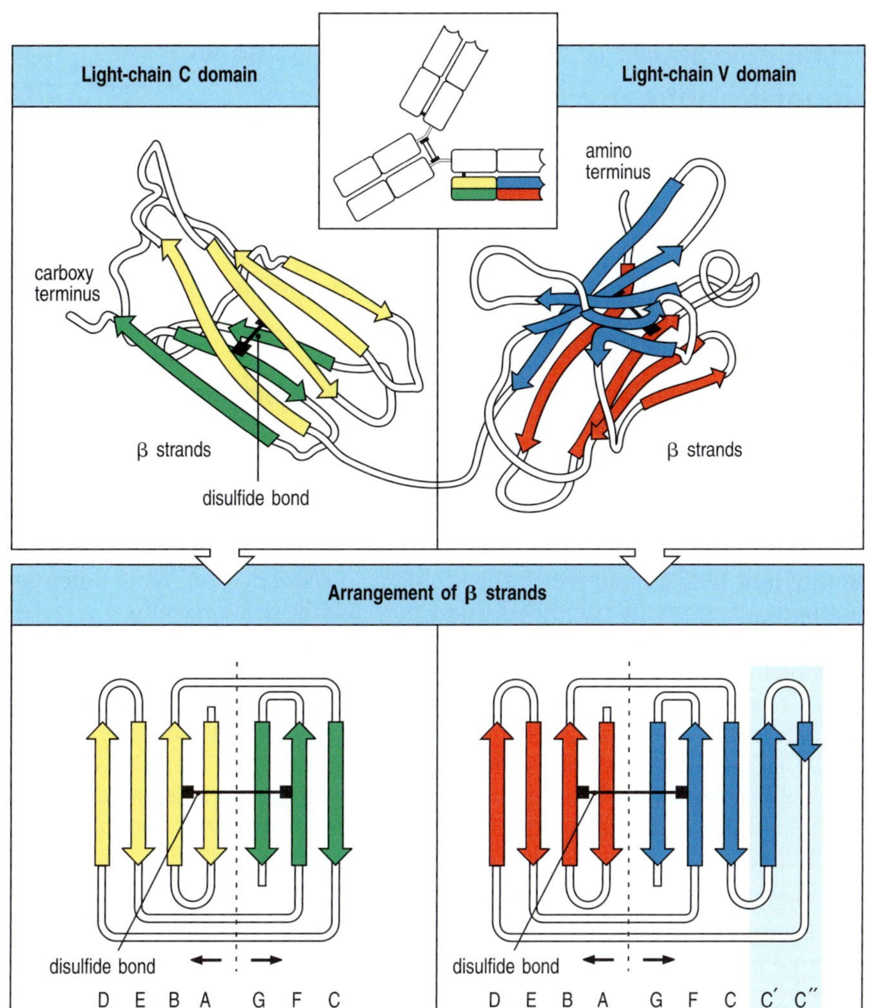

Fig. 4.5 The structure of immunoglobulin constant and variable domains. The upper panels show schematically the folding pattern of the constant (C) and variable (V) domains of an immunoglobulin light chain. Each domain is a barrel-shaped structure in which strands of polypeptide chain (β strands) running in opposite directions (antiparallel) pack together to form two β sheets (shown in yellow and green for the C domain and red and blue for the V domain), which are held together by a disulfide bond. The way in which the polypeptide chain folds to give the final structure can be seen more clearly when the sheets are opened out, as shown in the lower panels. The β strands are lettered sequentially with respect to the order of their occurrence in the amino acid sequence of the domains; the order in each β sheet is characteristic of immunoglobulin domains. The β strands C′ and C″ that are found in the V domains but not in the C domains are indicated by a blue shaded background. The characteristic four-strand plus three-strand (C-region type domain) or four-strand plus five-strand (V-region type domain) arrangements are typical immunoglobulin superfamily domain building blocks, found in a whole range of other proteins as well as antibodies and T-cell receptors.

Summary.

The IgG antibody molecule is made up of four polypeptide chains, comprising two identical light chains and two identical heavy chains, and can be thought of as forming a flexible Y-shaped structure. Each of the four chains has a variable (V) region at its amino terminus, which contributes to the antigen-binding site, and a constant (C) region, which determines the isotype. The isotype of the heavy chain determines the functional properties of the antibody. The light chains are bound to the heavy chains by many noncovalent interactions and by disulfide bonds, and the V regions of the heavy and light chains pair in each arm of the Y to generate two identical antigen-binding sites, which lie at the tips of the arms of the Y. The possession of two antigen-binding sites allows antibody molecules to cross-link antigens and to bind them much more stably and with higher avidity. The trunk of the Y, called the Fc fragment, is composed of the carboxy-terminal domains of the heavy chains. Joining the arms of the Y to the trunk are the flexible hinge regions. The Fc fragment and hinge regions differ in antibodies of different isotypes, thus determining their functional properties. However, the overall organization of the domains is similar in all isotypes.

The interaction of the antibody molecule with specific antigen.

In this part of the chapter we look at the antigen-binding site of an immunoglobulin molecule in more detail. We discuss the different ways in which antigens can bind to antibody, and address the question of how variation in the sequences of the antibody V domains determines the specificity for antigen.

4-6 Localized regions of hypervariable sequence form the antigen-binding site.

The V regions of any given antibody molecule differ from those of every other. Sequence variability is not, however, distributed evenly throughout the V region but is concentrated in certain segments, as is clearly seen in a **variability plot** (Fig. 4.6), in which the amino acid sequences of many different antibody V regions are compared. Three particularly variable segments can be identified in both the V_H and V_L domains. They are designated **hypervariable regions** and are denoted HV1, HV2, and HV3. In the heavy chains they run roughly from residues 30 to 36, 49 to 65, and 95 to 103, respectively, while in the light chains they are located roughly at residues 28 to 35, 49 to 59, and 92 to 103, respectively. The most variable part of the domain is in the HV3 region. The regions between the hypervariable regions, which comprise the rest of the V domain, show less variability and are termed the **framework regions**. There are four such regions in each V domain, designated FR1, FR2, FR3, and FR4.

The framework regions form the β sheets that provide the structural framework of the domain, whereas the hypervariable sequences correspond to three loops at the outer edge of the β barrel, which are juxtaposed in the folded domain (Fig. 4.7). Thus, not only is diversity concentrated in particular parts of the V domain sequence, but it is also localized to a particular region on the surface of the molecule. When the V_H and V_L immunoglobulin domains are paired in the antibody molecule, the three hypervariable loops

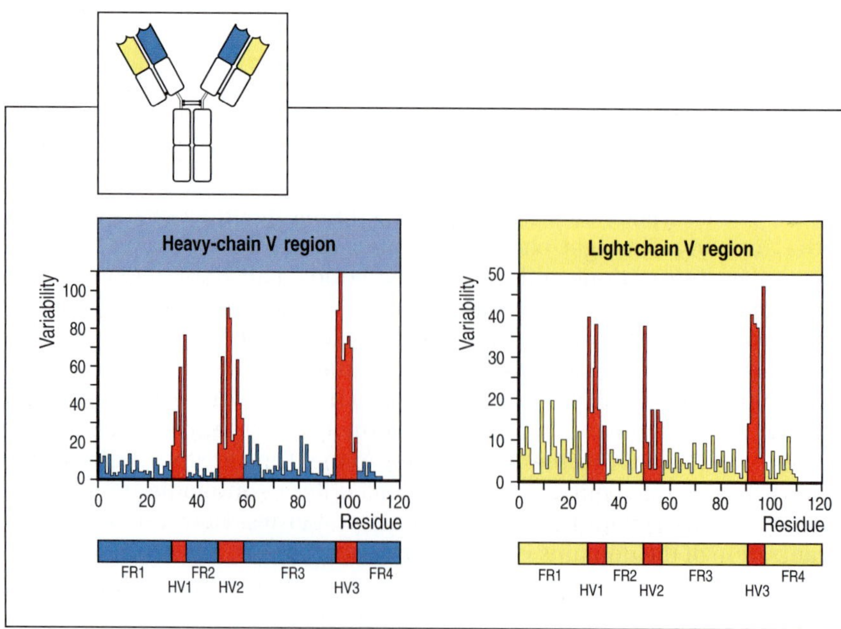

Fig. 4.6 There are discrete regions of hypervariability in V domains. A variability plot derived from comparison of the amino acid sequences of several dozen heavy-chain and light-chain V domains is shown. At each amino acid position the degree of variability is the ratio of the number of different amino acids seen in all of the sequences together to the frequency of the most common amino acid. Three hypervariable regions (HV1, HV2, and HV3) are indicated in red. They are flanked by less variable framework regions (FR1, FR2, FR3, and FR4, shown in blue or yellow).

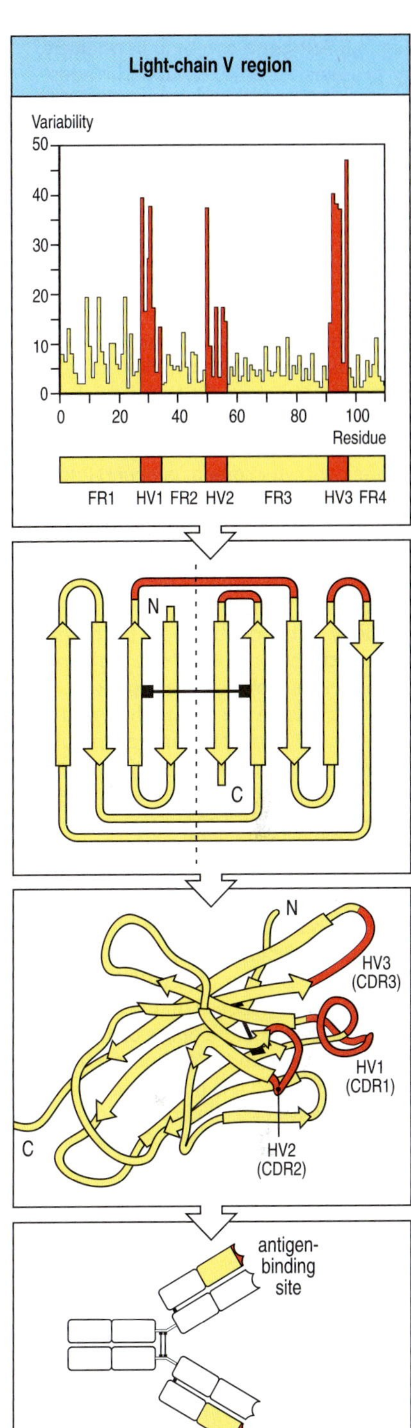

Fig. 4.7 The hypervariable regions lie in discrete loops of the folded structure. When the hypervariable regions are positioned on the structure of a V domain it can be seen that they occur in loops that are brought together in the folded structure. In the antibody molecule, the pairing of a heavy chain and a light chain brings together the hypervariable loops from each chain to create a single hypervariable surface, which forms the antigen-binding site at the tip of each arm. Because they are complementary to the antigen surface, the hypervariable regions are more commonly known as the complementarity-determining regions (CDRs). C, carboxy terminus; N, amino terminus.

from each domain are brought together, creating a single hypervariable site at the tip of each arm of the molecule. This is the **antigen-binding site** or **antibody combining site**, which determines the antigen specificity of the antibody. These six hypervariable loops are more commonly termed the **complementarity-determining regions**, or **CDRs**, because the surface they form is complementary to that of the antigen they bind. There are three CDRs from each of the heavy and light chains, namely CDR1, CDR2, and CDR3. Because CDRs from both V_H and V_L domains contribute to the antigen-binding site, it is the combination of the heavy and the light chain, and not either alone, that determines the final antigen specificity (see Fig. 4.6). Thus, one way in which the immune system is able to generate antibodies of different specificities is by generating different combinations of heavy-chain and light-chain V regions. This is known as **combinatorial diversity**; we will encounter a second form of combinatorial diversity when we consider in Chapter 5 how the genes encoding the heavy-chain and light-chain V regions are created from smaller segments of DNA.

4-7 Antibodies bind antigens via contacts with amino acids in CDRs, but the details of binding depend upon the size and shape of the antigen.

In early investigations of antigen binding to antibodies, the only available sources of large quantities of a single type of antibody molecule were tumors of antibody-secreting cells. The antigen specificities of these antibodies were unknown, so many compounds had to be screened to identify ligands that could be used to study antigen binding. In general, the substances found to bind to these antibodies were haptens (see Section 4-4) such as phosphocholine or vitamin K_1. Structural analysis of complexes of antibodies with their hapten ligands provided the first direct evidence that the hypervariable regions form the antigen-binding site, and demonstrated the structural basis of specificity for the hapten. Subsequently, with the discovery of methods of generating **monoclonal antibodies** (see Appendix I, Section A-12), it became possible to make large amounts of pure antibody specific for a given antigen. This has provided a more general picture of how antibodies interact with their antigens, confirming and extending the view of antibody–antigen interactions derived from the study of haptens.

The surface of the antibody molecule formed by the juxtaposition of the CDRs of the heavy and light chains is the site to which an antigen binds. Clearly, as the amino acid sequences of the CDRs are different in different antibodies, so too are the shapes and properties of the surfaces created by these CDRs. As a general principle, antibodies bind ligands whose surfaces are complementary to that of the antigen-binding site. A small antigen, such as a hapten or a short peptide, generally binds in a pocket or groove lying between the heavy-chain and light-chain V domains (Fig. 4.8a and b). Some antigens, such as proteins, can be the same size as, or larger than, the antibody itself. In these cases, the interface between antigen and antibody is often an extended surface that involves all the CDRs and, in some cases, part of the framework region as well (see Fig. 4.8c). This surface need not be concave but can be flat, undulating, or even convex. In some cases, antibody molecules with elongated CDR3 loops can protrude a 'finger' into recesses in the surface of the antigen, as shown in

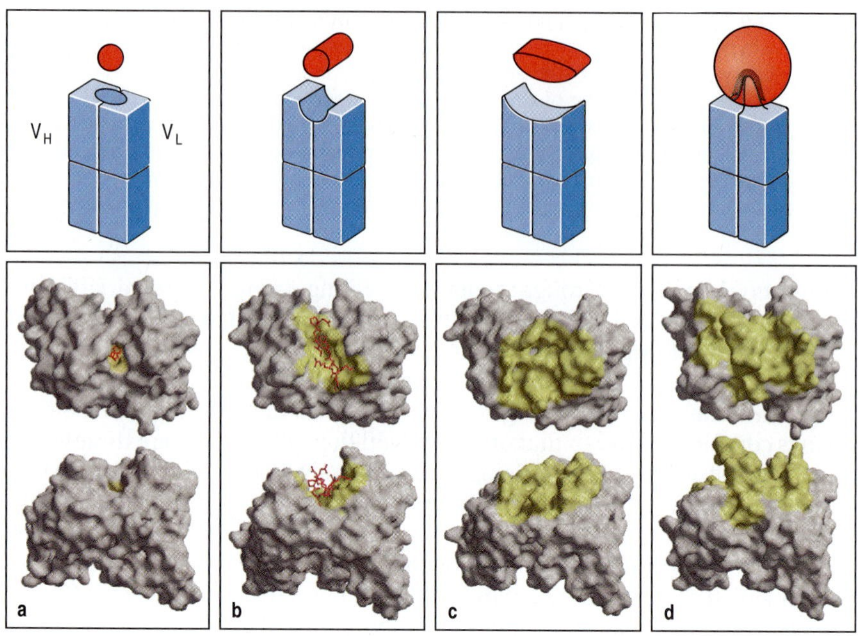

Fig. 4.8 Antigens can bind in pockets, or grooves, or on extended surfaces in the binding sites of antibodies. The panels in the top row show schematic representations of the different types of binding sites in a Fab fragment of an antibody: first panel, pocket; second panel, groove; third panel, extended surface; and fourth panel, protruding surface. Below are examples of each type. Panel a: the top image shows the molecular surface of the interaction of a small hapten with the complementarity-determining regions (CDRs) of a Fab fragment as viewed looking into the antigen-binding site. The ferrocene hapten, shown in green, is bound into the antigen-binding pocket (yellow). In the bottom image (and in those of panels b, c, and d) the molecule has been rotated by about 90° to give a side-on view of the binding site. Panel b: in a complex of an antibody with a peptide from the human immunodeficiency virus (HIV), the peptide (green) binds along a groove (yellow) formed between the heavy-chain and light-chain V domains. Panel c: complex between hen egg-white lysozyme and the Fab fragment of its corresponding antibody (HyHel5). The surface on the antibody that comes into contact with the lysozyme is colored yellow. All six CDRs of the antigen-binding site are involved in the binding. Panel d: an antibody molecule against the HIV gp120 antigen has an elongated CDR3 loop that protrudes into a recess on the side of the antigen. The structure of the complex between this antibody and gp120 has been solved, and it is now known that only the heavy chain interacts with gp120. Structures courtesy of R.L. Stanfield and I.A. Wilson.

Fig. 4.8d, where an antibody binding to the HIV gp120 antigen projects a long loop against its target.

4-8 Antibodies bind to conformational shapes on the surfaces of antigens.

The biological function of antibodies is to bind to pathogens and their products, and to facilitate their removal from the body. An antibody generally recognizes only a small region on the surface of a large molecule such as a polysaccharide or protein. The structure recognized by an antibody is called an **antigenic determinant** or **epitope**. Some of the most important pathogens have polysaccharide coats, and antibodies that recognize epitopes formed by the sugar subunits of these molecules are essential in providing immune protection against such pathogens. In many cases, however, the antigens that provoke an immune response are proteins. For example, protective antibodies against viruses recognize viral coat proteins. In all such cases, the structures recognized by the antibody are located on the surface of the protein. Such sites are likely to be composed of amino acids from different parts of the polypeptide chain that have been brought together by protein folding. Antigenic determinants of this kind are known as **conformational** or **discontinuous epitopes** because the structure recognized is composed of segments of the protein that are discontinuous in the amino acid sequence of the antigen but are brought together in the three-dimensional structure. In contrast, an epitope composed of a single segment of polypeptide chain is termed a **continuous** or **linear epitope**. Although most antibodies raised against intact, fully folded proteins recognize discontinuous epitopes, some will bind to peptide fragments of the protein. Conversely, antibodies raised against peptides of a protein or against synthetic peptides corresponding to part of its sequence are occasionally found to bind to the natural folded protein. This makes it possible, in some cases, to use synthetic peptides in vaccines that aim at raising antibodies against a pathogen protein.

4-9 Antigen–antibody interactions involve a variety of forces.

The interaction between an antibody and its antigen can be disrupted by high salt concentrations, by extremes of pH, by detergents, and sometimes by competition with high concentrations of the pure epitope itself. The binding is therefore a reversible noncovalent interaction. The forces, or bonds, involved in these noncovalent interactions are outlined in Fig. 4.9.

Electrostatic interactions occur between charged amino acid side chains, as in salt bridges. Most antibody–antigen interactions involve at least one electrostatic interaction. Interactions also occur between electric dipoles, as in hydrogen bonds, or can involve short-range van der Waals forces. High salt concentrations and extremes of pH disrupt antigen–antibody binding by weakening electrostatic interactions and/or hydrogen bonds. This principle is employed in the purification of antigens by using affinity columns of immobilized antibodies, and vice versa for antibody purification (see Appendix I, Section A-5). Hydrophobic interactions occur when two hydrophobic surfaces come together to exclude water. The strength of a hydrophobic interaction is proportional to the surface area that is hidden from water, and for some antigens, hydrophobic interactions probably account for most of the binding energy. In some cases, water molecules are trapped in pockets in the interface between antigen and antibody. These trapped water molecules, especially those between polar amino acid residues, may also contribute to binding and hence to the specificity of the antibody.

The contribution of each of these forces to the overall interaction depends on the particular antibody and antigen involved. A striking difference between antibody interactions with protein antigens and most other natural protein–protein interactions is that antibodies often have many aromatic amino acids in their antigen-binding sites. These amino acids participate mainly in van der Waals and hydrophobic interactions, and sometimes in hydrogen bonds. Tyrosine, for example, can take part in both hydrogen bonding and hydrophobic interactions; it is therefore particularly suitable for providing diversity in antigen recognition and is over-represented in antigen-binding sites. In general, the hydrophobic and van der Waals forces operate over very short ranges and serve to pull together two surfaces that are complementary in shape: hills on one surface must fit into valleys on the other for good binding to occur. In contrast, electrostatic interactions between charged side chains, and hydrogen bonds bridging oxygen and/or nitrogen atoms, accommodate specific features or reactive groups while strengthening the interaction overall.

An example of a reaction involving a specific amino acid in the antigen can be seen in the complex of hen egg-white lysozyme with the antibody D1.3

Noncovalent forces	Origin	
Electrostatic forces	Attraction between opposite charges	$-NH_3^{+}\ \ ^{-}OOC-$
Hydrogen bonds	Hydrogen shared between electronegative atoms (N, O)	$>N^{\delta-}-H^{\delta+}--O^{\delta-}=C<$
Van der Waals forces	Fluctuations in electron clouds around molecules polarize neighboring atoms oppositely	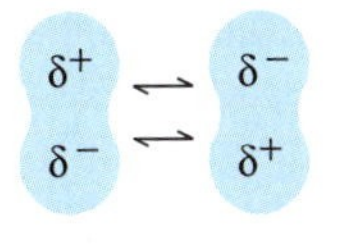 $\delta+ \rightarrow \delta-$; $\delta- \rightarrow \delta+$
Hydrophobic forces	Hydrophobic groups interact unfavorably with water and tend to pack together to exclude water molecules. The attraction also involves van der Waals forces	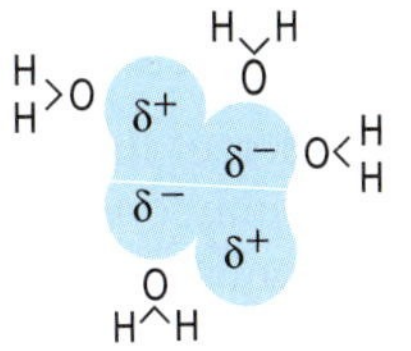 H_2O, $\delta+$, $\delta-$, $\delta-$, $\delta+$, H_2O

Fig. 4.9 The noncovalent forces that hold together the antigen:antibody complex. Partial charges found in electric dipoles are shown as δ^+ or δ^-. Electrostatic forces diminish as the inverse square of the distance separating the charges, whereas van der Waals forces, which are more numerous in most antigen–antibody contacts, fall off as the sixth power of the separation and therefore operate only over very short ranges. Covalent bonds never occur between antigens and naturally produced antibodies.

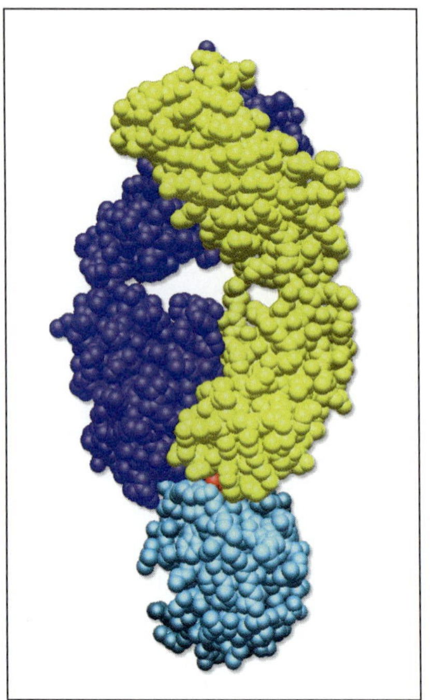

Fig. 4.10 The complex of lysozyme with the antibody D1.3. The interaction of the Fab fragment of D1.3 with hen egg-white lysozyme is shown, with the lysozyme in blue, the heavy chain in purple and the light chain in green. A glutamine residue of lysozyme, shown in red, protrudes between the two V domains of the antigen-binding site and makes hydrogen bonds that are important to the antigen–antibody binding. Courtesy of R.J. Poljak.

(Fig. 4.10), where strong hydrogen bonds are formed between the antibody and a particular glutamine in the lysozyme molecule that protrudes between the V_H and V_L domains. Lysozymes from partridge and turkey have another amino acid in place of the glutamine and do not bind to this antibody. In the high-affinity complex of hen egg-white lysozyme with another antibody, HyHel5 (see Fig. 4.8c), two salt bridges between two basic arginines on the surface of the lysozyme interact with two glutamic acids, one each from the V_H CDR1 and CDR2 loops. Lysozymes that lack one of the two arginine residues show a 1000-fold decrease in affinity for HyHel5. Overall surface complementarity must have an important role in antigen–antibody interactions, but in most antibodies that have been studied at this level of detail only a few residues make a major contribution to the binding energy and hence to the final specificity of the antibody. Although many antibodies naturally bind their ligands with high affinity (in the nanomolar range), genetic engineering by site-directed mutagenesis can tailor an antibody to bind even more strongly to its epitope.

Summary.

X-ray crystallographic analyses of antigen:antibody complexes have shown that the hypervariable loops (complementarity-determining regions, CDRs) of immunoglobulin V regions determine the binding specificity of an antibody. Contact between an antibody molecule and a protein antigen usually occurs over a broad area of the antibody surface that is complementary to the surface recognized on the antigen. Electrostatic interactions, hydrogen bonds, van der Waals forces, and hydrophobic interactions can all contribute to binding. Depending on the size of the antigen, amino acid side chains in most or all of the CDRs make contact with antigen and determine both the specificity and the affinity of the interaction. Other parts of the V region normally play little part in the direct contact with the antigen, but they provide a stable structural framework for the CDRs and help to determine their position and conformation. Antibodies raised against intact proteins usually bind to the surface of the protein and make contact with residues that are discontinuous in the primary structure of the molecule; they may, however, occasionally bind peptide fragments of the protein, and antibodies raised against peptides derived from a protein can sometimes be used to detect the native protein molecule. Peptides binding to antibodies usually bind in a cleft or pocket between the V regions of the heavy and light chains, where they make specific contact with some, but not necessarily all, of the CDRs. This is also the usual mode of binding for carbohydrate antigens and small molecules such as haptens.

Antigen recognition by T cells.

In contrast to the immunoglobulins, which interact with pathogens and their toxic products in the extracellular spaces of the body, T cells recognize foreign antigens only when they are displayed on the surfaces of the body's own cells. These antigens can derive from pathogens such as viruses or intracellular bacteria, which replicate within cells, or from pathogens or their products that have been internalized by endocytosis from the extracellular fluid.

T cells detect the presence of an intracellular pathogen because the infected cells display on their surface peptide fragments of the pathogen's proteins. These foreign peptides are delivered to the cell surface by specialized host-cell glycoproteins—the MHC molecules. These are encoded in a large cluster of genes that were first identified by their powerful effects on the immune response to transplanted tissues. For that reason, the gene complex was called the major histocompatibility complex (MHC) and the peptide-binding glycoproteins are known as MHC molecules. The recognition of antigen as a small peptide fragment bound to an MHC molecule and displayed at the cell surface is one of the most distinctive features of T cells, and will be the focus of this part of the chapter. How the peptide fragments of antigen are generated and become associated with MHC molecules will be described in Chapter 6.

We describe here the structure and properties of the T-cell receptor (TCR). As might be expected from their function as highly variable antigen-recognition structures, the genes for T-cell receptors are closely related to those for immunoglobulins. There are, however, important differences between T-cell receptors and immunoglobulins that reflect the special features of antigen recognition by T cells.

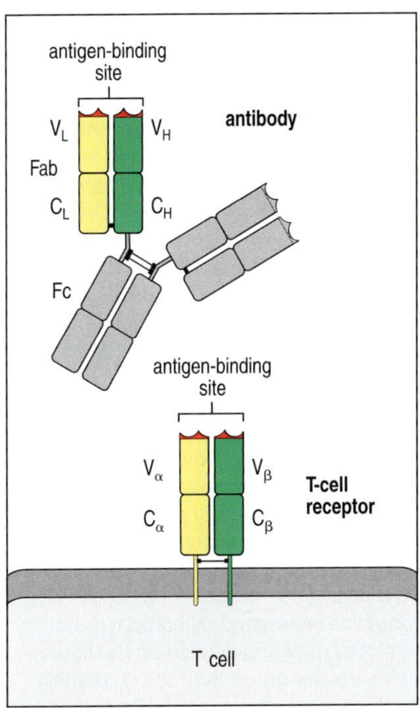

Fig. 4.11 The T-cell receptor resembles a membrane-bound Fab fragment. The Fab fragment of an antibody molecule is a disulfide-linked heterodimer, each chain of which contains one immunoglobulin C domain and one V domain; the juxtaposition of the V domains forms the antigen-binding site (see Section 4-6). The T-cell receptor is also a disulfide-linked heterodimer, with each chain containing an immunoglobulin C-like domain and an immunoglobulin V-like domain. As in the Fab fragment, the juxtaposition of the V domains forms the site for antigen recognition.

4-10 The T-cell receptor is very similar to a Fab fragment of immunoglobulin.

T-cell receptors were first identified by using monoclonal antibodies that bound to a single cloned T-cell line: such antibodies either specifically inhibit antigen recognition by the clone or specifically activate it by mimicking the antigen (see Appendix I, Section A-19). These **clonotypic** antibodies were then used to show that each T cell bears about 30,000 identical antigen receptors on its surface, each receptor consisting of two different polypeptide chains, termed the **T-cell receptor α** (**TCRα**) and **β** (**TCRβ**) chains, linked by a disulfide bond. The **α:β heterodimers** are very similar in structure to the Fab fragment of an immunoglobulin molecule (Fig. 4.11), and they account for antigen recognition by most T cells. A minority of T cells bear an alternative, but structurally similar, receptor made up of a different pair of polypeptide chains designated γ and δ. The **γ:δ T-cell receptors** seem to have different antigen-recognition properties from the **α:β T-cell receptors**, and the function of γ:δ T cells in immune responses is not yet entirely clear (see Section 3-24). In the rest of this chapter and elsewhere in the book we use the term T-cell receptor to mean the α:β receptor, except where specified otherwise. Both types of T-cell receptors differ from the membrane-bound immunoglobulin that serves as the B-cell receptor in two main ways. A T-cell receptor has only one antigen-binding site, whereas a B-cell receptor has two, and T-cell receptors are never secreted, whereas immunoglobulin can be secreted as antibody.

Further insights into the structure and function of the α:β T-cell receptor came from studies of cloned cDNA encoding the receptor chains. The amino acid sequences predicted from the cDNA showed that both chains of the T-cell receptor have an amino-terminal variable (V) region with homology to an immunoglobulin V domain, a constant (C) region with homology to an immunoglobulin C domain, and a short stalk segment containing a cysteine residue that forms the interchain disulfide bond (Fig. 4.12). Each chain spans the lipid bilayer by a hydrophobic transmembrane domain, and ends in a short cytoplasmic tail. These close similarities of T-cell receptor chains to

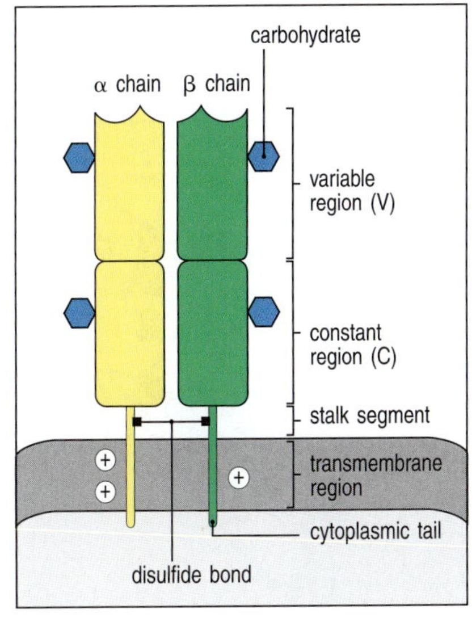

Fig. 4.12 Structure of the T-cell receptor. The T-cell receptor heterodimer is composed of two transmembrane glycoprotein chains, α and β. The extracellular portion of each chain consists of two domains, resembling immunoglobulin V and C domains, respectively. Both chains have carbohydrate side chains attached to each domain. A short stalk segment, analogous to an immunoglobulin hinge region, connects the Ig-like domains to the membrane and contains the cysteine residue that forms the interchain disulfide bond. The transmembrane helices of both chains are unusual in containing positively charged (basic) residues within the hydrophobic transmembrane segment. The α chain carries two such residues; the β chain has one.

the heavy and light immunoglobulin chains first enabled prediction of the structural resemblance of the T-cell receptor heterodimer to a Fab fragment of immunoglobulin.

The three-dimensional structure of the T-cell receptor has since been determined by X-ray crystallography, and the two structures are indeed similar: the T-cell receptor chains fold in much the same way as those of a Fab fragment (Fig. 4.13a), although the final structure appears a little shorter and wider. There are, however, some distinct structural differences between T-cell receptors and Fab fragments. The most striking is in the C_α domain, where the fold is unlike that of any other Ig-like domain. The half of the domain that is juxtaposed with the C_β domain forms a β sheet similar to that found in other Ig-like domains, but the other half of the domain is formed of loosely packed strands and a short segment of α helix (Fig. 4.13b). In a C_α domain the intramolecular disulfide bond, which in Ig-like domains normally joins two β strands, joins a β strand to this segment of α helix.

There are also differences in the way in which the domains interact. The interface between the V and C domains of both T-cell receptor chains is more extensive than in most antibodies. The interaction between the C_α and C_β domains is distinctive as it might be assisted by carbohydrate, with a sugar group from the C_α domain making a number of hydrogen bonds to the C_β domain (see Fig. 4.13b). Finally, a comparison of the variable binding sites shows that, although the CDR loops align fairly closely with those of antibody molecules, there is some relative displacement (see Fig. 4.13c). This is particularly marked in the V_α CDR2 loop, which is oriented at roughly right angles to the equivalent loop in antibody V domains, as a result of a shift in the β strand that anchors one end of the loop from one face of the domain to the other. A strand displacement also causes a change in the orientation of the V_β CDR2 loop in some V_β domains whose structures are known. As relatively few crystallographic structures have been solved to this level of resolution, it remains to be seen to what degree all T-cell receptors share these features, and whether there are more differences to be discovered.

4-11 A T-cell receptor recognizes antigen in the form of a complex of a foreign peptide bound to an MHC molecule.

Antigen recognition by T-cell receptors clearly differs from recognition by B-cell receptors and antibodies. The immunoglobulin on B cells binds directly to the intact antigen and, as discussed in Section 4-8, antibodies typically bind to the surface of protein antigens, contacting amino acids that are discontinuous in the primary structure but are brought together in the folded protein. T cells, in contrast, respond to short continuous amino acid

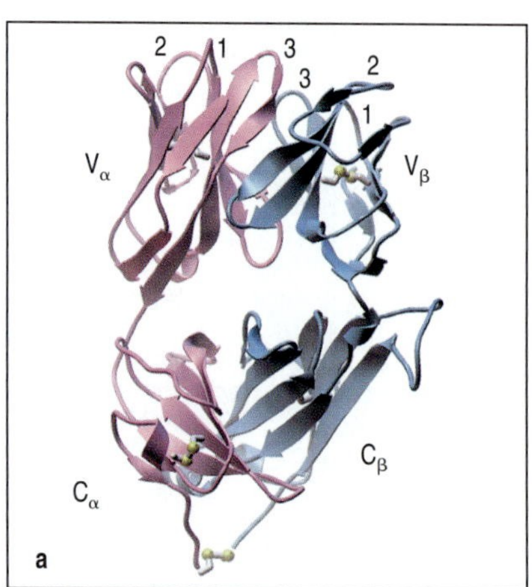

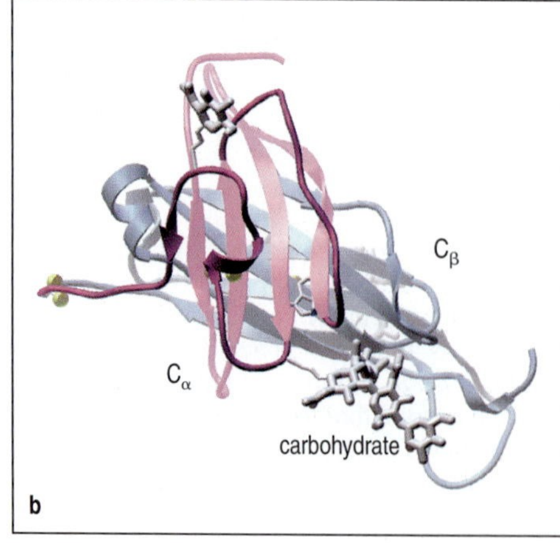

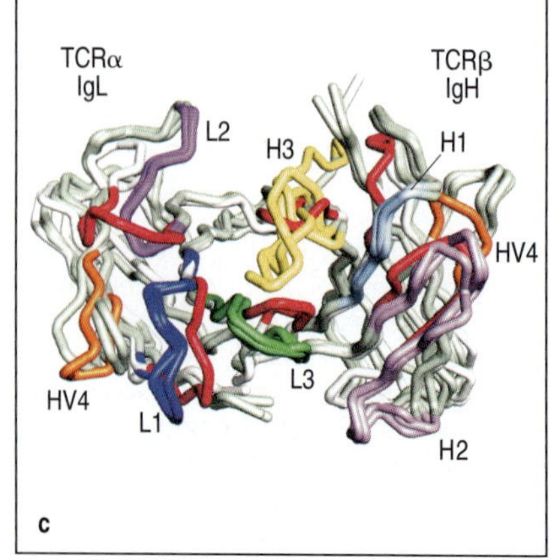

Fig. 4.13 The crystal structure of an α:β T-cell receptor resolved at 2.5 Å. In panels a and b the α chain is shown in pink and the β chain in blue. Disulfide bonds are shown in green. In panel a, the T-cell receptor is viewed from the side as it would sit on a cell surface, with the CDR loops that form the antigen-binding site (labeled 1, 2, and 3) arrayed across its relatively flat top. In panel b, the C_α and C_β domains are shown. The C_α domain does not fold into a typical Ig-like domain; the face of the domain away from the C_β domain is mainly composed of irregular strands of polypeptide rather than β sheet. The intramolecular disulfide bond joins a β strand to this segment of α helix. The interaction between the C_α and C_β domains is assisted by carbohydrate (colored gray and labeled on the figure), with a sugar group from the C_α domain making hydrogen bonds to the C_β domain. In panel c, the T-cell receptor is shown aligned with the antigen-binding sites from three different antibodies. This view is looking down into the binding site. The V_α domain of the T-cell receptor is aligned with the V_L domains of the antigen-binding sites of the antibodies, and the V_β domain is aligned with the V_H domains. The CDRs of the T-cell receptor and immunoglobulin molecules are colored, with CDRs 1, 2, and 3 of the TCR shown in red and the HV4 loop in orange. For the immunoglobulin V domains, the CDR1 loops of the heavy chain (H1) and light chain (L1) are shown in light and dark blue, respectively, and the CDR2 loops (H2, L2) in light and dark purple, respectively. The heavy-chain CDR3 loops (H3) are in yellow; the light-chain CDR3s (L3) are in bright green. The HV4 loops of the TCR (orange) have no hypervariable counterparts in immunoglobulins. Model structures courtesy of I.A. Wilson.

sequences. These sequences are often buried within the native structure of the protein and thus cannot be recognized directly by T-cell receptors unless the protein is unfolded and processed into peptide fragments (Fig. 4.14). We shall see in Chapter 6 how this occurs.

The nature of the antigen recognized by T cells became clear with the realization that the peptides that stimulate T cells are recognized only when bound to an MHC molecule. The ligand recognized by the T cell is thus a complex of peptide and MHC molecule. The evidence for involvement of the MHC in T-cell recognition of antigen was at first indirect, but it has been proved conclusively by stimulating T cells with purified peptide:MHC complexes. The T-cell receptor interacts with this ligand by making contacts with both the MHC molecule and the antigen peptide.

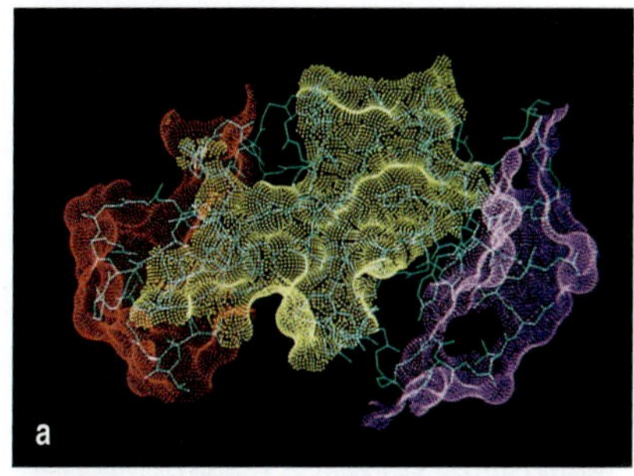

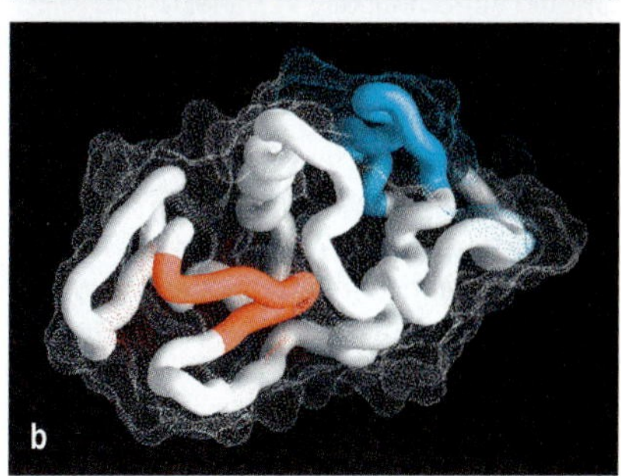

Fig. 4.14 Differences in the recognition of hen egg-white lysozyme by immunoglobulins and T-cell receptors. Antibodies can be shown by X-ray crystallography to bind epitopes on the surface of proteins, as shown in panel a, where the epitopes for three antibodies are shown in different colors on the surface of hen egg-white lysozyme (see also Fig. 4.10). In contrast, the epitopes recognized by T-cell receptors need not lie on the surface of the molecule, because the T-cell receptor recognizes not the antigenic protein itself but a peptide fragment of the protein. The peptides corresponding to two T-cell epitopes of lysozyme are shown in panel b. One epitope, shown in blue, lies on the surface of the protein but a second, shown in red, lies mostly within the core and is inaccessible in the folded protein. For this residue to be accessible to the T-cell receptor, the protein must be unfolded and processed. Panel a courtesy of S. Sheriff.

4-12 There are two classes of MHC molecules with distinct subunit compositions but similar three-dimensional structures.

There are two classes of MHC molecules—**MHC class I** and **MHC class II**—which differ in both their structure and in their expression pattern in the tissues of the body. As shown in Figs 4.15 and 4.16, MHC class I and MHC class II molecules are closely related in overall structure but differ in their subunit compositions. In both classes, the two paired protein domains nearest to the membrane resemble immunoglobulin domains, whereas the two domains furthest away from the membrane fold together to create a long cleft, or groove, which is the site at which a peptide binds. Purified peptide:MHC class I and peptide:MHC class II complexes have been characterized structurally, allowing us to describe in detail both the MHC molecules themselves and the way in which they bind peptides.

MHC class I molecules (see Fig. 4.15) consist of two polypeptide chains. One chain, the α chain, is encoded in the MHC (on chromosome 6 in humans) and is noncovalently associated with a smaller chain, **β_2-microglobulin**, which is not polymorphic and is encoded on a different chromosome—chromosome 15 in humans. Only the class I α chain spans the membrane. The complete molecule has four domains, three formed from the MHC-encoded α chain, and one contributed by β_2-microglobulin. The α_3 domain and β_2-microglobulin closely resemble Ig-like domains in their folded structure. The folded α_1 and α_2 domains form the walls of a cleft on the surface of the molecule; this is where the peptide binds and is known as the **peptide-binding cleft** or **peptide-binding groove**. The MHC molecules are highly polymorphic and the major differences between the different forms are located in the peptide-binding cleft, influencing which peptides will bind and thus the specificity of the dual antigen presented to T cells.

An MHC class II molecule consists of a noncovalent complex of two chains, α and β, both of which span the membrane (see Fig. 4.16). The MHC class II α chain is a different protein from the class I α chain. The MHC class II α and β chains are both encoded within the MHC. The crystallographic structure of the MHC class II molecule shows that it is folded very much like the MHC class I molecule, but the peptide-binding cleft is formed by two domains from different chains, the α_1 and β_1 domains. The major differences lie at the ends of the peptide-binding cleft, which are more open in MHC class II molecules than in MHC class I molecules. Consequently, the ends of a peptide bound to an MHC class I molecule are substantially buried within the molecule, whereas the ends of peptides bound to MHC class II molecules are not. In both MHC class I and class II molecules, bound peptides are sandwiched between the two α-helical segments of the MHC molecule (Fig. 4.17). The T-cell receptor interacts with this compound ligand, making contacts with both the MHC molecule and the peptide antigen. The sites of major polymorphism in MHC class II molecules are again located in the peptide-binding cleft.

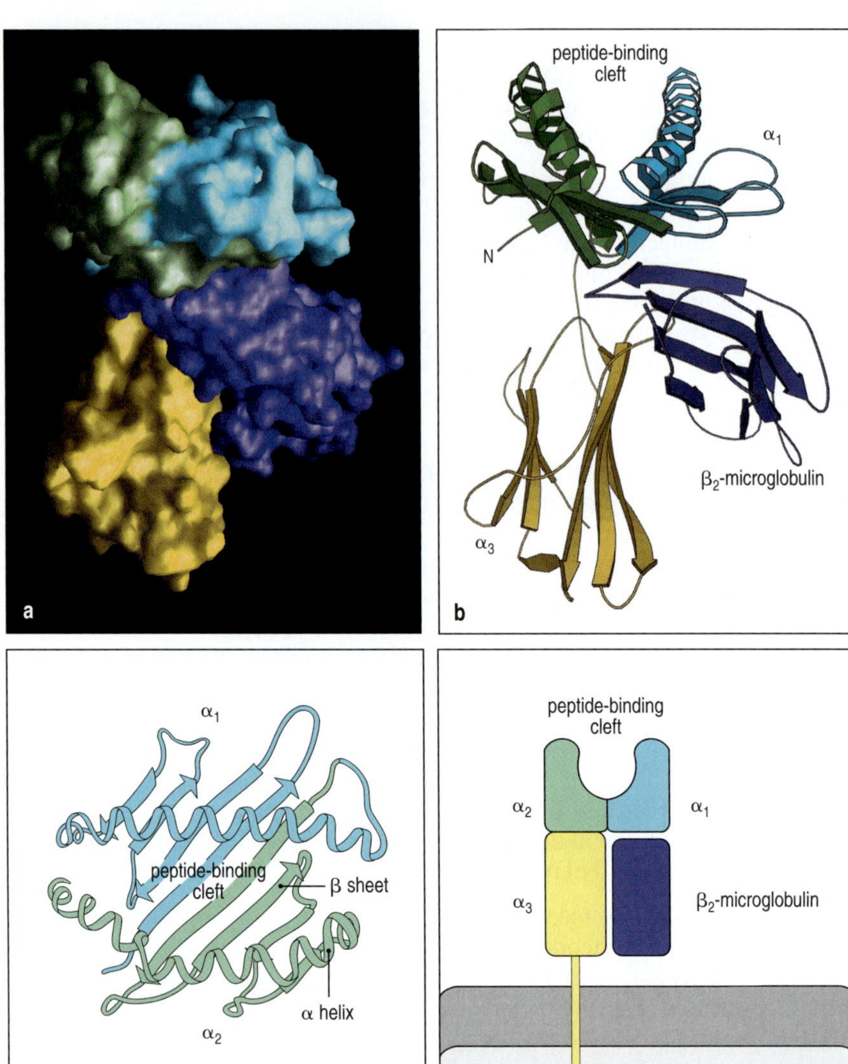

Fig. 4.15 The structure of an MHC class I molecule determined by X-ray crystallography. Panel a shows a computer graphic representation of a human MHC class I molecule, HLA-A2, which has been cleaved from the cell surface by the enzyme papain. The surface of the molecule is shown, colored according to the domains shown in panels b–d and described below. Panels b and c show a ribbon diagram of that structure. Shown schematically in panel d, the MHC class I molecule is a heterodimer of a membrane-spanning α chain (molecular weight 43 kDa) bound noncovalently to β_2-microglobulin (12 kDa), which does not span the membrane. The α chain folds into three domains: α_1, α_2, and α_3. The α_3 domain and β_2-microglobulin show similarities in amino acid sequence to immunoglobulin C domains and have similar folded structures, whereas the α_1 and α_2 domains fold together into a single structure consisting of two separated α helices lying on a sheet of eight antiparallel β strands. The folding of the α_1 and α_2 domains creates a long cleft or groove, which is the site at which peptide antigens bind to the MHC molecules. The transmembrane region and the short stretch of peptide that connects the external domains to the cell surface are not seen in panels a and b because they have been removed by the digestion with papain. As can be seen in panel c, looking down on the molecule from above, the sides of the cleft are formed from the inner faces of the two α helices; the β-pleated sheet formed by the pairing of the α_1 and α_2 domains creates the floor of the cleft.

4-13 Peptides are stably bound to MHC molecules, and also serve to stabilize the MHC molecule on the cell surface.

An individual can be infected by a wide variety of pathogens, whose proteins will not generally have peptide sequences in common. If T cells are to be alerted to all possible infections, the MHC molecules on each cell (both class I and class II) must be able to bind stably to many different peptides. This behavior is quite distinct from that of other peptide-binding receptors, such as those for peptide hormones, which usually bind only a single type of peptide. The crystal structures of peptide:MHC complexes have helped to show how a single binding site can bind peptides with high affinity while retaining the ability to bind a wide variety of different peptides.

An important feature of the binding of peptides to MHC molecules is that the peptide is bound as an integral part of the MHC molecule's structure, and MHC molecules are unstable when peptides are not bound. Stable peptide binding is important, because otherwise peptide exchanges occurring at the cell surface would prevent peptide:MHC complexes from being reliable indicators of infection or of uptake of a specific antigen. When MHC molecules are purified from cells, their stably bound peptides co-purify with them, and this has enabled the peptides bound by particular MHC molecules to be

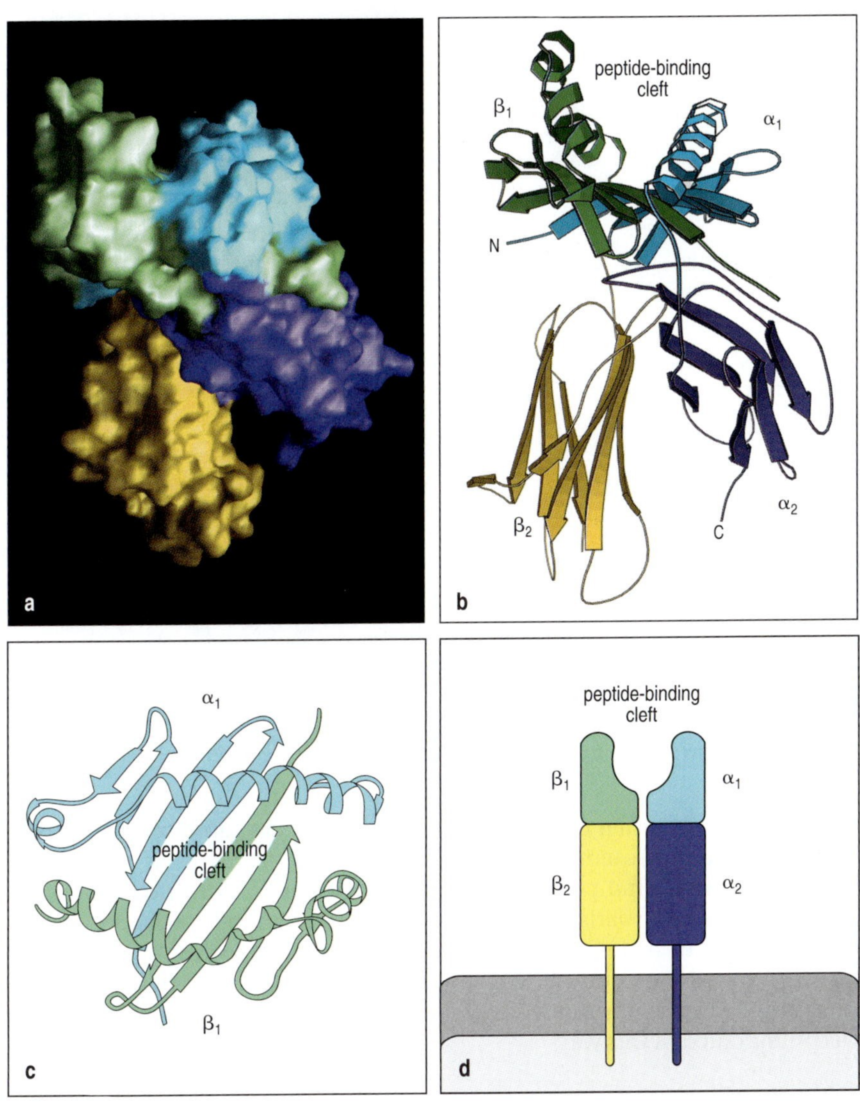

Fig. 4.16 MHC class II molecules resemble MHC class I molecules in overall structure. The MHC class II molecule is composed of two transmembrane glycoprotein chains, α (34 kDa) and β (29 kDa), as shown schematically in panel d. Each chain has two domains, and the two chains together form a compact four-domain structure similar to that of the MHC class I molecule (compare with panel d of Fig. 4.15). Panel a shows a computer graphic representation of the surface of the MHC class II molecule, in this case the human protein HLA-DR1, and panel b shows the equivalent ribbon diagram. N, amino terminus; C, carboxy terminus. The α_2 and β_2 domains, like the α_3 and β_2-microglobulin domains of the MHC class I molecule, have amino acid sequence and structural similarities to immunoglobulin C domains; in the MHC class II molecule the two domains forming the peptide-binding cleft are contributed by different chains and are therefore not joined by a covalent bond (see panels c and d). Another important difference, not apparent in this diagram, is that the peptide-binding groove of the MHC class II molecule is open at both ends.

analyzed. Peptides are released from the MHC molecules by denaturing the complex in acid, and they are then purified and sequenced. Pure synthetic peptides can also be incorporated into empty MHC molecules and the structure of the complex determined, revealing details of the contacts between the MHC molecule and the peptide. From such studies a detailed picture of the binding interactions has been built up. We first discuss the peptide-binding properties of MHC class I molecules.

4-14 MHC class I molecules bind short peptides of 8–10 amino acids by both ends.

Binding of a peptide to an MHC class I molecule is stabilized at both ends of the peptide-binding cleft by contacts between atoms in the free amino and carboxy termini of the peptide and invariant sites that are found at each end of the cleft in all MHC class I molecules (Fig. 4.18). These are thought to be the main stabilizing contacts for peptide:MHC class I complexes, because synthetic peptide analogs lacking terminal amino and carboxyl groups fail to bind stably to MHC class I molecules. Other residues in the peptide serve as additional anchors. Peptides that bind to MHC class I molecules are usually 8–10 amino acids long. Longer peptides are thought to bind, however, particularly

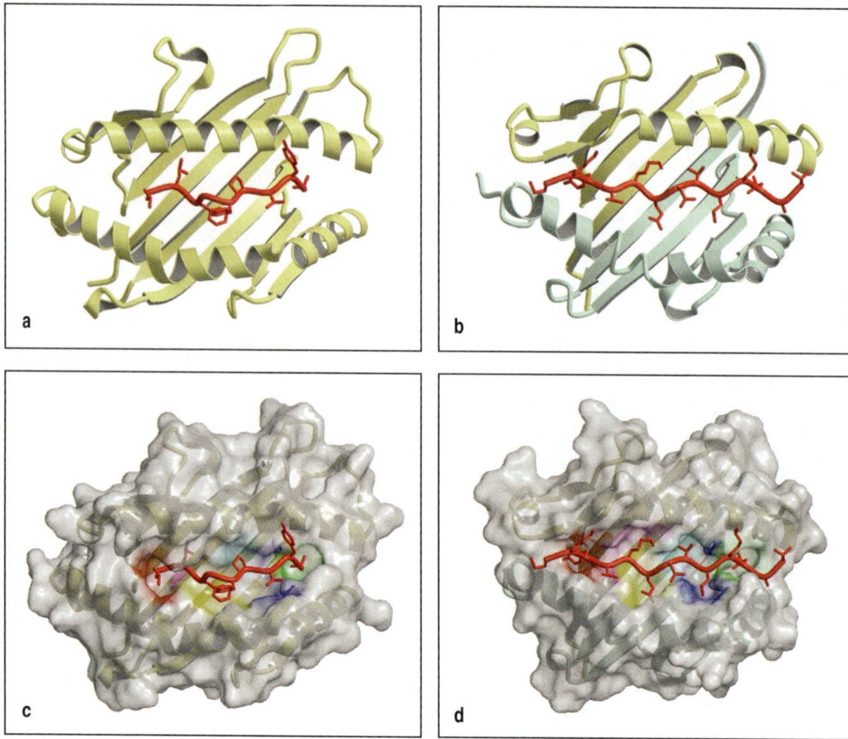

Fig. 4.17 MHC molecules bind peptides tightly within the cleft. When MHC molecules are crystallized with a single synthetic peptide antigen, the details of peptide binding are revealed. In MHC class I molecules (panels a and c) the peptide is bound in an elongated conformation with both ends tightly bound at either end of the cleft. In MHC class II molecules (panels b and d) the peptide is also bound in an elongated conformation but the ends of the peptide are not tightly bound and the peptide extends beyond the cleft. The upper surface of the peptide:MHC complex is recognized by T cells, and is composed of residues of the MHC molecule and the peptide. In representations c and d, the electrostatic potential of the MHC molecule surface is shown, with blue areas indicating a positive potential and red a negative potential. Structures courtesy of R.L. Stanfield and I.A. Wilson.

if they can bind at their carboxy terminus, but they are subsequently cleaved by exopeptidases present in the endoplasmic reticulum, which is where MHC class I molecules bind peptides. The peptide lies in an elongated conformation along the cleft; variations in peptide length seem to be accommodated, in most cases, by a kinking in the peptide backbone. However, two examples of MHC class I molecules in which the peptide is able to extend out of the cleft at the carboxy terminus suggest that some length variation can also be accommodated in this way.

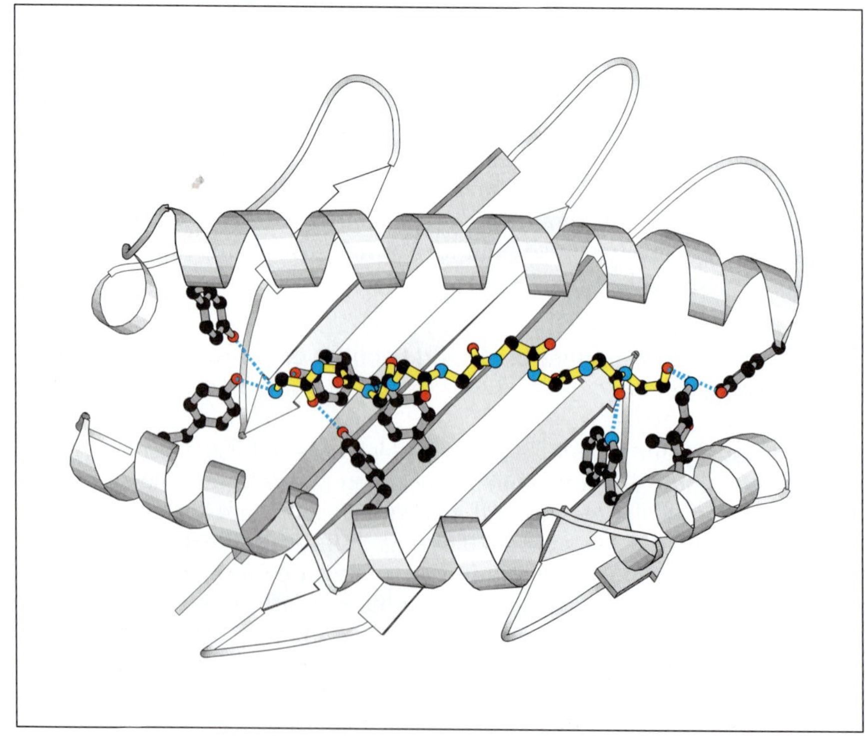

Fig. 4.18 Peptides are bound to MHC class I molecules by their ends. MHC class I molecules interact with the backbone of a bound peptide (shown in yellow) through a series of hydrogen bonds and ionic interactions (shown as dotted blue lines) at each end of the peptide. The amino terminus of the peptide is to the left, the carboxy terminus to the right. Black circles are carbon atoms; red are oxygen; blue are nitrogen. The amino acid residues in the MHC molecule that form these bonds are common to all MHC class I molecules, and their side chains are shown in full (in gray) on a ribbon diagram of the MHC class I groove. A cluster of tyrosine residues common to all MHC class I molecules forms hydrogen bonds to the amino terminus of the bound peptide, while a second cluster of residues forms hydrogen bonds and ionic interactions with the peptide backbone at the carboxy terminus and with the carboxy terminus itself.

These interactions give MHC class I molecules a broad peptide-binding specificity. In addition, MHC molecules are highly polymorphic. There are hundreds of different versions, or **alleles**, of the MHC class I genes in the human population, and each individual carries only a small selection. The main differences between allelic MHC variants are found at certain sites in the peptide-binding cleft, resulting in different amino acids in key peptide-interaction sites. In consequence, the different MHC variants preferentially bind different peptides. The peptides that can bind to a given MHC variant have the same or very similar amino acid residues at two or three particular positions along the peptide sequence. The amino acid side chains at these positions insert into pockets in the MHC molecule that are lined by the polymorphic amino acids. Because this binding anchors the peptide to the MHC molecule, the peptide residues involved are called the **anchor residues**. Both the position and identity of these anchor residues can vary, depending on the particular MHC class I variant that is binding the peptide. However, most peptides that bind to MHC class I molecules have a hydrophobic (or sometimes basic) anchor residue at the carboxy terminus (Fig. 4.19). Whereas changing an anchor residue will in most cases prevent the peptide from binding, not every synthetic peptide of suitable length that contains these anchor residues will bind the appropriate MHC class I molecule, and so the overall binding must also depend on the nature of the amino acids at other positions in the peptide. In some cases, particular amino acids are preferred in certain positions, whereas in others the presence of particular amino acids prevents binding. These additional amino acid positions are called 'secondary anchors.' These features of peptide binding enable an individual MHC class I molecule to bind a wide variety of different peptides, yet allow different MHC class I allelic variants to bind different sets of peptides.

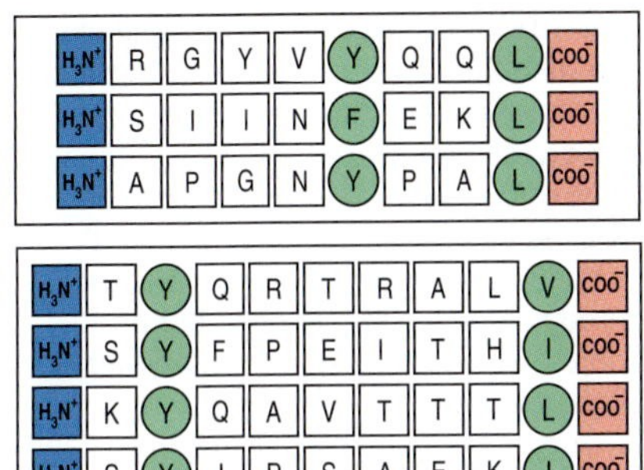

Fig. 4.19 Peptides bind to MHC molecules through structurally related anchor residues. Peptides eluted from two different MHC class I molecules are shown in the upper and lower panels, respectively. The anchor residues (green) differ for peptides that bind different alleles of MHC class I molecules but are similar for all peptides that bind to the same MHC molecule. The anchor residues that bind a particular MHC molecule need not be identical, but are always related (for example, phenylalanine (F) and tyrosine (Y) are both aromatic amino acids, whereas valine (V), leucine (L), and isoleucine (I) are all large hydrophobic amino acids). Peptides also bind to MHC class I molecules through their amino (blue) and carboxy (red) termini.

4-15 The length of the peptides bound by MHC class II molecules is not constrained.

Peptide binding to MHC class II molecules has also been analyzed by elution of bound peptides and by X-ray crystallography, and differs in several ways from peptide binding to MHC class I molecules. Natural peptides that bind to MHC class II molecules are at least 13 amino acids long and can be much longer. The clusters of conserved residues that bind the two ends of a peptide in MHC class I molecules are not found in MHC class II molecules, and the ends of the peptide are not bound. Instead, the peptide lies in an extended conformation along the peptide-binding cleft. It is held there both by peptide side chains that protrude into shallow and deep pockets lined by polymorphic residues and by interactions between the peptide backbone and side chains of conserved amino acids that line the peptide-binding cleft in all MHC class II molecules (Fig. 4.20). Structural data show that amino acid side chains at residues 1, 4, 6, and 9 of an MHC class II-bound peptide can be held in these binding pockets.

The binding pockets of MHC class II molecules accommodate a greater variety of side chains than those of MHC class I molecules, making it more difficult to define anchor residues and to predict which peptides will be able to bind a particular MHC class II variant (Fig. 4.21). Nevertheless, by comparing the sequences of known binding peptides it is usually possible to detect patterns of amino acids that permit binding to different MHC class II variants, and to model how the amino acids of this peptide sequence motif will interact with the amino acids of the peptide-binding cleft. Because the peptide is bound by its backbone and allowed to emerge from both ends of the binding groove, there is, in principle, no upper limit to the length of peptides that could bind to MHC class II molecules. However, it seems that longer peptides bound to MHC class II molecules are trimmed by peptidases to a length of around 13–17 amino acids in most cases. Like MHC class I molecules, MHC class II molecules that lack bound peptide are unstable.

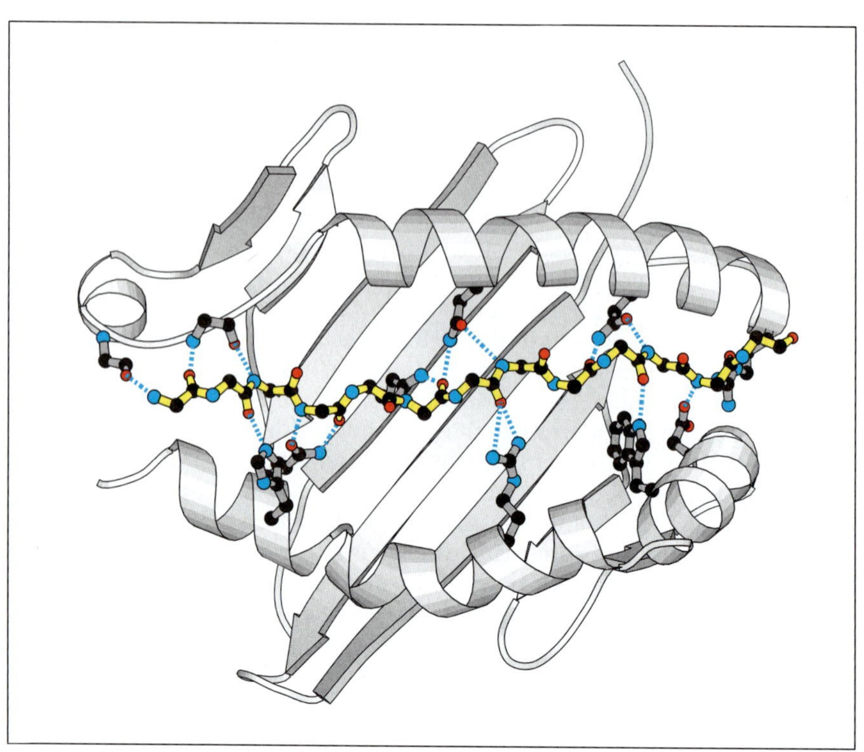

Fig. 4.20 Peptides bind to MHC class II molecules by interactions along the length of the binding groove. A peptide (yellow; shown as the peptide backbone only, with the amino terminus to the left and the carboxy terminus to the right) is bound by an MHC class II molecule through a series of hydrogen bonds (dotted blue lines) that are distributed along the length of the peptide. The hydrogen bonds toward the amino terminus of the peptide are made with the backbone of the MHC class II polypeptide chain, whereas throughout the peptide's length bonds are made with residues that are highly conserved in MHC class II molecules. The side chains of these residues are shown in gray on the ribbon diagram of the MHC class II groove.

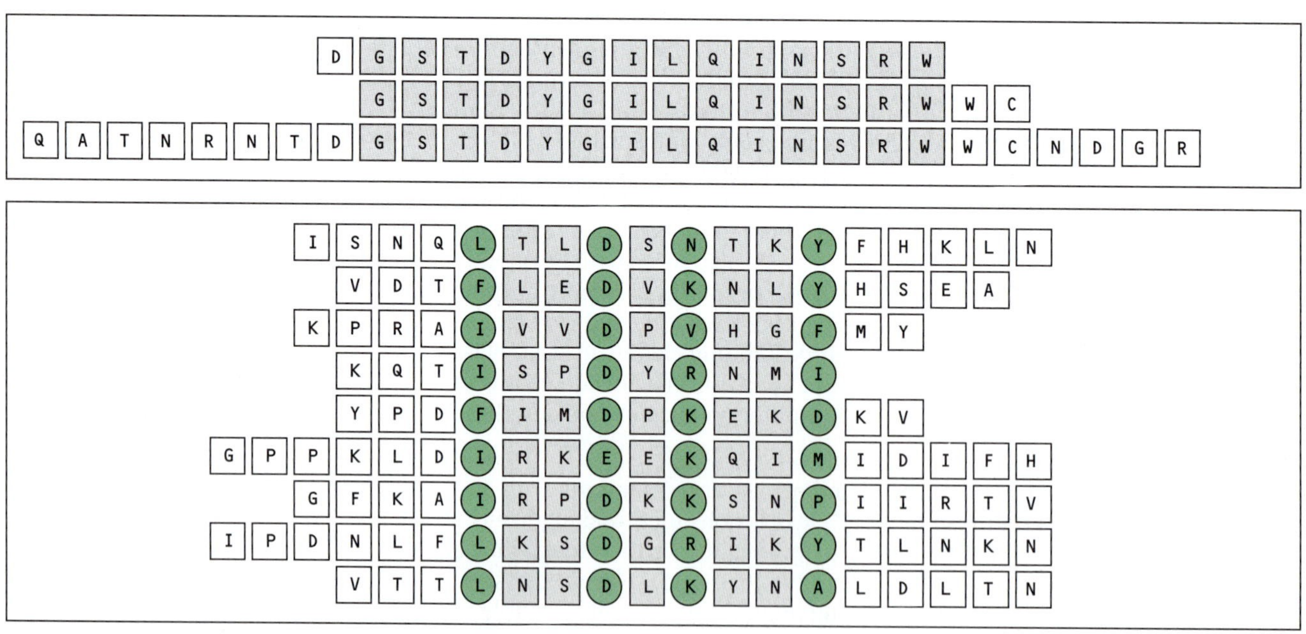

Fig. 4.21 Peptides that bind MHC class II molecules are variable in length and their anchor residues lie at various distances from the ends of the peptide. The sequences of a set of peptides that bind to the mouse MHC class II A^k allele are shown in the upper panel. All contain the same core sequence (shaded) but differ in length. In the lower panel, different peptides binding to the human MHC class II allele HLA-DR3 are shown. Anchor residues are shown as green circles. The lengths of these peptides can vary, and so by convention the first anchor residue is denoted as residue 1. Note that all of the peptides share a hydrophobic residue in position 1, a negatively charged residue (aspartic acid (D) or glutamic acid (E)) in position 4, and a tendency to have a basic residue (lysine (K), arginine (R), histidine (H), glutamine (Q), or asparagine (N)) in position 6 and a hydrophobic residue (for example, tyrosine (Y), leucine (L), phenylalanine (F)) in position 9.

4-16 The crystal structures of several peptide:MHC:T-cell receptor complexes show a similar orientation of the T-cell receptor over the peptide:MHC complex.

At the time that the first X-ray crystallographic structure of a T-cell receptor was published, a structure of the same T-cell receptor bound to a peptide:MHC class I ligand was also produced. This structure (Fig. 4.22), which had been predicted by site-directed mutagenesis of the MHC class I molecule, showed the T-cell receptor aligned diagonally over the peptide and the peptide-binding cleft, with the TCRα chain lying over the α_2 domain and the amino-terminal end of the bound peptide, the TCRβ chain lying over the α_1 domain and the carboxy-terminal end of the peptide, and the CDR3 loops of both TCRα and TCRβ chains meeting over the central amino acids of the peptide. The T-cell receptor is threaded through a valley between the two high points on the two surrounding α helices that form the walls of the peptide-binding cleft.

Analysis of other peptide:MHC class I:T-cell receptor complexes and of peptide:MHC class II:T-cell receptor complexes (Fig. 4.23) shows that all have a very similar orientation, particularly for the V_α domain, although some variability does occur in the location and orientation of the V_β domain. In this orientation, the V_α domain makes contact primarily with the amino-terminal half of the bound peptide, whereas the V_β domain contacts primarily the carboxy-terminal half. Both chains also interact with the α helices of the MHC class I molecule (see Fig. 4.22). The T-cell receptor contacts are not symmetrically distributed over the MHC molecule: whereas the V_α CDR1 and CDR2 loops are in close contact with the helices of the peptide:MHC complex around the amino terminus of the bound peptide, the β-chain CDR1 and CDR2 loops, which interact with the complex at the carboxy terminus of the bound peptide, have variable contributions to the binding.

Comparison of the three-dimensional structure of an unliganded T-cell receptor and the same T-cell receptor complexed to its peptide:MHC ligand shows that the binding results in some degree of conformational change, or 'induced fit,' particularly within the V_α CDR3 loop. It has also been shown that subtly different peptides can have strikingly different effects on the recognition of an otherwise identical peptide:MHC ligand by the same T cell. The flexibility in the CDR3 loop demonstrated by these two structures helps to explain how the T-cell receptor can adopt conformations that recognize related, but different, ligands.

From an examination of the available structures it is hard to predict whether the main binding energy is contributed by T-cell receptor contacts with the bound peptide or by T-cell receptor contacts with the MHC molecule.

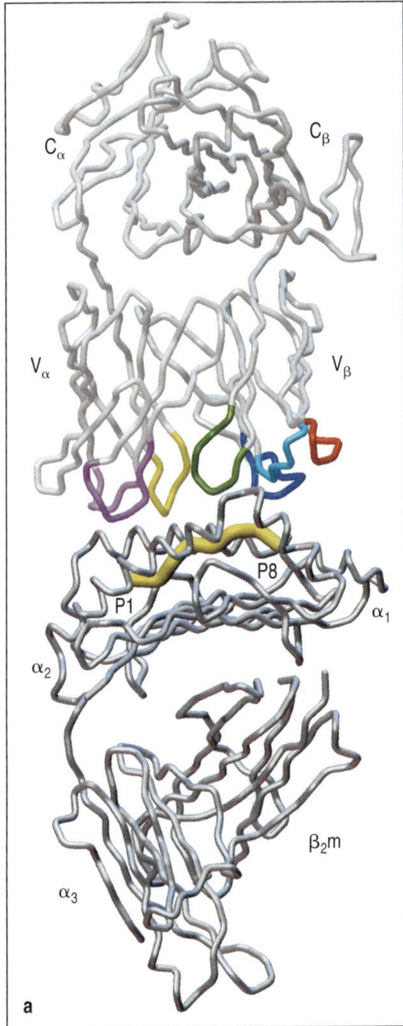

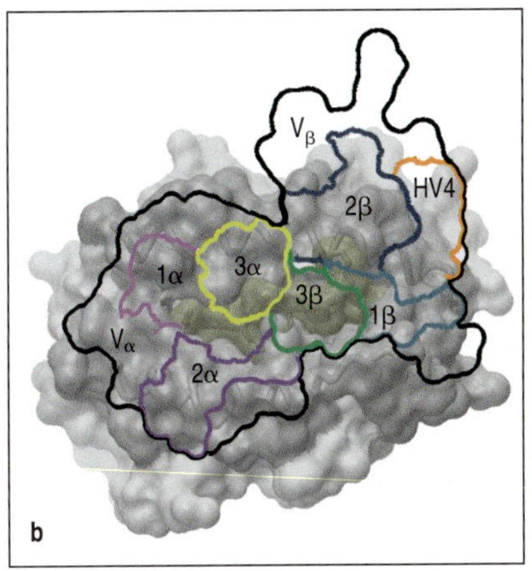

Fig. 4.22 The T-cell receptor binds to the peptide:MHC complex. Panel a: the T-cell receptor binds to the top of the peptide:MHC complex, straddling, in the case of the MHC class I molecule shown here, both the α_1 and α_2 domain helices. The CDRs of the T-cell receptor are indicated in color: the CDR1 and CDR2 loops of the β chain in light and dark blue, respectively; and the CDR1 and CDR2 loops of the α chain in light and dark purple, respectively. The α-chain CDR3 loop is in yellow, and the β-chain CDR3 loop is in green. The β-chain HV4 loop is in red. The thick yellow line P1–P8 is the bound peptide. Panel b: the outline of the T-cell receptor's antigen-binding site (thick black line) is superimposed on the top surface of the peptide:MHC complex (the peptide is shaded dull yellow). The T-cell receptor lies diagonally across the peptide:MHC complex, with the α and β CDR3 loops of the T-cell receptor (3α, 3β, yellow and green, respectively) contacting the center of the peptide. The α-chain CDR1 and CDR2 loops (1α, 2α, light and dark purple, respectively) contact the MHC helices at the amino terminus of the bound peptide, whereas the β-chain CDR1 and CDR2 loops (1β, 2β, light and dark blue, respectively) make contact with the helices at the carboxy terminus of the bound peptide. Courtesy of I.A. Wilson.

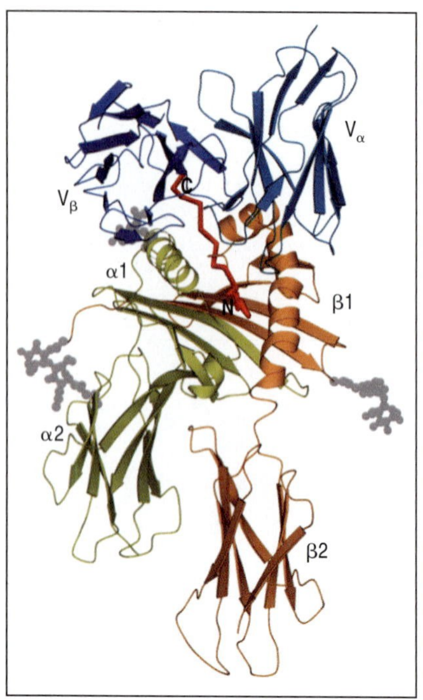

Fig. 4.23 The T-cell receptor interacts with MHC class I and MHC class II molecules in a similar fashion. The structure of a T-cell receptor binding to an MHC class II molecule has been determined, and shows the T-cell receptor binding to an equivalent site, and in an equivalent orientation, to the way in which T-cell receptors bind to MHC class I molecules (see Fig. 4.22). Only the V_α and V_β domains of the T-cell receptor are shown, colored in blue. The peptide is colored red, and carbohydrate residues are indicated in gray. The T-cell receptor sits in a shallow saddle formed between the α-helical regions of the MHC class II α (yellow-green) and β chain (orange), at roughly 90° to the long axis of the MHC class II molecule and the bound peptide. Courtesy of E.L. Reinherz and J-H. Wang.

Measurements of the kinetics of T-cell receptor binding to peptide:MHC ligands suggest that the interactions between the T-cell receptor and the MHC molecule might predominate at the start of the contact, guiding the receptor into the correct position where a second, more detailed, interaction with the peptide as well as the MHC molecule dictates the final outcome of the interaction—binding or dissociation. As with antibody–antigen interactions, only a few amino acids at the interface might provide the essential contacts that determine the specificity and strength of binding. Simply changing a leucine to isoleucine in the peptide, for example, is sufficient to alter a T-cell response from strong killing to no response at all. Mutations of single residues in the presenting MHC molecules can have the same effect. Thus, the specificity of T-cell recognition involves both the peptide and its presenting MHC molecule. This dual specificity underlies the MHC restriction of T-cell responses, a phenomenon that was observed long before the peptide-binding properties of MHC molecules were known. We will recount the story of how MHC restriction was discovered when we return to the issue of how MHC polymorphism affects antigen recognition by T cells in Chapter 6. Another consequence of this dual specificity is a need for T-cell receptors to be able to interact appropriately with the antigen-presenting surface of MHC molecules. It seems that some inherent specificity for MHC molecules is encoded in the T-cell receptor genes, and there is selection during T-cell development for a repertoire of receptors able to interact appropriately with the particular MHC molecules present in that individual (discussed in Chapter 8).

4-17 The CD4 and CD8 cell-surface proteins of T cells are required to make an effective response to antigen.

As well as engaging a peptide:MHC ligand via its antigen receptor, a T cell makes additional interactions with the MHC molecule that stabilize the interaction and are required for the cell to respond effectively to antigen. T cells fall into two major classes, which have different effector functions and are distinguished by the expression of the cell-surface proteins **CD4** and **CD8**. CD8 is carried by cytotoxic T cells, while CD4 is carried by T cells whose function is to activate other cells (see Section 1-19). CD4 and CD8 were known as markers for these functional sets for some time before it became clear that the distinction was based on the ability of T cells to recognize different classes of MHC molecules: CD8 recognizes MHC class I molecules and CD4 recognizes MHC class II. During antigen recognition, CD4 or CD8 (depending on the type of T cell) associates on the T-cell surface with the T-cell receptor and binds to invariant sites on the MHC portion of the composite peptide:MHC ligand, away from the peptide-binding site. This binding is required for the T cell to make an effective response, and so CD4 and CD8 are called **co-receptors**.

CD4 is a single-chain protein composed of four Ig-like domains (Fig. 4.24). The first two domains (D1 and D2) are packed tightly together to form a rigid rod about 60 Å long, which is joined by a flexible hinge to a similar rod formed by the third and fourth domains (D3 and D4). The MHC-binding region on CD4 is located mainly on a lateral face of the D1 domain, and CD4 binds to a hydrophobic crevice formed at the junction of the α_2 and β_2 domains of the MHC class II molecule. This site is well away from the site where the T-cell receptor binds (Fig. 4.25a), and so the CD4 molecule and the T-cell receptor can bind simultaneously to the same peptide:MHC class II complex. The intracellular portion of CD4 interacts strongly with a cytoplasmic tyrosine kinase called Lck, and brings Lck close to the intracellular signaling components associated with the T-cell receptor. This results in enhancement of the signal that is generated when the T-cell receptor binds its ligand, as we discuss in Chapter 7. When CD4 and the T-cell receptor bind simultaneously to the same MHC class II:peptide complex, the T cell is about a hundredfold more sensitive to the antigen than if CD4 were absent.

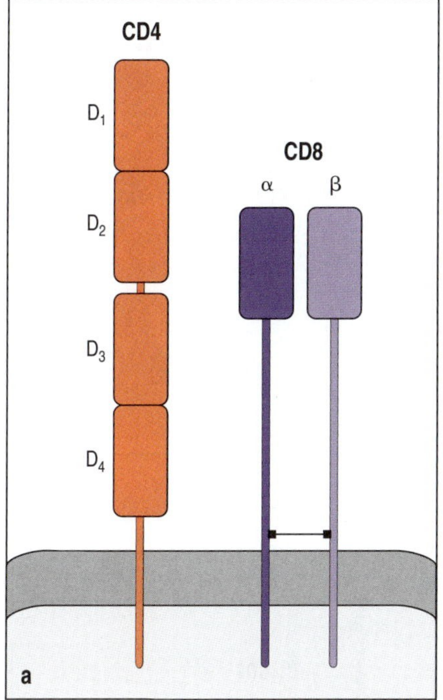

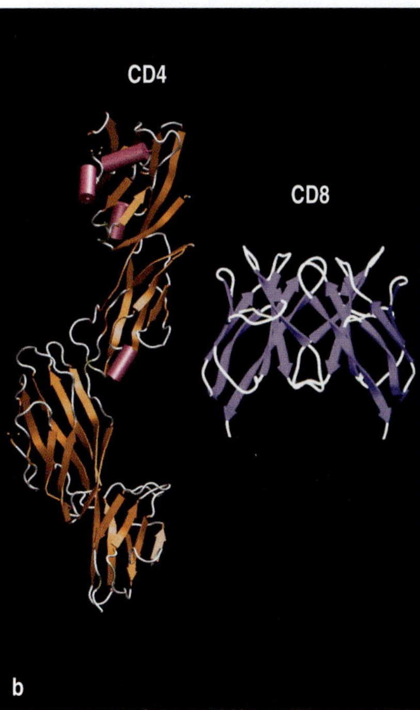

Fig. 4.24 The structures of the CD4 and CD8 co-receptor molecules. The CD4 molecule contains four Ig-like domains, shown in schematic form in panel a and as a ribbon diagram of the crystal structure in panel b. The amino-terminal domain, D_1, is similar in structure to an immunoglobulin V domain. The second domain, D_2, although clearly related to an immunoglobulin domain, is different from both V and C domains and has been termed a C2 domain. The first two domains of CD4 form a rigid rod-like structure that is linked to the two carboxy-terminal domains by a flexible link. The binding site for MHC class II molecules is thought to involve mainly the D_1 domain. The CD8 molecule is a heterodimer of an α and a β chain covalently linked by a disulfide bond; an alternative form of CD8 exists as a homodimer of α chains. The heterodimer is depicted in panel a, whereas the ribbon diagram in panel b is of the homodimer. CD8α and CD8β chains have very similar structures, each having a single domain resembling an immunoglobulin V domain and a stretch of polypeptide chain, believed to be in a relatively extended conformation, that anchors the V-like domain to the cell membrane.

The structure of CD8 is quite different. It is a disulfide-linked dimer of two different chains, called α and β, each containing a single Ig-like domain linked to the membrane by a segment of extended polypeptide (see Fig. 4.24). This segment is extensively glycosylated, which is thought to maintain it in an extended conformation and protect it from cleavage by proteases. CD8α chains can form homodimers, although these are not found when the CD8β chains are present. The CD8α homodimer may have a specific function in recognizing a specialized subset of nonclassical MHC class I molecules that we describe in Chapter 6.

CD8 binds weakly to an invariant site in the α_3 domain of an MHC class I molecule (see Fig. 4.25b). Although only the interaction of the CD8α homodimer with MHC class I is known in detail, it shows that the binding site on the CD8 α:β heterodimer is formed by the interaction of the CD8α and β chains. In addition, CD8 interacts (most probably through its α chain) with residues in the base of the α_2 domain of the MHC class I molecule. The strength of binding of CD8 to the MHC class I molecule is influenced by the glycosylation state of the CD8 molecule; increased numbers of sialic acid residues added to the CD8 carbohydrate structures decrease the strength of the interaction. The pattern of sialylation of CD8 changes during the maturation of T cells and also on activation, and this is likely to have a role in modulating antigen recognition.

By binding to the membrane-proximal domains of the MHC class I and class II molecules, the co-receptors leave the upper surface of the MHC molecule exposed and free to interact with a T-cell receptor, as shown for CD8 in Fig. 4.26. Both CD4 and CD8 bind Lck—in the case of the CD8α:β heterodimer through the cytoplasmic tail of the α chain—and bring it into close proximity to the T-cell receptor. As with CD4, the presence of CD8 increases the sensitivity of T cells to antigen presented by MHC class I molecules about a hundredfold. Thus, CD4 and CD8 have similar functions and bind to the same approximate location in MHC class I and MHC class II molecules, even though the structures of the two co-receptor proteins are only distantly related.

Fig. 4.25 The binding sites for CD4 and CD8 on MHC class II and class I molecules lie in the Ig-like domains. The binding sites for CD4 and CD8 on the MHC class II and class I molecules, respectively, lie in the Ig-like domains nearest to the membrane and distant from the peptide-binding cleft. The binding of CD4 to an MHC class II molecule is shown as a structure graphic in panel a and schematically in panel c. The α chain of the MHC class II molecule is shown in pink, and the β chain in white, while CD4 is in gold. Only the D_1 and D_2 domains of the CD4 molecule are shown in panel a. The binding site for CD4 lies at the base of the β_2 domain of an MHC class II molecule, in the hydrophobic crevice between the β_2 and α_2 domains. The binding of CD8 to an MHC class I molecule is shown in panel b and schematically in panel d. The class I heavy chain and β_2-microglobulin are shown in white and pink, respectively, and the two chains of the CD8 dimer are shown in light and dark purple. The structure is actually of the binding of the CD8α homodimer, but the CD8α:β heterodimer is believed to bind in a similar way. The binding site for CD8 on the MHC class I molecule lies in a similar position to that of CD4 in the MHC class II molecule, but CD8 binding also involves the base of the α_1 and α_2 domains, and thus the binding of CD8 to MHC class I is not completely equivalent to the binding of CD4 to MHC class II.

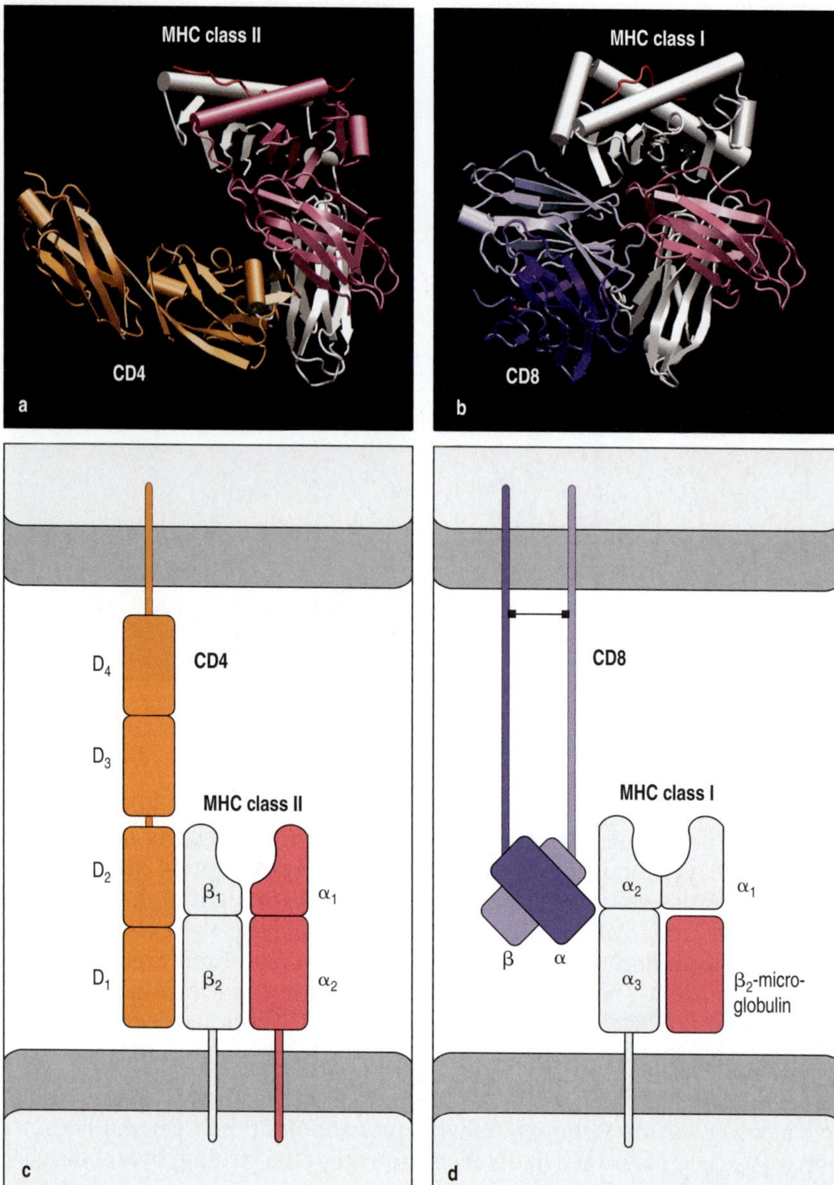

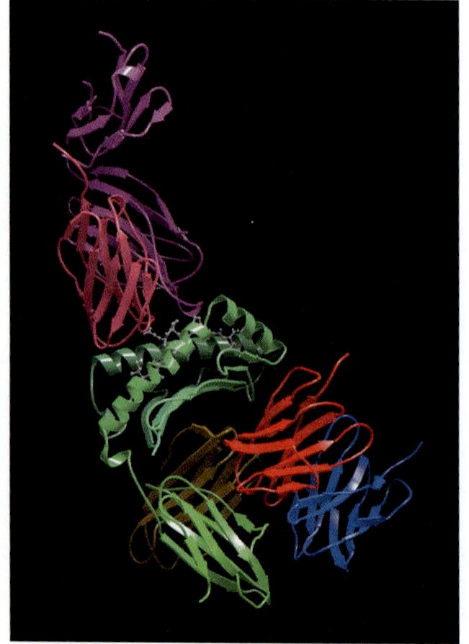

4-18 The two classes of MHC molecules are expressed differentially on cells.

MHC class I and MHC class II molecules have distinct distributions among cells, and these reflect the different effector functions of the T cells that recognize them (Fig. 4.27). MHC class I molecules present peptides from pathogens, commonly viruses, to CD8 cytotoxic T cells, which are specialized

Fig. 4.26 CD8 binds to a site on MHC class I molecules distant from that to which the T-cell receptor binds. The relative positions of the T-cell receptor and CD8 molecules bound to the same MHC class I molecule can be seen in this hypothetical reconstruction of the interaction of an MHC class I molecule (the α chain is shown in green; β_2-microglobulin (dull yellow) can be seen faintly in the background) with a T-cell receptor and CD8. The α and β chains of the T-cell receptor are shown in pink and purple, respectively. The CD8 structure is that of a CD8α homodimer, but is colored to represent the likely orientation of the subunits in the heterodimer, with the CD8β subunit in red and the CD8α subunit in blue. Courtesy of G. Gao.

to kill any cell that they specifically recognize. Because viruses can infect any nucleated cell, almost all such cells express MHC class I molecules, although the level of constitutive expression varies from one cell type to the next. For example, cells of the immune system express abundant MHC class I on their surface, whereas liver cells (hepatocytes) express relatively low levels (see Fig. 4.27). Nonnucleated cells, such as mammalian red blood cells, express little or no MHC class I, and thus the interior of red blood cells is a site in which an infection can go undetected by cytotoxic T cells. Because red blood cells cannot support viral replication, this is of no great consequence for viral infection, but it might be the absence of MHC class I that allows the *Plasmodium* parasites that cause malaria to live in this privileged site.

In contrast, the main function of the CD4 T cells that recognize MHC class II molecules is to activate other effector cells of the immune system. Thus, MHC class II molecules are normally found on B lymphocytes, dendritic cells, and macrophages—cells that participate in immune responses—but not on other tissue cells (see Fig. 4.27). When CD4 T cells recognize peptides bound to MHC class II molecules on B cells, they stimulate the B cells to produce antibody. Similarly, CD4 T cells recognizing peptides bound to MHC class II molecules on macrophages activate these cells to destroy the pathogens in their vesicles. We shall see in Chapter 9 that MHC class II molecules are also expressed on specialized antigen-presenting cells, the dendritic cells, in lymphoid tissues where naive T cells encounter antigen and are first activated.

The expression of both MHC class I and MHC class II molecules is regulated by cytokines, in particular interferons, released in the course of immune responses. Interferon-α (IFN-α) and IFN-β increase the expression of MHC class I molecules on all types of cells, whereas IFN-γ increases the expression of both MHC class I and MHC class II molecules, and can induce the expression of MHC class II molecules on certain cell types that do not normally express them. Interferons also enhance the antigen-presenting function of MHC class I molecules by inducing the expression of key components of the intracellular machinery that enables peptides to be loaded onto the MHC molecules.

Tissue	MHC class I	MHC class II
Lymphoid tissues		
T cells	+++	+*
B cells	+++	+++
Macrophages	+++	++
Dendritic cells	+++	+++
Epithelial cells of the thymus	+	+++
Other nucleated cells		
Neutrophils	+++	–
Hepatocytes	+	–
Kidney	+	–
Brain	+	–†
Nonnucleated cells		
Red blood cells	–	–

Fig. 4.27 The expression of MHC molecules differs between tissues. MHC class I molecules are expressed on all nucleated cells, although they are most highly expressed in hematopoietic cells. MHC class II molecules are normally expressed only by a subset of hematopoietic cells and by thymic stromal cells, although they may be expressed by other cell types on exposure to the inflammatory cytokine IFN-γ.
*In humans, activated T cells express MHC class II molecules, whereas in mice all T cells are MHC class II-negative.
†In the brain, most cell types are MHC class II-negative, but microglia, which are related to macrophages, are MHC class II-positive.

4-19 A distinct subset of T cells bears an alternative receptor made up of γ and δ chains.

During the search for the gene for the TCRα chain, another T-cell receptor-like gene was unexpectedly discovered. This gene was named TCRγ, and its discovery led to a search for further T-cell receptor genes. Another receptor chain was identified by using antibody against the predicted sequence of the γ chain and was called the δ chain. It was soon discovered that a minority population of T cells bore a distinct type of T-cell receptor made up of γ:δ heterodimers rather than α:β heterodimers. The development of these cells is described in Sections 8-11 and 8-12.

The crystallographic structure of a γ:δ T-cell receptor reveals that, as expected, it is similar in shape to α:β T-cell receptors (Fig. 4.28). γ:δ T-cell receptors may be specialized to bind certain kinds of ligands, including heat-shock proteins and nonpeptide ligands such as phosphorylated ligands or mycobacterial lipid antigens. It seems likely that γ:δ T-cell receptors are not restricted by the 'classical' MHC class I and class II molecules. They may bind the free antigen, much as immunoglobulins do, and/or they may bind to peptides or other antigens presented by nonclassical MHC-like molecules. These are proteins that resemble MHC class I molecules but are relatively nonpolymorphic and are described in Chapter 6. We still know little about how γ:δ T-cell receptors bind antigen and thus how these cells function, and what their role is in immune responses. The structure and rearrangement of the genes for γ:δ T-cell receptors are covered in Sections 5-11 and 8-12.

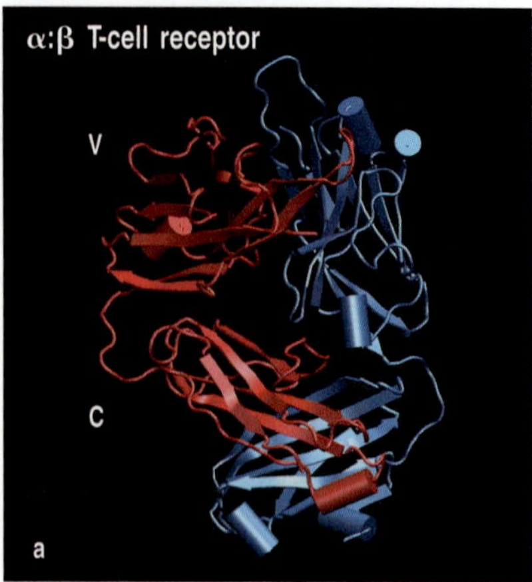

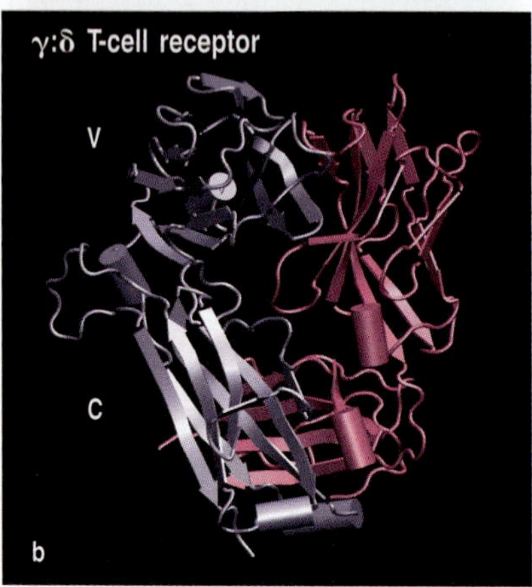

Fig. 4.28 Structures of α:β and γ:δ T-cell receptors. The structures of the α:β and the γ:δ T-cell receptors have both been determined by X-ray crystallography. The α:β T-cell receptor is shown in panel a, with the α chain colored red and the β chain blue. Panel b shows the γ:δ receptor, with the γ chain colored purple and the δ chain pink. The receptors have very similar structures, somewhat resembling that of a Fab fragment of an immunoglobulin molecule. The C_δ domain is more like an immunoglobulin domain than is the corresponding C_α domain of the α:β T-cell receptor.

Summary.

The receptor for antigen on most T cells, the α:β T-cell receptor, is composed of two protein chains, TCRα and TCRβ, and resembles in many respects a single Fab fragment of immunoglobulin. T-cell receptors are always membrane-bound. α:β T-cell receptors do not recognize antigen in its native state, in contrast with the immunoglobulin receptors of B cells, but recognize a composite ligand of a peptide antigen bound to an MHC molecule. MHC molecules are highly polymorphic glycoproteins encoded by genes in the major histocompatibility complex (MHC). Each MHC molecule binds a wide variety of different peptides, but the different variants each preferentially recognize sets of peptides with particular sequence and physical features. The peptide antigen is generated intracellularly, and is bound stably in a peptide-binding cleft on the surface of the MHC molecule. There are two classes of MHC molecules, and these are bound in their nonpolymorphic domains by CD8 and CD4 molecules that distinguish two different functional classes of α:β T cells. CD8 binds MHC class I molecules and can bind simultaneously to the same peptide:MHC class I complex being recognized by a T-cell receptor, thus acting as a co-receptor and enhancing the T-cell response; CD4 binds MHC class II molecules and acts as a co-receptor for T-cell receptors that recognize peptide:MHC class II ligands. A T-cell receptor interacts directly both with the antigenic peptide and with polymorphic features of the MHC molecule that displays it, and this dual specificity underlies the MHC restriction of T-cell responses. A second type of T-cell receptor, composed of a γ and a δ chain, is structurally similar to the α:β T-cell receptor, but it seems to bind different ligands, including nonpeptide ligands. It is thought not to be MHC restricted and is found on a minority population of T cells, the γ:δ T cells.

Summary to Chapter 4.

B cells and T cells use different, but structurally similar, molecules to recognize antigen. The antigen-recognition molecules of B cells are immunoglobulins, and are made both as a membrane-bound receptor for antigen, the B-cell receptor, and as secreted antibodies that bind antigens and elicit humoral effector functions. The antigen-recognition molecules of T cells, in contrast, are made only as cell-surface receptors. Immunoglobulins and T-cell receptors are highly variable molecules, with the variability concentrated in that part of the molecule, the variable (V) region, that binds to antigen. Immunoglobulins bind a wide variety of chemically different antigens, whereas the major α:β type of T-cell receptor predominantly recognizes peptide fragments of foreign proteins bound to the MHC molecules that are ubiquitous on cell surfaces.

Binding of antigen by immunoglobulins has chiefly been studied with antibodies. The binding of antibody to its corresponding antigen is highly specific, and this specificity is determined by the shape and physicochemical properties of the antigen-binding site. The part of the antibody that elicits effector functions, once the variable part has bound an antigen, is located at the other end of the molecule from the antigen-binding sites, and is termed the constant region. There are five main functional classes of antibodies, each encoded by a different type of constant region. As we will see in Chapter 10, these interact with different components of the immune system to incite an inflammatory response and eliminate the antigen.

T-cell receptors differ in several respects from the B-cell immunoglobulins. One is the absence of a secreted form of the receptor. This reflects the functional differences between T cells and B cells. B cells deal with pathogens and their protein products circulating within the body; secretion of a soluble antigen-recognition molecule by the activated B cell after antigen has been encountered enables them to mop up antigen effectively throughout the extracellular spaces of the body. T cells, in contrast, are specialized for cell–cell interactions. They either kill cells that are infected with intracellular pathogens and that bear foreign antigenic peptides on their surface, or interact with cells of the immune system that have taken up foreign antigen and are displaying it on the cell surface. T-cell recognition does not require a soluble, secreted receptor.

The second distinctive feature of the T-cell receptor is that it recognizes a composite ligand made up of the foreign peptide bound to a self MHC molecule. This means that T cells can interact only with a body cell displaying the antigen, not with the intact pathogen or protein. Each T-cell receptor is specific for a particular combination of peptide and a self MHC molecule.

MHC molecules are encoded by a family of highly polymorphic genes; although each individual expresses several of these genes, this represents only a small selection of all possible variants. During T-cell development, the T-cell receptor repertoire is selected so that the T cells of each individual recognize antigen only in conjunction with their own MHC molecules. Expression of multiple variant MHC molecules, each with a different peptide-binding repertoire, helps to ensure that T cells from an individual will be able to recognize at least some peptides generated from nearly every pathogen.

Questions.

4.1 *The immunoglobulin superfamily is one of the most abundant families of protein domain structures. (a) What are the characteristics of an immunoglobulin domain, and how do the various subtypes of these domains differ? (b) What regions of the V-type immunoglobulin domain contribute to its complementarity-determining regions (CDRs), and how do the V-type and C-type immunoglobulin domains differ in those regions?*

4.2 *How do antibodies, which all have the same basic shape, recognize antigens of a wide variety of different shapes?*

4.3 *Although the antigen receptors on B cells and T cells are structurally related, there are important differences between them. (a) Describe the similarities and differences in the antigen-recognition properties of B-cell and T-cell antigen receptors. (b) How do these differences influence which antigens are recognized by B cells and T cells? (c) Given these differences, what would you say is the essential difference in the function of B cells and T cells?*

4.4 *There are two kinds of MHC molecules: class I and class II. (a) What role do MHC molecules have in the activation of antigen-specific T cells? (b) Explain how the peptide-binding region of MHC class I and class II molecules can be so similar, even though one is encoded by a single gene and the other is encoded by two different genes. (c) Besides interactions with the T-cell receptors, what additional interactions made by MHC molecules with T cells help to functionally distinguish between antigens presented by MHC class I and MHC class II molecules?*

General references.

Ager, A., Callard, R., Ezine, S., Gerard, C., and Lopez-Botet, M.: **Immune receptor Supplement.** *Immunol. Today* 1996, 17.

Davies, D.R., and Chacko, S.: **Antibody structure.** *Acc. Chem. Res.* 1993, **26**:421–427.

Frazer, K., and Capra, J.D.: **Immunoglobulins: structure and function**, in Paul W.E. (ed): *Fundamental Immunology*, 4th ed. New York, Raven Press, 1998.

Germain, R.N.: **MHC-dependent antigen processing and peptide presentation: providing ligands for T lymphocyte activation.** *Cell* 1994, **76**:287–299.

Honjo, T., and Alt, F.W. (eds): *Immunoglobulin Genes*, 2nd ed. London, Academic Press, 1996.

Moller, G. (ed): **Origin of major histocompatibility complex diversity.** *Immunol. Rev.* 1995, **143**:5–292.

Poljak, R.J.: **Structure of antibodies and their complexes with antigens.** *Mol. Immunol.* 1991, **28**:1341–1345.

Rudolph, M.G., Stanfield, R.L., and Wilson, I.A: **How TCRs bind MHCs, peptides, and coreceptors.** *Annu. Rev. Immunol.* 2006, **24**:419–466.

Section references.

4-1 IgG antibodies consist of four polypeptide chains.

Edelman, G.M.: **Antibody structure and molecular immunology.** *Scand. J. Immunol.* 1991, **34**:4–22.

Faber, C., Shan, L., Fan, Z., Guddat, L.W., Furebring, C., Ohlin, M., Borrebaeck, C.A.K., and Edmundson, A.B.: **Three-dimensional structure of a human Fab with high affinity for tetanus toxoid.** *Immunotechnology* 1998, **3**:253–270.

Harris, L.J., Larson, S.B., Hasel, K.W., Day, J., Greenwood, A., and McPherson, A.: **The three-dimensional structure of an intact monoclonal antibody for canine lymphoma.** *Nature* 1992, **360**:369–372.

4-2 Immunoglobulin heavy and light chains are composed of constant and variable regions.
&
4-3 The antibody molecule can readily be cleaved into functionally distinct fragments.

Porter, R.R.: **Structural studies of immunoglobulins.** *Scand. J. Immunol.* 1991, **34**:382–389.

Yamaguchi, Y., Kim, H., Kato, K., Masuda, K., Shimada, I., and Arata, Y.: **Proteolytic fragmentation with high specificity of mouse IgG—mapping of proteolytic cleavage sites in the hinge region.** *J. Immunol. Meth.* 1995, **181**:259–267.

4-4 The immunoglobulin molecule is flexible, especially at the hinge region.

Gerstein, M., Lesk, A.M., and Chothia, C.: **Structural mechanisms for domain movements in proteins.** *Biochemistry* 1994, **33**:6739–6749.

Jimenez, R., Salazar, G., Baldridge, K.K., and Romesberg, F.E.: **Flexibility and molecular recognition in the immune system.** *Proc. Natl Acad. Sci. USA* 2003, **100**:92–97.

Saphire, E.O., Stanfield, R.L., Crispin, M.D., Parren, P.W., Rudd, P.M., Dwek, R.A., Burton, D.R., and Wilson, I.A.: **Contrasting IgG structures reveal extreme asymmetry and flexibility.** *J. Mol. Biol.* 2002, **319**:9–18.

4-5 The domains of an immunoglobulin molecule have similar structures.

Barclay, A.N., Brown, M.H., Law, S.K., McKnight, A.J., Tomlinson, M.G., and van der Merwe, P.A. (eds): *The Leukocyte Antigen Factsbook*, 2nd ed. London, Academic Press, 1997.

Brummendorf, T., and Lemmon, V.: **Immunoglobulin superfamily receptors: *cis*-interactions, intracellular adapters and alternative splicing regulate adhesion.** *Curr. Opin. Cell Biol.* 2001, **13**:611–618.

Marchalonis, J.J., Jensen, I., and Schluter, S.F.: **Structural, antigenic and evolutionary analyses of immunoglobulins and T cell receptors.** *J. Mol. Recog.* 2002, **15**:260–271.

Ramsland, P.A., and Farrugia, W.: **Crystal structures of human antibodies: a detailed and unfinished tapestry of immunoglobulin gene products.** *J. Mol. Recog.* 2002, **15**:248–259.

4-6 Localized regions of hypervariable sequence form the antigen-binding site.

Chitarra, V., Alzari, P.M., Bentley, G.A., Bhat, T.N., Eiselé, J.-L., Houdusse, A., Lescar, J., Souchon, H., and Poljak, R.J.: **Three-dimensional structure of a heteroclitic antigen-antibody cross-reaction complex.** *Proc. Natl Acad. Sci. USA* 1993, **90**:7711–7715.

Decanniere, K., Muyldermans, S., and Wyns, L.: **Canonical antigen-binding loop structures in immunoglobulins: more structures, more canonical classes?** *J. Mol. Biol.* 2000, **300**:83–91.

Gilliland, L.K., Norris, N.A., Marquardt, H., Tsu, T.T., Hayden, M.S., Neubauer, M.G., Yelton, D.E., Mittler, R.S., and Ledbetter, J.A.: **Rapid and reliable cloning of antibody variable regions and generation of recombinant single-chain antibody fragments.** *Tissue Antigens* 1996, **47**:1–20.

Johnson, G., and Wu, T.T.: **Kabat Database and its applications: 30 years after the first variability plot.** *Nucleic Acids Res.* 2000, **28**:214–218.

Wu, T.T., and Kabat, E.A.: **An analysis of the sequences of the variable regions of Bence Jones proteins and myeloma light chains and their implications for antibody complementarity.** *J. Exp. Med.* 1970, **132**:211–250.

Xu, J., Deng, Q., Chen, J., Houk, K.N., Bartek, J., Hilvert, D., and Wilson, I.A.: **Evolution of shape complementarity and catalytic efficiency from a primordial antibody template.** *Science* 1999, **286**:2345–2348.

4-7 Antibodies bind antigens via contacts with amino acids in CDRs, but the details of binding depend upon the size and shape of the antigen.
&
4-8 Antibodies bind to conformational shapes on the surfaces of antigens.

Ban, N., Day, J., Wang, X., Ferrone, S., and McPherson, A.: **Crystal structure of an anti-anti-idiotype shows it to be self-complementary.** *J. Mol. Biol.* 1996, **255**:617–627.

Davies, D.R., and Cohen, G.H.: **Interactions of protein antigens with antibodies.** *Proc. Natl Acad. Sci. USA* 1996, **93**:7–12.

Decanniere, K., Desmyter, A., Lauwereys, M., Ghahroudi, M.A., Muyldermans, S., and Wyns, L.: **A single-domain antibody fragment in complex with RNase A: non-canonical loop structures and nanomolar affinity using two CDR loops.** *Structure Fold. Des.* 1999, **7**:361–370.

Padlan, E.A.: **Anatomy of the antibody molecule.** *Mol. Immunol.* 1994, **31**:169–217.

Saphire, E.O., Parren, P.W., Pantophlet, R., Zwick, M.B., Morris, G.M., Rudd, P.M., Dwek, R.A., Stanfield, R.L., Burton, D.R., and Wilson, I.A.: **Crystal structure of a neutralizing human IGG against HIV-1: a template for vaccine design.** *Science* 2001, **293**:1155–1159.

Stanfield, R.L., and Wilson, I.A.: **Protein–peptide interactions.** *Curr. Opin. Struct. Biol.* 1995, **5**:103–113.

Tanner, J.J., Komissarov, A.A., and Deutscher, S.L.: **Crystal structure of an antigen-binding fragment bound to single-stranded DNA.** *J. Mol. Biol.* 2001, **314**:807–822.

Wilson, I.A., and Stanfield, R.L.: **Antibody–antigen interactions: new structures and new conformational changes.** *Curr. Opin. Struct. Biol.* 1994, **4**:857–867.

4-9 Antigen–antibody interactions involve a variety of forces.

Braden, B.C., and Poljak, R.J.: **Structural features of the reactions between antibodies and protein antigens.** *FASEB J.* 1995, **9**:9–16.

Braden, B.C., Goldman, E.R., Mariuzza, R.A., and Poljak, R.J.: **Anatomy of an antibody molecule: structure, kinetics, thermodynamics and mutational studies of the antilysozyme antibody D1.3.** *Immunol. Rev.* 1998, **163**:45–57.

Ros, R., Schwesinger, F., Anselmetti, D., Kubon, M., Schäfer, R., Plückthun, A., and Tiefenauer, L.: **Antigen binding forces of individually addressed single-chain Fv antibody molecules.** *Proc. Natl Acad. Sci. USA* 1998, **95**:7402–7405.

4-10 The T-cell receptor is very similar to a Fab fragment of immunoglobulin.

Al-Lazikani, B., Lesk, A.M., and Chothia, C.: **Canonical structures for the hypervariable regions of T cell αβ receptors.** *J. Mol. Biol.* 2000, **295**:979–995.

Kjer-Nielsen, L., Clements, C.S., Brooks, A.G., Purcell, A.W., McCluskey, J., and Rossjohn, J.: **The 1.5 Å crystal structure of a highly selected antiviral T cell receptor provides evidence for a structural basis of immunodominance.** *Structure (Camb.)* 2002, **10**:1521–1532.

Machius, M., Cianga, P., Deisenhofer, J., and Ward, E.S.: **Crystal structure of a T cell receptor Vα11 (AV11S5) domain: new canonical forms for the first and second complementarity determining regions.** *J. Mol. Biol.* 2001, **310**:689–698.

4-11 A T-cell receptor recognizes antigen in the form of a complex of a foreign peptide bound to an MHC molecule.

Garcia, K.C., and Adams, E.J.: **How the T cell receptor sees antigen—a structural view.** *Cell* 2005, **122**:333–336.

Hennecke, J., and Wiley, D.C.: **Structure of a complex of the human αβ T cell receptor (TCR) HA1.7, influenza hemagglutinin peptide, and major histocompatibility complex class II molecule, HLA-DR4 (DRA*0101 and DRB1*0401): insight into TCR cross-restriction and alloreactivity.** *J. Exp. Med.* 2002, **195**:571–581.

Luz, J.G., Huang, M., Garcia, K.C., Rudolph, M.G., Apostolopoulos, V., Teyton, L., and Wilson, I.A.: **Structural comparison of allogeneic and syngeneic T cell receptor–peptide–major histocompatibility complex complexes: a buried alloreactive mutation subtly alters peptide presentation substantially increasing V_β interactions.** *J. Exp. Med.* 2002, **195**:1175–1186.

Reinherz, E.L., Tan, K., Tang, L., Kern, P., Liu, J., Xiong, Y., Hussey, R.E., Smolyar, A., Hare, B., Zhang, R., *et al.* **The crystal structure of a T cell receptor in complex with peptide and MHC class II.** *Science* 1999, **286**:1913–1921.

Rudolph, M.G., Stanfield, R.L., and Wilson, I.A.: **How TCRs bind MHCs, peptides, and coreceptors.** *Annu. Rev. Immunol.* 2006, **24**:419–466.

4-12 There are two classes of MHC molecules with distinct subunit compositions but similar three-dimensional structures.
&
4-13 Peptides are stably bound to MHC molecules, and also serve to stabilize the MHC molecule on the cell surface.

Bouvier, M.: **Accessory proteins and the assembly of human class I MHC molecules: a molecular and structural perspective.** *Mol. Immunol.* 2003, **39**:697–706.

Dessen, A., Lawrence, C.M., Cupo, S., Zaller, D.M., and Wiley, D.C.: **X-ray crystal structure of HLA-DR4 (DRA*0101, DRB1*0401) complexed with a peptide from human collagen II.** *Immunity* 1997, **7**:473–481.

Fremont, D.H., Hendrickson, W.A., Marrack, P., and Kappler, J.: **Structures of an MHC class II molecule with covalently bound single peptides.** *Science* 1996, **272**:1001–1004.

Fremont, D.H., Matsumura, M., Stura, E.A., Peterson, P.A. and Wilson, I.A.: **Crystal structures of two viral peptides in complex with murine MHC class 1 H-2K^b.** *Science* 1992, **257**:919–927.

Fremont, D.H., Monnaie, D., Nelson, C.A., Hendrickson, W.A., and Unanue, E.R.: **Crystal structure of I-Ak in complex with a dominant epitope of lysozyme.** *Immunity* 1998, **8**:305–317.

Macdonald, W.A., Purcell, A.W., Mifsud, N.A., Ely, L.K., Williams, D.S., Chang, L., Gorman, J.J., Clements, C.S., Kjer-Nielsen, L., Koelle, D.M., *et al.*: **A naturally selected dimorphism within the HLA-B44 supertype alters class I structure, peptide repertoire, and T cell recognition.** *J. Exp. Med.* 2003, **198**:679–691.

Zhu, Y., Rudensky, A.Y., Corper, A.L., Teyton, L., and Wilson, I.A.: **Crystal structure of MHC class II I-Ab in complex with a human CLIP peptide: prediction of an I-Ab peptide-binding motif.** *J. Mol. Biol.* 2003, **326**:1157–1174.

4-14 MHC class I molecules bind short peptides of 8–10 amino acids by both ends.

Bouvier, M., and Wiley, D.C.: **Importance of peptide amino and carboxyl termini to the stability of MHC class I molecules.** *Science* 1994, **265**:398–402.

Govindarajan, K.R., Kangueane, P., Tan, T.W., and Ranganathan, S.: **MPID: MHC-Peptide Interaction Database for sequence–structure–function information on peptides binding to MHC molecules.** *Bioinformatics* 2003, **19**:309–310.

Saveanu, L., Fruci, D., and van Endert, P.: **Beyond the proteasome: trimming, degradation and generation of MHC class I ligands by auxiliary proteases.** *Mol. Immunol.* 2002, **39**:203–215.

Weiss, G.A., Collins, E.J., Garboczi, D.N., Wiley, D.C., and Schreiber, S.L.: **A tricyclic ring system replaces the variable regions of peptides presented by three alleles of human MHC class I molecules.** *Chem. Biol.* 1995, **2**:401–407.

4-15 The length of the peptides bound by MHC class II molecules is not constrained.

Conant, S.B., and Swanborg, R.H.: **MHC class II peptide flanking residues of exogenous antigens influence recognition by autoreactive T cells.** *Autoimmun. Rev.* 2003, **2**:8–12.

Guan, P., Doytchinova, I.A., Zygouri, C., and Flower, D.R.: **MHCPred: a server for quantitative prediction of peptide–MHC binding.** *Nucleic Acids Res.* 2003, **31**:3621–3624.

Lippolis, J.D., White, F.M., Marto, J.A., Luckey, C.J., Bullock, T.N., Shabanowitz, J., Hunt, D.F., and Engelhard, V.H.: **Analysis of MHC class II antigen processing by quantitation of peptides that constitute nested sets.** *J. Immunol.* 2002, **169**:5089–5097.

Park, J.H., Lee, Y.J., Kim, K.L., and Cho, E.W.: **Selective isolation and identification of HLA-DR-associated naturally processed and presented epitope peptides.** *Immunol. Invest.* 2003, **32**:155–169.

Rammensee, H.G.: **Chemistry of peptides associated with MHC class I and class II molecules.** *Curr. Opin. Immunol.* 1995, **7**:85–96.

Rudensky, A.Y., Preston-Hurlburt, P., Hong, S.C., Barlow, A., and Janeway, C.A., Jr: **Sequence analysis of peptides bound to MHC class II molecules.** *Nature* 1991, **353**:622–627.

Sercarz, E.E., and Maverakis, E.: **MHC-guided processing: binding of large antigen fragments.** *Nat. Rev. Immunol.* 2003, **3**:621–629.

Sinnathamby, G., and Eisenlohr, L.C.: **Presentation by recycling MHC class II molecules of an influenza hemagglutinin-derived epitope that is revealed in the early endosome by acidification.** *J. Immunol.* 2003, **170**:3504–3513.

4-16 The crystal structures of several peptide:MHC:T-cell receptor complexes show a similar orientation of the T-cell receptor over the peptide:MHC complex.

Buslepp, J., Wang, H., Biddison, W.E., Appella, E., and Collins, E.J.: **A correlation between TCR Vα docking on MHC and CD8 dependence: implications for T cell selection.** *Immunity* 2003, **19**:595–606.

Ding, Y.H., Smith, K.J., Garboczi, D.N., Utz, U., Biddison, W.E., and Wiley, D.C.: **Two human T cell receptors bind in a similar diagonal mode to the HLA-A2/Tax peptide complex using different TCR amino acids.** *Immunity* 1998, **8**:403–411.

Kjer-Nielsen, L., Clements, C.S., Purcell, A.W., Brooks, A.G., Whisstock, J.C., Burrows, S.R., McCluskey, J., and Rossjohn, J.: **A structural basis for the selection of dominant αβ T cell receptors in antiviral immunity.** *Immunity* 2003, **18**:53–64.

Garcia, K.C., Degano, M., Pease, L.R., Huang, M., Peterson, P.A., Leyton, L., and Wilson, I.A.: **Structural basis of plasticity in T cell receptor recognition of a self peptide-MHC antigen.** *Science* 1998, **279**:1166–1172.

Reiser, J.B., Darnault, C., Gregoire, C., Mosser, T., Mazza, G., Kearney, A., van der Merwe, P.A., Fontecilla-Camps, J.C., Housset, D., and Malissen, B.: **CDR3 loop flexibility contributes to the degeneracy of TCR recognition.** *Nat. Immunol.* 2003, **4**:241–247.

Sant'Angelo, D.B., Waterbury, G., Preston-Hurlburt, P., Yoon, S.T., Medzhitov, R., Hong, S.C., and Janeway, C.A., Jr: **The specificity and orientation of a TCR to its peptide-MHC class II ligands.** *Immunity* 1996, **4**:367–376.

Teng, M.K., Smolyar, A., Tse, A.G.D., Liu, J.H., Liu, J., Hussey, R.E., Nathenson, S.G., Chang, H.C., Reinherz, E.L., and Wang, J.H.: **Identification of a common docking topology with substantial variation among different TCR–MHC–peptide complexes.** *Curr. Biol.* 1998, **8**:409–412.

4-17 The CD4 and CD8 cell-surface proteins of T cells are required to make an effective response to antigen.

Chang, H. C., Tan, K., Ouyang, J., Parisini, E., Liu, J. H., Le, Y., Wang, X., Reinherz, E. L., and Wang, J. H. **Structural and mutational analyses of CD8αβ heterodimer and comparison with the CD8αα homodimer.** *Immunity* 2005, **6**: 661–671.

Gao, G.F., Tormo, J., Gerth, U.C., Wyer, J.R., McMichael, A.J., Stuart, D.I., Bell, J.I., Jones, E.Y., and Jakobsen, B.Y.: **Crystal structure of the complex between human CD8αα and HLA-A2.** *Nature* 1997, **387**:630–634.

Gaspar, R., Jr, Bagossi, P., Bene, L., Matko, J., Szollosi, J., Tozser, J., Fesus, L., Waldmann, T.A., and Damjanovich, S.: **Clustering of class I HLA oligomers with CD8 and TCR: three-dimensional models based on fluorescence resonance energy transfer and crystallographic data.** *J. Immunol.* 2001, **166**:5078–5086.

Kim, P.W., Sun, Z.Y., Blacklow, S.C., Wagner, G., and Eck, M.J.: **A zinc clasp structure tethers Lck to T cell coreceptors CD4 and CD8.** *Science* 2003, **301**:1725–1728.

Moody, A.M., North, S.J., Reinhold, B., Van Dyken, S.J., Rogers, M.E., Panico, M., Dell, A., Morris, H.R., Marth, J.D., and Reinherz, E.L.: **Sialic acid capping of CD8β core 1-*O*-glycans controls thymocyte-major histocompatibility complex class I interaction.** *J. Biol. Chem.* 2003, **278**:7240–7260.

Wang, J.H., and Reinherz, E.L.: **Structural basis of T cell recognition of peptides bound to MHC molecules.** *Mol. Immunol.* 2002, **38**:1039–1049.

Wu, H., Kwong, P.D., and Hendrickson, W.A.: **Dimeric association and segmental variability in the structure of human CD4.** *Nature* 1997, **387**:527–530.

Zamoyska, R.: **CD4 and CD8: modulators of T cell receptor recognition of antigen and of immune responses?** *Curr. Opin. Immunol.* 1998, **10**:82–86.

4-18 The two classes of MHC molecules are expressed differentially on cells.

Steimle, V., Siegrist, C.A., Mottet, A., Lisowska-Grospierre, B., and Mach, B.: **Regulation of MHC class II expression by interferon-γ mediated by the transactivator gene CIITA.** *Science* 1994, **265**:106–109.

4-19 A distinct subset of T cells bears an alternative receptor made up of γ and δ chains.

Allison, T.J., and Garboczi, D.N.: **Structure of γδ T cell receptors and their recognition of non-peptide antigens.** *Mol. Immunol.* 2002, **38**:1051–1061.

Allison, T.J., Winter, C.C., Fournie, J.J., Bonneville, M., and Garboczi, D.N.: **Structure of a human γδ T-cell antigen receptor.** *Nature* 2001, **411**:820–824.

Carding, S.R., and Egan, P.J.: **γδ T cells: functional plasticity and heterogeneity.** *Nat. Rev. Immunol.* 2002, **2**:336–345.

Das, H., Wang, L., Kamath, A., and Bukowski, J.F.: **$V_\gamma 2V_\delta 2$ T-cell receptor-mediated recognition of aminobisphosphonates.** *Blood* 2001, **98**:1616–1618.

Wilson, I.A., and Stanfield, R.L.: **Unraveling the mysteries of γδ T cell recognition.** *Nat. Immunol.* 2001, **2**:579–581.

Wu, J., Groh, V., and Spies, T.: **T cell antigen receptor engagement and specificity in the recognition of stress-inducible MHC class I-related chains by human epithelial γδ T cells.** *J. Immunol.* 2002, **169**:1236–1240.

The Generation of Lymphocyte Antigen Receptors

5

Lymphocyte antigen receptors, in the form of immunoglobulins on B cells and T-cell receptors on T cells, are the means by which lymphocytes sense the presence of antigens in their environment. Individual lymphocytes bear numerous copies of a single antigen receptor with a unique antigen-binding site, which determines the antigens that the lymphocyte can bind. Because each person possesses billions of lymphocytes, these cells collectively enable a response to a great variety of antigens. The wide range of antigen specificities in the antigen-receptor repertoire is due to variation in the amino acid sequence at the antigen-binding site, which is made up from the variable (V) regions of the receptor protein chains. In each chain the V region is linked to an invariant constant (C) region, which provides effector or signaling functions.

Given the importance of a diverse repertoire of lymphocyte receptors in the defense against infection, it is not surprising that a complex and elegant genetic mechanism has evolved for generating these highly variable proteins. Each receptor-chain variant cannot be encoded in full in the genome, as this would require more genes for antigen receptors than there are genes in the entire genome. Instead, we will see that the V regions of the receptor chains are encoded in several pieces—so-called gene segments. These are assembled in the developing lymphocyte by somatic DNA recombination to form a complete V-region sequence, a mechanism known generally as **gene rearrangement**. A fully assembled V-region sequence is made up of two or three types of gene segment, each of which is present in multiple copies in the germline genome. The selection of a gene segment of each type during gene rearrangement occurs at random, and the large number of possible combinations accounts for much of the diversity of the receptor repertoire.

In the first and second parts of this chapter we describe the intrachromosomal gene rearrangements that generate the primary repertoire of V regions of immunoglobulin and T-cell receptor genes. The mechanism of gene rearrangement is common to both B cells and T cells, and its evolution was probably critical to the evolution of the vertebrate adaptive immune system. The antigen receptors expressed after these primary gene rearrangements provide the repertoire of diverse antigen specificities of naive B cells and T cells.

Immunoglobulins can be synthesized as either transmembrane receptors or secreted antibodies, unlike T-cell receptors, which only exist as transmembrane receptors. In the third part of the chapter we shall see how the transition from the production of transmembrane immunoglobulins by activated B cells to the production of secreted antibodies by plasma cells is achieved. The C regions of antibodies have important effector functions in an immune response, and we also briefly consider here the different types of antibody C regions and their properties, a topic that we shall return to in more detail in Chapter 10.

Next we consider two kinds of secondary modifications that can take place in rearranged immunoglobulin genes in B cells but do not occur in T cells. These all provide further diversity in the antibody repertoire that helps make the antibody response more effective over time. One is a process known as somatic hypermutation, which introduces point mutations into the V regions of rearranged immunoglobulin genes in activated B cells, producing some variants that bind more strongly to the antigen. This leads to the phenomenon of affinity maturation, in which the affinity of antibodies for the antigen increases as the immune response progresses. The second modification is the limited, but functionally important, sequential expression of different immunoglobulin C regions in activated B cells by a process called class switching, which enables antibodies with the same antigen specificity but different functional properties to be produced. We end the chapter with a brief look at the evolution of adaptive immunity and the different ways in which diversity is achieved in different species.

Primary immunoglobulin gene rearrangement.

Virtually any substance can be the target of an antibody response, and the response to even a single epitope comprises many different antibody molecules, each with a subtly different specificity for the epitope and a unique affinity, or binding strength. The total number of antibody specificities available to an individual is known as the **antibody repertoire** or **immunoglobulin repertoire**, and in humans is at least 10^{11} and probably several orders of magnitude greater. The number of antibody specificities present at any one time is, however, limited by the total number of B cells in an individual, as well as by each individual's previous encounters with antigens.

Before it was possible to examine the immunoglobulin genes directly, there were two main hypotheses for the origin of this diversity. The **germline theory** held that there is a separate gene for each different immunoglobulin chain and that the antibody repertoire is largely inherited. In contrast, **somatic diversification theories** proposed that the observed repertoire is generated from a limited number of inherited V-region sequences that undergo alteration within B cells during the individual's lifetime. Cloning of the immunoglobulin genes revealed that elements of both theories were correct and that the DNA sequence encoding each V region is generated by rearrangements of a relatively small group of inherited gene segments. Diversity is further enhanced by the process of somatic hypermutation in mature activated B cells. Thus the somatic diversification theory was essentially correct, although the concept of multiple germline genes embodied in the germline theory also proved true.

5-1 Immunoglobulin genes are rearranged in antibody-producing cells.

In nonlymphoid cells, the gene segments encoding the greater part of the V region of an immunoglobulin chain are a considerable distance away from the sequence encoding the C region. In mature B lymphocytes, however, the assembled V-region sequence lies much nearer the C region, as a consequence of gene rearrangement. Rearrangement within the immunoglobulin genes was originally discovered about 30 years ago, when the techniques of restriction enzyme analysis first made it possible to study the organization of the immunoglobulin genes in both B cells and nonlymphoid cells. Such experiments showed that segments of genomic DNA within the immunoglobulin genes are rearranged in cells of the B-lymphocyte lineage, but not in other cells. This process of rearrangement is known as **somatic recombination**, to distinguish it from the meiotic recombination that takes place during the production of gametes.

5-2 Complete genes that encode a variable region are generated by the somatic recombination of separate gene segments.

The V region, or V domain, of an immunoglobulin heavy or light chain is encoded by more than one gene segment. For the light chain, the V domain is encoded by two separate DNA segments. The first encodes the first 95–101 amino acids, the greater part of the domain, and is called a **variable** or **V gene segment**. The second encodes the remainder of the domain (up to 13 amino acids) and is called a **joining** or **J gene segment**.

The rearrangements that produce a complete immunoglobulin light-chain gene are shown in Fig. 5.1 (center panel). The joining of a V and a J gene segment creates an exon that encodes the whole light-chain V region. In the

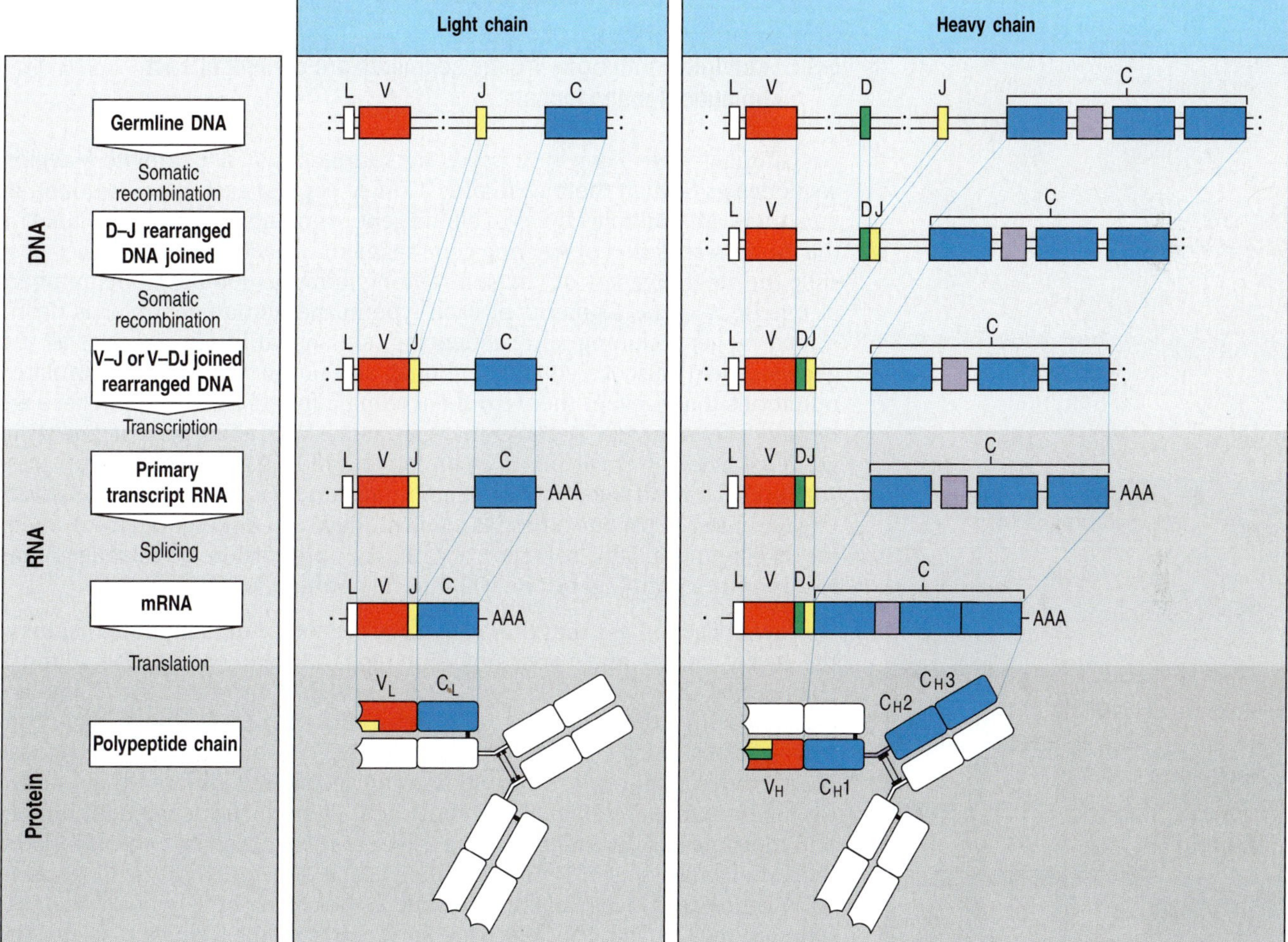

Fig. 5.1 V-region genes are constructed from gene segments. Light-chain V-region genes are constructed from two segments (center panel). A variable (V) and a joining (J) gene segment in the genomic DNA are joined to form a complete light-chain V-region exon. Immunoglobulin chains are extracellular proteins and the V gene segment is preceded by an exon encoding a leader peptide (L), which directs the protein into the cell's secretory pathways and is then cleaved. The light-chain C region is encoded in a separate exon and is joined to the V-region exon by splicing of the light-chain RNA to remove the L-to-V and the J-to-C introns. Heavy-chain V regions are constructed from three gene segments (right panel). First, the diversity (D) and J gene segments join, then the V gene segment joins to the combined DJ sequence, forming a complete V_H exon. A heavy-chain C-region gene is encoded by several exons. The C-region exons, together with the leader sequence, are spliced to the V-domain sequence during processing of the heavy-chain RNA transcript. The leader sequence is removed after translation, and the disulfide bonds that link the polypeptide chains are formed. The hinge region is shown in purple.

unrearranged DNA, the V gene segments are located relatively far away from the C region. The J gene segments are located close to the C region, however, and the joining of a V gene segment to a J gene segment also brings the V gene segment close to a C-region sequence. The J gene segment of the rearranged V region is separated from a C-region sequence only by a short intron. To make a complete immunoglobulin light-chain messenger RNA, the V-region exon is joined to the C-region sequence by RNA splicing after transcription (see Fig. 5.1).

A heavy-chain V region is encoded in three gene segments. In addition to the V and J gene segments (denoted V_H and J_H to distinguish them from the light-chain V_L and J_L), there is a third gene segment called the **diversity** or **D_H gene segment**, which lies between the V_H and J_H gene segments. The recombination process that generates a complete heavy-chain V region is shown in Fig. 5.1 (right panel), and occurs in two separate stages. In the first, a D_H gene segment is joined to a J_H gene segment; then a V_H gene segment rearranges to DJ_H to make a complete V_H-region exon. As with the light-chain genes, RNA splicing joins the assembled V-region sequence to the neighboring C-region gene.

5-3 Multiple contiguous V gene segments are present at each immunoglobulin locus.

For simplicity we have discussed the formation of a complete V-region sequence as though there were only a single copy of each gene segment. In fact, there are multiple copies of all the gene segments in germline DNA. It is the random selection of just one gene segment of each type that makes possible the great diversity of V regions among immunoglobulins. The numbers of functional gene segments of each type in the human genome, as determined by gene cloning and sequencing, are shown in Fig. 5.2. Not all the gene segments discovered are functional, as a proportion have accumulated mutations that prevent them from encoding a functional protein. These are termed 'pseudogenes.' Because there are many V, D, and J gene segments in germline DNA, no single one is essential. This reduces the evolutionary pressure on each gene segment to remain intact, and has resulted in a relatively large number of pseudogenes. As some of these can undergo rearrangement just like a normal gene segment, a significant proportion of rearrangements incorporate a pseudogene and will thus be nonfunctional.

Number of functional gene segments in human immunoglobulin loci

Segment	Light chains		Heavy chain
	κ	λ	H
Variable (V)	34–38	29–33	38–46
Diversity (D)	0	0	23
Joining (J)	5	4–5	6
Constant (C)	1	4–5	9

Fig. 5.2 The numbers of functional gene segments for the V regions of human heavy and light chains. These numbers are derived from exhaustive cloning and sequencing of DNA from one individual and exclude all pseudogenes (mutated and nonfunctional versions of a gene sequence). As a result of genetic polymorphism, the numbers will not be the same for all people.

We saw in Section 4-1 that there are three sets of immunoglobulin chains—the heavy chains and two equivalent types of light chains, the κ and λ chains. The immunoglobulin gene segments that encode these chains are organized into three clusters or **genetic loci**—the κ, λ, and heavy-chain loci—each of which can assemble a complete V-region sequence. Each locus is on a different chromosome and is organized slightly differently, as shown for the human loci in Fig. 5.3. At the λ light-chain locus, located on human chromosome 22, a cluster of V_λ gene segments is followed by four (or in some individuals five) sets of J_λ gene segments each linked to a single C_λ gene. In the κ light-chain locus, on chromosome 2, the cluster of V_κ gene segments is followed by a cluster of J_κ gene segments, and then by a single C_κ gene. The organization of the heavy-chain locus, on chromosome 14, resembles that of the κ locus, with separate clusters of V_H, D_H, and J_H gene segments and of C_H genes. The heavy-chain locus differs in one important way: instead of a single C region, it contains a series of C regions arrayed one after the other, each of which corresponds to a different isotype. B cells initially express the heavy-chain isotypes μ and δ (see Section 4-1), which is accomplished by alternative mRNA splicing and leads to the expression of immunoglobulins IgM and IgD, as we shall see in Section 5-14. The expression of other isotypes, such as γ (giving IgG), occurs at a later stage, through DNA rearrangements referred to as class switching, which is described in Section 5-19.

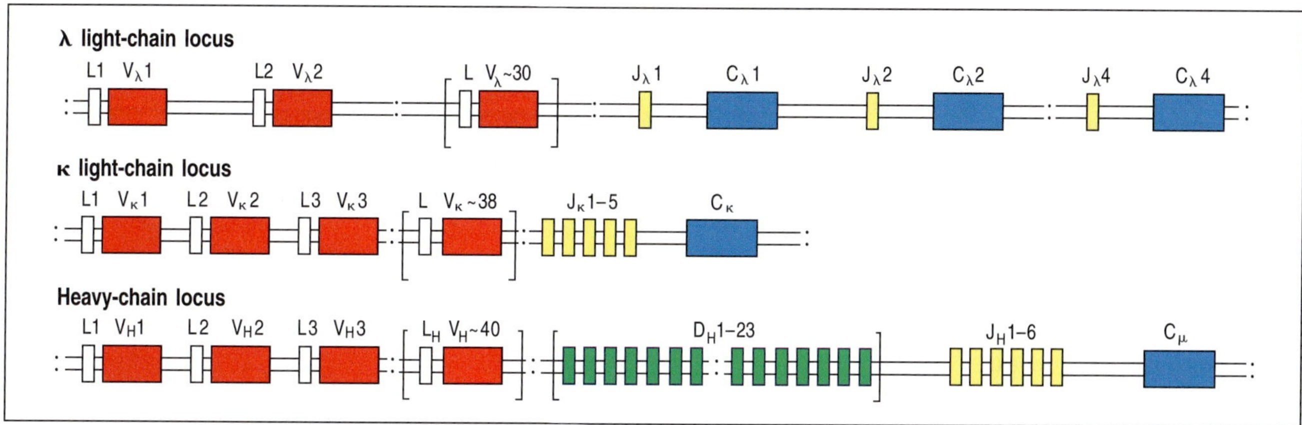

Fig. 5.3 The germline organization of the immunoglobulin heavy- and light-chain loci in the human genome. The genetic locus for the λ light chain (chromosome 22) has between 29 and 33 functional V_λ gene segments and four or five pairs of functional J_λ gene segments and C_λ genes, depending on variation between individuals. The κ locus (chromosome 2) is organized in a similar way, with about 38 functional V_κ gene segments accompanied by a cluster of five J_κ gene segments but with a single C_κ gene. In approximately 50% of individuals, the entire cluster of V_κ gene segments has undergone an increase by duplication (not shown, for simplicity). The heavy-chain locus (chromosome 14) has about 40 functional V_H gene segments and a cluster of around 23 D_H segments lying between these V_H gene segments and six J_H gene segments. The heavy-chain locus also contains a large cluster of C_H genes that are shown in Fig. 5.16. For simplicity, all V gene segments have been shown in the same chromosomal orientation, only the first C_H gene (for Cμ) is shown, without illustrating its separate exons, and all pseudogenes have been omitted. This diagram is not to scale: the total length of the heavy-chain locus is more than 2 megabases (2 million bases), whereas some of the D gene segments are only six bases long.

The human V gene segments can be grouped into families in which each member shares at least 80% DNA sequence identity with all others in the family. Both the heavy-chain and κ-chain V gene segments can be subdivided into seven families, and there are eight families of V_λ gene segments. The families can be grouped into clans, made up of families that are more similar to each other than to families in other clans. Human V_H gene segments fall into three clans. All the V_H gene segments identified from amphibians, reptiles, and mammals also fall into the same three clans, suggesting that these clans existed in a common ancestor of these modern animal groups. Thus, the V gene segments that we see today have arisen by a series of gene duplications and diversification through evolutionary time.

5-4 Rearrangement of V, D, and J gene segments is guided by flanking DNA sequences.

For a complete immunoglobulin or T-cell receptor chain to be expressed, DNA rearrangements must take place at the correct locations relative to the V, D, or J gene segment coding regions. In addition, joins must be regulated such that a V gene segment joins to a D or J and not to another V. DNA rearrangements are guided by conserved noncoding DNA sequences that are found adjacent to the points at which recombination takes place and are called **recombination signal sequences** (**RSSs**). An RSS consists of a conserved block of seven nucleotides—the **heptamer** 5′CACAGTG3′—which is always contiguous with the coding sequence, followed by a nonconserved region known as the **spacer**, which is either 12 or 23 base pairs (bp) long, followed by a second conserved block of nine nucleotides—the **nonamer** 5′ACAAAAACC3′ (Fig. 5.4). The sequences given here are the consensus sequences and can vary slightly from individual to individual. The spacers vary in sequence but their conserved lengths correspond to one turn (12 bp) or two turns (23 bp) of the DNA double helix. This is thought to bring the heptamer and nonamer sequences to the same side of the DNA helix to allow interactions with proteins catalyzing recombination, but this idea still lacks structural proof. The heptamer–spacer–nonamer sequence motif—the RSS—is always found directly adjacent to the coding sequence of V, D, or J gene segments. Recombination normally occurs between gene segments located on the same chromosome. A gene segment flanked by an RSS with a 12-bp spacer typically can be joined only to one flanked by a 23-bp spacer RSS. This is known as the **12/23 rule**. Thus, for the heavy chain, a D_H gene segment can be joined to a J_H gene segment and a V_H gene segment to a D_H gene segment, but V_H gene segments cannot be joined to J_H gene segments directly, as both V_H and J_H gene segments are flanked by 23-bp spacers and the D_H gene segments have 12-bp spacers on both sides (Fig. 5.4).

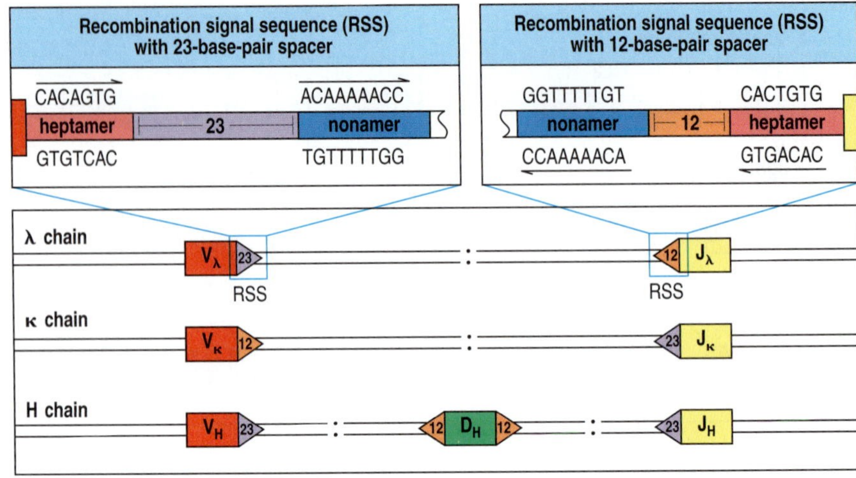

Fig. 5.4 Recombination signal sequences are conserved heptamer and nonamer sequences that flank the gene segments encoding the V, D, and J regions of immunoglobulins. Recombination signal sequences (RSSs) are composed of heptamer (CACAGTG) and nonamer (ACAAAAACC) sequences that are separated either by 12 bp or approximately 23 bp nucleotides. The heptamer–12-bp spacer–nonamer motif is depicted here as an orange arrowhead; the motif that includes the 23-bp spacer is depicted as a purple arrowhead. Joining of gene segments almost always involves a 12-bp and a 23-bp RSS—the 12/23 rule. The arrangement of RSSs in the V (red), D (green), and J (yellow) gene segments of heavy (H) and light (λ and κ) chains of immunoglobulin is shown here. Note that according to the 12/23 rule, the arrangement of RSSs in the immunoglobulin heavy-chain gene segments precludes direct V-to-J joining.

Recall from Section 4-6 that the antigen-binding region of an immunoglobulin is formed by three hypervariable regions. The first two, CDR1 and CDR2, are encoded in the V gene segment itself. The third, CDR3, is encoded by the additional DNA sequence that is created by the joining of the V and J gene segments for the light chain, and the V, D, and J gene segments for the heavy chain. Additional diversity in the antibody repertoire can result from the generation of CDR3 regions that seem to result from the joining of one D gene segment to another D gene segment. Although infrequent, such D–D joining would seem to violate the 12/23 rule, and it is not clear how these rare rearrangements are generated. In humans, D–D joining is found in approximately 5% of antibodies and is the major mechanism accounting for the unusually long CDR3 loops found in some heavy chains.

Movie 5.1 

The mechanism of DNA rearrangement is similar for the heavy- and light-chain loci, although only one joining event is needed to generate a light-chain gene but two are required for a heavy-chain gene. When two gene segments are in the same transcriptional orientation in the DNA, rearrangement involves the looping-out and deletion of the DNA between them (Fig. 5.5, left panels), but if the gene segments have opposite transcriptional orientations the intervening DNA meets a different fate (see Fig. 5.5, right panels). In the latter case, the intervening DNA is retained in the chromosome in an inverted orientation. This mode of recombination is less common, but it accounts for about half of all V_κ to J_κ joins in humans because the orientation of half the V_κ gene segments is opposite to that of the J_κ gene segments.

5-5 The reaction that recombines V, D, and J gene segments involves both lymphocyte-specific and ubiquitous DNA-modifying enzymes.

The molecular mechanism of V-region rearrangement, or **V(D)J recombination**, is illustrated in Fig. 5.6. The two RSSs are brought together by interactions between proteins that specifically recognize the length of spacer and thus enforce the 12/23 rule for recombination. The DNA molecule is then broken in two places and rejoined in a different configuration. The ends of the heptamer sequences are joined precisely in a head-to-head fashion to form a **signal joint**; when the joining segments are in the same orientation, the signal joint is in a circular piece of extrachromosomal DNA (see Fig. 5.5, left panels), which is lost from the genome when the cell divides. The V and J gene segments, which remain on the chromosome, join to form what is called the **coding joint**. In the case of rearrangement by inversion (see Fig. 5.5, right panels), the signal joint is also retained within the chromosome,

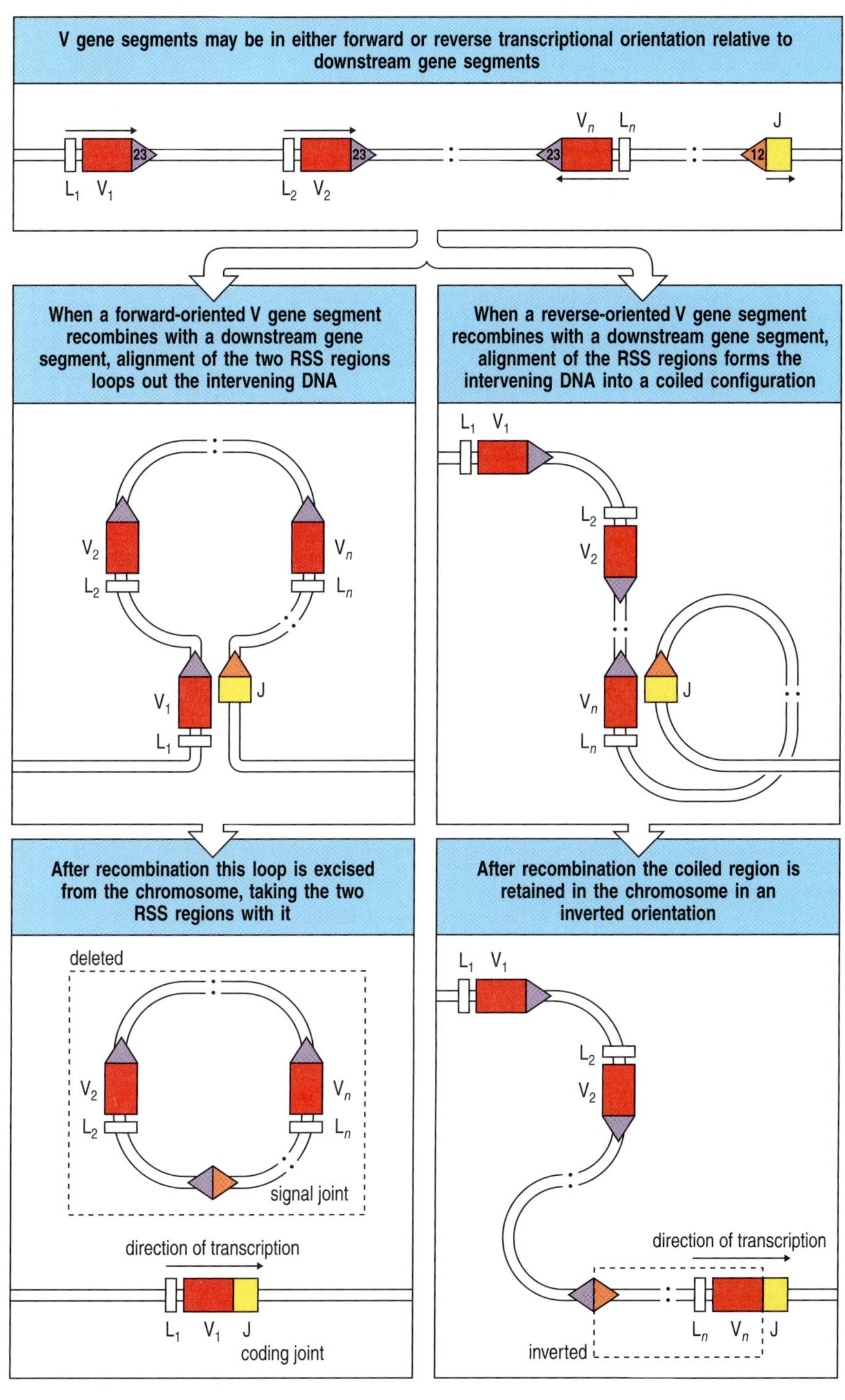

V gene segments may be in either forward or reverse transcriptional orientation relative to downstream gene segments
When a forward-oriented V gene segment recombines with a downstream gene segment, alignment of the two RSS regions loops out the intervening DNA
When a reverse-oriented V gene segment recombines with a downstream gene segment, alignment of the RSS regions forms the intervening DNA into a coiled configuration
After recombination this loop is excised from the chromosome, taking the two RSS regions with it
After recombination the coiled region is retained in the chromosome in an inverted orientation
deleted
signal joint
direction of transcription
coding joint
inverted

Fig. 5.5 V-region gene segments are joined by recombination. In every V-region recombination event, the recombination signal sequences (RSSs) flanking the gene segments are brought together to allow recombination to take place. The 12-bp-spaced RSSs are shown in orange, the 23-bp-spaced RSSs in purple. For simplicity, the recombination of a light-chain gene is illustrated; for a heavy-chain gene, two separate recombination events are required to generate a functional V region. In most cases, the two segments undergoing rearrangement (the V and J gene segments in this example) are arranged in the same transcriptional orientation in the chromosome (left panels), and juxtaposition of the RSSs results in the looping out of the intervening DNA. Recombination occurs at the ends of the heptamer sequences in the RSSs, creating the so-called signal joint and releasing the intervening DNA in the form of a closed circle. Subsequently, the joining of the V and J gene segments creates the coding joint in the chromosomal DNA. In other cases, illustrated in the right panels, the V and J gene segments are initially oriented in opposite transcriptional directions. In this case, alignment of the RSSs requires the coiled topology shown, rather than a simple loop, so that joining the ends of the two heptamer sequences now results in the inversion and integration of the intervening DNA into a new position on the chromosome. Again, the joining of the V and J segments creates a functional V-region exon.

and the region of DNA between the V gene segment and the RSS of the J gene segment is inverted to form the coding joint. As we shall see later, the coding joint junction is imprecise, and consequently it generates much additional variability in the V-region sequence.

The complex of enzymes that act in concert to carry out somatic V(D)J recombination is termed the **V(D)J recombinase**. The lymphoid-specific components of the recombinase are called **RAG-1** and **RAG-2**, and they are encoded by two recombination-activating genes, *RAG1* and *RAG2*. This pair of genes is expressed in developing lymphocytes only while they are engaged in assembling their antigen receptors, as described in more detail in Chapter 8, and they are essential for V(D)J recombination. Indeed, the *RAG* genes

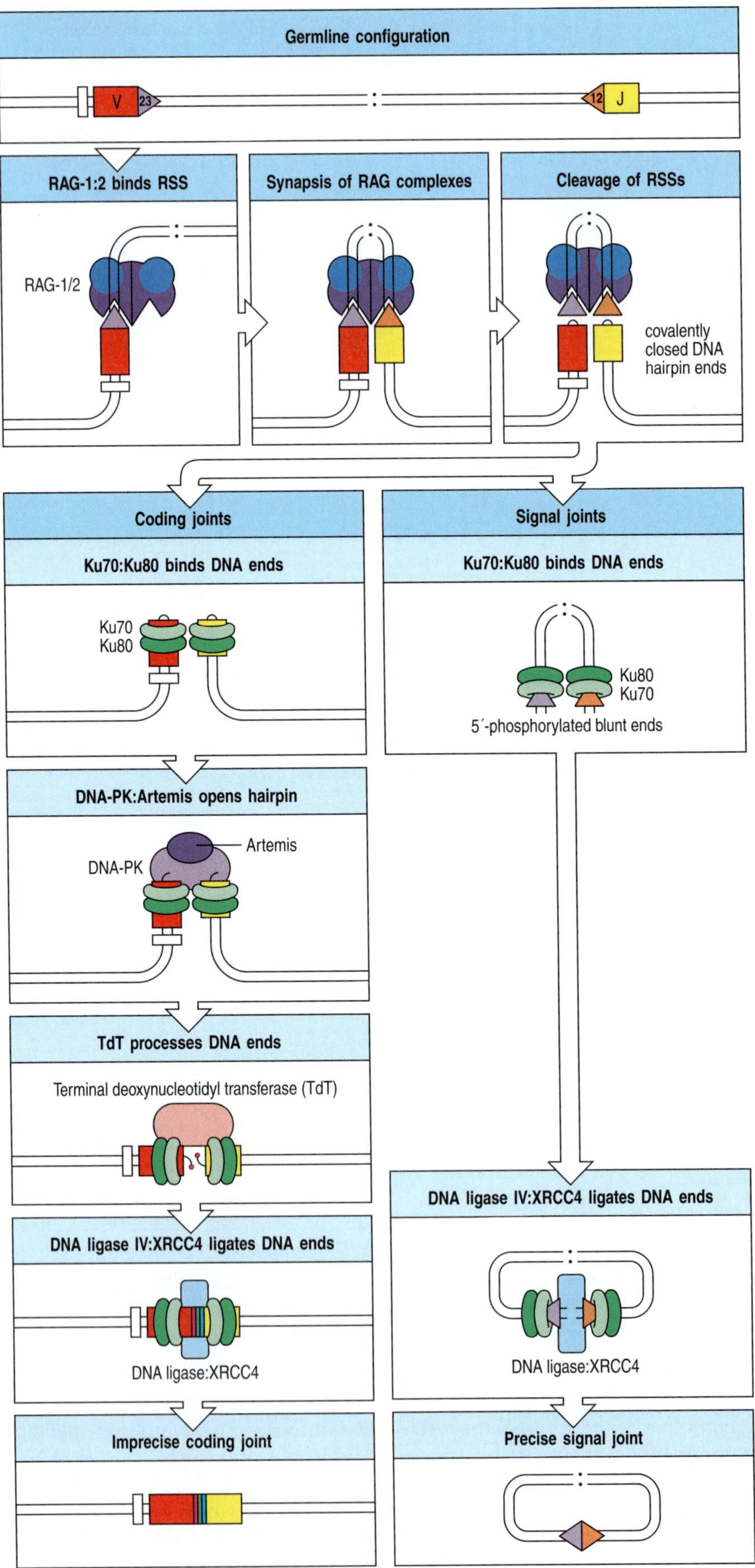

Fig. 5.6 Enzymatic steps in RAG-dependent V(D)J rearrangement. Recombination of gene segments containing recombination signal sequences (RSSs) (triangles) begins with the binding of a complex of RAG-1, RAG-2 (blue and purple), and high-mobility group (HMG) proteins (not shown) to one of the RSSs flanking the coding sequences to be joined (second row). The RAG complex then recruits the other RSS. In the cleavage step, the endonuclease activity of RAG makes single-stranded cuts in the DNA backbone precisely between each coding segment and its RSS. At each cutting point this creates a 3′-OH group, which then reacts with a phosphodiester bond on the opposite DNA strand to generate a hairpin, leaving a blunt double-stranded break at the end of the RSS. These two types of DNA ends are resolved in different ways. At the coding ends (left panels) essential repair proteins such as Ku70:Ku80 (green) bind to the hairpin. The DNA-PK:Artemis complex (purple) then joins the complex and its endonuclease activity opens the DNA hairpin at a random site, yielding either two flush-ended DNA strands or a single-strand extension. The cut end is then modified by terminal deoxynucleotidyl transferase (TdT) (pink) and exonuclease, which randomly add and remove nucleotides, respectively (this step is shown in more detail in Fig. 5.7). The two coding ends are finally ligated by DNA ligase IV in association with XRCC4 (turquoise). At the signal ends (right panels), Ku70:Ku80 binds to the RSS but the ends are not further modified. Instead, a complex of DNA ligase IV:XRCC4 joins the two ends precisely to form the signal joint.

expressed together can confer on nonlymphoid cells such as fibroblasts the capacity to rearrange exogenous segments of DNA containing the appropriate RSSs; this is how RAG-1 and RAG-2 were initially discovered.

The other proteins in the recombinase complex are mainly ubiquitous DNA-modifying proteins that are involved in the repair of DNA double-strand breaks and the modification of the ends of broken DNA strands. One is **Ku**, which is a heterodimer (Ku70:Ku80); this forms a ring around the DNA and associates tightly with a protein kinase catalytic subunit, DNA-PKcs, to form the **DNA-dependent protein kinase** (**DNA-PK**). Another is the protein **Artemis**, which has nuclease activity. The DNA ends are finally joined together by the enzyme **DNA ligase IV**, which forms a complex with the DNA repair protein XRCC4. DNA polymerases μ and λ participate in DNA-end fill-in synthesis. In addition, polymerase μ can add nucleotides in a template-independent manner.

V(D)J recombination is a multistep enzymatic process in which the first reaction is an endonucleolytic cleavage that requires the coordinated activity of both RAG proteins. Initially, a complex of RAG-1 and RAG-2 proteins, together with high-mobility group chromatin proteins, recognize and align the two RSSs that are the target of the cleavage reaction (see Fig. 5.6). RAG-1 is thought to specifically recognize the nonamer of the RSS. At this stage, the 12/23 rule is established through mechanisms that are still poorly understood. The endonuclease activity of the RAG protein complex, which is thought to reside in RAG-1, then makes two single-strand DNA breaks at sites just 5′ of each bound RSS, leaving a free 3′-OH group at the end of each coding segment. The 3′-OH group then attacks the phosphodiester bond on the other strand, creating a DNA 'hairpin' at the end of the gene segment coding region and a flush double-strand break at the ends of the two heptamer sequences. The DNA ends do not float apart, however, but are held tightly in the complex until the join has been completed. The blunt ends are precisely joined by a complex of DNA ligase IV and XRCC4 to form the signal joint.

Formation of the coding joint is more complex. The DNA ends with the hairpins are bound by Ku, which recruits the DNA-PKcs subunit. Artemis recruited to the complex is activated by phosphorylation by DNA-PK and then opens the DNA hairpins by making a single-strand nick in the DNA. This nicking can happen at various points along the hairpin, which leads to sequence variability in the final joint. The DNA repair enzymes in the complex modify the opened hairpins by removing nucleotides, while at the same time the lymphoid-specific enzyme **terminal deoxynucleotidyl transferase** (**TdT**), which is also part of the recombinase complex, adds nucleotides randomly to the single-strand ends. Addition and deletion of nucleotides can occur in any order, and one does not necessarily precede the other. Finally, DNA ligase IV joins the processed ends together, thus reconstituting a chromosome that includes the rearranged gene. This repair process creates diversity in the joint between gene segments while ensuring that the RSS ends are ligated without modification and that unintended genetic damage such as a chromosome break is avoided. Despite the use of some ubiquitous mechanisms of DNA repair, adaptive immunity based on the RAG-mediated generation of antigen receptors by somatic recombination seems to be unique to the jawed vertebrates, and its evolution is discussed in the last part of this chapter.

The *in vivo* roles of the enzymes involved in V(D)J recombination have been established through natural or artificially induced mutations. Mice lacking TdT do not add extra nucleotides to the joints between gene segments. Mice in which either of the *RAG* genes have been knocked out, or which lack DNA-PKcs, Ku, or Artemis, suffer a complete block in lymphocyte development at the gene-rearrangement stage or make only trivial numbers of B and T cells. They are said to suffer from **severe combined immune deficiency** (**SCID**).

The original ***scid*** mutation was discovered some time before the components of the recombination pathway were identified and was subsequently identified as a mutation in DNA-PKcs. In humans, mutations in *RAG1* or *RAG2* that result in partial V(D)J recombinase activity are responsible for an inherited disorder called **Omenn syndrome**, which is characterized by an absence of circulating B cells and an infiltration of skin by activated oligoclonal T lymphocytes. Mice deficient in components of ubiquitous DNA repair pathways, such as DNA-PKcs, Ku, or Artemis, are defective in double-strand break repair generally and are therefore also hypersensitive to ionizing radiation (which produces double-strand breaks). Defects in Artemis in humans, for example, produce a combined immunodeficiency of B and T cells that is associated with increased radiosensitivity. SCID caused by mutations in DNA repair pathways is called **irradiation-sensitive SCID** (**IR-SCID**) to distinguish it from SCID due to lymphocyte-specific defects.

Another genetic condition in which radiosensitivity is associated with some degree of immunodeficiency is **ataxia telangiectasia**, which is due to mutations in the *ATM* gene. This encodes a protein kinase (ataxia telangiectasia mutated) of the same family as DNA-PKcs and which has a known role in general cellular DNA repair of double-strand breaks. What part ATM might play in V(D)J recombination is under investigation. It seems that at least some V(D)J recombination can occur in the absence of ATM, as the immune deficiencies seen in ataxia telangiectasia (low numbers of B and T cells and/or a deficiency in antibody class switching) are variable in their severity and are less severe than in SCID.

5-6 The diversity of the immunoglobulin repertoire is generated by four main processes.

The gene rearrangement that combines gene segments to form a complete V-region exon generates diversity in two ways. First, there are multiple different copies of each type of gene segment, and different combinations of gene segments can be used in different rearrangement events. This **combinatorial diversity** is responsible for a substantial part of the diversity of V regions. Second, **junctional diversity** is introduced at the joints between the different gene segments as a result of the addition and subtraction of nucleotides by the recombination process. A third source of diversity is also combinatorial, arising from the many possible different combinations of heavy- and light-chain V regions that pair to form the antigen-binding site in the immunoglobulin molecule. The two means of generating combinatorial diversity alone could give rise, in theory, to approximately 1.9×10^6 different antibody molecules (see Section 5-7). Coupled with junctional diversity, it is estimated that at least 10^{11} different receptors could make up the repertoire of receptors expressed by naive B cells, and diversity could be several orders of magnitude greater, depending on how one calculates junctional diversity. Finally, **somatic hypermutation**, which we discuss later in this chapter, introduces point mutations into the rearranged V-region genes of activated B cells, creating further diversity that can be selected for enhanced binding to antigen.

5-7 The multiple inherited gene segments are used in different combinations.

There are multiple copies of the V, D, and J gene segments, each of which can contribute to an immunoglobulin V region. Many different V regions can therefore be made by selecting different combinations of these segments. For human κ light chains, there are approximately 40 functional V_κ gene segments and 5 J_κ gene segments, and thus potentially 200 different V_κ regions. For λ light chains there are approximately 30 functional V_λ gene segments and

4 J_λ gene segments, yielding 120 possible V_λ regions. So, in all, 320 different light chains can be made as a result of combining different light-chain gene segments. For the heavy chains of humans, there are 40 functional V_H gene segments, approximately 25 D_H gene segments, and 6 J_H gene segments, and thus around 6000 different possible V_H regions ($40 \times 25 \times 6 = 6000$). During B-cell development, rearrangement at the heavy-chain gene locus to produce a heavy chain is followed by several rounds of cell division before light-chain gene rearrangement takes place, resulting in the same heavy chain being paired with different light chains in different cells. Because both the heavy- and the light-chain V regions contribute to antibody specificity, each of the 320 different light chains could be combined with each of the approximately 6000 heavy chains to give around 1.9×10^6 different antibody specificities.

This theoretical estimate of combinatorial diversity is based on the number of germline V gene segments contributing to functional antibodies (see Fig. 5.2); the total number of V gene segments is larger, but the additional gene segments are pseudogenes and do not appear in expressed immunoglobulin molecules. In practice, combinatorial diversity is likely to be less than one might expect from the calculations above. One reason is that not all V gene segments are used at the same frequency; some are common in antibodies, while others are found only rarely. Also, not every heavy chain can pair with every light chain: certain combinations of V_H and V_L regions will not form a stable molecule. Cells in which heavy and light chains fail to pair may undergo further light-chain gene rearrangement until a suitable chain is produced or they will be eliminated. Nevertheless, it is thought that most heavy and light chains can pair with each other, and that this type of combinatorial diversity has a major role in forming an immunoglobulin repertoire with a wide range of specificities.

5-8 Variable addition and subtraction of nucleotides at the junctions between gene segments contributes to the diversity of the third hypervariable region.

As noted earlier, of the three hypervariable loops in an immunoglobulin chain, CDR1 and CDR2 are encoded within the V gene segment. CDR3, however, falls at the joint between the V gene segment and the J gene segment, and in the heavy chain it is partly encoded by the D gene segment. In both heavy and light chains, the diversity of CDR3 is significantly increased by the addition and deletion of nucleotides at two steps in the formation of the junctions between gene segments. The added nucleotides are known as P-nucleotides and N-nucleotides, and their addition is illustrated in Fig. 5.7.

P-nucleotides are so called because they make up palindromic sequences added to the ends of the gene segments. As described in Section 5-5, the RAG proteins generate DNA hairpins at the coding ends of the V, D, or J segments, after which Artemis catalyzes a single-stranded cleavage at a random point within the coding sequence but near the original point at which the hairpin was first formed. When this cleavage occurs at a different point from the initial break induced by the RAG1/2 complex, a single-stranded tail is formed from a few nucleotides of the coding sequence plus the complementary nucleotides from the other DNA strand (see Fig. 5.7). In most light-chain gene rearrangements, DNA repair enzymes then fill in complementary nucleotides on the single-stranded tails, which would leave short palindromic sequences (the P-nucleotides) at the joint if the ends were rejoined without any further exonuclease activity.

In heavy-chain gene rearrangements and in a proportion of human light-chain gene rearrangements, however, **N-nucleotides** are added by a quite different mechanism before the ends are rejoined. N-nucleotides are so called

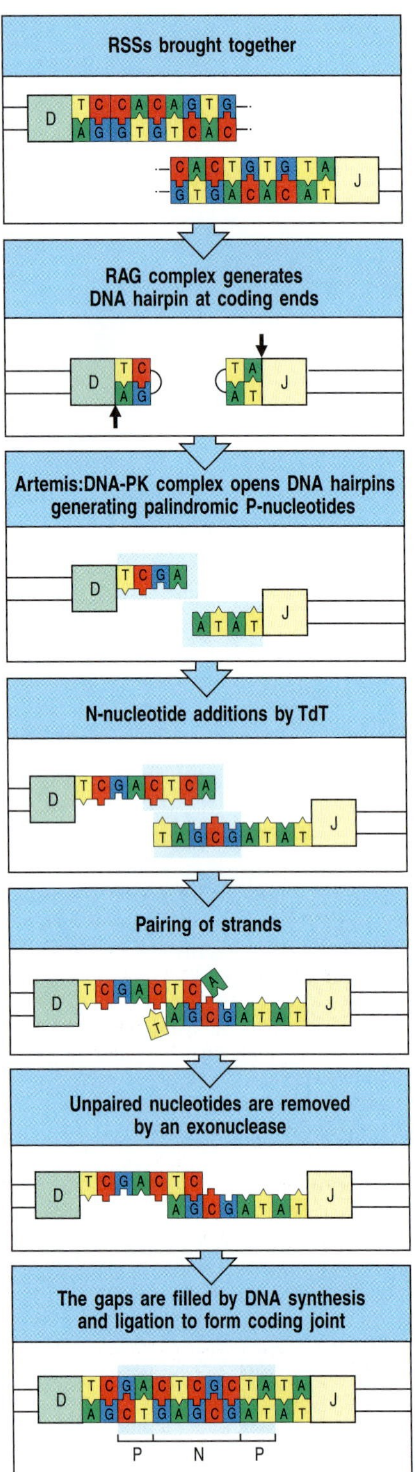
RSSs brought together
RAG complex generates DNA hairpin at coding ends
Artemis:DNA-PK complex opens DNA hairpins generating palindromic P-nucleotides
N-nucleotide additions by TdT
Pairing of strands
Unpaired nucleotides are removed by an exonuclease
The gaps are filled by DNA synthesis and ligation to form coding joint
D
J
P
N
P

Fig. 5.7 The introduction of P- and N-nucleotides diversifies the joints between gene segments during immunoglobulin gene rearrangement. The process is illustrated for a D_H to J_H rearrangement (first panel); however, the same steps occur in V_H to D_H and in V_L to J_L rearrangements. After formation of the DNA hairpins (second panel), the two heptamer sequences are ligated to form the signal joint (not shown here), while the Artemis:DNA-PK complex cleaves the DNA hairpin at a random site (indicated by the arrows) to yield a single-stranded DNA end (third panel). Depending on the site of cleavage, this single-stranded DNA may contain nucleotides that were originally complementary in the double-stranded DNA and which therefore form short DNA palindromes, such as TCGA and ATAT, as indicated by the blue-shaded box. Such stretches of nucleotides that originate from the complementary strand are known as P-nucleotides. For example, the sequence GA at the end of the D segment shown is complementary to the preceding sequence TC. Where the enzyme terminal deoxynucleotidyl transferase (TdT) is present, nucleotides are added at random to the ends of the single-stranded segments (fourth panel), indicated by the shaded box surrounding these nontemplated, or N, nucleotides. The two single-stranded ends then pair (fifth panel). Exonuclease trimming of unpaired nucleotides (sixth panel) and repair of the coding joint by DNA synthesis and ligation (bottom panel) leaves both the P- and N-nucleotides present in the final coding joint (indicated by light blue shading). The randomness of insertion of P- and N-nucleotides makes an individual P–N region virtually unique and a valuable marker for following an individual B-cell clone as it develops, for instance in studies of somatic hypermutation.

because they are non-template-encoded. They are added by the enzyme TdT to the single-stranded ends of the coding DNA after hairpin cleavage. After the addition of up to 20 nucleotides, the two single-stranded stretches form complementary base pairs. Repair enzymes then trim off nonmatching nucleotides, synthesize complementary DNA to fill in the remaining single-stranded gaps, and ligate the new DNA to the palindromic region (see Fig. 5.7). TdT is maximally expressed during the period in B-cell development when the heavy-chain gene is being assembled, and so N-nucleotides are common in heavy chain V–D and D–J junctions. N-nucleotides are less common in light-chain genes, which undergo rearrangement after heavy-chain genes (see Chapter 8).

Nucleotides can also be deleted at gene segment junctions. This is accomplished by exonucleases, and although these have not yet been identified, Artemis has dual endonuclease and exonuclease activity and so could well be involved in this step. Thus, a heavy-chain CDR3 can be shorter than even the smallest D segment. In some instances it is difficult, if not impossible, to recognize the D segment that contributed to CDR3 formation because of the excision of most of its nucleotides. Deletions may also erase the traces of P-nucleotide palindromes introduced at the time of hairpin opening. For this reason, many completed VDJ joins do not show obvious evidence of P-nucleotides. As the total number of nucleotides added by these processes is random, the added nucleotides often disrupt the reading frame of the coding sequence beyond the joint. Such frameshifts will lead to a nonfunctional protein, and DNA rearrangements leading to such disruptions are known as **nonproductive rearrangements**. As roughly two in every three rearrangements will be nonproductive, many B-cell progenitors never succeed in producing functional immunoglobulin and therefore never become mature B cells. Thus, junctional diversity is achieved only at the expense of considerable cell wastage. We discuss this further in Chapter 8.

Summary.

The extraordinary diversity of the immunoglobulin repertoire is achieved in several ways. Perhaps the most important factor enabling this diversity is that V regions are encoded by separate gene segments (V, D, and J gene segments), which are brought together by a somatic recombination process (V(D)J recombination) to produce a complete V-region exon. Many different gene segments are present in the genome of an individual, thus providing a heritable source of diversity that this combinatorial mechanism can use. Unique lymphocyte-specific recombinases, the RAG proteins, are absolutely required to catalyze this rearrangement, and the evolution of RAG proteins coincided with the appearance of the modern vertebrate adaptive immune system. Another substantial fraction of the functional diversity of immunoglobulins comes from the joining process itself. Variability at the joints between gene segments is generated by the insertion of random numbers of P- and N-nucleotides and by the variable deletion of nucleotides at the ends of some segments. The association of different light- and heavy-chain V regions to form the antigen-binding site of an immunoglobulin molecule contributes further diversity. The combination of all these sources of diversity generates a vast primary repertoire for antibody specificities. Additional changes in the rearranged V regions, introduced by somatic hypermutation (discussed later in the chapter), add even greater diversity to this primary repertoire.

T-cell receptor gene rearrangement.

The mechanism by which B-cell antigen receptors are generated is such a powerful means of creating diversity that it is not surprising that the antigen receptors of T cells bear structural resemblances to immunoglobulins and are generated by the same mechanism. In this part of the chapter we describe the organization of the T-cell receptor loci and the generation of the genes for the individual T-cell receptor chains.

5-9 The T-cell receptor gene segments are arranged in a similar pattern to immunoglobulin gene segments and are rearranged by the same enzymes.

Like immunoglobulin light and heavy chains, T-cell receptor α and β chains each consist of a variable (V) amino-terminal region and a constant (C) region (see Section 4-10). The organization of the TCRα and TCRβ loci is shown in Fig. 5.8. The organization of the gene segments is broadly homologous to that of the immunoglobulin gene segments (see Sections 5-2 and 5-3). The TCRα locus, like those for the immunoglobulin light chains, contains V and J gene segments (V_{α} and J_{α}). The TCRβ locus, like that for the immunoglobulin heavy chain, contains D gene segments in addition to V_{β} and J_{β} gene segments. The T-cell receptor gene segments rearrange during T-cell development to form complete V-domain exons (Fig. 5.9). T-cell receptor gene rearrangement

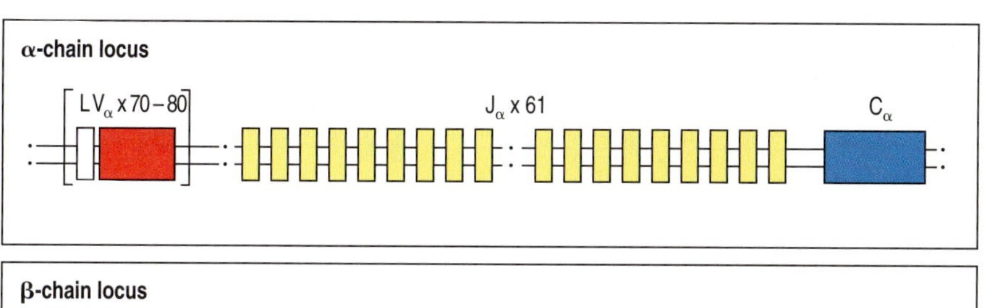

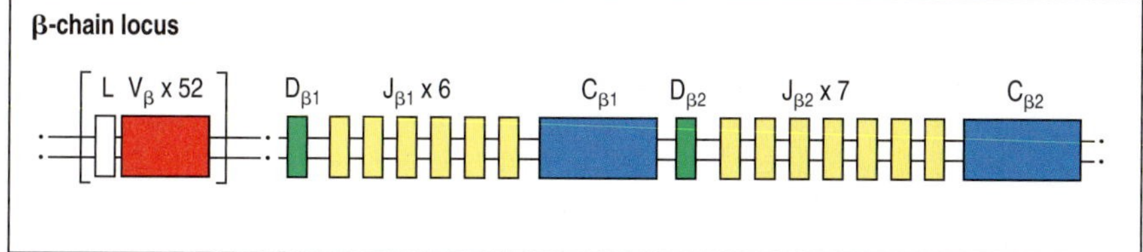

Fig. 5.8 The germline organization of the human T-cell receptor α and β loci. The arrangement of the gene segments resembles that at the immunoglobulin loci, with separate variable (V), diversity (D), and joining (J) gene segments, and constant (C) genes. The TCRα locus (chromosome 14) consists of 70–80 V_{α} gene segments, each preceded by an exon encoding the leader sequence (L). How many of these V_{α} gene segments are functional is not known exactly. A cluster of 61 J_{α} gene segments is located a considerable distance from the V_{α} gene segments. The J_{α} gene segments are followed by a single C gene, which contains separate exons for the constant and hinge domains and a single exon encoding the transmembrane and cytoplasmic regions (not shown). The TCRβ locus (chromosome 7) has a different organization, with a cluster of 52 functional V_{β} gene segments located distantly from two separate clusters each containing a single D gene segment together with six or seven J gene segments and a single C gene. Each TCRβ C gene has separate exons encoding the constant domain, the hinge, the transmembrane region, and the cytoplasmic region (not shown). The TCRα locus is interrupted between the J and V gene segments by another T-cell receptor locus—the TCRδ locus (not shown here; see Fig. 5.13).

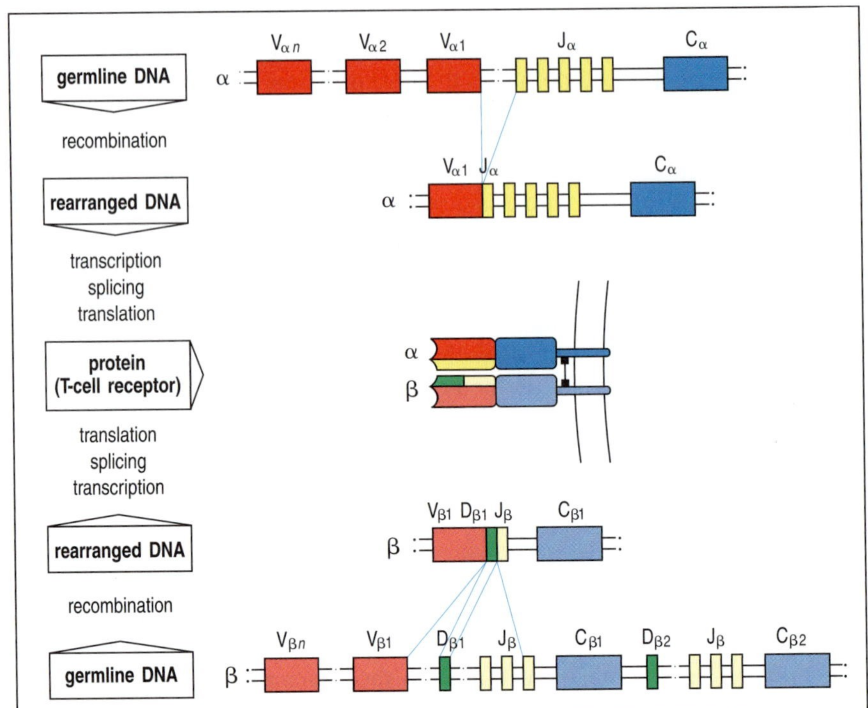

Fig. 5.9 T-cell receptor α- and β-chain gene rearrangement and expression. The TCRα- and β-chain genes are composed of discrete segments that are joined by somatic recombination during development of the T cell. Functional α- and β-chain genes are generated in the same way that complete immunoglobulin genes are created. For the α chain (upper part of figure), a V_α gene segment rearranges to a J_α gene segment to create a functional V-region exon. Transcription and splicing of the VJ_α exon to C_α generates the mRNA that is translated to yield the T-cell receptor α-chain protein. For the β chain (lower part of figure), like the immunoglobulin heavy chain, the variable domain is encoded in three gene segments, V_β, D_β, and J_β. Rearrangement of these gene segments generates a functional VDJ_β V-region exon that is transcribed and spliced to join to C_β; the resulting mRNA is translated to yield the T-cell receptor β chain. The α and β chains pair soon after their synthesis to yield the α:β T-cell receptor heterodimer. Not all J gene segments are shown, and the leader sequences preceding each V gene segment are omitted for simplicity.

takes place in the thymus; the order and regulation of the rearrangements are dealt with in detail in Chapter 8. Essentially, however, the mechanics of gene rearrangement are similar for B and T cells. The T-cell receptor gene segments are flanked by 12-bp and 23-bp spacer recombination signal sequences (RSSs) that are homologous to those flanking immunoglobulin gene segments (Fig. 5.10, and see Section 5-4) and are recognized by the same enzymes. The DNA circles resulting from gene rearrangement (see Fig. 5.5) are known as T-cell receptor excision circles (TRECs) and are used as markers for T cells that have recently emigrated from the thymus. All known defects in genes that control V(D)J recombination affect T cells and B cells equally, and animals with these genetic defects lack functional lymphocytes altogether (see Section 5-5). A further shared feature of immunoglobulin and T-cell receptor gene rearrangement is the presence of P- and N-nucleotides in the junctions between the V, D, and J gene segments of the rearranged TCRβ gene. In T cells, P- and N-nucleotides are also added between the V and J gene segments of all rearranged TCRα genes, whereas only about half the V–J joints in immunoglobulin light-chain genes are modified by N-nucleotide addition, and these are often left without any P-nucleotides as well (Fig. 5.11, and see Section 5-8).

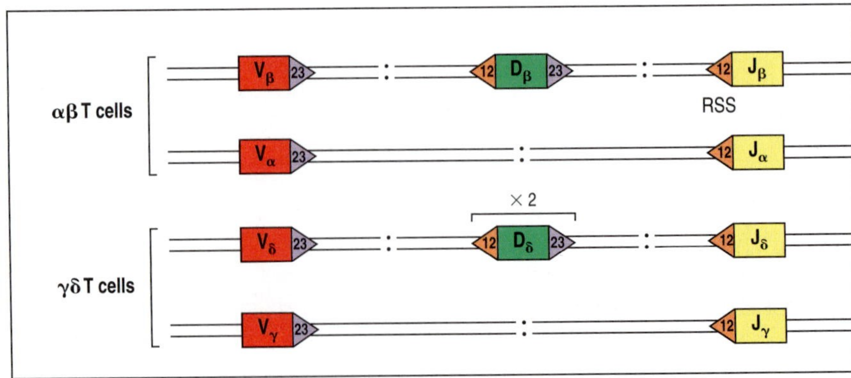

Fig. 5.10 Recombination signal sequences flank T-cell receptor gene segments. As in the immunoglobulin gene loci (see Fig. 5.4), the individual gene segments at the TCRα and TCRβ loci are flanked by heptamer–spacer–nonamer recombination signal sequences (RSSs). RSS motifs containing 12-bp spacers are depicted here as orange arrowheads, and those containing 23-bp spacers are shown in purple. Joining of gene segments almost always follows the 12/23 rule. Because of the disposition of heptamer and nonamer RSSs in the TCRβ and TCRδ loci, direct V_β to J_β joining is in principle allowed by the 12/23 rule (unlike in the immunoglobulin heavy-chain gene), although this occurs very rarely owing to other types of regulation that occur.

Element	Immunoglobulin		α:β T-cell receptors	
	H	κ+λ	β	α
Variable segments (V)	~40	~70	52	~70
Diversity segments (D)	23	0	2	0
D segments read in three frames	rarely	–	often	–
Joining segments (J)	6	5(κ) 4(λ)	13	61
Joints with N- and P-nucleotides	2	50% of joints	2	1
Number of V gene pairs	1.9×10^6		5.8×10^6	
Junctional diversity	$\sim 3 \times 10^7$		$\sim 2 \times 10^{11}$	
Total diversity	$\sim 5 \times 10^{13}$		$\sim 10^{18}$	

Fig. 5.11 The numbers of human T-cell receptor gene segments and the sources of T-cell receptor diversity compared with those of immunoglobulins. Note that only about half of human κ chains contain N-nucleotides. Somatic hypermutation as a source of diversity is not included in this figure because it does not occur in T cells.

The main differences between the immunoglobulin genes and those encoding T-cell receptors reflect the fact that all the effector functions of B cells depend upon secreted antibodies whose different heavy-chain C-region isotypes trigger distinct effector mechanisms. The effector functions of T cells, in contrast, depend upon cell–cell contact and are not mediated directly by the T-cell receptor, which serves only for antigen recognition. Thus, the C regions of the TCRα and TCRβ loci are much simpler than those of the immunoglobulin heavy-chain locus. There is only one Cα gene, and although there are two Cβ genes they are very closely homologous and there is no known functional distinction between their products. The T-cell receptor C-region genes encode only transmembrane polypeptides.

5-10 T-cell receptors concentrate diversity in the third hypervariable region.

The three-dimensional structure of the antigen-recognition site of a T-cell receptor looks much like that of an antibody molecule (see Sections 4-10 and 4-7, respectively). In an antibody, the center of the antigen-binding site is formed by the CDR3s of the heavy and light chains. The structurally equivalent third hypervariable loops (CDR3s) of the T-cell receptor α and β chains, to which the D and J gene segments contribute, also form the center of the antigen-binding site of a T-cell receptor; the periphery of the site consists of the CDR1 and CDR2 loops, which are encoded within the germline V gene segments for the α and β chains. The extent and pattern of variability in T-cell receptors and immunoglobulins reflect the distinct nature of their ligands. Whereas the antigen-binding sites of immunoglobulins must conform to the surfaces of an almost infinite variety of different antigens, and thus come in a wide variety of shapes and chemical properties, the ligand for the major class of human T-cell receptor (α:β) is always a peptide bound to an MHC molecule. As a group, the antigen-recognition sites of T-cell receptors would therefore be predicted to have a less variable shape, with most of the variability focused on the bound antigenic peptide occupying the center of the surface in contact with the receptor. Indeed, the less variable CDR1 and CDR2 loops of a T-cell receptor will mainly contact the relatively less variable MHC component of the ligand, whereas the highly variable CDR3 regions will mainly contact the unique peptide component (Fig. 5.12).

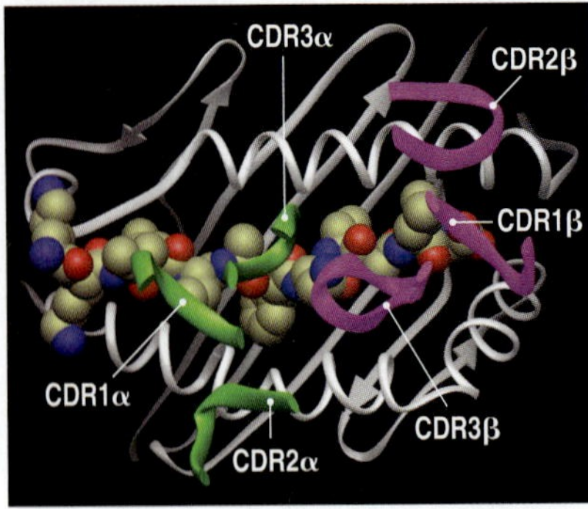

Fig. 5.12 The most variable parts of the T-cell receptor interact with the peptide bound to an MHC molecule. The positions of the CDR loops of a T-cell receptor are shown as colored tubes, which in this figure are superimposed on the peptide:MHC complex (MHC, gray; peptide, yellow-green with O atoms in red and N atoms in blue). The CDR loops of the α chain are in green, while those of the β chain are in magenta. The CDR3 loops lie in the center of the interface between the TCR and the peptide:MHC complex, and make direct contacts with the antigenic peptide.

The structural diversity of T-cell receptors is attributable mainly to combinatorial and junctional diversity generated during the process of gene rearrangement. It can be seen from Fig. 5.11 that most of the variability in T-cell receptor chains is in the junctional regions, which are encoded by V, D, and J gene segments and modified by P- and N-nucleotides. The TCRα locus contains many more J gene segments than either of the immunoglobulin light-chain loci: in humans, 61 J_α gene segments are distributed over about 80 kb of DNA, whereas immunoglobulin light-chain loci have only 5 J gene segments at most (see Fig. 5.11). Because the TCRα locus has so many J gene segments, the variability generated in this region is even greater for T-cell receptors than for immunoglobulins. Thus, most of the diversity resides in the CDR3 loops that contain the junctional region and form the center of the antigen-binding site.

5-11 γ:δ T-cell receptors are also generated by gene rearrangement.

A minority of T cells bear T-cell receptors composed of γ and δ chains (see Section 4-19). The organization of the TCRγ and TCRδ loci (Fig. 5.13) resembles that of the TCRα and TCRβ loci, although there are important differences. The cluster of gene segments encoding the δ chain is found entirely within the TCRα locus, between the V_α and the J_α gene segments. V_δ genes are interspersed with the V_α genes but are located primarily in the 3′ region of the locus. Because all V_α gene segments are oriented such that rearrangement will delete the intervening DNA, any rearrangement at the α locus results in the loss of the δ locus (Fig. 5.14). There are substantially fewer V gene segments at the TCRγ and TCRδ loci than at either the TCRα or TCRβ loci or any of the immunoglobulin loci. Increased junctional variability in the δ chains may compensate for the small number of V gene segments and has the effect of focusing almost all the variability in the γ:δ receptor in the junctional region. As we have seen for the α:β T-cell receptors, the amino acids encoded by the junctional regions lie at the center of the T-cell receptor binding site.

T cells bearing γ:δ receptors are a distinct lineage of T cells whose functions are at present unclear. The ligands for these receptors are also largely unknown. Some γ:δ T-cell receptors seem able to recognize antigen directly,

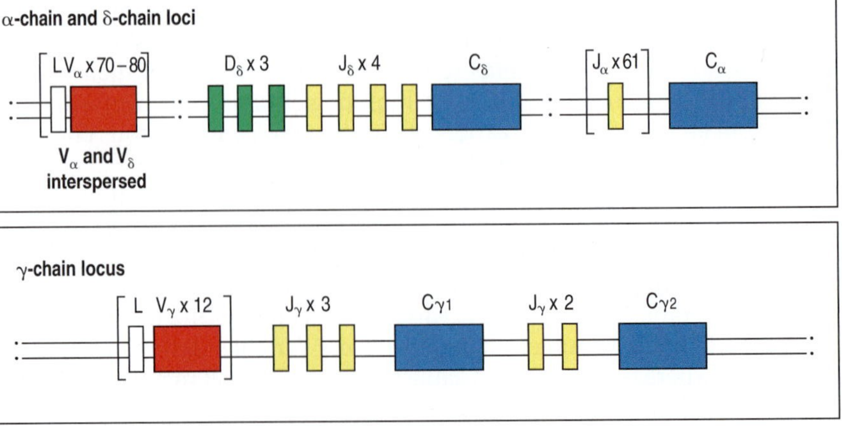

Fig. 5.13 The organization of the T-cell receptor γ- and δ-chain loci in humans. The TCRγ and TCRδ loci, like the TCRα and TCRβ loci, have discrete V, D, and J gene segments, and C genes. Uniquely, the locus encoding the δ chain is located entirely within the α-chain locus. The three D_δ gene segments, four J_δ gene segments, and the single δ C gene lie between the cluster of V_α gene segments and the cluster of J_α gene segments. There are two V_δ gene segments located near the δ C gene, one just upstream of the D regions and one in inverted orientation just downstream of the C gene (not shown). In addition, there are six V_δ gene segments interspersed among the V_α gene segments. Five are shared with V_α and can be used by either locus, and one is unique to the δ locus. The human TCRγ locus resembles the TCRβ locus in having two C genes each with its own set of J gene segments. The mouse γ locus (not shown) has a more complex organization and there are three functional clusters of γ gene segments, each containing V and J gene segments and a C gene. Rearrangement at the γ and δ loci proceeds as for the other T-cell receptor loci, with the exception that during TCRδ rearrangement two D segments can be used in the same gene. The use of two D segments greatly increases the variability of the δ chain, mainly because extra N-region nucleotides can be added at the junction between the two D gene segments as well as at the V–D and D–J junctions.

much as antibodies do, without presentation by an MHC molecule or antigen processing. Detailed analysis of the rearranged V regions of γ:δ T-cell receptors shows that they resemble the V regions of antibody molecules more than those of α:β T-cell receptors.

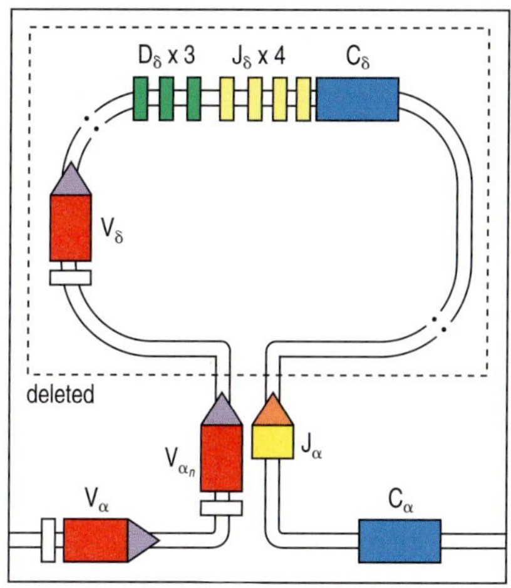

Fig. 5.14 Deletion of the TCRδ locus is induced by rearrangement of a V_α to J_α gene segment. The TCRδ locus is entirely contained within the chromosomal region containing the TCRα locus. When any V region in the V_α/V_δ region rearranges to any one of the J_α segments, the intervening region, and the entire V_δ locus, is deleted. Thus, V_α rearrangement prevents any continued expression of a V_δ gene and precludes lineage development down the γ:δ pathway.

Summary.

T-cell receptors are structurally similar to immunoglobulins and are encoded by homologous genes. T-cell receptor genes are assembled by somatic recombination from sets of gene segments in the same way that the immunoglobulin genes are. Diversity is, however, distributed differently in immunoglobulins and T-cell receptors: the T-cell receptor loci have roughly the same number of V gene segments but more J gene segments, and there is greater diversification of the junctions between gene segments during gene rearrangement. Moreover, functional T-cell receptors are not known to diversify their V genes after rearrangement through somatic hypermutation. This leads to a T-cell receptor in which the highest diversity is in the central part of the receptor, which in the case of α:β T-cell receptors contacts the bound peptide fragment of the ligand. Most of the diversity among γ:δ T-cell receptors is also in CDR3, but how this affects ligand binding is less clear because γ:δ T cells directly recognize poorly characterized ligands which in some cases are independent of MHC molecules.

Structural variation in immunoglobulin constant regions.

So far in this chapter we have focused on the structural variation in antigen receptors that results from the assembly of the V regions. We now turn to the C regions. The C regions of T-cell receptors have no functional purpose beyond supporting the V regions and anchoring the molecule in the membrane, and we shall not discuss them further here. Immunoglobulins, in contrast, can be made as both a transmembrane receptor and a secreted antibody, and the C domains of antibodies are crucial to their diverse effector functions.

Immunoglobulins are made in several different classes, which are distinguished by their heavy chains. Different heavy chains are produced in a given clone of B cells by linking different heavy-chain C regions (C_H) to the rearranged V_H gene. All classes of immunoglobulins produced by a clone of B cells thus have the same basic V region, although, as we shall see, this will become modified by somatic hypermutation. In the heavy-chain locus, the different C regions are encoded in separate genes located downstream of the V-region segments. Initially, naive B cells use only the first two of these, the C_μ and C_δ genes, which are expressed along with the associated assembled V-region sequence to produce transmembrane IgM and IgD on the surface of the naive B cell. During the course of an antibody response, activated B cells can switch to the expression of C_H genes other than C_μ and C_δ by a type of somatic recombination known as class switching. Along with other mechanisms that further diversify immunoglobulins, class switching will be discussed in the last part of this chapter. In contrast to heavy-chain C regions, light-chain C regions (C_L) do not provide specific effector function other than structural attachment for V regions and do not undergo class switching, and there seem to be no functional differences between λ and κ light chains.

In this part of the chapter we consider the structural features that distinguish the C_H regions of antibodies of the five major classes, and discuss some of their special properties. The functions of the different antibody classes are considered in more detail in Chapter 10. We also explain how the same antibody gene can generate both membrane-bound immunoglobulin and secreted immunoglobulin through alternative mRNA splicing.

5-12 Different classes of immunoglobulins are distinguished by the structure of their heavy-chain constant regions.

The five main classes of immunoglobulin are IgM, IgD, IgG, IgE, and IgA, all of which can occur as transmembrane antigen receptors or secreted antibodies. In humans, IgG can be further subdivided into four subclasses (IgG1, IgG2, IgG3, and IgG4), and IgA antibodies are found as two subclasses (IgA1 and IgA2). The IgG subclasses in humans are named in order of the abundance of the antibodies in serum, with IgG1 being the most abundant. The different heavy chains that define these classes are known as isotypes and are designated by the lower-case Greek letters μ, δ, γ, ε, and α, as shown in Fig. 5.15, which lists the major physical and functional properties of the different human antibody classes.

The functions of the immunoglobulin classes are discussed in detail in Chapter 10, in the context of the humoral immune response, and here we will just touch on them briefly. IgM is the first class of immunoglobulin produced after activation of a B cell, and the IgM antibody is secreted as a pentamer (see Fig. 5.19). This accounts for its high molecular weight and the fact that it is normally present in the bloodstream but not in tissues. Being a pentamer also increases the avidity of IgM for antigens before its affinity is increased through the process of affinity maturation (see Section 5-18).

IgG isotypes produced during an immune response are found in the bloodstream and in the extracellular spaces in tissues. IgM and most IgG isotypes

Fig. 5.15 The physical properties of the human immunoglobulin isotypes. IgM is so called because of its size: although monomeric IgM is only 190 kDa, it normally forms pentamers, known as macroglobulin (hence the M), of very large molecular weight (see Fig. 5.19). IgA dimerizes to give a molecular weight of around 390 kDa in secretions. IgE antibody is associated with immediate-type hypersensitivity. When fixed to tissue mast cells, IgE has a much longer half-life than its half-life in plasma shown here.

	Immunoglobulin								
	IgG1	IgG2	IgG3	IgG4	IgM	IgA1	IgA2	IgD	IgE
Heavy chain	γ_1	γ_2	γ_3	γ_4	μ	α_1	α_2	δ	ϵ
Molecular weight (kDa)	146	146	165	146	970	160	160	184	188
Serum level (mean adult mg/ml)	9	3	1	0.5	1.5	3.0	0.5	0.03	5×10^{-5}
Half-life in serum (days)	21	20	7	21	10	6	6	3	2
Classical pathway of complement activation	++	+	+++	–	++++	–	–	–	–
Alternative pathway of complement activation	–	–	–	–	–	+	–	–	–
Placental transfer	+++	+	++	–/+	–	–	–	–	–
Binding to macrophage and phagocyte Fc receptors	+	–	+	–/+	–	+	+	–	+
High-affinity binding to mast cells and basophils	–	–	–	–	–	–	–	–	+++
Reactivity with staphylococcal Protein A	+	+	–/+	+	–	–	–	–	–

can interact with the complement component C1 to activate the classical complement pathway (described in Section 2-7). IgA and IgE do not activate complement. IgA can be found in the bloodstream, but it also acts in the defense of mucosal surfaces; it is secreted into the gut and respiratory tract, and also into mother's milk. IgE is particularly involved in defense against multicellular parasites (for example schistosomes) but is also the antibody involved in common allergic diseases such as allergic asthma. IgG and IgE are always monomers, but IgA can be secreted as either a monomer or a dimer.

Sequence differences in the constant regions of the immunoglobulin heavy chain produce the distinct characteristics of each antibody isotype. These characteristics include the number and location of interchain disulfide bonds, the number of attached oligosaccharide moieties, the number of C domains, and the length of the hinge region (Fig. 5.16). IgM and IgE heavy chains contain an extra C domain that replaces the hinge region found in γ, δ, and α chains. The absence of the hinge region does not imply that IgM and IgE molecules lack flexibility; electron micrographs of IgM molecules binding to ligands show that the Fab arms can bend relative to the Fc portion. However, such a difference in structure may have functional consequences that are not yet characterized. Different isotypes and subtypes also differ in their ability to engage various effector functions, as described later. The distinct properties of the different C regions are encoded by different immunoglobulin C_H genes that are present in a cluster located 3′ of the J_H segments. We describe the rearrangement process by which the V region becomes associated with a different C_H gene in Section 5-19.

5-13 The constant region confers functional specialization on the antibody.

Antibodies can protect the body in a variety of ways. In some cases it is enough for the antibody simply to bind antigen. For instance, by binding tightly to a toxin or virus, an antibody can prevent it from recognizing its receptor on a host cell (see Fig. 1.24). The antibody V regions are sufficient for this activity. The C region is essential, however, for recruiting the help of other cells and molecules to destroy and dispose of pathogens to which the antibody has bound.

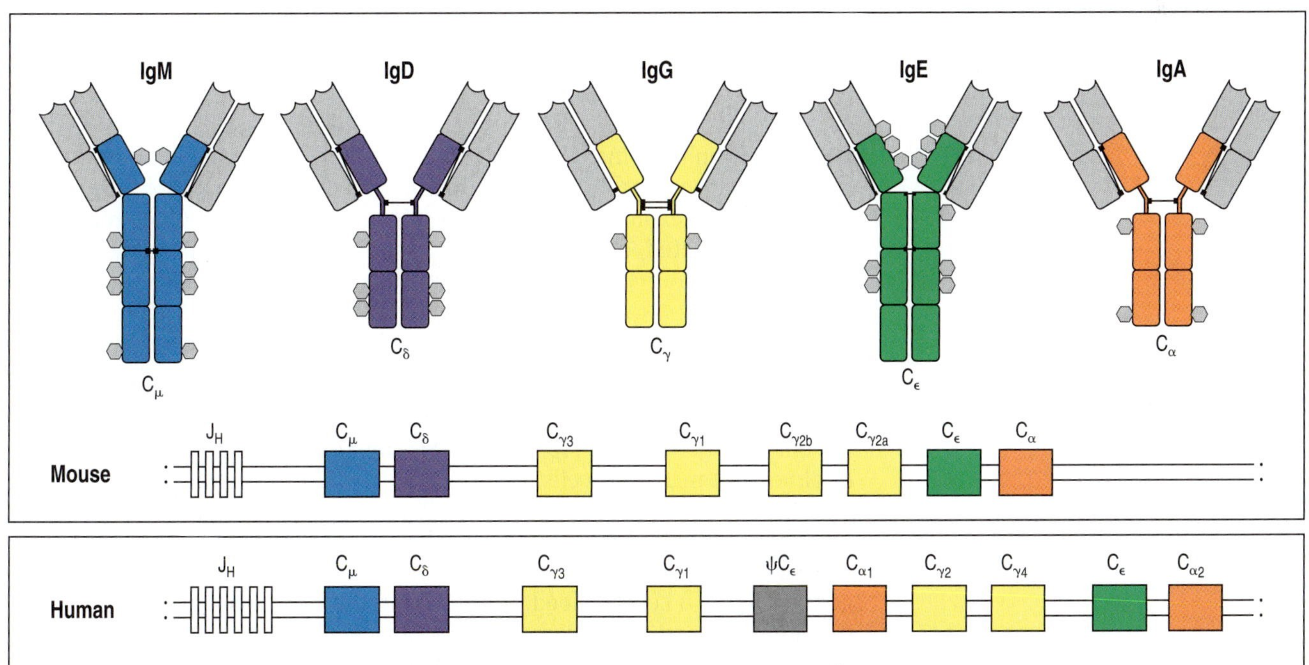

Fig. 5.16 The immunoglobulin isotypes are encoded by a cluster of immunoglobulin heavy-chain C-region genes. The general structure of the main immunoglobulin isotypes (upper panel) is indicated, with the immunoglobulin domain indicated as a rectangle. These are encoded by separate heavy-chain C-region genes arranged in a cluster in both mouse and human (lower panel). The constant region of the heavy chain for each istoype is indicated by the same color as the C-region gene segment that encodes it. IgM and IgE lack a hinge region but each contains an extra heavy-chain domain. Note the differences in the numbers and locations of the disulfide bonds (black lines) linking the chains. The isotypes also differ in the distribution of *N*-linked carbohydrate groups, shown as hexagons. In humans, the cluster shows evidence of evolutionary duplication of a unit consisting of two γ genes, an ε gene, and an α gene. One of the ε genes is a pseudogene (ψ); hence only one subtype of IgE is expressed. For simplicity, other pseudogenes are not illustrated, and the exon details within each C gene are not shown. The classes of immunoglobulins found in mice are called IgM, IgD, IgG1, IgG2a, IgG2b, IgG3, IgA, and IgE.

The C regions (Fc portions) of antibodies have three main effector functions. First, the Fc portions of certain isotypes are recognized by specialized **Fc receptors** expressed by immune effector cells. Fcγ receptors present on the surface of phagocytic cells such as macrophages and neutrophils bind the Fc portions of IgG1 and IgG3 antibodies, thus facilitating the phagocytosis of pathogens coated with these antibodies. The Fc portion of IgE binds to a high-affinity Fcε receptor on mast cells, basophils, and activated eosinophils, enabling these cells to respond to the binding of specific antigen by releasing inflammatory mediators. Second, the Fc portions of antigen:antibody complexes can bind to the C1q complement protein (see Section 2-7) and initiate the classical complement cascade, which recruits and activates phagocytes to engulf and destroy pathogens. Third, the Fc portion can deliver antibodies to places they would not reach without active transport. These include mucous secretions, tears, and milk (IgA), and the fetal blood circulation by transfer from the pregnant mother (IgG). In both cases, the Fc portion of IgA or IgG engages a specific receptor, the neonatal Fc receptor (FcRn), that actively transports the immunoglobulin through cells to reach different body compartments. Podocytes in the kidney glomerulus express FcRn to help remove IgG that has been filtered from the blood and accumulated at the glomerular basement membrane. We will return to the Fc receptors in Chapter 10.

The role of the Fc portion in these effector functions has been demonstrated by studying immunoglobulins that have had one or other Fc domain cleaved off enzymatically (see Section 4-3) or, more recently, by genetic engineering, which permits detailed mapping of the amino acid residues within the Fc that are needed for particular functions. Many microorganisms have responded to the destructive potential of the Fc portion by evolving proteins that either bind it or cleave it, and so prevent the Fc region from working; examples are Protein A and Protein G of *Staphylococcus* and Protein D of *Haemophilus*. Researchers have exploited these proteins to help map the Fc region and as immunological reagents (see Appendix I, Section A-10). Not all immunoglobulin classes have the same capacity to engage each of the effector functions. The different functional properties of each heavy-chain isotype are summarized in Fig. 5.15. For example, IgG1 and IgG3 have a higher affinity than IgG2 for the most common type of Fc receptor.

5-14 Mature naive B cells express both IgM and IgD at their surface.

The immunoglobulin C_H genes form a large cluster spanning about 200 kb to the 3′ side of the J_H gene segments (see Fig. 5.16). Each C_H gene is split into several exons (not shown in the figure), each corresponding to an individual immunoglobulin domain in the folded C region. The gene encoding the μ C region lies closest to the J_H gene segments, and therefore closest to the assembled V_H-region exon (VDJ exon) after DNA rearrangement. Once rearrangement is completed, a complete μ heavy-chain transcript is produced. Any J_H gene segments remaining between the assembled V gene and the C_μ gene are removed during RNA processing to generate the mature mRNA. μ heavy chains are therefore the first to be expressed, and IgM is the first immunoglobulin to be produced during B-cell development.

Immediately 3′ to the μ gene lies the δ gene, which encodes the C region of the IgD heavy chain (see Fig. 5.16). IgD is coexpressed with IgM on the surface of almost all mature B cells, although this isotype is secreted in only small amounts by plasma cells. The unique function of IgD is still unclear. Because IgD has hinge regions that are more flexible than those in IgM, IgD has been suggested to be an auxiliary receptor that may facilitate the binding of antigens by naive B cells. Indeed, mice lacking the C_δ exons show normal B-cell development and can generate largely normal antibody responses, but show a delay in the process of affinity maturation, which we describe in the next part of the chapter and which may involve the uptake of antigens by B cells for presentation to helper T cells.

B cells expressing IgM and IgD have not undergone class switching, which, as we will see, entails an irreversible change in the DNA. Instead, these cells produce a long primary mRNA transcript that is differentially cleaved and spliced to yield either of two distinct mRNA molecules. In one, the VDJ exon is linked to the C_μ exons to encode a μ heavy chain, while in the other the VDJ exon is linked to the C_δ exons to encode a δ heavy chain (Fig. 5.17). The processing of the long mRNA transcript is developmentally regulated, with immature B cells making mostly the μ transcript and mature B cells making mostly δ along with some μ. When a B cell is activated it ceases to coexpress IgD with IgM, either because μ and δ sequences have been removed as a consequence of a class switch or, in IgM-secreting plasma cells, because transcription from the V_H promoter no longer extends through the C_δ exons.

5-15 Transmembrane and secreted forms of immunoglobulin are generated from alternative heavy-chain transcripts.

Immunoglobulins of all classes can be produced either as a membrane-bound receptor or as secreted antibodies. All B cells initially express the transmembrane form of IgM; after stimulation by antigen, some of their progeny differentiate into plasma cells producing IgM antibodies, whereas others undergo class switching to express transmembrane immunoglobulins of a different class followed by the production of secreted antibody of the new class. The membrane forms of all immunoglobulin classes are monomers comprising two heavy and two light chains: IgM and IgA polymerize only when they have been secreted. In its membrane-bound form, the immunoglobulin heavy chain has a hydrophobic transmembrane domain of about 25 amino acid residues at the carboxy terminus, which anchors it to the surface of the B lymphocyte. This domain is absent from the secreted form, whose carboxy terminus is a hydrophilic secretory tail. The carboxy termini of the transmembrane and secreted forms of immunoglobulin heavy chains are encoded in two different exons, and production of the two forms is achieved by alternative RNA processing (Fig. 5.18). The last two exons of each C_H gene contain the sequences encoding the secreted and the transmembrane regions, respectively; if the primary transcript is cleaved and polyadenylated at a site downstream of these exons, the sequence encoding the carboxy terminus of the secreted form is removed by splicing, and the cell-surface form of immunoglobulin is produced. Alternatively, if the primary transcript is cleaved at the polyadenylation site located before the last two exons, only the secreted

Fig. 5.17 Coexpression of IgD and IgM is regulated by RNA processing. In mature B cells, transcription initiated at the V_H promoter extends through both C_μ and C_δ exons. This long primary transcript is then processed by cleavage and polyadenylation (AAA), and by splicing. Cleavage and polyadenylation at the μ site (pA1) and splicing between C_μ exons yields an mRNA encoding the μ heavy chain (left panel). Cleavage and polyadenylation at the δ site (pA2) and a different pattern of splicing that removes the C_μ exons yields mRNA encoding the δ heavy chain (right panel). For simplicity we have not shown all the individual C-region exons.

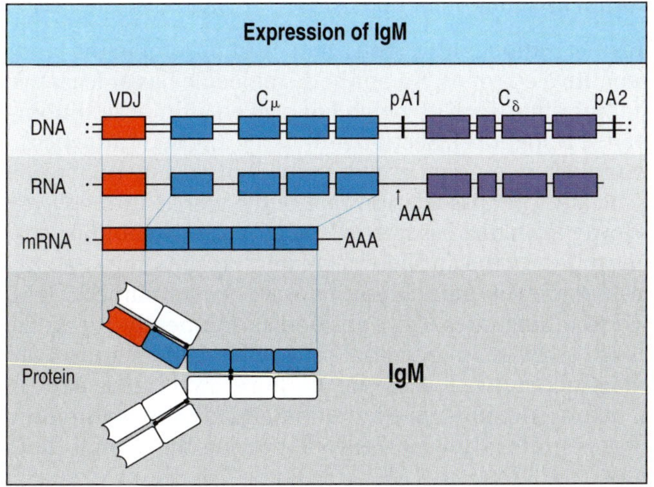

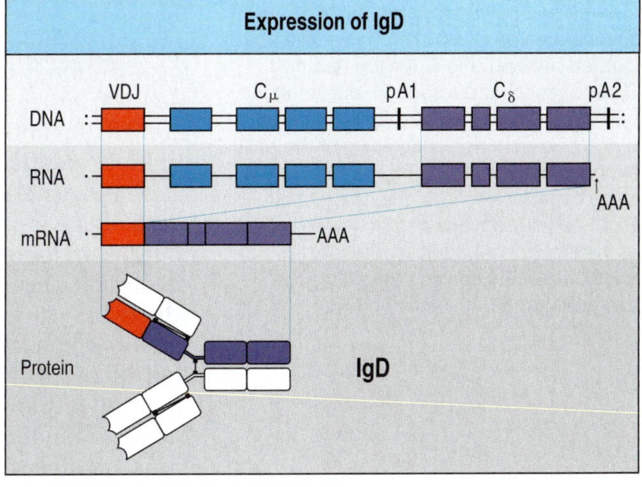

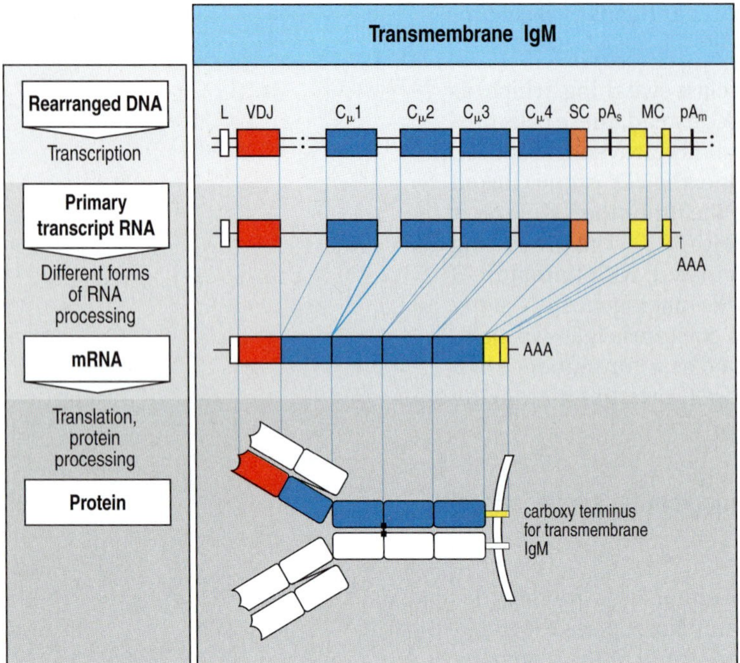

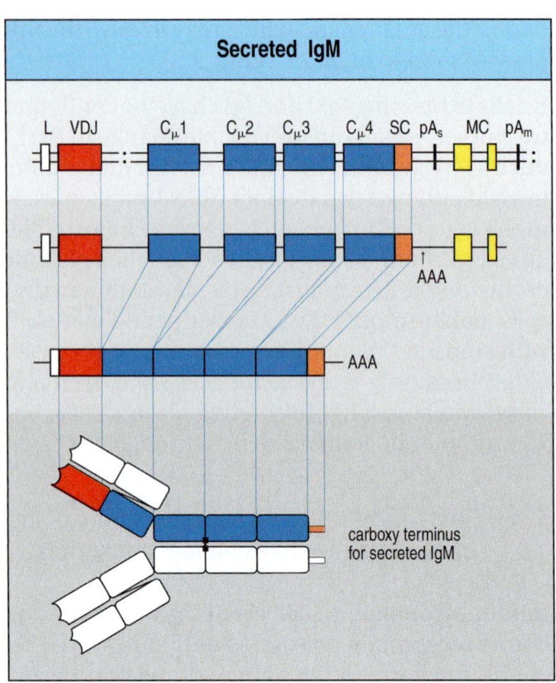

Fig. 5.18 Transmembrane and secreted forms of immunoglobulins are derived from the same heavy-chain sequence by alternative RNA processing. Each heavy-chain C gene has two exons (membrane-coding (MC), yellow) that encode the transmembrane region and cytoplasmic tail of the transmembrane form, and a secretion-coding (SC) sequence (orange) that encodes the carboxy terminus of the secreted form. In the case of IgD, the SC sequence is present on a separate exon (not shown), but for the other isotypes, including IgM as shown here, the SC sequence is contiguous with the last C-domain exon. The events that dictate whether a heavy-chain RNA will result in a secreted or transmembrane immunoglobulin occur during processing of the initial transcript. Each heavy-chain C gene has two potential polyadenylation sites (shown as pA_s and pA_m). Left panel: the transcript is cleaved and polyadenylated (AAA) at the second site (pA_m). Splicing between a site located between the $C_\mu 4$ exon and the SC sequence, and a second site at the 5′ end of the MC exons, results in removal of the SC sequence and joining of the MC exons to the $C_\mu 4$ exon. This generates the transmembrane form of the heavy chain. Right panel: cleavage and polyadenylation at the first poly(A) addition site (pA_s) and subsequent splicing generates the secreted form of the heavy chain.

molecule can be produced. This differential RNA processing is illustrated for C_μ in Fig. 5.18, but it occurs in the same way for all isotypes. In activated B cells that differentiate to become antibody-secreting plasma cells, most of the transcripts are spliced to yield the secreted rather than the transmembrane form of whichever heavy-chain isotype the B cell is expressing.

5-16 IgM and IgA can form polymers.

Although all immunoglobulin molecules are constructed from a basic unit of two heavy and two light chains, both IgM and IgA can form multimers of these basic units (Fig. 5.19). C regions include a 'tailpiece' of 18 amino acids that contains a cysteine residue essential for polymerization. A separate 15 kDa polypeptide chain called the J chain promotes polymerization by linking to the cysteines of this tailpiece, which is found only in the secreted forms of the μ and α chains. (This J chain should not be confused with the immunoglobulin J region encoded by a J gene segment; see Section 5-2.) In the case of IgA, dimerization is required for transport through epithelia, as we discuss in Chapter 10. IgM molecules are found as pentamers, and occasionally hexamers (without J chain), in plasma, whereas IgA is found mainly as a dimer in mucous secretions but as a monomer in plasma.

Immunoglobulin polymerization is also thought to be important in the binding of antibody to repetitive epitopes. An antibody molecule has at least two identical antigen-binding sites, each of which has a given affinity, or binding strength, for antigen (see Appendix I, Section A-9). If the antibody attaches to multiple identical epitopes on a target antigen, it will dissociate only when all binding sites dissociate. The dissociation rate of the whole antibody will therefore be much slower than the dissociation rate for a single binding site; multiple binding sites thus give the antibody a greater total binding strength, or avidity. This consideration is particularly relevant for pentameric IgM, which has 10 antigen-binding sites. IgM antibodies frequently recognize repetitive epitopes such as those on bacterial cell-wall polysaccharides, but individual binding sites are often of low affinity because IgM is made early in immune responses, before somatic hypermutation and affinity maturation. Multisite binding makes up for this, markedly improving the overall functional binding strength.

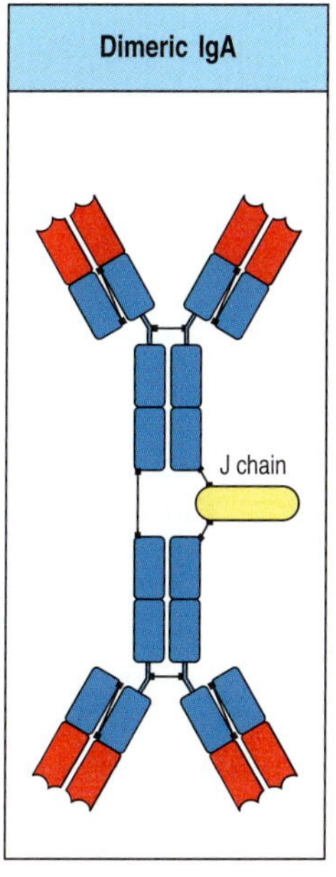

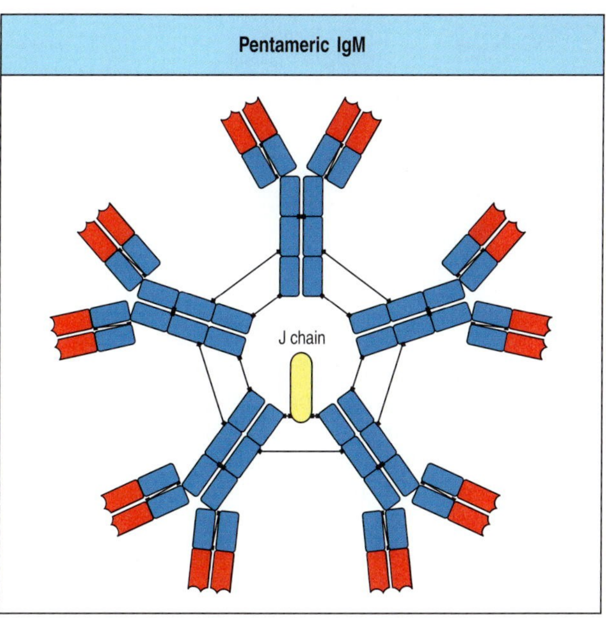

Fig. 5.19 The IgM and IgA molecules can form multimers. IgM and IgA are usually synthesized as multimers in association with an additional polypeptide chain, the J chain. In dimeric IgA (left panel), the monomers have disulfide bonds to the J chain as well as to each other. In pentameric IgM (right panel), the monomers are cross-linked by disulfide bonds to each other and to the J chain. IgM can also form hexamers that lack a J chain (not shown).

Summary.

The classes of immunoglobulins are defined by their heavy-chain C regions, with the different heavy-chain isotypes being encoded by different C-region genes. The heavy-chain C-region genes lie in a cluster 3′ to the V and J gene segments. A productively rearranged V-region exon is initially expressed in association with μ and δ C_H genes, which are coexpressed in naive B cells by alternative splicing of an mRNA transcript that contains both the μ and δ C_H exons. In addition, B cells can express any class of immunoglobulin as a membrane-bound antigen receptor or as secreted antibody. This is achieved by differential splicing of mRNA to include exons that encode either a hydrophobic membrane anchor or a secretable tailpiece. The antibody that a B cell secretes upon activation thus recognizes the antigen that initially activated the B cell via its antigen receptor. The same V-region exon can subsequently be associated with any one of the other isotypes to direct the production of antibodies of different classes. This process of class switching is described in the next part of the chapter.

Secondary diversification of the antibody repertoire.

The RAG-mediated V(D)J recombination described in the first part of the chapter is responsible for the initial antibody repertoire of B cells developing in the bone marrow. These somatic mutations—in the form of gene rearrangements—assemble the genes that produce the primary repertoire of immunoglobulins, and this takes place without interaction of B cells with antigen. Although this primary repertoire is large, further diversification can occur that enhances both the ability of immunoglobulins to recognize and bind to foreign antigens and the effector capacities of the expressed antibodies. This secondary phase of diversification occurs in activated B cells and is largely driven by antigen. Diversification is achieved through three mechanisms—**somatic hypermutation**, **class switching** or **class switch recombination**, and **gene conversion**—which alter the sequence of the secreted immunoglobulin in distinct ways (Fig. 5.20).

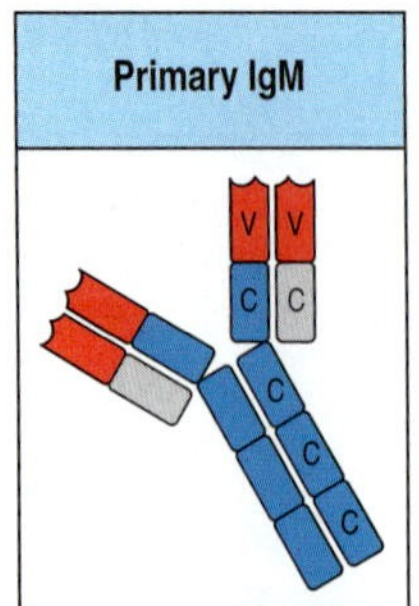

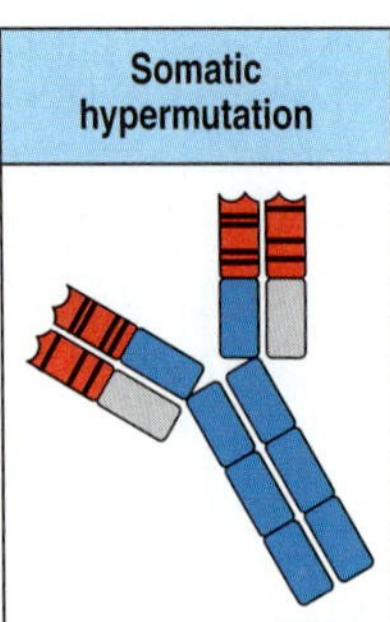

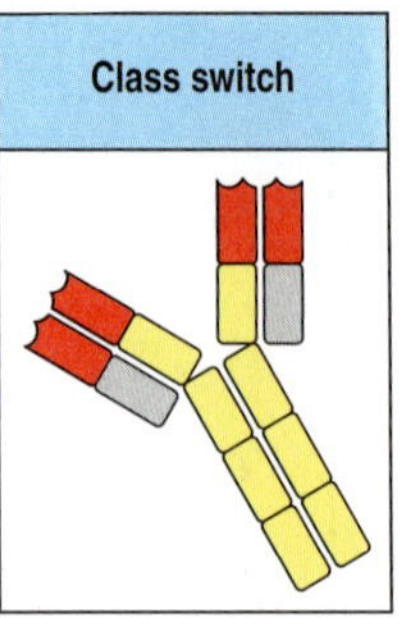

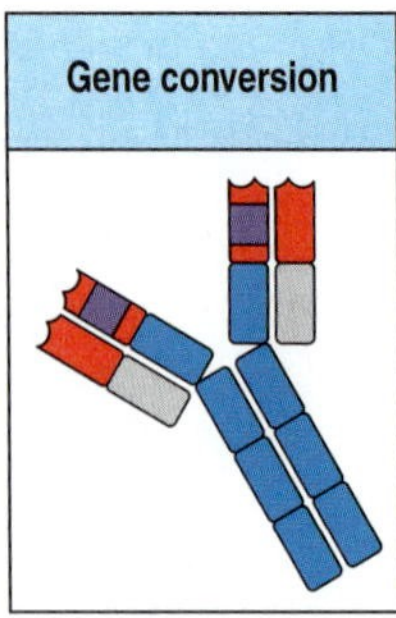

Fig. 5.20 The primary antibody repertoire is diversified by three processes that modify the rearranged immunoglobulin gene. First panel: The primary antibody repertoire is initially composed of IgM containing variable regions (red) produced by V(D)J recombination and constant regions from the μ gene segment (blue). The range of reactivity of this primary repertoire can be further modified by somatic hypermutation, class switch recombination at the immunoglobulin loci, and in some species by gene conversion (not shown). Second panel: somatic hypermutation results in mutations (shown as blue lines) being introduced into the heavy-chain and light-chain V regions (red), altering the affinity of the antibody for its antigen. Third panel: in class switch recombination, the initial μ heavy-chain C regions (blue) are replaced by heavy-chain regions of another isotype (shown as yellow), modifying the effector activity of the antibody but not its antigen specificity. Fourth panel: in gene conversion, the rearranged V region is modified by the introduction of sequences derived from V gene segment pseudogenes, creating additional antibody specificities.

Somatic hypermutation affects the V region and diversifies the antibody repertoire by introducing point mutations into the V regions of both chains, which alters the affinity of the antibody for antigen. Class switch recombination involves the C region only: it replaces the original C_{μ} heavy-chain C region with an alternative C region, thereby increasing the functional diversity of the immunoglobulin repertoire. Gene conversion diversifies the primary antibody repertoire in some animals, replacing blocks of sequence in the V regions with sequences derived from the V regions of pseudogenes. Like RAG-mediated V(D)J recombination, all these processes result in irreversible somatic mutation of the immunoglobulin genes, but unlike V(D)J recombination they are initiated by an enzyme called **activation-induced cytidine deaminase** (**AID**), which is expressed specifically in activated B cells; these processes do not occur in T-cell receptor genes. The initiation mechanism underlying all these processes is similar and so we will start with a general description of the enzymes involved.

5-17 Activation-induced cytidine deaminase (AID) introduces mutations into genes transcribed in B cells.

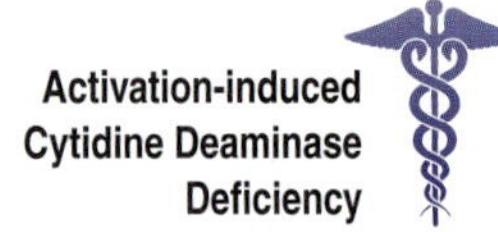

The enzyme AID was initially identified as a gene that is expressed specifically upon activation of B cells. Its importance for antibody diversification was revealed by disrupting AID expression in mice, which caused a defect in both somatic hypermutation and class switch recombination. People with mutations in the *AID* gene have been identified resulting in a lack of functional enzyme—**activation-induced cytidine deaminase deficiency** (**AID deficiency**). They lack class switching and somatic hypermutation, leading to the production of predominantly IgM and the absence of affinity maturation, a syndrome known as hyper IgM type 2 immunodeficiency (discussed in Chapter 13). The sequence of AID is related to that of a protein known for short as APOBEC1 (apolipoprotein B mRNA editing catalytic polypeptide 1), which converts cytosine in apolipoprotein B mRNA to uracil by deamination. Current evidence suggests that AID acts as a cytidine deaminase on DNA rather than RNA. When AID deaminates cytidine residues in the immunoglobulin V regions, somatic hypermutation is initiated; when cytidine residues in switch regions are deaminated, class switch recombination is initiated.

AID can bind to and deaminate single-stranded DNA but not double-stranded DNA. Thus, the DNA double helix must be temporarily unwound locally for AID to act, explaining why AID seems to target only genes that are being transcribed. By analogy with other cytidine deaminases, it is thought that AID initiates a nucleophilic attack on the pyrimidine ring of the exposed cytidine to generate uridine (Fig. 5.21). AID is expressed only in activated B cells, and so the targeted change of cytidine to uridine in the immunoglobulin genes only takes place in these cells. This uridine represents a dual lesion in DNA; not only is uridine foreign to normal DNA, but it is now a mismatch with the guanosine nucleoside on the opposite DNA strand.

The presence of uridine in DNA can trigger several types of DNA repair—including the **mismatch repair** and the **base-excision repair** pathways—which

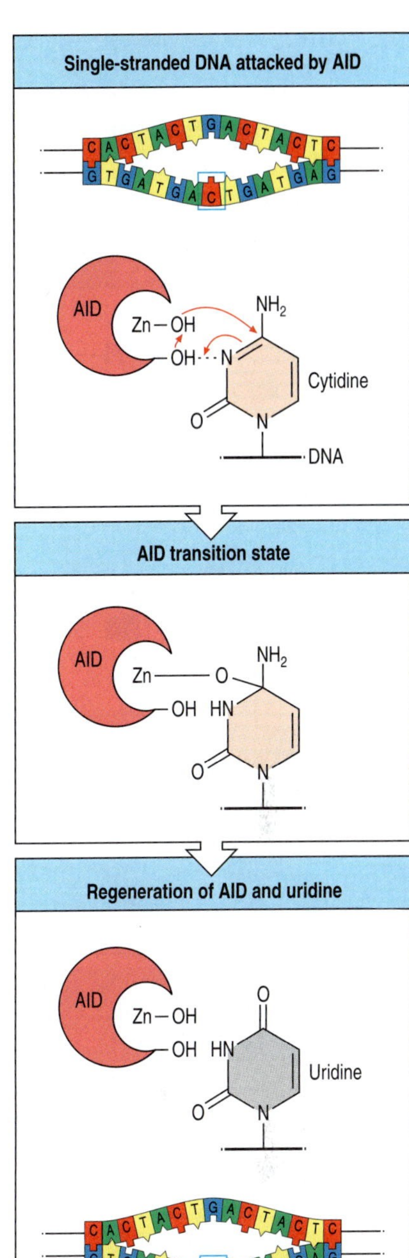

Fig. 5.21 Activation-induced cytidine deaminase (AID) is the initiator of mutations in somatic hypermutation, gene conversion, and class switching. The activity of AID, which is expressed only in B cells, requires access to the cytidine side chain of a single-stranded DNA molecule (first panel), which is normally prevented by the hydrogen bonding in double-stranded DNA. AID initiates a nucleophilic attack on the cytosine ring (second panel), which is resolved by the deamination of the cytidine to form uridine (third panel).

further alter the DNA sequence. The various repair processes lead to different mutational outcomes (Fig. 5.22). In the mismatch repair pathway, the presence of uridine is detected by the mismatch repair proteins MSH2 and MSH6 (MSH2/6). They recruit nucleases that remove the complete uridine nucleotide along with several adjacent nucleotides from the damaged DNA strand. This is followed by a fill-in 'patch repair' by a DNA polymerase; in B cells this DNA synthesis is error-prone and tends to introduce mutations, including mutations at nearby A:T base pairs.

The initial steps in the base-excision repair pathway are shown in Fig. 5.23. In this pathway, the enzyme **uracil-DNA glycosylase** (**UNG**) removes the uracil base from the uridine to create an abasic site in the DNA. If no further

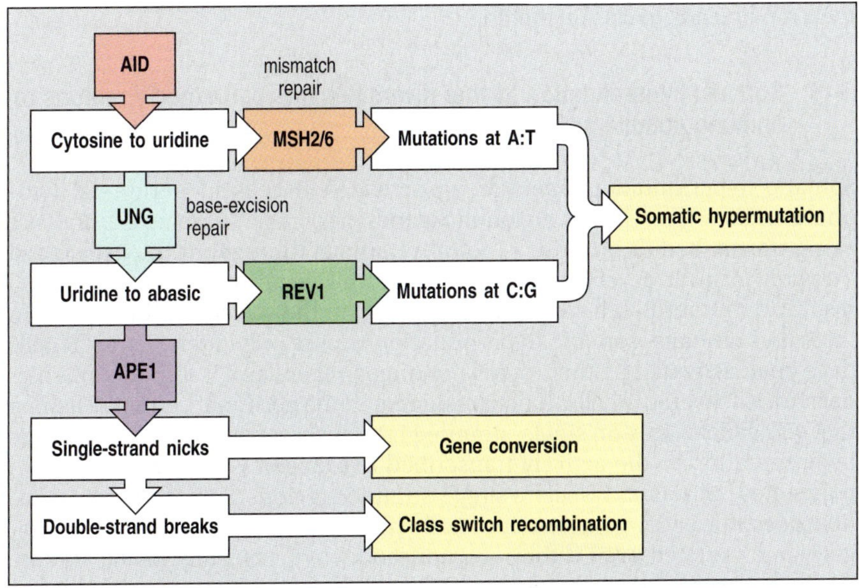

Fig. 5.22 AID initiates DNA lesions whose repair leads to somatic hypermutation, class switch recombination, or gene conversion. When AID converts a cytidine (C) to uridine (U) in the DNA of an immunoglobulin gene, the final mutation produced depends on which repair pathways are used. Somatic hypermutation can result from either mismatch repair (MSH2/6) or base-excision repair (UNG) pathways. Acting together, these can generate point mutations at and around the site of the original C:G pair. REV1 is a DNA repair enzyme that can synthesize DNA over the abasic sites in damaged DNA, causing a random nucleotide to be inserted at C:G residues where AID initially acted. Both class switch recombination and gene conversion require the formation of a single-strand break in the DNA. A single-strand break is formed when apurinic/apyrimidinic endonuclease 1 (APE1) removes a damaged residue from the DNA as part of the repair process (see Fig. 5.22, bottom panel). In class-switch recombination, single-strand breaks made in two of the so-called switch regions flanking the C-region genes are converted to double-strand breaks. The cell's machinery for repairing double-strand breaks, which is very similar to the later stages of V(D)J recombination, then rejoins the DNA ends in a way that leads to a recombination event in which a different C-region gene is brought adjacent to the rearranged V region. Gene conversion results from the broken DNA strand using homologous sequences flanking the immunoglobulin gene as a template for repair DNA synthesis, thus replacing part of the gene with new sequence.

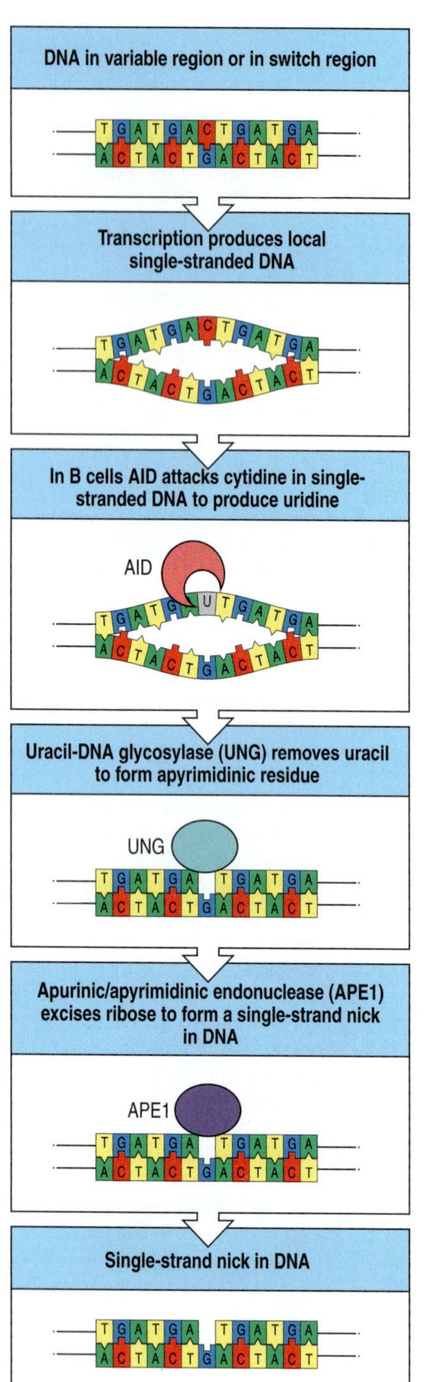
DNA in variable region or in switch region
Transcription produces local single-stranded DNA
In B cells AID attacks cytidine in single-stranded DNA to produce uridine
AID
Uracil-DNA glycosylase (UNG) removes uracil to form apyrimidinic residue
UNG
Apurinic/apyrimidinic endonuclease (APE1) excises ribose to form a single-strand nick in DNA
APE1
Single-strand nick in DNA

Fig. 5.23 The triggering of the base-excision repair pathway can generate single-strand nicks in DNA by the sequential action of AID, uracil-DNA-glycosylase (UNG), and apurinic/apyrimidinic endonuclease 1 (APE1). Double-stranded DNA (first panel) can be made accessible to AID by transcription that unwinds the DNA helix locally (second panel). AID, which is specifically expressed in activated B cells, converts cytidine residues to uridines (third panel). The ubiquitous base-excision repair enzyme UNG can then remove the uracil ring from uridine, creating an abasic site (fourth panel). The repair exonuclease APE1 can then excise the abasic residue from the DNA strand (fifth panel), leading to the formation of a single-strand nick in the DNA (sixth panel).

modification is made, this will result at the next round of DNA replication in the random insertion of a nucleotide opposite the abasic site by DNA polymerase, leading to mutation (see Fig. 5.22). The action of UNG may, however, be followed by the action of another enzyme, **apurinic/apyrimidinic endonuclease 1** (**APE1**), which excises the abasic residue to create a single-strand discontinuity (known as a single-strand nick) in the DNA at the site of the original cytidine (see Fig. 5.23). Repair of the single-strand nick by a process of homologous recombination results in gene conversion. Gene conversion is not used in the diversification of immunoglobulin genes in humans and mice but, as discussed later in this chapter, it is of great importance in some other mammals and in birds. In some circumstances, the generation of a single-strand nick leads to double-strand breaks in the DNA. When these occur as staggered breaks in specific locations in the immunoglobulin C-region genes, their repair leads to class switching.

5-18 Somatic hypermutation further diversifies the rearranged V regions of immunoglobulin genes.

Somatic hypermutation operates on activated B cells in peripheral lymphoid organs after immunoglobulin genes have been assembled and are being transcribed. It introduces point mutations throughout the rearranged V-region exon at a very high rate—hence the term 'hypermutation'—and results in mutant B-cell receptors appearing on the surface of the B cells. In mice and humans, somatic hypermutation occurs only after mature B cells have been activated by their corresponding antigen, and it also requires signals from activated T cells. It occurs mainly in the germinal centers. The fact that AID can act only on single-stranded DNA restricts the process of somatic hypermutation to the actively transcribed rearranged V regions where DNA polymerase generates transient single-stranded regions. Somatic hypermutation does not occur in transcriptionally inactive loci. Rearranged V_H and V_L genes are mutated even if they are 'nonproductive' rearrangements and are not being expressed as protein, as long as they are being transcribed. Some actively transcribed genes in B cells besides those for immunoglobulins can also be affected by the somatic mutation process, but at a much lower level.

Somatic hypermutation in functional V-region genes can have various consequences. Detrimental mutations that alter amino acid sequences in the conserved framework regions (see Fig. 4.6) will tend to disrupt basic immunoglobulin structure and are selected against, because in the germinal center clones of B cells are competing with each other for interaction with antigen via their B-cell receptors. Favorable mutations make changes that increase the affinity of the B-cell receptor for its antigen, and B-cell clones producing receptors with the highest affinity for antigen are favored for survival. Some of the mutant immunoglobulins bind antigen better than the original B-cell receptors, and B cells expressing them are preferentially selected to mature into antibody-secreting cells (Fig. 5.24). This gives rise to a phenomenon called **affinity maturation** of the antibody population, which we discuss in more detail in Chapters 10 and 11. The result of selection for enhanced binding to antigen is that the nucleotide changes that alter amino acid sequences,

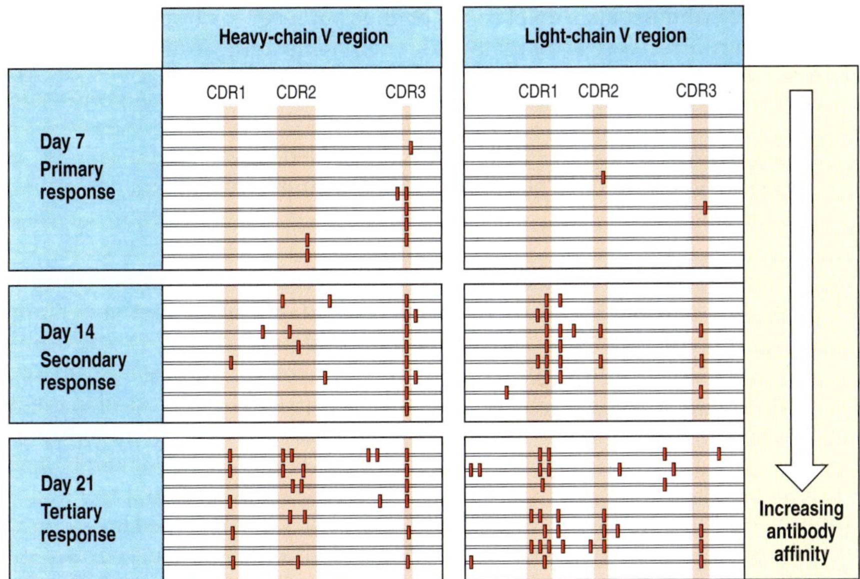

Fig. 5.24 Somatic hypermutation introduces mutations into the rearranged immunoglobulin variable (V) regions that improve antigen binding. It is possible to follow the process of somatic hypermutation by sequencing immunoglobulin V regions from hybridomas (clones of antibody-producing cells; see Appendix I, Section A-13) established at different time points after the experimental immunization of mice. The result of one experiment is depicted here. Each V region sequenced is represented by a horizontal line, on which the positions of the complementarity-determining regions CDR1, CDR2, and CDR3 are shown by pink shading. Mutations that change the amino acid sequence are represented by red bars. Within a few days of immunization, the V regions within a particular clone of responding B cells begin to acquire mutations, and over the course of the next week more mutations accumulate (top panels). B cells whose V regions have accumulated deleterious mutations and can no longer bind antigen die. B cells whose V regions have acquired mutations that improve the affinity of the B-cell receptor for antigen are able to compete more effectively for antigen, and receive signals that drive their proliferation and expansion. The antibodies they produce also have this improved affinity. This process of mutation and selection can continue in the lymph node germinal center through multiple cycles in response to secondary and tertiary immune responses elicited by further immunization with the same antigen (center and bottom panels). In this way, the antigen-binding efficiency of the antibody response is improved over time.

and thus protein structure, tend to be clustered in the CDRs of the immunoglobulin V-region genes (see Fig. 5.24), whereas silent, or neutral, mutations that preserve amino acid sequence and do not alter protein structure are scattered throughout the V region.

Somatic hypermutation involves both mutation at the original cytidines targeted by AID and mutation at nearby non-cytidine nucleotides. If the original U:G mismatch is recognized by UNG, then an abasic site will be generated in the DNA (see Fig. 5.22). If no further modification is made to this site, it can be replicated without instructive base pairing from the template strand by a class of error-prone 'translesion' DNA polymerases that normally repair gross damage to DNA, such as that caused by ultraviolet (UV) radiation. These polymerases can incorporate any nucleotide into the new DNA strand opposite the abasic site, and after a further round of DNA replication this can result in a stable mutation at the site of the original C:G base pair.

Mutations at A:T base pairs near the original cytidine arise by mismatch repair. Here, the U:G mismatch is recognized by MSH2/6, which recruits nucleases that remove the complete uridine residue along with a number of adjacent residues on the same strand. For reasons that are still unclear, during somatic hypermutation in B cells (but not during routine mismatch repair in other cell types) this DNA lesion is then repaired by error-prone translesion DNA polymerases rather than by more accurate polymerases that faithfully copy the undamaged template strand. Individuals with a defect in the translesion polymerase Polη have relatively fewer mutations than usual at A:T, but not at C:G, in their hypermutated immunoglobulin V regions, indicating that Polη is the repair polymerase involved in this pathway of somatic hypermutation. These individuals also have a form of xeroderma pigmentosum, a condition resulting from the inability of their cells to repair DNA damage caused by UV radiation.

In contrast to B cells, all the diversity in T-cell receptors is generated during gene rearrangement, and somatic hypermutation of rearranged V regions does not occur in T cells. This means that variability in the CDR1 and CDR2 regions is limited to that of the germline V gene segments, and that most diversity is focused on the CDR3 regions. The strongest argument for why T cells lack somatic hypermutation is that hypermutation is simply an adaptive specialization for B cells to make very high-affinity secreted antibodies that will efficiently carry out their effector functions. Because T cells do not need this capacity, and because deleterious changes in receptor-binding specificities in mature T cells are potentially more damaging to the immune response than are those in B cells, somatic hypermutation in T cells has never evolved.

5-19 Class switching enables the same assembled V_H exon to be associated with different C_H genes in the course of an immune response.

The V_H-region exons expressed by any given B cell are determined during its early differentiation in the bone marrow and, although they may subsequently be modified by somatic hypermutation, no further V(D)J recombination occurs. All the progeny of that B cell will therefore express the same V_H gene. In contrast, several different C-region isotypes can be expressed in the B cell's progeny as the cells mature and proliferate in the course of an immune response. The first antigen receptors expressed by B cells are IgM and IgD, and the first antibody produced in an immune response is always IgM. Later in the immune response, the same assembled V region may be expressed in IgG, IgA, or IgE antibodies. This change is known as class switching (or **isotype switching**), and, unlike the expression of IgD, it involves irreversible DNA recombination. It is stimulated in the course of an immune response by external signals such as cytokines released by T cells or mitogenic signals delivered by pathogens, as we discuss further in Chapter 10. Here we are concerned with the molecular basis of the class switch.

Movie 5.2 

Switching from IgM to the other immunoglobulin classes occurs only after B cells have been stimulated by antigen. It is achieved through class switch recombination, which is a type of nonhomologous DNA recombination that is guided by stretches of repetitive DNA known as **switch regions**. Switch regions lie in the intron between the J_H gene segments and the C_μ gene, and at equivalent sites upstream of the genes for each of the other heavy-chain isotypes, with the exception of the δ gene, the expression of which is not dependent on DNA rearrangement (Fig. 5.25, first panel). When a B cell switches from the coexpression of IgM and IgD to the expression of another subtype, DNA recombination occurs between S_μ and the S region immediately upstream of the gene for that isotype. In such a recombination event the C_δ coding regions and all of the intervening DNA between it and the S region undergoing rearrangement are deleted. Fig. 5.25 illustrates switching from C_μ to C_ε in the mouse. All switch recombination events produce genes that can encode a functional protein, because the switch sequences lie in introns and therefore cannot cause frameshift mutations.

As noted in Section 5-17, AID can act only on single-stranded DNA. It is known that transcription through the switch regions is required for efficient class switching, and this transcription is presumably necessary to open up the DNA and allow AID access to cytidine residues in the switch regions. The sequences of the switch regions have characteristics that may promote accessibility of the unwound DNA to AID when they are being transcribed. First, the non-template strand is G-rich. S_μ consists of about 150 repeats of the sequence $(GAGCT)_n(GGGGGT)$, where n is usually 3 but can be as many as 7. The sequences of the other switch regions (S_γ, S_α, and S_ε) differ in detail, but all contain repeats of the GAGCT and GGGGGT sequences. It is thought that transcription produces bubble-like structures, called **R-loops**, that are formed when the transcribed RNA displaces the non-template strand of the DNA double helix (see Fig. 5.25). The switch region is now a good substrate for AID, which initiates the formation of single-strand nicks at the sites of C residues. In addition, particular sequences, such as AGCT, may be particularly good substrates for AID, and because they are palindromic they may allow AID to act on the cytidine residues of both strands concurrently, introducing multiple single-strand nicks on both strands that eventually lead to a double-strand break in the DNA.

Whatever the precise mechanism, transcription through the switch regions seems to induce the generation of double-strand breaks in these regions. Cellular mechanisms for repairing double-strand breaks could then lead to the nonhomologous recombination between switch regions that results in class switching, with the ends to be joined being brought together by the alignment of repetitive sequences common to the different switch regions. Rejoining of the DNA ends would then lead to excision of all DNA between the two switch regions and the formation of a chimeric region at the junction.

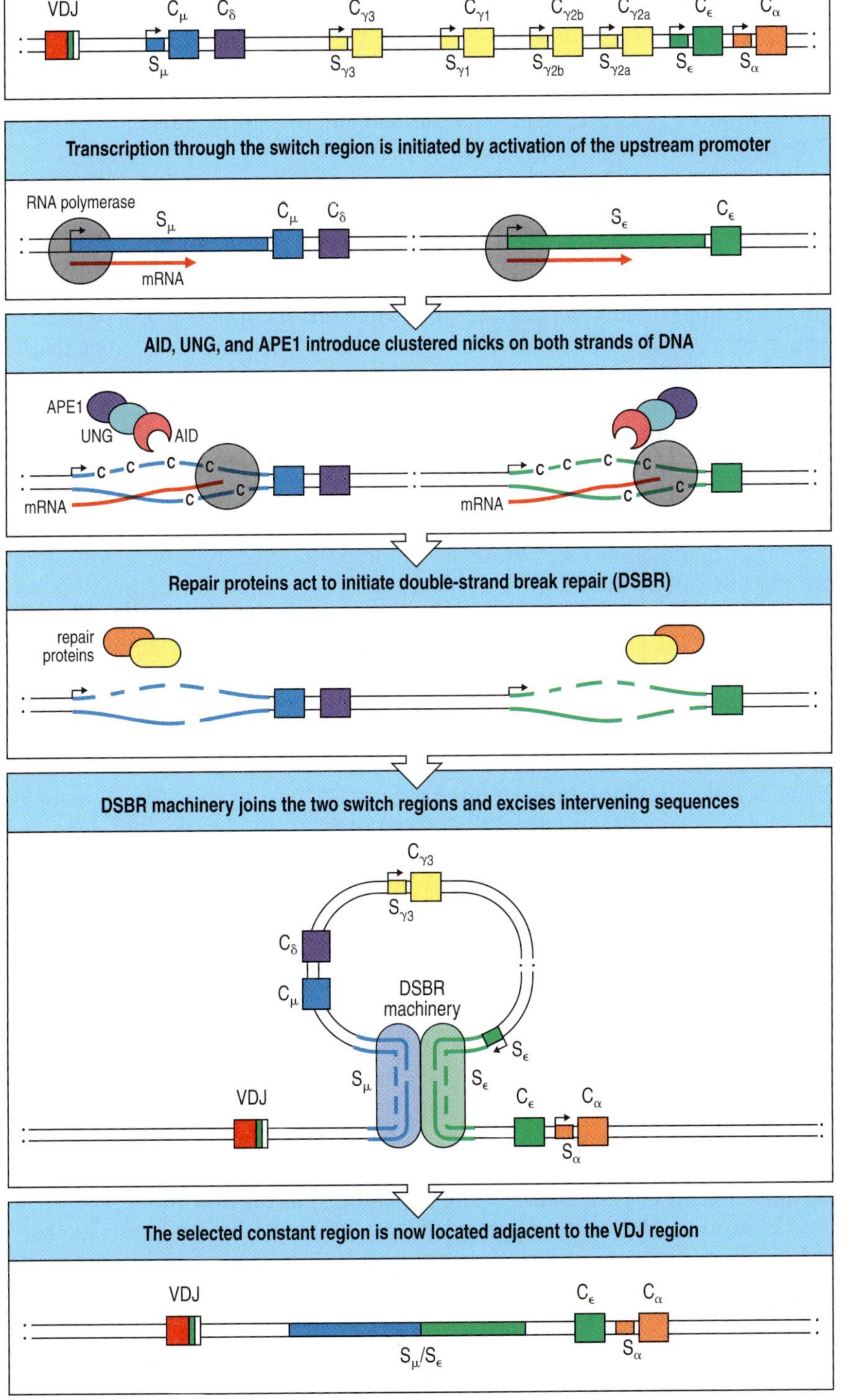

Fig. 5.25 Class switching involves recombination between specific switch signals. Top panel: organization of a rearranged immunoglobulin heavy-chain locus before class switching. Switching between the μ and ε isotypes in the mouse heavy-chain locus is illustrated in this figure. Switch regions (S), repetitive DNA sequences that guide class switching, are found upstream of each of the immunoglobulin C-region genes, with the exception of the δ gene. Switching is guided by the initiation of transcription by RNA polymerase through these regions from promoters (shown as arrows) located upstream of each S. Because of the nature of the repetitive sequences, transcription through S regions generates R-loops (extended regions of single-stranded DNA formed by the non-template strand), which serve as substrates for AID, and subsequently for UNG and APE1. These activities introduce a high density of single-strand nicks into the non-template DNA strand, and presumably a smaller number of nicks into the template strand. Staggered nicks are converted to double-strand breaks by a mechanism that is not yet understood. These breaks are then recognized by the cell's double-strand break repair machinery, which involves DNA-PKcs and other repair proteins. The two switch regions, in this case S_μ and S_ε, are brought together by the repair proteins, and class switching is completed by excision of the intervening region of DNA (including C_μ, C_δ) and ligation of the S_μ and S_ε regions.

While a lack of AID completely blocks class switching, deficiency of UNG in both mouse and humans severely impairs class switching, further suggesting the sequential actions of AID and UNG indicated in Section 5-17. Joining of DNA ends is probably mediated by classic nonhomologous end-joining (as in V(D)J recombination) as well as by a poorly understood alternative end-joining pathway. Class switching is sometimes impaired in the disease ataxia telangiectasia, which is caused by mutations in the DNA-PKcs-family kinase ATM, a known DNA repair protein. The role of ATM in class switching is not yet entirely clear, however.

Activation-induced Cytidine Deaminase Deficiency

Although they both involve DNA rearrangement and some of the same enzymatic machinery, class switch recombination is unlike V(D)J recombination in several ways. First, all class switch recombination is productive; second, it uses different recombination signal sequences and does not require the RAG enzymes; third, it happens after antigen stimulation and not during B-cell development in the bone marrow; and fourth, the switching process is not random but is directed by external signals such as those provided by T cells, as discussed in Chapter 10.

Summary.

Immunoglobulin genes rearranged by V(D)J recombination can be further diversified by somatic hypermutation, gene conversion, and class switching, which all rely on DNA repair and recombination processes initiated by the enzyme activation-induced cytidine deaminase (AID). Unlike V(D)J recombination, this secondary diversification occurs only in B cells and, in the case of somatic hypermutation and class switching, only after B-cell activation by antigen. Somatic hypermutation diversifies the V region by the introduction of point mutations. Where this results in a greater affinity for the antigen, activated B cells producing the mutated immunoglobulin are selected for survival, which in turn results in an increase in the affinity of antibodies for the antigen as the immune response proceeds. Class switching does not affect the V region but increases the functional diversity of immunoglobulins by replacing the C_{μ} region in the immunoglobulin gene first expressed with another heavy-chain C region to produce IgG, IgA, or IgE antibodies. Class switching provides antibodies with the same antigen specificity but distinct effector capacities. Gene conversion is the main mechanism used to provide a diverse immunoglobulin repertoire in animals in which only limited diversity can be generated from the germline genes by V(D)J recombination. It involves the replacement of segments of the rearranged V region by sequences derived from pseudogenes. The changes in immunoglobulin and T-cell receptor genes that occur during B-cell and T-cell development are summarized in Fig. 5.26.

Evolution of the adaptive immune response.

Classical adaptive immunity depends on the action of the RAG-1/RAG-2 recombinase to generate an enormously diverse clonally distributed repertoire of immunoglobulins and T-cell receptors, and it is found only in the jawed vertebrates (the gnathostomes), which split off from the other vertebrates around 500 million years ago. Adaptive immunity seems to have arisen abruptly in evolution. Even the cartilaginous fish, the earliest group of jawed fishes to survive to the present day, have organized lymphoid tissue, T-cell receptors and immunoglobulins, and the ability to mount adaptive immune responses. The diversity generated within the vertebrate adaptive immune system was once viewed as unique among animal immune systems.

Event	Process	Nature of change	Process occurs in:	
			B cells	T cells
V-region assembly	Somatic recombination of DNA	Irreversible	Yes	Yes
Junctional diversity	Imprecise joining, N-sequence insertion in DNA	Irreversible	Yes	Yes
Transcriptional activation	Activation of promoter by proximity to the enhancer	Irreversible but regulated	Yes	Yes
Switch recombination	Somatic recombination of DNA	Irreversible	Yes	No
Somatic hypermutation	DNA point mutation	Irreversible	Yes	No
IgM, IgD expression on surface	Differential splicing of RNA	Reversible, regulated	Yes	No
Membrane vs secreted form	Differential splicing of RNA	Reversible, regulated	Yes	No

Fig. 5.26 Changes in immunoglobulin and T-cell receptor genes that occur during B-cell and T-cell development and differentiation. Those changes that establish immunological diversity are all irreversible, as they involve changes in B-cell or T-cell DNA. Certain changes in the organization of DNA or in its transcription are unique to B cells. Somatic hypermutation has not been observed in functional T-cell receptors. The B-cell-specific processes, such as switch recombination, allow the same V region to be attached to several functionally distinct heavy-chain C regions, and thereby create functional diversity in an irreversible manner. By contrast, the expression of IgM versus IgD, and of membrane-bound versus secreted forms of all immunoglobulin types, can in principle be reversibly regulated.

But, as discussed briefly here, we now know that organisms as different as insects, echinoderms, and molluscs use a variety of genetic mechanisms to increase their repertoires of pathogen-detecting molecules, although they do not achieve true adaptive immunity. Nearer to home, it has been found that the surviving species of jawless vertebrates (agnathans)—the lampreys and hagfish—have a form of adaptive or 'anticipatory' immunity that is based on non-immunoglobulin 'antibody'-like proteins, and which involves a system of somatic gene rearrangement that is quite distinct from RAG-dependent V(D)J rearrangement. So we should now view our adaptive immune system as only one solution, albeit the most powerful, to the problem of generating highly diverse systems for pathogen recognition.

5-20 Some invertebrates generate extensive diversity in a repertoire of immunoglobulin-like genes.

Until very recently, it was thought that invertebrate immunity was limited to an innate system that had a very restricted diversity in recognizing pathogens. This idea was based on the knowledge that innate immunity in vertebrates relied on around 10 distinct Toll-like receptors and a similar number of other receptors that also recognize PAMPs, and also on the assumption that there were no greater numbers in invertebrates. Recent studies have, however, uncovered at least two invertebrate examples of extensive diversification of an immunoglobulin superfamily member, which could potentially provide an extended range of recognition of pathogens.

In *Drosophila*, fat-body cells and hemocytes act as part of the immune system. Fat-body cells secrete proteins, such as the antimicrobial defensins (see Chapters 2 and 3), into the hemolymph. Another protein found in hemolymph is the **Down syndrome cell adhesion molecule (Dscam)**, a member of the immunoglobulin superfamily. Dscam was originally discovered in the fly as a protein involved in the specification of neuronal wiring. It is also made in fat-body cells and hemocytes, which can secrete it into the hemolymph. Here it is thought to opsonize invading bacteria and aid their engulfment by phagocytes.

The Dscam protein contains multiple, usually 10, immunoglobulin-like domains. The gene that encodes Dscam has, however, evolved to contain a large number of alternative exons for several of these domains (Fig. 5.27).

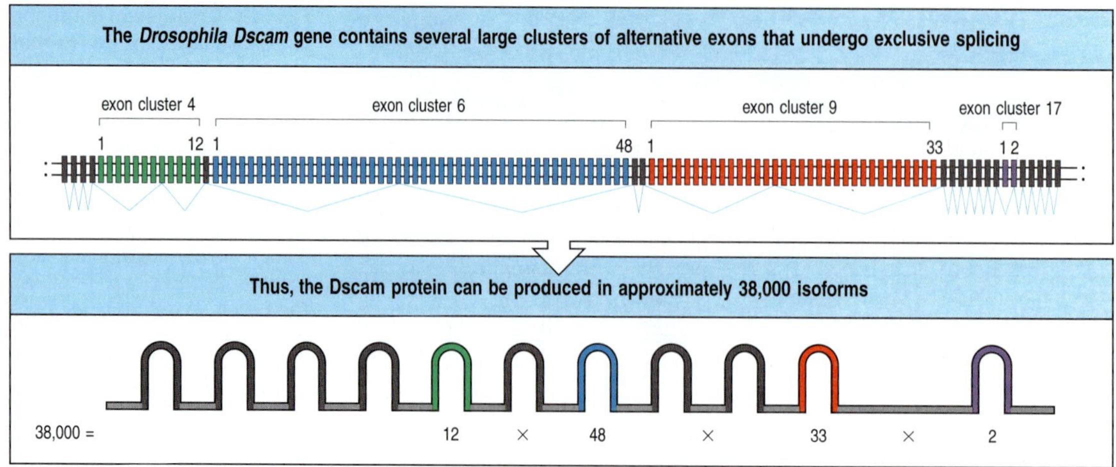

Fig. 5.27 The Dscam protein of *Drosophila* innate immunity contains multiple immunoglobulin domains and is highly diversified through alternative splicing. The gene encoding Dscam in *Drosophila* contains several large clusters of alternative exons. The clusters encoding exon 4 (green), exon 6 (light blue), exon 9 (red) and exon 17 (dark blue) contain 12, 48, 33, and 2 alternative exons, respectively. For each of these clusters, only one alternative exon is used in the complete *Dscam* mRNA. There is some differential usage of exons in neurons, fat-body cells, and hemocytes. All three cell types use the entire range of alternative exons for exons 4 and 6. For exon 9, there is a restricted use of alternative exons in hemocytes and fat-body cells. The combinatorial use of alternative exons in the *Dscam* gene makes it possible to generate more than 38,000 protein isoforms. Adapted from Anastassiou, D.: *Genome Biol.* 2006, 7:R2.

Exon 4 of the Dscam protein can be encoded by any one of 12 different exons, each specifying an immunoglobulin domain of differing sequence. Exon cluster 6 has 48 alternative exons, cluster 9 another 33, and cluster 17 a further 2: it is estimated that the Dscam gene could encode around 38,000 protein isoforms. A role for Dscam in immunity was proposed when it was found that *in vitro* phagocytosis of *Escherichia coli* by isolated hemocytes lacking Dscam was less efficient than normal. These observations suggest that at least some of this extensive repertoire of alternative exons may have evolved to diversify insects' ability to recognize pathogens. This role for Dscam has been confirmed in the mosquito *Anopheles gambiae*, in which silencing of the Dscam homolog AgDscam has been shown to weaken the mosquito's normal resistance to bacteria and to the malaria parasite *Plasmodium*. There is also evidence from the mosquito that some Dscam exons have specificity for particular pathogens.

Another invertebrate, this time a mollusc, uses a different strategy to diversify an immunoglobulin superfamily protein for use in immunity. The freshwater snail *Biomphalaria glabrata* expresses a small family of **fibrinogen-related proteins** (**FREPs**) thought to have a role in innate immunity. FREPs are produced by hemocytes and secreted into the hemolymph. Their concentration increases when the snail is infected by parasites—it is the intermediate host for the parasitic schistosomes that cause human schistosomiasis. FREPs have one or two immunoglobulin domains at their amino-terminal end and a fibrinogen domain at their carboxy-terminal end. The immunoglobulin domains may interact with pathogens, while the fibrinogen domain may confer on the FREP lectin-like properties that help precipitate the complex.

The *B. glabrata* genome contains many copies of FREP genes that can be divided into approximately 13 subfamilies. A study of the sequences of expressed FREP3 subfamily members has revealed that the FREPs expressed in an individual organism are extensively diversified compared with the germline genes. There are fewer than five genes in the FREP3 subfamily, but an individual snail was found to generate more than 45 distinct FREP3 proteins, all with slightly different sequences. An analysis of the protein sequences suggested that this diversification was due to the accumulation of point mutations in one of the germline FREP3 genes. Although the precise mechanism of this diversification, and the cell type in which it occurs, is not yet known, it does suggest some similarity to somatic hypermutation.

Both the insect and *Biomphalaria* examples seem to represent a way of diversifying molecules involved in immune defense, but although they resemble in some ways the strategy of an adaptive immune response, there is no evidence of clonal selection—the cornerstone of true adaptive immunity.

5-21 Agnathans possess an adaptive immune system that uses somatic gene rearrangement to diversify receptors built from LRR domains.

It had been known for many years that the hagfish and the lamprey could mount a form of accelerated rejection of transplanted skin grafts and exhibit a kind of immunological delayed-type hypersensitivity. Their serum also seemed to contain an activity that behaved as a specific agglutinin, increasing in titer after secondary immunizations, in a similar way to an antibody response in higher vertebrates. Although these phenomena seemed reminiscent of adaptive immunity, there was no evidence of a thymus or of immunoglobulins, but these animals did have cells that could be considered to be genuine lymphocytes on the basis of morphological and molecular analysis. Analysis of the genes expressed by lymphocytes of the sea lamprey *Petromyzon marinus* revealed none related to T-cell receptor or immunoglobulin genes. However, these cells expressed large amounts of mRNAs from genes encoding proteins with multiple LRR domains, the same protein domain from which the pathogen-recognizing Toll-like receptors (TLRs) are built (see Section 3-5).

This might simply have meant that these cells are specialized for recognizing and reacting to pathogens, but the LRR proteins expressed had some surprises in store. Instead of being present in a relatively few forms (like the invariant TLRs), they had highly variable amino acid sequences, with a large number of variable LRR units placed between less variable amino-terminal and carboxy-terminal LRR units. These LRR-containing proteins, called **variable lymphocyte receptors** (**VLRs**), have an invariant stalk region connecting them to the plasma membrane by a glycosylphosphatidylinositol linkage, and can either be tethered to the cell or, at other times, like antibodies, they can be secreted into the blood.

Analysis of the expressed lamprey VLR genes indicates that they are assembled by a process of somatic gene rearrangement (Fig. 5.28). In the germline

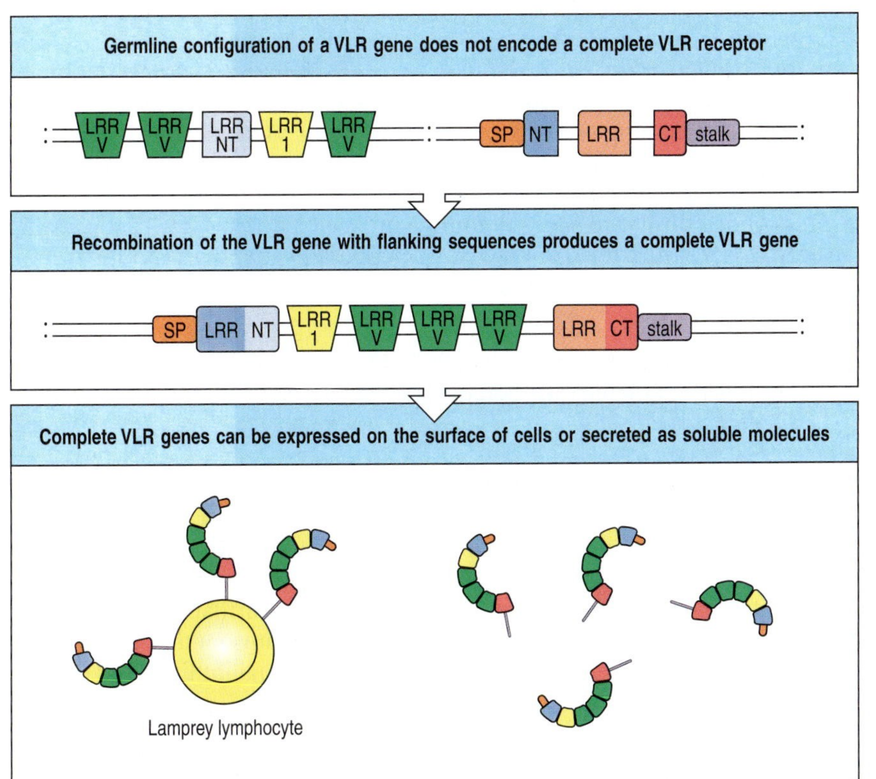

Fig. 5.28 Somatic recombination of an incomplete germline VLR gene generates a diverse repertoire of complete VLR genes in the lamprey. Top panel: an incomplete germline copy of a lamprey VLR gene contains a framework for the complete gene: the signal peptide (SP), part of an amino-terminal LRR unit (NT, dark blue), and a carboxy-terminal LRR unit (red) that is split into two parts (LRR and CT) by intervening noncoding DNA sequences. Nearby flanking regions contain multiple copies of VLR gene-'cassettes' with single or double copies of variable LRR domains (green) and cassettes that encode part of the amino-terminal LRR domains (light blue and yellow). Middle panel: somatic recombination causes various LRR units to be copied into the original VLR gene. This creates a complete VLR gene that contains the assembled amino-terminal LRR cassette (LRR NT) and first LRR (yellow) followed by several variable LRR units (green) and the completed carboxy-terminal LRR unit. The cytidine deaminases PmCDA1 and PmCDA2 from the lamprey *P. marinus* are candidates for enzymes that may initiate this rearrangement. After rearrangement, the complete receptor is attached to the cell membrane by glycosylphosphatidylinositol (GPI) linkage of the stalk region (purple). Bottom panel: an individual lymphocyte undergoes somatic gene rearrangement to produce a unique VLR receptor. These receptors can be tethered to the surface of the lymphocyte via the GPI linkage or can be secreted into the blood. Unique somatic rearrangement events in each developing lymphocyte generate a repertoire of VLR receptors of differing specificities. Adapted from Pancer, Z. and Cooper, M.D.: *Annu. Rev. Immunol.* 2006, 24:497–518.

configuration, there are two incomplete VLR genes, *VLRA* and *VLRB*. These each encode a signal peptide, a partial amino-terminal LRR unit and a partial carboxy-terminal LRR unit, but these three blocks of coding sequence are separated by noncoding DNA that contains no typical signals for RNA splicing nor the RSSs present in immunoglobulin genes (see Section 5-4). Instead, the regions flanking the incomplete VLR genes include a large number of DNA 'cassettes' that contain LRR units—one, two, or three LRR domains at a time. Each mature lamprey lymphocyte expresses a complete and unique VLR gene, either *VLRA* or *VLRB*, which has undergone recombination of these flanking regions with the germline VLR gene.

The creation of a complete VLR gene is currently thought to occur during replication of lamprey lymphocyte DNA by a 'copy-choice' mechanism that is similar, but not identical, to gene conversion (described in Section 5-23). During DNA replication, LRR units flanking the VLR gene are copied into the VLR gene—presumably when a DNA strand being synthesized switches template and copies sequences from one of these LRR units. Although final proof is still lacking, this template-switching mechanism may be triggered by enzymes of the AID-APOBEC family that are expressed by lamprey lymphocytes, and whose cytidine deaminase activity could cause the single-strand DNA breaks that can start the copy-choice process. Lampreys possess two such enzymes, one of which is expressed in *VLRA*-lineage lymphocytes, and the other in *VLRB*-lineage lymphocytes. The final VLR gene contains a complete amino-terminal capping LRR subunit, followed by the addition of up to seven internal LRR domains, each 24 amino acids long, and the removal of the internal noncoding regions to complete the formation of the carboxy-terminal LRR domain (see Fig. 5.28).

It is estimated that this somatic rearrangement mechanism can generate as much diversity in the VLR proteins as is possible for immunoglobulins. Indeed, the crystal structure of a VLR protein shows that the concave surface formed by the series of LRR repeats interacts with a variable insert in the carboxy-terminal LRR to form a surface capable of interacting with a great diversity of antigens. Thus, the diversity of the anticipatory repertoire of agnathans may be limited not by the numbers of possible receptors they can generate but by the number of lymphocytes present in any individual, as in the adaptive immune system of their evolutionary cousins, the gnathostomes. As noted above, each lamprey lymphocyte rearranges only one of the two germline VLR genes, expressing either a complete VLRA or VLRB protein. These two cell populations seem to have some characteristics of mammalian T and B lymphocytes, respectively. For example, VLRA-expressing lymphocytes also express genes similar to some mammalian T-cell cytokine genes, suggesting an even closer similarity to our own RAG-dependent adaptive immune system than was previously appreciated.

5-22 RAG-dependent adaptive immunity based on a diversified repertoire of immunoglobulin-like genes appeared abruptly in the cartilaginous fishes.

Within the vertebrates, we can trace the development of immune functions from the agnathans through the cartilaginous fishes (sharks, skates, and rays) to the bony fishes, then to the amphibians, to reptiles and birds, and finally to mammals. RAG-dependent V(D)J recombination has not been found in agnathans, other chordates, or any invertebrate. But cartilaginous fishes, the earliest group of jawed vertebrates to survive to the present day, have organized lymphoid tissue, T-cell receptors and immunoglobulins, and mount adaptive immune responses.

The origins of RAG-dependent adaptive immunity are now becoming clearer as the genome sequences of many more animals become available. The first clue was that RAG-dependent recombination shares many features with the transposition mechanism of DNA transposons—mobile genetic elements that encode their own transposase, an enzymatic activity that allows them to excise from one site in the genome and reinsert themselves elsewhere. The mammalian RAG complex can act as a transposase *in vitro*, and even the structure of the *RAG* genes, which lie close together in the chromosome and lack the usual introns of mammalian genes, is reminiscent of a transposon.

All this provoked speculation that the origin of RAG-dependent adaptive immunity was the invasion of a DNA transposon into a gene similar to an immunoglobulin or a T-cell receptor V-region gene, an event that would have occurred in some ancestor of the jawed vertebrates (Fig. 5.29). DNA transposons carry inverted repeated sequences at either end, which are bound by the transposase for transposition to occur. These terminal repeats are considered to be the ancestors of the RSSs in present-day antigen-receptor genes (see Section 5-4), while the RAG-1 protein is believed to have evolved from a transposase. Subsequent duplication, reduplication, and recombination of the immune-receptor gene and its inserted RSSs eventually led to the separation of the RAG genes from the rest of the relic transposon and to the multi-segmented immunoglobulin and T-cell receptor loci of present-day vertebrates.

The ultimate origins of the RSSs and the RAG-1 catalytic core are now thought to lie in the Transib superfamily of DNA transposons, and genome sequencing has led to the discovery of sequences related to *RAG-1* in animals as distantly related to vertebrates as the sea anemone *Nematostella*. The origin of *RAG-2* is more obscure, but a *RAG1–RAG2*-related gene cluster was recently discovered in sea urchins, invertebrate relatives of the chordates. Sea urchins themselves show no evidence of immunoglobulins, T-cell receptors or adaptive immunity, but the proteins expressed by the sea-urchin *RAG* genes form a complex with each other and with RAG proteins from the bull shark (*Carcharias leucas*), a primitive jawed vertebrate (but not with those of mammals). This suggests that these proteins could indeed be related to the vertebrate RAGs, and that RAG-1 and RAG-2 were already present in a common ancestor of chordates and echinoderms (the group to which sea urchins belong), presumably fulfilling some other cellular function.

The origin of somatic gene rearrangement in the excision of a transposable element makes sense of an apparent paradox in the rearrangement of

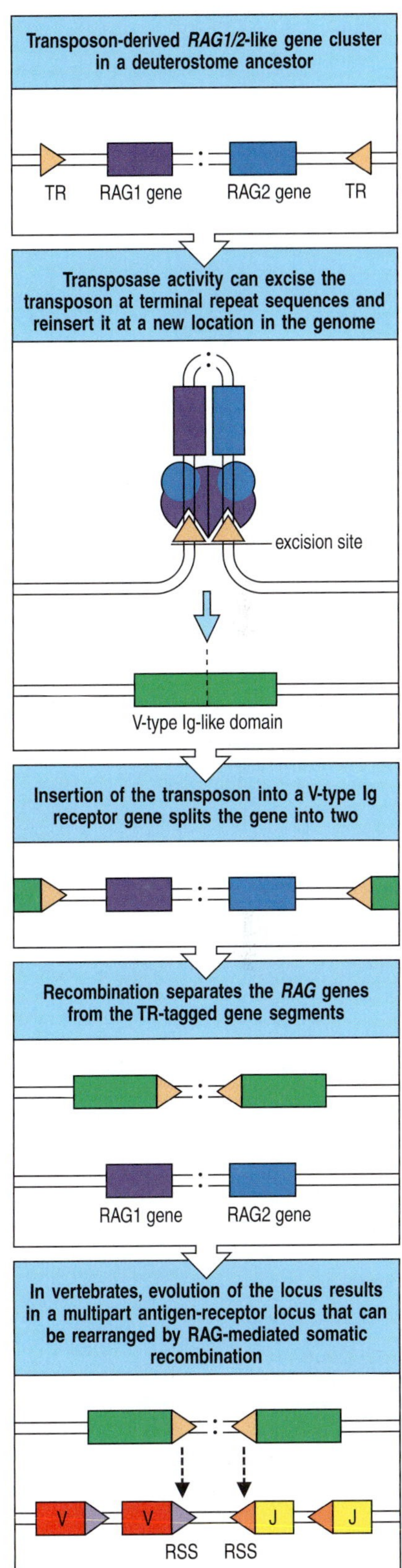

Fig. 5.29 Integration of a transposon into a V-type immunoglobulin receptor gene is thought to have given rise to the T-cell receptor and immunoglobulin genes. Top panel: a DNA transposon in an ancestor of the deuterostomes (the large group of phyla to which the chordates belong) is thought to have had genes related to *RAG1* and *RAG2*—prototype *RAG1* (purple) and *RAG2* (blue), which acted as its transposase. DNA transposons are bounded by terminal inverted repeat (TR) sequences. Second panel: to excise a transposon from DNA, the transposase proteins (purple and blue) bind the TRs, bringing them together, and the transposase enzymatic activity cuts the transposon out of the DNA, leaving a footprint in the host DNA that resembles the TRs. Third panel: after excision from one site, the transposon reinserts elsewhere in the genome, in this case into a V-type immunoglobulin receptor (green). The enzymatic activity of the transposase enables the transposon to insert into DNA in a reaction that is the reverse of the excision reaction. Fourth panel: the integration of the *RAG1/2*-like transposon into the middle of the gene for a V-type immunoglobulin receptor splits the V exon into two parts. Fifth panel: in the evolution of the immunoglobulin and T-cell receptor (TCR) genes, the initial integration event has been followed by DNA rearrangements that separated the transposase genes (now known as the *RAG1* and *RAG2* genes) from the transposon TRs, which we now term the recombination signal sequences (RSSs). The purple sea urchin (an invertebrate deuterostome) has a *RAG1/2*-like gene cluster (not shown) and expresses proteins similar to RAG-1 and RAG-2 proteins, but does not have immunoglobulins, T-cell receptors, or adaptive immunity. The RAG-like proteins presumably retain some other cellular function (so far unknown) in this animal.

immune-system genes. This is that the RSSs are joined precisely in the excised DNA (see Section 5-5), which has no further function and whose fate is irrelevant to the cell, whereas the cut ends in the genomic DNA, which form part of the immunoglobulin or T-cell receptor gene, are joined by an error-prone process, which could be viewed as a disadvantage. However, when looked at from the transposon's point of view this makes sense, because the transposon preserves its integrity by this excision mechanism, whereas the fate of the DNA it leaves behind is of no significance to it. As it turned out, the error-prone joining in the primitive immunoglobulin gene generated useful diversity in molecules used for antigen recognition and was strongly selected for. The RAG-based rearrangement system also provided something else that mutation could not—a means of rapidly modifying the size of the coding region, not just its diversity.

The next question is what sort of gene the transposon inserted into. Proteins containing Ig-like domains are ubiquitous throughout the plant, animal, and bacterial kingdoms, making this one of the most abundant protein superfamilies; in species whose genomes have been fully sequenced, the immunoglobulin superfamily is one of the largest families of protein domains in the genome. The functions of the members of this superfamily are very disparate, and they are a striking example of natural selection taking a useful structure—the basic Ig-domain fold—and adapting it to different purposes.

The immunoglobulin superfamily domains can be divided into four families on the basis of differences in structure and sequence. These are V (resembling an immunoglobulin variable domain), C1 and C2 (resembling constant-region domains), and the more diverse I domains. The target of the RSS-containing element is likely to have been a gene encoding a cell-surface receptor containing an Ig-like V domain, most probably a type similar to present-day VJ domains. These domains are found in some invariant receptor proteins and are so called because of the resemblance of one of the strands to a J segment. It is possible to imagine how transposon movement into such a gene could produce separate V and J gene segments (see Fig. 5.29). On the basis of phylogenetic analysis, a multigene family found in hagfish and lamprey called APARs (agnathan paired receptors resembling Ag receptors) are currently the best candidates for relatives of the ancestor of the antigen receptor. Their DNA sequences predict single-pass transmembrane proteins with a single extracellular VJ domain and a cytoplasmic region containing signaling modules. APARs are expressed in leukocytes.

5-23 Different species generate immunoglobulin diversity in different ways.

Most of the vertebrates we are familiar with generate a large part of their antigen receptor diversity in the same way as mice and humans, by putting together gene segments in different combinations. There are exceptions, however, even within the mammals. Some animals use gene rearrangement to always join together the same V and J gene segment initially, and then diversify this recombined V region. In birds, rabbits, cows, pigs, sheep, and horses, there is little or no germline diversity in the V, D, and J gene segments that are rearranged to form the genes for the initial B-cell receptors, and the rearranged V-region sequences are identical or similar in most immature B cells. These immature B cells migrate to specialized microenvironments—the bursa of Fabricius in chickens, and another intestinal lymphoid organ in rabbits. Here, B cells proliferate rapidly, and their rearranged immunoglobulin genes undergo further diversification.

In birds and rabbits this occurs mainly by gene conversion, a process by which short sequences in the expressed rearranged V-region gene are replaced with sequences from an upstream V gene segment pseudogene. The germline arrangement of the chicken heavy-chain locus is a single set of rearranging

V, J, D, and C gene segments and multiple copies of V-segment pseudogenes. Diversity in this system is created by gene conversion in which sequences from the V_H pseudogenes are copied into the single rearranged V_H gene (Fig. 5.30). It seems that gene conversion is related to somatic hypermutation in its mechanism, because gene conversion in a chicken B-cell line has been shown to require AID. The single-strand nicks generated by the endonuclease APE1 subsequent to cytosine deamination (see Section 5-17) are thought to be the signal that initiates a homology-directed repair process in which a homologous V gene segment is used as the template for the DNA replication that repairs the V-region gene.

In sheep and cows, immunoglobulin diversification is the result of somatic hypermutation, which occurs in an organ known as the ileal Peyer's patch. Somatic hypermutation, independent of T cells and a particular driving antigen, also contributes to immunoglobulin diversification in birds, sheep, and rabbits.

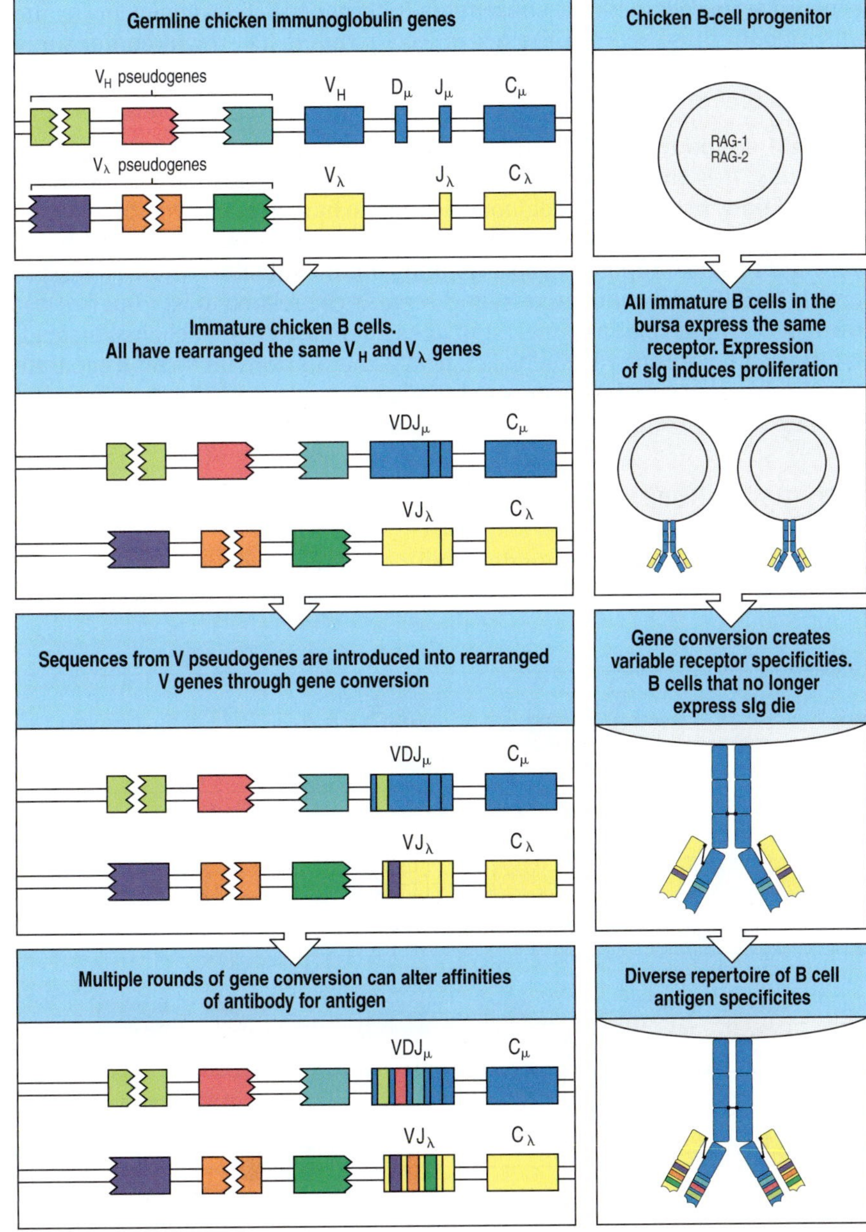

Fig. 5.30 The diversification of chicken immunoglobulins occurs through gene conversion. In chickens, the immunoglobulin diversity that can be created by V(D)J recombination is extremely limited. Initially, there is only one active V, one J gene segment and 15 D gene segments at the chicken heavy-chain locus and one active V and one J gene segment at the single light-chain locus (top left panel). Primary gene rearrangement can thus produce only a very limited number of receptor specificities (second panels). Immature B cells expressing this receptor migrate to the bursa of Fabricius, where the cross-linking of surface immunoglobulin (sIg) induces cell proliferation (second panels). Gene conversion events introduce sequences from adjacent V gene segment pseudogenes into the expressed gene, creating diversity in the receptors (third panel). Some of these gene conversions will inactivate the previously expressed gene (not shown). If a B cell can no longer express sIg after such a gene conversion, it is eliminated. Repeated gene conversion events can continue to diversify the repertoire (bottom panels).

A more fundamentally different organization of immunoglobulin genes is found in the cartilaginous fish, the most primitive jawed vertebrates. Sharks have multiple copies of discrete V_L–J_L–C_L and V_H–D_H–J_H–C_H cassettes, and activate rearrangement within individual cassettes (Fig. 5.31). Although this is somewhat different from the kind of combinatorial gene rearrangement of higher vertebrates, in most cases there is still a requirement for a RAG-mediated somatic rearrangement event. As well as rearranging genes, cartilaginous fish have multiple 'rearranged' V_L regions (and sometimes rearranged V_H regions) in the germline genome (see Fig. 5.31) and apparently generate diversity by activating the transcription of different copies. Even here, some diversity is also contributed by combinatorial means by the subsequent pairing of heavy and light chains.

This 'germline-joined' organization of the light-chain loci is unlikely to represent an intermediate evolutionary stage, because in that case the heavy-chain and light-chain genes would have had to independently acquire the capacity for rearrangement by convergent evolution. It is much more likely that, after the divergence of the cartilaginous fishes, some immunoglobulin loci became rearranged in the germline of various ancestors through activation of the *RAG* genes in germ cells, with the consequent inheritance of rearranged loci by the offspring. In these species, the rearranged germline loci might confer some advantages, such as ensuring rapid responses to common pathogens by producing a preformed set of immunoglobulin chains.

The IgM antibody isotype is thought to go back to the origins of adaptive immunity. It is the predominant form of immunoglobulin in cartilaginous fish and bony fish. The cartilaginous fishes also have at least two other heavy-chain isotypes not found in more recently evolved species. One, IgW, has a constant region composed of six immunoglobulin domains, whereas the second, IgNAR (for new antigen receptor) seems to be related to IgW but has lost the first constant-region domain and does not pair with light chains. Instead, it forms a homodimer in which each heavy-chain V domain forms a separate antigen-binding site. IgW seems to be related to IgD (which is first found in bony fish) and, like IgM, to go back to the origin of adaptive immunity.

5-24 Both α:β and γ:δ T-cell receptors are present in cartilaginous fish.

Neither the T-cell receptors nor the immunoglobulins have been found in any species evolutionarily earlier than the cartilaginous fishes. What is surprising

Fig. 5.31 The organization of immunoglobulin genes is different in different species, but all can generate a diverse repertoire of receptors. The organization of the immunoglobulin heavy-chain genes in mammals, in which there are separated clusters of repeated V, D, and J gene segments, is not the only solution to the problem of generating a diverse repertoire of receptors. Other vertebrates have found alternative solutions. In 'primitive' groups, such as the sharks, the locus consists of multiple repeats of a basic unit composed of a V gene segment, one or two D gene segments, a J gene segment, and a C gene segment. A more extreme version of this organization is found in the κ-like light-chain locus of some cartilaginous fishes such as the rays and the carcharine sharks, in which the repeated unit consists of already rearranged VJ-C genes, from which a random choice is made for expression. In chickens, there is a single rearranging set of gene segments at the heavy-chain locus but multiple copies of V-segment pseudogenes. Diversity in this system is created by gene conversion, in which sequences from the V_H pseudogenes are copied onto the single rearranged V_H gene.

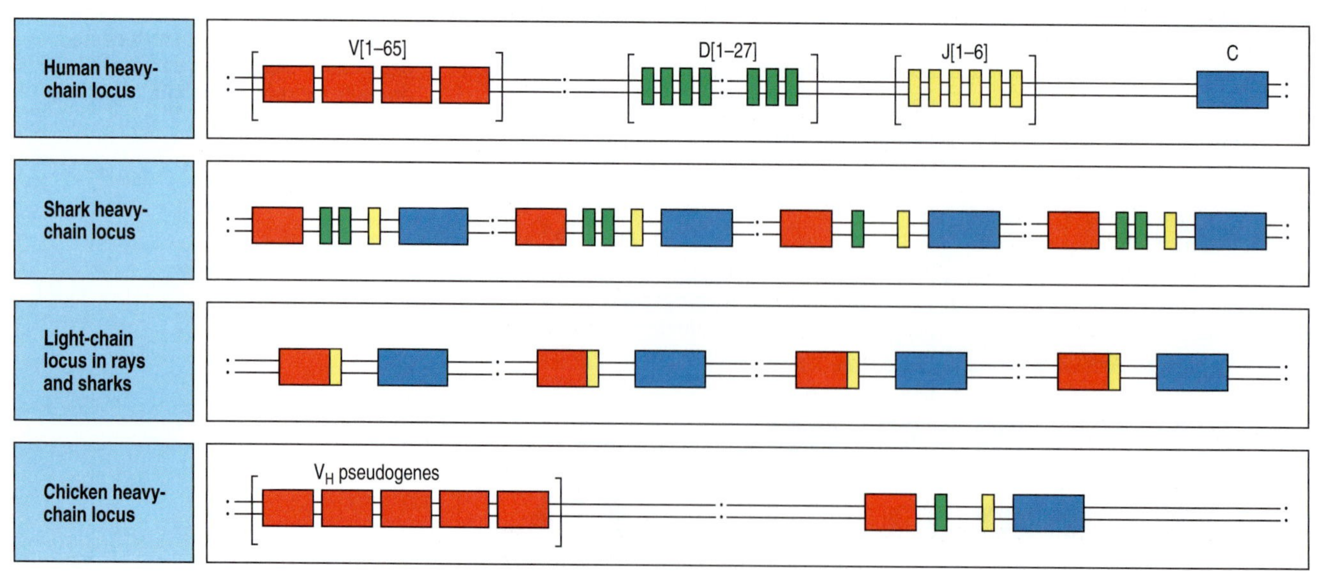

is that by the time we first observe them, they have essentially the same form that we see in mammals. The identification of TCRβ-chain and δ-chain homologs from sharks, and of distinct TCRα, β, γ, and δ chains from a skate, show that even at the earliest time that these adaptive immune system receptors can be identified, they had already diversified into at least two recognition systems. Moreover, each lineage shows diversity resulting from combinatorial somatic rearrangement. Although we still do not fully understand the role of γ:δ T cells in the mammalian adaptive immune system, the very early divergence of the two classes of T-cell receptors and their conservation through subsequent evolution suggests an early separation of functions.

5-25 MHC class I and class II molecules are also first found in the cartilaginous fishes.

One would expect to see the specific ligands of T-cell receptors, the MHC molecules, emerge at around the same time in evolution. Indeed, MHC molecules are present in the cartilaginous fishes and in all higher vertebrates but, like the T-cell receptors, they have not been found in agnathans or invertebrates. Both MHC class I and class II α-chain and β-chain genes are present in sharks, and their products seem to function in an identical way to mammalian MHC molecules. The key residues of the peptide-binding cleft that interact with the ends of the peptide, in MHC class I molecules, or with the central region of the peptide, in MHC class II molecules, are conserved in shark MHC molecules.

Moreover, the MHC genes are also polymorphic in sharks, with multiple alleles of class I and class II loci. In some species, more than 20 MHC class I alleles have been identified so far. For the shark MHC class II molecules, both the class II α and the class II β chains are polymorphic. Thus, not only has the function of the MHC molecules in selecting peptides for presentation evolved during the divergence of the agnathans and the cartilaginous fishes, but the continuous selection imposed by pathogens has also resulted in the polymorphism that is a characteristic feature of the MHC.

The MHC class I genes can be classified into the classical MHC class I genes (sometimes called class Ia) and the nonclassical MHC class Ib genes (discussed in Section 6-18). This is also the case in cartilaginous fishes, because the class I genes of sharks include some that resemble mammalian class Ib molecules. However, it is thought that the shark class Ib genes are not the direct ancestors of the mammalian class Ib genes. For the class I genes, it seems that within each of the five major vertebrate lineages studied (cartilaginous fishes, lobe-finned fishes, ray-finned fishes, amphibians, and mammals) these genes have independently separated into classical and nonclassical loci.

Thus, the characteristic features of the MHC molecules are all present when these molecules are first encountered, and there are no intermediate forms to guide our understanding of their evolution. Although we can trace the evolution of the components of the innate immune system (see Chapters 2 and 3), the mystery of the origin of the adaptive immune system still largely persists. But although we may not have a sure answer to the question of what selective forces led to RAG-dependent elaboration of adaptive immunity, it has never been clearer that, as Charles Darwin remarked about evolution in general, "from so simple a beginning endless forms most beautiful and most wonderful have been, and are being, evolved."

Summary.

Evolution of RAG-dependent adaptive immunity in jawed vertebrates was once considered a wholly unique and inexplicable 'immunological Big Bang.' However, we now understand that true adaptive immunity has also evolved independently at least one other time during evolution. Our close vertebrate cousins, the jawless fishes, have evolved an adaptive immune system built on a completely different basis—the diversification of LRR domains rather than immunoglobulin domains—but which otherwise seems to have the essential features of clonal selection and immunological memory of a true adaptive immune system. Also, we now understand that evolution of the RAG-dependent adaptive immune system is probably related to the chance insertion of a transposon into a member of the immunoglobulin gene superfamily. This event must have taken place in a germline cell in an ancestor of the vertebrates. By chance, the transposon terminal sequences, the forerunners of the RSSs, were placed in an appropriate location within this primordial antigen-receptor gene to enable intramolecular somatic recombination, thus paving the way for the full-blown somatic gene rearrangement seen in present-day immunoglobulin and T-cell receptor genes. The MHC family that are ligands for T-cell receptors first appears in the cartilaginous fishes, suggesting co-evolution with RAG-dependent adaptive immunity. The transposase genes (the *RAG* genes) could have already been present and active in some other function in the genome of this ancestor. *RAG1* seems to be of very ancient origin, as RAG1-related sequences have been found in a wide variety of animal genomes.

Summary to Chapter 5.

Lymphocyte receptors are remarkably diverse, and developing B cells and T cells use the same basic mechanism to achieve this diversity. In each cell, functional genes for the immunoglobulin and T-cell receptor chains are assembled by somatic recombination from sets of separate gene segments that together encode the V region. The substrates for the joining process are arrays of V, D, and J gene segments, which are similar in all the antigen-receptor gene loci, although there are some important differences in the details of their arrangement. The lymphoid-specific proteins RAG-1 and RAG-2 direct the V(D)J recombination process in both T and B cells. These proteins function in concert with ubiquitous DNA-modifying enzymes and at least one other lymphoid-specific enzyme, TdT, to complete the gene rearrangements. As each type of gene segment is present in multiple, slightly different, versions, the random selection of one gene segment from each set for assembly is the source of substantial potential diversity. During the process of assembly, a high degree of functionally important diversity is introduced at the gene segment junctions through imprecise joining mechanisms. This diversity is concentrated in the DNA encoding the CDR3 loops of the receptors, which lie at the center of the antigen-binding site. The independent association of the two chains of immunoglobulins or T-cell receptors to form a complete antigen receptor multiplies the overall diversity available. In addition, mature B cells that are activated by antigen initiate a process of somatic point mutation of the V-region DNA, which creates numerous variants of the original assembled V regions. An important difference between immunoglobulins and T-cell receptors is that immunoglobulins exist in both membrane-bound forms (B-cell receptors) and secreted forms (antibodies). The ability to express both a secreted and a membrane-bound form of the same molecule is due to differential splicing of the heavy-chain mRNA to include exons that encode different forms of the carboxy terminus. Heavy-chain C regions contain three or four immunoglobulin domains, whereas the T-cell receptor chains have only one. Finally, B cells are able to increase the diversity of immunoglobulins by three mechanisms that involve AID-dependent somatic mutation of the primary repertoire—somatic hypermutation, class switching and gene conversion. Somatic hypermutation and gene conversion increase diversity by changes to the V regions of immunoglobulin genes. Class switching diversifies the effector functions of antibodies by providing alternative heavy-chain C regions, but with the same V region and thus same specificity. In this way, the progeny of a single B cell can express several different antibody classes, thus maximizing the possible effector functions of a given antigen-specific antibody. Adaptive immunity in jawed vertebrates appears to have arisen by the integration of a retrotransposon that encoded prototype *RAG1/2* genes into a preexisting V-type immunoglobulin-like gene which subsequently diversified to generate T and B cell receptor genes.

Questions.

5.1 *(a) What are the two kinds of somatic rearrangements of DNA that occur in the immunoglobulin gene loci? (b) Compare and contrast the mechanisms that generate these types of rearrangements. (c) Which one of these types of rearrangements also occurs in the loci that encode the T-cell receptor? (d) What might be a consequence of AID activity occurring in T cells?*

5.2 *The complete V(D)J recombination process uses enzymatic activities that are present in all cell types as well as lymphocyte-specific enzymes. (a) Identify two ubiquitous enzymatic activities that are required for the completion of V(D)J recombination and discuss their functions. (b) Why do these activities not result in inappropriate V(D)J DNA rearrangements in other cell types?*

5.3 *(a) Discuss the four main processes that generate the diversity in the lymphocyte repertoire. (b) Which of these processes is not shared by both B and T cells? (c) How does this difference relate to the kinds of DNA rearrangements that occur in B cells and T cells? (d) What other processes occur in B cells that do not occur in T cells, and why?*

5.4 *What are the physiologic functions of immunoglobulin class switching, and how does this type of diversification differ from that offered by somatic hypermutation?*

5.5 *Mutations produced during affinity maturation are initially generated randomly by the action of AID, and yet eventually amino acid changes to the antibody are clustered to the three CDRs in the V regions. (a) What is the relationship between the location of the CDRs and the V(D)J junctions? (b) How might randomly generated mutations in DNA lead to an enrichment of mutations that are restricted to the CDRs after an immune response?*

5.6 *Variable LRR receptors are produced by somatic rearrangements of incomplete VLR genes in some agnathan species. (a) What are the essential features of an adaptive immune system? (b) Sea lamprey lymphocytes seem to produce two populations of lymphocytes (with different rearranged VLR genes) with different properties. Do you think it more likely that the adaptive immune system of jawed vertebrates evolved directly from the VLR system of agnathans or that the VLR system and the immunoglobulin and T-cell system represent examples of convergent evolution? Discuss the reasons for your answer. (c) How might the presence of two distinct cytidine deaminase-related enzymes in the sea lamprey be of importance?*

General references.

Chaudhuri, J., Basu, U., Zarrin, A., Yan, C., Franco, S., Perlot, T., Vuong, B., Wang, J., Phan, R.T., Datta, A., *et al.*: **Evolution of the immunoglobulin heavy chain class switch recombination mechanism.** *Adv. Immunol.* 2007, **94**:157–214.

Fugmann, S.D., Lee, A.I., Shockett, P.E., Villey, I.J., and Schatz, D.G.: **The RAG proteins and V(D)J recombination: complexes, ends, and transposition.** *Annu. Rev. Immunol.* 2000, **18**:495–527.

Jung, D., Giallourakis, C., Mostoslavsky, R., and Alt, F.W.: **Mechanism and control of V(D)J recombination at the immunoglobulin heavy chain locus.** *Annu. Rev. Immunol.* 2006, **24**:541–570.

Longerich, S., Basu, U., Alt, F., and Storb, U.: **AID in somatic hypermutation and class switch recombination.** *Curr. Opin. Immunol.* 2006, **18**:164–174.

Odegard, V.H., and Schatz, D.G.: **Targeting of somatic hypermutation.** *Nat. Rev. Immunol.* 2006, **6**:573–583.

Schatz, D.G.: **Antigen receptor genes and the evolution of a recombinase.** *Semin. Immunol.* 2004, **16**:245–256.

Schatz, D.G.: **V(D)J recombination.** *Immunol. Rev.* 2004, **200**:5–11.

Section references.

5-1 Immunoglobulin genes are rearranged in antibody-producing cells.

Hozumi, N., and Tonegawa, S.: **Evidence for somatic rearrangement of immunoglobulin genes coding for variable and constant regions.** *Proc. Natl Acad. Sci. USA* 1976, **73**:3628–3632.

Tonegawa, S., Brack, C., Hozumi, N., and Pirrotta, V.: **Organization of immunoglobulin genes.** *Cold Spring Harbor Symp. Quant. Biol.* 1978, **42**:921–931.

5-2 Complete genes that encode a variable region are generated by the somatic recombination of separate gene segments.

Early, P., Huang, H., Davis, M., Calame, K., and Hood, L.: **An immunoglobulin heavy chain variable region gene is generated from three segments of DNA: V_H, D and J_H.** *Cell* 1980, **19**:981–992.

Tonegawa, S., Maxam, A.M., Tizard, R., Bernard, O., and Gilbert, W.: **Sequence of a mouse germ-line gene for a variable region of an immunoglobulin light chain.** *Proc. Natl Acad. Sci. USA* 1978, **75**:1485–1489.

5-3 Multiple contiguous V gene segments are present at each immunoglobulin locus.

Maki, R., Traunecker, A., Sakano, H., Roeder, W., and Tonegawa, S.: **Exon shuffling generates an immunoglobulin heavy chain gene.** *Proc. Natl Acad. Sci. USA* 1980, **77**:2138–2142.

Matsuda, F., and Honjo, T.: **Organization of the human immunoglobulin heavy-chain locus.** *Adv. Immunol.* 1996, **62**:1–29.

Thiebe, R., Schable, K.F., Bensch, A., Brensing-Kuppers, J., Heim, V., Kirschbaum, T., Mitlohner, H., Ohnrich, M., Pourrajabi, S., Roschenthaler, F., *et al.*: **The variable genes and gene families of the mouse immunoglobulin kappa locus.** *Eur. J. Immunol.* 1999, **29**:2072–2081.

5-4 Rearrangement of V, D, and J gene segments is guided by flanking DNA sequences.

Grawunder, U., West, R.B., and Lieber, M.R.: **Antigen receptor gene rearrangement.** *Curr. Opin. Immunol.* 1998, **10**:172–180.

Lieber, M. R.: **The mechanism of human nonhomologous DNA end joining.** *J. Biol. Chem.* 2008, **283**:1–5.

Sakano, H., Huppi, K., Heinrich, G., and Tonegawa, S.: **Sequences at the somatic recombination sites of immunoglobulin light-chain genes.** *Nature* 1979, **280**:288–294.

5-5 The reaction that recombines V, D, and J gene segments involves both lymphocyte-specific and ubiquitous DNA-modifying enzymes.

Agrawal, A., and Schatz, D.G.: **RAG1 and RAG2 form a stable postcleavage synaptic complex with DNA containing signal ends in V(D)J recombination.** *Cell* 1997, **89**:43–53.

Ahnesorg, P., Smith, P., and Jackson, S.P.: **XLF interacts with the XRCC4-DNA ligase IV complex to promote nonhomologous end-joining.** *Cell* 2006, **124**:301–313.

Blunt, T., Finnie, N.J., Taccioli, G.E., Smith, G.C.M., Demengeot, J., Gottlieb, T.M., Ma, Y., Pannicke, U., Schwarz, K., and Lieber, M.R.: **Hairpin opening and overhang processing by an Artemis:DNA-PKcs complex in V(D)J recombination and in nonhomologous end joining.** *Cell* 2002, **108**:781–794.

Buck, D., Malivert, L., deChasseval, R., Barraud, A., Fondaneche, M.-C., Xanal, O., Plebani, A., Stephan, J.-L., Hufnagel, M., LeDiest, F., *et al.*: **Cernunnos, a novel nonhomologous end-joining factor, is mutated in human immunodeficiency with microcephaly.** *Cell* 2006, **124**:287–299.

Jung, D., Giallourakis, C., Mostoslavsky, R., and Alt, F.W.: **Mechanism and control of V(D)J recombination at the immunoglobulin heavy chain locus.** *Annu. Rev. Immunol.* 2006, **24**:541–570.

Li, Z.Y., Otevrel, T., Gao, Y.J., Cheng, H.L., Seed, B., Stamato, T.D., Taccioli, G.E., and Alt, F.W.: **The XRCC4 gene encodes a novel protein involved in DNA double-strand break repair and V(D)J recombination.** *Cell* 1995, **83**:1079–1089.

Mizuta, R., Varghese, A.J., Alt, F.W., Jeggo, P.A., and Jackson, S.P.: **Defective DNA-dependent protein kinase activity is linked to V(D)J recombination and DNA-repair defects associated with the murine-scid mutation.** *Cell* 1995, **80**:813–823.

Moshous, D., Callebaut, I., de Chasseval, R., Corneo, B., Cavazzana-Calvo, M., Le Deist, F., Tezcan, I., Sanal, O., Bertrand, Y., Philippe, N., *et al.*: **Artemis, a novel DNA double-strand break repair/V(D)J recombination protein, is mutated in human severe combined immune deficiency.** *Cell* 2001, **105**:177–186.

Oettinger, M.A., Schatz, D.G., Gorka, C., and Baltimore, D.: **RAG-1 and RAG-2, adjacent genes that synergistically activate V(D)J recombination.** *Science* 1990, **248**:1517–1523.

Villa, A., Santagata, S., Bozzi, F., Giliani, S., Frattini, A., Imberti, L., Gatta, L.B., Ochs, H.D., Schwarz, K., Notarangelo, L.D., *et al.*: **Partial V(D)J recombination activity leads to Omenn syndrome.** *Cell* 1998, **93**:885–896.

5-6 The diversity of the immunoglobulin repertoire is generated by four main processes.

Weigert, M., Perry, R., Kelley, D., Hunkapiller, T., Schilling, J., and Hood, L.: **The joining of V and J gene segments creates antibody diversity.** *Nature* 1980, **283**:497–499.

5-7 The multiple inherited gene segments are used in different combinations.

Lee, A., Desravines, S., and Hsu, E.: **IgH diversity in an individual with only one million B lymphocytes.** *Dev. Immunol.* 1993, **3**:211–222.

5-8 Variable addition and subtraction of nucleotides at the junctions between gene segments contributes to the diversity in the third hypervariable region.

Gauss, G.H., and Lieber, M.R.: **Mechanistic constraints on diversity in human V(D)J recombination.** *Mol. Cell. Biol.* 1996, **16**:258–269.

Gilfillan, S., Dierich, A., Lemeur, M., Benoist, C., and Mathis, D.: **Mice lacking TdT: mature animals with an immature lymphocyte repertoire.** *Science* 1993, **261**:1755–1759.

Komori, T., Okada, A., Stewart, V., and Alt, F.W.: **Lack of N regions in antigen receptor variable region genes of TdT-deficient lymphocytes.** *Science* 1993, **261**:1171–1175.

Weigert, M., Gatmaitan, L., Loh, E., Schilling, J., and Hood, L.: **Rearrangement of genetic information may produce immunoglobulin diversity.** *Nature* 1978, **276**:785–790.

5-9 The T-cell receptor gene segments are arranged in a similar pattern to immunoglobulin gene segments and are rearranged by the same enzymes.

Bertocci, B., DeSmet, A., Weill, J.-C., and Reynaud, C.A. **Non-overlapping functions of polX family DNA polymerases, pol μ, pol λ, and TdT, during immunoglobulin V(D)J recombination *in vivo*.** *Immunity* 2006, **25**:31–41.

Lieber, M.R.: **The polymerases for V(D)J recombination.** *Immunity* 2006, **25**:7–9.

Rowen, L., Koop, B.F., and Hood, L.: **The complete 685-kilobase DNA sequence of the human β T cell receptor locus.** *Science* 1996, **272**:1755–1762.

Shinkai, Y., Rathbun, G., Lam, K.P., Oltz, E.M., Stewart, V., Mendelsohn, M., Charron, J., Datta, M., Young, F., Stall, A.M., *et al.*: **RAG-2 deficient mice lack mature lymphocytes owing to inability to initiate V(D)J rearrangement.** *Cell* 1992, **68**:855–867.

5-10 T-cell receptors concentrate diversity in the third hypervariable region.

Davis, M.M., and Bjorkman, P.J.: **T-cell antigen receptor genes and T-cell recognition.** *Nature* 1988, **334**:395–402.

Garboczi, D.N., Ghosh, P., Utz, U., Fan, Q.R., Biddison, W.E., and Wiley, D.C.: **Structure of the complex between human T-cell receptor, viral peptide and HLA-A2.** *Nature* 1996, **384**:134–141.

Hennecke, J., and Wiley, D.C.: **T cell receptor–MHC interactions up close.** *Cell* 2001, **104**:1–4.

Hennecke, J., Carfi, A., and Wiley, D.C.: **Structure of a covalently stabilized complex of a human αβ T-cell receptor, influenza HA peptide and MHC class II molecule, HLA-DR1.** *EMBO J.* 2000, **19**:5611–5624.

Jorgensen, J.L., Esser, U., Fazekas de St. Groth, B., Reay, P.A., and Davis, M.M.: **Mapping T-cell receptor–peptide contacts by variant peptide immunization of single-chain transgenics.** *Nature* 1992, **355**:224–230.

5-11 γ:δ T-cell receptors are also generated by gene rearrangement.

Chien, Y.H., Iwashima, M., Kaplan, K.B., Elliott, J.F., and Davis, M.M.: **A new T-cell receptor gene located within the alpha locus and expressed early in T-cell differentiation.** *Nature* 1987, **327**:677–682.

Lafaille, J.J., DeCloux, A., Bonneville, M., Takagaki, Y., and Tonegawa, S.: **Junctional sequences of T cell receptor gamma delta genes: implications for gamma delta T cell lineages and for a novel intermediate of V-(D)-J joining.** *Cell* 1989, **59**:859–870.

Tonegawa, S., Berns, A., Bonneville, M., Farr, A.G., Ishida, I., Ito, K., Itohara, S., Janeway, C.A., Jr, Kanagawa, O., Kubo, R., *et al.*: **Diversity, development, ligands, and probable functions of gamma delta T cells.** *Adv. Exp. Med. Biol.* 1991, **292**:53–61.

5-12 Different classes of immunoglobulins are distinguished by the structure of their heavy-chain constant regions.

Davies, D.R., and Metzger, H.: **Structural basis of antibody function.** *Annu. Rev. Immunol.* 1983, **1**:87–117.

5-13 The constant region confers functional specialization on the antibody.

Helm, B.A., Sayers, I., Higginbottom, A., Machado, D.C., Ling, Y., Ahmad, K., Padlan, E.A., and Wilson, A.P.M.: **Identification of the high affinity receptor binding region in human IgE.** *J. Biol. Chem.* 1996, **271**:7494–7500.

Jefferis, R., Lund, J., and Goodall, M.: **Recognition sites on human IgG for Fcγ receptors—the role of glycosylation.** *Immunol. Lett.* 1995, **44**:111–117.

Sensel, M.G., Kane, L.M., and Morrison, S.L.: **Amino acid differences in the N-terminus of C_H2 influence the relative abilities of IgG2 and IgG3 to activate complement.** *Mol. Immunol.* **34**:1019–1029.

5-14 Mature naive B cells express both IgM and IgD at their surface.

Abney, E.R., Cooper, M.D., Kearney, J.F., Lawton, A.R., and Parkhouse, R.M.: **Sequential expression of immunoglobulin on developing mouse B lymphocytes: a systematic survey that suggests a model for the generation of immunoglobulin isotype diversity.** *J. Immunol.* 1978, **120**:2041–2049.

Blattner, F.R. and Tucker, P.W.: **The molecular biology of immunoglobulin D.** *Nature* 1984, **307**:417–422.

Goding, J.W., Scott, D.W., and Layton, J.E.: **Genetics, cellular expression and function of IgD and IgM receptors.** *Immunol. Rev.* 1977, **37**:152–186.

5-15 Transmembrane and secreted forms of immunoglobulin are generated from alternative heavy-chain transcripts.

Early, P., Rogers, J., Davis, M., Calame, K., Bond, M., Wall, R., and Hood, L.: **Two mRNAs can be produced from a single immunoglobulin μ gene by alternative RNA processing pathways.** *Cell* 1980, **20**:313–319.

Peterson, M.L., Gimmi, E.R., and Perry, R.P.: **The developmentally regulated shift from membrane to secreted μ mRNA production is accompanied by an increase in cleavage-polyadenylation efficiency but no measurable change in splicing efficiency.** *Mol. Cell. Biol.* 1991, **11**:2324–2327.

Rogers, J., Early, P., Carter, C., Calame, K., Bond, M., Hood, L., and Wall, R.: **Two mRNAs with different 3′ ends encode membrane-bound and secreted forms of immunoglobulin μ chain.** *Cell* 1980, **20**:303–312.

5-16 IgM and IgA can form polymers.

Hendrickson, B.A., Conner, D.A., Ladd, D.J., Kendall, D., Casanova, J.E., Corthesy, B., Max, E.E., Neutra, M.R., Seidman, C.E., and Seidman, J.G.: **Altered hepatic transport of IgA in mice lacking the J chain.** *J. Exp. Med.* 1995, **182**:1905–1911.

Niles, M.J., Matsuuchi, L., and Koshland, M.E.: **Polymer IgM assembly and secretion in lymphoid and nonlymphoid cell-lines—evidence that J chain is required for pentamer IgM synthesis.** *Proc. Natl Acad. Sci. USA* 1995, **92**:2884–2888.

5-17 Activation-induced cytidine deaminase (AID) introduces mutations into genes transcribed in B cells.

Bransteitter, R., Pham, P., Scharff, M.D., and Goodman, M.F.: **Activation-induced cytidine deaminase deaminates deoxycytidine on single-stranded DNA but requires the action of RNase.** *Proc. Natl Acad. Sci. USA* 2003, **100**:4102–4107.

Muramatsu, M., Kinoshita, K., Fagarasan, S., Yamada, S., Shinkai, Y., and Honjo, T.: **Class switch recombination and hypermutation require activation-induced cytidine deaminase (AID), a potential RNA editing enzyme.** *Cell* 2000, **102**:553–563.

Petersen-Mahrt, S.K., Harris, R.S., and Neuberger, M.S.: **AID mutates *E. coli* suggesting a DNA deamination mechanism for antibody diversification.** *Nature* 2002, **418**:99–103.

Pham, P., Bransteitter, R., Petruska, J., and Goodman, M.F.: **Processive AID-catalyzed cytosine deamination on single-stranded DNA stimulates somatic hypermutation.** *Nature* 2003, **424**:103–107.

Yu, K., Huang, F.T., and Lieber, M.R.: **DNA substrate length and surrounding sequence affect the activation-induced deaminase activity at cytidine.** *J. Biol. Chem.* 2004, **279**:6496–6500.

5-18 Somatic hypermutation further diversifies the rearranged V regions of immunoglobulin genes.

Basu, U., Chaudhuri, J., Alpert, C., Dutt, S., Ranganath, S., Li, G., Schrum, J.P., Manis, J.P., and Alt, F.W.: **The AID antibody diversification enzyme is regulated by protein kinase A phosphorylation.** *Nature* 2005, **438**:508–511.

Betz, A.G., Rada, C., Pannell, R., Milstein, C., and Neuberger, M.S.: **Passenger transgenes reveal intrinsic specificity of the antibody hypermutation mechanism: clustering, polarity, and specific hot spots.** *Proc. Natl Acad. Sci. USA* 1993, **90**:2385–2388.

Chaudhuri, J., Khuong, C., and Alt, F.W.: **Replication protein A interacts with AID to promote deamination of somatic hypermutation targets.** *Nature* 2004, **430**:992–998.

Di Noia, J. and Neuberger, M.S.: **Altering the pathway of immunoglobulin hypermutation by inhibiting uracil-DNA glycosylase.** *Nature* 2002, **419**:43–48.

McKean, D., Huppi, K., Bell, M., Straudt, L., Gerhard, W., and Weigert, M.: **Generation of antibody diversity in the immune response of BALB/c mice to influenza virus hemagglutinin.** *Proc. Natl Acad. Sci. USA* 1984, **81**:3180–3184.

Weigert, M.G., Cesari, I.M., Yonkovich, S.J., and Cohn, M.: **Variability in the lambda light chain sequences of mouse antibody.** *Nature* 1970, **228**:1045–1047.

5-19 Class switching enables the same assembled V_H exon to be associated with different C_H genes in the course of an immune response.

Chaudhuri, J., and Alt, F.W.: **Class-switch recombination: interplay of transcription, DNA deamination and DNA repair.** *Nat. Rev. Immunol.* 2004, **4**:541–552.

Jung, S., Rajewsky, K., and Radbruch, A.: **Shutdown of class switch recombination by deletion of a switch region control element.** *Science* 1993, **259**:984.

Revy, P., Muto, T., Levy, Y., Geissmann, F., Plebani, A., Sanal, O., Catalan, N., Forveille, M., Dufourcq-Lagelouse, R., Gennery, A., *et al.*: **Activation-induced cytidine deaminase (AID) deficiency causes the autosomal recessive form of the hyper-IgM syndrome (HIGM2).** *Cell* 2000, **102**:565–575.

Sakano, H., Maki, R., Kurosawa, Y., Roeder, W., and Tonegawa, S.: **Two types of somatic recombination are necessary for the generation of complete immunoglobulin heavy-chain genes.** *Nature* 1980, **286**:676–683.

Shinkura, R., Tian, M., Smith, M., Chua, K., Fujiwara, Y., and Alt, F.W.: **The influence of transcriptional orientation on endogenous switch region function.** *Nat. Immunol.* 2003, **4**:435–441.

Yu, K., Chedin, F., Hsieh, C.-L., Wilson, T.E., and Lieber, M.R.: **R-loops at immunoglobulin class switch regions in the chromosomes of stimulated B cells.** *Nat. Immunol.* 2003, **4**:442–451.

5-20 Some invertebrates generate extensive diversity in a repertoire of immunoglobulin-like genes.

Dong, Y., Taylor, H.E., and Dimopoulos, G.: **AgDscam, a hypervariable immunoglobulin domain-containing receptor of the *Anopheles gambiae* innate immune system.** *PLoS Biol.* 2006, **4**:e229.

Loker, E.S., Adema, C.M., Zhang, S.M., and Kepler, T.B.: **Invertebrate immune systems—not homogeneous, not simple, not well understood.** *Immunol. Rev.* 2004, **198**:10–24.

Watson, F.L., Puttmann-Holgado, R., Thomas, F., Lamar, D.L., Hughes, M., Kondo, M., Rebel, V.I., and Schmucker, D.: **Extensive diversity of Ig-superfamily proteins in the immune system of insects.** *Science* 2005, **309**:1826–1827.

Zhang, S.M., Adema, C.M., Kepler, T.B., and Loker, E.S.: **Diversification of Ig superfamily genes in an invertebrate.** *Science* 2004, **305**:251–254.

5-21 Agnathans possess an adaptive immune system that uses somatic gene rearrangement to diversify receptors built from LRR domains.

Cooper, M.D., and Alder, M.N.: **The evolution of adaptive immune systems.** *Cell* 2006, **124**:815–822.

Guo, P., Hirano, M., Herrin, B.R., Li, J., Yu, C., Sadlonova, A., and Cooper, M.D.: **Dual nature of the adaptive immune system in lampreys.** *Nature* 2009, **459**:796–801. [Erratum: *Nature* 2009, **460**:1044.]

Han, B.W., Herrin, B.R., Cooper, M.D., and Wilson, I.A.: **Antigen recognition by variable lymphocyte receptors.** *Science* 2008, **321**:1834–1837.

Litman, G.W., Finstad, F.J., Howell, J., Pollara, B.W., and Good, R.A.: **The evolution of the immune response. 3. Structural studies of the lamprey immuoglobulin.** *J. Immunol.* 1970, **105**:1278–1285.

Nagawa, F., Kishishita, N., Shimizu, K., Hirose, S., Miyoshi, M., Nezu, J., Nishimura, T., Nishizumi, H., Takahashi, Y., Hashimoto, S., *et al.*: **Antigen-receptor genes of the agnathan lamprey are assembled by a process involving copy choice.** *Nat. Immunol.* 2007, **8**:206–213.

Rogozin, I.B., Iyer, L.M., Liang, L., Glazko, G.V., Liston, V.G., Pavlov, Y.I., Aravind, L., and Pancer, Z.: **Evolution and diversification of lamprey antigen receptors: evidence for involvement of an AID-APOBEC family cytosine deaminase.** *Nat. Immunol.* 2007, **8**:647–656.

5-22 RAG-dependent adaptive immunity based on a diversified repertoire of immunoglobulin-like genes appeared abruptly in the cartilaginous fishes.

Fugmann, S.D., Messier, C., Novack, L.A., Cameron, R.A., and Rast, J.P.: **An ancient evolutionary origin of the *Rag1/2* gene locus.** *Proc. Natl Acad. Sci. USA* 2006, **103**:3728–3733.

Kapitonov, V.V., and Jurka J.: **RAG1 core and V(D)J recombination signal sequences were derived from Transib transposons.** *PLoS Biol.* 2005, **3**:e181.

Suzuki, T., Shin-I, T., Fujiyama, A., Kohara, Y., and Kasahara, M.: **Hagfish leukocytes express a paired receptor family with a variable domain resembling those of antigen receptors.** *J. Immunol.* 2005, **174**:2885–2891.

van den Berg, T.K., Yoder, J.A., and Litman, G.W.: **On the origins of adaptive immunity: innate immune receptors join the tale.** *Trends Immunol.* 2004, **25**:11–16.

5-23 Different species generate immunoglobulin diversity in different ways.

Knight, K.L., and Crane, M.A.: **Generating the antibody repertoire in rabbit.** *Adv. Immunol.* 1994, **56**:179–218.

Reynaud, C.A., Bertocci, B., Dahan, A., and Weill, J.C.: **Formation of the chicken B-cell repertoire—ontogeny, regulation of Ig gene rearrangement, and diversification by gene conversion.** *Adv. Immunol.* 1994, **57**:353–378.

Reynaud, C.A., Garcia, C., Hein, W.R., and Weill, J.C.: **Hypermutation generating the sheep immunoglobulin repertoire is an antigen independent process.** *Cell* 1995, **80**:115–125.

Vajdy, M., Sethupathi, P., and Knight, K.L.: **Dependence of antibody somatic diversification on gut-associated lymphoid tissue in rabbits.** *J. Immunol.* 1998, **160**:2725–2729.

5-24 Both α:β and γ:δ T-cell receptors are present in cartilaginous fish.

Rast, J.P., and Litman, G.W.: **T-cell receptor gene homologs are present in the most primitive jawed vertebrates.** *Proc. Natl Acad. Sci. USA* 1994, **91**:9248–9252.

Rast, J.P., Anderson, M.K., Strong, S.J., Luer, C., Litman, R.T., and Litman, G.W.: **α, β, γ, and δ T-cell antigen receptor genes arose early in vertebrate phylogeny.** *Immunity* 1997, **6**:1–11.

5-25 MHC class I and class II molecules are also first found in the cartilaginous fishes.

Hashimoto, K., Okamura, K., Yamaguchi, H., Ototake, M., Nakanishi, T., and Kurosawa, Y.: **Conservation and diversification of MHC class I and its related molecules in vertebrates.** *Immunol. Rev.* 1999, **167**:81–100.

Kurosawa, Y., and Hashimoto, K.: **How did the primordial T cell receptor and MHC molecules function initially?** *Immunol. Cell Biol.* 1997, **75**:193–196.

Ohta, Y., Okamura, K., McKinney, E.C., Bartl, S., Hashimoto, K., and Flajnik, M.F.: **Primitive synteny of vertebrate major histocompatibility complex class I and class II genes.** *Proc. Natl Acad. Sci. USA* 2000, **97**:4712–4717.

Okamura, K., Ototake, M., Nakanishi, T., Kurosawa, Y., and Hashimoto, K.: **The most primitive vertebrates with jaws possess highly polymorphic MHC class I genes comparable to those of humans.** *Immunity* 1997, **7**:777–790.

Antigen Presentation to T Lymphocytes

6

In an adaptive immune response, antigen is recognized by two distinct sets of highly variable receptor molecules—the immunoglobulins that serve as antigen receptors on B cells and the antigen-specific receptors of T cells. As we saw in Chapter 4, T cells only recognize antigens that are displayed on cell surfaces. These antigens can be derived from pathogens that replicate within cells, such as viruses or intracellular bacteria, or from pathogens or their products that cells have taken up from the extracellular fluid. In either case, cells display on their surface peptide fragments derived from the pathogens' proteins, and in this way the presence of infected cells and foreign antigens can be detected by T cells. The pathogen-derived peptides are delivered to the cell surface by specialized glycoproteins, the **MHC molecules**, whose structure and function were also introduced in Chapter 4. The MHC molecules are encoded within a large cluster of genes that were first identified by their powerful effects on the immune response to transplanted tissues. For that reason, the gene complex is called the **major histocompatibility complex** (**MHC**).

We begin by discussing the mechanisms by which protein antigens are degraded into peptides inside cells and the peptides are carried to the cell surface bound to MHC molecules. We will see that the two different classes of MHC molecules, known as MHC class I and MHC class II, obtain peptides at different cellular locations. Peptides derived from the cytosol are transported in the endoplasmic reticulum where they are placed into newly synthesized MHC class I molecules; when transported to the cell surface, these peptide:MHC complexes are recognized by CD8 T cells. Peptides generated by the degradation of proteins in intracellular endosomal vesicles are bound to MHC class II molecules and are recognized by CD4 T cells. CD8 and CD4 T cells have very different activities, which are effective against different types of pathogens. The class of MHC molecule presenting a pathogen peptide is therefore crucial in ensuring recognition by a T cell whose actions will be able to eliminate that pathogen.

Whether an infection is present or absent, MHC molecules routinely bind peptides from self proteins, which are continually being broken down in the cell, and display these at the cell surface. The tolerance mechanisms that prevent the immune system from reacting against self tissues normally prevent these self peptides from initiating an immune response. If tolerance breaks down, however, self peptides displayed on cell surfaces can elicit autoimmune responses, as discussed in Chapter 15.

The second part of this chapter focuses on the MHC class I and II genes and their remarkable variability. There are several different MHC molecules in each class and each of their genes is highly polymorphic, with many variants present in the population. MHC polymorphism has a profound effect on antigen recognition by T cells, and the combination of multiple genes and polymorphism greatly extends the range of peptides that can be presented to T cells by each individual and by populations as a whole, thus enabling them to respond to the wide range of potential pathogens they will encounter. The MHC also contains genes other than those for the MHC molecules, and the products of many of these genes are involved in the production of peptide:MHC complexes. We shall also consider a group of proteins, encoded both within and outside the MHC, that are similar to MHC class I molecules but have limited polymorphism. They have various functions, such as activating T cells and NK cells by binding NKG2D (see Section 3-22) and the presentation of microbial lipid antigens to a special subset of α:β T cells known as iNKT cells, which express a restricted repertoire of T-cell receptors.

The generation of T-cell receptor ligands.

The protective function of T cells depends on their ability to recognize cells that are harboring pathogens or that have internalized pathogens or their products. As we saw in Chapter 4, the ligand recognized by a T-cell receptor is a peptide bound to an MHC molecule and displayed on a cell surface. The generation of peptides from native proteins is commonly referred to as **antigen processing**, while the display of the peptide at the cell surface by the MHC molecule is referred to as **antigen presentation.** We have already described the structure of MHC molecules and seen how they bind peptide antigens in a cleft on their outer surface (see Sections 4-13 to 4-16). In this chapter we look at how peptides are generated from pathogen proteins present in two distinct intracellular compartments—the cytosol and the vesicular compartment (Fig. 6.1)—and how they are loaded onto MHC class I or MHC class II molecules, respectively.

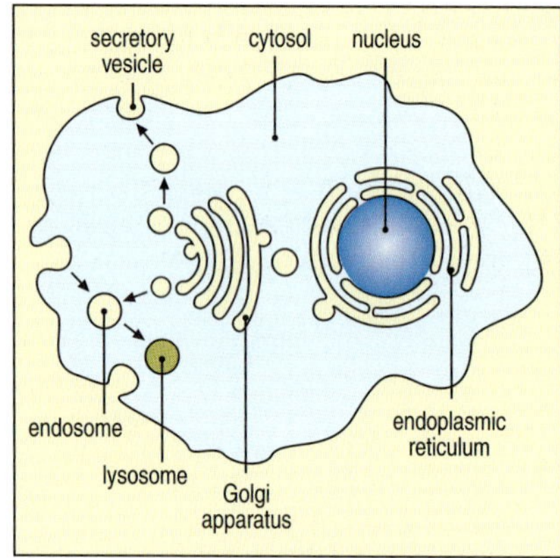

Fig. 6.1 There are two major intracellular compartments, separated by membranes. One compartment is the cytosol, which also communicates with the nucleus via the nuclear pores in the nuclear membrane. The other is the vesicular system, which comprises the endoplasmic reticulum, Golgi apparatus, endosomes, lysosomes, and other intracellular vesicles. The vesicular system can be thought of as continuous with the extracellular fluid. Secretory vesicles bud off from the endoplasmic reticulum and are transported via fusion with Golgi membranes to move vesicular contents out of the cell, whereas extracellular material is taken up by endocytosis or phagocytosis into endosomes or phagosomes, respectively. The inward and outward pathways can be linked by the fusion of incoming and outgoing vesicles, which is important both for pathogen destruction in specialized cells such as neutrophils and for antigen presentation, as we shall see in this chapter.

6-1 The MHC class I and class II molecules deliver peptides to the cell surface from two intracellular compartments.

Antigens from infectious agents get into the cytosolic or vesicular compartments of cells by various routes. Viruses and some bacteria replicate in the cytosol or in the contiguous nuclear compartment (Fig. 6.2, first panel). Microbial antigens enter the vesicular compartment in either of two ways. Certain pathogenic bacteria and protozoan parasites survive ingestion by macrophages and replicate inside the intracellular vesicles of the endosomal–lysosomal system (Fig. 6.2, second panel). Other pathogenic bacteria proliferate outside cells, where they cause tissue damage by secreting toxins and other proteins. These bacteria and their toxic products can be internalized by phagocytosis, receptor-mediated endocytosis, or macropinocytosis into endosomes and lysosomes, where they are broken down by digestive enzymes; illustrated here is the case of receptor-mediated endocytosis by B cells, which are very efficient at taking up and internalizing extracellular antigens by means of their antigen-specific B-cell receptors (Fig. 6.2, third panel). Virus particles and parasite antigens in extracellular fluids can also be taken up by these routes and degraded, and their peptides presented to T cells.

Peptides originating in the different compartments are delivered to the cell surface by different classes of MHC molecules. MHC class I molecules deliver peptides originating in the cytosol, whereas MHC class II molecules deliver peptides originating in the vesicular system. The immune system has different strategies for eliminating pathogens replicating in the cytosol or in the endosomal system, and the MHC molecules ensure that the appropriate strategy is activated. Cells infected with viruses or cytosolic bacteria are detected and eliminated by **cytotoxic T cells**; these effector T cells are distinguished by the co-receptor molecule CD8, which binds to MHC class I molecules (see Section 4-17). The function of CD8 T cells is to kill the cells they recognize;

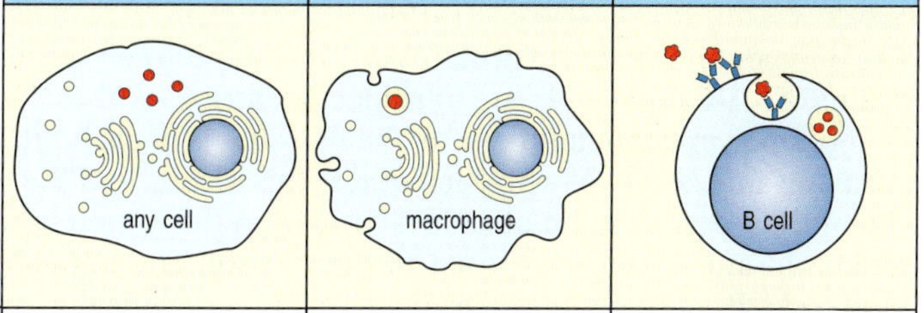

	Cytosolic pathogens	Intravesicular pathogens	Extracellular pathogens and toxins
	any cell	macrophage	B cell
Degraded in	Cytosol	Endocytic vesicles (low pH)	Endocytic vesicles (low pH)
Peptides bind to	MHC class I	MHC class II	MHC class II
Presented to	Effector CD8 T cells	Effector CD4 T cells	Effector CD4 T cells
Effect on presenting cell	Cell death	Activation to kill intravesicular bacteria and parasites	Activation of B cells to secrete Ig to eliminate extracellular bacteria/toxins

Fig. 6.2 Pathogens and their products can be found in either the cytosolic or the vesicular compartment of cells. First panel: all viruses and some bacteria replicate in the cytosolic compartment. Their antigens are presented by MHC class I molecules to CD8 T cells. Second panel: other bacteria and some parasites are taken up into endosomes, usually by specialized phagocytic cells such as macrophages. Here they are killed and degraded, or in some cases are able to survive and proliferate within the vesicle. Their antigens are presented by MHC class II molecules to CD4 T cells. Third panel: proteins derived from extracellular pathogens may enter the intracellular vesicular system by binding to cell-surface receptors followed by endocytosis. This is illustrated here for antigens bound by the surface immunoglobulin (B-cell receptor) of B cells (the endoplasmic reticulum and Golgi apparatus have been omitted for simplicity). The B cells present these antigens to CD4 helper T cells, which can then stimulate the B cells to produce antibody. Other types of cells that bear receptors for the Fc regions of antibody molecules can also internalize antigens in this way and are able to activate T cells.

this is an important means of eliminating sources of new viral particles and bacteria that must live in the cytosol, and so freeing the host from infection.

Pathogens and their products in the vesicular compartment are detected by a different class of T cell, distinguished by the co-receptor molecule CD4, which binds to MHC class II molecules (see Section 4-17). **Effector CD4 T cells** comprise several different subsets, each of which has a distinct activity that helps to eliminate the pathogens against which it is targeted. Intravesicular pathogens are adapted to resist intracellular killing, and the macrophages in which they live have to be given an extra boost of activation to kill the pathogen: this is one of the roles of the T_H1 subset of CD4 T cells. Other CD4 T-cell subsets have roles in regulating other aspects of the immune response, and some CD4 T cells even have cytotoxic activity. The different activities of CD8 and CD4 T cells can largely be viewed as being adapted to deal with pathogens found in different cellular compartments; however, as we shall see later, there is significant cross-talk between these two pathways.

MHC class I and MHC class II molecules have distinct distributions on the cells of the body, and these reflect the different effector functions of the T cells that recognize them. MHC class I molecules are expressed by virtually all the cells of the body (except red blood cells), whereas the expression of MHC class II molecules is mainly restricted to immune-system cells—dendritic cells, macrophages, B cells, and T cells (the last in humans, but not in mice)—and thymic cortical epithelial cells, as described in Section 4-18 and Fig. 4.27. The cells illustrated in Fig. 6.2 are targets for already activated effector T cells. To initiate an adaptive immune response, however, antigen must be presented to naive T cells by specialized antigen-presenting cells, primarily conventional dendritic cells (see Section 3.12). Dendritic cells are highly specialized for this task and activate both CD8 and CD4 naive T cells. Macrophages and B cells can also act as antigen-presenting cells, albeit in a more limited way. Macrophages take up particulate material by phagocytosis and so mainly present pathogen-derived peptides on MHC class II molecules, activating the CD4 T cells that are needed to act back on the macrophage to clear pathogens living in its vesicles. By efficiently endocytosing specific antigen via their surface immunoglobulin and presenting the antigen-derived peptides on MHC class II molecules, B cells can activate CD4 T cells that will in turn serve as helper T cells for the production of antibodies against that antigen.

6-2 Peptides that bind to MHC class I molecules are actively transported from the cytosol to the endoplasmic reticulum.

The polypeptide chains of proteins destined for the cell surface, including the chains of MHC molecules, are translocated during synthesis into the lumen of the endoplasmic reticulum. Here, the two chains of each MHC molecule fold correctly and assemble with each other. This means that the peptide-binding site of the MHC class I molecule is formed in the lumen of the endoplasmic reticulum and is never exposed to the cytosol. The antigen fragments that bind to MHC class I molecules, however, are typically derived from proteins made in the cytosol. This raised the question: How are these peptides able to bind to MHC class I molecules and be delivered to the cell surface?

The answer is that peptides are continually being transported from the cytosol into the endoplasmic reticulum. The first clues to this delivery mechanism came from mutant cells with a defect in antigen presentation by MHC class I molecules. Although both chains of MHC class I molecules were synthesized normally, there were far fewer MHC class I proteins than normal on the cell surface. The defect could be corrected by the addition of synthetic peptides to the culture medium, suggesting that it was the supply of peptides to the MHC class I molecules that was being affected. This defect was also the first indication that MHC molecules are unstable in the absence of bound peptide and that peptide binding is required for the appearance and maintenance of MHC class I molecules at the cell surface. Analysis of the DNA of the mutant cells showed that the genes affected encoded members of the ATP-binding cassette (ABC) family of proteins and that these proteins were absent or non-functional in the mutant cells. ABC proteins mediate the ATP-dependent transport of ions, sugars, amino acids, and peptides across membranes. The two ABC proteins missing from the mutant cells are normally associated with the endoplasmic reticulum membrane and are called ***t*ransporters associated with *a*ntigen *p*rocessing-1** and **-2** (**TAP1** and **TAP2**). Transfection of the mutant cells with the missing genes restored the presentation of peptides by the cell's MHC class I molecules. The two TAP proteins form a heterodimer in the membrane (Fig. 6.3), and mutations in either TAP gene can prevent antigen presentation by MHC class I molecules. The genes *TAP1* and *TAP2* map within the MHC (see Section 6-11) and are inducible by interferons, which are produced in response to viral infection; indeed, viral infection increases the delivery of cytosolic peptides into the endoplasmic reticulum.

In assays *in vitro* using fractions from nonmutant cells, microsomal vesicles that mimic the endoplasmic reticulum will internalize peptides, which bind to MHC class I molecules present in the microsome lumen. Vesicles from TAP1- or TAP2-deficient cells do not take up peptides. Peptide transport into normal microsomes requires ATP hydrolysis, confirming that the TAP1:TAP2 complex is an ATP-dependent peptide transporter. Similar experiments in

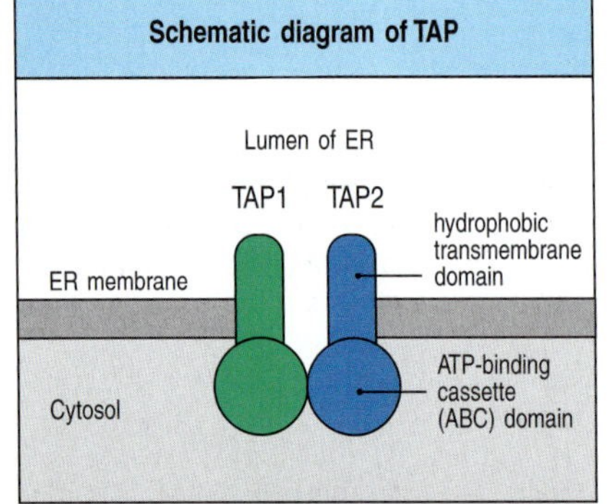

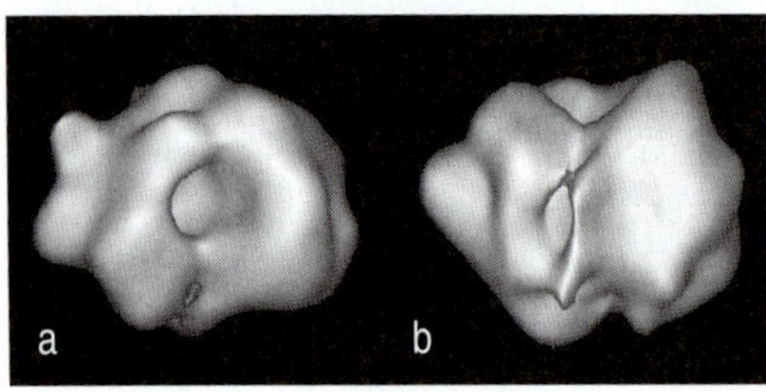

Fig. 6.3 TAP1 and TAP2 form a peptide transporter in the endoplasmic reticulum membrane. Upper panel: TAP1 and TAP2 are individual polypeptide chains, each with one hydrophobic and one ATP-binding domain. The two chains assemble into a heterodimer to form a four-domain transporter typical of the ATP-binding cassette (ABC) family. The hydrophobic transmembrane domains have multiple transmembrane regions (not shown here). The ATP-binding domains lie within the cytosol, whereas the hydrophobic domains project through the membrane into the lumen of the endoplasmic reticulum (ER) to form a channel through which peptides can pass. Lower panel: electron microscopic reconstruction of the structure of the TAP1:TAP2 heterodimer. Panel a shows the surface of the TAP transporter as seen from the lumen of the ER, looking down onto the top of the transmembrane domains, while panel b shows a lateral view of TAP in the plane of the membrane. The ATP-binding domains form two lobes beneath the transmembrane domains; the bottom edges of these lobes are just visible at the back of the lateral view. TAP structures courtesy of G. Velarde.

human cells show that the TAP complex has some specificity for the peptides it will transport. It prefers peptides of between 8 and 16 amino acids in length, with hydrophobic or basic residues at the carboxy terminus—the precise features of peptides that bind MHC class I molecules (see Section 4-14)—and has a bias against proline in the first three amino-terminal residues. The discovery of TAP explained how viral peptides from proteins synthesized in the cytosol gain access to the lumen of the endoplasmic reticulum and are bound by MHC class I molecules, but left open the question of how these peptides are generated.

6-3 Peptides for transport into the endoplasmic reticulum are generated in the cytosol.

Proteins in cells are continually being degraded and replaced with newly synthesized proteins. Much cytosolic protein degradation is carried out by a large, multicatalytic protease complex called the **proteasome** (Fig. 6.4). A typical proteasome is composed of one 20S catalytic core and two 19S regulatory caps, one at each end; both the core and the caps are multisubunit complexes of proteins. The 20S core is a large cylindrical complex of some 28 subunits, arranged in four stacked rings of seven subunits each. It has a hollow core lined by the active sites of the proteolytic subunits. Proteins are often tagged for degradation by the attachment of molecules of the protein ubiquitin. This modification targets them to the 19S cap, which recognizes ubiquitin and unfolds the tagged protein so that it can be introduced into the proteasome's catalytic core. There the protein chain is broken down into short peptides, which are subsequently released into the cytosol.

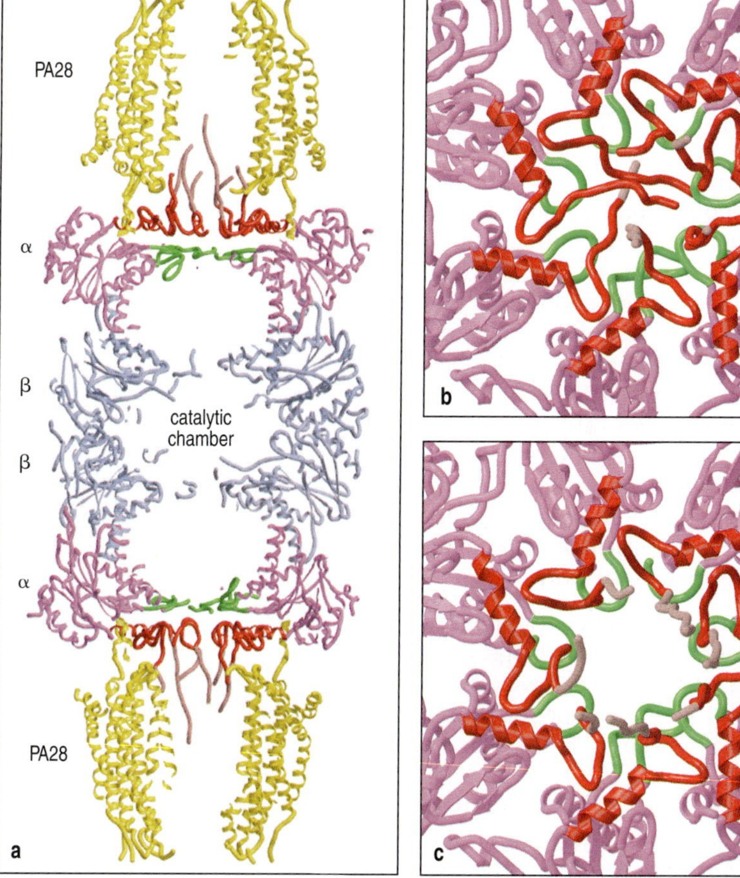

Fig. 6.4 The PA28 proteasome activator binds to either end of the proteasome. Panel a: the heptamer rings of the PA28 proteasome activator (yellow) interact with the α subunits (pink) at either end of the core proteasome (the β subunits that make up the catalytic cavity of the core are in blue). Within this region is the α-annulus (green), a narrow ring-like opening that is normally blocked by other parts of the α subunits (shown in red). Panel b: close-up view of the α-annulus. Panel c: binding of PA28 (not shown, for simplicity) to the proteasome changes the conformation of the α subunits, moving those parts of the molecule that block the α-annulus, and opening the end of the cylinder. Structures courtesy of F. Whitby.

Various lines of evidence implicate the proteasome in the production of peptide ligands for MHC class I molecules. Experimentally tagging proteins with ubiquitin results in more efficient presentation of their peptides by MHC class I molecules, and inhibitors of the proteolytic activity of the proteasome inhibit antigen presentation by MHC class I molecules. Whether the proteasome is the only cytosolic protease capable of generating peptides for transport into the endoplasmic reticulum is not known.

The two inner rings of the 20S proteasome core are composed of constitutively expressed proteolytic subunits called β1, β2, and β5, which form the catalytic chamber (see Fig. 6.4). These constitutive subunits are sometimes displaced by three alternative catalytic subunits, called LMP2 (PSMB9) and LMP7 (PSMB8), which are encoded within the MHC near *TAP1* and *TAP2*, and MECL-1, which is not encoded within the MHC. LMP2, LMP7, and MECL-1 are induced by interferons, similarly to MHC class I and TAP proteins. Thus, the proteasome can exist in two forms—as a constitutive proteasome present in all cells, and as the **immunoproteasome**, present in cells stimulated with interferons. The replacement of the β subunits by their interferon-inducible counterparts alters the enzymatic specificity of the proteasome such that there is increased cleavage of polypeptides after hydrophobic residues, and decreased cleavage after acidic residues. This produces peptides with carboxy-terminal residues that are preferred anchor residues for binding to most MHC class I molecules (see Chapter 4) and are also the preferred structures for transport by TAP.

The production of antigenic peptides of the right length is increased by a further modification of the proteasome induced by interferon-γ (IFN-γ). This is the binding to the proteasome of a protein complex called the PA28 proteasome-activator complex. PA28 is a six- or seven-membered ring composed of two proteins, PA28α and PA28β, both of which are induced by IFN-γ. A PA28 ring can bind to either end of the 20S proteasome core in place of the 19S regulatory cap, and acts to increase the rate at which peptides are released (see Fig. 6.4). In addition to simply providing more peptides, the increased rate of flow will allow potentially antigenic peptides to escape additional processing that might destroy their antigenicity.

Translation of self or pathogen-derived mRNAs in the cytoplasm generates not only properly folded proteins but also a significant quantity—possibly up to 30%—of peptides and proteins that are known as ***d*efective *ri*bosomal *prod*ucts (DRiPs)**. These include peptides translated from introns in improperly spliced mRNAs, translations of frameshifts, and improperly folded proteins. DRiPs are recognized and tagged by ubiquitin for rapid degradation by the proteasome. This seemingly wasteful process ensures that both self proteins and proteins derived from pathogens generate abundant peptides for eventual presentation by MHC class I proteins. The proteasome can also increase the pool of peptides through an excision–splicing mechanism, in which an internal segment of a protein is removed and the surrounding noncontiguous polypeptide segments are joined and used as the peptide presented by MHC class I. It is not yet clear how frequently excision–splicing occurs, but there are several examples of melanoma-specific CD8 T cells that recognize peptide antigens formed in this way.

The proteasome produces peptides that are ready for delivery into the endoplasmic reticulum. At this stage, cellular chaperones such as the TCP-1 ring complex (TRiC) protect these peptides from complete degradation in the cytoplasm. Many of the peptides are, however, longer than can be bound by MHC class I molecules. Thus, cleavage in the proteasome may not be the only processing step for cytosolic peptides. There is good evidence that the carboxy-terminal ends of peptide antigens are produced by cleavage in the proteasome, but the amino termini may be produced by another mechanism. Peptides too long to bind MHC class I molecules can still be transported into the endoplasmic reticulum, where their amino termini can be trimmed by an aminopeptidase called the ***e*ndoplasmic *r*eticulum *a*minopeptidase associated with *a*ntigen *p*rocessing** (**ERAAP**). Like other components of the antigen-processing pathway, ERAAP is upregulated by IFN-γ. Mice lacking ERAAP have an altered repertoire of peptides loaded onto MHC class I molecules. Although the loading of some peptides is not affected by the absence of ERAAP, other peptides fail to load normally, and many unstable and immunogenic peptides not normally present are found in complexes with MHC molecules on the cell surface. This causes cells from ERAAP-deficient mice to be immunogenic for T cells from wild-type mice, demonstrating that ERAAP is an important editor of the normal peptide:MHC repertoire.

6-4 Newly synthesized MHC class I molecules are retained in the endoplasmic reticulum until they bind a peptide.

Binding a peptide is an important step in the assembly of a stable MHC class I molecule. When the supply of peptides into the endoplasmic reticulum is disrupted, as in *TAP*-mutant cells, newly synthesized MHC class I molecules are held in the endoplasmic reticulum in a partly folded state. This explains why cells with mutations in *TAP1* or *TAP2* lack MHC class I molecules at the cell surface. The folding and assembly of a complete MHC class I molecule (see Fig. 4.17) depends on the association of the MHC class I α chain first with β_2-microglobulin and then with peptide, and this process involves a number of accessory proteins with chaperone-like functions. Only after peptide has bound is the MHC class I molecule released from the endoplasmic reticulum and allowed to travel to the cell surface. This explains why the rare human patients that have been identified with immunodeficiency due to defects in *TAP-1* and *TAP-2* have few MHC class I molecules on their cell surfaces, a condition known as **MHC class I deficiency**.

Newly synthesized MHC class I α chains that enter the endoplasmic reticulum membranes bind to the chaperone protein **calnexin**, which retains the MHC class I molecule in a partly folded state (Fig. 6.5). Calnexin also associates with

Fig. 6.5 MHC class I molecules do not leave the endoplasmic reticulum unless they bind peptides. Newly synthesized MHC class I α chains assemble in the endoplasmic reticulum (ER) with the membrane-bound protein calnexin. When this complex binds β_2-microglobulin (β_2m), the MHC class I α:β_2m dimer dissociates from calnexin, and the partly folded MHC class I molecule then binds to the peptide transporter TAP by interacting with one molecule of the TAP-associated protein tapasin. The chaperone molecules ERp57, which forms a heterodimer with tapasin, and calreticulin also bind to form the MHC class I peptide-loading complex. The MHC class I molecule is retained within the ER until released by the binding of a peptide, which completes the folding of the MHC molecule. Even in the absence of infection, there is a continual flow of peptides from the cytosol into the ER. Defective ribosomal products (DRiPs) and old proteins marked for destruction are degraded in the cytoplasm by the proteasome to generate peptides that are transported into the lumen of the endoplasmic reticulum by TAP. Some of these peptides will bind to MHC class I molecules. The aminopeptidase ERAAP trims the peptides at their amino termini, allowing peptides that are too long to bind, increasing the repertoire of potential peptides for presentation. Once a peptide has bound to the MHC molecule, the peptide:MHC complex leaves the endoplasmic reticulum and is transported through the Golgi apparatus and finally to the cell surface.

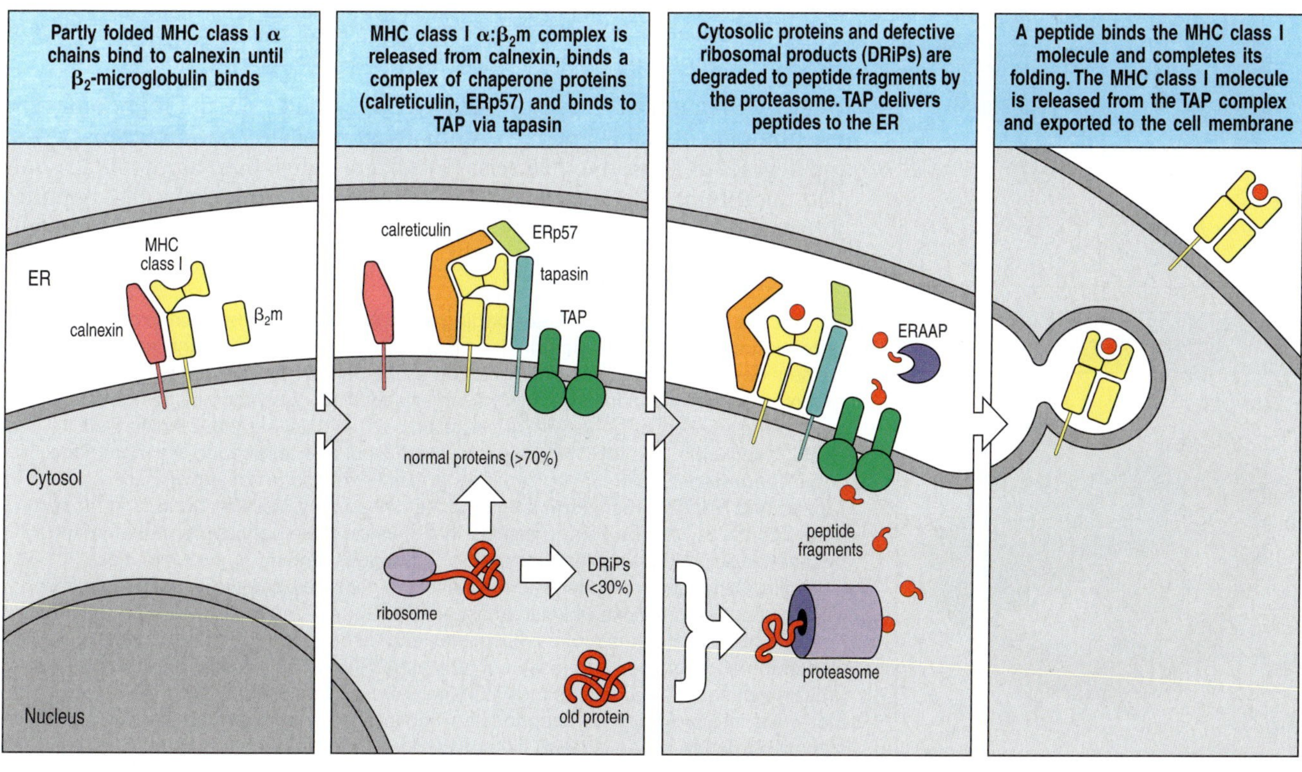

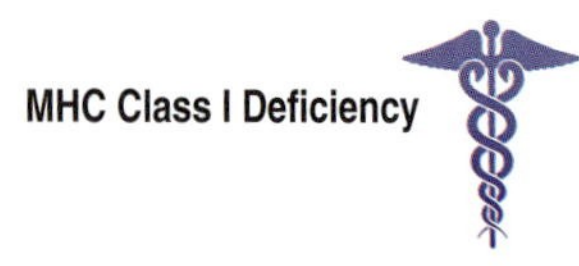

Movie 6.1

partly folded T-cell receptors, immunoglobulins, and MHC class II molecules, and so has a central role in the assembly of many immunological proteins. When β_2-microglobulin binds to the α chain, the partly folded MHC class I α:β_2-microglobulin heterodimer dissociates from calnexin and now binds to an assembly of proteins called the MHC class I **peptide-loading complex (PLC)**. One component of the PLC—**calreticulin**—is similar to calnexin and probably has a similar chaperone function. A second component of the complex is the TAP-associated protein **tapasin**, encoded by a gene within the MHC. Tapasin forms a bridge between MHC class I molecules and TAP, allowing the partly folded α:β_2-microglobulin heterodimer to await the transport of a suitable peptide from the cytosol. A third component of this complex is the chaperone **ERp57**, a thiol oxidoreductase that may have a role in breaking and reforming the disulfide bond in the MHC class I α_2 domain during peptide loading (Fig. 6.6). ERp57 forms a stable disulfide-linked heterodimer with tapasin. Calnexin, ERp57, and calreticulin bind various other glycoproteins assembling in the endoplasmic reticulum and seem to be part of the cell's general quality-control machinery.

The final component of the PLC is the peptide-transporting TAP molecule itself. The other components seem to be essential to maintain the MHC class I molecule in a state receptive to a peptide and also to enable the exchange of low-affinity peptides bound to the MHC molecule for peptides of higher affinity, a process called **peptide editing**. *In vitro* binding studies suggest that the ERp57:tapasin heterodimer functions in editing peptides binding to MHC class I. Cells defective in either calreticulin or tapasin show defects in the assembly of MHC class I molecules, and those molecules that reach the cell surface are bound to suboptimal, low-affinity peptides. The binding of a peptide to the partly folded MHC class I molecule finally releases it from the loading complex. The fully folded MHC molecule and its bound peptide can now leave the endoplasmic reticulum and be transported to the cell surface. It is not yet clear whether the PLC actively loads peptides onto MHC class I molecules or whether binding to the PLC simply allows the MHC class I molecule to scan the peptides transported by TAP before they diffuse away. Most of the peptides transported by TAP will not bind to the MHC molecules in that cell and are rapidly cleared out of the endoplasmic reticulum; there is evidence that they are transported back into the cytosol by an ATP-dependent transport complex distinct from TAP, known as the Sec61 complex.

In cells with mutant *TAP* genes, the MHC class I molecules in the endoplasmic reticulum are unstable and are eventually translocated back into the cytosol and degraded. Thus, the MHC class I molecule must bind a peptide to complete its folding and be transported onward. In uninfected cells, peptides derived from self proteins fill the peptide-binding cleft of the mature MHC

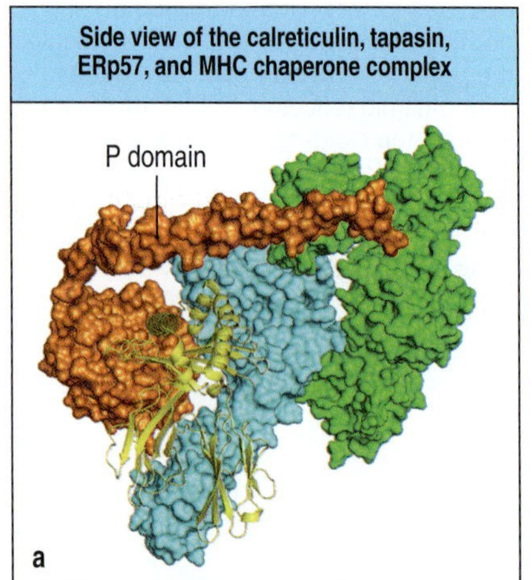

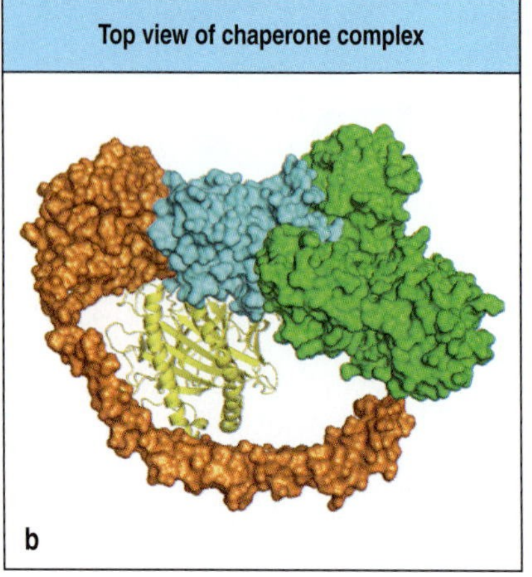

Fig. 6.6 The MHC class I peptide-loading complex includes the chaperones calreticulin, ERp57, and tapasin. This model shows a side view (a) and top view (b) of the peptide-loading complex (PLC) oriented as it extends from the luminal surface of the endoplasmic reticulum. The newly synthesized MHC class I and β_2-microglobulin are shown as yellow ribbons, with the α helices of the MHC peptide-binding cleft clearly identifiable. The MHC and tapasin (cyan) would be tethered to the membrane of the endoplasmic reticulum by carboxy-terminal extensions not shown here. Tapasin and ERp57 (green) form a heterodimer linked by a disulfide bond, and tapasin makes contacts with the MHC molecule that stabilize the empty conformation of the peptide-binding cleft; they function in editing peptides binding to the MHC class I molecule. Calreticulin (orange), like the calnexin it replaces (see Fig. 6.5), binds to the monoglucosylated *N*-linked glycan at asparagine 86 of the immature MHC molecule. The long, flexible P domain of calreticulin extends around the top of the peptide-binding cleft of the MHC molecule to make contact with ERp57. The transmembrane region of tapasin (not shown) associates the PLC with TAP (see Fig. 6.5), bringing the empty MHC molecules into proximity with peptides arriving into the endoplasmic reticulum from the cytosol. Structure based on PDB file provided by Karin Reinisch and Peter Cresswell.

class I molecules and are carried to the cell surface. In normal cells, MHC class I molecules are retained in the endoplasmic reticulum for some time, which suggests that they are present in excess of peptide. This is important for the immunological function of MHC class I molecules, which must be immediately available to transport viral peptides to the cell surface if the cell becomes infected.

6-5 Many viruses produce immunoevasins that interfere with antigen presentation by MHC class I molecules.

The presentation of viral peptides by MHC class I molecules at a cell surface signals CD8 T cells to kill the infected cell. Some viruses evade immune recognition by producing proteins called **immunoevasins**, which prevent the appearance of peptide:MHC class I complexes on the infected cell (Fig. 6.7). Some immunoevasins block peptide entry into the endoplasmic reticulum by targeting the TAP transporter (Fig. 6.8, top panel). The herpes simplex virus produces the protein ICP47, which binds to the cytosolic surface of TAP and prevents peptides from entering the transporter, whereas the US6 protein from human cytomegalovirus binds to the luminal face of TAP and prevents peptides being transported by inhibiting TAP ATPase activity. The UL49.5 protein from bovine herpes virus inhibits TAP peptide transport by blocking conformational changes in TAP that are required for peptide translocation, and also by targeting TAP proteins for proteasomal degradation.

Virus proteins can also prevent peptide:MHC complexes from reaching the cell surface by retaining MHC class I molecules in the endoplasmic reticulum (Fig. 6.8, middle panel). The adenovirus E19 protein interacts with certain MHC class I proteins and contains a motif that retains the protein complex in the endoplasmic reticulum. E19 also prevents the tapasin–TAP interaction required for peptide loading onto the MHC class I molecule. Several viral proteins catalyze the degradation of newly synthesized MHC class I molecules

Virus	Protein	Category	Mechanism
Herpes simplex virus 1	ICP47	Blocks peptide entry to endoplasmic reticulum	Blocks peptide binding to TAP
Human cytomegalovirus (HCMV)	US6		Inhibits TAP ATPase activity and blocks peptide release into endoplasmic reticulum
Bovine herpes virus	UL49.5		Inhibits TAP peptide transport
Adenovirus	E19	Retention of MHC class I in endoplasmic reticulum	Competitive inhibitor of tapasin
HCMV	US3		Blocks tapasin function
Murine cytomegalovirus (CMV)	M152		Unknown
HCMV	US2	Degradation of MHC class I (dislocation)	Transports some newly synthesized MHC class I molecules into cytosol
Murine gamma herpes virus 68	mK3		E3-ubiquitin ligase activity
Murine CMV	m4	Binds MHC class I at cell surface	Interferes with recognition by cytotoxic lymphocytes by an unknown mechanism

Fig. 6.7 Immunoevasins produced by viruses interfere with the processing of antigens that bind to MHC class I molecules.

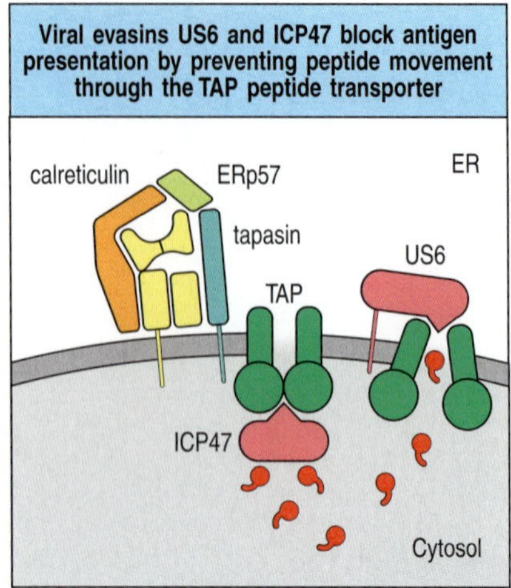

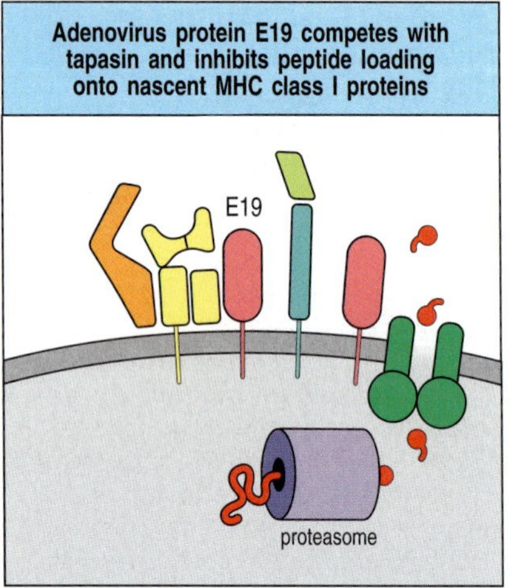

Adenovirus protein E19 competes with tapasin and inhibits peptide loading onto nascent MHC class I proteins
E19
proteasome

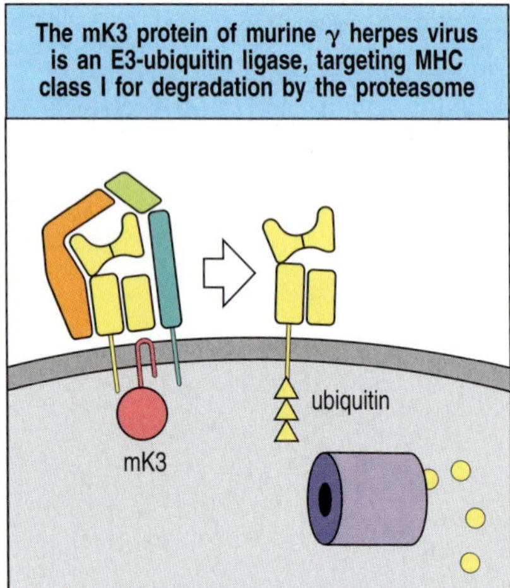

The mK3 protein of murine γ herpes virus is an E3-ubiquitin ligase, targeting MHC class I for degradation by the proteasome
ubiquitin
mK3

Fig. 6.8 The peptide-loading complex in the endoplasmic reticulum is targeted by viral immunoevasins. The top panel shows blockade of peptide entry to the endoplasmic reticulum (ER). The cytosolic ICP47 protein from the virus HSV-1 (see Fig. 6.7) prevents peptides from binding to TAP in the cytosol, whereas the US6 protein from human CMV interferes with the ATP-dependent transfer of peptides through TAP. The center panel shows the retention of MHC class I molecules in the ER by the adenovirus E19 protein. This binds certain MHC molecules and retains them in the ER through an ER-retention motif, at the same time competing with tapasin to prevent association with TAP and peptide loading. The bottom panel shows how the murine herpes virus mK3 protein, an E3-ubiquitin ligase, targets newly synthesized MHC class I molecules. mK3 associates with tapasin:TAP complexes and directs the addition of ubiquitin subunits with K48 linkages (see Section 7-5) to the cytoplasmic tail of the MHC class I molecule (not shown). The polyubiquitination of the cytoplasmic tail of MHC initiates the process of degradation of the MHC molecule by the proteasomal pathway.

by a process known as **dislocation**, which initiates the pathway normally used to degrade misfolded endoplasmic reticulum proteins by directing them back into the cytosol. For example, the US11 protein of human cytomegalovirus binds nascent MHC class I molecules and in conjunction with derlin, a ubiquitous protein of endoplasmic reticulum membranes, delivers them into the cytosol, where they are degraded. The mK3 protein of the murine gamma herpes virus 68 has two transmembrane regions that direct its association with the tapasin:TAP complex (Fig. 6.8, bottom panel). Its E3-ubiquitin ligase activity adds ubiquitin chains to the tails of MHC class I molecules, leading to degradation of MHC by the proteasome. Most viral immunoevasins are found in DNA viruses, such as the herpes viruses, which have large genomes and can typically persist in a latent or quiescent form in their hosts.

6-6 Peptides presented by MHC class II molecules are generated in acidified endocytic vesicles.

Several types of pathogens, including the protozoan parasite *Leishmania* and the mycobacteria that cause leprosy and tuberculosis, replicate inside intracellular vesicles in macrophages. Because they reside in membrane-enclosed vesicles, the proteins of these pathogens are not usually accessible to proteasomes in the cytosol. Instead, after activation of the macrophage, the pathogens are degraded by activated intravesicular proteases into peptide fragments that can bind to MHC class II molecules, which pass through this compartment on their way from the endoplasmic reticulum to the cell surface. Like all membrane proteins, MHC class II molecules are first delivered into the endoplasmic reticulum membrane, and are transported onward as part of membrane-enclosed vesicles that bud off the endoplasmic reticulum. Complexes of peptides and MHC class II molecules are then delivered to the cell surface, where they can be recognized by CD4 T cells. Extracellular pathogens and proteins that are internalized into endocytic vesicles are also processed and presented by this pathway (Fig. 6.9).

Most of what we know about protein processing in the endocytic pathway has come from experiments in which simple proteins are fed to macrophages and are taken up by endocytosis; in this way, the processing of added antigen can be quantified. Proteins that bind to surface immunoglobulin on B cells and are internalized by receptor-mediated endocytosis are processed by the same pathway. Proteins that enter cells through endocytosis are delivered to endosomes, which become increasingly acidic as they progress into the interior of the cell, eventually fusing with lysosomes. The endosomes and lysosomes contain proteases, known as acid proteases, that are activated at low pH and eventually degrade the protein antigens contained in the vesicles. Larger particulate material, such as whole cells, internalized by phagocytosis or macropinocytosis, is also handled by this pathway of antigen processing (see Chapter 3).

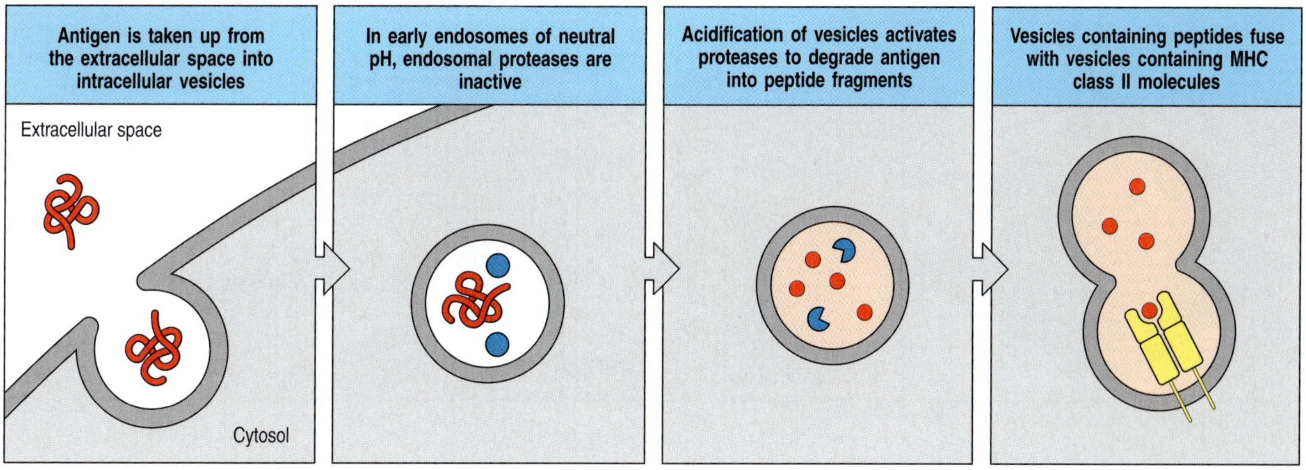

Antigen is taken up from the extracellular space into intracellular vesicles	In early endosomes of neutral pH, endosomal proteases are inactive	Acidification of vesicles activates proteases to degrade antigen into peptide fragments	Vesicles containing peptides fuse with vesicles containing MHC class II molecules

Extracellular space

Cytosol

Fig. 6.9 Peptides that bind to MHC class II molecules are generated in acidified endocytic vesicles. In the case illustrated here, extracellular foreign antigens, such as bacteria or bacterial antigens, have been taken up by an antigen-presenting cell such as a macrophage or immature dendritic cell. In other cases, the source of the peptide antigen may be bacteria or parasites that have invaded the cell to replicate in intracellular vesicles. In both cases the antigen-processing pathway is the same. The pH of the endosomes containing the engulfed pathogens decreases progressively, activating proteases within the vesicles to degrade the engulfed material. At some point on their pathway to the cell surface, newly synthesized MHC class II molecules pass through such acidified vesicles and bind peptide fragments of the antigen, transporting the peptides to the cell surface.

Drugs such as chloroquine that raise the pH of endosomes, making them less acidic, inhibit the presentation of intravesicular antigens, suggesting that acid proteases are responsible for processing internalized antigen. These proteases include the cysteine proteases known as cathepsins B, D, S, and L, of which L is the most active. Antigen processing can be mimicked to some extent by the digestion of proteins with these enzymes *in vitro* at acid pH. Cathepsins S and L may be the predominant proteases in the processing of vesicular antigens; mice that lack cathepsin B or cathepsin D process antigens normally, whereas mice with no cathepsin S show some deficiencies. It is likely that the overall repertoire of peptides produced within the vesicular pathway reflects the activities of the many proteases present in endosomes and lysosomes.

Disulfide bonds, particularly intramolecular disulfide bonds, may need to be reduced before proteins that contain them can be digested in endosomes. An IFN-γ-induced thiol reductase present in endosomes—IFN-γ-induced lysosomal thiol reductase (GILT)—carries out this role in the antigen-processing pathway.

6-7 The invariant chain directs newly synthesized MHC class II molecules to acidified intracellular vesicles.

The immunological function of MHC class II molecules is to bind peptides generated in the intracellular vesicles of macrophages, immature dendritic cells, B cells, and other antigen-presenting cells and to present these peptides to CD4 T cells. The biosynthetic pathway for MHC class II molecules, like that of other cell-surface glycoproteins, starts with their translocation into the endoplasmic reticulum, and they must therefore be prevented from binding prematurely to peptides transported into the endoplasmic reticulum lumen or to the cell's own newly synthesized polypeptides. The endoplasmic reticulum is full of unfolded and partly folded polypeptide chains, and so a general mechanism is needed to prevent these from binding in the open-ended peptide-binding groove of the MHC class II molecule.

Binding is prevented by the assembly of newly synthesized MHC class II molecules with a membrane protein known as the MHC class II-associated **invariant chain (Ii,** CD74). Ii forms trimers, with each Ii subunit binding non-covalently to an MHC class II α:β heterodimer (Fig. 6.10). An Ii subunit binds to an MHC class II molecule with part of the chain lying within the peptide-binding groove, thus blocking the groove and preventing the binding of either peptides or partly folded proteins. While this complex is being assembled in the endoplasmic reticulum, its component parts are associated with calnexin. Only when a nine-chain complex has been assembled is this complex released

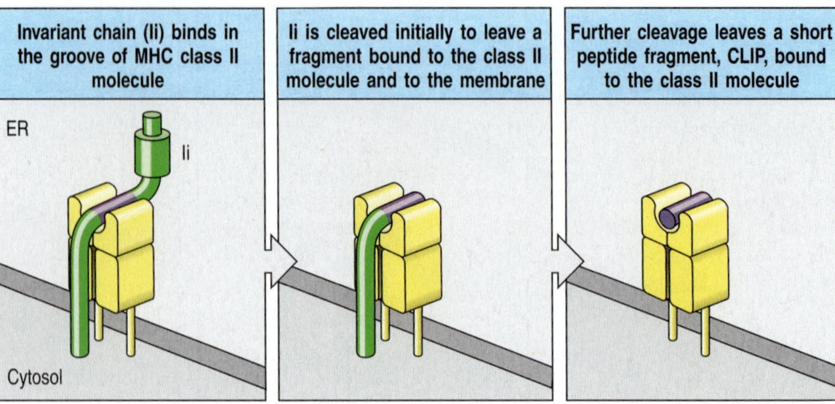
Invariant chain (Ii) binds in the groove of MHC class II molecule
Ii is cleaved initially to leave a fragment bound to the class II molecule and to the membrane
Further cleavage leaves a short peptide fragment, CLIP, bound to the class II molecule
ER
Ii
Cytosol

Fig. 6.10 The invariant chain is cleaved to leave a peptide fragment, CLIP, bound to the MHC class II molecule. A model of the trimeric invariant chain bound to MHC class II α:β heterodimers is shown on the left. The CLIP portion is shown in purple, the rest of the invariant chain is shown in green, and the MHC class II molecules are shown in yellow. In the endoplasmic reticulum, the invariant chain (Ii) binds to MHC class II molecules with the CLIP section of its polypeptide chain lying along the peptide-binding groove (model and left of three panels). After transport into an acidified vesicle, Ii is cleaved, initially just at one side of the MHC class II molecule (center panel). The remaining portion of Ii (known as the leupeptin-induced peptide or LIP fragment) retains the transmembrane and cytoplasmic segments that contain the signals that target Ii:MHC class II complexes to the endosomal pathway. Subsequent cleavage (right panel) of LIP leaves only a short peptide still bound by the class II molecule; this peptide is the CLIP fragment. Model structure courtesy of P. Cresswell.

from calnexin for transport out of the endoplasmic reticulum. As part of the nine-chain complex, the MHC class II molecules cannot bind peptides or unfolded proteins, so that peptides present in the endoplasmic reticulum are not usually presented by MHC class II molecules. There is evidence that in the absence of Ii many MHC class II molecules are retained in the endoplasmic reticulum as complexes with misfolded proteins.

Ii has a second function, which is to target delivery of the MHC class II molecules to a low-pH endosomal compartment where peptide loading can occur. The complex of MHC class II α:β heterodimers with Ii trimers is retained for 2–4 hours in this compartment. During this time, each Ii molecule is cleaved by acid proteases such as cathepsin S in several steps, as shown in Fig. 6.10. The initial cleavages generate a truncated form of Ii that remains bound to the MHC class II molecule and retains it within the proteolytic compartment. A subsequent cleavage releases the MHC class II molecule from the membrane-associated Ii, leaving a short fragment of Ii called **CLIP** (for ***c**lass **II**-associated **i**nvariant-chain **p**eptide*) bound to the MHC molecule. MHC class II molecules associated with CLIP cannot bind other peptides. CLIP must either dissociate or be displaced to allow a peptide to bind to the MHC molecule and enable the complex to be delivered to the cell surface. Cathepsin S cleaves Ii in most MHC class II-positive cells, including the antigen-presenting cells. The exception are thymic cortical epithelial cells, which seem to use cathepsin L.

The endosomal compartment in which Ii is cleaved and MHC class II molecules encounter peptide is not yet clearly defined. Newly synthesized MHC class II molecules are brought toward the cell surface in vesicles, most of which at some point fuse with incoming endosomes. However, it also seems likely that some MHC class II:Ii complexes are first transported to the cell surface and then reinternalized into endosomes. In either case, MHC class II:Ii complexes enter the endosomal pathway and there encounter and bind peptides derived from either internalized pathogen proteins or self proteins. Immunoelectron-microscopy using antibodies tagged with gold particles to localize Ii and MHC class II molecules within cells suggests that Ii can be cleaved and peptides bind to MHC class II molecules in a particular endosomal compartment called the **MIIC** (***M**HC class **II** **c**ompartment*), late in the endosomal pathway (Fig. 6.11). MHC class II molecules that do not bind peptide after dissociation from the invariant chain are unstable in the acidic pH after fusion with lysosomes, and they are rapidly degraded.

Movie 6.3

6-8 A specialized MHC class II-like molecule catalyzes loading of MHC class II molecules with peptides.

Another component of the vesicular antigen-processing pathway was revealed by analysis of mutant human B-cell lines with a defect in antigen presentation.

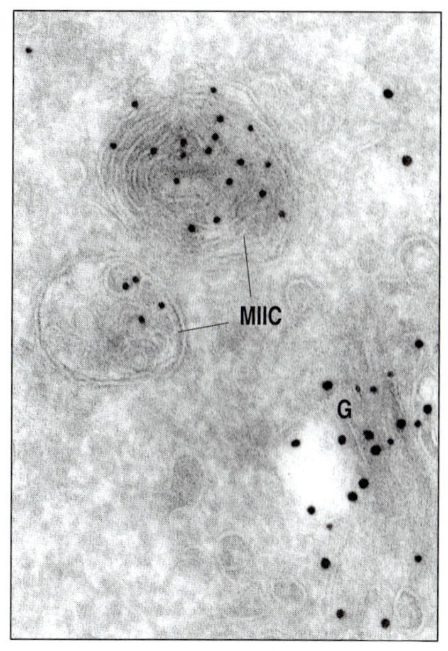

Fig. 6.11 MHC class II molecules are loaded with peptide in a specialized intracellular compartment. MHC class II molecules are transported from the Golgi apparatus (labeled G in this electron micrograph of an ultrathin section of a B cell) to the cell surface via specialized intracellular vesicles called the MHC class II compartment (MIIC). These have a complex morphology, showing internal vesicles and sheets of membrane. Antibodies labeled with different-sized gold particles identify the presence of both MHC class II molecules (small gold particles) and the invariant chain (large gold particles) in the Golgi, whereas only MHC class II molecules are detectable in the MIIC. This compartment is therefore thought to be the one in which the invariant chain is cleaved and peptide loading occurs. Photograph (×135,000) courtesy of H.J. Geuze.

MHC class II molecules in these cell lines assemble correctly with the Ii and seem to follow the normal vesicular route. However, they fail to bind peptides derived from internalized proteins and often arrive at the cell surface with the CLIP peptide still bound.

The defect in these cells lies in an MHC class II-like molecule called **HLA-DM** in humans (H-2M in mice). The HLA-DM genes are found near the TAP and LMP (now also known as PSMB) genes in the MHC class II region (see Fig. 6.13); they encode an α chain and a β chain that closely resemble those of other MHC class II molecules. The HLA-DM molecule is not present at the cell surface, however, but is found predominantly in the MIIC. HLA-DM binds to and stabilizes empty MHC class II molecules that would otherwise aggregate; in addition, it catalyzes both the release of the CLIP fragment from MHC class II:CLIP complexes and the binding of other peptides to the empty MHC class II molecule (Fig. 6.12). The HLA-DM molecule itself does not bind peptides, and the open groove found in other MHC class II molecules is closed in HLA-DM.

HLA-DM also catalyzes the release of unstably bound peptides from MHC class II molecules. In the presence of a mixture of peptides capable of binding to MHC class II molecules, as occurs in the MIIC, HLA-DM continuously

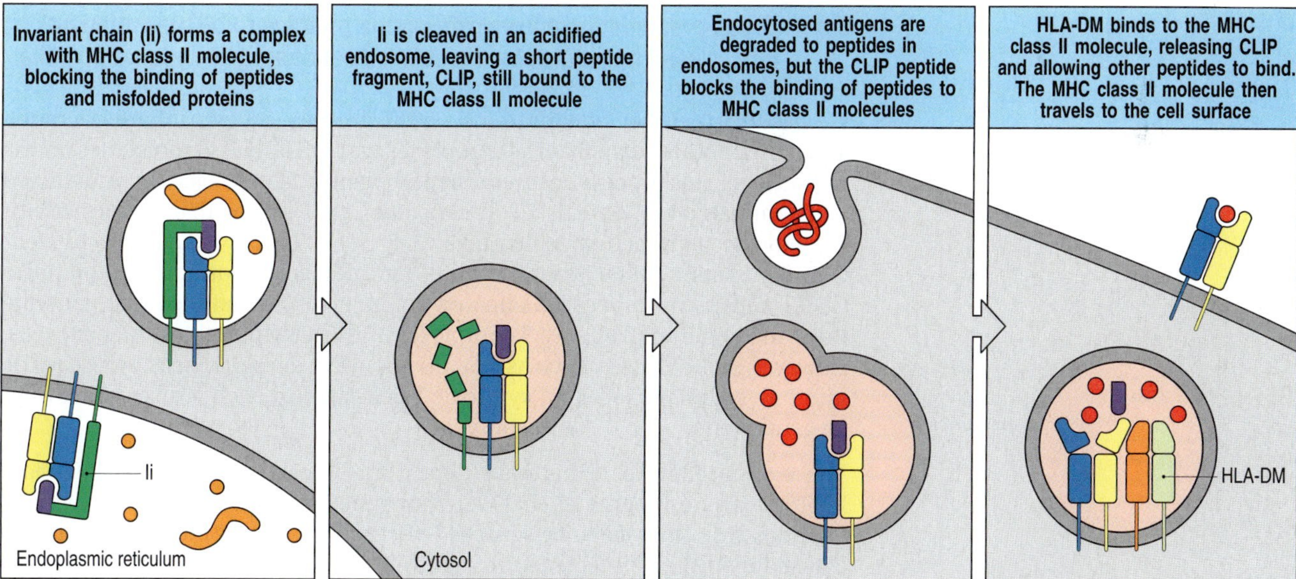

Invariant chain (Ii) forms a complex with MHC class II molecule, blocking the binding of peptides and misfolded proteins
Ii is cleaved in an acidified endosome, leaving a short peptide fragment, CLIP, still bound to the MHC class II molecule
Endocytosed antigens are degraded to peptides in endosomes, but the CLIP peptide blocks the binding of peptides to MHC class II molecules
HLA-DM binds to the MHC class II molecule, releasing CLIP and allowing other peptides to bind. The MHC class II molecule then travels to the cell surface
Ii
Endoplasmic reticulum
Cytosol
HLA-DM

Fig. 6.12 HLA-DM facilitates the loading of antigenic peptides onto MHC class II molecules. The invariant chain (Ii) (shown here schematically) binds to newly synthesized MHC class II molecules and blocks the binding of peptides and unfolded proteins in the endoplasmic reticulum and during the transport of the MHC class II molecule into acidified endocytic vesicles (first panel). In such vesicles, proteases cleave the invariant chain, leaving the CLIP peptide bound to the MHC class II molecule (second panel). Pathogens and their proteins are broken down into peptides within acidified endosomes, but these peptides cannot bind to MHC class II molecules that are occupied by CLIP (third panel). The class II-like molecule HLA-DM binds to MHC class II:CLIP complexes, catalyzing the release of CLIP and the binding of antigenic peptides (fourth panel).

binds and rebinds to peptide:MHC class II complexes, removing weakly bound peptides and allowing other peptides to replace them. Antigens presented by MHC class II molecules may have to persist on the surface of antigen-presenting cells for some days before encountering T cells able to recognize them. The ability of HLA-DM to remove unstably bound peptides, sometimes called peptide editing, ensures that the peptide:MHC class II complexes displayed on the surface of the antigen-presenting cell survive long enough to stimulate the appropriate CD4 cells.

A second atypical MHC class II molecule, called **HLA-DO** (H-2O in mice), is produced in thymic epithelial cells, B cells, and dendritic cells. This molecule is a heterodimer of the HLA-DOα chain and the HLA-DOβ chain (see Fig. 6.13). HLA-DO is not present at the cell surface, being found only in intracellular vesicles, and it does not seem to bind peptides. Instead, it acts as a negative regulator of HLA-DM, binding to it and inhibiting both the HLA-DM-catalyzed release of CLIP from, and the binding of other peptides to, MHC class II molecules. Expression of the HLA-DOβ chain is not increased by IFN-γ, whereas the expression of HLA-DM is. Thus, during inflammatory responses, in which IFN-γ is produced by T cells and NK cells, the increased expression of HLA-DM is able to overcome the inhibitory effects of HLA-DO. Why the antigen-presenting ability of thymic epithelial cells and B cells should be regulated in this way is not known; in thymic epithelial cells the function may be to select developing CD4 T cells by using a repertoire of self peptides different from those to which they will be exposed as mature T cells, as discussed further in Chapter 8. The role of HLA-DM in facilitating the binding of peptides to MHC class II molecules parallels that of TAP in facilitating peptide binding to MHC class I molecules. Thus, it seems likely that specialized mechanisms of delivering peptides have coevolved with the MHC molecules themselves. It is also likely that pathogens have evolved strategies to inhibit the loading of peptides onto MHC class II molecules, much as viruses have found ways of subverting antigen processing and presentation through the MHC class I molecules.

6-9 Cross-presentation allows exogenous proteins to be presented on MHC class I molecules by a restricted set of antigen-presenting cells.

We described earlier how proteins synthesized in the cytosol are degraded and their peptides displayed as complexes with MHC class I molecules on the cell surface. This process ensures that pathogen-infected cells can be detected and eliminated by cytotoxic T cells, effector CD8 T cells with cytolytic activity. But how do pathogen-specific naive CD8 T cells initially become activated? Our explanation so far would require dendritic cells that activate the naive CD8 T cells to have become infected by the cytosolic pathogen, to generate the same peptide:MHC class I complexes as those displayed by infected cells. However, some viruses, such as HIV, do not infect dendritic cells. How can the immune system ensure that cytotoxic T cells against such viruses can always be generated?

The answer seems to be that some dendritic cells can form peptide:MHC class I complexes from peptides that were not generated within their own cytosol. Peptides from extracellular sources—such as viruses, bacteria, and phagocytosed dying cells infected with cytosolic pathogens—can be presented on MHC class I molecules. This process is called **cross-presentation**. It appears that the capacity for cross-presentation is not equally distributed across all antigen-presenting cells, but is most efficiently performed by a subset of dendritic cells present in mice and humans. We describe the properties of these dendritic cells in greater detail in Section 9-5. The biochemical mechanisms enabling cross-presentation are still unclear, and there may be several different pathways. What is clear is that material that is captured by receptors and taken into endosomes can be partly, but not completely, degraded and eventually allowed to enter the pathway for loading antigen onto MHC class I molecules (Fig. 6.13, upper panel).

Cross-presentation was first recognized in the mid-1970s from studies of minor histocompatibility antigens. These are protein antigens that elicit responses between mice of different genetic backgrounds (such as B10 and BALB) but are not encoded by genes of the MHC. When spleen cells from B10 mice of MHC type H-2^{b} were injected into BALB mice of MHC type H-2$^{b\times d}$ (which express both b and d MHC types), the BALB mice generated cytotoxic T cells reactive against minor antigens of the B10 background. Some of these cytotoxic T cells recognized minor antigens presented by the H-2^{b} B10 cells used for immunization, as one might expect from direct priming of T cells by the B10 antigen-presenting cells. However, others recognized minor B10 antigens that are normally presented only by B10 background cells but, surprisingly, only when they were presented by cells of the H-2^{d} type. This meant that the *in vivo* priming of the BALB H2$^{b\times d}$ naive CD8 T cells had involved the presentation of minor B10 antigens by the BALB host's H-2^{d} molecules, and not only the direct recognition of antigens presented by the H-2^{b} B10 cells. Thus, minor histocompatibility antigens must have become transferred from the original immunizing B10 cells to the BALB host's dendritic cells and processed for MHC class I presentation. We now know that cross-presentation in MHC class I molecules occurs not only for antigens on tissue or cell grafts, as in the original experiment described above, but also for viral and bacterial antigens.

Another exception to the normal rules of antigen presentation is the loading of cytosol-derived peptides on MHC class II molecules. A significant number of the self peptides bound to MHC class II molecules arise from common proteins that are cytosolic in location, such as actin and ubiquitin. The most likely way in which cytosolic proteins are processed for MHC class II presentation is by the natural process of protein turnover known as **autophagy**, in which damaged organelles and cytosolic proteins are delivered to lysosomes for degradation. Here their peptides could encounter MHC class II molecules present in the lysosome membranes, and the resulting peptide:MHC class II complex could be transported to the cell surface via endolysosomal tubules (Fig. 6.13, lower panel). Autophagy is constitutive, but it is increased by cellular stresses such as starvation, when the cell catabolizes intracellular proteins to obtain energy. In **microautophagy**, cytosol is continuously internalized into the vesicular system by lysosomal invaginations, whereas in **macroautophagy**, which is induced by starvation, a double-membraned autophagosome engulfs cytosol and fuses with lysosomes. A third autophagic pathway uses the heat-shock cognate protein 70 (Hsc70) and the lysosome-associated membrane protein-2 (LAMP-2) to transport cytosolic proteins to lysosomes. Autophagy has been shown to be involved in the processing of the Epstein–Barr virus nuclear antigen 1 (EBNA-1) for presentation on MHC class II molecules. Such presentation enables cytotoxic CD4 T cells to recognize and kill B cells infected with Epstein–Barr virus.

6-10 Stable binding of peptides by MHC molecules provides effective antigen presentation at the cell surface.

For MHC molecules to perform their essential function of signaling intracellular infection, the peptide:MHC complex must be stable at the cell surface. If the complex were to dissociate too readily, the pathogen in the infected cell could escape detection. In addition, MHC molecules on uninfected cells could pick up peptides released by MHC molecules on infected cells and falsely signal to cytotoxic T cells that a healthy cell is infected, triggering its unwarranted destruction. The tight binding of peptides by MHC molecules makes both of these undesirable outcomes unlikely.

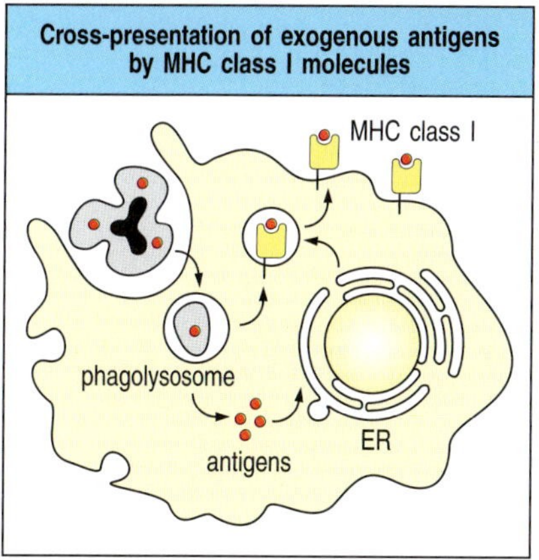

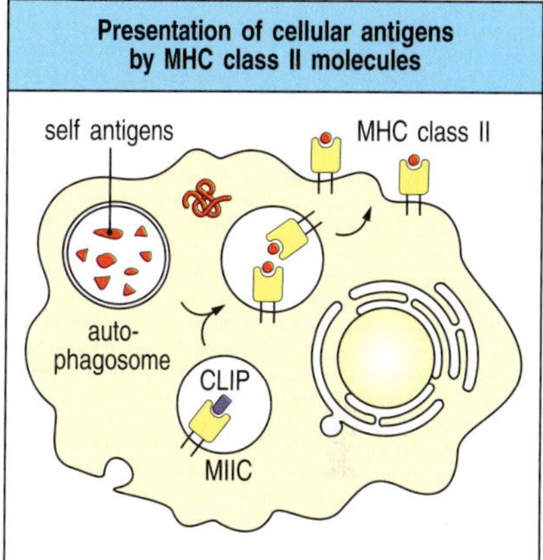

Fig. 6.13 Cross-presentation and autophagy of antigens. Top panel: cross-presentation of extracellular antigens by MHC class I molecules. The molecular pathway of cross-presentation is still unclear. One possible route could involve the translocation of ingested proteins from the phagolysosome into the cytosol, where they can undergo degradation by the proteasome and enter the endoplasmic reticulum through TAP, where they are loaded onto MHC class I molecules in the usual way. Another possible route could involve the transport of antigens directly from the phagolysosome into a vesicular compartment, without passage through the cytosol, where peptides are allowed to be bound to mature MHC class I molecules. Bottom panel: autophagy of cytosolic self antigens permits their presentation by MHC class II molecules. Autophagy is the phenomenon by which a cell's own cytoplasm is taken into specialized vesicles, autophagosomes, in a process requiring a cascade of protein–protein interactions. The autophagosomes are fused with lysosomes and their contents are catabolized, from which some of the resulting peptides can be bound to and presented by MHC class II molecules, on the cell surface.

The persistence of a peptide:MHC complex on a cell can be measured by its ability to stimulate T cells, while the fate of the MHC molecules themselves can be followed directly by specific staining. Such experiments have shown that specific peptide:MHC complexes on living cells are lost from the surface and reinternalized as part of natural protein turnover at a similar rate to that of the MHC molecules themselves, suggesting that most peptide complexes are long-lived. Such stable binding enables even rare peptides to be transported efficiently to the cell surface by MHC molecules for prolonged display of these complexes on the surface of the infected cell. Particularly for MHC class I, which display peptides derived largely from cytosolic proteins, it is important that dissociation of a peptide from a cell-surface MHC molecule does not allow extracellular peptides to bind in the empty peptide-binding site. In fact, removal of the peptide from a purified MHC class I molecule has been shown to require denaturation of the protein. When peptide dissociates from an MHC class I molecule at the surface of a living cell, the molecule changes conformation, the β_2-microglobulin dissociates, and the α chain is internalized and rapidly degraded. Thus, most empty MHC class I molecules are quickly lost from the cell surface. In this manner, MHC class I molecules are largely prevented from acquiring peptides directly from the surrounding extracellular fluid, ensuring that T cells act selectively on infected cells while sparing surrounding healthy cells.

At neutral pH, empty MHC class II molecules are more stable than empty MHC class I molecules, yet empty MHC class II molecules are also removed from the cell surface. They aggregate readily, and internalization of such aggregates may account for their removal. Moreover, peptide loss from MHC class II molecules is most likely when the molecules are transiting through acidified endosomes as part of the normal process of cell-membrane recycling. At acidic pH, MHC class II molecules are able to bind peptides that are present in the vesicles, but those that fail to do so are rapidly degraded.

Some binding of extracellular peptides to MHC molecules at the cell surface can occur, however, as the addition of peptides to chemically fixed cells can generate peptide:MHC complexes that are recognized by T cells specific for those peptides. This has been readily demonstrated for many peptides that bind MHC class II molecules. Whether this phenomenon is due to the presence of empty class II proteins on the cells or to peptide exchange is not clear. Nevertheless it happens, and it may have some biological importance in the recognition of small bacterial toxins, or portions of bacterial toxins that are relatively unstructured.

Summary.

The most distinctive feature of antigen recognition by T cells is the form of the ligand recognized by the T-cell receptor. This comprises a peptide derived from the degradation of pathogen or self antigens and bound to an MHC molecule. MHC molecules are cell-surface glycoproteins with a peptide-binding groove that can bind a wide variety of different peptides. The MHC molecule binds the peptide in an intracellular location and delivers it to the cell surface, where the combined ligand can be recognized by a T cell. There are two classes of MHC molecules, MHC class I and MHC class II, which acquire peptides at different intracellular sites, and which activate CD8 and CD4 T cells, respectively. Cells presenting peptides derived from virus replication in the cytosol can thus be recognized by CD8 cytotoxic T cells, which are specialized to kill any cells displaying foreign antigens. MHC class I molecules are synthesized in the endoplasmic reticulum and acquire their peptides at this location. The peptides loaded onto MHC class I are derived from proteins degraded in the cytosol by a multicatalytic protease, the proteasome. Peptides generated by proteasomes are transported into the endoplasmic reticulum by a heterodimeric ATP-binding protein called TAP; they are further processed by the aminopeptidase ERAAP and are then available for binding by partly folded MHC class I molecules. Peptide binding is an integral part of MHC class I molecule assembly, and must occur before the MHC class I molecule can complete its folding and leave the endoplasmic reticulum for the cell surface. Some dendritic cells are able to obtain exogenous antigens from pathogen-infected cells and load peptides from these antigens onto MHC class I molecules, a process called cross-presentation, which is important for the generation of effective immune responses.

In contrast to peptide loading of MHC class I, MHC class II molecules do not acquire their peptide ligands in the endoplasmic reticulum because of their early association with the invariant chain (Ii), which binds to and blocks their peptide-binding groove. They are targeted by Ii to an acidic endosomal compartment where—in the presence of active proteases, in particular cathepsin S, and with the help of HLA-DM, a specialized MHC class II-like molecule that catalyzes peptide loading—Ii is released and other peptides are bound. MHC class II molecules thus bind peptides from proteins that are degraded in endosomes. There they can capture peptides from pathogens that have entered the vesicular system of macrophages, or from antigens internalized by immature dendritic cells or the immunoglobulin receptors of B cells. The process of autophagy can deliver cytosolic proteins to the vesicular system for presentation by MHC class II. The CD4 T cells that recognize peptide:MHC class II complexes have a variety of specialized effector activities. Subsets of CD4 T cells activate macrophages to kill the intravesicular pathogens they harbor, help B cells to secrete immunoglobulins against foreign molecules, and regulate immune responses.

The major histocompatibility complex and its function.

The function of MHC molecules is to bind peptide fragments derived from pathogens and display them on the cell surface for recognition by the appropriate T cells. The consequences are almost always deleterious to the pathogen—virus-infected cells are killed, macrophages are activated to kill bacteria living in their intracellular vesicles, and B cells are activated to produce antibodies that eliminate or neutralize extracellular pathogens. Thus, there is strong selective pressure in favor of any pathogen that has mutated in such a way that it escapes presentation by an MHC molecule.

Two separate properties of the major histocompatibility complex (MHC) make it difficult for pathogens to evade immune responses in this way. First, the MHC is **polygenic**: it contains several different MHC class I and MHC class II genes, so that every individual possesses a set of MHC molecules with different ranges of peptide-binding specificities. Second, the MHC is highly **polymorphic**; that is, there are multiple variants, or alleles, of each gene within the population as a whole. The MHC genes are, in fact, the most polymorphic genes known. In this section we describe the organization of the genes in the MHC and discuss how the variation in MHC molecules arises. We will also see how the effect of polygeny and polymorphism on the range of peptides that can be bound contributes to the ability of the immune system to respond to the multitude of different and rapidly evolving pathogens.

6-11 Many proteins involved in antigen processing and presentation are encoded by genes within the MHC.

The MHC is located on chromosome 6 in humans and chromosome 17 in the mouse and extends over at least 4 million base pairs. In humans it contains more than 200 genes. As work continues to define the genes within and around the MHC, it becomes difficult to establish precise boundaries for this locus, which is now thought to span as many as 7 million base pairs. The genes encoding the α chains of MHC class I molecules and the α and β chains of MHC class II molecules are linked within the complex; the genes for β_2-microglobulin and the invariant chain are on different chromosomes (chromosomes 15 and 5, respectively, in humans, and chromosomes 2 and 18 in the mouse). Figure 6.14 shows the general organization of the MHC class I and II genes in human and mouse. In humans these genes are called ***h**uman **l**eukocyte **a**ntigen* or **HLA** genes, because they were first discovered through antigenic differences between white blood cells from different individuals; in the mouse they are known as the **H-2** genes. The mouse MHC class II genes were in fact first identified as genes that controlled whether an immune response was made to a given antigen and were originally called **Ir** (Immune response) genes. Because of this, the mouse MHC class II *A* and *E* genes were in the past referred to as *I-A* and *I-E*; but this terminology could be confused with MHC class I genes and it is no longer used.

There are three class I α-chain genes in humans, called *HLA-A*, *-B*, and *-C*. There are also three pairs of MHC class II α- and β-chain genes, called *HLA-DR*, *-DP*, and *-DQ*. In many people, however, the HLA-DR cluster contains an extra β-chain gene whose product can pair with the DRα chain. This means that the three sets of genes can give rise to four types of MHC class II molecules. All the MHC class I and class II molecules can present peptides to T cells, but each protein binds a different range of peptides (see Sections 4-14 and 4-15). Thus, the presence of several different genes for each MHC class means that any one individual is equipped to present a much broader range of peptides than if only one MHC molecule of each class were expressed at the cell surface.

Figure 6.15 shows a more detailed map of the human MHC region. An inspection of this map shows that many genes within this locus participate in antigen processing or antigen presentation, or have other functions related to the innate or adaptive immune response. The two *TAP* genes lie in the MHC class II region, in close association with the *LMP* genes, whereas the gene encoding tapasin (*TAPBP*), a protein that binds to both TAP and empty MHC class I molecules, lies at the edge of the MHC nearest the centromere. The genetic

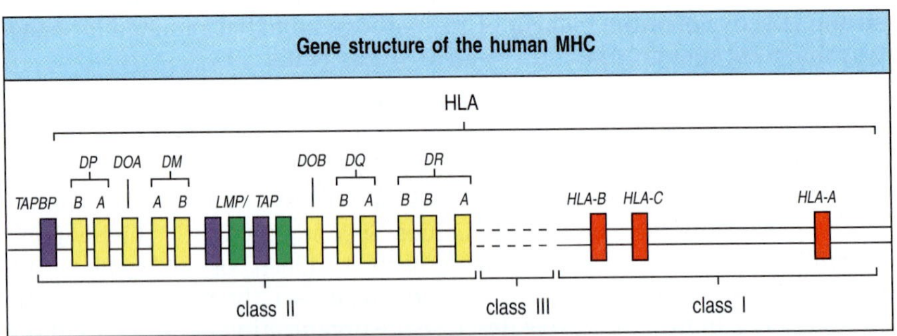

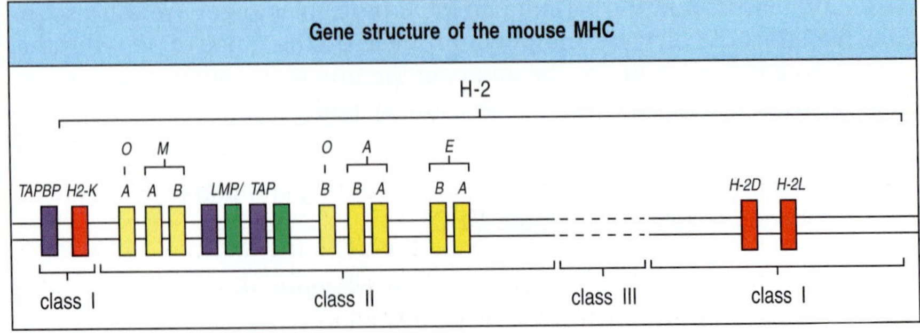

Fig. 6.14 The genetic organization of the major histocompatibility complex (MHC) in human and mouse. The organization of the principal MHC genes is shown for both humans (where the MHC is called HLA and is on chromosome 6) and mice (in which the MHC is called H-2 and is on chromosome 17). The organization of the MHC genes is similar in both species. There are separate clusters of MHC class I genes (red) and MHC class II genes (yellow), although in the mouse an MHC class I gene (*H-2K*) seems to have translocated relative to the human MHC so that the class I region in mice is split in two. In both species there are three main class I genes, which are called *HLA-A*, *HLA-B*, and *HLA-C* in humans, and *H2-K*, *H2-D*, and *H2-L* in the mouse. Each of these encodes the α chain of the respective MHC class I protein (HLA-A, HLA-B, etc.). The other subunit of an MHC class I molecule, β_2-microglobulin, is encoded by a gene located on a different chromosome—chromosome 15 in humans and chromosome 2 in the mouse. The class II region includes the genes for the α and β chains (designated *A* and *B*, respectively, in the gene names) of the MHC class II molecules HLA-DR, -DP, and -DQ (H-2A and -E in the mouse). In addition, the genes for the TAP1:TAP2 peptide transporter, the *LMP* genes that encode proteasome subunits, the genes encoding the DMα and DMβ chains (*DMA* and *DMB*), the genes encoding the α and β chains of the DO molecule (*DOA* and *DOB*, respectively), and the gene encoding tapasin (*TAPBP*) are also in the MHC class II region. The so-called class III genes encode various other proteins with functions in immunity (see Fig. 6.15).

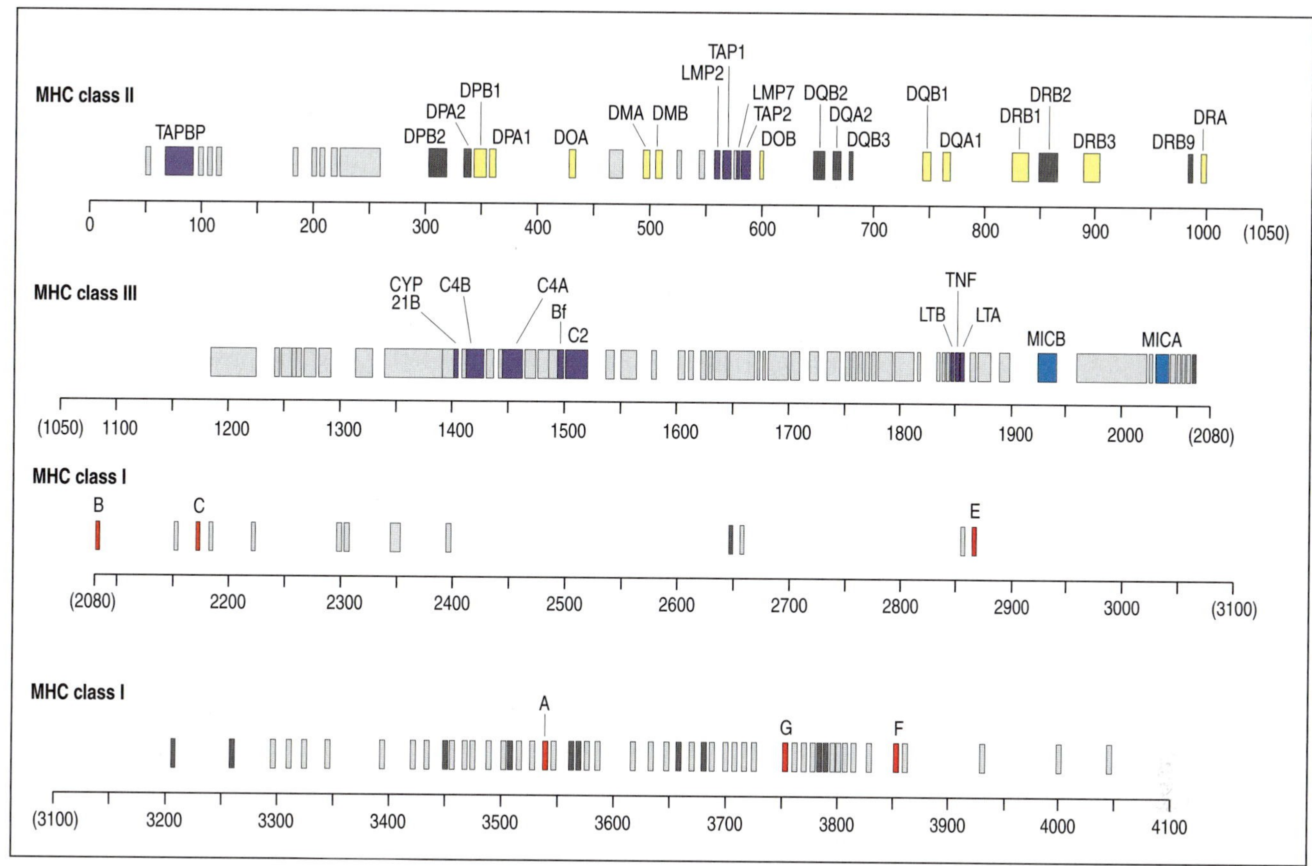

Fig. 6.15 Detailed map of the human MHC. The organization of the class I, class II, and class III regions of the human MHC is shown, with approximate genetic distances given in thousands of base pairs. Most of the genes in the class I and class II regions are mentioned in the text. The additional genes indicated in the class I region (for example, E, F, and G) are class I-like genes, encoding class Ib molecules; the additional class II genes are pseudogenes. The genes shown in the class III region encode the complement proteins C4 (two genes, shown as C4A and C4B), C2, and factor B (shown as Bf) as well as genes that encode the cytokines tumor necrosis factor-α (TNF) and lymphotoxin (LTA, LTB). Closely linked to the C4 genes is the gene encoding 21-hydroxylase (shown as CYP 21B), an enzyme involved in steroid biosynthesis. Immunologically important functional protein-coding genes mentioned in the text are color coded, with the MHC class I genes being shown in red, except for the MIC genes, which are shown in blue; these are distinct from the other class I-like genes and are under different transcriptional control. The MHC class II genes are shown in yellow. Genes in the MHC region that have immune functions but are not related to the MHC class I and class II genes are shown in purple. Genes in dark gray are pseudogenes related to immune-function genes. Unnamed genes unrelated to immune function are shown in light gray.

linkage of the MHC class I genes (whose products deliver cytosolic peptides to the cell surface) with the *TAP*, tapasin, and proteasome (*LMP*) genes (whose products deliver cytosolic peptides into the endoplasmic reticulum) suggests that the entire MHC has been selected during evolution for antigen processing and presentation.

When cells are treated with the interferons IFN-α, -β, or -γ, there is a marked increase in the transcription of MHC class I α-chain and β_2-microglobulin genes and of the proteasome, tapasin, and *TAP* genes. Interferons are produced early in viral infections as part of the innate immune response, as described in Chapter 3, so this effect increases the ability of all cells to process viral proteins and present the resulting virus-derived peptides at the cell surface (all cells except red blood cells express MHC class I molecules, whereas the distribution of class II is more restricted; see Fig. 4.27). This helps to activate the appropriate T cells and initiate the adaptive immune response in response to the virus. The coordinated regulation of the genes encoding these components may be facilitated by the linkage of many of them in the MHC.

The *HLA-DM* genes, which encode the DM molecule whose function is to catalyze peptide binding to MHC class II molecules (see Section 6-8), are clearly related to the MHC class II genes. The *DOA* and *DOB* genes, which encode the DOα and DOβ subunits of the DO molecule, a negative regulator of DM, are also clearly related to the MHC class II genes. Expression of the classical MHC class II genes, along with the invariant-chain gene (which is located on another chromosome, chromosome 5 in humans) and the genes encoding DMα, DMβ, and DOα, but not DOβ, can be coordinately increased by IFN-γ. This distinct regulation of MHC class II genes by IFN-γ, which is made by activated T cells of the T_H1 type as well as by activated CD8 T cells and NK cells, allows T cells responding to bacterial infections to increase the expression of those molecules concerned with the processing and presentation of intravesicular antigens. Expression of all these molecules is induced by IFN-γ (but not by IFN-α or -β), via the production of a protein known as **MHC class *II* *t*rans*a*ctivator** (**CIITA**), which acts as a positive transcriptional co-activator of MHC class II genes. An absence of CIITA causes severe immunodeficiency due to the nonproduction of MHC class II molecules—**MHC class II deficiency**. Finally, the MHC region contains many so-called 'nonclassical' MHC genes, which resemble MHC class I genes in structure. We will return to these genes, often called MHC class Ib genes, in Section 6-18, after we complete our discussion of the classical MHC genes.

6-12 The protein products of MHC class I and class II genes are highly polymorphic.

Because of the polygeny of the MHC, every person expresses at least three different MHC class I molecules and three (or sometimes four) MHC class II molecules on his or her cells. In fact, the number of different MHC molecules expressed by most people is greater because of the extreme polymorphism of the MHC (Fig. 6.16).

The term **polymorphism** comes from the Greek *poly*, meaning many, and *morphe*, meaning shape or structure. As used here, it means within-species variation at a gene locus, and thus in its protein product; the variant genes that can occupy the locus are termed **alleles**. There are more than 800 alleles of some human MHC class I and class II genes in the human population, far more than the number of alleles for other genes found within the MHC region. Each MHC class I and class II allele is relatively frequent in the population, so there is only a small chance that the corresponding gene loci on both homologous chromosomes of an individual will have the same allele;

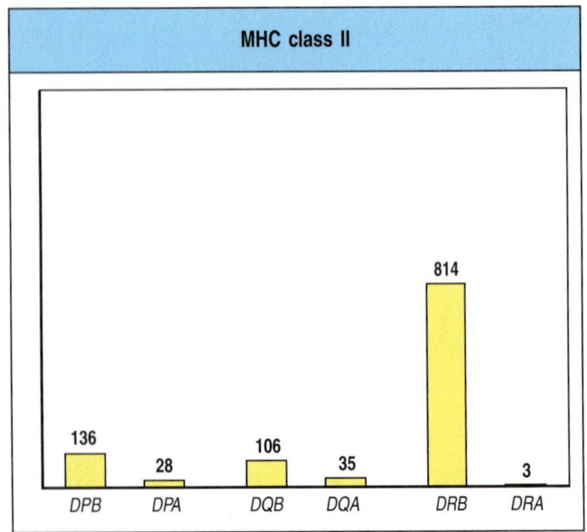

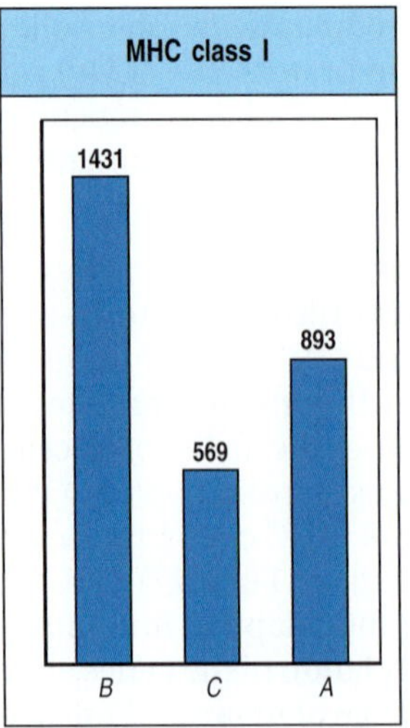

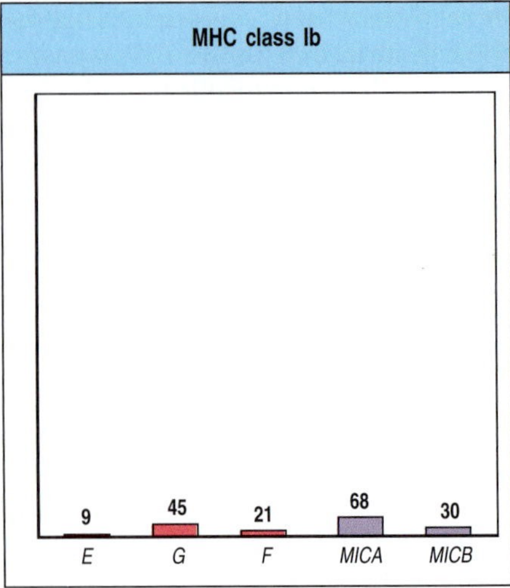

Fig. 6.16 Human MHC genes are highly polymorphic. With the notable exception of the DRα locus, which is functionally monomorphic, each gene locus has many alleles. The number of functional proteins encoded is less than the total number of alleles. Shown in this figure as the heights of the bars are the number of different HLA alleles currently assigned by the WHO Nomenclature Committee for Factors of the HLA System as of January 2010.

most individuals will be **heterozygous** for the genes encoding MHC class I and class II molecules. The particular combination of MHC alleles found on a single chromosome is known as an **MHC haplotype**. Expression of MHC alleles is **codominant**, meaning that the protein products of both the alleles at a locus are expressed equally in the cell, and both gene products can present antigens to T cells. The number of MHC alleles discovered that do not code for a functional protein is remarkably small. The extensive polymorphism at each locus thus has the potential to double the number of different MHC molecules expressed in an individual and thereby increase the diversity already available through polygeny (Fig. 6.17).

For the MHC class II genes, the number of different MHC molecules may be increased still further by the combination of α and β chains encoded by different chromosomes (so that two α chains and two β chains can give rise to four different proteins, for example). In mice it has been shown that not all combinations of α and β chains can form stable dimers and so, in practice, the exact number of different MHC class II molecules expressed depends on which alleles are present on each chromosome.

Because most individuals are heterozygous, most matings will produce offspring that receive one of four possible combinations of the parental MHC haplotypes. Thus siblings are also likely to differ in the MHC alleles they express, there being one chance in four that an individual will share both haplotypes with a sibling. One consequence of this is the difficulty of finding suitable donors for tissue transplantation, even among siblings.

All MHC class I and II proteins are polymorphic to a greater or lesser extent, with the exception of the DRα chain and its mouse homolog Eα. These chains do not vary in sequence between different individuals and are said to be **monomorphic**. This might indicate a functional constraint that prevents variation in the DRα and Eα proteins, but no such special function has been found. Many mice, both domestic and wild, have a mutation in the Eα gene that prevents synthesis of the Eα protein. They thus lack cell-surface H-2E molecules, so if H2-E does have a special function it is unlikely to be essential.

MHC polymorphisms at individual MHC genes seem to have been strongly selected by evolutionary pressures. Several genetic mechanisms contribute to the generation of new alleles. Some new alleles arise by point mutations and others by gene conversion, a process in which a sequence in one gene is replaced, in part, by sequences from a different gene (Fig. 6.18). The effects of selective pressure in favor of polymorphism can be seen clearly in the pattern of point mutations in the MHC genes. Point mutations can be classified as replacement substitutions, which change an amino acid, or silent substitutions, which simply change the codon but leave the amino acid the same. Replacement substitutions occur within the MHC at a higher frequency relative to silent substitutions than would be expected, providing evidence that polymorphism has been actively selected for in the evolution of the MHC.

Fig. 6.17 Polymorphism and polygeny both contribute to the diversity of MHC molecules expressed by an individual. The high polymorphism of the classical MHC genes ensures a diversity in MHC gene expression in the population as a whole. However, no matter how polymorphic a gene, no individual can express more than two alleles at a single gene locus. Polygeny, the presence of several different related genes with similar functions, ensures that each individual produces a number of different MHC molecules. Polymorphism and polygeny combine to produce the diversity of MHC molecules seen both within an individual and in the population at large.

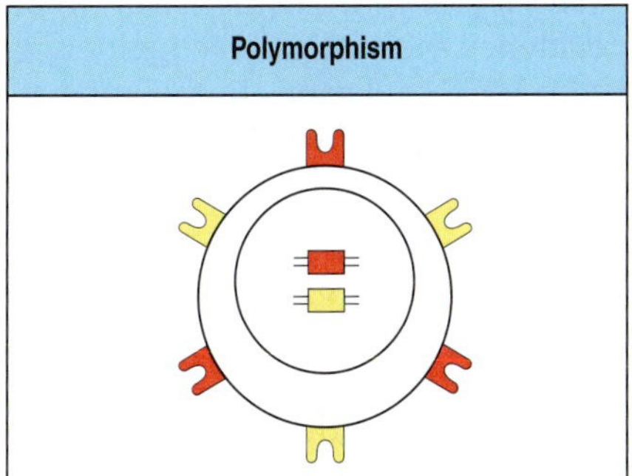

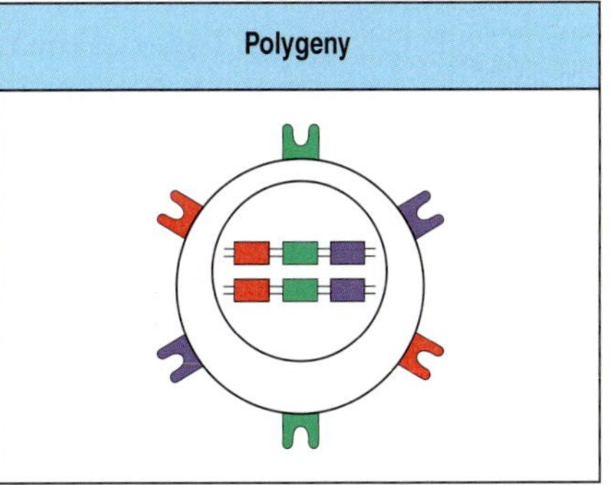

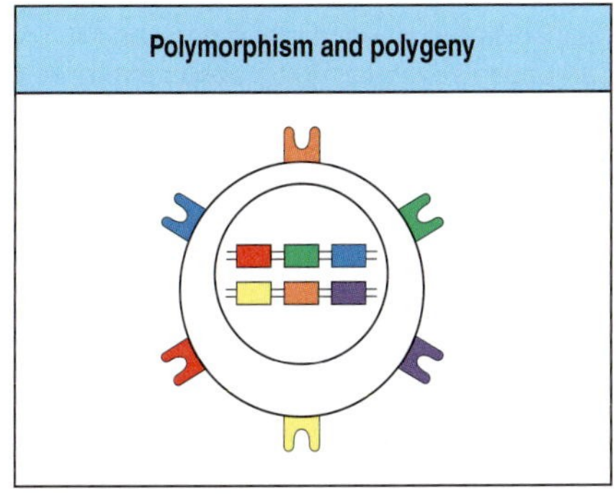

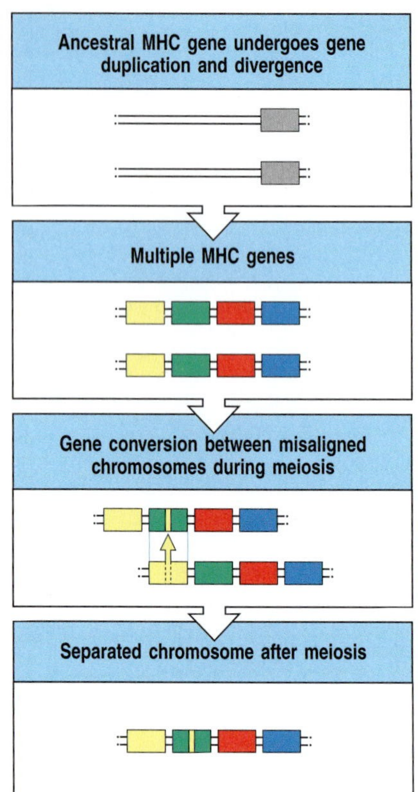

Fig. 6.18 Gene conversion can create new alleles by copying sequences from one MHC gene to another. Multiple MHC genes of generally similar structure were derived over evolutionary time by duplication of an unknown ancestral MHC gene (gray) followed by genetic divergence. Further interchange between these genes can occur by a process known as gene conversion, in which sequences can be transferred from part of one gene to a similar gene. For this to happen, the two genes must become apposed during meiosis. This can occur as a consequence of the misalignment of the two paired homologous chromosomes when there are many copies of similar genes arrayed in tandem—somewhat like buttoning in the wrong buttonhole. During the process of crossing-over and DNA recombination, a DNA sequence from one chromosome is sometimes copied to the other, replacing the original sequence. In this way, several nucleotide changes can be inserted all at once into a gene and can cause several simultaneous amino acid changes between the new gene sequence and the original gene. Because of the similarity of the MHC genes to each other and their close linkage, gene conversion has occurred many times in the evolution of MHC alleles.

6-13 MHC polymorphism affects antigen recognition by T cells by influencing both peptide binding and the contacts between T-cell receptor and MHC molecule.

The next few sections describe how MHC polymorphisms benefit the immune response and how pathogen-driven selection can account for the large number of MHC alleles. The products of individual MHC alleles, often known as protein **isoforms**, can differ from one another by up to 20 amino acids, making each variant protein quite distinct. Most of the differences are localized to exposed surfaces of the extracellular domain furthest from the membrane, and to the peptide-binding cleft in particular (Fig. 6.19). We have seen that peptides bind to MHC class I and class II molecules through the interaction of specific anchor residues with peptide-binding pockets in the peptide-binding cleft (see Sections 4-14 and 4-15). Many of the polymorphisms in MHC molecules alter the amino acids that line these pockets and thus change the pockets' binding specificities. This in turn changes the anchor residues of peptides that can bind to each MHC isoform. The set of anchor residues that allows binding to a given isoform of an MHC class I or class II molecule is called a **sequence motif**, and this can be used to predict peptides within a protein that might bind that variant (Fig. 6.20). Such predictions may be very important in designing peptide vaccines.

In rare cases, processing of a protein will not generate any peptides with a suitable sequence motif for binding to any of the MHC molecules expressed by an individual. This individual fails to respond to the antigen. Such failures in responsiveness to simple antigens were first reported in inbred animals, where they were called **immune response (Ir) gene** defects. These defects were mapped to genes within the MHC long before the structure or function of MHC molecules was understood, and they were the first clue to the antigen-presenting function of MHC molecules. We now understand that Ir gene defects are common in inbred strains of mice because they are homozygous at all their MHC genes, which limits the range of peptides they can present to T cells. Ordinarily, MHC polymorphism guarantees a sufficient number of different MHC molecules in a single individual to make this type of nonresponsiveness unlikely, even to relatively simple antigens such as small toxins. This has obvious importance for host defense.

Initially, the only evidence linking Ir gene defects to the MHC was genetic—mice of one MHC genotype could make antibody in response to a particular antigen, whereas mice of a different MHC genotype, but otherwise genetically identical, could not. The MHC genotype was somehow controlling the ability of the immune system to detect or respond to specific antigens, but it was not clear at the time that direct recognition of MHC molecules was involved.

Later experiments showed that the antigen specificity of T-cell recognition was controlled by MHC molecules. The immune responses affected by the Ir genes were known to depend on T cells, and this led to a series of experiments in mice to ascertain how MHC polymorphism might control T-cell responses. The earliest of these experiments showed that T cells could be activated only by macrophages or B cells that shared MHC alleles with the mouse in which the T cells originated. This was the first evidence that antigen recognition by T cells depends on the presence of specific MHC molecules in the antigen-presenting cell—the phenomenon we now know as **MHC restriction**, as noted in Chapter 4.

The clearest example of this feature of T-cell recognition came, however, from studies of virus-specific cytotoxic T cells, for which Peter Doherty and Rolf Zinkernagel were awarded the Nobel Prize in 1996. When mice are infected with a virus, they generate cytotoxic T cells that kill cells infected with the virus, while sparing uninfected cells or cells infected with unrelated viruses. The cytotoxic T cells are thus virus-specific. The additional and striking outcome of their experiments was the demonstration that the ability of cytotoxic

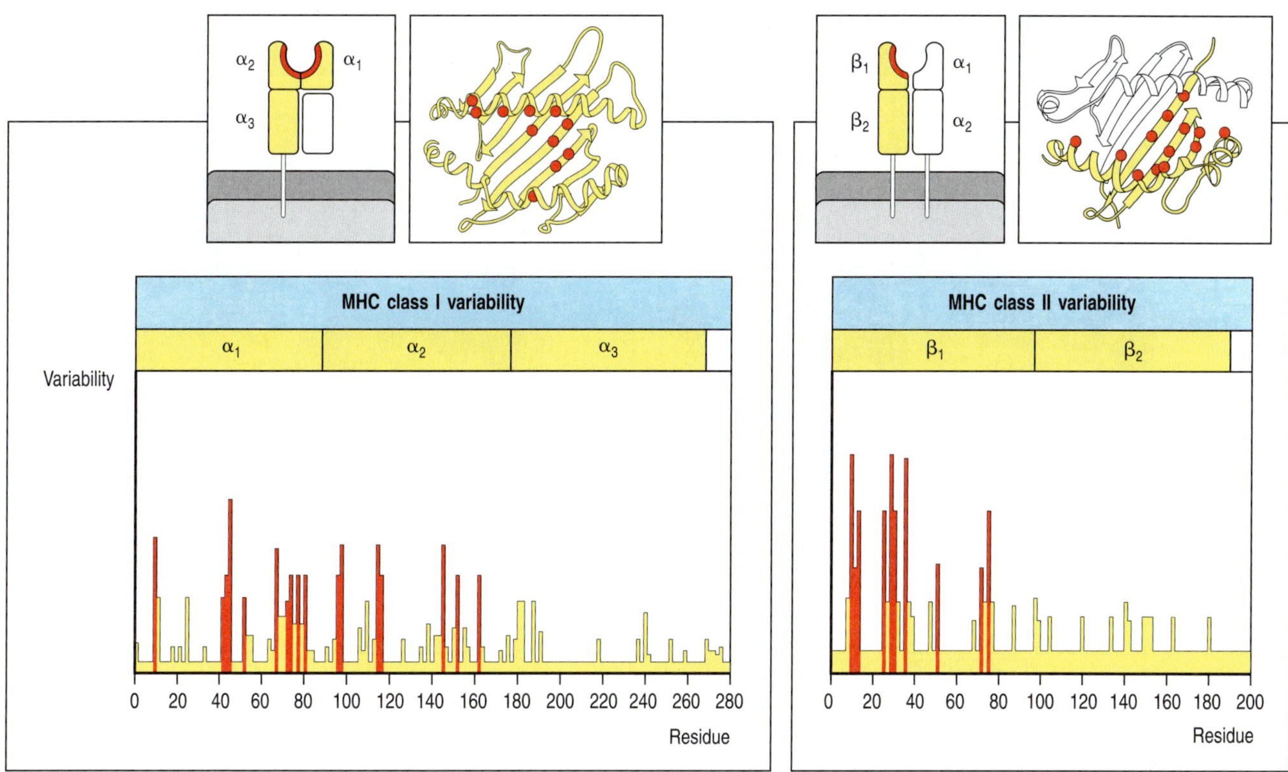

Fig. 6.19 Allelic variation occurs at specific sites within MHC molecules. Variability plots of the amino acid sequences of MHC molecules show that the variation arising from genetic polymorphism is restricted to the amino-terminal domains (α_1 and α_2 domains of MHC class I molecules, and α_1 and β_1 domains of MHC class II molecules). These are the domains that form the peptide-binding cleft. Moreover, allelic variability is clustered in specific sites within the amino-terminal domains, lying in positions that line the peptide-binding cleft, either on the floor of the groove or directed inward from the walls. For the MHC class II molecule, the variability of the HLA-DR alleles is shown. For HLA-DR, and its homologs in other species, the α chain is essentially invariant and only the β chain shows significant polymorphism.

T cells to kill virus-infected cells was also affected by the polymorphism of MHC molecules. Cytotoxic T cells induced by viral infection in mice of MHC genotype a (MHC^a) would kill any MHC^a cell infected with that virus but would not kill cells of MHC genotype b, or c, and so on, even if they were infected with the same virus. In other words, cytotoxic T cells kill cells infected by virus only if those cells express self MHC. Because the MHC genotype 'restricts' the antigen specificity of the T cells, this effect was called MHC restriction. Together with the earlier studies on both B cells and macrophages, this work showed that MHC restriction is a critical feature of antigen recognition by all functional classes of T cells.

We now know that MHC restriction is due to the fact that the binding specificity of an individual T-cell receptor is not for its peptide antigen alone but for the complex of peptide and MHC molecule (see Chapter 4). MHC restriction is explained in part by the fact that different MHC molecules bind different peptides. In addition, some of the polymorphic amino acids in MHC molecules are located in the α helices that flank the peptide-binding groove, but have side chains oriented toward the exposed surface of the peptide:MHC complex that can directly contact the T-cell receptor (see Figs. 6.19 and 4.22). It is therefore not surprising that T cells can readily distinguish between a peptide bound to MHC^a and the same peptide bound to MHC^b. This restricted recognition may sometimes be caused both by differences in the conformation of the bound peptide imposed by the different MHC molecules and by direct recognition of polymorphic amino acids in the MHC molecule itself. Thus, the specificity of a T-cell receptor is defined both by the peptide it recognizes and by the MHC molecule bound to it (Fig. 6.21).

6-14 Alloreactive T cells recognizing nonself MHC molecules are very abundant.

The discovery of MHC restriction also helped to explain the otherwise puzzling phenomenon of recognition of nonself MHC in the rejection of organs and tissues transplanted between members of the same species. Transplanted

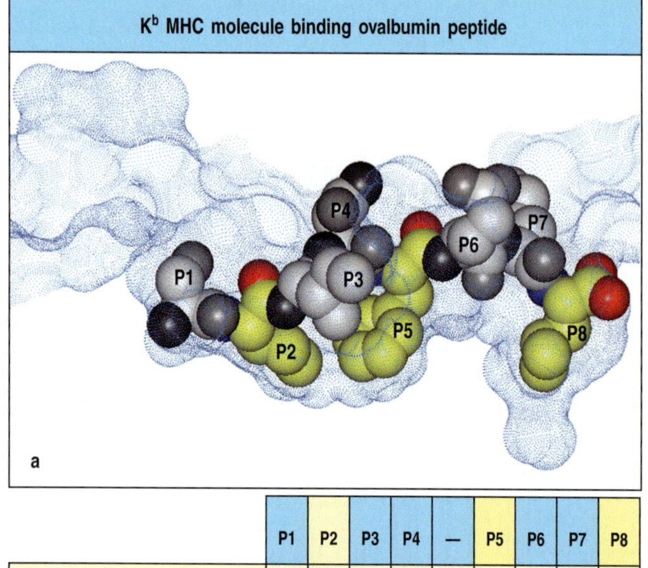

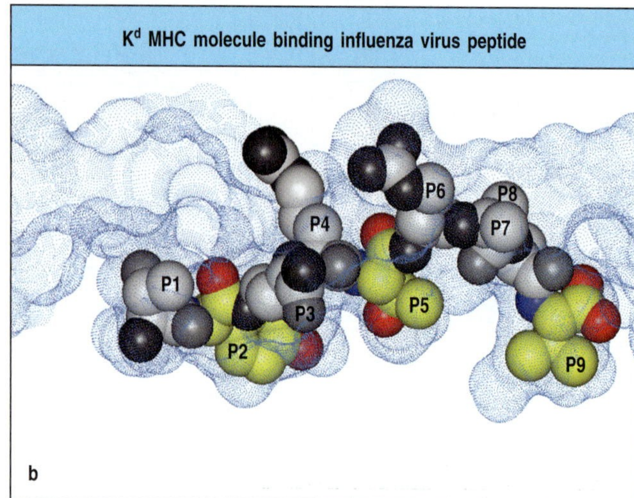

	P1	P2	P3	P4	—	P5	P6	P7	P8
Ovalbumin (257–264)	S	I	I	N		F	E	K	L
HBV surface antigen (208–215)	I	L	S	P		F	L	P	L
Influenza NS2 (114–121)	R	T	F	S		F	Q	L	I
LCMV NP (205–212)	Y	T	V	K		Y	P	N	L
VSV NP (52–59)	R	G	Y	V		Y	Q	G	L
Sendai virus NP (324–332)	F	A	P	G	N	Y	P	A	L

	P1	P2	P3	P4	P5	P6	P7	P8	P9
Influenza NP (147–155)	T	Y	Q	R	T	R	A	L	V
ERK4 (136–144)	Q	Y	I	H	S	A	N	V	L
P198 (14–22)	K	Y	Q	A	V	T	T	T	L
P. yoelii CSP (280–288)	S	Y	V	P	S	A	E	Q	I
P. berghei CSP (25)	G	Y	I	P	S	A	E	K	I
JAK1 (367–375)	S	Y	F	P	E	I	T	H	I

Fig. 6.20 Different allelic variants of an MHC class I molecule bind different peptides. Shown are cutaway views of (a) ovalbumin peptide bound to the mouse H2-K^b MHC class I molecule and (b) influenza nucleoprotein (NP) peptide bound to the H2-K^d MHC class I molecule, respectively. The solvent-accessible surface of the MHC molecules is shown as a blue dotted surface. Class I MHC molecules typically have six pockets in the peptide-binding groove, which are conventionally called A–F. The bound peptides, shown as space-filling models, fit into the peptide-binding groove, with side chains from the anchor residues extending to fill the pockets. H2-K^b is binding SIINFEKL (single-letter amino acid code), a peptide of eight residues (P1–8) from ovalbumin, and H2-K^d is binding TYQRTRALV, a peptide of nine residues (P1–9) from the influenza nucleoprotein (NP). Anchor residues (shown in yellow) may be primary or secondary in their influence on peptide binding. For H2-K^b, the sequence motif is determined by two primary anchors, P5 and P8; the C pocket binds the P5 side chain of the peptide (a tyrosine (Y) or a phenylalanine (F)), and the F pocket binds the P8 residue (a non-aromatic hydrophobic side chain from leucine (L), isoleucine (I), methionine (M), or valine (V)). The B pocket binds P2, a secondary anchor residue in H-2K^b. For H2-K^d, the sequence motif is primarily determined by the two primary anchors, P2 and P9. The B pocket accommodates a tyrosine side chain. The F pocket binds leucine, isoleucine, or valine. Beneath the structures are shown sequence motifs from peptides that are known to bind to the MHC molecule. CSP, circumsporozoite antigen; ERK4, extracellular signal-related kinase 4; HBV, hepatitis B virus; JAK1, Janus-associated kinase 1; LCMV, lymphocytic choriomeningitis virus; NS2, NS2 protein; P198, modified tumor-cell antigen; *P. berghei*, *Plasmodium berghei*; *P. yoelii*, *Plasmodium yoelii*; VSV, vesicular stomatitis virus. An extensive collection of motifs can be found at http://www.syfpeithi.de. Structures courtesy of V.E. Mitaksov and D. Fremont.

organs from donors bearing MHC molecules that differ from those of the recipient—even by as little as one amino acid—are rapidly rejected owing to the presence in any individual of large numbers of T cells that react to nonself, or **allogeneic**, MHC molecules. Early studies on T-cell responses to allogeneic MHC molecules used the **mixed lymphocyte reaction**, in which T cells from one individual are mixed with lymphocytes from a second individual. If the T cells of this individual recognize the other individual's MHC molecules as 'foreign,' the T cells will divide and proliferate. (The lymphocytes from

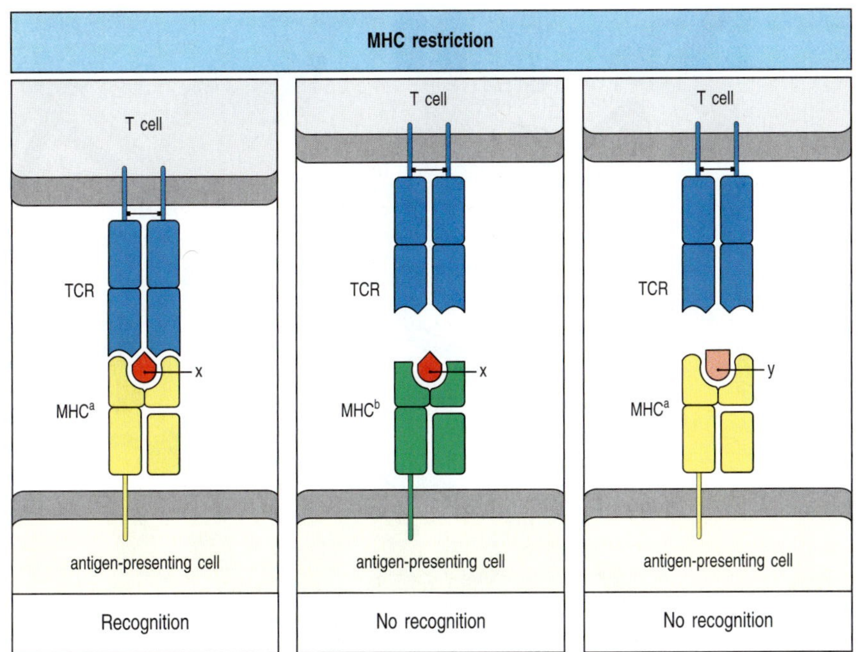

Fig. 6.21 T-cell recognition of antigens is MHC restricted. The antigen-specific T-cell receptor (TCR) recognizes a complex of an antigenic peptide and a self MHC molecule. One consequence of this is that a T cell specific for peptide x and a particular MHC allele, MHC^a (left panel), will not recognize the complex of peptide x with a different MHC allele, MHC^b (center panel), or the complex of peptide y with MHC^a (right panel). The co-recognition of a foreign peptide and MHC molecule is known as MHC restriction because the MHC molecule is said to restrict the ability of the T cell to recognize antigen. This restriction may either result from direct contact between the MHC molecule and T-cell receptor or be an indirect effect of MHC polymorphism on the peptides that bind or on their bound conformation.

the second individual are usually prevented from dividing by irradiation or treatment with the cytostatic drug mitomycin C.) Such studies have shown that roughly 1–10% of all T cells in an individual will respond to stimulation by cells from another, unrelated, member of the same species. This type of T-cell response is called an **alloreaction** or **alloreactivity**, because it represents the recognition of allelic polymorphism in allogeneic MHC molecules. The phenomenon of alloreactivity in the context of organ transplantation is discussed in more detail in Chapter 15.

Before the role of the MHC molecules in antigen presentation was understood, it was a mystery why so many T cells should recognize nonself MHC molecules, as there is no reason that the immune system should have evolved a defense against tissue transplants. Once it was realized that T-cell receptors have evolved to recognize foreign peptides in combination with polymorphic MHC molecules, alloreactivity became easier to explain. We now know of at least two processes that can contribute to the high frequency of alloreactive T cells. T cells developing in the thymus go through a process of positive selection that favors the survival of cells whose T-cell receptors interact weakly with the self MHC molecules expressed in the thymus (this is described in detail in Chapter 8). It is thought that selecting T-cell receptors for their interaction with one type of MHC molecule increases the likelihood that they will cross-react with other (nonself) MHC variants.

It seems that alloreactivity is also promoted by an inherent ability of T-cell receptor genes to recognize MHC molecules. This was implied by observations that T cells artificially driven to mature in animals lacking MHC class I and class II, and in which positive selection in the thymus cannot occur, still display frequent alloreactivity. In agreement with this idea, specific amino acid residues within the germline-encoded region of certain TCRβ genes have been shown to promote the general recognition of MHC molecules (Fig. 6.22). Given the large number of variable-region sequences in T-cell receptors, it may be that each T-cell receptor has its own idiosyncratic way of binding MHC molecules.

In principle, alloreactive T cells might depend on recognizing either a foreign peptide antigen or the nonself MHC molecule to which it is bound for their reactivity against nonself MHC; these options have been called

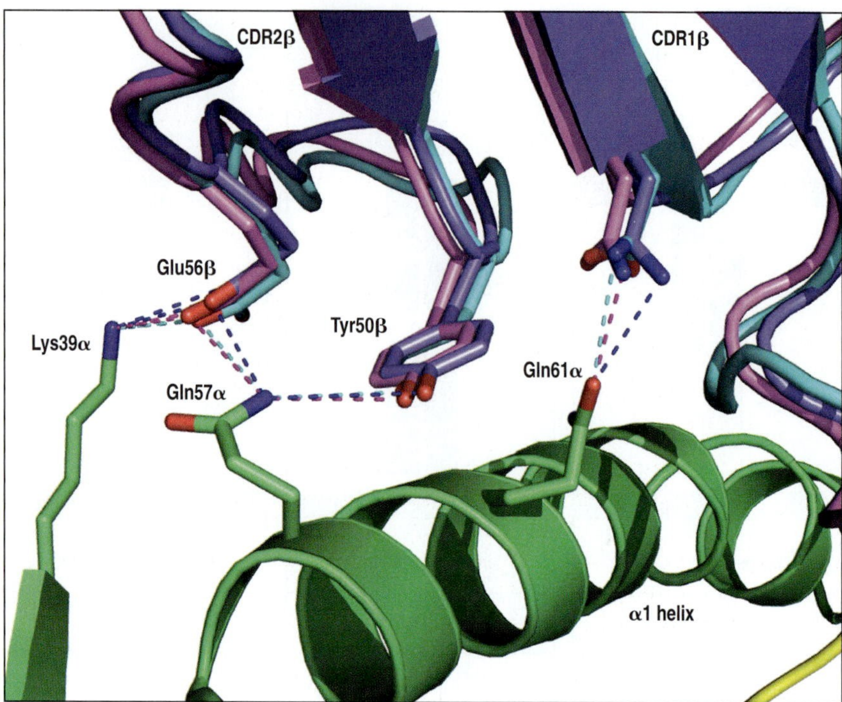

Fig. 6.22 Germline-encoded residues in the CDRs of Vβ and Vα genes confer the affinity of T-cell receptor for MHC molecules. Shown is the structure for several T-cell receptors bound to a class II MHC molecule. Conserved residues (Lys39, Gln57, and Gln61) within the α1 helix of the MHC (green) make an extended hydrogen-bonded network with germline-encoded and nonpolymorphic residues (Glu56, Tyr50, and Asn31) that are located in the CDR1 and CDR2 regions of the Vβ 8.2 gene. Courtesy of Chris Garcia.

peptide-dependent and peptide-independent allorecognition. But as the number of individual alloreactive T-cell clones studied has increased, it seems that most alloreactive T cells actually recognize both; that is, most individual alloreactive T-cell clones respond to a foreign MHC molecule only when a particular peptide is bound to it. In this sense, the structural basis of allorecognition may be quite similar to normal MHC-restricted peptide recognition, being dependent on contacts with both peptide and MHC molecule (see Fig. 6.21, left panel), but in this case a foreign MHC molecule. In practice, alloreactive responses against a transplanted organ are likely to represent the combined activity of many alloreactive T cells, and it is not possible to determine what peptides from the donor might be involved in recognition by the alloreactive T cells.

6-15 Many T cells respond to superantigens.

Superantigens are a distinct class of antigens that stimulate a primary T-cell response similar in magnitude to a response to allogeneic MHC molecules. Such responses were first observed in mixed lymphocyte reactions using lymphocytes from strains of mice that were MHC identical but otherwise genetically distinct. The antigens provoking this reaction were originally designated **minor lymphocyte stimulating (Mls) antigens**, and it seemed reasonable to suppose that they might be functionally similar to the MHC molecules themselves. We now know that this is not true. The Mls antigens in these mouse strains are encoded by retroviruses, such as the mouse mammary tumor virus, that have become stably integrated at various sites in the mouse chromosomes. Superantigens are produced by many different pathogens, including bacteria, mycoplasmas, and viruses, and the responses they provoke are helpful to the pathogen rather than the host.

Mls proteins act as superantigens because they have a distinctive mode of binding to both MHC and T-cell receptor molecules that enables them to stimulate very large numbers of T cells. Superantigens are unlike other protein antigens, in that they are recognized by T cells without being processed into peptides that are captured by MHC molecules. Indeed, fragmentation of a superantigen destroys its biological activity, which depends on binding as an intact protein to the outside surface of an MHC class II molecule that has already bound peptide. In addition to binding MHC class II molecules, superantigens are able to bind the V_β region of many T-cell receptors (Fig. 6.23). Bacterial superantigens bind mainly to the V_β CDR2 loop, and to a smaller extent to the V_β CDR1 loop and an additional loop called the hypervariable 4 or HV4 loop. The HV4 loop is the predominant binding site for viral superantigens, at least for the Mls antigens encoded by the endogenous mouse mammary tumor viruses. Thus, the α-chain V region and the CDR3 of the β chain of the T-cell receptor have little effect on superantigen recognition, which is determined largely by the germline-encoded V gene segments that encode the expressed V_β chain. Each superantigen is specific for one or a few of the different V_β gene products, of which there are 20–50 in mice and humans; a superantigen can thus stimulate 2–20% of all T cells.

This mode of stimulation does not prime an adaptive immune response specific for the pathogen. Instead, it causes a massive production of cytokines by CD4 T cells, the predominant responding population of T cells. These cytokines have two effects on the host: systemic toxicity and suppression of the adaptive immune response. Both these effects contribute to microbial pathogenicity. Among the bacterial superantigens are the **staphylococcal enterotoxins** (**SEs**), which cause food poisoning, and the **toxic shock syndrome toxin-1** (**TSST-1**) of *Staphylococcus aureus*, the etiologic principle in **toxic shock syndrome**, which can be caused by a localized infection with toxin-producing strains of the bacterium. The role of viral superantigens in human disease is less clear.

6-16 MHC polymorphism extends the range of antigens to which the immune system can respond.

Most polymorphic genes encode proteins that vary by only one or a few amino acids, whereas the allelic variants of MHC proteins differ from each other by up to 20 amino acids. The extensive polymorphism of the MHC proteins has almost certainly evolved to outflank the evasive strategies of pathogens. The requirement that pathogen antigens must be presented by an MHC molecule provides two possible ways in which pathogens could evolve to evade detection. One is through mutations that eliminate from the pathogen's proteins all peptides able to bind MHC molecules. The Epstein–Barr virus provides an example. There are small isolated populations in south-east China and Papua New Guinea in which about 60% of the people carry the HLA-A11 allele. Many isolates of the Epstein–Barr virus obtained from these populations have mutations in a dominant peptide epitope normally presented by HLA-A11; the mutant peptides no longer bind to HLA-A11 and cannot be recognized by HLA-A11-restricted T cells. This strategy is clearly much less successful if there are many different MHC molecules, and the polygeny of the MHC may have evolved in response.

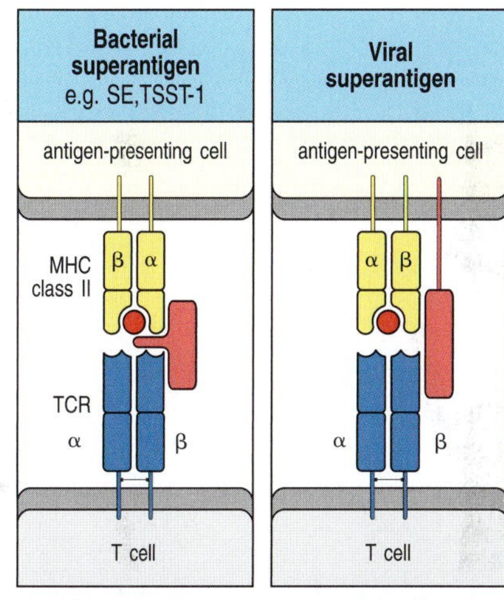

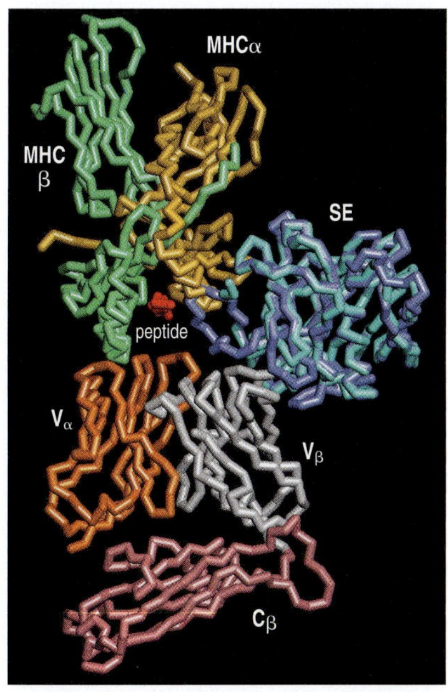

Fig. 6.23 Superantigens bind directly to T-cell receptors and to MHC molecules. Superantigens can bind independently to MHC class II molecules and to T-cell receptors, binding to the V_β domain of the T-cell receptor (TCR), away from the complementarity-determining regions, and to the outer faces of the MHC class II molecule, outside the peptide-binding site (top panels). The bottom panel shows a reconstruction of the interaction between a T-cell receptor, an MHC class II molecule and a staphylococcal enterotoxin (SE) superantigen, produced by superimposing separate structures of an enterotoxin:MHC class II complex onto an enterotoxin:T-cell receptor complex. The two enterotoxin molecules (actually SEC3 and SEB) are shown in turquoise and blue, binding to the α chain of the MHC class II molecule (yellow) and to the β chain of the T-cell receptor (colored gray for the V_β domain and pink for the C_β domain). Molecular model courtesy of H.M. Li, B.A. Fields, and R.A. Mariuzza.

In addition, in large outbred populations, polymorphism at each locus can potentially double the number of different MHC molecules expressed by an individual, as most individuals will be heterozygotes. Polymorphism has the additional advantage that individuals in a population will differ in the combinations of MHC molecules that they express and will therefore present different sets of peptides from each pathogen. This makes it unlikely that all individuals in a population will be equally susceptible to a given pathogen, and its spread will therefore be limited. The fact that exposure to pathogens over an evolutionary timescale can select for particular MHC alleles is indicated by the strong association of the HLA-B53 allele with recovery from a potentially lethal form of malaria. This allele is very common in people from West Africa, where malaria is endemic, and rare elsewhere, where lethal malaria is uncommon.

Similar arguments apply to a second strategy for evading recognition. Pathogens that can block the presentation of their peptides by MHC molecules can avoid the adaptive immune response. Adenoviruses encode a protein that binds to MHC class I molecules in the endoplasmic reticulum and prevents their transport to the cell surface, thus preventing the recognition of viral peptides by CD8 cytotoxic T cells. This MHC-binding protein interacts with a polymorphic region of the MHC class I molecule, as some allelic variants are retained in the endoplasmic reticulum by the adenoviral protein, whereas others are not. Increasing the variety of MHC molecules expressed reduces the likelihood that a pathogen will be able to block presentation by all of them and completely evade an immune response.

These arguments raise a question: If having three MHC class I loci is better than having one, why are there not far more? The probable explanation is that each time a distinct MHC molecule is added to the repertoire, all T cells that can respond to the self peptide bound to that MHC molecule must be removed to maintain self tolerance. It seems that the number of MHC genes present in humans and mice is about optimal to balance out the advantages of presenting an increased range of foreign peptides and the disadvantages of a loss of T cells from the repertoire.

6-17 A variety of genes with specialized functions in immunity are also encoded in the MHC.

In addition to the highly polymorphic 'classical' MHC class I and class II genes, there are many 'nonclassical' MHC genes in the MHC region encoding MHC class I-type molecules that show comparatively little polymorphism; many of these have yet to be assigned a function. They are linked to the class I region of the MHC, and their exact number varies greatly between species and even between members of the same species. These genes have been termed **MHC class Ib** genes; like MHC class I genes, many, but not all, associate with β_2-microglobulin when expressed on the cell surface. Their expression on cells is variable, both in the amount present at the cell surface and in tissue distribution. The characteristics of several MHC class Ib genes are shown in Fig. 6.24.

Fig. 6.24 Mouse and human MHC class Ib proteins and their functions. MHC class Ib proteins are encoded both within the MHC region and on other chromosomes. The functions of some MHC class Ib proteins are unrelated to the adaptive immune response, but many have a role in innate immunity by interacting with receptors on NK cells (see the text and Section 3-21). HLA-C, which is a classical MHC molecule (class Ia), is included here because, as well as its role in presenting peptides to T-cell receptors in adaptive immunity, all HLA-C isoforms interact with the KIR class of NK-cell receptors to regulate NK-cell function in the innate immune response. CTL, cytotoxic T lymphocyte.

One mouse MHC class Ib molecule, H2-M3, can present peptides with *N*-formylated amino termini, which is of interest because all bacteria initiate protein synthesis with *N*-formylmethionine. Cells infected with cytosolic bacteria can be killed by CD8 T cells that recognize *N*-formylated bacterial peptides bound to H2-M3. Whether an equivalent MHC class Ib molecule exists in humans is not known.

	Class 1b molecule						Receptors or interacting proteins			
	Human	Mouse	Expression pattern	Associates with β_2m	Poly-morphism	Ligand	T-cell receptor	NK receptor	Other	Biological function
MHC encoded	HLA-C (class 1a)		Ubiquitous	Yes	High	Peptide	TCR	KIRs		Activate T cells Inhibit NK cells
		H2-M3	Limited	Yes	Low	fMet peptide	TCR			Activate CTLs with bacterial peptides
		T22 T10	Splenocytes	Yes	Low	None	γ:δ TCR			Regulation of activated splenocytes
	HLA-E	Qa-1	Ubiquitous	Yes	Low	MHC leader peptides (Qdm)		NKG2A NKG2C		NK cell inhibition
	HLA-F		Widely expressed	Yes	Low	Peptide?		LILRB1 LILRB2		Unknown
	HLA-G		Maternal/fetal interface	Yes	Low	Peptide	TCR	LILRB1		Modulate maternal/ fetal interaction
	MIC-A MIC-B		GI tract, widely expressed	No	Moderate	None		NKG2D		Stress-induced activation of NK and CD8 cells
		TL	Small intestine epithelium	Yes	Low	None	CD8α:α			Potential modulation of T-cell activation
		M10	Vomeronasal neurons	Yes	Low	Unknown			Vomeronasal receptor V2R	Pheromone detection
Non-MHC encoded	ULBPs	MULT1 H60, Rae1	Limited	No	Low	None		NKG2D		Induced NK-cell-activating ligand
	MR1	MR1	Ubiquitous	Yes	None	Unknown		LILRB2		Control of inflammatory response
	CD1a–CD1e	CD1d	Limited	Yes	None	Lipids glycolipids	α:β TCR			Activate T cells against bacterial lipids
		Mill1 Mill2	Ubiquitous	Yes?	Low	Unknown	Unknown			Unknown
	HFE	HFE	Liver and gut	Yes	Low	None			Transferrin receptor	Iron homeostasis
	FcRn	FcRn	Maternal/fetal interface	Yes	Low	None			Fc (IgG)	Shuttle maternal IgG to fetus (passive immunity)
	ZAG	ZAG	Bodily fluid	No	None	Fatty acid				Lipid homeostasis

Two other closely related mouse MHC class Ib genes, *T22* and *T10*, are expressed by activated lymphocytes and are recognized by a subset of γ:δ T cells. The exact function of the T22 and T10 proteins is unclear, but it has been proposed that this interaction allows the regulation of activated lymphocytes by γ:δ T cells.

The other genes that map within the MHC include some that encode complement components (for example, C2, C4, and factor B) and some that encode cytokines—for example, tumor necrosis factor-α (TNF-α) and lymphotoxin—all of which have important functions in immunity. These genes lie in the so-called 'MHC class III' region (see Fig. 6.15).

Many studies have established associations between susceptibility to certain diseases and particular alleles of MHC genes (see Chapter 15), and we now have considerable insight into how polymorphism in the classical MHC class I and class II genes can affect resistance or susceptibility. Most MHC-influenced traits or diseases are known or suspected to have an immunological cause, but this is not so for all of them: many genes lying within the MHC have no known or suspected immunological function. For example, the class Ib gene *M10* encodes a protein that is recognized by pheromone receptors in the vomeronasal organ. *M10* could thus influence mating preference, a trait that has been linked to the MHC region in rodents.

The gene for HLA-H, which has been renamed *HFE* (see Fig. 6.15), lies some 3 million base pairs from HLA-A. Its protein product is expressed on cells in the intestinal tract and has a function in iron metabolism, regulating the uptake of dietary iron into the body, most probably through interactions with the transferrin receptor that decrease its affinity for iron-loaded transferrin. Individuals defective for this gene have an iron-storage disease, hereditary hemochromatosis, in which an abnormally high level of iron is retained in the liver and other organs. Mice lacking β_2-microglobulin, and hence defective in the expression of all class I molecules, have a similar iron overload. Another MHC gene with a nonimmune function encodes the enzyme 21-hydroxylase, which, when deficient, causes congenital adrenal hyperplasia and, in severe cases, salt-wasting syndrome. Even where a disease-related gene is clearly homologous to immune-system genes, as is the case with *HFE*, the disease mechanism may not be immune related. Disease associations mapping to the MHC must therefore be interpreted with caution, in the light of a detailed understanding of its genetic structure and the functions of its individual genes. Much remains to be learned about the latter and about the significance of all the genetic variation localized within the MHC. For instance, in humans the complement component C4 comes in two versions, C4A and C4B, and different individuals have variable numbers of the gene for each type in their genomes, but the adaptive significance of this genetic variability is not well understood.

6-18 Specialized MHC class I molecules act as ligands for the activation and inhibition of NK cells.

Some MHC class Ib genes, for example the members of the **MIC** gene family, are under a different regulatory control from the classical MHC class I genes and are induced in response to cellular stress (such as heat shock). There are seven MIC genes, but only two—*MICA* and *MICB*—are expressed and produce protein products (see Fig. 6.24). They are expressed in fibroblasts and epithelial cells, particularly in intestinal epithelial cells, and have a role in innate immunity or in the induction of immune responses in circumstances in which interferons are not produced. The MICA and MICB proteins are recognized by the NKG2D receptor expressed by NK cells, γ:δ T cells, and some CD8 T cells, and can activate these cells to kill MIC-expressing targets. NKG2D is an 'activating' member of the NKG2 family of NK-cell receptors;

its cytoplasmic domain lacks the inhibitory sequence motif found in other members of this family, which act as inhibitory receptors (see Sections 3-21 and 3-22). NKG2D is coupled to the adaptor protein DAP10, which transmits the signal into the interior of the cell by interacting with and activating intracellular phosphatidylinositol 3-kinase.

Even more distantly related to MHC class I genes are a small family of proteins known in humans as the UL16-binding proteins (ULBPs) or the RAET1 proteins (see Fig. 6.24); the homologous proteins in mice are known as Rae1 (retinoic acid early inducible 1) and H60. These proteins also bind NKG2D, as described in Section 3-21. They seem to be expressed under conditions of cellular stress, such as when cells are infected with pathogens (UL16 is a human cytomegalovirus protein) or have undergone transformation to tumor cells. By expressing ULBPs, stressed or infected cells can activate NKG2D on NK cells, γ:δ T cells, and CD8 cytotoxic α:β T cells, and so be recognized and eliminated.

The human MHC class Ib molecule HLA-E and its mouse counterpart Qa-1 (see Fig. 6.24) have a specialized role in cell recognition by NK cells. HLA-E and Qa-1 bind a very restricted subset of nonpolymorphic peptides, called Qa-1 determinant modifiers (Qdm), that are derived from the leader peptides of other HLA class I molecules. These peptide:HLA-E complexes can bind to the receptor NKG2A, which is present on NK cells in a complex with the cell-surface molecule CD94 (see Section 3-21). NKG2A is an inhibitory member of the NKG2 family, and upon HLA-E engagement it inhibits the cytotoxic activity of the NK cell. Thus, a cell that expresses either HLA-E or Qa-1 is not killed by NK cells.

The classical class Ia MHC molecules HLA-A, -B and -C, which present a diverse repertoire of peptides to CD8 T cells, are also recognized by members of the killer cell immunoglobulin-like receptors (KIRs) expressed by NK cells (see Section 3-21). KIRs interact with the same face of the MHC class I molecule as do T-cell receptors, but KIRs bind over only one end, and not over the whole area recognized by the T-cell receptor. Like MHC molecules, KIRs themselves are highly polymorphic, and they have undergone rapid evolution in humans. Only a few alleles of HLA-A and HLA-B bind KIRs, but all HLA-C alleles bind KIRs, indicating a specialization of HLA-C for regulating NK cells in humans.

Two other MHC class Ib molecules, HLA-F and HLA-G (see Fig. 6.24), can also inhibit cell killing by NK cells. HLA-G is expressed on fetus-derived placental cells that migrate into the uterine wall. These cells express no classical MHC class I molecules and cannot be recognized by CD8 T cells but, unlike other cells lacking such proteins, they are not killed by NK cells. This seems to be because HLA-G is recognized by another inhibitory receptor on NK cells, the leukocyte immunoglobulin-like receptor subfamily B member 1 (LILRB1), also called ILT-2 or LIR-1, which prevents the NK cell from killing the placental cells. HLA-F is expressed in a variety of tissues, although it is usually not detected at the cell surface except, for example, on some monocyte cell lines or on virus-transformed lymphoid cells. HLA-F is also thought to interact with LILRB1.

6-19 The CD1 family of MHC class I-like molecules is encoded outside the MHC and presents microbial lipids to CD1-restricted T cells.

Some MHC class I-like genes map outside the MHC region. One small family of such genes is called **CD1** and is expressed on dendritic cells, monocytes, and some thymocytes. Humans have five CD1 genes, CD1a through e, whereas mice express only two highly homologous versions of CD1d, namely CD1d1 and CD1d2. CD1 proteins can present antigens to T cells, but they have two features that distinguish them from classical MHC class I molecules. The first is that CD1, although similar to an MHC class I molecule in its subunit organization and association with β_2-microglobulin, behaves like an MHC class II molecule. It is not retained within the endoplasmic reticulum by association with the TAP complex but is targeted to vesicles, where it binds its ligand. The second unusual feature is that, unlike MHC class I, CD1 molecules have a hydrophobic channel that is specialized for binding hydrocarbon alkyl chains. This confers on CD1 molecules an ability to bind and present a variety of glycolipids.

CD1 molecules are classified into group 1, comprising CD1a, CD1b, and CD1c, and group 2, containing only CD1d; CD1e is considered intermediate. Group 1 molecules bind microbially derived glycolipids, phospholipids, and lipopeptide antigens, such as the mycobacterial membrane components mycolic acid, glucose monomycolate, phosphoinositol mannosides, and lipoarabinomannan. Group 2 CD1 molecules are thought to bind mainly self lipid antigens such as sphingolipids and diacylglycerols. Structural studies show that the CD1 molecule has a deep binding groove into which the glycolipid antigens bind (Fig. 6.25). Unlike the binding of peptide to MHC, in which the peptide takes on a linear, extended conformation, Group 1 CD1 molecules bind their antigens by anchoring the alkyl chains in the hydrophobic groove, which orients the variable carbohydrate headgroups or other hydrophilic parts of these molecules protruding from the end of the binding groove, allowing recognition by the T-cell receptors on CD1-restricted T cells.

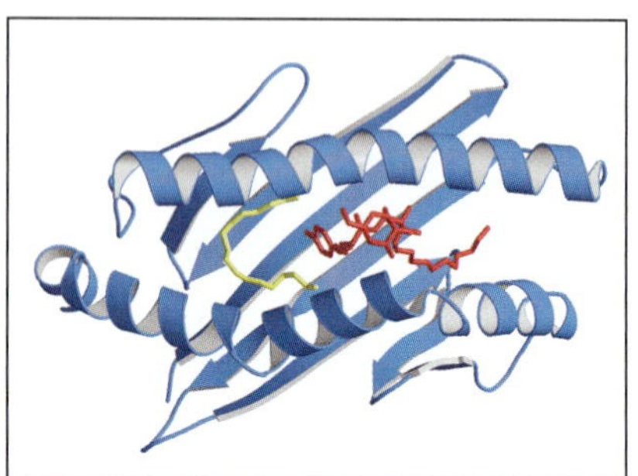

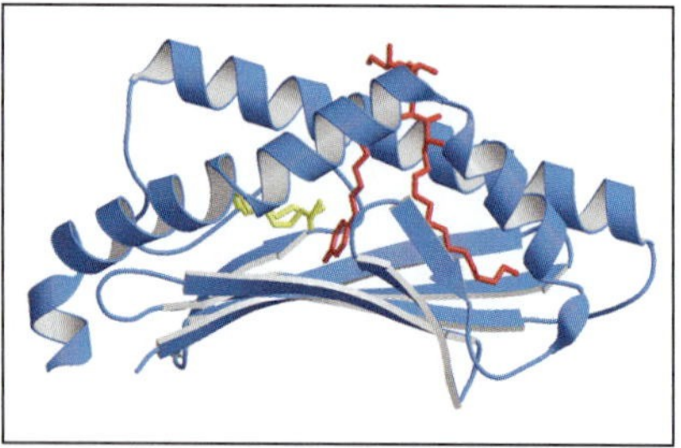

Fig. 6.25 Structure of CD1 binding to a lipid antigen. Shown are top and side views of the structure of mouse CD1d bound to C8PhF, a synthetic lipid that is an analog of α-GalCer. The helical side chains of CD1d (blue) form a binding pocket generally similar in shape to MHC class I and II molecules. However, the C8PhF (red) ligand binds to CD1 molecules in a distinctly different conformation from that of peptides. The two long alkyl side chains extend deep inside the binding groove, where they make contacts with hydrophobic residues. This orientation of the alkyl side chains places the carbohydrate component of α-GalCer to the outer surface of CD1, where it can be recognized by the T-cell receptor. In addition, the CD1 molecule contains an endogenous lipid molecule (yellow), derived from cellular sources, that binds to a distinct region within the groove and prevents a large pocket adjacent to the α-GalCer binding region from collapsing. The ability to incorporate additional ligands into the binding groove may provide flexibility to CD1d in accommodating a variety of exogenous glycosphingolipids from microorganisms. Courtesy of I.A. Wilson.

Whereas T cells that recognize antigen presented by MHC class I and class II molecules express CD8 and CD4, respectively, the T cells that recognize lipids presented by CD1 molecules express neither CD4 nor CD8. Most of the T cells recognizing lipids presented by group 1 CD1 molecules have a diverse repertoire of α:β receptors, and respond to these lipids presented by CD1a, CD1b, and CD1c. In contrast, CD1d-restricted T cells are less diverse, many using the same TCRα chain ($V_\alpha 24$–$J_\alpha 18$ in humans), but they also express NK-cell receptors. This CD1d-restricted population has been called **invariant NKT (iNKT) cells**.

One recognized ligand for CD1d molecules is α-galactoceramide (α-GalCer), which was isolated from an extract of marine sponge. When α-galactoceramide is bound to CD1d, it forms a structure that is recognized by many iNKT cells. It seems that iNKT cells have an important role in bridging the gap between innate immunity and adaptive immunity. Their ability to recognize different glycolipid constituents from microorganisms presented by CD1d molecules places them in an 'innate' category; their possession of a fully rearranged T cell receptor, despite its relatively limited repertoire, makes them 'adaptive.'

It seems that the CD1 proteins have evolved as a separate lineage of antigen-presenting molecules able to present microbial lipids and glycolipids to T cells. Just as peptides are loaded onto classical MHC proteins at various cellular locations, the various CD1 proteins are transported differently through the endoplasmic reticulum and endocytic compartments, providing access to different lipid antigens. Transport is regulated by an amino acid sequence motif at the terminus of the cytoplasmic domain of the CD1 protein, which controls interaction with adaptor-protein (AP) complexes. CD1a lacks this binding motif and moves to the cell surface, where it is transported only through the early endocytic compartment. CD1c and CD1d have motifs that interact with the adaptor AP-2 and are transported through early and late endosomes; CD1d is also targeted to lysosomes. CD1b and mouse CD1d bind AP-2 and AP-3 and can be transported through late endosomes, lysosomes, and the MIIC. CD1 proteins can thus bind lipids delivered into and processed within the endocytic pathway, such as by the internalization of mycobacteria or the ingestion of mycobacterial lipoarabinomannans mediated by mannose receptors.

From an evolutionary perspective it is interesting to note that some class Ib genes, notably CD1 and some with functions distinct from antigen presentation, seem to have evolved early, before the divergence of the cartilaginous fishes from the vertebrate line, and are likely to have homologs in all vertebrates. It seems that the other class I genes have independently evolved into classical and nonclassical loci within the vertebrate lineages that have been studied (for example, cartilaginous fishes, lobe-finned fishes, ray-finned fishes, amphibians, and mammals).

Summary.

The major histocompatibility complex (MHC) of genes consists of a linked set of genetic loci encoding many of the proteins involved in antigen presentation to T cells, most notably the MHC class I and class II glycoproteins (the MHC molecules) that present peptides to the T-cell receptor. The outstanding feature of the MHC molecules is their extensive polymorphism. This polymorphism is of critical importance in antigen recognition by T cells. A T cell recognizes antigen as a peptide bound by a particular allelic variant of an MHC molecule, and will not recognize the same peptide bound to other MHC molecules. This behavior of T cells is called MHC restriction. Most MHC alleles differ from one another by multiple amino acid substitutions, and these differences are focused on the peptide-binding site and surface-exposed regions that make direct contact with the T-cell receptor. At least three properties of MHC molecules are affected by MHC polymorphism: the range of peptides bound; the conformation of the bound peptide; and the direct interaction of the MHC molecule with the T-cell receptor. Thus, the highly polymorphic nature of the MHC has functional consequences, and the evolutionary selection for this polymorphism suggests that it is critical to the role of the MHC molecules in the immune response. Powerful genetic mechanisms generate the variation that is seen among MHC alleles, and a compelling argument can be made that selective pressure to maintain a wide variety of MHC molecules in the population comes from infectious agents. As a consequence, the immune system is highly individualized—each individual responds differently to a given antigen.

Within the MHC region there are also a large number of genes whose structure is closely related to the MHC class I molecules—the so-called nonclassical, or class Ib, MHC. Although some of these genes serve purposes that are unrelated to the immune system, many are involved in recognition by activating and inhibitory receptors expressed by NK cells, γ:δ T cells, and α:β T cells. MHC class Ib proteins called CD1 molecules are encoded outside the MHC region, and bind lipids and glycolipid antigens for presentation to T cells.

Summary to Chapter 6.

The T-cell antigen receptor recognizes antigen in the form of a peptide bound to an MHC molecule. During an infection, pathogen-derived peptides become bound to MHC molecules and are displayed on the cell surface, where they can be recognized by T cells specific for that combination. In the absence of infection, MHC molecules are occupied by self peptides but do not normally provoke a T-cell response because the immune system is normally tolerant of the body's own antigens. There are two classes of MHC molecules—MHC class I molecules, which bind stably to peptides derived from proteins synthesized and degraded in the cytosol, and MHC class II molecules, which bind stably to peptides derived from proteins degraded in endocytic vesicles. In addition to being bound by the T-cell receptor, the two classes of MHC molecules are differentially recognized by the two co-receptor molecules, CD8 and CD4, which characterize the two major subsets of T cells. CD8 T cells recognize peptide:MHC class I complexes and are activated to kill cells displaying foreign peptides derived from cytosolic pathogens, such as viruses. Exogenous antigens, such as those obtained during phagocytosis of viral antigens by dendritic cells, can be delivered from the vesicular system to the cytosol, a process known as cross-presentation, for loading and presentation by MHC class I molecules. This pathway is important in the initial activation of CD8 T cells by dendritic cells. CD4 T cells recognize peptide:MHC class II complexes and are specialized to activate other immune effector cells, for example B cells or macrophages, to act against the foreign antigens or pathogens that they have taken up.

There are several genes for each class of MHC molecule, arranged in clusters within a larger region known as the major histocompatibility complex (MHC). Within the MHC, the genes for the MHC molecules are closely linked to genes involved in the degradation of proteins into peptides, the formation of the complex of peptide and MHC molecule, and the transport of these complexes to the cell surface. Because the several different genes for the MHC class I and class II molecules are highly polymorphic and are expressed in a codominant fashion, each individual expresses a number of different MHC class I and class II molecules. Each different MHC molecule can bind stably to a range of different peptides, and thus the MHC repertoire of each individual can recognize and bind many different peptide antigens. Because the T-cell receptor binds a combined peptide:MHC ligand, T cells show MHC-restricted antigen recognition, such that a given T cell is specific for a particular peptide bound to a particular MHC molecule.

Questions.

6.1 *MHC class I and class II molecules are structurally and functionally homologous, yet have dissimilar pathways of assembly and delivery to the cell surface. (a) Describe how these differences in assembly and delivery are integrated with the different functions of class I and class II molecules. (b) How do these functions relate to the source from which class I or class II MHC receive peptides? (c) Given that the processes of cross-presentation and autophagy can redirect antigens from various sources for processing by alternative pathways, how do these processes alter your answer to (b)?*

6.2 *Viral pathogens have acquired diverse mechanisms to evade the immune response. (a) Describe the steps at which viruses can prevent recognition of viral antigens by CD8 T cells, and provide a specific example for each. (b) Of the examples of viral evasion presented in this chapter, most were concerned with antigens presented by class I MHC. Why might there be more examples of viral inhibition of antigen presentation by class I MHC than by class II MHC? (c) Suggest a reason that large DNA viruses might use these mechanisms more than small RNA viruses.*

6.3 *Many of the proteins encoded within the MHC exist in the population in multiple forms, or allelic variants. (a) What genetic events give rise to this variation and what are its functional consequences? (b) In some cases, particular combinations of alleles of the different MHC genes are found to be present at a far higher frequency than chance would predict. What are the possible mechanisms that might explain this finding?*

6.4 *Many genes outside the MHC region encode proteins that are structurally and functionally related to class I MHC proteins. (a) Discuss the cell types that recognize various 'nonclassical' MHC proteins and what their functions are. (b) Discuss the kinds of ligand(s), if any, that are presented by various members of these proteins.*

General references.

Bodmer, J.G., Marsh, S.G.E., Albert, E.D., Bodmer, W.F., DuPont, B., Erlich, H.A., Mach, B., Mayr, W.R., Parham, P., Saszuki, T., *et al.*: **Nomenclature for factors of the HLA system, 1991.** *Tissue Antigens* 2000, **56**:289–290.

Germain, R.N.: **MHC-dependent antigen processing and peptide presentation: providing ligands for T lymphocyte activation.** *Cell* 1994, **76**:287–299.

Klein, J.: *Natural History of the Major Histocompatibility Complex.* New York, J. Wiley & Sons, 1986.

Moller, G. (ed.): **Origin of major histocompatibility complex diversity.** *Immunol. Rev.* 1995, **143**:5–292.

Section references.

6-1 The MHC class I and class II molecules deliver peptides to the cell surface from two intracellular compartments.

Brocke, P., Garbi, N., Momburg, F., and Hammerling, G.J.: **HLA-DM, HLA-DO and tapasin: functional similarities and differences.** *Curr. Opin. Immunol.* 2002, **14**:22–29.

Gromme, M., and Neefjes, J.: **Antigen degradation or presentation by MHC class I molecules via classical and non-classical pathways.** *Mol. Immunol.* 2002, **39**:181–202.

Villadangos, J.A.: **Presentation of antigens by MHC class II molecules: getting the most out of them.** *Mol. Immunol.* 2001, **38**:329–346.

Williams, A., Peh, C.A., and Elliott, T.: **The cell biology of MHC class I antigen presentation.** *Tissue Antigens* 2002, **59**:3–17.

6-2 Peptides that bind to MHC class I molecules are actively transported from the cytosol to the endoplasmic reticulum.

Gorbulev, S., Abele, R., and Tampe, R.: **Allosteric crosstalk between peptide-binding, transport, and ATP hydrolysis of the ABC transporter TAP.** *Proc. Natl Acad. Sci. USA* 2001, **98**:3732–3737.

Lankat-Buttgereit, B., and Tampe, R.: **The transporter associated with antigen processing: function and implications in human diseases.** *Physiol. Rev.* 2002, **82**:187–204.

Townsend, A., Ohlen, C., Foster, L., Bastin, J., Lunggren, H.G., and Karre, K.: **A mutant cell in which association of class I heavy and light chains is induced by viral peptides.** *Cold Spring Harbor Symp. Quant. Biol.* 1989, **54**:299–308.

Uebel, S., and Tampe, R.: **Specificity of the proteasome and the TAP transporter.** *Curr. Opin. Immunol.* 1999, **11**:203–208.

6-3 Peptides for transport into the endoplasmic reticulum are generated in the cytosol.

Cascio, P., Call, M., Petre, B.M., Walz, T., and Goldberg, A.L.: **Properties of the hybrid form of the 26S proteasome containing both 19S and PA28 complexes.** *EMBO J.* 2002, **21**:2636–2645.

Goldberg, A.L., Cascio, P., Saric, T., and Rock, K.L.: **The importance of the proteasome and subsequent proteolytic steps in the generation of antigenic peptides.** *Mol. Immunol.* 2002, **39**:147–164.

Hammer, G.E., Gonzalez, F., Champsaur, M., Cado, D., and Shastri, N.: **The aminopeptidase ERAAP shapes the peptide repertoire displayed by major histocompatibility complex class I molecules.** *Nat. Immunol.* 2006, **7**:103–112.

Hammer, G.E., Gonzalez, F., James, E., Nolla, H., and Shastri, N.: **In the absence of aminopeptidase ERAAP, MHC class I molecules present many unstable and highly immunogenic peptides.** *Nat. Immunol.* 2007, **8**:101–108.

Schubert, U., Anton, L.C., Gibbs, J., Norbury, C.C., Yewdell, J.W., and Bennink, J.R.: **Rapid degradation of a large fraction of newly synthesized proteins by proteasomes.** *Nature* 2000, **404**:770–774.

Serwold, T., Gonzalez, F., Kim, J., Jacob, R., and Shastri, N.: **ERAAP customizes peptides for MHC class I molecules in the endoplasmic reticulum.** *Nature* 2002, **419**:480–483.

Shastri, N., Schwab, S., and Serwold, T.: **Producing nature's gene-chips: the generation of peptides for display by MHC class I molecules.** *Annu. Rev. Immunol.* 2002, **20**:463–493.

Sijts, A., Sun, Y., Janek, K., Kral, S., Paschen, A., Schadendorf, D., and Kloetzel, P.M.: **The role of the proteasome activator PA28 in MHC class I antigen processing.** *Mol. Immunol.* 2002, **39**:165–169.

Vigneron, N., Stroobant, V., Chapiro, J., Ooms, A., Degiovanni, G., Morel, S., van der Bruggen, P., Boon, T., and Van den Eynde, B.J.: **An antigenic peptide produced by peptide splicing in the proteasome.** *Science* 2004, **304**:587–590.

6-4 Newly synthesized MHC class I molecules are retained in the endoplasmic reticulum until they bind a peptide.

Bouvier, M.: **Accessory proteins and the assembly of human class I MHC molecules: a molecular and structural perspective.** *Mol. Immunol.* 2003, **39**:697–706.

Gao, B., Adhikari, R., Howarth, M., Nakamura, K., Gold, M.C., Hill, A.B., Knee, R., Michalak, M., and Elliott, T.: **Assembly and antigen-presenting function of MHC class I molecules in cells lacking the ER chaperone calreticulin.** *Immunity* 2002, **16**:99–109.

Grandea, A.G. III, and Van Kaer, L.: **Tapasin: an ER chaperone that controls MHC class I assembly with peptide.** *Trends Immunol.* 2001, **22**:194–199.

Van Kaer, L.: **Accessory proteins that control the assembly of MHC molecules with peptides.** *Immunol. Res.* 2001, **23**:205–214.

Williams, A., Peh, C.A., and Elliott, T.: **The cell biology of MHC class I antigen presentation.** *Tissue Antigens* 2002, **59**:3–17.

Williams, A.P., Peh, C.A., Purcell, A.W., McCluskey, J., and Elliott, T.: **Optimization of the MHC class I peptide cargo is dependent on tapasin.** *Immunity* 2002, **16**:509–520.

6-5 Many viruses produce immunoevasins that interfere with antigen presentation by MHC class I molecules.

Lilley, B.N., and Ploegh, H.L.: **A membrane protein required for dislocation of misfolded proteins from the ER.** *Nature* 2004, **429**:834–840.

Lilley, B.N., and Ploegh, H.L.: **Viral modulation of antigen presentation: manipulation of cellular targets in the ER and beyond.** *Immunol. Rev.* 2005, **207**:126–144.

Lybarger, L., Wang, X., Harris, M., and Hansen, T.H.: **Viral immune evasion molecules attack the ER peptide-loading complex and exploit ER-associated degradation pathways.** *Curr. Opin. Immunol.* 2005, **17**:79–87.

Verweij, M.C., Koppers-Lalic, D., Loch, S., Klauschies, F., de la Salle, H., Quinten, E., Lehner, P.J., Mulder, A., Knittler, M.R., Tampé, R., *et al.*: **The varicellovirus UL49.5 protein blocks the transporter associated with antigen processing (TAP) by inhibiting essential conformational transitions in the 6+6 transmembrane TAP core complex.** *J. Immunol.* 2008, **181**:4894–4907.

6-6 Peptides presented by MHC class II molecules are generated in acidified endocytic vesicles.

Godkin, A.J., Smith, K.J., Willis, A., Tejada-Simon, M.V., Zhang, J., Elliott, T., and Hill, A.V.: **Naturally processed HLA class II peptides reveal highly conserved immunogenic flanking region sequence preferences that reflect antigen processing rather than peptide–MHC interactions.** *J. Immunol.* 2001, **166**:6720–6727.

Hiltbold, E.M., and Roche, P.A.: **Trafficking of MHC class II molecules in the late secretory pathway.** *Curr. Opin. Immunol.* 2002, **14**:30–35.

Hsieh, C.S., deRoos, P., Honey, K., Beers, C., and Rudensky, A.Y.: **A role for cathepsin L and cathepsin S in peptide generation for MHC class II presentation.** *J. Immunol.* 2002, **168**:2618–2625.

Lennon-Duménil, A.M., Bakker, A.H., Wolf-Bryant, P., Ploegh, H.L., and Lagaudrière-Gesbert, C.: **A closer look at proteolysis and MHC-class-II-restricted antigen presentation.** *Curr. Opin. Immunol.* 2002, **14**:15–21.

Maric, M., Arunachalam, B., Phan, U.T., Dong, C., Garrett, W.S., Cannon, K.S., Alfonso, C., Karlsson, L., Flavell, R.A., and Cresswell, P.: **Defective antigen processing in GILT-free mice.** *Science* 2001, **294**:1361–1365.

Pluger, E.B., Boes, M., Alfonso, C., Schroter, C.J., Kalbacher, H., Ploegh, H.L., and Driessen, C.: **Specific role for cathepsin S in the generation of antigenic peptides in vivo.** *Eur. J. Immunol.* 2002, **32**:467–476.

6-7 The invariant chain directs newly synthesized MHC class II molecules to acidified intracellular vesicles.

Gregers, T.F., Nordeng, T.W., Birkeland, H.C., Sandlie, I., and Bakke, O.: **The cytoplasmic tail of invariant chain modulates antigen processing and presentation.** *Eur. J. Immunol.* 2003, **33**:277–286.

Hiltbold, E.M., and Roche, P.A.: **Trafficking of MHC class II molecules in the late secretory pathway.** *Curr. Opin. Immunol.* 2002, **14**:30–35.

Kleijmeer, M., Ramm, G., Schuurhuis, D., Griffith, J., Rescigno, M., Ricciardi-Castagnoli, P., Rudensky, A.Y., Ossendorp, F., Melief, C.J., Stoorvogel, W., *et al.*: **Reorganization of multivesicular bodies regulates MHC class II antigen presentation by dendritic cells.** *J. Cell Biol.* 2001, **155**:53–63.

van Lith, M., van Ham, M., Griekspoor, A., Tjin, E., Verwoerd, D., Calafat, J., Janssen, H., Reits, E., Pastoors, L., and Neefjes, J.: **Regulation of MHC class II antigen presentation by sorting of recycling HLA-DM/DO and class II within the multivesicular body.** *J. Immunol.* 2001, **167**:884–892.

6-8 A specialized MHC class II-like molecule catalyzes loading of MHC class II molecules with peptides.

Pathak, S.S., Lich, J.D., and Blum, J.S.: **Cutting edge: editing of recycling class II:peptide complexes by HLA-DM.** *J. Immunol.* 2001, **167**:632–635.

Qi, L., and Ostrand-Rosenberg, S.: **H2-O inhibits presentation of bacterial superantigens, but not endogenous self antigens.** *J. Immunol.* 2001, **167**:1371–1378.

Zarutskie, J.A., Busch, R., Zavala-Ruiz, Z., Rushe, M., Mellins, E.D., and Stern, L.J.: **The kinetic basis of peptide exchange catalysis by HLA-DM.** *Proc. Natl Acad. Sci. USA* 2001, 98:12450–12455.

6-9 Cross-presentation allows exogenous proteins to be presented on MHC class I molecules by a restricted set of antigen-presenting cells.

Ackerman, A.L., and Cresswell, P.: **Cellular mechanisms governing cross-presentation of exogenous antigens.** *Nat. Immunol.* 2004, **5**:678–684.

Bevan, M.J.: **Minor H antigens introduced on H-2 different stimulating cells cross-react at the cytotoxic T cell level during in vivo priming.** *J. Immunol.* 1976, **117**:2233–2238.

Bevan, M.J.: **Helping the CD8$^+$ T cell response.** *Nat. Rev. Immunol.* 2004, **4**:595–602.

Li, P., Gregg, J.L., Wang, N., Zhou, D., O'Donnell, P., Blum, J.S., and Crotzer, V.L.: **Compartmentalization of class II antigen presentation: contribution of cytoplasmic and endosomal processing.** *Immunol. Rev.* 2005, **207**:206–217.

6-10 Stable binding of peptides by MHC molecules provides effective antigen presentation at the cell surface.

Apostolopoulos, V., McKenzie, I.F., and Wilson, I.A.: **Getting into the groove: unusual features of peptide binding to MHC class I molecules and implications in vaccine design.** *Front. Biosci.* 2001, **6**:D1311–D1320.

Buslepp, J., Zhao, R., Donnini, D., Loftus, D., Saad, M., Appella, E., and Collins, E.J.: **T cell activity correlates with oligomeric peptide-major histocompatibility complex binding on T cell surface.** *J. Biol. Chem.* 2001, **276**:47320–47328.

Hill, J.A., Wang, D., Jevnikar, A.M., Cairns, E., and Bell, D.A.: **The relationship between predicted peptide-MHC class II affinity and T-cell activation in a HLA-DRβ1*0401 transgenic mouse model.** *Arthritis Res. Ther.* 2003, **5**:R40–R48.

Su, R.C., and Miller, R.G.: **Stability of surface H-2K^b, H-2D^b, and peptide-receptive H-2K^b on splenocytes.** *J. Immunol.* 2001, **167**:4869–4877.

6-11 Many proteins involved in antigen processing and presentation are encoded by genes within the MHC.

Aguado, B., Bahram, S., Beck, S., Campbell, R.D., Forbes, S.A., Geraghty, D., Guillaudeux, T., Hood, L., Horton, R., Inoko, H., *et al.* (The MHC Sequencing Consortium): **Complete sequence and gene map of a human major histocompatibility complex.** *Nature* 1999, **401**:921–923.

Chang, C.H., Gourley, T.S., and Sisk, T.J.: **Function and regulation of class II transactivator in the immune system.** *Immunol. Res.* 2002, **25**:131–142.

Kumnovics, A., Takada, T., and Lindahl, K.F.: **Genomic organization of the mammalian MHC.** *Annu. Rev. Immunol.* 2003, **21**:629–657.

Lefranc, M.P.: **IMGT, the international ImMunoGeneTics database.** *Nucleic Acids Res.* 2003, **31**:307–310.

6-12 The protein products of MHC class I and class II genes are highly polymorphic.

Gaur, L.K., and Nepom, G.T.: **Ancestral major histocompatibility complex DRB genes beget conserved patterns of localized polymorphisms.** *Proc. Natl Acad. Sci. USA* 1996, **93**:5380–5383.

Marsh, S.G.: **Nomenclature for factors of the HLA system, update December 2002.** *Eur. J. Immunogenet.* 2003, **30**:167–169.

Robinson, J., and Marsh, S.G.: **HLA informatics. Accessing HLA sequences from sequence databases.** *Methods Mol. Biol.* 2003, **210**:3–21.

6-13 MHC polymorphism affects antigen recognition by T cells by influencing both peptide binding and the contacts between T-cell receptor and MHC molecule.

Falk, K., Rotzschke, O., Stevanovic, S., Jung, G., and Rammensee, H.G.: **Allele-specific motifs revealed by sequencing of self-peptides eluted from MHC molecules.** *Nature* 1991, **351**:290–296.

Garcia, K.C., Degano, M., Speir, J.A., and Wilson, I.A.: **Emerging principles for T cell receptor recognition of antigen in cellular immunity.** *Rev. Immunogenet.* 1999, **1**:75–90.

Katz, D.H., Hamaoka, T., Dorf, M.E., Maurer, P.H., and Benacerraf, B.: **Cell interactions between histoincompatible T and B lymphocytes. IV. Involvement of immune response (Ir) gene control of lymphocyte interaction controlled by the gene.** *J. Exp. Med.* 1973, **138**:734–739.

Kjer-Nielsen, L., Clements, C.S., Brooks, A.G., Purcell, A.W., Fontes, M.R., McCluskey, J., and Rossjohn, J.: **The structure of HLA-B8 complexed to an immunodominant viral determinant: peptide-induced conformational changes and a mode of MHC class I dimerization.** *J. Immunol.* 2002, **169**:5153–5160.

Wang, J.H., and Reinherz, E.L.: **Structural basis of T cell recognition of peptides bound to MHC molecules.** *Mol. Immunol.* 2002, **38**:1039–1049.

Zinkernagel, R.M., and Doherty, P.C.: **Restriction of *in vivo* T-cell mediated cytotoxicity in lymphocytic choriomeningitis within a syngeneic or semiallogeneic system.** *Nature* 1974, **248**:701–702.

6-14 Alloreactive T cells recognizing nonself MHC molecules are very abundant.

Felix, N.J., and Allen, P.M.: **Specificity of T-cell alloreactivity.** *Nat. Rev. Immunol.* 2007, **7**:942–953.

Feng, D., Bond, C.J., Ely, L.K., Maynard, J., and Garcia, K.C.: **Structural evidence for a germline-encoded T cell receptor–major histocompatibility complex interaction 'codon.'** *Nat. Immunol.* 2007, **8**:975–993.

Hennecke, J., and Wiley, D.C.: **Structure of a complex of the human α/β T cell receptor (TCR) HA1.7, influenza hemagglutinin peptide, and major histocompatibility complex class II molecule, HLA-DR4 (DRA*0101 and DRB1*0401): insight into TCR cross-restriction and alloreactivity.** *J. Exp. Med.* 2002, **195**:571–581.

Jankovic, V., Remus, K., Molano, A., and Nikolich-Zugich, J.: **T cell recognition of an engineered MHC class I molecule: implications for peptide-independent alloreactivity.** *J. Immunol.* 2002, **169**:1887–1892.

Nesic, D., Maric, M., Santori, F.R., and Vukmanovic, S.: **Factors influencing the patterns of T lymphocyte allorecognition.** *Transplantation* 2002, **73**:797–803.

Reiser, J.B., Darnault, C., Guimezanes, A., Gregoire, C., Mosser, T., Schmitt-Verhulst, A.M., Fontecilla-Camps, J.C., Malissen, B., Housset, D., and Mazza, G.: **Crystal structure of a T cell receptor bound to an allogeneic MHC molecule.** *Nat. Immunol.* 2000, **1**:291–297.

Speir, J.A., Garcia, K.C., Brunmark, A., Degano, M., Peterson, P.A., Teyton, L., and Wilson, I.A.: **Structural basis of 2C TCR allorecognition of H-2Ld peptide complexes.** *Immunity* 1998, **8**:553–562.

6-15 Many T cells respond to superantigens.

Acha-Orbea, H., Finke, D., Attinger, A., Schmid, S., Wehrli, N., Vacheron, S., Xenarios, I., Scarpellino, L., Toellner, K.M., MacLennan, I.C., *et al.*: **Interplays between mouse mammary tumor virus and the cellular and humoral immune response.** *Immunol. Rev.* 1999, **168**:287–303.

Kappler, J.W., Staerz, U., White, J., and Marrack, P.: **T cell receptor Vb elements which recognize Mls-modified products of the major histocompatibility complex.** *Nature* 1988, **332**:35–40.

Rammensee, H.G., Kroschewski, R., and Frangoulis, B.: **Clonal anergy induced in mature Vβ6⁺ T lymphocytes on immunizing Mls-1b mice with Mls-1a expressing cells.** *Nature* 1989, **339**:541–544.

Sundberg, E.J., Li, H., Llera, A.S., McCormick, J.K., Tormo, J., Schlievert, P.M., Karjalainen, K., and Mariuzza, R.A.: **Structures of two streptococcal superantigens bound to TCR β chains reveal diversity in the architecture of T cell signaling complexes.** *Structure* 2002, **10**:687–699.

Torres, B.A., Perrin, G.Q., Mujtaba, M.G., Subramaniam, P.S., Anderson, A.K., and Johnson, H.M.: **Superantigen enhancement of specific immunity: antibody production and signaling pathways.** *J. Immunol.* 2002, **169**:2907–2914.

White, J., Herman, A., Pullen, A.M., Kubo, R., Kappler, J.W., and Marrack, P.: **The V_β-specific super antigen staphylococcal enterotoxin B: stimulation of mature T cells and clonal deletion in neonatal mice.** *Cell* 1989, **56**:27–35.

6-16 MHC polymorphism extends the range of antigens to which the immune system can respond.

Hill, A.V., Elvin, J., Willis, A.C., Aidoo, M., Allsopp, C.E.M., Gotch, F.M., Gao, X.M., Takiguchi, M., Greenwood, B.M., Townsend, A.R.M., *et al.*: **Molecular analysis of the association of B53 and resistance to severe malaria.** *Nature* 1992, **360**:435–440.

Martin, M.P., and Carrington, M.: **Immunogenetics of viral infections.** *Curr. Opin. Immunol.* 2005, **17**:510–516.

Messaoudi, I., Guevara Patino, J.A., Dyall, R., LeMaoult, J., and Nikolich-Zugich, J.: **Direct link between *mhc* polymorphism, T cell avidity, and diversity in immune defense.** *Science* 2002, **298**:1797–1800.

Potts, W.K., and Slev, P.R.: **Pathogen-based models favouring MHC genetic diversity.** *Immunol. Rev.* 1995, **143**:181–197.

6-17 A variety of genes with specialized functions in immunity are also encoded in the MHC.

Alfonso, C., and Karlsson, L.: **Nonclassical MHC class II molecules.** *Annu. Rev. Immunol.* 2000, **18**:113–142.

Allan, D.S., Lepin, E.J., Braud, V.M., O'Callaghan, C.A., and McMichael, A.J.: **Tetrameric complexes of HLA-E, HLA-F, and HLA-G.** *J. Immunol. Methods* 2002, **268**:43–50.

Gao, G.F., Willcox, B.E., Wyer, J.R., Boulter, J.M., O'Callaghan, C.A., Maenaka, K., Stuart, D.I., Jones, E.Y., Van Der Merwe, P.A., Bell, J.I., *et al.*: **Classical and nonclassical class I major histocompatibility complex molecules exhibit subtle conformational differences that affect binding to CD8αα.** *J. Biol. Chem.* 2000, **275**:15232–15238.

Powell, L.W., Subramaniam, V.N., and Yapp, T.R.: **Haemochromatosis in the new millennium.** *J. Hepatol.* 2000, **32**:48–62.

6-18 Specialized MHC class I molecules act as ligands for the activation and inhibition of NK cells.

Borrego, F., Kabat, J., Kim, D.K., Lieto, L., Maasho, K., Pena, J., Solana, R., and Coligan, J.E.: **Structure and function of major histocompatibility complex (MHC) class I specific receptors expressed on human natural killer (NK) cells.** *Mol. Immunol.* 2002, **38**:637–660.

Boyington, J.C., Riaz, A.N., Patamawenu, A., Coligan, J.E., Brooks, A.G., and Sun, P.D.: **Structure of CD94 reveals a novel C-type lectin fold: implications for the NK cell-associated CD94/NKG2 receptors.** *Immunity* 1999, **10**:75–82.

Braud, V.M., and McMichael, A.J.: **Regulation of NK cell functions through interaction of the CD94/NKG2 receptors with the nonclassical class I molecule HLA-E.** *Curr. Top. Microbiol. Immunol.* 1999, **244**:85–95.

Lanier, L.L.: **NK cell recognition.** *Annu. Rev. Immunol.* 2005, **23**:225–274.

Lopez-Botet, M., and Bellon, T.: **Natural killer cell activation and inhibition by receptors for MHC class I.** *Curr. Opin. Immunol.* 1999, **11**:301–307.

Lopez-Botet, M., Bellon, T., Llano, M., Navarro, F., Garcia, P., and de Miguel, M.: **Paired inhibitory and triggering NK cell receptors for HLA class I molecules.** *Hum. Immunol.* 2000, **61**:7–17.

Lopez-Botet, M., Llano, M., Navarro, F., and Bellon, T.: **NK cell recognition of non-classical HLA class I molecules.** *Semin. Immunol.* 2000, 12:109–119.

Rodgers, J.R., and Cook, R.G.: **MHC class Ib molecules bridge innate and acquired immunity.** *Nat. Rev. Immunol.* 2005, **5**:459–471.

6-19 The CD1 family of MHC class I-like molecules is encoded outside the MHC and presents microbial lipids to CD1-restricted T cells.

Gadola, S.D., Zaccai, N.R., Harlos, K., Shepherd, D., Castro-Palomino, J.C., Ritter, G., Schmidt, R.R., Jones, E.Y., and Cerundolo, V.: **Structure of human CD1b with bound ligands at 2.3 Å, a maze for alkyl chains.** *Nat. Immunol.* 2002, **3**:721–726.

Gendzekhadze, K., Norman, P.J., Abi-Rached, L., Graef, T., Moesta, A.K., Layrisse, Z., and Parham, P.: **Co-evolution of KIR2DL3 with HLA-C in a human population retaining minimal essential diversity of KIR and HLA class I ligands.** *Proc. Natl Acad. Sci. USA* 2009, **106**:18692–18697.

Godfrey, D.I., Stankovic, S., and Baxter, A.G.: **Raising the NKT cell family.** *Nat. Immunol.* 2010, **11**:197–206.

Hava, D.L., Brigl, M., van den Elzen, P., Zajonc, D.M., Wilson, I.A., and Brenner, M.B.: **CD1 assembly and the formation of CD1-antigen complexes.** *Curr. Opin. Immunol.* 2005, **17**:88–94.

Jayawardena-Wolf, J., and Bendelac, A.: **CD1 and lipid antigens: intracellular pathways for antigen presentation.** *Curr. Opin. Immunol.* 2001, **13**:109–113.

Moody, D.B., and Besra, G.S.: **Glycolipid targets of CD1-mediated T-cell responses.** *Immunology* 2001, **104**:243–251.

Moody, D.B., and Porcelli, S.A.: **CD1 trafficking: invariant chain gives a new twist to the tale.** *Immunity* 2001, **15**:861–865.

Moody, D.B., and Porcelli, S.A.: **Intracellular pathways of CD1 antigen presentation.** *Nat. Rev. Immunol.* 2003, **3**:11–22.

Schiefner, A., Fujio, M., Wu, D., Wong, C.H., and Wilson, I.A.: **Structural evaluation of potent NKT cell agonists: implications for design of novel stimulatory ligands.** *J. Mol. Biol.* 2009, **394**:71–82.

PART III THE DEVELOPMENT OF MATURE LYMPHOCYTE RECEPTOR REPERTOIRES

Signaling Through Immune-System Receptors

7

Cells of the immune system use a variety of cell-surface receptors to sense their environment and to communicate with other cells. Among these, the lymphocyte antigen receptors have historically been the most studied, and they are the focus of this chapter. The operation of many other receptors on lymphocytes and other immune cells are now well understood, and some of these are included here. We have already described the signaling pathways used by G-protein-coupled receptors and Toll-like receptors (TLRs) involved in innate immunity in Chapter 3. Cell-surface receptors convey information received from the extracellular environment across the plasma membrane to generate intracellular biochemical events, which are transmitted along **intracellular signaling pathways** composed of proteins that interact with each other in a variety of ways. The signals are converted into different biochemical forms—the process known as **signal transduction**—distributed to different sites in the cell, and sustained and regulated as they proceed toward their destinations.

The final destinations of most of the signaling pathways we shall consider in this chapter are the nucleus and the cytoskeleton. Signals reaching the nucleus alter gene expression, leading to the synthesis of new proteins such as cytokines, chemokines, and cell-adhesion molecules. Signals reaching the nucleus can also induce cell division and differentiation, thus expanding lymphocyte populations during an immune response, or they may induce cell death after the immune response has occurred. Other signals can affect the cytoskeleton to alter cell shape, size, and motility. Such changes can occur without new protein synthesis and allow immediate effector activities in differentiated cells.

We begin by discussing some general principles of intracellular signaling, and then outline the pathways activated when a naive lymphocyte encounters its specific antigen. Next, we briefly discuss the co-stimulatory signaling that is necessary to activate naive T cells and, in most cases, naive B cells. In the last part of the chapter we look at a selection of other signaling pathways used by immune-system cells, including those from some cytokine receptors and from the 'death receptors' that stimulate apoptosis.

General principles of signal transduction and propagation.

In this part of the chapter we review briefly some general principles of receptor action and signal transduction that are common to many of the pathways discussed here. We start with the cell-surface receptors through which cells receive extracellular signals.

7-1 Transmembrane receptors convert extracellular signals into intracellular biochemical events.

All cell-surface receptors that have a signaling function are either transmembrane proteins themselves or form parts of protein complexes that link the exterior and interior of the cell. Different classes of receptors transduce extracellular signals in a variety of ways: a common theme among the receptors covered in this chapter is that ligand binding results in the activation of an intracellular enzymatic activity. The enzymes most commonly associated with receptor activation are the **protein kinases**. This large group of enzymes catalyzes the covalent attachment of a phosphate group to a protein in the reversible process known as **protein phosphorylation**. For receptors that use protein kinases, the binding of ligand to the extracellular part of the receptor allows the receptor-associated protein kinase to become 'active'—that is, to phosphorylate its intracellular substrate—and thus to propagate the signal. As we shall see, receptor-associated kinases can become activated in various ways, such as by undergoing modifications to the kinase itself that alter its intrinsic catalytic efficiency, or by changes in subcellular localization that increase access to its biochemical substrates.

In animals, protein kinases phosphorylate proteins on three amino acids—tyrosine, serine, or threonine. Most of the enzyme-linked receptors we discuss in detail in this chapter activate **tyrosine protein kinases**. Tyrosine kinases are specific for tyrosine residues, whereas serine/threonine kinases phosphorylate serine and threonine residues. In general cell biology, protein tyrosine phosphorylation is a much rarer modification than serine/threonine phosphorylation, and is found mainly in signaling pathways. One large group of receptors—the so-called **receptor tyrosine kinases**—carry a kinase activity within the cytoplasmic region of the receptor itself (Fig. 7.1, top panel). This group contains receptors for many growth factors; lymphocyte receptors of this type include Kit and FLT3, which are expressed on developing lymphocytes and are discussed in Chapter 8. The receptor for transforming growth factor-β (TGF-β), an important regulatory cytokine produced by many cells, is a **receptor serine/threonine kinase**.

Even more important to the function of mature lymphocytes are receptors that have no intrinsic enzymatic activity themselves but associate with intracellular tyrosine kinases. The antigen receptors on B lymphocytes and T lymphocytes are of this type, as are the receptors for some types of cytokines. Ligand binding to the extracellular domain of such receptors causes particular amino acid residues in their cytoplasmic domains to become phosphorylated by specific cytoplasmic tyrosine kinases (Fig. 7.1, bottom panel). These **nonreceptor kinases** can either be constitutively associated with the cytoplasmic domains of the receptors, as with many cytokine receptors, or they may become associated with the receptor when it binds a ligand, as is the case for the antigen receptors.

For many cytokine receptors, ligand binding causes dimerization or clustering of individual receptor molecules, bringing the associated kinases together and enabling them to phosphorylate the cytoplasmic tail of adjacent receptors—thus initiating an intracellular signal. In the case of the lymphocyte antigen receptors, association with cytoplasmic tyrosine kinases occurs after ligand binding but is unlikely to be due to a simple clustering mechanism. Instead, the actions of co-receptors are required: these bring cytoplasmic tyrosine kinases into proximity with the cytoplasmic regions of the antigen receptor, a complex process that we will describe later.

Signaling is usually not a simple 'on or off' switch, but frequently involves quantitative aspects of thresholds, amplitude, duration, and regulation, which are influenced by affinity and by the spatial and temporal abundance of

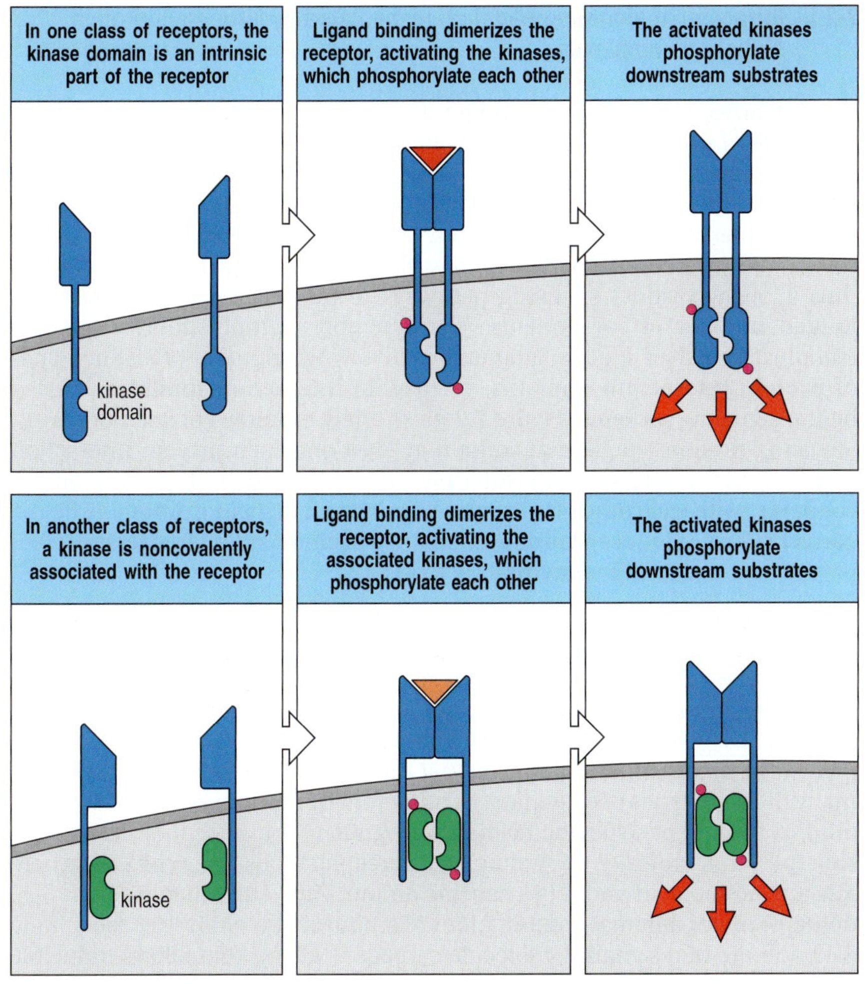

In one class of receptors, the kinase domain is an intrinsic part of the receptor
Ligand binding dimerizes the receptor, activating the kinases, which phosphorylate each other
The activated kinases phosphorylate downstream substrates
kinase domain
In another class of receptors, a kinase is noncovalently associated with the receptor
Ligand binding dimerizes the receptor, activating the associated kinases, which phosphorylate each other
The activated kinases phosphorylate downstream substrates
kinase

Fig. 7.1 Enzyme-associated receptors of the immune system can use intrinsic or associated protein kinases to signal. These receptors convey information that a ligand has bound to its extracellular portion by activating a protein kinase on the cytoplasmic side of the membrane. Receptor tyrosine kinases (top panels) contain the kinase activity as part of the receptor itself. Ligand binding results in clustering of the receptor, activation of catalytic activity, and the consequent phosphorylation of the receptor tails and other substrates, transmitting the signal onward. Receptors that lack intrinsic kinase activity associate with nonreceptor kinases (bottom panels). Receptor dimerization or clustering after ligand binding activates the associated enzyme. In all receptors of these types encountered in this chapter, the enzyme is a tyrosine kinase.

ligand. These features are often merged into the simple term 'signal strength,' but it is important to keep them in mind when considering the impact of receptor signaling on gene expression and cell function.

The role of protein kinases in cell signaling is not confined to receptor activation, and they act at many different stages in intracellular signaling pathways. Protein kinases figure largely in cell signaling because phosphorylation and **dephosphorylation**—the removal of a phosphate group—are the means of regulating the activity of many enzymes, transcription factors, and other proteins. Equally important to the workings of signaling pathways is the fact that phosphorylation generates sites on proteins to which other signaling proteins can bind.

Phosphate groups are removed from proteins by a large class of enzymes called **protein phosphatases**. Different classes of protein phosphatases remove phosphate groups from phosphotyrosine or from phosphoserine/phosphothreonine. Specific dephosphorylation by phosphatases is one important means of regulating signaling pathways by resetting a protein to its original state and thus switching signaling off. Dephosphorylation does not always inhibit a protein's activity. In many instances the removal of a particular phosphate group by a specific phosphatase is needed to activate an enzyme. In other cases, the level of phosphorylation on an enzyme determines its activity, and represents a balance between the activity of kinases and phosphatases.

7-2 Intracellular signal propagation is mediated by large multiprotein signaling complexes.

As we learned in Chapter 3, binding of a ligand to its receptor can initiate a cascade of events involving intracellular proteins that sequentially convey signaling information onward. The unique enzymes and other components assembled into a particular multiprotein receptor complex will determine the character of the signal it generates. These components may be shared by several receptor pathways, or they may be exclusive to one receptor pathway, thus allowing distinct signaling pathways to be built up from a relatively limited number of components. The assembly of multisubunit signaling complexes involves specific interactions involving a number of distinct types of **protein-interaction domains**, or **protein-interaction modules**, carried by the signaling proteins. Figure 7.2 gives a few examples of such domains. Signaling proteins in general contain at least one such protein-interaction domain, but many contain multiple domains. These protein modules cooperate with each other, for example, to organize signaling proteins into the correct subcellular localizations, to enable specific binding between protein partners, and to modify enzymatic activity.

For the pathways considered in this chapter, the most important mechanism underlying the formation of signaling complexes is the phosphorylation of protein tyrosine residues. Phosphotyrosines are binding sites for a number of protein-interaction domains, including the **SH2 (Src homology 2) domain** (see Fig. 7.2). This module, built from approximately 100 amino acids, is present in many intracellular signaling proteins, where it is associated with many different types of enzymatic or other functional domains. SH2 domains bind in a sequence-specific fashion, recognizing the phosphorylated tyrosine (pY) and, typically, the amino acid three positions away (pYXXZ, where X is any amino acid and Z is a specific amino acid), such that different SH2 domains prefer different combinations of amino acids. In this way, the unique SH2 domain of a signaling molecule can act as a 'key' that allows inducible and specific association with a particular protein containing the appropriate pY-containing amino acid sequence.

Tyrosine kinase-associated receptors can assemble multiprotein signaling complexes by using proteins called **scaffolds** and **adaptors**. Scaffolds and

Fig. 7.2 Signaling proteins interact with each other and with lipid signaling molecules via modular protein domains. A few of the most common protein domains used by immune-system signaling proteins are listed, together with some proteins containing the domain that are mentioned in this chapter or elsewhere in the book, and the general class of ligand bound by the interaction domain. The right-hand column lists specific examples of a protein motif bound (in single-letter amino acid code) or, for the phosphoinositide-binding domains, the particular phosphoinositide they bind. All these domains are used in many other nonimmune signaling pathways as well. C termini, carboxy termini.

Protein domain	Found in	Ligand class	Example of ligand
SH2	Lck, ZAP-70, Fyn, Src, Grb2, PLC-γ, STAT, Cbl, Btk, Itk, SHIP, Vav, SAP, PI3K	phosphotyrosine	pYXXZ
SH3	Lck, Fyn, Src, Grb2, Btk, Itk, Tec, Fyb, Nck, GADS	proline	PXXP
PH	Tec, PLC-γ, Akt, Btk, Itk, SOS	phosphoinositides	PIP_3
PX	$P40^{phox}$, $P47^{phox}$, PLD	phosphoinositides	PIP_2
PDZ	CARMA1	C termini of proteins	IESDV, VETDV
C1	RasGRP, PKC-θ	membrane lipid	diacylglycerol (DAG) phorbol ester
NZF	TAB2	polyubiquitin (K63-linked)	polyubiquitinated RIP, TRAF-6, or NEMO

adaptors lack enzymatic activity, and they function by recruiting other proteins into a signaling complex so that these proteins can interact with each other. Scaffolds are relatively large proteins that can, for example, become tyrosine phosphorylated on multiple sites and so can recruit many different proteins (Fig. 7.3, top panel). By specifying which proteins are recruited, scaffolds can define the character of a particular signaling response. This function of tyrosine phosphorylation in generating binding sites may explain why it is so commonly used in signaling pathways.

Adaptors are smaller proteins with usually no more than two or three signaling modules that serve to link two other proteins together. The adaptor proteins Grb2 and Gads, for example, each contain an SH2 domain and two copies of another module called the SH3 domain (see Fig. 7.2). This arrangement of modules can be used to link tyrosine phosphorylation of a receptor to molecules acting in the next stage of signaling. For example, the SH2 domain of Grb2 binds to a phosphotyrosine residue on a receptor (or another scaffold), while its two SH3 domains bind to proline-rich motifs on other signaling proteins (Fig. 7.3, bottom panel), such as Sos, which we discuss in the next section.

7-3 Small G proteins act as molecular switches in many different signaling pathways.

Monomeric GTP-binding proteins known as **small G proteins** or **small GTPases** are important in the signaling pathways leading from many tyrosine kinase-associated receptors. The small GTPases are distinct from the larger heterotrimeric G proteins associated with G-protein-coupled receptors such

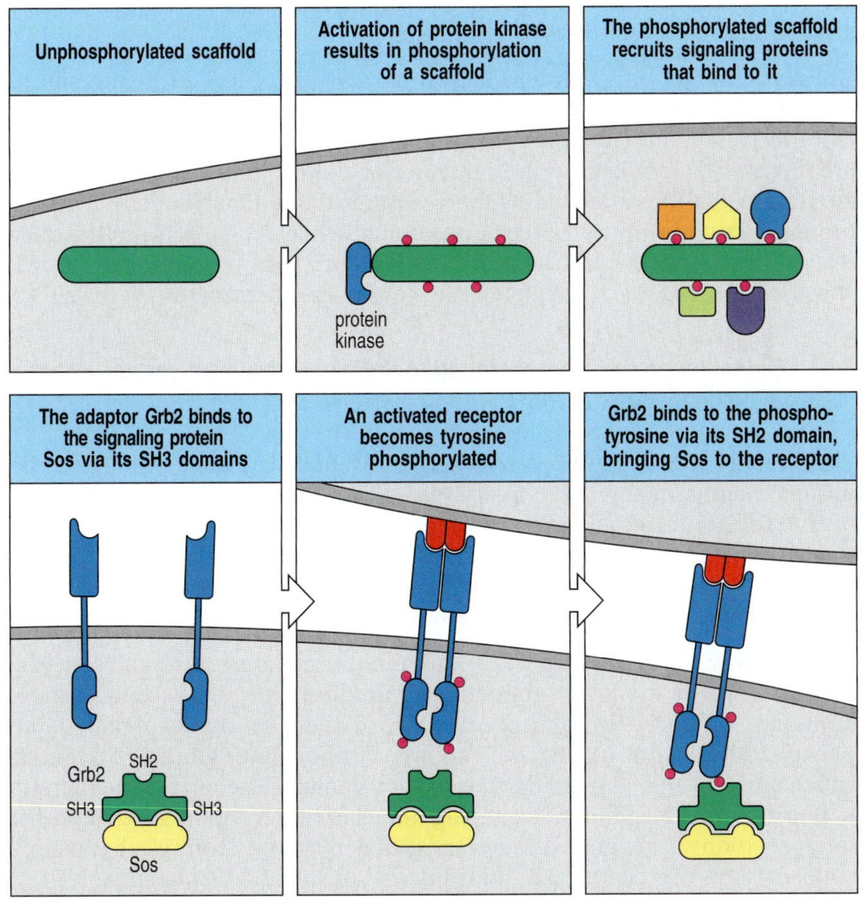

Fig. 7.3 Assembly of signaling complexes is mediated by scaffold and adaptor proteins. Assembly of signaling complexes is an important aspect of signal transduction. This is often achieved through scaffold and adaptor proteins. Scaffolds function to bring many different signaling proteins together (top panel). They generally have numerous sites of tyrosine phosphorylation that, after phosphorylation, can recruit many different proteins that contain, for example, SH2 domains. These proteins will determine the character of the signaling response. An adaptor protein functions to bring two different proteins together (bottom panel). The adaptor protein (green) shown here contains two SH3 domains and an SH2 domain. With the SH3 domains it can, for example, bind proline-rich sites on an intracellular signaling molecule (orange). Activation and tyrosine phosphorylation of a receptor generates a binding site for the SH2 domain of the adaptor, resulting in the recruitment of the signaling molecule to the activated receptor.

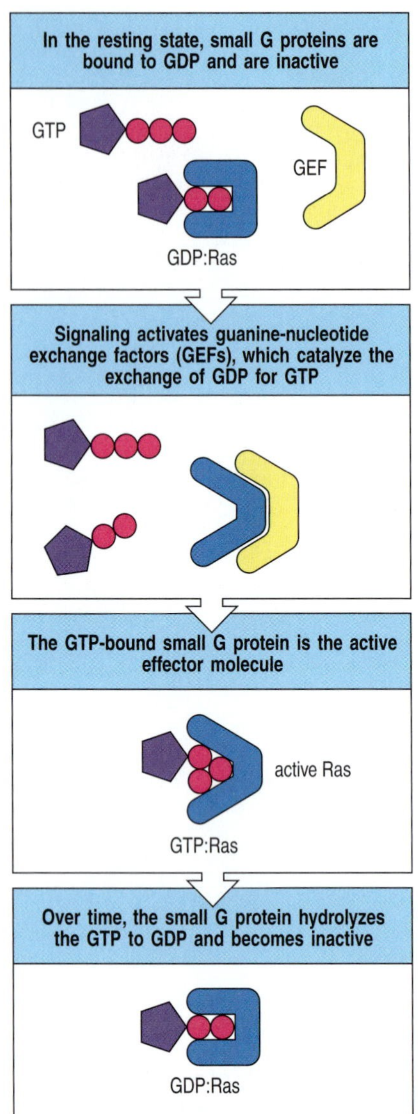
In the resting state, small G proteins are bound to GDP and are inactive
GTP
GEF
GDP:Ras
Signaling activates guanine-nucleotide exchange factors (GEFs), which catalyze the exchange of GDP for GTP
The GTP-bound small G protein is the active effector molecule
active Ras
GTP:Ras
Over time, the small G protein hydrolyzes the GTP to GDP and becomes inactive
GDP:Ras

Fig. 7.4 Small G proteins are switched from inactive to active states by guanine-nucleotide exchange factors (GEFs) and the binding of GTP. Ras is a small GTP-binding protein with intrinsic GTPase activity. In its resting state, Ras is bound to GDP. Receptor signaling activates guanine-nucleotide exchange factors (GEFs), which can bind to small G proteins such as Ras and displace GDP, allowing GTP to bind in its place (center panels). The GTP-bound form of Ras can then bind to a large number of effectors, recruiting them to the membrane. Over time, the intrinsic GTPase activity of Ras will result in the hydrolysis of GTP to GDP. GTPase-activating proteins (GAPs) can accelerate the hydrolysis of GTP to GDP, thus shutting off the signal more rapidly.

as the chemokine receptors discussed in Chapter 3. The superfamily of small GTPases comprises more than 100 different proteins, and many are important in lymphocyte signaling. One of these, **Ras**, is involved in many different pathways leading to cell proliferation. Other small GTPases include Rac, Rho, and Cdc42, which control changes in an effector T cell's actin cytoskeleton caused by signals received through the T-cell receptor. We will describe their actions in Chapter 9 when we examine the functions of effector T cells.

Small GTPases exist in two states, depending on whether they are bound to GTP or to GDP. The GDP-bound form is inactive but is converted into the active form by exchange of the GDP for GTP. This reaction is mediated by proteins known as **guanine-nucleotide exchange factors**, or **GEFs**, which cause the GTPase to release GDP and to bind the more abundant GTP (Fig. 7.4). Sos, which is recruited to signaling pathways by the adaptor Grb2 (see Section 7-2), is one of the GEFs for Ras. The binding of GTP induces a conformational change in the small GTPase that enables it to bind to and induce effector activity in a variety of target proteins. Thus, GTP binding functions like an on/off switch for small GTPases.

This GTP-bound form does not remain permanently active but is eventually converted into the inactive GDP-bound form by the intrinsic GTPase activity in the G protein, which removes a phosphate group from the bound GTP. Regulatory cofactors known as **GTPase-activating proteins** (**GAPs**) accelerate the conversion of GTP to GDP, thus rapidly downregulating the activity of the small GTPase. Because of GAP activity, small GTPases are usually present in the inactive GDP-bound state and are activated only transiently in response to a signal from an activated receptor. *RAS* is commonly found to be mutated in cancer cells, and the mutated Ras protein is thought to be a significant contributor to the cancerous state. The importance of GAPs in signaling regulation is indicated by the fact that some mutations in Ras found in cancer act by preventing the ability of GAP to mediate nucleotide exchange, thereby locking Ras into the active GTP-bound state.

GEFs are the key to G-protein activation and are recruited to the site of receptor activation at the cell membrane by binding to adaptor proteins. Once recruited, they are able to activate Ras or other small G proteins, which are themselves localized to the inner surface of the plasma membrane via fatty acids that are attached to the G protein post-translationally. Thus, G proteins act as molecular switches, becoming switched on when a cell-surface receptor is activated and then being switched off. Each G protein has its own specific GEFs and GAPs, which helps to confer specificity on the pathway.

7-4 Signaling proteins are recruited to the membrane by a variety of mechanisms.

We have seen how receptors can recruit intracellular signaling proteins to the plasma membrane through tyrosine phosphorylation of the receptor itself or of an associated scaffold, followed by recruitment of SH2-domain-containing signaling proteins or adaptors, such as Grb2 (Fig. 7.5). Subsequent recruitment of GEFs, such as Sos, to the plasma membrane can activate membrane-associated small GTPases, such as Ras, which can act on downstream targets.

Another way in which receptors can recruit signaling molecules to the membrane is by the local production of modified membrane lipids. These lipids are produced by phosphorylation of the membrane phospholipid phosphatidylinositol by enzymes known as **phosphatidylinositol kinases**, which are activated as a result of receptor signaling. The inositol headgroup of phosphatidylinositol is a sugar ring that can be phosphorylated at one or more positions to generate a wide variety of derivatives. The ones that mainly concern us in this chapter are phosphatidylinositol 3,4-bisphosphate (**PIP$_2$**)

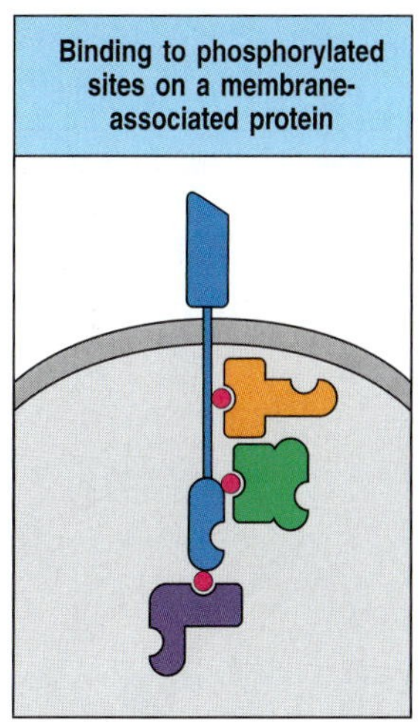

Binding to phosphorylated sites on a membrane-associated protein

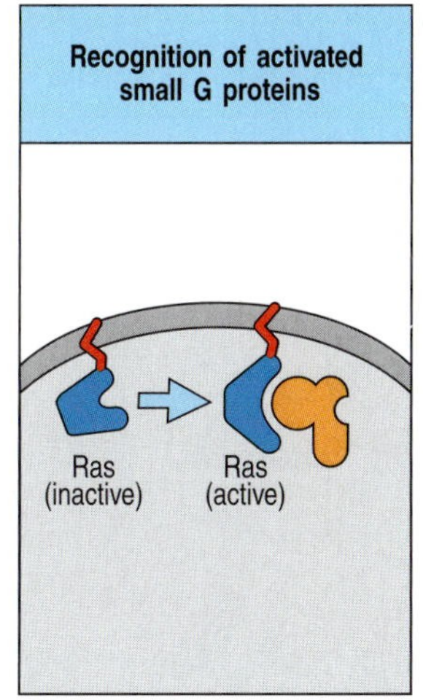

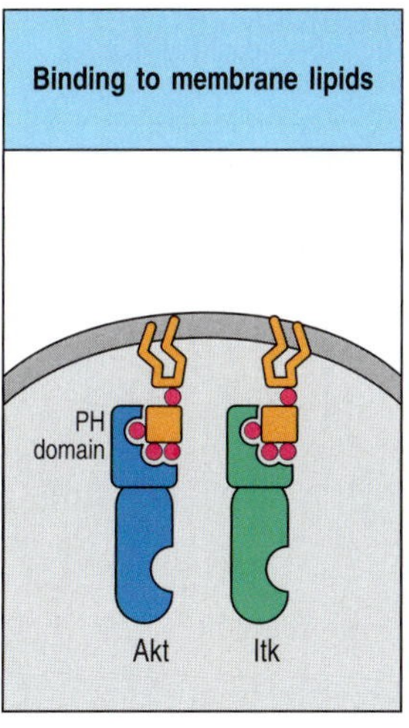

Fig. 7.5 Signaling proteins can be recruited to the membrane in a variety of ways. Recruitment of signaling proteins to the membrane is important in signal propagation because this is where receptors are usually located. Left panel: tyrosine phosphorylation of membrane-associated proteins, such as receptors themselves, recruits phosphotyrosine-binding proteins. This can also protect the receptor tail from dephosphorylation by tyrosine phosphatases, which inhibit signaling. Center panel: small G proteins such as Ras can associate with the membrane by having lipid modifications (shown in red). When activated, they can bind a variety of signaling proteins. Right panel: modifications to the membrane itself that result from receptor activation can recruit signaling proteins. In this example the membrane lipid PIP_3 has been produced in the membrane by the phosphorylation of PIP_2 by PI 3-kinase (not shown). PIP_3 is recognized by the PH domains of signaling proteins such as the kinases Akt and Itk.

and phosphatidylinositol 3,4,5-trisphosphate (**PIP_3**), which is generated from PIP_2 by the enzyme **phosphatidylinositol 3-kinase** (**PI 3-kinase**) (see Fig. 7.5). PI 3-kinase is recruited by binding of the SH2 domain of its regulatory subunit to phosphotyrosines in a receptor tail, bringing its catalytic subunit into proximity with inositol substrates in the membrane. In this way, membrane phosphoinositides such as PIP_3 are rapidly produced after receptor activation and are short-lived, which makes them ideal signaling molecules. PIP_3 is recognized specifically by proteins containing a pleckstrin homology (PH) domain or a PX domain (see Fig. 7.2), and one of its functions is to recruit such proteins to the membrane and in some cases contribute to the activation of enzymatic activity.

7-5 Ubiquitin conjugation of proteins can both activate and inhibit signaling responses.

A general mechanism of signal termination is protein degradation, most commonly initiated by the covalent attachment of one or more molecules of the small protein **ubiquitin**. Ubiquitin is attached by its carboxy-terminal glycine to lysine residues of target proteins by enzymes known as **ubiquitin ligases**. These enzymes can continue to add ubiquitin molecules to form polyubiquitin. Importantly, different ubiquitin ligases add the carboxy terminus of one ubiquitin molecule to different lysine residues of the conjugated ubiquitin, typically either lysine 48 (K48) or lysine 63 (K63). These different forms of polyubiquitin produce divergent outcomes for signaling pathways.

When polyubiquitin chains are formed using K48 linkages, the outcome is to target the protein for degradation by the proteasome. An important ubiquitin ligase of this kind in lymphocytes is Cbl, which selects its targets via its SH2 domain. Cbl can thus bind to specific tyrosine-phosphorylated targets, causing them to become ubiquitinated via K48 linkages. Proteins that recognize this form of polyubiquitin then target the ubiquitinated proteins to degradative pathways via the proteasome. Membrane proteins such as receptors can be tagged by single ubiquitin molecules or by di-ubiquitin. These are recognized not by the proteasome but rather by specific ubiquitin-binding proteins, and targeted for degradation in lysosomes (Fig. 7.6). Thus, ubiquitination of proteins can inhibit signaling, somewhat like phosphatases, except that the inhibition by ubiquitin is more permanent, whereas the dephosphorylation by phosphatases is reversible.

Ubiquitination can also be used to activate signaling pathways. We have already discussed this aspect in Section 3-7 in connection with the NFκB

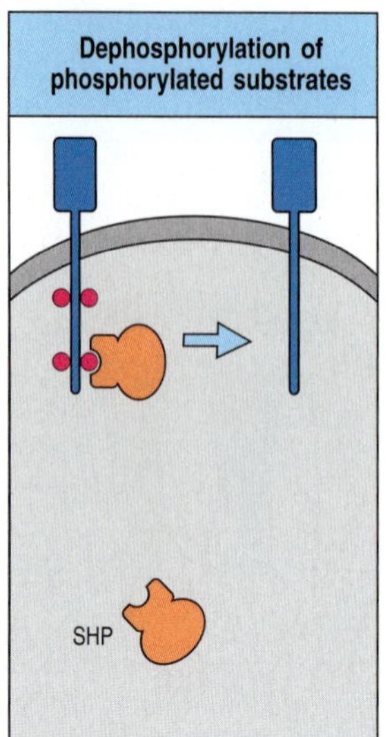
Dephosphorylation of phosphorylated substrates
SHP

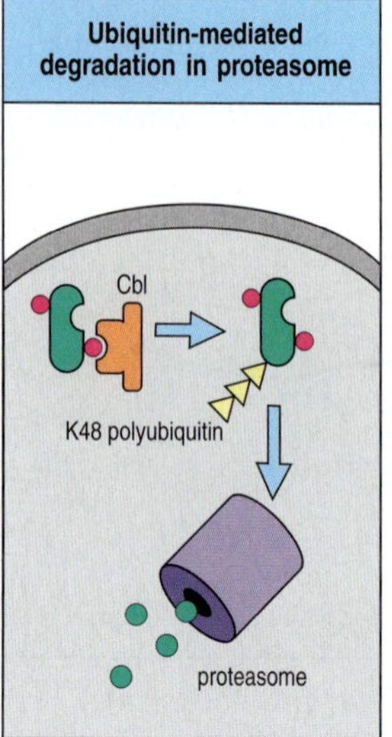
Ubiquitin-mediated degradation in proteasome
Cbl
K48 polyubiquitin
proteasome

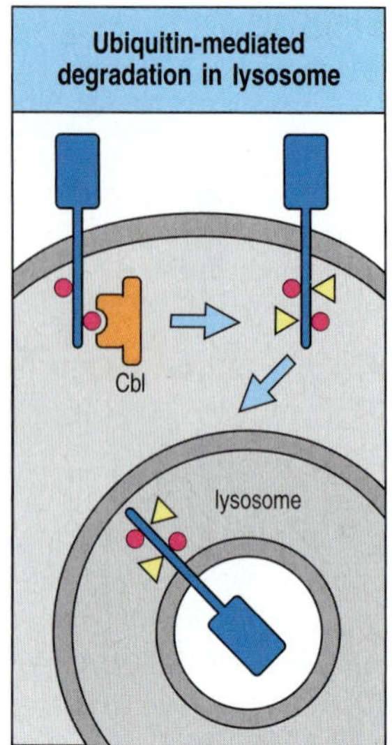
Ubiquitin-mediated degradation in lysosome
Cbl
lysosome

Fig. 7.6 Signaling must be turned off as well as turned on. The inability to terminate a signaling pathway can result in serious diseases such as autoimmunity or cancer. As a significant proportion of signaling events depend on protein phosphorylation, protein phosphatases, such as SHP, have an important part in shutting down signaling pathways (left panel). Another common mechanism for terminating signaling is regulated protein degradation (center and right panels). Phosphorylated proteins recruit ubiquitin ligases, such as Cbl, that add the small protein ubiquitin to proteins, thus targeting them for degradation. Cytoplasmic proteins are targeted for destruction in the proteasome by the addition of polyubiquitin chains, linked through lysine 48 (K48) of ubiquitin (center panel). Membrane receptors that become ubiquitinated by individual ubiquitin molecules or di-ubiquitin are internalized and transported to the lysosome for destruction (right panel).

signaling pathway from TLRs. There, the ubiquitin ligase TRAF-6 produces K63-linked polyubiquitin chains on TRAF-6 and NEMO. This form of polyubiquitin is recognized by specific domains in signaling proteins that recruit additional signaling molecules to the pathway (see Fig. 3.13).

7-6 The activation of some receptors generates small-molecule second messengers.

After an initial intracellular signal has been generated, the information is then transmitted to the intracellular targets that will carry out the appropriate cellular response. In many cases, the signaling pathway involves the activation of enzymes that produce small-molecule biochemical mediators known as **second messengers** (Fig. 7.7). These mediators can diffuse throughout the cell, enabling the signal to activate a variety of target proteins. The enzymatic production of second messengers also serves the purpose of achieving concentrations sufficient to activate the next stage of the pathway. The second messengers generated by receptors that signal via tyrosine kinases include calcium ions (Ca^{2+}) and a variety of membrane lipids and their soluble derivatives. Although some of these lipid messengers are confined to membranes, they can move within them. A second messenger binding to its target protein typically induces a conformational change that allows the protein to be activated.

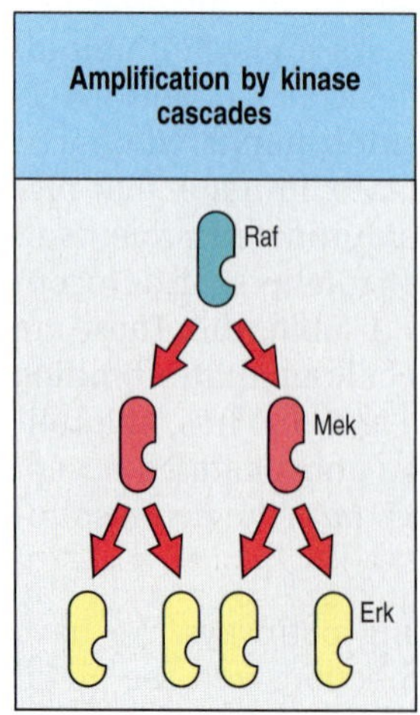
Amplification by kinase cascades
Raf
Mek
Erk

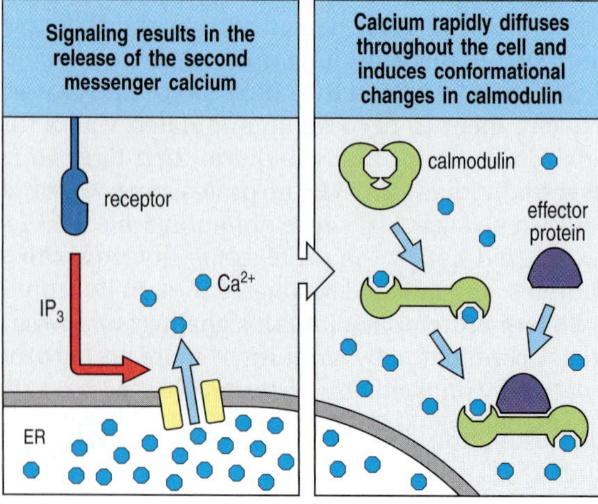
Signaling results in the release of the second messenger calcium
receptor
IP3
Ca2+
ER
Calcium rapidly diffuses throughout the cell and induces conformational changes in calmodulin
calmodulin
effector protein

Fig. 7.7 Signaling pathways amplify the initial signal. Amplification of the initial signal is an important element of most signal transduction pathways. One means of amplification is a kinase cascade (left panel), in which protein kinases successively phosphorylate and activate each other. In this example, taken from a commonly used kinase cascade (see Fig. 7.17), activation of the kinase Raf results in the phosphorylation and activation of a second kinase, Mek, that phosphorylates yet another kinase, Erk. As each kinase can phosphorylate many different substrate molecules, the signal is amplified at each step, resulting in a huge amplification of the initial signal. Another method of signal amplification is the generation of second messengers (center and right panels). In this example, signaling results in the release of the second messenger calcium (Ca^{2+}) from intracellular stores into the cytosol or its influx from the extracellular environment. Ca^{2+} release from the endoplasmic reticulum (ER) is shown here. The sharp increase in free Ca^{2+} in the cytoplasm can potentially activate many downstream signaling molecules, such as the calcium-binding protein calmodulin. Calcium binding induces a conformational change in calmodulin, which allows it to bind to and regulate a variety of effector proteins.

Summary.

Cell-surface receptors serve as the front line of a cell's interaction with its environment, sensing extracellular events and converting them into biochemical signals for the cell. As most receptors sit in the plasma membrane, a critical step in the transduction of extracellular signals to the interior of the cell is recruitment of intracellular proteins to the membrane and changes in the composition of the membrane surrounding the receptor. Many immune receptors operate by activating tyrosine kinases to transmit their signals onward, often using scaffolds and adaptors to form large multiprotein signaling complexes. Both qualitative and quantitative changes take place in the composition of these signaling complexes that determine the character of response and biological outcomes. Formation of signaling complexes is mediated by the wide variety of protein-interaction domains, or modules, including the SH2, SH3, and PH domains found in proteins. In many cases, the increase in enzymatically produced small-molecule signaling intermediates called second messengers regulates and amplifies the signaling cascade. Termination of signaling involves protein dephosphorylation as well as ubiquitin-mediated protein degradation.

Antigen receptor signaling and lymphocyte activation.

Movie 7.1

The ability of T cells and B cells to recognize and respond to their specific antigen is central to adaptive immunity. As described in Chapters 4 and 5, the B-cell antigen receptor (BCR) and the T-cell antigen receptor (TCR) are made up of antigen-binding chains—the heavy and light immunoglobulin chains in the B-cell receptor, and the TCRα and TCRβ chains in the T-cell receptor. These variable antigen-binding chains have exquisite specificity for antigen but no intrinsic signaling capacity. In the fully functional antigen receptor complex they are associated with invariant accessory proteins that initiate signaling when the receptors bind antigen. Assembly with these accessory proteins is also essential for transport of the receptor to the cell surface. In this part of the chapter we describe the structure of the antigen receptor complexes on T cells and B cells, and the signaling pathways that lead from them. In addition, because signaling from the antigen receptors is not on its own sufficient to activate a naive lymphocyte, we shall look at signaling from the co-receptors and co-stimulatory receptors that help achieve full lymphocyte activation.

7-7 Antigen receptors consist of variable antigen-binding chains associated with invariant chains that carry out the signaling function of the receptor.

In T cells, the highly variable TCRα:β heterodimer (see Chapter 5) is not sufficient on its own to make up a complete cell-surface antigen receptor. When cells were transfected with cDNAs encoding the TCRα and TCRβ chains, the heterodimers formed were degraded and did not appear on the cell surface. This implied that other molecules are required for the T-cell receptor to be expressed on the cell surface. In the T-cell receptor, the other required molecules are the **CD3γ**, **CD3δ**, and **CD3ε** chains, which together form the **CD3 complex**, and the **ζ chain**, which is present as a disulfide-linked homodimer. The CD3 proteins have an extracellular immunoglobulin-like domain, whereas the ζ chain is distinct in having only a short extracellular domain. Throughout the remainder of the chapter, we will use the term T-cell receptor to refer to the entire T-cell receptor complex including these associated signaling subunits.

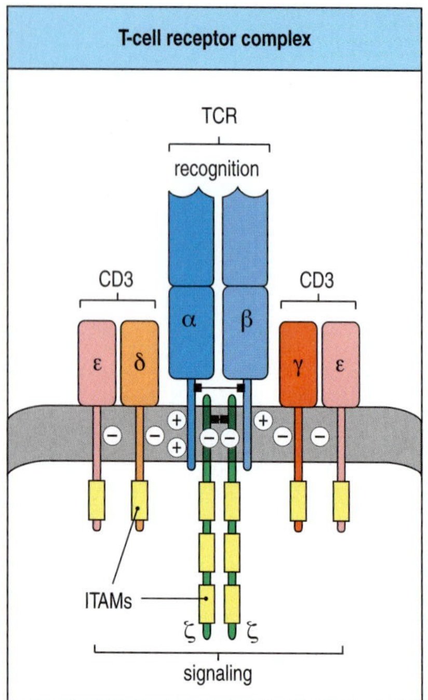

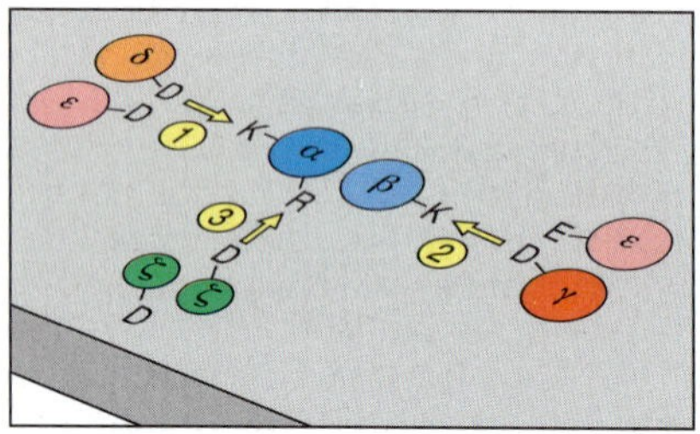

Fig. 7.8 The T-cell receptor complex is made up of variable antigen-recognition proteins and invariant signaling proteins. Upper panel: the functional T-cell receptor (TCR) complex is composed of the antigen-binding TCRα:β heterodimer associated with four signaling chains (two ε, one δ, and one γ) collectively called CD3, which are required for the cell-surface expression of the antigen-binding chains and for signaling. A homodimer of ζ chains is also associated with the receptor. Each CD3 chain has one immunoreceptor tyrosine-based activation motif (ITAM), shown as a yellow segment, whereas each ζ chain has three. The transmembrane regions of each chain have either positive or negative charges as shown. Lower panel: the transmembrane regions of the various TCR subunits are represented in cross-section. It is thought that one of the positive charges from a lysine (K) of the α chain interacts with the two negative charges of aspartic acid (D) of the CD3δ:ε dimer, while the other positive charge of arginine (R) interacts with aspartic acids of the ζ homodimer. The positive arginine (K) charge of the β chain interacts with the negative charges of aspartic acid and glutamic acid (E) in the CD3γ:ε dimer.

Although the exact stoichiometry of the complete T-cell receptor is not definitively established, it is thought that the receptor α chain interacts with one CD3δ:CD3ε dimer and one ζ dimer, while the receptor β chain interacts with one CD3γ:CD3ε dimer (Fig. 7.8). These interactions are mediated by reciprocal charge interactions between basic and acidic intramembrane amino acids of the receptor subunits. There are two positive charges in the TCRα transmembrane region and one in the TCRβ transmembrane domain. Negative charges in the CD3 and ζ transmembrane domains interact with the positive charges in α and β. Assembly of CD3 with the α:β heterodimer stabilizes the dimer during its production in the endoplasmic reticulum and allows the complex to be transported to the plasma membrane. These associations ensure that all T-cell receptors present on the plasma membrane are properly assembled. Recent evidence suggests that the composition of the T-cell receptor signaling chains is dynamic and can change after stimulation of the receptor by its ligand.

Signaling from the T-cell receptor is initiated by tyrosine phosphorylation within cytoplasmic regions in the CD3ε, γ, δ, and ζ chains called **immunoreceptor tyrosine-based activation motifs** (**ITAMs**). CD3γ, δ, and ε each contain one ITAM, and each ζ chain contains three, giving the T-cell receptor a total of 10 ITAMs. This motif is also present in the signaling chains of the B-cell receptor and in the NK-cell receptors described in Chapter 3, as well as in the receptors for the immunoglobulin constant region (Fc receptors) that are present on mast cells, macrophages, monocytes, neutrophils, and NK cells.

Each ITAM contains two tyrosine residues that become phosphorylated by specific protein tyrosine kinases when the receptor binds its ligand, providing sites for the recruitment of the SH2 domains of signaling proteins as described earlier in the chapter. Two YXXL/I motifs are separated by about six to nine amino acids within each ITAM, so the canonical ITAM sequence is ...YXX[L/I]X_{6-9}YXX[L/I]..., where Y is tyrosine, L is leucine, I is isoleucine, and X represents any amino acid. The two tyrosines of the ITAM make it particularly efficient in recruiting signaling proteins that contain two tandem SH2 domains (Fig. 7.9).

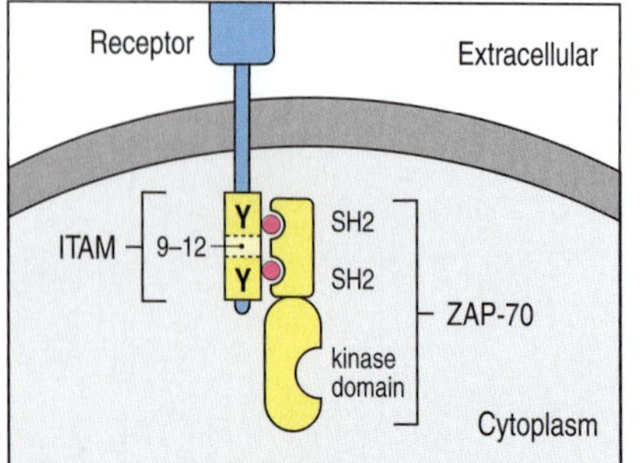

Fig. 7.9 ITAMs recruit signaling proteins that have tandem SH2 domains. The ITAMs of the TCR and BCR contain tyrosine residues contained within the motif ...YXX[L/I] X_{6-9}YXX[L/I].... The spacing between the tyrosines is important in binding to tandem SH2-containing proteins such as Syk and ZAP-70. After phosphorylation of both tyrosines within one ITAM, a signaling protein containing properly oriented tandem SH2 domains can dock cooperatively to both phosphotyrosines, as shown here for ZAP-70. By being recruited into the active signaling complex, ZAP-70 can itself be phosphorylated, so that it becomes an active kinase that can then phosphorylate its substrates.

As in the T-cell receptor, the antigen-binding portion of the B-cell receptor has no signaling function. On the cell surface, the antigen-binding immunoglobulin is associated with invariant protein chains, called **Igα** and **Igβ**, that are required both for its transport to the surface and for the signaling function of the B-cell receptor (Fig. 7.10). Igα and Igβ are single-chain proteins composed of an extracellular immunoglobulin-like domain connected by a transmembrane domain to a cytoplasmic tail. They form a disulfide-linked heterodimer that becomes associated with immunoglobulin heavy chains and enables their transport to the cell surface, thus ensuring that only fully assembled B-cell receptors are present on the cell. The Igα:Igβ dimer associates with the B-cell receptor through hydrophilic interactions, not charged interactions, between their transmembrane regions. The complete B-cell receptor is thought to be a complex of six chains—two identical light chains, two identical heavy chains, and one associated heterodimer of Igα and Igβ. Like CD3 and the ζ chains of the T-cell receptor, Igα and Igβ each have one ITAM, and these are essential for the ability of the B-cell receptor to signal.

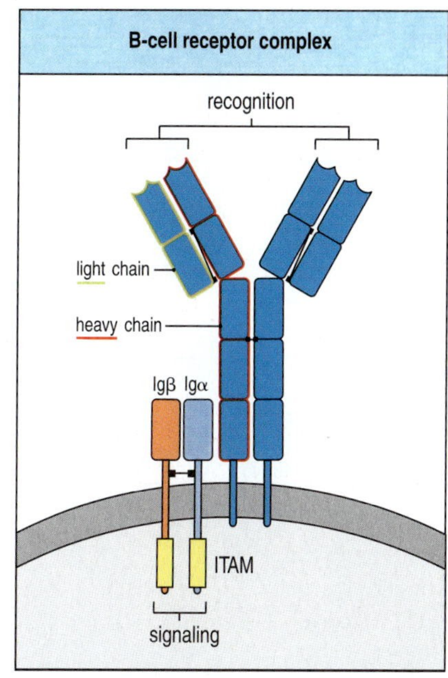

Fig. 7.10 The B-cell receptor complex is made up of cell-surface immunoglobulin with one each of the invariant signaling proteins Igα and Igβ. The immunoglobulin recognizes and binds antigen but cannot itself generate a signal. It is associated with antigen-nonspecific signaling molecules—Igα and Igβ. These each have a single ITAM (yellow segment) in their cytosolic tails that enables them to signal when the B-cell receptor is ligated with antigen. Igα and Igβ form a disulfide-linked heterodimer that is associated with the heavy chains, but it is not known which binds to the heavy chain.

7-8 Antigen recognition by the T-cell receptor and its co-receptors leads to phosphorylation of ITAMs by Src-family kinases.

To make an effective immune response, T cells and B cells must be able to respond to their specific antigen even when it is present at extremely low levels. This is especially important for T cells, as the antigen-presenting cell will display many different peptides from both self and foreign proteins on its surface, and so the number of peptide:MHC complexes specific for a particular T-cell receptor is likely to be very low. Some estimates suggest that a naive CD4 T cell can become activated by fewer than 50 antigenic peptide:MHC complexes displayed by an antigen-presenting cell, and effector CD8 cytotoxic T cells may be even more sensitive. B cells become activated when about 20 B-cell receptors are engaged. These estimates based on *in vitro* studies may not be precise for cells *in vivo*, but it is clear that the antigen receptors on T cells and B cells confer remarkable sensitivity to antigen.

For a peptide:MHC complex to activate T cells it must bind directly to the T-cell receptor (Fig. 7.11, upper panel, and see Fig. 4.22). However, it is still unclear precisely how this extracellular recognition event is transmitted across the T-cell membrane to initiate signaling. The essential unknowns are the stoichiometry and arrangement of T-cell receptors bound to peptide:MHC complexes that initiates the signaling cascade. We will discuss this area of active research only briefly before moving on to explain well-understood intracellular events that occur after antigen recognition. Recall that ligand-induced receptor dimerization explains signaling by TLRs and many cytokine receptors owing to the reciprocal activation of receptor-associated tyrosine kinases. T-cell receptor signaling seems more complex, however. For one thing it involves the CD4 or CD8 co-receptors that recognize nonpolymorphic sites on the MHC molecule (see Section 4-17), and whose involvement will be discussed in more detail later.

One suggestion is that signaling is initiated by T-cell receptor dimerization through formation of **'pseudo-dimeric' peptide:MHC complexes** containing one antigen peptide:MHC molecule and one self-peptide:MHC molecule on the surface of the antigen-presenting cell. This model relies on a weak interaction occurring between the T-cell receptor and self-peptide:MHC complexes, but it would explain signaling induced by low densities of antigenic peptides.

It has also been suggested that signaling might involve receptor oligomerization, or clustering, as antibodies that bind to and cross-link T-cell receptors can activate T cells. However, antigenic peptides are vastly outnumbered by other peptides displayed on the surface of the antigen-binding cell, and it is unlikely that physiologic amounts of antigen induce conventional

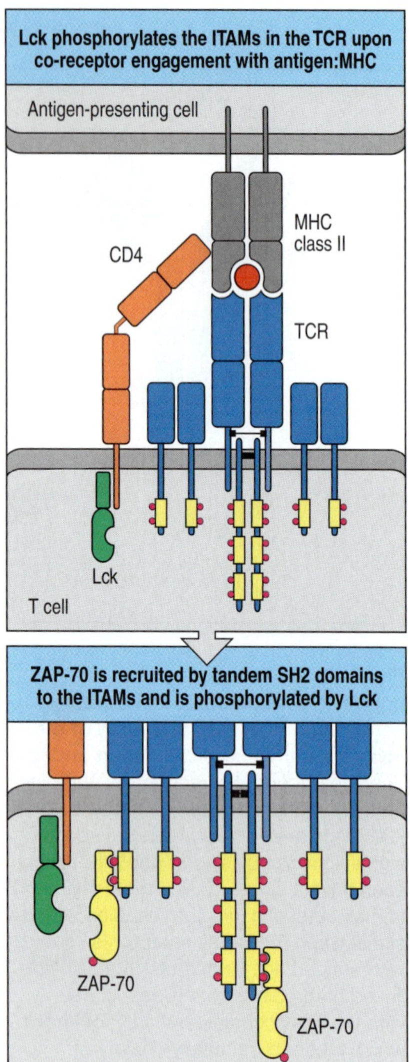

Lck phosphorylates the ITAMs in the TCR upon co-receptor engagement with antigen:MHC
Antigen-presenting cell
MHC class II
CD4
TCR
Lck
T cell
ZAP-70 is recruited by tandem SH2 domains to the ITAMs and is phosphorylated by Lck
ZAP-70
ZAP-70

Fig. 7.11 Engagement of co-receptors with the T-cell receptor enhances phosphorylation of the ITAMs. Upper panel: for simplicity, we show the CD4 co-receptor engaging the same MHC molecule as the T-cell receptor (TCR), although signaling within receptor microclusters may differ from this arrangement. When T-cell receptors and co-receptors are brought together by binding to peptide:MHC complexes on the surface of an antigen-presenting cell, recruitment of the co-receptor-associated kinase Lck leads to phosphorylation of ITAMs in CD3γ, δ, and ε, and in the ζ chain. Lower panel: the tyrosine kinase ZAP-70 binds to phosphorylated ITAMs through its SH2 domain, enabling ZAP-70 to be phosphorylated and activated by Lck. ZAP-70 then phosphorylates other intracellular signaling molecules.

oligomerization as observed with antibodies. However, assemblies of small numbers of T-cell receptors called **microclusters** have been observed in the zone of contact between the T cell and the antigen-presenting cell. Currently, the molecular composition of microclusters and their role in signaling is undefined. Finally, it has been suggested that the antigenic peptide:MHC complex induces a conformational change in the T-cell receptor, or a change in its composition, that somehow generates the signal; however, no direct structural evidence yet supports this model.

The first intracellular signal generated after the T cell has detected its specific antigen is the phosphorylation of both tyrosines in the ITAMs of the T-cell receptor. It is the CD4 and CD8 co-receptors, which bind to MHC class II molecules and class I molecules, respectively (see Section 4-17), that help initiate T-cell receptor signaling by their association with nonreceptor kinases. The Src family kinase **Lck** is constitutively associated with the cytoplasmic domains of CD4 and CD8 and is thought to be the kinase primarily responsible for phosphorylation of the ITAMs of the T-cell receptor (see Fig. 7.11). It is thought that binding of the co-receptor to the peptide:MHC complex that binds the T-cell receptor allows Lck to phosphorylate the T-cell receptor ITAMs. For CD4, engagement within a pseudo-dimer has been proposed, as noted earlier, such that CD4 might engage a different MHC molecule than the one bound to the T-cell receptor. But no matter how Lck becomes associated with the antigen receptor, the importance of this event is demonstrated by the profound reduction in T-cell development in Lck-deficient mice. This indicates the essential role of Lck in T-cell receptor signaling during the selection of developing T cells in the thymus (discussed in Chapter 8). Lck is important for T-cell receptor signaling in naive T cells and effector T cells, but is less important for the activation or maintenance of memory CD8 T cells by their specific antigen. A related tyrosine kinase **Fyn** is weakly associated with the ITAMs of the T-cell receptor, and may have some role in signaling. Whereas mice lacking Fyn develop normal CD4 and CD8 T cells that respond essentially normally to antigen, mice lacking both Lck and Fyn show a more complete loss of T-cell development than mice lacking Lck alone.

Another role of the co-receptors in T-cell receptor signaling may be to stabilize interactions between the receptor and the peptide:MHC complex. Affinities of individual receptors for their specific peptide:MHC complexes are in the micromolar range, which means that the T-cell receptor:peptide:MHC complexes have half lives of less than 1 second and dissociate rapidly. The additional binding of a co-receptor to the MHC molecule is thought to stabilize the interaction by increasing its duration, thereby giving time for an intracellular signal to be generated.

The Lck bound to the cytoplasmic tails of CD4 or CD8 is brought near to its substrate ITAMs in the T-cell receptor when the co-receptor binds the receptor:peptide:MHC complex (see Fig. 7.11, lower panel). Lck's activity is also regulated allosterically by phosphorylation of a tyrosine in its carboxy terminus by the **C-terminal Src kinase** (**Csk**). The resulting phosphotyrosine interacts with Lck's SH2 domain and causes a conformational change that keeps Lck in the catalytically inactive state (Fig. 7.12). The absence of Csk during T-cell development causes T cells to mature autonomously in the thymus without needing to bind peptide:MHC, presumably as a result of abnormal activation of TCR signaling in Csk-deficient thymocytes. This suggests that Csk normally acts to reduce Lck activity and to attenuate TCR signaling. Dephosphorylation of the tyrosine or engagement of the SH2 or SH3 domains by their ligands releases Lck from its inactive conformation. Full activation also requires phosphorylation of a tyrosine in Lck's catalytic domain. In lymphocytes, the **tyrosine phosphatase CD45**, which can dephosphorylate both the tyrosine phosphorylation sites, has an important role in maintaining Src family kinases such as Lck in a partly active, dephosphorylated state.

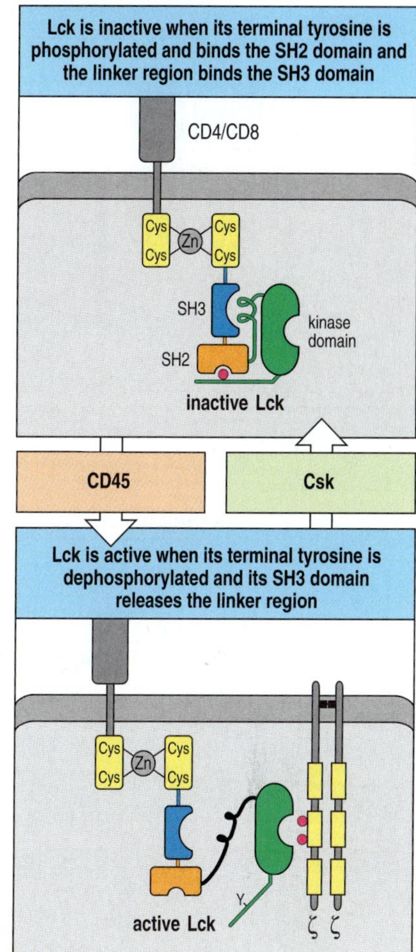

Fig. 7.12 Lck activity is regulated by tyrosine phosphorylation and dephosphorylation. Src kinases such as Lck contain SH3 (blue) and SH2 (orange) domains preceding the kinase domain (green). Lck also contains a unique amino-terminal motif (yellow) with two cysteine residues that bind a Zn ion that is also bound to a similar motif in the cytoplasmic domain of CD4 or CD8. Upper panel: in inactive Lck, the two lobes of the kinase domain are constrained by interactions with both the SH2 and SH3 domains. The SH2 domain interacts with a phosphorylated tyrosine (red) at the carboxy-terminal end of the kinase domain. The SH3 domain interacts with a proline-rich sequence in the linker between the SH2 domain and the kinase domain. Lower panel: dephosphorylation of the carboxy-terminal tyrosine by the phosphatase CD45 (not shown) releases the SH2 domain and results in kinase activation. Binding of other ligands to the SH3 region can facilitate release of the linker region (not shown). Active Lck can then phosphorylate ITAMs in the signaling chains of the nearby T-cell receptor. Rephosphorylation of the carboxy-terminal tyrosine by the C-terminal Src kinase (Csk) or loss of the SH3 ligand returns Lck to the inactive state.

7-9 Phosphorylated ITAMs recruit and activate the tyrosine kinase ZAP-70, which phosphorylates scaffold proteins that recruit the phospholipase PLC-γ.

The precise spacing of the two YXXL/I motifs in an ITAM suggests that the ITAM is a binding site for a signaling protein with two SH2 domains. In the case of the T-cell receptor, this protein is the tyrosine kinase **ZAP-70 (ζ-chain-associated protein)**, which carries the signal onward. ZAP-70 has two tandem SH2 domains that can be simultaneously engaged by the two phosphorylated tyrosines in an ITAM (see Fig. 7.9). The affinity of the phosphorylated YXXL sequence for a single SH2 domain is low; binding of both SH2 domains to the ITAM is significantly stronger and confers specificity on ZAP-70 binding. Thus, when Lck has sufficiently phosphorylated an ITAM in the T-cell receptor, ZAP-70 binds to it, which allows ZAP-70 also to be phosphorylated and activated by Lck. ZAP-70 can also be activated by autophosphorylation.

Once ZAP-70 has been recruited to the receptor complex and activated, its proximity to the cell membrane allows it to phosphorylate the scaffold protein **LAT (linker of activated T cells)**, a transmembrane protein with a large cytoplasmic domain (Fig. 7.13). ZAP-70 also phosphorylates another adaptor protein, **SLP-76**. LAT and SLP-76 can be linked by the adaptor protein Gads; this linking seems to be important for their function, because mice lacking Gads have defects in T-cell activation. The next step in the pathway is the activation of the key signaling protein **phospholipase C-γ (PLC-γ)** to membrane. First, PLC-γ is brought to the inner face of the plasma membrane by the binding of its PH domain to PIP_3 that has been formed by the phosphorylation of PIP_2 by PI 3-kinase; it then binds to phosphorylated LAT and SLP-76, where it can be activated by the membrane-associated tyrosine kinase Itk (see the next section).

The actions of PLC-γ produce three different second messengers that activate three distinct terminal branches of the T-cell receptor pathway. PLC-γ is therefore the gatekeeper to the final steps of T-cell activation and, in line with this crucial role, its own activation is controlled at several different levels. The next section will discuss these in some greater detail.

7-10 The activation of PLC-γ requires a co-stimulatory signal.

On recruitment to the membrane, PLC-γ is not yet activated, and its activation requires phosphorylation by Itk, a member of the Tec family of cytoplasmic tyrosine kinases, as described above. This requirement provides a further level of control of PLC-γ activation. Tec kinases contain PH, SH2, and SH3 domains and are recruited to the plasma membrane by their PH domain,

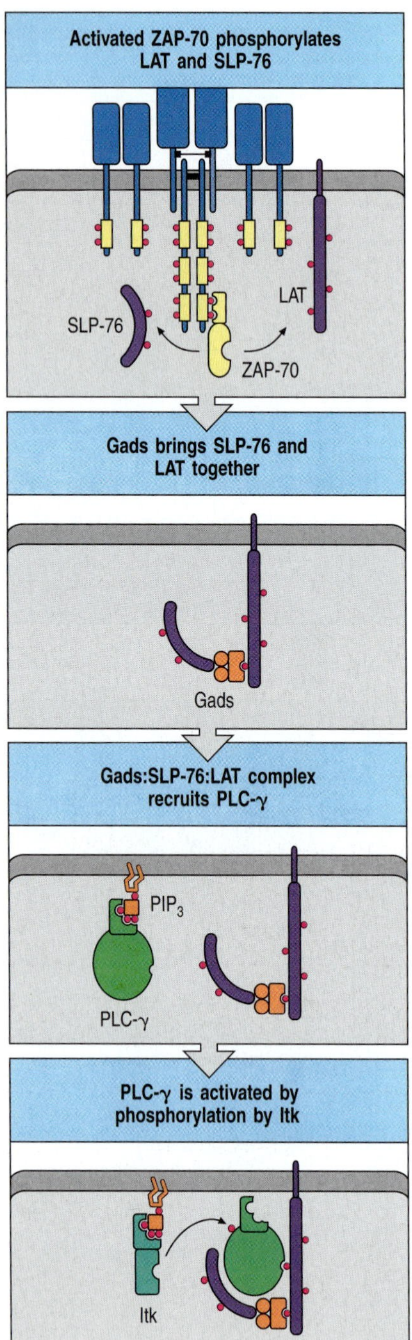
Activated ZAP-70 phosphorylates LAT and SLP-76
LAT
SLP-76
ZAP-70
Gads brings SLP-76 and LAT together
Gads
Gads:SLP-76:LAT complex recruits PLC-γ
PIP$_3$
PLC-γ
PLC-γ is activated by phosphorylation by Itk
Itk

Fig. 7.13 The recruitment of phospholipase C-γ by LAT and SLP-76, and its phosphorylation and activation by the protein kinase Itk, are crucial steps in T-cell activation. ZAP-70 phosphorylates and recruits the scaffold proteins LAT and SLP-76 to the activated T-cell receptor complex. An adaptor protein, Gads, holds the tyrosine-phosphorylated LAT and SLP-76 together. Phospholipase C-γ (PLC-γ) is recruited to the membrane by its PH domain binding to PIP_3 (formed by the phosphorylation of PIP_2 by PI 3-kinase), and then binds to phosphorylated sites in both LAT and SLP-76. To be activated, PLC-γ must be phosphorylated by the Tec-family kinase Itk. Itk is recruited to the membrane by interaction of its PH domain with PIP_3, and by interactions with phosphorylated SLP-76. Once phosphorylated by Itk, phospholipase C-γ is active.

which interacts with PIP_3 on the inner face of the membrane (see Fig. 7.13). PIP_3 is generated by the actions of PI 3-kinase (see Section 7-4), however, and it is not clear that T-cell receptor signaling directly activates PI 3-kinase. An additional stimulatory signal required to activate PI-3-kinase, and thus PLC-γ, is delivered through the cell-surface receptor CD28. This process is called **co-stimulation** (see Sections 1-7 and 1-16). Naive T cells need co-stimulatory signals as well as signals delivered through the antigen receptor to become fully activated and to differentiate into effector T cells.

Of the four Tec kinases expressed in lymphoid cells, **Itk** is the one chiefly expressed in T lymphocytes. When CD28 engagement activates PI 3-kinase to produce PIP_3, Itk is recruited by its PH domain to the membrane and is phosphorylated by Lck. Activated Itk is then recruited to the phosphorylated LAT/SLP-76 scaffolds by its SH2 and SH3 domains, and can then phosphorylate and activate PLC-γ (see Fig. 7.13). Thus, the full activation of PLC-γ requires signals emanating from both the T-cell receptor and CD28. The generation of the CD28 co-stimulatory signal is discussed in more detail later in the chapter.

7-11 Activated PLC-γ generates the second messengers diacylglycerol and inositol trisphosphate.

Once PLC-γ has been recruited to the inner face of the plasma membrane and has been activated, it can catalyze the breakdown of the membrane lipid PIP_2 (see Section 7-4) to generate two products, the membrane lipid **diacylglycerol** (**DAG**) and the diffusible second messenger **inositol 1,4,5-trisphosphate** (**IP_3**) (not to be confused with the membrane lipid PIP_3) (Fig. 7.14). DAG is confined to the membrane, but diffuses in the plane of the membrane and serves as a molecular target that recruits other signaling molecules to the membrane. IP_3 diffuses into the cytosol and binds to IP_3 receptors on the endoplasmic reticulum (ER) membrane. These receptors are Ca^{2+} channels, which open and release all the calcium stored in the ER into the cytosol. The consequent low levels of calcium in the ER then cause the transmembrane protein **STIM1** to cluster within the ER membrane. Through an unknown mechanism, STIM1 clustering triggers the opening of calcium channels in the cell's plasma membrane; these are known as **CRAC channels** (***c**alcium **r**elease-**a**ctivated **c**alcium channels*), and they allow extracellular calcium to flow into the cell to activate further signaling pathways and to replenish ER calcium stores. CRAC channels are formed at least partly by the product of the ***ORAI1*** gene, which is mutated in some cases of **severe combined immunodeficiency** (**SCID**).

The activation of PLC-γ marks an important step in T-cell activation, because after this point the antigen signaling pathway splits into three distinct branches—the stimulation of Ca^{2+} entry, and the activation of both Ras and **protein kinase C-θ** (**PKC-θ**)—each of which ends in the activation of a different transcription factor. The signaling pathways from the T-cell receptor are summarized in Fig. 7.15. These signaling pathways are not exclusive to lymphocytes but are versions of pathways used in many different cell types. Their importance in T-cell activation is shown by the observation that treatment of T cells with phorbol myristate acetate (an analog of DAG) and ionomycin (a

Fig. 7.14 Phospholipase C-γ cleaves inositol phospholipids to generate two important signaling molecules. Top panel: phosphatidylinositol bisphosphate (PIP_2) is a component of the inner leaflet of the plasma membrane. When PLC-γ is activated by phosphorylation, it cleaves PIP_2 into two parts: inositol trisphosphate (IP_3), which diffuses away from the membrane into the cytosol, and diacylglycerol (DAG), which stays in the membrane. Both these molecules are important in signaling. Middle panel: there are two phases of calcium release. IP_3 first binds to a receptor in the endoplasmic reticulum (ER) membrane, opening calcium channels (yellow) and allowing the early phase of calcium ions (Ca^{2+}) to enter the cytosol from the ER. The depletion of Ca^{2+} stores in the ER causes the aggregation of an ER sensor, STIM1. Bottom panel: aggregated STIM1 stimulates the second phase of calcium entry by opening channels in the plasma membrane, called CRAC channels. This further increases cytosolic calcium and restores ER Ca^{2+} stores. DAG binds and recruits signaling proteins to the membrane, most importantly the Ras-GEF called RasGRP and a serine/threonine kinase called protein kinase C-θ (PKC-θ). Recruitment of RasGRP to the plasma membrane activates Ras, and PKC-θ activation results in the activation of the transcription factor NFκB.

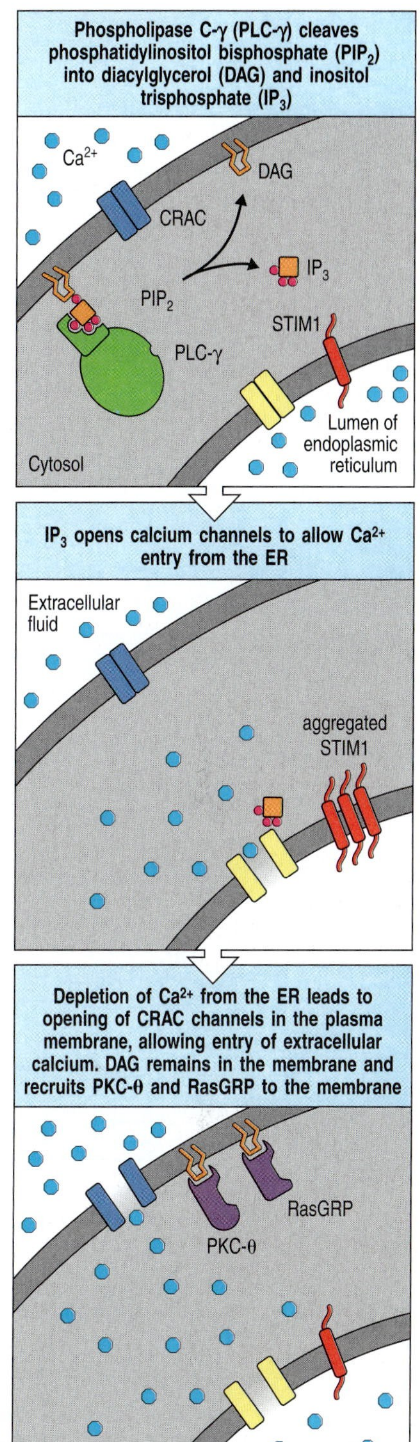

pore-forming drug that allows extracellular calcium to flow into the cell) can largely reconstitute the effects of T-cell activation.

7-12 Ca^{2+} entry activates the transcription factor NFAT.

An important outcome of the increased cytosolic Ca^{2+} resulting from T-cell receptor signaling is the activation of a family of transcription factors called **NFAT** (**nuclear factor of activated T cells**). NFAT is something of a misnomer, because the five members of this family are expressed in many different tissues. NFAT is present in the cytosol of resting T cells, and in the absence of activating signals it is kept in the cytosol by phosphorylation by serine/threonine kinases, including glycogen synthase kinase 3 (GSK3) and casein kinase 2 (CK2). This phosphorylation blocks entry of NFAT into the nucleus by preventing its nuclear localization sequence from being recognized by nuclear transporters (Fig. 7.16).

The cytoplasmic Ca^{2+} resulting from T-cell receptor signaling binds a protein called **calmodulin** and induces a conformational change in this protein that allows it to bind to and activate a wide range of different target enzymes. In T cells, an important target of calmodulin is **calcineurin**, a protein phosphatase that acts on NFAT. Dephosphorylation of NFAT by calcineurin allows the nuclear localization sequence to be recognized by nuclear transporters, and NFAT enters the nucleus (see Fig. 7.16). There it functions in turning on many of the genes crucial for T-cell activation, such as the gene for the cytokine interleukin-2 (IL-2).

The importance of NFAT in T-cell activation is illustrated by the effects of selective inhibitors of calcineurin called **cyclosporin A** (**CsA**) and **tacrolimus** (also known as FK506). CsA forms a complex with the protein cyclophilin A, and this complex inhibits calcineurin. Tacrolimus binds a different protein, FK-binding protein (FKBP), making a complex that similarly inhibits calcineurin. By inhibiting calcineurin, these drugs prevent the formation of active NFAT. T cells express low levels of calcineurin, so they are more sensitive to inhibition of this pathway than are many other cell types. Both cyclosporin A and tacrolimus thus act as effective immunosuppressants with only limited side-effects, and are widely used to prevent the rejection of organ transplants (discussed in Chapter 16).

7-13 Ras activation stimulates the mitogen-activated protein kinase (MAPK) relay and induces expression of the transcription factor AP-1.

The next step in this pathway is the activation of the small GTPase Ras. This can occur by various means. The DAG generated by PLC-γ diffuses in the plasma membrane and activates a variety of proteins. One of these is the protein RasGRP, which is a guanine-nucleotide exchange factor that specifically activates Ras. RasGRP contains a protein-interaction module called a C1 domain that binds to DAG. This interaction recruits RasGRP to the membrane near active signaling complexes (see Fig. 7.14), where it activates Ras by promoting the exchange of GDP for GTP. Ras is also activated in the T-cell receptor signaling pathway by the guanine-exchange factor Sos, which is recruited by

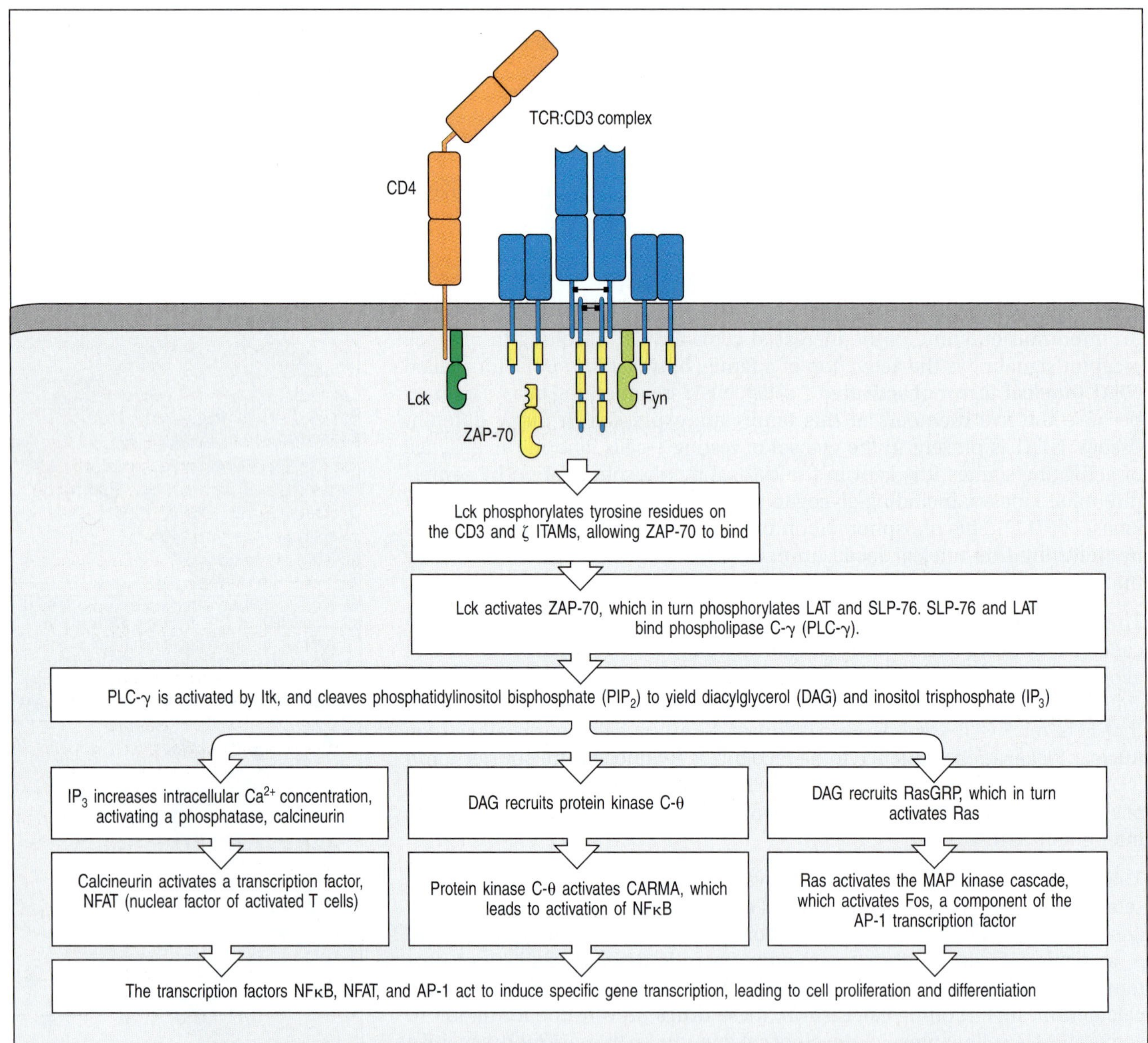

Fig. 7.15 Simplified outline of the intracellular signaling pathways initiated by the T-cell receptor and its co-receptor. The T-cell receptor and co-receptor (in this example the CD4 molecule) are associated with Src-family protein kinases Fyn and Lck, respectively. It is thought that binding of a peptide:MHC ligand to the T-cell receptor and co-receptors brings together Lck with the ITAMs in the receptor. Phosphorylation of the ITAMs in CD3ε, γ, and δ and the ζ chain enables them to bind the cytosolic tyrosine kinase ZAP-70. ZAP-70 recruited to the T-cell receptor is phosphorylated and activated by Lck. Activated ZAP-70 phosphorylates the adaptor proteins LAT and SLP-76, which in turn leads to membrane recruitment of PLC-γ and its phosphorylation and activation by Tec kinases. Activated PLC-γ initiates three important signaling pathways that culminate in the activation of transcription factors in the nucleus. Together, NFκB, NFAT, and AP-1 act in the nucleus to initiate gene transcription that results in the differentiation, proliferation, and effector actions of T cells. This diagram is a highly simplified version of the pathways, showing the main events only.

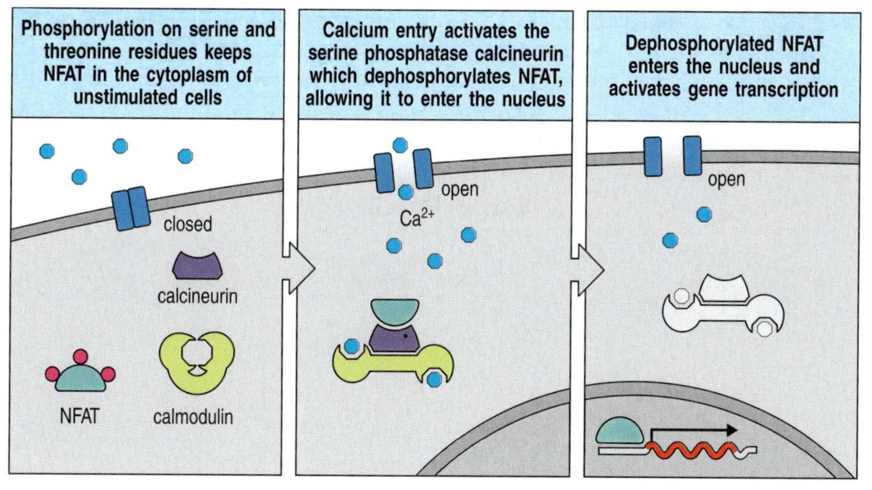

Phosphorylation on serine and threonine residues keeps NFAT in the cytoplasm of unstimulated cells
closed
calcineurin
NFAT
calmodulin
Calcium entry activates the serine phosphatase calcineurin which dephosphorylates NFAT, allowing it to enter the nucleus
open
Ca2+
Dephosphorylated NFAT enters the nucleus and activates gene transcription
open

Fig. 7.16 The transcription factor NFAT is regulated by calcium signaling. Left panel: NFAT is maintained in the cytoplasm by phosphorylation on serine and threonine residues. Center panel: after antigen receptor stimulation, calcium enters the cytosol, first from the endoplasmic reticulum (not shown) as described in Fig. 7.14, and later from the extracellular space (shown). After entering the cytosol, calcium binds to calmodulin, and the Ca^{2+}:calmodulin complex binds to the serine/threonine phosphatase calcineurin, activating it to dephosphorylate NFAT. Right panel: once dephosphorylated, NFAT moves into the nucleus, where it binds to promoter elements and activates the transcription of various genes.

the adaptor protein Grb2 (see Sections 7-2 and 7-3), which is itself recruited by binding to phosphorylated LAT/SLP-76 (see Fig. 7.13).

Activated Ras then triggers a three-kinase relay that ends in the activation of a serine/threonine kinase known as a **mitogen-activated protein kinase** or **MAP kinase** (**MAPK**) (Fig. 7.17). In the case of antigen receptor signaling, the first member of the relay is a MAPK kinase kinase (MAP3K) called **Raf**. Raf is a serine/threonine kinase that phosphorylates the next member of the series, a MAPK kinase (MAP2K) called **MEK1**. MEK1 is a dual-specificity protein kinase that phosphorylates a tyrosine and a threonine residue on the last member of the relay, a MAPK which in T cells and B cells is **Erk** (**extracellular signal-related kinase**).

Signaling by MAPK cascades is facilitated by specialized scaffold proteins that bind to all three kinases in a particular MAPK relay, thereby accelerating their interactions. The scaffold protein **kinase suppressor of Ras** (**KSR**) functions in the Raf/MEK1/Erk pathway. During T-cell receptor signaling, KSR associates with Raf, MEK1, and Erk and localizes itself and its cargo to the membrane. In that location, activated Ras can engage with the Raf bound to KSR and trigger the kinase relay (see Fig. 7.17).

An important function of MAPKs is to phosphorylate and activate transcription factors that can then induce new gene expression. Erk acts indirectly to generate the transcription factor **AP-1**, which is a heterodimer composed of one monomer each from the Fos and Jun families of transcription factors (Fig. 7.18). Active Erk phosphorylates the transcription factor Elk-1, which

Fig. 7.17 MAPK cascades activate transcription factors. All MAPK cascades are initiated by the activation of small G proteins, such as Ras in this example. Ras is switched from an inactive state (first panel) to an active state (second panel) by a guanine-nucleotide exchange factor (GEF) RasGRP, which is recruited to the membrane by DAG. Ras activates the first enzyme of the cascade, a protein kinase called Raf, a MAPK kinase kinase (MAP3K) (third panel). Raf phosphorylates Mek, a MAP2K, which in turn phosphorylates and activates Erk, a MAPK. The scaffold protein KSR associates with Raf, Mek, and Erk to facilitate their efficient interactions. Phosphorylation and activation of Erk releases it from the complex so that it can diffuse within the cell and enter the nucleus (fourth panel). Phosphorylation of transcription factors by Erk results in new gene transcription.

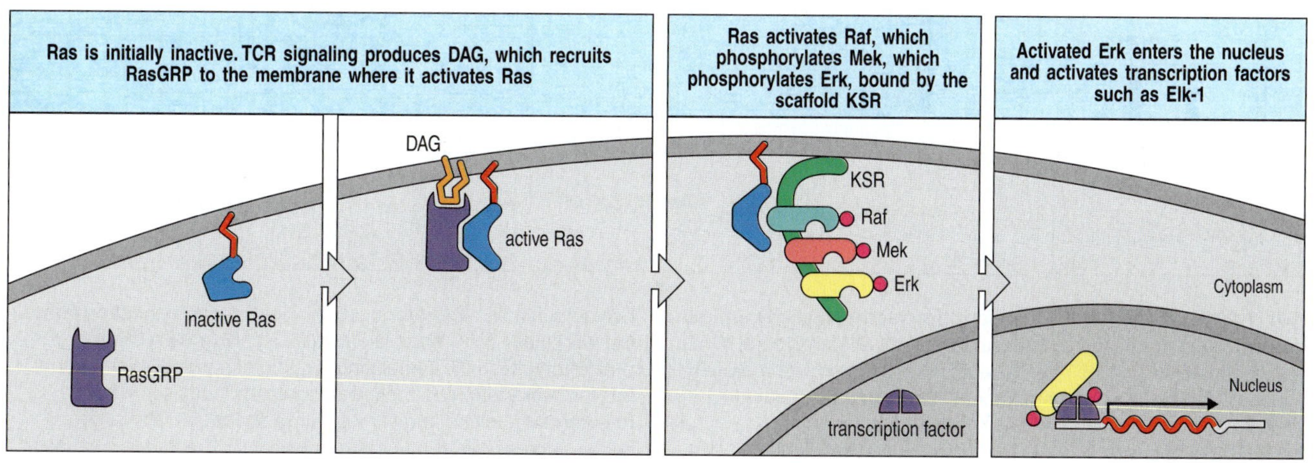

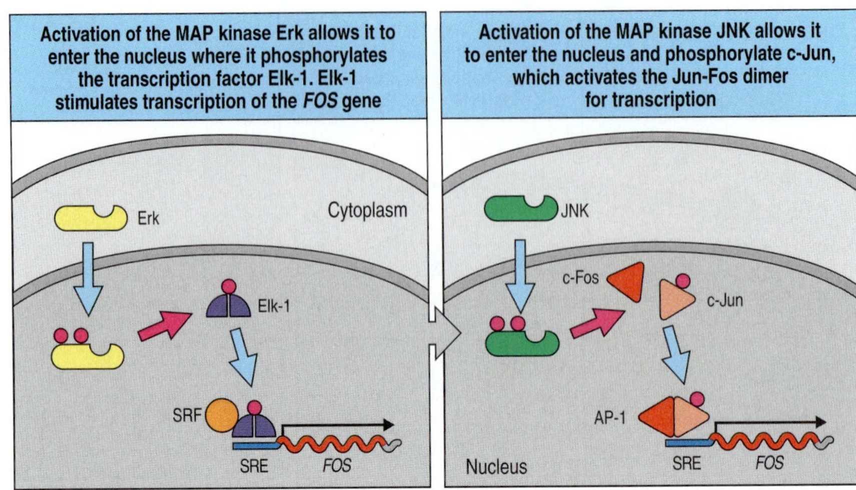

Fig. 7.18 The transcription factor AP-1 is formed as a result of the Ras/MAPK signaling pathway. Left panel: phosphorylation of the MAPK Erk activated as a result of the Ras–MAPK cascade allows Erk to enter the nucleus, where it phosphorylates the transcription factor Elk-1, which along with serum response factor (SRF) binds to the serum response element (SRE) in the promoter of the gene (*FOS*) for the transcription factor c-Fos, stimulating its transcription. Right panel: the protein kinase PKC-θ can induce the phosphorylation of another MAPK called Jun kinase (JNK), which enables it to enter the nucleus and phosphorylate the transcription factor c-Jun, which forms a dimer with c-Fos. The phosphorylated c-Jun/Fos dimer is an active AP-1 transcription factor.

cooperates with a transcription factor called serum response factor to initiate transcription of the *FOS* gene. Fos protein then associates with Jun to form the AP-1 heterodimer, but this remains transcriptionally inactive until another MAPK called **Jun kinase** (**JNK**) phosphorylates Jun.

7-14 Protein kinase C activates the transcription factors NFκB and AP-1.

The third signaling pathway leading from PLC-γ results in the activation of **PKC-θ**, an isoform of protein kinase C that is restricted to T cells and muscle. Mice lacking PKC-θ develop T cells in the thymus, but their mature T cells have a defect in the activation of two crucial transcription factors, NFκB and AP-1, in response to signaling by the T-cell receptor and CD28. This makes PKC-θ an important component of T-cell activation.

PKC-θ has a C1 domain and is recruited to the membrane when DAG is generated by activated PLC-γ (see Fig. 7.14). In this location, the kinase activity of PKC-θ initiates a series of steps that results in the activation of NFκB (Fig. 7.19). PKC-θ phosphorylates the large membrane-localized scaffold protein CARMA1, causing it to oligomerize and form a multisubunit complex

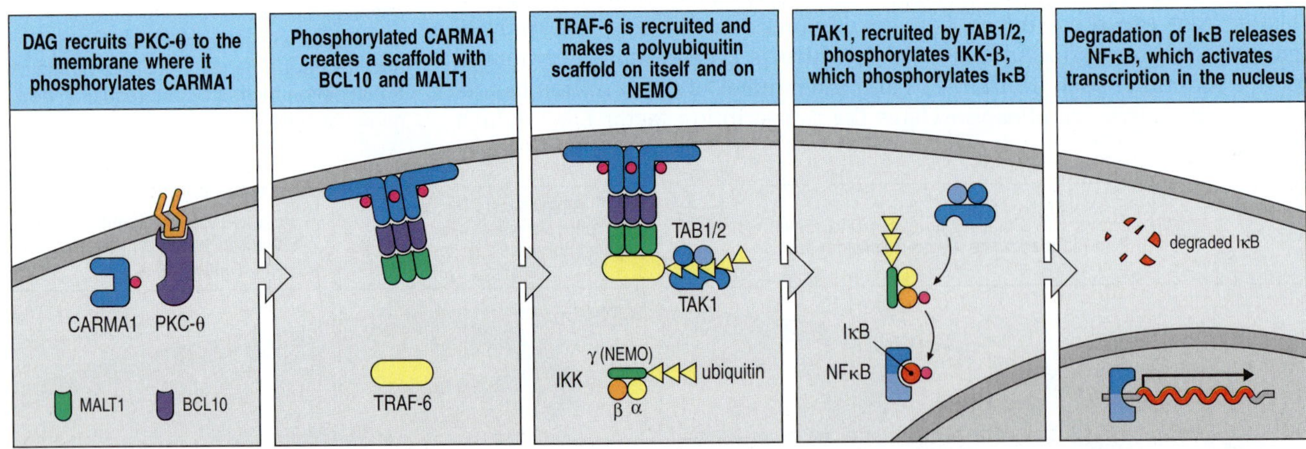

Fig. 7.19 Activation of the transcription factor NFκB by antigen receptors is mediated by protein kinase C. Diacylglycerol (DAG) produced as a result of T-cell receptor signaling recruits a protein kinase C (PKC-θ) to the membrane, where it phosphorylates a scaffold protein called CARMA1. This forms a complex with BCL10 and MALT1 that recruits the E3 ligase TRAF-6. As described in Fig. 3.13, the kinase TAK1 is recruited by the polyubiquitin scaffold produced by TRAF-6 and phosphorylates the IκB kinase (IKK) complex (IKKα:IKKβ:IKKγ (NEMO)). IKK phosphorylates IκB, stimulating its ubiquitination, which targets IκB for degradation by the proteasome and releases NFκB to enter the nucleus and stimulate transcription of its target genes. A defect in NEMO that prevents NFκB activation causes immunodeficiency, among other symptoms.

with other proteins. This complex recruits and activates TRAF-6, the same protein that we encountered in Chapter 3 in its role in activating NFκB in the TLR signaling pathway (see Fig. 3.13).

NFκB is the general name for a member of a family of homo- and heterodimeric transcription factors made up of the Rel family of proteins. The most common NFκB activated in lymphocytes is a heterodimer of p50:p65Rel. The dimer is held in an inactive state in the cytoplasm by binding to an inhibitory protein called inhibitor of κB (IκB). As described for TLR signaling (see Fig. 3.13), TRAF-6 stimulates the degradation of IκB by first activating the kinase TAK1, which activates a complex of serine kinases, IκB kinase (IKK). IKK phosphorylates IκB, causing its ubiquitination and subsequent degradation, and the consequent release and entry into the nucleus of active NFκB. Inherited deficiency of the IKKγ subunit (also called **NEMO**) leads to a syndrome known as **X-linked hypohidrotic ectodermal dysplasia and immunodeficiency**, which is characterized by developmental defects in ectodermal structures such as skin and teeth, as well as immunodeficiency.

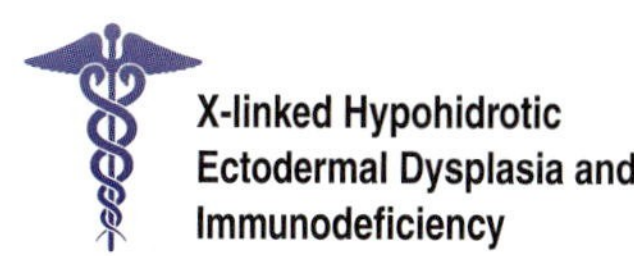

PKC-θ can also activate JNK, and might be able to activate the transcription factor AP-1 by this route. However, T cells lacking PKC-θ have a defect in AP-1 activation in addition to their defect in NFκB activation, but no defect in JNK activation, indicating that our understanding of this pathway is still incomplete.

7-15 The cell-surface protein CD28 is a co-stimulatory receptor for naive T cells.

The signaling through the T-cell receptor complex described in the previous sections is not by itself sufficient to activate a naive T cell. As noted in Chapter 1, antigen-presenting cells that can activate naive T cells bear cell-surface proteins known as **co-stimulatory molecules** or co-stimulatory ligands. These interact with cell-surface receptors, known as **co-stimulatory receptors**, on the naive T cell to transmit a signal that is required, along with antigen stimulation, for T-cell activation—this signal is sometimes called 'signal 2.' We discuss the immunological consequences of this requirement for co-stimulation in detail in Chapter 9. The best understood of these co-stimulatory receptors is the cell-surface protein **CD28**.

CD28 is present on the surface of all naive T cells and binds the co-stimulatory ligands **B7.1** (**CD80**) and **B7.2** (**CD86**), which are expressed mainly on specialized antigen-presenting cells such as dendritic cells (Fig. 7.20). To become

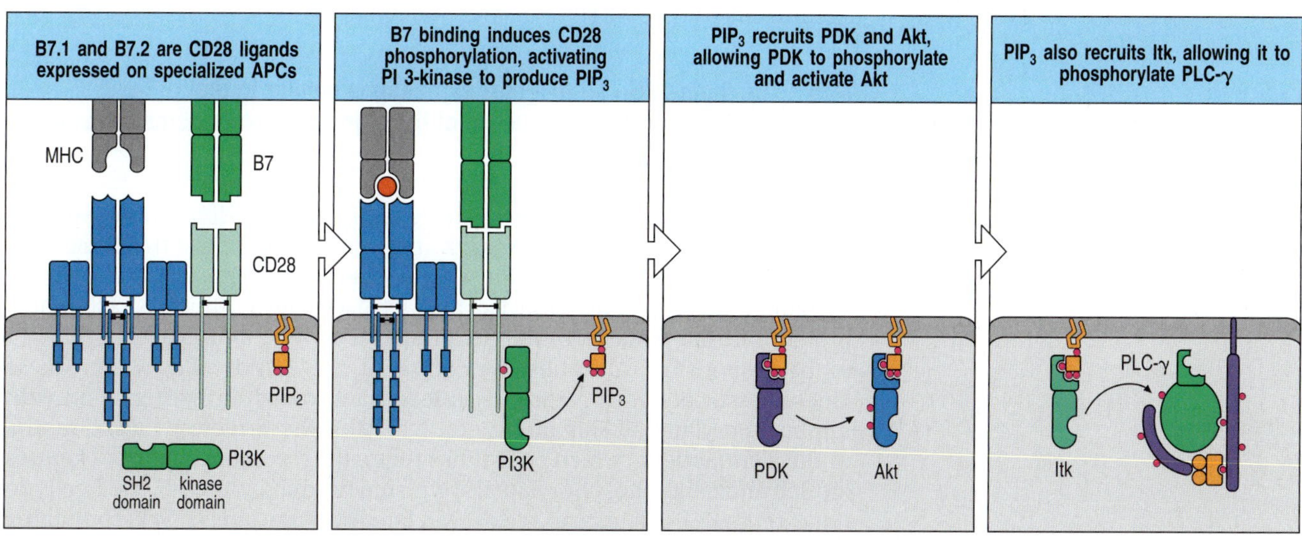

Fig. 7.20 The co-stimulatory protein CD28 transduces a variety of different signals. The ligands for CD28, namely B7.1 and B7.2, are expressed only on specialized antigen-presenting cells (APCs) such as dendritic cells (first panel). Engagement of CD28 induces its tyrosine phosphorylation, which activates PI 3-kinase (PI3K), with subsequent production of PIP_3 that recruits several enzymes via their PH domains, thus bringing them together with their substrates in the membrane. The protein kinase Akt, which becomes phosphorylated by phosphoinositide-dependent protein kinase-1 (PDK-1), is activated and enhances cell survival and upregulates cell metabolism. Recruitment of the kinase Itk to the membrane is critical for the full activation of PLC-γ (see Fig. 7.13).

Movie 7.3

activated, the naive lymphocyte must engage both antigen and a co-stimulatory ligand on the same antigen-presenting cell. The requirement for CD28 signaling thus means that naive T cells can be activated only by dedicated antigen-presenting cells expressing this molecule and not by other bystander cells that might happen to carry antigen on their surface. Because co-stimulatory ligands are induced on antigen-presenting cells by infection (see Chapter 3), this also helps ensure that T cells are activated only in response to infection. It is thought that CD28 signaling aids antigen-dependent T-cell activation mainly by promoting T-cell proliferation, cytokine production, and cell survival. All these effects are mediated by signaling motifs present in the cytoplasmic domain of CD28.

After engagement by B7 molecules, CD28 becomes tyrosine phosphorylated by Lck in its cytoplasmic domain on tyrosine residues in a YXN motif that can recruit the adaptor protein Grb2, and in a non-ITAM motif YMNM. The cytoplasmic tail of CD28 also carries a proline-rich motif (PXXP) that binds the SH3 domains of Lck and Itk. Although details are uncertain, the effect of CD28 phosphorylation is to activate PI 3-kinase to generate PIP_3, which recruits Itk to the membrane where Lck can phosphorylate it. Activated Itk can then phosphorylate and activate PLC-γ (see Fig. 7.13). Thus, signaling through both the T-cell receptor and CD28 contribute to PLC-γ activation.

Another effect of PIP_3 is to recruit the serine/threonine kinase Akt (also known as protein kinase B) to the membrane via Akt's PH domain (see Fig. 7.20). In that location, Akt becomes activated and can then phosphorylate a variety of downstream proteins. One of its effects is to promote cell survival by inhibiting the cell-death pathway that we discuss later in this chapter; another is to stimulate the cell's metabolism by increasing the utilization of glucose. PIP_3 can also recruit guanine-nucleotide exchange factors of the **Vav** family, which regulate the cytoskeleton and additional signaling pathways.

In T cells, one of the major functions of NFAT, AP-1, and NFκB is to act together to stimulate expression of the gene for the cytokine IL-2, which is essential for promoting T-cell proliferation and differentiation into effector cells. The promoter for the *IL-2* gene contains multiple regulatory elements that must be bound by transcription factors to initiate *IL-2* expression. Some control sites are already bound by transcription factors, such as Oct1, that are produced constitutively in lymphocytes, but this is not sufficient to switch on *IL-2*. Only when AP-1, NFAT, and NFκB are all activated and are bound to their control sites in the *IL-2* promoter is the gene expressed. NFAT and AP-1 bind to the promoter cooperatively and with higher affinity by forming a heterotrimer of NFAT, Jun, and Fos. Thus, the *IL-2* promoter integrates signals from the different signaling pathways to ensure that IL-2 is produced only in appropriate circumstances (Fig. 7.21).

7-16 The logic of B-cell receptor signaling is similar to that of T-cell receptor signaling, but some of the signaling components are specific to B cells.

There are many similarities between signaling from T-cell receptors and signaling from B-cell receptors. As with the T-cell receptor, the B-cell receptor is composed of antigen-specific chains associated with ITAM-containing signaling chains, in this case Igα and Igβ (see Fig. 7.10). In B cells, three protein tyrosine kinases of the Src family—Fyn, Blk, and Lyn—are thought to be responsible for phosphorylation of the ITAMs (Fig. 7.22). These kinases associate with resting receptors via a low-affinity interaction with unphosphorylated ITAMs in Igα and Igβ. After the receptors have bound a multivalent antigen, which cross-links them, the receptor-associated kinases are activated and phosphorylate the tyrosine residues in the ITAMs. B cells do not express ZAP-70; instead, a closely related tyrosine kinase, **Syk**, containing

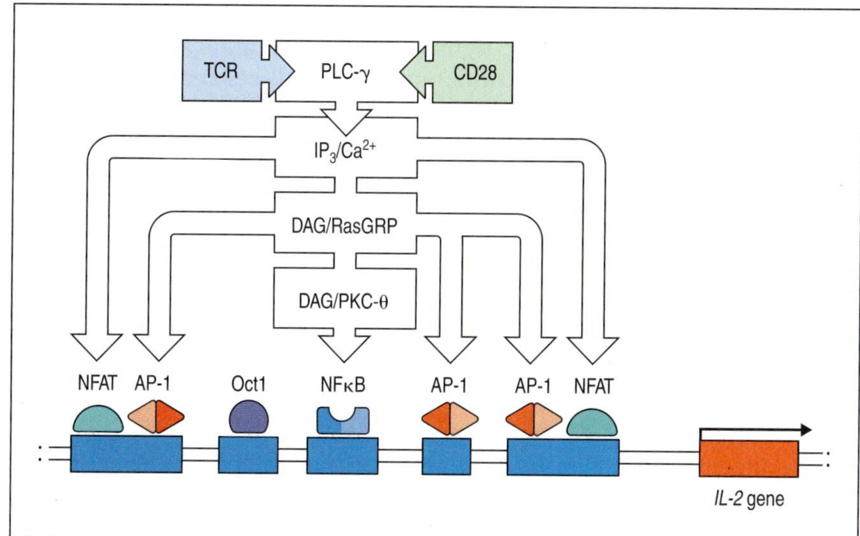

Fig. 7.21 Multiple signaling pathways converge on the *IL-2* promoter. AP-1, NFAT, and NFκB binding to the promoter of the *IL-2* gene integrate multiple signaling pathways into a single output, the production of the cytokine IL-2. The MAPK pathway activates AP-1; calcium activates NFAT; and protein kinase C activates NFκB. All three pathways are required to stimulate *IL-2* transcription. Activation of the gene requires both the binding of NFAT and AP-1 to a specific promoter element, and the additional binding of AP-1 on its own to another site. Oct1 is a transcription factor that is required for *IL-2* transcription. Unlike the other transcription factors, it is constitutively bound to the promoter and is therefore not regulated by T-cell receptor signaling.

two SH2 domains, is recruited to the phosphorylated ITAM. In contrast to ZAP-70, which requires additional Lck phosphorylation for activation, Syk is activated simply by its binding to the phosphorylated site.

The B-cell equivalent to the co-receptors CD4 and CD8 is a complex of cell-surface proteins—**CD19**, **CD21**, and **CD81**—which is known as the **B-cell co-receptor** (Fig. 7.23). As with T cells, antigen-dependent signaling from the B-cell receptor is enhanced if the B-cell co-receptor is simultaneously bound by its ligand and clusters with the antigen receptor. CD21 (also known as complement receptor 2, CR2) is a receptor for the C3dg fragment of complement. This means that antigens such as bacterial pathogens on which C3dg is bound (see Fig. 7.23) can cross-link the B-cell receptor with the CD21:CD19:CD81 complex. This induces phosphorylation of the cytoplasmic tail of CD19 by B-cell receptor-associated tyrosine kinases, which in turn leads to the binding of Src-family kinases, the augmentation of signaling through the B-cell receptor itself, and the recruitment of PI 3-kinase (see Section 7-4). PI 3-kinase initiates an additional signaling pathway to that leading from the B-cell receptor (see Fig. 7.23). Thus, the B-cell co-receptor serves to strengthen the signal resulting from antigen recognition. The role of the third component of the B-cell receptor complex, CD81 (TAPA-1), is as yet unknown.

Once activated, Syk phosphorylates the scaffold protein **BLNK** (also known as SLP-65). Like LAT in T cells, BLNK has multiple sites for tyrosine phosphorylation and recruits a variety of SH2-containing proteins, including enzymes and adaptor proteins, to form several distinct multiprotein signaling complexes that can act in concert. As in T cells, a key signaling protein is the phospholipase PLC-γ, which is activated with the aid of the B-cell-specific Tec kinase

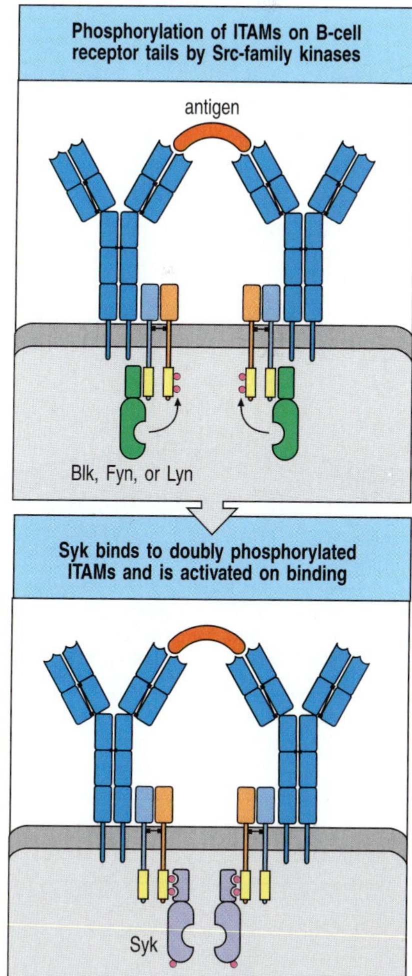

Fig. 7.22 Src-family kinases are associated with B-cell receptors and phosphorylate the tyrosines in ITAMs to create binding sites for Syk and Syk activation via transphosphorylation. The membrane-bound Src-family kinases Fyn, Blk, and Lyn associate with the B-cell antigen receptor by binding to ITAMs, either (as shown in the figure) through their amino-terminal domains or by binding a single phosphorylated tyrosine through their SH2 domains. After ligand binding and receptor clustering, the associated kinases phosphorylate tyrosines in the ITAMs on the cytoplasmic tails of Igα and Igβ. Subsequently, Syk binds to the phosphorylated ITAMs of the Igβ chain. Because there are at least two receptor complexes in each cluster, Syk molecules become bound in close proximity and can activate each other by transphosphorylation, thus initiating further signaling.

Bruton's tyrosine kinase (**Btk**) and hydrolyzes PIP_2 to form DAG and IP_3. As discussed for the T-cell receptor, signaling by calcium and DAG leads to the activation of downstream transcription factors. The B-cell receptor signaling pathway is summarized in Fig. 7.24. A deficiency in Btk (which is encoded by a gene on the X chromosome) prevents the development and functioning of B cells, resulting in the disease **X-linked agammaglobulinemia**, which is characterized by a lack of antibodies. Besides Btk, mutations in other signaling molecules in B cells, including receptor chains and BLNK, have been linked to B-cell immunodeficiencies.

7-17 ITAMs are also found in other receptors on leukocytes that signal for cell activation.

Other immune-system receptors also use ITAM-containing accessory chains to transduce activating signals (Fig. 7.25). One example is **FcγRIII** (CD16); this is a receptor for IgG that triggers antibody-dependent cell-mediated cytotoxicity (ADCC) by NK cells, which we consider in Chapter 11; CD16 is also found on macrophages and neutrophils, where it facilitates the uptake and destruction of antibody-bound pathogens. To signal, FcγRIII must associate with either the ζ chain found also in the T-cell receptor, or with another member of the same protein family known as the Fcγ chain. The Fcγ chain is also the signaling component of another Fc receptor—the Fcε receptor I (FcεRI) on mast cells. As we discuss in Chapter 14, this receptor binds IgE antibodies, and on cross-linking by allergens it triggers the degranulation of mast cells. Lastly, many activating receptors on NK cells are associated with DAP12, another ITAM-containing protein (see Section 3-21).

Several viral pathogens seem to have acquired ITAM-containing receptors from their hosts. These include the Epstein–Barr virus (EBV), whose *LMP2A* gene encodes a membrane protein with a cytoplasmic tail containing an ITAM. This enables EBV to trigger B-cell proliferation by the signaling pathways discussed in Section 7-15 and the preceding sections. Another virus that expresses an ITAM-containing protein is the Kaposi sarcoma herpes virus (KSHV or HHV8), which also causes malignant transformation and proliferation of the cells it infects.

7-18 Inhibitory receptors on lymphocytes help regulate immune responses.

CD28 is one of a structurally related family of receptors that are expressed by lymphocytes and bind B7-family ligands. Some, such as the receptor ICOS, which is discussed in Chapter 9, act as activating receptors, but others inhibit signaling by the antigen receptors, can stimulate apoptosis, and are important in regulating the immune response. Inhibitory receptors related to CD28 and expressed by T cells include **CTLA-4** (CD152) and **PD-1** (**programmed death-1**), while the **B and T lymphocyte attenuator** (**BTLA**) is expressed by both T cells and B cells. Of these, CTLA-4 seems to be the most important: mice lacking CTLA-4 die at a young age from an uncontrolled proliferation of

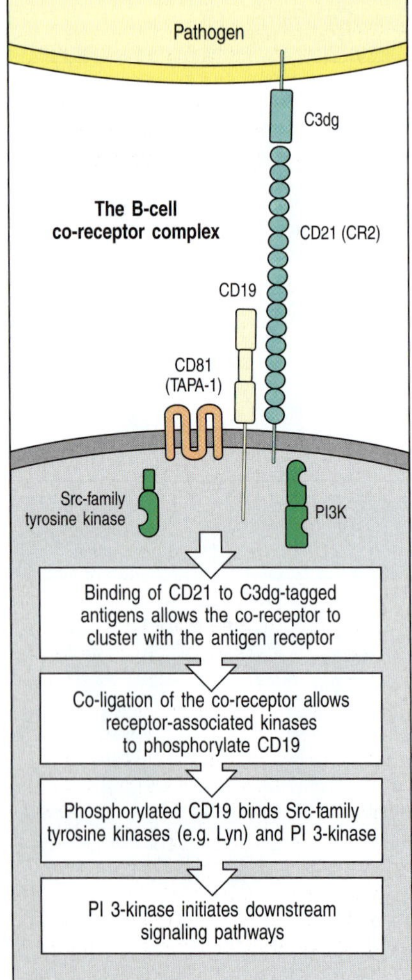

Fig. 7.23 B-cell antigen receptor signaling is modulated by a co-receptor complex of at least three cell-surface molecules, CD19, CD21, and CD81. Cleavage of the antigen-bound complement component C3 to C3dg (see Fig. 2.28) allows the tagged antigen to bind to both the B-cell receptor and the cell-surface protein CD21 (complement receptor 2, CR2), a component of the B-cell co-receptor complex. Cross-linking and clustering of the co-receptor with the antigen receptor results in the phosphorylation of tyrosine residues in the cytoplasmic domain of CD19 by protein kinases associated with the B-cell receptor; other Src-family kinases can bind to phosphorylated CD19 and so augment signaling through the B-cell receptor. Phosphorylated CD19 can also bind PI 3-kinase.

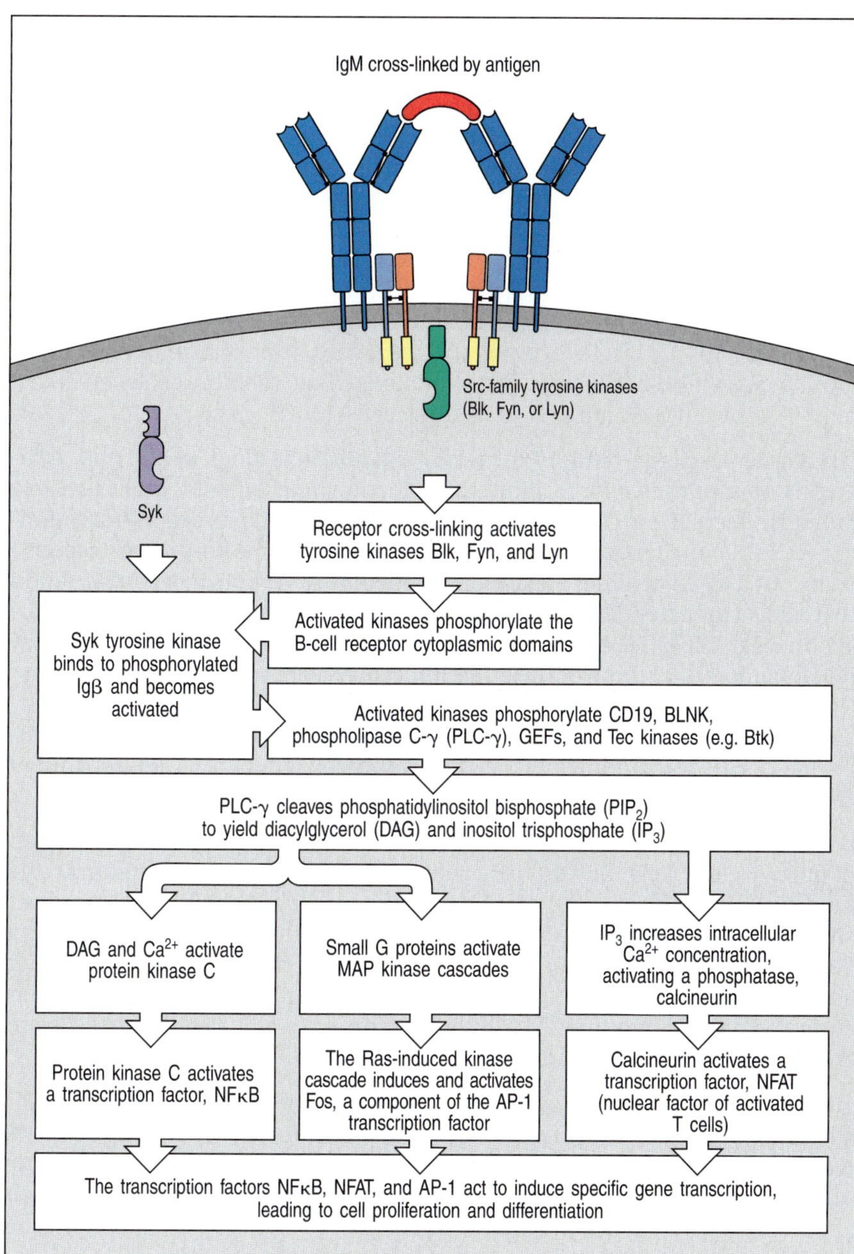

Fig. 7.24 Simplified outline of the intracellular signaling pathways initiated by cross-linking of B-cell receptors by antigen. Cross-linking of surface immunoglobulin molecules activates the receptor-associated Src-family protein tyrosine kinases Blk, Fyn, and Lyn. The receptor-associated kinases phosphorylate the ITAMs in the receptor complex, which bind and activate the cytosolic protein kinase Syk, whose activation is described in Fig. 7.22. Syk then phosphorylates other targets, including the adaptor protein BLNK, which helps to recruit Tec kinases that in turn phosphorylate and activate the enzyme PLC-γ. PLC-γ cleaves the membrane phospholipid PIP_2 into IP_3 and DAG, thus initiating two of the three main signaling pathways to the nucleus. IP_3 releases Ca^{2+} from intracellular and extracellular sources, and Ca^{2+}-dependent enzymes are activated, whereas DAG activates protein kinase C with the help of Ca^{2+}. The third main signaling pathway is initiated by guanine-nucleotide exchange factors (GEFs) that become associated with the receptor and activate small GTP-binding proteins such as Ras. These in turn trigger protein kinase cascades (MAP kinase cascades) that lead to the activation of MAP kinases that move into the nucleus and phosphorylate proteins that regulate gene transcription. This scheme is a simplification of the events that actually occur during signaling, showing only the main events and pathways.

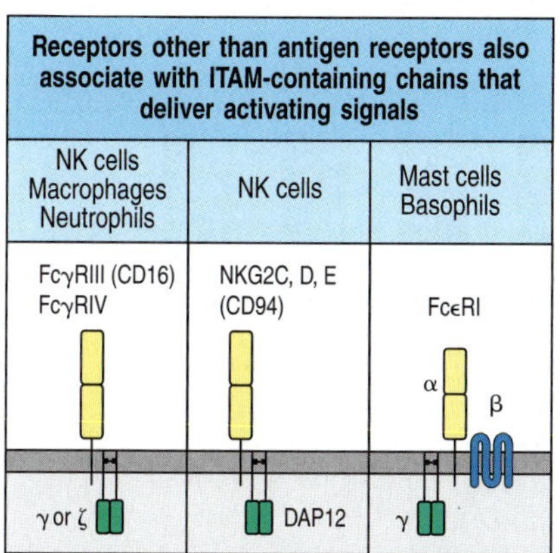

Fig. 7.25 Other receptors that pair with ITAM-containing chains can deliver activating signals. Cells other than B and T cells have receptors that pair with accessory chains containing ITAMs, which are phosphorylated when the receptor is cross-linked. These receptors deliver activating signals. The Fcγ receptor III (FcγRIII, or CD16) is found on NK cells, macrophages, and neutrophils. Binding of IgG to this receptor activates the killing function of the NK cell, leading to the process known as antibody-dependent cell-mediated cytotoxicity (ADCC). Activating receptors on NK cells, such as NKG2C, NKG2D, and NKG2E, also associate with ITAM-containing signaling chains. The Fcε receptor (FcεRI) is found on mast cells and basophils. The α subunit binds to IgE antibodies with very high affinity. The β subunit is a four-spanning transmembrane protein. When antigen subsequently binds to the IgE, the mast cell is triggered to release granules containing inflammatory mediators. The γ chain associated with the Fc receptors, and the DAP12 chain that associates with the NK killer-activating receptors, also contain one ITAM per chain and are present as homodimers.

T cells in multiple organs, whereas loss of PD-1 or BTLA causes less marked changes of a quantitative rather than a qualitative nature.

CTLA-4 is induced on activated T cells and binds to the same co-stimulatory ligands (B7.1 and B7.2) as CD28, but CTLA-4 engagement is inhibitory for T-cell activation, rather than enhancing it. The function of CTLA-4 is controlled largely by regulation of its surface expression. Initially, CTLA-4 resides on intracellular membranes but moves to the cell surface after T-cell receptor signaling. The surface expression of CTLA-4 is controlled by phosphorylation of the tyrosine-based motif GVYVKM in its cytoplasmic tail. When this motif is not phosphorylated, it is able to bind to the clathrin adaptor molecule AP-2, which removes CTLA-4 from the surface. When it is phosphorylated, this motif cannot bind AP-2, and CTLA-4 remains in the membrane, where it can bind B7 molecules on antigen-presenting cells.

CTLA-4 has a higher affinity for its B7 ligands than does CD28, and, apparently of importance for its inhibitory function, it engages B7 molecules in a different orientation. CD28, CTLA-4, and B7.1 are all expressed as homodimers. A CD28 dimer engages one B7.1 dimer in a direct one-to-one correspondence, but a CTLA-4 dimer engages two different B7 dimers in a configuration that allows for extended cross-linkages that confer high avidity on the interaction (Fig. 7.26). CTLA-4 was once presumed to act by recruiting inhibitory phosphatases, like some of the other inhibitory receptors described later, but this is no longer thought to be so. It is still not clear whether CTLA-4 directly activates inhibitory signaling pathways. Instead, its actions may result in part from blocking the binding of CD28 to B7, thereby reducing CD28-dependent co-stimulation.

CTLA-4-expressing T cells can also exert an inhibitory effect on the activation of other T cells. How they do this is not yet clear, but it might result from the binding up of B7 molecules on antigen-presenting cells by CTLA-4, in effect stealing the CD28 required by the other T cells. Direct actions of CTLA-4 on T cells have not been excluded, however. Notably, the regulatory T cells that are needed to suppress autoimmunity express high levels of CTLA-4 on their surface, and they require CTLA-4 to function normally. Regulatory cells are described in detail in Chapter 9.

Some other receptors that can inhibit T-cell activation possess motifs in their cytoplasmic regions that are known as the **immunoreceptor tyrosine-based inhibitory motif** (**ITIM**, consensus sequence [I/V]XYXX[L/I], where X is any amino acid) (Fig. 7.27) or the related **immunoreceptor tyrosine-based switch**

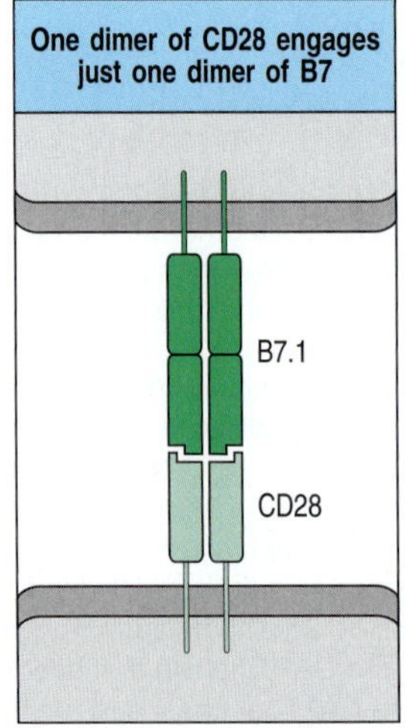

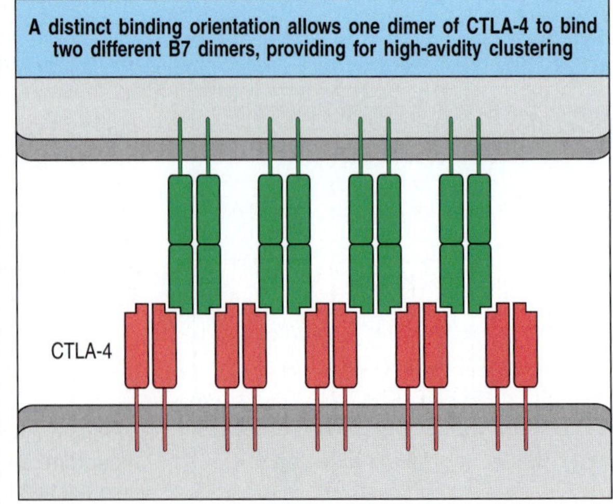

A distinct binding orientation allows one dimer of CTLA-4 to bind two different B7 dimers, providing for high-avidity clustering
CTLA-4

Fig. 7.26 CTLA-4 has a higher affinity than CD28 for B7 and engages it in a multivalent orientation. CD28 and CTLA-4 are both expressed as dimers on the cell surface and both bind to the ligands B7.1, which is a dimer, and B7.2, which is not. However, the orientation of the B7 binding of CD28 and CTLA-4 differ in a way that contributes to the inhibitory action of CTLA-4. One dimer of CD28 engages just one dimer of B7.1. But one dimer of CTLA-4 binds in such a way that two different dimers of B7.1 are engaged at once, allowing these molecules to cluster into complexes of high avidity. This, and the higher affinity of CTLA-4 for B7 molecules, may give it an advantage in competing for available B7 molecules on an antigen-presenting cell, providing one mechanism by which it could block the co-stimulation of T cells.

motif (**ITSM**, consensus sequence, TXYXX[V/I]). When the tyrosine in an ITIM or ITSM is phosphorylated, it can recruit either of two inhibitory phosphatases, called **SHP** (SH2-containing phosphatase) and **SHIP** (SH2-containing inositol phosphatase), via their SH2 domains. SHP is a protein tyrosine phosphatase and removes phosphate groups added by tyrosine kinases to a variety of proteins. SHIP is an inositol phosphatase and removes the phosphate from PIP_3 to give PIP_2, thus reversing the recruitment of proteins such as Tec kinases and Akt to the cell membrane and thereby inhibiting signaling.

One ITIM-containing receptor is PD-1 (see Fig. 7.27), which is induced transiently on activated T cells, B cells, and myeloid cells. It can bind to the B7-family ligands **PD-L1** (programmed death ligand-1, B7-H1) and **PD-L2** (programmed death ligand-2, B7-DC). Despite their names, we now understand that these proteins function as ligands for the inhibitory receptor PD-1, rather than acting directly in cell death. PD-L1 is constitutively expressed by a wide variety of cells, whereas PD-L2 expression is induced on antigen-presenting cells during inflammation. Because PD-L1 is expressed constitutively, regulation of its expression could have a critical role in controlling T-cell responses. For example, signaling by pro-inflammatory cytokines can repress PD-1, thus enhancing the T-cell response. Mice lacking PD-1 gradually develop autoimmunity, presumably because of an inability to regulate T-cell activation. In chronic infections, the widespread expression of PD-1 reduces the effector activity of T cells; this helps to limit potential damage to bystander cells, but at the expense of pathogen clearance.

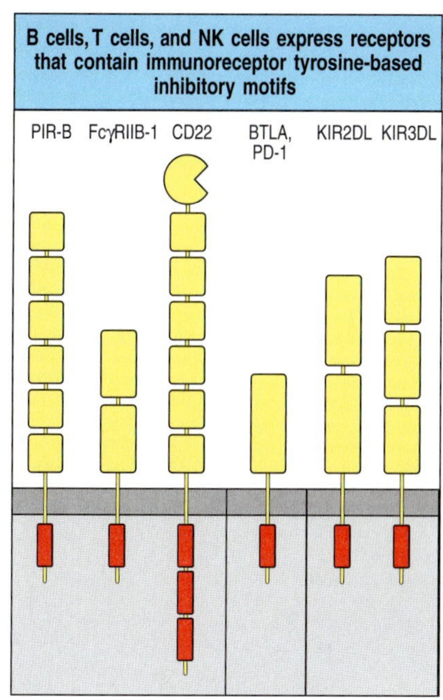
B cells, T cells, and NK cells express receptors that contain immunoreceptor tyrosine-based inhibitory motifs
PIR-B
FcγRIIB-1
CD22
BTLA, PD-1
KIR2DL
KIR3DL

Fig. 7.27 Some lymphocyte cell-surface receptors contain motifs involved in downregulating activation. Several receptors that transduce signals that inhibit lymphocyte or NK-cell activation contain one or more ITIMs (immunoreceptor tyrosine-based inhibitory motifs) in their cytoplasmic tails. ITIMs bind to various phosphatases that, when activated, inhibit signals derived from ITAM-containing receptors.

BTLA contains an ITIM and an ITSM and is expressed on activated T cells and B cells, as well as on some cells of the innate immune system. Unlike other CD28-family members, however, BTLA does not interact with B7 ligands but binds a member of the tumor necrosis factor (TNF) receptor family called the **herpes virus entry molecule** (**HVEM**), which is highly expressed on resting T cells and immature dendritic cells.

Other structural types of receptors on B cells and T cells also contain ITIMs and can inhibit cell activation when ligated along with the antigen receptors. One example is the receptor **FcγRIIB-1** on B cells, which binds the Fc region of IgG antibodies. It has long been known that the activation of naive B cells in response to antigen can be inhibited by soluble IgG antibodies that recognize the same antigen and therefore co-ligate the B-cell receptor with this Fc receptor. The ITIM in FcγRIIB-1 recruits SHIP into a complex with the B-cell receptor to block the actions of PI 3-kinase. Another inhibitory receptor on B cells is the transmembrane protein **CD22**, which contains an ITIM that interacts with SHP, which can dephosphorylate adaptors such as BLNK that associate with CD22, thereby inhibiting signaling from the B-cell receptor.

The ITIM motif is also an important motif in signaling by receptors on NK cells that inhibit the killer activity of these cells (see Section 3-21). These inhibitory receptors recognize MHC class I molecules and transmit signals that inhibit the release of the NK cell's cytotoxic granules when the NK cell recognizes a healthy uninfected cell.

Summary.

The antigen receptors on the surface of lymphocytes are multiprotein complexes in which the antigen-binding chains interact with additional proteins that are responsible for signaling from the receptor. These protein chains carry tyrosine-containing signaling motifs known as ITAMs. Activation of the receptors by antigen results in phosphorylation of the ITAMs by Src-family kinases. The phosphorylated ITAM then recruits another tyrosine kinase known as ZAP-70 in T cells and Syk in B cells. Activation of ZAP-70 or Syk results in the phosphorylation of scaffolds called LAT and SLP-76 in T cells, and BLNK in B cells. The most important of the signaling proteins recruited and activated by these phosphorylated scaffolds is phospholipase C-γ, which when activated generates inositol trisphosphate (IP_3) and diacylglycerol (DAG). IP_3 has an important role in inducing changes in intracellular calcium concentrations, and DAG is involved in activating protein kinase C-θ and the small G protein Ras. With contributions from CD28 signaling, these pathways ultimately result in the activation of three transcription factors, namely AP-1, NFAT, and NFκB, which together induce transcription of the cytokine IL-2, which is essential for the proliferation and further differentiation of the activated lymphocyte. Signaling by antigen receptors is facilitated by co-receptors that become engaged as a result of receptor–antigen binding. These co-receptors are the MHC-binding CD4 and CD8 transmembrane proteins in T cells and the complement-binding B-cell co-receptor complex in B cells. An important secondary signaling system in naive T cells is provided by the CD28 family of co-stimulatory proteins, which bind members of the B7 family of proteins. Activating members of the CD28 family provide co-stimulatory signals that amplify the signal from the T-cell receptor and are important in ensuring the activation of naive T cells by the appropriate target cell. Inhibitory members of this and other receptor families function to attenuate or completely block signaling by activating receptors. The regulated expression of activating and inhibitory receptors and their ligands generates a sophisticated level of control of immune responses that is only beginning to be understood.

Other receptors and signaling pathways.

Lymphocytes are normally studied in terms of their responsiveness to antigen. However, they and other immune-system cells bear many other receptors that make them aware of events occurring both in their immediate neighborhood and at distant sites in the body. In this part of the chapter, we focus on the mechanism of signal transduction by three classes of receptors: receptors for the hematopoietin family of cytokines, receptors for the TNF cytokine family, and the 'death receptors' that signal for apoptosis. The receptor for the cytokine IL-1 signals via a pathway almost identical to that of TLR-4 (see Chapter 3).

7-19 Cytokines and their receptors fall into distinct families of structurally related proteins.

One major way in which cells of the immune system communicate with each other and with the other cells of the body is through a class of small secreted proteins known as cytokines, some of which were introduced in Chapter 3. Cytokines are usually secreted in response to an extracellular stimulus and may act on the cells that produce them, on other cells in the immediate vicinity, or on cells at a distance after being transported in blood or tissue fluids. They can influence the growth, development, functional differentiation, and activation of lymphocytes and other leukocytes. Cytokines produce immediate responses in the cells they affect, and the signaling properties of the cytokine receptors reflect this, causing rapid changes in gene expression in the nucleus.

Cytokines can be grouped by structure into families—the **hematopoietins**, the **interferons**, and the **TNF family**—and their receptors can likewise be grouped (Fig. 7.28). We have already encountered members of all of these families in Chapter 3. One large structurally related class of cytokine receptors, the hematopoietin receptor family, are tyrosine kinase-associated receptors that

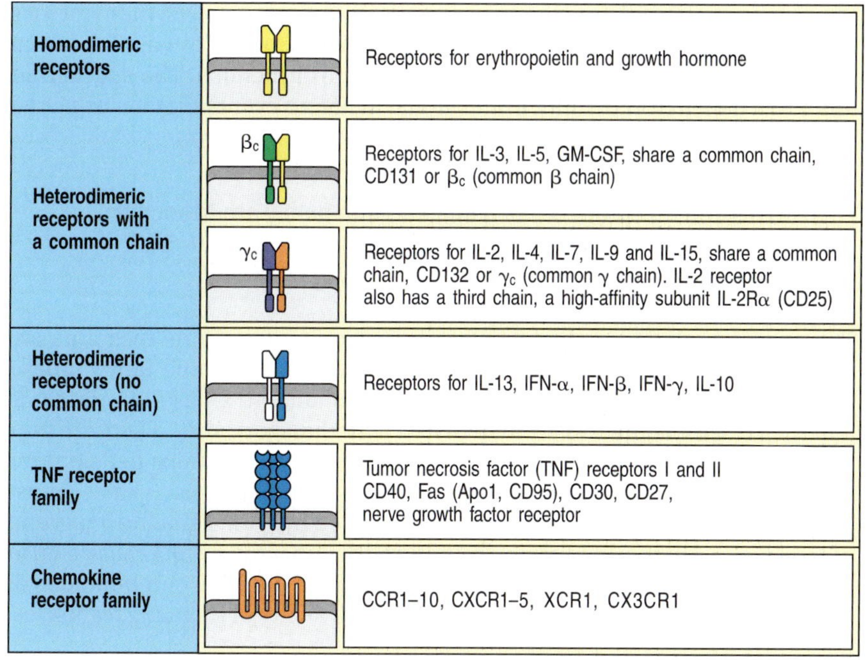

Homodimeric receptors
Receptors for erythropoietin and growth hormone
Heterodimeric receptors with a common chain
βc
Receptors for IL-3, IL-5, GM-CSF, share a common chain, CD131 or βc (common β chain)
γc
Receptors for IL-2, IL-4, IL-7, IL-9 and IL-15, share a common chain, CD132 or γc (common γ chain). IL-2 receptor also has a third chain, a high-affinity subunit IL-2Rα (CD25)
Heterodimeric receptors (no common chain)
Receptors for IL-13, IFN-α, IFN-β, IFN-γ, IL-10
TNF receptor family
Tumor necrosis factor (TNF) receptors I and II CD40, Fas (Apo1, CD95), CD30, CD27, nerve growth factor receptor
Chemokine receptor family
CCR1–10, CXCR1–5, XCR1, CX3CR1

Fig. 7.28 Cytokine receptors belong to families of receptor proteins, each with a distinctive structure. Many cytokines signal through receptors of the hematopoietin receptor superfamily, named after its first member, the erythropoietin receptor. The hematopoietin receptor superfamily includes homodimeric and heterodimeric receptors, which are subdivided into families on the basis of protein sequence and structure. Examples of these are given in the first three rows. The α chain often defines the ligand specificity of the receptor, whereas the β or γ chain confers the intracellular signaling function. A smaller number of receptors fall into the class II cytokine receptor superfamily, such as receptors for interferons or interferon-like cytokines (fourth row). Other superfamilies of cytokine receptors are the tumor necrosis factor receptor (TNFR) family and the chemokine receptor family, the latter belonging to the very large family of G-protein-coupled receptors. For the TNFR family, the ligands act as trimers and may be associated with the cell membrane rather than being secreted.

form dimers when their cytokine ligand binds. Dimerization initiates intracellular signaling from the tyrosine kinases associated with the cytoplasmic domains of the receptor. Some types of cytokine receptors are composed of two identical subunits, but others have two different subunits. An important feature of cytokine signaling is the large variety of different receptor subunit combinations that occur.

Many of the soluble cytokines made by activated T cells are members of the hematopoietin family. These cytokines and their receptors can be further divided into subfamilies characterized by functional similarities and genetic linkage. For instance, IL-3, IL-4, IL-5, IL-13, and GM-CSF are related structurally, their genes are closely linked in the genome, and they are often produced together by the same kinds of cells. In addition, they bind to closely related receptors, which belong to the family of class I cytokine receptors. The IL-3, IL-5, and GM-CSF receptors form a subgroup that shares a common β chain. Another subgroup of class I cytokine receptors is defined by the use of the γ chain of the IL-2 receptor. This chain is shared by receptors for the cytokines IL-2, IL-4, IL-7, IL-9, and IL-15; it is called the **common γ chain** (γ_c) and is encoded by a gene located on the X chromosome. Mutations that inactivate γ_c cause an **X-linked severe combined immunodeficiency** (**X-linked SCID**) due to inactivation of the signaling pathways for several cytokines—IL-7, IL-15, and IL-2—that are required for normal lymphocyte development. More distantly related, the receptor for IFN-γ is a member of a small family of heterodimeric cytokine receptors with some similarities to the hematopoietin receptor family. These so-called class II cytokine receptors include the receptor for IFN-α and IFN-β, and the IL-10 receptor.

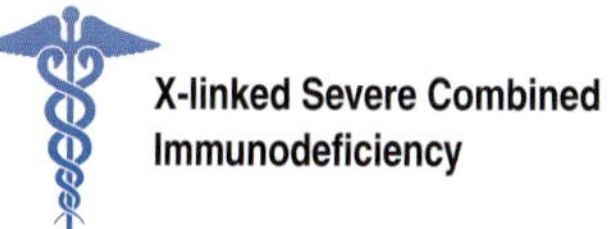

The second class of cytokine receptors contains those for cytokines of the TNF family. These are structurally unrelated to the receptors described above but also have to cluster to become activated. TNF-family cytokines, such as TNF-α and lymphotoxin, are produced as trimers, and binding of these cytokines induces the clustering of three identical receptor subunits. Many cytokines of this family are transmembrane proteins or proteins that remain associated with the cell surface. Nevertheless, they share some important properties with the soluble T-cell cytokines, because they are also synthesized *de novo* by T cells after antigen recognition and affect the behavior of the target cell.

Overall, the structural, functional, and genetic relations between the cytokines and their receptors suggest that they may have diversified in parallel during the evolution of increasingly specialized effector functions. The specialized functions depend on distinct intracellular signaling events that are triggered by the cytokines binding to their specific receptors.

7-20 Cytokine receptors of the hematopoietin family are associated with the JAK family of tyrosine kinases, which activate STAT transcription factors.

Movie 7.4

The signaling chains of the hematopoietin family of cytokine receptors are noncovalently associated with protein tyrosine kinases of the **Janus kinase (JAK) family**—so called because they have two tandem kinase-like domains and thus resemble the two-headed mythical Roman god Janus. There are four members of the JAK family: Jak1, Jak2, Jak3, and Tyk2. As mice deficient for individual JAK family members show different phenotypes, each kinase must have a distinct function. For example, Jak3 is used by γ_c for signaling by several of the cytokines described above. Mutations that inactivate Jak3 cause a form of **SCID** that is not X-linked. Presumably, the use of different combinations of JAKs by different cytokine receptors enables a diversity of signaling responses.

The dimerization or clustering of the signaling chains allows the JAKs to cross-phosphorylate each other on tyrosine, thus stimulating their kinase activity. These activated JAKs then phosphorylate their associated receptors on specific tyrosine residues to generate binding sites for proteins with SH2 domains (Fig. 7.29). Some of the tyrosine-phosphorylated sites recruit SH2-containing transcription factors known as **signal transducers and activators of transcription (STATs)**, which reside in the cytoplasm in an inactive form until activated by cytokine receptors.

There are seven STATs (1–4, 5a, 5b, and 6). The specificity of a particular STAT for a particular receptor is determined by the recognition of the distinctive phosphotyrosine sequence on the activated receptor by the SH2 domain of the STAT. Recruitment of a STAT to the activated receptor brings the STAT close to an activated JAK, which can then phosphorylate it on a conserved tyrosine in its carboxy terminus. This leads to a conformational change: two STATs form a dimer in which the phosphotyrosine on each STAT protein binds to the SH2 domain of the other. STATs predominantly form homodimers, with

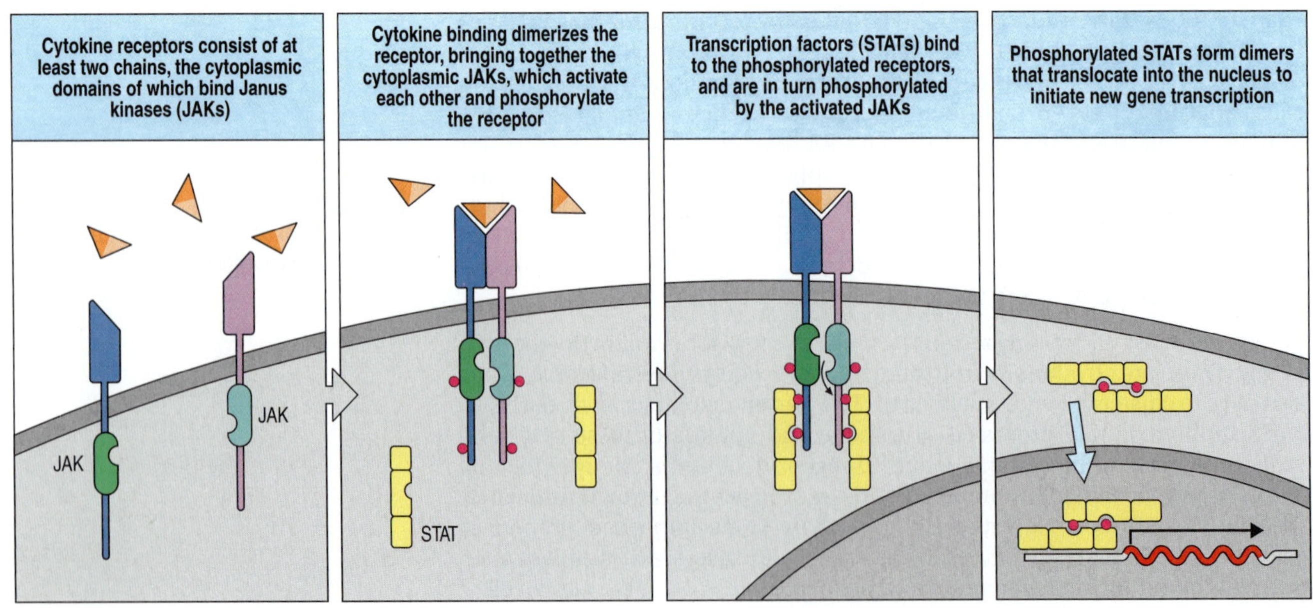

Cytokine receptors consist of at least two chains, the cytoplasmic domains of which bind Janus kinases (JAKs)
JAK
JAK
Cytokine binding dimerizes the receptor, bringing together the cytoplasmic JAKs, which activate each other and phosphorylate the receptor
STAT
Transcription factors (STATs) bind to the phosphorylated receptors, and are in turn phosphorylated by the activated JAKs
Phosphorylated STATs form dimers that translocate into the nucleus to initiate new gene transcription

Fig. 7.29 Many cytokine receptors signal using a rapid pathway called the JAK–STAT pathway. Many cytokines act via receptors that are associated with cytoplasmic Janus kinases (JAKs). The receptor consists of at least two chains, each associated with a specific JAK (first panel). Binding of dimeric ligand results in dimerization of the receptor chains, bringing together the JAKs, which can phosphorylate and activate each other. The activated JAKs then phosphorylate tyrosines in the receptor tails (second panel). Members of the STAT (signal transducers and activators of transcription) family of proteins (which contain SH2 domains) bind to the tyrosine-phosphorylated receptors and are themselves phosphorylated by the JAKs (third panel). After phosphorylation, STAT proteins form a dimer by the binding of their SH2 domains to phosphotyrosine residues on the other STAT, and translocate to the nucleus (last panel), where they bind to and activate the transcription of a variety of genes important for adaptive immunity.

a cytokine typically activating one type of STAT. For example, IFN-γ activates STAT1 and generates STAT1 homodimers, whereas IL-4 activates STAT6, generating STAT6 homodimers. Other cytokine receptors can activate several STATs, and some STAT heterodimers can be formed. The phosphorylated STAT dimer enters the nucleus, where it acts as a transcription factor to initiate the expression of selected genes. Genes regulated by STATs include ones that contribute to the growth and differentiation of particular subsets of lymphocytes. An example of the specificity of STAT-mediated transcription is that STAT1 and STAT4 are essential for T_H1 cell development, whereas STAT6 is required for T_H2 cell development.

STAT-mediated transcription is not the only pathway that can be initiated by cytokine receptors. Cytokine receptors can, for example, activate the Ras–MAP kinase pathway and the phosphatidylinositide pathway. Relatively little is known about how cytokine receptors activate these pathways, but it is possible that the ability of closely related cytokines to induce distinct biological responses may result from the selective activation of different combinations of multiple possible signaling pathways.

7-21 Cytokine signaling is terminated by a negative feedback mechanism.

As cytokines have so many and such powerful effects, the activation of cytokine signaling pathways must be tightly controlled; breakdown in control can lead to significant pathological effects. A variety of cytokine-specific inhibitory mechanisms ensure that cytokine signaling pathways can be efficiently terminated. As cytokine receptor signaling depends on tyrosine phosphorylation, dephosphorylation of the receptor complex by tyrosine phosphatases is one important means of termination. A variety of tyrosine phosphatases have been implicated in the dephosphorylation of cytokine receptors, JAKs, and STATs; these include SHP, CD45, and the T-cell phosphatase (TCPTP).

Cytokine signaling can also be terminated by negative feedback involving specific inhibitors that are induced by cytokine activation. The suppressor of cytokine signaling (SOCS) proteins are a class of inhibitors that terminate the signaling of many cytokine and hormone receptors. SOCS proteins are induced by STAT activation, and thus inhibit receptor signaling after the cytokine has had its effect. SOCS proteins contain an SH2 domain that can recruit them to the phosphorylated JAK kinase or receptor, and can inhibit JAK kinases directly, compete for the receptor, and direct the ubiquitination and subsequent degradation of JAKs and STATs. Their importance can be seen in SOCS1-deficient mice, which develop a multiorgan inflammatory infiltrate caused by increased signaling from interferon receptors, γ_c-containing receptors, and TLRs. Another class of inhibitory proteins consists of the protein inhibitors of activated STAT (PIAS) proteins, which also seem to be involved in promoting the degradation of receptors and pathway components.

7-22 The receptors that induce apoptosis activate specialized intracellular proteases called caspases.

Programmed cell death or **apoptosis** (see Section 1-14) is a normal process that is crucial to the proper development and function of the immune system. In particular, it has an important role in the termination of immune responses by getting rid of cells that are no longer needed after an infection has been cleared. It also has a key role in lymphocyte development in removing developing lymphocytes that fail to generate functional antigen receptors (see Chapter 5) or that have produced potentially autoreactive receptors, as discussed in Chapter 8. Apoptosis is a regulated process that is induced by specific extracellular signals (or in some cases by the lack of signals required for survival) and proceeds by a series of cellular events that include plasma membrane blebbing, changes in the distribution of membrane lipids, and enzymatic fragmentation of chromosomal DNA.

Two general pathways are involved in signaling cell death. One, called the **extrinsic pathway of apoptosis**, is mediated by the activation of so-called **death receptors** by extracellular ligands. Engagement of the ligand stimulates apoptosis in the receptor-bearing cell. The other pathway is known as the **intrinsic** or **mitochondrial pathway of apoptosis** and mediates apoptosis in response to noxious stimuli including ultraviolet irradiation, chemotherapeutic drugs, starvation, or lack of the growth factors required for survival. Common to both pathways is the activation of specialized proteases called aspartic-acid-specific cysteine proteases or **caspases**, which were introduced in Chapter 3 for their role in processing the cytokines IL-1 and IL-18 to their mature forms.

Like many other proteases, caspases are synthesized as inactive pro-caspases, in which the catalytic domain is inhibited by an adjacent pro-domain. Pro-caspases are activated by other caspases that cleave the protein to release the inhibitory pro-domain. There are two classes of caspases involved in the apoptotic pathway: **initiator caspases** promote apoptosis by cleaving and activating other caspases, and the **effector caspases** are the ones that initiate the cellular changes associated with apoptosis. The extrinsic pathway uses two related initiator caspases, caspase 8 and caspase 10, whereas the intrinsic pathway uses caspase 9. Both pathways use caspases 3, 6, and 7 as effector caspases. The effector caspases cleave a variety of proteins that are critical for cellular integrity and also activate enzymes that promote the death of the cell. For example, they cleave and degrade nuclear proteins, such as lamin B, that are required for the structural integrity of the nucleus, and activate the endonucleases that fragment the chromosomal DNA.

Movie 7.5

We shall first consider the apoptosis pathway leading from death receptors, as these are involved in many immune-system functions. The activation of caspase 8 is the critical step in the apoptosis pathway and starts with the recruitment of this initiator pro-caspase to the activated death receptor.

The death receptors are members of the large TNF receptor family but are distinguished from other receptors in this family by having a cytoplasmic **death domain** (**DD**), which we introduced in Section 3-7 in the light of its role in MyD88-dependent signaling from TLRs. Of the death receptors expressed in immune-system cells, **Fas** (CD95) and **TNFR-I**, a receptor for the cytokine TNF-α, are the best understood. TNFR-II, a second receptor for TNF expressed primarily on T cells, lacks a DD and activates the NFκB pathway, promoting cell survival rather than cell death. Fas and its ligand FasL are expressed widely, not only in the immune system. Fas-mediated cell death occurs in numerous contexts, including the protection of immunologically privileged sites (see Chapter 13) and the regulation and termination of immune responses (see Chapter 9). Loss-of-function mutations in Fas lead to the increased survival of lymphocytes and are one cause of the disease **autoimmune lymphoproliferative syndrome** (**ALPS**). This disease can also be due to mutations in FasL and in caspase 10, an enzyme involved in Fas-mediated apoptosis. The signaling pathway resulting from the stimulation of Fas by FasL is shown in Fig. 7.30.

Autoimmune Lymphoproliferative Syndrome (ALPS)

The first step in Fas-mediated apoptosis is the binding of the trimeric FasL, which results in the trimerization of Fas. This causes the death domains of Fas to bind to the death domains of an adaptor protein FADD (Fas-associated via death domain), which we introduced in Section 3-9 for its role in RLH signaling. FADD contains a death domain and an additional domain called a death effector domain (DED) that can bind to DEDs present in other proteins. When FADD is recruited to Fas, the DED of FADD then recruits the **initiator caspases**, pro-caspase 8 and pro-caspase 10, via interaction with a DED in

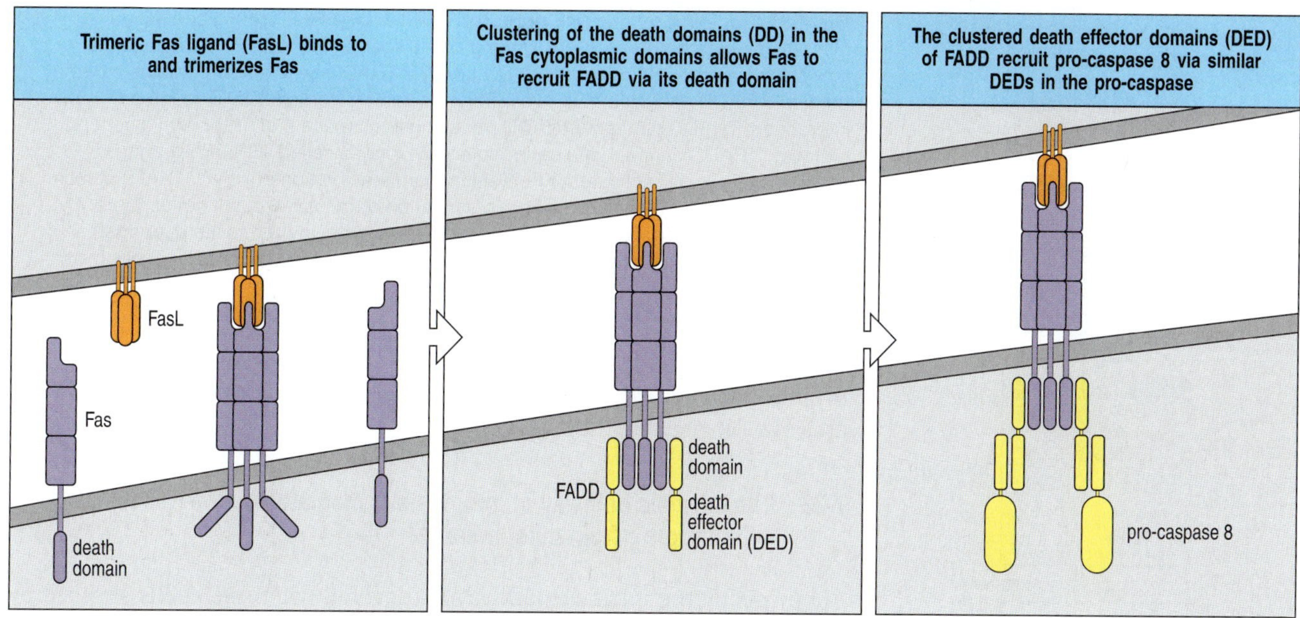

Trimeric Fas ligand (FasL) binds to and trimerizes Fas
FasL
Fas
death domain
Clustering of the death domains (DD) in the Fas cytoplasmic domains allows Fas to recruit FADD via its death domain
FADD
death domain
death effector domain (DED)
The clustered death effector domains (DED) of FADD recruit pro-caspase 8 via similar DEDs in the pro-caspase
pro-caspase 8

Fig. 7.30 Binding of Fas ligand to Fas initiates the extrinsic pathway of apoptosis. The cell-surface receptor Fas contains a so-called death domain in its cytoplasmic tail. When Fas ligand (FasL) binds Fas, this trimerizes the receptor (left panel). The adaptor protein FADD (also known as MORT-1) also contains a death domain and can bind to the clustered death domains of Fas (center panel). FADD also contains a domain called a death effector domain (DED) that allows it to recruit the pro-caspase 8 (which also contains a DED domain) (right panel). Clustered pro-caspase 8 activates itself to release an active caspase into the cytoplasm (not shown).

the pro-caspases. The high local concentration of these caspases around the receptors allows the caspases to cleave themselves, resulting in their activation. Once activated, caspases 8 and 10 are released from the receptor complex and can activate the downstream effector caspases.

A slightly different pathway is used by TNFR-I when stimulated by its ligand TNF-α. In some cells, TNFR-I signaling induces apoptosis; in other cells, TNFR-I signaling induces the induction of pro-inflammatory response genes. What determines whether apoptosis or new gene transcription is activated is not known. The current hypothesis suggests that the two different responses are regulated by two different signaling complexes that can be assembled by TNFR-I. In both cases, the DD in the cytoplasmic domain of the receptor recruits a DD-containing adaptor protein called TRADD, and then the pathways diverge. When TRADD binds FADD, the pathway proceeds as it did for the Fas pathway, leading to pro-caspase activation and apoptosis (Fig. 7.31). In other conditions, however, TRADD recruits an adaptor called TRAF-2 (TNF receptor associated factor-2) and a serine/threonine kinase called RIP (receptor-interacting protein). TRAF-2, like TRAF-6, is an E3 ubiquitin ligase, and it adds K63-linked polyubiquitin onto RIP (see Section 7-5). The K63 polyubiquitin scaffold attracts the TAB1:TAB2:TAK1 complex (see Section 3-7), allowing RIP to phosphorylate TAK1, which, as we have seen, activates IKK followed by activation of NFκB (see Figure 3.13). There are several different signaling routes to NFκB activation. Although T cells lacking PKC-θ (see Section 7-14) have defective activation of NFκB on stimulation through the antigen receptor, they activate NFκB normally through the above pathway in response to TLR signaling and inflammatory cytokines such as TNF-α. TRAF-2 also stimulates a MAPK signaling pathway that results in the activation of JNK, which phosphorylates Jun and thus activates AP-1. This pathway also activates the MAPK **p38**, which is important in the production of many inflammatory mediators.

The *Drosophila* Imd protein, which is part of the pathogen-recognition system of the fly, recruits the *Drosophila* homolog of FADD and activates a pathway very similar to the TNF signaling pathway. *Drosophila* FADD activates DREDD, a homolog of mammalian caspase-8. *Drosophila* dTAK1 may be the homolog of mammalian TAK1, which activates the NFκ B pathway in mammals (see Section 3-7). In *Drosophila* it activates IKK, resulting in the activation of Relish (the *Drosophila* version of NFκB), yet another example of

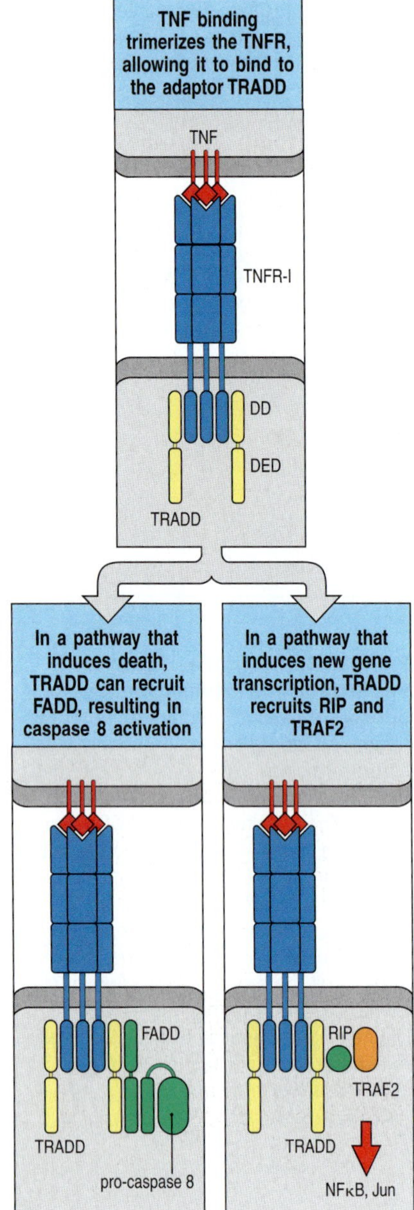

TNF binding trimerizes the TNFR, allowing it to bind to the adaptor TRADD
TNF
TNFR-I
DD
DED
TRADD
In a pathway that induces death, TRADD can recruit FADD, resulting in caspase 8 activation
FADD
TRADD
pro-caspase 8
In a pathway that induces new gene transcription, TRADD recruits RIP and TRAF2
RIP
TRAF2
TRADD
NFκB, Jun

Fig. 7.31 Signaling by the TNF receptor TNFR-I. Like Fas, TNFR-I contains a cytoplasmic death domain (DD). Binding of the ligand TNF to TNFR-I induces the receptor's death domains to recruit the adaptor TRADD, which also contains a death domain. TRADD can assemble two different signaling complexes. Through a DD–DD interaction, TRADD can recruit FADD, resulting in caspase 8 activation. In a second pathway, TRADD can recruit a serine/threonine kinase called RIP and an adaptor protein TRAF-2. RIP activates IKK, resulting in the activation of NFκB. TRAF2 stimulates the JNK signaling pathway, resulting in activation of the AP-1 transcription factor by the phosphorylation of Jun. It is not clear how one pathway is chosen over another. Caspase 8 activation can lead to cell death, but this is blocked in healthy cells by activation of NFκB.

evolution finding different roles for different proteins within the same broad area of biological function.

7-23 The intrinsic pathway of apoptosis is mediated by the release of cytochrome *c* from mitochondria.

Apoptosis by the intrinsic pathway is triggered when the cell is stressed by exposure to noxious stimuli or does not receive extracellular signals that are required for cell survival. The critical step is the release of cytochrome *c* from mitochondria, which triggers the activation of caspases. Once in the cytoplasm, cytochrome *c* binds to a protein called Apaf-1 (apoptotic protease activating factor-1), stimulating its oligomerization. The Apaf-1 oligomer then recruits an initiator caspase, pro-caspase 9. Aggregation of caspase 9 permits its self-cleavage, and frees it to stimulate the activation of effector caspases as in the death receptor pathways (Fig. 7.32).

The release of cytochrome *c* is controlled by interactions between members of the Bcl-2 family of proteins. The **Bcl-2 family** of proteins is defined by the presence of one or more Bcl-2 homology (BH) domains and can be divided into two general groups: members that promote apoptosis and members that inhibit apoptosis (Fig. 7.33). Some pro-apoptotic Bcl-2 family members, such as Bax, Bak, and Bok (referred to as executioners), bind to mitochondrial membranes and can directly cause cytochrome *c* release. How they do this is still not known, but they may form pores in the membranes.

The anti-apoptotic Bcl-2 family members are induced by stimuli that promote cell survival. The best known of the anti-apoptotic proteins is Bcl-2 itself. The *Bcl-2* gene was first identified as an oncogene in a B-cell lymphoma, and its overexpression in tumors makes the cells more resistant to apoptotic stimuli and thus more likely to progress to an invasive cancer and difficult to kill. Other members of the inhibitory family include Bcl-X_L and Bcl-W. Anti-apoptotic proteins function by binding to the mitochondrial membrane to block the release of cytochrome *c*. The precise mechanism of inhibition is not clear, but they may function by directly blocking the function of the pro-apoptotic family members.

Movie 7.6 

A second family of pro-apoptotic Bcl-2 family members are sentinels and are activated by apoptotic stimuli. Once activated, these proteins, which include Bad, Bid, and PUMA, can either act to block the activity of the anti-apoptotic proteins or act directly to stimulate the activity of the executioner pro-apoptotic proteins.

Summary.

Many different signals govern lymphocyte behavior, only some of which are delivered via the antigen receptor. Lymphocyte development, activation, and longevity are clearly influenced by the antigen receptor, but these processes

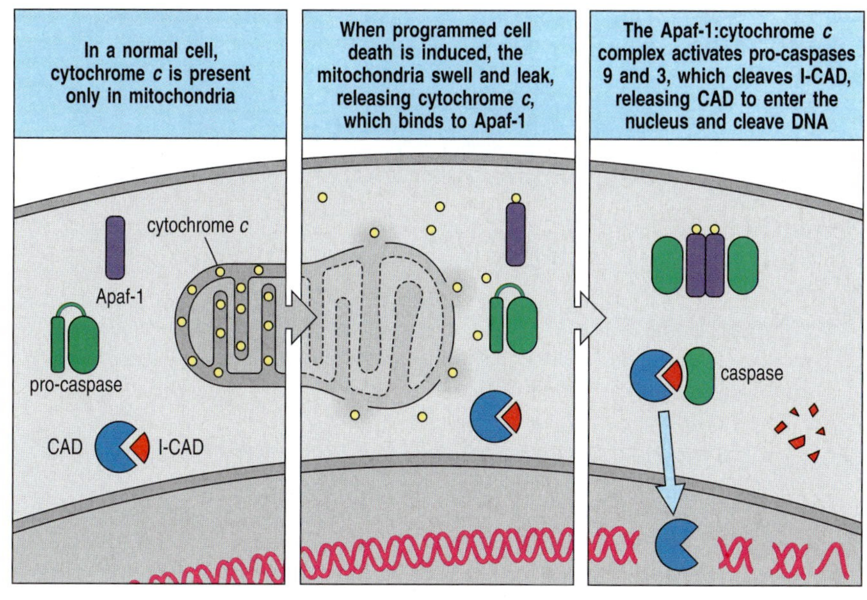

In a normal cell, cytochrome c is present only in mitochondria
When programmed cell death is induced, the mitochondria swell and leak, releasing cytochrome c, which binds to Apaf-1
The Apaf-1:cytochrome c complex activates pro-caspases 9 and 3, which cleaves I-CAD, releasing CAD to enter the nucleus and cleave DNA
cytochrome c
Apaf-1
pro-caspase
CAD
I-CAD
caspase

Fig. 7.32 In the intrinsic pathway, cytochrome *c* release from mitochondria induces programmed cell death. In normal cells, cytochrome *c* is confined to the mitochondria (first panel). However, during stimulation of the intrinsic pathway, the mitochondria swell, allowing the cytochrome *c* to leak out into the cytosol (second panel). There it interacts with the protein Apaf-1, forming a cytochrome *c*:Apaf-1 complex which recruits pro-caspase 9. Clustering of pro-caspase 9 activates it, allowing it to cleave downstream caspases, such as caspase 3, resulting in the activation of enzymes such as I-CAD, which can cleave DNA (third panel).

are also regulated by other extracellular signals. Other signals are delivered in a variety of ways. Most cytokines signal through an express pathway that links receptor-associated JAK kinases to preformed STAT transcription proteins, which after phosphorylation dimerize through their SH2 domains and head for the nucleus. Activated lymphocytes are programmed to die when the Fas receptor that they express binds the Fas ligand. This transmits a death signal, which activates a protease cascade that triggers apoptosis. Lymphocyte apoptosis is inhibited by some members of the intracellular Bcl-2 family and promoted by others. Working out the complete picture of the signals processed by lymphocytes as they develop, circulate, respond to antigen, and die is an immensely exciting prospect.

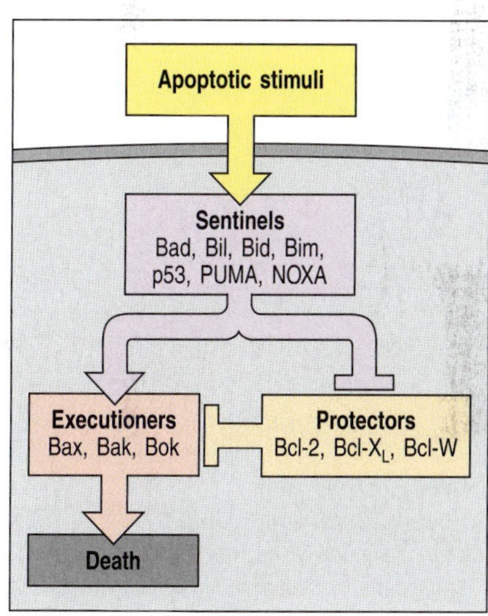

Fig. 7.33 General scheme of intrinsic pathway regulation by Bcl-2-family proteins. Extracellular apoptotic stimuli activate a group of pro-apoptotic (sentinel) proteins. Sentinel proteins can function either to block the protection provided by pro-survival, protector proteins or to directly activate pro-apoptotic, executioner proteins. In mammalian cells, apoptosis is mediated by the executioner proteins Bax, Bak, and Bok. In normal cells, these proteins are prevented from acting by the protector proteins (Bcl-2, Bcl-X_L, and Bcl-W). The release of activated executioner proteins causes the release of cytochrome *c* and subsequent cell death, as shown in Fig. 7.32.

Summary to Chapter 7.

Signaling by cell-surface receptors of many different sorts is crucial to the ability of the immune system to respond appropriately to foreign pathogens. The importance of these signaling pathways is demonstrated by the many diseases that are due to aberrant signaling, which include both immunodeficiency diseases and autoimmune diseases. Common features of many signaling pathways are the generation of second messengers such as calcium and phosphoinositides and the activation of both serine/threonine and tyrosine kinases. An important concept in the initiation of signaling pathways by receptor proteins is the recruitment of signaling proteins to the plasma membrane and the assembly of multiprotein signaling complexes. In many cases, signal transduction leads to the activation of transcription factors that lead directly or indirectly to the proliferation, differentiation, and effector function of activated lymphocytes. Other roles of signal transduction are to mediate changes in the cytoskeleton important for cell functions such as migration and shape changes.

While we are beginning to understand the basic circuitry of signal transduction pathways, it is important to keep in mind that we do not yet understand why these pathways are so complex. The complexity of signaling pathways could have roles in properties such as amplification, robustness, diversity, and efficiency of signaling responses. An important goal for the future will be to understand how the design of each signaling pathway contributes to the particular quality and sensitivity needed for specific signaling responses.

Questions.

7.1 *Compare and contrast the roles of inducible tyrosine phosphorylation and protein ubiquitination in signal transduction.*

7.2 *Describe different mechanisms used (1) to recruit signaling molecules to the plasma membrane and (2) to terminate the activity of these molecules.*

7.3 *What are some of the advantages of using complexes of many signaling proteins for signal transduction? Suggest some reasons why signaling pathways are so complicated.*

7.4 *How are small G proteins such as Ras regulated? How do they exert their effect?*

7.5 *Describe how phospholipase C-γ is activated by T-cell receptor signaling. Where does the CD28 pathway intersect with the T-cell receptor pathway to control this important step in T-cell activation?*

7.6 *Describe three different pathways used by immune-system cells to activate NFκB.*

7.7 *Name at least three differences between T-cell and B-cell receptor signaling.*

7.8 *Speculate why CD28 family members include both positive and negative regulators of T-cell activation.*

7.9 *Compare and contrast the intrinsic pathway versus the extrinsic pathway of apoptosis.*

General references.

Alberts, B., Johnson, A., Lewis, J., Raff, M., Roberts, K., and Walter, P.: *Molecular Biology of the Cell*, 5th ed. New York: Garland Science, 2008.

Gomperts, B., Kramer, I., and Tatham, P.: *Signal Transduction.* San Diego: Elsevier, 2002.

Marks, F., Klingmüller, U., and Müller-Decker, K.: Cellular *Signal Processing*, New York: Garland Science, 2009.

Section references.

7-1 Transmembrane receptors convert extracellular signals into intracellular biochemical events.

Lin, J., and Weiss, A.: **T cell receptor signalling.** *J. Cell Sci.* 2001, **114**:243–244.

Smith-Garvin, J.E., Koretzky, G.A., and Jordan, M.S.: **T cell activation**. *Annu. Rev. Immunol.* 2009, **27**:591–619.

7-2 Intracellular signal propagation is mediated by large multiprotein signaling complexes.

Jordan, M.S., Singer, A.L., and Koretzky, G.A.: **Adaptors as central mediators of signal transduction in immune cells.** *Nat. Immunol.* 2003, **4**:110–116.

Pawson, T.: **Specificity in signal transduction: from phosphotyrosine-SH2 domain interactions to complex cellular systems.** *Cell* 2004, **116**:191–203.

Pawson, T., and Nash, P.: **Assembly of cell regulatory systems through protein interaction domains.** *Science* 2003, **300**:445–452.

7-3 Small G proteins act as molecular switches in many different signaling pathways.

Cantrell, D.A.: **GTPases and T-cell activation.** *Immunol. Rev.* 2003, **192**:122–130.

Etienne-Manneville, S., and Hall, A.: **Rho GTPases in cell biology.** *Nature* 2002, **420**:629–635.

Mitin, N., Rossman, K.L., and Der, C.J.: **Signaling interplay in Ras superfamily function.** *Curr. Biol.* 2005, **15**:R563–R574.

7-4 Signaling proteins are recruited to the membrane by a variety of mechanisms.

Buday, L.: **Membrane-targeting of signaling molecules by SH2/SH3 domain-containing adaptor proteins.** *Biochim. Biophys. Acta* 1999, **1422**:187–204.

Kanai, F., Liu, H., Field, S.J., Akbary, H., Matsuo, T., Brown, G.E., Cantley, L.C., and Yaffe, M.B.: **The PX domains of p47phox and p40phox bind to lipid products of PI(3)K.** *Nat. Cell Biol.* 2001, **3**:675–678.

Kholodenko, B.N., Hoek, J.B., and Westerhoff, H.V.: **Why cytoplasmic signaling proteins should be recruited to cell membranes.** *Trends Cell Biol.* 2000, **10**:173–178.

Lemmon, M.A.: **Phosphoinositide recognition domains.** *Traffic* 2003, **4**:201–213.

7-5 Ubiquitin conjugation of proteins can both activate and inhibit signaling responses.

Ciechanover, A.: **Proteolysis: from the lysosome to ubiquitin and the proteasome.** *Nat. Rev. Mol. Cell Biol.* 2005, **6**:79–87.

Hurley, J.H., Lee, S., and Prag, G.: **Ubiquitin-binding domains.** *Biochem. J.* 2006, **399**:361–372.

Kanayama, A., Seth, R.B., Sun, L., Ea, C.K., Hong, M., Shaito, A., Chiu, Y.H., Deng, L., and Chen, Z.J.: **TAB2 and TAB3 activate the NF-κB pathway through binding to polyubiquitin chains.** *Mol. Cell* 2004, **15**:535–548.

Katzmann, D.J., Odorizzi, G., and Emr, S.D.: **Receptor downregulation and multivesicular-body sorting.** *Nat. Rev. Mol. Cell Biol.* 2002, **3**:893–905.

Liu, Y.C., Penninger, J., and Karin, M.: **Immunity by ubiquitylation: a reversible process of modification.** *Nat. Rev. Immunol.* 2005, **5**:941–952.

7-6 The activation of some receptors generates small-molecule second messengers.

Kresge, N., Simoni, R.D., and Hill, R.L.: **Earl W. Sutherland's discovery of cyclic adenine monophosphate and the second messenger system.** *J. Biol. Chem.* 2005, **280**:39–40.

Rall, T.W., and Sutherland, E.W.: **Formation of a cyclic adenine ribonucleotide by tissue particles.** *J. Biol. Chem.* 1958, **232**:1065–1076.

7-7 Antigen receptors consist of variable antigen-binding chains associated with invariant chains that carry out the signaling function of the receptor.

Call, M.E., Pyrdol, J., Wiedmann, M., and Wucherpfennig, K.W.: **The organizing principle in the formation of the T cell receptor-CD3 complex.** *Cell* 2002, **111**:967–979.

Exley, M., Terhorst, C., and Wileman, T.: **Structure, assembly and intracellular transport of the T cell receptor for antigen.** *Semin. Immunol.* 1991, **3**:283–297.

7-8 Antigen recognition by the T-cell receptor and its co-receptors leads to phosphorylation of ITAMs by Src-family kinases.

Irving, B.A., and Weiss, A.: **The cytoplasmic domain of the T cell receptor ζ chain is sufficient to couple to receptor-associated signal transduction pathways.** *Cell* 1991, **64**:891–901.

Letourneur, F., and Klausner, R.D.: **Activation of T cells by a tyrosine kinase activation domain in the cytoplasmic tail of CD3 epsilon.** *Science* 1992, **255**:79–82.

Li, Q.J., Dinner, A.R., Qi, S., Irvine, D.J., Huppa, J.B., Davis, M.M., and Chakraborty, A.K.: **CD4 enhances T cell sensitivity to antigen by coordinating Lck accumulation at the immunological synapse.** *Nat. Immunol.* 2004, **5**:791–799.

7-9 Phosphorylated ITAMs recruit and activate the tyrosine kinase ZAP-70, which phosphorylates scaffold proteins that recruit the phospholipase PLC-γ.

Chan, A.C., Dalton, M., Johnson, R., Kong, G.H., Wang, T., Thoma, R., and Kurosaki, T.: **Activation of ZAP-70 kinase activity by phosphorylation of tyrosine 493 is required for lymphocyte antigen receptor function.** *EMBO J.* 1995, **14**:2499–2508.

Chan, A.C., Iwashima, M., Turck, C.W., and Weiss, A.: **ZAP-70: a 70 kd protein-tyrosine kinase that associates with the TCR ζ chain.** *Cell* 1992, **71**:649–662.

Iwashima, M., Irving, B.A., van Oers, N.S., Chan, A.C., and Weiss, A.: **Sequential interactions of the TCR with two distinct cytoplasmic tyrosine kinases.** *Science* 1994, **263**:1136–1139.

Samelson, L.E.: **Signal transduction mediated by the T cell antigen receptor: the role of adapter proteins.** *Annu. Rev. Immunol.* 2002, **20**:371–394.

7-10 The activation of PLC-γ requires a co-stimulatory signal.
&

7-11 Activated PLC-γ generates the second messengers diacylglycerol and inositol trisphosphate.

Berg, L.J., Finkelstein, L.D., Lucas, J.A., and Schwartzberg, P.L.: **Tec family kinases in T lymphocyte development and function.** *Annu. Rev. Immunol.* 2005, **23**:549–600.

Lewis, C.M., Broussard, C., Czar, M.J., and Schwartzberg, P.L.: **Tec kinases: modulators of lymphocyte signaling and development.** *Curr. Opin. Immunol.* 2001, **13**:317–325.

7-12 Ca^{2+} entry activates the transcription factor NFAT.

Hogan, P.G., Chen, L., Nardone, J., and Rao, A.: **Transcriptional regulation by calcium, calcineurin, and NFAT.** *Genes Dev.* 2003, **17**:2205–2232.

Macian, F., Lopez-Rodriguez, C., and Rao, A.: **Partners in transcription: NFAT and AP-1.** *Oncogene* 2001, **20**:2476–2489.

Picard, C., McCarl, C.A., Papolos, A., Khalil, S., Lüthy, K., Hivroz, C., LeDeist, F., Rieux-Laucat, F., Rechavi, G., Rao, A., *et al.*: **STIM1 mutation associated with a syndrome of immunodeficiency and autoimmunity.** *N. Engl. J. Med.* 2009, **360**:1971–1980.

Prakriya, M., Feske, S., Gwack, Y., Srikanth, S., Rao, A., and Hogan, P.G.: **Orai1 is an essential pore subunit of the CRAC channel.** *Nature* 2006, **443**:230–233.

7-13 Ras activation stimulates the mitogen-activated protein kinase (MAPK) relay and induces expression of the transcription factor AP-1.

Downward, J., Graves, J.D., Warne, P.H., Rayter, S., and Cantrell, D.A.: **Stimulation of p21ras upon T-cell activation.** *Nature* 1990, **346**:719–723.

Leevers, S.J., and Marshall, C.J.: **Activation of extracellular signal-regulated kinase, ERK2, by p21ras oncoprotein.** *EMBO J.* 1992, **11**:569–574.

Shaw, A.S., and Filbert, E.L.: **Scaffold proteins and immune-cell signalling.** *Nat. Rev. Immunol.* 2009, **9**:47–56.

7-14 Protein kinase C activates the transcription factors NFκB and AP-1.

Matsumoto, R., Wang, D., Blonska, M., Li, H., Kobayashi, M., Pappu, B., Chen, Y., Wang, D., and Lin, X.: **Phosphorylation of CARMA1 plays a critical role in T cell receptor-mediated NF-κB activation.** *Immunity* 2005, **23**:575–585.

Rueda, D., and Thome, M.: **Phosphorylation of CARMA1: the link(er) to NF-κB activation.** *Immunity* 2005, **23**:551–553.

Sommer, K., Guo, B., Pomerantz, J.L., Bandaranayake, A.D., Moreno-Garcia, M.E., Ovechkina, Y.L., and Rawlings, D.J.: **Phosphorylation of the CARMA1 linker controls NF-κB activation.** *Immunity* 2005, **23**:561–574.

7-15 The cell-surface protein CD28 is a co-stimulatory receptor for naive T cells.

Acuto, O., and Michel, F.: **CD28-mediated co-stimulation: a quantitative support for TCR signalling.** *Nat. Rev. Immunol.* 2003, **3**:939–951.

Frauwirth, K.A., Riley, J.L., Harris, M.H., Parry, R.V., Rathmell, J.C., Plas, D.R., Elstrom, R.L., June, C.H., and Thompson, C.B.: **The CD28 signaling pathway regulates glucose metabolism.** *Immunity* 2002, **16**:769–777.

7-16 The logic of B-cell receptor signaling is similar to that of T-cell receptor signaling, but some of the signaling components are specific to B cells.

Cambier, J.C., Pleiman, C.M., and Clark, M.R.: **Signal transduction by the B cell antigen receptor and its coreceptors.** *Annu. Rev. Immunol.* 1994, **12**:457–486.

DeFranco, A.L., Richards, J.D., Blum, J.H., Stevens, T.L., Law, D.A., Chan, V.W., Datta, S.K., Foy, S.P., Hourihane, S.L., Gold, M.R., *et al.*: **Signal transduction by the B-cell antigen receptor.** *Ann. NY Acad. Sci.* 1995, **766**:195–201.

Kurosaki, T.: **Functional dissection of BCR signaling pathways.** *Curr. Opin. Immunol.* 2000, **12**:276–281.

7-17 ITAMs are also found in other receptors on leukocytes that signal for cell activation.

Daeron, M.: **Fc receptor biology.** *Annu. Rev. Immunol.* 1997, **15**:203–234.

Lanier, L.L., and Bakker, A.B.: **The ITAM-bearing transmembrane adaptor DAP12 in lymphoid and myeloid cell function.** *Immunol. Today* 2000, **21**:611–614.

7-18 Inhibitory receptors on lymphocytes help regulate immune responses.

Chen, L.: **Co-inhibitory molecules of the B7-CD28 family in the control of T-cell immunity.** *Nat. Rev. Immunol.* 2004, **4**:336–347.

Rudd, C.E., and Schneider, H.: **Unifying concepts in CD28, ICOS and CTLA4 co-receptor signalling.** *Nat. Rev. Immunol.* 2003, **3**:544–556.

Rudd, C.E., Taylor, A., and Schneider, H.: **CD28 and CTLA-4 coreceptor expression and signal transduction.** *Immunol. Rev.* 2009, **229**:12–26.

Sharpe, A.H., and Freeman, G.J.: **The B7-CD28 superfamily.** *Nat. Rev. Immunol.* 2002, **2**:116–126.

7-19 Cytokines and their receptors fall into distinct families of structurally related proteins.

Basler, C.F., and Garcia-Sastre, A.: **Viruses and the type I interferon antiviral system: induction and evasion.** *Int. Rev. Immunol.* 2002, **21**:305–337.

Boulay, J.L., O'Shea, J.J., and Paul, W.E.: **Molecular phylogeny within type I cytokines and their cognate receptors.** *Immunity* 2003, **19**:159–163.

Collette, Y., Gilles, A., Pontarotti, P., and Olive, D.: **A co-evolution perspective of the TNFSF and TNFRSF families in the immune system.** *Trends Immunol.* 2003, **24**:387–394.

Ihle, J.N.: **Cytokine receptor signalling.** *Nature* 1995, **377**:591–594.

Proudfoot, A.E.: **Chemokine receptors: multifaceted therapeutic targets.** *Nat. Rev. Immunol.* 2002, **2**:106–115.

Taniguchi, T., and Takaoka, A.: **The interferon-α/β system in antiviral responses: a multimodal machinery of gene regulation by the IRF family of transcription factors.** *Curr. Opin. Immunol.* 2002, **14**:111–116.

7-20 Cytokine receptors of the hematopoietin family are associated with the JAK family of tyrosine kinases, which activate STAT transcription factors.

Fu, X.Y.: **A transcription factor with SH2 and SH3 domains is directly activated by an interferon α-induced cytoplasmic protein tyrosine kinase(s).** *Cell* 1992, **70**:323–335.

Leonard, W.J., and O'Shea, J.J.: **Jaks and STATs: biological implications.** *Annu. Rev. Immunol.* 1998, **16**:293–322.

Levy, D.E., and Darnell, J.E., Jr: **Stats: transcriptional control and biological impact.** *Nat. Rev. Mol. Cell Biol.* 2002, **3**:651–662.

Pesu, M., Candotti, F., Husa, M., Hofmann, S.R., Notarangelo, L.D., and O'Shea, J.J.: **Jak3, severe combined immunodeficiency, and a new class of immunosuppressive drugs.** *Immunol Rev.* 2005, **203**:127–142.

Schindler, C., Shuai, K., Prezioso, V.R., and Darnell, J.E., Jr: **Interferon-dependent tyrosine phosphorylation of a latent cytoplasmic transcription factor.** *Science* 1992, **257**:809–813.

7-21 Cytokine signaling is terminated by a negative feedback mechanism.

Krebs, D.L., and Hilton, D.J.: **SOCS proteins: negative regulators of cytokine signaling.** *Stem Cells* 2001, **19**:378–387.

Rytinki, M.M., Kaikkonen, S., Pehkonen, P., Jääskeläinen, T., and Palvimo, J.J.: **PIAS proteins: pleiotropic interactors associated with SUMO.** *Cell. Mol. Life Sci.* 2009, **66**:3029–3041.

Shuai, K., and Liu, B.: **Regulation of JAK-STAT signalling in the immune system.** *Nat. Rev. Immunol.* 2003, **3**:900–911.

Yasukawa, H., Sasaki, A., and Yoshimura, A.: **Negative regulation of cytokine signaling pathways.** *Annu. Rev. Immunol.* 2000, **18**:143–164.

7-22 The receptors that induce apoptosis activate specialized intracellular proteases called caspases.

Aggarwal, B.B.: **Signalling pathways of the TNF superfamily: a double-edged sword.** *Nat. Rev. Immunol.* 2003, **3**:745–756.

Bishop, G.A.: **The multifaceted roles of TRAFs in the regulation of B-cell function.** *Nat. Rev. Immunol.* 2004, **4**:775–786.

Siegel, R.M.: **Caspases at the crossroads of immune-cell life and death.** *Nat. Rev. Immunol.* 2006, **6**:308–317.

7-23 The intrinsic pathway of apoptosis is mediated by the release of cytochrome *c* from mitochondria.

Borner, C.: **The Bcl-2 protein family: sensors and checkpoints for life-or-death decisions.** *Mol. Immunol.* 2003, **39**:615–647.

Hildeman, D.A., Zhu, Y., Mitchell, T.C., Kappler, J., and Marrack, P.: **Molecular mechanisms of activated T cell death *in vivo*.** *Curr. Opin. Immunol.* 2002, **14**:354–359.

Strasser, A.: **The role of BH3-only proteins in the immune system.** *Nat. Rev. Immunol.* 2005, **5**:189–200.

The Development and Survival of Lymphocytes

8

The antigen receptors carried by B lymphocytes and T lymphocytes are immensely variable in their antigen specificity, enabling an individual to make immune responses against the wide range of pathogens encountered during a lifetime. This diverse repertoire of B-cell receptors and T-cell receptors is generated during the development of B cells and T cells from their uncommitted precursors. The production of new lymphocytes, or **lymphopoiesis**, takes place in specialized lymphoid tissues—the **central lymphoid tissues**—which are the bone marrow for most B cells and the thymus for most T cells. Lymphocyte precursors originate in the bone marrow, but whereas B cells complete most of their development there, the precursors of most T cells migrate to the thymus, where they develop into mature T cells. B cells also originate and develop in the fetal liver and the neonatal spleen. Some T cells that form specialized populations within the gut epithelium may migrate as immature precursors from the bone marrow to develop in sites called 'cryptopatches' just under the intestinal epithelial crypts. We will mainly focus here on the development of bone marrow derived B cells and thymus-derived T cells.

In the fetus and the juvenile, the central lymphoid tissues are the sources of large numbers of new lymphocytes, which migrate to populate the **peripheral lymphoid tissues** such as lymph nodes, spleen, and mucosal lymphoid tissue (these tissues are also known as secondary lymphoid tissues). In mature individuals, the development of new T cells in the thymus slows down, and T-cell numbers are maintained through long-lived individual T cells together with the division of mature T cells outside the central lymphoid organs. New B cells, in contrast, are continually produced from the bone marrow, even in adults.

The structure of the antigen receptor genes expressed by B cells and T cells, and the mechanisms by which a complete antigen receptor is assembled, were described in Chapters 4 and 5. This chapter builds on that foundation to explain how B and T lymphocytes themselves develop from a common progenitor through a series of stages, and how each of these stages tests for the proper assembly of antigen receptors.

Once an antigen receptor has been formed, rigorous testing is required to select lymphocytes that carry useful antigen receptors—that is, antigen receptors that can recognize pathogens and yet will not react against an individual's own cells. Given the incredible diversity of receptors that the rearrangement process can generate, it is important that those lymphocytes that mature are likely to be useful in recognizing and responding to foreign antigens, especially as an individual can express only a small fraction of the total possible receptor repertoire in his or her lifetime. We describe how the specificity and affinity of the receptor for self ligands are tested to determine whether the immature lymphocyte will survive into the mature repertoire, or die. In general, it seems that developing lymphocytes whose receptors interact weakly with self antigens, or that bind self antigens in a particular way, receive a signal that enables them to survive; this type of selection is known as **positive selection**. Positive selection is particularly critical in the development of α:β T cells, which recognize composite antigens consisting of peptides bound to MHC molecules, because it ensures that an individual's T cells will be able to respond to peptides bound to his or her MHC molecules.

In contrast, lymphocytes with strongly self-reactive receptors must be eliminated to prevent autoimmune reactions; this process of **negative selection** is one of the ways in which the immune system is made self-tolerant. The default fate of developing lymphocytes, in the absence of any signal being received from the receptor, is death, and as we will see, the vast majority of developing lymphocytes die before emerging from the central lymphoid organs or before completing maturation in the peripheral lymphoid organs.

In this chapter we describe the different stages of the development of B cells and T cells in mice and humans from the uncommitted stem cell up to the mature, functionally specialized lymphocyte with its unique antigen receptor ready to respond to a foreign antigen. The final stages in the life history of a mature lymphocyte, in which an encounter with its antigen activates it to become an effector or memory lymphocyte, are discussed in Chapters 9–11. The chapter is divided into four parts. The first two describe B-cell and T-cell development, respectively. We then look at the positive and negative selection of T cells in the thymus. Next, we describe the fate of newly generated lymphocytes as they leave the central lymphoid organs and migrate to the peripheral lymphoid tissues, where further maturation occurs. Mature lymphocytes continually recirculate between the blood and peripheral lymphoid tissues (see Chapter 1) and, in the absence of infection, their numbers remain relatively constant, despite the continual production of new lymphocytes. Finally, we look at the factors that govern the survival of naive lymphocytes in the peripheral lymphoid organs, and the maintenance of lymphocyte **homeostasis**—the regulation of their numbers and turnover.

Development of B lymphocytes.

The main phases of a B lymphocyte's life history are shown in Fig. 8.1. The stages in both B-cell and T-cell development are defined mainly by the successive steps in the assembly and expression of functional antigen-receptor genes. At each step of lymphocyte development, the progress of gene rearrangement is monitored, and the major recurring theme is that successful gene rearrangement leads to the production of a protein chain that serves as a signal for the cell to progress to the next stage. We will see that a developing B cell is presented with opportunities for multiple rearrangements that increase the likelihood of expressing a functional antigen receptor, but there are also checkpoints that reinforce the requirement that each B cell expresses just one receptor specificity. We will start by looking at how the earliest recognizable cells of the B-cell lineage develop from the multipotent hematopoietic stem cells in the bone marrow, and at what point the B-cell and T-cell lineages diverge.

8-1 Lymphocytes derive from hematopoietic stem cells in the bone marrow.

The cells of the lymphoid lineage—B cells, T cells, and NK cells—are all derived from common lymphoid progenitor cells, which themselves derive from the multipotent **hematopoietic stem cells** (**HSCs**) that give rise to all blood cells

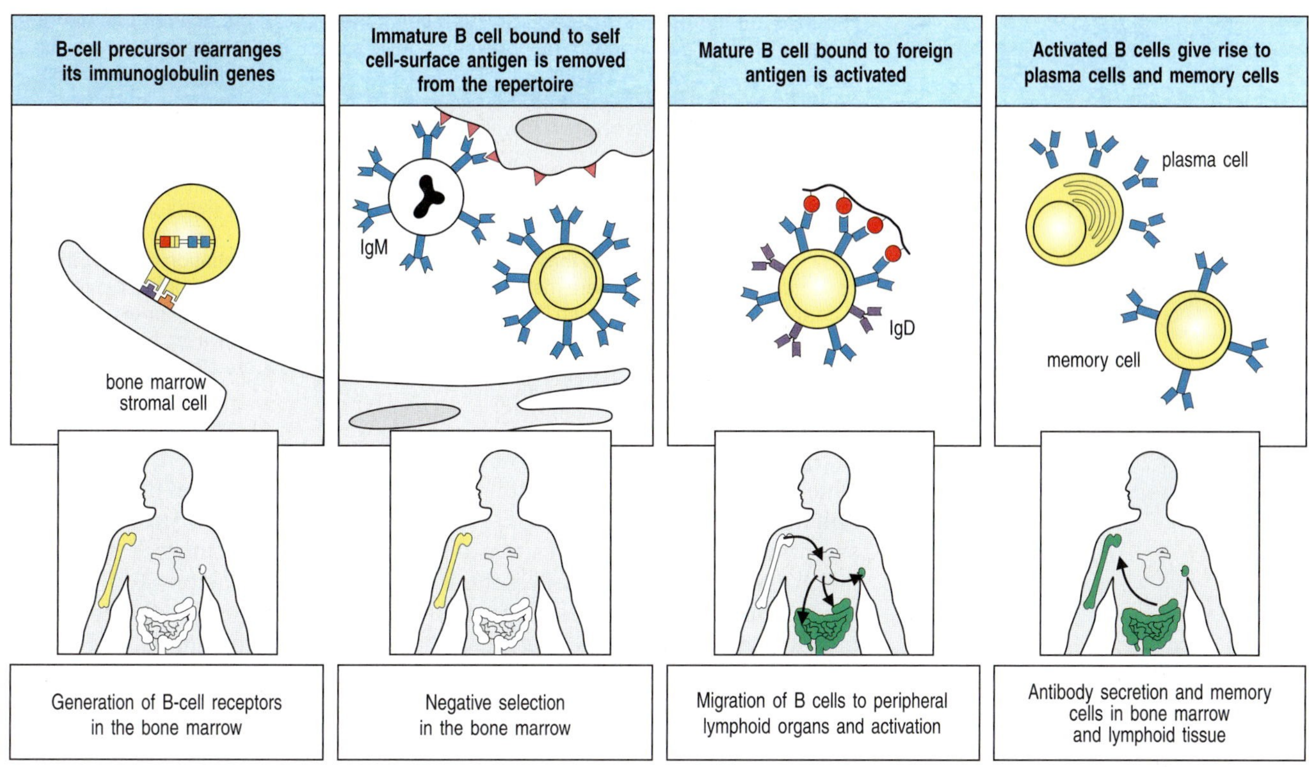
B-cell precursor rearranges its immunoglobulin genes
Immature B cell bound to self cell-surface antigen is removed from the repertoire
Mature B cell bound to foreign antigen is activated
Activated B cells give rise to plasma cells and memory cells
bone marrow stromal cell
IgM
IgD
plasma cell
memory cell
Generation of B-cell receptors in the bone marrow
Negative selection in the bone marrow
Migration of B cells to peripheral lymphoid organs and activation
Antibody secretion and memory cells in bone marrow and lymphoid tissue

Fig. 8.1 B cells develop in the bone marrow and migrate to peripheral lymphoid organs, where they can be activated by antigens. In the first phase of development, progenitor B cells in the bone marrow rearrange their immunoglobulin genes. This phase is independent of antigen but is dependent on interactions with bone marrow stromal cells (first panels). It ends in an immature B cell that carries an antigen receptor in the form of cell-surface IgM and can now interact with antigens in its environment. Immature B cells that are strongly stimulated by antigen at this stage either die or are inactivated in a process of negative selection, thus removing many self-reactive B cells from the repertoire (second panels). In the third phase of development, the surviving immature B cells emerge into the periphery and mature to express IgD as well as IgM. They can now be activated by encounter with their specific foreign antigen in a peripheral lymphoid organ (third panels). Activated B cells proliferate, and differentiate into antibody-secreting plasma cells and long-lived memory cells (fourth panels).

(see Fig. 1.3). Development from the precursor stem cell into cells that are committed to becoming B cells or T cells follows the basic principles of cell differentiation. Properties that are essential for the function of the mature cell are gradually acquired, along with the loss of properties that are more characteristic of the immature cell. In the case of lymphocyte development, cells become committed first to the lymphoid, as opposed to the myeloid, lineage, and then to either the B-cell or the T-cell lineages (Fig. 8.2).

The specialized microenvironment of the bone marrow provides signals both for the development of lymphocyte progenitors from hematopoietic stem cells and for the subsequent differentiation of B cells. Such signals act on the developing lymphocytes to switch on key genes that direct the developmental program and are produced by the network of specialized non-lymphoid connective-tissue **stromal cells** that are in intimate contact with the developing lymphocytes (Fig. 8.3). The contribution of the stromal cells is twofold. First, they form specific adhesive contacts with the developing lymphocytes by interactions between cell-adhesion molecules and their ligands. Second, they provide soluble and membrane-bound cytokines and chemokines that control lymphocyte differentiation and proliferation.

The hematopoietic stem cell first differentiates into **multipotent progenitor cells** (**MPPs**), which can produce both lymphoid and myeloid cells but are no longer self-renewing stem cells. Multipotent progenitors express a cell-surface receptor tyrosine kinase known as FLT3 (originally called stem-cell kinase 1 (STK1) in humans and Flt3/Flk2 in mice) that binds the membrane-bound FLT3 ligand on stromal cells. Signaling through FLT3 is needed for differentiation to the next stage, the **common lymphoid progenitor** (**CLP**). This cell gives rise to the earliest committed B-lineage cell, the **pro-B cell** (see Fig. 8.3). The common lymphoid progenitor is so called because it was originally thought to give rise to both the B-cell and T-cell lineages. But although it can do so in culture, it is not yet clear whether it does so *in vivo*.

The production of lymphocyte progenitors from the multipotent progenitor cell is accompanied by expression of the receptor for interleukin-7 (IL-7),

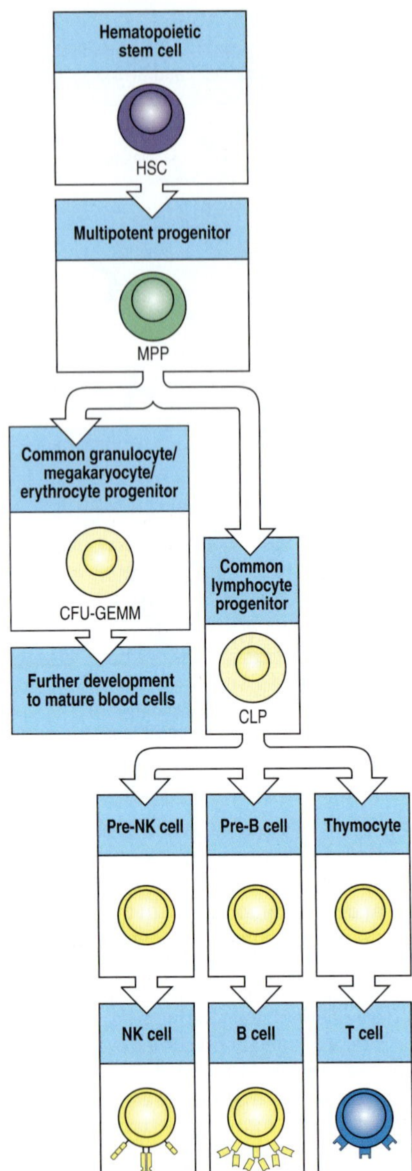

Fig. 8.2 A multipotent hematopoietic stem cell generates all the cells of the immune system. In the bone marrow or other hematopoietic sites, the multipotent stem cell gives rise to cells with progressively more limited potential. A simplified progression is shown here. The multipotent progenitor (MPP), for example, has lost its stem-cell properties. The first branch leads to cells with myeloid and erythroid potential on the one hand (CFU-GEMMs) and on the other to the common lymphoid progenitors (CLPs), with lymphoid potential. The former give rise to all nonlymphoid cellular blood elements, including circulating monocytes and granulocytes and the macrophages and dendritic cells that reside in tissues and peripheral lymphoid organs (not shown). The CLP can give rise to NK cells, T cells, or B cells through successive stages of differentiation in either the bone marrow or thymus. There may be considerable plasticity in these pathways, in that in certain circumstances progenitor cells may switch their commitment. For example, a progenitor cell may give rise to either B cells or macrophages; however, for simplicity these alternative pathways are not shown. Some dendritic cells are also thought to be derived from the lymphoid progenitor.

which is induced by FLT3 signaling together with the activity of the transcription factor PU.1. The cytokine IL-7, secreted by bone marrow stromal cells, is essential for the growth and survival of developing B cells in mice (but possibly not in humans). Another essential factor is stem-cell factor (SCF), a membrane-bound cytokine present on stromal cells that stimulates the growth of hematopoietic stem cells and the earliest B-lineage progenitors. SCF interacts with the receptor tyrosine kinase Kit on the precursor cells (see Fig. 8.3). The chemokine CXCL12 (stromal cell-derived factor 1, SDF-1) is also essential for the early stages of B-cell development. It is produced constitutively by bone marrow stromal cells, and one of its roles may be to retain developing B-cell precursors in the marrow microenvironment. Thymic stroma-derived lymphopoietin (**TSLP**) resembles IL-7 and binds a receptor sharing the common γ chain of the IL-7 receptor. Despite its name, TSLP may promote B-cell development in the embryonic liver and, in the perinatal period at least, in the mouse bone marrow.

A definitive B-cell fate, the pro-B cell, is specified by induction of the B-lineage-specific transcription factor E2A, which is present as two alternatively spliced forms called E12 and E47, and the early B-cell factor (EBF). It is not clear what initiates the expression of E2A in some progenitors, but it is known that the transcription factors PU.1 and Ikaros are required for E2A expression. E2A then induces the expression of EBF. IL-7 signaling promotes the survival of these committed progenitors, while E2A and EBF act together to drive the expression of proteins that determine the pro-B cell state.

As B-lineage cells mature, they migrate within the marrow, remaining in contact with the stromal cells. The earliest stem cells lie in a region called the **endosteum**, which lines the inner cavity of the long bones. Developing B-lineage cells make contact with reticular stromal cells in the trabecular spaces, and as they mature they move toward the central sinus of the marrow cavity. The final stages of development of **immature B cells** into **mature B cells** occur in peripheral lymphoid organs such as the spleen, which we describe in the fourth part of this chapter.

8-2 B-cell development begins by rearrangement of the heavy-chain locus.

The stages of B-cell development are, in the order that they occur: **early pro-B cell**, **late pro-B cell**, **large pre-B cell**, **small pre-B cell**, and **mature B cell** (Fig. 8.4). Only one gene locus is rearranged at a time, in a fixed sequence. The first rearrangement to take place involves the locus that contains D gene segments—the immunoglobulin heavy-chain (IgH) locus in the case of B cells—and the joining of a D gene segment to a J segment. D to J_H rearrangement takes place mostly in the early pro-B-cell stage, but can be seen as early as

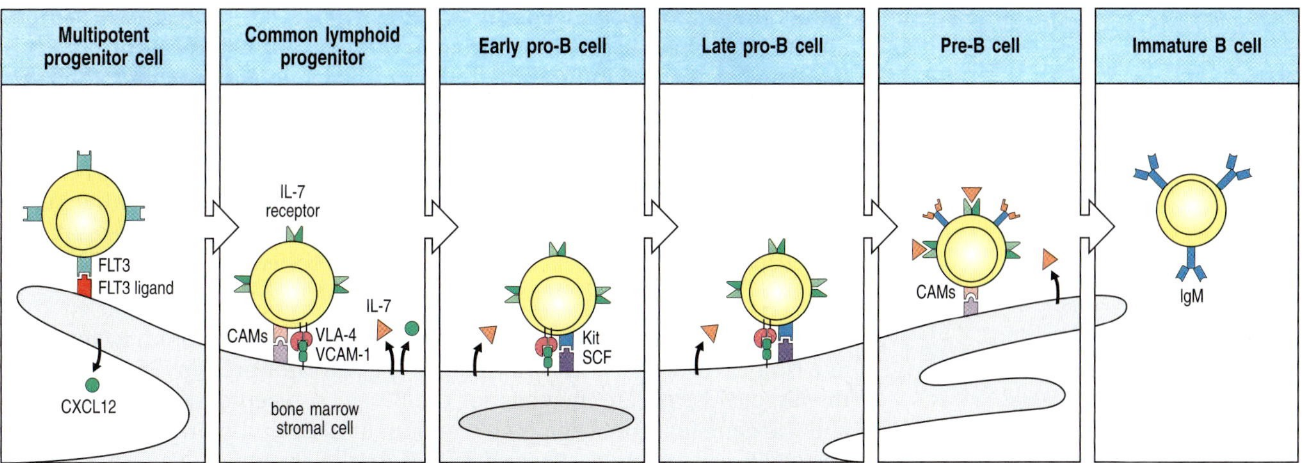

Fig. 8.3 The early stages of B-cell development are dependent on bone marrow stromal cells. Interaction of B-cell progenitors with bone marrow stromal cells is required for development to the immature B-cell stage. The designations pro-B cell and pre-B cell refer to defined phases of B-cell development as described in Fig. 8.4. Multipotent progenitor cells express the receptor tyrosine kinase FLT3, which binds to its ligand on stromal cells. Signaling through FLT3 is required for differentiation to the next stage, the common lymphoid progenitor. The chemokine CXCL12 (SDF-1) acts to retain stem cells and lymphoid progenitors to appropriate stromal cells in the bone marrow. The receptor for interleukin-7 (IL-7) is present from this stage, and IL-7 produced by stromal cells is required for the development of B-lineage cells. Progenitor cells bind to the adhesion molecule VCAM-1 on stromal cells through the integrin VLA-4 and also interact through other cell-adhesion molecules (CAMs). The adhesive interactions promote the binding of the receptor tyrosine kinase Kit (CD117) on the surface of the pro-B cell to stem-cell factor (SCF) on the stromal cell, which activates the kinase and induces the proliferation of B-cell progenitors.

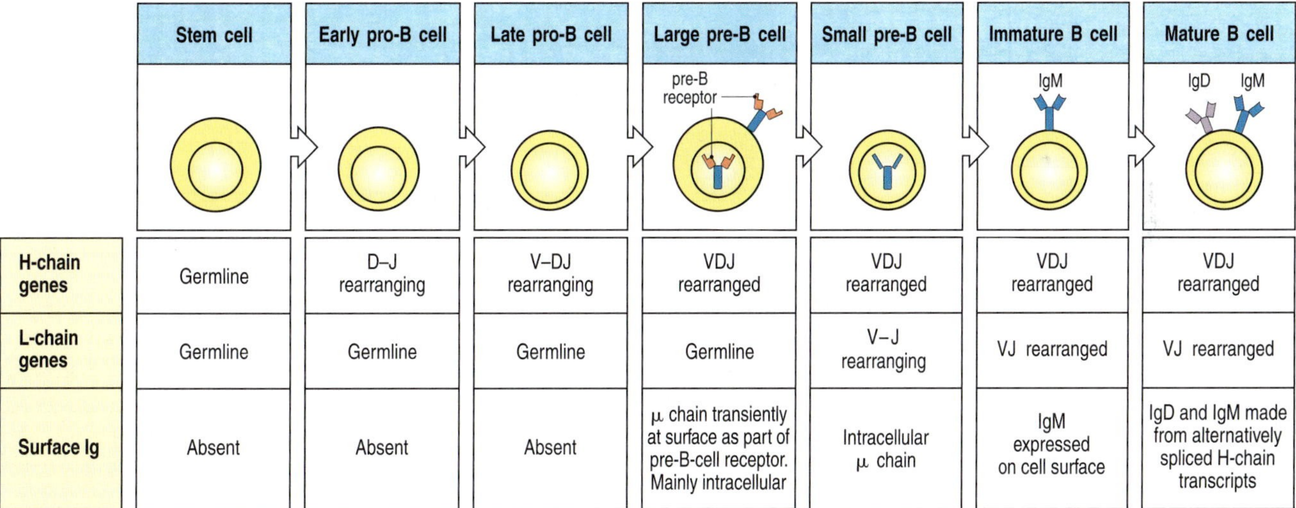

	Stem cell	Early pro-B cell	Late pro-B cell	Large pre-B cell	Small pre-B cell	Immature B cell	Mature B cell
H-chain genes	Germline	D–J rearranging	V–DJ rearranging	VDJ rearranged	VDJ rearranged	VDJ rearranged	VDJ rearranged
L-chain genes	Germline	Germline	Germline	Germline	V–J rearranging	VJ rearranged	VJ rearranged
Surface Ig	Absent	Absent	Absent	μ chain transiently at surface as part of pre-B-cell receptor. Mainly intracellular	Intracellular μ chain	IgM expressed on cell surface	IgD and IgM made from alternatively spliced H-chain transcripts

Fig. 8.4 The development of a B-lineage cell proceeds through several stages marked by the rearrangement and expression of the immunoglobulin genes. The stem cell has not yet begun to rearrange its immunoglobulin (Ig) gene segments; they are in the germline configuration as found in all nonlymphoid cells. The heavy-chain (H-chain) locus rearranges first. Rearrangement of a D gene segment to a J_H gene segment starts in the common lymphoid progenitor and occurs mostly in early pro-B cells, generating late pro-B cells in which V_H to DJ_H rearrangement occurs. A successful VDJ_H rearrangement leads to the expression of a complete immunoglobulin heavy chain as part of the pre-B-cell receptor, which is found mainly in the cytoplasm and to some degree on the surface of the cell. Once this occurs, the cell is stimulated to become a large pre-B cell, which proliferates. Large pre-B cells then cease dividing and become small resting pre-B cells, at which point they cease expression of the surrogate light chains and express the μ heavy chain alone in the cytoplasm. Small pre-B cells reexpress the RAG proteins and start to rearrange the light-chain (L-chain) genes. Upon successfully assembling a light-chain gene, a cell becomes an immature B cell that expresses a complete IgM molecule at the cell surface. Mature B cells produce a δ heavy chain as well as a μ heavy chain, by a mechanism of alternative mRNA splicing (see Fig. 5.17), and are marked by the additional appearance of IgD on the cell surface.

the common lymphoid progenitor. As shown in Fig. 8.4, expression of a functional heavy chain allows the formation of the **pre-B-cell receptor**, which is the signal to the cell to proceed to the next stage of development, the rearrangement of a light-chain gene. The transcription factors E2A and EBF in the early pro-B cell induce the expression of several key proteins that enable gene rearrangement to occur, including the RAG-1 and RAG-2 components of the V(D)J recombinase (see Chapter 5). Thus, E2A and EBF allow the initiation of V(D)J recombination at the heavy-chain locus and the expression of a heavy chain. In the absence of E2A or EBF, even the earliest identifiable stage in B-cell development, D to J_H joining, fails to occur.

Another key protein induced by E2A and EBF is the transcription factor Pax5, one isoform of which is known as the B-cell activator protein (BSAP). Among the targets of Pax5 are the gene for the B-cell co-receptor component CD19 and the gene for Igα, a signaling component of both the pre-B-cell receptor and the B-cell receptor (see Section 7-7). In the absence of Pax5, pro-B cells fail to develop further down the B-cell pathway but can be induced to give rise to T cells and myeloid cell types, indicating that Pax5 is required for commitment of the pro-B cell to the B-cell lineage. Pax5 also induces the expression of the B-cell linker protein (BLNK), an SH2-containing scaffold protein that is required for further development of the pro-B cell and for signaling from the mature B-cell antigen receptor (see Section 7-16). The temporal expression of some of the surface proteins, receptors, and transcription factors required for B-cell development are listed in Fig. 8.5.

Although the V(D)J recombinase system operates in both B- and T-lineage cells and uses the same core enzymes, rearrangements of T-cell receptor

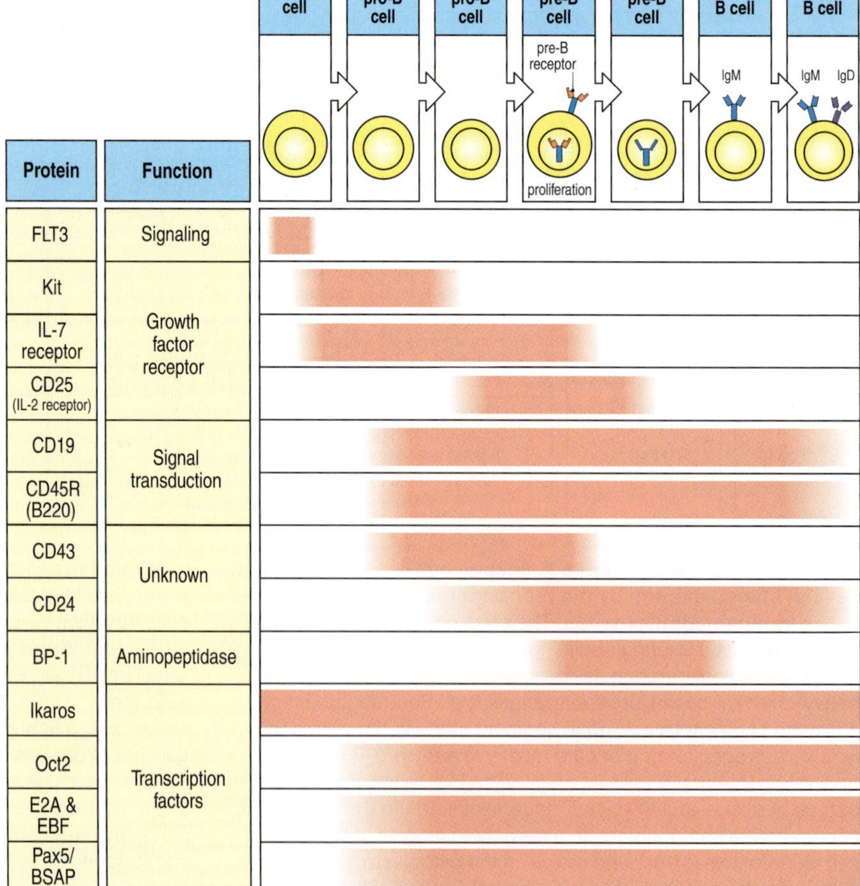

Fig. 8.5 Expression of surface proteins, receptors, and transcription factors in B-cell development. The stages of B-cell development corresponding to those shown in Fig. 8.4 are shown at the top of the figure. The receptor FLT3 is expressed on hematopoietic stem cells and the common lymphoid progenitor. The earliest B-lineage surface markers are CD19 and CD45R (B220 in the mouse), which are expressed throughout B-cell development. A pro-B cell is also distinguished by the expression of CD43 (a marker of unknown function), Kit (CD117), and the IL-7 receptor. A late pro-B cell starts to express CD24 (a marker of unknown function) and the IL-2 receptor α chain CD25. A pre-B cell is phenotypically distinguished by the expression of the enzyme BP-1, whereas Kit and the IL-7 receptor are no longer expressed. The actions of the listed transcription factors in B-cell development are discussed in the text, with the exception of the octamer transcription factor, Oct2, which binds the octamer ATGCAAAT found in the heavy-chain promoter and elsewhere.

genes do not occur in B-lineage cells, nor do complete rearrangements of immunoglobulin genes occur in T cells. The ordered rearrangement events that do occur are associated with lineage-specific low-level transcription of the gene segments about to be joined.

The initial D to J_H rearrangements in the immunoglobulin heavy-chain locus (Fig. 8.6) typically occur on both alleles, at which point the cell becomes a late pro-B cell. Most D to J_H joins in humans are potentially useful, because most human D gene segments can be translated in all three reading frames without encountering a stop codon. Thus, there is no need of a special mechanism for distinguishing successful D to J_H joins, and at this early stage there is also no need to ensure that only one allele undergoes rearrangement. Indeed, given the likely rate of failure at later stages, starting off with two successfully rearranged D–J_H sequences is an advantage.

To produce a complete immunoglobulin heavy chain, the late pro-B cell now proceeds with a rearrangement of a V_H gene segment to a DJ_H sequence. In contrast to D to J_H rearrangement, V_H to DJ_H rearrangement occurs first on only one chromosome. A successful rearrangement leads to the production of

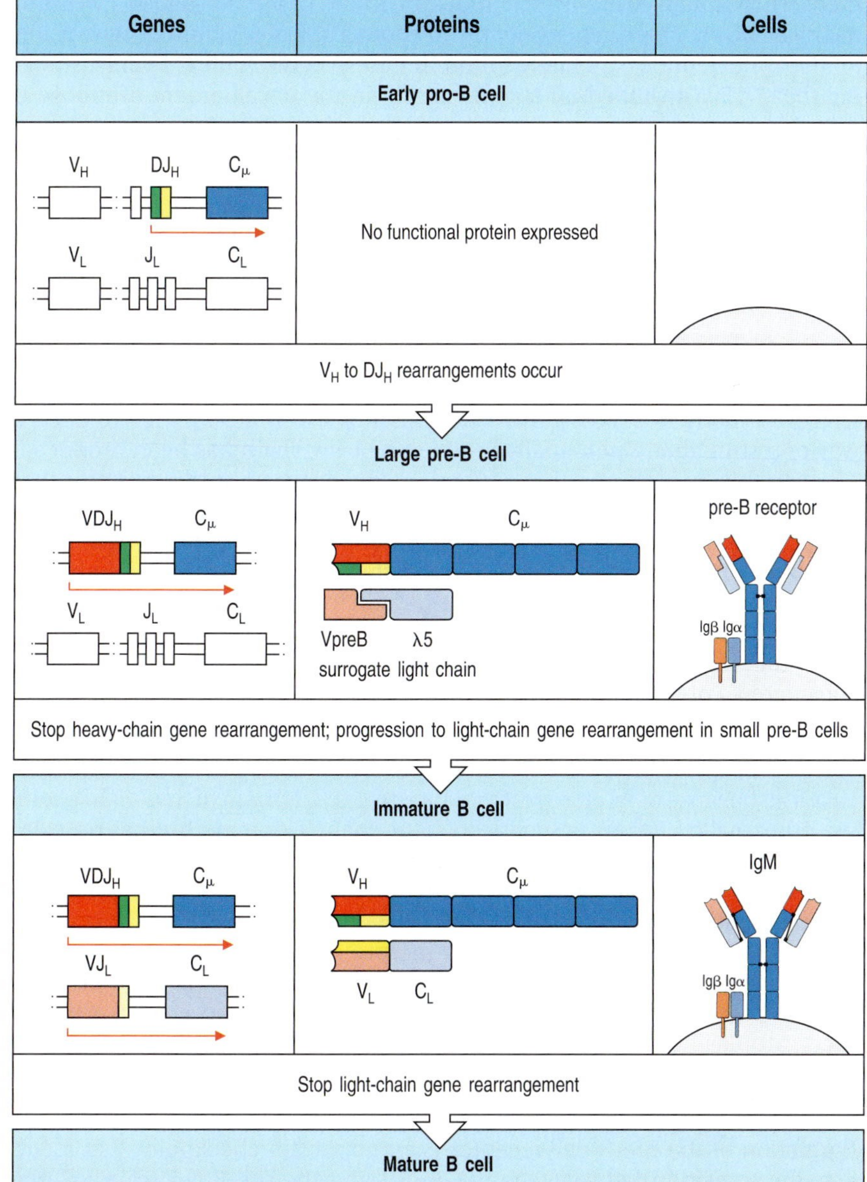

Fig. 8.6 A productively rearranged immunoglobulin gene is immediately expressed as a protein by the developing B cell. In early pro-B cells, heavy-chain gene rearrangement is not yet complete and no functional μ protein is expressed, although transcription occurs (red arrow), as shown in the top panel. As soon as a productive heavy-chain gene rearrangement takes place, μ chains are expressed by the cell in a complex with two other chains, λ5 and VpreB, which together make up a surrogate light chain. The whole immunoglobulin-like complex is known as the pre-B-cell receptor (center panel). It is associated with two other protein chains, Igα and Igβ, which signal the B cell to halt heavy-chain gene rearrangement, which drives the transition to the large pre-B cell stage by inducing proliferation. The progeny of large pre-B cells stop dividing and become small pre-B cells, in which light-chain gene rearrangements commence. Successful light-chain gene rearrangement results in the production of a light chain that binds the μ chain to form a complete IgM molecule, which is expressed together with Igα and Igβ at the cell surface, as shown in the bottom panel. Signaling via this surface receptor complex is thought to trigger the cessation of light-chain gene rearrangement.

intact μ heavy chains, after which V_H to DJ_H rearrangement ceases and the cell becomes a pre-B cell. Pro-B cells that do not produce a μ chain are eliminated, and at least 45% of pro-B cells are lost at this stage. In at least two out of three cases the first V_H to DJ_H rearrangement is nonproductive, and rearrangement then occurs on the other chromosome, again with a theoretical two in three chance of failure. A rough estimate of the chance of generating a pre-B cell is thus 55% ($1/3 + (2/3 \times 1/3) = 0.55$). The actual frequency is somewhat lower, because the V gene segment repertoire contains pseudogenes that can rearrange yet have major defects that prevent the expression of a functional protein. An initial nonproductive rearrangement need not signal the immediate failure of development of the pro-B cell, because it is possible for most loci to undergo successive rearrangements on the same chromosome, and where that fails, the locus on the other chromosome will rearrange.

The diversity of the B-cell antigen-receptor repertoire is enhanced at this stage by the enzyme terminal deoxynucleotidyl transferase (TdT). TdT is expressed by the pro-B cell and adds nontemplated nucleotides (N-nucleotides) at the joints between rearranged gene segments (see Section 5-8). In adult humans it is expressed in pro-B cells during heavy-chain gene rearrangement, but its expression declines at the pre-B-cell stage during light-chain gene rearrangement. This explains why N-nucleotides are found in the V–D and D–J joints of nearly all heavy-chain genes but only in about a quarter of human light-chain joints. N-nucleotides are rarely found in mouse light-chain V–J joints, showing that TdT is switched off slightly earlier in the development of mouse B cells. In fetal development, when the peripheral immune system is first being supplied with T and B lymphocytes, TdT is expressed at low levels, if at all.

8-3 The pre-B-cell receptor tests for successful production of a complete heavy chain and signals for the transition from the pro-B cell to pre-B cell stage.

The imprecise nature of V(D)J recombination is a double-edged sword. Although it produces increased diversity in the antibody repertoire, it also results in many unsuccessful rearrangements. Pro-B cells therefore need a way of testing that a potentially functional heavy chain has been produced. They do this by incorporating a functional heavy chain into a receptor that can signal its successful production. This test takes place in the absence of light chains, whose loci have not yet rearranged. Instead, pro-B cells make two invariant 'surrogate' proteins that together have a structural resemblance to the light chain and can pair with the μ chain to form the pre-B-cell receptor (pre-BCR) (see Fig. 8.6). The assembly of a pre-B-cell receptor signals to the B cell that a productive rearrangement has been made, and the cell is then considered a pre-B cell.

The surrogate chains are encoded by nonrearranging genes separate from the antigen-receptor loci, and their expression is induced by E2A and EBF. One is called **λ5** because of its close resemblance to the C domain of the λ light chain; the other, called **VpreB**, resembles a light-chain V domain but has an extra region at the amino-terminal end. Other proteins expressed by pro-B cells and pre-B cells are also required for the formation of a functional receptor complex and are essential for B-cell development. The invariant proteins Igα and Igβ are components of both the pre-B-cell receptor and the B-cell receptor complexes on the cell surface. Igα and Igβ transduce signals from these receptors by interacting with intracellular tyrosine kinases through their cytoplasmic tails (see Section 7-7). Igα and Igβ are expressed from the pro-B-cell stage until the death of the cell or until its terminal differentiation into an antibody-secreting plasma cell.

Formation of the pre-B-cell receptor is an important checkpoint that mediates the transition between the pro-B cell and the pre-B cell. In mice that either lack λ5 or have mutant heavy-chain genes that cannot produce the transmembrane domain, the pre-B-cell receptor cannot be formed and B-cell development is blocked after heavy-chain gene rearrangement. In normal B-cell development, the pre-B-cell receptor complex is expressed transiently, perhaps because the production of λ5 mRNA stops as soon as pre-B-cell receptors begin to be formed. Although present at only low levels on the cell surface, the pre-B-cell receptor generates signals required for the transition from pro-B cell to pre-B cell. No antigen or other external ligand seems to be involved in signaling by the receptor. Instead, pre-B-cell receptors are thought to interact with each other, forming dimers or oligomers that generate signals as described in Section 7-16. Dimerization involves 'unique' regions in the amino termini of λ5 and VpreB proteins that are not present in other immunoglobulin-like domains and which mediate the cross-linking of adjacent pre-B-cell receptors on the cell surface (Fig. 8.7). The signaling generated by receptor clustering halts further rearrangement of the heavy-chain locus and allows the pro-B cell to become sensitive to IL-7. This induces cell proliferation, initiating the transition to the large pre-B cell. Pre-B-cell receptor signaling requires the scaffold protein BLNK and Bruton's tyrosine kinase (Btk), an intracellular Tec-family tyrosine kinase (see Section 7-16). In humans and mice, deficiency of BLNK leads to a block in B-cell development at the pro-B-cell stage. In humans, mutations in the *BTK* gene cause a profound B-lineage-specific immune deficiency, Bruton's X-linked agammaglobulinemia (XLA), in which no mature B cells are produced. The block in B-cell development caused by mutations in *BTK* is almost total, interrupting the transition from pre-B cell to immature B cell. A similar, but less severe, defect called **X-linked immunodeficiency** or xid arises from mutations in the *Btk* gene in mice.

X-linked
Agammaglobulinemia

8-4 Pre-B-cell receptor signaling inhibits further heavy-chain locus rearrangement and enforces allelic exclusion.

Successful rearrangements at both heavy-chain alleles could result in a B cell producing two receptors of different antigen specificities. To prevent this, signaling by the pre-B-cell receptor enforces **allelic exclusion**, the state in which only one of the two alleles of a gene is expressed in a diploid cell. Allelic exclusion, which occurs at both the heavy-chain locus and the light-chain loci, was

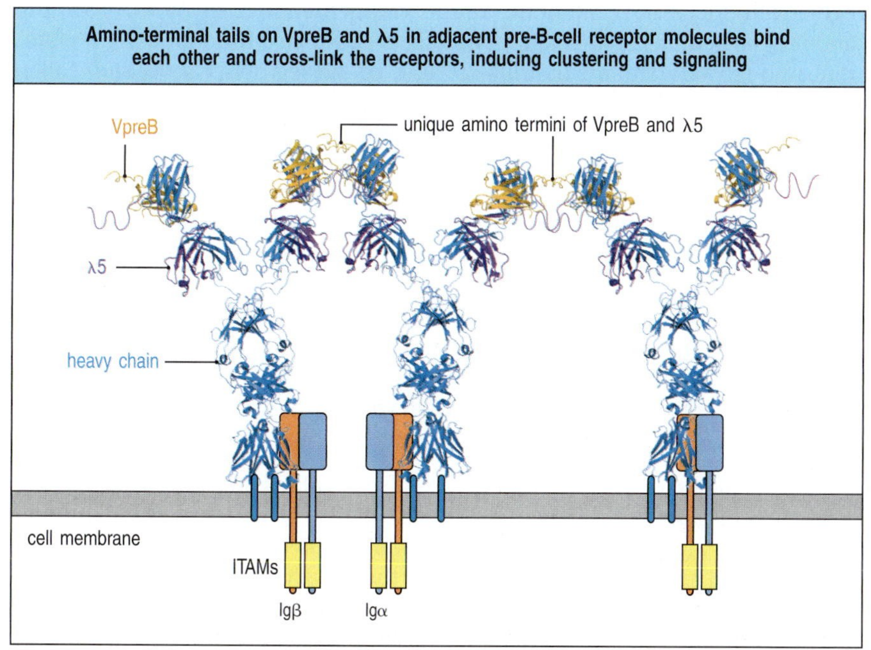
Amino-terminal tails on VpreB and λ5 in adjacent pre-B-cell receptor molecules bind each other and cross-link the receptors, inducing clustering and signaling
VpreB
unique amino termini of VpreB and λ5
λ5
heavy chain
cell membrane
ITAMs
Igβ
Igα

Fig. 8.7 The pre-B-cell receptor initiates signaling through spontaneous dimerization induced by the unique regions of VpreB and λ5. Two surrogate protein chains, VpreB (orange) and λ5 (purple), substitute for a light chain and bind to a heavy chain, thus allowing its surface expression. VpreB substitutes for the light-chain V region in this surrogate interaction, while λ5 takes the part of the light-chain constant region. Both VpreB and λ 5 contain 'unique' amino-terminal regions that are not present in other immunoglobulin-like domains, shown here as unstructured tails extending out from the globular domains. These amino-terminal regions associated with one pre-B-cell receptor can interact with the corresponding regions on the adjacent pre-B cell receptor, promoting the spontaneous formation of pre-B-cell receptor dimers on the cell surface. Dimerization generates signaling from the pre-B-cell receptor that is dependent on the presence of the ITAM-containing signaling chains Igα and Igβ. The signals cause the inhibition of RAG-1 and RAG-2 expression and the proliferation of the large pre-B cell. Courtesy of Chris Garcia.

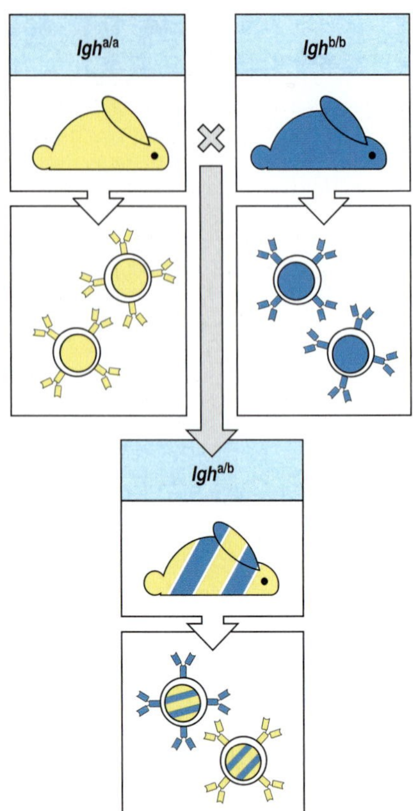

Fig. 8.8 Allelic exclusion in individual B cells. Most species have genetic polymorphisms of the constant regions of their immunoglobulin heavy-chain and light-chain genes; these are known as allotypes (see Appendix I, Section A-10). In rabbits, for example, all of the B cells in an individual homozygous for the *a* allele of the immunoglobulin heavy-chain locus ($Igh^{a/a}$) will express immunoglobulin of allotype a, whereas in an individual homozygous for the *b* allele ($Igh^{b/b}$) all the B cells make immunoglobulin of allotype b. In a heterozygous animal ($Igh^{a/b}$), which carries the *a* allele at one of the Igh loci and the *b* allele on the other, individual B cells can be shown to express surface immunoglobulin of either the a-allotype or the b-allotype, but not both. This allelic exclusion reflects the productive rearrangement of only one of the two *Igh* alleles in the B cell, because the production of a successfully rearranged immunoglobulin heavy chain forms a pre-B-cell receptor, which signals the cessation of further heavy-chain gene rearrangement.

discovered more than 30 years ago and provided one of the original pieces of experimental support for the theory that one lymphocyte expresses one type of antigen receptor (Fig. 8.8).

Signaling from the pre-B-cell receptor promotes heavy-chain allelic exclusion in three ways. First, it reduces V(D)J recombinase activity by directly reducing the expression of the *RAG-1* and *RAG-2* genes. Second, it further reduces levels of RAG-2 by indirectly causing this protein to be targeted for degradation, which occurs when RAG-2 is phosphorylated in response to the entry of the pro-B cell into S phase (the DNA synthesis phase) of the cell cycle. Finally, pre-B-cell receptor signaling reduces access of the heavy-chain locus to the recombinase machinery, although the precise details are not clear. At a later stage of B-cell development, RAG proteins will again be expressed in order to carry out light-chain locus rearrangement, but at that point the heavy-chain locus does not undergo further rearrangement. In the absence of pre-B-cell receptor signaling, allelic exclusion of the heavy-chain locus does not occur. For example, in λ5 knockout mice, in which the pre-B-cell receptor is not formed and the signal for V_H to DJ_H rearrangement to stop is not given, rearrangements of the heavy-chain genes are found on both chromosomes in all B-cell precursors, so that about 10% of the cells have two productive VDJ_H rearrangements.

8-5 Pre-B cells rearrange the light-chain locus and express cell-surface immunoglobulin.

The transition from the pro-B cell to the large pre-B-cell stage is accompanied by several rounds of cell division, expanding the population of cells with successful in-frame joins by about 30–60-fold before they become resting small pre-B cells. A large pre-B cell with a particular rearranged heavy-chain gene therefore gives rise to numerous small pre-B cells. RAG proteins are produced again in the small pre-B cells, and rearrangement of the light-chain locus begins. Each of these cells can make a different rearranged light-chain gene and so cells with many different antigen specificities are generated from a single pre-B cell, which makes an important contribution to overall B-cell receptor diversity.

Light-chain rearrangement also exhibits allelic exclusion. Rearrangements at the light-chain locus generally take place at only one allele at a time. The light-chain loci lack D segments and rearrangement occurs by V to J joining, and if a particular VJ rearrangement fails to produce a functional light chain, repeated rearrangements of unused V and J gene segments at the same allele can occur (Fig. 8.9). Several attempts at productive rearrangement of a light-chain gene can therefore be made on one chromosome before initiating any rearrangements on the second chromosome. This greatly increases the chances of eventually generating an intact light chain, especially as there are two different light-chain loci. As a result, many cells that reach the pre-B-cell stage succeed in generating progeny that bear intact IgM molecules and can be classified as **immature B cells**. Figure 8.10 lists some of the proteins involved in V(D)J recombination and shows how their expression is regulated throughout B-cell development. Figure 8.11 summarizes the stages of B-cell development up to the point of assembly of a complete surface immunoglobulin, indicating the points at which developing B cells can be lost as a result of failure to produce a productive join.

As well as allelic exclusion, light chains also display **isotypic exclusion**; that is, the expression of only one type of light chain—κ or λ—by an individual B cell. In mice and humans, the κ light-chain locus tends to rearrange before the λ locus. This was first deduced from the observation that myeloma cells secreting λ light chains generally have both their κ and λ light-chain genes rearranged, whereas in myelomas secreting κ light chains, generally only the

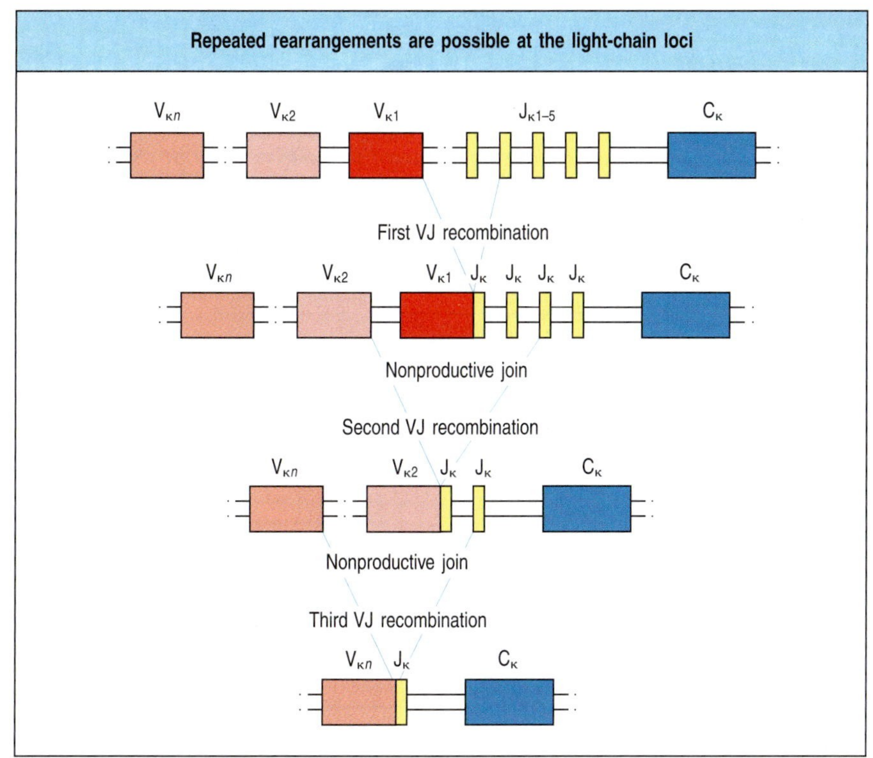

Fig. 8.9 Nonproductive light-chain gene rearrangements can be rescued by further rearrangement. The organization of the light-chain loci in mice and humans offers many opportunities for the rescue of pre-B cells that initially make an out-of-frame rearrangement. Light-chain rescue is illustrated here at the human κ locus. If the first rearrangement is nonproductive, a 5′ V_κ gene segment can recombine with a 3′ J_κ gene segment to remove the out-of-frame join located between them and to replace it with a new rearrangement. In principle, this can happen up to five times on each chromosome, because there are five functional J_κ gene segments in humans. If all rearrangements of κ-chain genes fail to yield a productive light-chain join, λ-chain gene rearrangement may succeed (not shown; see Fig. 8.11).

κ genes are rearranged. This order is occasionally reversed, however, and λ gene rearrangement does not absolutely require the previous rearrangement of the κ genes. The ratios of κ-expressing versus λ-expressing mature B cells vary from one extreme to the other in different species. In mice and rats it is 95% κ to 5% λ, in humans it is typically 65%:35%, and in cats it is 5%:95%, the opposite of that in mice. These ratios correlate most strongly with the number of functional V_κ and V_λ gene segments in the genome of the species. They also reflect the kinetics and efficiency of gene segment rearrangements. The κ:λ ratio in the mature lymphocyte population is useful in clinical diagnostics, because an aberrant κ:λ ratio indicates the dominance of one clone and the presence of a lymphoproliferative disorder, which may be malignant.

8-6 Immature B cells are tested for autoreactivity before they leave the bone marrow.

Once a rearranged light chain has paired with a μ chain, IgM can be expressed on the cell surface (sIgM) and the pre-B cell becomes an **immature B cell**. At this stage, the antigen receptor is first tested for reactivity to self antigens, or autoreactivity. The elimination or inactivation of autoreactive B cells ensures that the B-cell population as a whole will be tolerant of self antigens. The **tolerance** produced at this stage of B-cell development is known as **central tolerance** because it arises in a central lymphoid organ, the bone marrow. As we shall see later in the chapter and in Chapter 15, self-reactive B cells that escape this test and go on to mature may still be removed from the repertoire after they have left the bone marrow, a process that produces **peripheral tolerance**.

sIgM associates with Igα and Igβ to form a functional B-cell receptor complex, and the fate of an immature B cell in the bone marrow depends on signals delivered from this receptor complex on its interaction with ligands in the environment. Igα signaling is particularly important in dictating the

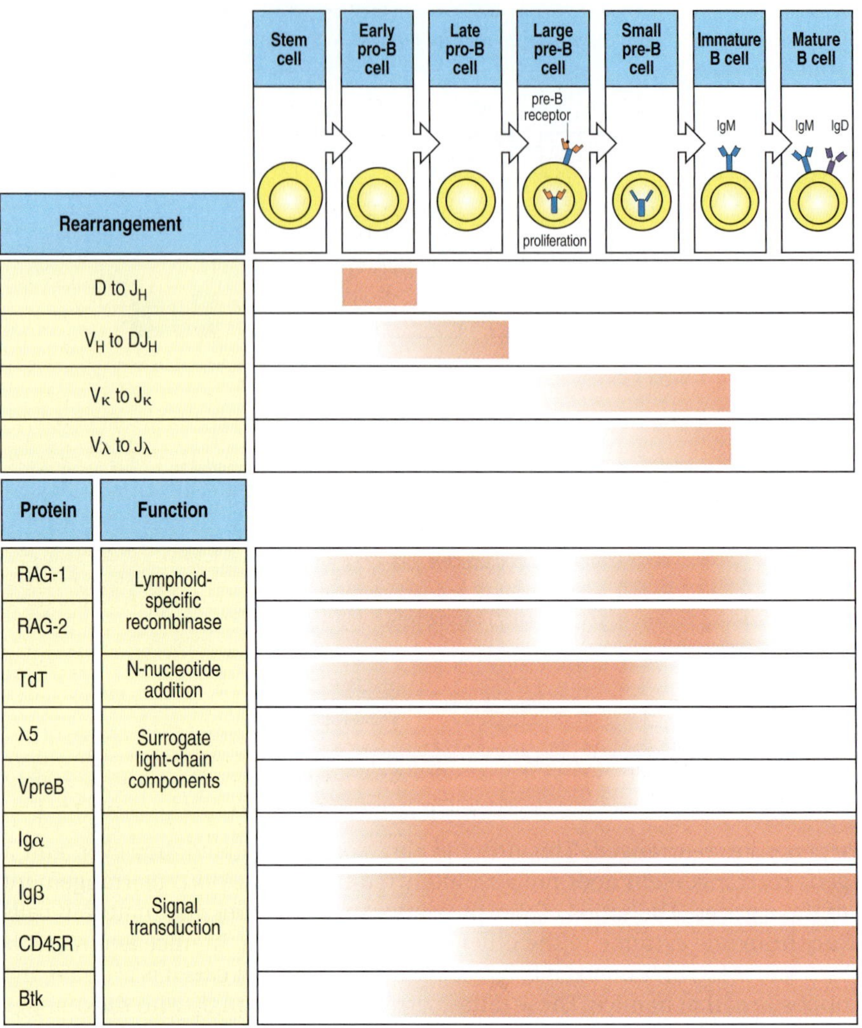

Fig 8.10 Expression of proteins involved in gene rearrangement and the production of pre-B-cell and B-cell receptors. The proteins listed here have been included because of their proven importance in the developmental sequence, largely on the basis of studies in mice. Their individual contributions to B-cell development are discussed in the text. Also shown is the temporal sequence of gene rearrangement. Signaling proteins and transcription factors involved in early B-lineage development are listed in Fig. 8.5.

emigration of B cells from the bone marrow and/or their survival in the periphery: mice that express Igα with a truncated cytoplasmic domain that cannot signal show a fourfold reduction in the number of immature B cells in the marrow, and a hundredfold reduction in the number of peripheral B cells.

Immature B cells that have no strong reactivity to self antigens are allowed to mature (Fig. 8.12, first panel). They leave the marrow via sinusoids that enter the central sinus and are carried by the venous blood supply to the spleen. If, however, the newly expressed receptor encounters a strongly cross-linking antigen in the bone marrow—that is, if the B cell is strongly self-reactive—development is arrested and the cell will not mature. This was first demonstrated by experiments in which antigen receptors on immature B cells were experimentally stimulated *in vivo* using anti-μ chain antibodies (see Appendix I, Section A-10); the outcome was elimination of the immature B cells.

More recent experiments using mice expressing transgenes that enforce the expression of self-reactive B-cell receptors confirmed these earlier findings, but also showed that immediate elimination is not the only possible outcome of binding to a self antigen. There are four possible fates for self-reactive immature B cells, depending on the nature of the ligand they recognize (Fig. 8.12, last three panels). These fates are: cell death by apoptosis or clonal deletion; the production of a new receptor by a process known as receptor editing; the induction of a permanent state of unresponsiveness, or anergy, to antigen; and a state of immunological ignorance.

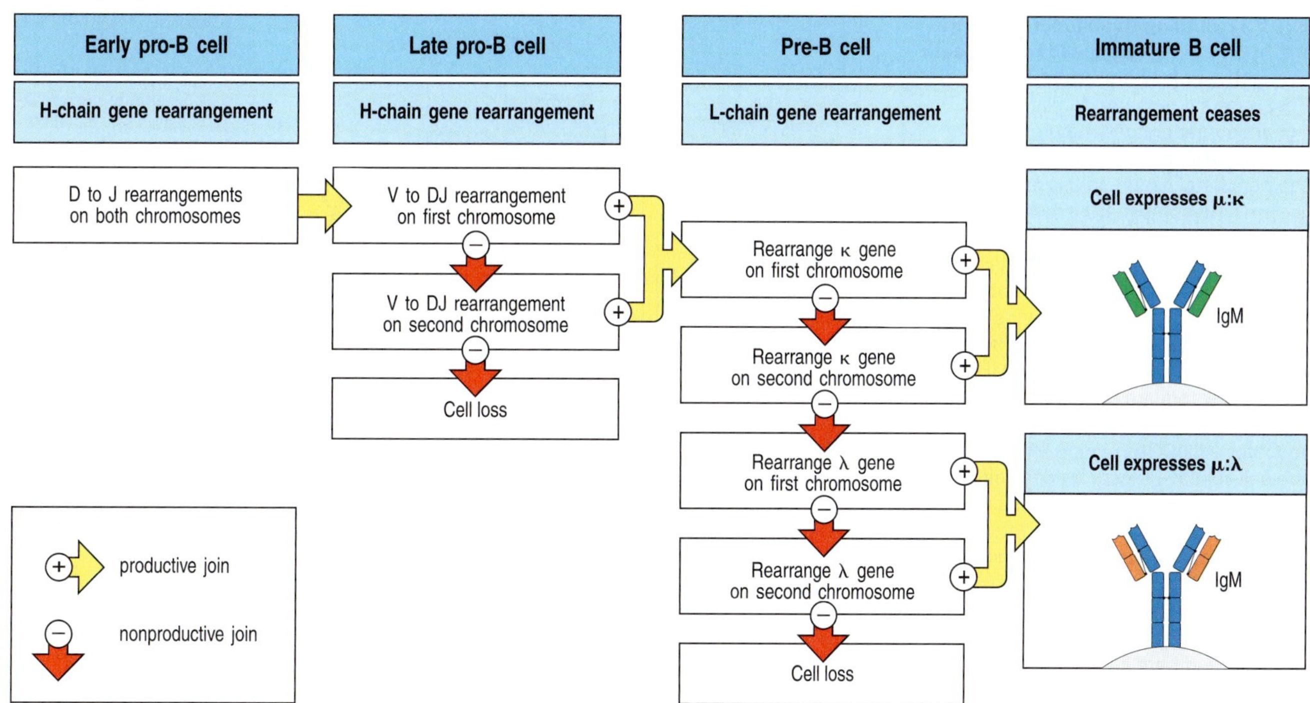

Fig. 8.11 The steps in immunoglobulin gene rearrangement at which developing B cells can be lost. The developmental program rearranges the heavy-chain (H-chain) locus first and then the light-chain (L-chain) loci. Cells are allowed to progress to the next stage when a productive rearrangement has been achieved. Each rearrangement has about a one in three chance of being successful, but if the first attempt is nonproductive, development is suspended and there is a chance for one or more further attempts, so by simple mathematics four in nine rearrangements give rise to a heavy chain. The scope for repeated rearrangements is greater at the light-chain loci (see Fig. 8.9), so that fewer cells are lost between the pre-B and immature B-cell stages than in the pro-B to pre-B transition.

Clonal deletion, or the removal of cells of a particular antigen specificity from the repertoire, seems to predominate when the interacting self antigen is multivalent. The effect of an encounter with a multivalent antigen was tested in mice transgenic for both chains of an immunoglobulin specific for H-2K^b MHC class I molecules. In such mice, nearly all the B cells that develop bear the anti-MHC immunoglobulin as sIgM. If the transgenic mouse does not express H-2K^b, normal numbers of B cells develop, all bearing transgene-encoded anti-H-2K^b receptors. However, in mice expressing both H-2K^b and the immunoglobulin transgenes, B-cell development is blocked. Normal numbers of pre-B cells and immature B cells are found, but B cells expressing the anti-H-2K^b immunoglobulin as sIgM never mature to populate the spleen and lymph nodes; instead, most of these immature B cells die in the bone marrow by apoptosis.

Not all lymphocytes with strongly autoreactive receptors undergo clonal deletion. There is an interval before cell death during which the autoreactive B cell can be rescued by further gene rearrangements that replace the autoreactive receptor with a new receptor that is not self-reactive. This mechanism is termed **receptor editing** (Fig. 8.13). When an immature B cell first produces sIgM, RAG protein is still being made. If the receptor is not self-reactive, the absence of sIgM cross-linking allows gene rearrangement to cease and B-cell development continues, with RAG proteins eventually disappearing. For an autoreactive receptor, however, an encounter with the self antigen results in strong cross-linking of sIgM, RAG expression continues, and light-chain gene rearrangement can continue, as described in Fig. 8.9. These secondary rearrangements can rescue immature self-reactive B cells by deleting the self-reactive light-chain gene and replacing it with another sequence. If the

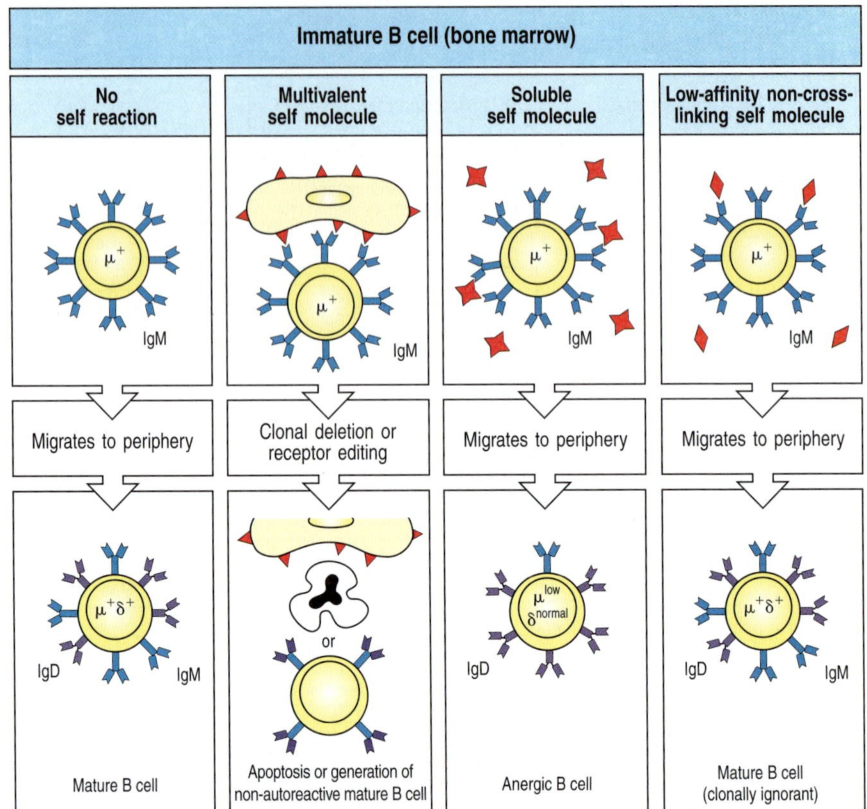

Fig. 8.12 Binding to self molecules in the bone marrow can lead to the death or inactivation of immature B cells. First panels: immature B cells that do not encounter antigen mature normally; they migrate from the bone marrow to the peripheral lymphoid tissues, where they may become mature recirculating B cells bearing both IgM and IgD on their surface. Second panels: when developing B cells express receptors that recognize multivalent ligands, for example ubiquitous self cell-surface molecules such as those of the MHC, these receptors are deleted from the repertoire. The B cells either undergo receptor editing (see Fig. 8.13), so that the self-reactive receptor specificity is deleted, or the cells themselves undergo programmed cell death or apoptosis (clonal deletion). Third panels: immature B cells that bind soluble self antigens able to cross-link the B-cell receptor are rendered unresponsive to the antigen (anergic) and bear little surface IgM. They migrate to the periphery, where they express IgD but remain anergic; if in competition with other B cells in the periphery, they are rapidly lost. Fourth panels: immature B cells whose antigen is inaccessible to them, or which bind monovalent or soluble self antigens with low affinity, do not receive any signal and mature normally. Such cells are potentially self-reactive, however, and are said to be clonally ignorant because their ligand is present but is unable to activate them.

new light chain is not autoreactive, the B cell continues normal development. If the receptor remains autoreactive, rearrangement continues until a non-autoreactive receptor is produced or light-chain V and J gene segments are exhausted. Cells that remain autoreactive undergo apoptosis.

Receptor editing has been shown definitively in mice bearing transgenes for autoantibody heavy and light chains that have been placed within the immunoglobulin loci by homologous recombination (see Appendix I, Section A-46, for details of this method). The transgene imitates a primary gene rearrangement and is surrounded by unused endogenous gene segments. In mice that express the antigen recognized by the transgene-encoded receptor, the mature B cells that emerge into the periphery have used these surrounding gene segments for rearrangements that replace the autoreactive light-chain transgene with a non-autoreactive gene.

It is not clear whether receptor editing occurs at the heavy-chain locus. There are no available D segments at a rearranged heavy-chain locus, so a new rearrangement cannot simply occur by the normal mechanism and remove the preexisting one. Instead, a process of V_H replacement may use embedded recombination signal sequences in a recombination event that displaces the V gene segment from the self-reactive rearrangement and replaces it with a new V gene segment. This has been observed in some B-cell tumors, but whether it occurs during normal B-cell development in response to signals from autoreactive B-cell receptors is not certain.

It was originally thought that the successful production of a heavy chain and a light chain caused the almost instantaneous shutdown of light-chain locus rearrangement and that this ensured both allelic and isotypic exclusion. The unexpected ability of self-reactive B cells to continue to rearrange their light-chain genes, even after having made a productive rearrangement, has raised questions about this supposed mechanism of allelic exclusion. This

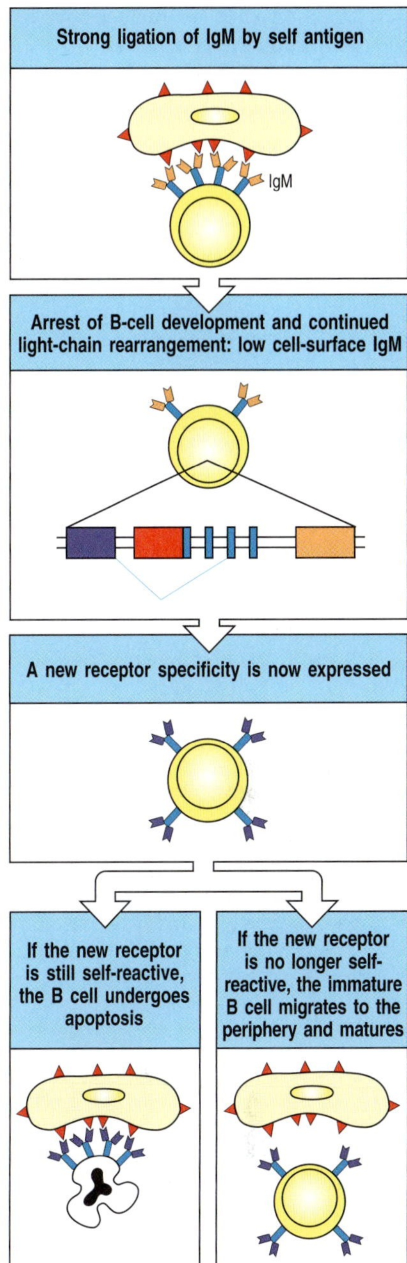

Fig. 8.13 Replacement of light chains by receptor editing can rescue some self-reactive B cells by changing their antigen specificity. When a developing B cell expresses antigen receptors that are strongly cross-linked by multivalent self antigens such as MHC molecules on cell surfaces (top panel), its development is arrested. The cell decreases surface expression of IgM and does not turn off the *RAG* genes (second panel). Continued synthesis of RAG proteins allows the cell to continue light-chain gene rearrangement. This usually leads to a new productive rearrangement and the expression of a new light chain, which combines with the previous heavy chain to form a new receptor (receptor editing; third panel). If this new receptor is not self-reactive, the cell is 'rescued' and continues normal development much like a cell that had never reacted with self (bottom right panel). If the cell remains self-reactive it may be rescued by another cycle of rearrangement, but if it continues to react strongly with self it will undergo programmed cell death or apoptosis and be deleted from the repertoire (clonal deletion; bottom left panel).

suggests that the fall in the level of RAG protein that follows a successful non-autoreactive rearrangement could be the principal, but not necessarily the sole, mechanism behind allelic exclusion at the light-chain locus. It seems that allelic exclusion is not absolute as there are rare B cells that express two different light chains.

We have so far discussed the fate of newly formed B cells that undergo multivalent cross-linking of their sIgM. Immature B cells that encounter more weakly cross-linking self antigens of low valence, such as small soluble proteins, respond differently. In this situation, self-reactive B cells tend to be inactivated and enter a state of permanent unresponsiveness, or **anergy**, but do not immediately die (see Fig. 8.12). Anergic B cells cannot be activated by their specific antigen even with help from antigen-specific T cells. Again, this phenomenon was elucidated using transgenic mice. Hen egg-white lysozyme (HEL) was expressed in soluble form from a transgene in mice that were also transgenic for high-affinity anti-HEL immunoglobulin. The HEL-specific B cells matured but could not respond to antigen. Anergic cells retain their IgM within the cell and transport little to the surface. In addition, they develop a partial block in signal transduction so that, despite normal levels of HEL-binding sIgD, the cells cannot be stimulated by cross-linking this receptor. The signaling defect may involve the action of two ubiquitin ligases (see Section 7-5), Cbl and Cbl-b. These proteins seem to target Igα and Syk for degradation, reducing sIgM expression and BLNK phosphorylation (see Section 7-16). In mice lacking these proteins, B cells that would normally become anergic go on to develop into mature B cells with normal surface expression of IgM.

The migration of anergic B cells within peripheral lymphoid organs is impaired, and their life-span and their ability to compete with immuno-competent B cells are compromised. In normal circumstances, in which B cells binding a soluble self antigen are rare, self-reactive anergic B cells are detained in the T-cell areas of peripheral lymphoid tissues and are excluded from lymphoid follicles. Anergic B cells cannot be activated by T cells, because all the T cells will be tolerant to soluble antigens. Instead they die relatively quickly, presumably because they fail to get survival signals from T cells, which ensures that the long-lived pool of peripheral B cells is purged of potentially self-reactive cells.

The fourth potential fate of self-reactive immature B cells is that nothing happens to them; they remain in a state of **immunological ignorance** of their self antigen (see Fig. 8.12). Immunologically ignorant cells have affinity for a self antigen but for various reasons do not sense and respond to it. The antigen may not be accessible to developing B cells in the bone marrow or spleen, or may be in low concentration, or may bind so weakly to the B-cell receptor that it does not generate an activating signal. Because some ignorant cells can be (and in fact are) activated under certain conditions such as inflammation or when the self antigen becomes available or reaches an unusually high concentration, they should not be considered inert, and they are fundamentally different from cells with non-autoreactive receptors that could never be activated by self antigens.

The fact that central tolerance is not perfect and some self-reactive B cells are allowed to mature reflects the balance that the immune system strikes between purging all self-reactivity and maintaining the ability to respond to pathogens. If the elimination of self-reactive cells were too efficient, the receptor repertoire might become too limited and thus unable to recognize a wide variety of pathogens. Some autoimmune disease is the price of this balance: we shall see in Chapter 15 that ignorant self-reactive lymphocytes can be activated and cause disease under certain circumstances. Normally, however, ignorant B cells are held in check by a lack of T-cell help, the continued inaccessibility of the self antigen, or the tolerance that can be induced in mature B cells, which is described later in this chapter.

Summary.

So far we have followed B-cell development from the earliest progenitors in the bone marrow to the immature B cell that is ready to emerge into the peripheral lymphoid tissue. The heavy-chain locus is rearranged first and, if this is successful, a μ heavy chain is produced that combines with surrogate light chains to form the pre-B-cell receptor; this is the first checkpoint in B-cell development. Production of the pre-B-cell receptor signals successful heavy-chain gene rearrangement, and causes cessation of this rearrangement, thus enforcing allelic exclusion. It also initiates pre-B-cell proliferation, generating numerous progeny in which subsequent light-chain rearrangement can be attempted. If the initial light-chain gene rearrangement is productive, a complete immunoglobulin B-cell receptor is formed, gene rearrangement again ceases, and the B cell continues its development. If the first light-chain gene rearrangement is unsuccessful, rearrangement continues until either a productive rearrangement is made or all available J regions are used up. If no productive rearrangement is made, the developing B cell dies. In the next part of the chapter, we turn to T-cell development.

The development of T lymphocytes in the thymus.

Like B cells, T lymphocytes derive from the multipotent hematopoietic stem cells in the bone marrow. However, their progenitor cells migrate from the bone marrow via the blood to the thymus, where they mature (Fig. 8.14); this is the reason for the name thymus-dependent (T) lymphocytes or T cells. T-cell development parallels that of B cells in many ways, including the orderly and stepwise rearrangement of antigen-receptor genes, the sequential testing for successful gene rearrangement, and the eventual assembly of a heterodimeric antigen receptor. Nevertheless, T-cell development in the thymus has some features not seen for B cells, such as the generation of two distinct lineages of T cells, the γ:δ lineage and the α:β lineage, which express distinct antigen-receptor genes. Developing T cells, which are known generally as **thymocytes**, also undergo rigorous selection that depends on interactions with thymic cells and that shapes the mature repertoire of T cells to ensure self-MHC restriction as well as self-tolerance. We begin with a general overview of the stages of thymocyte development and its relationship to thymic anatomy before considering gene rearrangement and the mechanisms of selection.

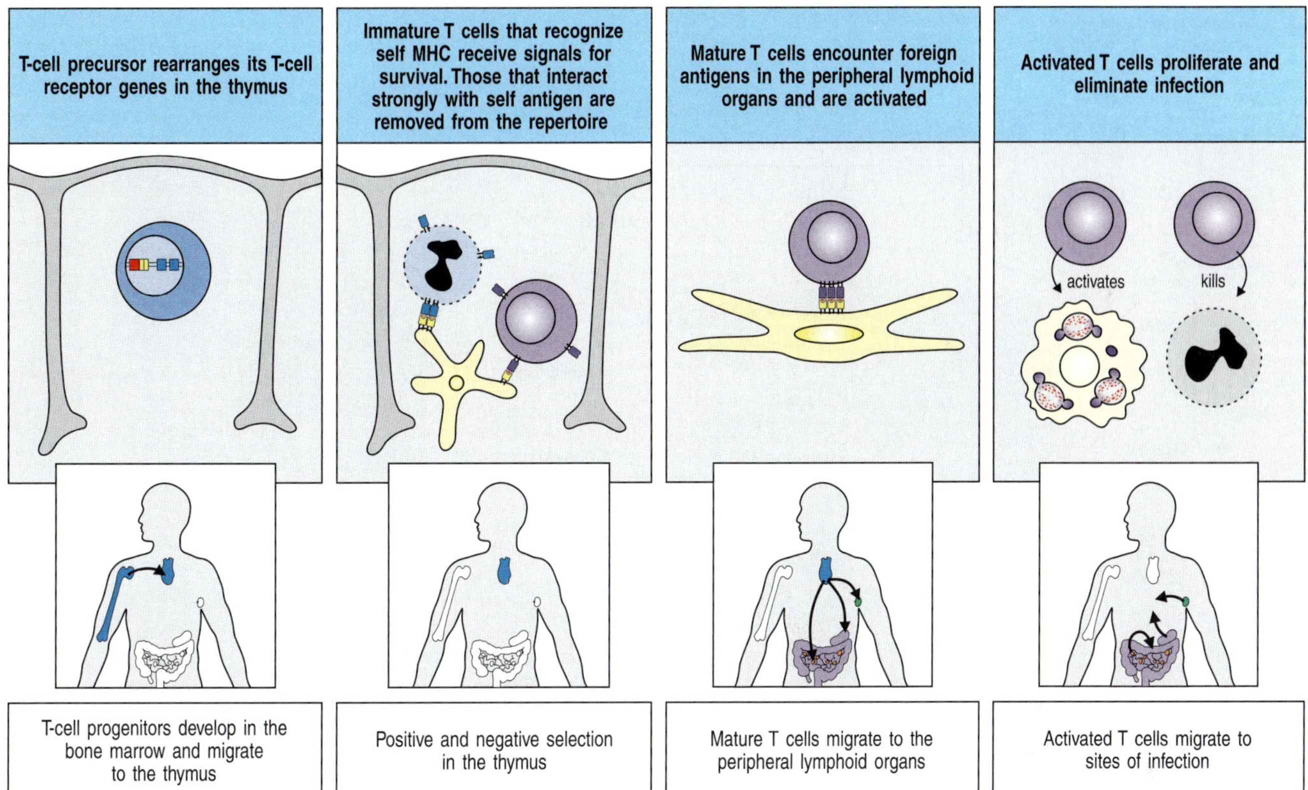
T-cell precursor rearranges its T-cell receptor genes in the thymus
Immature T cells that recognize self MHC receive signals for survival. Those that interact strongly with self antigen are removed from the repertoire
Mature T cells encounter foreign antigens in the peripheral lymphoid organs and are activated
Activated T cells proliferate and eliminate infection
activates
kills
T-cell progenitors develop in the bone marrow and migrate to the thymus
Positive and negative selection in the thymus
Mature T cells migrate to the peripheral lymphoid organs
Activated T cells migrate to sites of infection

Fig. 8.14 T cells undergo development in the thymus and migrate to the peripheral lymphoid organs, where they are activated by foreign antigens. T-cell precursors migrate from the bone marrow to the thymus, where the T-cell receptor genes are rearranged (first panels); α:β T-cell receptors that are compatible with self-MHC molecules transmit a survival signal on interacting with thymic epithelium, leading to positive selection of the cells that bear them. Self-reactive receptors transmit a signal that leads to cell death, and cells bearing them are removed from the repertoire in a process of negative selection (second panels). T cells that survive selection mature and leave the thymus to circulate in the periphery; they repeatedly leave the blood to migrate through the peripheral lymphoid organs, where they may encounter their specific foreign antigen and become activated (third panels). Activation leads to clonal expansion and differentiation into effector T cells. These are attracted to sites of infection, where they can kill infected cells or activate macrophages (fourth panels); others are attracted into B-cell areas, where they help to activate an antibody response (not shown).

8-7 T-cell progenitors originate in the bone marrow, but all the important events in their development occur in the thymus.

The thymus is situated in the upper anterior thorax, just above the heart. It consists of numerous lobules, each clearly differentiated into an outer cortical region—the **thymic cortex**—and an inner **medulla** (Fig. 8.15). In young individuals, the thymus contains large numbers of developing T-cell precursors embedded in a network of epithelia known as the **thymic stroma**. This provides a unique microenvironment for T-cell development analogous to that provided for B cells by the stromal cells of the bone marrow.

T lymphocytes develop from a lymphoid progenitor in the bone marrow that also gives rise to B lymphocytes. Some of these progenitors leave the bone marrow and migrate to the thymus. In the thymus, the progenitor cell receives a signal, most probably from stromal cells, that is transduced through a receptor called Notch1 to switch on specific genes. Notch signaling is widely used in animal development to specify tissue differentiation; in lymphocyte development, the Notch signal instructs the precursor to commit to the T-cell lineage rather than the B-cell lineage. Notch signaling is required throughout T-cell development and is also thought to help regulate other T-cell lineage choices, including the α:β versus γ:δ choice and the CD4 versus CD8 decision.

The thymic epithelium arises early in embryonic development from endoderm-derived structures known as the third pharyngeal pouches. These epithelial tissues form a rudimentary thymus, or **thymic anlage**. This is colonized by cells of hematopoietic origin that give rise to large numbers of thymocytes, which are committed to the T-cell lineage, and to **intrathymic dendritic cells**. Thymocytes are not simply passengers within the thymus: they influence the arrangement of the thymic epithelial cells on which they depend for survival, inducing the formation of a reticular epithelial structure that surrounds the developing thymocytes (Fig. 8.16). The thymus is independently colonized by numerous macrophages, also of bone marrow origin.

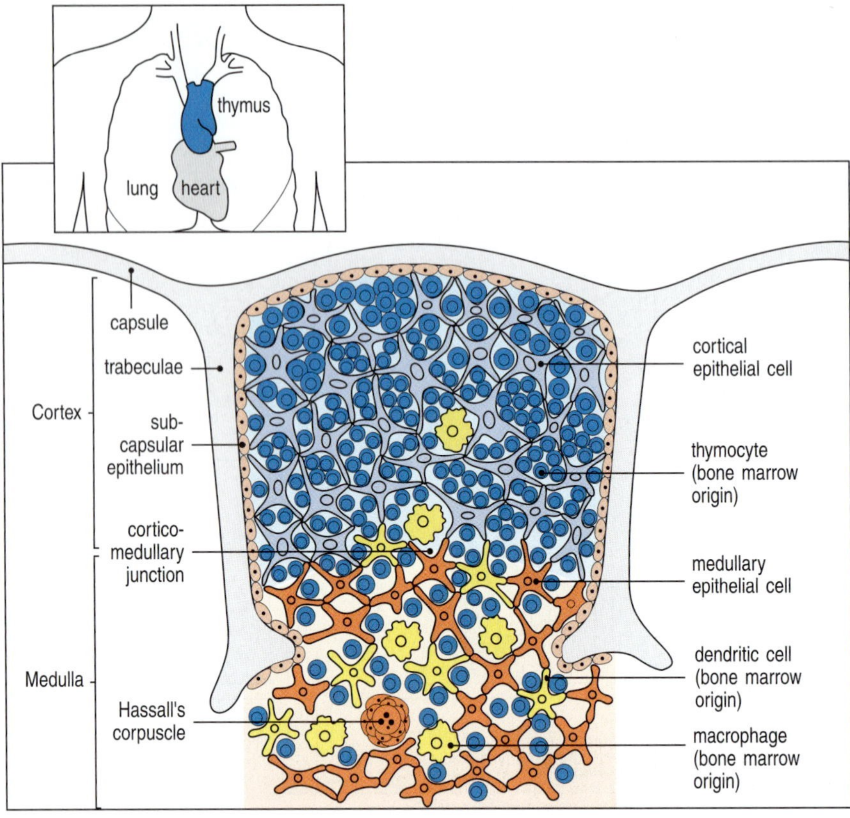

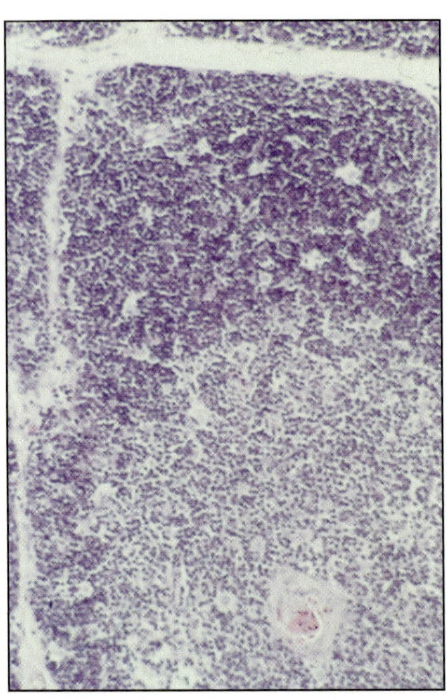

Fig. 8.15 The cellular organization of the human thymus. The thymus, which lies in the midline of the body, above the heart, is made up of several lobules, each of which contains discrete cortical (outer) and medullary (central) regions. As shown in the diagram on the left, the cortex consists of immature thymocytes (dark blue), branched cortical epithelial cells (pale blue), with which the immature cortical thymocytes are closely associated, and scattered macrophages (yellow), which are involved in clearing apoptotic thymocytes. The medulla consists of mature thymocytes (dark blue) and medullary epithelial cells (orange), along with macrophages (yellow) and dendritic cells (yellow) of bone marrow origin. Hassall's corpuscles are probably also sites of cell destruction. The thymocytes in the outer cortical cell layer are proliferating immature cells, whereas the deeper cortical thymocytes are mainly immature T cells undergoing thymic selection. The photograph shows the equivalent section of a human thymus, stained with hematoxylin and eosin. The cortex is darkly staining, whereas the medulla is lightly stained. The large body in the medulla is a Hassall's corpuscle. Photograph courtesy of C.J. Howe.

The cellular architecture of the human thymus is illustrated in Fig. 8.15. Bone marrow derived cells are differentially distributed between the thymic cortex and medulla. The cortex contains only immature thymocytes and scattered macrophages, whereas more mature thymocytes, along with dendritic cells and macrophages, are found in the medulla. This reflects the different developmental events that occur in these two compartments.

The importance of the thymus in immunity was first discovered through experiments on mice; indeed, most of our knowledge of T-cell development in the thymus comes from the mouse. It was found that surgical removal of the thymus (**thymectomy**) at birth resulted in immunodeficient mice, focusing interest on this organ at a time when the difference between T and B lymphocytes in mammals had not been defined. Much evidence, including observations in immunodeficient children, has since confirmed the importance of the thymus in T-cell development. In **DiGeorge syndrome** in humans and in mice with the ***nude*** mutation, the thymus does not form

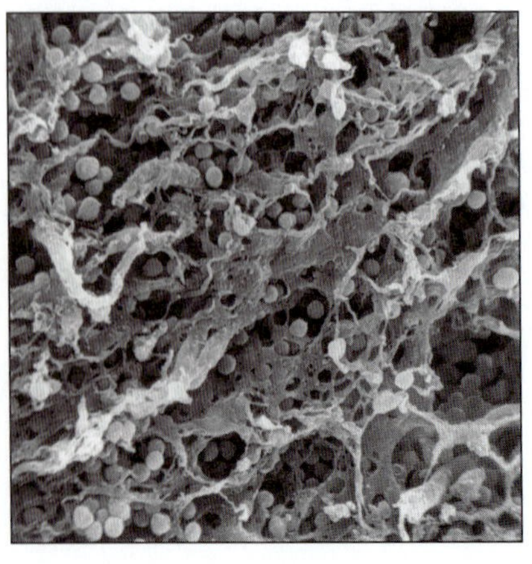

Fig. 8.16 The epithelial cells of the thymus form a network surrounding developing thymocytes. In this scanning electron micrograph of the thymus, the developing thymocytes (the spherical cells) occupy the interstices of an extensive network of epithelial cells. Photograph courtesy of W. van Ewijk.

and the affected individual produces B lymphocytes but few T lymphocytes. DiGeorge syndrome is a complex combination of cardiac, facial, endocrine, and immune defects associated with deletions of chromosome 22q11. The *nude* mutation in mice is due to a defect in the gene for Foxn1, a transcription factor required for terminal epithelial cell differentiation; the name *nude* was given to this mutation because it also causes hairlessness. Rare cases of a defect in the human *FOXN1* gene (which is on chromosome 17) have been associated with T-cell immunodeficiency, absence of a thymus, congenital alopecia, and nail dystrophy.

The crucial role of the thymic stroma in inducing the differentiation of the bone marrow derived T-cell progenitors can be demonstrated by the results of tissue grafts between two mutant mice, each lacking mature T cells for a different reason. In *nude* mice the thymic epithelium fails to differentiate, whereas in *scid* mice B and T lymphocytes fail to develop because of a defect in antigen-receptor gene rearrangement (see Section 5-5). Reciprocal grafts of thymus and bone marrow between these immunodeficient strains show that *nude* bone marrow precursors develop normally in a *scid* thymus (Fig. 8.17).

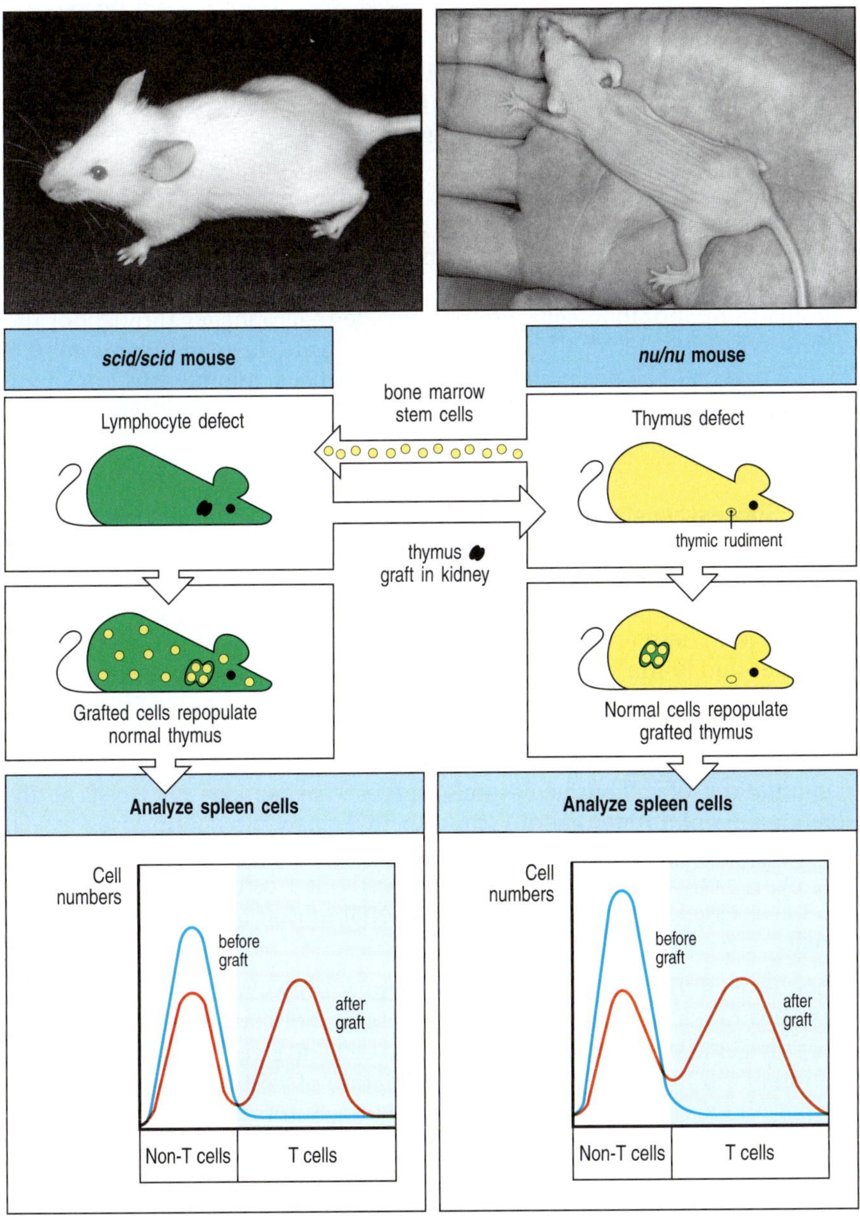

Fig. 8.17 The thymus is critical for the maturation of bone marrow derived cells into T cells. Mice with the *scid* mutation (upper left photograph) have a defect that prevents lymphocyte maturation, whereas mice with the *nude* mutation (upper right photograph) have a defect that affects the development of the cortical epithelium of the thymus. T cells do not develop in either strain of mouse: this can be demonstrated by staining spleen cells with antibodies specific for mature T cells and analyzing them in a flow cytometer (see Appendix I, Section A-22), as represented by the blue line in the graphs in the bottom panels. Bone marrow cells from *nude* mice can restore T cells to *scid* mice (red line in graph on left), showing that, in the correct environment, the *nude* bone marrow cells are intrinsically normal and capable of producing T cells. Thymic epithelial cells provided by a thymus graft from *scid* mice into the kidney of *nude* mice can induce the maturation of T cells in *nude* mice (red line in graph on right), demonstrating that the thymus provides the essential microenvironment for T-cell development.

Thus, the defect in *nude* mice is in the thymic stromal cells. Transplanting a *scid* thymus into a *nude* mouse leads to T-cell development. However, *scid* bone marrow cannot develop T cells, even in a wild-type recipient.

In mice, the thymus continues to develop for 3–4 weeks after birth, whereas in humans it is fully developed at birth. The rate of T-cell production by the thymus is greatest before puberty. After puberty, the thymus begins to shrink and the production of new T cells in adults is lower, although it does continue throughout life. In both mice and humans, removal of the thymus after puberty is not accompanied by any notable loss of T-cell function or numbers. Thus, it seems that once the T-cell repertoire is established, immunity can be sustained without the production of significant numbers of new T cells; the pool of peripheral T cells is instead maintained by long-lived T cells and also by some division of mature T cells.

8-8 T-cell precursors proliferate extensively in the thymus, but most die there.

T-cell precursors arriving in the thymus from the bone marrow spend up to a week differentiating there before they enter a phase of intense proliferation. In a young adult mouse the thymus contains about 10^8 to 2×10^8 thymocytes. About 5×10^7 new cells are generated each day; however, only about 10^6 to 2×10^6 (roughly 2–4%) of these leave the thymus each day as mature T cells. Despite the disparity between the number of T cells generated in the thymus and the number leaving, the thymus does not continue to grow in size or cell numbers. This is because about 98% of the thymocytes that develop in the thymus also die in the thymus by apoptosis (see Section 1-14). Changes in the plasma membrane of cells undergoing apoptosis lead to their rapid phagocytosis, and apoptotic bodies, which are the residual condensed chromatin of apoptotic cells, are seen inside macrophages throughout the thymic cortex (Fig. 8.18). This apparently profligate waste of thymocytes is a crucial part of T-cell development because it reflects the intensive screening that each thymocyte undergoes for the ability to recognize self-peptide:self-MHC complexes and for self-tolerance.

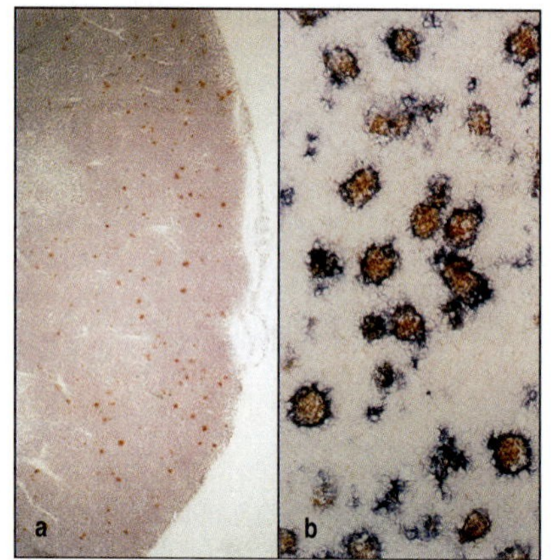

Fig. 8.18 Developing T cells that undergo apoptosis are ingested by macrophages in the thymic cortex. Panel a shows a section through the thymic cortex and part of the medulla in which cells have been stained for apoptosis with a red dye. Thymic cortex is to the right of the photograph. Apoptotic cells are scattered throughout the cortex but are rare in the medulla. Panel b shows a section of thymic cortex at higher magnification that has been stained red for apoptotic cells and blue for macrophages. The apoptotic cells can be seen within macrophages. Magnifications: panel a, ×45; panel b, ×164. Photographs courtesy of J. Sprent and C. Surh.

8-9 Successive stages in the development of thymocytes are marked by changes in cell-surface molecules.

Like developing B cells, developing thymocytes pass through a series of distinct stages. These are marked by changes in the status of the T-cell receptor genes and in the expression of the T-cell receptor, and by changes in the expression of cell-surface proteins such as the CD3 complex (see Section 7-7) and the co-receptor proteins CD4 and CD8 (see Section 4-17). These surface changes reflect the state of functional maturation of the cell, and particular combinations of cell-surface proteins are used as markers for T cells at different stages of differentiation. The principal stages are summarized in Fig. 8.19. Two distinct lineages of T cells—$\alpha{:}\beta$ and $\gamma{:}\delta$, which have different types of T-cell receptor chains—are produced early in T-cell development. Later, $\alpha{:}\beta$ T cells develop into two distinct functional subsets—CD4 T cells and CD8 T cells.

When progenitor cells first enter the thymus from the bone marrow, they lack most of the surface molecules characteristic of mature T cells, and their receptor genes are not rearranged. These cells give rise to the major population of $\alpha{:}\beta$ T cells and the minor population of $\gamma{:}\delta$ T cells. If injected into the peripheral circulation, these lymphoid progenitors can even give rise to B cells and NK cells. Interactions with the thymic stroma trigger an initial phase of differentiation along the T-cell lineage pathway followed by cell proliferation and the expression of the first cell-surface molecules specific for T cells,

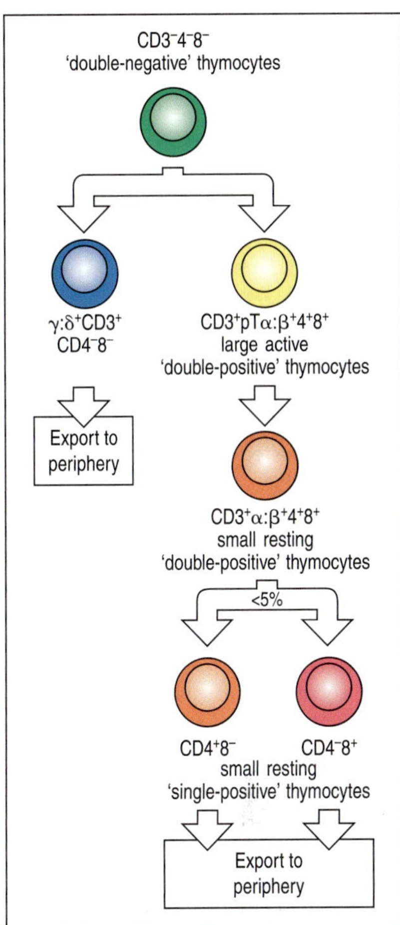

Fig. 8.19 Two distinct lineages of thymocytes are produced in the thymus. CD4, CD8, and T-cell receptor complex molecules (CD3, and the T-cell receptor α and β chains) are important cell-surface molecules for identifying thymocyte subpopulations. The earliest cell population in the thymus does not express any of these proteins, and because they do not express CD4 or CD8 they are called 'double-negative' thymocytes. These cells include precursors that give rise to two T-cell lineages: the minority population of γ:δ T cells (which lack CD4 or CD8 even when mature), and the majority α:β T-cell lineage. The development of prospective α:β T cells proceeds through stages in which both CD4 and CD8 are expressed by the same cell; these are known as 'double-positive' thymocytes. These cells enlarge and divide. Later, they become small resting double-positive cells that express low levels of the T-cell receptor. Most thymocytes die within the thymus after becoming small double-positive cells, but those cells whose receptors can interact with self-peptide:self-MHC molecular complexes lose expression of either CD4 or CD8 and increase the level of expression of the T-cell receptor. The outcome of this process is the 'single-positive' thymocytes, which, after maturation, are exported from the thymus as mature single-positive CD4 or CD8 T cells.

for example CD2 and (in mice) Thy-1. At the end of this phase, which can last about a week, the thymocytes bear distinctive markers of the T-cell lineage but do not express any of the three cell-surface markers that define mature T cells. These are the CD3:T-cell receptor complex and the co-receptors CD4 or CD8. Because of the absence of CD4 and CD8, such cells are called **double-negative thymocytes** (see Fig. 8.19).

In the fully developed thymus, immature double-negative T cells constitute about 60% of the thymocytes that lack both CD4 and CD8. This pool (about 5% of all thymocytes) also includes two populations of more mature T cells that belong to minority lineages. One of these, representing about 20% of the double-negative cells, comprises cells that have rearranged and are expressing the genes encoding the γ:δ T-cell receptor; we will return to these cells in Section 8-12. The second, representing another 20% of all double-negative thymocytes, includes cells bearing α:β T-cell receptors of very limited diversity. These cells also express the NK1.1 receptor commonly found on NK cells; they are therefore known as **invariant NKT cells (iNKT cells)**. iNKT cells are activated as part of the early response to many infections; they differ from the major lineage of α:β T cells in recognizing CD1 molecules rather than MHC class I or MHC class II molecules (see Section 6-19) and they are not shown in Fig. 8.19. In this and subsequent discussions, we reserve the term 'double-negative thymocytes' for the immature thymocytes that do not yet express a complete T-cell receptor molecule. These cells give rise to both γ:δ and α:β T cells (see Fig. 8.19), although most of them develop along the α:β pathway.

The α:β pathway is shown in more detail in Fig. 8.20. The double-negative stage can be further subdivided into four stages on the basis of expression of the adhesion molecule CD44, CD25 (the α chain of the IL-2 receptor), and Kit, the receptor for SCF (see Section 8-1). At first, double-negative thymocytes express Kit and CD44 but not CD25 and are called **DN1** cells; in these cells, the genes encoding both chains of the T-cell receptor are in the germline configuration. As thymocytes mature, they begin to express CD25 on their surface and are called **DN2** cells; later, expression of CD44 and Kit is reduced, and they are called **DN3** cells.

Rearrangement of the T-cell receptor β-chain locus begins in DN2 cells with some D_β to J_β rearrangements and continues in DN3 cells with V_β to DJ_β rearrangement. Cells that fail to make a successful rearrangement of the β-chain locus remain at the DN3 ($CD44^{low}CD25^{+}$) stage and soon die, whereas cells that make productive β-chain gene rearrangements and express the β chain lose expression of CD25 once again and progress to the **DN4** stage, in which they proliferate. The functional significance of the transient expression of CD25 is unclear: T cells develop normally in mice in which the IL-2 gene has been deleted by gene knockout (see Appendix I, Section A-46). By contrast,

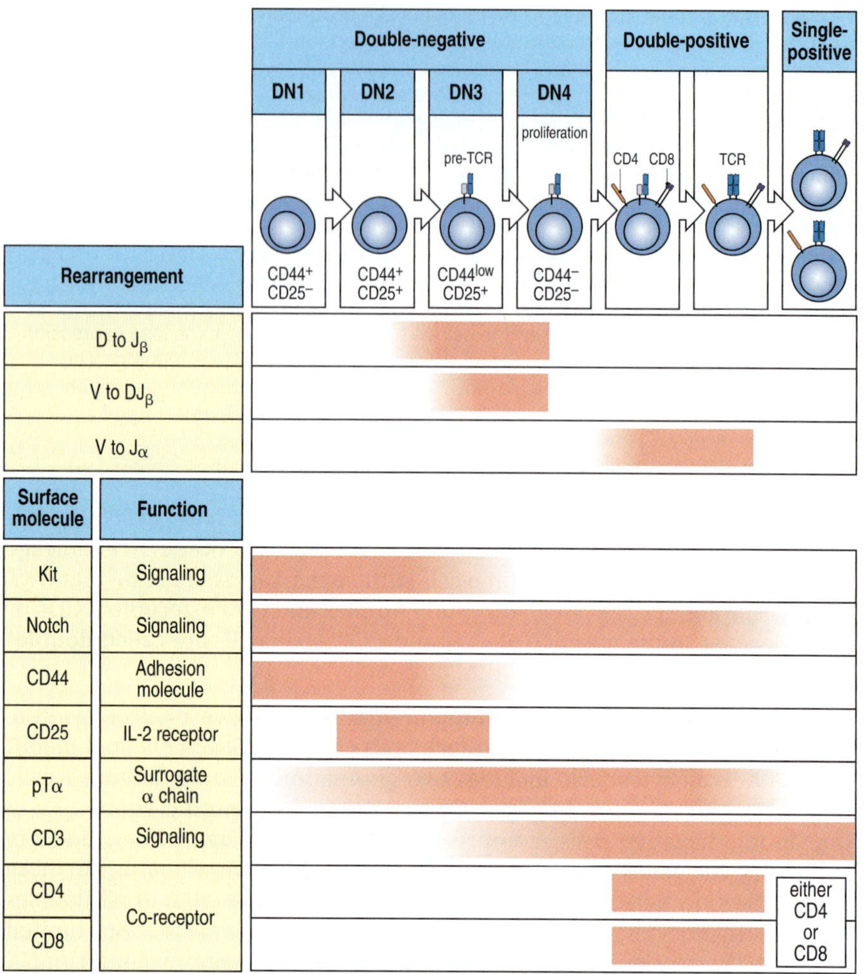

Fig. 8.20 The correlation of stages of α:β T-cell development in the mouse thymus with the program of gene rearrangement and the expression of cell-surface proteins. Lymphoid precursors are triggered to proliferate and become thymocytes committed to the T-cell lineage through interactions with the thymic stroma. These double-negative (DN1) cells express CD44 and Kit, and at a later stage (DN2) express CD25, the α chain of the IL-2 receptor. After this, the DN2 ($CD44^+CD25^+$) cells begin to rearrange the β-chain locus, becoming $CD44^{low}$ and Kit^{low} as this occurs, and become DN3 cells. The DN3 cells are arrested in the $CD44^{low}CD25^+$ stage until they productively rearrange the β-chain locus; the in-frame β chain then pairs with a surrogate chain called pTα to form the pre-T-cell receptor (pre-TCR) and is expressed on the cell surface, which triggers entry into the cell cycle. Expression of small amounts of pTα:β on the cell surface in association with CD3 signals the cessation of β-chain gene rearrangement and triggers rapid cell proliferation, which causes the loss of CD25. The cells are then known as DN4 cells. Eventually, the DN4 cells cease to proliferate and CD4 and CD8 are expressed. The small $CD4^+CD8^+$ double-positive cells begin efficient rearrangement at the α-chain locus. The cells then express low levels of an α:β T-cell receptor and the associated CD3 complex and are ready for selection. Most cells die by failing to be positively selected or as a consequence of negative selection, but some are selected to mature into CD4 or CD8 single-positive cells and eventually to leave the thymus. Expression of some other cell-surface proteins is depicted with respect to the stages of thymocyte development. Their individual contributions to T-cell development are discussed in the text.

Kit is quite important for the development of the earliest double-negative thymocytes, in that mice lacking Kit have a much smaller number of double-negative T cells. In addition, IL-7 produced by the thymic stroma is essential for early T-cell development, because there is a severe block to development when the IL-7 receptor is defective. Finally, continuous Notch signaling is important for progression through each stage of T-cell development.

In DN3 thymocytes, the expressed β chains pair with a surrogate pre-T-cell receptor α chain called **pTα** (pre-T-cell α), which allows the assembly of a complete **pre-T-cell receptor** (pre-TCR) that is analogous in structure and function to the pre-B-cell receptor. The pre-TCR is expressed on the cell surface in a complex with the CD3 molecules that provide the signaling components of T-cell receptors (see Section 7-7). As with the pre-B-cell receptor, the assembly of the CD3:pre-T-cell receptor complex causes constitutive signaling that does not require interaction with a ligand. Recent structural evidence shows that the pre-TCR forms dimers in a manner similar to pre-BCR dimerization. The pTα Ig domain makes two important contacts. It associates with the constant-region Ig domain of the Vβ subunit to form the pre-TCR itself. A distinct surface of the pre-Tα then binds to a Vβ domain from another pre-TCR molecule, forming a bridge between two different pre-TCRs. Importantly, the region of contact with the Vβ involves residues that are highly conserved across many Vβ families. In this way, expression of the pre-TCR induces ligand-independent dimerization, which leads to cell proliferation, the arrest of further β-chain gene rearrangement, and the expression of both CD8 and CD4. These **double-positive thymocytes** make up the vast majority of thymocytes. Once the large double-positive thymocytes have ceased to proliferate and have become small double-positive cells, the α-chain locus begins to rearrange. As we will see later in this chapter, the structure of the α locus (see Section 5-9) allows multiple successive attempts at rearrangement, so that it is successfully rearranged in most developing thymocytes. Thus, most double-positive cells produce an α:β T-cell receptor during their relatively short life-span.

Small double-positive thymocytes initially express low levels of the T-cell receptor. Most of these receptors cannot recognize self-peptide:self-MHC molecular complexes; they will fail positive selection and the cells will die. In contrast, those double-positive cells that recognize self-peptide:self-MHC complexes, and can therefore be positively selected, go on to mature, and express high levels of the T-cell receptor. At the same time they cease to express one or other of the two co-receptor molecules, becoming either CD4 or CD8 **single-positive thymocytes**. Thymocytes also undergo negative selection during and after the double-positive stage, which eliminates those cells capable of responding to self antigens. About 2% of the double-positive thymocytes survive this dual screening and mature as single-positive T cells that are gradually exported from the thymus to form the peripheral T-cell repertoire. The time between the entry of a T-cell progenitor into the thymus and the export of its mature progeny is estimated to be about 3 weeks in the mouse.

8-10 Thymocytes at different developmental stages are found in distinct parts of the thymus.

The thymus is divided into two main regions, a peripheral cortex and a central medulla (see Fig. 8.15). Most T-cell development takes place in the cortex; only mature single-positive thymocytes are seen in the medulla. Initially, progenitors from the bone marrow enter at the cortico-medullary junction and migrate to the outer cortex (Fig. 8.21). At the outer edge of the cortex, in the subcapsular region of the thymus, large immature double-negative thymocytes proliferate vigorously; these cells are thought to represent committed thymocyte progenitors and their immediate progeny and will give rise to all subsequent thymocyte populations. Deeper in the cortex, most of the thymocytes are small double-positive cells. The cortical stroma is composed of epithelial cells with long branching processes that express both MHC class II and MHC class I molecules on their surface. The thymic cortex is

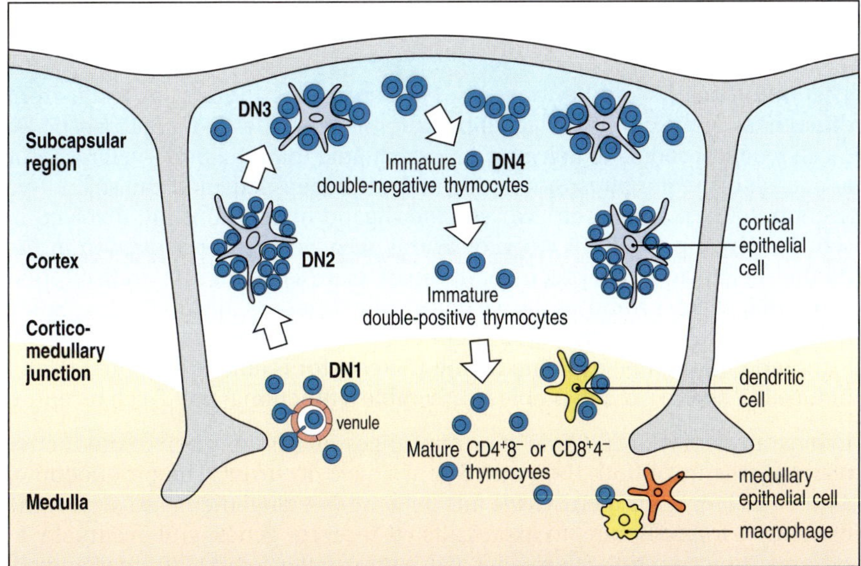

Fig. 8.21 Thymocytes at different developmental stages are found in distinct parts of the thymus. The earliest precursor thymocytes enter the thymus from the bloodstream via venules near the cortico-medullary junction. Ligands that interact with the receptor Notch1 are expressed in the thymus and act on the immigrant cells to commit them to the T-cell lineage. As these cells differentiate through the early $CD4^-CD8^-$ double-negative (DN) stages described in the text, they migrate through the cortico-medullary junction and to the outer cortex. DN3 cells reside near the subcapsular region. As the progenitor matures further to the $CD4^+CD8^+$ double-positive stage, it migrates back through the cortex. Finally, the medulla contains only mature single-positive T cells, which eventually leave the thymus.

densely packed with thymocytes, and the branching processes of the thymic cortical epithelial cells make contact with almost all cortical thymocytes (see Fig. 8.16). Contact between the MHC molecules on thymic cortical epithelial cells and the receptors of developing T cells has a crucial role in positive selection, as we will see later in this chapter.

After positive selection, developing T cells migrate from the cortex to the medulla. The medulla contains fewer lymphocytes, and those present are predominantly the newly matured single-positive T cells that will eventually leave the thymus. The medulla plays a role in negative selection. The antigen-presenting cells in this environment include dendritic cells that express co-stimulatory molecules, which are generally absent from the cortex. In addition, specialized medullary epithelial cells present peripheral antigens for the negative selection of T cells reactive for these self antigens.

8-11 T cells with α:β or γ:δ receptors arise from a common progenitor.

T cells bearing γ:δ receptors differ from α:β T cells in that they are found primarily in epithelial and mucosal sites and lack expression of the CD4 and CD8 co-receptors; in comparison with α:β T cells, relatively little is known about the ligands they recognize and the γ:δ receptor is thought not to be MHC restricted (see Sections 3-24 and 4-19). Recall from Section 5-11 that different genetic loci are used to make these two types of T-cell receptors. The β, γ, and δ loci begin to undergo rearrangement almost simultaneously in developing thymocytes, and the two cell lineages diverge from a common precursor only after certain gene rearrangements have already occurred (Fig. 8.22). This can be deduced from the pattern of gene rearrangements found in thymocytes and in mature γ:δ and α:β T cells. Mature γ:δ T cells can contain rearranged β-chain genes, although 80% of these are nonproductive, and mature α:β T cells often contain rearranged, but mostly out-of-frame, γ-chain genes.

The decision of a precursor to commit to the γ:δ or the α:β lineage is thought to depend on which type of receptor—a functional γ:δ receptor or the pre-T-cell receptor (β:pTα)—is expressed first at the DN stage of thymocyte development. Formation of a pre-T-cell receptor is more likely to occur first, as formation of a γ:δ receptor requires functional rearrangements at both the γ and the δ loci, whereas formation of a pre-T-cell receptor requires only a functional β locus rearrangement, to produce a β chain that pairs with pTα (see Section 8-9). If a γ:δ receptor is expressed first, the double-negative precursor commits to the γ:δ lineage, fails to express CD4 and CD8 co-receptors and migrates from the thymus to the periphery without undergoing positive or negative selection. If the pre-T-cell receptor is expressed first, the cell commits to the α:β lineage and goes on to express CD4 and CD8.

This difference in fate between the two lineages is thought to result from differences in the quality of signaling from the two types of receptors. The γ:δ T-cell receptor seems to deliver a stronger signal than the pre-T-cell receptor as a result of greater phosphorylation of the mitogen-activated kinase Erk (see Section 7-13). It is not yet clear whether ligand interactions are involved in signaling by the γ:δ T-cell receptor at this step. Strong Erk activation in the double-negative T-cell precursor induces expression of the transcription factor Id3, which promotes the γ:δ lineage and opposes the α:β lineage. Some mature γ:δ cells actually carry productive rearrangements at the β-chain locus, and this can be explained if the γ:δ receptor is able to activate Erk and induce Id3 before the cell is able to assemble a functional pre-T-cell receptor.

In most thymocytes a β-chain gene rearranges successfully before productive rearrangements of both the γ and δ genes have occurred. The production of a pre-T-cell receptor then arrests further gene rearrangement and signals the thymocyte to proliferate, to express its co-receptor genes, and eventually to start rearranging the α-chain genes. Once the α-chain locus starts rearranging,

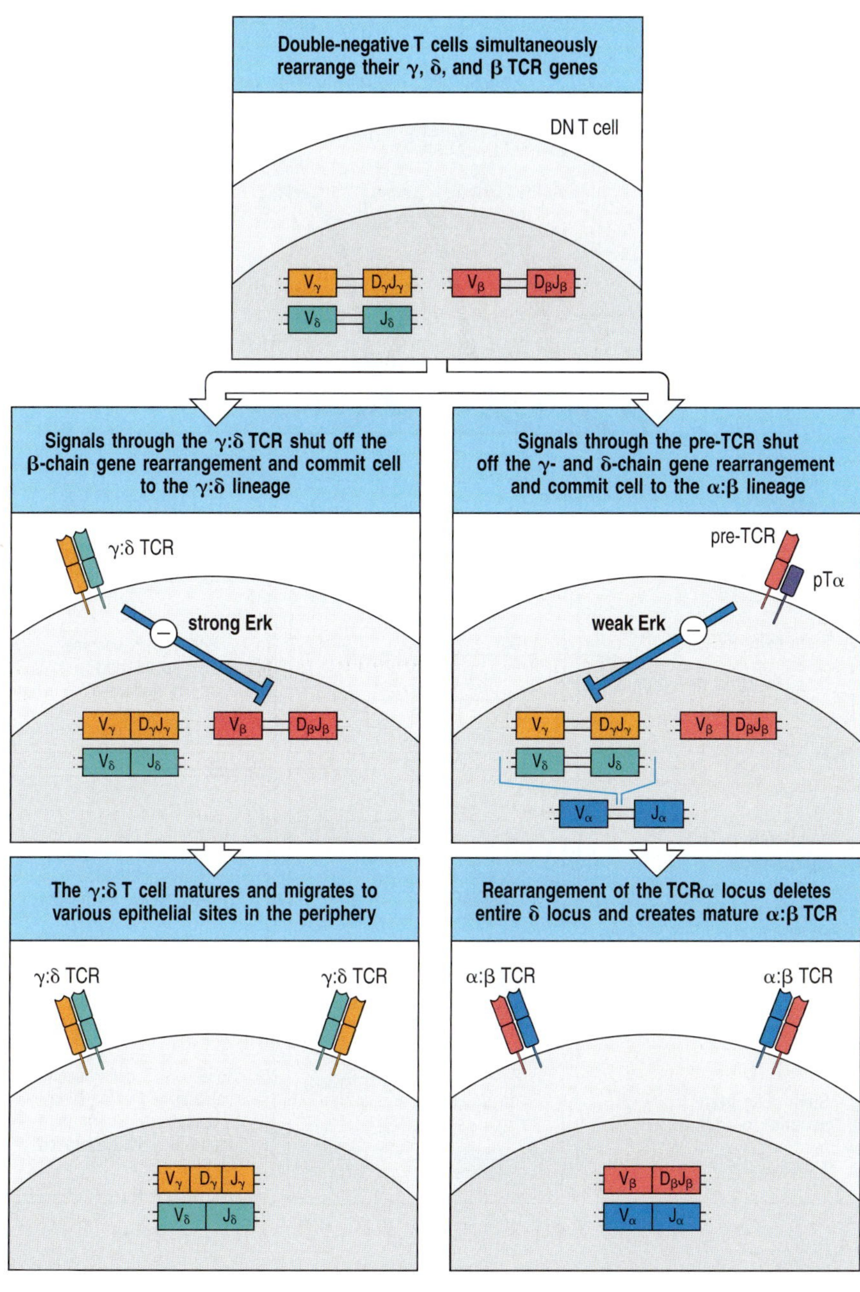

Double-negative T cells simultaneously rearrange their γ, δ, and β TCR genes
DN T cell
Signals through the γ:δ TCR shut off the β-chain gene rearrangement and commit cell to the γ:δ lineage
γ:δ TCR
strong Erk
Signals through the pre-TCR shut off the γ- and δ-chain gene rearrangement and commit cell to the α:β lineage
pre-TCR
pTα
weak Erk
The γ:δ T cell matures and migrates to various epithelial sites in the periphery
γ:δ TCR
γ:δ TCR
Rearrangement of the TCRα locus deletes entire δ locus and creates mature α:β TCR
α:β TCR
α:β TCR

Fig. 8.22 Signals through the γ:δ T-cell receptor and the pre-T-cell receptor compete to determine the fate of thymocytes. During the development of T cells in the thymus, double-negative (DN) thymocytes begin to rearrange the γ, δ, and β T-cell receptor loci simultaneously (top panel). If a complete γ:δ T-cell receptor is formed before a successful β-chain gene rearrangement has led to the production of the pre-T-cell receptor (left panels), the thymocyte receives signals through the γ:δ receptor, which shuts off further rearrangement of the β-chain gene. Signaling by the γ:δ receptor induces strong activation of Erk (see Section 7-13) that commits the cell to the γ:δ lineage. This cell then matures into a γ:δ T cell and migrates out of the thymus into the peripheral circulation (bottom left panel). If a functional β chain is formed before a complete γ:δ receptor, it pairs with the pTα to generate a pre-T-cell receptor (right panels). In this case, the developing thymocyte receives a signal through the pre-T-cell receptor that shuts off rearrangements of the γ and δ loci. Weaker Erk activation via the pre-T-cell receptor in comparison with the γ:δ receptor leads to commitment to the α:β lineage. The thymocyte passes from the DN3 stage through the proliferating DN4 stage into the double-positive stage, at which the TCRα-chain locus rearranges and a mature α:β T-cell receptor is produced (bottom right panel). α-chain locus rearrangement deletes the δ genes, thus precluding the production of a γ:δ receptor on the same cell.

the δ-chain gene segments located within the α-chain locus are deleted as an extrachromosomal circle. This ensures that cells committed to the α:β lineage cannot make a complete γ:δ receptor.

8-12 T cells expressing particular γ- and δ-chain V regions arise in an ordered sequence early in life.

During embryonic development, the generation of the various types of T cells—even the particular V region assembled in γ:δ cells—is developmentally controlled. The first T cells to appear carry γ:δ T-cell receptors (Fig 8.23). In the mouse, γ:δ T cells first appear in discrete waves or bursts in the fetus, with the T cells in each wave populating distinct sites in the adult animal.

The first wave of γ:δ T cells populates the epidermis; the T cells become wedged among the keratinocytes and adopt a dendritic-like form that has given them

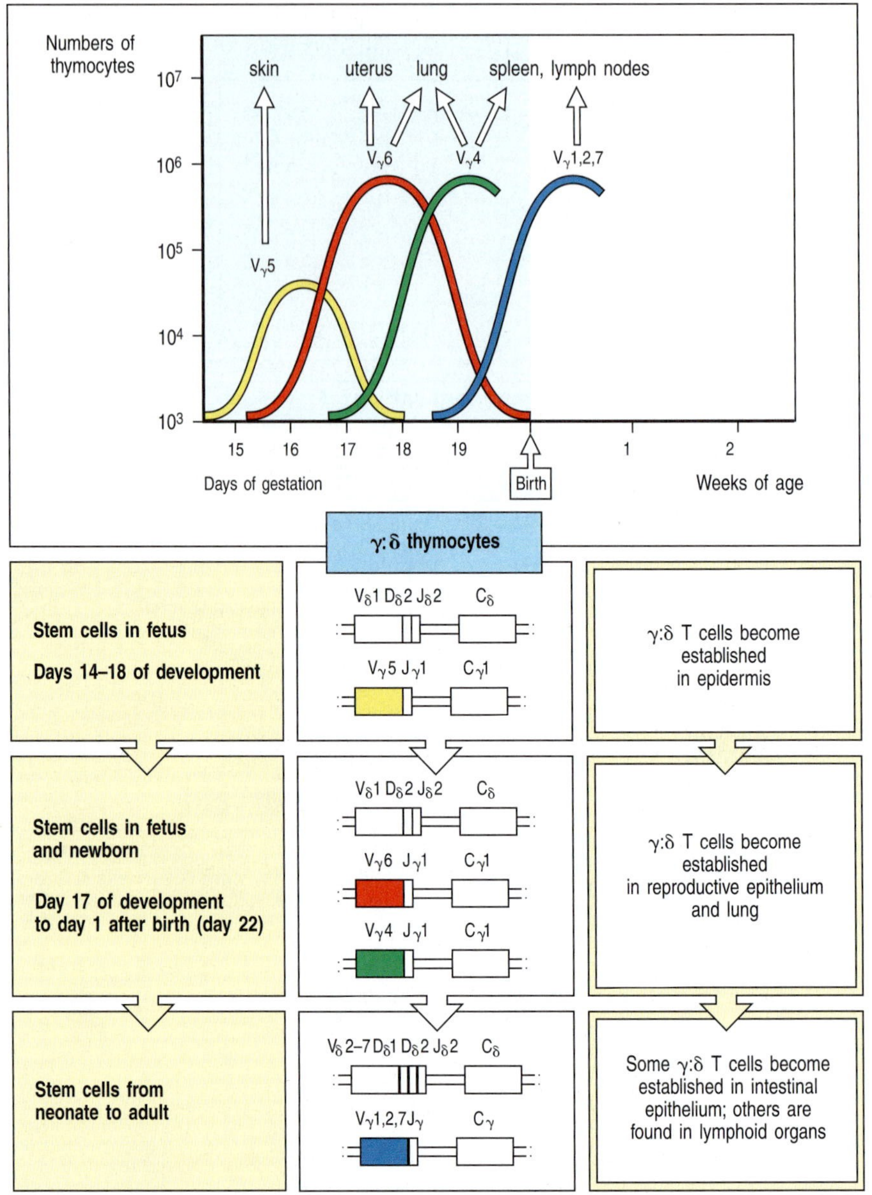

Fig. 8.23 The rearrangement of T-cell receptor γ and δ genes in the mouse proceeds in waves of cells expressing different V_γ and V_δ gene segments. At about 2 weeks of gestation, the $C\gamma_1$ locus is expressed with its closest V gene ($V\gamma_5$). After a few days, $V\gamma_5$-bearing cells decline (upper panel) and are replaced by cells expressing the next most proximal gene, $V\gamma_6$. Both these rearranged γ chains are expressed with the same rearranged δ-chain gene, as shown in the lower panels, and there is little junctional diversity in either the Vγ or the Vδ chain. As a consequence, most of the γ:δ T cells produced in each of these early waves have the same specificity, although the antigen recognized in each case is not known. The $V\gamma_5$-bearing cells become established selectively in the epidermis, whereas the $V\gamma_6$-bearing cells become established in the epithelium of the reproductive tract. After birth, the α:β T-cell lineage becomes dominant and, although γ:δ T cells are still produced, they are a much more heterogeneous population, bearing receptors with a great deal of junctional diversity. Note that Vγ gene segments are described using the system proposed by Tonegawa.

the name of **dendritic epidermal T cells (dETCs)** (Fig. 8.24). The second wave homes to the epithelia of the reproductive tract. Remarkably, given the large number of theoretically possible rearrangements, the receptors expressed by these early waves of γ:δ T cells are essentially invariant. All the cells in each wave assemble the same V_γ and V_δ regions, but each wave uses a different set of V, D, and J gene segments. Why certain V, D, and J gene segments are selected for rearrangement at particular times during embryonic development is still poorly understood. There are no N-nucleotides contributing additional diversity at the junctions between V, D, and J gene segments, reflecting the absence of the enzyme TdT in these fetal γ:δ T cells.

After the initial waves, T cells are produced continuously and α:β T cells predominate, making up more than 95% of thymocytes. The γ:δ T cells produced at this stage are different from those produced earlier. They have much more diverse receptors, for which several different V gene segments have been used, and the receptor sequences have abundant N-nucleotide additions. Most of these γ:δ T cells, like α:β T cells, migrate to peripheral lymphoid tissues rather than to epithelia.

The developmental changes in V gene segment usage and N-nucleotide addition in murine γ:δ T cells parallel changes in B-cell populations during fetal development, which are discussed later (see Section 8-28). Their functional significance is unclear, however, and not all of these changes in the pattern of receptors expressed by γ:δ T cells occur in humans. The dETCs, for example, do not seem to have exact human counterparts, although there are γ:δ T cells in the epithelial linings of the human reproductive and gastrointestinal tracts. The mouse dETCs may serve as sentinel cells that are activated upon local tissue damage or as cells that regulate inflammatory processes.

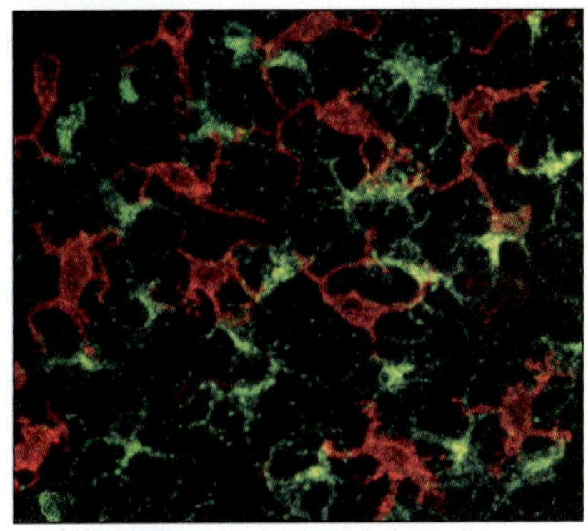

Fig. 8.24 Dendritic epidermal T cells reside within the epithelial layer, forming an interdigitating network with Langerhans cells. This face-on view of a murine epidermal sheet shows Langerhans cells (green) and dendritic epidermal T cells (dETCs) (red) forming an interdigitating network within the layers of the epidermis. The epidermal epithelial cells are not visible in this fluorescence image. The branching dendritic-like form of these γ:δ T cells is the source of their name. Although the ligands for all γ:δ T-cell receptors are not known, some γ:δ T cells recognize nonclassical MHC molecules (see Section 6-17), which can be induced in epithelia by stresses such as UV damage or pathogens. Thus, dETCs may serve as sentinels of such damage, producing cytokines that activate the innate immune response, and in turn, adaptive immunity. Courtesy of Adrian Hayday.

8-13 Successful synthesis of a rearranged β chain allows the production of a pre-T-cell receptor that triggers cell proliferation and blocks further β-chain gene rearrangement.

We now return to the development of α:β T cells. The rearrangement of the β- and α-chain loci closely parallels the rearrangement of immunoglobulin heavy-chain and light-chain loci during B-cell development (see Sections 8-2 and 8-5). As shown in Fig. 8.25, the β-chain gene segments rearrange first, with the D_β gene segments rearranging to J_β gene segments, and this is followed by V_β to DJ_β rearrangement. If no functional β chain can be synthesized from this rearrangement, the cell will not be able to produce a pre-T-cell receptor and will die unless it makes successful rearrangements at both the γ and the δ loci (see Fig. 8.22). However, unlike B cells with nonproductive heavy-chain gene rearrangements, thymocytes with nonproductive β-chain VDJ rearrangements can be rescued by further rearrangement, which is possible because of the two clusters of D_β and J_β gene segments upstream of two C_β genes (see Fig. 5.9). The likelihood of a productive VDJ join at the β locus is therefore somewhat higher than the 55% chance of a productive immunoglobulin heavy-chain gene arrangement.

Once a productive β-chain gene rearrangement has occurred, the β chain is expressed together with the invariant pTα and the CD3 molecules (see Fig. 8.25) and is transported in this complex to the cell surface. The β:pTα complex is a functional pre-T-cell receptor analogous to the μ:VpreB:λ5 pre-B-cell receptor complex in B-cell development (see Section 8-3). Expression of the pre-T-cell receptor at the DN3 stage of thymocyte development induces signals that cause the phosphorylation and degradation of RAG-2, thus halting β-chain gene rearrangement and ensuring allelic exclusion at the β locus. These signals also induce the DN4 stage in which rapid cell proliferation occurs, and eventually the co-receptor proteins CD4 and CD8 are expressed. The pre-T-cell receptor signals constitutively via the cytoplasmic protein kinase Lck, a Src-family tyrosine kinase (see Fig. 7.12), but seems not to require a ligand on the thymic epithelium. Lck subsequently associates with the co-receptor proteins. In mice genetically deficient in Lck, T-cell development is arrested before the CD4CD8 double-positive stage, and no α-chain gene rearrangements can be made.

The role of the expressed β chain in suppressing further β-locus rearrangement can be demonstrated in mice containing a rearranged TCRβ transgene: these mice express the transgenic β chain on virtually 100% of their T cells, and rearrangement of their endogenous β-chain genes is strongly suppressed. The importance of pTα has been shown in mice deficient in pTα, in which there is a hundredfold decrease in α:β T cells and an absence of allelic exclusion at the β locus.

During the proliferation of DN4 cells triggered by expression of the pre-T-cell receptor, the *RAG-1* and *RAG-2* genes are repressed. Hence, no rearrangement of the α-chain locus occurs until the proliferative phase ends, when *RAG-1* and *RAG-2* are transcribed again, and the functional RAG-1:RAG-2 complex accumulates. This ensures that each cell in which a β-chain gene has been successfully rearranged gives rise to many CD4CD8 thymocytes. Once the cells stop dividing, each of them can independently rearrange its α-chain genes, so that a single functional β chain can be associated with many different α chains in the progeny cells. During the period of α-chain gene rearrangement, α:β T-cell receptors are first expressed and selection by self-peptide:self-MHC complexes on the thymus cells can begin.

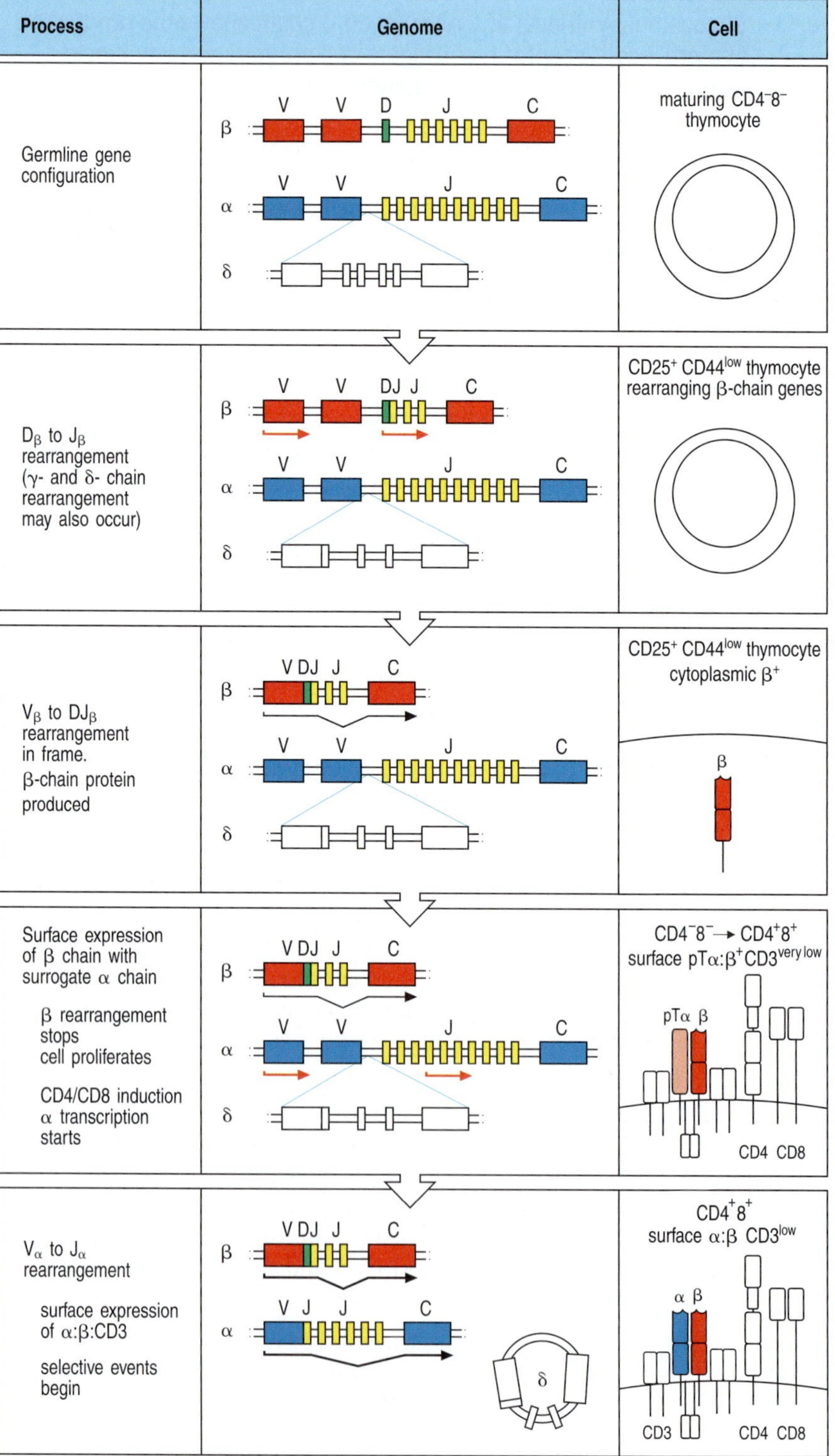

Fig. 8.25 The stages of gene rearrangement in α:β T cells. The sequence of gene rearrangements is shown, together with an indication of the stage at which the events take place and the nature of the cell-surface receptor molecules expressed at each stage. The β-chain locus rearranges first, in $CD4^-CD8^-$ double-negative thymocytes expressing CD25 and low levels of CD44. As with immunoglobulin heavy-chain genes, D to J gene segments rearrange before V gene segments rearrange to DJ (second and third panels). It is possible to make up to four attempts to generate a productive rearrangement at the β-chain locus, because there are four D gene segments with two sets of J gene segments associated with each TCR β chain locus (not shown). The productively rearranged gene is expressed initially within the cell and then at low levels on the cell surface. It associates with pTα, a surrogate 33 kDa α chain that is equivalent to λ5 in B-cell development, and this pTα:β heterodimer forms a complex with the CD3 chains (fourth panel). The expression of the pre-T-cell receptor signals the developing thymocytes to halt β-chain gene rearrangement and to undergo multiple cycles of division. At the end of this proliferative burst, the CD4 and CD8 molecules are expressed, the cell ceases cycling, and the α chain is now able to undergo rearrangement. The first α-chain gene rearrangement deletes all δ D, J, and C gene segments on that chromosome, although these are retained as a circular DNA, proving that these are nondividing cells (bottom panel). This permanently inactivates the δ-chain gene. Rearrangements at the α-chain locus can proceed through several cycles, because of the large number of V_α and J_α gene segments, so that productive rearrangements almost always occur. When a functional α chain is produced that pairs efficiently with the β chain, the $CD3^{low}CD4^+CD8^+$ thymocyte is ready to undergo selection for its ability to recognize self peptides in association with self-MHC molecules.

The progression of thymocytes from the double-negative to the double-positive, and finally to the single-positive, stage is accompanied by a distinct pattern of expression of proteins involved in DNA rearrangement, signaling, and T-cell-specific gene expression (Fig. 8.26). TdT, the enzyme responsible for the insertion of N-nucleotides, is expressed throughout T-cell receptor gene rearrangement; N-nucleotides are found at the junctions of all rearranged α and β genes. Lck and another tyrosine kinase, ZAP-70, are both expressed from an early stage in thymocyte development. As well as its key role in signaling from the pre-T-cell receptor, Lck is also important for γ:δ T-cell development. In contrast, gene knockout studies (see Appendix I, Section A-46) show that ZAP-70, although expressed from the double-negative stage onward, has an essential role later: it promotes the development of single-positive thymocytes from double-positive thymocytes. Fyn, a Src-family kinase similar to Lck, is expressed at increasing levels from the double-positive stage onward. It is not essential for the development of α:β thymocytes as long as Lck is present, but is required for the development of iNKT cells.

Finally, several transcription factors have been identified that guide the development of thymocytes from one stage to the next. Ikaros and GATA3 are expressed in early T-cell progenitors; in the absence of either, T-cell development is generally disrupted. Moreover, these factors also have roles in the normal functioning of mature T cells. In contrast, Ets1, though also expressed in early progenitors, is not essential for T-cell development, although mice

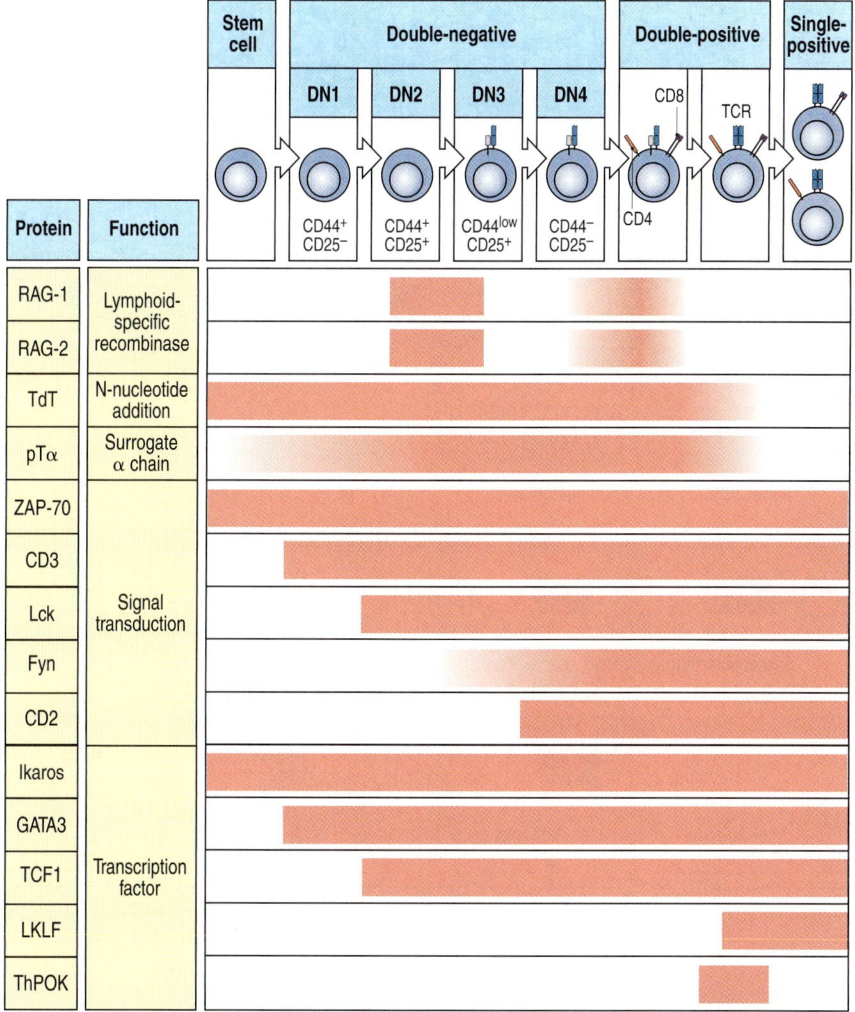

Fig. 8.26 The temporal pattern of expression of some cellular proteins important in early T-cell development. Expression is depicted with regard to the stages of thymocyte development as determined by cell-surface marker expression. The proteins listed are a selection of those known to be associated with early T-lineage development and have been included because of their proven importance in the developmental sequence, largely on the basis of studies in mice. Some of these proteins are involved in gene rearrangement and signaling through receptors, and their individual contributions are discussed in the text. Several transcription factors have been identified that guide the development of thymocytes from one stage to the next by regulating gene expression. Ikaros and GATA3 are expressed in early T-cell progenitors; in the absence of either, T-cell development is generally disrupted. These proteins also have roles in mature T cells. In the absence of TCF1 (T-cell factor-1), double-negative T cells that make productive β-chain gene rearrangements do not proliferate in response to the pre-T-cell receptor signal, thus preventing the efficient production of double-positive thymocytes. LKLF (lung Kruppel-like factor) is first expressed at the single-positive stage; if it is absent, thymocytes exhibit a defect in emigration to populate peripheral lymphoid tissues, due in part to their failure to express receptors involved in trafficking, such as the sphingosine-1-phosphate (S1P) receptor, $S1P_1$ (see Chapter 9). The transcription factor Ets1 (not shown on this figure) is not essential for T-cell development, but mice lacking this factor do not make NK cells.

lacking this factor do not make NK cells. TCF1 (T-cell factor-1) is first expressed during the double-negative stage. In its absence, double-negative T cells that make productive β-chain gene rearrangements do not proliferate, preventing the efficient production of double-positive thymocytes. Thus, transcription factors expressed at various developmental stages control normal thymocyte development by controlling the expression of appropriate genes.

8-14 T-cell α-chain genes undergo successive rearrangements until positive selection or cell death intervenes.

The T-cell receptor α-chain genes are comparable to the immunoglobulin κ and λ light-chain genes in that they do not have D gene segments and are rearranged only after their partner receptor chain has been expressed. As with the light-chain genes, repeated attempts at α-chain gene rearrangement are possible, as illustrated in Fig. 8.27. The presence of multiple V_α gene segments, and about 60 J_α gene segments spread over some 80 kilobases of DNA, allows many successive V_α to J_α rearrangements to take place at both α-chain alleles. This means that T cells with an initial nonproductive α-gene rearrangement are much more likely to be rescued by a subsequent rearrangement than are B cells with a nonproductive light-chain gene rearrangement.

One key difference between B and T cells is that the final assembly of an immunoglobulin leads to the cessation of gene rearrangement and initiates the further differentiation of the B cell, whereas rearrangement of the V_α gene segments continues in T cells unless there is signaling by a self-peptide:self-MHC complex that positively selects the receptor. This means that many T cells have in-frame rearrangements on both chromosomes and so can produce two types of α chains. This is possible because expression of the T-cell receptor is not in itself sufficient to shut off gene rearrangement. Continued rearrangements on both chromosomes can allow several different α chains to be produced successively as well as simultaneously in each developing T cell and to be tested for self-peptide:self-MHC recognition in partnership with

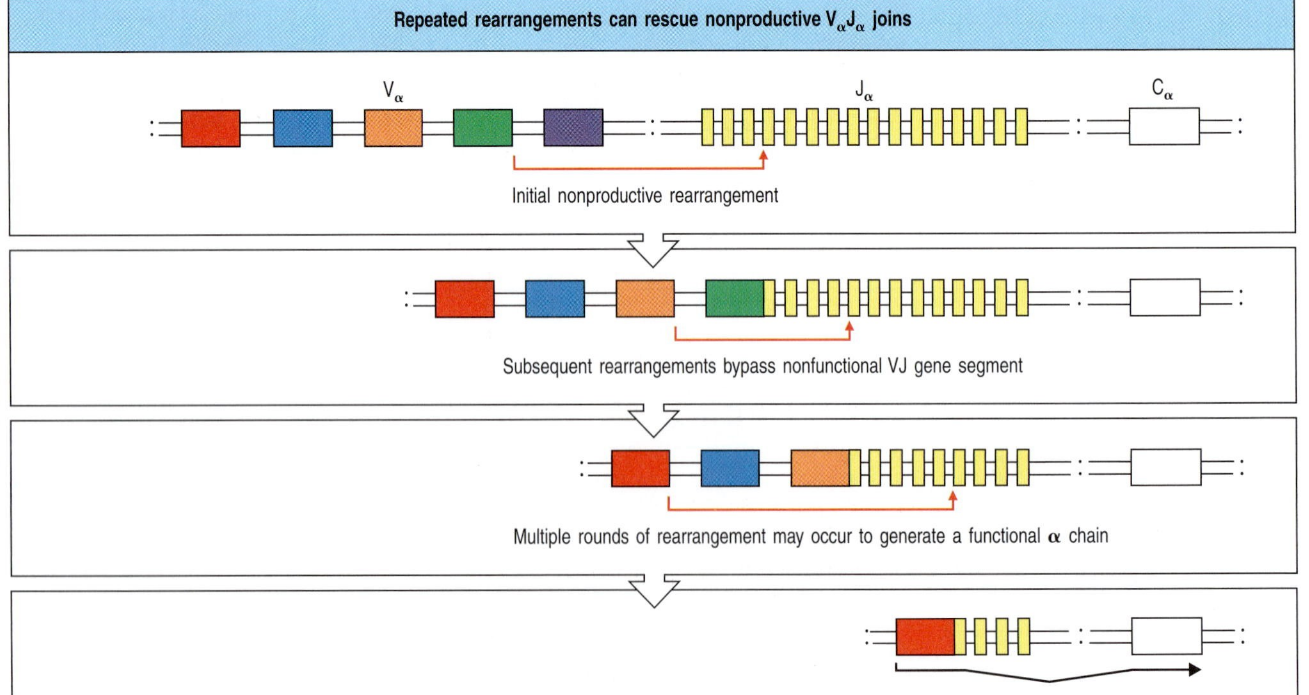

Fig. 8.27 Multiple successive rearrangement events can rescue nonproductive T-cell receptor α-chain gene rearrangements. The multiplicity of V and J gene segments at the α-chain locus allows successive rearrangement events to 'leapfrog' over previously rearranged VJ segments, deleting any intervening gene segments. The α-chain rescue pathway resembles that of the immunoglobulin κ light-chain genes (see Section 8-5), but the number of possible successive rearrangements is greater. α-chain gene rearrangement continues until either a productive rearrangement leads to positive selection or the cell dies.

the same β chain. This phase of gene rearrangement lasts for 3 or 4 days in the mouse and ceases only when positive selection occurs as a consequence of receptor engagement, or when the cell dies. One can predict that if the frequency of positive selection is sufficiently low, roughly one in three mature T cells will express two productively rearranged α chains at the cell surface. This has been confirmed for both human and mouse T cells. Thus, in the strict sense, T-cell receptor α-chain genes are not subject to allelic exclusion.

T cells with dual specificity might be expected to give rise to inappropriate immune responses if the cell is activated through one receptor yet can act upon target cells recognized by the second receptor. However, only one of the two receptors is likely to be able to recognize peptide presented by a self-MHC molecule, and so the T cell will have only a single functional specificity. This is because once a thymocyte has been positively selected by self-peptide:self-MHC recognition, α-chain gene rearrangement ceases. Thus, the existence of cells with two α-chain genes productively rearranged and two α chains expressed at the cell surface does not truly challenge the idea that a single functional specificity is expressed by each cell.

Summary.

The thymus provides a specialized and architecturally organized microenvironment for the development of mature T cells. Precursors of T cells migrate from the bone marrow to the thymus, where they interact with environmental cues, such as ligands for the Notch receptor, that drive commitment to the T lineage. Developing thymocytes develop along one of three T-cell lineages: γ:δ T cells, α:β T cells, and iNKT cells (which are T cells with α:β receptors of very limited diversity).

T-cell progenitors make a choice between the γ:δ and the α:β T-cell lineages. Early in ontogeny, the production of γ:δ T cells predominates over α:β T cells, and these cells populate several peripheral tissues, including the skin, reproductive epithelium, and intestine. Later, more than 90% of thymocytes express α:β T-cell receptors. In developing thymocytes, the γ, δ, and β genes rearrange virtually simultaneously; signaling by a functional γ:δ receptor commits the precursor toward the γ:δ lineage; these cells halt further gene rearrangement and do not express CD4 and CD8 co-receptors. Production of a functionally rearranged β-chain gene and signaling by the pre-T-cell receptor commits the precursor to the α:β lineage.

T cells of the conventional α:β lineage pass through a series of stages distinguished by rearrangement of the α-chain gene and the production of an α:β T-cell receptor, the differential expression of CD44 and CD25, CD3:T-cell receptor complex proteins, and the co-receptors CD4 and CD8. Most steps in T-cell development take place in the thymic cortex, whereas the medulla contains mainly mature T cells.

Positive and negative selection of T cells.

Up to the stage at which an α:β receptor is produced, T-cell development has been independent of antigen. From this point onward, developmental decisions in the α:β T-cell lineage depend on the interaction of the receptor with peptide:MHC ligands it encounters in the thymus, and we now consider this phase of T-cell development.

T-cell precursors committed to the α:β lineage at the DN3 stage undergo vigorous proliferation in the subcapsular region—the DN4 stage. These cells then rapidly transit through an immature CD8-single-positive stage and then into double-positive cells that express low levels of the T-cell receptor and both the CD4 and CD8 co-receptors and move deeper into the thymic cortex. These double-positive cells have a life-span of only about 3–4 days unless they are rescued by engagement of their T-cell receptor. The rescue of double-positive cells from programmed cell death and their maturation into CD4 or CD8 single-positive cells is the process known as positive selection. Only about 10–30% of the T-cell receptors generated by gene rearrangement will be able to recognize self-peptide:self-MHC complexes and thus function in self-MHC-restricted responses to foreign antigens (see Chapter 4); those that have this capability are selected for survival in the thymus. Double-positive cells also undergo negative selection: T cells whose receptors recognize self-peptide:self-MHC complexes too strongly undergo apoptosis, thus eliminating potentially self-reactive cells. In this section we examine the interactions between developing double-positive thymocytes and different thymic components and examine the mechanisms by which these interactions shape the mature T-cell repertoire.

8-15 The MHC type of the thymic stroma selects a repertoire of mature T cells that can recognize foreign antigens presented by the same MHC type.

Positive selection was first demonstrated in classic experiments using mice whose bone marrow had been completely replaced by bone marrow from a mouse that had a different MHC genotype but was otherwise genetically identical. These mice are known as **bone marrow chimeras** (see Appendix I, Section A-42). The recipient mouse is first irradiated to destroy all its own lymphocytes and bone marrow progenitor cells; after bone marrow transplantation, all bone marrow derived cells will be of the donor genotype. These will include all lymphocytes, as well as the antigen-presenting cells they interact with. The rest of the animal's tissues, including the nonlymphoid stromal cells of the thymus, will be of the recipient MHC genotype.

In the experiments that demonstrated positive selection (Fig. 8.28), the donor mice were F_1 hybrids derived from MHC^a and MHC^b parents and thus were of the $MHC^{a \times b}$ genotype. The irradiated recipients were one of the parental strains, either MHC^a or MHC^b. Because of MHC restriction, individual T cells recognize either MHC^a or MHC^b, but not both. Normally, roughly equal numbers of the $MHC^{a \times b}$ T cells from $MHC^{a \times b}$ F_1 hybrid mice will recognize antigen presented by MHC^a or MHC^b. However, in bone marrow chimeras in which T cells of $MHC^{a \times b}$ genotype develop in an MHC^a thymus, T cells immunized to a particular antigen turn out to recognize that antigen mainly, if not exclusively, when it is presented by MHC^a molecules, even though the antigen-presenting cells display antigen bound to both MHC^a and MHC^b. These experiments demonstrated that the MHC molecules present in the environment in which T cells develop determine the MHC restriction of the mature T-cell receptor repertoire.

A similar kind of experiment, using grafts of thymic tissue, showed that the radioresistant cells of the thymic stroma are responsible for positive selection. In these experiments, the recipient animals were athymic *nude* or thymectomized mice of $MHC^{a \times b}$ genotype that were given thymic stromal grafts of MHC^a genotype. Thus, all their cells except the thymic stroma carried both MHC^a and MHC^b. The $MHC^{a \times b}$ bone marrow cells of these mice matured into T cells that recognized antigens presented by MHC^a but not by MHC^b. This result showed that it is the MHC molecules expressed by the thymic stromal cells that determine what mature T cells consider to be self MHC. These results also argued that the MHC restriction seen in the immunized bone marrow chimeras could originate in the thymus, presumably by the selection of T cells as they developed.

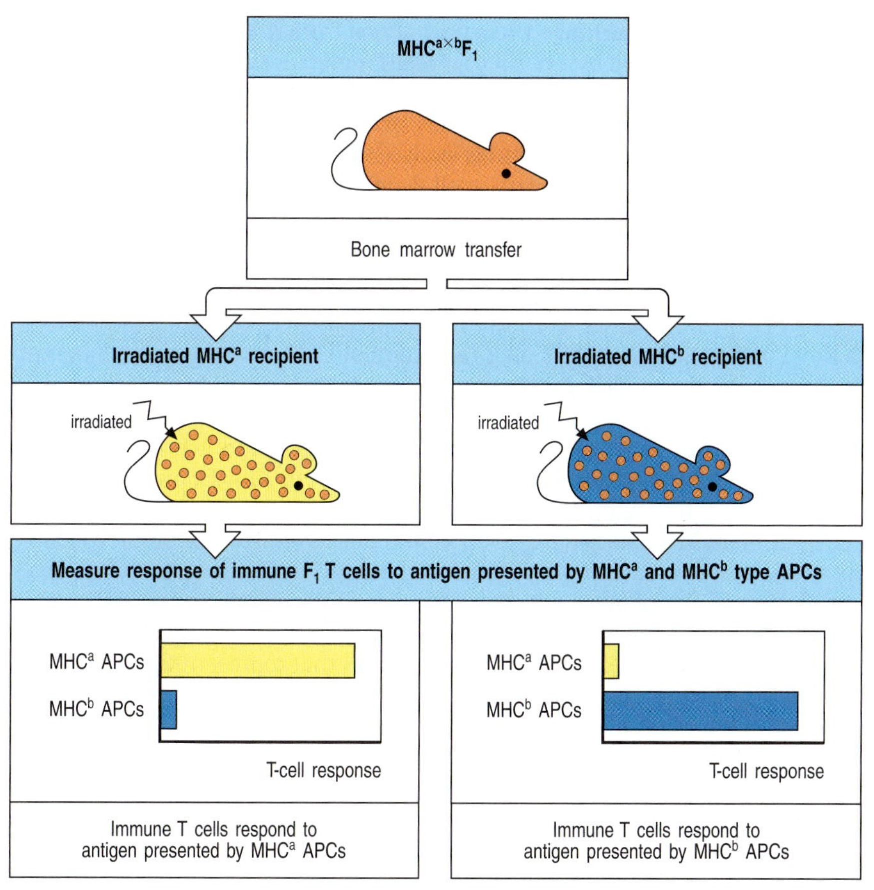

Fig. 8.28 Positive selection is revealed by bone marrow chimeric mice. As shown in the top two sets of panels, bone marrow from an MHC$^{a\times b}$ F$_1$ hybrid mouse is transferred to a lethally irradiated recipient mouse of either parental MHC type (MHC^a or MHC^b). When these chimeric mice are immunized with antigen, the antigen can be presented by the bone marrow derived MHC$^{a\times b}$ antigen-presenting cells (APCs) in association with both MHC^a and MHC^b molecules. The T cells from an MHC$^{a\times b}$ F$_1$ mouse include cells that respond to antigen presented by APCs from MHC^a mice and cells that respond to APCs from MHC^b mice (not shown). But when immunized T cells from the chimeric animals are tested *in vitro* with APCs bearing MHC^a or MHC^b only, they respond far better to antigen presented by the MHC molecules of the recipient MHC type, as shown in the bottom panels. This shows that the T cells have undergone positive selection for MHC restriction in the recipient thymus.

8-16 Only thymocytes whose receptors interact with self-peptide: self-MHC complexes can survive and mature.

Bone marrow chimeras and thymic grafting provided evidence that MHC molecules in the thymus influence the MHC-restricted T-cell repertoire. However, mice transgenic for rearranged T-cell receptor genes provided the first conclusive evidence that the interaction of the T cell with self-peptide:self-MHC complexes is necessary for the survival of immature T cells and their maturation into naive CD4 or CD8 T cells. For these experiments, the rearranged α- and β-chain genes were cloned from a T-cell clone (see Appendix I, Section A-24) whose origin, antigen specificity, and MHC restriction were known. When such genes are introduced into the mouse genome, they are expressed early during thymocyte development and the rearrangement of endogenous T-cell receptor genes is inhibited; endogenous β-chain gene rearrangement is inhibited completely but that of α-chain genes is inhibited only incompletely. The result is that most of the developing thymocytes express the T-cell receptor encoded by the transgenes.

By introducing T-cell receptor transgenes specific for a known MHC genotype, the effect of MHC molecules on the maturation of thymocytes with receptors of known specificity can be studied directly without the need for immunization and analysis of effector function. Such studies showed that thymocytes bearing a particular T-cell receptor could develop to the double-positive stage in thymuses that expressed different MHC molecules from those in which the cell bearing the T-cell receptor originally developed. However, these transgenic thymocytes only developed beyond the double-positive stage and became mature T cells if the thymus expressed the same self-MHC molecule as that on which the original T-cell clone was selected (Fig. 8.29). Such experiments also discovered the fate of T cells that fail

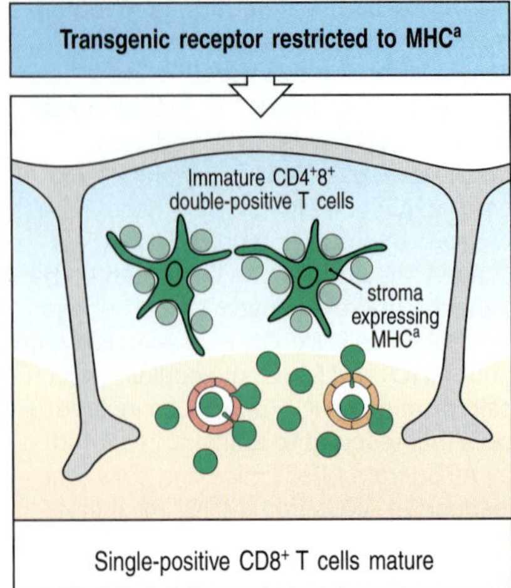

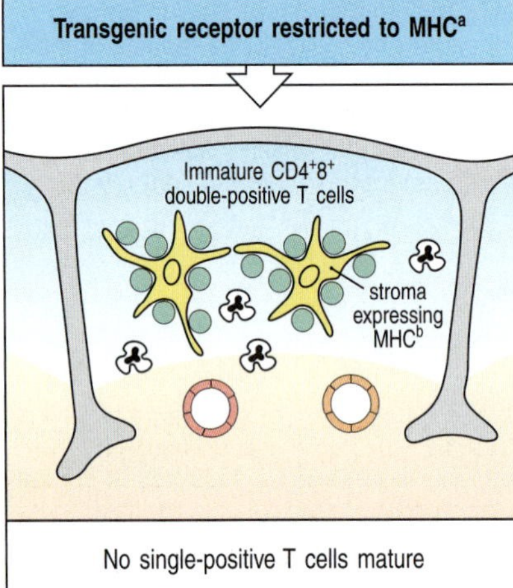

Fig. 8.29 Positive selection is demonstrated by the development of T cells expressing rearranged T-cell receptor transgenes. In mice transgenic for rearranged α:β T-cell receptor genes, the maturation of T cells depends on the MHC haplotype expressed in the thymus. If the transgenic mice express the same MHC haplotype in their thymic stromal cells as the mouse from which the rearranged TCRα-chain and TCRβ-chain genes originally developed (both MHC^a, top panels), then the T cells expressing the transgenic T-cell receptor will develop from the double-positive stage (pale green) into mature T cells (dark green), in this case mature $CD8^+$ single-positive cells. If the MHC^a-restricted TCR transgenes are genetically crossed into a different MHC background (MHC^b, yellow) (bottom panel), then developing T cells expressing the transgenic receptor will progress to the double-positive stage but will fail to mature further. This failure is due to the absence of an interaction between the transgenic T-cell receptor with MHC molecules on the thymic cortex, and thus no signal for positive selection is delivered, leading to apoptotic death by neglect.

positive selection. Rearranged receptor genes from a mature T cell specific for a peptide presented by a particular MHC molecule were introduced into a recipient mouse lacking that MHC molecule, and the fate of thymocytes was investigated by staining with antibodies specific for the transgenic receptor. Antibodies against other molecules such as CD4 and CD8 were used at the same time to mark the stages of T-cell development. It was found that cells that fail to recognize the MHC molecules present on the thymic epithelium never progress further than the double-positive stage and die in the thymus within 3 or 4 days of their last division.

8-17 Positive selection acts on a repertoire of T-cell receptors with inherent specificity for MHC molecules.

Positive selection acts on a repertoire of T-cell receptors whose specificity is determined by randomly generated combinations of V, D, and J gene segments (see Section 5-7). Despite this, T-cell receptors exhibit a bias toward recognition of MHC molecules even before positive selection. If the specificity of the unselected repertoire were completely random, only a very small proportion of thymocytes would be expected to recognize any MHC molecule. However, an inherent specificity of T-cell receptors for MHC molecules has been detected by examining mature T cells that represent the unselected receptor repertoire. Such T cells can be produced *in vitro* from fetal thymuses that lack expression of MHC class I and MHC class II molecules by triggering generalized 'positive selection' using antibodies that bind to the V_β chain of T-cell receptors and to the CD4 co-receptor. When such antibody-selected CD4 T cells are examined, roughly 5% can respond to any one MHC class II genotype. Because these cells developed without selection by MHC molecules, this reactivity must reflect an inherent MHC-specificity encoded in the germline V gene segments. This specificity should significantly increase the proportion of receptors that can be positively selected by any individual's MHC molecules.

The germline-encoded reactivity seems to be due to specific amino acids in the CDR1 and CDR2 regions of T-cell receptor V_β and V_α regions. The CDR1 and CDR2 regions are encoded in the germline V gene segments and are highly variable (see Section 5-8). But among this variability, certain amino acids are conserved and common to many V segments. Analysis of numerous crystal structures has revealed that when the T-cell receptor binds a peptide:MHC complex, specific amino acids of the V_β region interact with a particular part of the MHC protein. For example, in many human and mouse V_β regions, the CRD2 has a tyrosine at position 48, and this interacts with a region in the middle of the α1 helix of MHC class I and class II proteins (see Fig. 6.22). Two other amino acids commonly found in other V_β regions (tyrosine at 46 and glutamic acid at 54) interact with the same region of MHC. T cells expressing V_β genes with mutations at any of these positions showed reduced positive selection, demonstrating that the interaction of such V regions with MHC molecules contributes to T-cell development.

8-18 Positive selection coordinates the expression of CD4 or CD8 with the specificity of the T-cell receptor and the potential effector functions of the T cell.

At the time of positive selection, the thymocyte expresses both CD4 and CD8 co-receptor molecules. By the end of thymic selection, mature α:β T cells ready for export to the periphery have stopped expressing one of these co-receptors and belong to one of three categories: conventional CD4 or CD8 T cells, or a subset of regulatory T cells expressing CD4 and high levels of CD25. Moreover, almost all mature T cells that express CD4 have receptors that recognize peptides bound to self-MHC class II molecules and are programmed to become cytokine-secreting helper T cells. In contrast, most of the cells that express CD8 have receptors that recognize peptides bound to self-MHC class I molecules and are programmed to become cytotoxic effector cells. Thus, positive selection also determines the cell-surface phenotype and functional potential of the mature T cell, selecting the appropriate co-receptor for efficient antigen recognition and the appropriate program for the T cell's eventual functional differentiation in an immune response.

Experiments with mice transgenic for rearranged T-cell receptor genes show clearly that the specificity of the T-cell receptor for self-peptide:self-MHC molecule complexes determines which co-receptor a mature T cell will express. If the transgenes encode a T-cell receptor specific for antigen presented by self-MHC class I molecules, mature T cells that express the transgenic receptor are CD8 T cells. Similarly, in mice made transgenic for a receptor that recognizes antigen with self-MHC class II molecules, mature T cells that express the transgenic receptor are CD4 T cells (Fig. 8.30).

The importance of MHC molecules in this selection is illustrated by the class of human immunodeficiency diseases known as **bare lymphocyte syndromes**, which are caused by mutations that lead to an absence of MHC molecules on lymphocytes and thymic epithelial cells. People who lack MHC class II molecules have CD8 T cells but only a few, highly abnormal, CD4 T cells; a similar result has been obtained in mice in which MHC class II expression has been eliminated by targeted gene disruption (see Appendix I, Section A-46). Likewise, mice and humans that lack MHC class I molecules lack CD8 T cells. Thus, MHC class II molecules are absolutely required for CD4 T-cell development, whereas MHC class I molecules are similarly required for CD8 T-cell development.

In mature T cells, the co-receptor functions of CD8 and CD4 depend on their respective abilities to bind invariant sites on MHC class I and MHC class II molecules (see Section 4-17). Co-receptor binding to an MHC molecule is also required for normal positive selection, as shown for CD4 in the experiment discussed in the next section. Thus, positive selection depends on engagement of both the antigen receptor and co-receptor by an MHC molecule, and determines the survival of single-positive cells that express only the appropriate co-receptor. Commitment to either the CD4 or CD8 lineage is coordinated with receptor specificity and it seems that the developing thymocyte integrates signals from both the antigen receptor and the co-receptor. Co-receptor-associated Lck signals are most effectively delivered when CD4 rather than CD8 is engaged as a co-receptor, and these Lck signals play a large part in the decision to become a mature CD4 cell. When a T cell receives a signal inducing positive selection, it first downregulates both CD4 and CD8, but then reexpresses CD4, making a $CD4^+CD8^{low}$ cell regardless of whether the T-cell receptor has been engaged by MHC class I or MHC class II molecules (Fig. 8.31). It appears that the strength of signaling in this $CD4^+CD8^{low}$ cell is what determines the lineage choice. If the cell is being selected by MHC class II, the reexpression of CD4 provides a stronger or more sustained signal, mediated in part by Lck, and this is responsible for further differentiation along the CD4 pathway, with the complete loss of CD8. If the cell is being

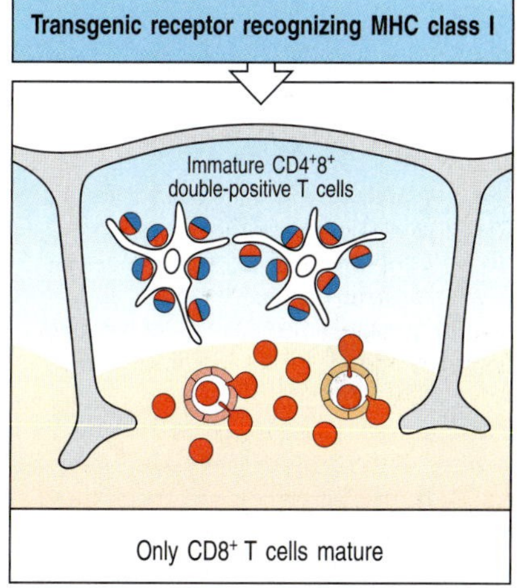

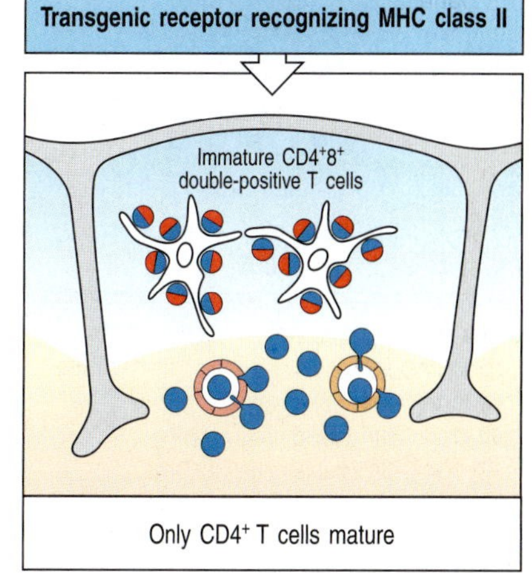

Fig. 8.30 The MHC molecules that induce positive selection determine co-receptor specificity. In mice transgenic for T-cell receptors restricted by an MHC class I molecule (top panel), the mature T cells that develop all have the CD8 (red) phenotype. In mice transgenic for receptors restricted by an MHC class II molecule (bottom panel), all mature T cells have the CD4 (blue) phenotype. In both cases, normal numbers of immature, double-positive thymocytes (half blue, half red) are found. The specificity of the T-cell receptor determines the outcome of the developmental pathway, ensuring that the only T cells that mature are those equipped with a co-receptor that is able to bind the same self-MHC molecule as the T-cell receptor.

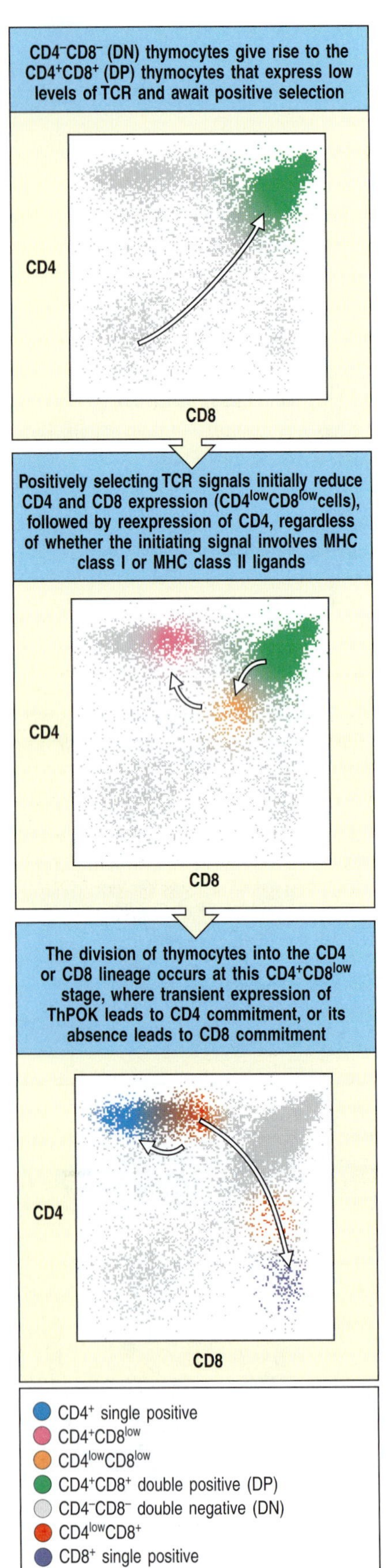

CD4⁻CD8⁻ (DN) thymocytes give rise to the CD4⁺CD8⁺ (DP) thymocytes that express low levels of TCR and await positive selection
CD4
CD8
Positively selecting TCR signals initially reduce CD4 and CD8 expression (CD4^low CD8^low cells), followed by reexpression of CD4, regardless of whether the initiating signal involves MHC class I or MHC class II ligands
CD4
CD8
The division of thymocytes into the CD4 or CD8 lineage occurs at this CD4⁺CD8^low stage, where transient expression of ThPOK leads to CD4 commitment, or its absence leads to CD8 commitment
CD4
CD8
CD4⁺ single positive
CD4⁺CD8^low
CD4^low CD8^low
CD4⁺CD8⁺ double positive (DP)
CD4⁻CD8⁻ double negative (DN)
CD4^low CD8⁺
CD8⁺ single positive

selected by MHC class I, reexpression of CD4 will not lead to further signaling via Lck; this weaker signal in turn determines CD8 commitment, with a subsequent loss of CD4 expression and the later reexpression of CD8.

T-cell receptor signaling regulates this choice of the CD4 versus the CD8 lineage by controlling the expression of two transcription factors, ThPOK and Runx3 (see Fig. 8.31). ThPOK was identified through a naturally occurring loss-of-function mutation in mice that lacked CD4 T-cell development. In mice lacking ThPOK, MHC class II-restricted thymocytes are redirected toward the CD8 lineage. ThPOK is not expressed in double-positive thymocytes, but strong T-cell receptor signaling at the $CD4^{+}CD8^{low}$ stage of development induces its expression. ThPOK in turn represses expression of Runx3; together, the expression of ThPOK and absence of Runx3 lead to CD4 commitment and the ability to express cytokine genes typical of CD4 cells. If T-cell signaling is of insufficient strength or duration, however, ThPOK is not induced, and Runx3 is allowed to be expressed. This leads to silencing of CD4 expression, reexpression of CD8 and the ability to express genes typical of CD8 T cells that encode proteins involved in target-cell killing.

The majority of double-positive thymocytes that undergo positive selection develop into either CD4 or CD8 single-positive T cells. The thymus also generates a smaller population of single-positive CD4 T cells that represent a distinct lineage known as **natural regulatory T cells (T_{reg} cells)**. Like CD4 and CD8 cells, T_{reg} cells arise at the $CD4^{+}CD8^{low}$ stage of development. Commitment to the T_{reg} lineage is determined by expression of a Forkhead-family transcription factor called **FoxP3** and seems to be caused by T-cell receptor signaling that is stronger than that inducing conventional CD4 commitment but is not strong enough to induce deletion (discussed in Section 8-20). T_{reg} cells express high levels of the surface proteins CD25 and CTLA-4 (see Section 7-18). It is thought that the T-cell receptors expressed by T_{reg} cells strongly recognize self antigens expressed in the thymus, which triggers stable expression of FoxP3.

8-19 Thymic cortical epithelial cells mediate positive selection of developing thymocytes.

The thymus transplantation studies described in Section 8-15 suggested that stromal cells were important for positive selection. These cells form a web of cell processes that make close contacts with the double-positive T cells undergoing positive selection (see Fig. 8.16), and T-cell receptors can be seen clustering with MHC molecules at the sites of contact. Direct evidence that thymic cortical epithelial cells mediate positive selection comes from

Fig. 8.31 Stages in the positive selection of $\alpha{:}\beta$ T cells as identified by FACS analysis. The diagram represents FACS analysis (see Appendix I, Fig. A.25) of various stages in thymocyte development with reference to the co-receptor molecules CD4 and CD8. Each dot represents an individual cell, and its position on the axes represents the quantity of CD4 and CD8 expression that was determined at the time of analysis. Double-negative (DN) cells are located in the bottom left corner of the plot. DN cells that have successfully rearranged a β chain receive a signal through pre-TCR and undergo proliferation, and then induce expression of the CD8 and CD4 co-receptors to become double-positive (DP) cells (green). Rearrangement of the α-chain locus occurs in these DP cells, with expression of a T-cell receptor on the cell surface first at low and then at intermediate levels. In these cells, signaling is co-receptor dependent. If the expressed T-cell receptor interacts successfully with MHC molecules on thymic stroma to induce positive selection, the cell initially reduces CD8 and CD4 expression (orange), followed by a subsequent increased expression of CD4 to generate the $CD4^{+}CD8^{low}$ population (red). If selection was provided by an MHC class II molecule, signaling in the $CD4^{+}CD8^{low}$ T cell is of a longer duration and commitment to CD4 occurs, with maintenance of CD4 and loss of CD8 expression (blue). If the selection was provided by an MHC class I molecule, signaling in the $CD4^{+}CD8^{low}$ T cell will be of shorter duration, and this leads to commitment to the CD8 lineage, with reexpression of CD8 and loss of CD4 (purple).

an ingenious manipulation of mice whose MHC class II genes have been eliminated by targeted gene disruption (Fig. 8.32). Mutant mice that lack MHC class II molecules do not normally produce CD4 T cells. To test the role of the thymic epithelium in positive selection, an MHC class II gene was placed under the control of a promoter that restricted its expression to thymic cortical epithelial cells. This was then introduced as a transgene into the CD4-mutant mice, and CD4 T-cell development was restored. A variant of this experiment showed that, to promote the development of CD4 T cells, the MHC class II molecule on the thymic cortical epithelium must be able to interact effectively with CD4. Thus, when the MHC class II transgene expressed in the thymus contains a mutation that prevents its binding to CD4, very few CD4 T cells develop. Equivalent studies of CD8 interaction with MHC class I molecules showed that co-receptor binding is also necessary for the positive selection of CD8 cells.

The critical role of the thymic cortical epithelium in positive selection raises the question of whether there is anything distinctive about the antigen-presenting properties of these cells. The thymic stromal cells may simply be in closest proximity to the developing thymocytes, as there are very few macrophages and dendritic cells in the cortex. In addition, however, thymic epithelium differs from other tissues in the expression of key proteases that are involved in MHC class I and II antigen processing (see Section 6-8). Cortical epithelial cells express cathepsin L as opposed to the more widely expressed cathepsin S, and mice deficient in cathepsin L have severely impaired CD4 T-cell development. Thymic epithelial cells from mice lacking cathepsin L exhibit a relatively high density of MHC class II molecules on their surface that retain the invariant chain-associated peptide (CLIP) (see Fig. 6.10). Cortical epithelial cells also express a unique proteasome subunit, β5T, whereas other cells express β5 or β5i. Mice deficient in β5T have severely impaired CD8 T-cell development. Because mice that lack either cathepsin L or β5 still have normal levels of MHC on the surface of their thymic cortical cells, it would seem that it is the peptide repertoire displayed by the MHC molecules on cortical epithelial cells that is responsible for altering CD8 T-cell development, although the mechanism is still unclear.

Fig. 8.32 Thymic cortical epithelial cells mediate positive selection. In the thymus of normal mice (first panels), which express MHC class II molecules on epithelial cells in the thymic cortex (blue) as well as on medullary epithelial cells (orange) and bone marrow derived cells (yellow), both CD4 (blue) and CD8 (red) T cells mature. Double-positive thymocytes are shown as half red and half blue. The second panels represent mutant mice in which MHC class II expression has been eliminated by targeted gene disruption; in these mice, few CD4 T cells develop, although CD8 T cells develop normally. In MHC class II-negative mice containing an MHC class II transgene engineered so that it is expressed only on the epithelial cells of the thymic cortex (third panels), normal numbers of CD4 T cells mature. In contrast, if a mutant MHC class II molecule with a defective CD4-binding site is expressed (fourth panel), positive selection of CD4 T cells does not take place. This indicates that the cortical epithelial cells are the critical cell type mediating positive selection and that the MHC class II molecule needs to be able to interact with the CD4 protein.

8-20 T cells that react strongly with ubiquitous self antigens are deleted in the thymus.

When the T-cell receptor of a mature naive T cell is strongly ligated by a peptide:MHC complex displayed on a specialized antigen-presenting cell in a

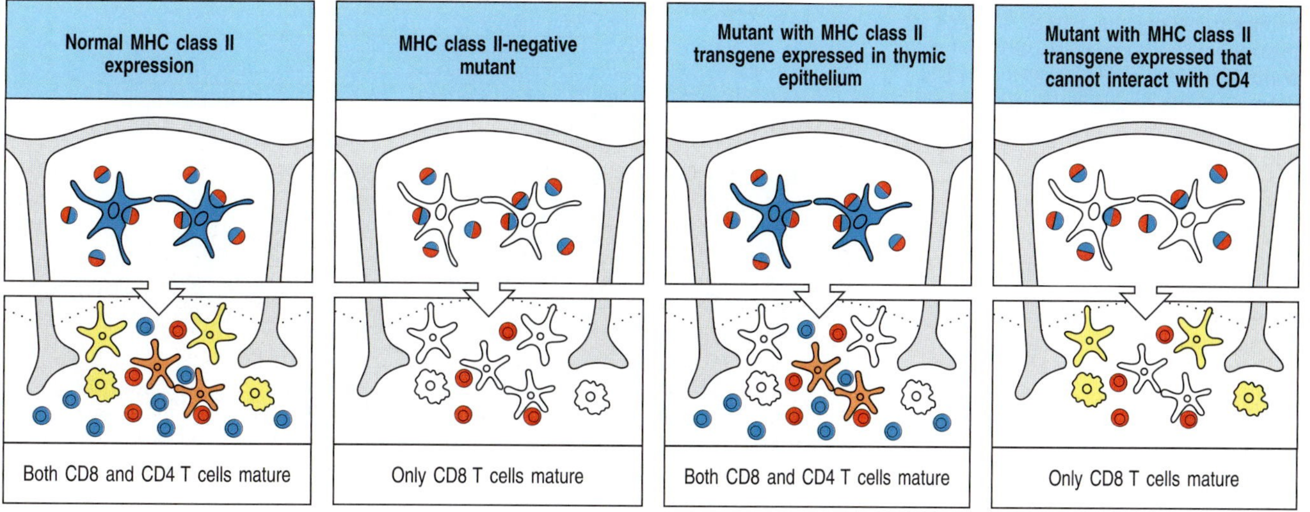
Normal MHC class II expression
MHC class II-negative mutant
Mutant with MHC class II transgene expressed in thymic epithelium
Mutant with MHC class II transgene expressed that cannot interact with CD4
Both CD8 and CD4 T cells mature
Only CD8 T cells mature
Both CD8 and CD4 T cells mature
Only CD8 T cells mature

peripheral lymphoid organ, the T cell is activated to proliferate and produce effector T cells. In contrast, when the T-cell receptor of a developing thymocyte is similarly ligated by a self-peptide:MHC complex in the thymus, it dies by apoptosis. The response of immature T cells to stimulation by antigen is the basis of negative selection. Elimination of immature T cells in the thymus prevents their potentially harmful activation later, should they encounter the same self peptides when they are mature T cells.

Negative selection has been demonstrated by the use of both artificial and naturally occurring self peptides. Negative selection of thymocytes reactive to an artificial self peptide was demonstrated by using TCR-transgenic mice in which the majority of thymocytes expressed a T-cell receptor specific for a peptide of ovalbumin bound to an MHC class II molecule. When the ovalbumin peptide was injected into these mice, most of the CD4CD8 double-positive thymocytes in the thymic cortex died by apoptosis (Fig. 8.33). Negative selection by a natural self peptide was observed by using TCR-transgenic mice expressing T-cell receptors specific for self peptides normally expressed only in male mice. Thymocytes bearing these receptors disappear from the developing T-cell population in male mice at the double-positive stage of development, and no single-positive cells bearing the transgenic receptors mature. By contrast, in female mice, which lack the male-specific peptide, T cells bearing the transgenic receptors mature normally. Negative selection to male-specific peptides has also been demonstrated in normal mice and also occurs by deletion of T cells.

TCR transgenic mice were very useful for the classic experiments above, but they express a functional T-cell receptor earlier during development than normal mice and have a very high frequency of cells reactive to any particular peptide. A more realistic system for evaluating negative selection involves the transgenic expression of only the β chain of a T-cell receptor reactive to a given peptide antigen. In such mice, the β chain pairs with endogenous α chains, yet the frequency of peptide-reactive T cells is sufficient for detection using peptide:MHC tetramers (see Appendix I, Section A-28). These and other

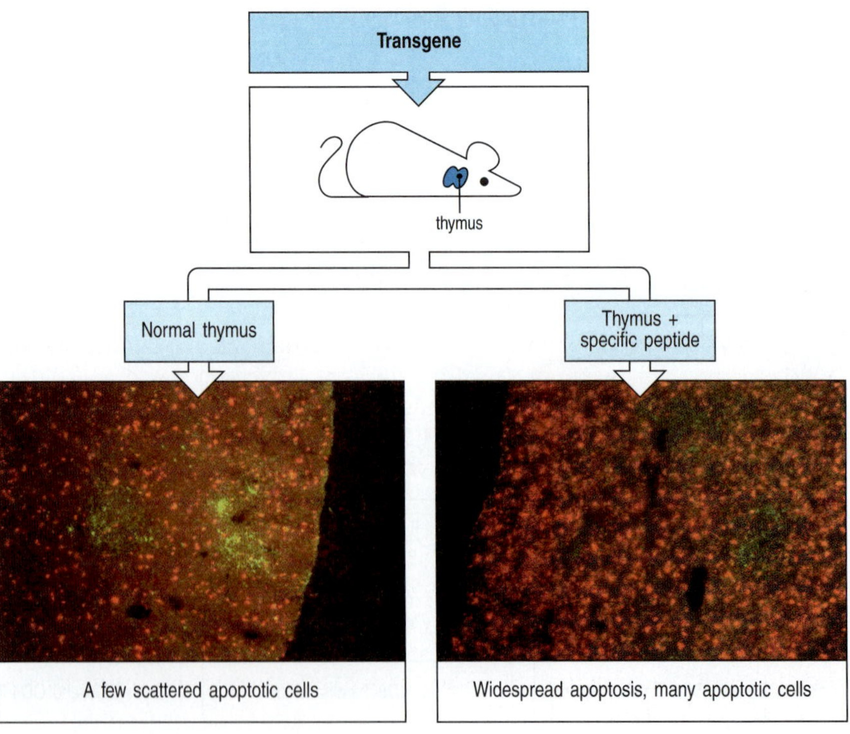

Fig. 8.33 T cells specific for self antigens are deleted in the thymus. In mice transgenic for a T-cell receptor that recognizes a known peptide antigen complexed with self MHC, all the T cells have the same specificity. In the absence of the peptide, most thymocytes mature and emigrate to the periphery. This can be seen in the bottom left panel, where a normal thymus is stained with antibody to identify the medulla (in green), and by the TUNEL technique (see Appendix I, Section A-32) to identify apoptotic cells (in red). If the mice are injected with the peptide that is recognized by the transgenic T-cell receptor, massive cell death occurs in the thymus, as shown by the increased numbers of apoptotic cells in the right-hand bottom panel. Photographs courtesy of A. Wack and D. Kioussis.

more physiologic approaches showed that clonal deletion can occur at either the double-positive or single-positive stage, presumably depending on where the T cell encounters the antigen that causes deletion.

These experiments illustrate the principle that self-peptide:self-MHC complexes encountered in the thymus purge the mature T-cell repertoire of cells bearing self-reactive receptors. One obvious problem with this scheme is that many tissue-specific proteins, such as pancreatic insulin, would not be expected to be expressed in the thymus. However, it is now clear that many such 'tissue-specific' proteins are expressed by certain stromal cells in the thymic medulla; thus, intrathymic negative selection could apply even to proteins that are otherwise restricted to tissues outside the thymus. The expression of some, but not all, tissue-specific proteins in the thymic medulla is controlled by a gene called ***AIRE*** (**autoimmune regulator**). *AIRE* is expressed in medullary stromal cells (Fig. 8.34), interacts with many proteins involved in transcription, and seems to lengthen transcripts from promoters that would otherwise terminate. Mutations in *AIRE* give rise to the human autoimmune disease known as **autoimmune polyendocrinopathy–candidiasis–ectodermal dystrophy** (**APECED**) or **autoimmune polyglandular syndrome type I**, highlighting the important role of intrathymic expression of tissue-specific proteins in maintaining tolerance to self. Negative selection of developing T cells involves interactions with ubiquitous self antigens and tissue-restricted self antigens, and can take place in both the thymic cortex and the thymic medulla.

It is unlikely that all possible self proteins are expressed in the thymus. Thus, negative selection in the thymus may not remove all T cells reactive to self antigens that appear exclusively in other tissues or are expressed at different stages in development. There are, however, several mechanisms operating in the periphery that can prevent mature T cells from responding to tissue-specific antigens; these are discussed in Chapter 15, when we consider the problem of autoimmune responses and their avoidance.

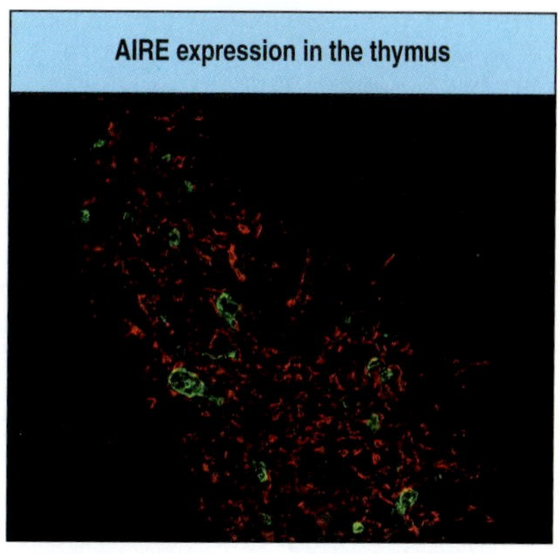

Fig. 8.34 AIRE is expressed in the medulla of the thymus and promotes the expression of proteins normally expressed in peripheral tissues. Expression of AIRE by thymic medullary cells is limited to the medullary region of the thymus, where it is expressed in a subset of epithelial-like cells. The expression of the thymic medullary epithelial marker MTS10 is shown in red. AIRE expression is shown in green by immunofluorescence, and is present in only a fraction of the medullary epithelial cells. Photograph courtesy of R.K. Chin and Y.-X. Fu.

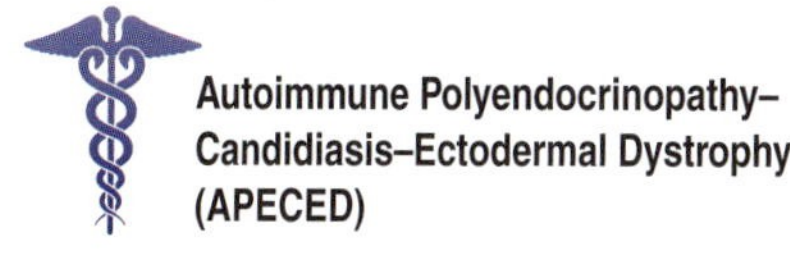

8-21 Negative selection is driven most efficiently by bone marrow derived antigen-presenting cells.

As discussed above, negative selection occurs throughout thymocyte development, both in the thymic cortex and in the medulla, and so is likely to be mediated by antigen presentation by several different cell types. There does seem to be a hierarchy in the effectiveness of cells in mediating negative selection. At the top are bone marrow derived dendritic cells and macrophages. These are antigen-presenting cells that also activate mature T cells in peripheral lymphoid tissues, as we shall see in Chapter 9. The self antigens presented by these cells are therefore the most important source of potential autoimmune responses, and T cells responding to such self peptides must be eliminated in the thymus.

Experiments using bone marrow chimeric mice have clearly shown the role of thymic macrophages and dendritic cells in negative selection (Fig. 8.35). In these experiments, $MHC^{a \times b}$ F_1 bone marrow is grafted into one of the parental strains (MHC^a in Fig. 8.35). The $MHC^{a \times b}$ T cells developing in the grafted animals are thus exposed to MHC^a thymic epithelium. Bone marrow derived dendritic cells and macrophages will, however, express both MHC^a and MHC^b. The bone marrow chimeras will tolerate skin grafts from either MHC^a or MHC^b animals, and from the acceptance of both grafts we can infer that the developing T cells are not self-reactive for either of the two MHC antigens. The only cells that could present self-peptide:MHC^b complexes to thymocytes, and thus induce tolerance to MHC^b, are the bone marrow derived cells. The dendritic cells and macrophages are therefore assumed to have a crucial role in negative selection.

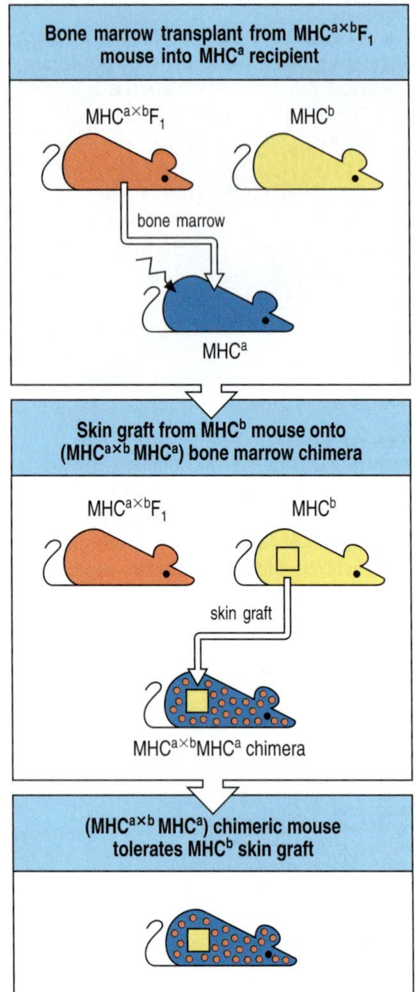

Fig. 8.35 Bone marrow derived cells mediate negative selection in the thymus. When MHCa$^{a \times b}$ F$_1$ bone marrow is injected into an irradiated MHC^a mouse, the T cells mature on thymic epithelium expressing only MHC^a molecules. Nevertheless, the chimeric mice are tolerant to skin grafts expressing MHC^b molecules (provided that these grafts do not present skin-specific peptides that differ between strains a and b). This implies that the T cells whose receptors recognize self antigens presented by MHC^b have been eliminated in the thymus. Because the transplanted MHC$^{a \times b}$ F$_1$ bone marrow cells are the only source of MHC^b molecules in the thymus, bone marrow derived cells must be able to induce negative selection.

In addition, both the thymocytes themselves and the thymic epithelial cells can cause the deletion of self-reactive cells. Such reactions may normally be of secondary significance compared with the dominant role of bone marrow derived cells. In patients undergoing bone marrow transplantation from an unrelated donor, however, where all the thymic macrophages and dendritic cells are of donor type, negative selection mediated by thymic epithelial cells can assume a special importance in maintaining tolerance to the recipient's own antigens.

8-22 The specificity and/or the strength of signals for negative and positive selection must differ.

T cells undergo both positive selection for self-MHC restriction and negative selection for self-tolerance by interacting with self-peptide:self-MHC complexes expressed on stromal cells in the thymus. An unresolved issue is how the interaction of the T-cell receptor with self-peptide:self-MHC complexes distinguishes between these opposite outcomes. First, more receptor specificities must be positively selected than are negatively selected. Otherwise, all the cells that were positively selected in the thymic cortex would be eliminated by negative selection, and no T cells would ever be produced (Fig. 8.36, left panels). Second, the consequences of the interactions that lead to positive and negative selection must differ: cells that recognize self-peptide:self-MHC complexes on cortical epithelial cells are induced to mature, whereas those whose receptors might confer strong and potentially damaging autoreactivity are induced to die.

Currently, the choice between positive and negative selection is thought to hinge on the strength of self-peptide:MHC binding by the T-cell receptor, an idea known as the **affinity hypothesis** (Fig. 8.36, right panels). Low-affinity interactions rescue the cell from death by neglect, leading to positive selection; high-affinity interactions induce apoptosis and thus negative selection. Because more complexes are likely to bind weakly than to bind strongly, this model explains the positive selection of a larger repertoire of cells than are negatively selected. Using T-cell receptor transgenic thymocytes, it was shown that variants of the antigenic peptide could induce positive selection in thymic organ cultures or *in vivo*. Peptide variants that induced positive selection had a lower affinity for the T-cell receptor than did antigenic peptide. How this quantitative difference in receptor affinity translates into a qualitatively distinct cell fate is still an area of active investigation. Many of the biochemical signals induced by low-affinity interactions are weaker or of shorter duration than those from high-affinity interactions. However, low-affinity interactions lead to increased activation of the protein kinase Erk, whereas high-affinity interactions lead to only transient activation of Erk, suggesting that differential activation of this or other MAPKs might determine the outcome of thymic selection.

Finally, other populations besides α:β T cells emerge from the thymus; they are numerically minor but functionally important. These include the T$_{reg}$ cells discussed earlier (Section 8-18), the iNKT cells described in Section 8-9 and in Section 3-24, and CD8α:α intraepithelial lymphocytes (IELs) (see Section 12-10); each has unique developmental requirements. T$_{reg}$ cells require signaling downstream of the IL-2 receptor, whereas other T cells do not. iNKT cells require a T-cell receptor interaction with nonclassical MHC molecules expressed on thymocytes and signal through the adaptor protein SAP, unlike other T cells. They have a memory/activated phenotype and have been suggested to develop in response to 'agonist' signaling—in other words, interactions of a T-cell receptor with a self peptide that would normally activate a T cell. Such interactions are also known to cause clonal deletion, and at this point it is not clear which activating interactions lead to clonal deletion in the thymus and which lead to selection of the nonconventional iNKT cells.

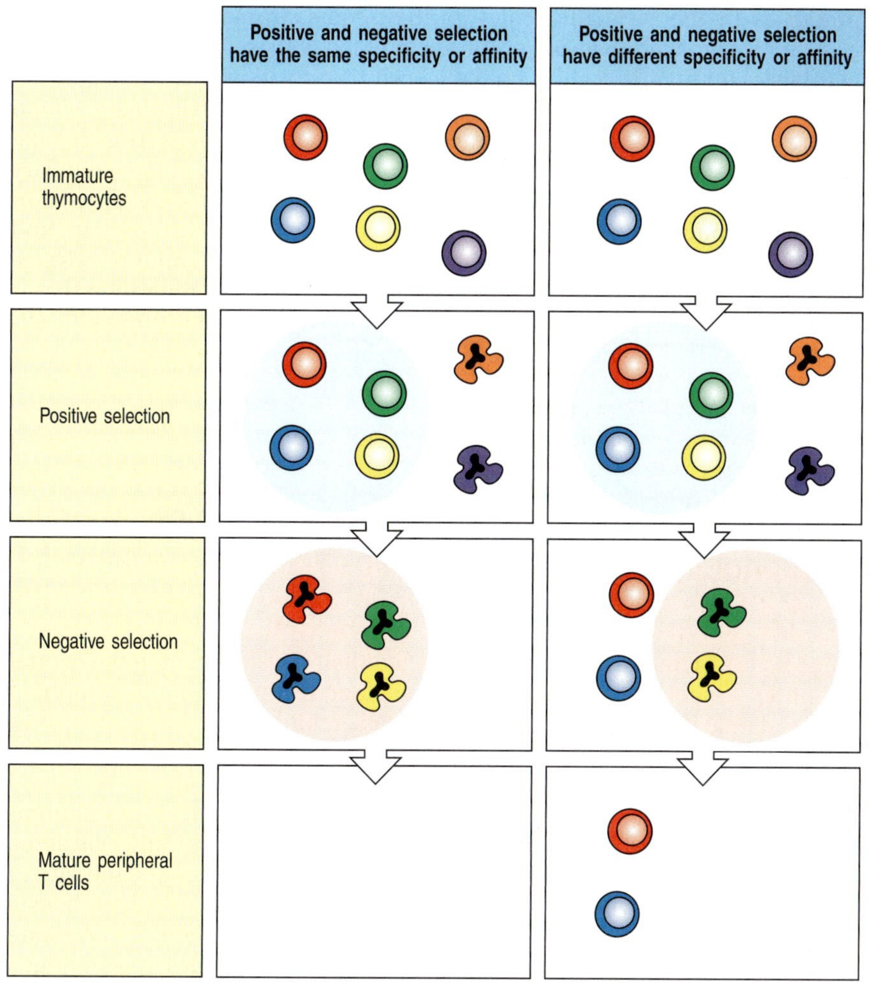
Positive and negative selection have the same specificity or affinity
Positive and negative selection have different specificity or affinity
Immature thymocytes
Positive selection
Negative selection
Mature peripheral T cells

Fig. 8.36 The specificity or affinity of positive selection must differ from that of negative selection. Immature T cells are positively selected in such a way that only those thymocytes whose receptors can engage the peptide:MHC complexes on thymic epithelium mature, giving rise to a population of thymocytes restricted for self MHC. Negative selection removes those thymocytes whose receptors can be activated by self peptides complexed with self-MHC molecules, giving a self-tolerant population of thymocytes. If the specificity and affinity of positive and negative selection were the same (left panels), all the T cells that survive positive selection would be deleted during negative selection. Only if the specificity and affinity of negative selection are different from those of positive selection (right panels) can thymocytes mature into T cells.

After surviving positive and negative selection, thymocytes complete their maturation in the thymic medulla and then emigrate to peripheral lymphoid organs. Their final maturation results in changes to the T-cell receptor signaling machinery. Whereas an immature double-positive or single-positive thymocyte stimulated through the T-cell receptor will undergo apoptosis, a mature single-positive thymocyte responds by proliferating. The final maturation stage takes less than 4 days, and functionally competent T cells then emigrate from the thymus into the bloodstream. Emigration requires recognition of the lipid molecule sphingosine 1-phosphate (S1P) by the G-protein-coupled receptor $S1P_1$, which is expressed by thymocytes during their final maturation. S1P is present in high concentration in blood and lymph, and mature thymocytes seem to be drawn toward it. Mature thymocytes also express CD62L (L-selectin), a lymph-node homing receptor that facilitates the localization of mature naive T cells to peripheral lymphoid organs after their emigration from the thymus.

Summary.

The stages of thymocyte development up to the expression of the pre-T-cell receptor—including the decision between commitment to either the α:β or the γ:δ lineage—are all independent of interactions with peptide:MHC antigens. With the successful rearrangement of α-chain genes and expression of the T-cell receptor, α:β thymocytes undergo further development that is determined by the interactions of their T-cell receptors with self peptides presented by the MHC molecules on the thymic stroma. CD4CD8 double-positive thymocytes whose receptors interact with self-peptide:self-MHC complexes on thymic cortical epithelial cells are positively selected, and will eventually mature into CD4 or CD8 single-positive cells. T cells that react too strongly with self antigens are deleted in the thymus, a process driven most efficiently by bone marrow derived antigen-presenting cells in the medullary region of the thymus. Certain tissue-restricted proteins can be expressed in medullary epithelial cells as a result of the actions of the AIRE protein, and so can mediate negative selection in the thymus. The outcome of positive and negative selection is the generation of a mature T-cell repertoire that is both MHC-restricted and self-tolerant. Precisely how the recognition of self-peptide:self-MHC ligands by the T-cell receptor leads either to positive or negative selection remains an unsolved problem.

Survival and maturation of lymphocytes in peripheral lymphoid tissues.

Once B and T lymphocytes have completed their development in the central lymphoid tissues, they are carried in the blood to the peripheral lymphoid tissues. These tissues have a highly organized architecture, with distinct areas of B cells and T cells, and organization that is determined by interactions between lymphocytes and the other cell types that make up the lymphoid tissues. Before considering the factors governing the survival and maturation of newly formed lymphocytes in the periphery, we will briefly look at the development and organization of peripheral lymphoid tissues and the signals that guide lymphocytes to their correct locations. Normally, a lymphocyte will leave peripheral lymphoid tissue and recirculate via lymph and blood (see Section 1-15), continually reentering lymphoid tissues until antigen is encountered or the lymphocyte dies. When it meets its antigen, the lymphocyte stops recirculating, proliferates, and differentiates, as described in Chapters 9–11. When a lymphocyte dies, its place is taken by a newly formed lymphocyte; this enables a turnover of the receptor repertoire and ensures that lymphocyte numbers remain constant.

8-23 Different lymphocyte subsets are found in particular locations in peripheral lymphoid tissues.

The various peripheral lymphoid organs are organized roughly along the same lines, as discussed in Chapter 1, with distinct areas of B cells and T cells, and they also contain macrophages, dendritic cells, and non-leukocyte stromal cells. The lymphoid tissue of the spleen is the white pulp, whose overall architecture is illustrated in Fig. 1.19. Each area of white pulp is demarcated by a **marginal sinus**, a vascular network that branches from the central arteriole. The **marginal zone** of the white pulp, the outer border of which is the edge of the marginal sinus, is a highly organized region of cells whose function is poorly understood. It has few T cells but is rich in macrophages and contains a unique population of B cells, the **marginal zone B cells**, which do not recirculate. Pathogens reaching the bloodstream are efficiently trapped in the marginal zone by macrophages, and it could be that marginal zone B cells are adapted to provide the first responses to such pathogens.

The white pulp contains clearly separated areas of T cells and B cells. T cells are clustered around the central arteriole, and the globular B-cell areas or follicles are located farther out. Some follicles may contain **germinal centers**, in which B cells involved in an adaptive immune response are proliferating and undergoing somatic hypermutation (see Section 5-18). In follicles with germinal centers, the resting B cells that are not part of the immune response are pushed outward to make up the **mantle zone** around the proliferating lymphocytes. The antigen-driven production of germinal centers will be described in detail when we consider B-cell responses in Chapter 10.

Other types of cells are found within the B-cell and T-cell areas. The B-cell zone contains a network of **follicular dendritic cells** (**FDCs**), which are concentrated mainly in the area of the follicle most distant from the central arteriole. FDCs have long processes, from which they get their name, and these are in contact with B cells. FDCs are a distinct type of cell from the dendritic cells we have encountered previously (see Section 1-3), in that they are not leukocytes and are not derived from bone marrow precursors; in addition, they are not phagocytic and do not express MHC class II proteins. FDCs seem to be specialized to capture antigen in the form of immune complexes—complexes of antigen, antibody, and complement. The immune complexes are not internalized but remain intact on the surface of the FDC, where the antigen can be recognized by B cells. FDCs are also important in the development of B-cell follicles.

T-cell zones contain a network of bone marrow derived dendritic cells, sometimes known as **interdigitating dendritic cells** from the way in which their processes interweave among the T cells. There are two subtypes of these dendritic cells, distinguished by characteristic cell-surface proteins; one expresses the α chain of CD8, whereas the other is CD8 negative but expresses CD11b:CD18, an integrin that is also expressed by macrophages.

As in the spleen, the T cells and B cells in lymph nodes are organized into discrete T-cell and B-cell areas (see Fig. 1.18). B-cell follicles have a similar structure and composition to those in the spleen and are located just under the outer capsule of the lymph node. T-cell zones surround the follicles in the paracortical areas. Unlike the spleen, lymph nodes have connections to both the blood system and the lymphatic system. Lymph enters into the subcapsular space, which is also known as the marginal sinus, and brings in antigen and antigen-bearing dendritic cells from the tissues.

The **mucosa-associated lymphoid tissues** (**MALT**) are associated with the body's epithelial surfaces that provide physical barriers against infection. Peyer's patches are part of the MALT and are lymph node-like structures interspersed at intervals just beneath the gut epithelium. They have B-cell follicles and T-cell zones (see Fig. 1.20), and the epithelium overlying them contains specialized M cells that are adapted to channel antigens and pathogens from the gut lumen to the Peyer's patch (see Section 1-15 and Chapter 12). Peyer's patches and similar tissue present in the tonsils provide specialized sites where B cells can become committed to synthesizing IgA. The stromal cells of the MALT secrete the cytokine TGF-β, which has been shown to induce IgA secretion by B cells in culture. In addition, as discussed in Section 8-12, during fetal development waves of γ:δ T cells with specific γ- and δ-gene rearrangements leave the thymus and migrate to these epithelial barriers. The mucosal immune system is discussed in more detail in Chapter 12.

8-24 The development of peripheral lymphoid tissues is controlled by lymphoid tissue inducer cells and proteins of the tumor necrosis factor family.

Before discussing how lymphocytes find their way to their respective zones in the peripheral lymphoid organs, we shall briefly look at how these organs and the zones within them develop in the first place. Lymphatic vessels are formed during embryonic development from endothelial cells that originate in blood vessels. Some endothelial cells in the early venous system begin to express the homeobox transcription factor Prox1. These cells bud from the vein, migrate away, and reassociate to form a parallel network of lymphatic vessels. Mice lacking Prox1 have normal arteries and veins, but fail to form a lymphatic system, showing this factor to be critical in establishing the identity of lymphatic endothelium. As the lymphatic vessels form, blood-lineage cells called **lymphoid tissue inducer (LTi) cells** arise in the fetal liver and are carried in the bloodstream to sites of prospective lymph nodes and Peyer's patches. LTi cells initiate the formation of lymph nodes and Peyer's patches by interacting with stromal cells and inducing the production of cytokines and chemokines, which recruit other lymphoid cells to these sites. Members of the tumor necrosis factor (TNF)/TNF receptor (TNFR) family of cytokines turn out to be critically involved in the interactions between LTi cells and stromal cells.

The role of this family of cytokines in the formation of peripheral lymphoid organs has been demonstrated in a series of knockout mice in which either the TNF family ligand or its receptor was inactivated (Fig. 8.37). These knockouts have complicated phenotypes, which is partly due to the fact that individual TNF-family proteins can bind to multiple receptors and, conversely, many receptors can bind more than one protein. In addition, it seems clear that there is some overlapping function or cooperation between TNF-family proteins. Nonetheless, some general conclusions can be drawn.

Lymph-node development depends on the expression of TNF-family proteins known as the **lymphotoxins (LTs)**, and different types of lymph nodes depend on signals from different LTs. **LT-α_3**, a soluble homotrimer of the LT-α chain, supports the development of cervical and mesenteric lymph nodes, and possibly lumbar and sacral lymph nodes. All these lymph nodes drain mucosal sites. LT-α_3 probably exerts its effects by binding to TNFR-I. The membrane-bound heterotrimer comprising LT-α and the distinct protein LT-β (**LT-α_2:β_1**), often known as **LT-β**, binds only to the LT-β receptor and supports the development of all the other lymph nodes. Peyer's patches also do not form in the absence of LT-β. The effects of the LT knockouts are not reversible in adult animals and there are certain critical developmental periods during which the absence or inhibition of these LT-family proteins will permanently prevent the development of lymph nodes and Peyer's patches.

LTi cells express LT-β, which engages the LT-β receptors on stromal cells in the prospective lymphoid site, activating the NFκB pathway (see Section 7-22). This induces the stromal cells to express adhesion molecules and chemokines such as CXCL13 (B-lymphocyte chemokine, BLC), which in turn recruits more LTi cells, which have receptors for these molecules, eventually generating large clusters of cells that will become lymph nodes or Peyer's patches. The chemokines also attract cells such as lymphocytes and other hematopoietic-lineage cells with appropriate receptors to populate the forming lymphoid organ. The principles, and even some of the molecules, underlying the development of peripheral lymphoid organs in the fetus are

Fig. 8.37 Normal architecture of the peripheral lymphoid organs requires TNF family members and their receptors. The role of TNF family members in the development of peripheral lymphoid organs has been deduced mainly from the study of knockout mice deficient in one or more TNF-family ligands or receptors. Some receptors bind more than one ligand, and some ligands bind more than one receptor, complicating the effects of their deletion. (Note that receptors are named for the first ligand known to bind them.) The defects are organized here with respect to the two main receptors, TNFR-I and the LT-β receptor and their ligands, TNF-α and the lymphotoxins (LTs). In some cases, the loss of ligands that bind the same receptor leads to different phenotypes. This is due to the ability of the ligand to bind another receptor, as indicated in the figure. The LT-α protein chain contributes to two distinct ligands, the trimer LT-α_3 and the heterodimer LT-α_2:β_1, each of which acts through a distinct receptor. In general, signaling through the LT-β receptor is required for lymph node and FDC development and normal splenic architecture, whereas signaling through TNFR-I is also required for FDCs and normal splenic architecture but not for lymph-node development.

		Effects seen in knockout (KO) mice				
Receptor	Ligands	Spleen	Peripheral lymph node	Mesenteric lymph node	Peyer's patch	Follicular dendritic cells
TNFR-I	TNF-α LT-α_3	Distorted architecture	Present in TNF-α KO Absent in LT-α KO owing to lack of LT-β signals	Present	Reduced	Absent
LT-β receptor	TNF-α LT-α_2/β_1	Distorted No marginal zones	Absent	Present in LT-β KO Absent in LT-β receptor KO	Absent	Absent

very similar to those that maintain the organization of lymphoid organs in the adult, as we shall see in the next section.

Although the spleen develops in mice deficient in any of the known TNF or TNFR family members, its architecture is abnormal in many of these mutants (see Fig. 8.37). LT (most probably the membrane-bound LT-β) is required for the normal segregation of T-cell and B-cell zones in the spleen. TNF-α, binding to TNFR-I, also contributes to the organization of the white pulp: when TNF-α signals are disrupted, B cells surround T-cell zones in a ring rather than forming discrete follicles. In addition, the marginal zones are not well defined when TNF-α or its receptor is absent.

Perhaps the most important role of TNF-α and TNFR-I in lymphoid organ development is in the development of FDCs, as these cells are lacking in mice with knockouts of either TNF-α or TNFR-I (see Fig. 8.37). The knockout mice do have lymph nodes and Peyer's patches, because they express LTs, but these structures lack FDCs. LT-β is also required for FDC development: mice that cannot form LT-β or signal through it lack normal FDCs in the spleen and any residual lymph nodes. Unlike the disruption of lymph-node development, the disorganized lymphoid architecture in the spleen is reversible if the missing TNF-family member is restored. B cells are the likely source of the LT-β, because normal B cells can restore FDCs and follicles when transferred to RAG-deficient recipients (which lack lymphocytes).

8-25 The homing of lymphocytes to specific regions of peripheral lymphoid tissues is mediated by chemokines.

Newly formed lymphocytes enter the spleen via the blood, exiting first in the marginal sinus, from which they migrate to the appropriate areas of the white pulp. Lymphocytes that survive their passage through the spleen most probably leave via venous sinuses in the red pulp. In lymph nodes, lymphocytes enter from the blood through the walls of specialized blood vessels, the high endothelial venules (HEVs), which are located within the T-cell zones. Naive B cells migrate through the HEVs in the T-cell area and come to rest in the follicle where, unless they encounter their specific antigen and become activated, they remain for about a day. B cells and T cells leave in the lymph via the efferent lymphatic, which returns them eventually to the blood. The precise location of B cells, T cells, macrophages, and dendritic cells in peripheral lymphoid tissue is controlled by chemokines, which are produced by both stromal cells and bone marrow derived cells (Fig. 8.38).

B cells constitutively express the chemokine receptor CXCR5 and are attracted to the follicles by the ligand for this receptor, the chemokine CXCL13. The

Fig. 8.38 The organization of a lymphoid organ is orchestrated by chemokines. The cellular organization of lymphoid organs is initiated by stromal cells and vascular endothelial cells, which secrete the chemokine CCL21 (first panel). Dendritic cells with a receptor for CCL21, CCR7, are attracted to the site of the developing lymph node by CCL21 (second panel); it is not known whether at the earliest stages of lymph node development immature dendritic cells enter from the bloodstream or via the lymphatics, as they do later in life. Once in the lymph node, the dendritic cells express the chemokines CCL18 (also called DC-CK1) and CCL19, for which T cells express receptors. Together, the chemokines secreted by stromal cells and dendritic cells attract T cells to the developing lymph node (third panel). The same combination of chemokines also attracts B cells into the developing lymph node (fourth panel). The B cells are able to either induce the differentiation of the non-leukocyte FDCs (which are a distinct lineage from the bone marrow derived dendritic cells) or direct their recruitment into the lymph node. Once present, the FDCs secrete a chemokine, CXCL13, which is a chemoattractant for B cells. The production of CXCL13 drives the organization of B cells into discrete B-cell areas (follicles) around the FDCs and contributes to the further recruitment of B cells from the circulation into the lymph node (fifth panel).

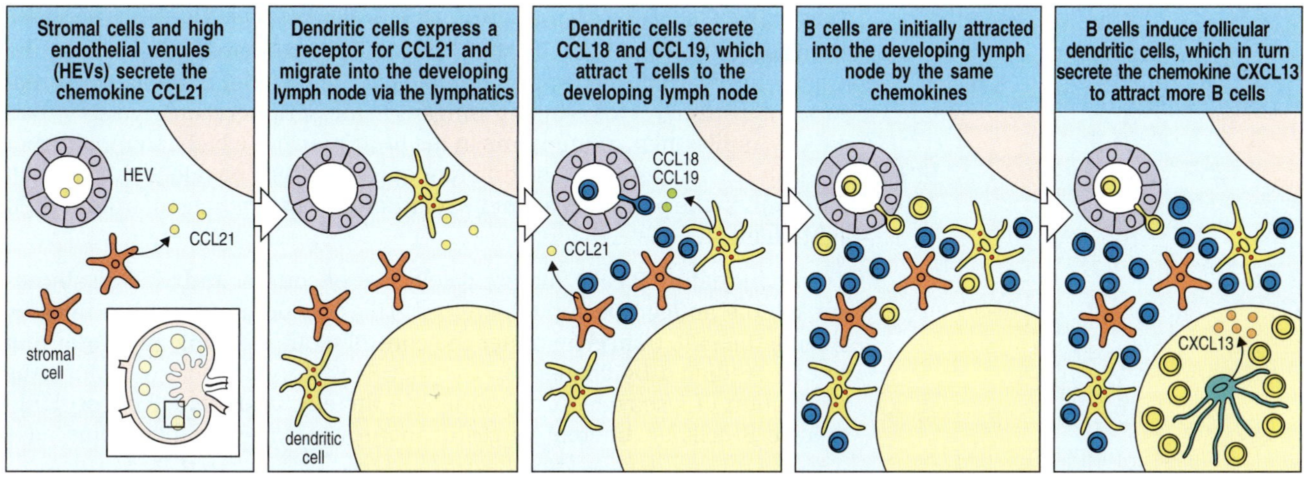
Stromal cells and high endothelial venules (HEVs) secrete the chemokine CCL21
HEV
CCL21
stromal cell
Dendritic cells express a receptor for CCL21 and migrate into the developing lymph node via the lymphatics
dendritic cell
Dendritic cells secrete CCL18 and CCL19, which attract T cells to the developing lymph node
CCL18
CCL19
CCL21
B cells are initially attracted into the developing lymph node by the same chemokines
B cells induce follicular dendritic cells, which in turn secrete the chemokine CXCL13 to attract more B cells
CXCL13

most likely source of CXCL13 is the FDC, possibly along with other follicular stromal cells. This is reminiscent of the expression of CXCL13 by stromal cells during the formation of the lymph node. B cells are, in turn, the source of the LT that is required for the development of FDCs, which is reminiscent of LTi cells expressing the LT required to activate stromal cells. The reciprocal dependence of B cells and FDCs, and LTis and stromal cells, illustrates the complex web of interactions that organizes peripheral lymphoid tissues. T cells can also express CXCR5, although at a lower level, and this may explain how T cells are able to enter B-cell follicles, which they do when activated, to participate in the formation of the germinal center.

Movie 8.2

T-cell localization to the T zones involves two chemokines, CCL19 (MIP-3β) and CCL21 (secondary lymphoid chemokine, SLC). Both of these bind the receptor CCR7, which is expressed by T cells; mice that lack CCR7 do not form normal T-cell zones and have impaired primary immune responses. CCL21 is produced by stromal cells of the T-cell zone in spleen, and by the endothelial cells of HEVs in lymph nodes and Peyer's patches. Another source of CCL19 and CCL21 is the interdigitating dendritic cells, which are also prominent in the T zones. Indeed, dendritic cells themselves express CCR7 and will localize to T zones even in RAG-deficient mice. Thus, in lymph-node development, the T zone might be organized first through the attraction of dendritic cells and T cells by CCL21 produced by stromal cells. This organization would then be reinforced by CCL21 and CCL19 secreted by resident mature dendritic cells, which in turn attract more T cells and immature dendritic cells.

8-26 Lymphocytes that encounter sufficient quantities of self antigens for the first time in the periphery are eliminated or inactivated.

Large numbers of autoreactive lymphocytes have been purged from the population of new lymphocytes in the central lymphoid organs (discussed in Section 8-20); however, this only involves lymphocytes specific for autoantigens that are expressed in or can reach these organs. Not all potential self antigens are expressed in the central lymphoid organs. Some, like the thyroid product thyroglobulin, are highly tissue specific, or are compartmentalized so that little if any is available in the circulation. Therefore, newly emigrated self-reactive lymphocytes that encounter their specific autoantigen for the first time in the periphery must be eliminated or inactivated. This tolerance mechanism is known as peripheral tolerance. Like self-reactive lymphocytes in the central lymphoid organs, lymphocytes that encounter self antigens *de novo* in the periphery can have several fates: deletion, anergy, or survival.

In the absence of an infection, newly mature B cells that encounter a strongly cross-linking antigen in the periphery will undergo clonal deletion. This was elegantly shown in studies of B cells expressing B-cell receptors specific for H-2K^b MHC class I molecules. These cells are deleted even when, in transgenic animals, the expression of the H-2K^b molecule is restricted to the liver by the use of a liver-specific gene promoter. There is no receptor editing: B cells that encounter strongly cross-linking antigens in the periphery undergo apoptosis directly, unlike their counterparts in the bone marrow, which attempt further receptor rearrangements. The different outcomes may be due to the fact that the B cells in the periphery are more mature and can no longer rearrange their light-chain loci.

As with immature B cells, mature B cells that encounter and bind an abundant soluble antigen become anergized. This was demonstrated in mice by placing the *HEL* transgene under the control of an inducible promoter that can be regulated by changes in the diet. It is thus possible to induce the production of lysozyme at any time and thereby study its effects on HEL-specific B cells at different stages of maturation. These experiments have shown that both mature and immature B cells are inactivated when they are chronically exposed to soluble antigen.

The situation is similar for T cells. Again, our understanding of the fates of autoreactive T cells in the periphery comes mainly from the study of mice transgenic for self-reactive T-cell receptors. In some cases, T cells reacting to self antigens in the periphery are eliminated. This usually follows a brief period of activation and cell division, and so is known as **activation-induced cell death**. In other cases, the self-reactive cells may be rendered anergic. When studied *in vitro*, these anergic T cells prove refractory to signals delivered through the T-cell receptor.

The question immediately arises: if the encounter of a mature naive lymphocyte with a self antigen leads to cell death or anergy, why does this not also happen to a mature lymphocyte that recognizes a pathogen-derived antigen? The answer is that infection sets up inflammation, which induces the expression of co-stimulatory molecules on the antigen-presenting dendritic cells and the production of cytokines promoting lymphocyte activation. The outcome of an encounter with antigen in these conditions is the activation, proliferation, and differentiation of the lymphocyte to effector-cell status. In the absence of infection or inflammation, dendritic cells still process and present self antigens, but in the absence of co-stimulatory and other signals, any interaction of a mature lymphocyte with its specific antigen seems to result in a tolerance-inducing (**tolerogenic**) signal from the antigen receptor.

8-27 Immature B cells arriving in the spleen turn over rapidly and require cytokines and positive signals through the B-cell receptor for maturation and survival.

When B cells emerge from bone marrow into the periphery, they are still functionally immature. Immature B cells express high levels of sIgM but little sIgD, whereas mature B cells express low levels of IgM and high levels of IgD. Most immature B cells leaving the bone marrow will not survive to become fully mature B cells. Figure 8.39 shows the possible fates of newly produced B cells

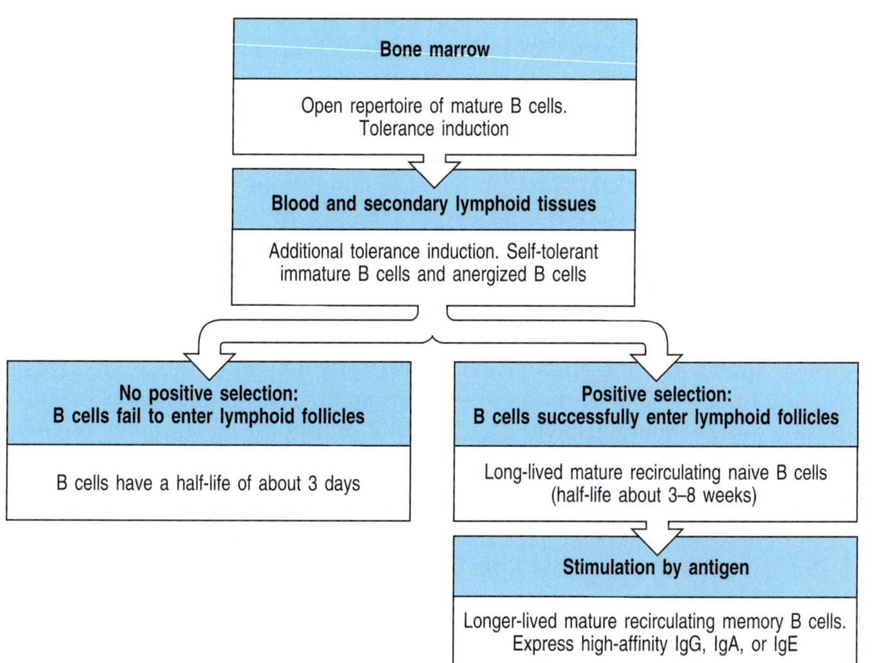

Fig. 8.39 Proposed population dynamics of conventional B cells. B cells are produced as receptor-positive immature B cells in the bone marrow. The most avidly self-reactive B cells are removed at this stage. B cells then migrate to the periphery, where they enter the peripheral lymphoid tissues. It is estimated that $10–20 \times 10^6$ B cells are produced by the bone marrow and exported each day in a mouse, and an equal number are lost from the periphery. There seem to be two classes of peripheral B cells: long-lived B cells and rapidly turning-over B cells. The short-lived B cells are, by definition, recently formed B cells. Most of the turnover of short-lived B cells might result from B cells that fail to enter lymphoid follicles. In some cases this is a consequence of being rendered anergic by binding to soluble self antigen; for the remaining immature B cells, entry into lymphoid follicles is thought to entail some form of positive selection. Thus, the remainder of the short-lived B cells fail to join the long-lived pool because they are not positively selected. About 90% of all peripheral B cells are relatively long-lived mature B cells that seem to have undergone positive selection in the periphery. These mature naive B cells recirculate through peripheral lymphoid tissues and have a half-life of 6–8 weeks in mice. Memory B cells, which have been activated previously by antigen and T cells, are thought to have a longer life.

that enter the periphery. The daily output of new B cells from the bone marrow is roughly 5–10% of the total B-lymphocyte population in the steady-state peripheral pool. The size of this pool seems to remain constant in unimmunized animals, a regulatory process known as homeostasis, and so the stream of new B cells needs to be balanced by the removal of an equal number of peripheral B cells. However, the majority of peripheral mature B cells are long-lived, and only 1–2% of these die each day. Most of the B cells that die are in the rapidly turning-over immature peripheral B-cell population, of which more than 50% die every 3 days. The failure of most newly formed B cells to survive for more than a few days in the periphery seems to be due to competition between peripheral B cells for access to the follicles in peripheral lymphoid tissues. If newly produced immature B cells do not enter a follicle, their passage through the periphery is halted and they eventually die. The limited number of lymphoid follicles cannot accommodate all of the B cells generated each day and so there is continual competition for entry.

The follicle provides signals necessary for B-cell survival. In particular, the TNF-family member **BAFF** (for B-cell activating factor belonging to the TNF family) is made by several cell types, but is produced abundantly by the FDCs. B cells express three different receptors for BAFF, namely BAFF-R, BCMA, and TACI. The BAFF-R is the most important for follicular B-cell survival, because mutants lacking BAFF-R have mainly immature B cells and few long-lived peripheral B cells. BCMA and TACI also bind a related TNF family cytokine APRIL, which is not required for the survival of immature B cells but is important for IgA antibody production, as we shall see in Chapter 10. Immature B cells proceed through two defined **transitional stages** in the spleen, called T1 and T2, defined by the absence or presence of the B-cell co-receptor component CD21 (complement receptor 2) (see Section 2-13). In mice lacking BAFF, immature B cells progress to the T1 stage in the spleen but fail to express CD21, and the mice lack mature B cells.

Peripheral B cells also include memory B cells, which are produced in addition to antibody-producing plasma cells from mature B cells after their first encounter with antigen; we will return to B-cell memory in Chapter 11. Competition for follicular entry favors mature B cells that are already established in the relatively long-lived and stable peripheral B-cell pool. Mature B cells have undergone phenotypic changes that might make their access to the follicles easier; for example, they express CXCR5, the receptor for CXCL13, which is expressed by FDCs (see Fig. 8.38). They also have increased expression of CD21, which enhances the signaling capacity of the B cell.

The B-cell receptor plays a positive role in the maturation and continued recirculation of peripheral B cells. Continuous expression of the B-cell receptor is required for B-cell survival, because conditional gene deletion of the B-cell receptor in mature B cells leads to the loss of these B cells. Mice that lack the tyrosine kinase Syk, which is involved in signaling from the B-cell receptor (see Section 7-16), have immature B cells but fail to develop mature B cells. Thus, a Syk-transduced signal may be required for final B-cell maturation or for the survival of mature B cells. Although each B-cell receptor has a unique specificity, antigen-specific interactions may not induce the signals used for final B-cell maturation; the receptor could, for example, be responsible for 'tonic' signaling, in which a weak but important signal is generated by the assembly of the receptor complex and infrequently triggers some or all of the downstream signaling events.

8-28 B-1 cells and marginal zone B cells are distinct B-cell subtypes with unique antigen receptor specificity.

The receptor specificity is important in shaping the peripheral B-cell pools that derive from immature B cells that reach the spleen. This is shown most clearly in the role of the B-cell receptor and antigen in the selection of two subsets of B cells that do not reside in B-cell follicles: the so-called **B-1 cells** and the marginal zone B cells of the spleen.

B-1 cells are a unique subset comprising about 5% of all B cells in mice and humans, and are the major population in rabbits. B-1 cells have high levels of sIgM and little sIgD, opposite to the pattern of follicular B cells. They are called B-1 because they are the first to appear during fetal development (Fig. 8.40), before the conventional B cells whose development we have discussed up to now—which are called **B-2 cells**. B-1 cells are found primarily in the peritoneal and pleural cavity fluid. Certain autoantigens and environmental antigens encountered in the periphery are thought to drive the expansion and maintenance of B-1 cells. Some of these antigens, such as phosphocholine, are encountered on the surface of bacteria that colonize the gut.

It is not clear whether B-1 cells arise as a distinct lineage from a unique precursor cell or differentiate to the B-1 phenotype from a precursor cell that could also give rise to B-2 cells. In the mouse, fetal liver produces mainly B-1 cells, whereas adult bone marrow generates predominantly B-2 cells, and this has been interpreted as support for the unique precursor hypothesis. However, the weight of evidence favors the idea that commitment to the B-1 or B-2 subset is due to a selection step, rather than their being distinct lineages like the γ:δ and α:β T cells.

Marginal zone B cells, so called because they reside in the marginal sinus of the white pulp in the spleen, are another unique subset of B cells. They seem to be resting mature B cells, yet they have a different set of surface proteins from those present on the major follicular B-cell population. For example, they express lower levels of CD23 (the low-affinity receptor for IgE), and high

Property	B-1 cells	Conventional B-2 cells	Marginal zone B cells
When first produced	Fetus	After birth	After birth
N-regions in VDJ junctions	Few	Extensive	Yes
V-region repertoire	Restricted	Diverse	Partly restricted
Primary location	Body cavities (peritoneal, pleural)	Secondary lymphoid organs	Spleen
Mode of renewal	Self-renewing	Replaced from bone marrow	Long-lived
Spontaneous production of immunoglobulin	High	Low	Low
Isotypes secreted	IgM >> IgG	IgG > IgM	IgM > IgG
Response to carbohydrate antigen	Yes	Maybe	Yes
Response to protein antigen	Maybe	Yes	Yes
Requirement for T-cell help	No	Yes	Sometimes
Somatic hypermutation	Low to none	High	?
Memory development	Little or none	Yes	?

Fig. 8.40 A comparison of the properties of B-1 cells, conventional B cells (B-2 cells), and marginal zone B cells. B-1 cells can develop in unusual sites in the fetus, such as the omentum, in addition to the liver. B-1 cells predominate in the young animal, although they probably can be produced throughout life. Being produced mainly during fetal and neonatal life, their rearranged variable-region sequences contain few N-nucleotides. In contrast, marginal zone B cells accumulate after birth and do not reach peak levels in the mouse until 8 weeks of age. B-1 cells are best thought of as a partly activated self-renewing pool of lymphocytes that are selected by ubiquitous self and foreign antigens. Because of this selection, and possibly because the cells are produced early in life, the B-1 cells have a restricted repertoire of variable regions and antigen-binding specificities. Marginal zone B cells also have a restricted repertoire that may be selected by a set of antigens similar to those that select B-1 cells. B-1 cells seem to be the major population of B cells in certain body cavities, most probably because of exposure at these sites to antigens that drive B-1 cell proliferation. Marginal zone B cells remain in the marginal zone of the spleen and are not thought to recirculate. Partial activation of B-1 cells leads to the secretion of mainly IgM antibody; B-1 cells contribute much of the IgM that circulates in the blood. The limited diversity of both the B-1 and marginal zone B-cell repertoire and the propensity of these cells to react with common bacterial carbohydrate antigens suggest that they carry out a more primitive, less adaptive, immune response than conventional B cells (B-2 cells). In this regard they are comparable to γ:δ T cells.

levels of both the MHC class I-like molecule CD1 (see Section 6-19) and two receptors for the C3 fragment of complement, CR1 (CD35) and CR2 (CD21). Marginal zone B cells have restricted antigen specificities, biased toward self antigens and common bacterial antigens, and may be adapted to provide a quick response if the latter enter the bloodstream. They may not require T-cell help to become activated. Functionally and phenotypically, marginal zone B cells resemble B-1 cells; recent experiments suggest that they are positively selected for survival by certain self antigens, much as B-1 cells are.

The functions of B-1 cells and marginal zone B cells are being clarified. Their locations suggest a role for B-1 cells in defending the body cavities and a role for marginal zone B cells in defense against bacteria that penetrate the bloodstream. The restricted repertoire of receptors in both cell types seems to equip them for a function in the early, nonadaptive phase of an immune response (see Section 3-24). Indeed, the V gene segments that are used to encode the receptors of B-1 and marginal zone B cells might have evolved by natural selection to recognize common bacterial antigens, thus allowing them to contribute to the very early phases of the adaptive immune response. In practice, it is found that B-1 cells make little contribution to adaptive immune responses to most protein antigens, but contribute strongly to some antibody responses against carbohydrate antigens. Moreover, a large proportion of the IgM that normally circulates in the blood of unimmunized mice derives from B-1 cells. The existence of these so-called **natural antibodies**, which are highly cross-reactive and bind with low affinity to both microbial and self antigens, supports the view that B-1 cells are partly activated because they are selected for self-renewal by ubiquitous self and environmental antigens.

8-29 T-cell homeostasis in the periphery is regulated by cytokines and self-MHC interactions.

When T cells have expressed their receptors and co-receptors, and matured within the thymus for a further week or so, they emigrate to the periphery. Unlike B cells emigrating from bone marrow, only relatively small numbers of T cells are exported from the thymus, roughly $1–2 \times 10^6$ per day in the mouse. As with B cells, the size and composition of the peripheral pool of naive T cells are also regulated by homeostatic mechanisms that maintain it at a roughly constant size and a composition of diverse but potentially functional T-cell receptors. These mechanisms involve both cytokines and signals received through the T-cell receptor in response to its interaction with self-MHC molecules.

A requirement for the cytokine IL-7 and interactions with self-peptide:self-MHC complexes for T-cell survival in the periphery has been shown experimentally. If T cells are transferred from their normal environment to recipients lacking MHC molecules, or lacking the 'correct' MHC molecules that originally selected the T cells, they do not survive long. In contrast, if T cells are transferred into recipients that have the correct MHC molecules, they survive. Contact with the appropriate self-peptide:self-MHC complex as they circulate through peripheral lymphoid organs leads mature naive T cells to undergo infrequent cell division. This slow increase in T-cell numbers must be balanced by a slow loss of T cells, such that the number of T cells remains roughly constant. Most probably, this loss occurs among the daughters of the dividing naive T cells.

Where do the mature naive CD4 and CD8 T cells encounter their positively selecting ligands? Current evidence favors self-MHC molecules on dendritic cells resident in the T-cell zones of peripheral lymphoid tissues. These cells are similar to the dendritic cells that migrate to the lymph nodes from other tissues but lack sufficient co-stimulatory potential to induce full T-cell activation. The study of peripheral positive selection is in its infancy, however, and a clear picture has yet to emerge. Memory T cells are also part of the peripheral T-cell pool, and we return to their regulation in Chapter 11.

Summary.

The formation and organization of the peripheral lymphoid tissues is controlled by proteins of the TNF family and their receptors (TNFRs). LTi cells expressing LT-β interact with stromal cells expressing the receptor TNFR-I in the developing embryo to induce chemokine production, which in turn initiates formation of the lymph nodes and Peyer's patches. Similar interactions between lymphotoxin-expressing B cells and TNFR-I-expressing follicular dendritic cells (FDCs) establishes the normal architecture of the spleen and lymph nodes. The homing of B and T cells to distinct areas of lymphoid tissue involves attraction by specific chemokines. B and T lymphocytes that survive selection in the bone marrow and thymus are exported to the peripheral lymphoid organs. Most of the newly formed B cells that emigrate from the bone marrow die soon after their arrival in the periphery, thus keeping the number of circulating B cells fairly constant. A small number mature and become longer-lived naive B cells. T cells leave the thymus as fully mature cells and are produced in smaller numbers than B cells. The fate of mature lymphocytes in the periphery is still controlled by their antigen receptors. In the absence of an encounter with their specific foreign antigen, naive lymphocytes require some tonic signaling through their antigen receptors for long-term survival.

T cells are generally long-lived and are thought to be slowly self-renewing in the peripheral lymphoid tissues, being maintained by repeated contacts with self-peptide:self-MHC complexes that can be recognized by the T-cell receptor but do not cause T-cell activation, in combination with signals derived from IL-7. The evidence for receptor-mediated survival signals is clearest for T cells, but they also seem to be needed for B-1 cells and marginal zone B cells, in which case they may promote differentiation, expansion, and survival, and most probably also for B-2 cells, in which case they promote survival without expansion. The lymphoid follicle, through which B cells must circulate to survive, seems to provide signals for their maturation and survival. A few ligands that select B-1 and marginal zone B cells are known, but in general the ligands involved in B-cell selection are unknown. The distinct minority subpopulations of lymphocytes, such as the B-1 cells, marginal zone B cells, γ:δ T cells, and iNKT cells, have different developmental histories and functional properties from those of conventional B-2 cells and α:β T cells and are likely to be regulated independently of these majority B-cell and T-cell populations.

Summary to Chapter 8.

In this chapter we have learned about the formation of the B-cell and T-cell lineages from an uncommitted hematopoietic stem cell. The somatic gene rearrangements that generate the highly diverse repertoire of antigen receptors—immunoglobulin for B cells, and the T-cell receptor for T cells—occur in the early stages of the development of T cells and B cells from a common bone marrow derived lymphoid progenitor. Mammalian B-cell development takes place in fetal liver and, after birth, in the bone marrow; T cells also originate in the bone marrow but undergo most of their development in the thymus. Much of the somatic recombination machinery, including the RAG proteins that are an essential part of the V(D)J recombinase, is common to both B and T cells. In both B and T cells, gene rearrangements begin with the loci that contain D gene segments, and proceed successively at each locus. The first step in B-cell development is the rearrangement of the locus for the immunoglobulin heavy chain, and for T cells the β chain. In each case, the developing cell is allowed to proceed to the next stage of development only if the rearrangement has produced an in-frame sequence that can be translated into a protein expressed on the cell surface: either the pre-B-cell receptor or the pre-T cell receptor. Cells that do not generate successful rearrangements for both receptor chains die by apoptosis. The course of conventional B-cell development is summarized in Fig. 8.45, and that of α:β T cells in Fig. 8.46.

Once a functional antigen receptor has appeared on the cell surface, the lymphocyte is tested in two ways. Positive selection tests for the potential usefulness of the antigen receptor, whereas negative selection removes self-reactive cells from the lymphocyte repertoire, rendering it tolerant to the antigens of the body. Positive selection is particularly important for T cells, because it ensures that only cells bearing T-cell receptors that can recognize antigen in combination with self-MHC molecules will continue to mature. Positive selection also coordinates the choice of co-receptor expression. CD4 becomes expressed by T cells harboring MHC class II restricted receptors, and CD8 by cells harboring MHC class I restricted receptors. This ensures the optimal use of these receptors in responses to pathogens. For B cells, positive selection seems to occur at the final transition from immature to mature B cells, which occurs in peripheral lymphoid tissues. Tolerance to self antigens is enforced by negative selection at different stages throughout the development of both B and T cells, and positive selection likewise seems to represent a continuous process.

B and T cells surviving development in the central lymphoid organs emigrate to the periphery, where they home to occupy specific sites. The formation of peripheral lymphoid organs begins during embryonic development by the interaction of lymphoid tissue inducer cells expressing TNF family cytokines with stromal cells. The organization of the peripheral lymphoid organs, such as spleen and lymph nodes, also involves interactions between cells expressing TNF and TNFR family proteins. The homing of B and T cells to different parts of these peripheral tissues involves their expression of distinct chemokine receptors and the secretion of specific chemokines by various stromal elements. Maturation and survival of B and T lymphocytes in these peripheral tissues involves other specific factors. Naive B cells receive survival signals in the follicle through interaction with BAFF. Naive T cells require the cytokines IL-7 and IL-15 for survival, along with signals received through the T-cell receptor interacting with self-MHC molecules.

Stage	B cells		Heavy-chain genes	Light-chain genes	Intracellular proteins	Surface marker proteins	Location
ANTIGEN INDEPENDENT	Stem cell		Germline	Germline		CD34 CD45 AA4.1	BONE MARROW
ANTIGEN INDEPENDENT	Early pro-B cell		D–J rearranged	Germline	RAG-1 RAG-2 TdT λ5, VpreB	CD34 CD45R AA4.1, IL-7R MHC class II CD10, CD19 CD38	BONE MARROW
ANTIGEN INDEPENDENT	Late pro-B cell		V–DJ rearranged	Germline	TdT λ5, VpreB	CD45R AA4.1, IL-7R MHC class II CD10, CD19 CD38, CD20 CD40	BONE MARROW
ANTIGEN INDEPENDENT	Large pre-B cell	pre-B receptor	VDJ rearranged	Germline	λ5, VpreB	CD45R AA4.1, IL-7R MHC class II pre-B-R CD19, CD38 CD20, CD40	BONE MARROW
ANTIGEN INDEPENDENT	Small pre-B cell	cytoplasmic μ	VDJ rearranged	V–J rearrangement	μ RAG-1 RAG-2	CD45R AA4.1 MHC class II CD19, CD38 CD20, CD40	BONE MARROW
ANTIGEN DEPENDENT	Immature B cell	IgM	VDJ rearranged. μ heavy chain produced in membrane form	VJ rearranged		CD45R AA4.1 MHC class II IgM CD19, CD20 CD40	BONE MARROW
ANTIGEN DEPENDENT	Mature naive B cell	IgD IgM	VDJ rearranged. μ chain produced in membrane form. Alternative splicing yields μ + δ mRNA			CD45R MHC class II IgM, IgD CD19, CD20 CD21, CD40	PERIPHERY
ANTIGEN DEPENDENT	Lympho-blast	IgM	Alternative splicing yields secreted μ chains		Ig	CD45R MHC class II CD19, CD20 CD21, CD40	PERIPHERY
ANTIGEN DEPENDENT	Memory B cell	IgG	Isotype switch to Cγ, Cα, or Cε. Somatic hypermutation	Somatic hypermutation		CD45R MHC class II IgG, IgA CD19, CD20 CD21, CD40	PERIPHERY
TERMINAL DIFFERENTIATION	Plasma blast and plasma cell	IgG	Alternative splicing yields both membrane and secreted Ig	VJ rearranged	Ig	CD135 Plasma cell antigen-1 CD38	PERIPHERY

Fig. 8.45 A summary of the development of human conventional B-lineage cells. The state of the immunoglobulin genes, the expression of some essential intracellular proteins, and the expression of some cell-surface molecules are shown for successive stages of B-2-cell development. The immunoglobulin genes undergo further changes during antigen-driven B-cell differentiation, such as class switching and somatic hypermutation (see Chapter 5), which are evident in the immunoglobulins produced by memory cells and plasma cells. These antigen-dependent stages are described in more detail in Chapter 9.

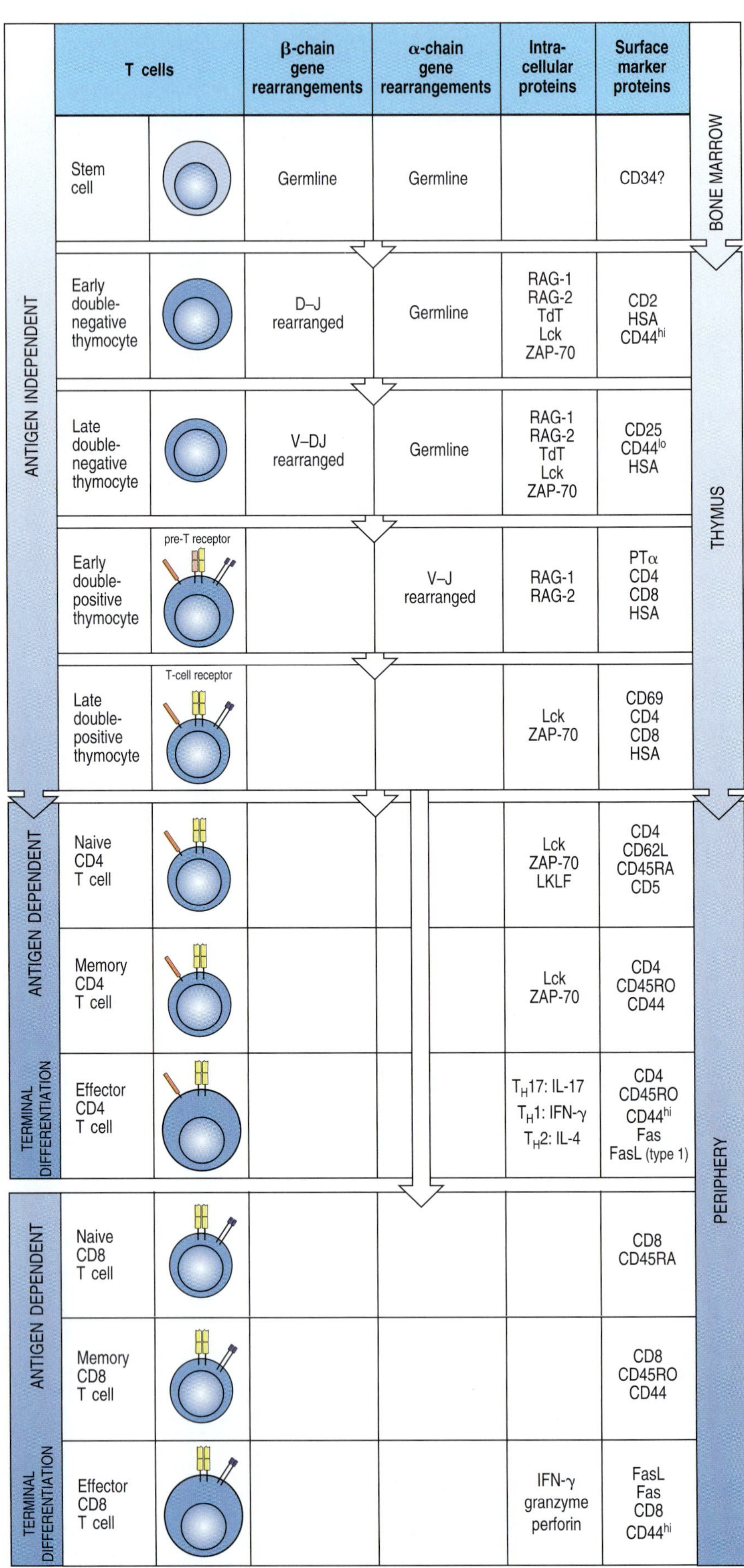

T cells	β-chain gene rearrangements	α-chain gene rearrangements	Intra-cellular proteins	Surface marker proteins
Stem cell	Germline	Germline		CD34?
Early double-negative thymocyte	D–J rearranged	Germline	RAG-1 RAG-2 TdT Lck ZAP-70	CD2 HSA CD44^{hi}
Late double-negative thymocyte	V–DJ rearranged	Germline	RAG-1 RAG-2 TdT Lck ZAP-70	CD25 CD44^{lo} HSA
Early double-positive thymocyte		V–J rearranged	RAG-1 RAG-2	PTα CD4 CD8 HSA
Late double-positive thymocyte			Lck ZAP-70	CD69 CD4 CD8 HSA
Naive CD4 T cell			Lck ZAP-70 LKLF	CD4 CD62L CD45RA CD5
Memory CD4 T cell			Lck ZAP-70	CD4 CD45RO CD44
Effector CD4 T cell			T_H17: IL-17 T_H1: IFN-γ T_H2: IL-4	CD4 CD45RO CD44^{hi} Fas FasL (type 1)
Naive CD8 T cell				CD8 CD45RA
Memory CD8 T cell				CD8 CD45RO CD44
Effector CD8 T cell			IFN-γ granzyme perforin	FasL Fas CD8 CD44^{hi}

Fig. 8.46 A summary of the development of human α:β T cells. The state of the T-cell receptor genes, the expression of some essential intracellular proteins, and the expression of some cell-surface molecules are shown for successive stages of α:β T-cell development. Note that because the T-cell receptor genes do not undergo further changes during antigen-driven development, only the phases during which they are actively undergoing rearrangement in the thymus are indicated. The antigen-dependent phases of CD4 and CD8 cells are depicted separately, and are detailed in Chapter 9.

Questions.

8.1 *B-cell development in the bone marrow shares many features with T-cell development in the thymus. (a) What are the two major goals of lymphocyte development? (b) Discuss the ordered steps of receptor rearrangement in B and T cells, drawing the parallels between the two cell types. (c) What is the function of the pre-B-cell receptor and the pre-T-cell receptor? (d) Why do T cells develop in the thymus and B cells develop in the bone marrow?*

8.2 *Lymphocyte development is notable for huge cell losses at several steps. (a) What are the major reasons that lymphocytes die without progressing beyond the pre-T-cell or pre-B-cell stage? (b) What is the major reason that lymphocytes die after reaching the immature stage of expressing a complete TCR or BCR?*

8.3 *Discuss the process of positive selection of T cells in the thymus. (a) Where does it take place? (b) What are the ligands? (c) When (at what stage) during T-cell development does positive selection occur? (d) Describe how the choice between expression of the co-receptor—CD4 or CD8—occurs, and identify any known regulators of this process.*

8.4 *Peripheral lymphoid tissues become organized through communication between several kinds of cells and several kinds of receptor interactions. (a) What families of molecules are critical for the proper organization of peripheral lymphoid tissues? (b) Which are important for organizing the B-cell zones? (c) Which are important for organizing the T-cell zone?*

8.5 *There are three main subsets of B cells: follicular, marginal zone, and B-1. Compare and contrast their development and functions, covering at least five different categories.*

General references.

Loffert, D., Schaal, S., Ehlich, A., Hardy, R.R., Zou, Y.R., Muller, W., and Rajewsky, K.: **Early B-cell development in the mouse—insights from mutations introduced by gene targeting.** *Immunol. Rev.* 1994, **137**:135–153.

Melchers, F., ten Boekel, E., Seidl, T., Kong, X.C., Yamagami, T., Onishi, K., Shimizu, T., Rolink, A.G., and Andersson, J.: **Repertoire selection by pre-B-cell receptors and B-cell receptors, and genetic control of B-cell development from immature to mature B cells.** *Immunol. Rev.* 2000, **175**:33–46.

Starr, T.K., Jameson, S.C., and Hogquist, K.A.: **Positive and negative selection of T cells.** *Annu. Rev. Immunol.* 2003, **21**:139–176.

von Boehmer, H.: **The developmental biology of T lymphocytes.** *Annu. Rev. Immunol.* 1993, **6**:309–326.

Weinberg, R.A.: *The Biology of Cancer*, 1st ed. New York:Garland Science, 2007.

Section references.

8-1 Lymphocytes derive from hematopoietic stem cells in the bone marrow.

Busslinger, M.: **Transcriptional control of early B cell development.** *Annu. Rev. Immunol.* 2004, **22**:55–79.

Chao, M.P., Seita, J., and Weissman, I.L.: **Establishment of a normal hematopoietic and leukemia stem cell hierarchy.** *Cold Spring Harbor Symp. Quant. Biol.* 2008, **73**:439–449.

Funk, P.E., Kincade, P.W., and Witte, P.L.: **Native associations of early hematopoietic stem-cells and stromal cells isolated in bone-marrow cell aggregates.** *Blood* 1994, **83**:361–369.

Jacobsen, K., Kravitz, J., Kincade, P.W., and Osmond, D.G.: **Adhesion receptors on bone-marrow stromal cells—in vivo expression of vascular cell adhesion molecule-1 by reticular cells and sinusoidal endothelium in normal and γ-irradiated mice.** *Blood* 1996, **87**:73–82.

Kiel, M.J., and Morrison, S.J.: **Uncertainty in the niches that maintain haematopoietic stem cells.** *Nat. Rev. Immunol.* 2008, **8**:290–301.

8-2 B-cell development begins by rearrangement of the heavy-chain locus.

Allman, D., Li, J., and Hardy, R.R.: **Commitment to the B lymphoid lineage occurs before DH-JH recombination.** *J. Exp. Med.* 1999, **189**:735–740.

Allman, D., Lindsley, R.C., DeMuth, W., Rudd, K., Shinton, S.A., and Hardy, R.R.: **Resolution of three nonproliferative immature splenic B cell subsets reveals multiple selection points during peripheral B cell maturation.** *J. Immunol.* 2001, **167**:6834–6840.

Hardy, R.R., Carmack, C.E., Shinton, S.A., Kemp, J.D., and Hayakawa, K.: **Resolution and characterization of pro-B and pre-pro-B cell stages in normal mouse bone marrow.** *J. Exp. Med.* 1991, **173**:1213–1225.

Osmond, D.G., Rolink, A., and Melchers, F.: **Murine B lymphopoiesis: towards a unified model.** *Immunol. Today* 1998, **19**:65–68.

8-3 The pre-B-cell receptor tests for successful production of a complete heavy chain and signals for the transition from the pro-B cell to pre-B cell stage.

Bankovich, A.J., Raunser, S., Juo, Z.S., Walz, T., Davis, M.M., and Garcia, K.C., **Structural insight into pre-B cell receptor function.** *Science* 2007, **316**:291–294.

Grawunder, U., Leu, T.M.J., Schatz, D.G., Werner, A., Rolink, A.G., Melchers, F., and Winkler, T.H.: **Down-regulation of Rag1 and Rag2 gene expression in pre-B cells after functional immunoglobulin heavy-chain rearrangement.** *Immunity* 1995, **3**:601–608.

Monroe, J.G.: **ITAM-mediated tonic signalling through pre-BCR and BCR complexes.** *Nat. Rev. Immunol.* 2006, **6**:283–294.

8-4 Pre-B-cell receptor signaling inhibits further heavy-chain locus rearrangement and enforces allelic exclusion.

Geier, J.K., and Schlissel, M.S.: **Pre-BCR signals and the control of Ig gene rearrangements.** *Semin. Immunol.* 2006, **18**:31–39.

Loffert, D., Ehlich, A., Muller, W., and Rajewsky, K.: **Surrogate light-chain expression is required to establish immunoglobulin heavy-chain allelic exclusion during early B-cell development.** *Immunity* 1996, **4**:133–144.

Melchers, F., ten Boekel, E., Yamagami, T., Andersson, J., and Rolink, A.: **The roles of preB and B cell receptors in the stepwise allelic exclusion of mouse IgH and L chain gene loci.** *Semin. Immunol.* 1999, **11**:307–317.

8-5 Pre-B cells rearrange the light-chain locus and express cell-surface immunoglobulin.

Arakawa, H., Shimizu, T., and Takeda, S.: **Reevaluation of the probabilities for productive rearrangements on the κ-loci and λ-loci.** *Int. Immunol.* 1996, **8**:91–99.

Gorman, J.R., van der Stoep, N., Monroe, R., Cogne, M., Davidson, L., and Alt, F.W.: **The Igk 3′ enhancer influences the ratio of Igκ versus Igλ B lymphocytes.** *Immunity* 1996, **5**:241–252.

Hesslein, D.G., and Schatz, D.G.: **Factors and forces controlling V(D)J recombination.** *Adv. Immunol.* 2001, **78**:169–232.

Kee, B.L., and Murre, C.: **Transcription factor regulation of B lineage commitment.** *Curr. Opin. Immunol.* 2001, **13**:180–185.

Sleckman, B.P., Gorman, J.R., and Alt, F.W.: **Accessibility control of antigen receptor variable region gene assembly—role of *cis*-acting elements.** *Annu. Rev. Immunol.* 1996, **14**:459–481.

Takeda, S., Sonoda, E., and Arakawa, H.: **The κ–λ ratio of immature B cells.** *Immunol. Today* 1996, **17**:200–201.

8-6 Immature B cells are tested for autoreactivity before they leave the bone marrow.

Casellas, R., Shih, T.A., Kleinewietfeld, M., Rakonjac, J., Nemazee, D., Rajewsky, K., and Nussenzweig, M.C.: **Contribution of receptor editing to the antibody repertoire.** *Science* 2001, **291**:1541–1544.

Chen, C., Nagy, Z., Radic, M.Z., Hardy, R.R., Huszar, D., Camper, S.A., and Weigert, M.: **The site and stage of anti-DNA B-cell deletion.** *Nature* 1995, **373**:252–255.

Cornall, R.J., Goodnow, C.C., and Cyster, J.G.: **The regulation of self-reactive B cells.** *Curr. Opin. Immunol.* 1995, **7**:804–811.

Huang, F., and Gu, H.: **Negative regulation of lymphocyte development and function by the Cbl family of proteins.** *Immunol. Rev.* 2008, **224**:229–238.

Melamed, D., Benschop, R.J., Cambier, J.C., and Nemazee, D.: **Developmental regulation of B lymphocyte immune tolerance compartmentalizes clonal selection from receptor selection.** *Cell* 1998, **92**:173–182.

Nemazee, D.: **Receptor editing in lymphocyte development and central tolerance.** *Nat. Rev. Immunol.* 2006, **6**:728–740.

Prak, E.L., and Weigert, M.: **Light-chain replacement—a new model for antibody gene rearrangement.** *J. Exp. Med.* 1995, **182**:541–548.

8-7 T-cell progenitors originate in the bone marrow, but all the important events in their development occur in the thymus.

Anderson, G., Moore, N.C., Owen, J.J.T., and Jenkinson, E.J.: **Cellular interactions in thymocyte development.** *Annu. Rev. Immunol.* 1996, **14**:73–99.

Carlyle, J.R., and Zúñiga-Pflücker, J.C.: **Requirement for the thymus in α:β T lymphocyte lineage commitment.** *Immunity* 1998, **9**:187–197.

Ciofani, M., Knowles, G., Wiest, D., von Boehmer, H., and Zúñiga-Pflücker, J.: **Stage-specific and differential Notch dependency at the α:β and γ:δ T lineage bifurcation.** *Immunity* 2006, **25**:105–116.

Gordon, J., Wilson, V.A., Blair, N.F., Sheridan, J., Farley, A., Wilson, L., Manley, N.R., and Blackburn, C.C.: **Functional evidence for a single endodermal origin for the thymic epithelium.** *Nat. Immunol.* 2004, **5**:546–553.

Nehls, M., Kyewski, B., Messerle, M., Waldschütz, R., Schüddekopf, K., Smith, A.J.H., and Boehm, T.: **Two genetically separable steps in the differentiation of thymic epithelium.** *Science* 1996, **272**:886–889.

Rodewald, H.R.: **Thymus organogenesis.** *Annu. Rev. Immunol.* 2008, **26**:355–388.

van Ewijk, W., Hollander, G., Terhorst, C., and Wang, B.: **Stepwise development of thymic microenvironments *in vivo* is regulated by thymocyte subsets.** *Development* 2000, **127**:1583–1591.

8-8 T-cell precursors proliferate extensively in the thymus, but most die there.

Shortman, K., Egerton, M., Spangrude, G.J., and Scollay, R.: **The generation and fate of thymocytes.** *Semin. Immunol.* 1990, **2**:3–12.

Surh, C.D., and Sprent, J.: **T-cell apoptosis detected *in situ* during positive and negative selection in the thymus.** *Nature* 1994, **372**:100–103.

8-9 Successive stages in the development of thymocytes are marked by changes in cell-surface molecules.

Borowski, C., Martin, C., Gounari, F., Haughn, L., Aifantis, I., Grassi, F., and von Boehmer, H.: **On the brink of becoming a T cell.** *Curr. Opin. Immunol.* 2002, **14**:200–206.

Pang, S.S., Berry, R., Chen, Z., Kjer-Nielsenm, L., Perugini, M.A., King, G.F., Wang, C., Chew, S.H., La Gruta, N.L., Williams, N.K., *et al.*: **The structural basis for autonomous dimerization of the pre-T-cell antigen receptor.** *Nature* 2010, **467**:844–848.

Saint-Ruf, C., Ungewiss, K., Groetrrup, M., Bruno, L., Fehling, H.J., and von Boehmer, H.: **Analysis and expression of a cloned pre-T-cell receptor gene.** *Science* 1994, **266**:1208–1212.

Shortman, K., and Wu, L.: **Early T lymphocyte progenitors.** *Annu. Rev. Immunol.* 1996, **14**:29–47.

8-10 Thymocytes at different developmental stages are found in distinct parts of the thymus.

Benz, C., Heinzel, K., and Bleul, C.C.: **Homing of immature thymocytes to the subcapsular microenvironment within the thymus is not an absolute requirement for T cell development.** *Eur. J. Immunol.* 2004, **34**:3652–3663.

Bleul, C.C., and Boehm, T.: **Chemokines define distinct microenvironments in the developing thymus.** *Eur. J. Immunol.* 2000, **30**:3371–3379.

Nitta, T., Murata, S., Ueno, T., Tanaka, K., and Takahama, Y.: **Thymic microenvironments for T-cell repertoire formation.** *Adv. Immunol.* 2008, **99**:59–94.

Ueno, T., Saito F., Gray, D.H.D., Kuse, S., Hieshima, K., Nakano, H., Kakiuchi, T., Lipp, M., Boyd, R.L., and Takahama, Y.: **CCR7 signals are essential for cortex–medulla migration of developing thymocytes.** *J. Exp. Med.* 2004, **200**:493–505.

8-11 T cells with α:β or γ:δ receptors arise from a common progenitor.

Fehling, H.J., Gilfillan, S., and Ceredig, R.: **α β/γ δ lineage commitment in the thymus of normal and genetically manipulated mice.** *Adv. Immunol.* 1999, **71**:1–76.

Hayday, A.C., Barber, D.F., Douglas, N., and Hoffman, E.S.: **Signals involved in**

$\gamma\delta$ T cell versus $\alpha\beta$ T cell lineage commitment. *Semin. Immunol.* 1999, **11**:239–249.

Hayes, S.M., and Love, P.E.: **Distinct structure and signaling potential of the $\gamma\delta$ TCR complex.** *Immunity* 2002, **16**:827–838.

Kang, J., and Raulet, D.H.: **Events that regulate differentiation of $\alpha\beta$ TCR$^+$ and $\gamma\delta$ TCR$^+$ T cells from a common precursor.** *Semin. Immunol.* 1997, **9**:171–179.

Kreslavsky, T., Garbe, A.I., Krueger, A., and von Boehmer, H.: **T cell receptor-instructed $\alpha\beta$ versus $\gamma\delta$ lineage commitment revealed by single-cell analysis.** *J. Exp. Med.* 2008, **205**:1173–1186.

Lauritsen, J.P., Haks, M.C., Lefebvre, J.M., Kappes, D.J., and Wiest, D.L.: **Recent insights into the signals that control $\alpha\beta/\gamma\delta$-lineage fate.** *Immunol. Rev.* 2006, **209**:176–190.

Livak, F., Petrie, H.T., Crispe, I.N., and Schatz, D.G.: **In-frame TCRδ gene rearrangements play a critical role in the $\alpha\beta/\gamma\delta$ T cell lineage decision.** *Immunity* 1995, **2**:617–627.

Xiong, N., and Raulet, D.H.: **Development and selection of $\gamma\delta$ T cells.** *Immunol, Rev.* 2007, **215**:15–31.

8-12 T cells expressing particular γ- and δ-chain V regions arise in an ordered sequence early in life.

Carding, S.R., and Egan, P.J.: **$\gamma\delta$ T cells: functional plasticity and heterogeneity.** *Nat. Rev. Immunol.* 2002, **2**:336–345.

Ciofani, M., Knowles, G.C., Wiest, D.L., von Boehmer, H., and Zúñiga-Pflücker, J.C.: **Stage-specific and differential notch dependency at the $\alpha{:}\beta$ and $\gamma{:}\delta$ T lineage bifurcation.** *Immunity* 2006, **25**:105–116.

Dunon, D., Courtois, D., Vainio, O., Six, A., Chen, C.H., Cooper, M.D., Dangy, J.P., and Imhof, B.A.: **Ontogeny of the immune system: $\gamma{:}\delta$ and $\alpha{:}\beta$ T cells migrate from thymus to the periphery in alternating waves.** *J. Exp. Med.* 1997, **186**:977–988.

Haas, W., Pereira, P., Tonegawa, S.: **Gamma/delta cells.** *Annu. Rev. Immunol.* 1993, **11**:637–685.

Lewis, J.M., Girardi, M., Roberts, S.J., Barbee, S.D., Hayday, A.C., and Tigelaar, R.E.: **Selection of the cutaneous intraepithelial $\gamma\delta^+$ T cell repertoire by a thymic stromal determinant.** *Nat. Immunol.* 2006, **7**:843–850.

Strid, J., Tigelaar, R.E., and Hayday, A.C.: **Skin immune surveillance by T cells—a new order?** *Semin. Immunol.* 2009, **21**:110–120.

8-13 Successful synthesis of a rearranged β chain allows the production of a pre-T-cell receptor that triggers cell proliferation and blocks further β-chain gene rearrangement.

Borowski, C., Li, X., Aifantis, I., Gounari, F., and von Boehmer, H.: **Pre-TCRα and TCRα are not interchangeable partners of TCRβ during T lymphocyte development.** *J. Exp. Med.* 2004, **199**:607–615.

Dudley, E.C., Petrie, H.T., Shah, L.M., Owen, M.J., and Hayday, A.C.: **T-cell receptor β chain gene rearrangement and selection during thymocyte development in adult mice.** *Immunity* 1994, **1**:83–93.

Philpott, K.I., Viney, J.L., Kay, G., Rastan, S., Gardiner, E.M., Chae, S., Hayday, A.C., and Owen, M.J.: **Lymphoid development in mice congenitally lacking T cell receptor $\alpha\beta$-expressing cells.** *Science* 1992, **256**:1448–1453.

von Boehmer, H., Aifantis, I., Azogui, O., Feinberg, J., Saint-Ruf, C., Zober, C., Garcia, C., and Buer, J.: **Crucial function of the pre-T-cell receptor (TCR) in TCRβ selection, TCRβ allelic exclusion and $\alpha{:}\beta$ versus $\gamma{:}\delta$ lineage commitment.** *Immunol. Rev.* 1998, **165**:111–119.

8-14 T-cell α-chain genes undergo successive rearrangements until positive selection or cell death intervenes.

Buch, T., Rieux-Laucat, F., Förster, I., and Rajewsky, K.: **Failure of HY-specific thymocytes to escape negative selection by receptor editing.** *Immunity* 2002, **16**:707–718.

Hardardottir, F., Baron, J.L., and Janeway, C.A., Jr: **T cells with two functional antigen-specific receptors.** *Proc. Natl Acad. Sci. USA* 1995, **92**:354–358.

Huang, C.-Y., Sleckman, B.P., and Kanagawa, O.: **Revision of T cell receptor α chain genes is required for normal T lymphocyte development.** *Proc. Natl Acad. Sci. USA* 2005, **102**:14356–14361.

Marrack, P., and Kappler, J.: **Positive selection of thymocytes bearing $\alpha{:}\beta$ T cell receptors.** *Curr. Opin. Immunol.* 1997, **9**:250–255.

Padovan, E., Casorati, G., Dellabona, P., Meyer, S., Brockhaus, M., and Lanzavecchia, A.: **Expression of two T-cell receptor α chains: dual receptor T cells.** *Science* 1993, **262**:422–424.

Petrie, H.T., Livak, F., Schatz, D.G., Strasser, A., Crispe, I.N., and Shortman, K.: **Multiple rearrangements in T-cell receptor α-chain genes maximize the production of useful thymocytes.** *J. Exp. Med.* 1993, **178**:615–622.

8-15 The MHC type of the thymic stroma selects a repertoire of mature T cells that can recognize foreign antigens presented by the same MHC type.

Fink, P.J., and Bevan, M.J.: **H-2 antigens of the thymus determine lymphocyte specificity.** *J. Exp. Med.* 1978, **148**:766–775.

Zinkernagel, R.M., Callahan, G.N., Klein, J., and Dennert, G.: **Cytotoxic T cells learn specificity for self H-2 during differentiation in the thymus.** *Nature* 1978, **271**:251–253.

8-16 Only thymocytes whose receptors interact with self-peptide:self-MHC complexes can survive and mature.

Hogquist, K.A., Tomlinson, A.J., Kieper, W.C., McGargill, M.A., Hart, M.C., Naylor, S., and Jameson, S.C.: **Identification of a naturally occurring ligand for thymic positive selection.** *Immunity* 1997, **6**:389–399.

Huessman, M., Scott, B., Kisielow, P., and von Boehmer, H.: **Kinetics and efficacy of positive selection in the thymus of normal and T-cell receptor transgenic mice.** *Cell* 1991, **66**:533–562.

Stefanski, H.E., Mayerova, D., Jameson, S.C., and Hogquist, K.A.: **A low affinity TCR ligand restores positive selection of CD8$^+$ T cells *in vivo*.** *J. Immunol.* 2001, **166**:6602–6607.

8-17 Positive selection acts on a repertoire of T-cell receptors with inherent specificity for MHC molecules.

Marrack, P., Scott-Browne, J.P., Dai, S., Gapin, L., and Kappler, J.W.: **Evolutionarily conserved amino acids that control TCR-MHC interaction.** *Annu. Rev. Immunol.* 2008, **26**:171–203.

Merkenschlager, M., Graf, D., Lovatt, M., Bommhardt, U., Zamoyska, R., and Fisher, A.G.: **How many thymocytes audition for selection?** *J. Exp. Med.* 1997, **186**:1149–1158.

Scott-Browne, J.P., White, J., Kappler, J.W., Gapin, L., and Marrack, P.: **Germline-encoded amino acids in the $\alpha\beta$ T-cell receptor control thymic selection.** *Nature* 2009, **458**:1043–1046.

Zerrahn, J., Held, W., and Raulet, D.H.: **The MHC reactivity of the T cell repertoire prior to positive and negative selection.** *Cell* 1997, **88**:627–636.

8-18 Positive selection coordinates the expression of CD4 or CD8 with the specificity of the T-cell receptor and the potential effector functions of the T cell.

Egawa, T., and Littman, D.R.: **ThPOK acts late in specification of the helper T cell lineage and suppresses Runx-mediated commitment to the cytotoxic T cell lineage.** *Nat. Immunol.* 2008, **9**:1131–1139.

He, X., Park, K., and Kappes, D.J.: **The role of ThPOK in control of CD4/CD8 lineage commitment.** *Annu. Rev. Immunol.* 2010, **28**:295–320.

He, X., Xi, H., Dave, V.P., Zhang, Y., Hua, X., Nicolas, E., Xu, W., Roe, B.A., and Kappes, D.J.: **The zinc finger transcription factor Th-POK regulates CD4 versus CD8 T-cell lineage commitment.** *Nature* 2005, **433**:826–833.

Lundberg, K., Heath, W., Kontgen, F., Carbone, F.R., and Shortman, K.: **Intermediate steps in positive selection: differentiation of CD4$^+$8^{int}TCR^{int} thymocytes into CD4$^-$8$^+$TCR^{hi} thymocytes.** *J. Exp. Med.* 1995, **181**:1643–1651.

Singer, A., Adoro, S., and Park, J.H.: **Lineage fate and intense debate: myths, models and mechanisms of CD4- versus CD8-lineage choice.** *Nat. Rev. Immunol.* 2008, **8**:788–801.

von Boehmer, H., Kisielow, P., Lishi, H., Scott, B., Borgulya, P., and Teh, H.S.: **The expression of CD4 and CD8 accessory molecules on mature T cells is not random but correlates with the specificity of the α:β receptor for antigen.** *Immunol. Rev.* 1989, **109**:143–151.

Zheng, Y., and Rudensky, A.Y.: **Foxp3 in control of the regulatory T cell lineage.** *Nat. Immunol.* 2007, **8**:457–462.

8-19 Thymic cortical epithelial cells mediate positive selection of developing thymocytes.

Cosgrove, D., Chan, S.H., Waltzinger, C., Benoist, C., and Mathis, D.: **The thymic compartment responsible for positive selection of CD4+ T cells.** *Int. Immunol.* 1992, **4**:707–710.

Ernst, B.B., Surh, C.D., and Sprent, J.: **Bone marrow-derived cells fail to induce positive selection in thymus reaggregation cultures.** *J. Exp. Med.* 1996, **183**:1235–1240.

Fowlkes, B.J., and Schweighoffer, E.: **Positive selection of T cells.** *Curr. Opin. Immunol.* 1995, **7**:188–195.

Nakagawa, T., Roth, W., Wong, P., Nelson, A., Farr, A., Deussing, J., Villadangos, J.A., Ploegh, H., Peters, C., and Rudensky, A.Y.: **Cathepsin L: critical role in Ii degradation and CD4 T cell selection in the thymus.** *Science* 1998, **280**:450–453.

8-20 T cells that react strongly with ubiquitous self antigens are deleted in the thymus.

Kishimoto, H., and Sprent, J.: **Negative selection in the thymus includes semimature T cells.** *J. Exp. Med.* 1997, **185**:263–271.

Mathis, D., and Benoist, C.: **Aire.** *Annu. Rev. Immunol.* 2009, **27**:287–312.

Zal, T., Volkmann, A., and Stockinger, B.: **Mechanisms of tolerance induction in major histocompatibility complex class II-restricted T cell specific for a blood-borne self antigen.** *J. Exp. Med.* 1994, **180**:2089–2099.

8-21 Negative selection is driven most efficiently by bone marrow derived antigen-presenting cells.

McCaughtry, T.M., Baldwin, T.A., Wilken, M.S., and Hogquist, K.A.: **Clonal deletion of thymocytes can occur in the cortex with no involvement of the medulla.** *J. Exp. Med.* 2008, **205**:2575–2584.

Sprent, J., and Webb, S.R.: **Intrathymic and extrathymic clonal deletion of T cells.** *Curr. Opin. Immunol.* 1995, **7**:196–205.

Webb, S.R., and Sprent, J.: **Tolerogenicity of thymic epithelium.** *Eur. J. Immunol.* 1990, **20**:2525–2528.

8-22 The specificity and/or the strength of signals for negative and positive selection must differ.

Alberola-Ila, J., Hogquist, K.A., Swan, K.A., Bevan, M.J., and Perlmutter, R.M.: **Positive and negative selection invoke distinct signaling pathways.** *J. Exp. Med.* 1996, **184**:9–18.

Ashton-Rickardt, P.G., Bandeira, A., Delaney, J.R., Van Kaer, L., Pircher, H.P., Zinkernagel, R.M., and Tonegawa, S.: **Evidence for a differential avidity model of T-cell selection in the thymus.** *Cell* 1994, **76**:651–663.

Bommhardt, U., Basson, M.A., Krummrei, U., and Zamoyska, R.: **Activation of the extracellular signal-related kinase/mitogen-activated protein kinase pathway discriminates CD4 versus CD8 lineage commitment in the thymus.** *J. Immunol.* 1999, **163**:715–722.

Bommhardt, U., Scheuring, Y., Bickel, C., Zamoyska, R., and Hunig, T.: **MEK activity regulates negative selection of immature CD4+CD8+ thymocytes.** *J. Immunol.* 2000, **164**:2326–2337.

Hogquist, K.A., Jameson, S.C., Heath, W.R., Howard, J.L., Bevan, M.J., and Carbone, F.R.: **T-cell receptor antagonist peptides induce positive selection.** *Cell* 1994, **76**:17–27.

8-23 Different lymphocyte subsets are found in particular locations in peripheral lymphoid tissues.

Liu, Y.J.: **Sites of B lymphocyte selection, activation, and tolerance in spleen.** *J. Exp. Med.* 1997, **186**:625–629.

Loder, F., Mutschler, B., Ray, R.J., Paige, C.J., Sideras, P., Torres, R., Lamers, M.C., and Carsetti, R.: **B cell development in the spleen takes place in discrete steps and is determined by the quality of B cell receptor-derived signals.** *J. Exp. Med.* 1999, **190**:75–89.

Mebius, R.E.: **Organogenesis of lymphoid tissues.** *Nat. Rev. Immunol.* 2003, **3**:292–303.

8-24 The development of peripheral lymphoid tissues is controlled by lymphoid tissue inducer cells and proteins of the tumor necrosis factor family.

Douni, E., Akassoglou, K., Alexopoulou, L., Georgopoulos, S., Haralambous, S., Hill, S., Kassiotis, G., Kontoyiannis, D., Pasparakis, M., Plows, D., *et al.*: **Transgenic and knockout analysis of the role of TNF in immune regulation and disease pathogenesis.** *J. Inflamm.* 1996, **47**:27–38.

Fu, Y.X., and Chaplin, D.D.: **Development and maturation of secondary lymphoid tissues.** *Annu. Rev. Immunol.* 1999, **17**:399–433.

Mariathasan, S., Matsumoto, M., Baranyay, F., Nahm, M.H., Kanagawa, O., and Chaplin, D.D.: **Absence of lymph nodes in lymphotoxin-α (LTα)-deficient mice is due to abnormal organ development, not defective lymphocyte migration.** *J. Inflamm.* 1995, **45**:72–78.

Mebius, R.E., Rennert, P., and Weissman, I.L.: **Developing lymph nodes collect CD4+CD3− LTβ+ cells that can differentiate to APC, NK cells, and follicular cells but not T or B cells.** *Immunity* 1997, **7**:493–504.

Wigle, J.T., and Oliver, G.: **Prox1 function is required for the development of the murine lymphatic system.** *Cell* 1999, **98**:769–778.

8-25 The homing of lymphocytes to specific regions of peripheral lymphoid tissues is mediated by chemokines.

Ansel, K.M., and Cyster, J.G.: **Chemokines in lymphopoiesis and lymphoid organ development.** *Curr. Opin. Immunol.* 2001, **13**:172–179.

Cyster, J.G.: **Chemokines and cell migration in secondary lymphoid organs.** *Science* 1999, **286**:2098–2102.

Cyster, J.G.: **Leukocyte migration: scent of the T zone.** *Curr. Biol.* 2000, **10**:R30–R33.

Cyster, J.G., Ansel, K.M., Reif, K., Ekland, E.H., Hyman, P.L., Tang, H.L., Luther, S.A., and Ngo, V.N.: **Follicular stromal cells and lymphocyte homing to follicles.** *Immunol. Rev.* 2000, **176**:181–193.

8-26 Lymphocytes that encounter sufficient quantities of self antigens for the first time in the periphery are eliminated or inactivated.

Cyster, J.G., Hartley, S.B., and Goodnow, C.C.: **Competition for follicular niches excludes self-reactive cells from the recirculating B-cell repertoire.** *Nature* 1994, **371**:389–395.

Goodnow, C.C., Crosbie, J., Jorgensen, H., Brink, R.A., and Basten, A.: **Induction of self-tolerance in mature peripheral B lymphocytes.** *Nature* 1989, **342**:385–391.

Lam, K.P., Kuhn, R., and Rajewsky, K.: **In vivo ablation of surface immunoglobulin on mature B cells by inducible gene targeting results in rapid cell death.** *Cell* 1997, **90**:1073–1083.

Russell, D.M., Dembic, Z., Morahan, G., Miller, J.F.A.P., Burki, K., and Nemazee, D.: **Peripheral deletion of self-reactive B cells.** *Nature* 1991, **354**:308–311.

Steinman, R.M., and Nussenzweig, M.C.: **Avoiding horror autotoxicus: the importance of dendritic cells in peripheral T cell tolerance.** *Proc. Natl Acad. Sci. USA* 2002, **99**:351–358.

8-27 Immature B cells arriving in the spleen turn over rapidly and require cytokines and positive signals through the B-cell receptor for maturation and survival.

Allman, D.M., Ferguson, S.E., Lentz, V.M., and Cancro, M.P.: **Peripheral B cell maturation. II. Heat-stable antigen^{hi} splenic B cells are an immature developmental intermediate in the production of long-lived marrow-derived B cells.** *J. Immunol.* 1993, **151**:4431–4444.

Harless, S.M., Lentz, V.M., Sah, A.P., Hsu, B.L., Clise-Dwyer, K., Hilbert, D.M., Hayes, C.E., and Cancro, M.P.: **Competition for BLyS-mediated signaling through Bcmd/BR3 regulates peripheral B lymphocyte numbers.** *Curr. Biol.* 2001, **11**:1986–1989.

Levine, M.H., Haberman, A.M., Sant'Angelo, D.B., Hannum, L.G., Cancro, M.P., Janeway, C.A., Jr, and Shlomchik, M.J.: **A B-cell receptor-specific selection step governs immature to mature B cell differentiation.** *Proc. Natl Acad. Sci. USA* 2000, **97**:2743–2748.

Rolink, A.G., Tschopp, J., Schneider, P., and Melchers, F.: **BAFF is a survival and maturation factor for mouse B cells.** *Eur. J. Immunol.* 2002, **32**:2004–2010.

Schiemann, B., Gommerman, J.L., Vora, K., Cachero, T.G., Shulga-Morskaya, S., Dobles, M., Frew, E., and Scott, M.L.: **An essential role for BAFF in the normal development of B cells through a BCMA-independent pathway.** *Science* 2001, **293**:2111–2114.

8-28 B-1 cells and marginal zone B cells are distinct B-cell subtypes with unique antigen receptor specificity.

Clarke, S.H., and Arnold, L.W.: **B-1 cell development: evidence for an uncommitted immunoglobulin (Ig)M^{+} B cell precursor in B-1 cell differentiation.** *J. Exp. Med.* 1998, **187**:1325–1334.

Hardy, R.R., and Hayakawa, K.: **A developmental switch in B lymphopoiesis.** *Proc. Natl Acad. Sci. USA* 1991, **88**:11550–11554.

Hayakawa, K., Asano, M., Shinton, S.A., Gui, M., Allman, D., Stewart, C.L., Silver, J., and Hardy, R.R.: **Positive selection of natural autoreactive B cells.** *Science* 1999, **285**:113–116.

Martin, F., and Kearney, J.F.: **Marginal-zone B cells.** *Nat. Rev. Immunol.* 2002, **2**:323–335.

8-29 T-cell homeostasis in the periphery is regulated by cytokines and self-MHC interactions.

Judge, A.D., Zhang, X., Fujii, H., Surh, C.D., and Sprent, J.: **Interleukin 15 controls both proliferation and survival of a subset of memory-phenotype CD8^{+} T cells.** *J. Exp. Med.* 2002, **196**:935–946.

Kassiotis, G., Garcia, S., Simpson, E., and Stockinger, B.: **Impairment of immunological memory in the absence of MHC despite survival of memory T cells.** *Nat. Immunol.* 2002, **3**:244–250.

Ku, C.C., Murakami, M., Sakamoto, A., Kappler, J., and Marrack, P.: **Control of homeostasis of CD8^{+} memory T cells by opposing cytokines.** *Science* 2000, **288**:675–678.

Murali-Krishna, K., Lau, L.L., Sambhara, S., Lemonnier, F., Altman, J., and Ahmed, R.: **Persistence of memory CD8 T cells in MHC class I-deficient mice.** *Science* 1999, **286**:1377–1381.

Seddon, B., Tomlinson, P., and Zamoyska, R.: **IL-7 and T cell receptor signals regulate homeostasis of CD4 memory cells.** *Nat. Immunol.* 2003, **4**:680–686.

PART IV THE ADAPTIVE IMMUNE RESPONSE

T Cell-Mediated Immunity

9

An adaptive immune response is induced when an infection overwhelms innate defense mechanisms. The pathogen continues to replicate and antigen accumulates. Together with the changed cellular environment produced by innate immunity, this triggers the adaptive immune response. Some infections may be dealt with solely by innate immunity, as discussed in Chapters 2 and 3, but most pathogens, almost by definition, can overcome the innate immune system, and adaptive immunity is essential for defense against them. This is shown by the immunodeficiency syndromes that are associated with failure of particular parts of the adaptive immune response; these will be discussed in Chapter 13. In the next three chapters, we will learn how the adaptive immune response involving the antigen-specific T cells and B cells is initiated and deployed. We will consider T cell-mediated immune responses first, in this chapter, and humoral immunity—the antibody response produced by B cells—in Chapter 10. In Chapter 11 we will combine what we have learned to present a dynamic view of adaptive immune responses to pathogens, including a discussion of one of its most important features—immunological memory.

Once T cells have completed their development in the thymus, they enter the bloodstream. On reaching a peripheral lymphoid organ, they leave the blood to migrate through the lymphoid tissue, returning via the lymphatics to the bloodstream to recirculate between blood and peripheral lymphoid tissues. Mature recirculating T cells that have not yet encountered their specific antigens are known as **naive T cells**. To participate in an adaptive immune response, a naive T cell must meet its specific antigen, present to it as a peptide:MHC complex on the surface of an antigen-presenting cell, and be induced to proliferate and differentiate into cells that have acquired new activities that contribute to removing the antigen. These cells are called **effector T cells** and, unlike naive T cells, perform their function as soon as they encounter their specific antigen on other cells. Because of their requirement to recognize peptide antigens presented by MHC molecules, all effector T cells act on other host cells, not on the pathogen itself. The cells on which effector T cells act will be referred to as their **target cells**.

On antigen recognition, naive T cells differentiate into several functional classes of effector T cells that are specialized for different activities. CD8 T cells recognize pathogen peptides presented by MHC class I molecules, and naive CD8 T cells differentiate into cytotoxic effector T cells that recognize and kill infected cells. CD4 T cells have a more flexible repertoire of effector activities. After recognition of pathogen peptides presented by MHC class II molecules, naive CD4 T cells can differentiate down distinct pathways that generate effector subsets with different immunological functions. The main CD4 effector subsets currently distinguished are T_H1, T_H2, T_H17, and T_{FH}, which activate their target cells, and several regulatory T-cell subsets with inhibitory activity that limits the extent of immune activation (Fig. 9.1).

The activation of naive T cells in response to antigen, and their subsequent proliferation and differentiation into effector cells, constitute a **primary cell-mediated immune response**. Effector T cells differ in many ways from their

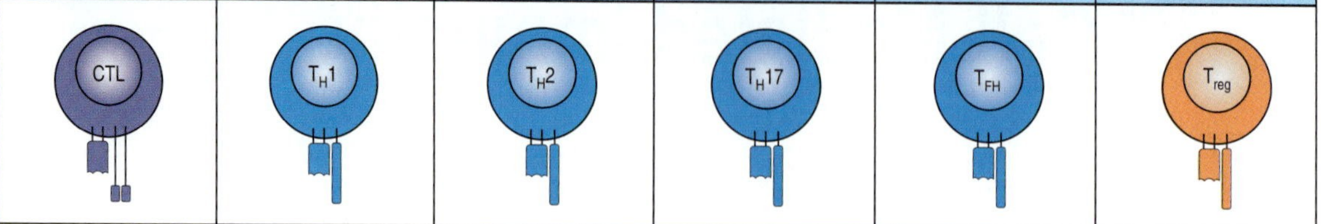

	CD8 cytotoxic T cells	CD4 T_H1 cells	CD4 T_H2 cells	CD4 T_H17 cells	T_{FH} cells	CD4 regulatory T cells (various types)
Types of effector T cell	CTL	T_H1	T_H2	T_H17	T_{FH}	T_{reg}
Main functions in adaptive immune response	Kill virus-infected cells	Activate infected macrophages Provide help to B cells for antibody production	Provide help to B cells for antibody production, especially switching to IgE	Enhance neutrophil response Promote barrier integrity (skin, intestine)	B-cell help Isotype switching Antibody production	Suppress T-cell responses
Pathogens targeted	Viruses (e.g. influenza, rabies, vaccinia) Some intracellular bacteria	Microbes that persist in macrophage vesicles (e.g. mycobacteria, *Listeria*, *Leishmania donovani*, *Pneumocystis carinii*) Extracellular bacteria	Helminth parasites	*Klebsiella pneumoniae* Fungi (*Candida albicans*)	All types	

Fig. 9.1 The roles of effector T cells in cell-mediated and humoral immune responses. Cell-mediated immune responses are directed principally at intracellular pathogens. They involve the destruction of infected cells by cytotoxic CD8 T cells, or the destruction of intracellular pathogens in macrophages activated by CD4 T_H1 cells. CD4 T_H17 cells help to recruit neutrophils to sites of infection early in the adaptive immune response, which is also a response aimed mainly at extracellular pathogens. CD4 T_H2 cells induce the switch to production of IgE antibodies, which are involved in the activation of effector responses aimed against extracellular multicellular parasites such as helminth worms (discussed in detail in Chapter 10). T_{FH} cells contribute to humoral immunity by stimulating the production of antibodies by B cells and inducing class switching, and can produce cytokines characteristic of either T_H1 or T_H2 cells. All classes of antibody contribute to humoral immunity, which is directed principally at extracellular pathogens. Both cell-mediated and humoral immunity are involved in many infections. Regulatory T cells tend to suppress the adaptive immune response and are important in preventing immune responses from becoming uncontrolled and in preventing autoimmunity.

naive precursors, and these changes equip them to respond quickly and efficiently when they encounter specific antigen on target cells. In this chapter we will describe the specialized mechanisms of T cell-mediated cytotoxicity and of macrophage activation by effector T cells, which make up the major components of **cell-mediated immunity**. The other main function of effector T cells is to provide help to B cells to trigger antibody production. We will only touch on this in this chapter and will discuss it in detail in Chapter 10. At the same time as providing effector T cells, the primary T-cell response also generates **memory T cells**, long-lived cells that give an enhanced response to antigen, which yields protection from subsequent challenge by the same pathogen. We will discuss T-cell and B-cell immunological memory together in Chapter 11.

In this chapter, we will see how naive T cells are activated to proliferate and produce effector T cells the first time they encounter their specific antigen. The activation and clonal expansion of a naive T cell on its initial encounter with antigen is often called **priming**, to distinguish it from the responses of effector T cells to antigen on their target cells and the responses of primed memory T cells. The initiation of adaptive immunity is one of the most compelling narratives in immunology. As we will learn, the activation of naive T cells is controlled by a variety of signals: in the nomenclature used in this book these are called signal 1, signal 2, and signal 3. A naive T cell recognizes antigen in the form of a peptide:MHC complex on the surface of a specialized antigen-presenting cell, as discussed in Chapter 6. Antigen-specific activation of the T-cell receptor delivers signal 1; interaction of co-stimulatory molecules on antigen-presenting cells with ligands on T cells delivers signal 2; and cytokines that control differentiation into different types of effector cells deliver signal 3. All these events are set in motion by much earlier signals that arise from the initial detection of the pathogens by the innate immune system. These signals are delivered to cells of the innate immune system by receptors such as the Toll-like receptors (TLRs), which recognize pathogen-associated molecular patterns that signify the presence of nonself (see Chapters 2 and 3). As we will see in this chapter, those signals are essential to activate antigen-presenting cells so that they are able, in turn, to activate naive T cells.

By far the most important antigen-presenting cells in the activation of naive T cells are the highly specialized **dendritic cells**, whose major function is to ingest and present antigen. Tissue dendritic cells take up antigen at sites of infection and are activated as part of the innate immune response. This induces their migration to local lymphoid tissue and their maturation into cells that are highly effective at presenting antigen to recirculating naive T cells. In the first part of this chapter we shall see how naive T cells and dendritic cells meet in the peripheral lymphoid organs, and how dendritic cells become activated to full antigen-presenting cell status.

Entry of naive T cells and antigen-presenting cells into peripheral lymphoid organs.

Adaptive immune responses are initiated in the peripheral lymphoid organs—lymph nodes, spleen, and the mucosa-associated lymphoid tissues such as the Peyer's patches in the gut. This means that for a T-cell immune response to be induced, the rare naive T cells specific for the appropriate antigens must meet dendritic cells presenting those antigens in a peripheral lymphoid organ. An infection can originate in virtually any site in the body, however, and so the pathogen's antigens must be brought from these sites to peripheral lymphoid organs. In this part of the chapter we shall see how dendritic cells pick up antigen and travel to local lymphoid organs, where they mature into cells that can both present antigen to T cells and activate them. Free antigens, such as bacteria and virus particles, also travel through lymphatics and in the blood directly to lymphoid organs, where they can be taken up and presented by antigen-presenting cells. As we learned in Chapter 1, naive T cells are continuously recirculating through the peripheral lymphoid tissues, surveying the antigen-presenting cells for foreign antigens. We shall look first at how this cellular traffic is orchestrated by chemotactic cytokines (chemokines) and adhesion molecules, which direct naive T cells out of the blood and into the lymphoid organs.

9-1 Naive T cells migrate through peripheral lymphoid tissues, sampling the peptide:MHC complexes on dendritic cell surfaces.

Naive T cells circulate from the bloodstream into lymph nodes, spleen, and mucosa-associated lymphoid tissues and back to the blood (see Fig. 1.17 for the overall circulation in respect of a lymph node). This enables them to make contact with thousands of dendritic cells in the lymphoid tissues every day and sample the peptide:MHC complexes on the surfaces of the dendritic cells. Each T cell thus has a high probability of encountering antigens derived from any pathogen that has set up an infection in whatever location (Fig. 9.2). Naive T cells that do not encounter their specific antigen exit from the lymphoid tissue via the efferent lymphatics, eventually reenter the bloodstream, and continue recirculating. When a naive T cell recognizes its specific antigen on the surface of a mature dendritic cell, however, it ceases to migrate. It proliferates for several days, undergoing **clonal expansion** and differentiation, and gives rise to effector T cells and memory cells of identical antigen specificity. At the end of this period, the effector T cells exit into the efferent lymphatics and reenter the bloodstream, through which they migrate to the sites of infection. The exception to this type of recirculation is the spleen, which has no connection with the lymphatic system; all cells enter the spleen from the blood and exit directly back into it.

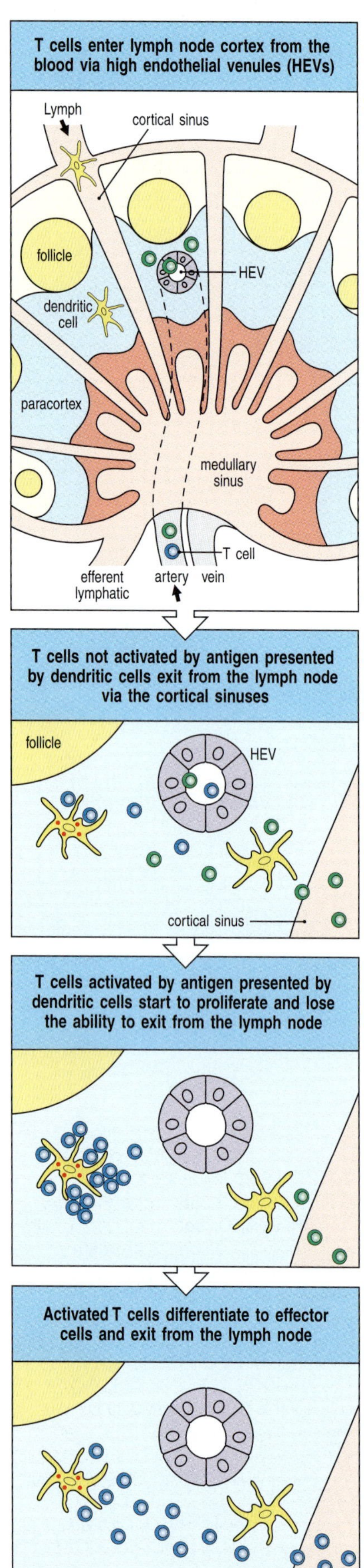
T cells enter lymph node cortex from the blood via high endothelial venules (HEVs)
Lymph
cortical sinus
follicle
HEV
dendritic cell
paracortex
medullary sinus
T cell
efferent lymphatic
artery
vein
T cells not activated by antigen presented by dendritic cells exit from the lymph node via the cortical sinuses
follicle
HEV
cortical sinus
T cells activated by antigen presented by dendritic cells start to proliferate and lose the ability to exit from the lymph node
Activated T cells differentiate to effector cells and exit from the lymph node

Fig. 9.2 Naive T cells encounter antigen during their recirculation through peripheral lymphoid organs. Naive T cells recirculate through peripheral lymphoid organs, such as the lymph node shown here, entering from the arterial blood via the specialized vascular endothelium of high endothelial venules (HEVs). Entry into the lymph node is regulated by chemokines (not shown) that direct the T cells' migration through the HEV wall and into the paracortical areas, where they encounter mature dendritic cells (top panel). Those T cells shown in green do not encounter their specific antigen; they receive a survival signal through their interaction with self-peptide:self-MHC complexes and IL-7, and leave the lymph node through the lymphatics to return to the circulation (second panel). T cells shown in blue encounter their specific antigen on the surface of mature dendritic cells; they lose their ability to exit from the node and become activated to proliferate and to differentiate into effector T cells (third panel). After several days, these antigen-specific effector T cells regain the expression of receptors needed to exit from the node, leave via the efferent lymphatics, and enter the circulation in greatly increased numbers (bottom panel).

The efficiency with which T cells screen each antigen-presenting cell in lymph nodes is very high, as can be seen by the rapid trapping of antigen-specific T cells in a single lymph node containing antigen: all of the antigen-specific T cells in a sheep were trapped in one lymph node within 48 hours of antigen deposition (Fig. 9.3). Such efficiency is crucial for the initiation of an adaptive immune response, as only one naive T cell in 10^4–10^6 is likely to be specific for a particular antigen, and adaptive immunity depends on the activation and expansion of these rare cells.

9-2 Lymphocyte entry into lymphoid tissues depends on chemokines and adhesion molecules.

Migration of naive T cells into peripheral lymphoid tissues depends on their binding to high endothelial venules (HEVs) through cell–cell interactions that are not antigen-specific but are governed by cell-adhesion molecules. The main classes of adhesion molecules involved in lymphocyte interactions are the selectins, the integrins, members of the immunoglobulin superfamily, and some mucin-like molecules (see Fig. 3.24). Entry of lymphocytes into lymph nodes occurs in distinct stages that include initial rolling of lymphocytes along the endothelial surface, activation of integrins, firm adhesion, and transmigration or **diapedesis** across the endothelial layer into the paracortical areas, the T-cell zones (Fig. 9.4). These stages are regulated by a coordinated interplay of adhesion molecules and chemokines. Adhesion molecules have fairly broad roles in immune responses, being involved not only in lymphocyte migration but also in interactions between naive T cells and antigen-presenting cells, between effector T cells and their targets, and between other types of leukocytes and endothelium (such as the entry of monocytes and neutrophils into infected tissue discussed in Chapter 3).

The selectins (Fig. 9.5) are important for specifically guiding leukocytes to particular tissues, a phenomenon known as leukocyte **homing**. **L-selectin** (CD62L) is expressed on leukocytes, whereas P-selectin (CD62P) and E-selectin (CD62E) are expressed on vascular endothelium (see Section 3-15). L-selectin on naive T cells guides their exit from the blood into peripheral lymphoid tissues by initiating a light attachment to the wall of the HEV that results in the T cells rolling along the endothelium surface (see Fig. 9.4). P-selectin and E-selectin are expressed on the vascular endothelium at sites of infection, and serve to recruit effector cells into the infected tissue. Selectins are cell-surface molecules with a common core structure, distinguished from each other by the presence of different lectin-like domains in their extracellular portion. The lectin domains bind to particular sugar groups, and each selectin binds to a cell-surface carbohydrate. L-selectin binds to the carbohydrate moiety—sulfated sialyl-Lewis^x—of mucin-like molecules called **vascular**

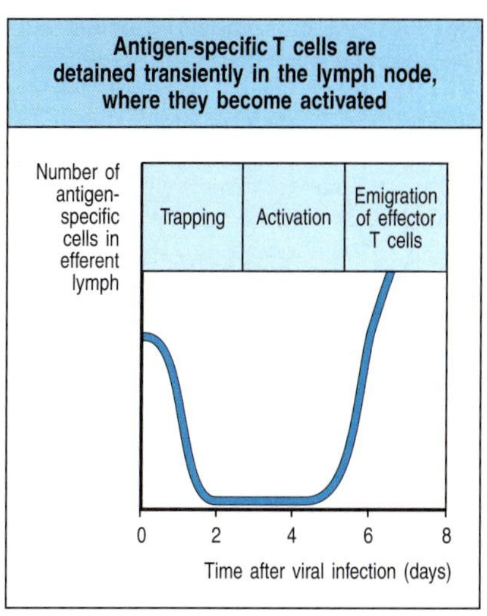

Fig. 9.3 Trapping and activation of antigen-specific naive T cells in lymphoid tissue. Naive T cells entering the lymph node from the blood encounter antigen-presenting dendritic cells in the lymph node cortex. T cells that recognize their specific antigen bind stably to the dendritic cells and are activated through their T-cell receptors, resulting in the production of effector T cells. By 5 days after the arrival of antigen, activated effector T cells are leaving the lymph node in large numbers via the efferent lymphatics. Lymphocyte recirculation and recognition are so effective that all the naive T cells in the peripheral circulation specific for a particular antigen can be trapped by that antigen in one node within 2 days.

addressins, which are expressed on the surface of vascular endothelial cells. Two of these addressins, **CD34** and **GlyCAM-1** (see Fig. 9.5), are expressed on high endothelial venules in lymph nodes. A third, **MAdCAM-1** (see Fig. 9.5), is expressed on endothelium in mucosae, and guides lymphocyte entry into mucosal lymphoid tissue such as the Peyer's patches in the gut.

The interaction between L-selectin and the vascular addressins is responsible for the specific homing of naive T cells to lymphoid organs. On its own, however, it does not enable the cell to cross the endothelial barrier into the lymphoid tissue. This requires the concerted action of chemokines and integrins.

9-3 Activation of integrins by chemokines is responsible for the entry of naive T cells into lymph nodes.

Entry of naive T cells into peripheral lymphoid organs requires two additional types of cell-adhesion molecules—the integrins and members of the immunoglobulin superfamily. These proteins also have a crucial role in the subsequent interactions of lymphocytes with antigen-presenting cells and later with their target cells. Integrins bind tightly to their ligands after receiving signals that induce a change in their conformation. Signaling by chemokines activates integrins on leukocytes to bind tightly to the vascular wall in preparation for their migration into sites of inflammation (see Section 3-14). Similarly, chemokines present at the luminal surface of the HEV activate integrins expressed on naive T cells during their migration into lymphoid organs (see Fig. 9.4).

An integrin molecule consists of a large α chain that pairs noncovalently with a smaller β chain. There are several integrin subfamilies, broadly defined by their common β chains. We will be concerned here chiefly with the **leukocyte integrins**, which have a common β_2 chain paired with distinct α chains (Fig. 9.6). All T cells express the integrin $\alpha_L:\beta_2$ (CD11a:CD18), better known as leukocyte functional antigen-1 (**LFA-1**). This integrin is also present on

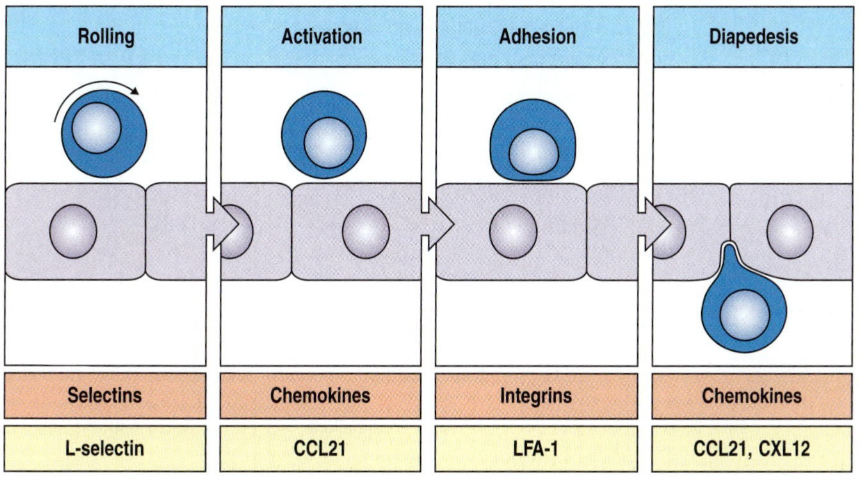

Fig. 9.4 Lymphocyte entry into a lymph node from the blood occurs in distinct stages involving the activity of adhesion molecules, chemokines, and chemokine receptors. Naive T cells are induced to roll along the surface of a high endothelial venule (HEV) by the interactions of selectins expressed by the T cells with vascular addressins on the endothelial cell membranes. Chemokines present at the HEV surface activate receptors on the T cell, and chemokine receptor signaling leads to an increase in the affinity of integrins on the T cell for the adhesion molecules expressed on the HEV. This induces strong adhesion. After adhesion, the T cells follow gradients of chemokines to pass through the HEV wall into the paracortical region of the lymph node.

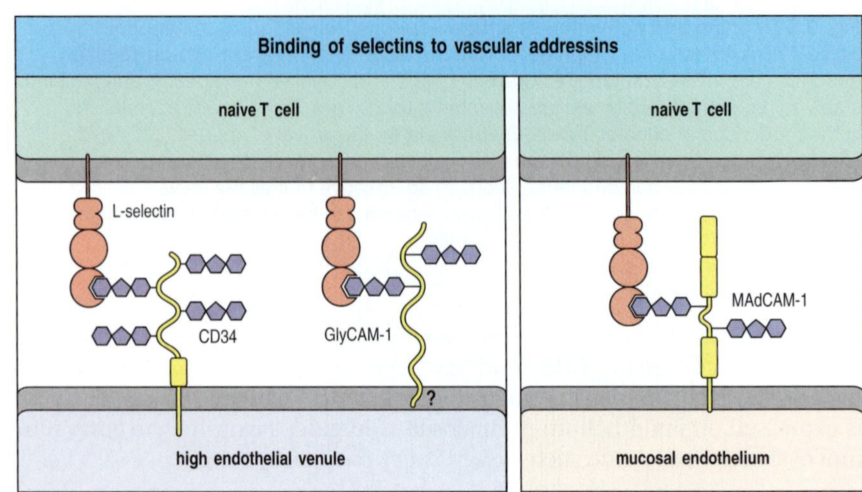

Fig. 9.5 L-selectin binds to mucin-like vascular addressins. L-selectin is expressed on naive T cells and recognizes carbohydrate motifs. Its binding to sulfated sialyl-Lewis^x moieties on the vascular addressins CD34 and GlyCAM-1 on HEVs binds the lymphocyte weakly to the endothelium. The relative importance of CD34 and GlyCAM-1 in this interaction is unclear. CD34 has a transmembrane anchor and is expressed in appropriately glycosylated form only on HEV cells, although it is found in other forms on other endothelial cells. GlyCAM-1 is expressed on HEVs but has no transmembrane region and may be secreted into the HEVs. The addressin MAdCAM-1 is expressed on mucosal endothelium and guides lymphocytes to mucosal lymphoid tissue. The icon shown represents mouse MAdCAM-1, which contains an IgA-like domain closest to the cell membrane; human MAdCAM-1 has an elongated mucin-like domain and lacks the IgA-like domain.

macrophages and neutrophils, and is involved in their recruitment to sites of infection (see Section 3-15). LFA-1 has a similar role in both naive and effector T cells in enabling their migration out of the blood.

LFA-1 is also important in the adhesion of both naive and effector T cells to their target cells. Nevertheless, T-cell responses can be normal in individuals genetically lacking the β_2 integrin chain and hence all β_2 integrins, including LFA-1. This is probably because T cells also express other adhesion molecules, including the immunoglobulin superfamily member CD2 and β_1 integrins, that could compensate for the absence of LFA-1. Expression of the β_1 integrins increases significantly at a late stage in T-cell activation, and they are thus often called **VLAs**, for **very late activation antigens**; they are important in directing effector T cells to inflamed target tissues.

At least five members of the immunoglobulin superfamily are especially important in T-cell activation (Fig. 9.7). Three very similar **intercellular adhesion molecules (ICAMs)**—**ICAM-1, ICAM-2**, and **ICAM-3**—all bind to the T-cell integrin LFA-1. ICAM-1 and ICAM-2 are expressed on endothelium as well as on antigen-presenting cells, and binding to these molecules enables lymphocytes to migrate through blood-vessel walls. ICAM-3 is expressed only on naive T cells and is thought to have an important role in the adhesion of T cells to antigen-presenting cells by binding to LFA-1 expressed on dendritic cells. The two remaining immunoglobulin superfamily adhesion molecules, **CD58** (formerly known as LFA-3) on the antigen-presenting cell and **CD2** on

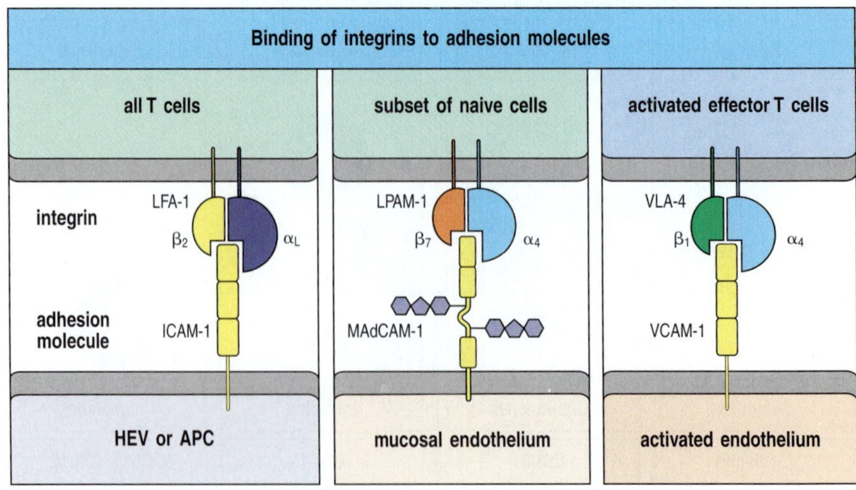

Fig. 9.6 Integrins are important in T-lymphocyte adhesion. Integrins are heterodimeric proteins containing a β chain, which defines the class of integrin, and an α chain, which defines the different integrins within a class. The α chain is larger than the β chain and contains binding sites for divalent cations that may be important in signaling. LFA-1 (integrin $\alpha_L:\beta_2$) is expressed on all leukocytes. It binds ICAMs and is important in cell migration and in the interactions of T cells with antigen-presenting cells (APCs) or target cells; it is expressed at higher levels on effector T cells than on naive T cells. Lymphocyte Peyer's patch adhesion molecule (LPAM-1 or integrin $\alpha_4:\beta_7$) is expressed by a subset of naive T cells and contributes to lymphocyte entry into mucosal lymphoid tissues by supporting adhesive interactions with vascular addressin MAdCAM-1. VLA-4 (integrin $\alpha_4:\beta_1$) is expressed strongly after T-cell activation. It binds to VCAM-1 on activated endothelium and is important for recruiting effector T cells into sites of infection.

Immunoglobulin superfamily	Name	Tissue distribution	Ligand
ICAM1/3, VCAM1 CD58 CD2	CD2 (LFA-2)	T cells	CD58 (LFA-3)
	ICAM-1 (CD54)	Activated vessels, lymphocytes, dendritic cells	LFA-1, Mac-1
	ICAM-2 (CD102)	Resting vessels	LFA-1
	ICAM-3 (CD50)	Naive T cells	LFA-1
	LFA-3 (CD58)	Lymphocytes, antigen-presenting cells	CD2
	VCAM-1 (CD106)	Activated endothelium	VLA-4

Fig. 9.7 Immunoglobulin superfamily adhesion molecules involved in leukocyte interactions. Adhesion molecules of the immunoglobulin superfamily bind to adhesion molecules of various types, including integrins (LFA-1 and VLA-4) and other immunoglobulin superfamily members (the CD2–CD58 (LFA-3) interaction). These interactions have a role in lymphocyte migration, homing, and cell–cell interactions; other molecules listed here have been introduced in Fig. 3.24.

the T cell, bind to each other; this interaction synergizes with that of ICAM-1 or ICAM-2 with LFA-1.

As in phagocyte migration, naive T cells are specifically attracted into the lymph node by chemokines secreted by cells in the lymph node. The chemokines bind to proteoglycans in the extracellular matrix and high endothelial venule wall, forming a chemical gradient, and are recognized by receptors on the naive T cell. The extravasation of naive T cells is prompted by the chemokine **CCL21** (secondary lymphoid tissue chemokine, SLC). CCL21 is expressed by vascular high endothelial cells and the stromal cells of lymphoid tissues, and binds to the chemokine receptor **CCR7** on naive T cells, stimulating activation of the intracellular receptor-associated G-protein subunit $G\alpha_i$. The resulting intracellular signaling rapidly increases the affinity of integrin binding (see Section 3-15).

The entry of a naive T cell into a lymph node is shown in detail in Fig. 9.8. Initial rolling along the high endothelial venule surface is mediated by L-selectin. Contact of naive T cells with CCL21 in the high endothelial venule causes the integrin LFA-1 on the naive T cell to become activated, increasing its affinity for ICAM-2 and ICAM-1. ICAM-2 is expressed constitutively on all endothelial cells, whereas in the absence of inflammation, ICAM-1 is expressed only on the high endothelial cells of peripheral lymphoid tissues. The mobility of

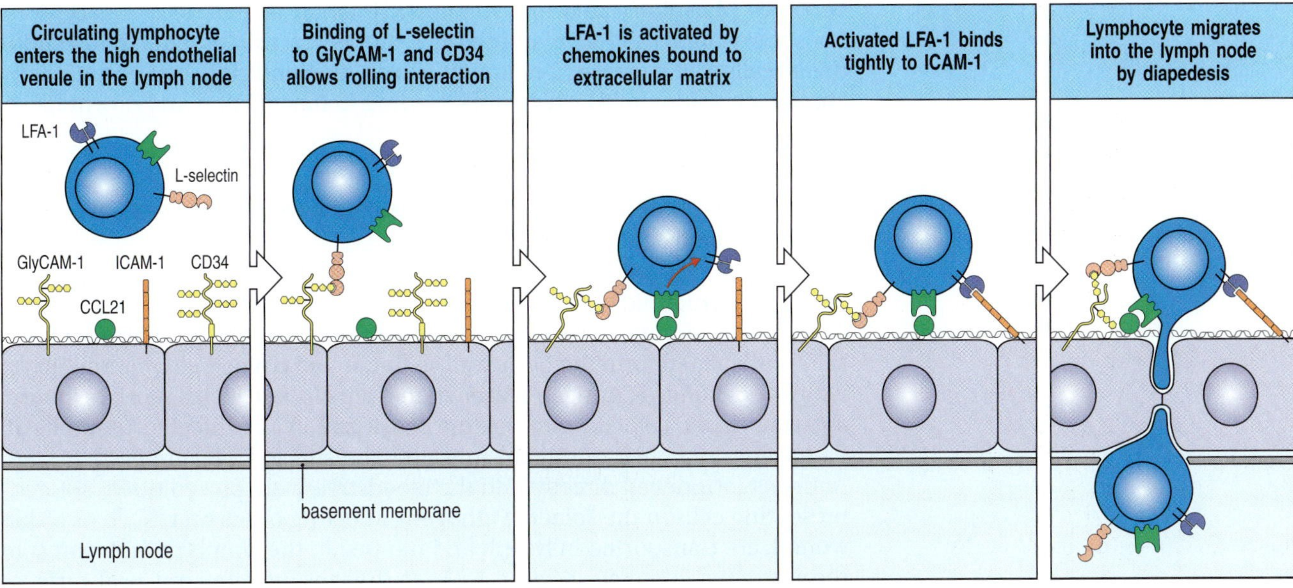

Fig. 9.8 Lymphocytes in the blood enter lymphoid tissue by crossing the walls of high endothelial venules. The first step is the binding of L-selectin on the lymphocyte to sulfated carbohydrates (sulfated sialyl-Lewis[x]) of GlyCAM-1 and CD34 on the HEV. Local chemokines such as CCL21 bound to a proteoglycan matrix on the HEV surface stimulate chemokine receptors on the T cell, leading to the activation of LFA-1. This causes the T cell to bind tightly to ICAM-1 on the endothelial cell, allowing migration across the endothelium. As in the case of neutrophil migration (see Fig. 3.26), matrix metalloproteinases on the lymphocyte surface (not shown) enable it to penetrate the basement membrane.

integrin in the T-cell membrane is also increased by chemokine stimulation, so that integrin molecules migrate into the area of cell–cell contact. This produces stronger binding, which arrests the T cell on the endothelial surface and thus enables it to enter the lymphoid tissue.

The interplay of chemokines and cell-adhesion molecules, together with the architecture of the peripheral lymphoid organs (see Figs 1.18–1.20), virtually guarantees the contact of foreign antigen with the T-cell receptors specific for it. Once naive T cells have arrived in the T-cell zone via the high endothelial venules, CCR7 directs their retention in this location, as stromal cells in the T-cell zone produce CCL21 and **CCL19**, another ligand for CCR7. The T-cell zone is also rich in mature dendritic cells, which produce CCL19 and **CCL18** (**DC-CK**), which also attracts naive T cells. The naive T cells scan the surfaces of dendritic cells for specific peptide:MHC complexes and if they find their antigen and bind to it, they are trapped in the lymph node. If they are not activated by antigen, naive T cells soon leave the lymph node (see Fig. 9.2).

T cells exit from a lymph node via the cortical sinuses, which lead into the medullary sinus and hence into the efferent lymphatic vessel. The egress of T cells from peripheral lymphoid organs involves the lipid molecule **sphingosine 1-phosphate** (**S1P**). This lipid has chemotactic activity and signaling properties similar to those of chemokines, in that the receptors for S1P are G-protein-coupled receptors; S1P signaling activates $G\alpha_i$. S1P is produced by phosphorylation of sphingosine, and can be degraded by S1P lyases or by S1P phosphatases. There seems to be a S1P concentration gradient between the lymphoid tissues and lymph or blood, such that naive T cells expressing a S1P receptor are drawn away from the lymphoid tissues and back into circulation.

T cells activated by antigen in the lymphoid organs downregulate the surface expression of the S1P receptor, **$S1P_1$**, for several days. This loss of $S1P_1$ surface expression is caused by CD69, a surface protein that is induced by T-cell receptor signaling and which acts to internalize $S1P_1$. During this period, T cells cannot respond to the S1P gradient and do not exit the lymphoid organ. After several days of proliferation, as T-cell activation wanes, CD69 expression decreases and $S1P_1$ reappears on the cell's surface, allowing the effector T cell to migrate. The regulation of the exit of both naive and effector lymphocytes from peripheral lymphoid organs by S1P is the basis for a new kind of potential immunosuppressive drug, FTY720 (fingolimod). FTY720 inhibits immune responses in animal models of transplantation and autoimmunity by preventing lymphocytes from returning to the circulation, causing rapid onset of lymphopenia (a lack of lymphocytes in the blood). *In vivo*, FTY720 becomes phosphorylated and mimics S1P as an agonist at S1P receptors. Phosphorylated FTY720 may inhibit lymphocyte exit by effects on endothelial cells that increase tight junction formation and close exit portals, or by chronic activation of S1P receptors, leading to inactivation and downregulation of the receptor.

9-4 T-cell responses are initiated in peripheral lymphoid organs by activated dendritic cells.

Peripheral lymphoid organs were first shown to be important in the initiation of adaptive immune responses by ingenious experiments in which a flap of skin was isolated from the body wall so that it had a blood circulation but no lymphatic drainage. Antigen placed in the flap did not elicit a T-cell response, showing that T cells do not become sensitized in the infected tissue itself. Pathogens and their products must be transported to lymphoid tissues. Antigens introduced directly into the bloodstream are picked up by antigen-presenting cells in the spleen. Pathogens infecting other sites, such as a skin wound, are transported in lymph and trapped in the lymph nodes nearest to the site of infection (see Section 1-15). Pathogens infecting mucosal surfaces are transported directly across the mucosa into lymphoid tissues such as the tonsils or Peyer's patches.

In this chapter we will focus on T-cell activation by dendritic cells as it occurs in organs of the systemic immune system—lymph nodes and spleen. The activation of T cells by dendritic cells in the mucosal immune system follows the same principles, but differs in some details, such as the route by which antigen is delivered and the subsequent circulation patterns of the effector cells, which are described in Chapter 11. The delivery of antigen from an infection to lymphoid tissue is actively aided by the innate immune response. One effect of innate immunity is an inflammatory reaction at the site of infection that increases the rate of entry of blood plasma into the infected tissues and thus increases the drainage of extracellular fluid into the lymph, taking with it free antigen that is then carried to lymphoid tissues. Even more important for the initiation of the adaptive response is the induced maturation of tissue dendritic cells that have taken up particulate and soluble antigens at the site of infection (Fig. 9.9). Immature dendritic cells in those tissues can

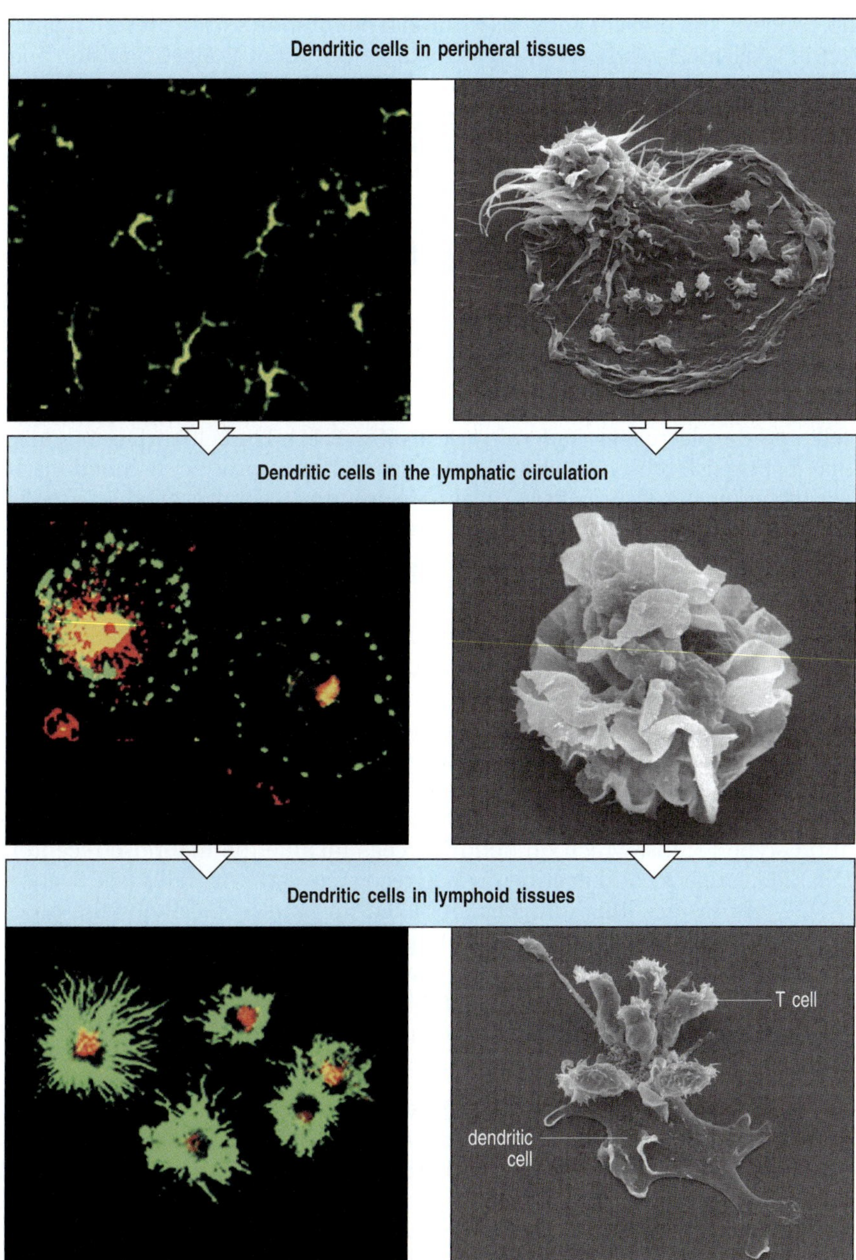

Fig. 9.9 Dendritic cells in different stages of maturation. The left panels show fluorescence micrographs of dendritic cells stained for MHC class II molecules in green and for a lysosomal protein in red. The right panels show scanning electron micrographs of single dendritic cells. Immature dendritic cells (top panels) have many long processes, or dendrites, from which the cells get their name. The cell bodies are difficult to distinguish in the left panel, but the cells contain many endocytic vesicles that stain both for MHC class II molecules and for the lysosomal protein; when these two colors overlap they give rise to a yellow fluorescence. The immature cells are activated and leave the tissues to migrate through the lymphatics to secondary lymphoid tissues. During this migration their morphology changes. The dendritic cells stop phagocytosing antigen, and the staining for lysosomal proteins is beginning to be distinct from that for MHC class II molecules (center left panel). The dendritic cell now has many folds of membrane (right panel), which gave these cells their original name of 'veil' cells. Finally, in the lymph nodes, they become mature dendritic cells that express high levels of peptide:MHC complexes and co-stimulatory molecules, and are very good at stimulating naive CD4 and naive CD8 T cells. These cells do not phagocytose, and the red staining of lysosomal proteins is quite distinct from the green-stained MHC class II molecules displayed at high density on many dendritic processes (bottom left panel). The typical morphology of a mature dendritic cell is shown on the right, as it interacts with a T cell. Fluorescent micrographs courtesy of I. Mellman, P. Pierre, and S. Turley. Scanning electron micrographs courtesy of K. Dittmar.

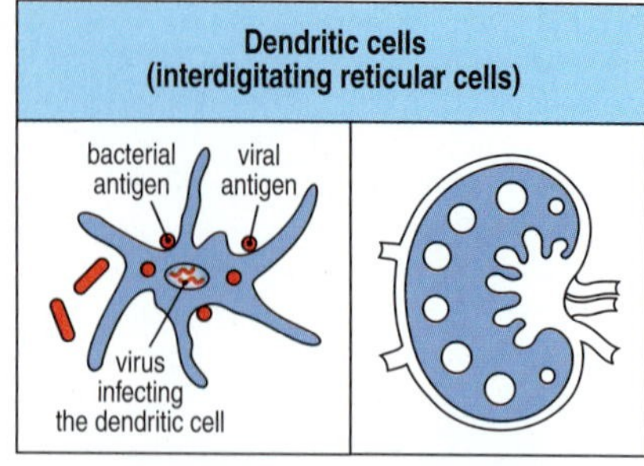

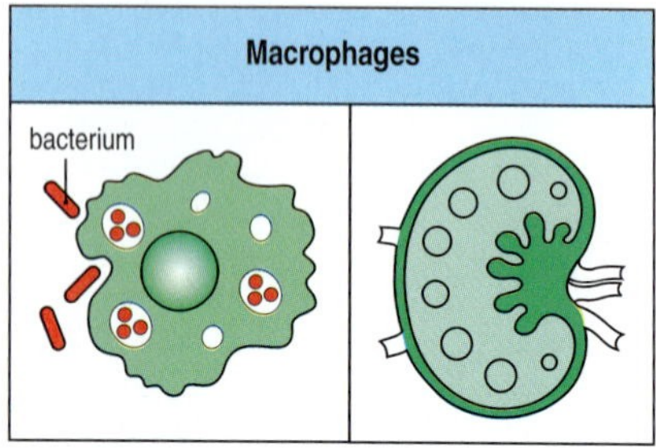

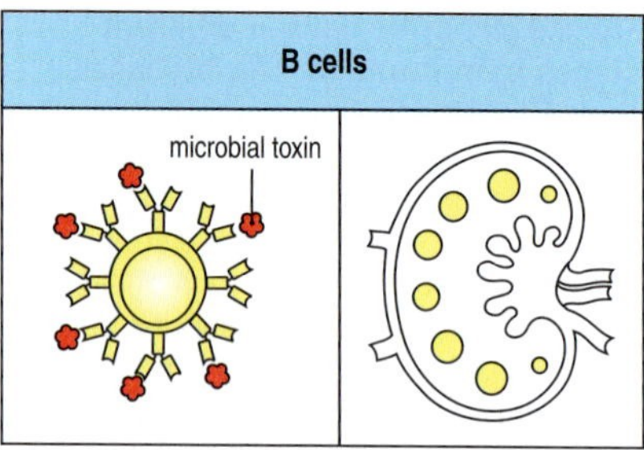

Fig. 9.10 Antigen-presenting cells are distributed differentially in the lymph node. Dendritic cells are found throughout the cortex of the lymph node in the T-cell areas. Mature dendritic cells are by far the strongest activators of naive T cells, and can present antigens from many types of pathogens, such as bacteria or viruses as shown here. Macrophages are distributed throughout the lymph node but are concentrated mainly in the marginal sinus, where the afferent lymph collects before percolating through the lymphoid tissue, and also in the medullary cords, where the efferent lymph collects before passing via the efferent lymphatics into the blood. B cells are found mainly in the follicles and can contribute to neutralizing soluble antigens such as toxins.

be activated via their TLRs and other pathogen-recognition receptors (see Chapter 3), or by tissue damage, or by cytokines produced during the inflammatory response. Activated dendritic cells migrate to the lymph node and express the co-stimulatory molecules that are required, in addition to antigen, for the activation of naive T cells. In the lymphoid tissues, these mature dendritic cells present antigen to naive T lymphocytes and activate any antigen-specific T cells to divide and mature into effector cells that reenter the circulation.

Macrophages, which are found in most tissues including lymphoid tissue, and B cells, which are located primarily in lymphoid tissue, can be similarly induced through the same pathogen-recognition receptors to express co-stimulatory molecules and act as antigen-presenting cells. The distribution of dendritic cells, macrophages, and B cells in a lymph node is shown schematically in Fig. 9.10. Only these three cell types express the specialized co-stimulatory molecules required to activate naive T cells, and they only express these molecules when activated in the context of infection. However, these cells activate T-cell responses in distinct ways. Dendritic cells can take up, process, and present antigens from all types of sources, are present mainly in the T-cell areas, and overwhelmingly drive the initial clonal expansion and differentiation of naive T cells into effector T cells. By contrast, macrophages and B cells specialize in processing and presenting antigens from intracellular pathogens and soluble antigens, respectively, and interact mainly with already primed effector CD4 T cells.

9-5 Dendritic cells process antigens from a wide array of pathogens.

Dendritic cells arise from both myeloid and lymphoid progenitors within the bone marrow; they emerge from the bone marrow to migrate via the blood to tissues throughout the body, and also directly to peripheral lymphoid organs. There are at least two classes of dendritic cells: the conventional dendritic cells and the plasmacytoid dendritic cells (Fig. 9.11). The cell-surface markers that distinguish these two classes, and the interferon-producing functions of plasmacytoid dendritic cells in the innate immune response, are discussed in Sections 3-15 and 3-19. In this chapter we shall focus on the role of conventional dendritic cells in the adaptive immune response—presenting antigens to and activating naive T cells.

Conventional dendritic cells are found under most surface epithelia, and in most solid organs such as the heart and kidneys. There they have an immature phenotype that is associated with low levels of MHC proteins and B7 co-stimulatory molecules, and so are not yet equipped to stimulate naive T cells. Like macrophages, immature dendritic cells are very active in ingesting antigens by phagocytosis using complement receptors and Fc receptors (which recognize the constant regions of antibodies in antigen:antibody complexes), and C-type lectins, which on dendritic cells include the mannose receptor, DEC 205, langerin, and dectin-1, which recognize carbohydrates (see Section 3-1). Other extracellular antigens are taken up nonspecifically by the process of **macropinocytosis**, in which large volumes of surrounding fluid are engulfed. In addition, dendritic cells can detect pathogens by using signaling receptors such as TLRs that recognize pathogen-associated molecular patterns and respond by secreting cytokines that influence the course of both innate and adaptive immune responses.

This versatility enables dendritic cells to present antigens from virtually any type of pathogen, including fungi, parasites, viruses, and bacteria (Fig. 9.12). Phagocytic receptors that between them recognize a wide variety of bacteria and viruses take up extracellular pathogens into the endocytic pathway, where they are processed and presented on MHC class II molecules (see Chapter 6) for recognition by CD4 T cells. Some microbes have evolved thick

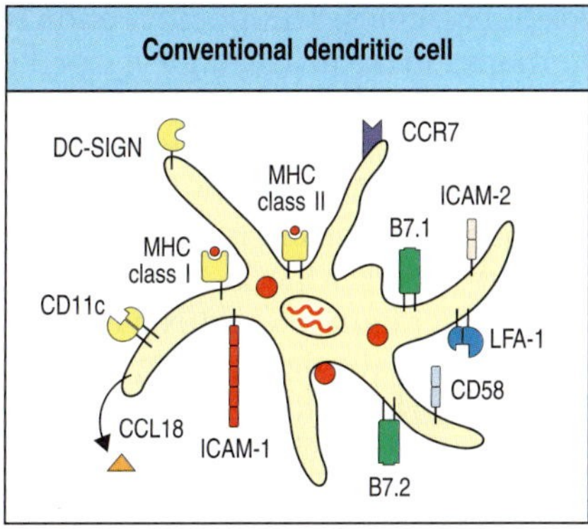

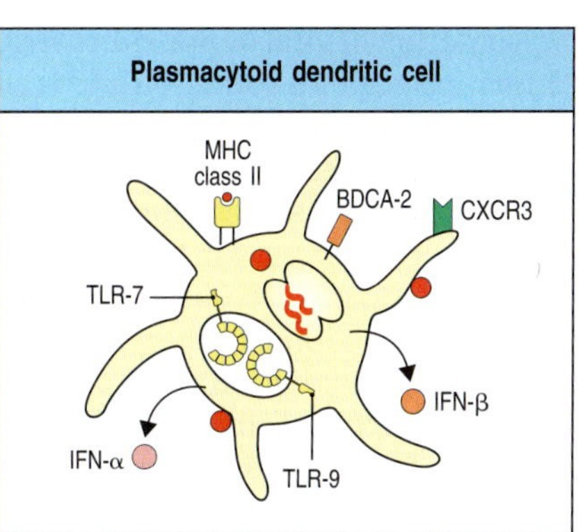

Fig. 9.11 Conventional and plasmacytoid dendritic cells have different roles in the immune response. Mature conventional dendritic cells (left panel) are primarily concerned with the activation of naive T cells. There are several subsets of conventional dendritic cells, but these all process antigen efficiently, and when they are mature they express MHC proteins and co-stimulatory molecules for priming naive T cells. The cell-surface proteins expressed by the mature dendritic cell are described in the text. Immature dendritic cells lack many of the cell-surface molecules shown here but have numerous surface receptors that recognize pathogen molecules, including most of the Toll-like receptors (TLRs). Plasmacytoid dendritic cells (right panel) are sentinels primarily for viral infections, and secrete large amounts of class I interferons. This category of dendritic cell is less efficient in priming naive T cells, but they express the intracellular receptors TLR-7 and TLR-9 for sensing viral infections.

polysaccharide capsules to escape recognition by phagocytic receptors, but these pathogens can be taken up by macropinocytosis and enter the endocytic pathway that way (see Fig. 9.12).

A second route is entry directly into the cytosol, for example through viral infection. Dendritic cells are particularly important in stimulating T-cell responses to viruses, which fail to induce co-stimulatory activity in the other types of antigen-presenting cells. Dendritic cells are susceptible to infection by quite a large number of viruses, which enter the cytoplasm by binding to cell-surface proteins that act as entry receptors. Viral proteins synthesized in the cytoplasm are processed in the proteasome and presented as peptides bound to MHC class I molecules, as in any other type of virus-infected cell (see Chapter 6). This enables dendritic cells to present antigen to and activate naive CD8 T cells, which will differentiate into cytotoxic effector CD8 T cells that recognize and kill virus-infected cells.

Uptake of extracellular virus particles or virus-infected cells by phagocytosis or macropinocytosis into the endocytic pathway can also result in the

Routes of antigen processing and presentation by dendritic cells	Receptor-mediated phagocytosis	Macropinocytosis	Viral infection	Cross-presentation after phagocytic or macropinocytic uptake	Transfer from incoming dendritic cell to resident dendritic cell
					?
Type of pathogen presented	Extracellular bacteria	Extracellular bacteria, soluble antigens, virus particles	Viruses	Viruses	Viruses
MHC molecules loaded	MHC class II	MHC class II	MHC class I	MHC class I	MHC class I
Type of naive T cell activated	CD4 T cells	CD4 T cells	CD8 T cells	CD8 T cells	CD8 T cells

Fig. 9.12 The different routes by which dendritic cells can take up, process, and present protein antigens. Uptake of antigens into the endocytic system, either by receptor-mediated phagocytosis or by macropinocytosis, is considered to be the major route for delivering peptides to MHC class II molecules for presentation to CD4 T cells (first two panels). Production of antigens in the cytosol, for example as a result of viral infection, is thought to be the major route for delivering peptides to MHC class I molecules for presentation to CD8 T cells (third panel). It is possible, however, for exogenous antigens taken into the endocytic pathway to be delivered into the cytosol for eventual delivery to MHC class I molecules for presentation to CD8 T cells, a process called cross-presentation (fourth panel). Finally, it seems that antigens can be transmitted from one dendritic cell to another, particularly for presentation to CD8 T cells, although the details of this route are still unclear (fifth panel).

presentation of viral peptides on MHC class I molecules. This phenomenon is known as cross-presentation and is an alternative to the usual endocytic pathway for MHC class I antigen processing described in Section 6-3. The result is that viruses that do not infect dendritic cells can stimulate the activation of CD8 T cells. Cross-presentation is performed most efficiently by a subset of dendritic cells present in both mice and humans. This subset is characterized by expression of CD8α in the mouse, and by the marker BDCA-3 in humans. Any viral infection can therefore lead to the generation of cytotoxic effector CD8 T cells. In addition, viral peptides presented on the dendritic cell's MHC class II molecules activate naive CD4 T cells, which leads to the production of effector CD4 T cells that stimulate the production of antiviral antibodies by B cells and produce cytokines that enhance the immune response.

Movie 9.2

In some cases, such as infections with herpes simplex or influenza viruses, the dendritic cells that migrate to the lymph nodes from peripheral tissues may not be the same cells that finally present antigen to naive T cells. In herpes simplex infection, for example, immature dendritic cells resident in the skin capture antigen and transport it to the draining lymph nodes (Fig. 9.13). There, some antigen is transferred to a CD9-positive subset of dendritic cells resident in the lymph node, which seem to be the dominant dendritic cells responsible for priming naive CD8 T cells in this disease. This type of transfer means that antigens from viruses that infect but rapidly kill dendritic cells can still be presented by uninfected dendritic cells that have been activated via their TLRs and can take up the dying dendritic cells and cross-present this material.

Langerhans cells are a type of immature conventional dendritic cells that reside in the skin. They are actively phagocytic and contain large granules known as Birbeck granules—which are an endosomal recycling compartment formed by the accumulation of **langerin**, a transmembrane lectin with mannose-binding specificity. In the presence of an infection in the skin, Langerhans cells will pick up antigens from pathogens by any of the routes outlined above. Encounter with pathogens also triggers their migration to the regional lymph nodes (see Fig. 9.13). Here they rapidly lose the ability to take up antigen but briefly increase the synthesis of MHC molecules. On arriving

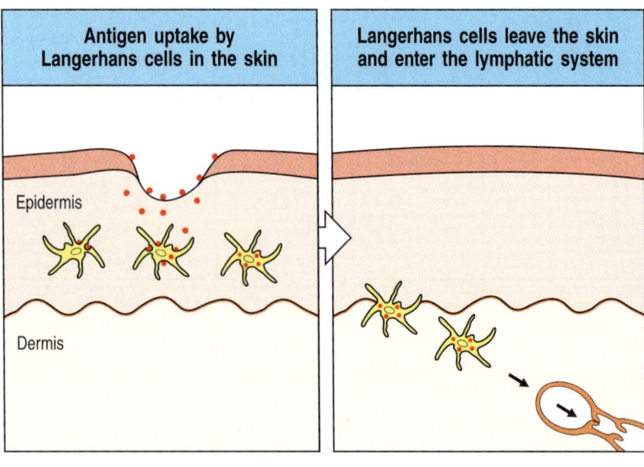

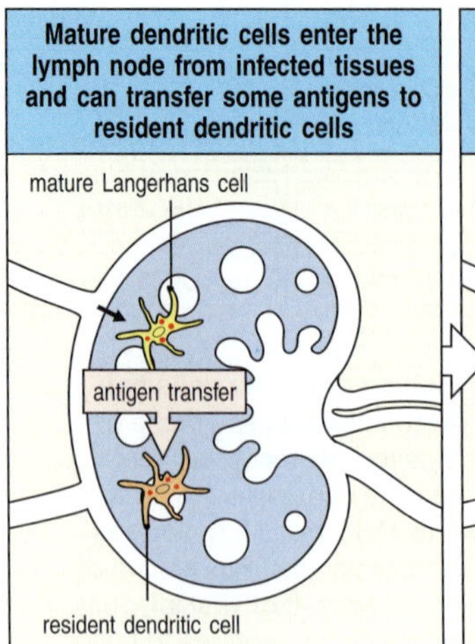

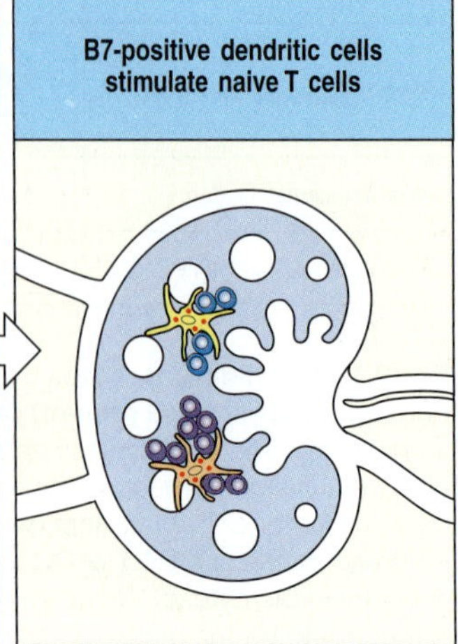

Fig. 9.13 Langerhans cells take up antigen in the skin, migrate to the peripheral lymphoid organs, and present foreign antigens to T cells. Langerhans cells (yellow) are one type of immature dendritic cell that resides in the epidermis. They ingest antigen in various ways but have no co-stimulatory activity (first panel). In the presence of infection, they take up antigen locally and then migrate to the lymph nodes (second panel). There they differentiate into mature dendritic cells that can no longer ingest antigen but have co-stimulatory activity. Now they can prime both naive CD8 and CD4 T cells. In the case of some viral infections, for example with herpes simplex virus, some dendritic cells arriving from the site of infection seem able to transfer antigen to resident dendritic cells (orange) in the lymph nodes (third panel) for presentation of class I MHC-restricted antigens to naive CD8 T cells (fourth panel).

in the lymph node they also express co-stimulatory B7 molecules and large numbers of adhesion molecules, which enable them to interact with antigen-specific T cells. In this way the Langerhans cells capture antigens from invading pathogens and differentiate into mature dendritic cells that are uniquely fitted for presenting these antigens to and activating naive T cells.

Immature dendritic cells resident in the spleen are ideally suited to sample antigens from infectious agents present in the blood, such as malaria parasites or bacteria during sepsis. Dendritic cells also present alloantigens derived from transplanted organs, thus triggering graft rejection (discussed in Chapter 15), and present the environmental protein antigens that cause allergies (discussed in Chapter 14). In principle, any nonself antigen will be immunogenic if it is taken up and subsequently presented by an activated dendritic cell. The normal physiology of dendritic cells is to migrate, and this is increased by stimuli, such as transplantation, that activate the linings of the lymphatics; this is why dendritic cells are so potent at stimulating reactions against transplanted tissues.

9-6 Pathogen-induced TLR signaling in immature dendritic cells induces their migration to lymphoid organs and enhances antigen processing.

We shall now look at dendritic cell maturation in more detail. Working together in ways that are not yet completely understood, TLR signaling and signals received from chemokines convert the immature tissue dendritic cell into a mature dendritic cell arriving in the lymphoid tissues. When an infection occurs, dendritic cells can capture pathogens by means of phagocytic receptors such as DEC 205, and then activate responses to these pathogens through pattern recognition receptors such as TLRs (Fig. 9.14, top panel). In addition to the pattern recognition receptors described in Chapter 3, dendritic cells express various lectins that can recognize and signal in response to pathogens. For example, the lectin DC-SIGN binds mannose and fucose residues present on a wide range of pathogens; likewise, dectin-1, which is also expressed by macrophages and neutrophils, recognizes β-1,3-linked glucans found in fungal cell walls (see Fig. 3.2). These signals are of key importance in determining whether an adaptive immune response will be initiated. Several members of the TLR family are expressed on tissue dendritic cells and are thought to be involved in detecting and signaling the presence of the various classes of pathogens (see Fig. 3.16). In humans, conventional dendritic cells

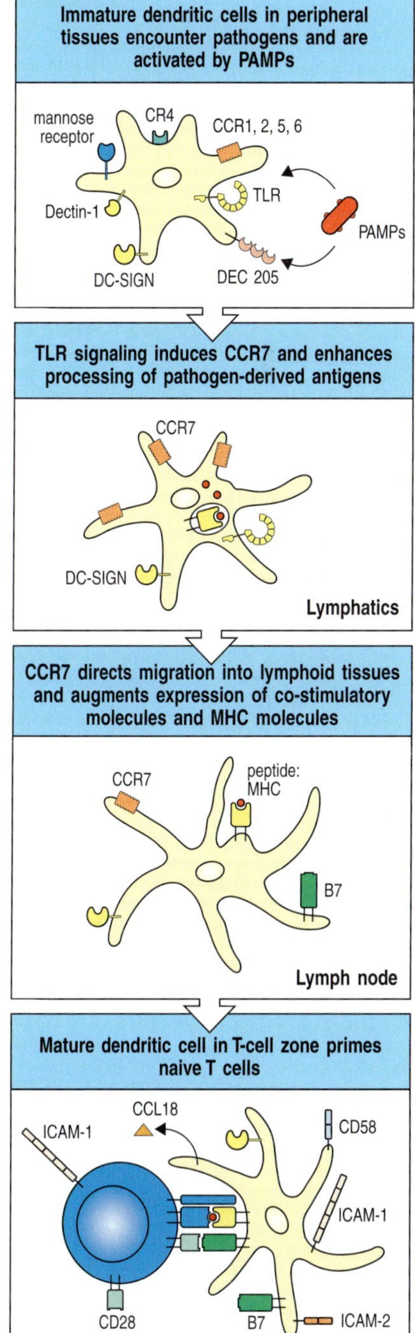

Fig. 9.14 Conventional dendritic cells mature through at least two definable stages to become potent antigen-presenting cells in peripheral lymphoid tissue. Immature dendritic cells originate from bone marrow progenitors and migrate via the blood, from which they enter and populate most tissues, including some direct entry into peripheral lymphoid tissues. Entry to particular tissues is based on the particular chemokine receptors they express: CCR1, CCR2, CCR5, CCR6, CXCR1, and CXCR2 (not all shown here, for simplicity). Immature dendritic cells in tissues are highly phagocytic via receptors such as dectin-1, DEC 205, DC-SIGN, and langerin, and are actively macropinocytic, but they do not express co-stimulatory molecules. They carry most of the different types of Toll-like receptors (TLRs) (see the text). At sites of infection, immature dendritic cells are exposed to pathogens, leading to activation of their TLRs (top panel). TLR signaling causes the dendritic cells to become licensed and begin to undergo maturation, which involves induction of the chemokine receptor CCR7. TLR signaling also increases the processing of antigens taken up into phagosomes (second panel). Dendritic cells expressing CCR7 are sensitive to CCL19 and CCL21, which directs them to the draining lymphoid tissues. CCL19 and CCL21 provide further maturation signals, which result in higher levels of co-stimulatory B7 molecules and MHC molecules (third panel). In the draining lymph node, mature conventional dendritic cells have become powerful activators of naive T cells but are no longer phagocytic. They express B7.1, B7.2, and high levels of MHC class I and class II molecules, as well as high levels of the adhesion molecules ICAM-1, ICAM-2, LFA-1, and CD58 (bottom panel).

express all known TLRs except for TLR-9, which is, however, expressed by plasmacytoid dendritic cells along with TLR-1 and TLR-7, and other TLRs to a lesser degree. Other receptors that can bind pathogens, such as receptors for complement, or phagocytic receptors such as the mannose receptor, may contribute to dendritic-cell activation as well as to phagocytosis.

TLR signaling results in a significant alteration in the chemokine receptors expressed by dendritic cells, which facilitates their entry into peripheral lymphoid tissues (Fig. 9.14, second panel). This change in dendritic cell behavior is often called **licensing**, as the cells are now embarked on the program of differentiation that will enable them to activate T cells. TLR signaling induces expression of the receptor CCR7, which makes the activated dendritic cells sensitive to the chemokine CCL21 produced by lymphoid tissue and induces their migration through the lymphatics and into the local lymphoid tissues. Whereas T cells have to cross the high endothelial venule wall to leave the blood and reach the T-cell zones, dendritic cells entering via the afferent lymphatics can migrate directly into the T-cell zones from the marginal sinus.

Pathogen-derived proteins that enter the immature dendritic cell via phagocytosis are processed in the endocytic compartment for presentation by MHC class II molecules (see Fig. 9.14, second panel). The efficiency of antigen processing by this endocytic compartment may be augmented by concurrent TLR signaling activated by encounter with pathogens within phagosomes. Although the precise details of this mechanism are unclear, this may preferentially help to deliver pathogen-derived peptides into the pool of peptide:MHC complexes that are transported to the surface of the dendritic cell, where they can be presented to naive T cells in the context of co-stimulation.

CCL21 signaling through CCR7 not only induces the migration of dendritic cells into lymphoid tissue, but it also contributes to their maturation (Fig. 9.14, third panel). By the time mature dendritic cells arrive within lymphoid tissues, they are no longer able to engulf antigens by phagocytosis or macropinocytosis. They now express very high levels of long-lived MHC class I and MHC class II molecules, which enables them to stably present peptides from pathogens already taken up and processed. Equally importantly, by this time they also have high levels of co-stimulatory molecules on their surface. These are two structurally related transmembrane glycoproteins called B7.1 (CD80) and B7.2 (CD86), which deliver co-stimulatory signals by interacting with receptors on naive T cells (see Section 7-15). Mature dendritic cells also express very high levels of adhesion molecules, including DC-SIGN, and they secrete the chemokine CCL18, which specifically attracts naive T cells. Together, these properties enable the dendritic cell to stimulate strong responses in naive T cells (Fig. 9.14, bottom panel).

Despite their enhanced presentation of pathogen-derived antigens, mature dendritic cells also present some self peptides, which could present a problem for the maintenance of self-tolerance. The T-cell receptor repertoire has, however, been purged of receptors that recognize self peptides presented in the thymus (see Chapter 8), so that T-cell responses against most ubiquitous self antigens are avoided. In addition, dendritic cells in the lymphoid tissues that have not been activated by infection will bear self-peptide:MHC complexes on their surface, derived from the breakdown of their own proteins and tissue proteins present in the extracellular fluid. Because these cells do not express the appropriate co-stimulatory molecules, however, they do not have the same capacity to activate naive T cells as do activated, mature dendritic cells. Although the details are still unclear, the presentation of self peptides by immature, or 'unlicensed,' dendritic cells instead induces a state of unresponsiveness in naive T cells to these antigens.

Intracellular degradation of pathogens is thought to reveal pathogen components, other than peptides, that trigger dendritic cell activation. For example, bacterial or viral DNA containing unmethylated CpG dinucleotide motifs induces the rapid activation of plasmacytoid dendritic cells, probably as a consequence of recognition of the DNA by TLR-9, which is present in intracellular vesicles (see Fig. 3.10). Exposure to bacterial DNA activates NFκB and mitogen-activated protein kinase (MAPK) signaling pathways (see Fig. 7.17), leading to the production of cytokines such as IL-6, IL-12, IL-18, and interferon (IFN)-α and IFN-β by dendritic cells. In turn, these cytokines act on the dendritic cells themselves to augment the expression of co-stimulatory molecules. Heat-shock proteins are another internal bacterial constituent that can activate the antigen-presenting function of dendritic cells. Some viruses are thought to be recognized by TLRs inside the dendritic cell via double-stranded RNA produced in the course of their replication. As discussed in Section 3-9, viral infection also induces the production of IFN-α and IFN-β by all types of infected cells; both of these interferons can further activate dendritic cells to increase the expression of co-stimulatory molecules.

The induction of co-stimulatory activity in antigen-presenting cells by common microbial constituents is believed to allow the immune system to distinguish antigens borne by infectious agents from antigens associated with innocuous proteins, including self proteins. Indeed, many foreign proteins do not induce an immune response when injected on their own, presumably because they fail to induce co-stimulatory activity in antigen-presenting cells. When such protein antigens are mixed with bacteria, however, they become immunogenic, because the bacteria induce the essential co-stimulatory activity in cells that ingest the protein. Bacteria or bacterial components used in this way are known as adjuvants (see Appendix I, Section A-4). We will see in Chapter 15 how self proteins mixed with bacterial adjuvants can induce autoimmune disease, illustrating the crucial importance of the regulation of co-stimulatory activity in the discrimination of self from nonself.

9-7 Plasmacytoid dendritic cells produce abundant type I interferons and may act as helper cells for antigen presentation by conventional dendritic cells.

Plasmacytoid dendritic cells are thought to act as sentinels in early defense against viral infection on the basis of their expression of TLRs and the intracellular nucleic-acid-sensing RIG-I-like helicases, and their production of antiviral type I interferons (see Sections 3-9 and 3-19). For several reasons, they are not thought to be involved in a major way in the antigen-specific activation of naive T cells. Plasmacytoid dendritic cells express fewer MHC class II and co-stimulatory molecules on their surface, and they process antigens less efficiently than conventional dendritic cells. In addition, unlike conventional dendritic cells, plasmacytoid dendritic cells do not cease the synthesis and recycling of MHC class II molecules after being activated. This means that they rapidly recycle their surface MHC II molecules and so cannot present pathogen-derived peptide:MHC complexes to T cells for extended periods, as conventional dendritic cells do.

Plasmacytoid dendritic cells may, however, act as helper cells for antigen presentation by conventional dendritic cells. This activity was revealed by studies in mice infected with the intracellular bacterium *Listeria monocytogenes*. Normally, IL-12 made by conventional dendritic cells induces CD4 T-cells to produce abundant IFN-γ, which helps macrophages kill the bacteria. When plasmacytoid dendritic cells were experimentally eliminated, IL-12 production by conventional dendritic cells decreased, and the mice become susceptible to *Listeria*. It seems that the plasmacytoid dendritic cells interact with conventional dendritic cells to sustain IL-12 production. Activation of plasmacytoid dendritic cells through TLR-9 induces the expression of **CD40 ligand** (CD40L or CD154), a TNF-family transmembrane cytokine, which binds to **CD40**, a TNF-family receptor that is expressed by activated conventional dendritic cells. This interaction enables conventional dendritic cells to sustain production of the pro-inflammatory cytokine IL-12, strengthening the IL-12-induced production of IFN-γ by T cells. Human and mouse plasmacytoid dendritic cells can also produce IL-12 themselves, although in smaller amounts than conventional dendritic cells do. Finally, the interferons produced by plasmacytoid dendritic cells can promote the development of conventional dendritic cells from blood monocytes.

9-8 Macrophages are scavenger cells that can be induced by pathogens to present foreign antigens to naive T cells.

The two other cell types that can act as antigen-presenting cells to T cells are macrophages and B cells, but there is an important distinction between the function of antigen presentation by these cells as compared with dendritic cells. It is likely that macrophages and B cells do not present antigen mainly to activate naive T cells, but rather to make use of the effector functions of T cells that have been previously primed by conventional dendritic cells. As we learned in Chapter 3, many of the microorganisms that enter the body are engulfed and destroyed by phagocytes, which provide an innate, antigen-nonspecific first line of defense against infection. However, pathogens have developed many mechanisms to avoid elimination by innate immunity, such as resisting the killing properties of phagocytes. Macrophages that have ingested microorganisms but have failed to destroy them can use antigen presentation to recruit the adaptive immune response to enhance their microbicidal capacities, as we will see later in this chapter.

As well as being resident in tissues, macrophages are found in lymphoid organs (see Fig. 9.10). They are present in many areas of the lymph node, especially in the marginal sinus, where the afferent lymph enters the lymphoid tissue, and in the medullary cords, where the efferent lymph collects before flowing into the blood (see Fig. 1.18). Their main role is to ingest microbes and particulate antigens and so prevent them from entering the blood. Although macrophages do process ingested microbes and antigens and display peptide:MHC class II antigens on their surface in conjunction with co-stimulatory molecules, it is thought that their main function in lymphoid tissues is as scavengers of pathogens and of apoptotic lymphocytes.

Resting macrophages have few or no MHC class II molecules on their surface and do not express B7. The expression of both MHC class II molecules and B7 is induced by the ingestion of microorganisms and recognition of their foreign molecular patterns. Macrophages, like tissue dendritic cells, have a variety of receptors that recognize microbial surface components (see Chapter 3). Receptors such as dectin-1, scavenger receptors, and complement receptors take up microorganisms into phagosomes, where they are degraded, thus producing peptides for presentation, while recognition of pathogen components via TLRs triggers intracellular signaling that contributes to the expression of MHC class II molecules and B7. Macrophages may also take up soluble antigens into the endocytic pathway through the process of pinocytosis, although this seems to be less efficient than receptor-mediated endocytosis.

Macrophages continuously scavenge dead or dying cells, which are rich sources of self antigens, so it is particularly important that they should not activate naive T cells in the absence of microbial infection. The Kupffer cells of the liver sinusoids and the macrophages of the splenic red pulp, in particular, remove large numbers of dying cells from the blood daily. Kupffer cells express little MHC class II and no TLR-4, the receptor that signals the presence of bacterial LPS. Thus, although they generate large amounts of self peptides in their endosomes, these macrophages are not likely to elicit an autoimmune response.

At present, there is very little evidence that macrophages ever initiate T-cell immunity, so it is likely that their expression of co-stimulatory molecules is more important for expanding primary or secondary responses already initiated by dendritic cells. This might be envisaged to be important for the maintenance and functioning of effector or memory T cells that enter the site of infection.

9-9 B cells are highly efficient at presenting antigens that bind to their surface immunoglobulin.

B cells are uniquely adapted to bind specific soluble molecules through their cell-surface immunoglobulin, and will internalize the bound molecules by receptor-mediated endocytosis. If the antigen contains a protein component, the B cell will process the internalized protein to peptide fragments and then display peptide fragments as peptide:MHC class II complexes. This mechanism of antigen uptake is extremely efficient, concentrating the specific antigen in the endocytic pathway. B cells also constitutively express high levels of MHC class II molecules, and so high levels of specific peptide:MHC class II complexes appear on the B-cell surface (Fig. 9.15). This pathway of antigen presentation allows B cells to be targeted by antigen-specific CD4 T cells that have been previously activated, and which drive their differentiation, as we will see in Chapter 10.

B cells do not constitutively express co-stimulatory activity but, as with dendritic cells and macrophages, they can be induced by various microbial constituents to express B7 molecules. In fact, B7.1 was first identified as a protein on B cells activated by LPS, and B7.2 is predominantly expressed by B cells *in vivo*. Soluble protein antigens are not abundant during infections; most natural antigens, such as bacteria and viruses, are particulate, and soluble bacterial toxins act by binding to cell surfaces and so are present only at low concentrations in solution. Some natural immunogens enter the body as soluble molecules, however; examples are insect toxins, anticoagulants injected by blood-sucking insects, snake venoms, and many allergens. Nevertheless, it is unlikely that B cells are important in priming naive T cells to soluble antigens in natural immune responses. Tissue dendritic cells can take up soluble antigens by macropinocytosis, and although they could not concentrate these antigens as antigen-specific B cells do, dendritic cells are more likely to encounter a naive T cell with the appropriate antigen specificity than are the limited number of antigen-specific B cells. The chances of a B cell encountering a T cell that can recognize the peptide antigens it displays are greatly increased once a naive T cell has been detained in lymphoid tissue by finding its antigen on the surface of a dendritic cell and has undergone clonal expansion.

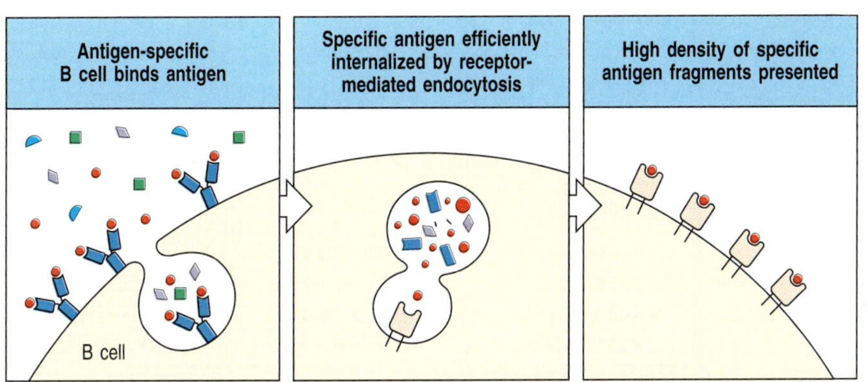
Antigen-specific B cell binds antigen
Specific antigen efficiently internalized by receptor-mediated endocytosis
High density of specific antigen fragments presented
B cell

Fig. 9.15 B cells can use their surface immunoglobulin to present specific antigen very efficiently to T cells. Surface immunoglobulin allows B cells to bind and internalize specific antigen very efficiently, especially if the antigen is present as a soluble protein, as most toxins are. The internalized antigen is processed in intracellular vesicles where it binds to MHC class II molecules. The vesicles are transported to the cell surface where the foreign-peptide:MHC class II complexes can be recognized by T cells. When the protein antigen is not specific for the B-cell receptor, its internalization is inefficient and only a few fragments of such proteins are subsequently presented at the B-cell surface (not shown).

The three types of antigen-presenting cell are compared in Fig. 9.16. In each of these cell types the expression of co-stimulatory activity is controlled so as to provoke responses against pathogens while avoiding immunization against self.

Summary.

An adaptive immune response is generated when naive T cells contact mature, activated antigen-presenting cells in the peripheral lymphoid organs. To ensure that rare antigen-specific T cells survey the body effectively for equally rare pathogen-bearing antigen-presenting cells, T cells continuously recirculate through the lymphoid organs and thus can sample antigens brought in by antigen-presenting cells from many different sites of infection. The migration of naive T cells into lymphoid organs is guided by the chemokine receptor CCR7, which binds the chemokine CCL21 produced by stromal cells in the T-cell zones of peripheral lymphoid organs. L-selectin expressed by naive T cells initiates their rolling along the specialized surfaces of high endothelial venules, and contact with CCL21 there induces a switch in the integrin LFA-1 expressed by T cells to a configuration with affinity for the ICAM-1 expressed on the venule endothelium. This initiates strong adhesion, diapedesis, and migration of the T cells into the T-cell zone. There, naive T cells meet antigen-bearing dendritic cells. There are two main populations of dendritic cells: CD11c-positive conventional dendritic cells and plasmacytoid dendritic cells. Conventional dendritic cells continuously survey peripheral tissues for invading pathogens and are the dendritic cells responsible for activating naive T cells. Contact with pathogens delivers signals to dendritic cells via TLRs and other receptors that accelerate antigen processing and the production of

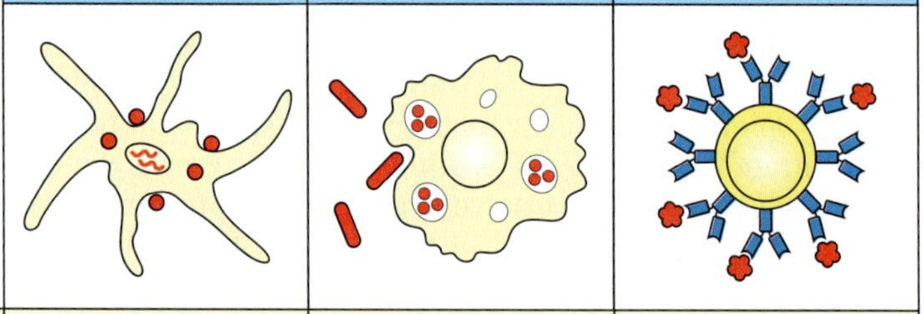

	Dendritic cells	Macrophages	B cells
Antigen uptake	+++ Macropinocytosis and phagocytosis by tissue dendritic cells	+++ Macropinocytosis +++ Phagocytosis	Antigen-specific receptor (Ig) ++++
MHC expression	Low on immature dendritic cells High on dendritic cells in lymphoid tissues	Inducible by bacteria and cytokines – to +++	Constitutive Increases on activation +++ to ++++
Co-stimulation delivery	Constitutive by mature, nonphagocytic lymphoid dendritic cells ++++	Inducible – to +++	Inducible – to +++
Location	Ubiquitous throughout the body	Lymphoid tissue Connective tissue Body cavities	Lymphoid tissue Peripheral blood
Effect	Results in activation of naive T cells	Results in activation of macrophages	Results in delivery of help to B cell

Fig. 9.16 The properties of the various antigen-presenting cells. Dendritic cells, macrophages, and B cells are the main cell types involved in the presentation of foreign antigens to T cells. These cells vary in their means of antigen uptake, MHC class II expression, co-stimulator expression, the type of antigen they present effectively, their locations in the body, and their surface adhesion molecules (not shown). Antigen presentation by dendritic cells is primarily involved in activating naive T cells for expansion and differentiation. Macrophages and B cells present antigen primarily to receive specific help from effector T cells in the form of cytokines or surface molecules.

foreign-peptide:self-MHC complexes. TLR signaling also induces expression of CCR7 by the dendritic cells, which directs their migration to T-cell zones of peripheral lymphoid organs where they encounter and activate naive T cells.

Macrophages and B cells can also process particulate or soluble antigens from pathogens to be presented as peptide:MHC complexes to T cells. Whereas antigen presentation by dendritic cells is exclusively to activate native T cells, antigen presentation by macrophages and B cells enables them to make use of the effector activities of previously activated antigen-specific T cells. For example, by presenting antigens of ingested pathogens, macrophages gain help from IFN-γ-producing CD4 T cells to augment their intracellular killing of these pathogens. Presentation of antigens by B cells recruits help from T cells to stimulate antibody production and class switching, as we shall discuss in Chapter 10. In all three types of antigen-presenting cells, the expression of co-stimulatory molecules is activated in response to signals from receptors that also function in innate immunity to signal the presence of infectious agents.

Priming of naive T cells by pathogen-activated dendritic cells.

T-cell responses are initiated when a mature naive CD4 or CD8 T cell encounters a properly activated antigen-presenting cell displaying the appropriate peptide:MHC ligand. We will now describe the generation of effector T cells from naive T cells. The activation and differentiation of naive T cells, often called priming, is distinct from the later responses of effector T cells to antigen on their target cells, and from the responses of primed memory T cells to subsequent encounters with the same antigen. Priming of naive CD8 T cells generates cytotoxic T cells capable of directly killing pathogen-infected cells. CD4 cells develop into a diverse array of effector cell types depending on the nature of the signals they receive during priming. CD4 effector activity can include cytotoxicity, but more frequently it involves the secretion of a set of cytokines that directs the target cell to make a particular response.

9-10 Cell-adhesion molecules mediate the initial interaction of naive T cells with antigen-presenting cells.

As they migrate through the cortical region of the lymph node, naive T cells bind transiently to each antigen-presenting cell that they encounter. Mature dendritic cells bind naive T cells very efficiently through interactions between LFA-1 and CD2 on the T cell, and ICAM-1, ICAM-2, and CD58 on the dendritic cell (Fig. 9.17). Perhaps because of this synergy, the precise role of each adhesion molecule has been difficult to distinguish. People lacking LFA-1 can have normal T-cell responses, and this also seems to be true for genetically engineered mice lacking CD2. It would not be surprising if there were enough redundancy in the molecules mediating T-cell adhesive interactions to enable immune responses to occur in the absence of any one of them; such molecular redundancy has been observed in other complex biological processes.

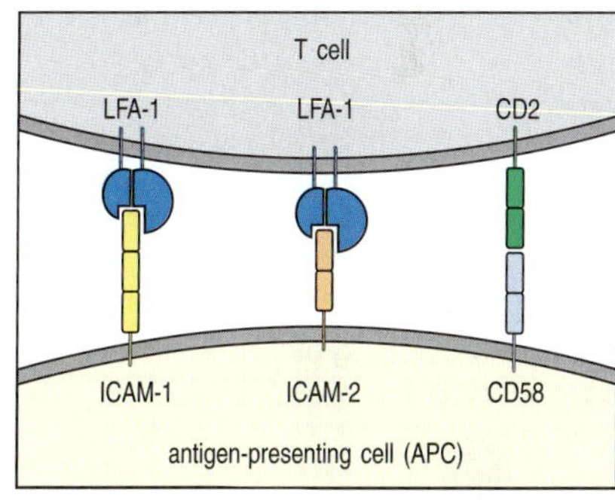

Fig. 9.17 Cell-surface molecules of the immunoglobulin superfamily are important in the interactions of lymphocytes with antigen-presenting cells. In the initial encounter of T cells with antigen-presenting cells, CD2 binding to CD58 on the antigen-presenting cell synergizes with LFA-1 binding to ICAM-1 and ICAM-2. LFA-1 is the $\alpha_L:\beta_2$ integrin heterodimer CD11a:CD18. ICAM-1 and ICAM-2 are also known as CD54 and CD102, respectively.

The transient binding of naive T cells to antigen-presenting cells is crucial in providing time for T cells to sample large numbers of MHC molecules on each antigen-presenting cell for the presence of a specific peptide. In those rare cases in which a naive T cell recognizes a peptide:MHC ligand, signaling through the T-cell receptor induces a conformational change in LFA-1 that greatly increases its affinity for ICAM-1 and ICAM-2. This conformational change is the same as that induced by signaling through chemokine receptors

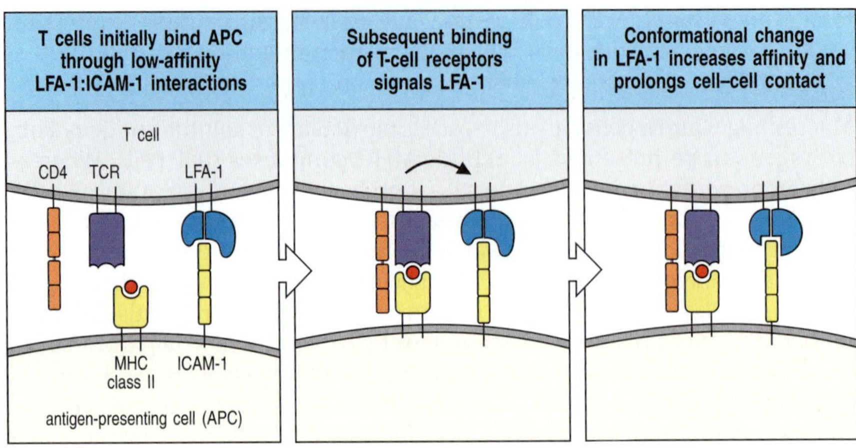

Fig. 9.18 Transient adhesive interactions between T cells and antigen-presenting cells are stabilized by specific antigen recognition. When a T cell binds to its specific ligand on an antigen-presenting cell, intracellular signaling through the T-cell receptor (TCR) induces a conformational change in LFA-1 that causes it to bind with higher affinity to ICAMs on the antigen-presenting cell. The T cell shown here is a CD4 T cell.

during the migration of naive T cells into a peripheral lymphoid organ (see Section 9-2). The change in LFA-1 stabilizes the association between the antigen-specific T cell and the antigen-presenting cell (Fig. 9.18). The association can persist for several days, during which time the naive T cell proliferates and its progeny, which also adhere to the antigen-presenting cell, differentiate into effector T cells.

Most encounters of T cells with antigen-presenting cells do not, however, result in the recognition of an antigen. In this case, the T cell must be able to separate efficiently from the antigen-presenting cell so that it can continue to migrate through the lymph node, eventually leaving via the efferent lymphatic vessels to reenter the blood and continue circulating. Dissociation, like stable binding, may also involve signaling between the T cell and the antigen-presenting cells, but little is known of its mechanism.

9-11 Antigen-presenting cells deliver three kinds of signals for the clonal expansion and differentiation of naive T cells.

When discussing the activation of naive T cells, it is useful to consider three different types of signals (Fig. 9.19). Signal 1 comprises those antigen-specific signals derived from the interaction of a specific peptide:MHC complex with the T-cell receptor. Engagement of the T-cell receptor with its peptide antigen is essential for activating a naive T cell, but even if the co-receptor—CD4 or CD8—is also ligated, this does not on its own stimulate the T cell to fully proliferate and differentiate into effector T cells. Expansion and differentiation of naive T cells involves at least two other kinds of signals, which are generally delivered by the same antigen-presenting cell. These additional signals can be divided into the co-stimulatory signals that promote the survival and

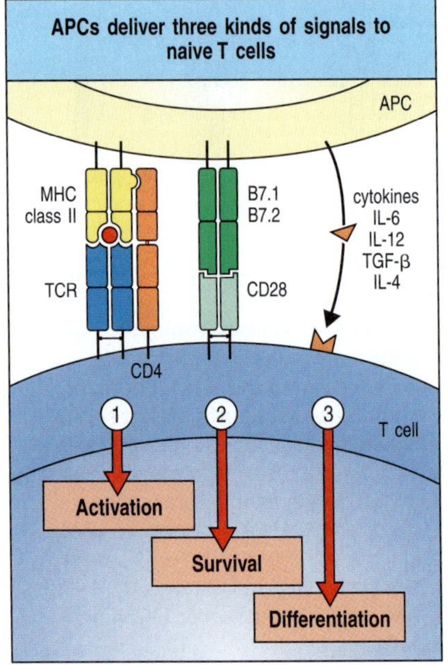

Fig. 9.19 Three kinds of signals are involved in activation of naive T cells by antigen-presenting cells. Binding of the foreign-peptide:self-MHC complex by the T-cell receptor and, in this example, a CD4 co-receptor, transmits a signal (arrow 1) to the T cell that antigen has been encountered. Effective activation of naive T cells requires a second signal (arrow 2), the co-stimulatory signal, to be delivered by the same antigen-presenting cell (APC). In this example, CD28 on the T cell encountering B7 molecules on the antigen-presenting cell delivers signal 2, whose net effect is the increased survival and proliferation of the T cell that has received signal 1. ICOS and various members of the TNF receptor family may also provide co-stimulatory signals. For CD4 T cells in particular, different pathways of differentiation produce subsets of effector T cells that carry out different effector responses, depending on the nature of a third signal (arrow 3) delivered by the antigen-presenting cell. Cytokines are commonly, but not exclusively, involved in directing this differentiation.

expansion of the T cells (signal 2), and those that are involved in directing T-cell differentiation into the different subsets of effector T cells (signal 3).

The best-characterized co-stimulatory molecules that deliver signal 2 are the B7 molecules. These homodimeric members of the immunoglobulin superfamily are found exclusively on the surfaces of cells, such as dendritic cells, that stimulate naive T-cell proliferation (see Section 9-6). Their role in co-stimulation has been demonstrated by transfecting fibroblasts that express a T-cell ligand with genes encoding B7 molecules and showing that the fibroblasts could then stimulate the clonal expansion of naive T cells. The receptor for B7 molecules on the T cell is **CD28**, a member of the immunoglobulin superfamily (see Section 7-15). Ligation of CD28 by B7 molecules or by anti-CD28 antibodies is necessary for the optimal clonal expansion of naive T cells, whereas anti-B7 antibodies, which inhibit the binding of B7 molecules to CD28, have been shown experimentally to inhibit T-cell responses.

9-12 CD28-dependent co-stimulation of activated T cells induces expression of the T-cell growth factor interleukin-2 and the high-affinity IL-2 receptor.

Naive T cells are found as small resting cells with condensed chromatin and scanty cytoplasm, and synthesize little RNA or protein. On activation, they reenter the cell cycle and divide rapidly to produce the large numbers of progeny that will differentiate into effector T cells. Their proliferation and differentiation are driven by the cytokine **interleukin-2** (**IL-2**), which is produced by the activated T cell itself.

The initial encounter with specific antigen in the presence of a co-stimulatory signal triggers entry of the T cell into the G_1 phase of the cell cycle; at the same time, it also induces the synthesis of IL-2 along with the α chain of the IL-2 receptor (also known as CD25). The IL-2 receptor is composed of three chains: α, β, and γ (Fig. 9.20). Resting T cells express a form of the receptor composed of β and γ chains only, which binds IL-2 with moderate affinity, allowing resting T cells to respond to very high concentrations of IL-2. Association of the α chain with the β and γ heterodimer creates a receptor with a much higher affinity for IL-2, allowing the cell to respond to very low concentrations of IL-2. Binding of IL-2 to the high-affinity receptor then triggers progression through the rest of the cell cycle (Fig. 9.21). T cells activated in this way can divide two or three times a day for several days, allowing one cell to give rise to a clone of thousands of cells that all bear the same receptor for antigen. IL-2 is a survival factor for these cells, and also allows their differentiation into effector T cells. The removal of IL-2 from activated T cells results in their death.

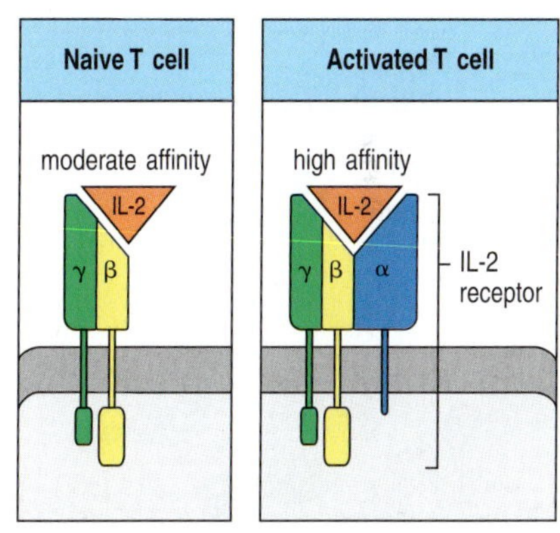

Fig. 9.20 High-affinity IL-2 receptors are three-chain structures that are present only on activated T cells. On resting T cells, the β and γ chains are expressed constitutively. They bind IL-2 with moderate affinity. Activation of T cells induces the synthesis of the α chain and the formation of the high-affinity heterotrimeric receptor. The β and γ chains show similarities in amino acid sequence to cell-surface receptors for growth hormone and prolactin, each of which also regulates cell growth and differentiation.

Antigen recognition by the T-cell receptor induces the synthesis or activation of the transcription factors NFAT, AP-1, and NFκB, which bind to the promoter region of the IL-2 gene and are essential to activate its transcription (see Section 7-15). Co-stimulation through CD28 contributes to the production of IL-2 in at least three ways. First, CD28 signaling activates PI 3-kinase, which increases production of the AP-1 and NFκB transcription factors, thereby increasing the transcription of IL-2 mRNA. However, the mRNAs of many cytokines, including IL-2, are very short-lived because of an 'instability' sequence (AUUUAUUUA) in the 3′ untranslated region. CD28 signaling prolongs the lifetime of an IL-2 mRNA molecule by inducing the expression of proteins that block the activity of the instability sequence, resulting in increased translation and more IL-2 protein. Finally, PI 3-kinase helps activate the protein kinase Akt (see Section 7-15), which generally promotes cell growth and survival, increasing the total production of IL-2 by activated T cells.

The central importance of IL-2 in initiating adaptive immune responses is exploited by drugs commonly used to suppress undesirable immune responses

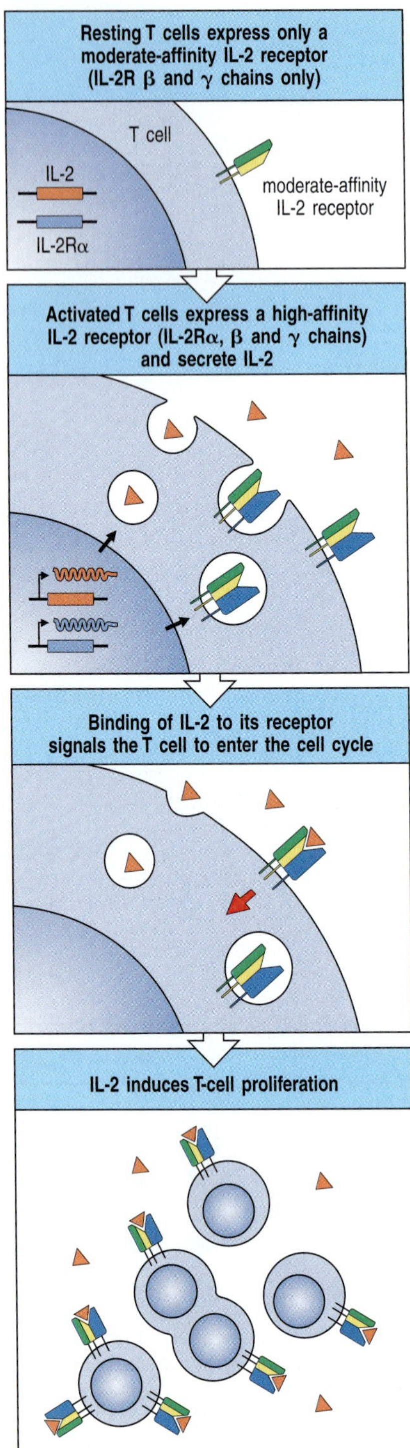

Resting T cells express only a moderate-affinity IL-2 receptor (IL-2R β and γ chains only)
T cell
IL-2
moderate-affinity IL-2 receptor
IL-2Rα
Activated T cells express a high-affinity IL-2 receptor (IL-2Rα, β and γ chains) and secrete IL-2
Binding of IL-2 to its receptor signals the T cell to enter the cell cycle
IL-2 induces T-cell proliferation

Fig. 9.21 Activated T cells secrete and respond to IL-2. Activation of naive T cells in the presence of co-stimulation through CD28 signaling induces the expression and secretion of IL-2 and the expression of high-affinity IL-2 receptors. IL-2 binds to the high-affinity IL-2 receptors to promote T-cell growth in an autocrine fashion.

such as transplant rejection. The immunosuppressive drugs cyclosporin A and FK506 (tacrolimus or fujimycin) inhibit IL-2 production by disrupting signaling through the T-cell receptor, whereas rapamycin (sirolimus) inhibits signaling through the IL-2 receptor. Cyclosporin A and rapamycin act synergistically to inhibit immune responses by preventing the IL-2-driven clonal expansion of T cells. The mode of action of these drugs will be considered in detail in Chapter 16.

9-13 Signal 2 can be modified by additional co-stimulatory pathways.

Once a naive T cell is activated, it expresses a number of proteins in addition to CD28 that contribute to sustaining or modifying the co-stimulatory signal. These other co-stimulatory receptors generally belong to either the CD28 family of receptors or the tumor necrosis factor (TNF)/TNF receptor families.

CD28-related proteins are induced on activated T cells and modify the co-stimulatory signal as the T-cell response develops. One is the inducible co-stimulator (**ICOS**), which binds a ligand known as **ICOSL** (**ICOS ligand** or **B7-H2**), a structural relative of B7.1 and B7.2. ICOSL is produced on activated dendritic cells, monocytes, and B cells. Although ICOS resembles CD28 in driving T-cell proliferation, it does not induce IL-2 but seems to regulate the expression of other cytokines made by the CD4 T-cell subsets, such as IL-4 and IFN-γ. ICOS is particularly important for enabling CD4 T cells to function as helper cells for B-cell responses such as isotype switching. ICOS is expressed on T cells in germinal centers within lymphoid follicles, and mice lacking ICOS fail to develop germinal centers and have severely diminished antibody responses.

Another receptor for B7 molecules is **CTLA-4** (CD152), which is related in sequence to CD28. CTLA-4 binds B7 molecules about 20 times more avidly than does CD28, but its effect is to inhibit, rather than activate, the T cell (Fig. 9.22). CTLA-4 does not contain an ITIM motif, and it is suggested to inhibit T-cell activation by competing with CD28 for interaction with B7 molecules expressed by antigen-presenting cells. Activation of naive T cells induces the surface expression of CTLA-4, making activated T cells less sensitive than naive T cells to stimulation by the antigen-presenting cell, thereby restricting IL-2 production. Thus, binding of CTLA-4 to B7 molecules is essential for limiting the proliferative response of activated T cells to antigen and B7. This was confirmed by producing mice with a disrupted CTLA-4 gene; such mice develop a fatal disorder characterized by a massive overgrowth of lymphocytes. Antibodies that block CTLA-4 from binding to B7 molecules markedly increase T-cell dependent immune responses.

Several different TNF-family molecules acting through their receptors can also deliver co-stimulatory signals. These all seem to function by activation of NFκB through a TRAF-dependent pathway (see Section 7-22). **CD70** on dendritic cells binding to its receptor **CD27** constitutively expressed on naive T cells delivers a potent co-stimulatory signal to T cells early in the activation process. The receptor CD40 on dendritic cells (see Section 9-7) binds to CD40 ligand expressed on T cells, initiating two-way signaling that transmits activating signals to the T cell and also induces the dendritic cell to express B7, thus stimulating further T-cell proliferation. The role of the CD40–CD40 ligand pair in sustaining a T-cell response is demonstrated in mice lacking CD40 ligand; when these mice are immunized, the clonal expansion of responding T cells is curtailed at an early stage. The T-cell molecule **4-1BB** (CD137) and its ligand **4-1BBL**, which is expressed on activated dendritic cells, macrophages, and B cells, make up another pair of TNF-family co-stimulators. The effects of this interaction are also bidirectional, with both the T cell and the antigen-presenting cell receiving activating signals; this type of interaction is sometimes referred to as the T-cell:antigen-presenting cell dialog. Another co-stimulatory receptor and its ligand, **OX40** and **OX40L**, are expressed on activated T cells and dendritic cells respectively. Mice deficient in OX40 show reduced CD4 T-cell proliferation in response to viral infection, indicating a role in sustaining ongoing T-cell responses by enhancing T-cell survival and proliferation.

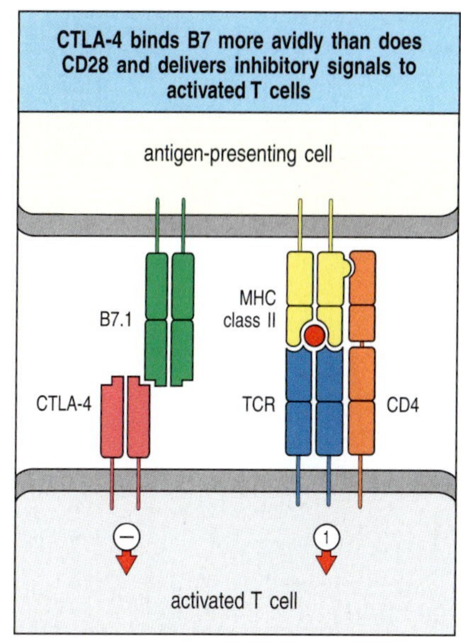

Fig. 9.22 CTLA-4 is an inhibitory receptor for B7 molecules. Naive T cells express CD28, which delivers a co-stimulatory signal on binding B7 molecules (see Fig. 9.19), driving their survival and expansion. Activated T cells express increased levels of CTLA-4 (CD152). CTLA-4 has a higher affinity than CD28 for B7 molecules and thus binds most or all of the B7 molecules, serving to regulate the proliferative phase of the response.

9-14 Antigen recognition in the absence of co-stimulation leads to functional inactivation or clonal deletion of peripheral T cells.

Despite the deletion of many self-reactive T cells in the thymus (see Section 8-20), some T cells specific for self antigens do survive and enter peripheral tissues. This is evident from diseases in which such T cells become activated and cause tissue-specific autoimmunity. Self peptides can be presented by MHC class I molecules expressed on peripheral tissues, and by MHC class I and class II molecules expressed on dendritic cells, which bring self proteins into the lymphoid organs (see Section 9-6). However, peripheral tissues do not express co-stimulatory molecules, and in the absence of infection, dendritic cells express very few co-stimulatory molecules. For these reasons, naive T cells recognizing self peptides are normally not activated, and instead meet alternative fates. Some may be converted into regulatory T cells (see Section 9-19); others are thought to be clonally deleted or to enter a state of anergy (Fig. 9.23).

Anergy was first demonstrated for clones of CD4 T cells maintained *in vitro*. When stimulated first with antigen in the absence of co-stimulation, such T cells became refractory to subsequent activation by specific antigen even when the antigen was presented by antigen-presenting cells expressing co-stimulatory molecules. This state is called anergy. A similar state of unresponsiveness has been demonstrated *in vivo* using T cells transgenic for a T-cell receptor of known antigen specificity. When specific antigen was delivered in soluble form without adjuvants to induce co-stimulatory activity, T cells remained viable but were much less responsive to subsequent stimulation, even in the presence of co-stimulation. By becoming anergic, self-reactive T cells are prevented from undergoing clonal expansion and acquiring effector functions that could be directed against self tissues.

Anergy involves an acquired blockade in the T-cell receptor signaling pathway, which may be due to the induction of E3 ubiquitin ligases in anergic T cells

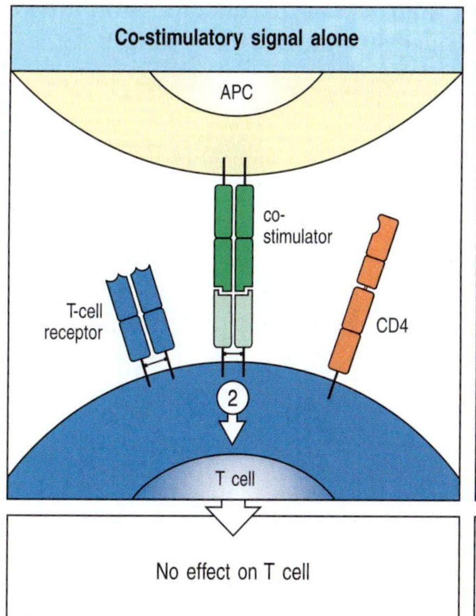

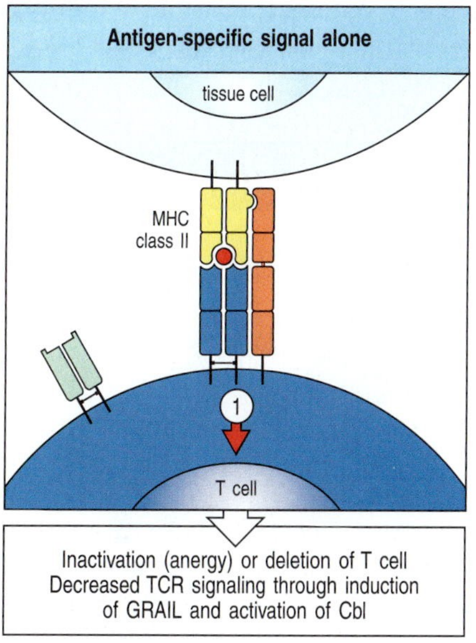

Fig. 9.23 T-cell tolerance to antigens expressed on tissue cells results from antigen recognition in the absence of co-stimulation. An antigen-presenting cell (APC) will neither activate nor inactivate a T cell if the appropriate antigen is not present on the APC surface, even if it expresses a co-stimulatory molecule (left panel). However, when a T cell recognizes antigen (signal 1) in the absence of co-stimulatory molecules (right panel), the T cell becomes anergic and may be deleted. Anergic cells have a blockade in signaling by the T-cell receptor caused by the induction of the E3 ubiquitin ligases GRAIL and/or Cbl, which target components of the T-cell receptor signaling pathway, such as CD3ζ, for degradation.

that target components of the CD3 signaling complex for degradation. The E3 ligase GRAIL (gene related to anergy in lymphocytes) has been linked to anergy. Like the E3 ligase Cbl (see Section 7-5), GRAIL can target components of CD3, most probably CD3ζ, for degradation by the proteasome, thus blocking T-cell receptor signaling. In mice deficient in GRAIL, antigen-stimulated T cells proliferate and produce cytokines even without co-stimulation, which is consistent with a role for GRAIL in establishing anergy in the absence of co-stimulation. CD28 signaling prevents the induction of GRAIL, so that no blockade in the T-cell pathway occurs in these circumstances.

However, anergy does not explain all the peripheral tolerance to potentially self-reactive T cells. Peripheral tolerance also requires regulatory T cells (T_{reg} cells) expressing FoxP3 (see Section 8-18). Experimentally, this has been shown by deleting FoxP3-expressing regulatory T cells in normal adult mice, which leads to the rapid onset of autoimmunity. We will return to a discussion of regulatory T cells later in this chapter.

9-15 Proliferating T cells differentiate into effector T cells that do not require co-stimulation to act.

After 4–5 days of rapid proliferation induced by IL-2, activated T cells differentiate into effector T cells that can synthesize all the molecules required for their specialized helper or cytotoxic functions. In addition, effector T cells have undergone changes that distinguish them from naive T cells. One of the most important is in their activation requirements: once a T cell has differentiated into an effector cell, encounter with its specific antigen results in immune attack without the need for co-stimulation (Fig. 9.24). This distinction is particularly easy to understand for CD8 cytotoxic T cells, which must be able to act on any cell infected with a virus, whether or not the infected cell can express co-stimulatory molecules. However, this feature is also important for the effector function of CD4 cells, as effector CD4 T cells must be able to activate B cells and macrophages that have taken up antigen even if these cells are not initially expressing co-stimulatory molecules.

Changes are also seen in the cell-adhesion molecules and receptors expressed by effector T cells. They express higher levels of LFA-1 and CD2 than do naive T cells, but lose cell-surface L-selectin and therefore cease to recirculate through lymph nodes. Instead, they express the integrin VLA-4, which allows them to bind to vascular endothelium bearing the adhesion molecule **VCAM-1**, which is expressed at sites of inflammation. This allows effector T cells to enter sites of infection and put their armory of effector proteins to good use. These changes in the T-cell surface are summarized in Fig. 9.25.

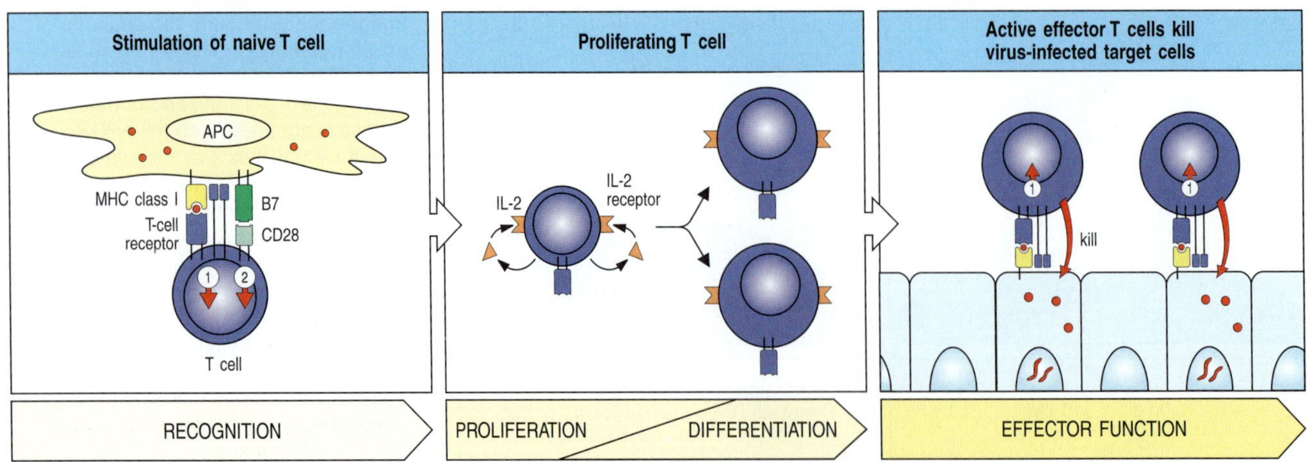

Fig. 9.24 Effector T cells can respond to their target cells without co-stimulation. A naive T cell that recognizes antigen on the surface of an antigen-presenting cell and receives the required two signals (arrows 1 and 2, left panel) becomes activated, and both secretes and responds to IL-2. IL-2-driven clonal expansion (center panel) is followed by the differentiation of the T cells to effector cell status. Once the cells have differentiated into effector T cells, any encounter with specific antigen triggers their effector actions without the need for co-stimulation. Thus, as illustrated here, a cytotoxic T cell can kill virus-infected target cells that express only the peptide:MHC ligand and not co-stimulatory signals (right panel).

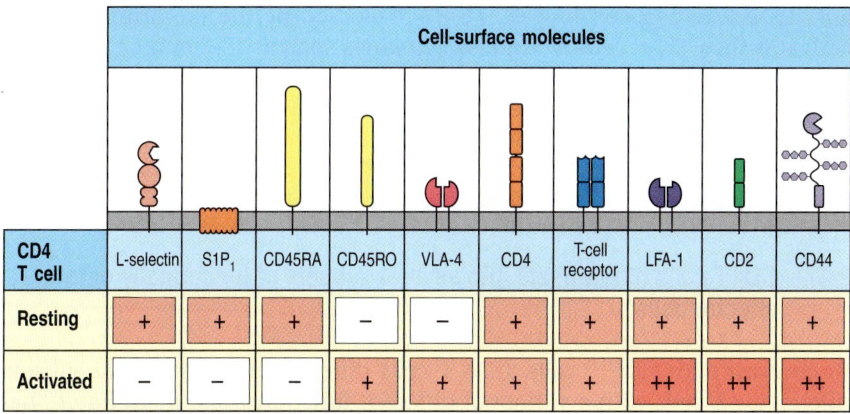

Cell-surface molecules										
CD4 T cell	L-selectin	$S1P_1$	CD45RA	CD45RO	VLA-4	CD4	T-cell receptor	LFA-1	CD2	CD44
Resting	+	+	+	−	−	+	+	+	+	+
Activated	−	−	−	+	+	+	+	++	++	++

Fig. 9.25 Activation of T cells changes the expression of several cell-surface molecules. The example here is a CD4 T cell. Resting naive T cells express L-selectin, through which they home to lymph nodes, but express relatively low levels of other adhesion molecules such as CD2 and LFA-1. Upon activation, expression of L-selectin ceases and, instead, increased amounts of the integrin LFA-1 are produced, which is activated to bind its ligands ICAM-1 and ICAM-2. A newly expressed integrin called VLA-4, which acts as a homing receptor for vascular endothelium at sites of inflammation, ensures that activated T cells enter peripheral tissues at sites where they are likely to encounter infection. Activated T cells also have a higher density of the adhesion molecule CD2 on their surface, increasing the avidity of the interaction with potential target cells, and a higher density of the adhesion molecule CD44. There is a change in the isoform of CD45 that is expressed, by alternative splicing of the RNA transcript of the CD45 gene, so that activated T cells express the CD45RO isoform, which associates with the T-cell receptor and CD4. This change makes the T cell more sensitive to stimulation by lower concentrations of peptide:MHC complexes. Finally, the sphingosine 1-phosphate receptor ($S1P_1$) is expressed by resting naive T cells, allowing the egress of cells that do not become activated. Downregulation of $S1P_1$ for several days after activation prevents T-cell egress during the period of proliferation and differentiation. After several days, it is expressed again, allowing effector cells to exit from the lymphoid tissues.

9-16 CD8 T cells can be activated in different ways to become cytotoxic effector cells.

Naive T cells fall into two large classes, one of which carries the co-receptor CD8 on its surface and the other bears the co-receptor CD4. CD8 T cells all differentiate into **CD8 cytotoxic T cells** (sometimes called **cytotoxic lymphocytes** or **CTLs**), which kill their target cells (Fig. 9.26). They are important in the defense against intracellular pathogens, especially viruses. Virus-infected cells display fragments of viral proteins as peptide:MHC class I complexes on their surface, and these are recognized by cytotoxic T lymphocytes.

Perhaps because the effector actions of these cells are so destructive, naive CD8 T cells require more co-stimulatory activity to drive them to become activated effector cells than do naive CD4 T cells. This requirement can be met in two ways. The simplest is activation by mature dendritic cells, which have high intrinsic co-stimulatory activity. In some viral infections, dendritic cells become sufficiently activated to directly induce CD8 T cells to produce the IL-2 required for their proliferation and differentiation, without help by CD4 effector cells. This property of dendritic cells has been exploited to generate cytotoxic T-cell responses against tumors, as we will see in Chapter 16.

In the majority of viral infections, however, CD8 T-cell activation requires additional help, which is provided by CD4 effector T cells. Effector CD4 T cells that recognize related antigens presented by the antigen-presenting cell can amplify the activation of naive CD8 T cells by further activating the

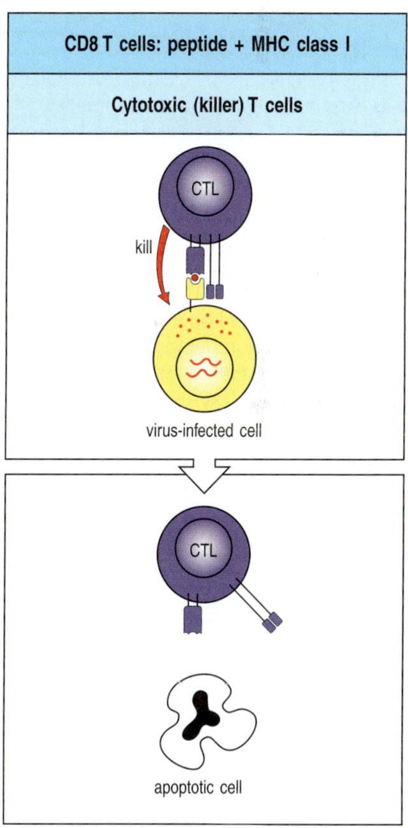

Fig. 9.26 CD8 cytotoxic T cells are specialized to kill cells infected with intracellular pathogens. CD8 cytotoxic cells kill target cells that display peptide fragments of cytosolic pathogens, most notably viruses, bound to MHC class I molecules at the cell surface.

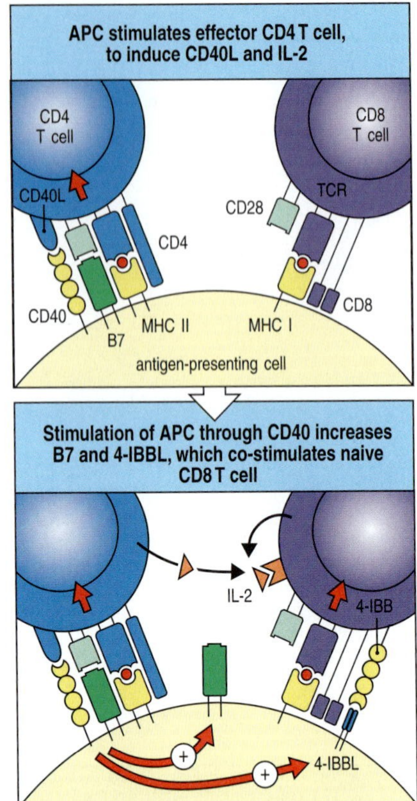
APC stimulates effector CD4 T cell, to induce CD40L and IL-2
CD4 T cell
CD8 T cell
CD40L
TCR
CD28
CD4
CD40
CD8
MHC II
MHC I
B7
antigen-presenting cell
Stimulation of APC through CD40 increases B7 and 4-IBBL, which co-stimulates naive CD8 T cell
IL-2
4-IBB
4-IBBL

Fig. 9.27 Most CD8 T-cell responses require CD4 T cells. CD8 T cells recognizing antigen on weakly co-stimulatory cells may become activated only in the presence of CD4 T cells interacting with the same antigen-presenting cell (APC). This happens mainly by an effector CD4 T cell recognizing antigen on the antigen-presenting cell and being triggered to induce increased levels of co-stimulatory activity on the antigen-presenting cell. The CD4 T cells also produce abundant IL-2 and thus help drive CD8 T-cell proliferation. This may in turn activate the CD8 T cell to make its own IL-2.

antigen-presenting cell (Fig. 9.27). B7 expressed by the dendritic cell first activates the CD4 T cells to express cytokines such as IL-2 and CD40 ligand (see Section 9-7). CD40 ligand binds CD40 on the dendritic cell, delivering an additional signal that increases the expression of B7 and 4-1BBL by the dendritic cell, which in turn provides additional co-stimulation to the naive CD8 T cell. The IL-2 produced by effector CD4 T cells also acts as a growth factor to promote CD8 T-cell differentiation.

9-17 CD4 T cells differentiate into several subsets of functionally different effector cells.

In contrast with CD8 T cells, CD4 T cells differentiate into several subsets of effector T cells with a variety of different functions. The main functional classes are **T_H1, T_H2, T_H17**, and the **regulatory T cells**. A recently recognized class specialized for providing help to B cells in the lymphoid follicles is called the **T follicular helper cell**, or **T_{FH}**. The subsets, particularly T_H1, T_H2, and T_H17, are defined on the basis of the different combinations of cytokines that they secrete (Fig. 9.28). The first to be distinguished were the T_H1 and T_H2 subsets, hence their names.

T_H1 cells help control bacteria that can set up intravesicular infections in macrophages, such as the mycobacteria that cause tuberculosis and leprosy. These bacteria are taken up by macrophages in the usual way but can evade the killing mechanisms described in Chapter 3. If a T_H1 cell recognizes bacterial antigens displayed on the surface of an infected macrophage, it will interact with the infected cell to activate it further, stimulating the macrophage's microbicidal activity to enable it to kill its resident bacteria. We shall describe the macrophage-activating functions of T_H1 cells later in this chapter. By contrast, T_H2 cells help to control infections by parasites, particularly helminths, rather than intracellular bacteria or viruses, through promoting responses mediated by eosinophils, mast cells, and the IgE antibody isotype. In particular, cytokines produced by T_H2 cells are required for the switching of B cells to produce the IgE class of antibody, the primary role of which is to fight parasite infections, as we shall see in Chapter 10. IgE is also the antibody responsible for allergies, and thus T_H2 differentiation is of additional medical interest, as discussed in Chapter 14. A third major effector subset of CD4 T cells are the T_H17 cells. They seem to be induced early in adaptive immune responses, and their main function seems to be to help protect against extracellular bacteria and fungi through stimulating the neutrophil response that helps to clear such pathogens (see Fig. 9.28). As fully differentiated effector cells, T_H1, T_H2, and T_H17 cells can function outside the lymphoid tissues at sites of infection to activate macrophages or aid in recruiting cells such as eosinophils and neutrophils.

Another crucial function of CD4 T cells is in providing help to B cells for antibody production (see Section 1-4). A point of confusion in the past has been whether both T_H1 and T_H2 cells could provide B-cell help, and it was often inaccurately implied that this was the function of T_H2 cells alone. The current view is that the T_{FH} cell, rather than either T_H1 or T_H2 cells, is the effector T cell that mostly provides B-cell help for high-affinity antibody production in lymphoid follicles. T_{FH} cells are identified mainly by their location and by the expression of certain markers, such as CXCR5 and ICOS (Section 9-13), and have been identified in both mice and humans. Their crucial feature in relation to help for antibody production is that they can secrete cytokines characteristic of either T_H1 or T_H2 cells. This explains how, in the course of an infection, B cells could receive help in a follicle to switch to IgE, through the presence of 'T_H2' cytokines, and to switch to other isotypes such as IgG2a, through the presence of 'T_H1' cytokines. The existence of the T_{FH} subset

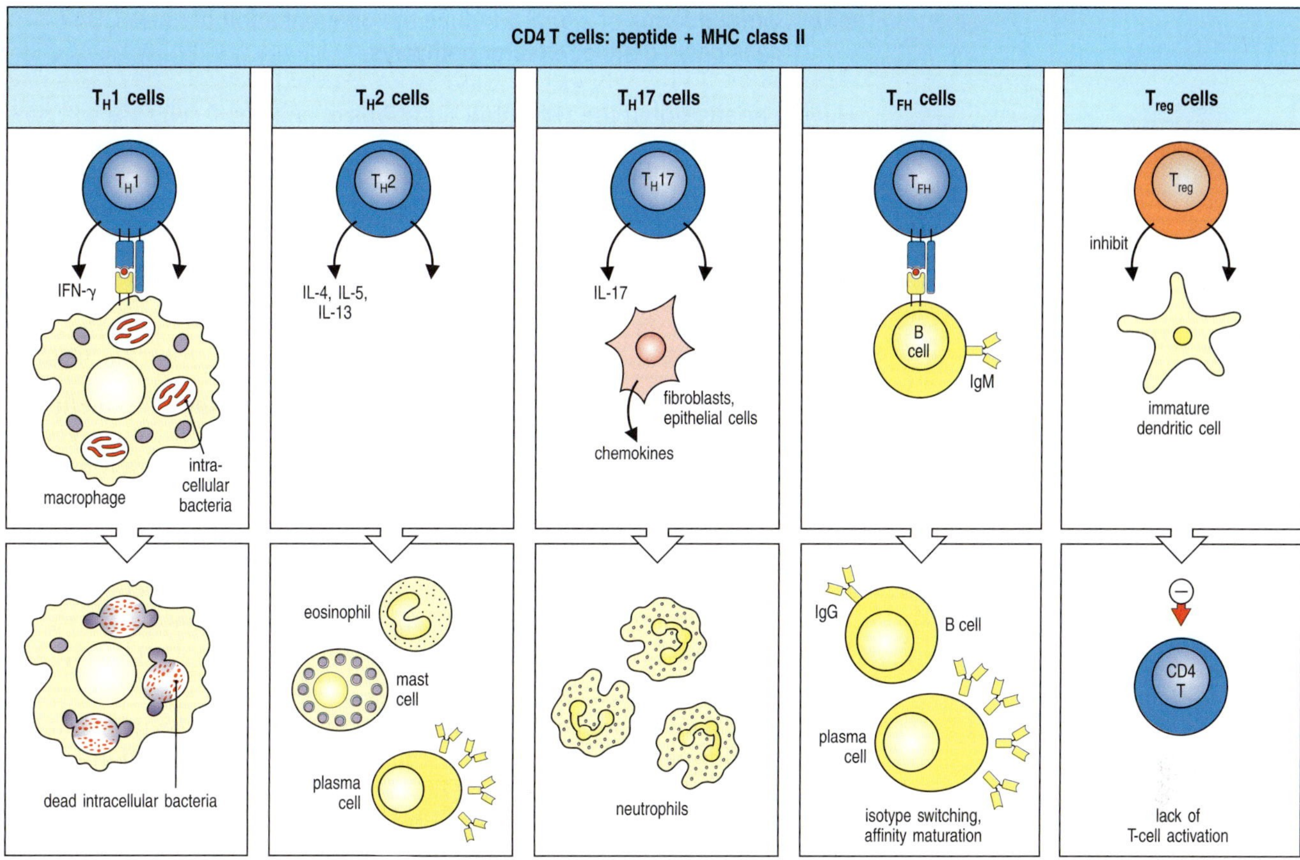

Fig. 9.28 Subsets of CD4 effector T cells are specialized to provide help for different classes of pathogens. T_H1 cells (first panels) produce cytokines that activate macrophages, enabling them to destroy intracellular microorganisms more efficiently. T_H2 cells (second panels) produce cytokines that recruit and activate eosinophils, mast cells, and basophils, and promote barrier immunity at mucosal surfaces. T_H17 cells (third panels) secrete IL-17-family cytokines that induce local epithelial and stromal cells to produce chemokines that recruit neutrophils to sites of infection early in the adaptive immune response. T_{FH} cells are a subset that localizes in B-cell follicles, but that produce cytokines characteristic of other subsets. T_{FH} cells producing IFN-γ activate B cells to produce strongly opsonizing antibodies belonging to certain IgG subclasses (IgG1 and IgG3 in humans, and their homologs IgG2a and IgG2b in the mouse). Those T_{FH} cells producing IL-4 drive B cells to differentiate and produce immunoglobulins of other types, especially IgE. Regulatory T cells (right panels) are a heterogeneous class of cells that suppress T-cell activity and help prevent the development of autoimmunity during immune responses.

could also explain the mass of previous research over many years indicating that cells with cytokine profiles characteristic of T_H2 and T_H1 are absolutely required for the complete range of antibody production, even though when the development of the T_H1 and T_H2 subsets is prevented by knockout of crucial transcription factors, most antibody classes are still produced. The identification of T_{FH} cells does not, however, mean that the other CD4 T-cell subsets have no role at all in antibody production or in influencing the class of the antibodies produced. T_H2 cells and the cytokines they produce are, for example, important in driving IgE production in response to parasites and in allergic responses. The developmental relationship of T_{FH} to other CD4 subsets is still a matter of active research, but they seem to represent a distinct branch of effector T cells that remain within the lymphoid tissues and are specialized for providing B-cell help. We will return to the helper functions of T_{FH} cells in more detail in Chapter 10.

All the effector T cells described above are involved in activating their target cells to make responses that help clear the pathogen from the body. The other CD4 T cells found in the periphery have a different function. These are the regulatory T cells, whose function is to suppress T-cell responses rather than activate them. They are involved in limiting the immune response and preventing autoimmune responses. Two main groups of regulatory T cells are currently recognized. One subset becomes committed to a regulatory fate while still in the thymus; they are known as the **natural regulatory T cells**, which we introduced in Section 8-18. Other subsets of CD4 regulatory T cells with different phenotypes have been recognized more recently and are thought to differentiate from naive CD4 T cells in the periphery under the influence of particular environmental conditions. This group is known as **induced regulatory T cells** (or adaptive regulatory T cells).

9-18 Various forms of signal 3 induce the differentiation of naive CD4 T cells down distinct effector pathways.

Having briefly noted the types and functions of effector T cells, we will now consider how they are derived from naive T cells. The fate of the progeny of a naive CD4 T cell is largely decided during the initial priming period and is regulated by signals provided by the local environment, particularly the priming antigen-presenting cell. These are the signals we will call signal 3. The five effector subtypes—T_H1, T_H2, T_H17, T_{FH}, and the induced regulatory T cells (iT_{reg} cells)—are associated with distinct signals that induce their formation, different transcription factors that drive their differentiation, and unique cytokines and surface markers that define their identity (Fig. 9.29).

The T_H1 and T_H2 subsets are distinguished principally by their production of specific cytokines, such as **interferon (IFN)-γ** and IL-2 by T_H1 cells, and **IL-4** and **IL-5** by T_H2 cells. One or the other of these subsets will often become predominant in chronic immune responses, such as persistent infections, autoimmunity, or allergies. In most acute responses to infection, however, it is likely that both T_H1 and T_H2 cells are involved in making an effective response. The decision to differentiate into T_H1 or T_H2 cells occurs early in the immune response, and one important determinant of this decision is the mix of cytokines produced by cells of the innate immune system in response to pathogens.

T_H1 development is induced when signal 3 is composed of the cytokines IFN-γ and IL-12 during the early stages in T-cell activation. As described in Section 7-20, many key cytokines, including IFN-γ and IL-12, stimulate the JAK–STAT intracellular signaling pathway, resulting in the activation of specific genes. JAKs (Janus tyrosine kinases) and STATs (signal-transducing activators of transcription) are present as families of proteins, which can be activated by different cytokines to achieve different effects. For T_H1 development, STAT1 and STAT4 are important and are activated by cytokines produced by innate immune cells during early infection. Activated NK cells may be an important source of IFN-γ, because the IFN-γ gene in resting naive CD4 T cells is switched off. STAT1 induces the expression of another transcription factor, T-bet, in the activated CD4 T cells, which switches on the genes for IFN-γ and the IL-12 receptor. These T cells are now committed to becoming T_H1 cells. The cytokine IL-12, produced by dendritic cells and macrophages, can then activate STAT4, which promotes the expansion and differentiation of the committed T_H1 cells.

These effector T_H1 cells will generate copious amounts of IFN-γ when they recognize antigen on a target cell, thus reinforcing the signal for the differentiation of more T_H1 cells. In this way, recognition of a particular type of pathogen by the innate immune system initiates a chain reaction that links the innate response to the adaptive immune response. For example, bacterial

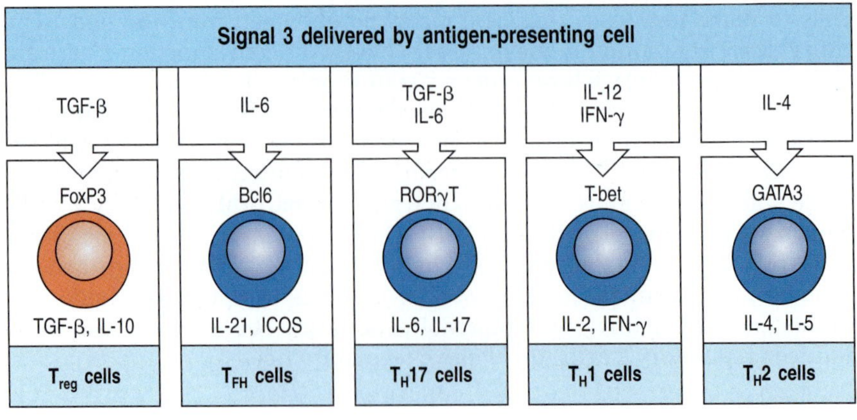

Fig. 9.29 Variation in signal 3 causes naive CD4 T cells to acquire several distinct types of effector functions. Antigen-presenting cells, principally dendritic cells, provide signal 3 in the form of various cytokines or express surface proteins that induce the development of CD4 T cells into distinct types of effector cells. The environmental conditions, such as the exposure to various pathogens, determine which signal the antigen-presenting cell will produce. When pathogens are absent, an abundance of TGF-β and the lack of IL-6, IFN-γ, and IL-12 favor the development of FoxP3-expressing induced T_{reg} cells. Early in infection, IL-6 produced by dendritic cells acts with transforming growth factor-β (TGF-β) to induce T_H17 cells expressing the transcription factor RORγT, which are amplified by IL-23. T_{FH} cells, which require IL-6 and the transcription factor Bcl-6 for their function, provide help to B cells in the form of cytokines such as IL-21, and the surface molecule ICOS. Later, dendritic cells and other antigen-presenting cells produce cytokines that promote either T_H1 (IFN-γ and IL-12) or T_H2 (IL-4) and suppress T_H17 development. T_H1 and T_H2 cells express the T-bet and GATA3 transcription factors, respectively.

infections induce dendritic cells and macrophages to produce IL-12, favoring the emergence of T_H1 effector cells. These promote effector functions such as macrophage activation, which is required to clear infections caused by mycobacteria and *Listeria*, for example, and the provision of help for antibody production against extracellular bacteria.

T_H2 development is favored by a different signal 3, in this case IL-4 (see Fig. 9.29). This cytokine is the most powerful trigger for inducing T_H2 development from naive CD4 T cells. If IL-4 is encountered while the naive T cells are being activated by antigen, the IL-4 receptor activates STAT6, which promotes expression of the transcription factor GATA3 in the T cell. GATA3 is a powerful activator of the genes for several cytokines characteristically produced by T_H2 cells, such as IL-4 and IL-13. GATA3 also induces its own expression, helping to stabilize T_H2 differentiation. A longstanding question both for the induction of allergy and the response to infection has been the initial source of the IL-4 that triggers the T_H2 response. Eosinophils, basophils, and mast cells are an attractive possibility because they can produce abundant IL-4, and recent evidence indicates that these cells can be activated by chitin, a polysaccharide present in fungal and helminth parasites, as well as in insects and in crustaceans. In mice treated with chitin, eosinophils and basophils were recruited into tissues and were activated to produce IL-4.

T_H17 cells arise when the cytokines IL-6 and transforming growth factor (TGF)-β are present but IL-4 and IL-12 are absent (see Fig. 9.29); they are distinguished by their ability to produce cytokines of the IL-17 family, but not IFN-γ or IL-4. Development as T_H17 involves the initial production by the T cell of the cytokine IL-21, which acts in an autocrine manner to activate STAT3, a transcription factor required for development as T_H17. The signature transcription factor expressed by differentiated T_H17 cells is RORγT, an orphan nuclear hormone receptor that drives the expression of characteristic T_H17-cell cytokines.

T_H17 cells express the receptor for the cytokine IL-23, rather than the IL-12 receptor typical of T_H1 cells, and the expansion and further development of T_H17 effector activity seem to require IL-23, in the same way that effective T_H1 responses require IL-12. T_H17 cells promote inflammation indirectly. The IL-17 secreted by T_H17 cells acts on receptors on local tissue cells such as stromal cells or epithelium; these respond by producing chemokines, such as IL-8, that recruit innate effector cells, particularly neutrophils. T_H17 cells also make IL-22, which acts on receptors expressed in the gut, skin, and lungs to promote local innate defenses to pathogens.

Induced regulatory T cells are distinguished by expression of the transcription factor FoxP3 and cell-surface CD4 and CD25, and are produced when naive T cells are activated in the presence of the cytokine **transforming growth factor-β (TGF-β)**alone and in the absence of IL-6 and other pro-inflammatory cytokines. Differentiated T_{reg} cells themselves produce TGF-β and **IL-10**, which act in an inhibitory manner to suppress immune responses and inflammation. Thus, it is the presence or absence of IL-6 that decides between the development of immunosuppressive T_{reg} cells or of T_H17 cells, which promote inflammation and the generation of immunity (see Fig. 9.29). The generation of IL-6 by innate immune cells is regulated by the presence or absence of pathogens, with pathogen products tending to stimulate its production. In the absence of pathogens, IL-6 production is low, favoring differentiation of the immunosuppressive T_{reg} cells and so preventing unwanted immune responses.

T_{FH} cells, unlike the four subsets described above, have not been produced efficiently *in vitro*, and so the requirements for their differentiation are not yet clearly established. IL-6 seems to be required for T_{FH} development, but much remains to be learned about the control of this subset. One transcription factor important for T_{FH} development is Bcl6, which is required for the expression of CXCR5, the receptor for the chemokine CXCL13, which is produced by the stromal cells of the B-cell follicle. This receptor is essential for T_{FH} localization in follicles, and the other effector T-cell subsets do not express it to any extent. T_{FH} cells also express ICOS, whose ligand is expressed abundantly by B cells. ICOS seems crucial for the helper activity of T_{FH} cells, because mice lacking ICOS show a severe defect in T-cell dependent antibody responses.

The consequences of inducing the development of these various CD4 subsets are profound. On the one hand, the selective production of T_H1 cells leads to cell-mediated immunity, and the cytokines they produce help promote the switch of antibody production to opsonizing antibody classes (predominantly IgG). On the other hand the production of predominantly T_H2 cells results in the presence of cytokines that favor humoral immunity and the production of IgM, IgA, and IgE. T_H17 cells seem to be important in the recruitment of neutrophils to control the early stages of an infection, and the regulatory T-cell subsets restrain inflammation and maintain tolerance.

A striking example of the difference that different T-cell subsets can make to the outcome of infection is seen in leprosy, a disease caused by infection with *Mycobacterium leprae. M. leprae*, like *M. tuberculosis*, grows in macrophage vesicles, and effective host defense requires macrophage activation by T_H1 cells. In patients with **tuberculoid leprosy**, in which T_H1 cells are preferentially induced, few live bacteria are found, little antibody is produced, and, although skin and peripheral nerves are damaged by the inflammatory responses associated with macrophage activation, the disease progresses slowly and the patient usually survives. However, when T_H2 cells are preferentially induced, the main response is humoral, the antibodies produced cannot reach the intracellular bacteria, and the patients develop **lepromatous leprosy**, in which *M. leprae* grows abundantly in macrophages, causing gross tissue destruction that is eventually fatal.

9-19 Regulatory CD4 T cells are involved in controlling adaptive immune responses.

The regulatory T cells found in the periphery are a heterogeneous group of cells with different developmental origins. The natural T regulatory cells (natural T_{reg} cells) that develop in the thymus (see Section 8-18) are CD4-positive cells that also express CD25 and high levels of the L-selectin receptor CD62L and of CTLA-4, and represent about 10–15% of the CD4 T cells in the human circulation. The induced T_{reg} cells that arise in the periphery from naive CD4 T cells also express CD25 (see Section 9-18). A hallmark of both natural and induced T_{reg} cells is expression of the transcription factor FoxP3, which interferes with the interaction between AP-1 and NFAT at the IL-2 gene promoter, preventing transcriptional activation of the gene and production of IL-2. Some T-cell subsets in the periphery have also been described that lack FoxP3 expression but produce the immunosuppressive cytokines characteristic of T_{reg} cells.

Natural T_{reg} cells are potentially self-reactive T cells that express conventional α:β T-cell receptors and seem to be selected in the thymus by high-affinity binding to MHC molecules containing self peptides. It is currently not known whether they are activated to express their regulatory function in the periphery by the same self ligands that selected them in the thymus or by other self or nonself antigens. Once activated, they may mediate their effects in a contact-dependent fashion. The high levels of CTLA-4 on the surface of natural T_{reg} cells is necessary for their regulatory activity. One possible mechanism for contact-dependent inhibition by natural T_{reg} cells is that the CTLA-4 on their surface competes for B7 expressed by antigen-presenting cells, and prevents adequate co-stimulation of naive T cells. Other evidence suggests that natural T_{reg} cells can secrete IL-10 and TGF-β, cytokines that inhibit T-cell proliferation (see Fig. 9.29). IL-10 also affects the differentiation of dendritic cells, inhibiting their secretion of IL-12 and thus impairing their ability to promote T-cell activation and T_H1 differentiation. Failure of natural T_{reg}-cell function is known to be involved in several autoimmune syndromes and is described in more detail in Chapter 15. In addition to their ability to prevent autoimmune disease *in vivo*, natural T_{reg} cells have been shown to suppress antigen-specific T-cell proliferation and T-cell proliferation in response to allogeneic cells *in vitro*.

The other well-characterized class of regulatory T cells in peripheral tissues comprises the FoxP3-expressing induced CD4 T_{reg} cells. These are found in both the systemic and the mucosal immune systems, and their importance in the routine prevention of unwanted immune responses is described in more detail in Chapters 12, 14, and 15. Pre-dating the discovery of the FoxP3-expressing induced T_{reg} cells, other subsets of regulatory T cells were described that produce inhibitory cytokines. One subset, called **T_H3**, is found predominantly in the mucosal immune system, and they are characterized by their production of IL-4, IL-10, and TGF-β; the production of TGF-β distinguishes them from T_H2 cells. T_H3 cells seem to function to suppress or control immune responses in the mucosae. T_H3 cells were described before the role of FoxP3 in the development of induced T_{reg} cells was recognized, and although there is considerable overlap in the characteristics of T_H3 cells and FoxP3-expressing induced T_{reg} cells, a distinct transcriptional signature for T_H3 cells has not yet been established.

Another type of T cell with regulatory activity described previously is called **T_R1**. These cells have been differentiated *in vitro*, and were defined largely by their production of the cytokines IL-10 and TGF-β, but not IL-4, which distinguished them from T_H3 cells. We now recognize that many different cells, including T_H1, T_H2, T_H17, and B cells, can produce IL-10 under certain circumstances, such as high antigen dose, so the uniqueness of T_R1 cells is somewhat uncertain.

Whatever the source, IL-10 is important in immune regulation, because it suppresses the T-cell production of IL-2, TNF-α, and IL-5, and inhibits antigen-presenting cells by reducing the expression of MHC molecules and co-stimulatory molecules. TGF-β similarly blocks T-cell cytokine production, cell division, and killing ability. Not all the effects of IL-10 and TGF-β are immunosuppressive, however: IL-10 can enhance B-cell survival and maturation into plasma cells and increase the activity of CD8 T cells. Nevertheless, the dominant effects *in vivo* of both IL-10 and TGF-β are immunosuppressive, as shown by the fact that mice lacking either cytokine are prone to autoimmune disease.

Summary.

The crucial first step in adaptive immunity is the activation or priming of naive antigen-specific T cells by antigen-presenting cells within the lymphoid tissues and organs through which they are constantly passing. The most distinctive feature of antigen-presenting cells is the expression of cell-surface co-stimulatory molecules, of which the B7 molecules are the most important in natural responses to infection. Naive T cells will only respond to antigen when the antigen-presenting cell presents both a specific antigen to the T-cell receptor (signal 1) and a B7 molecule to CD28 on the T cell (signal 2).

The activation of naive T cells leads to their proliferation and the differentiation of their progeny into effector T cells. Proliferation and differentiation depend on the production of cytokines, in particular IL-2, which binds to a high-affinity receptor on the activated T cell. T cells whose antigen receptors are ligated in the absence of co-stimulatory signals fail to make IL-2 and instead become anergic or die. This dual requirement for both receptor ligation and co-stimulation by the same antigen-presenting cell helps to prevent naive T cells from responding to self antigens on tissue cells, which lack co-stimulatory activity.

Antigen-stimulated proliferating T cells develop into effector T cells, the critical event in most adaptive immune responses. Various combinations of cytokines provide signal 3 to regulate the type of effector T cell that develops in response to an infection. In turn, the cytokines present during primary T-cell activation are influenced by the innate immune system as it first recognizes the pathogen. Once an expanded clone of T cells achieves effector function, its progeny can act on any target cell that displays antigen on its surface. Effector T cells have a variety of functions. CD8 cytotoxic T cells recognize virus-infected cells and kill them. T_H1 effector cells promote the activation of macrophages, and together they make up cell-mediated immunity. T_H2 cells promote mucosal barrier immunity against pathogens such as helminths requiring the effector activities of cells such as eosinophils and mast cells for their elimination. T_H17 cells enhance the acute inflammatory response to infection by recruiting neutrophils to sites of infection. T_{FH} cells can make cytokines characteristic of other subsets, but localize to the B-cell follicle and germinal centers, where they interact with B cells to provide help with antibody production and isotype switching. Regulatory CD4 T-cell subsets restrain the immune response by producing inhibitory cytokines, sparing surrounding tissues from collateral damage.

General properties of effector T cells and their cytokines.

All T-cell effector functions involve the interaction of an effector T cell with a target cell displaying specific antigen. The effector proteins released by the T cells are focused on the target by mechanisms that are activated by antigen recognition. The focusing mechanism is common to all types of effector T cells, whereas their effector actions depend on the array of membrane and secreted proteins that they express or release upon ligation of their antigen receptors. The different types of effector T cells are specialized to deal with different types of pathogens, and the effector molecules that they are programmed to produce cause distinct and appropriate effects on the target cell.

9-20 Effector T-cell interactions with target cells are initiated by antigen-nonspecific cell-adhesion molecules.

Once an effector T cell has completed its differentiation in the lymphoid tissue, it must find target cells that are displaying the peptide:MHC complex that it recognizes. T_{FH} cells encounter their B-cell targets without leaving the lymphoid tissue. However, most of the effector T cells emigrate from their site of activation in lymphoid tissues and enter the blood via the thoracic duct. Because of the cell-surface changes that have occurred during differentiation, they can now migrate into tissues, particularly at sites of infection. They are guided to these sites by changes in the adhesion molecules expressed on the endothelium of the local blood vessels as a result of infection, and by local chemotactic factors.

The initial binding of an effector T cell to its target, like that of a naive T cell to an antigen-presenting cell, is an antigen-nonspecific interaction mediated by LFA-1 and CD2. The levels of LFA-1 and of CD2 are two to fourfold higher on effector T cells than on naive T cells, and so effector T cells can bind efficiently to target cells that have less ICAM and CD58 on their surface than do antigen-presenting cells. This interaction is transient unless recognition of antigen on the target cell by the T-cell receptor triggers an increase in the affinity of the T-cell's LFA-1 for its ligands. The T cell then binds more tightly to its target and remains bound for long enough to release its effector molecules. CD4 effector T cells, which activate macrophages or induce B cells to secrete antibody, have to switch on new genes and synthesize new proteins to carry out their effector actions and so must maintain contact with their targets for relatively long periods. Cytotoxic T cells, by contrast, can be observed under the microscope attaching to and dissociating from successive targets relatively rapidly as they kill them (Fig. 9.30). Killing of the target, or some local change in the T cell, allows the effector T cell to detach and address new targets. How CD4 effector T cells disengage from their antigen-negative targets is not known, although evidence suggests that CD4 binding to MHC class II molecules without engagement of the T-cell receptor provides a signal for the cell to detach.

Movie 9.5

9-21 An immunological synapse forms between effector T cells and their targets to regulate signaling and to direct the release of effector molecules.

When binding to their specific antigenic peptide:self-MHC complexes or to self-peptide:self-MHC complexes, the T-cell receptors and their associated co-receptors cluster at the site of cell–cell contact, forming what is called the **supramolecular activation complex** (**SMAC**) or the **immunological synapse**. Other cell-surface molecules also cluster here. For example, the tight binding of LFA-1 to ICAM-1 induced by ligation of the T-cell receptor creates a molecular seal that surrounds the T-cell receptor and its co-receptor (Fig. 9.31). In some cases, the contact surface organizes into two zones: a central zone known as the central supramolecular activation complex (c-SMAC) and an outer zone known as the peripheral supramolecular activation complex (p-SMAC). The c-SMAC contains most of the signaling proteins known to be important in T-cell activation. The p-SMAC is notable mainly for the presence of the integrin LFA-1 and the cytoskeletal protein talin, which connects integrin to the actin cytoskeleton (see Section 3-15). The immunological synapse is not necessarily a static structure as implied by Fig. 9.31, but is actually quite dynamic. T-cell receptors move from the periphery into the c-SMAC, where they undergo endocytosis through ubiquitin-mediated degradation involving the E3 ligase Cbl (see Section 7-5). Because T-cell receptors are being degraded in the c-SMAC, signaling is actually weaker there than in the peripheral contact areas, where microclusters of T-cell receptors are being formed and are highly active (see Section 7-8).

Clustering of the T-cell receptors signals a reorientation of the cytoskeleton that polarizes the effector cell and focuses the release of effector molecules at the site of contact with the target cell. This is illustrated for a cytotoxic T cell in Fig. 9.32. An important intermediary in the effect of T-cell signaling on the cytoskeleton is the Wiskott–Aldrich syndrome protein (WASP), defects in which result in the inability of T cells to become polarized, among other

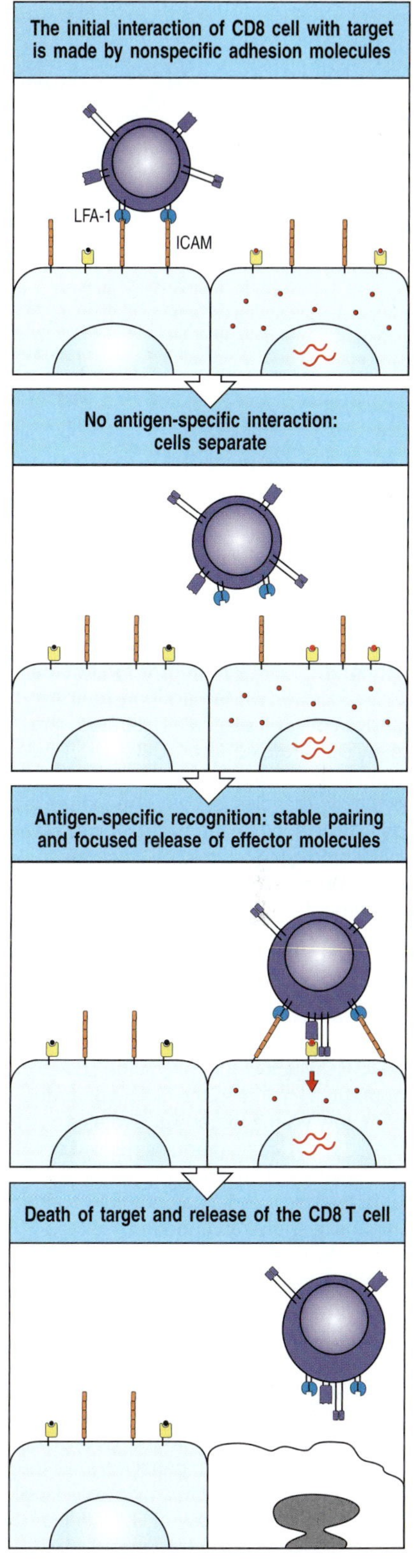
The initial interaction of CD8 cell with target is made by nonspecific adhesion molecules
LFA-1
ICAM
No antigen-specific interaction: cells separate
Antigen-specific recognition: stable pairing and focused release of effector molecules
Death of target and release of the CD8 T cell

Fig. 9.30 Interactions of T cells with their targets initially involve nonspecific adhesion molecules. The major initial interaction is between LFA-1 on the T cell, illustrated here as a cytotoxic CD8 T cell, and ICAM-1 or ICAM-2 on the target cell (top panel). This binding allows the T cell to remain in contact with the target cell and to scan its surface for the presence of specific peptide:MHC complexes. If the target cell does not carry the specific antigen, the T cell disengages (second panel) and can scan other potential targets until it finds the specific antigen (third panel). Signaling through the T-cell receptor increases the strength of the adhesive interactions, prolonging the contact between the two cells and stimulating the T cell to deliver its effector molecules. The T cell then disengages (bottom panel).

Outer ring (red) pSMAC	Inner circle (green) cSMAC
LFA-1:ICAM-1 talin	TCR, CD4, CD28 MHC:peptide CD8, PKC-θ

Fig. 9.31 The area of contact between an effector T cell and its contact forms an immunological synapse. A confocal fluorescence micrograph of the area of contact between a CD4 T cell and a B cell (as viewed through one of the cells) is shown. Proteins in the contact area between the T cell and the antigen-presenting cell form a structure called the immunological synapse, also known as the supramolecular activation complex (SMAC), which is organized into two distinct regions: the outer, or peripheral SMAC (pSMAC), indicated by the red ring; and the inner, or central SMAC (cSMAC), indicated in bright green. The cSMAC is enriched in the T-cell receptor (TCR), CD4, CD8, CD28, CD2, and PKC-θ. The pSMAC is enriched for the integrin LFA-1 and the cytoskeletal protein talin. Photograph courtesy of A. Kupfer.

Movie 9.6 

effects, and cause an immune deficiency syndrome for which the protein is named (see Section 13-15). WASP is activated via T-cell receptor signaling through several pathways, for example by an adaptor protein called Nck or by the small GTP-binding proteins Cdc42 and Rac1 (see Section 7-3), which are activated by the adaptor protein Vav (see Section 7-15). Polarization starts with the local reorganization of the cortical actin cytoskeleton at the site of contact; this in turn leads to the reorientation of the microtubule-organizing center (MTOC), the center from which the microtubule cytoskeleton is produced, and of the Golgi apparatus (GA), through which most proteins destined for secretion travel. In the cytotoxic T cell, the cytoskeletal reorientation focuses exocytosis of the preformed cytotoxic granules at the site of contact with its target cell. The polarization of a T cell also focuses the secretion of soluble effector molecules whose synthesis is induced *de novo* by ligation of the T-cell receptor. For example, the secreted cytokine IL-4, which is the principal effector molecule of T_H2 cells, is confined and concentrated at the site of contact with the target cell (see Fig. 10.6).

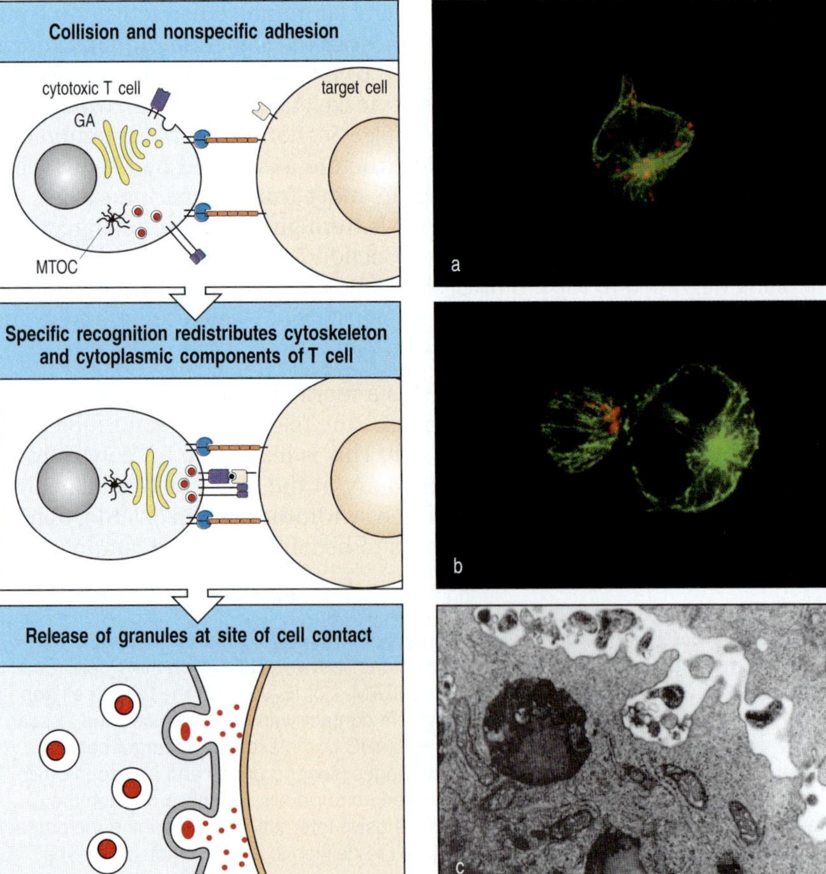

Fig. 9.32 The cellular polarization of T cells during specific antigen recognition allows effector molecules to be focused on the antigen-bearing target cell. The example illustrated here is a CD8 cytotoxic T cell. Cytotoxic T cells contain specialized lysosomes called cytotoxic granules (shown in red in the left panels), which contain cytotoxic proteins. Initial binding to a target cell through adhesion molecules does not have any effect on the location of the cytotoxic granules. Binding of the T-cell receptor causes the T cell to become polarized: reorganization within the cortical actin cytoskeleton at the site of contact aligns the microtubule-organizing center (MTOC), which in turn aligns the secretory apparatus, including the Golgi apparatus (GA), toward the target cell. Proteins stored in cytotoxic granules derived from the Golgi are then directed specifically onto the target cell. The photomicrograph in panel a shows an unbound, isolated cytotoxic T cell. The microtubule cytoskeleton is stained in green and the cytotoxic granules in red. Note how the granules are dispersed throughout the T cell. Panel b depicts a cytotoxic T cell bound to a (larger) target cell. The granules are now clustered at the site of cell–cell contact in the bound T cell. The electron micrograph in panel c shows the release of granules from a cytotoxic T cell. Panels a and b courtesy of G. Griffiths. Panel c courtesy of E. Podack.

Movie 9.7

Thus, the antigen-specific T-cell receptor controls the delivery of effector signals in three ways: it induces the tight binding of effector cells to their target cells to create a narrow space in which effector molecules can be concentrated; it focuses their delivery at the site of contact by inducing a reorientation of the secretory apparatus of the effector cell; and it triggers their synthesis and/or release. All these mechanisms contribute to targeting the action of effector molecules onto the cell bearing a specific antigen. Effector T-cell activity is thus highly selective for the appropriate target cells, even though the effector molecules themselves are not antigen-specific.

9-22 The effector functions of T cells are determined by the array of effector molecules that they produce.

The effector molecules produced by effector T cells fall into two broad classes: **cytotoxins**, which are stored in specialized cytotoxic granules and released by CD8 cytotoxic T cells (see Fig. 9.32), and cytokines and related membrane-associated proteins, which are synthesized *de novo* by all effector T cells. Cytotoxins are the principal effector molecules of cytotoxic T cells and will be discussed in Section 9-28. Their release in particular must be tightly regulated because they are not specific: they can penetrate the lipid bilayer and trigger apoptosis in any cell. By contrast, CD4 effector T cells act mainly through the production of cytokines and membrane-associated proteins, and their actions are restricted to cells bearing MHC class II molecules and expressing receptors for these proteins.

The main effector molecules of T cells are summarized in Fig. 9.33. The cytokines are a diverse group of proteins and we will review them briefly before discussing the T-cell cytokines and their actions. Soluble cytokines and membrane-associated molecules often act in combination to mediate these effects.

CD8 T cells: peptide + MHC class I		CD4 T cells: peptide + MHC class II							
Cytotoxic (killer) T cells		T_H1 cells		T_H2 cells		T_H17 cells		T_{reg} cells	
Cytotoxic effector molecules	Others	Macrophage-activating effector molecules	Others	Barrier immunity activating effector molecules	Others	Neutrophil recruitment	Others	Suppressive cytokines	Others
Perforin Granzymes Granulysin Fas ligand	IFN-γ LT-α TNF-α	IFN-γ GM-CSF TNF-α CD40 ligand Fas ligand	IL-3 LT-α CXCL2 (GROβ)	IL-4 IL-5 IL-13 CD40 ligand	IL-3 GM-CSF IL-10 TGF-β CCL11 (eotaxin) CCL17 (TARC)	IL-17A IL-17F IL-6	TNF CXCL1 (GROα)	IL-10 TGF-β	GM-CSF

Fig. 9.33 The different types of effector T cell subsets produce different effector molecules. CD8 T cells are predominantly killer T cells that recognize peptide:MHC class I complexes. They release perforin (which helps deliver granzymes into the target cell) and granzymes (which are pro-proteases that are activated intracellularly to trigger apoptosis in the target cell), and often also produce the cytokine IFN-γ. They also carry the membrane-bound effector molecule Fas ligand (CD178). When this binds to Fas (CD95) on a target cell it activates apoptosis in the Fas-bearing cell. The various functional subsets of CD4 T cells recognize peptide:MHC class II complexes. T_H1 cells are specialized to activate macrophages that are infected by or have ingested pathogens; they secrete IFN-γ to activate the infected cell, as well as other effector molecules. They can express membrane-bound CD40 ligand and/or Fas ligand. CD40 ligand triggers activation of the target cell, whereas Fas ligand triggers the death of Fas-bearing targets, and so which molecule is expressed strongly influences T_H1 function. T_H2 cells are specialized for promoting immune responses to parasites and also promote allergic responses. They provide help in B-cell activation and secrete the B-cell growth factors IL-4, IL-5, IL-9, and IL-13. The principal membrane-bound effector molecule expressed by T_H2 cells is CD40 ligand, which binds to CD40 on B cells and induces B-cell proliferation and isotype switching (see Chapter 10). T_H17 cells produce members of the IL-17 family and IL-6, and promote acute inflammation by helping to recruit neutrophils to sites of infection. T_{reg} cells, of which there are several types, produce inhibitory cytokines such as IL-10 and TGF-β and exert inhibitory actions through unknown mechanisms that are dependent on cell contact.

9-23 Cytokines can act locally or at a distance.

Cytokines are small soluble proteins secreted by cells that can alter the behavior or properties of the secreting cell itself or of another cell. They are produced by many cell types in addition to those of the immune system. We have already introduced the families of cytokines and their receptors important in innate and adaptive immunity in Chapters 3 and 7 (see Section 3-13 and Section 7-19). Here we are concerned mainly with the cytokines that mediate the effector functions of T cells. Cytokines produced by lymphocytes are often called **lymphokines**, but this nomenclature can be confusing because some lymphokines are also secreted by nonlymphoid cells; we will therefore use the generic term 'cytokine' for all of them. Most cytokines produced by T cells are given the name **interleukin** (**IL**) followed by a number: we have encountered several interleukins already in this chapter. The cytokines produced by T cells are shown in Fig. 9.34, and a more comprehensive list of cytokines of immunological interest is in Appendix III. Most cytokines have a multitude of different biological effects when tested at high concentration in biological assays *in vitro*, but targeted disruption of the genes for cytokines and cytokine receptors by gene knockout in mice (see Appendix I, Section A-46) has helped to clarify their physiological roles.

The main cytokine released by CD8 effector T cells is IFN-γ, which can block viral replication or even lead to the elimination of virus from infected cells without killing them. CD4 effector subsets release different, but overlapping, sets of cytokines, which define their distinct actions in immunity. T_H17 cells secrete IL-17, TNF-α, and the chemokine CXCL1, all of which act to recruit neutrophils to sites of infection early in the adaptive immune response. T_H1 cells secrete IFN-γ, which is the main macrophage-activating cytokine, and lymphotoxin-α (LT-α), which activates macrophages, inhibits B cells, and is directly cytotoxic for some cells. T_H2 cells secrete IL-4, IL-5, and IL-13, which stimulate eosinophils and mast cells and activate B cells, and IL-10, which inhibits the development of T_H1 cells and cytokine release from macrophages. During the earliest stages of T-cell activation, provided that co-stimulatory signals are present, the differentiating CD4 T cells produce IL-2, and only very small amounts of IL-4 and IFN-γ.

Binding of the T-cell receptor orchestrates the polarized release of these cytokines so that they are concentrated at the site of contact with the target cell (see Section 9-21). Furthermore, most of the soluble cytokines have local actions that synergize with those of the membrane-bound effector molecules. The effect of all these molecules is therefore combinatorial, and, because the membrane-bound effectors can bind only to receptors on an interacting cell, this is another mechanism by which selective effects of cytokines are focused on the target cell. The effects of some cytokines are further confined to target cells by tight regulation of their synthesis: the synthesis of IL-2, IL-4, and IFN-γ is controlled by mRNA instability (see Section 9-13), so that their secretion by T cells does not continue after the interaction with a target cell has ended.

Some cytokines have more distant effects. IL-3 and GM-CSF (see Fig. 9.34) are released by T_H1 and T_H2 cells and act on bone marrow cells to stimulate the production of macrophages and granulocytes, both of which are important nonspecific effector cells in both humoral and cell-mediated immunity. IL-3 and GM-CSF also stimulate the production of dendritic cells from bone marrow precursors. The predominant T cells activated in allergic reactions are T_H2 cells, and the IL-5 they produce stimulates the production of eosinophils, which contributes to the later phases of an allergic reaction (see Chapter 14). Whether a given cytokine effect is local or more distant is likely to reflect the amounts released, the degree to which this release is focused on the target cell, and the stability of the cytokine *in vivo*.

9-24 T cells express several TNF-family cytokines as trimeric proteins that are usually associated with the cell surface.

Most effector T cells express members of the TNF protein family as membrane-associated proteins on the cell surface. The most important in T-cell effector function are TNF-α, the lymphotoxins (LTs), **Fas ligand** (CD178), and **CD40 ligand**, the latter two always being cell-surface associated. TNF-α is made by T cells in soluble and membrane-associated forms and forms a homotrimer (see Fig. 7.28). Secreted LT-α is a homotrimer, but when it is in membrane-bound form, LT-α is linked to a third, transmembrane member of this family called LT-β to form heterotrimers, commonly called LT-β (see Section 7-19).

Fig. 9.34 The nomenclature and functions of well-defined T-cell cytokines. Each cytokine has multiple activities on different cell types. Major activities of effector cytokines are highlighted in red. The mixture of cytokines secreted by a given cell type produces many effects through what is called a 'cytokine network.' ↑, increase; ↓, decrease; CTL, cytotoxic lymphocyte; NK cells, natural killer cells; CSF, colony-stimulating factor; IBD, inflammatory bowel disease; NO, nitric oxide.

Cytokine	T-cell source	Effects on					Effect of gene knockout
		B cells	T cells	Macrophages	Hematopoietic cells	Other tissue cells	
Interleukin-2 (IL-2)	Naive, T_H1, some CD8	Stimulates growth and J-chain synthesis	Growth	–	Stimulates NK cell growth	–	↓T-cell responses IBD
Interferon-γ (IFN-γ)	T_H1, CTL	Differentiation IgG2a synthesis (mouse)	Inhibits T_H2 cell growth	Activation, ↑MHC class I and class II	Activates NK cells	Antiviral ↑MHC class I and class II	Susceptible to mycobacteria, some viruses
Lymphotoxin-α (LT-α, TNF-β)	T_H1, some CTL	Inhibits	Kills	Activates, induces NO production	Activates neutrophils	Kills fibroblasts and tumor cells	Absence of lymph nodes Disorganized spleen
Interleukin-4 (IL-4)	T_H2	Activation, growth IgG1, IgE ↑MHC class II induction	Growth, survival	Inhibits macrophage activation	↑Growth of mast cells	–	No T_H2
Interleukin-5 (IL-5)	T_H2	Mouse: Differentiation IgA synthesis	–	–	↑Eosinophil growth and differentiation	–	Reduced eosinophilia
Interleukin-10 (IL-10)	T_H2 (human: some T_H1), T_{reg}	↑MHC class II	Inhibits T_H1	Inhibits cytokine release	Co-stimulates mast cell growth	–	IBD
Interleukin-3 (IL-3)	T_H1, T_H2, some CTL	–	–	–	Growth factor for progenitor hematopoietic cells (multi-CSF)	–	–
Tumor necrosis factor-α (TNF-α)	T_H1, some T_H2, some CTL	–	–	Activates, induces NO production	–	Activates microvascular endothelium	Susceptibility to Gram –ve sepsis
Granulocyte-macrophage colony-stimulating factor (GM-CSF)	T_H1, some T_H2, some CTL	Differentiation	Inhibits growth?	Activation Differentiation to dendritic cells	↑Production of granulocytes and macrophages (myelopoiesis) and dendritic cells	–	–
Transforming growth factor-β (TGF-β)	CD4 T cells (T_{reg})	Inhibits growth IgA switch factor	Inhibits growth, promotes survival	Inhibits activation	Activates neutrophils	Inhibits/ stimulates cell growth	Death at ~10 weeks
Interleukin-17 (IL-17)	CD4 T cells (T_H17), macrophages	–	–	–	Stimulates neutrophil recruitment	Stimulates fibroblasts and epithelial cells to secrete chemokines	–

The receptors for these molecules, TNFR-I and TNFR-II, form homotrimers when bound to either TNF-α or LT-α. The trimeric structure is characteristic of all members of the TNF family, and the ligand-induced trimerization of their receptors seems to be the critical event in initiating signaling.

Fas ligand and CD40 ligand bind respectively to the transmembrane proteins **Fas** (CD95) and **CD40** on target cells. Fas contains a 'death' domain in its cytoplasmic tail, and binding of Fas by Fas ligand induces death by apoptosis in the Fas-bearing cell (see Fig. 7.29). Other TNFR-family members, including TNFR-I, are also associated with death domains and can also induce apoptosis. Thus, TNF-α and LT-α can induce apoptosis by binding to TNFR-I.

CD40 ligand is particularly important for CD4 T-cell effector function; it is induced on T_H1, T_H2, and T_{FH} cells, and delivers activating signals to B cells and macrophages through CD40. The cytoplasmic tail of CD40 lacks a death domain; instead, it seems to be linked downstream to proteins called TRAFs (TNF-receptor-associated factors). CD40 is involved in the activation of B cells and macrophages; the ligation of CD40 on B cells promotes growth and isotype switching, whereas CD40 ligation on macrophages induces them to secrete TNF-α and become receptive to much lower concentrations of IFN-γ. Deficiency in CD40 ligand expression is associated with immunodeficiency, as we will learn in Chapters 10 and 14.

Summary.

Interactions between effector T cells and their targets are initiated by transient antigen-nonspecific adhesion between the cells. T-cell effector functions are elicited only when peptide:MHC complexes on the surface of the target cell are recognized by the receptor on an effector T cell. This recognition event triggers the effector T cell to adhere more strongly to the antigen-bearing target cell and to release its effector molecules directly at the target cell, leading to the activation or death of the target. The immunological consequences of antigen recognition by an effector T cell are determined largely by the set of effector molecules that it produces on binding a specific target cell. CD8 cytotoxic T cells store preformed cytotoxins in specialized cytotoxic granules whose release is tightly focused at the site of contact with the infected target cell, thus killing it without killing any uninfected cells nearby. Cytokines and members of the TNF family of membrane-associated effector proteins are synthesized *de novo* by most effector T cells. T_H1 cells express effector proteins that activate macrophages, and cytokines that induce class switching to certain antibody classes. Cytokines made by T_H2 cells direct class switching to antibodies involved in anti-parasitic and allergic type responses. T_H17 cells secrete IL-17, which recruits acute inflammatory cells such as neutrophils to the site of infection. Membrane-associated effector molecules can deliver signals only to an interacting cell bearing the appropriate receptor, whereas soluble cytokines can act on cytokine receptors expressed locally on the target cell, or on hematopoietic cells at a distance. The actions of cytokines and membrane-associated effector molecules through their specific receptors, together with the effects of the cytotoxins released by CD8 cells, account for most of the effector functions of T cells.

T cell-mediated cytotoxicity.

All viruses, and some bacteria, multiply in the cytoplasm of infected cells; indeed, a virus is a highly sophisticated parasite that has no biosynthetic or metabolic apparatus of its own and, in consequence, can replicate only inside cells. Although susceptible to antibody before they enter cells, once they have done so these pathogens are not accessible to antibodies and can be eliminated only by the destruction or modification of the infected cells on which they depend. This role in host defense is largely filled by CD8 cytotoxic T cells, although CD4 T cells may also acquire cytotoxic capacities. The crucial role of cytotoxic T cells in limiting such infections is seen in the increased susceptibility of animals artificially depleted of these T cells, or of mice or humans that lack the MHC class I molecules that present antigen to CD8 T cells. The elimination of infected cells without the destruction of healthy tissue requires the cytotoxic mechanisms of CD8 T cells to be both powerful and accurately targeted.

9-25 Cytotoxic T cells can induce target cells to undergo programmed cell death.

Cells can die in various ways. Physical or chemical injury, such as the deprivation of oxygen that occurs in heart muscle during a heart attack or membrane damage with antibody and complement, leads to cell disintegration or necrosis. The dead or necrotic tissue is taken up and degraded by phagocytic cells, which eventually clear the damaged tissue and heal the wound. The other form of cell death is known as programmed cell death, which can be by apoptosis or by autophagic cell death. Apoptosis is a normal cellular response that is crucial in the tissue remodeling that occurs during development and metamorphosis in all multicellular animals. As we saw in Chapter 8, most thymocytes die by apoptosis when they fail positive selection. Early changes seen in apoptosis are nuclear blebbing, alteration in cell morphology, and, eventually, fragmentation of the DNA. The cell then destroys itself from within, shrinking by shedding membrane-bound vesicles, and degrading itself until little is left. A hallmark of apoptosis is the fragmentation of nuclear DNA into pieces 200 base pairs long through the activation of nucleases that cleave the DNA between nucleosomes. As we described in Chapter 6, autophagy is the process of degrading senescent or abnormal proteins and organelles. In autophagic programmed cell death, large vacuoles degrade cellular organelles before the condensation and destruction of the nucleus that is characteristic of apoptosis.

Cytotoxic T cells kill their targets by inducing them to undergo apoptosis (Fig. 9.35). When cytotoxic T cells are mixed with target cells and rapidly brought into contact by centrifugation, they can induce antigen-specific target cells to die within 5 minutes, although death can take hours to become fully evident. The rapidity of this response reflects the release of preformed effector molecules, which activate an apoptotic pathway within the target cell.

A mechanism for inducing apoptosis that does not depend on cytotoxic granules involves members of the TNF family, particularly Fas and Fas ligand. In contrast to the killing of infected tissue cells, this mechanism is used mainly to regulate lymphocyte numbers. Activated lymphocytes express both Fas and Fas ligand, and thus activated cytotoxic T cells can kill other lymphocytes through the activation of caspases, which induces apoptosis in the target lymphocyte. Thus, Fas–Fas ligand interactions are important in terminating lymphocyte proliferation after the pathogen initiating an immune response has been cleared. As well as cytotoxic T cells, T_H1 cells and some T_H2 cells have been shown to be able to kill cells by this pathway. The importance of Fas in maintaining lymphocyte homeostasis can be seen from the effects of mutations in the genes encoding Fas and Fas ligand. Mice and humans with a mutant form of Fas develop a lymphoproliferative disease associated with severe autoimmunity (**autoimmune lymphoproliferative syndrome** (**ALPS**)), which is described in Section 15-19. A mutation in the gene encoding Fas ligand in another mouse strain creates a nearly identical phenotype. These

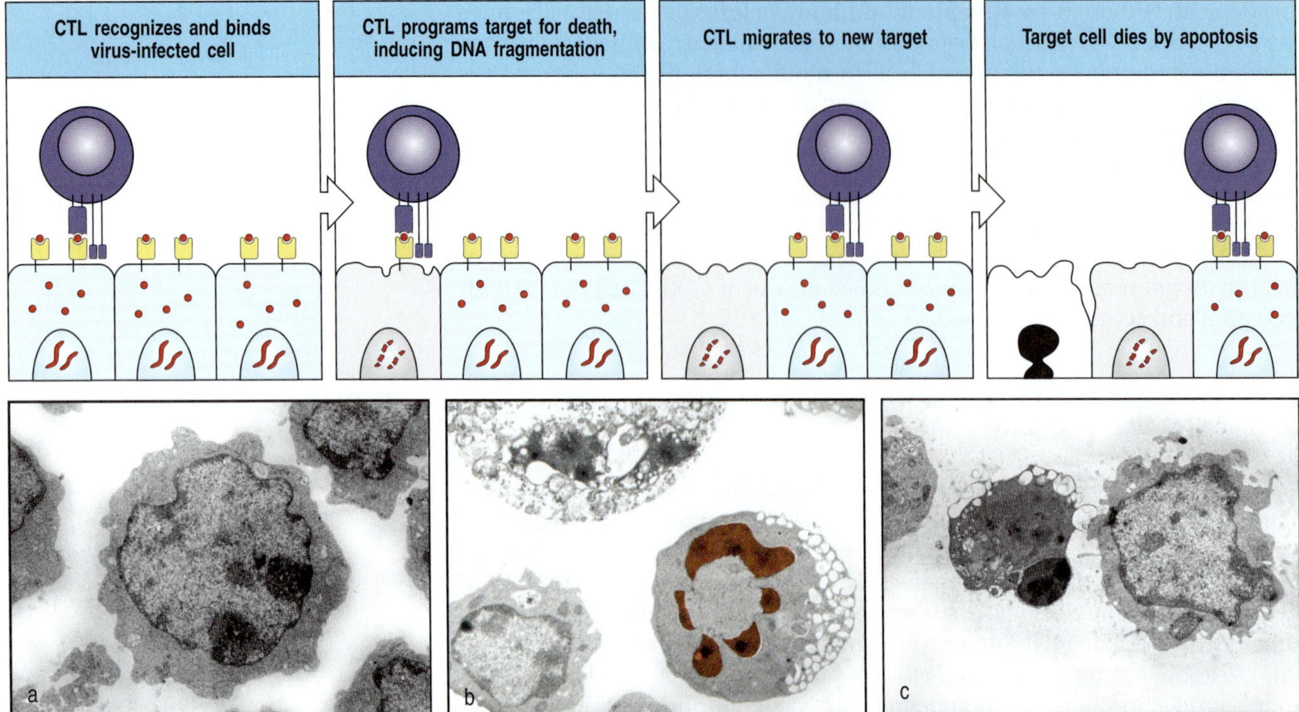

Fig. 9.35 Cytotoxic CD8 T cells can induce apoptosis in target cells. Specific recognition of peptide:MHC complexes on a target cell (top panels) by a cytotoxic CD8 T cell (CTL) leads to the death of the target cell by apoptosis. Cytotoxic T cells can recycle to kill multiple targets. Each killing requires the same series of steps, including receptor binding and the directed release of cytotoxic proteins stored in granules. The process of apoptosis is shown in the micrographs (bottom panels), where panel a shows a healthy cell with a normal nucleus. Early in apoptosis (panel b) the chromatin becomes condensed (red) and, although the cell sheds membrane vesicles, the integrity of the cell membrane is retained, in contrast to the necrotic cell in the upper part of the same field. In late stages of apoptosis (panel c), the cell nucleus (middle cell) is very condensed, no mitochondria are visible, and the cell has lost much of its cytoplasm and membrane through the shedding of vesicles. Photographs (×3500) courtesy of R. Windsor and E. Hirst.

mutant phenotypes represent the best-characterized examples of generalized autoimmunity caused by single-gene defects.

As well as killing the host cell, the apoptotic mechanism may also act directly on cytosolic pathogens. For example, the nucleases that are activated in apoptosis to destroy cellular DNA can also degrade viral DNA. This prevents the assembly of virions and the release of infectious virus, which could otherwise infect nearby cells. Other enzymes activated in the course of apoptosis may destroy nonviral cytosolic pathogens. Apoptosis is therefore preferable to necrosis as a means of killing infected cells; in cells dying by necrosis, intact pathogens are released from the dead cell and these can continue to infect healthy cells or parasitize the macrophages that ingest them.

Protein in granules of cytotoxic T cells	Actions on target cells
Perforin	Aids in delivering contents of granules into the cytoplasm of target cell
Granzymes	Serine proteases, which activate apoptosis once in the cytoplasm of the target cell
Granulysin	Has antimicrobial actions and can induce apoptosis

Fig. 9.36 Cytotoxic effector proteins released by cytotoxic T cells.

9-26 Cytotoxic effector proteins that trigger apoptosis are contained in the granules of CD8 cytotoxic T cells.

The principal mechanism of cytotoxic T-cell action is the calcium-dependent release of specialized **cytotoxic granules** upon recognition of antigen on the surface of a target cell. Cytotoxic granules are modified lysosomes that contain at least three distinct classes of cytotoxic effector proteins that are expressed specifically in cytotoxic T cells (Fig. 9.36). Such proteins are stored in the cytotoxic granules in an active form, but conditions within the granules prevent them from functioning until after their release. One of these cytotoxic proteins, known as **perforin**, acts in the delivery of the contents of cytotoxic granules to target-cell membranes. The importance of perforin in cytotoxicity is well illustrated in mice that have had their perforin gene knocked out. They are severely defective in their ability to mount a cytotoxic T-cell response to many, but not all, viruses. Another class of cytotoxic proteins comprises a family of serine proteases, called **granzymes**, of which there are 5 in humans and 10 in the mouse. The third cytotoxic protein, **granulysin**, which is expressed in humans but not in mice, has antimicrobial activity and at high concentrations is also able to induce apoptosis in target cells. Granules that store perforin, granzymes, and granulysin can be seen in CD8 cytotoxic effector cells in infected tissue. The granules also contain the proteoglycan **serglycin**, which acts as a scaffold, forming a complex with perforin and the granzymes.

Both perforin and granzymes are required for effective cell killing. Their separate roles have been investigated in experiments that rely on similarities between the cytotoxic granules of CD8 T cells and the granules of mast cells, which are more easily studied. The release of mast-cell granules occurs on cross-linking of a cell-surface receptor for IgE, just as the release of cytotoxic granules from T cells occurs after the aggregation of T-cell receptors at the immunological synapse. The mechanism of signaling for granule release is thought to be the same or similar in both cases, because both the IgE receptor and the T-cell receptor have ITAM motifs in their cytoplasmic domains, and their cross-linking leads to tyrosine phosphorylation of the ITAMs (see Chapter 7). When a mast-cell line is transfected with the genes for perforin or a granzyme, the gene products are stored in mast-cell granules, and when the cell is activated these granules are released. When transfected with the gene encoding perforin alone, mast cells can kill other cells, but large numbers of transfected cells are needed because the killing is very inefficient. In contrast, mast cells transfected with the gene encoding granzyme B alone are unable to kill other cells. However, when perforin-transfected mast cells are also transfected with the gene encoding granzyme B, the cells or their purified granules become as effective at killing targets as granules from cytotoxic cells. It is generally thought that perforin acts by causing a pore to form in the target cell plasma membrane, through which granzymes enter.

The granzymes trigger apoptosis in the target cell by activating caspases. Granzyme B cleaves and activates caspase 3, which is a cysteine protease that cuts after aspartic acid residues (hence the name caspase). Caspase 3 activates a caspase proteolytic cascade, which eventually activates the caspase-activated deoxyribonuclease (CAD) by cleaving an inhibitory protein (ICAD) that binds to and inactivates CAD. This nuclease is believed to be the enzyme that degrades the DNA (Fig. 9.37). Granzyme B also activates other pathways of cell death. One important target is the protein BID (for BH3-interacting domain death agonist protein). When BID is cleaved, either directly by granzyme B or indirectly by activated caspase 3, the mitochondrial

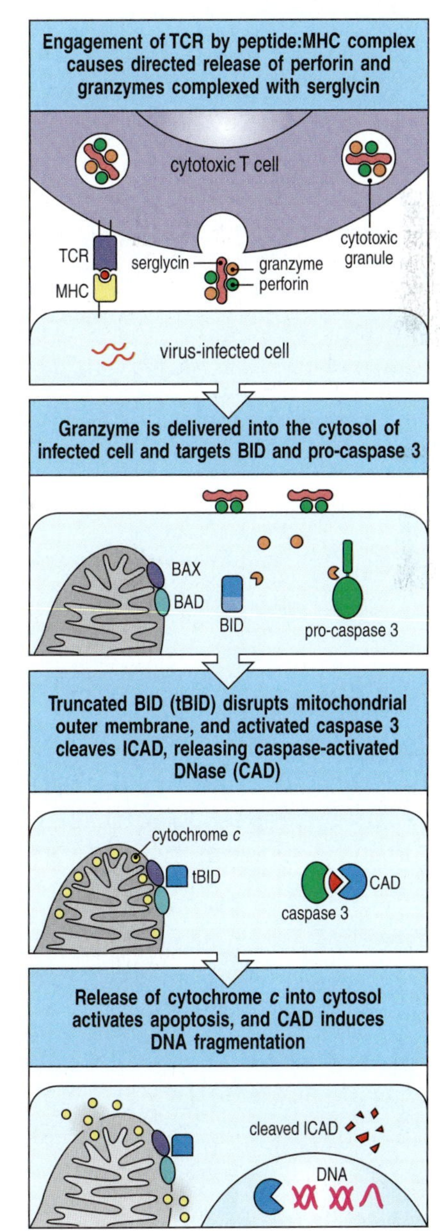

Fig. 9.37 Perforin, granzymes, and serglycin are released from cytotoxic granules and deliver granzymes into the cytosol of target cells to induce apoptosis. Recognition of its antigen on a virus-infected cell by a cytotoxic CD8 T cell induces the release of the contents of its cytotoxic granules in a directed fashion. Perforin and granzymes, complexed with the proteoglycan serglycin, are delivered as a complex to the membrane of the target cell (top panel). By an unknown mechanism, perforin directs the entry of the granule contents into the cytosol of the target cell without apparent pore formation, and the introduced granzymes then act on specific intracellular targets such as the proteins BID and pro-caspase 3. Either directly or indirectly, the granzymes cause the cleavage of BID into truncated BID (tBID) and the cleavage of pro-caspase 3 to an active caspase (second panel). tBID acts on mitochondria to release cytochrome *c* into the cytosol, and activated caspase 3 targets ICAD to release caspase-activated DNase (CAD) (third panel). Cytochrome *c* in the cytosol promotes apoptosis, and CAD fragments the DNA (bottom panel).

outer membrane becomes disrupted, causing the release from the mitochondrial intermembrane space of pro-apoptotic molecules such as cytochrome *c*. Other granzymes are thought to promote apoptosis by targeting different cellular components.

Cells undergoing programmed cell death are rapidly ingested by phagocytic cells, which recognize a change in the cell membrane: phosphatidylserine, which is normally found only in the inner leaflet of the membrane, replaces phosphatidylcholine as the predominant phospholipid in the outer leaflet. The ingested cell is broken down and completely digested by the phagocyte without the induction of co-stimulatory proteins. Thus, apoptosis is normally an immunologically 'quiet' process; that is, apoptotic cells do not normally contribute to or stimulate immune responses.

9-27 Cytotoxic T cells are selective and serial killers of targets expressing a specific antigen.

When cytotoxic T cells are offered a mixture of equal amounts of two target cells, one bearing a specific antigen and the other not, they kill only the target cell bearing the specific antigen. The 'innocent bystander' cells and the cytotoxic T cells themselves are not killed. The cytotoxic T cells are probably not killed because release of the cytotoxic effector molecules is highly polarized. As we saw in Fig. 9.32, cytotoxic T cells orient their Golgi apparatus and microtubule-organizing center to focus secretion on the point of contact with a target cell. Granule movement toward the point of contact is shown in Fig. 9.38 Cytotoxic T cells attached to several different target cells reorient their secretory apparatus toward each cell in turn and kill them one by one, strongly suggesting that the mechanism whereby cytotoxic mediators are released allows attack at only one point of contact at any one time. The narrowly focused action of CD8 cytotoxic T cells allows them to kill single infected cells in a tissue without creating widespread tissue damage (Fig. 9.39) and is of crucial importance in tissues where cell regeneration does not occur, as with the neurons of the central nervous system, or is very limited, as in the pancreatic islets.

Cytotoxic T cells can kill their targets rapidly because they store preformed cytotoxic proteins in forms that are inactive in the environment of the cytotoxic granule. Cytotoxic proteins are synthesized and loaded into the granules during the first encounter of a naive cytotoxic precursor T cell with its specific antigen. Ligation of the T-cell receptor similarly induces *de novo* synthesis of perforin and granzymes in effector CD8 T cells, so that the supply of cytotoxic granules is replenished. This makes it possible for a single CD8 T cell to kill a series of targets in succession.

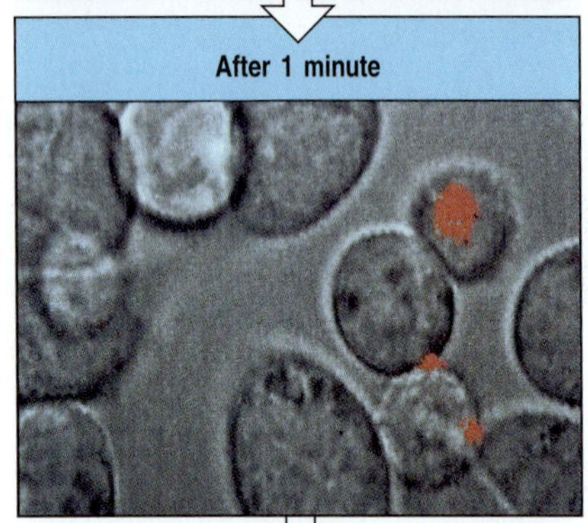

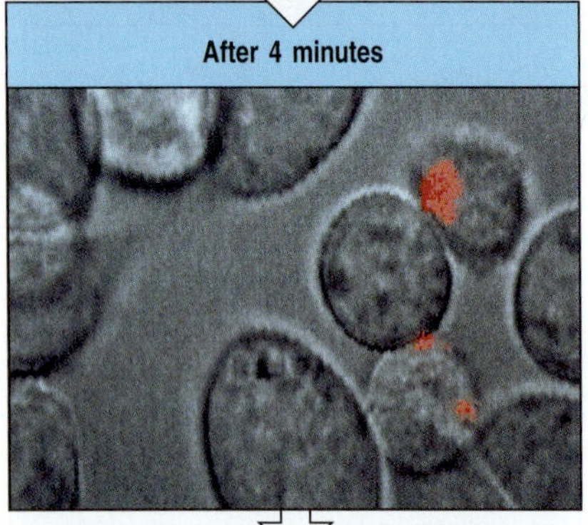

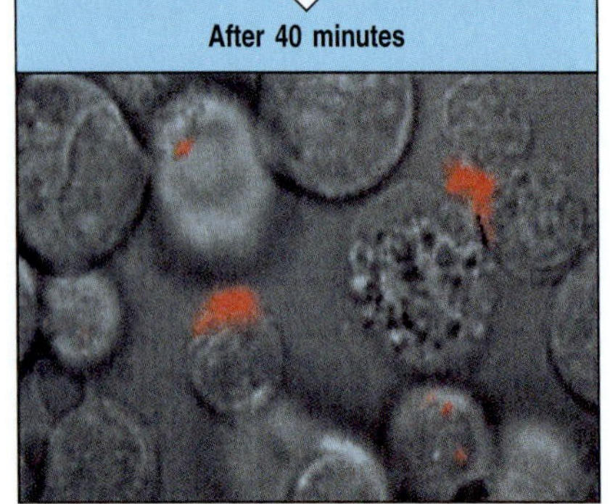

Fig. 9.38 Effector molecules are released from T-cell granules in a highly polar fashion. The granules of cytotoxic T cells can be labeled with fluorescent dyes, allowing them to be seen under the microscope, and their movements can be followed by time-lapse photography. Here we show a series of pictures taken during the interaction of a cytotoxic T cell with a target cell, which is eventually killed. In the top panel, at time 0, the T cell (upper right) has just made contact with a target cell (diagonally below). At this time, the granules of the T cell, labeled with a red fluorescent dye, are distant from the point of contact. In the second panel, after 1 minute has elapsed, the granules have begun to move toward the target cell, a move that has essentially been completed in the third panel, after 4 minutes. After 40 minutes, in the last panel, the granule contents have been released into the space between the T cell and the target, which has begun to undergo apoptosis (note the fragmented nucleus). The T cell will now disengage from the target cell and can recognize and kill other targets. Photographs courtesy of G. Griffiths.

9-28 Cytotoxic T cells also act by releasing cytokines.

Inducing apoptosis in target cells is the main way in which CD8 cytotoxic T cells eliminate infection. However, most CD8 cytotoxic T cells also release the cytokines IFN-γ, TNF-α, and LT-α, which contribute to host defense in several ways. IFN-γ inhibits viral replication directly, and induces the increased expression of MHC class I molecules and of other proteins that are involved in peptide loading of these newly synthesized MHC class I molecules in infected cells. This increases the chance that infected cells will be recognized as target cells for cytotoxic attack. IFN-γ also activates macrophages, recruiting them to sites of infection both as effector cells and as antigen-presenting cells. TNF-α and LT-α can synergize with IFN-γ in macrophage activation and in killing some target cells through their interaction with TNFR-I, which induces apoptosis (see Section 9-24). Thus, effector CD8 cytotoxic T cells act in a variety of ways to limit the spread of cytosolic pathogens. The relative importance of each of these mechanisms is being rapidly determined through gene knockouts in mice.

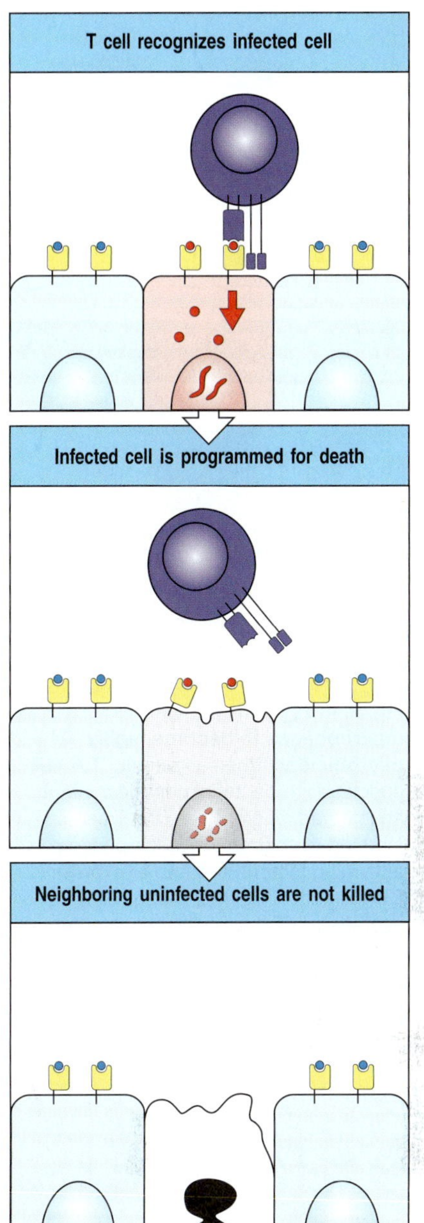
T cell recognizes infected cell
Infected cell is programmed for death
Neighboring uninfected cells are not killed

Fig. 9.39 Cytotoxic T cells kill target cells bearing specific antigen while sparing neighboring uninfected cells. All the cells in a tissue are susceptible to killing by the cytotoxic proteins of armed effector CD8 T cells, but only infected cells are killed. Specific recognition by the T-cell receptor identifies which target cell to kill, and the polarized release of the cytotoxic granules (not shown) ensures that neighboring cells are spared.

Summary.

Effector CD8 cytotoxic T cells are essential in host defense against pathogens that live in the cytosol: most commonly these will be viruses. These cytotoxic T cells can kill any cell harboring such pathogens by recognizing foreign peptides that are transported to the cell surface bound to MHC class I molecules. CD8 cytotoxic T cells perform their killing function by releasing three types of preformed cytotoxic proteins: the granzymes, which seem able to induce apoptosis in any type of target cell; perforin, which acts in the delivery of granzymes into the target cell; and granulysin. These properties allow the cytotoxic T cell to attack and destroy virtually any cell infected with a cytosolic pathogen. The membrane-bound Fas ligand, expressed by CD8 and some CD4 T cells, may also induce apoptosis by binding to Fas on some target cells, but this pathway is probably more important in removing Fas-bearing activated lymphocytes after an infection has been cleared and in maintaining lymphocyte homoeostasis. CD8 cytotoxic T cells also produce IFN-γ, which inhibits viral replication and is an important inducer of MHC class I molecule expression and macrophage activation. Cytotoxic T cells kill infected targets with great precision, sparing adjacent normal cells. This precision is crucial in minimizing tissue damage while allowing the eradication of infected cells.

Macrophage activation by T_H1 cells.

Some microorganisms, most notably mycobacteria, are intracellular pathogens that grow primarily in the phagosomes of macrophages, shielded from the effects of both antibodies and cytotoxic T cells. Peptides derived from such microorganisms can be displayed on the macrophage surface by MHC class II molecules. When these peptide:MHC complexes are recognized by the T-cell receptor of antigen-specific effector T_H1 cells, the T cell is stimulated to synthesize membrane-associated proteins and soluble cytokines that stimulate the macrophage and enable it to eliminate the pathogen. This boost to antimicrobial mechanisms is known as **macrophage activation**. T_H1 cells similarly activate macrophages to increase the destruction of recently ingested pathogens. This coordination between T_H1 cells and macrophages underlies the formation of the immunological reaction called the granuloma, in which microbes are held in check within a central area of macrophages surrounded by activated lymphocytes.

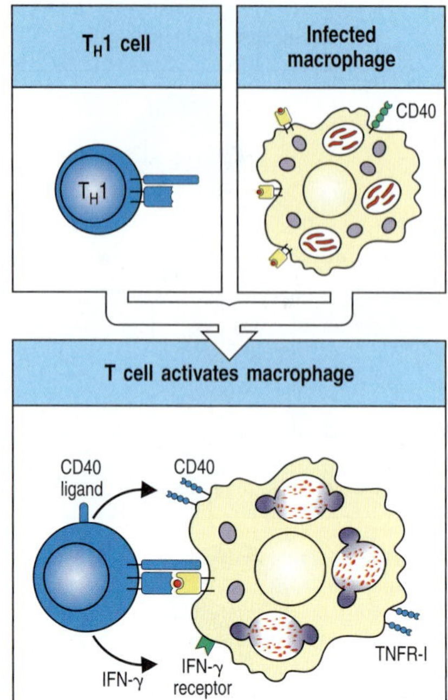

Fig. 9.40 T_H1 cells activate macrophages to become highly microbicidal. When an effector T_H1 cell specific for a bacterial peptide contacts an infected macrophage, the T cell is induced to secrete the macrophage-activating factor IFN-γ and to express CD40 ligand. Together these newly synthesized T_H1 proteins activate the macrophage.

9-29 T_H1 cells have a central role in macrophage activation.

Pathogens of all types are ingested by macrophages from the extracellular fluid, and are often destroyed without the need for additional macrophage activation, as we saw in Chapter 3. In several clinically important infections, however, ingested pathogens are not killed, and can even set up a chronic infection in the macrophage and incapacitate it. Such microorganisms maintain themselves in the usually hostile environment of the phagosomes by inhibiting the fusion of phagosomes and lysosomes, or by preventing the acidification required to activate lysosomal proteases. They can be eliminated when the infected macrophage is recognized by a T_H1 cell, which provides additional signals that further stimulate the macrophage's intracellular antimicrobial defenses.

Macrophages require two signals for activation, and effector T_H1 cells can deliver both. One signal is the cytokine IFN-γ; the other sensitizes the macrophage to respond to IFN-γ. T_H1 cells interacting with their target cells characteristically secrete IFN-γ, while the CD40 ligand expressed by the T_H1 cell delivers the sensitizing signal by contacting CD40 on the macrophage (Fig. 9.40). T_H1 cells also secrete lymphotoxin-alpha (LT-α), which can substitute for CD40 ligand in macrophage activation. When T_H1 cells stimulate macrophages through these molecules, the macrophage secretes TNF-α, further stimulating macrophages through the TNFR-1, the same receptor activated by LT-α. This TNF receptor seems to be required to maintain the viability of the macrophage in this setting; in mice lacking the TNFR-I (see Section 7-22), infection by *Mycobacterium avium* leads to an excess apoptosis of macrophages that is dependent on T_H1 cells, leading to disintegration of granulomas and dissemination of the pathogen.

CD8 T cells also produce IFN-γ and can activate macrophages presenting antigens derived from cytosolic proteins on MHC class I molecules. Macrophages can also be made more sensitive to IFN-γ by very small amounts of bacterial LPS, and this latter pathway may be particularly important when CD8 T cells are the primary source of the IFN-γ. T_H2 cells are inefficient macrophage activators because they produce IL-10, a cytokine that can deactivate macrophages, and they also do not produce IFN-γ. They do express CD40 ligand, however, and can deliver the contact-dependent signal required to sensitize macrophages to respond to IFN-γ.

After a T_H1 cell encounters its specific antigen, the expression of genes for effector cytokines and cell-surface molecules starts within an hour of contact, and their production and secretion require several hours. T_H1 cells must therefore adhere to their target cells for far longer than do cytotoxic T cells. As in cytotoxic T cells, the T_H1 cell's secretory machinery becomes polarized and the newly synthesized cytokines are secreted at the site of contact between T cell and macrophage (see Fig. 9.32). CD40 ligand also seems to be delivered to the cell surface in this polarized fashion. So although all macrophages have receptors for IFN-γ, the infected macrophage presenting antigen to the T_H1 cell is far more likely to become activated than nearby uninfected macrophages.

9-30 Activation of macrophages by T_H1 cells promotes microbial killing and must be tightly regulated to avoid tissue damage.

Activation converts the macrophage into a potent antimicrobial effector cell, as illustrated in Fig. 9.41. Phagosomes fuse with lysosomes, and microbicidal reactive oxygen and nitrogen species are generated as described in Section 3-2. Because activated macrophages are extremely effective in destroying pathogens, one might ask why they are not simply maintained in a state of constant activation. Besides the fact that macrophages consume huge quantities of energy to maintain the activated state, macrophage activation *in vivo* is usually associated with localized tissue destruction resulting from the release of oxygen radicals, NO, and proteases, which are toxic to host cells as well as to pathogens.

The release of toxic mediators enables macrophages to attack large extracellular pathogens that they cannot ingest, such as parasitic worms, but it comes at the price of tissue damage. Antigen-specific macrophage activation by T_H1 cells is a means of deploying this powerful defensive mechanism to maximum effect while minimizing local tissue damage and energy consumption. Macrophage activation is inhibited by cytokines such as TGF-β and IL-10, which are produced by CD4 T_H2 cells and various regulatory cells, and so the induction of these types of CD4 T cells is important for limiting macrophage activation.

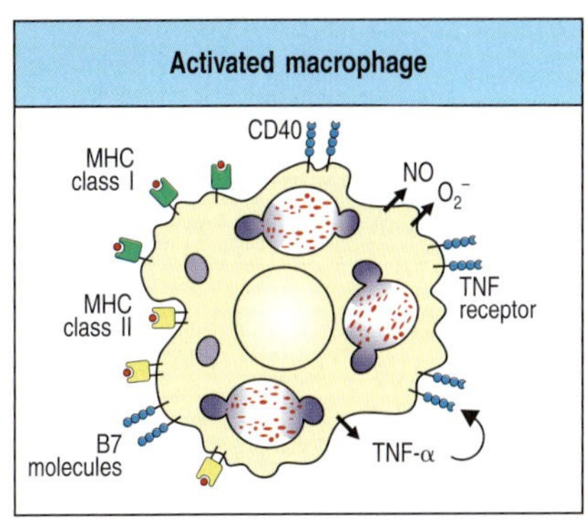

Fig. 9.41 Activated macrophages undergo changes that greatly increase their antimicrobial effectiveness and amplify the immune response. Activated macrophages increase their expression of CD40 and of TNF receptors, and are stimulated to secrete TNF-α. This autocrine stimulus synergizes with IFN-γ secreted by T_H1 cells to increase the antimicrobial action of the macrophage, in particular by inducing the production of nitric oxide (NO) and superoxide (O_2^-). The macrophage also upregulates its B7 molecules in response to binding to CD40 ligand on the T cell, and increases its expression of MHC class II molecules, thus allowing further activation of resting CD4 T cells.

9-31 T_H1 cells coordinate the host response to intracellular pathogens.

The activation of macrophages by T_H1 cells is central to the host response to pathogens that proliferate in macrophage vesicles. As well as increased intracellular killing, other changes occur in activated macrophages that help to amplify the adaptive immune response against those pathogens. The number of B7 molecules, CD40, MHC class II molecules, and TNF receptors on the macrophage surface increases (see Fig. 9.41), making the cell more effective at presenting antigen to T cells, and more responsive to CD40 ligand and TNF-α. In addition, activated macrophages secrete IL-12, which increases the amount of IFN-γ produced by T_H1 cells and also promotes the differentiation of activated naive CD4 T cells into T_H1 effector cells (see Section 9-18). Cytokines and chemokines secreted by activated macrophages are also important in stimulating the production of antibodies and in recruiting other immune cells to sites of infection.

In mice whose gene for IFN-γ or CD40 ligand has been destroyed by targeted gene disruption, the production of antimicrobial agents by macrophages is impaired, and the animals succumb to sublethal doses of *Mycobacterium* and *Leishmania* species. Macrophage activation is also crucial in controlling vaccinia virus. Mice lacking TNF receptors are more susceptible to these pathogens. However, although IFN-γ and CD40 ligand are probably the most important effector molecules synthesized by T_H1 cells, the immune response to pathogens that proliferate in macrophage vesicles is complex, and other cytokines secreted by T_H1 cells may also be crucial in coordinating these responses (Fig. 9.42). For example, macrophages that are chronically infected with intracellular bacteria may lose the ability to become activated, and such cells could provide a reservoir of infection shielded from immune attack. Activated T_H1 cells can also express Fas ligand and thus kill a limited range of target cells that express Fas, including macrophages, thereby destroying these infected cells. Certain intravesicular bacteria, including some mycobacteria and *Listeria monocytogenes*, escape from cell vesicles and enter the cytoplasm, where they are not susceptible to macrophage activation. Their presence can, however, be detected by CD8 cytotoxic T cells. The pathogens released when macrophages are killed either by T_H1 cells or by CD8 cytotoxic T cells can be taken up by freshly recruited macrophages still capable of activation to antimicrobial activity.

The depletion of CD4 T cells in people with HIV/AIDS can lead to microbes that are normally cleared by macrophages becoming a problem and causing disease. This is the case with the opportunist fungal pathogen *Pneumocystis jirovecii* (formerly known as *P. carinii*). The lungs of healthy people are kept clear of *P. jirovecii* by phagocytosis and intracellular killing by alveolar macrophages. Pneumonia caused by *P. jirovecii* is, however, a frequent cause of death in people with AIDS. In the absence of CD4 T cells, phagocytosis of

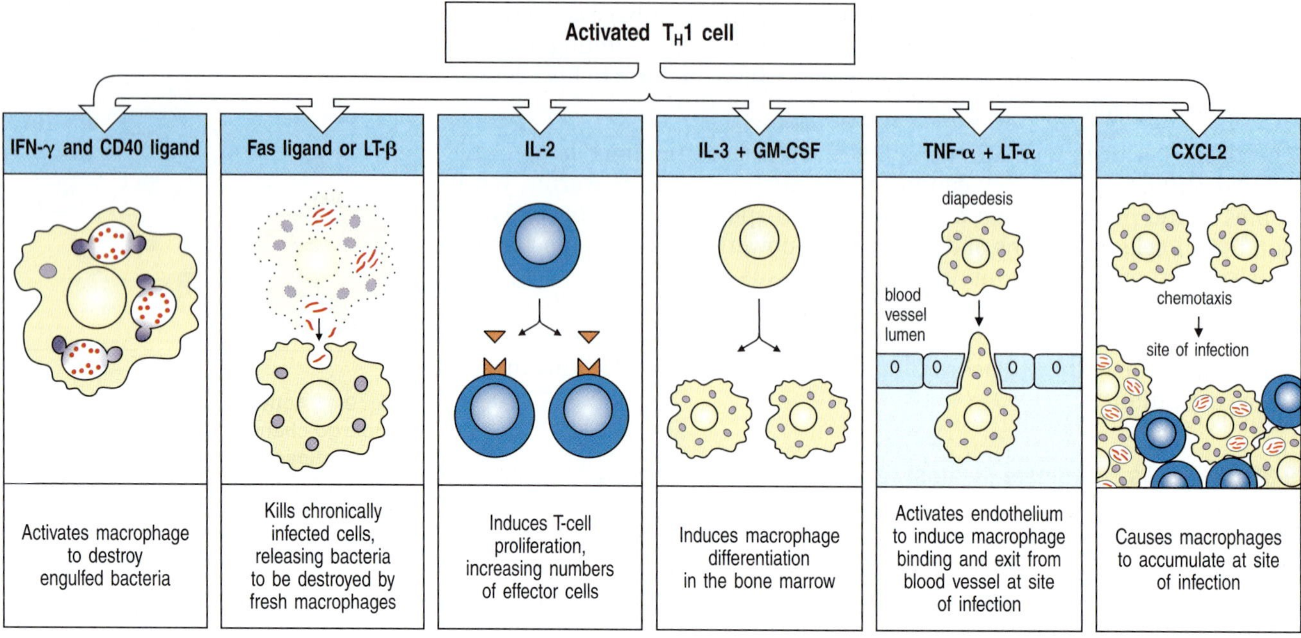

Fig. 9.42 The immune response to intracellular bacteria is coordinated by activated T_H1 cells. The activation of T_H1 cells by infected macrophages results in the synthesis of cytokines that both activate the macrophage and coordinate the immune response to intracellular pathogens. IFN-γ and CD40 ligand synergize in activating the macrophage, which allows it to kill engulfed pathogens. Chronically infected macrophages lose the ability to kill intracellular bacteria, and membrane-bound Fas ligand or LT-β produced by the T_H1 cell can kill these macrophages, releasing the engulfed bacteria, which are taken up and killed by fresh macrophages. In this way, IFN-γ and LT-β synergize in the removal of intracellular bacteria. IL-2 produced by T_H1 cells induces T-cell proliferation and potentiates the release of other cytokines. IL-3 and GM-CSF stimulate the production of new macrophages by acting on hematopoietic stem cells in the bone marrow. New macrophages are recruited to the site of infection by the actions of secreted TNF-α, LT-α, and other cytokines on vascular endothelium, which signal macrophages to leave the bloodstream and enter the tissues. A chemokine with macrophage chemotactic activity (CXCL2) signals macrophages to migrate into sites of infection and accumulate there. Thus, the T_H1 cell coordinates a macrophage response that is highly effective in destroying intracellular infectious agents.

P. jirovecii and intracellular killing by lung macrophages are impaired, and the pathogen colonizes the surface of the lung epithelium and invades lung tissue. The requirement for CD4 T cells seems to be due, at least in part, to a requirement for the macrophage-activating cytokines IFN-γ and TNF-α.

Another important function of T_H1 cells is the recruitment of phagocytic cells to sites of infection. T_H1 cells recruit macrophages by two mechanisms. First, they make the hematopoietic growth factors IL-3 and GM-CSF, which stimulate the production of new phagocytic cells in the bone marrow. Second, the TNF-α and LT-α secreted by T_H1 cells at sites of infection change the surface properties of endothelial cells so that phagocytes adhere to them. Chemokines such as CXCL2, which is produced by T_H1 cells in the inflammatory response, direct the migration of monocytes through the vascular endothelium and into the infected tissue (see Section 3-14).

When microbes effectively resist the microbicidal effects of activated macrophages, chronic infection with inflammation can develop. This often has a characteristic pattern, consisting of a central area of macrophages surrounded by activated lymphocytes. This pathological pattern is called a granuloma (Fig. 9.43). Giant cells consisting of fused macrophages can form in the center of these granulomas. A granuloma serves to 'wall off' pathogens that resist destruction. T_H2 cells seem to participate in granulomas along with T_H1 cells, perhaps by regulating their activity and preventing widespread tissue damage. In tuberculosis, the centers of large granulomas can become isolated and the cells there die, probably from a combination of lack of oxygen and the cytotoxic effects of activated macrophages. As the dead tissue in the center resembles cheese, this process is called caseation necrosis. Thus, the activation of T_H1 cells can cause significant pathology. Their non-activation, however, leads to the more serious consequence of death from disseminated infection, which is now seen frequently in patients with AIDS and concomitant mycobacterial infection.

Summary.

CD4 T cells that can activate macrophages have a critical role in host defense against those intracellular and extracellular pathogens that resist killing after

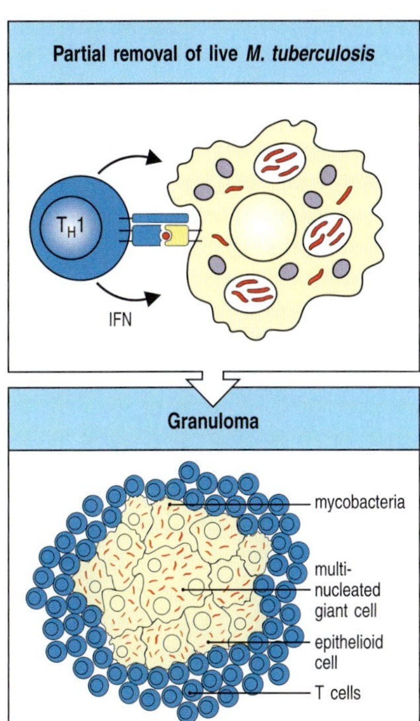

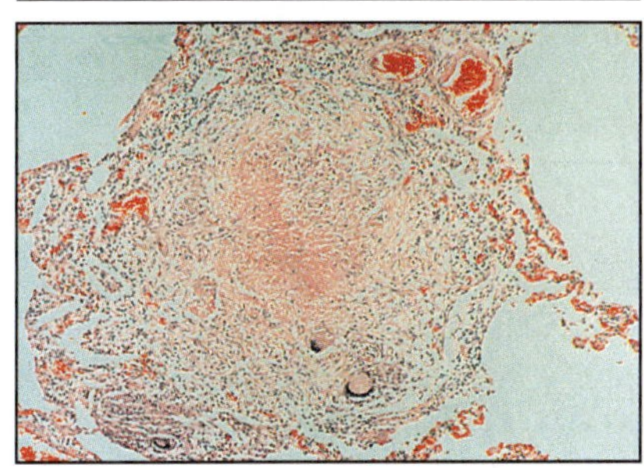

Fig. 9.43 Granulomas form when an intracellular pathogen or its constituents cannot be totally eliminated. When mycobacteria (red) resist the effects of macrophage activation, a characteristic localized inflammatory response called a granuloma develops. This consists of a central core of infected macrophages. The core may include multinucleate giant cells, which are fused macrophages, surrounded by large macrophages often called epithelioid cells, but in granulomas caused by mycobacteria the core usually becomes necrotic. Mycobacteria can persist in the cells of the granuloma. The central core is surrounded by T cells, many of which are CD4-positive. The exact mechanisms by which this balance is achieved, and how it breaks down, are unknown. Granulomas, as seen in the bottom panel, also form in the lungs and elsewhere in a disease known as sarcoidosis, which may be caused by inapparent mycobacterial infection. Photograph courtesy of J. Orrell.

being engulfed by macrophages. Macrophages are activated by membrane-bound signals delivered by activated T_H1 cells and by the potent macrophage-activating cytokine IFN-γ, which is secreted by activated T cells. Once activated, the macrophage can kill intracellular and ingested bacteria. Activated macrophages can also cause local tissue damage, and this explains why their activity is strictly regulated by antigen-specific T cells. T_H1 cells produce a range of cytokines, chemokines, and surface molecules that not only activate infected macrophages but also kill chronically infected senescent macrophages, stimulate the production of new macrophages in bone marrow, and recruit fresh macrophages to sites of infection. Thus, T_H1 cells control and coordinate host defense against certain intracellular pathogens. It is likely that the absence of this function explains the preponderance of infections with intracellular pathogens in adult AIDS patients.

Summary to Chapter 9.

An adaptive immune response is initiated when naive T cells encounter specific antigen on the surface of an antigen-presenting cell that also expresses the co-stimulatory molecules B7.1 and B7.2. In most cases, the antigen-presenting cells responsible for activating naive T cells, and inducing their clonal expansion, are conventional dendritic cells. Conventional dendritic cells—the subset expressing high levels of CD11c—not only reside in lymphoid tissues, but they also survey the periphery, where they encounter pathogens, take up antigen at sites of infection, become activated through innate recognition, and migrate to local lymphoid tissue. The dendritic cell may mature to become a potent direct activator of naive T cells, or it may transfer antigen to dendritic cells resident in peripheral lymphoid organs for cross-presentation to naive CD8 T cells. Plasmacytoid dendritic cells contribute to rapid responses against viruses by the production of type I interferons. Activated T cells produce IL-2, which drives them to proliferate; various other signals drive the differentiation of several types of effector T cells, which act by releasing mediators directly onto their target cells. This triggering of effector T cells by peptide:MHC complexes is independent of co-stimulation, so that any infected target cell can be activated or destroyed by an effector T cell. CD8 cytotoxic T cells kill target cells infected with cytosolic pathogens, removing sites of pathogen replication. CD4 T cells can become specialized effectors that promote phagocytic/inflammatory (T_H1), allergic and mucosal/barrier immunity (T_H2), or acute inflammatory (T_H17) responses to pathogens, or that provide help to B cells (T_{FH}). CD4 T_H1 cells activate macrophages to kill intracellular parasites. CD4 T cells are also essential in the activation of B cells to secrete the antibodies that mediate humoral immune responses directed against extracellular pathogens. T_H17 cells help enhance the neutrophil response to extracellular pathogens. Thus, effector T cells control virtually all known effector mechanisms of the adaptive immune response. In addition, subsets of CD4 regulatory T cells are produced that help control and limit immune responses by suppressing T-cell activity.

Questions.

9.1 Dendritic cells migrate through tissues, providing a surveillance mechanism for infection by pathogens. (a) What types of dendritic cells are there? (b) Describe how dendritic cells identify the presence of infection in peripheral tissues and initiate an immune response to it in the lymph nodes or secondary lymphoid tissues. (c) What mechanisms prevent dendritic cells from initiating immune responses to self antigens?

9.2 Activation of a naive T cell requires interaction with an antigen-presenting cell, such as a dendritic cell. (a) Which molecules on T cells are involved in this process, and what do they interact with on the antigen-presenting cell? (b) What consequences would you expect if these molecules were deficient in an individual? (c) What scope do these molecules offer for the design of anti-inflammatory or immunosuppressive drugs?

9.3 In some particle-physics experiments, coincidence detection—the simultaneous measurement of the same event by two separate detectors—is used to discriminate real events from spurious fluctuations in the detector systems. How do the requirements for T-cell activation follow the same principle in (a) the recognition of pathogens, or (b) the prevention of autoimmune reactions?

9.4 Consider the claim "T-cell effector functions are primarily mediated by secreted products." (a) To what extent is this statement true for CD4 cells and for CD8 T cells? (b) Describe the roles of T-cell membrane-bound effector molecules in activating macrophages.

9.5 CD4 T cells can develop into several types of effector cells, which have sometimes been considered separate lineages. (a) Describe the known CD4 subsets and correlate their immunological functions with their specific effector mechanisms. (b) What types of properties would determine whether or not these subsets are distinct lineages of cells? (c) Describe the role of antigen-presenting cells and pathogens in generating each subset. (d) Discuss how antigen-presenting cells and CD4 T-cell subsets are related to the maintenance of tolerance.

General references.

Dustin, M.L.: **Coordination of T-cell activation and migration through formation of the immunological synapse.** *Ann. NY Acad. Sci.* 2003, **987**:51–59.

Heath, W.R., and Carbone, F.R.: **Dendritic cell subsets in primary and secondary T cell responses at body surfaces.** *Nat. Immunol.* 2009, **10**:1237–1244.

Korn, T, Bettelli, E., Oukka, M., and Kuchroo, V.K.: **IL-17 and Th17 cells.** *Annu. Rev. Immunol.* 2009, **27**:485–517.

Mosmann, T.R., Li, L., Hengartner, H., Kagi, D., Fu, W., and Sad, S.: **Differentiation and functions of T cell subsets.** *Ciba Found. Symp.* 1997, **204**:148–154; discussion 154–158.

Snyder, J.E., and Mosmann, T.R.: **How to 'spot' a real killer.** *Trends Immunol.* 2003, **24**:231–232.

Springer, T.A.: **Traffic signals for lymphocyte recirculation and leukocyte emigration: the multistep paradigm.** *Cell* 1994, **76**:301–314.

Tseng, S.Y., and Dustin, M.L.: **T-cell activation: a multidimensional signaling network.** *Curr. Opin. Cell Biol.* 2002, **14**:575–580.

Section references.

9-1 Naive T cells migrate through peripheral lymphoid tissues, sampling the peptide:MHC complexes on dendritic cell surfaces.

Caux, C., Ait-Yahia, S., Chemin, K., de Bouteiller, O., Dieu-Nosjean, M.C., Homey, B., Massacrier, C., Vanbervliet, B., Zlotnik, A., and Vicari, A.: **Dendritic cell biology and regulation of dendritic cell trafficking by chemokines.** *Springer Semin. Immunopathol.* 2000, **22**:345–369.

Itano, A.A., and Jenkins, M.K.: **Antigen presentation to naive CD4 T cells in the lymph node.** *Nat. Immunol.* 2003, **4**:733–739.

Mackay, C.R., Kimpton, W.G., Brandon, M.R., and Cahill, R.N.: **Lymphocyte subsets show marked differences in their distribution between blood and the afferent and efferent lymph of peripheral lymph nodes.** *J. Exp. Med.* 1988, **167**:1755–1765.

Picker, L.J., and Butcher, E.C.: **Physiological and molecular mechanisms of lymphocyte homing.** *Annu. Rev. Immunol.* 1993, **10**:561–591.

Steptoe, R.J., Li, W., Fu, F., O'Connell, P.J., and Thomson, A.W.: **Trafficking of APC from liver allografts of Flt3L-treated donors: augmentation of potent allostimulatory cells in recipient lymphoid tissue is associated with a switch from tolerance to rejection.** *Transpl. Immunol.* 1999, **7**:51–57.

Yoshino, M., Yamazaki, H., Nakano, H., Kakiuchi, T., Ryoke, K., Kunisada, T., and Hayashi, S.: **Distinct antigen trafficking from skin in the steady and active states.** *Int. Immunol.* 2003, **15**:773–779.

9-2 Lymphocyte entry into lymphoid tissues depends on chemokines and adhesion molecules.

Hogg, N., Henderson, R., Leitinger, B., McDowall, A., Porter, J., and Stanley, P.: **Mechanisms contributing to the activity of integrins on leukocytes.** *Immunol. Rev.* 2002, **186**:164–171.

Kunkel, E.J., Campbell, D.J., and Butcher, E.C.: **Chemokines in lymphocyte trafficking and intestinal immunity.** *Microcirculation* 2003, **10**:313–323.

Madri, J.A., and Graesser, D.: **Cell migration in the immune system: the evolving interrelated roles of adhesion molecules and proteinases.** *Dev. Immunol.* 2000, **7**:103–116.

Rasmussen, L.K., Johnsen, L.B., Petersen, T.E., and Sørensen, E.S.: **Human GlyCAM-1 mRNA is expressed in the mammary gland as splicing variants and encodes various aberrant truncated proteins.** *Immunol. Lett.* 2002, **83**:73–75.

Rosen, S.D.: **Ligands for L-selectin: homing, inflammation, and beyond.** *Annu. Rev. Immunol.* 2004, **22**:129–156.

von Andrian, U.H., and Mempel, T.R.: **Homing and cellular traffic in lymph nodes.** *Nat. Rev. Immunol.* 2003, **3**:867–878.

9-3 Activation of integrins by chemokines is responsible for the entry of naive T cells into lymph nodes.

Cyster, J.G.: **Chemokines, sphingosine-1-phosphate, and cell migration in secondary lymphoid organs.** *Annu. Rev. Immunol.* 2005, **23**:127–159.

Laudanna, C., Kim, J.Y., Constantin, G., and Butcher, E.: **Rapid leukocyte integrin activation by chemokines.** *Immunol. Rev.* 2002, **186**:37–46.

Lo, C.G., Lu, T.T., and Cyster, J.G.: **Integrin-dependence of lymphocyte entry into the splenic white pulp.** *J. Exp. Med.* 2003, **197**:353–361.

Luo, B.H., Carman, C.V., and Springer, T.A.: **Structural basis of integrin regulation and signaling.** *Annu. Rev. Immunol.* 2007, **25**:619–647.

Rosen, H., and Goetzl, E.J.: **Sphingosine 1-phosphate and its receptors: an autocrine and paracrine network.** *Nat. Rev. Immunol.* 2005, **5**:560–570.

9-4 T-cell responses are initiated in peripheral lymphoid organs by activated dendritic cells.

Germain, R.N., Miller, M.J., Dustin, M.L., and Nussenzweig, M.C.: **Dynamic imaging of the immune system: progress, pitfalls and promise.** *Nat. Rev. Immunol.* 2006, **6**:497–507.

Miller, M.J., Wei, S.H., Cahalan, M.D., and Parker, I.: **Autonomous T cell trafficking examined *in vivo* with intravital two-photon microscopy.** *Proc. Natl Acad. Sci. USA* 2003, **100**:2604–2609.

Schlienger, K., Craighead, N., Lee, K.P., Levine, B.L., and June, C.H.: **Efficient priming of protein antigen-specific human CD4⁺ T cells by monocyte-derived dendritic cells.** *Blood* 2000, **96**:3490–3498.

Thery, C., and Amigorena, S.: **The cell biology of antigen presentation in dendritic cells.** *Curr. Opin. Immunol.* 2001, **13**:45–51.

9-5 Dendritic cells process antigens from a wide array of pathogens.

Belz, G.T., Carbone, F.R., and Heath, W.R.: **Cross-presentation of antigens by dendritic cells.** *Crit. Rev. Immunol.* 2002, **22**:439–448.

Guermonprez, P., Valladeau, J., Zitvogel, L., Thery, C., and Amigorena, S.: **Antigen presentation and T cell stimulation by dendritic cells.** *Annu. Rev. Immunol.* 2002, **20**:621–667.

Shortman, K., and Heath, W.R.: **The CD8⁺ dendritic cell subset.** *Immunol. Rev.* 2010, **234**:18–31.

Shortman, K., and Naik, S.H.: **Steady-state and inflammatory dendritic-cell development.** *Nat. Rev. Immunol.* 2007, **7**:19–30.

9-6 Pathogen-induced TLR signaling in immature dendritic cell induces their migration to lymphoid organs and enhances antigen processing.

Allan, R.S., Waithman, J., Bedoui, S., Jones, C.M., Villadangos, J.A., Zhan, Y., Lew, A.M., Shortman, K., Heath, W.R., and Carbone, F.R.: **Migratory dendritic cells transfer antigen to a lymph node-resident dendritic cell population for efficient CTL priming.** *Immunity* 2006, **25**:153–162.

Bachman, M.F., Kopf, M., and Marsland, B.J.: **Chemokines: more than just road signs.** *Nat. Rev. Immunol.* 2006, **6**:159–164.

Blander, J.M., and Medzhitov, R.: **Toll-dependent selection of microbial antigens for presentation by dendritic cells.** *Nature* 2006, **440**:808–812.

Reis e Sousa, C.: **Toll-like receptors and dendritic cells: for whom the bug tolls.** *Semin. Immunol.* 2004, **16**:27–34.

9-7 Plasmacytoid dendritic cells produce abundant type I interferons and may act as helper cells for antigen presentation by conventional dendritic cells.

Asselin-Paturel, C., and Trinchieri, G.: **Production of type I interferons: plasmacytoid dendritic cells and beyond.** *J. Exp. Med.* 2005, **202**:461–465.

Krug, A., Veeraswamy, R., Pekosz, A., Kanagawa, O., Unanue, E.R., Colonna, M., and Cella, M.: **Interferon-producing cells fail to induce proliferation of naive T cells but can promote expansion and T helper 1 differentiation of antigen-experienced unpolarized T cells.** *J. Exp. Med.* 2003, **197**:899–906.

Kuwajima, S., Sato, T., Ishida, K., Tada, H., Tezuka, H., and Ohteki, T.: **Interleukin 15-dependent crosstalk between conventional and plasmacytoid dendritic cells is essential for CpG-induced immune activation.** *Nat. Immunol.* 2006, **7**:740–746.

Swiecki, M., and Colonna, M.: **Unraveling the functions of plasmacytoid dendritic cells during viral infections, autoimmunity, and tolerance.** *Immunol. Rev.* 2010, **234**:142–162.

9-8 Macrophages are scavenger cells that can be induced by pathogens to present foreign antigens to naive T cells.

Barker, R.N., Erwig, L.P., Hill, K.S., Devine, A., Pearce, W.P., and Rees, A.J.: **Antigen presentation by macrophages is enhanced by the uptake of necrotic, but not apoptotic, cells.** *Clin. Exp. Immunol.* 2002, **127**:220–225.

Underhill, D.M., Bassetti, M., Rudensky, A., and Aderem, A.: **Dynamic interactions of macrophages with T cells during antigen presentation.** *J. Exp. Med.* 1999, **190**:1909–1914.

Zhu, F.G., Reich, C.F., and Pisetsky, D.S.: **The role of the macrophage scavenger receptor in immune stimulation by bacterial DNA and synthetic oligonucleotides.** *Immunology* 2001, **103**:226–234.

9-9 B cells are highly efficient at presenting antigens that bind to their surface immunoglobulin.

Guermonprez, P., England, P., Bedouelle, H., and Leclerc, C.: **The rate of dissociation between antibody and antigen determines the efficiency of antibody-mediated antigen presentation to T cells.** *J. Immunol.* 1998, **161**:4542–4548.

Shirota, H., Sano, K., Hirasawa, N., Terui, T., Ohuchi, K., Hattori, T., and Tamura, G.: **B cells capturing antigen conjugated with CpG oligodeoxynucleotides induce Th1 cells by elaborating IL-12.** *J. Immunol.* 2002, **169**:787–794.

Zaliauskiene, L., Kang, S., Sparks, K., Zinn, K.R., Schwiebert, L.M., Weaver, C.T., and Collawn, J.F.: **Enhancement of MHC class II-restricted responses by receptor-mediated uptake of peptide antigens.** *J. Immunol.* 2002, **169**:2337–2345.

9-10 Cell-adhesion molecules mediate the initial interaction of naive T cells with antigen-presenting cells.

Dustin, M.L.: **T-cell activation through immunological synapses and kinapses.** *Immunol. Rev.* 2008, **221**:77–89.

Friedl, P., and Brocker, E.B.: **TCR triggering on the move: diversity of T-cell interactions with antigen-presenting cells.** *Immunol. Rev.* 2002, **186**:83–89.

Gunzer, M., Schafer, A., Borgmann, S., Grabbe, S., Zanker, K.S., Brocker, E.B., Kampgen, E., and Friedl, P.: **Antigen presentation in extracellular matrix: interactions of T cells with dendritic cells are dynamic, short lived, and sequential.** *Immunity* 2000, **13**:323–332.

Montoya, M.C., Sancho, D., Vicente-Manzanares, M., and Sanchez-Madrid, F.: **Cell adhesion and polarity during immune interactions.** *Immunol. Rev.* 2002, **186**:68–82.

Wang, J., and Eck, M.J.: **Assembling atomic resolution views of the immunological synapse.** *Curr. Opin. Immunol.* 2003, **15**:286–293.

9-11 Antigen-presenting cells deliver three kinds of signals for the clonal expansion and differentiation of naive T cells.

Bour-Jordan, H., and Bluestone, J.A.: **CD28 function: a balance of costimulatory and regulatory signals.** *J. Clin. Immunol.* 2002, **22**:1–7.

Gonzalo, J.A., Delaney, T., Corcoran, J., Goodearl, A., Gutierrez-Ramos, J.C., and Coyle, A.J.: **Cutting edge: the related molecules CD28 and inducible costimulator deliver both unique and complementary signals required for optimal T-cell activation.** *J. Immunol.* 2001, **166**:1–5.

Kapsenberg, M.L.: **Dendritic-cell control of pathogen-driven T-cell polarization.** *Nat. Rev. Immunol.* 2003, **3**:984–993.

Wang, S., Zhu, G., Chapoval, A.I., Dong, H., Tamada, K., Ni, J., and Chen, L.: **Costimulation of T cells by B7-H2, a B7-like molecule that binds ICOS.** *Blood* 2000, **96**:2808–2813.

9-12 CD28-dependent co-stimulation of activated T cells induces expression of the T-cell growth factor interleukin-2 and the high-affinity IL-2 receptor.

Acuto, O., and Michel, F.: **CD28-mediated co-stimulation: a quantitative support for TCR signalling.** *Nat. Rev. Immunol.* 2003, **3**:939–951.

Gaffen, S.L.: **Signaling domains of the interleukin 2 receptor.** *Cytokine* 2001, **14**:63–77.

Seko, Y., Cole, S., Kasprzak, W., Shapiro, B.A., and Ragheb, J.A.: **The role of cytokine mRNA stability in the pathogenesis of autoimmune disease.** *Autoimmun. Rev.* 2006, **5**:299–305.

Zhou, X.Y., Yashiro-Ohtani, Y., Nakahira, M., Park, W.R., Abe, R., Hamaoka, T., Naramura, M., Gu, H., and Fujiwara, H.: **Molecular mechanisms underlying differential contribution of CD28 versus non-CD28 costimulatory molecules to IL-2 promoter activation.** *J. Immunol.* 2002, **168**:3847–3854.

9-13 Signal 2 can be modified by additional co-stimulatory pathways.

Greenwald, R.J., Freeman, G.J., and Sharpe, A.H.: **The B7 family revisited.** *Annu. Rev. Immunol.* 2005, **23**:515–548.

Watts, T.H.: **TNF/TNFR family members in costimulation of T cell responses.** *Annu. Rev. Immunol.* 2005, **23**:23–68.

9-14 Antigen recognition in the absence of co-stimulation leads to functional inactivation or clonal deletion of peripheral T cells.

Lin, A.E., and Mak, T.W.: **The role of E3 ligases in autoimmunity and the regulation of autoreactive T cells.** *Curr. Opin. Immunol.* 2007, **19**:665–673.

Nurieva, R.I., Zheng, S., Jin, W., Chung, Y., Zhang, Y., Martinez, G.J., Reynolds, J.M., Wang, S.L., Lin, X., Sun, S.C., *et al*: **The E3 ubiquitin ligase GRAIL regulates T cell tolerance and regulatory T cell function by mediating T cell receptor-CD3 degradation.** *Immunity* 2010, **32**:670–680.

Schwartz, R.H.: **T cell anergy.** *Annu. Rev. Immunol.* 2003, **21**:305–334.

Wekerle, T., Blaha, P., Langer, F., Schmid, M., and Muehlbacher, F.: **Tolerance through bone marrow transplantation with costimulation blockade.** *Transpl. Immunol.* 2002, **9**:125–133.

9-15 Proliferating T cells differentiate into effector T cells that do not require co-stimulation to act.

Gudmundsdottir, H., Wells, A.D., and Turka, L.A.: **Dynamics and requirements of T cell clonal expansion *in vivo* at the single-cell level: effector function is linked to proliferative capacity.** *J. Immunol.* 1999, **162**:5212–5223.

London, C.A., Lodge, M.P., and Abbas, A.K.: **Functional responses and costimulator dependence of memory CD4⁺ T cells.** *J. Immunol.* 2000, **164**:265–272.

Schweitzer, A.N., and Sharpe, A.H.: **Studies using antigen-presenting cells lacking expression of both B7-1 (CD80) and B7-2 (CD86) show distinct requirements for B7 molecules during priming versus restimulation of Th2 but not Th1 cytokine production.** *J. Immunol.* 1998, **161**:2762–2771.

9-16 CD8 T cells can be activated in different ways to become cytotoxic effector cells.

Andreasen, S.O., Christensen, J.E., Marker, O., and Thomsen, A.R.: **Role of CD40 ligand and CD28 in induction and maintenance of antiviral CD8⁺ effector T cell responses.** *J. Immunol.* 2000, **164**:3689–3697.

Blazevic, V., Trubey, C.M., and Shearer, G.M.: **Analysis of the costimulatory requirements for generating human virus-specific *in vitro* T helper and effector responses.** *J. Clin. Immunol.* 2001, **21**:293–302.

Croft, M.: **Co-stimulatory members of the TNFR family: keys to effective T-cell immunity?** *Nat. Rev. Immunol.* 2003, **3**:609–620.

Liang, L., and Sha, W.C.: **The right place at the right time: novel B7 family members regulate effector T cell responses.** *Curr. Opin. Immunol.* 2002, **14**:384–390.

Seder, R.A., and Ahmed, R.: **Similarities and differences in CD4⁺ and CD8⁺ effector and memory T cell generation.** *Nat. Immunol.* 2003, **4**:835–842.

Weninger, W., Manjunath, N., and von Andrian, U.H.: **Migration and differentiation of CD8⁺ T cells.** *Immunol. Rev.* 2002, **186**:221–233.

9-17 CD4 T cells differentiate into several subsets of functionally different effector cells.

Breitfeld, D., Ohl, L., Kremmer, E., Ellwart, J., Sallusto, F., Lipp, M., and Förster, R.: **Follicular B helper T cells express CXC chemokine receptor 5, localize to B cell follicles, and support immunoglobulin production.** *J. Exp. Med.* 2000, **192**:1545–1552.

Bluestone, J.A., and Abbas, A.K.: **Natural versus adaptive regulatory T cells.** *Nat. Rev. Immunol.* 2003, **3**:253–257.

King, C.: **New insights into the differentiation and function of T follicular helper cells.** *Nat. Rev. Immunol.* 2009, **9**:757–766.

Littman, D.R., and Rudensky, A.Y.: **Th17 and regulatory T cells in mediating and restraining inflammation.** *Cell* 2010, **140**:845–858.

Murphy, K.M., and Reiner, S.L.: **The lineage decisions of helper T cells.** *Nat. Rev. Immunol.* 2002, **2**:933–944.

Nurieva, R.I., and Chung, Y.: **Understanding the development and function of T follicular helper cells.** *Cell. Mol. Immunol.* 2010, **7**:190–197.

Schaerli, P., Willimann, K., Lang, A.B., Lipp, M., Loetscher, P., and Moser, B.: **CXC chemokine receptor 5 expression defines follicular homing T cells with B cell helper function.** *J. Exp. Med.* 2000, **192**:1553–1562.

9-18 Various forms of signal 3 induce the differentiation of naive CD4 T cells down distinct effector pathways.

Johnston, R.J., Poholek, A.C., DiToro, D., Yusuf, I., Eto, D., Barnett, B., Dent, A.L., Craft, J., and Crotty, S.: **Bcl6 and Blimp-1 are reciprocal and antagonistic regulators of T follicular helper cell differentiation.** *Science* 2009, **325**:1006–1010.

Nath, I., Vemuri, N., Reddi, A.L., Jain, S., Brooks, P., Colston, M.J., Misra, R.S., and Ramesh, V.: **The effect of antigen presenting cells on the cytokine profiles of stable and reactional lepromatous leprosy patients.** *Immunol. Lett.* 2000, **75**:69–76.

O'Shea, J.J., and Paul, W.E.: **Mechanisms underlying lineage commitment** and plasticity of helper CD4⁺ T cells. *Science* 2010, **327**:1098–1102.

Reese, T.A., Liang, H.E., Tager, A.M., Luster, A.D., Van Rooijen, N., Voehringer, D., and Locksley, R.M.: **Chitin induces the accumulation in tissue of innate immune cells associated with allergy.** *Nature* 2007, **447**:92–96.

Szabo, S.J., Sullivan, B.M., Peng, S.L., and Glimcher, L.H.: **Molecular mechanisms regulating Th1 immune responses.** *Annu. Rev. Immunol.* 2003, **21**:713–758.

Weaver, C.T., Harrington, L.E., Mangan, P.R., Gavrieli, M., and Murphy, K.M.: **Th17: an effector CD4 lineage with regulatory T cell ties.** *Immunity* 2006, **24**:677–688.

9-19 Regulatory CD4 T cells are involved in controlling adaptive immune responses.

Fontenot, J.D., and Rudensky, A.Y.: **A well adapted regulatory contrivance: regulatory T cell development and the forkhead family transcription factor Foxp3.** *Nat. Immunol.* 2005, **6**:331–337.

Roncarolo, M.G., Bacchetta, R., Bordignon, C., Narula, S., and Levings, M.K.: **Type 1 T regulatory cells.** *Immunol. Rev.* 2001, **182**:68–79.

Sakaguchi, S.: **Naturally arising Foxp3-expressing $CD25^+CD4^+$ regulatory T cells in immunological tolerance to self and non-self.** *Nat. Immunol.* 2005, **6**:345–352.

Sakaguchi, S., Ono, M., Setoguchi, R., Yagi, H., Hori, S., Fehervari, Z., Shimizu, J., Takahashi, T., and Nomura, T.: **$Foxp3^+$ $CD25^+$ $CD4^+$ natural regulatory T cells in dominant self-tolerance and autoimmune disease.** *Immunol. Rev.* 2006, **212**:8–27.

Saraiva, M., and O'Garra, A.: **The regulation of IL-10 production by immune cells.** *Nat. Rev. Immunol.* 2010, **10**:170–181.

9-20 Effector T-cell interactions with target cells are initiated by antigen-nonspecific cell-adhesion molecules.

Dustin, M.L.: **T-cell activation through immunological synapses and kinases.** *Immunol. Rev.* 2008, **221**:77–89.

van der Merwe, P.A., and Davis, S.J.: **Molecular interactions mediating T cell antigen recognition.** *Annu. Rev. Immunol.* 2003, **21**:659–684.

9-21 An immunological synapse forms between effector T cells and their targets to regulate signaling and to direct the release of effector molecules.

Bossi, G., Trambas, C., Booth, S., Clark, R., Stinchcombe, J., and Griffiths, G.M.: **The secretory synapse: the secrets of a serial killer.** *Immunol. Rev.* 2002, **189**:152–160.

Montoya, M.C., Sancho, D., Vicente-Manzanares, M., and Sanchez-Madrid, F.: **Cell adhesion and polarity during immune interactions.** *Immunol. Rev.* 2002, **186**:68–82.

Trambas, C.M., and Griffiths, G.M.: **Delivering the kiss of death.** *Nat. Immunol.* 2003, **4**:399–403.

9-22 The effector functions of T cells are determined by the array of effector molecules that they produce.

&

9-23 Cytokines can act locally or at a distance.

Basler, C.F., and Garcia-Sastre, A.: **Viruses and the type I interferon antiviral system: induction and evasion.** *Int. Rev. Immunol.* 2002, **21**:305–337.

Boulay, J.L., O'Shea, J.J., and Paul, W.E.: **Molecular phylogeny within type I cytokines and their cognate receptors.** *Immunity* 2003, **19**:159–163.

Guidotti, L.G., and Chisari, F.V.: **Cytokine-mediated control of viral infections.** *Virology* 2000, **273**:221–227.

Harty, J.T., Tvinnereim, A.R., and White, D.W.: **$CD8^+$ T cell effector mechanisms in resistance to infection.** *Annu. Rev. Immunol.* 2000, **18**:275–308.

Proudfoot, A.E.: **Chemokine receptors: multifaceted therapeutic targets.** *Nat. Rev. Immunol.* 2002, **2**:106–115.

9-24 T cells express several TNF family cytokines as trimeric proteins that are usually associated with the cell surface.

Bekker, L.G., Freeman, S., Murray, P.J., Ryffel, B., and Kaplan, G.: **TNF-alpha controls intracellular mycobacterial growth by both inducible nitric oxide synthase-dependent and inducible nitric oxide synthase-independent pathways.** *J. Immunol.* 2001, **166**:6728–6734.

Hehlgans, T., and Mannel, D.N.: **The TNF–TNF receptor system.** *Biol. Chem.* 2002, **383**:1581–1585.

Ware, C.F.: **Network communications: lymphotoxins, LIGHT, and TNF.** *Annu. Rev. Immunol.* 2005, **23**:787–819.

9-25 Cytotoxic T cells can induce target cells to undergo programmed cell death.

Ashton-Rickardt, P.G.: **The granule pathway of programmed cell death.** *Crit. Rev. Immunol.* 2005, **25**:161–182.

Green, D.R., Droin, N., and Pinkoski, M.: **Activation-induced cell death in T cells.** *Immunol. Rev.* 2003, **193**:70–81.

Russell, J.H., and Ley, T.J.: **Lymphocyte-mediated cytotoxicity.** *Annu. Rev. Immunol.* 2002, **20**:323–370.

Wallin, R.P., Screpanti, V., Michaelsson, J., Grandien, A., and Ljunggren, H.G.: **Regulation of perforin-independent NK cell-mediated cytotoxicity.** *Eur. J. Immunol.* 2003, **33**:2727–2735.

9-26 Cytotoxic effector proteins that trigger apoptosis are contained in the granules of CD8 cytotoxic T cells.

Barry, M., Heibein, J.A., Pinkoski, M.J., Lee, S.F., Moyer, R.W., Green, D.R., and Bleackley, R.C.: **Granzyme B short-circuits the need for caspase 8 activity during granule-mediated cytotoxic T-lymphocyte killing by directly cleaving Bid.** *Mol. Cell Biol.* 2000, **20**:3781–3794.

Grossman, W.J., Revell, P.A., Lu, Z.H., Johnson, H., Bredemeyer, A.J., and Ley, T.J.: **The orphan granzymes of humans and mice.** *Curr. Opin. Immunol.* 2003, **15**:544–552.

Yasukawa, M., Ohminami, H., Arai, J., Kasahara, Y., Ishida, Y., and Fujita, S.: **Granule exocytosis, and not the Fas/Fas ligand system, is the main pathway of cytotoxicity mediated by alloantigen-specific $CD4^+$ as well as $CD8^+$ cytotoxic T lymphocytes in humans.** *Blood* 2000, **95**:2352–2355.

9-27 Cytotoxic T cells are selective and serial killers of targets expressing a specific antigen.

Stinchcombe, J.C., and Griffiths, G.M.: **Secretory mechanisms in cell-mediated cytotoxicity.** *Annu. Rev. Cell Dev. Biol.* 2007, **23**:495–517.

Veugelers, K., Motyka, B., Frantz, C., Shostak, I., Sawchuk, T., and Bleackley, R.C.: **The granzyme B-serglycin complex from cytotoxic granules requires dynamin for endocytosis.** *Blood* 2004, **103**:3845–3853.

9-28 Cytotoxic T cells also act by releasing cytokines.

Amel-Kashipaz, M.R., Huggins, M.L., Lanyon, P., Robins, A., Todd, I., and Powell, R.J.: **Quantitative and qualitative analysis of the balance between type 1 and type 2 cytokine-producing $CD8^-$ and $CD8^+$ T cells in systemic lupus erythematosus.** *J. Autoimmun.* 2001, **17**:155–163.

Dobrzanski, M.J., Reome, J.B., Hollenbaugh, J.A., and Dutton, R.W.: **Tc1 and Tc2 effector cell therapy elicit long-term tumor immunity by contrasting mechanisms that result in complementary endogenous type 1 antitumor responses.** *J. Immunol.* 2004, **172**:1380–1390.

Prezzi, C., Casciaro, M.A., Francavilla, V., Schiaffella, E., Finocchi, L., Chircu, L.V., Bruno, G., Sette, A., Abrignani, S., and Barnaba, V.: **Virus-specific $CD8^+$ T cells with type 1 or type 2 cytokine profile are related to different disease activity in chronic hepatitis C virus infection.** *Eur. J. Immunol.* 2001, **31**:894–906.

9-29 T_H1 cells have a central role in macrophage activation.

Bekker, L.G., Freeman, S., Murray, P.J., Ryffel, B., and Kaplan, G.: **TNF-alpha controls intracellular mycobacterial growth by both inducible nitric oxide synthase-dependent and inducible nitric oxide synthase-independent pathways.** *J. Immunol.* 2001, **166**:6728–6734.

Ehlers, S., Kutsch, S., Ehlers, E.M., Benini, J., and Pfeffer, K.: **Lethal granuloma disintegration in mycobacteria-infected TNFRp55$^{-/-}$ mice is dependent on T cells and IL-12.** *J. Immunol.* 2000, **165**:483–492.

Muñoz-Fernández, M.A., Fernández, M.A., and Fresno, M.: **Synergism between tumor necrosis factor-α and interferon-γ on macrophage activation for the killing of intracellular *Trypanosoma cruzi* through a nitric oxide-dependent mechanism.** *Eur. J. Immunol.* 1992, **22**:301–307.

Stout, R.D., Suttles, J., Xu, J., Grewal, I.S., and Flavell, R.A.: **Impaired T cell-mediated macrophage activation in CD40 ligand-deficient mice.** *J. Immunol.* 1996, **156**:8–11.

9-30 Activation of macrophages by T_H1 cells promotes microbial killing and must be tightly regulated to avoid tissue damage.

Duffield, J.S.: **The inflammatory macrophage: a story of Jekyll and Hyde.** *Clin. Sci.* 2003, **104**:27–38.

James, D.G.: **A clinicopathological classification of granulomatous disorders.** *Postgrad. Med. J.* 2000, **76**:457–465.

Labow, R.S., Meek, E., and Santerre, J.P.: **Model systems to assess the destructive potential of human neutrophils and monocyte-derived macrophages during the acute and chronic phases of inflammation.** *J. Biomed. Mater. Res.* 2001, **54**:189–197.

Wigginton, J.E., and Kirschner, D.: **A model to predict cell-mediated immune regulatory mechanisms during human infection with *Mycobacterium tuberculosis*.** *J. Immunol.* 2001, **166**:1951–1967.

9-31 T_H1 cells coordinate the host response to intracellular pathogens.

Berberich, C., Ramirez-Pineda, J.R., Hambrecht, C., Alber, G., Skeiky, Y.A., and Moll, H.: **Dendritic cell (DC)-based protection against an intracellular pathogen is dependent upon DC-derived IL-12 and can be induced by molecularly defined antigens.** *J. Immunol.* 2003, **170**:3171–3179.

Biedermann, T., Zimmermann, S., Himmelrich, H., Gumy, A., Egeter, O., Sakrauski, A.K., Seegmuller, I., Voigt, H., Launois, P., Levine, A.D., *et al.*: **IL-4 instructs T_H1 responses and resistance to *Leishmania major* in susceptible BALB/c mice.** *Nat. Immunol.* 2001, **2**:1054–1060.

Kaplan, M.H., Whitfield, J.R., Boros, D.L., and Grusby, M.J.: **Th2 cells are required for the *Schistosoma mansoni* egg-induced granulomatous response.** *J. Immunol.* 1998, **160**:1850–1856.

Koguchi, Y., and Kawakami, K.: **Cryptococcal infection and Th1-Th2 cytokine balance.** *Int. Rev. Immunol.* 2002, **21**:423–438.

Neighbors, M., Xu, X., Barrat, F.J., Ruuls, S.R., Churakova, T., Debets, R., Bazan, J.F., Kastelein, R.A., Abrams, J.S., and O'Garra, A.: **A critical role for interleukin 18 in primary and memory effector responses to *Listeria monocytogenes* that extends beyond its effects on interferon gamma production.** *J. Exp. Med.* 2001, **194**:343–354.

The Humoral Immune Response

10

Many of the bacteria that cause infectious disease in humans multiply in the extracellular spaces of the body, and most intracellular pathogens spread by moving from cell to cell through the extracellular fluids. The extracellular spaces are protected by the **humoral immune response**, in which antibodies produced by B cells cause the destruction of extracellular microorganisms and prevent the spread of intracellular infections. Activation of naive B cells is triggered by antigen and usually requires helper T cells, such as the T follicular helper (T_{FH}) cells introduced in Section 9-17; the activated B cells then differentiate into antibody-secreting **plasma cells** (Fig. 10.1) and memory B cells. In this chapter we use the general term **helper T cell** to mean any effector CD4 T cell that can activate a B cell.

Antibodies contribute to immunity in three main ways (see Fig. 10.1). The first is known as **neutralization**. To enter cells, viruses and intracellular bacteria bind to specific molecules on the target cell surface. Antibodies that bind to the pathogen can prevent this and are said to neutralize the pathogen. Neutralization by antibodies is also important in preventing bacterial toxins from entering cells. Second, antibodies protect against bacteria that multiply outside cells, and do this mainly by facilitating uptake of the pathogen by phagocytes. Coating the surface of a pathogen to enhance phagocytosis is called **opsonization**. Antibodies bound to the pathogen are recognized by phagocytic cells by means of receptors called Fc receptors that bind to the antibody constant region (C region). Third, antibodies coating a pathogen can activate the proteins of the complement system by the classical pathway, as described in Chapter 2. Complement proteins bound to the pathogen surface opsonize the pathogen by binding complement receptors on phagocytes. Other complement components recruit phagocytic cells to the site of infection, and the terminal components of complement can lyse certain microorganisms directly by forming pores in their membranes. Which effector mechanisms are engaged in a particular response is determined by the heavy-chain isotype of the antibodies produced, which determines their class (see Section 5-12).

In the first part of this chapter we describe the interactions of naive B cells with antigen and with helper T cells that lead to the activation of B cells and antibody production. Some important microbial antigens can provoke antibody production without the help of T cells, and we shall also consider these responses here. Most antibody responses undergo a process called affinity maturation, in which antibodies of greater affinity for their target antigen are produced by the somatic hypermutation of antibody variable-region (V-region) genes. The molecular mechanism of somatic hypermutation was described in Chapter 5, and here we look at its immunological consequences. We also revisit class switching (see Section 5-19), which produces antibodies of different functional classes and confers functional diversity on the antibody response. Both affinity maturation and class switching occur only in B cells and require T-cell help. In the rest of the chapter we look in detail at the various effector mechanisms by which antibodies contain and eliminate infections. Like the T-cell response, the humoral immune response produces immunological memory, and this will be discussed in Chapter 11.

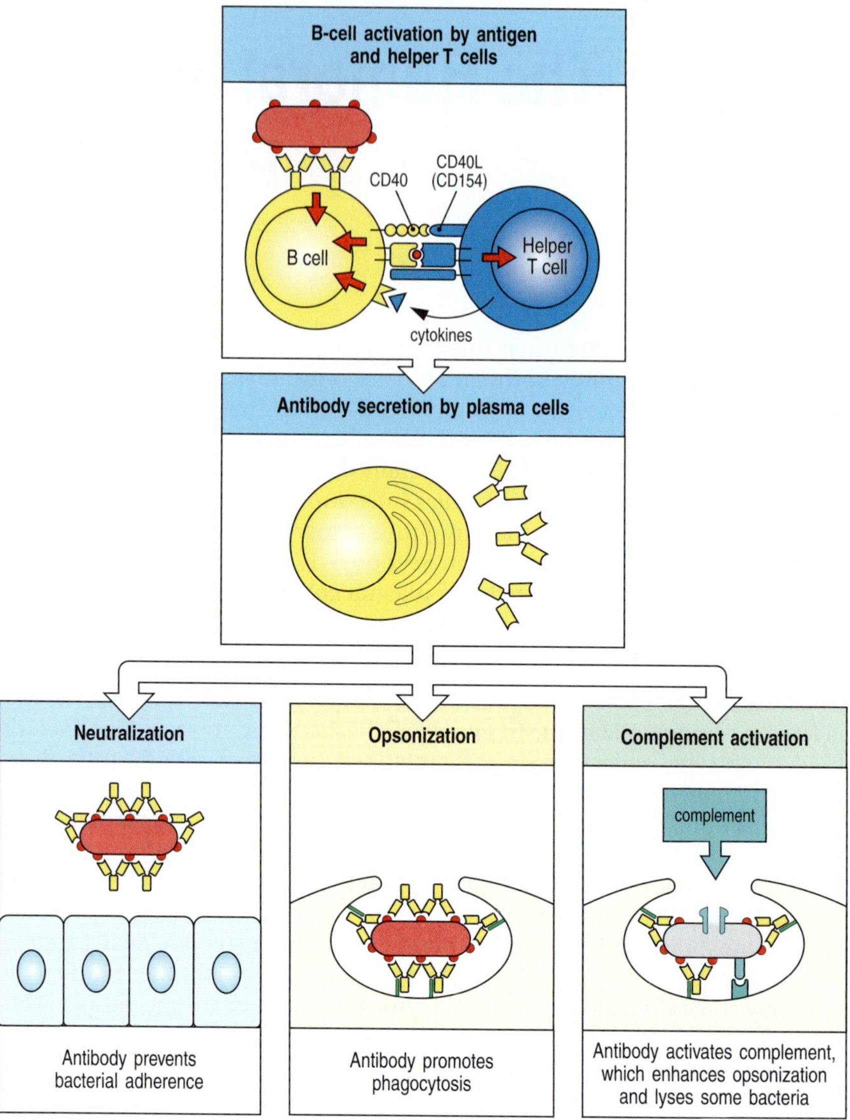

Fig. 10.1 The humoral immune response is mediated by antibodies secreted by plasma cells. Antigen that binds to the B-cell antigen receptor both sends a signal into the B cell and is internalized and processed into peptides that activate effector helper T cells. Signals from the bound antigen and from the helper T cell (in the form of the co-stimulatory CD40 ligand (CD40L) and cytokines) induce the B cell to proliferate and differentiate into plasma cells secreting specific antibody (top two panels). Antibodies protect the host from infection in three main ways. First, they can inhibit the toxic effects or infectivity of pathogens by binding to them, which is termed neutralization (bottom left panel). Second, by coating the pathogens they can enable accessory cells that recognize the Fc portions of arrays of antibodies to ingest and kill the pathogen, a process called opsonization (bottom center panel). Third, antibodies can trigger the activation of the complement system. Complement proteins strongly enhance opsonization, and can directly kill certain bacterial cells (bottom right panel).

B-cell activation by helper T cells.

The surface immunoglobulin that serves as the **B-cell receptor** (**BCR**) for antigen can bind a vast variety of chemical structures. In the context of natural infections it binds native proteins, glycoproteins, and polysaccharides, as well as whole virus particles and bacterial cells, by recognizing epitopes on their surfaces. It has two roles in B-cell activation. First, like the antigen receptor on T cells, it signals to the cell's interior when antigen is bound (see Chapter 7). Second, the B-cell antigen receptor delivers the bound antigen to intracellular sites, where it can be degraded to give peptides that are returned to the B-cell surface bound to MHC class II molecules (see Chapter 6). These peptide:MHC class II complexes are recognized by antigen-specific helper T cells that have already differentiated in response to the same pathogen, as described in Chapter 9. The effector T cells make cytokines that cause the B cell to proliferate and its progeny to differentiate into antibody-secreting cells and into memory B cells. Some microbial antigens can activate B cells directly in the absence of T-cell help, and the ability of B cells to respond directly to these antigens provides a rapid response to many important pathogens.

However, the fine tuning of antibody responses to increase the affinity of the antibody for the antigen and the switching to most immunoglobulin classes other than IgM depend on the interaction of antigen-stimulated B cells with helper T cells and other cells in the peripheral lymphoid organs. Thus, antibodies induced by microbial antigens alone tend to have lower affinity and to be less functionally versatile than those induced with T-cell help.

10-1 The humoral immune response is initiated when B cells that bind antigen are signaled by helper T cells or by certain microbial antigens alone.

It is a general rule in adaptive immunity that naive antigen-specific lymphocytes are difficult to activate by antigen alone. As we saw in Chapter 9, priming of naive T cells requires a co-stimulatory signal from professional antigen-presenting cells; naive B cells also require accessory signals that can come either from a helper T cell or, in some cases, directly from microbial constituents.

Antibody responses to protein antigens require antigen-specific T-cell help. These antigens are unable to induce antibody responses in animals or humans who lack T cells, and they are therefore known as **thymus-dependent** or **TD antigens**. To receive T-cell help, the B cell must be displaying antigen on its surface in a form a T cell can recognize. This occurs when antigen bound by surface immunoglobulin on the B cell is internalized and returned to the cell surface as peptides bound to MHC class II molecules. Helper T cells that recognize the peptide:MHC complex then deliver activating signals to the B cell (Fig. 10.2, top two panels). Thus, protein antigens binding to B cells both provide a specific signal to the B cell by cross-linking its antigen receptors and allow the B cell to attract antigen-specific T-cell help. When an activated helper T cell recognizes and binds to a peptide:MHC class II complex on the B-cell surface it induces the B cell to proliferate and differentiate into antibody-producing plasma cells (Fig. 10.3). The requirement for T-cell help means that before a B cell can be induced to make antibody against the molecules of an infecting pathogen, CD4 T cells specific for peptides from this pathogen must be activated to produce helper T cells. This occurs when naive T cells interact with dendritic cells presenting the appropriate peptides, as described in Chapter 9.

Although peptide-specific helper T cells are required for B-cell responses to protein antigens, some microbial constituents, such as bacterial polysaccharides, can induce antibody production in the absence of helper T cells. These microbial antigens are known as **thymus-independent** or **TI antigens** because they can induce antibody responses in individuals who have no T lymphocytes. The second signal required to activate antibody production against TI antigens is provided either directly by recognition of a common microbial constituent (see Fig. 10.2, bottom panel) or by extensive cross-linking of B-cell receptors, which would occur when a B cell binds repeating epitopes on the bacterial cell. Thymus-independent antibody responses provide some protection against extracellular bacteria, and we will return to them later.

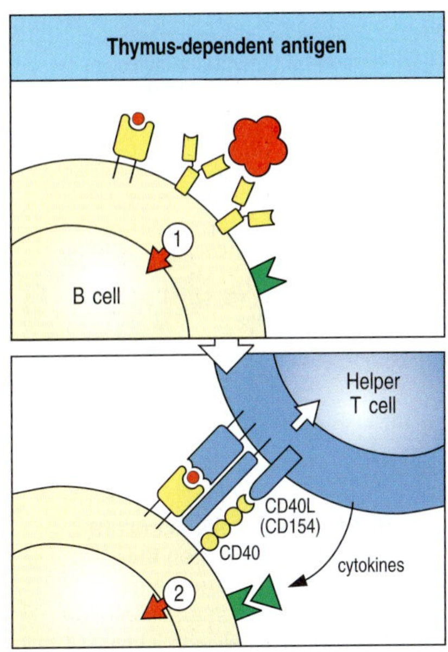

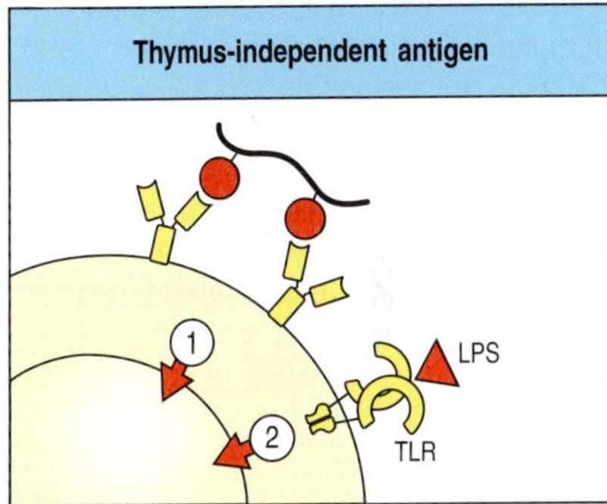

Fig. 10.2 A second signal is required for B-cell activation by either thymus-dependent or thymus-independent antigens. The first signal (indicated as 1 in the figure) required for B-cell activation is delivered through its antigen receptor (top panel). For thymus-dependent antigens, the second signal (indicated as 2) is delivered by a helper T cell that recognizes degraded fragments of the antigen as peptides bound to MHC class II molecules on the B-cell surface (center panel); the interaction between CD40 ligand (CD40L, also called CD154) on the T cell and CD40 on the B cell contributes an essential part of this second signal. For thymus-independent antigens, a second signal can be delivered along with the antigen itself, through Toll-like receptors that recognize antigen-associated TLR ligands, such as bacterial lipopolysaccharide (LPS) or bacterial DNA (bottom panel).

10-2 B-cell responses are enhanced by co-ligation of the B-cell receptor and B-cell co-receptor by antigen and complement fragments on microbial surfaces.

We have already described the **B-cell co-receptor complex** as being composed of three cell-surface proteins: CD19, CD21, and CD81 (see Fig. 7.23). CD21 (also known as complement receptor 2, CR2) is a receptor for the complement fragments C3d and C3dg (see Section 2-13). These complement

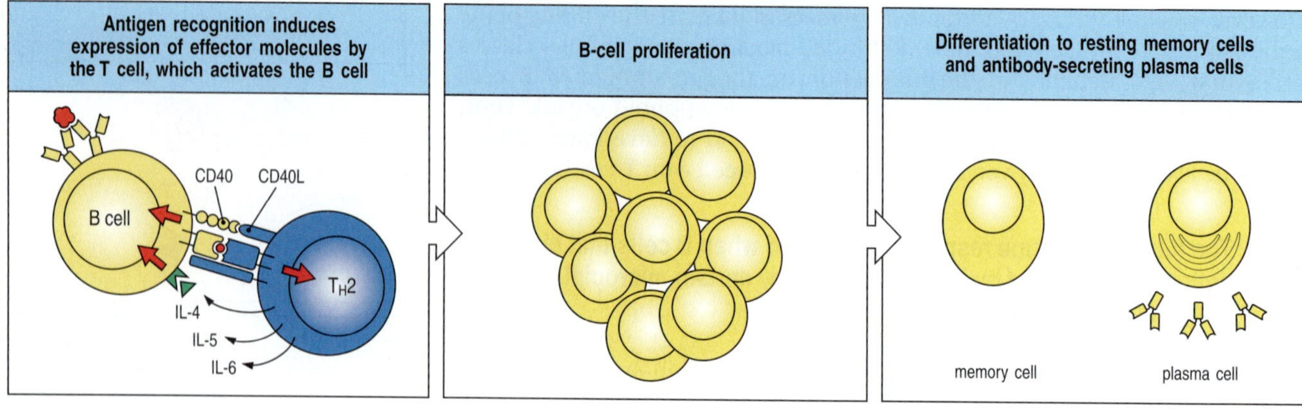
Antigen recognition induces expression of effector molecules by the T cell, which activates the B cell
B cell
CD40
CD40L
TH2
IL-4
IL-5
IL-6
B-cell proliferation
Differentiation to resting memory cells and antibody-secreting plasma cells
memory cell
plasma cell

Fig. 10.3 Helper T cells stimulate the proliferation and then the differentiation of antigen-binding B cells. The specific interaction of an antigen-binding B cell with a helper T cell leads to the expression of the B-cell stimulatory molecule CD40 ligand on the helper T-cell surface and to the secretion by the T cell of the B-cell stimulatory cytokines IL-4, IL-5, and IL-6, which drive the proliferation and differentiation of the B cell into antibody-secreting plasma cells. Alternatively, an activated B cell can become a memory cell.

fragments are deposited on the surface of pathogens, such as a bacterial cell, when the complement pathway is activated either by the innate pathways or by antibody bound to the antigen itself. When the B-cell receptor binds to antigen epitopes on surfaces that have also bound C3d or C3dg, CD21 binds to the complement fragments and brings the B-cell receptor together with the co-receptor. This generates signals through CD19 that activate a PI 3-kinase signaling pathway and co-stimulate the B-cell response (see Fig. 7.23) by enhancing proliferation, differentiation, and antibody production.

Simultaneous ligation (co-ligation) of both the antigen receptor and the B-cell co-receptor powerfully amplifies B-cell activation and antibody production. This effect is shown dramatically when mice are immunized with hen egg-white lysozyme coupled to three linked molecules of C3dg. In this case the dose of modified lysozyme needed to induce antibody in the absence of added adjuvant is as little as 1/10,000 of that needed with the unmodified lysozyme.

10-3 Helper T cells activate B cells that recognize the same antigen.

A given B cell can only be activated by helper T cells that respond to the same antigen; this is called **linked recognition**. However, the specific peptide recognized by the helper T cell may be quite distinct from the protein epitope recognized by the B cell's antigen receptor. Indeed, most complex natural antigens, such as viruses and bacteria, are composed of multiple proteins and carry both protein and carbohydrate epitopes. For linked recognition to occur, the peptide recognized by the T cell must be physically associated with the antigen recognized by the B cell, so that the B cell can internalize the antigen through its B-cell receptors and display the appropriate peptide to the T cell. For example, by recognizing an epitope on a viral protein coat, a B cell can bind and internalize a complete virus particle. It then can degrade either coat proteins or internal viral proteins into peptides for display on MHC class II molecules on the B-cell surface. Helper T cells that have been primed earlier in the infection by dendritic cells presenting these peptides can then activate the B cell to make antibodies that recognize the coat protein (Fig. 10.4).

The specific activation of the B cell by its **cognate** T cell—that is, a helper T cell primed by the same antigen—depends on the ability of the antigen-specific B cell to concentrate the appropriate peptide on its surface MHC class II molecules. B cells that bind a particular antigen are up to 10,000 times more efficient at displaying peptide fragments of that antigen on their MHC class II molecules than are B cells that do not bind the antigen. A B cell most efficiently receives help from a T cell that recognizes a peptide that is part of the antigen bound by the B cell.

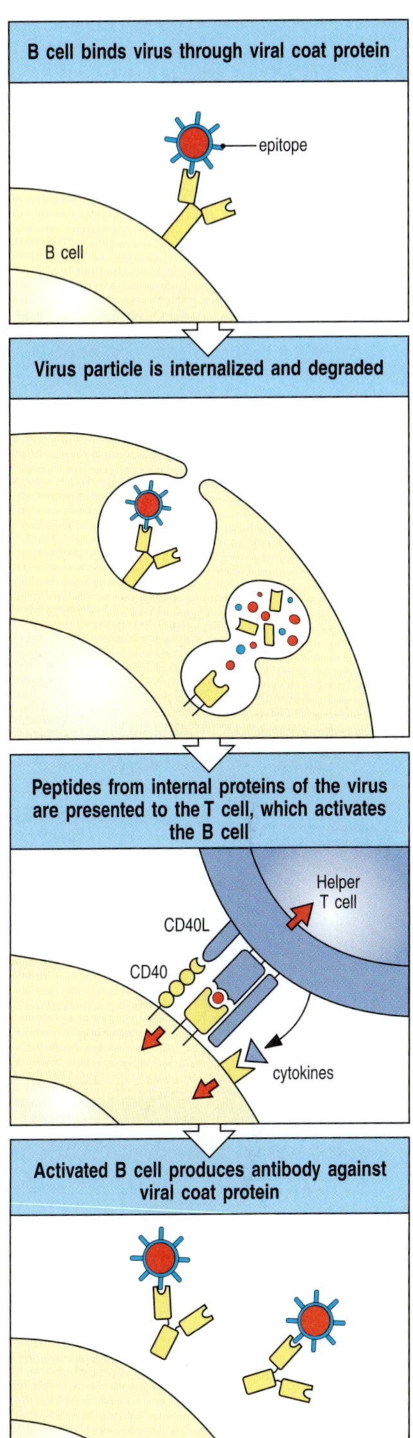
B cell binds virus through viral coat protein
epitope
B cell
Virus particle is internalized and degraded
Peptides from internal proteins of the virus are presented to the T cell, which activates the B cell
Helper T cell
CD40L
CD40
cytokines
Activated B cell produces antibody against viral coat protein

Fig. 10.4 B cells and helper T cells must recognize epitopes of the same molecular complex in order to interact. An epitope on a viral coat protein is recognized by the surface immunoglobulin on a B cell, and the virus is internalized and degraded. Peptides derived from viral proteins, including internal proteins, are returned to the B-cell surface bound to MHC class II molecules (see Chapter 6). Here, these complexes are recognized by helper T cells, which help to activate the B cells to produce antibody against the coat protein. This process is known as linked recognition.

Linked recognition was originally discovered through studies of the production of antibodies against haptens (see Appendix I, Section A-1). Haptens are small chemical groups that cannot elicit antibody responses on their own because they cannot cross-link B-cell receptors and do not recruit T-cell help. When coupled to a carrier protein, however, they become immunogenic, because the protein will carry multiple hapten groups that can now cross-link B-cell receptors. In addition, T-cell dependent responses are possible because T cells can be primed to respond to peptides derived from the protein. Accidental coupling of a hapten to a protein is responsible for the allergic responses shown by many people to the antibiotic penicillin, which reacts with host proteins to form a coupled hapten that can stimulate an antibody response, as we will learn in Chapter 14.

The requirement for linked recognition helps to ensure self tolerance, because it means that an autoimmune response will occur only if both a self-reactive T cell and a self-reactive B cell are present at the same time. This is discussed further in Chapter 15. Vaccine design can take advantage of linked recognition, as in the vaccine used to immunize infants against *Haemophilus influenzae* type b. This bacterial pathogen can infect the covering of the brain, called the meninges, causing meningitis. In adults, protective immunity to *H. influenzae* is due to a strong thymus-independent antibody response to the capsular polysaccharide. Infants, however, make weak responses to these polysaccharide antigens. To make a vaccine that is effective in infants, the polysaccharide is linked chemically to tetanus toxoid, a foreign protein against which infants are routinely and successfully vaccinated (see Chapter 16). B cells that bind the polysaccharide component of the vaccine are activated by helper T cells specific for peptides of the linked toxoid (Fig. 10.5).

10-4 T cells make membrane-bound and secreted molecules that activate B cells.

Recognition of peptide:MHC class II complexes on B cells triggers helper T cells to synthesize both cell-bound and secreted effector molecules that synergize in activating the B cell. As described in Section 9-13, CD40 ligand is expressed by helper T cells and binds to CD40 expressed by B cells; this interaction sustains T-cell growth and differentiation. Reciprocally, the engagement of CD40 on B cells increases B-cell proliferation, immunoglobulin class switching, and somatic hypermutation. Binding of CD40 by CD40 ligand helps to drive the resting B cell into the cell cycle and is essential for B-cell responses to thymus-dependent antigens. It also causes the B cell to increase its expression of co-stimulatory molecules, especially those of the B7 family.

T cells provide additional signals to B cells in the form of secreted cytokines that regulate B-cell proliferation and antibody production. In the B-cell follicle, it is the follicular helper T cell (T_{FH}) (see Section 9-17) that secretes these cytokines. The IL-4 produced by T cells is an important cytokine in B-cell activation. It is made by some T_{FH} cells and by T_H2 cells when they recognize their specific ligand on the B-cell surface, and IL-4 and CD40 ligand are thought to synergize in driving the clonal expansion of B cells that precedes antibody

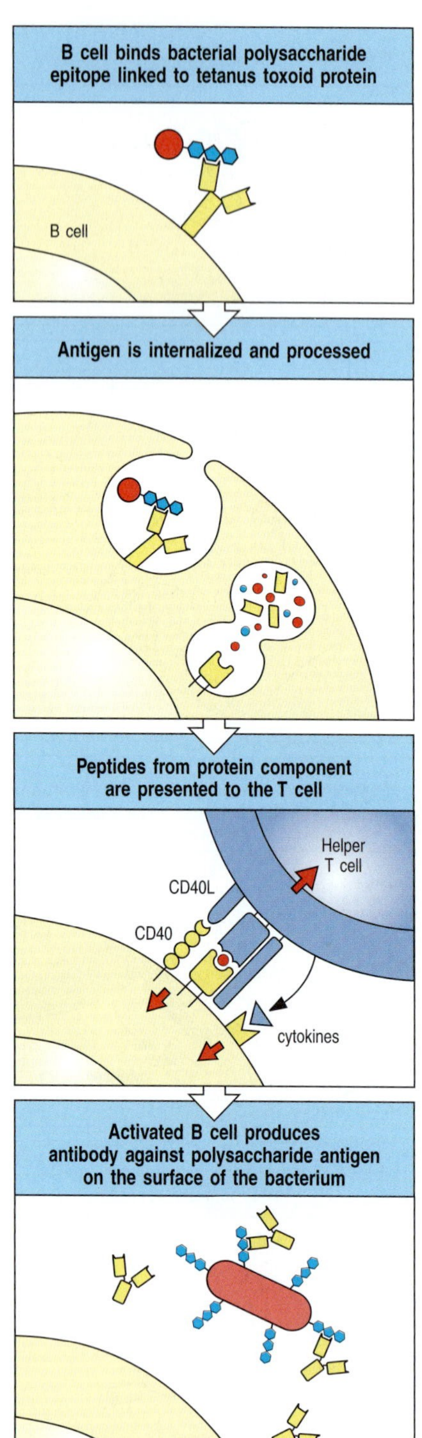
B cell binds bacterial polysaccharide epitope linked to tetanus toxoid protein
B cell
Antigen is internalized and processed
Peptides from protein component are presented to the T cell
Helper T cell
CD40L
CD40
cytokines
Activated B cell produces antibody against polysaccharide antigen on the surface of the bacterium

Fig. 10.5 Linked recognition can be exploited in the design of vaccines that boost B-cell responses against polysaccharide antigens. The Hib vaccine against *Haemophilus influenzae* type b is a conjugate of bacterial polysaccharide and the tetanus toxoid protein. The B cell recognizes and binds the polysaccharide, internalizes and degrades the whole conjugate and then displays toxoid-derived peptides on surface MHC class II molecules. Helper T cells generated in response to earlier vaccination against the toxoid recognize the complex on the B-cell surface and activate the B cell to produce anti-polysaccharide antibody. This antibody can then protect against infection with *H. influenzae* type b.

production *in vivo*. IL-4 is secreted in a polar fashion by T_H2 cells and is focused at the site of contact of the T cell with its target B cell (Fig. 10.6), so that it acts selectively on the antigen-specific target cell. Once a T-cell response is under way, however, the cytokines abundantly secreted by helper T cells can also help to activate nearby B cells not in contact with the T cell. After several rounds of proliferation, B cells differentiate into antibody-secreting plasma cells. Two additional cytokines, IL-5 and IL-6, both secreted by helper T cells, contribute to these later stages of B-cell activation.

Members of the TNF receptor/TNF family other than the CD40–CD40 ligand pair are involved in B-cell activation. Activated T cells express **CD30 ligand** (**CD30L**), which binds to **CD30** present on B cells and has been shown to promote B-cell activation. Mice lacking CD30 show reduced proliferation of activated B cells in lymphoid follicles and weaker secondary humoral responses than normal, both of which could be due to the inability of their B cells to respond to signals from T cells expressing CD30 ligand. The soluble TNF-family cytokine BAFF is secreted by dendritic cells and macrophages and acts as a survival factor for differentiating B cells (see Section 8-27).

10-5 B cells that encounter their antigens migrate toward the boundaries between B-cell and T-cell areas in secondary lymphoid tissues.

The frequency of naive lymphocytes specific for any given antigen is estimated to be between 1 in 10,000 and 1 in 1,000,000. Thus, the chance of an encounter between a T lymphocyte and a B lymphocyte that recognize the same antigen should be between 1 in 10^8 and 1 in 10^{12}. Also, as T cells and B cells mostly occupy two quite distinct zones in peripheral lymphoid tissues—the **T-cell areas** and the **primary lymphoid follicles** (also called B-cell areas or B-cell zones), respectively (see Figs 1.18–1.20), it is remarkable that B cells are able to find and interact with T cells of similar antigen specificity. Thus, linked recognition requires precise regulation of the migration of activated B and T cells into specific locations within the lymphoid tissues.

When circulating naive B cells migrate into lymphoid tissues, they enter the primary lymphoid follicles, attracted by the chemokine CXCL13 (Fig. 10.7, first panel). Within the follicle, stromal cells and a specialized cell type, the **follicular dendritic cell** (**FDC**), secrete CXCL13, while naive B cells express CXCR5, the receptor for this chemokine. The FDC is a nonphagocytic cell of nonhematopoietic origin that bears numerous long processes (see Section 8-23). Antigens derived from microorganisms are transported to lymph nodes via the lymph, and to the spleen via the blood.

Complement activation and the deposition of C3b onto microbial and viral antigens contribute to their efficient transport into and accumulation in the follicles. Antigens coated with C3b or C3dg can enter the follicle directly and be trapped there by the complement receptors CR1 and CR2, which are present on the dense network of FDC processes. In mice, intravital two-photon microscopy has indicated that opsonized particulate antigen entering lymph nodes in the afferent lymph (or entering the spleen via the blood), can also

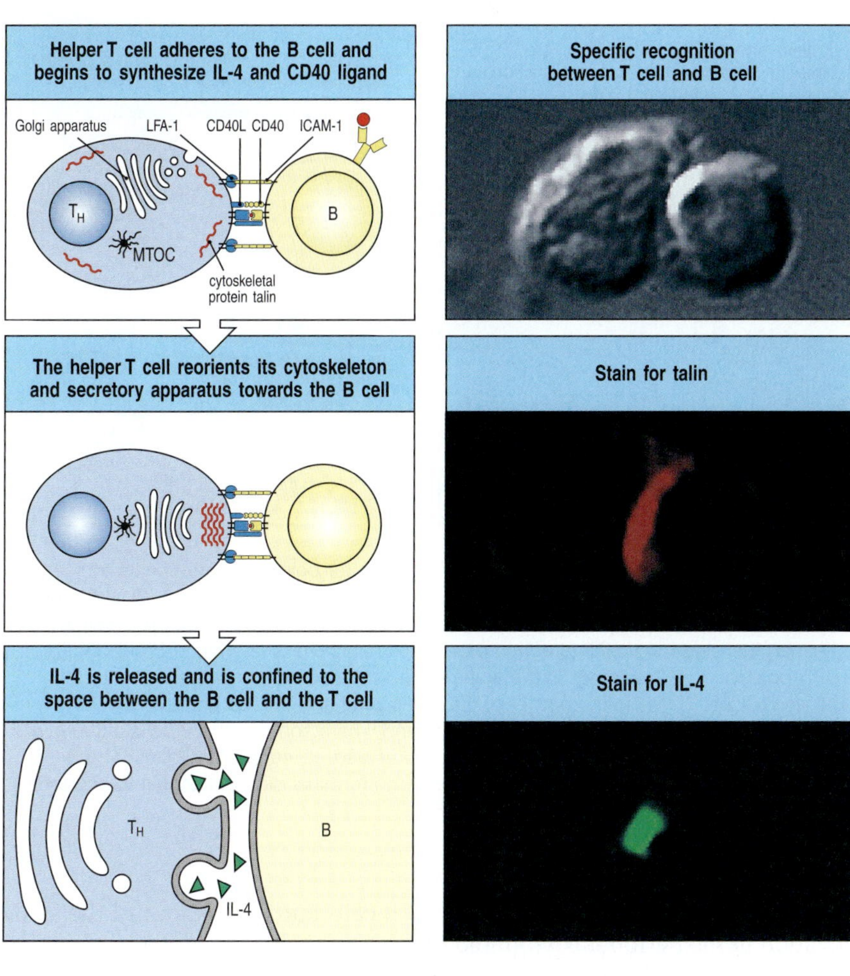

Fig. 10.6 Helper T cells stimulate B cells through binding of CD40 and directed cytokine secretion. When the T-cell receptor binds antigen presented by a B cell, CD40 ligand is induced on the T cell, which then binds to CD40 on the B cell (upper panels). The integrin LFA-1 on the T cell interacts with the adhesion molecule ICAM-1 on the B cell in the periphery of this immunological synapse (see Fig. 9.31). The cytoskeletal protein talin (stained red in the right center panel) becomes relocated to the point of cell–cell contact, and the secretory apparatus (the Golgi apparatus) is reoriented by the cytoskeleton toward the point of contact with the B cell. Cytokines are released at the point of contact (bottom panels). In this example, IL-4 (stained green) is confined to the space between the B cell and the helper T cell. MTOC, microtubule-organizing center. Photographs courtesy of A. Kupfer.

be taken up by specialized macrophages residing in the subcapsular sinus of lymph nodes and the marginal sinus of the spleen, parts of which are adjacent to the follicles (Fig. 10.8). These macrophages seem to retain the antigen on their surface rather than ingesting it and breaking it down. The antigen can then be sampled and picked up by antigen-specific follicular B cells. B cells of any antigen specificity could also acquire antigen from these macrophages via their complement receptors and transport it within the follicle.

When a naive B cell in the follicle encounters its specific antigen, either on specialized macrophages or displayed by FDCs, this induces expression of the chemokine receptor CCR7, and expression of CXCR5 is also retained. The B cell then moves toward the boundary with the T-cell area, where chemokine ligands for CCR7, such as CCL21, are highly expressed by stromal cells and dendritic cells (see Section 9-3). As with T cells, activation of B cells leads to a decrease in the sphingosine receptor $S1P_1$ on the B-cell surface, which retains the cells in the lymphoid tissues. Naive T cells express CCR7, but not CXCR5, and so are localized to the T-cell areas. When a naive T cell encounters its cognate peptide antigen presented by a dendritic cell, expression of CXCR5 is induced as the T cell begins to proliferate. Some T cells differentiate into effector cells and exit the lymphoid tissue, but others become T_{FH} cells and migrate to the boundary between the T-cell area and a follicle, where they can encounter activated B cells (see Fig. 10.7, second panel). When a T_{FH} cell recognizes a peptide displayed by MHC class II molecules on the surface of an activated B cell, it increases the expression of cell-surface molecules and secreted cytokines that promote B-cell activation—this is the basis for the phenomenon of linked recognition described earlier (see Section 10-3).

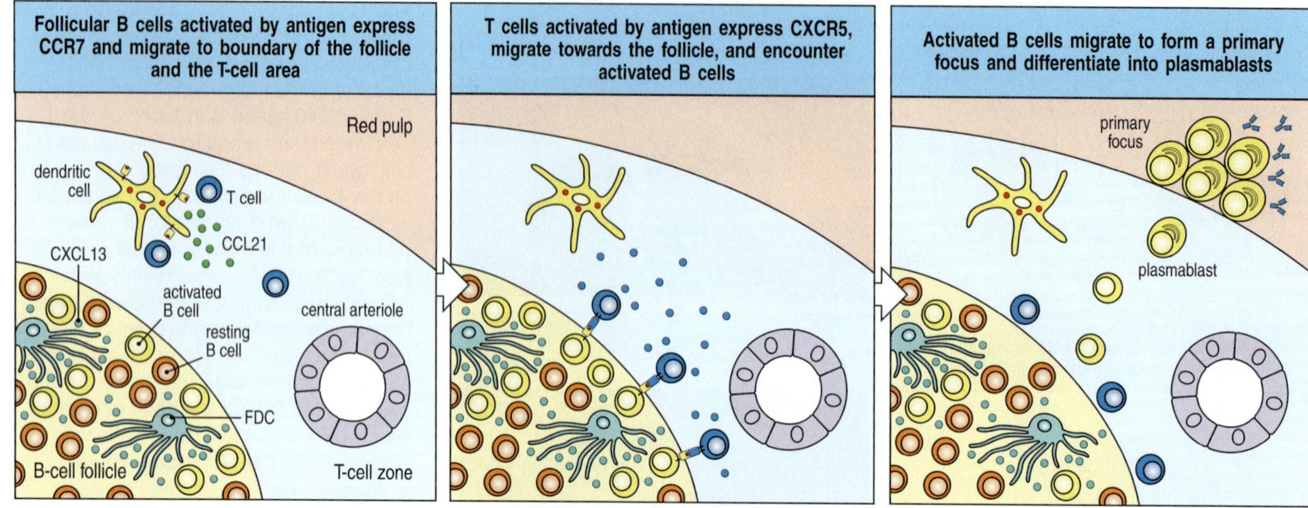
Follicular B cells activated by antigen express CCR7 and migrate to boundary of the follicle and the T-cell area
Red pulp
dendritic cell
T cell
CCL21
CXCL13
activated B cell
resting B cell
central arteriole
FDC
B-cell follicle
T-cell zone
T cells activated by antigen express CXCR5, migrate towards the follicle, and encounter activated B cells
Activated B cells migrate to form a primary focus and differentiate into plasmablasts
primary focus
plasmablast

Fig. 10.7 Antigen-binding B cells meet T cells at the border between the T-cell area and a B-cell follicle in secondary lymphoid tissue. B-cell activation in the spleen is shown here. On entry into the spleen from the blood through the marginal sinus (not shown), naive CCR7-positive T cells and CXCR5-positive B cells home to distinct regions where the chemokines CCL21 and CXCL13, respectively, are being produced (first panel). If a B cell encounters its antigen, either on a follicular dendritic cell (FDC) or a macrophage, it migrates toward the border between the follicle and the T-cell area. There it may encounter a T cell that has migrated to this border after being activated by its antigen on the surface of an antigen-presenting dendritic cell in the T-cell area (second panel). Through linked recognition, the T-cell–B-cell interaction produces an initial proliferation of B cells (third panel). In the spleen, the activated lymphocytes then migrate to the border of the T-cell zone and the red pulp, where they continue to proliferate and where some of the B cells differentiate into plasmablasts, forming a so-called primary focus. These plasmablasts then undergo terminal differentiation into antibody-secreting plasma cells.

The coordinated migration of activated B cells and T cells toward the same location in the peripheral lymphoid organ helps solve the problem of getting B cells together with their appropriate helper T cells. Activated B cells bearing peptide:MHC complexes end up in precisely the location to maximize

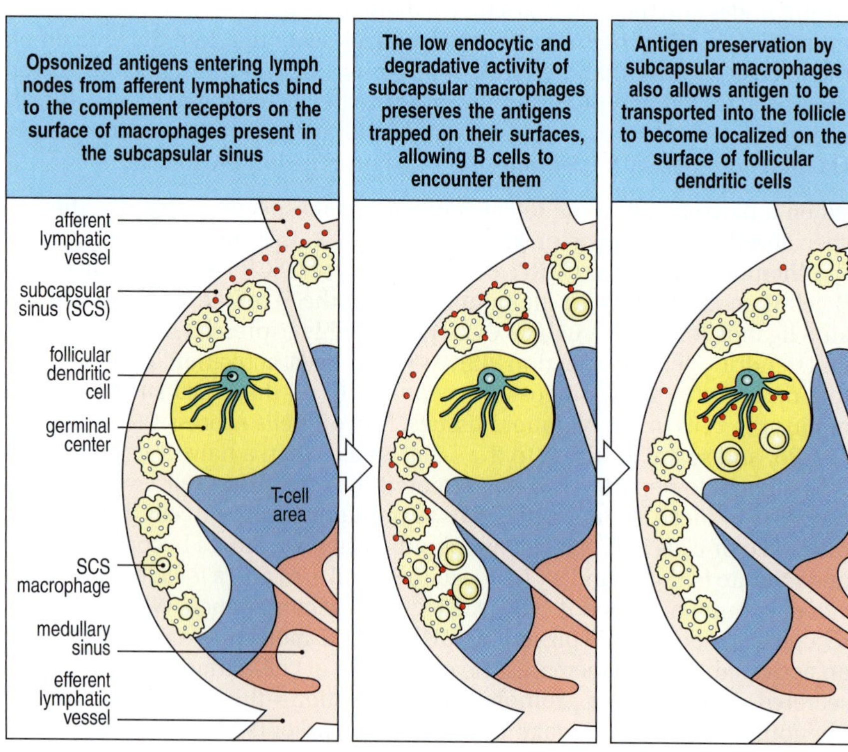
Opsonized antigens entering lymph nodes from afferent lymphatics bind to the complement receptors on the surface of macrophages present in the subcapsular sinus
The low endocytic and degradative activity of subcapsular macrophages preserves the antigens trapped on their surfaces, allowing B cells to encounter them
Antigen preservation by subcapsular macrophages also allows antigen to be transported into the follicle to become localized on the surface of follicular dendritic cells
afferent lymphatic vessel
subcapsular sinus (SCS)
follicular dendritic cell
germinal center
T-cell area
SCS macrophage
medullary sinus
efferent lymphatic vessel

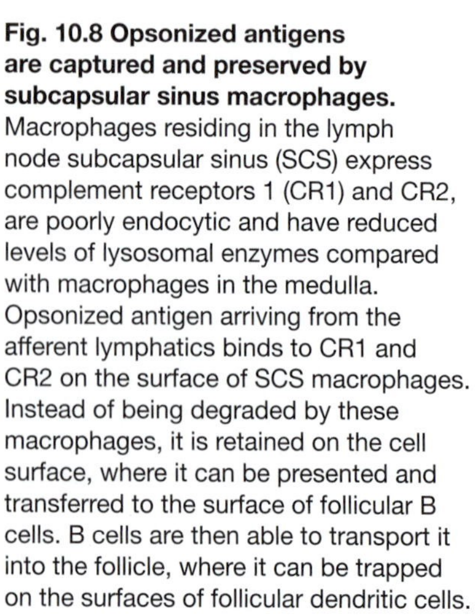

Fig. 10.8 Opsonized antigens are captured and preserved by subcapsular sinus macrophages. Macrophages residing in the lymph node subcapsular sinus (SCS) express complement receptors 1 (CR1) and CR2, are poorly endocytic and have reduced levels of lysosomal enzymes compared with macrophages in the medulla. Opsonized antigen arriving from the afferent lymphatics binds to CR1 and CR2 on the surface of SCS macrophages. Instead of being degraded by these macrophages, it is retained on the cell surface, where it can be presented and transferred to the surface of follicular B cells. B cells are then able to transport it into the follicle, where it can be trapped on the surfaces of follicular dendritic cells.

their chance of encountering helper T cells that can activate them. Antigen-stimulated B cells that fail to interact with T cells that recognize the same antigen die within 24 hours. After their initial encounter, B cells that have received T-cell help migrate from the follicle border to continue their proliferation and differentiation. In the spleen, these cells move to the border between the T-cell area and the red pulp. Here they establish a **primary focus** of clonal expansion (see Fig. 10.7, third panel). In lymph nodes, the primary focus is located in the medullary cords, where lymph drains out of the node. Primary foci appear about 5 days after an infection or immunization with an antigen not previously encountered, which correlates with the time needed for helper T cells to differentiate.

10-6 Antibody-secreting plasma cells differentiate from activated B cells.

Both T cells and B cells proliferate in the primary focus for several days, and this constitutes the first phase of the primary humoral immune response. Some of these proliferating B cells differentiate into antibody-synthesizing **plasmablasts** in the primary focus. Others may migrate into the lymphoid follicle and differentiate further there before becoming plasma cells, as we will describe later. Plasmablasts are cells that have begun to secrete antibody, yet are still dividing and express many of the characteristics of activated B cells that allow their interaction with T cells. After a few more days, the plasmablasts in the primary focus stop dividing and either die or undergo terminal differentiation into plasma cells. Some of the plasma cells remain in the lymphoid organs, where they are short lived, while others migrate to the bone marrow and continue antibody production there.

The properties of resting B cells, plasmablasts, and plasma cells are compared in Fig. 10.9. The differentiation of a B cell into a plasma cell is accompanied by many morphological changes that reflect its commitment to the production of large amounts of secreted antibody, which can make up to 20% of all the protein synthesized by a plasma cell. Plasmablasts and plasma cells have a prominent perinuclear Golgi apparatus and an extensive rough endoplasmic reticulum that is rich in immunoglobulin molecules being synthesized and exported into the lumen of the endoplasmic reticulum for secretion (see Fig. 1.23). Plasmablasts still express B7 co-stimulatory molecules and MHC class II molecules, but plasma cells do not, and so can no longer present antigen to helper T cells. Nevertheless, T cells still provide important signals for plasma-cell differentiation and survival, such as IL-6 and CD40 ligand. Plasmablasts also bear relatively large numbers of B-cell receptors on the cell surface,

	Intrinsic properties			Inducible by antigen stimulation		
B-lineage cell	Surface Ig	Surface MHC class II	High-rate Ig secretion	Growth	Somatic hyper-mutation	Class switch
Resting B cell	High	Yes	No	Yes	Yes	Yes
Plasmablast	High	Yes	Yes	Yes	Unknown	Yes
Plasma cell	Low	No	Yes	No	No	No

Fig. 10.9 Plasma cells secrete antibody at a high rate but can no longer respond to antigen or helper T cells. Resting naive B cells have membrane-bound immunoglobulin (usually IgM and IgD) and MHC class II molecules on their surface. Although their V genes do not carry somatic mutations, B cells can take up antigen and present it to helper T cells. The T cells in return induce the B cells to proliferate and to undergo isotype switching and somatic hypermutation, but B cells do not secrete significant amounts of antibody during this period. Plasmablasts have an intermediate phenotype. They secrete antibody but retain substantial surface immunoglobulin and MHC class II molecules and so can continue to take up and present antigen to T cells. Plasmablasts early in the immune response and those activated by T-independent antigens have usually not undergone somatic hypermutation and class switching, and therefore secrete IgM. Plasma cells are terminally differentiated plasmablasts that secrete antibodies. They can no longer interact with helper T cells because they have very low levels of surface immunoglobulin and lack MHC class II molecules. Early in the immune response they differentiate from unswitched plasmablasts and secrete IgM; later in the response they derive from plasmablasts that have undergone class switching and somatic hypermutation. Plasma cells have lost the ability to change the class of their antibody or undergo further somatic hypermutation.

whereas plasma cells have many fewer. These low levels of surface immunoglobulin may, however, be physiologically important, because the survival of plasma cells seems to be determined in part by their ability to continue to bind antigen. Plasma cells have a range of life spans. Some survive for only days to a few weeks after their final differentiation, whereas others are very long lived and account for the persistence of antibody responses.

Movie 10.1

10-7 The second phase of a primary B-cell immune response occurs when activated B cells migrate into follicles and proliferate to form germinal centers.

Some of the B cells that are activated and start to proliferate early in the immune response take a more circuitous route before they become plasma cells. Together with their associated T cells, they move into a primary lymphoid follicle (Fig. 10.10), where they continue to proliferate and ultimately form a **germinal center**; follicles with germinal centers are called **secondary lymphoid follicles**.

Germinal centers are composed mainly of proliferating B cells, but antigen-specific T cells make up about 10% of germinal center lymphocytes and provide indispensable help to the B cells. The germinal center is essentially an island of cell division that sets up amid a sea of resting B cells in the primary follicle. Proliferating germinal center B cells displace the resting B cells toward the periphery of the follicle, forming a **mantle zone** of resting cells around the two distinguishable areas of activated B cells, called the **light zone** and the **dark zone** (Fig. 10.11, top panel). The germinal center grows in size as the immune response proceeds, and then shrinks and finally disappears when the infection is cleared. Germinal centers are present for about 3–4 weeks after initial antigen exposure.

The events in the primary focus lead to the prompt secretion of specific antibody, mostly of the IgM isotype, that immediately protects the infected individual. In contrast, the germinal center reaction provides for a more effective humoral response later, should the pathogen establish a chronic infection or the host become reinfected. B cells undergo several important modifications in the germinal center that produce a more effective antibody response. These include somatic hypermutation, which alters the V regions of immunoglobulin genes (see Section 5-18) and enables a process called affinity maturation, which selects for the survival of mutated B cells with high affinity for the antigen. In addition, class switching allows the selected B cells to express a variety of effector functions in the form of antibodies of different classes (see Section 5-19). These B cells will differentiate either into memory B cells,

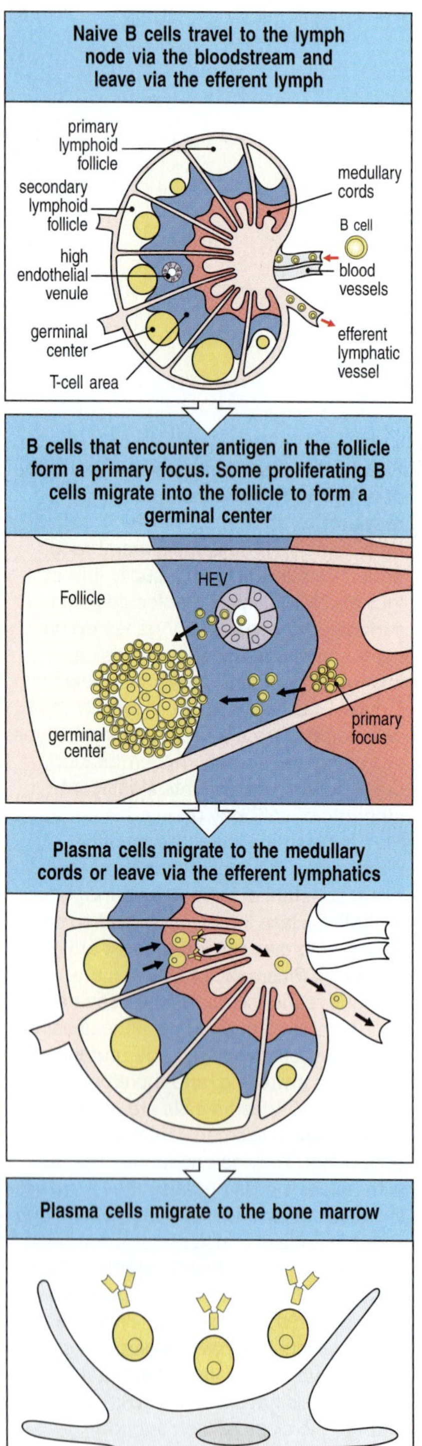

Fig. 10.10 Activated B cells form germinal centers in lymphoid follicles. Activation of B cells in a lymph node is shown here. Top panel: naive circulating B cells enter lymph nodes from the blood via high endothelial venules and are attracted by chemokines into the primary lymphoid follicle; if they do not encounter antigen in the follicle, they leave via the efferent lymphatic vessel. Second panel: B cells that have bound antigen move to the border with the T-cell area where they may encounter activated helper T cells specific for the same antigen, which interact with the B cells and activate them to start proliferation and differentiation into plasmablasts. Some B cells activated at the T-cell–B-cell border migrate to form a primary focus of antibody-secreting plasmablasts in the medullary cords, whereas others move back into the follicle, where they continue to proliferate and form a germinal center. Germinal centers are sites of sustained B-cell proliferation and differentiation. Follicles in which germinal centers have formed are known as secondary follicles. Within the germinal center, B cells begin their differentiation into either antibody-secreting plasma cells or memory B cells. Third and fourth panels: plasma cells leave the germinal center and migrate to the medullary cords, or leave the lymph node altogether via the efferent lymphatics and migrate to the bone marrow.

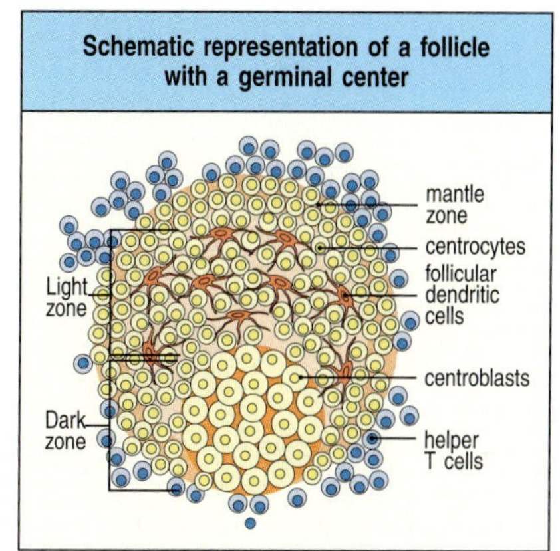

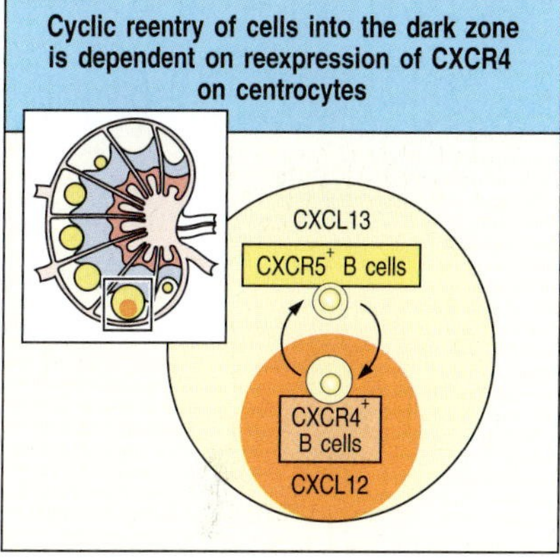

Fig. 10.11 The structure of a germinal center. The germinal center is a specialized microenvironment in which B-cell proliferation, somatic hypermutation, and selection for strength of antigen binding all occur. Closely packed centroblasts, which express CXCR4 and CXCR5, form the so-called 'dark zone' of the germinal center; the less densely packed 'light zone' contains centrocytes, which express only CXCR5. Cells in the dark zone produce CXCL12, which attracts the CXCR4-expressing centroblasts. Cyclic reentry describes the process by which B cells can lose and gain expression of CXCR4 and thus move from the light zone to the dark zone and back again, a process described in detail in Section 10-8.

whose function is described in Chapter 11, or into plasma cells that secrete higher-affinity and class-switched antibody during the latter part of the primary immune response.

B cells in the germinal center proliferate, dividing every 6–8 hours. Initially, these rapidly proliferating B cells, called **centroblasts**, express the chemokine receptors CXCR4 and CXCR5 but markedly reduce their expression of surface immunoglobulin, particularly of IgD. Centroblasts proliferate in the dark zone of the germinal center, named for its densely packed appearance (Fig. 10.12). Stromal cells in the dark zone produce CXCL12 (SDF-1), a ligand for CXCR4 that acts to retain centroblasts in this region. As time goes on, some centroblasts reduce their rate of cell division and enter the growth phase, while in the G_1 phase of the cell cycle they stop expressing CXCR4 and begin to produce higher levels of surface immunoglobulin. These B cells are termed

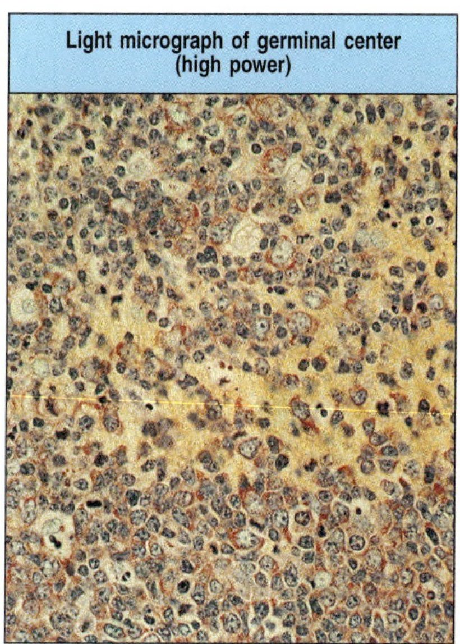

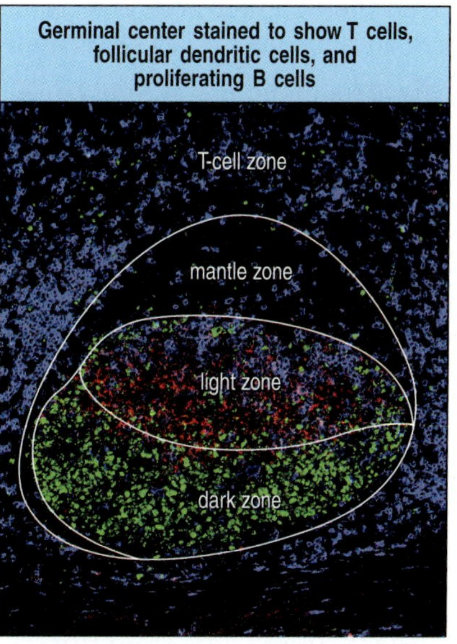

Fig. 10.12 Germinal centers are sites of intense cell proliferation and cell death. The photomicrograph (left panel) shows a high-power view of a section through a human tonsillar germinal center. Closely packed centroblasts seen in the lower part of this photomicrograph form the so-called dark zone of the germinal center. Above this region is the less densely packed light zone. The right panel shows immunofluorescent staining of a germinal center. B cells are found in the dark zone, light zone, and mantle zone. Proliferating cells are stained green for Ki67, an antigen expressed in nuclei of dividing cells, revealing the rapidly proliferating centroblasts in the dark zone. The dense network of follicular dendritic cells, stained red, mainly occupies the light zone. Centrocytes in the light zone proliferate to a lesser degree than centroblasts. Small recirculating B cells occupy the mantle zone at the edge of the B-cell follicle. Large masses of CD4 T cells, stained blue, can be seen in the T-cell zones, which separate the follicles. There are also significant numbers of T cells in the light zone of the germinal center; CD4 staining in the dark zone is associated mainly with CD4-positive phagocytes. Photographs courtesy of I. MacLennan.

centrocytes. The loss of CXCR4 allows centrocytes to move into the light zone, a less densely packed area containing abundant FDCs that produce the chemokine CXCL13 (BLC), a ligand for CXCR5 (see Fig. 10.11, bottom panels). The B cells proliferate in the light zone, but to a lesser extent than in the dark zone.

10-8 Germinal center B cells undergo V-region somatic hypermutation, and cells with mutations that improve affinity for antigen are selected.

The original diversification of antigen receptors produced by DNA rearrangement generates clones of B cells with radically differing B-cell receptors, as described in Sections 5-1 to 5-6. In contrast, the secondary diversification of immunoglobulin genes by somatic hypermutation introduces individual point mutations that change a single, or just a few, amino acids in the resulting immunoglobulin, producing closely related clones of B cells that differ subtly from each other in their specificity and affinity for antigen. When undergoing somatic hypermutation, immunoglobulin V-region genes accumulate mutations at a rate of about one base pair change per 10^3 base pairs per cell division. These mutations are targeted by the action of the enzyme AID (activation-induced cytidine deaminase) to the rearranged V genes (see Section 5-18 for the molecular detail of this process). The mutation rate in the rest of the DNA is much lower: around one base pair change per 10^{10} base pairs per cell division. Somatic hypermutation also affects some DNA flanking the rearranged V gene, but the mutations generally do not extend into the C-region exons. Each of the expressed heavy-chain and light-chain V-region genes is encoded by about 360 base pairs, and about three out of every four base changes results in an altered amino acid. This means that there is about a 50% chance at each cell division that a B cell will acquire a mutation in its receptor.

The point mutations accumulate in a stepwise manner as the descendants of each B cell proliferate in the germinal center to form B-cell clones. An altered receptor can affect the ability of a B cell to bind antigen and thus will affect the fate of the B cell in the germinal center (Fig. 10.13). Most mutations have a negative impact on the ability of the B-cell receptor to bind the original antigen, by preventing the correct folding of the immunoglobulin molecule or by blocking the complementarity-determining regions from binding antigen. Cells that harbor such mutations are eliminated by apoptosis, either because they can no longer make a functional B-cell receptor or because they cannot compete with sibling cells that bind antigen more strongly (see Fig. 10.13, left panels). Germinal centers are filled with apoptotic B cells that are quickly engulfed by macrophages, giving rise to the characteristic **tingible body macrophages**. These contain dark-staining nuclear debris in their cytoplasm and are a long-recognized histologic feature of germinal centers. Less frequently, mutations improve the affinity of a B-cell receptor for antigen. Cells that harbor these mutations are efficiently selected and expanded (see Fig. 10.13, right panels). Expansion seems to be due to the prevention of apoptosis, and thus increased survival compared with low-affinity cells, rather than an increased rate of proliferation.

Selection occurs in increments. It is thought that somatic hypermutation occurs in the centroblasts in the dark zone; when a centroblast reduces its rate of proliferation and becomes a centrocyte, it increases the number of B-cell receptors on its surface and moves to the light zone, where there are abundant FDCs bearing antigen. There, the centrocyte's ability to bind antigen is tested, in competition with the other clonally related centrocytes harboring different mutations. After each round of mutation, the centrocyte begins to express the new antigen receptor. If the new receptor can bind antigen, the centrocyte receives a signal through its receptor, which stimulates it to undergo additional rounds of division and to reexpress CXCR4, in effect becoming a centroblast again. The B cell reenters the dark zone, where the process of mutation and

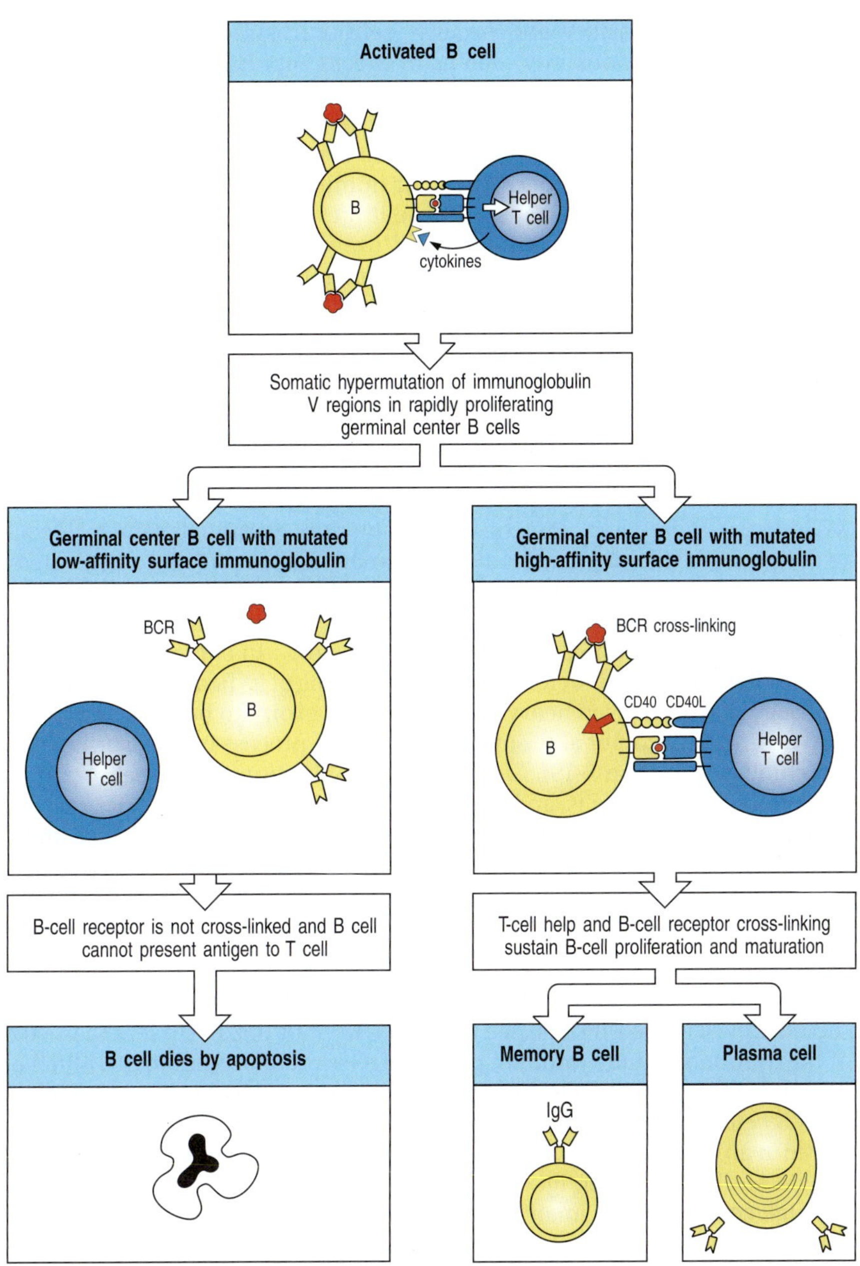

Fig. 10.13 Activated B cells undergo rounds of mutation and selection for higher-affinity mutants in the germinal center, resulting in high-affinity antibody-secreting plasma cells and high-affinity memory B cells. B cells are first activated at the follicle border by a combination of antigen and helper T cells (top panel). They migrate to germinal centers, where the remaining events occur. Somatic hypermutation can result in amino acid replacements in immunoglobulin V regions that affect the fate of the B cell. Mutations that result in a B-cell receptor (BCR) of lower or no affinity for the antigen (left panels) will prevent the B cell from being activated as efficiently, because both B-cell receptor cross-linking and the ability of the B cell to present peptide antigen to T cells are reduced. This results in such B cells dying by apoptosis, which purges low-affinity B cells from the germinal center. Most mutations are either negative or neutral, and thus the germinal center is a site of massive B-cell death as well as of proliferation. Some mutations, however, will improve the ability of the B-cell receptor to bind antigen. This increases the B cell's chance of interacting with T cells, and thus of proliferating and surviving (right panels). Surviving cells undergo repeated cycles of mutation and selection during which some of the progeny B cells undergo differentiation to either memory B cells or plasma cells (bottom right panels) and leave the germinal center. The signals that control these differentiation decisions are unknown.

selection is repeated. Being able to bind antigen will also allow the centrocyte to take up, process, and present antigen to T_{FH} cells, which will then provide survival signals to the centrocyte. This widely accepted model of the dynamics of B-cell migration within the germinal center is known as the **cyclic reentry model** (see Fig. 10.11, bottom right panel). The length of time that B cells can survive in the light zone is limited: they must reenter the dark zone or exit the germinal center within a few hours, or else they undergo apoptosis.

In this way, the affinity and specificity of positively selected B cells are continually refined during the germinal center response, the process known as **affinity maturation**. This leads to the average affinity of the population of responding B cells for its antigen increasing over time. The selection process can be quite stringent: although 50–100 B cells may seed the germinal center, most of them leave no progeny, and by the time the germinal center reaches maximum size, it is typically composed of the descendants of only one or a few B cells.

Evidence of both positive and negative selection is seen in the pattern of somatic hypermutations in V regions of B cells that have survived passage through the germinal center (see Section 5-18). The existence of negative selection is shown by the relative scarcity of amino acid replacements in the framework regions, reflecting the loss of cells that had mutated any one of the many residues that are critical for immunoglobulin V-region folding. Negative selection is important, because it prevents rapidly dividing B cells from expanding to numbers that would overwhelm the lymphoid tissues. Positive selection is evident in the accumulation of numerous amino acid replacements in the complementarity-determining regions, which determine antibody specificity and affinity (see Fig. 5.24).

10-9 Class switching in thymus-dependent antibody responses requires expression of CD40 ligand by helper T cells and is directed by cytokines.

As well as undergoing somatic hypermutation, germinal center B cells undergo class switching. Antibodies are remarkable not only for the diversity of their antigen-binding sites but also for their versatility as effector molecules. The antigen specificity of an antibody is determined by the variable domains of the immunoglobulin chains—V_H and V_L. In contrast, the effector action of the antibody is determined by the isotype of its heavy-chain C region (see Section 4-1). A given heavy-chain V domain exon can become associated with a C-region exon of any isotype through the process of class, or isotype, switching (see Section 5-19). Class switching only starts after B cells become activated by helper T cells, and can occur in B cells in the primary foci as well as in a proportion of the B cells in the germinal center. The DNA rearrangements that underlie class switching and confer this functional diversity on the humoral immune response are directed by cytokines, especially those released by effector CD4 T cells. We will see later in this chapter how antibodies of each class contribute to the elimination of pathogens.

All naive B cells express cell-surface IgM and IgD. IgM is the first antibody secreted by activated B cells (see Section 5-15), but makes up less than 10% of the immunoglobulin found in plasma; IgG is the most abundant. Much of the antibody in plasma has therefore been produced by plasma cells derived from B cells that have undergone class switching. Little IgD antibody is produced at any time, so the early stages of the antibody response are dominated by IgM antibodies. Later, IgG and IgA are the predominant antibody classes, with IgE contributing a small but biologically important part of the response. The overall predominance of IgG is also due in part to its longer lifetime in the plasma (see Fig. 5.15).

CD40 Ligand Deficiency

Productive interactions between B cells and helper T cells are essential for class switching to occur. This is demonstrated by people who have a genetic deficiency of CD40 ligand, which is required for those interactions. Class switching is greatly reduced in such individuals and they have abnormally high levels of IgM in their plasma, a condition known as **hyper IgM syndrome**, which is characterized by a lack of antibodies of classes other than IgM and severe humoral immunodeficiency, shown by repeated infections with common bacterial pathogens. Other defects that interfere with class switching, such as a deficiency of CD40, or of the enzyme AID, which is essential for the class-switch recombination process, also result in forms of hyper IgM syndrome (discussed in Chapter 13). Much of the IgM in hyper IgM syndromes may be induced by thymus-independent antigens on the pathogens that chronically infect these patients. Nevertheless, people with CD40 ligand deficiency can make IgM antibodies in response to thymus-dependent antigens, which indicates that in the B-cell response, CD40L–CD40 interactions are most important in enabling a sustained response that includes class switching and affinity maturation, rather than in the initial activation of B cells.

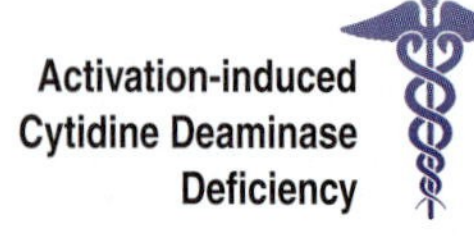

The molecular mechanisms that underlie class-switch recombination were outlined in Section 5-19. In this process, the rearranged V region of the heavy-chain gene is translocated from its original location upstream of the C_μ constant region and placed in front of a different C region, with deletion of the intervening chromosomal DNA (see Fig. 5.25). The selection of a particular C region for recombination is not random but is regulated by the cytokines being produced by helper T cells and other cells during the immune response; in this way it ensures that antibodies most useful in dealing with the particular infectious agent are produced. Much of what is known about the regulation of class switching by cytokines has come from experiments *in vitro* in which mouse B cells are exposed to various nonspecific stimuli, such as bacterial lipopolysaccharide (LPS), along with purified cytokines (Fig. 10.14). These experiments showed that different cytokines preferentially induce switching to different isotypes. In the mouse, IL-4 preferentially induces switching to IgG1 ($C\gamma_1$) and IgE ($C\varepsilon$), transforming growth factor (TGF)-β induces switching to IgG2b ($C\gamma_{2b}$) and IgA ($C\alpha$), IL-5 promotes switching to IgA, and interferon (IFN)-γ preferentially induces switching to IgG2a and IgG3 (Fig. 10.15). These cytokines can be produced by the CD4 T-cell subsets described in Section 9-17—for example, T_H2 cells make IL-4 and TGF-β, and T_H1 cells produce IFN-γ. However, T_{FH} cells are likely to regulate most B-cell switching in the germinal center, and they seem to be able to secrete cytokines characteristic of whichever T-cell subset is dominant within the effector T-cell response. Thus, helper T cells both regulate the production of antibody by B cells and also influence the heavy-chain isotype that determines the effector function of the antibody.

Cytokines induce class switching in part by inducing the production of RNA transcripts through the switch regions that lie 5′ to each heavy-chain C gene segment (see Fig. 10.14). When activated B cells are exposed to IL-4, for example, transcription from promoters that lie upstream of the switch regions of $C\gamma_1$ and $C\varepsilon$ can be detected a day or two before switching occurs. Interestingly, each of the cytokines that induces switching seems to induce transcription from the switch regions of two different heavy-chain C genes, but specific recombination occurs in only one of these genes.

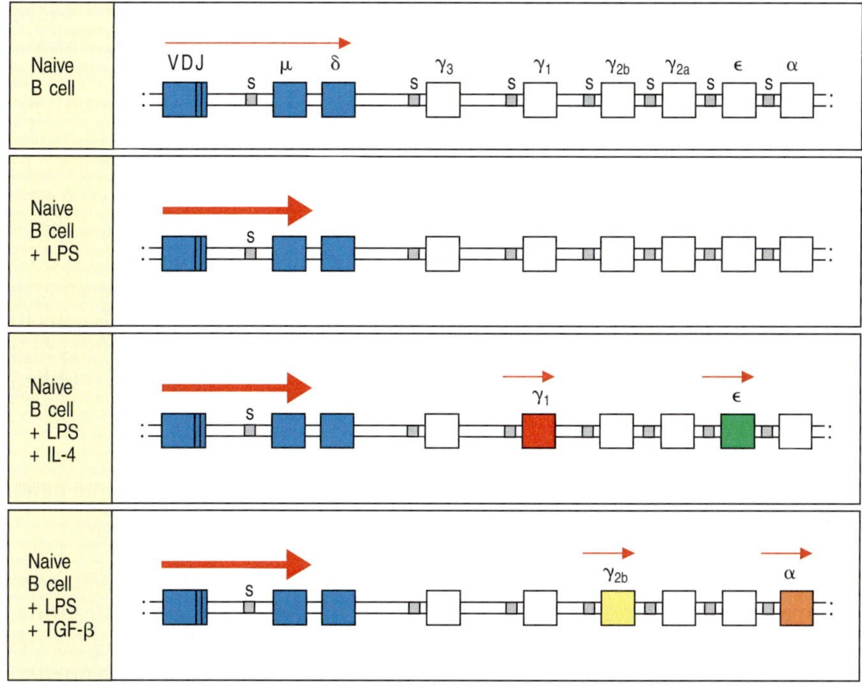

Fig. 10.14 Class switching is preceded by transcriptional activation of heavy-chain C-region genes. Resting naive B cells transcribe the genes for the heavy-chain isotypes μ and δ at a low rate, giving rise to surface IgM and IgD (first row). Bacterial lipopolysaccharide (LPS), which can activate B cells independently of antigen, induces IgM secretion (second row). In the presence of IL-4, however, transcripts of $C_{\gamma1}$ and C_ε are initiated at a low rate from promoters that are located at the 5′ end of each switch region (third row). These do not code for a protein but drive switching to IgG1 or IgE, respectively (see Section 5-19). Similarly, TGF-β gives rise to $C_{\gamma2b}$ and C_α transcripts and drives switching to IgG2b or IgA (fourth row). It is not known what determines which of the two transcriptionally activated heavy-chain C genes undergoes switching in a given event. Red arrows indicate transcription. The figure shows class switching in the mouse.

Fig. 10.15 Different cytokines induce switching to different antibody classes. The individual cytokines induce (violet) or inhibit (red) the production of certain antibody classes. Much of the inhibitory effect is probably the result of directed switching to a different class. These data are drawn from experiments with mouse cells.

Role of cytokines in regulating expression of antibody classes							
Cytokines	IgM	IgG3	IgG1	IgG2b	IgG2a	IgE	IgA
IL-4	Inhibits	Inhibits	Induces		Inhibits	Induces	
IL-5							Augments production
IFN-γ	Inhibits	Induces	Inhibits		Induces	Inhibits	
TGF-β	Inhibits	Inhibits		Induces			Induces

10-10 Ligation of CD40 and prolonged contact with T follicular helper cells is required to sustain germinal center B cells.

Germinal center B cells are inherently prone to dying, and to survive they must receive specific signals. It was originally discovered *in vitro* that resting B cells could be kept alive by simultaneously cross-linking their B-cell receptors and ligating their cell-surface CD40. *In vivo* these signals are delivered by antigen and by T_{FH} cells, respectively. The precise source of antigen in the germinal center has been a matter of some controversy. Antigen can be trapped and stored for long periods in the form of immune complexes on FDCs (Figs 10.16 and 10.17). It has been assumed that this is the antigen that sustains germinal center B-cell proliferation; other cells, however, may present antigen as well, and this issue remains an active area of research.

T_{FH} cells and germinal center B cells interact to deliver signals that are important for both cells. B cells express ICOS ligand (ICOSL), which co-stimulates T_{FH} cells through ICOS (see Section 9-13). Mice that lack ICOS are deficient in the germinal center reaction and have severely reduced switched antibody responses—in this case as a result of defective T_{FH}-cell function rather than any defect in the B cells. In turn, T_{FH} cells express CD40 ligand, which binds to CD40 on B cells, increasing their expression of a protein called Bcl-X_L, a relative of Bcl-2, which promotes B-cell survival by inhibiting apoptosis (see Section 7-23). Another important interaction involves receptors of the SLAM (signaling lymphocyte activation molecule) family, which are members of the immunoglobulin superfamily. SLAM-family receptors bind homotypically—that is, a SLAM receptor on one cell binds to a SLAM receptor on the other cell. Two members of the SLAM family, CD84 and Ly108, are involved in promoting prolonged contact between cognate T cells and B cells in the germinal center. Intravital microscopy has revealed that mice lacking CD84 have reduced numbers of conjugates between antigen-specific T cells and B cells in germinal centers, and these mice also have a reduced humoral response to antigen. Thus, it seems that T_{FH} cells make prolonged contact with B cells in germinal centers through several different receptor–ligand interactions that deliver signals in both directions. This cellular dialog is the basis for linked recognition.

10-11 Surviving germinal center B cells differentiate into either plasma cells or memory cells.

Some B cells eventually exit from the light zone and start to differentiate into plasma cells that produce large amounts of antibody. In B cells, the transcription factors Pax5 and Bcl6 inhibit the expression of transcription factors required for plasma-cell differentiation, and Pax5 and Bcl6 are downregulated

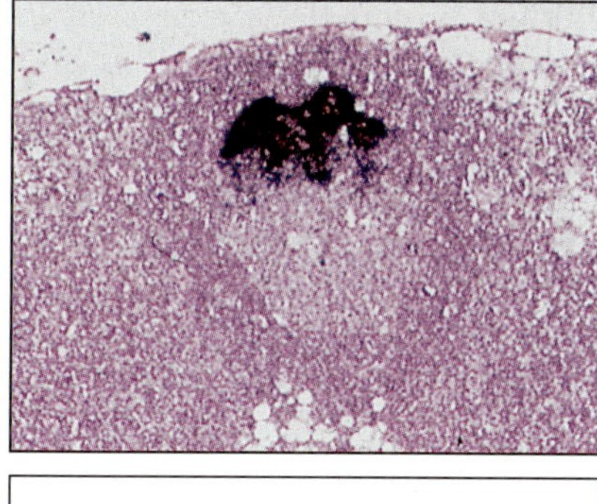

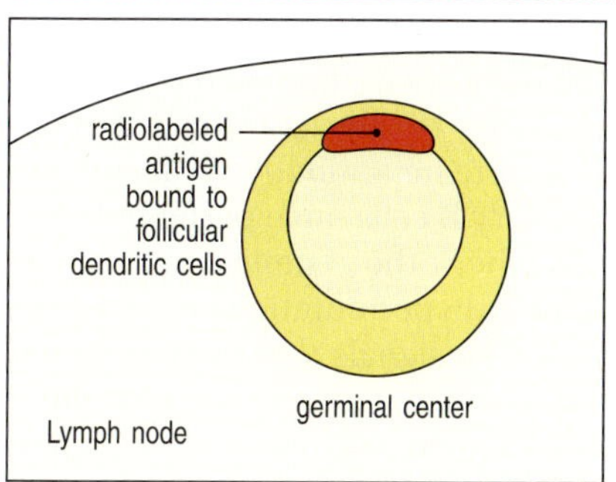

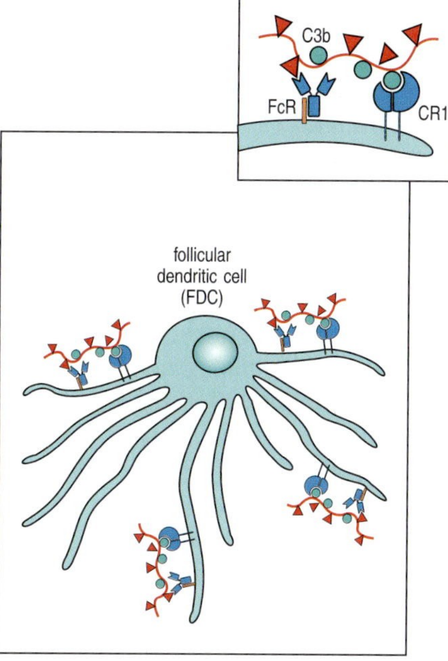

Fig. 10.16 Antigens are trapped in immune complexes that bind to the surface of follicular dendritic cells. Radiolabeled antigen localizes to, and persists in, lymphoid follicles of draining lymph nodes (see the light micrograph and the schematic representation below, showing a germinal center in a lymph node). Radiolabeled antigen has been injected 3 days previously and its localization in the germinal center is shown by the intense dark staining. The antigen is in the form of antigen:antibody:complement complexes bound to Fc receptors and to complement receptors CR1 or CR2 on the surface of the follicular dendritic cell (FDC), as depicted schematically in the right panel and inset. These complexes are not internalized. Antigen can persist in this form for long periods. Photograph courtesy of J. Tew.

when the B cell starts differentiation. The transcription factor IRF4 then induces the expression of BLIMP-1, a transcriptional repressor that switches off genes required for B-cell proliferation, class switching, and affinity maturation. B cells in which BLIMP-1 is induced become plasma cells; they cease proliferating, increase the synthesis and secretion of immunoglobulins, and change their cell-surface properties. They downregulate the chemokine receptor CXCR5, and upregulate CXCR4 and α_4:β_1 integrins, so that the plasma cells can now leave the germinal centers and home to peripheral tissues.

Some plasma cells deriving from germinal centers in lymph nodes or spleen migrate to the bone marrow, where a subset lives for a long period, whereas others migrate to the medullary cords in lymph nodes or splenic red pulp.

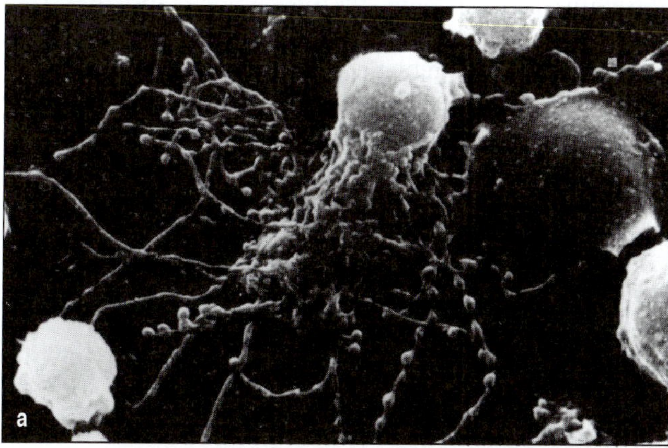

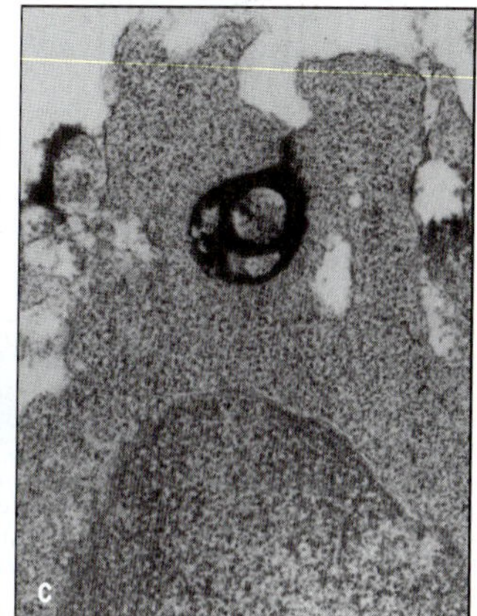

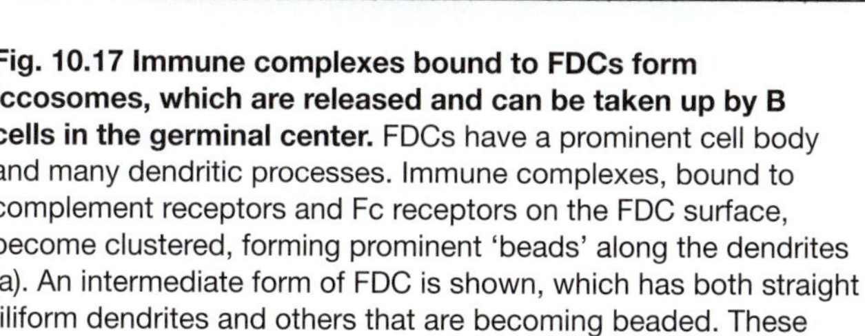

Fig. 10.17 Immune complexes bound to FDCs form iccosomes, which are released and can be taken up by B cells in the germinal center. FDCs have a prominent cell body and many dendritic processes. Immune complexes, bound to complement receptors and Fc receptors on the FDC surface, become clustered, forming prominent 'beads' along the dendrites (a). An intermediate form of FDC is shown, which has both straight filiform dendrites and others that are becoming beaded. These beads are shed from the cell as iccosomes (immune complex-coated bodies), which can bind to a B cell in the germinal center (b) and be taken up by it (c). In panels b and c, the iccosome has been formed with immune complexes containing horseradish peroxidase, which is electron-dense and therefore appears dark in the transmission electron micrographs. Photographs courtesy of A.K. Szakal.

The transcription factor XBP1 (X-box binding protein 1) is expressed in plasma cells and helps to regulate their secretory capacity; XBP1 is also required for plasma cells to colonize bone marrow successfully. B cells that have been activated in germinal centers in mucosal tissues, and which are predominantly switched to IgA production, stay within the mucosal system (discussed in Chapter 12). Plasma cells in bone marrow receive signals from stromal cells that are essential for their survival and they can be very long lived, whereas plasma cells in the medullary cords or red pulp are not long lived. Plasma cells in the bone marrow are the source of long-lasting high-affinity class-switched antibody.

Other germinal center B cells differentiate into **memory B cells**. Memory B cells are long-lived descendants of cells that were once stimulated by antigen and had proliferated in the germinal center. They divide very slowly if at all; they express surface immunoglobulin but secrete no antibody, or do so only at a low rate. Because the precursors of memory B cells once participated in a germinal center reaction, memory B cells inherit the genetic changes that occur in germinal center cells, including somatic hypermutation and the gene rearrangements that result in a class switch. The signals that control which differentiation path a B cell takes, and even whether at any given point the B cell continues to divide instead of differentiating, are still being investigated. We discuss memory B cells in Chapter 11.

10-12 Some bacterial antigens do not require T-cell help to induce B-cell responses.

Although antibody responses to most protein antigens are dependent on helper T cells, humans and mice with T-cell deficiencies nevertheless make antibodies against many bacterial antigens. This is because some bacterial polysaccharides, polymeric proteins, and lipopolysaccharides are able to stimulate naive B cells in the absence of T-cell help. The nonprotein bacterial products cannot elicit classical T-cell responses, yet they induce antibody responses in normal individuals. Such antigens are known as thymus-independent antigens (TI antigens).

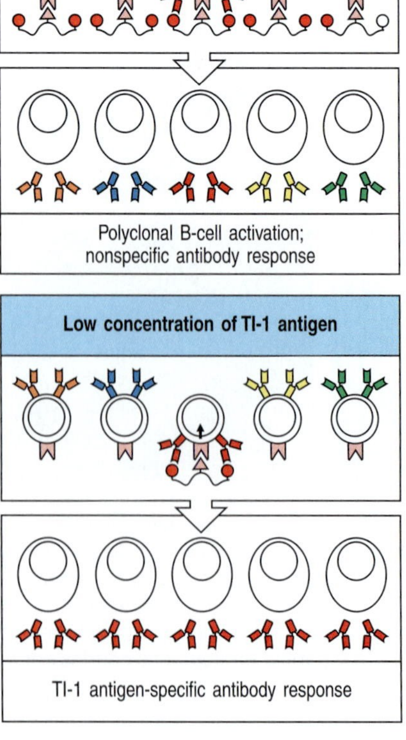

Thymus-independent antigens fall into two classes, which activate B cells by two different mechanisms. **TI-1 antigens** rely on activity that can directly induce B-cell division without T-cell help. We now understand that TI-1 antigens contain molecules that cause the proliferation and differentiation of most B cells regardless of their antigen specificity; this is known as **polyclonal activation** (Fig. 10.18, top panels). TI-1 antigens are therefore often called **B-cell mitogens**, a mitogen being a substance that induces cells to undergo mitosis. For example, LPS and bacterial DNA are both TI-1 antigens because they activate TLRs expressed by B cells (see Section 3-5) and can act as a mitogen. Naive murine B cells express most TLRs constitutively, but naive human B cells do not express high levels of most TLRs until they receive stimulation through the B-cell receptor. So, by the time a B cell has been stimulated by antigen through its B-cell receptor, it is likely to express several TLRs and be responsive to stimulation by TLR ligands that accompany the antigens. Thus, when B cells are exposed to concentrations of TI-1 antigens

Fig. 10.18 TI-1 antigens induce polyclonal B-cell responses at high concentrations, and antigen-specific antibody responses at low concentrations. At high concentration, the signal delivered by the B-cell-activating moiety of TI-1 antigens is sufficient to induce proliferation and antibody secretion by B cells in the absence of specific antigen binding to surface immunoglobulin. Thus, all B cells respond (top panels). At low concentration, only B cells specific for the TI-1 antigen bind enough of it to focus its B-cell activating properties onto the B cell; this gives a specific antibody response to epitopes on the TI-1 antigen (lower panels).

that are 10^3–10^5 times lower than those used for polyclonal activation, only those B cells whose B-cell receptors specifically bind the TI-1 antigen become activated. At these low concentrations, amounts of TI-1 antigen sufficient for B-cell activation can only be concentrated on the B-cell surface with the aid of this specific binding (Fig. 10.18, bottom panels).

The small amounts of TI-1 antigens present during the early stages of an infection are therefore likely to activate antigen-specific B cells only and to induce antibodies specific for the TI-1 antigen. Such responses may be important in defense against several extracellular pathogens: they arise earlier than thymus-dependent responses because they do not require the previous priming and clonal expansion of helper T cells. However, TI-1 antigens are inefficient inducers of affinity maturation and memory B cells, both of which require antigen-specific T-cell help.

Responses to TI antigens can, however, be influenced by T cells and natural killer (NK) cells, if such cells become activated in the immune response. In particular, these cells secrete cytokines that can affect the isotype of the antibody secreted. iNKT cells (see Section 8-9) are particularly intriguing as cells that might influence the TI response to nonprotein antigens. Because the T-cell receptors on these cells recognize certain polysaccharides bound to unconventional MHC class I or class I-like molecules such as CD1 (see Section 6-19), they could become activated by the same TI antigens as those that are activating the B cell, and so provide B-cell help in a manner similar to linked recognition.

10-13 B-cell responses to bacterial polysaccharides do not require peptide-specific T-cell help.

The second class of thymus-independent antigens consists of molecules such as bacterial capsular polysaccharides that have highly repetitive structures. These thymus-independent antigens, called **TI-2 antigens**, contain no intrinsic B-cell-stimulating activity. Whereas TI-1 antigens can activate both immature and mature B cells, TI-2 antigens can activate only mature B cells; immature B cells, as we saw in Section 8-6, are inactivated by encounter with repetitive epitopes. Infants and young children up to about 5 years of age do not make fully effective antibody responses against polysaccharide antigens, and this might be because most of their B cells are immature.

Responses to several TI-2 antigens are made prominently by B-1 cells (also known as CD5 B cells), which comprise an autonomously replicating subpopulation of nonconventional B cells (see Sections 3-24 and 8-28), and by marginal zone B cells, another unique subset of nonrecirculating B cells that line the border of the splenic white pulp (see Section 8-28). Marginal zone B cells are rare at birth and accumulate with age; they might therefore be responsible for most physiological TI-2 responses, which increase in efficiency with age.

TI-2 antigens most probably act by simultaneously cross-linking a critical number of B-cell receptors on the surface of antigen-specific mature B cells (Fig. 10.19, left panels). There is also evidence that dendritic cells and macrophages provide co-stimulatory signals for the initial activation of B cells by TI-2 antigens, signals that are necessary for the survival of the antigen-specific B cell and its differentiation into a plasmablast secreting IgM. One of these co-stimulatory signals is the TNF-family cytokine BAFF, which is secreted by the dendritic cell and interacts with the receptor TACI on the B cell (Fig. 10.19, right panels).

Excessive cross-linking of B-cell receptors renders mature B cells unresponsive or anergic, just as it does immature B cells. Thus, the density of TI-2 antigen epitopes presented to the B cell is critical. If it is too low, receptor cross-linking is insufficient to activate the cell; if too high, the B cell becomes anergic.

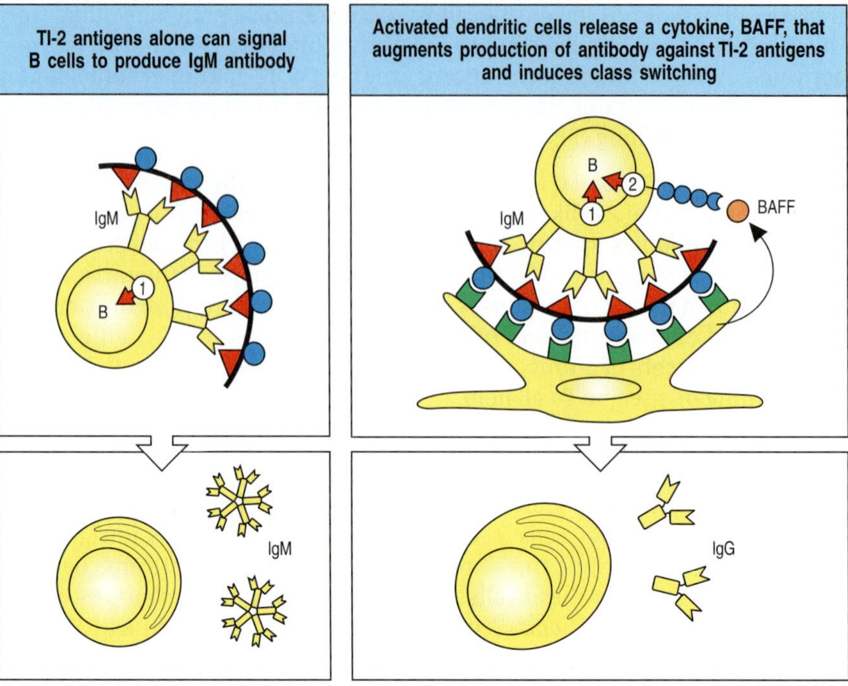

Fig. 10.19 B-cell activation by TI-2 requires, or is greatly enhanced by, cytokines. Multiple cross-linking of the B-cell receptor by TI-2 antigens can lead to IgM antibody production (left panels), but there is evidence that in addition, cytokines greatly augment these responses and lead to isotype switching as well (right panels). It is not clear where such cytokines are produced, but one possibility is that dendritic cells, which may be able to bind the antigen through innate immune system receptors on their surface and so present it to the B cells, secrete a soluble TNF-family cytokine called BAFF, which can activate class switching by the B cell.

B-cell responses to TI-2 antigens provide a prompt and specific response to an important class of pathogen—capsulated bacteria. Many common extracellular bacterial pathogens are surrounded by a polysaccharide capsule that enables them to resist ingestion by phagocytes. The bacteria not only escape direct destruction by phagocytes but also avoid stimulating T-cell responses in response to bacterial peptides presented by macrophages. IgM antibodies rapidly produced against the capsular polysaccharide independently of peptide-specific T-cell help will coat the bacteria, promoting their ingestion and destruction by phagocytes early in the infection.

As well as producing IgM, thymus-independent responses can include switching to certain other antibody classes, such as IgG3 in the mouse. This is probably the result of help from dendritic cells (see Fig. 10.19, right panels), which provide secreted cytokines such as BAFF and membrane-bound signals to proliferating plasmablasts as they respond to TI antigens.

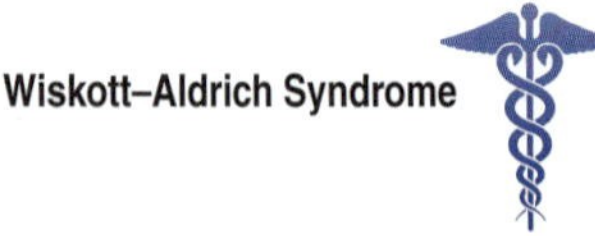

Not all antibodies against bacterial polysaccharides are produced strictly through this TI-2 mechanism. We mentioned earlier the importance of antibodies against the capsular polysaccharide of *Haemophilus influenzae* type b in protective immunity to this bacterium. The immunodeficiency disease Wiskott–Aldrich syndrome is caused by defects in T cells that impair their interaction with B cells (described in more detail in Chapter 13). Patients with Wiskott–Aldrich syndrome respond poorly to protein antigens, but, unexpectedly, also fail to make IgM and IgG antibody against polysaccharide antigens and are highly susceptible to infection with encapsulated bacteria such as *H. influenzae*. The failure to make IgM seems to be due in part to greatly reduced development of the marginal zone of the spleen, which contains B cells responsible for making much of the 'natural' IgM antibody against ubiquitous carbohydrate antigens. Thus, IgM and IgG antibodies induced by TI-2 antigens are likely to be an important part of the humoral immune response in many bacterial infections, and in humans at least, the production of class-switched antibodies to TI-2 antigens might normally rely on some degree of T-cell help. The distinguishing features of thymus-dependent, TI-1, and TI-2 antibody responses are summarized in Fig. 10.20.

	TD antigen	TI-1 antigen	TI-2 antigen
Antibody response in infants	Yes	Yes	No
Antibody production in congenitally athymic individual	No	Yes	Yes
Antibody response in absence of all T cells	No	Yes	No
Primes T cells	Yes	No	No
Polyclonal B-cell activation	No	Yes	No
Requires repeating epitopes	No	No	Yes
Examples of antigen	Diphtheria toxin Viral hemagglutinin Purified protein derivative (PPD) of *Mycobacterium tuberculosis*	Bacterial lipopoly-saccharide *Brucella abortus*	Pneumococcal polysaccharide *Salmonella* polymerized flagellin Dextran Hapten-conjugated Ficoll (polysucrose)

Fig. 10.20 Properties of different classes of antigen that elicit antibody responses.

Summary.

B-cell activation by many antigens requires both binding of the antigen by the B-cell surface immunoglobulin—the B-cell receptor—and interaction of the B cell with antigen-specific helper T cells. Helper T cells recognize peptide fragments derived from the antigen internalized by the B cell and displayed by the B cells as peptide:MHC class II complexes. Follicular helper T cells stimulate B cells by prolonged conjugation in germinal centers, with binding of CD40 ligand on the T cell to CD40 on the B cell, and by the directed release of cytokines. Activated B cells also provide signals to T cells, for example via B7-family molecules, that promote the T cells' continued activation. The initial interaction occurs at the border of the T-cell and B-cell areas of secondary lymphoid tissue, to which antigen-activated helper T cells and B cells migrate in response to chemokines. Further interactions between T cells and B cells continue after migration into the follicle and the formation of a germinal center.

Helper T cells induce a phase of vigorous B-cell proliferation in the germinal center reaction and direct the differentiation of clonally expanded B cells into either antibody-secreting plasma cells or memory B cells. The switching to different antibody isotypes is regulated by cytokines released from helper T cells. Somatic hypermutation and selection for high-affinity binding occur in the germinal centers. Helper T cells control these processes by selectively activating B cells that have retained their specificity for the antigen and by inducing proliferation and differentiation into plasma cells and memory B cells. Some nonprotein antigens stimulate B cells in the absence of linked recognition by peptide-specific helper T cells. Responses to these thymus-independent antigens are accompanied by only limited class switching and do not induce memory B cells. However, such responses have a crucial role in host defense against pathogens whose surface antigens cannot elicit peptide-specific T-cell responses.

The distributions and functions of immunoglobulin classes.

Extracellular pathogens can find their way to most sites in the body, and antibodies must be equally widely distributed to combat them. Most classes of antibodies are distributed by diffusion from their site of synthesis, but specialized transport mechanisms are required to deliver antibodies to epithelial surfaces lining the lumina of organs such as the lungs and intestine. The distribution of antibodies is determined by their heavy-chain isotype, which can limit their diffusion or enable them to engage specific transporters that deliver them across epithelia. In this part of the chapter we describe the mechanisms by which antibodies of different classes are directed to the compartments of the body in which their particular effector functions are appropriate, and discuss the protective functions of antibodies that result solely from their binding to pathogens. In the last part of the chapter we discuss the effector cells and molecules that are specifically engaged by different antibody classes.

10-14 Antibodies of different classes operate in distinct places and have distinct effector functions.

Pathogens most commonly enter the body across the epithelial barriers of the mucosa lining the respiratory, digestive, and urogenital tracts, or through damaged skin. Less often, insects, wounds, or hypodermic needles introduce microorganisms directly into the blood. Antibodies protect all the body's mucosal surfaces, tissues, and blood from such infections; these antibodies serve to neutralize the pathogen or promote its elimination before it can establish a significant infection. Antibodies of different classes are adapted to function in different compartments of the body. Because a given V region can become associated with any C region through class switching (see Section 10-9), the progeny of a single B cell can produce antibodies that share the same specificity yet provide all of the protective functions appropriate for each body compartment.

IgM antibodies are produced first in a humoral immune response, because they can be expressed without class switching (see Fig. 5.17). These early IgM antibodies tend to be of low affinity, being produced before B cells have undergone somatic hypermutation. However, IgM molecules form pentamers with 10 antigen-binding sites, conferring higher overall avidity when binding to multivalent antigens such as bacterial capsular polysaccharides, and so compensating for the low affinity of the IgM monomers. Because of the large size of the pentamers, IgM is found mainly in the bloodstream and, to a lesser extent, in the lymph, rather than in intercellular spaces within tissues. The pentameric structure of IgM makes it especially effective in activating the complement system, as we will see in the last part of this chapter. Infection of the bloodstream has serious consequences unless it is controlled quickly, and the rapid production of IgM and its efficient activation of the complement system are important in controlling such infections. Some IgM is produced in secondary and subsequent responses, and also from B cells that have not undergone class switching during somatic hypermutation, although other classes dominate the later phases of the antibody response. Much IgM is also produced by the nonconventional B-1 cells that reside in the peritoneal cavity and the pleural spaces and by the marginal zone B cells of the spleen. These cells secrete antibodies against commonly encountered carbohydrate antigens, including those of bacteria, and do not require T-cell help; they therefore provide a preformed repertoire of IgM antibodies in blood and body cavities that can recognize invading pathogens (see Sections 3-24 and 8-28).

Antibodies of the other classes—IgG, IgA, and IgE—are smaller, and diffuse easily out of the blood into the tissues. IgA can form dimers, as we saw in Chapter 4, but IgG and IgE are always monomeric. The affinity of the individual antigen-binding sites for their antigen is therefore critical for the effectiveness of these antibodies, and most of the B cells expressing these classes have been selected in the germinal centers for their increased affinity for antigen after somatic hypermutation. IgG4 is the least abundant of the IgG subclasses, but has the unusual ability to form hybrid antibodies. One IgG4 heavy chain and attached light chain can split from the original heavy-chain dimer and reassociate with a different IgG4 heavy chain–light chain pair, forming a bivalent IgG4 antibody with two distinct antigen specificities.

IgG is the principal class of antibody in blood and extracellular fluid, whereas IgA is the principal class in secretions, the most important being those from the epithelia lining the intestinal and respiratory tracts. IgG efficiently opsonizes pathogens for engulfment by phagocytes and activates the complement system, but IgA is a less potent opsonin and a weak activator of complement. IgG operates mainly in the tissues, where accessory cells and molecules are available, whereas dimeric IgA operates mainly on epithelial surfaces, where complement and phagocytes are not normally present; therefore IgA functions chiefly as a neutralizing antibody. IgA is also produced by plasma cells that differentiate from class-switched B cells in lymph nodes and spleen, and acts as a neutralizing antibody in extracellular spaces and in the blood. This IgA is monomeric and is predominantly of the subclass IgA1; the ratio of IgA1 to IgA2 in the blood is 10:1. The IgA antibodies produced by plasma cells in the gut are dimeric and predominantly of subclass IgA2; the ratio of IgA2 to IgA1 in the gut is 3:2.

Finally, IgE antibody is present only at very low levels in blood or extracellular fluid, but is bound avidly by receptors on mast cells that are found just beneath the skin and mucosa and along blood vessels in connective tissue. Antigen binding to this cell-associated IgE triggers mast cells to release powerful chemical mediators that induce reactions, such as coughing, sneezing, and vomiting, that in turn can expel infectious agents, as discussed later in this chapter. The distribution and main functions of antibodies of the different classes are summarized in Fig. 10.21.

10-15 Transport proteins that bind to the Fc regions of antibodies carry particular isotypes across epithelial barriers.

In the mucosal immune system, IgA-secreting plasma cells are found predominantly in the lamina propria, which lies immediately below the basement membrane of many surface epithelia. From there the IgA antibodies can be transported across the epithelium to its external surface, for example to the lumen of the gut or of the bronchi (Fig. 10.22). IgA antibody synthesized in the lamina propria is secreted as a dimeric IgA molecule associated with a single J chain (see Fig. 5.19). This polymeric form of IgA binds specifically to a receptor called the **polymeric immunoglobulin receptor** (**pIgR**), which is present on the basolateral surfaces of the overlying epithelial cells. When the pIgR has bound a molecule of dimeric IgA, the complex is internalized and carried through the cytoplasm of the epithelial cell in a transport vesicle to its luminal surface. This process is called transcytosis. IgM also binds to the pIgR and can be secreted into the gut by the same mechanism. Upon reaching the luminal surface of the enterocyte, the antibody is released into the mucous layer covering the gut lining by proteolytic cleavage of the extracellular domain of the pIgR. The cleaved extracellular domain of the pIgR is known as secretory component (frequently abbreviated to SC) and remains associated with the antibody. Secretory component is bound to the part of the Fc region of IgA that contains the binding site for the Fcα receptor I, which

Fig. 10.21 Each human immunoglobulin class has specialized functions and a unique distribution. The major effector functions of each class (+++) are shaded in dark red, whereas lesser functions (++) are shown in dark pink, and very minor functions (+) in pale pink. The distributions are marked similarly, with actual average levels in serum being shown in the bottom row. IgA has two subclasses, IgA1 and IgA2. The IgA column refers to both. *IgG2 can act as an opsonin in the presence of an Fc receptor of the appropriate allotype, found in about 50% of white people.

Functional activity	IgM	IgD	IgG1	IgG2	IgG3	IgG4	IgA	IgE
Neutralization	+	–	++	++	++	++	++	–
Opsonization	+	–	+++	*	++	+	+	–
Sensitization for killing by NK cells	–	–	++	–	++	–	–	–
Sensitization of mast cells	–	–	+	–	+	–	–	+++
Activates complement system	+++	–	++	+	+++	–	+	–

Distribution	IgM	IgD	IgG1	IgG2	IgG3	IgG4	IgA	IgE
Transport across epithelium	+	–	–	–	–	–	+++ (dimer)	–
Transport across placenta	–	–	+++	+	++	+/–	–	–
Diffusion into extravascular sites	+/–	–	+++	+++	+++	+++	++ (monomer)	+
Mean serum level (mg ml^{-1})	1.5	0.04	9	3	1	0.5	2.1	3×10^{-5}

is why secretory IgA does not bind to this receptor. Secretory component serves several physiological roles. It binds to mucins in mucus, acting as 'glue' to bind secreted IgA to the mucous layer on the luminal surface of the gut epithelium, where the antibody binds and neutralizes gut pathogens and their toxins (see Fig. 10.22). Secretory component also protects the antibodies against cleavage by gut enzymes.

The principal sites of IgA synthesis and secretion are the gut, the respiratory epithelium, the lactating breast, and various other exocrine glands such as the salivary and tear glands. It is believed that the primary functional role of IgA antibodies is to protect epithelial surfaces from infectious agents, just as IgG antibodies protect the extracellular spaces inside tissues. By binding bacteria, virus particles, and toxins, IgA antibodies prevent the attachment of bacteria and viruses to epithelial cells and the uptake of toxins, and provide

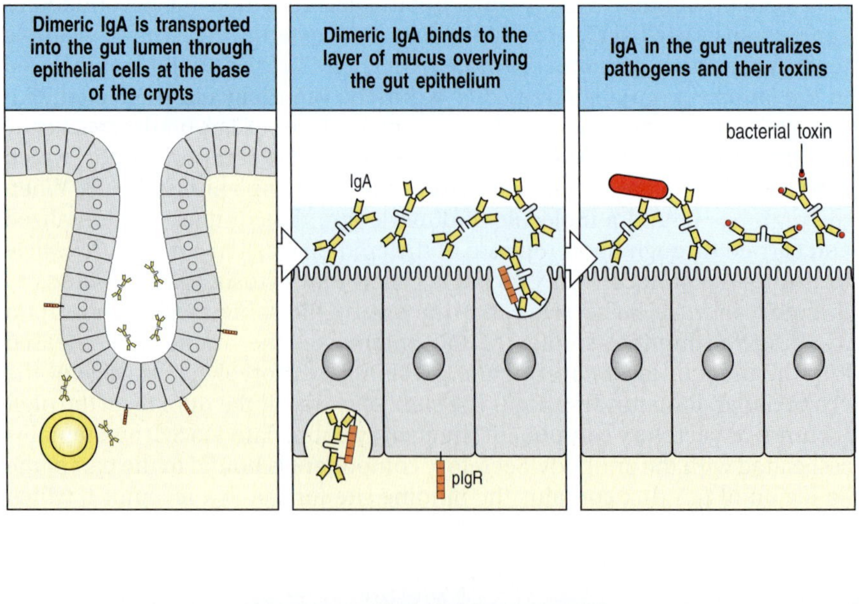

Fig. 10.22 The major class of antibody present in the lumen of the gut is secretory dimeric IgA. This is synthesized by plasma cells in the lamina propria and transported into the lumen of the gut through epithelial cells at the base of the crypts. Dimeric IgA binds to the layer of mucus overlying the gut epithelium and acts as an antigen-specific barrier to pathogens and toxins in the gut lumen.

the first line of defense against a wide variety of pathogens. IgA is also thought to have an additional role in the gut, that of regulating the gut microbiota (see Chapter 12).

Newborn infants are especially vulnerable to infection, having had no previous exposure to the microbes in the environment they enter at birth. IgA antibodies are secreted in breast milk and are thereby transferred to the gut of the newborn infant, where they provide protection from newly encountered bacteria until the infant can synthesize its own protective antibody. IgA is not the only protective antibody that a mother passes on to her baby. Maternal IgG is transported across the placenta directly into the bloodstream of the fetus during intrauterine life; human babies at birth have as high levels of plasma IgG as their mothers, and with the same range of antigen specificities. The selective transport of IgG from mother to fetus is due to an IgG transport protein in the placenta, **FcRn** (**neonatal Fc receptor**), which is closely related in structure to MHC class I molecules. Despite this similarity, FcRn binds IgG quite differently from the binding of peptide to MHC class I molecules, because its peptide-binding groove is occluded. It binds to the Fc portion of IgG molecules (Fig. 10.23). Two molecules of FcRn bind one molecule of IgG, bearing it across the placenta. In some rodents, FcRn also delivers IgG to the circulation of the neonate from the gut lumen. Maternal IgG is also ingested by the newborn animal in its mother's milk and colostrum, the protein-rich fluid secreted by the early postnatal mammary gland. In this case, FcRn transports the IgG from the lumen of the neonatal gut into the blood and tissues. Interestingly, FcRn is also found in adults in the gut and liver and on endothelial cells. Its function in adults is to maintain the levels of IgG in plasma, which it does by binding antibody, endocytosing it, and recycling it to the blood, thus preventing its excretion from the body.

By means of these specialized transport systems, mammals are supplied from birth with antibodies against pathogens common in their environments. As they mature and make their own antibodies of all isotypes, these are distributed selectively to different sites in the body (Fig. 10.24). Thus, throughout life, class switching and the distribution of antibody classes throughout the body provide effective protection against infection in extracellular spaces.

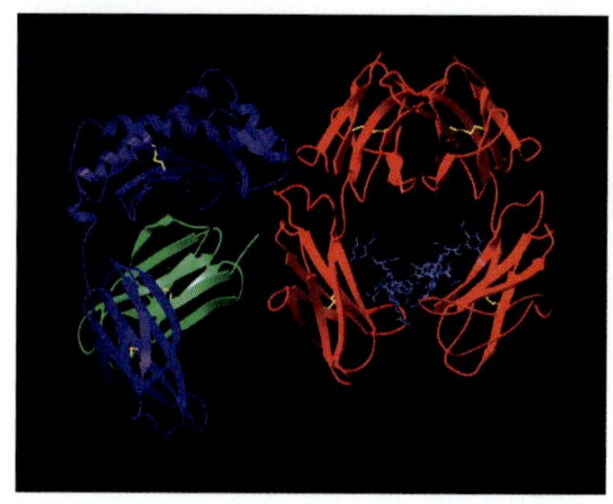

Fig. 10.23 The neonatal Fc receptor (FcRn) binds to the Fc portion of IgG. The structure of a molecule of FcRn (blue and green) is shown bound to one chain of the Fc portion of IgG (red), at the interface of the $C_{\gamma2}$ and $C_{\gamma3}$ domains, with the $C_{\gamma2}$ region at the top. The β_2-microglobulin component of the FcRn is green. The dark-blue structure attached to the Fc portion of IgG is a carbohydrate chain, reflecting glycosylation. FcRn transports IgG molecules across the placenta in humans and also across the gut in rats and mice. It also has a role in maintaining the levels of IgG in adults. Although only one molecule of FcRn is shown binding to the Fc portion, it is thought that it takes two molecules of FcRn to capture one molecule of IgG. Courtesy of P. Björkman.

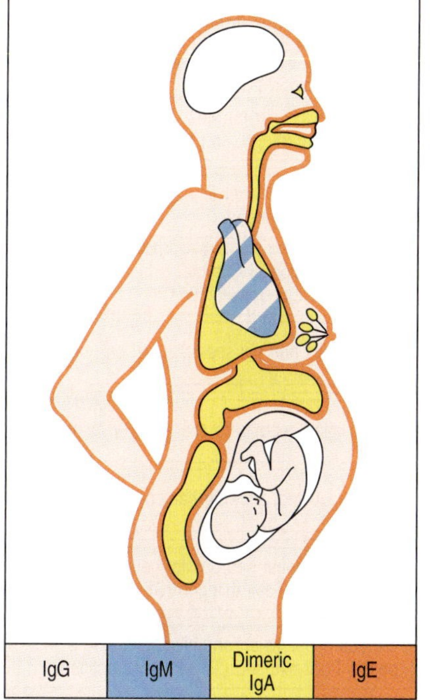

Fig. 10.24 Immunoglobulin classes are selectively distributed in the body. IgG and IgM predominate in blood (shown here for simplicity by IgM and IgG in the heart), whereas IgG and monomeric IgA are the major antibodies in extracellular fluid within the body. Dimeric IgA predominates in secretions across epithelia, including breast milk. The fetus receives IgG from the mother by transplacental transport. IgE is found mainly associated with mast cells just beneath epithelial surfaces (especially of the respiratory tract, gastrointestinal tract, and skin). The brain is normally devoid of immunoglobulin.

10-16 High-affinity IgG and IgA antibodies can neutralize bacterial toxins.

Many bacteria cause disease by secreting proteins called toxins, which damage or disrupt the function of the host's cells (Fig. 10.25). To have an effect, a toxin must interact specifically with a molecule that serves as a receptor on the surface of the target cell. In many toxins the receptor-binding domain is on one polypeptide chain but the toxic function is carried by a second chain. Antibodies that bind to the receptor-binding site on the toxin molecule can prevent the toxin from binding to the cell and thus protect the cell from attack (Fig. 10.26). Antibodies that act in this way to neutralize toxins are referred to as neutralizing antibodies.

Most toxins are active at nanomolar concentrations: a single molecule of diphtheria toxin can kill a cell. To neutralize toxins, therefore, antibodies must be able to diffuse into the tissues and bind the toxin rapidly and with high affinity. The ability of IgG antibodies to diffuse easily throughout the extracellular fluid, and their high affinity for antigen once affinity maturation has taken place, make them the principal antibodies that neutralize toxins in tissues. High-affinity IgA antibodies similarly neutralize toxins at the mucosal surfaces of the body.

Diphtheria and tetanus toxins are two bacterial toxins in which the toxic and receptor-binding functions are on separate protein chains. It is therefore possible to immunize individuals, usually as infants, with modified toxin molecules in which the toxic chain has been denatured. These modified toxins,

Fig. 10.25 Many common diseases are caused by bacterial toxins. These toxins are all exotoxins—proteins secreted by the bacteria. High-affinity IgG and IgA antibodies protect against these toxins. Bacteria also have nonsecreted endotoxins, such as lipopolysaccharide, which are released when the bacterium dies. The endotoxins are also important in the pathogenesis of disease, but there the host response is more complex because the innate immune system has receptors for some endotoxins (see Chapter 3).

Disease	Organism	Toxin	Effects *in vivo*
Tetanus	*Clostridium tetani*	Tetanus toxin	Blocks inhibitory neuron action, leading to chronic muscle contraction
Diphtheria	*Corynebacterium diphtheriae*	Diphtheria toxin	Inhibits protein synthesis, leading to epithelial cell damage and myocarditis
Gas gangrene	*Clostridium perfringens*	Clostridial toxin	Phospholipase activation, leading to cell death
Cholera	*Vibrio cholerae*	Cholera toxin	Activates adenylate cyclase, elevates cAMP in cells, leading to changes in intestinal epithelial cells that cause loss of water and electrolytes
Anthrax	*Bacillus anthracis*	Anthrax toxic complex	Increases vascular permeability, leading to edema, hemorrhage, and circulatory collapse
Botulism	*Clostridium botulinum*	Botulinum toxin	Blocks release of acetylcholine, leading to paralysis
Whooping cough	*Bordetella pertussis*	Pertussis toxin	ADP-ribosylation of G proteins, leading to lymphoproliferation
		Tracheal cytotoxin	Inhibits cilia and causes epithelial cell loss
Scarlet fever	*Streptococcus pyogenes*	Erythrogenic toxin	Vasodilation, leading to scarlet fever rash
		Leukocidin Streptolysins	Kill phagocytes, allowing bacterial survival
Food poisoning	*Staphylococcus aureus*	Staphylococcal enterotoxin	Acts on intestinal neurons to induce vomiting. Also a potent T-cell mitogen (SE superantigen)
Toxic-shock syndrome	*Staphylococcus aureus*	Toxic-shock syndrome toxin	Causes hypotension and skin loss. Also a potent T-cell mitogen (TSST-1 superantigen)

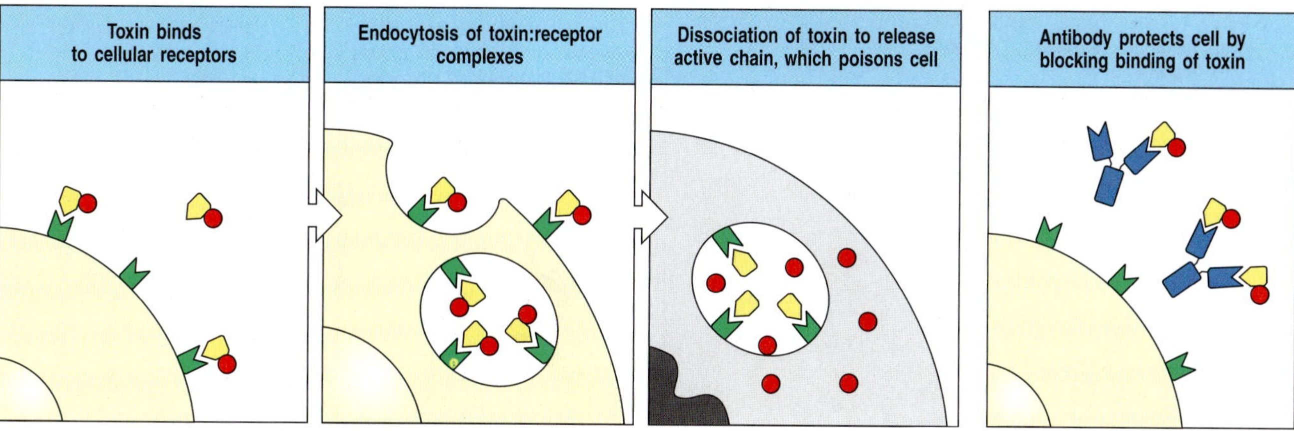
Toxin binds to cellular receptors
Endocytosis of toxin:receptor complexes
Dissociation of toxin to release active chain, which poisons cell
Antibody protects cell by blocking binding of toxin

Fig. 10.26 Neutralization of toxins by IgG antibodies protects cells from their damaging action. The damaging effects of many bacteria are due to the toxins they produce (see Fig. 10.25). These toxins are usually composed of several distinct moieties. One part of the toxin molecule binds a cell-surface receptor, which enables the molecule to be internalized. Another part of the toxin molecule then enters the cytoplasm and poisons the cell. Antibodies that inhibit toxin binding can prevent, or neutralize, these effects.

called toxoids, lack toxic activity but retain the receptor-binding site. Thus, immunization with the toxoid induces neutralizing antibodies that protect against the native toxin.

Some insect or animal venoms are so toxic that a single exposure can cause severe tissue damage or death, and for these the adaptive immune response is too slow to be protective. Exposure to these venoms is a rare event, and protective vaccines have not been developed for use in humans. Instead, neutralizing antibodies are generated by immunizing other species, such as horses, with insect and snake venoms to produce anti-venom antibodies (antivenins). The antivenins are injected into exposed individuals to protect them against the toxic effects of the venom. Transfer of antibodies in this way is known as passive immunization (see Appendix I, Section A-36).

10-17 High-affinity IgG and IgA antibodies can inhibit the infectivity of viruses.

Animal viruses infect cells by binding to a particular cell-surface receptor, often a cell-type-specific protein that determines which cells they can infect. The hemagglutinin of influenza virus, for example, binds to terminal sialic acid residues on the carbohydrates of glycoproteins present on epithelial cells of the respiratory tract. It is known as hemagglutinin because it recognizes and binds to similar sialic acid residues on chicken red blood cells and agglutinates these red blood cells. Antibodies against the hemagglutinin can prevent infection by the influenza virus. Such antibodies are called virus-neutralizing antibodies and, as with the neutralization of toxins, high-affinity IgA and IgG antibodies are particularly important.

Many antibodies that neutralize viruses do so by directly blocking the binding of virus to surface receptors (Fig. 10.27). However, viruses are sometimes successfully neutralized when only a single molecule of antibody is bound to a virus particle that has many receptor-binding sites on its surface. In these cases, the antibody must cause some change in the virus that disrupts its structure. This might prevent it from interacting with its receptor, or it might interfere with the fusion of the virus with the plasma membrane after the virus has engaged its receptor.

10-18 Antibodies can block the adherence of bacteria to host cells.

Many bacteria have cell-surface molecules called **adhesins** that enable them to bind to the surfaces of host cells. This adherence is crucial to the ability of these bacteria to cause disease, whether they subsequently enter the cell, as do *Salmonella* species, or remain attached to the cell surface as extracellular

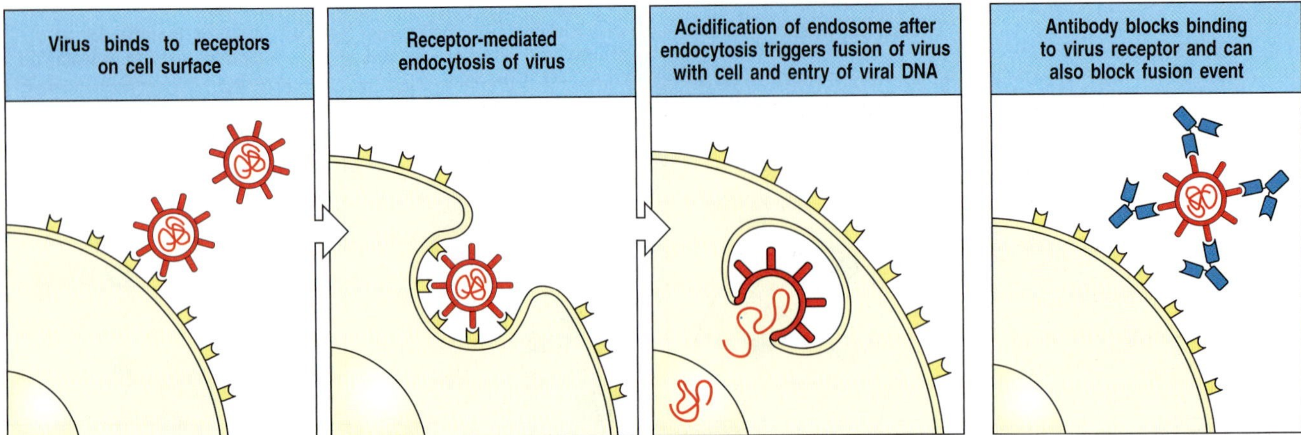
Virus binds to receptors on cell surface
Receptor-mediated endocytosis of virus
Acidification of endosome after endocytosis triggers fusion of virus with cell and entry of viral DNA
Antibody blocks binding to virus receptor and can also block fusion event

Fig. 10.27 Viral infection of cells can be blocked by neutralizing antibodies. For a virus to multiply within a cell, it must introduce its genes into the cell. The first step in entry is usually the binding of the virus to a receptor on the cell surface. For enveloped viruses, as shown in the figure, entry into the cytoplasm requires fusion of the viral envelope and the cell membrane. For some viruses this fusion event takes place on the cell surface (not shown); for others it can occur only within the more acidic environment of endosomes, as shown here. Non-enveloped viruses must also bind to receptors on cell surfaces, but they enter the cytoplasm by disrupting endosomes. Antibodies bound to viral surface proteins neutralize the virus, inhibiting either its initial binding to the cell or its subsequent entry.

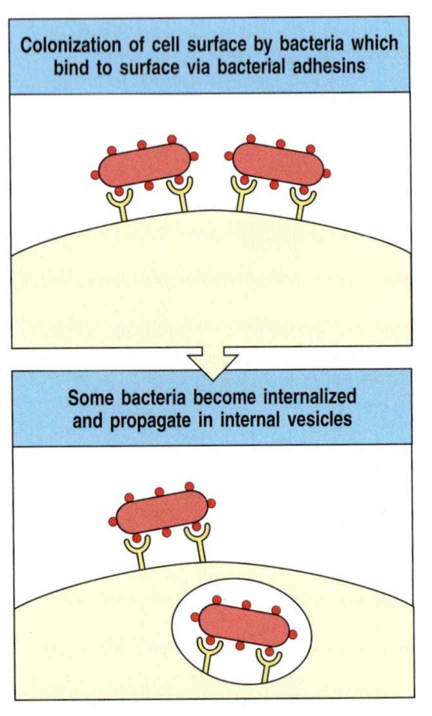
Colonization of cell surface by bacteria which bind to surface via bacterial adhesins
Some bacteria become internalized and propagate in internal vesicles

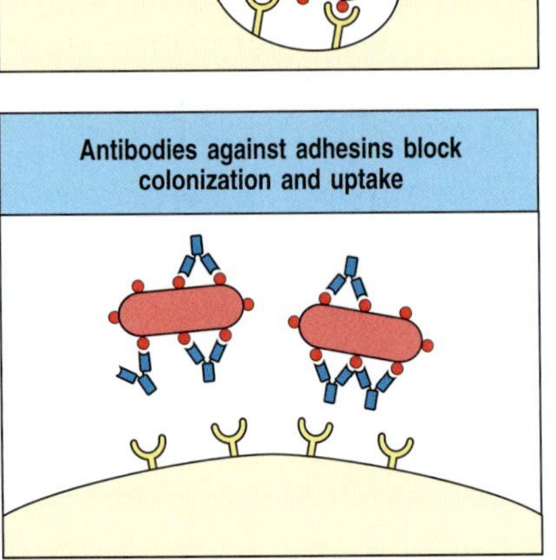
Antibodies against adhesins block colonization and uptake

pathogens (Fig. 10.28). *Neisseria gonorrhoeae*, the causative agent of the sexually transmitted disease gonorrhea, has a cell-surface protein known as pilin, which enables the bacterium to adhere to the epithelial cells of the urinary and reproductive tracts and is essential to its infectivity. Antibodies against pilin can inhibit this adhesive reaction and prevent infection.

IgA antibodies secreted onto the mucosal surfaces of the intestinal, respiratory, and reproductive tracts are particularly important in inhibiting the colonization of these surfaces by pathogens and in preventing infection of the epithelial cells. Adhesion of bacteria to cells within tissues can also contribute to pathogenesis, and IgG antibodies against adhesins protect tissues from damage in much the same way as IgA antibodies protect mucosal surfaces.

10-19 Antibody:antigen complexes activate the classical pathway of complement by binding to C1q.

Another way in which antibodies can protect against infection is by activation of complement. We described the complement system in Chapter 2 because it is first activated in the absence of antibody as part of an innate immune response. Complement activation proceeds via a cascade of proteolytic cleavage reactions, in which inactive complement proteins in the plasma are cleaved to form proteases that then cleave and activate the next enzyme in the series. Three pathways of complement activation converge to coat pathogen surfaces or antigen:antibody complexes with covalently attached complement fragment C3b, which acts as an opsonin to promote uptake and removal by phagocytes. In addition, the terminal complement components can form a membrane-attack complex that damages some bacteria.

Fig. 10.28 Antibodies can prevent the attachment of bacteria to cell surfaces. Many bacterial infections require an interaction between the bacterium and a cell-surface receptor. This is particularly true for infections of mucosal surfaces. The attachment process involves very specific molecular interactions between bacterial adhesins and their receptors on host cells; antibodies against bacterial adhesins can block such infections.

Antibodies initiate complement activation by a pathway known as the classical pathway because it was the first pathway of complement activation to be discovered. The full details of this pathway, and of the other two known pathways of complement activation, are given in Chapter 2. There we focused on how the classical pathway can be activated in innate immunity in the absence of specific antibody, but here we describe how antibody formed in the adaptive immune response initiates the classical pathway.

The first component of the classical pathway of complement activation is C1, which is a complex of three proteins called C1q, C1r, and C1s (see Fig. 2.17). Recall that C1r and C1s are serine proteases, and two molecules each of C1r and C1s are bound to each molecule of C1q. Complement activation is initiated when antibodies attached to the surface of a pathogen bind C1 via C1q (Fig. 10.29). C1q can be bound by either IgM or IgG antibodies but, because of the structural requirements of binding to C1q, neither of these antibody classes can activate complement in solution; the complement reactions are initiated only when the antibodies are already bound to multiple sites on a cell surface, normally that of a pathogen.

Each globular head of a C1q molecule can bind to one Fc region, and binding of two or more heads activates the C1 complex. In plasma, the pentameric IgM molecule has a planar conformation that does not bind C1q (Fig. 10.30, left panel); however, binding to the surface of a pathogen deforms the IgM pentamer so that it looks like a staple (Fig. 10.30, right panel), and this distortion exposes binding sites for the C1q heads. Although C1q binds with low affinity to some subclasses of IgG in solution, the binding energy required for C1q activation is achieved only when a single molecule of C1q can bind two or more IgG molecules that are held within 30–40 nm of each other as a result of binding antigen. This requires that multiple molecules of IgG be bound to a single pathogen or to an antigen in solution. For this reason, IgM is much more efficient than IgG in activating complement. The binding of

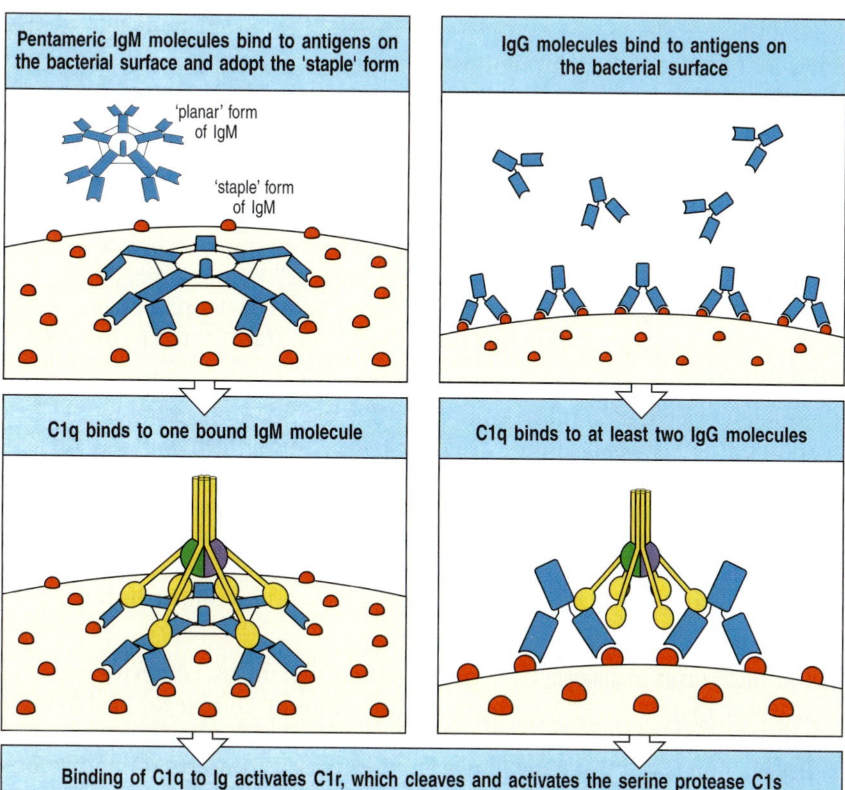
Pentameric IgM molecules bind to antigens on the bacterial surface and adopt the 'staple' form
'planar' form of IgM
'staple' form of IgM
IgG molecules bind to antigens on the bacterial surface
C1q binds to one bound IgM molecule
C1q binds to at least two IgG molecules
Binding of C1q to Ig activates C1r, which cleaves and activates the serine protease C1s

Fig. 10.29 The classical pathway of complement activation is initiated by the binding of C1q to antibody on a pathogen surface. When a molecule of IgM binds several identical epitopes on a pathogen surface, it is bent into the 'staple' conformation, which allows the globular heads of C1q to bind to its Fc pieces (left panels). Multiple molecules of IgG bound on the surface of a pathogen allow the binding of a single molecule of C1q to two or more Fc pieces (right panels). In both cases, the binding of C1q activates the associated C1r, which becomes an active enzyme that cleaves the pro-enzyme C1s, generating a serine protease that initiates the classical complement cascade (see Chapter 2).

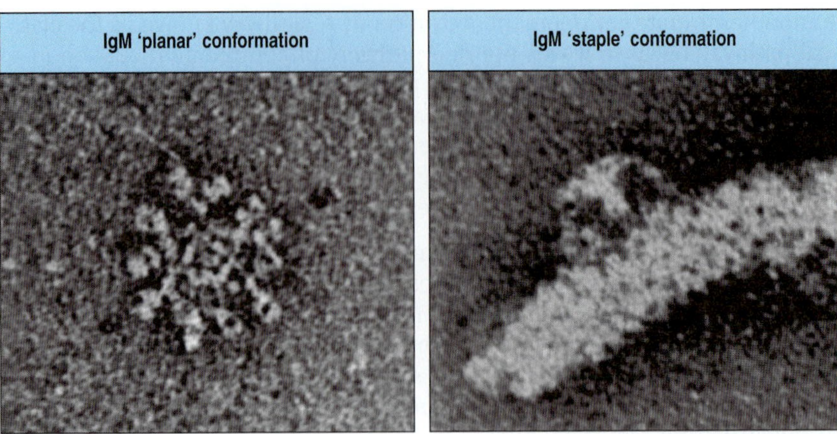

Fig. 10.30 The two conformations of IgM. The left panel shows the planar conformation of soluble IgM; the right panel shows the 'staple' conformation of IgM bound to a bacterial flagellum. Photographs (×760,000) courtesy of K.H. Roux.

C1q to a single bound IgM molecule, or to two or more bound IgG molecules (see Fig. 10.29), leads to activation of the protease activity of C1r, triggering the complement cascade (see Section 2-7 for a full description of the subsequent reactions). This translates antibody binding into the activation of the complement cascade.

10-20 Complement receptors are important in the removal of immune complexes from the circulation.

Many small soluble antigens form antibody:antigen complexes (immune complexes) that contain too few molecules of IgG to be efficiently bound by the Fcγ receptors that we discuss in the next part of the chapter. These antigens include toxins bound by neutralizing antibodies, and debris from dead host cells and microorganisms. Such immune complexes are found after most infections and are removed from the circulation through the action of complement. The soluble immune complexes trigger their own removal by activating complement, again through the binding of C1q, leading to the covalent binding of the activated complement fragments C4b and C3b to the complex, which is then cleared from the circulation by the binding of C4b and C3b to complement receptor 1 (CR1) on the surface of erythrocytes (see Section 2-13 for a description of the different types of complement receptors). The erythrocytes transport the bound complexes of antigen, antibody, and complement to the liver and spleen. Here, macrophages bearing CR1 and Fc receptors remove the complexes from the erythrocyte surface without destroying the cell, and then degrade them (Fig. 10.31). Even larger aggregates of particulate antigen, such as bacteria, viruses, and cell debris, can be coated with complement, picked up by erythrocytes, and transported to the spleen for destruction.

Complement-coated immune complexes that are not removed from the circulation tend to deposit in the basement membranes of small blood vessels, most notably those of the renal glomerulus, where the blood is filtered to form urine. Immune complexes that pass through the basement membrane of the glomerulus bind to CR1 present on the renal podocytes, cells that lie beneath the basement membrane. The functional significance of these receptors in the kidney is unknown; however, they have an important role in the pathology of some autoimmune diseases.

In the autoimmune disease systemic lupus erythematosus, which we describe in Chapter 15, excessive levels of circulating immune complexes lead to their deposition in large amounts on the podocytes, damaging the glomerulus; kidney failure is the principal danger in this disease. Antigen:antibody complexes can also be a cause of pathology in patients with deficiencies in the

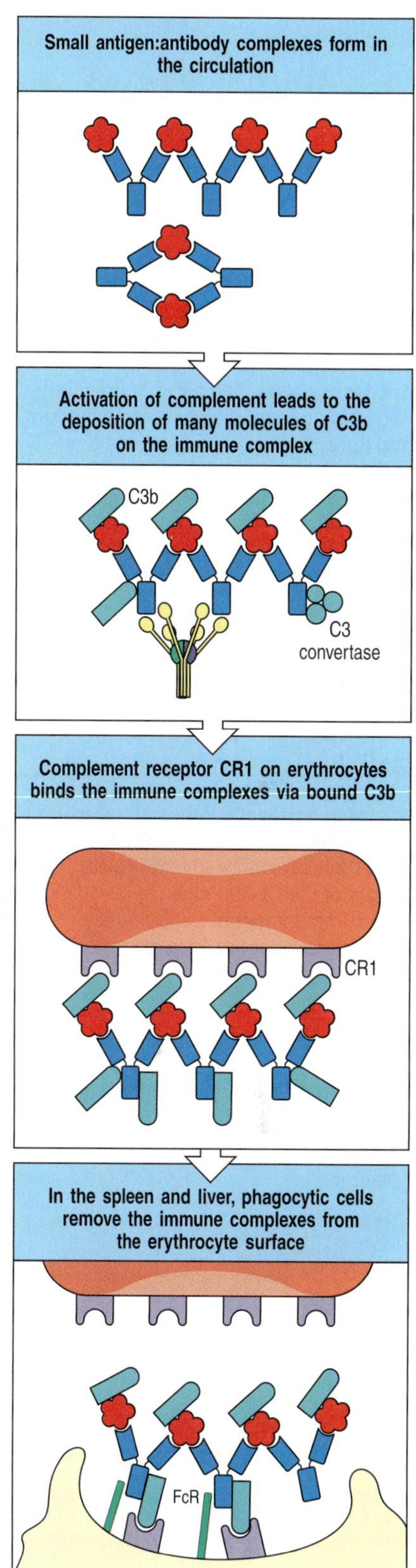

Small antigen:antibody complexes form in the circulation
Activation of complement leads to the deposition of many molecules of C3b on the immune complex
C3b
C3 convertase
Complement receptor CR1 on erythrocytes binds the immune complexes via bound C3b
CR1
In the spleen and liver, phagocytic cells remove the immune complexes from the erythrocyte surface
FcR

Fig. 10.31 Erythrocyte CR1 helps to clear immune complexes from the circulation. CR1 on the erythrocyte surface has an important role in the clearance of immune complexes from the circulation. Immune complexes bind to CR1 on erythrocytes, which transport them to the liver and spleen, where they are removed by macrophages expressing receptors for both Fc and bound complement components.

early components of complement (C1, C2, and C4). The classical complement pathway is not activated properly, and immune complexes are not cleared effectively because they do not become tagged with complement. These patients also suffer tissue damage as a result of immune-complex deposition, especially in the kidneys.

Summary.

The T-cell dependent antibody response begins with IgM secretion but quickly progresses to the production of additional antibody classes. Each class is specialized both in its localization in the body and in the functions it can perform. IgM antibodies are found mainly in blood; they are pentameric in structure. IgM is specialized to activate complement efficiently upon binding antigen and to compensate for the low affinity of a typical IgM antigen-binding site. IgG antibodies are usually of higher affinity and are found in blood and in extracellular fluid, where they can neutralize toxins, viruses, and bacteria, opsonize them for phagocytosis, and activate the complement system. IgA antibodies are synthesized as monomers, which enter blood and extracellular fluids, or as dimeric molecules by plasma cells in the lamina propria of various mucosal tissues. IgA dimers are selectively transported across the epithelial layer into sites such as the lumen of the gut, where they neutralize toxins and viruses and block the entry of bacteria across the intestinal epithelium. Most IgE antibody is bound to the surface of mast cells that reside mainly just below body surfaces; antigen binding to this IgE triggers local defense reactions. Antibodies can defend the body against extracellular pathogens and their toxic products in several ways. The simplest is by direct interactions with pathogens or their products, for example by binding to the active sites of toxins and neutralizing them or by blocking their ability to bind to host cells through specific receptors. When antibodies of the appropriate isotype bind to antigens, they can activate the classical pathway of complement, which leads to the elimination of the pathogen by the various mechanisms described in Chapter 2. Soluble immune complexes of antigen and antibody also fix complement and are cleared from the circulation via complement receptors on red blood cells.

The destruction of antibody-coated pathogens via Fc receptors.

The neutralization of toxins, viruses, or bacteria by high-affinity antibodies can protect against infection but does not, on its own, solve the problem of how to remove the pathogens and their products from the body. Moreover, many pathogens cannot be neutralized by antibody and must be destroyed by other means. Many pathogen-specific antibodies do not bind to neutralizing targets on pathogen surfaces and thus need to be linked to other effector mechanisms to play their part in host defense. We have already seen how the binding of antibody to antigen can activate complement. Another important defense mechanism is the activation of a variety of **accessory effector**

cells bearing receptors called **Fc receptors** because they are specific for the Fc portion of antibodies. These receptors facilitate the phagocytosis of antibody-bound extracellular pathogens by macrophages, dendritic cells, and neutrophils. Other, nonphagocytic, cells of the immune system—NK cells, eosinophils, basophils, and mast cells (see Fig. 1.4)—are triggered to secrete stored mediators when their Fc receptors are engaged by antibody-coated pathogens. These mechanisms maximize the effectiveness of all antibodies regardless of where they bind.

10-21 The Fc receptors of accessory cells are signaling receptors specific for immunoglobulins of different classes.

The Fc receptors are a family of cell-surface molecules that bind the Fc portion of immunoglobulins. Each member of the family recognizes immunoglobulin of one or a few closely related heavy-chain isotypes through a recognition domain on the α chain of the Fc receptor. Most Fc receptors are themselves members of the immunoglobulin gene superfamily. Different cell types bear different sets of Fc receptors, and the isotype of the antibody thus determines which types of cells will be engaged in a given response. The different Fc receptors, the cells that express them, and their specificities for different antibody classes are shown in Fig. 10.32.

Most Fc receptors function as part of a multisubunit complex. Only the α chain is required for antibody recognition; the other chains are required for transport of the receptor to the cell surface and for signal transduction when an Fc region is bound. Some Fcγ receptors, the Fcα receptor I, and the high-affinity receptor for IgE (FcεRI) all use a γ chain for signaling. This chain, which is closely related to the ζ chain of the T-cell receptor complex (see Section 7-7), associates noncovalently with the Fc-binding α chain. The human FcγRII-A is a single-chain receptor in which the cytoplasmic domain of the α chain replaces the function of the γ chain. FcγRII-B1 and FcγRII-B2 are also single-chain receptors but function as inhibitory receptors because

Fig. 10.32 Distinct receptors for the Fc region of the different immunoglobulin classes are expressed on different accessory cells. The subunit structure and binding properties of these receptors and the cell types expressing them are shown. The exact chain composition of any receptor can vary from one cell type to another. For example, FcγRIII in neutrophils is expressed as a molecule with a glycosylphosphatidylinositol membrane anchor, without γ chains, whereas in NK cells it is a transmembrane molecule associated with γ chains. The FcγRII-B1 differs from the FcγRII-B2 by the presence of an additional exon in the intracellular region. This exon prevents the FcγRII-B1 from being internalized after cross-linking. The binding affinities are taken from data on human receptors. *Only some allotypes of FcγRII-A bind IgG2. †In eosinophils, the molecular weight of the CD89α chain is 70–100 kDa. All are immunoglobulin superfamily members except FcεRII, which is a lectin and can form trimers.

Receptor	FcγRI (CD64)	FcγRII-A (CD32)	FcγRII-B2 (CD32)	FcγRII-B1 (CD32)	FcγRIII (CD16)	FcεRI	FcεRII (CD23)	FcαRI (CD89)	Fcα/μR
Structure	α 72 kDa γ	α 40 kDa γ-like domain	ITIM	ITIM	α 50–70 kDa or γ or ζ	α 45 kDa β 33 kDa γ 9 kDa	lectin domain trimer N	α 55–75 kDa γ 9 kDa	α 70 kDa
Binding	IgG1 10^8 M^{-1}	IgG1 2 × 10^6 M^{-1}	IgG1 2 × 10^6 M^{-1}	IgG1 2 × 10^6 M^{-1}	IgG1 5 × 10^5 M^{-1}	IgE 10^{10} M^{-1}	IgE 2–7 × 10^7 M^{-1} (trimer) 2–7 × 10^6 M^{-1} (monomer)	IgA1, IgA2 10^7 M^{-1}	IgA, IgM 3 × 10^9 M^{-1}
Order of affinity	1) IgG1=IgG3 2) IgG4 3) IgG2	1) IgG1 2) IgG3=IgG2* 3) IgG4	1) IgG1=IgG3 2) IgG4 3) IgG2	1) IgG1=IgG3 2) IgG4 3) IgG2	IgG1=IgG3			IgA1=IgA2	1) IgM 2) IgA
Cell type	Macrophages Neutrophils Eosinophils Dendritic cells	Macrophages Neutrophils Eosinophils Platelets Langerhans cells	Macrophages Neutrophils Eosinophils	B cells Mast cells	NK cells Eosinophils Macrophages Neutrophils Mast cells	Mast cells Basophils	Eosinophils B cells	Macrophages Eosinophils† Neutrophils	Macrophages B cells
Effect of ligation	Uptake Stimulation Activation of respiratory burst Induction of killing	Uptake Granule release (eosinophils)	Uptake Inhibition of stimulation	No uptake Inhibition of stimulation	Induction of killing (NK cells)	Secretion of granules	Degranulation	Uptake Induction of killing	Uptake

they contain an ITIM that engages the inositol 5′-phosphatase SHIP (see Section 7-18). The most prominent function of Fc receptors is the activation of accessory cells to attack pathogens, but they also contribute in other ways to immune responses. For example, FcγRII-B receptors negatively regulate the activities of B cells, mast cells, macrophages, and neutrophils by adjusting the threshold at which immune complexes will activate these cells. Fc receptors expressed by dendritic cells enable them to ingest antigen:antibody complexes efficiently and so be able to process these antigens and present their peptides to T cells.

10-22 Fc receptors on phagocytes are activated by antibodies bound to the surface of pathogens and enable the phagocytes to ingest and destroy pathogens.

The most important Fc-bearing cells in humoral immune responses are the phagocytic cells of the monocytic and myelocytic lineages, particularly macrophages and neutrophils. Many bacteria are directly recognized, ingested, and destroyed by phagocytes, and these bacteria are not pathogenic in normal individuals. Bacterial pathogens with polysaccharide capsules resist direct engulfment by phagocytes, however, and such bacteria only become susceptible to phagocytosis when they are coated with antibodies and complement that engage the Fcγ or Fcα receptors and the complement receptor CR1 on phagocytic cells, triggering bacterial uptake (Fig. 10.33). The stimulation of phagocytosis by complement-coated antigens binding to complement receptors is particularly important early in the immune response, before isotype-switched antibodies have been made. Capsular polysaccharides belong to the TI-2 class of thymus-independent antigens (see Section 10-13) and can therefore stimulate the early production of IgM antibodies, which are very effective at activating the complement system. IgM binding to encapsulated bacteria thus triggers the opsonization of these bacteria by complement and their prompt ingestion and destruction by phagocytes bearing complement receptors. Recently, Fcα/μR was discovered as a receptor that binds both IgA and IgM. Fcα/μR is expressed primarily on macrophages and B cells in the lamina propria of the intestine and in germinal centers. It is thought to have a role in the endocytosis of IgM antibody complexed with bacteria such as *Staphylococcus aureus*.

Phagocyte activation can initiate an inflammatory response that causes tissue damage, and so Fc receptors on phagocytes must be able to distinguish antibody molecules bound to a pathogen from the much larger number of free antibody molecules that are not bound to anything. This distinction is made possible by the aggregation of antibodies that occurs when they bind

Fig. 10.33 Fc and complement receptors on phagocytes trigger the uptake and degradation of antibody-coated bacteria. Many bacteria resist phagocytosis by macrophages and neutrophils. Antibodies bound to these bacteria, however, enable them to be ingested and degraded through the interaction of the multiple Fc domains arrayed on the bacterial surface with Fc receptors on the phagocyte surface. Antibody coating also induces activation of the complement system and the binding of complement components to the bacterial surface. These can interact with complement receptors (for example CR1) on the phagocyte. Fc receptors and complement receptors synergize in inducing phagocytosis. Bacteria coated with IgG antibody and complement are therefore more readily ingested than those coated with IgG alone. Binding of Fc and complement receptors signals the phagocyte to increase the rate of phagocytosis, to fuse lysosomes with phagosomes, and to increase its bactericidal activity.

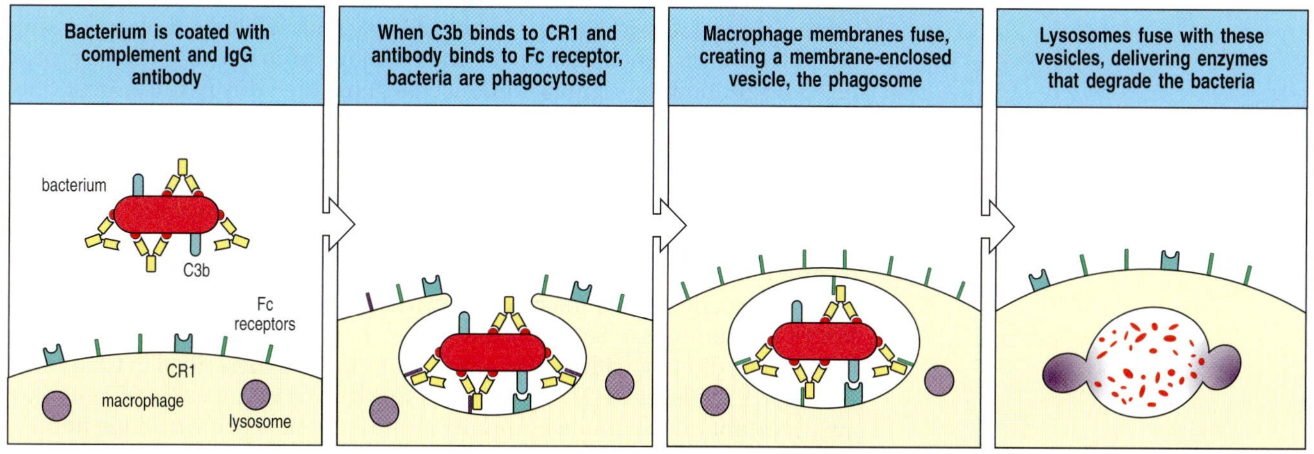
Bacterium is coated with complement and IgG antibody
bacterium
C3b
Fc receptors
CR1
macrophage
lysosome
When C3b binds to CR1 and antibody binds to Fc receptor, bacteria are phagocytosed
Macrophage membranes fuse, creating a membrane-enclosed vesicle, the phagosome
Lysosomes fuse with these vesicles, delivering enzymes that degrade the bacteria

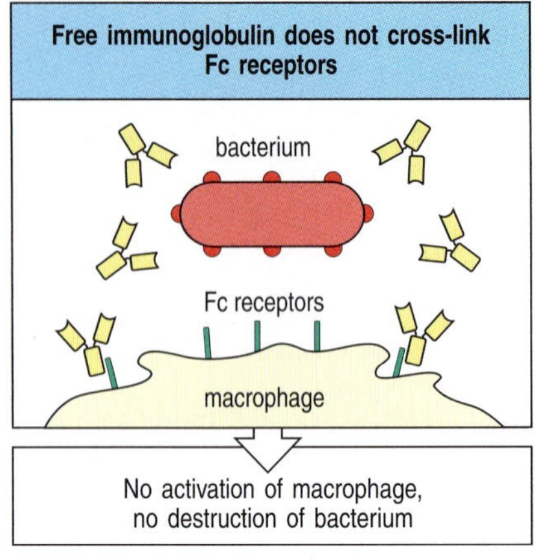
Free immunoglobulin does not cross-link Fc receptors
bacterium
Fc receptors
macrophage
No activation of macrophage, no destruction of bacterium

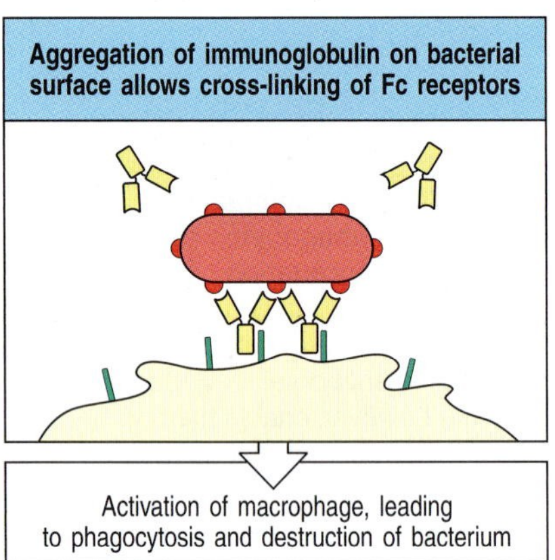
Aggregation of immunoglobulin on bacterial surface allows cross-linking of Fc receptors
Activation of macrophage, leading to phagocytosis and destruction of bacterium

Fig. 10.34 Bound antibody is distinguishable from free immunoglobulin by its state of aggregation. Free immunoglobulin molecules bind most Fc receptors with very low affinity and cannot cross-link Fc receptors. Antigen-bound immunoglobulin, however, binds to Fc receptors with high avidity because several antibody molecules that are bound to the same surface bind to multiple Fc receptors on the surface of the accessory cell. This Fc receptor cross-linking sends a signal to activate the cell bearing it. With Fc receptors that have ITIMs, the result is inhibition.

to multimeric antigens or to multivalent particulate antigens such as viruses and bacteria. Individual Fc receptors on a cell surface bind monomers of free antibody with low affinity, but when presented with an antibody-coated particle, the simultaneous binding by multiple Fc receptors results in binding of high avidity, and this is the principal mechanism by which bound antibodies are distinguished from free immunoglobulin (Fig. 10.34). The result is that Fc receptors enable cells to detect pathogens via the antibody molecules bound to them. Fc receptors therefore give phagocytic cells that lack intrinsic specificity the ability to identify and remove pathogens and their products from the extracellular spaces.

Phagocytosis is greatly enhanced by interactions between the molecules coating an opsonized microorganism and receptors on the phagocyte surface. When an antibody-coated pathogen binds to Fcγ receptors, for example, the cell surface extends around the surface of the particle through successive binding of the Fcγ receptors to the antibody Fc regions bound to the pathogen. This is an active process that is triggered by the stimulation of the Fcγ receptors. Phagocytosis leads to enclosure of the particle in an acidified cytoplasmic vesicle—the phagosome. This then fuses with one or more lysosomes to generate a phagolysosome; lysosomal enzymes are released into the vesicle interior, where they destroy the bacterium (see Fig. 10.33). The process of intracellular killing by phagocytes is described in more detail in Chapter 2.

Some particles are too large for a phagocyte to ingest; parasitic worms are one example. In this case the phagocyte attaches to the surface of the antibody-coated parasite via its Fcγ, Fcα, or Fcε receptors, and the contents of the secretory granules or lysosomes of the phagocyte are released by exocytosis. The contents are discharged directly onto the surface of the parasite and damage it. Thus, stimulation of Fcγ and Fcα receptors can trigger either the internalization of external particles by phagocytosis or the externalization of internal vesicles by exocytosis. The principal leukocytes involved in the destruction of bacteria are macrophages and neutrophils, but large parasites such as helminths are more usually attacked by eosinophils (Fig. 10.35), nonphagocytic cells that can bind antibody-coated parasites via several different Fc receptors, including the low-affinity Fcε receptor for IgE, CD23 (Fig. 10.32). Cross-linking of these receptors by antibody-coated surfaces activates the eosinophil to release its granule contents, which include proteins toxic to parasites (see Fig. 14.13). Cross-linking by antigen of IgE bound to the high-affinity FcεRI on mast cells and basophils also results in exocytosis of their granule contents, as we describe later in this chapter.

10-23 Fc receptors activate NK cells to destroy antibody-coated targets.

Virus-infected cells are usually destroyed by T cells that recognize virus-derived peptides bound to cell-surface MHC molecules. Cells infected by some viruses also signal the presence of intracellular infection by expressing on their surfaces proteins, such as viral envelope proteins, that can be recognized by antibodies originally produced against the virus particle. Host cells with antibodies bound to them can be killed by a specialized non-T, non-B cell of the lymphoid lineage called a natural killer cell (NK cell), which we met in Chapter 3. NK cells are large cells with prominent intracellular granules and make up a small fraction of peripheral blood lymphocytes. Although belonging to the lymphoid linage, NK cells express a limited repertoire of invariant receptors recognizing a range of ligands that are induced on abnormal cells, such as those infected with viruses; they are considered as part of innate immunity (see Section 3-21). On recognition of a ligand, the NK cell kills the cell directly, without the need for antibody, as described in Chapter 3. Although first discovered for their ability to kill some tumor cells, NK cells play an important role in innate immunity in the early stages of virus infection.

As well as this innate function, NK cells can recognize and destroy antibody-coated target cells in a process called **antibody-dependent cell-mediated cytotoxicity** (**ADCC**). This is triggered when antibody bound to the surface of a cell interacts with Fc receptors on the NK cell (Fig. 10.36). NK cells express the receptor FcγRIII (CD16), which recognizes the IgG1 and IgG3 subclasses. The killing mechanism is analogous to that of cytotoxic T cells, involving the release of cytoplasmic granules containing perforin and granzymes (see Section 9-27). ADCC has been shown to have a role in the defense against infection by viruses, and represents another mechanism by which antibodies can direct an antigen-specific attack by an effector cell that itself lacks specificity for antigen.

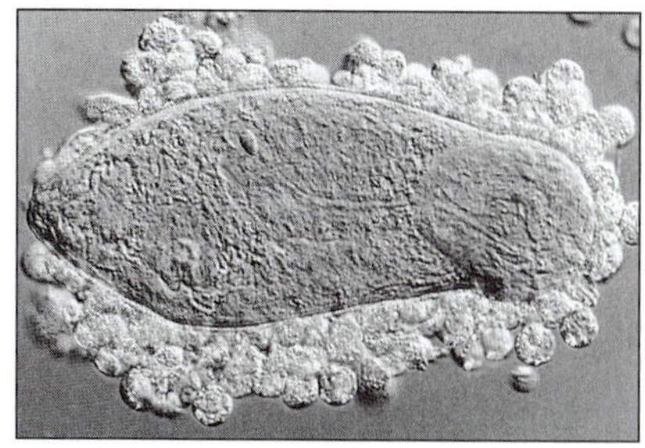

Fig. 10.35 Eosinophils attacking a schistosome larva in the presence of serum from an infected patient. Large parasites, such as worms, cannot be ingested by phagocytes; however, when the worm is coated with antibody, eosinophils can attack it through binding via their Fc receptors for IgG and IgA. Similar attacks can be mounted by other Fc receptor-bearing cells on large targets. These cells release the toxic contents of their granules directly onto the target, a process known as exocytosis. Photograph courtesy of A. Butterworth.

10-24 Mast cells and basophils bind IgE antibody via the high-affinity Fcε receptor.

When pathogens cross epithelial barriers and establish a local focus of infection, the host must mobilize its defenses and direct them to the site of pathogen growth. One way in which this is achieved is to activate the cells known as **mast cells**. Mast cells are large cells containing distinctive cytoplasmic granules that contain a mixture of chemical mediators, including histamine, that act rapidly to make local blood vessels more permeable. Mast cells have a distinctive appearance after staining with the dye toluidine blue that makes them readily identifiable in tissues (see Fig. 1.4). They are found in particularly high concentrations in vascularized connective tissues just beneath epithelial surfaces, including the submucosal tissues of the gastrointestinal and respiratory tracts and the dermis of the skin.

Mast cells have Fc receptors specific for IgE (FcεRI) and IgG (FcγRIII) and can be activated to release their granules, and to secrete lipid inflammatory mediators and cytokines, via antibody bound to these receptors. Most Fc receptors bind stably to the Fc regions of antibodies only when the antibodies have themselves bound antigen, and cross-linking of multiple Fc receptors is needed for strong binding. In contrast, FcεRI binds IgE antibody monomers with a very high affinity—approximately 10^{10} M^{-1}. Thus, even at the low levels of circulating IgE present in normal individuals, a substantial portion of the total IgE is bound to the FcεRI on mast cells in tissues and on circulating basophils.

Although mast cells are usually stably associated with bound IgE, this on its own does not activate them, nor will the binding of monomeric antigen to the IgE. Mast-cell activation occurs only when the bound IgE is cross-linked by multivalent antigens. This signal activates the mast cell to release the

Fig. 10.36 Antibody-coated target cells can be killed by NK cells in antibody-dependent cell-mediated cytotoxicity (ADCC). NK cells (see Chapter 3) are large granular non-T, non-B lymphoid cells that have FcγRIII (CD16) on their surface. When these cells encounter cells coated with IgG antibody, they rapidly kill the target cell. ADCC is only one way in which NK cells can contribute to host defense.

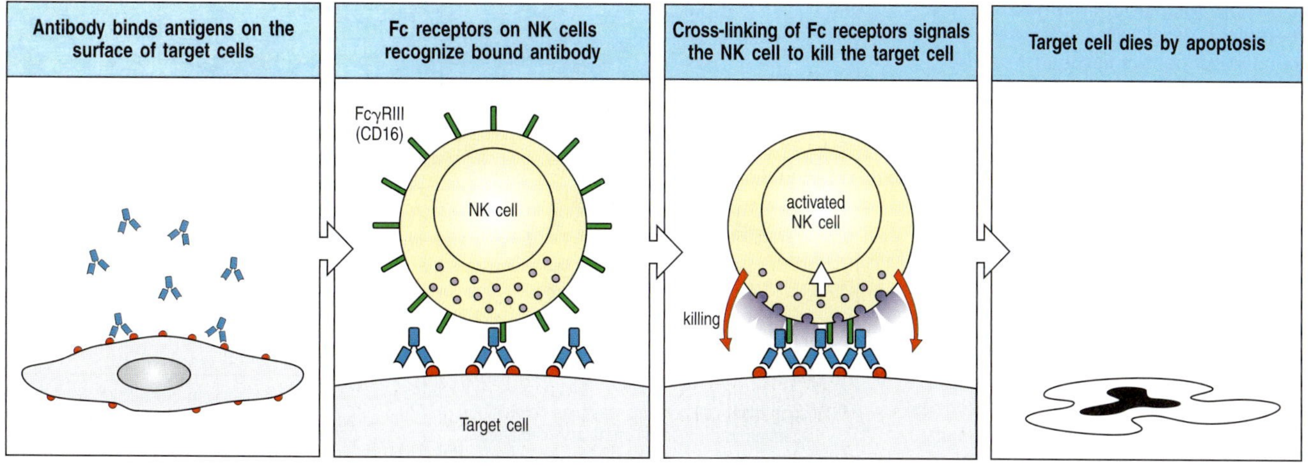
Antibody binds antigens on the surface of target cells
Fc receptors on NK cells recognize bound antibody
FcγRIII (CD16)
NK cell
Target cell
Cross-linking of Fc receptors signals the NK cell to kill the target cell
activated NK cell
killing
Target cell dies by apoptosis

contents of its granules, which occurs in seconds (Fig. 10.37), to synthesize and release lipid mediators such as prostaglandin D_2 and leukotriene C4, and to secrete cytokines such as TNF-α, thereby initiating a local inflammatory response. Degranulation releases stored histamine, which increases local blood flow and vascular permeability, which quickly leads to an accumulation of fluid and blood proteins, including antibodies, in the surrounding tissue. Shortly afterward there is an influx of blood-borne cells such as neutrophils and, later, macrophages, eosinophils, and effector lymphocytes. This influx can last from a few minutes to a few hours and produces an inflammatory response at the site of infection. Thus, mast cells are part of the front-line host defenses against pathogens that enter the body across epithelial barriers. They are also of additional medical importance because of their involvement in IgE-mediated allergic responses, which are discussed in Chapter 14. In allergic responses, mast cells are activated in the way described above by exposure to normally innocuous antigens (allergens) such as pollen, to which the individual has previously mounted a sensitizing immune response that produces allergen-specific IgE.

10-25 IgE-mediated activation of accessory cells has an important role in resistance to parasite infection.

Mast cells are thought to serve at least three important functions in host defense. First, their location near body surfaces allows them to recruit both pathogen-specific elements, such as antigen-specific lymphocytes, and nonspecific effector elements, such as neutrophils, macrophages, basophils, and eosinophils, to sites where infectious agents are most likely to enter the

Resting mast cell

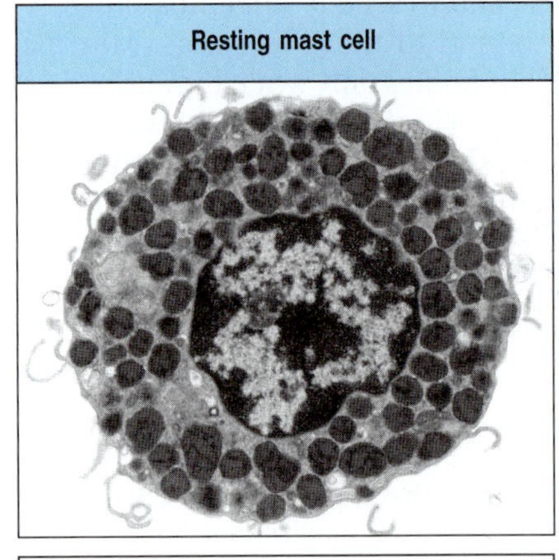

Activated mast cell

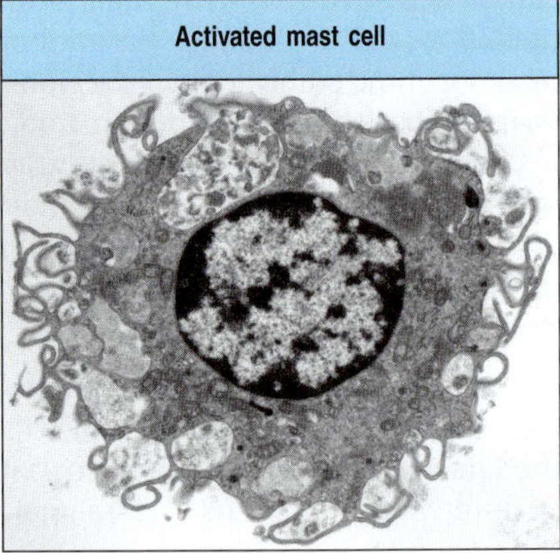

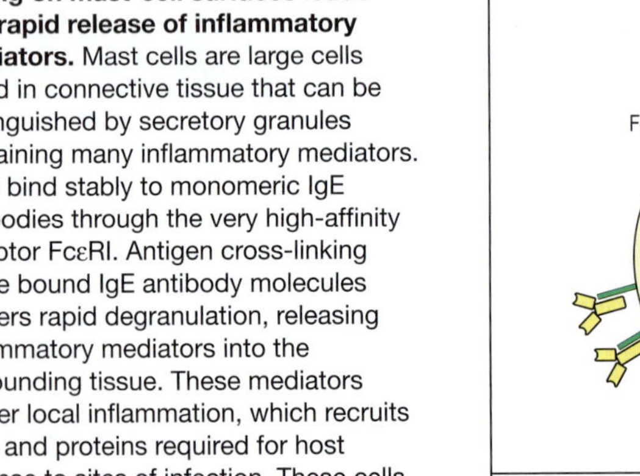

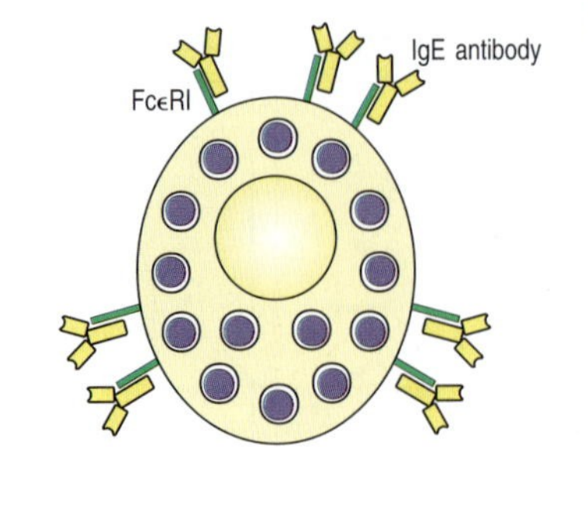

Resting mast cell contains granules containing histamine and other inflammatory mediators

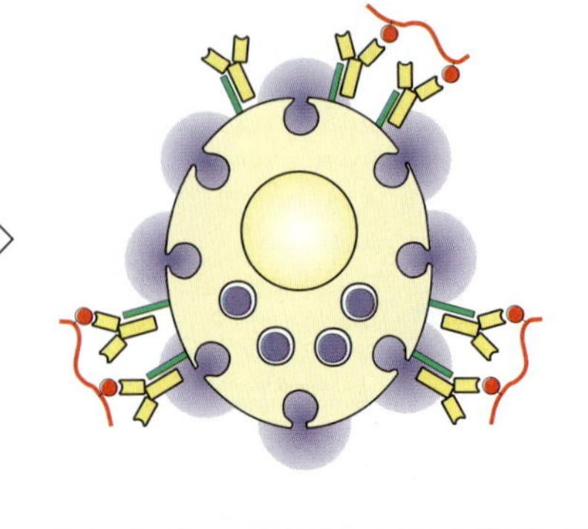

Multivalent antigen cross-links bound IgE antibody, causing release of granule contents

Fig. 10.37 IgE antibody cross-linking on mast-cell surfaces leads to a rapid release of inflammatory mediators. Mast cells are large cells found in connective tissue that can be distinguished by secretory granules containing many inflammatory mediators. They bind stably to monomeric IgE antibodies through the very high-affinity receptor FcεRI. Antigen cross-linking of the bound IgE antibody molecules triggers rapid degranulation, releasing inflammatory mediators into the surrounding tissue. These mediators trigger local inflammation, which recruits cells and proteins required for host defense to sites of infection. These cells are also triggered during allergic reactions when allergens bind to IgE on mast cells. Photographs courtesy of A.M. Dvorak.

internal milieu. Second, the inflammation they cause increases the flow of lymph from sites of antigen deposition to the regional lymph nodes, where naive lymphocytes are first activated. Third, the ability of mast-cell products to trigger muscular contraction can contribute to the physical expulsion of pathogens from the lungs or the gut. Mast cells respond rapidly to the binding of antigen to surface-bound IgE antibodies, and their activation leads to the initiation of an inflammatory response and the recruitment and activation of basophils and eosinophils, which contribute further to the inflammatory response (see Chapter 14). There is increasing evidence that such IgE-mediated responses are crucial to defense against parasite infestation.

A role for mast cells in the clearance of parasites is suggested by the accumulation of mast cells in the intestine, known as **mastocytosis**, that accompanies helminth infection, and by observations in *W*/*W*^{V} mutant mice, which have a profound mast-cell deficiency caused by mutation of the gene c-*kit*. These mutant mice show impaired clearance of the intestinal nematodes *Trichinella spiralis* and *Strongyloides* species. Clearance of *Strongyloides* is even more impaired in *W*/*W*^{V} mice that lack IL-3 and so also fail to produce basophils. Thus, both mast cells and basophils seem to contribute to defense against these helminth parasites.

Other evidence points to the importance of IgE antibodies and eosinophils in defense against parasites. Infection with certain types of multicellular parasites, particularly helminths, is strongly associated with the production of IgE antibodies and the presence of abnormally large numbers of eosinophils (eosinophilia) in blood and tissues. Furthermore, experiments in mice show that depletion of eosinophils by polyclonal anti-eosinophil antisera increases the severity of infection with the parasitic helminth *Schistosoma mansoni*. Eosinophils seem to be directly responsible for helminth destruction; examination of infected tissues shows degranulated eosinophils adhering to helminths, and experiments *in vitro* have shown that eosinophils can kill *S. mansoni* in the presence of anti-schistosome IgG or IgA antibodies (see Fig. 10.35).

The role of IgE, mast cells, basophils, and eosinophils can also be seen in resistance to the feeding of blood-sucking ixodid ticks. Skin at the site of a tick bite has degranulated mast cells and an accumulation of degranulated basophils and eosinophils, an indicator of recent activation. Subsequent resistance to feeding by these ticks develops after the first exposure, suggesting a specific immunological mechanism. Mice deficient in mast cells show no such acquired resistance to ticks, and in guinea pigs the depletion of either basophils or eosinophils by specific polyclonal antibodies also reduces resistance to tick feeding. Finally, experiments in mice showed that resistance to ticks is mediated by specific IgE antibody. Thus, many clinical studies and experiments support a role for this system of IgE bound to the high-affinity FcεRI in host resistance to pathogens that enter across epithelia or exoparasites such as ticks that breach it.

Summary.

Antibody-coated pathogens are recognized by effector cells through Fc receptors that bind to an array of constant regions (Fc portions) provided by the pathogen-bound antibodies. Binding activates the cell and triggers destruction of the pathogen, through either phagocytosis or granule release, or through both. Fc receptors comprise a family of proteins, each of which recognizes immunoglobulins of particular isotypes. Fc receptors on macrophages and neutrophils recognize the constant regions of IgG or IgA antibodies bound to a pathogen and trigger the engulfment and destruction of bacteria coated with IgG or IgA. Binding to the Fc receptor also induces the production of microbicidal agents in the intracellular vesicles of the phagocyte. Eosinophils are important in the elimination of parasites too large to be engulfed: they bear Fc receptors specific for the constant region of IgG, as well as receptors for IgE; aggregation of these receptors triggers the release of toxic substances onto the surface of the parasite. NK cells, tissue mast cells, and blood basophils also release their granule contents when their Fc receptors are engaged. The high-affinity receptor for IgE is expressed constitutively by mast cells and basophils. It differs from other Fc receptors in that it can bind free monomeric antibody, thus enabling an immediate response to pathogens at their site of first entry into the tissues. When IgE bound to the surface of a mast cell is aggregated by binding to antigen, it triggers the release of histamine and many other mediators that increase the blood flow to sites of infection; it thereby recruits antibodies and effector cells to these sites. Mast cells are found principally below epithelial surfaces of the skin and beneath the basement membrane of the digestive and respiratory tracts. Their activation by innocuous substances is responsible for many of the symptoms of acute allergic reactions, as will be described in Chapter 14.

Summary to Chapter 10.

The humoral immune response to infection involves the production of antibody by plasma cells derived from B lymphocytes, the binding of this antibody to the pathogen, and the elimination of the pathogen by phagocytic cells and molecules of the humoral immune system. The production of antibody usually requires the action of helper T cells specific for a peptide fragment of the antigen recognized by the B cell, a phenomenon called linked recognition. An activated B cell first moves to the T-zone–B-zone boundary in secondary lymphoid tissues, where it may encounter its cognate T cell and begin to proliferate. Some B cells become plasmablasts, and others will move to the germinal center, where somatic hypermutation and class switch recombination take place. B cells generated there that bind antigen most avidly are selected for survival and further differentiation, leading to affinity maturation of the antibody response. Cytokines made by helper T cells direct class switching, leading to the production of antibody of various classes that can be distributed to various body compartments.

IgM antibodies are produced early in an infection by conventional, or B-2, cells, and is also made in the absence of infection by subsets of nonconventional B cells in particular locations (natural antibodies). IgM has a major role in protecting against infection in the bloodstream, whereas isotypes secreted later in an adaptive immune response, such as IgG, diffuse into the tissues. Antigens with highly repeating antigenic determinants and that contain mitogens—called TI antigens—can elicit IgM and some IgG independently of T-cell help, and this provides an early protective immune response. Multimeric IgA is produced in the lamina propria and is transported across epithelial surfaces, whereas IgE is made in small amounts and binds avidly to receptors on the surface of mast cells.

Antibodies that bind with high affinity to critical sites on toxins, viruses, and bacteria can neutralize them. However, pathogens and their products are destroyed and removed from the body largely through uptake into phagocytes and degradation inside these cells. Antibodies that coat pathogens bind to Fc receptors on phagocytes, which are thereby triggered to engulf and destroy the pathogen. Binding of antibody C regions to Fc receptors on other cells leads to the exocytosis of stored mediators; this is particularly important in parasite infections, in which Fcε-expressing mast cells are triggered by the binding of antigen to IgE antibody to release inflammatory mediators directly onto parasite surfaces. Antibodies can also initiate the destruction of pathogens by activating the complement system. Complement components can opsonize pathogens for uptake by phagocytes, recruit phagocytes to sites of infection, and directly destroy pathogens by creating pores in their cell membrane. Receptors for complement components and Fc receptors often synergize in activating the uptake and destruction of pathogens and immune complexes. Thus, the humoral immune response is targeted to the infecting pathogen through the production of specific antibody; however, the effector actions of that antibody are determined by its heavy-chain isotype, which determines its class, and are the same for all pathogens bound by antibody of a particular class.

Questions.

10.1 *Describe the requirements for the activation of naive B cells by a thymus-dependent antigen. By what mechanisms do T cells provide help to B cells in the humoral response?*

10.2 *What is meant by the term linked recognition? What are the advantages of this process for immune tolerance? What are the advantages for specificity?*

10.3 *Compare and contrast the properties and functions of antibodies of the IgM and IgG classes.*

10.4 *In the initiation of a thymus-dependent antibody response, B cells and T cells change their locations. What determined the location of T and B cells at different stages of their activation?*

10.5 *In the germinal center reaction, B cells (centroblasts and centrocytes) cycle between the light and dark zones. What factors regulate this movement and localization?*

10.6 *Describe the process responsible for the phenomenon of affinity maturation of the antibody response. Where does affinity maturation mainly take place?*

10.7 *Assuming a rate of one mutation in 10^3 base pairs within the V regions per cell division and that about three out of four changes will result in an altered amino acid, explain how the estimate was arrived at of a 50% chance of mutation in the B cell's antigen receptor per division during the germinal center reaction.*

10.8 *Which of the antibody classes mainly activates mast cells? How does it do so and what are the results? Which type of pathogen is this class of antibody mainly directed against? What unwanted reaction is this antibody also responsible for?*

10.9 *How do antibodies interact with the complement system to rid the body of pathogens?*

10.10 *Which classes of maternal antibodies would you expect to find in a breast-fed newborn infant and how have they got there?*

General references.

Batista, F.D., and Harwood, N.E.: **The who, how and where of antigen presentation to B cells.** *Nat. Rev. Immunol.* 2009, **9**:15–27.

Nimmerjahn, F., and Ravetch, J.V.: **Fcγ receptors as regulators of immune responses.** *Nat. Rev. Immunol.* 2008, **8**:34–47.

Rajewsky, K.: **Clonal selection and learning in the antibody system.** *Nature* 1996, **381**:751–758.

Section references.

10-1 The humoral immune response is initiated when B cells that bind antigen are signaled by helper T cells or by certain microbial antigens alone.

Gulbranson-Judge, A., and MacLennan, I.: **Sequential antigen-specific growth of T cells in the T zones and follicles in response to pigeon cytochrome c.** *Eur. J. Immunol.* 1996, **26**:1830–1837.

10-2 B-cell responses are enhanced by co-ligation of the B-cell receptor and B-cell co-receptor by antigen and complement fragments on microbial surfaces.

Barrington, R.A., Zhang, M., Zhong, X., Jonsson, H., Holodick, N., Cherukuri, A., Pierce, S.K., Rothstein, T.L., and Carroll, M.C.: **CD21/CD19 coreceptor signaling promotes B cell survival during primary immune responses.** *J. Immunol.* 2005, **175**:2859–2867.

Fearon, D.T., and Carroll, M.C.: **Regulation of B lymphocyte responses to foreign and self-antigens by the CD19/CD21 complex.** *Annu. Rev. Immunol.* 2000, **18**:393–422.

O'Rourke, L., Tooze, R., and Fearon, D.T.: **Co-receptors of B lymphocytes.** *Curr. Opin. Immunol.* 1997, **9**:324–329.

Rickert, R.C.: **Regulation of B lymphocyte activation by complement C3 and the B cell coreceptor complex.** *Curr. Opin. Immunol.* 2005, **17**:237–243.

10-3 Helper T cells activate B cells that recognize the same antigen.

Eskola, J., Peltola, H., Takala, A.K., Kayhty, H., Hakulinen, M., Karanko, V., Kela, E., Rekola, P., Ronnberg, P.R., Samuelson, J.S., *et al.*: **Efficacy of *Haemophilus influenzae* type b polysaccharide-diphtheria toxoid conjugate vaccine in infancy.** *N. Engl. J. Med.* 1987, **317**:717–722.

MacLennan, I.C.M., Gulbranson-Judge, A., Toellner, K.M., Casamayor-Palleja, M., Chan, E., Sze, D.M.Y., Luther, S.A., and Orbea, H.A.: **The changing preference of T and B cells for partners as T-dependent antibody responses develop.** *Immunol. Rev.* 1997, **156**:53–66.

McHeyzer-Williams, L.J., Malherbe, L.P., and McHeyzer-Williams, M.G.: **Helper T cell-regulated B cell immunity.** *Curr. Top. Microbiol. Immunol.* 2006, **311**:59–83.

Parker, D.C.: **T cell-dependent B-cell activation.** *Annu. Rev. Immunol.* 1993, **11**:331–340.

10-4 T cells make membrane-bound and secreted molecules that activate B cells.

Gaspal, F.M., Kim, M.Y., McConnell, F.M., Raykundalia, C., Bekiaris, V., and Lane, P.J.: **Mice deficient in OX40 and CD30 signals lack memory antibody responses because of deficient CD4 T cell memory.** *J. Immunol.* 2005, **174**:3891–3896.

Jaiswal, A.I., and Croft, M.: **CD40 ligand induction on T cell subsets by peptide-presenting B cells.** *J. Immunol.* 1997, **159**:2282–2291.

Kalled, S.L.: **Impact of the BAFF/BR3 axis on B cell survival, germinal center maintenance and antibody production.** *Semin. Immunol.* 2006, **18**:290–296.

Mackay, F., and Browning, J.L.: **BAFF: a fundamental survival factor for B cells.** *Nat. Rev. Immunol.* 2002, **2**:465–475.

Shanebeck, K.D., Maliszewski, C.R., Kennedy, M.K., Picha, K.S., Smith, C.A., Goodwin, R.G., and Grabstein, K.H.: **Regulation of murine B cell growth and differentiation by CD30 ligand.** *Eur. J. Immunol.* 1995, **25**:2147–2153.

Yoshinaga, S.K., Whoriskey, J.S., Khare, S.D., Sarmiento, U., Guo, J., Horan, T., Shih, G., Zhang, M., Coccia, M.A., Kohno, T. *et al.*: **T-cell co-stimulation through B7RP-1 and ICOS.** *Nature* 1999, **402**:827–832.

10-5 B cells that encounter their antigens migrate toward the boundaries between B-cell and T-cell areas in secondary lymphoid tissues.

Cahalan, M.D., and Parker, I.: **Close encounters of the first and second kind: T-DC and T-B interactions in the lymph node.** *Semin. Immunol.* 2005, **17**:442–451.

Fang, Y., Xu, C., Fu, Y.X., Holers, V.M., and Molina, H.: **Expression of complement receptors 1 and 2 on follicular dendritic cells is necessary for the generation of a strong antigen-specific IgG response.** *J. Immunol.* 1998, **160**:5273–5279.

Garside, P., Ingulli, E., Merica, R.R., Johnson, J.G., Noelle, R.J., and Jenkins, M.K.: **Visualization of specific B and T lymphocyte interactions in the lymph node.** *Science* 1998, **281**:96–99.

Okada, T., and Cyster, J.G.: **B cell migration and interactions in the early phase of antibody responses.** *Curr. Opin. Immunol.* 2006, **18**:278–285.

Pape, K.A., Kouskoff, V., Nemazee, D., Tang, H.L., Cyster, J.G., Tze, L.E., Hippen, K.L., Behrens, T.W., and Jenkins, M.K.: **Visualization of the genesis and fate of isotype-switched B cells during a primary immune response.** *J. Exp. Med.* 2003, **197**:1677–1687.

Phan, T.G., Gray, E.E., and Cyster, J.G.: **The microanatomy of B cell activation.** *Curr. Opin. Immunol.* 2009, **21**:258–265.

10-6 Antibody-secreting plasma cells differentiate from activated B cells.

Moser, K., Tokoyoda, K., Radbruch, A., MacLennan, I., and Manz, R.A.: **Stromal niches, plasma cell differentiation and survival.** *Curr. Opin. Immunol.* 2006, **18**:265–270.

Radbruch, A., Muehlinghaus, G., Luger, E.O., Inamine, A., Smith, K.G., Dorner, T., and Hiepe, F.: **Competence and competition: the challenge of becoming a long-lived plasma cell.** *Nat. Rev. Immunol.* 2006, **6**:741–750.

Sciammas, R., and Davis, M.M.: **Blimp-1; immunoglobulin secretion and the switch to plasma cells.** *Curr. Top. Microbiol. Immunol.* 2005, **290**:201–224.

Shapiro-Shelef, M, and Calame, K.: **Regulation of plasma-cell development.** *Nat. Rev. Immunol.* 2005, **5**:230–242.

10-7 The second phase of a primary B-cell immune response occurs when activated B cells migrate into follicles and proliferate to form germinal centers.

Allen, C.D., Okada, T., and Cyster, J.G.: **Germinal-center organization and cellular dynamics.** *Immunity* 2007, **27**:190–202.

Cozine, C.L., Wolniak, K.L., and Waldschmidt, T.J.: **The primary germinal center response in mice.** *Curr. Opin. Immunol.* 2005, **17**:298–302.

Jacob, J., Przylepa, J., Miller, C., and Kelsoe, G.: **In situ studies of the primary immune response to (4-hydroxy-3-nitrophenyl)acetyl. III. The kinetics of V region mutation and selection in germinal center B cells.** *J. Exp. Med.* 1993, **178**:1293–1307.

Kelsoe, G.: **The germinal center: a crucible for lymphocyte selection.** *Semin. Immunol.* 1996, **8**:179–184.

Kunkel, E.J., and Butcher, E.C.: **Plasma-cell homing.** *Nat. Rev. Immunol.* 2003, **3**:822–829.

MacLennan, I.C.: **Germinal centers still hold secrets.** *Immunity* 2005, **22**:656–657.

10-8 Germinal center B cells undergo V-region somatic hypermutation, and cells with mutations that improve affinity for antigen are selected.

Anderson, S.M., Khalil, A., Uduman, M., Hershberg, U., Louzoun, Y., Haberman, A.M., Kleinstein, S.H., and Shlomchik, M.J.: **Taking advantage: high-affinity B cells in the germinal center have lower death rates, but similar rates of division, compared to low-affinity cells.** *J. Immunol.* 2009, **183**:7314–7325.

Jacob, J., Kelsoe, G., Rajewsky, K., and Weiss, U.: **Intraclonal generation of antibody mutants in germinal centres.** *Nature* 1991, **354**:389–392.

Li, Z., Woo, C.J., Iglesias-Ussel, M.D., Ronai, D., and Scharff, M.D.: **The generation of antibody diversity through somatic hypermutation and class switch recombination.** *Genes Dev.* 2004, **18**:1–11.

Odegard, V.H., and Schatz, D.G.: **Targeting of somatic hypermutation.** *Nat. Rev. Immunol.* 2006, **6**:573–583.

Pereira, J.P., Kelly, L.M., and Cyster, J.G.: **Finding the right niche: B-cell migration in the early phases of T-dependent antibody responses.** *Int. Immunol.* 2010, **22**:413–419.

10-9 Class switching in thymus-dependent antibody responses requires expression of CD40 ligand by helper T cells and is directed by cytokines.

Francke, U., and Ochs, H.D.: **The CD40 ligand, gp39, is defective in activated T cells from patients with X-linked hyper-IgM syndrome.** *Cell* 1993, **72**:291–300.

Jumper, M., Splawski, J., Lipsky, P., and Meek, K.: **Ligation of CD40 induces sterile transcripts of multiple Ig H chain isotypes in human B cells.** *J. Immunol.* 1994, **152**:438–445.

Litinskiy, M.B., Nardelli, B., Hilbert, D.M., He, B., Schaffer, A., Casali, P., and Cerutti, A.: **DCs induce CD40-independent immunoglobulin class switching through BLyS and APRIL.** *Nat. Immunol.* 2002, **3**:822–829.

MacLennan, I.C., Toellner, K.M., Cunningham, A.F., Serre, K., Sze, D.M., Zuniga, E., Cook, M.C., and Vinuesa, C.G.: **Extrafollicular antibody responses.** *Immunol. Rev.* 2003, **194**:8–18.

Snapper, C.M., Kehry, M.R., Castle, B.E., and Mond, J.J.: **Multivalent, but not divalent, antigen receptor cross-linkers synergize with CD40 ligand for induction of Ig synthesis and class switching in normal murine B cells.** *J. Immunol.* 1995, **154**:1177–1187.

Stavnezer, J.: **Immunoglobulin class switching.** *Curr. Opin. Immunol.* 1996, **8**:199–205.

10-10 Ligation of CD40 and prolonged contact with T follicular helper cells is required to sustain germinal center B cells.

Banchereau, J., de Paoli, P., Vallé, A., Garcia, E., and Rousset, F.: **Long-term human B cell lines dependent on interleukin-4 and antibody to CD40.** *Science* 1991, **251**:70–72.

Cannons, J.L., Qi, H., Lu, K.T., Dutta, M., Gomez-Rodriguez, J., Cheng, J., Wakeland, E.K., Germain, R.N., and Schwartzberg, P.L.: **Optimal germinal center responses require a multistage T cell:B cell adhesion process involving integrins, SLAM-associated protein, and CD84.** *Immunity* 2010, **32**:253–265.

Hannum, L.G., Haberman, A.M., Anderson, S.M., and Shlomchik, M.J.: **Germinal center initiation, variable gene region hypermutation, and mutant B cell selection without detectable immune complexes on follicular dendritic cells.** *J. Exp. Med.* 2000, **192**:931–942.

Liu, Y.J., Joshua, D.E., Williams, G.T., Smith, C.A., Gordon, J., and MacLennan, I.C.M.: **Mechanism of antigen-driven selection in germinal centres.** *Nature* 1989, **342**:929–931.

Wang, Z., Karras, J.G., Howard, R.G., and Rothstein, T.L.: **Induction of bcl-x by CD40 engagement rescues sIg-induced apoptosis in murine B cells.** *J. Immunol.* 1995, **155**:3722–3725.

10-11 Surviving germinal center B cells differentiate into either plasma cells or memory cells.

Hu, C.C., Dougan, S.K., McGehee, A.M., Love, J.C., and Ploegh, H.L.: **XBP-1 regulates signal transduction, transcription factors and bone marrow colonization in B cells.** *EMBO J.* 2009, **28**:1624–1636.

Nera, K.P., and Lassila, O.: **Pax5—a critical inhibitor of plasma cell fate.** *Scand. J. Immunol.* 2006, **64**:190–199.

Omori, S.A., Cato, M.H., Anzelon-Mills, A., Puri, K.D., Shapiro-Shelef, M., Calame, K., and Rickert, R.C.: **Regulation of class-switch recombination and plasma cell differentiation by phosphatidylinositol 3-kinase signaling.** *Immunity* 2006, **25**:545–557.

Radbruch, A., Muehlinghaus, G., Luger, E.O., Inamine, A., Smith, K.G., Dorner, T., and Hiepe, F.: **Competence and competition: the challenge of becoming a long-lived plasma cell.** *Nat. Rev. Immunol.* 2006, **6**:741–750.

Schebesta, M., Heavey, B., and Busslinger, M.: **Transcriptional control of B-cell development.** *Curr. Opin. Immunol.* 2002, **14**:216–223.

10-12 Some bacterial antigens do not require T-cell help to induce B-cell responses.

Anderson, J., Coutinho, A., Lernhardt, W., and Melchers, F.: **Clonal growth and maturation to immunoglobulin secretion in vitro of every growth-inducible B lymphocyte.** *Cell* 1977, **10**:27–34.

Bekeredjian-Ding, I., and Jego, G.: **Toll-like receptors—sentries in the B-cell response.** *Immunology* 2009, **128**:311–323.

Garcia De Vinuesa, C., Gulbranson-Judge, A., Khan, M., O'Leary, P., Cascalho, M., Wabl, M., Klaus, G.G., Owen, M.J., and MacLennan, I.C.: **Dendritic cells associated with plasmablast survival.** *Eur. J. Immunol.* 1999, **29**:3712–3721.

Ruprecht, C.R., and Lanzavecchia, A.: **Toll-like receptor stimulation as a third signal required for activation of human naive B cells.** *Eur. J. Immunol.* 2006, **36**:810–816.

10-13 B-cell responses to bacterial polysaccharides do not require peptide-specific T-cell help.

Balazs, M., Martin, F., Zhou, T., and Kearney, J.: **Blood dendritic cells interact with splenic marginal zone B cells to initiate T-independent immune responses.** *Immunity* 2002, **17**:341–352.

Craxton, A., Magaletti, D., Ryan, E.J., and Clark, E.A.: **Macrophage- and dendritic cell-dependent regulation of human B-cell proliferation requires the TNF family ligand BAFF.** *Blood* 2003, **101**:4464–4471.

Fagarasan, S., and Honjo, T.: **T-independent immune response: new aspects of B cell biology.** *Science* 2000, **290**:89–92.

MacLennan, I., and Vinuesa, C.: **Dendritic cells, BAFF, and APRIL: innate players in adaptive antibody responses.** *Immunity* 2002, **17**:341–352.

Mond, J.J., Lees, A., and Snapper, C.M.: **T cell-independent antigens type 2.** *Annu. Rev. Immunol.* 1995, **13**:655–692.

Snapper, C.M., Shen, Y., Khan, A.Q., Colino, J., Zelazowski, P., Mond, J.J., Gause, W.C., and Wu, Z.Q.: **Distinct types of T-cell help for the induction of a humoral immune response to *Streptococcus pneumoniae*.** *Trends Immunol.* 2001, **22**:308–311.

10-14 Antibodies of different classes operate in distinct places and have distinct effector functions.

Clark, M.R.: **IgG effector mechanisms.** *Chem. Immunol.* 1997, **65**:88–110.

Herrod, H.G.: **IgG subclass deficiency.** *Allergy Proc.* 1992, **13**:299–302.

Rispens, T., den Bleker, T.H., and Aalberse, R.C.: **Hybrid IgG4/IgG4 Fc antibodies form upon 'Fab-arm' exchange as demonstrated by SDS-PAGE or size-exclusion chromatography.** *Mol. Immunol.* 2010, **47**:1592–1594.

Suzuki, K., Meek, B., Doi, Y., Muramatsu, M., Chiba, T., Honjo, T., and Fagarasan, S.: **Aberrant expansion of segmented filamentous bacteria in IgA-deficient gut.** *Proc. Natl Acad. Sci. USA* 2004, **101**:1981–1986.

Ward, E.S., and Ghetie, V.: **The effector functions of immunoglobulins: implications for therapy.** *Ther. Immunol.* 1995, **2**:77–94.

10-15 Transport proteins that bind to the Fc regions of antibodies carry particular isotypes across epithelial barriers.

Burmeister, W.P., Gastinel, L.N., Simister, N.E., Blum, M.L., and Bjorkman, P.J.: **Crystal structure at 2.2 Å resolution of the MHC-related neonatal Fc receptor.** *Nature* 1994, **372**:336–343.

Corthesy, B., and Kraehenbuhl, J.P.: **Antibody-mediated protection of mucosal surfaces.** *Curr. Top. Microbiol. Immunol.* 1999, **236**:93–111.

Ghetie, V., and Ward, E.S.: **Multiple roles for the major histocompatibility complex class I-related receptor FcRn.** *Annu. Rev. Immunol.* 2000, **18**:739–766.

Lamm, M.E.: **Current concepts in mucosal immunity. IV. How epithelial transport of IgA antibodies relates to host defense.** *Am. J. Physiol.* 1998, **274**:G614–G617.

Mostov, K.E.: **Transepithelial transport of immunoglobulins.** *Annu. Rev. Immunol.* 1994, **12**:63–84.

10-16 High-affinity IgG and IgA antibodies can neutralize bacterial toxins.
&
10-17 High-affinity IgG and IgA antibodies can inhibit the infectivity of viruses.

Brandtzaeg, P.: **Role of secretory antibodies in the defence against infections.** *Int. J. Med. Microbiol.* 2003, **293**:3–15.

Mandel, B.: **Neutralization of polio virus: a hypothesis to explain the mechanism and the one hit character of the neutralization reaction.** *Virology* 1976, **69**:500–510.

Roost, H.P., Bachmann, M.F., Haag, A., Kalinke, U., Pliska, V., Hengartner, H., and Zinkernagel, R.M.: **Early high-affinity neutralizing anti-viral IgG responses without further overall improvements of affinity.** *Proc. Natl Acad. Sci. USA* 1995, **92**:1257–1261.

Sougioultzis, S., Kyne, L., Drudy, D., Keates, S., Maroo, S., Pothoulakis, C., Giannasca, P.J., Lee, C.K., Warny, M., Monath, T.P., *et al.*: ***Clostridium difficile*** **toxoid vaccine in recurrent *C. difficile*-associated diarrhea.** *Gastroenterology* 2005, **128**:764–770.

10-18 Antibodies can block the adherence of bacteria to host cells.

Fischetti, V.A., and Bessen, D.: **Effect of mucosal antibodies to M protein in colonization by group A streptococci.** In Switalski, L., Hook, M., and Beachery, E. (eds): *Molecular Mechanisms of Microbial Adhesion.* New York, Springer, 1989.

Wizemann, T.M., Adamou, J.E., and Langermann, S.: **Adhesins as targets for vaccine development.** *Emerg. Infect. Dis.* 1999, **5**:395–403.

10-19 Antibody:antigen complexes activate the classical pathway of complement by binding to C1q.

Cooper, N.R.: **The classical complement pathway. Activation and regulation of the first complement component.** *Adv. Immunol.* 1985, **37**:151–216.

Perkins, S.J., and Nealis, A.S.: **The quaternary structure in solution of human complement subcomponent $C1r_2C1s_2$.** *Biochem. J.* 1989, **263**:463–469.

10-20 Complement receptors are important in the removal of immune complexes from the circulation.

Nash, J.T., Taylor, P.R., Botto, M., Norsworthy, P.J., Davies, K.A., and Walport, M.J.: **Immune complex processing in C1q-deficient mice.** *Clin. Exp. Immunol.* 2001, **123**:196–202.

Nash, J.T., Taylor, P.R., Botto, M., Norsworthy, P.J., Davies, K.A., Walport, M.J., Schifferli, J.A., and Taylor, J.P.: **Physiologic and pathologic aspects of circulating immune complexes.** *Kidney Int.* 1989, **35**:993–1003.

Schifferli, J.A., Ng, Y.C., and Peters, D.K.: **The role of complement and its receptor in the elimination of immune complexes.** *N. Engl. J. Med.* 1986, **315**:488–495.

Walport, M.J., Davies, K.A., and Botto, M.: **C1q and systemic lupus erythematosus.** *Immunobiology* 1998, **199**:265–285.

10-21 The Fc receptors of accessory cells are signaling receptors specific for immunoglobulins of different classes.

Kinet, J.P., and Launay, P.: **Fc α/microR: single member or first born in the family?** *Nat. Immunol.* 2000, **1**:371–372.

Ravetch, J.V., and Bolland, S.: **IgG Fc receptors.** *Annu. Rev. Immunol.* 2001, **19**:275–290.

Ravetch, J.V., and Clynes, R.A.: **Divergent roles for Fc receptors and complement *in vivo*.** *Annu. Rev. Immunol.* 1998, **16**:421–432.

Shibuya, A., Sakamoto, N., Shimizu, Y., Shibuya, K., Osawa, M., Hiroyama, T., Eyre, H.J., Sutherland, G.R., Endo, Y., Fujita, T., *et al.*: **Fc α/μ receptor mediates endocytosis of IgM-coated microbes.** *Nat. Immunol.* 2000, **1**:441–446.

Stefanescu, R.N., Olferiev M., Liu, Y., and Pricop, L.: **Inhibitory Fc gamma receptors: from gene to disease.** *J. Clin. Immunol.* 2004, **24**:315–326.

10-22 Fc receptors on phagocytes are activated by antibodies bound to the surface of pathogens and enable the phagocytes to ingest and destroy pathogens.

Dierks, S.E., Bartlett, W.C., Edmeades, R.L., Gould, H.J., Rao, M., and Conrad, D.H.: **The oligomeric nature of the murine Fc epsilon RII/CD23. Implications for function.** *J. Immunol.* 1993, **150**:2372–2382.

Hogan, S.P., Rosenberg, H.F., Moqbel, R., Phipps, S., Foster, P.S., Lacy, P., Kay, A.B., and Rothenberg, M.E.: **Eosinophils: biological properties and role in health and disease.** *Clin. Exp. Allergy* 2008, **38**:709–750.

Karakawa, W.W., Sutton, A., Schneerson, R., Karpas, A., and Vann, W.F.: **Capsular antibodies induce type-specific phagocytosis of capsulated *Staphylococcus aureus* by human polymorphonuclear leukocytes.** *Infect. Immun.* 1986, **56**:1090–1095.

10-23 Fc receptors activate NK cells to destroy antibody-coated targets.

Chung, A.W., Rollman, E., Center, R.J., Kent, S.J., and Stratov, I.: **Rapid degranulation of NK cells following activation by HIV-specific antibodies.** *J. Immunol.* 2009, **182**:1202–1210.

Lanier, L.L., and Phillips, J.H.: **Evidence for three types of human cytotoxic lymphocyte.** *Immunol. Today* 1986, **7**:132.

Leibson, P.J.: **Signal transduction during natural killer cell activation: inside the mind of a killer.** *Immunity* 1997, **6**:655–661.

Sulica, A., Morel, P., Metes, D., and Herberman, R.B.: **Ig-binding receptors on human NK cells as effector and regulatory surface molecules.** *Int. Rev. Immunol.* 2001, **20**:371–414.

Takai, T.: **Multiple loss of effector cell functions in FcR γ-deficient mice.** *Int. Rev. Immunol.* 1996, **13**:369–381.

10-24 Mast cells and basophils bind IgE antibody via the high-affinity Fcε receptor.

Beaven, M.A., and Metzger, H.: **Signal transduction by Fc receptors: the FcεRI case.** *Immunol. Today* 1993, **14**:222–226.

Kalesnikoff, J., Huber, M., Lam, V., Damen, J.E., Zhang, J., Siraganian, R.P., and Krystal, G.: **Monomeric IgE stimulates signaling pathways in mast cells that lead to cytokine production and cell survival.** *Immunity* 2001, **14**:801–811.

Sutton, B.J., and Gould, H.J.: **The human IgE network.** *Nature* 1993, **366**:421–428.

10-25 IgE-mediated activation of accessory cells has an important role in resistance to parasite infection.

Capron, A., Riveau, G., Capron, M., and Trottein, F.: **Schistosomes: the road from host-parasite interactions to vaccines in clinical trials.** *Trends Parasitol.* 2005, **21**:143–149.

Grencis, R.K.: **Th2-mediated host protective immunity to intestinal nematode infections.** *Phil. Trans. R. Soc. Lond. B* 1997, **352**:1377–1384.

Grencis, R.K., Else, K.J., Huntley, J.F., and Nishikawa, S.I.: **The in vivo role of stem cell factor (c-kit ligand) on mastocytosis and host protective immunity to the intestinal nematode *Trichinella spiralis* in mice.** *Parasite Immunol.* 1993, **15**:55–59.

Kasugai, T., Tei, H., Okada, M., Hirota, S., Morimoto, M., Yamada, M., Nakama, A., Arizono, N., and Kitamura, Y.: **Infection with *Nippostrongylus brasiliensis* induces invasion of mast cell precursors from peripheral blood to small intestine.** *Blood* 1995, **85**:1334–1340.

Ushio, H., Watanabe, N., Kiso, Y., Higuchi, S., and Matsuda, H.: **Protective immunity and mast cell and eosinophil responses in mice infested with larval *Haemaphysalis longicornis* ticks.** *Parasite Immunol.* 1993, **15**:209–214.

Dynamics of Adaptive Immunity

11

Throughout this book we have examined the separate ways in which the innate and the adaptive immune responses protect the individual from invading microorganisms. In this chapter, we consider how the cells and molecules of the immune system work as an integrated defense system to eliminate or control an infectious agent and how the adaptive immune system provides long-lasting protective immunity. This is the first of several chapters that consider how the immune system functions as a whole in health and disease. The next chapter describes the role and specializations of the mucosal immune system, which forms the front-line defense against most pathogens. Subsequent chapters examine how immune defenses can fail (Chapter 13) or unwanted immune responses occur (Chapters 14 and 15), and how the immune response can be manipulated to benefit the individual (Chapter 16).

In Chapters 2 and 3, we saw how innate immunity is brought into play in the earliest phases of an infection and is probably sufficient to prevent colonization of the body by most of the microorganisms encountered in the environment. However, pathogenic microorganisms, by definition, have developed strategies that allow them to elude or overcome innate immune defenses and to establish a focus of infection from which they can spread. In these circumstances, the innate immune response sets the scene for the induction of an adaptive immune response. In the **primary immune response**, which occurs against a pathogen encountered for the first time, several days are required for the clonal expansion and differentiation of naive lymphocytes into effector T cells and antibody-secreting B cells, as described in Chapters 9 and 10. In most cases, these cells and antibodies will effectively target the pathogen for elimination (Fig. 11.1).

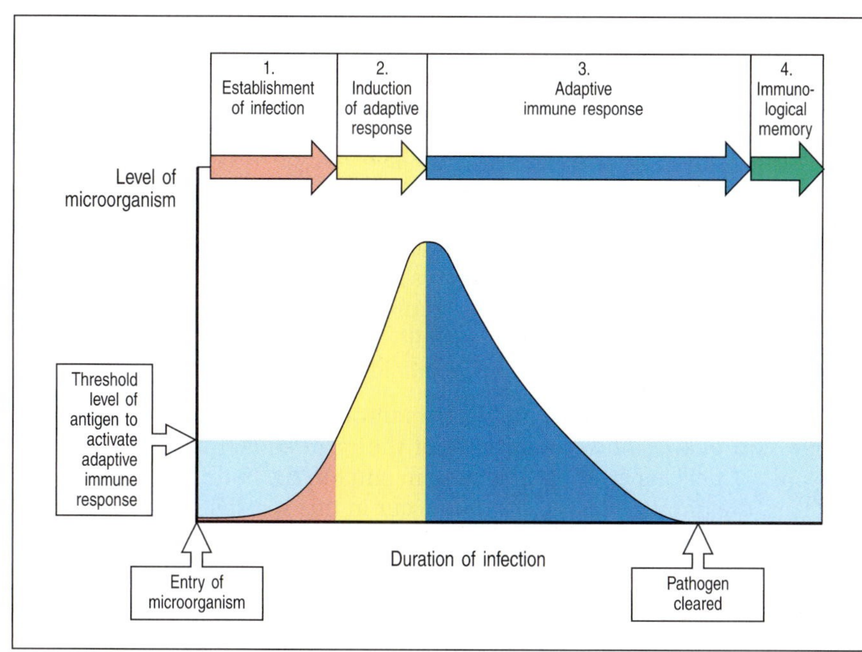

Fig. 11.1 The course of a typical acute infection that is cleared by an adaptive immune reaction. 1. The level of infectious agent increases as the pathogen replicates. 2. When numbers of the pathogen exceed the threshold dose of antigen required for an adaptive response, the response is initiated; the pathogen continues to grow, retarded only by responses of the innate immune system. At this stage, immunological memory also starts to be induced. 3. After 4–7 days, effector cells and molecules of the adaptive response start to clear the infection. 4. When the infection has been cleared and the dose of antigen has fallen below the response threshold, the response ceases, but antibody, residual effector cells, and immunological memory provide lasting protection against reinfection in most cases.

During this period, specific **immunological memory** is also established. This ensures a rapid reinduction of antigen-specific antibody and effector T cells on subsequent encounters with the same pathogen, thus providing long-lasting and often lifelong protection against it. Immunological memory is discussed in the last part of the chapter. Memory responses differ in several ways from primary responses. We discuss the reasons for this, and what is known about how immunological memory is maintained.

The course of the immune response to infection.

The immune response is a dynamic process, and both its nature and intensity change over time. It starts with the relatively nonspecific responses of innate immunity and becomes both more focused on the pathogen and more powerful as the adaptive immune response is initiated and rapidly develops. In this part of the chapter, we discuss how the different phases of an immune response are orchestrated in space and time, how the response develops in both strength and precision, how changes in specialized cell-surface molecules and chemokines guide effector lymphocytes to the appropriate site of action, and how these cells are regulated during the different stages.

An innate immune response is an essential prerequisite to a primary adaptive immune response, because the co-stimulatory molecules induced on cells of the innate immune system during their interaction with microorganisms are essential for the activation of the antigen-specific lymphocytes (see Chapter 9). Cells of the innate immune system hand on other important signals in the form of secreted cytokines that influence the characteristics of the adaptive response and tailor it to the type of pathogen encountered. For this to happen, cells from different locations must engage to coordinate the specific activation of naive T cells and B cells, and the migration of cells to precise locations within lymphoid tissues is thus critical for the coordination of an adaptive response.

11-1 The course of an infection can be divided into several distinct phases.

An infection can be broken down into various stages (see Fig. 2.5), but in Chapters 2 and 3 we considered in detail only the responses of innate immunity. In this chapter, we return to the various stages of an infection but will now integrate the adaptive immune response into the picture.

In the first stage of infection with a pathogen, a new host is exposed to infectious particles either shed by an infected individual or present in the environment. The numbers, route, mode of transmission, and stability of an infectious agent outside the host determine its infectivity. Some pathogens, such as the anthrax bacterium, are spread by spores that are highly resistant to heat and drying, whereas others, such as the human immunodeficiency virus (HIV), are spread only by the exchange of bodily fluids or tissues because they are unable to survive as infectious agents outside the body.

The first contact with a new host occurs through an epithelial surface, such as the skin or the mucosal surfaces of the respiratory, gastrointestinal, or urogenital tracts. As most pathogens gain entry to the body through mucosal surfaces, the immune responses that occur in this specialized compartment of the immune system are of great importance and are considered in detail in Chapter 12. After making contact, an infectious agent must establish a focus of infection. It must either adhere to the epithelial surface and colonize it, or penetrate it to replicate in the tissues (Fig. 11.2, first two panels). Wounds

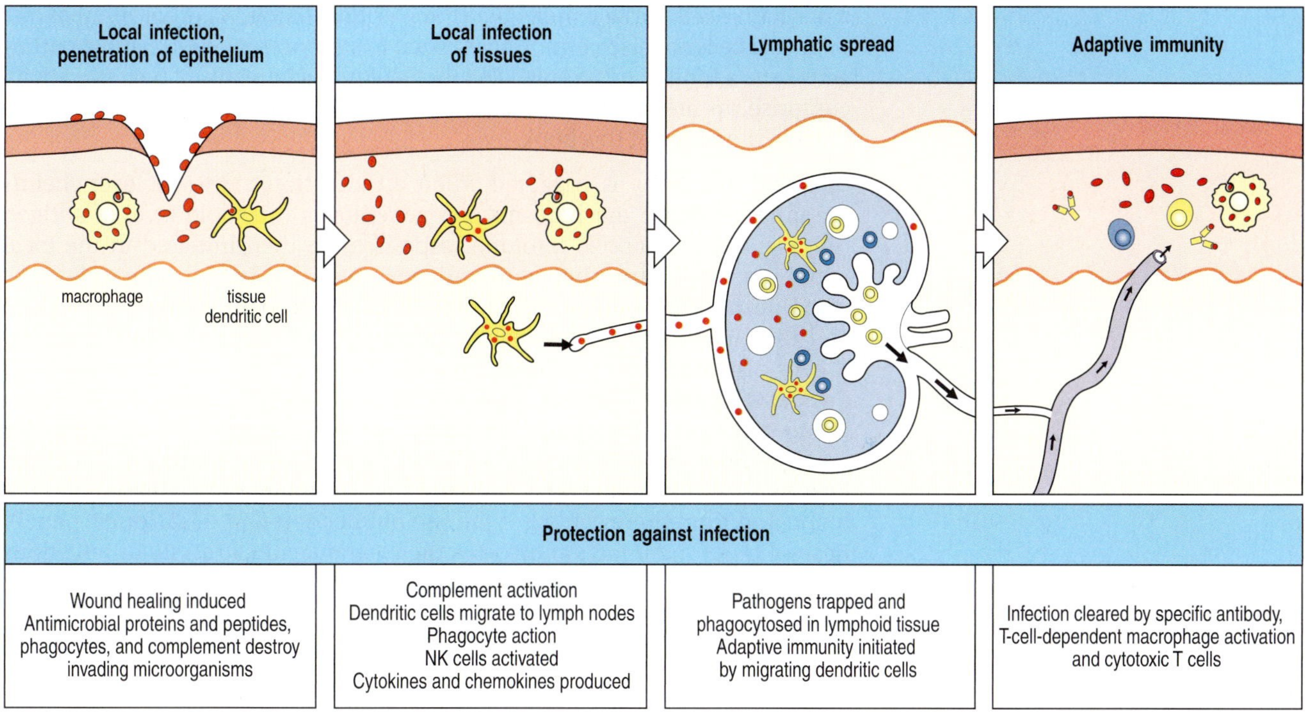
Local infection, penetration of epithelium
Local infection of tissues
Lymphatic spread
Adaptive immunity
macrophage
tissue dendritic cell
Protection against infection
Wound healing induced
Antimicrobial proteins and peptides, phagocytes, and complement destroy invading microorganisms
Complement activation
Dendritic cells migrate to lymph nodes
Phagocyte action
NK cells activated
Cytokines and chemokines produced
Pathogens trapped and phagocytosed in lymphoid tissue
Adaptive immunity initiated by migrating dendritic cells
Infection cleared by specific antibody, T-cell-dependent macrophage activation and cytotoxic T cells

Fig. 11.2 Infections and the responses to them can be divided into a series of stages. These are illustrated here for a pathogenic microorganism (red) entering across a wound in an epithelium. The microorganism first adheres to epithelial cells and then invades beyond the epithelium into underlying tissues (first panel). A local innate immune response helps to contain the infection, and delivers antigen and antigen-loaded dendritic cells to lymphatics (second panel) and thence to local lymph nodes (third panel). This leads to an adaptive immune response in the lymph node that involves the activation and further differentiation of B cells and T cells with the eventual production of antibody and effector T cells, which clear the infection (fourth panel).

and insect and tick bites that breach the epidermal barrier help some microorganisms to get through the skin. Many microorganisms are repelled or kept in check at this stage by innate defenses, which include the induced innate immune response triggered by stimulation of the various germline-encoded receptors (such as TLRs and NK-cell receptors) that discriminate between foreign microbial and self host-cell surfaces or between infected and normal cells (see Chapters 2 and 3). These responses are not as effective as adaptive immune responses, which can afford to be more powerful because they are antigen specific and thus target the pathogen precisely. However, they can prevent an infection from being established or, failing that, can contain it and prevent the spread of a pathogen into the bloodstream while an adaptive immune response develops.

Only when a microorganism has successfully established a focus of infection in the host does disease occur. With the possible exception of lung infections such as tuberculosis, and diarrhea-inducing gut infections such as cholera, in which the primary infection can cause life-threatening disease, little damage will be caused unless the agent spreads from the original focus or secretes toxins that can spread to other parts of the body. Extracellular pathogens spread by direct extension of the infection through the lymphatics or the bloodstream. Usually, spreading into the bloodstream occurs only after the lymphatic system has been overwhelmed. Obligate intracellular pathogens spread from cell to cell; they do so either by direct transmission from one cell to the next or by release into the extracellular fluid and reinfection of both adjacent and distant cells. In contrast, some of the bacteria that cause gastroenteritis exert their effects without spreading into the tissues. They establish a site of infection on the luminal surface of the epithelium lining the gut and cause no direct pathology themselves, but secrete toxins that cause damage either *in situ* or after crossing the epithelial barrier and entering the circulation.

Most infectious agents show a significant degree of host specificity, causing disease in only one or a few related species. What determines host specificity for every agent is not known, but the requirement for attachment to a particular cell-surface molecule is one critical factor. As other interactions with host cells are also commonly needed to support replication, most pathogens have a limited host range. The molecular mechanisms of host specificity comprise an area of research known as molecular pathogenesis, which is outside the scope of this book.

Adaptive immunity is triggered when an infection eludes or overwhelms the innate defense mechanisms and generates a threshold level of antigen (see Fig. 11.1). Adaptive immune responses are then initiated in the local lymphoid tissue, in response to antigens presented by dendritic cells activated during the course of the innate immune response (Fig. 11.2, second and third panels). Antigen-specific effector T cells and antibody-secreting B cells are generated by clonal expansion and differentiation over several days, as described in greater detail in Chapters 9 and 10. During this time, the induced responses of innate immunity, such as the acute-phase responses and interferon production (see Sections 3-7 and 3-18), continue to function. Eventually, antigen-specific T cells and then antibodies are released into the blood and from there can enter the site of infection (Fig. 11.2, fourth panel). Resolution of the infection involves the clearance of extracellular infectious particles by opsonizing antibodies and phagocytes (see Chapter 10) and the clearance of intracellular residues of infection through the actions of effector T cells (see Chapter 9).

After many types of infection, little or no residual pathology follows an effective primary adaptive response. In some cases, however, the infection or the response to it causes significant tissue damage. In yet other cases, such as infection with cytomegalovirus or *Mycobacterium tuberculosis*, the pathogen is contained but not eliminated, and can persist in a latent form. If the adaptive immune response is later weakened, as it is in acquired immune deficiency syndrome (AIDS), these pathogens may resurface to cause virulent systemic infections. We will focus on the strategies used by certain pathogens to evade or subvert adaptive immunity, and thereby establish a persistent, or chronic, infection, in the first part of Chapter 13. In addition to clearing the infectious agent, an effective adaptive immune response prevents reinfection. For some infectious agents, this protection is essentially absolute, whereas for others infection is only reduced or attenuated on reexposure to the pathogen.

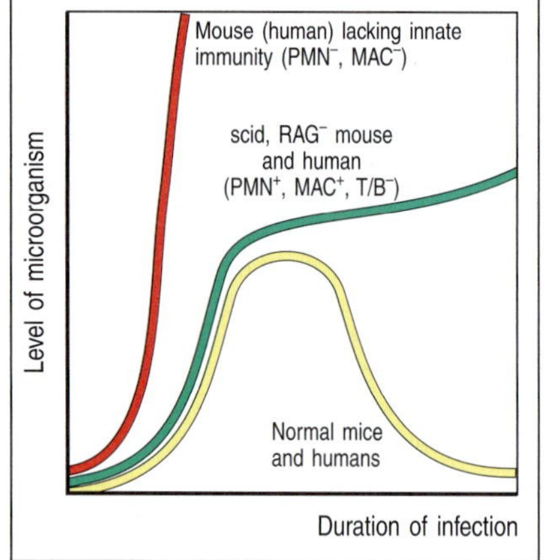

Fig. 11.3 The time course of infection in normal and immunodeficient mice and humans. The red curve shows the rapid growth of microorganisms in the absence of innate immunity, when macrophages (MAC) and polymorphonuclear leukocytes (PMN) are lacking. The green curve shows the course of infection in mice and humans that have innate immunity but have no T or B lymphocytes and so lack adaptive immunity. The yellow curve shows the normal course of an infection in immunocompetent mice or humans.

It is not known how many infections are dealt with solely by the nonadaptive mechanisms of innate immunity, because such infections are eliminated early and produce little in the way of symptoms or pathology. Naturally occurring deficiencies in nonadaptive defenses are rare, so it has seldom been possible to study their consequences. Innate immunity does, however, seem to be essential for effective host defense, as shown by the progression of infection in mice that lack components of innate immunity but have an intact adaptive immune system (Fig. 11.3). In humans, for example, mutations in the Toll-like receptor TLR-3 have been associated with increased susceptibility to encephalitis due to the herpes simplex virus, which more usually causes self-limiting cold sores on the skin. Adaptive immunity is also essential, as shown by the immunodeficiency syndromes associated with defects in various components of the adaptive immune response (discussed in Chapter 13).

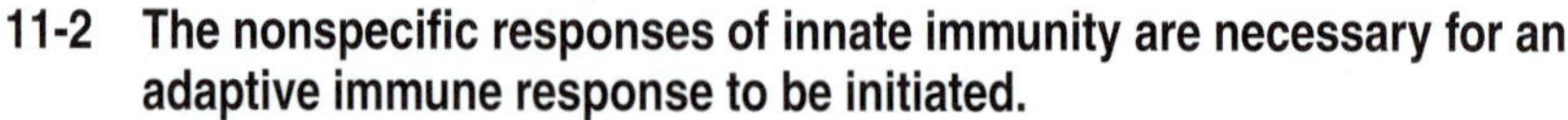

11-2 The nonspecific responses of innate immunity are necessary for an adaptive immune response to be initiated.

The establishment of a focus of infection in tissues and the response of the innate immune system produce changes in the immediate environment. Many of these changes have been described in earlier chapters, but we review them briefly here to provide a cohesive framework for the induction of adaptive immunity.

In a bacterial infection, the first thing that usually happens is that the infected tissue becomes inflamed. This is initially the result of the activation of the resident macrophages by bacterial components such as lipopolysaccharide (LPS) acting through Toll-like receptors (TLRs) on the macrophage. The cytokines and chemokines secreted by the activated macrophages, especially the cytokine tumor necrosis factor-α (TNF-α), induce numerous changes in the endothelial cells of nearby blood capillaries, a process known as endothelial cell activation. Inflammation also results from the activation of complement, resulting in the production of the anaphylatoxins C3a and C5a, which are able to activate vascular endothelium. In a primary infection, complement is activated mainly via the alternative and lectin pathways (see Fig. 2.12).

Activation of the vascular endothelium causes the release of the contents of Weibel–Palade bodies (the cell adhesion molecule P-selectin and von Willebrand factor) within the endothelial cells to the cell surface (see Section 3-15). Activation also induces the expression of E-selectin, which then also appears on the endothelial cell surface. These two selectins cause neutrophils, monocytes, and other leukocytes to adhere to and roll on the endothelial surface. Cytokine activation of the endothelium also induces the production of the adhesion molecule ICAM-1. By binding to adhesion molecules, such as LFA-1, on neutrophils and monocytes, ICAM-1 strengthens the interaction of these cells with the endothelium and aids their entry in large numbers into the infected tissue to form an inflammatory focus (see Fig. 3.26). As monocytes mature into tissue macrophages and become activated in their turn, additional inflammatory cells are attracted into the infected tissue, and the inflammatory response is maintained and reinforced. The inflammatory response can be thought of as putting up a flag on the endothelial cells to signal the presence of infection, but as yet the response is nonspecific for the pathogen.

A second crucial effect of infection is the activation of specialized antigen-presenting cells, the dendritic cells residing in most tissues, as described in Sections 9-4 to 9-6. Dendritic cells take up antigen in the infected tissues and, like macrophages, they are activated through innate immune receptors that respond to common constituents of pathogens, such as TLRs (Section 3-7) and NOD proteins (Section 3-8). Activated dendritic cells increase their synthesis of MHC class II molecules and, most importantly, begin to express the co-stimulatory molecules B7.1 and B7.2 on their surface. As described in Chapter 9, these antigen-presenting cells migrate away from the infected tissue through the lymphatics, along with their antigen cargo, to enter peripheral lymphoid tissues, where they initiate the adaptive immune response. They arrive in large numbers at the draining lymph nodes, or other nearby lymphoid tissue, attracted by the chemokines CCL19 and CCL21 produced by the lymph-node stroma (see Section 9-3).

When dendritic cells arrive in the lymphoid tissues, they seem to have reached their final destination. They activate antigen-specific naive T lymphocytes in these tissues, after which they die. Naive lymphocytes are continually passing through the lymph nodes, which they enter from the blood across the walls of high endothelial venules (see Fig. 9.4). Those naive T cells that recognize antigen on the surface of dendritic cells are activated, and divide and mature into effector cells that reenter the circulation. Where there is a local infection, the changes induced by inflammation in the walls of nearby venules induce these effector T cells to leave the blood vessel and migrate into the site of infection.

Thus, the local release of cytokines and chemokines at the site of infection has far-reaching consequences. In addition to recruiting neutrophils and macrophages, which are not specific for antigen, the changes induced in the blood vessel walls also enable newly activated effector T lymphocytes to enter infected tissue, as discussed in more detail later in this chapter.

11-3 Cytokines made during infection can direct differentiation of CD4 T cells toward the T_H17 subset.

In Chapter 9 we described transcriptional mechanisms that control how specific cytokines direct the differentiation of naive CD4 T cells into distinct classes of CD4 effector T cells—T_H17, T_H1, or T_H2, or regulatory subsets (see Fig. 9.29). The cytokines that are produced during the progression of an infection depend on how the microorganism influences the behavior of innate immune cells and antigen-presenting cells. The conditions produced by these interactions have a major impact on how T cells differentiate during their initial contact with antigen-presenting cells, thus determining the types of T cells that are generated (see Chapter 9). In turn, these T-cell subsets influence the nature of the effector responses that are recruited, such as the extent of macrophage activation, neutrophil and eosinophil recruitment to the site of infection, and which classes of antibody will predominate.

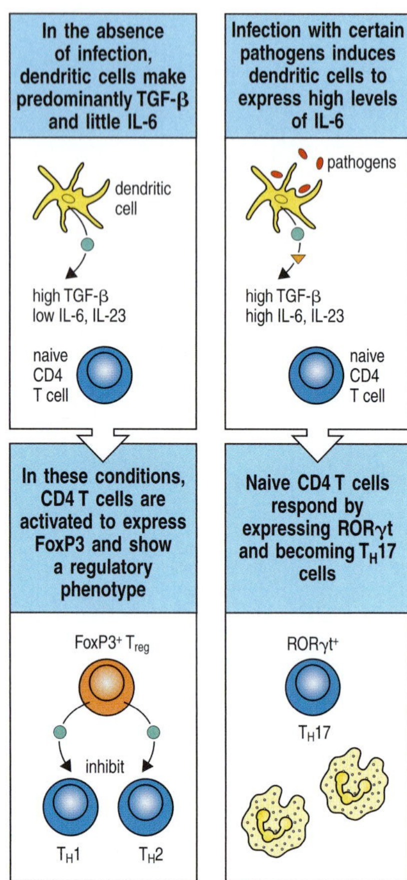

Fig. 11.4 Cytokines produced by dendritic cells regulate the balance of regulatory T-cell development and T_H17 differentiation. The balance between TGF-β and IL-6 production acts to induce either the transcription factor FoxP3, which is characteristic of regulatory T cells, or RORγt (an 'orphan' member of the nuclear receptor family), which is characteristic of T_H17 cells. In the absence of infection, IL-6 production by dendritic cells is low, and TGF-β production dominates. In these conditions, T cells that do encounter their cognate antigen will be induced to express FoxP3 and predominantly acquire a regulatory phenotype, whereas those that do not encounter antigen remain naive. Infection by certain bacteria and fungi induce dendritic cells to produce abundant IL-6 and IL-23, but less IL-12 (see Fig. 11.5); under these conditions, naive T cells will express RORγt and become T_H17 cells. The cytokines produced by this T-cell subset, IL-17 and IL-17F, induce cells such as epithelium to secrete chemokines that attract inflammatory cells such as neutrophils.

The subset of effector T cells generated in response to infection by extracellular bacteria and fungi is often T_H17. At the beginning of an infection, dendritic cells are not yet fully activated; in this state they produce TGF-β, but little IL-6 or other cytokines that direct T-cell differentiation. Upon encountering such pathogens, dendritic cells become activated and are induced to synthesize IL-6 along with IL-23. As there is no source of IL-4 or IL-12 at this time, naive CD4 T cells will differentiate into T_H17 cells rather than T_H1 or T_H2 cells (Fig. 11.4). For example, activation of dendritic cells through Dectin-1, a receptor that recognizes carbohydrates common to yeast and other fungi, induces dendritic cells to produce abundant IL-23 but little IL-12. The effect is that dendritic cells activated by these organisms will tend to promote T_H17 differentiation.

When the T_H17 cells leave the lymph node and migrate to the sites of infection, they encounter pathogen antigens and are stimulated to synthesize and release cytokines, which include various members of the IL-17 family such as IL-17A and IL-17E (also known as IL-25). The receptor for IL-17 is expressed ubiquitously on cells such as fibroblasts, epithelial cells, and keratinocytes. IL-17 induces these cells to secrete various cytokines, including IL-6, the chemokines CXCL8 and CXCL2, and the hematopoietic factors granulocyte colony-stimulating factor (G-CSF) and granulocyte–macrophage colony-stimulating factor (GM-CSF). These chemokines can act directly to recruit neutrophils, whereas G-CSF and GM-CSF act to augment neutrophil and macrophage production in the bone marrow or increase the differentiation of local monocytes into macrophages.

Thus, one important action of IL-17 at sites of infection is to induce local cells to secrete cytokines and chemokines that attract neutrophils. T_H17 cells also produce IL-22, a cytokine related to IL-10. IL-22 acts cooperatively with IL-17 to induce the expression of antimicrobial peptides, such as β-defensins, by the keratinocytes of the epidermis. In this way, the presence of pathogen-specific T_H17 cells serves as an efficient amplifier of an acute inflammatory response by the innate immune system at sites of early infection. CD4 T cells that acquire the T_H17 phenotype are not the only cells that can produce IL-17 in response to infections. CD8 T cells have also been shown to produce abundant IL-17.

The cytokine environment is also influential in preventing the immune system from making inappropriate responses to self antigens or those of commensal microorganisms, the microorganisms that normally inhabit the body. Even in the absence of infection, dendritic cells take up self and environmental antigens and eventually carry them to peripheral lymphoid tissues, where they may meet antigen-specific naive T cells. In such circumstances, the pro-inflammatory signals are not present; dendritic cells are not activated and so produce the cytokine TGF-β but not the other cytokines that can affect CD4 T-cell differentiation. In this setting, dendritic cells seem to actively generate tolerance to the antigens that the naive T cells encounter (Fig. 11.4, left panels). TGF-β on its own inhibits the proliferation and differentiation of T_H17, T_H1, and T_H2 cells. When a naive CD4 T cell encounters its cognate peptide:MHC ligand in the presence of TGF-β alone, it acquires the phenotype of a regulatory T cell, in that it can inhibit the activation of other T cells. Regulatory T cells induced in this way outside the thymus are called adaptive regulatory T cells and some express the transcription factor FoxP3 (see Section 9-18). The regulatory T cells, in theory, should not be specific for pathogen antigens—which they have not yet encountered—but should instead be specific for either self antigens or peptides from commensal organisms. Other FoxP3-expressing regulatory CD4 T cells, the natural regulatory T cells, seem to acquire their regulatory phenotype in the thymus (see Section 8-18).

The reciprocal pathways for the development of T_H17 cells and regulatory T cells seem to be based on an evolutionarily ancient system of activation and inactivation, because proteins similar to TGF-β and IL-17 are present in invertebrates that possess primitive intestinal immune systems. This might suggest that the dichotomy between T_H17 cells and regulatory T cells is largely concerned with maintaining the lymphocyte balance in tissues exposed to large numbers of potential pathogens, such as the mucosae of the gut and the lungs, where a rapid response to infection is critical. For example, IL-17-producing T cells have an important role in mice in resistance to infections of the lung by Gram-negative bacteria such as *Klebsiella pneumoniae.* Mice lacking the receptor for IL-17 are significantly more susceptible than normal mice to lung infection by this pathogen, and they show decreased production of G-CSF and CXCL2 and poorer recruitment of neutrophils to infected lungs. T_H17 cells also promote resistance to the gut nematode *Nippostrongylus brasiliensis.* This effect seems to be due to the induction or recruitment by IL-17E of a population of non-T non-B leukocytes, perhaps similar to basophils, that secrete the T_H2 cytokines IL-4, IL-5, and IL-13. These cytokines, particularly IL-13, promote resistance to *N. brasiliensis* by, for example, inducing its expulsion from the gut by augmenting the production of mucus (see Chapter 12).

11-4 T_H1 and T_H2 cells are induced by cytokines generated in response to different pathogens.

T_H1 responses tend to be induced by viruses and by bacterial and protozoan pathogens that can survive inside macrophage intracellular vesicles. In the case of viruses, the T_H1 response is generally involved in helping to activate the CD8 cytotoxic T cells that will recognize virus-infected cells and destroy them (see Chapter 9). T_H1 cells also induce the production of some subclasses of IgG antibodies, which neutralize virus particles in the blood and extracellular fluid. In the case of mycobacteria, and of protozoa such as *Leishmania* and *Toxoplasma,* which all take up residence inside macrophages, the role of T_H1 cells is to activate macrophages to a degree that will destroy the invaders.

Experiments *in vitro* have shown that naive CD4 T cells initially stimulated in the presence of IL-12 and IFN-γ tend to develop into T_H1 cells (Fig. 11.5, left panels). In part this is because these cytokines induce or activate the transcription factors leading to T_H1 development, and in part it is because IFN-γ inhibits the proliferation of T_H2 cells, as described in Chapter 9. NK cells and CD8 cells are also both activated in response to infections with viruses and some other intracellular pathogens, as discussed in Chapters 3 and 9, and both produce abundant IFN-γ. Dendritic cells and macrophages produce IL-12. Thus, CD4 T-cell responses in these infections tend eventually to be dominated by T_H1 cells.

Toll-like receptor (TLR) signaling is of major importance in driving dendritic cells to produce IL-12. This has been shown in mice lacking the adaptor

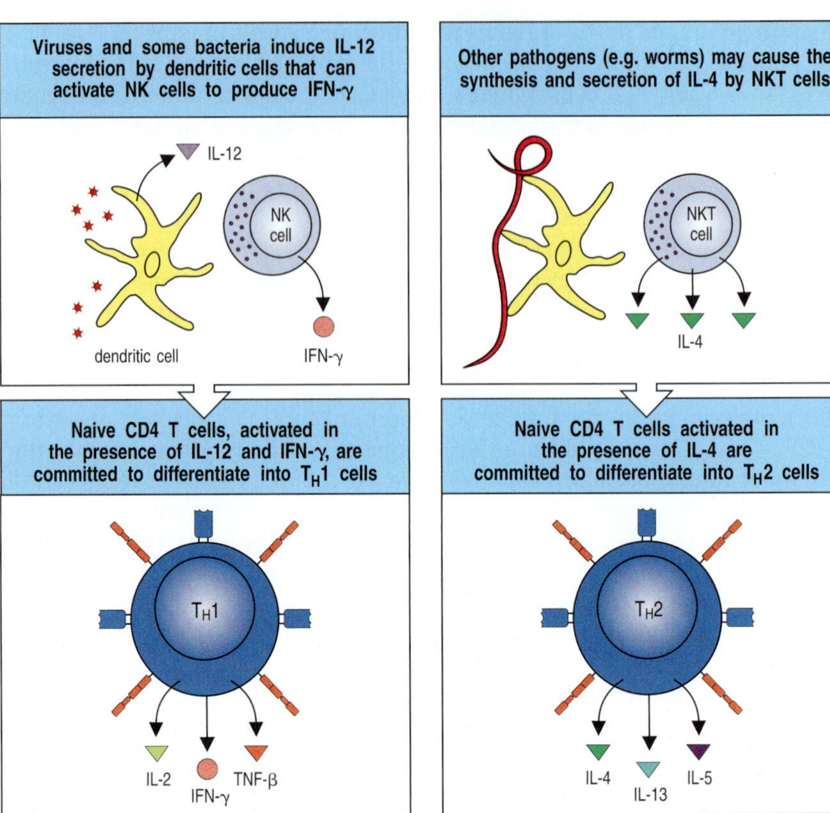

Fig. 11.5 The differentiation of naive CD4 T cells into subclasses of effector T cells is influenced by cytokines elicited by the pathogen. Left panels: many pathogens, especially viruses and intracellular bacteria, activate dendritic cells to produce IL-12 and NK cells to produce IFN-γ. These cytokines cause proliferating CD4 T cells to differentiate into T_H1 cells. NK cells can be induced by certain stimuli and adjuvants to migrate into lymph nodes, where they could promote T_H1 responses. Right panels: IL-4, which can be produced by various cells, is made in response to parasitic worms and some other pathogens, and acts on proliferating CD4 T cells to cause them to become T_H2 cells. An iNKT cell is shown as a source of IL-4 here, but these cells are not the only source of IL-4 that can promote T_H2 responses (see the text). The mechanisms by which these cytokines induce the selective differentiation of CD4 T cells are discussed in Section 9-18 and Fig. 9.29. Selective induction of transcription factors induced by cytokine binding to cytokine receptors leads to the activation of these two different fates.

protein MyD88, a component of an intracellular signaling pathway activated by the stimulation of some TLRs (see Section 3-7). Mice deficient in MyD88 do not survive a challenge with *Toxoplasma gondii*, which normally elicits a strong T_H1 response. Dendritic and other cells from mice lacking MyD88 failed to produce IL-12 in response to parasite antigens, and the animals failed to mount a T_H1 response (Fig. 11.6). In contrast to dendritic cells stimulated through Dectin-1, dendritic cells stimulated through TLR-9 produce abundant IL-12 and support T_H1 differentiation. It is important to note that many immune responses to pathogens produce a mixed response, with both T_H17 and T_H1 cells being generated. The basis for such mixed responses, and for potential plasticity between these subsets, is an area of active research.

The killer lymphocytes of innate immunity, the NK cells, may contribute to T_H1 development (see Fig. 11.5). NK cells are not normally found within lymph nodes, but injection of mice with certain adjuvants, or with mature dendritic cells, can induce their recruitment to lymph nodes via expression of the chemokine receptor CXCR3 by the NK cell. As NK cells produce abundant IFN-γ, but little IL-4, they may act in lymph nodes during infections to direct the development of T_H1 cells.

For T_H2 responses, the mechanisms linking innate immunity to regulation of the adaptive T_H2 response are somewhat less clear (see Section 9-18). Naive CD4 T cells activated in the presence of IL-4, especially when IL-6 is also present, tend to differentiate into T_H2 cells (see Fig. 11.5, right panels). Some pathogens, such as helminths and other extracellular parasites, consistently induce development of T_H2 responses *in vivo*, and do so in a manner that requires IL-4 signaling by T cells. But it is still uncertain how these pathogens are initially sensed by the immune system and how this triggers commitment of the naive T cell to the T_H2 subset. But once some T cells have differentiated into effector T_H2 cells, their production of IL-4 can strongly reinforce the development of more T_H2 cells. Many cells have been proposed to act as

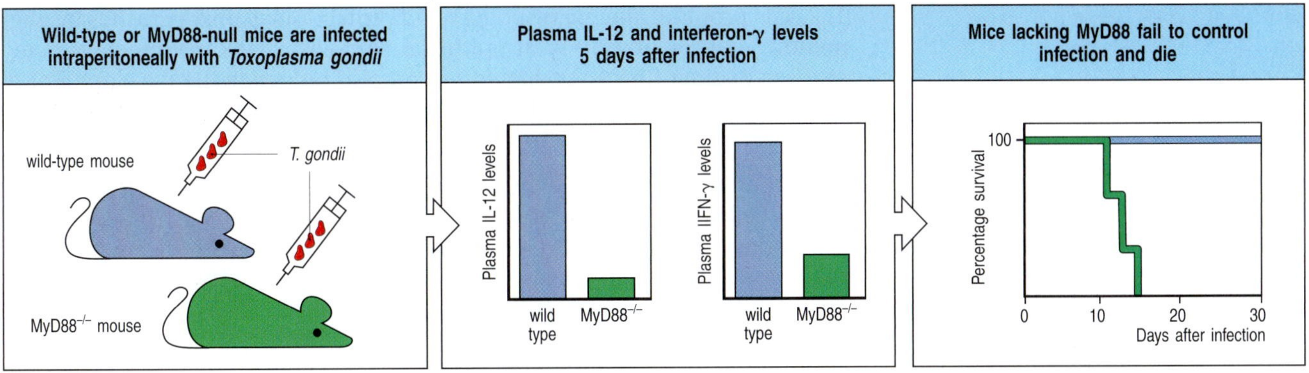

Fig. 11.6 Infection may trigger T_H1 polarization through Toll-like receptor signaling pathways. The adaptor protein MyD88 is a key component of Toll-like receptor signaling. Wild-type mice and mice deficient in MyD88 were infected intraperitoneally with the protozoan parasite *Toxoplasma gondii* (left panel). Five days after infection, mice lacking MyD88 showed a severe reduction in the amount of IL-12 in plasma compared with wild-type mice (center panel), and dendritic cells from the spleens of these animals failed to produce IL-12 on stimulation with antigens from *T. gondii*. The mice lacking MyD88 also failed to produce a large IFN-γ response (center panel) and a T_H1 response to the infection, and died about 2 weeks after infection (right panel, green line). In contrast, wild-type mice produced a strong IL-12, IFN-γ, and T_H1 response, controlled the infection, and survived (right panel, blue line). Sensing of *T. gondii* in mice may involve TLR-11 (see Section 3-5).

an initial and rapid source of IL-4, including invariant NKT cells (iNKT cells), mast cells, and basophils.

iNKT cells are a class of innate-like lymphocytes (see Section 3-24) that express the cell-surface marker NK1.1 normally associated with NK cells. Like other NKT cells (see Section 8-9), development of iNKT cells in the thymus does not involve selection by MHC class I or class II molecules, but instead depends on interactions with the MHC class IB molecules known as CD1 proteins (see Section 6-19), which bind self lipids. iNKT cells have a nearly invariant repertoire of α:β T-cell receptor composed of gene segments $V_\alpha14$–$J_\alpha28$ in mice, or the equivalent gene segments, $V_\alpha24$–$J_\alpha18$, in humans. Expression of CD1 proteins can be induced at sites of infection, which may allow the presentation of microbial lipids to T cells. At least some iNKT cells recognize specific glycolipid antigens presented by CD1d. On activation, iNKT cells secrete very large amounts of IL-4 and IFN-γ, and may provide an initial source of cytokines that polarize a T-cell response, particularly in the direction of T_H2 cells. iNKT cells are not the only T cells that recognize antigens presented by CD1 molecules. CD1b presents the bacterial lipid mycolic acid to α:β T cells, and other CD1 molecules are recognized by γ:δ T cells. However, a particular role for iNKT cells in T_H2 responses has emerged from studies of mice in which the genes encoding $J_\alpha18$ or CD1d have been inactivated, and so iNKT cells fail to develop. These mice also fail to develop antigen-induced airway hypersensitivity that mimics a condition similar to asthma in humans and which is strongly associated with a T_H2 response.

Mast cells and basophils are also potent IL-4 producers and can migrate to peripheral lymphoid organs, making them possible candidates for an early source of IL-4 in response to infection. However, transgenic mice engineered to lack basophils still developed T_H2 responses and IgE production when infected by the helminth *Nippostrongylus brasiliensis*, but basophils were required for protection against secondary infection. This suggests that their role is mainly as an effector, not an initiator, of T_H2-type responses. Some evidence suggests that certain TLR ligands can cause dendritic cells to produce cytokines favoring T_H2 development rather than T_H1. For example, dendritic cells make more IL-10 and less IL-12 when stimulated by some ligands for TLR-2, including bacterial lipoproteins, peptidoglycans, and zymosan, a carbohydrate derivative of yeast cell walls, compared with other TLR ligands, but these can also stimulate production of IL-23, which favors T_H17 development.

11-5 CD4 T-cell subsets can cross-regulate each other's differentiation.

The various subsets of CD4 T cells—regulatory T cells (T_{reg}), T_H17, T_H1, and T_H2—each have very different functions. T_{reg} cells maintain tolerance and limit immunopathology. T_H17 cells amplify acute inflammation at sites of infection. T_H1 cells are crucial for cell-mediated immunity due to phagocytes.

T_H2 cells promote allergic responses and protection against parasites by augmenting barrier immunity at epithelial surfaces. Cytokines mediate many of these distinct functions, but some can also influence the development of these T-cell subsets themselves. Thus, there is a complex pattern of cross-regulation during the development of CD4 T-cell subsets.

For example, T_H17 cells are induced by IL-6 and TGF-β, but IFN-γ (produced by T_H1 cells) or IL-4 (produced by T_H2 cells) can inhibit T_H17 development, promoting T_H1 or T_H2 development, respectively (Fig. 11.7). Furthermore, there is some degree of plasticity between these subsets, at least *in vitro*. Treatment with IFN-γ and IL-12 can convert established T_H17 cells into IFN-γ-producing T_H1 cells, and the combination of IFN-α, IFN-γ, and IL-12 can even induce IFN-γ production from T_H2 cells. There is also cross-regulation between T_H1 and T_H2 cells. IL-4 and IL-10, which are products of T_H2 cells, can inhibit the production of IL-12 by dendritic cells, and thereby inhibit T_H1 development. Conversely, IFN-γ, a product of T_H1 cells, can inhibit the proliferation of T_H2 cells (see Fig. 11.7). IL-10 is not made exclusively by T_H2 cells: it can also be made by T_H1 cells under conditions of strong stimulation of the T-cell receptor. In this way, antigen concentration can potentially modulate T_H1 and T_H2 differentiation by changing the overall balance of different cytokines such as IL-10.

It is sometimes possible to shift the balance between T_H1 and T_H2 cells by administering appropriate cytokines or antibodies, although studying their effects on human CD4 T-cells poses obvious difficulties. IL-2 and IFN-γ have been used to stimulate cell-mediated immunity in human diseases such as lepromatous leprosy and can cause both a local resolution of the lesion and a systemic change in T-cell responses. However, the links between cytokine action and disease have been explored mainly in mouse models. Such studies indicate that the appropriate choice of the CD4 T cell can sometimes be crucial for clearance of infection, and show that subtle differences in CD4 T-cell responses can have a significant impact on the outcome of infection.

One clear example of this is the murine model of infection by the protozoan parasite *Leishmania major*, which requires a T_H1 response and activation of

Fig. 11.7 The subsets of CD4 T cells each produce cytokines that can negatively regulate the development or effector activity of other subsets. In the absence of infection, in homeostatic conditions, the TGF-β produced by T_{reg} cells can inhibit the activation of naive T cells, thus preventing the development of a T_H17, T_H1, or T_H2 response (upper panels). During an infection, development of T_H17 cells occurs in response to IL-6 produced by dendritic cells. However, if signals are present to induce T_H1 or T_H2 cells, the cytokines IFN-γ or IL-4 produced by them can override the effect of IL-6 and inhibits T_H17 development (lower center panel). T_H2 cells can produce IL-10, which inhibits the production of IL-12 by macrophages, reducing T_H1 cells, and TGF-β, which inhibits T_H1 growth (left panels). T_H1 cells produce IFN-γ, which blocks the growth of T_H2 cells (right panels).

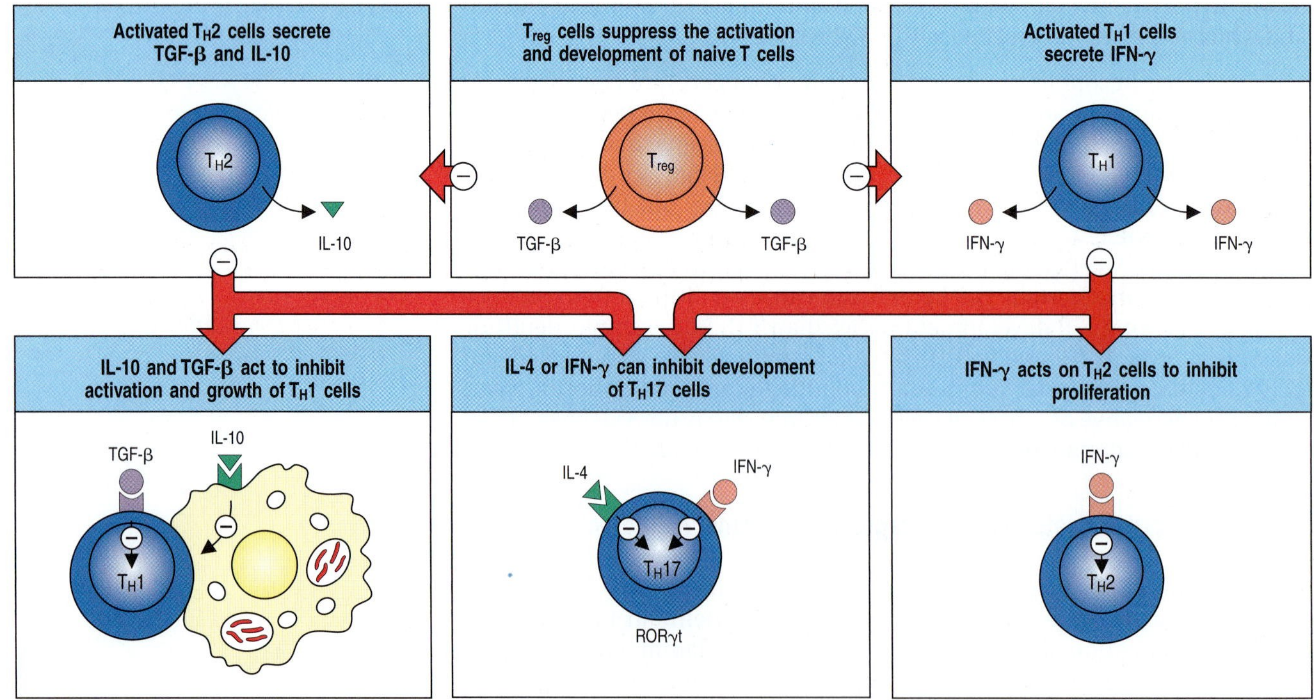

macrophages for clearance. C57BL/6 mice produce T_H1 cells that protect the animal by activating infected macrophages to kill *L. major*. In BALB/c mice infected with *L. major*, however, CD4 T cells fail to differentiate into T_H1 cells; instead, they become T_H2 cells, which are unable to activate macrophages to inhibit leishmanial growth. This difference seems to result from a population of memory T cells that are specific for gut-derived antigens but cross-react with an antigen, LACK (*Leishmania* analog of the receptors of activated C kinase), expressed by the *Leishmania* parasite. These memory cells are present in both strains of mice, but for unknown reasons they produce IL-4 in BALB/c mice but not in C57BL/6 mice. In BALB/c mice, the small amount of IL-4 secreted by these memory cells that are activated during *Leishmania* infection drives new *Leishmania*-specific CD4 T cells to become T_H2 cells, which leads eventually to a failure to eliminate the pathogen, and to death. The preferential development of T_H2 rather than T_H1 cells in BALB/c mice can be reversed if IL-4 is blocked in the first days of infection by injecting anti-IL-4 antibody, but this treatment is ineffective after a week or so of infection, demonstrating the crucial importance of the early exposure to cytokines for decisions by naive T cells (Fig. 11.8).

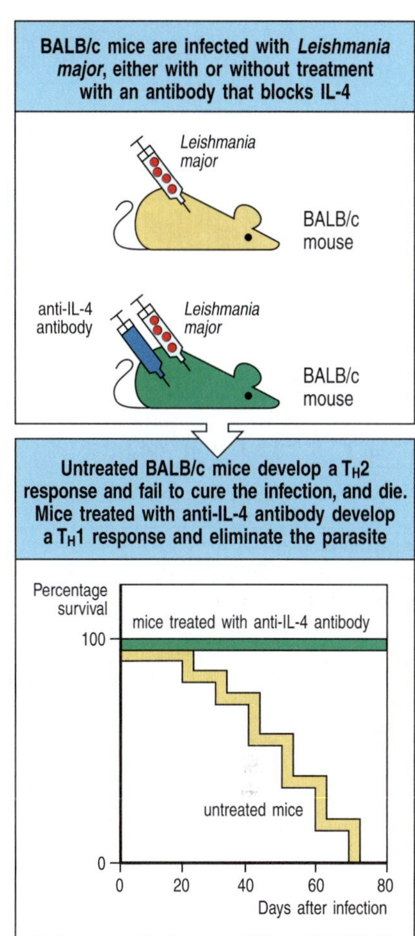

Fig. 11.8 The development of CD4 T-cell subsets can be manipulated by altering the cytokines acting during the early stages of infection. Elimination of infection with the intracellular protozoan parasite *Leishmania major* requires a T_H1 response, because IFN-γ is needed to activate the macrophages that provide protection. BALB/c mice are normally susceptible to *L. major* because they generate a T_H2 response to the pathogen. This is because they produce IL-4 early during infection and this induces naive T cells into the T_H2 lineage (see the text). Treatment of BALB/c mice with neutralizing anti-IL-4 antibodies at the beginning of infection inhibits this IL-4 and prevents the diversion of naive T cells toward the T_H2 lineage; these mice develop a protective T_H1 response.

CD8 T cells are also able to regulate the immune response by producing cytokines. Effector CD8 T cells can, in addition to their familiar cytotoxic function, also respond to antigen by secreting cytokines typical of either T_H1 or T_H2 cells. Such CD8 T cells, called T_C1 or T_C2 by analogy with the T_H subsets, seem to be responsible for the development of leprosy in its lepromatous rather than its tuberculoid form. As we saw in Chapter 9, lepromatous leprosy is due to the predominance of a T_H2 cell response, which does not clear the bacteria. Patients with the less destructive tuberculoid leprosy make T_C1 cells whose cytokines induce T_H1 cells, which can activate macrophages to rid the body of its burden of leprosy bacilli. Patients with lepromatous leprosy have CD8 T cells that suppress the T_H1 response by making IL-10 and TGF-β. The expression of these cytokines could explain the suppression of CD4 T cells by CD8 T cells that has been observed in various situations.

Another factor that may possibly influence the differentiation of CD4 T cells into various subsets is the amount, or sequence, of the antigenic peptide that initiates the response. In some cases, large amounts of antigenic peptide, or peptides that interact strongly with the T-cell receptor, favor T_H1 responses, whereas a low density of peptide or peptides that bind weakly tend to elicit T_H2 responses. These effects do not seem to be due to differences in signaling through the T-cell receptor but may involve changes in the overall balance of different cytokines produced by the cells involved in activating naive T cells. It is not clear whether this effect is universal, but it could be important in certain circumstances. For instance, allergic reactions are caused by the production of IgE antibody, which requires high levels of IL-4 but does not occur in the presence of IFN-γ, a powerful inhibitor of IL-4-driven class switching to IgE. Antigens that elicit IgE-mediated allergy are generally delivered in minute doses and elicit T_H2 cells that make IL-4 and no IFN-γ. Also, it is possible that many allergens do not activate cytokine production from innate immune responses, which might override the effects of antigen dose. Finally, allergens are delivered to humans in minute doses across a thin mucosa, such as that of the lung, and something about this route of sensitization seems to favor T_H2 responses.

11-6 Effector T cells are guided to sites of infection by chemokines and newly expressed adhesion molecules.

The full activation of naive T cells takes 4–5 days and is accompanied by marked changes in their homing behavior. Effector cytotoxic CD8 T cells must travel from the peripheral lymphoid tissue in which they have been activated to attack and destroy infected cells. Effector CD4 T_H1 cells must also leave

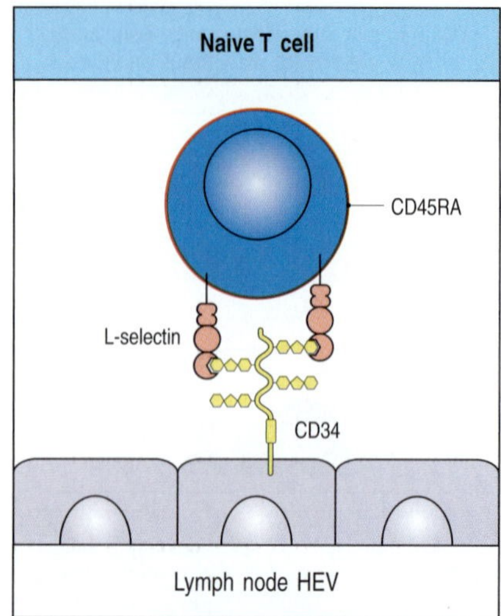

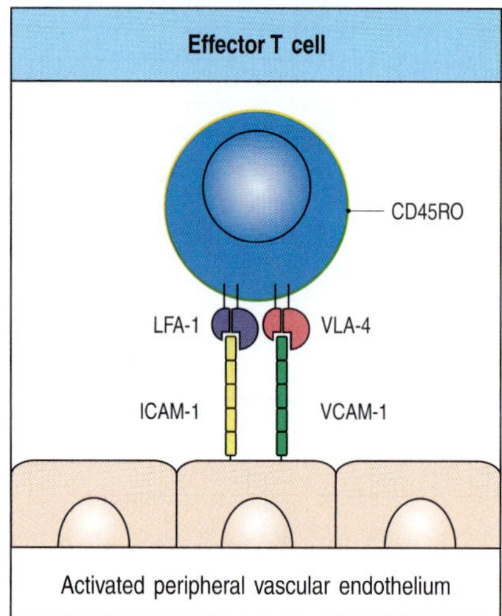

Fig. 11.9 Effector T cells change their surface molecules, allowing them to home to sites of infection. Naive T cells home to lymph nodes through the binding of L-selectin to sulfated carbohydrates displayed by various proteins, such as CD34 and GlyCAM-1, on the high endothelial venule (HEV, upper panel). After encounter with antigen, many of the differentiated effector T cells lose expression of L-selectin, leave the lymph node about 4–5 days later, and now express the integrin VLA-4 and increased levels of LFA-1. These bind to VCAM-1 and ICAM-1, respectively, on peripheral vascular endothelium at sites of inflammation (lower panel). On differentiating into effector cells, T cells also alter their splicing of the mRNA encoding the cell-surface protein CD45. The CD45RO isoform expressed by effector T cells lacks one or more exons that encode extracellular domains present in the CD45RA isoform expressed by naive T cells, and somehow makes effector T cells more sensitive to stimulation by specific antigen.

the lymphoid tissues to activate macrophages at the site of infection. Exit of effector T cells from lymphoid tissues is regulated by changes in their responsiveness to the chemotactic lipid sphingosine 1-phosphate, as described in Section 9-3. Most effector T cells cease production of L-selectin, which mediates homing to the lymph nodes, whereas the expression of other adhesion molecules is increased (Fig. 11.9). One important change is a marked increase in synthesis of the integrin $\alpha_4{:}\beta_1$, also known as VLA-4. This binds to the adhesion molecule VCAM-1, a member of the immunoglobulin superfamily that is induced on activated endothelial cell surfaces and initiates the extravasation of the effector T cells. Thus, if the innate immune response has already activated the endothelium at the site of infection, as described in Section 11-2, effector T cells will be rapidly recruited.

In the early stage of an immune response, only a few of the effector T cells that enter the infected tissue will be expected to be specific for pathogen, because any effector T cell specific for any antigen will also be able to enter. Specificity of the reaction is maintained, however, because only those effector T cells that recognize pathogen antigens will carry out their function, destroying infected cells or specifically activating pathogen-loaded macrophages. By the peak of an adaptive immune response, after several days of clonal expansion and differentiation, most of the recruited T cells will be specific for the infecting pathogen.

Not all infections trigger innate immune responses that activate local endothelial cells, and it is not so clear how effector T cells are guided to the sites of infection in these cases. However, activated T cells seem to enter all tissues in very small numbers, perhaps via adhesive interactions such as the binding of P-selectin on the endothelial cells to its ligand **P-selectin glycoprotein ligand-1** (**PSGL-1**), which is expressed by activated T cells, and so effector T cells could encounter their antigens even in the absence of a previous inflammatory response.

By the means just described, one or a few specific effector T cells encountering antigen in a tissue can initiate or augment a potent local inflammatory response that recruits many more effector lymphocytes and nonspecific inflammatory cells to that site. Effector T cells that recognize pathogen antigens in the tissues produce cytokines such as TNF-α, which activates endothelial cells to express E-selectin, VCAM-1, and ICAM-1, and chemokines such as CCL5, which acts on effector T cells to activate their adhesion molecules. VCAM-1 and ICAM-1 on endothelial cells bind VLA-4 and LFA-1, respectively, on effector T cells, recruiting more of these cells into tissues that contain the antigen. At the same time, monocytes and polymorphonuclear leukocytes are recruited by adhesion to E-selectin. TNF-α and IFN-γ released by activated T cells also act synergistically to change the shape of endothelial cells, allowing increased blood flow, increased vascular permeability, and increased emigration of leukocytes, fluid, and protein into the infected tissue. Thus, at the later stages of an infection, the protective effects of macrophages secreting TNF-α and other pro-inflammatory cytokines at the infected site (see Section 3-17) are reinforced by the actions of effector T cells.

By contrast, effector T cells that enter tissues but do not recognize their antigen are rapidly lost. They either enter the lymph from the tissues and eventually return to the bloodstream, or undergo apoptosis. Most of the T cells in the afferent lymph that drains tissues are memory or effector T cells, which characteristically express the CD45RO isoform of the cell-surface molecule CD45 and lack L-selectin (see Fig. 11.9). Effector T cells and memory T cells have similar phenotypes, as we discuss later, and both seem to be committed to migration through potential sites of infection. In addition to allowing effector T cells to clear all sites of infection, this pattern of migration allows them to contribute, along with memory cells, to protecting the host against reinfection with the same pathogen.

Expression of particular adhesion molecules can direct different subsets of effector T cells to specific sites. As we shall see in Chapter 12, the peripheral immune system is compartmentalized, such that different populations of lymphocytes migrate through different lymphoid compartments and—after activation—through the different tissues they serve. This is achieved by the selective expression of adhesion molecules that bind selectively to tissue-specific addressins. In this context, the adhesion molecules are often known as **homing receptors**. For example, some activated T cells specifically populate the skin. They are induced during activation to express the adhesion molecule **cutaneous lymphocyte antigen (CLA)** (Fig. 11.10). This is a specifically glycosylated isoform of PSGL-1 that binds to E-selectin on cutaneous vascular endothelium. CLA-expressing T lymphocytes also produce the chemokine receptor CCR4. This binds the chemokine CCL17 (TARC), which is present at high levels on the endothelium of cutaneous blood vessels. Interaction of CLA with E-selectin causes the T cell to roll against the wall of the vessel, and the signal delivered by endothelial CCL17 is thought to cause the arrest of lymphocytes and to bring about their adhesion to the wall, probably by inducing tight binding of integrin, as described for the action of CCL21 on naive T cells (see Section 9-3). The condition known as leukocyte adhesion deficiency is found in people who lack integrin β subunits, and such people suffer from recurrent infections with pyogenic bacteria and problems with wound healing. In addition to CCR4, skin-homing T cells carry the chemokine receptor CCR10 (GPR-2), which binds the chemokine CCL27 (CTACK) expressed by keratinocytes, the epithelial cells of the skin. Different combinations of chemokines and receptors are used by T cells that home to the mucosal immune system of the gut, and these are discussed in Chapter 12.

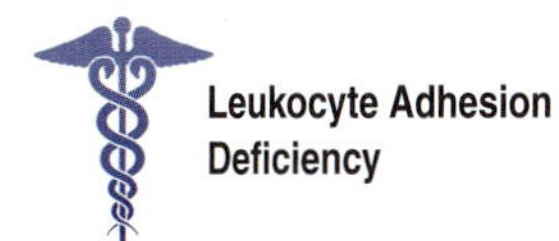

11-7 Differentiated effector T cells are not a static population but continue to respond to signals as they carry out their effector functions.

The commitment of CD4 T cells to become distinct lineages of effector cells begins in peripheral lymphoid tissues, such as lymph nodes, as described in Sections 11-3 and 11-4. However, the effector activities of these cells once they enter sites of infection are not defined simply by the signals received in the lymphoid tissues. Instead, evidence suggests that there is continuous regulation of the expansion and the effector activities of differentiated CD4 cells, in particular of T_H17 and T_H1 cells.

As noted in Chapter 9, commitment of naive T cells to become T_H17 cells is triggered by exposure to TGF-β and IL-6; commitment to T_H1 cells is initially triggered by IFN-γ. These initial conditions are not, however, sufficient to

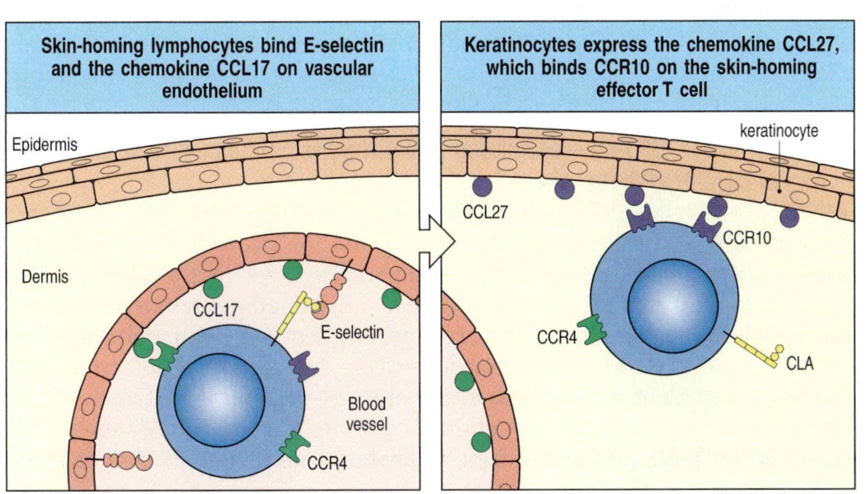

Fig. 11.10 Skin-homing T cells use specific combinations of integrins and chemokines to migrate specifically to the skin. Left panel: a skin-homing lymphocyte binds to the endothelium lining a cutaneous blood vessel by interactions between cutaneous lymphocyte antigen (CLA) and constitutively expressed E-selectin on the endothelial cells. The adhesion is strengthened by an interaction between lymphocyte chemokine receptor CCR4 and the endothelial chemokine CCL17. Right panel: once through the endothelium, effector T lymphocytes are attracted to keratinocytes of the epidermis by the chemokine CCL27, which binds to the receptor CCR10 on lymphocytes.

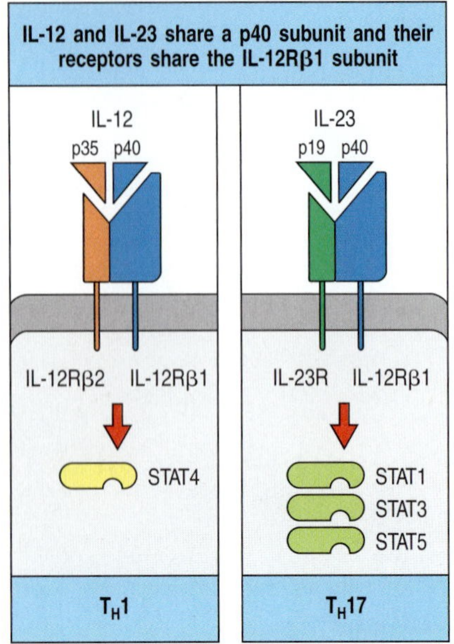

Fig. 11.11 The cytokines IL-12 and IL-23 share subunits, and their receptors have a common component. The dimeric cytokines IL-12 and IL-23 share the p40 subunit, and the receptors for IL-12 and IL-23 have the IL-12Rβ1 subunit in common. IL-12 signaling activates the transcriptional activators STAT1, STAT3, and STAT4, but its action in increasing IFN-γ production is due to STAT4. IL-23 activates several other STATs, but activates STAT4 only weakly. Both cytokines augment the activity and proliferation of the CD4 subsets that express receptors for them; T_H1 cells express IL-12R, and T_H17 cells express IL-23R. Mice deficient in p40 lack expression of both of these cytokines, and manifest immune defects as a result of deficiencies in both T_H1 and T_H17 activities.

generate complete or effective T_H17 or T_H1 responses. In addition, each T-cell subset also requires stimulation by another cytokine: IL-23 in the case of T_H17 cells, and IL-12 in the case of T_H1 cells. IL-23 and IL-12 are closely related in structure; each is a heterodimer and they share a subunit. IL-23 is composed of one p40 and one p19 subunit, whereas IL-12 has the p40 subunit and a unique p35 subunit. Committed T_H17 cells express a receptor for IL-23, whereas T_H1 cells express a receptor for IL-12. The receptors for IL-12 and IL-23 are also related, having a common subunit (Fig. 11.11).

IL-23 and IL-12 amplify the activities of T_H17 and T_H1 cells, respectively. Like many other cytokines, they both act through the JAK–STAT intracellular signaling pathway (see Fig. 7.29). IL-23 signaling activates the intracellular transcriptional activators STAT1, STAT3, and STAT5, but activates STAT4 very weakly. In contrast, IL-12 activates STAT1 and STAT3 and also strongly activates STAT4. IL-23 does not initiate the commitment of naive CD4 T cells to T_H17 cells, but it does stimulate their expansion. Many *in vivo* responses that depend on IL-17 are diminished in the absence of IL-23. For example, mice lacking the IL-23-specific subunit p19 show decreased production of IL-17 and IL-17F in the lung after infection by *Klebsiella pneumoniae.*

Mice lacking the p40 subunit, which is shared by IL-12 and IL-23, are deficient in both IL-23 and IL-12. This was the source of some confusion before the separate role of IL-23 in T_H17 activity was understood. It had been assumed that inflammation in the brains of mice with experimental autoimmune encephalomyelitis (EAE) was due to T_H1 cells and IFN-γ, because EAE could not be induced in p40-deficient mice. But later it was found that EAE could be induced in p35-deficient mice, which lack IL-12 but maintain IL-23, and that EAE cannot be induced in p19-deficient mice. It turns out that the brain inflammation in EAE is largely a result of the activity of T_H17 cells.

IL-12 regulates the effector activity of committed T_H1 cells at sites of infection, but other cytokines such as IL-18 may also be involved. Studies of two different pathogens have shown that the initial differentiation of T_H1 cells is not sufficient for protection, and that continuous signals are required. Mice deficient in p40 can resist initial infection by *T. gondii* as long as IL-12 is administered continuously. If IL-12 is administered during the first 2 weeks of infection, p40-deficient mice survive the initial infection and establish a latent chronic infection characterized by cysts containing the pathogen. When IL-12 administration is stopped, however, these mice gradually reactivate the latent cysts and the animals eventually die of toxoplasmic encephalitis. IFN-γ production by pathogen-specific T cells decreases in the absence of IL-12 but could be restored by IL-12 administration. Similarly, the adoptive transfer of differentiated T_H1 cells from mice cured of *L. major* protects RAG-deficient mice infected by *L. major*, but cannot protect p40-deficient mice (Fig. 11.12). Together, these experiments suggest that T_H1 cells continue to respond to signals during an infection, and that continuous IL-12 is needed to sustain the effectiveness of differentiated T_H1 cells against at least some pathogens.

11-8 Primary CD8 T-cell responses to pathogens can occur in the absence of CD4 T-cell help.

Many CD8 T-cell responses require help from CD4 T cells (see Section 9-16). This is typically the case where the antigen recognized by the CD8 T cells is derived from an agent that does not cause inflammation on initial infection. In such circumstances, CD4 T-cell help is required to activate dendritic cells to become able to stimulate a complete CD8 T-cell response, an activity that has been described as licensing of the antigen-presenting cell (see Section 9-6). Licensing involves the induction of co-stimulatory molecules such as B7, CD40, and 4-1BBL on the dendritic cell, which can then deliver signals that

Fig. 11.12 Continuous IL-12 is required for resistance to pathogens requiring T_H1 responses. Mice that have eliminated an infection with *Leishmania major* and have generated T_H1 cells specific to the pathogen were used as a source of T cells that were adoptively transferred into either RAG2-deficient mice, which lack T cells and B cells and cannot control *L. major* infection but can produce IL-12, or into mice lacking p40, which cannot produce IL-12. On subsequent infection of the RAG2-deficient mice, lesions did not enlarge because the transferred T_H1 cells conferred immunity. But despite the fact that the transferred cells were already differentiated T_H1 cells, they did not confer resistance to IL-12 p40-deficient mice in which a continuous source of IL-12 was not present.

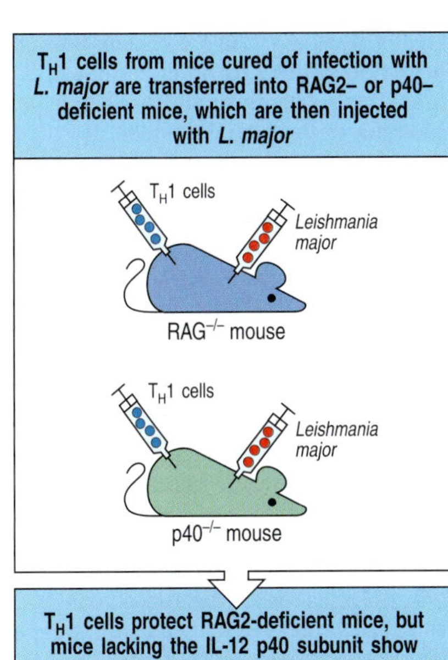
TH1 cells from mice cured of infection with L. major are transferred into RAG2– or p40–deficient mice, which are then injected with L. major
TH1 cells
Leishmania major
RAG−/− mouse
TH1 cells
Leishmania major
p40−/− mouse

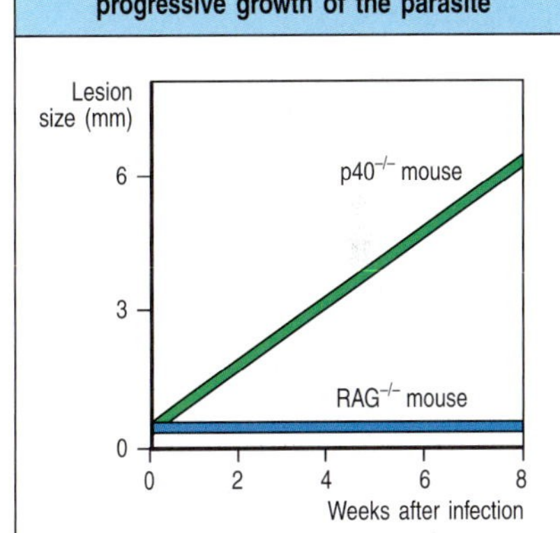
TH1 cells protect RAG2-deficient mice, but mice lacking the IL-12 p40 subunit show progressive growth of the parasite
Lesion size (mm)
6
3
0
p40−/− mouse
RAG−/− mouse
0
2
4
6
8
Weeks after infection

fully activate naive CD8 T cells (see Fig. 9.27). Licensing enforces a requirement for dual recognition of an antigen by the immune system by both CD4 and CD8 T cells, which provides a useful safeguard against autoimmunity. Dual recognition is also seen in the cooperation between T cells and B cells for antibody generation (see Chapter 10). However, not all CD8 T-cell responses require such help.

Some infectious agents, such as the intracellular Gram-positive bacterium *Listeria monocytogenes* and the Gram-negative bacterium *Burkholderia pseudomallei*, provide the inflammatory environment required to license dendritic cells and thus can induce primary CD8 T-cell responses without help from CD4 T cells. These pathogens carry a number of immunostimulatory signals, such as ligands for TLRs, and so can directly activate antigen-presenting cells to express the co-stimulatory molecules B7 and CD40. Thus, fully activated dendritic cells presenting *Listeria* or *Burkholderia* antigens can activate naive antigen-specific CD8 T cells without the help of CD4 T cells, and can induce them to undergo clonal expansion (Fig. 11.13). The activated dendritic cell also secretes cytokines such as IL-12 and IL-18, which act on naive CD8 T cells in a so-called 'bystander' effect to induce them to produce IFN-γ, which in turn induces other protective effects (see Fig. 11.13).

Primary CD8 T-cell responses to *L. monocytogenes* were examined in mice that were genetically deficient in MHC class II molecules and thus lacked CD4 T cells (see Section 8-18). The numbers of CD8 T cells specific for a particular antigen expressed by the pathogen were measured by using MHC tetramers (see Appendix I, Section A-28). On day 7 after infection, wild-type mice and mice lacking CD4 T cells show equivalent expansion, and equivalent cytotoxic capacity, of pathogen-specific CD8 T cells. Mice lacking CD4 T cells cleared the initial infection by *L. monocytogenes* as effectively as wild-type mice. These experiments clearly show that protective responses can be generated by pathogen-specific CD8 T cells without CD4 T-cell help. However, as we will see later, the nature of the CD8 memory response is different and is diminished in the absence of CD4 T-cell help.

Naive CD8 T cells can also undergo 'bystander' activation by IL-12 and IL-18 to produce IFN-γ very early during infection (see Fig. 11.13). Mice infected with *L. monocytogenes* or *B. pseudomallei* rapidly produce a strong IFN-γ response, which is essential for their survival. The source of this IFN-γ seems to be both the NK cells of innate immunity and naive CD8 T cells, which begin to secrete it within the first few hours after infection. This is believed to be too soon for any significant expansion of pathogen-specific CD8 T cells, which would initially be too rare to contribute in an antigen-specific manner. The production of IFN-γ by both NK and CD8 T cells at this early time can be blocked experimentally by antibodies against IL-12 and IL-18, suggesting that these cytokines are responsible. The source of the IL-12 and IL-18 was not identified in this experiment, but they are produced by macrophages and dendritic cells in response to activation via TLRs. These experiments indicate that naive CD8 T cells can contribute nonspecifically in a kind of innate defense, not requiring CD4 T cells, in response to early signals of infection.

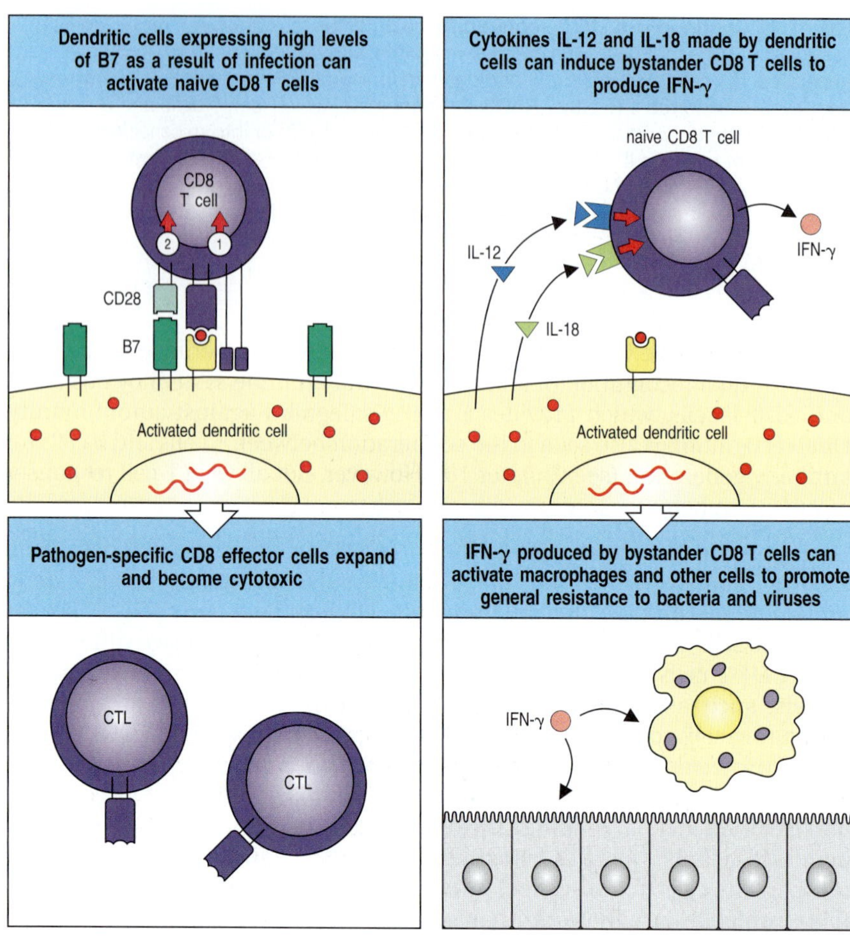
Dendritic cells expressing high levels of B7 as a result of infection can activate naive CD8 T cells
CD8 T cell
CD28
B7
Activated dendritic cell
Cytokines IL-12 and IL-18 made by dendritic cells can induce bystander CD8 T cells to produce IFN-γ
naive CD8 T cell
IL-12
IL-18
IFN-γ
Activated dendritic cell
Pathogen-specific CD8 effector cells expand and become cytotoxic
CTL
CTL
IFN-γ produced by bystander CD8 T cells can activate macrophages and other cells to promote general resistance to bacteria and viruses
IFN-γ

Fig. 11.13 Naive CD8 T cells can be activated directly by potent antigen-presenting cells through their T-cell receptor or through the action of cytokines. Left panels: naive CD8 T cells that encounter peptide:MHC class I complexes on the surface of dendritic cells expressing high levels of co-stimulatory molecules as a result of the inflammatory environment produced by some pathogens (upper left panel) are activated to proliferate in response, eventually differentiating into cytotoxic CD8 T cells (lower left panel). Right panels: activated dendritic cells also produce the cytokines IL-12 and IL-18, whose combined effect on CD8 T cells rapidly induces the production of IFN-γ (upper right panel). This activates macrophages for the destruction of intracellular bacteria and can promote antiviral responses in other cells (lower right panel).

11-9 Antibody responses develop in lymphoid tissues under the direction of T_{FH} cells.

As described in Chapter 10, B cells specific for a protein antigen cannot be activated to proliferate, form germinal centers, or differentiate into plasma cells until they encounter a helper T cell that is specific for one of the peptides derived from that antigen. Recall that naive antigen-binding B cells first enter the follicle, but when they encounter antigen, CCR7 is expressed and they move toward the boundary of the B-cell and T-cell zones (see Section 10-5). Similarly, in the T-cell zones, some naive T cells that are activated by antigen-bearing dendritic cells become follicular T helper (T_{FH}) cells that express CXCR5; these move toward the boundary of the B-cell and T-cell zones, where they can interact with recently activated B cells (see Fig. 10.5). Through linked recognition (see Section 10-3), this initial T cell–B cell interaction gives rise to the primary focus after about 5 days, and later can generate a germinal center reaction in the follicle that facilitates B cells to undergo isotype switching and affinity maturation (see Sections 5-18 and 10-10). Some B cells activated in the primary focus migrate to the medullary cords of the lymph node, or to those parts of the red pulp that are next to the T-cell zones of the spleen, where they become plasma cells and secrete specific antibody for a few days (see Fig. 10.7). The final products of the germinal center response are high-affinity memory B cells and long-lived plasma cells, which are important for the maintenance of serum antibody levels and long-lasting immunity.

Antigen is retained for very long periods in lymphoid follicles in the form of antigen:antibody complexes on the surface of the local follicular dendritic cells. The antigen:antibody complexes, which become coated with fragments of C3, are held on the cell by receptors for the complement fragments (CR1, CR2, and CR3) as well as by a nonphagocytic Fc receptor (see Fig. 10.16). The function of this long-lived antigen is unclear, because there is evidence that it is not absolutely required for the stimulation of B cells in the germinal center (see Section 10-10), but it may regulate the long-term antibody response.

11-10 Antibody responses are sustained in medullary cords and bone marrow.

The B cells activated in primary foci migrate either to adjacent follicles or to local extrafollicular sites of proliferation. B cells grow exponentially in these sites for 2–3 days and undergo six or seven cell divisions before the progeny come out of the cell cycle and form antibody-producing plasma cells *in situ* (Fig. 11.14, upper panel). Most of these plasma cells have a life-span of 2–4 days, after which they undergo apoptosis. About 10% of plasma cells in these extrafollicular sites live longer; their origin and ultimate fate are unknown. Surviving B cells that have undergone class switching and affinity maturation in the germinal center will either become memory cells or can leave the germinal center as plasmablasts (pre-plasma cells) to become relatively long-lived antibody-producing cells in other locations (see Sections 10-7 to 10-9).

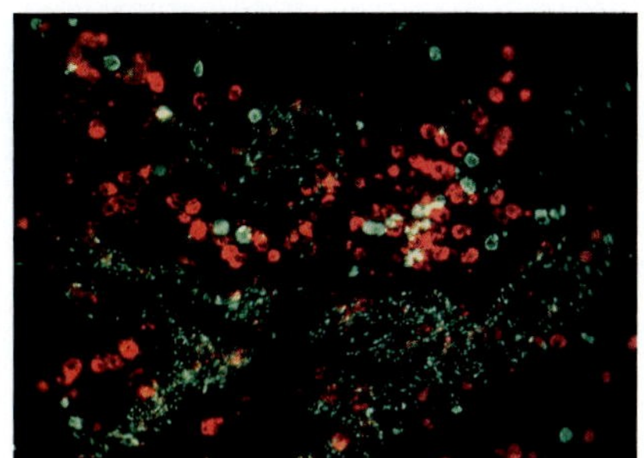

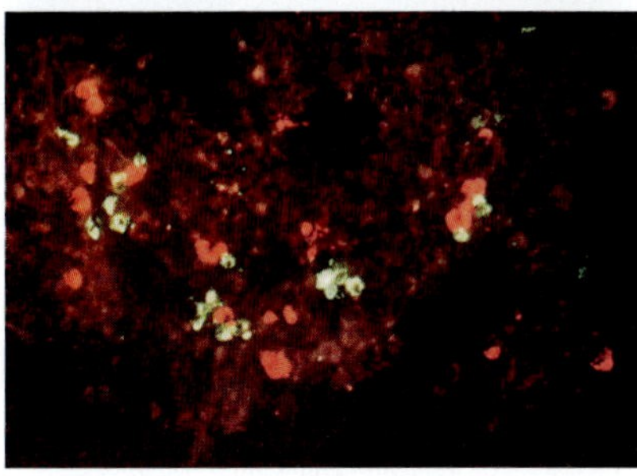

Fig. 11.14 Plasma cells are dispersed in medullary cords and bone marrow. In these sites they secrete antibody at high rates directly into the blood for distribution to the rest of the body. In the upper micrograph, plasma cells in lymph node medullary cords are stained green (with fluorescein anti-IgA) if they are secreting IgA, and red (with rhodamine anti-IgG) if they are secreting IgG. The plasma cells in these local extrafollicular sites are short lived (2–4 days). The lymphatic sinuses are outlined by green granular staining selective for IgA. In the lower micrograph, longer-lived plasma cells (3 weeks to 3 months or more) in the bone marrow are revealed with antibodies specific for light chains (fluorescein anti-λ and rhodamine anti-κ stain). Plasma cells secreting immunoglobulins containing λ light chains show on this micrograph as yellow. Those secreting immunoglobulins containing κ light chains stain red. Photographs courtesy of P. Brandtzaeg.

Plasmablasts originating in the follicles of Peyer's patches and mesenteric lymph nodes migrate via lymph to the blood and then enter the lamina propria of the gut and other epithelial surfaces. Those originating in peripheral lymph node or splenic follicles migrate to the bone marrow (Fig. 11.14, lower panel). In these distant sites of antibody production, the plasmablasts differentiate into plasma cells that have a life-span of months to years. These cells are thought to provide the antibody that can be present in the blood for years after the initial immune response. Whether this supply of plasma cells is replenished by the continual but occasional differentiation of memory cells is not yet known. Studies of responses to nonreplicating antigens show that germinal centers are present for only 3–4 weeks after initial exposure to antigen. However, small numbers of B cells continue to proliferate in the follicles for months and may be the precursors of antigen-specific plasma cells in the mucosa and bone marrow throughout the subsequent months and years.

11-11 The effector mechanisms used to clear an infection depend on the infectious agent.

Most infections engage both the cell-mediated and the humoral aspects of immunity, and in many cases both are helpful in clearing or containing the pathogen and setting up protective immunity, as shown in Fig. 11.15, although the relative importance of the different effector mechanisms, and the effective classes of antibody involved, vary with different pathogens. As we learned in Chapter 9, cytotoxic T cells are important in destroying virus-infected cells, and in some viral diseases they are the predominant class of lymphocytes present in the blood during a primary infection. Nevertheless, the role of antibodies in clearing viruses from the body and preventing them from getting a hold should not be forgotten. Ebola virus causes a hemorrhagic fever and is one of the most lethal viruses known, but some patients do survive and some people even become infected but remain asymptomatic. In both cases, a strong antiviral IgG response early in infection seems to be essential for survival. The antibody response seems to clear the virus from the bloodstream and gives the patient time to activate cytotoxic T cells. In contrast, this antibody response did not occur in infections that proved fatal; the virus continued to replicate, and even though there was some activation of T cells, the disease progressed.

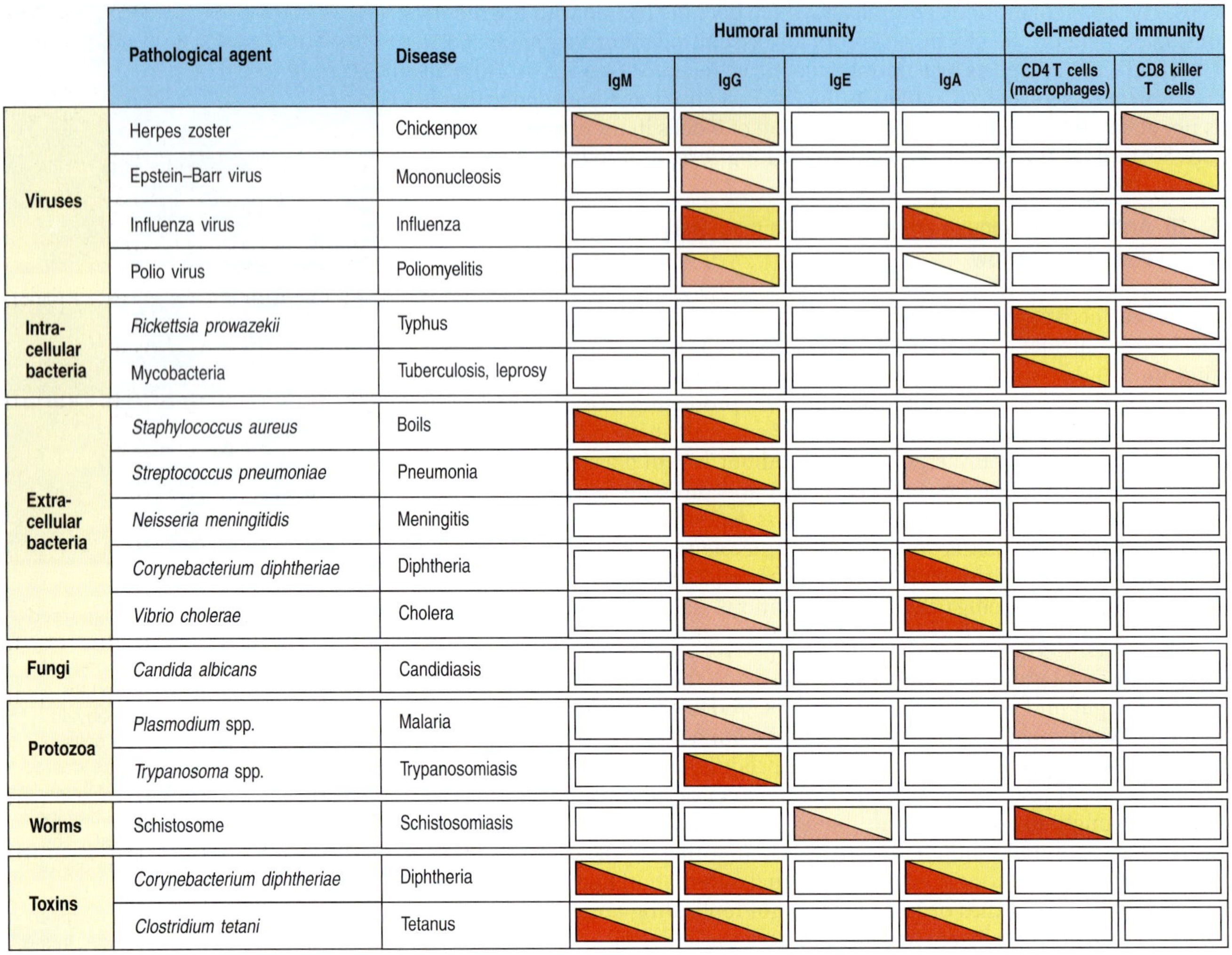

	Pathological agent	Disease	Humoral immunity				Cell-mediated immunity	
			IgM	IgG	IgE	IgA	CD4 T cells (macrophages)	CD8 killer T cells
Viruses	Herpes zoster	Chickenpox						
	Epstein–Barr virus	Mononucleosis						
	Influenza virus	Influenza						
	Polio virus	Poliomyelitis						
Intra-cellular bacteria	*Rickettsia prowazekii*	Typhus						
	Mycobacteria	Tuberculosis, leprosy						
Extra-cellular bacteria	*Staphylococcus aureus*	Boils						
	Streptococcus pneumoniae	Pneumonia						
	Neisseria meningitidis	Meningitis						
	Corynebacterium diphtheriae	Diphtheria						
	Vibrio cholerae	Cholera						
Fungi	*Candida albicans*	Candidiasis						
Protozoa	*Plasmodium* spp.	Malaria						
	Trypanosoma spp.	Trypanosomiasis						
Worms	Schistosome	Schistosomiasis						
Toxins	*Corynebacterium diphtheriae*	Diphtheria						
	Clostridium tetani	Tetanus						

Fig. 11.15 Different effector mechanisms are used to clear primary infections with different classes of pathogens and to protect against reinfection. The defense mechanisms used to clear a primary infection are identified by red shaded boxes. Yellow shading indicates a role in protective immunity. Paler shades indicate less well established mechanisms. It is clear that classes of pathogens elicit similar protective immune responses, reflecting similarities in their lifestyles. The CD4 responses indicated on this diagram refer only to those involved in macrophage activation. In addition, in virtually all diseases, helper CD4 T-cell responses will be involved in stimulating antibody production, class switching, and the production of memory cells.

Cytotoxic T cells are also required for the destruction of cells infected with some intracellular bacterial pathogens, such as *Rickettsia*, the causative agent of typhus. In contrast, mycobacteria, which live inside macrophage vesicles, are mainly kept in check by CD4 T_H1 cells, which activate infected macrophage to kill the bacteria. Antibodies are the principal immune reactants that clear primary infections with common extracellular bacteria such as *Staphylococcus aureus* and *Streptococcus pneumoniae*. The IgM and IgG antibodies produced against components of the bacterial surface coat opsonize the bacteria and make them more susceptible to phagocytosis.

Figure 11.15 also indicates the mechanisms involved in immunity to reinfection, or protective immunity, against the pathogens listed. Inducing protective immunity is the goal of vaccine development, which we discuss more detail in Chapter 16. To achieve protective immunity requires a vaccine to induce an adaptive immune response that has both the antigen specificity and the appropriate functional elements to combat the particular pathogen concerned. Pathogens carry multiple epitopes for both B cells and T cells, and thus generate diverse antibody and T-cell responses, but not all of these will be equally effective in clearing the disease. Protective immunity consists of two components—immune reactants, such as antibody or effector T cells generated in the initial infection or by vaccination, and long-lived immunological memory (Fig. 11.16).

The type of antibody or effector T cell that offers protection depends on the infectious strategy and lifestyle of the pathogen. Thus, when opsonizing

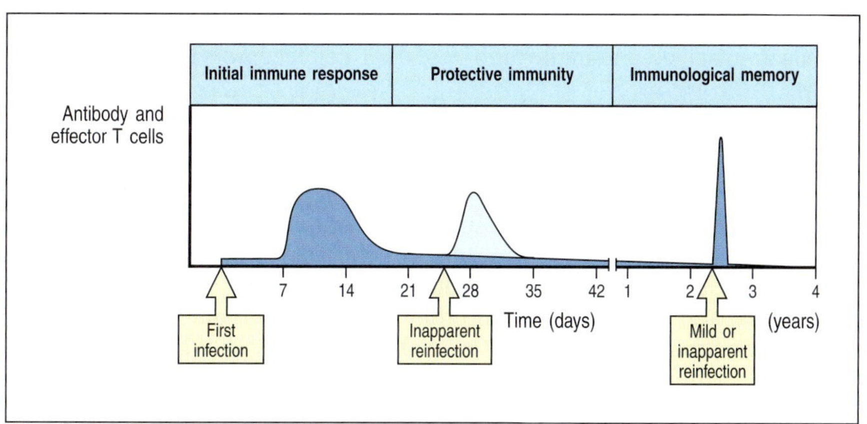

Fig. 11.16 Protective immunity consists of preformed immune reactants and immunological memory. The first time a particular pathogen is encountered, pathogen-specific antibody and effector T cells are produced. After the infection has been cleared, their levels gradually decline. An early reinfection with the same pathogen is rapidly cleared by these preformed immune reactants. There are few symptoms, but levels of immune reactants are found to increase temporarily (light blue peak). Reinfection years afterward leads to an immediate and rapid increase in pathogen-specific antibody and effector T cells as a result of immunological memory, and disease symptoms are mild or inapparent.

antibodies such as IgG1 are present (see Section 10-14), opsonization and phagocytosis of extracellular pathogens will be more efficient. If specific IgE is present, pathogens will trigger activation of mast cells, rapidly initiating an inflammatory response through the release of histamine and leukotrienes. In many cases the most efficient protective immunity is mediated by neutralizing antibody that can prevent pathogens from establishing an infection, and most of the established vaccines against acute childhood viral infections work primarily by inducing protective antibodies. Effective immunity against polio virus, for example, requires preexisting antibody (see Fig. 11.15), because the virus rapidly infects motor neurons and destroys them unless it is immediately neutralized by antibody and prevented from spreading within the body. In polio, specific IgA on mucosal epithelial surfaces also neutralizes the virus before it enters the tissues. Thus, protective immunity can involve effector mechanisms (IgA in this case) that do not operate in the elimination of the primary infection.

When a primary adaptive immune response is successful in halting an infection, it will often clear the primary infection from the body by the effector mechanisms discussed in Chapters 9 and 10. However, as we discuss in Chapter 13, many pathogens evade complete clearance and persist for the life of the host. The virus herpes zoster, which causes chickenpox on primary infection, then lies latent in the body for years without causing disease, but can later in life, or if the body is stressed, become reactivated and cause shingles.

11-12 Resolution of an infection is accompanied by the death of most of the effector cells and the generation of memory cells.

When an infection is effectively repelled by the adaptive immune system, two things occur. The actions of effector cells remove the specific stimulus that originally recruited them. In the absence of this stimulus, the cells then undergo 'death by neglect,' removing themselves by apoptosis. The dying cells are rapidly cleared by phagocytes and other cells, which recognize the membrane lipid phosphatidylserine. This lipid is normally found only on the inner surface of the plasma membrane, but in apoptotic cells it rapidly redistributes to the outer surface, where it can be recognized by specific receptors on many cells. Thus, not only does the ending of infection lead to the removal of the pathogen, it also leads to the loss of most of the pathogen-specific effector cells.

However, some of the effector cells are retained, and these provide the raw material for memory T-cell and B-cell responses. These are crucially important to the operation of the adaptive immune system. Memory T cells, in particular, are retained virtually forever. The mechanisms underlying the decision to induce apoptosis in most effector cells and to retain only a few are only now being discovered and are still not well understood. It seems likely that the answers will lie in the cytokines produced by the environment and by the T cells themselves, and in the affinity of the T-cell receptors for their antigens.

Summary.

The adaptive immune response is required for effective protection of the host against pathogenic microorganisms. The response of the innate immune system to pathogens helps to initiate the adaptive immune response. Pathogens cause the activation of dendritic cells to full antigen-presenting cell status, and interactions with other cells of the innate immune system lead to the production of cytokines that direct the quality of the CD4 T-cell response. Pathogen antigens are transported to local lymphoid organs by the migrating dendritic cells and are presented to antigen-specific naive T cells that continuously recirculate through the lymphoid organs. T-cell priming and the differentiation of effector T cells occur here, and the effector T cells either leave the lymphoid organ to provide cell-mediated immunity in sites of infection in the tissues or remain in the lymphoid organ to participate in humoral immunity by activating antigen-binding B cells. Distinct types of CD4 T cells develop in response to infection by different types of pathogens. T_H17 responses provide for the generation of acute inflammation at sites of infection by robust recruitment of neutrophils. T_H1 responses activate the phagocytic pathways to protect against intracellular pathogens. T_H2 responses are directed against infections by parasites such as helminths by promoting barrier and mucosal immunity, IgE production, and recruitment of eosinophils to sites of infection. CD8 T cells have an important role in protective immunity, especially in protecting the host against infection by viruses and intracellular infections by *Listeria* and other microbial pathogens that have special means for entering the host cell's cytoplasm. Primary CD8 T-cell responses to pathogens usually require CD4 T-cell help, but can occur in response to some pathogens without such help. CD4-independent responses can lead either to the generation and expansion of antigen-specific cytotoxic T cells or to the nonspecific activation of naive CD8 T cells to secrete IFN-γ, which in turn contributes to host protection. Ideally, the adaptive immune response eliminates the infectious agent and provides the host with a state of protective immunity against reinfection with the same pathogen.

Immunological memory.

Having considered how an appropriate primary immune response is mounted, we now turn to how long-lasting protective immunity is generated. The establishment of immunological memory is perhaps the most important consequence of an adaptive immune response, because it enables the immune system to respond more rapidly and effectively to pathogens that have been encountered previously, and prevents them from causing disease. Memory responses, which are called **secondary immune responses**, **tertiary immune responses**, and so on, depending on the number of exposures to antigen, also differ qualitatively from primary responses. This is particularly clear in the antibody response, in which the characteristics of antibodies produced in secondary and subsequent responses are distinct from those produced in the primary response to the same antigen. Memory T-cell responses can also be distinguished qualitatively from the responses of naive or effector T cells. The principal focus of this part of the chapter is the altered character of memory responses, although we also discuss emerging explanations of how immunological memory persists after exposure to antigen.

11-13 Immunological memory is long-lived after infection or vaccination.

Most children in developed countries are now vaccinated against measles virus; before vaccination was widespread, most were naturally exposed to this virus and developed an acute, unpleasant, and potentially dangerous illness. Whether through vaccination or infection, children exposed to the virus acquire long-term protection from measles, lasting for most people for the whole of their life. The same is true of many other acute infectious diseases (see Chapter 16): this state of protection is a consequence of immunological memory.

The basis of immunological memory has been hard to explore experimentally. Although the phenomenon was first recorded by the ancient Greeks and has been exploited routinely in vaccination programs for more than 200 years, it is only now becoming clear that this memory reflects a small population of specialized **memory cells** formed during the adaptive immune response that can persist in the absence of the antigen that originally induced them. This mechanism of maintaining memory is consistent with the finding that only individuals who were themselves previously exposed to a given infectious agent are immune, and that memory is not dependent on repeated exposure to infection as a result of contacts with other infected individuals. This was established by observations made of populations on remote islands, where a virus such as measles can cause an epidemic, infecting all people living on the island at that time, after which the virus disappears for many years. On reintroduction from outside the island, the virus does not affect the original population but causes disease in those people born since the first epidemic.

Studies have attempted to determine the duration of immunological memory by evaluating responses in people who received vaccinia, the virus used to immunize against smallpox (Fig. 11.17). Because smallpox was eradicated in 1978, it is presumed that their responses represent true immunological memory and are not due to restimulation from time to time by the smallpox virus. The study found strong vaccinia-specific CD4 and CD8 T-cell memory responses as long as 75 years after the original immunization, and from the strength of these responses it was estimated that the memory response had an approximate half-life of between 8 and 15 years. Half-life represents the time that a response takes to reduce to 50% of its original strength. Titers of antivirus antibody remained stable, without measurable decline.

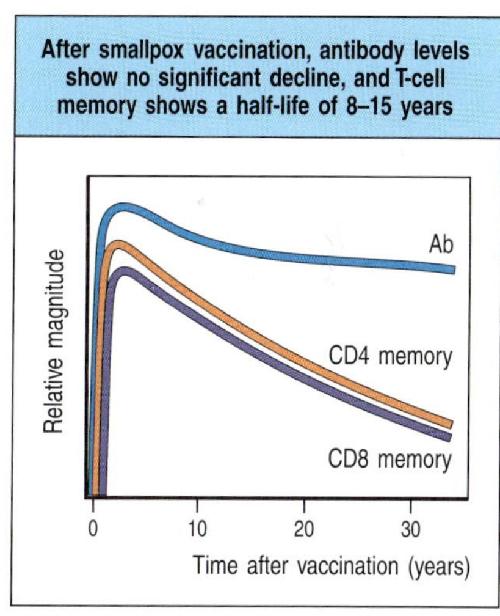

Fig. 11.17 Antiviral immunity after smallpox vaccination is long lived. Because smallpox has been eradicated, recall responses measured in people who were vaccinated for smallpox can be taken to represent true memory in the absence of reinfection. After smallpox vaccination, antibody levels show an early peak with a period of rapid decay, which is followed by long-term maintenance that shows no significant decay. CD4 and CD8 T-cell memory is long-lived but gradually decays, with a half-life in the range 8–15 years.

These findings show that immunological memory need not be maintained by repeated exposure to infectious virus. Instead, it is likely that memory is sustained by long-lived antigen-specific lymphocytes that were induced by the original exposure and that persist until a second encounter with the pathogen. Although most of the memory cells are in a resting state, careful studies have shown that a small percentage of these cells are dividing at any one time. What stimulates this infrequent cell division is unclear, but it is likely that cytokines produced either constitutively or during antigen-specific immune responses directed at other, non-cross-reactive, antigens are responsible. The number of memory cells for a given antigen is highly regulated, remaining practically constant during the memory phase, which reflects a control mechanism that maintains a balance between cell proliferation and cell death.

Immunological memory can be measured experimentally in various ways. Adoptive transfer assays (see Appendix I, Section A-36) of lymphocytes from animals immunized with simple, nonliving antigens have been favored for such studies, because the antigen cannot proliferate. In these experiments, the existence of memory cells is measured purely in terms of the transfer of specific responsiveness from an immunized, or 'primed,' animal to a nonimmunized recipient, as tested by a subsequent immunization with the antigen. Animals that received memory cells have a faster and more robust response to antigen challenge than do controls that did not receive cells, or that received cells from nonimmune donors.

Experiments like these have shown that when an animal is first immunized with a protein antigen, functional helper T-cell memory against that antigen appears abruptly and reaches a maximum after 5 days or so. Functional antigen-specific B-cell memory appears some days later, then enters a phase of proliferation and selection in lymphoid tissue. By 1 month after immunization, memory B cells are present at their maximum level. These levels of memory cells are then maintained, with little alteration, for the lifetime of the animal. It is important to recognize that the functional memory elicited in these experiments can be due to the precursors of memory cells as well as the memory cells themselves. These precursors are probably activated T cells and B cells, some of whose progeny will later differentiate into memory cells. Thus, precursors to memory can appear very shortly after immunization, even though resting memory-type lymphocytes may not yet have developed.

In the following sections we look in more detail at the changes that occur in lymphocytes after antigen priming that lead to the development of resting memory lymphocytes, and discuss the mechanisms that might account for these changes.

11-14 Memory B-cell responses differ in several ways from those of naive B cells.

Immunological memory in B cells can be examined quite conveniently *in vitro* by isolating B cells from immunized or unimmunized mice and restimulating them with antigen in the presence of helper T cells specific for the same antigen (Fig. 11.18). B cells from immunized mice produce responses that differ both quantitatively and qualitatively compared with naive B cells from unimmunized mice. B cells that respond to the antigen increase in frequency by up to 100-fold after their initial priming in the primary immune response, and, as a result of the process of affinity maturation (see Section 10-8), produce antibody of higher affinity than unprimed B lymphocytes. The response observed from immunized mice is due to **memory B cells**, which we introduced in Section 11-9 as B cells that arise from the germinal center reaction. These cells populate the spleen and lymph nodes as well as circulating through the blood, and express some markers that distinguish them from naive B cells and plasma cells. In the human, a marker of memory B cells is CD27, a member of the TNF receptor family that is also expressed by naive T cells and binds the TNF-family ligand CD70, which is expressed by dendritic cells (see Section 9-13).

	Source of B cells	
	Unimmunized donor Primary response	Immunized donor Secondary response
Frequency of antigen-specific B cells	1:10^4 – 1:10^5	1:10^2 – 1:10^3
Isotype of antibody produced	IgM > IgG	IgG, IgA
Affinity of antibody	Low	High
Somatic hypermutation	Low	High

Fig. 11.18 The generation of secondary antibody responses from memory B cells is distinct from the generation of the primary antibody response. These responses can be studied and compared by isolating B cells from immunized and unimmunized donor mice, and stimulating them in culture in the presence of antigen-specific effector T cells. The primary response usually consists of antibody molecules made by plasma cells derived from a quite diverse population of precursor B cells specific for different epitopes of the antigen and with receptors with a range of affinities for the antigen. The antibodies are of relatively low affinity overall, with few somatic mutations. The secondary response derives from a far more limited population of high-affinity B cells, which have, however, undergone significant clonal expansion. Their receptors and antibodies are of high affinity for the antigen and show extensive somatic mutation. The overall effect is that although there is usually only a 10–100-fold increase in the frequency of activatable B cells after priming, the quality of the antibody response is radically altered, in that these precursors induce a far more intense and effective response.

A primary antibody response is characterized by the initial rapid production of IgM, accompanied by an IgG response, due to class switching, which lags slightly behind it (Fig. 11.19). The secondary antibody response is characterized in its first few days by the production of relatively small amounts of IgM antibody and much larger amounts of IgG antibody, with some IgA and IgE. At the beginning of the secondary response, the source of these antibodies is memory B cells that were generated in the primary response and that have already switched from IgM to another isotypes, and express either IgG, IgA, or IgE on their surface, as well as a somewhat higher level of MHC class II molecules and B7.1 than is typical of naive B cells. The average affinity of IgG antibodies increases throughout the primary response and continues to increase during the ongoing secondary and subsequent responses (see Fig. 11.19). The higher affinity of memory B cells for antigen and their higher levels of cell-surface MHC class II molecules facilitate antigen uptake and presentation, which together with an increased expression of co-stimulatory molecules allows memory B cells to initiate their crucial interactions with helper T cells at lower doses of antigen than do naive B cells. This means that B-cell differentiation and antibody production start earlier after antigen stimulation than in the primary response. The secondary response is characterized by a more vigorous and earlier generation of plasma cells than in the primary response, thus accounting for the almost immediate abundant production of IgG (see Fig. 11.19).

The antibodies made in primary and secondary responses can be clearly distinguished in cases in which the primary response is dominated by antibodies that are closely related to each other and show little somatic hypermutation. This occurs in some inbred mouse strains in which certain haptens are recognized by a limited set of naive B cells. For example, in C57BL/6 mice, antibodies against the hapten nitrophenol (NP) are encoded by the same V_H (VH186.2) and V_L (λ1) genes in all animals of the strain. As a result of this uniformity in the primary response, changes in the antibody molecules produced in secondary responses to the same antigens are easy to observe. These differences include not only numerous somatic hypermutations in antibodies containing the dominant V regions (see Fig. 5.24) but also the addition of antibodies containing V_H and V_L gene segments not detected in the primary response. These are thought to derive from B cells that were activated at low frequency during the primary response, and thus were not detected, and which differentiated into memory B cells.

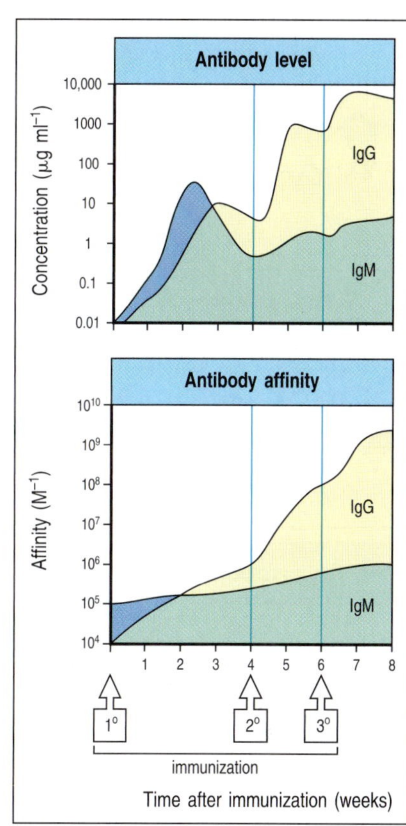

Fig. 11.19 Both the affinity and the amount of antibody increase with repeated immunization. The upper panel shows the increase in antibody concentration with time after a primary (1°), followed by a secondary (2°) and a tertiary (3°), immunization; the lower panel shows the increase in the affinity of the antibodies (affinity maturation). Affinity maturation is seen largely in IgG antibody (as well as in IgA and IgE, which are not shown) coming from mature B cells that have undergone isotype switching and somatic hypermutation to yield higher-affinity antibodies. The blue shading represents IgM on its own; the yellow shading IgG, and the green shading the presence of both IgG and IgM. Although some affinity maturation occurs in the primary antibody response, most arises in later responses to repeated antigen injections. Note that these graphs are on a logarithmic scale; it would otherwise be impossible to represent the overall increase of around a million-fold in the concentration of specific IgG antibody from its initial level.

11-15 Repeated immunization leads to increasing affinity of antibody due to somatic hypermutation and selection by antigen in germinal centers.

In secondary and subsequent immune responses, any antibodies persisting from previous responses are immediately available to bind the newly introduced pathogen. These antibodies divert the antigen to phagocytes for degradation and disposal (see Section 10-22), and if there is sufficient antibody to clear or inactivate the pathogen completely, it is possible that no secondary immune response will ensue. If the levels of pathogen overwhelm the amount of circulating antibody, excess antigens will bind to receptors on B cells and initiate a secondary B-cell response in the peripheral lymphoid organs. B cells with the highest avidity for antigen are the first to be recruited into this secondary response, and so memory B cells, which have already been selected for their avidity to antigen, make up a substantial part of the cells that contribute to the secondary response.

Secondary B-cell responses begin with the proliferation of B cells and T cells at the interface between the T-cell and B-cell zones, as in primary responses. Memory T cells reside in lymphoid tissues, but can also enter nonlymphoid tissues (see Section 11-6). Memory B cells, in contrast, continue to recirculate

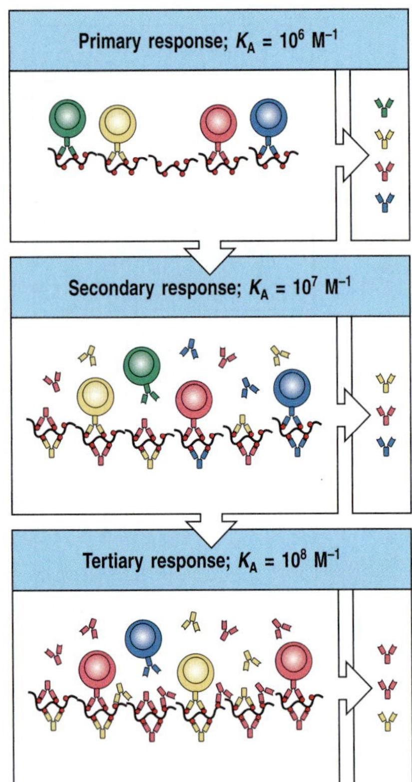

Fig. 11.20 The mechanism of affinity maturation in an antibody response. At the beginning of a primary response, B cells with receptors of a wide variety of affinities (K_A), most of which will bind antigen with low affinity, take up antigen, present it to helper T cells, and become activated to produce antibody of varying and relatively low affinity (top panel). These antibodies then bind and clear antigen, so that only those B cells with receptors of the highest affinity can continue to capture antigen and interact effectively with helper T cells. Such B cells will therefore be selected to undergo further expansion and clonal differentiation, and the antibodies they produce will dominate a secondary response (middle panel). These higher-affinity antibodies will in turn compete for antigen and select for the activation of B cells bearing receptors of still higher affinity in the tertiary response (bottom panel).

through the same secondary lymphoid compartments as naive B cells, principally the follicles of the spleen, lymph nodes, and the Peyer's patches of the gut mucosa. Memory B cells that have picked up antigen are able to present peptide:MHC class II complexes to their cognate helper T cells surrounding the follicle. Contact between the antigen-presenting B cells and helper T cells leads to the rapid proliferation of both the B cells and T cells. As the higher-affinity memory B cells compete most effectively for antigen, these B cells are most efficiently stimulated in the secondary immune response. Reactivated B cells that have not yet undergone differentiation into plasma cells migrate into the follicle and become germinal center B cells. There, they enter a second round of proliferation, during which the DNA encoding their immunoglobulin V domains undergoes somatic hypermutation, before differentiating into antibody-secreting plasma cells (see Section 10-8). The affinity of the antibodies produced rises progressively and rapidly, because B cells with the highest-affinity antigen receptors produced by somatic hypermutation bind antigen most efficiently, and will be selected to proliferate by their interactions with antigen-specific helper T cells in the germinal center (Fig. 11.20).

Memory B cells may not produce all antibodies in the secondary response. On secondary exposure to antigen, preexisting antibody can sometimes permit the formation of immune complexes, which do not form immediately in the primary response. Recent studies have shown that such immune complexes can bind to and activate signaling by Fc receptors on antigen-specific naive B cells, which acts to accelerate their kinetics of response. The formation of immune complexes, however, requires equivalent levels of antigen and antibody, which may not always occur, for example if antibody is present in great excess. Still, the shorter lag phase in antibody production in the secondary response may arise not only from the faster intrinsic responses of memory B cells, but also from accelerated naive B-cell responses if antigen:antibody complexes can form.

11-16 Memory T cells are increased in frequency compared with naive T cells specific for the same antigen, and have distinct activation requirements and cell-surface proteins that distinguish them from effector T cells.

Because the T-cell receptor does not undergo class switching or somatic hypermutation, it is not as easy to identify a memory T cell unequivocally as it is to identify a memory B cell. After immunization, the number of T cells reactive to a given antigen increases markedly as effector T cells are produced, and then falls back to persist at a level 100–1000-fold above the initial frequency for the rest of the life of the animal or person (Fig. 11.21). These persisting cells are designated **memory T cells**. They are long-lived cells with a particular set of cell-surface proteins, responses to stimuli, and expression of genes that control cell survival. Overall, their cell-surface proteins are similar to those of effector cells, but there are some distinctive differences (Fig. 11.22). In B cells, there is an obvious distinction between effector and memory cells, because effector B cells are terminally differentiated plasma cells that have already been activated to secrete antibody until they die.

A major problem in experiments aimed at establishing the existence of memory T cells is that many assays for T-cell effector function take several days, during which the putative memory T cells are reinduced to effector cell status. Thus, assays requiring several days do not distinguish preexisting effector cells from memory T cells, because memory cells can acquire effector activity during the period of the assay. This problem does not apply to cytotoxic T cells, however, because cytotoxic effector T cells can program a target cell for lysis in 5 minutes, whereas memory CD8 T cells need more time

Fig. 11.21 Generation of memory T cells after a virus infection. After an infection, in this case a reactivation of latent cytomegalovirus (CMV), the number of T cells specific for viral antigen increases dramatically and then falls back to give a sustained low level of memory T cells. The upper panel shows the numbers of T cells (orange); the lower panel shows the course of the virus infection (blue), as estimated by the amount of viral DNA in the blood. Data courtesy of G. Aubert.

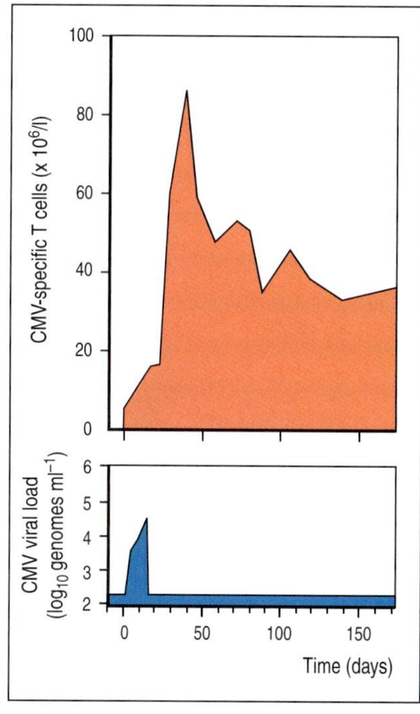

than this to be reactivated to become cytotoxic. Thus, their cytotoxic actions will appear later than those of any preexisting effector cells, even though they can become activated without undergoing DNA synthesis, as shown by studies conducted in the presence of mitotic inhibitors.

Recently it has become possible to track particular clones of antigen-specific CD8 T cells by staining them with tetrameric peptide:MHC complexes (see Appendix I, Section A-28). It has been found that the number of antigen-specific CD8 T cells increases markedly during an infection, and then decreases by up to 100-fold; nevertheless, this final level is distinctly higher than before priming. These cells continue to express some markers characteristic of activated cells, such as CD44, but stop expressing other activation markers, such

Protein	Naive	Effector	Memory	Comments
CD44	+	+++	+++	Cell-adhesion molecule
CD45RO	+	+++	+++	Modulates T-cell receptor signaling
CD45RA	+++	+	+++	Modulates T-cell receptor signaling
CD62L	+++	–	Some +++	Receptor for homing to lymph node
CCR7	+++	+/–	Some +++	Chemokine receptor for homing to lymph node
CD69	–	+++	–	Early activation antigen
Bcl-2	++	+/–	+++	Promotes cell survival
Interferon-γ	–	+++	+++	Effector cytokine; mRNA present and protein made on activation
Granzyme B	–	+++	+/–	Effector molecule in cell killing
FasL	–	+++	+	Effector molecule in cell killing
CD122	+/–	++	++	Part of receptor for IL-15 and IL-2
CD25	–	++	–	Part of receptor for IL-2
CD127	++	–	+++	Part of receptor for IL-7
Ly6C	+	+++	+++	GPI-linked protein
CXCR4	+	+	++	Receptor for chemokine CXCL12; controls tissue migration
CCR5	+/–	++	Some +++	Receptor for chemokines CCL3 and CCL4; tissue migration

Fig. 11.22 Expression of many proteins alters when naive T cells become memory T cells. Proteins that are differently expressed in naive T cells, effector T cells, and memory T cells include adhesion molecules, which govern interactions with antigen-presenting cells and endothelial cells; chemokine receptors, which affect migration to lymphoid tissues and sites of inflammation; proteins and receptors that promote the survival of memory cells; and proteins that are involved in effector functions, such as granzyme B. Some changes also increase the sensitivity of the memory T cell to antigen stimulation. Many of the changes that occur in memory T cells are also seen in effector cells, but some, such as expression of the cell-surface proteins CD25 and CD69, are specific to effector T cells; others, such as expression of the survival factor Bcl-2, are limited to long-lived memory T cells. This list represents a general picture that applies to both CD4 and CD8 T cells in mice and humans, but some details that may differ between these sets of cells have been omitted for simplicity.

as CD69. In addition they express more Bcl-2, a protein that promotes cell survival and may be responsible for the long half-life of memory CD8 cells.

The α subunit of the IL-7 receptor (IL-7Rα or CD127) may be a good marker for activated T cells that will become long-lived memory cells (see Fig. 11.22). Naive T cells express IL-7Rα, but it is rapidly lost upon activation and is not expressed by most effector T cells. For example, during the peak of the effector response against lymphocytic choriomeningitis virus (LCMV) in mice, around day 7 of the infection, a small population of approximately 5% of CD8 effector T cells expressed high levels of IL-7Rα. Adoptive transfer of these cells, but not the effector T cells expressing low levels of IL-7Rα, could provide functional CD8 T-cell memory to uninfected mice (Fig. 11.23). This experiment suggests that the early maintenance, or the reexpression, of IL-7Rα identifies effector CD8 T cells that generate memory T cells, although it is still not known whether, and how, this process is regulated. Memory T cells are more sensitive to restimulation by antigen than are naive T cells, and more quickly and more vigorously produce several cytokines such as IFN-γ, TNF-α, and IL-2 in response to such stimulation. A similar progression occurs for T cells in humans after immunization with a vaccine against yellow fever virus.

CD4 T-cell memory has been more difficult to study, in part because their responses are smaller than those of CD8 T cells and also because, until recently, there were no peptide:MHC class II reagents similar to the peptide:MHC class I tetramers. Nevertheless, the transfer and priming of naive T cells carrying T-cell receptor transgenes that give the T cells a known peptide:MHC specificity has made it possible to visualize memory CD4 T cells. They appear as a long-lived population of cells that share some surface characteristics of activated effector T cells but are distinct from effector T cells in that they require additional restimulation before acting on target cells. Changes in three cell-surface proteins—L-selectin, CD44, and CD45—that occur on the putative memory CD4 T cells after exposure to antigen are particularly significant. L-selectin is lost on most memory CD4 T cells, whereas CD44 levels are increased on all memory T cells; these changes contribute to directing the migration of memory T cells from the blood into the tissues rather than directly into lymphoid tissue. The isoform of CD45 changes because of alternative splicing of exons that encode the extracellular domain of CD45, leading to isoforms, such as CD45RO, that are smaller and more readily associated with the T-cell receptor and facilitate antigen recognition (see Fig. 11.22). These changes are characteristic of cells that have been activated to become effector T cells, yet some of the cells on which these changes have occurred have many characteristics of resting CD4 T cells, suggesting that they represent memory CD4 T cells. Only after reexposure to antigen on an antigen-presenting cell do they achieve effector T-cell status and acquire all the characteristics of T_H2 or T_H1 cells, secreting IL-4 and IL-5, or IFN-γ, respectively.

Fig. 11.23 Expression of the IL-7 receptor (IL-7R) indicates which CD8 effector T cells can generate robust memory responses. Mice expressing a T-cell receptor (TCR) transgene specific for a viral antigen from lymphocytic choriomeningitis virus (LCMV) were infected with the virus and effector cells were collected on day 11. Effector CD8 T cells expressing high levels of IL-7R (IL-7R^{hi}, blue) were separated and transferred into one group of naive mice, and effector CD8 T cells expressing low IL-7R (IL7R^{lo}, green) were transferred into another group. Three weeks after transfer, the mice were challenged with a bacterium engineered to express the original viral antigen, and the numbers of responding transferred T cells (detected by their expression of the transgenic TCR) were measured at various times after challenge. Only the transferred IL-7R^{hi} effector cells could generate a robust expansion of CD8 T cells after the secondary challenge.

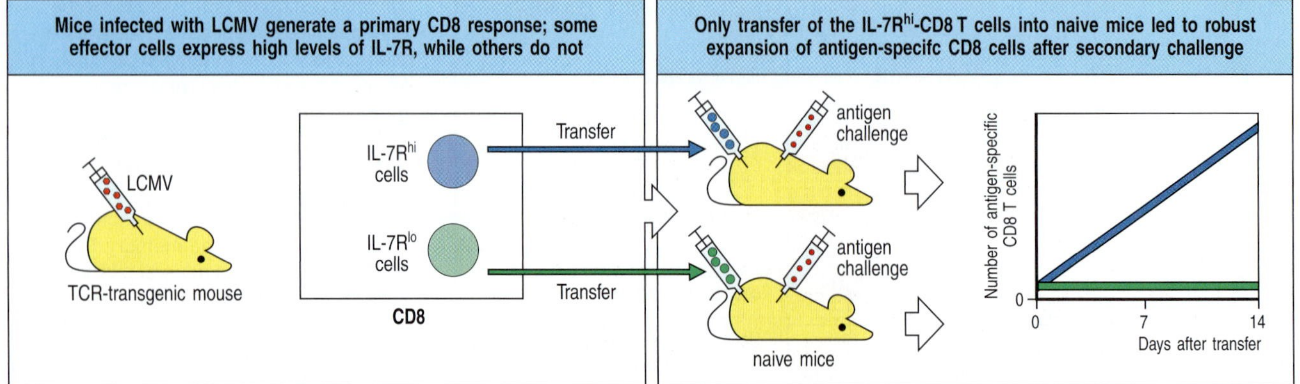

It therefore seems reasonable to designate these cells as memory CD4 T cells and to surmise that naive CD4 T cells differentiate into effector T cells, some of which then become memory cells. As with memory CD8 T cells, direct staining of CD4 T cells with peptide:MHC class II tetramers can identify antigen-specific CD4 T cells, and analysis using intracellular cytokine staining (see Appendix I, Section A-27) can determine whether they are T_H1, T_H17, or T_H2 cells. These improvements in the identification and phenotyping of CD4 T cells will rapidly increase our knowledge in this area and could contribute valuable comparative information on naive, memory, and effector CD4 T cells.

The homeostatic mechanisms governing the survival of memory T cells differ from those for naive T cells. Memory T cells divide more frequently than naive T cells, and their expansion is controlled by a balance between proliferation and cell death. As with naive cells, the survival of memory T cells requires stimulation by the cytokines IL-7 and IL-15. IL-7 is required for the survival of both CD4 and CD8 memory T cells, but in addition, IL-15 is critical for the long-term survival and proliferation of CD8 memory T cells under normal conditions. For memory CD4 T cells, the role of IL-15 is still under investigation.

In addition to cytokine stimulation, naive T cells also require contact with self-peptide:self-MHC complexes for their long-term survival in the periphery (see Section 8-29), but it seems that memory T cells do not have this requirement. It has been found, however, that memory T cells surviving after transfer to MHC-deficient hosts have some defects in typical T-cell memory functions, indicating that stimulation by self-peptide:self-MHC complexes may be required for their continued proliferation and optimal function (Fig. 11.24).

11-17 Memory T cells are heterogeneous and include central memory and effector memory subsets.

CD4 and CD8 T cells can differentiate into two types of memory cells with distinct activation characteristics (Fig. 11.25). One type is called an **effector memory cell** because it can rapidly mature into an effector T cell and secrete large amounts of IFN-γ, IL-4, and IL-5 early after restimulation. These cells lack the chemokine receptor CCR7 but express high levels of β_1 and β_2 integrins, as well as receptors for inflammatory chemokines. This profile suggests that these effector memory cells are specialized for rapidly entering inflamed tissues. The other type is called a **central memory cell**. It expresses CCR7 and would therefore be expected to recirculate more easily to the T-cell zones of peripheral lymphoid tissues, as do naive T cells. Central memory cells are very sensitive to cross-linking of their T-cell receptors and rapidly express CD40 ligand in response; however, they take longer than effector memory cells to differentiate into effector T cells and thus do not secrete such large amounts of cytokines early after restimulation.

The distinction between central memory cells and effector memory cells has been made both in humans and in the mouse. However, this general

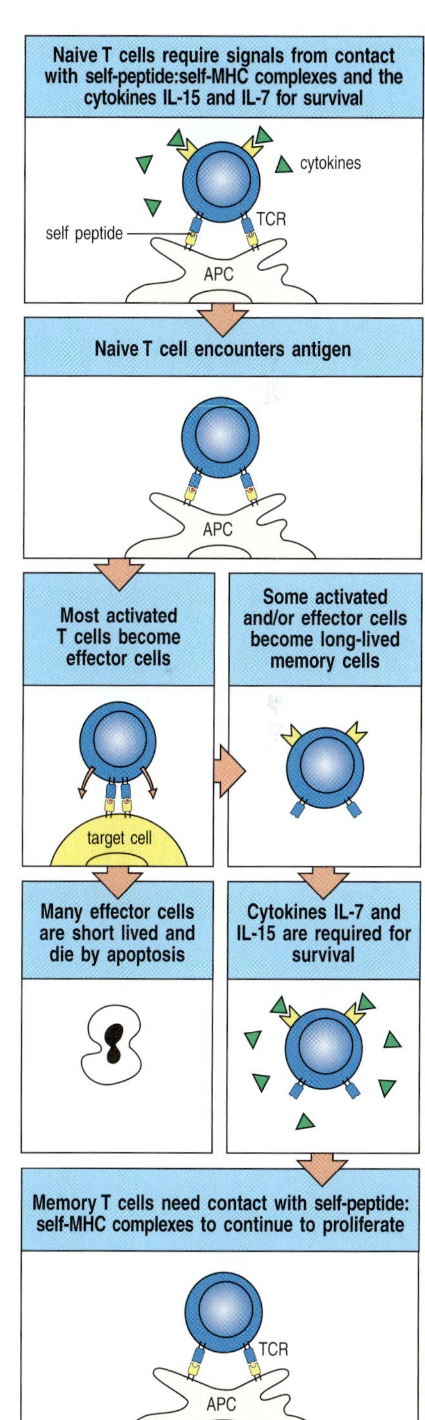

Fig. 11.24 Naive T cells and memory T cells have different requirements for survival. For their survival in the periphery, naive T cells require periodic stimulation with the cytokines IL-7 and IL-15 and with self antigens presented by MHC molecules. On priming with its specific antigen, a naive T cell divides and differentiates. Most of the progeny differentiate into relatively short-lived effector cells, but some effector cells become long-lived memory T cells, which need to be sustained by cytokines but do not require contact with self-peptide:self-MHC complexes purely for survival. However, contact with self antigens does seem necessary for memory T cells to continue to proliferate and thus keep up their numbers in the memory pool.

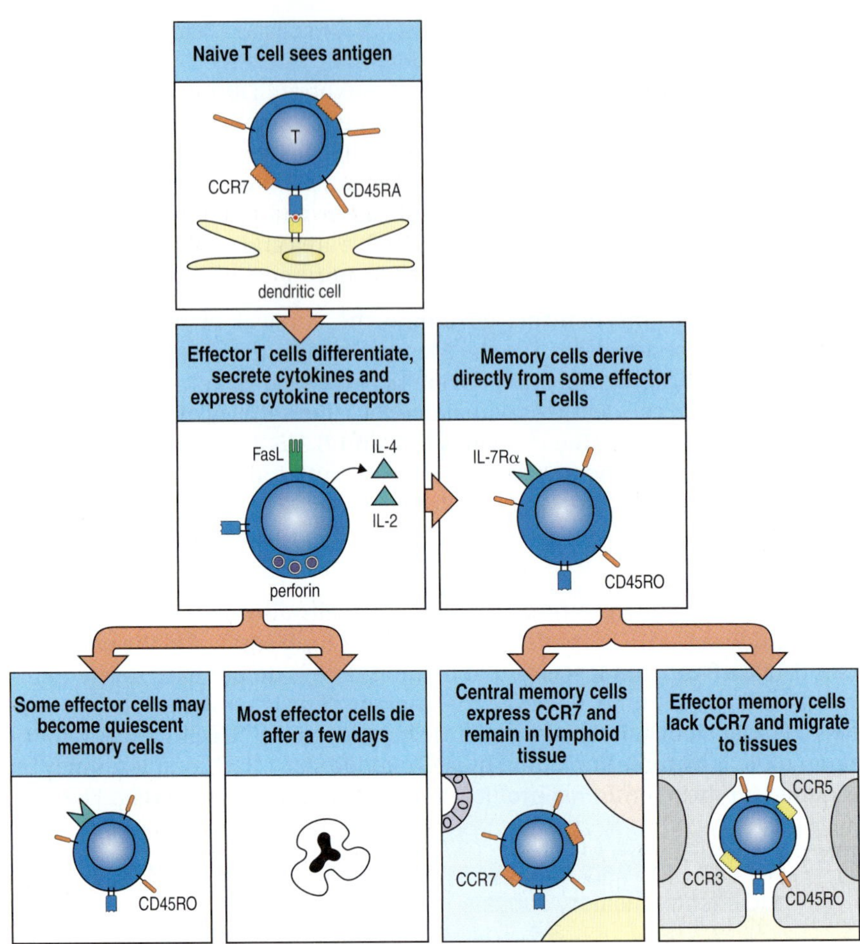

Fig. 11.25 T cells differentiate into central memory and effector memory subsets distinguished by expression of the chemokine receptor CCR7. Quiescent memory cells bearing the characteristic CD45RO surface protein can arise from activated effector cells (right half of diagram) or directly from activated naive T cells (left half of diagram). Two types of quiescent memory T cells can derive from the primary T-cell response. Central memory cells express CCR7 and remain in peripheral lymphoid tissues after restimulation. Memory cells of the other type—effector memory cells—mature rapidly into effector T cells after restimulation, and secrete large amounts of IFN-γ, IL-4, and IL-5. They do not express the receptor CCR7, but express receptors (CCR3 and CCR5) for inflammatory chemokines.

distinction does not imply that each subset is a uniform population. Within the CCR7-expressing central memory subset there are extensive differences in the expression of other markers, particularly receptors for other chemokines. For example, a subset of the CCR7-positive central memory cells also expresses CXCR5, similarly to T_{FH} cells, although it is not yet clear whether these memory cells can provide help to B cells in the germinal center.

On stimulation by antigen, central memory cells rapidly lose expression of CCR7 and differentiate into effector memory cells. Effector memory cells are also heterogeneous in the chemokine receptors they express, and have been classified according to chemokine receptors typical of T_H1 cells, such as CCR5, and of T_H2 cells, such as CCR4. Central memory cells are not yet committed to particular effector lineages, and even effector memory cells are not fully committed to the T_H1 or the T_H2 lineage, although there is some correlation between their eventual output of T_H1 or T_H2 cells and the chemokine receptors expressed. Further stimulation with antigen seems to drive the differentiation of effector memory cells gradually into the distinct effector T-cell lineages.

11-18 CD4 T-cell help is required for CD8 T-cell memory and involves CD40 and IL-2 signaling.

In Section 11-8 we saw how primary CD8 T-cell responses to *Listeria monocytogenes* can occur in mice that lack CD4 T cells. After 7 days of infection, wild-type mice and mice lacking CD4 T cells both showed equivalent expansion and activity of pathogen-specific CD8 effector T cells. However, they are not equally able to generate memory CD8 T cells. Mice that lack CD4 T cells as the result of a deficiency in MHC class II were found to generate much weaker secondary responses, characterized by many fewer expanding memory CD8 T cells specific for the pathogen. In this experiment, the *Listeria* carried a gene for the protein ovalbumin, and it was the response to this protein that was measured as a marker for CD8 T-cell memory (Fig. 11.26). CD4 T cells in these mice are lacking both during the primary response and in any secondary challenge, so the requirement for CD4 T cells could be either in the initial programming of CD8 T cells during their primary activation to enable memory development, or, alternatively, in providing help only during the secondary memory response.

Other experiments indicate that this CD4 T-cell help is necessary for programming naive CD8 T cells to be able to generate memory cells capable of robust secondary expansion. Memory CD8 T cells that developed in the absence of CD4 help were transferred into wild-type mice. After transfer, the recipient mice were challenged again. In mice that received CD8 memory T cells that developed without CD4 T-cell help, the CD8 T cells showed a reduced ability to proliferate even though the recipient mice expressed MHC class II. This result indicates that CD4 T-cell help is required during the priming of CD8 T cells and not simply at the time of secondary responses. This requirement for CD4 help in CD8 memory generation has also been demonstrated by experiments in which CD4 T cells were depleted by treatment with antibody or in which mice were deficient in the CD4 gene.

The mechanism underlying this requirement for CD4 T cells is not completely understood. It may involve two types of signals received by the CD8 T cell—those received through CD40 and those received through the IL-2 receptor. CD8 T cells that do not express CD40 are unable to generate memory T cells. Although many cells could potentially express the CD40 ligand needed to stimulate CD40, it is most likely that CD4 T cells are the source of this signal.

The requirement for IL-2 signaling in programming CD8 memory was discovered by using CD8 T cells with a genetic deficiency in the IL-2Rα subunit, which were therefore unable to respond to IL-2. Because IL-2Rα signaling is required for the development of T_{reg} cells, mice lacking IL-2Rα develop a lymphoproliferative disorder. However, this disorder does not develop in mice that are mixed bone marrow chimeras harboring both wild-type and IL-2Rα-deficient cells, and these chimeras can be used to study the behavior of IL-2Rα-deficient cells. When these chimeric mice were infected with LCMV and their responses were tested, memory CD8 responses were found to be defective specifically in the T cells lacking IL-2Rα.

CD4 T cells also provide help in maintaining the numbers of CD8 memory T cells that is distinct from their effect in programming naive CD8 T cells to become memory cells (Fig. 11.27). When CD8 memory T cells are transferred

Fig. 11.26 CD4 T cells are required for the development of functional CD8 memory T cells. Mice that do not express MHC class II molecules (MHC $II^{-/-}$) fail to develop CD4 T cells. Wild-type and MHC $II^{-/-}$ mice were infected with *Listeria monocytogenes* expressing the model antigen ovalbumin (LM-OVA). After 7 days, the numbers of OVA-specific CD8 T cells can be measured by using specific MHC tetramers that contain an OVA peptide, and therefore bind to T-cell receptors that react with this antigen. After 7 days of infection, mice lacking CD4 T cells have the same number of OVA-specific CD8 T cells as wild-type mice do. However, when mice are allowed to recover for 60 days, during which time memory T cells develop, and are then rechallenged with LM-OVA, the mice lacking CD4 T cells fail to expand CD8 memory cells specific to OVA, whereas there is a strong CD8 memory response in the wild-type mice.

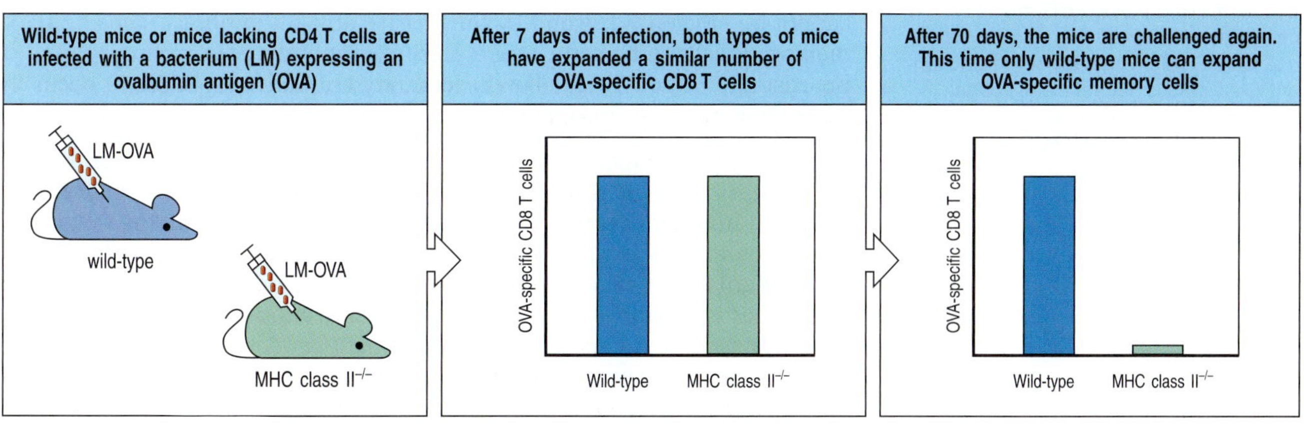

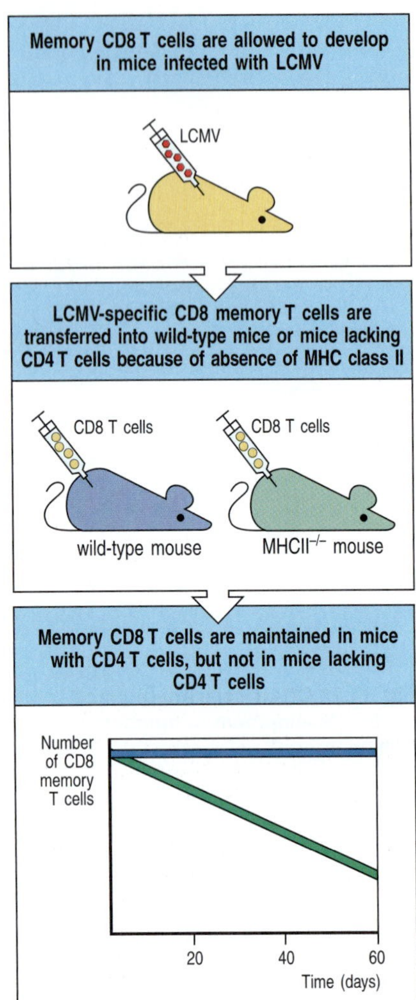

Fig. 11.27 CD4 T cells promote the maintenance of CD8 memory cells. The dependence of memory CD8 T cells on CD4 T cells is shown by the different lifetimes after transfer into host mice that either have normal CD4 T cells (wild-type), or lack CD4 T cells (MHC II$^{-/-}$). In the absence of MHC class II proteins, CD4 T cells fail to develop in the thymus. When CD8 memory T cells specific for LCMV were isolated from donor mice 35 days after infection with the virus and transferred into these hosts, memory cells were maintained only in mice that had CD4 T cells. The basis for this action of CD4 T cells is not yet clear, but has implications for conditions such as HIV-AIDS in which numbers of CD4 T cells are diminished.

into immunologically naive mice, the presence or absence of CD4 T cells in the recipient influences the maintenance of the CD8 memory cells. Transfer of CD8 memory cells into mice lacking CD4 T cells is followed by a gradual decrease in the number of memory cells in comparison with a similar transfer into wild-type mice. In addition, CD8 effector cells transferred into mice lacking CD4 T cells had a relative impairment of CD8 effector functions. These experiments show that the CD4 T cells activated during an immune response have a significant impact on the quantity and quality of the CD8 T-cell response, even when they are not needed for the initial CD8 T-cell activation. CD4 T cells help to program naive CD8 T cells to be able to generate memory T cells, help to promote efficient effector activity, and help to maintain memory T-cell numbers.

11-19 In immune individuals, secondary and subsequent responses are mainly attributable to memory lymphocytes.

In the normal course of an infection, a pathogen proliferates to a level sufficient to elicit an adaptive immune response and then stimulates the production of antibodies and effector T cells that eliminate the pathogen from the body. Most of the effector T cells then die, and antibody levels gradually decline, because the antigens that elicited the response are no longer present at the level needed to sustain it. We can think of this as feedback inhibition of the response. Memory T and B cells remain, however, and maintain a heightened ability to mount a response to a recurrence of infection with the same pathogen.

Antibody and memory lymphocytes remaining in an immunized individual can have the effect of reducing the activation of naive B and T cells on a subsequent encounter with the same antigen. In fact, passively transferring antibody to a naive recipient can be used to inhibit naive B cell responses to that same antigen. This phenomenon has been put to practical use to prevent Rh$^-$ mothers from making an immune response to a Rh$^+$ fetus, which can result in hemolytic disease of the newborn (see Appendix I, Section A-11). If anti-Rh antibody is given to the mother before she is first exposed to her child's Rh$^+$ red blood cells, her response will be inhibited. The mechanism of this suppression is likely to involve the antibody-mediated clearance and destruction of fetal red blood cells that have entered the mother, thus preventing naive B cells and T cells from mounting an immune response. Presumably, anti-Rh antibody is in excess over antigen, so that not only is antigen eliminated, but immune complexes are not formed to stimulate naive B cells through Fc receptors. Memory B-cell responses are, however, not inhibited by antibody, so the Rh$^-$ mothers at risk must be identified and treated before a primary response has occurred. Because of their high affinity for antigen and alterations in their B-cell receptor signaling requirements, memory B cells are much more sensitive to the small amounts of antigen that cannot be efficiently cleared by the passive anti-Rh antibody. The ability of memory B cells to be activated to produce antibody, even when exposed to preexisting antibody, also allows secondary antibody responses to occur in individuals who are already immune.

These suppressive mechanisms might also explain the phenomenon known as **original antigenic sin**. This term was coined to describe the tendency of people to make antibodies only against epitopes expressed on the first influenza virus variant to which they are exposed, even in subsequent infections with variants that bear additional, highly immunogenic, epitopes (Fig. 11.28). Antibodies against the original virus will tend to suppress responses of naive B cells specific for the new epitopes. This might benefit the host by using only those B cells that can respond most rapidly and effectively to the virus. This pattern is broken only if the person is exposed to an influenza virus that lacks

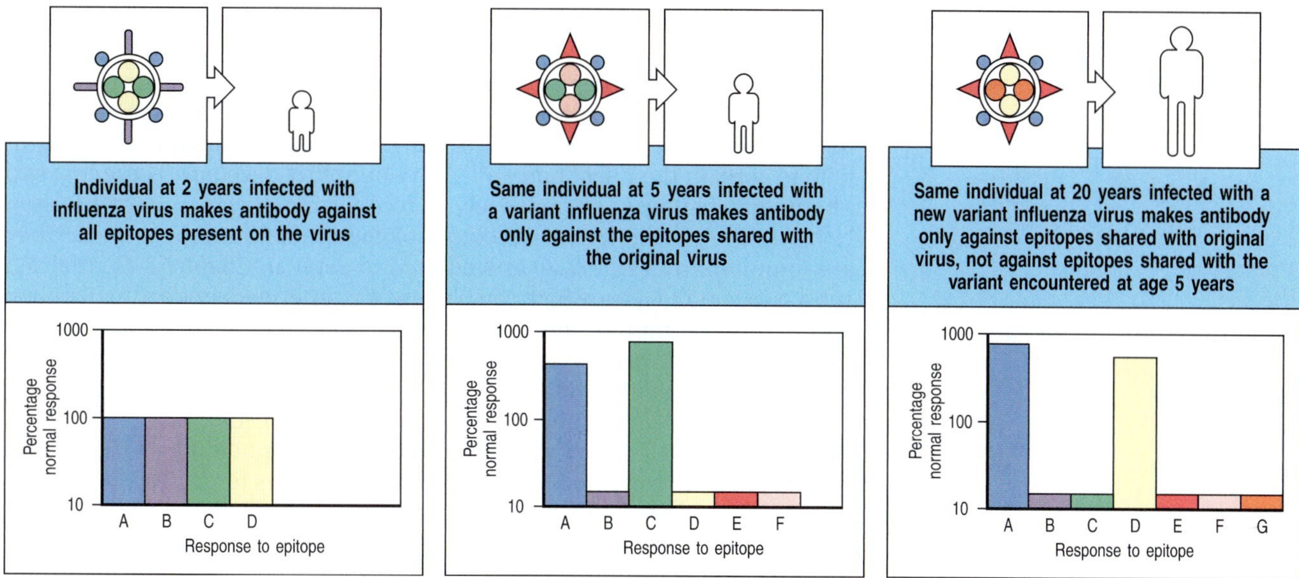

Fig. 11.28 When individuals who have been infected with one variant of influenza virus are infected with a second or third variant, they make antibodies only against epitopes that were present on the initial virus. A child infected for the first time with an influenza virus at 2 years of age makes a response to all epitopes (left panel). At age 5 years, the same child exposed to a different influenza virus responds preferentially to those epitopes shared with the original virus, and makes a smaller than normal response to new epitopes on the virus (center panel). Even at age 20 years, this commitment to respond to epitopes shared with the original virus, and the subnormal response to new epitopes, is retained (right panel). This phenomenon is called 'original antigenic sin.'

all epitopes seen in the original infection, because now no preexisting antibodies bind the virus, and naive B cells are able to respond.

A similar suppressive effect of antigen-specific memory T cells on naive T-cell responses has been observed in some settings, such as infection by lymphocytic choriomeningitis virus (LCMV) in the mouse or dengue virus in humans. For example, mice that were primed with one strain of LCMV responded to a subsequent infection with a variant LCMV by using CD8 T cells directed against antigens specific for the first, rather than the new, variant. However, this effect was not observed when responses to variable ovalbumin antigenic epitopes were examined in the setting of recurrent infections using the bacterial pathogen *Listeria monocytogenes*, suggesting that original antigenic sin may not be a universal occurrence in all immune responses.

Summary.

Protective immunity against reinfection is one of the most important consequences of adaptive immunity. Protective immunity depends not only on preformed antibody and effector T cells but most importantly on the establishment of a population of lymphocytes that mediate long-lived immunological memory. The capacity of memory cells to respond rapidly to restimulation with the same antigen can be transferred to naive recipients by primed B and T cells. The precise changes that distinguish naive, effector, and memory lymphocytes include the regulation of expression of receptors for cytokines, such as IL-7 and IL-15, that help to maintain these cells, and the regulation of chemokine receptors, such as CCR7, that distinguish between functional subsets of memory cells. The advent of receptor-specific reagents—MHC tetramers—has allowed an analysis of the relative contributions of clonal expansion and differentiation to the memory phenotype. Memory B cells can be distinguished by changes in their immunoglobulin genes because of isotype switching and somatic hypermutation, and secondary and subsequent immune responses are characterized by antibodies with increasing affinity for the antigen. As for T-cell memory, there is a complex interplay between CD4 and CD8 T cells that is only partly understood. Although CD8 T cells can generate effective primary responses in the absence of help from CD4 T cells, it is becoming clear that CD4 T cells have an integral role in regulating CD8 T-cell memory. These issues will be critical in understanding, for example, how to design effective vaccines for diseases such as HIV/AIDS.

Summary to Chapter 11.

Vertebrates resist infection by pathogenic microorganisms in several ways. The innate defenses can act immediately and may succeed in repelling the infection, but if not they are followed by a series of induced early responses that help to contain the infection as adaptive immunity develops. These first two phases of the immune response rely on recognizing the presence of infection by using the nonclonotypic receptors of the innate immune system. They are summarized in Fig. 11.29 and covered in detail in Chapter 3. Specialized subsets of lymphocytes, which can be viewed as intermediates between innate and adaptive immunity, include iNKT cells, which can help to bias the CD4 T-cell response toward a T_H1 or a T_H2 phenotype, and NK cells, which can be recruited to lymph nodes and secrete IFN-γ, and thus promote a T_H1 response. The third phase of an immune response is the adaptive immune response (see Fig. 11.29), which is mounted in the peripheral lymphoid tissue that serves the particular site of infection and takes several days to develop, because T and B lymphocytes must encounter their specific antigen, proliferate, and differentiate into effector cells. T-cell dependent B-cell responses cannot be initiated until antigen-specific T_{FH} cells have had a chance to proliferate and differentiate. Once an adaptive immune response has occurred, the antibodies and effector T cells are dispersed via the circulation and recruited into the infected tissues; the infection is usually controlled and the pathogen is contained or eliminated. The final effector mechanisms used to clear an infection depend on the type of infectious agent, and in most cases they are the same as those employed in the early phases of immune defense; only the recognition mechanism changes and is more selective (see Fig. 11.29).

An effective adaptive immune response leads to a state of protective immunity. This state consists of the presence of effector cells and molecules produced in the initial response, and of immunological memory. Immunological memory

	Phases of the immune response		
	Immediate (0–4 hours)	Early (4–96 hours)	Late (96–100 hours)
	Nonspecific Innate No memory No specific T cells	Nonspecific + specific Inducible No memory No specific T cells	Specific Inducible Memory Specific T cells
Barrier functions	Skin, epithelia, mucins, acid	Local inflammation (C5a) Local TNF-α	IgA antibody in luminal spaces IgE antibody on mast cells Local inflammation
Response to extracellular pathogens	Phagocytes Alternative and MBL complement pathway Lysozyme Lactoferrin Peroxidase Defensins	Mannan-binding lectin C-reactive protein T-independent B-cell antibody Complement	IgG antibody and Fc receptor-bearing cells IgG, IgM antibody + classical complement pathway
Response to intracellular bacteria	Macrophages	Activated NK-dependent macrophage activation IL-1, IL-6, TNF-α, IL-12	T-cell activation of macrophages by IFN-γ
Response to virus-infected cells	Natural killer (NK) cells	IFN-α and IFN-β IL-12-activated NK cells	Cytotoxic T cells IFN-γ

Fig. 11.29 The components of the three phases of the immune response against different classes of microorganisms. The mechanisms of innate immunity that operate in the first two phases of the immune response are described in Chapters 2 and 3, and thymus-independent (T-independent) B-cell responses are covered in Chapter 10. The early phases contribute to the initiation of adaptive immunity, and they influence the functional character of the antigen-specific effector T cells and antibodies that appear on the scene in the late phase of the response. There are striking similarities in the effector mechanisms at each phase of the response; the main change is in the recognition structure used.

is manifested as a heightened ability to respond to pathogens that have previously been encountered and successfully eliminated. Memory T and B lymphocytes have the property of being able to transfer immune memory to naive recipients. The mechanisms that maintain immunological memory include certain cytokines, such as IL-7 and IL-15, as well as homeostatic interactions between the T cell receptors on memory cells with self-peptide:self-MHC complexes. The artificial induction of protective immunity, which includes immunological memory, by vaccination is the most outstanding accomplishment of immunology in the field of medicine. The understanding of how this is accomplished is now catching up with its practical success. However, as we will see in Chapter 13, many pathogens do not induce protective immunity that completely eliminates the pathogen, so we will need to learn what prevents this before we can prepare effective vaccines against these pathogens.

Questions.

11.1 *Communication is critical in any large enterprise. (a) How is the body alerted to an invasion by microbes, and (b) how does it ensure that its responses reach the site of infection?*

11.2 *The immune system responds to particular classes of pathogens in different ways. What properties of viruses and bacteria are used to induce T_H1 responses to them, and what host cells provide the information about the type of pathogen present?*

11.3 *Differentiated T cells require continued signals to maintain their function. (a) What signals do T_H1 cells need? (b) What advantages might the requirement for continued signals have? What disadvantages?*

11.4 *One could question the need for immunological memory. Invertebrates get by without adaptive immunity or memory. After all, if you survive the first infection without memory, you should be able to survive it the second time without memory. And if you fail to survive the first infection, memory is of no help. (a) What are the advantages of immunological memory that counter this argument? What features of pathogens might have driven the evolution of immunological memory? (b) Innate immune responses seem to lack memory, but may be augmented for a time after infection. What features of immunological memory provided by adaptive immunity are of greater value than simply an augmented innate response? In what way could these features be a disadvantage? Give an example.*

11.5 *Memory responses differ from primary immune responses in several important properties. Name three ways in which they differ, and describe the underlying mechanism(s) involved in each case.*

11.6 *(a) Discuss the relative roles of cytokine signals and signals received through the T-cell receptor in the survival and function of memory T cells. (b) Compare and contrast their requirements and responses to such signals with those of naive T cells.*

11.7 *You are swimming in freshwater and are infected by a parasite that entered your body through your skin. Outline how your body would generate an immune response against this pathogen. Would the response be T_H1 or T_H2 polarized? Include all relevant cell types and molecules that would be involved.*

Section references.

11-1 The course of an infection can be divided into several distinct phases.

Mandell, G., Bennett, J., and Dolin, R. (eds): *Principles and Practice of Infectious Diseases*, 5th ed. New York, Churchill Livingstone, 2000.

Zhang, S.Y., Jouanguy, E., Sancho-Shimizu, V., von Bernuth, H., Yang, K., Abel, L., Picard, C., Puel, A., and Casanova, J.L.: **Human Toll-like receptor-dependent induction of interferons in protective immunity to viruses.** *Immunol. Rev.* 2007, **220**:225–236.

11-2 The nonspecific responses of innate immunity are necessary for an adaptive immune response to be initiated.

Fearon, D.T., and Carroll, M.C.: **Regulation of B lymphocyte responses to foreign and self-antigens by the CD19/CD21 complex.** *Annu. Rev. Immunol.* 2000, **18**:393–422.

Fearon, D.T., and Locksley, R.M.: **The instructive role of innate immunity in the acquired immune response.** *Science* 1996, **272**:50–53.

Janeway, C.A., Jr: **The immune system evolved to discriminate infectious nonself from noninfectious self.** *Immunol. Today* 1992, **13**:11–16.

11-3 Cytokines made during infection can direct differentiation of CD4 T cells toward the T_H17 subset.

Dillon, S., Agrawal, A., Van Dyke, T., Landreth, G., McCauley, L., Koh, A., Maliszewski, C., Akira, S., and Pulendran, B.: **A Toll-like receptor 2 ligand stimulates Th2 responses *in vivo*, via induction of extracellular signal-regulated kinase mitogen-activated protein kinase and c-Fos in dendritic cells.** *J. Immunol.* 2004, **172**:4733–4743.

Fallon, P.G., Ballantyne, S.J., Mangan, N.E., Barlow, J.L., Dasvarma, A., Hewett, D.R., McIlgorm, A., Jolin, H.E., and McKenzie, A.N.J.: **Identification of an interleukin (IL)-25-dependent cell population that provides IL-4, IL-5, and IL-13 at the onset of helminth expulsion.** *J. Exp. Med.* 2006, **203**:1105–1116.

Fossiez, F., Djossou, O., Chomarat, P., Flores-Romo, L., Ait-Yahia, S., Maat, C., Pin, J.J., Garrone, P., Garcia, E., Saeland, S., *et al.*: **T cell interleukin-17 induces stromal cells to produce proinflammatory and hematopoietic cytokines.** *J. Exp. Med.* 1996, **183**:2593–2603.

Happel, K.I., Zheng, M., Young, E., Quinton, L.J., Lockhart, E., Ramsay, A.J., Shellito, J.E., Schurr, J.R., Bagby, G.J., Nelson, S., *et al.*: **Cutting edge: roles of Toll-like receptor 4 and IL-23 in IL-17 expression in response to *Klebsiella pneumoniae* infection.** *J. Immunol.* 2003, **170**:4432–4436.

LeibundGut-Landmann, S., Gross, O., Robinson, M.J., Osorio, F., Slack, E.C., Tsoni, S.V., Schweighoffer, E., Tybulewicz, V., Brown, G.D., Ruland, J., *et al.*: **Syk- and CARD9-dependent coupling of innate immunity to the induction of T helper cells that produce interleukin 17.** *Nat. Immunol.* 2007, **8**:630–638.

Tato, C.M., and O'Shea, J.J.: **What does it mean to be just 17?** *Nature* 2006, **441**:166–168.

Ye, P., Rodriguez, F.H., Kanaly, S., Stocking, K.L., Schurr, J., Schwarzenberger, P., Oliver, P., Huang, W., Zhang, P., Zhang, J., *et al.*: **Requirement of interleukin 17 receptor signaling for lung CXC chemokine and granulocyte colony-stimulating factor expression, neutrophil recruitment, and host defense.** *J. Exp. Med.* 2001, **194**:519–527.

11-4 T_H1 and T_H2 cells are induced by cytokines generated in response to different pathogens.

Amsen, D., Blander, J.M., Lee, G.R., Tanigaki, K., Honjo, T., and Flavell, R.A.: **Instruction of distinct CD4 T helper cell fates by different Notch ligands on antigen-presenting cells.** *Cell* 2004, **117**:515–526.

Bendelac, A., Rivera, M.N., Park, S.H., and Roark, J.H.: **Mouse CD1-specific NK1 T cells: development, specificity, and function.** *Annu. Rev. Immunol.* 1997, **15**:535–562.

Finkelman, F.D., Shea-Donohue, T., Goldhill, J., Sullivan, C.A., Morris, S.C., Madden, K.B., Gauser, W.C., and Urban, J.F., Jr: **Cytokine regulation of host defense against parasitic intestinal nematodes.** *Annu. Rev. Immunol.* 1997, **15**:505–533.

Godfrey, D.I., MacDonald, H.R., Kronenberg, M., Smyth, M.J., and Van Kaer, L.: **NKT cells: what's in a name?** *Nat. Rev. Immunol.* 2004, **4**:231–237.

Hammad, H., Plantinga, M., Deswarte, K., Pouliot, P., Willart. M.A., Kool, M., Muskens, F., and Lambrecht, B.N.: **Inflammatory dendritic cells—not basophils—are necessary and sufficient for induction of Th2 immunity to inhaled house dust mite allergen.** *J. Exp. Med.* 2010, **207**:2097–2111.

Hsieh, C.S., Macatonia, S.E., Tripp, C.S., Wolf, S.F., O'Garra, A., and Murphy, K.M.: **Development of T_H1 CD4$^+$ T cells through IL-12 produced by *Listeria*-induced macrophages.** *Science* 1993, **260**:547–549.

Jankovic, D., Sher, A., and Yap, G.: **Th1/Th2 effector choice in parasitic infection: decision making by committee.** *Curr. Opin. Immunol.* 2001, **13**:403–409.

Moser, M., and Murphy, K.M.: **Dendritic cell regulation of T_H1-T_H2 development.** *Nat. Immunol.* 2000, **1**:199–205.

Pulendran, B., and Ahmed, R.: **Translating innate immunity into immunological memory: implications for vaccine development.** *Cell* 2006, **124**:849–863.

Ohnmacht, C., Schwartz, C., Panzer, M., Schiedewitz, I., Naumann, R., and Voehringer, D.: **Basophils orchestrate chronic allergic dermatitis and protective immunity against helminths.** *Immunity* 2010, **33**:364–374.

11-5 CD4 T-cell subsets can cross-regulate each other's differentiation.

Constant, S.L., and Bottomly, K.: **Induction of Th1 and Th2 CD4$^+$ T cell responses: the alternative approaches.** *Annu. Rev. Immunol.* 1997, **15**:297–322.

Croft, M., Carter, L., Swain, S.L., and Dutton, R.W.: **Generation of polarized antigen-specific CD8 effector populations: reciprocal action of interleukin-4 and IL-12 in promoting type 2 versus type 1 cytokine profiles.** *J. Exp. Med.* 1994, **180**:1715–1728.

Grakoui, A., Donermeyer, D.L., Kanagawa, O., Murphy, K.M., and Allen, P.M.: **TCR-independent pathways mediate the effects of antigen dose and altered peptide ligands on Th cell polarization.** *J. Immunol.* 1999, **162**:1923–1930.

Harrington, L.E., Hatton, R.D., Mangan, P.R., Turner, H., Murphy, T.L., Murphy, K.M., and Weaver, C.T.: **Interleukin 17-producing CD4$^+$ effector T cells develop via a lineage distinct from the T helper type 1 and 2 lineages.** *Nat. Immunol.* 2005, **6**:1123–1132.

Julia, V., McSorley, S.S., Malherbe, L., Breittmayer, J.P., Girard-Pipau, F., Beck, A., and Glaichenhaus, N.: **Priming by microbial antigens from the intestinal flora determines the ability of CD4$^+$ T cells to rapidly secrete IL-4 in BALB/c mice infected with *Leishmania major*.** *J. Immunol.* 2000, **165**:5637–5645.

Lee, Y.K., Turner, H., Maynard, C.L., Oliver, J.R., Chen, D., Elson, C.O., and Weaver, C.T.: **Late developmental plasticity in the T helper 17 lineage.** *Immunity* 2009, **30**:92–107.

Martin-Fontecha, A., Thomsen, L.L., Brett, S., Gerard, C., Lipp, M., Lanzavecchia, A., and Sallusto, F.: **Induced recruitment of NK cells to lymph nodes provides IFN-γ for T_H1 priming.** *Nat. Immunol.* 2004, **5**:1260–1265.

Nakamura, T., Kamogawa, Y., Bottomly, K., and Flavell, R.A.: **Polarization of IL-4- and IFN-γ-producing CD4$^+$ T cells following activation of naive CD4$^+$ T cells.** *J. Immunol.* 1997, **158**:1085–1094.

Seder, R.A., and Paul, W.E.: **Acquisition of lymphokine producing phenotype by CD4$^+$ T cells.** *Annu. Rev. Immunol.* 1994, **12**:635–673.

Wang, L.F., Lin, J.Y., Hsieh, K.H., and Lin, R.H.: **Epicutaneous exposure of protein antigen induces a predominant T_H2-like response with high IgE production in mice.** *J. Immunol.* 1996, **156**:4079–4082.

11-6 Effector T cells are guided to sites of infection by chemokines and newly expressed adhesion molecules.

MacKay, C.R., Marston, W., and Dudler, L.: **Altered patterns of T-cell migration through lymph nodes and skin following antigen challenge.** *Eur. J. Immunol.* 1992, **22**:2205–2210.

Romanic, A.M., Graesser, D., Baron, J.L., Visintin, I., Janeway, C.A., Jr, and Madri, J.A.: **T cell adhesion to endothelial cells and extracellular matrix is modulated upon transendothelial cell migration.** *Lab. Invest.* 1997, **76**:11–23.

Sallusto, F., Kremmer, E., Palermo, B., Hoy, A., Ponath, P., Qin, S., Forster, R., Lipp, M., and Lanzavecchia, A.: **Switch in chemokine receptor expression upon TCR stimulation reveals novel homing potential for recently activated T cells.** *Eur. J. Immunol.* 1999, **29**:2037–2045.

11-7 Differentiated effector T cells are not a static population but continue to respond to signals as they carry out their effector functions.

Cua, D.J., Sherlock, J., Chen, Y., Murphy, C.A., Joyce, B., Seymour, B., Lucian, L., To, W., Kwan, S., Churakova, T., *et al.*: **Interleukin-23 rather than interleukin-12 is the critical cytokine for autoimmune inflammation of the brain.** *Nature* 2003, **421**:744–748.

Ghilardi, N., Kljavin, N., Chen, Q., Lucas, S., Gurney, A.L., and De Sauvage, F.J.: **Compromised humoral and delayed-type hypersensitivity responses in IL-23-deficient mice.** *J. Immunol.* 2004, **172**:2827–2833.

Gran, B., Zhang, G.X., Yu, S., Li, J., Chen, X.H., Ventura, E.S., Kamoun, M., and Rostami, A.: **IL-12p35-deficient mice are susceptible to experimental autoimmune encephalomyelitis: evidence for redundancy in the IL-12 system in the induction of central nervous system autoimmune demyelination.** *J. Immunol.* 2002, **169**:7104–7110.

Parham, C., Chirica, M., Timans, J., Vaisberg, E., Travis, M., Cheung, J., Pflanz, S., Zhang, R., Singh, K.P., Vega, F., *et al.*: **A receptor for the heterodimeric cytokine IL-23 is composed of IL-12Rβ1 and a novel cytokine receptor subunit, IL-23R.** *J. Immunol.* 2002, **168**:5699–5708.

Park, A.Y., Hondowics, B.D., and Scott, P.: **IL-12 is required to maintain a Th1 response during *Leishmania major* infection.** *J. Immunol.* 2000, **165**:896–902.

Stobie, L., Gurunathan, S., Prussin, C., Sacks, D.L., Glaichenhaus, N., Wu, C.Y., and Seder, R.A.: **The role of antigen and IL-12 in sustaining Th1 memory cells *in vivo*: IL-12 is required to maintain memory/effector Th1 cells sufficient to mediate protection to an infectious parasite challenge.** *Proc. Natl Acad. Sci. USA* 2000, **97**:8427–8432.

Yap, G., Pesin, M., and Sher, A.: **Cutting edge: IL-12 is required for the maintenance of IFN-γ production in T cells mediating chronic resistance to the intracellular pathogen *Toxoplasma gondii*.** *J. Immunol.* 2000, **165**:628–631.

11-8 Primary CD8 T-cell responses to pathogens can occur in the absence of CD4 T-cell help.

Lertmemongkolchai, G., Cai, G., Hunter, C.A., and Bancroft, G.J.: **Bystander activation of CD8 T cells contributes to the rapid production of IFN-γ in response to bacterial pathogens.** *J. Immunol.* 2001, **166**:1097–1105.

Rahemtulla, A., Fung-Leung, W.P., Schilham, M.W., Kundig, T.M., Sambhara, S.R., Narendran, A., Arabian, A., Wakeham, A., Paige, C.J., Zinkernagel, R.M., *et al.*: **Normal development and function of CD8$^+$ cells but markedly decreased helper cell activity in mice lacking CD4.** *Nature* 1991, **353**:180–184.

Schoenberger, S.P., Toes, R.E., van der Voort, E.I., Offringa, R., and Melief, C.J.: **T-cell help for cytotoxic T lymphocytes is mediated by CD40–CD40L interactions.** *Nature* 1998, **393**:480–483.

Sun, J.C., and Bevan, M.J.: **Defective CD8 T cell memory following acute infection without CD4 T-cell help.** *Science* 2003, **300**:339–349.

11-9 Antibody responses develop in lymphoid tissues under the direction of T_{FH} cells.

Garside, P., Ingulli, E., Merica, R.R., Johnson, J.G., Noelle, R.J., and Jenkins, M.K.: **Visualization of specific B and T lymphocyte interactions in the lymph node.** *Science* 1998, **281**:96–99.

Jacob, J., Kassir, R., and Kelsoe, G.: ***In situ* studies of the primary immune response to (4-hydroxy-3-nitrophenyl)acetyl. I. The architecture and dynamics of responding cell populations.** *J. Exp. Med.* 1991, **173**:1165–1175.

Kelsoe, G., and Zheng, B.: **Sites of B-cell activation *in vivo*.** *Curr. Opin. Immunol.* 1993, **5**:418–422.

King, C.: **New insights into the differentiation and function of T follicular helper cells.** *Nat. Rev. Immunol.* 2009, **9**:757–766.

Liu, Y.J., Zhang, J., Lane, P.J., Chan, E.Y., and MacLennan, I.C.: **Sites of specific B cell activation in primary and secondary responses to T cell-dependent and T cell-independent antigens.** *Eur. J. Immunol.* 1991, **21**:2951–2962.

MacLennan, I.C.M.: **Germinal centres.** *Annu. Rev. Immunol.* 1994, **12**:117–139.

Okada, T., and Cyster, J.G.: **B cell migration and interactions in the early phase of antibody responses.** *Curr. Opin. Immunol.* 2006, **18**:278–285.

Victora, G.D., Schwickert, T.A., Fooksman, D.R., Kamphorst, A.O., Meyer-Hermann, M., Dustin, M.L., and Nussenzweig, M.C.: **Germinal center dynamics revealed by multiphoton microscopy with a photoactivatable fluorescent reporter.** *Cell* 2010, **143**:592–605.

11-10 Antibody responses are sustained in medullary cords and bone marrow.

Benner, R., Hijmans, W., and Haaijman, J.J.: **The bone marrow: the major source of serum immunoglobulins, but still a neglected site of antibody formation.** *Clin. Exp. Immunol.* 1981, **46**:1–8.

Manz, R.A., Thiel, A., and Radbruch, A.: **Lifetime of plasma cells in the bone marrow.** *Nature* 1997, **388**:133–134.

Slifka, M.K., Antia, R., Whitmire, J.K., and Ahmed, R.: **Humoral immunity due to long-lived plasma cells.** *Immunity* 1998, **8**:363–372.

Takahashi, Y., Dutta, P.R., Cerasoli, D.M., and Kelsoe, G.: ***In situ* studies of the primary immune response to (4-hydroxy-3-nitrophenyl)acetyl. V. Affinity maturation develops in two stages of clonal selection.** *J. Exp. Med.* 1998, **187**:885–895.

11-11 The effector mechanisms used to clear an infection depend on the infectious agent.

Baize, S., Leroy, E.M., Georges-Courbot, M.C., Capron, M., Lansoud-Soukate, J., Debre, P., Fisher-Hoch, S.P., McCormick, J.B., and Georges, A.J.: **Defective humoral responses and extensive intravascular apoptosis are associated with fatal outcome in Ebola virus-infected patients.** *Nat. Med.* 1999, **5**:423–426.

Kaufmann, S.H.E., Sher, A., and Ahmed, R. (eds): *Immunology of Infectious Diseases*. Washington DC, ASM Press, 2002.

Mims, C.A.: *The Pathogenesis of Infectious Disease*, 5th ed. London, Academic Press, 2000.

11-12 Resolution of an infection is accompanied by the death of most of the effector cells and the generation of memory cells.

Murali-Krishna, K., Altman, J.D., Suresh, M., Sourdive, D.J., Zajac, A.J., Miller, J.D., Slansky, J., and Ahmed, R.: **Counting antigen-specific CD8 T cells: a reevaluation of bystander activation during viral infection.** *Immunity* 1998, **8**:177–187.

Webb, S., Hutchinson, J., Hayden, K., and Sprent, J.: **Expansion/deletion of mature T cells exposed to endogenous superantigens *in vivo*.** *J. Immunol.* 1994, **152**:586–597.

11-13 Immunological memory is long-lived after infection or vaccination.

Black, F.L., and Rosen, L.: **Patterns of measles antibodies in residents of Tahiti and their stability in the absence of re-exposure.** *J. Immunol.* 1962, **88**:725–731.

Hammarlund, E., Lewis, M.W., Hansen, S.G., Strelow, L.I., Nelson, J.A., Sexton, G.J., Hanifin, J.M., and Slifka, M.K.: **Duration of antiviral immunity after smallpox vaccination.** *Nat. Med.* 2003, **9**:1131–1137.

Kassiotis, G., Garcia, S., Simpson, E., and Stockinger, B.: **Impairment of immunological memory in the absence of MHC despite survival of memory T cells.** *Nat. Immunol.* 2002, **3**:244–250.

Ku, C.C., Murakami, M., Sakamoto, A., Kappler, J., and Marrack, P.: **Control of homeostasis of CD8$^+$ memory T cells by opposing cytokines.** *Science* 2000, **288**:675–678.

Murali-Krishna, K., Lau, L.L., Sambhara, S., Lemonnier, F., Altman, J., and Ahmed, R.: **Persistence of memory CD8 T cells in MHC class I-deficient mice.** *Science* 1999, **286**:1377–1381.

Seddon, B., Tomlinson, P., and Zamoyska, R.: **Interleukin 7 and T cell receptor signals regulate homeostasis of CD4 memory cells.** *Nat. Immunol.* 2003, **4**:680–686.

11-14 Memory B-cell responses differ in several ways from those of naive B cells.

Berek, C., and Milstein, C.: **Mutation drift and repertoire shift in the maturation of the immune response.** *Immunol. Rev.* 1987, **96**:23–41.

Cumano, A., and Rajewsky, K.: **Clonal recruitment and somatic mutation in the generation of immunological memory to the hapten NP.** *EMBO J.* 1986, **5**:2459–2468.

Goins, C.L., Chappell, C.P., Shashidharamurthy, R., Selvaraj, P., and Jacob, J.: **Immune complex-mediated enhancement of secondary antibody responses.** *J. Immunol.* 2010, **184**:6293–6298.

Klein, U., Tu, Y., Stolovitzky, G.A., Keller, J.L., Haddad, J., Jr, Miljkovic, V., Cattoretti, G., Califano, A., and Dalla-Favera, R.: **Transcriptional analysis of the B cell germinal center reaction.** *Proc. Natl Acad. Sci. USA* 2003, **100**:2639–2644.

11-15 Repeated immunization leads to increasing affinity of antibody due to somatic hypermutation and selection by antigen in germinal centers.

Berek, C., Jarvis, J.M., and Milstein, C.: **Activation of memory and virgin B cell clones in hyperimmune animals.** *Eur. J. Immunol.* 1987, **17**:1121–1129.

Liu, Y.J., Zhang, J., Lane, P.J., Chan, E.Y., and MacLennan, I.C.: **Sites of specific B cell activation in primary and secondary responses to T cell-dependent and T cell-independent antigens.** *Eur. J. Immunol.* 1991, **21**:2951–2962.

Siskind, G.W., Dunn, P., and Walker, J.G.: **Studies on the control of antibody synthesis: II. Effect of antigen dose and of suppression by passive antibody on the affinity of antibody synthesized.** *J. Exp. Med.* 1968, **127**:55–66.

11-16 Memory T cells are increased in frequency compared with naive T cells specific for the same antigen, and have distinct activation requirements and cell-surface proteins that distinguish them from effector T cells.

Akondy, R.S., Monson, N.D., Miller, J.D., Edupuganti, S., Teuwen, D., Wu, H., Quyyumi, F., Garg, S., Altman, J.D., Del Rio, C., *et al.*: **The yellow fever virus vaccine induces a broad and polyfunctional human memory CD8⁺ T cell response.** *J. Immunol.* 2009, **183**:7919–7930.

Bradley, L.M., Atkins, G.G., and Swain, S.L.: **Long-term CD4⁺ memory T cells from the spleen lack MEL-14, the lymph node homing receptor.** *J. Immunol.* 1992, **148**:324–331.

Hataye, J., Moon, J.J., Khoruts, A., Reilly, C., and Jenkins, M.K.: **Naive and memory CD4⁺ T cell survival controlled by clonal abundance.** *Science* 2006, **312**:114–116.

Kaech, S.M., Hemby, S., Kersh, E., and Ahmed, R.: **Molecular and functional profiling of memory CD8 T cell differentiation.** *Cell* 2002, **111**:837–851.

Kaech, S.M., Tan, J.T., Wherry, E.J., Konieczny, B.T., Surh, C.D., and Ahmed, R.: **Selective expression of the interleukin 7 receptor identifies effector CD8 T cells that give rise to long-lived memory cells.** *Nat. Immunol.* 2003, **4**:1191–1198.

Rogers, P.R., Dubey, C., and Swain, S.L.: **Qualitative changes accompany memory T cell generation: faster, more effective responses at lower doses of antigen.** *J. Immunol.* 2000, **164**:2338–2346.

Wherry, E.J., Teichgraber, V., Becker, T.C., Masopust, D., Kaech, S.M., Antia, R., von Andrian, U.H., and Ahmed, R.: **Lineage relationship and protective immunity of memory CD8 T cell subsets.** *Nat. Immunol.* 2003, **4**:225–234.

11-17 Memory T cells are heterogeneous and include central memory and effector memory subsets.

Lanzavecchia, A., and Sallusto, F.: **Understanding the generation and function of memory T cell subsets.** *Curr. Opin. Immunol.* 2005, **17**:326–332.

Sallusto, F., Geginat, J., and Lanzavecchia, A.: **Central memory and effector memory T cell subsets: function, generation, and maintenance.** *Annu. Rev. Immunol.* 2004, **22**:745–763.

Sallusto, F., Lenig, D., Forster, R., Lipp, M., and Lanzavecchia, A.: **Two subsets of memory T lymphocytes with distinct homing potentials and effector functions.** *Nature* 1999, **401**:708–712.

11-18 CD4 T-cell help is required for CD8 T-cell memory and involves CD40 and IL-2 signaling.

Bourgeois, C., and Tanchot, C.: **CD4 T cells are required for CD8 T cell memory generation.** *Eur. J. Immunol.* 2003, **33**:3225–3231.

Bourgeois, C., Rocha, B., and Tanchot, C.: **A role for CD40 expression on CD8 T cells in the generation of CD8 T cell memory.** *Science* 2002, **297**:2060–2063.

Janssen, E.M., Lemmens, E.E., Wolfe, T., Christen, U., von Herrath, M.G., and Schoenberger, S.P.: **CD4 T cells are required for secondary expansion and memory in CD8 T lymphocytes.** *Nature* 2003, **421**:852–856.

Shedlock, D.J., and Shen, H.: **Requirement for CD4 T cell help in generating functional CD8 T cell memory.** *Science* 2003, **300**:337–339.

Sun, J.C., Williams, M.A., and Bevan, M.J.: **CD4 T cells are required for the maintenance, not programming, of memory CD8 T cells after acute infection.** *Nat. Immunol.* 2004, **5**:927–933.

Tanchot, C., and Rocha, B.: **CD8 and B cell memory: same strategy, same signals.** *Nat. Immunol.* 2003, **4**:431–432.

Williams, M.A., Tyznik, A.J., and Bevan, M.J.: **Interleukin-2 signals during priming are required for secondary expansion of CD8 memory T cells.** *Nature* 2006, **441**:890–893.

11-19 In immune individuals, secondary and subsequent responses are mainly attributable to memory lymphocytes.

Fazekas de St Groth, B., and Webster, R.G.: **Disquisitions on original antigenic sin. I. Evidence in man.** *J. Exp. Med.* 1966, **140**:2893–2898.

Fridman, W.H.: **Regulation of B cell activation and antigen presentation by Fc receptors.** *Curr. Opin. Immunol.* 1993, **5**:355–360.

Klenerman, P., and Zinkernagel, R.M.: **Original antigenic sin impairs cytotoxic T lymphocyte responses to viruses bearing variant epitopes.** *Nature* 1998, **394**:482–485.

Mongkolsapaya, J., Dejnirattisai, W., Xu, X.N., Vasanawathana, S., Tangthawornchaikul, N., Chairunsri, A., Sawasdivorn, S., Duangchinda, T., Dong, T., Rowland-Jones, S., *et al.*: **Original antigenic sin and apoptosis in the pathogenesis of dengue hemorrhagic fever.** *Nat. Med.* 2003, **9**:921–927.

Pollack, W., Gorman, J.G., Freda, V.J., Ascari, W.Q., Allen, A.E., and Baker, W.J.: **Results of clinical trials of RhoGAm in women.** *Transfusion* 1968, **8**:151–153.

Zehn, D., Turner, M.J., Lefrançois, L., and Bevan, M.J.: **Lack of original antigenic sin in recall CD8⁺ T cell responses.** *J. Immunol.* 2010, **184**:6320–6326.

The Mucosal Immune System

12

A series of anatomically distinct compartments can be distinguished within the immune system, each of which is specially adapted to generate a response to antigens encountered in a particular set of tissues. In previous chapters we have mainly discussed adaptive immune responses that are initiated in lymph nodes and spleen—the peripheral lymphoid tissues that respond to antigens that have entered the body via the skin, are present in the internal organs, or have spread into the blood. These are the immune responses most studied by immunologists, as they are the responses evoked when antigens are administered by injection. There is, however, an additional compartment of the adaptive immune system, of even greater size, located near the surfaces where most pathogens actually invade. This is the **mucosal immune system**—the subject of this chapter.

The organization of the mucosal system.

The thin layer of mucosal epithelium lining internal body surfaces is the only physical barrier against invasion of the underlying tissues by potential pathogens and the body's own commensal microorganisms, which are present in vast numbers at most mucosal surfaces. These surfaces therefore require continual protection against invaders. The epithelium can be breached relatively easily and so its barrier function needs to be supplemented by defenses provided by the cells and molecules of the mucosal immune system. The innate defenses of mucosal tissues, such as antimicrobial peptides and cells bearing invariant pathogen-recognition receptors, are described in Chapters 2 and 3. In this chapter we concentrate on the adaptive mucosal immune system. Many of the anatomical and immunological principles underlying the mucosal immune system apply to all its constituent tissues; here we will use the intestine as our example.

12-1 The mucosal immune system protects the internal surfaces of the body.

The mucosal immune system comprises the body surfaces lined by mucus-secreting epithelium—the gastrointestinal tract, the upper and lower respiratory tract, and the urogenital tract. It also includes the exocrine glands associated with these organs, such as the conjunctivae and lachrymal glands of the eye, the salivary glands, and the lactating breast (Fig. 12.1). The mucosal surfaces represent an enormous area to be protected. The human small intestine, for instance, has a surface area of almost 400 m^2, which is 200 times

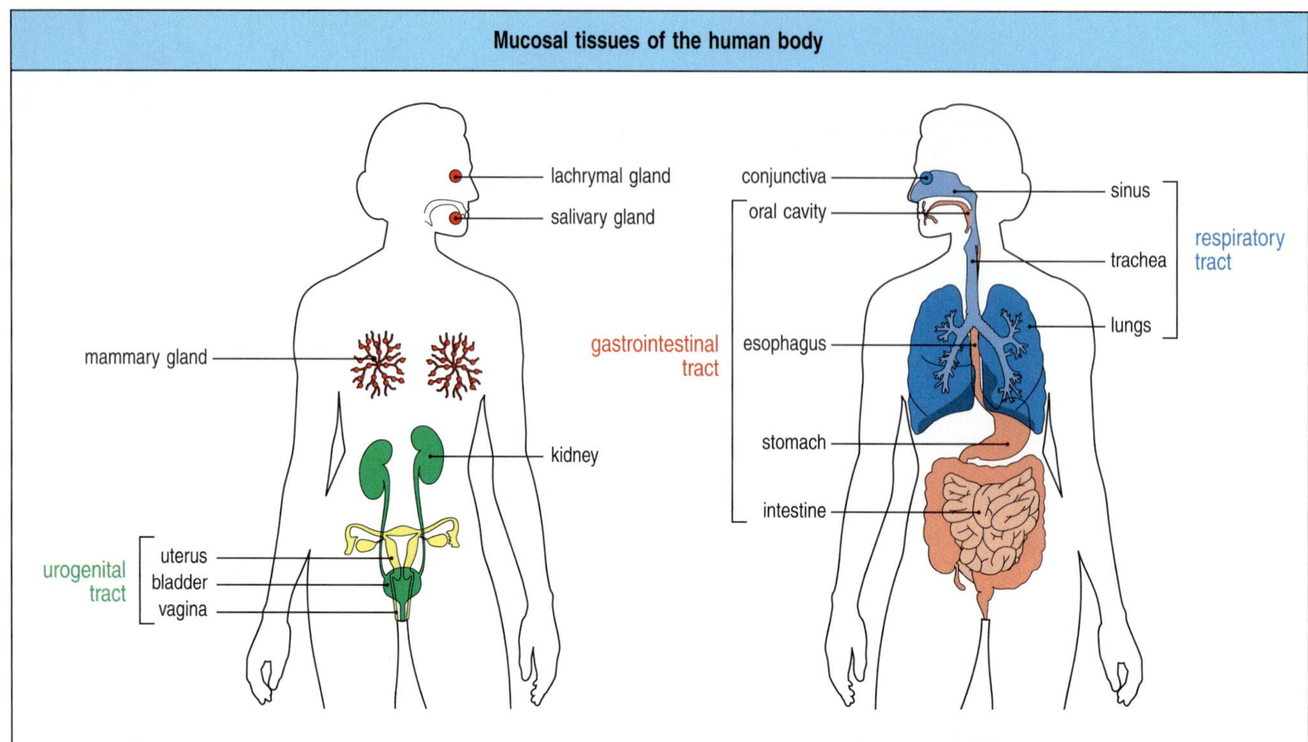

Fig. 12.1 The mucosal immune system. The tissues of the mucosal immune system are the lymphoid organs associated with the intestine, respiratory tract, and urogenital tract, as well as the oral cavity and pharynx and the glands associated with these tissues, such as the salivary glands and lachrymal glands. The lactating breast is also part of the mucosal immune system.

that of the skin. Because of their physiological functions in gas exchange (the lungs), food absorption (the gut), sensory activities (eyes, nose, mouth, and throat), and reproduction (uterus and vagina), the mucosal surfaces are thin and permeable barriers to the interior of the body. The importance of these tissues to life means that effective defense mechanisms are essential to protect them from invasion. Equally, their fragility and permeability create obvious vulnerability to infection, and it is not surprising that the vast majority of infectious agents invade the human body by these routes (Fig. 12.2). Diarrheal diseases, acute respiratory infections, pulmonary tuberculosis, measles, whooping cough, and worm infestations continue to be the major causes of death throughout the world, especially in infants in developing countries. To these must be added the human immunodeficiency virus (HIV), a pathogen whose natural route of entry via a mucosal surface is often overlooked, as well as other sexually transmitted infections such as syphilis.

The mucosal surfaces are also portals of entry for a vast array of foreign antigens that are not pathogenic. This is best seen in the gut, which is exposed to enormous quantities of food proteins—an estimated 30–35 kg per year per person. At the same time, the healthy large intestine is colonized by at least a thousand species of bacteria that live in symbiosis with their host and are known as **commensal microorganisms**, or the **microbiota**. These bacteria are present at levels of at least 10^{12} organisms per milliliter in the colon contents, making them the most numerous cells in the body by a factor of 10. In normal circumstances they do no harm and are beneficial to their host in many ways. Many of the other mucosal surfaces have equivalent, if smaller, populations of resident commensal organisms (Fig. 12.3).

As food proteins and the microbiota contain many foreign antigens, they are capable of being recognized by the adaptive immune system. Generating protective immune responses against these harmless agents would, however, be inappropriate and wasteful. Indeed, aberrant immune responses of this kind are now believed to be the cause of some relatively common diseases, including celiac disease (caused by a response to the wheat protein gluten) and inflammatory bowel diseases such as Crohn's disease (a response

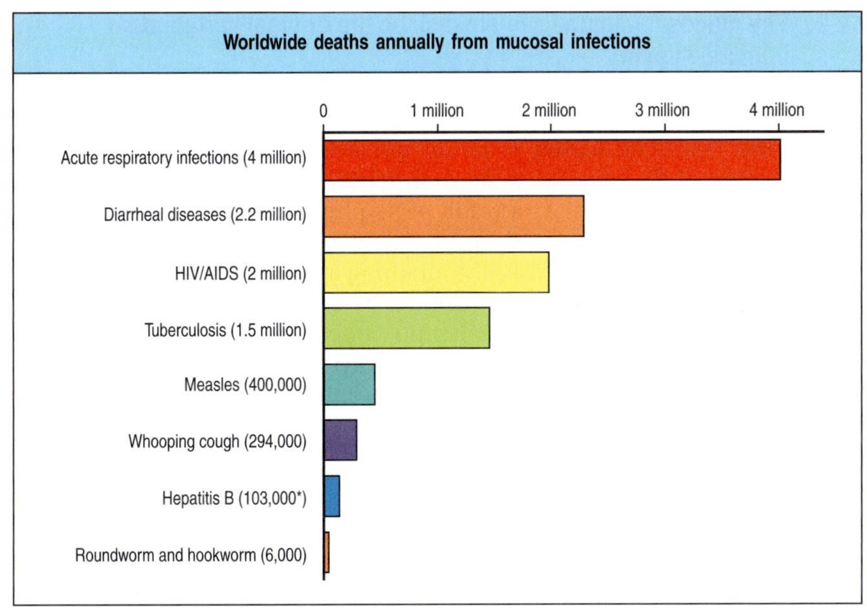

Fig. 12.2 Mucosal infections are one of the biggest health problems worldwide. Most of the pathogens that cause the deaths of large numbers of people are those of mucosal surfaces or enter the body through these routes. Respiratory infections are caused by numerous bacteria (such as *Streptococcus pneumoniae* and *Haemophilus influenzae*, which cause pneumonia, and *Bordetella pertussis*, the cause of whooping cough) and viruses (such as influenza and respiratory syncytial virus). Diarrheal diseases are caused by both bacteria (such as the cholera bacterium *Cholera vibrio*) and viruses (such as rotaviruses). The human immunodeficiency virus (HIV) that causes AIDS enters through the mucosa of the urogenital tract or is secreted into breast milk and is passed from mother to child in this way. The bacterium *Mycobacterium tuberculosis*, which causes tuberculosis, also enters through the respiratory tract. Measles manifests itself as a systemic disease, but it originally enters via the oral/respiratory route. Hepatitis B is also a sexually transmitted virus. Finally, parasitic worms inhabiting the intestine cause chronic debilitating disease and premature death. Most of these deaths, especially those from acute respiratory and diarrheal diseases, occur in children under 5 years old in the developing world, and there are still no effective vaccines against many of these pathogens. Numbers shown are the most recent estimated figures available (*The Global Burden of Disease: 2004 Update*. World Health Organization, 2008). *Does not include deaths from liver cancer or cirrhosis resulting from chronic infection.

to commensal bacteria). As we shall see, the intestinal mucosal immune system has evolved means of distinguishing harmful pathogens from antigens in food and the natural gut microbiota. Similar issues are faced at other mucosal surfaces, such as the respiratory tract and female genital tract. Here, protective immunity against pathogens is essential but many of the antigens entering these tissues are also harmless, being derived from commensal organisms, pollen, other innocuous environmental material, and, in the lower urogenital tract, sperm.

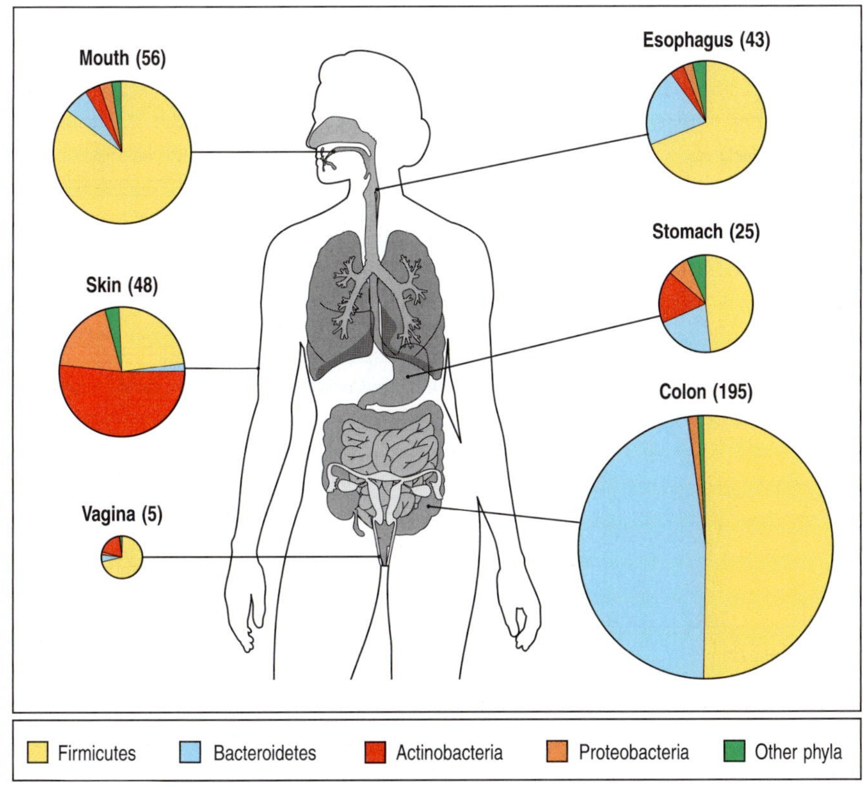

Fig. 12.3 Composition of the commensal microbiota at different mucosal surfaces in healthy humans. The different sizes of the pie-charts for different sites reflect the numbers of distinct bacterial 'species' typically present at that site. The colon contains the greatest number of different species (around 195 as estimated from individual surveys). The color key indicates the four bacterial phyla that contain the majority of commensal species. Ubiquitous commensal bacteria include *Lactobacillus* spp. (Firmicutes), *Bifidobacterium* spp. (Actinobacteria), *Bacteroides fragilis* (Bacteroidetes), and *Escherichia coli* (Proteobacteria). Adapted from Dethlefsen, L. *et al.*: *Nature* 2007, 449:811–818.

12-2 The mucosal immune system may be the original vertebrate immune system.

From the point of view of traditional immunology, the mucosal immune system has been considered to be an unusual and relatively minor sub-compartment of the immune system. In terms of size and function, this is an inaccurate description. As a result of its physiologically critical role and extent of exposure to antigens, the mucosal immune system forms the largest part of the body's immune tissues, containing approximately three-quarters of all lymphocytes and producing the majority of immunoglobulin in healthy individuals. When compared with lymph nodes and spleen (which in this chapter we will call the **systemic immune system**), the mucosal immune system has many unique and unusual features. The main distinctive features are listed in Fig. 2.4.

The mucosal immune system, in particular that of the gut, may well have been the first part of the vertebrate adaptive immune system to evolve, and it has been proposed that its evolution could be linked to the need to deal with the vast populations of commensal bacteria that coevolved with the vertebrates. Organized lymphoid tissues and immunoglobulin antibodies are first found in vertebrates in the gut of primitive cartilaginous fishes, and two important central lymphoid organs—the thymus and the avian bursa of Fabricius—derive from the embryonic intestine. It has therefore been suggested that the mucosal immune system represents the original vertebrate immune system, and that the spleen and lymph nodes are later specializations.

12-3 Cells of the mucosal immune system are located both in anatomically defined compartments and scattered throughout mucosal tissues.

Lymphocytes and other immune-system cells such as macrophages and dendritic cells are found throughout the intestinal tract, both in organized tissues and scattered throughout the surface epithelium of the mucosa and an underlying layer of connective tissue called the **lamina propria**. The

Fig.12.4 Distinctive features of the mucosal immune system. The mucosal immune system is bigger, encounters a wider range of antigens, and encounters them much more frequently, than the rest of the immune system—what we call in this chapter the systemic immune system. This is reflected in distinctive anatomical features, specialized mechanisms for the uptake of antigen, and unusual effector and regulatory responses that are designed to prevent unwanted immune responses to food and other innocuous antigens.

Distinctive features of the mucosal immune system	
Anatomical features	Intimate interactions between mucosal epithelia and lymphoid tissues
	Discrete compartments of diffuse lymphoid tissue and more organized structures such as Peyer's patches, isolated lymphoid follicles, and tonsils
	Specialized antigen-uptake mechanisms, e.g. M cells in Peyer's patches, adenoids, and tonsils
Effector mechanisms	Activated/memory T cells predominate even in the absence of infection
	Multiple activated 'natural' effector/regulatory T cells present
	Secretory IgA antibodies
	Presence of distinctive microbiota
Immunoregulatory environment	Active downregulation of immune responses (e.g. to food and other innocuous antigens) predominates
	Inhibitory macrophages and tolerance-inducing dendritic cells

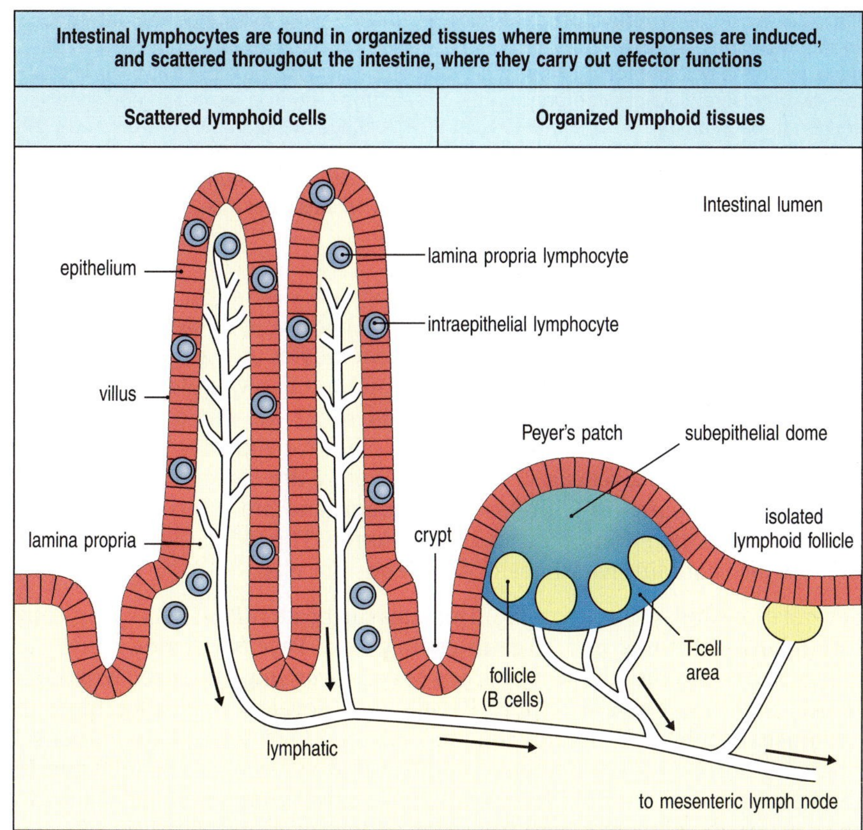

Fig. 12.5 Gut-associated lymphoid tissues and lymphocyte populations. The intestinal mucosa of the small intestine is made up of finger-like processes (villi) covered by a thin layer of epithelial cells (red) that are responsible for digestion of food and absorption of nutrients. These epithelial cells are replaced continually by new cells that derive from stem cells in the crypts. The tissue layer underlying the epithelium is called the lamina propria, and will be colored pale yellow throughout this chapter. Lymphocytes are found in several discrete compartments in the intestine, with the organized lymphoid tissues such as Peyer's patches and isolated lymphoid follicles forming what is known as the gut-associated lymphoid tissues (GALT). These tissues lie in the wall of the intestine itself, separated from the contents of the intestinal lumen by the single layer of epithelium. The draining lymph nodes for the gut are the mesenteric lymph nodes (see Fig. 12.12), which are connected to Peyer's patches and the intestinal mucosa by lymphatic vessels and are the largest lymph nodes in the body. Together, these organized tissues are the sites of antigen presentation to T cells and B cells and are responsible for the induction phase of immune responses. Peyer's patches and mesenteric lymph nodes contain discrete T-cell areas (blue) and B-cell follicles (yellow), while the isolated follicles comprise mainly B cells. Many lymphocytes are found scattered throughout the mucosa outside the organized lymphoid tissues: these are effector cells—effector T cells and antibody-secreting plasma cells. Effector lymphocytes are found both in the epithelium and in the lamina propria. Lymphatics also drain from the lamina propria to the mesenteric lymph nodes.

organized secondary lymphoid tissues in the gut comprise a group of organs known as the **gut-associated lymphoid tissues (GALT)**, together with the draining **mesenteric lymph nodes** (Fig. 12.5). The GALT and the mesenteric lymph nodes have the anatomically compartmentalized structure typical of peripheral lymphoid organs, and are sites at which immune responses are initiated. The cells scattered throughout the epithelium and the lamina propria comprise the effector cells of the local immune response.

The GALT comprises the **Peyer's patches**, which are present in the small intestine, **isolated lymphoid follicles**, which are found throughout the intestine, and the appendix (in humans). The **palatine tonsils**, **adenoids**, and **lingual tonsils** are large aggregates of lymphoid tissue covered by a layer of squamous epithelium and form a ring, known as Waldeyer's ring, at the back of the mouth at the entrance of the gut and airways (Fig. 12.6). They often become extremely enlarged in childhood because of recurrent infections, and in the past were victims of a vogue for surgical removal. A reduced IgA response to oral polio vaccination has been seen in individuals who have had their tonsils and adenoids removed.

The Peyer's patches of the small intestine, the lymphoid tissue of the appendix (which is another frequent victim of the surgeon's knife), and the isolated lymphoid follicles are located within the intestinal wall. Peyer's patches are extremely important sites for the initiation of immune responses in the gut. Visible to the naked eye, they have a distinctive appearance, forming dome-like aggregates of lymphoid cells that project into the intestinal lumen (Fig. 12.7). There are 100–200 Peyer's patches in the human small intestine. They are much richer in B cells than the systemic peripheral lymphoid organs, each consisting of a large number of B-cell follicles with germinal centers, with small T-cell areas between and immediately below the follicles. The subepithelial dome area lies immediately beneath the epithelium and is rich in dendritic cells, T cells, and B cells. Separating the lymphoid tissues from the

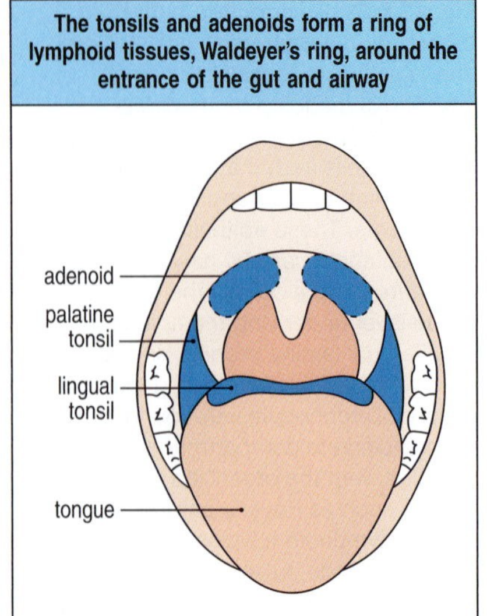

The tonsils and adenoids form a ring of lymphoid tissues, Waldeyer's ring, around the entrance of the gut and airway
adenoid
palatine tonsil
lingual tonsil
tongue

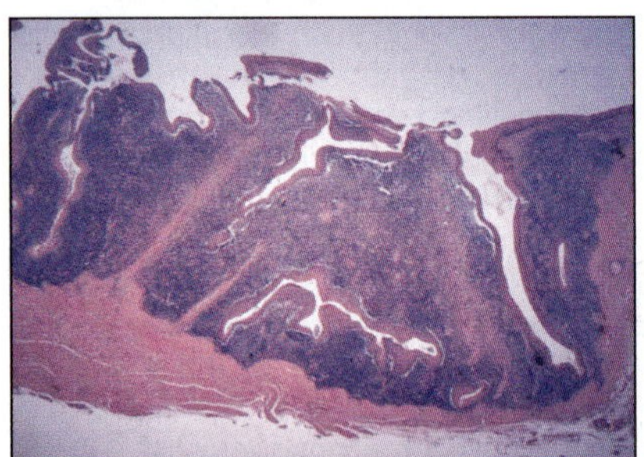

Fig. 12.6 A ring of lymphoid organs called Waldeyer's ring surrounds the entrance to the intestine and respiratory tract. The adenoids lie at either side of the base of the nose, while the palatine tonsils lie at either side of the back of the oral cavity. The lingual tonsils are discrete lymphoid organs on the base of the tongue. The micrograph shows a section through an inflamed human tonsil. In the absence of inflammation, the tonsils and adenoids normally comprise areas of organized tissue with both B-cell and T-cell areas, covered by a layer of squamous epithelium (at top of photo). The surface contains deep crevices (crypts) that increase the surface area but can easily become sites of infection. Hematoxylin and eosin staining. Magnification ×100.

gut lumen is a layer of **follicle-associated epithelium**. This contains conventional intestinal epithelial cells known as enterocytes and a smaller number of specialized epithelial cells called **microfold cells** (**M cells**), which have a folded luminal surface instead of the microvilli present on enterocytes. Unlike enterocytes, M cells do not secrete digestive enzymes or mucus and lack a thick surface glycocalyx. They are therefore directly exposed to microorganisms and particles within the gut lumen and are the route by which antigen enters the Peyer's patch from the lumen. The follicle-associated epithelium also contains lymphocytes and dendritic cells.

Several thousand isolated lymphoid follicles can be identified microscopically throughout the small and large intestines. Like Peyer's patches, these are composed of an epithelium containing M cells overlying organized lymphoid tissue, but they contain mainly B cells and develop only after birth in response to antigen stimulation due to colonization of the gut by commensal microorganisms. Peyer's patches, in contrast, are present in the fetal gut. In the gut, isolated lymphoid follicles seem to arise from small aggregates in the intestinal wall called **cryptopatches**, which contain dendritic cells and lymphoid tissue inducer cells (see Section 8-24). Peyer's patches and isolated lymphoid follicles are connected by lymphatics to the draining mesenteric lymph nodes, which are located in the connective tissue that tethers the intestine to the rear wall of the abdomen. These are the largest lymph nodes in

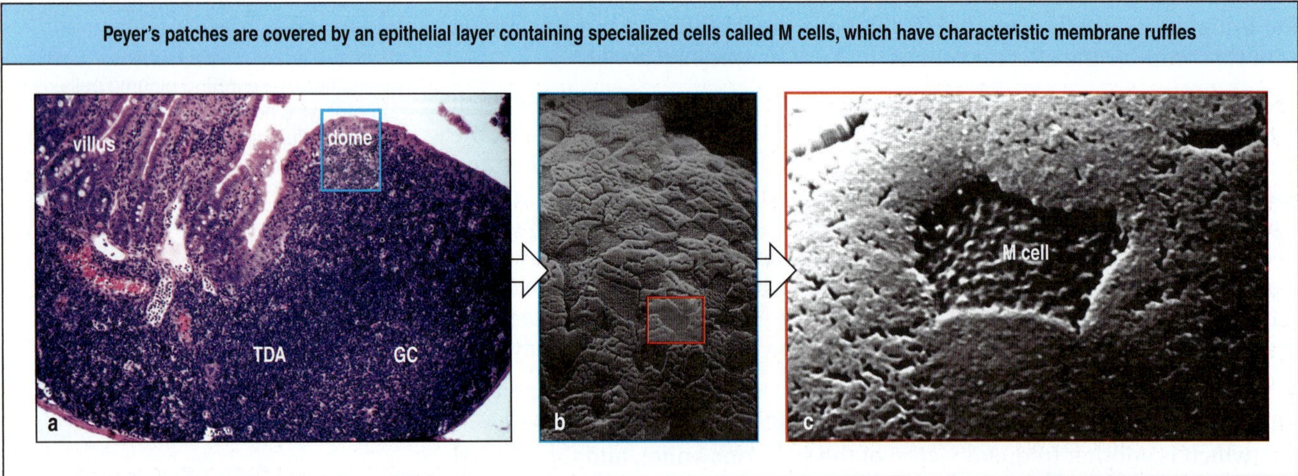

Fig. 12.7 A Peyer's patch and its specialized surface epithelium. Panel a: Peyer's patches are organized lymphoid tissues lying in the submucosal layer of the intestinal wall. Each comprises numerous, highly active B-cell follicles with germinal centres (GC), as well as intervening T-cell dependent areas (TDA) and a layer between the surface epithelium and the follicles known as the subepithelial dome, which is rich in dendritic cells, T cells, and B cells (see Figs 12.5 and 1.20 for schematic views of a Peyer's patch). The surface epithelium is known as the follicle-associated epithelium and is a single layer of columnar epithelial cells. Panel b: scanning electron micrograph of the follicle-associated epithelium of the mouse Peyer's patch shown boxed in (a) reveals microfold (M) cells, which lack the microvilli and the layer of mucus that is present on normal epithelial cells. Each M cell appears as a sunken area on the epithelial surface. Panel c: a higher-magnification view of the boxed area in (b) shows the characteristic ruffled surface of an M cell. M cells are the portal of entry for many pathogens and other particles. (a) Hematoxylin and eosin staining. Magnification ×100. (b) ×5000. (c) ×23,000. Source: Mowat, A., Viney, J.: *Immunol. Rev.* 1997, **156**:145–166.

the body and play a crucial role in initiating and shaping immune responses to intestinal antigens.

The immune responses generated when antigen is recognized in one of the tissues of the GALT are quite distinct from those stimulated in lymph nodes or spleen when antigen is introduced into the skin, muscle, or bloodstream. This is because the microenvironment of the GALT has its own characteristic content of lymphoid cells, hormones, and other immunomodulatory factors. The mesenteric lymph nodes and Peyer's patches differentiate independently of the systemic immune system during fetal development, with the involvement of specific chemokines and receptors of the tumor necrosis factor (TNF) family (see Fig. 12.8; see also Section 8-24). The differences between the GALT and the systemic lymphoid organs are thus imprinted early in life and are independent of exposure to antigen.

In some species such as mice, isolated lymphoid follicles are also found in the lining of the nose, where they are called **nasal-associated lymphoid tissues** (**NALT**), and in the wall of the upper respiratory tract, when they are known as **bronchus-associated lymphoid tissues** (**BALT**). The term **mucosa-associated lymphoid tissues** (**MALT**) is sometimes used to refer collectively to all such tissues found in mucosal organs, although defined organized lymphoid tissues are not found in the nose or respiratory tract in humans unless infection is present.

Control of development of the GALT compared with systemic lymphoid tissues

	Protein required for tissue development									
Tissue	TNFR-I	LT-α	LT-β	LTβR	TRANCE	IL-7R	$\alpha_4{:}\beta_7$ integrin	L-selectin	CXCR5	NFκB2
Peyer's patch	+	+	+	+	–	+	+/–	–	+/–	+
Isolated lymphoid follicle	+	+	+	+	–	+	–	–	+	+
Mesenteric lymph node	–	+	–	+	+	–	+/–	+/–	–	–
Systemic lymph node	+/–	+	+/–	+	+	–	–	+	–	+/–

Fig. 12.8 The fetal development of intestinal lymphoid tissues is controlled by a specific set of cytokines. Experiments in knockout mice show that the mesenteric lymph nodes and Peyer's patches differ from each other, and from lymph nodes in other parts of the body, in the signals that are required for their development in fetal and early neonatal life. The development of all these lymphoid tissues requires an interchange of signals between lymphoid-tissue inducer cells and local stromal cells. Signals from the stromal cells induce the lymphoid-tissue inducer cells to express lymphotoxin (LT)-α and -β subunits. These can form homotrimers (LT-α_3) or heterotrimers (LT-$\alpha_1{:}\beta_2$); LT-$\alpha_1{:}\beta_2$ acts on local stromal cells via the LT-β receptor, and this receptor is required for the development of all the lymphoid tissues considered here, as is the production of the LT-α subunit. Stimulation of stromal cells via the LT-β receptor leads to the expression of adhesion molecules such as VCAM-1 and the production of chemokines such as CCL19, CCL21, and CXCL13, all of which recruit lymphocytes into the developing organ, as well as more lymphoid-tissue inducer cells. Mesenteric lymph nodes are the first lymphoid tissues to develop in the fetus. Lymphoid-tissue inducer cells in these sites produce LT-$\alpha_1{:}\beta_2$ in response to the TNF-family cytokine TRANCE produced by the stromal cells, but knockout experiments in mice show that the LT-β subunit is not essential for mesenteric lymph node development and that it can be replaced by another TNF-family molecule, LIGHT, which can also bind the LT-β receptor. The development of Peyer's patches is absolutely dependent on the presence of both LT-α and LT-β subunits, which are produced by lymphoid-tissue inducer cells in response to IL-7 produced by stromal cells. Lymphoid-tissue inducer cells are also uniquely recruited to Peyer's patches via their CXCR5 receptors, and the TNF receptor TNFR-I is also involved in the development of Peyer's patches but not of the other tissues shown here. In respect of LT signals, the requirements of the peripheral lymph nodes are more similar to those of the mesenteric lymph node. The differences in the requirements for LT subunits and receptors probably reflect subtle differences in the signaling pathways used in the different sites. Adhesion molecules are also involved in lymphoid tissue development. Peyer's patches develop normally in the absence of L-selectin but are partly dependent on the integrin $\alpha_4{:}\beta_7$ and are entirely absent if both these proteins are lacking. Mesenteric lymph nodes also require either L-selectin or $\alpha_4{:}\beta_7$ integrin, but develop normally in the absence of either. Systemic lymph nodes require only L-selectin for their development.

12-4 The intestine has distinctive routes and mechanisms of antigen uptake.

Antigens present at mucosal surfaces must be transported across an epithelial barrier before they can stimulate the mucosal immune system. Peyer's patches and isolated lymphoid follicles are highly adapted for the uptake of antigen from the intestinal lumen, particularly those present on bacteria and viruses. The M cells in the follicle-associated epithelium are continually taking up molecules and particles from the gut lumen by endocytosis or phagocytosis (Fig. 12.9). In the case of bacteria this may involve specific recognition of the bacterial FimH protein found in type 1 pili by a glycoprotein (GP2) on the M cell. This material is transported through the interior of the cell in membrane-enclosed vesicles to the basal cell membrane, where it is released into the extracellular space—a process known as **transcytosis**. Because M cells are much more accessible than enterocytes, a number of pathogens target M cells to gain access to the subepithelial space, even though they then find themselves in the heart of the intestinal adaptive immune system (see Section 12-3).

The basal cell membrane of an M cell is extensively folded, forming a pocket that encloses lymphocytes and dendritic cells. The dendritic cells take up the transported material released from the M cells and process it for presentation to T lymphocytes. These dendritic cells are in a particularly favorable position to acquire gut antigens, and they are recruited to the follicle-associated epithelium in response to chemokines that are released constitutively by the epithelial cells. The chemokines include CCL20 (MIP-3α) and CCL9 (MIP-1γ), which bind to the receptors CCR6 and CCR1, respectively, on the dendritic cell (see Appendix IV for a listing of chemokines and their receptors). The antigen-loaded dendritic cells then migrate from the dome region to the T-cell areas of the Peyer's patch, where they meet naive, antigen-specific T cells. Together, the dendritic cells and primed T cells then activate B cells and initiate class switching to IgA. All these processes—the uptake of antigen by M cells, the migration of dendritic cells into the epithelial layer, the production of chemokines, and the subsequent migration of dendritic cells into T-cell areas—are markedly enhanced in the presence of pathogenic organisms and their products.

Fig. 12.9 Uptake and transport of antigens by M cells. The first three panels show uptake via M cells in the follicle-associated epithelium of Peyer's patches. These have convoluted basal membranes that form 'pockets' within the epithelial layer, allowing close contact with lymphocytes and other cells. This favors the local transport of antigens that have been taken up from the intestine by the M cells and their delivery to dendritic cells for antigen presentation. The micrograph of part of a Peyer's patch on the right shows epithelial cells (dark blue), some of which are M cells that form pockets where T cells (red) and B cells (green) accumulate. The cells have been stained with fluorescently labeled antibodies specific for individual cell types. Micrograph from Brandtzaeg, P., *et al.*: *Immunol. Today* 1999, 20:141–151.

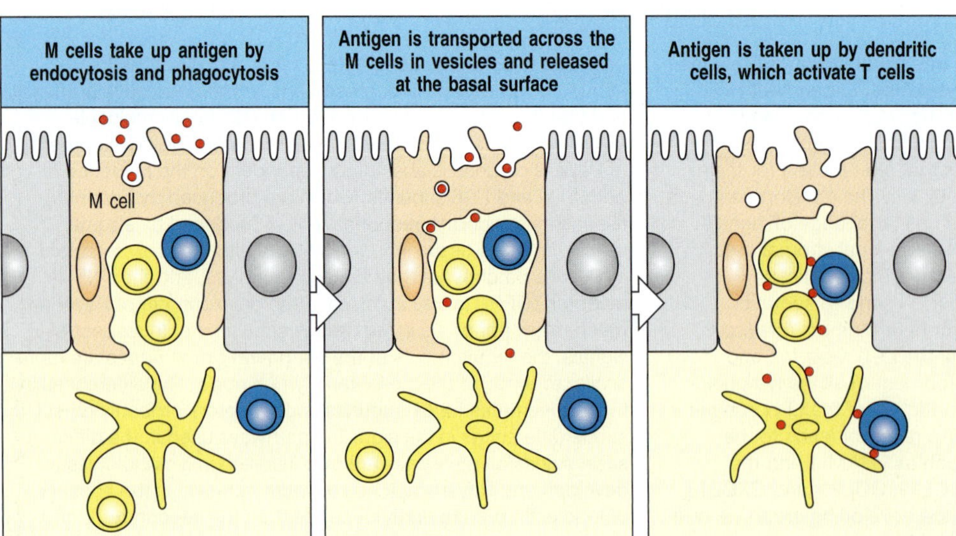

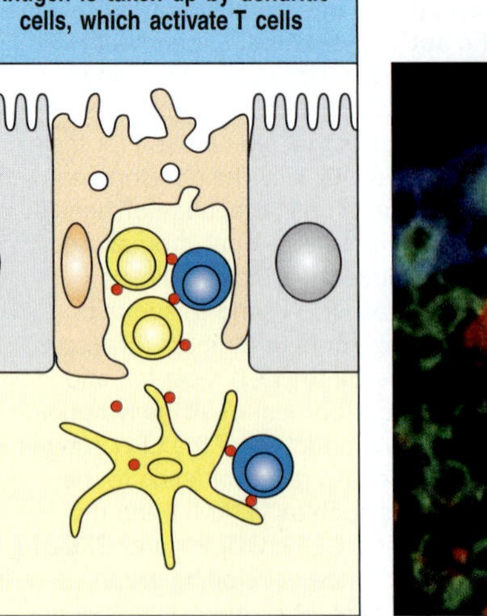

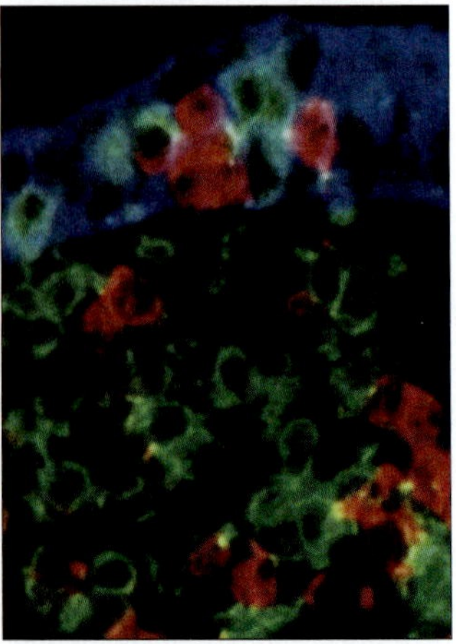

12-5 The mucosal immune system contains large numbers of effector lymphocytes even in the absence of disease.

In addition to the organized lymphoid organs, a mucosal surface contains enormous numbers of lymphocytes and other leukocytes scattered throughout the tissue. Most of the scattered lymphocytes have the appearance of cells that have been activated by antigen, and they comprise the effector T cells and plasma cells of the mucosal immune system. In the intestine, effector cells are found in two main compartments: the epithelium and the lamina propria (Fig. 12.10). These tissues are quite distinct in immunological terms, despite being separated by only a thin layer of basement membrane. The epithelium contains mainly lymphocytes, which in the small intestine are virtually all CD8 T cells. The lamina propria is much more heterogeneous, with large numbers of CD4 T cells and CD8 T cells, as well as plasma cells, macrophages, dendritic cells, and occasional eosinophils and mast cells. Neutrophils are rare in the healthy intestine, although their numbers increase rapidly during inflammatory disease or infection. The total number of lymphocytes in the epithelium and lamina propria probably exceeds that of most other parts of the body.

The healthy intestinal mucosa therefore displays many characteristics of a chronic inflammatory response—namely, the presence of numerous effector lymphocytes and other leukocytes in the tissues. The presence of such large numbers of effector cells is not dependent on infection by a pathogen and is unusual for a healthy, nonlymphoid tissue. It is the result of the local responses that are continually being made to the myriad of innocuous antigens that are bombarding the mucosal surfaces. As we shall see, this is a physiological process that is essential for maintaining the beneficial symbiosis between the host and its intestinal contents, and involves a balanced generation of effector and regulatory T cells. When required, however, the mucosal

Fig. 12.10 The lamina propria and epithelium of the intestinal mucosa are discrete lymphoid compartments. The lamina propria contains a heterogeneous mixture of IgA-producing plasma cells, lymphocytes with a 'memory' phenotype (see Chapter 10), conventional CD4 and CD8 effector T cells, dendritic cells, macrophages, and mast cells. T cells in the lamina propria of the small intestine express the integrin $\alpha_4{:}\beta_7$ and the chemokine receptor CCR9, which attracts them into the tissue from the bloodstream. Intraepithelial lymphocytes express CCR9 and the integrin $\alpha_E{:}\beta_7$, which binds to E-cadherin on epithelial cells. They are mostly CD8 T cells, some of which express the conventional $\alpha{:}\beta$ form of CD8 and others the CD8$\alpha{:}\alpha$ homodimer. CD4 T cells predominate in the lamina propria, whereas CD8 T cells predominate in the epithelium.

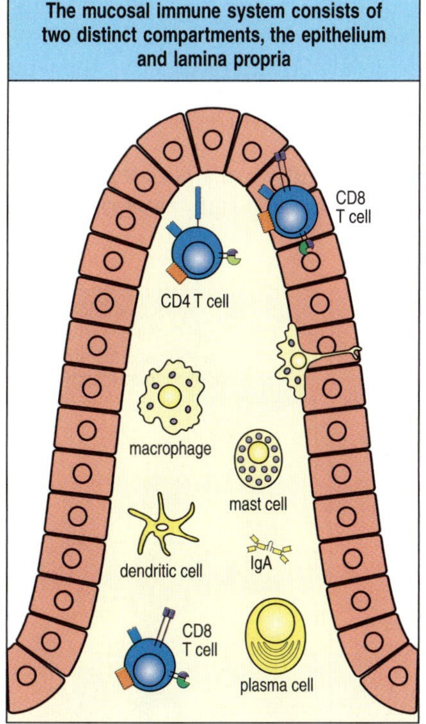

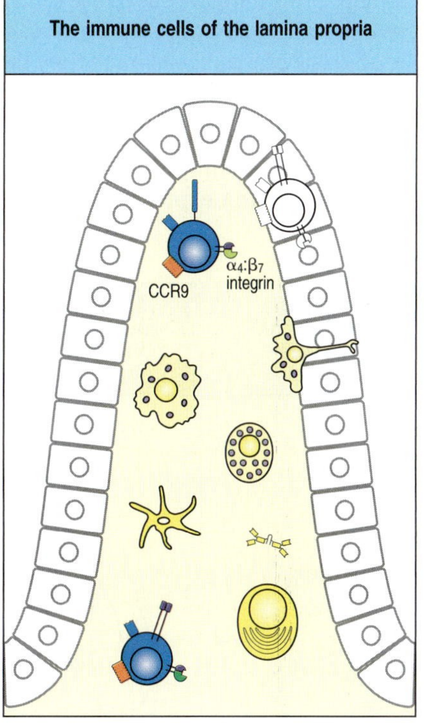

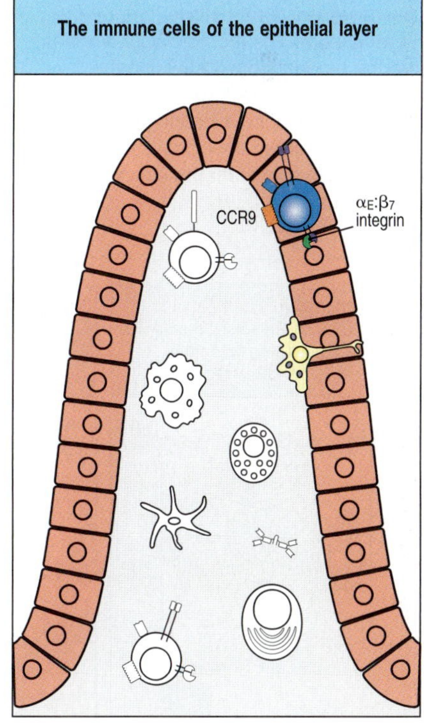

immune system can be refocused on producing a full adaptive immune response to invading pathogens.

12-6 The circulation of lymphocytes within the mucosal immune system is controlled by tissue-specific adhesion molecules and chemokine receptors.

The arrival of effector lymphocytes in the mucosal surface layer is the outcome of a series of events in which the homing characteristics of lymphocytes change as they become activated. The life history of mucosal lymphocytes starts with the emergence of naive T cells and B cells from the thymus and bone marrow, respectively. At this point, the naive lymphocytes circulating in the bloodstream are not predetermined as to which compartment of the immune system they will end up in. Naive lymphocytes arriving at Peyer's patches and mesenteric lymph nodes enter them through high endothelial venules (Fig. 12.11). As in the systemic immune system, entry to peripheral lymphoid organs is controlled by the chemokines CCL21 and CCL19, which are released from the lymphoid tissues and bind the receptor CCR7 on naive lymphocytes. If the naive lymphocytes do not see their antigen, they exit from the lymphoid organ via the lymphatics and return to the bloodstream. If they encounter antigen in the GALT, the lymphocytes become activated and lose expression of CCR7 and L-selectin. This means that that they have lost their ability to home to secondary lymphoid organs, because they cannot enter them via the high endothelial venules (see Section 9-3).

Although primed lymphocytes leave the mucosal lymphoid organs in which they were activated, they travel back to the mucosa as effector cells. T and B lymphocytes initially activated in Peyer's patches, for example, leave via the lymphatics before they differentiate fully into effector cells, pass through mesenteric lymph nodes, and eventually end up in the thoracic duct. From there they circulate in the bloodstream (see Fig. 12.11) and selectively reenter the lamina propria via small blood vessels. Antigen-specific B cells are primed as IgM-producing B cells in the follicular areas of Peyer's patches and undergo switching to IgA production there, but they only differentiate fully into IgA-producing plasma cells once they have recirculated and returned to the lamina propria. Plasma cells are rarely found in Peyer's patches, and this is also true of effector T cells.

Gut-specific homing by antigen-stimulated T and B cells is determined in large part by the expression of the adhesion molecule **α_4:β_7 integrin** on the

Fig. 12.11 Priming of naive T cells and the redistribution of effector T cells in the intestinal immune system. Naive T cells carry the chemokine receptor CCR7 and L-selectin, which direct their entry into Peyer's patches via high endothelial venules (HEV). In the T-cell area they encounter antigen that has been transported into the lymphoid tissue by M cells and is presented by local dendritic cells. During activation, and under the selective control of gut-derived dendritic cells, the T cells lose L-selectin and acquire the chemokine receptor CCR9 and the integrin α_4:β_7. After activation, but before full differentiation, the primed T cells exit from the Peyer's patch via the draining lymphatics, passing through the mesenteric lymph node to enter the thoracic duct. The thoracic duct empties into the bloodstream, delivering the activated T cells back to the wall of the intestine. Here T cells bearing CCR9 and α_4:β_7 are attracted specifically to leave the bloodstream and enter the lamina propria of the villus.

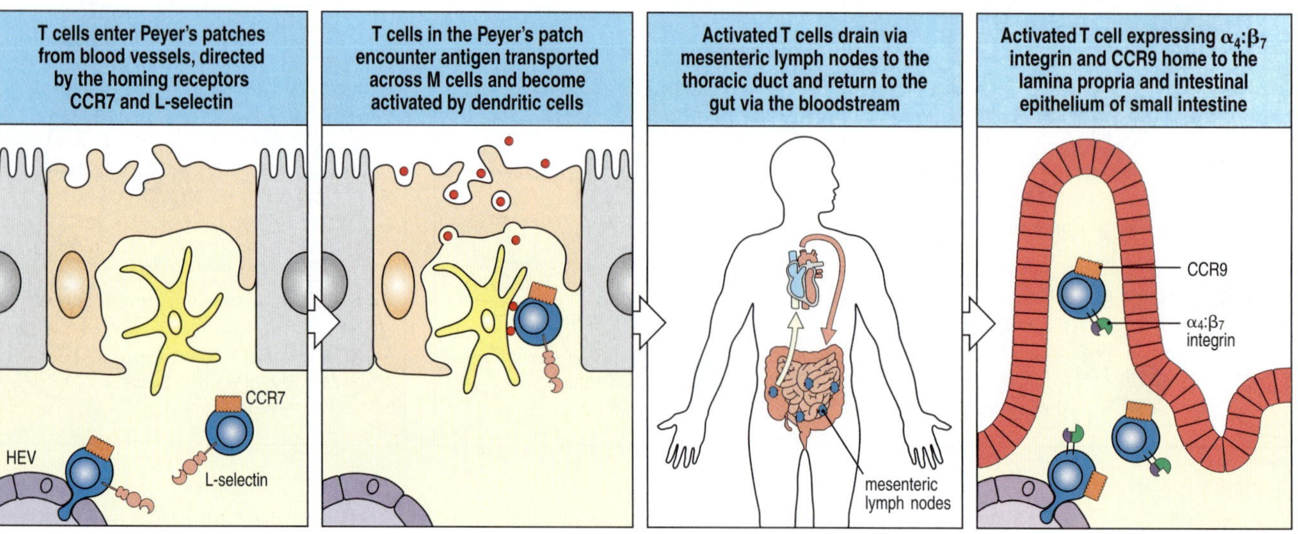
T cells enter Peyer's patches from blood vessels, directed by the homing receptors CCR7 and L-selectin
CCR7
HEV
L-selectin
T cells in the Peyer's patch encounter antigen transported across M cells and become activated by dendritic cells
Activated T cells drain via mesenteric lymph nodes to the thoracic duct and return to the gut via the bloodstream
mesenteric lymph nodes
Activated T cell expressing α4:β7 integrin and CCR9 home to the lamina propria and intestinal epithelium of small intestine
CCR9
α4:β7 integrin

lymphocytes. This binds to the mucosal vascular addressin **MAdCAM-1**, which is found mainly on the endothelial cells that line the blood vessels within the gut wall (Fig. 12.12). Lymphocytes originally primed in the gut are also lured back as a result of tissue-specific expression of chemokines by the gut epithelium. **CCL25** (TECK) is expressed by the epithelium of the small intestine and is a ligand for the receptor **CCR9** expressed on gut-homing T cells and B cells. Within the intestine there seems to be regional specialization of chemokine expression, as CCL25 is not expressed outside the small intestine and CCR9 is not required for migration of lymphocytes to the colon. However, the colon, lactating mammary gland, and salivary glands express **CCL28** (MEC, mucosal epithelial chemokine), which is a ligand for the receptor **CCR10** on gut-primed lymphocytes and attracts IgA-producing B lymphoblasts to these tissues.

Under most normal circumstances, only lymphocytes that first encounter antigen in a gut-associated secondary lymphoid organ are induced to express gut-specific homing receptors and integrins. As we shall see in the next section, these molecules are induced or 'imprinted' on T lymphocytes by intestinal dendritic cells during antigen presentation and activation. In contrast, dendritic cells from nonmucosal lymphoid tissues induce activated lymphocytes to express other adhesion molecules and chemokine receptors—for example α_4:β_1 integrin, cutaneous lymphocyte antigen (CLA), and the chemokine receptor CCR4—which direct them to tissues such as the skin (see Section 11-6). The tissue-specific consequences of lymphocyte priming in the GALT explain why vaccination against intestinal infections requires immunization by a mucosal route, because other routes, such as subcutaneous or intramuscular immunization, do not involve dendritic cells with the correct imprinting properties.

12-7 Priming of lymphocytes in one mucosal tissue can induce protective immunity at other mucosal surfaces.

Not all parts of the mucosal immune system exploit the same tissue-specific chemokines (see Section 12-6); this allows the compartmentalization of lymphocyte recirculation within the system. Thus, effector T and B cells primed in lymphoid organs draining the small intestine (mesenteric lymph nodes and Peyer's patches) are most likely to return to the small intestine; similarly, those primed in the respiratory tract migrate most efficiently back to the respiratory mucosa. This homing is obviously useful in returning antigen-specific effector

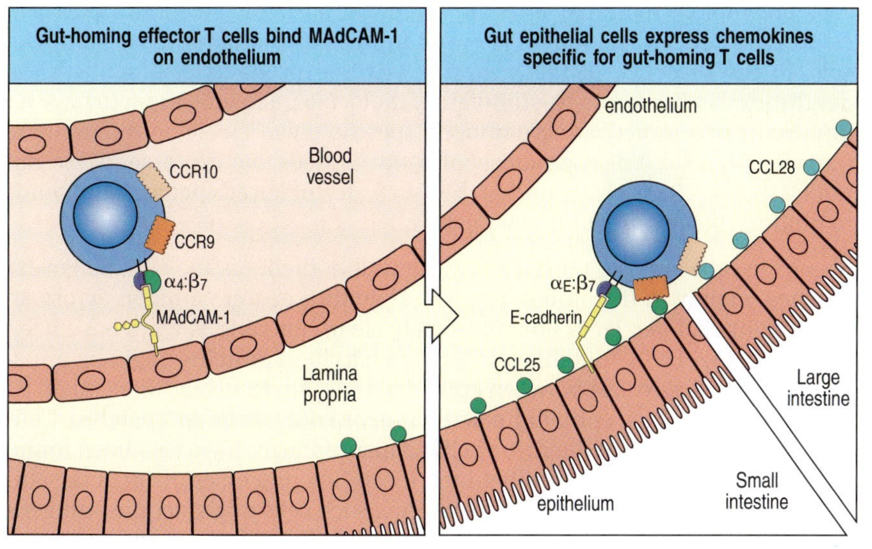

Fig. 12.12 Molecular control of intestine-specific homing of lymphocytes. Left panel: T and B lymphocytes primed by antigen in the Peyer's patches or mesenteric lymph nodes arrive as effector lymphocytes in the bloodstream supplying the intestinal wall (see Fig. 12.11). The lymphocytes express the integrin α_4:β_7, which binds specifically to MAdCAM-1 expressed selectively on the endothelium of blood vessels in mucosal tissues. This provides the adhesion signal needed for the emigration of cells into the lamina propria. Right panel: if primed in the small intestine, the effector lymphocytes also express the chemokine receptor CCR9, which allows them to respond to CCL25 (green circles) produced by epithelial cells of the small intestine; this enhances selective recruitment. Effector lymphocytes that have been primed in the large intestine do not express CCR9 but do express CCR10. This may respond to CCL28 (blue circles) produced by colon epithelial cells to fulfill a similar function. Lymphocytes that will enter the epithelial layer stop expressing the α_4:β_7 integrin and instead express the α_E:β_7 integrin. The receptor for this is E-cadherin on the epithelial cells. These interactions may help keep lymphocytes in the epithelium once they have entered it.

cells to the mucosal organ in which they will be most effective in fighting an infection or in controlling immune responses against foreign proteins and commensals. Nevertheless, the adhesion molecule MAdCAM-1 is present on the vasculature in all mucosae. As a result, some lymphocytes that have been primed in the GALT, for example, can also recirculate as effector cells to other mucosal tissues such as the respiratory tract, urogenital tract, and lactating breast. This overlap between mucosal recirculation routes gave rise to the idea of a **common mucosal immune system**, which is distinct from other parts of the immune system. Although this is now understood to be an oversimplification, it does have important implications for vaccine development, because it enables immunization by one mucosal route to be used to protect against infection at another mucosal surface. This phenomenon has been illustrated in many experimental models, the most interesting being the ability of nasal immunization to prime immune responses in the urogenital tract against HIV. In addition, the induction of IgA antibody production in the lactating breast by natural infection or vaccination at mucosal surfaces elsewhere, such as the intestine, is an important means of generating protective immunity that is transmitted to infants by passive transfer of the antibodies in milk.

12-8 Unique populations of dendritic cells control mucosal immune responses.

As in any compartment of the immune system, dendritic cells are the most important cell type for initiating and shaping adaptive immune responses in mucosal tissues. Dendritic cells are abundant in all mucosal tissues, being located both in the lymphoid organs and scattered throughout the mucosal surfaces. Several different subpopulations of mucosal dendritic cells have been described, many of which are distinct from their counterparts in other parts of the immune system. The nature of mucosal dendritic cells is very much determined by their local environment; in turn these cells play the major role in defining the unusual features of mucosal immune responses.

Dendritic cells are found in the Peyer's patches in two main areas. One population is found in the subepithelial dome region, and acquires antigen from M cells (see Section 12-4). Peyer's patch dendritic cells are mostly CD11b (α_M integrin)-positive and CD8α-negative, and express CCR6, the receptor for the chemokine CCL20 produced by follicle-associated epithelial cells. In resting conditions, these dendritic cells remain mostly beneath the epithelium and produce the cytokine IL-10 in response to antigen uptake, which tends to prevent the priming of T cells to become pro-inflammatory T cells.

However, during infection by a pathogen such as *Salmonella*, these dendritic cells are rapidly recruited into the epithelial layer of the Peyer's patch in response to the CCL20 that is released in increased quantities by epithelial cells in the presence of the bacteria. Bacterial products also activate the dendritic cells to express co-stimulatory molecules, allowing them to have an important role in activating pathogen-specific naive T cells to differentiate into effector cells. A distinct subset of dendritic cells that expresses CD8α but not C11b or CCR6 is also found in the T-cell area of Peyer's patches and tends to produce the pro-inflammatory cytokine IL-12.

Dendritic cells are also abundant in the wall of the intestine outside Peyer's patches, mainly in the lamina propria. How they acquire antigen across an intact epithelial barrier is not completely clear, but various ways have been proposed (Fig. 12.13). Soluble antigens such as food proteins might be transported directly across or between epithelial cells. Alternatively, there might be M cells in the surface epithelium of the mucosa outside Peyer's patches. Cells with the appearance of dendritic cells or macrophages have also been found making their way into the epithelium, or sending processes through the epithelial layer without disturbing its integrity. Such cells might acquire antigens such as bacteria from the lumen before returning with them to the lamina

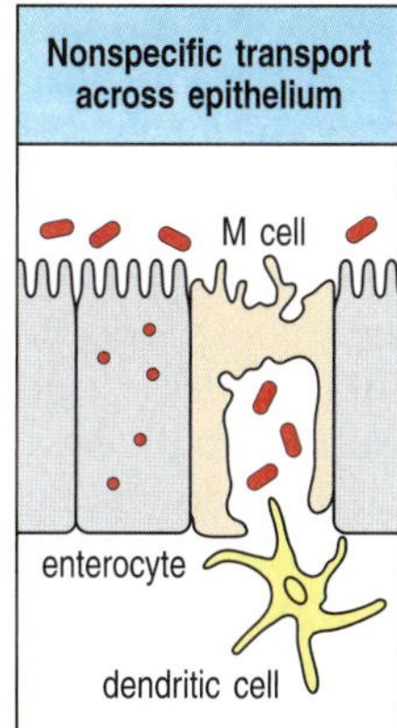

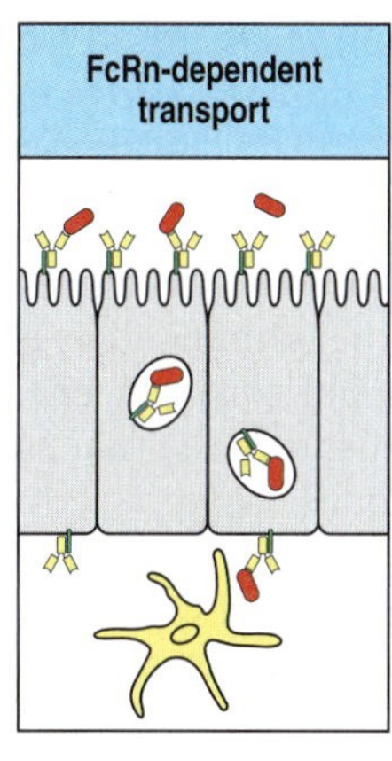

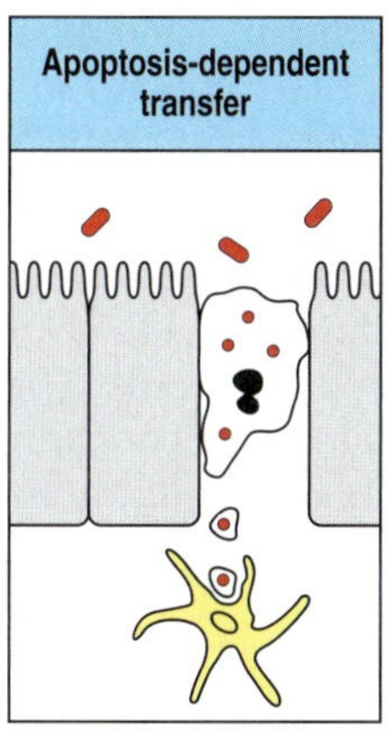

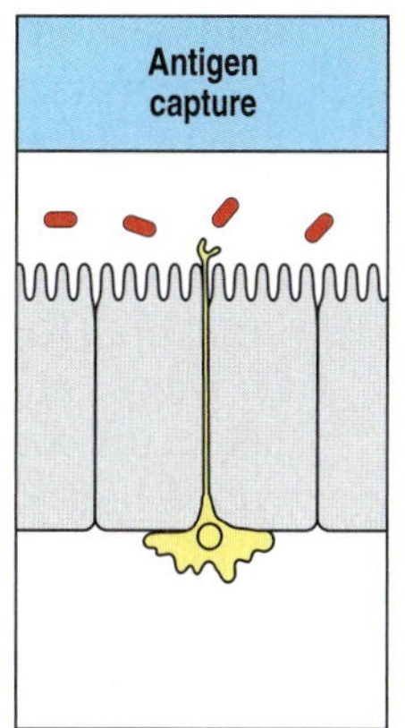

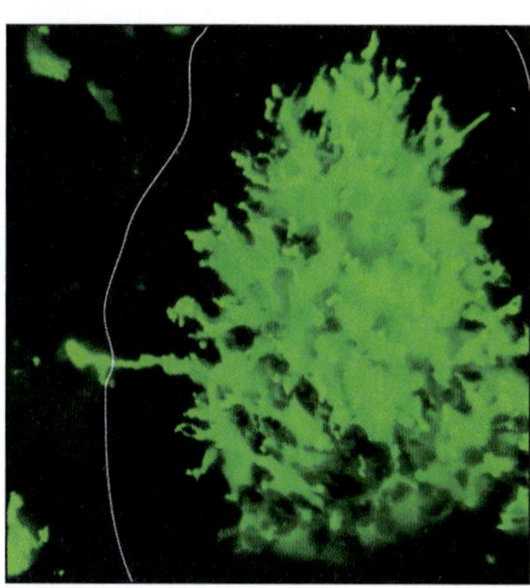

Fig. 12.13 Capture of antigens from the intestinal lumen by mononuclear cells in the lamina propria. First panel: soluble antigens such as food proteins might be transported directly across or between enterocytes, or there might be M cells in the surface epithelium outside Peyer's patches (see Fig. 12.9 for details of transport across M cells). Second panel: enterocytes can capture and internalize antigen:antibody complexes by means of the FcRn on their surface and transport them across the epithelium by transcytosis. At the basal face of the epithelium, lamina propria dendritic cells expressing FcRn and other Fc receptors pick up and internalize the complexes. Third panel: an enterocyte infected with an intracellular pathogen undergoes apoptosis and its remains are phagocytosed by the dendritic cell. Fourth panel: mononuclear cells have been seen extending processes between the cells of the epithelium without disturbing its integrity. The cell process could pick up and internalize antigen from the gut lumen and then retract. The micrograph shows mononuclear cells, which may be dendritic cells or macrophages, (stained green with a fluorescent tag on the CD11c molecule) in the lamina propria of a villus of mouse small intestine. The epithelium is not stained and appears black, but its luminal (outer) surface is shown by the white line. A cell process has squeezed between two epithelial cells and its tip is present in the lumen of the intestine. Magnification ×200. Micrograph from Niess, J.H., *et al.*: *Science* 2005, 307:254–258.

propria. Other routes by which material from the lumen can be delivered to lamina propria dendritic cells include the uptake of antibody-coated antigens by epithelial cells expressing the neonatal Fc receptor (FcRn), or the phagocytosis of antigen-containing apoptotic epithelial cells by dendritic cells (see Fig. 12.13).

The dendritic cells in the lamina propria comprise a subset with several unique properties, and they make important contributions to maintaining tolerance to harmless antigens in the intestine, especially those derived from food proteins. Most lamina propria dendritic cells express the integrin α_E:β_7 (**CD103**), and once loaded with antigen, CD103 dendritic cells leave the mucosa and migrate to the T-cell areas of mesenteric lymph nodes via the afferent lymphatics that drain the intestinal wall. In the lymph nodes they interact with naive T cells and induce the gut-homing properties that enable the T cells to efficiently return to the intestinal wall as differentiated effector T cells. The migration of CD103 dendritic cells depends on their expression of the chemokine receptor CCR7, but this expression is constitutive and does not require the presence of pathogens or other inflammatory stimuli, although it is enhanced by such agents. It is estimated that 5–10% of the mucosal dendritic cell population emigrates to the mesenteric lymph node every day in the resting intestine.

Another unusual property of CD103 dendritic cells is their production of the nonprotein signaling molecule **retinoic acid**, which is derived from the metabolism of dietary vitamin A through the action of retinal dehydrogenases. Retinoic acid production endows these dendritic cells with the unique ability to induce the expression of the gut-homing molecules CCR9 and integrin α_4:β_7 in T and B cells. Lamina propria dendritic cells respond poorly to inflammatory stimuli such as the microbial ligands for TLRs, and they produce IL-10 rather than inflammatory cytokines such as IL-12. As a result, when CD103 dendritic cells arrive in the mesenteric lymph node under resting conditions, they promote the generation of FoxP3-positive regulatory T cells (induced T_{reg} cells) from antigen-specific naive CD4 T cells (see Chapter 9 for details of the differentiation and functions of induced regulatory T cells). This process also depends on the retinoic acid produced by the dendritic cells and is assisted by transforming growth factor-β (TGF-β), a cytokine produced in abundance by intestinal cells and which promotes T_{reg} cells. Intestinal dendritic cells also produce indoleamine dioxygenase (IDO), an enzyme that catabolizes and depletes tryptophan from the environment and produces kynurenine metabolites. By some mechanism not yet understood, these actions of IDO favor the development of induced T_{reg} cells.

The anti-inflammatory behavior of mucosal CD103 dendritic cells in the healthy gut is actively promoted by factors that are constitutively produced in the mucosal environment. These include thymic stromal lymphopoietin (TSLP), retinoic acid, and TGF-β released from epithelial cells, as well as lipid-derived mediators such as the prostaglandin PGE_2, which is produced by stromal cells. Macrophages in the mucosa also produce IL-10 constitutively and this helps keep dendritic cells in a quiescent state, as well as maintaining the local population of regulatory T cells.

Once arrived at the mesenteric lymph node, the immunomodulatory properties of CD103 dendritic cells are further enhanced by retinoic acid produced by lymph-node stromal cells. The rapid turnover of CD103 dendritic cells allows the constant delivery of antigens from the mucosal surface to the local lymph nodes, while their immunomodulatory properties make them crucial in maintaining tolerance to harmless food proteins in the small intestine (discussed in Section 12-14). CD103 dendritic cells are also present in the large intestine and are believed to play similar roles in controlling immune responses to commensal bacteria (see Section 12-15), although this has not yet been studied in detail. Retinoic-acid-producing dendritic cells that stimulate the appearance of gut-homing molecules on T cells are also found in Peyer's patches, and may also be important for generating regulatory T cells that recognize commensal bacteria. However, CD103 dendritic cells themselves are rare in this tissue. Migratory populations of dendritic cells that continuously take up local antigens in the tissue and transport them to the draining lymph nodes are found in the lungs and at other mucosal surfaces.

Despite the overall bias toward tolerance, pro-inflammatory cells of the myeloid lineage are also present in resting lamina propria. These cells produce cytokines such as IL-6, IL-23, tumor necrosis factor-α (TNF-α), and nitric oxide, and can drive the differentiation of effector T_H17 cells and IgA switching in B cells. It has been suggested that these cells are a distinct population of dendritic cells that have been stimulated via the receptor TLR-5; they express CX3CR1, the receptor for the chemokine fractalkine (see Fig. 3.22), and do not express CD103. However, recent work indicates that these CX3CR1-positive cells in the intestine do not migrate to lymph nodes, cannot present antigen to naive T cells, and do not produce retinoic acid, and so may not be classical dendritic cells. Instead, they could be a population of macrophage-like cells whose main role is not to shape naive T-cell responses but to produce pro-inflammatory mediators during an ongoing inflammation.

12-9 The intestinal lamina propria contains antigen-experienced T cells and populations of unusual innate-type lymphocytes.

As noted above, most of the T cells in the lamina propria seem to be antigen-experienced effector cells, which have been activated by the dendritic cells discussed in the previous section. They have markers associated with effector or memory T cells, such as CD45RO in humans, and express the gut-homing markers CCR9 and $\alpha_4{:}\beta_7$ integrin and receptors for pro-inflammatory chemokines such as CCL5 (RANTES). The T-cell population of the lamina propria has a ratio of CD4 to CD8 T cells of 3:1 or more, similar to that in systemic lymphoid tissues.

Lamina propria effector CD4 T cells proliferate poorly when stimulated by mitogens or antigen, but in the healthy intestine they secrete large amounts of cytokines such as interferon (IFN)-γ, IL-5, IL-17, and IL-10, even in the absence of inflammation. It is particularly noteworthy that the colon and the ileum are the only locations in the healthy body where effector T_H17 cells are found, perhaps reflecting the high bacterial load in these sites. Indeed, studies in mice show that the presence of mucosal T_H17 cells is dependent on commensal microbes, with the *Clostridium*-related segmented filamentous bacterium (SFB) playing a dominant and specific role in driving T_H17 differentiation. Cytokine-producing T_H1 and T_H2 cells are also present in the healthy lamina propria, reflecting the constant state of immune recognition of the microbiota. These lamina propria effector T cells help maintain the mutualistic response to commensals under resting conditions by producing cytokines such as IL-4, IL-5, IL-6, IL-21, and TGF-β, which can assist IgA production, or IL-22, which induces the secretion of antimicrobial peptides and promotes epithelial repair.

In any other situation, the presence of such large numbers of apparent effector cells would suggest the presence of a pathogen and would be likely to lead to inflammation. The fact that it does not in the healthy lamina propria is because the generation of T_H1, T_H2, and T_H17 cells is balanced by the presence of IL-10-producing, FoxP3-positive or FoxP3-negative T_{reg} cells (see Section 12-8).

During infection by a pathogen, the balance tips in favor of effector cell accumulation, either because of increased recruitment of these cells, or because the differentiation and/or functions of regulatory T cells are diminished. Now, the effector T_H1, T_H2, and T_H17 cells in the lamina propria can act without restriction to provide appropriate forms of protective immunity against the pathogen. Effector CD8 T cells are also present in the lamina propria and are capable of both cytokine production and cytotoxic activity during a protective immune response to a pathogen. Because of the pro-inflammatory cytokines they produce, T_H1 cells, T_H17 cells, and cytotoxic T cells are also the principal effectors of inflammatory disorders such as celiac disease and inflammatory bowel diseases.

The healthy lamina propria also contains unusual 'natural' effector lymphocytes. In addition to the CD1-restricted iNKT cells that are also present in the systemic immune system (see Section 6-19), there is a small subset of **mucosal invariant T cells (MAIT)**, which express an invariant TCRα chain and respond to antigens presented by the nonclassical class I MHC molecule MR1 on B cells. MAITs are active producers of a variety of cytokines in response to unknown antigens, probably derived from commensal bacteria.

Very recently, additional and distinctive populations of non-T lymphocytes with innate effector functions have been described in human and mouse intestinal mucosa. These cells produce large amounts of IL-22, and some express the NK-cell receptors NKp44 and NKp46. They all express the transcription factor RORγT, which among other things controls the development of lymphoid tissue inducer cells. Thus they combine some of the characteristics of NK cells (see Section 3-22) and lymphoid tissue inducer cells (see Section 8-24), but seem to be distinct from both. These unusual non-T lymphocytes seem to be a major innate source of IL-22 in response to commensal microbiota. IL-22 drives the production of antimicrobial peptides in the gut and helps maintain the function of the epithelial barrier.

12-10 The intestinal epithelium is a unique compartment of the immune system.

The lymphocytes found in the intestinal epithelium—the **intraepithelial lymphocytes (IELs)**—are quite distinct in character from the lymphocyte population in the lamina propria or from lymphocyte populations in the systemic immune system (Fig. 12.14). There are 10–15 lymphocytes for every 100 epithelial cells in the healthy small intestine; given the enormous surface area of the mucosa, this makes the IELs one of the single largest populations of lymphocytes in the body. More than 90% of small intestinal IELs are T cells, and around 80% of these carry CD8, in complete contrast to the lymphocytes in the lamina propria. IELs are also present in the large intestine, although there are fewer relative to the number of epithelial cells and a greater proportion of CD4 T cells in comparison with the small intestine.

Like the lymphocytes in the lamina propria, most IELs have an activated appearance even in the absence of infection by a pathogen, and they contain

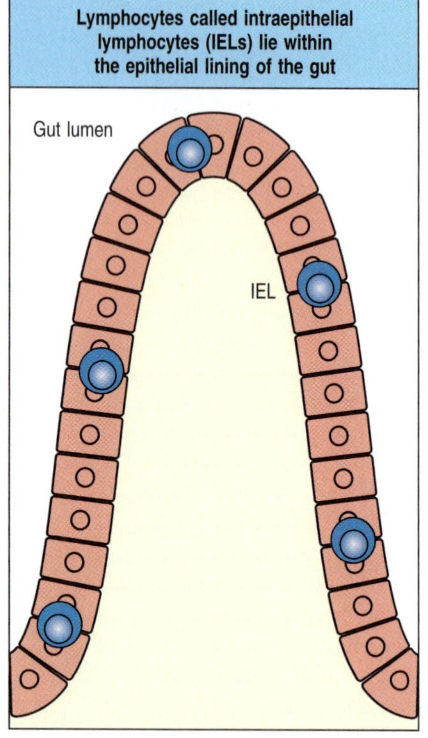
Lymphocytes called intraepithelial lymphocytes (IELs) lie within the epithelial lining of the gut
Gut lumen
IEL

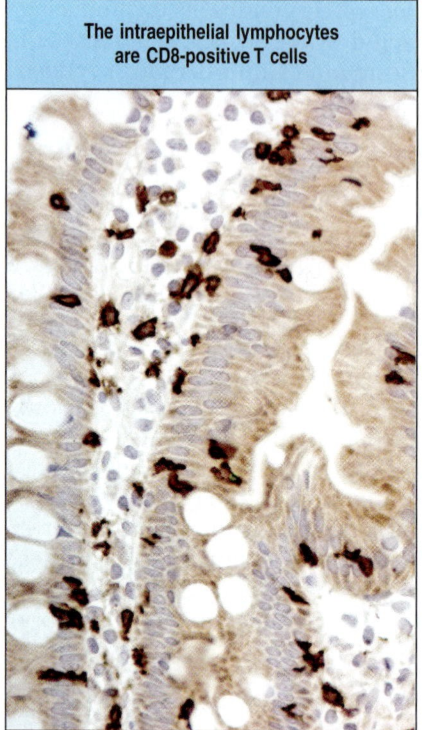
The intraepithelial lymphocytes are CD8-positive T cells

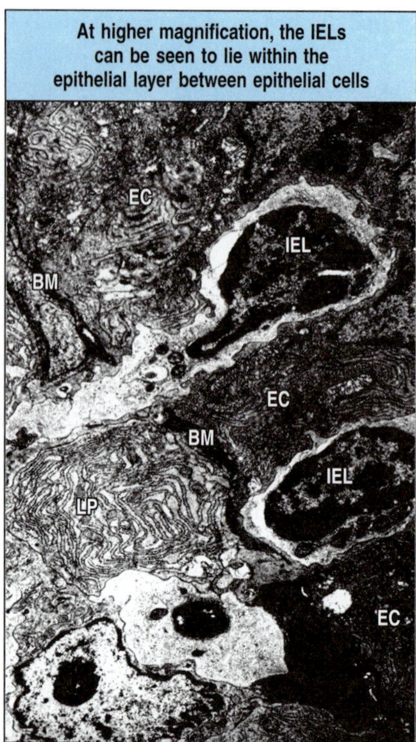
At higher magnification, the IELs can be seen to lie within the epithelial layer between epithelial cells
EC
IEL
BM
EC
BM
IEL
LP
EC

Fig. 12.14 Intraepithelial lymphocytes. The epithelium of the small intestine contains a large population of lymphocytes known as intraepithelial lymphocytes (IELs) (left panel). The micrograph in the center is of a section through human small intestine in which CD8 T cells have been stained brown with a peroxidase-labeled monoclonal antibody. Most of the lymphocytes in the epithelium are CD8 T cells. Magnification ×400. The electron micrograph on the right shows that the IELs lie between epithelial cells (EC) on the basement membrane (BM) separating the lamina propria (LP) from the epithelium. One IEL can be seen having crossed the basement membrane into the epithelium, leaving a trail of cytoplasm in its wake. Magnification ×8000.

intracellular granules containing perforin and granzymes, like those in conventional effector CD8 cytotoxic T cells. However, the T-cell receptors of most CD8 IELs show a restricted use of V(D)J gene segments, indicating that they may expand locally in response to a relatively small number of antigens. The IELs of the small intestine express the chemokine receptor CCR9, but have the α_E:β_7 integrin (CD103) on their surface instead of the α_4:β_7 integrin found on other gut-homing T cells. The receptor for α_E:β_7 integrin is E-cadherin on the surface of epithelial cells, and this interaction may assist IELs to remain in the epithelium (see Fig. 12.12).

The origin and functions of IELs are controversial. In young animals and the adults of some species, this T-cell population contains an unusually large number of γ:δ T cells. In normal adult mice and humans, however, γ:δ T cells are found in similar numbers in the gut epithelium and the bloodstream. In mice, the CD8 intraepithelial T cells can be divided into two subsets—type a and type b—depending on which form of CD8 is expressed. The relative proportions of the subsets vary with the age, mouse strain, and the number of bacteria in the intestine. Type a IELs are conventional T cells bearing α:β T-cell receptors and the CD8α:β heterodimer. They are derived from naive CD8 T cells activated by antigen in the Peyer's patches, and function as conventional class I MHC-restricted cytotoxic T cells, killing virus-infected cells, for example (Fig. 12.15, top panels). They also secrete effector cytokines such as IFN-γ and are involved in protective immunity against parasites such as *Toxoplasma gondii*.

Type b CD8 IELs comprise T cells expressing the CD8α homodimer (CD8α:α) and either an α:β or a γ:δ T-cell receptor. Some of the α:β receptors on this cell population bind nonconventional ligands, including those presented by MHC class Ib molecules (see Sections 6-17 and 6-18). All IELs express high levels of the activating C-type lectin NK receptor NKG2D (see Sections 3-21 and 3-23). This receptor can bind to two MHC-like molecules—MIC-A and MIC-B—that are expressed on intestinal epithelial cells in response to cellular injury, stress, or ligation of TLRs. The injured cells can then be recognized and killed by the IELs, a process that is enhanced by the production of IL-15

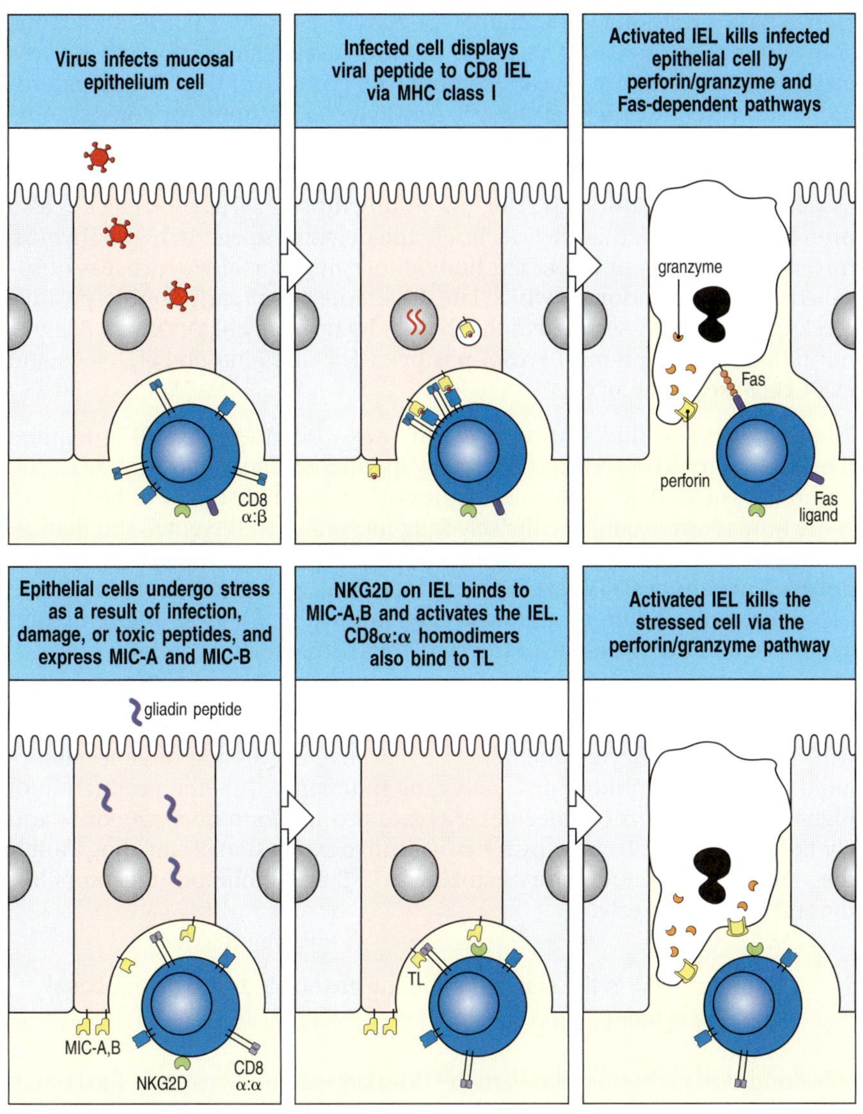

Virus infects mucosal epithelium cell
Infected cell displays viral peptide to CD8 IEL via MHC class I
Activated IEL kills infected epithelial cell by perforin/granzyme and Fas-dependent pathways
CD8
α:β
granzyme
Fas
perforin
Fas ligand
Epithelial cells undergo stress as a result of infection, damage, or toxic peptides, and express MIC-A and MIC-B
NKG2D on IEL binds to MIC-A,B and activates the IEL. CD8α:α homodimers also bind to TL
Activated IEL kills the stressed cell via the perforin/granzyme pathway
gliadin peptide
MIC-A,B
NKG2D
CD8
α:α
TL

Fig. 12.15 Effector functions of intraepithelial lymphocytes. There are two main types of intraepithelial lymphocyte (IEL). As shown in the top panels, one type (type a) are conventional CD8 cytotoxic T cells that recognize peptides derived from viruses or other intracellular pathogens bound to classical MHC class I molecules on infected epithelial cells. The activated IEL recognizes specific peptide:MHC complexes by using its α:β T-cell receptor, with the CD8α:β heterodimer as co-receptor. The IEL releases perforin and granzyme, which kill the infected cell. Apoptosis of epithelial cells can also be induced by the binding of Fas ligand on the T cell to Fas on the epithelial cell. In the bottom panels, epithelial cells that have been stressed by infection or altered cell growth, or by a toxic peptide from the protein α-gliadin (a component of gluten), upregulate expression of the nonclassical MHC class I molecules MIC-A and MIC-B and produce IL-15. Neighboring type b IELs are activated by IL-15 and recognize MIC-A and MIC-B using the receptor NKG2D (see Section 3-23). They also kill the epithelial cells by releasing perforin and granzyme. These IELs carry the CD8α:α homodimer, and this protein may also contribute to their recognition of infected cells by binding directly to the nonclassical MHC class I molecule TL, encoded in the T region of the MHC, which is present on epithelial cells.

by the damaged epithelial cells. IELs can thus be considered in evolutionary terms as being at the interface between innate and adaptive immunity. Their role in the gut may be the rapid recognition and elimination of epithelial cells that express an abnormal phenotype as a result of stress or infection (Fig. 12.15, bottom panels). Gut IELs are also thought to be important in aiding the repair of the mucosa after inflammatory damage: they stimulate the release of antimicrobial peptides, thus helping to remove the source of the inflammation; they release cytokines such as keratinocyte growth factor that promote epithelial barrier function; and they suppress inflammation directly by producing cytokines such as TGF-β, which can inhibit all aspects of immune function. These repair functions have been associated particularly with the γ:δ subset of IELs, which have a similar role in skin repair.

Inappropriate or excess activation of IELs can, however, give rise to disease. MIC-A-dependent cytotoxic activity of intraepithelial T cells contributes to intestinal damage in celiac disease, which is caused by an abnormal response to the wheat protein gluten (see Section 14-18). Certain components of gluten can stimulate the production of IL-15 by epithelial cells, and celiac disease is associated with increased numbers of IELs.

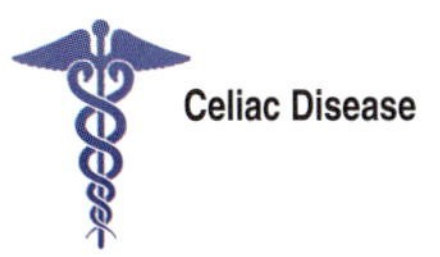

Until recently, it was believed that type b IELs developed entirely within the gut. But it now seems that all IELs require the thymus for their development.

Unlike type a IELs, however, many type b IELs seem not to have undergone conventional positive and negative selection (see Chapter 8), and express apparently autoreactive T-cell receptors. The absence of the CD8α:β heterodimer, however, means that these T cells have low affinity for conventional peptide:MHC complexes, because the CD8β chain binds more strongly than the CD8α chain to classical MHC molecules. Type b IELs therefore cannot act as self-reactive effector cells. The low affinity for self-MHC molecules is also probably the reason that these cells escape negative selection in the thymus. Instead, expression of the CD8α homodimer might enable a process of so-called **agonist selection**, in which late double-negative/early double-positive T cells are positively selected in the thymus by relatively high-affinity ligands, not unlike the selection of FoxP3-positive CD4 CD25 'natural' T_{reg} cells and iNKT cells (see Chapter 8).

The precursors of IELs exit from the thymus while they are still immature; they mature in the intestine, which may involve additional positive selection on nonclassical MHC molecules expressed on the intestinal epithelium. In some mouse strains, one of the selecting molecules in the gut is the thymus leukemia antigen (TL), which is a nonclassical MHC class I molecule that does not present peptides. TL is expressed by intestinal epithelial cells and binds the CD8α homodimer directly and with high affinity. These local differentiation events require the presence of the cytokine IL-15, which is 'trans-presented' to IELs in a complex with the IL-15 receptor present on epithelial cells.

In addition to agonist selection, type b IELs share several other properties with innate lymphocytes, including constitutive expression of genes associated with both inhibitory and activating functions. Thus the production of high levels of cytotoxic molecules, NO and pro-inflammatory cytokines and chemokines is kept in check by the constitutive expression of signaling inhibitors—the immunomodulatory cytokine TGF-β and inhibitory receptors like those found on NK cells.

12-11 Secretory IgA is the class of antibody associated with the mucosal immune system.

The dominant class of antibody in the mucosal immune system is **IgA**, which is produced locally by plasma cells present in the mucosal wall. The nature of IgA differs between the two main compartments in which it is found—the blood and mucosal secretions. IgA in the blood is mainly in the form of a monomer (mIgA) and has been produced in the bone marrow by plasma cells derived from B cells activated in lymph nodes. In mucosal tissues, IgA is produced almost exclusively as a polymer, usually as a dimer, in which the two immunoglobulin monomers are linked by a J chain (see Fig. 5.19).

The naive B-cell precursors of the IgA-secreting mucosal plasma cells are activated in Peyer's patches and mesenteric lymph nodes. Class switching of activated B cells to IgA is controlled by the cytokine TGF-β. In the human gut, class switching is T-cell dependent and occurs only in the organized lymphoid tissues, using the same molecular mechanisms as in lymph nodes and spleen (the molecular mechanisms of class switching are discussed in detail in Chapter 5, and the general consequences for immune responses in Chapter 10). The subsequent expansion and differentiation of IgA-switched B cells are driven by IL-5, IL-6, IL-10, and IL-21. Upward of 75,000 IgA-producing plasma cells are present in the normal human intestine, and 3–4 g of IgA is secreted by the mucosal tissues each day, considerably exceeding the production of all other immunoglobulin classes. This continuous production of large quantities of IgA occurs in the absence of pathogenic invasion and is driven almost entirely by recognition of the resident microbiota.

In humans, monomeric and dimeric IgA are both found as two isotypes, IgA1 and IgA2. The ratio of IgA1 to IgA2 varies markedly depending on the tissue, being about 10:1 in blood and upper respiratory tract, about 3:2 in the small intestine and 2:3 in colon. Some common pathogens of respiratory mucosa (such as *Haemophilus influenzae*) and genital mucosa (such as *Neisseria gonorrhoeae*) produce proteolytic enzymes that can cleave IgA1, whereas IgA2 is much more resistant to cleavage. How this is related to the prevalence of IgA1 or IgA2 is uncertain. The higher proportion of plasma cells secreting IgA2 in the large intestine might be because the high density of commensal microorganisms at this site drives the production of cytokines that cause selective class switching.

After activation and differentiation, the resulting IgA-expressing B lymphoblasts express the mucosal homing integrin $\alpha_4{:}\beta_7$, as well as the chemokine receptors CCR9 and CCR10, and localize to mucosal tissues by the mechanisms discussed in Section 12-6. Once in the lamina propria, the B cells undergo final differentiation into plasma cells, which synthesize IgA dimers and secrete them into the subepithelial space (Fig. 12.16). To reach its target antigen in the gut lumen, the IgA has to be transported across the epithelium. This is done via immature epithelial cells located at the base of the intestinal crypts, which constitutively express the **polymeric immunoglobulin receptor** (**pIgR**), which is located on their basolateral surfaces. This receptor has a high affinity for J-chain-linked polymeric immunoglobulins such as dimeric IgA and pentameric IgM, and transports the antibody by transcytosis to the luminal surface of the epithelium, where it is released by proteolytic cleavage of the extracellular domain of the receptor (see Fig. 12.16). Part of the cleaved pIgR remains associated with the IgA and is known as **secretory component** (frequently abbreviated to SC). The resulting antibody is referred to as **secretory IgA** (**SIgA**).

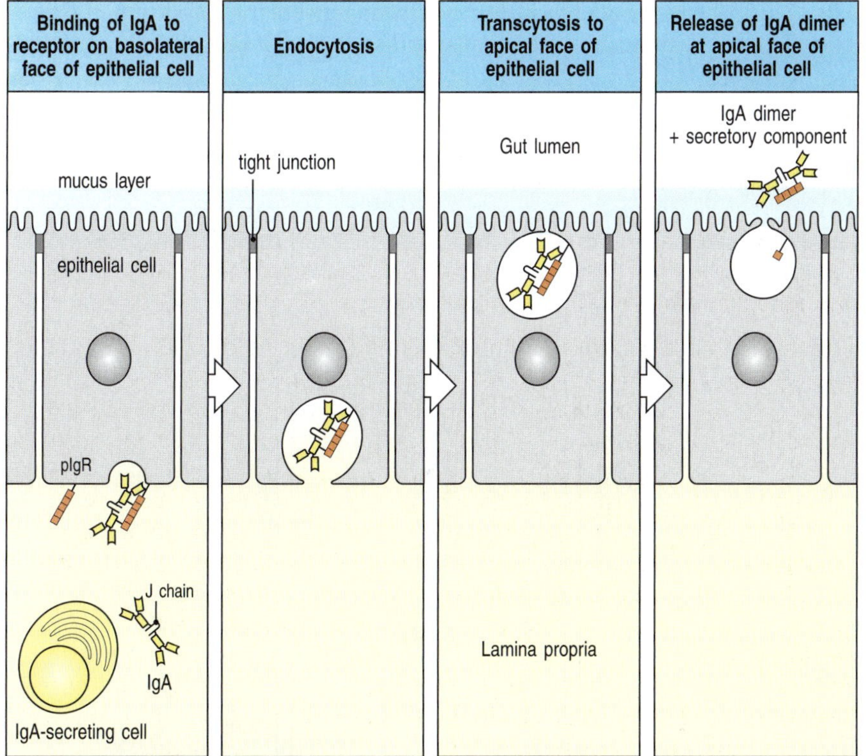

Fig. 12.16 Transcytosis of IgA antibody across epithelia is mediated by the polymeric Ig receptor (pIgR), a specialized transport protein. Most IgA antibody is synthesized in plasma cells lying just beneath epithelial basement membranes of the gut, the respiratory epithelia, the tear and salivary glands, and the lactating mammary gland. The IgA dimer linked by a J chain diffuses across the basement membrane and is bound by the pIgR on the basolateral surface of the epithelial cell. The bound complex undergoes transcytosis, by which it is transported in a vesicle across the cell to the apical surface, where the pIgR is cleaved to leave the extracellular IgA-binding component bound to the IgA molecule as the so-called secretory component. Carbohydrate on the secretory component binds to mucins in mucus and holds the IgA at the epithelial surface. The residual piece of the pIgR is nonfunctional and is degraded. IgA is transported across epithelia in this way into the lumina of several organs that are in contact with the external environment.

In some animals there is a second route of IgA secretion into the intestine—the **hepatobiliary route**. Dimeric IgA that has not bound pIgR is taken up into the portal veins in the lamina propria, which drain intestinal blood to the liver. In the liver these small veins (sinusoids) are lined by an endothelium that allows the antibodies access to underlying hepatocytes, which have pIgR on their surface. IgA is taken up into the hepatocytes and transported by transcytosis into an adjacent bile duct. In this way, secretory IgA can be delivered directly into the upper small intestine via the common bile duct. This hepatobiliary route allows dimeric IgA to eliminate antigens that have invaded the lamina propria and have been bound there by IgA. Although highly efficient in rats, rabbits, and chickens, this route does not seem to be of great significance in humans and other primates, in whom hepatocytes do not express pIgR.

IgA secreted into the gut lumen binds to the layer of mucus coating the epithelial surface via carbohydrate determinants in secretory component. There it is involved in preventing invasion by pathogenic organisms and, just as importantly, it also has a crucial role in maintaining the homeostatic balance between the host and the commensal microbiota. IgA inhibits microbial adherence to the epithelium, an effect that is assisted by the unusually wide and flexible angle between the Fab pieces of both IgA isotypes, allowing very efficient bivalent binding to large antigens such as bacteria. Secretory IgA can also neutralize microbial toxins or enzymes.

In addition to its activities in the lumen, IgA can neutralize bacterial lipopolysaccharide and viruses it encounters inside epithelial cells, and after bacteria and viruses have penetrated across the epithelial barrier into the lamina propria. The resulting IgA:antigen complexes are then reexported into the gut lumen, from where they are excreted from the body (Fig. 12.17). Complexes containing dimeric IgA formed in the lamina propria can also be excreted via the hepatobiliary route described above. In addition to enabling the elimination of antigens, the formation of IgA:antigen complexes can enhance the uptake and transcytosis of luminal antigen by M cells and facilitate its uptake by Peyer's patch dendritic cells via their Fcα receptors (see Fig. 12.13). Secretory IgA has little capacity to activate the classical pathway of complement or to act as an opsonin, and so does not induce inflammation.

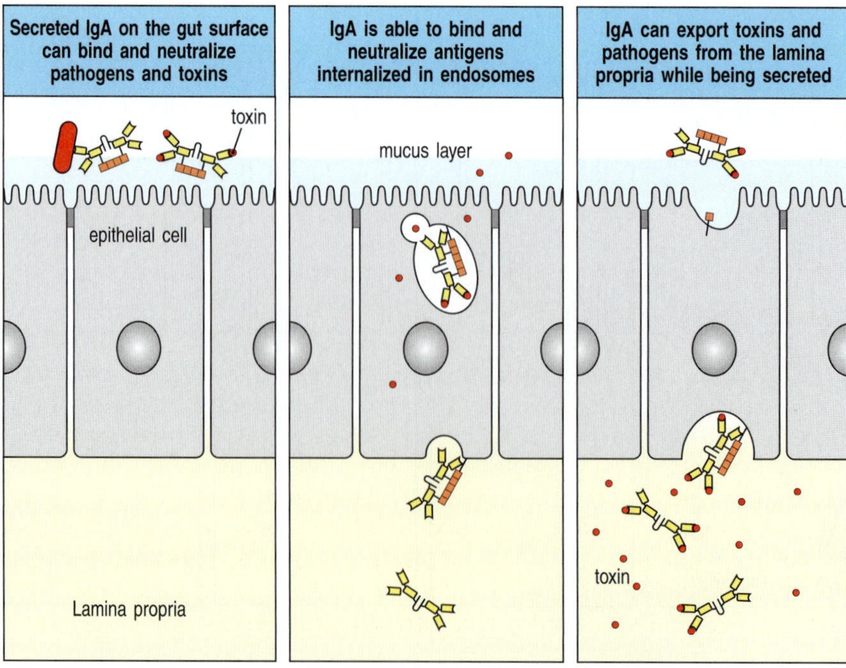

Fig. 12.17 Mucosal IgA has several functions in epithelial surfaces. First panel: IgA adsorbs on the layer of mucus covering the epithelium, where it can neutralize pathogens and their toxins, preventing their access to tissues and inhibiting their functions. Second panel: antigen internalized by the epithelial cell can meet and be neutralized by IgA in endosomes. Third panel: toxins or pathogens that have reached the lamina propria encounter pathogen-specific IgA in the lamina propria, and the resulting complexes are reexported into the lumen across the epithelial cell as the dimeric IgA is secreted.

This means it can limit the penetration of microbes into the mucosa without risking inflammatory damage to these fragile tissues, something that is beneficial during infection by intestinal pathogens. In addition, by restricting uptake of microbes from the intestinal lumen and by facilitating antigen uptake by noninflammatory dendritic cells in the GALT, secretory IgA is crucial to the beneficial symbiosis between an individual and their gut commensal bacteria (see Section 12-15).

In mice, unlike humans, a significant proportion of intestinal IgA is derived from T-cell-independent B-cell activation and class switching (see Section 10-1), which depends on activation of the innate immune system by the products of commensal microbes and may result from the direct interaction between B cells and dendritic cells in solitary lymphoid follicles. This antibody production seems to involve lymphocytes of the B-1 subset (see Section 8-28), which arise from precursor B cells in the peritoneal cavity and migrate to the intestinal wall in response to microbial constituents such as lipopolysaccharide. Once in the mucosa, TGF-β-dependent class switching to IgA occurs under the influence of factors present in the local environment. In addition to TGF-β, these include IL-6, retinoic acid, and the TNF-family members BAFF (B-cell activating factor of the TNF family) and APRIL (a proliferation-inducing ligand), which bind to TACI (transmembrane activator and calcium-modulator and cyclophilin ligand interactor) on B cells, substituting for signals otherwise supplied by CD4 helper T cells (see Section 10-4). Intestinal epithelial cells can produce BAFF and APRIL, but the major source seems to be a specialized population of local dendritic cells that also produce nitric oxide (NO) and TNF-α, both of which assist in the processing and activation of TGF-β.

The IgA antibodies produced in these T-cell-independent responses are of limited diversity and of generally low affinity, with little evidence of somatic hypermutation. They are nevertheless an important source of 'natural' antibodies directed at commensal bacteria. As yet, there is little evidence for this source of IgA in humans, in whom all secretory IgA responses involve somatic hypermutation and seem to be T-cell dependent. The enzyme activation-induced cytidine deaminase (AID), which is essential for class switching (see Chapter 5), cannot be detected in human intestinal lamina propria, indicating that class switching is unlikely to occur there. Nevertheless, its occurrence in lamina propria B cells in mice may offer a glimpse into the evolutionary history of specific antibody responses in the mucosa, and might indicate pathways that could be activated when T-cell dependent IgA production is compromised in humans, as it is in AIDS.

12-12 IgA deficiency is common in humans but may be compensated for by secretory IgM.

Selective deficiency of IgA production is the commonest primary immune deficiency in humans, occurring in about 1 in 500 to 1 in 700 individuals in populations of Caucasian origin, although it is somewhat rarer in other ethnic groups. A slightly higher incidence of respiratory infections, atopy (a tendency to make allergic reactions to harmless environmental antigens), and autoimmune disease has been reported in older people with IgA deficiency. However, most individuals with IgA deficiency are not overly susceptible to infections unless there is also a deficiency in IgG2 production. The dispensability of IgA probably reflects the ability of IgM to replace IgA as the predominant antibody in secretions, and increased numbers of IgM-producing plasma cells are indeed found in the intestinal mucosa of IgA-deficient people. Because IgM is a J-chain-linked polymer, IgM produced in the gut mucosa is bound efficiently by the pIgR and is transported across epithelial cells into the gut lumen as secretory IgM. The importance of this back-up mechanism has been shown in knockout mice. Animals lacking IgA alone have a normal phenotype, but those lacking the pIgR are susceptible to mucosal infections. They also show increased penetration of commensal bacteria into tissues and a consequent systemic immune response to these bacteria. Genetic absence of the pIgR has never been reported in humans, suggesting that such a defect is lethal.

Summary.

The mucosal tissues of the body such as the intestine and respiratory tract are exposed continuously to enormous amounts of different antigens, which can be either pathogenic invaders or harmless materials such as foods and commensal organisms. Potential immune responses to this antigen load are controlled by a distinct compartment of the immune system, the mucosal immune system, which is the largest in the body and possesses many unique features. These include distinctive routes and processes for the uptake and presentation of antigens, exploiting M cells to transport antigens across the epithelium of Peyer's patches, and a unique subset of retinoic acid-producing, CD103-positive dendritic cells that imprint the T and B cells they activate with gut-homing properties. They also favor the generation of FoxP3-positive T_{reg} cells in the normal gut. Lymphocytes primed in the mucosa-associated lymphoid tissues acquire specific homing receptors, allowing them to redistribute preferentially back to mucosal surfaces as effector cells. Exposure to antigen outside the mucosal immune system cannot reproduce these effects. The mucosa-associated lymphoid tissues also generate different effector responses from those in other parts of the body, including unique forms of innate immunity. The adaptive immune response in mucosal tissues is characterized by the production of secretory dimeric IgA, and by the presence of distinctive populations of effector T cells whose functional and phenotypic properties are highly influenced by their anatomical location.

The mucosal response to infection and regulation of mucosal immune responses.

The major role of the mucosal immune response is defense against infectious agents, which include all forms of microorganisms from viruses to multicellular parasites. This means that the host must be able to generate a wide spectrum of immune responses tailored to meet the challenge of individual pathogens, and it is equally unsurprising that many microbes have evolved means of adapting to and subverting the host response. To ensure an adequate response to pathogens, the mucosal immune system needs to be able to recognize and respond to any foreign antigen, but it must not produce the same effector response to a harmless antigen (from food or commensals) as it would to a pathogen. A major role of the mucosal immune system is to balance these competing demands, and how it does this will be the focus of this part of the chapter.

12-13 Enteric pathogens cause a local inflammatory response and the development of protective immunity.

Despite the array of innate immune mechanisms in the gut, and stiff competition from the indigenous microbiota, the gut is the most frequent site of infection by pathogenic organisms. These include many viruses, enteric bacteria such as *Helicobacter pylori*, *Salmonella* and *Shigella* species, protozoans such as *Entamoeba histolytica*, and multicellular helminth parasites such as tapeworms and pinworms (Fig. 12.18). These pathogens cause disease in many ways, but certain common features of infection are crucial to understanding how they stimulate a productive immune response by the host. The key to this in the gut, as elsewhere in the body, is the activation of the innate immune system, one of whose functions is to stimulate the expression of co-stimulatory molecules on local dendritic cells, thus overcoming the inherent unresponsiveness that characterizes these cells in the healthy intestine and favoring the generation of an appropriate adaptive immune response.

The effector mechanisms of the innate immune system can themselves eliminate most intestinal infections rapidly and without significant spread beyond

Intestinal pathogens and human disease	
Bacteria	
Salmonella enterica Typhi	Typhoid fever
Salmonella enterica Paratyphi	Enteric fever (paratyphoid)
Salmonella enterica Enteritidis	Food poisoning
Vibrio cholera	Cholera
Shigella dysenteriae, flexneri, sonnei	Dysentery
Enteropathogenic *E. coli* (EPEC)	Gastroenteritis, systemic infection
Enterohemolytic *E. coli* (EHEC)	Gastroenteritis, systemic infection
Enterotoxigenic *E. coli* (ETEC)	Gastroenteritis, 'travelers diarrhea'
Enteroaggregative *E. coli* (EAEC)	Gastroenteritis, systemic infection
Yersinia enterocolitica	Gastroenteritis, systemic infection
Clostridium difficile	Necrotizing enterocolitis
Campylobacter jejuni	Gastroenteritis
Staphylococcus aureus	Gastroenteritis
Bacillus cereus	Gastroenteritis
Clostridium perfringens	Gastroenteritis
Helicobacter pylori	Gastritis, peptic ulcer, gastric cancer
Mycobacterium tuberculosis	Intestinal tuberculosis
Listeria monocytogenes	Foodborne infection
Viruses	
Rotaviruses	Gastroenteritis
Noroviruses	'Winter vomiting' disease, Epidemic gastroenteritis
Astroviruses	'Winter vomiting' disease
Adenoviruses	'Winter vomiting' disease
Parasites	
Protozoa	
Giardia lamblia	Gastroenteritis
Blastocystis hominis	Gastroenteritis (esp. in immunocompromised hosts)
Toxoplasma gondii	Gastroenteritis, systemic disease (esp. in immunocompromised hosts)
Cryptosporidium parvum	Gastroenteritis (esp. in immunocompromised hosts)
Entamoeba histolytica	Amebic dysentery + liver abscesses
Microsporidium species	Diarrheal disease
Helminths	
Ascaris lumbricoides	Roundworm infection of small intestine
Necator americanus	Hookworm infection of small intestine
Strongyloides species	Roundworm infection of small intestine
Enterobius species	Pinworm infection of large intestine
Trichinella spiralis	Trichinosis
Trichuris trichiura	Whipworm infection of large intestine
Taenia species	Tapeworm infections
Schistosoma species	Schistosomiasis: enteritis, mesenteric vein infection

Fig. 12.18 Intestinal pathogens and infectious disease in humans. Many species of bacteria, viruses, and parasites can cause disease in the human intestine.

the intestine. Pattern-recognition receptors such as the TLRs are important in this process (see Chapter 3), and are expressed on both inflammatory cells and intestinal epithelial cells. Epithelial cells bear TLRs on both their apical and basal surfaces, which allows them to sense bacteria in the gut lumen and those that have penetrated across the epithelium. For example, TLR-5 on gut epithelial cells enables them to recognize flagellin (the protein of which bacterial flagella are made), and mutant mice that lack this receptor show increased susceptibility to infection by *Salmonella*. Epithelial cells also carry TLRs in intracellular vacuoles that can detect extracellular pathogens and their products that have been internalized by endocytosis (Fig. 12.19).

Epithelial cells also have intracellular sensors that can react to microorganisms or their products that enter the cytoplasm (see Fig. 12.19). These sensors include the nucleotide-binding oligomerization domain proteins NOD1 and NOD2, which are related to the TLRs and induce activation of NFκB (the details of this pathway are discussed in Section 3-8 and Figs 3.13 and 3.15). NOD1 recognizes a diaminopimelic acid-containing peptide that is found only in the cell walls of Gram-negative bacteria; NOD2 recognizes a muramyl dipeptide found in the peptidoglycans of most bacteria, and epithelial cells defective in NOD2 are less resistant to infection by intracellular bacteria. Mice lacking NOD2 also show increased translocation of bacteria across the epithelium and out of Peyer's patches. The Paneth cells of the small intestinal epithelium respond very effectively to TLR and NOD2 stimulation by producing antimicrobial peptides such as defensins (see Section 2-4), and mice lacking NOD2 or defensins are more susceptible to developing intestinal inflammation. A defect in recognition of the commensal microbiota by NOD2 also seems to be important in Crohn's disease, as up to 25% of patients carry a mutation in the *NOD2* gene that renders the NOD2 protein nonfunctional (see Section 15-23).

Ligation of TLRs or NOD proteins in epithelial cells stimulates the release of antimicrobial peptides, the production of cytokines such as IL-1 and IL-6, and the production of chemokines. The chemokines include CXCL8, which is a potent neutrophil chemoattractant (see Fig. 3.22), and CCL2, CCL3, CCL4, and CCL5, which attract monocytes, eosinophils, and T cells out of the blood. Stimulated epithelial cells also increase their production of the chemokine CCL20, which attracts immature dendritic cells resident in the tissue (see Section 12-4). In this way, the onset of infection triggers an influx of inflammatory cells and lymphocytes into the mucosa from the bloodstream, and the positioning of dendritic cells where they can capture antigen, all of which aids in the induction of a pathogen-specific adaptive immune response.

Epithelial cells also express NLRP3, a member of the NOD-like receptor family. NLRP3 is activated in response to various TLR ligands and by-products of cell injury such as ATP and reactive oxygen species, and forms part of the inflammasome, a protein complex in which caspase-1 becomes activated

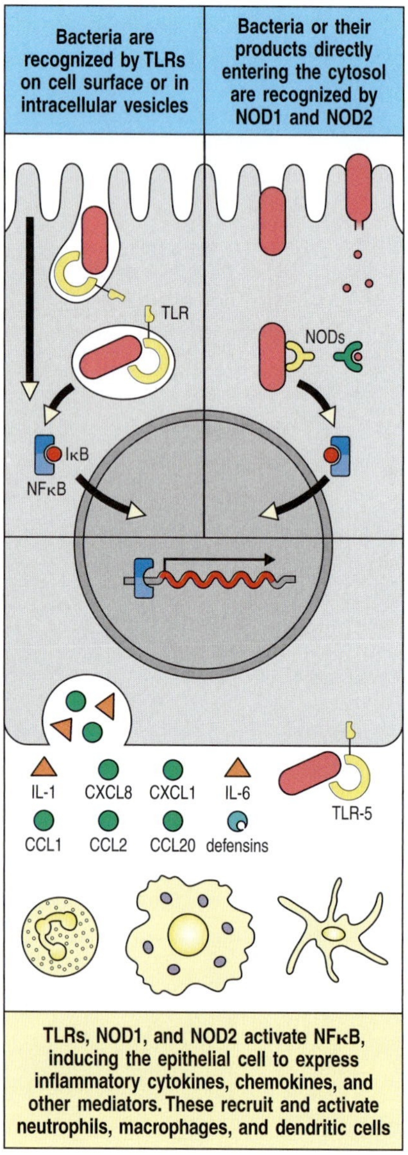

Bacteria are recognized by TLRs on cell surface or in intracellular vesicles

Bacteria or their products directly entering the cytosol are recognized by NOD1 and NOD2

TLRs, NOD1, and NOD2 activate NFκB, inducing the epithelial cell to express inflammatory cytokines, chemokines, and other mediators. These recruit and activate neutrophils, macrophages, and dendritic cells

Fig. 12.19 Epithelial cells have a crucial role in innate defense against pathogens. TLRs are present in intracellular vesicles or on the basolateral or apical surfaces of epithelial cells, where they recognize different components of invading bacteria. NOD1 and NOD2 pattern-recognition receptors are found in the cytoplasm and recognize cell-wall peptides from bacteria. Both TLRs and NODs activate the NFκB pathway (see Fig. 3.13), leading to the generation of pro-inflammatory responses by epithelial cells. These include the production of chemokines such as CXCL8, CXCL1 (GROα), CCL1, and CCL2, which attract neutrophils and macrophages, and CCL20 and β-defensin, which attract immature dendritic cells in addition to possessing antimicrobial properties. The cytokines IL-1 and IL-6 are also produced and activate macrophages and other components of the acute inflammatory response. The epithelial cells also express MIC-A and MIC-B and other stress-related nonclassical MHC molecules, which can be recognized by cells of the innate immune system. IκB, inhibitor of NFκB.

(see Section 3-8). Caspase-1 cleaves pro-IL-1 and pro-IL-18 to produce the active cytokines (see Fig. 3.16), both of which contribute to epithelial defense against bacterial invasion and can cause tissue damage if present for long periods. In acute inflammation, however, IL-18 has an important protective role, stimulating the renewal and repair of epithelial cells.

Injury and stress to the enterocytes lining the gut, and also ligation of TLR-3 by viral RNA, stimulate the expression of nonclassical MHC molecules, such as MIC-A and MIC-B (see Fig. 12.15). As discussed earlier, these proteins can be recognized by the receptor NKG2D on local cytotoxic lymphocytes, which are then activated to kill the infected epithelial cells.

12-14 The outcome of infection by intestinal pathogens is determined by a complex interplay between the microorganism and the host immune response.

Many enteric pathogens exploit host mechanisms of antigen uptake via M cells and inflammation as part of their invasive strategy. Poliovirus, reoviruses, and some retroviruses are transported through M cells by transcytosis, and initiate infection in tissues distant from the intestine after delivery into the subepithelial space. HIV may use a similar route into the lymphoid tissue of the rectal mucosa, where it first encounters and infects dendritic cells. Prions such as the causal agent of scrapie follow the same entry routes. Many of the most important enteric bacterial pathogens also gain entry through M cells. These include *Salmonella enterica* Typhi, the causative agent of typhoid fever; other *Salmonella enterica* serotypes, which are major causes of bacterial food poisoning; *Shigella* species that cause dysentery; *Yersinia pestis*, which causes plague; and *H. pylori*, which causes gastritis (inflammation of the intestine). After entry into the M cell, these bacteria produce proteins that reorganize the M-cell cytoskeleton in a manner that encourages their transcytosis. Some of the entry mechanisms used by salmonellae are shown in Fig. 12.20, and those of shigellae in Fig. 12.21.

M cells are not the only port of entry into the mucosa. Some intestinal bacteria such as *Clostridium difficile* or *Vibrio cholerae* secrete large amounts of protein toxins, enabling them to cause disease without the need to invade the epithelium. Other bacteria, such as enteropathogenic and enterohemolytic strains of *E. coli*, have specialized means of attaching to and invading epithelial cells, allowing them to cause intestinal damage and produce harmful toxins from an intracellular location. Some enteric viruses, such as rotaviruses, which cause diarrheal disease, are specialized to infect enterocytes directly.

Once delivered into the subepithelial space, pathogenic bacteria and viruses can cause a more widespread infection. Paradoxically, the host inflammatory response is an additional and often essential part of the invasive process. Many enteric bacteria stimulate the production of chemokines such as CCL20 by the epithelial cells, which recruits dendritic cells into the epithelial layer. Bacteria transcytosed through M cells are free to interact with TLRs on inflammatory cells such as macrophages and on the basal surfaces of adjacent epithelial cells. In addition, after being ingested by phagocytes, many of these microbes can be recognized by intracellular NOD1 or NLRP3 and induce NFκB and caspase-1 activation, respectively (see Section 3-8). All this stimulates the production of a cascade of inflammatory mediators, among which IL-1β and TNF-α drastically loosen the tight junctions between epithelial cells. This removes the normal barrier to bacterial invasion, allowing microorganisms to flood into the intestinal tissue from the lumen and extend the infection.

Despite their apparent benefit to the invader, it is important to remember that the principal role of the mediators and cells induced by the innate immune

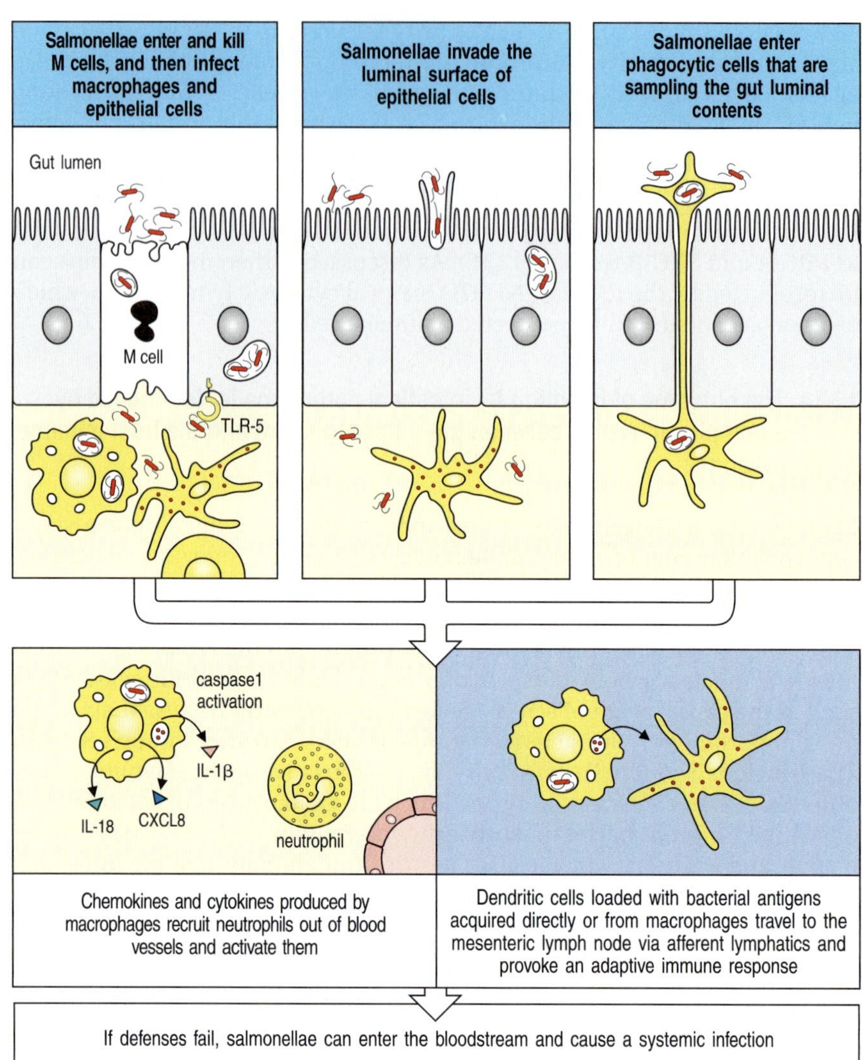

Fig. 12.20 *Salmonella enterica* serovar Typhimurium, an important cause of food poisoning, can penetrate the gut epithelial layer by three routes. In the first route (top left panel), *Salmonella* Typhimurium adheres to and enters M cells, which it then kills by causing apoptosis. Having penetrated the epithelium, it infects macrophages and gut epithelial cells. Epithelial cells express TLR-5 on their basal membrane; this binds flagellin on the salmonella flagella, activating an inflammatory response via the NFκB pathway. After uptake by macrophages in the lamina propria, invasive *Salmonella* induces caspase 1 activation, promoting production of IL-1 and IL-8. CXCL8 is also produced by the infected macrophages, and together these mediators recruit and activate neutrophils (lower left panel). Dendritic cells in Peyer's patches acquire organisms directly from M cells or from infected macrophages and present the antigens to T cells in the Peyer's patch (not shown). Salmonellae can also invade gut epithelial cells directly by adherence of their fimbriae (fine threadlike protrusions) to the luminal epithelial surface (middle panel). In the third route of entry, mononuclear phagocytes sampling the gut lumen extrude cell processes between epithelial cells. These effectively breach the epithelial layer and can be infected by salmonellae in the lumen (top right panel). Dendritic cells in villus lamina propria can acquire organisms either directly or from infected macrophages and carry them to the draining mesenteric lymph node to prime the adaptive immune response (lower right panel). If these local containment processes fail, *Salmonella* can invade beyond the intestine and its lymphoid tissues, enter the bloodstream and cause systemic infection.

response is to help initiate the adaptive immune response that will ultimately eliminate the microbe. Central to these protective effects are the cytokines IL-12, IL-18, and IL-23 produced by activated macrophages and dendritic cells, as well as IL-1, IL-6, and IL-8 produced by macrophages, dendritic cells, and epithelial cells. In infections by extracellular bacteria, IL-1, IL-6, and IL-23 drive the differentiation of T_H17 cells, which activate and recruit neutrophils; T_H17 cells also produce IL-22, which stimulates Paneth cells to produce antimicrobial peptides. In response to intracellular infections, such as those due to *Salmonella* or *Mycobacterium* species, IL-18, assisted by IL-12, drives IFN-γ production by antigen-specific T_H1 cells. In turn, this enhances the ability of the macrophage to kill bacteria it has ingested (see Section 9-29). Thus, the innate immune response to enteric bacteria has apparently opposing effects. It orchestrates a series of potent effector mechanisms aimed at eliminating infection, but these mechanisms can also be exploited by the invading organism. The fact that the protective immune response wins out in most cases is testimony to the efficiency and adaptability of the mucosal immune system.

The host–pathogen interaction is complicated further by the ability of many enteric microbes to modulate the host inflammatory response. For instance, *Yersinia* species produce Yop proteins, which can both inhibit the inflammatory response and block phagocytosis and intracellular killing of microbes by phagocytes, as well as the virulence antigen LcrV, which stimulates IL-10

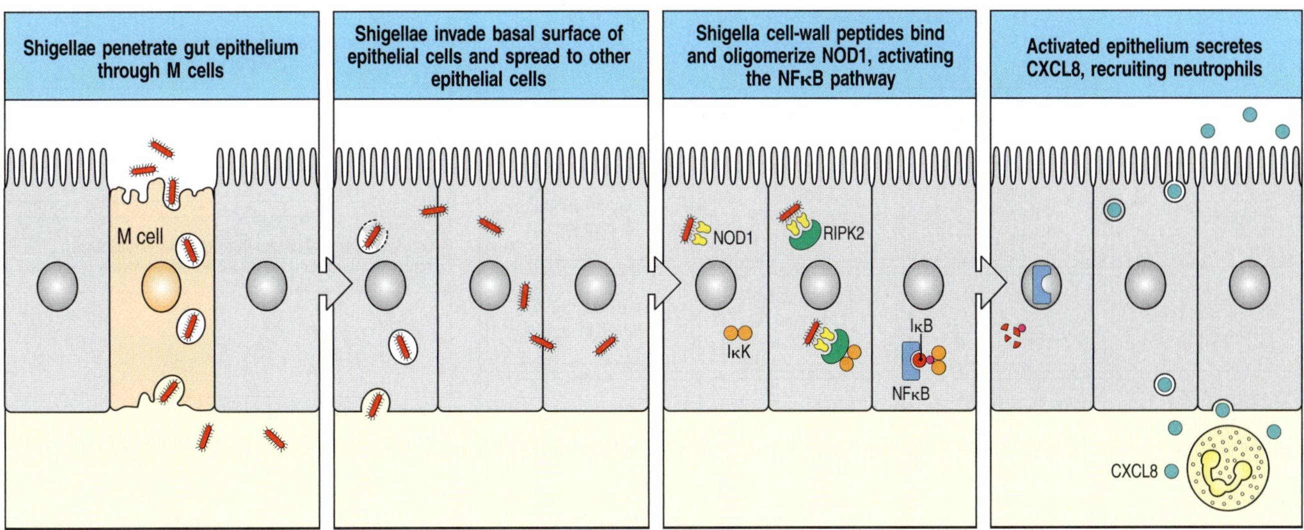
Shigellae penetrate gut epithelium through M cells
Shigellae invade basal surface of epithelial cells and spread to other epithelial cells
Shigella cell-wall peptides bind and oligomerize NOD1, activating the NFκB pathway
Activated epithelium secretes CXCL8, recruiting neutrophils
M cell
NOD1
RIPK2
IκK
IκB
NFκB
CXCL8

Fig. 12.21 *Shigella flexneri*, a cause of bacterial dysentery, infects intestinal epithelial cells, triggering activation of the NFκB pathway. *Shigella flexneri* binds to M cells and is translocated beneath the gut epithelium (first panel). The bacteria infect intestinal epithelial cells from their basal surface and are released into the cytoplasm (second panel). Muramyl tripeptides containing diaminopimelic acid in the cell walls of the shigellae bind to and oligomerize the protein NOD1. Oligomerized NOD1 binds the serine/threonine kinase RIPK2, which triggers activation of the NFκB pathway (see Fig. 3.15), leading to the transcription of genes for chemokines and cytokines (third panel). Activated epithelial cells release the chemokine CXCL8, which acts as a neutrophil chemoattractant (fourth panel). IκB, inhibitor of NFκB; IκK, IκB kinase.

production by binding to TLR-2. Enteropathogenic *E. coli* can inject inhibitory molecules into dendritic cells via its type III secretory system. *Salmonella enterica* Typhi creates its own safe haven within phagosomes by modifying the phagosome membrane and preventing intracellular killing (see Section 3-2). *Shigella*, in contrast, resides in the cytoplasm of epithelial cells, where it remodels the actin cytoskeleton, creating a molecular machinery that allows the bacterium to spread directly from cell to cell without exposure to the immune system. All these microorganisms also induce apoptosis of the cells they infect via the activation of caspases, and can produce factors that inhibit the activation of NFκB, thus disabling important arms of the inflammatory response as well as enhancing the spread of the pathogen (see Sections 3-7 and 3-8 for discussion of these signaling pathways). The immunomodulatory molecules produced by these bacteria are frequently essential to their ability to cause disease.

12-15 The mucosal immune system must maintain a balance between protective immunity and homeostasis to a large number of different foreign antigens.

The majority of antigens encountered by the normal intestinal immune system are not derived from pathogens, but come from food and commensal bacteria. These antigens are not only harmless but are in fact also highly beneficial to the host. They normally do not induce an inflammatory immune response, despite the fact that, like any other foreign antigens, there will be no central tolerance to them because they were not present in the thymus during lymphocyte development (see Chapter 8). The mucosal immune system has developed sophisticated means of discriminating between pathogens and innocuous antigens.

Contrary to popular belief, food proteins are not digested completely in the intestine; significant amounts are absorbed into the body in an immunologically relevant form. The default response to oral administration of a protein antigen is the development of a phenomenon known as **oral tolerance**. This is a form of **peripheral tolerance** that renders the systemic and mucosal immune systems relatively unresponsive to the same antigen. It can be demonstrated in mice by feeding them a foreign protein such as ovalbumin (Fig. 12.22). When the animals are then challenged with the antigen by a nonmucosal route, such as injection into the skin, the immune response one would expect is blunted. This suppression of systemic immune responses is long

	Protective immunity	Mucosal tolerance	
Antigen	Invasive bacteria, viruses, toxins	Food proteins	Commensal bacteria
Primary Ig production	Intestinal IgA Specific Ab present in serum	Some local IgA Low or no Ab in serum	Local IgA No Ab in serum
Primary T-cell response	Local and systemic effector and memory T cells	No local effector T-cell response	No local effector T-cell response
Response to antigen reexposure	Enhanced (memory) response	Low or no response	Low or no mucosal or systemic response

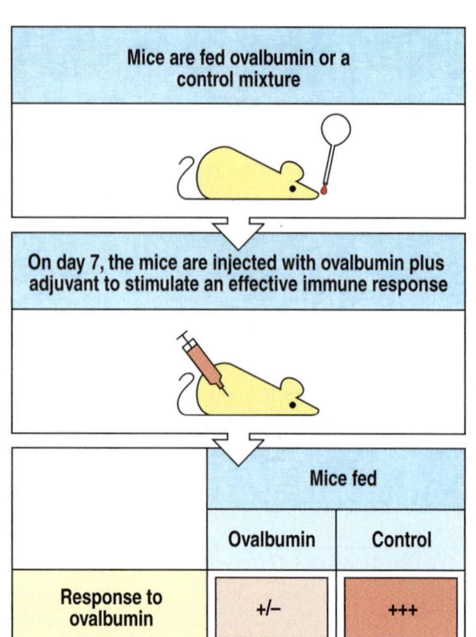

Fig. 12.22 Immune priming and tolerance are different outcomes of intestinal exposure to antigen. Left panel, the intestinal immune system generates protective immunity against antigens that are a threat to the host, such as pathogenic organisms and their products. IgA antibodies are produced locally, serum IgG and IgA are made, and the appropriate effector T cells are activated in the intestine and elsewhere. When the antigen is encountered again, there is effective memory, ensuring rapid protection. Harmless antigens such as food proteins or antigens from commensal bacteria induce tolerance, either locally, or both locally and systemically. They lack the danger signals needed to activate local antigen-presenting cells, or do not invade sufficiently to cause inflammation. In the case of food proteins, there is little or no local IgA antibody production or primary systemic antibody response, nor are effector T cells activated. Subsequent local and systemic immune responses to challenge are also specifically suppressed. This phenomenon is known as 'oral tolerance' and is shown in the right-hand panels, where it has been induced by feeding a protein such as ovalbumin to a normal mouse. First, mice are fed either ovalbumin or a different protein as a control. Seven days later, the mice are immunized subcutaneously with ovalbumin and an adjuvant; 2 weeks later, systemic immune responses such as serum antibodies and T-cell function are measured. Mice fed ovalbumin have a lower ovalbumin-specific systemic immune response than those fed the control protein. In the case of commensal bacteria, there is evidence of local IgA antibody production, but no primary systemic antibody responses, and effector T cells are not activated in either site. Profound tolerance of T-cell responses is retained thereafter, but the systemic immune system remains ignorant of commensals unless they gain access to the circulation, when a primary systemic immune response will result.

lasting and is antigen specific: responses to other antigens are not affected. A similar suppression of subsequent immune responses is observed after the administration of proteins into the respiratory tract, giving rise to the concept of **mucosal tolerance** as the usual response to such antigens delivered via a mucosal surface.

All aspects of the peripheral immune response can be affected by oral tolerance, although T-cell dependent effector responses and IgE production tend to be more inhibited than serum IgG antibody responses. Similar inhibition of systemic T-cell responses has been found in humans fed protein antigens that they have not previously encountered, although here, serum antibody responses can be completely unaffected. Thus, the systemic immune responses most susceptible to oral tolerance are those that are usually associated with tissue inflammation. Mucosal immune responses to the antigen are also downregulated. Again, this mainly affects effector T-cell responses, and low levels of secretory IgA antibodies directed at food proteins can be found in healthy humans. However, as discussed above (see Section 12-11), antibodies of this class will not lead to inflammation in the intestine. A breakdown in oral tolerance is believed to occur in celiac disease. In this condition, genetically susceptible individuals generate IFN-γ-producing, CD4 T-cell responses against the wheat protein gluten, and the resulting inflammation destroys the upper small intestine (see Section 14-18). Defective generation of oral tolerance has also been demonstrated in children with IgE-mediated food allergies.

Various mechanisms are likely to account for oral tolerance to protein antigens, including anergy, deletion of antigen-specific T cells, and the generation of regulatory T cells (see Chapters 8 and 9). Current evidence suggests that although antigen-specific FoxP3-positive T_{reg} cells can be generated in Peyer's patches after feeding protein antigens to experimental animals, the mesenteric lymph node plays the dominant role in their generation. This is because protein antigens seem to be mostly taken up by the retinoic-acid-producing CD103 dendritic cells of the lamina propria that migrate to a mesenteric lymph node and induce the generation of gut-homing T_{reg} cells (see Section 12-8). The events in the mesenteric lymph node are also essential for the suppression of systemic immune responses, but the mechanisms responsible for this are not yet understood.

As discussed in Chapters 9 and 11, regulatory T cells can act in various ways, but their production of TGF-β has been particularly associated with oral tolerance. TGF-β has many immunosuppressive properties and also stimulates B-cell switching to IgA. Together, these properties could help prevent inflammatory responses against food proteins by ensuring tolerance of effector T cells and by favoring the production of noninflammatory IgA antibodies. IL-10-producing T_{reg} cells may also be involved in oral tolerance; they are important in the equivalent tolerance that occurs to antigens introduced via the respiratory route. Other cytokines that can contribute to tolerance include IFN-γ, which is particularly important in preventing potentially harmful T_H2-driven allergic responses such as eosinophilia and IgE antibody production.

In addition to its physiological role in preventing inappropriate food-related immune responses, mucosal tolerance has proved effective as a way of preventing inflammatory disease in experimental animal models. Oral or intranasal administration of appropriate antigens has been found to prevent or even treat type 1 diabetes mellitus, experimental arthritis, encephalomyelitis, and other autoimmune diseases in animals. Clinical trials using mucosal tolerance to treat the equivalent diseases in humans have been less successful, possibly because it is much more difficult to inhibit inflammatory responses that are already under way; however, oral tolerance remains a potentially attractive means of inducing antigen-specific tolerance in clinical situations.

12-16 The healthy intestine contains large quantities of bacteria but does not generate potentially harmful immune responses against them.

We each harbor more than 1000 species of commensal bacteria in our intestine, and they are present in greatest numbers in the colon and lower ileum. Despite the fact that there are at least 10^{14} of these bacterial cells and they collectively weigh about 1 kg, for most of the time we cohabit with our intestinal microbiota in a happy symbiotic relationship known as **mutualism**. Nevertheless, they do represent a potential threat, as is shown when the integrity of the intestinal epithelium is damaged, allowing large numbers of commensal bacteria to enter the mucosa. This can occur when the blood flow to the gut is compromised by trauma, infection, or blood vessel disease, for example, or by endotoxic shock (see Sections 3-6 and 3-17). In these circumstances, normally innocuous gut bacteria, such as nonpathogenic *E. coli*, can cross the mucosa, invade the bloodstream, and cause fatal systemic infection.

The normal gut microbiota has coevolved with vertebrates over many millennia and has an essential role in maintaining health. Its members assist in the metabolism of dietary constituents such as cellulose, as well as degrading toxins and producing essential cofactors such as vitamin K_1 and short-chain fatty acids. By having direct effects on epithelial cells, the commensal bacteria are also essential for maintaining the normal barrier function of the epithelium. Another important property of commensal organisms is that they interfere with the ability of pathogenic bacteria to colonize and invade the gut. Commensals do this partly by competing for space and nutrients, but they can also directly inhibit the pro-inflammatory signaling pathways that pathogens stimulate in epithelial cells and that are needed for invasion.

The protective role of the commensal microbiota is dramatically illustrated by the adverse effects of broad-spectrum antibiotics. These antibiotics can kill large numbers of commensal gut bacteria, thereby creating an ecological niche for bacteria that would not otherwise be able to compete successfully. One example of a bacterium that grows in the antibiotic-treated gut and can cause a severe infection is *Clostridium difficile*; this produces two toxins, which can cause severe bloody diarrhea associated with mucosal injury (Fig. 12.23).

Several host factors help prevent commensal bacteria from invading beyond the intestine (Fig. 12.24). One of these is TLR signaling in epithelial cells and phagocytes, which is also important in protecting against intestinal inflammation. Mice lacking TLR-2, TLR-5, TLR-9, or the TLR adaptor protein MyD88 are much more susceptible to the induction of experimental inflammatory bowel diseases. The protective effects of TLR stimulation involve the epithelial cells being made more resistant to inflammation-induced damage, as well as tuning of the inflammatory responses themselves.

Commensal bacteria and their products are also essential for the normal development and function of the immune system. The scale of this effect is illustrated by **germ-free** (or **gnotobiotic**) animals, in which there is no colonization of the gut by microorganisms. These animals have marked reductions in the size of all lymphoid organs, low serum immunoglobulin levels, fewer mature T cells, and reduced immune responses, especially of the T_H1 and T_H17 types. In the intestine, Peyer's patches do not develop normally, isolated lymphoid follicles are absent, and there are severely reduced numbers of T lymphocytes in both the lamina propria and epithelium. There is also defective production of several of the mediators that normally regulate local immunity, such as antimicrobial peptides, IL7, IL-25, IL-33, and TSLP.

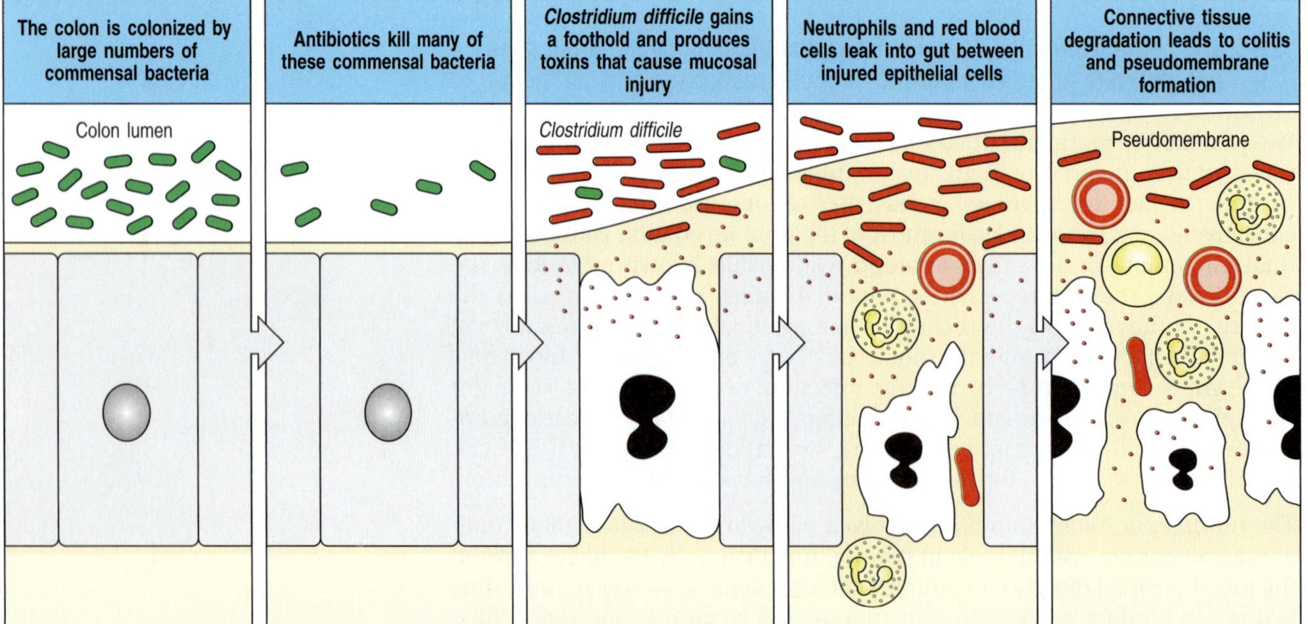

Fig. 12.23 Infection by *Clostridium difficile*. Treatment with antibiotics causes massive death of the commensal bacteria that normally colonize the colon. This allows pathogenic bacteria to proliferate and to occupy an ecological niche that is normally occupied by harmless commensal bacteria. *Clostridium difficile* is an example of a pathogen producing toxins that can cause severe bloody diarrhea in patients treated with antibiotics.

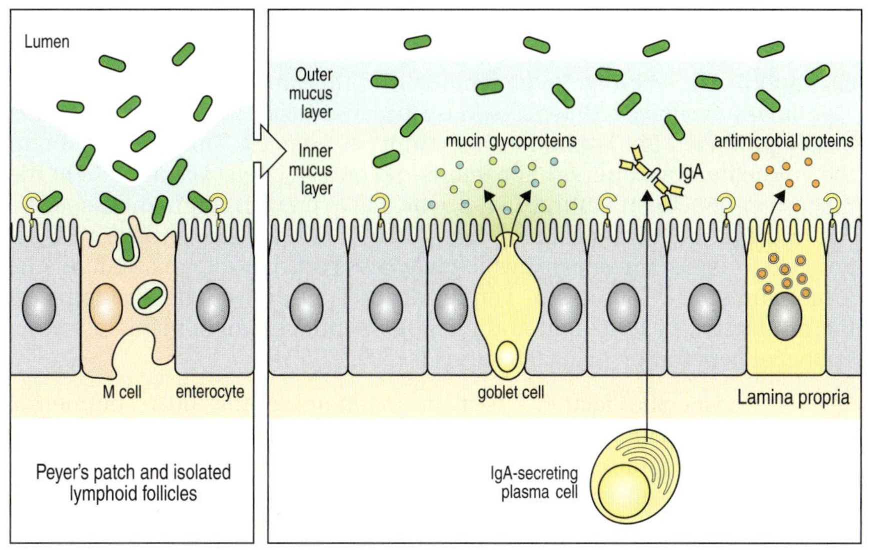

Fig. 12.24 Local responses to commensals. Several local processes ensure peaceful coexistence between the microbiota and the host, allowing the commensal organisms to be recognized by the immune system without inducing inflammation or an immune response that would eliminate them. Commensal bacteria in the lumen gain access to the immune system via M cells in Peyer's patches and isolated follicles (left panel). Uptake and presentation of these noninvasive organisms by resting dendritic cells generates IgA-switched B cells that localize in the lamina propria as IgA-producing plasma cells (right panel). The secretory IgA that is produced limits the access of commensals to the epithelium and helps prevent their penetration. This is assisted further by the presence of thick layers of mucus, which also contain mucin glycoproteins that have antibacterial properties. In addition, stimulation of pattern-recognition receptors on epithelial cells and local leukocytes induces the production of antimicrobial peptides such as defensins.

How the microbiota affects immune-system development is still being worked out; many different mechanisms are probably involved. Ligation of TLRs is likely to be important, and bacterial species have recently been identified with specific functions. Thus, polysaccharide A (PSA) from *Bacteroides fragilis* has been shown to drive the differentiation of several subsets of effector CD4 T cells, including IL-10-producing T_{reg} cells, whereas segmented filamentous bacteria (SFB) induce the appearance of T_H1, T_H17, and FoxP3-positive T cells in the mouse intestine. The effects of commensal microbes extend far beyond the intestine; for example, the incidence of several autoimmune diseases in experimental animals can be either reduced or increased in germ-free animals. In the case of type 1 diabetes, the germ-free state greatly increases the severity of the disease, perhaps because of defective development of regulatory T cells.

Commensal bacteria stimulate local IgA antibody production, and in the healthy gut there is also active suppression of local effector T-cell responses. The intestinal secretions of healthy animals contain high levels of secretory IgA directed at commensal bacteria (see Fig. 12.24). Experiments in mice that lack secretory antibodies show the importance of this response, because these animals have increased numbers of commensal bacteria penetrating the intestinal mucosa and its draining lymphoid tissues. In addition, normal individuals contain T cells that can recognize commensal bacteria, which probably accounts for the large numbers of fully differentiated T_H1 and T_H17 cells present in the resting intestine. However, as we have discussed (see Section 12-14), these effector T cells are held in check by the regulatory T cells present, producing a balanced state often called **physiological inflammation**. When strong effector T-cell responses are made against commensal bacteria, inflammatory disease such as Crohn's disease can develop (see Section 15-23).

Commensal bacteria do not induce a state of systemic immune unresponsiveness similar to the oral tolerance induced by protein antigens, and when such bacteria get into the bloodstream they stimulate a normal primary systemic immune response. They are normally confined to the gut because, unlike pathogenic bacteria, they do not possess the virulence factors necessary for penetrating the mucus layer or intact epithelium and cannot disseminate throughout the body. As a result, the only route of entry into the body for gut commensal bacteria is via M cells in Peyer's patches, with subsequent transfer to local dendritic cells. These gut-derived dendritic cells are specialized to activate naive B cells to become IgA-expressing B lymphocytes and to drive the generation of FoxP3-positive T_{reg} cells, which also assist the production of IgA antibodies via their secretion of TGF-β. In addition, the influence of these dendritic cells ensures that the activated T cells and B cells will migrate preferentially to the lamina propria (Fig. 12.25). This anatomically restricted response has two main outcomes. First, the locally produced IgA antibodies help limit adherence to and penetration of the epithelium by commensal bacteria. Second, T_{reg} cells accumulate in the mucosa, producing IL-10 and TGF-β, which help prevent commensal-specific T_H1 and T_H17 cells from causing overt inflammation. These effector cells do, however, retain some activity, which may also contribute to maintaining the local symbiotic relationship. For instance, T_H17-derived IL-17 stimulates the production of IL-22 and antimicrobial peptides, which helps to restrict epithelial penetration of local bacteria (see Fig. 12.24).

Several nonspecific factors assist in maintaining the host–commensal balance. Unlike pathogens, commensal bacteria do not stimulate full-blown inflammatory responses that loosen the epithelial barrier. Structural differences have been observed between the molecular patterns of commensal bacteria and of pathogens, meaning that they may induce different kinds of

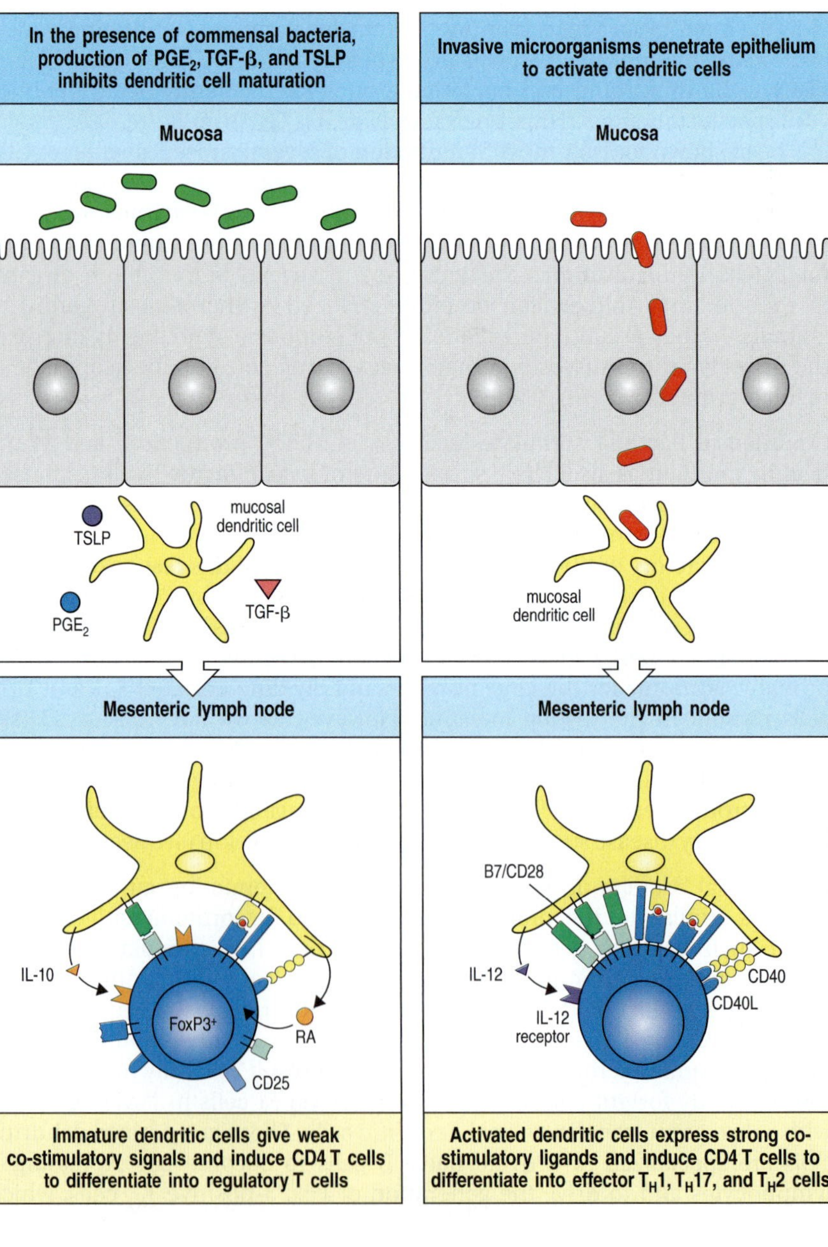

Fig. 12.25 Mucosal dendritic cells regulate the induction of tolerance and immunity in the intestine. Under normal conditions (left panels), dendritic cells are present in the mucosa underlying the epithelium and can acquire antigens from foods or commensal organisms. They take these antigens to the draining mesenteric lymph node, where they present them to naive CD4 T cells. There is, however, constitutive production by epithelial cells and mesenchymal cells of molecules such as TGF-β, thymic stromal lymphopoietin (TSLP), and prostaglandin E_2 (PGE_2), which maintain the local dendritic cells in a quiescent state with low levels of co-stimulatory molecules, so that when they present antigen to naive CD4 T cells, anti-inflammatory or regulatory T cells are generated. These recirculate back to the intestinal wall and maintain tolerance to the harmless antigens. Invasion by pathogens or a massive influx of commensal bacteria (right panels) overcomes these homeostatic mechanisms, resulting in full activation of local dendritic cells and their expression of co-stimulatory molecules and pro-inflammatory cytokines such as IL-12. Presentation of antigen to naive CD4 T cells in the mesenteric lymph node by these dendritic cells causes differentiation into effector T_H1 and T_H2 cells, leading to a full immune response.

response through the same receptor. So although commensal bacteria are continually triggering TLRs and NODs in epithelial cells and leukocytes, they probably trigger different signaling pathways from those induced by pathogens. Structural differences may also account for the fact that the endotoxin present on commensal bacteria seems unusually sensitive to neutralization by gut enzymes such as alkaline phosphatase. In addition to these inherent differences, commensal bacteria also actively inhibit the pro-inflammatory, NFκB-mediated responses induced in epithelial cells by pathogenic bacteria. This inhibition involves strategies for preventing the activation of NFκB by, for example, inhibiting the degradation of IκB (the inhibitory protein that keeps NFκB in a complex in the cytoplasm) or promoting the export of NFκB from the nucleus via the peroxisome proliferator activated receptor-γ (PPARγ) (Fig. 12.26). Finally, if commensal bacteria do cross the epithelium in small numbers, their lack of virulence factors means they cannot resist uptake and killing by phagocytic cells, and they are rapidly destroyed.

Commensal organisms can therefore remain associated with the mucosal surface without invading it or provoking inflammation and a consequent adaptive immune response. In parallel, the lack of tolerance to commensals in the systemic immune system means that the immune system will be able to generate protective immunity if they do manage to enter the body through a damaged intestine.

12-17 Full immune responses to commensal bacteria provoke intestinal disease.

It is now generally accepted that potentially aggressive T cells that can respond to commensal bacteria are present in normal animals but are usually kept in check by active regulation. If these regulatory mechanisms fail, unrestricted immune responses to commensals lead to **inflammatory bowel diseases** such as Crohn's disease (see Section 15-23). This has been demonstrated in animals with defects in immunoregulatory m echanisms involving IL-10 and TGF-β, or in which a disrupted epithelial barrier allows large numbers of commensal bacteria to penetrate. In these conditions, systemic immune responses are generated against commensal bacterial antigens such as flagellin. Strong

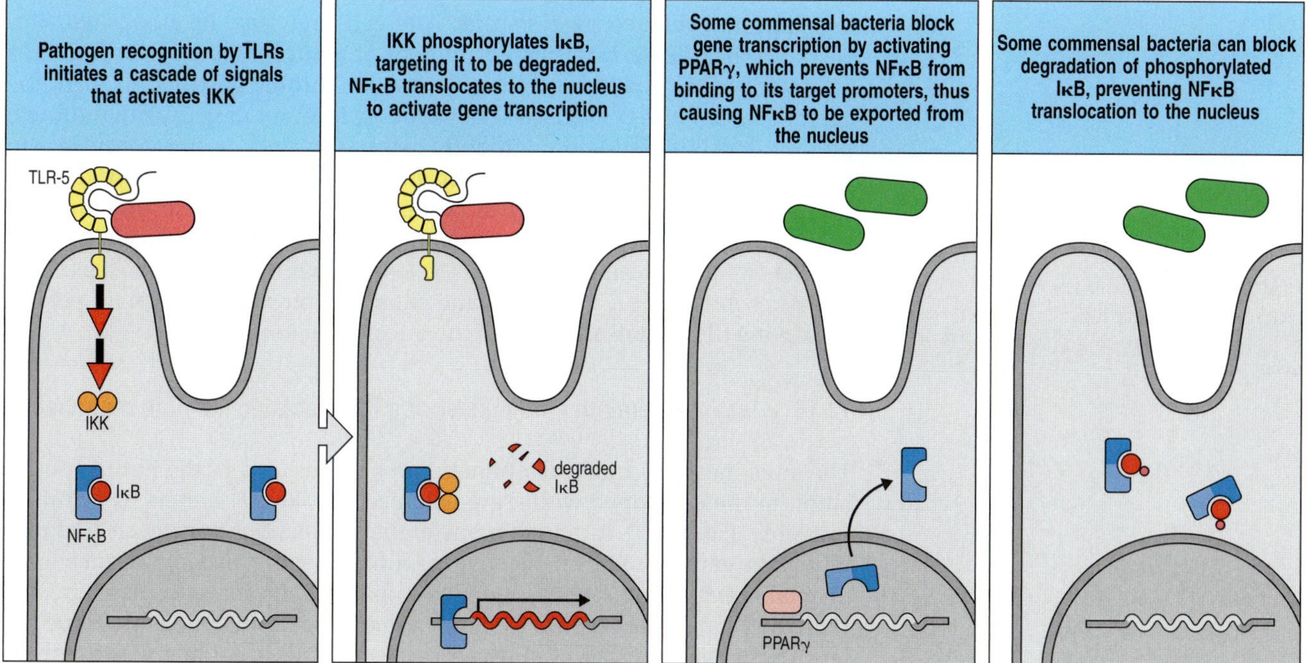

Fig. 12.26 Commensal bacteria can prevent inflammatory responses in the intestine. The pro-inflammatory transcription factor NFκB pathway is activated in epithelial cells via the ligation of TLRs by pathogens (first two panels). Commensal bacteria have been found to inhibit this pathway and thus prevent inflammation. One way is by activation of the nuclear receptor PPARγ, leading to the export of NFκB from the nucleus (third panel). Another is by blocking the degradation of the inhibitor IκB and thus retaining NFκB in the cytoplasm (fourth panel).

inflammatory T-cell responses are also generated in the mucosa, leading to severe intestinal damage. IL-23 has a particularly important role in this process, probably mostly via its ability to drive the differentiation of T_H17 effector cells, which produce IL-17, IL-21, and IL-22. IL-23 can also assist inflammatory T_H1 responses in the intestine. These experimental results are consistent with clinical evidence for a linkage between polymorphisms in the IL-23 receptor and Crohn's disease in humans. TNF-α is also heavily involved as a final common mediator, and antibody-mediated neutralization of this cytokine is effective in treating Crohn's disease (see Fig. 12.24, right panels). In all cases, these disorders are entirely dependent on the presence of commensal bacteria, because they can be prevented by treatment with antibiotics and do not occur in germ-free animals. It is not known whether all, or only some, commensal species can provoke inflammation.

An excellent example of an inappropriate immune response against a bacterium is the gastritis associated with chronic infection by *Helicobacter pylori*. Although, strictly speaking, this is a pathogenic bacterium, it has coevolved with humans over millennia and is present in the stomachs of a large proportion of the world's population. In developing countries, more than 90% of adults may be infected for most of their lives, but even in North America and Europe up to 50% of adults are infected. *H. pylori* achieves this in the face of an ongoing host immune response, making it one of the most successful persistent infections of humans. Curiously for such a successful inhabitant of humans, *H. pylori* induces a chronic inflammatory reaction characterized by high levels of cytokines and chemokines such as IL-1, IL-6, IL-8, IL-12, IL-23, and TNF-α. Production of these cytokines is stimulated by the bacterium's virulence factors, including cytotoxin-associated gene A (CagA) and vacuolating cytotoxin A (VacA). It is unclear whether the host obtains any benefit from the persistent colonization with *H. pylori*; indeed, the chronic infection causes gastritis, and also duodenal and gastric ulcers, and is one of the leading causes of stomach cancer in the world. It also predisposes to the development of lymphoma in mucosa-associated lymphoid tissues. Although studies in animals indicate that it is possible to induce sterilizing immunity against *H. pylori*, persistently infected humans clearly fail to induce the appropriate protective response, allowing chronic stimulation of the immune system.

The exact nature of the protective and inflammatory responses against *H. pylori* is not yet clear. Innate responses such as mucus production are key. Current evidence suggests that T_H17 cells may also be associated with clearance of *H. pylori* via the recruitment of neutrophils, whereas T_H1 cells producing IFN-γ may be responsible for the chronic gastritis; in turn, the gastritis favors bacterial survival by reducing levels of gastric acid. Although present, local and systemic antibody responses seem not to contribute to either form of immunity. *H. pylori* also actively manipulates the host immune response to favor its own survival. Not only does it employ several immune evasion strategies, but it also drives the development of induced T_{reg} cells *in vivo* by mechanisms that are as yet unknown. The T_{reg} cells selectively inhibit protective T_H17 responses and allow chronic inflammation to persist, preserving the organism's niche in the infected host.

12-18 Intestinal helminths provoke strong T_H2-mediated immune responses.

The intestines of virtually all animals and humans, except the human population of the developed world, are colonized by large numbers of helminth parasites (Fig. 12.27). Although many of these infections may be cleared rapidly by the generation of an effective response, they are also important causes of debilitating and chronic disease in humans and animals. In these circumstances, the parasite persists for long periods apparently undisturbed by the host's attempts to expel it, and causes disease by competing with the host

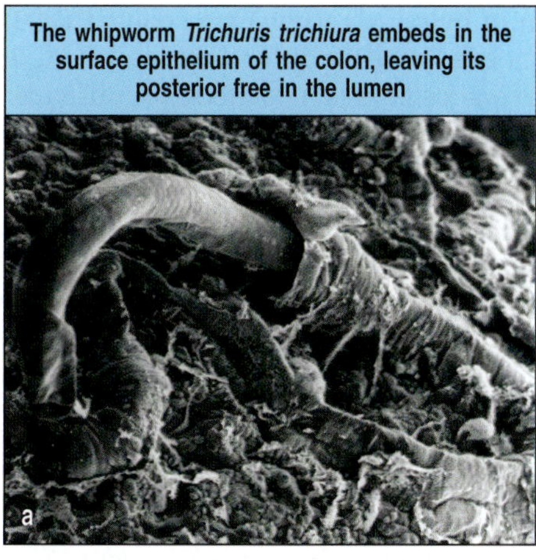

The whipworm Trichuris trichiura embeds in the surface epithelium of the colon, leaving its posterior free in the lumen
a

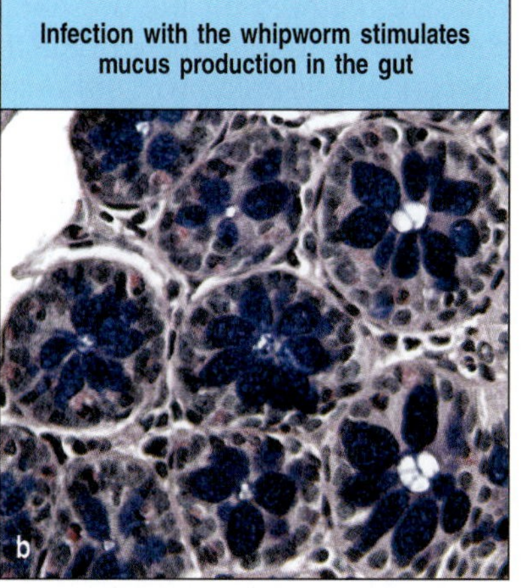

Infection with the whipworm stimulates mucus production in the gut
b

Fig. 12.27 Intestinal helminth infection. Panel a: the whipworm *Trichuris trichiura* is a helminth parasite that lives partly embedded in intestinal epithelial cells. This scanning electron micrograph of mouse colon shows the head of the parasite buried in an epithelial cell and its posterior lying free in the lumen. Panel b: a cross-section of crypts from the colon of a mouse infected with *T. trichiura* shows the markedly increased production of mucus by goblet cells in the intestinal epithelium. The mucus is seen as large droplets in vesicles inside the goblet cells and stains dark blue with periodic acid–Schiff reagent. Magnification ×400.

for nutrients, or by causing local damage to epithelial cells or blood vessels. In addition, the host immune response against these parasites can produce many harmful effects.

The exact nature of the host–pathogen interaction in helminth infections depends very much on the type of parasite involved. Some remain within the gut lumen, whereas others invade and colonize epithelial cells; others invade beyond the intestine and spend part of their life cycle in other tissues, such as liver, lung, or muscle; some are found only the small intestine, whereas others inhabit the large intestine. In virtually all cases, the protective immune response is generated by CD4 T_H2 cells, which, among other things, induce B cells to switch to the IgE isotype (Fig. 12.28), whereas a T_H1 response does not clear the pathogen and tends to produce an inflammatory reaction that damages the mucosa (Fig. 12.29, last two panels). A T_H1 response is generated when the activating dendritic cells express IL-12.

The T_H2 response is polarized by worm products acting on a variety of different cells such as basophils, dendritic cells, and epithelial cells. Dendritic

Fig. 12.28 Protective responses to intestinal helminths. Most intestinal helminths induce both protective and pathological immune responses by CD4 T cells. T_H2 responses tend to be protective, creating an unfriendly environment for the parasite, and leading to its expulsion and the generation of protective immunity (see the text for details). M2 macrophage, alternatively activated macrophage.

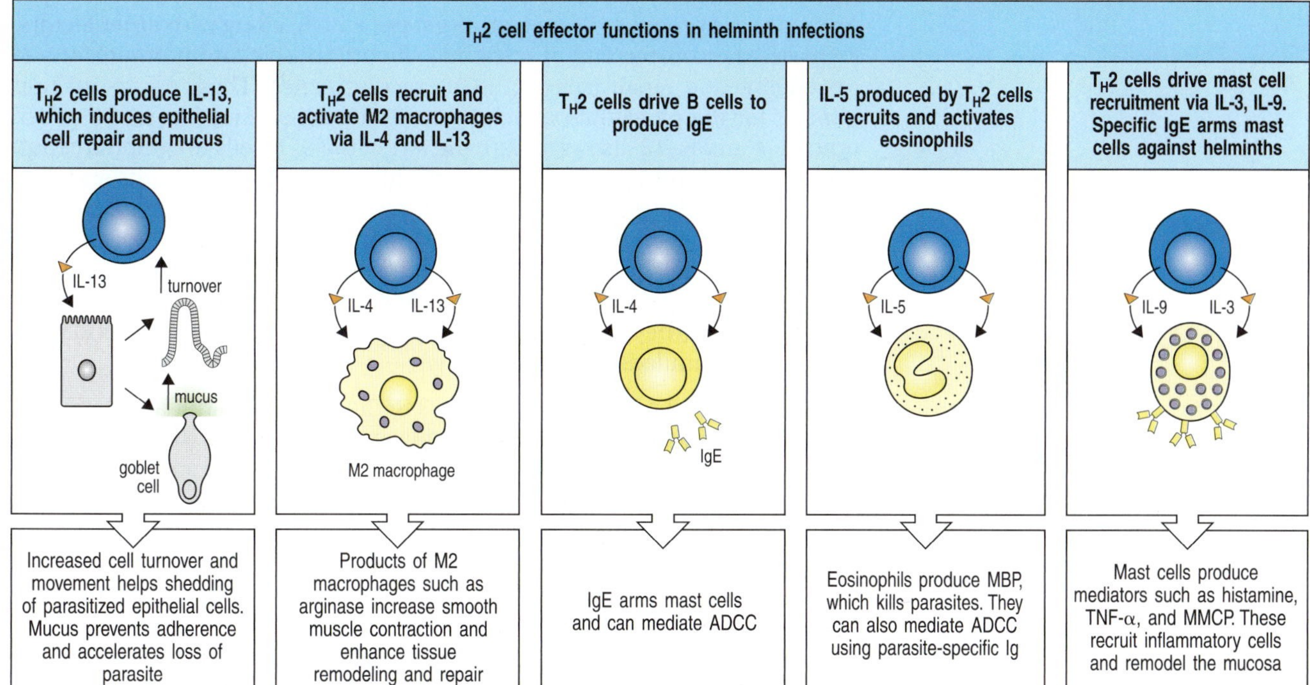

T_H2 cell effector functions in helminth infections
T_H2 cells produce IL-13, which induces epithelial cell repair and mucus
IL-13
turnover
mucus
goblet cell
Increased cell turnover and movement helps shedding of parasitized epithelial cells. Mucus prevents adherence and accelerates loss of parasite
T_H2 cells recruit and activate M2 macrophages via IL-4 and IL-13
IL-4
IL-13
M2 macrophage
Products of M2 macrophages such as arginase increase smooth muscle contraction and enhance tissue remodeling and repair
T_H2 cells drive B cells to produce IgE
IL-4
IgE
IgE arms mast cells and can mediate ADCC
IL-5 produced by T_H2 cells recruits and activates eosinophils
IL-5
Eosinophils produce MBP, which kills parasites. They can also mediate ADCC using parasite-specific Ig
T_H2 cells drive mast cell recruitment via IL-3, IL-9. Specific IgE arms mast cells against helminths
IL-9
IL-3
Mast cells produce mediators such as histamine, TNF-α, and MMCP. These recruit inflammatory cells and remodel the mucosa

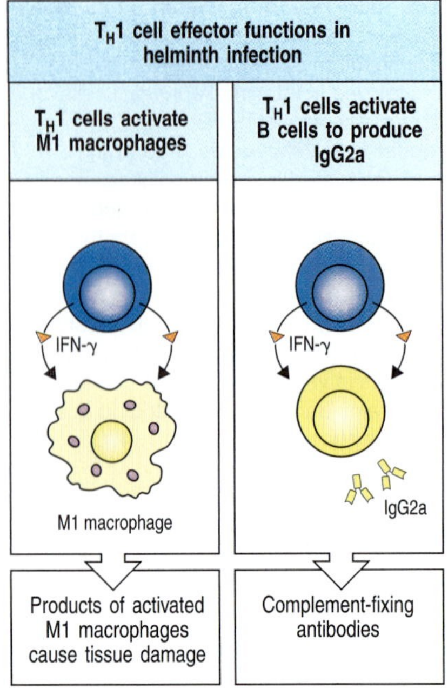
T_H1 cell effector functions in helminth infection
T_H1 cells activate M1 macrophages
T_H1 cells activate B cells to produce IgG2a
IFN-γ
IFN-γ
M1 macrophage
IgG2a
Products of activated M1 macrophages cause tissue damage
Complement-fixing antibodies

Fig. 12.29 Pathological responses to intestinal helminths. If the CD4 T-cell response to a helminth parasite is polarized to produce predominantly T_H1 effector T cells (for example, by the production of IL-12 by dendritic cells), it does not clear the pathogen (see the text for details). If not balanced by a protective T_H2 response (see Fig. 12.28), the T_H1 response leads to persistent infection and chronic intestinal pathology. M1 macrophage, classically activated macrophage.

cells are required for the presentation of helminth antigens to naive CD4 T cells, but the initial source of the IL-4 needed to drive differentiation into T_H2 effector cells is unknown. Although several cell types have been proposed as the source, including iNKT cells, mast cells, and basophils, none of these has been proved to be essential. The helminth-induced factors that drive the differentiation of naive CD4 T cells into T_H2 cells are also still a matter of debate, but seem to include the cytokines IL-25, perhaps produced by basophils, and IL-33 and TSLP produced by epithelial cells and other nonhematopoietic cells. These may act directly on the differentiating T cells, or indirectly via cells such as mast cells. Worm products and IL-25 further favor T_H2 responses by preventing IL-12 production and the generation of T_H1 cells.

The typical mucosal response to helminths shows high levels of IgE antibody, the activation of mast cells and eosinophils (see Section 10-24), activated macrophages, and tissue remodeling (see Fig. 12.28), although the precise involvement of each response depends on the type of parasite. The antihelminth response is mediated by cytokines produced by T_H2 cells. IL-13 directly enhances the production of mucus by goblet cells and increases the migration and turnover of epithelial cells (see Fig. 12.28, first panel). This is a critical component of the host response to parasitic worms. It helps to eliminate parasites that have attached to the epithelium and decreases the surface area available for colonization. Turnover increases partly because epithelial cells in the crypt respond to the loss of damaged cells from the surface epithelium by dividing more rapidly in an attempt to repair the damage. Increased epithelial cell turnover is also a direct and specific effect of the IL-13 produced by T cells, NK cells, and iNKT cells in the presence of infection. Although it makes life difficult for the parasite, increased epithelial turnover also compromises intestinal function, because the newly produced epithelial cells are immature and so are deficient in absorptive and digestive activity. Thus, the host immune response in intestinal helminths has to tread an extremely fine line, because the most efficient aspects of the protective response are also likely to be deleterious in the local environment.

IL-4 and IL-13 produced by T_H2 cells result in the differentiation of **alternatively activated macrophages** (also called **M2-type macrophages**). Unlike conventional pro-inflammatory macrophages (**M1-type macrophages**), which tend to differentiate after interaction with T_H1 cells (see Fig. 12.29), M2-type macrophages produce the enzyme arginase, which, along with other factors, increases the contractility of intestinal smooth muscle and promotes tissue remodeling and repair (see Fig. 12.28, second panel). These effects make it difficult for the parasite to attach to the epithelial surface for any length of time. IL-4 and IL-13 also stimulate the switching of B cells to IgE production (see Fig. 12.28, third panel). The IL-5 produced by T_H2 cells recruits and activates eosinophils (see Fig. 12.28, fourth panel), which can have direct toxic effects on pathogens by releasing cytotoxic molecules such as major basic protein (MBP). Eosinophils bear Fc receptors for IgG and can show antibody-dependent cell-mediated cytotoxicity (ADCC) against IgG-coated parasites (see Fig. 10.23). They also express the Fcα receptor (CD89) and degranulate in response to stimulation by secretory IgA.

The IL-3 and IL-9 produced by T_H2 cells in the mucosa recruit and activate a specialized population of mast cells known as **mucosal mast cells** (see Fig. 12.28, fifth panel). The innate cytokines IL-25 and IL-33 also activate mucosal mast cells early in a response to helminths. Mucosal mast cells differ from their counterparts in other tissues by having only small numbers of IgE receptors and producing very little histamine. When activated by cytokines, or by the binding of antigen to receptor-bound IgE, mucosal mast cells produce large amounts of preformed inflammatory mediators such as prostaglandins, leukotrienes, and several proteases, including the mucosal mast cell protease (MMCP-1) (see Chapter 10). This enzyme can remodel intestinal mucosal tissues by digesting the basement membrane between the epithelium and the lamina propria, and may also have direct effects on parasites. Together, the mast-cell-derived mediators increase vascular permeability, induce leukocyte recruitment, increase intestinal motility, and stimulate the production of mucus by goblet cells, all of which help to create a hostile microenvironment for the parasite. Mast cells also produce large quantities of TNF-α, which may assist in killing parasites and infected epithelial cells. TNF-α is, however, also an important cause of the inflammation and intestinal damage that occurs in such infections.

Many intestinal helminths are the ultimately adapted chronic infectious agents, having evolved sophisticated methods of persisting for long periods in the face of an ongoing immune response. They modulate the host immune response in several ways, including the production of mediators that dampen the innate inflammatory response, and the expression of decoy receptors for inflammatory cytokines and chemokines. In addition, several molecules secreted by helminths modify the differentiation of T cells, selectively encouraging the generation of T_{reg} cells at the expense of effector cells. TGF-β induced by IL-13 is crucial to this, suppressing many inflammatory responses and favoring the development of IL-10-producing T_{reg} cells. The overall effect is to limit the production and inflammatory potential of cytokines such as IFN-γ and TNF-α and to modulate effector T_H2 responses, which produces a state of persistent infection in which damage to the host is limited. These opposing immunological processes operate simultaneously in many parasitic infections, rather as we saw in the response to commensal bacteria but to a more exaggerated extent. This can result in an intestine that seems badly inflamed but can retain some physiological function, despite harboring large numbers of live, multicellular parasites.

12-19 Other eukaryotic parasites provoke protective immunity and pathology in the gut.

In addition to multicellular worms, the intestinal immune system has to contend with a variety of unicellular eukaryotic parasites, including protozoans such as *Giardia lamblia*, *Cryptosporidium parvum*, *Entamoeba histolytica*, and *Toxoplasma gondii*. *Giardia lamblia* is a widespread non-invasive waterborne microorganism that is an important cause of intestinal inflammation. Protective immunity to *G. lamblia* is associated with the production of local antibodies and infiltration of the mucosa by effector T cells, including IELs, but immunity can be inefficient, leading to chronic disease. *Cryptosporidium parvum* and *T. gondii* are usually opportunistic infections, being most commonly found in people with immune deficiencies such as AIDS. They are intracellular pathogens that require both CD4 T_H1 cells and CD8 T cells to clear them. Chronic infection is associated with marked pathology caused by the overproduction of IFN-γ and TNF-α by T cells and macrophages, respectively.

12-20 The mucosal immune system has to compromise between suppression and activation of an immune response.

The immune system in the normal intestine and other mucosal surfaces is biased to avoid making active immune responses against the majority of antigens encountered (see Section 12-15). But the antigens are still recognized, and potent protective immune responses must be, and are, generated against pathogens when required, as we saw in preceding sections. How can these apparently opposing needs be met without compromising the health of the host? The answer seems to be in the interactions between local dendritic cells and factors in the mucosal microenvironment (see Fig. 12.25).

These interactions normally result in a population of noninflammatory dendritic cells, as discussed earlier in this chapter (see Section 12-8). Fortunately for our health, however, this predominantly inhibitory microenvironment can be changed by the presence of invasive pathogens or of adjuvants, allowing dendritic cells to be fully activated and productive immunity to be induced when required (see Fig. 12.25). The ability of mucosal dendritic cells to change their behavior rapidly and with high sensitivity probably reflects the fact that, even in the absence of overt infection, both the inflammatory and regulatory components of the immune response are likely to be operating simultaneously in the mucosa. The term physiological inflammation is used to describe the appearance of the normal intestine, which contains large numbers of lymphocytes and other cells that are normally associated with chronic inflammation and are generally not present in other organs in the absence of disease. This 'inflammation' is driven mainly by the presence of commensal bacteria and to a lesser extent by food antigens, and it is essential for the normal function of both the intestine and the mucosal immune system. It probably also ensures that dendritic cells are always in a state of high readiness to respond appropriately to changes in their local environment.

In addition to combating infection, these regulatory interactions may have had a wider influence on the evolution of the gut and the immune system, being one of the factors underlying the **hygiene hypothesis** (see Section 14-4). According to this idea, the human immune system has evolved in the face of continued exposure to ubiquitous pathogens and commensals, whose immunomodulatory products have helped to condition the polarization of responses to other foreign antigens. With the increasing cleanliness of the human environment, our immune system is no longer exposed to this influence during the critical early period of life, allowing hypersensitivity reactions of all kinds to develop unchecked against self antigens and harmless environmental substances. As the main source of exposure to environmental microbes, the intestine is heavily involved in these processes. In particular, there is clear evidence that the increasing incidence of disorders such as type 1 diabetes, Crohn's disease, and atopy correlates specifically with the eradication of immunomodulatory organisms such as helminths and *H. pylori* from the Western world.

Summary.

The immune system in the mucosa has to distinguish between potential pathogens and harmless antigens, generating strong effector responses to pathogens but remaining unresponsive to foods and commensals. Pathogenic microorganisms such as enteric bacteria use several strategies to invade, often exploiting the host's antigen-uptake and inflammatory mechanisms as well as modulating different components of the immune response. The strong T_H1 or T_H17 reactions they provoke normally result in elimination of the infection. In contrast, food proteins induce an active form of immunological tolerance in the systemic and mucosal immune systems, which may be mediated by regulatory T cells producing IL-10 and/or TGF-β. Commensal bacteria are also recognized by the immune system, but this is limited to the mucosa and its draining lymphoid tissues, because they are presented to T cells by dendritic cells that migrate from the intestinal wall and lodge in the draining mesenteric lymph node. This results in active mucosal tolerance and the production of local IgA antibodies that restrict colonization by the microorganisms, but 'ignorance' of these antigens by the systemic immune system. Because commensal bacteria have many beneficial effects for the host, these immunoregulatory processes are important in allowing the bacteria to coexist with the immune system.

Intestinal helminths frequently produce chronic infections, partly because they produce several factors that can modulate the host immune system. The dominant protective response against helminths is T_H2 mediated, with the involvement of IgE, mast cells, and eosinophils and the production of TNF-α. Such a response can also damage the intestine, and the immune system maintains a balance between protective immunity and immunopathology. A similar situation occurs with *H. pylori* infection of the stomach, where this bacterium manipulates the host response to ensure the presence of a chronic inflammatory state that promotes its growth and prevents protective immunity. The absence of helminth- and *Helicobacter*-derived immunomodulatory factors may contribute to the increased incidence of allergic and inflammatory diseases in developed countries.

The key factor deciding between the generation of protective immunity and immune tolerance in the gut mucosa is local dendritic cells. Normally, intestinal dendritic cells do not respond to pro-inflammatory stimuli and will polarize the T-cell response toward the differentiation of gut-homing regulatory T cells. Nevertheless, the dendritic cells can still respond fully to invading organisms and inflammatory signals when required, allowing the priming of T cells to effector status. When the normal regulatory processes break down, inflammatory diseases can occur. As a consequence of these competing, but interacting, needs of the immune response, the intestine normally has the appearance of physiological inflammation, which helps maintain normal function of the gut and immune system.

Summary to Chapter 12.

The mucosal immune system is a large and complex apparatus that has a crucial role in health, not just by protecting physiologically vital organs but also by helping to regulate the tone of the entire immune system and prevent disease. The peripheral lymphoid organs focused on by most immunologists may be a recent specialization of an original template that evolved in mucosal tissues. The mucosal surfaces of the body are highly vulnerable to infection and possess a complex array of innate and adaptive mechanisms of immunity. The adaptive immune system of the mucosa-associated lymphoid tissues differs from that of the rest of the peripheral lymphoid system in several respects: the immediate juxtaposition of mucosal epithelium and lymphoid tissue; diffuse lymphoid tissue as well as more organized lymphoid organs; specialized antigen uptake mechanisms and dendritic cells; the predominance of activated/memory lymphocytes even in the absence of infection; the production of dimeric secretory IgA as the predominant antibody; and the downregulation of immune responses to innocuous antigens such as food antigens and commensal microorganisms. No systemic immune response can normally be detected to these antigens. In contrast, pathogenic microorganisms induce strong protective responses. The key factor in the decision between tolerance and the development of powerful adaptive immune responses is the context in which antigen is presented to T lymphocytes in the mucosal immune system. When there is no inflammation, presentation of antigen to T cells by specialized CD103-positive dendritic cells induces the differentiation of regulatory T cells. By contrast, pathogenic microorganisms crossing the mucosa induce an inflammatory response in the tissues, which stimulates the maturation of antigen-presenting cells and their expression of co-stimulatory molecules, thus favoring a protective T-cell response.

Questions.

12.1 *Describe the processes that allow a specific CD4 T cell to be primed against antigen in the intestine, and discuss how the resulting effector T cells can return to the intestinal surface.*

12.2 *We are exposed to foreign antigens in large quantities in the food that we eat. (a) Why do we not mount effective immune responses against these food antigens? (b) How does the immune system distinguish between food antigens and antigens that are potentially harmful?*

12.3 *Discuss how dendritic cells in the intestine are distinct from their counterparts in other tissues, and describe how their functions are adapted to the local challenges faced by the immune system in the gut.*

12.4 *Describe how different aspects of the host immune response may produce either protective immunity or tissue damage during infection by an intestinal worm.*

12.5 *An unusual aspect of the intestine is the close interaction between epithelial cells and cells of the immune system. Outline how these interactions help shape immune responses to the different kinds of antigen encountered by the gut.*

General references.

Brandtzaeg, P.: **Mucosal immunity: induction, dissemination, and effector functions.** *Scand. J. Immunol.* 2009, **70**:505–515.

Kohlmeier, J.E., and Woodland, D.L.: **Immunity to respiratory viruses.** *Annu. Rev. Immunol.* 2009, **27**:61–82.

Macpherson, A.J., McCoy, K.D., Johansen, F.E., and Brandtzaeg, P.: **The immune geography of IgA induction and function.** *Mucosal Immunol.* 2008, **1**:11–22.

Mowat, A.M.: **Anatomical basis of tolerance and immunity to intestinal antigens.** *Nat. Rev. Immunol.* 2003, **3**:331–341.

Lee, Y.K., and Mazmanian, S.K.: **Has the microbiota played a critical role in the evolution of the adaptive immune system?** *Science* 2010, **330**:1768–1773.

Section references.

12-1 The mucosal immune system protects the internal surfaces of the body.

Brandtzaeg, P.: **Induction of secretory immunity and memory at mucosal surfaces.** *Vaccine* 2007, **25**:5467–5484.

Kunisawa, J., Nochi, T., and Kiyono, H.: **Immunological commonalities and distinctions between airway and digestive immunity.** *Trends Immunol.* 2008, **29**:505–513.

Wira, C.R., Fahey, J.V., Sentman, C.L., Pioli, P.A., and Shen, L.: **Innate and adaptive immunity in female genital tract: cellular responses and interactions.** *Immunol. Rev.* 2005, **206**:306–335.

World Health Organization: *The Global Burden of Disease 2004 Update.* Geneva, World Health Organization, 2008.

12-2 The mucosal immune system may be the original vertebrate immune system.

Fagarasan, S., Kawamoto, S., Kanagawa, O., and Suzuki, K.: **Adaptive immune regulation in the gut: T cell-dependent and T cell-independent IgA synthesis.** *Annu. Rev. Immunol.* 2010, **28**:243–273.

Matsunaga, T., and Rahman, A.: **In search of the origin of the thymus: the thymus and GALT may be evolutionarily related.** *Scand. J. Immunol.* 2001, **53**:1–6.

Rocha, B.: **The extrathymic T-cell differentiation in the murine gut.** *Immunol. Rev.* 2007, **215**:166–177.

12-3 Cells of the mucosal immune system are located both in anatomically defined compartments and scattered throughout mucosal tissues.

Bouskra, D., Brézillon, C., Bérard, M., Werts, C., Varona, R., Boneca, I.G., and Eberl, G.: **Lymphoid tissue genesis induced by commensals through NOD1 regulates intestinal homeostasis.** *Nature* 2008, **456**:507–510.

Brandtzaeg, P., Kiyono, H., Pabst, R., and Russell, M.W.: **Terminology: nomenclature of mucosa-associated lymphoid tissue.** *Mucosal Immunol.* 2008, **1**:31–37.

Eberl, G., and Lochner, M.: **The development of intestinal lymphoid tissues at the interface of self and microbiota.** *Mucosal Immunol.* 2009, **2**:478–485.

Eberl, G., and Sawa, S.: **Opening the crypt: current facts and hypotheses on the function of cryptopatches.** *Trends Immunol.* 2010, **31**:50–55.

Tsuji, M., Suzuki, K., Kitamura, H., Maruya, M., Kinoshita, K., Ivanov, I.I., Itoh, K., Littman, D.R., and Fagarasan, S.: **Requirement for lymphoid tissue-inducer cells in isolated follicle formation and T cell-independent immunoglobulin A generation in the gut.** *Immunity* 2008, **29**:261–271.

van de Pavert, S.A., and Mebius, R.E.: **New insights into the development of lymphoid tissues.** *Nat. Rev. Immunol.* 2010, **10**:664–674.

12-4 The intestine has distinctive routes and mechanisms of antigen uptake.

Anosova, N.G., Chabot, S., Shreedhar, V., Borawski, J.A., Dickinson, B.L., and Neutra, M.R.: **Cholera toxin, *E. coli* heat-labile toxin, and non-toxic derivatives induce dendritic cell migration into the follicle-associated epithelium of Peyer's patches.** *Mucosal Immunol.* 2008, **1**:59–67.

Chieppa, M., Rescigno, M., Huang, A.Y., and Germain, R.N.: **Dynamic imaging of dendritic cell extension into the small bowel lumen in response to epithelial cell TLR engagement.** *J. Exp. Med.* 2006, **203**:2841–2852.

Hase, K., Kawano, K., Nochi, T., Pontes, G.S., Fukuda, S., Ebisawa, M., Kadokura, K., Tobe, T., Fujimura, Y., Kawano, S., *et al.*: **Uptake through glycoprotein 2 of FimH+ bacteria by M cells initiates mucosal immune response.** *Nature* 2009, **462**:226–230.

Knoop, K.A., Kumar, N., Butler, B.R., Sakthivel, S.K., Taylor, R.T., Nochi, T., Akiba, H., Yagita, H., Kiyono, H., and Williams, I.R.: **RANKL is necessary and sufficient to initiate development of antigen-sampling M cells in the intestinal epithelium.** *J. Immunol.* 2009, **183**:5738–5747.

Lelouard, H., Henri, S., De Bovis, B., Mugnier, B., Chollat-Namy, A., Malissen, B., Méresse, S., and Gorvel, J.P.: **Pathogenic bacteria and dead cells are internalized by a unique subset of Peyer's patch dendritic cells that express lysozyme.** *Gastroenterology* 2010, **138**:173–184.

Salazar-Gonzalez, R.M., Niess, J.H., Zammit, D.J., Ravindran, R., Srinivasan, A., Maxwell, J.R., Stoklasek, T., Yadav, R., Williams, I.R., Gu, X., *et al.*: **CCR6-mediated dendritic cell activation of pathogen-specific T cells in Peyer's patches.** *Immunity* 2006, **24**:623–632.

12-5 The mucosal immune system contains large numbers of effector lymphocytes even in the absence of disease.

Gaboriau-Routhiau, V., Rakotobe, S., Lécuyer, E., Mulder, I., Lan, A., Bridonneau, C., Rochet, V., Pisi, A., De Paepe, M., Brandi, G., *et al.*: **The key role of segmented filamentous bacteria in the coordinated maturation of gut helper T cell responses.** *Immunity* 2009, **31**:677–689.

Macdonald, T.T., and Monteleone, G.: **Immunity, inflammation, and allergy in the gut.** *Science* 2005, **307**:1920–1925.

Maynard, C.L., and Weaver, C.T.: **Intestinal effector T cells in health and disease.** *Immunity* 2009, **31**:389–400.

Niess, J.H., Leithäuser, F., Adler, G., and Reimann, J.: **Commensal gut flora drives the expansion of proinflammatory CD4 T cells in the colonic lamina propria under normal and inflammatory conditions.** *J. Immunol.* 2008, **180**:559–568.

12-6 The circulation of lymphocytes within the mucosal immune system is controlled by tissue-specific adhesion molecules and chemokine receptors.

Agace, W.: **Generation of gut-homing T cells and their localization to the small intestinal mucosa.** *Immunol. Lett.* 2010, **128**:21–23.

Brandtzaeg, P.: **Mucosal immunity: induction, dissemination, and effector functions.** *Scand. J. Immunol.* 2009, **70**:505–515.

Hammerschmidt, S.I., Ahrendt, M., Bode, U., Wahl, B., Kremmer, E., Förster, R., and Pabst, O.: **Stromal mesenteric lymph node cells are essential for the generation of gut-homing T cells *in vivo*.** *J. Exp. Med.* 2008, **205**:2483–2490.

Iwata, M., Hirakiyama, A., Eshima, Y., Kagechika, H., Kato, C., and Song, S.Y.: **Retinoic acid imprints gut-homing specificity on T cells.** *Immunity* 2004, **21**:527–538.

Mora, J.R., and von Andrian, U.H.: **Differentiation and homing of IgA-secreting cells.** *Mucosal Immunol.* 2008, **1**:96–109.

Morteau, O., Gerard, C., Lu, B., Ghiran, S., Rits, M., Fujiwara, Y., Law, Y., Distelhorst, K., Nielsen, E.M., Hill, E.D., *et al.*: **An indispensable role for the chemokine receptor CCR10 in IgA antibody-secreting cell accumulation.** *J. Immunol.* 2008, **181**:6309–6315.

12-7 Priming of lymphocytes in one mucosal tissue can induce protective immunity at other mucosal surfaces.

Brandtzaeg, P.: **Induction of secretory immunity and memory at mucosal surfaces.** *Vaccine* 2007, **25**:5467–5484.

Johansen, F.E., Baekkevold, E.S., Carlsen, H.S., Farstad, I.N., Soler, D., and Brandtzaeg, P.: **Regional induction of adhesion molecules and chemokine receptors explains disparate homing of human B cells to systemic and mucosal effector sites: dispersion from tonsils.** *Blood* 2005, **106**:593–600.

12-8 Unique populations of dendritic cells control mucosal immune responses.

Annacker, O., Coombes, J.L., Malmstrom, V., Uhlig, H.H., Bourne, T., Johansson-Lindbom, B., Agace, W.W., Parker, C.M., and Powrie, F.: **Essential role for CD103 in the T cell-mediated regulation of experimental colitis.** *J. Exp. Med.* 2005, **202**:1051–1061.

Iwasaki, A.: **Mucosal dendritic cells.** *Annu. Rev. Immunol.* 2007, **25**:381–418.

Lambrecht, B.N., and Hammad, H.: **Biology of lung dendritic cells at the origin of asthma.** *Immunity* 2009, **31**:412–424.

Milling, S.W., Yrlid, U., Jenkins, C., Richards, C.M., Williams, N.A., and MacPherson, G.: **Regulation of intestinal immunity: effects of the oral adjuvant *Escherichia coli* heat-labile enterotoxin on migrating dendritic cells.** *Eur. J. Immunol.* 2007, **37**:87–99.

Persson, E.K., Jaensson, E., and Agace, W.W.: **The diverse ontogeny and function of murine small intestinal dendritic cell/macrophage subsets.** *Immunobiology* 2010, **215**:692–697.

Rescigno, M., and Di Sabatino, A.: **Dendritic cells in intestinal homeostasis and disease.** *J. Clin. Invest.* 2009, **119**:2441–2450.

Sato, A., Hashiguchi, M., Toda, E., Iwasaki, A., Hachimura, S., and Kaminogawa, S.: **CD11b+ Peyer's patch dendritic cells secrete IL-6 and induce IgA secretion from naive B cells.** *J. Immunol.* 2003, **171**:3684–3690.

Schulz, O., Jaensson, E., Persson, E.K., Liu, X., Worbs, T., Agace, W.W., and Pabst, O.: **Intestinal CD103+, but not CX3CR1+, antigen sampling cells migrate in lymph and serve classical dendritic cell functions.** *J. Exp. Med.* 2009, **206**:3101–3114.

Uematsu, S., Fujimoto, K., Jang, M.H., Yang, B.G., Jung, Y.J., Nishiyama, M., Sato, S., Tsujimura, T., Yamamoto, M., Yokota, Y., *et al.*: **Regulation of humoral and cellular gut immunity by lamina propria dendritic cells expressing Toll-like receptor 5.** *Nat. Immunol.* 2008, **9**:769–776.

12-9 The intestinal lamina propria contains antigen-experienced T cells and populations of unusual innate-type lymphocytes.

Buonocore, S., Ahern, P.P., Uhlig, H.H., Ivanov, I.I., Littman, D.R., Maloy, K.J., and Powrie, F.: **Innate lymphoid cells drive interleukin-23-dependent innate intestinal pathology.** *Nature* 2010, **464**:1371–1375.

Colonna, M.: **Interleukin-22-producing natural killer cells and lymphoid tissue inducer-like cells in mucosal immunity.** *Immunity* 2009, **31**:15–23.

Cua, D.J., and Tato, C.M.: **Innate IL-17-producing cells: the sentinels of the immune system.** *Nat. Rev. Immunol.* 2010, **10**:479–489.

Khader, S.A., Gaffen, S.L., and Kolls, J.K.: **Th17 cells at the crossroads of innate and adaptive immunity against infectious diseases at the mucosa.** *Mucosal Immunol.* 2009, **2**:403–411.

Maynard, C.L., and Weaver, C.T.: **Intestinal effector T cells in health and disease.** *Immunity* 2009, **31**:389–400.

Meresse, B., and Cerf-Bensussan, N.: **Innate T cell responses in human gut.** *Semin. Immunol.* 2009, **21**:121–129.

Middendorp, S., and Nieuwenhuis, E.E.: **NKT cells in mucosal immunity.** *Mucosal Immunol.* 2009, **2**:393–402.

Sawa, S., Cherrier, M., Lochner, M., Satoh-Takayama, N., Fehling, H.J., Langa,

F., Di Santo, J.P., and Eberl, G.: **Lineage relationship analysis of RORγt$^+$ innate lymphoid cells.** *Science* 2010, **330**:665–669.

van Wijk, F., and Cheroutre, H.: **Intestinal T cells: facing the mucosal immune dilemma with synergy and diversity.** *Semin. Immunol.* 2009, **21**:130–138.

12-10 The intestinal epithelium is a unique compartment of the immune system.

Eberl, G., and Littman, D.R.: **Thymic origin of intestinal $\alpha\beta$ T cells revealed by fate mapping of RORγt$^+$ cells.** *Science* 2004, **305**:248–251.

Eberl, G., and Sawa, S.: **Opening the crypt: current facts and hypotheses on the function of cryptopatches.** *Trends Immunol.* 2010, **31**:50–55.

Rocha, B.: **The extrathymic T-cell differentiation in the murine gut.** *Immunol Rev* 2007, **215**:166–177.

Staton, T.L., Habtezion, A., Winslow, M.M., Sato, T., Love, P.E., and Butcher, E.C.: **CD8$^+$ recent thymic emigrants home to and efficiently repopulate the small intestine epithelium.** *Nat. Immunol.* 2006, **7**:482–488.

van Wijk, F., and Cheroutre, H.: **Intestinal T cells: facing the mucosal immune dilemma with synergy and diversity.** *Semin. Immunol.* 2009, **21**:130–138.

12-11 Secretory IgA is the class of antibody associated with the mucosal immune system.

Cerutti, A.: **Immunology. IgA changes the rules of memory.** *Science* 2010, **328**:1646–1647.

Cerutti, A., and Rescigno, M.: **The biology of intestinal immunoglobulin A responses.** *Immunity* 2008, **28**:740–750.

Fagarasan, S., Kawamoto, S., Kanagawa, O., and Suzuki, K.: **Adaptive immune regulation in the gut: T cell-dependent and T cell-independent IgA synthesis.** *Annu. Rev. Immunol.* 2010, **28**:243–273.

Spencer, J., Barone, F., and Dunn-Walters, D.: **Generation of Immunoglobulin diversity in human gut-associated lymphoid tissue.** *Semin. Immunol.* 2009, **21**:139–146.

12-12 IgA deficiency is common in humans but may be compensated for by secretory IgM.

Karlsson, M.R., Johansen, F.E., Kahu, H., Macpherson, A., and Brandtzaeg, P.: **Hypersensitivity and oral tolerance in the absence of a secretory immune system.** *Allergy* 2010, **65**:561–570.

Yel, L.: **Selective IgA deficiency.** *J. Clin. Immunol.* 2010, **30**:10–16.

12-13 Enteric pathogens cause a local inflammatory response and the development of protective immunity.

Dubin, P.J., and Kolls, J.K.: **Th17 cytokines and mucosal immunity.** *Immunol. Rev.* 2008, **226**:160–171.

Fritz, J.H., Le Bourhis, L., Magalhaes, J.G., and Philpott, D.J.: **Innate immune recognition at the epithelial barrier drives adaptive immunity: APCs take the back seat.** *Trends Immunol.* 2008, **29**:41–49.

Lavelle, E.C., Murphy, C., O'Neill, L.A., and Creagh, E.M.: **The role of TLRs, NLRs, and RLRs in mucosal innate immunity and homeostasis.** *Mucosal Immunol.* 2010, **3**:17–28.

Ouellette, A.J.: **Paneth cells and innate mucosal immunity.** *Curr. Opin. Gastroenterol.* 2010, **26**:547–553.

Philpott, D.J., and Girardin, S.E.: **Nod-like receptors: sentinels at host membranes.** *Curr. Opin. Immunol.* 2010, **22**:428–434.

Sansonetti, P.J.: **To be or not to be a pathogen: that is the mucosally relevant question.** *Mucosal Immunol.* 2011, **4**:8–14.

Santos, R.L., Raffatellu, M., Bevins, C.L., Adams, L.G., Tükel, C., Tsolis, R.M., and Bäumler, A.J.: **Life in the inflamed intestine, *Salmonella* style.** *Trends Microbiol.* 2009, **17**:498–506.

Sellge, G., Magalhaes, J.G., Konradt, C., Fritz, J.H., Salgado-Pabon, W., Eberl, G., Bandeira, A., Di Santo, J.P., Sansonetti, P.J., and Phalipon, A.: **Th17 cells are the dominant T cell subtype primed by *Shigella flexneri* mediating protective immunity.** *J. Immunol.* 2010, **184**:2076–2085.

Tam, M.A., Rydström, A., Sundquist, M., and Wick, M.J.: **Early cellular responses to *Salmonella* infection: dendritic cells, monocytes, and more.** *Immunol. Rev.* 2008, **225**:140–162.

Vaishnava, S., Behrendt, C.L., Ismail, A.S., Eckmann, L., and Hooper, L.V.: **Paneth cells directly sense gut commensals and maintain homeostasis at the intestinal host–microbial interface.** *Proc. Natl Acad. Sci. USA* 2008, **105**:20858–20863.

12-14 The outcome of infection by intestinal pathogens is determined by a complex interplay between the microorganism and the host immune response.

Carneiro, L.A., Travassos, L.H., Soares, F., Tattoli, I., Magalhaes, J.G., Bozza, M.T., Plotkowski, M.C., Sansonetti, P.J., Molkentin, J.D., Philpott, D.J., *et al.*: ***Shigella* induces mitochondrial dysfunction and cell death in nonmyleoid cells.** *Cell Host Microbe* 2009, **5**:123–136.

Cossart, P., and Sansonetti, P.J.: **Bacterial invasion: the paradigms of enteroinvasive pathogens.** *Science* 2004, **304**:242–248.

Pédron, T., and Sansonetti, P.: **Commensals, bacterial pathogens and intestinal inflammation: an intriguing ménage à trois.** *Cell Host Microbe* 2008, **3**:344–347.

Trosky, J.E., Liverman, A.D., and Orth, K.: ***Yersinia* outer proteins: Yops.** *Cell Microbiol.* 2008, **10**:557–565.

12-15 The mucosal immune system must maintain a balance between protective immunity and homeostasis to a large number of different foreign antigens.

Barnes, M.J., and Powrie, F.: **Regulatory T cells reinforce intestinal homeostasis.** *Immunity* 2009, **31**:401–411.

Cario, E.: **Innate immune signalling at intestinal mucosal surfaces: a fine line between host protection and destruction.** *Curr. Opin. Gastroenterol.* 2008, **24**:725–732.

Dubin, P.J., and Kolls, J.K.: **Th17 cytokines and mucosal immunity.** *Immunol. Rev.* 2008, **226**:160–171.

Hand, T., and Belkaid, Y.: **Microbial control of regulatory and effector T cell responses in the gut.** *Curr. Opin. Immunol.* 2010, **22**:63–72.

Macdonald, T.T., and Monteleone, G.: **Immunity, inflammation, and allergy in the gut.** *Science* 2005, **307**:1920–1925.

Sansonetti, P.J., and Medzhitov, R.: **Learning tolerance while fighting ignorance.** *Cell* 2009, **138**:416–420.

Strobel, S., and Mowat, A.M.: **Oral tolerance and allergic responses to food proteins.** *Curr. Opin. Allergy Clin. Immunol.* 2006, **6**:207–213.

Worbs, T., Bode, U., Yan, S., Hoffmann, M.W., Hintzen, G., Bernhardt, G., Forster, R., and Pabst, O.: **Oral tolerance originates in the intestinal immune system and relies on antigen carriage by dendritic cells.** *J. Exp. Med.* 2006, **203**:519–527.

12-16 The healthy intestine contains large quantities of bacteria but does not generate potentially harmful immune responses against them.

Abreu, M.T.: **Toll-like receptor signalling in the intestinal epithelium: how bacterial recognition shapes intestinal function.** *Nat. Rev. Immunol.* 2010, **10**:131–144.

Asquith, M.J., Boulard, O., Powrie, F., and Maloy, K.J.: **Pathogenic and protective roles of MyD88 in leukocytes and epithelial cells in mouse models of inflammatory bowel disease.** *Gastroenterology* 2010, **139**:519–529.

Barnes, M.J., and Powrie, F.: **Regulatory T cells reinforce intestinal homeostasis.** *Immunity* 2009, **31**:401–411.

Bouskra, D., Brézillon, C., Bérard, M., Werts, C., Varona, R., Boneca, I.G., and Eberl, G.: **Lymphoid tissue genesis induced by commensals through NOD1 regulates intestinal homeostasis.** *Nature* 2008, **456**:507–510.

Cerf-Bensussan, N., and Gaboriau-Routhiau, V.: **The immune system and the gut microbiota: friends or foes?** *Nat. Rev. Immunol.* 2010, **10**:735–744.

Duerkop, B.A., Vaishnava, S., and Hooper, L.V.: **Immune responses to the**

microbiota at the intestinal mucosal surface. *Immunity* 2009, **31**:368–376.

Garrett, W.S., Gordon, J.I., and Glimcher, L.H.: **Homeostasis and inflammation in the intestine.** *Cell* 2010, **140**:859–870.

Platt, A.M., and Mowat, A.M.: **Mucosal macrophages and the regulation of immune responses in the intestine.** *Immunol. Lett.* 2008, **119**:22–31.

Rakoff-Nahoum, S., Hao, L., and Medzhitov, R.: **Role of toll-like receptors in spontaneous commensal-dependent colitis.** *Immunity* 2006, **25**:319–329.

Round, J.L., and Mazmanian, S.K.: **Inducible Foxp3⁺ regulatory T-cell development by a commensal bacterium of the intestinal microbiota.** *Proc. Natl Acad. Sci. USA* 2010, **107**:12204–12209.

Siegmund, B.: **Interleukin-18 in intestinal inflammation: friend and foe?** *Immunity* 2010, **32**:300–302.

Slack, E., Hapfelmeier, S., Stecher, B., Velykoredko, Y., Stoel, M., Lawson, M.A., Geuking, M.B., Beutler, B., Tedder, T.F., Hardt, W.D., *et al.*: **Innate and adaptive immunity cooperate flexibly to maintain host–microbiota mutualism.** *Science* 2009, **325**:617–620.

12-17 Full immune responses to commensal bacteria provoke intestinal disease.

Asquith, M.J., Boulard, O., Powrie, F., and Maloy, K.J.: **Pathogenic and protective roles of MyD88 in leukocytes and epithelial cells in mouse models of inflammatory bowel disease.** *Gastroenterology* 2010, **139**:519–529.

Buonocore, S., Ahern, P.P., Uhlig, H.H., Ivanov, I.I., Littman, D.R., Maloy, K.J., and Powrie, F.: **Innate lymphoid cells drive interleukin-23-dependent innate intestinal pathology.** *Nature* 2010, **464**:1371–1375.

Fischer, W., Prassl, S., and Haas, R.: **Virulence mechanisms and persistence strategies of the human gastric pathogen *Helicobacter pylori*.** *Curr. Top. Microbiol. Immunol.* 2009, **337**:129–171.

Kaser, A., Zeissig, S., and Blumberg, R.S.: **Inflammatory bowel disease.** *Annu. Rev. Immunol.* 2010, **28**:573–621.

Lodes, M.J., Cong, Y., Elson, C.O., Mohamath, R., Landers, C.J., Targan, S.R., Fort, M., and Hershberg, R.M.: **Bacterial flagellin is a dominant antigen in Crohn's disease.** *J. Clin. Invest.* 2004, **113**:1296–1306.

Sartor, R.B.: **Key questions to guide a better understanding of host–commensal microbiota interactions in intestinal inflammation.** *Mucosal Immunol.* 2011 Jan 19.

Siegmund, B.: **Interleukin-18 in intestinal inflammation: friend and foe?** *Immunity* 2010, **32**:300–302.

Watanabe, N., Kiriya, K., and Chiba, T.: **Small intestine Peyer's patches are major induction sites of the *Helicobacter*-induced host immune responses.** *Gastroenterology* 2008, **134**:642–643.

12-18 Intestinal helminths provoke strong T_H2-mediated immune responses.

Artis, D., and Grencis, R.K.: **The intestinal epithelium: sensors to effectors in nematode infection.** *Mucosal Immunol.* 2008, **1**:252–264.

Bischoff, S.C.: **Physiological and pathophysiological functions of intestinal mast cells.** *Semin. Immunopathol.* 2009, **31**:185–205.

Hasnain, S.Z., Wang, H., Ghia, J.E., Haq, N., Deng, Y., Velcich, A., Grencis, R.K., Thornton, D.J., and Khan, W.I.: **Mucin gene deficiency in mice impairs host resistance to an enteric parasitic infection.** *Gastroenterology* 2010, **138**:1763–1771.

Humphreys, N.E., Xu, D., Hepworth, M.R., Liew, F.Y., and Grencis, R.K.: **IL-33, a potent inducer of adaptive immunity to intestinal nematodes.** *J. Immunol.* 2008, **180**:2443–2449.

Ierna, M.X., Scales, H.E., Saunders, K.L., and Lawrence, C.E.: **Mast cell production of IL-4 and TNF may be required for protective and pathological responses in gastrointestinal helminth infection.** *Mucosal Immunol.* 2008, **1**:147–155.

Maizels, R.M., Pearce, E.J., Artis, D., Yazdanbakhsh, M., and Wynn, T.A.: **Regulation of pathogenesis and immunity in helminth infections.** *J. Exp. Med.* 2009, **206**:2059–2066.

Neill, D.R., Wong, S.H., Bellosi, A., Flynn, R.J., Daly, M., Langford, T.K., Bucks, C., Kane, C.M., Fallon, P.G., Pannell, R., *et al.*: **Nuocytes represent a new innate effector leukocyte that mediates type-2 immunity.** *Nature* 2010, **464**:1367–1370.

Saenz, S.A., Noti, M., and Artis, D.: **Innate immune cell populations function as initiators and effectors in Th2 cytokine responses.** *Trends Immunol.* 2010, **31**:407–413.

Specht, S., Saeftel, M., Arndt, M., Endl, E., Dubben, B., Lee, N.A., Lee, J.J., and Hoerauf, A.: **Lack of eosinophil peroxidase or major basic protein impairs defense against murine filarial infection.** *Infect. Immun.* 2006, **74**:5236–5243.

Taylor, B.C., Zaph, C., Troy, A.E., Du, Y., Guild, K.J., Comeau, M.R., and Artis, D.: **TSLP regulates intestinal immunity and inflammation in mouse models of helminth infection and colitis.** *J. Exp. Med.* 2009, **206**:655–667.

12-19 Other eukaryotic parasites provoke protective immunity and pathology in the gut.

Dalton, J.E., Cruickshank, S.M., Egan, C.E., Mears, R., Newton, D.J., Andrew, E.M., Lawrence, B., Howell, G., Else, K.J., Gubbels, M.J., *et al.*: **Intraepithelial $\gamma\delta^+$ lymphocytes maintain the integrity of intestinal epithelial tight junctions in response to infection.** *Gastroenterology* 2006, **131**:818–829.

Eckmann, L.: **Mucosal defences against *Giardia*.** *Parasite Immunol.* 2003, **25**:259–270.

Petry, F., Jakobi, V., and Tessema, T.S.: **Host immune response to *Cryptosporidium parvum* infection.** *Exp. Parasitol.* 2010, **126**:304–309.

Schulthess, J., Fourreau, D., Darche, S., Meresse, B., Kasper, L., Cerf-Bensussan, N., and Buzoni-Gatel, D.: **Mucosal immunity in *Toxoplasma gondii* infection.** *Parasite* 2008, **15**:389–395.

12-20 The mucosal immune system has to compromise between suppression and activation of an immune response.

Abt, M.C., and Artis, D.: **The intestinal microbiota in health and disease: the influence of microbial products on immune cell homeostasis.** *Curr. Opin. Gastroenterol.* 2009, **25**:496–502.

Altmann, D.M.: **Review series on helminths, immune modulation and the hygiene hypothesis: nematode coevolution with adaptive immunity, regulatory networks and the growth of inflammatory diseases.** *Immunology* 2009, **126**:1–2.

Cario, E.: **Innate immune signalling at intestinal mucosal surfaces: a fine line between host protection and destruction.** *Curr. Opin. Gastroenterol.* 2008, **24**:725–732.

Dunne, D.W., and Cooke, A.: **A worm's eye view of the immune system: consequences for evolution of human autoimmune disease.** *Nat. Rev. Immunol.* 2005, **5**:420–426.

Gaboriau-Routhiau, V., Rakotobe, S., Lécuyer, E., Mulder, I., Lan, A., Bridonneau, C., Rochet, V., Pisi, A., De Paepe, M., Brandi, G., *et al.*: **The key role of segmented filamentous bacteria in the coordinated maturation of gut helper T cell responses.** *Immunity* 2009, **31**:677–689.

Okada, H., Kuhn, C., Feillet, H., and Bach, J.F.: **The 'hygiene hypothesis' for autoimmune and allergic diseases: an update.** *Clin. Exp. Immunol.* 2010, **160**:1–9.

Round, J.L., and Mazmanian, S.K.: **Inducible Foxp3⁺ regulatory T-cell development by a commensal bacterium of the intestinal microbiota.** *Proc. Natl Acad. Sci. USA* 2010, **107**:12204–12209.

Saenz, S.A., Taylor, B.C., and Artis, D.: **Welcome to the neighborhood: epithelial cell-derived cytokines license innate and adaptive immune responses at mucosal sites.** *Immunol. Rev.* 2008, **226**:172–190.

Zaph, C., Troy, A.E., Taylor, B.C., Berman-Booty, L.D., Guild, K.J., Du, Y., Yost, E.A., Gruber, A.D., May, M.J., Greten, F.R., *et al.*: **Epithelial-cell-intrinsic IKK-β expression regulates intestinal immune homeostasis.** *Nature* 2007, **446**:552–556.

PART V

THE IMMUNE SYSTEM IN HEALTH AND DISEASE

Failures of Host Defense Mechanisms

13

In the normal course of an infection, the infectious agent first triggers an innate immune response. The foreign antigens of the infectious agent, enhanced by signals from innate immune cells, then induce an adaptive immune response that clears the infection and establishes a state of protective immunity. This does not always happen, however, and in this chapter we examine three circumstances in which there are failures of host defense against pathogens: the avoidance or subversion of a normal immune response by the pathogen; inherited failures of immune defenses because of gene defects; and the acquired immune deficiency syndrome (AIDS), a generalized susceptibility to infection that is itself due to the failure of the host to control and eliminate the human immunodeficiency virus (HIV).

To propagate itself, a pathogen must replicate in the infected host and spread to new hosts. Common pathogens must therefore grow without activating too vigorous an immune response, but they must not kill the host too quickly. The most successful pathogens persist either because they do not elicit an immune response or because they evade the response once it has occurred. Over millions of years of coevolution with their hosts, pathogens have developed various strategies for avoiding destruction by the immune system, and these are examined in the first part of this chapter.

In the second part of the chapter, we turn to the **immunodeficiency diseases**, in which host defense fails. In most of these diseases, a defective gene results in the elimination of one or more components of the immune system, leading to heightened susceptibility to infection with particular classes of pathogens. Immunodeficiency diseases caused by defects in T- or B-lymphocyte development, phagocyte function, and complement components have all been discovered. In the last part of the chapter, we consider how the persistent infection of immune system cells by HIV leads to AIDS, an example of an acquired immunodeficiency. The study of all these immunodeficiencies has already contributed greatly to our understanding of host defense mechanisms and, in the longer term, might help to provide new methods of controlling or preventing infectious diseases, including AIDS.

Evasion and subversion of immune defenses.

Just as vertebrates have developed many different defenses against pathogens, so pathogens have evolved numerous ways of overcoming them. These range from resisting phagocytosis to avoiding recognition by the adaptive immune system and even actively suppressing immune responses. We start by looking at how some pathogens keep one step ahead of the adaptive immune response.

13-1 Antigenic variation allows pathogens to escape from immunity.

One way in which an infectious agent can evade surveillance by the immune system is by altering its antigens; this is known as **antigenic variation** and is particularly important for extracellular pathogens, which are generally eliminated by antibodies against their surface structures (see Chapter 10). There are three main forms of antigenic variation. First, many infectious agents exist in a wide variety of antigenic types. There are, for example, 84 known types of *Streptococcus pneumoniae*, an important cause of bacterial pneumonia, which differ in the structure of their polysaccharide capsules. The different types are distinguished by using specific antibodies as reagents in serological tests and so are often known as **serotypes**. Infection with one serotype can lead to type-specific immunity, which protects against reinfection with that type but not with a different serotype. Thus, from the point of view of the adaptive immune system, each serotype of *S. pneumoniae* represents a distinct organism, with the result that essentially the same pathogen can cause disease many times in the same individual (Fig. 13.1).

A second, more dynamic, mechanism of antigenic variation is an important feature of the influenza virus. At any one time, a single virus type is responsible for most cases of influenza throughout the world. The human population gradually develops protective immunity to this type, chiefly through neutralizing antibody directed against the viral hemagglutinin, the main surface protein of the influenza virus. Because the virus is rapidly cleared from immune individuals, it might be in danger of running out of potential hosts if it had not evolved two distinct ways of changing its antigenic type (Fig. 13.2).

Movie 13.1

The first of these is called **antigenic drift** and is caused by point mutations in the genes encoding hemagglutinin and a second surface protein, neuraminidase. Every 2–3 years a variant flu virus arises with mutations that allow it to evade neutralization by the antibodies present in the population. Other mutations affect epitopes in the proteins that are recognized by T cells, in

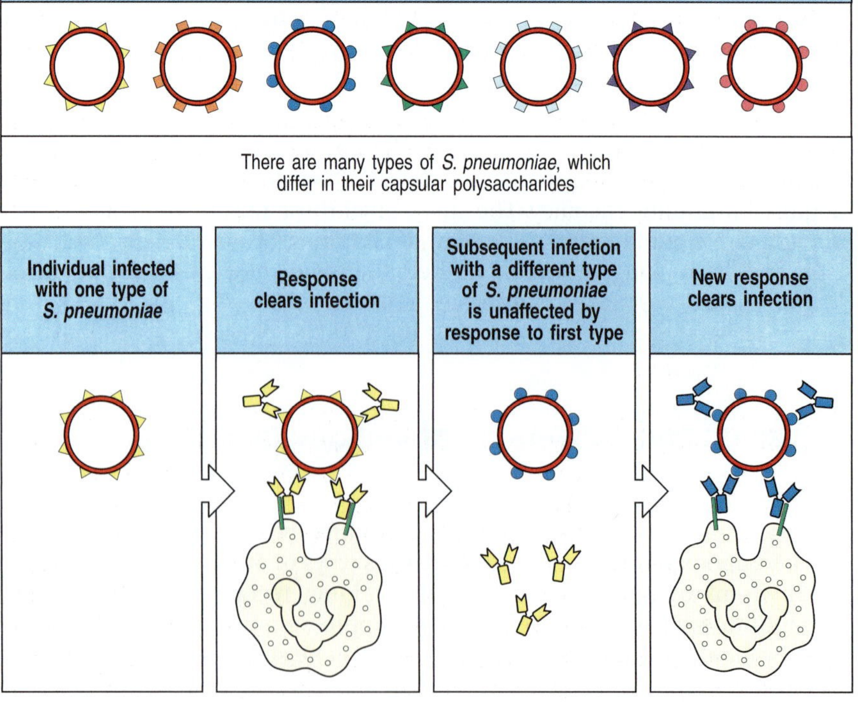

Fig. 13.1 Host defense against *Streptococcus pneumoniae* is type specific. The different strains of *S. pneumoniae* have antigenically distinct capsular polysaccharides. The capsule prevents effective phagocytosis until the bacterium is opsonized by specific antibody and complement, allowing phagocytes to destroy it. Antibody against one type of *S. pneumoniae* does not cross-react with the other types, so an individual immune to one type has no protective immunity to a subsequent infection with a different type. An individual must generate a new adaptive immune response each time he or she is infected with a different type of *S. pneumoniae*.

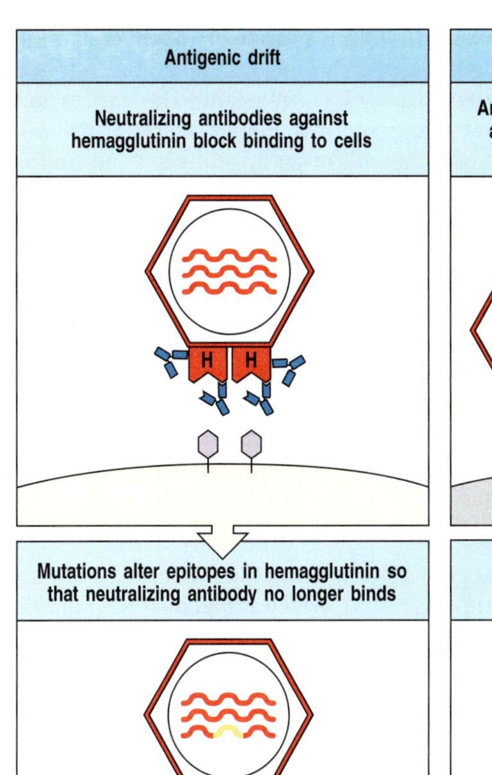
Antigenic drift
Neutralizing antibodies against hemagglutinin block binding to cells
H
H
Mutations alter epitopes in hemagglutinin so that neutralizing antibody no longer binds

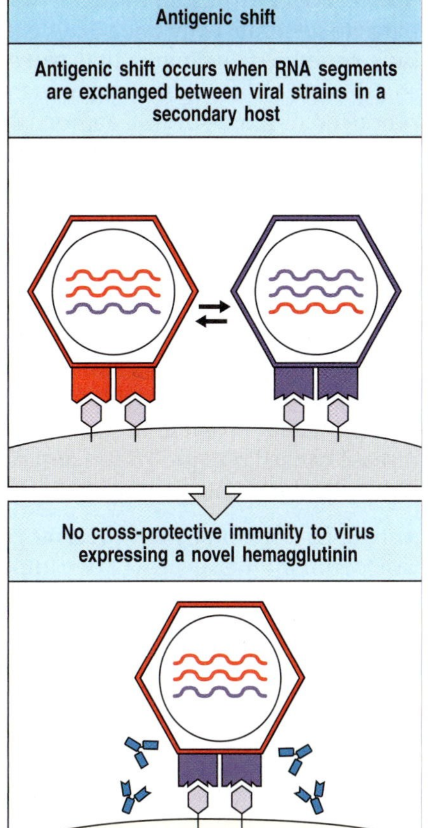
Antigenic shift
Antigenic shift occurs when RNA segments are exchanged between viral strains in a secondary host
No cross-protective immunity to virus expressing a novel hemagglutinin

Fig. 13.2 Two types of variation allow repeated infection with type A influenza virus. Neutralizing antibody that mediates protective immunity is directed at the viral surface protein hemagglutinin (H), which is responsible for viral binding to and entry into cells. Antigenic drift (left panels) involves the emergence of point mutants with altered binding sites for protective antibodies on the hemagglutinin. The new virus can grow in a host that is immune to the previous strain of virus, but as T cells and some antibodies can still recognize epitopes that have not been altered, the new variants cause only mild disease in previously infected individuals. Antigenic shift (right panels) is a rare event involving the reassortment of the segmented RNA viral genomes of two different influenza viruses, probably in a bird or a pig. These antigen-shifted viruses have large changes in their hemagglutinin, and therefore T cells and antibodies produced in earlier infections are not protective. These shifted strains cause severe infection that spreads widely, causing the influenza pandemics that occur every 10–50 years. There are eight RNA molecules in each viral genome, but for simplicity only three are shown.

particular CD8 cytotoxic T cells, and so cells infected with the mutant virus also escape destruction. People immune to the old flu virus are thus susceptible to the new variant, but because the changes in the viral proteins are relatively minor, there is still some cross-reaction with antibodies and memory T cells produced against the previous variant, and most of the population still has some level of immunity (see Fig. 10.27). An epidemic resulting from antigenic drift is usually relatively mild.

The other type of antigenic change in influenza virus is known as **antigenic shift** and is due to major changes in the hemagglutinin of the virus. Antigenic shifts cause global pandemics of severe disease, often with substantial mortality, because the new hemagglutinin is recognized poorly, if at all, by antibodies and T cells directed against the previous variant. Antigenic shift is due to reassortment of the segmented RNA genome of the human influenza virus and animal influenza viruses in an animal host, in which the hemagglutinin gene from the animal virus replaces the hemagglutinin gene in the human virus.

Movie 13.2

The third mechanism of antigenic variation in pathogens involves programmed gene rearrangements. The most striking example occurs in African trypanosomes, where changes in the major surface antigen occur repeatedly within the same infected host. African trypanosomes are insect-borne protozoan parasites that replicate in the extracellular spaces of tissues and cause the disease known as trypanosomiasis or sleeping sickness. The trypanosome is coated with a single type of glycoprotein, the variant-specific glycoprotein (VSG), which elicits a potent protective antibody response that rapidly clears most of the parasites. The trypanosome genome, however, contains about 1000 VSG genes, each encoding a protein with distinct antigenic properties.

A VSG gene is expressed by being placed into the active expression site in the parasite genome. Only one VSG is expressed at a time, and it can be changed by a gene rearrangement that places a new VSG gene into the expression site (Fig. 13.3). So, by using gene rearrangement to change the VSG protein produced, trypanosomes keep one step ahead of an immune system that is capable of generating many distinct antibodies by gene rearrangement. The few trypanosomes with changed surface glycoproteins are not affected by the antibodies previously made by the host, and these variants multiply and cause a recurrence of disease (see Fig. 13.3, bottom panel). Antibodies are then made against the new VSG, and the whole cycle is repeated. The chronic cycles of antigen clearance lead to immune-complex damage and inflammation, and eventually to neurological damage, resulting finally in the coma that gives sleeping sickness its name. The cycles of evasive action make trypanosome infections very difficult for the immune system to defeat, and they are a major health problem in Africa. Malaria is another serious and widespread disease caused by a protozoan parasite that varies its antigens to avoid elimination by the immune system.

Antigenic variation by DNA rearrangement also occurs in bacteria and helps to account for the success of two important bacterial pathogens—*Salmonella enterica* serotype Typhimurium, a common cause of salmonella food poisoning, and *Neisseria gonorrhoeae*, which causes gonorrhea, a major sexually transmitted disease and an increasing public-health problem in the United States. *Salmonella* Typhimurium regularly alternates two versions of its surface flagellin protein. Inversion of a segment of DNA containing the promoter for one flagellin gene turns off expression of the gene and allows the expression of a second flagellin gene that encodes an antigenically distinct protein. *N. gonorrhoeae* has several variable antigens, the most important of which is the pilin protein, which is responsible for adherence of the bacterium to a mucosal surface. Like the VSGs of the African trypanosome, there is more than one pilin gene variant, only one of which is active at any given time. From time to time, a different pilin gene replaces the active gene under the control of the pilin promoter. All these mechanisms of antigenic variation help the pathogen to evade an otherwise specific and effective immune response.

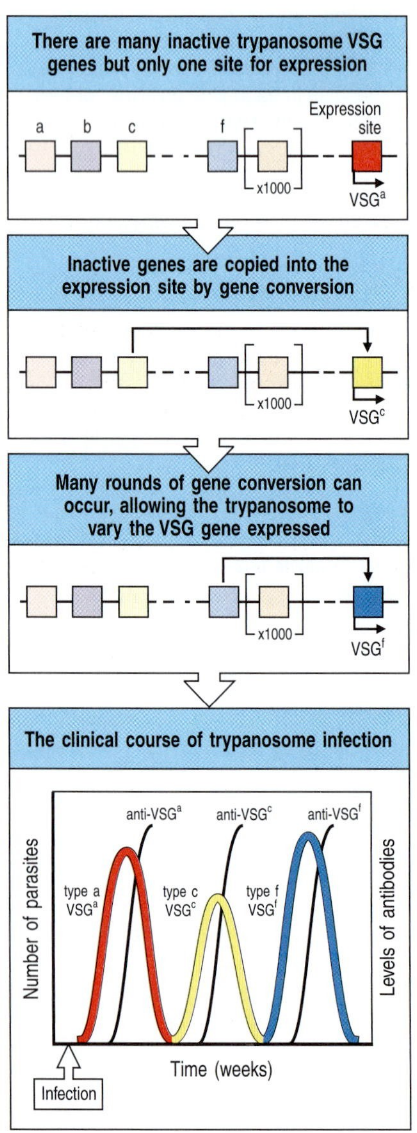
There are many inactive trypanosome VSG genes but only one site for expression
a
b
c
f
Expression site
x1000
VSG^a
Inactive genes are copied into the expression site by gene conversion
x1000
VSG^c
Many rounds of gene conversion can occur, allowing the trypanosome to vary the VSG gene expressed
x1000
VSG^f
The clinical course of trypanosome infection
anti-VSG^a
anti-VSG^c
anti-VSG^f
type a VSG^a
type c VSG^c
type f VSG^f
Number of parasites
Levels of antibodies
Time (weeks)
Infection

13-2 Some viruses persist *in vivo* by ceasing to replicate until immunity wanes.

Once they have entered cells, viruses usually betray their presence to the immune system by directing the synthesis of viral proteins, fragments of which are displayed on the MHC molecules of the infected cell where they can be detected by T lymphocytes. To replicate, a virus must make viral proteins, and rapidly replicating viruses that produce acute illnesses are readily detected by T cells, which normally control them. Some viruses, however, can enter a state known as **latency**, in which the virus is not being replicated. In the latent state, the virus does not cause disease; however, because there are

Fig. 13.3 Antigenic variation in trypanosomes allows them to escape immune surveillance. The surface of a trypanosome is covered with a variant-specific glycoprotein (VSG). Each trypanosome has about 1000 genes encoding different VSGs, but only the gene in a specific expression site within the telomere at one end of the chromosome is active. Although several genetic mechanisms have been observed for changing the VSG gene expressed, the usual mechanism is gene conversion. An inactive gene, which is not at the telomere, is copied and transposed into the telomeric expression site, where it becomes active. When an individual is first infected, antibodies are raised against the VSG initially expressed by the trypanosome population. A small number of trypanosomes spontaneously switch their VSG gene to a new type, and although the host antibody eliminates the initial variant, the new variant is unaffected. As the new variant grows, the whole sequence of events is repeated.

no viral peptides to signal its presence it cannot be eliminated. Latent infections can be reactivated, and this results in recurrent illness.

A major class of viral agents that cause latent infections in humans are the herpesviruses, which are characterized by their ability to establish lifelong infections. An example is herpes simplex virus (HSV), the cause of cold sores, which infects epithelial cells and spreads to sensory neurons serving the infected area. An effective immune response controls the epithelial infection, but the virus persists in a latent state in the sensory neurons. Factors such as sunlight, bacterial infection, or hormonal changes reactivate the virus, which then travels down the axons of the sensory neuron and reinfects the epithelial tissues (Fig. 13.4). At this point, the immune response again becomes active and controls the local infection by killing the epithelial cells, producing a new sore. This cycle can be repeated many times.

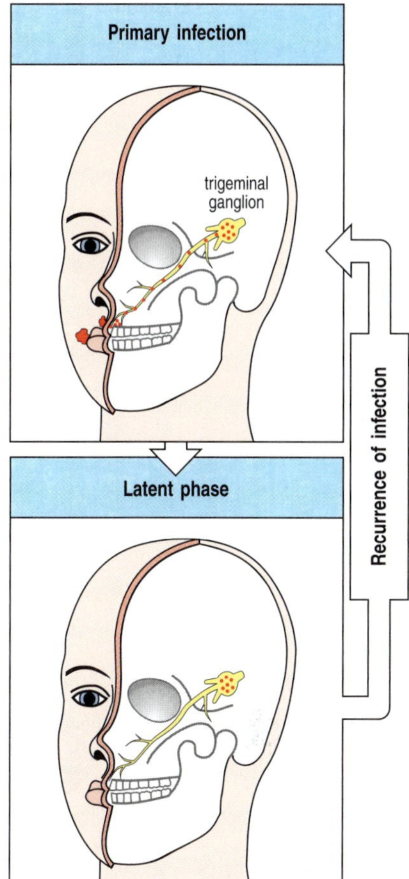

Fig. 13.4 Persistence and reactivation of herpes simplex virus infection. The initial infection in the skin is cleared by an effective immune response, but residual infection persists in sensory neurons such as those of the trigeminal ganglion, whose axons innervate the lips. When the virus is reactivated, usually by some environmental stress and/or alteration in immune status, the skin in the area served by the nerve is reinfected from virus in the ganglion and a new cold sore results. This process can be repeated many times.

There are two reasons why the sensory neuron remains infected: first, the virus is quiescent and generates few virus-derived peptides to present on MHC class I molecules; second, neurons carry very low levels of MHC class I molecules, which makes it harder for CD8 cytotoxic T cells to recognize infected neurons and attack them. The low level of MHC class I expression might be beneficial, because it reduces the risk that neurons, which do not regenerate or do so only very slowly, will be attacked inappropriately by cytotoxic T cells. It does, however, make neurons attractive cellular reservoirs that are unusually vulnerable to persistent infections. Herpesviruses often enter latency: herpes zoster (varicella zoster), which causes chickenpox, remains latent in one or a few dorsal root ganglia after the acute illness is over, and can be reactivated by stress or immunosuppression. It then spreads down the nerve and reinfects the skin to cause the disease **shingles**, which is marked by the reappearance of the classic varicella rash in the area of skin served by the infected dorsal root. Unlike herpes simplex, in which reactivation occurs frequently, herpes zoster usually reactivates only once in a lifetime in an immunocompetent host.

Yet another herpesvirus, the Epstein–Barr virus (EBV), establishes a persistent infection in most individuals. EBV enters latency in B cells after a primary infection that often passes without being diagnosed. In a minority of infected individuals, the initial acute infection of B cells is more severe, causing the disease known as **infectious mononucleosis** or glandular fever. EBV infects B cells by binding to CR2 (CD21), a component of the B-cell co-receptor complex, and to MHC class II molecules. In the primary infection, most of the infected cells proliferate and produce virus, leading in turn to the proliferation of antigen-specific T cells and the excess of mononuclear white cells in the blood that gives the disease its name. Virus is released from the B cells, destroying them in the process, and virus can be recovered from saliva. The infection is eventually controlled by virus-specific CD8 cytotoxic T cells, which kill the infected proliferating B cells. A fraction of memory B lymphocytes become latently infected, however, and EBV remains quiescent in these cells.

These two forms of infection are accompanied by quite different patterns of expression of viral genes. EBV has a large DNA genome encoding more than 70 proteins. Many of these are required for viral replication and are expressed by the replicating virus, providing a source of viral peptides by which infected cells can be recognized. In a latent infection, in contrast, the virus survives within the host B cells without replicating, and a very limited set of viral proteins is expressed. One of these is the Epstein–Barr nuclear antigen 1 (EBNA-1), which is needed to maintain the viral genome. EBNA-1 interacts with the proteasome (see Section 6-3) to prevent its own degradation into peptides that would otherwise elicit a T-cell response.

Latently infected B cells can be isolated by culturing B cells from individuals who have apparently cleared their EBV infection: in the absence of T cells, latently infected cells retaining the EBV genome become transformed into so-called immortal cell lines, the equivalent of tumorigenesis *in vitro*. Infected B cells occasionally undergo malignant transformation *in vivo*, giving rise to a B-cell lymphoma called Burkitt's lymphoma. In this lymphoma, expression of the peptide transporters TAP-1 and TAP-2 is downregulated (see Section 6-2), and so cells are unable to process endogenous antigens for presentation on HLA class I molecules (the human MHC class I). This deficiency provides one explanation for how these tumors escape attack by CD8 cytotoxic T cells. Patients with acquired and inherited immunodeficiencies of T-cell function have an increased risk of developing EBV-associated lymphomas, presumably as a result of a failure of immune surveillance.

13-3 Some pathogens resist destruction by host defense mechanisms or exploit them for their own purposes.

Some pathogens induce a normal immune response but have evolved specialized mechanisms for resisting its effects. For instance, some bacteria that are engulfed by macrophages have evolved means of avoiding destruction by these phagocytes and instead use macrophages as their primary host cell. *Mycobacterium tuberculosis*, for example, is taken up by macrophages but prevents the fusion of the phagosome with the lysosome, protecting itself from the bactericidal actions of the lysosomal contents.

Movie 13.3 

Other microorganisms, such as the bacterium *Listeria monocytogenes*, escape from the phagosome into the cytoplasm of the macrophage, where they multiply. They then spread to adjacent cells in the tissue without emerging into the extracellular environment. They do this by hijacking the cytoskeletal protein actin, which assembles into filaments at the rear of the bacterium. The actin filaments drive the bacteria forward into vacuolar projections to adjacent cells; the vacuoles are then lysed by the *Listeria*, releasing the bacteria into the cytoplasm of the adjacent cell. In this way *Listeria* avoids attack by antibodies, but the infected cells are still susceptible to killing by cytotoxic T cells. The protozoan parasite *Toxoplasma gondii* generates its own vesicle, which does not fuse with any cellular vesicle and thus isolates the parasite from the rest of the cell. This might render peptides derived from *T. gondii* less available for loading onto MHC molecules.

The spirochete bacterium *Treponema pallidum*, the cause of syphilis, can avoid elimination by antibodies and establish a persistent and extremely damaging infection in tissues. *T. pallidum* is believed to avoid recognition by antibodies by coating its surface with host proteins until it has invaded tissues such as the central nervous system, where it is less accessible to antibodies. Another spirochete, the tick-borne *Borrelia burgdorferi*, is the cause of Lyme disease, which occurs as a result of chronic infection by the bacterium. Some strains of *B. burgdorferi* may avoid lysis by complement by coating themselves in the complement-inhibitory protein factor H made by the host (see Section 2-16), which binds to receptor proteins in the bacterium's outer membrane.

Finally, many viruses subvert particular parts of the immune system. The mechanisms used include the capture of cellular genes for cytokines or chemokines and their receptors, the synthesis of complement-regulatory molecules, the inhibition of MHC class I molecule synthesis or assembly (as observed in EBV infections), and the production of decoy proteins that mimic the TIR domains that are part of the TLR/IL-1 receptor signaling pathway (see Fig. 3.13). The human cytomegalovirus (CMV), another herpesvirus, produces a protein called UL18, which is homologous to an HLA class I molecule. By the interaction of UL18 with the receptor protein LIR-1, an inhibitory receptor on NK cells, the virus is thought to provide an inhibitory signal to the innate immune response. CMV also impairs antiviral responses by producing a homolog of the cytokine IL-10, called cmvIL-10, which downregulates the production of several pro-inflammatory cytokines by immune cells, including IFN-γ, IL-12 and IL-23, IL-1, IL-6, and TNF-α, to promote tolerogenic rather than immunogenic adaptive responses to viral antigens. Several viruses also produce molecules that interfere with chemokine responses, either by producing decoy chemokine receptors or chemokine homologs that interfere with natural ligand-induced signaling through chemokine receptors. Subversion of immune responses is one of the most rapidly expanding areas of research into host–pathogen relationships. Examples of how members of the herpesvirus and poxvirus families subvert host responses are shown in Fig. 13.5.

13-4 Immunosuppression or inappropriate immune responses can contribute to persistent disease.

Many pathogens suppress immune responses in general. For example, staphylococci produce toxins, such as the **staphylococcal enterotoxins** and **toxic shock syndrome toxin-1**, that act as superantigens. Superantigens are proteins that bind the antigen receptors of very large numbers of T cells (see Section 6-15), stimulating them to produce cytokines that cause a severe

Viral strategy	Specific mechanism	Result	Virus examples
Inhibition of humoral immunity	Virally encoded Fc receptor	Blocks effector functions of antibodies bound to infected cells	Herpes simplex Cytomegalovirus
	Virally encoded complement receptor	Blocks complement-mediated effector pathways	Herpes simplex
	Virally encoded complement control protein	Inhibits complement activation by infected cell	Vaccinia
Inhibition of inflammatory response	Virally encoded chemokine receptor homolog, e.g., β-chemokine receptor	Sensitizes infected cells to effects of β-chemokine; advantage to virus unknown	Cytomegalovirus
	Virally encoded soluble cytokine receptor, e.g., IL-1 receptor homolog, TNF receptor homolog, interferon-γ receptor homolog	Blocks effects of cytokines by inhibiting their interaction with host receptors	Vaccinia Rabbit myxoma virus
	Viral inhibition of adhesion molecule expression, e.g., LFA-3 ICAM-1	Blocks adhesion of lymphocytes to infected cells	Epstein–Barr virus
	Protection from NFκB activation by short sequences that mimic TLRs	Blocks inflammatory responses elicited by IL-1 or bacterial pathogens	Vaccinia
Blocking of antigen processing and presentation	Inhibition of MHC class I expression	Impairs recognition of infected cells by cytotoxic T cells	Herpes simplex Cytomegalovirus
	Inhibition of peptide transport by TAP	Blocks peptide association with MHC class I	Herpes simplex
Immunosuppression of host	Virally encoded cytokine homolog of IL-10	Inhibits T_H1 lymphocytes Reduces interferon-γ production	Epstein–Barr virus

Fig. 13.5 Mechanisms used by viruses of the herpes and pox families to subvert the host immune system.

inflammatory illness—**toxic shock**. The stimulated T cells proliferate and then rapidly undergo apoptosis, leaving a generalized immunosuppression together with the deletion of certain families of peripheral T cells.

Bacillus anthracis, the cause of anthrax, also suppresses immune responses through the release of a toxin. Anthrax is contracted by inhalation of, contact with, or ingestion of *B. anthracis* endospores and is often fatal if the endospores become disseminated throughout the body. *B. anthracis* produces a toxin called anthrax lethal toxin, which is a complex of two proteins—lethal factor and protective antigen. The main role of the protective antigen is to route the lethal factor into the host-cell cytosol. Lethal factor is a metalloproteinase with a unique specificity for MAPK kinases, components of many intracellular signaling pathways, and induces apoptosis of infected macrophages and abnormal maturation of dendritic cells. This results in the disruption of the immunological effector pathways that might otherwise delay bacterial growth.

The viruses hepatitis B (HBV, a DNA virus) and hepatitis C (HCV, an RNA virus) infect the liver and cause acute and chronic hepatitis, liver cirrhosis, and in some cases hepatocellular carcinoma. Immune responses probably have an important role in the clearance of both types of hepatitis infection, but in many cases HBV and HCV set up a chronic infection. Although HCV mainly infects the liver during the early stage of a primary infection, the virus subverts the adaptive immune response by interfering with dendritic-cell activation and maturation. This leads to inadequate activation of CD4 T cells and a consequent lack of T_H1 cell differentiation, which is thought to be responsible for the infection becoming chronic, most probably because of the lack of CD4 T-cell help to activate naive CD8 cytotoxic T cells. There is evidence that the decrease in levels of viral antigen seen after antiviral treatment improves CD4 T-cell help and allows the restoration of cytotoxic CD8 T-cell function and memory CD8 T-cell function. The delay in dendritic-cell maturation caused by HCV is thought to synergize with another property of the virus that helps it to evade an immune response. The RNA polymerase that the virus uses to replicate its genome lacks proofreading capacity. This contributes to a high viral mutation rate and thus a change in its antigenicity, which allows it to evade adaptive immunity.

In the well-characterized mouse model of viral infection caused by lymphocytic choriomeningitis virus (LCMV), certain strains promote chronic infection that is associated with 'exhaustion' of antiviral CD8 T cells. CD8 T cells induced in this setting are characterized by expression of an inhibitory receptor of the CD28 superfamily, the programmed death-1 (PD-1) receptor (see Section 7-18), activation of which by its ligand PD-L1 suppresses CD8 T-cell effector function. Blockade of the PD-L1–PD-1 interaction restores antiviral CD8 effector function and decreases the viral load, indicating that ongoing activation of this pathway is involved in impaired viral clearance. A similar mechanism has been implicated in chronic infections in humans caused by HBV, HCV, and HIV.

Leprosy, which we discussed in Section 9-18, is a more complex example of immunosuppression by an infection. In lepromatous leprosy, cell-mediated immunity is profoundly depressed, cells infected with *Mycobacterium leprae* are present in great profusion, and cellular immune responses to many other antigens are suppressed (Fig. 13.6). This leads to a state called anergy, which in this context specifically means the absence of delayed-type hypersensitivity reactions (see Chapter 14) in tests with a wide range of antigens unrelated to *M. leprae* (see Section 9-14 for the more general definition of anergy in use in other contexts). In tuberculoid leprosy, in contrast, there is strong cell-mediated immunity with macrophage activation that controls, but does not eradicate, infection. Most of the pathology in tuberculoid leprosy is caused by the ongoing localized inflammatory response to the mycobacteria that persist.

Many other pathogens cause a mild or transient immunosuppression during acute infection. These forms of suppressed immunity are poorly understood but important, as they often make the host susceptible to secondary infections by common environmental microorganisms. The measles virus can cause a relatively long-lasting immunosuppression after an infection, which is a particular problem in malnourished or undernourished children. In spite of the widespread availability of an effective vaccine, measles still accounts for 10% of the global mortality of children under 5 years old and is the eighth leading cause of death worldwide. Malnourished children are the main victims, and the cause of death is usually a secondary bacterial infection,

Infection with *Mycobacterium leprae* can result in different clinical forms of leprosy

There are two polar forms, tuberculoid and lepromatous leprosy, but several intermediate forms also exist

Tuberculoid leprosy	Lepromatous leprosy
Organisms present at low to undetectable levels	Organisms show florid growth in macrophages
Low infectivity	High infectivity
Granulomas and local inflammation. Peripheral nerve damage	Disseminated infection. Bone, cartilage, and diffuse nerve damage
Normal serum immunoglobulin levels	Hypergammaglobulinemia
Normal T-cell responsiveness. Specific response to *M. leprae* antigens	Low or absent T-cell responsiveness. No response to *M. leprae* antigens

Cytokine patterns in leprosy lesions

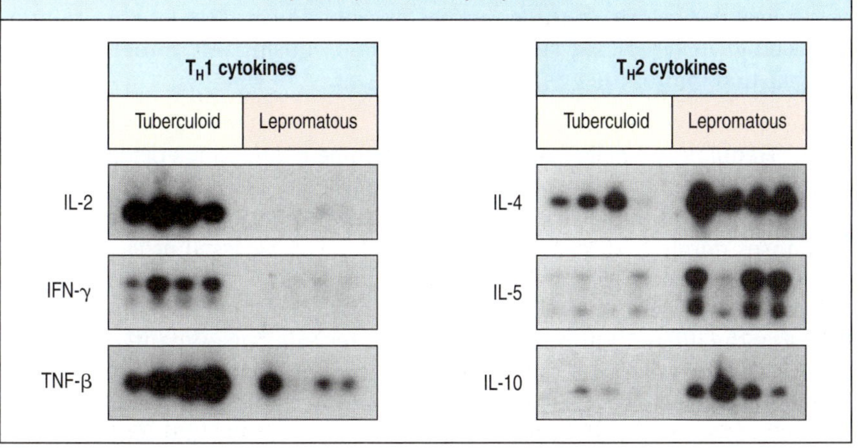

Fig. 13.6 T-cell and macrophage responses to *Mycobacterium leprae* are sharply different in the two polar forms of leprosy. Infection with *M. leprae*, whose cells stain as small dark red dots in the photographs, can lead to two very different forms of disease (top panels). In tuberculoid leprosy (left), growth of the organism is well controlled by T_H1-like cells that activate infected macrophages. The tuberculoid lesion contains granulomas and is inflamed, but the inflammation is local and causes only local effects, such as peripheral nerve damage. In lepromatous leprosy (right), infection is widely disseminated and the bacilli grow uncontrolled in macrophages; in the late stages of disease there is major damage to connective tissues and to the peripheral nervous system. There are several intermediate stages between these two polar forms. The lower panel shows Northern blots demonstrating that the cytokine patterns in the two polar forms of the disease are sharply different, as shown by the analysis of RNA isolated from lesions of four patients with lepromatous leprosy and four patients with tuberculoid leprosy. Cytokines typically produced by T_H2 cells (IL-4, IL-5, and IL-10) dominate in the lepromatous form, whereas cytokines produced by T_H1 cells (IL-2, IFN-γ, and TNF-β) dominate in the tuberculoid form. It therefore seems that T_H1-like cells predominate in tuberculoid leprosy, and T_H2-like cells in lepromatous leprosy. IFN-γ would be expected to activate macrophages, enhancing the killing of *M. leprae*, whereas IL-4 can actually inhibit the induction of bactericidal activity in macrophages. Photographs courtesy of G. Kaplan; cytokine patterns courtesy of R.L. Modlin.

particularly pneumonia, caused by measles-induced immunosuppression. This immunosuppression can last for several months after the disease is over and is associated with reduced T- and B-cell function. An important factor in measles-induced immunosuppression is the infection of dendritic cells by the measles virus. The infected dendritic cells render T lymphocytes generally unresponsive to antigen by mechanisms that are not yet understood, and it seems likely that this is the immediate cause of the immunosuppression.

13-5 Immune responses can contribute directly to pathogenesis.

Tuberculoid leprosy is just one example of an infection in which the pathology is caused largely by the immune response, the phenomenon known as **immunopathology**. This is true to some degree in most infections; for example, the fever that accompanies a bacterial infection is caused by the release of cytokines by macrophages. One medically important example of immunopathology is the wheezy bronchiolitis caused by **respiratory syncytial virus** (**RSV**) infection. Bronchiolitis caused by RSV is the major cause of admission of young children to hospital in the Western world, with as many as 90,000 admissions and 4500 deaths each year in the United States alone. The first indication that the immune response to the virus might have a role in the pathogenesis of this disease came from the observation that infants vaccinated with an alum-precipitated killed virus preparation had a more severe illness than children who did not receive the vaccine. This occurred because the vaccine failed to induce neutralizing antibodies but succeeded in producing effector T_H2 cells. When the vaccinated children encountered the virus, the T_H2 cells released interleukins IL-3, IL-4, and IL-5, which induced bronchospasm, increased the secretion of mucus, and increased tissue eosinophilia. Mice can be infected with RSV and develop a disease similar to that seen in humans.

Another example of a pathogenic immune response is the response to the eggs of schistosomes (blood flukes). These helminth parasites lay their eggs in the hepatic portal vein. Some reach the intestine and are shed in the feces, spreading the infection; other eggs lodge in the portal circulation of the liver, where they elicit a potent immune response leading to chronic inflammation, hepatic fibrosis, and eventually liver failure. This process reflects the excessive activation of T_H1 cells and can be modulated by T_H2 cells, IL-4, or CD8 T cells, which can also produce IL-4.

Unremitting T_H17 responses to chronic bacterial infections of the respiratory tract can lead to injury and dilation of the conducting airways, or bronchi, a condition called **bronchiectasis**. This is especially common in patients with cystic fibrosis, who have an inherited defect that impairs mucociliary clearance in the lungs. This defect leads to colonization of the airways with biofilms composed of bacteria such as *Pseudomonas aeruginosa* that provoke T_H17 immune responses. As discussed in Chapter 15, unrestrained T_H17 responses to bacteria of the intestinal microbiota can also lead to immunopathology in genetically susceptible patients, causing **inflammatory bowel disease** that injures the intestinal tissues.

13-6 Regulatory T cells can affect the outcome of infectious disease.

Some pathogens may avoid eradication by promoting adaptive immune responses dominated by regulatory T (T_{reg}) cells (discussed in Section 9-19) rather than effector T cells. 'Natural' FoxP3$^+$ T_{reg} cells arise in the thymus and migrate to the periphery, where they help to maintain tolerance by suppressing the differentiation of lymphocytes recognizing autoantigens. Other FoxP3$^+$ CD4 regulatory T cells, called 'induced' or 'adaptive' T_{reg} cells, differentiate from naive CD4 T cells in the periphery. Pathogen-specific induced T_{reg} cells can be elicited in response to infectious agents and may normally curb effector responses as a mechanism to control immunopathology and restore homeostasis as an infection is cleared. However, in some infections, the induction of T_{reg} cells is promoted by the pathogen, which thus avoids clearance and can set up a chronic infection. This mechanism seems to contribute to chronic liver infections caused by HBV and HCV, and perhaps to HIV persistence. Patients infected with HBV and HCV have elevated numbers of FoxP3$^+$ T_{reg} cells in the circulation and in the liver, and *in vitro* depletion of T_{reg} cells enhances cytotoxic lymphocyte responses against the virus. During infections with the protozoan parasite *Leishmania major*, T_{reg} cells accumulate in the dermis, where they impair the ability of effector T cells to eliminate pathogens from this site.

In contrast, studies in both humans and mice have shown that the inflammation occurring during ocular infections with HSV is limited by the presence of T_{reg} cells. If these cells are depleted from mice before HSV infection, a more severe disease results, even when smaller doses of virus are used to cause infection. T_{reg} cells also restrain inflammation in the pulmonary disease that occurs in immunodeficient mice infected with the opportunistic yeast-like fungal pathogen *Pneumocystis jirovecii* (formerly known as *Pneumocystis carinii*), which is a common pathogen in immunodeficient humans.

Summary.

Infectious agents can cause recurrent or persistent disease by avoiding normal host defense mechanisms or by subverting them to promote their own replication. There are many different ways of evading or subverting the immune response. Antigenic variation, latency, resistance to immune effector mechanisms, and suppression of the immune response all contribute to persistent and medically important infections. In some cases the immune response is part of the problem: some pathogens use immune activation to spread infection, and others would not cause disease if it were not for the immune response. Each of these mechanisms teaches us something about the nature of the immune response and its weaknesses, and each requires a different medical approach to prevent or to treat infection.

Immunodeficiency diseases.

Immunodeficiencies occur when one or more components of the immune system are defective; immunodeficiencies are classified as primary (or congenital) or secondary. **Primary immunodeficiencies** are caused by inherited mutations in any of a large number of genes that are involved in or control immune responses. Well over 100 primary immunodeficiencies have been described that affect the development of immune cells, their function, or both. Clinical features of these disorders are therefore highly variable, although a common feature is recurrent and often overwhelming infections in very young children. Allergy, abnormal proliferation of lymphocytes, autoimmunity, and certain types of cancer can also occur. In contrast, **secondary immunodeficiencies** are acquired as a consequence of other diseases, or are secondary to environmental factors such as starvation, or are an adverse consequence of medical intervention.

Primary immunodeficiencies can be classified on the basis of the component of the immune system involved. Adaptive immune defects include combined immunodeficiencies that compromise T- and B-cell immunity, or those limited to antibody deficiencies alone. Innate immune defects include deficiencies of complement, phagocytes, and TLR signaling.

By examining which infectious diseases accompany a particular inherited (or acquired) immunodeficiency, we gain insights into components of the immune system that are important in the response to particular agents. The inherited immunodeficiencies also reveal how interactions between different cell types contribute to the immune response and to the development of T and B lymphocytes. Finally, these inherited diseases can lead us to the defective gene, often revealing new information about the molecular basis of immune processes and providing the necessary information for diagnosis, for genetic counseling, and eventually the possibility of gene therapy.

13-7 A history of repeated infections suggests a diagnosis of immunodeficiency.

Patients with immune deficiency are usually detected clinically by a history of recurrent infection with the same or similar pathogens. The type of infection is a guide to which part of the immune system is deficient. Recurrent infection by pyogenic, or pus-forming, bacteria suggests a defect in antibody, complement, or phagocyte function, reflecting the role of these parts of the immune system in defense against such infections. In contrast, a history of persistent fungal skin infection, such as cutaneous candidiasis, or recurrent viral infections is more suggestive of a defect in host defense mediated by T lymphocytes.

13-8 Primary immunodeficiency diseases are caused by inherited gene defects.

Before the advent of antibiotics, it is likely that most individuals with inherited immune defects died in infancy or early childhood because of their susceptibility to particular classes of pathogens. Such cases were not easily identified, because many normal infants also died of infection. Most of the gene defects that cause inherited immunodeficiencies are recessive, and many are caused by mutations in genes on the X chromosome. As males have only one X chromosome, all males who inherit an X chromosome carrying a defective gene will be affected by the disease. In contrast, female carriers with one defective X chromosome are usually healthy. Immunodeficiency diseases that affect various steps in B- and T-lymphocyte development have been described, as have defects in surface or signaling molecules that are important for T- or B-cell function. Defects in phagocytic cells, in complement, in cytokines, in cytokine receptors, and in molecules that mediate effector responses also occur. Thus, immunodeficiency can be caused by defects in either the adaptive or the innate immune system.

Gene knockout techniques in mice (see Appendix I, Section A-46) have created many immunodeficient states that are adding rapidly to our knowledge of the contribution of individual proteins to normal immune function. Nevertheless, human immunodeficiency diseases remain the best source of insight into the normal pathways of defense against infectious diseases. For example, a deficiency of antibody, of complement, or of phagocytic function each increases the risk of infection by certain pyogenic bacteria. This shows that the normal pathway of host defense against such bacteria is the binding of antibody followed by the fixation of complement, which allows the opsonized bacteria to be taken up by phagocytic cells and killed. Breaking any of the links in this chain of events causes a similar immunodeficient state.

Immunodeficiencies also teach us about the redundancy of defense mechanisms against infectious disease. For example, the first two people to be discovered with a hereditary deficiency of complement were healthy immunologists who used their own blood in their experiments. This teaches us that there are multiple protective immune mechanisms against infection, such that a defect in one component of immunity might be compensated for by other components. Thus, although there is abundant evidence that complement deficiency increases susceptibility to pyogenic infection, not every human with complement deficiency suffers from recurrent infections.

Examples of immunodeficiency diseases are listed in Fig. 13.7. None is very common (a selective deficiency in IgA being the most frequently reported), and some are extremely rare. Some of these diseases are described in subsequent sections, and we have grouped the diseases according to where the

Name of deficiency syndrome	Specific abnormality	Immune defect	Susceptibility
Severe combined immune deficiency	See text and Fig. 13.8		General
DiGeorge's syndrome	Thymic aplasia	Variable numbers of T cells	General
MHC class I deficiency	TAP mutations	No CD8 T cells	Chronic lung and skin inflammation
MHC class II deficiency	Lack of expression of MHC class II	No CD4 T cells	General
Wiskott–Aldrich syndrome	X-linked; defective WASP gene	Defective anti-polysaccharide antibody, impaired T-cell activation responses, and T_{reg} dysfunction	Encapsulated extracellular bacteria Herpesvirus infections (e.g., HSV, EBV)
X-linked agammaglobulinemia	Loss of Btk tyrosine kinase	No B cells	Extracellular bacteria, viruses
Hyper IgM syndrome	CD40 ligand deficiency CD40 deficiency NEMO (IKK) deficiency	No isotype switching and/or no somatic hypermutation plus T-cell defects	Extracellular bacteria *Pneumocystis jirovecii* *Cryptosporidium parvum*
Hyper IgM syndrome—B-cell intrinsic	AID deficiency UNG deficiency	No isotype switching +/– normal somatic hypermutation	Extracellular bacteria
Hyper IgE syndrome (Job's syndrome)	Defective STAT3	Block in T_H17 cell differentiation Elevated IgE	Extracellular bacteria and fungi
Common variable immunodeficiency	ICOS deficiency, other unknown	Defective IgA and IgG production	Extracellular bacteria
Selective IgA	Unknown; MHC-linked	No IgA synthesis	Respiratory infections
Phagocyte deficiencies	Many different	Loss of phagocyte function	Extracellular bacteria and fungi
Complement deficiencies	Many different	Loss of specific complement components	Extracellular bacteria especially *Neisseria* spp.
X-linked lymphoproliferative syndrome	SAP (SH2D1A) mutant	Inability to control B-cell growth	EBV-driven B-cell tumors Fatal infectious mononucleosis
Ataxia telangiectasia	Mutation of kinase domain of ATM	T cells reduced	Respiratory infections
Bloom's syndrome	Defective DNA helicase	T cells reduced Reduced antibody levels	Respiratory infections

Fig. 13.7 Human immunodeficiency syndromes. The specific gene defect, the consequence for the immune system, and the resulting disease susceptibilities are listed for some common and some rare human immunodeficiency syndromes. Severe combined immunodeficiency (SCID) can be due to many different defects, as summarized in Fig. 13.8 and described in the text. AID, activation-induced cytidine deaminase; ATM, ataxia telangiectasia-mutated protein; EBV, Epstein–Barr virus; IKKγ, γ subunit of the kinase IKK; STAT3, signal transducer and activator of transcription 3; TAP, transporters associated with antigen processing; UNG, uracil-DNA glycosylase; WASP, Wiskott–Aldrich syndrome protein.

specific causal defects lie along the development and activation pathways of the T- and B-cell lineages.

13-9 Defects in T-cell development can result in severe combined immunodeficiencies.

The developmental pathways leading to circulating naive T cells and B cells are summarized in Fig. 13.8. Patients with defects in T-cell development are highly susceptible to a broad range of infectious agents. This demonstrates the central role of T-cell differentiation and maturation in adaptive immune responses to virtually all antigens. Because such patients make neither T-cell dependent antibody responses nor cell-mediated immune responses, and thus cannot develop immunological memory, they are said to suffer from **severe combined immunodeficiency** (**SCID**).

X-linked SCID (**XSCID**) is the most frequent form of SCID and is caused by mutations in the gene *IL2RG* on the human X chromosome, which encodes the interleukin-2 receptor (IL-2R) common gamma chain (γ_c). γ_c is required in all receptors of the IL-2 cytokine family (IL-2, IL-4, IL-7, IL-9, IL-15, and

Fig. 13.8 Defects in T-cell and B-cell development that cause immunodeficiency. The pathways leading to circulating naive T cells and B cells are shown here. Mutations in genes that encode the proteins (indicated in red boxes) are known to cause human immunodeficiency diseases. BCR, B-cell receptor; CLP, common lymphoid progenitor; DP T cell, double-positive T cell (see Chapter 8); HSC, hematopoietic stem cell; MZ B cell, marginal zone B cell; pre-BCR, pre-B-cell receptor; pre-TCR, pre T-cell receptor; RSCID, radiation-sensitive SCID; SCID, severe combined immunodeficiency; TCR, T-cell receptor; XSCID, X-linked SCID. See the text for details. Immunodeficiency can also be caused by mutations in genes in the thymic epithelium that impair thymic development, and thus T-cell development (see text).

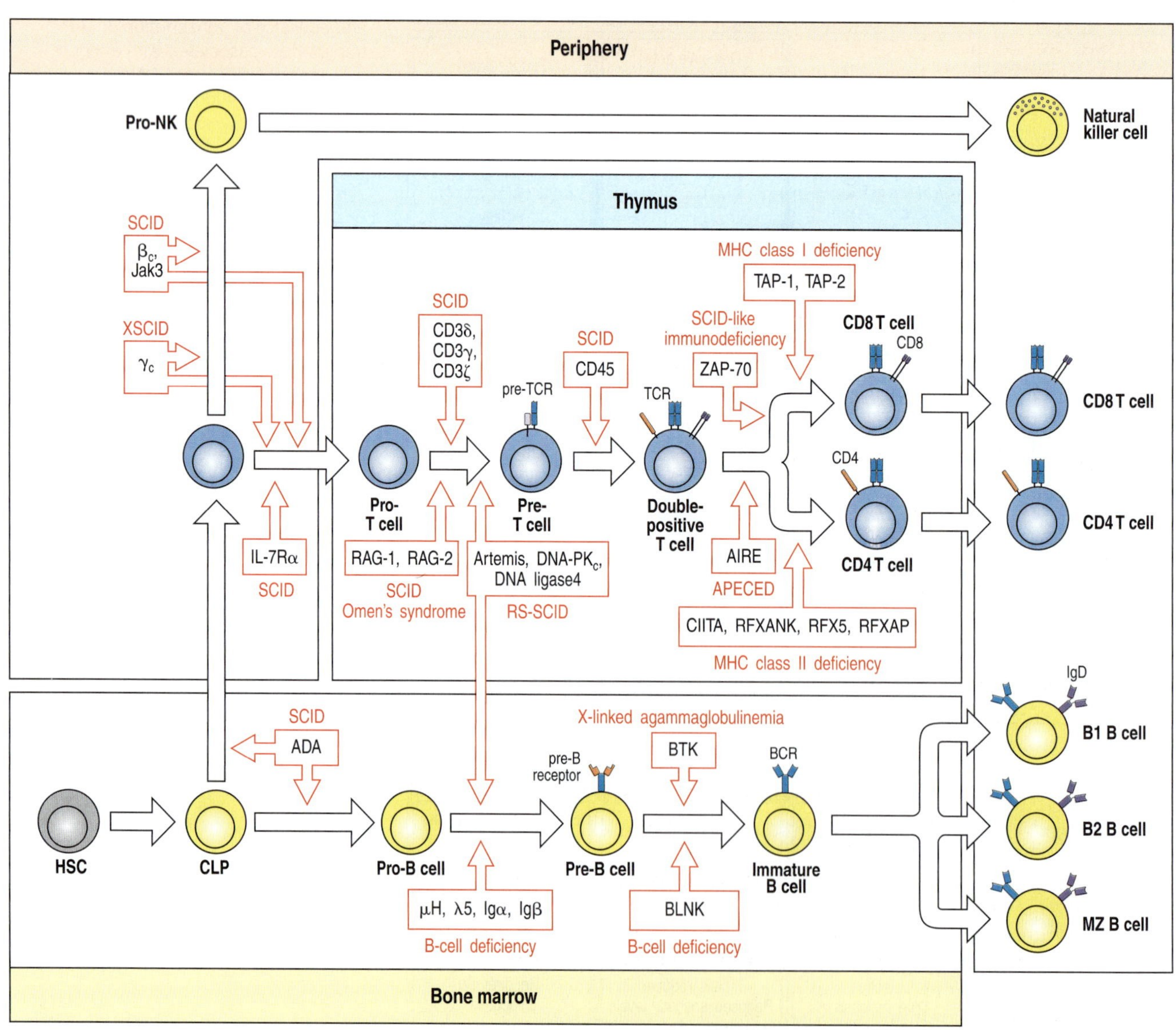

IL-21), and two of these cytokines (and their receptors) are essential for the early development of T-cell progenitors (IL-7) or NK cells (IL-15) (see Fig. 13.8). Patients with XSCID, who are overwhelmingly likely to be boys, thus have defects in signaling of all IL-2-family cytokines and, owing to the defects in IL-7 and IL-15, T cells and NK cells fail to develop normally, whereas B-cell numbers, but not function, are normal. XSCID is also known as the 'bubble boy disease' after a boy with XSCID who lived in a protective bubble for more than a decade before he died after an unsuccessful bone marrow transplant. A clinically and immunologically indistinguishable type of SCID is associated with an inactivating mutation in the kinase Jak3 (see Section 7-20), which physically associates with γ_c and transduces signaling through γ_c-chain cytokine receptors. This autosomal recessive mutation also impairs the development of T and NK cells, but the development of B cells is unaffected.

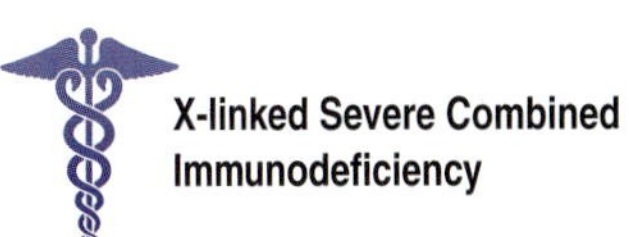

Other immunodeficiencies in humans and mice have pinpointed more precisely the roles of individual cytokines and their receptors in T-cell and NK-cell development. For example, a child was reported with SCID who lacked NK cells and T cells but had normal genes for γ_c and Jak3 kinase. It transpired that he had a deficiency of the common beta chain (β_c) shared by the IL-2 and IL-15 receptors. This child, and mice with targeted mutations in the β_c gene (*IL2RB*), defined a key role for IL-15 as a growth factor for the development of NK cells (see Fig. 13.8), as well as a role for the cytokine in T-cell maturation and trafficking. Mice with targeted mutations in IL-15 itself or the α chain of its receptor have no NK cells and relatively normal T-cell development, but they show reduced T-cell homing to peripheral lymphoid tissues and a decrease in the number of CD8-positive T cells.

Humans with a deficiency of the IL-7 receptor α chain have no T cells but normal levels of NK cells, illustrating that IL-7 signaling is not essential for the development of NK cells (see Fig. 13.8). In humans and mice whose T cells show defective production of IL-2 after receptor stimulation, T-cell development itself is normal. The more limited effects of individual cytokine signaling defects are in contrast to the global defects in T- and NK-cell development in patients with XSCID.

As in all serious T-cell deficiencies, patients with XSCID do not make effective antibody responses to most antigens, although their B cells seem normal. Most, but not all, naive IgM-positive B cells from female carriers of XSCID have inactivated the defective X chromosome rather than the normal one (see Section 13-14), showing that B-cell development is affected by, but is not wholly dependent on, the γ_c chain. Mature memory B cells that have undergone class switching have inactivated the defective X chromosome almost without exception. This might reflect the fact that the γ_c chain is also part of the receptors for IL-4 and IL-21. Thus, B cells that lack this chain have defective IL-4 and IL-21 receptors and do not proliferate in T-cell-dependent antibody responses (see Section 10-4).

13-10 SCID can also be due to defects in the purine salvage pathway.

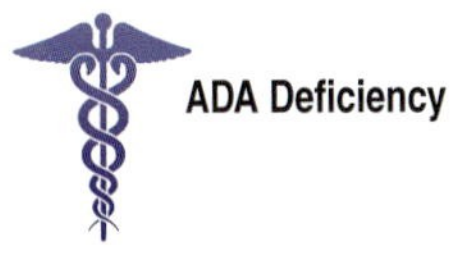

Variants of autosomal recessive SCID that arise from defects in enzymes of the salvage pathway of purine synthesis include **adenosine deaminase (ADA) deficiency** and **purine nucleotide phosphorylase (PNP) deficiency** (see Fig. 13.8). ADA catalyzes the conversion of adenosine and deoxyadenosine to inosine and deoxyinosine, respectively, and its deficiency results in the accumulation of deoxyadenosine and its precursor, *S*-adenosylhomocysteine, which are toxic to developing T and B cells. PNP catalyzes the conversion of inosine and guanosine to hypoxanthine and guanine, respectively. PNP deficiency, which is a rarer form of SCID, also causes the accumulation of toxic precursors but affects developing T cells more severely than B cells. In both diseases, the development of lymphopenia is progressive after birth, resulting in profound lymphopenia within the first few years of life. Because both enzymes are housekeeping proteins expressed by many cell types, the immune deficiency associated with each of these inherited defects is part of a broader clinical syndrome.

13-11 Defects in antigen receptor gene rearrangement can result in SCID.

Another set of autosomally inherited defects leading to SCID is that caused by failures of DNA rearrangement in developing lymphocytes. Mutations in either the *RAG1* or *RAG2* gene that result in nonfunctional proteins cause arrest of lymphocyte development at the pro- to pre-T cell and B-cell transitions because of a failure of V(D)J recombination (see Fig. 13.8). Thus there is a complete lack of T cells and B cells in these patients. Because the effects of RAG deficiencies are limited to lymphocytes that undergo antigen gene rearrangement, NK-cell development is not impaired in these patients. There are other children with **hypomorphic** mutations (which cause reduced, but not absent, function) in either *RAG1* or *RAG2* who can make a small amount of functional RAG protein, allowing limited V(D)J recombination. This latter group suffers from a distinctive and severe disease called **Omenn syndrome**, which, in addition to increased susceptibility to multiple opportunistic infections, has clinical features very similar to graft-versus-host disease (see Section 15-36), with rashes, eosinophilia, diarrhea, and enlargement of the lymph nodes. Normal or increased numbers of T cells, all of which are activated, are found in these unfortunate children. A possible explanation for this phenotype is that very low levels of *RAG* activity allow some limited T-cell receptor gene recombination. No B cells are found, however, suggesting that B cells have more stringent requirements for *RAG* activity. The T cells produced in patients with Omenn syndrome have an abnormal and highly restricted receptor repertoire, both in the thymus and in the periphery, where they have undergone activation and clonal expansion. The clinical features strongly suggest that these peripheral T cells are autoreactive and are responsible for the graft-versus-host phenotype.

A subset of patients with autosomal recessive SCID are characterized by an abnormal sensitivity to ionizing radiation. They produce very few mature B and T cells because there is a failure of DNA rearrangement in their developing lymphocytes; only rare VJ or VDJ joints are seen, and most of these are abnormal. This type of SCID is due to defects in ubiquitous DNA repair proteins involved in repairing DNA double-strand breaks, which are generated not only during antigen receptor gene rearrangement (see Section 5-5) but also by ionizing radiation. Owing to the increased radiosensitivity in these patients, this class of SCID is called **radiation-sensitive SCID** (**RS-SCID**) to distinguish it from SCID due to lymphocyte-specific defects. Defects in the genes for Artemis, DNA protein-kinase catalytic subunit (DNA-PKcs), and DNA ligase IV cause RS-SCID (see Fig. 13.8). Because defects in repair of DNA breaks increase the risk of translocations during cell division that can lead to malignant transformation, patients with these SCID variants can also be more likely to develop cancer.

13-12 Defects in signaling from T-cell antigen receptors can cause severe immunodeficiency.

Several gene defects have been described that interfere with signaling through the T-cell receptor (TCR), and thus block the activation of T cells early in thymic development. After productive rearrangement of the VDJ genes of the TCRβ-chain locus in pro-T cells (double-negative thymocytes), the TCRβ chain must complex with the pTα chain and accessory components of the CD3 complex (CD3γ, CD3δ, and CD3ε) to be transported to the cell surface as the pre-T-cell receptor (pre-TCR) (see Section 8-9). Assembly and expression of the pre-TCR, and its signaling through these CD3 chains and the CD3ζ dimer, represent a critical checkpoint in T-cell development and promote transition to the pre-T-cell stage that is characterized by coexpression of CD4 and CD8 (double-positive thymocytes) and initiation of TCRα-chain gene rearrangements. Patients with mutations in the CD3δ, CD3ε, or CD3ζ chains of the CD3 complex have defective pre-TCR signaling and fail to progress to the double-positive stage of thymic development (see Fig. 13.8), resulting in SCID. Another lymphocyte signaling defect that leads to severe immunodeficiency is caused by mutations in the tyrosine phosphatase CD45. Humans and mice with CD45 deficiency show a marked reduction in peripheral T-cell numbers and also abnormal B-cell maturation.

Although not strictly classifiable as SCID, severe immunodeficiency occurs in patients who make a defective form of the cytosolic protein tyrosine kinase ZAP-70, which transmits signals from the T-cell receptor (see Section 7-9). CD4 T cells emerge from the thymus in normal numbers, whereas CD8 T cells are absent. However, the CD4 T cells that mature fail to respond to stimuli that normally activate the cells through the T-cell receptor.

Wiskott–Aldrich syndrome (**WAS**) has shed new light on the molecular basis of signaling and immune synapse formation between various cells in the immune system. The disease also affects platelets and was first described as a blood-clotting disorder, but it is also associated with immunodeficiency due to impaired lymphocyte function, leading to reduced T-cell numbers, defective NK-cell cytotoxicity, and a failure of antibody responses (see Section 10-13). WAS is caused by a defective gene on the X chromosome, encoding a protein called WAS protein (WASP). WASP is expressed in all hematopoietic cell lineages and is likely to be a key regulator of lymphocyte and platelet development and to function through its effects on the actin cytoskeleton, which is critical for immune synapse formation and the polarization of effector T cells (see Section 9-21). It has also recently been suggested that WASP is required for the suppressive function of natural T_{reg} cells, and this may help explain why patients with WASP are susceptible to autoimmune diseases. WASP has a key role in transducing signals to the cytoskeletal framework of cells because it activates the Arp2/3 complex, which is essential for initiating actin polymerization. In patients with WAS, and in mice whose *WASP* gene has been knocked out, T cells fail to respond normally to T-cell receptor cross-linking. Several signaling pathways leading from the T-cell receptor are known to activate WASP. One involves the scaffold protein SLP-76, which serves as the binding site for an adaptor protein, Nck, which in turn binds WASP. WASP can also be activated by small GTP-binding proteins, notably Cdc42 and Rac1, which can themselves be activated via the T-cell receptor signaling through the adaptor protein Vav (see Chapter 7).

13-13 Genetic defects in thymic function that block T-cell development result in severe immunodeficiencies.

A disorder of thymic development, associated with SCID and a lack of body hair, has been known for many years in mice; the mutant strain is descriptively named ***nude*** (see Section 8-7). A small number of children have been described with the same phenotype. In both mice and humans this syndrome is caused by mutations in the gene *FOXN1* (also known as *WHN*), which encodes a transcription factor selectively expressed in skin and thymus. FOXN1 is necessary for the differentiation of thymic epithelium and the formation of a functional thymus. In patients with a mutation in *FOXN1*, the lack of thymic function prevents normal T-cell development. In many cases, B-cell development is normal in individuals with the mutation, yet the response to nearly all pathogens is profoundly impaired because of the lack of T cells.

DiGeorge Syndrome

MHC Class II Deficiency

MHC Class I Deficiency

DiGeorge syndrome is another disorder in which the thymic epithelium fails to develop normally, resulting in SCID. The genetic abnormality underlying this complex developmental disorder is a deletion within one copy of chromosome 22. The deletion varies between 1.5 and 5 megabases in size, with the smallest deletion that causes the syndrome containing approximately 24 genes. The relevant gene within this interval is *TBX1*, which encodes the transcription factor T-box 1. DiGeorge syndrome is caused by the deletion of a single copy of this gene, such that patients with this disorder are **haploinsufficient** for *TBX1*. Without the proper inductive thymic environment, T cells cannot mature, and both cell-mediated immunity and T-cell dependent antibody production are impaired. Patients with this syndrome have normal levels of serum immunoglobulin but an absence of, or incomplete development of, the thymus and parathyroid glands, with varying degrees of T-cell immunodeficiency.

Defects in the expression of MHC molecules can lead to severe immunodeficiency as a result of effects on the positive selection of T cells in the thymus (see Fig. 13.8). Individuals with **bare lymphocyte syndrome** lack expression of all MHC class II molecules, and the disease is now called **MHC class II deficiency**. Because the thymus lacks MHC class II molecules, CD4 T cells cannot be positively selected and few develop. The antigen-presenting cells in these individuals also lack MHC class II molecules and so the few CD4 T cells that do develop cannot be stimulated by antigen. MHC class I expression is normal, and CD8 T cells develop normally. However, such people suffer from severe immunodeficiency, illustrating the central importance of CD4 T cells in adaptive immunity to most pathogens.

MHC class II deficiency is caused not by mutations in the MHC genes themselves but by mutations in one of several genes encoding gene-regulatory proteins that are required for the transcriptional activation of MHC class II promoters. Four complementing gene defects (known as groups A, B, C, and D) have been defined in patients who fail to express MHC class II molecules, indicating that the products of at least four different genes are required for the normal expression of these proteins. Genes corresponding to each complementation group have been identified: the *MHC class II transactivator*, or *CIITA*, is mutated in group A, and the genes *RFXANK*, *RFX5*, and *RFXAP* are mutated in groups B, C, and D, respectively (see Fig. 13.8). These last three encode proteins that are components of a multimeric complex, RFX, which is involved in the control of gene transcription. RFX binds a DNA sequence named an X-box, which is present in the promoter region of all MHC class II genes.

A more limited immunodeficiency, associated with chronic respiratory bacterial infections and skin ulceration with vasculitis, has been observed in a small number of patients who have almost no cell-surface MHC class I molecules—a condition known as **MHC class I deficiency**. Affected individuals have normal levels of mRNA encoding MHC class I molecules and normal production of MHC class I proteins, but very few of the proteins reach the cell surface. This condition is due to mutations in either *TAP1* or *TAP2*, which encode the subunits of the peptide transporter responsible for transporting peptides generated in the cytosol into the endoplasmic reticulum, where they are loaded into nascent MHC I molecules. This defect is similar to that in the *TAP* mutant cells mentioned in Section 6-2. By analogy with MHC class II deficiency, the absence of MHC class I molecules on the surfaces of thymic epithelial cells results in a lack of CD8 T cells expressing the $\alpha{:}\beta$ T-cell receptor (see Fig. 13.8), but patients do have $\gamma{:}\delta$ CD8 T cells, certain subsets of which develop independently of the thymus. People with MHC class I deficiency are not abnormally susceptible to viral infections, which is surprising given the key role of MHC class I presentation and of cytotoxic CD8 T cells in combating viral infections. There is, however, evidence for TAP-independent pathways for the presentation of certain peptides by MHC class I molecules, and the clinical phenotype of *TAP1*- and *TAP2*-deficient patients indicates that these pathways may be sufficient to allow viruses to be controlled.

Some defects in thymic cells lead to a phenotype with other effects besides those of immunodeficiency. The gene *AIRE* encodes a transcription factor that enables thymic epithelial cells to express many proteins and so to mediate efficient negative selection. Defects in *AIRE* lead to a complex syndrome called **APECED**, characterized by immunodeficiency, autoimmunity, and developmental defects (see Section 8-20).

13-14 Defects in B-cell development result in deficiencies in antibody production that cause an inability to clear extracellular bacteria.

In addition to inherited defects in proteins that are crucial to both T-cell and B-cell development, such as RAG-1 and RAG-2, defects in proteins that are specific to B-cell development have also been identified (see Fig. 13.8). Patients with these defects are characterized by an inability to cope with extracellular bacteria and some viruses whose efficient clearance requires specific antibodies. Pyogenic bacteria, such as staphylococci and streptococci, have polysaccharide capsules that are not directly recognized by the receptors on macrophages and neutrophils that stimulate phagocytosis. The bacteria escape elimination by the innate immune response and are successful extracellular pathogens, but can be cleared by an adaptive immune response. In this, opsonization by antibody and complement enables phagocytes to ingest and destroy the bacteria (see Section 10-22). The principal effect of deficiencies in antibody production is therefore a failure to control infections by pyogenic bacteria. Susceptibility to some viral infections, notably those caused by enteroviruses, is also increased because of the importance of antibodies in neutralizing viruses that enter the body through the gut.

The first description of an immunodeficiency disease was Ogden C. Bruton's account, in 1952, of the failure of a male child to produce antibody. Because inheritance of this condition is X-linked and is characterized by the absence of immunoglobulin in the serum (**agammaglobulinemia**), it was called **Bruton's X-linked agammaglobulinemia (XLA)** (see Fig. 13.8). Since then, many more defects of antibody production have been described. Infants with these diseases are usually identified as a result of recurrent infections with pyogenic bacteria such as *Streptococcus pneumoniae* and chronic infections with viruses such as HBV, HCV, poliovirus, and ECHO virus. In this regard, it should be noted that normal infants have a transient deficiency in immunoglobulin production in the first 3–12 months of life. The newborn infant has antibody levels comparable to those of the mother because of the transplacental transport of maternal IgG (see Section 10-15). As this IgG is catabolized, antibody levels gradually decrease until the infant begins to produce significant amounts of its own IgG at about 6 months (Fig. 13.9). Thus, IgG levels are

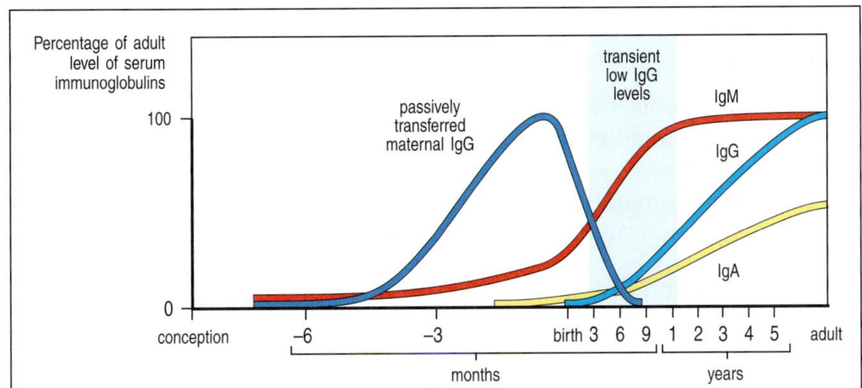

Fig. 13.9 Immunoglobulin levels in newborn infants fall to low levels at about 6 months of age. Babies are born with high levels of IgG, which is actively transported across the placenta from the mother during gestation. After birth, the production of IgM starts almost immediately; the production of IgG, however, does not begin for about 6 months, during which time the total level of IgG falls as the maternally acquired IgG is catabolized. Thus, IgG levels are low from about the age of 3 months to 1 year, which can lead to susceptibility to disease.

quite low between the ages of 3 months and 1 year. This can lead to a period of heightened susceptibility to infection, especially in premature babies, who begin with lower levels of maternal IgG and also reach immune competence later after birth. Because of the transient protection afforded newborn infants by maternal antibodies, XLA is typically detected several months after birth, when maternal antibody levels in the infant have declined.

The defective gene in XLA encodes a protein tyrosine kinase called BTK (Bruton's tyrosine kinase), which is a member of the Tec family of kinases that transduce signals through the pre-B-cell receptor (pre-BCR) (see Section 7-16). As discussed in Section 8-3, the pre-BCR is composed of successfully rearranged μ heavy chains complexed with the surrogate light chain composed of λ5 and VpreB, and with the signal-transducing subunits Igα and Igβ. Stimulation of the pre-BCR recruits cytoplasmic proteins, including BTK, which convey signals required for the proliferation and differentiation of pre-B cells. In the absence of BTK function, B-cell maturation is largely arrested at the pre-B-cell stage (see Fig. 13.8; see also Section 8-3), resulting in profound B-cell deficiency and agammaglobulinemia. Some B cells do mature, however, perhaps as a result of compensation by other Tec kinases.

During embryonic development, females randomly inactivate one of their two X chromosomes. Because BTK is required for B-lymphocyte development, only cells in which the normal allele of *BTK* is active develop into mature B cells. Thus, in all B cells in female carriers of a mutant *BTK* gene, the active X chromosome is the normal one and the abnormal X chromosome is inactivated. This fact allowed female carriers of XLA to be identified even before the nature of the BTK protein was known. In contrast, the active X chromosomes in the T cells and macrophages of carriers are an equal mixture of normal and *BTK* mutant X chromosomes. Nonrandom X inactivation only in B cells shows conclusively that the *BTK* gene is required for the development of B cells but not of other cell types, and that BTK must act in the B cells themselves rather than in stromal or other cells required for B-cell development (Fig. 13.10).

Autosomal recessive deficiencies in other components of the pre-BCR also block early B-cell development and cause severe B-cell deficiency and congenital agammaglobulinemia similar to that of XLA (see Fig. 13.8). These disorders are much rarer than XLA, and include mutations in the genes that encode the μ heavy chain (*IGHM*), which is the second most common cause of agammaglobulinemia, λ5 (*IGLL1*), and Igα (*CD79A*) and Igβ (*CD79B*). Mutations that cripple the B-cell receptor signaling adaptor, B-cell linker protein (encoded by *BLNK*), also cause the arrest of early B-cell development that results in selective B-cell deficiency.

Patients with pure B-cell defects resist many pathogens other than pyogenic bacteria. Fortunately, the latter can be suppressed with antibiotics and with monthly infusions of human immunoglobulin collected from a large pool of donors. Because there are antibodies against many common pathogens in this pooled immunoglobulin, it serves as a fairly successful shield against infection.

13-15 Immune deficiencies can be caused by defects in B-cell or T-cell activation and function.

After their development in the bone marrow or thymus, B and T cells require antigen-driven activation and differentiation to mount effective immune responses. Analogous to defects in early T-cell development, defects in T-cell activation and differentiation that occur after thymic selection have an impact on both cell-mediated immunity and antibody responses as a result of deficient help for B-cell class switching (Fig. 13.11). Defects specific to the

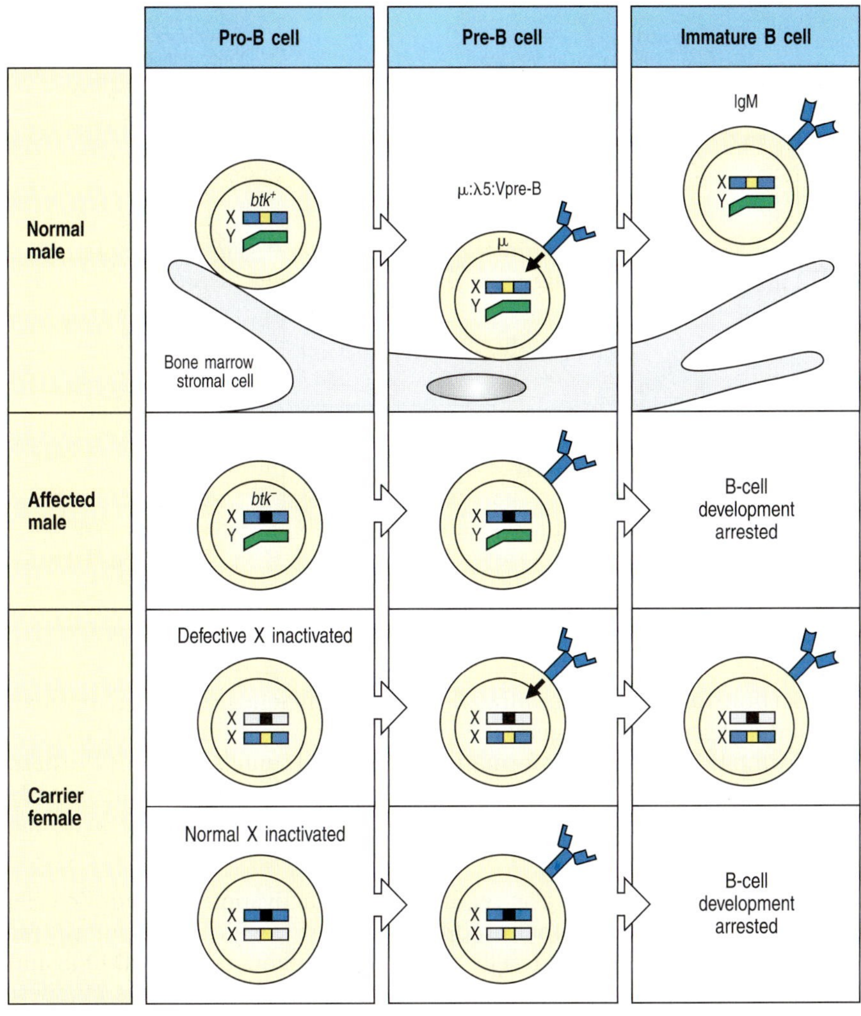

Fig. 13.10 The product of the *BTK* gene is important for B-cell development. In X-linked agammaglobulinemia (XLA), a protein tyrosine kinase of the Tec family called Btk, which is encoded on the X chromosome, is defective. In normal individuals, B-cell development proceeds through a stage in which the pre-B-cell receptor, consisting of μ:λ5:VpreB (see Section 8-3), transduces a signal via Btk, triggering further B-cell development. In males with XLA, no signal can be transduced and, although the pre-B-cell receptor is expressed, the B cells develop no further. In female mammals, including humans, one of the two X chromosomes in each cell is permanently inactivated early in development. Because the choice of which chromosome to inactivate is random, half of the pre-B cells in a carrier female will have inactivated the chromosome with the wild-type *BTK* gene, meaning that they can express only the defective *btk* gene and cannot develop further. Therefore, in the carrier, mature B cells always have the nondefective X chromosome active. This is in sharp contrast to all other cell types, which have the nondefective X chromosome active in only half of their cells. Nonrandom X-chromosome inactivation in a particular cell lineage is a clear indication that the product of the X-linked gene is required for the development of cells of that lineage. It is also sometimes possible to identify the stage at which the gene product is required, by detecting the point in development at which X-chromosome inactivation develops bias. Using this kind of analysis, one can identify carriers of X-linked traits such as XLA without needing to know the nature of the mutant gene.

activation and differentiation of B cells can impair their ability to undergo class switching to IgG, IgA, and IgE while leaving cell-mediated immunity largely intact. Depending on where in the process of T- or B-cell differentiation these defects occur, the characteristics of the immune deficiency that results can be either profound or relatively circumscribed.

A common feature of patients with defects that affect B-cell class switching is **hyper IgM syndrome** (see Fig. 13.11). These patients have normal B- and T-cell development and normal or high serum levels of IgM, but make very limited antibody responses against antigens that require T-cell help. Thus immunoglobulin isotypes other than IgM and IgD are produced only in trace amounts. This renders these patients highly susceptible to infection with extracellular pathogens. Several causes of hyper IgM syndrome have been distinguished, and these have helped to elucidate the pathways that are essential for normal class-switch recombination and somatic hypermutation in B cells. Defects have been found in both T-cell helper function and in the B cells themselves.

The most common form of hyper IgM syndrome is **X-linked hyper IgM syndrome**, or **CD40 ligand deficiency**, which is caused by mutations in the gene encoding CD40 ligand (CD154) (see Fig. 13.11). CD40 ligand is normally expressed on activated T cells, enabling them to engage the CD40 protein on antigen-presenting cells, including B cells, dendritic cells, and macrophages

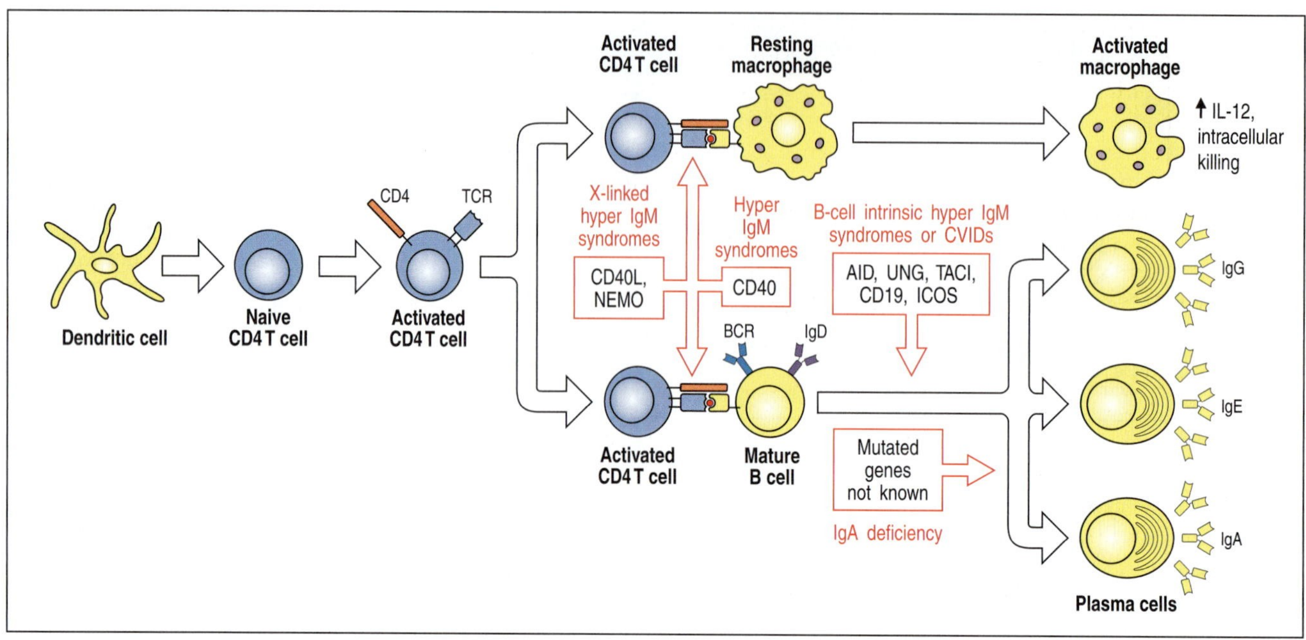

Fig. 13.11 Defects in T-cell and B-cell activation and differentiation cause immunodeficiencies. The pathways leading to activation and differentiation of naive T cells and B cells are shown here. Genes known to be mutated in the relevant human immunodeficiency diseases are indicated in red boxes. BCR, B-cell receptor; CVIDs, common variable immunodeficiencies; TCR, T-cell receptor. Note that the defect in cytoskeletal function in Wiskott–Aldrich syndrome (WAS) affects immune-cell function at many steps in this schema, and is not included in the figure for the sake of clarity. See the text for details.

(see Section 10-4). In males with CD40 ligand deficiency, B cells are normal, but in the absence of engagement of CD40, their B cells do not undergo isotype switching or initiate the formation of germinal centers (Fig. 13.12). These patients therefore have severe reductions in circulating levels of all antibody isotypes except IgM and are highly susceptible to infections by pyogenic extracellular bacteria.

Because CD40 signaling is also required for the activation of dendritic cells and macrophages for optimal production of IL-12, which is important for the production of IFN-γ by T cells and NK cells, patients with CD40 ligand deficiency also have defects in cell-mediated immunity and thus manifest a form of combined immunodeficiency. Inadequate cross-talk between T cells and dendritic cells via CD40L–CD40 interaction can lead to lower levels of co-stimulatory molecules on dendritic cells, thus impairing their ability to stimulate naive T cells (see Section 9-14). These patients are therefore susceptible to infections by extracellular pathogens that require class-switched antibodies, such as pyogenic bacteria, but also have defects in the clearance of intracellular pathogens, such as mycobacteria, and are particularly prone to opportunistic infections by *P. jirovecii*, which is normally killed by activated macrophages.

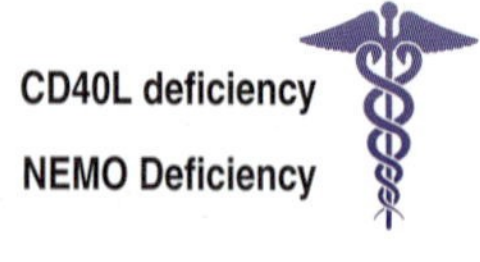

A very similar syndrome has been identified in patients with mutations in two other genes (see Fig. 13.11). Not unexpectedly, one is the gene encoding CD40 on chromosome 20, mutations in which have been found in several patients with an autosomal recessive variant of hyper IgM syndrome. In another form of X-linked hyper IgM syndrome, known as **NEMO deficiency**, mutations occur in the gene encoding the protein NEMO (also known as IKKγ, a subunit of the kinase IKK), which is an essential component of the intracellular signaling pathway from CD40 that leads to activation of the transcription factor NFκB (see Fig. 3.13). This group of hyper IgM syndromes shows that mutations at different points in the CD40L–CD40 signaling pathway result in a similar combined immunodeficiency syndrome.

Other variants of hyper IgM syndrome are due to intrinsic defects in the process of B-cell class switch recombination. These patients are susceptible to severe extracellular bacterial infections, but because T-cell differentiation and function are spared, they do not show increased susceptibility to intracellular pathogens or opportunistic agents such as *P. jirovecii*. One class-switching defect is due to mutations in the gene for activation-induced cytidine deaminase (AID), which is required for both somatic hypermutation and class switching (see Section 5-17). Patients with autosomally inherited defects in the AID gene (*AICDA*) fail to switch antibody isotype and also have greatly reduced somatic hypermutation (see Fig. 13.11). Immature B cells accumulate in abnormal germinal centers, causing enlargement of the lymph nodes and spleen. Another variant of B-cell-intrinsic hyper IgM syndrome was identified recently in a small number of patients with an autosomal recessive defect in the DNA repair enzyme uracil-DNA glycosylase (UNG) (see Section 5-17), which is also involved in class switching; these patients have normal AID function and normal somatic hypermutation, but defective class switching.

Other examples of predominantly humoral immunodeficiencies include the most common forms of primary immunodeficiency, referred to as **common variable immunodeficiencies** (**CVIDs**). CVIDs are a clinically and genetically heterogeneous group of disorders that typically do not come to attention until late childhood or adulthood, because the immune deficiency is relatively mild. Unlike other causes of immunoglobulin deficiency, patients with CVID can have defects in immunoglobulin production that are limited to one or more isotypes (see Fig. 13.11). **IgA deficiency**, the most common primary immunodeficiency, exists in both sporadic and familial forms, and both autosomal recessive and autosomal dominant inheritance has been described. The etiology of IgA deficiency in most patients is not understood, and these patients are asymptomatic. In IgA-deficient patients who do develop recurrent infections, an associated defect in one of the IgG subclasses is often found.

A small minority of CVID patients with IgA deficiency have a genetic defect in the transmembrane protein TACI (TNF-like receptor transmembrane activator and CAML interactor) encoded by the gene *TNFRSF13B*. TACI is the receptor for the cytokines BAFF and APRIL, which are produced by T cells, dendritic cells, and macrophages, and which can provide co-stimulatory and survival signals for B-cell activation and class switching (see Section 10-13). Patients with selective deficiencies in IgG subclasses have also been described. B-cell numbers are typically normal in these patients, but serum levels of the affected immunoglobulin isotype are depressed. Although some of these patients have recurrent bacterial infections, as in IgA deficiency, many are asymptomatic. CVID patients with other defects that affect immunoglobulin class switching have been identified. Included in this group are patients with inherited defects in CD19, which is a component of the B-cell co-receptor (see Fig. 13.11). A genetic defect that has been linked to a small percentage of people with CVID is deficiency of the co-stimulatory molecule ICOS. As described in Section 9-13, ICOS is upregulated on T cells when they are activated. The effects of ICOS deficiency have confirmed its essential role in T-cell help for the later stages of B-cell differentiation, including class switching and the formation of memory cells.

The final immunodeficiency to be considered in this section is **hyper IgE syndrome** (**HIES**), also called **Job's syndrome**. This disease is characterized by recurrent skin and pulmonary infections caused by pyogenic bacteria and fungi, very high serum concentrations of IgE, and chronic eczematous dermatitis or skin rash. HIES is inherited in an autosomal recessive or dominant pattern, with the latter manifesting skeletal and dental abnormalities not found in the recessive variant. The inherited defect is in the transcription factor STAT3, which is activated downstream of several cytokine receptors, including those for IL-6 and IL-23, and which is central to the differentiation of T_H17 cells and activation of innate immune responses at skin and mucosal barriers (see Sections 9-18 and 12-11). Because differentiation of T_H17 cells is deficient in these patients, the recruitment of neutrophils normally orchestrated by the

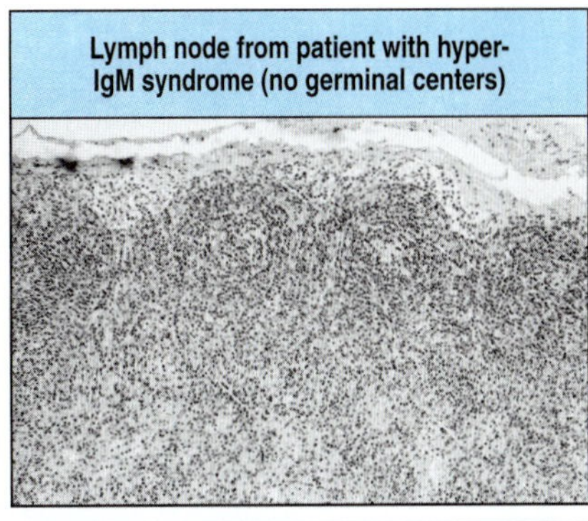

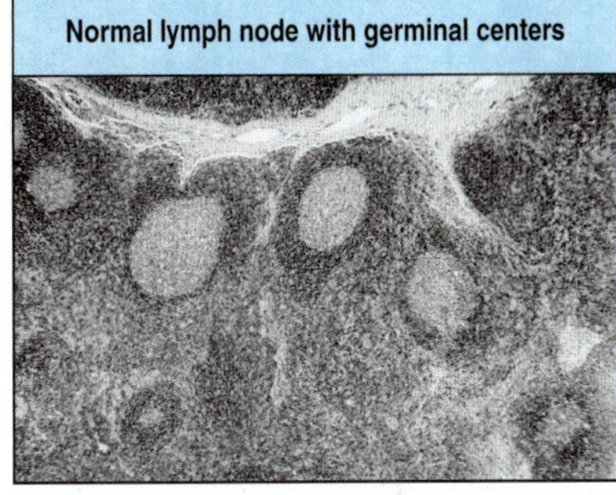

Fig. 13.12 Patients with CD40 ligand deficiency are unable to activate their B cells fully. Lymphoid tissues in patients with CD40 ligand deficiency, which manifests as a hyper IgM syndrome, are devoid of germinal centers (top panel), unlike a normal lymph node (bottom panel). B-cell activation by T cells is required both for isotype switching and for the formation of germinal centers, where extensive B-cell proliferation takes place. Photographs courtesy of R. Geha and A. Perez-Atayde.

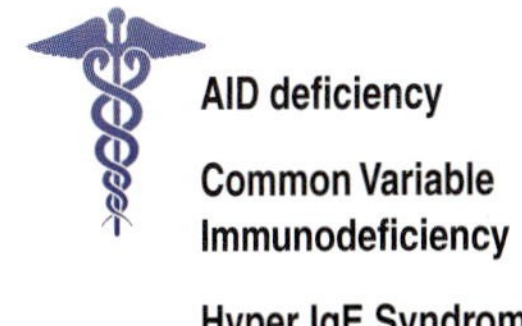

AID deficiency

Common Variable Immunodeficiency

Hyper IgE Syndrome

T_H17 response is also defective, and this is thought to underlie the impaired defense against extracellular bacteria and fungi. The cause of the elevated IgE is not understood, but it might be due to an abnormal accentuation of skin and mucosal T_H2 responses as a result of T_H17 deficiency.

13-16 Defects in complement components and complement-regulatory proteins cause defective humoral immune function and tissue damage.

The diseases discussed so far are mainly due to disturbances of the adaptive immune system. In the next few sections we look at some immunodeficiency diseases that affect cells and molecules of the innate immune system. We start with the complement system, which can be activated by any of three pathways that converge on the cleavage and activation of complement component C3, allowing it to bind covalently to pathogen surfaces where it acts as an opsonin (discussed in Chapter 2). Not surprisingly, the spectrum of infections associated with complement deficiencies overlaps substantially with that seen in patients with deficiencies in antibody production. In particular, there is increased susceptibility to extracellular bacteria that require opsonization by antibody and/or complement for efficient clearance by phagocytes (Fig. 13.13). Defects in the activation of C3 by any of the three pathways, as well as defects in C3 itself, are associated with increased susceptibility to infection by a range of pyogenic bacteria, including *S. pneumoniae*, emphasizing the role of C3 as a central effector that promotes the phagocytosis and clearance of capsulated bacteria.

In contrast, defects in the membrane-attack components of complement (C5–C9) downstream of C3 activation have more limited effects, and result almost exclusively in susceptibility to *Neisseria* species. A similar susceptibility phenotype is found in patients with defects in the alternative complement pathway components, factor D and properdin, which are activated by *Neisseria* species. This indicates that defense against these bacteria, which can survive intracellularly, is mediated by extracellular lysis by the membrane-attack

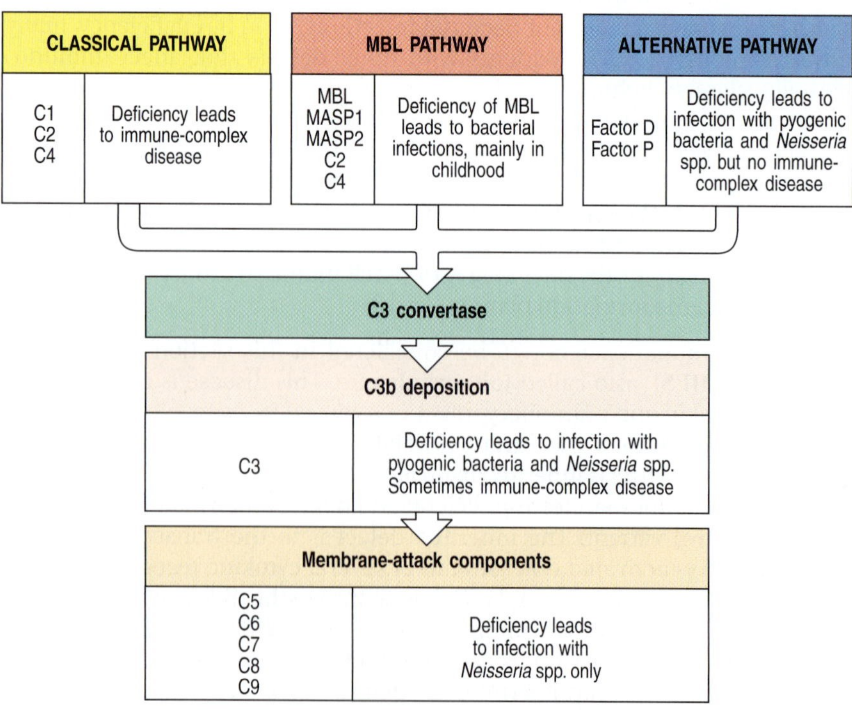

Fig. 13.13 Defects in complement components are associated with susceptibility to certain infections and the accumulation of immune complexes. Defects in the early components of the alternative pathway and in C3 lead to susceptibility to extracellular pathogens, particularly pyogenic bacteria. Defects in the early components of the classical pathway predominantly affect the processing of immune complexes and the clearance of apoptotic cells, leading to immune-complex disease. Deficiency of mannose-binding lectin (MBL), the recognition molecule of the mannose-binding lectin pathway, is associated with bacterial infections, mainly in early childhood. Defects in the membrane-attack components are associated only with susceptibility to strains of *Neisseria* species, the causative agents of meningitis and gonorrhea, implying that the effector pathway is important chiefly in defense against these organisms.

complex. Data from large population studies in Japan, where endemic *N. meningitidis* infection is rare, show that the risk each year of infection with this organism is approximately 1 in 2,000,000 to a normal person. This compares with a risk of 1 in 200 to a person in the same population with an inherited deficiency of one of the membrane-attack complex proteins—a 10,000-fold increase in risk.

The early components of the classical complement pathway are particularly important for the elimination of immune complexes and apoptotic cells, which can cause significant pathology in autoimmune diseases such as systemic lupus erythematosus. This aspect of inherited complement deficiency is discussed in Chapter 15. Deficiencies in mannose-binding lectin (MBL), which initiates complement activation in innate immunity (see Section 2-6), are relatively common (5% of the population). MBL deficiency may be associated with a mild immunodeficiency with an increased incidence of bacterial infection in early childhood. A similar phenotype is found in patients with defects in the gene that encodes the MBL-associated serine-protease-2 (*MASP2*).

Disease	Complement control protein deficient or affected
Paroxysmal nocturnal hemoglobinuria	DAF or CD59
Factor I deficiency	Factor I
Atypical hemolytic uremic syndrome	MCP, factor I, or factor H
Age-related macular degeneration	Polymorphisms in factor H
Hereditary angioedema	C1 inhibitor

Fig. 13.14 Defects in complement control proteins are associated with a range of diseases.

Another set of complement-related diseases is caused by defects in complement control proteins (Fig. 13.14). Deficiencies in decay-accelerating factor (DAF) or protectin (CD59), membrane-associated control proteins that protect the surfaces of the body's cells from complement activation, lead to destruction of red blood cells, resulting in the disease **paroxysmal nocturnal hemoglobinuria**, as discussed in Section 2-16. Deficiencies in soluble complement-regulatory proteins such as factor I and factor H have various outcomes. Homozygous **factor I deficiency** is a rare defect that results in uncontrolled activity of the alternative pathway C3 convertase, leading to a *de facto* C3 deficiency (see Section 2-16). Deficiencies in MCP, factor I, or factor H can also cause a condition known as **atypical hemolytic uremic syndrome**. A polymorphism in factor H has quite recently been associated with exacerbation of the eye disease age-related macular degeneration. It is thought that this association is due to defective complement-mediated clearance of debris in the eye in the absence of factor H.

A striking consequence of the loss of a complement-regulatory protein is seen in patients with C1-inhibitor defects, which cause the syndrome known as **hereditary angioedema** (**HAE**) (see Section 2-16). Deficiency of C1 inhibitor leads to a failure to regulate both the blood clotting and complement activation pathways, leading to excessive production of vasoactive mediators that cause fluid accumulation in the tissues (edema) and local laryngeal swelling that can result in suffocation.

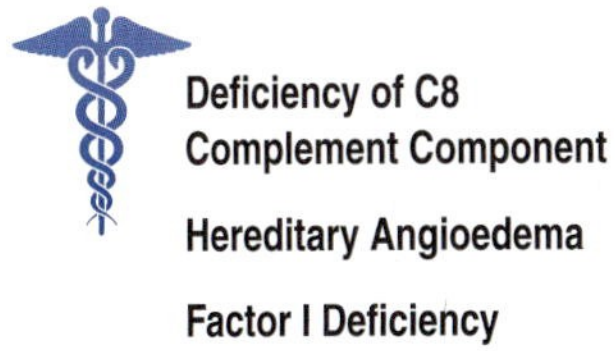

Deficiency of C8 Complement Component
Hereditary Angioedema
Factor I Deficiency

13-17 Defects in phagocytic cells permit widespread bacterial infections.

Deficiencies in phagocyte numbers or function can be associated with severe immunodeficiency; indeed, a total absence of neutrophils is incompatible with survival in a normal environment. Phagocyte immunodeficiencies can be grouped into four general types: deficiencies in phagocyte production, phagocyte adhesion, phagocyte activation, and phagocyte killing of microorganisms (Fig. 13.15). We consider each in turn.

Inherited deficiencies of neutrophil production (**neutropenias**) are classified either as **severe congenital neutropenia** (**SCN**, or **Kostmann disease**) or **cyclic neutropenia**. In severe congenital neutropenia, which can be inherited as a dominant or recessive trait, the neutrophil count is persistently less than 0.5×10^9 per liter of blood (normal numbers are 3×10^9–5.5×10^9 per liter). Cyclic neutropenia is characterized by neutrophil numbers that fluctuate from near normal to very low or undetectable, with an approximate cycle time of 21 days that results in periodicity of infectious risk. The most common

Type of defect/name of syndrome	Associated infections or other diseases
Congenital neutropenias (e.g., elastase 2 deficiency)	Widespread pyogenic bacterial infections
Leukocyte adhesion deficiency	Widespread pyogenic bacterial infections
TLR signaling defects (e.g., MyD88 or IRAK4)	Severe cold pyogenic bacterial infections
Chronic granulomatous disease	Intracellular and extracellular infection, granulomas
G6PD deficiency	Defective respiratory burst, chronic infection
Myeloperoxidase deficiency	Defective intracellular killing, chronic infection
Chediak–Higashi syndrome	Intracellular and extracellular infection, granulomas

Fig. 13.15 Defects in phagocytic cells are associated with persistence of bacterial infection. Defects in neutrophil development caused by congenital neutropenias result in profound defects in anti-bacterial defense. Impairment of the leukocyte integrins with a common β_2 subunit (CD18) or defects in the selectin ligand, sialyl-Lewis^x, prevent phagocytic cell adhesion and migration to sites of infection (leukocyte adhesion deficiency). Inability to transmit signals through Toll-like receptors (TLRs), such as those resulting from defects in MyD88 or IRAK4, impairs the proximal sensing of many infectious agents by innate immune cells. The respiratory burst is defective in chronic granulomatous disease, glucose-6-phosphate dehydrogenase (G6PD) deficiency, and myeloperoxidase deficiency. In chronic granulomatous disease, infections persist because macrophage activation is defective, leading to chronic stimulation of CD4 T cells and hence to granulomas. Vesicle fusion in phagocytes is defective in Chediak–Higashi syndrome. These diseases illustrate the critical role of phagocytes in removing and killing pathogenic bacteria.

causes of SCN are sporadic or autosomal dominant mutations of the gene that encodes neutrophil elastase (*ELA2*), a component of the azurophilic granules involved in the degradation of phagocytosed microbes. Inappropriate targeting of defective elastase 2 causes apoptosis of developing myelocytes and a developmental block at the promyelocyte–myelocyte stage. Some mutations of *ELA2* cause cyclic neutropenia. How the mutant elastase causes a 21-day cycle in neutropenia and the effects on other bone marrow cell types is still a mystery.

A rare autosomal dominant form of SCN is caused by mutations in the oncogene *GFI1*, which encodes a transcriptional repressor. This finding arose from the unexpected observation that mice lacking the protein Gfi1 are neutropenic. Closer analysis revealed that mutation in mouse *Gfi1* affects the expression of *Ela2*, providing a link between these two genes in a common pathway of myeloid cell differentiation.

Autosomal recessive forms of SCN have also been identified. Deficiency of the mitochondrial protein HAX1 leads to increased apoptosis in developing myeloid cells, resulting in neutropenia. The heightened sensitivity of developing neutrophils to apoptosis is highlighted by SCN associated with genetic defects in glucose metabolism. Patients with recessive mutations in the genes encoding the glucose-6-phosphatase catalytic subunit 3 (*G6PC3*) or the glucose-6-phosphate translocase 1 (*SLC37A4*) also demonstrate increased apoptosis during granulocyte development that results in neutropenia. In addition to the immune deficiency, patients with severe congenital or cyclic neutropenias have an increased risk of developing myelodysplasia or myeloid leukemia, often preceded by somatic mutations in the gene that encodes the granulocyte colony-stimulating factor (G-CSF) receptor (*CSF3R*). This may be aggravated by the chronic administration to these patients of G-CSF, which is now the mainstay of treatment for congenital neutropenias. Acquired neutropenia associated with chemotherapy, malignancy, or aplastic anemia is also associated with a similar spectrum of severe pyogenic bacterial infections.

Defects in the migration of phagocytic cells to extravascular sites of infection can cause serious immunodeficiency. Leukocytes reach such sites by emigrating from blood vessels in a tightly regulated process (see Fig. 3.25). The first stage is the rolling adherence of leukocytes to endothelial cells, through the binding of a fucosylated tetrasaccharide ligand known as sialyl-Lewis^x on the leukocyte to E-selectin and P-selectin on endothelium. The second stage is signaling through G-protein-coupled chemokine receptors on the leukocyte, which encounters chemokines tethered to the activated endothelium while rolling. This induces the activation of a high-affinity state of binding of leukocyte β_2 integrins such as CD11b:CD18 (Mac-1:CR3) that initiates tight adherence of the leukocytes to counter-receptors on endothelial cells. The final stage is the transmigration of leukocytes through the endothelium along gradients of chemokines originating from the site of tissue injury.

Deficiencies in the molecules involved in each of these stages can prevent neutrophils and macrophages from penetrating infected tissues, and are referred to as **leukocyte adhesion deficiencies** (**LADs**). Deficiency in the leukocyte integrin common β_2 subunit CD18, which is a component of LFA-1, MAC-1, and p150:95, prevents the migration of leukocytes into an infected site by abolishing the cells' ability to adhere tightly to the endothelium. Because it was the first LAD to be characterized, it is now referred to as **type 1 LAD**, or **LAD-1**. Reduced rolling adhesion has been described in rare patients who lack the sialyl-Lewis^x antigen owing to a deficiency in the GDP-fucose-specific transporter that is involved in the biosynthesis of sialyl-Lewis^x and other fucosylated ligands for the selectins. This is referred to as **type 2 LAD** or **LAD-2**. **LAD-3** results from deficiency of Kindlin-3, a protein involved in the induction of the high-affinity binding state of β integrins required for firm adhesion (see Section 3-15).

Each variant of LAD has an autosomal recessive pattern of inheritance and causes severe, life-threatening bacterial or fungal infections early in life that are characterized by impaired wound healing and, in pyogenic bacterial infections, the absence of pus formation. The infections that occur in these patients are resistant to antibiotic treatment and persist despite an apparently effective cellular and humoral adaptive immune response. LAD-3 is also associated with defects in platelet aggregation that cause increased bleeding.

A key step in the activation of innate immune cells, including phagocytes, involves their recognition of pathogen-associated molecular patterns through Toll-like receptors (TLRs) (see Section 3-5). Several primary immunodeficiencies caused by defects in intracellular signaling components of the TLR pathways have been described, although, remarkably, only a single immunodeficiency has so far been linked to a specific TLR receptor—TLR-3. A deficiency of this receptor has been linked to **recurrent herpes simplex encephalitis**.

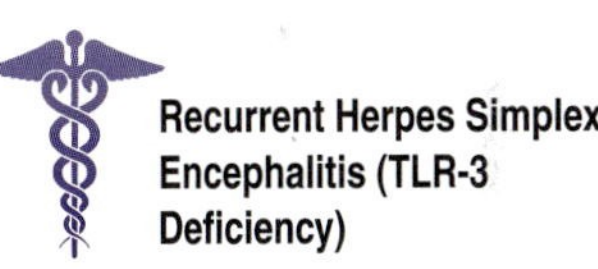
Recurrent Herpes Simplex Encephalitis (TLR-3 Deficiency)

Autosomal recessive mutations in the genes encoding the TLR-associated adaptor protein MyD88 or the kinase IRAK4, which activate the ubiquitous NFκB and MAPK pathways (see Section 3-7), have a similar phenotype that is characterized by recurrent, severe peripheral, and invasive infections by pyogenic bacteria that elicit little inflammation, a situation known as a 'cold' infection. Hemizygous (dominant) and homozygous (recessive) mutations in the genes for TLR-3 and the TLR-transport protein UNC93B1, respectively, also have a shared phenotype characterized by impaired type I interferon responses and herpes simplex virus-1 (HSV-1) infections of the central nervous system (herpes simplex encephalitis). UNC93B1 is required for the transport of intracellular TLRs from the endoplasmic reticulum to the endolysosome. Remarkably, these patients show only a limited predisposition to other viral infections, implying redundancy for immune protection against most other types of viral infection. Note that NEMO deficiency, which impairs B-cell class switching (see Section 13-15), also impairs TLR signaling though its block of normal NFκB activation downstream of all TLRs. Immunodeficiency associated with defects in NEMO therefore affects both adaptive and innate immune function.

Most of the other known defects in phagocytic cells affect their ability to kill intracellular bacteria or ingest extracellular bacteria (see Fig. 13.15). Patients with **chronic granulomatous disease** (**CGD**) are highly susceptible to bacterial and fungal infections and form granulomas as a result of an inability to kill bacteria ingested by phagocytes (see Fig. 9.43). The defect in this case is in the production of reactive oxygen species (ROS) such as the superoxide anion (see Section 3-2). Discovery of the molecular defect in this disease gave weight to the idea that these agents killed bacteria directly; this notion has since been challenged by the finding that the generation of ROS is not itself sufficient to kill target microorganisms. It is now thought that ROS cause an influx of K^+ ions into the phagocytic vacuole, increasing the pH to the optimal level for the action of microbicidal peptides and proteins, which are the key agents in killing the invading microorganism.

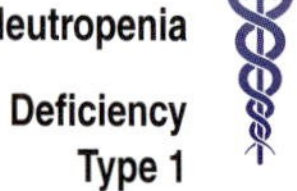

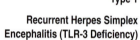

Severe Congenital Neutropenia

Leukocyte Adhesion Deficiency Type 1

Recurrent Herpes Simplex Encephalitis (TLR-3 Deficiency)

IRAK4 Deficiency

Chronic Granulomatous Disease

Chediak–Higashi Syndrome

Genetic defects affecting any of the constituent proteins of the NADPH oxidase expressed in neutrophils and monocytes (see Section 3-2) can cause chronic granulomatous disease. Patients with the disease have chronic bacterial infections, which in some cases lead to the formation of granulomas. Deficiencies in glucose-6-phosphate dehydrogenase and myeloperoxidase also impair intracellular bacterial killing and lead to a similar, although less severe, phenotype. Finally, in **Chediak–Higashi syndrome**, a complex syndrome characterized by partial albinism, abnormal platelet function, and severe immunodeficiency, a defect in a protein called CHS1, which is involved in intracellular vesicle formation and trafficking, causes a failure of lysosomes and phagosomes to fuse properly; the phagocytes in these patients have enlarged granules and impaired intracellular killing ability. This defect also impairs the general secretory pathway. The consequences of this are described in Section 13-21.

13-18 Mutation in the molecular regulators of inflammation can cause uncontrolled inflammatory responses that result in 'autoinflammatory disease.'

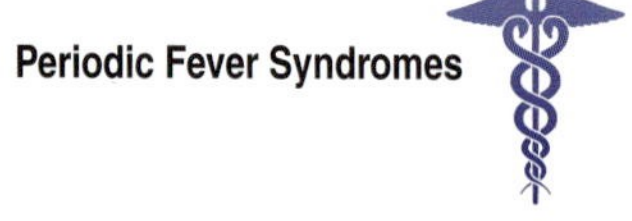

There are a small number of diseases in which mutations in genes that control the life, death, and activities of inflammatory cells are associated with severe inflammatory disease. We have included them in this chapter, although they do not lead to immunodeficiency, because they are single-gene defects affecting a crucial aspect of innate immunity—the inflammatory response. These conditions represent a failure of the normal mechanisms that limit inflammation, and are known as **autoinflammatory diseases**: they can lead to inflammation even in the absence of infection (Fig. 13.16). **Familial Mediterranean fever** (**FMF**) is characterized by episodic attacks of severe inflammation, which can occur at various sites throughout the body and are associated with fever, an acute-phase response (see Section 3-18), and severe malaise. The pathogenesis of FMF was a mystery until its cause was discovered to be mutations in the gene *MEFV*, which encodes a protein called pyrin, to reflect its association with fever. Pyrin and pyrin-domain-containing proteins are involved in pathways that lead to the apoptosis of inflammatory cells, and in pathways that inhibit the secretion of pro-inflammatory cytokines such as IL-1β. It is proposed that an absence of functional pyrin leads to unregulated cytokine activity and defective apoptosis, resulting in a failure to control inflammation. In mice, an absence of pyrin causes increased sensitivity to lipopolysaccharide and a defect in macrophage apoptosis. A disease with similar clinical manifestations, known as **TNF-receptor associated periodic syndrome** (**TRAPS**) is caused by mutations in quite a different gene, that encoding the TNF-α receptor TNFR-I. Patients with TRAPS have reduced levels of TNFR-1, which leads to increased levels of pro-inflammatory TNF-α in the circulation because it is not mopped up by binding to the receptors. The disease responds to therapeutic blockade with anti-TNF agents such as etanercept, a soluble TNF receptor developed primarily to treat patients with rheumatoid

Disease (common abbreviation)	Clinical features	Inheritance	Mutated gene	Protein (alternative name)
Familial Mediterranean fever (FMF)	Periodic fever, serositis (inflammation of the pleural and/or peritoneal cavity), arthritis, acute-phase response	Autosomal recessive	*MEFV*	Pyrin
TNF-receptor associated periodic syndrome (TRAPS) (also known as familial Hibernian fever)	Periodic fever, myalgia, rash, acute-phase response	Autosomal dominant	*TNFRSF1A*	TNF-α 55 kDa receptor (TNFR-I)
Pyogenic arthritis, pyoderma gangrenosum, and acne (PAPA)		Autosomal dominant	*PTSTPIP*	CD2-binding protein 1
Muckle–Wells syndrome	Periodic fever, urticarial rash, joint pains, conjunctivitis, progressive deafness	Autosomal dominant	*NLRP3*	Cryopyrin
Familial cold autoinflammatory syndrome (FCAS) (familial cold urticaria)	Cold-induced periodic fever, urticarial rash, joint pains, conjunctivitis			
Chronic infantile neurologic cutaneous and articular syndrome (CINCA)	Neonatal onset recurrent fever, urticarial rash, chronic arthropathy, facial dysmorphia, neurologic involvement			
Hyper-IgD syndrome (HIDS)	Periodic fever, elevated IgD levels, lymphadenopathy	Autosomal recessive	*MVK*	Mevalonate synthase
Blau syndrome	Granulomatous inflammation of skin, eye, and joints	Autosomal dominant	*NOD2*	NOD2

Fig. 13.16 The autoinflammatory diseases.

arthritis (see Section 16-8). Mutations in the gene encoding PSTPIP1 (proline-serine-threonine phosphatase-interacting protein 1), which interacts with pyrin, are associated with another dominantly inherited autoinflammatory syndrome—**pyogenic arthritis, pyoderma gangrenosum, and acne** (**PAPA**). The mutations accentuate the interaction between pyrin and PSTPIP1, and it has been proposed that the interaction sequesters pyrin and limits its normal regulatory function.

The episodic autoinflammatory diseases **Muckle–Wells syndrome** and **familial cold autoinflammatory syndrome** (**FCAS**) are clearly linked to the inappropriate stimulation of inflammation, because they are due to mutations in NLRP3, a component of the 'inflammasome' that normally senses cell damage and stress as a result of infection (see Section 3-8). The mutations lead to the activation of NLRP3 in the absence of such stimuli and the unregulated production of pro-inflammatory cytokines. These dominantly inherited syndromes present with episodes of fever—which is induced by exposure to cold in the case of FCAS—as well as urticarial rash, joint pains, and conjunctivitis. Mutations in *NLRP3* are also associated with the autoinflammatory disorder **chronic infantile neurologic cutaneous and articular syndrome** (**CINCA**), in which short recurrent fever episodes are common, although severe arthropathic, neurologic, and dermatologic symptoms predominate. Both pyrin and NLRP3 are predominantly expressed in leukocytes and in cells that act as innate barriers to pathogens, such as intestinal epithelial cells. The stimuli that modulate pyrin and related molecules include inflammatory cytokines and stress-related changes in cells. Indeed, Muckle–Wells syndrome responds dramatically to the drug anakinra, an antagonist of the receptor for IL-1.

Not all autoinflammatory diseases are caused by mutations in genes involved in the regulation of apoptosis or cytokine production. **Hyper IgD syndrome** (**HIDS**), which is associated with attacks of fever starting in infancy, high levels of IgD in serum, and lymphadenopathy, is caused by mutations that result in a partial deficiency of mevalonate kinase, an enzyme in the pathway for the synthesis of isoprenoids and cholesterol. It is not yet clear how this enzyme deficiency causes the autoinflammatory disease.

13-19 The normal pathways for host defense against intracellular bacteria are pinpointed by genetic deficiencies of IFN-γ and IL-12 and their receptors.

A small number of families have been identified that contain several individuals who suffer from persistent and eventually fatal attacks by intracellular pathogens, especially mycobacteria and salmonellae. These people typically suffer from the ubiquitous, environmental, nontuberculous strains of mycobacteria, such as *Mycobacterium avium*. They may also develop disseminated infection after vaccination with *Mycobacterium bovis* bacillus Calmette–Guérin (BCG), the strain of *M. bovis* that is used as a live vaccine against *M. tuberculosis*. Susceptibility to these infections is conferred by a variety of mutations that abolish the function of any of the following: the cytokine IL-12; the IL-12 receptor; or the receptor for interferon (IFN)-γ and its signaling pathway. Mutations have been found in the p40 subunit of IL-12, the IL-12 receptor β_1 chain, and the two subunits (R1 and R2) of the IFN-γ receptor. p40 is shared by IL-12 and IL-23, and so p40 deficiency can cause both IL-12 and IL-23 deficiency. A mutation in STAT1, a protein in the signaling pathway activated after ligation of the IFN-γ receptor, is also associated in humans with increased susceptibility to mycobacterial infections. Similar susceptibility to intracellular bacterial infection is seen in mice with induced mutations in these same genes, and also in mice lacking tumor necrosis factor (TNF)-α or the TNF p55 receptor gene. Why tuberculosis itself is not seen more often in patients with these defects, especially since *M. tuberculosis* is more virulent than *M. avium* and *M. bovis*, remains unclear.

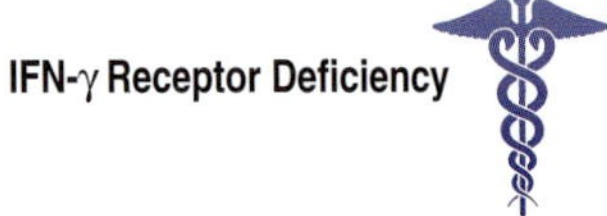

Mycobacteria and salmonellae enter dendritic cells and macrophages, where they can reproduce and multiply. At the same time they provoke an immune response that occurs in several stages and eventually controls the infection with the help of CD4 T cells. First, lipoproteins and peptidoglycans on the surface of the bacteria ligate receptors on macrophages and dendritic cells as they enter the cells. These receptors include the Toll-like receptors (TLRs) (see Section 3-5), particularly TLR-2, and the mannose receptor, and their ligation stimulates nitric oxide (NO) production within the cells, which is toxic to bacteria. Signaling by the TLRs stimulates the release of IL-12, which in turn drives NK cells to produce IFN-γ in the early phase of the immune response. IL-12 also stimulates antigen-specific CD4 T cells to release IFN-γ and TNF-α. These cytokines activate and recruit more macrophages to the site of infection, resulting in the formation of granulomas (see Section 9-29).

The key role of IFN-γ in activating macrophages to kill intracellular bacteria is dramatically illustrated by the failure to control infection in patients who are genetically deficient in either of the two subunits of its receptor. In the total absence of IFN-γ receptor expression, granuloma formation is much reduced, showing a role for this receptor in granuloma development. In contrast, if the underlying mutation is associated with the presence of low levels of functional receptor, granulomas form but the macrophages within them are not sufficiently activated to be able to control the division and spread of the mycobacteria. It is important to remember that this cascade of cytokine reactions is occurring in the context of cognate interactions between antigen-specific CD4 T cells and the macrophages and dendritic cells harboring the intracellular bacteria. T-cell receptor ligation and co-stimulation of the phagocyte by the interaction between CD40 and CD40 ligand, for example (Section 13-15), sends signals that help activate infected phagocytes to kill the intracellular bacteria. Thus, as described above, inherited deficiencies in the CD40–CD40L pathway encompass defects in the clearance of intracellular bacteria in addition to the defects in B-cell maturation that cause hyper IgM syndrome.

Atypical mycobacterial infections have been reported in several patients with NEMO deficiency, and are due to impaired NFκB activation and its effects on many cellular responses, including those to TLR ligands and TNF-α. The conclusion to be drawn from these diseases is that pathways controlled by TLRs and NFκB seem to be important in immune responses against a collection of unrelated pathogens, whereas the IL-12/IFN-γ pathway is especially important for immunity to mycobacteria and salmonellae but not to other pathogens.

13-20 X-linked lymphoproliferative syndrome is associated with fatal infection by Epstein–Barr virus and with the development of lymphomas.

The Epstein–Barr virus (EBV) we encountered earlier in the chapter (see Section 13-2) can transform B lymphocytes and is used to immortalize clones of B cells in the laboratory. Transformation does not normally occur *in vivo* because EBV infection is actively controlled and the virus maintained in a latent state by the actions of NK cells, NKT cells, and cytotoxic T cells with specificity for B cells expressing EBV antigens. In the presence of certain types of immunodeficiency, however, this control can break down, resulting in overwhelming EBV infection (infectious mononucleosis) that is accompanied by unrestrained proliferation of EBV-infected B cells and cytotoxic T cells, hypogammaglobulinemia (low levels of circulating immunoglobulins), and the potential for the development of lethal, non-Hodgkin B-cell lymphomas. These occur in the rare immunodeficiency **X-linked lymphoproliferative** (**XLP**) **syndrome**, which results from mutations in one of two X-linked genes: the SH2-domain containing gene 1A (*SH2D1A*) or the X-linked inhibitor of apoptosis gene (*XIAP*).

In **XLP1**, which accounts for approximately 80% of patients with this syndrome, the defect is in the protein SAP (signaling lymphocyte activation molecule (SLAM)-associated protein), which is encoded by *SH2D1A*. SAP links signaling through the SLAM family of immune-cell receptors to the Src-family tyrosine kinase Fyn in T cells and NK cells. SLAM family members interact through homotypic or heterotypic binding to modulate the outcome of interactions between T cells and antigen-presenting cells and between NK cells and their target cells. In the absence of SAP, defective EBV-specific cytotoxic T-cell and NK-cell responses are made to EBV, indicating that SAP has a vital, nonredundant role in the normal control of EBV infection. **XLP2** is due to defects in the XIAP protein, which normally binds the TNF-receptor-associated factors TRAF-1 and TRAF-2 (see Section 7-22) and inhibits the activation of apoptosis-inducing caspases. The lack of XIAP results in the enhanced death of T cells and NKT cells, creating a deficiency in these cells. The exact reason for the impaired control of EBV infection in these distinct inherited immune deficiencies remains to be defined, but the apparent targeting of this particular virus might simply be due to the very high prevalence of EBV infection in humans rather than a specific immune defect against the virus.

13-21 Genetic abnormalities in the secretory cytotoxic pathway of lymphocytes cause uncontrolled lymphoproliferation and inflammatory responses to viral infections.

A small group of inherited immunodeficiency diseases also affect skin pigmentation, causing albinism. The link between these two apparently unrelated phenotypes is a defect in the regulated secretion of lysosomes. In response to specific stimuli, many cell types derived from the bone marrow, including lymphocytes, granulocytes, and mast cells, exocytose secretory lysosomes that contain specialized collections of proteins. Other cell types are also capable of regulated secretion of lysosomes, in particular the melanocytes, the pigment cells of the skin. The contents of the secretory lysosomes differ depending on cell type. In melanocytes, melanin is the major component, whereas in cytotoxic T cells, secretory lysosomes contain the cytolytic proteins perforin, granulysin, and granzymes (see Section 9-26). Although the contents of the granules differ between cell types, the fundamental mechanisms for their secretion do not, and this explains how inherited disorders affecting the regulated secretion of lysosomes can cause the combination of albinism and immunodeficiency.

We learned in Section 13-20 that X-linked lymphoproliferative syndrome is associated with uncontrolled inflammation in response to EBV infection. In that respect it is very similar to a group of diseases known as the **hemophagocytic syndromes**, in which there is a dysregulated expansion of CD8 cytotoxic lymphocytes that is associated with macrophage activation. The clinical manifestations of the disease are due to an inflammatory response caused by an increased release of pro-inflammatory cytokines such as IFN-γ, TNF, IL-6, IL-10, and macrophage colony-stimulating factor (M-CSF). These mediators are secreted by activated T lymphocytes and macrophages that infiltrate all tissues, causing tissue necrosis and organ failure. The activated macrophages phagocytose blood cells, including erythrocytes and leukocytes, which gives the syndromes their name. Some hemophagocytic syndromes are inherited, and these can be classified into two types according to the nature of the gene defect. In the first type, the effects of the mutation are confined to lymphocytes or other cells of the immune system because the mutated protein is located in the granules of NK and cytotoxic T cells. In the second type, the genetic abnormality is located in the regulated secretory pathway of lysosomes and affects all cell types that use this pathway; in these cases albinism may also result.

The disease **familial hemophagocytic lymphohistiocytosis** (**FHL**) is caused by an inherited deficiency of the cytotoxic protein perforin. This is a lymphocyte-specific disorder, in which polyclonal CD8-positive T cells accumulate in lymphoid tissue and other organs, in association with activated hemophagocytic macrophages. The progressive inflammation is lethal unless checked by immunosuppressive therapy. In mice that lack perforin, no immediate defect is observed, but when the mice are infected with LCMV or other viruses, a disease resembling human TRAPS develops, driven by an uncontrolled virus-specific T-cell response. This rare syndrome powerfully demonstrates a role for CD8-positive lymphocytes in limiting T-cell immune responses, for example in response to viral infection, by perforin-dependent cytotoxic mechanisms. When this mechanism fails, uncontrolled activated T cells kill their host. Perforin is also critical for NK-cell cytotoxicity, which is impaired in TRAPS.

Familial Hemophagocytic Lymphohistiocytosis

Griscelli Syndrome

Chediak-Higashi Syndrome

Examples of inherited diseases that affect the regulated secretion of lysosomes are Chediak–Higashi syndrome (see Fig. 13.15 and Section 13-17), caused by mutations in a protein, CHS1, that regulates lysosomal trafficking, and **Griscelli syndrome**, caused by mutations in a small GTPase, Rab27a, that controls the movement of vesicles within cells. Two other types of Griscelli syndrome have been identified, in which patients have pigmentary changes only and no immunological deficiency. In Chediak–Higashi syndrome, abnormal giant lysosomes accumulate in melanocytes, neutrophils, lymphocytes, eosinophils, and platelets. The hair is typically a metallic silver color, vision is poor because of abnormalities in retinal pigment cells, and platelet dysfunction causes increased bleeding. Children with the syndrome suffer from recurrent severe infections because of a failure of T-cell, neutrophil, and NK-cell function. After a few years, hemophagocytic lymphohistiocytosis develops, with fatal consequences if untreated. Antibiotics are needed to treat and to prevent infections, and immunosuppression is required to deal with the uncontrolled inflammation; only bone marrow transplantation offers any real hope to these patients.

13-22 Hematopoietic stem cell transplantation or gene therapy can be useful to correct genetic defects.

It is frequently possible to correct the defects in lymphocyte development that lead to SCID and some other immunodeficiency phenotypes by replacing the defective component, generally by hematopoietic stem cell (HSC) transplantation. The main difficulties in these therapies result from MHC polymorphism. To be useful, the graft must share some MHC alleles with the host. As we learned in Section 8-15, the MHC alleles expressed by the thymic epithelium determine which T cells can be positively selected. When HSCs are used to restore immune function to individuals with a normal thymic stroma, both the T cells and the antigen-presenting cells are derived from the graft. Therefore, unless the graft shares at least some MHC alleles with the recipient, the T cells that are selected on host thymic epithelium cannot be activated by graft-derived antigen-presenting cells (Fig. 13.17). There is also a danger that mature, post-thymic T cells that contaminate donor HSCs prepared from the peripheral blood or bone marrow might recognize the host as foreign and attack it, causing **graft-versus-host disease** (**GVHD**) (Fig. 13.18, top panel). This can be overcome by depleting the donor graft of mature T cells. For immunodeficiency diseases such as XLP, in which there is residual T-cell or NK-cell function, myeloablative treatment of recipients (destruction of the bone marrow, typically using cytotoxic drugs) is carried out before transplantation, both to generate space for engraftment of the transplanted HSCs and to minimize the threat of **host-versus-graft disease** (**HVGD**) (Fig. 13.18, third panel). In SCID patients, however, there is little problem with the host response to the transplanted HSCs because the patient is immunodeficient, and transplants can be performed without bone marrow destruction.

Now that many specific gene defects have been identified, an alternative approach to correcting inherited immune deficiencies is through **somatic gene therapy**. This strategy involves the isolation of HSCs from the patient's bone marrow or peripheral blood, introduction of a normal copy of the defective gene with the use of a retroviral vector, and reinfusion of the stem cells into the patient. Although in principle this is an attractive approach, in practice it has met with severe complications. Trials using this approach to treat X-linked SCID and ADA deficiency complemented the defects, but led to malignancies in a high proportion of treated patients: five of the ten children with XSCID whose immunodeficiency was corrected by this gene therapy developed leukemia due to insertion of the retrovirus within a proto-oncogene. The inability to control the site in the genome in which retrovirally encoded genes insert is therefore problematic. Recently, a technique for the generation of **induced pluripotent stem cells** (**iPS cells**) from a patient's own somatic cells has been demonstrated by forced expression of a set of pluripotency transcription factors. This approach offers the promise of 'repairing' specific defective genes in patient-derived stem cells by gene targeting *ex vivo*

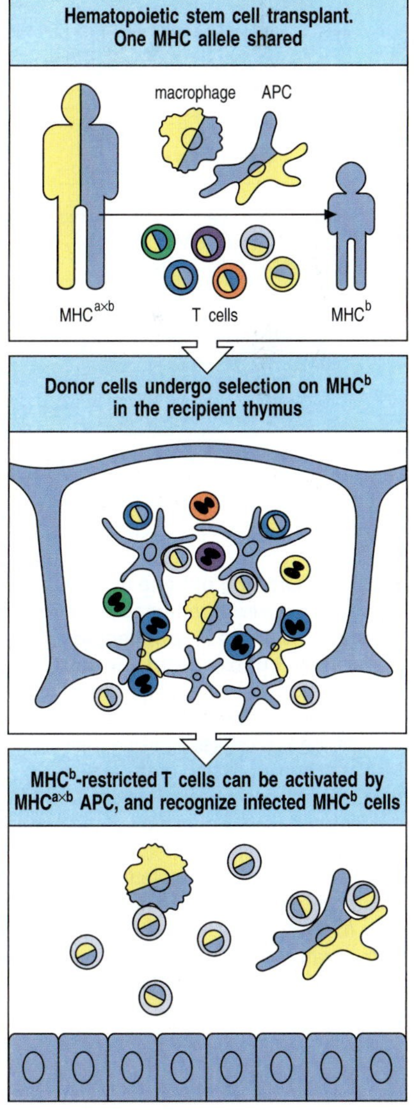

Hematopoietic stem cell transplant. One MHC allele shared
macrophage
APC
$MHC^{a \times b}$
T cells
MHC^b
Donor cells undergo selection on MHC^b in the recipient thymus
MHC^b-restricted T cells can be activated by $MHC^{a \times b}$ APC, and recognize infected MHC^b cells

Fig. 13.17 The donor and the recipient of a hematopoietic stem cell (HSC) graft must share at least some MHC molecules to restore immune function. An HSC transplant from a genetically different donor is illustrated in which the donor marrow cells share some MHC molecules with the recipient. The shared MHC type is designated 'b' and illustrated in blue; the MHC type of the donor HSCs that is not shared is designated 'a' and shown in yellow. In the recipient, developing donor lymphocytes are positively selected on MHC^b on thymic epithelial cells and negatively selected by the recipient's stromal epithelial cells and at the cortico-medullary junction by encounter with dendritic cells derived from both the donor HSCs and residual recipient dendritic cells. The negatively selected cells are shown as apoptotic cells. The donor-derived antigen-presenting cells (APCs) in the periphery can activate T cells that recognize MHC^b molecules; the activated T cells can then recognize the recipient's infected MHC^b-bearing cells.

before reinfusion, but this technique is not yet established. Until better methods for introduction of corrected genes into self-renewing stem cells are identified, allogeneic HSC transplantation will remain the mainstay of treatment for many primary immunodeficiencies.

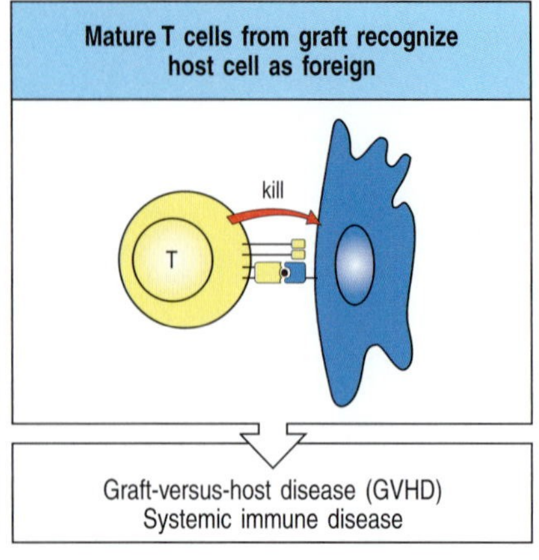

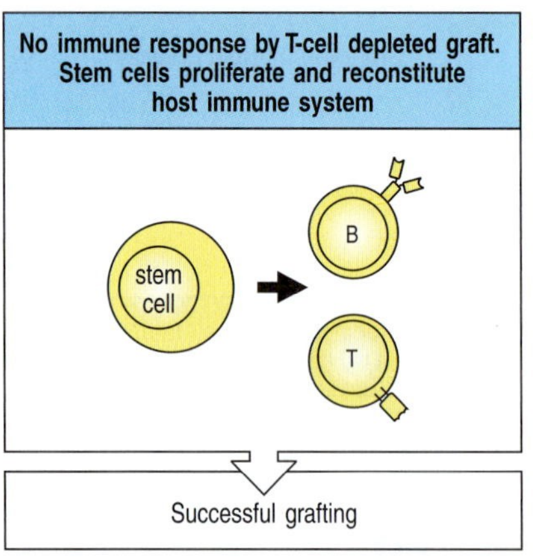

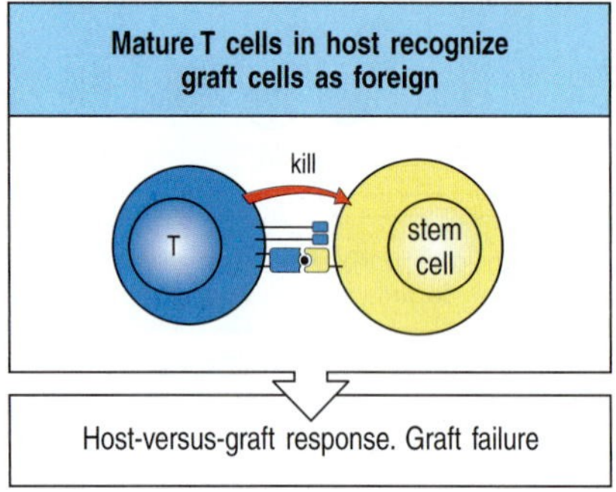

Fig. 13.18 Grafting of bone marrow can be used to correct immunodeficiencies caused by defects in lymphocyte maturation, but two problems can arise. First, if there are mature T cells in the bone marrow, they can attack cells of the host by recognizing their MHC antigens, causing graft-versus-host disease (top panel). This can be prevented by T-cell depletion of the donor bone marrow (center panel). Second, if the recipient has competent T cells, these can attack the bone marrow stem cells (bottom panel). This causes failure of the graft by the usual mechanism of transplant rejection (see Chapter 15).

13-23 Secondary immunodeficiencies are major predisposing causes of infection and death.

The primary immunodeficiencies have taught us much about the biology of specific proteins of the immune system. Fortunately, these conditions are rare. In contrast, secondary immunodeficiency is extremely common and important in everyday medical practice. Malnutrition devastates many populations around the world, and a major feature of malnutrition is secondary immunodeficiency. This particularly affects cell-mediated immunity, and death in famines is frequently caused by infection. Measles, which itself causes immunosuppression (see Section 13-4), is an important cause of death in malnourished children. In the developed world, measles is an unpleasant illness but major complications are uncommon. In contrast, measles in the malnourished has a high mortality. Tuberculosis is another important infection in the malnourished. In mice, protein starvation causes immunodeficiency by affecting antigen-presenting cell function, but in humans it is not clear how malnourishment specifically affects immune responses. Links between the endocrine and immune systems may provide part of the answer. Adipocytes (fat cells) produce the hormone leptin, and leptin levels are related directly to the amount of fat present in the body; leptin levels fall in starvation. Both mice and humans with genetic leptin deficiency have reduced T-cell responses, and in mice the thymus atrophies. In both starved mice and those with inherited leptin deficiency, these abnormalities can be reversed by the administration of leptin.

Secondary immunodeficiency states are also associated with hematopoietic tumors such as leukemia and lymphomas. Myeloproliferative diseases, such as leukemia, can be associated with deficiencies of neutrophils (neutropenia) or an excess of immature myeloid progenitors that lack functional properties of mature neutrophils, either of which increases susceptibility to bacterial and fungal infections, as described in Section 13-17. Destruction or invasion of peripheral lymphoid tissue by lymphomas or metastases from other cancers can promote opportunistic infections. Surgical removal of the spleen, or destruction of spleen function by certain diseases, is associated with a lifelong predisposition to overwhelming infection by *S. pneumoniae*, graphically illustrating the role of mononuclear phagocytic cells within the spleen in the clearance of this organism from blood. Patients who have lost spleen function should be vaccinated against pneumococcal infection and are often recommended to take prophylactic antibiotics throughout their life.

Unfortunately, a major complication of the cytotoxic drugs used to treat cancer is immunosuppression and increased susceptibility to infection. These drugs kill all dividing cells, and cells of the bone marrow and lymphoid systems are frequently unwanted targets of these agents. Infection is thus one of the major side effects of cytotoxic drug therapy. This is also the case when these and similar drugs are used therapeutically as immunosuppressants. Another undesirable side-effect of medical intervention is the increased risk of infection around implanted medical devices such as catheters, artificial heart valves, and artificial joints. These act as privileged sites for the development of infections that resist easy elimination by antibiotics. These implanted materials lack the innate defensive mechanisms of normal body tissues and act as a 'protected' matrix for the growth of bacteria and fungi. Catheters used for peritoneal dialysis or for the infusion of drugs or fluids into the circulation can also act as a conduit for bacteria to bypass the normal defensive barrier of the skin.

Finally, medical treatments that aim at suppressing immune function for therapeutic ends are a major cause of secondary immunodeficiencies. Immune suppression to induce host tolerance of solid organ allografts, such as kidney or heart transplants, carries a substantial increased risk for infection and even for malignancy. The recent introduction of biologic therapies for some forms of autoimmunity has led to an increased risk of infection because of their immunosuppressive effects. For example, administration of antibodies that block TNF-α in patients with rheumatoid arthritis or other forms of autoimmunity has been associated with infrequent, but increased, instances of infectious complications.

Summary.

Genetic defects can occur in almost any molecule involved in the immune response. These defects give rise to characteristic deficiency diseases, which, although rare, provide much information about the development and functioning of the immune system in normal humans. Inherited immunodeficiencies illustrate the vital role of the adaptive immune response and T cells in particular, without which both cell-mediated and humoral immunity fails. They have provided information about the separate roles of B lymphocytes in humoral immunity and of T lymphocytes in cell-mediated immunity, the importance of phagocytes and complement in humoral and innate immunity, and the specific functions of a growing number of cell-surface or signaling molecules in the adaptive immune response. There are also some inherited immune disorders whose causes we still do not understand. The study of these diseases will undoubtedly teach us more about the normal immune response and its control. Acquired defects in the immune system, the secondary immunodeficiencies, are much commoner than the primary inherited immunodeficiencies. In the next section we consider the pandemic of acquired immune deficiency syndrome caused by infection with the HIV virus.

Acquired immune deficiency syndrome.

The most extreme case of immune suppression caused by a pathogen is the **acquired immune deficiency syndrome** (**AIDS**) caused by infection with the **human immunodeficiency virus** (**HIV**). HIV infection leads to a gradual loss of immune competence, allowing infection with organisms that are not normally pathogenic. The earliest documented case of HIV infection in humans was reported in a sample of serum from Kinshasa (Democratic Republic of Congo) that was stored in 1959. It was not until 1981, however, that the first cases of AIDS were officially reported. The disease is characterized by a susceptibility to infection with opportunistic pathogens or by the occurrence of an aggressive form of Kaposi's sarcoma or B-cell lymphoma, accompanied by a profound decrease in the number of CD4 T cells.

Because the disease seemed to be spread by contact with body fluids, the cause was suspected to be a new virus, and by 1983 the virus responsible, HIV, was isolated and identified. It is now clear that there are at least two types of HIV—HIV-1 and HIV-2— that are closely related. HIV-2 is endemic in West Africa and is now spreading in India. Most AIDS worldwide, however, is caused by the more virulent HIV-1. Both viruses seem to have spread to humans from other primate species, and the best evidence from nucleotide sequence relationships suggests that HIV-1 has passed to humans on at least three independent occasions from the chimpanzee, *Pan troglodytes*, whereas HIV-2 originated in the sooty mangabey, *Cercocebus atys*.

HIV-1 shows marked genetic variability and is classified on the basis of nucleotide sequence into three major groups named M (main), O (outlier), and N (non-M, non-O). These are only distantly related to each other and are thought to have passed into humans by independent transmissions from chimpanzees. The M group is the major cause of AIDS worldwide and is genetically diversified into subtypes, sometimes known as clades, that are designated by letters A to K; in different parts of the world different subtypes predominate. The ancestry of the M group of HIV-1 to a common parent has been traced by phylogenetic analysis and it is most likely that the subtypes of the M virus have evolved within humans after the original transmission of the virus from a chimpanzee, which harbors the related simian immunodeficiency virus (SIV_{cpz}). It has been estimated that the common ancestor of the M group may date back as far as 1915–1941; if correct, this means that HIV-1 has been infecting humans in central Africa for far longer than had been thought.

HIV infection does not immediately cause AIDS, and the mechanisms of how it does, and whether all HIV-infected patients will progress to overt disease, are incompletely understood. Nevertheless, accumulating evidence clearly implicates the growth of the virus in CD4 T cells, and the immune response to it, as the central keys to the puzzle of AIDS. HIV is now a worldwide pandemic, and although great strides are being made in understanding the pathogenesis and epidemiology of the disease, the number of infected people around the world continues to grow at an alarming rate, presaging the death of many people from AIDS for many years to come. Estimates from the World Health Organization are that more than 25 million people have died from AIDS since the beginning of the epidemic, and that there are currently around 33.5 million people living with HIV infection (Fig. 13.19). Most of these are living in sub-Saharan Africa, where approximately 5.2% of adults are infected. In some countries within this region, such as Swaziland and Botswana, around 25% of adults are infected. There are growing epidemics of HIV infection and AIDS in China and in India, where surveys in several states have shown a 1–2% prevalence of HIV infection in pregnant women. The incidence of HIV infection is rising faster in Eastern Europe and Central Asia than in the rest of the world. About one-third of those infected with HIV are aged between 15 and 24 years, and most are unaware that they carry the virus.

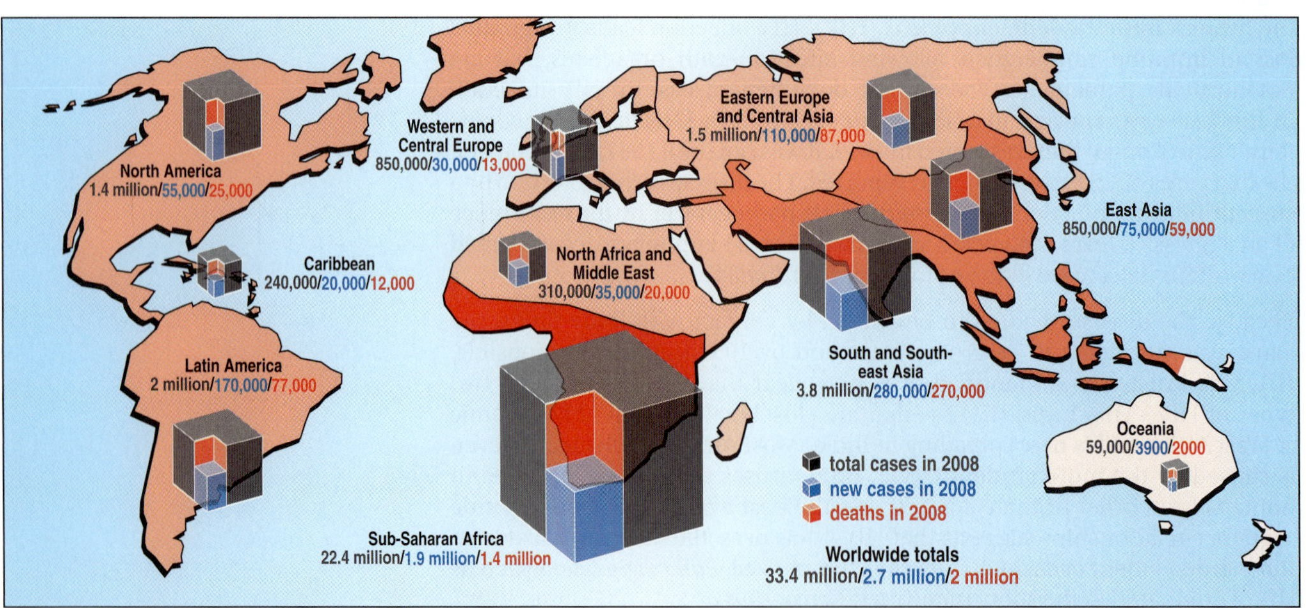

Fig. 13.19 The incidence of new HIV infection is increasing more slowly in many regions of the world, but AIDS is still a major disease burden. The number of individuals living with HIV/AIDS is large and continues to grow, but the number of new infections in 2008 has decreased by 30% compared with the apparent peak of the epidemic in 1996. Worldwide, in 2008 there were around 33.5 million individuals infected with HIV, including some 2.7 million new cases, and more than 2 million deaths from AIDS. Data are estimated numbers of adults and children living with HIV/AIDS at the end of 2008 (AIDS Epidemic Update, UNAIDS/World Health Organization, 2009).

13-24 Most individuals infected with HIV progress over time to AIDS.

Many viruses cause an acute but limited infection that induces lasting protective immunity. Others, such as herpesviruses, set up a latent infection that is not eliminated but is chronically controlled by an adaptive immune response (see Section 13-2). Infection with HIV, however, seems rarely to lead to an immune response that can prevent ongoing replication of the virus. Although the initial acute infection does seem to be controlled by the immune system, HIV continues to replicate and infect new cells.

Infection with HIV generally occurs after the transfer of body fluids from an infected person to an uninfected one. HIV infection is most commonly spread by sexual intercourse, contaminated needles used for intravenous drug delivery, and the therapeutic use of infected blood or blood products, although this last route of transmission has largely been eliminated in the developed world, where blood products are routinely screened for the presence of HIV. An important route of virus transmission is from an infected mother to her baby at birth or through breast milk. Rates of transmission from an infected mother to a child range from as low as 11% to as high as 60% depending on the severity of the infection (as judged by the viral load) and the frequency of breastfeeding. Antiretroviral drugs such as zidovudine (AZT) or nevirapine administered during pregnancy significantly reduce the amount of virus passed to the newborn infant, thus reducing the frequency of transmission.

The virus is carried mainly in infected cells that express CD4, which acts as the receptor for the virus, along with a co-receptor, usually the chemokine receptors CCR5 or CXCR4. Free virus is also present in blood, semen, vaginal fluid, or mother's milk. The gastrointestinal and genital mucosae are the dominant sites of primary infection. HIV virions seem to establish infection initially in a small number of mucosal CD4 T cells, where they replicate locally before spreading to lymph nodes draining the mucosa. The lymphoid compartment of mucosal tissues is enriched for memory CD4 T cells that express CCR5, and so viral replication is favored in these sites and the lymph nodes that drain them. After accelerated replication in regional lymph nodes, the virus disseminates widely via the bloodstream, and in particular gains broad access to the gut-associated lymphoid tissues (GALT), where the highest number of CD4 T cells in the body reside (see Chapter 12). Because of the important role of CCR5 as the predominant co-receptor for viral entry early in the course of HIV infection, new experimental drugs, called entry inhibitors, have been designed to block the interaction between HIV virus and CCR5, as discussed further below.

The **acute phase** of HIV infection is clinically characterized by an influenza-like illness in up to 80% of cases, with an abundance of virus (viremia) in the peripheral blood and a marked decrease in the numbers of circulating CD4 T cells, the latter due largely to the extensive death of CD4 T cells in the GALT that are killed directly by viral cytopathic effects or indirectly by activation-induced apoptosis (Fig. 13.20). Diagnosis at this stage is usually missed unless there is a high index of suspicion. The acute viremia is associated in virtually all patients with the activation of CD8 T cells, which kill HIV-infected cells, and subsequently with antibody production, or **seroconversion**. The cytotoxic T-cell response is thought to be important in controlling virus levels, which peak and then decline, as the CD4 T-cell counts rebound to around 800 cells μl^{-1} (the normal value is around 1200 cells μl^{-1}).

By 3–4 months after infection, the symptoms of acute viremia have usually passed. The level of virus that persists in the blood plasma at this stage of infection, referred to as the **viral set point**, is usually the best indicator of future disease progression. There is then a period of apparent quiescence of the disease that is known as clinical latency or the **asymptomatic phase** (see Fig. 13.20). This period is not silent, however, for there is persistent replication

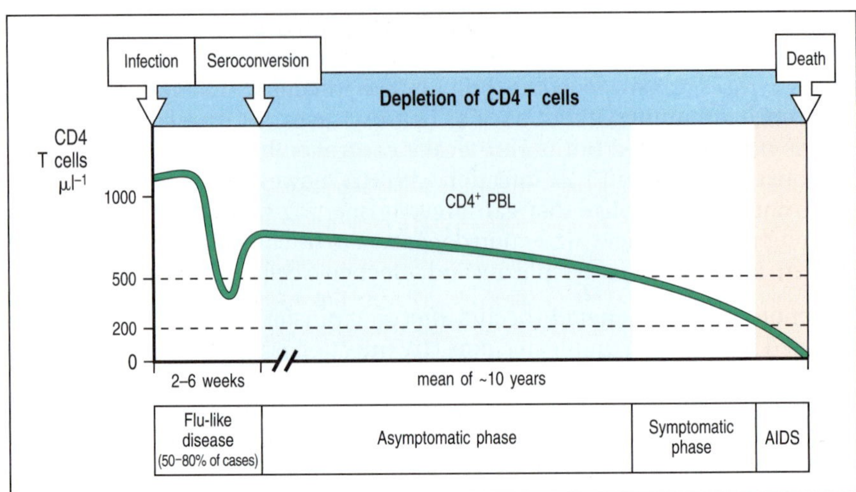

Fig. 13.20 The typical course of untreated infection with HIV. The first few weeks are typified by an acute influenza-like viral illness, sometimes called seroconversion disease, with high titers of virus in the blood. An adaptive immune response follows, which controls the acute illness and largely restores levels of CD4 T cells (CD4⁺ PBL) but does not eradicate the virus. This is the asymptomatic phase. Opportunistic infections and other symptoms become more frequent as the CD4 T-cell count falls, starting at about 500 cells μl^{-1}. The disease then enters the symptomatic phase. When CD4 T-cell counts fall below 200 cells μl^{-1}, the patient is said to have AIDS. Note that CD4 T-cell counts are measured for clinical purposes in cells per microliter (cells μl^{-1}), rather than cells per milliliter (cells ml^{-1}), the unit used elsewhere in this book.

of the virus, and a gradual decline in the function and numbers of CD4 T cells until eventually such patients have few CD4 T cells left. At this point, which can occur anywhere between 6 months and 20 years or more after the primary infection, the period of clinical latency ends and opportunistic infections begin to appear, marking the onset of AIDS. Almost all people who are infected with HIV will eventually develop AIDS, unless viral replication is suppressed for life by antiretroviral drugs (Fig. 13.21).

There are at least three dominant mechanisms that cause the loss of CD4 T cells in HIV infection. First, infected cells can be killed directly by the virus; second, infected cells are more likely to undergo apoptosis; and third, infected CD4 T cells are killed by CD8 cytotoxic lymphocytes that recognize viral peptides. In addition, regeneration of new T cells is also defective in infected people, suggesting infection and destruction of the progenitors of CD4 T cells within the thymus. This deficiency in regeneration might provide an explanation for the rapid progression of disease in infants.

Figure 13.20 illustrates the typical course of an infection with HIV. However, it is increasingly clear that the course of the disease can vary widely. Most people infected with HIV and remaining untreated go on to develop AIDS and ultimately die of opportunistic infections or cancer. But not all do. A small

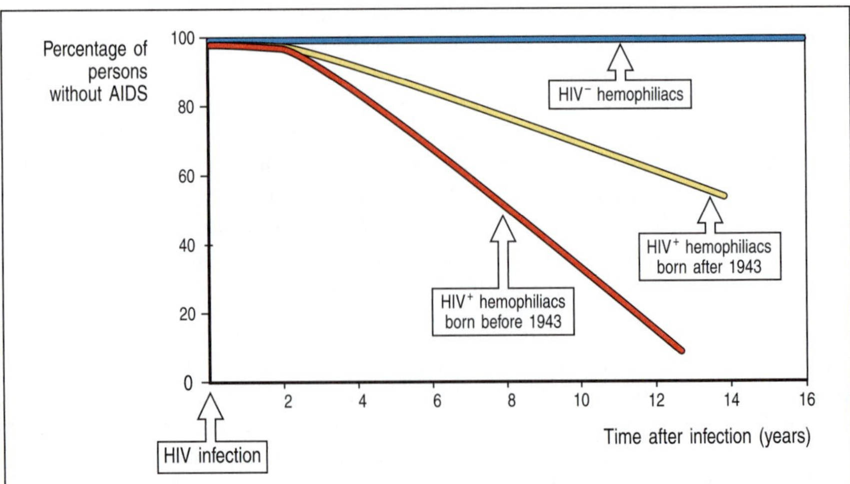

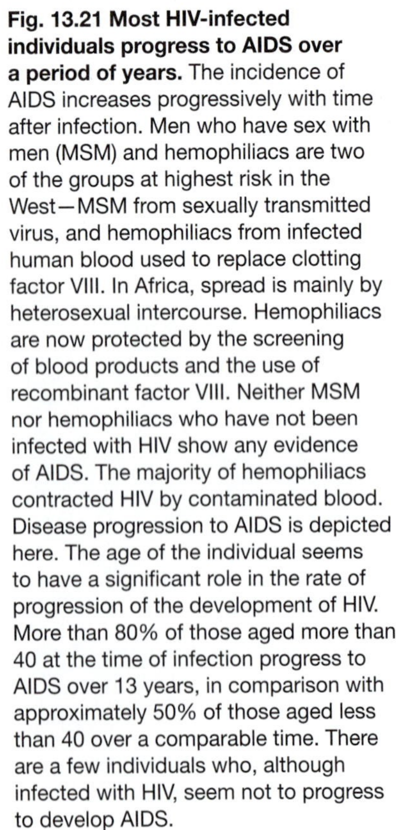

Fig. 13.21 Most HIV-infected individuals progress to AIDS over a period of years. The incidence of AIDS increases progressively with time after infection. Men who have sex with men (MSM) and hemophiliacs are two of the groups at highest risk in the West—MSM from sexually transmitted virus, and hemophiliacs from infected human blood used to replace clotting factor VIII. In Africa, spread is mainly by heterosexual intercourse. Hemophiliacs are now protected by the screening of blood products and the use of recombinant factor VIII. Neither MSM nor hemophiliacs who have not been infected with HIV show any evidence of AIDS. The majority of hemophiliacs contracted HIV by contaminated blood. Disease progression to AIDS is depicted here. The age of the individual seems to have a significant role in the rate of progression of the development of HIV. More than 80% of those aged more than 40 at the time of infection progress to AIDS over 13 years, in comparison with approximately 50% of those aged less than 40 over a comparable time. There are a few individuals who, although infected with HIV, seem not to progress to develop AIDS.

percentage of people exposed to the virus seroconvert, making antibodies against many HIV proteins, but do not seem to have progressive disease, in that their CD4 T-cell counts and other measures of immune competence are maintained. These long-term non-progressors have unusually low levels of circulating virus and are being studied intensively to discover how they are able to control their HIV infection. A second group consists of seronegative people who have been highly exposed to HIV yet remain disease-free and virus-negative. Some of these people have specific cytotoxic lymphocytes and T_H1 lymphocytes directed against infected cells, which confirms that they have been exposed to HIV or possibly to noninfectious HIV antigens. It is not clear whether this immune response accounts for clearing the infection, but it is of considerable interest for the development and design of vaccines.

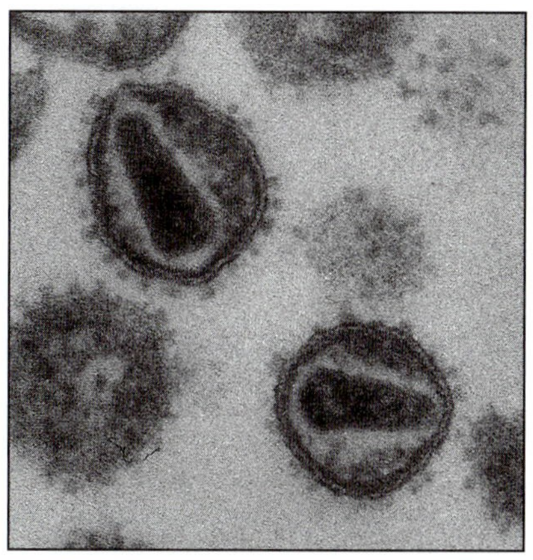

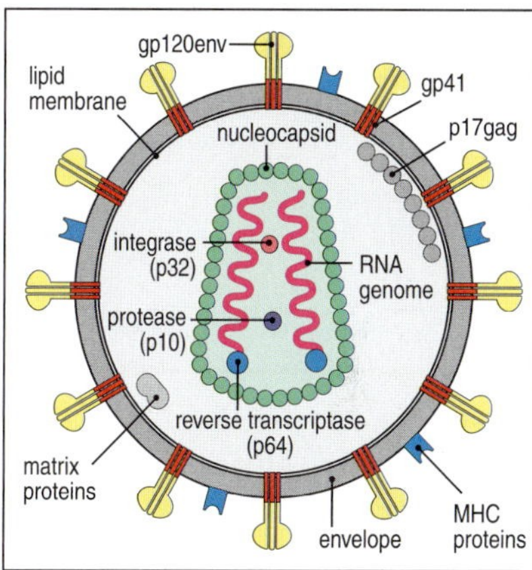

Fig. 13.22 The virion of human immunodeficiency virus (HIV). The virus illustrated is HIV-1, the leading cause of AIDS. The reverse transcriptase, integrase, and viral protease enzymes are packaged in the virion and are shown schematically in the viral capsid. In reality, many molecules of these enzymes are contained in each virion. Photograph courtesy of H. Gelderblom.

13-25 HIV is a retrovirus that infects CD4 T cells, dendritic cells, and macrophages.

HIV is an enveloped retrovirus whose structure is shown in Fig. 3.22. Each virus particle, or virion, contains two copies of an RNA genome and numerous copies of essential enzymes that are required for the initial steps of infection and genome replication, before new viral proteins are produced. The viral genome is reverse transcribed into DNA in the infected cell by the viral **reverse transcriptase**, and the DNA is integrated into the host-cell chromosomes with the aid of the viral **integrase**. RNA transcripts are produced from the integrated viral DNA and serve both as mRNAs to direct the synthesis of viral proteins and later as the RNA genomes of new viral particles. These escape from the cell by budding from the plasma membrane, each enclosed in a membrane envelope. HIV belongs to a group of retroviruses called the **lentiviruses**, from the Latin *lentus*, meaning slow, because of the gradual course of the diseases that they cause. These viruses persist and continue to replicate for many years before causing overt signs of disease.

The ability of HIV to enter particular cell types, known as its cellular **tropism**, is determined by the expression of specific receptors for the virus on the surfaces of those cells. HIV enters cells by means of a complex of two noncovalently associated viral glycoproteins, gp120 and gp41, in the viral envelope. The gp120 portion of the glycoprotein complex binds with high affinity to the cell-surface molecule CD4. The virus thus binds to CD4 T cells and to dendritic cells and macrophages, which also express some CD4. Before fusion and entry of the virus, gp120 must also bind to a co-receptor in the membrane of the host cell. Several chemokine receptors can serve as co-receptors for HIV entry. The major co-receptors are CCR5, which is predominantly expressed on memory CD4 T cells, dendritic cells, macrophages, and CXCR4, which is expressed on activated T cells. After binding of gp120 to the receptor and co-receptor, gp41 then causes fusion of the viral envelope with the cell's plasma membrane, allowing the viral genome and associated viral proteins to enter the cytoplasm. This fusion process has become a target for drug therapy, and active development of **fusion inhibitors** is providing new alternatives for combination therapy against HIV. Peptide analogs of the carboxy-terminal peptide of gp41 inhibit fusion of the viral envelope and plasma membrane, and administration of one such peptide, called T-20, to patients with HIV infection caused an approximately 20-fold decline in HIV RNA levels in plasma.

After a period of clonal homogeneity during the acute phase of infection, HIV begins to mutate rapidly as the antiviral CD8 T-cell response to infection progresses and selects for **escape mutants** that are no longer detected by the adaptive immune response. This gives rise to many different variants in a single infection and to even broader variation within the population as a whole. Different variants infect different cell types, and their tropism is determined to a large degree by which chemokine receptor the virus uses as co-receptor. The co-receptor used by variants of HIV that are associated with primary infections is CCR5, which binds the CC chemokines CCL3, CCL4, and CCL5, and these variants require only a low level of CD4 on the cells they infect. The variants of HIV that use CCR5 infect dendritic cells, macrophages, and T cells *in vivo*, and are now usually designated as 'R5' viruses, reflecting their chemokine receptor usage. In contrast, 'X4' viruses preferentially infect CD4 T cells and use CXCR4 (the receptor for chemokine CXCL12) as a co-receptor.

It seems that R5 isolates of HIV are preferentially transmitted by sexual contact, because they are the dominant viral phenotype found in newly infected individuals and target the CCR5-expressing cells that are enriched in mucosal lymphoid tissues. Because the intestinal and genital mucosae are constantly exposed to commensal microbes, they harbor a large number of activated immune cells in which HIV can readily replicate. Infection occurs across two types of epithelium. The mucosae of the vagina, penis, cervix, and anus are covered by stratified squamous epithelium, which is an epithelium composed of several layers of cells. A second type of epithelium, composed of a single layer of cells, is present in the rectum and in the endocervix.

A complex ferrying mechanism seems to transfer HIV picked up by dendritic cells in squamous epithelia to CD4 T cells in lymphoid tissue. *In vitro* studies have shown that HIV attaches to monocyte-derived dendritic cells by the binding of viral gp120 to C-type lectin receptors such as langerin (CD207), the mannose receptor (CD206), and DC-SIGN. A portion of the bound virus is rapidly taken up into vacuoles, where it can remain for days in an infectious state. In this way the virus is protected and remains stable until it encounters a susceptible CD4 T cell, whether in the local mucosal environment or after being carried to draining lymphoid tissue (Fig. 13.23). The existence of this transport mechanism confirms the suggestion that HIV can infect CD4 cells either directly or via the immunological synapse formed between dendritic cells and CD4 T cells.

Epithelial cells from the single-layer epithelium covering rectum and endocervix express CCR5 and another HIV-binding molecule, glycosphingolipid galactosyl ceramide, and have been shown to selectively translocate R5 HIV variants, but not X4, through the epithelial monolayer. This allows HIV to bind to and infect submucosal CD4 T cells and dendritic cells. Infection of CD4 T cells via CCR5 occurs early in the course of infection and continues, with activated CD4 T cells accounting for the major production of HIV

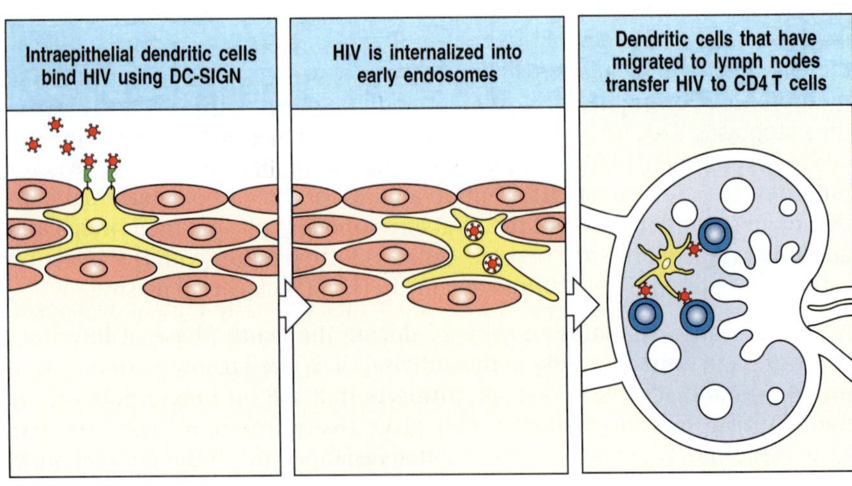

Intraepithelial dendritic cells bind HIV using DC-SIGN
HIV is internalized into early endosomes
Dendritic cells that have migrated to lymph nodes transfer HIV to CD4 T cells

Fig. 13.23 Dendritic cells initiate infection by transporting HIV from mucosal surfaces to lymphoid tissue. HIV adheres to the surface of intraepithelial dendritic cells by the binding of viral gp120 to DC-SIGN (left panel). It gains access to dendritic cells at sites of mucosal injury or possibly to dendritic cells that have protruded between epithelial cells to sample the external world. Dendritic cells internalize HIV into mildly acidic early endosomes and migrate to lymphoid tissue (center panel). The HIV is translocated back to the cell surface, and when the dendritic cell encounters CD4 T cells in a secondary lymphoid tissue, the HIV is transferred to the T cell (right panel).

throughout infection. Late in infection, in approximately 50% of cases, the viral phenotype switches to the X4 type that infects T cells via CXCR4 co-receptors, and this is followed by a rapid decline in CD4 T-cell count and progression to AIDS.

13-26 Genetic variation in the host can alter the rate of progression of disease.

The rate of progression of HIV infection toward AIDS can be modified by the genetic makeup of the infected person. Genetic variation in HLA type is one factor: alleles HLA-B57 and HLA-B27 are associated with a better prognosis, and HLA-B35 with more rapid disease progression. Homozygosity of HLA class I (HLA-A, HLA-B, and HLA-C) is associated with more rapid progression, presumably because the T-cell response to infection is less diverse. Certain polymorphisms of the killer cell immunoglobulin-like receptors (KIRs) present on NK cells (see Section 3-21), in particular the receptor KIR-3DS1 in combination with certain alleles of HLA-B, delay the progression to AIDS.

The clearest case of host genetic variation affecting HIV infection is a mutant allele of CCR5 that when homozygous effectively blocks HIV-1 infection, and when heterozygous slows AIDS progression. This is discussed in more detail in the next section. Mutations that affect the production of cytokines such as IL-10 and IFN-γ have also been implicated in the restriction of HIV progression. Genes that influence the progression to AIDS are listed in Fig. 13.24.

13-27 A genetic deficiency of the co-receptor CCR5 confers resistance to HIV infection *in vivo*.

Evidence for the importance of chemokine receptors in HIV infection has come from studies in a small group of individuals with a high risk of exposure to HIV-1 but who remain seronegative. Cultures of lymphocytes and macrophages from these people were relatively resistant to HIV infection and were found to secrete high levels of the chemokines CCL3, CCL4, and CCL5 in response to inoculation with HIV. The resistance of these rare individuals to HIV infection has now been explained by the discovery that they are homozygous for an allelic, nonfunctional variant of CCR5 called Δ32, which is caused by a 32-base-pair deletion from the coding region that leads to a frameshift mutation and a truncated protein. The gene frequency of this mutant allele in caucasoid populations is quite high, at 0.09 (that is, about 10% of the population are heterozygous carriers of the allele and about 1% are homozygous). The mutant allele has not been found in Japanese or in black Africans from Western or Central Africa. Heterozygous deficiency of CCR5 might provide some protection against sexual transmission of HIV infection and a modest reduction in the rate of progression of the disease. In addition to the structural polymorphism of the gene, variation in the promoter region of the CCR5 gene has been found in both Caucasian and African Americans. Different promoter variants were associated with different rates of progression of disease.

These results provide dramatic confirmation that CCR5 is the major macrophage and T-lymphocyte co-receptor used by HIV to establish primary infection *in vivo*, and offers the possibility that primary infection might be blocked by antagonists of the CCR5 receptor. Indeed, there is preliminary evidence that low molecular weight inhibitors of this receptor can block infection of macrophages by HIV *in vitro*. Such inhibitors might be the precursors of useful drugs that could be taken by mouth to prevent infection. Such drugs are very unlikely to provide complete protection, however, because a very small number of individuals who are homozygous for the nonfunctional variant of CCR5 are infected with HIV. These individuals seem to have suffered from primary infection by X4 strains of the virus.

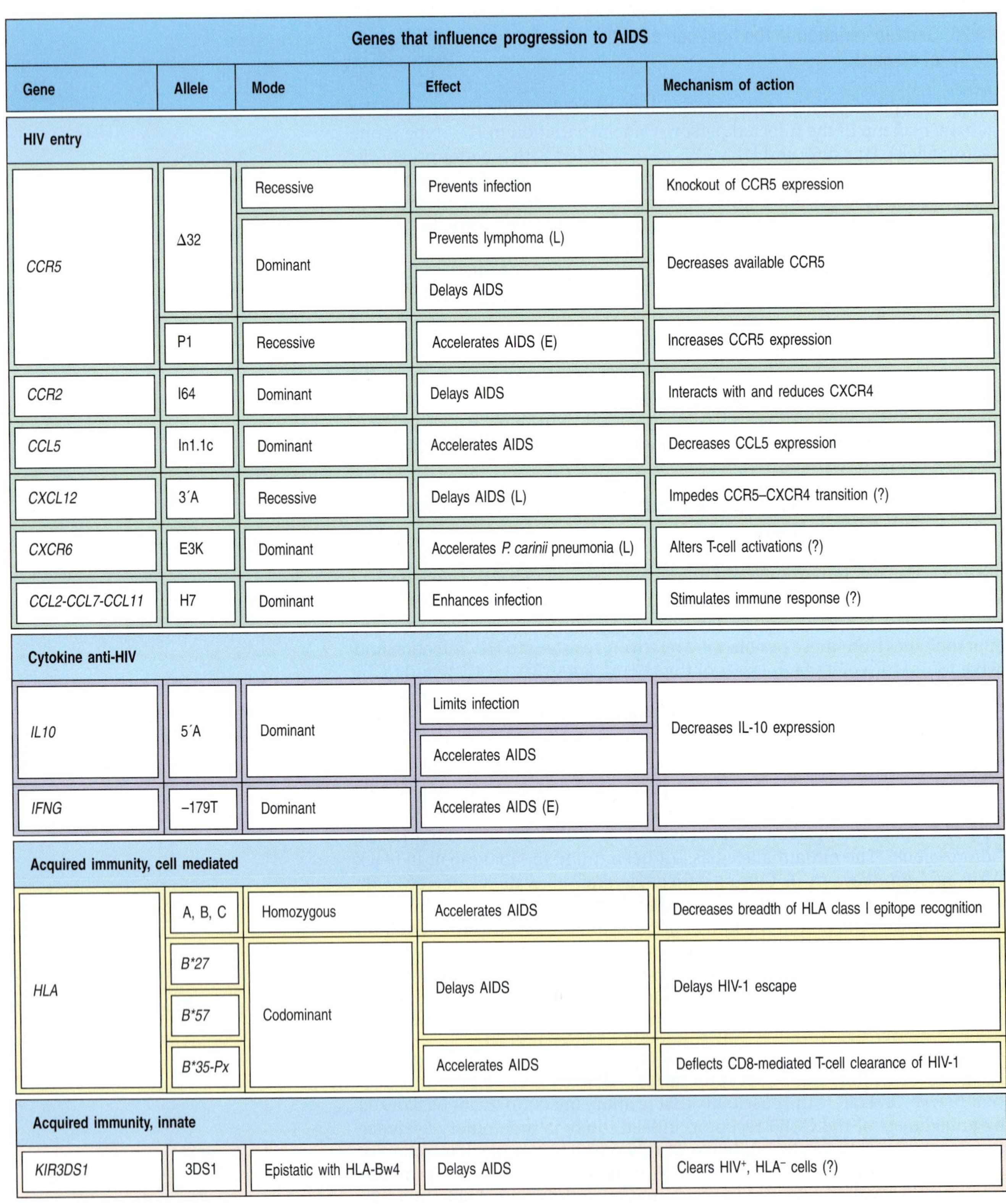

Genes that influence progression to AIDS				
Gene	**Allele**	**Mode**	**Effect**	**Mechanism of action**
HIV entry				
CCR5	Δ32	Recessive	Prevents infection	Knockout of CCR5 expression
		Dominant	Prevents lymphoma (L)	Decreases available CCR5
			Delays AIDS	
	P1	Recessive	Accelerates AIDS (E)	Increases CCR5 expression
CCR2	I64	Dominant	Delays AIDS	Interacts with and reduces CXCR4
CCL5	In1.1c	Dominant	Accelerates AIDS	Decreases CCL5 expression
CXCL12	3´A	Recessive	Delays AIDS (L)	Impedes CCR5–CXCR4 transition (?)
CXCR6	E3K	Dominant	Accelerates *P. carinii* pneumonia (L)	Alters T-cell activations (?)
CCL2-CCL7-CCL11	H7	Dominant	Enhances infection	Stimulates immune response (?)
Cytokine anti-HIV				
IL10	5´A	Dominant	Limits infection	Decreases IL-10 expression
			Accelerates AIDS	
IFNG	–179T	Dominant	Accelerates AIDS (E)	
Acquired immunity, cell mediated				
HLA	A, B, C	Homozygous	Accelerates AIDS	Decreases breadth of HLA class I epitope recognition
	*B*27*	Codominant	Delays AIDS	Delays HIV-1 escape
	*B*57*			
	*B*35-Px*		Accelerates AIDS	Deflects CD8-mediated T-cell clearance of HIV-1
Acquired immunity, innate				
KIR3DS1	3DS1	Epistatic with HLA-Bw4	Delays AIDS	Clears HIV$^+$, HLA$^-$ cells (?)

Fig. 13.24 Genes that influence progression to AIDS in humans. E, an effect that acts early in progression to AIDS; L, acts late in AIDS progression; ?, plausible mechanism of action with no direct support. Reprinted with permission from Macmillan Publishers Ltd: S.J. O'Brien, G.W. Nelson, Nat. Genet. 36:565–574, © 2004.

13-28 HIV RNA is transcribed by viral reverse transcriptase into DNA that integrates into the host-cell genome.

Once the virus has entered cells, it replicates in the same way as other retroviruses. One of the proteins carried in the virus particle is the viral reverse transcriptase, which transcribes the viral RNA into a complementary DNA (cDNA) copy. The viral cDNA is then integrated into the host-cell genome by the viral integrase, which also enters the cell with the viral RNA. The integrated cDNA copy is known as the **provirus**. The entire infectious cycle is shown in Fig. 13.25. In activated CD4 T cells, virus replication is initiated by transcription of the provirus, as we discuss in the next section. However, HIV can, like other retroviruses, establish a latent infection in which the provirus remains quiescent. This seems to occur in memory CD4 T cells and in dormant macrophages, and these cells are thought to be important reservoirs of infection.

The HIV genome consists of nine genes flanked by long terminal repeat (LTR) sequences. The latter are required for the integration of the provirus into the host-cell DNA and contain binding sites for gene regulatory proteins that control the expression of the viral genes. Like other retroviruses, HIV has three major genes—*gag*, *pol*, and *env* (Fig. 13.26). The *gag* gene encodes the structural proteins of the viral core, *pol* encodes the enzymes involved in viral replication and integration, and *env* encodes the viral envelope glycoproteins. The *gag* and *pol* mRNAs are translated to give polyproteins—long polypeptide chains that are then cleaved by the **viral protease** (also encoded by *pol*) into individual functional proteins. The product of the *env* gene, gp160, has to be cleaved by a host-cell protease into gp120 and gp41, which are then assembled as trimers into the viral envelope. As shown in Fig. 13.26, HIV has six other, smaller, genes encoding proteins that affect viral replication and infectivity in various ways. Two of these, Tat and Rev, perform regulatory functions that are essential for viral replication. The remaining four—Nef, Vif, Vpr, and Vpu—are essential for efficient virus production *in vivo*.

13-29 Replication of HIV occurs only in activated T cells.

The production of infectious virus particles from an HIV provirus in CD4 T cells is stimulated by T-cell activation. This induces the transcription factors NFκB and NFAT, which bind to promoters in the viral LTR, thereby initiating the transcription of viral RNA by the cellular RNA polymerase II. This transcript is spliced in various ways to produce mRNAs for the viral proteins. The Gag and Gag–Pol proteins are translated from unspliced mRNA; Vif, Vpr, Vpu, and Env are translated from singly spliced viral mRNA; Tat, Rev, and Nef are translated from multiply spliced mRNA. Tat greatly enhances the transcription of viral RNA from the provirus by the RNA polymerase II complex. It binds to the transcriptional activation region (TAR) in the provirus 5′LTR in a complex with the cellular cyclin T1 and its partner, cyclin-dependent kinase 9 (CDK9), to form a complex that phosphorylates RNA polymerase and stimulates its RNA elongation activity. The expression of the cyclin T1–CDK9 complex is greatly increased in activated T cells compared with quiescent ones. Together with the increased expression of NFκB and NFAT in activated T cells, this may explain the ability of HIV to lie dormant in resting T cells and replicate in activated T cells.

Eukaryotic cells have mechanisms to prevent the export from the cell nucleus of incompletely spliced mRNA transcripts. This could pose a problem for a retrovirus that is dependent on the export of unspliced, singly spliced, and multiply spliced mRNA species to translate the full complement of viral proteins. The Rev protein is the viral solution to this problem. Export from the nucleus and translation of the three HIV proteins encoded by the fully spliced mRNA transcripts—Tat, Nef, and Rev—occurs early after viral infection by means of the normal host cellular mechanisms of mRNA export. The expressed Rev protein then enters the nucleus and binds to a specific viral RNA sequence, the Rev response element (RRE). In the presence of *rev*, RNA is exported from

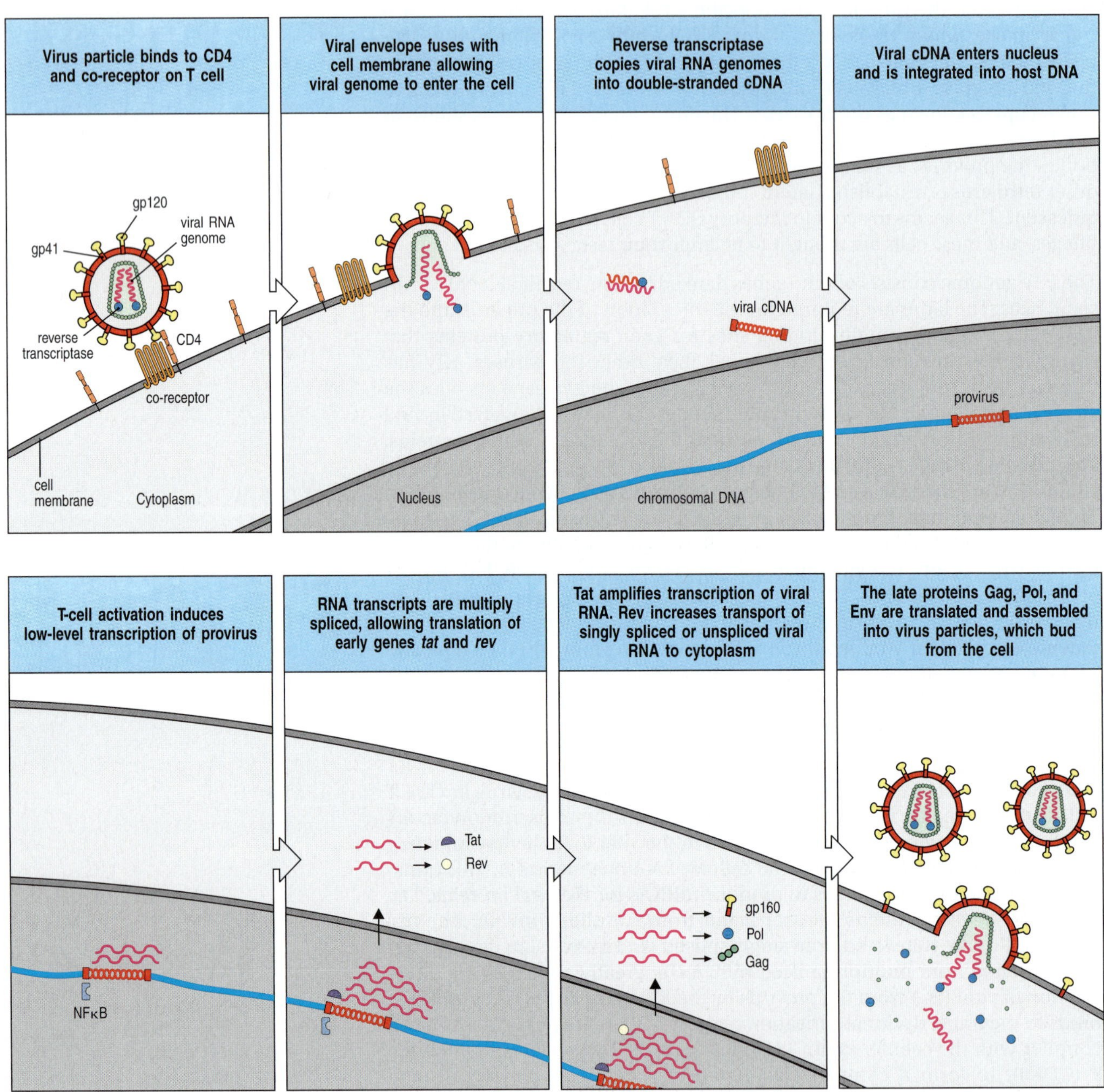

Fig. 13.25 The life cycle of HIV. Top row: the virus binds to CD4 using gp120, which is altered by CD4 binding so that it now also binds a chemokine receptor that acts as a co-receptor for viral entry. This binding releases gp41, which causes fusion of the viral envelope with the cell membrane and release of the viral core into the cytoplasm. Once in the cytoplasm, the viral core releases the RNA genome, which is reverse-transcribed into double-stranded cDNA using the viral reverse transcriptase. The double-stranded cDNA migrates to the nucleus in association with the viral integrase and the Vpr protein and is integrated into the cell genome, becoming a provirus. Bottom row: activation of CD4 T cells induces the expression of the transcription factors NFκB and NFAT, which bind to the proviral LTR and initiate transcription of the HIV genome. The first viral transcripts are extensively processed, producing spliced mRNAs encoding several regulatory proteins, including Tat and Rev. Tat both enhances transcription from the provirus and binds to the RNA transcripts, stabilizing them in a form that can be translated. Rev binds the RNA transcripts and transports them to the cytosol. As levels of Rev increase, less extensively spliced and unspliced viral transcripts are transported out of the nucleus. The singly spliced and unspliced transcripts encode the structural proteins of the virus, and unspliced transcripts, which are also the new viral genomes, are packaged with these proteins to form many new virus particles.

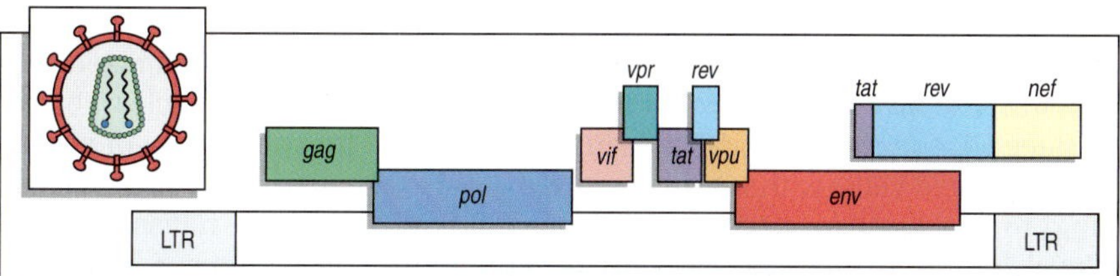

Gene		Gene product/function
gag	Group-specific antigen	Core proteins and matrix proteins
pol	Polymerase	Reverse transcriptase, protease, and integrase enzymes
env	Envelope	Transmembrane glycoproteins. gp120 binds CD4 and CCR5; gp41 is required for virus fusion and internalization
tat	Transactivator	Positive regulator of transcription
rev	Regulator of viral expression	Allows export of unspliced and partially spliced transcripts from nucleus
vif	Viral infectivity	Affects particle infectivity
vpr	Viral protein R	Transport of DNA to nucleus. Augments virion production. Cell-cycle arrest
vpu	Viral protein U	Promotes intracellular degradation of CD4 and enhances release of virus from cell membrane
nef	Negative-regulation factor	Augments viral replication *in vivo* and *in vitro*. Decreases CD4, MHC class I and II expression

Fig. 13.26 The genomic organization of HIV. Like all retroviruses, HIV-1 has an RNA genome flanked by long terminal repeats (LTR) involved in viral integration and in regulation of transcription of the viral genome. The genome can be read in three frames, and several of the viral genes overlap in different reading frames. This allows the virus to encode many proteins in a small genome. The three main protein products—Gag, Pol, and Env—are synthesized by all infectious retroviruses. The known functions of the different genes and their products are listed. The products of *gag*, *pol*, and *env* are known to be present in the mature viral particle, together with the viral RNA. The mRNAs for Tat, Rev, and Nef proteins are produced by splicing of viral transcripts, so their genes are split in the viral genome. In the case of Nef, only one exon, shown in yellow, is translated.

the nucleus before it can be spliced, so that the structural proteins and RNA genome can be produced. Rev also binds to a cellular transport protein named Crm1, which engages a host pathway for exporting mRNA species through nuclear pores into the cytoplasm.

When the provirus is first activated, Rev levels are low, the transcripts are translocated slowly from the nucleus, and thus multiple splicing events can occur. In this way, more Tat and Rev are produced, and Tat in turn ensures that more viral transcripts are made. Later, when Rev levels have increased, the transcripts are translocated rapidly from the nucleus unspliced or only singly spliced. These unspliced or singly spliced transcripts are translated to produce the structural components of the viral core and envelope, together with the reverse transcriptase, the integrase, and the viral protease, all of which are needed to make new viral particles. The complete, unspliced transcripts that are exported from the nucleus late in the infectious cycle are required for the translation of *gag* and *pol* and are also destined to be packaged with the proteins as the RNA genomes of the new virus particles.

The success of virus replication also depends on the proteins Nef, Vif, Vpr, and Vpu. Vif (viral infectivity factor) is an RNA-binding protein that accumulates in the cytoplasm and on the plasma membrane of infected cells. Vif acts to overcome a natural cellular defense against retroviruses. Cells express a cytidine deaminase called APOBEC, which can be incorporated into virions. This enzyme, which belongs to the same protein family as the activation-induced cytidine deaminase (see Section 13-15), catalyzes the conversion of deoxycytidine to deoxyuridine in the first strand of reverse-transcribed viral cDNA, thus destroying its ability to encode viral proteins. Vif induces the transport of APOBEC to proteasomes, where it is degraded. The expression of Nef (negative regulation factor) early in the viral life cycle induces T-cell activation and the establishment of a persistent state of HIV infection. Nef is also important in the downregulation of expression of two critical immune molecules—MHC class I and CD4. By inhibiting the expression of MHC class I molecules on infected cells, Nef assists in immune evasion by making actively infected cells less likely to be killed by cytotoxic T cells. Nef also promotes the clearance of surface CD4 molecules, which seems to be critical for the release of virions from infected cells; without this function, CD4 expressed on the host cell can bind to the virion during budding and interfere with virion release. Finally, Nef also inhibits the MHC class II-restricted presentation of peptides to CD4 T cells, further inhibiting the generation of an antiviral immune response. The function of Vpr (viral protein R) is not fully understood, but it has various activities that enhance viral production and release. Vpu (viral protein U) is unique to HIV-1 and variants of SIV, and is required for the maturation of progeny virions and their efficient release.

13-30 Lymphoid tissue is the major reservoir of HIV infection.

Although HIV load and turnover are usually measured in terms of the RNA present in virions in the blood, a major reservoir of HIV infection is lymphoid tissue, in which infected CD4 T cells, monocytes, macrophages, and dendritic cells are found. In addition, HIV is trapped in the form of immune complexes on the surface of follicular dendritic cells in the germinal center. These cells are not themselves infected but may act as a reservoir of infective virions. Several other potential reservoirs for HIV-1 that may contribute to its long-term persistence are infected cells in the central nervous system, the gastrointestinal system, and the male urogenital tract.

From studies of patients receiving drug treatment, it is estimated that more than 95% of the virus that can be detected in the plasma is derived from productively infected CD4 T cells that have a very short half-life of about 2 days. Virus-producing CD4 T cells are found in the T-cell areas of lymphoid tissues, and these are thought to succumb to infection while being activated in an immune response. Latently infected CD4 memory T cells that become reactivated by antigen also produce virus that can spread to other activated CD4 T cells. Unfortunately, latently infected CD4 memory T cells have an extremely long mean half-life of around 44 months. This means that drug therapy may never be able to eliminate an HIV infection and therefore needs to be administered throughout life. In addition to cells that are productively or latently infected, a further large population of cells are infected by defective proviruses, which do not produce infectious virus.

Macrophages and dendritic cells seem to be able to harbor replicating HIV without necessarily being killed by it, and they are believed to be an important reservoir of infection; they also serve as a means of spreading virus to other tissues, such as the brain. Although the function of macrophages as antigen-presenting cells does not seem to be compromised by HIV infection, it is thought that the virus causes abnormal patterns of cytokine secretion that could account for the wasting that commonly occurs in AIDS patients late in their disease.

13-31 An immune response controls but does not eliminate HIV.

Infection with HIV generates an immune response that contains the virus but only very rarely, if ever, eliminates it. The time course of various elements in the adaptive immune response to HIV is shown, together with the levels of infectious virus in plasma, in Fig. 13.27. The initial acute phase that occurs as the adaptive immune response develops is followed by the chronic, semi-stable phase that eventually culminates in AIDS. Current thinking is that virus-mediated cytopathicity is very important during early infection and that this results in a substantial depletion of CD4 T cells, particularly in the mucosal tissues where the largest number of T cells normally reside. After the acute

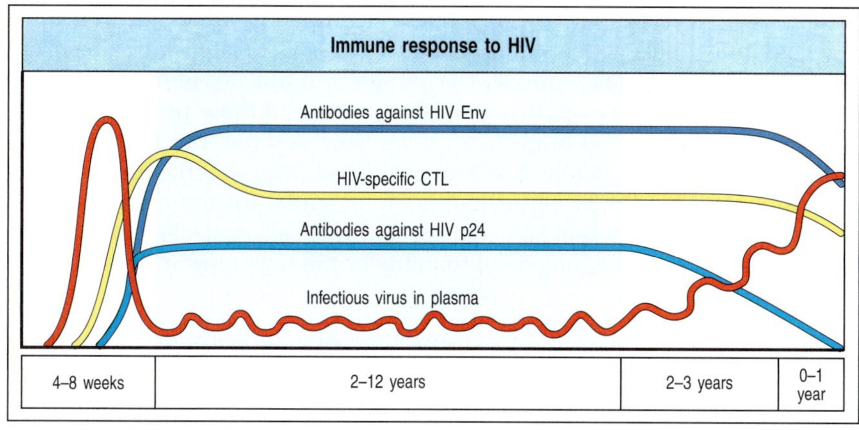

Fig. 13.27 The immune response to HIV. Infectious virus is present at relatively low levels in the peripheral blood of infected individuals during a prolonged asymptomatic phase, during which the virus is replicated persistently in lymphoid tissues. During this period, CD4 T-cell counts gradually decline, although antibodies and CD8 cytotoxic T cells directed against the virus remain at high levels. Two different antibody responses are shown in the figure, one to the envelope protein (Env) of HIV, and one to the core protein p24. Eventually, the levels of antibody and HIV-specific cytotoxic T lymphocytes (CTLs) also decline, and there is a progressive increase in infectious HIV in the peripheral blood.

phase there is a good initial recovery, but cytotoxic lymphocytes directed against HIV-infected cells, immune activation (direct and indirect), viral cytopathic effects, and insufficient T-cell regeneration combine to establish the chronic state, during which immunodeficiency develops. In this section we consider in turn the roles of CD8 cytotoxic T cells, CD4 T cells, antibodies, and soluble factors in the immune response to HIV infection that ultimately fails to contain the infection.

Studies of peripheral blood cells from infected individuals reveal cytotoxic T cells specific for viral peptides that can kill infected cells *in vitro. In vivo,* cytotoxic T cells can be seen to invade sites of HIV replication and they could, in theory, be responsible for killing many productively infected cells before any infectious virus is released, thereby containing viral load at the quasi-stable levels that are characteristic of the asymptomatic period. Evidence for the clinical importance of the control of HIV-infected cells by CD8 cytotoxic T cells comes from studies relating the numbers and activity of CD8 T cells to viral load. An inverse correlation was found between the number of CD8 T cells carrying a receptor specific for an HLA-A2-restricted HIV peptide and the amount of viral RNA in the plasma. Similarly, patients with high levels of HIV-specific CD8 T cells showed slower progression of disease than those with low levels. There is also direct evidence from experiments in macaques infected with SIV that CD8 cytotoxic T cells control retrovirus-infected cells *in vivo*. Treatment of infected animals with monoclonal antibodies that remove CD8 T cells was followed by a large increase in viral load.

A variety of factors produced by CD4, CD8, and NK cells are important in antiviral immunity. Evidence for a noncytotoxic suppressor activity of CD8 cells on HIV-1 came from the observation that peripheral blood mononuclear cells (PMBCs) from seropositive asymptomatic individuals failed to replicate HIV-1 *in vitro*, but that depletion of CD8 T cells, but not other cells (for example, NK cells) from this PMBC fraction led to an increase in viral replication. The inhibition is now known to be mediated by secreted proteins. Chemokines such as CCL5, CCL3, and CCL4 are released at the site of infection and inhibit virus spread (without killing the cell) by competing with R5 strains of HIV-1 for the engagement of co-receptor CCR5, whereas factors still unknown compete with R4 strains for binding to CXCR4. Cytokines such as IFN-α and IFN-γ may also be involved in controlling virus spread, but a mechanism for this is not clear.

In addition to being a major target for HIV infection, three pieces of evidence show that CD4 T cells also have an important role in the host response to HIV-infected cells. First, an inverse correlation is found between the strength of CD4 T-cell proliferative responses to HIV antigen and viral load. Second, some patients who did not progress to AIDS long after infection by HIV showed strong CD4 T-cell proliferative responses. Third, early treatment of acutely infected individuals with antiretroviral drugs was associated with a recovery in CD4 proliferative responses to HIV antigens. If this antiretroviral therapy was stopped, the CD4 responses persisted in some of these people and were associated with reduced levels of viremia. However, the infection persisted in all patients and it is likely that immunological control of the infection will ultimately fail. If CD4 T-cell responses are essential for the control of HIV infection, then the fact that HIV is tropic for these cells and kills them may be the explanation for the long-term inability of the host immune response to control the infection.

Antibodies against gp120 and gp41 envelope viral antigens are produced in response to infection but, as with T cells, are unable to clear the infection. The antibodies react well with purified antigens *in vitro* and with viral debris, but bind poorly to intact enveloped virions or to infected cells. This suggests that the native conformation of these antigens, which are heavily glycosylated, is not accessible to naturally produced antibodies. The evidence is strong that antibodies cannot significantly modify established disease. Nevertheless, the passive administration of antibodies against HIV can protect experimental animals from mucosal infection by HIV, and this offers hope that an effective vaccine might be developed that could prevent new infections.

The mutations that occur as HIV replicates can allow the resulting virus variants to escape recognition by neutralizing antibody or cytotoxic T cells and contribute to the long-term failure of the immune system to contain the infection. An immune response is often dominated by T cells specific for particular epitopes—the **immunodominant** epitopes—and mutations in immunodominant HIV peptides presented by MHC class I molecules have been found. Mutant peptides have been found to inhibit T cells responsive to the wild-type epitope, thus allowing both the mutant and wild-type viruses to survive. Inhibitory mutant peptides have also been reported in hepatitis B virus infections, and similar mutant immunodominant peptides might contribute to the persistence of other viral infections.

An exciting development in the study of HIV immunity is the identification of a number of cellular proteins that can target HIV replication. The enzyme APOBEC (see Section 13-29) causes extensive mutation of newly formed HIV cDNA, thus destroying its coding and replicative capacity. APOBEC is active in resting CD4 T cells but is degraded in infected CD4 T cells, providing yet another reason that resting CD4 T cells are resistant to infection. The powerful antiretroviral action of APOBEC has provoked considerable interest in finding small molecules that interfere with its virus-induced degradation. Another cytoplasmic protein, TRIM 5α, limits HIV-1 infections in rhesus monkeys, probably by targeting the viral capsid and preventing the uncoating and release of viral RNA.

13-32 The destruction of immune function as a result of HIV infection leads to increased susceptibility to opportunistic infection and eventually to death.

When CD4 T-cell numbers decline below a critical level, cell-mediated immunity is lost, and infections with a variety of opportunistic microbes appear (Fig. 13.28). Typically, resistance is lost early to oral *Candida* species and to *M. tuberculosis*, which is manifested as an increased prevalence of thrush (oral candidiasis) and tuberculosis, respectively. Later, patients suffer from shingles, caused by the activation of latent herpes zoster, from aggressive EBV-induced B-cell lymphomas, and from Kaposi's sarcoma, a tumor of endothelial cells that probably represents a response both to cytokines produced in the infection and to a herpesvirus called Kaposi's sarcoma-associated herpesvirus (KSHV, or HHV8) that was identified in these lesions. Pneumonia caused by

Infections	
Parasites	*Toxoplasma* spp. *Cryptosporidium* spp. *Leishmania* spp. *Microsporidium* spp.
Intracellular bacteria	*Mycobacterium tuberculosis* *Mycobacterium avium intracellulare* *Salmonella* spp.
Fungi	*Pneumocystis jirovecii* *Cryptococcus neoformans* *Candida* spp. *Histoplasma capsulatum* *Coccidioides immitis*
Viruses	Herpes simplex Cytomegalovirus Herpes zoster

Malignancies
Kaposi's sarcoma – (HHV8) Non-Hodgkin's lymphoma, including EBV-positive Burkitt's lymphoma Primary lymphoma of the brain

Fig. 13.28 A variety of opportunistic pathogens and cancers can kill AIDS patients. Infections are the major cause of death in AIDS, with respiratory infection with *P. jirovecii* and mycobacteria being the most prominent. Most of these pathogens require effective macrophage activation by CD4 T cells or effective cytotoxic T cells for host defense. Opportunistic pathogens are present in the normal environment but cause severe disease primarily in immunocompromised hosts, such as AIDS patients and cancer patients. AIDS patients are also susceptible to several rare cancers, such as Kaposi's sarcoma (associated with human herpesvirus 8 (HHV8)) and various lymphomas, suggesting that immune surveillance of the causative herpesviruses by T cells can normally prevent such tumors (see Chapter 16).

P. jirovecii is common and was often fatal before effective antifungal therapy was introduced. Co-infection by hepatitis C virus is common and associated with more rapid progression of hepatitis. In the final stages of AIDS, infection with cytomegalovirus or a member of the *Mycobacterium avium* group of bacteria is more prominent. It is important to note that not all patients with AIDS get all these infections or tumors, and there are other tumors and infections that are less prominent but still significant. Figure 13.28 lists the commonest opportunistic infections and tumors, most of which are normally controlled by robust CD4 T cell-mediated immunity that wanes as the CD4 T-cell counts drop toward zero (see Fig. 13.20).

13-33 Drugs that block HIV replication lead to a rapid decrease in titer of infectious virus and an increase in CD4 T cells.

Studies with powerful drugs that completely block the cycle of HIV replication indicate that the virus is replicating rapidly at all phases of infection, including the asymptomatic phase. Two viral proteins in particular have been the target of drugs aimed at arresting viral replication. These are the viral reverse transcriptase, which is required for synthesis of the provirus, and the viral protease, which cleaves the viral polyproteins to produce the virion proteins and viral enzymes. The reverse transcriptase is inhibited by nucleoside analogs such as zidovudine (AZT), which was the first anti-HIV drug to be licensed in the United States. Inhibitors of the reverse transcriptase and the protease prevent the establishment of further infection in uninfected cells. Cells that are already infected can continue to produce virions because, once the provirus is established, reverse transcriptase is not needed to make new virus particles, while the viral protease acts at a very late maturation step of the virus, and inhibition of the protease does not prevent virus from being released. However, in both cases, the released virions are not infectious and further cycles of infection and replication are prevented.

The introduction of combination therapy with a cocktail of viral protease inhibitors and nucleoside analogs, also known as **highly active antiretroviral therapy** (**HAART**), dramatically reduced mortality and morbidity in patients with advanced HIV infection in the United States between 1995 and 1997 (Fig. 13.29). Many patients treated with HAART show a rapid and dramatic reduction in viremia, eventually maintaining levels of HIV RNA close to the limit of detection (50 copies per ml of plasma) for a long period (Fig. 13.30). It is unclear how the virus particles are removed so rapidly from the circulation after the initiation of HAART therapy. It seems most likely that they are opsonized by specific antibody and complement and removed by cells of the mononuclear phagocyte system. Opsonized HIV particles can also be

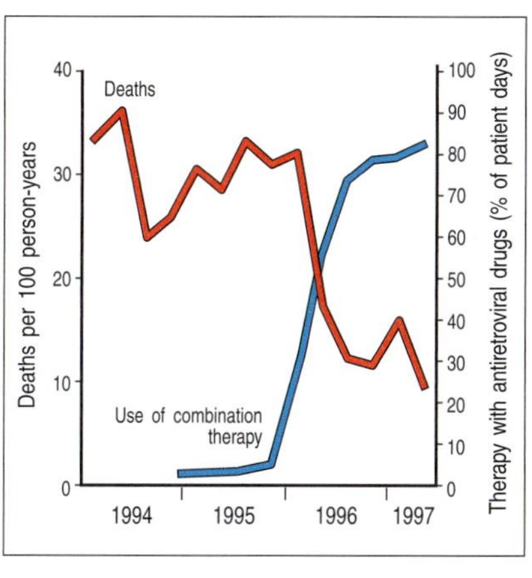

Fig. 13.29 The mortality in advanced HIV infection fell in the United States in parallel with the introduction of combination antiretroviral drug therapy. The graph shows the number of deaths, expressed each calendar quarter as the deaths per 100 person-years. Figure based on data from F. Palella.

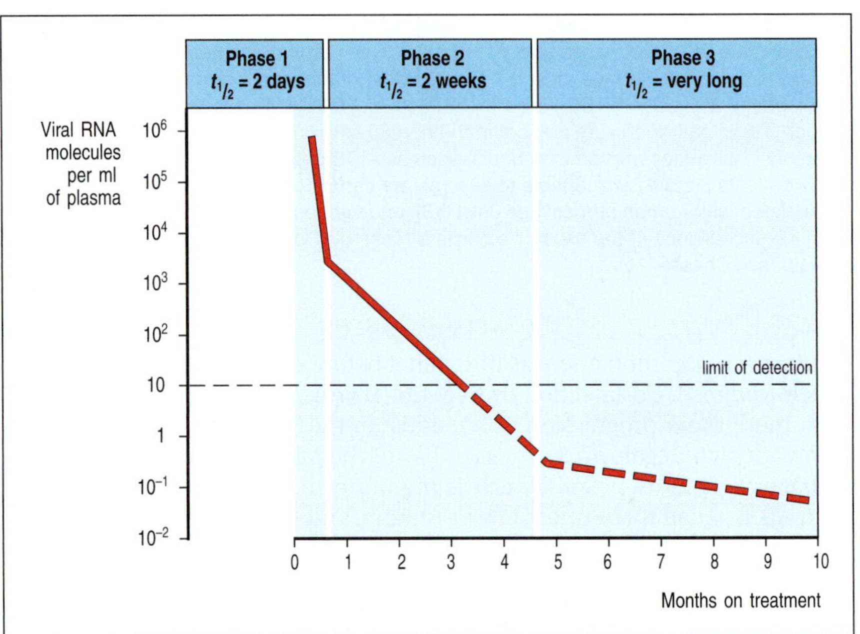

Fig. 13.30 The time-course of the reduction of HIV circulating in the blood on drug treatment. The production of new HIV particles can be arrested for prolonged periods by combinations of protease inhibitors and viral reverse-transcriptase inhibitors. After the initiation of such treatment, virus production is curtailed as these cells die and no new cells are infected. The half-life of virus decay occurs in three phases. The first phase has a half-life of about 2 days, reflecting the half-life of productively infected CD4 T cells, and lasts for about 2 weeks, during which time viral production declines as the lymphocytes that were productively infected at the onset of treatment die. Released virus is rapidly cleared from the circulation, where it has a half-life ($t_{1/2}$) of 6 hours, and there is a decrease in virus levels in plasma of more than 95% during this first phase. The second phase lasts for about 6 months and the virus has a half-life of about 2 weeks. During this phase, virus is released from infected macrophages and from resting, latently infected CD4 T cells stimulated to divide and develop productive infection. It is thought that there is then a third phase of unknown length that results from the reactivation of integrated provirus in memory T cells and other long-lived reservoirs of infection. This reservoir of latently infected cells might remain present for many years. Measurement of this phase of viral decay is impossible at present because viral levels in plasma are below detectable levels (dotted line). Data courtesy of G.M. Shaw.

trapped on the surface of follicular dendritic cells in lymphoid follicles, which are known to capture antigen:antibody complexes and retain them for long periods.

HAART therapy is also accompanied by a slow but steady increase in CD4 T cells, despite the fact that many other compartments of the immune system remain compromised. What is the source of the new CD4 T cells that appear once treatment has started? Three complementary mechanisms have been established for the recovery in CD4 T-cell numbers. The first of these is a redistribution of CD4 T memory cells from lymphoid tissues into the circulation as viral replication is controlled; this occurs within weeks of starting treatment. The second is the reduction in the abnormal levels of immune activation as the HIV infection is controlled, associated with reduced cytotoxic T lymphocyte killing of infected CD4 T cells. The third is much slower and is caused by the emergence of new naive T cells from the thymus. Although the thymus involutes with aging, evidence that these later-arriving cells are indeed of thymic origin is provided by the observation that they contain T-cell receptor excision circles (TRECs) (see Section 5-9).

Although HAART is effective at inhibiting HIV replication, and thus prevents the progression to AIDS and greatly decreases the transmission of HIV, it is ineffective at eradicating all viral stores. Cessation of HAART therefore leads to a rapid rebound of virus multiplication, so that patients require treatment indefinitely. This, coupled with the serious side-effects and high cost of HAART, have stimulated investigation into other targets to block viral replication (Fig. 13.31) as well as ways of flushing out viral reservoirs to eradicate infection permanently. New classes of anti-HIV replication drugs include **viral entry inhibitors**, which block the binding of gp120 to CCR5 or block viral fusion by inhibiting gp41, and **viral integrase inhibitors**, which block the insertion of the reverse-transcribed viral genome into the host DNA. Strategies to induce viral replication in cells latently harboring the virus, and thus to facilitate the actions of replication inhibitors such as HAART, are also being considered. Examples include the administration of cytokines such as IL-2, IL-6, and TNF-α that favor viral transcription and replication. IL-2 is one of the few T-cell-activating cytokines that has been trialed in the treatment of AIDS to boost the depleted immune system. Despite its lack of effect on

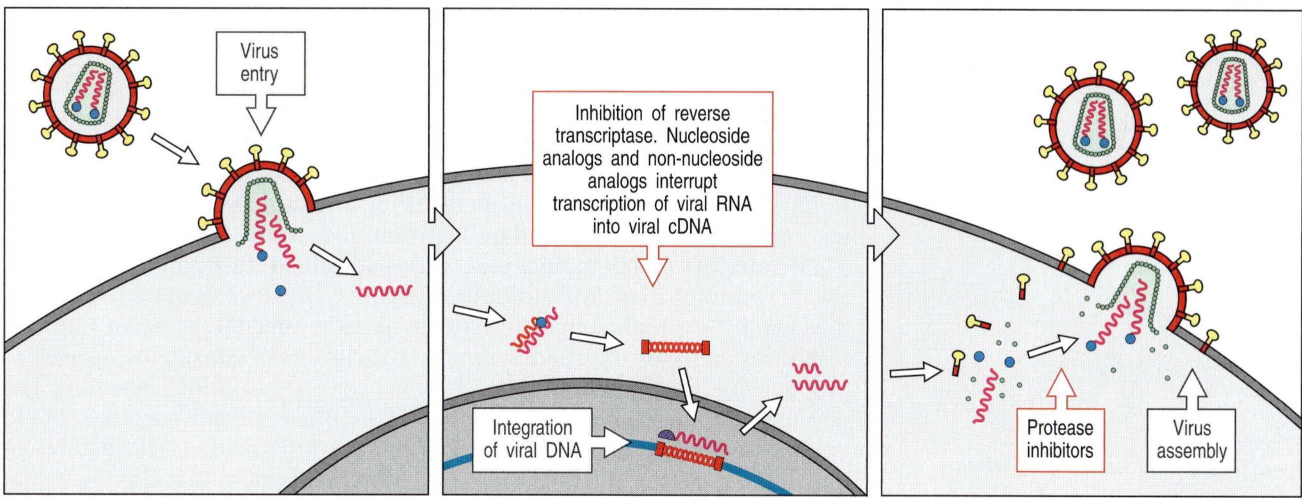

Fig. 13.31 Possible targets for interference with the HIV life cycle. In principle, HIV could be attacked by therapeutic drugs at multiple points in its life cycle: virus entry, reverse transcription of viral RNA, insertion of viral cDNA into cellular DNA by the viral integrase, cleavage of viral polyproteins by the viral protease, and assembly and budding of infectious virions. As yet, only drugs that inhibit reverse transcriptase and protease action have been developed. There are eight nucleoside analog inhibitors and three non-nucleoside inhibitors of reverse transcriptase available, and seven protease inhibitors. Combination therapy using different kinds of drugs is more effective than using a single drug.

the clearance of HIV-1 RNA, IL-2 treatment induces an approximately sixfold increase in the CD4 T-cell count when administered in combination with antiretroviral therapy, with the increase being predominantly in naive T cells rather than memory T cells. Whether IL-2 will have clinical benefit remains to be tested, particularly in view of the associated side-effects, including flu-like symptoms, sinus congestion, low blood pressure, and liver toxicity.

13-34 HIV accumulates many mutations in the course of infection, and drug treatment is soon followed by the outgrowth of drug-resistant variants.

The rapid replication of HIV, with the generation of 10^9 to 10^{10} virions every day, is coupled with a mutation rate of approximately 3×10^{-5} per nucleotide base per cycle of replication, and thus leads to the generation of many variants of HIV in a single infected patient in the course of a day. This high mutation rate arises from the error-prone nature of retroviral replication. Reverse transcriptase lacks the proofreading mechanisms associated with cellular DNA polymerases, and the RNA genomes of retroviruses are copied into DNA with relatively low fidelity. The transcription of the proviral DNA into RNA by RNA polymerase II is also a low-fidelity process. A rapidly replicating persistent virus that is going through these two steps repeatedly in the course of an infection can accumulate many mutations, and numerous variants of HIV, sometimes called **quasi-species**, are found within a single infected individual. This phenomenon was first recognized in HIV and has since proved to be common to all lentiviruses.

As a consequence of its high variability, HIV rapidly develops resistance to antiviral drugs. When a drug is administered, variants of the virus with mutations conferring resistance to the drug emerge and multiply until the previous levels of virus are regained. Resistance to some of the protease inhibitors requires only a single mutation and appears after only a few days (Fig. 13.32); resistance to some of the inhibitors of reverse transcriptase develops in a similarly short time. In contrast, resistance to the nucleoside analog zidovudine takes months to develop, as it requires three or four mutations to occur in the viral reverse transcriptase. As a result of the relatively rapid appearance of resistance to all known anti-HIV drugs, successful drug treatment depends on combination therapy (see Section 13-33). It might also be important to begin treatment early in the course of an infection, thereby reducing the chance that a variant virus will have accumulated all the mutations necessary to resist the entire cocktail.

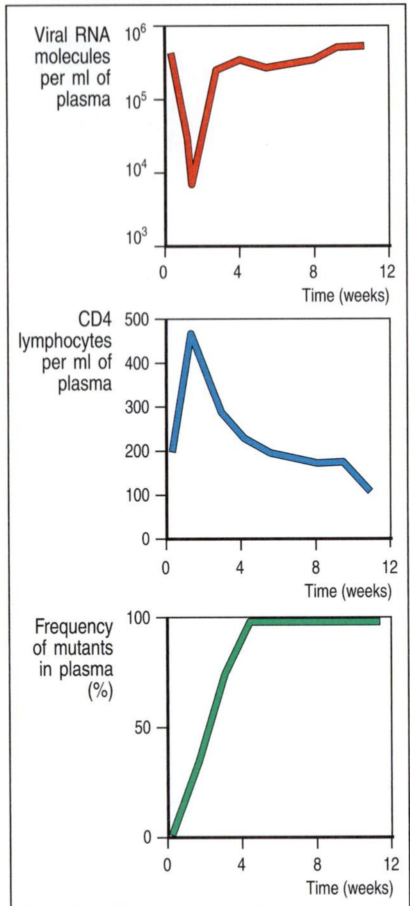

Fig. 13.32 Resistance of HIV to protease inhibitors develops rapidly. After the administration of a single protease inhibitor drug to a patient with HIV there is a precipitous fall in viral RNA levels in plasma, with a half-life of about 2 days (top panel). This is accompanied by an initial rise in the number of CD4 T cells in peripheral blood (center panel). Within days of starting the drug, mutant drug-resistant variants can be detected in plasma (bottom panel) and in peripheral blood lymphocytes. After only 4 weeks of treatment, viral RNA levels and CD4 lymphocyte levels have returned to their original pre-drug levels, and 100% of plasma HIV is present as the drug-resistant mutant.

13-35 Vaccination against HIV is an attractive solution but poses many difficulties.

A safe and effective vaccine for the prevention of HIV infection and AIDS is the ultimate goal, but its attainment is fraught with difficulties that have not been faced in developing vaccines against other diseases. The main problem is the nature of the infection itself, featuring a virus that directly undermines the central component of adaptive immunity—the CD4 T cell—and that proliferates extremely rapidly and causes sustained infection in the face of strong cytotoxic T cell and antibody responses. The development of vaccines that could be administered to patients already infected, to boost immune responses and prevent progression to AIDS, has been considered, as well as prophylactic vaccines that would be given to prevent initial infection. The development of therapeutic vaccination in those already infected would be extremely difficult. As discussed in the previous section, HIV evolves in individual patients by the selective proliferation of mutant viruses that escape recognition by antibodies and cytotoxic T cells. The ability of the virus to persist in latent form as a transcriptionally silent provirus invisible to the immune system might also prevent even an immunized person from clearing an infection once it has been established.

There has been more hope for prophylactic vaccination to prevent new infection. But even here, the lack of effect of the normal immune response and the sheer scale of sequence diversity among HIV strains in the population as a whole is a significant challenge. Patients infected with one strain of virus do not seem to be resistant to closely related strains, ruling out a universal vaccine. For example, a patient infected with HIV-1 clade AE was successfully treated for 28 months, but 3 months after ceasing treatment he contracted an infection with a clade B HIV-1 as a result of sexual encounters in Brazil, where this clade is endemic. Cases of superinfection, where two strains simultaneously infect the same cell, have also been described. Key among the difficulties is our uncertainty over what form protective immunity to HIV might take. It is not known whether antibodies, responses by CD4 T cells, or responses by CD8 cytotoxic T cells, or all three, are necessary to achieve protective immunity, and which epitopes might provide the targets for protective immunity.

However, against this pessimistic background, there are grounds for hope that successful vaccines can be developed. Of particular interest are rare groups of people who have been exposed often enough to HIV to make it virtually certain that they should have become infected but who have not developed the disease. In some cases this is due to an inherited deficiency in the chemokine receptor used as co-receptor for HIV entry (see Section 13-27). However, this mutant chemokine receptor does not occur in Africa, where one such group has been identified. A small group of Gambian and Kenyan sex workers who are estimated to have been exposed to many HIV-infected male partners each month for up to 5 years were found to lack antibody responses but to have cytotoxic T-cell responses to a variety of peptide epitopes from HIV. These women seem to have been naturally immunized against HIV. Follow-up of a number of them showed that about 10% subsequently acquired HIV infection. Paradoxically, HIV infection was found more frequently in women who had reduced their sex work and thereby their regular exposure to the virus. A possible explanation is that the absence of repeated exposure to HIV antigens led to a loss of the cytotoxic T-cell response, thus rendering the women susceptible to infection.

Various strategies are being tried in an attempt to develop vaccines against HIV. Many successful vaccines against other viral diseases contain a live attenuated strain of the virus, which raises an immune response but does not cause disease (see Section 16-23). There are substantial difficulties in the development of live attenuated vaccines against HIV, not least the worry of recombination between vaccine strains and wild-type viruses, which would lead to a reversion to virulence. An alternative approach is the use of DNA vaccination, a technique that we discuss in Section 16-27. DNA vaccination against HIV followed by the administration of a recombinant modified vaccinia boost, containing HIV antigens, has been piloted in primate experiments and was successful in preventing infection from an intrarectal challenge given 7 months after the booster vaccination. For every success in the route to HIV vaccination there is a setback, however. Rhesus monkeys were vaccinated with a DNA vaccine against SIV together with an IL-2 fusion protein and then challenged with a pathogenic SIV–HIV hybrid. Six months after the challenge, one of the monkeys developed an AIDS-like illness that was associated with the emergence of a virus carrying a point mutation in an immunodominant Gag epitope that is recognized by cytotoxic T cells. This is a beautiful but depressing example of the ability of HIV to escape immune control under the pressure of a cytotoxic T-cell response.

Subunit vaccines, which induce a response against only some proteins in the virus, have also been made. One such vaccine made from the envelope protein gp120 has been tested on chimpanzees. This vaccine proved to be specific to the precise strain of virus used to make it and was therefore useless in protection against natural infection. Subunit vaccines are also less efficient at inducing prolonged cytotoxic T-cell responses. In spite of the results in chimpanzees, a recombinant gp120 protein vaccine has been trialed in uninfected human volunteers. Immunized volunteers developed antibodies against gp120, but they were nonneutralizing, and small number of those immunized subsequently contracted HIV-1 infection, the course of which was not modified by the previous vaccination. Similarly disappointing results were obtained in a large recent trial of a subunit vaccine that used individual attenuated adenoviral vectors to deliver the HIV Gag, Pol, and Nef proteins to volunteers with documented HIV infection or who were HIV-negative but at high risk for exposure. Early results from the trial indicated the induction of heightened CD8 responses to the viral proteins, but lack of efficacy in protecting against infection or disease progression.

In addition to the biological obstacles to developing effective HIV vaccines, there are difficult ethical issues. It would be unethical to conduct a vaccine trial without trying at the same time to minimize the exposure of a vaccinated population to the virus itself. The effectiveness of a vaccine can, however, only be assessed in a population in which the exposure rate to the virus is high enough to assess whether vaccination protects against infection. This means that initial vaccine trials might have to be conducted in countries where the incidence of infection is very high and public health measures have not yet succeeded in reducing the spread of HIV.

13-36 Prevention and education are one way in which the spread of HIV and AIDS can be controlled.

The one way in which we know we can protect against infection with HIV is by avoiding contact with body fluids, such as semen, blood, blood products, or milk, from people who are infected. Indeed, it has been repeatedly demonstrated that this precaution, simple enough in the developed world, is sufficient to prevent infection, because healthcare workers can take care of AIDS patients for long periods without seroconversion or signs of infection.

For this strategy to work, however, one must be able to periodically test people at risk of infection with HIV, so that they can take the steps necessary to avoid passing the virus to others. This, in turn, requires strict confidentiality and mutual trust. A barrier to the control of HIV is the reluctance of individuals to find out whether or not they are infected, especially as one of the consequences of a positive HIV test is stigmatization by society. As a result, infected individuals can unwittingly infect many others. Balanced against this is the success of combination drug therapy (see Section 13-33), which provides an incentive for potentially infected people to identify the presence of infection and gain the benefits of treatment. Responsibility is at the heart of AIDS prevention, and a law guaranteeing the rights of people infected with HIV might go a long way toward encouraging responsible behavior. The rights of HIV-infected people are protected in a few countries. The problem in the less-developed nations, where elementary health precautions are extremely difficult to establish, is more profound.

Summary.

Infection with the human immunodeficiency virus (HIV) is the cause of acquired immune deficiency syndrome (AIDS). This worldwide epidemic is now spreading at an alarming rate, especially through heterosexual contact in less-developed countries. HIV is an enveloped retrovirus that replicates in cells of the immune system. Viral entry requires the presence of CD4 and a particular chemokine receptor, and the viral cycle is dependent on transcription factors found in activated T cells. Infection with HIV causes a loss of CD4 T cells and an acute viremia that rapidly subsides as cytotoxic T-cell responses develop, but HIV infection is not eliminated by this immune response. Uninfected cells become activated and also subsequently die, which is a key feature distinguishing HIV from nonpathogenic natural infections of African primates with various SIVs. HIV establishes a state of persistent infection in which the virus is continually replicating in newly infected cells. Current treatment consists of combinations of viral protease inhibitors together with nucleoside analogs that inhibit the reverse transcriptase, causing a rapid decrease in virus levels and a slower increase in CD4 T-cell counts. The main effect of HIV infection is the destruction of CD4 T cells, which occurs through the direct cytopathic effects of HIV infection and through killing by CD8 cytotoxic T cells. As the CD4 T-cell counts wane, the body becomes progressively more susceptible to opportunistic infection. Eventually, most HIV-infected individuals develop AIDS and die; however, a small minority (3–7%) remain healthy for many years with no apparent ill effects of infection. We hope to be able to learn from these people how infection with HIV can be controlled. The existence of these people, and of others who seem to have been naturally immunized against infection, gives hope that it will be possible to develop effective vaccines against HIV.

Summary to Chapter 13.

Whereas most infections elicit protective immunity, most successful pathogens have developed some means of at least partly resisting the immune response, and they can result in serious, persistent disease. Some individuals have inherited deficiencies in different components of the immune system, making them highly susceptible to certain classes of infectious agents. Persistent infection and the inherited immunodeficiency diseases illustrate the importance of innate and adaptive immunity in an effective host defense against infection, and present huge challenges for future immunological research. The human immunodeficiency virus (HIV), which leads to acquired autoimmune disease syndrome (AIDS), combines the characteristics of a persistent infectious agent with the ability to create immunodeficiency in its human host, a combination that is usually slowly lethal to the patient. The key to fighting new pathogens such as HIV is to develop our understanding of the basic properties of the immune system and its role in combating infection more fully.

Questions.

13.1 *List the different ways in which viruses can evade the immune system. Which of these strategies lead to chronic infection, and why?*

13.2 *Compare and contrast the host response to intracellular versus extracellular bacterial pathogens.*

13.3 *List examples of immunodeficiencies that primarily affect T lymphocytes. Why do these generally affect immune responses more severely than deficiencies involving B cells only?*

13.4 *How are primary immune deficiencies that affect either the complement system or B-cell development similar and different?*

13.5 *What do people with inherited and acquired immunodeficiencies teach us about the normal mechanism of host protection against tuberculosis?*

13.6 *How does infection by HIV cause AIDS?*

13.7 *Why can HIV infection not be cured by current drug therapies?*

13.8 *Compare antigenic drift and antigenic shift in the context of influenza infections and pandemics. How do these different types of antigenic variation occur?*

13.9 *Discuss the factors that allow herpesviruses to maintain latent infections in the host, and how reactivation occurs so that the virus can spread from one host to another.*

13.10 *From what you have learned about Leishmania infection in other chapters (for example Chapters 9 and 11), discuss how the accumulation of T_{reg} cells in the dermis is likely to impair the elimination of the pathogen from this site.*

13.11 *Hepatitis C virus can cause an acute or chronic infection. What mechanisms does HCV use to evade clearance by the host and cause a chronic infection?*

13.12 *How do HIV accessory proteins protect the virus from immune clearance?*

13.13 *Why is it difficult to make an HIV vaccine?*

General references.

Chapel, H., Geha, R., and Rosen, F.: **Primary immunodeficiency diseases: an update.** *Clin. Exp. Immunol.* 2003, **132**:9–15.

De Cock, K.M., Mbori-Ngacha, D., and Marum, E.: **Shadow on the continent: public health and HIV/AIDS in Africa in the 21st century.** *Lancet* 2002, **360**:67–72.

Fischer, A., Cavazzana-Calvo, M., De Saint Basile, G., DeVillartay, J.P., Di Santo, J.P., Hivroz, C., Rieux-Laucat, F., and Le Deist, F.: **Naturally occurring primary deficiencies of the immune system.** *Annu. Rev. Immunol.* 1997, **15**:93–124.

Hill, A.V.: **The immunogenetics of human infectious diseases.** *Annu. Rev. Immunol.* 1998, **16**:593–617.

Korber, B., Muldoon, M., Theiler, J., Gao, F., Gupta, R., Lapedes, A., Hahn, B.H., Wolinsky, S., and Bhattacharya, T.: **Timing the ancestor of the HIV-1 pandemic strains.** *Science* 2000, **288**:1789–1796.

Lederberg, J.: **Infectious history.** *Science* 2000, **288**:287–293.

Notarangelo, L.D.: **Primary immunodeficiencies.** *J. Allergy Clin. Immunol.* 2010, **125**:S182–S194.

Royce, R.A., Sena, A., Cates, W., Jr, and Cohen, M.S.: **Sexual transmission of HIV.** *N. Engl. J. Med.* 1997, **336**:1072–1078.

Xu, X.N., Screaton, G.R., and McMichael, A.J.: **Virus infections: escape, resistance, and counterattack.** *Immunity* 2001, **15**:867–870.

Section references.

13-1 Antigenic variation allows pathogens to escape from immunity.

Clegg, S., Hancox, L.S., and Yeh, K.S.: ***Salmonella typhimurium* fimbrial phase variation and FimA expression.** *J. Bacteriol.* 1996, **178**:542–545.

Cossart, P.: **Host/pathogen interactions. Subversion of the mammalian cell cytoskeleton by invasive bacteria.** *J. Clin. Invest.* 1997, **99**:2307–2311.

Donelson, J.E., Hill, K.L., and El-Sayed, N.M.: **Multiple mechanisms of immune evasion by African trypanosomes.** *Mol. Biochem. Parasitol.* 1998, **91**:51–66.

Gibbs, M.J., Armstrong, J.S., and Gibbs, A.J.: **Recombination in the hemagglutinin gene of the 1918 'Spanish flu.'** *Science* 2001, **293**:1842–1845.

Hatta, M., Gao, P., Halfmann, P., and Kawaoka, Y.: **Molecular basis for high virulence of Hong Kong H5N1 influenza A viruses.** *Science* 2001, **293**:1840–1842.

Kuppers, R.: **B cells under the influence: transformation of B cells by Epstein–Barr virus.** *Nat. Rev. Immunol.* 2003, **3**:801–812.

Laver, G., and Garman, E.: **Virology. The origin and control of pandemic influenza.** *Science* 2001, **293**:1776–1777.

Ressing, M.E., Keating, S.E., van Leeuwen, D., Koppers-Lalic, D., Pappworth, I.Y., Wiertz, E.J., and Rowe, M.: **Impaired transporter associated with antigen processing-dependent peptide transport during productive EBV infection.** *J. Immunol.* 2005, **174**:6829–6838.

Seifert, H.S., Wright, C.J., Jerse, A.E., Cohen, M.S., and Cannon, J.G.: **Multiple gonococcal pilin antigenic variants are produced during experimental human infections.** *J. Clin. Invest.* 1994, **93**:2744–2749.

Webster, R.G.: **Virology. A molecular whodunit.** *Science* 2001, **293**:1773–1775.

13-2 Some viruses persist *in vivo* by ceasing to replicate until immunity wanes.

Cohen, J.I.: **Epstein–Barr virus infection.** *N. Engl. J. Med.* 2000, **343**:481–492.

Ehrlich, R.: **Selective mechanisms utilized by persistent and oncogenic viruses to interfere with antigen processing and presentation.** *Immunol. Res.* 1995, **14**:77–97.

Garcia Blanco, M.A., and Cullen, B.R.: **Molecular basis of latency in pathogenic human viruses.** *Science* 1991, **254**:815–820.

Hahn, G., Jores, R., and Mocarski, E.S.: **Cytomegalovirus remains latent in a common precursor of dendritic and myeloid cells.** *Proc. Natl Acad. Sci. USA* 1998, **95**:3937–3942.

Ho, D.Y.: **Herpes simplex virus latency: molecular aspects.** *Prog. Med. Virol.* 1992, **39**:76–115.

Macsween, K.F., and Crawford, D.H.: **Epstein–Barr virus—recent advances.** *Lancet Infect. Dis.* 2003, **3**:131–140.

Mitchell, B.M., Bloom, D.C., Cohrs, R.J., Gilden, D.H., and Kennedy, P.G.: **Herpes simplex virus-1 and varicella-zoster virus latency in ganglia.** *J. Neurovirol.* 2003, **9**:194–204.

Nash, A.A.: **T cells and the regulation of herpes simplex virus latency and reactivation.** *J. Exp. Med.* 2000, **191**:1455–1458.

Wensing, B., and Farrell, P.J.: **Regulation of cell growth and death by Epstein–Barr virus.** *Microbes Infect.* 2000, **2**:77–84.

Yewdell, J.W., and Hill, A.B.: **Viral interference with antigen presentation.** *Nat. Immunol.* 2002, **2**:1019–1025.

13-3 Some pathogens resist destruction by host defense mechanisms or exploit them for their own purposes.

Alcami, A.: **Viral mimicry of cytokines, chemokines and their receptors.** *Nat. Rev. Immunol.* 2003, **3**:36–50.

Arvin, A.M.: **Varicella-zoster virus: molecular virology and virus–host interactions.** *Curr. Opin. Microbiol.* 2001, **4**:442–449.

Brander, C., and Walker, B.D.: **Modulation of host immune responses by clinically relevant human DNA and RNA viruses.** *Curr. Opin. Microbiol.* 2000, **3**:379–386.

Connolly, S.E., and Benach, J.L.: **The versatile roles of antibodies in *Borrelia* infections.** *Nat. Rev. Microbiol.* 2005, **3**:411–420.

Cooper, S.S., Glenn, J., and Greenberg, H.B.: **Lessons in defense: hepatitis C, a case study.** *Curr. Opin. Microbiol.* 2000, **3**:363–365.

Cosman, D., Fanger, N., Borges, L., Kubin, M., Chin, W., Peterson, L., and Hsu, M.L.: **A novel immunoglobulin superfamily receptor for cellular and viral MHC class I molecules.** *Immunity* 1997, **7**:273–282.

Hadler, J.L.: **Learning from the 2001 anthrax attacks: immunological characteristics.** *J. Infect. Dis.* 2007, **195**:163–164.

Lauer, G.M., and Walker, B.D.: **Hepatitis C virus infection.** *N. Engl. J. Med.* 2001, **345**:41–52.

McFadden, G., and Murphy, P.M.: **Host-related immunomodulators encoded by poxviruses and herpesviruses.** *Curr. Opin. Microbiol.* 2000, **3**:371–378.

Miller, J.C., and Stevenson, B.: ***Borrelia burgdorferi erp* genes are expressed at different levels within tissues of chronically infected mammalian hosts.** *Int. J. Med. Microbiol.* 2006, **296** Suppl. 40:185–194.

Park, J.M., Greten, F.R., Li, Z.W., and Karin, M.: **Macrophage apoptosis by anthrax lethal factor through p38 MAP kinase inhibition.** *Science* 2002, **297**:2048–2051.

Radolf, J.D.: **Role of outer membrane architecture in immune evasion by *Treponema pallidum* and *Borrelia burgdorferi*.** *Trends Microbiol.* 1994, **2**:307–311.

Sinai, A.P., and Joiner, K.A.: **Safe haven: the cell biology of nonfusogenic pathogen vacuoles.** *Annu. Rev. Microbiol.* 1997, **51**:415–462.

13-4 Immunosuppression or inappropriate immune responses can contribute to persistent disease.

Auffermann-Gretzinger, S., Keeffe, E.B., and Levy, S.: **Impaired dendritic cell maturation in patients with chronic, but not resolved, hepatitis C virus infection.** *Blood* 2001, **97**:3171–3176.

Bhardwaj, N.: **Interactions of viruses with dendritic cells: a double-edged sword.** *J. Exp. Med.* 1997, **186**:795–799.

Bloom, B.R., Modlin, R.L., and Salgame, P.: **Stigma variations: observations on suppressor T cells and leprosy.** *Annu. Rev. Immunol.* 1992, **10**:453–488.

Kanto, T., Hayashi, N., Takehara, T., Tatsumi, T., Kuzushita, T., Ito, A., Sasaki, Y., Kasahara, A., and Hori, M.: **Impaired allostimulatory capacity of peripheral blood dendritic cells recovered from hepatitis C virus-infected individuals.** *J. Immunol.* 1999, **162**:5584–5591.

Lerat, H., Rumin, S., Habersetzer, F., Berby, F., Trabaud, M.A., Trepo, C., and Inchauspe, G.: ***In vivo* tropism of hepatitis C virus genomic sequences in hematopoietic cells: influence of viral load, viral genotype, and cell phenotype.** *Blood* 1998, **91**:3841–3849.

Salgame, P., Abrams, J.S., Clayberger, C., Goldstein, H., Convit, J., Modlin, R.L., and Bloom, B.R.: **Differing lymphokine profiles of functional subsets of human CD4 and CD8 T cell clones.** *Science* 1991, **254**:279–282.

Swartz, M.N.: **Recognition and management of anthrax—an update.** *N. Engl. J. Med.* 2001, **345**:1621–1626.

13-5 Immune responses can contribute directly to pathogenesis.

Cheever, A.W., and Yap, G.S.: **Immunologic basis of disease and disease regulation in schistosomiasis.** *Chem. Immunol.* 1997, **66**:159–176.

Doherty, P.C., Topham, D.J., Tripp, R.A., Cardin, R.D., Brooks, J.W., and Stevenson, P.G.: **Effector CD4$^+$ and CD8$^+$ T-cell mechanisms in the control of respiratory virus infections.** *Immunol. Rev.* 1997, **159**:105–117.

Varga, S.M., Wang, X., Welsh, R.M., and Braciale, T.J.: **Immunopathology in RSV infection is mediated by a discrete oligoclonal subset of antigen-specific CD4$^+$ T cells.** *Immunity* 2001, **15**:637–646.

13-6 Regulatory T cells can affect the outcome of infectious disease.

Belkaid, Y., and Tarbell, K.: **Regulatory T cells in the control of host–microorganism interactions.** *Annu. Rev. Immunol.* 2009, **27**:551–589.

Rouse, B.T., Sarangi, P.P., and Suvas, S.: **Regulatory T cells in virus infections.** *Immunol. Rev.* 2006, **212**:272–286.

Waldmann, H., Adams, E., Fairchild, P., and Cobbold, S.: **Infectious tolerance and the long-term acceptance of transplanted tissue.** *Immunol. Rev.* 2006, **212**:301–313.

13-7 A history of repeated infections suggests a diagnosis of immunodeficiency.

Carneiro-Sampaio, M., and Coutinho, A.: **Immunity to microbes: lessons from primary immunodeficiencies.** *Infect. Immun.* 2007, **75**:1545–1555.

Cunningham-Rundles, C., and Ponda, P.P.: **Molecular defects in T- and B-cell primary immunodeficiency diseases.** *Nat. Rev. Immunol.* 2005, **5**:880–892.

Rosen, F.S., Cooper, M.D., and Wedgwood, R.J.: **The primary immunodeficiencies.** *N. Engl. J. Med.* 1995, **333**:431–440.

13-8 Primary immunodeficiency diseases are caused by inherited gene defects.

Cunningham-Rundles, C., and Ponda, P.P.: **Molecular defects in T- and B-cell primary immunodeficiency diseases.** *Nat. Rev. Immunol.* 2005, **5**:880–892.

Kokron, C.M., Bonilla, F.A., Oettgen, H.C., Ramesh, N., Geha, R.S., and Pandolfi, F.: **Searching for genes involved in the pathogenesis of primary immunodeficiency diseases: lessons from mouse knockouts.** *J. Clin. Immunol.* 1997, **17**:109–126.

Marodi, L., and Notarangelo, L.D.: **Immunological and genetic bases of new primary immunodeficiencies.** *Nat. Rev. Immunol.* 2007, **7**:851–861.

Smart, B.A., and Ochs, H.D.: **The molecular basis and treatment of primary immunodeficiency disorders.** *Curr. Opin. Pediatr.* 1997, **9**:570–576.

Smith, C.I., and Notarangelo, L.D.: **Molecular basis for X-linked immunodeficiencies.** *Adv. Genet.* 1997, **35**:57–115.

13-9 Defects in T-cell development can result in severe combined immunodeficiencies.

Buckley, R.H., Schiff, R.I., Schiff, S.E., Markert, M.L., Williams, L.W., Harville, T.O., Roberts, J.L., and Puck, J.M.: **Human severe combined immunodeficiency: genetic, phenotypic, and functional diversity in one hundred eight infants.** *J. Pediatr.* 1997, **130**:378–387.

Leonard, W.J.: **The molecular basis of X linked severe combined immunodeficiency.** *Annu. Rev. Med.* 1996, **47**:229–239.

Leonard, W.J.: **Cytokines and immunodeficiency diseases.** *Nat. Rev. Immunol.* 2001, **1**:200–208.

Stephan, J.L., Vlekova, V., Le Deist, F., Blanche, S., Donadieu, J., De Saint-Basile, G., Durandy, A., Griscelli, C., and Fischer, A.: **Severe combined immunodeficiency: a retrospective single-center study of clinical presentation and outcome in 117 patients.** *J. Pediatr.* 1993, **123**:564–572.

13-10 SCID can also be due to defects in the purine salvage pathway.

Hirschhorn, R.: **Adenosine deaminase deficiency: molecular basis and recent developments.** *Clin. Immunol. Immunopathol.* 1995, **76**:S219–S227.

13-11 Defects in antigen receptor gene rearrangement can result in SCID.

Bosma, M.J., and Carroll, A.M.: **The SCID mouse mutant: definition, characterization, and potential uses.** *Annu. Rev. Immunol.* 1991, **9**:323–350.

Fugmann, S.D.: **DNA repair: breaking the seal.** *Nature* 2002, **416**:691–694.

Gennery, A.R., Cant, A.J., and Jeggo, P.A.: **Immunodeficiency associated with DNA repair defects.** *Clin. Exp. Immunol.* 2000, **121**:1–7.

Lavin, M.F., and Shiloh, Y.: **The genetic defect in ataxia-telangiectasia.** *Annu. Rev. Immunol.* 1997, **15**:177–202.

Moshous, D., Callebaut, I., de Chasseval, R., Corneo, B., Cavazzana-Calvo, M., Le Deist, F., Tezcan, I., Sanal, O., Bertrand, Y., Philippe, N., *et al.*: **Artemis, a novel DNA double-strand break repair/V(D)J recombination protein, is mutated in human severe combined immune deficiency.** *Cell* 2001, **105**:177–186.

13-12 Defects in signaling from T-cell antigen receptors can cause severe immunodeficiency.

Castigli, E., Pahwa, R., Good, R.A., Geha, R.S., and Chatila, T.A.: **Molecular basis of a multiple lymphokine deficiency in a patient with severe combined immunodeficiency.** *Proc. Natl Acad. Sci. USA* 1993, **90**:4728–4732.

DiSanto, J.P., Keever, C.A., Small, T.N., Nicols, G.L., O'Reilly, R.J., and Flomenberg, N.: **Absence of interleukin 2 production in a severe combined immunodeficiency disease syndrome with T cells.** *J. Exp. Med.* 1990, **171**:1697–1704.

DiSanto, J.P., Rieux Laucat, F., Dautry Varsat, A., Fischer, A., and de Saint Basile, G.: **Defective human interleukin 2 receptor γ chain in an atypical X chromosome-linked severe combined immunodeficiency with peripheral T cells.** *Proc. Natl Acad. Sci. USA* 1994, **91**:9466–9470.

Gilmour, K.C., Fujii, H., Cranston, T., Davies, E.G., Kinnon, C., and Gaspar, H.B.: **Defective expression of the interleukin-2/interleukin-15 receptor β subunit leads to a natural killer cell-deficient form of severe combined immunodeficiency.** *Blood* 2001, **98**:877–879.

Humblet-Baron, S., Sather, B., Anover, S., Becker-Herman, S., Kasprowicz, D.J., Khim, S., Nguyen, T., Hudkins-Loya, K., Alpers, C.E., Ziegler, S.F., *et al.*: **Wiskott-Aldrich syndrome protein is required for regulatory T cell homeostasis.** *J. Clin. Invest.* 2007, **117**:407–418.

Kung, C., Pingel, J.T., Heikinheimo, M., Klemola, T., Varkila, K., Yoo, L.I., Vuopala, K., Poyhonen, M., Uhari, M., Rogers, M., *et al.*: **Mutations in the tyrosine phosphatase CD45 gene in a child with severe combined immunodeficiency disease.** *Nat. Med.* 2000, **6**:343–345.

Roifman, C.M., Zhang, J., Chitayat, D., and Sharfe, N.: **A partial deficiency of interleukin-7R α is sufficient to abrogate T-cell development and cause severe combined immunodeficiency.** *Blood* 2000, **96**:2803–2807.

13-13 Genetic defects in thymic function that block T-cell development result in severe immunodeficiencies.

Adriani, M., Martinez-Mir, A., Fusco, F., Busiello, R., Frank, J., Telese, S., Matrecano, E., Ursini, M.V., Christiano, A.M., and Pignata, C.: **Ancestral founder mutation of the nude (FOXN1) gene in congenital severe combined immunodeficiency associated with alopecia in Southern Italy population.** *Ann. Hum. Genet.* 2004, **68**:265–268.

Coffer, P.J., and Burgering, B.M.: **Forkhead-box transcription factors and their role in the immune system.** *Nat. Rev. Immunol.* 2004, **4**:889–899.

Gadola, S.D., Moins-Teisserenc, H.T., Trowsdale, J., Gross, W.L., and Cerundolo, V.: **TAP deficiency syndrome.** *Clin. Exp. Immunol.* 2000, **121**:173–178.

Grusby, M.J., and Glimcher, L.H.: **Immune responses in MHC class II-deficient mice.** *Annu. Rev. Immunol.* 1995, **13**:417–435.

Masternak, K., Barras, E., Zufferey, M., Conrad, B., Corthals, G., Aebersold, R., Sanchez, J.C., Hochstrasser, D.F., Mach, B., and Reith, W.: **A gene encoding a novel RFX-associated transactivator is mutated in the majority of MHC class II deficiency patients.** *Nat. Genet.* 1998, **20**:273–277.

Pignata, C., Gaetaniello, L., Masci, A.M., Frank, J., Christiano, A., Matrecano, E., and Racioppi, L.: **Human equivalent of the mouse Nude/SCID phenotype: long-term evaluation of immunologic reconstitution after bone marrow transplantation.** *Blood* 2001, **97**:880–885.

Steimle, V., Reith, W., and Mach, B.: **Major histocompatibility complex class II deficiency: a disease of gene regulation.** *Adv. Immunol.* 1996, **61**:327–340.

13-14 Defects in B-cell development result in deficiencies in antibody production that cause an inability to clear extracellular bacteria.

Bruton, O.C.: **Agammaglobulinemia.** *Pediatrics* 1952, **9**:722–728.

Conley, M.E.: **Genetics of hypogammaglobulinemia: what do we really know?** *Curr. Opin. Immunol.* 2009, **21**:466–471.

Desiderio, S.: **Role of Btk in B cell development and signaling.** *Curr. Opin. Immunol.* 1997, **9**:534–540.

Fuleihan, R., Ramesh, N., and Geha, R.S.: **X-linked agammaglobulinemia and immunoglobulin deficiency with normal or elevated IgM: immunodeficiencies of B cell development and differentiation.** *Adv. Immunol.* 1995, **60**:37–56.

Lee, M.L., Gale, R.P., and Yap, P.L.: **Use of intravenous immunoglobulin to prevent or treat infections in persons with immune deficiency.** *Annu. Rev. Med.* 1997, **48**:93–102.

Notarangelo, L.D.: **Immunodeficiencies caused by genetic defects in protein kinases.** *Curr. Opin. Immunol.* 1996, **8**:448–453.

Ochs, H.D., and Wedgwood, R.J.: **IgG subclass deficiencies.** *Annu. Rev. Med.* 1987, **38**:325–340.

Preud'homme, J.L., and Hanson, L.A.: **IgG subclass deficiency.** *Immunodefic. Rev.* 1990, **2**:129–149.

13-15 Immune deficiencies can be caused by defects in B-cell or T-cell activation and function.

Burrows, P.D., and Cooper, M.D.: **IgA deficiency.** *Adv. Immunol.* 1997, **65**:245–276.

Doffinger, R., Smahi, A., Bessia, C., Geissmann, F., Feinberg, J., Durandy, A., Bodemer, C., Kenwrick, S., Dupuis-Girod, S., Blanche, S., *et al.*: **X-linked anhidrotic ectodermal dysplasia with immunodeficiency is caused by impaired NF-κB signaling.** *Nat. Genet.* 2001, **27**:277–285.

Durandy, A., and Honjo, T.: **Human genetic defects in class-switch recombination (hyper-IgM syndromes).** *Curr. Opin. Immunol.* 2001, **13**:543–548.

Ferrari, S., Giliani, S., Insalaco, A., Al Ghonaium, A., Soresina, A.R., Loubser, M., Avanzini, M.A., Marconi, M., Badolato, R., Ugazio, A.G., *et al.*: **Mutations of CD40 gene cause an autosomal recessive form of immunodeficiency with hyper IgM.** *Proc. Natl Acad. Sci. USA* 2001, **98**:12614–12619.

Harris, R.S., Sheehy, A.M., Craig, H.M., Malim, M.H., and Neuberger, M.S.: **DNA deamination: not just a trigger for antibody diversification but also a mechanism for defense against retroviruses.** *Nat. Immunol.* 2003, **4**:641–643.

Minegishi, Y.: **Hyper-IgE syndrome.** *Curr. Opin. Immunol.* 2009, **21**:487–492.

Park, M.A., Li, J.T., Hagan, J.B., Maddox, D.E., and Abraham, R.S.: **Common variable immunodeficiency: a new look at an old disease.** *Lancet* 2008, **372**:489–503.

Thrasher, A.J., and Burns, S.O.: **WASP: a key immunological multitasker.** *Nat. Rev. Immunol.* 2010, **10**:182–192.

Yel, L.: **Selective IgA deficiency.** *J. Clin. Immunol.* 2010, **30**:10–16.

Yong, P.F., Salzer, U., and Grimbacher, B.: **The role of costimulation in antibody deficiencies: ICOS and common variable immunodeficiency.** *Immunol. Rev.* 2009, **229**:101–113.

13-16 Defects in complement components and complement-regulatory proteins cause defective humoral immune function and tissue damage.

Botto, M., Dell'Agnola, C., Bygrave, A.E., Thompson, E.M., Cook, H.T., Petry, F., Loos, M., Pandolfi, P.P., and Walport, M.J.: **Homozygous C1q deficiency causes glomerulonephritis associated with multiple apoptotic bodies.** *Nat. Genet.* 1998, **19**:56–59.

Colten, H.R., and Rosen, F.S.: **Complement deficiencies.** *Annu. Rev. Immunol.* 1992, **10**:809–834.

Dahl, M., Tybjaerg-Hansen, A., Schnohr, P., and Nordestgaard, B.G.: **A population-based study of morbidity and mortality in mannose-binding lectin deficiency.** *J. Exp. Med.* 2004, **199**:1391–1399.

Walport, M.J.: **Complement. First of two parts.** *N. Engl. J. Med.* 2001, **344**:1058–1066.

Walport, M.J.: **Complement. Second of two parts.** *N. Engl. J. Med.* 2001, **344**:1140–1144.

13-17 Defects in phagocytic cells permit widespread bacterial infections.

Ambruso, D.R., Knall, C., Abell, A.N., Panepinto, J., Kurkchubasche, A., Thurman, G., Gonzalez-Aller, C., Hiester, A., deBoer, M., Harbeck, R.J., *et al.*: **Human neutrophil immunodeficiency syndrome is associated with an inhibitory Rac2 mutation.** *Proc. Natl Acad. Sci. USA* 2000, **97**:4654–4659.

Andrews, T., and Sullivan, K.E.: **Infections in patients with inherited defects in phagocytic function.** *Clin. Microbiol. Rev.* 2003, **16**:597–621.

Aprikyan, A.A., and Dale, D.C.: **Mutations in the neutrophil elastase gene in cyclic and congenital neutropenia.** *Curr. Opin. Immunol.* 2001, **13**:535–538.

Ellson, C.D., Davidson, K., Ferguson, G.J., O'Connor, R., Stephens, L.R., and Hawkins, P.T.: **Neutrophils from *p40phox*$^{-/-}$ mice exhibit severe defects in NADPH oxidase regulation and oxidant-dependent bacterial killing.** *J. Exp. Med.* 2006, **203**:1927–1937.

Etzioni, A.: **Genetic etiologies of leukocyte adhesion defects.** *Curr. Opin. Immunol.* 2009, **21**:481–486.

Fischer, A., Lisowska Grospierre, B., Anderson, D.C., and Springer, T.A.: **Leukocyte adhesion deficiency: molecular basis and functional consequences.** *Immunodefic. Rev.* 1988, **1**:39–54.

Goldblatt, D., and Thrasher, A.J.: **Chronic granulomatous disease.** *Clin. Exp. Immunol.* 2000, **122**:1–9.

Klein, C., and Welte, K.: **Genetic insights into congenital neutropenia.** *Clin. Rev. Allergy Immunol.* 2010, **38**:68–74.

Ku, C.L., Yang, K., Bustamante, J., Puel, A., von Bernuth, H., Santos, O.F., Lawrence, T., Chang, H.H., Al-Mousa, H., Picard, C., *et al.*: **Inherited disorders of human Toll-like receptor signaling: immunological implications.** *Immunol. Rev.* 2005, **203**:10–20.

Luhn, K., Wild, M.K., Eckhardt, M., Gerardy-Schahn, R., and Vestweber, D.: **The gene defective in leukocyte adhesion deficiency II encodes a putative GDP-fucose transporter.** *Nat. Genet.* 2001, **28**:69–72.

Malech, H.L., and Nauseef, W.M.: **Primary inherited defects in neutrophil function: etiology and treatment.** *Semin. Hematol.* 1997, **34**:279–290.

Rotrosen, D., and Gallin, J.I.: **Disorders of phagocyte function.** *Annu. Rev. Immunol.* 1987, **5**:127–150.

Spritz, R.A.: **Genetic defects in Chediak–Higashi syndrome and the beige mouse.** *J. Clin. Immunol.* 1998, **18**:97–105.

Suhir, H., and Etzioni, A.: **The role of Toll-like receptor signaling in human immunodeficiencies.** *Clin. Rev. Allergy Immunol.* 2010, **38**:11–19.

13-18 Mutation in the molecular regulators of inflammation can cause uncontrolled inflammatory responses that result in 'autoinflammatory disease.'

Chae, J.J., Komarow, H.D., Cheng, J., Wood, G., Raben, N., Liu, P.P., and Kastner, D.L.: **Targeted disruption of pyrin, the FMF protein, causes heightened sensitivity to endotoxin and a defect in macrophage apoptosis.** *Mol. Cell* 2003, **11**:591–604.

Delpech, M., and Grateau, G.: **Genetically determined recurrent fevers.** *Curr. Opin. Immunol.* 2001, **13**:539–542.

Dinarello, C.A.: **Immunological and inflammatory functions of the interleukin-1 family.** *Annu. Rev. Immunol.* 2009, **27**:519–550.

Drenth, J.P., and van der Meer, J.W.: **Hereditary periodic fever.** *N. Engl. J. Med.* 2001, **345**:1748–1757.

Hoffman, H.M., Mueller, J.L., Broide, D.H., Wanderer, A.A., and Kolodner, R.D.: **Mutation of a new gene encoding a putative pyrin-like protein causes familial cold autoinflammatory syndrome and Muckle–Wells syndrome.** *Nat. Genet.* 2001, **29**:301–305.

Houten, S.M., Frenkel, J., Rijkers, G.T., Wanders, R.J., Kuis, W., and Waterham, H.R.: **Temperature dependence of mutant mevalonate kinase activity as a pathogenic factor in hyper-IgD and periodic fever syndrome.** *Hum. Mol. Genet.* 2002, **11**:3115–3124.

Kastner, D.L., and O'Shea, J.J.: **A fever gene comes in from the cold.** *Nat. Genet.* 2001, **29**:241–242.

McDermott, M.F., Aksentijevich, I., Galon, J., McDermott, E.M., Ogunkolade, B.W., Centola, M., Mansfield, E., Gadina, M., Karenko, L., Pettersson, T., *et al.*: **Germline mutations in the extracellular domains of the 55 kDa TNF receptor, TNFR1, define a family of dominantly inherited autoinflammatory syndromes.** *Cell* 1999, **97**:133–144.

Stehlik, C., and Reed, J.C.: **The PYRIN connection: novel players in innate immunity and inflammation.** *J. Exp. Med.* 2004, **200**:551–558.

Wise, C.A., Gillum, J.D., Seidman, C.E., Lindor, N.M., Veile, R., Bashiardes, S., and Lovett, M.: **Mutations in CD2BP1 disrupt binding to PTP PEST and are responsible for PAPA syndrome, an autoinflammatory disorder.** *Hum. Mol. Genet.* 2002, **11**:961–969.

13-19 The normal pathways for host defense against intracellular bacteria are pinpointed by genetic deficiencies of IFN-γ and IL-12 and their receptors.

Casanova, J.L., and Abel, L.: **Genetic dissection of immunity to mycobacteria: the human model.** *Annu. Rev. Immunol.* 2002, **20**:581–620.

Dupuis, S., Dargemont, C., Fieschi, C., Thomassin, N., Rosenzweig, S., Harris, J., Holland, S.M., Schreiber, R.D., and Casanova, J.L.: **Impairment of mycobacterial but not viral immunity by a germline human STAT1 mutation.** *Science* 2001, **293**:300–303.

Keane, J., Gershon, S., Wise, R.P., Mirabile-Levens, E., Kasznica, J., Schwieterman, W.D., Siegel, J.N., and Braun, M.M.: **Tuberculosis associated with infliximab, a tumor necrosis factor α-neutralizing agent.** *N. Engl. J. Med.* 2001, **345**:1098–1104.

Lammas, D.A., Casanova, J.L., and Kumararatne, D.S.: **Clinical consequences of defects in the IL-12-dependent interferon-γ (IFN-γ) pathway.** *Clin. Exp. Immunol.* 2000, **121**:417–425.

Newport, M.J., Huxley, C.M., Huston, S., Hawrylowicz, C.M., Oostra, B.A., Williamson, R., and Levin, M.: **A mutation in the interferon-γ-receptor gene and susceptibility to mycobacterial infection.** *N. Engl. J. Med.* 1996, **335**:1941–1949.

Shtrichman, R., and Samuel, C.E.: **The role of γ interferon in antimicrobial immunity.** *Curr. Opin. Microbiol.* 2001, **4**:251–259.

Van de Vosse, E., Hoeve, M.A., and Ottenhoff, T.H.: **Human genetics of intracellular infectious diseases: molecular and cellular immunity against mycobacteria and salmonellae.** *Lancet Infect. Dis.* 2004, **4**:739–749.

13-20 X-linked lymphoproliferative syndrome is associated with fatal infection by Epstein–Barr virus and with the development of lymphomas.

Latour, S., Gish, G., Helgason, C.D., Humphries, R.K., Pawson, T., and Veillette, A.: **Regulation of SLAM-mediated signal transduction by SAP, the X-linked lymphoproliferative gene product.** *Nat. Immunol.* 2001, **2**:681–690.

Milone, M.C., Tsai, D.E., Hodinka, R.L., Silverman, L.B., Malbran, A., Wasik, M.A., and Nichols, K.E.: **Treatment of primary Epstein–Barr virus infection in patients with X-linked lymphoproliferative disease using B-cell-directed therapy.** *Blood* 2005, **105**:994–996.

Morra, M., Howie, D., Grande, M.S., Sayos, J., Wang, N., Wu, C., Engel, P., and Terhorst, C.: **X-linked lymphoproliferative disease: a progressive immunodeficiency.** *Annu. Rev. Immunol.* 2001, **19**:657–682.

Nichols, K.E., Koretzky, G.A., and June, C.H.: **SAP: natural inhibitor or grand SLAM of T-cell activation?** *Nat. Immunol.* 2001, **2**:665–666.

Rigaud, S., Fondaneche, M.C., Lambert, N., Pasquier, B., Mateo, V., Soulas, P., Galicier, L., Le Deist, F., Rieux-Laucat, F., Revy, P., *et al.*: **XIAP deficiency in humans causes an X-linked lymphoproliferative syndrome.** *Nature* 2006, **444**:110–114.

Satterthwaite, A.B., Rawlings, D.J., and Witte, O.N.: **DSHP: a 'power bar' for sustained immune responses?** *Proc. Natl Acad. Sci. USA* 1998, **95**:13355–13357.

13-21 Genetic abnormalities in the secretory cytotoxic pathway of lymphocytes cause uncontrolled lymphoproliferation and inflammatory responses to viral infections.

de Saint, B.G., and Fischer, A.: **The role of cytotoxicity in lymphocyte homeostasis.** *Curr. Opin. Immunol.* 2001, **13**:549–554.

Dell'Angelica, E.C., Mullins, C., Caplan, S., and Bonifacino, J.S.: **Lysosome-related organelles.** *FASEB J.* 2000, **14**:1265–1278.

Huizing, M., Anikster, Y., and Gahl, W.A.: **Hermansky–Pudlak syndrome and Chediak–Higashi syndrome: disorders of vesicle formation and trafficking.** *Thromb. Haemost.* 2001, **86**:233–245.

Menasche, G., Pastural, E., Feldmann, J., Certain, S., Ersoy, F., Dupuis, S., Wulffraat, N., Bianchi, D., Fischer, A., Le Deist, F., *et al.*: **Mutations in RAB27A cause Griscelli syndrome associated with haemophagocytic syndrome.** *Nat. Genet.* 2000, **25**:173–176.

Stinchcombe, J.C., and Griffiths, G.M.: **Normal and abnormal secretion by haemopoietic cells.** *Immunology* 2001, **103**:10–16.

13-22 Hematopoietic stem cell transplantation or gene therapy can be useful to correct genetic defects.

Anderson, W.F.: **Human gene therapy.** *Nature* 1998, **392**:25–30.

Candotti, F., and Blaese, R.M.: **Gene therapy of primary immunodeficiencies.** *Springer Semin. Immunopathol.* 1998, **19**:493–508.

Fischer, A., Hacein-Bey, S., and Cavazzana-Calvo, M.: **Gene therapy of severe combined immunodeficiencies.** *Nat. Rev. Immunol.* 2002, **2**:615–621.

Fischer, A., Haddad, E., Jabado, N., Casanova, J.L., Blanche, S., Le Deist, F., and Cavazzana-Calvo, M.: **Stem cell transplantation for immunodeficiency.** *Springer Semin. Immunopathol.* 1998, **19**:479–492.

Fischer, A., Le Deist, F., Hacein-Bey-Abina, S., Andre-Schmutz, I., de Saint, B.G., de Villartay, J.P., and Cavazzana-Calvo, M.: **Severe combined immunodeficiency. A model disease for molecular immunology and therapy.** *Immunol. Rev.* 2005, **203**:98–109.

Hacein-Bey-Abina, S., Le Deist, F., Carlier, F., Bouneaud, C., Hue, C., De Villartay, J.P., Thrasher, A.J., Wulffraat, N., Sorensen, R., Dupuis-Girod, S., *et al.*: **Sustained correction of X-linked severe combined immunodeficiency by *ex vivo* gene therapy.** *N. Engl. J. Med.* 2002, **346**:1185–1193.

Hacein-Bey-Abina, S., Von Kalle, C., Schmidt, M., McCormack, M.P., Wulffraat, N., Leboulch, P., Lim, A., Osborne, C.S., Pawliuk, R., Morillon, E., *et al.*: **LMO2-associated clonal T cell proliferation in two patients after gene therapy for SCID-X1.** *Science* 2003, **302**:415–419.

Onodera, M., Ariga, T., Kawamura, N., Kobayashi, I., Ohtsu, M., Yamada, M., Tame, A., Furuta, H., Okano, M., Matsumoto, S., *et al.*: **Successful peripheral T-lymphocyte-directed gene transfer for a patient with severe combined immune deficiency caused by adenosine deaminase deficiency.** *Blood* 1998, **91**:30–36.

Pesu, M., Candotti, F., Husa, M., Hofmann, S.R., Notarangelo, L.D., and O'Shea, J.J.: **Jak3, severe combined immunodeficiency, and a new class of immunosuppressive drugs.** *Immunol. Rev.* 2005, **203**:127–142.

Rosen, F.S.: **Successful gene therapy for severe combined immunodeficiency.** *N. Engl. J. Med.* 2002, **346**:1241–1243.

13-23 Secondary immunodeficiencies are major predisposing causes of infection and death.

Chandra, R.K.: **Nutrition, immunity and infection: from basic knowledge of dietary manipulation of immune responses to practical application of ameliorating suffering and improving survival.** *Proc. Natl Acad. Sci. USA* 1996, **93**:14304–14307.

Lord, G.M., Matarese, G., Howard, J.K., Baker, R.J., Bloom, S.R., and Lechler, R.I.: **Leptin modulates the T-cell immune response and reverses starvation-induced immunosuppression.** *Nature* 1998, **394**:897–901.

13-24 Most individuals infected with HIV progress over time to AIDS.

Baltimore, D.: **Lessons from people with nonprogressive HIV infection.** *N. Engl. J. Med.* 1995, **332**:259–260.

Barre-Sinoussi, F.: **HIV as the cause of AIDS.** *Lancet* 1996, **348**:31–35.

Gao, F., Bailes, E., Robertson, D.L., Chen, Y., Rodenburg, C.M., Michael, S.F., Cummins, L.B., Arthur, L.O., Peeters, M., Shaw, G.M., *et al.*: **Origin of HIV-1 in the chimpanzee *Pan troglodytes troglodytes*.** *Nature* 1999, **397**:436–441.

Heeney, J.L., Dalgleish, A.G., and Weiss, R.A.: **Origins of HIV and the evolution of resistance to AIDS.** *Science* 2006, **313**:462–466.

Kirchhoff, F., Greenough, T.C., Brettler, D.B., Sullivan, J.L., and Desrosiers, R.C.: **Brief report: absence of intact nef sequences in a long-term survivor with non-progressive HIV-1 infection.** *N. Engl. J. Med.* 1995, **332**:228–232.

Pantaleo, G., Menzo, S., Vaccarezza, M., Graziosi, C., Cohen, O.J., Demarest, J.F., Montefiori, D., Orenstein, J.M., Fox, C., Schrager, L.K., *et al.*: **Studies in subjects with long-term nonprogressive human immunodeficiency virus infection.** *N. Engl. J. Med.* 1995, **332**:209–216.

Peckham, C., and Gibb, D.: **Mother-to-child transmission of the human immunodeficiency virus.** *N. Engl. J. Med.* 1995, **333**:298–302.

Rosenberg, P.S., and Goedert, J.J.: **Estimating the cumulative incidence of HIV infection among persons with haemophilia in the United States of America.** *Stat. Med.* 1998, **17**:155–168.

Volberding, P.A.: **Age as a predictor of progression in HIV infection.** *Lancet* 1996, **347**:1569–1570.

13-25 HIV is a retrovirus that infects CD4 T cells, dendritic cells, and macrophages.

Bomsel, M., and David, V.: **Mucosal gatekeepers: selecting HIV viruses for early infection.** *Nat. Med.* 2002, **8**:114–116.

Cammack, N.: **The potential for HIV fusion inhibition.** *Curr. Opin. Infect. Dis.* 2001, **14**:13–16.

Chan, D.C., and Kim, P.S.: **HIV entry and its inhibition.** Cell 1998, **93**:681–684.

Connor, R.I., Sheridan, K.E., Ceradini, D., Choe, S., and Landau, N.R.: **Change in coreceptor use correlates with disease progression in HIV-1—infected individuals.** *J. Exp. Med.* 1997, **185**:621–628.

Farber, J.M., and Berger, E.A.: **HIV's response to a CCR5 inhibitor: I'd rather tighten than switch!** *Proc. Natl Acad. Sci. USA* 2002, **99**:1749–1751.

Grouard, G., and Clark, E.A.: **Role of dendritic and follicular dendritic cells in HIV infection and pathogenesis.** *Curr. Opin. Immunol.* 1997, **9**:563–567.

Kilby, J.M., Hopkins, S., Venetta, T.M., DiMassimo, B., Cloud, G.A., Lee, J.Y., Alldredge, L., Hunter, E., Lambert, D., Bolognesi, D., *et al.*: **Potent suppression of HIV-1 replication in humans by T-20, a peptide inhibitor of gp41-mediated virus entry.** *Nat. Med.* 1998, **4**:1302–1307.

Kwon, D.S., Gregorio, G., Bitton, N., Hendrickson, W.A., and Littman, D.R.: **DC-SIGN-mediated internalization of HIV is required for trans-enhancement of T cell infection.** *Immunity* 2002, **16**:135–144.

Moore, J.P., Trkola, A., and Dragic, T.: **Co-receptors for HIV-1 entry.** *Curr. Opin. Immunol.* 1997, **9**:551–562.

Pohlmann, S., Baribaud, F., and Doms, R.W.: **DC-SIGN and DC-SIGNR: helping hands for HIV.** *Trends Immunol.* 2001, **22**:643–646.

Root, M.J., Kay, M.S., and Kim, P.S.: **Protein design of an HIV-1 entry inhibitor.** *Science* 2001, **291**:884–888.

Sol-Foulon, N., Moris, A., Nobile, C., Boccaccio, C., Engering, A., Abastado, J.P., Heard, J.M., van Kooyk, Y., and Schwartz, O.: **HIV-1 Nef-induced upregulation of DC-SIGN in dendritic cells promotes lymphocyte clustering and viral spread.** *Immunity* 2002, **16**:145–155.

Unutmaz, D., and Littman, D.R.: **Expression pattern of HIV-1 coreceptors on T cells: implications for viral transmission and lymphocyte homing.** *Proc. Natl Acad. Sci. USA* 1997, **94**:1615–1618.

Wyatt, R., and Sodroski, J.: **The HIV-1 envelope glycoproteins: fusogens, antigens, and immunogens.** *Science* 1998, **280**:1884–1888.

13-26 Genetic variation in the host can alter the rate of progression of disease.

Bream, J.H., Ping, A., Zhang, X., Winkler, C., and Young, H.A.: **A single nucleotide polymorphism in the proximal IFN-gamma promoter alters control of gene transcription.** *Genes Immun.* 2002, **3**:165–169.

Martin, M.P., Gao, X., Lee, J.H., Nelson, G.W., Detels, R., Goedert, J.J., Buchbinder, S., Hoots, K., Vlahov, D., Trowsdale, J., *et al.*: **Epistatic interaction between KIR3DS1 and HLA-B delays the progression to AIDS.** *Nat. Genet.* 2002, **31**:429–434.

Shin, H.D., Winkler, C., Stephens, J.C., Bream, J., Young, H., Goedert, J.J., O'Brien, T.R., Vlahov, D., Buchbinder, S., Giorgi, J., *et al.*: **Genetic restriction of HIV-1 pathogenesis to AIDS by promoter alleles of IL10.** *Proc. Natl Acad. Sci. USA* 2000, **97**:14467–14472.

13-27 A genetic deficiency of the co-receptor CCR5 confers resistance to HIV infection *in vivo*.

Berger, E.A., Murphy, P.M., and Farber, J.M.: **Chemokine receptors as HIV-1 coreceptors: roles in viral entry, tropism, and disease.** *Annu. Rev. Immunol.* 1999, **17**:657–700.

Gonzalez, E., Kulkarni, H., Bolivar, H., Mangano, A., Sanchez, R., Catano, G., Nibbs, R.J., Freedman, B.I., Quinones, M.P., Bamshad, M.J., *et al.*: **The influence of CCL3L1 gene-containing segmental duplications on HIV-1/AIDS susceptibility.** *Science* 2005, **307**:1434–1440.

Lehner, T.: **The role of CCR5 chemokine ligands and antibodies to CCR5 coreceptors in preventing HIV infection.** *Trends Immunol.* 2002, **23**:347–351.

Littman, D.R.: **Chemokine receptors: keys to AIDS pathogenesis?** *Cell* 1998, **93**:677–680.

Liu, R., Paxton, W.A., Choe, S., Ceradini, D., Martin, S.R., Horuk, R., Macdonald, M.E., Stuhlmann, H., Koup, R.A., and Landau, N.R.: **Homozygous defect in HIV-1 coreceptor accounts for resistance of some multiply exposed individuals to HIV 1 infection.** *Cell* 1996, **86**:367–377.

Murakami, T., Nakajima, T., Koyanagi, Y., Tachibana, K., Fujii, N., Tamamura, H., Yoshida, N., Waki, M., Matsumoto, A., Yoshie, O., *et al.*: **A small molecule CXCR4 inhibitor that blocks T cell line-tropic HIV-1 infection.** *J. Exp. Med.* 1997, **186**:1389–1393.

Samson, M., Libert, F., Doranz, B.J., Rucker, J., Liesnard, C., Farber, C.M., Saragosti, S., Lapoumeroulie, C., Cognaux, J., Forceille, C., *et al.*: **Resistance to HIV-1 infection in Caucasian individuals bearing mutant alleles of the CCR 5 chemokine receptor gene.** *Nature* 1996, **382**:722–725.

Yang, A.G., Bai, X., Huang, X.F., Yao, C., and Chen, S.: **Phenotypic knockout of HIV type 1 chemokine coreceptor CCR-5 by intrakines as potential therapeutic approach for HIV-1 infection.** *Proc. Natl Acad. Sci. USA* 1997, **94**:11567–11572.

13-28 HIV RNA is transcribed by viral reverse transcriptase into DNA that integrates into the host-cell genome.

Andrake, M.D., and Skalka, A.M.R.: **Retroviral integrase, putting the pieces together.** *J. Biol. Chem.* 1995, **271**:19633–19636.

Baltimore, D.: **The enigma of HIV infection.** *Cell* 1995, **82**:175–176.

McCune, J.M.: **Viral latency in HIV disease.** *Cell* 1995, **82**:183–188.

Wei, P., Garber, M.E., Fang, S.M., Fischer, W.H., and Jones, K.A.: **A novel CDK9-associated C-type cyclin interacts directly with HIV-1 Tat and mediates its high-affinity, loop-specific binding to TAR RNA.** *Cell* 1998, **92**:451–462.

13-29 Replication of HIV occurs only in activated T cells.

Cullen, B.R.: **HIV-1 auxiliary proteins: making connections in a dying cell.** *Cell* 1998, **93**:685–692.

Cullen, B.R.: **Connections between the processing and nuclear export of** mRNA: evidence for an export license? *Proc. Natl Acad. Sci. USA* 2000, **97**:4–6.

Emerman, M., and Malim, M.H.: **HIV-1 regulatory/accessory genes: keys to unraveling viral and host cell biology.** *Science* 1998, **280**:1880–1884.

Fujinaga, K., Taube, R., Wimmer, J., Cujec, T.P., and Peterlin, B.M.: **Interactions between human cyclin T, Tat, and the transactivation response element (TAR) are disrupted by a cysteine to tyrosine substitution found in mouse cyclin T.** *Proc. Natl Acad. Sci. USA* 1999, **96**:1285–1290.

Kinoshita, S., Su, L., Amano, M., Timmerman, L.A., Kaneshima, H., and Nolan, G.P.: **The T-cell activation factor NF-ATc positively regulates HIV-1 replication and gene expression in T cells.** *Immunity* 1997, **6**:235–244.

Pollard, V.W., and Malim, M.H.: **The HIV-1 Rev protein.** *Annu. Rev. Microbiol.* 1998, **52**:491–532.

Subbramanian, R.A., and Cohen, E.A.: **Molecular biology of the human immuno-deficiency virus accessory proteins.** *J. Virol.* 1994, **68**:6831–6835.

Trono, D.: **HIV accessory proteins: leading roles for the supporting cast.** *Cell* 1995, **82**:189–192.

13-30 Lymphoid tissue is the major reservoir of HIV infection.

Burton, G.F., Masuda, A., Heath, S.L., Smith, B.A., Tew, J.G., and Szakal, A.K.: **Follicular dendritic cells (FDC) in retroviral infection: host/pathogen perspectives.** *Immunol. Rev.* 1997, **156**:185–197.

Chun, T.W., Carruth, L., Finzi, D., Shen, X., DiGiuseppe, J.A., Taylor, H., Hermankova, M., Chadwick, K., Margolick, J., Quinn, T.C., *et al.*: **Quantification of latent tissue reservoirs and total body viral load in HIV-1 infection.** *Nature* 1997, **387**:183–188.

Clark, E.A.: **HIV: dendritic cells as embers for the infectious fire.** *Curr. Biol.* 1996, **6**:655–657.

Emerman, M., and Malim, M.H.: **HIV-1 regulatory/accessory genes: keys to unraveling viral and host cell biology.** *Science* 1998, **280**:1880–1884.

Finzi, D., Blankson, J., Siliciano, J.D., Margolick, J.B., Chadwick, K., Pierson, T., Smith, K., Lisziewicz, J., Lori, F., Flexner, C., *et al.*: **Latent infection of CD4⁺ T cells provides a mechanism for lifelong persistence of HIV-1, even in patients on effective combination therapy.** *Nat. Med.* 1999, **5**:512–517.

Fujinaga, K., Taube, R., Wimmer, J., Cujec, T.P., and Peterlin, B.M.: **Interactions between human cyclin T, Tat, and the transactivation response element (TAR) are disrupted by a cysteine to tyrosine substitution found in mouse cyclin T.** *Proc. Natl Acad. Sci. USA* 1999, **96**:1285–1290.

Haase, A.T.: **Population biology of HIV-1 infection: viral and CD4⁺ T cell demographics and dynamics in lymphatic tissues.** *Annu. Rev. Immunol.* 1999, **17**:625–656.

Kinoshita, S., Su, L., Amano, M., Timmerman, L.A., Kaneshima, H., and Nolan, G.P.: **The T-cell activation factor NF-ATc positively regulates HIV-1 replication and gene expression in T cells.** *Immunity* 1997, **6**:235–244.

Orenstein, J.M., Fox, C., and Wahl, S.M.: **Macrophages as a source of HIV during opportunistic infections.** *Science* 1997, **276**:1857–1861.

Palella, F.J., Jr, Delaney, K.M., Moorman, A.C., Loveless, M.O., Fuhrer, J., Satten, G.A., Aschman, D.J., and Holmberg, S.D.: **Declining morbidity and mortality among patients with advanced human immunodeficiency virus infection. HIV Outpatient Study Investigators.** *N. Engl. J. Med.* 1998, **338**:853–860.

Pierson, T., McArthur, J., and Siliciano, R.F.: **Reservoirs for HIV-1: mechanisms for viral persistence in the presence of antiviral immune responses and antiretroviral therapy.** *Annu. Rev. Immunol.* 2000, **18**:665–708.

Pollard, V.W., and Malim, M.H.: **The HIV-1 Rev protein.** *Annu. Rev. Microbiol.* 1998, **52**:491–532.

Subbramanian, R.A., and Cohen, E.A.: **Molecular biology of the human immuno-deficiency virus accessory proteins.** *J. Virol.* 1994, **68**:6831–6835.

Trono, D.: **HIV accessory proteins: leading roles for the supporting cast.** *Cell* 1995, **82**:189–192.

13-31 An immune response controls but does not eliminate HIV.

Barouch, D.H., and Letvin, N.L.: **CD8⁺ cytotoxic T lymphocyte responses to lentiviruses and herpesviruses.** *Curr. Opin. Immunol.* 2001, **13**:479–482.

Chiu, Y.L., Soros, V.B., Kreisberg, J.F., Stopak, K., Yonemoto, W., and Greene, W.C.: **Cellular APOBEC3G restricts HIV-1 infection in resting CD4⁺ T cells.** *Nature* 2005, **435**:108–114.

Evans, D.T., O'Connor, D.H., Jing, P., Dzuris, J.L., Sidney, J., da Silva, J., Allen, T.M., Horton, H., Venham, J.E., Rudersdorf, R.A., *et al.*: **Virus-specific cytotoxic T-lymphocyte responses select for amino-acid variation in simian immunodeficiency virus Env and Nef.** *Nat. Med.* 1999, **5**:1270–1276.

Haase, A.T.: **Targeting early infection to prevent HIV-1 mucosal transmission.** *Nature* 2010, **464**:217–223.

Johnson, W.E., and Desrosiers, R.C.: **Viral persistance: HIV's strategies of immune system evasion.** *Annu. Rev. Med.* 2002, **53**:499–518.

McMichael, A.J., Borrow, P., Tomaras, G.D., Goonetilleke, N., and Haynes, B.F.: **The immune response during acute HIV-1 infection: clues for vaccine development.** *Nat. Rev. Immunol.* 2010, **10**:11–23.

Poignard, P., Sabbe, R., Picchio, G.R., Wang, M., Gulizia, R.J., Katinger, H., Parren, P.W., Mosier, D.E., and Burton, D.R.: **Neutralizing antibodies have limited effects on the control of established HIV-1 infection *in vivo*.** *Immunity* 1999, **10**:431–438.

Price, D.A., Goulder, P.J., Klenerman, P., Sewell, A.K., Easterbrook, P.J., Troop, M., Bangham, C.R., and Phillips, R.E.: **Positive selection of HIV-1 cytotoxic T lymphocyte escape variants during primary infection.** *Proc. Natl Acad. Sci. USA* 1997, **94**:1890–1895.

Schmitz, J.E., Kuroda, M.J., Santra, S., Sasseville, V.G., Simon, M.A., Lifton, M.A., Racz, P., Tenner-Racz, K., Dalesandro, M., Scallon, B.J., *et al.*: **Control of viremia in simian immunodeficiency virus infection by CD8⁺ lymphocytes.** *Science* 1999, **283**:857–860.

Stremlau, M., Owens, C.M., Perron, M.J., Kiessling, M., Autissier, P., and Sodroski, J.: **The cytoplasmic body component TRIM5α restricts HIV-1 infection in Old World monkeys.** *Nature* 2004, **427**:848–583.

13-32 The destruction of immune function as a result of HIV infection leads to increased susceptibility to opportunistic infection and eventually to death.

Badley, A.D., Dockrell, D., Simpson, M., Schut, R., Lynch, D.H., Leibson, P., and Paya, C.V.: **Macrophage-dependent apoptosis of CD4⁺ T lymphocytes from HIV-infected individuals is mediated by FasL and tumor necrosis factor.** *J. Exp. Med.* 1997, **185**:55–64.

Ho, D.D., Neumann, A.U., Perelson, A.S., Chen, W., Leonard, J.M., and Markowitz, M.: **Rapid turnover of plasma virions and CD4 lymphocytes in HIV-1 infection.** *Nature* 1995, **373**:123–126.

Kedes, D.H., Operskalski, E., Busch, M., Kohn, R., Flood, J., and Ganem, D.R.: **The seroepidemiology of human herpesvirus 8 (Kaposi's sarcoma associated herpesvirus): distribution of infection in KS risk groups and evidence for sexual transmission.** *Nat. Med.* 1996, **2**:918–924.

Kolesnitchenko, V., Wahl, L.M., Tian, H., Sunila, I., Tani, Y., Hartmann, D.P., Cossman, J., Raffeld, M., Orenstein, J., Samelson, L.E., *et al.*: **Human immunodeficiency virus 1 envelope-initiated G2-phase programmed cell death.** *Proc. Natl Acad. Sci. USA* 1995, **92**:11889–11893.

Lauer, G.M., and Walker, B.D.: **Hepatitis C virus infection.** *N. Engl. J. Med.* 2001, **345**:41–52.

Miller, R.: **HIV-associated respiratory diseases.** *Lancet* 1996, **348**:307–312.

Pantaleo, G., and Fauci, A.S.: **Apoptosis in HIV infection.** *Nat. Med.* 1995, **1**:118–120.

Zhong, W.D., Wang, H., Herndier, B., and Ganem, D.R.: **Restricted expression of Kaposi sarcoma associated herpesvirus (human herpesvirus 8) genes in Kaposi sarcoma.** *Proc. Natl Acad. Sci. USA* 1996, **93**:6641–6646.

13-33 Drugs that block HIV replication lead to a rapid decrease in titer of infectious virus and an increase in CD4 T cells.

Boyd, M., and Reiss, P.: **The long-term consequences of antiretroviral therapy: a review.** *J. HIV Ther.* 2006, **11**:26–35.

Carcelain, G., Debre, P., and Autran, B.: **Reconstitution of CD4⁺ T lymphocytes

in HIV-infected individuals following antiretroviral therapy. *Curr. Opin. Immunol.* 2001, **13**:483–488.

Chun, T.W., and Fauci, A.S.: **Latent reservoirs of HIV: obstacles to the eradication of virus.** *Proc. Natl Acad. Sci. USA* 1999, **96**:10958–10961.

Ho, D.D.: **Perspectives series: host/pathogen interactions. Dynamics of HIV-1 replication *in vivo*.** *J. Clin. Invest.* 1997, **99**:2565–2567.

Lipsky, J.J.: **Antiretroviral drugs for AIDS.** *Lancet* 1996, **348**:800–803.

Lundgren, J.D., and Mocroft, A.: **The impact of antiretroviral therapy on AIDS and survival.** *J. HIV Ther.* 2006, **11**:36–38.

Palella, F.J., Jr, Delaney, K.M., Moorman, A.C., Loveless, M.O., Fuhrer, J., Satten, G.A., Aschman, D.J., and Holmberg, S.D.: **Declining morbidity and mortality among patients with advanced human immunodeficiency virus infection. HIV Outpatient Study Investigators.** *N. Engl. J. Med.* 1998, **338**:853–860.

Pau, A.K., and Tavel, J.A.: **Therapeutic use of interleukin-2 in HIV-infected patients.** *Curr. Opin. Pharmacol.* 2002, **2**:433–439.

Perelson, A.S., Essunger, P., Cao, Y.Z., Vesanen, M., Hurley, A., Saksela, K., Markowitz, M., and Ho, D.D.: **Decay characteristics of HIV-1-infected compartments during combination therapy.** *Nature* 1997, **387**:188–191.

Smith, D.: **The long-term consequences of antiretroviral therapy.** *J. HIV Ther.* 2006, **11**:24–25.

Smith, K.A.: **To cure chronic HIV infection, a new therapeutic strategy is needed.** *Curr. Opin. Immunol.* 2001, **13**:617–624.

Wei, X., Ghosh, S.K., Taylor, M.E., Johnson, V.A., Emini, E.A., Deutsch, P., Lifson, J.D., Bonhoeffer, S., Nowak, M.A., Hahn, B.H., *et al.*: **Viral dynamics in human immunodeficiency virus type 1 infection.** *Nature* 1995, **373**:117–122.

13-34 HIV accumulates many mutations in the course of infection, and drug treatment is soon followed by the outgrowth of drug-resistant variants.

Bonhoeffer, S., May, R.M., Shaw, G.M., and Nowak, M.A.: **Virus dynamics and drug therapy.** *Proc. Natl Acad. Sci. USA* 1997, **94**:6971–6976.

Condra, J.H., Schleif, W.A., Blahy, O.M., Gabryelski, L.J., Graham, D.J., Quintero, J.C., Rhodes, A., Robbins, H.L., Roth, E., Shivaprakash, M., *et al.*: ***In vivo* emergence of HIV-1 variants resistant to multiple protease inhibitors.** *Nature* 1995, **374**:569–571.

Finzi, D., and Siliciano, R.F.: **Viral dynamics in HIV-1 infection.** *Cell* 1998, **93**:665–671.

Katzenstein, D.: **Combination therapies for HIV infection and genomic drug resistance.** *Lancet* 1997, **350**:970–971.

Moutouh, L., Corbeil, J., and Richman, D.D.: **Recombination leads to the rapid emergence of HIV 1 dually resistant mutants under selective drug pressure.** *Proc. Natl Acad. Sci. USA* 1996, **93**:6106–6111.

13-35 Vaccination against HIV is an attractive solution but poses many difficulties.

Amara, R.R., Villinger, F., Altman, J.D., Lydy, S.L., O'Neil, S.P., Staprans, S.I., Montefiori, D.C., Xu, Y., Herndon, J.G., Wyatt, L.S., *et al.*: **Control of a mucosal challenge and prevention of AIDS by a multiprotein DNA/MVA vaccine.** *Science* 2001, **292**:69–74.

Baba, T.W., Liska, V., Hofmann-Lehmann, R., Vlasak, J., Xu, W., Ayehunie, S., Cavacini, L.A., Posner, M.R., Katinger, H., Stiegler, G., *et al.*: **Human neutralizing monoclonal antibodies of the IgG1 subtype protect against mucosal simian–human immunodeficiency virus infection.** *Nat. Med.* 2000, **6**:200–206.

Barouch, D.H., Kunstman, J., Kuroda, M.J., Schmitz, J.E., Santra, S., Peyerl, F.W., Krivulka, G.R., Beaudry, K., Lifton, M.A., Gorgone, D.A., *et al.*: **Eventual AIDS vaccine failure in a rhesus monkey by viral escape from cytotoxic T lymphocytes.** *Nature* 2002, **415**:335–339.

Burton, D.R.: **A vaccine for HIV type 1: the antibody perspective.** *Proc. Natl Acad. Sci. USA* 1997, **94**:10018–10023.

Kaul, R., Rowland-Jones, S.L., Kimani, J., Dong, T., Yang, H.B., Kiama, P., Rostron, T., Njagi, E., Bwayo, J.J., MacDonald, K.S., *et al.*: **Late seroconversion in HIV-resistant Nairobi prostitutes despite pre-existing HIV-specific CD8$^+$ responses.** *J. Clin. Invest.* 2001, **107**:341–349.

Letvin, N.L.: **Progress and obstacles in the development of an AIDS vaccine.** *Nat. Rev. Immunol.* 2006, **6**:930–939.

Letvin, N.L., and Walker, B.D.: **HIV versus the immune system: another apparent victory for the virus.** *J. Clin. Invest.* 2001, **107**:273–275.

MacQueen, K.M., Buchbinder, S., Douglas, J.M., Judson, F.N., McKirnan, D.J., and Bartholow, B.: **The decision to enroll in HIV vaccine efficacy trials: concerns elicited from gay men at increased risk for HIV infection.** *AIDS Res. Hum. Retroviruses* 1994, **10** Suppl. 2:S261–S264.

Mascola, J.R., and Nabel, G.J.: **Vaccines for the prevention of HIV-1 disease.** *Curr. Opin. Immunol.* 2001, **13**:489–495.

Robert-Guroff, M.: **IgG surfaces as an important component in mucosal protection.** *Nat. Med.* 2000, **6**:129–130.

Shiver, J.W., Fu, T.M., Chen, L., Casimiro, D.R., Davies, M.E., Evans, R.K., Zhang, Z.Q., Simon, A.J., Trigona, W.L., Dubey, S.A., *et al.*: **Replication-incompetent adenoviral vaccine vector elicits effective anti-immunodeficiency-virus immunity.** *Nature* 2002, **415**:331–335.

13-36 Prevention and education are one way in which the spread of HIV and AIDS can be controlled.

Coates, T.J., Aggleton, P., Gutzwiller, F., Des-Jarlais, D., Kihara, M., Kippax, S., Schechter, M., and van-den-Hoek, J.A.: **HIV prevention in developed countries.** *Lancet* 1996, **348**:1143–1148.

Decosas, J., Kane, F., Anarfi, J.K., Sodji, K.D., and Wagner, H.U.: **Migration and AIDS.** *Lancet* 1995, **346**:826–828.

Dowsett, G.W.: **Sustaining safe sex: sexual practices, HIV and social context.** *AIDS* 1993, **7** Suppl. 1:S257–S262.

Kimball, A.M., Berkley, S., Ngugi, E., and Gayle, H.: **International aspects of the AIDS/HIV epidemic.** *Annu. Rev. Public. Health* 1995, **16**:253–282.

Kirby, M.: **Human rights and the HIV paradox.** *Lancet* 1996, **348**:1217–1218.

Nelson, K.E., Celentano, D.D., Eiumtrakol, S., Hoover, D.R., Beyrer, C., Suprasert, S., Kuntolbutra, S., and Khamboonruang, C.: **Changes in sexual behavior and a decline in HIV infection among young men in Thailand.** *N. Engl. J. Med.* 1996, **335**:297–303.

Weniger, B.G., and Brown, T.: **The march of AIDS through Asia.** *N. Engl. J. Med.* 1996, **335**:343–345.

Allergy and Allergic Diseases

14

The adaptive immune response is a critical component of host defense against infection and is essential for normal health. Adaptive immune responses are sometimes elicited by antigens not associated with infectious agents, and this can cause disease. One circumstance in which this occurs is when harmful immunologically mediated hypersensitivity reactions known generally as **allergic reactions** are made in response to inherently harmless 'environmental' antigens such as pollen, food, and drugs.

Hypersensitivity reactions due to immunological responses were classified into four broad types by Coombs and Gell (Fig. 14.1). Type I hypersensitivity reactions in this classification are immediate-type allergic reactions mediated by **IgE** antibodies, but many of the allergic diseases that are initiated by IgE antibodies, such as allergic asthma, have chronic features characteristic of other types of immune response, particularly of T_H2 cell-mediated type IV hypersensitivity (see Fig. 14.1). In most **allergies**, such as those to food, pollen, and house dust, reactions occur because the individual has become **sensitized** to an innocuous antigen—the **allergen**—by producing IgE antibodies against it. Subsequent exposure to the allergen triggers the activation of IgE-binding cells, chiefly mast cells and basophils, in the exposed tissue, leading to a series of responses that are characteristic of this type of allergic reaction. In hay fever (allergic rhinoconjunctivitis), for example, symptoms occur when allergenic proteins leached out of grass pollen grains come into contact with the mucous membrane of the nose and eyes. Other allergic diseases, such as serum sickness, allergic contact dermatitis, and celiac disease, do not involve IgE and are due to type II, III, or T_H1- or CD8 T-cell type IV hypersensitivity reactions (see Fig. 14.1).

Although everybody is exposed to common environmental allergens, most of the population does not develop allergic reactions to them. In surveys, up to 40% of the population shows an exaggerated tendency to become sensitized to a wide variety of common environmental allergens. A predisposition to become IgE-sensitized to environmental allergens is called **atopy**, and later in the chapter we discuss the various factors—both genetic and environmental—that may contribute to predisposition. The importance of genetic factors in predisposing to IgE-mediated allergic disease is shown by the fact that if both parents are atopic, a child has a 40–60% chance of developing an IgE-mediated allergy, whereas the risk for a child neither of whose parents is atopic is much lower, of the order of 10%, although this percentage is increasing, as we discuss later in the chapter.

The main biological role of IgE is thought to be in adaptive immunity to infection with parasitic worms (see Chapter 10), which are prevalent in less developed countries. In the industrialized countries, IgE-mediated allergic responses to innocuous antigens predominate and are an important cause of disease (Fig. 14.2). Almost half the population of North America and Europe

	Type I	Type II		Type III	Type IV		
Immune reactant	IgE	IgG		IgG	T_H1 cells	T_H2 cells	CTL
Antigen	Soluble antigen	Cell- or matrix-associated antigen	Cell-surface receptor	Soluble antigen	Soluble antigen	Soluble antigen	Cell-associated antigen
Effector mechanism	Mast-cell activation	Complement, FcR^+ cells (phagocytes, NK cells)	Antibody alters signaling	Complement, phagocytes	Macrophage activation	IgE production, eosinophil activation, mastocytosis	Cytotoxicity
	Ag	platelets + complement		immune complex, blood vessel, + complement	IFN-γ, T_H1, chemokines, cytokines, cytotoxins	IL-4, IL-5, T_H2, eotaxin, cytotoxins, inflammatory mediators	CTL
Example of hypersensitivity reaction	Allergic rhinitis, allergic asthma, atopic eczema, systemic anaphylaxis, some drug allergies	Some drug allergies (e.g. penicillin)	Chronic urticaria (antibody against FcεRI alpha chain)	Serum sickness, Arthus reaction	Allergic contact dermatitis, tuberculin reaction	Chronic asthma, chronic allergic rhinitis	Graft rejection, allergic contact dermatitis to poison ivy

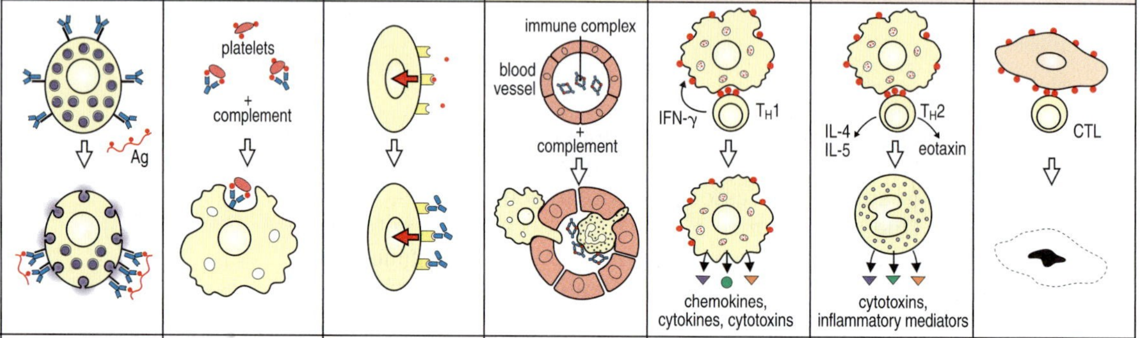

Fig. 14.1 Immunological hypersensitivity reactions, or allergic reactions, are mediated by immune reactions that cause tissue damage. Four types of allergic reaction are generally recognized. Types I–III are antibody-mediated and are distinguished by the different types of antigen recognized and the different classes of antibody involved. Type I responses are mediated by IgE, which induces mast-cell activation, whereas types II and III are mediated by IgG, which can engage complement-mediated and phagocytic effector mechanisms to varying degrees, depending on the subclass of IgG and the nature of the antigen involved. Type II responses are directed against cell-surface or matrix antigens, whereas type III responses are directed against soluble antigens, and the tissue damage involved is caused by responses triggered by immune complexes. A special category of type II responses involves IgG antibodies against cell-surface receptors that disrupt the normal functions of the receptor, either by causing uncontrollable activation or by blocking receptor function. Type IV hypersensitivity reactions are mediated by T cells and can be subdivided into three groups. In the first group, tissue damage is caused by the activation of macrophages by T_H1 cells, which results in an inflammatory response. In the second, damage is caused by the activation by T_H2 cells of inflammatory responses in which eosinophils predominate; in the third, damage is caused directly by cytotoxic T cells (CTLs).

is sensitized to one or more common environmental antigens and, although rarely life-threatening, allergic diseases initiated by contact with a specific allergen can cause much distress and lost time from school and work. The burden of allergic diseases in the Western world is considerable, with a more than doubling in prevalence in the past 15 years or so, and so most clinical and scientific attention has been paid to the role of IgE in allergic disease rather than in its protective capacity. Until a few years ago, developing countries in Africa and the Middle East reported a relatively low prevalence of allergy (although this situation is rapidly changing as a result of Western-style modernization).

In this chapter we first consider the mechanisms that favor the sensitization of an individual to an allergen through the production of IgE. We then describe the IgE-mediated allergic reaction itself—the pathological consequences of the interaction between allergen and the IgE bound to the high-affinity Fcε receptor on mast cells and basophils. Finally, we consider the causes and consequences of other types of immunological hypersensitivity reaction.

IgE-mediated allergic reactions			
Reaction or disease	Common allergens	Route of entry	Response
Systemic anaphylaxis	Drugs Venoms Food, e.g. peanuts Serum	Intravenous (either directly or following oral absorption into the blood after oral intake)	Edema Increased vascular permeability Laryngeal edema Circulatory collapse Death
Acute urticaria (wheal-and-flare)	Animal hair Insect bites Allergy testing	Through skin Systemic	Local increase in blood flow and vascular permeability Edema
Seasonal rhinoconjunctivitis (hay fever)	Pollens (ragweed, trees, grasses) Dust-mite feces	Contact with conjunctiva of eye and nasal mucosa	Edema of conjunctiva and nasal mucosa Sneezing
Asthma	Danders (cat) Pollens Dust-mite feces	Inhalation leading to contact with mucosal lining of lower airways	Bronchial constriction Increased mucus production Airway inflammation
Food allergy	Peanuts Tree nuts Shellfish Fish Milk Eggs Soy Wheat	Oral	Vomiting Diarrhea Pruritis (itching) Urticaria (hives) Anaphylaxis (rarely)

Fig. 14.2 IgE-mediated reactions to extrinsic antigens. All IgE-mediated responses involve mast-cell degranulation, but the symptoms experienced by the patient can be very different depending, for example, on whether the allergen is injected directly into the bloodstream, is eaten, or comes into contact with the mucosa of the respiratory tract.

IgE and IgE-mediated allergic diseases.

Type I hypersensitivity reactions are those allergic reactions due to the production of IgE against innocuous antigens. IgE is produced both by plasma cells in lymph nodes draining the site of antigen entry and by plasma cells at the site of the allergic reaction—typically a mucosal tissue or the skin. In mucosal tissues, germinal centers develop within the inflamed tissue. IgE differs from other antibody isotypes in being predominantly localized in the tissues, where it is tightly bound to the surface of mast cells, and some other cell types, through the high-affinity IgE receptor **FcεRI** (see Section 10-24). Binding of antigen to IgE cross-links these receptors, causing the release of chemical mediators from the mast cells that can lead to allergic disease (Fig. 14.3). How an initial antibody response to environmental antigens comes to be dominated by IgE production in atopic individuals is still being worked out. In this part of the chapter we describe the current understanding of the factors that contribute to this process.

14-1 Sensitization involves class switching to IgE production on first contact with an allergen.

To produce an allergic reaction against a given antigen, an individual has first to be exposed to the antigen and become sensitized to it by producing IgE antibodies. Atopic individuals often develop multiple types of allergic disease to multiple allergens—for example, atopic eczema developing in childhood in response to sensitization to food antigens is followed in a sizable proportion of these individuals by the development of allergic rhinitis and/or asthma

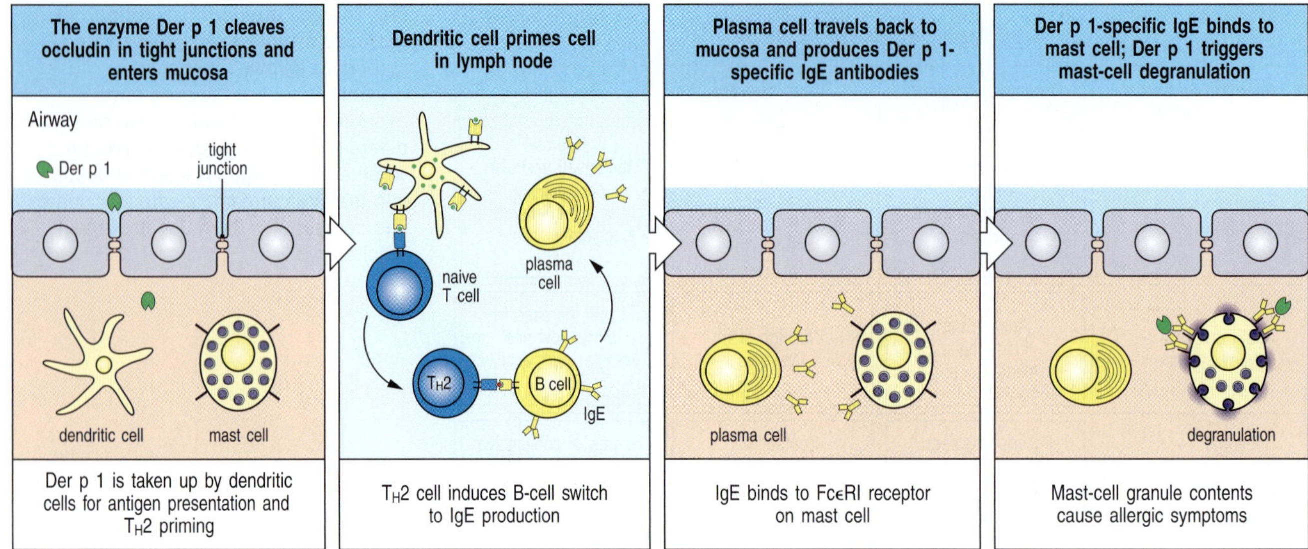
The enzyme Der p 1 cleaves occludin in tight junctions and enters mucosa
Airway
Der p 1
tight junction
dendritic cell
mast cell
Der p 1 is taken up by dendritic cells for antigen presentation and T_H2 priming
Dendritic cell primes cell in lymph node
naive T cell
plasma cell
T_H2
B cell
IgE
T_H2 cell induces B-cell switch to IgE production
Plasma cell travels back to mucosa and produces Der p 1-specific IgE antibodies
plasma cell
IgE binds to FcεRI receptor on mast cell
Der p 1-specific IgE binds to mast cell; Der p 1 triggers mast-cell degranulation
degranulation
Mast-cell granule contents cause allergic symptoms

Fig. 14.3 Sensitization to an inhaled allergen. A common respiratory allergen is the protein Der p 1, found in fecal pellets of the house dust mite. On a first encounter with Der p 1 in an atopic individual, T_H2 cells specific for Der p 1 may be produced (first and second panels). Interaction of these T cells with Der p 1-specific B cells leads to the production of class-switched plasma cells producing Der p 1-specific IgE in the mucosal tissues (third panel), and this IgE becomes bound to Fc receptors on resident submucosal mast cells. On a subsequent encounter with Der p 1, the allergen binds to the mast-cell-bound IgE, triggering mast-cell activation and the release of mast-cell granule contents, which cause the symptoms of the allergic reaction (last panel). Der p 1 is a protease that cleaves occludin, a protein that helps to maintain the tight junctions; the enzymatic activity of Der p 1 is thought to help it to pass through the epithelium.

caused by airborne allergens. Allergic reactions in non-atopic people, in contrast, are predominantly due to sensitization to one specific allergen, such as bee venom or a drug such as penicillin, and can develop at any time of life. It is important to remember, however, that not all encounters with a potential allergen will lead to sensitization, and not all sensitizations will lead to a symptomatic allergic response, even in atopic individuals.

The immune response leading to IgE production in response to antigen is driven by two main groups of signals. The first consists of signals that favor the differentiation of naive T cells to a T_H2 phenotype. The second comprises the action of cytokines and co-stimulatory signals from T_H2 cells that stimulate B cells to switch to the production of IgE. The fate of a naive CD4 T cell responding to a peptide presented by a dendritic cell is determined by the cytokines it is exposed to before and during this response, and by the intrinsic properties of the antigen, the antigen dose, and the route of presentation. Exposure to IL-4, IL-5, IL-9, and IL-13 favors the development of T_H2 cells, whereas IFN-γ and IL-12 (and its relatives IL-23 and IL-27) favor T_H1-cell development (see Section 9-18).

Immune defenses against multicellular parasites are found mainly at the sites of parasite entry, namely under the skin and in the mucosal tissues of the airways and the gut; cells of the innate and adaptive immune systems at these sites are specialized to secrete cytokines that promote a T_H2-cell response to parasite infection. In the presence of infection, dendritic cells taking up antigens in these tissues migrate to regional lymph nodes, where they tend to drive antigen-specific naive CD4 T cells to become effector T_H2 cells. T_H2 cells themselves secrete IL-4, IL-5, IL-9, and IL-13, thus maintaining an environment in which further differentiation of T_H2 cells is favored. The cytokine IL-33, which can be produced by activated mast cells, also seems to be important in amplifying the T_H2 response. Allergic responses against common environmental antigens are normally avoided by the propensity of mucosal dendritic cells in the absence of infection to induce the production of antigen-specific regulatory T cells (T_{reg} cells) from naive CD4 T cells. The T_{reg} cells suppress T-cell responses and produce a state of tolerance to the antigen (see Section 12-8).

The cytokines and chemokines produced by T_H2 cells both amplify the T_H2 response and stimulate the class switching of B cells to IgE production. As we saw in Chapter 10, IL-4 or IL-13 provides the first signal that switches B cells to IgE production. Cytokines IL-4 and IL-13 activate the Janus-family tyrosine kinases Jak1 and Jak3 (see Section 7-20), which ultimately leads to phosphorylation of the transcriptional regulator STAT6 in T and B lymphocytes. Mice lacking functional IL-4, IL-13, or STAT6 have impaired T_H2 responses and impaired IgE switching, demonstrating the key importance of these cytokines and their signaling pathways. The second signal is a co-stimulatory interaction between CD40 ligand on the T-cell surface and CD40 on the B-cell surface. This interaction is essential for all antibody class switching: patients with a genetic deficiency of CD40 ligand produce no IgG, IgA, or IgE, and display a hyper IgM syndrome phenotype (see Section 13-15).

CD40 Ligand Deficiency

Mast cells and basophils can also drive IgE production by B cells (Fig. 14.4). Mast cells and basophils express FcεRI, and when they are activated by antigen cross-linking their FcεRI-bound IgE, they express cell-surface CD40 ligand and secrete IL-4. Like T_H2 cells, therefore, they can drive class switching and IgE production by B cells. The interaction between mast cells or basophils and B cells can occur at the site of the allergic reaction, because B cells are observed to form germinal centers at inflammatory foci. One goal of therapy for allergies is to block this amplification process and thus prevent allergic reactions from becoming self-sustaining.

In humans, the IgE response, once initiated, can also be amplified by the capture of IgE by Fcε receptors on dendritic cells. Some populations of human immature dendritic cells—for example, the Langerhans cells of the skin—express surface FcεRI in an inflammatory setting, and once anti-allergen IgE antibodies have been produced they can bind to these receptors. The bound IgE forms a highly effective trap for allergen, which is then efficiently processed by the dendritic cell for presentation to naive T cells, thus maintaining and reinforcing the T_H2 response to the allergen. Eosinophils have also been reported to express IgE receptors, but this is still controversial. However, eosinophils may act as antigen-presenting cells to T cells in a standard fashion after upregulation of eosinophil MHC class II molecules and co-stimulatory molecules, although this probably occurs in sites where activated T cells have migrated rather than in lymph nodes where naive T cells are primed by dendritic cells.

14-2 Allergens are usually delivered transmucosally at low dose, a route that favors IgE production.

Most airborne allergens are relatively small, highly soluble proteins that are carried on dry particles such as pollen grains or mite feces (Fig. 14.5). On contact with the mucus-covered epithelia of the eyes, nose, or airways, the soluble allergen is eluted from the particle and diffuses into the mucosa, where it can be picked up by dendritic cells and provoke sensitization (see Fig. 14.3). Allergens are typically presented to the immune system at low concentrations. It has been estimated that the maximum exposure of a person to the common pollen allergens in ragweed (*Ambrosia* species) does not exceed 1 μg

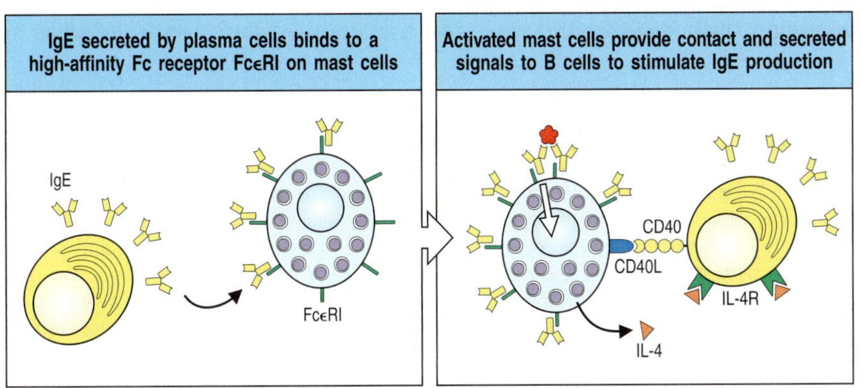
IgE secreted by plasma cells binds to a high-affinity Fc receptor FcεRI on mast cells
IgE
FcεRI
Activated mast cells provide contact and secreted signals to B cells to stimulate IgE production
CD40
CD40L
IL-4R
IL-4

Fig. 14.4 Antigen binding to IgE on mast cells or basophils leads to amplification of IgE production. Left panel: IgE secreted by plasma cells binds to the high-affinity IgE receptor on mast cells (illustrated here) and basophils. Right panel: when the surface-bound IgE is cross-linked by antigen, these cells express CD40 ligand (CD40L) and secrete IL-4, which in turn binds to IL-4 receptors (IL-4R) on the activated B cell, stimulating class switching by B cells and the production of more IgE. These interactions can occur *in vivo* at the site of allergen-triggered inflammation, for example in bronchus-associated lymphoid tissue.

Features of airborne allergens that may promote the priming of T_H2 cells that drive IgE responses	
Protein, often with carbohydrate side chains	Only proteins induce T-cell responses
Low dose	Favors activation of IL-4-producing CD4 T cells
Low molecular weight	Allergen can diffuse out of particles into the mucosa
Highly soluble	Allergen can be readily eluted from particle
Stable	Allergen can survive in desiccated particle
Contains peptides that bind host MHC class II	Required for T-cell priming

Fig. 14.5 Properties of inhaled allergens. The typical characteristics of inhaled allergens are described in this table.

per year. Yet these minute doses of allergen can provoke irritating and even life-threatening T_H2-driven IgE antibody responses in atopic individuals.

It seems likely that presenting an antigen across a mucosal epithelium and at very low doses is a particularly efficient way of inducing T_H2-driven IgE responses. In mice, IgE antibody production requires help from T_H2 cells that produce interleukin-4 (IL-4) and IL-13, and it can be inhibited by T_H1 cells that produce interferon-γ (IFN-γ) (see Fig. 9.28). Low doses of antigen can favor the activation of T_H2 cells over T_H1 cells, and many common allergens are delivered in low doses to the respiratory mucosa. In the mucosa these allergens encounter dendritic cells that take up and process protein antigens very efficiently. In some circumstances, mast cells, basophils, and eosinophils can also present allergen-derived antigen to activated T cells that have already been primed by dendritic cells, further promoting the responses of T_H2 cells.

Many parasitic worms invade their hosts by secreting proteolytic enzymes that break down connective tissue and allow the parasite access to internal tissues, and it has been proposed that these enzymes are particularly active at promoting T_H2 responses. One ubiquitous protease allergen is the cysteine protease Der p 1 present in the feces of the house dust mite (*Dermatophagoides pteronyssimus*), which provokes allergic reactions in about 20% of the North American population. This enzyme has been found to cleave occludin, a protein component of intercellular tight junctions. This reveals one possible reason for the allergenicity of certain enzymes. By destroying the integrity of the tight junctions between epithelial cells, Der p 1 may gain abnormal access to subepithelial antigen-presenting cells (see Fig. 14.3). The tendency of proteases to induce IgE production is highlighted by individuals with Netherton's syndrome (Fig. 14.6), which is characterized by high levels of IgE and multiple allergies. The defect in this disease is the lack of a protease inhibitor called SPINK5, which is thought to inhibit the proteases released by bacteria such as *Staphylococcus aureus*, thus raising the possibility that protease inhibitors might be novel therapeutic targets in some allergic disorders. The cysteine protease papain, derived from the papaya fruit, is used as a meat tenderizer and causes allergic reactions in workers preparing the enzyme; such allergies are called **occupational allergies**. Not all allergens are enzymes, however; for example, two allergens identified from filarial worms are enzyme inhibitors, and, in general, allergenic pollen-derived proteins do not seem to possess enzymatic activity. Thus, there seems to be no systematic association between enzymatic activity and allergenicity.

Knowledge of the identity of allergenic proteins can be important to public health and can have economic significance, as illustrated by the following cautionary tale. Some years ago, the gene for a protein from brazil nuts that is rich in methionine and cysteine was transferred by genetic engineering into soy beans intended for animal feed. This was done to improve the nutritional value of soy beans, which are intrinsically poor in these sulfur-containing amino acids. This experiment led to the discovery that the protein, 2S albumin, was the major brazil nut allergen. Injection of extracts of the genetically modified soy beans into the epidermis triggered an allergic skin response in

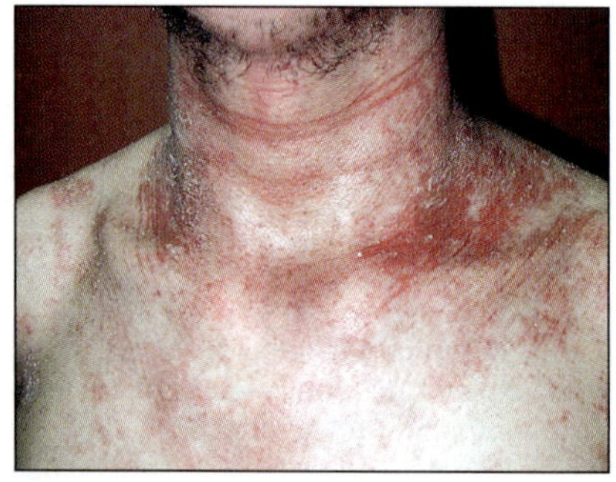

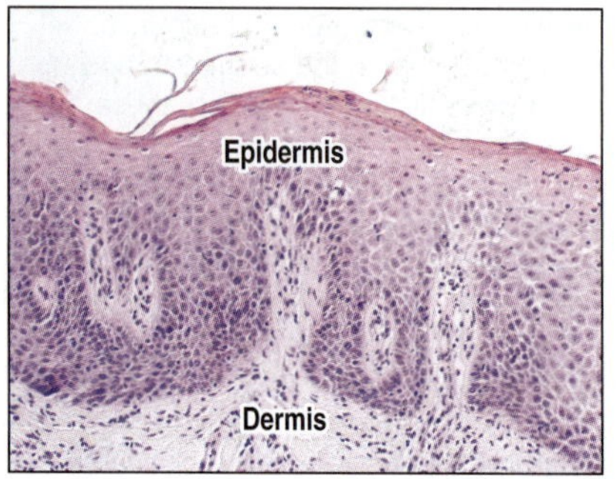

Fig. 14.6 Netherton's syndrome illustrates the association of proteases with the development of high levels of IgE and allergy. This 26-year-old man with Netherton's syndrome, caused by a deficiency in the protease inhibitor SPINK5, had persistent erythroderma, recurrent infections of the skin and elsewhere, and multiple food allergies associated with high serum IgE levels. In the top photograph, large erythematous plaques covered with scales and erosions are visible over the upper trunk. The lower panel shows a section through the skin of the same patient. Note the psoriasis-like hyperplasia of the epidermis. Neutrophils are also present in the epidermis. In the dermis, a perivascular infiltrate containing both mononuclear cells and neutrophils is evident. Source: Sprecher, E., *et al.*: *Clin. Exp. Dermatol.* 2004, 29:513–517.

people allergic to brazil nuts. As there could be no guarantee that the modified soy beans could be kept out of the human food chain if they were produced on a large scale, development of this genetically modified food was abandoned.

14-3 Genetic factors contribute to the development of IgE-mediated allergic disease.

The risk of developing allergic disease has both genetic and environmental components. In studies performed in Western industrialized countries, up to 40% of the test population shows an exaggerated tendency to mount IgE responses to a wide variety of common environmental allergens. This is the state called atopy. It has a strong familial basis and is known to be influenced by multiple genetic loci. Atopic individuals have higher total levels of IgE in the circulation and higher levels of eosinophils than their non-atopic counterparts and are more susceptible to developing allergic diseases such as allergic rhinoconjunctivitis, allergic asthma, or atopic eczema.

Genome-wide linkage scans have uncovered several distinct susceptibility genes for the allergic skin condition atopic eczema (also known as atopic dermatitis) and for allergic asthma, although there is little overlap between the two sets of genes, suggesting that the genetic predisposition differs somewhat (Fig. 14.7). In addition, there are many ethnic differences in the susceptibility genes for a given allergic disease. Several of the chromosome regions associated with allergy or asthma are also associated with the inflammatory disease psoriasis and with autoimmune diseases, suggesting the presence of genes that are involved in exacerbating inflammation (see Fig. 14.7).

One candidate susceptibility gene for both allergic asthma and atopic eczema, at chromosome 11q12–13, encodes the β subunit of the high-affinity IgE receptor FcεRI. Another region of the genome associated with allergic disease, 5q31–33, contains at least four types of candidate gene that might

Fig. 14.7 Susceptibility loci identified by genome screens for asthma, atopic dermatitis, and other immune disorders. Only loci with significant linkages are indicated. Clustering of disease-susceptibility genes is found for the MHC on chromosome 6p21, and also in several other genomic regions. There is in fact little overlap between susceptibility genes for asthma and atopic dermatitis, suggesting that specific genetic factors are involved in both. There is also some overlap between susceptibility genes for asthma and those for autoimmune diseases, and between those for the inflammatory skin disease psoriasis and atopic dermatitis. Adapted from Cookson, W.: *Nat. Rev. Immunol.* 2004, 4:978–988.

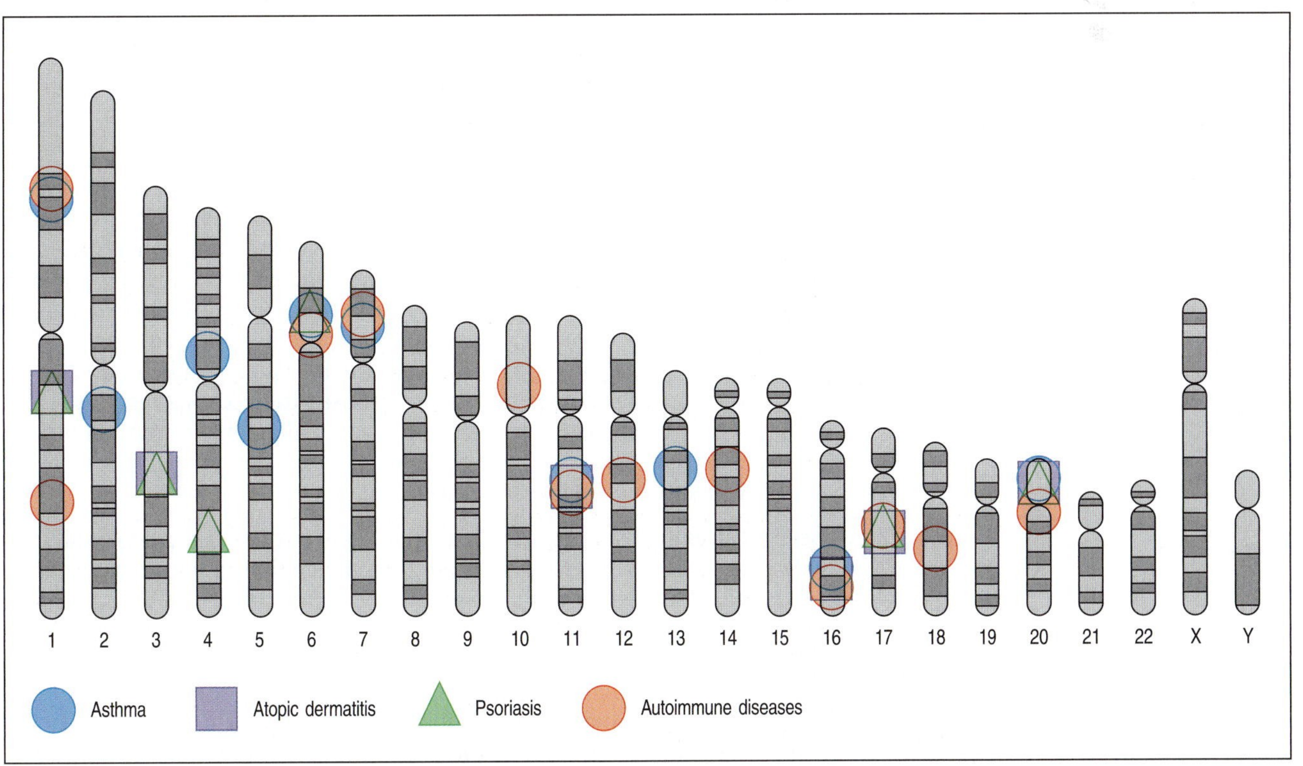

be responsible for increased susceptibility. First, there is a cluster of tightly linked genes for cytokines that enhance IgE class switching, eosinophil survival, and mast-cell proliferation, all of which help to produce and maintain an IgE-mediated allergic response. This cluster includes the genes for IL-3, IL-4, IL-5, IL-9, IL-13, and granulocyte–macrophage colony-stimulating factor (GM-CSF). In particular, genetic variation in the promoter region of the gene encoding IL-4 has been associated with raised IgE levels in atopic individuals. The variant promoter directs increased expression of a reporter gene in experimental systems and thus might produce increased IL-4 *in vivo*. Atopy has also been associated with a gain-of-function mutation of the α subunit of the IL-4 receptor that causes increased signaling after ligation of the receptor.

A second set of genes in this region of chromosome 5 is the TIM family (for *T* cell, *i*mmunoglobulin domain, and *m*ucin domain), which encode T-cell-surface proteins. In mice, Tim-3 protein is specifically expressed on T_H1 cells and negatively regulates T_H1 responses, whereas Tim-2 (and to a lesser extent Tim-1) is preferentially expressed in, and negatively regulates, T_H2 cells. Mouse strains that carry different variants of the *Tim* genes differ both in their susceptibility to allergic inflammation of the airways and in the production of IL-4 and IL-13 by their T cells. Inherited variation in the *TIM* genes in humans has been correlated with levels of airway **hyperreactivity** or **hyperresponsiveness**. In this condition, contact not only with allergen but also with nonspecific irritants causes airway narrowing with wheezy breathlessness similar to that seen in asthma. The third candidate susceptibility gene in this part of the genome encodes p40, one of the two subunits of IL-12. This cytokine promotes T_H1 responses, and genetic variation in p40 expression that could cause reduced production of IL-12 was found to be associated with more severe asthma. A fourth candidate susceptibility gene, that encoding the β-adrenergic receptor, is also encoded in this region. Variation in this receptor might be associated with alteration in smooth-muscle responsiveness to endogenous and pharmacological ligands.

This complexity illustrates a common challenge in identifying the genetic basis of complex disease traits. Relatively small regions of the genome, identified as containing genes for altered disease susceptibility, may contain many good candidates, judging by their known physiological activities. Identifying the correct gene, or genes, may require studies of several very large populations of patients and controls. For chromosome 5q31–33, for example, it is still too early to know how important each of the different polymorphisms is in the complex genetics of atopy.

A second type of inherited variation in IgE responses is linked to the HLA class II region (the human MHC class II region) and affects responses to specific allergens, rather than a general susceptibility to atopy. IgE production in response to particular allergens is associated with certain HLA class II alleles, implying that particular peptide:MHC combinations might favor a strong T_H2 response; for example, IgE responses to several ragweed pollen allergens are associated with haplotypes containing the HLA class II allele *DRB1*1501*. Many people are therefore generally predisposed to make T_H2 responses and are specifically predisposed to respond to some allergens more than others. However, allergic responses to drugs such as penicillin show no association with HLA class II or with the presence or absence of atopy.

There are also likely to be genes that affect only particular aspects of allergic disease. In asthma, for example, there is evidence that different genes affect at least three aspects of the disease—IgE production, the inflammatory response, and clinical responses to particular treatments. Polymorphism of the gene on chromosome 20 encoding ADAM33, a metalloproteinase expressed by bronchial smooth muscle cells and lung fibroblasts, has been associated with asthma and bronchial hyperreactivity. This is likely to be an example of genetic variation in the pulmonary inflammatory response and in the pathological anatomical changes that occur in the airways (airway remodeling).

Some of the many genes that have been associated with asthma are shown in Fig. 14.8, where they are grouped into several categories of immune response in which they participate.

14-4 Environmental factors may interact with genetic susceptibility to cause allergic disease.

Studies of susceptibility suggest that environmental factors and genetic variation each account for about 50% of the risk of developing a disease such as allergic asthma. The prevalence of atopic allergic diseases, and of asthma in particular, is increasing in economically advanced regions of the world, and this is likely to be due to changing environmental factors.

The main candidate environmental factors for the increase in allergy are changes in exposure to infectious diseases in early childhood; the change from 'traditional' rural societies that has meant less early exposure to animal microorganisms and microorganisms in the soil, for example; and changes in the intestinal microbiota, which performs an important immunomodulatory function (discussed in Chapter 12). Changes in exposure to ubiquitous microorganisms as a possible cause of the increase in allergy has received much attention since the idea was first mooted in 1989, and this is known as the **hygiene hypothesis** (Fig. 14.9). The original proposition was that less hygienic environments, specifically environments that predispose to infections early in childhood, help to protect against the development of atopy and allergic asthma. It was originally proposed that the protective effect might be due to mechanisms that skewed immune responses away from the production of T_H2 cells and their associated cytokines, which dispose toward IgE production, and toward the production of T_H1 cells, whose cytokines do not induce class switching to IgE.

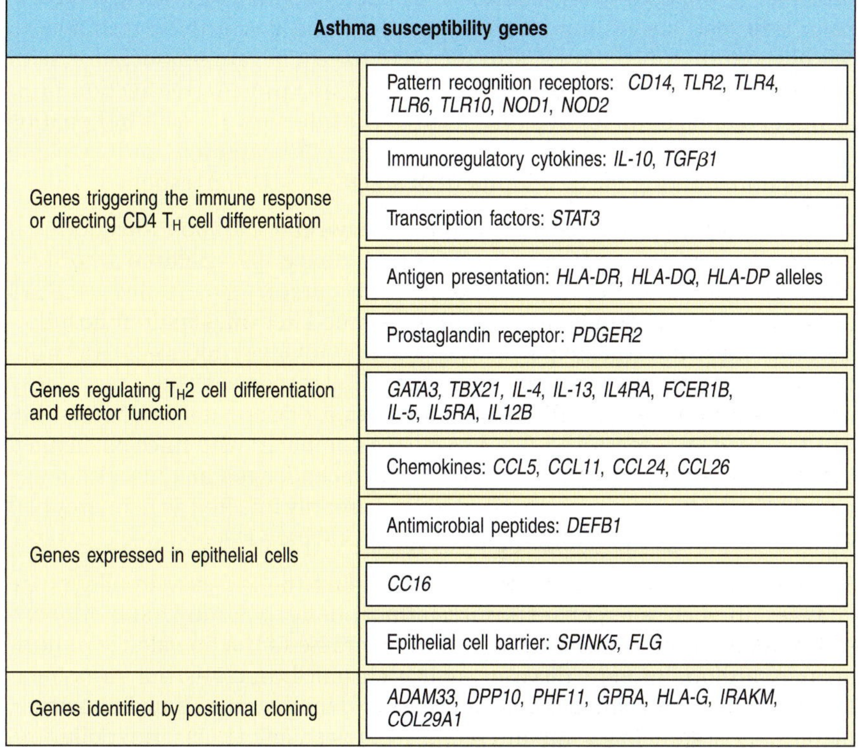

Asthma susceptibility genes	
Genes triggering the immune response or directing CD4 T_H cell differentiation	Pattern recognition receptors: *CD14, TLR2, TLR4, TLR6, TLR10, NOD1, NOD2*
	Immunoregulatory cytokines: *IL-10, TGFβ1*
	Transcription factors: *STAT3*
	Antigen presentation: *HLA-DR, HLA-DQ, HLA-DP* alleles
	Prostaglandin receptor: *PDGER2*
Genes regulating T_H2 cell differentiation and effector function	*GATA3, TBX21, IL-4, IL-13, IL4RA, FCER1B, IL-5, IL5RA, IL12B*
Genes expressed in epithelial cells	Chemokines: *CCL5, CCL11, CCL24, CCL26*
	Antimicrobial peptides: *DEFB1*
	CC16
	Epithelial cell barrier: *SPINK5, FLG*
Genes identified by positional cloning	*ADAM33, DPP10, PHF11, GPRA, HLA-G, IRAKM, COL29A1*

Fig. 14.8 Candidate susceptibility genes for asthma.

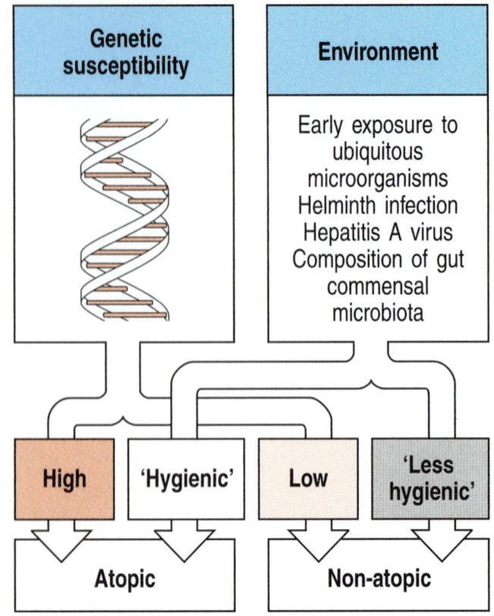

Fig. 14.9 Genes, the environment, and atopic allergic diseases. Both inherited and environmental factors are important determinants of the likelihood of developing atopic allergic disease. Some genes known to influence the development of asthma are shown in Fig. 14.8. The postulate of the 'hygiene hypothesis' or 'counter-regulation hypothesis' is that exposure to some infections and to common environmental microorganisms in infancy and childhood drives the immune system toward a general state of non-atopy. In contrast, children with genetic susceptibility to atopy and who live in an environment with low exposure to infectious disease and environmental microorganisms are thought not to develop efficient immunoregulatory mechanisms and to be most susceptible to the development of atopic allergic disease.

The biggest drawback to this interpretation, however, was the strong negative correlation between infection by helminths (such as hookworm and schistosomes) and the development of allergic disease. A study in Venezuela showed that children treated for a prolonged period with antihelminthic agents had a higher prevalence of atopy than did untreated and heavily parasitized children. As helminths provoke a strong T_H2-mediated IgE response, this seemed to run counter to the hygiene hypothesis.

These observations led to a modification of the hypothesis known as the **counter-regulation hypothesis**. This proposes that all types of infection might protect against the development of atopy by driving the production of cytokines such as IL-10 and transforming growth factor (TGF)-β, which downregulate both T_H1 and T_H2 responses (see Sections 9-18 and 9-19). A large proportion of allergic reactions are initiated by antigens that enter though mucosal surfaces such as the respiratory or intestinal epithelium. As described in Chapter 12, the human mucosal immune system has evolved mechanisms of regulating responses to commensal flora and environmental antigens (such as food antigens) that involve the generation of IL-10/TGF-β-producing regulatory T cells. The idea underlying the current version of the hygiene hypothesis is that decreased early exposure to common microbial pathogens and commensals in some way makes the body less efficient at producing these regulatory T cells, thus increasing the risk of making an allergic response to a common environmental antigen.

In support of the counter-regulation hypothesis is evidence that exposure to certain types of childhood infection, with the important exception of some respiratory infections that we consider below, helps to protect against the development of allergic disease. Younger children from families with three or more older siblings, and children aged less than 6 months who are exposed to other children in daycare facilities—situations linked to a greater exposure to infections—are somewhat protected against atopy and asthma. Furthermore, early colonization of the gut by commensal bacteria such as lactobacilli and bifidobacteria, or infection by gut pathogens such as *Toxoplasma gondii* or *Helicobacter pylori*, is associated with a reduced prevalence of allergic disease.

A history of infection with measles or hepatitis A virus, or a positive tuberculin skin test (suggesting previous exposure and an immune response to *Mycobacterium tuberculosis*), also seems to have a negative association with atopy. The human counterpart of the murine Tim-1 protein (see Section 14-3) is the cellular receptor for hepatitis A virus. The infection of T cells by hepatitis A virus could thus directly influence their differentiation and cytokine production, limiting the development of an IgE-generating response.

In contrast to these negative associations between childhood infection and the development of atopy and asthma is evidence that children who have had attacks of bronchiolitis associated with respiratory syncytial virus (RSV) infection are more prone to developing asthma later on. Children hospitalized with RSV infection have a skewed ratio of cytokine production away from IFN-γ toward IL-4, the cytokine that induces T_H2 responses. This effect of RSV may depend on age at first infection. Infection of neonatal mice with RSV was followed by a decreased IFN-γ response compared with mice challenged at 4 or 8 weeks of age. When the mice were rechallenged at 12 weeks of age with RSV infection, animals that had been primarily infected as neonates had more severe lung inflammation than animals infected at 4 or 8 weeks of age.

Other environmental factors that might explain the increase in allergy are changes in diet, allergen exposure, atmospheric pollution, and tobacco smoke. Pollution has been blamed for an increase in the prevalence of non-allergic cardiopulmonary diseases such as chronic bronchitis, but an association with allergic disease has been less easy to demonstrate. There is, however, increasing evidence for an interaction between allergens and pollution, particularly in genetically susceptible individuals. Diesel exhaust particles are the best-studied pollutant in this context; they increase IgE production 20–50-fold when combined with allergen, with an accompanying shift to T_H2 cytokine production. Reactive oxidant chemicals such as ozone are generated as a result of such pollution, and individuals less able to deal with this onslaught may be at increased risk of allergic disease.

Genes that might be governing this aspect of susceptibility are *GSTP1* and *GSTM1*, members of the glutathione-*S*-transferase superfamily that are important in preventing oxidant stress. Individuals who were allergic to ragweed pollen and who carried particular variant alleles of these genes showed an increased airway hyperreactivity when challenged with the allergen plus diesel exhaust particles, compared with the allergen alone. A study in Mexico City on the effects of atmospheric ozone levels on atopic children with allergic asthma also found that the children carrying the null allele of *GSTM1* were more susceptible to airway hyperreactivity than were noncarriers when exposed to given levels of ozone. Indeed, genetic factors such as these, and the complexity of genetic and environmental interactions, may explain why the epidemiological evidence for an association between pollution and allergy remains moderate at best.

14-5 Regulatory T cells can control allergic responses.

Peripheral blood mononuclear cells (PBMCs) from atopic individuals have a tendency to secrete T_H2 cytokines after nonspecific stimulation via the T-cell receptor, whereas those from non-atopic individuals do not. This has led to the suggestion that regulatory mechanisms have an important role in preventing IgE responses to allergens. Regulatory T cells, in particular, are receiving considerable attention with regard to all types of immunologically mediated disease. The different types of regulatory T cells (see Section 9-19) may all have a role in modulating allergy. Circulating CD4 CD25 T_{reg} cells from atopic individuals are defective in suppressing T_H2 cytokine production compared with those from non-atopic individuals, and this defect is even more pronounced during the pollen season. More evidence comes from mice deficient in the transcription factor FoxP3, the master switch for producing both natural (thymus-derived) and some induced T_{reg} cells. These mice develop manifestations of allergic disease including eosinophilia, hyper IgE, and allergic airway inflammation, suggesting that these symptoms result from the absence of regulatory T cells. This syndrome could be partly reversed by a concomitant deficiency in STAT6, which independently prevents the development of a T_H2 response (see Section 14-1).

Regulatory T cells can also be induced by the actions of the anti-inflammatory enzyme indoleamine 2,3-dioxygenase (IDO), which is expressed in a variety of cell types in response to stimulation by certain cytokines, such as IFN-γ, or by unmethylated CpG DNA acting via the receptor TLR-9. IDO activity in resident dendritic cells in the lung stimulated in this way has been shown to ameliorate experimental asthma in mice. IDO breaks down tryptophan to metabolites called kynurenines, which are thought to be the active agents in the various effects reported for IDO on immune system functions.

Summary.

Type I allergic reactions are the result of the production of specific IgE antibody against common, innocuous antigens. Allergens are small antigens that commonly provoke an IgE antibody response. Such antigens normally enter the body at very low doses by diffusion across mucosal surfaces and thus trigger a T_H2 response. The differentiation of naive allergen-specific T cells into T_H2 cells is also favored by cytokines such as IL-4 and IL-13. Allergen-specific

T_H2 cells producing IL-4 and IL-13 drive allergen-specific B cells to produce IgE. The specific IgE produced in response to the allergen binds to the high-affinity receptor for IgE on mast cells and basophils. IgE production can be amplified by these cells because, upon activation, they produce IL-4 and CD40 ligand. The tendency to IgE overproduction is influenced by both genetic and environmental factors. Once IgE has been produced in response to an allergen, reexposure to the allergen triggers an allergic response. Immunoregulation is critical in the control of allergic disease through a variety of mechanisms, including regulatory T cells. We describe the mechanism and pathology of the allergic responses themselves in the next part of the chapter.

Effector mechanisms in IgE-mediated allergic reactions.

Allergic reactions are triggered when allergens cross-link preformed IgE bound to the high-affinity receptor FcεRI on mast cells. Mast cells line external mucosal surfaces and serve to alert the immune system to local infection. Once activated, they induce inflammatory reactions by secreting pharmacological mediators such as histamine stored in preformed granules and by synthesizing prostaglandins, leukotrienes, and platelet-activating factor from the plasma membrane. They also release various cytokines and chemokines after activation. In the case of an allergic reaction, they provoke unpleasant reactions to innocuous antigens that are not associated with invading pathogens that need to be expelled. The consequences of IgE-mediated mast-cell activation depend on the dose of antigen and its route of entry; symptoms range from the swollen eyes and rhinitis associated with contact of pollen with the conjunctiva of the eye and the nasal epithelium, to the life-threatening circulatory collapse that occurs in anaphylaxis (Fig. 14.10). The immediate reaction caused by mast-cell degranulation is followed, to a greater or lesser

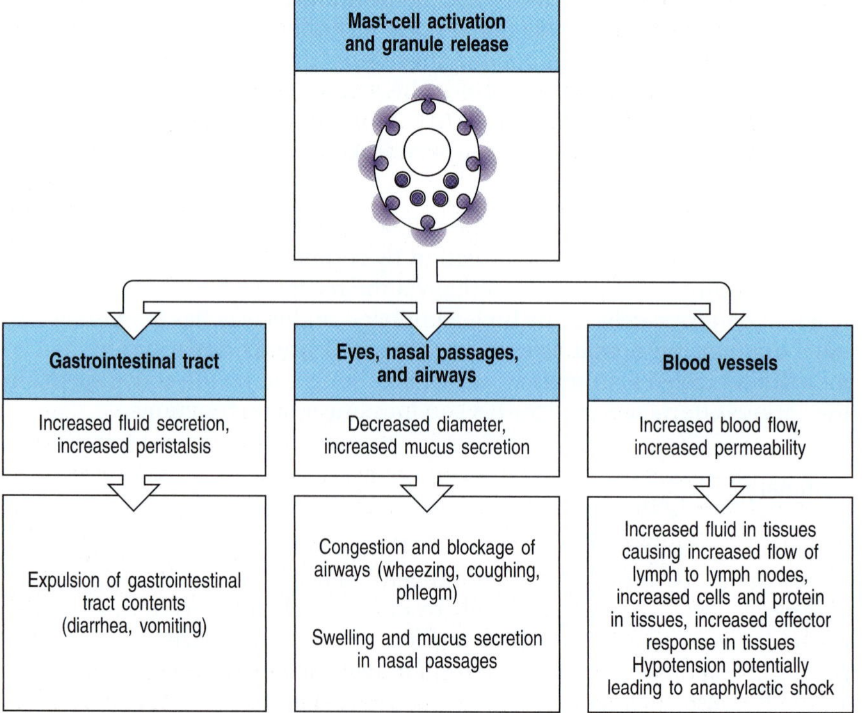

Fig. 14.10 Mast-cell activation has different effects on different tissues.

extent depending on the disease, by a more sustained inflammation, which is due to the recruitment of other effector leukocytes, notably T_H2 lymphocytes, eosinophils, and basophils.

14-6 Most IgE is cell-bound and engages effector mechanisms of the immune system by different pathways from those of other antibody isotypes.

Antibodies engage effector cells such as mast cells by binding to receptors specific for the Fc constant regions. Most antibodies engage Fc receptors only after the antibody variable region has bound specific antigen, forming an immune complex of antigen and antibody. IgE is an exception, because it is captured by the high-affinity Fcε receptor (FcεRI) in the absence of bound antigen. This means that, unlike other antibodies, which are found mainly in body fluids, IgE is mostly found fixed on cells that carry this receptor—mast cells in tissues and basophils in the circulation. The ligation of the cell-bound IgE antibody by specific antigen triggers the activation of these cells at the sites of antigen entry into the tissues. The release of inflammatory lipid mediators, cytokines, and chemokines at sites of IgE-triggered reactions recruits eosinophils and basophils to augment the type I hypersensitivity response. It also recruits T_H2 cells, which can then mount a local T_H2 type IV hypersensitivity response.

There are two types of IgE-binding Fc receptor. The first, FcεRI, is a high-affinity receptor of the immunoglobulin superfamily that binds IgE on mast cells and basophils (see Section 10-24). When the cell-bound IgE is cross-linked by specific antigen, FcεRI transduces an activating signal. High levels of IgE, such as those that exist in people with allergic diseases or parasite infections, can result in a marked increase in FcεRI on the surface of mast cells, an enhanced sensitivity of such cells to activation by low concentrations of specific antigen, and a markedly increased IgE-dependent release of chemical mediators and cytokines.

The second IgE receptor, FcεRII, usually known as **CD23**, is a C-type lectin and is structurally unrelated to FcεRI; it binds IgE with low affinity. CD23 is present on many cell types, including B cells, activated T cells, monocytes, eosinophils, platelets, follicular dendritic cells, and some thymic epithelial cells. This receptor was thought to be crucial for the regulation of IgE levels, but mouse strains in which the gene for CD23 has been inactivated still develop relatively normal polyclonal IgE responses. Nevertheless, CD23 does seem to be involved in enhancing IgE antibody levels in some situations. Responses against a specific antigen are known to be increased in the presence of the antigen complexed with IgE, but such enhancement fails to occur in mice that lack the gene for CD23. This has been interpreted to indicate that CD23 on antigen-presenting cells has a role in the capture of antigen complexed with IgG.

14-7 Mast cells reside in tissues and orchestrate allergic reactions.

Mast cells were described by Ehrlich in the mesentery of rabbits and named *Mastzellen* ('fattened cells'). Like basophils, mast cells contain granules rich in acidic proteoglycans that take up basic dyes. Mast cells are derived from hematopoietic stem cells but mature locally, often residing near surfaces exposed to pathogens and allergens, such as mucosal tissues and the connective tissues surrounding blood vessels. Mucosal mast cells differ in some of their properties from submucosal or connective tissue mast cells, but both can be involved in allergic reactions.

The major factors for mast-cell growth and development include stem-cell factor (the ligand for the receptor tyrosine kinase Kit), IL-3, and T_H2-associated cytokines such as IL-4 and IL-9. Mice with defective Kit lack differentiated mast cells, and although they produce IgE they cannot make IgE-mediated inflammatory responses. This shows that such responses depend almost exclusively on mast cells. Mast-cell activation depends on the activation of phosphatidylinositol 3-kinase (PI 3-kinase) in mast cells by Kit, and pharmacological inactivation of the p110δ isoform of PI 3-kinase has been shown to protect mice against allergic responses.

Mast cells express FcεRI constitutively on their surface and are activated when antigens cross-link IgE bound to these receptors (see Fig. 10.37). A relatively low level of allergen is sufficient to trigger degranulation. There are many mast-cell precursors in tissues, which can rapidly differentiate to mature mast cells in conditions of allergic inflammation, thus aiding the continuation of the allergic response. Mast-cell degranulation begins within seconds of antigen binding, releasing an array of preformed and newly generated inflammatory mediators (Fig. 14.11). Granule contents include the short-lived vasoactive amine **histamine**, serine esterases, and proteases such as chymase and tryptase.

Human mast cells are classified on the basis of their protease content. Mast cells of one class (MC_T) predominantly express tryptase only, and these predominate in mucosal epithelia, whereas those of another type (MC_{CT}) express tryptase, chymase, carboxypeptidase A, and cathepsin G and predominate in the submucosa and other connective tissues. Histamine acts via H_1 receptors on local blood vessels to cause an immediate increase in local blood flow and vessel permeability. It also has immunomodulatory and inflammatory activity. Acting through the H_1 receptor on dendritic cells, histamine can increase antigen-presenting capacity and T_H1 priming; acting through H_1 on T cells, it can enhance T_H1 proliferation and IFN-γ production.

Fig. 14.11 Molecules released by mast cells on activation. Mast cells produce a wide variety of biologically active proteins and other chemical mediators. The enzymes and toxic mediators listed in the first two rows are released from the preformed granules. The cytokines, chemokines, and lipid mediators are synthesized after activation.

Class of product	Examples	Biological effects
Enzyme	Tryptase, chymase, cathepsin G, carboxypeptidase	Remodel connective tissue matrix
Toxic mediator	Histamine, heparin	Toxic to parasites Increase vascular permeability Cause smooth muscle contraction Anticoagulation
Cytokine	IL-4, IL-13	Stimulate and amplify T_H2-cell response
	IL-3, IL-5, GM-CSF	Promote eosinophil production and activation
	TNF-α (some stored preformed in granules)	Promotes inflammation, stimulates cytokine production by many cell types, activates endothelium
Chemokine	CCL3	Attracts monocytes, macrophages, and neutrophils
Lipid mediator	Prostaglandins D_2, E_2 Leukotrienes C4, D4, E4	Smooth muscle contraction Chemotaxis of eosinophils, basophils, and T_H2 cells Increase vascular permeability Stimulate mucus secretion Bronchoconstriction
	Platelet-activating factor	Attracts leukocytes Amplifies production of lipid mediators Activates neutrophils, eosinophils, and platelets

The proteases released by the mast cells activate matrix metalloproteinases, which break down extracellular matrix proteins, causing tissue disintegration and damage. Large amounts of the cytokine tumor necrosis factor (TNF)-α are also released by mast cells after activation. Some comes from stores in the granules; some is newly synthesized by the activated mast cells. TNF-α activates endothelial cells, resulting in increased expression of adhesion molecules, which in turn promotes the influx of pro-inflammatory leukocytes and lymphocytes into the affected tissue (see Chapter 3).

On activation, mast cells also synthesize and release chemokines, cytokines, and lipid mediators—prostaglandins, leukotrienes, thromboxanes (collectively called eicosanoids), and platelet-activating factor. Mucosal and submucosal mast cells, for example, produce the cytokine IL-4, which helps perpetuate the T_H2 response. These secreted products contribute to both acute and chronic inflammatory responses. The lipid mediators, in particular, act rapidly to cause smooth muscle contraction, increased vascular permeability, and the secretion of mucus, and also induce the influx and activation of leukocytes, which contribute to the allergic inflammation.

Eicosanoids derive mainly from the fatty acid arachidonic acid. This is cleaved from membrane phospholipids by phospholipase A2, which is activated at the plasma membrane as a result of cell activation. Arachidonic acid can be modified by either of two pathways to give rise to lipid mediators. Modification via the cyclooxygenase pathway produces the prostaglandins and thromboxanes, whereas leukotrienes are produced via the lipoxygenase pathway. Prostaglandin D_2 is the major prostaglandin produced by mast cells and recruits T_H2 cells, eosinophils, and basophils, all of which express its receptor (PTGDR). Prostaglandin D_2 is critical to the development of allergic diseases such as asthma, and polymorphisms in the *PTGDR* gene have been linked to an increased risk of developing asthma. The leukotrienes, especially C4, D4, and E4, are also important in sustaining inflammatory responses in tissues. Nonsteroidal anti-inflammatory drugs such as aspirin and ibuprofen exert their effects by preventing prostaglandin production. They inhibit the cyclooxygenases that act on arachidonic acid to form the ring structure present in prostaglandins.

IgE-mediated activation of mast cells thus orchestrates an important inflammatory cascade that is amplified by the recruitment of several cell types including eosinophils, basophils, T_H2 lymphocytes, and B cells. The physiological importance of this reaction is as a defense against parasite infection (see Section 10-25). In an allergic reaction, however, the acute and chronic inflammatory reactions triggered by mast-cell activation have important pathophysiological consequences, as seen in the diseases associated with allergic responses to environmental antigens. The role of mast cells is not limited to IgE-driven pro-inflammatory responses, however. Increasingly, mast cells are also considered to have a role in immunoregulation. They can secrete the immunosuppressive cytokine IL-10, and interaction with regulatory T cells may prevent degranulation.

14-8 Eosinophils and basophils cause inflammation and tissue damage in allergic reactions.

Eosinophils are granulocytic leukocytes that originate in bone marrow. They are so called because their granules, which contain arginine-rich basic proteins, are colored bright orange by the acidic stain eosin. Only very small numbers of these cells are normally present in the circulation; most eosinophils are found in tissues, especially in the connective tissue immediately underneath respiratory, gut, and urogenital epithelium, implying a likely role for these cells in defense against invading organisms. They possess numerous cell-surface receptors, including receptors for cytokines (such as IL-5), Fcγ and Fcα receptors, and the complement receptor C3, through which they can be activated and stimulated to degranulate. For example, parasites coated with IgG, C3b, or IgA can cause eosinophil degranulation. In allergic tissue reactions, the large concentrations of IL-5/IL-3 and GM-CSF that are typically present are likely inducers of degranulation.

Eosinophils have two kinds of effector function. First, on activation they release highly toxic granule proteins and free radicals, which can kill micro-organisms and parasites but also cause significant tissue damage in allergic reactions (Fig. 14.12). Second, activation induces the synthesis of chemical mediators such as prostaglandins, leukotrienes, and cytokines. These amplify the inflammatory response by activating epithelial cells and by recruiting and activating more eosinophils and leukocytes. Eosinophils also secrete proteins involved in airway tissue remodeling.

What were later to be defined as eosinophils were observed in the 19th century in the first pathological description of fatal status asthmaticus, but the precise role of these cells in allergic disease generally is still unclear. In allergic tissue reactions, for example, those that lead to chronic asthma, mast-cell degranulation and T_H2 activation cause eosinophils to accumulate in large numbers and to become activated. Among other things, eosinophils secrete T_H2-type cytokines and *in vitro* can promote the apoptosis of T_H1 cells by their expression of IDO and consequent production of kynurenine, which acts on the T_H1 cells. Their apparent promotion of T_H2-cell expansion may thus be partly due to a relative reduction in T_H1-cell numbers. The continued

Fig. 14.12 Eosinophils secrete a range of highly toxic granule proteins and other inflammatory mediators.

Class of product	Examples	Biological effects
Enzyme	Eosinophil peroxidase	Toxic to targets by catalyzing halogenation Triggers histamine release from mast cells
	Eosinophil collagenase	Remodels connective tissue matrix
	Matrix metalloproteinase-9	Matrix protein degradation
Toxic protein	Major basic protein	Toxic to parasites and mammalian cells Triggers histamine release from mast cells
	Eosinophil cationic protein	Toxic to parasites Neurotoxin
	Eosinophil-derived neurotoxin	Neurotoxin
Cytokine	IL-3, IL-5, GM-CSF	Amplify eosinophil production by bone marrow Eosinophil activation
	TGF-α, TGF-β	Epithelial proliferation, myofibroblast formation
Chemokine	CXCL8 (IL-8)	Promotes influx of leukocytes
Lipid mediator	Leukotrienes C4, D4, E4	Smooth muscle contraction Increase vascular permeability Increase mucus secretion Bronchoconstriction
	Platelet-activating factor	Attracts leukocytes Amplifies production of lipid mediators Activates neutrophils, eosinophils, and platelets

presence of eosinophils is characteristic of chronic allergic inflammation, and eosinophils are thought to be major contributors to tissue damage.

The activation and degranulation of eosinophils is strictly regulated, because their inappropriate activation would be harmful to the host. The first level of control acts on the production of eosinophils by the bone marrow. Few eosinophils are produced in the absence of infection or other immune stimulation. But when T_H2 cells are activated, cytokines such as IL-5 are released that increase the production of eosinophils in the bone marrow and their release into the circulation. However, transgenic animals overexpressing IL-5 have increased numbers of eosinophils (**eosinophilia**) in the circulation but not in their tissues, indicating that the migration of eosinophils from the circulation into tissues is regulated separately, by a second set of controls. The key molecules in this case are CC chemokines. Most of these cause chemotaxis of several types of leukocyte, but three are particularly important in attracting and activating eosinophils, and have been named the **eotaxins**: CCL11 (eotaxin 1), CCL24 (eotaxin 2), and CCL26 (eotaxin 3).

The eotaxin receptor on eosinophils, CCR3, is quite promiscuous and binds other CC chemokines, including CCL7, CCL13, and CCL5, which also induce eosinophil chemotaxis and activation. Identical or similar chemokines stimulate mast cells and basophils. For example, eotaxins attract basophils and cause their degranulation, and CCL2, which binds to CCR2, similarly activates mast cells in both the presence and the absence of antigen. CCL2 can also promote the differentiation of naive T cells to T_H2 cells; T_H2 cells also carry the receptor CCR3 and migrate toward eotaxins. It is striking that these interactions between different chemokines and their receptors show a high degree of overlap and redundancy; we do not understand the significance of this complexity. However, these findings show that families of chemokines, as well as cytokines, can coordinate certain kinds of immune response.

Basophils are also present at the site of an inflammatory reaction, and growth factors for basophils are very similar to those for eosinophils; they include IL-3, IL-5, and GM-CSF. There is evidence for reciprocal control of the maturation of the stem-cell population into basophils or eosinophils. For example, TGF-β in the presence of IL-3 suppresses eosinophil differentiation and enhances that of basophils. Basophils are normally present in very low numbers in the circulation and seem to have a similar role to that of eosinophils in defense against pathogens. Like eosinophils, they are recruited to the sites of IgE-mediated allergic reactions. Basophils express high-affinity FcεRI on the cell surface and so have IgE bound. On activation by antigen binding to IgE or by cytokines, they release histamine from the basophilic granules after which they are named; they also produce IL-4 and IL-13.

Eosinophils, mast cells, and basophils can interact with each other. Eosinophil degranulation releases **major basic protein** (see Fig. 14.12), which in turn causes the degranulation of mast cells and basophils. This effect is augmented by any of the cytokines that affect eosinophil and basophil growth, differentiation, and activation, such as IL-3, IL-5, and GM-CSF.

14-9 IgE-mediated allergic reactions have a rapid onset but can also lead to chronic responses.

Under laboratory conditions, the clinical response of a sensitized individual to challenge by intradermal allergen or inhalation of allergen can be divided into an 'immediate reaction' and a 'late-phase reaction' (Fig. 14.13). The **immediate reaction** is due to IgE-mediated mast-cell activation and starts within seconds of allergen exposure. It is the result of the actions of histamine, prostaglandins, and other preformed or rapidly synthesized mediators released by mast cells. These cause a rapid increase in vascular permeability,

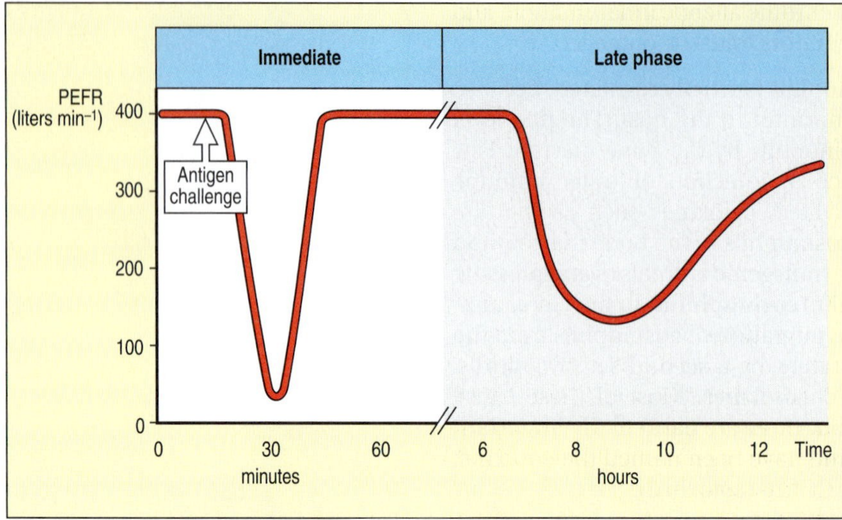

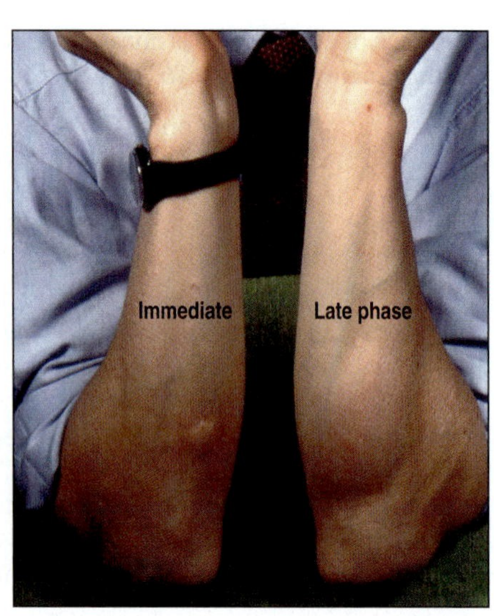

Fig. 14.13 Allergic reactions in response to test antigens can be divided into an immediate response and a late-phase response. Left panel: the response to an inhaled antigen can be divided into early and late responses. An asthmatic response in the lungs with narrowing of the airways caused by the constriction of bronchial smooth muscle and development of edema can be measured as a fall in the peak expiratory flow rate (PEFR). The immediate response peaks within minutes after antigen inhalation and then subsides. Six to eight hours after antigen challenge, there is a late-phase response that also results in a fall in the PEFR. The immediate response is caused by the direct effects on blood vessels and smooth muscle of rapidly metabolized mediators such as histamine and lipid mediators released by mast cells. The late-phase response is caused by the continued production of these mediators and by the production of vasoactive compounds that dilate blood vessels and produce edema. Right panel: a wheal-and-flare allergic reaction develops within a minute or two of intradermal injection of antigen and lasts for up to 30 minutes. The more widespread edematous response characteristic of the late phase develops approximately 6 hours later and can persist for some hours. The photograph shows an intradermal skin challenge with allergen showing a 15-minute wheal-and-flare (early-phase) reaction (left) and a 6-hour late-phase reaction (right). The allergen was grass pollen extract. Photograph courtesy of S.R. Durham.

resulting in visible edema and reddening of the skin (in a skin response) and airway narrowing as result of edema and the constriction of smooth muscle (in an airway response). In the skin, histamine acting on H_1 receptors on local blood vessels causes an immediate increase in vascular permeability, which leads to extravasation of fluid and edema. Histamine also acts on H_1 receptors on local nerve endings, leading to vasodilation of cutaneous blood vessels and local reddening of the skin. The resulting skin lesion is called a **wheal-and-flare reaction**.

Whether a **late-phase reaction** occurs is dependent on allergen dose; at doses that are deemed safe in tests of asthmatic reactions, for example, a late reaction occurs in about 50% of individuals who show an immediate response (see Fig. 14.13, left panel). The late reaction peaks between 3 and 9 hours after antigen challenge and in skin tests becomes obvious as a much increased area and degree of edema (see Fig. 14.13, right panel). The late-phase reaction is caused by the continued synthesis and release of inflammatory mediators by mast cells, especially vasoactive mediators such as calcitonin gene related peptide (CGRP) and vascular endothelial growth factor (VEGF), which cause vasodilation and vascular leakage resulting in edema. In experimental responses to inhaled allergens, the late-phase reactions are associated with a second phase of airway narrowing with sustained edema.

Allergists take advantage of the immediate response as an aid to the assessment and confirmation of sensitization, in the light of a clinical history of allergic disease, and to determine which allergens are responsible. Minute amounts of potential allergens are introduced into the skin by skin prick—one site for each allergen—and if the individual is sensitive to any of the allergens tested, a wheal-and-flare reaction will occur at the site within a few minutes (see Fig. 14.13). Although the reaction after the administration of such small amounts of allergen is usually very localized, there is a still a small risk of inducing anaphylaxis. Another standard test for allergy is to measure circulating concentrations of IgE antibody specific for a particular allergen in a sandwich ELISA (see Appendix I, Section A-6).

The late-phase reaction described above occurs under controlled experimental conditions to a single relatively high dose of allergen and so does not reflect all the effects of long-term natural exposure. In IgE-mediated allergic diseases, a long-term sequel of allergen exposure is **chronic allergic inflammation**, which is in essence a T_H2 type IV hypersensitivity reaction (see Fig. 14.1). These chronic reactions contribute to much serious long-term illness, such as chronic asthma. Inflammatory mediators released by mast cells and basophils recruit other leukocytes, mainly T_H2 cells and eosinophils, to the site of inflammation. In chronic asthma, for example, the cytokines released by T_H2 cells and the effector molecules released by eosinophils (see Fig. 14.12) result in persistent edema, which narrows the airways, and in **airway tissue remodeling**, a change in the bronchial tissue due to smooth muscle hypertrophy (an increase in size due to cell growth) and hyperplasia (an increase in the number of cells). The chronic phase of asthma is characterized by the presence of both T_H1 cytokines (such as IFN-γ) and T_H2 cytokines, although the latter seem to predominate.

In the natural situation, the clinical symptoms produced by an IgE-mediated allergic reaction depend critically on several variables: the amount of allergen-specific IgE present, the route by which the allergen is introduced, the dose of allergen, and most probably some underlying defect in barrier function in the particular tissue or organ affected. The outcomes produced by different combinations of allergen dose and route of entry are summarized in Fig. 14.14. When exposure to allergen in a sensitized individual triggers an allergic reaction, both the immediate and the chronic effects are focused on the site at which mast-cell degranulation occurs.

14-10 Allergen introduced into the bloodstream can cause anaphylaxis.

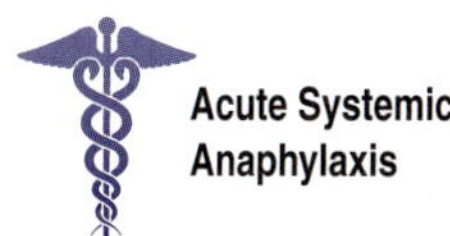

If allergen is introduced directly into the bloodstream, for example by a bee or wasp sting, or is rapidly absorbed into the bloodstream from the gut in a sensitized individual, connective-tissue mast cells associated with blood vessels throughout the body can become immediately activated, resulting in a widespread release of histamine and other mediators that causes a reaction called **anaphylaxis**. The symptoms of anaphylaxis can range in severity from a mild **urticaria** (hives) to potentially fatal **anaphylactic shock** (see Fig. 14.14, first and last panels). Acute urticaria is a response to ingested allergens entering the bloodstream and reaching the skin. Histamine released by mast cells activated by allergen in the skin causes large, itchy, red swellings all over the body, a disseminated version of the wheal-and-flare reaction. Although acute urticaria is commonly caused by an IgE-mediated reaction against an allergen, the causes of chronic urticaria, in which the urticarial rash recurs over long periods, are mostly unknown. It seems likely that from one-third to one-half of the cases of chronic urticaria are caused by autoantibodies against either the α chain of FcεRI or against IgE itself, and are thus due to autoimmunity. Interaction of the autoantibody with the receptor triggers mast-cell degranulation, with resulting urticaria. This is an example of a type II hypersensitivity reaction (see Fig. 14.1).

In anaphylactic shock, a widespread increase in vascular permeability resulting from a massive release of histamine leads to a catastrophic loss of blood pressure, resulting in shock; airways constrict, causing difficulty in breathing; and swelling of the epiglottis can cause suffocation. The major causes of anaphylaxis are wasp and bee stings, or allergic responses to foods in sensitized individuals—anaphylaxis in response to peanuts is relatively common. Severe anaphylactic shock can be rapidly fatal if untreated, but can usually be controlled by the immediate injection of epinephrine, which reverses the action of histamine at the H_1 receptor, thus relaxing the bronchial smooth muscle, and inhibits the immediately life-threatening cardiovascular effects. Sensitization to, and anaphylactic reactions to, insect venom, drugs, and food such as peanuts are not particularly associated with atopy, even when they are IgE-mediated. This may be because these antigens are delivered at high doses compared with the very low doses of airborne antigens such as pollen.

A frequent IgE-mediated allergic reaction to drugs occurs to penicillin and its relatives. In people with IgE antibodies against penicillin, injection of the drug can cause anaphylaxis and even death. Penicillin acts as a hapten (see

Appendix I, Section A-1); it is a small molecule with a highly reactive β-lactam ring that is crucial for its antibacterial activity. This ring reacts with amino groups on host proteins to form covalent conjugates. When penicillin is ingested or injected, it forms conjugates with self proteins, and the penicillin-modified self peptides provoke a T_H2 response in some individuals. These T_H2 cells then activate penicillin-binding B cells to produce IgE antibody against the penicillin hapten. Thus, penicillin acts both as the B-cell antigen and,

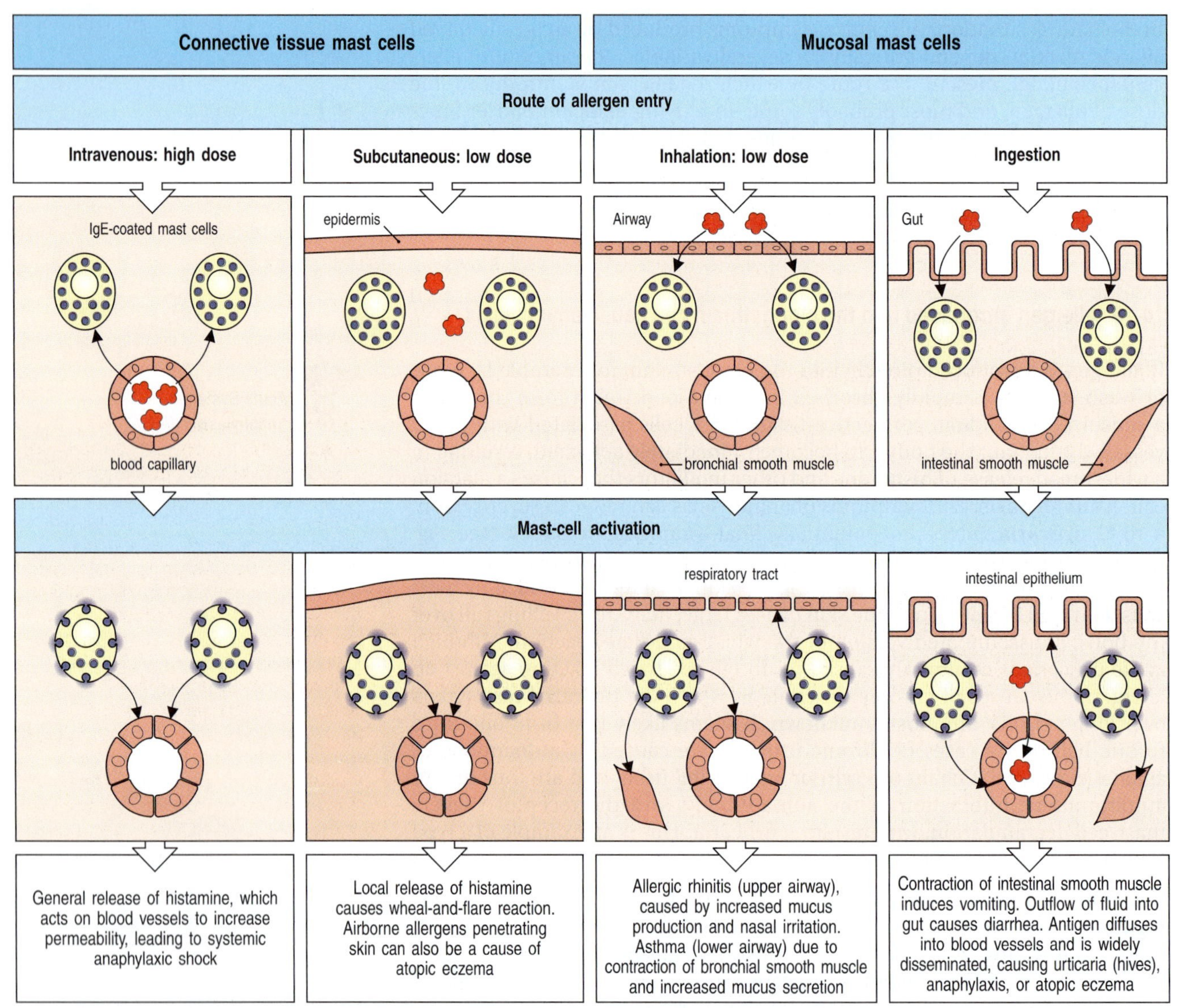

Fig. 14.14 The dose and route of allergen administration determine the type of IgE-mediated allergic reaction that results. There are two main anatomical distributions of mast cells: those associated with vascularized connective tissues, called connective tissue mast cells, and those found in submucosal layers of the gut and respiratory tract, called mucosal mast cells. In an allergic individual, all of these mast cells are loaded with IgE directed against specific allergens. The response to an allergen then depends on which mast cells are activated. Allergen in the bloodstream (intravenous) activates connective tissue mast cells throughout the body, resulting in the systemic release of histamine and other mediators. Entry of allergen through the skin activates local connective tissue mast cells, leading to a local inflammatory reaction. In an experimental skin prick with allergen or after some types of insect bite, this manifests as a wheal-and-flare reaction. In atopic individuals, airborne allergens penetrating the skin may lead to atopic eczema. Inhaled allergen, penetrating respiratory mucosal epithelia, activates mainly mucosal mast cells, causing increased secretion of mucus by the mucosal epithelium and irritation in the nasal mucosa, leading to allergic rhinitis, or to the additional constriction of smooth muscle in the lower airways, leading to asthma. Ingested allergen penetrates the gut epithelium, causing vomiting due to intestinal smooth muscle contraction and diarrhea due to outflow of fluid across the gut epithelium. Food allergens can also be disseminated in the bloodstream, causing a widespread urticaria (hives) when the allergen reaches the skin, or they may cause eczema.

by modifying self peptides, as the T-cell antigen. When penicillin is injected intravenously into an allergic individual, the penicillin-modified proteins can cross-link IgE molecules on tissue mast cells and circulating basophils and thus cause anaphylaxis.

Another drug that is known to provoke anaphylaxis is the anesthetic hexamethonium. Great care should be taken to avoid giving a drug to patients with a past history of allergy to that drug or a close structural relative.

14-11 Allergen inhalation is associated with the development of rhinitis and asthma.

Allergic Asthma

The respiratory tract is an important route of allergen entry (see Fig. 14.14, third panels). Many atopic people react to airborne allergens with an IgE-mediated allergic reaction known as **allergic rhinitis**. This results from the activation of mucosal mast cells beneath the nasal epithelium by allergens such as pollens that release their soluble protein contents, which then diffuse across the mucous membranes of the nasal passages. Allergic rhinitis is characterized by intense itching and sneezing, local edema leading to blocked nasal passages, a nasal discharge, which is typically rich in eosinophils, and irritation of the nasal mucosa as a result of histamine release. A similar reaction to airborne allergens deposited on the conjunctiva of the eye is called **allergic conjunctivitis**. Allergic rhinitis and conjunctivitis are commonly caused by environmental allergens that are present only during certain seasons of the year. For example, hay fever (known clinically as **seasonal allergic rhinoconjunctivitis**) is caused by a variety of allergens, including certain grass and tree pollens. Summer and autumnal symptoms may be caused by weed pollen, such as that of ragweed, or the spores of fungi such as *Alternaria*. Ubiquitous allergens such as cat dander and house dust mites can be a cause of year-round allergic rhinoconjunctivitis.

A more serious IgE-mediated respiratory disease is **allergic asthma**, which is triggered by allergen-induced activation of submucosal mast cells in the lower airways (Fig. 14.15). This leads within seconds to bronchial constriction and an increased secretion of fluid and mucus, making breathing more difficult by trapping inhaled air in the lungs. Patients with allergic asthma usually need treatment, and asthmatic attacks can be life threatening. The same allergens that cause allergic rhinitis and conjunctivitis commonly cause asthma attacks. For example, respiratory arrest caused by severe attacks of asthma in the summer or autumn has been associated with the inhalation of *Alternaria* spores.

Fig. 14.15 The acute response in allergic asthma leads to T_H2-mediated chronic inflammation of the airways. In sensitized individuals, cross-linking of specific IgE on the surface of mast cells by an inhaled allergen triggers them to secrete inflammatory mediators, causing increased vascular permeability, contraction of bronchial smooth muscle, and increased secretion of mucus. There is an influx of inflammatory cells, including eosinophils and T_H2 cells, from the blood. Activated mast cells and T_H2 cells secrete cytokines that augment eosinophil activation and degranulation, which causes further tissue injury and the entry of more inflammatory cells. The result is chronic inflammation, which can cause irreversible damage to the airways.

Acute responses

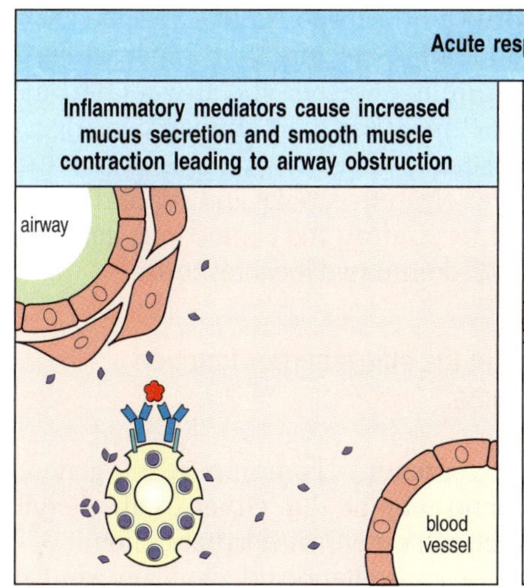
Inflammatory mediators cause increased mucus secretion and smooth muscle contraction leading to airway obstruction
airway
blood vessel

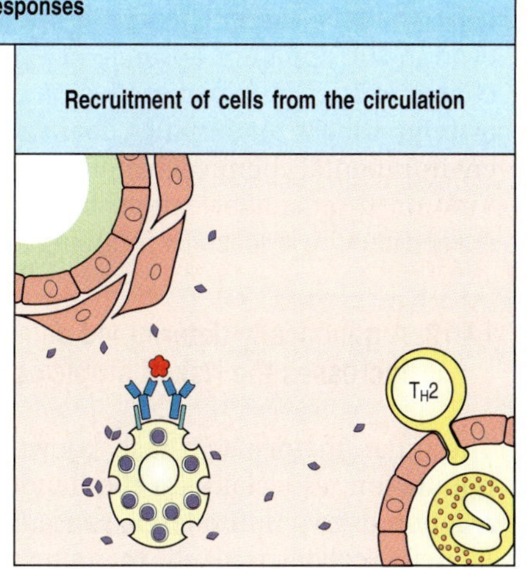
Recruitment of cells from the circulation
T_H2

Chronic response

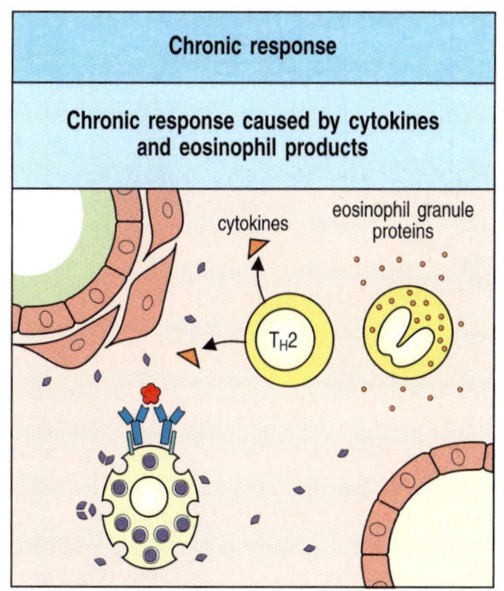
Chronic response caused by cytokines and eosinophil products
cytokines
eosinophil granule proteins
T_H2

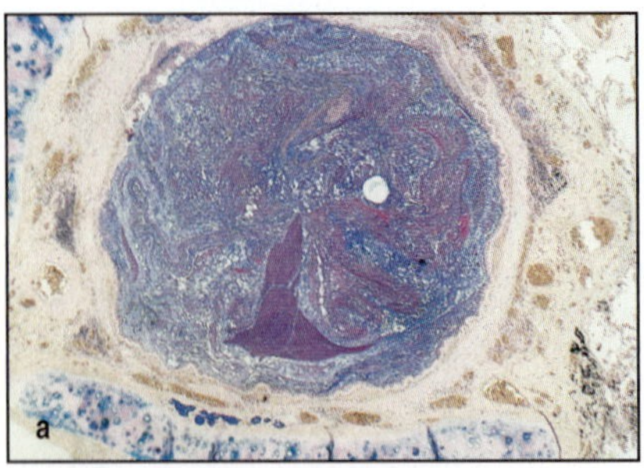

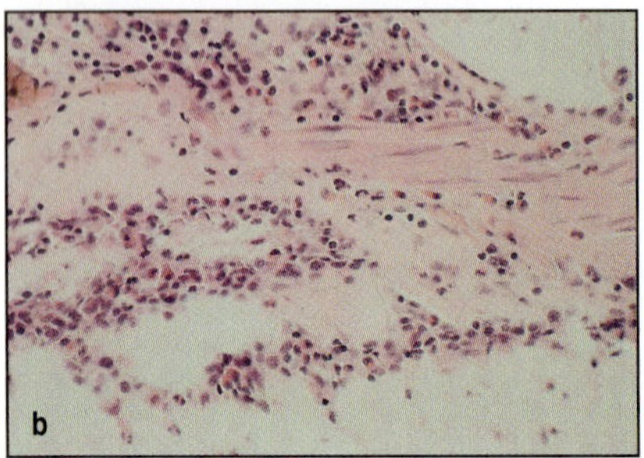

Fig. 14.16 Morphological evidence of chronic inflammation in the airways of an asthmatic patient. Panel a shows a section through a bronchus of a patient who died of asthma; there is almost total occlusion of the airway by a mucus plug. In panel b, a close-up view of the bronchial wall shows injury to the epithelium lining the bronchus, accompanied by a dense inflammatory infiltrate that includes eosinophils, neutrophils, and lymphocytes. Photographs courtesy of T. Krausz.

An important feature of asthma is chronic inflammation of the airways, which is characterized by the continued presence of increased numbers of T_H2 lymphocytes, eosinophils, neutrophils, and other leukocytes (Fig. 14.16). The concerted actions of these cells cause airway remodeling—a thickening of the airway walls due to hyperplasia and hypertrophy of the smooth muscle layer, with the eventual development of fibrosis. This remodeling leads to a permanent narrowing of the airways, and is responsible for many of the clinical manifestations of chronic allergic asthma. In chronic asthmatics, a general hyperreactivity of the airways to nonimmunological stimuli also often develops. Bronchial epithelial cells can produce at least two of the chemokine ligands—CCL5 and CCL11—of the receptor CCR3 expressed on T_H2 cells, macrophages, eosinophils, and basophils. These chemokines enhance the T_H2 response by attracting more T_H2 cells and eosinophils to the damaged lungs. The direct effects of T_H2 cytokines and chemokines on airway smooth muscle and fibroblasts lead to the apoptosis of epithelial cells and airway remodeling, induced in part by the production of TGF-β, which has numerous effects on the epithelium, ranging from inducing apoptosis to stimulating cell proliferation. The direct action of T_H2 cytokines such as IL-9 and IL-13 on airway epithelial cells may also have a dominant role in another major feature of chronic allergic asthma, the induction of goblet-cell metaplasia, which is the increased differentiation of epithelial cells as goblet cells, and a consequent increase in mucus secretion. CD1d-restricted invariant NKT cells (iNKTs, a type of innate-like lymphocyte; see Sections 3-24 and 8-9), also seem to have an important role in the development of airway hyperreactivity, whether allergen-induced or nonspecific. Animal models of asthma have shown that airway hyperreactivity requires the presence of iNKT cells.

Mice do not naturally develop asthma, but a disease resembling human asthma develops in mice that lack the transcription factor T-bet, which is required for T_H1 differentiation (see Section 9-18), and in which T-cell responses are thought to be skewed to T_H2. These mice have increased levels of the T_H2 cytokines IL-4, IL-5, and IL-13, and develop airway inflammation involving lymphocytes and eosinophils (Fig. 14.17). They also develop nonspecific airway hyperreactivity to nonimmunological stimuli, similar to that seen in human asthma. These changes occur in the absence of any exogenous inflammatory stimulus and show that, in extreme circumstances, a genetic imbalance toward T_H2 responses can cause allergic disease. The involvement of eosinophils in asthma seems somewhat different in humans and in mice. In human asthma patients, the number of eosinophils is directly associated with the severity of asthma. In mice deficient in eosinophils, however, the only consistent finding relevant to asthma pathophysiology is a reduction in airway remodeling without a reduction in airway hyperreactivity.

Although allergic asthma is initially driven by a response to a specific allergen, the subsequent chronic inflammation seems to be perpetuated even in the apparent absence of exposure to allergen. The airways become characteristically hyperreactive, and factors other than antigen can trigger asthma attacks. Asthmatics characteristically show hyperresponsiveness to environmental chemical irritants such as cigarette smoke and sulfur dioxide. Viral or, to a smaller extent, bacterial respiratory tract infections can also exacerbate the disease by inducing a T_H2-dominated local response.

14-12 A genetically determined defect in the skin's barrier function increases the risk of atopic eczema.

The inflammatory skin rash known as **eczema** is common in the general population, especially on the hands, and may be due either to an allergic reaction or to a non-allergic cause (such as contact with irritant chemicals). Allergic eczema is likely to represent a constellation of clinically similar

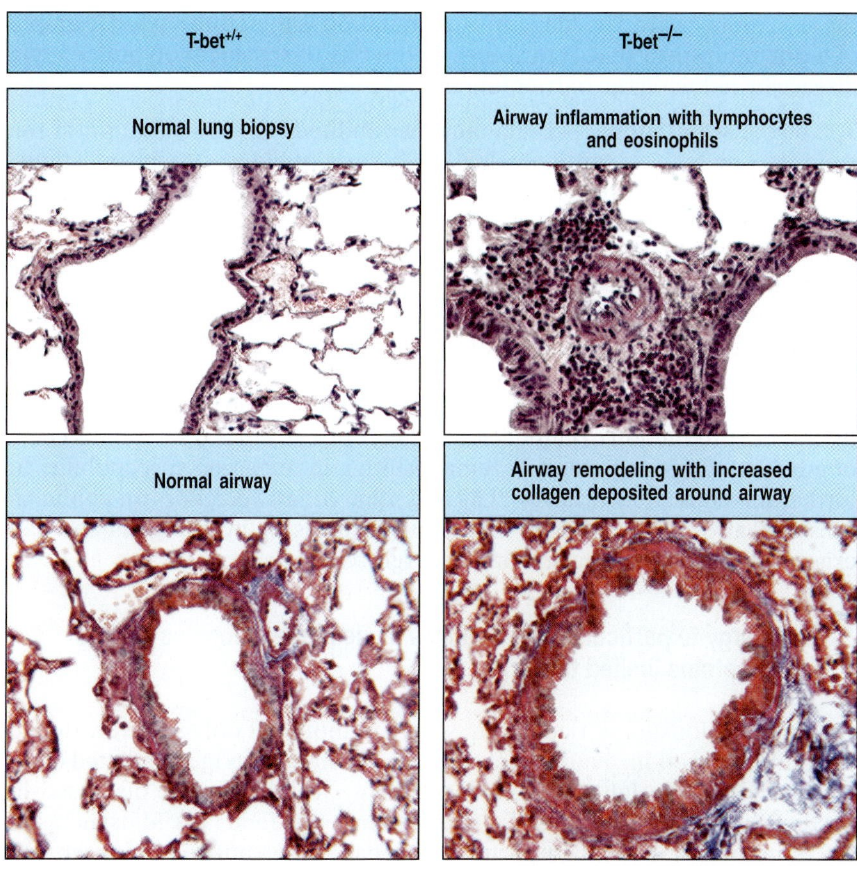

Fig. 14.17 Mice lacking the transcription factor T-bet develop asthma and T-cell responses polarized toward T_H2. T-bet binds to the promoter of the gene encoding IL-2 and is present in T_H1 but not T_H2 cells. Mice with a gene-targeted deletion of T-bet (T-bet$^{-/-}$) developed a spontaneous asthma-like phenotype in the lungs. Left-hand panels: lung and airways in normal mice. Right-hand panels: T-bet-deficient mice showed lung inflammation, with lymphocytes and eosinophils around the airway and blood vessels (top) and airway remodeling with increased collagen around the airway (bottom). Photographs courtesy of L. Glimcher.

conditions with various underlying allergic mechanisms, not all of which predominantly involve IgE. Although allergy is often considered solely in the context of a T_H2 phenotype, both T_H1 and T_H2 cytokines can contribute to the immunopathogenesis. About one-third of patients with eczema show minimal, if any, rise in circulating IgE, and T_H1-cell development is preferentially observed in the lesions of people with a history of persistent disease. However, early-onset and/or chronic eczema occurring in young children, often in response to food antigens such as those in cow's milk, is often the first indication of atopy, and in such cases sensitization to an allergen and IgE-mediated allergic reactions are usually eventually demonstrated. The terms **atopic eczema** or **atopic dermatitis** are now recommended for use only in such cases. Atopic eczema is the result of a chronic inflammatory response with features of tissue remodeling and fibrosis similar to those seen in the bronchial walls of patients with asthma. In both atopic and non-atopic allergic eczema, the apoptosis of keratinocytes induced by T-cell-produced IFN-γ and TNF-α contributes to the pathophysiology.

A recently discovered association between a deficiency in the epidermal protein **filaggrin** and atopic eczema is throwing new light on the condition. Filaggrin binds to keratin fibers in epidermal cells, contributing to the physical barrier at the skin surface that keeps skin waterproof and prevents the entry of airborne allergens. Many patients with atopic eczema have mutations in the filaggrin gene that lead to the nonproduction of filaggrin, and have a 'leaky skin,' which is more liable to become dry owing to increased water loss. A recent survey in Denmark showed a positive association between the filaggrin-null mutation and the occurrence of eczema in atopic patients with a history of the disease, whereas there was no significant association between the occurrence of hand eczema and filaggrin deficiency in non-atopic people. It is proposed that the filaggrin defect facilitates the sensitization of atopic people to airborne allergens by allowing the allergens to penetrate the skin more easily. The filaggrin-null mutation is therefore a good example of a polymorphism that contributes to the risk of sensitization and allergic disease by reducing the barrier function of a tissue.

T_H2 responses in atopic eczema may lead indirectly to exacerbation of the condition by making the individual more susceptible to certain infections. For example, at the time when children were routinely vaccinated against smallpox with a live vaccinia virus vaccine, it was observed that this often led to a disseminated vaccinia infection in the skin of children with atopic eczema; atopic individuals are now excluded from voluntary smallpox vaccination. The antimicrobial peptide cathelicidin (see Chapter 2), which is produced by keratinocytes in reponse to infection, can inhibit vaccinia virus. The spread of the virus in atopic skin seems to be due to the actions of the T_H2 cytokines IL-4 and IL-13, which are overexpressed in atopic eczema and inhibit the production of cathelicidin. Thus, one can envisage a vicious circle of infection triggering atopic eczema resulting in increased susceptibility to further infection. Activation of TLRs and other innate receptors on epithelial cells can also occur either by microbial antigens or by allergens and can exacerbate atopic eczema and other allergic reactions.

14-13 Allergy to particular foods causes systemic reactions as well as symptoms limited to the gut.

Adverse reactions to particular foods are common, but only some are due to an immune reaction. 'Food allergy' can be classified into IgE-mediated allergic reactions, non-IgE-mediated food allergy (celiac disease, discussed in Section 14-18), idiosyncrasies, food intolerance, and food fads. Idiosyncrasies are abnormal responses to particular foods whose cause is unknown but which can provoke symptoms resembling those of an allergic response. Food intolerances are nonimmune adverse reactions due mainly to metabolic deficits, such as intolerance of cow's milk due to the inability to digest lactose.

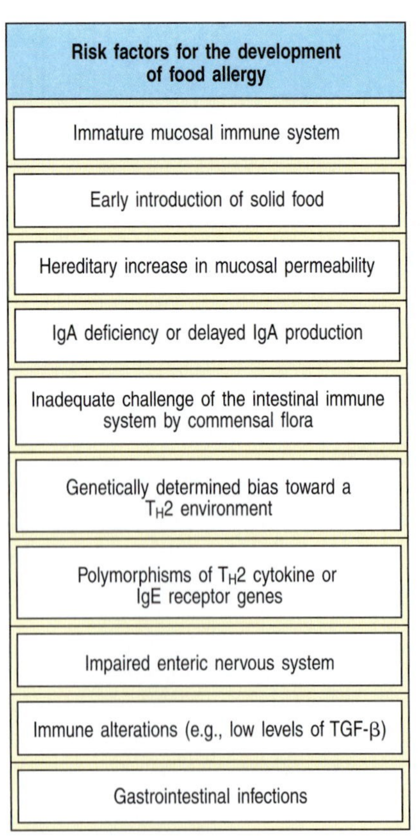

Risk factors for the development of food allergy
Immature mucosal immune system
Early introduction of solid food
Hereditary increase in mucosal permeability
IgA deficiency or delayed IgA production
Inadequate challenge of the intestinal immune system by commensal flora
Genetically determined bias toward a T_H2 environment
Polymorphisms of T_H2 cytokine or IgE receptor genes
Impaired enteric nervous system
Immune alterations (e.g., low levels of TGF-β)
Gastrointestinal infections

Fig. 14.18 Risk factors for the development of food allergy.

IgE-mediated food allergies affect about 1–4% of American and European adults, with allergies being slightly more frequent in children (around 5%). About 25% of this is to peanuts, and peanut allergy is increasing in incidence. Figure 14.18 illustrates the risk factors for the development of IgE-mediated food allergy. IgE-mediated food allergy can manifest itself in a variety of ways, ranging from a swelling of the lips and oral tissue on contact with the allergen, to gastrointestinal symptoms, urticaria, asthma, and in the most severe cases a severe anaphylactic reaction leading to cardiovascular collapse (see Section 14-10). Local gastrointestinal symptoms are due to activation of mucosal mast cells, leading to transepithelial fluid loss and smooth muscle contraction, causing diarrhea and vomiting. Food allergens that subsequently reach the bloodstream can lead to urticaria or to a severe anaphylactic reaction. Certain foods, most importantly peanuts, tree nuts, and shellfish, are particularly associated with severe anaphylaxis. Around 150 deaths occur each year in the United States as a result of a severe allergic reaction to food, with peanut and tree nut allergies accounting for most of the deaths. Peanut allergy is a significant public-health problem, especially in school: children may be unwittingly exposed to peanuts, which are present in many foods.

One of the characteristic features of food allergens is a high degree of resistance to digestion by pepsin in the stomach. This allows them to reach the mucosal surface of the small intestine as intact allergens. Cases of IgE-mediated food allergies arising in previously unaffected adults who were taking antacids or proton-pump inhibitors for ulcers or acid reflux have been proposed to be due to impaired digestion of potential allergens in the less-acid stomach conditions produced by these medications.

14-14 IgE-mediated allergic disease can be treated by inhibiting the effector pathways that lead to symptoms or by desensitization techniques that aim at restoring tolerance to the allergen.

Most of the current drug treatments for allergic disease either treat only the symptoms—examples of such drugs are antihistamines and β-antagonists—or are general anti-inflammatory drugs such as the corticosteroids (Fig. 14.19). Treatment is largely palliative, rather than curative, and often needs to be taken throughout life. Anaphylactic reactions are treated with epinephrine, which stimulates the re-formation of endothelial tight junctions, promotes the relaxation of constricted bronchial smooth muscle, and stimulates the heart. Antihistamines, which block the H_1 receptor, reduce the symptoms that follow the release of histamine from mast cells in allergic rhinoconjunctivitis and IgE-triggered urticaria. In urticaria, for example, the relevant H_1 receptors include those on blood vessels and unmyelinated nerve fibers in the skin.

Treatments for allergic disease

Target step	Mechanism of treatment	Specific approach
In clinical use		
Mediator action	Inhibit effects of mediators on specific receptors Inhibit synthesis of specific mediators	Antihistamines, β-blockers Lipoxygenase inhibitors
Chronic inflammatory reactions	General anti-inflammatory effects	Corticosteroids
T_H2 response	Induction of regulatory T cells	Desensitization therapy by injections of specific antigen
IgE binding to mast cell	Bind to IgE Fc region and prevent IgE binding to Fc receptors on mast cells	Anti-IgE antibodies (omalizumab)
Proposed or under investigation		
T_H2 activation	Induction of regulatory T cells	Injection of specific antigen peptides Administration of cytokines, e.g., IFN-γ, IL-10, IL-12, TGF-β Use of adjuvants such as CpG oligodeoxynucleotides to stimulate T_H1 response
Activation of B cell to produce IgE	Block co-stimulation Inhibit T_H2 cytokines	Inhibit CD40L Inhibit IL-4 or IL-13
Mast-cell activation	Inhibit effects of IgE binding to mast cell	Blockade of IgE receptor
Eosinophil-dependent inflammation	Block cytokine and chemokine receptors that mediate eosinophil recruitment and activation	Inhibit IL-5 Block CCR3

Fig. 14.19 Approaches to the treatment of allergic disease. Examples of treatments in current clinical use for allergic reactions are listed in the top half of the table, with approaches under investigation listed below.

Antileukotriene drugs act as antagonists of leukotriene receptors on smooth muscle, endothelial cells, and mucous-gland cells, and are also used to relieve the symptoms of allergic rhinoconjunctivitis and in asthma. Inhaled bronchodilators that act on β-adrenergic receptors to relax constricted muscle relieve acute asthma attacks. In chronic allergic disease it is extremely important to treat and prevent the chronic inflammatory injury to tissues, and regular use of inhaled corticosteroids is now recommended in persistent asthma to help suppress inflammation. Topical corticosteroids are used to suppress the chronic inflammatory changes seen in eczema.

The treatments noted above have been in use for many years. Ways are still being sought to more precisely and effectively suppress the T-cell response to a given allergen through immunological approaches. Two types of immunotherapy are already used in the clinic. One relies on active immunotherapy—known as **desensitization immunotherapy** or **allergen-specific immunotherapy**—that manipulates the immune response itself. The other relies on passive immunotherapy, such as the blockade of IgE by anti-IgE antibodies. Several other approaches are still in the experimental or clinical trial stage.

In desensitization immunotherapy the aim is to restore tolerance to the allergen by reducing its tendency to induce IgE production. Patients are desensitized by injection with escalating doses of allergen, starting with tiny amounts, an injection schedule that gradually decreases the IgE-dominated response. The mechanisms underlying desensitization therapy are complex, but the key to success seems to be the induction of regulatory T cells secreting IL-10 and/or TGF-β, which skew the response away from IgE production (see Section 14-3). For example, beekeepers exposed to repeated stings are often naturally protected from severe allergic reactions such as anaphylaxis through a mechanism that involves IL-10-secreting T cells. Similarly, specific allergen immunotherapy for sensitivity to insect venom and airborne allergens induces the increased production of IL-10 and in some cases TGF-β, as well as the induction of IgG isotypes, particularly IgG4, an isotype selectively promoted by IL-10. Recent evidence shows that desensitization is also associated with a reduction in the numbers of inflammatory cells at the site of the allergic reaction. A potential complication of the desensitization approach is the risk of inducing an IgE-mediated allergic response, and it is contraindicated in patients with severe asthma. When a patient is allergic to a drug that is essential for their treatment (such as an antibiotic, insulin, or a chemotherapeutic agent), a high-risk procedure called **acute** or **rapid desensitization** is sometimes used to induce temporary tolerance. In this case the drug is introduced in repeated increasing subthreshold doses over a short period of time, from hours to days, until a total cumulative therapeutic dose is reached. The tolerance lasts only for as long as the medication continues.

An alternative, and still experimental, approach to desensitization is vaccination with peptides derived from common allergens. This procedure induces T_{reg} cells, with the accompanying production of IL-10. An IgE-mediated allergic reaction is not induced by the peptide because it is too short to cross-link IgE. A potential difficulty with this approach is that an individual's response to peptide antigens is restricted by their MHC class II alleles (see Section 6-14); patients with different MHC class II molecules therefore respond to different allergen-derived peptides. One possible solution is the use of peptides that contain short sequences with multiple overlapping MHC-binding motifs that would provide coverage for most of the population. A vaccination strategy that shows promise in experimental animal models of allergy is the use of oligodeoxynucleotides rich in unmethylated CpG as adjuvants for desensitization regimes. These oligonucleotides mimic the CpG motifs in bacterial

DNA and strongly promote T_H1 responses, probably through the stimulation of TLR-9 in dendritic cells and the suppression of T_H2 responses. The mechanism of action of adjuvants is discussed in Appendix I, Section A-4.

An alternative approach to manipulating the T-cell response is to administer cytokines that promote T_H1-type responses. IFN-γ, IFN-α, and IL-12 have each been shown to reduce IL-4-stimulated IgE synthesis *in vitro*, and IFN-γ and IFN-α have been shown to reduce IgE synthesis *in vivo*. Administration of IL-12 to patients with mild allergic asthma caused a decrease in the number of eosinophils in blood and sputum but had no effect on immediate or late-phase responses to inhaled allergen. The treatment with IL-12 was accompanied by quite severe flu-like symptoms in most patients, which is likely to limit its possible therapeutic value. Cytokines that enhance the switching of B cells to IgE production are also potential targets for therapy with cytokine inhibitors. Inhibitors of IL-4, IL-5, and IL-13 would be predicted to reduce the production of IgE, but redundancy in the activities of these cytokines might make this approach difficult to implement in practice.

One successful approach to passive immunotherapy has been the development of anti-IgE antibodies that bind the Fc region of free IgE and prevent it from binding to FcεRI. The humanized mouse monoclonal anti-IgE antibody omalizumab is mainly used in cases of chronic allergic asthma in which other treatments have failed to control the disease. In clinical trials, omalizumab treatment reduced circulating IgE levels by more than 95%, which was accompanied by downregulation of the numbers of high-affinity IgE receptors on basophils and mast cells. Omalizumab also seems to exert its therapeutic effect in chronic allergic asthma by reducing IgE-mediated antigen trapping and presentation by dendritic cells, thus preventing the activation of new allergen-specific T_H2 cells.

A further approach to the treatment of allergic disease would be to block the recruitment of eosinophils to sites of allergic inflammation. The eotaxin receptor CCR3 is a potential target in this context. The production of eosinophils in bone marrow and their exit into the circulation might also be reduced by a blockade of IL-5 action. Anti-IL-5 antibody (mepolizumab) is of benefit in treating **hypereosinophilic syndrome**, in which chronic overproduction of eosinophils causes severe organ damage. Clinical trials of anti-IL-5 treatment in asthma, however, show that any beneficial effect is in practice likely to be limited to a small subset of asthma patients with prednisone-dependent eosinophilic asthma, in which IL-5 seems to reduce the number of exacerbations of asthma when the corticosteroid dose is reduced.

Summary.

The allergic response to innocuous antigens reflects the pathophysiological aspects of a defensive immune response whose physiological role is to protect against helminth parasites. It is triggered by the binding of antigen to IgE antibodies bound to the high-affinity IgE receptor FcεRI on mast cells. Mast cells are strategically distributed beneath the mucosal surfaces of the body and in connective tissue. Antigen cross-linking the IgE on their surface causes them to release large amounts of inflammatory mediators. The resulting inflammation can be divided into early events, characterized by short-lived mediators such as histamine, and later events that involve leukotrienes, cytokines, and chemokines, which recruit and activate eosinophils and basophils. This response can evolve into chronic inflammation, characterized by the presence of effector T cells and eosinophils, which is most clearly seen in chronic allergic asthma.

Non-IgE-mediated allergic diseases.

In this part of the chapter we focus on immunological hypersensitivity responses involving IgG antibodies (type II and type III responses in Fig. 14.1) and type IV responses involving antigen-specific T_H1 cells or CD8 T cells (the type IV T_H2 responses that are characteristic of chronic IgE-initiated allergic disease were discussed in the previous part of the chapter). These effector arms of the immune response occasionally react with noninfectious antigens to produce acute or chronic allergic reactions. Although the mechanisms initiating the various forms of hypersensitivity are different, much of the pathology is due to the same immunological effector mechanisms.

14-15 Innocuous antigens can cause type II hypersensitivity reactions in susceptible individuals by binding to the surfaces of circulating blood cells.

Antibody-mediated destruction of red blood cells (hemolytic anemia) or platelets (thrombocytopenia) can be caused by some drugs, including the antibiotics penicillin and cephalosporin. These are examples of **type II hypersensitivity reactions** in which the drug binds to the cell surface and is a target for anti-drug IgG antibodies that cause destruction of the cell (see Fig. 14.1). The anti-drug antibodies are made in only a minority of people, and it is not clear why these individuals make them. The cell-bound antibody triggers the clearance of the cell from the circulation, predominantly by tissue macrophages in the spleen, which bear Fcγ receptors.

14-16 Systemic disease caused by immune-complex formation can follow the administration of large quantities of poorly catabolized antigens.

Type III hypersensitivity reactions can arise with soluble antigens (see Fig. 14.1). The pathology is caused by the deposition of antigen:antibody aggregates, or **immune complexes**, in particular tissues and sites. Immune complexes are generated in all antibody responses, but their pathogenic potential is determined, in part, by their size and by the amount, affinity, and isotype of the responding antibody. Larger aggregates fix complement and are readily cleared from the circulation by the mononuclear phagocyte system. However, the small complexes that form when antigen is in excess tend to be deposited in blood vessel walls. There they can ligate Fc receptors on leukocytes, leading to leukocyte activation and tissue injury.

A local type III hypersensitivity reaction called an **Arthus reaction** (Fig. 14.20) can be triggered in the skin of sensitized individuals who possess IgG antibodies against the sensitizing antigen. When antigen is injected into the skin, circulating IgG antibody that has diffused into the skin forms immune complexes locally. The immune complexes bind Fc receptors such as FcγRIII on mast cells and other leukocytes, generating a local inflammatory response and increased vascular permeability. Fluid and cells, especially polymorphonuclear leukocytes, then enter the site of inflammation from local blood vessels. The immune complexes also activate complement, leading to the production of the complement fragment C5a. This is a key participant in the inflammatory reaction because it interacts with C5a receptors on leukocytes to activate these cells and attract them to the site of inflammation (see Section 2-5). Both C5a and FcγRIII have been shown to be required for the experimental induction of an Arthus reaction in the lung by macrophages in the walls of the alveoli, and they are probably required for the same reaction induced by mast cells in the skin and the linings of joints (synovia).

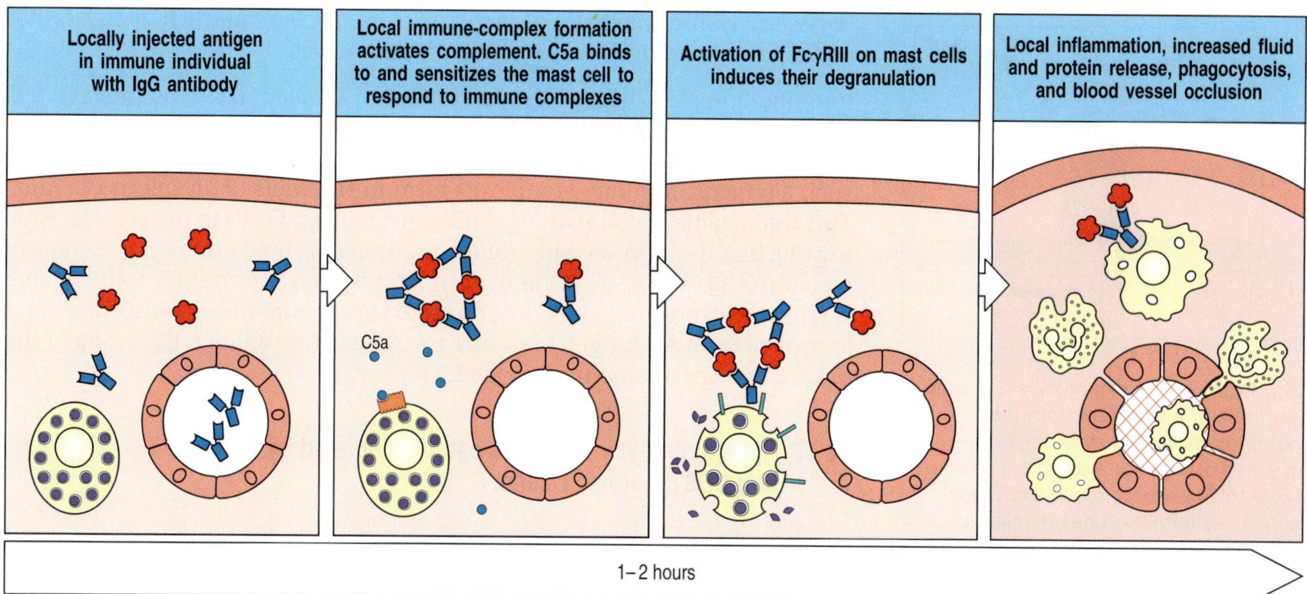

Locally injected antigen in immune individual with IgG antibody
Local immune-complex formation activates complement. C5a binds to and sensitizes the mast cell to respond to immune complexes
Activation of FcγRIII on mast cells induces their degranulation
Local inflammation, increased fluid and protein release, phagocytosis, and blood vessel occlusion
C5a
1–2 hours

Fig. 14.20 The deposition of immune complexes in tissues causes a local inflammatory response known as an Arthus reaction (type III hypersensitivity reaction). In individuals who have already made IgG antibody against an antigen, the same antigen injected into the skin forms immune complexes with IgG antibody that has diffused out of the capillaries. Because the dose of antigen is low, the immune complexes are only formed close to the site of injection, where they activate mast cells bearing Fcγ receptors (FcγRIII). The complement component C5a seems to be important in sensitizing the mast cell to respond to immune complexes. As a result of mast-cell activation, inflammatory cells invade the site, and blood vessel permeability and blood flow are increased. Platelets also accumulate inside the vessel at the site, ultimately leading to vessel occlusion.

A systemic type III hypersensitivity reaction, known as **serum sickness**, can result from the injection of large quantities of a poorly catabolized foreign antigen. This illness was so named because it frequently followed the administration of therapeutic horse antiserum. In the pre-antibiotic era, antiserum made by immunizing horses was often used to treat pneumococcal pneumonia; the specific anti-pneumococcal antibodies in the horse serum would help the patient to clear the infection. In much the same way, **antivenin** (serum from horses immunized with snake venoms) is still used today as a source of neutralizing antibodies to treat people suffering from the bites of poisonous snakes. The increasing use of monoclonal antibodies in the treatment of disease (for example, anti-TNF-α in rheumatoid arthritis) has led to the development of serum sickness in a small minority of patients.

Serum sickness occurs 7–10 days after the injection of horse serum, an interval that corresponds to the time required to mount an IgG-switched primary immune response against the foreign antigens. The clinical features of serum sickness are chills, fever, rash, arthritis, and sometimes glomerulonephritis (inflammation of the glomeruli of the kidneys). Urticaria is a prominent feature of the rash, implying a role for histamine derived from mast-cell degranulation. In this case, the mast-cell degranulation is triggered by the ligation of cell-surface FcγRIII by IgG-containing immune complexes.

The course of serum sickness is illustrated in Fig. 14.21. The onset of disease coincides with the development of antibodies against the abundant soluble proteins in the foreign serum; these antibodies form immune complexes with their antigens throughout the body. These immune complexes fix complement and can bind to and activate leukocytes bearing Fc and complement receptors; these in turn cause widespread tissue damage. The formation of immune complexes causes clearance of the foreign antigen, so serum sickness is usually a self-limiting disease. Serum sickness after a second dose of antigen follows the kinetics of a secondary antibody response (see Section 10-14), with symptoms typically appearing within a day or two.

Pathological immune-complex deposition is seen in other situations in which antigen persists. One is when an adaptive antibody response fails to clear the infecting pathogen, as occurs in subacute bacterial endocarditis or chronic viral hepatitis. In these situations, the replicating pathogen is continuously generating new antigen in the presence of a persistent antibody response, with the consequent formation of abundant immune complexes. These are

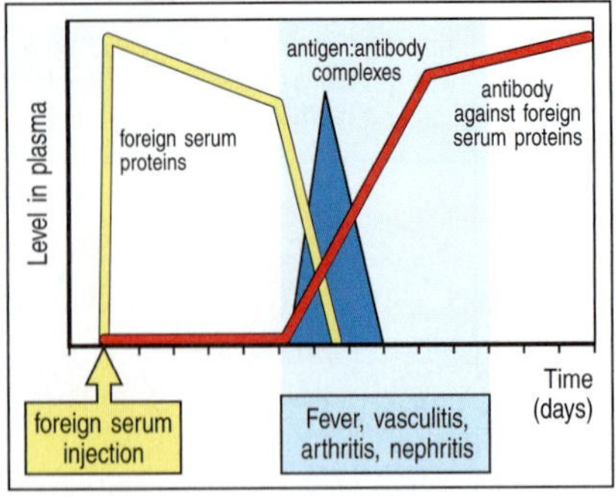

Fig. 14.21 Serum sickness is a classic example of a transient immune complex-mediated syndrome. An injection of a foreign protein or proteins leads to an antibody response. These antibodies form immune complexes with the circulating foreign proteins. The complexes are deposited in small vessels and activate complement and phagocytes, inducing fever and the symptoms of vasculitis, nephritis, and arthritis. All these effects are transient and resolve when the foreign protein is cleared.

deposited within small blood vessels, with consequent injury in many tissues and organs, including the skin, kidneys, and nerves.

Immune-complex disease also occurs when inhaled allergens provoke IgG rather than IgE antibody responses, perhaps because they are present at relatively high levels in the air. When a person is reexposed to high doses of such allergens, immune complexes form in the walls of alveoli in the lung. This leads to the accumulation of fluid, protein, and cells in the alveolar wall, slowing blood–gas interchange and compromising lung function. This type of reaction is more likely to occur in occupations such as farming, in which there is repeated exposure to hay dust or mold spores, and the resulting disease is known as **farmer's lung**. If exposure to antigen is sustained, the lining of the lungs can be permanently damaged.

14-17 Hypersensitivity reactions can be mediated by T_H1 cells and CD8 cytotoxic T cells.

Unlike the immediate hypersensitivity reactions (type I and type II), which are mediated by antibodies, **type IV hypersensitivity reactions** or **delayed-type hypersensitivity** are mediated by antigen-specific effector T cells. We have already seen the involvement of T_H2 effector cells and the cytokines they produce in the chronic response in IgE-initiated allergic reactions. Here we consider the allergic diseases caused by type IV hypersensitivity reactions mediated by T_H1 and CD8 cytotoxic T cells (Fig. 14.22). These cells function in essentially the same way as they do in response to a pathogen, as described in Chapter 9, and the responses can be transferred between experimental animals by purified T cells or cloned T-cell lines. Much of the chronic inflammation seen in some of the allergic diseases described earlier is due to type IV hypersensitivity reactions mediated by antigen-specific T_H1 cells as well as by T_H2 cells.

The prototypic delayed-type hypersensitivity reaction is the tuberculin test (see Appendix I, Section A-38). This is a T_H1-type IV hypersensitivity reaction (see Fig. 14.1) that is used to determine whether an individual has previously been infected with *M. tuberculosis*. In the Mantoux test, small amounts of tuberculin—a complex mixture of peptides and carbohydrates derived from *M. tuberculosis*—are injected intradermally. In people who have been exposed to the bacterium, either by infection or by immunization with the BCG vaccine (an attenuated form of *M. tuberculosis*), a local T cell-mediated

Type IV hypersensitivity reactions are mediated by antigen-specific effector T cells

Syndrome	Antigen	Consequence
Delayed-type hypersensitivity	Proteins: Insect venom Mycobacterial proteins (tuberculin, lepromin)	Local skin swelling: Erythema Induration Cellular infiltrate Dermatitis
Contact hypersensitivity	Haptens: Pentadecacatechol (poison ivy) DNFB Small metal ions: Nickel Chromate	Local epidermal reaction: Erythema Cellular infiltrate Vesicles Intraepidermal abscesses
Gluten-sensitive enteropathy (celiac disease)	Gliadin	Villous atrophy in small bowel Malabsorption

Fig. 14.22 Type IV hypersensitivity responses. These reactions are mediated by T cells and all take some time to develop. They can be grouped into three syndromes, according to the route by which antigen passes into the body. In delayed-type hypersensitivity the antigen is injected into the skin; in contact hypersensitivity it is absorbed into the skin; and in gluten-sensitive enteropathy it is absorbed by the gut. DNFB, dinitrofluorobenzene.

inflammatory reaction evolves over 24–72 hours. The response is caused by T_H1 cells, which enter the site of antigen injection, recognize complexes of peptide:MHC class II molecules on antigen-presenting cells, and release inflammatory cytokines such as IFN-γ and TNF-β. These stimulate the expression of adhesion molecules on endothelium and increase local blood vessel permeability, allowing plasma and accessory cells to enter the site, thus causing a visible swelling (Fig. 14.23). Each of these phases takes several hours and so the fully developed response only appears 24–48 hours after challenge. The cytokines produced by the activated T_H1 cells and their actions are shown in Fig. 14.24.

Very similar reactions are observed in **allergic contact dermatitis**, which is an immune-mediated local inflammatory reaction in the skin caused by direct skin contact with certain antigens—or sometimes by oral uptake of the antigen, when it is known as **systemic allergic contact dermatitis**. It is important to note that not all contact dermatitis is immune-mediated and allergic in nature; it can also be caused by direct damage to the skin by irritant or toxic chemicals.

Allergic contact dermatitis can be caused by the activation of CD4 or CD8 T cells, depending on the pathway by which the antigen is processed. Typical antigens that cause allergic contact dermatitis are highly reactive small molecules that can easily penetrate intact skin, especially if they cause itching that leads to scratching. These chemicals then react with self proteins, creating hapten:protein complexes that can be processed to hapten:peptide complexes capable of being presented by MHC molecules and recognized by T cells as foreign antigens. As with other allergic responses, there are two phases to a cutaneous allergic response: sensitization and elicitation. During the sensitization phase, Langerhans cells (dendritic cells) in the skin take up and process antigen, and migrate to regional lymph nodes, where they activate T cells (see Fig. 9.13) with the consequent production of memory T cells, which end up in the dermis. In the elicitation phase, a subsequent exposure to the sensitizing chemical leads to antigen presentation to memory T cells in the dermis, with the release of T-cell cytokines such as IFN-γ and IL-17. This stimulates the keratinocytes of the epidermis to release IL-1, IL-6, TNF-α, GM-CSF, the chemokine CXCL8, and the interferon-inducible chemokines CXCL11 (IP-9), CXCL10 (IP-10), and CXCL9 (Mig; monokine induced by IFN-γ). These cytokines and chemokines enhance the inflammatory response by inducing the migration of monocytes into the lesion and their maturation into macrophages, and by attracting more T cells (Fig. 14.25).

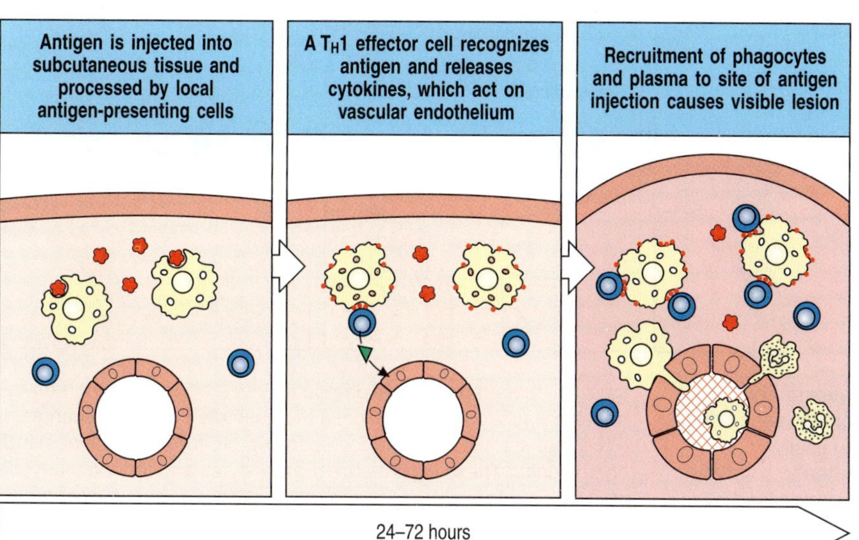

Antigen is injected into subcutaneous tissue and processed by local antigen-presenting cells
A T_H1 effector cell recognizes antigen and releases cytokines, which act on vascular endothelium
Recruitment of phagocytes and plasma to site of antigen injection causes visible lesion
24–72 hours

Fig. 14.23 The stages of a delayed-type hypersensitivity reaction. The first phase involves uptake, processing, and presentation of the antigen by local antigen-presenting cells. In the second phase, T_H1 cells that have been primed by a previous exposure to the antigen migrate into the site of injection and become activated. Because these specific cells are rare, and because there is little inflammation to attract cells into the site, it can take several hours for a T cell of the correct specificity to arrive. These cells release mediators that activate local endothelial cells, recruiting an inflammatory cell infiltrate dominated by macrophages and causing the accumulation of fluid and protein. At this point the lesion becomes apparent.

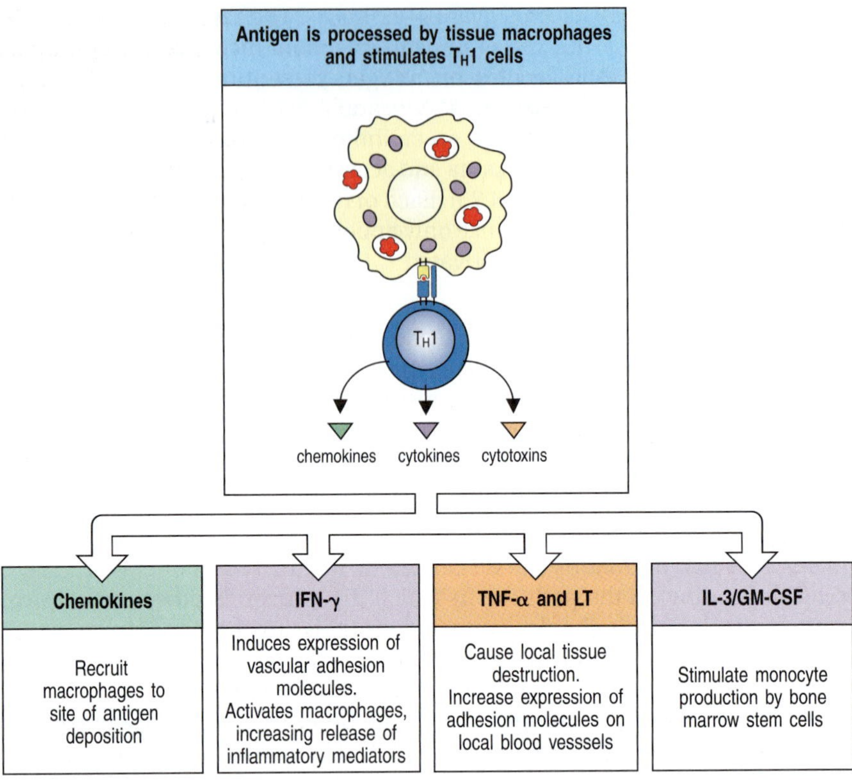

Fig. 14.24 The delayed-type (type IV) hypersensitivity response is directed by chemokines and cytokines released by antigen-stimulated T_H1 cells. Antigen in the local tissues is processed by antigen-presenting cells and presented on MHC class II molecules. Antigen-specific T_H1 cells that recognize the antigen locally at the site of injection release chemokines and cytokines that recruit macrophages to the site of antigen deposition. Antigen presentation by the newly recruited macrophages then amplifies the response. T cells can also affect local blood vessels through the release of TNF-α and lymphotoxin (LT), and stimulate the production of macrophages through the release of IL-3 and GM-CSF. T_H1 cells activate macrophages through the release of IFN-γ and TNF-α, and kill macrophages and other sensitive cells through the cell-surface expression of the Fas ligand.

The rash produced by contact with the poison ivy plant (Fig. 14.26) is a common example of allergic contact dermatitis in the United States and is caused by a CD8 T-cell response to urushiol oil (a mixture of pentadecacatechols) in the plant. These chemicals are lipid-soluble and so can cross the cell membrane and attach to intracellular proteins. The modified proteins generate modified peptides within the cytosol, which are translocated into

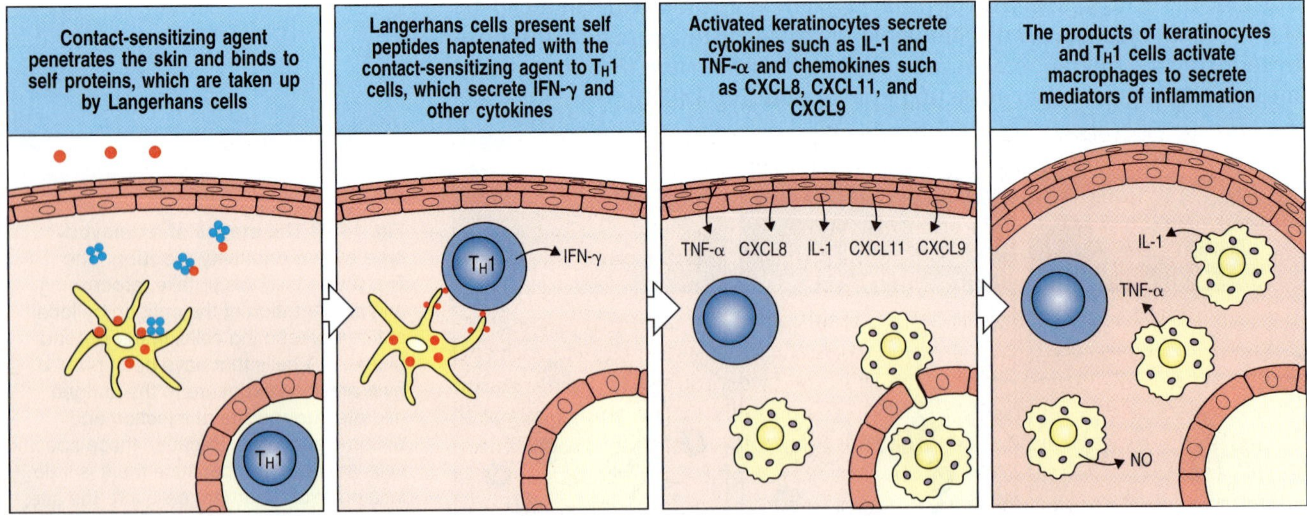

Fig. 14.25 Elicitation of a delayed-type hypersensitivity response to a contact-sensitizing agent. A contact-sensitizing agent is a small highly reactive molecule that can easily penetrate intact skin. It binds covalently as a hapten to a variety of endogenous proteins, which are taken up and processed by Langerhans cells, the major antigen-presenting cells of skin. These present haptenated peptides to effector T_H1 cells (which must have been previously primed in lymph nodes and then have traveled back to the skin). These then secrete cytokines such as IFN-γ that stimulate keratinocytes to secrete further cytokines and chemokines. These in turn attract monocytes and induce their maturation into activated tissue macrophages, which contribute to the inflammatory lesions depicted in Fig. 14.26. NO, nitric oxide.

the endoplasmic reticulum and delivered to the cell surface bound to MHC class I molecules. CD8 T cells recognizing the peptides cause damage either by killing the eliciting cell or by secreting cytokines such as IFN-γ. The well-studied chemical picryl chloride produces a CD4 T-cell allergic contact dermatitis. Picryl chloride modifies extracellular self proteins, which are then processed into modified self peptides that bind to self-MHC class II molecules and are recognized by T_H1 cells. When sensitized T_H1 cells recognize these complexes, they produce extensive inflammation by activating macrophages (see Fig. 14.25).

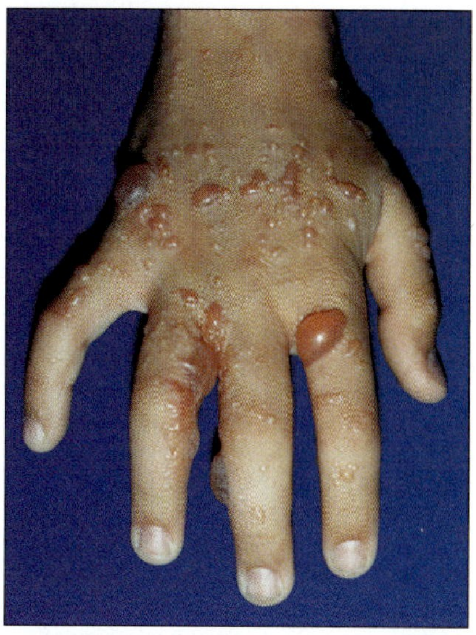

Fig. 14.26 Blistering skin lesions on the hand of a patient with allergic contact dermatitis caused by poison ivy. Photograph courtesy of R. Geha.

Some insect proteins also elicit a delayed-type hypersensitivity response. One example of this in the skin is a severe reaction to mosquito bites. Instead of a small itchy bump, people allergic to proteins in mosquito saliva can develop an immediate reaction such as urticaria and swelling or, much more rarely, anaphylactic shock (see Section 14-10). Some allergic individuals subsequently develop a delayed reaction to a bite in which the whole affected limb swells up.

Important delayed-type hypersensitivity responses to divalent cations such as nickel have also been observed. These divalent cations can alter the conformation or the peptide binding of MHC class II molecules, and thus provoke a T-cell response. In humans, nickel can also bind to the receptor TLR-4 and produce a pro-inflammatory signal. Sensitization to nickel is widespread as a result of prolonged contact with nickel-containing items such as jewelry, buttons, and clothing fasteners, but some countries now have standards that specify that such products must have non-nickel coatings or potentially release very low amounts of the metal, and this is reducing the prevalence of nickel allergy in those countries.

Finally, although this section has focused on the role of T_H1 and cytotoxic T cells in inducing delayed-type hypersensitivity reactions, there is evidence that antibody and complement might also have a role. Mice deficient in B cells, antibody, or complement show impaired contact hypersensitivity reactions. In particular, IgM antibodies (produced in part by B1 cells), which activate the complement cascade, facilitate the initiation of these reactions.

14-18 Celiac disease has features of both an allergic response and autoimmunity.

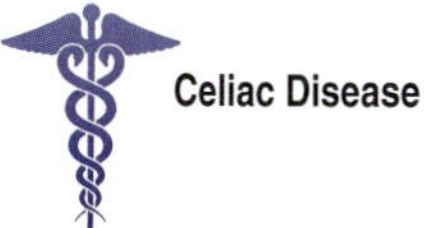

Celiac disease is a chronic condition of the upper small intestine caused by an immune response directed at gluten, a complex of proteins present in wheat, oats, and barley. Elimination of gluten from the diet restores normal gut function, but must be continued throughout life. The pathology of celiac disease is characterized by the loss of the slender, finger-like villi formed by the intestinal epithelium (a condition termed villous atrophy), together with an increase in the size of the sites in which epithelial cells are renewed (crypt hyperplasia) (Fig. 14.27). These pathological changes result in the loss of the mature epithelial cells that cover the villi and which normally absorb and digest food, and is accompanied by severe inflammation of the intestinal wall, with increased numbers of T cells, macrophages, and plasma cells in the lamina propria, as well as increased numbers of lymphocytes in the epithelial layer. Gluten seems to be the only food protein that provokes intestinal inflammation in this way, a property that reflects gluten's ability to stimulate both innate and specific immune responses in genetically susceptible individuals.

Celiac disease shows an extremely strong genetic predisposition, with more than 95% of patients expressing the HLA-DQ2 class II MHC allele, and there is an 80% concordance in monozygotic twins (that is, if one twin develops it, there is an 80% probability that the other will) but only a 10% concordance in dizygotic twins. Nevertheless, most individuals expressing HLA-DQ2 do not

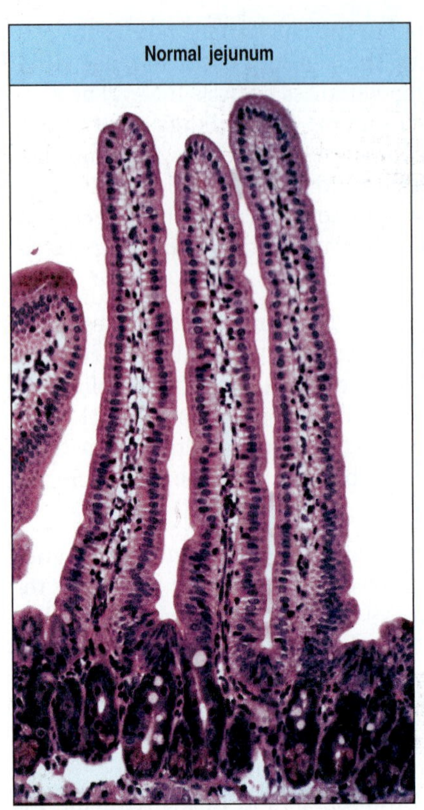

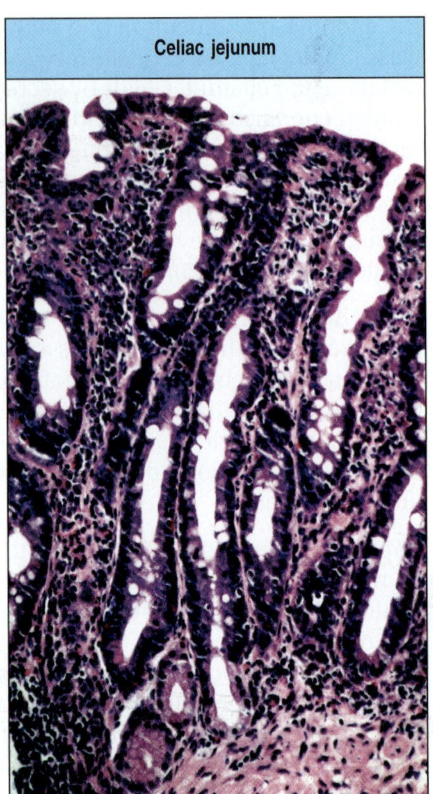

Fig. 14.27 The pathological features of celiac disease. Left: the surface of the normal small intestine is folded into finger-like villi, which provide an extensive surface for nutrient absorption. Right: the local immune response against the food protein α-gliadin provokes destruction of the villi. In parallel, there is lengthening and increased mitotic activity in the underlying crypts where new epithelial cells are produced. There is also a marked inflammatory infiltrate in the intestinal mucosa, with increased numbers of lymphocytes in the epithelial layer and accumulation of CD4 T cells, plasma cells, and macrophages in the deeper layer, the lamina propria. Because the villi contain all the mature epithelial cells that digest and absorb foodstuffs, their loss results in life-threatening malabsorption and diarrhea. Photographs courtesy of Allan Mowat.

develop celiac disease despite the almost universal presence of gluten in the Western diet. Thus, other genetic factors must make important contributions to susceptibility.

Most evidence indicates that celiac disease requires the aberrant priming of IFN-γ-producing CD4 T cells by antigenic peptides present in α-gliadin, one of the major proteins in gluten. It is generally accepted that only a limited number of peptides can provoke an immune response leading to celiac disease. This is likely to be due to the unusual structure of the peptide-binding groove of the HLA-DQ2 molecule. The key step in the immune recognition of α-gliadin is the deamidation of its peptides by the enzyme tissue transglutaminase (tTG), which converts selected glutamine residues to negatively charged glutamic acid. Only peptides containing negatively charged residues in certain positions bind strongly to HLA-DQ2, and thus the transamination reaction promotes the formation of peptide:HLA-DQ2 complexes, which can activate antigen-specific CD4 T cells (Fig. 14.28). Multiple peptide epitopes can be generated from gliadin. Activated gliadin-specific CD4 T cells accumulate in the lamina propria, producing IFN-γ, a cytokine that leads to intestinal inflammation.

Celiac disease is entirely dependent on the presence of a foreign antigen (gluten) and is not associated with a specific immune response against antigens in the tissue—the intestinal epithelium—that is damaged during the immune response. But it does have some features of autoimmunity. Autoantibodies against tissue transglutaminase are found in all patients with celiac disease; indeed, the presence of serum IgA antibodies against this enzyme is used as a sensitive and specific test for the disease. Interestingly, no tTG-specific T cells have been found, and it has been proposed that gluten-reactive T cells provide help to B cells reactive to tissue transglutaminase. In support of this hypothesis, gluten can complex with the enzyme and therefore could be taken up by tTG-reactive B cells (Fig. 14.29). There is no evidence that these autoantibodies contribute to tissue damage, however.

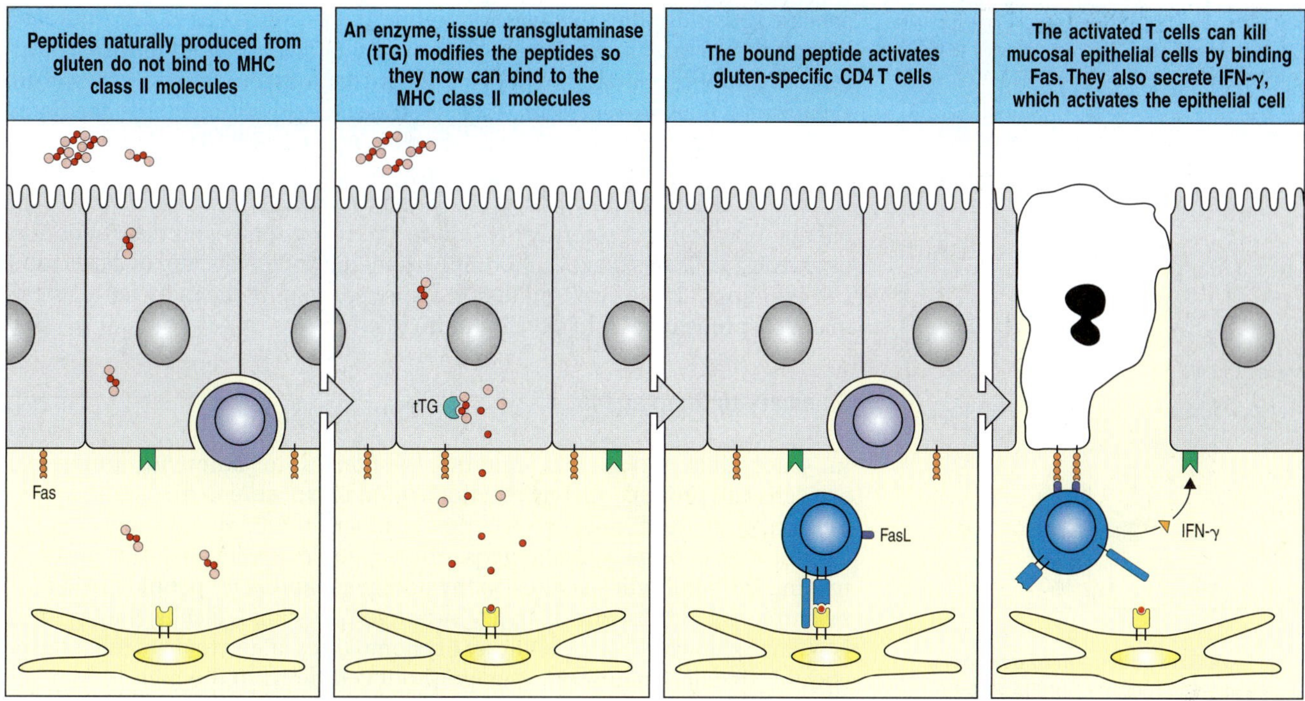

Peptides naturally produced from gluten do not bind to MHC class II molecules
An enzyme, tissue transglutaminase (tTG) modifies the peptides so they now can bind to the MHC class II molecules
The bound peptide activates gluten-specific CD4 T cells
The activated T cells can kill mucosal epithelial cells by binding Fas. They also secrete IFN-γ, which activates the epithelial cell
Fas
tTG
FasL
IFN-γ

Fig. 14.28 Molecular basis of immune recognition of gluten in celiac disease. After the digestion of gluten by gut digestive enzymes, deamidation of epitopes by tissue transglutaminase leads to their binding to HLA-DQ molecules and priming of the immune system.

Chronic T-cell responses against food proteins are normally prevented by the development of oral tolerance (see Section 12-13). Why this breaks down in patients with celiac disease is unknown. The properties of the HLA-DQ2 molecule provide a partial explanation, but there must be additional factors because most HLA-DQ2-positive individuals do not develop celiac disease, and the high concordance rates in monozygotic twins indicate a role for additional genetic factors. Polymorphisms in the gene for CTLA-4 or in other immunoregulatory genes may be associated with susceptibility. There could also be differences in how individuals digest gliadin in the intestine, so that differing amounts survive for deamidation and presentation to T cells.

The gluten protein also seems to have several properties that contribute to pathogenesis. As well as its relative resistance to digestion, there is mounting evidence that some gliadin-derived peptides stimulate the innate immune system by inducing the release of IL-15 by intestinal epithelial cells. This process is antigen-nonspecific and involves peptides that cannot be bound by HLA-DQ2 molecules or recognized by CD4 T cells. IL-15 release leads to the activation of dendritic cells in the lamina propria, as well as the upregulation of MIC-A expression by epithelial cells. CD8 T cells in the mucosal epithelium can be activated via their NKG2D receptors, which recognize MIC-A, and they can kill MIC-A-expressing epithelial cells via these same NKG2D receptors (Fig. 14.30). Triggering of these innate immune responses by α-gliadin may create some intestinal damage on its own and also induce some of the co-stimulatory events necessary for initiating an antigen-specific CD4 T-cell response to other parts of the α-gliadin molecule. The ability of gluten to stimulate both innate and adaptive immune responses may thus explain its unique ability to induce celiac disease.

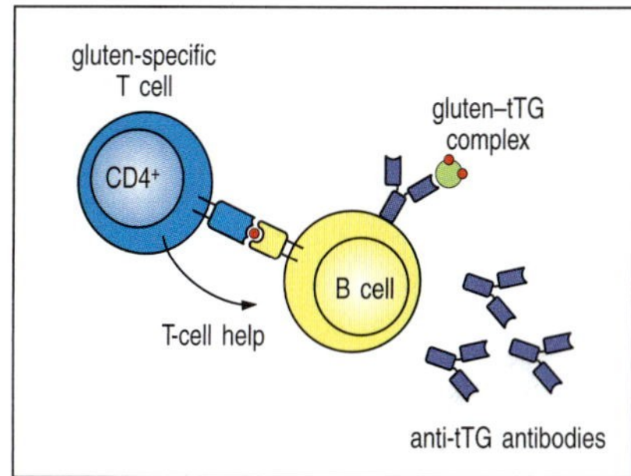

Fig. 14.29 A hypothesis to explain antibody production against tissue transglutaminase (tTG) in the absence of T cells specific for tTG in celiac patients. tTG-reactive B cells endocytose gluten–tTG complexes and present gluten peptides to the gluten-specific T cells. The stimulated T cells can now provide help to these B cells, which produce autoantibodies against tTG.

Summary.

Non-IgE-mediated immunological hypersensitivity also reflects normal immune mechanisms that are inappropriately directed against innocuous antigens or inflammatory stimuli. It comprises both immediate-type and delayed-type reactions. Immediate-type reactions are due to the binding of

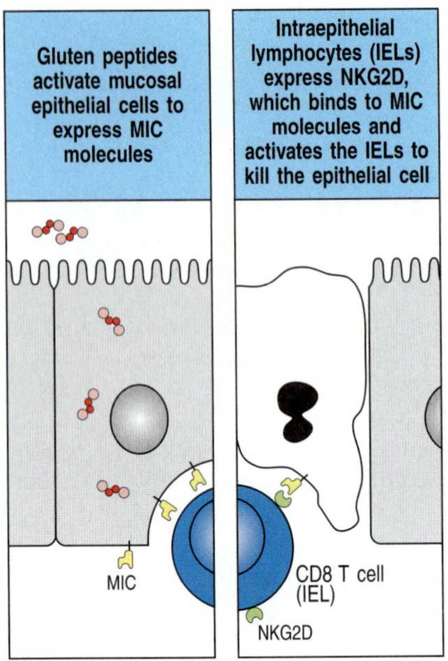
Gluten peptides activate mucosal epithelial cells to express MIC molecules
Intraepithelial lymphocytes (IELs) express NKG2D, which binds to MIC molecules and activates the IELs to kill the epithelial cell
MIC
CD8 T cell (IEL)
NKG2D

Fig. 14.30 The activation of cytotoxic T cells by the innate immune system in celiac disease. Gluten peptides can induce the expression of the MHC class Ib molecules MIC-A and MIC-B on gut epithelial cells. Intraepithelial lymphocytes (IELs), many of which are CD8 cytotoxic T cells, recognize these proteins via the receptor NKG2D, which activates the IELs to kill the MIC-bearing cells, leading to destruction of the gut epithelium.

specific IgG antibodies to allergen-modified cell surfaces, as in drug-induced hemolytic anemia (a type II reaction), or to the formation of immune complexes of antibodies bound to poorly catabolized antigens, as occurs in serum sickness (a type III reaction). Type IV hypersensitivity reactions mediated by T_H1 cells and cytotoxic T cells develop more slowly. The T_H1-mediated hypersensitivity reaction in the skin provoked by mycobacterial tuberculin is used to diagnose previous exposure to *Mycobacterium tuberculosis*. The allergic reaction to poison ivy is due to the recognition and destruction by cytotoxic T cells of skin cells modified by a plant molecule, and to cytotoxic T cell cytokines. These T cell-mediated responses require the induced synthesis of effector molecules and develop more slowly.

Summary to Chapter 14.

In susceptible individuals, immune responses to otherwise innocuous antigens can produce allergic reactions upon reexposure to the same antigen. Most allergic reactions involve the production of IgE antibody against common environmental allergens. Some people are intrinsically prone to making IgE antibodies against many allergens, and such people are said to be atopic. IgE production is driven by antigen-specific T_H2 cells; the response is polarized toward T_H2 by an array of chemokines and cytokines that engage specific signaling pathways. The IgE produced binds to the high-affinity IgE receptor FcεRI on mast cells and basophils. Specific effector T cells, mast cells, and eosinophils, in combination with T_H1 and T_H2 cytokines and chemokines, orchestrate chronic allergic inflammation, which is the major cause of the chronic morbidity of asthma. Failure to regulate these responses can occur at many levels of the immune system, including defects in regulatory T cells. Antibodies of other isotypes and antigen-specific effector T cells contribute to allergic hypersensitivity to other antigens.

Questions.

14.1 *List three allergic reactions that involve IgE and three that involve other mechanisms.*

14.2 *Describe how a person becomes sensitized to an allergen. Discuss the factors predisposing to the production of IgE.*

14.3 *What are the key features that differentiate acute and chronic allergic reactions?*

14.4 *How can the innate immune system contribute to allergy? How do infectious agents modulate allergy?*

14.5 *Which types of white blood cell participate in allergic responses, and what do they do?*

14.6 *Describe how an ingested food allergen can give rise to the allergic skin reaction urticaria.*

14.7 *How does desensitization therapy work?*

General references.

Johansson, S.G., Bieber, T., Dahl, R., Friedmann, P.S., Lanier, B.Q., Lockey, R.F., Motala, C., Ortega Martell, J.A., Platts-Mills, T.A., Ring, J., *et al.*: **Revised nomenclature for allergy for global use: report of the Nomenclature Review Committee of the World Allergy Organization, October 2003.** *J. Allergy Clin. Immunol.* 2004, **113**:832–836.

Kay, A.B.: *Allergy and Allergic Diseases.* Oxford, Blackwell Science, 1997.

Kay, A.B.: **Allergy and allergic diseases. First of two parts.** *N. Engl. J. Med.* 2001, **344**:30–37.

Kay, A.B.: **Allergy and allergic diseases. Second of two parts.** *N. Engl. J. Med.* 2001, **344**:109–113.

Kay, A.B.: **The role of T lymphocytes in asthma.** *Chem. Immunol. Allergy* 2006, **91**:59–75.

Maddox, L., and Schwartz, D.A.: **The pathophysiology of asthma.** *Annu. Rev. Med.* 2002, **53**:477–498.

Papageorgiou, P.S.: **Clinical aspects of food allergy.** *Biochem. Soc. Trans.* 2002, **30**:901–906.

Rosen, F.S.: **Urticaria, angioedema, and anaphylaxis.** *Pediatr. Rev.* 1992, **13**:387–390.

Section references.

14-1 Sensitization involves class switching to IgE production on first contact with an allergen.

Akuthota, P., Wang, H., and Weller, P.F.: **Eosinophils as antigen-presenting cells in allergic upper airway disease.** *Curr. Opin. Allergy Clin. Immunol.* 2010, **10**:14–19.

Bieber T: **The pro- and anti-inflammatory properties of human antigen-presenting cells expressing the high affinity receptor for IgE (FcεRI).** *Immunobiology* 2007, **212**:499–503.

Chen, Z., Lund, R., Aittokallio, T., Kosonen, M., Nevalainen, O., and Lahesmaa, R.: **Identification of novel IL-4/Stat6-regulated genes in T lymphocytes.** *J. Immunol.* 2003, **171**:3627–3635.

Geha, R.S., Jabara, H.H., and Brodeur, S.R.: **The regulation of immunoglobulin E class-switch recombination.** *Nat. Rev. Immunol.* 2003, **3**:721–732.

Holt, P.G.: **The role of airway dendritic cell populations in regulation of T-cell responses to inhaled antigens: atopic asthma as a paradigm.** *J. Aerosol Med.* 2002, **15**:161–168.

Mikhak, Z., and Luster, A.D.: **The emergence of basophils as antigen-presenting cells in Th2 inflammatory responses.** *J. Mol. Cell Biol.* 2009, **1**:69–71.

Romagnani, S.: **Cytokines and chemoattractants in allergic inflammation.** *Mol. Immunol.* 2002, **38**:881–885.

Spencer, L.A., and Weller, P.F.: **Eosinophils and Th2 immunity: contemporary insights.** *Immunol. Cell Biol.* 2010, **88**:250–256.

Urban, J.F., Jr, Noben-Trauth, N., Donaldson, D.D., Madden, K.B., Morris, S.C., Collins, M., and Finkelman, F.D.: **IL-13, IL-4Rα, and Stat6 are required for the expulsion of the gastrointestinal nematode parasite *Nippostrongylus brasiliensis.*** *Immunity* 1998, **8**:255–264.

14-2 Allergens are usually delivered transmucosally at low dose, a route that favors IgE production.

Grunstein, M.M., Veler, H., Shan, X., Larson, J., Grunstein, J.S., and Chuang, S.: **Proasthmatic effects and mechanisms of action of the dust mite allergen, Der p 1, in airway smooth muscle.** *J. Allergy Clin. Immunol.* 2005, **116**:94–101.

Kauffman, H.F., Tomee, J.F., van de Riet, M.A., Timmerman, A.J., and Borger, P.: **Protease-dependent activation of epithelial cells by fungal allergens leads to morphologic changes and cytokine production.** *J. Allergy Clin. Immunol.* 2000, **105**:1185–1193.

Nordlee, J.A., Taylor, S.L., Townsend, J.A., Thomas, L.A., and Bush, R.K.: **Identification of a Brazil-nut allergen in transgenic soybeans.** *N. Engl. J. Med.* 1996, **334**:688–692.

Pease, J.E.: **Asthma, allergy and chemokines.** *Curr. Drug Targets* 2006, **7**:3–12.

Sehgal, N., Custovic, A., and Woodcock, A.: **Potential roles in rhinitis for protease and other enzymatic activities of allergens.** *Curr. Allergy Asthma Rep.* 2005, **5**:221–226.

Sprecher, E., Tesfaye-Kedjela, A., Ratajczak, P., Bergman, R., and Richard, G.: **Deleterious mutations in SPINK5 in a patient with congenital ichthyosiform erythroderma: molecular testing as a helpful diagnostic tool for Netherton syndrome.** *Clin. Exp. Dermatol.* 2004, **29**:513–517.

Thomas, W.R., Smith, W., and Hales, B.J.: **House dust mite allergen characterisation: implications for T-cell responses and immunotherapy.** *Int. Arch. Allergy Immunol.* 1998, **115**:9–14.

Wan, H., Winton, H.L., Soeller, C., Tovey, E.R., Gruenert, D.C., Thompson, P.J., Stewart, G.A., Taylor, G.W., Garrod, D.R., Cannell, M.B., *et al.*: **Der p 1 facilitates transepithelial allergen delivery by disruption of tight junctions.** *J. Clin. Invest.* 1999, **104**:123–133.

14-3 Genetic factors contribute to the development of IgE-mediated allergic disease.

Cookson, W.: **The immunogenetics of asthma and eczema: a new focus on the epithelium.** *Nat. Rev. Immunol.* 2004, **4**:978–988.

Palmer, L.J., Silverman, E.S., Weiss, S.T., and Drazen, J.M.: **Pharmacogenetics of asthma.** *Am. J. Respir. Crit. Care Med.* 2002, **165**:861–866.

Shapiro, S.D., and Owen, C.A.: **ADAM-33 surfaces as an asthma gene.** *N. Engl. J. Med.* 2002, **347**:936–938.

Umetsu, D.T., McIntire, J.J., Akbari, O., Macaubas, C., and DeKruyff, R.H.: **Asthma: an epidemic of dysregulated immunity.** *Nat. Immunol.* 2002, **3**:715–720.

Van Eerdewegh, P., Little, R.D., Dupuis, J., Del Mastro, R.G., Falls, K., Simon, J., Torrey, D., Pandit, S., McKenny, J., Braunschweiger, K., *et al.*: **Association of the *ADAM33* gene with asthma and bronchial hyperresponsiveness.** *Nature* 2002, **418**:426–430.

Weiss, S.T., Raby, B.A., and Rogers, A.: **Asthma genetics and genomics.** *Curr. Opin. Genet. Dev.* 2009, **19**:279–282.

14-4 Environmental factors may interact with genetic susceptibility to cause allergic disease.

Culley, F.J., Pollott, J., and Openshaw, P.J.: **Age at first viral infection determines the pattern of T cell-mediated disease during reinfection in adulthood.** *J. Exp. Med.* 2002, **196**:1381–1386.

Dunne, D.W., and Cooke, A.: **Opinion: a worm's eye view of the immune system: consequences for evolution of human autoimmune disease.** *Nat. Rev. Immunol.* 2005, **5**:420–426.

Eder, W., and von Mutius, E.: **Hygiene hypothesis and endotoxin: what is the evidence?** *Curr. Opin. Allergy Clin. Immunol.* 2004, **4**:113–117.

Huang, L., Baban, B., Johnson, B.A. 3rd, and Mellor, A.L.: **Dendritic cells, indoleamine 2,3 dioxygenase and acquired immune privilege.** *Int. Rev. Immunol.* 2010, **29**:133–155.

Lynch, N.R., Hagel, I., Perez, M., Di Prisco, M.C., Lopez, R., and Alvarez, N.: **Effect of antihelminthic treatment on the allergic reactivity of children in a tropical slum.** *J. Allergy Clin. Immunol.* 1993, **92**:404–411.

Minelli, C., Granell, R., Newson, R., Rose-Zerilli, M.-J., Torrent, M., Ring, S.M., Holloway, J.W., Shaheen, S.O., and Henderson, J.A.: **Glutathione-*S*-transferase genes and asthma phenotypes: a Human Genome Epidemiology (HuGE) systematic review and meta-analysis including unpublished data.** *Int. J. Epidemiol.* 2010, **39**:539–562.

Morahan, G., Huang, D., Wu, M., Holt, B.J., White, G.P., Kendall, G.E., Sly, P.D., and Holt, P.G.: **Association of IL12B promoter polymorphism with severity of atopic and non-atopic asthma in children.** *Lancet* 2002, **360**:455–459.

Raitala, A., Karjalainen, J., Oja, S.S., Kosunen, T.U., and Hurme, M.: **Indoleamine 2,3-dioxygenase (IDO) activity is lower in atopic than in non-atopic individuals

and is enhanced by environmental factors protecting from atopy. *Mol. Immunol.* 2006, **43**:1054–1056.

Romieu, I., Ramirez-Aguilar, M., Sienra-Monge, J.J., Moreno-Macías, H., del Rio-Navarro, B.E., David, G., Marzec, J., Hernández-Avila, M., and London, S.: **GSTM1 and GSTP1 and respiratory health in asthmatic children exposed to ozone.** *Eur. Respir. J.* 2006, **28**:953–959.

Saxon, A., and Diaz-Sanchez, D. **Air pollution and allergy: you are what you breathe.** *Nat. Immunol.* 2005, **6**:223–226.

Summers, R.W., Elliott, D.E., Urban, J.F., Jr, Thompson, R.A., and Weinstock, J.V.: ***Trichuris suis* therapy for active ulcerative colitis: a randomized controlled trial.** *Gastroenterology* 2005, **128**:825–832.

Wills-Karp, M., Santeliz, J., and Karp, C.L.: **The germless theory of allergic disease: revisiting the hygiene hypothesis.** *Nat. Rev. Immunol.* 2001, **1**:69–75.

14-5 Regulatory T cells can control allergic responses.

Akdis, M., Blaser, K., and Akdis, C.A.: **T regulatory cells in allergy: novel concepts in the pathogenesis, prevention, and treatment of allergic diseases.** *J. Allergy Clin. Immunol.* 2005, **116**:961–968.

Hawrylowicz, C.M.: **Regulatory T cells and IL-10 in allergic inflammation.** *J. Exp. Med.* 2005, **202**:1459–1463.

Hayashi, T., Beck, L., Rossetto, C., Gong, X., Takikawa, O., Takabayashi, K., Broide, D.H., Carson, D.A., and Raz, E.: **Inhibition of experimental asthma by indoleamine 2,3-dioxygenase.** *J. Clin. Invest.* 2004, **114**:270–279.

Lin, W., Truong, N., Grossman, W.J., Haribhai, D., Williams, C.B., Wang, J., Martin, M.G., and Chatila, T.A.: **Allergic dysregulation and hyperimmunoglobulinemia E in Foxp3 mutant mice.** *J. Allergy Clin. Immunol.* 2005, **116**:1106–1115.

Mellor, A.L., and Munn, D.H.: **IDO expression by dendritic cells: tolerance and tryptophan catabolism.** *Nat. Rev. Immunol.* 2004, **4**:762–774.

14-6 Most IgE is cell-bound and engages effector mechanisms of the immune system by different pathways from those of other antibody isotypes.

Conner, E.R., and Saini, S.S.: **The immunoglobulin E receptor: expression and regulation.** *Curr. Allergy Asthma Rep.* 2005, **5**:191–196.

Gilfillan, A.M., and Tkaczyk, C.: **Integrated signalling pathways for mast-cell activation.** *Nat. Rev. Immunol.* 2006, **6**:218–230.

Heyman, B.: **Regulation of antibody responses via antibodies, complement, and Fc receptors.** *Annu. Rev. Immunol.* 2000, **18**:709–737.

Kinet, J.P.: **The high-affinity IgE receptor (FcεRI): from physiology to pathology.** *Annu. Rev. Immunol.* 1999, **17**:931–972.

14-7 Mast cells reside in tissues and orchestrate allergic reactions.

Ali, K., Bilancio, A., Thomas, M., Pearce, W., Gilfillan, A.M., Tkaczyk, C., Kuehn, N., Gray, A., Giddings, J., Peskett, E., *et al.*: **Essential role for the p110δ phosphoinositide 3-kinase in the allergic response.** *Nature* 2004, **431**:1007–1011.

Bingham, C.O., and Austen, K.F.: **Mast-cell responses in the development of asthma.** *J. Allergy Clin. Immunol.* 2000, **105**:S527–S534.

Galli, S.J., Nakae, S., and Tsai, M.: **Mast cells in the development of adaptive immune responses.** *Nat. Immunol.* 2005, **6**:135–142.

Gonzalez-Espinosa, C., Odom, S., Olivera, A., Hobson, J.P., Martinez, M.E., Oliveira-Dos-Santos, A., Barra, L., Spiegel, S., Penninger, J.M., and Rivera, J.: **Preferential signaling and induction of allergy-promoting lymphokines upon weak stimulation of the high affinity IgE receptor on mast cells.** *J. Exp. Med.* 2003,**197**:1453–1465.

Luster, A.D., and Tager, A.M.: **T-cell trafficking in asthma: lipid mediators grease the way.** *Nat Rev Immunol.* 2004, **4**:711–724.

Oguma, T., Palmer, L.J., Birben, E., Sonna, L.A., Asano, K., and Lilly, C.M.: **Role of prostanoid DP receptor variants in susceptibility to asthma.** *N. Engl. J. Med.* 2004, **351**:1752–1763.

Taube, C., Miyahara, N., Ott, V., Swanson, B., Takeda, K., Loader, J., Shultz, L.D., Tager, A.M., Luster, A.D., Dakhama, A., *et al.*: **The leukotriene B4 receptor (BLT1) is required for effector CD8+ T cell-mediated, mast cell-dependent airway hyperresponsiveness.** *J. Immunol.* 2006, **176**:3157–3164.

14-8 Eosinophils and basophils cause inflammation and tissue damage in allergic reactions.

Bisset, L.R., and Schmid-Grendelmeier, P.: **Chemokines and their receptors in the pathogenesis of allergic asthma: progress and perspective.** *Curr. Opin. Pulm. Med.* 2005, **11**:35–42.

Dvorak, A.M.: **Cell biology of the basophil.** *Int. Rev. Cytol.* 1998, **180**:87–236.

Hammad, H., Plantinga, M., Deswarte, K., Pouliot, P., Willart, M.A., Kool, M., Muskens, F., and Lambrecht, B.N.: **Inflammatory dendritic cells—not basophils—are necessary and sufficient for induction of Th2 immunity to inhaled house dust mite allergen.** *J. Exp. Med.* 2010, **207**:2097–2111.

Hogan, S.P., Rosenberg, H.F., Moqbel, R., Phipps, S., Foster, P.S., Lacy, P., Kay, A.B., and Rothenberg, M.E.: **Eosinophils: biological properties and role in health and disease.** *Clin. Exp. Allergy* 2008, **38**:709–750.

Lukacs, N.W.: **Role of chemokines in the pathogenesis of asthma.** *Nat. Rev. Immunol.* 2001, **1**:108–116.

MacGlashan, D., Jr, Gauvreau, G., and Schroeder, J.T.: **Basophils in airway disease.** *Curr. Allergy Asthma Rep.* 2002, **2**:126–132.

Odemuyiwa, S.O., Ghahary, A., Li, Y., Puttagunta, L., Lee, J.E., Musat-Marcu, S., and Moqbel, R.: **Cutting edge: human eosinophils regulate T cell subset selection through indoleamine 2,3-dioxygenase.** *J. Immunol.* 2004, **173**:5909–5913.

Ohnmacht, C., Schwartz, C., Panzer, M., Schiedewitz, I., Naumann, R., and Voehringer, D.: **Basophils orchestrate chronic allergic dermatitis and protective immunity against helminths.** *Immunity* 2010, **33**:364–374.

Plager, D.A., Stuart, S., and Gleich, G.J.: **Human eosinophil granule major basic protein and its novel homolog.** *Allergy* 1998, **53**:33–40.

14-9 IgE-mediated allergic reactions have a rapid onset but can also lead to chronic responses.

deShazo, R.D., and Kemp, S.F.: **Allergic reactions to drugs and biologic agents.** *JAMA* 1997, **278**:1895–1906.

Macfarlane, A.J., Kon, O.M., Smith, S.J., Zeibecoglou, K., Khan, L.N., Barata, L.T., McEuen, A.R., Buckley, M.G., Walls, A.F., Meng, Q., *et al.*: **Basophils, eosinophils, and mast cells in atopic and nonatopic asthma and in late-phase allergic reactions in the lung and skin.** *J. Allergy Clin. Immunol.* 2000, **105**:99–107.

Pearlman, D.S.: **Pathophysiology of the inflammatory response.** *J. Allergy Clin. Immunol.* 1999, **104**:S132–S137.

Taube, C., Duez, C., Cui, Z.H., Takeda, K., Rha, Y.H., Park, J.W., Balhorn, A., Donaldson, D.D., Dakhama, A., and Gelfand, E.W.: **The role of IL-13 in established allergic airway disease.** *J. Immunol.* 2002, **169**:6482–6489.

14-10 Allergen introduced into the bloodstream can cause anaphylaxis.

Fernandez, M., Warbrick, E.V., Blanca, M., and Coleman, J.W.: **Activation and hapten inhibition of mast cells sensitized with monoclonal IgE anti-penicillin antibodies: evidence for two-site recognition of the penicillin derived determinant.** *Eur. J. Immunol.* 1995, **25**:2486–2491.

Finkelman, F.D., Rothenberg, M.E., Brandt, E.B., Morris, S.C., and Strait, R.T.: **Molecular mechanisms of anaphylaxis: lessons from studies with murine models.** *J. Allergy Clin. Immunol.* 2005, **115**:449–457.

Kemp, S.F., Lockey, R.F., Wolf, B.L., and Lieberman, P.: **Anaphylaxis. A review of 266 cases.** *Arch. Intern. Med.* 1995, **155**:1749–1754.

Oettgen, H.C., Martin, T.R., Wynshaw Boris, A., Deng, C., Drazen, J.M., and Leder, P.: **Active anaphylaxis in IgE-deficient mice.** *Nature* 1994, **370**:367–370.

Padovan, E.: **T-cell response in penicillin allergy.** *Clin. Exp. Allergy* 1998, **28** Suppl. **4**:33–36.

Weltzien, H.U., and Padovan, E.: **Molecular features of penicillin allergy.** *J. Invest. Dermatol.* 1998, **110**:203–206.

14-11 Allergen inhalation is associated with the development of rhinitis and asthma.

Bousquet, J., Jeffery, P.K., Busse, W.W., Johnson, M., and Vignola, A.M.: **Asthma. From bronchoconstriction to airways inflammation and remodeling.** *Am. J. Respir. Crit. Care Med.* 2000, **161**:1720–1745.

Boxall, C., Holgate, S.T., and Davies, D.E.: **The contribution of transforming growth factor-β and epidermal growth factor signalling to airway remodelling in chronic asthma.** *Eur. Respir. J.* 2006, **27**:208–229.

Busse, W.W., and Lemanske, R.F., Jr: **Asthma.** *N. Engl. J. Med.* 2001, **344**:350–362.

Dakhama, A., Park, J.W., Taube, C., Joetham, A., Balhorn, A., Miyahara, N., Takeda, K., and Gelfand, E.W.: **The enhancement or prevention of airway hyperresponsiveness during reinfection with respiratory syncytial virus is critically dependent on the age at first infection and IL-13 production.** *J. Immunol.* 2005, **175**:1876–1883.

Day, J.H., Ellis, A.K., Rafeiro, E., Ratz, J.D., and Briscoe, M.P.: **Experimental models for the evaluation of treatment of allergic rhinitis.** *Ann. Allergy Asthma Immunol.* 2006, **96**:263–277; quiz 277–268, 315.

Finotto, S., Neurath, M.F., Glickman, J.N., Qin, S., Lehr, H.A., Green, F.H., Ackerman, K., Haley, K., Galle, P.R., Szabo, S.J., *et al.*: **Development of spontaneous airway changes consistent with human asthma in mice lacking T-bet.** *Science* 2002, **295**:336–338.

Haselden, B.M., Kay, A.B., and Larche, M.: **Immunoglobulin E-independent major histocompatibility complex-restricted T cell peptide epitope-induced late asthmatic reactions.** *J. Exp. Med.* 1999, **189**:1885–1894.

Kuperman, D.A., Huang, X., Koth, L.L., Chang, G.H., Dolganov, G.M., Zhu, Z., Elias, J.A., Sheppard, D., and Erle, D.J.: **Direct effects of interleukin-13 on epithelial cells cause airway hyperreactivity and mucus overproduction in asthma.** *Nat. Med.* 2002, **8**:885–889.

Lambrecht, B.N., and Hammad, H.: **The role of dendritic and epithelial cells as master regulators of allergic airway inflammation.** *Lancet* 2010, **376**:835–843.

Lloyd, C.M., and Hawrylowicz, C.M.: **Regulatory T cells in asthma.** *Immunity* 2009, **31**:438–449.

Meyer, E.H., DeKruyff, R.H., and Umetsu, D.T.: **T cells and NKT cells in the pathogenesis of asthma.** *Annu. Rev. Med.* 2008, **59**:281–292.

Platts-Mills, T.A.: **The role of allergens in allergic airway disease.** *J. Allergy Clin. Immunol.* 1998, **101**:S364–S366.

Robinson, D.S.: **Regulatory T cells and asthma.** *Clin. Exp. Allergy* 2009, **39**:1314–1323.

Szabo, S.J., Sullivan, B.M., Stemmann, C., Satoskar, A.R., Sleckman, B.P., and Glimcher, L.H.: **Distinct effects of T-bet in TH1 lineage commitment and IFN-γ production in CD4 and CD8 T cells.** *Science* 2002, **295**:338–342.

Wills-Karp, M.: **Interleukin-13 in asthma pathogenesis.** *Immunol. Rev.* 2004, **202**:175–190.

Zureik, M., Neukirch, C., Leynaert, B., Liard, R., Bousquet, J., and Neukirch, F.: **Sensitisation to airborne moulds and severity of asthma: cross sectional study from European Community respiratory health survey.** *BMJ* 2002, **325**:411–414.

14-12 A genetically determined defect in the skin's barrier function increases the risk of atopic eczema.

Howell, M.D., Gallo, R.L., Boguniewicz, M., Jones, J.F., Wong, C., Streib, J.E., and Leung, D.Y.: **Cytokine milieu of atopic dermatitis skin subverts the innate immune response to vaccinia virus.** *Immunity* 2006, **24**:341–348.

Thyssen, J.P., Carlsen, B.C., Menné, T., Linneberg, A., Nielsen, N.H., Meldgaard, M., Szecsi, P.B., Stender, S., and Johansen, J.D.: **Filaggrin null-mutations increase the risk and persistence of hand eczema in subjects with atopic dermatitis: results from a general population study.** *Br. J. Dermatol.* 2010, **163**:115–120.

Tsutsui, H., Yoshimoto, T., Hayashi, N., Mizutani, H., and Nakanishi, K.: **Induction of allergic inflammation by interleukin-18 in experimental animal models.** *Immunol. Rev.* 2004, **202**:115–138.

Van den Oord, R.A.H.M., and Sheikh, A.: **Filaggrin gene defects and risk of developing allergic sensitisation and allergic disorders: systematic review and meta-analysis.** *BMJ* 2009, **339**: b2433.

Verhagen, J., Akdis, M., Traidl-Hoffmann, C., Schmid-Grendelmeier, P., Hijnen, D., Knol, E.F., Behrendt, H., Blaser, K., and Akdis, C.A.: **Absence of T-regulatory cell expression and function in atopic dermatitis skin.** *J. Allergy Clin. Immunol.* 2006, **117**:176–183.

14-13 Allergy to particular foods causes systemic reactions as well as symptoms limited to the gut.

Ewan, P.W.: **Clinical study of peanut and nut allergy in 62 consecutive patients: new features and associations.** *BMJ* 1996, **312**:1074–1078.

Lee, L.A., and Burks, A.W.: **Food allergies: prevalence, molecular characterization, and treatment/prevention strategies.** *Annu. Rev. Nutr.* 2006, **26**:539–565.

14-14 IgE-mediated allergic disease can be treated by inhibiting the effector pathways that lead to symptoms or by desensitization techniques that aim at restoring tolerance to the allergen.

Adkinson, N.F., Jr, Eggleston, P.A., Eney, D., Goldstein, E.O., Schuberth, K.C., Bacon, J.R., Hamilton, R.G., Weiss, M.E., Arshad, H., Meinert, C.L., *et al.*: **A controlled trial of immunotherapy for asthma in allergic children.** *N. Engl. J. Med.* 1997, **336**:324–331.

Ali, F.R., Kay, A.B., and Larche, M.: **The potential of peptide immunotherapy in allergy and asthma.** *Curr. Allergy Asthma Rep.* 2002, **2**:151–158.

Bryan, S.A., O'Connor, B.J., Matti, S., Leckie, M.J., Kanabar, V., Khan, J., Warrington, S.J., Renzetti, L., Rames, A., Bock, J.A., *et al.*: **Effects of recombinant human interleukin-12 on eosinophils, airway hyper-responsiveness, and the late asthmatic response.** *Lancet* 2000, **356**:2149–2153.

Campbell, J.D., Buckland, K.F., McMillan, S.J., Kearley, J., Oldfield, W.L.G., Stern, L.J., Gronlund, H., van Hage, M., Reynolds, C.J., Boyton, R.J., *et al.*: **Peptide immunotherapy in allergic asthma generates IL-10-dependent immunological tolerance associated with linked epitope suppression.** *J. Exp. Med.* 2009, **206**:1535–1547.

D'Amato, G.: **Role of anti-IgE monoclonal antibody (omalizumab) in the treatment of bronchial asthma and allergic respiratory diseases.** *Eur. J. Pharmacol.* 2006, **533**:302–307.

Haldar, P., Brightling, C.E., Hargadon, B., Gupta, S., Monteiro, W., Sousa, A., Marshall, R.P., Bradding, P., Green, R.H., Wardlaw A.J., *et al.*: **Mepolizumab and exacerbations of refractory eosinophilic asthma.** *N. Engl. J. Med.* 2009, **360**:973–984.

Leckie, M.J., ten Brinke, A., Khan, J., Diamant, Z., O'Connor, B.J., Walls, C.M., Mathur, A.K., Cowley, H.C., Chung, K.F., Djukanovic, R., *et al.*: **Effects of an interleukin-5 blocking monoclonal antibody on eosinophils, airway hyper-responsiveness, and the late asthmatic response.** *Lancet* 2000, **356**:2144–2148.

Nair, P., Pizzichini, M.M.M., Kjarsgaard, M., Inman, M.D., Efthimiadis, A., Pizzichini, E., Hargreave, F.E., and O'Byrne, P.M.: **Mepolizumab for prednisone-dependent asthma with sputum eosinophilia.** *N. Engl. J. Med.* 2009, **360**:985–993.

Peters-Golden, M., and Henderson, W.R., Jr: **The role of leukotrienes in allergic rhinitis.** *Ann. Allergy Asthma Immunol.* 2005, **94**:609–618; quiz 618–620, 669.

Roberts, G., Hurley, C., Turcanu, V., and Lack, G.: **Grass pollen immunotherapy as an effective therapy for childhood seasonal allergic asthma.** *J. Allergy Clin. Immunol.* 2006, **117**:263–268.

Verhagen, J., Taylor, A., Blaser, K., Akdis, M., and Akdis, C.A.: **T regulatory cells in allergen-specific immunotherapy.** *Int. Rev. Immunol.* 2005, **24**:533–548.

Verhoef, A., Alexander, C., Kay, A.B., and Larche, M.: **T cell epitope immunotherapy induces a $CD4^+$ T cell population with regulatory activity.** *PLoS Med.* 2005, **2**:e78.

Zhu, D., Kepley, C.L., Zhang, K., Terada, T., Yamada, T., and Saxon, A.: **A chimeric human–cat fusion protein blocks cat-induced allergy.** *Nat. Med.* 2005, **11**:446–449.

14-15 Innocuous antigens can cause type II hypersensitivity reactions in susceptible individuals by binding to the surfaces of circulating blood cells.

Arndt, P.A., and Garratty, G.: **The changing spectrum of drug-induced immune hemolytic anemia.** *Semin. Hematol.* 2005, **42**:137–144.

Greinacher, A., Potzsch, B., Amiral, J., Dummel, V., Eichner, A., and Mueller Eckhardt, C.: **Heparin-associated thrombocytopenia: isolation of the antibody and characterization of a multimolecular PF4–heparin complex as the major antigen.** *Thromb. Haemost.* 1994, **71**:247–251.

Semple, J.W., and Freedman, J.: **Autoimmune pathogenesis and autoimmune hemolytic anemia.** *Semin. Hematol.* 2005, **42**:122–130.

14-16 Systemic disease caused by immune-complex formation can follow the administration of large quantities of poorly catabolized antigens.

Bielory, L., Gascon, P., Lawley, T.J., Young, N.S., and Frank, M.M.: **Human serum sickness: a prospective analysis of 35 patients treated with equine anti-thymocyte globulin for bone marrow failure.** *Medicine (Baltimore)* 1988, **67**:40–57.

Davies, K.A., Mathieson, P., Winearls, C.G., Rees, A.J., and Walport, M.J.: **Serum sickness and acute renal failure after streptokinase therapy for myocardial infarction.** *Clin. Exp. Immunol.* 1990, **80**:83–88.

Gamarra, R.M., McGraw, S.D., Drelichman, V.S., and Maas, L.C.: **Serum sickness-like reactions in patients receiving intravenous infliximab.** *J. Emerg. Med.* 2006, **30**:41–44.

Schifferli, J.A., Ng, Y.C., and Peters, D.K.: **The role of complement and its receptor in the elimination of immune complexes.** *N. Engl. J. Med.* 1986, **315**:488–495.

Schmidt, R.E., and Gessner, J.E.: **Fc receptors and their interaction with complement in autoimmunity.** *Immunol. Lett.* 2005, **100**:56–67.

Skokowa, J., Ali, S.R., Felda, O., Kumar, V., Konrad, S., Shushakova, N., Schmidt, R.E., Piekorz, R.P., Nurnberg, B., Spicher, K., *et al.*: **Macrophages induce the inflammatory response in the pulmonary Arthus reaction through $G\alpha_{i2}$ activation that controls C5aR and Fc receptor cooperation.** *J. Immunol.* 2005, **174**:3041–3050.

14-17 Hypersensitivity reactions can be mediated by T_H1 cells and CD8 cytotoxic T cells.

Bernhagen, J., Bacher, M., Calandra, T., Metz, C.N., Doty, S.B., Donnelly, T., and Bucala, R.: **An essential role for macrophage migration inhibitory factor in the tuberculin delayed-type hypersensitivity reaction.** *J. Exp. Med.* 1996, **183**:277–282.

Kalish, R.S., Wood, J.A., and LaPorte, A.: **Processing of urushiol (poison ivy) hapten by both endogenous and exogenous pathways for presentation to T cells *in vitro*.** *J. Clin. Invest.* 1994, **93**:2039–2047.

Kimber, I., and Dearman, R.J.: **Allergic contact dermatitis: the cellular effectors.** *Contact Dermatitis* 2002, **46**:1–5.

Mark, B.J., and Slavin, R.G.: **Allergic contact dermatitis.** *Med. Clin. North Am.* 2006, **90**:169–185.

Muller, G., Saloga, J., Germann, T., Schuler, G., Knop, J., and Enk, A.H.: **IL-12 as mediator and adjuvant for the induction of contact sensitivity *in vivo*.** *J. Immunol.* 1995, **155**:4661–4668.

Schmidt, M., Raghavan, B., Müller, V., Vogl, T., Fejer, G., Tchaptchet, S., Keck, S., Kalis, C., Nielsen, P.J., Galanos, C., *et al.*: **Crucial role for human Toll-like receptor 4 in the development of contact allergy to nickel.** *Nat. Immunol.* 2010, **11**:814–819.

Vollmer, J., Weltzien, H.U., and Moulon, C.: **TCR reactivity in human nickel allergy indicates contacts with complementarity-determining region 3 but excludes superantigen-like recognition.** *J. Immunol.* 1999, **163**:2723–2731.

14-18 Celiac disease has features of both an allergic response and autoimmunity.

Ciccocioppo, R., Di Sabatino, A., and Corazza, G.R.: **The immune recognition of gluten in celiac disease.** *Clin. Exp. Immunol.* 2005, **140**:408–416.

Koning, F.: **Celiac disease: caught between a rock and a hard place.** *Gastroenterology* 2005, **129**:1294–1301.

Shan, L., Molberg, O., Parrot, I., Hausch, F., Filiz, F., Gray, G.M., Sollid, L.M., and Khosla, C.: **Structural basis for gluten intolerance in celiac sprue.** *Science* 2002, **297**:2275–2279.

Sollid, L.M.: **Celiac disease: dissecting a complex inflammatory disorder.** *Nat. Rev. Immunol.* 2002, **2**:647–655.

Autoimmunity and Transplantation

15

We have already seen how undesirable adaptive immune responses can be elicited by environmental antigens, and how this can cause serious disease in the form of allergic and hypersensitivity reactions (discussed in Chapter 14). In this chapter we examine unwanted responses to other medically important categories of antigens—those expressed on the body's own cells and tissues, on transplanted organs, or on the commensal microbiota that populate the intestinal tract. The responses to self antigens or antigens associated with the commensal microbiota are called **autoimmunity** and can lead to **autoimmune diseases** that are characterized by tissue damage. The response to nonself antigens on transplanted organs is called **allograft rejection**.

The gene rearrangements that occur during lymphocyte development in the central lymphoid organs are random, and thus inevitably result in the generation of some lymphocytes with affinity for self antigens. Such lymphocytes are normally removed from the repertoire or held in check by a variety of mechanisms, many of which we have already encountered in Chapter 8. These generate a state of **self-tolerance** in which an individual's immune system does not attack the normal tissues of the body. Autoimmunity represents a breakdown or failure of the mechanisms of self-tolerance. We therefore first revisit the mechanisms that keep the lymphocyte repertoire self-tolerant and see how these may fail. We then discuss a selection of autoimmune diseases that illustrate the various pathogenic mechanisms by which autoimmunity can damage the body. How genetic and environmental factors predispose to or trigger autoimmunity are then considered. In the remaining part of the chapter, we discuss the adaptive immune responses to nonself tissue antigens that cause transplant rejection.

The making and breaking of self-tolerance.

To generate self-tolerance, the immune system must be able to distinguish self-reactive from nonself-reactive lymphocytes as they develop. As we learned in Chapter 8, the immune system takes advantage of surrogate markers of self and nonself to identify and delete potentially self-reactive lymphocytes. Despite this, some self-reactive lymphocytes escape elimination, leave the thymus, and can subsequently be activated to cause autoimmune disease. In part, autoreactivity occurs because the recognition of self-reactivity is indirect and therefore imperfect. In addition, many lymphocytes with some degree of self-reactivity can also make an immune response to foreign antigens; therefore, if all weakly self-reactive lymphocytes were eliminated, the function of the immune system would be impaired.

15-1 A critical function of the immune system is to discriminate self from nonself.

The immune system has very powerful effector mechanisms that can eliminate a wide variety of pathogens. Early in the study of immunity it was realized that these could, if turned against the host, cause severe tissue damage. The concept of autoimmunity was first presented at the beginning of the 20th century by **Paul Ehrlich**, who described it as '*horror autotoxicus*.' Autoimmune responses resemble normal immune responses to pathogens in that they are specifically activated by antigens, in this case self antigens or **autoantigens**, and give rise to autoreactive effector cells and to antibodies, called **autoantibodies**, against the self antigen. When reactions to self tissues do occur and are then improperly regulated, they cause a variety of chronic syndromes called autoimmune diseases. These syndromes are quite varied in their severity, in the tissues affected, and in the effector mechanisms that are most important in causing damage (Fig. 15.1).

Although individual autoimmune diseases are uncommon, collectively they affect approximately 5% of the populations of Western countries, and their incidence is on the rise. Nevertheless, their relative rarity indicates that the immune system has evolved multiple mechanisms to prevent damage to self tissues. The most fundamental principle underlying these mechanisms is the discrimination of self from nonself, but this discrimination is not easy to achieve. B cells recognize the three-dimensional shape of an epitope on an antigen, and a pathogen epitope can be indistinguishable from one in humans. Similarly, the short peptides derived from the processing of pathogen antigens can be identical to self peptides. So how does a lymphocyte know what 'self' really is if there are no unique molecular signatures of self?

Fig. 15.1 Some common autoimmune diseases. The diseases listed are among the commonest autoimmune diseases and will be used as examples in this part of the chapter. They are listed in order of prevalence. A more comprehensive listing and discussion of autoimmune diseases is given later in the chapter.

Disease	Disease mechanism	Consequence	Prevalence
Psoriasis	Autoreactive T cells against skin-associated antigens	Inflammation of skin with formation of scaly patches or plaques	1 in 50
Rheumatoid arthritis	Autoreactive T cells against antigens of joint synovium	Joint inflammation and destruction causing arthritis	1 in 100
Graves' disease	Autoantibodies against the thyroid-stimulating-hormone receptor	Hyperthyroidism: overproduction of thyroid hormones	1 in 100
Hashimoto's thyroiditis	Autoantibodies and autoreactive T cells against thyroid antigens	Destruction of thyroid tissue leading to hypothyroidism: underproduction of thyroid hormones	1 in 200
Systemic lupus erythematosus	Autoantibodies and autoreactive T cells against DNA, chromatin proteins, and ubiquitous ribonucleoprotein antigens	Glomerulonephritis, vasculitis, rash	1 in 200
Sjögren's syndrome	Autoantibodies and autoreactive T cells against ribonucleoprotein antigens	Lymphocyte infiltration of exocrine glands, leading to dry eyes and/or dry mouth; other organs may be involved, leading to systemic disease	1 in 300
Crohn's disease	Autoreactive T cells against intestinal flora antigens	Intestinal inflammation and scarring	1 in 500
Multiple sclerosis	Autoreactive T cells against brain antigens	Formation of sclerotic plaques in brain with destruction of myelin sheaths surrounding nerve cell axons, leading to muscle weakness, ataxia, and other symptoms	1 in 700
Type 1 diabetes (insulin-dependent diabetes mellitus, IDDM)	Autoreactive T cells against pancreatic islet cell antigens	Destruction of pancreatic islet β cells leading to nonproduction of insulin	1 in 800

The first mechanism proposed for distinguishing between self and nonself was that recognition of antigen by an immature lymphocyte leads to a negative signal that causes lymphocyte death or inactivation. Thus, 'self' was thought to comprise those molecules recognized by a lymphocyte shortly after it has begun to express its antigen receptor. Indeed, this is an important mechanism of inducing self-tolerance in lymphocytes developing in the thymus and bone marrow. The tolerance induced at this stage is known as **central tolerance** and is discussed in detail in Chapter 8. Newly formed lymphocytes are especially sensitive to inactivation by strong signals through their antigen receptors, whereas the same signals would activate a mature lymphocyte.

Another antigenic quality that correlates with self is a sustained, high concentration of the antigen. Many self proteins are expressed by multiple cell types in the body or are abundant in connective tissues. These can provide strong signals to lymphocytes, and even mature lymphocytes can be made tolerant to an antigen, or **tolerized**, by strong and constant signals through their antigen receptors. In contrast, pathogens and other foreign antigens are introduced to the immune system suddenly, and the concentrations of their antigens increase rapidly and exponentially as the pathogens replicate in the early stages of an infection. Naive mature lymphocytes are tuned to respond by activation to a sudden increase in antigen-receptor signals.

A third mechanism for discriminating between self and nonself relies on the innate immune system, which provides signals that are crucial in enabling the activation of an adaptive immune response to infection (see Chapter 3). In the absence of infection, these signals, which include pro-inflammatory cytokines (for example IL-6 or IL-12) and co-stimulatory molecules (for example B7.1) that are expressed by activated antigen-presenting cells, are not generated. In these circumstances, the encounter of a naive lymphocyte with a self antigen tends to lead to a negative inactivating signal, rather than no signal at all (see Section 8-26), or can promote the development of regulatory lymphocytes that suppress the development of effector responses that might injure tissues. This tolerance mechanism is particularly important for antigens that are encountered outside the thymus and bone marrow. Tolerance induced in the mature lymphocyte repertoire once cells have left the central lymphoid organs is known as **peripheral tolerance**.

Thus, several clues are used by lymphocytes to distinguish self ligands from nonself ligands: encounter with the ligand when the lymphocyte is still immature; a high and constant concentration of ligand; and binding the ligand in the absence of pro-inflammatory cytokine or co-stimulatory signals. All these mechanisms are error-prone, however, because none of them particularly distinguishes a self ligand from a foreign one at the molecular level. The immune system therefore has several additional mechanisms for controlling autoimmune responses should they start.

15-2 Multiple tolerance mechanisms normally prevent autoimmunity.

The mechanisms that normally prevent autoimmunity may be considered as a succession of checkpoints. Each checkpoint is partly effective in preventing anti-self responses, and together they act synergistically to provide efficient protection against autoimmunity without inhibiting the ability of the immune system to mount effective responses to pathogens. Central tolerance mechanisms eliminate newly formed strongly autoreactive lymphocytes. On the other hand, mature self-reactive lymphocytes that do not sense self strongly in the central lymphoid organs—because their cognate self antigens are not expressed there, for example—may be killed or inactivated in the periphery. The principal mechanisms of peripheral tolerance are anergy (functional unresponsiveness), suppression by regulatory T cells, induction of regulatory T-cell development instead of effector T-cell development (functional deviation), and deletion of lymphocytes from the repertoire due to activation-induced cell death. In addition, some antigens are sequestered in organs that are not normally accessible to the immune system (Fig. 15.2).

Each checkpoint strikes a balance between preventing autoimmunity and not impairing immunity too greatly, and in combination they provide an effective overall defense against autoimmune disease. It is relatively easy to find isolated breakdowns of one or even more layers of protection, even in healthy individuals. Thus, activation of autoreactive lymphocytes does not necessarily equal autoimmune disease. In fact, a low level of autoreactivity is physiological and crucial to normal immune function. Autoantigens help to form the repertoire of mature lymphocytes, and the survival of naive T cells and B cells in the periphery requires continuous exposure to autoantigens (see Chapter 8). Autoimmune disease develops only if enough of the safeguards are overcome to lead to a sustained reaction to self that includes the generation of effector cells and molecules that destroy tissues. Although the mechanisms by which this occurs are not completely known, autoimmunity is thought to result from a combination of genetic susceptibility, breakdown in natural tolerance mechanisms, and environmental triggers such as infections (Fig. 15.3).

15-3 Central deletion or inactivation of newly formed lymphocytes is the first checkpoint of self-tolerance.

Central tolerance mechanisms, which remove strongly autoreactive lymphocytes, are the first and most important checkpoints in self-tolerance and are covered in detail in Chapter 8. Without them, the immune system would be strongly self-reactive, and lethal autoimmunity would most certainly be present from birth. It is unlikely that the other, later-acting, mechanisms of tolerance would be sufficient to compensate for the failure to remove self-reactive lymphocytes during their primary development. Indeed, there are

Fig. 15.2 Self-tolerance depends on the concerted action of a variety of mechanisms that operate at different sites and stages of development. The different ways in which the immune system prevents activation of and damage caused by autoreactive lymphocytes are listed, along with the specific mechanism and where such tolerance predominantly occurs.

Layers of self-tolerance		
Type of tolerance	**Mechanism**	**Site of action**
Central tolerance	Deletion Editing	Thymus Bone marrow
Antigen segregation	Physical barrier to self-antigen access to lymphoid system	Peripheral organs (e.g. thyroid, pancreas)
Peripheral anergy	Cellular inactivation by weak signaling without co-stimulus	Secondary lymphoid tissue
Regulatory T cells	Suppression by cytokines, intercellular signals	Secondary lymphoid tissue and sites of inflammation
Functional deviation	Differentiation of regulatory T cells that limit inflammatory cytokine secretion	Secondary lymphoid tissue and sites of inflammation
Activation-induced cell death	Apoptosis	Secondary lymphoid tissue and sites of inflammation

no known autoimmune diseases that are attributable to a complete failure of these basic mechanisms, although some are associated with a partial failure of central tolerance.

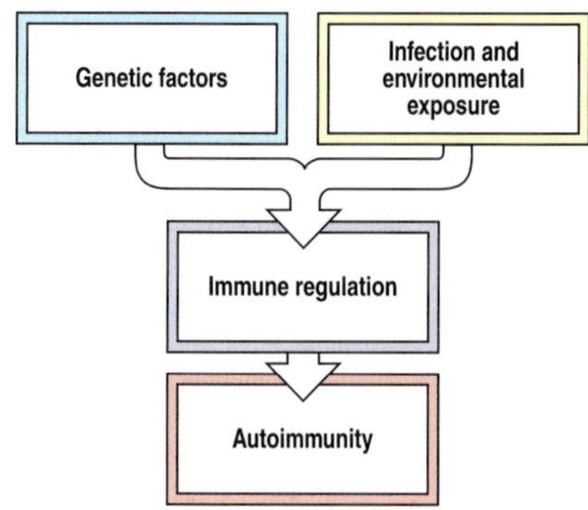

Fig. 15.3 Requirements for the development of autoimmune disease. In genetically predisposed individuals, autoimmunity may be triggered as a result of the failure of intrinsic tolerance mechanisms and/or environmental triggers such as infection.

Self-tolerance generated in the central lymphoid organs is effective, but for a long time it was thought that many self antigens were not expressed in the thymus or bone marrow, and that peripheral mechanisms must be the only way of generating tolerance to them. It is now clear, however, that many (but not all) tissue-specific antigens, such as insulin, are actually expressed in the thymus by a subset of dendritic-like cells, and thus self-tolerance against these antigens can be generated centrally. How these 'peripheral' genes are turned on ectopically in the thymus is not yet completely worked out, but an important clue has been found. A single transcription factor, AIRE (for *a*uto*i*mmune *re*gulator), is thought to be responsible for turning on many peripheral genes in the thymus (see Section 8-20). The *AIRE* gene is defective in patients with a rare inherited form of autoimmunity—**APECED** (**autoimmune polyendocrinopathy–candidiasis–ectodermal dystrophy**)—that leads to the destruction of multiple endocrine tissues, including insulin-producing pancreatic islets. This disease is also known as autoimmune polyglandular syndrome 1 (APS-1). Mice that have been engineered to lack the *AIRE* gene have a similar syndrome, although they do not seem to be susceptible to fungal infections such as candidiasis. Most importantly, these mice no longer express many of the peripheral genes in the thymus. This links the AIRE protein to the expression of these genes as well as suggesting that an inability to express these genes in the thymus leads to autoimmune disease (Fig. 15.4). The autoimmunity that accompanies AIRE deficiency takes time to develop and does not always affect all potential organ targets. So as well as emphasizing the importance of central tolerance, the disease shows that other layers of tolerance control have important roles.

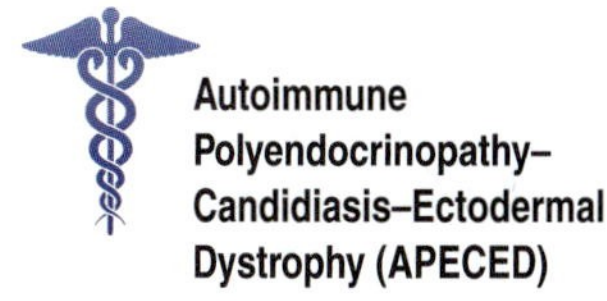

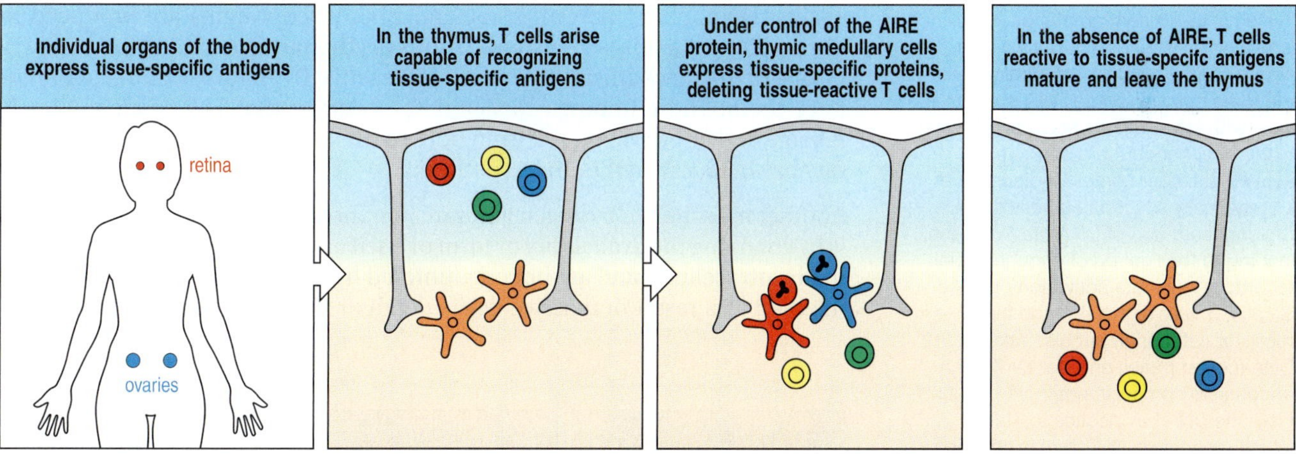

Fig. 15.4 The 'autoimmune regulator' gene *AIRE* promotes the expression of some tissue-specific antigens in thymic medullary cells, causing the deletion of immature thymocytes that can react to these antigens. Although the thymus expresses many genes, and thus self proteins, common to all cells, it is not obvious how antigens that are specific to specialized tissues, such as retina or ovary (first panel), gain access to the thymus to promote the negative selection of immature autoreactive thymocytes. It is now known that a gene called *AIRE* promotes the expression of many tissue-specific proteins in thymic medullary cells. Some developing thymocytes will be able to recognize these tissue-specific antigens (second panel). Peptides from these proteins are presented to the developing thymocytes as they undergo negative selection in the thymus (third panel), causing deletion of these cells. In the absence of *AIRE*, this deletion does not occur; instead, the autoreactive thymocytes mature and are exported to the periphery (fourth panel), where they could cause autoimmune disease. Indeed, people and mice that lack expression of *AIRE* develop an autoimmune syndrome called APECED, or autoimmune polyendocrinopathy–candidiasis–ectodermal dystrophy.

15-4 Lymphocytes that bind self antigens with relatively low affinity usually ignore them but in some circumstances become activated.

Most circulating lymphocytes have a low affinity for self antigens but make no response to them, and may be considered as 'ignorant' of self (see Section 8-6). Such ignorant but latently self-reactive cells can be recruited into autoimmune responses if their threshold for activation is lowered by co-activating factors. One such stimulus could be infection. Naive T cells with low affinity for a ubiquitous self antigen can become activated if they encounter an activated dendritic cell presenting that antigen and expressing high levels of co-stimulatory signals or pro-inflammatory cytokines as a result of the presence of infection.

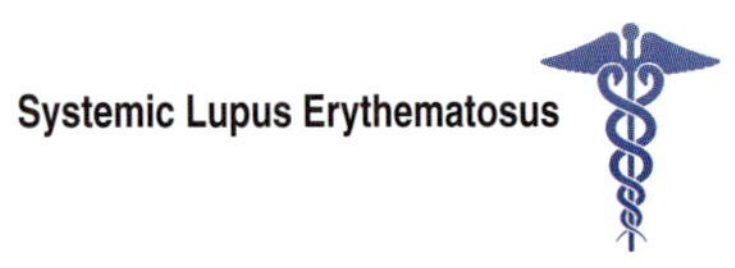

A particular situation in which ignorant lymphocytes may be activated is where their autoantigens are also the ligands for Toll-like receptors (TLRs). These receptors are usually considered to be pattern-recognition receptors specific for pathogen-associated molecular patterns (see Section 3-7), but these patterns are not exclusive to pathogens and can be found among self molecules. An example of this type of potential autoantigen is unmethylated CpG sequences in DNA that are recognized by TLR-9. Unmethylated CpG is normally much more common in bacterial DNA than in mammalian DNA but is enriched in mammalian cells undergoing apoptosis. In a scenario of extensive cell death coupled with inadequate clearance of apoptotic fragments (possibly as a result of infection), B cells specific for components of chromatin can internalize the CpG sequences via their B-cell receptors. These sequences encounter their receptor, TLR-9, intracellularly, leading to a co-stimulatory signal that, together with the signal from the B-cell receptor, activates the previously ignorant anti-chromatin B cell (Fig. 15.5). B cells activated in this way will proceed to produce anti-chromatin autoantibodies and also can act as antigen-presenting cells for autoreactive T cells. Ribonucleoprotein complexes containing uridine-rich RNA have similarly been shown to activate naive B cells through binding of the RNA by TLR-7 or TLR-8. Autoantibodies against DNA, chromatin proteins, and ribonucleoproteins are produced in the autoimmune disease **systemic lupus erythematosus** (**SLE**), and this may be one of the mechanisms by which self-reactive B cells are stimulated to produce them. These findings challenge the concept that TLRs are completely reliable at distinguishing self from nonself; their proposed role in autoimmunity has been called the 'Toll hypothesis.'

Another mechanism by which ignorant lymphocytes can be drawn into action is by changing the availability or form of a self antigen. Some antigens are normally intracellular and are not encountered by lymphocytes, but they may be released as a result of massive tissue death or inflammation. These antigens

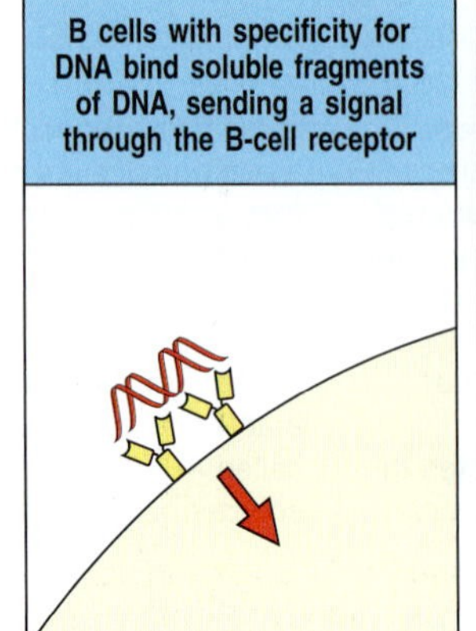

B cells with specificity for DNA bind soluble fragments of DNA, sending a signal through the B-cell receptor

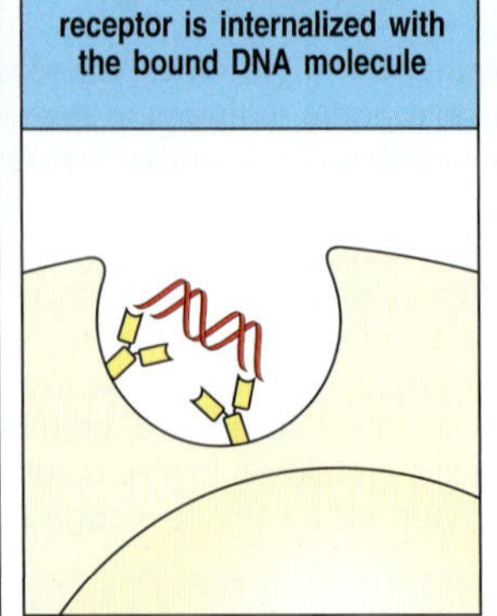

The cross-linked B-cell receptor is internalized with the bound DNA molecule

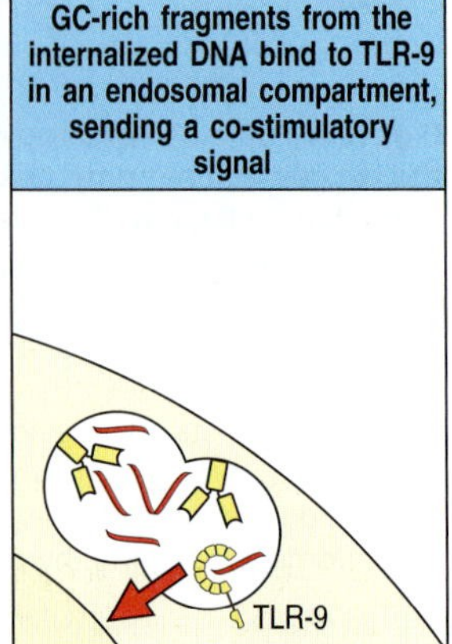

GC-rich fragments from the internalized DNA bind to TLR-9 in an endosomal compartment, sending a co-stimulatory signal
TLR-9

Fig. 15.5 Self antigens that are recognized by Toll-like receptors can activate autoreactive B cells by providing co-stimulation. The receptor TLR-9 promotes the activation of B cells specific for DNA, a common autoantibody in the autoimmune disease systemic lupus erythematosus (SLE) (see Fig. 15.1). Although B cells with strong affinity for DNA are eliminated in the bone marrow, some DNA-specific B cells with lower affinity escape and persist in the periphery but are not normally activated. Under some conditions and in genetically susceptible individuals, the concentration of DNA may increase, leading to the ligation of enough B-cell receptors to initiate activation of these B cells. B cells signal through their receptor (left panel) but also take up the DNA (center panel) and deliver it to an endosomal compartment (right panel). Here the DNA has access to TLR-9, which recognizes DNA that is enriched in unmethylated CpG DNA sequences. Such CpG-enriched sequences are much more common in microbial than eukaryotic DNA and normally this allows TLR-9 to distinguish pathogens from self. DNA in apoptotic mammalian cells is enriched in unmethylated CpG, however, and the DNA-specific B cell will also concentrate the self DNA in the endosomal compartment. This would provide sufficient ligand to activate TLR-9, potentiating the activation of the DNA-specific B cell and leading to the production of autoantibodies against DNA.

can then activate hitherto ignorant T and B cells, leading to autoimmunity. This can occur after myocardial infarction, when an autoimmune response is detectable some days after the release of cardiac antigens. Such reactions are typically transient and cease when the autoantigens have been removed; however, when clearance mechanisms are inadequate or genetically deficient, they can continue, causing clinical autoimmune disease.

Some autoantigens are present in great quantity but are usually in a nonimmunogenic form. IgG is a good example, because there are large quantities of it in blood and in other extracellular fluids. B cells specific for the IgG constant region are not usually activated because the IgG is monomeric and cannot cross-link the B-cell receptor. However, when immune complexes form after a severe infection or an immunization, enough IgG is in multivalent form to evoke a response from these otherwise ignorant B cells. The anti-IgG autoantibody they produce is known as **rheumatoid factor** because it is commonly present in rheumatoid arthritis. Again, this response is normally short lived, as long as the immune complexes are cleared rapidly.

A unique situation occurs in peripheral lymphoid organs when activated B cells undergo somatic hypermutation in germinal centers (see Section 10-7). This can result in some already activated B cells becoming self-reactive or increasing their affinity for a self antigen (Fig. 15.6). Like the ignorant lymphocytes discussed above, such self-reactive B cells would have bypassed all the other tolerance mechanisms but would now be a source of potentially pathogenic autoantibodies. There seems, however, to be a mechanism to control germinal-center B cells that have acquired affinity for self. In this case, the self antigen is likely to be present within the germinal center, whereas a pathogen is less likely to be. If a hypermutated self-reactive B cell encounters strong cross-linking of its B-cell receptor in the germinal center, it undergoes apoptosis rather than further proliferation.

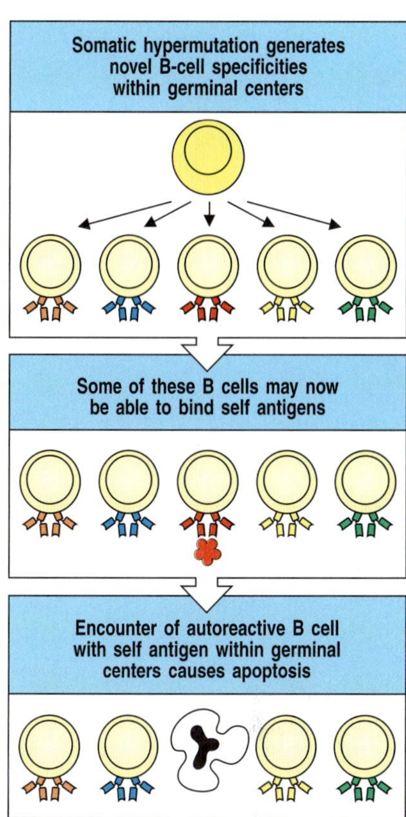

Fig. 15.6 Elimination of autoreactive B cells in germinal centers. During somatic hypermutation in germinal centers (top panel), B cells with autoreactive B-cell receptors can arise. Ligation of these receptors by soluble autoantigen (center panel) induces apoptosis of the autoreactive B cell by signaling through the B-cell antigen receptor in the absence of helper T cells (bottom panel).

15-5 Antigens in immunologically privileged sites do not induce immune attack but can serve as targets.

Tissue grafts placed in some sites in the body do not elicit immune responses. For instance, the brain and the anterior chamber of the eye are sites in which tissues can be grafted without inducing rejection. Such locations are termed **immunologically privileged sites** (Fig. 15.7). It was originally believed that immunological privilege arose from the failure of antigens to leave privileged sites and induce immune responses. Subsequent studies have shown that antigens do leave these sites and that they do interact with T cells. Instead of eliciting a destructive immune response, however, they induce tolerance or a response that is not destructive to the tissue.

Immunologically privileged sites seem to be unusual in three ways. First, the communication between the privileged site and the body is atypical in that extracellular fluid in these sites does not pass through conventional lymphatics, although proteins placed in these sites do leave them and can have immunological effects. Privileged sites are generally surrounded by tissue barriers that exclude naive lymphocytes. The brain, for example, is guarded by the blood–brain barrier. Second, soluble factors, presumably cytokines, that affect the course of an immune response are produced in privileged sites and leave them together with antigens. The anti-inflammatory transforming growth factor (TGF)-β seems to be particularly important in this regard. In

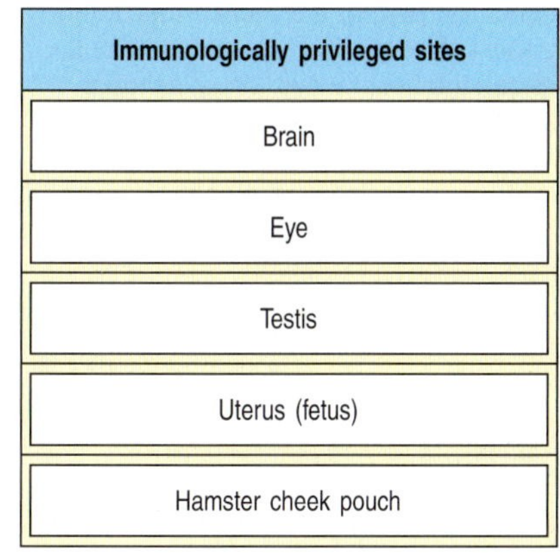

Fig. 15.7 Some sites in the body are immunologically privileged. Tissue grafts placed in these sites often last indefinitely, and antigens placed in these sites do not elicit destructive immune responses.

homeostatic conditions (that is, in the absence of infection and thus of pro-inflammatory signals), antigens recognized in concert with TGF-β tend to induce regulatory T-cell responses that do not damage tissues, rather than pro-inflammatory T_H17 responses, which are induced by TGF-β in the presence of IL-6 co-signaling (see Section 9-18). Third, the expression of Fas ligand by the tissues of immunologically privileged sites may provide a further level of protection by inducing the apoptosis of Fas-bearing effector lymphocytes that enter these sites.

Paradoxically, the antigens sequestered in immunologically privileged sites are often the targets of autoimmune attack; for example, brain autoantigens such as myelin basic protein are targeted in the autoimmune disease **multiple sclerosis**, a chronic inflammatory demyelinating disease of the central nervous system (see Fig. 15.1). It is therefore clear that the tolerance normally shown to this antigen cannot be due to previous deletion of the self-reactive T cells. In the condition **experimental autoimmune encephalomyelitis (EAE)**, a mouse model for multiple sclerosis, mice become diseased only when they are deliberately immunized with myelin basic protein, which causes substantial infiltration of the brain with antigen-specific T_H17 and T_H1 cells that cooperate to induce a local inflammatory response that damages nerve tissue.

This shows that at least some antigens expressed in immunologically privileged sites induce neither tolerance nor lymphocyte activation in normal circumstances, but if autoreactive lymphocytes are activated elsewhere, these autoantigens can become targets for autoimmune attack. It seems plausible that T cells specific for antigens sequestered in immunologically privileged sites are most likely to be in a state of immunological ignorance. Further evidence comes from the eye disease **sympathetic ophthalmia** (Fig. 15.8). If one eye is ruptured by a blow or other trauma, an autoimmune response to eye proteins can occur, although this happens only rarely. Once the response is induced, it often attacks both eyes. Immunosuppression—and, rarely, removal of the damaged eye, the source of antigen—is required to preserve vision in the undamaged eye.

It is not surprising that effector T cells can enter immunologically privileged sites: such sites can become infected, and effector cells must be able to enter these sites during infection. Effector T cells enter most or all tissues after

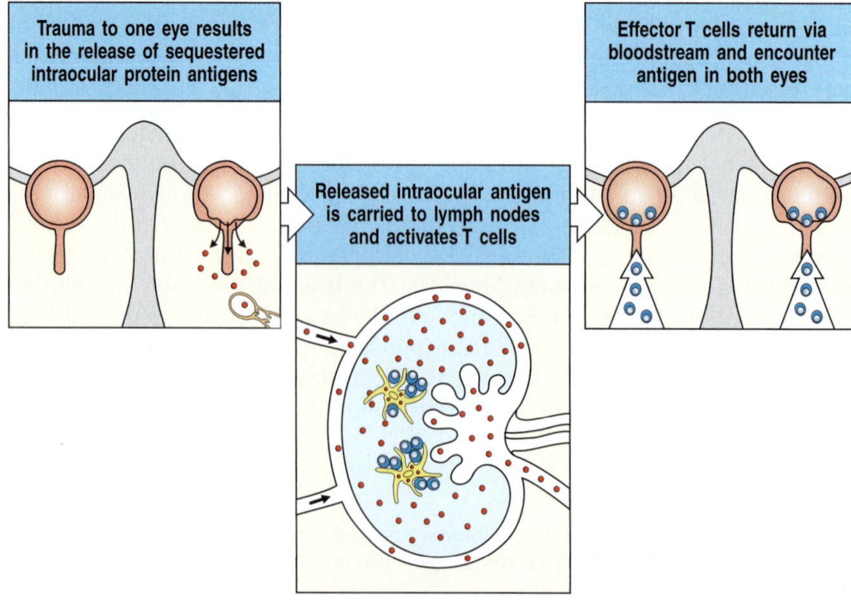

Trauma to one eye results in the release of sequestered intraocular protein antigens
Released intraocular antigen is carried to lymph nodes and activates T cells
Effector T cells return via bloodstream and encounter antigen in both eyes

Fig. 15.8 Damage to an immunologically privileged site can induce an autoimmune response. In the disease sympathetic ophthalmia, trauma to one eye releases the sequestered eye antigens into the surrounding tissues, making them accessible to T cells. The effector cells that are elicited attack the traumatized eye, and also infiltrate and attack the healthy eye. Thus, although the sequestered antigens do not induce a response by themselves, if a response is induced elsewhere they can serve as targets for attack.

activation (see Chapter 11), but accumulation of cells is seen only when antigen is recognized in the site, triggering the production of cytokines that alter tissue barriers.

15-6 Autoreactive T cells that express particular cytokines may be nonpathogenic or may suppress pathogenic lymphocytes.

We learned in Chapter 9 that, during the course of normal immune responses, CD4 T cells can differentiate into various types of effector cells, namely T_H1, T_H2, and T_H17. T_H1, T_H2, and T_H17 cells secrete different cytokines (interferon (IFN)-γ and tumor necrosis factor (TNF)-α for T_H1; interleukin (IL)-4, IL-5, IL-10, and IL-13 for T_H2; and IL-17 and IL-22 for T_H17). These effector cell subsets probably evolved to control different types of infectious threats and therefore orchestrate distinct types of immunity, which is reflected in their different effects on antigen-presenting cells, B cells, and innate effector cells such as macrophages, eosinophils, and neutrophils. A similar paradigm holds true for autoimmunity. In particular, certain T cell-mediated autoimmune diseases such as **type 1 diabetes mellitus** (also known as **insulin-dependent diabetes mellitus** or **IDDM**) (see Fig. 15.1) depend on T_H1 cells to cause disease, whereas others, such as psoriasis (an autoimmune disease of the skin), depend on T_H17 cells.

In murine models of diabetes, when cytokines were infused to influence T-cell differentiation or when knockout mice predisposed to T_H2 differentiation were studied, the development of diabetes was inhibited. In some cases, potentially pathogenic T cells specific for pancreatic islet-cell components, and expressing T_H2 instead of T_H1 cytokines, are actually suppressive of disease caused by T_H1 cells of the same specificity. So far, attempts to control human autoimmune disease by switching cytokine profiles from one effector cell type to another (for example T_H1 to T_H2), a procedure termed **immune modulation**, have not proved successful. Another important subset of CD4 T cells, the regulatory T cells, may prove to be more important in the natural prevention of autoimmune disease, and efforts to deviate effector T-cell responses to regulatory T-cell responses by immune modulation may have promise as a novel therapy for treatment of autoimmunity.

15-7 Autoimmune responses can be controlled at various stages by regulatory T cells.

Autoreactive cells that have escaped the tolerance-inducing mechanisms described previously can still be regulated so that they do not cause clinical disease. This regulation takes two forms: the first is extrinsic, coming from specific regulatory T cells that exert effects on activated T cells and on antigen-presenting cells. The second is intrinsic and has to do with limits on the size and duration of immune responses that are programmed into the lymphocytes themselves. We shall first discuss the role of regulatory T cells, which were introduced in Chapter 9.

Tolerance due to regulatory lymphocytes is distinguished from other forms of self-tolerance by the fact that a regulatory T (T_{reg}) cell has the potential to suppress self-reactive lymphocytes that recognize antigens different from those recognized by the T_{reg} cell (Fig. 15.9). This type of tolerance is therefore known as **regulatory tolerance** or **infectious tolerance**. The key feature of regulatory tolerance is that regulatory cells can suppress autoreactive lymphocytes that recognize a variety of different self antigens, as long as the antigens are from the same tissue or are presented by the same antigen-presenting cell. Two general types of regulatory T cells have been defined experimentally. One type, referred to as 'natural' T_{reg} cells, are moderately autoreactive CD4 CD25 T cells expressing the transcription factor FoxP3 that escape deletion in the

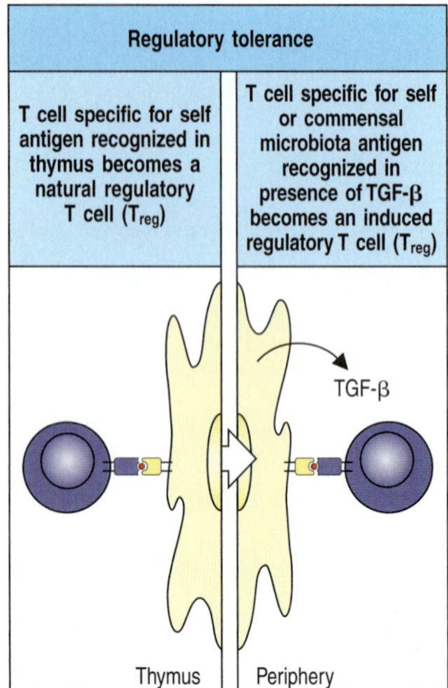
Regulatory tolerance
T cell specific for self antigen recognized in thymus becomes a natural regulatory T cell (T_{reg})
T cell specific for self or commensal microbiota antigen recognized in presence of TGF-β becomes an induced regulatory T cell (T_{reg})
TGF-β
Thymus
Periphery

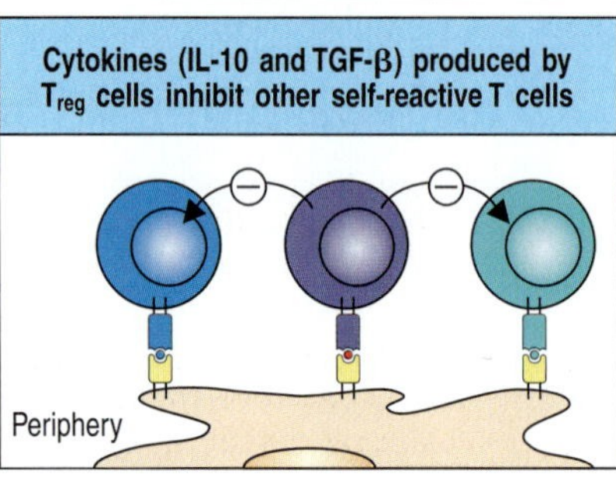
Cytokines (IL-10 and TGF-β) produced by T_{reg} cells inhibit other self-reactive T cells
Periphery

Fig. 15.9 Tolerance mediated by regulatory T cells can inhibit multiple autoreactive T cells that all recognize the same tissue. Specialized autoreactive natural regulatory T (T_{reg}) cells develop in the thymus in response to weak stimulation by self antigens that is not sufficient to cause deletion but is greater than that required for simple positive selection (upper left panel). Regulatory T cells can also be induced from naive self-reactive T cells in the periphery if the naive T cell recognizes its antigen and is activated in the presence of the cytokine TGF-β (upper right panel). The lower panel shows how regulatory T cells, both natural or induced, can inhibit other self-reactive T cells. If regulatory T cells encounter their self antigen on an antigen-presenting cell, they secrete inhibitory cytokines such as IL-10 and TGF-β that inhibit all surrounding autoreactive T cells, regardless of their precise autoantigen specificity.

thymus and when activated by self antigens in peripheral tissues do not differentiate into cells that can initiate an autoimmune response. Instead, they inhibit other self-reactive T cells that recognize antigens in the same tissue and prevent their differentiation into effector T cells or prevent their effector function. The second type, referred to as 'induced' or 'adaptive' T_{reg} cells, develop in peripheral immune tissues in response to antigens recognized on 'immature' dendritic cells that produce TGF-β in the absence of pro-inflammatory cytokines. Giving animals large amounts of self antigen orally, which induces so-called oral tolerance (see Section 12-15), can sometimes lead to unresponsiveness to these antigens when given by other routes, and can prevent autoimmune disease. Oral tolerance is routinely generated to antigens such as food antigens and is accompanied by the generation of induced T_{reg} cells in the gut-draining mesenteric lymph nodes. These cells are known to suppress immune responses to the given antigen in the gut itself, but how the suppression in the rest of the peripheral immune system is achieved is not yet known. Many investigators have hypothesized that T_{reg} cells could have therapeutic potential for the treatment of autoimmune disease if they could be isolated or induced to differentiate and then be infused into patients.

A common characteristic of all natural and many induced T_{reg} cells is the expression of CD4 and CD25 (the α chain of the IL-2 receptor) on their surface, and their expression of the transcription factor, FoxP3 (see Section 9-19). That FoxP3, and the T_{reg} cells whose development and function it controls, is important to the maintenance of immune tolerance is evident from the fact that humans and mice that carry mutations in the gene for FoxP3 rapidly develop severe, systemic autoimmunity (discussed in Section 15-20). A protective role for FoxP3-expressing T_{reg} cells has been demonstrated in respect of several autoimmune syndromes in mice, including diabetes, EAE, SLE, and inflammation of the large intestine, or colon (colitis). Experiments in mouse models of these diseases show that CD4 CD25 T_{reg} cells suppress disease when transferred *in vivo* and that depletion of these cells exacerbates or causes disease. A proposed model for the resolution of autoimmune colitis in mice by CD4 CD25 T_{reg} cells is shown in Fig. 15.10. These CD4 CD25 T_{reg} cells have also been shown to prevent or ameliorate other immunopathologic syndromes, such as graft-versus-host disease and graft rejection, which are described later in this chapter.

The importance of regulatory T cells has been demonstrated in several human autoimmune diseases. For example, in patients with multiple sclerosis or with autoimmune polyglandular syndrome type 2 (a rare syndrome in which two or more autoimmune diseases occur simultaneously), the suppressive activity of CD4 CD25 T_{reg} cells is defective, although their numbers are normal. A different picture emerges from studies of patients with active rheumatoid arthritis. Peripheral blood CD4 CD25 T_{reg} cells from these patients were found to be effective in suppressing the proliferation of the patients' own effector T cells *in vitro* but did not suppress the secretion of inflammatory cytokines by these cells. Thus, increasing evidence supports the notion that regulatory T cells normally have an important role in preventing autoimmunity, and that autoimmunity may be accompanied by a variety of functional defects in these cells.

FoxP3-expressing T_{reg} cells are not the only type of regulatory lymphocyte that has been found. Regulatory T cells that do not express FoxP3 have been identified *in vitro* and *in vivo*. These cells are characterized by their production of IL-10 and are enriched in the intestinal tissues, where they have been shown experimentally in mice to suppress inflammatory bowel disease (IBD), an autoinflammatory disease, through an IL-10-dependent mechanism.

Almost every type of lymphocyte has been shown to display regulatory activity in some circumstance. Even B cells can regulate experimentally induced

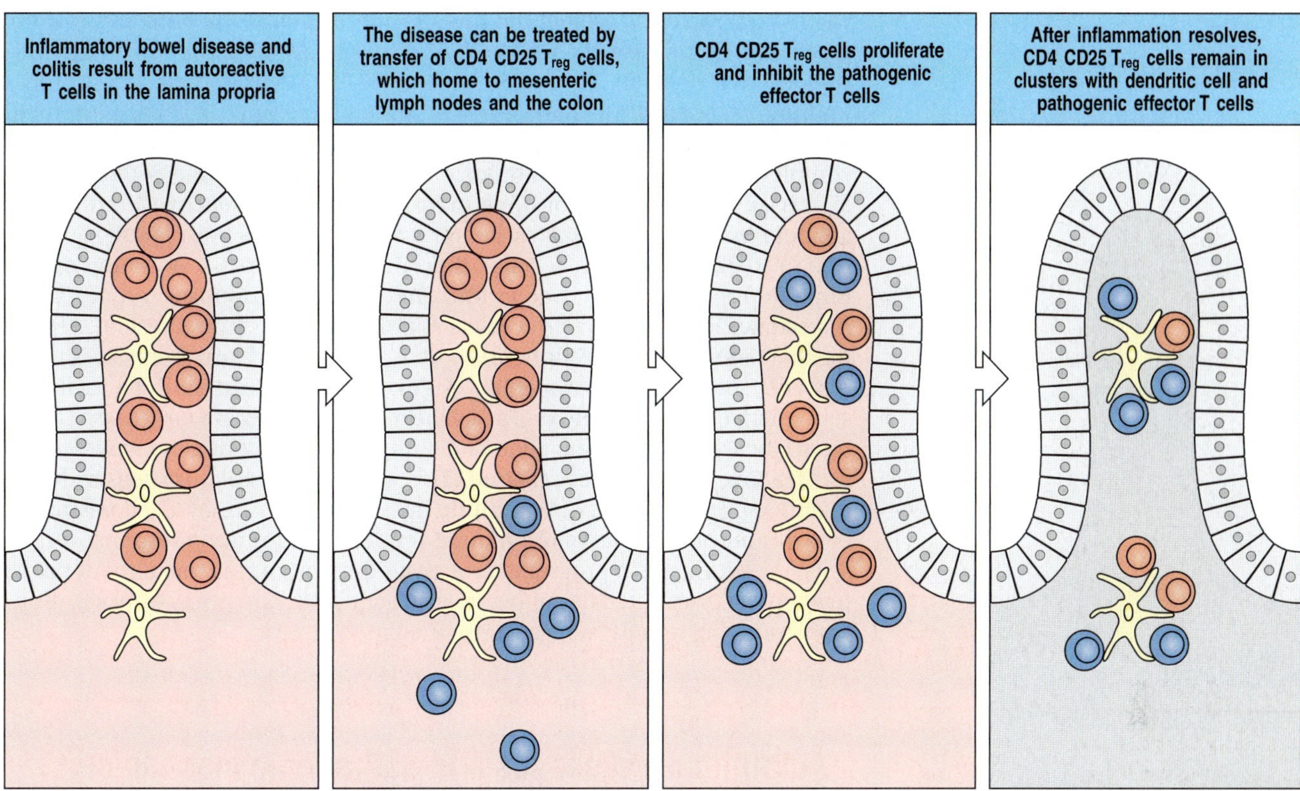

Inflammatory bowel disease and colitis result from autoreactive T cells in the lamina propria
The disease can be treated by transfer of CD4 CD25 T_{reg} cells, which home to mesenteric lymph nodes and the colon
CD4 CD25 T_{reg} cells proliferate and inhibit the pathogenic effector T cells
After inflammation resolves, CD4 CD25 T_{reg} cells remain in clusters with dendritic cell and pathogenic effector T cells

Fig. 15.10 CD4 CD25 regulatory T cells inhibit colitis by migrating to the colon and mesenteric lymph nodes, where they interact with dendritic and effector T cells. Naive T cells that contain some autoreactive clones (first panel, pink cells) cause colitis when transferred to T-cell-deficient mice. The naive population lacks CD4 CD25 T_{reg} cells, but if these are also transferred along with the naive T cells (second panel; blue cells are T_{reg} cells), colitis is blocked. The blocking mechanism includes migration of the T_{reg} cells to mesenteric lymph nodes (not shown) and later to the lamina propria of the colon. The T_{reg} cells proliferate and secrete regulatory cytokines (third panel), including IL-10, which is essential, and interact with both dendritic and autoreactive T cells, reducing activation (indicated by the smaller size of the pink cells) and ultimately reducing inflammation. Once inflammation has been quelled, regulatory T cells remain in the lamina propria (fourth panel). Based on a figure by F. Powrie.

autoimmune syndromes, including collagen-induced arthritis (CIA) and EAE in mice. This regulatory activity is probably mediated in a similar way to that of regulatory CD4 T cells, with the secretion of cytokines that inhibit T-cell proliferation and the differentiation of effector T cells being of major importance.

In addition to the extrinsic regulation of autoreactive T and B cells by regulatory cells, lymphocytes have intrinsic limits to proliferation and survival that can help to restrict autoimmune responses as well as normal immune responses (see Section 11-12). This is illustrated by the effects of mutations in the pathways that control apoptosis, such as the Bcl-2 pathway or the Fas pathway (see Section 7-23), that lead to spontaneous autoimmunity, as we shall see later in this chapter. This form of autoimmunity provides evidence that autoreactive cells are normally generated but are then controlled by apoptosis. This seems to be an important mechanism for both T- and B-cell tolerance.

Summary.

Discrimination between self and nonself is imperfect, partly because of its indirect nature and partly because a proper balance must be struck between preventing autoimmune disease and preserving immune competence. Self-reactive lymphocytes always exist in the natural immune repertoire but are not often activated. In autoimmune disease, however, these cells become activated by specific autoantigens. If activation persists, effector functions identical to those elicited in response to pathogens are generated and cause disease. The immune system has a remarkable set of mechanisms that work together to prevent autoimmune disease (see Fig. 15.2). This collective action means that each mechanism need not work perfectly nor apply to every possible self-reactive cell. Self-tolerance begins during lymphocyte development, when autoreactive T cells in the thymus and B cells in the bone marrow are deleted or, in the case of CD4 T cells, give rise to a subpopulation of self antigen-reactive, natural CD4 CD25 regulatory T (T_{reg}) cells that tend to suppress immune responses. Mechanisms of peripheral tolerance, such as peripheral anergy and deletion, or the extrathymic development of induced T_{reg} cells, complement these central tolerance mechanisms for antigens that are not expressed in the thymus or bone marrow. Weakly self-reactive lymphocytes are not removed at this stage; extending tolerance mechanisms such as deletion to weakly autoreactive cells would impose too great a limitation on the immune repertoire, resulting in impaired immune responses to pathogens. Instead, weakly self-reactive cells are suppressed only if they are activated, by mechanisms that include inhibition by T_{reg} cells, which are themselves autoreactive, although not pathogenic. Regulatory T cells can inhibit a variety of self-reactive lymphocytes, as long as the regulatory cells are targeting autoantigens located in the same general vicinity of the autoantigens to which the self-reactive lymphocytes respond. This allows the regulatory cells to home to and suppress sites of autoimmune inflammation. A final mechanism that controls autoimmunity is the natural tendency of immune responses to be self-limited: intrinsic programs in activated lymphocytes make them prone to apoptosis. Activated lymphocytes also acquire sensitivity to external apoptosis-inducing signals, such as those mediated by Fas.

Autoimmune diseases and pathogenic mechanisms.

Here we describe some of the more common clinical autoimmune syndromes, and the ways in which loss of self-tolerance and expansion of self-reactive lymphocytes lead to tissue damage. These mechanisms of pathogenesis resemble in many ways those that target invading pathogens. Damage by autoantibodies, mediated through the complement and Fc receptor systems, has an important role in some diseases, such as SLE. Similarly, cytotoxic T cells directed at self tissues destroy them much as they would virus-infected cells, and this is one way in which pancreatic β cells are destroyed in diabetes. However, self proteins cannot normally be eliminated, with rare exceptions such as those uniquely expressed by islet cells in the pancreas, and so the response continues. Some pathogenic mechanisms are unique to autoimmunity, such as antibodies against receptors on cell surfaces that affect their function, as in the disease myasthenia gravis, as well as hypersensitivity-type reactions. In this part of the chapter we describe the pathogenic mechanisms of some clinical syndromes of autoimmune disease.

15-8 Specific adaptive immune responses to self antigens can cause autoimmune disease.

In certain genetically susceptible strains of experimental animals, autoimmune disease can be induced artificially by the injection of 'self' tissues from a genetically identical animal that have been mixed with strong adjuvants containing bacteria (see Appendix I, Section A-4). This shows directly that autoimmunity can be provoked by inducing a specific adaptive immune response to self antigens. Such experimental systems highlight the importance of the activation of other components of the immune system, primarily dendritic cells, by the bacteria contained in the adjuvant. There are drawbacks to the use of such animal models for the study of autoimmunity, however. In humans and genetically autoimmune-prone animals, autoimmunity usually arises spontaneously: that is, we do not know what events initiate the immune response to self that leads to the autoimmune disease. By studying the patterns of autoantibodies and also the particular tissues affected, it has been possible to identify some of the self antigens that are targets of autoimmune disease, although it has still to be proved that the immune response was initiated in response to these same antigens.

Some autoimmune disorders may be triggered by infectious agents that express epitopes resembling self antigens and that lead to sensitization of the patient against that tissue. There is, however, also evidence from animal models of autoimmunity that many autoimmune disorders are caused by internal dysregulation of the immune system without the apparent participation of infectious agents.

15-9 Autoimmune diseases can be classified into clusters that are typically either organ-specific or systemic.

The classification of disease is an uncertain science, especially in the absence of a precise understanding of causative mechanisms. This is well illustrated by the difficulty in classifying the autoimmune diseases. From a clinical perspective it is often useful to distinguish between the following two major patterns of autoimmune disease: the diseases in which the expression of autoimmunity is restricted to specific organs of the body, known as 'organ-specific' autoimmune diseases; and those in which many tissues of the body are affected, the 'systemic' autoimmune diseases. In both types of autoimmunity, disease has a tendency to become chronic because, with a few notable exceptions (for example type 1 diabetes or Hashimoto's thyroiditis), the autoantigens are never cleared from the body. Some autoimmune diseases seem to be dominated by the pathogenic effects of a particular immune effector pathway, either autoantibodies or activated autoreactive T cells. However, both of these pathways often contribute to the overall pathogenesis of autoimmune disease.

In organ-specific diseases, autoantigens from one or a few organs only are targeted, and disease is therefore limited to those organs. Examples of organ-specific autoimmune diseases are **Hashimoto's thyroiditis** and **Graves' disease**, both predominantly affecting the thyroid gland, and type 1 diabetes, which is caused by immune attack on insulin-producing pancreatic β cells. Examples of systemic autoimmune disease are SLE and primary Sjögren's syndrome, in which tissues as diverse as the skin, kidneys, and brain may all be affected (Fig. 15.11).

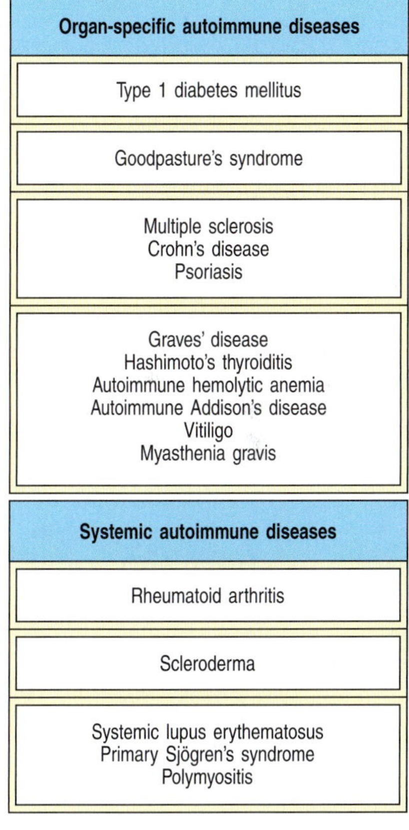

Fig. 15.11 Some common autoimmune diseases classified according to their 'organ-specific' or 'systemic' nature. Diseases that tend to occur in clusters are grouped in single boxes. Clustering is defined as more than one disease affecting a single patient or different members of a family. Not all autoimmune diseases can be classified according to this scheme. For example, autoimmune hemolytic anemia can occur in isolation or in association with systemic lupus erythematosus.

The autoantigens recognized in these two categories of disease are themselves organ-specific and systemic, respectively. Thus, Graves' disease is characterized by the production of antibodies against the thyroid-stimulating hormone (TSH) receptor, which is specific to the thyroid gland, Hashimoto's thyroiditis by antibodies against thyroid peroxidase, and type 1 diabetes by anti-insulin antibodies. By contrast, SLE is characterized by the presence of antibodies against antigens that are ubiquitous and abundant in every cell of the body, such as chromatin and the proteins of the pre-mRNA splicing machinery—the spliceosome complex.

An unusual, but prevalent, variant of chronic inflammatory disease is **inflammatory bowel disease** (**IBD**), which includes two distinct clinical entities—Crohn's disease (discussed later in this chapter) and ulcerative colitis (see Section 15-7). We discuss IBD in this chapter because it has many features of an autoimmune disease, even though it is not primarily targeted against self-tissue antigens. Instead, the targets of the dysregulated immune response in IBD are antigens derived from the commensal microbiota resident in the intestines. Strictly speaking, therefore, IBD is an outlier among autoimmune diseases in that the immune response is not directed against 'self' antigens; rather, it is directed against microbial antigens of the resident, or 'self,' microbiota. Nevertheless, features of immune tolerance breakdown characteristic of other autoimmune diseases are seen in IBD, and, in common with the organ-specific autoimmune diseases, the tissue destruction wrought by the aberrant immune response is primarily, although not exclusively, localized to a single organ—the intestines.

It is likely that the organ-specific and systemic autoimmune diseases have somewhat different etiologies, which provides a biological basis for their division into two broad categories. Evidence for the validity of this classification also comes from observations that different autoimmune diseases cluster within individuals and within families. The organ-specific autoimmune diseases frequently occur together in many combinations; for example, autoimmune thyroid disease and the autoimmune depigmenting disease vitiligo are often found in the same person. Similarly, SLE and Sjögren's syndrome can coexist within a single individual or among different members of a family.

These clusters of autoimmune diseases provide the most useful classification into different subtypes, each of which may turn out to have a distinct mechanism. The classification of autoimmune diseases given in Fig. 15.11 is based on such clustering. A strict separation of diseases into organ-specific and systemic categories does, however, break down to some extent, because not all autoimmune diseases can be usefully classified in this manner. For example, autoimmune hemolytic anemia, in which red blood cells are destroyed, sometimes occurs as a solitary entity and could be classified as an organ-specific disease. In other circumstances it can occur in conjunction with SLE as part of a systemic autoimmune disease.

15-10 Multiple components of the immune system are typically recruited in autoimmune disease.

Immunologists have long been concerned with the issue of which parts of the immune system are important in different autoimmune syndromes, because this can be useful in understanding how a disease is caused and how it is maintained, with the ultimate goal of finding effective therapies. In **myasthenia gravis**, for example, autoantibodies seem to have the main role in causing the disease symptoms. Antibodies produced against the acetylcholine receptor cause blocking of receptor function at the neuromuscular junction, resulting in a syndrome of muscle weakness. In other autoimmune conditions, antibodies in the form of immune complexes are deposited in tissues and cause tissue damage as a result of complement activation and ligation of Fc receptors on inflammatory cells, resulting in inflammation of the affected tissue.

Relatively common autoimmune diseases in which effector T cells seem to be the main destructive agents include type 1 diabetes, psoriasis, IBD, and multiple sclerosis. In these diseases, T cells recognize self peptides or peptides derived from antigens of the commensal microbiota complexed with self-MHC molecules. The damage in such diseases is caused by the T cells recruiting and activating myeloid cells of the innate immune system (for example macrophages and neutrophils) to cause a local inflammation, or by direct T-cell damage to tissue cells. Affected tissues are heavily infiltrated by T lymphocytes and activated myeloid cells.

When disease can be transferred from a diseased individual to a healthy one by transferring autoantibodies and/or self-reactive T cells, this both confirms that the disease is autoimmune in nature and also proves the involvement of the transferred material in the pathological process. In myasthenia gravis, serum from affected patients can transfer similar disease symptoms to animal recipients, thus proving the pathogenic role of the anti-acetylcholine autoantibodies (Fig. 15.12). Similarly, in the animal model disease EAE, T cells from affected animals can transfer disease to normal animals (Fig. 15.13).

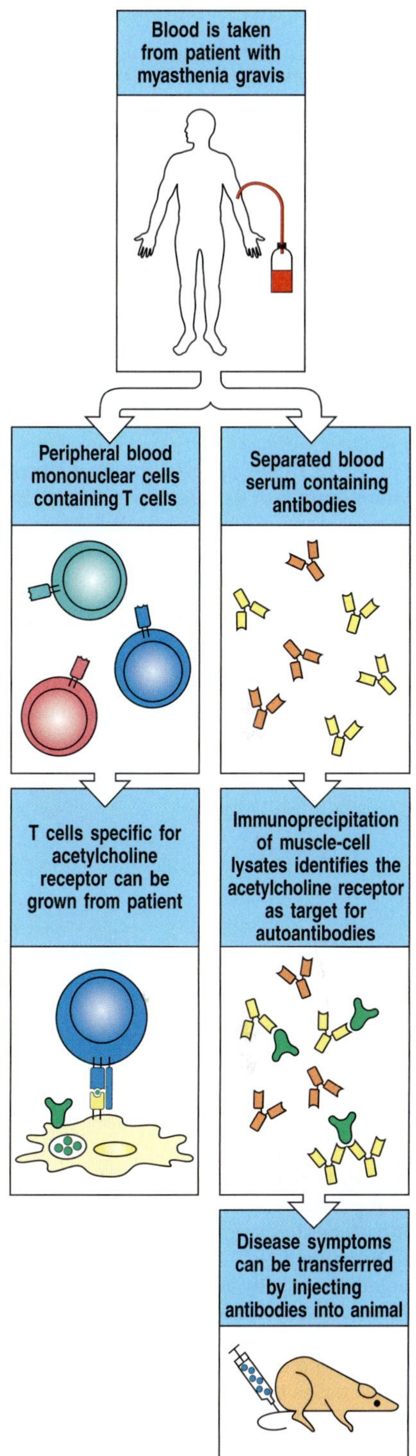

Fig. 15.12 Identification of autoantibodies that can transfer disease in patients with myasthenia gravis. Autoantibodies from the serum of patients with myasthenia gravis immunoprecipitate the acetylcholine receptor from lysates of skeletal muscle cells (right-hand panels). Because they can bind to both the murine and the human acetylcholine receptor, they can transfer disease when injected into mice (bottom panel). This experiment demonstrates that the antibodies are pathogenic. However, to be able to produce antibodies, the same patients should also have CD4 T cells that respond to a peptide derived from the acetylcholine receptor. To detect them, T cells from patients with myasthenia gravis are isolated and grown in the presence of the acetylcholine receptor plus antigen-presenting cells of the correct MHC type (left-hand panels). T cells specific for epitopes of the acetylcholine receptor are stimulated to proliferate and can thus be detected.

Pregnancy is an experiment of nature that can demonstrate a role for antibodies in the causation of disease. IgG antibodies, but not T cells, can cross the placenta (see Section 10-15). For some autoimmune diseases (Fig. 15.14), transmission of autoantibodies across the placenta leads to disease in the fetus or the neonate (Fig. 15.15). This provides proof in humans that such autoantibodies cause some of the symptoms of autoimmunity. The symptoms of disease in the newborn infant typically disappear rapidly as the maternal antibody is catabolized, but in some cases the antibodies cause chronic organ injury before they are removed, such as damage to the conducting tissue of the heart in babies of mothers with SLE or Sjögren's syndrome. Antibody clearance can be speeded up by exchange of the infant's blood or plasma (plasmapheresis), although this is of no clinical use after permanent injury has occurred, as in congenital heart block.

Although the diseases noted above are clear examples that a particular effector function, once established, can cause disease, the idea that most autoimmune diseases are caused solely by a single effector pathway of the immune system is an oversimplification. It is more useful to consider autoimmune responses, like immune responses to pathogens, as engaging the integrated immune system and therefore typically involving T cells, B cells, and innate immune cells. In the non-obese diabetic (NOD) mouse model of type 1 diabetes, for example—a disease that is usually considered to be T-cell mediated—B cells are required for disease initiation. In this case, the B cells are probably functioning as essential antigen-presenting cells for T cells, although the exact details are not clear. A selection of autoimmune diseases showing which parts of the immune response contribute to pathogenesis is given in Fig. 15.16.

15-11 Chronic autoimmune disease develops through positive feedback from inflammation, inability to clear the self antigen, and a broadening of the autoimmune response.

When normal immune responses are engaged to destroy a pathogen, the typical outcome is the elimination of the foreign invader, after which the immune response ceases, leaving only an expanded cohort of memory lymphocytes (see Chapter 11). In autoimmunity, however, the self antigen cannot easily be eliminated, because it is in vast excess or is ubiquitous, as with the SLE autoantigen, chromatin. Thus, a very important mechanism for limiting the extent of an immune response cannot apply to many autoimmune diseases. Instead, autoimmune diseases tend to evolve into a chronic state.

In general, autoimmune diseases are characterized by an early activation phase with the involvement of only a few autoantigens, followed by a chronic stage. The constant presence of autoantigen leads to chronic inflammation. This in turn leads to the release of more autoantigens as a result of tissue damage, and this breaks an important barrier to autoimmunity known

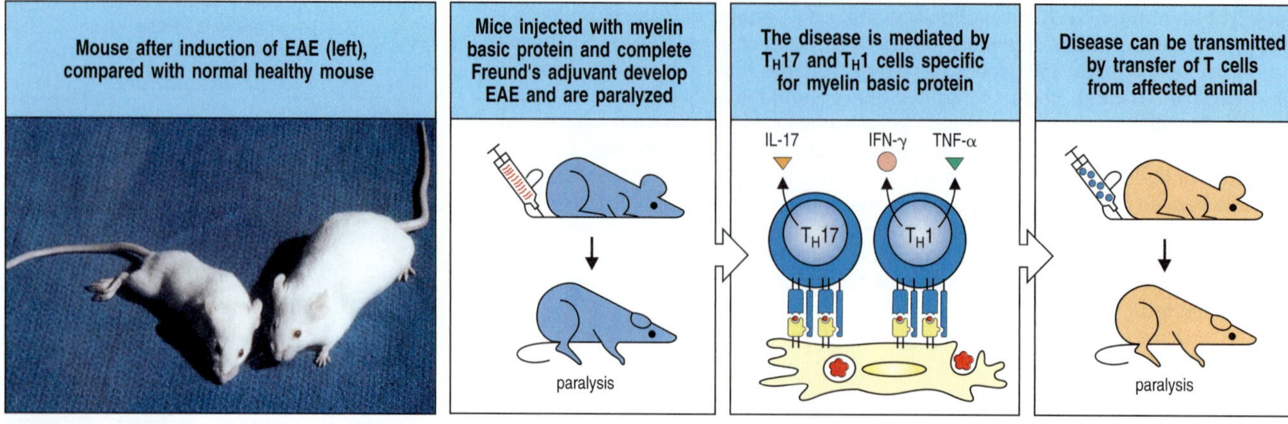

Fig. 15.13 T cells specific for myelin basic protein mediate inflammation of the brain in experimental autoimmune encephalomyelitis (EAE). This disease is produced in experimental animals by injecting them with isolated spinal cord homogenized in complete Freund's adjuvant. EAE is due to an inflammatory reaction in the brain that causes a progressive paralysis affecting first the tail and hind limbs (as shown in the mouse on the left of the photograph, compared with a healthy mouse on the right) before progressing to forelimb paralysis and eventual death. One of the autoantigens identified in the spinal cord homogenate is myelin basic protein (MBP). Immunization with MBP alone in complete Freund's adjuvant can also cause these disease symptoms. Inflammation of the brain and paralysis are mediated by T_H1 and T_H17 cells specific for MBP. Cloned MBP-specific T_H1 cells can transfer symptoms of EAE to naive recipients provided that the recipients carry the correct MHC allele. In this system it has therefore proved possible to identify the peptide:MHC complex recognized by the T_H1 clones that transfer disease. Other purified components of the myelin sheath can also induce the symptoms of EAE, so there is more than one autoantigen in this disease. Photograph from Wraith, D. *et al.*: *Cell* 1989, 59:247–255. With permission from Elsevier.

as 'sequestration,' by which many self antigens are normally kept apart from the immune system. It also leads to the attraction of nonspecific effector cells such as macrophages and neutrophils that respond to the release of cytokines and chemokines from injured tissues (Fig. 15.17). The result is a continuing and evolving self-destructive process.

The transition to the chronic stage is usually accompanied by an extension of the autoimmune response to new epitopes on the initiating autoantigen, and to new autoantigens. This phenomenon is known as **epitope spreading** and is important in perpetuating and amplifying the disease. As we saw in Chapter 10, activated B lymphocytes can efficiently internalize their cognate antigens by receptor-mediated endocytosis via their antigen receptor, process them and present the derived peptides to T cells. Epitope spreading can

Fig. 15.14 Some autoimmune diseases that can be transferred across the placenta by pathogenic IgG autoantibodies. These diseases are caused mostly by autoantibodies against cell-surface or tissue-matrix molecules. This suggests that an important factor determining whether an autoantibody that crosses the placenta causes disease in the fetus or newborn baby is the accessibility of the antigen to the autoantibody. Autoimmune congenital heart block is caused by fibrosis of the developing cardiac conducting tissue, which expresses abundant Ro antigen. Ro protein is a constituent of an intracellular small cytoplasmic ribonucleoprotein. It is not yet known whether it is expressed at the cell surface of cardiac conducting tissue to act as a target for autoimmune tissue injury. Nevertheless, autoantibody binding leads to tissue damage and results in slowing of the heart rate (bradycardia).

Autoimmune diseases transferred across the placenta to the fetus and newborn infant

Disease	Autoantibody	Symptom
Myasthenia gravis	Anti-acetylcholine receptor	Muscle weakness
Graves' disease	Anti-thyroid-stimulating-hormone (TSH) receptor	Hyperthyroidism
Thrombocytopenic purpura	Anti-platelet antibodies	Bruising and hemorrhage
Neonatal lupus rash and/or congenital heart block	Anti-Ro antibodies Anti-La antibodies	Photosensitive rash and/or bradycardia
Pemphigus vulgaris	Anti-desmoglein-3	Blistering rash

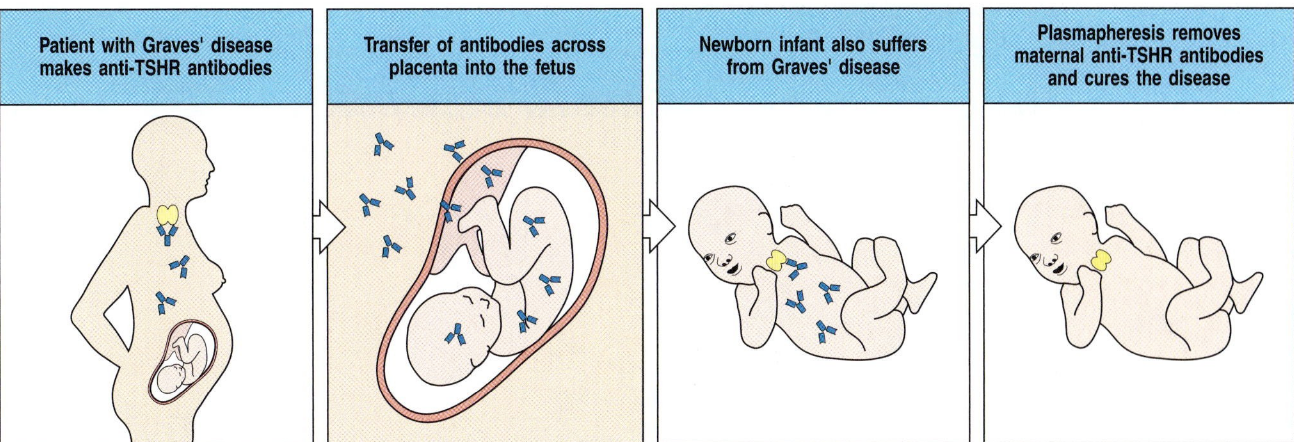

Fig. 15.15 Antibody-mediated autoimmune diseases can appear in the infants of affected mothers as a consequence of transplacental antibody transfer. In pregnant women, IgG antibodies cross the placenta and accumulate in the fetus before birth (see Fig. 10.24). Babies born to mothers with IgG-mediated autoimmune disease therefore frequently show symptoms similar to those of the mother in the first few weeks of life. Fortunately, there is little lasting damage because the symptoms disappear along with the maternal antibody. In Graves' disease, the symptoms are caused by antibodies against the thyroid-stimulating hormone receptor (TSHR). Children of mothers making thyroid-stimulating antibody are born with hyperthyroidism, but this can be corrected by replacing the plasma with normal plasma (plasmapheresis), thus removing the maternal antibody.

occur in several ways. Processing of the internalized autoantigen will reveal a variety of novel, previously hidden, peptide epitopes called **cryptic epitopes**, that the B cell can then present to T cells. Autoreactive T cells responding to these 'new' epitopes will provide help to any B cells presenting these peptides, recruiting additional B-cell clones to the autoimmune reaction, and resulting in the production of a greater variety of autoantibodies. In addition, on binding and internalizing specific antigen via their B-cell receptor, B cells will also internalize any other molecules closely associated with that antigen. By these routes, B cells can act as antigen-presenting cells for peptides derived from proteins completely different from the original autoantigen that initiated the autoimmune reaction.

The autoantibody response in SLE initiates these mechanisms of epitope spreading. In this disease, autoantibodies against both the protein and DNA components of chromatin are found. Figure 15.18 shows how autoreactive B cells specific for DNA can recruit autoreactive T cells specific for histone proteins, another component of chromatin, into the autoimmune response. In

Autoimmune diseases involve all aspects of the immune response			
Disease	**T cells**	**B cells**	**Antibody**
Systemic lupus erythematosus	Pathogenic Help for antibody	Present antigen to T cells	Pathogenic
Type 1 diabetes	Pathogenic	Present antigen to T cells	Present, but role unclear
Myasthenia gravis	Help for antibody	Antibody secretion	Pathogenic
Multiple sclerosis	Pathogenic	Present antigen to T cells	Present, but role unclear

Fig. 15.16 Autoimmune diseases involve all aspects of the immune response. Although some autoimmune diseases have traditionally been thought to be mediated by B cells or T cells, it is useful to consider that, typically, all aspects of the immune system have a role. For four important autoimmune diseases, the figure lists the roles of T cells, B cells, and antibody. In some diseases, such as SLE, T cells can have multiple roles such as helping B cells to make autoantibody and directly promoting tissue damage, whereas B cells can have two roles as well—presenting autoantigens to stimulate T cells and secreting pathogenic autoantibodies.

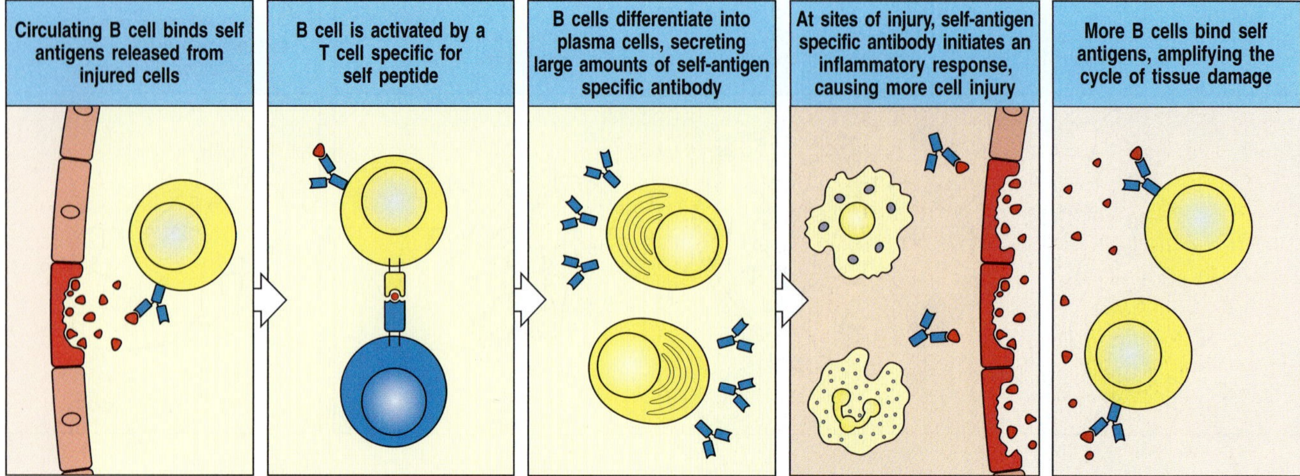
Circulating B cell binds self antigens released from injured cells
B cell is activated by a T cell specific for self peptide
B cells differentiate into plasma cells, secreting large amounts of self-antigen specific antibody
At sites of injury, self-antigen specific antibody initiates an inflammatory response, causing more cell injury
More B cells bind self antigens, amplifying the cycle of tissue damage

Fig. 15.17 Autoantibody-mediated inflammation can lead to the release of autoantigens from damaged tissues, which in turn promotes further activation of autoreactive B cells. Autoantigens, particularly intracellular ones that are targets in SLE, stimulate B cells only when released from dying cells (first panel). The result is the activation of autoreactive T and B cells and the eventual secretion of autoantibodies (second and third panels). These autoantibodies can mediate tissue damage through a variety of effector functions (see Chapter 10) and this results in the further death of cells (fourth panel). A positive feedback loop is established because these additional autoantigens recruit and activate additional autoreactive B cells (fifth panel). These in turn can start the cycle over again, as shown in the first panel.

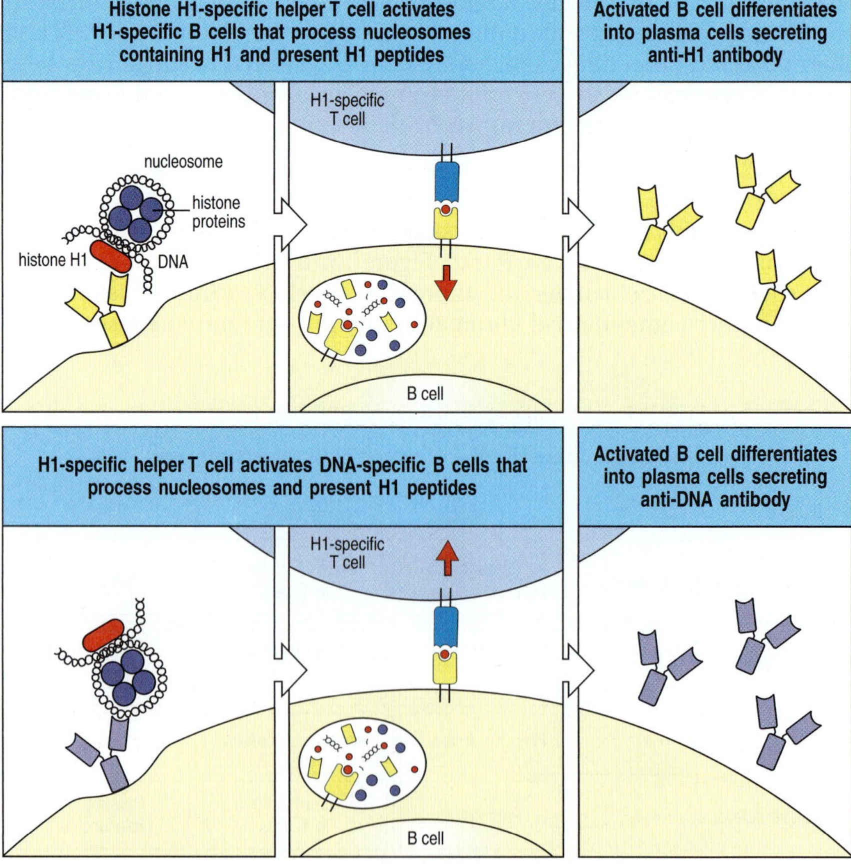
Histone H1-specific helper T cell activates H1-specific B cells that process nucleosomes containing H1 and present H1 peptides
Activated B cell differentiates into plasma cells secreting anti-H1 antibody
H1-specific T cell
nucleosome
histone proteins
histone H1
DNA
B cell
H1-specific helper T cell activates DNA-specific B cells that process nucleosomes and present H1 peptides
Activated B cell differentiates into plasma cells secreting anti-DNA antibody
H1-specific T cell
B cell

Fig. 15.18 Epitope spreading occurs when B cells specific for various components of a complex antigen are stimulated by an autoreactive helper T cell of a single specificity. In SLE, patients often produce autoantibodies against both the DNA and histone protein components of a nucleosome (a subunit of chromatin), or of some other complex antigen. The most likely explanation is that different autoreactive B cells have been activated by a single clone of autoreactive T cells specific for a peptide of one of the proteins in the complex. A B cell binding to any component of the complex through its surface immunoglobulin can internalize the whole complex, degrade it, and return peptides derived from the histone proteins to the cell surface bound to MHC class II molecules, where they stimulate helper T cells. These, in turn, activate the B cells. Thus, a T cell specific for the H1 histone protein of the nucleosome can activate both a B cell specific for H1 (upper panels) and a B cell specific for double-stranded DNA (lower panels). T cells of additional epitope specificities can also become recruited into the response in this way by antigen-presenting B cells bearing a variety of nucleosome-derived peptide:MHC complexes on their surface.

turn, these T cells provide help not only to the original DNA-specific B cells but also to histone-specific B cells, resulting in the production of both anti-DNA and anti-histone antibodies.

An autoimmune disease in which epitope spreading is linked to the progression of disease is **pemphigus vulgaris**, which is characterized by severe blistering of the skin and mucosal membranes. It is caused by autoantibodies against desmogleins, a type of cadherin present in cell junctions (desmosomes) that hold the cells of the epidermis together. Binding of autoantibodies to the extracellular domains of these adhesion molecules causes dissociation of the junctions and dissolution of the affected tissue. Pemphigus vulgaris usually starts with lesions in the oral and genital mucosa; only later does the skin become involved. In the mucosal stage, only autoantibodies against certain epitopes on desmoglein Dsg-3 are found, and these antibodies seem unable to cause skin blistering. Progression to the skin disease is associated both with epitope spreading within Dsg-3, which gives rise to autoantibodies that can cause deep skin blistering, and to another desmoglein, Dsg-1, which is more abundant in the epidermis. Dsg-1 is also the autoantigen in a less severe variant of the disease, pemphigus foliaceus. In that disease, the autoantibodies first produced against Dsg-1 cause no damage; disease appears only after autoantibodies are made against epitopes on parts of the protein involved in the adhesion of epidermal cells.

15-12 Both antibody and effector T cells can cause tissue damage in autoimmune disease.

The manifestations of autoimmune disease are caused by the effector mechanisms of the immune system being directed at the body's own tissues. As discussed previously, the response is usually amplified and maintained by the constant supply of new autoantigen. An important exception to this general rule is type 1 diabetes, in which the autoimmune response destroys the target organ completely. This leads to a failure to produce insulin—one of the major autoantigens in this disease—and it is the lack of insulin that is responsible for the disease symptoms.

The mechanisms of tissue injury in autoimmunity can be classified according to the scheme adopted for hypersensitivity reactions (Fig. 15.19; see also Fig. 14.1). It should be emphasized, however, that both B and T cells are involved in most autoimmune diseases, even in cases where a particular type of response predominates in causing tissue damage. The antigen, or group of antigens, against which the autoimmune response is directed, and the mechanism by which the antigen-bearing tissue is damaged, together determine the pathology and clinical expression of the disease.

Type I IgE-mediated hypersensitive responses play no major part in autoimmunity. By contrast, autoimmunity that damages tissues by mechanisms analogous to type II hypersensitivity reactions is quite common. In this form of autoimmunity, IgG or IgM responses to autoantigens located on cell surfaces or extracellular matrix cause the injury. In other autoimmune diseases, tissue damage is due primarily to type III responses involving the deposition of immune complexes (see Fig. 15.19); in autoimmunity, the immune complexes are composed of soluble autoantigens and their cognate autoantibodies. These autoimmune diseases are systemic and are characterized by autoimmune vasculitis—inflammation of blood vessels. In SLE, autoantibodies cause damage by both type II and type III mechanisms. Finally, several organ-specific autoimmune diseases are due to a type IV response in which T_H1 cells and/or cytotoxic T cells are directly involved in causing tissue damage.

In most autoimmune diseases, however, several mechanisms of immunopathogenesis operate. Notably, helper T cells are almost always required for the production of pathogenic autoantibodies. Reciprocally, B cells often have an important role in the maximal activation of T cells that mediate tissue damage or help autoantibody production (see Section 15-10). In type 1 diabetes and rheumatoid arthritis, for example, which are classed as T cell-mediated diseases, both T-cell and antibody-mediated pathways cause tissue injury. SLE is an example of an autoimmune disease that was previously thought to be mediated solely by antibodies and immune complexes but is now known to have a component of T cell-mediated pathogenesis as well. We will first examine how autoantibodies cause tissue damage, before considering self-reactive T-cell responses and their role in autoimmune disease.

Fig. 15.19 Mechanisms of tissue damage in autoimmune diseases. Autoimmune diseases can be grouped in the same way as hypersensitivity reactions, according to the predominant type of immune response and the mechanism by which it damages tissues. The immunopathological mechanisms given here are those illustrated for the hypersensitivity reactions in Fig. 14.1; type I IgE-mediated responses are not given here because they are not a known cause of autoimmune disease. Some additional autoimmune diseases in which the antigen is a cell-surface receptor, and in which the pathology is due to altered signaling, are listed later, in Fig. 15.23. In many autoimmune diseases, several immunopathogenic mechanisms operate in parallel. This is illustrated here for rheumatoid arthritis, which appears in more than one category of immunopathogenic mechanism.

Some common autoimmune diseases classified by immunopathogenic mechanism

Syndrome	Autoantigen	Consequence
Type II antibody against cell-surface or matrix antigens		
Autoimmune hemolytic anemia	Rh blood group antigens, I antigen	Destruction of red blood cells by complement and FcR^+ phagocytes, anemia
Autoimmune thrombocytopenic purpura	Platelet integrin GpIIb:IIIa	Abnormal bleeding
Goodpasture's syndrome	Noncollagenous domain of basement membrane collagen type IV	Glomerulonephritis, pulmonary hemorrhage
Pemphigus vulgaris	Epidermal cadherin	Blistering of skin
Acute rheumatic fever	Streptococcal cell-wall antigens. Antibodies cross-react with cardiac muscle	Arthritis, myocarditis, late scarring of heart valves
Type III immune-complex disease		
Mixed essential cryoglobulinemia	Rheumatoid factor IgG complexes (with or without hepatitis C antigens)	Systemic vasculitis
Rheumatoid arthritis	Rheumatoid factor IgG complexes	Arthritis
Type IV T-cell-mediated disease		
Type 1 diabetes	Pancreatic β-cell antigen	β-cell destruction
Rheumatoid arthritis	Unknown synovial joint antigen	Joint inflammation and destruction
Multiple sclerosis	Myelin basic protein, proteolipid protein, myelin oligodendrocyte glycoprotein	Brain invasion by CD4 T cells, muscle weakness, and other neurological symptoms
Crohn's disease	Antigens of intestinal microbiota	Regional intestinal inflammation and scarring
Psoriasis	Unknown skin antigens	Inflammation of skin with formation of plaques

15-13 Autoantibodies against blood cells promote their destruction.

IgG or IgM responses to antigens located on the surface of blood cells lead to the rapid destruction of these cells. An example of this is **autoimmune hemolytic anemia**, in which antibodies against self antigens on red blood cells trigger destruction of the cells, leading to anemia. This can occur in two ways (Fig. 15.20). Red cells with bound IgG or IgM antibody are rapidly cleared from the circulation by interaction with Fc or complement receptors, respectively, on cells of the fixed mononuclear phagocytic system; this occurs particularly in the spleen. Alternatively, the autoantibody-sensitized red cells are lysed by formation of the membrane-attack complex of complement. In **autoimmune thrombocytopenic purpura**, autoantibodies against the GpIIb:IIIa fibrinogen receptor or other platelet-specific surface antigens can cause thrombocytopenia (a depletion of platelets), which can in turn cause hemorrhage.

Lysis of nucleated cells by complement is less common because these cells are better defended by complement regulatory proteins, which protect cells against immune attack by interfering with the activation of complement components and their assembly into a membrane-attack complex (see Section 2-15). Nevertheless, nucleated cells targeted by autoantibodies are still destroyed by cells of the mononuclear phagocytic system. Autoantibodies against neutrophils, for example, cause neutropenia, which increases susceptibility to infection with pyogenic bacteria. In all these cases, accelerated clearance of autoantibody-sensitized cells is the cause of their depletion in the blood. One therapeutic approach to this type of autoimmunity is removal of the spleen, the organ in which the main clearance of red cells, platelets, and leukocytes occurs. Another is the administration of large quantities of nonspecific IgG (termed IVIG, for intravenous immunoglobulin), which among other mechanisms inhibits the Fc receptor-mediated uptake of antibody-coated cells.

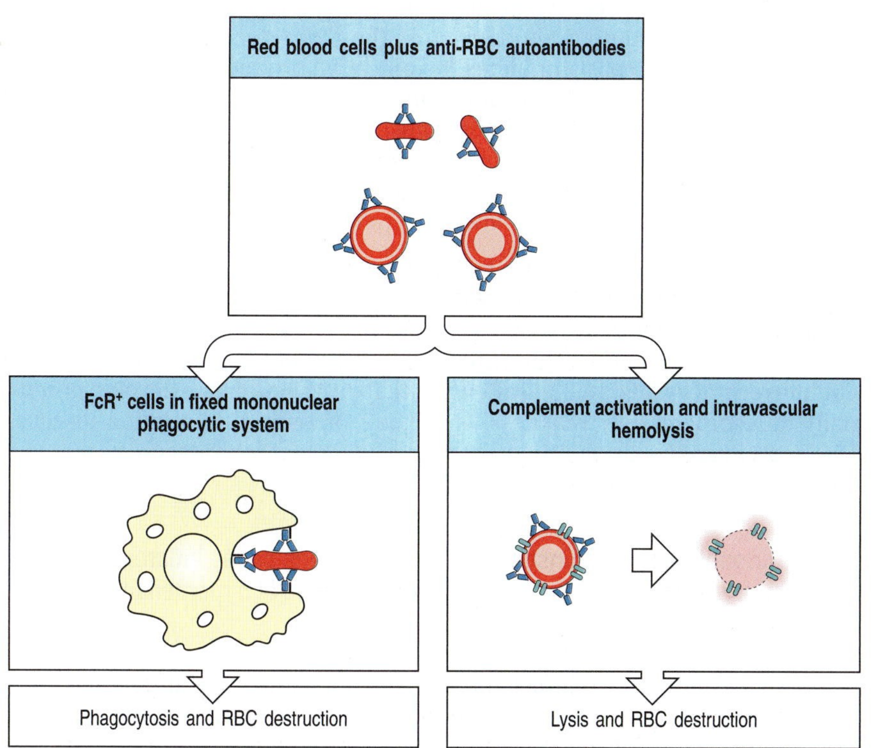

Fig. 15.20 Antibodies specific for cell-surface antigens can destroy cells. In autoimmune hemolytic anemias, red blood cells (RBCs) coated with IgG autoantibodies against a cell-surface antigen are rapidly cleared from the circulation by uptake by Fc receptor-bearing macrophages in the fixed mononuclear phagocytic system (left panel). Red cells coated with IgM autoantibodies fix C3 and are cleared by CR1- and CR3-bearing macrophages in the fixed mononuclear phagocytic system (not shown). Uptake and clearance by these mechanisms occurs mainly in the spleen. The binding of certain rare autoantibodies that fix complement extremely efficiently causes the formation of the membrane-attack complex on the red cells, leading to intravascular hemolysis (right panel).

15-14 The fixation of sublytic doses of complement to cells in tissues stimulates a powerful inflammatory response.

The binding of IgG and IgM antibodies to cells in tissues causes inflammatory injury by a variety of mechanisms. One of these is the fixation of complement. Although nucleated cells are relatively resistant to lysis by complement, the assembly of sublytic amounts of the membrane-attack complex on their surface provides a powerful activating stimulus. Depending on the type of cell, this interaction can cause cytokine release, the generation of a respiratory burst, or the mobilization of membrane phospholipids to generate arachidonic acid—the precursor of prostaglandins and leukotrienes, which are lipid mediators of inflammation.

Most cells in tissues are fixed in place, and innate and adaptive immune cells are attracted to them by chemoattractant molecules. One such molecule is the complement fragment C5a, which is released as a result of complement activation triggered by autoantibody binding. Other chemoattractants, such as leukotriene B4, can be released by the autoantibody-targeted cells. Inflammatory leukocytes are further activated by binding to autoantibody Fc regions and fixed complement C3 fragments on the tissue cells. Tissue injury can then result from the products of the activated leukocytes and by antibody-dependent cellular cytotoxicity mediated by NK cells (see Section 10-23).

A probable example of this type of autoimmunity is Hashimoto's thyroiditis, in which autoantibodies against tissue-specific antigens such as thyroid peroxidase and thyroglobulin are found at extremely high levels for prolonged periods. Direct T cell-mediated cytotoxicity, which we discuss later, is probably also important in this disease.

15-15 Autoantibodies against receptors cause disease by stimulating or blocking receptor function.

A special class of type II hypersensitivity reaction occurs when the autoantibody binds to a cell-surface receptor. Antibody binding to a receptor can either stimulate the receptor or block its stimulation by its natural ligand. In Graves' disease, autoantibody against the thyroid-stimulating hormone receptor on thyroid cells stimulates the excessive production of thyroid hormone. The production of thyroid hormone is normally controlled by feedback regulation; high levels of thyroid hormone inhibit the release of thyroid-stimulating hormone (TSH) by the pituitary. In Graves' disease, feedback inhibition fails because the autoantibody continues to stimulate the TSH receptor in the absence of TSH, and the patient becomes hyperthyroid (Fig. 15.21).

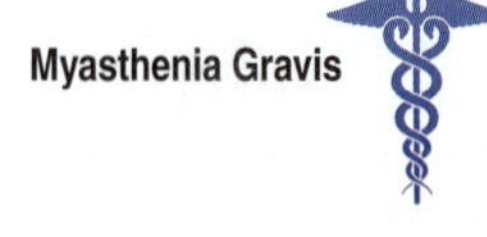

In myasthenia gravis, autoantibodies against the α chain of the nicotinic acetylcholine receptor, which is present on skeletal muscle cells at neuromuscular junctions, can block neuromuscular transmission. The antibodies are believed to drive the internalization and intracellular degradation of acetylcholine receptors (Fig. 15.22). Patients with myasthenia gravis develop potentially fatal progressive weakness as a result of their autoimmune disease. Diseases caused by autoantibodies that act as agonists or antagonists for cell-surface receptors are listed in Fig. 15.23.

15-16 Autoantibodies against extracellular antigens cause inflammatory injury by mechanisms akin to type II and type III hypersensitivity reactions.

Antibody responses to extracellular matrix molecules are infrequent, but they can be very damaging when they occur. In **Goodpasture's syndrome**, an example of a type II hypersensitivity reaction (see Fig. 14.1), antibodies are

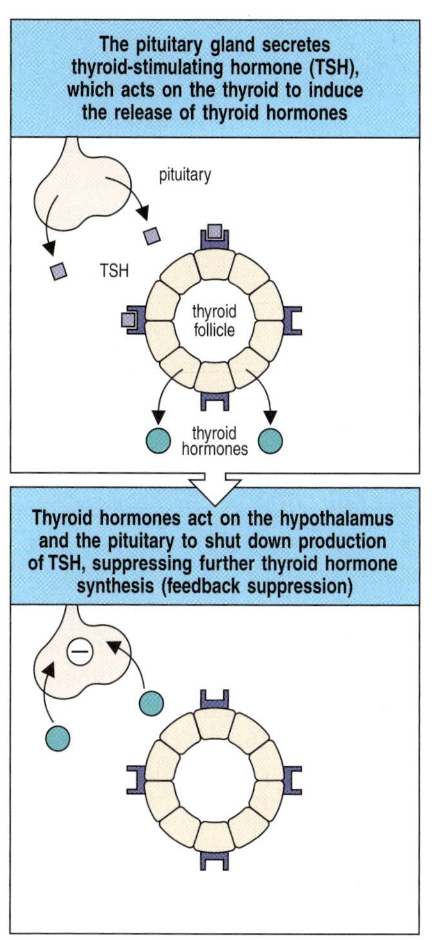

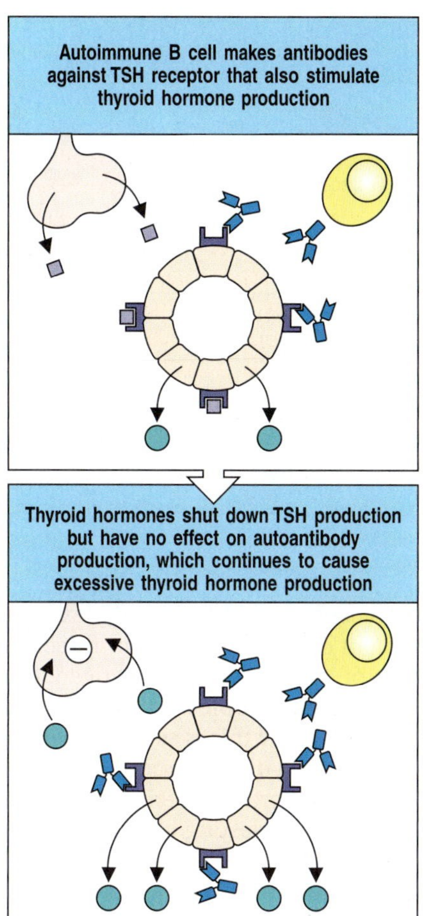

Fig. 15.21 Feedback regulation of thyroid hormone production is disrupted in Graves' disease. Graves' disease is caused by autoantibodies specific for the receptor for thyroid-stimulating hormone (TSH). Normally, thyroid hormones are produced in response to TSH and limit their own production by inhibiting the production of TSH by the pituitary (left panels). In Graves' disease, the autoantibodies are agonists for the TSH receptor and therefore stimulate the production of thyroid hormones (right panels). The thyroid hormones inhibit TSH production in the normal way but do not affect production of the autoantibody; the excessive thyroid hormone production induced in this way causes hyperthyroidism.

formed against the α_3 chain of basement membrane collagen (type IV collagen). These antibodies bind to the basement membranes of renal glomeruli (Fig. 15.24a) and, in some cases, to the basement membranes of pulmonary alveoli, causing a rapidly fatal disease if untreated. The autoantibodies bound to basement membrane ligate Fcγ receptors, leading to the activation of monocytes, neutrophils, and tissue basophils and mast cells. These release chemokines that attract a further influx of neutrophils into glomeruli, causing severe tissue injury (Fig. 15.24b). The autoantibodies also cause a local activation of complement, which may amplify the tissue injury.

Immune complexes are produced whenever there is an antibody response to a soluble antigen (see Appendix I, Section A-8). They are normally cleared efficiently by red blood cells bearing complement receptors and by phagocytes of the mononuclear phagocytic system that have both complement and

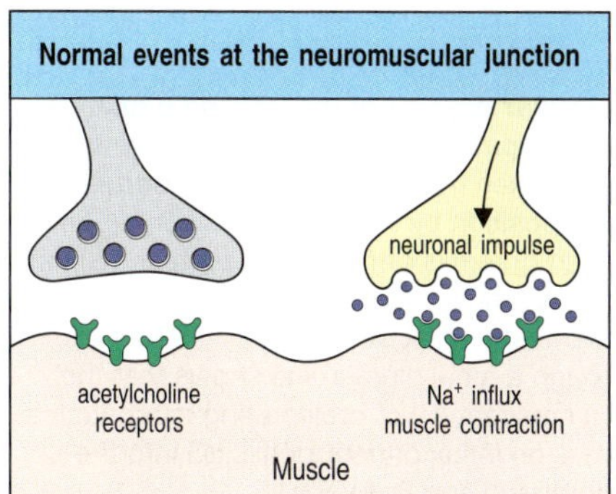

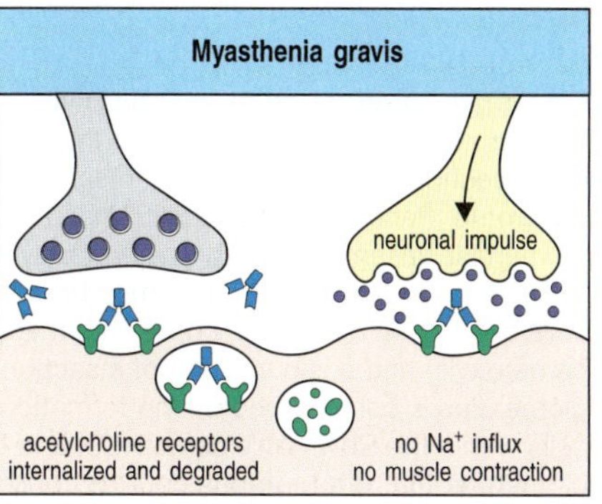

Fig. 15.22 Autoantibodies inhibit receptor function in myasthenia gravis. In normal circumstances, acetylcholine released from stimulated motor neurons at the neuromuscular junction binds to acetylcholine receptors on skeletal muscle cells, triggering muscle contraction (left panel). Myasthenia gravis is caused by autoantibodies against the α subunit of the receptor for acetylcholine. These autoantibodies bind to the receptor without activating it and also cause receptor internalization and degradation (right panel). As the number of receptors on the muscle is decreased, the muscle becomes less responsive to acetylcholine.

Fig. 15.23 Autoimmune diseases caused by autoantibodies against cell-surface receptors. These antibodies produce different effects depending on whether they are agonists (which stimulate the receptor) or antagonists (which inhibit it). Note that different autoantibodies against the insulin receptor can either stimulate or inhibit signaling.

Diseases mediated by autoantibodies against cell-surface receptors		
Syndrome	**Antigen**	**Consequence**
Graves' disease	Thyroid-stimulating hormone receptor	Hyperthyroidism
Myasthenia gravis	Acetylcholine receptor	Progressive weakness
Insulin-resistant diabetes (type 2 diabetes)	Insulin receptor (antagonist)	Hyperglycemia, ketoacidosis
Hypoglycemia	Insulin receptor (agonist)	Hypoglycemia
Chronic urticaria	Receptor-bound IgE or IgE receptor (agonist)	Persistent itchy rash

Fc receptors, and such complexes cause little tissue damage. This clearance system can, however, fail in three circumstances. The first follows the injection of large amounts of antigen, leading to the formation of large amounts of immune complexes that overwhelm the normal clearance mechanisms. An example of this is serum sickness (see Section 14-16), which is caused by the injection of large amounts of serum proteins or by small-molecule drugs binding to serum proteins and acting as haptens. Serum sickness is a transient disease, lasting only until the immune complexes have been cleared. The second circumstance is seen in chronic infections such as bacterial endocarditis, in which the immune response to bacteria lodged on a cardiac valve is incapable of clearing the infection. The persistent release of bacterial antigens from the valve infection in the presence of a strong antibacterial antibody response causes widespread immune-complex injury to small blood vessels in organs such as the kidney and the skin. Chronic infections, such as hepatitis C infection, can lead to the production of cryoglobulins and the condition **mixed essential cryoglobulinemia**, in which immune complexes are deposited in joints and tissues.

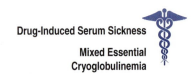

Drug-Induced Serum Sickness

Mixed Essential Cryoglobulinemia

Third, part of the pathogenesis of SLE can also be attributed to the failure to clear immune complexes. In SLE there is chronic IgG antibody production directed at ubiquitous self antigens present in all nucleated cells, leading to a wide range of autoantibodies against common cellular constituents. The main antigens are three types of intracellular nucleoprotein particles—the nucleosome subunits of chromatin, the spliceosome, and a small cytoplasmic ribonucleoprotein complex containing two proteins known as Ro and La (named after the first two letters of the surnames of the two patients in whom autoantibodies against these proteins were discovered). For these autoantigens to participate in immune-complex formation, they must become extracellular.

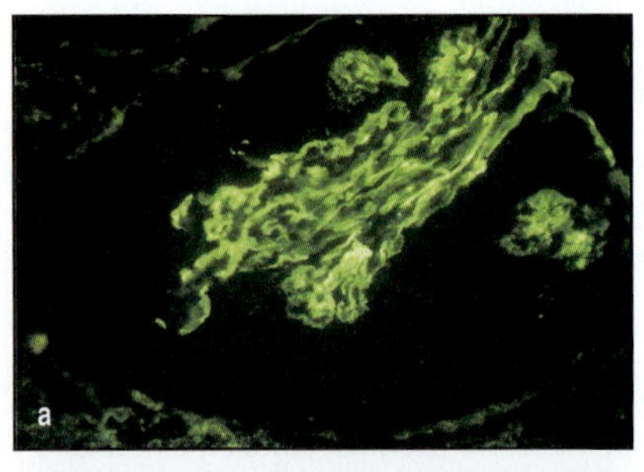

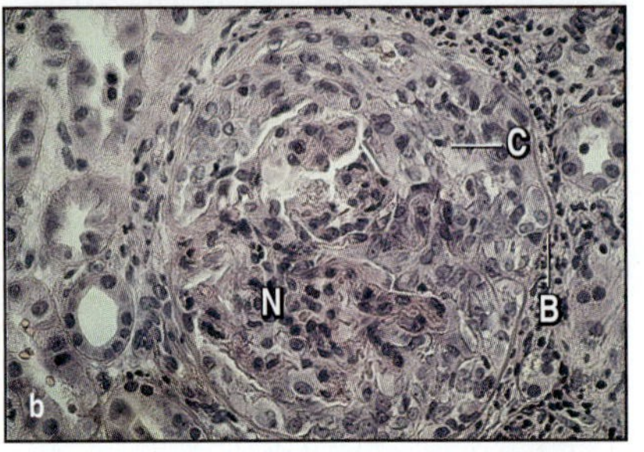

Fig. 15.24 Autoantibodies reacting with glomerular basement membrane cause the inflammatory glomerular disease known as Goodpasture's syndrome. The panels show sections of renal glomeruli in serial biopsies taken from patients with Goodpasture's syndrome. Panel a: glomerulus stained for IgG deposition by immunofluorescence. Anti-glomerular basement membrane antibody (stained green) is deposited in a linear fashion along the glomerular basement membrane. The autoantibody causes local activation of cells bearing Fc receptors, complement activation, and influx of neutrophils. Panel b: hematoxylin and eosin staining of a section through a renal glomerulus shows that the glomerulus is compressed by the formation of a crescent (C) of proliferating mononuclear cells within the Bowman's capsule (B) and there is an influx of neutrophils (N) into the glomerular tuft. Photographs courtesy of M. Thompson and D. Evans.

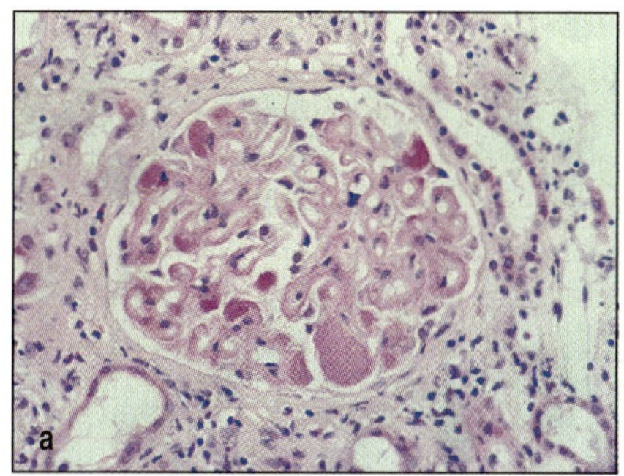

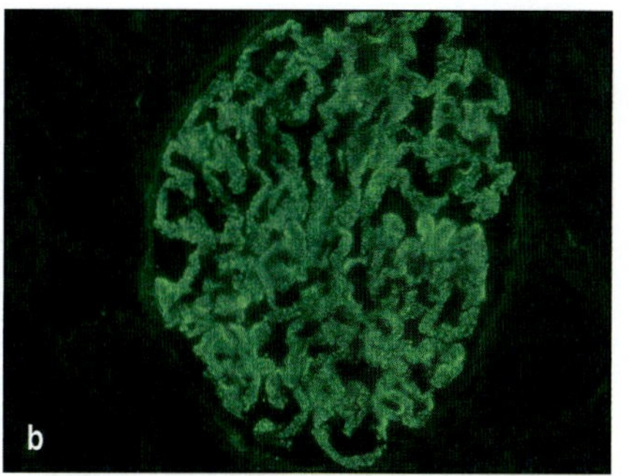

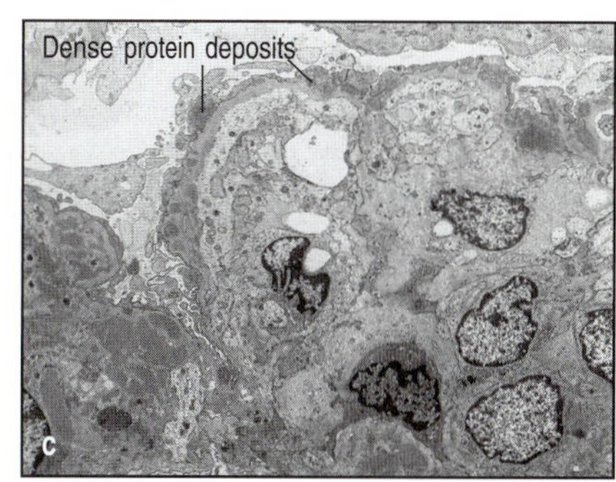

Fig. 15.25 Deposition of immune complexes in the renal glomerulus causes renal failure in systemic lupus erythematosus (SLE). Panel a: a section through a renal glomerulus from a patient with SLE, showing that the deposition of immune complexes has caused thickening of the glomerular basement membrane, seen as the clear 'canals' running through the glomerulus. Panel b: a similar section stained with fluorescent anti-immunoglobulin, revealing immunoglobulin deposits in the basement membrane. In panel c, the immune complexes are seen under the electron microscope as dense protein deposits between the glomerular basement membrane and the renal epithelial cells. Polymorphonuclear neutrophilic leukocytes are also present, attracted by the deposited immune complexes. Photographs courtesy of H.T. Cook and M. Kashgarian.

The autoantigens of SLE are exposed on dead and dying cells and are released from injured tissues. In SLE, large quantities of antigen are available, so large amounts of small immune complexes are produced continuously and are deposited in the walls of small blood vessels in the renal glomerulus, in glomerular basement membrane (Fig. 15.25), in joints, and in other organs. This leads to the activation of phagocytic cells through their Fc receptors. The consequent tissue damage releases more nucleoprotein complexes, which in turn form more immune complexes. During this process, autoreactive T cells also become activated, although much less is known about their specificity. The experimental animal models for SLE cannot be initiated without the help of T cells, and T cells can also be directly pathogenic, forming part of the cellular infiltrates in skin and the interstitial areas of the kidney. As we discuss in the next section, T cells contribute to autoimmune disease in two ways: by helping B cells to make antibodies, in an analogous manner to a normal T-dependent immune response, and by direct effector functions of T cells as they infiltrate and destroy target tissues such as skin, renal interstitium, and vessels. Eventually, the inflammation induced in these tissues can cause sufficient damage to kill the patient.

15-17 T cells specific for self antigens can cause direct tissue injury and sustain autoantibody responses.

It is much more difficult to demonstrate the existence of autoreactive T cells than it is to demonstrate the presence of autoantibodies. First, autoreactive human T cells cannot be used to transfer disease to experimental animals because T-cell recognition is MHC-restricted, and animals and humans have different MHC alleles. Second, it is difficult to identify the antigen recognized by a T cell; for example, autoantibodies can be used to stain self tissues to reveal the distribution of the autoantigen, whereas T cells cannot be used in the same way. Nevertheless, there is strong evidence for the involvement of autoreactive T cells in several autoimmune diseases. In type 1 diabetes, for example, the insulin-producing β cells of the pancreatic islets of Langerhans are selectively destroyed by specific cytotoxic T cells. In rare cases in which patients with diabetes were transplanted with half a pancreas from an identical twin donor, the β cells in the grafted tissue were rapidly and selectively destroyed by the recipient's T cells. Recurrence of disease can be prevented by the immunosuppressive drug cyclosporin A (see Chapter 16), which inhibits T-cell activation.

Autoantigens recognized by CD4 T cells can be identified by adding cells or tissues, containing autoantigens, to cultures of blood mononuclear cells and testing for recognition by CD4 cells derived from an autoimmune patient. If the autoantigen is present, it should be effectively presented, because phagocytes in the blood cultures can take up extracellular protein, degrade it in intracellular vesicles, and present the resulting peptides bound to MHC class II molecules. The identification of autoantigenic peptides is particularly difficult in autoimmune diseases in which CD8 T cells have a role, because autoantigens recognized by CD8 T cells are not effectively presented in such cultures. Peptides presented by MHC class I molecules must usually be made by the target cells themselves (see Chapter 6); intact cells of target tissue from the patient must therefore be used to study autoreactive CD8 T cells that cause tissue damage. Conversely, the pathogenesis of the disease can itself give clues to the identity of the antigen in some CD8 T cell-mediated diseases. For example, in type 1 diabetes, the insulin-producing β cells seem to be specifically targeted and destroyed by CD8 T cells (Fig. 15.26). This suggests that a protein unique to β cells is the source of the peptide recognized by the pathogenic CD8 T cells. Studies in the NOD mouse model of type 1 diabetes have shown that peptides from insulin itself are recognized by pathogenic CD8 cells, confirming insulin as one of the principal autoantigens in this diabetes model.

Multiple sclerosis is an example of a T cell-mediated chronic neurological disease that is caused by a destructive immune response against several brain antigens, including myelin basic protein (MBP), proteolipid protein (PLP), and myelin oligodendrocyte glycoprotein (MOG). It takes its name from the hard (sclerotic) lesions, or plaques, that develop in the white matter of the central nervous system. These lesions show dissolution of the myelin that normally sheathes nerve cell axons, along with inflammatory infiltrates of lymphocytes and macrophages, particularly along the blood vessels. Patients with multiple sclerosis develop a variety of neurological symptoms, including muscle weakness, ataxia, blindness, and paralysis of the limbs. Lymphocytes and other blood cells do not normally cross the blood–brain barrier, but if the

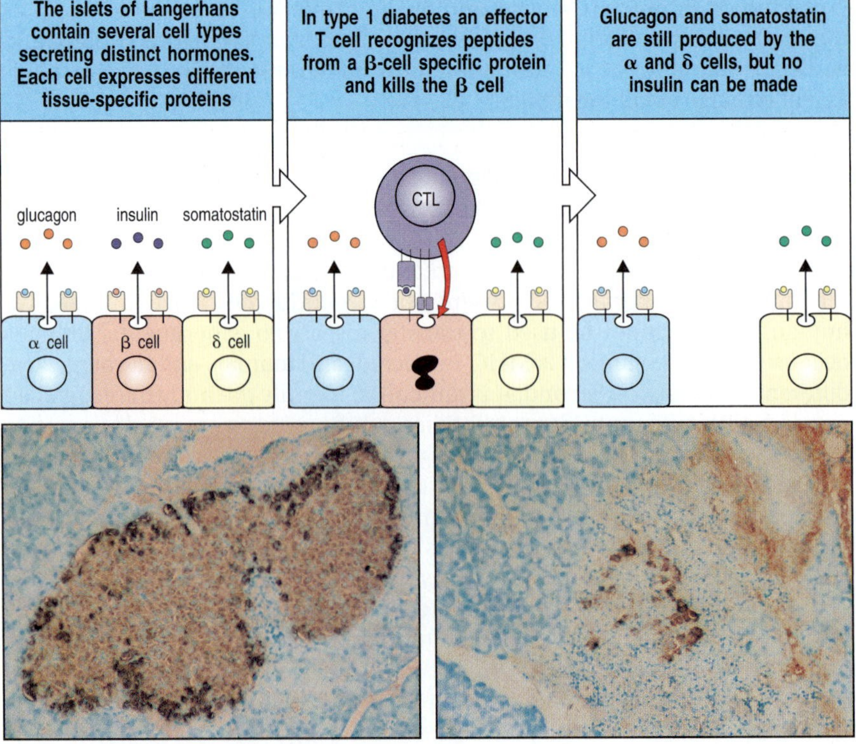

Fig. 15.26 Selective destruction of pancreatic β cells in type 1 diabetes indicates that the autoantigen is produced in β cells and recognized on their surface. In type 1 diabetes there is highly specific destruction of insulin-producing β cells in the pancreatic islets of Langerhans, sparing other islet cell types (α and δ). This is shown schematically in the upper panels. In the lower panels, islets from normal (left) and diabetic (right) mice are stained for insulin (brown), which shows the β cells, and for glucagon (black), which shows the α cells. Note the lymphocytes infiltrating the islet in the diabetic mouse (right) and the selective loss of the β cells (brown), whereas the α cells (black) are spared. The characteristic morphology of the islet is also disrupted with the loss of the β cells. Photographs courtesy of I. Visintin.

brain and its blood vessels become inflamed, the blood–brain barrier breaks down. When this happens, activated CD4 T cells autoreactive for brain antigens and expressing α_4:β_1 integrin can bind vascular cell adhesion molecules (VCAM) on the surface of the activated venule endothelium (see Section 11-6), enabling the T cells to migrate out of the blood vessel. There they reencounter their specific autoantigen presented by MHC class II molecules on microglial cells (Fig. 15.27). Microglia are phagocytic macrophage-like cells of the innate immune system resident in the central nervous system and, like macrophages, can act as antigen-presenting cells. Inflammation causes increased vascular permeability and the site becomes heavily infiltrated by T_H17 and T_H1 cells, which produce IL-17 and IFN-γ, respectively. Cytokines and chemokines produced by the infiltrating effector T cells in turn recruit and activate myeloid cells that exacerbate the inflammation, resulting in the further recruitment of T cells, B cells, and innate immune cells to the site of the lesion. Autoreactive B cells produce autoantibodies against myelin antigens with help from T cells. These combined activities lead to demyelination and interference with neuronal function.

Rheumatoid arthritis (**RA**) is a chronic disease characterized by inflammation of the synovium (the thin lining of a joint). As the disease progresses, the inflamed synovium invades and damages the cartilage, followed by erosion of the bone (Fig. 15.28). Patients with rheumatoid arthritis suffer chronic pain, loss of function, and disability. Rheumatoid arthritis was at first considered an autoimmune disease driven mainly by B cells producing anti-IgG autoantibodies called rheumatoid factor (see Section 15-4). However, the identification of rheumatoid factor in some healthy individuals, and its absence in some patients with rheumatoid arthritis, suggested that more complex mechanisms orchestrate this pathology. The discovery that rheumatoid arthritis has an association with particular class II HLA-DR genes of the MHC suggested that T cells were involved in the pathogenesis of this disease. In rheumatoid arthritis, as in multiple sclerosis, autoreactive CD4 T cells become activated by dendritic cells and by inflammatory cytokines produced by macrophages. Once activated, the autoreactive T cells provide help to B cells to differentiate into plasma cells producing arthritogenic antibodies. Autoantigens such as type II collagen, proteoglycans, aggrecan, cartilage link protein, and heat-shock proteins have been proposed as potential antigens because of their ability to induce arthritis in mice. Their pathogenic role in humans remains to be ascertained, however. The activated T cells produce cytokines, which in turn stimulate monocytes/macrophages, endothelial cells, and fibroblasts to produce more pro-inflammatory cytokines such as TNF-α, IL-1 and IFN-γ, or chemokines (CXCL8, CCL2), and finally matrix metalloproteinases, which

Fig. 15.27 The pathogenesis of multiple sclerosis. At sites of inflammation, activated T cells autoreactive for brain antigens can cross the blood–brain barrier and enter the brain, where they reencounter their antigens on microglial cells and secrete cytokines such as IFN-γ. The production of T-cell and macrophage cytokines exacerbates the inflammation and induces a further influx of blood cells (including macrophages, dendritic cells, and B cells) and blood proteins (such as complement) into the affected site. Mast cells also become activated. The individual roles of these components in demyelination and loss of neuronal function are still not well understood. CNS, central nervous system.

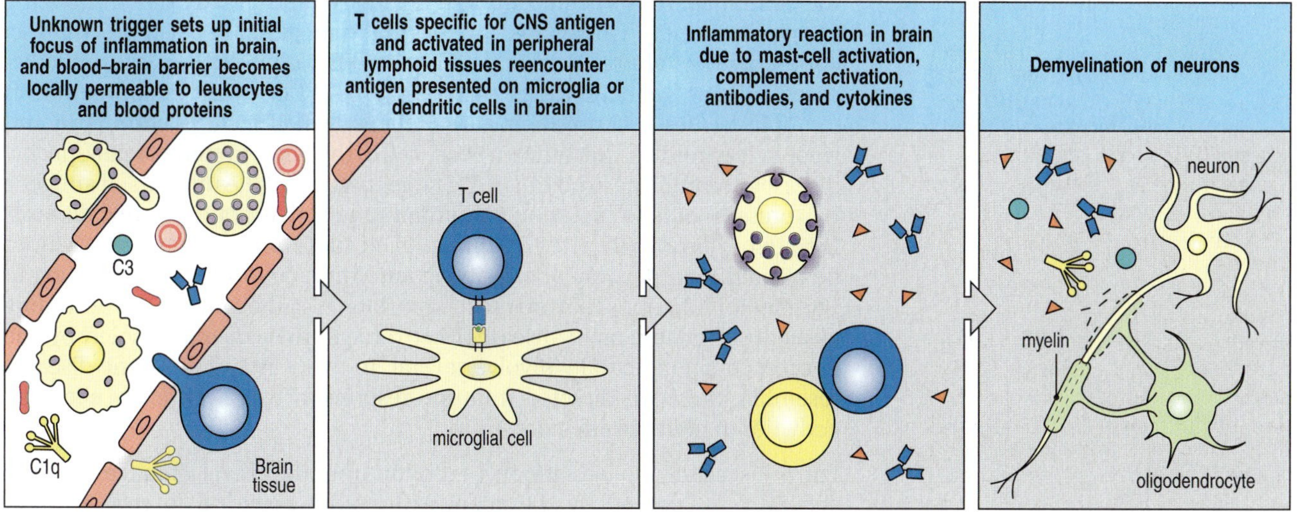

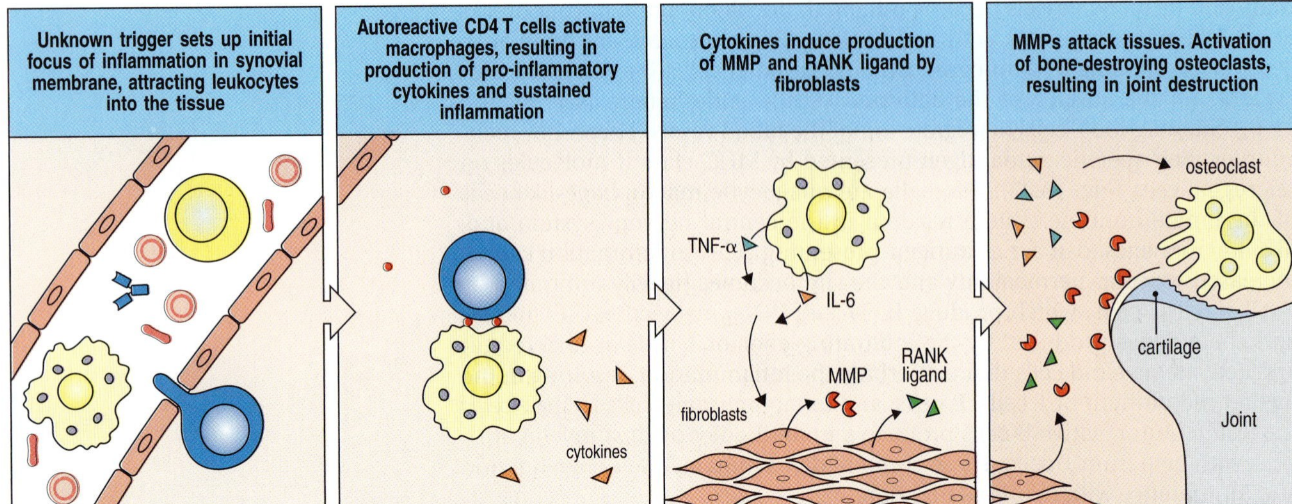

Fig. 15.28 The pathogenesis of rheumatoid arthritis. Inflammation of the synovial membrane, initiated by some unknown trigger, attracts autoreactive lymphocytes and macrophages to the inflamed tissue. Autoreactive effector CD4 T cells activate macrophages, with the production of pro-inflammatory cytokines such as IL-1, IL-6, IL-17, and TNF-α. Fibroblasts activated by cytokines produce matrix metalloproteinases (MMPs), which contribute to tissue destruction. The TNF family cytokine RANK ligand, expressed by T cells and fibroblasts in the inflamed joint, is the primary activator of bone-destroying osteoclasts. Antibodies against several joint proteins are also produced (not shown), but their role in pathogenesis is uncertain.

are responsible for tissue destruction. Therapeutic antibodies against TNF-α have been successful in treating the symptoms of the disease (discussed in Section 16-8). However, it needs to be realized that in rheumatoid arthritis, as in many other autoimmune diseases, we do not yet know how disease starts. Mouse models of rheumatoid arthritis teach us that both T cells and B cells are needed to initiate the disease, because mice lacking T cells or B cells are resistant to its development.

Rheumatoid Arthritis

Systemic Onset Juvenile Idiopathic Arthritis

Summary.

Autoimmune diseases can be broadly classified into those that affect a specific organ and those that affect tissues throughout the body. Organ-specific autoimmune diseases include diabetes, multiple sclerosis, psoriasis, Crohn's disease, myasthenia gravis, and Graves' disease. In each case the effector functions target autoantigens that are restricted to particular organs: insulin-producing β cells of the pancreas (diabetes), the myelin sheathing axons in the central nervous system (multiple sclerosis), and the thyroid-stimulating hormone receptor (Graves' disease), or, in the special case of Crohn's disease, components of the resident intestinal microbiota. In contrast, systemic diseases such as systemic lupus erythematosus (SLE) cause inflammation in multiple tissues because their autoantigens, which include chromatin and ribonucleoproteins, are found in every cell of the body. In some organ-specific diseases, immune destruction of the target tissue and the unique self antigens it expresses leads to cessation of autoimmune activity, but systemic diseases tend to be chronically active if untreated, because their autoantigens cannot be cleared. Another way of classifying autoimmune diseases is according to the effector functions that are most important in pathogenesis. It is becoming clear, however, that many diseases once thought to be mediated solely by one effector function actually involve several. In this way, autoimmune diseases resemble pathogen-directed immune responses, which typically elicit the activities of multiple effectors.

For a disease to be classified as autoimmune, the tissue damage must be shown to be caused by the adaptive immune response to self antigens.

Autoinflammatory reactions directed against the commensal microbiota of the intestines, such as those that cause inflammatory bowel disease such as Crohn's, are a special case in that the target antigens are not strictly 'self,' but are derived from the 'self' intestinal microbiota. The IBDs do, nevertheless, share immunopathogenic features with other autoimmune diseases. The most convincing proof that the immune response is causal in autoimmunity is the transfer of disease by transferring the active component of the immune response to an appropriate recipient. Autoimmune diseases are mediated by autoreactive lymphocytes and/or their soluble products, pro-inflammatory cytokines, and autoantibodies responsible for inflammation and tissue injury. A few autoimmune diseases are caused by antibodies that bind to cell-surface receptors, causing either excess activity or inhibition of receptor function. In these diseases, transplacental passage of natural IgG autoantibodies can cause disease in the fetus and in the neonate. T cells can be involved directly in inflammation or cellular destruction, and they are also required to sustain an autoantibody response. Similarly, B cells are important antigen-presenting cells for sustaining autoantigen-specific T-cell responses and for causing epitope spreading. In spite of our knowledge of the mechanisms of tissue damage and the therapeutic approaches that this information has engendered, the deeper, more important question is how the autoimmune response is induced.

The genetic and environmental basis of autoimmunity.

Given the complex and varied mechanisms that exist to prevent autoimmunity, it is not surprising that autoimmune diseases are the result of multiple factors, both genetic and environmental. We first discuss the genetic basis of autoimmunity, attempting to understand how genetic defects perturb the various tolerance mechanisms. Genetic defects alone are not, however, always sufficient to cause autoimmune disease. Environmental factors such as toxins, drugs, and infections also play a part, although these factors are poorly understood. As we shall see, genetic and environmental factors together can overcome tolerance mechanisms and result in autoimmune disease.

15-18 Autoimmune diseases have a strong genetic component.

Although the causes of autoimmunity are still being worked out, it is increasingly clear that some individuals are genetically predisposed to autoimmunity. Perhaps the clearest demonstration of this is found in the several inbred mouse strains that are prone to various types of autoimmune diseases. For example, mice of the NOD strain are very likely to get diabetes. The female mice become diabetic more quickly than the males (Fig. 15.29). Many autoimmune diseases are more common in females than in males (see Fig. 15.33 below), although occasionally the opposite is true. Autoimmune diseases in humans also have a genetic component. Some autoimmune diseases, including type 1 diabetes, run in families, suggesting a role for genetic susceptibility. Most convincingly, if one of two identical (monozygotic) twins is affected, then the other twin is quite likely to be as well, whereas concordance of disease is much less in nonidentical (dizygotic) twins.

Environmental influences are also clearly involved. For example, although most of a colony of NOD mice are destined to get diabetes, they will do so at different ages (see Fig. 15.29). Moreover, the timing of disease onset often differs from one investigator's animal colony to the next, even though all the mice are genetically identical. Thus, environmental variables must be, in part, determining the rate of development of diabetes; a few even escape

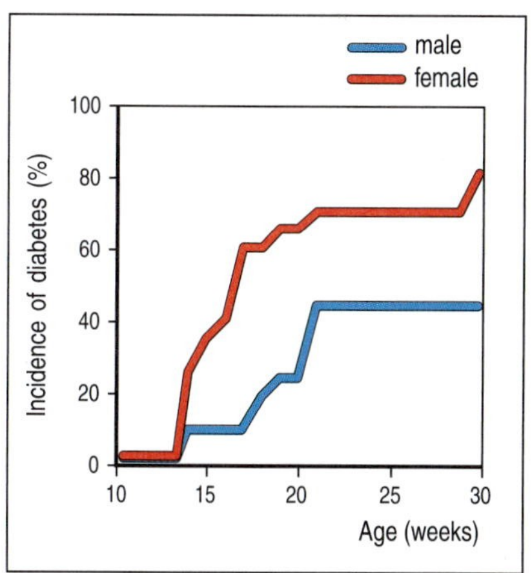

Fig. 15.29 Sex differences in the incidence of autoimmune disease. Many autoimmune diseases are more common in females than males, as illustrated here by the cumulative incidence of diabetes in a population of diabetes-prone NOD mice. Females (red line) get diabetes at a much younger age than do males, indicating their greater predisposition. Data kindly provided by S. Wong.

the disease entirely. Particularly striking is the importance of the intestinal microbiota in the development of inflammatory bowel disease in genetically susceptible mice. Treatment with broad-spectrum antibiotics that reduce or eliminate many components of the commensal flora can delay or eliminate disease onset, and re-derivation of susceptible mice under germ-free conditions eliminates disease. Identical twins tell a similar story. With Crohn's disease, although disease incidence in susceptible monozygotic twins is much higher than in dizygotic twins, the concordance rate is far from 100%. The explanation for the incomplete concordance could lie in variability in the composition of the intestinal microbiota, or it could simply be random.

15-19 Several approaches have given us insight into the genetic basis of autoimmunity.

Since the advent of gene knockout technology in mice (see Section A-47, Appendix I), many genes that encode proteins of the immune system have been experimentally disrupted. Several of these mutant mouse strains have signs of autoimmune disease, including autoantibodies and, in some cases, infiltration of organs by T cells. The study of these mice has greatly expanded our knowledge of the genetic pathways that can contribute to autoimmunity and that therefore might be candidates for naturally occurring mutations. A growing number of genes whose deletion or overexpression can contribute to the pathogenesis of autoimmunity have been identified. These encode cytokines, co-receptors, members of cytokine- or antigen-signaling cascades, co-stimulatory molecules, proteins involved in pathways that promote apoptosis and those that inhibit it, and proteins that clear antigen or antigen:antibody complexes. Some of the cytokines and signaling proteins implicated in autoimmune disease are listed in Fig. 15.30, and Fig. 15.31 lists some of the known associations for other categories of proteins.

Fig. 15.30 Defects in cytokine production or signaling that can lead to autoimmunity. Some of the signaling pathways involved in autoimmunity have been identified by genetic analysis, mainly in animal models. The effects of overexpression or underexpression of some of the cytokines and intracellular signaling molecules involved are listed here (see the text for further discussion).

Defects in cytokine production or signaling that can lead to autoimmunity		
Defect	**Cytokine or intracellular signal**	**Result**
Overexpression	TNF-α	Inflammatory bowel disease, arthritis, vasculitis
	IL-2, IL-7, IL-2R	Inflammatory bowel disease
	IL-3	Demyelinating syndrome
	IFN-γ	Overexpression in skin leads to SLE
	STAT4	Inflammatory bowel disease
Underexpression	TNF-α	SLE
	IL-1 receptor agonist	Arthritis
	IL-10, IL-10R, STAT3	Inflammatory bowel disease
	TGF-β	Ubiquitous underexpression leads to inflammatory bowel disease. Underexpression specifically in T cells leads to SLE

In humans, the association of autoimmunity with a particular gene or genetic region can be assessed by large-scale family studies, or by association studies in the general population (**genome-wide association studies**, or **GWASs**), that look for a correlation between disease frequency and variant alleles,

Proposed mechanism	Murine models	Disease phenotype	Human gene affected	Disease phenotype
Antigen clearance and presentation	C1q knockout	Lupus-like	*C1QA*	Lupus-like
	C4 knockout		*C2, C4*	
			Mannose-binding lectin	
	AIRE knockout	Multiorgan autoimmunity resembling APECED	*AIRE*	APECED
	Mer knockout	Lupus-like		
Signaling	SHP-1 knockout	Lupus-like		
	Lyn knockout			
	CD22 knockout			
	CD45 E613R point mutation			
	B cells deficient in all Src-family kinases (triple knockout)			
	FcγRIIB knockout (inhibitory signaling molecule)		*FCGR2A*	Lupus
Co-stimulatory molecules	CTLA-4 knockout (blocks inhibitory signal)	Lymphocyte infiltration into organs		
	PD-1 knockout (blocks inhibitory signal)	Lupus-like		
	BAFF overexpression (transgenic mouse)			
Apoptosis	Fas knockout (*lpr*)	Lupus-like with lymphocyte infiltrates	*FAS* and *FASL* mutations (ALPS)	Lupus-like with lymphocyte infiltrates
	FasL knockout (*gld*)			
	Bcl-2 overexpression (transgenic mouse)	Lupus-like		
	Pten heterozygous deficiency			
T_{reg} development/function	*scurfy* mouse	Multi-organ autoimmunity	*FOXP3*	IPEX
	foxp3 knockout			

Fig. 15.31 Categories of genetic defects that lead to autoimmune syndromes. Many genes have been identified in which mutations predispose to autoimmunity in humans and animal models. These are best understood by the type of process affected by the genetic defect. A list of such genes is given here, organized by process (see the text for further discussion). In some cases, the same gene has been identified in mice and humans. In other cases, different genes affecting the same mechanism are implicated in mice and humans. The smaller number of human genes identified so far undoubtedly reflects the difficulty of identifying the genes responsible in outbred human populations.

genetic markers, duplications or deletions, or **single-nucleotide polymorphisms** (**SNPs**), positions in the genome that differ by a single base between individuals. These studies have supported the concept that genetic susceptibility to autoimmune disease in humans is typically due to a combination of susceptibility alleles at multiple loci. For example, in large association studies looking at candidate susceptibility genes in humans, several of the commonest autoimmune diseases, including type 1 diabetes, Graves' disease, Hashimoto's thyroiditis, Addison's disease, rheumatoid arthritis, and multiple sclerosis, show genetic association with the *CTLA4* locus on chromosome 2. The cell-surface protein CTLA-4 is produced by activated T cells and is an inhibitory receptor for B7 co-stimulatory molecules (see Section 9-13). The effects of genetic variation in *CTLA4* on susceptibility to type 1 diabetes have been studied in mice. *CTLA4* is located on mouse chromosome 1 in a cluster with the genes for the other co-stimulatory receptors, CD28 and ICOS. When this genetic region in the diabetes-susceptible NOD mouse strain was replaced with the same region from the autoimmune-resistant B10 strain, it conferred diabetes resistance on the NOD mice. It seems that genetic variation in the splicing of the *CTLA4* mRNA may contribute to the difference in susceptibility. Splice variants of CTLA-4 that lack a portion essential for binding to its ligands B7.1 and B7.2 were still resistant to activation, and there was increased expression of this variant in the memory and regulatory T cells of diabetes-resistant mice.

15-20 Many genes that predispose to autoimmunity fall into categories that affect one or more of the mechanisms of tolerance.

The genes identified as predisposing to autoimmunity can be classified as follows: genes that affect autoantigen availability and clearance; those that affect apoptosis; those that affect signaling thresholds; those involved in cytokine gene expression or signaling; those affecting the expression of co-stimulatory molecules or their receptors; and those that affect the development or function of regulatory T cells (see Figs 15.30 and 15.31).

Genes that control antigen availability and clearance are important both centrally in the thymus, where they act to make self proteins available for inducing tolerance in developing lymphocytes, and in the periphery, where they control how self molecules are made available in an immunogenic form to peripheral lymphocytes. In the periphery, a hereditary deficiency of some complement proteins, specifically those for C1q, C2, and C4, is strongly associated with the development of SLE in humans. C1q, C2, and C4 are early components in the classical complement pathway, which is important in antibody-mediated clearance of apoptotic cells and immune complexes (see Chapter 2). If apoptotic cells and immune complexes are not cleared, the chance that their antigens will activate low-affinity self-reactive lymphocytes in the periphery is increased. Genes that control apoptosis, such as *Fas*, are important in regulating the duration and vigor of immune responses. Failure to regulate immune responses properly could cause excessive destruction of self tissues, releasing autoantigens. In addition, because clonal deletion and anergy are not absolute, immune responses can include some self-reactive cells. As long as their numbers are limited by apoptotic mechanisms, they may not be sufficient to cause autoimmune disease, but they could cause a problem if apoptosis is not properly regulated.

Perhaps the largest category of mutations associated with autoimmunity comprises those associated with signals that control lymphocyte activation. One subset contains mutations that inactivate negative regulators of lymphocyte activation and thus lead to the hyperproliferation of lymphocytes and exaggerated immune responses. These include mutations in CTLA-4 (as discussed in Section 15-19), in inhibitory Fc receptors, and in inhibitory receptors containing ITIMs, such as CD22 on B cells, which is a negative regulator of B-cell receptor signaling. Another subset contains mutations in proteins involved in signal transduction through the antigen receptor itself. Adjusting thresholds in either direction, to make signaling more or less sensitive, can result in autoimmunity, depending on the situation. A decrease in sensitivity in the thymus, for example, can lead to a failure of negative selection and thereby to autoreactivity in the periphery. In contrast, increasing receptor sensitivity in the periphery can lead to greater and prolonged activation, again resulting in an exaggerated immune response with the side effect of autoimmunity. Additionally, mutations that affect the expression of genes for cytokines and co-stimulatory molecules have been linked to autoimmunity. A final subset of mutations comprises those that predispose to autoimmunity by compromising regulatory T-cell development or function, as exemplified by the FoxP3 mutations that give rise to the IPEX syndrome (see Section 15-21).

15-21 A defect in a single gene can cause autoimmune disease.

Predisposition to most of the common autoimmune diseases is due to the combined effects of multiple genes, but there are very few known monogenic autoimmune diseases. In these, possession of the predisposing allele confers a very high risk of disease on the individual, but the overall impact on the population is minimal because these variants are rare (Fig. 15.32). The existence of monogenic autoimmune disease was first observed in mutant mice, in which the inheritance of an autoimmune syndrome followed a pattern consistent with a single-gene defect. Autoimmune disease alleles are usually recessive or X-linked. For example, the disease APECED, discussed in Section 15-3, is a recessive autoimmune disease caused by a defect in the gene *AIRE*.

Two monogenic autoimmune syndromes have been linked to defects in regulatory T cells. The X-linked recessive autoimmune syndrome **IPEX** (immune dysregulation, polyendocrinopathy, enteropathy, X-linked disease) is typically caused by missense mutations in the gene for the transcription factor FoxP3, which is a key factor in the differentiation of some types of T_{reg} cells (see Section 9-18). Also known as XLAAD (X-linked autoimmunity-allergic dysregulation syndrome), this disease is characterized by severe allergic inflammation, autoimmune polyendocrinopathy, secretory diarrhea, hemolytic

Fig. 15.32 Single-gene traits associated with autoimmunity. Listed are examples of monogenic disorders that cause autoimmunity in humans. Mice with targeted deletions (knockout) or spontaneous mutations (for example *lpr/lpr*) in homologous genes have similar disease characteristics and are useful models for the study of the pathogenic basis for these disorders. APECED, autoimmune polyendocrinopathy–candidiasis–ectodermal dystrophy; APS-1, autoimmune polyglandular syndrome 1; IPEX, immune dysregulation, polyendocrinopathy, enteropathy, X-linked syndrome; ALPS, autoimmune lymphoproliferative syndrome. The *lpr* mutation in mice affects the gene for Fas, whereas the *gld* mutation affects the gene for FasL. Reprinted from J.D. Rioux and A.K. Abbas: *Nature* 435:584–589, © 2005. With permission from Macmillan Publishers Ltd.

Single-gene traits associated with autoimmunity			
Gene	Human disease	Mouse mutant or knockout	Mechanism of autoimmunity
AIRE	APECED (APS-1)	Knockout	Decreased expression of self antigens in the thymus, resulting in defective negative selection of self-reactive T cells
CTLA4	Association with Graves' disease, type 1 diabetes and others	Knockout	Failure of T-cell anergy and reduced activation threshold of self-reactive T cells
FOXP3	IPEX	Knockout and mutation (*scurfy*)	Decreased function of CD4 CD25 regulatory T cells
FAS	ALPS	*lpr/lpr;gld/gld* mutants	Failure of apoptotic death of self-reactive B and T cells
C1q	SLE	Knockout	Defective clearance of immune complexes and apoptotic cells

Immune Dysregulation, Polyendocrinopathy, Enteropathy X-linked Disease (IPEX)

anemia, and thrombocytopenia, and usually leads to early death. Despite mutation of the *FOXP3* gene, the numbers of CD4 CD25 T_{reg} cells in the blood of individuals with IPEX were comparable with those in healthy individuals; however, the suppressive function normally displayed by cells with this phenotype was reduced. A spontaneous frameshift mutation in the mouse *Foxp3* gene (the *scurfy* mutation) that results in loss of the DNA-binding, forkhead domain of FoxP3, or the knockout of *Foxp3* lead to an analogous systemic autoimmune disease, in this case associated with the absence of CD4 CD25 T_{reg} cells.

A second instance of autoimmunity resulting from a genetic defect in T_{reg}-cell function has been identified in a single patient with a deficiency of CD25 as a result of a deletion in *CD25* and impaired peripheral tolerance. This patient suffered from multiple immunological deficiencies and autoimmune diseases and was highly susceptible to infections. These findings further confirm the important roles of CD25 CD4 T_{reg} cells in the regulation of the immune system.

An interesting case of a monogenic autoimmune disease is the systemic autoimmune syndrome caused by mutations in the gene for Fas, which is called **autoimmune lymphoproliferative syndrome** (**ALPS**) in humans. Fas is normally present on the surface of activated T and B cells, and when ligated by Fas ligand it signals the Fas-bearing cell to undergo apoptosis (see Section 9-25). In this way it functions to limit the extent of immune responses. Mutations that eliminate or inactivate Fas lead to a massive accumulation of lymphocytes, especially T cells, and in mice the production of large quantities of pathogenic autoantibodies. The resulting disease resembles SLE, although typical SLE in humans has not been associated with mutations in Fas. A mutation leading to this autoimmune syndrome was first observed in the mouse strain MRL and dubbed *lpr*, for lymphoproliferation; it was subsequently identified as a mutation in the *Fas* gene. Researchers studying a group of human patients with the rare autoimmune lymphoproliferative syndrome, a syndrome similar to that in the MRL/*lpr* mice, identified and cloned the mutant gene responsible for most of these cases, which also turned out to be *FAS* (see Fig. 15.32).

Autoimmune Lymphoproliferative Syndrome (ALPS)

Autoimmune diseases caused by single genes are not common. They are nonetheless of great interest, as the mutations that cause them identify some of the important pathways that normally prevent the development of autoimmune responses.

15-22 MHC genes have an important role in controlling susceptibility to autoimmune disease.

Among all the genetic loci that could contribute to autoimmunity, susceptibility to autoimmune disease has so far been most consistently associated with MHC genotype. Human autoimmune diseases that show associations with HLA (MHC) type are shown in Fig. 15.33. For most of these diseases, susceptibility is linked most strongly with MHC class II alleles, and thus with CD4 T cells, although in some cases there are strong associations with particular MHC class I alleles, implicating CD8 T cells. In some cases, class III alleles such as those for TNF-α or complement protein have been associated with disease. The development of experimental diabetes or arthritis in transgenic mice expressing specific human HLA antigens strongly suggests that particular MHC alleles can confer susceptibility to disease.

The association of MHC genotype with disease is assessed initially by comparing the frequency of different alleles in patients with their frequency in the normal population. For type 1 diabetes, this approach originally demonstrated an association with the HLA-DR3 and HLA-DR4 alleles identified by serotyping (Fig. 15.34). Such studies also showed that the MHC class II allele

Associations of HLA serotype with susceptibility to autoimmune disease			
Disease	**HLA allele**	**Relative risk**	**Sex ratio (♀:♂)**
Ankylosing spondylitis	B27	87.4	0.3
Acute anterior uveitis	B27	10	<0.5
Goodpasture's syndrome	DR2	15.9	~1
Multiple sclerosis	DR2	4.8	10
Graves' disease	DR3	3.7	4–5
Myasthenia gravis	DR3	2.5	~1
Systemic lupus erythematosus	DR3	5.8	10–20
Type 1 (insulin-dependent) diabetes mellitus	DR3/DR4 heterozygote	~25	~1
Rheumatoid arthritis	DR4	4.2	3
Pemphigus vulgaris	DR4	14.4	~1
Hashimoto's thyroiditis	DR5	3.2	4–5

Fig. 15.33 Associations of HLA serotype and sex with susceptibility to autoimmune disease. The 'relative risk' for an HLA allele in an autoimmune disease is calculated by comparing the observed number of patients carrying the HLA allele with the number that would be expected, given the prevalence of the HLA allele in the general population. For type 1 insulin-dependent diabetes mellitus, the association is in fact with the HLA-DQ gene, which is tightly linked to the DR genes but is not detectable by serotyping. Some diseases show a significant bias in the sex ratio; this is taken to imply that sex hormones are involved in pathogenesis. Consistent with this, the difference in the sex ratio in these diseases is greatest between the menarche and the menopause, when levels of such hormones are highest.

HLA-DR2 has a dominant protective effect: individuals carrying HLA-DR2, even in association with one of the susceptibility alleles, rarely develop diabetes. Another way of determining whether MHC genes are important in autoimmune disease is to study the families of affected patients; it has been shown that two siblings affected with the same autoimmune disease are far more likely than expected to share the same MHC haplotypes (Fig. 15.35). As HLA genotyping has become more exact through the DNA sequencing of HLA alleles, disease associations that were originally discovered through HLA serotyping have been defined more precisely. For example, the association between type 1 diabetes and the DR3 and DR4 alleles is now known to be due to their tight genetic linkage to DQβ alleles that actually confer susceptibility to the disease. Indeed, susceptibility is most closely associated with polymorphisms at a particular position in the DQβ amino acid sequence. The most common DQβ amino acid sequence has an aspartic acid residue at position 57 that is able to form a salt bridge across the end of the peptide-binding cleft of the DQ molecule. In contrast, patients with diabetes in Caucasoid populations mostly have valine, serine, or alanine at that position, and thus make DQ molecules that lack this salt bridge (Fig. 15.36). The NOD strain of mice, which

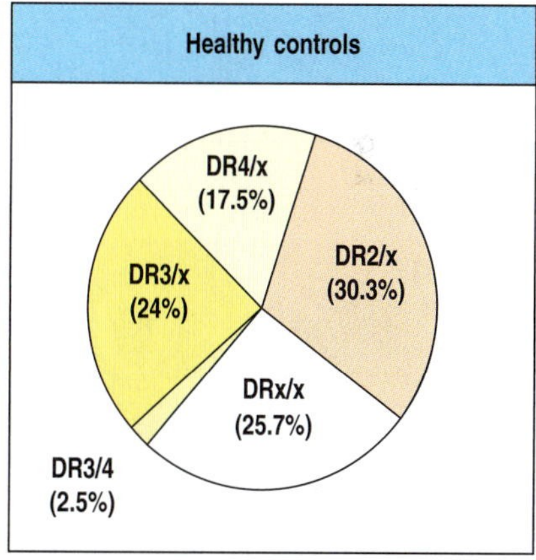

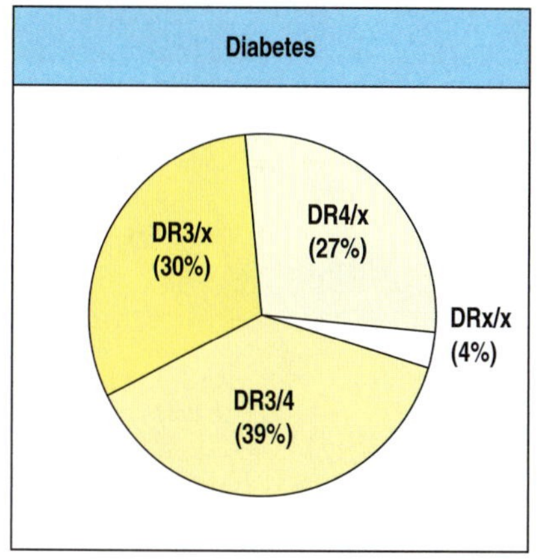

Fig. 15.34 Population studies show association of susceptibility to type 1 diabetes with HLA genotype. The HLA genotypes (determined by serotyping) of patients with diabetes (lower panel) are not representative of those found in the general population (upper panel). Almost all patients with diabetes express HLA-DR3 and/or HLA-DR4, and HLA-DR3/DR4 heterozygosity is greatly over-represented in diabetics compared with controls. These alleles are linked tightly to HLA-DQ alleles that confer susceptibility to type 1 diabetes. By contrast, HLA-DR2 protects against the development of diabetes and is found only extremely rarely in patients with diabetes. The small letter x represents any allele other than DR2, DR3, or DR4.

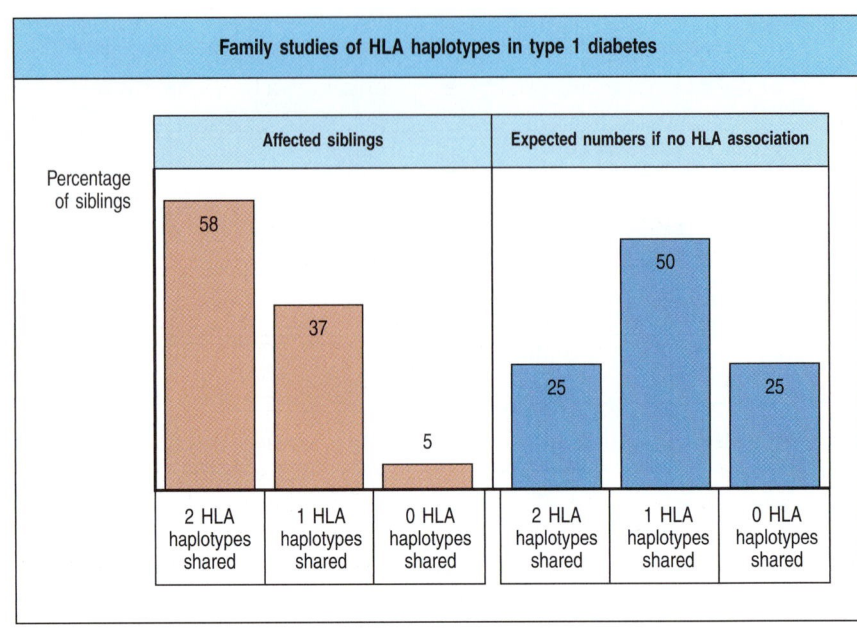

Fig. 15.35 Family studies show strong linkage of susceptibility to type 1 diabetes with HLA genotype. In families in which two or more siblings have type 1 diabetes, it is possible to compare the HLA genotypes of affected siblings. Affected siblings share two HLA haplotypes much more frequently than would be expected if the HLA genotype did not influence disease susceptibility.

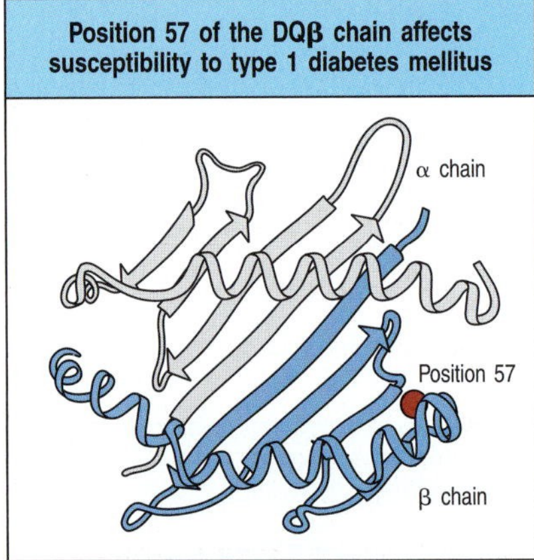

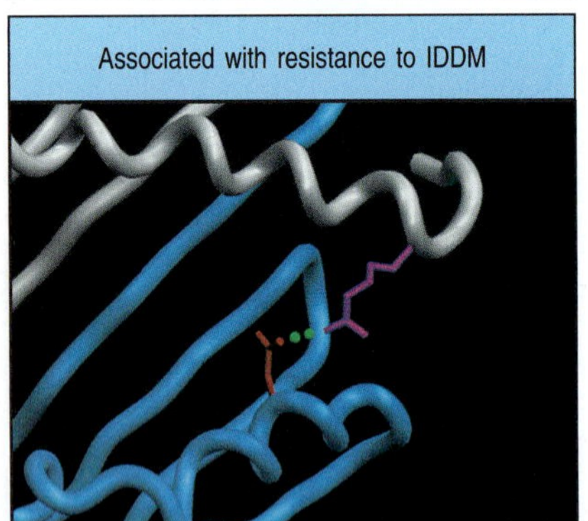

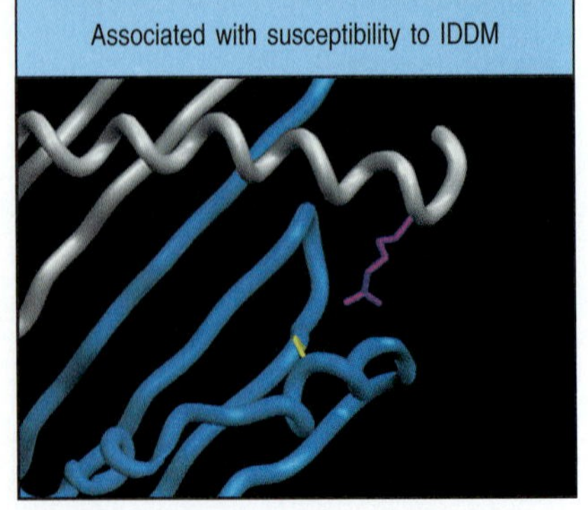

develops spontaneous diabetes, also has a serine residue at that position in the homologous mouse MHC class II molecule, known as I-A^{g7}.

The association of MHC genotype with autoimmune disease is not surprising, because autoimmune responses involve T cells, and the ability of T cells to respond to a particular antigen depends on MHC genotype. Thus, the associations can be explained by a simple model in which susceptibility to an autoimmune disease is determined by differences in the ability of different allelic variants of MHC molecules to present autoantigenic peptides to autoreactive T cells. This would be consistent with what we know of T-cell involvement in particular diseases. In diabetes, for example, there are associations with both MHC class I and MHC class II alleles, and this is consistent with the finding that both CD8 and CD4 T cells, which respond to antigens presented by MHC class I and MHC class II molecules, respectively, mediate the autoimmune response.

An alternative hypothesis for the association between MHC genotype and susceptibility to autoimmune diseases emphasizes the role of MHC alleles in shaping the T-cell receptor repertoire (see Chapter 8). This hypothesis proposes that self peptides associated with certain MHC molecules may drive the positive selection of developing thymocytes that are specific for particular autoantigens. Such autoantigenic peptides might be expressed at too low a level or bind too poorly to self MHC molecules to drive negative selection in

Fig. 15.36 Amino acid changes in the sequence of an MHC class II protein correlate with susceptibility to and protection from diabetes. The HLA-DQβ$_1$ chain contains an aspartic acid (Asp) residue at position 57 in most people; in Caucasoid populations, patients with type 1 diabetes more often have valine, serine, or alanine at this position instead, as well as other differences. Asp 57, shown in red on the backbone structure of the DQβ chain, forms a salt bridge (shown in green in the center panel) to an arginine residue (shown in pink) in the adjacent a chain (gray). The change to an uncharged residue (for example alanine, shown in yellow in the bottom panel) disrupts this salt bridge, altering the stability of the DQ molecule. The non-obese diabetic (NOD) strain of mice, which develops spontaneous diabetes, shows a similar substitution of serine for aspartic acid at position 57 of the homologous I-Aβ chain, and NOD mice transgenic for β chains with Asp 57 have a marked reduction in diabetes incidence. Courtesy of C. Thorpe.

the thymus, but might be present at a sufficient level or bind strongly enough to drive positive selection. This hypothesis is supported by observations that I-A^{g7}, the disease-associated MHC class II molecule in NOD mice, binds many peptides very poorly and may therefore be less effective in driving intrathymic negative selection of T cells that bind self peptides.

15-23 Genetic variants that impair innate immune responses can predispose to T cell-mediated chronic inflammatory disease.

As noted earlier in this chapter, a common chronic inflammatory disease is **Crohn's disease**, an intestinal disorder of the type known generally as inflammatory bowel disease, or IBD. The other main form of IBD is ulcerative colitis. Crohn's disease results from an abnormal over-responsiveness of CD4 T cells to antigens of the normal commensal gut microbiota, as opposed to true self antigens. Abnormal hyperactivity of T_H1 and T_H17 cells is thought to be pathogenic in this disease. This results from a failure of mucosal innate immune mechanisms to sequester luminal bacteria from the adaptive immune system, or is caused by intrinsic defects in cells of the adaptive immune system that result in heightened effector CD4 T-cell responses or failure of homeostatic T_{reg}-cell activity to suppress their development (Fig. 15.37). Patients with Crohn's disease have episodes of severe inflammation that commonly affect the terminal ileum, with or without involvement of the colon—hence the alternative name of regional ileitis for this disease—but any part of the gastrointestinal tract can be involved. The disease is characterized by a chronic inflammation of the mucosa and submucosa of the intestine that includes the prominent development of granulomatous lesions (Fig. 15.38) similar to those seen in the type IV hypersensitivity responses discussed in Section 14-17. Genetic analysis of patients with Crohn's disease and their families has identified a growing list of disease-susceptibility genes. One of the earliest to be identified was *NOD2* (also known as *CARD15*), which is expressed predominantly in monocytes, dendritic cells, and the Paneth cells of the small intestine, and which is involved in recognition of microbial antigens as part of the innate immune response (see Section 3-8). Mutations and uncommon polymorphic variants of the NOD2 protein are strongly associated with the

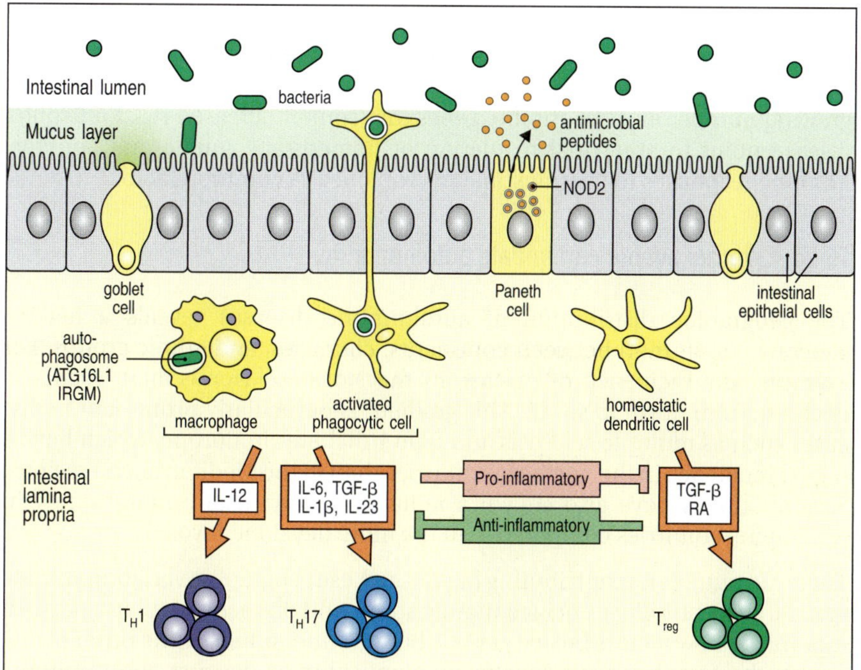

Fig. 15.37 Crohn's disease results from a breakdown of the normal homeostatic mechanisms that limit inflammatory responses to the gut microbiota. The innate and adaptive immune systems normally cooperate to limit inflammatory responses to intestinal bacteria through a combination of mechanisms: a mucus layer produced by goblet cells; tight junctions between the intestinal epithelial cells; antimicrobial peptides released from epithelial cells and Paneth cells; and induction of T_{reg} cells that inhibit effector CD4 T-cell development and promote the production of IgA antibodies that are transported into the intestinal lumen, where they inhibit translocation of intestinal bacteria (not shown). In individuals with impaired homeostatic mechanisms, dysregulated T_H1- and T_H17-cell responses to the intestinal microbiota can result, generating disease-causing chronic inflammation. Crohn's disease susceptibility genes of innate immunity include *NOD2* and the autophagy genes *ATG16L1* and *IRGM*. A major susceptibility gene that affects adaptive immune cells is *IL23R*, which is expressed by T_H17 cells.

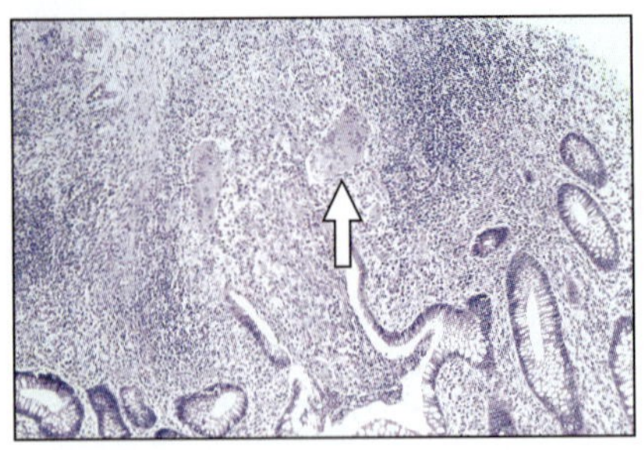

Fig. 15.38 Granulomatous inflammation in Crohn's disease. A section of bowel wall from a patient with Crohn's disease. The arrow marks a giant cell granuloma. There is a dense infiltrate of lymphocytes throughout the bowel submucosa. Photograph courtesy of H.T. Cook.

presence of Crohn's disease, with around 30% of patients carrying a loss-of-function mutation in *NOD2*. Mutations in the same gene are also the cause of a dominantly inherited granulomatous disease named **Blau syndrome**, in which granulomas typically develop in the skin, eyes, and joints. Whereas Crohn's disease results from a loss of function of NOD2, it is thought that Blau syndrome results from a gain of function.

NOD2 is an intracellular receptor for the muramyl dipeptide derived from bacterial peptidoglycan, and its stimulation leads to activation of the transcription factor NFκB and the induction of genes encoding pro-inflammatory cytokines and chemokines (see Section 3-8 and Fig. 12.19). In Paneth cells, which are specialized intestinal epithelial cells in the base of the small intestinal crypts, activation of NOD2 stimulates the release of granules containing antimicrobial proteins and peptides that contribute to the limitation of commensal bacteria to the intestinal lumen, away from cells of the adaptive immune system. Mutant forms of NOD2 that have lost this function limit the innate antibacterial response, thereby predisposing to heightened effector CD4 T-cell responses to the commensal microbiota that produce chronic intestinal inflammation (see Section 12-13).

In addition to NOD2, other deficiencies in innate immunity have been identified in patients with Crohn's disease, including defective CXCL8 production and defective neutrophil accumulation, which can synergize with defects in NOD2 to promote abnormal intestinal inflammation. Thus, compound defects in innate immunity and the regulation of inflammation may act synergistically to promote immunopathology in Crohn's disease. Genetic association studies have identified other susceptibility genes for Crohn's disease that may be linked to impaired innate immune functions. Two genes (*ATG161* and *IRGM*) that contribute to the process of autophagy have been linked to Crohn's disease, suggesting that other mechanisms that impair clearance of commensal bacteria might predispose to chronic intestinal inflammation. Autophagy, or the digestion of a cell's cytoplasm by its own lysosomes, is important in the turnover of damaged cellular organelles and proteins, and has a role in antigen processing and presentation (see Section 6-9), but is also thought to contribute to the clearance of some phagocytosed bacteria.

While defects in important pathways of the innate immune system contribute to Crohn's disease, genes that regulate the adaptive immune response have also been associated with susceptibility. Most notably, variants of the gene for the IL-23 receptor (*IL23R*) have been found that predispose to disease, consistent with heightened T_H17 responses found in diseased tissues. Collectively, the growing number of susceptibility genes that confer increased risk for Crohn's disease point to abnormal regulation of homeostatic innate and adaptive immune responses to the intestinal microbiota as a common factor.

15-24 External events can initiate autoimmunity.

The geographic distribution of autoimmune diseases reveals a heterogeneous distribution between continents, countries, and ethnic groups. For example, the incidence of disease in the Northern Hemisphere seems to decrease from north to south. This gradient is particularly prominent in diseases such as multiple sclerosis and type 1 diabetes in Europe, which have a higher incidence in the northern countries than in the Mediterranean regions. Several studies have also shown a reduced incidence of autoimmunity in developing countries compared with the more developed world.

There are numerous contributing factors to these geographic variations beside genetic susceptibility—socioeconomic status and diet seem to play a part. An example of how factors beside genetic background influence the onset of disease is the fact that even genetically identical mice develop autoimmunity at different rates and severity (see Fig. 15.29). In humans, exposure to infections and environmental toxins may be factors that help trigger autoimmunity. However, it should be noted that epidemiological and clinical studies over the past century have also shown a negative correlation between exposure to some types of infection in early life and the development of allergy and autoimmune diseases. This 'hygiene hypothesis' is discussed in detail in Section 14-4; it proposes that a lack of infection during childhood may affect the regulation of the immune system in later life, leading to a greater likelihood of allergic and autoimmune responses.

15-25 Infection can lead to autoimmune disease by providing an environment that promotes lymphocyte activation.

How might pathogens initiate or modulate autoimmunity? During an infection and the consequent immune response, the combination of the inflammatory mediators released from activated antigen-presenting cells and lymphocytes and the increased expression of co-stimulatory molecules can have effects on so-called bystander cells—lymphocytes that are not themselves specific for the antigens of the infectious agent. Self-reactive lymphocytes can become activated in these circumstances, particularly if tissue destruction by the infection leads to an increase in the availability of the self antigen (Fig. 15.39, first panel). Furthermore, pro-inflammatory cytokines, such as IL-1 and IL-6, impair the suppressive activity of regulatory T cells, allowing self-reactive naive T cells to become activated to differentiate into effector T cells that can initiate an autoimmune response.

In general, any infection will lead to an inflammatory response and recruitment of inflammatory cells to the site of the infection. The perpetuation, and even exacerbation, of autoimmune disease by viral or bacterial infections has been shown in experimental animal models. For example, the severity of type 1 diabetes in NOD mice is exacerbated by Coxsackie virus B4 infection, which leads to inflammation, tissue damage and the release of sequestered islet antigens, and the generation of autoreactive T cells.

We discussed earlier the ability of self ligands such as unmethylated CpG DNA sequences and RNA to directly activate ignorant autoreactive B cells via their TLRs and thus break tolerance to self (see Section 15-4). Microbial ligands for TLRs may also promote autoimmunity by stimulating dendritic cells and macrophages to produce large quantities of cytokines that cause local inflammation and help stimulate and maintain already activated autoreactive T cells and B cells. This mechanism might be relevant to the flare-ups of inflammation that follow infection in patients with autoimmune vasculitis associated with anti-neutrophil cytoplasmic antibodies.

One example of how exposure to TLR ligands can induce local inflammation derives from an animal model of arthritis in which injection of bacterial CpG DNA into the joints of healthy mice induces an aseptic arthritis characterized by macrophage infiltration. These macrophages express chemokine receptors on their surface and produce large amounts of CC chemokines, which promote leukocyte recruitment to the site of injection.

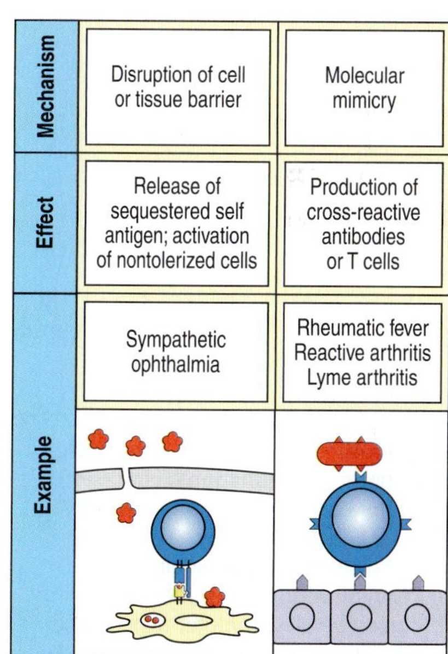

Fig. 15.39 Infectious agents could break self-tolerance in several different ways. Left panel: because some antigens are sequestered from the circulation, either behind a tissue barrier or within the cell, an infection that breaks cell and tissue barriers might expose hidden antigens. Right panel: molecular mimicry might result in infectious agents inducing either T- or B-cell responses that can cross-react with self antigens.

15-26 Cross-reactivity between foreign molecules on pathogens and self molecules can lead to anti-self responses and autoimmune disease.

Infection with certain pathogens is particularly associated with autoimmune sequelae. Some pathogens express protein or carbohydrate antigens that resemble host molecules, a phenomenon dubbed **molecular mimicry**. In such cases, antibodies produced against a pathogen epitope may cross-react with a self protein (see Fig. 15.39, second panel). Such structures do not necessarily have to be identical: it is sufficient if they are similar enough to be recognized by the same antibody. Molecular mimicry may also activate autoreactive naive or effector T cells if a processed peptide from a pathogen antigen is identical or similar to a host peptide, resulting in an attack on self tissues. A model system to demonstrate molecular mimicry has been generated by using transgenic mice that express a viral antigen in the pancreas. Normally, there is no response to this virus-derived 'self' antigen. But if the mice are infected with the virus that was the source of the transgenic antigen, they develop diabetes, because the virus activates T cells that are cross-reactive with the 'self' viral antigen and attack the pancreas (Fig. 15.40).

One might wonder why these self-reactive lymphocytes have not been deleted or inactivated by the usual mechanisms of self-tolerance. One reason, as discussed earlier in the chapter, is that lower-affinity self-reactive B and T cells are not removed efficiently and are present in the naive lymphocyte repertoire as ignorant lymphocytes (see Section 15-4). Second, the strong pro-inflammatory stimulus that accompanies an infection could be sufficient to activate even anergic T and B cells in the periphery, thus drawing into the response cells that would usually be silent. Third, pathogens may provide substantially higher local doses of the eliciting antigen in an immunogenic form, whereas normally it would be relatively unavailable to lymphocytes. Some examples of autoimmune syndromes thought to involve molecular mimicry are the rheumatic fever that sometimes follows streptococcal infection, and the reactive arthritis that can occur after enteric infection.

Autoimmmune Hemolytic Anemia

Once self-reactive lymphocytes have been activated by such a mechanism, their effector functions can destroy self tissues. Autoimmunity of this type is sometimes transient, and remits when the inciting pathogen is eliminated. This is the case in the autoimmune hemolytic anemia that follows mycoplasma infection, in which antibodies against the pathogen cross-react with an antigen on red blood cells, leading to hemolysis (see Section 15-13). The autoantibodies disappear when the patient recovers from the infection. Sometimes, however, the autoimmunity persists well beyond the initial infection. This is true in some cases of **rheumatic fever**, which occasionally follows sore throat or scarlet fever caused by *Streptococcus pyogenes*. The similarity of epitopes on streptococcal antigens to epitopes on some tissues leads to antibody-mediated, and possibly T cell-mediated, damage to a variety of tissues, including heart valves. Although rheumatic fever is often transient, especially with antibiotic treatment, it can sometimes become chronic. Similarly, Lyme

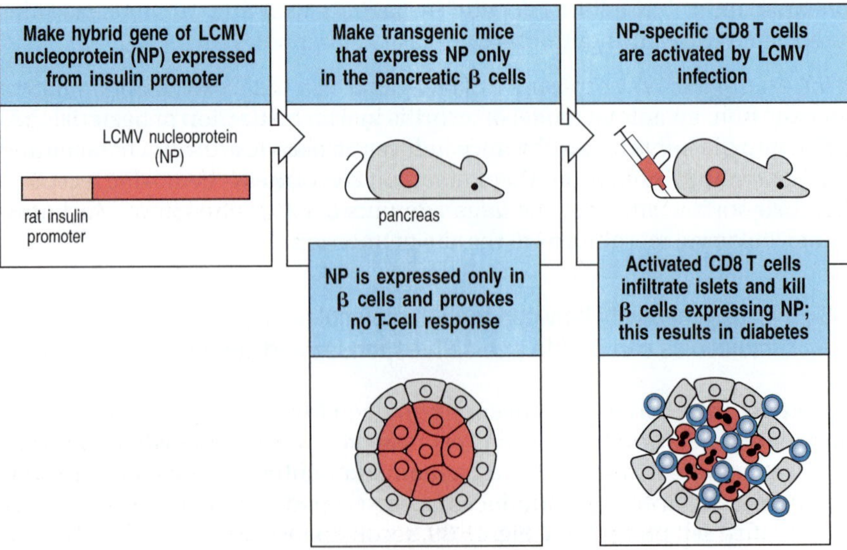

Fig. 15.40 Virus infection can break tolerance to a transgenic viral protein expressed in pancreatic β cells. Mice made transgenic for the lymphocytic choriomeningitis virus (LCMV) nucleoprotein under the control of the rat insulin promoter express the nucleoprotein in their pancreatic β cells but do not respond to this protein and therefore do not develop an autoimmune diabetes. However, if the transgenic mice are infected with LCMV, a potent antiviral cytotoxic T-cell response is elicited, and this kills the β cells, leading to diabetes. It is thought that infectious agents can sometimes elicit T-cell responses that cross-react with self peptides (a process known as molecular mimicry) and that this could cause autoimmune disease in a similar way.

disease, an infection with the spirochete *Borrelia burgdorferi*, is followed by later-developing autoimmunity, causing so-called Lyme arthritis. In this case, the mechanism is not entirely clear, but it is likely to involve cross-reactivity of pathogen and host components, leading to a self-perpetuating autoimmune reaction.

15-27 Drugs and toxins can cause autoimmune syndromes.

Perhaps some of the clearest evidence of external causative agents in human autoimmunity comes from the effects of certain drugs, which elicit autoimmune reactions as side effects in a small proportion of patients. Procainamide, a drug used to treat heart arrhythmias, is particularly notable for inducing autoantibodies similar to those in SLE, although these are rarely pathogenic. Several drugs are associated with the development of autoimmune hemolytic anemia, in which autoantibodies against surface components of red blood cells attack and destroy these cells (see Section 15-13). Toxins in the environment can also cause autoimmunity. When heavy metals, such as gold or mercury, are administered to genetically susceptible strains of mice, a predictable autoimmune syndrome, including the production of autoantibodies, ensues. The extent to which heavy metals promote autoimmunity in humans is debatable, but the animal models clearly show that environmental factors such as toxins could have key roles in certain syndromes.

The mechanisms by which drugs and toxins cause autoimmunity are uncertain. For some drugs it is thought that they react chemically with self proteins and form derivatives that the immune system recognizes as foreign. The immune response to these haptenated self proteins can lead to inflammation, complement deposition, destruction of tissue, and finally immune responses to the original underivatized self proteins.

15-28 Random events may be required for the initiation of autoimmunity.

Although scientists and physicians would like to attribute the onset of 'spontaneous' diseases to some specific cause, this may not always be possible. There may not be one virus or bacterium, or even any understandable pattern of events that precedes the onset of autoimmune disease. The chance encounter in the peripheral lymphoid tissues of a few autoreactive B and T cells that can interact with each other, at just the moment when an infection is providing pro-inflammatory signals, may be all that is needed. This could be a rare event, and in a genetically resistant individual it could even be brought under control. But in a susceptible individual such events could be more frequent and/or more difficult to control.

Thus, the onset or incidence of autoimmunity can seem to be random. Genetic predisposition represents, in part, an increased chance of occurrence of this random event. This view, in turn, may explain why many autoimmune diseases appear in early adulthood or later, after enough time has elapsed to permit low-frequency random events to occur. It may also explain why after certain kinds of experimental aggressive therapies for these diseases, such as bone marrow transplantation or B-cell depletion, the disease eventually recurs after a long interval of remission.

Summary.

The specific causes of most autoimmune diseases are in most cases not known. Genetic risk factors including particular alleles of MHC class II molecules and other genes have been identified, but many individuals with genetic variants that predispose to a particular autoimmune disease do not get the disease.

Epidemiological studies of genetically identical populations of animals have highlighted the role of environmental factors for the initiation of autoimmunity, but although environmental factors have at least as strong an influence on the outcome as genetics, they are even less well understood. Some toxins and drugs are known to cause autoimmune syndromes, but their role in the common varieties of autoimmune disease is unclear. Similarly, some autoimmune syndromes can follow viral or bacterial infections. Pathogens can promote autoimmunity by causing nonspecific inflammation and tissue damage. They can also sometimes elicit responses to self proteins if they express molecules that resemble self, a phenomenon known as molecular mimicry. Much more progress needs to be made to define environmental factors. It may be that there will not be a single, or even an identifiable, environmental factor that contributes to most diseases, and chance may have an important role in determining disease onset.

Responses to alloantigens and transplant rejection.

The transplantation of tissues to replace diseased organs is now an important medical therapy. In most cases, adaptive immune responses to the grafted tissues are the major impediment to successful transplantation. Rejection is caused by immune responses to alloantigens on the graft, which are proteins that vary from individual to individual within a species and are therefore perceived as foreign by the recipient. When tissues containing nucleated cells are transplanted, T-cell responses to the highly polymorphic MHC molecules almost always trigger a response against the grafted organ. Matching the MHC type of the donor and the recipient increases the success rate of grafts, but perfect matching is possible only when donor and recipient are related and, in these cases, genetic differences at other loci can still trigger rejection, although less severely. Nevertheless, advances in immunosuppression and transplantation medicine now mean that the precise matching of tissues for transplantation is no longer the major factor in graft survival. In blood transfusion, which was the earliest tissue transplant and is still the most common, MHC matching is not necessary for routine blood transfusions because red blood cells and platelets express only small amounts of MHC class I molecules and do not express MHC class II at all; thus, they are not usually targets for the T cells of the recipient. Antibodies made against platelet MHC class I can, however, be a problem when repeated transfusions of platelets are required. Blood must be matched for ABO and Rh blood group antigens to avoid the rapid destruction of mismatched red blood cells by antibodies in the recipient (see Appendix I, Section A-11). Because there are only four major ABO types and two Rh types, this is relatively easy. In this part of the chapter we examine the immune response to tissue grafts and also ask why such responses do not reject the one foreign tissue graft that is tolerated routinely—the mammalian fetus.

15-29 Graft rejection is an immunological response mediated primarily by T cells.

The basic rules of tissue grafting were first elucidated by skin transplantation between inbred strains of mice. Skin can be grafted with 100% success between different sites on the same animal or person (an **autograft**), or between genetically identical animals or people (a **syngeneic graft**). However, when skin is grafted between unrelated or **allogeneic** individuals (an **allograft**), the graft initially survives but is then rejected about 10–13 days after grafting (Fig. 15.41). This response is called an **acute rejection** and is quite

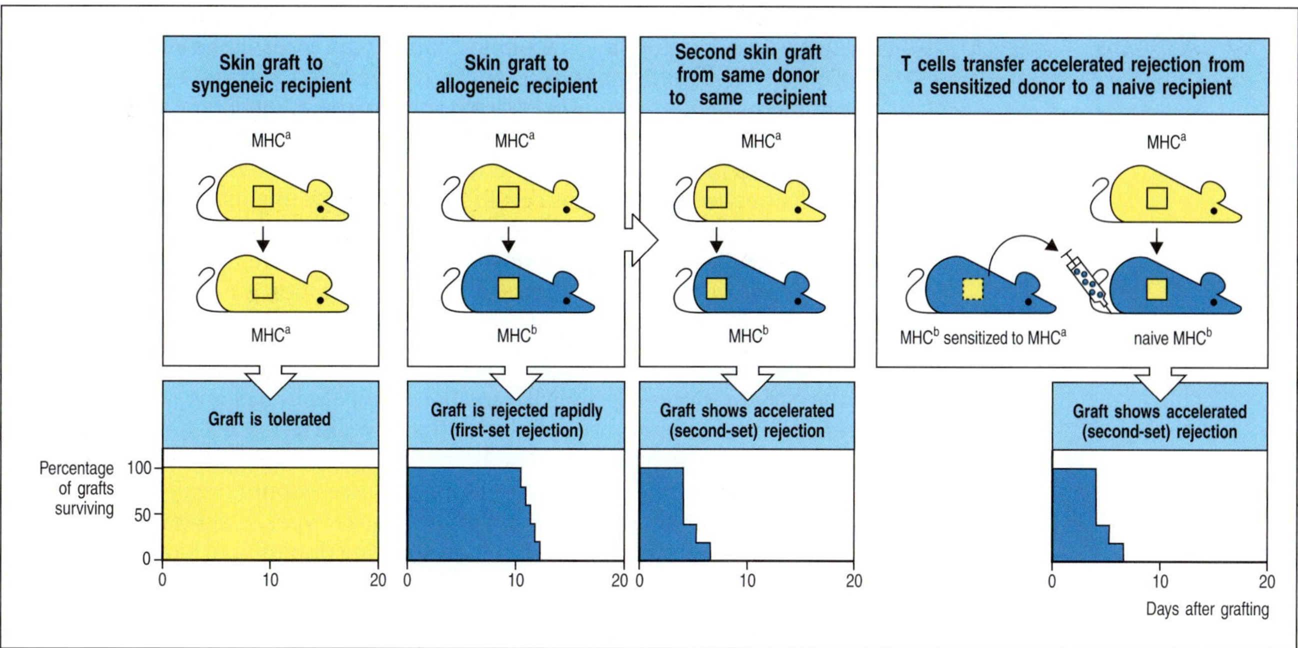

Fig. 15.41 Skin graft rejection is the result of a T cell-mediated anti-graft response. Grafts that are syngeneic are permanently accepted (first panels), but grafts differing at the MHC are rejected about 10–13 days after grafting (first-set rejection, second panels). When a mouse is grafted for a second time with skin from the same donor, it rejects the second graft faster (third panels). This is called a second-set rejection, and the accelerated response is MHC-specific; skin from a second donor of the same MHC type is rejected equally fast, whereas skin from an MHC-different donor is rejected in a first-set pattern (not shown). Naive mice that are given T cells from a sensitized donor behave as if they had already been grafted (final panels).

consistent. It depends on a T-cell response in the recipient, because skin grafted onto *nude* mice, which lack T cells, is not rejected. The ability to reject skin can be restored to *nude* mice by the adoptive transfer of normal T cells.

When a recipient that has previously rejected a graft is regrafted with skin from the same donor, the second graft is rejected more rapidly (6–8 days) in an **accelerated rejection** (see Fig. 15.41). Skin from a third-party donor grafted onto the same recipient at the same time does not show this faster response but follows a first-set rejection course. The rapid course of second-set rejection can also be transferred to normal or irradiated recipients by T cells from the initial recipient, showing that second-set rejection is caused by a memory-type immune response (see Chapter 11) from clonally expanded and primed T cells specific for the donor skin.

Immune responses are the major barrier to effective tissue transplantation, destroying grafted tissue by an adaptive immune response to its foreign proteins. These responses can be mediated by CD8 T cells, by CD4 T cells, or by both. Antibodies can also contribute to second-set rejection of tissue grafts.

15-30 Transplant rejection is caused primarily by the strong immune response to nonself MHC molecules.

Antigens that differ between members of the same species are known as **alloantigens**, and an immune response against such antigens is known as an **alloreactive** response. When donor and recipient differ at the MHC, an alloreactive immune response is directed at the nonself allogeneic MHC molecule or molecules present on the graft. In most tissues these will be predominantly MHC class I antigens. Once a recipient has rejected a graft of a particular MHC type, any further graft bearing the same nonself MHC molecule will be rapidly rejected in a second-set response. The frequency of T cells specific for any nonself MHC molecule is relatively high, making differences at MHC loci the most potent trigger of the rejection of initial grafts (see Section 6-14); indeed, the major histocompatibility complex was originally so named because of its central role in graft rejection.

Once it became clear that recognition of nonself MHC molecules was a major determinant of graft rejection, a considerable amount of effort was put into

MHC matching between recipient and donor. Today, with advances in immunosuppression that have facilitated solid organ transplants across MHC barriers, MHC matching has become largely irrelevant for most types of allograft, although it remains important for bone marrow transplantation, for reasons that will be discussed later. Even a perfect match at the MHC locus, known as the HLA locus in humans, does not in itself prevent rejection reactions. Grafts between HLA-identical siblings will invariably incite a rejection reaction, albeit more slowly than an unmatched graft, unless donor and recipient are identical twins. This reaction is the result of differences between antigens from non-MHC proteins that also vary between individuals.

Thus, unless donor and recipient are identical twins, all graft recipients must be given immunosuppressive drugs chronically to prevent rejection. Indeed, the current success of clinical transplantation of solid organs is more the result of advances in immunosuppressive therapy, discussed in Chapter 16, than of improved tissue matching. The limited supply of cadaveric organs, coupled with the urgency of identifying a recipient once a donor organ becomes available, means that accurate matching of tissue types is achieved only rarely, with the notable exception of matched sibling donation of kidneys.

15-31 In MHC-identical grafts, rejection is caused by peptides from other alloantigens bound to graft MHC molecules.

When donor and recipient are identical at the MHC but differ at other genetic loci, graft rejection is not as rapid, but left unchecked it will still destroy the graft (Fig. 15.42). This is, for example, the reason that grafts between HLA-identical siblings would be rejected without immunosuppressive treatment. MHC class I and class II molecules bind and present a large selection of peptides derived from self proteins made in the cell, and if these proteins are polymorphic, then different peptides will be produced from them in different members of a species. Such proteins can be recognized as **minor histocompatibility antigens** (Fig. 15.43). One set of proteins that induce minor histocompatibility responses are encoded on the male-specific Y chromosome. Responses induced by these proteins are known collectively as H-Y. As these Y-chromosome-specific genes are not expressed in females, female anti-male minor histocompatibility responses occur; however, male anti-female responses are not seen, because both males and females express

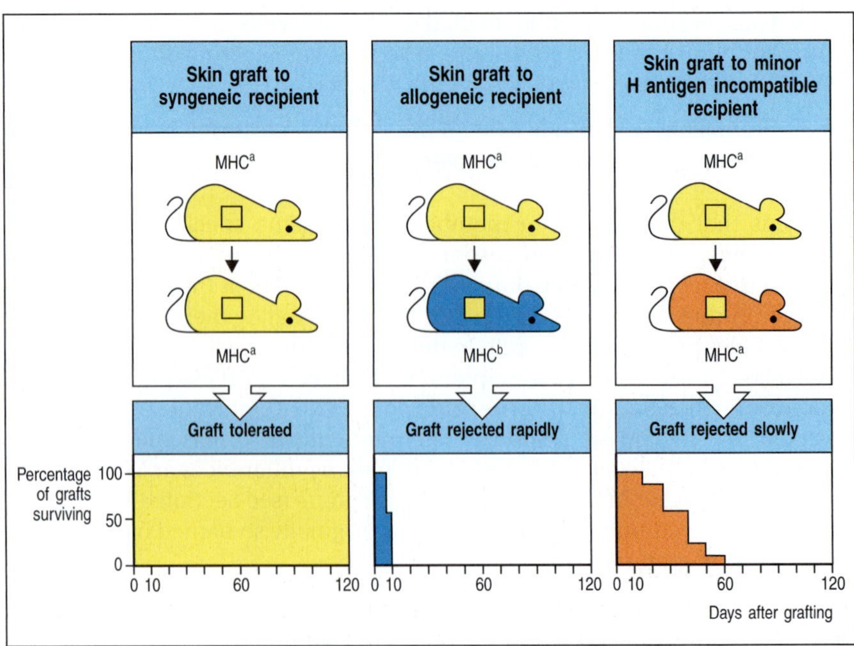

Fig. 15.42 Even complete matching at the MHC does not ensure graft survival. Although syngeneic grafts are not rejected (left panels), MHC-identical grafts from donors that differ at other loci (minor H antigen loci) are rejected (right panels), albeit more slowly than MHC-disparate grafts (center panels).

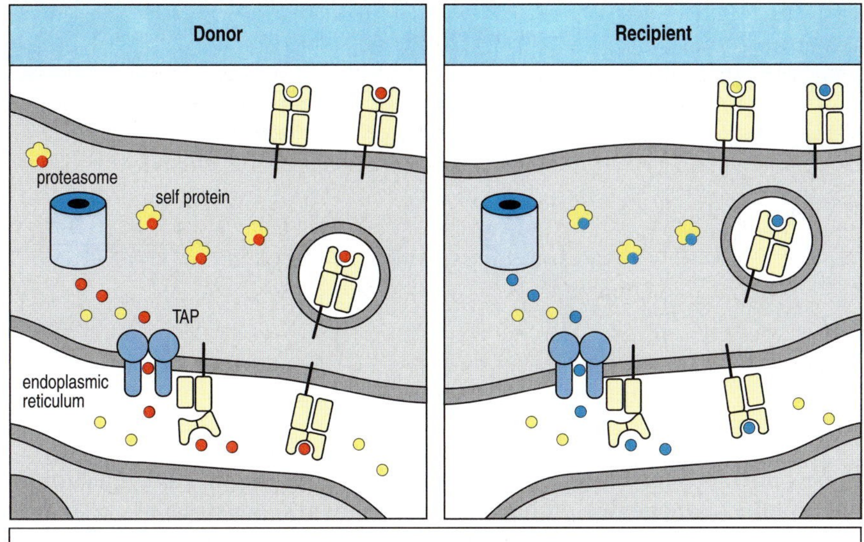
Donor
Recipient
proteasome
self protein
TAP
endoplasmic reticulum
Polymorphic self proteins that differ in amino acid sequence between individuals give rise to minor H antigen differences between donor and recipient

Fig. 15.43 Minor H antigens are peptides derived from polymorphic cellular proteins bound to MHC class I molecules. Self proteins are routinely digested by proteasomes within the cell's cytosol, and peptides derived from them are delivered to the endoplasmic reticulum, where they can bind to MHC class I molecules and be delivered to the cell surface. If a polymorphic protein differs between the graft donor (shown in red on the left) and the recipient (shown in blue on the right), it can give rise to an antigenic peptide (red on the donor cell) that can be recognized by the recipient's T cells as nonself and elicit an immune response. Such antigens are the minor H antigens.

X-chromosome genes. One H-Y antigen has been identified in mice and humans as peptides from a protein encoded by the Y-chromosome gene *Smcy.* An X-chromosome homolog of *Smcy*, called *Smcx*, does not contain these peptide sequences, which are therefore expressed uniquely in males. Most minor histocompatibility antigens are encoded by autosomal genes and their identity is largely unknown, although an increasing number have now been identified at the genetic level.

The response to minor histocompatibility antigens is in many ways analogous to the response to viral infection. However, whereas an antiviral response eliminates only infected cells, a large fraction of cells in the graft express minor histocompatibility antigens, and thus the graft is destroyed in the response against these antigens. Given the virtual certainty of mismatches in minor histocompatibility antigens between any two individuals, and the potency of the reactions they incite, it is understandable that successful transplantation requires the use of powerful immunosuppressive drugs.

15-32 There are two ways of presenting alloantigens on the transplanted donor organ to the recipient's T lymphocytes.

Before naive alloreactive T cells can develop into effector T cells that cause rejection, they must be activated by antigen-presenting cells that bear both the allogeneic MHC and co-stimulatory molecules. Organ grafts carry with them antigen-presenting cells of donor origin, sometimes called passenger leukocytes, and these are an important stimulus to alloreactivity. This route for sensitization of the recipient to a graft seems to involve donor antigen-presenting cells leaving the graft and migrating to secondary lymphoid tissues of the recipient, including the spleen and lymph nodes, where they can activate those host T cells that bear the corresponding T-cell receptors. Because the lymphatic drainage of solid organ allografts is interrupted by transplantation, migration of donor antigen-presenting cells occurs via the blood, not lymphatics. The activated alloreactive effector T cells can then circulate to the graft, which they attack directly (Fig. 15.44). This recognition pathway is known as **direct allorecognition** (Fig. 15.45, left panel). Indeed, if the grafted tissue is depleted of antigen-presenting cells by treatment with antibodies or by prolonged incubation, rejection occurs only after a much longer time.

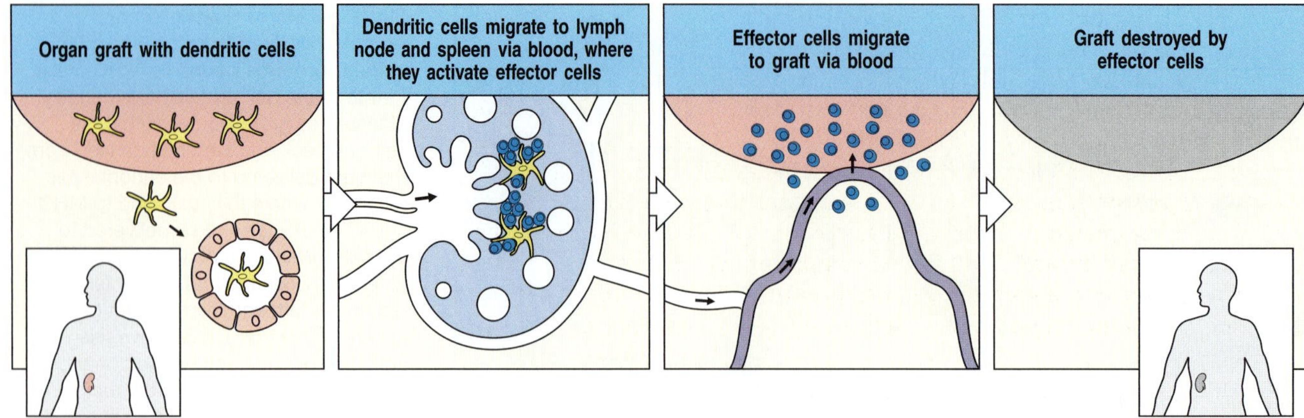

Organ graft with dendritic cells
Dendritic cells migrate to lymph node and spleen via blood, where they activate effector cells
Effector cells migrate to graft via blood
Graft destroyed by effector cells

Fig. 15.44 The initiation of graft rejection normally involves the migration of donor antigen-presenting cells from the graft to local lymph nodes. The example of an organ graft is illustrated here, in which dendritic cells are the antigen-presenting cells. They display peptides from the graft on their surface. After traveling to the spleen or a lymph node, these antigen-presenting cells encounter recirculating naive T cells specific for graft antigens, and stimulate these T cells to divide. The resulting activated effector T cells migrate via the thoracic duct to the blood and home to the grafted tissue, which they rapidly destroy. Destruction is highly specific for donor-derived cells, suggesting that it is mediated by direct cytotoxicity and not by nonspecific inflammatory processes.

A second mechanism of allograft recognition leading to graft rejection is the uptake of allogeneic proteins by the recipient's own antigen-presenting cells and their presentation to T cells by self-MHC molecules. This is known as **indirect allorecognition** (Fig. 15.45, right panel). Peptides from both the foreign MHC molecules themselves and from the minor histocompatibility antigens can be presented by indirect allorecognition.

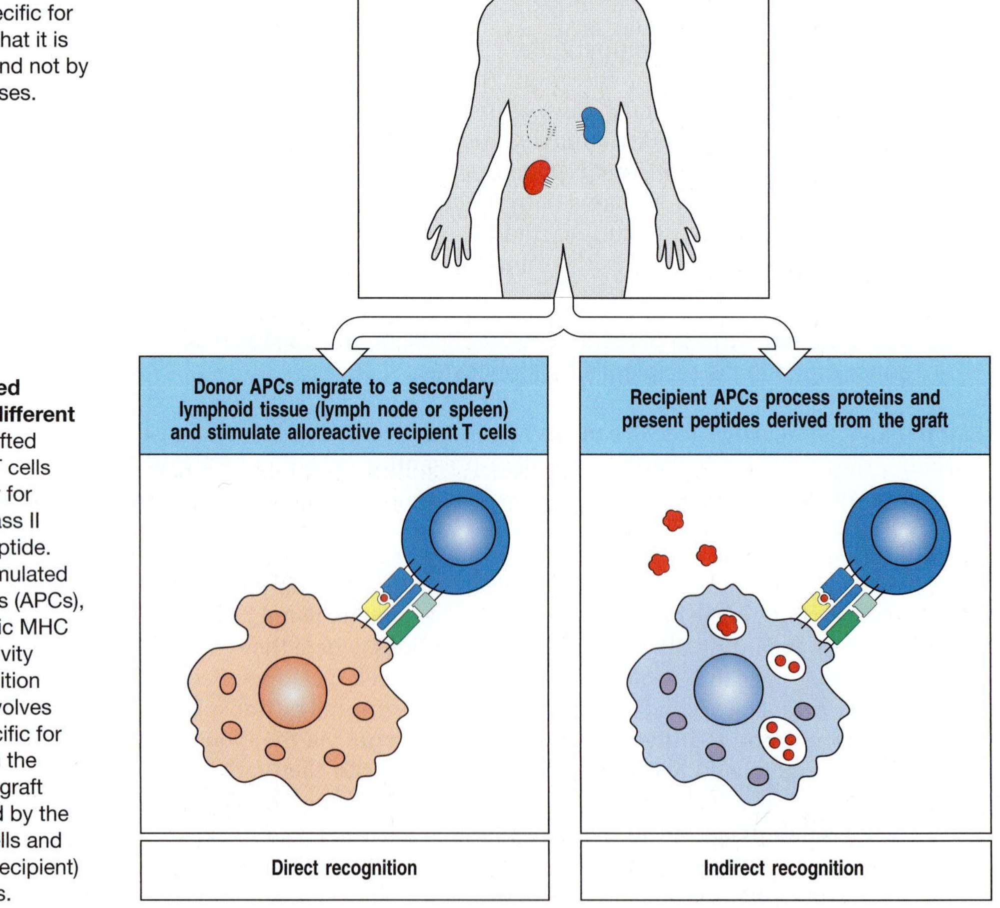

Donor APCs migrate to a secondary lymphoid tissue (lymph node or spleen) and stimulate alloreactive recipient T cells
Recipient APCs process proteins and present peptides derived from the graft
Direct recognition
Indirect recognition

Fig. 15.45 Alloantigens in grafted organs are recognized in two different ways. Direct recognition of a grafted organ (red in upper panel) is by T cells whose receptors have specificity for the allogeneic MHC class I or class II molecule in combination with peptide. These alloreactive T cells are stimulated by donor antigen-presenting cells (APCs), which express both the allogeneic MHC molecule and co-stimulatory activity (lower left panel). Indirect recognition of the graft (lower right panel) involves T cells whose receptors are specific for allogeneic peptides derived from the grafted organ. Proteins from the graft (red) are taken up and processed by the recipient's antigen-presenting cells and are therefore presented by self (recipient) MHC class I or class II molecules.

The relative contributions of direct and indirect allorecognition in graft rejection are not known. Direct allorecognition is thought to be largely responsible for acute rejection, especially when MHC mismatches mean that the frequency of directly alloreactive recipient T cells is high. Furthermore, a direct cytotoxic T-cell attack on graft cells can be made only by T cells that recognize the graft MHC molecules directly. Nonetheless, T cells with specificity for alloantigens presented on self MHC can contribute to graft rejection by activating macrophages, which cause tissue injury and fibrosis. T cells with indirect allospecificity are also likely to be important in the development of an antibody response to a graft. Antibodies produced against nonself antigens from another member of the same species are known as **alloantibodies**.

15-33 Antibodies that react with endothelium cause hyperacute graft rejection.

Antibody responses are an important potential cause of graft rejection. Preexisting alloantibodies against blood group antigens and polymorphic MHC antigens can cause the rapid rejection of transplanted organs in a complement-dependent reaction that can occur within minutes of transplantation. This type of reaction is known as **hyperacute graft rejection**. Most grafts that are transplanted routinely in clinical medicine are vascularized organ grafts linked directly to the recipient's circulation. In some cases the recipient might already have circulating antibodies against donor graft antigens. Antibodies of the ABO type can bind to all tissues, not just red blood cells. They are preformed and are relevant in all ABO-mismatched individuals. In addition, antibodies against other antigens can be produced in response to a previous transplant or a blood transfusion. All such preexisting antibodies can cause very rapid rejection of vascularized grafts because they react with antigens on the vascular endothelial cells of the graft and initiate the complement and blood clotting cascades. The vessels of the graft become blocked, or thrombosed, causing its immediate death. Such grafts become engorged and purple-colored from hemorrhaged blood, which becomes deoxygenated (Fig. 15.46). This problem can be avoided by ABO-matching as well as **cross-matching** donor and recipient. Cross-matching involves determining whether the recipient has antibodies that react with the white blood cells of the donor. If antibodies of this type are found, they have hitherto been considered a serious contraindication to transplantation of most solid organs, because in the absence of any treatment they lead to near-certain hyperacute rejection.

For reasons that are incompletely understood, some transplanted organs, particularly the liver, are less susceptible to this type of injury, and can be transplanted despite ABO incompatibilities. In addition, the presence of donor-specific MHC alloantibodies and a positive cross-match are no longer considered an absolute contraindication for transplantation, because the desensitization by treatment with intravenous immunoglobulin has been successful in a proportion of patients in whom antibodies against the donor tissue were already present.

A very similar problem prevents the routine use of animal organs—**xenografts**—in transplantation. If xenogeneic grafts could be used, it would circumvent the major limitation in organ replacement therapy, namely the

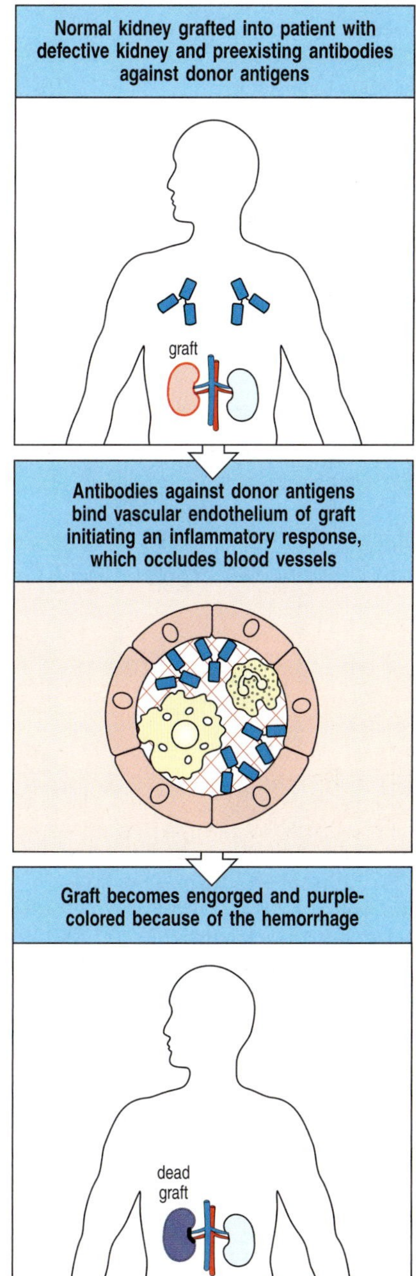

Fig. 15.46 Preexisting antibody against donor graft antigens can cause hyperacute graft rejection. In some cases, recipients already have antibodies against donor antigens, which are often blood group antigens. When the donor organ is grafted into such recipients, these antibodies bind to vascular endothelium in the graft, initiating the complement and clotting cascades. Blood vessels in the graft become obstructed by clots and leak, causing hemorrhage of blood into the graft. This becomes engorged and turns purple from the presence of deoxygenated blood.

severe shortage of donor organs. Pigs have been suggested as a potential source of organs for xenografting, because they are of a similar size to humans and are easily farmed. Most humans and other primates have natural antibodies that react with a ubiquitous cell-surface carbohydrate antigen (α-Gal) of other mammalian species, including pigs. When pig xenografts are placed in humans, these antibodies trigger hyperacute rejection by binding to the endothelial cells of the graft and initiating the complement and clotting cascades. The problem of hyperacute rejection is exacerbated in xenografts because complement-regulatory proteins such as CD59, DAF (CD55), and MCP (CD46) (see Section 2-16) work less efficiently across a species barrier, and so pig regulatory proteins, for example, cannot protect the graft from attack by human complement.

A recent step toward xenotransplantation has been the development of transgenic pigs expressing human DAF as well as pigs that lack α-Gal. These approaches might one day reduce or eliminate hyperacute rejection in xenotransplantation. However, hyperacute rejection is only the first barrier faced by a xenotransplanted organ. The T lymphocyte-mediated graft rejection mechanisms might be extremely difficult to overcome with the immunosuppressive regimes currently available.

15-34 Late failure of transplanted organs is caused by chronic injury to the graft.

The success of modern immunosuppression means that about 90% of cadaveric kidney grafts are still functioning a year after transplantation. There has, however, been little improvement in rates of long-term graft survival: the half-life for functional survival of renal allografts remains about 8 years. Although traditionally the late failure of a transplanted organ has been termed **chronic rejection**, it is typically difficult to determine whether the cause of chronic allograft injury involves specific immune alloreactivity, nonimmune injury, or both.

The dominant pattern of chronic injury to transplanted organs is variable, depending on the tissue. A major component of late failure of vascularized transplanted organs is a chronic reaction called **chronic allograft vasculopathy**, which is a prominent cause of injury in heart and kidney allografts. This is characterized by concentric arteriosclerosis of graft blood vessels, which leads to hypoperfusion of the graft and its eventual fibrosis and atrophy. Multiple mechanisms may contribute to this form of vascular injury, including recurring acute rejection events, allospecific antibodies reactive to the vascular endothelium of the graft, or some forms of immunosuppressive therapy (for example calcineurin inhibitors, such as cyclosporin). In transplanted livers, chronic rejection is associated with loss of bile ducts, the so-called ‘vanishing bile duct syndrome,’ whereas in transplanted lungs, the major cause of late organ failure is accumulation of scar tissue in the smallest airways, or bronchioles, which is termed bronchiolitis obliterans. Alloreactive responses can occur months to years after transplantation, and may be associated with gradual loss of graft function that is hard to detect clinically.

Other important causes of chronic graft dysfunction include ischemia–reperfusion injury, which occurs at the time of grafting but may have late adverse effects on the grafted organ, viral infections that emerge as a result of immunosuppression, and recurrence of the same disease in the allograft that destroyed the original organ. Irrespective of etiology, chronic allograft injury is typically irreversible and progressive, ultimately leading to complete failure of allograft function.

15-35 A variety of organs are transplanted routinely in clinical medicine.

Although the immune response makes organ transplantation difficult, there are few alternative therapies for organ failure. Three major advances have made it possible to use organ transplantation routinely in the clinic. First, surgical techniques for performing organ replacement have been advanced to the point at which they are now relatively routine in most major medical centers. Second, networks of transplantation centers have been organized to procure the few healthy organs that become available from cadaveric donors. Third, the use of powerful immunosuppressive drugs, especially cyclosporin A and tacrolimus (FK506) to inhibit T-cell activation (see Chapter 16), or blockade of IL-2 receptor signal with rapamycin (sirolimus), which provokes the apoptosis of allospecifically activated CD4 lymphocytes, has markedly increased graft survival rates. The different organs that are transplanted in the clinic are listed in Fig. 15.47. By far the most frequently transplanted solid organ is the kidney, the organ first successfully transplanted between identical twins in the 1950s. Transplantation of the cornea is even more frequent; this tissue is a special case because it is not vascularized, and corneal grafts between unrelated people are usually successful even without immunosuppression.

Many problems other than graft rejection are associated with organ transplantation. First, donor organs are difficult to obtain; this is especially a problem when the organ involved is a vital one, such as the heart or liver. Second, the disease that destroyed the patient's organ might also destroy the graft, as in the destruction of pancreatic β cells in autoimmune diabetes. Third, the immunosuppression required to prevent graft rejection increases the risk of cancer and infection. All of these problems need to be addressed before clinical transplantation can become commonplace. The problems most amenable to scientific solution are the development of more effective means of immunosuppression that prevent rejection with minimal impairment of more generalized immunity, the induction of graft-specific tolerance, and the development of xenografts as a practical solution to organ availability.

Tissue transplanted	No. of grafts in USA (2009)*	5-year graft survival
Kidney	17,683	81.4%#
Liver	6,320	68.3%
Heart	2,241	74.0%
Pancreas and pancreas/kidney	1,233	53.4%†
Lung	1,690	50.6%
Intestine	180	~48.4%
Cornea	~40,000	~70%
HSC transplants	15,000‡	40%/60%‡

Fig. 15.47 Tissues commonly transplanted in clinical medicine. The numbers of organ grafts performed in the United States in 2009 are shown. HSC, hematopoietic stem cells (includes bone marrow, peripheral blood HSCs, and cord blood transplants). *Number of grafts includes multiple organ grafts (for example kidney and pancreas, or heart and lung). For solid organs, 5-year survival is based on transplants performed between 2002 and 2007. Data from the United Network for Organ Sharing. #Kidney survival listed (81.4%) is for kidneys from living donors; 5-year survival for cadaveric donor transplants is 69.1%. †Pancreas survival listed (53.4%) is when transplanted alone; 5-year survival when transplanted with a kidney is 73.5%. ‡Latest data available. They refer to allogeneic transplants only; survival depends on disease and is 40% for patients with acute and 60% for patients with chronic forms of myelogenous leukemia. All grafts except corneal grafts require long-term immunosuppression.

15-36 The converse of graft rejection is graft-versus-host disease.

Transplantation of hematopoietic stem cells (HSCs) enriched from peripheral blood, bone marrow, or fetal cord blood is a successful therapy for some tumors derived from bone marrow precursor cells, such as certain leukemias and lymphomas. By replacing genetically defective stem cells with normal donor ones, HSC transplantation can also be used to cure some primary immunodeficiency diseases (see Chapter 13) and other inherited diseases due to defective blood cells, such as the severe forms of thalassemia. In leukemia therapy, the recipient's bone marrow, the source of the leukemia, must first be destroyed by a combination of irradiation and aggressive cytotoxic chemotherapy. One of the major complications of allogeneic HSC transplantation is **graft-versus-host disease** (**GVHD**), in which mature donor T cells that contaminate preparations of HSCs recognize the tissues of the recipient as foreign, causing a severe inflammatory disease characterized by rashes, diarrhea, and liver disease. GVHD is particularly virulent when there is mismatch of MHC class I or class II antigens. Most transplants are therefore undertaken only when the donor and recipient are HLA-matched siblings or, less frequently, when there is an HLA-matched unrelated donor. As in organ transplantation, GVHD also occurs in the context of disparities between minor histocompatibility antigens, so immunosuppression must also be used in every HSC transplant.

The presence of alloreactive donor T cells can easily be demonstrated experimentally by the **mixed lymphocyte reaction** (**MLR**), in which lymphocytes from a potential donor are mixed with irradiated lymphocytes from the potential recipient. If the donor lymphocytes contain naive T cells that recognize alloantigens on the recipient lymphocytes, they will respond by proliferating (Fig. 15.48). The MLR is sometimes used in the selection of donors for HSC transplants, when the lowest possible alloreactive response is essential. However, the limitation of the MLR in the selection of HSC donors is that the test does not accurately quantify alloreactive T cells. A more accurate test is a version of the limiting-dilution assay (see Appendix I, Section A-25), which precisely counts the frequency of alloreactive T cells.

Although GVHD is harmful to the recipient of a HSC transplant, it can have some beneficial effects that are crucial to the success of the therapy. Much of the therapeutic effect of HSC transplantation for leukemia can be due to a **graft-versus-leukemia effect**, in which the allogeneic preparations of HSCs contain donor T cells that recognize minor histocompatibility antigens or tumor-specific antigens expressed by the host leukemic cells, leading the donor cells to kill the leukemic cells. One of the treatment options for suppressing the development of GVHD is the elimination of mature T cells from the preparations of donor HSCs *in vitro* before transplantation, thereby removing alloreactive T cells. Those T cells that subsequently mature from the donor marrow *in vivo* in the recipient are tolerant to the recipient's antigens. Although the elimination of GVHD has benefits for the patient, there is an increased risk of leukemic relapse, which provides strong evidence in support of the graft-versus-leukemia effect.

Immunodeficiency is another complication of donor T-cell depletion. Because most of the recipient's T cells are destroyed by the combination of high-dose chemotherapy and irradiation used to treat the recipient before the transplant, donor T cells are the major source for reconstituting a mature T-cell repertoire after the transplant. This is particularly true in adults, who have poor residual thymic function. If too many T cells are depleted from the graft, therefore, transplant recipients experience, and often die from, numerous opportunistic infections. The need to balance the beneficial effects of

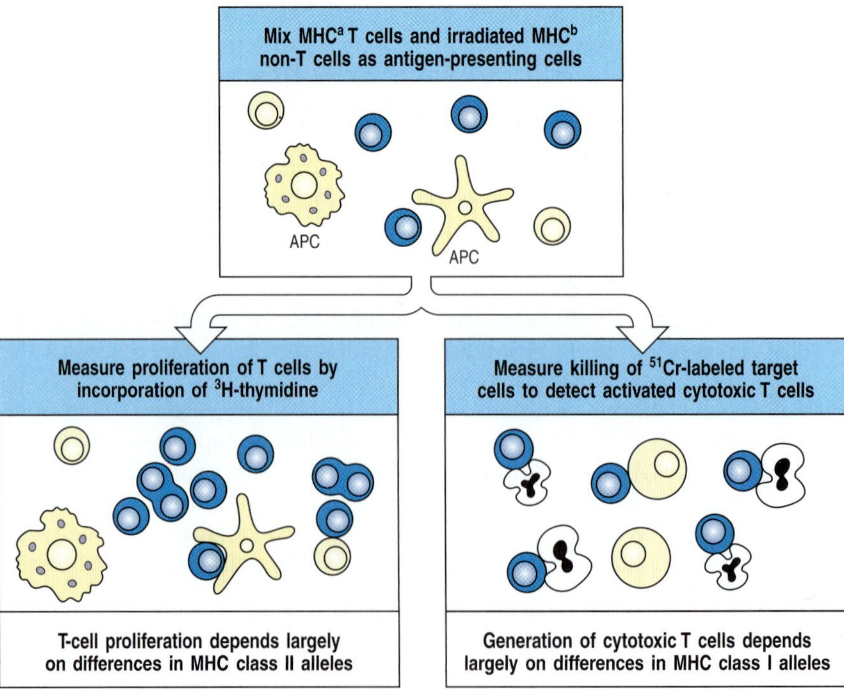

Fig. 15.48 The mixed lymphocyte reaction (MLR) can be used to detect histoincompatibility. Lymphocytes from the two individuals who are to be tested for compatibility are isolated from peripheral blood. The cells from one person (yellow), which will also contain antigen-presenting cells, are either irradiated or treated with mitomycin C; they will act as stimulator cells but cannot now respond by DNA synthesis and cell division to antigenic stimulation by the other person's cells. The cells from the two individuals are then mixed (upper panel). If the unirradiated lymphocytes (the responders, blue) contain alloreactive T cells, these will be stimulated to proliferate and differentiate to effector cells. Between 3 and 7 days after mixing, the culture is assessed for T-cell proliferation (lower left panel), which is mainly the result of CD4 T cells recognizing differences in MHC class II molecules, and for the generation of activated cytotoxic T cells (lower right panel), which respond to differences in MHC class I molecules.

the graft-versus-leukemia effect and immunocompetence with the adverse effects of GVHD caused by donor T cells has spawned much research. One particularly promising approach is to prevent donor T cells from reacting with recipient antigens that they could meet shortly after the transplant. This is accomplished by depleting the recipient's antigen-presenting cells, chiefly dendritic cells (Fig. 15.49). Evidently, in this situation, the donor T cells are not activated during the initial inflammation that accompanies the transplant, and thereafter they do not promote GVHD. However, it is unclear whether there would be a graft-versus-leukemia effect in this context.

15-37 Regulatory T cells are involved in alloreactive immune responses.

As in all immune responses, regulatory T cells are now thought to have an important immunoregulatory role in the alloreactive immune responses involved in graft rejection. Experiments on the transplantation of allogeneic HSCs in mice have thrown some light on this question. Here, depletion of either CD4 CD25 T_{reg} cells in the recipient or of the same class of T_{reg} cells in the HSC graft before transplantation accelerated the onset of GVHD and subsequent death. In contrast, supplementing the graft with fresh CD4 CD25 T_{reg} cells or such cells that had been activated and expanded *ex vivo* delayed, or even prevented, death from GVHD. Similar observations have been made in experimental mouse models of solid organ transplantation, where the transfer of either naturally occurring or induced CD4 CD25 T_{reg} cells significantly delays allograft rejection. These experiments suggest that enriching or generating T_{reg} cells in preparations of donor HSCs might provide a possible therapy for GVHD in the future.

Another class of regulatory T cells, $CD8^+$ $CD28^-$ T_{reg} cells, have an anergic phenotype and they are thought to maintain T-cell tolerance indirectly by inhibiting the capacity of antigen-presenting cells to activate helper T cells. Cells of this type have been isolated from transplant patients. They can be distinguished from alloreactive CD8 cytotoxic T cells because they do not display cytotoxic activity against donor cells and express high levels of the inhibitory killer receptor CD94 (see Section 3-21). This finding suggests the possibility that $CD8^+$ $CD28^-$ T_{reg} cells interfere with the activation of antigen-presenting cells and have a role in the maintenance of transplant tolerance.

15-38 The fetus is an allograft that is tolerated repeatedly.

All of the transplants discussed so far are artifacts of modern medical technology. However, one 'foreign' tissue that is repeatedly grafted and repeatedly

Fig. 15.49 Recipient type antigen-presenting cells are required for the efficient initiation of graft-versus-host disease (GVHD). T cells that accompany the hematopoietic stem cells from the donor (left panel) can recognize minor histocompatibility antigens of the recipient and start an immune response against the recipient's tissues. In stem-cell transplantation, minor antigens could be presented by either recipient- or donor-derived antigen-presenting cells, the latter deriving from the stem-cell graft and from precursor cells that differentiate after the transplant. Antigen-presenting cells are shown here as dendritic cells in a lymph node (middle panel). In mice, it has been possible to inactivate the host antigen-presenting cells by using gene knockouts. Such recipients are entirely resistant to GVHD mediated by donor CD8 T cells (right panel). Thus, cross-presentation of the recipient's minor histocompatibility antigens on donor dendritic cells is not sufficient to stimulate GVHD; those antigens endogenously synthesized and presented by the recipient's antigen-presenting cells are required to stimulate donor T cells. For this strategy to be useful for preventing GVHD in human patients, ways of depleting the recipient's antigen-presenting cells will be needed. These are the focus of research in several laboratories.

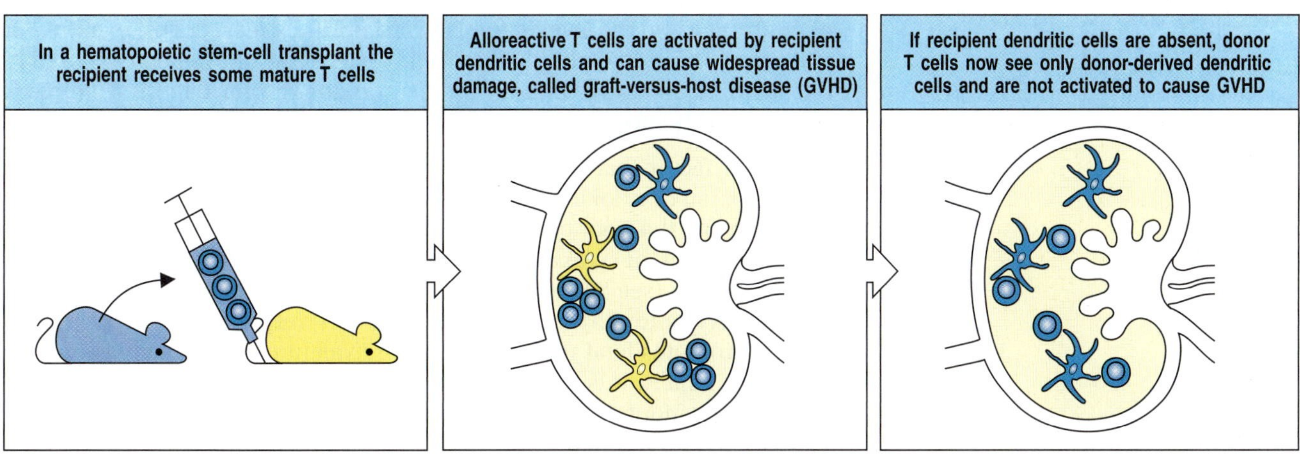

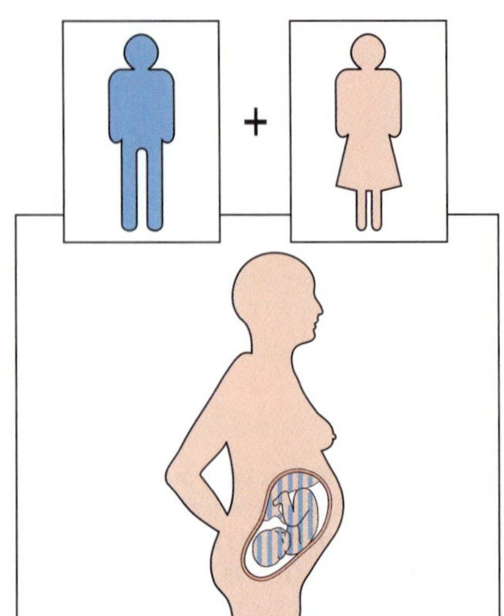

Fig. 15.50 The fetus is an allograft that is not rejected. Although the fetus carries MHC molecules derived from the father, and other foreign antigens, it is not rejected. Even when the mother bears several children to the same father, no sign of immunological rejection is seen.

tolerated is the mammalian fetus. The fetus carries paternal MHC and minor histocompatibility antigens that differ from those of the mother (Fig. 15.50), and yet a mother can successfully bear many children expressing the same nonself MHC proteins derived from the father. The mysterious lack of rejection of the fetus has puzzled generations of immunologists, and no comprehensive explanation has yet emerged. One problem is that acceptance of the fetal allograft is so much the norm that it is difficult to study the mechanism that prevents rejection; if the mechanism for rejecting the fetus is rarely activated, how can one analyze the mechanisms that control it?

Various hypotheses have been advanced to account for the tolerance shown to the fetus. It has been proposed that the fetus is simply not recognized as foreign. There is evidence against this hypothesis, because women who have borne several children usually make antibodies directed against the father's MHC proteins and red blood cell antigens; indeed, this is the best source of antibodies for human MHC typing. However, the placenta, which is a fetus-derived tissue, seems to sequester the fetus from the mother's T cells. The outer layer of the placenta, the interface between fetal and maternal tissues, is the trophoblast. This does not express classical MHC class I and class II proteins, making it resistant to recognition and attack by maternal T cells. Tissues lacking MHC class I expression are, however, vulnerable to attack by NK cells (see Section 3-21). The trophoblast might be protected from attack by NK cells by the expression of a nonclassical and minimally polymorphic HLA class I molecule—HLA-G. This protein has been shown to bind to the two major inhibitory NK receptors, KIR1 and KIR2, and to inhibit NK killing.

The placenta might also sequester the fetus from the mother's T cells by an active mechanism of nutrient depletion. The enzyme indoleamine 2,3-dioxygenase (IDO) is expressed at a high level by cells at the maternal–fetal interface. This enzyme catabolizes, and thereby depletes, the essential amino acid tryptophan at this site, and T cells starved of tryptophan show reduced responsiveness. Inhibition of IDO in pregnant mice, using the inhibitor 1-methyltryptophan, causes rapid rejection of allogeneic, but not syngeneic, fetuses. This supports the hypothesis that maternal T cells, alloreactive to paternal MHC proteins, might be held in check in the placenta by tryptophan depletion.

It is likely that fetal tolerance is a multifactorial process. The trophoblast does not act as an absolute barrier between mother and fetus, and fetal blood cells can cross the placenta and be detected in the maternal circulation, albeit in very low numbers. There is direct evidence from experiments in mice for specific T-cell tolerance against paternal MHC alloantigens. Pregnant female mice whose T cells bear a transgenic receptor specific for a paternal alloantigen showed reduced expression of this T-cell receptor during pregnancy. These same mice lost the ability to control the growth of an experimental tumor bearing the same paternal MHC alloantigen. After pregnancy, tumor growth was controlled and the level of the T-cell receptor increased. This experiment demonstrates that the maternal immune system must have been exposed to paternal MHC alloantigens and that the immune response to these antigens was temporarily suppressed.

Yet another factor that might contribute to maternal tolerance of the fetus is the secretion of cytokines at the maternal–fetal interface. Both the uterine epithelium and the trophoblast secrete cytokines, including TGF-β and IL-10. This combination of cytokines tends to suppress the development of effector T-cell responses (see Section 11-5). The induction or injection of cytokines such as IFN-γ and IL-12, which promote T_H1 responses in experimental animals, promotes fetal resorption, the equivalent of spontaneous abortion in humans. Finally, it is possible that regulatory T cells could have a role in suppressing responses to the fetus.

The fetus is thus tolerated for two main reasons: it occupies a site protected by a nonimmunogenic tissue barrier, and it promotes a local immunosuppressive response in the mother. Several sites in the body, such as the eye, have these characteristics and allow the prolonged acceptance of foreign tissue grafts. They are usually called immunologically privileged sites (see Section 15-5).

Summary.

Clinical transplantation is now an everyday reality, its success built on MHC matching, immunosuppressive drugs, and technical skill. However, even accurate MHC matching does not prevent graft rejection; other genetic differences between host and donor can result in allogeneic proteins whose peptides are presented as minor histocompatibility antigens by MHC molecules on the grafted tissue, and responses to these can lead to rejection. Because we lack the ability to specifically suppress the response to the graft without compromising host defense, most transplants require generalized immunosuppression of the recipient. This can be toxic and increases the risk of cancer and infection. The fetus is a natural allograft that must be accepted—it almost always is—or the species will not survive. Tolerance to the fetus might hold the key to inducing specific tolerance to grafted tissues, or it might be a special case not applicable to organ replacement therapy.

Summary to Chapter 15.

Ideally, the effector functions of the immune system would be targeted only to foreign pathogens and never to self tissues. In practice, because foreign and self proteins are chemically similar, strict discrimination between self and nonself is impossible. Yet the immune system maintains tolerance to self tissues. This is accomplished by layers of regulation, all of which use surrogate markers to distinguish self from nonself, thus properly directing the immune response. When these regulatory mechanisms break down, autoimmune disease can result. Minor breaches of single regulatory barriers probably occur every day but are quelled by the effects of other regulatory layers: thus, tolerance operates at the level of the overall immune system. For disease to occur, multiple layers of tolerance have to be overcome and the effect needs to be chronic. These layers begin with central tolerance in the bone marrow and thymus, and include peripheral mechanisms such as anergy, cytokine deviation, and regulatory T cells. Sometimes immune responses do not occur simply because the antigens are not available, as in immune sequestration.

Perhaps because of selective pressure to mount effective immune responses to pathogens, the damping of immune responses to promote self-tolerance is limited and prone to failure. Genetic predisposition has an important role in determining which individuals will develop an autoimmune disease. The MHC region has an important effect in many diseases. There are many other genes that contribute to immune regulation and thus, when defective, can cause or predispose to autoimmune disease. Environmental forces also have a significant role, because even identical twins are not always both affected by the same autoimmune disease. Influences from the environment could include infections, toxins, and chance events.

When self-tolerance is broken and autoimmune disease ensues, the effector mechanisms are quite similar to those used in pathogen responses. Although the details vary from disease to disease, both antibody and T cells can be involved. Much has been learned about immune responses made to tissue antigens by examining the response to nonself transplanted organs and tissues; lessons learned in the study of graft rejection apply to autoimmunity and vice versa. Transplantation of solid organs and hematopoietic stem cells has brought on syndromes of rejection that are in many ways similar to autoimmune disease, but the targets are either major or minor histocompatibility antigens. The latter come from polymorphic genes. T cells are the main effectors in graft rejection and graft-versus-host disease.

For each of the undesirable categories of response discussed here (along with allergy, discussed in Chapter 14), the question is how to control the response without adversely affecting protective immunity to infection. The answer might lie in a more complete understanding of the regulation of the immune response, especially the suppressive mechanisms that seem to be important in tolerance. The deliberate control of the immune response is examined further in Chapter 16.

Questions.

15.1 *What is the evidence that genetic predisposition has an important role in autoimmune disease? Give two examples, and for each explain why the example implicates genetics.*

15.2 *(a) Discuss one compelling piece of evidence that environment has a role in the development of autoimmunity. (b) Name two potential environmental factors, and for one of them describe in more detail how it might work to incite autoimmunity.*

15.3 *There are several different mechanisms that initiate autoimmunity. Provide and briefly discuss an example each for four of these. Include both antibody-dependent and T-cell dependent mechanisms.*

15.4 *Discuss potential mechanisms for and differences between hyperacute, acute, and chronic rejection.*

15.5 *(a) Describe the ways in which the fetus is protected from immune attack by the mother. (b) Nevertheless, newborn babies of mothers with some types of autoimmunity show symptoms of the same autoimmune disease. How does this happen and what can be done to alleviate it? Not all autoimmune diseases will affect a baby in this way; explain why.*

15.6 *What is the role of TNF-α in rheumatoid arthritis? Which cells does it come from?*

15.7 *Discuss the mechanisms leading to the production of donor antigen-reactive antibody in an allograft recipient and how such antibody could lead to injury to the graft.*

15.8 *What would be the consequences of depleting T cells from an allogeneic bone marrow transplant for a patient with leukemia? What explanation can you give for this effect?*

Section references.

15-1 A critical function of the immune system is to discriminate self from nonself.

Ehrlich, P., and Morgenroth, J.: **On haemolysins**, in Himmelweit, F. (ed): *The Collected Papers of Paul Ehrlich.* London, Pergamon, 1957: 246–255.

Janeway, C.A., Jr: **The immune system evolved to discriminate infectious nonself from noninfectious self.** *Immunol. Today* 1992, **13**:11–16.

Matzinger, P.: **The danger model: a renewed sense of self.** *Science* 2002, **296**:301–305.

15-2 Multiple tolerance mechanisms normally prevent autoimmunity.

Goodnow, C.C., Sprent, J., Fazekas de St Groth, B., and Vinuesa, C.G.: **Cellular and genetic mechanisms of self tolerance and autoimmunity.** *Nature* 2005, **435**:590–597.

Shlomchik, M.J.: **Sites and stages of autoreactive B cell activation and regulation.** *Immunity* 2008, **28**:18–28.

15-3 Central deletion or inactivation of newly formed lymphocytes is the first checkpoint of self-tolerance.

Hogquist, K.A., Baldwin, T.A., and Jameson, S.C.: **Central tolerance: learning self-control in the thymus.** *Nat. Rev. Immunol.* 2005, **5**:772–782.

Kappler, J.W., Roehm, N., and Marrack, P.: **T cell tolerance by clonal elimination in the thymus.** *Cell* 1987, **49**:273–280.

Kyewski, B., and Klein, L.: **A central role for central tolerance.** *Annu. Rev. Immunol.* 2006, **24**:571–606.

Nemazee, D.A., and Burki, K.: **Clonal deletion of B lymphocytes in a transgenic mouse bearing anti-MHC class-I antibody genes.** *Nature* 1989, **337**:562–566.

Schwartz, R.H.: **T cell anergy.** *Annu. Rev. Immunol.* 2003, **21**:305–334.

Steinman, R.M., Hawiger, D., and Nussenzweig, M.C.: **Tolerogenic dendritic cells.** *Annu. Rev. Immunol.* 2003, **21**:685–711.

15-4 Lymphocytes that bind self antigens with relatively low affinity usually ignore them but in some circumstances become activated.

Billingham, R.E., Brent, L., and Medawar, P.B.: **Actively acquired tolerance of foreign cells.** *Nature* 1953, **172**:603–606.

Hannum, L.G., Ni, D., Haberman, A.M., Weigert, M.G., and Shlomchik, M.J.: **A disease-related RF autoantibody is not tolerized in a normal mouse: implications for the origins of autoantibodies in autoimmune disease.** *J. Exp. Med.* 1996, **184**:1269–1278.

Kurts, C., Sutherland, R.M., Davey, G., Li, M., Lew, A.M., Blanas, E., Carbone, F.R., Miller, J.F., and Heath, W.R.: **CD8 T cell ignorance or tolerance to islet antigens depends on antigen dose.** *Proc. Natl Acad. Sci. USA* 1999, **96**:12703–12707.

Marshak-Rothstein, A.: **Toll-like receptors in systemic autoimmune disease.** *Nat. Rev. Immunol.* 2006, **6**:823–835.

Martin, D.A., and Elkon, K.B.: **Autoantibodies make a U-turn: the toll hypothesis for autoantibody specificity.** *J. Exp. Med.* 2005, **202**:1465–1469.

15-5 Antigens in immunologically privileged sites do not induce immune attack but can serve as targets.

Forrester, J.V., Xu, H., Lambe, T., and Cornall, R.: **Immune privilege or privileged immunity?** *Mucosal Immunol.* 2008, **1**:372–381.

Green, D.R., and Ware, C.F.: **Fas-ligand: privilege and peril.** *Proc. Natl Acad. Sci. USA* 1997, **94**:5986–5990.

Mellor, A.L., and Munn, D.H.: **Creating immune privilege: active local suppression that benefits friends, but protects foes.** *Nat. Rev. Immunol.* 2008, **8**:74–80.

Simpson, E.: **A historical perspective on immunological privilege.** *Immunol. Rev.* 2006, **213**:12–22.

15-6 Autoreactive T cells that express particular cytokines may be nonpathogenic or may suppress pathogenic lymphocytes.

von Herrath, M.G., and Harrison, L.C.: **Antigen-induced regulatory T cells in autoimmunity.** *Nat. Rev. Immunol.* 2003, **3**:223–232.

15-7 Autoimmune responses can be controlled at various stages by regulatory T cells.

Asano, M., Toda, M., Sakaguchi, N., and Sakaguchi, S.: **Autoimmune disease as a consequence of developmental abnormality of a T cell subpopulation.** *J. Exp. Med.* 1996, **184**:387–396.

Fillatreau, S., Sweenie, C.H., McGeachy, M.J., Gray, D., and Anderton, S.M.: **B cells regulate autoimmunity by provision of IL-10.** *Nat. Immunol.* 2002, **3**:944–950.

Fontenot, J.D., and Rudensky, A.Y.: **A well adapted regulatory contrivance: regulatory T cell development and the forkhead family transcription factor Foxp3.** *Nat. Immunol.* 2005, **6**:331–337.

Izcue, A., Coombes, J.L., and Powrie, F.: **Regulatory lymphocytes and intestinal inflammation.** *Annu. Rev. Immunol.* 2009, **27**:313–338.

Mayer, L., and Shao, L.: **Therapeutic potential of oral tolerance.** *Nat. Rev. Immunol.* 2004, **4**:407–419.

Maynard, C.L., and Weaver, C.T.: **Diversity in the contribution of interleukin-10 to T-cell-mediated immune regulation.** *Immunol. Rev.* 2008, **226**:219–233.

Sakaguchi, S.: **Naturally arising CD4$^+$ regulatory T cells for immunologic self-tolerance and negative control of immune responses.** *Annu. Rev. Immunol.* 2004, **22**:531–562.

Wildin, R.S., Ramsdell, F., Peake, J., Faravelli, F., Casanova, J.L., Buist, N., Levy-Lahad, E., Mazzella, M., Goulet, O., Perroni, L., *et al.*: **X-linked neonatal diabetes mellitus, enteropathy and endocrinopathy syndrome is the human equivalent of mouse scurfy.** *Nat. Genet.* 2001, **27**:18–20.

Yamanouchi, J., Rainbow, D., Serra, P., Howlett, S., Hunter, K., Garner, V.E.S., Gonzalez-Munoz, A., Clark, J., Veijola, R., Cubbon, R., *et al.*: **Interleukin-2 gene variation impairs regulatory T cell function and causes autoimmunity.** *Nat. Genet.* 2007, **39**:329–337.

15-8 Specific adaptive immune responses to self antigens can cause autoimmune disease.

Lotz, P.H.: **The autoantibody repertoire: searching for order.** *Nat. Rev. Immunol.* 2003, **3**:73–78.

Santamaria, P.: **The long and winding road to understanding and conquering type 1 diabetes.** *Immunity* 2010, **32**:437–445.

Steinman, L.: **Multiple sclerosis: a coordinated immunological attack against myelin in the central nervous system.** *Cell* 1996, **85**:299–302.

15-9 Autoimmune diseases can be classified into clusters that are typically either organ-specific or systemic.

Davidson, A., and Diamond, B.: **Autoimmune diseases.** *N. Engl. J. Med.* 2001, **345**:340–350.

D'Cruz, D.P., Khamashta, M.A., and Hughes, G.R.V.: **Systemic lupus erythematosus.** *Lancet* 2007, **369**:587–596.

Marrack, P., Kappler, J., and Kotzin, B.L.: **Autoimmune disease: why and where it occurs.** *Nat. Med.* 2001, **7**:899–905.

15-10 Multiple components of the immune system are typically recruited in autoimmune disease.

Drachman, D.B.: **Myasthenia gravis.** *N. Engl. J. Med.* 1994, **330**:1797–1810.

Firestein, G.S.: **Evolving concepts of rheumatoid arthritis.** *Nature* 2003, **423**:356–361.

Lehuen, A., Diana, J., Zaccone, P., and Cooke, A.: **Immune cell crosstalk in type 1 diabetes.** *Nat. Rev. Immunol.* 2010, **10**:501–513.

Shlomchik, M.J., and Madaio, M.P.: **The role of antibodies and B cells in**

the pathogenesis of lupus nephritis. *Springer Semin. Immunopathol.* 2003, **24**:363–375.

15-11 Chronic autoimmune disease develops through positive feedback from inflammation, inability to clear the self antigen, and a broadening of the autoimmune response.

Marshak-Rothstein, A.: **Toll-like receptors in systemic autoimmune disease.** *Nat. Rev. Immunol.* 2006, **6**:823–835.

Nagata, S., Hanayama, R., and Kawane, K.: **Autoimmunity and the clearance of dead cells.** *Cell* 2010, **140**:619–630.

Salato, V.K., Hacker-Foegen, M.K., Lazarova, Z., Fairley, J.A., and Lin, M.S.: **Role of intramolecular epitope spreading in pemphigus vulgaris.** *Clin. Immunol.* 2005, **116**:54–64.

Steinman, L.: **A few autoreactive cells in an autoimmune infiltrate control a vast population of nonspecific cells: a tale of smart bombs and the infantry.** *Proc. Natl Acad. Sci. USA* 1996, **93**:2253–2256.

Vanderlugt, C.L., and Miller, S.D.: **Epitope spreading in immune-mediated diseases: implications for immunotherapy.** *Nat. Rev. Immunol.* 2002, **2**:85–95.

15-12 Both antibody and effector T cells can cause tissue damage in autoimmune disease.

Naparstek, Y., and Plotz, P.H.: **The role of autoantibodies in autoimmune disease.** *Annu. Rev. Immunol.* 1993, **11**:79–104.

Vlahakos, D., Foster, M.H., Ucci, A.A., Barrett, K.J., Datta, S.K., and Madaio, M.P.: **Murine monoclonal anti-DNA antibodies penetrate cells, bind to nuclei, and induce glomerular proliferation and proteinuria *in vivo*.** *J. Am. Soc. Nephrol.* 1992, **2**:1345–1354.

15-13 Autoantibodies against blood cells promote their destruction.

Beardsley, D.S., and Ertem, M.: **Platelet autoantibodies in immune thrombocytopenic purpura.** *Transfus. Sci.* 1998, **19**:237–244.

Clynes, R., and Ravetch, J.V.: **Cytotoxic antibodies trigger inflammation through Fc receptors.** *Immunity* 1995, **3**:21–26.

Domen, R.E.: **An overview of immune hemolytic anemias.** *Cleveland Clin. J. Med.* 1998, **65**:89–99.

15-14 The fixation of sublytic doses of complement to cells in tissues stimulates a powerful inflammatory response.

Brandt, J., Pippin, J., Schulze, M., Hansch, G.M., Alpers, C.E., Johnson, R.J., Gordon, K., and Couser, W.G.: **Role of the complement membrane attack complex (C5b-9) in mediating experimental mesangioproliferative glomerulonephritis.** *Kidney Int.* 1996, **49**:335–343.

Hansch, G.M.: **The complement attack phase: control of lysis and non-lethal effects of C5b-9.** *Immunopharmacology* 1992, **24**:107–117.

15-15 Autoantibodies against receptors cause disease by stimulating or blocking receptor function.

Bahn, R.S., and Heufelder, A.E.: **Pathogenesis of Graves' ophthalmopathy.** *N. Engl. J. Med.* 1993, **329**:1468–1475.

Drachman, D.B.: **Myasthenia gravis.** *N. Engl. J. Med.* 1994, **330**:1797–1810.

Vincent, A., Lily, O., and Palace, J.: **Pathogenic autoantibodies to neuronal proteins in neurological disorders.** *J. Neuroimmunol.* 1999, **100**:169–180.

15-16 Autoantibodies against extracellular antigens cause inflammatory injury by mechanisms akin to type II and type III hypersensitivity reactions.

Casciola-Rosen, L.A., Anhalt, G., and Rosen, A.: **Autoantigens targeted in systemic lupus erythematosus are clustered in two populations of surface structures on apoptotic keratinocytes.** *J. Exp. Med.* 1994, **179**:1317–1330.

Clynes, R., Dumitru, C., and Ravetch, J.V.: **Uncoupling of immune complex formation and kidney damage in autoimmune glomerulonephritis.** *Science* 1998, **279**:1052–1054.

Kotzin, B.L.: **Systemic lupus erythematosus.** *Cell* 1996, **85**:303–306.

Lee, R.W., and D'Cruz, D.P.: **Pulmonary renal vasculitis syndromes.** *Autoimmun. Rev.* 2010, **9**:657–660.

Mackay, M., Stanevsky, A., Wang, T., Aranow, C., Li, M., Koenig, S., Ravetch, J.V., and Diamond, B.: **Selective dysregulation of the FcgIIB receptor on memory B cells in SLE.** *J. Exp. Med.* 2006, **203**:2157–2164.

Xiang, Z., Cutler, A.J., Brownlie, R.J., Fairfax, K., Lawlor, K.E., Severinson, E., Walker, E.U., Manz, R.A., Tarlinton, D.M., and Smith, K.G.: **FcgRIIb controls bone marrow plasma cell persistence and apoptosis.** *Nat. Immunol.* 2007, **8**:419–429.

15-17 T cells specific for self antigens can cause direct tissue injury and sustain autoantibody responses.

Feldmann, M., and Steinman, L.: **Design of effective immunotherapy for human autoimmunity.** *Nature* 2005, **435**:612–619.

Firestein, G.S.: **Evolving concepts of rheumatoid arthritis.** *Nature* 2003, **423**:356–361.

Frohman, E.M., Racke, M.K., and Raine, C.S.: **Multiple sclerosis—the plaque and its pathogenesis.** *N. Engl. J. Med.* 2006, **354**:942–955.

Zamvil, S., Nelson, P., Trotter, J., Mitchell, D., Knobler, R., Fritz, R., and Steinman, L.: **T-cell clones specific for myelin basic protein induce chronic relapsing paralysis and demyelination.** *Nature* 1985, **317**:355–358.

15-18 Autoimmune diseases have a strong genetic component.

Fernando, M.M.A., Stevens, C.R., Walsh, E.C., De Jager, P.L., Goyette, P., Plenge, R.M., Vyse, T.J., and Rioux, J.D.: **Defining the role of the MHC in autoimmunity: a review and pooled analysis.** *PLoS Genet.* 2008, **4**:e1000024.

Melanitou, E., Fain, P., and Eisenbarth, G.S.: **Genetics of type 1A (immune mediated) diabetes.** *J. Autoimmun.* 2003, **21**:93–98.

Rioux, J.D., and Abbas, A.K.: **Paths to understanding the genetic basis of autoimmune disease.** *Nature* 2005, **435**:584–589.

Wakeland, E.K., Liu, K., Graham, R.R., and Behrens, T.W.: **Delineating the genetic basis of systemic lupus erythematosus.** *Immunity* 2001, **15**:397–408.

15-19 Several approaches have given us insight into the genetic basis of autoimmunity.

Botto, M., Kirschfink, M., Macor, P., Pickering, M.C., Wurzner, R., and Tedesco, F.: **Complement in human diseases: lessons from complement deficiencies.** *Mol. Immunol.* 2009, **46**: 2774–2783.

Gregersen, P.K.: **Pathways to gene identification in rheumatoid arthritis: PTPN22 and beyond.** *Immunol. Rev.* 2005, **204**:74–86.

Kumar, K.R., Li, L., Yan, M., Bhaskarabhatla, M., Mobley, A.B., Nguyen, C., Mooney, J.M., Schatzle, J.D., Wakeland, E.K., and Mohan, C.: **Regulation of B cell tolerance by the lupus susceptibility gene *Ly108*.** *Science* 2006, **312**:1665–1669.

Nishimura, H., Nose, M., Hiai, H., Minato, N., and Honjo, T.: **Development of lupus-like autoimmune diseases by disruption of the PD-1 gene encoding an ITIM motif-carrying immunoreceptor.** *Immunity* 1999, **11**:141–151.

Xavier, R.J., and Rioux, J.D.: **Genome-wide association studies: a new window into immune-mediated diseases.** *Nat. Rev. Immunol.* 2008, **8**:631–643.

15-20 Many genes that predispose to autoimmunity fall into categories that affect one or more of the mechanisms of tolerance.

Goodnow, C.C.: **Polygenic autoimmune traits: Lyn, CD22, and SHP-1 are limiting elements of a biochemical pathway regulating BCR signaling and selection.** *Immunity* 1998, **8**:497–508.

Tivol, E.A., Borriello, F., Schweitzer, A.N., Lynch, W.P., Bluestone, J.A., and Sharpe, A.H.: **Loss of CTLA-4 leads to massive lymphoproliferation and fatal multiorgan tissue destruction, revealing a critical negative regulatory role of CTLA-4.** *Immunity* 1995, **3**:541–547.

Wakeland, E.K., Liu, K., Graham, R.R., and Behrens, T.W.: **Delineating the genetic basis of systemic lupus erythematosus.** *Immunity* 2001, **15**:397–408.

Walport, M.J.: **Lupus, DNase and defective disposal of cellular debris.** *Nat. Genet.* 2000, **25**:135–136.

15-21 A defect in a single gene can cause autoimmune disease.

Anderson, M.S., Venanzi, E.S., Chen, Z., Berzins, S.P., Benoist, C., and Mathis, D.: **The cellular mechanism of Aire control of T cell tolerance.** *Immunity* 2005, **23**:227–239.

Bacchetta, R., Passerini, L., Gambineri, E., Dai, M., Allan, S.E., Perroni, L., Dagna-Bricarelli, F., Sartirana, C., Matthes-Martin, S., Lawitschka, A., *et al.*: **Defective regulatory and effector T cell functions in patients with FOXP3 mutations.** *J. Clin. Invest.* 2006, **116**:1713–1722.

Ueda, H., Howson, J.M., Esposito, L., Heward, J., Snook, H., Chamberlain, G., Rainbow, D.B., Hunter, K.M., Smith, A.N., DiGenova, G., *et al.*: **Association of the T-cell regulatory gene CTLA4 with susceptibility to autoimmune disease.** *Nature* 2003, **423:** 506–511.

Wildin, R.S., Ramsdell, F., Peake, J., Faravelli, F., Casanova, J.L., Buist, N., Levy-Lahad, E., Mazzella, M., Goulet, O., Perroni, L., *et al.*: **X-linked neonatal diabetes mellitus, enteropathy and endocrinopathy syndrome is the human equivalent of mouse scurfy.** *Nat. Genet.* 2001, **27**:18–20.

15-22 MHC genes have an important role in controlling susceptibility to autoimmune disease.

Fernando, M.M.A., Stevens, C.R., Walsh, E.C., De Jager, P.L., Goyette, P., Plenge, R.M., Vyse, T.J., and Rioux, J.D.: **Defining the role of the MHC in autoimmunity: a review and pooled analysis.** *PLoS Genet.* 2008, **4**:e1000024.

McDevitt, H.O.: **Discovering the role of the major histocompatibility complex in the immune response.** *Annu. Rev. Immunol.* 2000, **18**:1–17.

15-23 Genetic variants that impair innate immune responses can predispose to T cell-mediated chronic inflammatory disease.

Cadwell, K., Liu, J.Y., Brown, S.L., Miyoshi, H., Loh, J., Lennerz, J.K., Kishi, C., Kc, W., Carrero, J.A., Hunt, S., *et al.*: **A key role for autophagy and the autophagy gene Atg16l1 in mouse and human intestinal Paneth cells.** *Nature* 2008, **456**:259–263.

Duerr, R.H., Taylor, K.D., Brant, S.R., Rioux, J.D., Silverberg, M.S., Daly, M.J., Steinhart, A.H., Abraham, C., Regueiro, M., Griffiths, A., *et al.*: **A genome-wide association study identifies IL23R as an inflammatory bowel disease gene.** *Science* 2006, **314**:1461–1463.

Eckmann, L., and Karin, M.: **NOD2 and Crohn's disease: loss or gain of function?** *Immunity* 2005, **22**:661–667.

Xavier, R.J., and Podolsky, D.K.: **Unravelling the pathogenesis of inflammatory bowel disease.** *Nature* 2007, **448**:427–434.

15-24 External events can initiate autoimmunity.

Klareskog, L., Padyukov, L., Ronnelid, J., and Alfredsson, L.: **Genes, environment and immunity in the development of rheumatoid arthritis.** *Curr. Opin. Immunol.* 2006 **18**:650–655.

Munger, K.L., Levin, L.I., Hollis, B.W., Howard, N.S., and Ascherio, A.: **Serum 25-hydroxyvitamin D levels and risk of multiple sclerosis.** *J. Am. Med. Assoc.* 2006, **296**:2832–2838.

15-25 Infection can lead to autoimmune disease by providing an environment that promotes lymphocyte activation.

Bach, J.F.: **Infections and autoimmune diseases.** *J. Autoimmunity* 2005, **25**:74–80.

Moens, U., Seternes, O.M., Hey, A.W., Silsand, Y., Traavik, T., Johansen, B., and Hober, D., and Sauter, P.: **Pathogenesis of type 1 diabetes mellitus: interplay between enterovirus and host.** *Nat. Rev. Endocrinol.* 2010, **6**:279–289.

Sfriso, P., Ghirardello, A., Botsios, C., Tonon, M., Zen, M., Bassi, N., Bassetto, F., and Doria, A.: **Infections and autoimmunity: the multifaceted relationship.** *J. Leukocyte Biol.* 2010, **87**:385–395.

Takeuchi, O., and Akira, S.: **Pattern recognition receptors and inflammation.** *Cell* 2010, **140**:805–820.

15-26 Cross-reactivity between foreign molecules on pathogens and self molecules can lead to anti-self responses and autoimmune disease.

Barnaba, V., and Sinigaglia, F.: **Molecular mimicry and T cell-mediated autoimmune disease.** *J. Exp. Med.* 1997, **185**:1529–1531.

Rose, N.R.: **Infection, mimics, and autoimmune disease.** *J. Clin. Invest.* 2001, **107**:943–944.

Rose, N.R., Herskowitz, A., Neumann, D.A., and Neu, N.: **Autoimmune myocarditis: a paradigm of post-infection autoimmune disease.** *Immunol. Today* 1988, **9**:117–120.

15-27 Drugs and toxins can cause autoimmune syndromes.
&
15-28 Random events may be required for the initiation of autoimmunity.

Bagenstose, L.M., Salgame, P., and Monestier, M.: **Murine mercury-induced autoimmunity: a model of chemically related autoimmunity in humans.** *Immunol. Res.* 1999, **20**:67–78.

Eisenberg, R.A., Craven, S.Y., Warren, R.W., and Cohen, P.L.: **Stochastic control of anti-Sm autoantibodies in MRL/Mp-lpr/lpr mice.** *J. Clin. Invest.* 1987, **80**:691–697.

Yoshida, S., and Gershwin, M.E.: **Autoimmunity and selected environmental factors of disease induction.** *Semin. Arthritis Rheum.* 1993, **22**:399–419.

15-29 Graft rejection is an immunological response mediated primarily by T cells.

Cornell, L.D., Smith, R.N., and Colvin, R.B.: **Kidney transplantation: mechanisms of rejection and acceptance.** *Annu. Rev. Pathol.* 2008, **3**:189–220.

Zelenika, D., Adams, E., Humm, S., Lin, C.Y., Waldmann, H., and Cobbold, S.P.: **The role of CD4⁺ T-cell subsets in determining transplantation rejection or tolerance.** *Immunol. Rev.* 2001, **182**:164–179.

15-30 Transplant rejection is caused primarily by the strong immune response to nonself MHC molecules.

Opelz, G.: **Factors influencing long-term graft loss. The Collaborative Transplant Study.** *Transplant. Proc.* 2000, **32**:647–649.

Opelz, G., and Wujciak, T.: **The influence of HLA compatibility on graft survival after heart transplantation. The Collaborative Transplant Study.** *N. Engl. J. Med.* 1994, **330**:816–819.

15-31 In MHC-identical grafts, rejection is caused by peptides from other alloantigens bound to graft MHC molecules.

den Haan, J.M., Meadows, L.M., Wang, W., Pool, J., Blokland, E., Bishop, T.L., Reinhardus, C., Shabanowitz, J., Offringa, R., Hunt, D.F., *et al.*: **The minor histocompatibility antigen HA-1: a diallelic gene with a single amino acid polymorphism.** *Science* 1998, **279**:1054–1057.

Mutis, T., Gillespie, G., Schrama, E., Falkenburg, J.H., Moss, P., and Goulmy, E.: **Tetrameric HLA class I-minor histocompatibility antigen peptide complexes demonstrate minor histocompatibility antigen-specific cytotoxic T lymphocytes in patients with graft-versus-host disease.** *Nat. Med.* 1999, **5**:839–842.

15-32 There are two ways of presenting alloantigens on the transplanted donor organ to the recipient's T lymphocytes.

Carbone, F.R., Kurts, C., Bennett, S.R., Miller, J.F., and Heath, W.R.: **Cross-presentation: a general mechanism for CTL immunity and tolerance.** *Immunol. Today* 1998, **19**:368–373.

Jiang, S., Herrera, O., and Lechler, R.I.: **New spectrum of allorecognition pathways: implications for graft rejection and transplantation tolerance.** *Curr. Opin. Immunol.* 2004, **16**:550–557.

Lakkis, F.G., Arakelov, A., Konieczny, B.T., and Inoue, Y.: **Immunologic ignorance of vascularized organ transplants in the absence of secondary lymphoid tissue.** *Nat. Med.* 2000, **6**:686–688.

15-33 Antibodies that react with endothelium cause hyperacute graft rejection.

Colvin, R.B., and Smith, R.N.: **Antibody-mediated organ-allograft rejection.** *Nat. Rev. Immunol.* 2005, **5**:807–817.

Kissmeyer Nielsen, F., Olsen, S., Petersen, V.P., and Fjeldborg, O.: **Hyperacute rejection of kidney allografts, associated with pre-existing humoral antibodies against donor cells.** *Lancet* 1966, **ii**:662–665.

Williams, G.M., Hume, D.M., Hudson, R.P., Jr, Morris, P.J., Kano, K., and Milgrom, F.: **'Hyperacute' renal-homograft rejection in man.** *N. Engl. J. Med.* 1968, **279**:611–618.

15-34 Late failure of transplanted organs is caused by chronic injury to the graft.

Colvin, R.B., and Smith, R.N.: **Antibody-mediated organ-allograft rejection.** *Nat. Rev. Immunol.* 2005, **5**:807–817.

Womer, K.L., Vella, J.P., and Sayegh, M.H.: **Chronic allograft dysfunction: mechanisms and new approaches to therapy.** *Semin. Nephrol.* 2000, **20**:126–147.

15-35 A variety of organs are transplanted routinely in clinical medicine.

Lechler, R.I., Sykes, M., Thomson, A.W., and Turka, L.A.: **Organ transplantation—how much of the promise has been realized?** *Nat. Med.* 2005, **11**:605–613.

Ricordi, C., and Strom, T.B.: **Clinical islet transplantation: advances and immunological challenges.** *Nat. Rev. Immunol.* 2004, **4**:259–268.

Yang, Y.-G., and Sykes, M.: **Xenotransplantation: current status and a perspective on the future.** *Nat. Rev. Immunol.* 2007, **7**:519–531.

15-36 The converse of graft rejection is graft-versus-host disease.

Dazzi, F., and Goldman, J.: **Donor lymphocyte infusions.** *Curr. Opin. Hematol.* 1999, **6**:394–399.

Kappel, L.W., Goldberg, G.L., King, C.G., Suh, D.Y., Smith, O.M., Ligh, C., Holland, A.M., Grubin, J., Mark, N.M., Liu, C., *et al.*: **IL-17 contributes to CD4-mediated graft-versus-host disease.** *Blood* 2009, **113**:945–952.

Porter, D.L., and Antin, J.H.: **The graft-versus-leukemia effects of allogeneic cell therapy.** *Annu. Rev. Med.* 1999, **50**:369–386.

Ratajczak, P., Janin, A., Peffault de Latour, R., Leboeuf, C., Desveaux, A., Keyvanfar, K., Robin, M., Clave, E., Douay, C., Quinquenel, A., *et al.*: **Th17/Treg ratio in human graft-versus-host disease.** *Blood* 2010, **116**:1165–1171.

Shlomchik, W.D.: **Graft-versus-host disease.** *Nat. Rev. Immunol.* 2007, **7**:340–352.

15-37 Regulatory T cells are involved in alloreactive immune responses.

Joffre, O., and van Meerwijk, J.P.: **CD4$^+$CD25$^+$ regulatory T lymphocytes in bone marrow transplantation.** *Semin. Immunol.* 2006, **18**:128–135.

Lu, L.F., Lind, E.F., Gondek, D.C., Bennett, K.A., Gleeson, M.W., Pino-Lagos, K., Scott, Z.A., Coyle, A.J., Reed, J.L., Van Snick, J., *et al.*: **Mast cells are essential intermediaries in regulatory T-cell tolerance.** *Nature* 2006, **31**:997–1002.

Wood, K.J., and Sakaguchi, S.: **Regulatory T cells in transplantation tolerance.** *Nat. Rev. Immunol.* 2003, **3**:199–210.

15-38 The fetus is an allograft that is tolerated repeatedly.

Carosella, E.D., Rouas-Freiss, N., Paul, P., and Dausset, J.: **HLA-G: a tolerance molecule from the major histocompatibility complex.** *Immunol. Today* 1999, **20**:60–62.

Mellor, A.L., and Munn, D.H.: **Immunology at the maternal–fetal interface: lessons for T cell tolerance and suppression.** *Annu. Rev. Immunol.* 2000, **18**:367–391.

Trowsdale, J., and Betz, A.G.: **Mother's little helpers: mechanisms of maternal–fetal tolerance.** *Nat. Immunol.* 2006, **7**:241–246.

Manipulation of the Immune Response

16

In this chapter we consider the ways in which the immune system can be manipulated or controlled, both to suppress unwanted immune responses in autoimmunity, allergy, and graft rejection, and to stimulate protective immune responses. It has long been felt that it should be possible to deploy the powerful and specific mechanisms of adaptive immunity to destroy tumors, and we discuss the present state of progress toward that goal. In the final section of the chapter we discuss current vaccination strategies and how a more rational approach to the design and development of vaccines promises to increase their efficacy and widen their usefulness and application.

Treatment of unwanted immune responses.

Unwanted immune responses occur in many settings, such as autoimmune disease, transplant rejection, and allergy, which present different therapeutic challenges. The goal of treatment in all cases is to avoid tissue damage and prevent the disruption of tissue function. Some unwanted immune responses can be anticipated so that preventive measures may be taken, as in the case of allograft rejection. Other unwanted responses may not be detected until after they are established, as is the case with autoimmune or allergic reactions. The relative difficulty of suppressing established immune responses is seen in animal models of autoimmunity, in which treatments that are able to prevent the induction of the disease generally fail to halt established disease.

Conventional immunosuppressive drugs—meaning natural or synthetic small-molecule compounds—can be divided into several different categories (Fig. 16.1). There are the powerful anti-inflammatory drugs of the corticosteroid family such as prednisone, the cytotoxic drugs such as azathioprine and cyclophosphamide, and the noncytotoxic fungal and bacterial derivatives such as cyclosporin A, tacrolimus (FK506 or fujimycin), and rapamycin (sirolimus), which inhibit intracellular signaling pathways within T lymphocytes. Finally, a recently introduced drug, fingolimod, interferes with signaling by the sphingosine-1-phosphate receptor that controls the egress of T cells from lymphoid organs, thus preventing effector lymphocytes from reaching the periphery. Most of these drugs exert broad inhibition of the immune system, and suppress helpful as well as harmful responses. Opportunistic infection is therefore a common complication of immunosuppressive drug therapy.

Newer treatments attempt to target aspects of the immune response that cause the tissue damage, such as cytokine action, while avoiding wholesale immunosuppression, but even these therapeutic agents can affect important components of the response to infectious disease. The most immediate way of inhibiting a particular part of the immune response is via highly specific antibodies, usually directed against specific proteins expressed and/or secreted by immune cells. Approaches of this type that were experimental at the time

Fig. 16.1 Conventional immunosuppressive drugs in clinical use.

Conventional immunosuppressive drugs in clinical use	
Immunosuppressive drug	**Mechanism of action**
Corticosteroids	Inhibit inflammation; inhibit many targets including cytokine production by macrophages
Azathioprine, cyclophosphamide, mycophenolate	Inhibit proliferation of lymphocytes by interfering with DNA synthesis
Cyclosporin A, tacrolimus (FK506)	Inhibit the calcineurin-dependent activation of NFAT; block IL-2 production and proliferation by T cells
Rapamycin (sirolimus)	Inhibits proliferation of effector T cells by blocking Rictor-dependent mTOR activation
Fingolimod (FTY270)	Blocks lymphocyte trafficking out of lymphoid tissues by interfering with signaling by the sphingosine-1-phosphate receptor

of previous editions of this book are now part of established medical practice. Anticytokine monoclonal antibodies, such as infliximab (anti-TNF-α), neutralize local excesses of cytokines or chemokines, or target natural cellular regulatory mechanisms to inhibit immune responses.

Corticosteroid therapy	
Effect on	**Physiological effects**
↓ IL-1, TNF-α, GM-CSF IL-3, IL-4, IL-5, CXCL8	↓ Inflammation caused by cytokines
↓ NOS	↓ NO
↓ Phospholipase A_2 ↓ Cyclooxygenase type 2 ↑ Annexin-1	↓ Prostaglandins ↓ Leukotrienes
↓ Adhesion molecules	Reduced emigration of leukocytes from vessels
↑ Endonucleases	Induction of apoptosis in lymphocytes and eosinophils

Fig. 16.2 Anti-inflammatory effects of corticosteroid therapy. Corticosteroids regulate the expression of many genes, with a net anti-inflammatory effect. First, they reduce the production of inflammatory mediators, including cytokines, prostaglandins, and nitric oxide (NO). Second, they inhibit inflammatory cell migration to sites of inflammation by inhibiting the expression of adhesion molecules. Third, corticosteroids promote the death by apoptosis of leukocytes and lymphocytes. The layers of complexity are illustrated by the actions of annexin-1 (originally identified as a factor induced by corticosteroids and named lipocortin), which has now been shown to participate in all of the effects of corticosteroids listed on the right. NOS, NO synthase.

16-1 Corticosteroids are powerful anti-inflammatory drugs that alter the transcription of many genes.

Corticosteroid drugs are powerful anti-inflammatory and immunosuppressant agents that are used widely to attenuate the harmful effects of immune responses of autoimmune or allergic origin (see Chapters 14 and 15), as well as those induced by organ transplants. **Corticosteroids** are derivatives of the glucocorticoid family of steroid hormones; one of the most widely used is **prednisone**, a synthetic version of the hormone cortisol. Corticosteroids cross the cell's plasma membrane and bind to intracellular receptors of the nuclear receptor family. Activated glucocorticoid receptors are transported to the nucleus, where they bind directly to DNA and interact with other transcription factors to regulate as many as 20% of the genes expressed in leukocytes. The response to steroid therapy is complex, given the large number of genes regulated in leukocytes and in other tissues. In respect of immunosuppression, corticosteroids exert multiple anti-inflammatory effects, which are briefly summarized in Fig. 16.2.

Corticosteroids target the pro-inflammatory functions of monocytes and macrophages and reduce the number of CD4 T cells. Corticosteroids can induce the expression of certain anti-inflammatory genes, such as *annexin I*, which encodes a protein inhibitor of phospholipase A2, an enzyme required for the generation of the pro-inflammatory prostaglandins and leukotrienes (see Sections 3-3 and 14-7). Conversely, corticosteroids also suppress the expression of some pro-inflammatory genes, including those encoding the cytokines IL-1β and TNF-α. Finally, corticosteroids directly regulate the activity of some proteins, such as phosphatidylinositol 3-kinase (see Section 7-4), which produces a more rapidly acting anti-inflammatory effect than could be explained by the induction of new gene transcription alone. The therapeutic effects of corticosteroid drugs are due to their presence at much higher concentrations than occur naturally, causing exaggerated responses with both toxic and beneficial effects; the drugs can also decrease in effectiveness over time. Adverse effects include fluid retention, weight gain, diabetes, bone mineral loss, and thinning of the skin, requiring a careful balance to be maintained between their beneficial and harmful effects. Inhaled corticosteroids, however, have proved highly beneficial in the treatment of chronic asthma (see Section 14-14). In the treatment of autoimmunity or allograft rejection, in which high doses of oral corticosteroids are needed to be effective, they are most often administered in combination with other immunosuppressant drugs to keep the corticosteroid dose and side-effects to a minimum. These drugs include cytotoxic agents that act as immunosuppressants by killing rapidly dividing lymphocytes, and drugs that more specifically target lymphocyte signaling pathways.

16-2 Cytotoxic drugs cause immunosuppression by killing dividing cells and have serious side-effects.

The three cytotoxic drugs most commonly used as immunosuppressants are **azathioprine**, **cyclophosphamide**, and **mycophenolate**. These drugs interfere with DNA synthesis, and their major pharmacological action is on tissues in which cells are continually dividing. Developed originally to treat cancer, these drugs were found to be immunosuppressive after observations that they were cytotoxic to dividing lymphocytes. Azathioprine also interferes with CD28 co-stimulation and blocks activation of the small GTPase Rac1, which in T cells promotes apoptosis (see Section 9-21). The use of these compounds is, however, limited by their toxic effects on all tissues in which cells are dividing, such as the skin, gut lining, and bone marrow. Effects include decreased immune function, as well as anemia, leukopenia, thrombocytopenia, damage to intestinal epithelium, hair loss, and fetal death or injury. As a result of their toxicity, these drugs are used at high doses only when the aim is to eliminate all dividing lymphocytes, as in the treatment of lymphoma and leukemia; in these cases, treated patients require subsequent bone marrow transplantation to restore their hematopoietic function. When used to treat unwanted immune responses such as autoimmune conditions, they are used at lower doses and in combination with other drugs such as corticosteroids.

Azathioprine is converted *in vivo* to the purine analog 6-thioguanine (6-TG), which is metabolized to 6-thioinosinic acid. This competes with inosine monophosphate, blocking the *de novo* synthesis of adenosine monophosphate and guanosine monophosphate, thus inhibiting DNA synthesis. 6-TG is also incorporated into the DNA in place of guanine, and accumulation of 6-TG increases the DNA's sensitivity to mutations induced by the ultraviolet radiation in sunlight. Thus, patients treated with azathioprine have the long-term side-effect of increased risk of skin cancer. Mycophenolate mofetil, the newest addition to the family of cytotoxic immunosuppressive drugs, works in a similar fashion to azathioprine. It is metabolized to mycophenolic acid, which inhibits the enzyme inosine monophosphate dehydrogenase, thus blocking the *de novo* synthesis of guanosine monophosphate.

Azathioprine and mycophenolate are less toxic than cyclophosphamide, which is metabolized to phosphoramide mustard, which alkylates DNA. Cyclophosphamide is a member of the nitrogen mustard family of compounds, which were originally developed as chemical weapons. It has a range of highly toxic effects including inflammation of, and hemorrhage from, the bladder, known as hemorrhagic cystitis, and induction of bladder neoplasia.

16-3 Cyclosporin A, tacrolimus (FK506), and rapamycin (sirolimus) are powerful immunosuppressive agents that interfere with T-cell signaling.

Three noncytotoxic alternatives to the cytotoxic drugs are available as immunosuppressants and are widely used to treat transplant recipients. These are **cyclosporin A**, **tacrolimus** (previously known as **FK506**), and **rapamycin** (also known as **sirolimus**). Cyclosporin A is a cyclic decapeptide derived from a soil fungus, *Tolypocladium inflatum*, from Norway. Tacrolimus is a macrolide compound from the filamentous bacterium *Streptomyces tsukabaensis*, found in Japan; macrolides are compounds that contain a many-membered lactone ring to which is attached one or more deoxysugars. Rapamycin, another *Streptomyces* macrolide, has become important in the prevention of transplant rejection; rapamycin is derived from *Streptomyces hygroscopicus*, found on Easter Island ('Rapa Nui' in Polynesian—hence the name of the drug). All three compounds exert their pharmacological effects by binding to members of a family of intracellular proteins known as the **immunophilins**, forming complexes that interfere with signaling pathways important for the clonal expansion of lymphocytes.

As explained in Section 7-12, cyclosporin A and tacrolimus block T-cell proliferation by inhibiting the phosphatase activity of the Ca^{2+}-activated protein phosphatase **calcineurin**, which is required for the activation of the transcription factor NFAT. Both drugs reduce the expression of several cytokine genes that are normally induced on T-cell activation (Fig. 16.3), including interleukin (IL)-2, which is an important growth factor for T cells (see Section 9-13). Cyclosporin A and tacrolimus inhibit T-cell proliferation in response to either specific antigens or allogeneic cells and are used extensively in medical practice to prevent the rejection of allogeneic organ grafts. Although the major immunosuppressive effects of both drugs are probably the result of inhibition of T-cell proliferation, they also act on other cells and have a large variety of other immunological effects (see Fig. 16.3).

These two drugs inhibit calcineurin by first binding to an immunophilin molecule; cyclosporin A binds to the cyclophilins, and tacrolimus to the FK-binding proteins (FKBP). Immunophilins are peptidyl-prolyl *cis-trans* isomerases, but their isomerase activity is not relevant to the immunosuppressive activity of the drugs that bind them. Rather, the immunophilin:drug complexes bind and inhibit the Ca^{2+}-activated serine/threonine phosphatase calcineurin. In a normal immune response, the increase in intracellular calcium ions in response to T-cell receptor signaling activates the calcium-binding protein calmodulin; calmodulin then activates calcineurin (see Fig. 7.16). Binding of the immunophilin:drug complex to calcineurin prevents its activation by calmodulin; the bound calcineurin is unable to dephosphorylate and activate NFAT (Fig. 16.4). Calcineurin is found in other cells besides T cells, but its levels in T cells are much lower than in other tissues. T cells are therefore particularly susceptible to the inhibitory effects of these drugs.

Fig. 16.3 Cyclosporin A and tacrolimus inhibit lymphocyte and some granulocyte responses.

Immunological effects of cyclosporin A and tacrolimus	
Cell type	**Effects**
T lymphocyte	Reduced expression of IL-2, IL-3, IL-4, GM-CSF, TNF-α Reduced proliferation following decreased IL-2 production Reduced Ca^{2+}-dependent exocytosis of granule-associated serine esterases Inhibition of antigen-driven apoptosis
B lymphocyte	Inhibition of proliferation secondary to reduced cytokine production by T lymphocytes Inhibition of proliferation following ligation of surface immunoglobulin Induction of apoptosis following B-cell activation
Granulocyte	Reduced Ca^{2+}-dependent exocytosis of granule-associated serine esterases

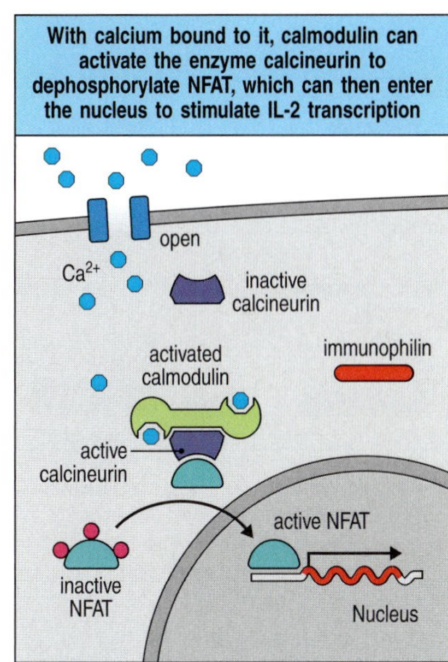
With calcium bound to it, calmodulin can activate the enzyme calcineurin to dephosphorylate NFAT, which can then enter the nucleus to stimulate IL-2 transcription
open
Ca^{2+}
inactive calcineurin
activated calmodulin
immunophilin
active calcineurin
active NFAT
inactive NFAT
Nucleus

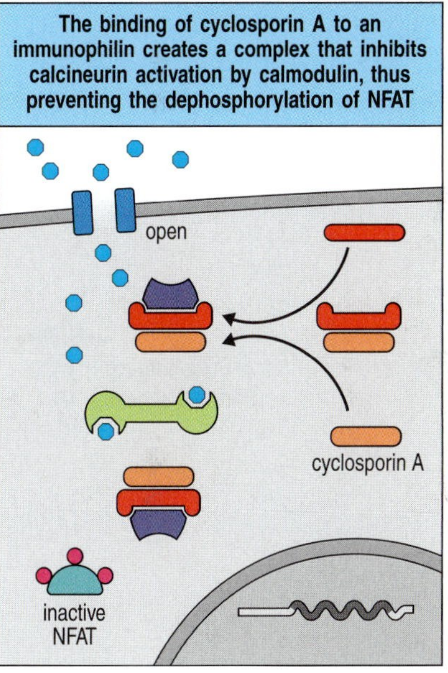
The binding of cyclosporin A to an immunophilin creates a complex that inhibits calcineurin activation by calmodulin, thus preventing the dephosphorylation of NFAT
open
cyclosporin A
inactive NFAT

Fig. 16.4 Cyclosporin A and tacrolimus inhibit T-cell activation by interfering with the serine/threonine-specific phosphatase calcineurin. As shown in the upper panel, signaling via T-cell receptor-associated tyrosine kinases leads to opening of CRAC channels in the plasma membrane. This increases the concentration of Ca^{2+} in the cytoplasm and promotes calcium binding to the regulatory protein calmodulin (see Fig. 7.16). Calmodulin is activated by binding Ca^{2+} and can then target many downstream effector proteins such as the phosphatase calcineurin. Binding by calmodulin activates calcineurin to dephosphorylate the transcription factor NFAT (see Section 7-12), which then enters the nucleus, where it activates genes required for T-cell activation to progress. As shown in the lower panel, when cyclosporin A or tacrolimus are present, they form complexes with their immunophilin targets, cyclophilin and FK-binding protein, respectively. These complexes bind to calcineurin, preventing it from becoming activated by calmodulin, and thereby preventing the dephosphorylation of NFAT.

Cyclosporin A and tacrolimus are effective immunosuppressants but are not problem-free. As with the cytotoxic agents, they affect all immune responses indiscriminately. Their immunosuppressive action is controlled by varying the dose; at the time of grafting, high doses are required but, once a graft is established, the dose can be decreased to allow useful protective immune responses while maintaining adequate suppression of the residual response to the grafted tissue. This balance is difficult to achieve. Furthermore, these drugs have effects on many different tissues, being toxic to kidney tubule epithelial cells, for example. Finally, treatment with these drugs is relatively expensive, because they are complex natural products that must be taken for long periods. Nevertheless, at present they are the immunosuppressants of choice in clinical transplantation, and they are also being tested in a variety of autoimmune diseases, especially those that, like graft rejection, are mediated by T cells.

Rapamycin has a different mode of action from either cyclosporin A or tacrolimus. Like tacrolimus it binds to the FKBP family of immunophilins, but the rapamycin:immunophilin complex does not inhibit calcineurin activity. Instead, it inhibits a serine/threonine kinase known as mTOR (mammalian target of rapamycin), which is involved in regulating cell growth and proliferation and can activate the PI 3-kinase/Akt (protein kinase B) signaling pathway, which is involved in preventing apoptosis and stimulating the cell's utilization of glucose (see Section 7-15). mTOR can form two distinct complexes, mTORC1 or mTORC2, whose activation is controlled by regulatory proteins called Raptor and Rictor, respectively. Rapamycin seems only to inhibit mTOR activation through the Raptor-dependent pathway, thus primarily inhibiting mTORC1 (Fig. 16.5). Blockade of this pathway markedly reduces T-cell proliferation, arresting cells in the G_1 phase of the cell cycle and promoting apoptosis. The drug similarly inhibits lymphocyte proliferation driven by growth factors such as IL-2, IL-4, and IL-6. Intriguingly, rapamycin increases the number of regulatory T cells, perhaps because these cells use different signaling pathways from those of effector T cells. In this same manner, rapamycin differentially influences the outgrowth of effector versus memory T cells, selectively reducing effector function while apparently enhancing the formation of memory T cells. Because of this, the use of rapamycin to augment T-cell memory induced by vaccines is being considered.

The most recently introduced small-molecule drug with the potential to damp down immune responses is one that prevents the migration of immune effector cells to the site of a graft or of autoimmune disease. In Section 9-3 we described how emigration of lymphocytes out of the lymphoid tissues requires recognition of the lipid molecule sphingosine 1-phosphate (S1P) by the G-protein-coupled receptor $S1P_1$. **Fingolimod** (FTY270), a sphingosine 1-phosphate analog, is a new drug that causes the retention of effector lymphocytes in peripheral lymphoid organs and inhibits the migration of dendritic cells, thus preventing these cells from mediating their effector

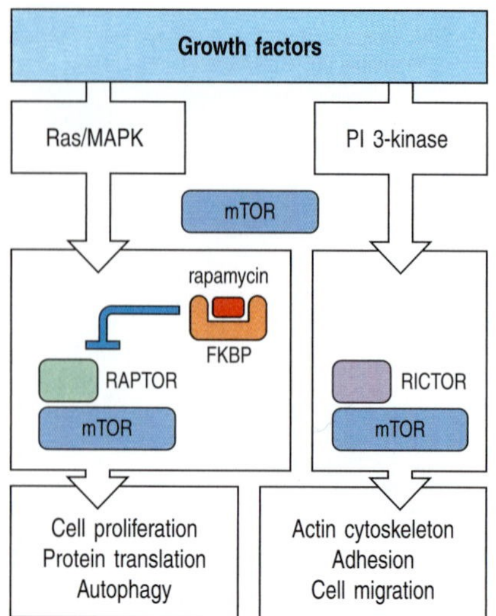

Fig. 16.5 Rapamycin inhibits cell growth and proliferation by selectively blocking activation of the kinase mTOR by RAPTOR. Rapamycin binds FK-binding protein (FKBP), the same immunophilin protein that binds to tacrolimus (FK506). The rapamycin:FKBP complex does not inhibit calcineurin, but instead blocks one of the two pathways that activate mTOR, a large kinase that regulates many metabolic pathways. mTOR can be activated by association with two proteins, RAPTOR (regulatory associated protein of mTOR) and RICTOR (rapamycin-insensitive companion of mTOR). RAPTOR, which is activated by the Ras/MAPK pathway, promotes cell proliferation, translation of proteins, and autophagy. RICTOR, which is activated downstream of PI 3-kinase, influences cell adhesion and migration by controlling the actin cytoskeleton. The rapamycin:FKBP complex acts to inhibit only the RAPTOR-mediated activation of mTOR, and thereby selectively reduces cell growth and proliferation.

activities in target tissues. Fingolimod was approved in 2010 for treatment of the autoimmune disease multiple sclerosis, and has shown promise in the treatment of kidney graft rejection and asthma.

16-4 Antibodies against cell-surface molecules can be used to eliminate lymphocyte subsets or to inhibit lymphocyte function.

All the drugs discussed so far exert a general inhibition on immune responses and can have severe side effects, but antibodies can act in a more specific manner and with less direct toxicity. The potential of antibodies to eliminate unwanted lymphocytes is demonstrated by **anti-lymphocyte globulin**, a preparation of polyclonal immunoglobulin from rabbits (and previously horses) immunized with human lymphocytes, which has been used for many years to treat episodes of acute graft rejection. Anti-lymphocyte globulin does not, however, discriminate between useful lymphocytes and those responsible for the unwanted responses. Foreign immunoglobulins are also highly antigenic in humans, and the large doses of anti-lymphocyte globulin used in therapy often cause a condition called serum sickness, resulting from the formation of immune complexes of the animal immunoglobulin and human antibodies against it (see Section 14-16).

Anti-lymphocyte globulin is nevertheless still used to treat acute rejection, and this has stimulated the quest for monoclonal antibodies (see Appendix I, Section A-12) that would achieve more specifically targeted effects. One such antibody is alemtuzumab (marketed as Campath-1H), which is directed at the cell-surface protein CD52 expressed by most lymphocytes. It has similar actions to anti-lymphocyte globulin, causing long-standing lymphopenia, and is used to eliminate cells in the treatment of chronic lymphocytic leukemia, and to deplete lymphocytes in transplantation.

Immunosuppressive monoclonal antibodies act by one of two general mechanisms. Some, such as alemtuzumab, trigger the destruction of lymphocytes *in vivo* and are referred to as **depleting antibodies**, whereas others are **non-depleting** and act by blocking the function of their target protein without killing the cell that bears it. Depleting monoclonal IgG antibodies tag lymphocytes and target them to macrophages and NK cells, which bear Fc receptors and kill the lymphocytes by phagocytosis and antibody-dependent cell-mediated cytotoxicity (ADCC), respectively. Complement-mediated lysis may also play a part in lymphocyte destruction.

16-5 Antibodies can be engineered to reduce their immunogenicity in humans.

The major impediment to therapy with monoclonal antibodies in humans is that these antibodies are most readily made using mouse cells (see Appendix I, Section A-12), and humans rapidly develop antibody responses to mouse antibodies. This not only blocks the actions of the mouse antibodies but also leads to allergic reactions, and if treatment is continued it can result in anaphylaxis (see Section 14-10). Once a patient has made a response to an antibody, it can no longer be used for future treatment. This problem can, in principle, be avoided by making antibodies that are not recognized as foreign by the human immune system, a process called **humanization**.

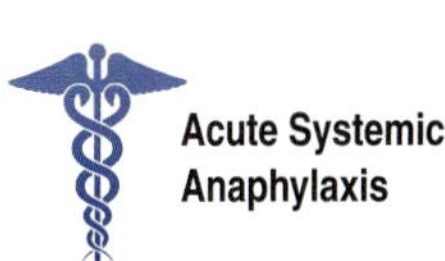

Various approaches have been tried to humanize antibodies. The variable regions encoding the antigen-recognition determinants from a murine antibody can be spliced onto the constant and Fc regions of human IgG by gene manipulation. Antibodies of this type are called **chimeric** antibodies. However, this approach leaves regions within the murine variable regions that could potentially induce immune responses (Fig. 16.6). Another approach is to clone human V regions into a phage display library and select for binding to human cells, as described in Appendix I (see Section A-13). In this way, monoclonal antibodies that are entirely human in origin can be obtained. Newer methods are aimed at generating fully human monoclonal antibodies directly from human cells through the use of viral transformation of human primary B-cell lines or antibody-secreting plasmablasts, or by generating human B-cell hybridomas.

Monoclonal antibodies belong to a new class of therapeutic compounds called **biologics**, which includes other natural proteins such as anti-lymphocyte globulin, cytokines, and fragments of proteins, and also the use of whole cells, as in the adoptive transfer of T cells in cancer immunotherapy. Many monoclonal antibodies have been, or are in the process of being, approved for clinical use by the US Food and Drug Administration (Fig. 16.7), and a systematic naming process identifies the type of antibody. Murine monoclonals are designated by the suffix -**omab**, such as muromomab (originally called OKT3), a murine antibody against CD3. Chimeric antibodies in which the entire variable region is spliced into human constant regions have the suffix -**ximab**, such as basiliximab, an anti-CD25 antibody approved for the treatment of transplantation rejection. Humanized antibodies in which the murine hypervariable regions have been spliced into a human antibody have the suffix -**zumab**, as in alemtuzumab and natalizumab (Tysabri). The latter is directed against the α_4 integrin subunit, and is used to treat multiple sclerosis and Crohn's disease. Antibodies derived entirely from human sequences have the suffix -**mumab**, as in adalimumab, an antibody derived from phage display that binds TNF-α; it is used to treat several autoimmune diseases.

16-6 Monoclonal antibodies can be used to prevent allograft rejection.

Antibodies specific for various physiological targets are being used, or are under investigation, to prevent the rejection of transplanted organs by inhibiting the development of harmful inflammatory and cytotoxic responses. For example, alemtuzumab (see Section 16-4) is licensed for the treatment of

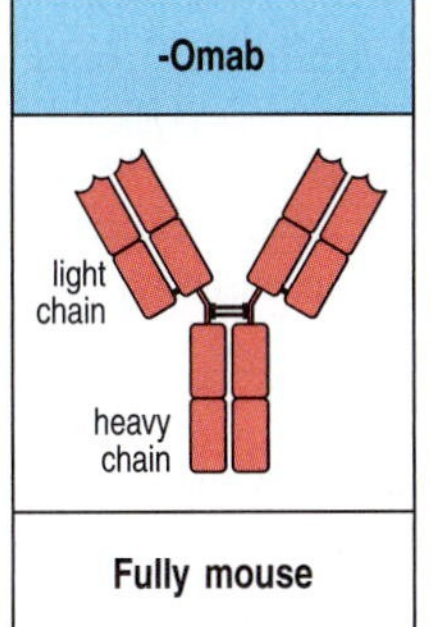

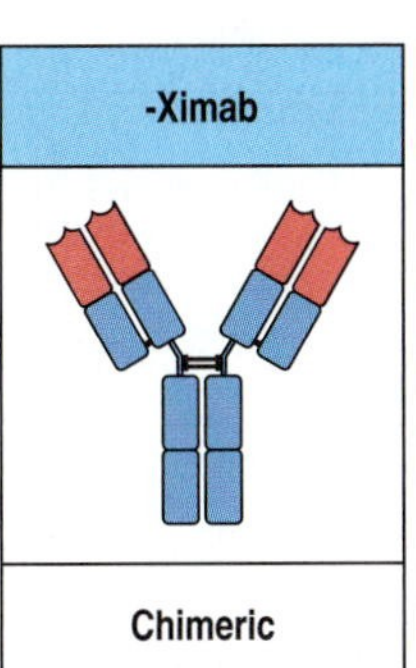

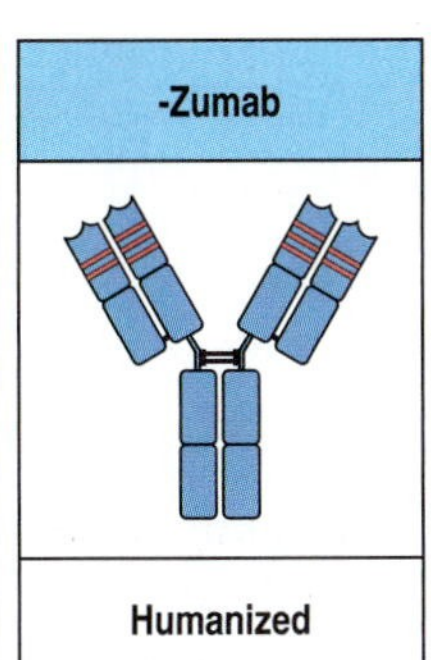

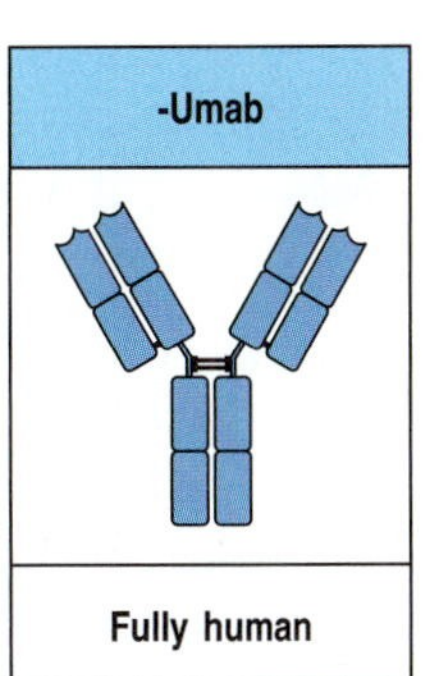

Fig. 16.6 Monoclonal antibodies used to treat human diseases can be engineered to decrease immunogenicity but maintain their antigen specificity. Antibodies that are derived fully from mouse, named with the suffix -omab, are immunogenic in humans. This causes patients to generate antibodies against them, limiting their usefulness over time. This immunogenicity can be reduced by making chimeric antibodies in which the V regions from the mouse are spliced onto human antibody constant regions, and named with the suffix -ximab. Humanization is the process of splicing just the complementarity-determining regions from the mouse antibody, further reducing immunogenicity; these are named with the suffix -zumab. New techniques are now allowing fully human (-umab) monoclonal antibodies to be derived, which have the minimum of immunogenicity.

certain leukemias but is also used in both solid organ and bone marrow transplantation. In solid-organ transplantation, alemtuzumab may be given to the recipient around the time of transplantation to remove mature T lymphocytes from the circulation. In bone marrow transplantation, in contrast,

Fig. 16.7 Monoclonal antibodies used to induce immunosuppression that are currently approved for use in humans.

Monoclonal antibodies developed for immunotherapy			
Generic name	**Specificity**	**Mechanism of action**	**Approved indication**
Rituximab	Anti-CD20	Eliminates B cells	Non-Hodgkin's lymphoma
Alemtuzumab (Campath-1H)	Anti-CD52	Eliminates lymphocytes	Chronic myeloid leukemia
Muromonab (OKT3)	Anti-CD3	Inhibits T-cell activation	Kidney transplantation
Daclizumab	Anti-IL2R	Reduces T-cell activation	
Basiliximab	Anti-IL2R	Reduces T-cell activation	
Infliximab	Anti-TNF-α	Inhibit inflammation induced by TNF-α	Crohn's disease
Certolizumab	Anti-TNF-α		Rheumatoid arthritis
Adalimumab	Anti-TNF-α		
Golimumab	Anti-TNF-α		
Tocilizumab	Anti-IL6R	Blocks inflammation induced by IL-6 signaling	
Canakinumab	Anti-IL1β	Blocks inflammation caused by IL-1	Muckle–Wells syndrome
Denosumab	Anti-RANK-L	Inhibits activation of osteoclasts by RANK-L	Bone loss
Ustekinumab	Anti-IL12/23	Inhibits inflammation caused by IL-12 and IL-23	Psoriasis
Efalizumab	Anti-CD11a (α_L integrin subunit)	Block lymphocyte trafficking	Psoriasis (withdrawn from use in United States and European Union)
Natalizumab	Anti-α4 integrin		Multiple sclerosis
Omalizumab	Anti-IgE	Removes IgE antibody	Chronic asthma
Belimumab	Anti-BLyS	Reduces B-cell responses	Systemic lupus erythematosus (pending approval)
Ipilimumab	Anti-CTLA-4	Increases CD4 T-cell responses	Metastatic melanoma
Raxibacumab	Anti-*B. anthracis* protective antigen (the cell-binding moiety of anthrax toxin)	Prevents action of anthrax toxins	Anthrax infection (pending approval)

alemtuzumab is used *in vitro* to deplete donor bone marrow of mature T cells before its infusion into a recipient. Elimination of mature T cells from donor bone marrow is very effective at reducing the incidence of graft-versus-host disease (see Section 15-36). In this disease, the T lymphocytes in the donor bone marrow recognize the recipient as foreign and mount a damaging response, causing rashes, diarrhea, and hepatitis, which can occasionally be fatal. It had been thought that elimination of mature donor T cells might not be so advantageous when the bone marrow graft was being given as a treatment for leukemia, because the antileukemic action of the donor T cells could be lost, but this seems not to be the case when alemtuzumab is used as the depleting agent.

Specific antibodies directed against T cells have been used to treat episodes of graft rejection that occur after transplantation. The murine antibody muromomab (OKT3) targets the CD3 complex and leads to T-cell immunosuppression by inhibiting signaling through the T-cell receptor. It has been used clinically in solid organ transplantation but is often associated with an unwanted stimulation of cytokine release, and its use is declining. The cytokine release is related to an intact Fc region, which when mutated (as in the antibody called OKT3g1(Ala-Ala)) no longer produces this potentially dangerous side-effect. The latter antibody retains the antigen-binding region of OKT3, but amino acids 234 and 235 of the human IgG1 Fc region have been changed to alanines, preventing the interactions that lead to cytokine release. Two other antibodies, daclizumab and basiliximab, approved for treating kidney transplant rejections, are directed against CD25 (a subunit of the IL-2 receptor) and reduce T-cell activation, presumably by blocking the growth-promoting signals delivered by IL-2.

A primate model of kidney transplant rejection showed promising effects for a humanized monoclonal antibody against the CD40 ligand expressed by T cells (see Section 9-7). A possible mechanism of protection by this antibody is blockade of the activation of dendritic cells by helper T cells that recognize donor antigens. Only preliminary studies of anti-CD40 ligand antibodies have been performed in humans. One antibody was associated with thromboembolic complications and was withdrawn; a different anti-CD40 ligand antibody was administered to patients with the autoimmune disease systemic lupus erythematosus (SLE) without significant complications, but also with little evidence of effectiveness.

In experimental models, monoclonal antibodies against other targets have also had some success in preventing graft rejection, including nondepleting antibodies that bind the CD4 co-receptor or co-stimulatory receptor CD28 on lymphocytes. Similarly, a soluble recombinant protein CTLA-4–Ig (abatacept), which binds to the co-stimulatory B7 molecules on antigen-presenting cells and prevents their interaction with CD28 on T cells, has allowed the long-term survival of certain experimentally grafted tissues.

16-7 Depletion of autoreactive lymphocytes can treat autoimmune disease.

In addition to their uses to prevent transplantation rejection, monoclonal antibodies can be used to treat certain autoimmune diseases, and the different immune mechanisms targeted are discussed in the next few sections. We start by discussing the use of depleting and nondepleting antibodies to remove lymphocytes nonspecifically. The anti-CD20 monoclonal antibody rituximab was originally developed to treat B-cell lymphomas, but has also been tried in certain autoimmune diseases. By ligating CD20, rituximab (Rituxan, MabThera) transduces a signal that induces lymphocyte apoptosis and depletes B cells for several months. Certain autoimmune diseases are believed to involve autoantibody-mediated pathogenesis. There is evidence for the efficacy of this antibody in some patients with autoimmune hemolytic

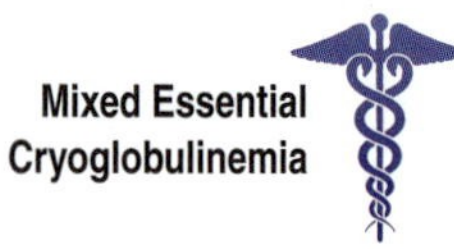

anemia, SLE, rheumatoid arthritis, or type II mixed cryoglobulinemia, in all of which autoantibodies are present. Although CD20 is not expressed on antibody-producing plasma cells, their B-cell precursors are targeted by anti-CD20, resulting in a substantial reduction in the short-lived, but not the long-lived, plasma-cell population.

Alemtuzumab, which is used in treating leukemia and in transplant rejection (see Section 16-5), has shown some beneficial effect in studies of small numbers of patients with multiple sclerosis. However, immediately after its infusion, most multiple sclerosis patients suffered a frightening, although fortunately brief, flare-up of their illness, illustrating another potential complication of antibody therapy. While alemtuzumab was binding and killing cells by complement- and Fc-dependent mechanisms, cytokines were released, including TNF-α, interferon (IFN)-γ, and IL-6. One of the effects of this was a transient blockade of nerve conduction in nerve fibers previously affected by demyelination, which caused the dramatic exacerbation of symptoms. Nevertheless, alemtuzumab could be useful at early stages of disease when the inflammatory response is maximal, but this has yet to be determined.

Treatment of patients with rheumatoid arthritis and multiple sclerosis using anti-CD4 antibodies has been tried, but with disappointing results. In controlled studies the antibodies showed only small therapeutic effects but caused depletion of T lymphocytes from peripheral blood for more than 6 years after treatment. The likely explanation for the failure seems to be that these antibodies failed to delete primed CD4 T_H1 cells secreting the pro-inflammatory cytokine IFN-γ, and may thus have missed their target. This cautionary tale shows that it is possible to deplete large numbers of lymphocytes and yet completely fail to kill the cells that matter.

16-8 Biologic agents that block TNF-α or IL-1 can alleviate autoimmune diseases.

Anti-inflammatory therapy can either attempt to eliminate an autoimmune response altogether, as with immunosuppressive drugs or depleting antibodies, or can try to reduce the tissue injury caused by the immune response. This second category of treatment is called **immunomodulatory therapy**, and is illustrated by the use of conventional anti-inflammatory agents such as aspirin, nonsteroidal anti-inflammatory drugs, or low-dose corticosteroids. A newer avenue of immunomodulatory therapy using biologics is illustrated by several FDA-approved antibodies that target the activity of the powerful pro-inflammatory cytokines, such as TNF-α, IL-1, and IL-6.

Anti-TNF therapy was the first specific biological therapeutic to enter the clinic. Anti-TNF-α antibodies have been found to induce striking remissions in rheumatoid arthritis (Fig. 16.8) and to reduce tissue inflammation in Crohn's disease, an inflammatory bowel disease (see Section 15-23). Two types of established biologic are used to antagonize TNF-α in clinical practice. The first type comprises the anti-TNF-α antibodies, such as infliximab and adalimumab, which bind to TNF-α and block its activity. The second type is a recombinant human TNF receptor (TNFR) subunit p75–Fc fusion protein called etanercept, which binds TNF-α, thereby neutralizing its activity. These are extremely potent anti-inflammatory agents, and the number of diseases in which they have been shown to be effective is growing as further clinical trials are performed. In addition to rheumatoid arthritis, the rheumatic diseases ankylosing spondylitis, psoriatic arthropathy, and juvenile idiopathic arthritis (other than the systemic-onset subset) respond well to blockade of TNF-α, and this treatment is now routine in many cases.

Rheumatoid Arthritis
Multiple Sclerosis
Crohn's Disease

An illustration of the importance of TNF-α in defense against infection is the observation that TNF-α blockade carries a small but increased risk to patients

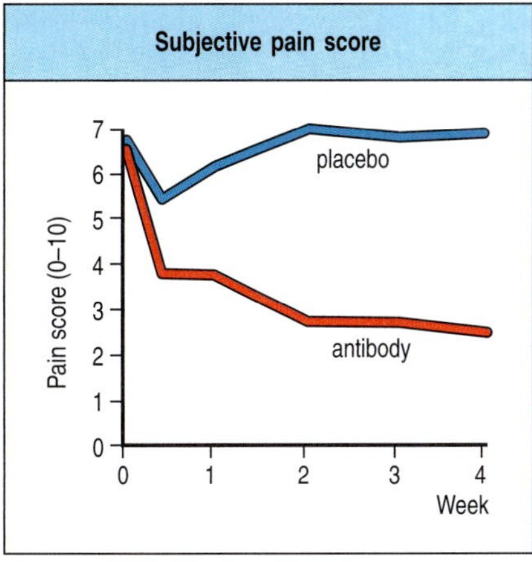

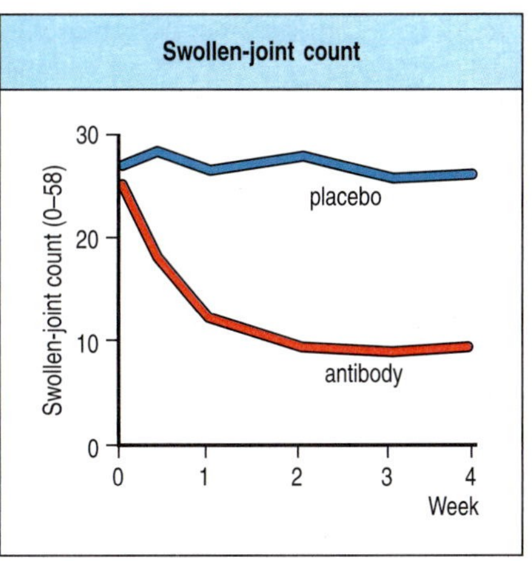

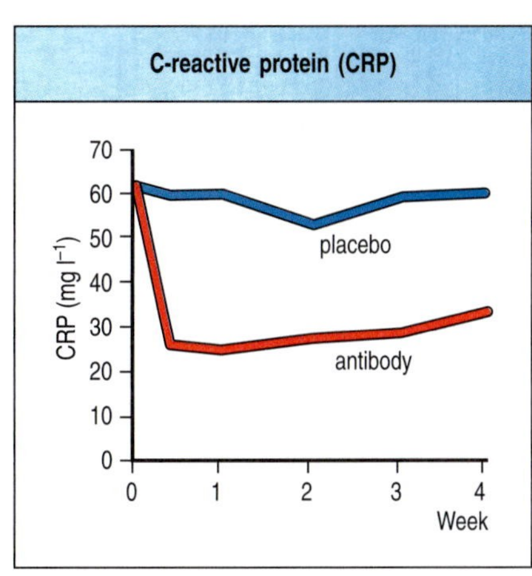

Fig. 16.8 Anti-inflammatory effects of anti-TNF-α therapy in rheumatoid arthritis. The clinical course of 24 patients was followed for 4 weeks after treatment with either a placebo or a monoclonal antibody against TNF-α at a dose of 10 mg kg^{-1}. The antibody therapy was associated with a reduction in both subjective and objective parameters of disease activity (as measured by pain score and swollen-joint count, respectively) and in the systemic inflammatory acute-phase response, measured as a fall in the concentration of the acute-phase C-reactive protein. Data courtesy of R.N. Maini.

of developing serious infections, including tuberculosis (see Section 3-17). Anti-TNF-α therapy has not been successful in all diseases. TNF-α blockade in experimental autoimmune encephalomyelitis (EAE, a mouse model of multiple sclerosis) led to amelioration of the disease, but in patients with multiple sclerosis treated with anti-TNF-α, relapses became more frequent, possibly because of an increase in T-cell activation. This illustrates the potential pitfalls of using animal models in the design of therapies for human disease.

Antibodies and recombinant proteins against the pro-inflammatory cytokine IL-1 and its receptor have not proved as effective as TNF-α blockade for treating rheumatoid arthritis in humans, despite being equally powerful in animal models of arthritis. An antibody against the cytokine IL-1 has been licensed for clinical use in the hereditary autoinflammatory disease Muckle–Wells syndrome (see Section 13-18), and blockade of the IL-1β receptor by the recombinant protein anakinra (Kineret) has also proved useful in controlling symptoms in autoinflammatory syndromes.

Another cytokine antagonist in clinical use is the humanized antibody tocilizumab, directed against the IL-6 receptor, which blocks the effects of the pro-inflammatory cytokine IL-6. This seems to be as effective as anti-TNF-α in patients with rheumatoid arthritis and also shows promise in treating systemic onset juvenile idiopathic arthritis, an autoinflammatory condition.

Interferon (IFN)-β (Avonex) is used to treat diseases of viral origin based on its ability to enhance immunity, but is also effective in treating multiple sclerosis, attenuating its course and severity and reducing the occurrence of relapses. Until recently, it was unclear how IFN-β could reduce rather than enhance immunity. In Section 3-8, we described the inflammasome, in which innate sensors of the NLR family activate caspase 1 to cleave the IL-1 pro-protein into the active form of the cytokine (see Fig. 3.16). We now know that IFN-β acts at two levels to reduce IL-1 production. It inhibits the activity of the NALP3 (NLRP3) and NLRP1 inflammasomes. Further, it reduces expression of the IL-1 pro-protein, reducing the substrate available to activated caspase 1. Thus, IFN-β limits the production of a powerful pro-inflammatory cytokine, which may explain its observed effects on the symptoms of multiple sclerosis.

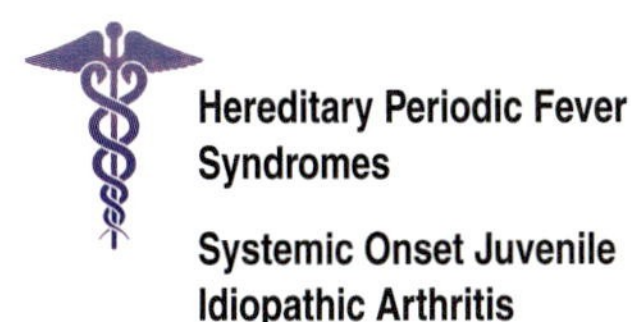

Hereditary Periodic Fever Syndromes
Systemic Onset Juvenile Idiopathic Arthritis

16-9 Biologic agents can block cell migration to sites of inflammation and reduce immune responses.

Effector lymphocytes expressing the integrin $\alpha_4{:}\beta_1$ (VLA-4) bind to VCAM-1 on endothelium in the central nervous system, while those expressing $\alpha_4{:}\beta_7$ (lamina propria-associated molecule 1) bind to MAdCAM-1 on endothelium in the gut. The humanized monoclonal antibody natalizumab is specific

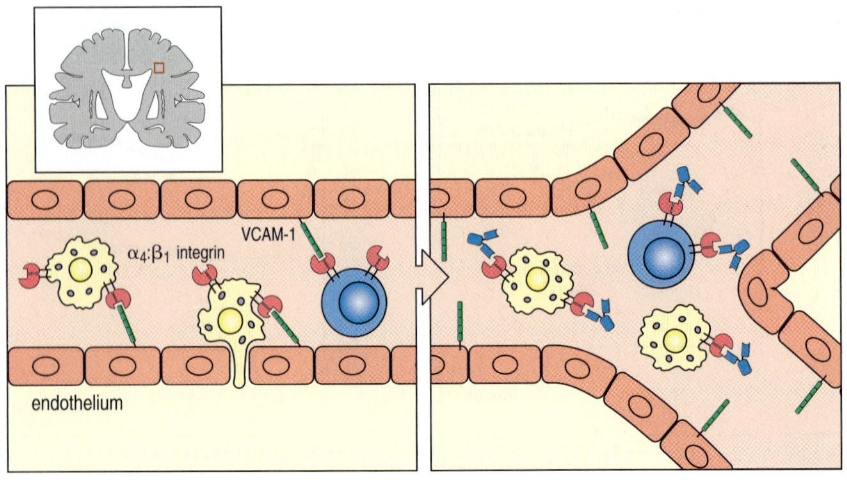

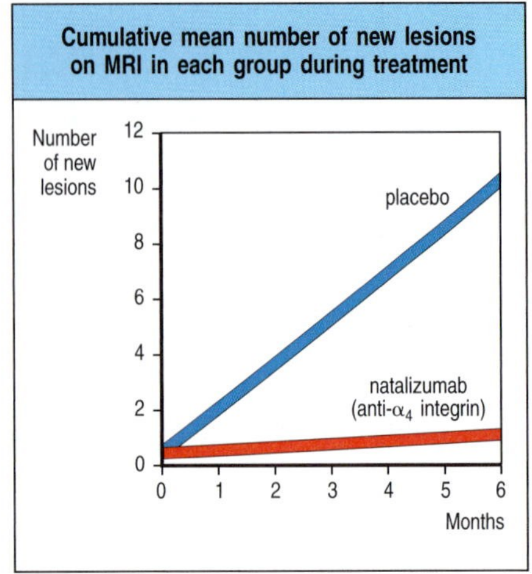

Fig. 16.9 Treatment with an anti-α_4 integrin humanized monoclonal antibody reduces relapses in multiple sclerosis. Left panel: interaction between α_4:β_1 integrin (VLA-4) on lymphocytes and macrophages and VCAM-1 expressed on endothelial cells permits the adhesion of these cells to brain endothelium. This facilitates the migration of these cells into the plaques of inflammation in multiple sclerosis. Center panel: the monoclonal antibody natalizumab (blue) binds to the α_4 chain of the integrin and blocks adhesive interactions between lymphocytes and monocytes and VCAM-1 on endothelial cells, thus preventing the cells from entering the tissue and exacerbating the inflammation. The future of this treatment is unclear because of the development of a rare infection as a side effect (see the text). Right panel: the number of new lesions detected on magnetic resonance imaging (MRI) of the brain is greatly reduced in patients treated with natalizumab compared with a placebo. Data from Miller, D.H. *et al.*: *N. Engl. J. Med.* 2003, 348:15–23.

for the α_4 integrin subunit and binds both VLA-4 and α_4:β_7, preventing their interaction with their ligands (Fig. 16.9). This antibody has shown therapeutic benefit in placebo-controlled trials in patients with Crohn's disease or with multiple sclerosis. The early signs that this treatment could be successful illustrate the fact that disease depends on the continuing emigration of lymphocytes, monocytes, and macrophages from the circulation into the tissues of the brain in multiple sclerosis, and into the gut wall in Crohn's disease. However, blockade of α_4:β_1 integrin is not specific and, like anti-TNF therapy, could lead to reduced defense against infection. Rare patients treated with natalizumab have developed progressive multifocal leukoencephalopathy, an opportunistic infection caused by the JC virus. This led to the temporary withdrawal of natalizumab from the market in 2005, but in June 2006 it was again allowed to be prescribed for multiple sclerosis and for Crohn's disease.

A similar problem with multifocal leukoencephalopathy led to the withdrawal from the market in the United States and Europe in 2009 of another anti-integrin antibody, efalizumab (which targets the α_L subunit CD11a), which had shown promise in treating psoriasis.

16-10 Blockade of co-stimulatory pathways that activate lymphocytes can be used to treat autoimmune disease.

The blocking of co-stimulatory pathways, noted above in connection with the prevention of transplantation rejection (see Section 16-6), has also been applied to autoimmune diseases. For example, CTLA-4–Ig (abatacept) blocks the interaction of B7, expressed by antigen-presenting cells, with CD28 expressed by T cells. This drug is approved for the treatment of rheumatoid arthritis, and also seems to be beneficial in treating psoriasis. Psoriasis is an inflammatory skin disease driven primarily by T cells, leading to the production of pro-inflammatory cytokines. When CTLA-4–Ig was given to patients with psoriasis, there was an improvement in the psoriatic rash and histological evidence of loss of activation of keratinocytes, T cells, and dendritic cells within the damaged skin. Another co-stimulatory pathway that has been targeted in psoriasis is the interaction between the adhesion molecules CD2 on T cells and CD58 (LFA-3) on antigen-presenting cells. A recombinant CD58–IgG1 fusion protein, called alefacept (Amevive), inhibits the interaction between CD2 and CD58, and is now a routine and effective treatment for psoriasis. Although memory T cells are targeted by this therapy, responses to vaccination such as antitetanus remain intact.

16-11 Some commonly used drugs have immunomodulatory properties.

Certain existing medications, such as the statins and angiotensin blockers widely used in the prevention and treatment of cardiovascular disease, can also modulate the immune response in experimental animals. Statins are very widely prescribed drugs that block the enzyme 3-hydroxy-3-methylglutaryl-coenzyme A (HMG-CoA) reductase, thereby reducing cholesterol levels. They also reduce the increased level of expression of MHC class II molecules in some autoimmune diseases. These effects may be due to an alteration in the cholesterol content of membranes, thereby influencing lymphocyte signaling. In animal models, these drugs also seem to cause T cells to switch from a more pathogenic T_H1 response to a more protective T_H2 response, although whether this occurs in human patients is not clear.

Vitamin D_3, an essential hormone for bone and mineral homeostasis, also exerts immunomodulatory effects. It decreases IL-12 production by dendritic cells and leads to a decrease in IL-2 and IFN-γ by CD4 T cells, and protective effects have been demonstrated in a variety of animal models of autoimmunity, such as EAE and diabetes, and in transplantation. The major drawback of vitamin D_3 is that its immunomodulatory effects are seen only at dosages that would lead to hypercalcemia and bone resorption in humans. There is a major search under way for structural analogs of vitamin D_3 that retain the immunomodulatory effects but do not cause hypercalcemia.

16-12 Controlled administration of antigen can be used to manipulate the nature of an antigen-specific response.

In some diseases, the target antigen of an unwanted immune response can be identified. It can then be possible to use the antigen itself, rather than drugs or antibodies, to treat the disease, because the manner of antigen presentation can alter the immune response and reduce or eliminate its pathogenic features. As discussed in Section 14-14, this principle has been applied with some success to the treatment of allergies caused by an IgE response to very low doses of antigen. Repeated treatment of allergic individuals with increasing doses of allergen seems to divert the allergic response to one dominated by T cells that favor the production of IgG and IgA antibodies. These antibodies are thought to desensitize the patient by binding the small amounts of allergen normally encountered and preventing it from binding to IgE.

There has been considerable interest in using peptide antigens to suppress pathogenic responses in T cell-mediated autoimmune disease. The type of CD4 T-cell response induced by a peptide depends on the way in which it is presented to the immune system (see Section 9-18). For instance, peptides given orally tend to prime regulatory T cells that make predominantly transforming growth factor (TGF)-β, without activating T_H1 cells or inducing a great deal of systemic antibody (see Section 12-14). Indeed, experiments in animals indicate that oral antigens can protect against induced autoimmune disease. Diseases resembling multiple sclerosis or rheumatoid arthritis can be induced in mice, by the injection of myelin basic protein (MBP) or collagen type II, respectively, in complete Freund's adjuvant. Oral administration of MBP or type II collagen inhibits the development of these diseases in animals, and has some beneficial effects in reducing the activity of already established disease. However, in people with multiple sclerosis or rheumatoid arthritis, the oral administration of the whole protein has shown only marginal therapeutic effects. Similarly, no protective effect was found in a large study to examine whether giving low-dose parenteral insulin could delay the onset of the disease in people at high risk of developing diabetes.

Other approaches using antigen to shift the autoimmune T-cell response to a less damaging T_H2 response have been more effective in humans. The peptide drug glatiramer acetate (Copaxone) is an approved drug for multiple sclerosis, reducing relapse rates by up to 30%. It mimics the amino-acid composition of MBP and induces a T_H2-type protective response. Another strategy uses **altered peptide ligands** (**APLs**), in which amino acid substitutions have been made in T-cell receptor contact positions in the antigenic peptide. APLs can be designed to act as partial agonists or antagonists, or even to induce the differentiation of regulatory T cells. Despite their success in ameliorating EAE in mice, the trial of these peptides for multiple sclerosis in some patients led to exacerbated disease or to allergic reaction associated with a vigorous T_H2 response. Whether such approaches can be effective in manipulating the established immune responses that drive human autoimmune diseases remains to be seen.

Summary.

Treatments for unwanted immune responses, such as graft rejection, autoimmunity, or allergic reactions, include conventional drugs—anti-inflammatory, cytotoxic, and immunosuppressive—as well as biologic agents such as monoclonal antibodies and immunomodulatory proteins. Anti-inflammatory drugs, of which the most potent are the corticosteroids, have a broad spectrum of actions and a wide range of toxic side-effects; their dose must be controlled carefully. They are therefore normally used in combination with either cytotoxic or immunosuppressive drugs. The cytotoxic drugs kill all dividing cells and thereby prevent lymphocyte proliferation, but they suppress all immune responses indiscriminately and also kill other types of dividing cell. The immunosuppressant drugs, such as cyclosporin A, act by intervening in the intracellular signaling pathways of T cells. They are generally less toxic and more expensive than the cytotoxic drugs, but still suppress the immune response indiscriminately.

Immunosuppression is used to suppress the immune response to the graft before it has become established. In contrast, autoimmune responses are already well established at the time of diagnosis and are consequently much more difficult to suppress. They are therefore less responsive to the immunosuppressant drugs, and for that reason autoimmune responses are usually controlled using a combination of corticosteroids and cytotoxic drugs.

Several types of biologic agent are now established in the clinic for treating transplant rejection and autoimmune diseases (Fig. 16.10). Many monoclonal antibodies have been approved for human use that deplete lymphocytes either generally or selectively, or inhibit lymphocyte activation through receptor blockade, or prevent lymphocyte migration into tissues. Immunomodulatory agents also include monoclonal antibodies or fusion proteins that inhibit the inflammatory actions of TNF-α, a triumph of immunotherapy.

Using the immune response to attack tumors.

Cancer is one of the three leading causes of death in industrialized nations, the others being infectious disease and cardiovascular disease. As treatments for infectious diseases and the prevention of cardiovascular disease continue to improve, and the average life expectancy increases, cancer is likely to become the most common fatal disease in these countries. Cancers are caused by the progressive growth of the progeny of a single transformed cell. Curing cancer

Therapeutic agents used to treat human autoimmune diseases				
Target	**Therapeutic agent**	**Disease**	**Disease outcome**	**Disadvantages**
Integrins	α_4:β_1 integrin-specific monoclonal antibody (mAb)	Relapsing/remitting multiple sclerosis (MS) Rheumatoid arthritis (RA) Inflammatory bowel disease	Reduction in relapse rate; delay in disease progression	Increased risk of infection; progressive multifocal encephalopathy
B cells	CD20-specific mAb	RA Systemic lupus erythematosus (SLE) MS	Improvement in arthritis, possibly in SLE	Increased risk of infection
HMG-coenzyme A reductase	Statins	MS	Reduction in disease activity	Hepatotoxicity; rhabdomyolysis
T cells	CD3-specific mAb	Type 1 diabetes mellitus	Reduced insulin use	Increased risk of infection
	CTLA4-immuno-globulin fusion protein	RA Psoriasis MS	Improvement in arthritis	
Cytokines	TNF-specific mAb and soluble TNFR fusion protein	RA Crohn's disease Psoriatic arthritis Ankylosing spondylitis	Improvement in disability; joint repair in arthritis	Increased risk of tuberculosis and other infections; slight increase in risk of lymphoma
	IL-1 receptor antagonist	RA	Improves disability	Low efficacy
	IL-15-specific mAb	RA	May improve disability	Increased risk of opportunistic infection
	IL-6 receptor-specific mAb	RA	Decreased disease activity	Increased risk of opportunistic infection
	Type I interferons	Relapsing/remitting MS	Reduction in relapse rate	Liver toxicity; influenza-like syndrome is common

Fig. 16.10 New therapeutic agents for human autoimmunity. The immunosuppressive agents listed in Figs 16.1 and 16.7 can act in one of three general ways. First (red), they can act by depleting cells from inflammatory sites, or cause global cell-specific depletion, or block integrin interactions, thereby inhibiting lymphocyte trafficking. Second (blue), agents may block specific cellular interactions or inhibit various co-stimulatory pathways. Third (green), agents may target the terminal effector mechanisms, such as the neutralization of various pro-inflammatory cytokines.

therefore requires that all the malignant cells be removed or destroyed without killing the patient. An attractive way of achieving this would be to induce an immune response against the tumor that would discriminate between the cells of the tumor and their normal cell counterparts, in the same way that vaccination against a viral or bacterial pathogen induces a specific immune response that provides protection only against that pathogen. Immunological approaches to the treatment of cancer have been attempted for more than a century, but it is only in the past decade that immunotherapy of cancer has shown real promise. An important conceptual advance has been the integration of conventional approaches such as surgery or chemotherapy, which substantially reduce tumor load, with immunotherapy.

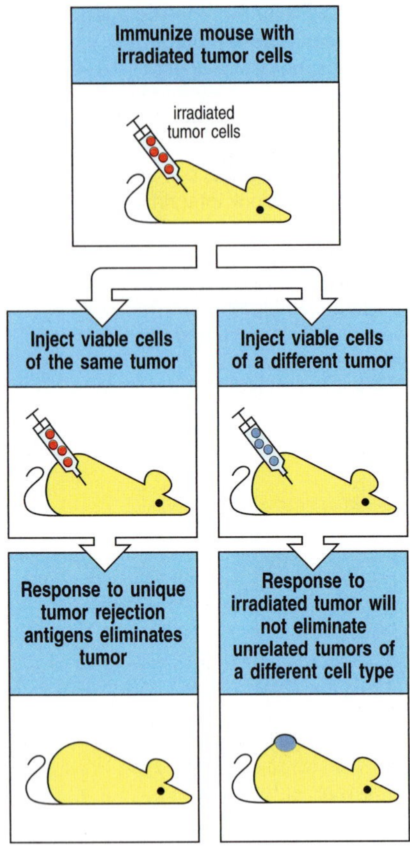

Fig. 16.11 Tumor rejection antigens are specific to individual tumors. Mice immunized with irradiated tumor cells and challenged with viable cells of the same tumor can, in some cases, reject a lethal dose of that tumor (left panels). This is the result of an immune response to tumor rejection antigens. If the immunized mice are challenged with viable cells of a different tumor, there is no protection and the mice die (right panels).

16-13 The development of transplantable tumors in mice led to the discovery of protective immune responses to tumors.

The finding that tumors could be induced in mice after treatment with chemical carcinogens or irradiation, coupled with the development of inbred strains of mice, made it possible to undertake the key experiments that led to the discovery of immune responses to tumors. These tumors could be transplanted between mice, and the experimental study of tumor rejection has generally been based on the use of such tumors. If they bear MHC molecules foreign to the mice into which they are transplanted, the tumor cells are readily recognized and destroyed by the immune system, a fact that was exploited to develop the first MHC-congenic strains of mice. Specific immunity to tumors must therefore be studied within inbred strains, so that host and tumor can be matched for their MHC type.

Transplantable tumors in mice exhibit a variable pattern of growth when injected into syngeneic recipients. Most tumors grow progressively and eventually kill the host. However, if mice are injected with irradiated tumor cells that cannot grow, they are frequently protected against subsequent injection with a normally lethal dose of viable cells of the same tumor (Fig. 16.11). There seems to be a spectrum of immunogenicity among transplantable tumors: injections of irradiated tumor cells seem to induce varying degrees of protective immunity against a challenge injection of viable tumor cells at a distant site. These protective effects are not seen in T-cell-deficient mice but can be conferred by adoptive transfer of T cells from immune mice, showing the need for T cells to mediate all these effects.

These observations indicate that the tumors express antigenic peptides that can become targets of a tumor-specific T-cell response that rejects the tumor. These **tumor rejection antigens** are expressed by experimentally induced murine tumors (in which they are often termed tumor-specific transplantation antigens), and are usually specific for an individual tumor. Thus, immunization with irradiated tumor cells from one tumor protects a syngeneic mouse from challenge with live cells from that same tumor, but not from challenge with a different syngeneic tumor (see Fig. 16.11).

16-14 Tumors are 'edited' by the immune system as they evolve and can escape rejection in many ways.

In the 1950s, Frank MacFarlane Burnet and Lewis Thomas formulated the '**immune surveillance**' hypothesis, in which cells of the immune system would detect and destroy tumor cells. Since then, it has become clear that the relationship between the immune system and cancer is considerably more complex, and this hypothesis has been modified to consider three phases of tumor growth. The first is the 'elimination phase,' in which the immune system recognizes and destroys potential tumor cells—the phenomenon previously called immune surveillance (Fig. 16.12). If elimination is not completely successful, what follows is an 'equilibrium phase,' in which tumor cells undergo changes or mutations that aid their survival as a result of the selection pressure imposed by the immune system. During the equilibrium phase, a process known as **cancer immunoediting** continuously shapes the properties of the tumor cells that survive. In the final 'escape phase,' tumor cells that have accumulated sufficient mutations elude the attentions of the immune system and grow unimpeded to become clinically detectable.

Mice with targeted gene deletions or treated with antibodies to remove specific components of innate and adaptive immunity have provided the best evidence that immune surveillance influences the development of certain types of tumor. For example, mice lacking perforin, part of the killing mechanism of NK cells and CD8 cytotoxic T cells (see Section 9-26), show an

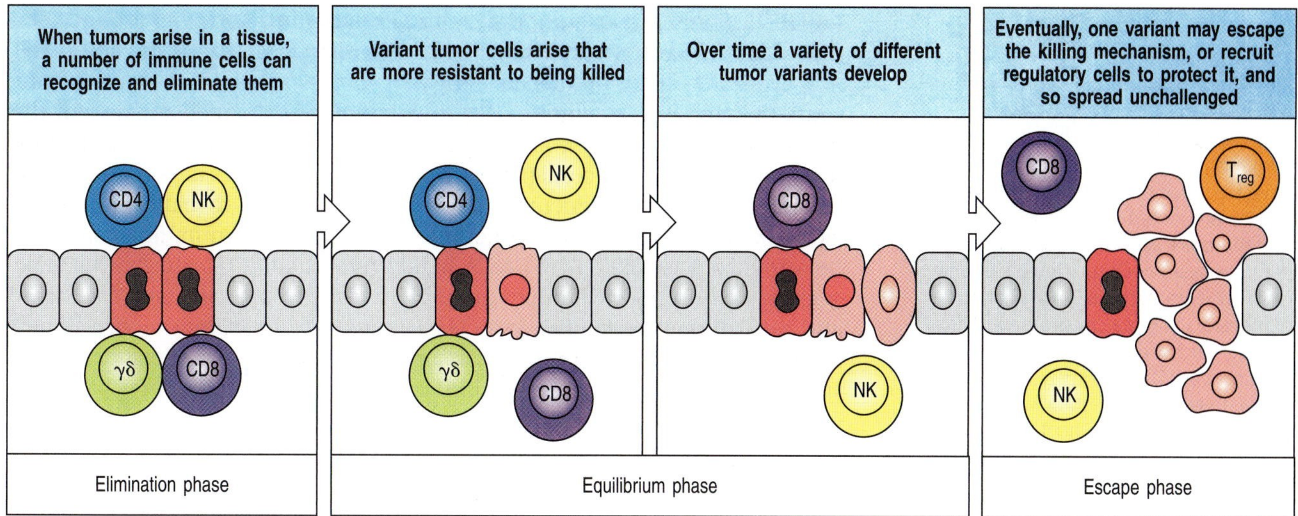

When tumors arise in a tissue, a number of immune cells can recognize and eliminate them
Variant tumor cells arise that are more resistant to being killed
Over time a variety of different tumor variants develop
Eventually, one variant may escape the killing mechanism, or recruit regulatory cells to protect it, and so spread unchallenged
CD4
NK
γδ
CD8
T_{reg}
Elimination phase
Equilibrium phase
Escape phase

Fig. 16.12 Malignant cells can be controlled by immune surveillance. Some types of tumor cell are recognized by a variety of immune-system cells, which can eliminate them (left panel). If the tumor cells are not completely eliminated, variants occur that eventually escape the immune system and proliferate to form a tumor.

increased frequency of lymphomas—tumors of the lymphoid system. Strains of mice lacking the RAG and STAT1 proteins, thus being deficient in both adaptive and certain innate immune mechanisms, develop gut epithelial and breast tumors. Mice lacking T lymphocytes expressing γ:δ receptors show markedly increased susceptibility to skin tumors induced by the topical application of carcinogens, illustrating a role for intraepithelial γ:δ T cells (see Section 12-13) in surveying and killing abnormal epithelial cells. Both IFN-γ and IFN-α are important in the elimination of tumor cells, either directly, or indirectly through their actions on other cells. Studies of the various effector cells of the immune system show that γ:δ T cells are a major source of IFN-γ, which may explain their importance in the removal of cancer cells.

According to the immunoediting hypothesis, those tumor cells that survive the equilibrium phase have acquired additional mutations that prevent their elimination by the immune system. In an immunocompetent individual, the equilibrium immune response continually removes tumor cells, delaying tumor growth; if the immune system is compromised, the equilibrium phase quickly turns into escape, as no tumor cells at all are removed. An excellent clinical example to support the presence of the equilibrium phase is the occurrence of cancer in recipients of organ transplants. One study reported the development of melanoma between 1 and 2 years after transplant in two patients who had received kidneys from the same donor, a patient who had had malignant melanoma, successfully treated, 16 years before her death. Presumably, melanoma cells, which are known to spread easily to other organs, were present in the donor kidneys at the time of transplantation but were in equilibrium phase with the immune system. If so, this would indicate that the melanoma cells are not killed off completely by the immune system but are held in check by an immunocompetent immune system. Because the recipients' immune systems were immunosuppressed, the melanoma cells were released from equilibrium and began to divide rapidly and spread to other parts of the body.

Another situation in which the breakdown of immune surveillance can lead to tumor development is in **post-transplant lymphoproliferative disorder**, which can occur when patients are immunosuppressed after, for example, solid organ transplantation. It usually takes the form of a B-cell expansion driven by Epstein–Barr virus (EBV) in which the B cells can undergo mutations and become malignant. Immune surveillance therefore seems to be critical for the control of virus-associated tumors.

Tumors can avoid stimulating an immune response, or can evade it when it occurs, by means of numerous mechanisms, which are summarized in

Fig. 16.13. Spontaneous tumors may initially lack mutations that produce new tumor-specific antigens that elicit T-cell responses (see Fig. 16.13, first panel). And even when a tumor-specific antigen is expressed and is taken up and presented by antigen-presenting cells, if co-stimulatory signals are absent this will tend to tolerize any antigen-specific naive T cells rather than activating them (see Fig. 16.13, second panel). How long such tumors are treated as 'self' is unclear. Recent sequencing of entire tumor genomes reveals that as many as 10–15 unique antigenic peptides may be generated by mutations that could be recognized as 'foreign' by T cells. In addition, cellular transformation is frequently associated with induction of MHC class 1b proteins (such as MIC-A and MIC-B) that are ligands for NKG2D, thus allowing tumor recognition by NK cells (see Section 6-18). But cancer cells tend to be genetically unstable, so that clones that are not recognized by an immune response may be able to escape elimination.

Some tumors, such as colon and cervical cancers, lose the expression of a particular MHC class I molecule, perhaps through immune selection by T cells specific for a peptide presented by that MHC class I molecule (see Fig. 16.13, third panel). In experimental studies, when a tumor loses expression of all MHC class I molecules (Fig. 16.14), it can no longer be recognized by cytotoxic T cells, although it might become susceptible to NK cells (Fig. 16.15). Tumors that lose only one MHC class I molecule might be able to avoid recognition by specific CD8 cytotoxic T cells while still remaining resistant to NK cells, conferring a selective advantage *in vivo*.

Tumors also seem to be able to evade immune attack by creating an environment that is generally immunosuppressive (see Fig. 16.13, fourth panel). Many tumors make immunosuppressive cytokines. Transforming growth factor-β (TGF-β) was first identified in the culture supernatant of a tumor (hence its name) and, as we have seen, it tends to suppress inflammatory T-cell responses and cell-mediated immunity, which are needed to control tumor growth. Recall that TGF-β induces the development of inducible regulatory T cells (T_{reg} cells) (see Section 9-18), which have been found in a variety of cancers and might expand specifically in response to tumor antigens. In mouse models, removal of T_{reg} cells increases resistance to cancer, whereas their transfer into a T_{reg}-negative recipient allows cancers to develop. The IL-2-induced expansion of T_{reg} cells may also explain the relatively low effectiveness

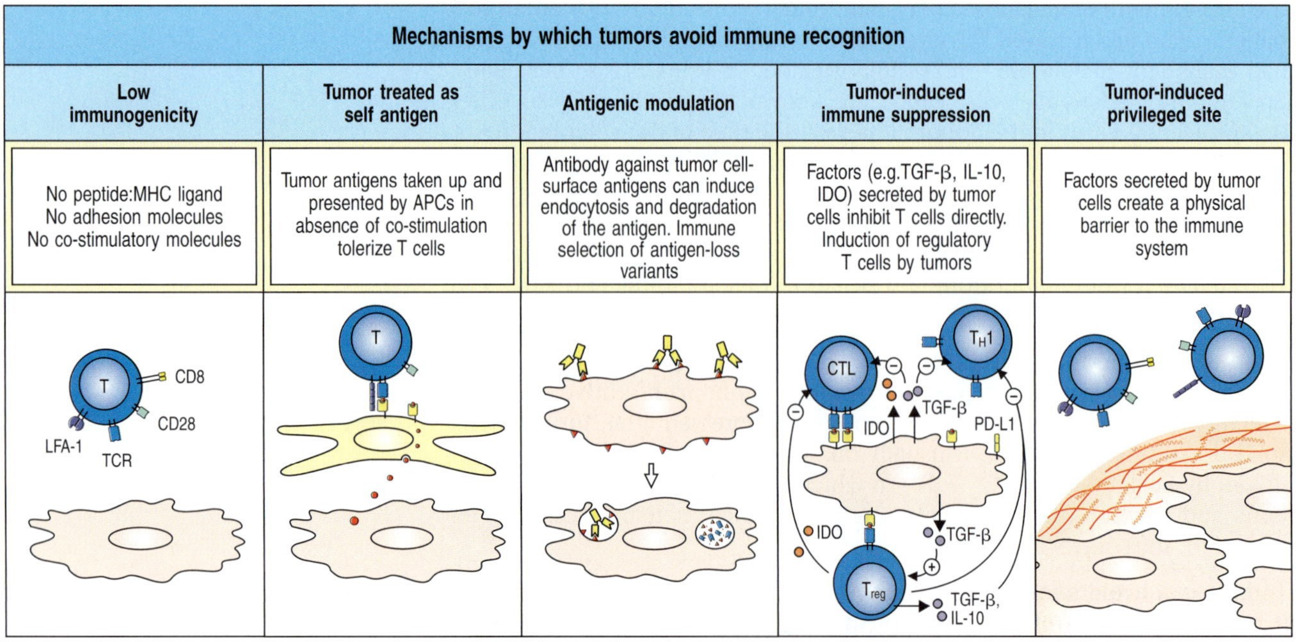

Mechanisms by which tumors avoid immune recognition

Low immunogenicity	Tumor treated as self antigen	Antigenic modulation	Tumor-induced immune suppression	Tumor-induced privileged site
No peptide:MHC ligand No adhesion molecules No co-stimulatory molecules	Tumor antigens taken up and presented by APCs in absence of co-stimulation tolerize T cells	Antibody against tumor cell-surface antigens can induce endocytosis and degradation of the antigen. Immune selection of antigen-loss variants	Factors (e.g.TGF-β, IL-10, IDO) secreted by tumor cells inhibit T cells directly. Induction of regulatory T cells by tumors	Factors secreted by tumor cells create a physical barrier to the immune system

Fig. 16.13 Tumors can avoid immune recognition in a variety of ways. First panel: tumors can have low immunogenicity. Some tumors do not have peptides of novel proteins that can be presented by MHC molecules, and therefore appear normal to the immune system. Others have lost one or more MHC molecules, and most do not express co-stimulatory proteins, which are required to activate naive T cells. Second panel: tumor antigens presented in the absence of co-stimulatory signals will make the responding T cells tolerant to that antigen. Third panel: tumors can initially express antigens to which the immune system responds but lose them by antibody-induced internalization or antigenic variation. The process of genetic instability leading to antigenic change is now considered to be part of an equilibrium phase, which can lead to outgrowth of the tumor when the immune system loses the race and is no longer able to adapt. When a tumor is attacked by cells responding to a particular antigen, any tumor cell that does not express that antigen will have a selective advantage. Fourth panel: tumors often produce molecules, such as TGF-β, IL-10, IDO, or PD-L1, that suppress immune responses directly or can recruit regulatory T cells that can themselves secrete immunosuppressive cytokines. Fifth panel: tumor cells can secrete molecules such as collagen that form a physical barrier around the tumor, preventing lymphocyte access. APC, antigen-presenting cell.

of IL-2 in increasing the immune response in melanoma. Although approved for clinical use, IL-2 leads to a long-term beneficial response in relatively few patients. Therefore, a possible additional therapy would be to deplete or inactivate T_{reg} cells together with the IL-2 administration. Many tumors seem to contain **myeloid-derived suppressor cells**, a heterogeneous population composed of both monocytic and polymorphonuclear cells that can inhibit T-cell activation within the tumor but are incompletely characterized at present. Several tumors of different tissue origins, such as melanoma, ovarian carcinoma, and B-cell lymphoma, have also been shown to produce the immunosuppressive cytokine IL-10, which can reduce dendritic cell activity and inhibit T-cell activation.

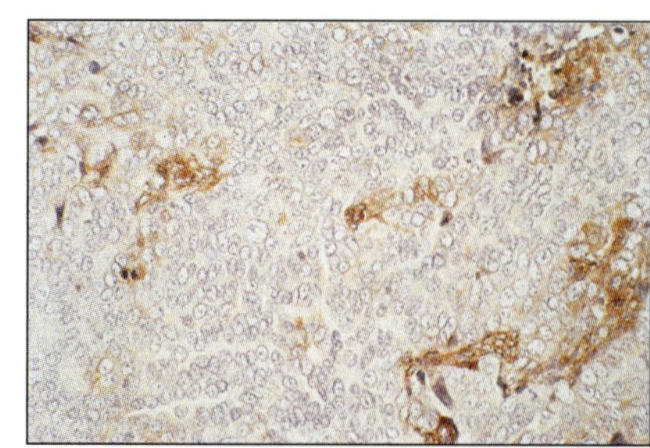

Fig. 16.14 Loss of MHC class I expression in a prostatic carcinoma. Some tumors can evade immune surveillance by a loss of expression of MHC class I molecules, preventing their recognition by CD8 T cells. A section of a human prostate cancer that has been stained with a peroxidase-conjugated antibody against HLA class I molecules is shown. The brown stain that represents HLA class I expression is restricted to infiltrating lymphocytes and tissue stromal cells. The tumor cells that occupy most of the section show no staining. Photograph courtesy of G. Stamp.

Some tumors express cell-surface proteins that directly inhibit immune responses (see Fig. 16.13, fourth panel). For example, many types of cancer express programmed death ligand-1 (PD-L1), a member of the B7 family and a ligand for the inhibitory receptor PD-1 expressed by activated T cells (see Section 7-18). Furthermore, tumors can produce enzymes that act to suppress local immune responses. The enzyme **indoleamine 2,3-dioxygenase** (IDO) catabolizes tryptophan, an essential amino acid, producing the immunosuppressive metabolite kynurenine. IDO seems normally to function in maintaining a balance between immune responses and tolerance during infections, but can be induced during the equilibrium phase of tumor development. Finally, tumor cells can produce materials such as collagen that create a physical barrier to interaction with cells of the immune system (see Fig. 16.13, last panel).

16-15 Tumor-specific antigens can be recognized by T cells and form the basis of immunotherapies.

The tumor rejection antigens recognized by the immune system are peptides of tumor-cell proteins that are presented to T cells by MHC molecules. These peptides become the targets of a tumor-specific T-cell response even though they can also be present on normal tissues. For instance, strategies to induce immunity to the relevant antigens in melanoma patients can induce vitiligo, an autoimmune destruction of pigmented cells in healthy skin. Several categories of tumor rejection antigens can be distinguished (Fig. 16.16). One comprises strictly tumor-specific antigens that result from point mutations or gene rearrangements occurring during oncogenesis. Point mutations may evoke a T-cell response either by allowing *de novo* binding of the mutant peptide to MHC class I molecules or by creating a new epitope for T cells by modification of a peptide that already binds class I molecules (Fig. 16.17). In B- and T-cell tumors, which are derived from single clones of lymphocytes, a special class of tumor-specific antigen comprises the idiotypes (see Appendix I, Section A-10) unique to the antigen receptor expressed by the clone. However, not all mutated peptides may be properly processed or be able to associate with MHC molecules and thus ensure that they stimulate an effective response.

The second category of tumor rejection antigens comprises the **cancer-testis antigens**. These are proteins encoded by genes that are normally expressed only in male germ cells in the testis. Male germ cells do not express MHC molecules, and therefore peptides from these molecules are not normally presented to T lymphocytes. Tumor cells show widespread abnormalities of gene expression, including the activation of genes encoding cancer-testis antigens, such as the MAGE antigens on melanomas (see Fig. 16.16). When expressed by tumor cells, peptides derived from these 'germ-cell' proteins can now be presented to T cells by tumor-cell MHC class I molecules; these proteins are therefore effectively tumor specific in their expression as antigens. Perhaps the cancer-testis antigen best characterized immunologically is NY-ESO-1 (New York, Esophageal Squamous Cell Carcinoma-1), which is

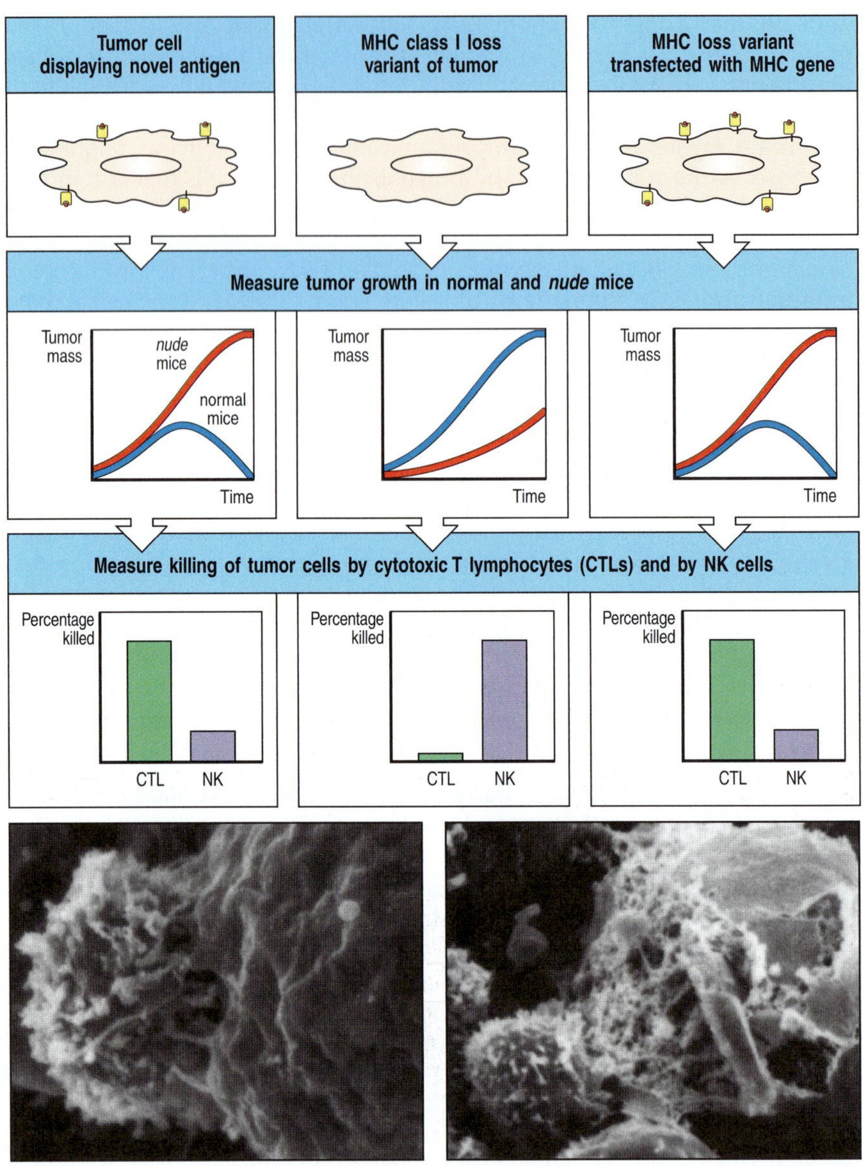

Fig. 16.15 Tumors that lose expression of all MHC class I molecules as a mechanism of escape from immune surveillance are more susceptible to being killed by NK cells. Regression of transplanted tumors is largely due to the actions of cytotoxic T lymphocytes (CTLs), which recognize novel peptides bound to MHC class I antigens on the surface of the cell (left panels). NK cells have inhibitory receptors that bind MHC class I molecules, so variants of the tumor that have low levels of MHC class I, although less sensitive to CD8 cytotoxic T cells, become susceptible to NK cells (center panels). *nude* mice lack T cells but have higher than normal levels of NK cells, and so tumors that are sensitive to NK cells grow less well in *nude* mice than in normal mice. Transfection with MHC class I genes can restore both resistance to NK cells and susceptibility to CD8 cytotoxic T cells (right panels). The bottom panels show scanning electron micrographs of NK cells attacking leukemia cells. Left panel: shortly after binding to the target cell, the NK cell has put out numerous microvillous extensions and established a broad zone of contact with the leukemia cell. The NK cell is the smaller cell on the left in both photographs. Right panel: 60 minutes after mixing, long microvillous processes can be seen extending from the NK cell (bottom left) to the leukemia cell and there is extensive damage to the leukemia cell; the plasma membrane has rolled up and fragmented. Photographs reprinted from Herberman, R., and Callewaert, D: *Mechanisms of Cytotoxicity by Natural Killer Cells*, 1985, with permission from Elsevier.

highly immunogenic and is expressed by a variety of human tumors, including melanoma.

The third category comprises 'differentiation antigens' encoded by genes that are expressed only in particular types of tissue. The best examples of these are the differentiation antigens expressed in melanocytes and melanoma cells; several of these antigens are proteins in the pathways that produce the black pigment melanin. The fourth category consists of antigens that are strongly overexpressed in tumor cells compared with their normal counterparts. An example is HER-2/neu (also known as c-Erb-2), which is a receptor tyrosine kinase homologous to the epidermal growth factor receptor. HER-2/neu is overexpressed in many adenocarcinomas, including breast and ovarian cancers, where it is associated with a poor prognosis. MHC class I-restricted, CD8-positive cytotoxic T lymphocytes have been found infiltrating solid tumors overexpressing HER-2/neu but are not capable of destroying such tumors *in vivo*. The fifth category of tumor rejection antigens comprises molecules that display abnormal post-translational modifications. An example is underglycosylated mucin, MUC-1, which is expressed by several tumors, including breast and pancreatic cancers. The sixth category comprises novel proteins that are generated when one or more introns are retained in the mRNA, which occurs

Potential tumor rejection antigens have a variety of origins			
Class of antigen	**Antigen**	**Nature of antigen**	**Tumor type**
Tumor-specific mutated oncogene or tumor suppressor	Cyclin-dependent kinase 4	Cell-cycle regulator	Melanoma
	β-Catenin	Relay in signal transduction pathway	Melanoma
	Caspase 8	Regulator of apoptosis	Squamous cell carcinoma
	Surface Ig/ Idiotype	Specific antibody after gene rearrangements in B-cell clone	Lymphoma
Cancer-testis antigens	MAGE-1 MAGE-3 NY-ESO-1	Normal testicular proteins	Melanoma Breast Glioma
Differentiation	Tyrosinase	Enzyme in pathway of melanin synthesis	Melanoma
Abnormal gene expression	HER-2/neu	Receptor tyrosine kinase	Breast Ovary
	Wilms' tumor	Transcription factor	Leukemia
Abnormal post-translational modification	MUC-1	Underglycosylated mucin	Breast Pancreas
Abnormal post-transcriptional modification	GP100 TRP2	Retention of introns in the mRNA	Melanoma
Oncoviral protein	HPV type 16, E6 and E7 proteins	Viral transforming gene products	Cervical carcinoma

Fig. 16.16 Proteins selectively expressed in human tumors are candidate tumor rejection antigens. The molecules listed here have all been shown to be recognized by cytotoxic T lymphocytes raised from patients with the tumor type listed.

in melanoma. Proteins encoded by viral oncogenes comprise the seventh category of tumor rejection antigen. These oncoviral proteins can have a critical role in the oncogenic process and, because they are foreign, they can evoke a T-cell response. Examples are the human papilloma virus type 16 proteins, E6 and E7, which are expressed in cervical carcinoma (see Section 16-17).

Although all types of tumor rejection antigen can evoke an antitumor response *in vitro* and *in vivo*, it is exceptional for such a response on its own to eliminate an established tumor. It is the goal of tumor immunotherapy to harness and augment such responses to treat cancer more effectively. The spontaneous remission occasionally observed in malignant melanoma and renal-cell carcinoma, even when disease is quite advanced, offers hope that this goal is achievable.

In melanoma, tumor-specific antigens were discovered by culturing irradiated tumor cells with autologous lymphocytes, a reaction known as the mixed lymphocyte–tumor cell culture. From such cultures, cytotoxic T cells were identified that were reactive against melanoma peptides and would kill tumor cells bearing the relevant tumor-specific antigen. Such studies have revealed that melanomas carry at least five different antigens that can be recognized by cytotoxic T lymphocytes. However, cytotoxic T lymphocytes reactive against melanoma antigens are not expanded *in vivo*, suggesting that these antigens are not normally immunogenic. Nonetheless, tumors can be selected *in vitro*—and possibly *in vivo*—for loss of these antigens by the presence of

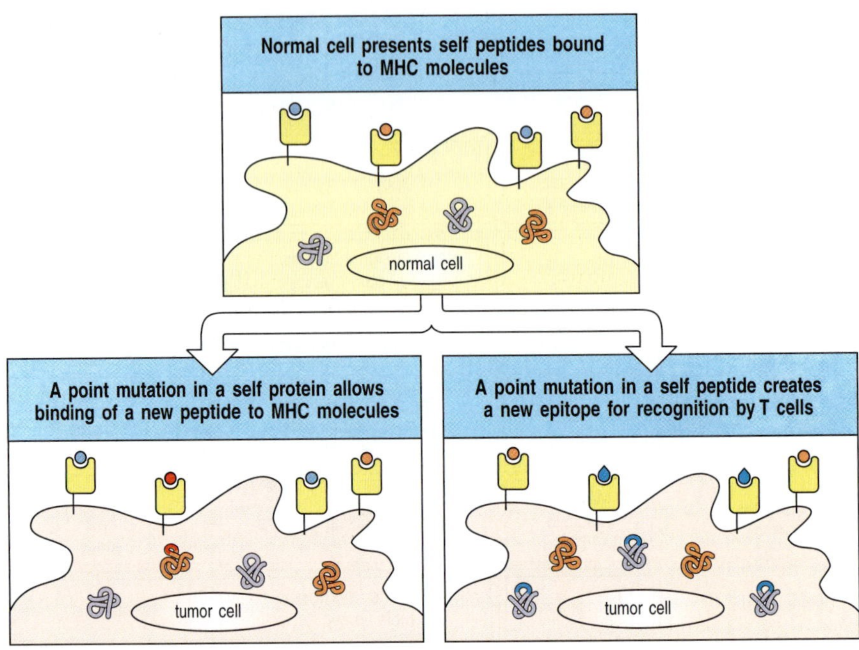

Fig. 16.17 Tumor rejection antigens may arise by point mutations in self proteins, which occur during the process of oncogenesis. In some cases a point mutation in a self protein may allow a new peptide to associate with MHC class I molecules (lower left panel). In other cases, a point mutation occurring within a self peptide that can bind self MHC proteins causes the expression of a new epitope for T-cell binding (lower right panel). In both cases, these mutated peptides will not have induced tolerance by the clonal deletion of developing T cells and can be recognized by mature T cells.

specific cytotoxic T cells, offering hope that they might be suitable targets for tumor immunotherapy.

Melanoma-specific T cells can be propagated from peripheral blood lymphocytes, from tumor-infiltrating lymphocytes, or by draining the lymph nodes of patients in whom the melanoma is growing. These T cells do not recognize proteins encoded by the mutant proto-oncogenes or tumor suppressor genes that caused the cancerous transformation of the cell. Instead, the T cells recognize antigens derived from other mutant genes or from normal proteins that are now displayed on tumor cells at levels detectable by T cells for the first time. Cancer-testis antigens such as the melanoma MAGE antigens discussed earlier probably represent early developmental antigens reexpressed in the process of tumorigenesis. Only a minority of melanoma patients have T cells reactive to the MAGE antigens, indicating that these antigens either are not expressed or are not immunogenic in most cases.

The most common melanoma antigens are peptides from the enzyme tyrosinase or from three other proteins—gp100, MART1, and gp75. These are differentiation antigens specific to the melanocyte lineage. It is likely that overexpression of these antigens in tumor cells leads to an abnormally high density of specific peptide:MHC complexes and it is this that makes them immunogenic. Although tumor rejection antigens are usually presented as peptides complexed with MHC class I molecules, the enzyme tyrosinase has been shown to stimulate CD4 T-cell responses in some melanoma patients by being ingested and presented by cells expressing MHC class II molecules. Both CD4 and CD8 T cells are likely to be important in achieving immunological control of tumors. CD8 cells can kill the tumor cells directly, while CD4 T cells have a role in the activation of CD8 cytotoxic T cells and the establishment of memory. CD4 T cells may also kill tumor cells by means of the cytokines, such as TNF-α, that they secrete.

Other potential tumor rejection antigens include the products of mutated cellular oncogenes or tumor suppressors, such as Ras and p53, and also fusion proteins, such as the Bcr–Abl tyrosine kinase that results from the chromosomal translocation (t9;22) found in chronic myeloid leukemia (CML). However, in each case, no specific cytotoxic T-cell response has been identified when the patient's lymphocytes are cultured with tumor cells bearing these mutated antigens.

When present on CML cells, the HLA class I molecule HLA-A*0301 can display a peptide derived from the fusion site between Bcr and Abl. This peptide was detected by a powerful technique known as 'reverse immunogenetics,' in which endogenous peptides are eluted from the MHC binding groove, and their sequence was determined by highly sensitive mass spectrometry. This technique has identified HLA-bound peptides from other tumor antigens, such as the MART1 and gp100 tumor antigens of melanomas, as well as candidate peptide sequences for vaccination against infectious diseases.

T cells specific for the Bcr–Abl fusion peptide can be identified in peripheral blood from patients with CML by using as specific ligands tetramers of HLA-A*0301 carrying the peptide (see Appendix I, Section A-28). Cytotoxic T lymphocytes specific for this and other tumor antigens can be selected *in vitro* by using peptides derived from the mutated or fused portions of these oncogenic proteins; these cytotoxic T cells are able to recognize and kill tumor cells.

After a bone marrow transplant to treat CML, mature lymphocytes from the bone marrow donor infused into the patient can help to eliminate any residual tumor. This technique is known as donor lymphocyte infusion (DLI). At present, it is not clear to what extent the clinical response is due to a graft-versus-host effect, in which the donor lymphocytes are responding to alloantigens expressed on the leukemia cells (see Section 15-36), or whether a specific antileukemic response is important. It is encouraging that it has been possible to separate T lymphocytes *in vitro* that mediate either a graft-versus-host effect or a graft-versus-leukemia effect. The ability to prime the donor cells against leukemia-specific peptides offers the prospect of enhancing the antileukemic effect while minimizing the risk of graft-versus-host disease.

There is now good reason to believe that T-cell immunotherapy against tumor antigens is a feasible clinical approach. Adoptive T-cell therapy involves the *ex vivo* expansion of tumor-specific T cells to large numbers and the infusion of those T cells into patients. Cells are expanded *in vitro* by culture with IL-2, anti-CD3 antibodies, and allogeneic antigen-presenting cells, which provide a co-stimulatory signal. Adoptive T-cell therapy is made more effective when patient is immunosuppressed before treatment and by the systemic administration of IL-2. T cells directed at malignancies expressing Epstein–Barr virus (EBV) antigens can also be expanded in an antigen-specific manner by using EBV-transformed B-lymphoblastoid cell lines from the patient. Another approach that has excited much interest is the use of retroviral vectors to transfer tumor-specific T-cell receptor genes into patients' T cells before reinfusion. This can have long-lasting effects as a result of the ability of T cells to become memory cells, and there is no requirement for histocompatibility because the transfused cells are derived from the patient.

16-16 Monoclonal antibodies against tumor antigens, alone or linked to toxins, can control tumor growth.

Using monoclonal antibodies to destroy tumors requires that a tumor-specific antigen be expressed on the tumor's cell surface, so that that antibody can direct the activity of a cytotoxic cell, toxin, or even radioactive nucleotide specifically to the tumor (Fig. 16.18). Some of the cell-surface molecules targeted in clinical trials are shown in Fig. 16.19, and some of these treatments have now been licensed. Striking improvements in survival have been reported for breast cancer patients treated with the monoclonal antibody trastuzumab (Herceptin), which targets the receptor HER-2/neu. This receptor is overexpressed in about one-quarter of breast cancer patients and is associated with a poorer prognosis. Herceptin is thought to act by blocking the binding of the natural ligand (so far unidentified) to this receptor, and by downregulating the level of expression of the receptor. The effects of this antibody can be enhanced when it is combined with conventional

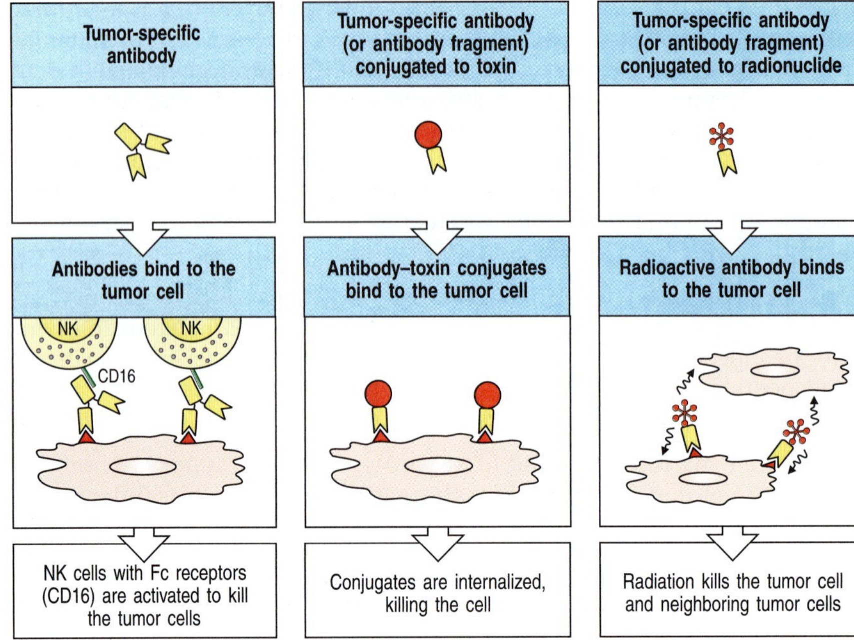

Fig. 16.18 Monoclonal antibodies that recognize tumor-specific antigens have been used to help eliminate tumors. Tumor-specific antibodies of the correct isotypes can lyse tumor cells by recruiting effector cells such as NK cells, activating the NK cells via their Fc receptors (left panels). Another strategy has been to couple the antibody to a powerful toxin (center panels). When the antibody binds to the tumor cell and is endocytosed, the toxin is released from the antibody and can kill the tumor cell. If the antibody is coupled to a radioisotope (right panels), binding of the antibody to a tumor cell will deliver a dose of radiation sufficient to kill the tumor cell. In addition, nearby tumor cells could also receive a lethal radiation dose, even though they do not bind the antibody. Antibody fragments have started to replace whole antibodies for coupling to toxins or radioisotopes.

chemotherapy. Beyond blocking a growth signal for tumor cells, experiments in mice suggest that some of trastuzumab's antitumor effects also involve innate and adaptive immune responses, such as directing ADCC or inducing antitumor T-cell responses. A monoclonal antibody that has yielded excellent results in the treatment of non-Hodgkin's B-cell lymphoma is the anti-CD20 antibody rituximab, which triggers the apoptosis of B cells on binding to CD20 on their surface (see Section 16-7).

Technical problems with monoclonal antibodies as therapeutic agents include inefficient killing of cells after binding the monoclonal antibody, inefficient penetration of the antibody into the tumor mass (which can be improved by

Fig. 16.19 Examples of tumor antigens that have been targeted by monoclonal antibodies in therapeutic trials. CEA, carcinoembryonic antigen.

Tumor tissue origin	Type of antigen	Antigen	Tumor type
Lymphoma/ leukemia	Differentiation antigen	CD5 Idiotype CD52 (CAMPATH1)	T-cell lymphoma B-cell lymphoma T- and B-cell lymphoma/ leukemia
	B-cell signaling receptor	CD20	Non-Hodgkin's B-cell lymphoma
Solid tumors	Cell-surface antigens Glycoprotein Carbohydrate	 CEA, mucin-1 Lewis^y CA-125	 Epithelial tumors (breast, colon, lung) Epithelial tumors Ovarian carcinoma
	Growth factor receptors	Epidermal growth factor receptor HER-2/neu IL-2 receptor Vascular endothelial growth factor (VEGF)	Lung, breast, head, and neck tumors Breast, ovarian tumors T- and B-cell tumors Colon cancer Lung, prostate, breast
	Stromal extracellular antigen	FAP-α Tenascin Metalloproteinases	Epithelial tumors Glioblastoma multiforme Epithelial tumors

using small antibody fragments), and soluble target antigens mopping up the antibody. The efficiency of killing can be enhanced by linking the antibody to a toxin, producing a reagent called an **immunotoxin** (see Fig. 16.18): two favored toxins are ricin A chain and *Pseudomonas* toxin. The antibody must be internalized to allow the cleavage of the toxin from the antibody in the endocytic compartment, allowing the toxin chain to penetrate and kill the cell. Toxins coupled to native antibodies have had limited success in cancer therapy, but fragments of antibodies such as single-chain Fv molecules (see Section 4-3) show more promise. An example of a successful immunotoxin is a recombinant Fv anti-CD22 antibody fused to a fragment of *Pseudomonas* toxin. This induced complete remissions in two-thirds of a group of patients with a type of B-cell leukemia known as hairy-cell leukemia, in which the disease was resistant to conventional chemotherapy.

Monoclonal antibodies can also be conjugated to chemotherapeutic drugs such as adriamycin or to radioisotopes. In the case of a drug-linked antibody, the binding of the antibody to a cell-surface antigen concentrates the drug to the site of the tumor. After internalization, the drug is released in the endosomes and exerts its cytostatic or cytotoxic effect. A variation on this approach is to link an antibody to an enzyme that metabolizes a nontoxic pro-drug to the active cytotoxic drug, a technique known as **antibody-directed enzyme/pro-drug therapy** (**ADEPT**). With this technique, a small amount of enzyme localized by the antibody can generate much larger amounts of active cytotoxic drug in the immediate vicinity of tumor cells than could be coupled directly to the targeting antibody. Monoclonal antibodies linked to radioisotopes (see Fig. 16.18) have been successfully used to treat refractory B-cell lymphoma, using anti-CD20 antibodies linked to yttrium-90 (ibritumomab tiuxetan). These approaches have the advantage of also killing neighboring tumor cells, because the released drug or radioactive emissions can affect cells adjacent to those that bind the antibody. Monoclonal antibodies coupled to γ-emitting radioisotopes have also been used successfully to image tumors for the purpose of diagnosis and monitoring tumor spread.

16-17 Enhancing the immune response to tumors by vaccination holds promise for cancer prevention and therapy.

A major breakthrough in anticancer vaccines occurred in 2005 with the completion of a clinical trial involving 12,167 women that tested a vaccine against human papilloma virus (HPV). This trial showed that a recombinant vaccine against HPV was 100% effective in preventing cervical cancer caused by two key strains, HPV-16 and HPV-18, which are associated with 70% of cervical cancers. The effect of the vaccine is most likely to be due to vaccine-induced anti-HPV antibodies preventing viral infection of cervical epithelium (Fig. 16.20). Although this trial showed the potential of vaccines to prevent cancer, attempts to use vaccines to treat existing tumors have been less effective. In

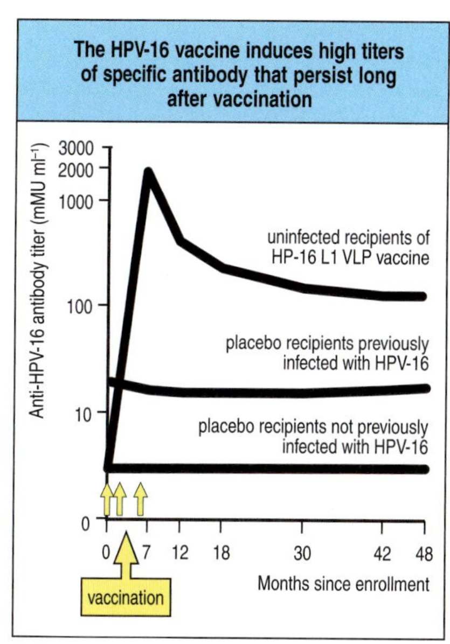

Fig. 16.20 An effective vaccine against human papilloma virus (HPV) induces antibodies that protect against HPV infection. Serotype 16 of HPV (HPV-16) is highly associated with the development of cervical cancer. In a clinical trial, 755 healthy uninfected women were immunized with a vaccine generated from highly purified non-infectious 'virus-like particles' (VLP) consisting of the capsid protein L1 of HPV-16 and formulated with an alum adjuvant (in this case aluminum hydroxyphosphate sulfate). In comparison with the very low titers of antibody in placebo-treated uninfected women (green line), or women previously infected with HPV that received placebo (blue line), the women treated with the virus-like particle vaccine (red line) developed high titers of antibody against the L1 capsid protein. None of these immunized women subsequently became infected by HPV-16. An anti-HPV vaccine marketed as Gardasil is now available and recommended for use in girls and young women as a protection from cervical cancer caused by HPV serotypes 6, 11, 16, and 18.

the case of HPV, certain types of vaccine that have increased immunogenicity for eliciting T-cell responses are beginning to show effectiveness in treating existing intraepithelial neoplasia caused by the virus.

Vaccines based on tumor antigens are, in principle, the ideal approach to T-cell-mediated cancer immunotherapy, but they are difficult to develop. For HPV, the relevant antigens are known. However, for most spontaneous tumors, relevant peptides of tumor rejection antigens may not be shared between different patients' tumors and may be presented only by particular MHC alleles. This requires that an effective tumor vaccine include a range of tumor antigens. MAGE-1 antigens, for example, are recognized only by T cells in melanoma patients expressing the HLA-A1 haplotype, but a range of MAGE-type proteins has now been characterized that encompass peptide epitopes presented by many different HLA class I and II molecules. It is clear that cancer vaccines for therapy should be used only where the tumor burden is low, such as after adequate surgery and chemotherapy.

Cell-based cancer vaccines have used the individual patient's tumor removed at surgery as a source of vaccine antigens. These were prepared by mixing either irradiated tumor cells or tumor extracts with killed bacteria such as Bacille Calmette–Guérin (BCG) or *Corynebacterium parvum*, which act as adjuvants to enhance their immunogenicity (see Appendix I, Section A-4). Although vaccination using BCG adjuvants has had variable results in the past, there is renewed interest as a result of a better understanding of their interaction with Toll-like receptors (TLRs). Stimulation of TLR-4 by BCG and other ligands has been tested in melanoma and other solid tumors. CpG DNA, which binds to TLR-9, has also been used to increase the immunogenicity of cancer vaccines.

In cases where candidate tumor rejection antigens have been identified, for example in melanoma, experimental vaccination strategies include the use of whole proteins, peptide vaccines based on sequences recognized by cytotoxic T lymphocytes and helper T lymphocytes (either administered alone or presented by the patient's own dendritic cells), and recombinant viruses encoding these peptide epitopes.

An experimental approach to tumor vaccination is based on the isolation of heat-shock proteins from tumor cells, because these proteins act as intracellular chaperones for antigenic peptides. There is evidence that dendritic cells express receptors that mediate the uptake of certain heat-shock proteins and can deliver the bound peptides into the antigen-processing pathways for presentation by MHC class I molecules. Although this approach does not require knowledge of the relevant tumor rejection antigens, heat-shock proteins bind many cellular peptides, and peptides from tumor rejection antigens would represent only a tiny fraction, limiting the effectiveness of vaccination.

16-18 Checkpoint blockade can augment immune responses to existing tumors.

Other approaches to tumor vaccination attempt to strengthen the natural immune response against a tumor by one of two means: by making the tumor itself more immunogenic, or by relieving the normal inhibitory mechanisms that regulate these responses. An example of the first approach shown to be effective in mice is the introduction of genes that encode co-stimulatory molecules or cytokines directly into tumor cells (Fig. 16.21). A tumor cell transfected with the gene encoding the co-stimulatory molecule B7 is implanted in a syngeneic animal. These B7-positive cells can activate tumor-specific naive T cells to become effector T cells able to reject the tumor cells. They are also able to stimulate further proliferation of the effector cells that reach the site of implantation. These T cells can then target the tumor

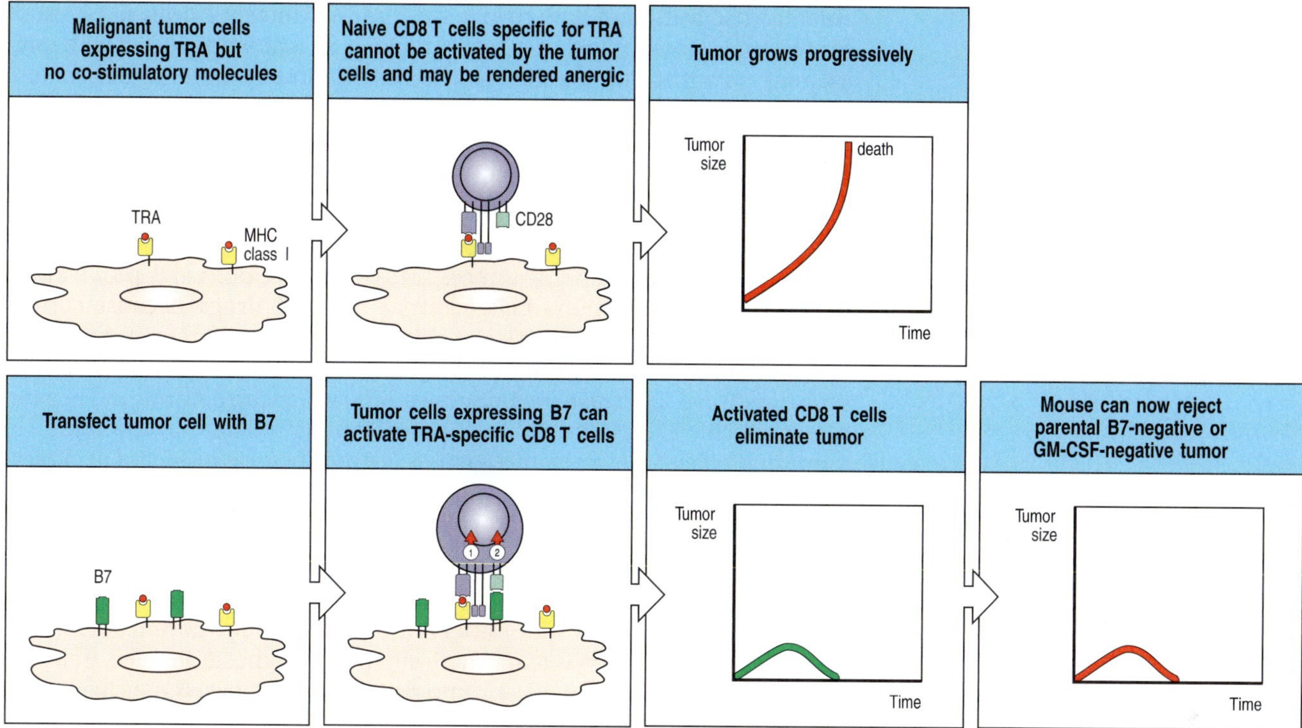

Fig. 16.21 Transfection of tumors with the gene for B7 enhances tumor immunogenicity. A tumor that does not express co-stimulatory molecules will not induce an immune response, even though it might express tumor rejection antigens (TRAs), because naive CD8 T cells specific for the TRA cannot be activated by the tumor. The tumor therefore grows progressively in normal mice and eventually kills the host (top panels). If such tumor cells are transfected with a co-stimulatory molecule, such as B7 (bottom panels), TRA-specific CD8 T cells now receive both signal 1 and signal 2 from the same cell and can therefore be activated. The same effect can be obtained by transfecting the tumor with the gene encoding GM-CSF, which attracts and stimulates the differentiation of dendritic cell precursors (not shown). Both these strategies have been tested in mice and shown to elicit memory T cells, although results with GM-CSF are more impressive. Because TRA-specific CD8 cells have now been activated, even the original B7-negative or GM-CSF-negative tumor cells can be rejected.

cells whether they express B7 or not; this can be shown by reimplanting nontransfected tumor cells, which are also rejected.

The same antitumor effect can be obtained by transfecting tumor cells with the gene encoding granulocyte–macrophage colony-stimulating factor (GM-CSF). This cytokine attracts and stimulates the differentiation of dendritic cell precursors and functions as an adjuvant to activate them. It is thought that these dendritic cells process the tumor antigens and migrate to the local lymph nodes, where they induce potent antitumor responses. In mice, B7-transfected cells seem less potent than the dendritic cells recruited by GM-CSF in inducing antitumor responses. Perhaps this is because dendritic cells express more of the molecules required to activate naive T cells than do B7-transfected tumor cells and are able to migrate into the T-cell areas of the lymph nodes, where they are optimally placed to interact with passing naive T cells (see Section 9-4). Treatment with GM-CSF alone has had limited success in patients because of the transient nature of the immune response it stimulates.

Another approach to cancer immunotherapy is called **checkpoint blockade**, which attempts to interfere with the normal inhibitory signals that regulate lymphocytes. Immune responses are controlled by several positive and negative immunological checkpoints. A positive checkpoint for T cells is controlled by the B7 co-stimulatory receptors expressed by professional antigen-presenting cells such as dendritic cells, as discussed earlier. Negative immunological checkpoints are provided by inhibitory receptors such as CTLA-4 and PD-1 (see Section 16-14). CTLA-4 imposes a critical checkpoint for potentially autoreactive T cells by binding to B7 molecules on dendritic cells and delivering a negative signal that must be overcome by other signals before T cells can become activated. In the absence of this critical checkpoint, self-reactive T cells that are normally held in check become activated instead, and produce multi-tissue autoimmune reactions, as seen in CTLA-4-deficient mice.

In checkpoint blockade directed at CTLA-4, anti-CTLA-4 antibodies disrupt its interaction with B7 and prevent it from delivering its inhibitory signal. This approach has shown some promise in treating melanoma, by causing enhanced activation of both helper T cells and cytotoxic T cells. In phase III clinical trials, patients with metastatic melanoma who were treated with ipilimumab, an anti-CTLA-4 antibody, showed an increase in the numbers and activity of T cells recognizing NY-ESO-1, a cancer-testis antigen expressed by melanoma. However, one side-effect of ipilimumab in these patients seemed to be an increased risk of autoimmune phenomena, as CTLA-4 is required to maintain the tolerance of self-reactive T cells.

Prospective treatments using checkpoint blockade rely on activation of the native immune response to tumors. This causes a potential problem for the design and evaluation of clinical trials for such drugs, because current guidelines for evaluation are based on the actions of chemotherapeutic drugs or radiation, which can kill tumor cells immediately. In checkpoint blockade, time is required to reverse the immune inhibition and to activate and expand tumor-specific T cells; in addition, the cells must migrate to the tumor to exert their effects. However, these issues are now being considered in clinical trials that are being conducted to examine checkpoint blockade used in combination with traditional anticancer therapies.

Other inhibitory receptors on lymphocytes are candidates for consideration in checkpoint blockade, including PD-1 and its ligands PD-L1 and PD-L2. PD-L1 is expressed on a wide variety of human tumors; in renal cell carcinoma, PD-L1 expression is associated with poor prognosis. In mice, transfection of the gene encoding PD-L1 into tumor cells increased their growth *in vivo* and reduced their susceptibility to lysis by cytotoxic T cells. These effects were reversed by an antibody against PD-L1.

The potency of dendritic cells in activating T-cell responses provides the rationale for yet another antitumor vaccination strategy. The use of antigen-loaded dendritic cells to stimulate therapeutically useful cytotoxic T-cell responses to tumors has been developed in animals, and there have been initial trials in humans with cancer. Other methods under trial include loading dendritic cells *ex vivo* with DNA encoding the tumor antigen or with mRNA derived from tumor cells, and the use of apoptotic or necrotic tumor cells as sources of antigens. Dendritic cell vaccination against tumors is a very active research field, and many variables are being explored in early-phase studies in patients.

Summary.

Some tumors elicit specific immune responses that suppress or modify their growth. A partly functioning immune system can lead to the outgrowth of tumors, suggesting that the immune system does have an important role in suppressing tumor development. Tumors evade or suppress the immune system in several ways, and regulatory T cells have received much attention in this area. Monoclonal antibodies have been successfully developed for tumor immunotherapy in several cases, such as anti-CD20 for B-cell lymphoma. Attempts are also being made to develop vaccines incorporating peptides designed to generate effective cytotoxic and helper T-cell responses. Checkpoint blockade strategies are being developed that use antibodies or other biologic agents to either stimulate a tumor-specific immune response or interfere with inhibitory mechanisms that tend to suppress immune responses against tumors. The efficiency of dendritic cells in presenting tumor antigens has been improved by pulsing the individual's dendritic cells *in vitro* with modified tumor cells or tumor antigens and then replacing them in the body. This approach has been extended in animal experiments to the transfection of tumor cells with genes encoding co-stimulatory molecules or cytokines that attract and activate dendritic cells. Current trends are attempting to incorporate immunotherapy with other traditional anticancer treatments to take advantage of the specificity and power of the immune system. The possibility of the near eradication of cervical cancer has been brought a step closer by the development of an effective vaccine against specific strains of the cancer-causing human papilloma virus.

Fighting infectious diseases with vaccination.

The two most important contributions to public health in the past 100 years—sanitation and vaccination—have markedly decreased deaths from infectious disease, and yet infectious diseases remain the leading cause of death worldwide. Modern immunology itself grew from the success of Jenner's and Pasteur's vaccines against smallpox and chicken cholera, respectively, and its greatest triumph has been the global eradication of smallpox, announced by the World Health Organization in 1979. A global campaign to eradicate polio is now well under way. With the past decade's tremendous progress in basic immunology, particularly in understanding innate immunity, there is now great hope that vaccines for other major infectious diseases, including malaria, tuberculosis, and HIV, are within reach. The vision of the current generation of vaccine scientists is to elevate their art to the level of modern drug design; to move it from an empiric practice to a true 'pharmacology of the immune system.'

The goal of vaccination is the generation of long-lasting and protective immunity. Throughout the course of this book, we have illustrated how the innate and the adaptive immune systems collaborate during infection to eliminate pathogens and generate protective immunity with immunological memory. Indeed, a single infection is often (but not always) sufficient to generate protective immunity to a pathogen. This important relationship was recognized long ago, and was recorded more than 2000 years ago in accounts of the Peloponnesian War, during which two successive outbreaks of plague struck Athens. The Greek historian Thucydides noted that people who had survived infection during the first outbreak were not susceptible to infection during the second.

The recognition of this type of relationship perhaps prompted the practice of **variolation** against smallpox, in which an inoculation of a small amount of dried material from a smallpox pustule was used to produce a mild infection that was then followed by long-lasting protection against reinfection. Smallpox itself has been recognized in medical literature for more that 1000 years; variolation seems to have been practiced in India and China many centuries before its introduction into the West (some time in the 1400s–1500s) and it was familiar to Jenner. However, infection after variolation was not always mild: fatal smallpox ensued in about 3% of cases, which would not meet modern criteria of safety. It seems there was some recognition that milkmaids exposed to a bovine virus similar to smallpox—cowpox—seemed protected from smallpox infection, and there is even one historical account suggesting that cowpox inoculation had been tried before Jenner. However, Jenner's achievement was not only the realization that infection with cowpox with would provide protective immunity against smallpox in humans without the risk of significant disease, but its experimental proof by the intentional variolation of people whom he had previously vaccinated. He named the process **vaccination** (from *vacca*, Latin for cow), and Pasteur, in his honor, extended the term to the stimulation of protection to other infectious agents. Humans are not a natural host of cowpox, which establishes only a brief and limited subcutaneous infection. But the cowpox virus contains antigens that stimulate an immune response that cross-reacts with smallpox antigens and thereby confers protection against the human disease. Since the early 20th century, the virus used to vaccinate against smallpox has been vaccinia virus, which is related to both cowpox and smallpox, but whose origin is obscure.

As we will see, many current vaccines offer protection by inducing the formation of neutralizing antibodies. However, that statement contains a hidden tautology; pathogens for which current vaccines are effective may also be pathogens for which antibodies are sufficient for protection. Several major pathogens are not so cooperative—malaria, tuberculosis, and HIV—and for these even a robust antibody response is not protective. The elimination of these pathogens requires additional effector activities, such as the generation of strong and durable cell-mediated immunity, which are not efficiently generated by current vaccine technologies. These are the issues that face modern vaccine science.

16-19 Vaccines can be based on attenuated pathogens or material from killed organisms.

Vaccine development in the early part of the 20th century followed two empirical approaches. The first was the search for **attenuated** organisms with reduced pathogenicity, which would stimulate protective immunity but not cause disease. This approach continues into the present with the design of genetically attenuated pathogens in which desirable mutations are introduced into the organism by recombinant DNA technologies. This idea is being applied to important pathogens, such as malaria, for which vaccines are currently unavailable, and may be important in the future for designing vaccines for influenza and HIV/AIDS.

The second approach was the development of vaccines based on killed organisms and, subsequently, on purified components of organisms that would be as effective as live whole organisms. Killed vaccines were desirable because any live vaccine, including vaccinia, can cause lethal systemic infection in immunosuppressed people. Evolving from this approach were vaccines based on the conjugation of purified antigens as described for *Haemophilus influenzae* (see Section 10-3). This approach continues with the addition of 'reverse immunogenetics' (see Section 16-15) to identify candidate peptide antigens for T cells and with strategies to use ligands that activate TLRs or other innate sensors as adjuvants to enhance responses to simple antigens.

Immunization is now considered so safe and so important that most states in the United States require all children to be immunized against measles, mumps, and polio viruses with live-attenuated vaccines as well as against tetanus (caused by *Clostridium tetani*), diphtheria (caused by *Corynebacterium diphtheriae*), and whooping cough (caused by *Bordetella pertussis*), with inactivated toxins or toxoids prepared from these bacteria. More recently, a vaccine has become available against *H. influenzae* type b (HiB), one of the causative agents of meningitis, as well as two vaccines for childhood diarrhea caused by rotaviruses, and, as described in Section 16-17, a vaccine for preventing HPV infection for protection against cervical cancer. Most vaccines are given to children within the first year of life. The vaccines against measles, mumps, and rubella (MMR), against chickenpox (varicella), and against influenza, when recommended, are usually given between the ages of 1 and 2 years.

Impressive as these accomplishments are, there are still many diseases for which we lack effective vaccines (Fig. 16.22). For many pathogens, natural infection does not seem to generate protective immunity, and infections become chronic or recurrent. In many infections of this type, antibodies are insufficient to prevent reinfection and to eliminate the pathogen, and cell-mediated immunity instead seems to be more important in limiting the pathogen, but is insufficient to provide full immunity, as in malaria, tuberculosis,

Some infections for which effective vaccines are not yet available

Disease	Estimated annual mortality
Malaria	889,000
Schistosomiasis	41,000
Intestinal worm infestation	6,000
Tuberculosis	1.5 million
Diarrheal disease	2.2 million
Respiratory infections	4 million
HIV/AIDS	2 million
Measles†	400,000

Fig. 16.22 Diseases for which effective vaccines are still needed. †Current measles vaccines are effective but heat-sensitive, which makes their use difficult in tropical countries; heat stability is being improved. Mortality data are the most recent estimated figures available (2004) (*The Global Burden of Disease: 2004 Update*. World Health Organization; 2008).

and HIV. It is not the absence of an immune response to the pathogen that is the problem, but rather that this response does not clear the pathogen, eliminate pathogenesis, or prevent reinfection.

Even when a vaccine such as measles can be used effectively in developed countries, technical and economic problems can prevent its widespread use in developing countries, where mortality from these diseases is still high. The development of vaccines therefore remains an important goal of immunology, and the latter half of the 20th century saw a shift to a more rational approach based on a detailed molecular understanding of microbial pathogenicity, analysis of the protective host response to pathogenic organisms, and an understanding of the regulation of the immune system to generate effective T- and B-lymphocyte responses.

16-20 Most effective vaccines generate antibodies that prevent the damage caused by toxins or that neutralize the pathogen and stop infection.

Although the requirements for generating protective immunity vary with the nature of the infecting organism, many effective vaccines currently work by inducing antibodies against the pathogen. For many pathogens, including extracellular organisms and viruses, antibodies can provide protective immunity. This is not the case for all pathogens, unfortunately, which may require additional cell-mediated immune responses such as CD8 T cells.

Effective protective immunity against some microorganisms requires the presence of preexisting antibody at the time of infection, either to prevent the damage caused by the pathogen or to prevent reinfection by the pathogen altogether. The first case is illustrated by vaccines to tetanus and diphtheria, in which clinical manifestations of infection are due to the effects of extremely powerful exotoxins (see Fig. 10.25). Preexisting antibody against the exotoxin is necessary to provide a defense against these diseases. Indeed, the tetanus exotoxin is so powerful that the tiny amount that can cause disease may be insufficient to lead to a protective immune response, such that even survivors of tetanus require vaccination to be protected against the risk of subsequent attack.

The second way in which antibodies can protect is by preventing secondary infection, as in the case of certain viral infections. This is called **neutralization**. The ability of an antibody to neutralize a pathogen may depend on its affinity, its isotype subclass, complement, and the activity of phagocytic cells. For example, preexisting antibodies are required to protect against the polio virus, which infects critical host cells within a short period after entering the body and is not easily controlled by T lymphocytes once intracellular infection has been established. Vaccines to seasonal influenza virus provide protection in this same manner, by inducing antibodies that limit reinfection. In the case of many viruses, antibodies produced by infection or vaccination are able to neutralize the virus, preventing further spread of infection, but this is not always the case. In HIV infection, despite the generation of antibodies that can bind to surface viral epitopes, most of these antibodies fail to neutralize virus. In addition, vaccines based on HIV proteins fail to induce antibodies that are broadly neutralizing.

Immune responses to infectious agents usually involve antibodies directed at multiple epitopes, and only some of these antibodies, if any, confer protection. The particular T-cell epitopes recognized can also affect the nature of the response. In Section 10-3, we described linked recognition, in which antigen-specific B cells and T cells provide mutually activating signals, leading to affinity maturation and isotype switching that may be required for neutralization. This process requires that an appropriate peptide epitope for T cells is presented by the B cells, and that the T-cell epitope typically must be contained

Features of effective vaccines	
Safe	Vaccine must not itself cause illness or death
Protective	Vaccine must protect against illness resulting from exposure to live pathogen
Gives sustained protection	Protection against illness must last for several years
Induces neutralizing antibody	Some pathogens (such as polio virus) infect cells that cannot be replaced (e.g. neurons). Neutralizing antibody is essential to prevent infection of such cells
Induces protective T cells	Some pathogens, particularly intracellular, are more effectively dealt with by cell-mediated responses
Practical considerations	Low cost per dose Biological stability Ease of administration Few side-effects

Fig. 16.23 There are several criteria for an effective vaccine.

within the region of protein epitope recognized by the B cell, a fact that must be considered in modern vaccine design. Indeed, as we discussed in Section 13-5, the predominant epitope recognized by T cells after vaccination with respiratory syncytial virus induces a vigorous inflammatory response but fails to elicit neutralizing antibodies and thus causes pathology without protection.

16-21 Effective vaccines must induce long-lasting protection while being safe and inexpensive.

A successful vaccine must possess several features in addition to its ability to provoke a protective immune response (Fig. 16.23). First, it must be safe. Vaccines must be given to huge numbers of people, relatively few of whom are likely to die of, or sometimes even catch, the disease that the vaccine is designed to prevent. This means that even a low level of toxicity is unacceptable. Second, the vaccine must be able to produce protective immunity in a very high proportion of the people to whom it is given. Third, particularly in poorer countries where it is impracticable to give regular 'booster' vaccinations to dispersed rural populations, a successful vaccine must generate long-lived immunological memory. This means that the vaccine must prime both B and T lymphocytes. Fourth, vaccines must be very cheap if they are to be administered to large populations. Vaccines are one of the most cost-effective measures in health care, but this benefit is eroded as the cost per dose rises.

Another benefit of an effective vaccination program is the 'herd immunity' that it confers on the general population. By lowering the number of susceptible members of a population, vaccination decreases the natural reservoir of infected individuals in that population and so reduces the probability of transmission of infection. Thus, even unvaccinated members will be protected because their individual chance of encountering the pathogen is decreased. However, the herd immunity effect is only seen at relatively high levels of vaccination within a population; for mumps it is estimated to be around 80%, and below this level sporadic epidemics can occur. This is illustrated by a marked increase in mumps in the United Kingdom in 2004–2005 in young adults as a result of the variable use in the mid-1990s of a measles/rubella vaccine, rather than the combined MMR, which was in short supply at that time.

16-22 Live-attenuated viral vaccines are usually more potent than 'killed' vaccines and can be made safer by the use of recombinant DNA technology.

Most antiviral vaccines currently in use consist of either live attenuated or inactivated viruses. Inactivated, or 'killed,' viral vaccines consist of viruses treated so that they are unable to replicate. Live-attenuated viral vaccines are generally far more potent: they elicit a greater number of effector mechanisms, including the activation of CD4 T cells and cytotoxic CD8 T cells. CD4 T cells help in shaping the antibody response, which is important for a vaccine's subsequent protective effect. Cytotoxic CD8 T cells would provide protection during infection by the virus itself, and if maintained, may contribute to protective memory. Inactivated viruses cannot produce proteins in the cytosol, so peptides from the viral antigens are not presented by MHC class I molecules. Thus, CD8 T cells are neither generated nor needed with killed virus vaccines. Attenuated viral vaccines are now in use for polio, measles, mumps, rubella, and varicella.

Traditionally, attenuation is achieved by growing the virus in cultured cells. Viruses are usually selected for preferential growth in nonhuman cells and, in the course of selection, become less able to grow in human cells (Fig. 16.24). Because these attenuated strains replicate poorly in human hosts, they induce immunity but not disease when given to people. Although attenuated

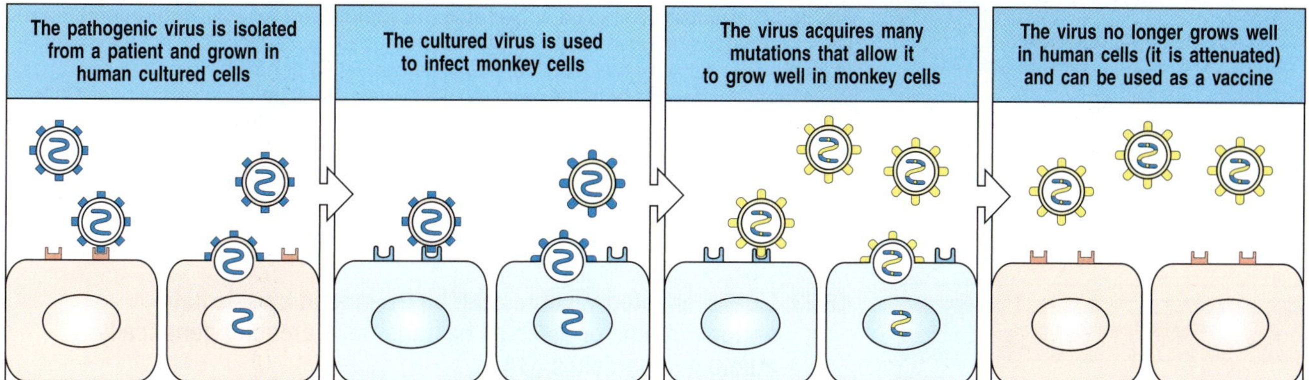

Fig. 16.24 Viruses are traditionally attenuated by selecting for growth in nonhuman cells. To produce an attenuated virus, the virus must first be isolated by growing it in cultured human cells. The adaptation to growth in cultured human cells can cause some attenuation in itself; the rubella vaccine, for example, was made in this way. In general, however, the virus is then adapted to growth in cells of a different species, until it grows only poorly in human cells. The adaptation is a result of mutation, usually a combination of several point mutations. It is usually hard to tell which of the mutations in the genome of an attenuated viral stock are critical to attenuation. An attenuated virus will grow poorly in the human host and will therefore produce immunity but not disease.

virus strains contain multiple mutations in genes encoding several of their proteins, it might be possible for a pathogenic virus strain to reemerge by a further series of mutations. For example, the type 3 Sabin polio vaccine strain differs from a wild-type progenitor strain at only 10 of 7429 nucleotides. On extremely rare occasions, reversion of the vaccine to a neurovirulent strain can occur, causing paralytic disease in the unfortunate recipient.

Attenuated viral vaccines can also pose particular risks to immunodeficient recipients, in whom they often behave as virulent opportunistic infections. Immunodeficient infants who are vaccinated with live attenuated polio before their inherited immunoglobulin deficiencies have been diagnosed are at risk because they cannot clear the virus from their gut, and there is therefore an increased chance that mutation of the virus associated with its continuing uncontrolled replication in the gut will lead to fatal paralytic disease.

An empirical approach to attenuation is still in use but might be superseded by two new approaches that use recombinant DNA technology. One is the isolation and *in vitro* mutagenesis of specific viral genes. The mutated genes are used to replace the wild-type gene in a reconstituted virus genome, and this deliberately attenuated virus can then be used as a vaccine (Fig. 16.25). The advantage of this approach is that mutations can be engineered so that reversion to wild type is virtually impossible.

Such an approach might be useful in developing live influenza vaccines. As described in Chapter 13, the influenza virus can reinfect the same host several times, because it undergoes antigenic shift and thus predominantly escapes the original immune response. A weak protection conferred by previous infections with a different subtype of influenza is observed in adults, but not in children, and is called heterosubtypic immunity. The current approach to vaccination against influenza is to use a killed virus vaccine that is reformulated annually on the basis of the prevalent strains of virus. The vaccine is moderately effective, reducing mortality in elderly people and illness in healthy adults. The ideal influenza vaccine would be an attenuated live organism that matched the prevalent virus strain. This could be created by first introducing a series of attenuating mutations into the gene encoding a viral polymerase protein, PB2. The mutated gene segment from the attenuated virus could then be substituted for the wild-type gene in a virus carrying the relevant hemagglutinin and neuraminidase antigen variants of the current epidemic or pandemic strain. This last procedure could be repeated as necessary to keep pace with the antigenic shift of the virus. Public attention has recently been directed toward the possibility of a flu pandemic caused by the H5N1 avian flu strain. This strain can be passed between birds and humans with a high mortality rate, but a pandemic would occur only if human-to-human transmission could occur. A live-attenuated vaccine would be used only if a pandemic occurred, because to give it beforehand would introduce new influenza virus genes that might recombine with existing influenza viruses.

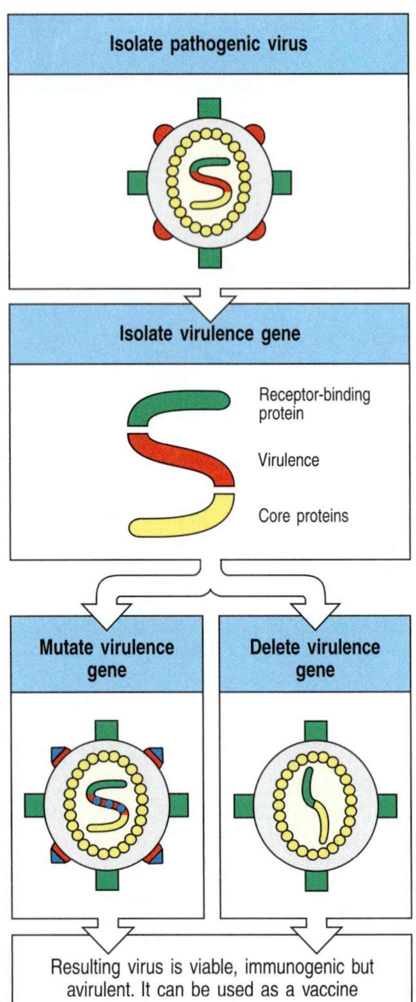
Isolate pathogenic virus
Isolate virulence gene
Receptor-binding protein
Virulence
Core proteins
Mutate virulence gene
Delete virulence gene
Resulting virus is viable, immunogenic but avirulent. It can be used as a vaccine

Fig. 16.25 Attenuation can be achieved more rapidly and reliably with recombinant DNA techniques. If a gene in the virus that is required for virulence but not for growth or immunogenicity can be identified, this gene can be either multiply mutated (left lower panel) or deleted from the genome (right lower panel) by using recombinant DNA techniques. This procedure creates an avirulent (nonpathogenic) virus that can be used as a vaccine. The mutations in the virulence gene are usually large, so that it is very difficult for the virus to revert to the wild type.

16-23 Live-attenuated vaccines can be developed by selecting nonpathogenic or disabled bacteria or by creating genetically attenuated parasites (GAPs).

Similar approaches have been used for bacterial vaccine development. The most important example of an attenuated vaccine is that of BCG, which is quite effective at protecting against serious disseminated tuberculosis in children, but is not protective against adult pulmonary disease. The current BCG vaccine, which remains the most widely used vaccine in the world, was obtained from a pathogenic isolate of *Mycobacterium bovis* and passaged in the laboratory at the beginning of the 20th century. Since then, several genetically diverse strains of BCG have evolved. The level of protection afforded by BCG is extremely variable, ranging from none in some countries, such as Malawi, to 50–80% in the UK.

Considering that tuberculosis remains one of the biggest killers worldwide, there is an urgent need for a new vaccine. Two recombinant BCG (rBCG) vaccines intended to prevent infection of unexposed individuals recently passed Phase I clinical trials. One was engineered to overexpress an immunodominant antigen of *M. tuberculosis*, to engender greater specificity toward the human pathogen. The second expressed the pore-forming protein listeriolysin from *L. monocytogenes* to induce the passage of BCG antigens from phagosomes into the cytoplasm and allow cross-presentation (see Section 6-9) on class I MHC, thereby stimulating BCG-specific cytotoxic T cells.

A similar approach is being used to generate new vaccines for malaria. Analysis of different stages of *Plasmodium falciparum*, the major cause of fatal malaria, identified genes that are selectively expressed in sporozoites within the mosquito's salivary gland, where they first become infectious for human hepatocytes. Deletion of two such genes from the *P. falciparum* genome rendered sporozoites incapable of establishing a blood-stage infection in mice, yet capable of inducing an immune response that protected mice from subsequent infection by wild-type *P. falciparum*. This protection was dependent on CD8 T cells, and to some extent on IFN-γ, indicating that cell-mediated immunity is important for protection against this parasite (Fig. 16.26). This highlights once again the importance of being able to generate vaccines that are capable of inducing strong cell-mediated immunity.

16-24 The route of vaccination is an important determinant of success.

The ideal vaccination induces host defense at the point of entry of the infectious agent. Stimulation of mucosal immunity is therefore an important goal of vaccination against those many organisms that enter through mucosal surfaces. Still, most vaccines are given by injection. This route has several disadvantages. Injections are painful and unpopular, reducing vaccine uptake, and they are expensive, requiring needles, syringes, and a trained injector. Mass vaccination by injection is laborious. There is also the immunological drawback that injection may not be the most effective way of stimulating an appropriate immune response because it does not mimic the usual route of entry of the majority of pathogens against which vaccination is directed.

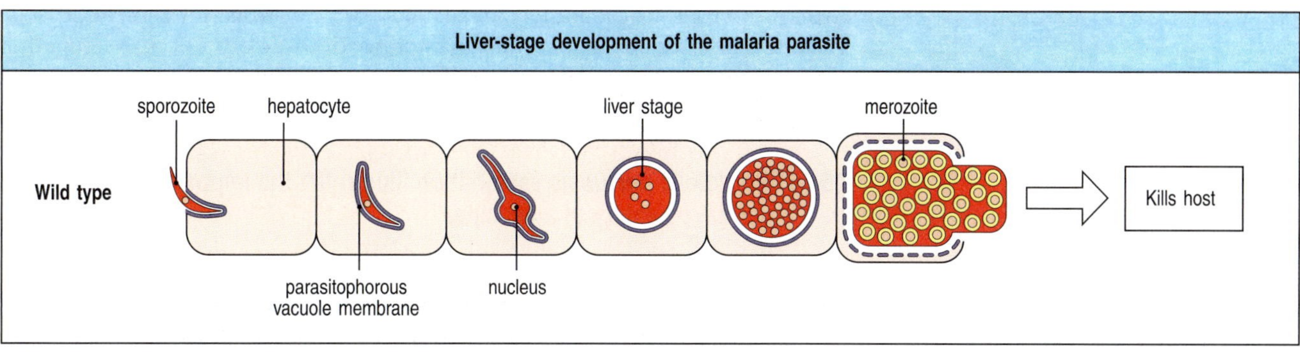

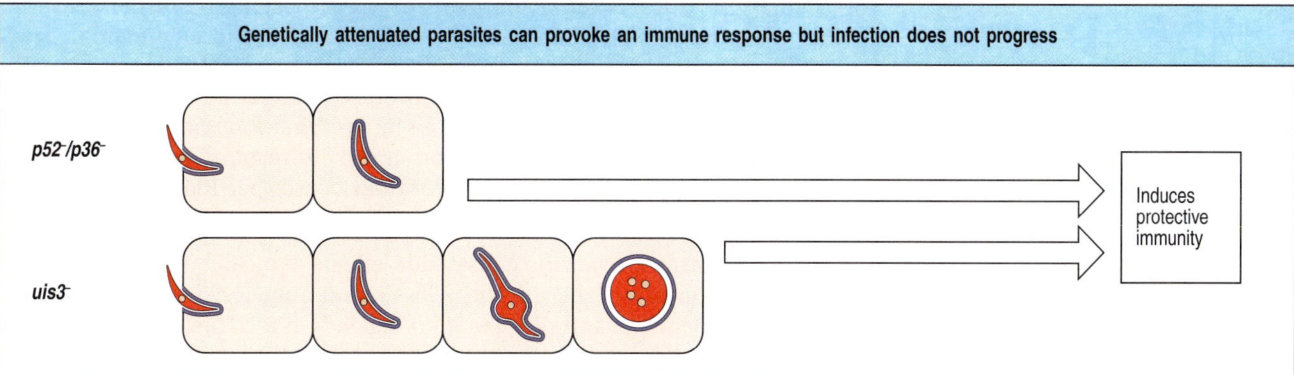

Fig. 16.26 Genetically attenuated parasites can be engineered as live vaccines to provide protective immunity. Top panel: wild-type *Plasmodium* sporozoites transmitted through the bite of an infected mosquito enter the bloodstream and are carried to the liver, where they infect hepatocytes. Each sporozoite multiplies in the liver, killing the infected cell and releasing thousands of merozoites, the next stage in infection. Bottom panels: in mice immunized with sporozoites with targeted disruption of key genes (for example, *p52* and *p36* (*p52*$^-$/*p36*$^-$), or *uis3* (*uis3*$^-$)), the sporozoites circulate in the bloodstream and mimic an early infection but cannot establish a productive infection in the liver. The mice do, however, produce an immune response against the sporozoites and are protected against a subsequent infection by wild-type sporozoites.

Many important pathogens infect mucosal surfaces or enter the body through mucosal surfaces. Examples include respiratory microorganisms such as *B. pertussis*, rhinoviruses, and influenza viruses, and enteric microorganisms such as *Vibrio cholerae*, *Salmonella typhi*, enteropathogenic *Escherichia coli*, and *Shigella*. Intranasally administered live-attenuated vaccine against influenza virus induces mucosal antibodies, which are more effective than systemic antibodies in the control of upper respiratory tract infection. However, the systemic antibodies induced by injection are effective in controlling lower respiratory tract disease, which is responsible for the severe morbidity and mortality in this disease. Thus, a realistic goal of any pandemic influenza vaccine is to prevent the lower respiratory tract disease but accept the fact that mild illness will not be prevented.

The power of the mucosal approach is illustrated by the effectiveness of live-attenuated polio vaccines. The Sabin oral polio vaccine consists of three attenuated polio virus strains and is highly immunogenic. Moreover, just as polio itself can be transmitted by fecal contamination of public swimming pools and other failures of hygiene, the vaccine can be transmitted from one individual to another by the fecal–oral route. Infection with *Salmonella* likewise stimulates a powerful mucosal and systemic immune response.

The rules of mucosal immunity are still poorly understood. Presentation of soluble protein antigens by the oral route often results in tolerance, which is important given the enormous load of foodborne and airborne antigens presented to the gut and respiratory tract (see Chapter 11). Nonetheless, the mucosal immune system responds to and eliminates mucosal infections, such as pertussis, cholera, and polio, that enter by the oral route. The proteins from these microorganisms that stimulate immune responses are therefore of special interest. One group of powerfully immunogenic proteins at mucosal surfaces is a group of bacterial toxins that have the property of binding to eukaryotic cells and are resistant to protease. A recent finding of potential practical importance is that certain of these proteins, such as the *E. coli* heat-labile toxin and pertussis toxin, have adjuvant properties that are retained even when the parent molecule has been engineered to eliminate its toxic properties. These molecules can be used as adjuvants for oral or nasal vaccines. In mice, nasal insufflation of either of these mutant toxins together with tetanus toxoid resulted in the development of protection against lethal challenge with tetanus toxin.

16-25 *Bordetella pertussis* vaccination illustrates the importance of the perceived safety of a vaccine.

The history of vaccination against the bacterium that causes whooping cough, *Bordetella pertussis*, illustrates the challenges of developing and disseminating an effective vaccine, as well as the public appeal of acellular conjugate vaccines over attenuated live organisms. At the beginning of the 20th century, whooping cough killed about 0.5% of American children under the age of 5 years. In the early 1930s, a trial of a killed, whole bacterial cell vaccine on the Faroe Islands provided evidence of a protective effect. In the United States, systematic use of a whole-cell vaccine in combination with diphtheria and tetanus toxoids (the DTP vaccine) from the 1940s resulted in a decline in the annual infection rate from 200 to fewer than 2 cases per 100,000 of the population. First vaccination with DTP was typically given at the age of 3 months.

Whole-cell pertussis vaccine causes side-effects, typically redness, pain, and swelling at the site of the injection; less commonly, vaccination is followed by high temperature and persistent crying. Very rarely, fits and a short-lived sleepiness or a floppy unresponsive state ensue. During the 1970s, widespread concern developed after several anecdotal observations that encephalitis leading to irreversible brain damage might very rarely follow pertussis vaccination. In Japan, in 1972, about 85% of children were given the pertussis vaccine, and fewer than 300 cases of whooping cough and no deaths were reported. As a result of two deaths after vaccination in Japan in 1975, the use of DTP was temporarily suspended and then reintroduced with the first vaccination at 2 years of age rather than at 3 months. In 1979 there were about 13,000 cases of whooping cough and 41 deaths. The possibility that pertussis vaccine very rarely causes severe brain damage has been studied extensively, and expert consensus is that pertussis vaccine is not a primary cause of brain injury. There is no doubt that there is greater morbidity from whooping cough than from the vaccine.

The public and medical perception that whole-cell pertussis vaccination might be unsafe provided a powerful incentive to develop safer pertussis vaccines. Study of the natural immune response to *B. pertussis* showed that infection induced antibodies against four components of the bacterium—pertussis toxin, filamentous hemagglutinin, pertactin, and fimbrial antigens. Immunization of mice with these antigens in purified form protected them against challenge with pertussis. This has led to the development of acellular pertussis vaccines, all of which contain purified pertussis toxoid—that is, toxin inactivated by chemical treatment, for example with hydrogen peroxide or formaldehyde—or more recently by genetic engineering of the toxin. Some pertussis vaccines also contain one or more of the filamentous hemagglutinin, pertactin, and fimbrial antigens. Current evidence shows that these are probably as effective as whole-cell pertussis vaccine and are free of the common minor side-effects of the whole-cell vaccine. The acellular vaccine is more expensive, however, thus restricting its use in poorer countries.

The history of pertussis vaccination illustrates that vaccines must first be extremely safe and free of side-effects; second, the public and medical profession must perceive the vaccine to be safe; and third, careful study of the nature of the protective immune response can lead to acellular vaccines that are safer than whole-cell vaccines but still as effective. Still, public concerns about vaccination remain high. Unwarranted fears of a link between the combined live-attenuated MMR vaccine and autism saw the uptake of MMR vaccine in England fall from a peak of 92% of children in 1995–1996 to 84% in 2001–2002. Small clustered outbreaks of measles during 2002 in London illustrate the importance of maintaining high uptake of vaccine to maintain herd immunity.

16-26 Conjugate vaccines have been developed as a result of understanding how T and B cells collaborate in an immune response.

Although acellular vaccines are inevitably safer than vaccines based on whole organisms, a fully effective vaccine cannot normally be made from a single isolated constituent of a microorganism, and it is now clear that this is because of the need to activate more than one cell type to initiate an immune response. One consequence of this insight has been the development of conjugate vaccines. We have already briefly described one of the most important of these, for *Haemophilus influenzae*, in Section 10-3.

Many bacteria, including *Neisseria meningitidis* (meningococcus), *Streptococcus pneumoniae* (pneumococcus), and *H. influenzae*, have an outer capsule composed of polysaccharides that are species- and type-specific for particular strains of the bacterium. The most effective defense against these microorganisms is opsonization of the polysaccharide coat with antibody. The aim of vaccination is therefore to elicit antibodies against the polysaccharide capsules of the bacteria.

Capsular polysaccharides can be harvested from bacterial growth medium and, because they are T-cell independent antigens, they can be used on their own as vaccines. However, young children under the age of 2 years cannot make good T-cell independent antibody responses and cannot be vaccinated effectively with polysaccharide vaccines. An efficient way of overcoming this problem (see Fig. 10.5) is to conjugate bacterial polysaccharides chemically to protein carriers, which provide peptides that can be recognized by antigen-specific T cells, thus converting a T-cell independent response into a T-cell dependent antipolysaccharide antibody response. Using this approach, various **conjugate vaccines** have been developed against *H. influenzae* type b, an important cause of serious childhood chest infections and meningitis, and against *N. meningitidis* serogroup C, an important cause of meningitis, and these are now widely applied. The success of the latter vaccine in the United Kingdom is illustrated in Fig. 16.27, which illustrates that the incidence of meningitis C has been markedly reduced in comparison with meningitis B, against which there is currently no vaccine.

16-27 Peptide-based vaccines can elicit protective immunity, but they require adjuvants and must be targeted to the appropriate cells and cell compartment to be effective.

Another vaccine-development strategy that does not require the whole organism, whether killed or attenuated, identifies the T-cell peptide epitopes that stimulate protective immunity. Candidate peptides can be identified in two ways: in one, overlapping peptides from immunogenic proteins are systematically synthesized and their ability to stimulate protective immunity is tested; alternatively, a reverse immunogenetic approach (see Section 16-15) can be used to predict potential peptide epitopes from a genome sequence. The latter approach has been applied to malaria by using the complete sequence of the *Plasmodium falciparum* genome. The starting point was the association between the human MHC class I molecule HLA-B53 and resistance to cerebral malaria, a relatively infrequent, but usually fatal, complication of infection. It was thought that HLA-B53 might protect from cerebral malaria because it could present peptides that are particularly good at activating naive cytotoxic T lymphocytes. Peptides eluted from HLA-B53 frequently contain a proline as

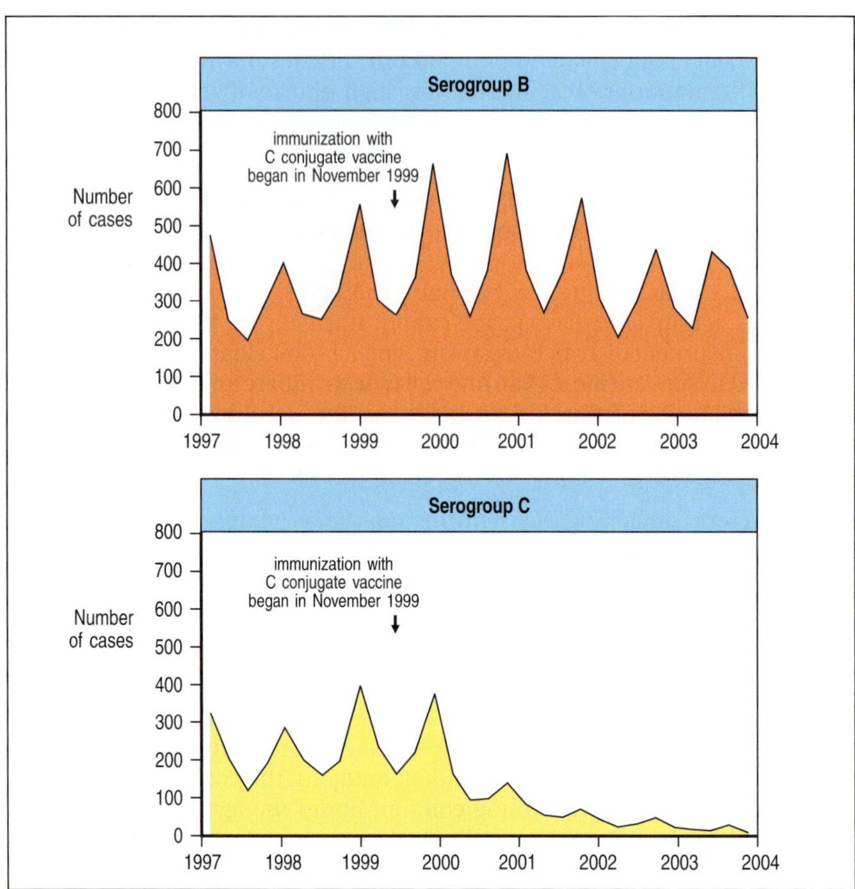

Fig. 16.27 The effect of vaccination against group C *Neisseria meningitidis* (meningococcus) on the number of cases of group B and group C meningococcal disease in England and Wales. Meningococcal infection affects roughly 5 in 100,000 people a year in the UK, with groups B and C meningococci accounting for almost all the cases. Before the introduction of the meningitis C vaccine, group C disease was the second most common cause of meningococcal disease, accounting for about 40% of cases. Group C disease now accounts for less than 10% of cases, with group B disease accounting for more than 80% of cases. After the introduction of the vaccine, there was a significant decrease in the number of laboratory-confirmed cases of group C disease in all age groups. The impact was greatest in the immunized groups, with reductions of more than 90% in all age groups. An impact has also been seen in the unimmunized age groups, with a reduction of about 70%, suggesting that this vaccine has had a herd immunity effect.

the second of their nine amino acids. On the basis of this information, reverse genetic analysis identified candidate protective peptides from four proteins of *P. falciparum* expressed in the early phase of hepatocyte infection—an important phase of infection to target in an effective immune response. One of the candidate peptides, from liver stage antigen-1, has been shown to be recognized by cytotoxic T cells when bound to HLA-B53.

Peptide-based vaccines, although promising, have several drawbacks, however. First, a particular peptide may not bind to all the MHC molecules present in the population. Because humans are highly polymorphic in the MHC (see Section 6-12), a large panel of protective peptides would need to be identified that would allow protective coverage of most individuals. Second, some direct exchange of short peptides on MHC molecules can occur without physiological antigen processing. If the required antigenic peptides load directly onto MHC molecules on cells other than dendritic cells, this may induce tolerance in T cells rather than stimulating immunity (see Section 8-26). Third, exogenous proteins and peptides delivered by a synthetic vaccine are efficiently processed for presentation by MHC class II molecules but do not enter the MHC class I processing pathway, and failure to present peptides on MHC class I molecules would severely restrict the activation of cytotoxic CD8 T cells. In certain dendritic cells, however, efficient 'cross-presentation' can occur, in which exogenously derived peptides are loaded onto MHC class I molecules (see Section 6-9), and directing peptide-based vaccines to such cells may enhance vaccine efficacy.

A recent development in peptide-based vaccine strategy seems to overcome several of these drawbacks and has already shown promise in humans. In clinical trials, patients with established vulvar intraepithelial neoplasia, an early form of vulvar cancer caused by human papilloma virus (HPV), were treated with a vaccine consisting of long peptides covering the entire length of two oncoproteins of HPV-16—E6 and E7—and delivered in an oil-in-water emulsion as adjuvant. By using very long peptides, around 100 amino acids in length, multiple candidate peptide epitopes can be delivered that may also be presented by different MHC alleles. It seems that these peptides are too long for direct exchange with peptides on cell surfaces and must instead be processed by professional antigen-presenting cells such as dendritic cells. This vaccine induced complete clinical remission in one-quarter of the patients, and about half of the treated patients showed significant clinical responses that correlated with *in vitro* evidence of enhanced cell-mediated immunity.

16-28 Adjuvants are important for enhancing the immunogenicity of vaccines, but few are approved for use in humans.

Another drawback of the peptide-based vaccines discussed above, as well as of vaccines based on highly purified protein components, is that they do not activate the innate immune system in the same way as a natural infection. Such vaccines require additional components to mimic how infections normally activate innate immunity, which induces dendritic cells to become optimally stimulatory for T cells (see Section 9-6). Such components of a vaccine are known as **adjuvants**, which are defined as substances that enhance the immunogenicity of antigens (see Appendix I, Section A-4). For example, tetanus toxoid is not immunogenic in the absence of adjuvants, and so tetanus toxoid vaccines contain inorganic aluminum salts (**alum**) in the form of noncrystalline gels, which bind polyvalently to the toxoid by ionic interactions. Pertussis toxin has adjuvant properties in its own right and, when given mixed as a toxoid with tetanus and diphtheria toxoids, not only protects against whooping cough but also acts as an additional adjuvant for the other two toxoids. This mixture makes up the DTP triple vaccine given to infants in the first year of life.

The antigenic components and adjuvants in a vaccine are not approved for use on their own; they are only approved in the context of the specific vaccine in which they are formulated. At present, alum is the only adjuvant that is approved by the FDA in the United States for use in marketed human vaccines, although some other adjuvant–vaccine combinations are undergoing clinical trials. Alum is the common name for certain inorganic aluminum salts, of which aluminum hydroxide and aluminum phosphate are most frequently used as adjuvants. In Europe, as well as the alum adjuvants, an oil (squalene)-in-water emulsion is used as an adjuvant in a formulation of influenza vaccine.

As we described in Section 3-8, alum seems to act as an adjuvant by stimulating one of the innate immune system's bacterial sensor mechanisms, NLRP3, thus activating the inflammasome and the inflammatory reactions that are a prerequisite for an effective adaptive immune response. Several other adjuvants are widely used experimentally in animals but are not approved for use in humans. Many of these are sterile constituents of bacteria, particularly of their cell walls. Freund's complete adjuvant is an oil–water emulsion containing killed mycobacteria. A complex glycolipid, muramyl dipeptide, which can be extracted from mycobacterial cell walls or synthesized, contains much of the adjuvant activity of whole killed mycobacteria. Other bacterial adjuvants include killed *B. pertussis*, bacterial polysaccharides, bacterial heat-shock proteins, and bacterial DNA. Many of these adjuvants cause quite marked inflammation and so are not suitable for use in vaccines for humans.

Many adjuvants seem to work by triggering the innate viral and bacterial sensor pathways, via TLRs and proteins of the NOD-like receptor family such as NLRP3 (see Chapter 3). Lipopolysaccharide (LPS), a TLR-4 agonist, has adjuvant effects, but these are limited by its toxicity. Small amounts of injected LPS can induce a state of shock and systemic inflammation that mimics Gram-negative sepsis, raising the question of whether its adjuvant effect can be separated from the toxic effects. Monophosphoryl lipid A, an LPS derivative and TLR-4 ligand, partly achieves this, retaining adjuvant effects but being associated with much lower toxicity than LPS. Unmethylated CpG DNA, which binds to and activates TLR-9, and imiquimod, a small-molecule drug that acts as a TLR-7 agonist, can both provide adjuvant activity experimentally, but neither is approved as an adjuvant in human vaccines. In natural infections, some bacterial proteins, for example cholera toxin, *E. coli* heat-labile enterotoxin, and pertussis toxin, act as adjuvants to stimulate mucosal immune responses, which are a particularly important defense against organisms entering through the digestive or respiratory tracts.

16-29 Protective immunity can be induced by DNA-based vaccination.

A more recent development in vaccination began with attempts to use nonreplicating bacterial plasmids encoding proteins for gene therapy. Surprisingly, proteins expressed *in vivo* from these plasmids were found to stimulate an immune response. When DNA encoding a viral immunogen is injected intramuscularly in mice, it leads to the development of antibody responses and cytotoxic T cells that allow the mice to reject a later challenge with whole virus. This response does not seem to damage the muscle tissue, is safe and effective, and, because it uses only a single microbial gene or a stretch of DNA encoding sets of antigenic peptides, does not carry the risk of active infection.

This procedure is termed **DNA vaccination**. DNA coated onto minute metal particles can be administered by a gene gun, so that particles penetrate the skin and, potentially, some underlying muscle. This technique has been shown to be effective in animals and might be suitable for mass immunization. One problem with DNA-based vaccines, however, is that they are comparatively weak. Mixing in plasmids that encode cytokines such as IL-12, IL-23, or GM-CSF makes immunization with genes encoding protective antigens much more effective. Another way to enhance DNA-based vaccines is to include genes that will express co-stimulatory molecules.

It seems that the antigens that stimulate immunity are produced in cells that are directly transfected, such as skin or muscle, but are transferred to dendritic cells for presentation to T cells. This means that for adjuvants expressed by the DNA vaccine to be able to enhance immune responses, their actions will also have to be transferred along with the antigen to dendritic cells. DNA vaccines for the prevention of malaria, influenza, and HIV infection are being tested in human trials.

16-30 The effectiveness of a vaccine can be enhanced by targeting it to sites of antigen presentation.

To be effective, vaccine antigens must be efficiently presented to the immune system by antigen-presenting cells. More efficient presentation can be achieved if proteolysis of the antigen on its way to the antigen-presenting cells is prevented, thereby preserving antigen structure. This is why many vaccines are given by injection rather than by the oral route, which exposes the vaccine to digestion in the gut. Other approaches can be taken to target the vaccine selectively to antigen-presenting cells once it has entered the body, and to engineer the selective uptake of vaccine antigens into the antigen-processing pathways within the cell. One is to coat the vaccine antigen with mannose to enhance uptake by mannose receptors on the cells. Alternatively, antigen can be delivered in the form of an immune complex, which takes advantage of antibody and complement binding by Fc and complement receptors.

A more complicated strategy is to target vaccine antigens selectively into the antigen-presenting pathways within the cell. Antigen can be coupled to specific antibodies that bind to surface proteins of dendritic cells so that receptor-mediated internalization of the antigen:antibody complex delivers the antigen into the MHC class II processing pathway and ensures that it is presented to T cells. Presentation on MHC class I molecules by such a route can be aided by targeting dendritic cells that perform efficient cross-presentation. For example, the receptor DEC205 is expressed by a subset of dendritic cells effective at cross-presentation, and linking an antigen to an antibody against DEC205 has been shown to increase the strength of immune responses to that antigen.

Antigen can also be targeted directly into antigen-processing pathways by other means. HPV E7 antigen has been coupled to the signal peptide that targets lysosome-associated membrane proteins to lysosomes and endosomes. The signal peptide therefore directs the E7 antigen directly to the intracellular compartments in which antigens are cleaved to peptides before binding to MHC class II molecules (see Section 6-7). In mice, a vaccinia virus incorporating this chimeric antigen induced a greater response to E7 antigen than did vaccinia incorporating wild-type E7 antigen alone.

16-31 An important question is whether vaccination can be used therapeutically to control existing chronic infections.

There are many chronic diseases in which infection persists because of a failure of the immune system to eliminate disease. Such infections can be divided into two groups: those in which there is an obvious immune response that fails to eliminate the organism, and those that seem to be invisible to the immune system and evoke a barely detectable immune response.

In the first category, the immune response is often partly responsible for the pathogenic effects. Infection by the helminth *Schistosoma mansoni* is associated with a powerful T_H2-type response, characterized by high levels of IgE, circulating and tissue eosinophilia, and a harmful fibrotic response to schistosome ova in the liver, leading to hepatic fibrosis. Other common parasites, such as *Plasmodium* and *Leishmania* species, also cause damage in many patients because they are not eliminated effectively by the immune response. The mycobacterial agents of tuberculosis and leprosy cause a persistent intracellular infection; a T_H1 response helps to contain these infections but also causes granuloma formation and tissue necrosis (see Fig. 9.43).

Among viruses, hepatitis B and hepatitis C infections are commonly followed by persistent viral carriage and hepatic injury, resulting in eventual death from hepatitis or from hepatocellular carcinoma. Infection with HIV, as we have seen in Chapter 13, also persists despite an ongoing immune response. In a preliminary trial involving HIV-infected patients, dendritic cells derived from the patients' own bone marrow were loaded with chemically inactivated HIV. After immunization with the loaded cells, a robust T-cell response to HIV was observed in some patients that was associated with the production of IL-2 and IFN-γ (Fig. 16.28). Viral load in these patients was reduced by 80%, and in almost half of these patients the suppression of viremia lasted for more than a year. Nonetheless, these responses were not sufficient to eliminate the HIV infection.

In the second category of chronic infection, which is predominantly viral, the immune response fails to clear infection because of the relative invisibility of the infectious agent to the immune system. A good example is herpes simplex type 2, which is transmitted venereally, becomes latent in nerve tissue, and causes genital herpes, which is frequently recurrent. This invisibility seems to be caused by a viral protein, ICP-47, that binds to the TAP complex (see Section 6-2) and inhibits peptide transport into the endoplasmic reticulum

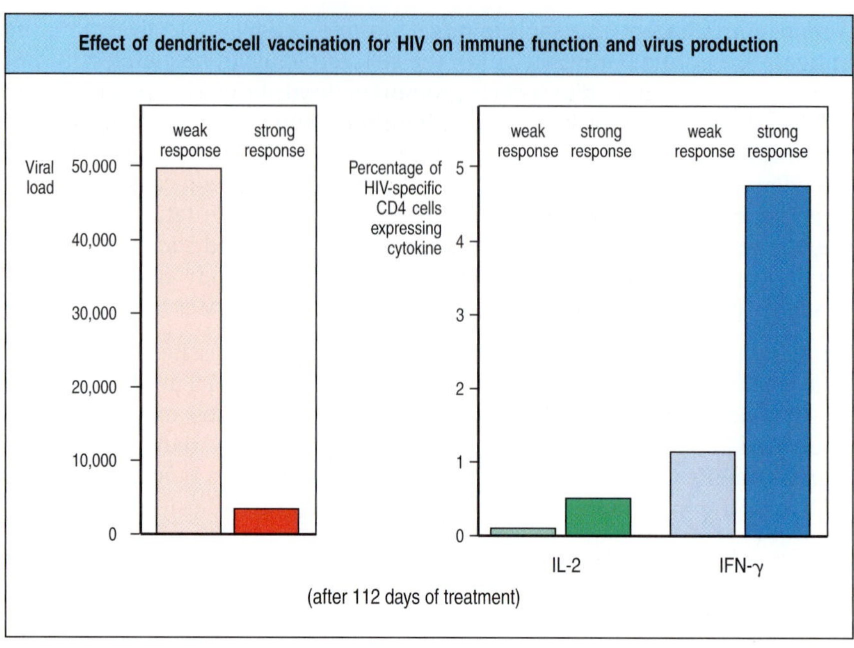

Fig. 16.28 Vaccination with dendritic cells loaded with HIV substantially reduces viral load and generates T-cell immunity. Left panel: viral load is shown for a weak and transient response to treatment (pink); the red bar represents individuals who made a strong and durable response. Right panel: CD4 T-cell IL-2 and interferon-γ production for individuals who made a weak or strong response. The production of both these cytokines, indicating T-cell activity, correlates with the response to treatment.

in infected cells. Thus, viral peptides are not presented to the immune system by MHC class I molecules. A similar example in this category of chronic infection is genital warts, caused by certain papilloma viruses that evoke very little immune response, particularly a cell-mediated response. As discussed previously, a recent clinical trial using long-peptide vaccines against HPV-16 was effective in increasing the strength of cell-mediated immune responses to viral antigens, and in reducing or eliminating pre-cancerous lesions associated with the HPV infection (see Section 16-27). This trial offers positive indications that vaccines directed at increasing cell-mediated responses to other pathogens may be similarly effective.

Summary.

The greatest triumphs of modern immunology have come from vaccination, which has eradicated or virtually eliminated several human diseases. It is the single most successful manipulation of the immune system so far, because it takes advantage of the immune system's natural specificity and inducibility. Nevertheless, there are important human infectious diseases for which there is still no effective vaccine. The most effective vaccines are based on attenuated live microorganisms, but these carry some risk and are potentially lethal to immunosuppressed or immunodeficient individuals. New techniques are being developed to generate genetically attenuated pathogens for use as vaccines, for malaria and tuberculosis in particular. Most current viral vaccines are based on live attenuated virus, but many bacterial vaccines are based on components of the microorganism, including components of the toxins that it produces. Protective responses to carbohydrate antigens, which in very young children do not provoke lasting immunity, can be enhanced by conjugation to a protein. Vaccines based on peptides, particularly very long peptides, are just emerging from the experimental stage and are beginning to be tested in humans. A vaccine's immunogenicity often depends on adjuvants that can help, directly or indirectly, to activate antigen-presenting cells that are necessary for the initiation of immune responses. Adjuvants activate these cells by engaging the innate immune system and providing ligands for TLRs and other receptors on antigen-presenting cells. The development of oral vaccines is particularly important for stimulating immunity to the many pathogens that enter through the mucosa.

Summary to Chapter 16.

One of the great future challenges in immunology is to be able to control the immune response so that unwanted immune responses can be suppressed and desirable responses elicited. Current methods of suppressing unwanted responses rely, to a great extent, on drugs that suppress adaptive immunity indiscriminately and are thus inherently flawed. We have seen in this book that the immune system can suppress its own responses in an antigen-specific manner and, by studying these endogenous regulatory events, it has been possible to devise strategies to manipulate specific responses while sparing general immune competence. New treatments, including many monoclonal antibodies, have emerged as clinically important therapies to selectively suppress the responses that lead to allergy, autoimmunity, or the rejection of grafted organs. Similarly, as we understand more about tumors and infectious agents, better strategies to mobilize the immune system against cancer and infection are becoming possible. To achieve all this, we need to learn more about the induction of immunity and the biology of the immune system, and to apply what we have learned to human disease.

Questions.

16.1 *Explain how the drug cyclosporin A, which binds to an immunophilin that is ubiquitous, can still exert selective inhibition in T cells.*

16.2 *If the drug rapamycin is immunosuppressive, why might it also be useful to augment the action of vaccines?*

16.3 *Explain the major risk, and its immunological basis, of anti-TNF-α antibodies used in the treatment of rheumatoid arthritis.*

16.4 *How do tumors evade the immune response?*

16.5 *What types of issue complicate the interpretation of clinical trials that use immunotherapy to treat cancer compared with trials based on cytotoxic drugs alone?*

16.6 *If the HPV vaccine can prevent infections that can cause cancer, should it be given to all women, or will this simply promote promiscuous behaviors? Should it be mandatory for boys as well?*

16.7 *Discuss the importance of herd immunity. What do you think should be done about people refusing vaccinations against important pathogens such as polio?*

16.8 *Discuss the benefits of immunotherapies that may eliminate cancers and their risks of inducing autoimmunity.*

General references.

Curtiss, R., III: **Bacterial infectious disease control by vaccine development.** *J. Clin. Invest.* 2002, **110**:1061–1066.

Feldmann, M.: **Translating molecular insights in autoimmunity into effective therapy.** *Annu. Rev. Immunol.* 2009, **27**:1–27.

Kappe, S.H., Vaughan, A.M., Boddey, J.A., and Cowman, A.F.: **That was then but this is now: malaria research in the time of an eradication agenda.** *Science* 2010, **328**:862–866.

Kaufmann, S.H.: **Future vaccination strategies against tuberculosis: thinking outside the box.** *Immunity* 2010, **33**:567–577.

Korman, A.J., Peggs, K.S., and Allison, J.P.: **Checkpoint blockade in cancer immunotherapy.** *Adv. Immunol.* 2006, **90**:297–339.

Plett, P.C.: **Peter Plett and other discoverers of cowpox vaccination before Edward Jenner.** *Sudhoffs Arch.* 2006, **90**:219–232. [in German]

Ulmer, J.B., and Liu, M.A.: **Ethical issues for vaccines and immunization.** *Nat. Rev. Immunol.* 2002, **2**:291–296.

Virgin, H.W., and Walker, B.D.: **Immunology and the elusive AIDS vaccine.** *Nature* 2010, **464**:224–231.

Section references.

16-1 Corticosteroids are powerful anti-inflammatory drugs that alter the transcription of many genes.

Galon, J., Franchimont, D., Hiroi, N., Frey, G., Boettner, A., Ehrhart-Bornstein, M., O'Shea, J.J., Chrousos, G.P., and Bornstein, S.R.: **Gene profiling reveals unknown enhancing and suppressive actions of glucocorticoids on immune cells.** *FASEB J.* 2002, **16**:61–71.

Kampa, M., and Castanas, E.: **Membrane steroid receptor signaling in normal and neoplastic cells.** *Mol. Cell. Endocrinol.* 2006, **246**:76–82.

Löwenberg, M., Verhaar, A.P., van den Brink, G.R., and Hommes, D.W.: **Glucocorticoid signaling: a nongenomic mechanism for T-cell immunosuppression.** *Trends Mol. Med.* 2007, **13**:158–163.

Rhen, T., and Cidlowski, J.A.: **Antiinflammatory action of glucocorticoids—new mechanisms for old drugs.** *N. Engl. J. Med.* 2005, **353**:1711–1723.

16-2 Cytotoxic drugs cause immunosuppression by killing dividing cells and have serious side-effects.

Aarbakke, J., Janka-Schaub, G., and Elion, G.B.: **Thiopurine biology and pharmacology.** *Trends Pharmacol. Sci.* 1997, **18**:3–7.

Allison, A.C., and Eugui, E.M.: **Mechanisms of action of mycophenolate mofetil in preventing acute and chronic allograft rejection.** *Transplantation* 2005, **80** Suppl. S181–S190.

O'Donovan, P., Perrett, C.M., Zhang, X., Montaner, B., Xu, Y.Z., Harwood, C.A., McGregor, J.M., Walker, S.L., Hanaoka, F., and Karran, P.: **Azathioprine and UVA light generate mutagenic oxidative DNA damage.** *Science* 2005, **309**:1871–1874.

Taylor, A.L., Watson, C.J., and Bradley, J.A.: **Immunosuppressive agents in solid organ transplantation: mechanisms of action and therapeutic efficacy.** *Crit. Rev. Oncol. Hematol.* 2005, **56**:23–46.

Zhu, L.P., Cupps, T.R., Whalen, G., and Fauci, A.S.: **Selective effects of cyclophosphamide therapy on activation, proliferation, and differentiation of human B cells.** *J. Clin. Invest.* 1987, **79**:1082–1090.

16-3 Cyclosporin A, tacrolimus (FK506), and rapamycin (sirolimus) are powerful immunosuppressive agents that interfere with T-cell signaling.

Araki, K., Turner, A.P., Shaffer, V.O., Gangappa, S., Keller, S.A., Bachmann, M.F., Larsen, C.P., and Ahmed, R.: **mTOR regulates memory CD8 T-cell differentiation.** *Nature* 2009, **460**:108–112.

Battaglia, M., Stabilini, A., and Roncarolo, M.G.: **Rapamycin selectively expands CD4$^+$CD25$^+$FoxP3$^+$ regulatory T cells.** *Blood* 2005, **105**:4743–4748.

Bierer, B.E., Mattila, P.S., Standaert, R.F., Herzenberg, L.A., Burakoff, S.J., Crabtree, G., and Schreiber, S.L.: **Two distinct signal transmission pathways in T lymphocytes are inhibited by complexes formed between an immunophilin and either FK506 or rapamycin.** *Proc. Natl Acad. Sci. USA* 1990, **87**:9231–9235.

Crabtree, G.R.: **Generic signals and specific outcomes: signaling through Ca^{2+}, calcineurin, and NF-AT.** *Cell* 1999, **96**:611–614.

Crespo, J.L., and Hall, M.N.: **Elucidating TOR signaling and rapamycin action: lessons from *Saccharomyces cerevisiae*.** *Microbiol. Mol. Biol. Rev.* 2002, **66**:579–591.

16-4 Antibodies against cell-surface molecules can be used to eliminate lymphocyte subsets or to inhibit lymphocyte function.

Graca, L., Le Moine, A., Cobbold, S.P., and Waldmann, H.: **Antibody-induced transplantation tolerance: the role of dominant regulation.** *Immunol. Res.* 2003, **28**:181–191.

Waldmann, H., and Hale, G.: **CAMPATH: from concept to clinic.** *Phil. Trans. R. Soc. Lond. B* 2005, **360**:1707–1711.

16-5 Antibodies can be engineered to reduce their immunogenicity in humans.

Kim, S.J., Park, Y., and Hong, H.J.: **Antibody engineering for the development of therapeutic antibodies.** *Mol. Cell* 2005, **20**:17–29.

Liu, X.Y., Pop, L.M., and Vitetta, E.S.: **Engineering therapeutic monoclonal antibodies.** *Immunol. Rev.* 2008, **222**:9–27.

Smith, K., Garman, L., Wrammert, J., Zheng, N.Y., Capra, J.D., Ahmed, R., and Wilson, P.C.: **Rapid generation of fully human monoclonal antibodies specific to a vaccinating antigen.** *Nat. Protocols* 2009, **4**:372–384.

Traggiai, E., Becker, S., Subbarao, K., Kolesnikova, L., Uematsu, Y., Gismondo, M.R., Murphy, B.R., Rappuoli, R., and Lanzavecchia, A.: **An efficient method to make human monoclonal antibodies from memory B cells: potent neutralization of SARS coronavirus.** *Nat. Med.* 2004, **10**:871–875.

Winter, G., Griffiths, A.D., Hawkins, R.E., and Hoogenboom, H.R.: **Making antibodies by phage display technology.** *Annu. Rev. Immunol.* 1994, **12**:433–455.

16-6 Monoclonal antibodies can be used to prevent allograft rejection.

Kirk, A.D., Burkly, L.C., Batty, D.S., Baumgartner, R.E., Berning, J.D., Buchanan, K., Fechner, J.H., Jr, Germond, R.L., Kampen, R.L., Patterson, N.B., *et al.*: **Treatment with humanized monoclonal antibody against CD154 prevents acute renal allograft rejection in nonhuman primates.** *Nat. Med.* 1999, **5**:686–693.

Li, X.C., Strom, T.B., Turka, L.A., and Wells, A.D.: **T-cell death and transplantation tolerance.** *Immunity* 2001, **14**:407–416.

Londrigan, S.L., Sutherland, R.M., Brady, J.L., Carrington, E.M., Cowan, P.J., d'Apice, A.J., O'Connell, P.J., Zhan, Y., and Lew, A.M.: **In situ protection against islet allograft rejection by CTLA4Ig transduction.** *Transplantation* 2010, **90**:951–957.

Pham, P.T., Lipshutz, G.S., Pham, P.T., Kawahji, J., Singer, J.S., and Pham, P.C.: **The evolving role of alemtuzumab (Campath-1H) in renal transplantation.** *Drug Des. Dev. Ther.* 2009, **3**:41–49.

Sageshima, J., Ciancio, G., Chen, L., and Burke, G.W.: **Anti-interleukin-2 receptor antibodies—basiliximab and daclizumab—for the prevention of acute rejection in renal transplantation.** *Biologics* 2009, **3**:319–336.

Waldmann, H.: **Therapeutic approaches for transplantation.** *Curr. Opin. Immunol.* 2001, **13**:606–610.

16-7 Depletion of autoreactive lymphocytes can treat autoimmune disease.

Coles, A., Deans, J., and Compston, A.: **Campath-1H treatment of multiple sclerosis: lessons from the bedside for the bench.** *Clin. Neurol. Neurosurg.* 2004, **106**:270–274.

Edwards, J.C., Leandro, M.J., and Cambridge, G.: **B lymphocyte depletion in rheumatoid arthritis: targeting of CD20.** *Curr. Dir. Autoimmun.* 2005, **8**:175–192.

Rep, M.H., van Oosten, B.W., Roos, M.T., Ader, H.J., Polman, C.H., and van Lier, R.A.: **Treatment with depleting CD4 monoclonal antibody results in a preferential** loss of circulating naive T cells but does not affect IFN-γ-secreting T_H1 cells in humans. *J. Clin. Invest.* 1997, **99**:2225–2231.

Singh, R., Robinson, D.B., and El-Gabalawy, H.S.: **Emerging biologic therapies in rheumatoid arthritis: cell targets and cytokines.** *Curr. Opin. Rheumatol.* 2005, **17**:274–279.

Willis, F., Marsh, J.C., Bevan, D.H., Killick, S.B., Lucas, G., Griffiths, R., Ouwehand, W., Hale, G., Waldmann, H., and Gordon-Smith, E.C.: **The effect of treatment with Campath-1H in patients with autoimmune cytopenias.** *Br. J. Haematol.* 2001, **114**:891–898.

Yazawa, N., Hamaguchi, Y., Poe, J.C., and Tedder, T.F.: **Immunotherapy using unconjugated CD19 monoclonal antibodies in animal models for B lymphocyte malignancies and autoimmune disease.** *Proc. Natl Acad. Sci. USA* 2005, **102**:15178–15783.

Zaja, F., De Vita, S., Mazzaro, C., Sacco, S., Damiani, D., De Marchi, G., Michelutti, A., Baccarani, M., Fanin, R., and Ferraccioli, G.: **Efficacy and safety of rituximab in type II mixed cryoglobulinemia.** *Blood* 2003, **101**:3827–3834.

16-8 Biologic agents that block TNF-α or IL-1 can alleviate autoimmune diseases.

Guarda, G., Braun, M., Staehli, F., Tardivel, A., Mattmann, C., Förster, I., Farlik, M., Decker, T., Du Pasquier, R.A., Romero, P., *et al.*: **Type I interferon inhibits interleukin-1 production and inflammasome activation.** *Immunity* 2011, **34**:213–223.

Feldmann, M., and Maini, R.N.: **Lasker Clinical Medical Research Award. TNF defined as a therapeutic target for rheumatoid arthritis and other autoimmune diseases.** *Nat. Med.* 2003, **9**:1245–1250.

Hallegua, D.S., and Weisman, M.H.: **Potential therapeutic uses of interleukin 1 receptor antagonists in human diseases.** *Ann. Rheum. Dis.* 2002, **61**:960–967.

Mackay, C.R.: **New avenues for anti-inflammatory therapy.** *Nat. Med.* 2002, **8**:117–118.

Sandborn, W.J., and Targan, S.R.: **Biologic therapy of inflammatory bowel disease.** *Gastroenterology* 2002, **122**:1592–1608.

16-9 Biologic agents can block cell migration to sites of inflammation and reduce immune responses.

Cyster, J.G.: **Chemokines, sphingosine-1-phosphate, and cell migration in secondary lymphoid organs.** *Annu. Rev. Immunol.* 2005, **23**:127–159.

Idzko, M., Hammad, H., van Nimwegen, M., Kool, M., Muller, T., Soullie, T., Willart, M.A., Hijdra, D., Hoogsteden, H.C., and Lambrecht, B.N.: **Local application of FTY720 to the lung abrogates experimental asthma by altering dendritic cell function.** *J. Clin. Invest.* 2006, **116**:2935–2944.

Kappos, L., Radue, E.W., O'Connor, P., Polman, C., Hohlfeld, R., Calabresi, P., Selmaj, K., Agoropoulou, C., Leyk, M., Zhang-Auberson, L., *et al.*: **A placebo-controlled trial of oral fingolimod in relapsing multiple sclerosis.** *N. Engl. J. Med.* 2010, **362**:387–401.

Miller, D.H., Khan, O.A., Sheremata, W.A., Blumhardt, L.D., Rice, G.P., Libonati, M.A., Willmer-Hulme, A.J., Dalton, C.M., Miszkiel, K.A., and O'Connor, P.W.: **A controlled trial of natalizumab for relapsing multiple sclerosis.** *N. Engl. J. Med.* 2003, **348**:15–23.

Podolsky, D.K.: **Selective adhesion-molecule therapy and inflammatory bowel disease—a tale of Janus?** *N. Engl. J. Med.* 2005, **353**:1965–1968.

16-10 Blockade of co-stimulatory pathways that activate lymphocytes can be used to treat autoimmune disease.

Abrams, J.R., Kelley, S.L., Hayes, E., Kikuchi, T., Brown, M.J., Kang, S., Lebwohl, M.G., Guzzo, C.A., Jegasothy, B.V., Linsley, P.S., *et al.*: **Blockade of T lymphocyte costimulation with cytotoxic T lymphocyte-associated antigen 4-immunoglobulin (CTLA4Ig) reverses the cellular pathology of psoriatic plaques, including the activation of keratinocytes, dendritic cells, and endothelial cells.** *J. Exp. Med.* 2000, **192**:681–694.

Ellis, C.N., and Krueger, G.G.: **Treatment of chronic plaque psoriasis by selective targeting of memory effector T lymphocytes.** *N. Engl. J. Med.* 2001, **345**:248–255.

Kraan, M.C., van Kuijk, A.W., Dinant, H.J., Goedkoop, A.Y., Smeets, T.J., de Rie, M.A., Dijkmans, B.A., Vaishnaw, A.K., Bos, J.D., and Tak, P.P.: **Alefacept treatment in psoriatic arthritis: reduction of the effector T cell population in peripheral blood and synovial tissue is associated with improvement of clinical signs of arthritis.** *Arthritis Rheum.* 2002, **46**:2776–2784.

Lowes, M.A., Chamian, F., Abello, M.V., Fuentes-Duculan, J., Lin, S.L., Nussbaum, R., Novitskaya, I., Carbonaro, H., Cardinale, I., Kikuchi, T., *et al.*: **Increase in TNF-α and inducible nitric oxide synthase-expressing dendritic cells in psoriasis and reduction with efalizumab (anti-CD11a).** *Proc. Natl Acad. Sci. USA* 2005, **102**:19057–19062.

Masharani, U.B., and Becker, J.: **Teplizumab therapy for type 1 diabetes.** *Expert Opin. Biol. Ther.* 2010, **10**:459–465.

16-11 Some commonly used drugs have immunomodulatory properties.

Baeke, F., Takiishi, T., Korf, H., Gysemans, C., and Mathieu, C.: **Vitamin D: modulator of the immune system.** *Curr. Opin. Pharmacol.* 2010, **10**:482–496.

Youssef, S., Stuve, O., Patarroyo, J.C., Ruiz, P.J., Radosevich, J.L., Hur, E.M., Bravo, M., Mitchell, D.J., Sobel, R.A., Steinman, L., *et al.*: **The HMG-CoA reductase inhibitor, atorvastatin, promotes a Th2 bias and reverses paralysis in central nervous system autoimmune disease.** *Nature* 2002, **420**:78–84.

16-12 Controlled administration of antigen can be used to manipulate the nature of an antigen-specific response.

Diabetes Prevention Trial: Type 1 Diabetes Study Group: **Effects of insulin in relatives of patients with type 1 diabetes mellitus.** *N. Engl. J. Med.* 2002, **346**:1685–1691.

Magee, C.C., and Sayegh, M.H.: **Peptide-mediated immunosuppression.** *Curr. Opin. Immunol.* 1997, **9**:669–675.

Mowat, A.M., Parker, L.A., Beacock-Sharp, H., Millington, O.R., and Chirdo, F.: **Oral tolerance: overview and historical perspectives.** *Ann. N.Y. Acad. Sci.* 2004, **1029**:1–8.

Steinman, L., Utz, P.J., and Robinson, W.H.: **Suppression of autoimmunity via microbial mimics of altered peptide ligands.** *Curr. Top. Microbiol. Immunol.* 2005, **296**:55–63.

Weiner, H.L.: **Oral tolerance for the treatment of autoimmune diseases.** *Annu. Rev. Med.* 1997, **48**:341–351.

16-13 The development of transplantable tumors in mice led to the discovery of protective immune responses to tumors.

Jaffee, E.M., and Pardoll, D.M.: **Murine tumor antigens: is it worth the search?** *Curr. Opin. Immunol.* 1996, **8**:622–627.

Klein, G.: **The strange road to the tumor-specific transplantation antigens (TSTAs).** *Cancer Immun.* 2001, **1**:6.

16-14 Tumors are 'edited' by the immune system as they evolve and can escape rejection in many ways.

Ahmadzadeh, M., and Rosenberg, S.A.: **IL-2 administration increases $CD4^+CD25^{hi}Foxp3^+$ regulatory T cells in cancer patients.** *Blood* 2006, **107**:2409–2414.

Belladonna, M.L., Puccetti, P., Orabona, C., Fallarino, F., Vacca, C., Volpi, C., Gizzi, S., Pallotta, M.T., Fioretti, M.C., and Grohmann, U.: **Immunosuppression via tryptophan catabolism: the role of kynurenine pathway enzymes.** *Transplantation* 2007, **84** Suppl. **1**:S17–S20.

Bodmer, W.F., Browning, M.J., Krausa, P., Rowan, A., Bicknell, D.C., and Bodmer, J.G.: **Tumor escape from immune response by variation in HLA expression and other mechanisms.** *Ann N.Y. Acad. Sci.* 1993, **690**:42–49.

Dunn, G.P., Old, L.J., and Schreiber, R.D.: **The immunobiology of cancer immunosurveillance and immunoediting.** *Immunity* 2004, **21**:137–148.

Gajewski,T.F., Meng, Y., Blank, C., Brown, I., Kacha, A., Kline, J., and Harlin, H.: **Immune resistance orchestrated by the tumor microenvironment.** *Immunol. Rev.* 2006, **213**:131–145.

Girardi, M., Oppenheim, D.E., Steele, C.R., Lewis, J.M., Glusac, E., Filler, R., Hobby, P., Sutton, B., Tigelaar, R.E., and Hayday, A.C.: **Regulation of cutaneous malignancy by γδ T cells.** *Science* 2001, **294**:605–609.

Ikeda, H., Lethe, B., Lehmann, F., van Baren, N., Baurain, J.F., de Smet, C., Chambost, H., Vitale, M., Moretta, A., Boon, T., *et al.*: **Characterization of an antigen that is recognized on a melanoma showing partial HLA loss by CTL expressing an NK inhibitory receptor.** *Immunity* 1997, **6**:199–208.

Koebel, C.M., Vermi, W., Swann, J.B., Zerafa, N., Rodig, S.J., Old, L.J., Smyth, M.J., and Schreiber, R.D.: **Adaptive immunity maintains occult cancer in an equilibrium state.** *Nature* 2007, **450**:903–907.

Koopman, L.A., Corver, W.E., van der Slik, A.R., Giphart, M.J., and Fleuren, G.J.: **Multiple genetic alterations cause frequent and heterogeneous human histocompatibility leukocyte antigen class I loss in cervical cancer.** *J. Exp. Med.* 2000, **191**:961–976.

Munn, D.H., and Mellor, A.L.: **Indoleamine 2,3-dioxygenase and tumor-induced tolerance.** *J. Clin. Invest.* 2007, **117**:1147–1154.

Ochsenbein, A.F., Sierro, S., Odermatt, B., Pericin, M., Karrer, U., Hermans, J., Hemmi, S., Hengartner, H., and Zinkernagel, R.M.: **Roles of tumour localization, second signals and cross priming in cytotoxic T-cell induction.** *Nature* 2001, **411**:1058–1064.

Peggs, K.S., Quezada, S.A., and Allison, J.P.: **Cell intrinsic mechanisms of T-cell inhibition and application to cancer therapy.** *Immunol. Rev.* 2008, **224**:141–165.

Peranzoni, E., Zilio, S., Marigo, I., Dolcetti, L., Zanovello, P., Mandruzzato, S., and Bronte, V.: **Myeloid-derived suppressor cell heterogeneity and subset definition.** *Curr. Opin. Immunol.* 2010, **22**:238–244.

Shroff, R., and Rees, L.: **The post-transplant lymphoproliferative disorder—a literature review.** *Pediatr. Nephrol.* 2004, **19**:369–377.

Tada, T., Ohzeki, S., Utsumi, K., Takiuchi, H., Muramatsu, M., Li, X.F., Shimizu, J., Fujiwara, H., and Hamaoka, T.: **Transforming growth factor-β-induced inhibition of T cell function. Susceptibility difference in T cells of various phenotypes and functions and its relevance to immunosuppression in the tumor-bearing state.** *J. Immunol.* 1991, **146**:1077–1082.

Wang, H.Y., Lee, D.A., Peng, G., Guo, Z., Li, Y., Kiniwa, Y., Shevach, E.M., and Wang, R.F.: **Tumor-specific human CD4⁺ regulatory T cells and their ligands: implications for immunotherapy.** *Immunity* 2004, **20**:107–118.

16-15 Tumor-specific antigens can be recognized by T cells and form the basis of immunotherapies.

Chaux, P., Vantomme, V., Stroobant, V., Thielemans, K., Corthals, J., Luiten, R., Eggermont, A.M., Boon, T., and van der Bruggen, P.: **Identification of MAGE-3 epitopes presented by HLA-DR molecules to CD4⁺ T lymphocytes.** *J. Exp. Med.* 1999, **189**:767–778.

Clark, R.E., Dodi, I.A., Hill, S.C., Lill, J.R., Aubert, G., Macintyre, A.R., Rojas, J., Bourdon, A., Bonner, P.L., Wang, L., *et al.*: **Direct evidence that leukemic cells present HLA-associated immunogenic peptides derived from the BCR-ABL b3a2 fusion protein.** *Blood* 2001, **98**:2887–2893.

Comoli, P., Pedrazzoli, P., Maccario, R., Basso, S., Carminati, O., Labirio, M., Schiavo, R., Secondino, S., Frasson, C., Perotti, C., *et al.*: **Cell therapy of Stage IV nasopharyngeal carcinoma with autologous Epstein–Barr virus-targeted cytotoxic T lymphocytes.** *J. Clin. Oncol.* 2005, **23**:8942–8949.

Disis, M.L., and Cheever, M.A.: **HER-2/neu oncogenic protein: issues in vaccine development.** *Crit. Rev. Immunol.* 1998, **18**:37–45.

Dudley, M.E., Wunderlich, J.R., Yang, J.C., Sherry, R.M., Topalian, S.L., Restifo, N.P., Royal, R.E., Kammula, U., White, D.E., Mavroukakis, S.A., *et al.*: **Adoptive cell transfer therapy following non-myeloablative but lymphodepleting chemotherapy for the treatment of patients with refractory metastatic melanoma.** *J. Clin. Oncol.* 2005, **23**:2346–2357.

Michalek, J., Collins, R.H., Durrani, H.P., Vaclavkova, P., Ruff, L.E., Douek, D.C., and Vitetta, E.S.: **Definitive separation of graft-versus-leukemia- and graft-versus-host-specific CD4⁺ T cells by virtue of their receptor β loci sequences.** *Proc. Natl Acad. Sci. USA* 2003, **100**:1180–1184.

Morris, E.C., Tsallios, A., Bendle, G.M., Xue, S.A., and Stauss, H.J.: **A critical role of T cell antigen receptor-transduced MHC class I-restricted helper T cells in tumor protection.** *Proc. Natl Acad. Sci. USA* 2005, **102**:7934–7939.

Van Der Bruggen, P., Zhang, Y., Chaux, P., Stroobant, V., Panichelli, C., Schultz, E.S., Chapiro, J., Van Den Eynde, B.J., Brasseur, F., and Boon, T.: **Tumor-specific shared antigenic peptides recognized by human T cells.** *Immunol. Rev.* 2002, **188**:51–64.

16-16 Monoclonal antibodies against tumor antigens, alone or linked to toxins, can control tumor growth.

Bagshawe, K.D., Sharma, S.K., Burke, P.J., Melton, R.G., and Knox, R.J.: **Developments with targeted enzymes in cancer therapy.** *Curr. Opin. Immunol.* 1999, **11**:579–583.

Cragg, M.S., French, R.R., and Glennie, M.J.: **Signaling antibodies in cancer therapy.** *Curr. Opin. Immunol.* 1999, **11**:541–547.

Fan, Z., and Mendelsohn, J.: **Therapeutic application of anti-growth factor receptor antibodies.** *Curr. Opin. Oncol.* 1998, **10**:67–73.

Hortobagyi, G.N.: **Trastuzumab in the treatment of breast cancer.** *N. Engl. J. Med.* 2005, **353**:1734–1736.

Houghton, A.N., and Scheinberg, D.A.: **Monoclonal antibody therapies 'constant' threat to cancer.** *Nat. Med.* 2000, **6**:373–374.

Kreitman, R.J., Wilson, W.H., Bergeron, K., Raggio, M., Stetler-Stevenson, M., FitzGerald, D.J., and Pastan, I.: **Efficacy of the anti-CD22 recombinant immunotoxin BL22 in chemotherapy-resistant hairy-cell leukemia.** *N. Engl. J. Med.* 2001, **345**:241–247.

Park, S., Jiang, Z., Mortenson, E.D., Deng, L., Radkevich-Brown, O., Yang, X., Sattar, H., Wang, Y., Brown, N.K., Greene, M., *et al.*: **The therapeutic effect of anti-HER2/neu antibody depends on both innate and adaptive immunity.** *Cancer Cell* 2010, **18**:160–170.

Tol, J., and Punt, C.J.: **Monoclonal antibodies in the treatment of metastatic colorectal cancer: a review.** *Clin. Ther.* 2010, **32**:437–453.

White, C.A., Weaver, R.L., and Grillo-Lopez, A.J.: **Antibody-targeted immunotherapy for treatment of malignancy.** *Annu. Rev. Med.* 2001, **52**:125–145.

16-17 Enhancing the immune response to tumors by vaccination holds promise for cancer prevention and therapy.

Kenter, G.G., Welters, M.J., Valentijn, A.R., Lowik, M.J., Berends-van der Meer, D.M., Vloon, A.P., Essahsah, F., Fathers, L.M., Offringa, R., Drijfhout, J.W., *et al.*: **Vaccination against HPV-16 oncoproteins for vulvar intraepithelial neoplasia.** *N. Engl. J. Med.* 2009, **361**:1838–1847.

Kugler, A., Stuhler, G., Walden, P., Zoller, G., Zobywalski, A., Brossart, P., Trefzer, U., Ullrich, S., Muller, C.A., Becker, V., *et al.*: **Regression of human metastatic renal cell carcinoma after vaccination with tumor cell-dendritic cell hybrids.** *Nat. Med.* 2000, **6**:332–336.

Mao, C., Koutsky, L.A., Ault, K.A., Wheeler, C.M., Brown, D.R., Wiley, D.J., Alvarez, F.B., Bautista, O.M., Jansen, K.U., and Barr, E.: **Efficacy of human papillomavirus-16 vaccine to prevent cervical intraepithelial neoplasia: a randomized controlled trial.** *Obstet. Gynecol.* 2006, **107**:18–27.

Murphy, A., Westwood, J.A., Teng, M.W., Moeller, M., Darcy, P.K., and Kershaw, M.H.: **Gene modification strategies to induce tumor immunity.** *Immunity* 2005, **22**:403–414.

Palucka, K., Ueno, H., Fay, J., and Banchereau, J.: **Dendritic cells and immunity against cancer.** *J. Intern. Med.* 2011, **269**:64–73.

Pardoll, D.M.: **Paracrine cytokine adjuvants in cancer immunotherapy.** *Annu. Rev. Immunol.* 1995, **13**:399–415.

Przepiorka, D., and Srivastava, P.K.: **Heat shock protein–peptide complexes as immunotherapy for human cancer.** *Mol. Med. Today* 1998, **4**:478–484.

Vambutas, A., DeVoti, J., Nouri, M., Drijfhout, J.W., Lipford, G.B., Bonagura, V.R., van der Burg, S.H., and Melief, C.J.: **Therapeutic vaccination with papillomavirus E6 and E7 long peptides results in the control of both established virus-induced lesions and latently infected sites in a pre-clinical cottontail rabbit papillomavirus model.** *Vaccine* 2005, **23**:5271–5280.

16-18 Checkpoint blockade can augment immune responses to existing tumors.

Bendandi, M., Gocke, C.D., Kobrin, C.B., Benko, F.A., Sternas, L.A., Pennington, R., Watson, T.M., Reynolds, C.W., Gause, B.L., Duffey, P.L., *et al.*: **Complete molecular remissions induced by patient-specific vaccination plus granulocyte-monocyte colony-stimulating factor against lymphoma.** *Nat. Med.* 1999, **5**:1171–1177.

Egen, J.G., Kuhns, M.S., and Allison, J.P.: **CTLA-4: new insights into its biological function and use in tumor immunotherapy.** *Nat. Immunol.* 2002, **3**:611–618.

Li, Y., Hellstrom, K.E., Newby, S.A., and Chen, L.: **Costimulation by CD48 and B7-1 induces immunity against poorly immunogenic tumors.** *J. Exp. Med.* 1996, **183**:639–644.

Phan, G.Q., Yang, J.C., Sherry, R.M., Hwu, P., Topalian, S.L., Schwartzentruber, D.J., Restifo, N.P., Haworth, L.R., Seipp, C.A., Freezer, L.J., *et al.*: **Cancer regression and autoimmunity induced by cytotoxic T lymphocyte-associated antigen 4 blockade in patients with metastatic melanoma.** *Proc. Natl Acad. Sci. USA* 2003, **100**:8372–8377.

Yuan, J., Gnjatic, S., Li, H., Powel, S., Gallardo, H.F., Ritter, E., Ku, G.Y., Jungbluth, A.A., Segal, N.H., Rasalan, T.S., *et al.*: **CTLA-4 blockade enhances polyfunctional NY-ESO-1 specific T cell responses in metastatic melanoma patients with clinical benefit.** *Proc. Natl Acad. Sci. USA* 2008, **105**:20410–20415.

16-19 Vaccines can be based on attenuated pathogens or material from killed organisms.

Anderson, R.M., Donnelly, C.A., and Gupta, S.: **Vaccine design, evaluation, and community-based use for antigenically variable infectious agents.** *Lancet* 1997, **350**:1466–1470.

Rabinovich, N.R., McInnes, P., Klein, D.L., and Hall, B.F.: **Vaccine technologies: view to the future.** *Science* 1994, **265**:1401–1404.

16-20 Most effective vaccines generate antibodies that prevent the damage caused by toxins or that neutralize the pathogen and stop infection.

Levine, M.M., and Levine, O.S.: **Influence of disease burden, public perception, and other factors on new vaccine development, implementation, and continued use.** *Lancet* 1997, **350**:1386–1392.

Mouque, H., Scheid, J.F., Z **of anti-HIV antibodies by heteroligation.** *Nature* 2010, **467**:591–595.

Nichol, K.L., Lind, A., Margolis, K.L., Murdoch, M., McFadden, R., Hauge, M., Palese, P., and Garcia-Sastre, A.: **Influenza vaccines: present and future.** *J. Clin. Invest.* 2002, **110**:9–13.

16-21 Effective vaccines must induce long-lasting protection while being safe and inexpensive.

Gupta, R.K., Best, J., and MacMahon, E.: **Mumps and the UK epidemic 2005.** *BMJ* 2005, **330**:1132–1135.

Magnan, S., and Drake, M.: **The effectiveness of vaccination against influenza in healthy, working adults.** *N. Engl. J. Med.* 1995, **333**:889–893.

16-22 Live-attenuated viral vaccines are usually more potent than 'killed' vaccines and can be made safer by the use of recombinant DNA technology.

Brochier, B., Kieny, M.P., Costy, F., Coppens, P., Bauduin, B., Lecocq, J.P., Languet, B., Chappuis, G., Desmettre, P., Afiademanyo, K., *et al.*: **Large-scale eradication of rabies using recombinant vaccinia–rabies vaccine.** *Nature* 1991, **354**:520–522.

Mueller, S.N., Langley, W.A., Carnero, E., García-Sastre, A., and Ahmed, R.: **Immunization with live attenuated influenza viruses that express altered NS1 proteins results in potent and protective memory CD8⁺ T-cell responses.** *J. Virol.* 2010, **84**:1847–1855.

Murphy, B.R., and Collins, P.L.: **Live-attenuated virus vaccines for respiratory syncytial and parainfluenza viruses: applications of reverse genetics.** *J. Clin. Invest.* 2002, **110**:21–27.

Parkin, N.T., Chiu, P., and Coelingh, K.: **Genetically engineered live attenuated influenza A virus vaccine candidates.** *J. Virol.* 1997, **71**:2772–2778.

Pena, L., Vincent, A.L., Ye, J., Ciacci-Zanella, J.R., Angel, M., Lorusso, A., Gauger, P.C., Janke, B.H., Loving, C.L., and Perez, D.R.: **Modifications in the polymerase genes of a swine-like triple-reassortant influenza virus to generate live attenuated vaccines against 2009 pandemic H1N1 viruses.** *J. Virol.* 2011, **85**:456–469.

Subbarao, K., Murphy, B.R., and Fauci, A.S.: **Development of effective vaccines against pandemic influenza.** *Immunity* 2006, **24**:5–9.

16-23 Live-attenuated vaccines can be developed by selecting nonpathogenic or disabled bacteria or by creating genetically attenuated parasites (GAPs).

Grode, L., Seiler, P., Baumann, S., Hess, J., Brinkmann, V., Nasser Eddine, A., Mann, P., Goosmann, C., Bandermann, S., Smith, D., *et al.*: **Increased vaccine efficacy against tuberculosis of recombinant *Mycobacterium bovis* bacille Calmette–Guérin mutants that secrete listeriolysin.** *J. Clin. Invest.* 2005, **115**:2472–2479.

Guleria, I., Teitelbaum, R., McAdam, R.A., Kalpana, G., Jacobs, W.R., Jr, and Bloom, B.R.: **Auxotrophic vaccines for tuberculosis.** *Nat. Med.* 1996, **2**:334–337.

Labaied, M., Harupa, A., Dumpit, R.F., Coppens, I., Mikolajczak, S.A., and Kappe, S.H.: ***Plasmodium yoelii* sporozoites with simultaneous deletion of P52 and P36 are completely attenuated and confer sterile immunity against infection.** *Infect. Immun.* 2007, **75**:3758–3768.

Martin, C.: **The dream of a vaccine against tuberculosis; new vaccines improving or replacing BCG?** *Eur. Respir. J.* 2005, **26**:162–167.

Thaiss, C.A., and Kaufmann, S.H.: **Toward novel vaccines against tuberculosis: current hopes and obstacles.** *Yale J. Biol. Med.* 2010, **83**:209–215.

Vaughan, A.M., Wang, R., and Kappe, S.H.: **Genetically engineered, attenuated whole-cell vaccine approaches for malaria.** *Hum. Vaccines* 2010, **6**:1–8.

16-24 The route of vaccination is an important determinant of success.

Amorij, J.P., Hinrichs, W.Lj., Frijlink, H.W., Wilschut, J.C., and Huckriede, A.: **Needle-free influenza vaccination.** *Lancet Infect. Dis.* 2010, **10**:699–711.

Belyakov, I.M., and Ahlers, J.D.: **What role does the route of immunization play in the generation of protective immunity against mucosal pathogens?** *J. Immunol.* 2009, **183**:6883–6892.

Douce, G., Fontana, M., Pizza, M., Rappuoli, R., and Dougan, G.: **Intranasal immunogenicity and adjuvanticity of site-directed mutant derivatives of cholera toxin.** *Infect. Immun.* 1997, **65**:2821–2828.

Dougan, G., Ghaem-Maghami, M., Pickard, D., Frankel, G., Douce, G., Clare, S., Dunstan, S., and Simmons, C.: **The immune responses to bacterial antigens encountered in vivo at mucosal surfaces.** *Phil. Trans. R. Soc. Lond. B* 2000, **355**:705–712.

Eriksson, K., and Holmgren, J.: **Recent advances in mucosal vaccines and adjuvants.** *Curr. Opin. Immunol.* 2002, **14**:666–672.

16-25 *Bordetella pertussis* vaccination illustrates the importance of the perceived safety of a vaccines.

Decker, M.D., and Edwards, K.M.: **Acellular pertussis vaccines.** *Pediatr. Clin. North Am.* 2000, **47**:309–335.

Madsen, K.M., Hviid, A., Vestergaard, M., Schendel, D., Wohlfahrt, J., Thorsen, P., Olsen, J., and Melbye, M.: **A population-based study of measles, mumps, and rubella vaccination and autism.** *N. Engl. J. Med.* 2002, **347**:1477–1482.

Mortimer, E.A.: **Pertussis vaccines**, in Plotkin, S.A., and Mortimer, E.A. (eds): *Vaccines*, 2nd ed. Philadelphia, W.B. Saunders Co., 1994.

Poland, G.A.: **Acellular pertussis vaccines: new vaccines for an old disease.** *Lancet* 1996, **347**:209–210.

16-26 Conjugate vaccines have been developed as a result of understanding how T and B cells collaborate in an immune response.

Bröker, M., Dull, P.M., Rappuoli, R., and Costantino, P.: **Chemistry of a new investigational quadrivalent meningococcal conjugate vaccine that is immunogenic at all ages.** *Vaccine* 2009, **27**:5574–5580.

Levine, O.S., Knoll, M.D., Jones, A., Walker, D.G., Risko, N., and Gilani, Z.: **Global status of *Haemophilus influenzae* type b and pneumococcal conjugate vaccines: evidence, policies, and introductions.** *Curr. Opin. Infect. Dis.* 2010, **23**:236–241.

Peltola, H., Kilpi, T., and Anttila, M.: **Rapid disappearance of *Haemophilus influenzae* type b meningitis after routine childhood immunisation with conjugate vaccines.** *Lancet* 1992, **340**:592–594.

Rappuoli, R.: **Conjugates and reverse vaccinology to eliminate bacterial meningitis.** *Vaccine* 2001, **19**:2319–2322.

16-27 Peptide-based vaccines can elicit protective immunity, but they require adjuvants and must be targeted to the appropriate cells and cell compartment to be effective.

Alonso, P.L., Sacarlal, J., Aponte, J.J., Leach, A., Macete, E., Aide, P., Sigauque, B., Milman, J., Mandomando, I., Bassat, Q., *et al.*: **Duration of protection with RTS,S/AS02A malaria vaccine in prevention of *Plasmodium falciparum* disease in Mozambican children: single-blind extended follow-up of a randomised controlled trial.** *Lancet* 2005, **366**:2012–2018.

Berzofsky, J.A.: **Epitope selection and design of synthetic vaccines. Molecular approaches to enhancing immunogenicity and cross-reactivity of engineered vaccines.** *Ann. N.Y. Acad. Sci.* 1993, **690**:256–264.

Davenport, M.P., and Hill, A.V.: **Reverse immunogenetics: from HLA-disease associations to vaccine candidates.** *Mol. Med. Today* 1996, **2**:38–45.

Hill, A.V.: **Pre-erythrocytic malaria vaccines: towards greater efficacy.** *Nat. Rev. Immunol.* 2006, **6**:21–32.

Hoffman, S.L., Rogers, W.O., Carucci, D.J., and Venter, J.C.: **From genomics to vaccines: malaria as a model system.** *Nat. Med.* 1998, **4**:1351–1353.

Ottenhoff, T.H., Doherty, T.M., Dissel, J.T., Bang, P., Lingnau, K., Kromann, I., and Andersen, P.: **First in humans: a new molecularly defined vaccine shows excellent safety and strong induction of long-lived *Mycobacterium tuberculosis*-specific Th1-cell like responses.** *Hum. Vaccin.* 2010, **6**:1007–1015.

Zwaveling, S., Ferreira Mota, S.C., Nouta, J., Johnson, M., Lipford, G.B., Offringa, R., van der Burg, S.H., and Melief, C.J.: **Established human papillomavirus type 16-expressing tumors are effectively eradicated following vaccination with long peptides.** *J. Immunol.* 2002, **169**:350–358.

16-28 Adjuvants are important for enhancing the immunogenicity of vaccines, but few are approved for use in humans.

Coffman, R.L., Sher, A., and Seder, R.A.: **Vaccine adjuvants: putting innate immunity to work.** *Immunity* 2010, **33**:492–503.

Hartmann, G., Weiner, G.J., and Krieg, A.M.: **CpG DNA: a potent signal for growth, activation, and maturation of human dendritic cells.** *Proc. Natl Acad. Sci. USA* 1999, **96**:9305–9310.

Palucka, K., Banchereau, J., and Mellman, I.: **Designing vaccines based on biology of human dendritic cell subsets.** *Immunity* 2010, **33**:464–478.

Persing, D.H., Coler, R.N., Lacy, M.J., Johnson, D.A., Baldridge, J.R., Hershberg, R.M., and Reed, S.G.: **Taking toll: lipid A mimetics as adjuvants and immunomodulators.** *Trends Microbiol.* 2002, **10**:S32–S37.

Pulendran, B.: **Modulating vaccine responses with dendritic cells and Toll-like receptors.** *Immunol. Rev.* 2004, **199**:227–250.

Takeda, K., Kaisho, T., and Akira, S.: **Toll-like receptors.** *Annu. Rev. Immunol.* 2003, **21**:335–376.

16-29 Protective immunity can be induced by DNA-based vaccination.

Donnelly, J.J., Ulmer, J.B., Shiver, J.W., and Liu, M.A.: **DNA vaccines.** *Annu. Rev. Immunol.* 1997, **15**:617–648.

Gurunathan, S., Klinman, D.M., and Seder, R.A.: **DNA vaccines: immunology, application, and optimization.** *Annu. Rev. Immunol.* 2000, **18**:927–974.

Nchinda, G., Kuroiwa, J., Oks, M., Trumpfheller, C., Park, C.G., Huang, Y., Hannaman, D., Schlesinger, S.J., Mizenina, O., Nussenzweig, M.C., *et al.*: **The efficacy of DNA vaccination is enhanced in mice by targeting the encoded protein to dendritic cells.** *J. Clin. Invest.* 2008, **118**:1427–1436.

Wolff, J.A., and Budker, V.: **The mechanism of naked DNA uptake and expression.** *Adv. Genet.* 2005, **54**:3–20.

16-30 The effectiveness of a vaccine can be enhanced by targeting it to sites of antigen presentation.

Bonifaz, L.C., Bonnyay, D.P., Charalambous, A., Darguste, D.I., Fujii, S., Soares, H., Brimnes, M.K., Moltedo, B., Moran, T.M., and Steinman, R.M.: ***In vivo* targeting of antigens to maturing dendritic cells via the DEC-205 receptor improves T cell vaccination.** *J. Exp. Med.* 2004, **199**:815–824.

Cheong, C., Choi, J.H., Vitale, L., He, L.Z., Trumpfheller, C., Bozzacco, L., Do, Y., Nchinda, G., Park, S.H., Dandamudi, D.B., *et al.*: **Improved cellular and humoral immune responses in vivo following targeting of HIV Gag to dendritic cells within human anti-human DEC205 monoclonal antibody.** *Blood* 2010, **116**:3828–3838.

Deliyannis, G., Boyle, J.S., Brady, J.L., Brown, L.E., and Lew, A.M.: **A fusion DNA vaccine that targets antigen-presenting cells increases protection from viral challenge.** *Proc. Natl Acad. Sci. USA* 2000, **97**:6676–6680.

Tan, M.C., Mommaas, A.M., Drijfhout, J.W., Jordens, R., Onderwater, J.J., Verwoerd, D., Mulder, A.A., van der Heiden, A.N., Scheidegger, D., Oomen, L.C., *et al.*: **Mannose receptor-mediated uptake of antigens strongly enhances HLA class II-restricted antigen presentation by cultured dendritic cells.** *Eur. J. Immunol.* 1997, **27**:2426–2435.

Thomson, S.A., Burrows, S.R., Misko, I.S., Moss, D.J., Coupar, B.E., and Khanna, R.: **Targeting a polyepitope protein incorporating multiple class II-restricted viral epitopes to the secretory/endocytic pathway facilitates immune recognition by CD4⁺ cytotoxic T lymphocytes: a novel approach to vaccine design.** *J. Virol.* 1998, **72**:2246–2252.

16-31 An important question is whether vaccination can be used therapeutically to control existing chronic infections.

Burke, R.L.: **Contemporary approaches to vaccination against herpes simplex virus.** *Curr. Top. Microbiol. Immunol.* 1992, **179**:137–158.

Grange, J.M., and Stanford, J.L.: **Therapeutic vaccines.** *J. Med. Microbiol.* 1996, **45**:81–83.

Hill, A., Jugovic, P., York, I., Russ, G., Bennink, J., Yewdell, J., Ploegh, H., and Johnson, D.: **Herpes simplex virus turns off the TAP to evade host immunity.** *Nature* 1995, **375**:411–415.

Lu, W., Arraes, L.C., Ferreira, W.T., and Andrieu, J.M.: **Therapeutic dendritic-cell vaccine for chronic HIV-1 infection.** *Nat. Med.* 2004, **10**:1359–1365.

Modlin, R.L.: **Th1–Th2 paradigm: insights from leprosy.** *J. Invest. Dermatol.* 1994, **102**:828–832.

Plebanski, M., Proudfoot, O., Pouniotis, D., Coppel, R.L., Apostolopoulos, V., and Flannery, G.: **Immunogenetics and the design of *Plasmodium falciparum* vaccines for use in malaria-endemic populations.** *J. Clin. Invest.* 2002, **110**:295–301.

Reiner, S.L., and Locksley, R.M.: **The regulation of immunity to *Leishmania major*.** *Annu. Rev. Immunol.* 1995, **13**:151–177.

Stanford, J.L.: **The history and future of vaccination and immunotherapy for leprosy.** *Trop. Geogr. Med.* 1994, **46**:93–107.

The Immunologist's Toolbox

APPENDIX I

Immunization.

Natural adaptive immune responses are normally directed at antigens borne by pathogenic microorganisms. The immune system can also be induced to respond to simple nonliving antigens, and experimental immunologists have focused on the responses to these simple antigens in developing our understanding of the immune response. The deliberate induction of an immune response is known as **immunization**. Experimental immunizations are routinely carried out by injecting the test antigen into the animal or human subject. The route, dose, and form in which antigen is administered can profoundly affect whether a response occurs and the type of response that is produced, and are considered in Sections A-1–A-4. The induction of protective immune responses against common microbial pathogens in humans is often called vaccination, although this term is correctly only applied to the induction of immune responses against smallpox by immunizing with the cross-reactive cowpox virus, vaccinia.

To determine whether an immune response has occurred and to follow its course, the immunized individual is monitored for the appearance of immune reactants directed at the specific antigen. Immune responses to most antigens elicit the production of both specific antibodies and specific effector T cells. Monitoring the antibody response usually involves the analysis of relatively crude preparations of **antiserum** (plural: **antisera**). The **serum** is the fluid phase of clotted blood, which, if taken from an immunized individual, is called antiserum because it contains specific antibodies against the immunizing antigen as well as other soluble serum proteins. To study immune responses mediated by T cells, blood lymphocytes or cells from lymphoid organs such as the spleen are tested; T-cell responses are more commonly studied in experimental animals than in humans.

Any substance that can elicit an immune response is said to be **immunogenic** and is called an **immunogen**. There is a clear operational distinction between an immunogen and an antigen. An antigen is defined as any substance that can bind to a specific antibody. All antigens therefore have the potential to elicit specific antibodies, but some need to be attached to an immunogen in order to do so. This means that although all immunogens are antigens, not all antigens are immunogenic. The antigens used most frequently in experimental immunology are proteins, and antibodies against proteins are of enormous utility in experimental biology and medicine. Purified proteins are, however, not always highly immunogenic, and to provoke an immune response they have to be administered with an adjuvant (see Section A-4). Carbohydrates, nucleic acids, and other types of molecules are all potential antigens, but will often only induce an immune response if attached to a protein carrier. Thus, the immunogenicity of protein antigens determines the outcome of virtually every immune response.

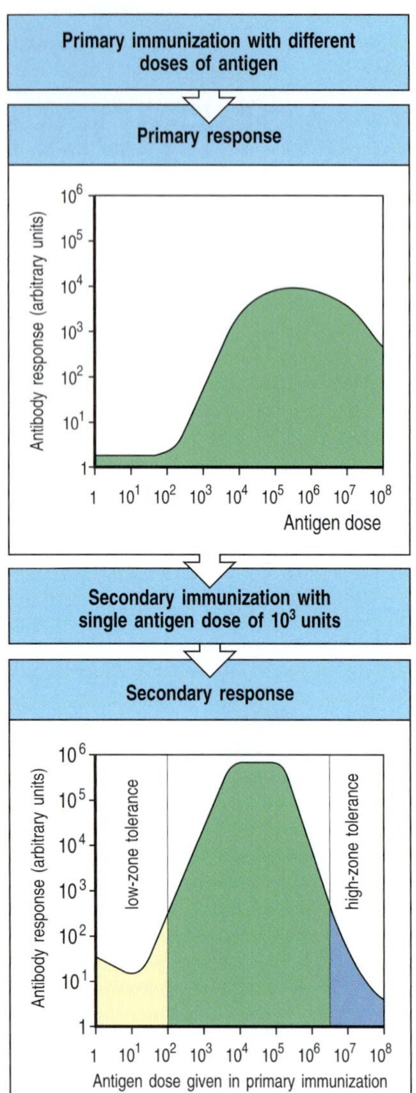

Fig. A.1 The dose of antigen used in an initial immunization affects the primary and secondary antibody response. The typical antigen dose–response curve shown here illustrates the influence of dose on both a primary antibody response (amounts of antibody produced expressed in arbitrary units) and the effect of the dose used for priming on a secondary antibody response elicited by a dose of antigen of 10^3 arbitrary mass units. Very low doses of antigen do not cause an immune response at all. Slightly higher doses seem to inhibit specific antibody production, an effect known as low-zone tolerance. Above these doses there is a steady increase in the response with antigen dose to reach a broad optimum. Very high doses of antigen also inhibit immune responsiveness to a subsequent challenge, a phenomenon known as high-zone tolerance.

Antisera generated by immunization with even the simplest antigen will contain many different antibody molecules that bind to the immunogen in slightly different ways. Some of the antibodies in an antiserum are cross-reactive. A **cross-reaction** is defined as the binding of an antibody to an antigen other than the immunogen; most antibodies cross-react with closely related antigens but, on occasion, some bind antigens with no clear relationship to the immunogen. These cross-reacting antibodies can create problems when the antiserum is used to detect a specific antigen. They can be removed from an antiserum by **absorption** with the cross-reactive antigen, leaving behind the antibodies that bind only to the immunogen. Absorption can be performed by affinity chromatography using immobilized antigen, a technique that is also used for the purification of antibodies or antigens (see Section A-5). Most problems of cross-reactivity can be avoided, however, by making monoclonal antibodies (see Section A-12).

Although almost any structure can be recognized by antibody as an antigen, usually only proteins elicit fully developed adaptive immune responses. This is because proteins have the ability to engage T cells, which contribute to inducing most antibody responses and are required for immunological memory. Proteins engage T cells because the T cells recognize antigens as peptide fragments of proteins bound to major histocompatibility complex (MHC) molecules. An adaptive immune response that includes immunological memory can be induced by nonpeptide antigens only when they are attached to a protein carrier that can engage the necessary T cells (see Section 10-3 and Fig. 10.5).

Immunological memory is produced as a result of the initial or **primary immunization**, which evokes the **primary immune response**. This is also known as **priming**, because the animal or person is now 'primed' like a pump to mount a more potent response to subsequent challenges with the same antigen. The response to each immunization is increasingly intense, so that **secondary**, **tertiary**, and subsequent responses are of increasing magnitude (Fig. A.1). Repetitive challenge with antigen to achieve a heightened state of immunity is known as **hyperimmunization**.

Certain properties of a protein that favor the priming of an adaptive immune response have been defined by studying antibody responses to simple natural proteins such as hen egg-white lysozyme and to synthetic polypeptide antigens (Fig. A.2). The larger and more complex a protein, and the more distant its relationship to self proteins, the more likely it is to elicit a response. This is because such responses depend on the proteins being degraded into peptides that can bind to MHC molecules, and on the subsequent recognition of these peptide:MHC complexes by T cells (see Chapter 6). The larger and more distinct the protein antigen, the more likely it is to contain such peptides. Particulate or aggregated antigens are more immunogenic because they are taken up more efficiently by the specialized antigen-presenting cells responsible for initiating a response. Indeed, small soluble proteins are unable to induce a response unless they are made to aggregate in some way. Many vaccines, for example, use aggregated protein antigens to potentiate the immune response.

A-1 Haptens.

Small organic molecules of simple structure, such as phenyl arsonates and nitrophenyls, do not provoke antibodies when injected by themselves. However, antibodies can be raised against them if the molecule is attached covalently, by simple chemical reactions, to a protein carrier. Such small molecules were termed **haptens** (from the Greek *haptein*, to fasten) by the immunologist Karl Landsteiner, who first studied them in the early 20th

Factors that influence the immunogenicity of proteins		
Parameter	**Increased immunogenicity**	**Decreased immunogenicity**
Size	Large	Small (MW<2500)
Dose	Intermediate	High or low
Route	Subcutaneous > intraperitoneal > intravenous or intragastric	
Composition	Complex	Simple
Form	Particulate	Soluble
	Denatured	Native
Similarity to self protein	Multiple differences	Few differences
Adjuvants	Slow release	Rapid release
	Bacteria	No bacteria
Interaction with host MHC	Effective	Ineffective

Fig. A.2 Intrinsic properties and extrinsic factors that affect the immunogenicity of proteins. MW, molecular weight in daltons.

century. He found that animals immunized with a hapten–carrier conjugate produced three distinct sets of antibodies (Fig. A.3). One set comprised hapten-specific antibodies that reacted with the same hapten on any carrier, as well as with free hapten. The second set of antibodies was specific for the carrier protein, as shown by their ability to bind both the hapten-modified and unmodified carrier protein. Finally, some antibodies reacted only with the specific conjugate of hapten and carrier used for immunization. Landsteiner studied mainly the antibody response to the hapten, as these small molecules could be synthesized in many closely related forms. He observed that antibodies raised against a particular hapten bind that hapten but, in general, fail to bind even very closely related chemical structures. The binding of haptens by anti-hapten antibodies has played an important part in defining the precision of antigen binding by antibody molecules. Anti-hapten antibodies are also important medically because they mediate allergic reactions to penicillin and other compounds that elicit antibody responses when they attach to self proteins (see Section 14-10).

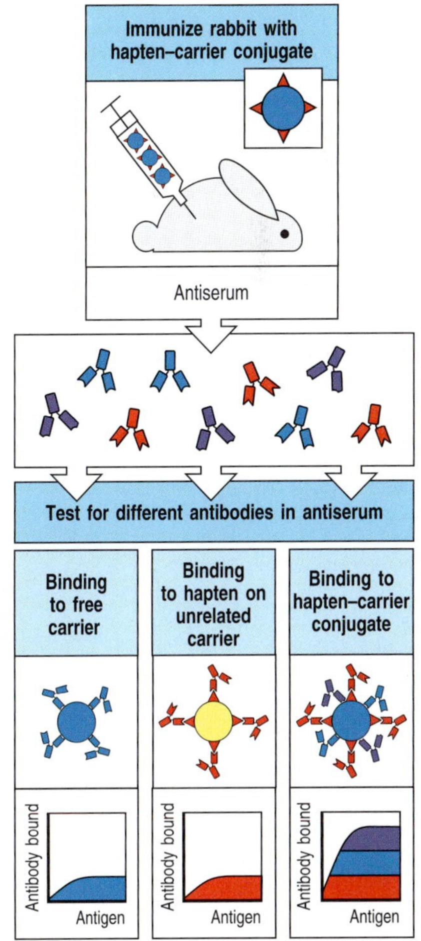

Fig. A.3 Antibodies can be elicited by small chemical groups called haptens only when the hapten is linked to an immunogenic protein carrier. Three types of antibodies are produced. One set (blue) binds the carrier protein alone and is called carrier-specific. One set (red) binds to the hapten on any carrier or to free hapten in solution and is called hapten-specific. One set (purple) only binds the specific conjugate of hapten and carrier used for immunization, apparently binding to sites at which the hapten joins the carrier, and is called conjugate-specific. The amount of antibody of each type in this serum is shown schematically in the graphs at the bottom; note that the original antigen binds more antibody than the sum of anti-hapten and anti-carrier antibodies as a result of the additional binding of conjugate-specific antibody.

A-2 Routes of immunization.

The route by which antigen is administered affects both the magnitude and the type of response obtained. The most common routes by which antigen is introduced experimentally or as a vaccine into the body are injection into tissue by **subcutaneous** (**s.c.**) injection into the fatty layer just below the dermis, or by **intradermal** (**i.d.**) or **intramuscular** (**i.m.**) injection; by **intravenous** (**i.v.**) injection or transfusion directly into the bloodstream; into the gastrointestinal tract by oral administration; and into the respiratory tract by **intranasal** (**i.n.**) administration or inhalation.

Antigens injected subcutaneously generally elicit the strongest responses, most probably because the antigen is taken up by Langerhans cells in the skin and efficiently presented in local lymph nodes, and so this is the method most commonly used when the object of the experiment is to elicit specific antibodies or T cells against a given antigen. Antigens injected or transfused directly into the bloodstream tend to induce immune unresponsiveness or tolerance unless they bind to host cells or are in the form of aggregates that are readily taken up by antigen-presenting cells.

Antigen administration via the gastrointestinal tract is used mostly in the study of allergy. It has distinctive effects, frequently eliciting a local antibody response in the intestinal lamina propria, while producing a systemic state of tolerance that manifests as a diminished response to the same antigen if subsequently administered in immunogenic form elsewhere in the body (see Chapter 12). This 'split tolerance' may be important in avoiding allergy to antigens in food, because the local response prevents food antigens from entering the body, while the inhibition of systemic immunity helps to prevent the formation of IgE antibodies, which are the cause of such allergies (see Chapter 14).

Introduction of antigen into the respiratory tract is also used mainly in the study of allergy. Protein antigens that enter the body through the respiratory epithelium tend to elicit allergic responses, for reasons that are not clear.

A-3 Effects of antigen dose.

The magnitude of the immune response depends on the dose of immunogen administered. Below a certain threshold dose, most proteins do not elicit any immune response. Above the threshold dose, there is a gradual increase in the response as the dose of antigen is increased, until a broad plateau level is reached, followed by a decline at very high antigen doses (see Fig. A.1). Because most infectious agents enter the body in small numbers, immune responses are generally elicited only by pathogens that multiply to a level sufficient to exceed the antigen dose threshold. The broad response optimum allows the system to respond to infectious agents across a wide range of doses. At very high antigen doses, the immune response is inhibited, which may be important in maintaining tolerance to abundant self proteins such as plasma proteins. In general, secondary and subsequent immune responses occur at lower antigen doses and achieve higher plateau values, which is a sign of immunological memory. However, under some conditions, very low or very high doses of antigen may induce specific unresponsive states, known respectively as acquired low-zone or high-zone tolerance.

A-4 Adjuvants.

Most proteins are poorly immunogenic or nonimmunogenic when administered by themselves. Strong adaptive immune responses to protein antigens almost always require that the antigen be injected in a mixture known as an

adjuvant. An adjuvant is any substance that enhances the immunogenicity of substances mixed with it. Adjuvants differ from protein carriers in that they do not form stable linkages with the immunogen. Furthermore, adjuvants are needed primarily for initial immunizations, whereas carriers are required to elicit not only primary but also subsequent responses to haptens. Commonly used adjuvants are listed in Fig. A.4.

Adjuvants can enhance immunogenicity in two different ways. First, adjuvants convert soluble protein antigens into particulate material, which is more readily ingested by antigen-presenting cells such as macrophages. For example, the antigen can be adsorbed on particles of the adjuvant (such as alum), made particulate by emulsification in mineral oils, or incorporated into the colloidal particles of immune stimulatory complexes (ISCOMs). This enhances immunogenicity somewhat, but such adjuvants are relatively weak unless they also contain bacteria or bacterial products. Such microbial constituents are the second means by which adjuvants enhance immunogenicity, and although their exact contribution to enhancing immunogenicity is unknown, they are clearly the more important component of an adjuvant. Microbial products may signal macrophages or dendritic cells to become more effective antigen-presenting cells (see Section 3-10). One of their effects is to induce the production of inflammatory cytokines and potent local inflammatory responses; this effect is probably intrinsic to their activity in enhancing responses, but precludes their use in humans.

Nevertheless, some human vaccines contain microbial antigens that can also act as effective adjuvants. For example, purified constituents of the bacterium *Bordetella pertussis*, which is the causative agent of whooping cough, are used as both antigen and adjuvant in the triplex DPT (diphtheria, pertussis, tetanus) vaccine against these diseases.

B cells contribute to adaptive immunity by secreting antibodies, and the response of B cells to an injected immunogen is usually measured by

Adjuvants that enhance immune responses		
Adjuvant name	**Composition**	**Mechanism of action**
Incomplete Freund's adjuvant	Oil-in-water emulsion	Delayed release of antigen; enhanced uptake by macrophages
Complete Freund's adjuvant	Oil-in-water emulsion with dead mycobacteria	Delayed release of antigen; enhanced uptake by macrophages; induction of co-stimulators in macrophages
Freund's adjuvant with MDP	Oil-in-water emulsion with muramyldipeptide (MDP), a constituent of mycobacteria	Similar to complete Freund's adjuvant
Alum (aluminum hydroxide)	Aluminum hydroxide gel	Delayed release of antigen; enhanced macrophage uptake
Alum plus *Bordetella pertussis*	Aluminum hydroxide gel with killed *B. pertussis*	Delayed release of antigen; enhanced uptake by macrophages; induction of co-stimulators
Immune stimulatory complexes (ISCOMs)	Matrix of Quil A containing viral proteins	Delivers antigen to cytosol; allows induction of cytotoxic T cells

Fig. A.4 Common adjuvants and their use. Adjuvants are mixed with the antigen and usually render it particulate, which helps to retain the antigen in the body and promotes uptake by macrophages. Most adjuvants include bacteria or bacterial components that stimulate macrophages, aiding in the induction of the immune response. ISCOMs are small micelles of the detergent Quil A; when viral proteins are placed in these micelles, they apparently fuse with the antigen-presenting cell, allowing the antigen to enter the cytosol. Thus, the antigen-presenting cell can stimulate a response to the viral protein, much as a virus infecting these cells would stimulate an antiviral response.

analyzing the specific antibody produced in a **humoral immune response**. This is most conveniently achieved by assaying the antibody that accumulates in the fluid phase of the blood or **plasma**; such antibodies are known as circulating antibodies. Circulating antibody is usually measured by collecting blood, allowing it to clot, and then isolating the serum from the clotted blood. The amount and characteristics of the antibody in the resulting antiserum are then determined using the assays we describe in Sections A-5–A-11.

The most important characteristics of an antibody response are the specificity, amount, isotype or class, and affinity of the antibodies produced. The **specificity** determines the ability of the antibody to distinguish the immunogen from other antigens. The amount of antibody can be determined in many different ways and is a function of the number of responding B cells, their rate of antibody synthesis, and the persistence of the antibody after production. The persistence of an antibody in the plasma and extracellular fluid bathing the tissues is determined mainly by its isotype or class (see Sections 5-12 and 10-14); each isotype has a different half-life *in vivo*. The isotypic composition of an antibody response also determines the biological functions these antibodies can perform and the sites in which antibody will be found. Finally, the strength of binding of the antibody to its antigen in terms of a single antigen-binding site binding to a monovalent antigen is termed its **affinity**; the total binding strength of a molecule with more than one binding site is called its **avidity**. Binding strength is important, because the higher the affinity of the antibody for its antigen, the less antibody is required to eliminate the antigen, because antibodies with higher affinity will bind at lower antigen concentrations. All these parameters of the humoral immune response help to determine the capacity of that response to protect the host from infection.

Antibody molecules are highly specific for their corresponding antigen, being able to detect one molecule of a protein antigen out of more than 10^8 similar molecules. This makes antibodies both easy to isolate and easy to study, and invaluable as probes of biological processes. Whereas standard chemistry would have great difficulty in distinguishing between two such closely related proteins as human and pig insulin, or two such closely related structures as *ortho-* and *para-*nitrophenyl, antibodies can be made that discriminate between these two structures absolutely. The value of antibodies as molecular probes has stimulated the development of many sensitive and highly specific techniques to measure their presence, to determine their specificity and affinity for a range of antigens, and to ascertain their functional capabilities. Many standard techniques used throughout biology exploit the specificity and stability of antigen binding by antibodies. Comprehensive guides to the conduct of these antibody assays are available in many books on immunological methodology; we will illustrate here only the most important techniques, especially those used in studying the immune response itself.

Some assays for antibody measure the direct binding of the antibody to its antigen. Such assays are based on **primary interactions**. Others determine the amount of antibody present by the changes it induces in the physical state of the antigen, such as the precipitation of soluble antigen or the clumping of antigenic particles; these are called **secondary interactions**. Both types of assay can be used to measure the amount and specificity of the antibodies produced after immunization, and both can be applied to a wide range of other biological questions.

Because assays for antibody were originally conducted using antisera from immune individuals, they are commonly referred to as **serological assays**, and the use of antibodies is often called **serology**. The amount of antibody is usually determined by antigen-binding assays after titration of the antiserum by serial dilution, and the point at which binding falls to 50% of the maximum is usually referred to as the **titer** of an antiserum.

The detection, measurement, and characterization of antibodies, and their use as research and diagnostic tools.

A-5 Affinity chromatography.

Specific antibody can be isolated from an antiserum by **affinity chromatography**, which exploits the specific binding of antibody to antigen held on a solid matrix (Fig. A.5). Antigen is bound covalently to small, chemically reactive beads, which are loaded into a column, and the antiserum is allowed to pass over the beads. The specific antibodies bind, while all the other proteins in the serum, including antibodies against other substances, can be washed away. The specific antibodies are then eluted, typically by lowering the pH to 2.5 or raising it to greater than 11. Antibodies bind stably under physiological conditions of salt concentration, temperature, and pH, but the binding is reversible because the bonds are noncovalent. Affinity chromatography can also be used to purify antigens from complex mixtures by using beads coated with specific antibody. The technique is known as affinity chromatography because it separates molecules on the basis of their affinity for one another.

A-6 Radioimmunoassay (RIA), enzyme-linked immunosorbent assay (ELISA), and competitive inhibition assay.

Radioimmunoassay (**RIA**) and **enzyme-linked immunosorbent assay** (**ELISA**) are direct binding assays for antibody (or antigen); both work on the same principle, but the means of detecting specific binding is different. Radioimmunoassays are commonly used to measure the levels of hormones in blood and tissue fluids, while ELISA assays are frequently used in viral diagnostics, for example in detecting cases of infection with the human immunodeficiency virus (HIV), which is the cause of AIDS. For both these methods one needs a pure preparation of a known antigen or antibody, or both, in order to standardize the assay. We will describe the assay with a sample of pure antibody, which is the more usual case, but the principle is similar if pure antigen is used instead. In RIA for an antigen, pure antibody against that antigen is radioactively labeled, usually with ^{125}I; for ELISA, an enzyme is linked chemically to the antibody. The unlabeled component, which in this case would be

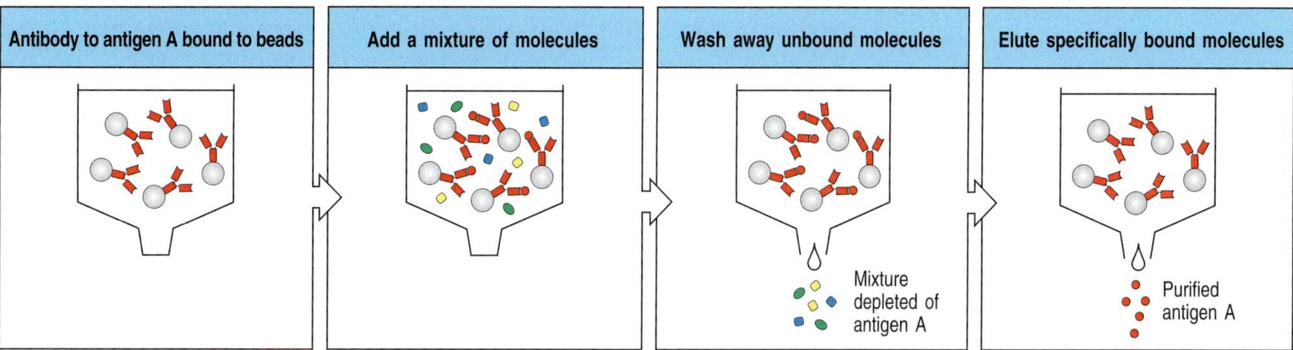

Fig. A.5 Affinity chromatography uses antigen–antibody binding to purify antigens or antibodies. To purify a specific antigen from a complex mixture of molecules, a monoclonal antibody is attached to an insoluble matrix, such as chromatography beads, and the mixture of molecules is passed over the matrix. The specific antibody binds the antigen of interest; other molecules are washed away. Specific antigen is then eluted by altering the pH, which can usually disrupt antibody–antigen bonds. Antibodies can be purified in the same way on beads coupled to antigen (not shown).

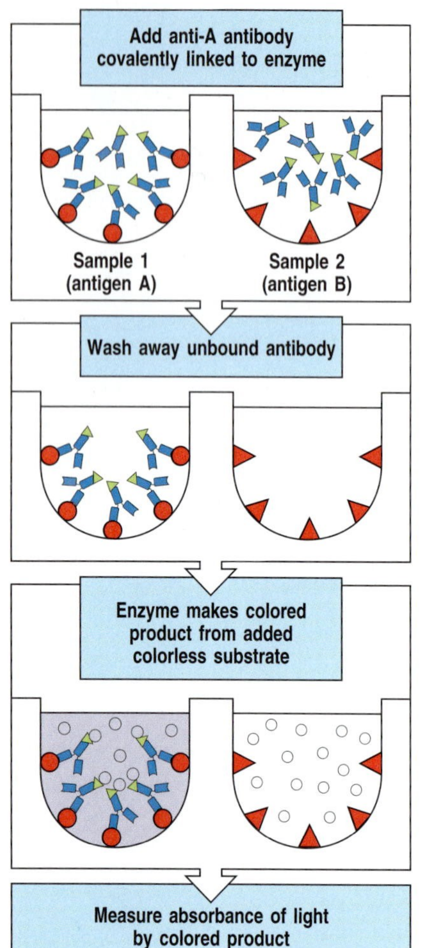

Fig. A.6 The principle of the enzyme-linked immunosorbent assay (ELISA). To detect antigen A, purified antibody specific for antigen A is linked chemically to an enzyme. The samples to be tested are coated onto the surface of plastic wells to which they bind nonspecifically; residual sticky sites on the plastic are blocked by adding irrelevant proteins (not shown). The labeled antibody is then added to the wells under conditions where nonspecific binding is prevented, so that only binding to antigen A causes the labeled antibody to be retained on the surface. Unbound labeled antibody is removed from all wells by washing, and bound antibody is detected by an enzyme-dependent color-change reaction. This assay allows arrays of wells known as microtiter plates to be read in fiberoptic multichannel spectrometers, greatly speeding the assay. Modifications of this basic assay allow antibody or antigen in unknown samples to be measured as shown in Figs A.7 and A.30 (see also Section A-10).

antigen, is attached to a solid support, such as the wells of a plastic multiwell plate, which will adsorb a certain amount of any protein.

The labeled antibody is allowed to bind to the unlabeled antigen, under conditions in which nonspecific adsorption is blocked, and any unbound antibody and other proteins are washed away. Antibody binding in RIA is measured directly in terms of the amount of radioactivity retained by the coated wells, whereas in ELISA the binding is detected by a reaction that converts a colorless substrate into a colored reaction product (Fig. A.6). The color change can be read directly in the reaction tray, making data collection very easy, and ELISA also avoids the hazards of radioactivity. This makes ELISA the preferred method for most direct-binding assays. Labeled anti-immunoglobulin antibodies (see Section A-10) can also be used in RIA or ELISA to detect the binding of unlabeled antibody to unlabeled antigen-coated plates. In this case, the labeled anti-immunoglobulin antibody is used in what is termed a 'second layer.' The use of such a second layer also amplifies the signal, because at least two molecules of the labeled anti-immunoglobulin antibody are able to bind to each unlabeled antibody. RIA and ELISA can also be carried out with unlabeled antibody stuck to the plates and labeled antigen added.

A modification of ELISA known as a **capture** or **sandwich ELISA** (or more generally as an **antigen-capture assay**) can be used to detect secreted products such as cytokines. Rather than the antigen being directly attached to a plastic plate, antigen-specific antibodies are bound to the plate. These are able to bind antigen with high affinity, and thus concentrate it on the surface of the plate, even with antigens that are present in very low concentrations in the initial mixture. A separate labeled antibody that recognizes a different epitope from that recognized by the immobilized first antibody is then used to detect the bound antigen.

These assays illustrate two crucial aspects of all serological assays. First, at least one of the reagents must be available in a pure, detectable form in order to obtain quantitative information. Second, there must be a means of separating the bound fraction of the labeled reagent from the unbound, free fraction so that the percentage of specific binding can be determined. Normally, this separation is achieved by having the unlabeled partner trapped on a solid support. Labeled molecules that do not bind can then be washed away, leaving just the labeled partner that has bound. In Fig. A.6, the unlabeled antigen is attached to the well and the labeled antibody is trapped by binding to it. The separation of bound from free is an essential step in every assay that uses antibodies.

RIA and ELISA do not allow one to measure directly the amount of antigen or antibody in a sample of unknown composition, because both depend on the binding of a pure labeled antigen or antibody. There are various ways around this problem, one of which is to use a **competitive inhibition assay**, as shown in Fig. A.7. In this type of assay, the presence and amount of a particular antigen in an unknown sample are determined by its ability to compete with a labeled reference antigen for binding to an antibody attached to a plastic well. A standard curve is first constructed by adding various amounts of a known, unlabeled standard preparation; the assay can then measure the amount of antigen in unknown samples by comparison with the standard. The competitive binding assay can also be used for measuring antibody in a sample of unknown composition by attaching the appropriate antigen to the plate and measuring the ability of the test sample to inhibit the binding of a labeled specific antibody.

A-7 Hemagglutination and blood typing.

The direct measurement of antibody binding to antigen is used in most quantitative serological assays. However, some important assays are based on the

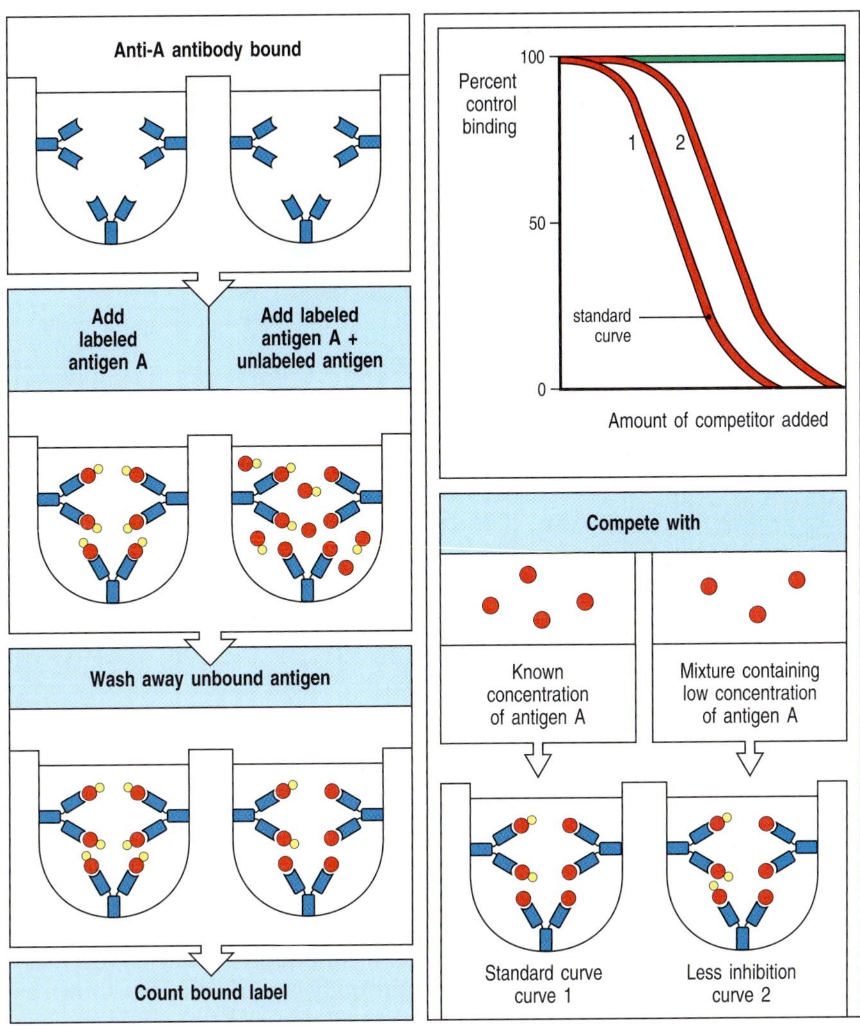

Fig. A.7 Competitive inhibition assay for antigen in unknown samples. A fixed amount of unlabeled antibody is attached to a set of wells, and a standard reference preparation of a labeled antigen is bound to it. Unlabeled standard or test samples are then added in various amounts and the displacement of labeled antigen is measured, generating characteristic inhibition curves. A standard curve is obtained by using known amounts of unlabeled antigen identical to that used as the labeled species, and comparison with this curve allows the amount of antigen in unknown samples to be calculated. The green line on the graph represents a sample lacking any substance that reacts with anti-A antibodies.

ability of antibody binding to alter the physical state of the antigen it binds to. These secondary interactions can be detected in a variety of ways. For instance, when the antigen is displayed on the surface of a large particle such as a bacterium, antibodies can cause the bacteria to clump or **agglutinate**. The same principle applies to the reactions used in blood typing, only here the target antigens are on the surface of red blood cells and the clumping reaction caused by antibodies against them is called **hemagglutination** (from the Greek *haima*, blood).

Hemagglutination is used to determine the **ABO blood group** of blood donors and transfusion recipients. Clumping or agglutination is induced by antibodies or agglutinins called anti-A or anti-B that bind to the A or B blood group substances, respectively (Fig. A.8). These blood group antigens are arrayed in many copies on the surface of the red blood cell, causing the cells to agglutinate when cross-linked by antibodies. Because hemagglutination involves the cross-linking of blood cells by the simultaneous binding of antibody molecules to identical antigens on different cells, this reaction also demonstrated that each antibody molecule must have at least two identical antigen-binding sites.

A-8 Precipitin reaction.

When sufficient quantities of antibody are mixed with soluble macromolecular antigens, a visible precipitate consisting of large aggregates of antigen cross-linked by antibody molecules can form. The amount of precipitate

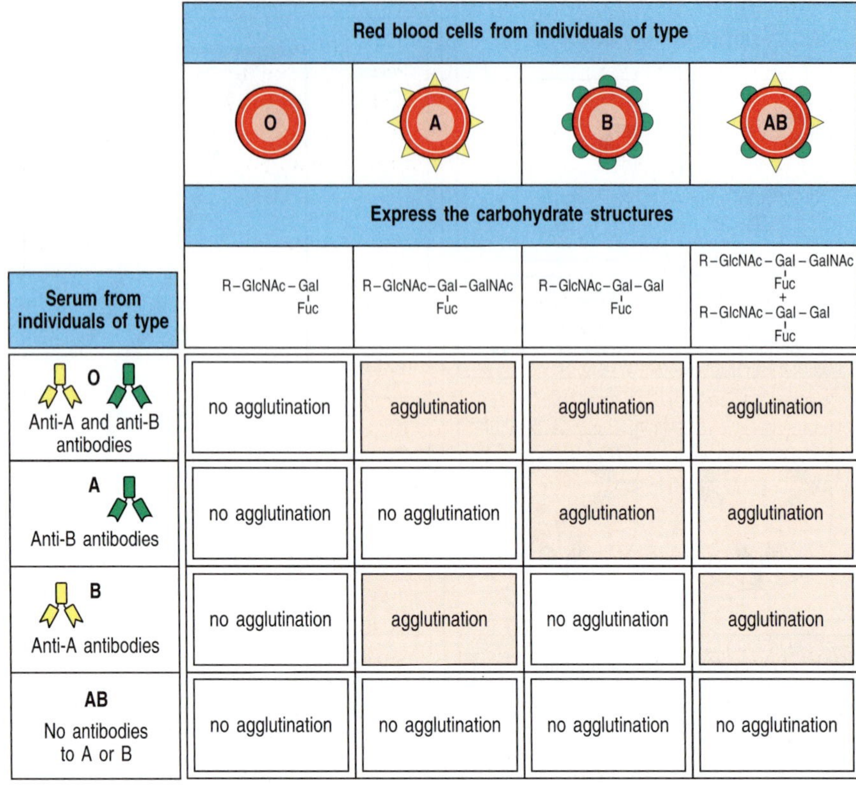

Serum from individuals of type	Red blood cells from individuals of type O	A	B	AB
Express the carbohydrate structures	R–GlcNAc–Gal, Fuc	R–GlcNAc–Gal–GalNAc, Fuc	R–GlcNAc–Gal–Gal, Fuc	R–GlcNAc–Gal–GalNAc, Fuc + R–GlcNAc–Gal–Gal, Fuc
O: Anti-A and anti-B antibodies	no agglutination	agglutination	agglutination	agglutination
A: Anti-B antibodies	no agglutination	no agglutination	agglutination	agglutination
B: Anti-A antibodies	no agglutination	agglutination	no agglutination	agglutination
AB: No antibodies to A or B	no agglutination	no agglutination	no agglutination	no agglutination

Fig. A.8 Hemagglutination is used to type blood groups and match compatible donors and recipients for blood transfusion. Common gut bacteria bear antigens that are similar or identical to blood group antigens, and these stimulate the formation of antibodies against these antigens in individuals who do not bear the corresponding antigen on their own red blood cells (left column); thus, type O individuals, who lack A and B, have both anti-A and anti-B antibodies, whereas type AB individuals have neither. The pattern of agglutination of the red blood cells of a transfusion donor or recipient with anti-A and anti-B antibodies reveals the individual's ABO blood group. Before transfusion, the serum of the recipient is also tested for antibodies that agglutinate the red blood cells of the donor, and vice versa, a procedure called a cross-match, which may detect potentially harmful antibodies against other blood groups that are not part of the ABO system.

depends on the amounts of antigen and antibody, and on the ratio between them (Fig. A.9). This **precipitin reaction** provided the first quantitative assay for antibody, but is now seldom used in immunology. However, it is important to understand the interaction of antigen with antibody that leads to this reaction, because the production of **antigen:antibody complexes**, also known as **immune complexes**, *in vivo* occurs in almost all immune responses and can occasionally cause significant pathology (see Chapters 14 and 15).

In the precipitin reaction, various amounts of soluble antigen are added to a fixed amount of serum containing antibody. As the amount of antigen added increases, the amount of precipitate generated also increases up to a maximum and then declines (see Fig. A.9). When small amounts of antigen are added, antigen:antibody complexes are formed under conditions of antibody excess so that each molecule of antigen is bound by antibody and cross-linked to other molecules of antigen. When large amounts of antigen are added, only small antigen:antibody complexes can form, and these are often soluble in this zone of antigen excess. Between these two zones, all of the antigen and antibody is found in the precipitate, generating a zone of equivalence. At equivalence, very large lattices of antigen and antibody are formed by cross-linking. Although all antigen:antibody complexes can potentially produce disease, the small, soluble immune complexes formed in the zone of antigen excess may persist and cause pathology *in vivo*.

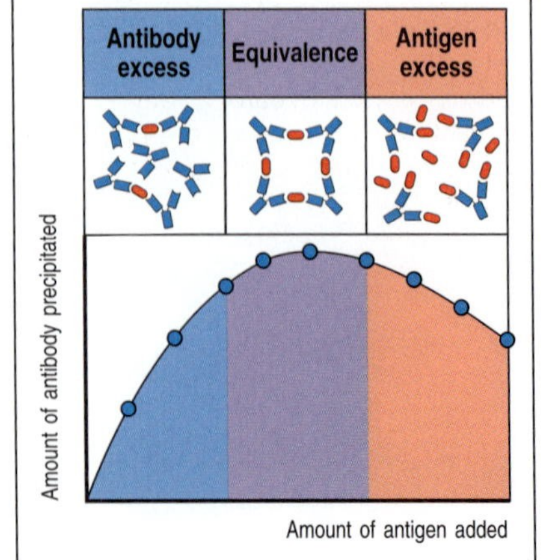

Fig. A.9 Antibody can precipitate soluble antigen. Analysis of the precipitate can generate a precipitin curve. Different amounts of antigen are added to a fixed amount of antibody, and precipitates form by antibody cross-linking of antigen molecules. The precipitate is recovered and the amount of precipitated antibody is measured; the supernatant is tested for residual antigen or antibody. This defines zones of antibody excess, equivalence, and antigen excess. At equivalence, the largest antigen:antibody complexes form. In the zone of antigen excess, some of the immune complexes are too small to precipitate. These soluble immune complexes can cause pathological damage to small blood vessels when they form *in vivo* (see Chapter 15).

The precipitin reaction is affected by the number of binding sites that each antibody has for antigen, and by the maximum number of antibodies that can be bound by an antigen molecule or particle at any one time. These quantities are defined as the **valence** of the antibody and the valence of the antigen: the valence of both the antibodies and the antigen must be two or greater before any precipitation can occur. The valence of an antibody depends on its structural class (see Section 5-12).

Antigen will be precipitated only if it has several antibody-binding sites. This condition is usually satisfied in macromolecular antigens, which have a complex surface with binding sites for several different antibodies. The site on an antigen to which each distinct antibody molecule binds is called an **antigenic determinant** or an **epitope**. Steric considerations limit the number of distinct antibody molecules that can bind to a single antigen molecule at any one time, however, because antibody molecules binding to epitopes that partly overlap will compete for binding. For this reason, the valence of an antigen is almost always less than the number of epitopes on the antigen (Fig. A.10).

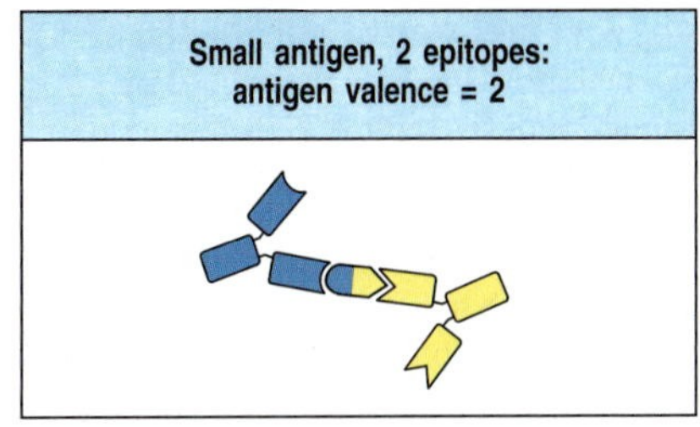

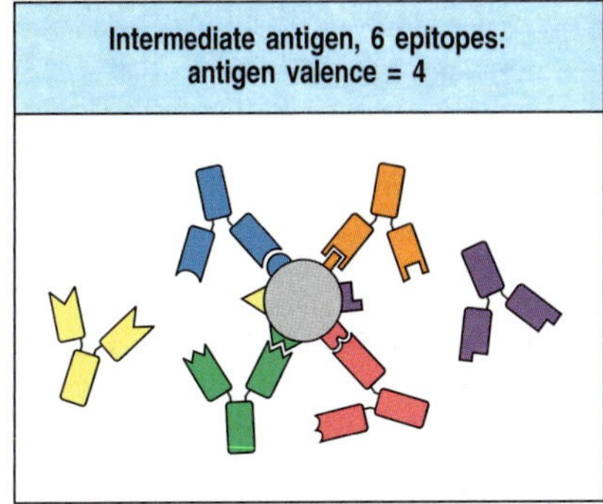

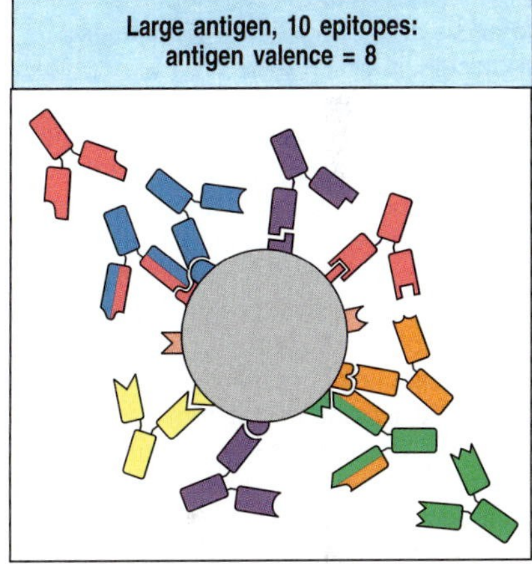

Fig. A.10 Different antibodies bind to distinct epitopes on an antigen molecule. The surface of an antigen possesses many potential antigenic determinants or epitopes, distinct sites to which an antibody can bind. The number of antibody molecules that can bind to a molecule of antigen at one time defines the antigen's valence. Steric considerations can limit the number of different antibodies that bind to the surface of an antigen at any one time (middle and bottom panels), so that the number of epitopes on an antigen is always greater than or equal to its valence.

A-9 Equilibrium dialysis: measurement of antibody affinity and avidity.

The affinity of an antibody is the strength of binding of a monovalent ligand to a single antigen-binding site on the antibody. The affinity of an antibody that binds small antigens, such as haptens, that can diffuse freely across a dialysis membrane can be determined directly by the technique of **equilibrium dialysis**. A known amount of antibody, whose molecules are too large to cross a dialysis membrane, is placed in a dialysis bag and offered various amounts of antigen. Molecules of antigen that bind to the antibody are no longer free to diffuse across the dialysis membrane, so only the unbound molecules of antigen equilibrate across it. By measuring the concentration of antigen inside the bag and in the surrounding fluid, one can determine the amount of the antigen that is bound as well as the amount that is free when equilibrium has been achieved. Given that the amount of antibody present is known, the affinity of the antibody and the number of specific binding sites for the antigen per molecule of antibody can be determined from this information. The data are usually analyzed using a **Scatchard analysis** (Fig. A.11); such analyses were used to demonstrate that a molecule of IgG antibody has two identical antigen-binding sites.

Whereas affinity measures the strength of binding of an antigenic determinant to a single antigen-binding site, an antibody reacting with an antigen that has multiple identical epitopes or with the surface of a pathogen will often bind the same molecule or particle with both of its antigen-binding sites. This increases the apparent strength of binding, because both binding sites must release at the same time for the two molecules to dissociate. This is often referred to as **cooperativity** in binding, but it should not be confused with the cooperative binding found in a protein such as hemoglobin, in which binding of ligand at one site enhances the affinity of a second binding site for its ligand. The overall strength of binding of an antibody molecule to an antigen or particle is called its avidity (Fig. A.12). For IgG antibodies, bivalent binding can significantly increase avidity; in IgM antibodies, which have 10 identical antigen-binding sites, the affinity of each site for a monovalent antigen is usually quite low, but the avidity of binding of the whole antibody to a surface such as a bacterium that displays multiple identical epitopes can be very high.

A-10 Anti-immunoglobulin antibodies.

A general approach to the detection of bound antibody that avoids the need to label each preparation of antibody molecules is to detect bound, unlabeled

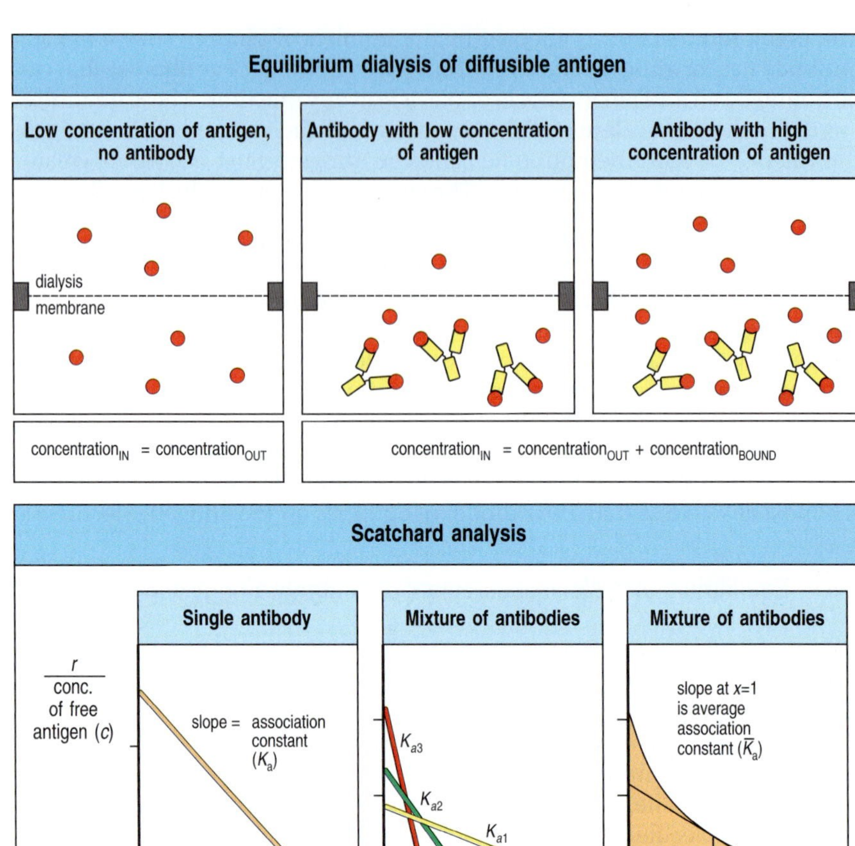

Fig. A.11 The affinity and valence of an antibody can be determined by equilibrium dialysis. A known amount of antibody is placed in the bottom half of a dialysis chamber and exposed to different amounts of a diffusible monovalent antigen, such as a hapten. At equilibrium, the concentration of free antigen will be the same on each side of the membrane, so that at each concentration of antigen added, the fraction of the antigen bound is determined from the difference in concentration of total antigen in the top and bottom chambers. This information can be transformed into a Scatchard plot as shown here. In Scatchard analysis, the ratio r/c (where r = moles of antigen bound per mole of antibody and c = molar concentration of free antigen) is plotted against r. The number of binding sites per antibody molecule can be determined from the value of r at infinite free-antigen concentration, where r/free = 0, in other words at the x-axis intercept. The analysis of a monoclonal IgG antibody molecule, in which there are two identical antigen-binding sites per molecule, is shown in the left panel. The slope of the line is determined by the affinity of the antibody molecule for its antigen; if all the antibody molecules in a preparation are identical, as for this monoclonal antibody, then a straight line is obtained whose slope is equal to $-K_a$, where K_a is the association (or affinity) constant, and the dissociation constant $K_d = 1/K_a$. However, antiserum raised even against a simple antigenic determinant such as a hapten contains a heterogeneous population of antibody molecules (see Section A-1). Each antibody molecule would, if isolated, make up part of the total and give a straight line whose x-axis intercept is less than two, because this antibody molecule contains only a fraction of the total binding sites in the population (middle panel). As a mixture, they give curved lines with an x-axis intercept of two for which an average affinity ($\overline{K}_a$) can be determined from the slope of this line at a concentration of antigen where 50% of the sites are bound, or at $x = 1$ (right panel). The association constant determines the equilibrium state of the reaction Ag + Ab = Ag:Ab, where Ag = antigen and Ab = antibody, and K_a = [Ag:Ab]/[Ag][Ab]. This constant reflects the 'on' and 'off' rates for antigen binding to the antibody; with small antigens such as haptens, binding is usually as rapid as diffusion allows, whereas differences in 'off' rates determine the affinity constant. However, with larger antigens the 'on' rate may also vary as the interaction becomes more complex.

antibody with a labeled antibody specific for immunoglobulins themselves. Immunoglobulins, like other proteins, are immunogenic when used to immunize individuals of another species. The majority of **anti-immunoglobulin antibodies** raised in this way recognize conserved features shared by all immunoglobulin molecules of the immunizing species. Others can be specific for immunoglobulin chains, such as the heavy or light chains, or for individual isotypes. Antibodies raised by immunization of goats with mouse IgG are commonly used in experimental immunology. Such goat anti-mouse IgG antibodies can be purified using affinity chromatography, then labeled and used as a general probe for bound IgG antibodies. Anti-immunoglobulin antisera have found many uses in clinical medicine and biological research since their introduction. Fluorescently labeled anti-immunoglobulin antibodies are now widely used both in immunology and other areas of biology as secondary reagents for detecting specific antibodies bound, for example, to cell structures (see Sections A-14 and A-16). Labeled anti-immunoglobulin antibodies can also be used in RIA or ELISA (see Section A-6) to detect the binding of unlabeled antibody to antigen-coated plates.

When an immunoglobulin is used as an antigen to immunize a different species of animal, it will be treated like any other foreign protein and will elicit an antibody response. Anti-immunoglobulin antibodies can be made that recognize the amino acids that characterize the isotype of the injected antibody. Such **anti-isotypic antibodies** recognize all immunoglobulins of the same isotype in all members of the species from which the injected antibody came.

It is also possible to raise antibodies that recognize differences in immunoglobulins from members of the same species that are due to the presence of multiple alleles of the individual C genes in the population (genetic polymorphism). Such allelic variants are called **allotypes**. In contrast to anti-isotypic antibodies, anti-allotypic antibodies will recognize immunoglobulin of a particular isotype in only some members of a species. Finally, because individual antibodies differ in their variable regions, one can raise antibodies against unique features of the antigen-binding site, which are called **idiotypes**.

A schematic picture of the differences between idiotypes, allotypes, and isotypes is given in Fig. A.13. Historically, the main features of immunoglobulins were defined by using isotypic and allotypic genetic markers identified by antisera raised in different species or in genetically distinct members of the same species. The independent segregation of allotypic and isotypic markers revealed the existence of separate heavy-chain, κ, and λ genes. Such anti-idiotypic, allotypic, and isotypic antibodies are still enormously useful in detecting antibodies and B cells in scientific experimentation and medical diagnostics.

Antibodies specific for individual immunoglobulin isotypes can be produced by immunizing an animal of a different species with a pure preparation of one isotype and then removing those antibodies that cross-react with immunoglobulins of other isotypes by using affinity chromatography (see Section A-5). Anti-isotype antibodies can be used to measure how much antibody of a particular isotype in an antiserum reacts with a given antigen. This reaction is particularly important for detecting small amounts of specific antibodies of the IgE isotype, which are responsible for most allergies. The presence in an individual's serum of IgE binding to an antigen correlates with allergic reactions to that antigen.

An alternative approach to detecting bound antibodies exploits bacterial proteins that bind to immunoglobulins with high affinity and specificity. One of these, **Protein A** from the bacterium *Staphylococcus aureus*, has been exploited widely in immunology for the affinity purification of immunoglobulin and for the detection of bound antibody. The use of standard second reagents such as labeled anti-immunoglobulin antibodies or Protein A to detect antibody bound specifically to its antigen allows great savings in reagent labeling costs, and also provides a standard detection system so that results in different assays can be compared directly.

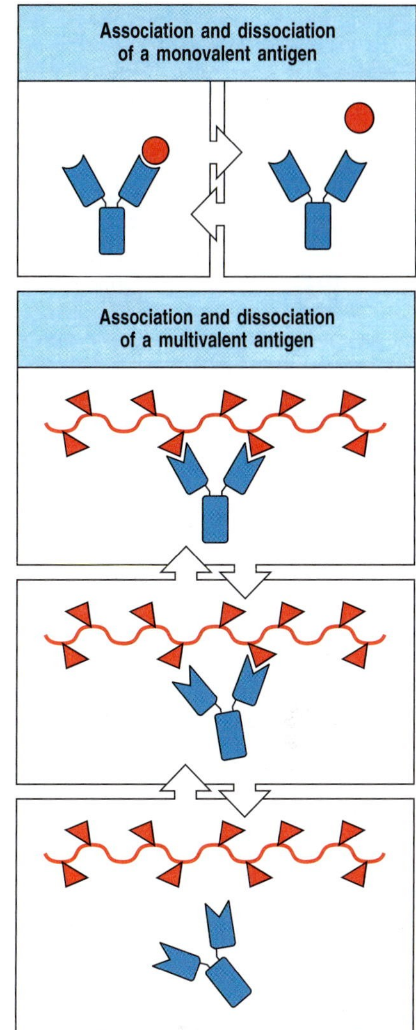

Fig. A.12 The avidity of an antibody is its strength of binding to intact antigen. When an IgG antibody binds a ligand with multiple identical epitopes, both binding sites can bind the same molecule or particle. The overall strength of binding, called avidity, is greater than the affinity, the strength of binding of a single site, because both binding sites must dissociate at the same time for the antibody to release the antigen. This property is very important in the binding of antibody to bacteria, which usually have multiple identical epitopes on their surfaces.

A-11 Coombs tests and the detection of Rhesus incompatibility.

These tests use anti-immunoglobulin antibodies (see Section A-10) to detect the antibodies that cause **hemolytic disease of the newborn**, or **erythroblastosis fetalis**. Anti-immunoglobulin antibodies were first developed by Robin Coombs, and the test for this disease is still called the Coombs test. Hemolytic disease of the newborn occurs when a mother makes IgG antibodies specific for the **Rhesus** or **Rh blood group antigen** expressed on the red blood cells of her fetus. Rh-negative mothers make these antibodies when they are exposed to Rh-positive fetal red blood cells bearing the paternally inherited Rh antigen. Maternal IgG antibodies are normally transported across the placenta to the fetus, where they protect the newborn infant against infection. However, IgG anti-Rh antibodies coat the fetal red blood cells, which are then destroyed by phagocytic cells in the liver, causing a hemolytic anemia in the fetus and newborn infant.

Rh antigens are widely spaced on the red blood cell surface, and so the IgG anti-Rh antibodies do not bind in the correct conformation to fix complement and so do not cause lysis of red blood cells *in vitro*. Furthermore, for reasons that are not fully understood, antibodies against Rh antigens do not agglutinate red blood cells, unlike antibodies against the ABO blood group antigens. Thus, detecting anti-Rh antibodies was difficult until anti-human immunoglobulin antibodies were developed. With these, maternal IgG

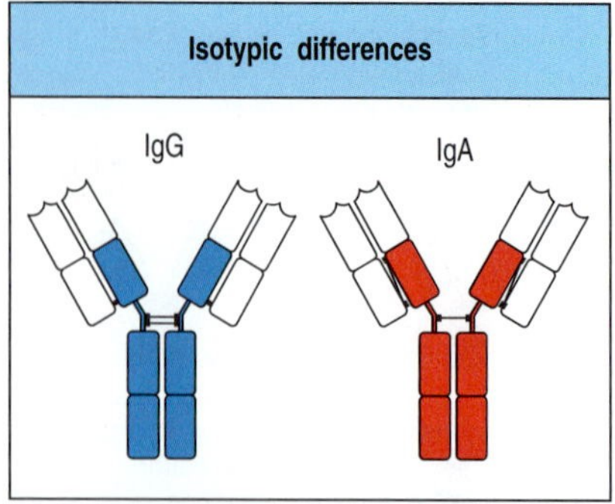

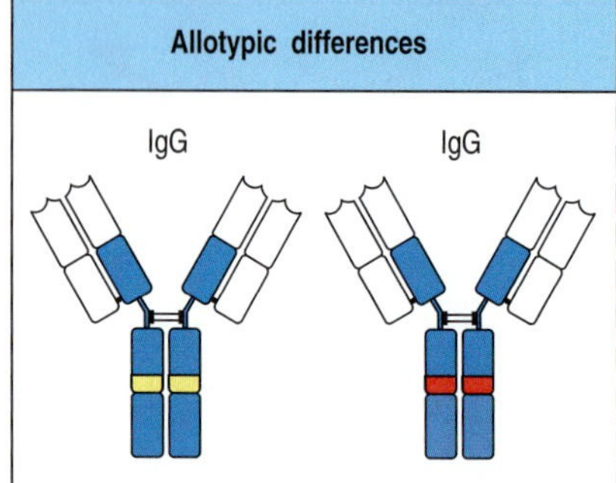

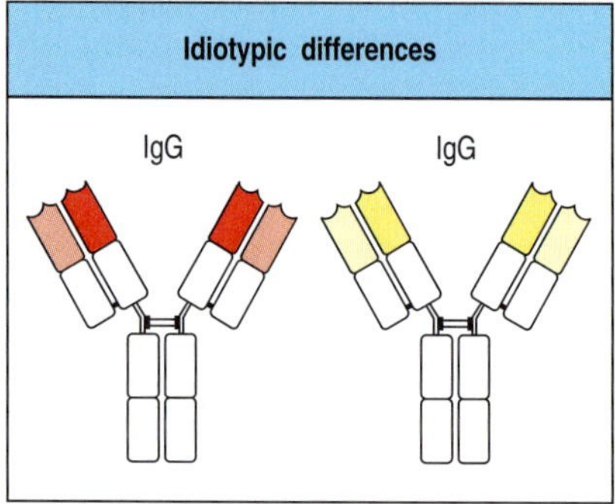

Fig. A.13 Different types of variation between immunoglobulins. Differences between constant regions due to the usage of different C-region genes are called isotypes; differences due to different alleles of the same C gene are called allotypes; differences due to particular rearranged V_H and V_L genes are called idiotypes.

antibodies bound to the fetal red blood cells can be detected after washing the cells to remove unbound immunoglobulin that is present in the fetal serum. Adding anti-human immunoglobulin antibodies against the washed fetal red blood cells agglutinates any cells to which maternal antibodies are bound. This is the **direct Coombs test** (Fig. A.14), so called because it directly detects antibody bound to the surface of the fetal red blood cells. An **indirect Coombs test** is used to detect nonagglutinating anti-Rh antibody in maternal serum: the serum is first incubated with Rh-positive red blood cells, which bind the anti-Rh antibody, after which the antibody-coated cells are washed to remove unbound immunoglobulin and are then agglutinated with anti-immunoglobulin antibody (see Fig. A.14). The indirect Coombs test allows Rh incompatibilities that might lead to hemolytic disease of the newborn to be detected, and this knowledge allows the disease to be prevented (see Section 11-19). The Coombs test is also commonly used to detect antibodies against drugs that bind to red blood cells and cause hemolytic anemia.

A-12 Monoclonal antibodies.

The antibodies generated in a natural immune response or after immunization in the laboratory are a mixture of molecules of different specificities and affinities. Some of this heterogeneity results from the production of antibodies that bind to different epitopes on the immunizing antigen, but even antibodies directed at a single antigenic determinant such as a hapten can be markedly heterogeneous, as shown by **isoelectric focusing**. In this technique, proteins are separated on the basis of their isoelectric point, the pH at which their net charge is zero. By electrophoresing proteins in a pH gradient for long enough, each molecule migrates along the pH gradient until it reaches the pH at which it is neutral and is thus concentrated (focused) at that point. When antiserum containing anti-hapten antibodies is treated in this way and then transferred to a solid support such as nitrocellulose paper, the anti-hapten antibodies can be detected by their ability to bind labeled hapten. The binding of antibodies of various isoelectric points to the hapten shows that even antibodies that bind the same antigenic determinant can be heterogeneous.

Antisera are valuable for many biological purposes but they have certain inherent disadvantages that relate to the heterogeneity of the antibodies they contain. First, each antiserum is different from all other antisera, even if raised in a genetically identical animal by using the identical preparation of antigen and the same immunization protocol. Second, antisera can be produced in only limited volumes, and thus it is impossible to use the identical serological reagent in a long or complex series of experiments or clinical tests. Finally, even antibodies purified by affinity chromatography (see Section A-5) can include minor populations of antibodies that give unexpected cross-reactions, which confound the analysis of experiments. To avoid these problems, and to harness the full potential of antibodies, it was necessary to develop a way of making an unlimited supply of antibody molecules of homogeneous structure and known specificity. This has been achieved through the production of monoclonal antibodies from cultures of hybrid antibody-forming cells or, more recently, by genetic engineering.

Biochemists in search of a homogeneous preparation of antibody that they could subject to detailed chemical analysis turned early to proteins produced by patients with multiple myeloma, a common tumor of plasma cells. It was known that antibodies are normally produced by plasma cells, and because this disease is associated with the presence of large amounts of a homogeneous gamma globulin called a **myeloma protein** in the patient's serum, it seemed likely that myeloma proteins would serve as models for normal antibody molecules. Thus, much of the early knowledge of antibody structure came from studies on myeloma proteins. These studies showed that

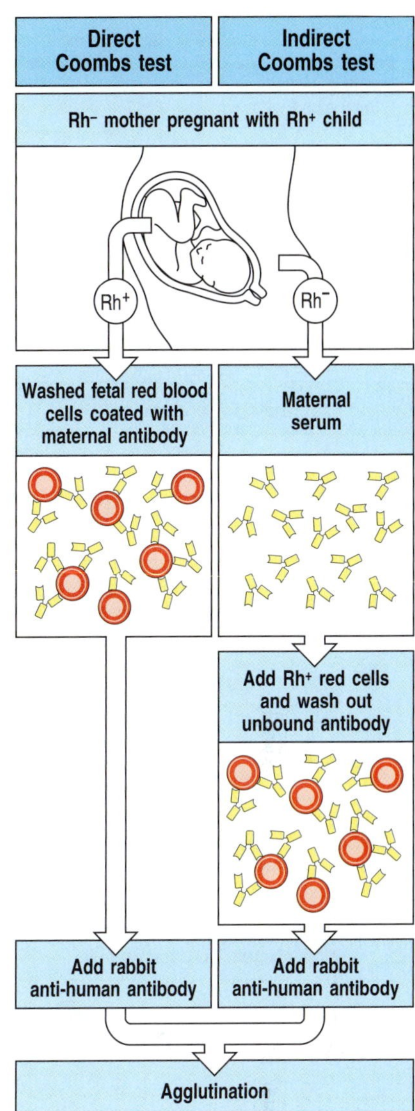

Fig. A.14 The Coombs direct and indirect anti-globulin tests for antibody against red blood cell antigens. A Rh⁻ mother of a Rh⁺ fetus can become immunized to fetal red blood cells that enter the maternal circulation at the time of delivery. In a subsequent pregnancy with a Rh⁺ fetus, IgG anti-Rh antibodies can cross the placenta and damage the fetal red blood cells. In contrast to anti-Rh antibodies, maternal anti-ABO antibodies are of the IgM isotype and cannot cross the placenta, and so do not cause harm. Anti-Rh antibodies do not agglutinate red blood cells, but their presence on the fetal red blood cell surface can be shown by washing away unbound immunoglobulin and then adding antibody against human immunoglobulin, which agglutinates the antibody-coated cells. Anti-Rh antibodies can be detected in the mother's serum in an indirect Coombs test; the serum is incubated with Rh⁺ red blood cells, and once the antibody has bound, the red blood cells are treated as in the direct Coombs test.

monoclonal antibodies could be obtained from immortalized plasma cells. However, the antigen specificity of most myeloma proteins was unknown, which limited their usefulness as objects of study or as immunological tools.

This problem was solved by Georges Köhler and César Milstein, who devised a technique for producing a homogeneous population of antibodies of known antigenic specificity. They did this by fusing spleen cells from an immunized mouse to cells of a mouse myeloma to produce hybrid cells that both proliferated indefinitely and secreted antibody specific for the antigen used to immunize the spleen cell donor. The spleen cell provides the ability to make specific antibody, while the myeloma cell provides the ability to grow indefinitely in culture and secrete immunoglobulin continuously. By using a myeloma cell partner that produces no antibody proteins itself, the antibody produced by the hybrid cells comes only from the immune spleen cell partner. After fusion, the hybrid cells are selected using drugs that kill the myeloma parental cell, while the unfused parental spleen cells have a limited life-span and soon die, so that only hybrid myeloma cell lines or **hybridomas** survive. Those hybridomas producing antibody of the desired specificity are then identified and cloned by regrowing the cultures from single cells (Fig. A.15). Because each hybridoma is a **clone** derived from fusion with a single B cell, all the antibody molecules it produces are identical in structure, including their antigen-binding site and isotype. Such antibodies are called **monoclonal antibodies**. This technology has revolutionized the use of antibodies by providing a limitless supply of antibody of a single and known specificity. Monoclonal antibodies are now used in most serological assays, as diagnostic probes, and as therapeutic agents. So far, however, only mouse monoclonals are routinely produced in this way, and efforts to use the same approach to make human monoclonal antibodies have met with limited success. 'Fully human' therapeutic monoclonal antibodies are currently made using either the phage display technology described in Section A-13 or transgenic mice carrying human antibody genes.

A-13 Phage display libraries for antibody V-region production.

This is a technique for producing antibody-like molecules. Gene segments encoding the antigen-binding variable, or V, domains of antibodies are fused to genes encoding the coat protein of a bacteriophage. Bacteriophages containing such gene fusions are used to infect bacteria, and the resulting phage particles have coats that express the antibody-like fusion protein, with the antigen-binding domain displayed on the outside of the bacteriophages. A collection of recombinant phages, each displaying a different antigen-binding domain on the surface, is known as a **phage display library**. In much the same way that antibodies specific for a particular antigen can be isolated from a complex mixture by affinity chromatography (see Section A-5), phages expressing antigen-binding domains specific for a particular antigen can be

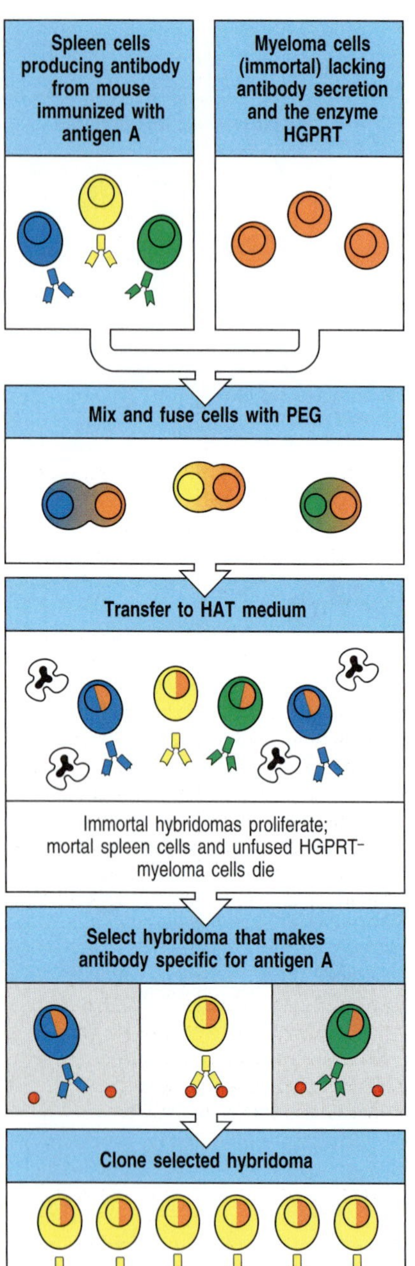

Fig. A.15 The production of monoclonal antibodies. Mice are immunized with antigen A and given an intravenous booster immunization 3 days before they are killed, in order to produce a large population of spleen cells secreting specific antibody. Spleen cells die after a few days in culture. To produce a continuous source of antibody they are fused with immortal myeloma cells by using polyethylene glycol (PEG) to produce a hybrid cell line called a hybridoma. The myeloma cells are selected beforehand to ensure that they are not secreting antibody themselves and that they are sensitive to the hypoxanthine–aminopterin–thymidine (HAT) medium that is used to select hybrid cells because they lack the enzyme hypoxanthine:guanine phosphoribosyl transferase (HGPRT). The HGPRT gene contributed by the spleen cell allows hybrid cells to survive in the HAT medium, and only hybrid cells can grow continuously in culture because of the malignant potential contributed by the myeloma cells. Unfused myeloma cells and unfused spleen cells therefore die in the HAT medium, as shown here by cells with dark, irregular nuclei. Individual hybridomas are then screened for antibody production, and cells that make antibody of the desired specificity are cloned by growing them up from a single antibody-producing cell. The cloned hybridoma cells are grown in bulk culture to produce large amounts of antibody. As each hybridoma is descended from a single cell, all the cells of a hybridoma cell line make the same antibody molecule, which is thus called a monoclonal antibody.

isolated by selecting the phages in the library for binding to that antigen. The phage particles that bind are recovered and used to infect fresh bacteria. Each phage isolated in this way will produce a monoclonal antigen-binding particle analogous to a monoclonal antibody (Fig. A.16). The genes encoding the antigen-binding site, which are unique to each phage, can then be recovered from the phage DNA and used to construct genes for a complete antibody molecule by joining them to parts of immunoglobulin genes that encode the invariant parts of an antibody. When these reconstructed antibody genes are introduced into a suitable host cell line, such as the nonantibody-producing myeloma cells used for hybridomas, the transfected cells can secrete antibodies with all the desirable characteristics of monoclonal antibodies produced from hybridomas.

In much the same way that a collection of phages can display a wide variety of potential antigen-binding sites, the phages can also be engineered to display a wide variety of antigens; such a library is known as an **antigen display library**. In such cases, the antigens displayed are often short peptides encoded by chemically synthesized DNA sequences that have mixtures of all four nucleotides in some positions, so that all possible amino acids are incorporated. It is not usual for every position in a peptide to be allowed to vary in this way, because the number of different peptide sequences increases dramatically with the number of variable positions; there are more than 2×10^{10} possible sequences of eight amino acids.

A-14 Microscopy and imaging.

Because antibodies bind stably and specifically to their corresponding antigen, they are invaluable as probes for identifying a particular molecule in cells, tissues, or biological fluids. Antibody molecules can be used to locate their target molecules accurately in single cells or in tissue sections by a variety of different labeling techniques. When the antibody itself, or the anti-immunoglobulin antibody used to detect it, is labeled with a fluorescent dye (a fluorochrome or fluorophore), the technique is known as **immunofluorescence microscopy**. As in all serological techniques, the antibody binds stably to its antigen, allowing unbound antibody to be removed by thorough washing. Because antibodies against proteins recognize the surface features of the native, folded protein, the native structure of the protein being sought usually needs to be preserved, either by using only the most gentle chemical fixation techniques or by using frozen tissue sections that are fixed only after the antibody reaction has been performed. Some antibodies, however, bind proteins

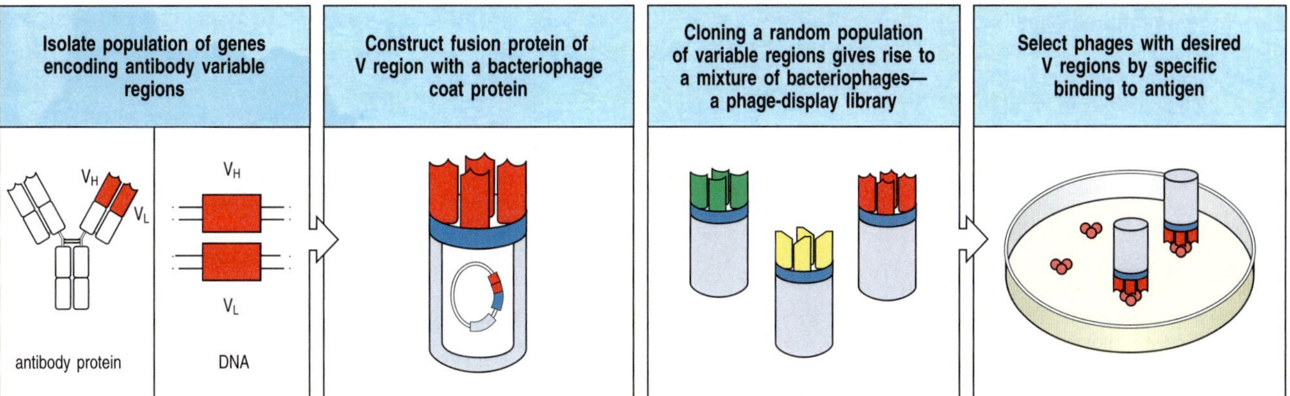
Isolate population of genes encoding antibody variable regions
Construct fusion protein of V region with a bacteriophage coat protein
Cloning a random population of variable regions gives rise to a mixture of bacteriophages—a phage-display library
Select phages with desired V regions by specific binding to antigen
VH
VL
VH
VL
antibody protein
DNA

Fig. A.16 The production of antibodies by genetic engineering. Short primers to consensus sequences in heavy- and light-chain variable (V) regions of immunoglobulin genes are used to generate a library of heavy- and light-chain V-region DNAs by PCR, with spleen DNA as the starting material. These heavy- and light-chain V-region genes are cloned randomly into filamentous phages such that each phage expresses one heavy-chain and one light-chain V region as a surface fusion protein with antibody-like properties. The resulting phage display library is multiplied in bacteria, and the phages are then bound to a surface coated with antigen. The unbound phages are washed away; the bound phages are recovered, multiplied in bacteria, and again bound to antigen. After a few cycles, only specific high-affinity antigen-binding phages are left. These can be used like antibody molecules, or their V genes can be recovered and engineered into antibody genes to produce genetically engineered antibody molecules (not shown). This technology may replace the hybridoma technology for producing monoclonal antibodies, and has the advantage that humans can be used as the source of DNA.

even if they are denatured, and such antibodies will bind specifically even to protein in fixed tissue sections.

The fluorescent dye can be covalently attached directly to the specific antibody; however, the bound antibody is more commonly detected by fluorescently labeled anti-immunoglobulin, a technique known as **indirect immunofluorescence**. The dyes chosen for immunofluorescence are excited by light of one wavelength, usually blue or green, and emit light of a different wavelength in the visible spectrum. The most commonly used fluorochromes are fluorescein, which emits green light, Texas Red and Peridinin chlorophyll protein (PerCP), which emit red light, and rhodamine and phycoerythrin (PE) which emit orange/red light (Fig. A.17). By using selective filters, only the light coming from the fluorochrome used is detected in the fluorescence microscope (Fig. A.18). Although Albert Coons first devised this technique to identify the plasma cell as the source of antibody, it can be used to detect the distribution of any protein. By attaching different dyes to different antibodies, the distribution of two or more molecules can be determined in the same cell or tissue section (see Fig. A.18).

The development of the **confocal fluorescent microscope**, which uses computer-aided techniques to produce ultrathin optical sections of a cell or tissue, gives very high resolution (sub-micrometer) fluorescence microscopy without the need for elaborate sample preparation. The light source for excitation (a laser) is focused onto a particular plane in the specimen, and the emitted light is refocused through a 'pinhole' so that only light from the desired plane reaches the detector, thus removing out-of-focus emissions from above or below the plane. This gives a sharper image than conventional fluorescence microscopy, and a three-dimensional picture can be built up from successive

Excitation and emission wavelengths of some commonly used fluorochromes		
Probe	Excitation (nm)	Emission (nm)
R-phycoerythrin (PE)	480; 565	578
Fluorescein	495	519
PerCP	490	675
Texas Red	589	615
Rhodamine	550	573

Fig. A.17 Excitation and emission wavelengths for common fluorochromes.

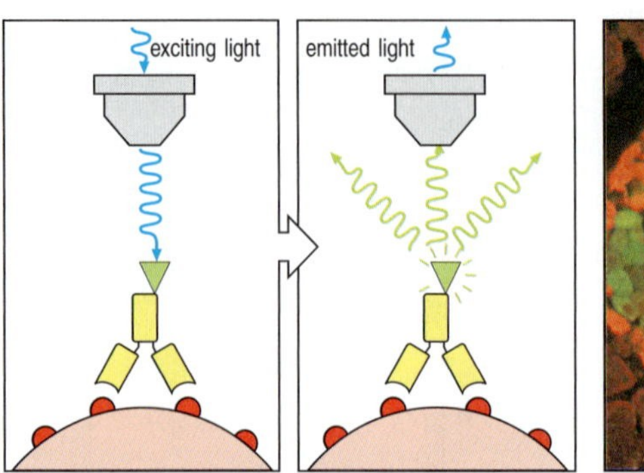

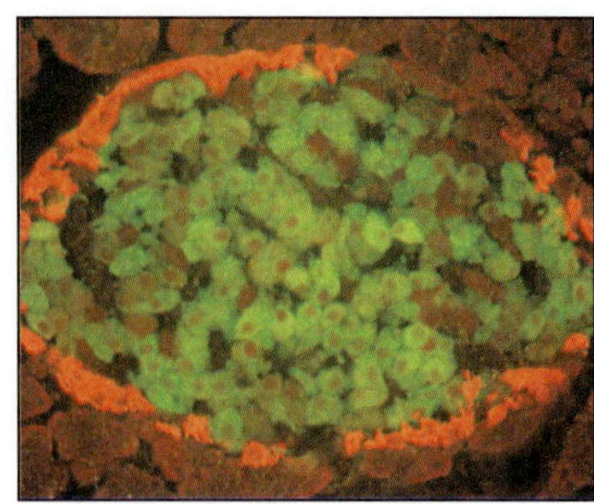

Fig. A.18 Immunofluorescence microscopy. Antibodies labeled with a fluorescent dye such as fluorescein (green triangle) are used to reveal the presence of their corresponding antigens in cells or tissues. The stained cells are examined in a microscope that exposes them to blue or green light to excite the fluorescent dye. The excited dye emits light at a characteristic wavelength, which is captured by viewing the sample through a selective filter. This technique is applied widely in biology to determine the location of molecules in cells and tissues. Different antigens can be detected in tissue sections by labeling antibodies with dyes of distinctive color. Here, antibodies against the protein glutamic acid decarboxylase (GAD) coupled to a green dye are shown to stain the β cells of pancreatic islets of Langerhans. The α cells do not make this enzyme and are labeled with antibodies against the hormone glucagon coupled with an orange fluorescent dye. GAD is an important antigen in diabetes, a disease in which the insulin-secreting β cells of the islets of Langerhans are destroyed by an immune attack on self tissues (see Chapter 15). Photograph courtesy of M. Solimena and P. De Camilli.

optical sections taken along the 'vertical' axis. Confocal microscopy can be used on fixed cells stained with fluorescently tagged antibodies or on living cells expressing proteins tagged with naturally fluorescent proteins. The first of these fluorescent proteins to come into wide use was green fluorescent protein (GFP), isolated from the jellyfish *Aequorea victoria*. The list of fluorescent proteins in routine use now includes those emitting red, blue, cyan, or yellow fluorescence. By using cells transfected with genes encoding different fusion proteins, it has been possible to visualize the redistribution of T-cell receptors, co-receptors, adhesion molecules, and other signaling molecules, such as CD45, that takes place when a T cell makes contact with a target cell (see Fig. 9.31).

Confocal microscopy, however, can only penetrate around 80 μm into a tissue, and at the wavelengths typically used for excitation, the source light will soon bleach the fluorescent label and damage the specimen. This means that the technique is not suitable for imaging a live specimen over a period of time sufficient, for example, to track the movements of cells in a tissue. The more recently developed technique of **two-photon scanning fluorescence microscopy** overcomes some of these limitations. Like confocal microscopy, two-photon microscopy produces thin optical sections from which a three-dimensional image can be built up. However, in this case, ultrashort pulses of laser light of much longer wavelength (and thus with photons of lower energy) are used for excitation, and two of these lower-energy photons arriving nearly simultaneously are required to excite the fluorophore. Excitation will therefore only occur in a very small region at the focus of the microscope, where the beam of light is most intense, and so fluorescence emission will be restricted to the plane of focus, producing a sharp, high-contrast image. The longer-wavelength light (typically in the near infrared) is also less damaging to living tissue than the blue and ultraviolet wavelengths typically used in confocal microscopy, and so imaging can be carried out over a longer period. More of the emitted light is collected than in confocal microscopy, and because single photons scattering within the tissue cannot cause fluorescence and consequent background haze, imaging to greater depths (several hundred micrometers) is possible.

To track the movements of molecules or cells over time, confocal or two-photon microscopy is combined with **time-lapse video imaging** using sensitive digital cameras. In immunology, time-lapse two-photon fluorescence imaging has been particularly valuable for tracking the movements of individual T cells and B cells expressing fluorescent proteins in intact lymphoid organs and observing where they interact (see Chapter 10).

A-15 Immunoelectron microscopy.

Antibodies can be used to detect the intracellular location of structures or particular proteins at high resolution by electron microscopy, a technique known as **immunoelectron microscopy**. Antibodies against the required antigen are labeled with gold particles and then applied to ultrathin sections, which are then examined in the transmission electron microscope. Antibodies labeled with gold particles of different diameters enable two or more proteins to be studied simultaneously (see Fig. 6.11). The difficulty with this technique is in staining the ultrathin section adequately, because few molecules of antigen are present in each section.

A-16 Immunohistochemistry.

An alternative to immunofluorescence (see Section A-14) for detecting a protein in tissue sections is **immunohistochemistry**, in which the specific antibody is chemically coupled to an enzyme that converts a colorless substrate into a colored reaction product *in situ*. The localized deposition of the colored product where antibody has bound can be directly observed under a light microscope. The antibody binds stably to its antigen, allowing unbound antibody to be removed by thorough washing. This method of detecting bound antibody is analogous to ELISA (see Section A-6) and frequently uses the same coupled enzymes, the difference in detection being primarily that in immunohistochemistry the colored products are insoluble and precipitate at the site where they are formed. Horseradish peroxidase and alkaline phosphatase are the two enzymes most commonly used in these applications. Horseradish peroxidase oxidizes the substrate diaminobenzidine to produce a brown precipitate, while alkaline phosphatase can produce red or blue dyes depending on the substrates used; a common substrate is 5-bromo-4-chloro-3-indolyl phosphate plus nitroblue tetrazolium (BCIP/NBT), which gives rise to a dark blue or purple stain. As with immunofluorescence, the native structure of the protein being sought usually needs to be preserved so that it will be recognized by the antibody. Tissues are fixed by the most gentle chemical fixation techniques, or frozen tissue sections are used that are fixed only after the antibody reaction has been performed.

A-17 Immunoprecipitation and co-immunoprecipitation.

To raise antibodies against membrane proteins and other cellular structures that are difficult to purify, mice are often immunized with whole cells or crude cell extracts. Antibodies against the individual molecules are then obtained by using these immunized mice to produce hybridomas making monoclonal antibodies (see Section A-12) that bind to the cell type used for immunization. To characterize the molecules identified by the antibodies, cells of the same type are labeled with radioisotopes and dissolved in nonionic detergents that disrupt cell membranes but do not interfere with antigen–antibody interactions. This allows the labeled protein to be isolated by binding to the antibody in a reaction known as **immunoprecipitation**. The antibody is usually attached to a solid support, such as the beads that are used in affinity chromatography (see Section A-5), or to Protein A. Cells can be labeled in two main ways for immunoprecipitation analysis. All the proteins in a cell can be labeled metabolically by growing the cell in radioactive amino acids that are incorporated into cellular protein (Fig. A.19). Alternatively, one can label only the cell-surface proteins by radioiodination under conditions that prevent iodine from crossing the plasma membrane and labeling proteins inside the cell, or by a reaction that labels only membrane proteins with biotin, a small molecule that is detected readily by labeled avidin, a protein found in egg whites that binds biotin with very high affinity.

Once the labeled proteins have been isolated by the antibody, they can be characterized in several ways. The most common is polyacrylamide gel electrophoresis (PAGE) of the proteins after they have been dissociated from

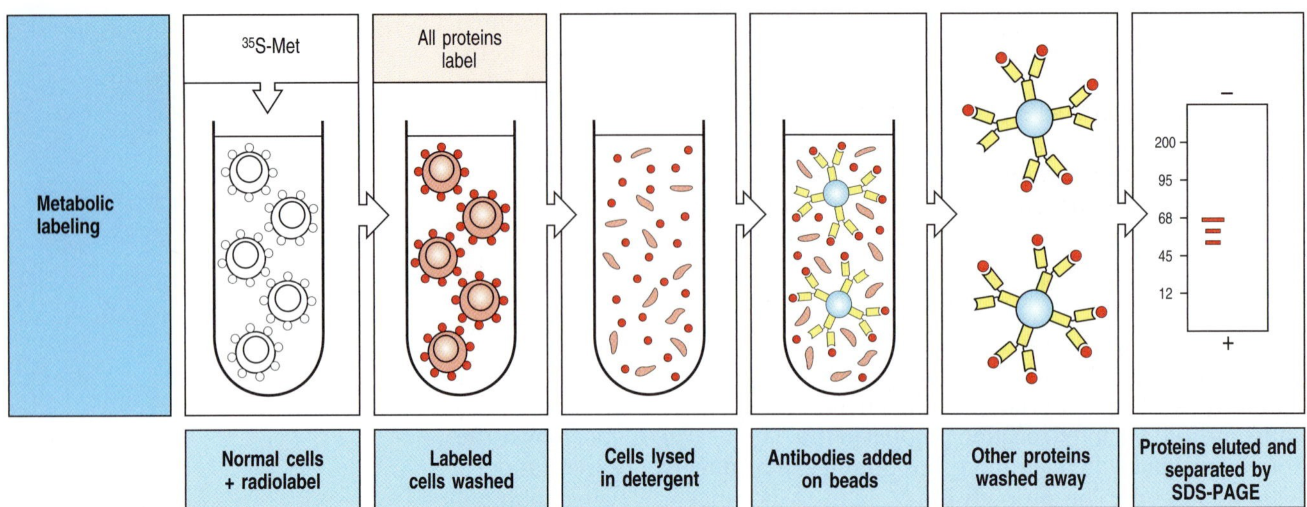

Fig. A.19 Cellular proteins reacting with an antibody can be characterized by immunoprecipitation of labeled cell lysates. All actively synthesized cellular proteins can be labeled metabolically by incubating cells with radioactive amino acids (shown here for methionine; ^{35}S-Met), or one can label just the cell-surface proteins by using radioactive iodine in a form that cannot cross the cell membrane or by a reaction with the small molecule biotin, detected by its reaction with labeled avidin (not shown). Cells are lysed with detergent and individual labeled cell-associated proteins can be precipitated with a monoclonal antibody attached to beads. After unbound proteins have been washed away, the bound protein is eluted in the detergent sodium dodecyl sulfate (SDS), which dissociates it from the antibody and also coats the protein with a strong negative charge, allowing it to migrate according to its size in polyacrylamide gel electrophoresis (PAGE). The positions of the labeled proteins are determined by autoradiography using X-ray film. This technique of SDS-PAGE can be used to determine the molecular weight and subunit composition of a protein. Patterns of protein bands observed with metabolic labeling are usually more complex than those revealed by radioiodination, owing to the presence of precursor forms of the protein (right panel). The mature form of a surface protein can be identified as being the same size as that detected by surface iodination or biotinylation (not shown).

antibody in the strong ionic detergent sodium dodecyl sulfate (SDS), a technique generally abbreviated as **SDS-PAGE**. SDS binds relatively homogeneously to proteins, conferring a charge that allows the electrophoretic field to drive protein migration through the gel. The rate of migration is controlled mainly by protein size (see Fig. A.19). Proteins of differing charges can be separated using isoelectric focusing (see Section A-12). This technique can be combined with SDS-PAGE in a procedure known as **two-dimensional gel electrophoresis**. For this, the immunoprecipitated protein is eluted in urea, a nonionic solubilizing agent, and run on an isoelectric focusing gel in a narrow tube of polyacrylamide. This first-dimensional isoelectric focusing gel is then placed across the top of an SDS-PAGE slab gel, which is then run vertically to separate the proteins by molecular weight (Fig. A.20). Two-dimensional gel electrophoresis is a powerful technique that allows many hundreds of proteins in a complex mixture to be distinguished from one another.

Immunoprecipitation and the related technique of immunoblotting (see Section A-18) are useful for determining the molecular weight and isoelectric point of a protein as well as its abundance, distribution, and whether, for

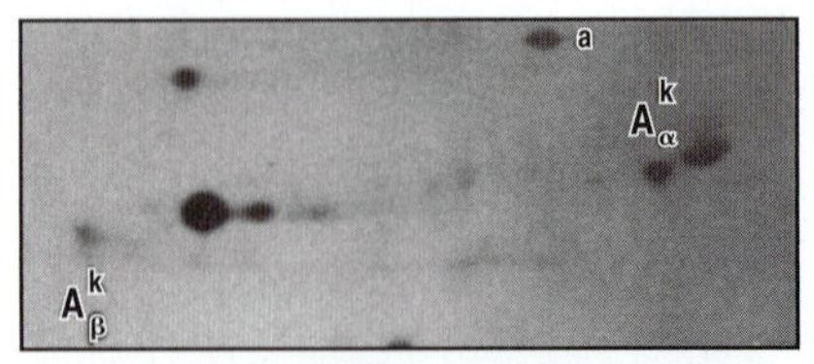

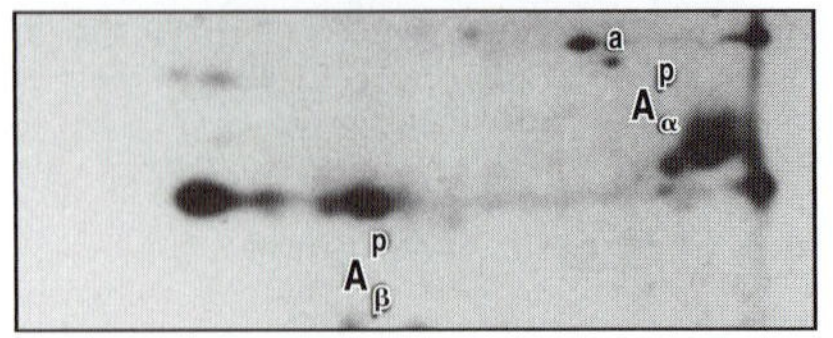

Fig. A.20 Two-dimensional gel electrophoresis of MHC class II molecules. Proteins in mouse spleen cells have been labeled metabolically (see Fig. A.19), precipitated with a monoclonal antibody against the mouse MHC class II molecule H2-A, and separated by isoelectric focusing in one direction and SDS-PAGE in a second direction at right angles to the first (hence the term two-dimensional gel electrophoresis). This allows one to distinguish molecules of the same molecular weight on the basis of their charge. The separated proteins are detected by autoradiography. The MHC class II molecules are composed of two chains, α and β, and in the different MHC class II molecules these have different isoelectric points (compare upper and lower panels). The MHC genotype of mice is indicated by lower-case superscripts (k, p). Actin, a common contaminant, is marked a. Photographs courtesy of J.F. Babich.

example, it undergoes changes in molecular weight and isoelectric point as a result of processing within the cell.

A-18 Immunoblotting (Western blotting).

Like immunoprecipitation (see Section A-17), **immunoblotting** is used for identifying the presence of a given protein in a cell lysate, but it avoids the problem of having to label large quantities of cells with radioisotopes. Unlabeled cells are placed in detergent to solubilize all cell proteins, and the lysate is run on SDS-PAGE to separate the proteins (see Section A-17). The size-separated proteins are then transferred from the gel to a stable support such as a nitrocellulose membrane. Specific proteins are detected by treatment with antibodies able to react with SDS-solubilized proteins (mainly those that react with denatured sequences); the bound antibodies are revealed by anti-immunoglobulin antibodies labeled with radioisotopes or an enzyme. The term **Western blotting** as a synonym for immunoblotting arose because the comparable technique for detecting specific DNA sequences is known as Southern blotting, after Ed Southern who devised it, which in turn provoked the name Northern for blots of size-separated RNA, and Western for blots of size-separated proteins. Western blots have many applications in basic research and clinical diagnosis. They are often used to test sera for the presence of antibodies against specific proteins, for example to detect antibodies against different constituents of HIV (Fig. A.21).

Co-immunoprecipitation is an extension of the immunoprecipitation technique that is used to determine whether a given protein interacts physically with another given protein. Cell extracts containing the presumed interaction complex are first immunoprecipitated with antibody against one of the proteins. The material identified by this means is then tested for the presence of the other protein by immunoblotting with a specific antibody.

A-19 Use of antibodies in the isolation and identification of genes and their products.

As a first step in isolating the gene that codes for a particular protein, antibodies specific for the protein can be used to isolate the purified protein from cells by affinity chromatography (see Section A-5). Small amounts of amino acid sequence can then be obtained from the protein's amino-terminal end or from peptide fragments generated by proteolysis. The information in these amino acid sequences is used to construct a set of synthetic oligonucleotides corresponding to the possible DNA sequences, which are then used as probes to isolate the gene encoding the protein from either a library of DNA sequences complementary to mRNA (a cDNA library) or a genomic DNA library (a library of chromosomal DNA fragments).

An alternative approach to gene identification uses antibodies to identify the protein produced by a gene that has been introduced into a cell that does not normally express it. This technique is most commonly applied to the

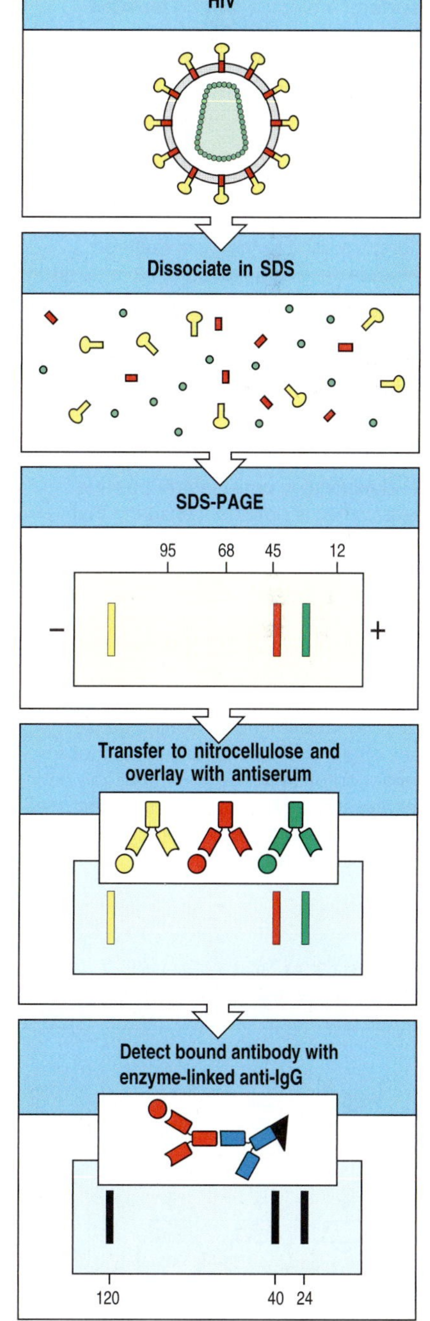

Fig. A.21 Western blotting is used to identify antibodies against the human immunodeficiency virus (HIV) in serum from infected individuals. The virus is dissociated into its constituent proteins by treatment with the detergent SDS, and its proteins are separated using SDS-PAGE. The separated proteins are transferred to a nitrocellulose sheet and reacted with the test serum. Anti-HIV antibodies in the serum bind to the various HIV proteins and are detected with enzyme-linked anti-human immunoglobulin, which deposits colored material from a colorless substrate. This general methodology will detect any combination of antibody and antigen and is used widely, although the denaturing effect of SDS means that the technique works most reliably with antibodies that recognize the antigen when it is denatured.

identification of genes encoding cell-surface proteins. A set of cDNA-containing expression vectors is first made from a cDNA library prepared from the total mRNA from a cell type that does express the protein of interest. The vectors are used to transfect a cell type that does not normally express the protein of interest, and the vector drives expression of the cDNA it contains without integrating into the host cell DNA. Cells expressing the required protein after transfection are then isolated by binding to specific antibodies that detect the presence of the protein on the cell surface. The vector containing the required gene can then be recovered from these cells (Fig. A.22).

The recovered vectors are then introduced into bacterial cells, where they replicate rapidly; these amplified vectors are used in a second round of transfection in mammalian cells. After several cycles of transfection, isolation, and amplification in bacteria, single colonies of bacteria are picked and the vectors prepared from cultures of each colony are used in a final transfection to identify a cloned vector carrying the cDNA of interest, which is then isolated and characterized. This methodology has been used to isolate many genes encoding cell-surface molecules.

The full amino acid sequence of the protein can be deduced from the nucleotide sequence of its cDNA, and this often gives clues to the nature of the protein and its biological properties. The nucleotide sequence of the gene and its regulatory regions can be determined from genomic DNA clones. The gene can be manipulated and introduced into cells by transfection for larger-scale production and functional studies. This approach has been used to characterize many immunologically important proteins, such as the MHC glycoproteins.

The converse approach is taken to identify the unknown protein product of a cloned gene. The gene sequence is used to construct synthetic peptides of 10–20 amino acids that are identical to part of the deduced protein sequence, and antibodies are then raised against these peptides by coupling them to carrier proteins; the peptides behave as haptens. These anti-peptide antibodies often bind the native protein and so can be used to identify its distribution in cells and tissues and to try to ascertain its function (Fig. A.23). This approach to identifying the function of a gene is often called 'reverse genetics' because it works from gene to phenotype rather than from phenotype to gene, which is the classical genetic approach. The great advantage of reverse genetics over the classical approach is that it does not require a detectable phenotypic genetic trait in order to identify a gene.

Antibodies can also be used in the determination of the function of gene products. Some antibodies are able to act as agonists, when the binding of the antibody to the molecule mimics the binding of the natural ligand and activates the function of the gene product. For example, antibodies against the CD3 molecule have been used to stimulate T cells, replacing the interaction of the

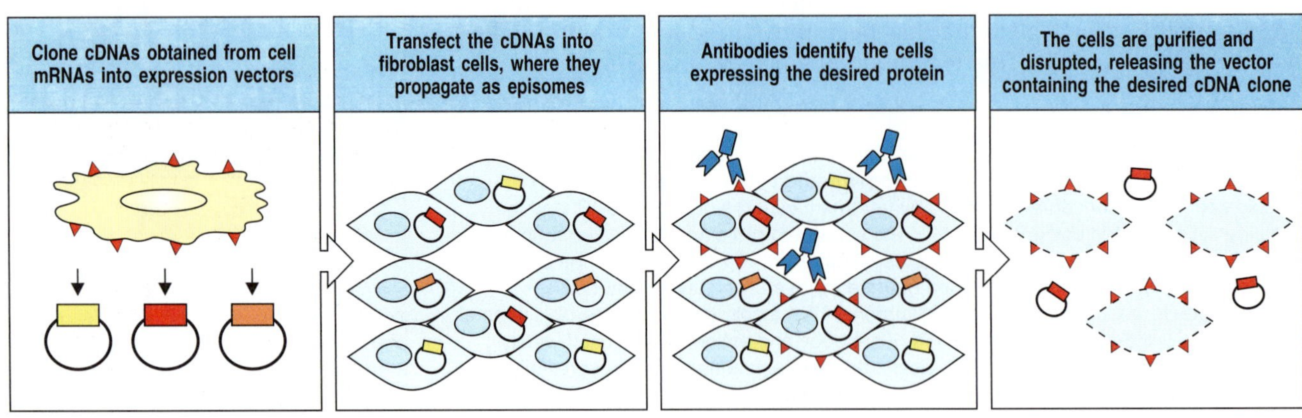

Clone cDNAs obtained from cell mRNAs into expression vectors
Transfect the cDNAs into fibroblast cells, where they propagate as episomes
Antibodies identify the cells expressing the desired protein
The cells are purified and disrupted, releasing the vector containing the desired cDNA clone

Fig. A.22 The gene encoding a cell-surface molecule can be isolated by expressing it in fibroblasts and detecting its protein product with monoclonal antibodies. Total mRNA from a cell line or tissue expressing the protein is isolated, converted into cDNA, and cloned as cDNAs in a vector designed to direct expression of the cDNA in fibroblasts. The entire cDNA library is used to transfect cultured fibroblasts. Fibroblasts that have taken up cDNA encoding a cell-surface protein express the protein on their surface; they can be isolated by binding a monoclonal antibody against that protein. The vector containing the gene is isolated from the cells that express the antigen and used for more rounds of transfection and reisolation until uniform positive expression is obtained, ensuring that the correct gene has been isolated. The cDNA insert can then be sequenced to determine the sequence of the protein it encodes and can also be used as the source of material for large-scale expression of the protein for analysis of its structure and function. The method illustrated is limited to cloning genes for single-chain proteins (that is, those encoded by only one gene) that can be expressed in fibroblasts. It has been used to clone many genes of immunological interest such as that for CD4.

Fig. A.23 **The use of antibodies to detect the unknown protein product of a known gene is called reverse genetics.** When the gene responsible for a genetic disorder such as Duchenne muscular dystrophy is isolated, the amino acid sequence of the unknown protein product of the gene can be deduced from the nucleotide sequence of the gene, and synthetic peptides representing parts of this sequence can be made. Antibodies are raised against these peptides and purified from the antiserum by affinity chromatography on a peptide column (see Fig. A.5). Labeled antibody is used to stain tissue from individuals with the disease and from unaffected individuals to determine differences in the presence, amount, and distribution of the normal gene product. The product of the dystrophin gene is present in normal mouse skeletal muscle cells, as shown in the bottom panel (red fluorescence), but is missing from the cells of mice bearing the mutation *mdx*, the mouse equivalent of Duchenne muscular dystrophy (not shown). Photograph (×15) courtesy of H.G.W. Lidov and L. Kunkel.

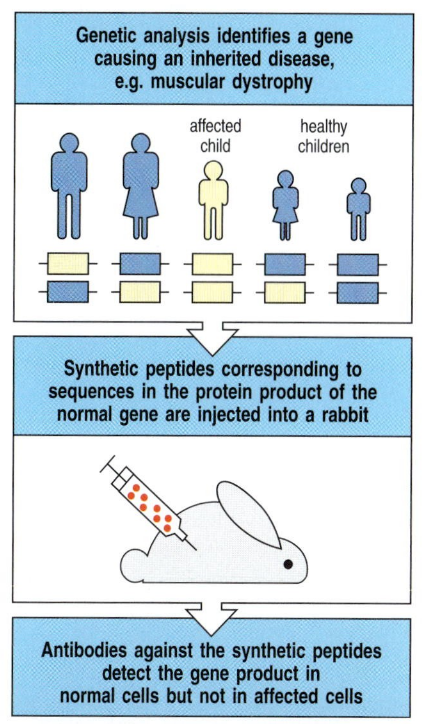

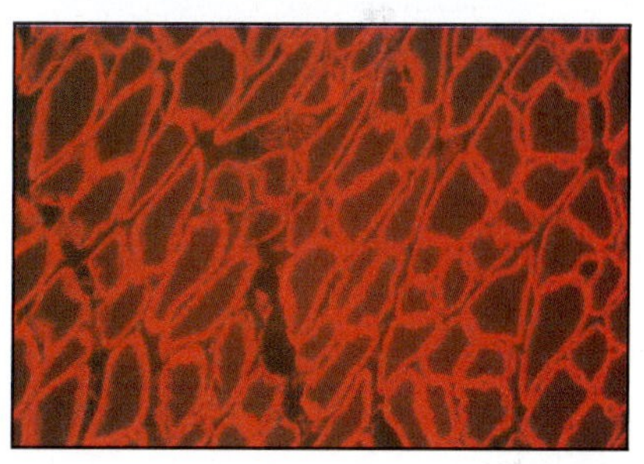

T-cell receptor with peptide:MHC antigens in cases where the specific peptide antigen is not known. Conversely, antibodies can function as antagonists, inhibiting the binding of the natural ligand and thus blocking the function.

Isolation of lymphocytes.

A-20 Isolation of peripheral blood lymphocytes by Ficoll-Hypaque™ gradient.

The first step in studying lymphocytes is to isolate them so that their behavior can be analyzed *in vitro*. Human lymphocytes can be isolated most readily from peripheral blood by density centrifugation over a step gradient consisting of a mixture of the carbohydrate polymer Ficoll-Hypaque™ and the dense iodine-containing compound metrizamide. This yields a population of mononuclear cells at the interface that has been depleted of red blood cells and most polymorphonuclear leukocytes or granulocytes (Fig. A.24). The resulting population, called **peripheral blood mononuclear cells (PBMCs)**, consists mainly of lymphocytes and monocytes. Although this population is readily accessible, it is not necessarily representative of the lymphoid system, because only recirculating lymphocytes can be isolated from blood.

A particular cell population can be isolated from a sample or culture by binding to antibody-coated plastic surfaces, a technique known as **panning**, or by removing unwanted cells by treatment with specific antibody and complement to kill them. Cells can also be passed over columns of antibody-coated,

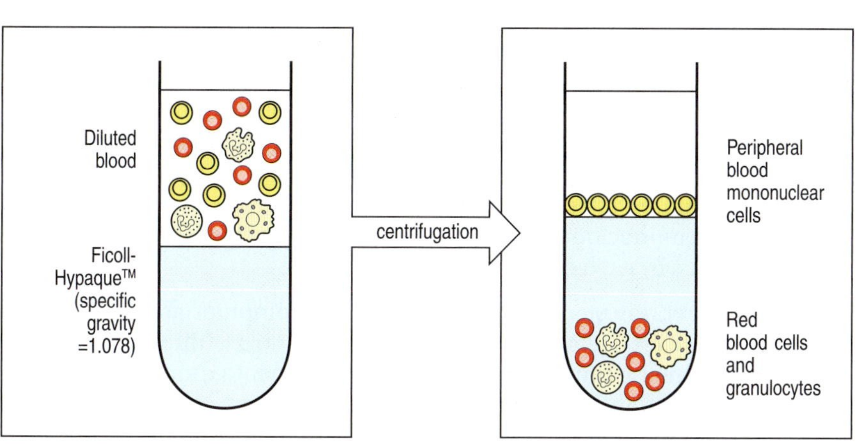

Fig. A.24 Peripheral blood mononuclear cells can be isolated from whole blood by Ficoll-Hypaque™ centrifugation. Diluted anticoagulated blood (left panel) is layered over Ficoll-Hypaque™ and centrifuged. Red blood cells and polymorphonuclear leukocytes or granulocytes are denser and travel through the Ficoll-Hypaque™, while mononuclear cells consisting of lymphocytes together with some monocytes band over it and can be recovered at the interface (right panel).

nylon-coated steel wool and different populations differentially eluted. This technique extends affinity chromatography to cells, and is now a very popular way to separate cells. All these techniques can also be used as a pre-purification step before sorting out highly purified populations by FACS (see Section A-22).

The 'normal' range of the numbers of the different types of white blood cells in blood, along with the normal ranges of concentrations of the various antibody classes, are given in Fig. A.42.

A-21 Isolation of lymphocytes from tissues other than blood.

In experimental animals, and occasionally in humans, lymphocytes are isolated from lymphoid organs, such as spleen, thymus, bone marrow, lymph nodes, or mucosal-associated lymphoid tissues, most commonly the palatine tonsils in humans (see Fig. 12.6). A specialized population of lymphocytes resides in surface epithelia; these cells are isolated by fractionating the epithelial layer after its detachment from the basement membrane. Finally, in situations where local immune responses are prominent, lymphocytes can be isolated from the site of the response itself. For example, in order to study the autoimmune reaction that is thought to be responsible for rheumatoid arthritis, an inflammatory response in joints, lymphocytes are isolated from the fluid aspirated from the inflamed joint space.

A-22 Flow cytometry and FACS analysis.

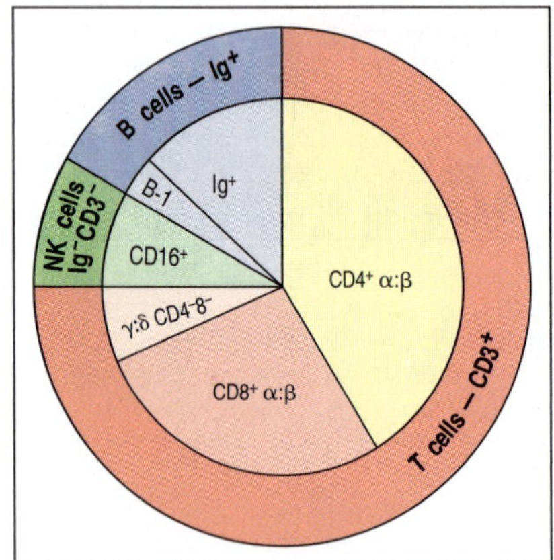

Fig. A.25 The distribution of lymphocyte subpopulations in human peripheral blood. As shown on the outside of the circle, lymphocytes can be divided into T cells bearing T-cell receptors (detected with anti-CD3 antibodies), B cells bearing immunoglobulin receptors (detected with anti-immunoglobulin antibodies), and null cells including natural killer (NK) cells, that label with neither. Further divisions of the T-cell and B-cell populations are shown inside. Using anti-CD4 and anti-CD8 antibodies, α:β T cells can be subdivided into two populations, whereas γ:δ T cells are identified with antibodies against the γ:δ T-cell receptor and mainly lack CD4 and CD8. A minority population of B cells, the so-called B-1 cells, have different characteristics from the majority population (B-2 cells) (see Section 3-24).

Resting lymphocytes are small round cells with a dense nucleus and little cytoplasm (see Fig. 1.7). However, these cells comprise many functional subpopulations, which are usually identified and distinguished from each other on the basis of their differential expression of cell-surface proteins, which can be detected with specific antibodies (Fig. A.25). B and T lymphocytes, for example, are identified unambiguously and separated from each other by antibodies against the constant regions of the B- and T-cell antigen receptors. T cells are further subdivided on the basis of expression of the co-receptor proteins CD4 and CD8.

An immensely powerful tool for defining and enumerating lymphocytes is the flow cytometer, which detects and counts individual cells passing in a stream through a laser beam. A flow cytometer equipped to separate the identified cells is called a **fluorescence-activated cell sorter (FACS)**. These instruments are used to study the properties of cell subsets identified using monoclonal antibodies against cell-surface proteins. Individual cells within a mixed population are first tagged by treatment with specific monoclonal antibodies labeled with fluorescent dyes, or by specific antibodies followed by labeled anti-immunoglobulin antibodies. The mixture of labeled cells is then forced with a much larger volume of saline through a nozzle, creating a fine stream of liquid containing cells spaced singly at intervals. As each cell passes through a laser beam it scatters the laser light, and any dye molecules bound to the cell will be excited and will fluoresce. Sensitive photomultiplier tubes detect both the scattered light, which gives information on the size and granularity of the cell, and the fluorescence emissions, which give information on the binding of the labeled monoclonal antibodies and hence on the expression of cell-surface proteins by each cell (Fig. A.26).

In the cell sorter, the signals passed back to the computer are used to generate an electric charge, which is passed from the nozzle through the liquid stream at the precise time that the stream breaks up into droplets, each containing no more than a single cell; droplets containing a charge can then be deflected from the main stream of droplets as they pass between plates of opposite charge, so that positively charged droplets are attracted to a negatively charged plate, and vice versa. In this way, specific subpopulations of cells, distinguished by the binding of the labeled antibody, can be purified from a mixed population of cells. Alternatively, to deplete a population of cells, the same fluorochrome can be used to label different antibodies

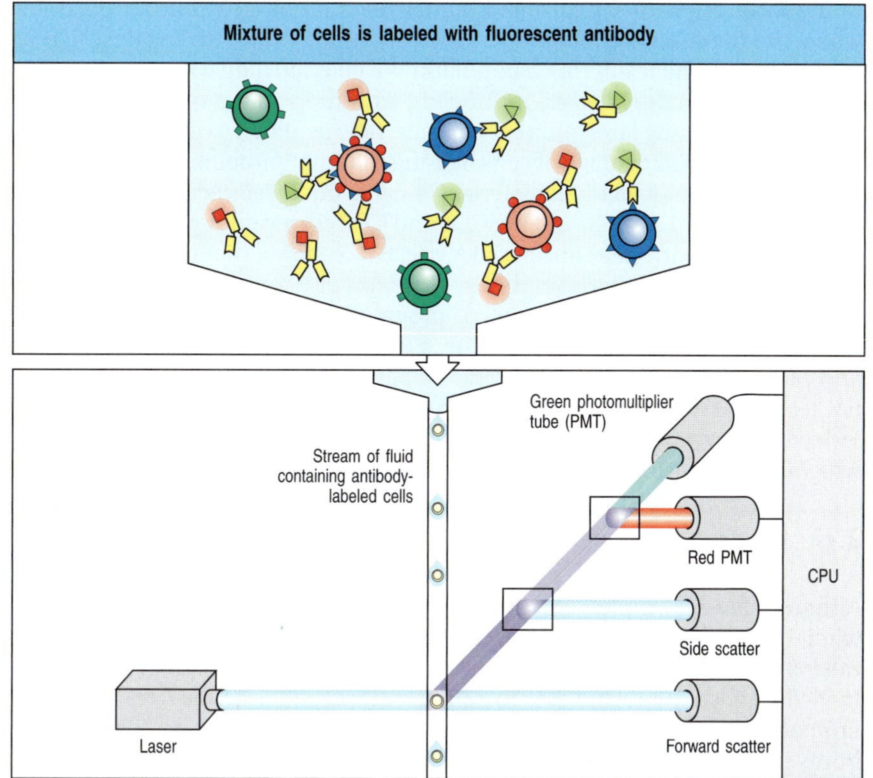

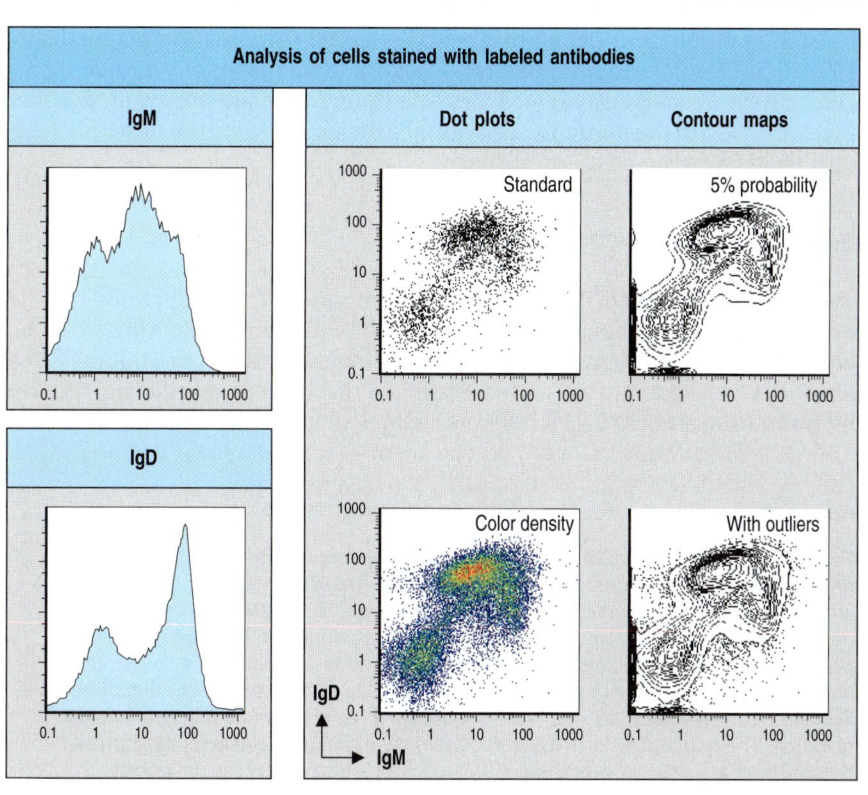

Fig. A.26 The FACS™ allows individual cells to be identified by their cell-surface antigens and to be sorted. Cells to be analyzed by flow cytometry are first labeled with fluorescent dyes (top panel). Direct labeling uses dye-coupled antibodies specific for cell-surface antigens (as shown here), while indirect labeling uses a dye-coupled immunoglobulin to detect unlabeled cell-bound antibody. The cells are forced through a nozzle in a single-cell stream that passes through a laser beam (second panel). Photomultiplier tubes (PMTs) detect the scattering of light, which is a sign of cell size and granularity, and emissions from the different fluorescent dyes. This information is analyzed by computer (CPU). By examining many cells in this way, the number of cells with a specific set of characteristics can be counted and levels of expression of various molecules on these cells can be measured. The lower part of the figure shows how these data can be represented, using the expression of two surface immunoglobulins, IgM and IgD, on a sample of B cells from a mouse spleen. The two immunoglobulins have been labeled with different colored dyes. When the expression of just one type of molecule is to be analyzed (IgM or IgD), the data are usually displayed as a histogram, as in the left-hand panels. Histograms display the distribution of cells expressing a single measured parameter (for example, size, granularity, fluorescence intensity). When two or more parameters are measured for each cell (IgM and IgD), various types of two-color plots can be used to display the data, as shown in the right-hand panel. All four plots represent the same data. The horizontal axis represents intensity of IgM fluorescence, and the vertical axis the intensity of IgD fluorescence. Two-color plots provide more information than histograms; they allow recognition, for example, of cells that are 'bright' for both colors, 'dull' for one and bright for the other, dull for both, negative for both, and so on. For example, the cluster of dots in the extreme lower left portions of the plots represents cells that do not express either immunoglobulin, and are mostly T cells. The standard dot plot (upper left) places a single dot for each cell whose fluorescence is measured. It is good for picking up cells that lie outside the main groups but tends to saturate in areas containing a large number of cells of the same type. A second means of presenting these data is the color dot plot (lower left), which uses color density to indicate high-density areas. A contour plot (upper right) draws 5% 'probability' contours, with 5% of the cells lying between each contour providing the best monochrome visualization of regions of high and low density. The lower right plot is a 5% probability contour map, which also shows outlying cells as dots.

directed at marker proteins expressed by the various undesired cell types. The cell sorter can be used to direct labeled cells to a waste channel, retaining only the unlabeled cells.

When cells are labeled with a single fluorescent antibody, the data from a flow cytometer are usually displayed in the form of a histogram of fluorescence intensity versus cell numbers. If two or more antibodies are used, each coupled to different fluorescent dyes, then the data are more usually displayed in the form of a two-dimensional scatter diagram or as a contour diagram, where the fluorescence of one dye-labeled antibody is plotted against that of a second, with the result that a population of cells labeling with one antibody can be further subdivided by its labeling with the second antibody (see Fig. A.26). By examining large numbers of cells, flow cytometry can give quantitative data on the percentage of cells bearing different proteins, such as surface immunoglobulin, which characterizes B cells, the T-cell receptor-associated molecules known as CD3, and the CD4 and CD8 co-receptor proteins that distinguish the major T-cell subsets. Likewise, FACS analysis has been instrumental in defining stages in the early development of B and T cells. As the power of the FACS technology has grown, progressively more antibodies labeled with distinct fluorescent dyes can be used at the same time. Three-, four-, and even five-color analyses can now be handled by very powerful machines. FACS analysis has been applied to a broad range of problems in immunology; indeed, it played a vital role in the early identification of AIDS as a disease in which T cells bearing CD4 are depleted selectively (see Chapter 13).

Heterogeneous population of lymphocytes is mixed with antibodies coupled to paramagnetic particles or beads and poured over an iron wool mesh

When a magnetic field is applied, the coupled cells stick to the iron wool; unlabeled cells are washed out

N

N

The magnetic field is removed, releasing the coupled cells

A-23 Lymphocyte isolation using antibody-coated magnetic beads.

Although FACS is superb for isolating small numbers of cells in pure form, when large numbers of lymphocytes must be prepared quickly, mechanical means of separating cells are preferable. A powerful and efficient way of isolating lymphocyte populations is to couple paramagnetic beads to monoclonal antibodies that recognize distinguishing cell-surface molecules. These antibody-coated beads are mixed with the cells to be separated, and run through a column containing material that attracts the paramagnetic beads when the column is placed in a strong magnetic field. Cells binding the magnetically labeled antibodies are retained; cells lacking the appropriate surface molecule can be washed away (Fig. A.27). The bound cells are positively selected for expression of the particular cell-surface molecule, while the unbound cells are negatively selected for its absence.

A-24 Isolation of homogeneous T-cell lines.

The analysis of specificity and effector function in T cells depends heavily on the study of monoclonal populations of T lymphocytes. These can be obtained in four distinct ways. First, by analogy with B-cell hybridomas (see Section A-12), normal T cells proliferating in response to specific antigen can be fused to malignant T-cell lymphoma lines to generate **T-cell hybrids**. The

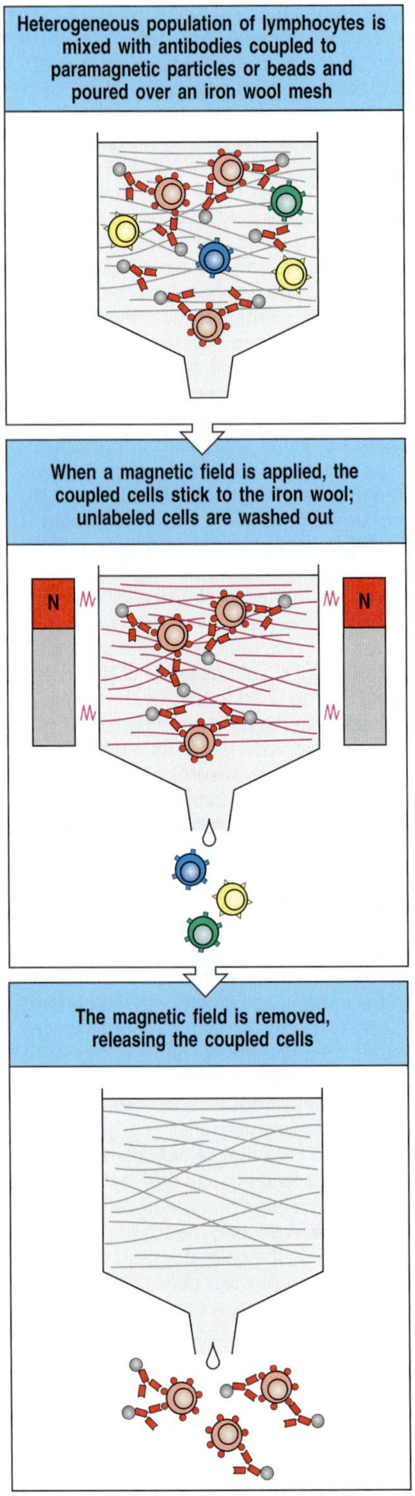

Fig. A.27 Lymphocyte subpopulations can be separated physically by using antibodies coupled to paramagnetic particles or beads. A mouse monoclonal antibody specific for a particular cell-surface molecule is coupled to paramagnetic particles or beads. It is mixed with a heterogeneous population of lymphocytes and poured over an iron wool mesh in a column. A magnetic field is applied so that the antibody-bound cells stick to the iron wool while cells that have not bound antibody are washed out; these cells are said to be negatively selected for lack of the molecule in question. The bound cells are released by removing the magnetic field; they are said to be positively selected for presence of the antigen recognized by the antibody.

hybrids express the receptor of the normal T cell, but proliferate indefinitely owing to the cancerous state of the lymphoma parent. T-cell hybrids can be cloned to yield a population of cells all having the same T-cell receptor. When stimulated by their specific antigen, these cells release cytokines such as the T-cell growth factor interleukin-2 (IL-2), and the production of cytokines is used as an assay to assess the antigen specificity of the T-cell hybrid.

T-cell hybrids are excellent tools for the analysis of T-cell specificity, because they grow readily in suspension culture. However, they cannot be used to analyze the regulation of specific T-cell proliferation in response to antigen because they are continually dividing. T-cell hybrids also cannot be transferred into an animal to test for function *in vivo* because they would give rise to tumors. Functional analysis of T-cell hybrids is also confounded by the fact that the malignant partner cell affects their behavior in functional assays. Therefore, the regulation of T-cell growth and the effector functions of T cells must be studied using **T-cell clones**. These are clonal cell lines of a single T-cell type and antigen specificity, which are derived from cultures of heterogeneous T cells, called **T-cell lines**, whose growth is dependent on periodic restimulation with specific antigen and, frequently, on the addition of T-cell growth factors (Fig. A.28). T-cell clones also require periodic restimulation with antigen and are more tedious to grow than T-cell hybrids but, because their growth depends on specific antigen recognition, they maintain antigen specificity, which is often lost in T-cell hybrids. Cloned T-cell lines can be used for studies of effector function both *in vitro* and *in vivo*. In addition, the proliferation of T cells, a critical aspect of clonal selection, can be characterized only in cloned T-cell lines, where such growth is dependent on antigen recognition. Thus, both types of monoclonal T-cell lines have valuable applications in experimental studies.

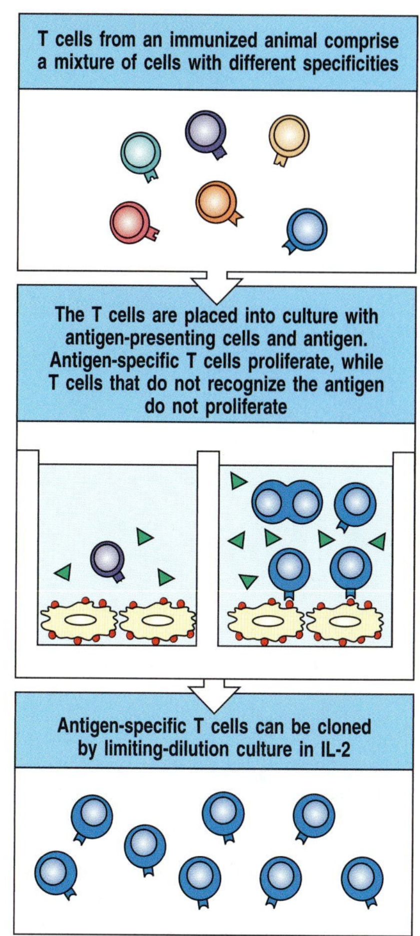

T cells from an immunized animal comprise a mixture of cells with different specificities
The T cells are placed into culture with antigen-presenting cells and antigen. Antigen-specific T cells proliferate, while T cells that do not recognize the antigen do not proliferate
Antigen-specific T cells can be cloned by limiting-dilution culture in IL-2

Fig. A.28 Production of cloned T-cell lines. T cells from an immune donor, comprising a mixture of cells with different specificities, are activated with antigen and antigen-presenting cells. Single responding cells are cultured by limiting dilution in the T-cell growth factor IL-2, which selectively stimulates the responding cells to proliferate. From these single cells, cloned lines specific for antigen are identified and can be propagated by culture with antigen, antigen-presenting cells, and IL-2.

Studies of human T cells have relied largely on T-cell clones because a suitable fusion partner for making T-cell hybrids has not been identified. However, a human T-cell lymphoma line, called Jurkat, has been characterized extensively because it secretes IL-2 when its antigen receptor is cross-linked with anti-receptor monoclonal antibodies. This simple assay system has yielded much information about signal transduction in T cells. One of the Jurkat cell line's most interesting features, shared with T-cell hybrids, is that it stops growing when its antigen receptor is cross-linked. This has allowed mutants lacking the receptor or having defects in signal transduction pathways to be selected simply by culturing the cells with anti-receptor antibody and selecting those that continue to grow. Thus, T-cell tumors, T-cell hybrids, and cloned T-cell lines all have valuable applications in experimental immunology.

Finally, primary T cells from any source can be isolated as single, antigen-specific cells by limiting dilution rather than by first establishing a mixed population of T cells in culture as a T-cell line and then deriving clonal subpopulations. During the growth of T-cell lines, particular T-cell clones can come to dominate the cultures and give a false picture of the number and specificities in the original sample. Direct cloning of primary T cells avoids this artifact.

Characterization of lymphocyte specificity, frequency, and function.

B cells are relatively easy to characterize because they have only one function—antibody production. T cells are more difficult to characterize because there are several different classes with different functions. It is also technically more difficult to study the membrane-bound T-cell receptors than the antibodies secreted in large amounts by B cells. All the methods in this part of the appendix can be used for T cells. Some are also used to detect and count B cells.

On many occasions it is important to know the frequency of antigen-specific lymphocytes, especially T cells, in order to measure the efficiency with which an individual responds to a particular antigen, for example, or the degree to which specific immunological memory has been established. There are a number of methods for doing this, either by detecting the cells directly by the specificity of their receptor, or by detecting activation of the cells to provide some particular function, such as cytokine secretion or cytotoxicity.

The first technique of this type to be established was the limiting-dilution culture (see Section A-25), in which the frequency of specific T or B cells responding to a particular antigen could be estimated by plating the cells into 96-well plates at increasing dilutions and measuring the number of wells in which there was no response. However, in this type of assay it became laborious to ask detailed questions about the phenotype of the responding cells, and to compare responses from different cell subpopulations.

A simpler assay for measuring the responses of T-cell populations has been developed from a variant of the antigen-capture ELISA method (see Section A-6), called the ELISPOT assay (see Section A-26). It assays T cells on the basis of cytokine production. In the ELISPOT assay, cytokine secreted by individual activated T cells is immobilized as discrete spots on a plastic plate, which are counted to give the number of activated T cells. The ELISPOT assay suffers from many of the same problems as the limiting-dilution assay in giving information about the nature of the activated cells, and it can be difficult to determine whether individual cells are capable of secreting mixtures of cytokines. It was therefore important to develop assays that could make these measurements on single cells. Measurements based on flow cytometry (see Section A-22) proved to be the answer, with the development of methods for detecting fluorescently labeled cytokines within activated T cells. The drawback of intracellular cytokine staining (see Section A-27) was that the T cells have to be killed and permeabilized by detergents to enable the cytokines to be detected. This led to the more sophisticated technique of capturing secreted labeled cytokines on the surfaces of the living T cells (see Section A-27).

Finally, methods for directly detecting T cells on the basis of the specificity of their receptor, using fluorochrome-tagged tetramers of specific peptide:MHC complexes (see Section A-28), have revolutionized the study of T-cell responses in a similar way to the use of monoclonal antibodies.

A-25 Limiting-dilution culture.

The response of a lymphocyte population is a measure of the overall response, but the frequency of lymphocytes able to respond to a given antigen can be determined only by **limiting-dilution culture**. This assay makes use of the Poisson distribution, a statistical function that describes how objects are distributed at random. For instance, when a sample of heterogeneous T cells is distributed equally into a series of culture wells, some wells will receive no T cells specific for a given antigen, some will receive one specific T cell, some two, and so on. The T cells in the wells are activated with specific antigen, antigen-presenting cells, and growth factors. After allowing several days for their growth and differentiation, the cells in each well are tested for a response to antigen, such as cytokine release or the ability to kill specific target cells (Fig. A.29). The assay is replicated with different numbers of T cells in the samples. The logarithm of the proportion of wells in which there is no response is plotted against the number of cells initially added to each well. If cells of one type, typically antigen-specific T cells because of their

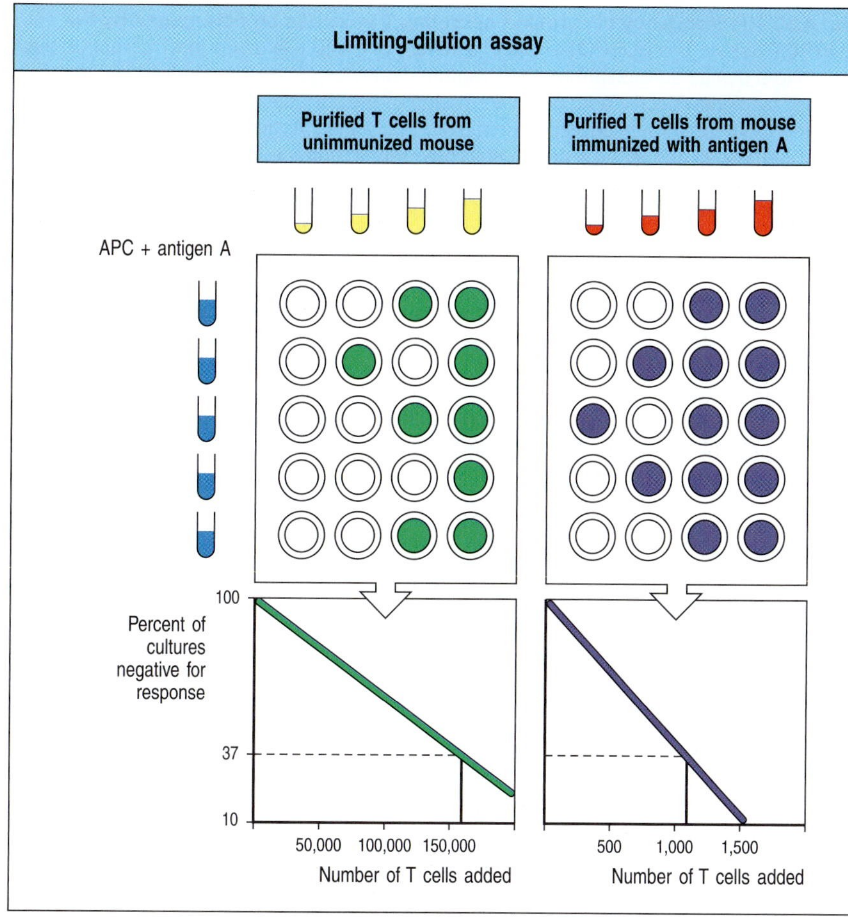

Fig. A.29 The frequency of specific lymphocytes can be determined using limiting-dilution assay. Various numbers of lymphoid cells from normal or immunized mice are added to individual culture wells and stimulated with antigen and antigen-presenting cells (APCs) or polyclonal mitogen and added growth factors. After several days, the wells are tested for a specific response to antigen, such as cytotoxic killing of target cells. Each well that initially contained a specific T cell will make a response to its target, and from the Poisson distribution one can determine that when 37% of the wells are negative, each well contained, on average, one specific T cell at the beginning of the culture. In the example shown, for the unimmunized mouse 37% of the wells are negative when 160,000 T cells have been added to each well; thus the frequency of antigen-specific T cells is 1 in 160,000. When the mouse is immunized, 37% of the wells are negative when only 1100 T cells have been added; hence the frequency of specific T cells after immunization is 1 in 1100, an increase in responsive cells of 150-fold.

rarity, are the only limiting factor for obtaining a response, then a straight line is obtained. From the Poisson distribution, it is known that there is, on average, one antigen-specific cell per well when the proportion of negative wells is 37%. Thus, the frequency of antigen-specific cells in the population equals the reciprocal of the number of cells added to each well when 37% of the wells are negative. After priming, the frequency of specific cells goes up substantially, reflecting the antigen-driven proliferation of antigen-specific cells. The limiting-dilution assay can also be used to measure the frequency of B cells that can make antibody against a given antigen.

A-26 ELISPOT assays.

A modification of the ELISA antigen-capture assay (see Section A-6), called the **ELISPOT assay**, has provided a powerful tool for measuring the frequency of T-cell responses. Populations of T cells are stimulated with the antigen of interest, and are then allowed to settle onto a plastic plate coated with antibodies against the cytokine that is to be assayed (Fig. A.30). If an activated T cell is secreting that cytokine, it is captured by the antibody on the plastic plate. After a period the cells are removed, and a second antibody against the cytokine is added to the plate to reveal a circle of bound cytokine surrounding the position of each activated T cell; counting each spot, and knowing the number of T cells originally added to the plate allows a simple calculation of the frequency of T cells secreting that particular cytokine, giving the ELISPOT assay its name. ELISPOT can also be used to detect specific antibody secretion by B cells, in this case by using antigen-coated surfaces to trap specific antibody and labeled anti-immunoglobulin to detect the bound antibody.

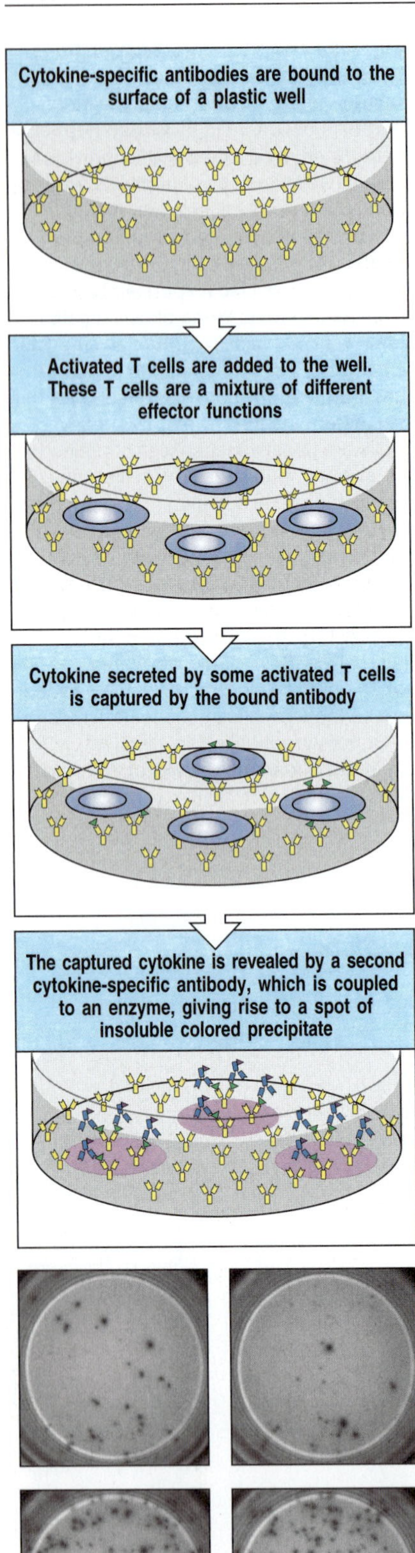

Cytokine-specific antibodies are bound to the surface of a plastic well
Activated T cells are added to the well. These T cells are a mixture of different effector functions
Cytokine secreted by some activated T cells is captured by the bound antibody
The captured cytokine is revealed by a second cytokine-specific antibody, which is coupled to an enzyme, giving rise to a spot of insoluble colored precipitate

Fig. A.30 The frequency of cytokine-secreting T cells can be determined by the ELISPOT assay. The ELISPOT assay is a variant of the ELISA assay in which antibodies bound to a plastic surface are used to capture cytokines secreted by individual T cells. Usually, cytokine-specific antibodies are bound to the surface of a plastic tissue-culture well and the unbound antibodies are removed (top panel). Activated T cells are then added to the well and settle onto the antibody-coated surface (second panel). If a T cell is secreting the appropriate cytokine, this will then be captured by the antibody molecules on the plate surrounding the T cell (third panel). After a period of time the T cells are removed, and the presence of the specific cytokine is detected using an enzyme-labeled second antibody specific for the same cytokine. Where this binds, a colored reaction product can be formed (fourth panel). Each T cell that originally secreted cytokine gives rise to a single spot of color, hence the name of the assay. The results of such an ELISPOT assay for T cells secreting IFN-γ in response to different stimuli are shown in the last panel. In this example, T cells from a stem-cell transplant recipient were treated with a control peptide (top two panels) or a peptide from cytomegalovirus (bottom two panels). You can see a greater number of spots in the bottom two panels, indicating clearly that the patient's T cells are able to respond to the viral peptide and produce IFN-γ. Photographs courtesy of S. Nowack.

A-27 Identification of functional subsets of T cells by staining for cytokines.

One problem with the detection of cytokine production on a single-cell level is that the cytokines are secreted by the T cells into the surrounding medium, and any association with the originating cell is lost. Two methods have been devised that allow the cytokine profile produced by individual cells to be determined. The first, that of **intracellular cytokine staining** (Fig. A.31), relies on the use of metabolic poisons that inhibit protein export from the cell. The cytokine thus accumulates within the endoplasmic reticulum and vesicular network of the cell. If the cells are subsequently fixed and rendered permeable by the use of mild detergents, antibodies can gain access to these intracellular compartments and detect the cytokine. The T cells can be stained for other markers simultaneously, and thus the frequency of IL-10-producing CD25$^+$ CD4 T cells, for example, can be easily obtained.

A second method, which has the advantage that the cells being analyzed are not killed in the process, is called **cytokine capture**. This technique uses hybrid antibodies, in which the two separate heavy- and light-chain pairs from different antibodies are combined to give a mixed antibody molecule in which the two antigen-binding sites recognize different ligands (Fig. A.32). In the bispecific antibodies used to detect cytokine production, one of the antigen-binding sites is specific for a T-cell surface marker, while the other is specific for the cytokine in question. The bispecific antibody binds to the T cells through the binding site for the cell-surface marker, leaving the cytokine-binding site free. If that T cell is secreting the particular cytokine, it is captured by the bound antibody before it diffuses away from the surface of the cell. It can then be detected by adding a fluorochrome-labeled second antibody specific for the cytokine to the cells.

A-28 Identification of T-cell receptor specificity using peptide:MHC tetramers.

For many years, the ability to identify antigen-specific T cells directly through their receptor specificity eluded immunologists. Foreign antigen could not be used directly to identify T cells because, unlike B cells, they do not recognize antigen alone but rather the complexes of peptide fragments of antigen bound to self MHC molecules. Moreover, the affinity of interaction between the T-cell receptor and the peptide:MHC complex was in practice so low that attempts to label T cells with their specific peptide:MHC complexes routinely failed. The breakthrough in labeling antigen-specific T cells came with the idea of making multimers of the peptide:MHC complex, so as to increase the avidity of the interaction.

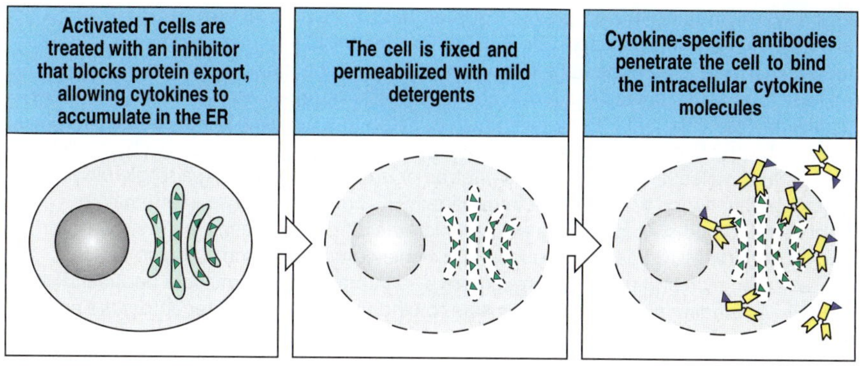

Fig. A.31 Cytokine-secreting cells can be identified by intracellular cytokine staining. The cytokines secreted by activated T cells can be determined by using fluorochrome-labeled antibodies to detect cytokine molecules that have been allowed to accumulate inside the cell. The accumulation of cytokine molecules, to allow them to reach a high enough concentration for efficient detection, is achieved by treating the activated T cells with inhibitors of protein export. In such treated cells, proteins destined to be secreted are instead retained within the endoplasmic reticulum (left panel). These treated cells are then fixed, to cross-link the proteins inside the cell and in the cell membranes, so that they are not lost when the cell is permeabilized by dissolving the cell membrane in a mild detergent (center panel). Fluorochrome-labeled antibodies can now enter the permeabilized cell and bind to the cytokines inside the cell (right panel). Cells labeled in this way can also be labeled with antibodies that bind to cell-surface proteins to determine which subsets of T cells are secreting particular cytokines.

Peptides can be biotinylated using the bacterial enzyme BirA, which recognizes a specific amino acid sequence. Recombinant MHC molecules containing this target sequence are used to make peptide:MHC complexes, which are then biotinylated. Avidin, or its bacterial counterpart streptavidin, contains four sites that bind biotin with extremely high affinity. Mixing the biotinylated peptide:MHC complex with avidin or streptavidin results in the formation of a **peptide:MHC tetramer**—four specific peptide:MHC complexes bound to a single molecule of streptavidin (Fig. A.33). Routinely, the streptavidin moiety is labeled with a fluorochrome to allow detection of those T cells capable of binding the peptide:MHC tetramer.

Peptide:MHC tetramers have been used to identify populations of antigen-specific T cells in, for example, patients with acute Epstein–Barr virus infections (infectious mononucleosis), showing that up to 80% of the peripheral T cells in infected individuals can be specific for a single peptide:MHC complex. They have also been used to follow responses over longer timescales in individuals with HIV or, in the example we show, cytomegalovirus infections. These reagents have also been important in identifying the cells responding, for example, to nonclassical class I molecules such as HLA-E or HLA-G, in

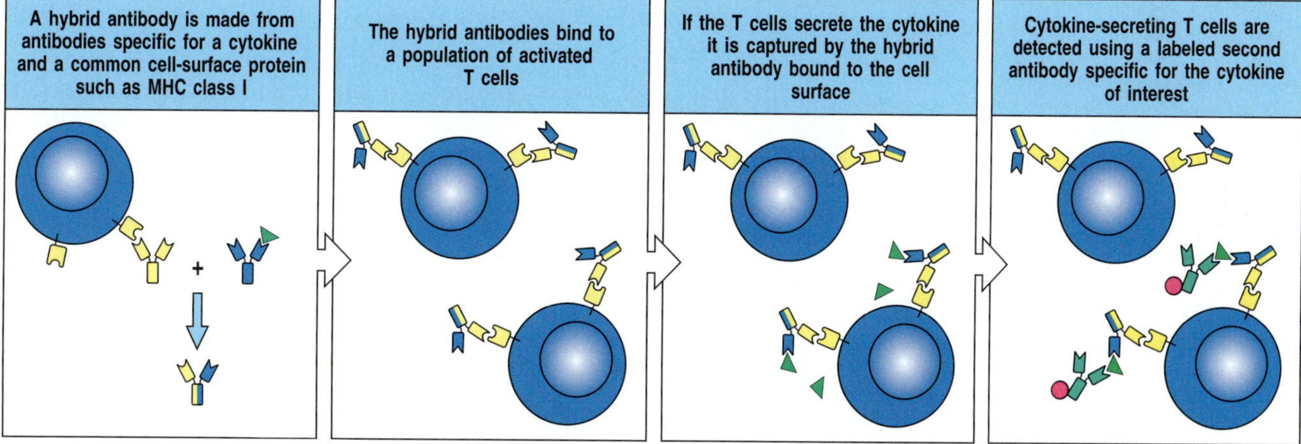

Fig. A.32 Hybrid antibodies containing cell-specific and cytokine-specific binding sites can be used to assay cytokine secretion by living cells and to purify cells secreting particular cytokines. Hybrid antibodies can be made by mixing together heavy- and light-chain pairs from antibodies of different specificities, for example an antibody against an MHC class I molecule and an antibody specific for a cytokine such as IL-4 (first panel). The hybrid antibodies are then added to a population of activated T cells, and bind to each cell via the MHC class I binding arm (second panel). If some of the cells in the population are secreting the appropriate cytokine, IL-4, this is captured by the cytokine-specific arm of the hybrid antibody (third panel). The presence of the cytokine can then be revealed, for example by using a fluorochrome-labeled second antibody specific for the same cytokine but binding to a different site to the one used for the hybrid antibody (last panel). The labeled cells are analyzed by flow cytometry or are isolated using a fluorescence-activated cell sorter. Alternatively, the second cytokine-specific antibody can be coupled to magnetic beads, and the cytokine-producing cells isolated magnetically.

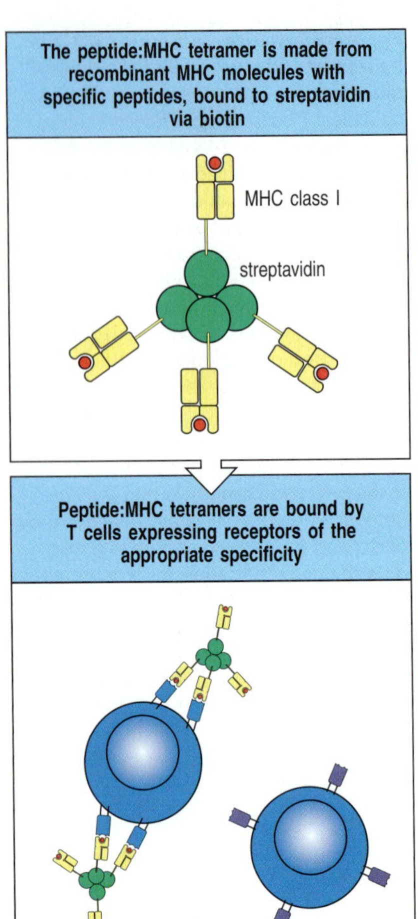

The peptide:MHC tetramer is made from recombinant MHC molecules with specific peptides, bound to streptavidin via biotin
MHC class I
streptavidin
Peptide:MHC tetramers are bound by T cells expressing receptors of the appropriate specificity

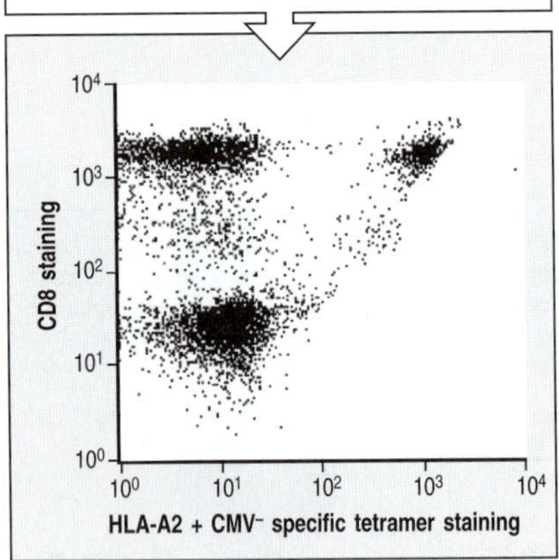

Fig. A.33 Peptide:MHC complexes coupled to streptavidin to form tetramers are able to stain antigen-specific T cells. Peptide:MHC tetramers are formed from recombinant refolded peptide:MHC complexes containing a single defined peptide epitope. The MHC molecules can be chemically derivatized to contain biotin, but more usually the recombinant MHC heavy chain is linked to a bacterial biotinylation sequence, a target for the *Escherichia coli* enzyme BirA, which is used to add a single biotin group to the MHC molecule. Streptavidin is a tetramer, each subunit having a single binding site for biotin; hence the streptavidin:peptide:MHC complex creates a tetramer of peptide:MHC complexes (top panel). Although the affinity between the T-cell receptor and its peptide:MHC ligand is too low for a single complex to bind stably to a T cell, the tetramer, by being able to make a more avid interaction with multiple peptide:MHC complexes binding simultaneously, is able to bind to T cells whose receptors are specific for the particular peptide:MHC complex (middle panel). Routinely, the streptavidin molecules are coupled to a fluorochrome, so that the binding to T cells can be monitored by flow cytometry. In the example shown in the bottom panel, T cells have been stained simultaneously with antibodies specific for CD3 and CD8, and with a tetramer of HLA-A2 molecules containing a cytomegalovirus peptide. Only the $CD3^+$ cells are shown, with the staining of CD8 displayed on the vertical axis and the tetramer staining displayed along the horizontal axis. The $CD8^-$ cells (mostly $CD4^+$) on the bottom left of the figure show no specific tetramer staining, while the bulk of the $CD8^+$ cells, on the top left, likewise show no tetramer staining. However, a discrete population of tetramer positive $CD8^+$ cells, at the top right of the panel, comprising some 5% of the total $CD8^+$ cells, can clearly be seen. Data courtesy of G. Aubert.

both cases showing that these nonclassical molecules are recognized by subsets of NK receptors.

A-29 Assessing the diversity of the T-cell repertoire by 'spectratyping.'

The extent of the diversity of the T-cell repertoire, either generally or during specific immune responses, is often of interest. In particular, as T cells do not undergo somatic hypermutation and affinity maturation in the same way that B cells do, the relationship between the repertoire of T cells making a primary response to antigen and the repertoire of T cells involved in secondary and subsequent responses to antigen has been difficult to determine. This information has usually been obtained through the laborious process of cloning the T cells involved in specific responses (see Section A-24), and the cloning and sequencing of their T-cell receptors.

It is possible, however, to estimate the diversity of T-cell responses by making use of the junctional diversity generated when T-cell receptors are created by somatic recombination, a technique known as **spectratyping**. Variability in the length of the CDR3 segments is created during the recombination process, both by variation in the exact positions at which the junctions between gene segments occur and by variation in the number of N-nucleotides added. Both these processes result in the length of the V_β CDR3 varying by up to nine amino acids. The problem in detecting this variability is that there are 24 families of V_β gene segments in humans and it is not possible to design a single oligonucleotide primer that will anneal to all of these families. Specific oligonucleotide primers can, however, be designed for each family, and these can be used in the polymerase chain reaction (PCR), together with a primer specific for the C_β region, to amplify, for each individual family, a segment of the mRNA for the T-cell receptor β chain that spans the CDR3 region. A population of TCR V_β genes will therefore show a distribution, or 'spectrum,' of CD3 lengths, and will give rise to PCR products of different lengths that can be resolved, usually by polyacrylamide gel electrophoresis (Fig. A.34). The deletion and addition of nucleotides during the generation of T-cell receptors by rearrangement is random, and so in a normal individual the CDR3 lengths follow a Gaussian distribution. Deviations from this Gaussian distribution, such as an excess of one particular CDR3 length, indicate the presence of clonal expansions of T cells, such as occurs during a T-cell response.

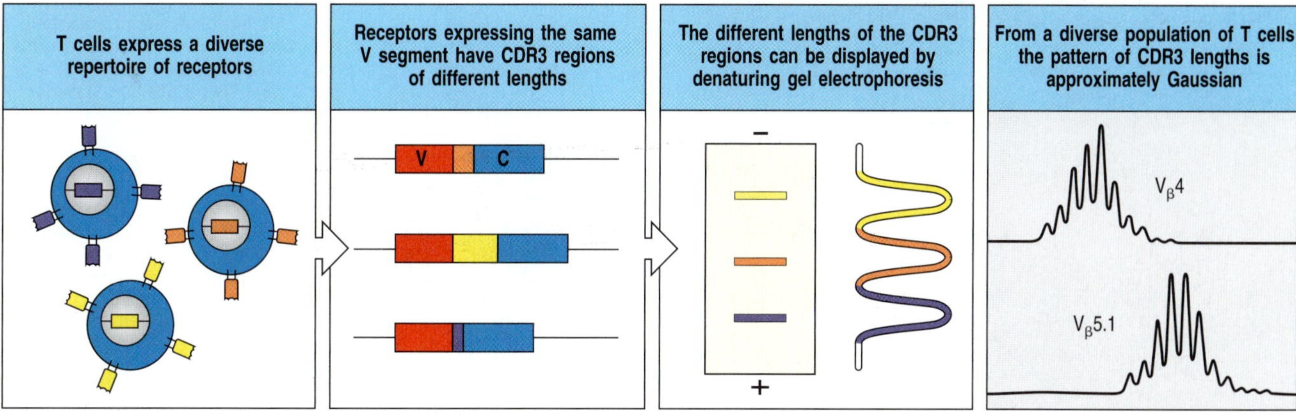

T cells express a diverse repertoire of receptors
Receptors expressing the same V segment have CDR3 regions of different lengths
V
C
The different lengths of the CDR3 regions can be displayed by denaturing gel electrophoresis
–
+
From a diverse population of T cells the pattern of CDR3 lengths is approximately Gaussian
Vβ4
Vβ5.1

Fig. A.34 The diversity of the T-cell receptor repertoire can be displayed by spectratyping, a PCR-based technique that separates different receptors on the basis of their CDR3 length. The process of generation of T-cell receptors is stochastic, giving rise to a population of mature T cells whose receptors are clonally distributed (first panel). In each of the cells expressing a particular V_β gene segment, all of the differences between the unique receptors are restricted to the CDR3 region, where there will be differences in length as well as sequence as a consequence of the imprecision of the rearrangement process (second panel). Using sets of primers for the PCR reaction that are specific for individual V_β gene segments at one end and for a conserved part of the C region at the other, it is possible to generate a set of DNA fragments that span the CDR3 region. If these are separated by denaturing acrylamide gel electrophoresis, a series of bands are formed or, because these fragments can be labeled with fluorochromes and analyzed by automated gel readers, a series of peaks corresponding to the different length fragments (third panel). The pattern of peaks obtained in this way is known as a spectratype. From a diverse population of cells, the distribution of fragment lengths is Gaussian, as shown in the last panel, in which the spectratypes of two different V_β regions from the same individual are shown. In this case, both of the patterns are approximately Gaussian; deviations from a Gaussian distribution may indicate expansion of particular clones of T cells, perhaps in response to antigenic challenge. Data courtesy of L. McGreavey.

A-30 Biosensor assays for measuring the rates of association and disassociation of antigen receptors for their ligands.

Two of the important questions that are always asked of any receptor–ligand interaction are: what is the strength of binding, or affinity, of the interaction, and what are the rates of association and disassociation? Traditionally, measurements of affinity were made by equilibrium binding measurements (see Section A-9), and measurements of rates of binding were difficult to obtain. In addition, equilibrium binding assays cannot be performed on T-cell receptors, which have large macromolecular ligands and cannot be isolated and purified in large quantity.

It is now possible to measure binding rates directly, by following the binding of ligands to receptors immobilized on gold-plated glass slides, using a phenomenon known as **surface plasmon resonance** (**SPR**) to detect the binding (Fig. A.35). A full explanation of surface plasmon resonance is beyond the scope of this textbook, as it is based on advanced physical and quantum mechanical principles. In brief, it relies on the total internal reflection of a beam of light from the surface of a gold-coated glass slide. As the light is reflected, some of its energy excites electrons in the gold coating and these excited electrons are in turn affected by the electric field of any molecules binding to the surface of the glass coating. The more molecules that bind to the surface, the greater is the effect on the excited electrons, and this in turn affects the reflected light beam. The reflected light thus becomes a sensitive measure of the number of atoms bound to the gold surface of the slide.

If a purified receptor is immobilized on the surface of the gold-coated glass slide, to make a biosensor 'chip,' and a solution containing the ligand is flowed over that surface, the binding of ligand to the receptor can be followed until it reaches equilibrium (see Fig. A.35). If the ligand is then washed out, dissociation of ligand from the receptor can easily be followed and the dissociation rate

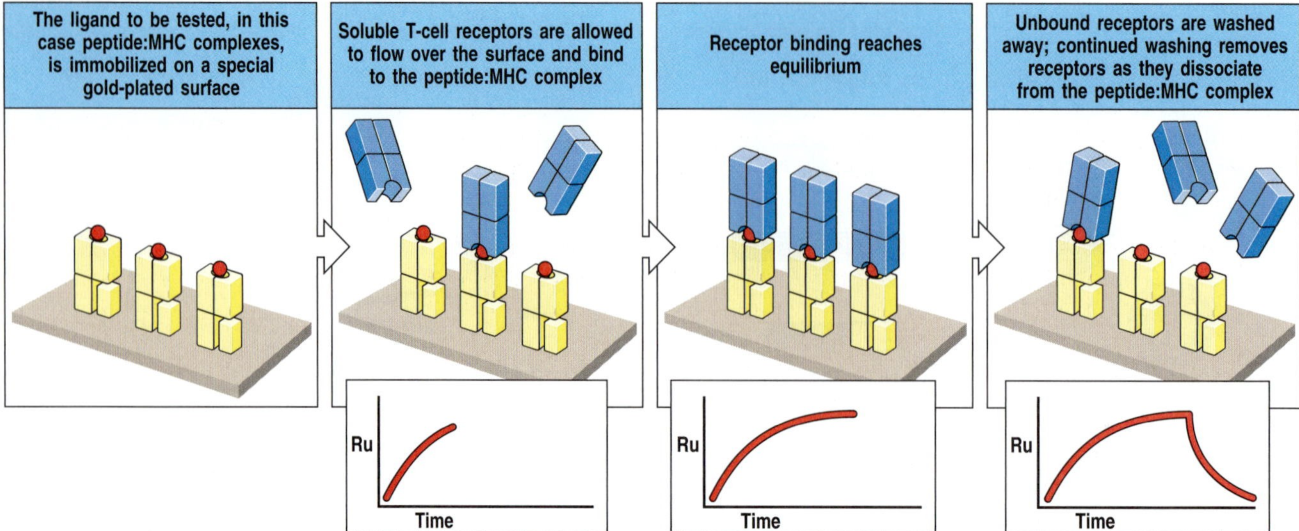

Fig. A.35 Measurement of receptor–ligand interactions can be made in real time using a biosensor. Biosensors are able to measure the binding of molecules on the surface of gold-plated glass chips through the indirect effects of the binding on the total internal reflection of a beam of polarized light at the surface of the chip. Changes in the angle and intensity of the reflected beam are measured in 'resonance units' (Ru) and plotted against time in what is termed a 'sensorgram.' Depending on the exact nature of the receptor–ligand pair to be analyzed, either the receptor or the ligand can be immobilized on the surface of the chip. In the example shown, peptide:MHC complexes are immobilized on such a surface (first panel). T-cell receptors in solution are now allowed to flow over the surface, and to bind to the immobilized peptide:MHC complexes (second panel). As the T-cell receptors bind, the sensorgram (inset panel below the main panel) reflects the increasing amount of protein bound. As the binding reaches either saturation or equilibrium (third panel), the sensorgram shows a plateau, as no more protein binds. At this point, unbound receptors can be washed away. With continued washing, bound receptors now start to dissociate and are removed in the flow of the washing solution (last panel). The sensorgram now shows a declining curve, reflecting the rate at which the receptor and ligand dissociation occurs.

calculated. A new solution of the ligand at a different concentration can then be flowed over the chip and the binding once again measured. The affinity of binding can be calculated in a number of ways in this type of assay. Most simply, the ratio of the rates of association and dissociation will give an estimate of the affinity, but more accurate estimates can be obtained from the measurements of the binding at different concentrations of ligand. From measurements of binding at equilibrium, a Scatchard plot (see Fig. A.11) will give a measurement of the affinity of the receptor–ligand interaction.

Mitogen	Responding cells
Phytohemagglutinin (PHA) (red kidney bean)	T cells
Concanavalin (ConA) (jack bean)	T cells
Pokeweed mitogen (PWM) (pokeweed)	T and B cells
Lipopolysaccharide (LPS) (*Escherichia coli*)	B cells (mouse)

Fig. A.36 Polyclonal mitogens, many of plant origin, stimulate lymphocyte proliferation in tissue culture. Many of these mitogens are used to test the ability of lymphocytes in human peripheral blood to proliferate.

A-31 Stimulation of lymphocyte proliferation by treatment with polyclonal mitogens or specific antigen.

To function in adaptive immunity, rare antigen-specific lymphocytes must proliferate extensively before they differentiate into functional effector cells in order to generate sufficient numbers of effector cells of a particular specificity. Thus, the analysis of induced lymphocyte proliferation is a central issue in their study. It is, however, difficult to detect the proliferation of normal lymphocytes in response to specific antigen because only a minute proportion of cells will be stimulated to divide. Great impetus was given to the field of lymphocyte culture by the finding that certain substances induce many or all lymphocytes of a given type to proliferate. These substances are referred to collectively as **polyclonal mitogens** because they induce mitosis in lymphocytes of many different specificities or clonal origins. T and B lymphocytes are stimulated by different polyclonal mitogens (Fig. A.36). Polyclonal mitogens seem to trigger essentially the same growth response mechanisms as antigen. Lymphocytes normally exist as resting cells in the G_0 phase of the cell cycle. When stimulated with polyclonal mitogens, they rapidly enter the G_1 phase and progress through the cell cycle. In most studies, lymphocyte proliferation is most simply measured by the incorporation of ^{3}H-thymidine into DNA. This assay is used clinically for assessing the ability of lymphocytes from patients with suspected immunodeficiencies to proliferate in response to a nonspecific stimulus.

Once lymphocyte culture had been optimized using the proliferative response to polyclonal mitogens as an assay, it became possible to detect antigen-specific T-cell proliferation in culture by measuring ^{3}H-thymidine uptake in response to an antigen to which the T-cell donor had previously been immunized (Fig. A.37). This is the assay most commonly used for assessing T-cell responses after immunization, but it reveals little about the functional capabilities of the responding T cells. These must be ascertained by functional assays, as outlined in Sections A-33 and A-34.

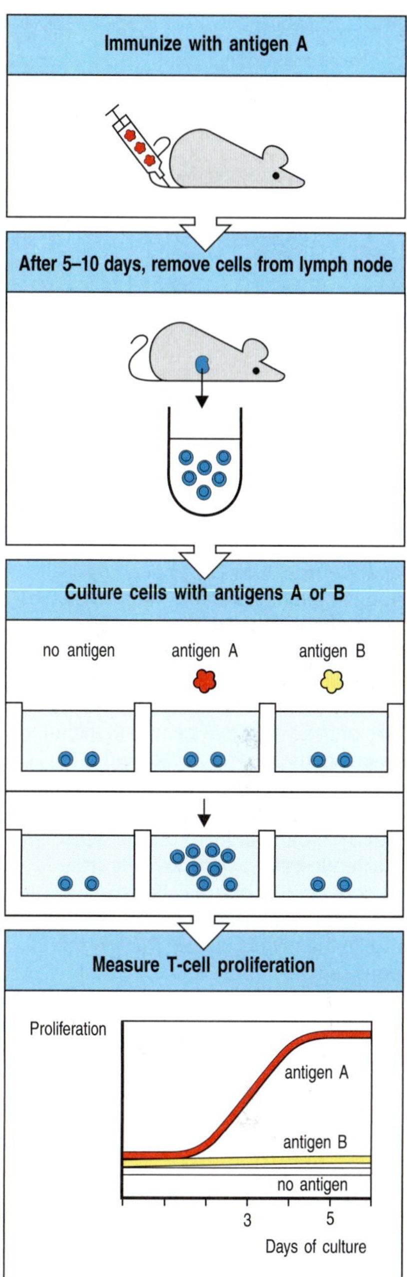

Fig. A.37 Antigen-specific T-cell proliferation is used frequently as an assay for T-cell responses. T cells from mice or humans that have been immunized with an antigen (A) proliferate when they are exposed to antigen A and antigen-presenting cells but not when cultured with unrelated antigens to which they have not been immunized (antigen B). Proliferation can be measured by incorporation of ^{3}H-thymidine into the DNA of actively dividing cells. Antigen-specific proliferation is a hallmark of specific CD4 T-cell immunity.

A-32 Measurements of apoptosis by the TUNEL assay.

Apoptotic cells can be detected by a procedure known as **TUNEL staining**. In this technique, the 3′ ends of the DNA fragments generated in apoptotic cells are labeled with biotin-coupled uridine by using the enzyme terminal deoxynucleotidyl transferase (TdT). The biotin label is then detected with enzyme-tagged streptavidin, which binds to biotin. When the colorless substrate of the enzyme is added to a tissue section or cell culture, it is reacted on to produce a colored precipitate only in cells that have undergone apoptosis (Fig. A.38). This technique has revolutionized the detection of apoptotic cells.

A-33 Assays for cytotoxic T cells.

Activated CD8 T cells generally kill any cells that display the specific peptide:MHC class I complex they recognize. Thus, CD8 T-cell function can be determined using the simplest and most rapid T-cell bioassay—the killing of a target cell by a cytotoxic T cell. This is usually detected in a ^{51}Cr-release assay. Live cells will take up, but do not spontaneously release, radioactively labeled sodium chromate, $Na_2{}^{51}CrO_4$. When these labeled cells are killed, the radioactive chromate is released and its presence in the supernatant of mixtures of target cells and cytotoxic T cells can be measured (Fig. A.39). In a similar assay, proliferating target cells such as tumor cells can be labeled with ^{3}H-thymidine, which is incorporated into the replicating DNA. On attack by a cytotoxic T cell, the DNA of the target cells is rapidly fragmented and retained in the filtrate, while large, unfragmented DNA is collected on a filter, and one can measure either the release of these fragments or the retention of ^{3}H-thymidine in chromosomal DNA. These assays provide a rapid, sensitive, and specific measure of the activity of cytotoxic T cells.

A-34 Assays for CD4 T cells.

CD4 T-cell functions usually involve the activation rather than the killing of cells bearing specific antigen. The activating effects of CD4 T cells on B cells or macrophages are mediated in large part by nonspecific mediator proteins called cytokines, which are released by the T cell when it recognizes antigen. Thus, CD4 T-cell function is usually studied by measuring the type and amount of these released proteins. Because different effector T cells release different amounts and types of cytokines, one can learn about the effector potential of that T cell by measuring the proteins it produces.

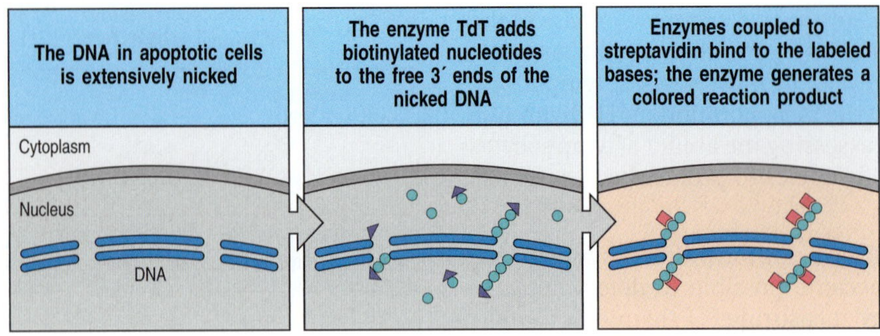
The DNA in apoptotic cells is extensively nicked
The enzyme TdT adds biotinylated nucleotides to the free 3′ ends of the nicked DNA
Enzymes coupled to streptavidin bind to the labeled bases; the enzyme generates a colored reaction product
Cytoplasm
Nucleus
DNA

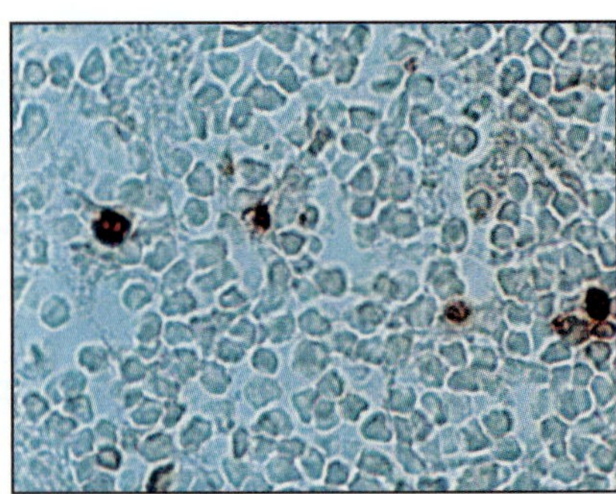

Fig. A.38 In the TUNEL assay, fragmented DNA is labeled by terminal deoxynucleotidyl transferase (TdT) to reveal apoptotic cells. When cells undergo programmed cell death, or apoptosis, their DNA becomes fragmented (first panel). The enzyme TdT is able to add nucleotides to the ends of DNA fragments; most commonly in this assay, biotin-labeled nucleotides (usually dUTP) are added (second panel). The biotinylated DNA can be detected by using streptavidin, which binds to biotin, coupled to enzymes that convert a colorless substrate into a colored insoluble product (third panel). Cells stained in this way can be detected by light microscopy, as shown in the photograph of apoptotic cells (stained red) in the thymic cortex. Photograph courtesy of R. Budd and J. Russell.

Cytokines can be detected by their activity in biological assays of cell growth, where they serve either as growth factors or as growth inhibitors. A more specific assay is a modification of ELISA known as a capture or sandwich ELISA (see Section A-6). In this assay, the cytokine is characterized by its ability to act as a bridge between two monoclonal antibodies reacting with different epitopes on the cytokine molecule. Cytokine-secreting cells can also be detected by ELISPOT (see Section A-26).

Sandwich ELISA and ELISPOT avoid a major problem of cytokine bioassays, namely the ability of different cytokines to stimulate the same response in a bioassay. Bioassays must always be confirmed by inhibition of the response with neutralizing monoclonal antibodies specific for the cytokine. Another way of identifying cells actively producing a given cytokine is to stain them with a fluorescently tagged anti-cytokine monoclonal antibody, and then identify and count them by FACS (see Section A-22).

A quite different approach to detecting cytokine production is to determine the presence and amount of the relevant cytokine mRNA in stimulated T cells. This can be done for single cells by *in situ* hybridization and for cell populations by **reverse transcriptase–polymerase chain reaction** (**RT-PCR**). Reverse transcriptase is an enzyme used by certain RNA viruses, such as HIV, to convert an RNA genome into a DNA copy, or cDNA. In RT–PCR, mRNA is isolated from cells and cDNA copies are made *in vitro* using reverse transcriptase. The desired cDNA is then selectively amplified by PCR by using sequence-specific primers. When the products of the reaction are subjected to electrophoresis on an agarose gel, the amplified DNA can be visualized as a band of a specific size. The amount of amplified cDNA sequence will be proportional to its representation in the mRNA; stimulated T cells actively producing a particular cytokine will produce large amounts of that particular mRNA and will thus give correspondingly large amounts of the selected cDNA on RT–PCR. The level of cytokine mRNA in the original tissue is usually determined by comparison with the outcome of RT–PCR on the mRNA produced by a so-called 'housekeeping gene' expressed by all cells.

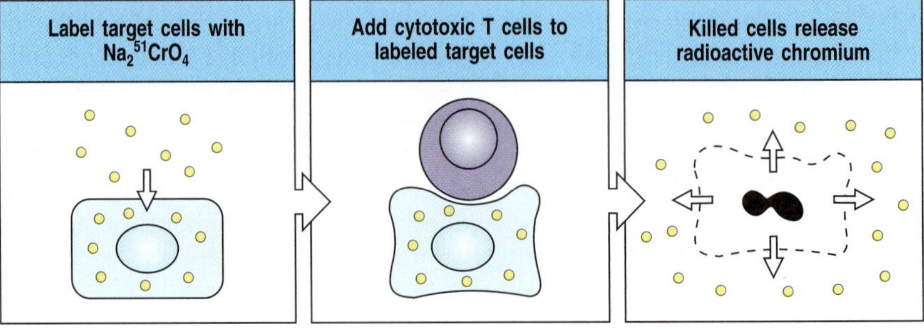
Label target cells with $Na_2{}^{51}CrO_4$
Add cytotoxic T cells to labeled target cells
Killed cells release radioactive chromium

Fig. A.39 Cytotoxic T-cell activity is often assessed by chromium release from labeled target cells. Target cells are labeled with radioactive chromium as $Na_2{}^{51}CrO_4$, washed to remove excess radioactivity, and exposed to cytotoxic T cells. Cell destruction is measured by the release of radioactive chromium into the medium, detectable within 4 hours of mixing target cells with T cells.

Detection of immunity *in vivo*.

A-35 Assessment of protective immunity.

An adaptive immune response against a pathogen often confers long-lasting immunity against infection with that pathogen; successful vaccination achieves the same end. The very first experiment in immunology, Jenner's successful vaccination against smallpox, is still the model for assessing the presence of such protective immunity. The assessment of protective immunity conferred by vaccination has three essential steps. First, an immune response is elicited by immunization with a candidate vaccine. Second, the immunized individuals, along with unimmunized controls, are challenged with the infectious agent (Fig. A.40). Finally, the prevalence and severity of infection in the immunized individual are compared with the course of the disease in the unimmunized controls. For obvious reasons, such experiments are usually performed first in animals, if a suitable animal model for the infection exists. However, eventually a trial must be conducted in humans. In this case, the infectious challenge is usually provided naturally by conducting the trial in a region where the disease is prevalent. The efficacy of the vaccine is determined by assessing the prevalence and severity of new infections in the immunized and control populations. Such studies necessarily give less precise results than a direct experiment; however, for most diseases they are the only way of assessing a vaccine's ability to induce protective immunity in humans.

A-36 Transfer of protective immunity.

The tests described in Section A-35 show that protective immunity has been established, but cannot show whether it involves humoral immunity, cell-mediated immunity, or both. When these studies are performed in inbred mice, the nature of protective immunity can be determined by transferring serum or lymphoid cells from an immunized donor animal to an unimmunized syngeneic recipient (that is, a genetically identical animal of the same inbred strain) (Fig. A.41). If protection against infection can be conferred by the transfer of serum, the immunity is provided by circulating antibodies and is called **humoral immunity**. Transfer of immunity by antiserum or purified antibodies provides immediate protection against many pathogens and against toxins such as those of tetanus and snake venom. However, although protection is immediate, it is temporary, lasting only so long as the transferred antibodies remain active in the recipient's body. This type of transfer is therefore called **passive immunization**. Only **active immunization** with antigen can provide lasting immunity. Moreover, the recipient may become immunized to the antiserum used to transfer immunity. Horse or sheep sera are the usual sources of anti-snake venoms used in humans, and repeated administration can lead either to serum sickness (see Section 14-16) or, if the recipient becomes allergic to the foreign serum, to anaphylaxis (see Section 14-10).

Protection against many diseases cannot be transferred with serum but can be transferred by lymphoid cells from immunized donors. The transfer of lymphoid cells from an immune donor to a normal syngeneic recipient is called **adoptive transfer** or **adoptive immunization**, and the immunity transferred is called **adoptive immunity**. Immunity that can be transferred only with lymphoid cells is called **cell-mediated immunity**. Such cell transfers must be between genetically identical donors and recipients, such as members of the same inbred strain of mouse, so that the donor lymphocytes are not rejected by the recipient and do not attack the recipient's tissues. Adoptive transfer of immunity is used clinically in humans in experimental approaches to cancer

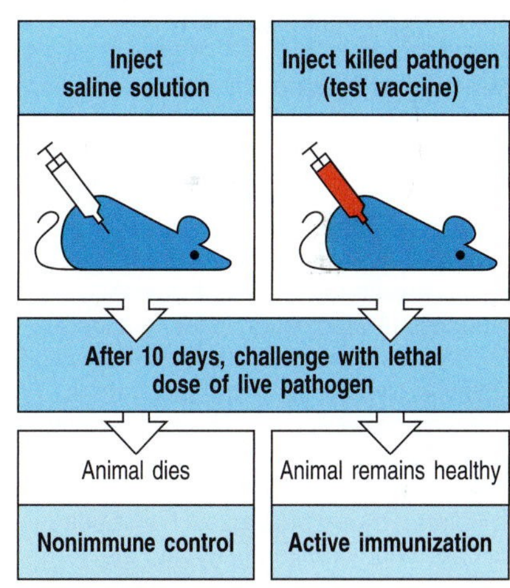

Fig. A.40 *In vivo* assay for the presence of protective immunity after vaccination in animals. Mice are injected with the test vaccine or a control such as saline solution. Different groups are then challenged with lethal or pathogenic doses of the test pathogen or with an unrelated pathogen as a specificity control (not shown). Unimmunized animals die or become severely infected. Successful vaccination is seen as specific protection of immunized mice against infection with the test pathogen. This is called active immunity, and the process is called active immunization.

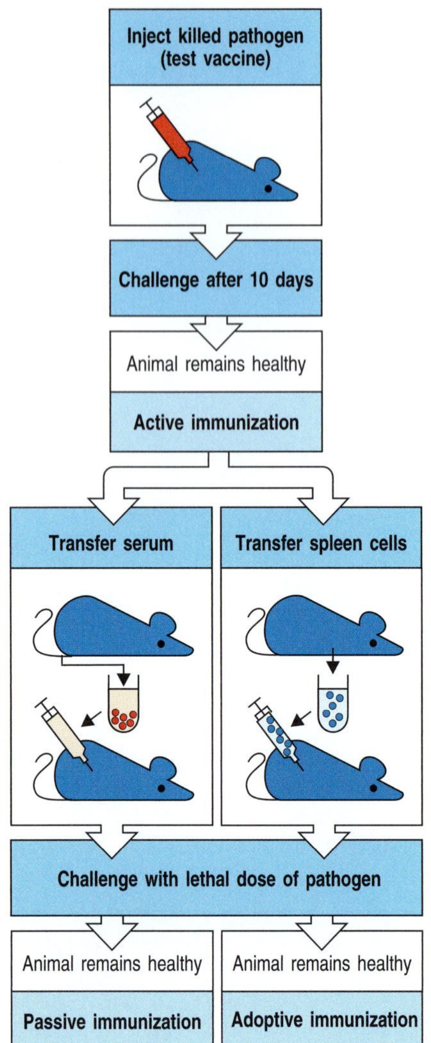

Fig. A.41 Immunity can be transferred by antibodies or by lymphocytes. Successful vaccination leads to a long-lived state of protection against the specific immunizing pathogen. If this immune protection can be transferred to a normal syngeneic recipient with serum from an immune donor, then immunity is mediated by antibodies; such immunity is called humoral immunity and the process is called passive immunization. If immunity can be transferred only by infusing lymphoid cells from the immune donor into a normal syngeneic recipient, then the immunity is called cell-mediated immunity and the transfer process is called adoptive transfer or adoptive immunization. Passive immunity is short-lived, because antibody is eventually catabolized, but adoptively transferred immunity is mediated by immune cells, which can survive and provide longer-lasting immunity.

therapy or as an adjunct to bone marrow transplantation; in these cases, the patient's own T cells, or the T cells of the bone marrow donor, are given.

A-37 The tuberculin test.

Local responses to antigen can indicate the presence of active immunity. Active immunity is often studied *in vivo*, especially in humans, by injecting antigens locally in the skin. If a reaction appears, this indicates the presence of antibodies or immune lymphocytes that are specific for that antigen; the **tuberculin test** is an example of this. When people have had tuberculosis they develop cell-mediated immunity that can be detected as a local response when their skin is injected with a small amount of tuberculin, an extract of *Mycobacterium tuberculosis*, the pathogen that causes tuberculosis. The response typically appears a day or two after the injection and consists of a raised, red, and hard (or indurated) area in the skin, which then disappears as the antigen is degraded.

A-38 Testing for allergic responses.

Local intracutaneous injections of minute doses of the antigens that cause allergies are used to determine which antigen triggers a patient's allergic reactions. Local responses that happen in the first few minutes after antigen injection in immune recipients are called **immediate hypersensitivity reactions**, and they can be of several forms, one of which is the wheal-and-flare response (see Fig. 14.13). Immediate hypersensitivity reactions are mediated by specific antibodies of the IgE class formed as a result of earlier exposures to the antigen. Responses that take hours to days to develop, such as the tuberculin test, are referred to as **delayed-type hypersensitivity responses** and are caused by preexisting immune T cells. This latter type of response was observed by Jenner when he tested vaccinated individuals with a local injection of vaccinia virus.

These tests work because the local deposit of antigen remains concentrated in the initial site of injection, eliciting responses in local tissues. They do not cause generalized reactions if sufficiently small doses of antigen are used. However, local tests carry a risk of systemic allergic reactions, and they should be used with caution in people with a history of hypersensitivity.

A-39 Assessment of immune responses and immunological competence in humans.

The methods used for testing immune function in humans are necessarily more limited than those used in experimental animals, but many different tests are available. They fall into several groups depending on the reason the patient is being studied.

Assessment of protective immunity in humans generally relies on tests conducted *in vitro*. To assess humoral immunity, specific antibody levels in the patient's serum are assayed by RIA or, more commonly, ELISA (see Section A-6), using the test microorganism or a purified microbial product as antigen. To test for humoral immunity against viruses, antibody production is often measured by the ability of serum to neutralize the infectivity of live virus for tissue culture cells. In addition to providing information about protective immunity, the presence of antibody against a particular pathogen indicates that the patient has been exposed to it, making such tests of crucial importance in epidemiology. At present, testing for antibody against HIV is the main screening test for infection with this virus, critical both for the patient and in blood banking, where blood from infected donors must be excluded from the supply. Essentially similar tests are used in investigating allergy, where allergens are used as the antigens in tests for specific IgE antibody by ELISA or RIA (see Section A-6), which may be used to confirm the results of skin tests.

Cell-mediated immunity—that is, immunity mediated by T cells—is technically more difficult to measure than humoral immunity. This is principally because T cells do not make a secreted antigen-binding product, so there is no simple binding assay for their antigen-specific responses. T-cell activity can be divided into an induction phase, in which T cells are activated to divide and differentiate, and an effector phase, in which their function is expressed. Both phases require that the T cell interact with another cell and that it recognize specific antigen displayed in the form of peptide:MHC complexes on the surface of this interacting cell. In the induction phase, the interaction must be with an antigen-presenting cell able to deliver co-stimulatory signals, whereas, in the effector phase, the nature of the target cell depends on the type of effector T cell that has been activated. Most commonly, the presence of T cells that have responded to a specific antigen is detected by their subsequent *in vitro* proliferation when reexposed to the same antigen (see Section A-31).

T-cell proliferation indicates only that cells able to recognize that antigen have been activated previously; it does not reveal what effector function they mediate. The effector function of a T cell is assayed by its effect on an appropriate target cell. Assays for cytotoxic CD8 T cells (see Section A-33) and for cytokine production by CD4 T cells (see Sections A-26, A-27, and A-34) are used to characterize the immune response. Cell-mediated immunity to infectious agents can also be tested by skin test with extracts of the pathogen, as in the tuberculin test (see Section A-37). These tests provide information about the exposure of the patient to the disease and also about the patient's ability to mount an adaptive immune response to it.

Patients with immune deficiency (see Chapter 13) are usually detected clinically by a history of recurrent infection. To determine the competence of the immune system in such patients, a battery of tests are usually conducted; these focus with increasing precision as the nature of the defect is narrowed down to a single element. The presence of the various cell types in blood is determined by routine hematology, often followed by FACS analysis of lymphocyte subsets (see Section A-22) and the measurement of serum immunoglobulins. Figure A.42 lists the main tests that are performed and the 'normal' ranges for leukocyte numbers and concentrations of each antibody class in peripheral blood. The phagocytic competence of freshly isolated polymorphonuclear leukocytes and monocytes is tested, and the efficiency of the complement system (see Chapters 2 and 9) is determined by testing the dilution of serum required for the lysis of 50% of antibody-coated red blood cells (this is denoted the CH_{50}).

In general, if such tests reveal a defect in one of the broad compartments of immune function, more specialized testing is then needed to determine the precise nature of the defect. Tests of lymphocyte function are often valuable, starting with the ability of polyclonal mitogens to induce T-cell proliferation and B-cell secretion of immunoglobulin in tissue culture (see Section A-31). These tests can eventually pinpoint the cellular defect in immunodeficiency.

In patients with autoimmune diseases (see Chapter 15), the same parameters are usually analyzed to determine whether there is a gross abnormality in the immune system. However, most patients with such diseases show few abnormalities in general immune function. To determine whether a patient is producing antibody against their own cellular antigens, the most informative test is to react their serum with tissue sections, which are then examined for bound antibody by indirect immunofluorescence using anti-human immunoglobulin labeled with fluorescent dye (see Section A-14). Most autoimmune diseases are associated with the production of broadly characteristic patterns of autoantibodies directed at self tissues. These patterns aid

Fig. A.42 Evaluation of immune function.

Evaluation of the cellular components of the human immune system	B cells	T cells	Phagocytes
Normal numbers ($\times 10^9$ per liter of blood)	Approximately 0.3	Total 1.0–2.5 CD4 0.5–1.6 CD8 0.3–0.9	Monocytes 0.15–0.6 Polymorphonuclear leukocytes Neutrophils 3.00–5.5 Eosinophils 0.05–0.25 Basophils 0.02
Measurement of function *in vivo*	Serum Ig levels Specific antibody levels	Skin test	—
Measurement of function *in vitro*	Induced antibody production in response to pokeweed mitogen	T-cell proliferation in response to phytohemagglutinin or to tetanus toxoid	Phagocytosis Nitro blue tetrazolium uptake Intracellular killing of bacteria

Evaluation of the humoral components of the human immune system	Immunoglobulins				Complement
Component	IgG	IgM	IgA	IgE	
Normal levels	600–1400 $mg\,dl^{-1}$	40–345 $mg\,dl^{-1}$	60–380 $mg\,dl^{-1}$	0–200 $IU\,ml^{-1}$	CH_{50} of 125–300 $IU\,ml^{-1}$

in the diagnosis of the disease and help to distinguish autoimmunity from tissue inflammation due to infectious causes.

It is also possible to investigate allergies by administering possible allergens by routes other than intracutaneous administration. Allergen may be given by inhalation to test for asthmatic allergic responses (see Fig. 14.13); this is done mainly for experimental purposes in studies of the mechanisms and treatment of asthma. Similarly, food allergens may be given by mouth. The administration of allergens is potentially very dangerous because of the risk of causing anaphylaxis, and must only be done by trained and experienced investigators in an environment in which full resuscitation facilities are available.

A-40 The Arthus reaction.

This inflammatory skin reaction (see Section 14-16) can be induced in animal models to study the formation of immune complexes in tissues and how they cause inflammation. The original reaction described by Maurice Arthus was induced by the repeated injection of horse serum into rabbits. Initial injections of horse serum into the skin induced no reaction, but later injections, after the production of antibodies against the proteins in horse serum, induced an inflammatory reaction at the site of injection after several hours, characterized by the presence of edema, hemorrhage, and neutrophil infiltration, which frequently progressed to tissue necrosis. Most investigators now use passive models of the Arthus reaction in which either antibody is infused systemically and antigen is given locally (passive Arthus reaction) or antigen is infused systemically and antibody is injected locally (reverse passive Arthus reaction).

A-41 Adoptive transfer of lymphocytes.

Ionizing radiation from X-ray or γ-ray sources kills lymphoid cells at doses that spare the other tissues of the body. This makes it possible to eliminate immune function in a recipient animal before attempting to restore immune function by adoptive transfer, and allows the effect of the adoptively transferred cells to be studied in the absence of other lymphoid cells. James Gowans originally used this technique to prove the role of the lymphocyte in immune responses. He showed that all active immune responses could be transferred to irradiated recipients by small lymphocytes from immunized donors. This technique can be refined by transferring only certain lymphocyte subpopulations, such as B cells or CD4 T cells. Even cloned T-cell lines have been tested for their ability to transfer immune function, and have been shown to confer adoptive immunity to their specific antigen. Such adoptive transfer studies are a cornerstone in the study of the intact immune system, as they can be performed rapidly, simply, and in any strain of mouse.

A-42 Hematopoietic stem-cell transfers.

All cells of hematopoietic origin can be eliminated by treatment with high doses of X rays, allowing replacement of the entire hematopoietic system, including lymphocytes, by transfusion of donor bone marrow or purified hematopoietic stem cells from another animal. The resulting animals are called **radiation bone marrow chimeras** from the Greek word *chimera*, a mythical animal that had the head of a lion, the tail of a serpent, and the body of a goat. This technique is used experimentally to examine the development of lymphocytes, as opposed to their effector functions, and has been particularly important in studying T-cell development. Essentially the same technique is used in humans to replace the hematopoietic system when it fails, as in aplastic anemia or after nuclear accidents, or to eradicate the bone marrow and replace it with normal marrow in the treatment of certain cancers. In humans, bone marrow is the main source of hematopoietic stem cells, but they are increasingly being obtained from peripheral blood after the donor has been treated with hematopoietic growth factors such as GM-CSF, or from umbilical cord blood, which is rich in such stem cells (see Chapter 15).

A-43 *In vivo* depletion of T cells.

The importance of T-cell function *in vivo* can be ascertained in mice with no T cells of their own. Under these conditions, the effect of a lack of T cells can be studied, and T-cell subpopulations can be restored selectively to analyze their specialized functions. T lymphocytes originate in the thymus, and neonatal **thymectomy**, the surgical removal of the thymus of a mouse at birth, prevents T-cell development from occurring because the export of most functionally mature T cells occurs only after birth in the mouse. Alternatively, adult mice can be thymectomized and then irradiated and reconstituted with bone marrow; such mice will develop all hematopoietic cell types except mature T cells.

The recessive *nude* mutation in mice is caused by a mutation in the gene for the transcription factor Wnt, and in homozygous form it causes hairlessness and absence of the thymus. Consequently, these animals fail to develop T cells from bone marrow progenitors. Grafting thymectomized or *nude*/*nude* mice with thymic epithelial elements depleted of lymphocytes allows the graft recipients to develop normal mature T cells. This procedure allows the role of the nonlymphoid thymic stroma to be examined; it has been crucial in determining the role of thymic stromal cells in T-cell development (see Chapter 8).

A-44 *In vivo* depletion of B cells.

There is no single site of B-cell development in mice, so techniques such as thymectomy cannot be applied to the study of B-cell function and development in rodents. However, bursectomy, the surgical removal of the bursa of

Fabricius in birds, can inhibit the development of B cells in these species. In fact, it was the effect of thymectomy versus bursectomy that led to the naming of T cells for thymus-derived lymphocytes and B cells for bursal-derived lymphocytes. There are no known spontaneous mutations (analogous to the *nude* mutation) in mice that produce animals with T cells but no B cells. However, such mutations exist in humans, leading to a failure to mount humoral immune responses or make antibody. The diseases produced by such mutations are called agammaglobulinemias because they were originally detected as the absence of gamma globulins. The genetic basis for several forms of this disease in humans have now been established (see Section 13-14), and some features of the disease can be reproduced in mice by targeted disruption of the corresponding gene (see Section A-45). Several different mutations in crucial regions of immunoglobulin genes have already been produced by gene targeting and have provided mice lacking B cells.

A-45 Transgenic mice.

The function of genes has traditionally been studied by observing the effects of spontaneous mutations in whole organisms and, more recently, by analyzing the effects of targeted mutations in cultured cells. The advent of gene cloning and *in vitro* mutagenesis now makes it possible to produce specific mutations in whole animals. Mice with extra copies or altered copies of a gene in their genome can be generated by **transgenesis**, which is now a well established procedure. To produce **transgenic mice**, a cloned gene is introduced into the mouse genome by microinjection into the male pronucleus of a fertilized egg, which is then implanted into the uterus of a pseudopregnant female mouse. In some of the eggs, the injected DNA becomes integrated randomly into the genome, giving rise to a mouse that has an extra genetic element of known structure, the transgene (Fig. A.43).

The transgene, to be studied in detail, needs to be introduced onto a stable, well-characterized genetic background. However, it is difficult to prepare transgenic embryos successfully in inbred strains of mice, and transgenic mice are routinely prepared in F_2 embryos (that is, the embryo formed after the mating of two F_1 animals). The transgene must then be bred onto a well-characterized genetic background; this requires 10 generations of back-crossing with an inbred strain to ensure that the integrated transgene is largely (more than 99%) free of heterogeneous genes from the founder mouse of the transgenic mouse line (Fig. A.44).

This technique allows one to study the impact of a newly discovered gene on development, to identify the regulatory regions of a gene required for its normal tissue-specific expression, to determine the effects of its overexpression or expression in inappropriate tissues, and to find out the impact of mutations on gene function. Transgenic mice have been particularly useful in studying the role of T-cell and B-cell receptors in lymphocyte development, as described in Chapter 8.

Female mouse is injected with follicle-stimulating hormone and chorionic gonadotropin to induce superovulation, and then mated
♀
×
♂
Fertilized eggs are removed from female. DNA containing the Eα gene is injected into the male pronucleus
Eα
♀
♂
Injected eggs are transferred into uterus of pseudopregnant female
♀
Some offspring will have incorporated the injected Eα gene (transgene)
♀
Eα⁻
Eα⁺
Eα⁻
Mate transgenic animal to Eα⁻ C57BL/6 mice to produce a strain expressing the Eα transgene

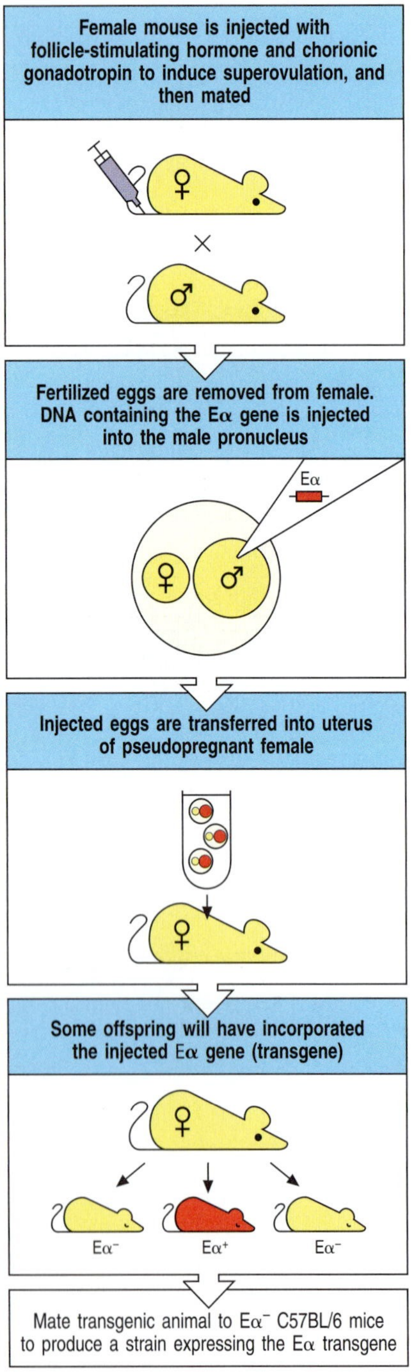

Fig. A.43 The function and expression of genes can be studied *in vivo* by using transgenic mice. DNA encoding a protein of interest, here the mouse MHC class II protein Eα, is purified and microinjected into the male pronuclei of fertilized eggs. The eggs are then implanted into pseudopregnant female mice. The resulting offspring are screened for the presence of the transgene in their cells, and positive mice are used as founders that transmit the transgene to their offspring, establishing a line of transgenic mice that carry one or more extra genes. The function of the Eα gene used here is tested by breeding the transgene into C57BL/6 mice that carry an inactivating mutation in their endogenous Eα gene.

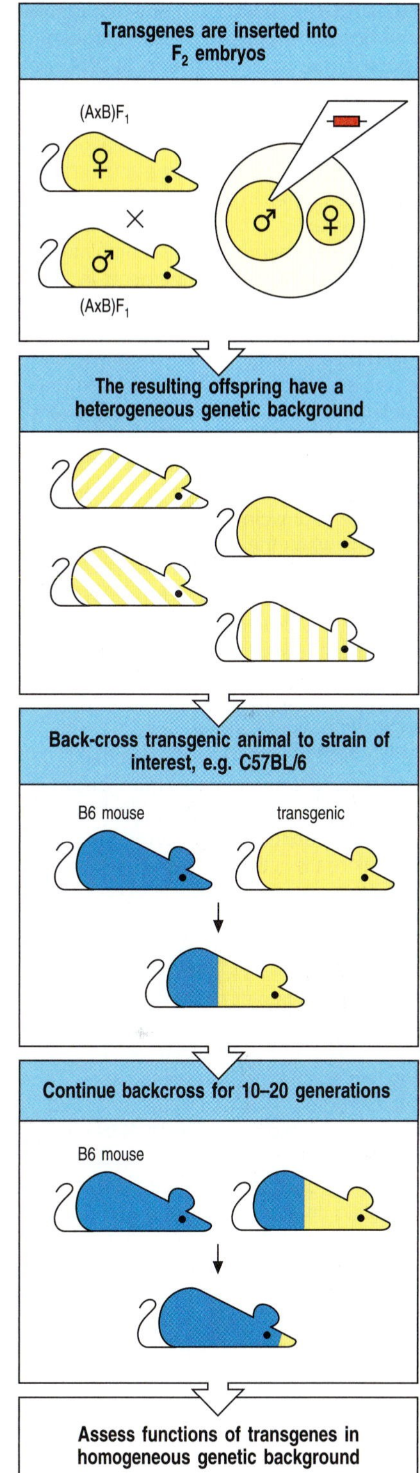

Transgenes are inserted into F_2 embryos
(AxB)F_1
♀
×
♂
(AxB)F_1
♂
♀
The resulting offspring have a heterogeneous genetic background
Back-cross transgenic animal to strain of interest, e.g. C57BL/6
B6 mouse
transgenic
Continue backcross for 10–20 generations
B6 mouse
Assess functions of transgenes in homogeneous genetic background

Fig. A.44 The breeding of transgenic co-isogeneic or congenic mouse strains. Transgenic mouse strains are routinely made in F_2 mice. To produce mice on an inbred background, the transgene is introgressively back-crossed onto a standard strain, usually C57BL/6 (B6). The presence of the transgene is tracked by performing PCR on genomic DNA extracted from the tail of young mice. After 10 generations of back-crossing, mice are more than 99% genetically identical, so that any differences observed between the mice are likely to be due to the transgene itself. The same technique can be used to breed a gene knockout into a standard strain of mice, as most gene knockouts are made in the 129 strain of mice (see Fig. A.46). The mice are then intercrossed and homozygous knockout mice are detected by an absence of an intact copy of the gene of interest (as determined by PCR).

A-46 Gene knockout by targeted disruption.

In many cases, the functions of a particular gene can only be fully understood if a mutant animal that does not express the gene can be obtained. Whereas genes used to be discovered through the identification of mutant phenotypes, it is now far more common to discover and isolate the normal gene and then determine its function by replacing it *in vivo* with a defective copy. This procedure is known as **gene knockout**, and it has been made possible by two fairly recent developments: a powerful strategy to select for targeted mutation by homologous recombination, and the development of continuously growing lines of **embryonic stem cells** (**ES cells**). These are embryonic cells that, on implantation into a blastocyst, can give rise to all cell lineages in a chimeric mouse.

The technique of **gene targeting** takes advantage of the phenomenon known as **homologous recombination** (Fig. A.45). Cloned copies of the target gene are altered to make them nonfunctional and are then introduced into the ES cell, where they recombine with the homologous gene in the cell's genome, replacing the normal gene with a nonfunctional copy. Homologous recombination is a rare event in mammalian cells, and thus a powerful selection strategy is required to detect those cells in which it has occurred. Most commonly, the introduced gene construct has its sequence disrupted by an inserted antibiotic-resistance gene such as that for neomycin resistance (*neo*^{r}). If this construct undergoes homologous recombination with the endogenous copy of the gene, the endogenous gene is disrupted but the antibiotic-resistance gene remains functional, allowing cells that have incorporated the gene to be selected in culture for resistance to the neomycin-like drug G418. However, antibiotic resistance on its own shows only that the cells have taken up and integrated the neomycin-resistance gene. To be able to select for those cells in which homologous recombination has occurred, the ends of the construct usually carry the thymidine kinase gene from the herpes simplex virus (*HSV-tk*). Cells that incorporate DNA randomly usually retain the entire DNA construct including *HSV-tk*, whereas homologous recombination between the construct and cellular DNA, the desired result, involves the exchange of homologous DNA sequences so that the nonhomologous *HSV-tk* genes at the ends of the construct are eliminated. Cells carrying *HSV-tk* are killed by the antiviral drug ganciclovir, and so cells with homologous recombinations have the unique feature of being resistant to both neomycin and ganciclovir, allowing them to be selected efficiently when these drugs are added to the cultures (see Fig. A.45).

This technique can be used to produce homozygous mutant cells in which the effects of knocking-out a specific gene can be analyzed. Diploid cells in which both copies of a gene have been mutated by homologous recombination can be selected after transfection with a mixture of constructs in which the gene to be targeted has been disrupted by one or other of two different antibiotic-resistance genes. Having obtained a mutant cell with a functional defect, the defect can be ascribed definitively to the mutated gene if the mutant phenotype can be reverted with a copy of the normal gene transfected into the

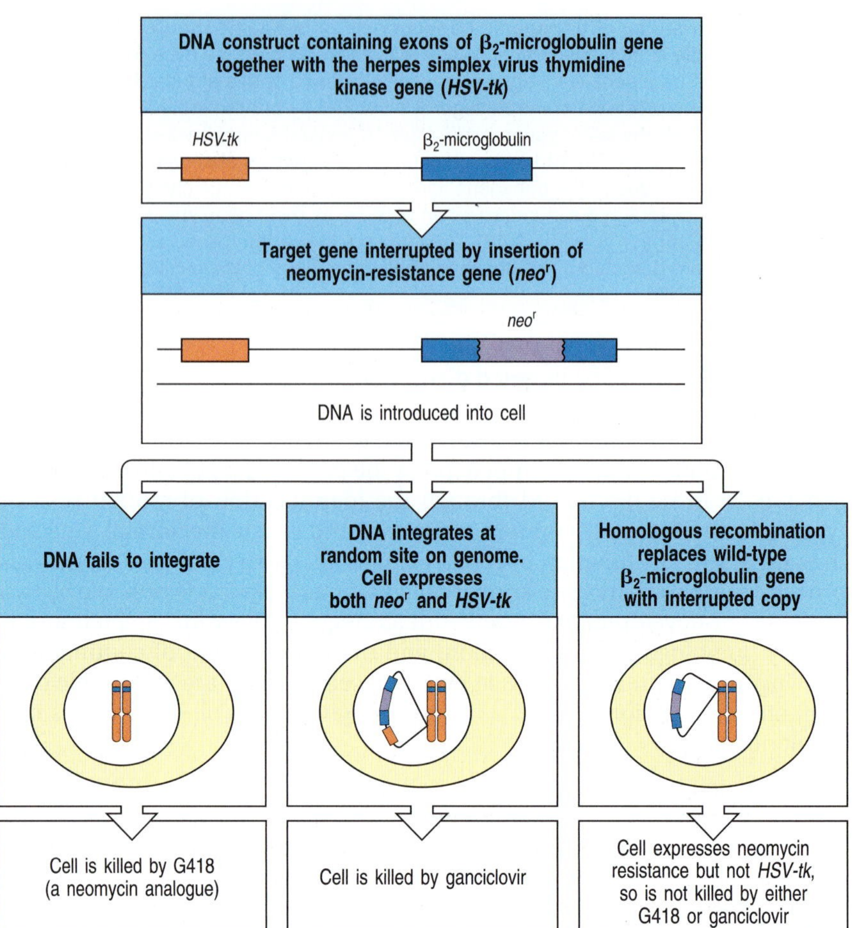

Fig. A.45 The deletion of specific genes can be accomplished by homologous recombination. When pieces of DNA are introduced into cells, they can integrate into cellular DNA in two different ways. If they randomly insert into sites of DNA breaks, the whole piece is usually integrated, often in several copies. However, extrachromosomal DNA can also undergo homologous recombination with the cellular copy of the gene, in which case only the central, homologous region is incorporated into cellular DNA. Inserting a selectable marker gene such as resistance to neomycin (*neo*^{r}) into the coding region of a gene does not prevent homologous recombination, and it achieves two goals. First, any cell that has integrated the injected DNA is protected from the neomycin-like antibiotic G418. Second, when the gene recombines with homologous cellular DNA, the *neo*^{r} gene disrupts the coding sequence of the modified cellular gene. Homologous recombinants can be discriminated from random insertions if the gene encoding herpes simplex virus thymidine kinase (*HSV-tk*) is placed at one or both ends of the DNA construct, which is often known as a 'targeting construct' because it targets the cellular gene. In random DNA integrations, *HSV-tk* is retained. HSV-tk renders the cell sensitive to the antiviral agent ganciclovir. However, as *HSV-tk* is not homologous to the target DNA, it is lost from homologous recombinants. Thus, cells that have undergone homologous recombination are uniquely resistant to both G418 and ganciclovir, and survive in a mixture of the two antibiotics. The presence of the disrupted gene has to be confirmed by Southern blotting or by PCR using primers in the *neo*^{r} gene and in cellular DNA lying outside the region used in the targeting construct. By using two different resistance genes one can disrupt the two cellular copies of a gene, making a deletion mutant (not shown).

mutant cell. Restoration of function means that the defect in the mutant gene has been complemented by the normal gene's function. This technique is very powerful because it allows the gene that is being transferred to be mutated in precise ways to determine which parts of the protein are required for function.

To knock out a gene *in vivo*, it is only necessary to disrupt one copy of the cellular gene in an ES cell. ES cells carrying the mutant gene are produced by targeted mutation (see Fig. A.45) and injected into a blastocyst, which is reimplanted into the uterus. The cells carrying the disrupted gene become incorporated into the developing embryo and contribute to all tissues of the resulting chimeric offspring, including those of the germline. The mutated gene can therefore be transmitted to some of the offspring of the original chimera, and further breeding of the mutant gene to homozygosity produces mice that completely lack the expression of that particular gene product (Fig. A.46). The effects of the absence of the gene's function can then be studied. In addition, the parts of the gene that are essential for its function can be identified by determining whether function can be restored by introducing different mutated copies of the gene back into the genome by transgenesis. The manipulation of the mouse genome by gene knockout and transgenesis is revolutionizing our understanding of the role of individual genes in lymphocyte development and function.

Because the most commonly used ES cells are derived from a poorly characterized strain of mice known as strain 129, the analysis of the function of a gene knockout often requires extensive back-crossing to another strain, just as in transgenic mice (see Fig. A.44). One can track the presence of the mutant copy of the gene by the presence of the *neo*^{r} gene. After sufficient back-crossing, the mice are intercrossed to produce mutants on a stable genetic background.

Fig. A.46 Gene knockout in embryonic stem cells enables mutant mice to be produced. Specific genes can be inactivated by homologous recombination in cultures of embryonic stem cells (ES cells). Homologous recombination is performed as described in Fig. A.45. In this example, the gene encoding β_2-microglobulin in ES cells is disrupted by homologous recombination with a targeting construct. Only a single copy of the gene needs to be disrupted. ES cells in which homologous recombination has taken place are injected into mouse blastocysts. If the mutant ES cells give rise to germ cells in the resulting chimeric mice (striped in the figure), the mutant gene can be transferred to their offspring. By breeding the mutant gene to homozygosity, a mutant phenotype is generated. These mutant mice are usually of the 129 strain because gene knockout is generally conducted in ES cells derived from the 129 strain of mice. In this case, the homozygous mutant mice lack MHC class I molecules on their cells, because MHC class I molecules have to pair with β_2-microglobulin for surface expression. The β_2-microglobulin-deficient mice can then be bred with mice transgenic for subtler mutants of the deleted gene, allowing the effect of such mutants to be tested *in vivo*.

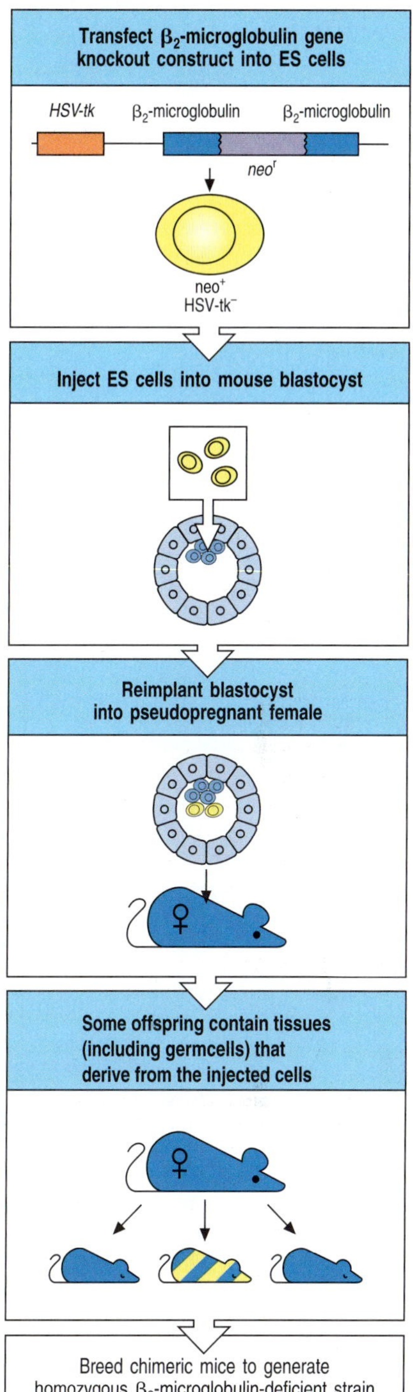

A problem with gene knockouts arises when the function of the gene is essential for the survival of the animal; in such cases the gene is termed a **recessive lethal gene**, and homozygous animals cannot be produced. However, by making chimeras with mice that are deficient in B and T cells, it is possible to analyze the function of recessive lethal genes in lymphoid cells. To do this, ES cells with homozygous lethal loss-of-function mutations are injected into blastocysts of mice lacking the ability to rearrange their antigen-receptor genes because of a mutation in their recombinase-activating genes (*RAG* knockout mice). As these chimeric embryos develop, the *RAG*-deficient cells can compensate for any developmental failure resulting from the gene knockout in the ES cells in all except the lymphoid lineage. As long as the mutated ES cells can develop into hematopoietic progenitors in the bone marrow, the embryos will survive and all of the lymphocytes in the resulting chimeric mouse will be derived from the mutant ES cells (Fig. A.47).

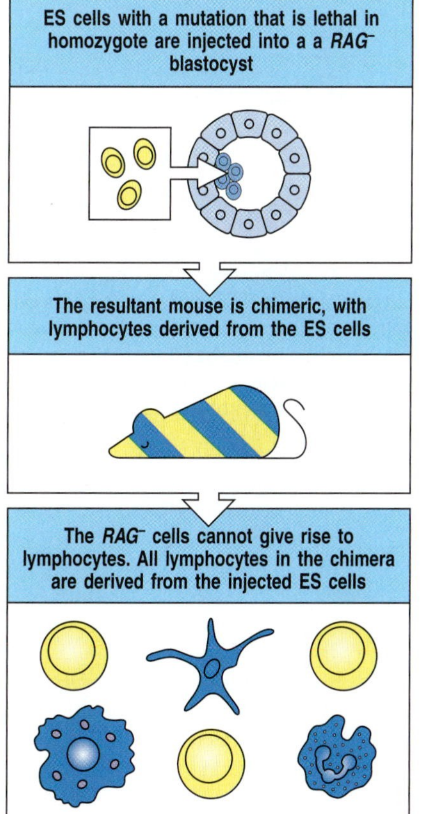

Fig. A.47 The role of recessive lethal genes in lymphocyte function can be studied using *RAG*-deficient chimeric mice. ES cells homozygous for the lethal mutation are injected into a *RAG*-deficient blastocyst (top panel). The *RAG*-deficient cells can give rise to all the tissues of a normal mouse except lymphocytes, and so can compensate for any deficiency in the developmental potential of the mutant ES cells (middle panel). If the mutant ES cells are capable of differentiating into hematopoietic stem cells—that is, if the gene function that has been deleted is not essential for this developmental pathway—then all the lymphocytes in the chimeric mouse will be derived from the ES cells (bottom panel), because *RAG*-deficient mice cannot make lymphocytes of their own.

A second powerful technique achieves tissue-specific or developmentally regulated gene deletion by employing the DNA sequences and enzymes used by bacteriophage P1 to excise itself from a host cell's genome. Integrated bacteriophage P1 DNA is flanked by recombination signal sequences called *loxP* sites. A recombinase, Cre, recognizes these sites, cuts the DNA and joins the two ends, thus excising the intervening DNA in the form of a circle. This mechanism can be adapted to allow the deletion of specific genes in a transgenic animal only in certain tissues or at certain times in development. First, *loxP* sites flanking a gene, or perhaps just a single exon, are introduced by homologous recombination (Fig. A.48). Usually, the introduction of these sequences into flanking or intronic DNA does not disrupt the normal function of the gene. Mice containing such *loxP* mutant genes are then mated with mice made transgenic for the Cre recombinase, under the control of a tissue-specific or inducible promoter. When the Cre recombinase is active, either in the appropriate tissue or when induced, it excises the DNA between the inserted *loxP* sites, thus inactivating the gene or exon. Thus, for example, using a T-cell specific promoter to drive expression of the Cre recombinase, a gene can be deleted only in T cells, while remaining functional in all other cells of the animal. This extremely powerful genetic technique was used to demonstrate the importance of B-cell receptors in B-cell survival.

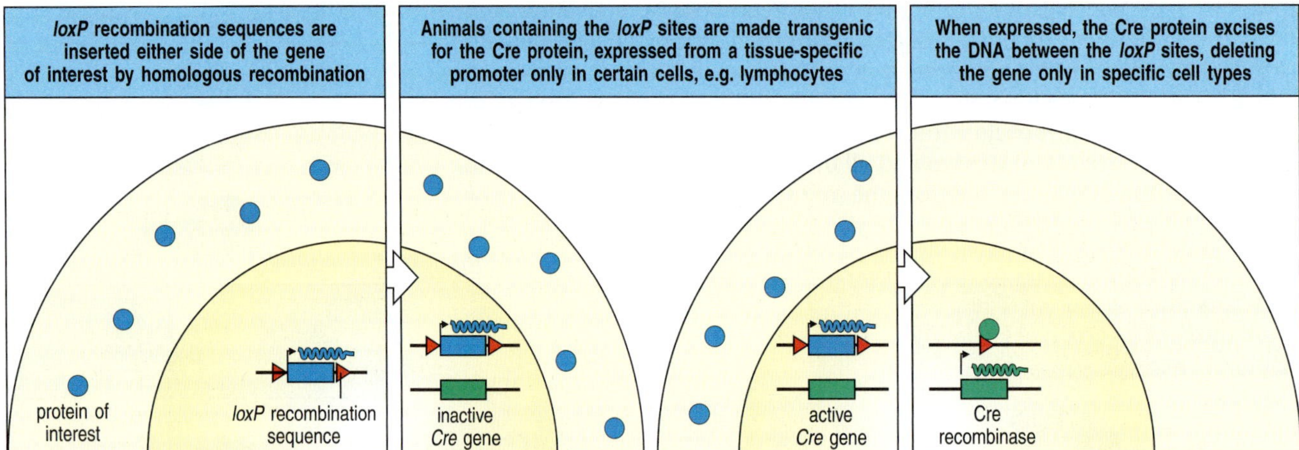

loxP recombination sequences are inserted either side of the gene of interest by homologous recombination
Animals containing the *loxP* sites are made transgenic for the Cre protein, expressed from a tissue-specific promoter only in certain cells, e.g. lymphocytes
When expressed, the Cre protein excises the DNA between the *loxP* sites, deleting the gene only in specific cell types
protein of interest
loxP recombination sequence
inactive *Cre* gene
active *Cre* gene
Cre recombinase

Fig. A.48 The P1 bacteriophage recombination system can be used to eliminate genes in particular cell lineages. The P1 bacteriophage protein Cre excises DNA that is bounded by recombination signal sequences called *loxP* sequences. These sequences can be introduced at either end of a gene by homologous recombination (left panel). Animals carrying genes flanked by *loxP* can also be made transgenic for the gene encoding the Cre protein, which is placed under the control of a tissue-specific promoter so that it is expressed only in certain cells or only at certain times during development (center panel). In the cells in which the Cre protein is expressed, it recognizes the *loxP* sequences and excises the DNA lying between them (right panel). Thus, individual genes can be deleted only in certain cell types or only at certain times. In this way, genes that are essential for the normal development of a mouse can be analyzed for their function in the developed animal and/or in specific cell types. Genes are shown as boxes, RNA as squiggles, and proteins as colored balls.

APPENDICES II–IV

Appendix II. CD antigens.					
CD antigen	**Cellular expression**	**Molecular weight (kDa)**	**Functions**	**Other names**	**Family relationships**
CD1a,b,c,d	Cortical thymocytes, Langerhans cells, dendritic cells, B cells (CD1c), intestinal epithelium, smooth muscle, blood vessels (CD1d)	43–49	MHC class I-like molecule, associated with β2-microglobulin. Has specialized role in presentation of lipid antigens		Immunoglobulin
CD2	T cells, thymocytes, NK cells	45–58	Adhesion molecule, binding CD58 (LFA-3). Binds Lck intracellularly and activates T cells	T11, LFA-2	Immunoglobulin
CD3	Thymocytes, T cells	γ: 25–28 δ: 20 ε: 20	Associated with the T-cell antigen receptor (TCR). Required for cell-surface expression of and signal transduction by the TCR	T3	Immunoglobulin
CD4	Thymocyte subsets, T_H1 and T_H2 T cells (about two thirds of peripheral T cells), monocytes, macrophages	55	Co-receptor for MHC class II molecules. Binds Lck on cytoplasmic face of membrane. Receptor for HIV-1 and HIV-2 gp120	T4, L3T4	Immunoglobulin
CD5	Thymocytes, T cells, subset of B cells	67		T1, Ly1	Scavenger receptor
CD6	Thymocytes, T cells, B cells in chronic lymphatic leukemia	100–130	Binds CD166	T12	Scavenger receptor
CD7	Pluripotential hematopoietic cells, thymocytes, T cells	40	Unknown, cytoplasmic domain binds PI 3-kinase on cross-linking. Marker for T cell acute lymphatic leukemia and pluripotential stem cell leukemias		Immunoglobulin
CD8	Thymocyte subsets, cytotoxic T cells (about one third of peripheral T cells)	α: 32–34 β: 32–34	Co-receptor for MHC class I molecules. Binds Lck on cytoplasmic face of membrane	T8, Lyt2,3	Immunoglobulin
CD9	Pre-B cells, monocytes, eosinophils, basophils, platelets, activated T cells, brain and peripheral nerves, vascular smooth muscle	24	Mediates platelet aggregation and activation via FcγRIIa, may play a role in cell migration		Tetraspanning membrane protein, also called transmembrane 4 (TM4)
CD10	B- and T-cell precursors, bone marrow stromal cells	100	Zinc metalloproteinase, marker for pre-B acute lymphatic leukemia (ALL)	Neutral endopeptidase, common acute lymphocytic leukemia antigen (CALLA)	
CD11a	Lymphocytes, granulocytes, monocytes and macrophages	180	αL subunit of integrin LFA-1 (associated with CD18); binds to CD54 (ICAM-1), CD102 (ICAM-2), and CD50 (ICAM-3)	LFA-1	Integrin α

CD antigen	Cellular expression	Molecular weight (kDa)	Functions	Other names	Family relationships
CD11b	Myeloid and NK cells	170	αM subunit of integrin CR3 (associated with CD18): binds CD54, complement component iC3b, and extracellular matrix proteins	Mac-1	Integrin α
CD11c	Myeloid cells	150	αX subunit of integrin CR4 (associated with CD18); binds fibrinogen	CR4, p150, 95	Integrin α
CD11d	Leukocytes	125	αD subunits of integrin; associated with CD18; binds to CD50		Integrin α
CDw12	Monocytes, granulocytes, platelets	90–120	Unknown		
CD13	Myelomonocytic cells	150–170	Zinc metalloproteinase	Aminopeptidase N	
CD14	Myelomonocytic cells	53–55	Receptor for complex of lipopoly-saccharide and lipopolysaccharide binding protein (LBP)		
CD15	Neutrophils, eosinophils, monocytes		Terminal trisaccharide expressed on glycolipids and many cell-surface glycoproteins	Lewis^x (Le^x)	
CD15s	Leukocytes, endothelium		Ligand for CD62E, P	Sialyl-Lewis^x (sLe^x)	poly-*N*-acetyl-lactosamine
CD15u			Sulfated CD15		Carbohydrate structures
CD16	Neutrophils, NK cells, macrophages	50–80	Component of low affinity Fc receptor, FcγRIII, mediates phagocytosis and antibody-dependent cell-mediated cytotoxicity	FcγRIII	Immunoglobulin
CDw17	Neutrophils, monocytes, platelets		Lactosyl ceramide, a cell-surface glycosphingolipid		
CD18	Leukocytes	95	Integrin β2 subunit, associates with CD11a, b, c, and d		Integrin β
CD19	B cells	95	Forms complex with CD21 (CR2) and CD81 (TAPA-1); co-receptor for B cells—cytoplasmic domain binds cytoplasmic tyrosine kinases and PI 3-kinase		Immunoglobulin
CD20	B cells	33–37	Oligomers of CD20 may form a Ca^{2+} channel; possible role in regulating B-cell activation		Contains 4 transmembrane segments
CD21	Mature B cells, follicular dendritic cells	145	Receptor for complement component C3d, Epstein–Barr virus. With CD19 and CD81, CD21 forms co-receptor for B cells	CR2	Complement control protein (CCP)
CD22	Mature B cells	α: 130 β: 140	Binds sialoconjugates	BL-CAM	Immunoglobulin
CD23	Mature B cells, activated macrophages, eosinophils, follicular dendritic cells, platelets	45	Low-affinity receptor for IgE, regulates IgE synthesis; ligand for CD19:CD21:CD81 co-receptor	FcεRII	C-type lectin
CD24	B cells, granulocytes	35–45	Unknown	Possible human homologue of mouse heat stable antigen (HSA)	
CD25	Activated T cells, B cells, and monocytes	55	IL-2 receptor α chain	Tac	CCP
CD26	Activated B and T cells, macrophages	110	Exopeptidase, cleaves N terminal X-Pro or X-Ala dipeptides from polypeptides	Dipeptidyl peptidase IV	Type II transmem-brane glycoprotein

CD antigen	Cellular expression	Molecular weight (kDa)	Functions	Other names	Family relationships
CD27	Medullary thymocytes, T cells, NK cells, some B cells	55	Binds CD70; can function as a co-stimulator for T and B cells		TNF receptor
CD28	T-cell subsets, activated B cells	44	Activation of naive T cells, receptor for co-stimulatory signal (signal 2) binds CD80 (B7.1) and CD86 (B7.2)	Tp44	Immunoglobulin and CD86 (B7.2)
CD29	Leukocytes	130	Integrin β1 subunit, associates with CD49a in VLA-1 integrin		Integrin β
CD30	Activated T, B, and NK cells, monocytes	120	Binds CD30L (CD153); cross-linking CD30 enhances proliferation of B and T cells	Ki-1	TNF receptor
CD31	Monocytes, platelets, granulocytes, T-cell subsets, endothelial cells	130–140	Adhesion molecule, mediating both leukocyte–endothelial and endothelial–endothelial interactions	PECAM-1	Immunoglobulin
CD32	Monocytes, granulocytes, B cells, eosinophils	40	Low affinity Fc receptor for aggregated immunoglobulin:immune complexes	FcγRII	Immunoglobulin
CD33	Myeloid progenitor cells, monocytes	67	Binds sialoconjugates		Immunoglobulin
CD34	Hematopoietic precursors, capillary endothelium	105–120	Ligand for CD62L (L-selectin)		Mucin
CD35	Erythrocytes, B cells, monocytes, neutrophils, eosinophils, follicular dendritic cells	250	Complement receptor 1, binds C3b and C4b, mediates phagocytosis	CR1	CCP
CD36	Platelets, monocytes, endothelial cells	88	Platelet adhesion molecule; involved in recognition and phagocytosis of apoptosed cells	Platelet GPIV, GPIIIb	
CD37	Mature B cells, mature T cells, myeloid cells	40–52	Unknown, may be involved in signal transduction; forms complexes with CD53, CD81, CD82, and MHC class II		Transmembrane 4
CD38	Early B and T cells, activated T cells, germinal center B cells, plasma cells	45	NAD glycohydrolase, augments B cell proliferation	T10	
CD39	Activated B cells, activated NK cells, macrophages, dendritic cells	78	Unknown, may mediate adhesion of B cells		
CD40	B cells, macrophages, dendritic cells, basal epithelial cells	48	Binds CD154 (CD40L); receptor for co-stimulatory signal for B cells, promotes growth, differentiation, and isotype switching of B cells, and cytokine production by macrophages and dendritic cells		TNF receptor
CD41	Platelets, megakaryocytes	Dimer: GPIIba: 125 GPIIbb: 22	αIIb integrin, associates with CD61 to form GPIIb, binds fibrinogen, fibronectin, von Willebrand factor, and thrombospondin	GPIIb	Integrin α
CD42a,b,c,d	Platelets, megakaryocytes	a: 23 b: 135, 23 c: 22 d: 85	Binds von Willebrand factor, thrombin; essential for platelet adhesion at sites of injury	a: GPIX b: GPIbα c: GPIbβ d: GPV	Leucine-rich repeat
CD43	Leukocytes, except resting B cells	115–135 (neutrophils) 95–115 (T cells)	Has extended structure, approx. 45 nm long and may be anti-adhesive	Leukosialin, sialophorin	Mucin
CD44	Leukocytes, erythrocytes	80–95	Binds hyaluronic acid, mediates adhesion of leukocytes	Hermes antigen, Pgp-1	Link protein

CD antigen	Cellular expression	Molecular weight (kDa)	Functions	Other names	Family relationships
CD45	All hematopoietic cells	180–240 (multiple isoforms)	Tyrosine phosphatase, augments signaling through antigen receptor of B and T cells, multiple isoforms result from alternative splicing (see below)	Leukocyte common antigen (LCA), T200, B220	Fibronectin type III
CD45RO	T-cell subsets, B-cell subsets, monocytes, macrophages	180	Isoform of CD45 containing none of the A, B, and C exons		Fibronectin type II
CD45RA	B cells, T-cell subsets (naive T cells), monocytes	205–220	Isoforms of CD45 containing the A exon		Fibronectin type II
CD45RB	T-cell subsets, B cells, monocytes, macrophages, granulocytes	190–220	Isoforms of CD45 containing the B exon	T200	Fibronectin type II
CD46	Hematopoietic and non-hematopoietic nucleated cells	56/66 (splice variants)	Membrane co-factor protein, binds to C3b and C4b to permit their degradation by Factor I	MCP	CCP
CD47	All cells	47–52	Adhesion molecule, thrombospondin receptor	IAP, MER6, OA3	Immunoglobulin superfamily
CD48	Leukocytes	40–47	Putative ligand for CD244	Blast-1	Immunoglobulin
CD49a	Activated T cells, monocytes, neuronal cells, smooth muscle	200	α1 integrin, associates with CD29, binds collagen, laminin-1	VLA-1	Integrin α
CD49b	B cells, monocytes, platelets, megakaryocytes, neuronal, epithelial and endothelial cells, osteoclasts	160	α2 integrin, associates with CD29, binds collagen, laminin	VLA-2, platelet GPIa	Integrin α
CD49c	B cells, many adherent cells	125, 30	α3 integrin, associates with CD29, binds laminin-5, fibronectin, collagen, entactin, invasin	VLA-3	Integrin α
CD49d	Broad distribution includes B cells, thymocytes, monocytes, granulocytes, dendritic cells	150	α4 integrin, associates with CD29, binds fibronectin, MAdCAM-1, VCAM-1	VLA-4	Integrin α
CD49e	Broad distribution includes memory T cells, monocytes, platelets	135, 25	α5 integrin, associates with CD29, binds fibronectin, invasin	VLA-5	Integrin α
CD49f	T lymphocytes, monocytes, platelets, megakaryocytes, trophoblasts	125, 25	α6 integrin, associates with CD29, binds laminin, invasin, merosine	VLA-6	Integrin α
CD50	Thymocytes, T cells, B cells, monocytes, granulocytes	130	Binds integrin CD11a/CD18	ICAM-3	Immunoglobulin
CD51	Platelets, megakaryocytes	125, 24	αV integrin, associates with CD61, binds vitronectin, von Willebrand factor, fibrinogen, and thrombospondin; may be receptor for apoptotic cells	Vitronectin receptor	Integrin α
CD52	Thymocytes, T cells, B cells (not plasma cells), monocytes, granulocytes, spermatozoa	25	Unknown, target for antibodies used therapeutically to deplete T cells from bone marrow	CAMPATH-1, HE5	
CD53	Leukocytes	35–42	Unknown	MRC OX44	Transmembrane 4
CD54	Hematopoietic and non-hematopoietic cells	75–115	Intercellular adhesion molecule (ICAM)-1 binds CD11a/CD18 integrin (LFA-1) and CD11b/CD18 integrin (Mac-1), receptor for rhinovirus	ICAM-1	Immunoglobulin
CD55	Hematopoietic and non-hematopoietic cells	60–70	Decay accelerating factor (DAF), binds C3b, disassembles C3/C5 convertase	DAF	CCP
CD56	NK cells	135–220	Isoform of neural cell-adhesion molecule (NCAM), adhesion molecule	NKH-1	Immunoglobulin
CD57	NK cells, subsets of T cells, B cells, and monocytes		Oligosaccharide, found on many cell-surface glycoproteins	HNK-1, Leu-7	

CD antigen	Cellular expression	Molecular weight (kDa)	Functions	Other names	Family relationships
CD58	Hematopoietic and non-hematopoietic cells	55–70	Leukocyte function-associated antigen-3 (LFA-3), binds CD2, adhesion molecule	LFA-3	Immunoglobulin
CD59	Hematopoietic and non-hematopoietic cells	19	Binds complement components C8 and C9, blocks assembly of membrane-attack complex	Protectin, Mac inhibitor	Ly-6
CD60a			Disialyl ganglioside D3 (GD3)		Carbohydrate structures
CD60b			9-*O*-acetyl-GD3		Carbohydrate structures
CD60c			7-*O*-acetyl-GD3		Carbohydrate structures
CD61	Platelets, megakaryocytes, macrophages	110	Intergrin β3 subunit, associates with CD41 (GPIIb/IIIa) or CD51 (vitronectin receptor)		Integrin β
CD62E	Endothelium	140	Endothelium leukocyte adhesion molecule (ELAM), binds sialyl-Lewis^x, mediates rolling interaction of neutrophils on endothelium	ELAM-1, E-selectin	C-type lectin, EGF, and CCP
CD62L	B cells, T cells, monocytes, NK cells	150	Leukocyte adhesion molecule (LAM), binds CD34, GlyCAM, mediates rolling interactions with endothelium	LAM-1, L-selectin, LECAM-1	C-type lectin, EGF, and CCP
CD62P	Platelets, megakaryocytes, endothelium	140	Adhesion molecule, binds CD162 (PSGL-1), mediates interaction of platelets with endothelial cells, monocytes and rolling leukocytes on endothelium	P-selectin, PADGEM	C-type lectin, EGF, and CCP
CD63	Activated platelets, monocytes, macrophages	53	Unknown, is lysosomal membrane protein translocated to cell surface after activation	Platelet activation antigen	Transmembrane 4
CD64	Monocytes, macrophages	72	High-affinity receptor for IgG, binds IgG3>IgG1>IgG4>>>IgG2, mediates phagocytosis, antigen capture, ADCC	FcγRI	Immunoglobulin
CD65	Myeloid cells		Oligosaccharide component of a ceramide dodecasaccharide		
CD66a	Neutrophils	160–180	Unknown, member of carcino-embryonic antigen (CEA) family (see below)	Biliary glyco-protein-1 (BGP-1)	Immunoglobulin
CD66b	Granulocytes	95–100	Unknown, member of carcino-embryonic antigen (CEA) family	Previously CD67	Immunoglobulin
CD66c	Neutrophils, colon carcinoma	90	Unknown, member of carcino-embryonic antigen (CEA) family	Nonspecific cross-reacting antigen (NCA)	Immunoglobulin
CD66d	Neutrophils	30	Unknown, member of carcino-embryonic antigen (CEA) family		Immunoglobulin
CD66e	Adult colon epithelium, colon carcinoma	180–200	Unknown, member of carcino-embryonic antigen (CEA) family	Carcino-embryonic antigen (CEA)	Immunoglobulin
CD66f	Unknown		Unknown, member of carcino-embryonic antigen (CEA) family	Pregnancy specific glycoprotein	Immunoglobulin
CD68	Monocytes, macrophages, neutrophils, basophils, large lymphocytes	110	Unknown	Macrosialin	Mucin

CD antigen	Cellular expression	Molecular weight (kDa)	Functions	Other names	Family relationships
CD69	Activated T and B cells, activated macrophages and NK cells	28, 32 homodimer	Unknown, early activation antigen	Activation inducer molecule (AIM)	C-type lectin
CD70	Activated T and B cells, and macrophages	75, 95, 170	Ligand for CD27, may function in co-stimulation of B and T cells	Ki-24	TNF
CD71	All proliferating cells, hence activated leukocytes	95 homodimer	Transferrin receptor	T9	
CD72	B cells (not plasma cells)	42 homodimer	Unknown	Lyb-2	C-type lectin
CD73	B-cell subsets, T-cell subsets	69	Ecto-5′-nucleotidase, dephosphorylates nucleotides to allow nucleoside uptake		
CD74	B cells, macrophages, monocytes, MHC class II positive cells	33, 35, 41, 43 (alternative initiation and splicing)	MHC class II-associated invariant chain	Ii, Iγ	
CD75	Mature B cells, T-cell subsets		Lactosamines, ligand for CD22, mediates B-cell–B-cell adhesion		
CD75s			α-2,6-sialylated lactosamines		Carbohydrate structures
CD77	Germinal center B cells		Neutral glycosphingolipid (Galα1$\rightarrow$4Galβ1$\rightarrow$4Glcβ1 $\rightarrow$ceramide), binds Shiga toxin, cross-linking induces apoptosis	Globotriaosyl-ceramide (Gb3) Pk blood group	
CD79α,β	B cells	α: 40–45 β: 37	Components of B-cell antigen receptor analogous to CD3, required for cell-surface expression and signal transduction	Igα, Igβ	Immunoglobulin
CD80	B-cell subset	60	Co-stimulator, ligand for CD28 and CTLA-4	B7 (now B7.1), BB1	Immunoglobulin
CD81	Lymphocytes	26	Associates with CD19, CD21 to form B cell co-receptor	Target of anti-proliferative antibody (TAPA-1)	Transmembrane 4
CD82	Leukocytes	50–53	Unknown	R2	Transmembrane 4
CD83	Dendritic cells, B cells, Langerhans cells	43	Unknown	HB15	Immunoglobulin
CDw84	Monocytes, platelets, circulating B cells	73	Unknown	GR6	Immunoglobulin
CD85	Dendritic cells		ILT/LIR family	GR4	Immunoglobulin superfamily
CD86	Monocytes, activated B cells, dendritic cells	80	Ligand for CD28 and CTLA4	B7.2	Immunoglobulin
CD87	Granulocytes, monocytes, macrophages, T cells, NK cells, wide variety of nonhematopoietic cell types	35–59	Receptor for urokinase plasminogen activator	uPAR	Ly-6
CD88	Polymorphonuclear leukocytes, macrophages, mast cells	43	Receptor for complement component C5a	C5aR	G protein-coupled receptor
CD89	Monocytes, macrophages, granulocytes, neutrophils, B-cell subsets, T-cell subsets	50–70	IgA receptor	FcαR	Immunoglobulin
CD90	CD34+ prothymocytes (human), thymocytes, T cells (mouse)	18	Unknown	Thy-1	Immunoglobulin
CD91	Monocytes, many non-hematopoietic cells	515, 85	α2-macroglobulin receptor		EGF, LDL receptor

CD antigen	Cellular expression	Molecular weight (kDa)	Functions	Other names	Family relationships
CD92	Neutrophils, monocytes, platelets, endothelium	70	Unknown	GR9	
CD93	Neutrophils, monocytes, endothelium	120	Unknown	GR11	
CD94	T-cell subsets, NK cells	43	Unknown	KP43	C-type lectin
CD95	Wide variety of cell lines, *in vivo* distribution uncertain	45	Binds TNF-like Fas ligand, induces apoptosis	Apo-1, Fas	TNF receptor
CD96	Activated T cells, NK cells	160	Unknown	T-cell activation increased late expression (TACTILE)	Immunoglobulin
CD97	Activated B and T cells, monocytes, granulocytes	75–85	Binds CD55	GR1	EGF, G protein-coupled receptor
CD98	T cells, B cells, natural killer cells, granulocytes, all human cell lines	80, 45 heterodimer	May be amino acid transporter	4F2, FRP-1	
CD99	Peripheral blood lymphocytes, thymocytes	32	Unknown	MIC2, E2	
CD100	Hematopoietic cells	150 homodimer	Unknown	GR3	Semaphorin
CD101	Monocytes, granulocytes, dendritic cells, activated T cells	120 homodimer	Unknown	BPC#4	Immunoglobulin
CD102	Resting lymphocytes, monocytes, vascular endothelium cells (strongest)	55–65	Binds CD11a/CD18 (LFA-1) but not CD11b/CD18 (Mac-1)	ICAM-2	Immunoglobulin
CD103	Intraepithelial lymphocytes, 2–6% peripheral blood lymphocytes	150, 25	αE integrin	HML-1, α6, αE integrin	Integrin α
CD104	$CD4^-$ $CD8^-$ thymocytes, neuronal, epithelial, and some endothelial cells, Schwann cells, trophoblasts	220	Integrin β4 associates with CD49f, binds laminins	β4 integrin	Integrin β
CD105	Endothelial cells, activated monocytes and macrophages, bone marrow cell subsets	90 homodimer	Binds TGF-β	Endoglin	
CD106	Endothelial cells	100–110	Adhesion molecule, ligand for VLA-4 ($\alpha_4\beta_1$ integrin)	VCAM-1	Immunoglobulin
CD107a	Activated platelets, activated T cells, activated neutrophils, activated endothelium	110	Unknown, is lysosomal membrane protein translocated to the cell surface after activation	Lysosomal associated membrane protein-1 (LAMP-1)	
CD107b	Activated platelets, activated T cells, activated neutrophils, activated endothelium	120	Unknown, is lysosomal membrane protein translocated to the cell surface after activation	LAMP-2	
CD108	Erythrocytes, circulating lymphocytes, lymphoblasts	80	Unknown	GR2, John Milton-Hagen blood group antigen	
CD109	Activated T cells, activated platelets, vascular endothelium	170	Unknown	Platelet activation factor, GR56	
CD110	Platelets		MPL, TPO R		
CD111	Myeloid cells		PPR1/Nectin1		
CD112	Myeloid cells		PRR2		
CD114	Granulocytes, monocytes	150	Granulocyte colony-stimulating factor (G-CSF) receptor		Immunoglobulin, fibronectin type III

CD antigen	Cellular expression	Molecular weight (kDa)	Functions	Other names	Family relationships
CD115	Monocytes, macrophages	150	Macrophage colony-stimulating factor (M-CSF) receptor	M-CSFR, c-fms	Immunoglobulin, tyrosine kinase
CD116	Monocytes, neutrophils, eosinophils, endothelium	70–85	Granulocyte-macrophage colony-stimulating factor (GM-CSF) receptor a chain	GM-CSFRα	Cytokine receptor, fibronectin type III
CD117	Hematopoietic progenitors	145	Stem-cell factor (SCF) receptor	c-Kit	Immunoglobulin, tyrosine kinase
CD118	Broad cellular expression		Interferon-α, β receptor	IFN-α, βR	
CD119	Macrophages, monocytes, B cells, endothelium	90–100	Interferon-γ receptor	IFN-γR	Fibronectin type III
CD120a	Hematopoietic and nonhematopoietic cells, highest on epithelial cells	50–60	TNF receptor, binds both TNF-α and LT	TNFR-I	TNF receptor
CD120b	Hematopoietic and nonhematopoietic cells, highest on myeloid cells	75–85	TNF receptor, binds both TNF-α and LT	TNFR-II	TNF receptor
CD121a	Thymocytes, T cells	80	Type I interleukin-1 receptor, binds IL-1α and IL-1β	IL-1R type I	Immunoglobulin
CDw121b	B cells, macrophages, monocytes	60–70	Type II interleukin-1 receptor, binds IL-1α and IL-1β	IL-1R type II	Immunoglobulin
CD122	NK cells, resting T-cell subsets, some B-cell lines	75	IL-2 receptor β chain	IL-2Rβ	Cytokine receptor, fibronectin type III
CD123	Bone marrow stem cells, granulocytes, monocytes, megakaryocytes	70	IL-3 receptor α chain	IL-3Rα	Cytokine receptor, fibronectin type III
CD124	Mature B and T cells, hematopoietic precursor cells	130–150	IL-4 receptor	IL-4R	Cytokine receptor, fibronectin type III
CD125	Eosinophils, basophils, activated B cells	55–60	IL-5 receptor	IL-5R	Cytokine receptor, fibronectin type III
CD126	Activated B cells and plasma cells (strong), most leukocytes (weak)	80	IL-6 receptor α subunit	IL-6Rα	Immunoglobulin, cytokine receptor, fibronectin type III
CD127	Bone marrow lymphoid precursors, pro-B cells, mature T cells, monocytes	68–79, possibly forms homodimers	IL-7 receptor	IL-7R	Fibronectin type III
CDw128	Neutrophils, basophils, T-cell subsets	58–67	IL-8 receptor	IL-8R	G protein-coupled receptor
CD129	Not yet assigned				
CD130	Most cell types, strong on activated B cells and plasma cells	130	Common subunit of IL-6, IL-11, oncostatin-M (OSM) and leukemia inhibitory factor (LIF) receptors	IL-6Rβ, IL-11Rβ, OSMRβ, LIFRβ, IFRβ	Immunoglobulin, cytokine receptor, fibronectin type III
CDw131	Myeloid progenitors, granulocytes	140	Common β subunit of IL-3, IL-5, and GM-CSF receptors	IL-3Rβ, IL-5Rbβ, GM-CSFRβ	Cytokine receptor, fibronectin type III
CD132	B cells, T cells, NK cells, mast cells, neutrophils	64	IL-2 receptor γ chain, common subunit of IL-2, IL-4, IL-7, IL-9, and IL-15 receptors		Cytokine receptor
CD133	Stem/progenitor cells		AC133		
CD134	Activated T cells	50	May act as adhesion molecule co-stimulator	OX40	TNF receptor
CD135	Multipotential precursors, myelomonocytic and B-cell progenitors	130, 155	Growth factor receptor	FLK2, STK-1	Immunoglobulin, tyrosine kinase
CDw136	Monocytes, epithelial cells, central and peripheral nervous system	180	Chemotaxis, phagocytosis, cell growth, and differentiation	MSP-R, RON	Tyrosine kinase

CD antigen	Cellular expression	Molecular weight (kDa)	Functions	Other names	Family relationships
CDw137	T and B lymphocytes, monocytes, some epithelial cells		Co-stimulator of T-cell proliferation	ILA (induced by lymphocyte activation), 4-1BB	TNF receptor
CD138	B cells		Heparan sulfate proteoglycan binds collagen type I	Syndecan-1	
CD139	B cells	209, 228	Unknown		
CD140a,b	Stromal cells, some endothelial cells	a: 180 b: 180	Platelet derived growth factor (PDGF) receptor α and β chains		
CD141	Vascular endothelial cells	105	Anticoagulant, binds thrombin, the complex then activates protein C	Thrombomodulin fetomodulin	C-type lectin, EGF
CD142	Epidermal keratinocytes, various epithelial cells, astrocytes, Schwann cells. Absent from cells in direct contact with plasma unless induced by inflammatory mediators	45–47	Major initiating factor of clotting. Binds Factor VIIa; this complex activates Factors VII, IX, and X	Tissue factor, thromboplastin	Fibronectin type III
CD143	Endothelial cells, except large blood vessels and kidney, epithelial cells of brush borders of kidney and small intestine, neuronal cells, activated macrophages and some T cells. Soluble form in plasma	170–180	Zn^{2+} metallopeptidase dipeptidyl peptidase, cleaves angiotensin I and bradykinin from precursor forms	Angiotensin converting enzyme (ACE)	
CD144	Endothelial cells	130	Organizes adherens junction in endothelial cells	Cadherin-5, VE-cadherin	Cadherin
CD145	Endothelial cells, some stromal cells	25, 90, 110	Unknown		
CD146	Endothelium	130	Potential adhesion molecule, localized at cell–cell junctions	MCAM, MUC18, S-ENDO	Immunoglobulin
CD147	Leukocytes, red blood cells, platelets, endothelial cells	55–65	Potential adhesion molecule	M6, neurothelin, EMMPRIN, basigin, OX-47	Immunoglobulin
CD148	Granulocytes, monocytes, dendritic cells, T cells, fibroblasts, nerve cells	240–260	Contact inhibition of cell growth	HPTPη	Fibronectin type III, protein tyrosine phosphatase
CD150	Thymocytes, activated lymphocytes	75–95	Unknown	SLAM	Immunoglobulin
CD151	Platelets, megakaryocytes, epithelial cells, endothelial cells	32	Associates with β1 integrins	PETA-3, SFA-1	Transmembrane 4
CD152	Activated T cells	33	Receptor for B7.1 (CD80), B7.2 (CD86); negative regulator of T-cell activation	CTLA-4	Immunoglobulin
CD153	Activated T cells, activated macrophages, neutrophils, B cells	38–40	Ligand for CD30, may co-stimulate T cells	CD30L	TNF
CD154	Activated CD4 T cells	30 trimer	Ligand for CD40, inducer of B cell proliferation and activation	CD40L, TRAP, T-BAM, gp39	TNF receptor
CD155	Monocytes, macrophages, thymocytes, CNS neurons	80–90	Normal function unknown; receptor for poliovirus	Poliovirus receptor	Immunoglobulin
CD156a	Neutrophils, monocytes	69	Unknown, may be involved in integrin leukocyte extravasation	MS2, ADAM 8 (A disintegrin and metalloproteinase)	
CD156b			TACE/ADAM17. Adhesion structures		
CD157	Granulocytes, monocytes, bone marrow stromal cells, vascular endothelial cells, follicular dendritic cells	42–45 (50 on monocytes)	ADP-ribosyl cyclase, cyclic ADP-ribose hydrolase	BST-1	
CD158	NK cells		KIR family		

CD antigen	Cellular expression	Molecular weight (kDa)	Functions	Other names	Family relationships
CD158a	NK-cell subsets	50 or 58	Inhibits NK cell cytotoxicity on binding MHC class I molecules	p50.1, p58.1	Immunoglobulin
CD158b	NK-cell subsets	50 or 58	Inhibits NK cell cytotoxicity on binding HLA-Cw3 and related alleles	p50.2, p58.2	Immunoglobulin
CD159a	NK cells		Binds CD94 to form NK receptor; inhibits NK cell cytotoxicity on binding MHC class I molecules	NKG2A	
CD160	T cells			BY55	
CD161	NK cells, T cells	44	Regulates NK cytotoxicity	NKRP1	C-type lectin
CD162	Neutrophils, lymphocytes, monocytes	120 homodimer	Ligand for CD62P	PSGL-1	Mucin
CD162R	NK cells			PEN5	
CD163	Monocytes, macrophages	130	Unknown	M130	
CD164	Epithelial cells, monocytes, bone marrow stromal cells	80	Unknown	MUC-24 (multi-glycosylated protein 24)	Mucin
CD165	Thymocytes, thymic epithelial cells, CNS neurons, pancreatic islets, Bowman's capsule	37	Adhesion between thymocytes and thymic epithelium	Gp37, AD2	
CD166	Activated T cells, thymic epithelium, fibroblasts, neurons	100–105	Ligand for CD6, involved integrin neurite extension	ALCAM, BEN, DM-GRASP, SC-1	Immunoglobulin
CD167a	Normal and transformed epithelial cells	63, 64 dimer	Binds collagen	DDR1, trkE, cak, eddr1	Receptor tyrosine kinase, discoidin-related
CD168	Breast cancer cells	Five isoforms: 58, 60, 64, 70, 84	Adhesion molecule. Receptor for hyaluronic acid-mediated motility—mediated cell migration	RHAMM	
CD169	Subsets of macrophages	185	Adhesion molecule. Binds sialylated carbohydrates. May mediate macrophage binding to granulocytes and lymphocytes	Sialoadhesin	Immunoglobulin superfamily, sialoadhesin family
CD170	Neutrophils	67 homodimer	Adhesion molecule. **S**ialic acid-binding **I**g-like **lec**tin (Siglec). Cytoplasmic tail contains ITIM motifs	Siglec-5, OBBP2, CD33L2	Immunoglobulin superfamily, sialoadhesin family
CD171	Neurons, Schwann cells, lymphoid and myelomonocytic cells, B cells, CD4 T cells (not CD8 T cells)	200–220, exact MW.varies with cell type	Adhesion molecule, binds CD9, CD24, CD56, also homophilic binding	L1, NCAM-L1	Immunoglobulin superfamily
CD172a		115–120	Adhesion molecule; the transmembrane protein is a substrate of activated receptor tyrosine kinases (RTKs) and binds to SH2 domains	SIRP, SHPS1, MYD-1, SIRP-α-1, protein tyrosine phosphatase, non-receptor type substrate 1 (PTPNS1)	Immunoglobulin superfamily
CD173	All cells		Blood group H type 2. Carbohydrate moiety		
CD174	All cells		Lewis y blood group. Carbohydrate moiety		
CD175	All cells		Tn blood group. Carbohydrate moiety		
CD175s	All cells		Sialyl-Tn blood group. Carbohydrate moiety		
CD176	All cells		TF blood group. Carbohydrate moiety		

CD antigen	Cellular expression	Molecular weight (kDa)	Functions	Other names	Family relationships
CD177	Myeloid cells	56–64	NB1 is a GPI-linked neutrophil-specific antigen, found on only a subpopulation of neutrophils present in NB1-positive adults (97% of healthy donors) NB1 is first expressed at the myelocyte stage of myeloid differentiation	NB1	
CD178	Activated T cells	38–42	Fas ligand; binds to Fas to induce apoptosis	FasL	TNF superfamily
CD179a	Early B cells	16–18	Immunoglobulin iota chain associates noncovalently with CD179b to form a surrogate light chain which is a component of the pre-B-cell receptor that plays a critical role in early B-cell differentiation	VpreB, IGVPB, Igι	Immunoglobulin superfamily
CD179b	B cells	22	Immunoglobulin λ-like polypeptide 1 associates noncovalently with CD179a to form a surrogate light chain that is selectively expressed at the early stages of B-cell development. Mutations in the CD179b gene have been shown to result in impairment of B-cell development and agammaglobulinemia in humans	IGLL1, λ5 (IGL5), IGVPB, 14.	Immunoglobulin superfamily
CD180	B cells	95–105	Type 1 membrane protein consisting of extracellular leucine-rich repeats (LRR). Is associated with a molecule called MD-1 and forms the cell-surface receptor complex, RP105/MD-1, which by working in concert with TLR4, controls B-cell recognition and signaling of lipopolysaccharide (LPS)	LY64, RP105	Toll-like receptors (TLR)
CD183	Particularly on malignant B cells from chronic lymphoproliferative disorders	46–52	CXC chemokine receptor involved in chemotaxis of malignant B lymphocytes. Binds INP10 and MIG[3]	CXCR3, G protein-coupled receptor 9 (GPR 9)	Chemokine receptors, G protein-coupled receptor superfamily
CD184	Preferentially expressed on the more immature CD34$^+$ hematopoietic stem cells	46–52	Binding to SDF-1 (LESTR/fusin); acts as a cofactor for fusion and entry of T-cell line; trophic strains of HIV-1	CXCR4, NPY3R, LESTR, fusin, HM89	Chemokine receptors, G protein-coupled receptor superfamily
CD195	Promyelocytic cells	40	Receptor for a CC type chemokine. Binds to MIP-1α, MIP-1β and RANTES. May play a role in the control of granulocytic lineage proliferation or differentiation. Acts as co-receptor with CD4 for primary macrophage-tropic isolates of HIV-1	CMKBR5, CCR5, CKR-5, CC-CKR-5, CKR5	Chemokine receptors, G protein-coupled receptor superfamily
CDw197	Activated B and T lymphocytes, strongly upregulated in B cells infected with EBV and T cells infected with HHV6 or 7	46–52	Receptor for the MIP-3β chemokine; probable mediator of EBV effects on B lymphocytes or of normal lymphocyte functions	CCR7. EBI1 (Epstein–Barr virus induced gene 1), CMKBR7, BLR2	Chemokine receptors, G protein-coupled receptor superfamily
CD200	Normal brain and B-cell lines	41 (rat thymocytes) 47 (rat brain)	Antigen identified by MoAb MRCOX-2. Nonlineage molecules. Function unknown	MOX-2, MOX-1	Immunoglobulin superfamily
CD201	Endothelial cells	49	Endothelial cell-surface receptor (EPCR) that is capable of high-affinity binding of protein C and activated protein C. It is downregulated by exposure of endothelium to tumor necrosis factor	EPCR	CD1major histocompatibility complex family
CD202b	Endothelial cells	140	Receptor tyrosine kinase, binds angiopoietin-1; important in angiogenesis, particularly for vascular network formation in endothelial cells. Defects in TEK are associated with inherited venous malformations; the TEK signaling pathway appears to be critical for endothelial cell–smooth muscle cell communication in venous morphogenesis	VMCM, TEK (tyrosine kinase, endothelial), TIE2 (tyrosine kinase with Ig and EGF homology domains), VMCM1	Immunoglobulin superfamily, tyrosine kinase

CD antigen	Cellular expression	Molecular weight (kDa)	Functions	Other names	Family relationships
CD203c	Myeloid cells (uterus, basophils, and mast cells)	101	Belongs to a series of ectoenzymes that are involved in hydrolysis of extracellular nucleotides. They catalyze the cleavage of phosphodiester and phosphosulfate bonds of a variety of molecules, including deoxynucleotides, NAD, and nucleotide sugars	NPP3, B10, PDNP3, PD-Iβ, gp130RB13-6	Type II transmembrane proteins, Ecto-nucleotide pyrophosphatase/phosphodiesterase (E-NPP) family
CD204	Myeloid cells	220	Mediate the binding, internalization, and processing of a wide range of negatively charged macromolecules. Implicated in the pathologic deposition of cholesterol in arterial walls during atherogenesis	Macrophage scavenger R (MSR1)	Scavenger receptor family, collagen-like
CD205	Dendritic cells	205	Lymphocyte antigen 75; putative antigen-uptake receptor on dendritic cells	LY75, DEC-205, GP200-MR6	Type I transmembrane protein
CD206	Macrophages, endothelial cells	175–190	Type I membrane glycoprotein; only known example of a C-type lectin that contains multiple C-type CRDs (carbohydrate-recognition domains); it binds high-mannose structures on the surface of potentially pathogenic viruses, bacteria, and fungi	Macrophage mannose receptor (MMR), MRC1	C-type lectin superfamily
CD207	Langerhans cells	40	Type II transmembrane protein; Langerhans cell specific C-type lectin; potent inducer of membrane superimposition and zippering leading to BG (Birbeck granule) formation	Langerin	C-type lectin superfamily
CD208	Interdigitating dendritic cells in lymphoid organs	70–90	Homologous to CD68, DC-LAMP is a lysosomal protein involved in remodeling of specialized antigen-processing compartments and in MHC class II-restricted antigen presentation. Upregulated in mature DCs induced by CD40L, TNF-α and LPS.	D lysosome-associated membrane protein, DC-LAMP	Major histocompatibility complex family
CD209	Dendritic cells	44	C-type lectin; binds ICAM3 and HIV-1 envelope glycoprotein gp120 enables T-cell receptor engagement by stabilization of the DC/T-cell contact zone, promotes efficient infection in *trans* cells that express CD4 and chemokine receptors; type II transmembrane protein	DC-SIGN (dendritic cell-specific ICAM3-grabbing non-integrin)	C-type lectin superfamily
CDw210	B cells, T helper cells, and cells of the monocyte/macrophage lineage	90–110	Interleukin 10 receptor α and β	IL-10Rα, IL-10RA, HIL-10R, IL-10Rβ, IL-10RB, CRF2-4, CRFB4	Class II cytokine receptor family
CD212	Activated CD4, CD8, and NK cells	130	IL-12 receptor β chain; a type I transmembrane protein involved in IL-12 signal transduction.	IL-12R IL-12RB	Hemopoietin cytokine receptor superfamily
CD213a1	B cells, monocytes, fibroblasts, endothelial cells	60–70	Receptor which binds Il-13 with a low affinity; together with IL-4Rα can form a functional receptor for IL-13, also serves as an alternate accessory protein to the common cytokine receptor gamma chain for IL-4 signaling	IL-13Rα 1, NR4, IL-13Ra	Hemopoietic cytokine receptor superfamily
CD213a2	B cells, monocytes, fibroblasts, endothelial cells		IL-13 receptor which binds as a monomer with high affinity to interleukin-13 (IL-13), but not to IL-4; human cells expressing IL-13RA2 show specific IL-13 binding with high affinity	IL-13Rα 2, IL-13BP	Hemopoietic cytokine receptor superfamily
CDw217	Activated memory T cells	120	Interleukin 17 receptor homodimer	IL-17R, CTLA-8	Chemokine/cytokine receptors

CD antigen	Cellular expression	Molecular weight (kDa)	Functions	Other names	Family relationships
CD220	Nonlineage molecules	α: 130 β: 95	Insulin receptor; integral transmembrane glycoprotein comprised of two α and two β subunits; this receptor binds insulin and has a tyrosine-protein kinase activity—autophosphorylation activates the kinase activity	Insulin receptor	Insulin receptor family of tyrosine-protein kinases, EGFR family
CD221	Nonlineage molecules	α: 135 β: 90	Insulin-like growth factor I receptor binds insulin-like growth factor with a high affinity. It has tyrosine kinase activity and plays a critical role in transformation events. Cleavage of the precursor generates α and β subunits.	IGF1R, JTK13	Insulin receptor family of tyrosine-protein kinases, EGFR family
CD222	Nonlineage molecules	250	Ubiquitously expressed multifunctional type I transmembrane protein. Its main functions include internalization of IGF-II, internalization or sorting of lysosomal enzymes and other M6P-containing proteins	IGF2R, CIMPR, CI-MPR, IGF2R, M6P-R (Mannose-6-phosphate receptor)	Mammalian lectins
CD223	Activated T and NK cells	70	Involved in lymphocyte activation; binds to HLA class-II antigens; role in down-regulating antigen specific response; close relationship of LAG3 to CD4	Lymphocyte-activation gene 3 LAG-3	Immunoglobulin superfamily
CD224	Nonlineage molecules	62 (unprocessed precursor)	Predominantly a membrane-bound enzyme; plays a key role in the γ-glutamyl cycle, a pathway for the synthesis and degradation of glutathione. This enzyme consists of two polypeptide chains, which are synthesized in precursor form from a single polypeptide	γ-glutamyl transferase, GGT1, D22S672 D22S732	γ-glutamyl transferase protein family
CD225	Leukocytes and endothelial cells	16–17	Interferon-induced transmembrane protein 1 is implicated in the control of cell growth. It is a component of a multimeric complex involved in the transduction of antiproliferative and homotypic adhesion signals	Leu 13, IFITM1, IFI17	IFN-induced transmembrane proteins
CD226	NK cells, platelets, monocytes, and a subset of T cells	65	Adhesion glycoprotein; mediates cellular adhesion to other cells bearing an unidentified ligand and cross-linking CD226 with antibodies causes cellular activation	DNAM-1 (PTA1), DNAX, TLiSA1	Immunoglobulin superfamily
CD227	Human epithelial tumors, such as breast cancer	122 (non-glycosylated)	Epithelial mucin containing a variable number of repeats with a length of twenty amino acids, resulting in many different alleles. Direct or indirect interaction with actin cytoskeleton.	PUM (peanut-reactive urinary mucin), MUC.1, mucin 1	Mucin
CD228	Predominantly in human melanomas	97	Tumor-associated antigen (melanoma) identified by monoclonal antibodies 133.2 and 96.5, involved in cellular iron uptake.	Melanotransferrin, P97	Transferrin superfamily
CD229	Lymphocytes	90–120	May participate in adhesion reactions between T lymphocytes and accessory cells by homophilic interaction	Ly9	Immunoglobulin superfamily (CD2 subfamily)
CD230	Expressed in both normal and infected cells	27–30	The function of PRP is not known. It is encoded in the host genome found in high quantity in the brain of humans and animals infected with neurodegenerative diseases known as transmissible spongiform encephalopathies or prion diseases (Creutzfeld–Jakob disease, Gerstmann–Strausler–Scheinker syndrome, fatal familial insomnia)	CJD, PRIP, prion protein (p27-30)	Prion family

CD antigen	Cellular expression	Molecular weight (kDa)	Functions	Other names	Family relationships
CD231	T-cell acute lymphoblastic leukemia, neuroblastoma cells and normal brain neuron	150	The function of CD231 is currently unknown. It is cell-surface glycoprotein which is a specific marker for T-cell acute lymphoblastic leukemia. Also found on neuroblastomas	TALLA-1, TM4SF2, A15, MXS1, CCG-B7	Transmembrane 4 superfamily (TM4SF also known as tetraspanins)
CD232	Nonlineage molecules	200	Receptor for an immunologically active semaphorin (virus-encoded semaphorin protein receptor)	VESPR, PLXN, PLXN-C1	Plexin family
CD233	Erythroid cells	93	Band 3 is the major integral glycoprotein of the erythrocyte membrane. It has two functional domains. Its integral domain mediates a 1:1 exchange of inorganic anions across the membrane, whereas its cytoplasmic domain provides binding sites for cytoskeletal proteins, glycolytic enzymes, and hemoglobin. Multifunctional transport protein	SLC4A1, Diego blood group, D1, AE1, EPB3	Anion exchanger family
CD234	Erythroid cells and nonerythroid cells	35	Fy-glycoprotein; Duffy blood group antigen; nonspecific receptor for many chemokines such as IL-8, GRO, RANTES, MCP-1 and TARC. It is also the receptor for the human malaria parasites *Plasmodium vivax* and *Plasmodium knowlesi* and plays a role in inflammation and in malaria infection	GPD, CCBP1, DARC (duffy antigen/receptor for chemokines)	Family 1 of G protein-coupled receptors, chemokine receptors superfamily
CD235a	Erythroid cells	31	Major carbohydrate rich sialoglycoprotein of human erythrocyte membrane which bears the antigenic determinants for the MN and Ss blood groups. The N-terminal glycosylated segment, which lies outside the erythrocyte membrane, has MN blood group receptors and also binds influenza virus	Glycophorin A, GPA, MNS	Glycophorin A family
CD235b	Erythroid cells	GYPD is smaller than GYPC (24 kDa vs 32 kDa)	This protein is a minor sialoglycoprotein in human erythrocyte membranes. Along with GYPA, GYPB is responsible for the MNS blood group system. The Ss blood group antigens are located on glycophorin B	Glycophorin B, MNS, GPB	Glycophorin A family
CD236	Erythroid cells	24	Glycophorin C (GPC) and glycophorin D (GPD) are closely related sialoglycoproteins in the human red blood cell (RBC) membrane. GPD is a ubiquitous shortened isoform of GPC, produced by alternative splicing of the same gene. The Webb and Duch antigens, also known as glycophorin D, result from single point mutations of the glycophorin C gene	Glycophorin D, GPD, GYPD	Type III membrane proteins
CD236R	Erythroid cells	32	Glycophorin C (GPC) is associated with the Gerbich (Ge) blood group deficiency. It is a minor red cell-membrane component, representing about 4% of the membrane sialoglycoproteins, but shows very little homology with the major red cell-membrane glycophorins A and B. It plays an important role in regulating the mechanical stability of red cells and is a putative receptor for the merozoites of *Plasmodium falciparum*	Glycophorin C, GYPC, GPC	Type III membrane proteins
CD238	Erythroid cells	93	KELL blood group antigen; homology to a family of zinc metalloglycoproteins with neutral endopeptidase activity, type II transmembrane glycoprotein	KELL	Belongs to peptidase family m13 (zinc metalloproteinase); also known as the neprilysin subfamily

CD antigen	Cellular expression	Molecular weight (kDa)	Functions	Other names	Family relationships
CD239	Erythroid cells	78	A type I membrane protein.The human F8/G253 antigen, B-CAM, is a cell surface glycoprotein that is expressed with restricted distribution pattern in normal fetal and adult tissues, and is upregulated following malignant transformation in some cell types. Its overall structure is similar to that of the human tumor marker MUC 18 and the chicken neural adhesion molecule SC1	B-CAM (B-cell adhesion molecule), LU, Lutheran blood group	Immunoglobulin superfamily
CD240CE	Erythroid cells	45.5	Rhesus blood group, CcEe antigens. May be part of an oligomeric complex which is likely to have a transport or channel function in the erythrocyte membrane. It is highly hydrophobic and deeply buried within the phospholipid bilayer	RHCE, RH30A, RHPI, Rh4	Rh family
CD240D	Erythroid cells	45.5 (product—30)	Rhesus blood group, D antigen. May be part of an oligomeric complex which is likely to have a transport or channel function in the erythrocyte membrane. Absent in the Caucasian RHD-negative phenotype	RhD, Rh4, RhPI, RhII, Rh30D	Rh family
CD241	Erythroid cells	50	Rhesus blood group-associated glycoprotein RH50, component of the RH antigen multisubunit complex; required for transport and assembly of the Rh membrane complex to the red blood cell surface. Highly homologous to RH, 30 kDa components. Defects in RhAg are a cause of a form of chronic hemolytic anemia associated with stomatocytosis, and spherocytosis, reduced osmotic fragility, and increased cation permeability	RhAg, RH50A	Rh family
CD242	Erythroid cells	42	Intercellular adhesion molecule 4, Landsteiner-Wiener blood group. LW molecules may contribute to the vaso-occlusive events associated with episodes of acute pain in sickle cell disease	ICAM-4, LW	Immunoglobulin superfamily, intercellular adhesion molecules (ICAMs)
CD243	Stem/progenitor cells	170	Multidrug resistance protein 1 (P-glycoprotein). P-gp has been shown to utilize ATP to pump hydrophobic drugs out of cells, thus increasing their intracellular concentration and hence their toxicity. The MDR 1 gene is amplified in multidrug resistant cell lines	MDR-1, p-170	ABC superfamily of ATP-binding transport proteins
CD244	NK cells	66	2B4 is a cell-surface glycoprotein related to CD2 and implicated in the regulation of natural killer and T-lymphocyte function. It appears that the primary function of 2B4 is to modulate other receptor–ligand interactions to enhance leukocyte activation	2B4, NK cell activation inducing ligand (NAIL)	Immunoglobulin superfamily
CD245	T cells	220–240	Cyclin E/Cdk2 interacting protein p220. NPAT is involved in a key S phase event and links cyclical cyclin E/Cdk2 kinase activity to replication-dependent histone gene transcription. NPAT gene may be essential for cell maintenance and may be a member of the housekeeping genes	NPAT	

CD antigen	Cellular expression	Molecular weight (kDa)	Functions	Other names	Family relationships
CD246	Expressed in the small intestine, testis, and brain but not in normal lymphoid cells	177 kDa; after glycosylation, produces a 200 kDa mature glyco-protein	Anaplastic ($CD30^+$ large cell) lymphoma kinase; plays an important role in brain development, involved in anaplastic nodal non Hodgkin lymphoma or Hodgkin's disease with translocation t(2;5)(p23;q35) or inv2(23;q35). Oncogenesis via the kinase function is activated by oligomerization of NPM1-ALK mediated by the NPM1 part	ALK	Insulin receptor family of tyrosine-protein kinases
CD247	T cells, NK cells	16	T-cell receptor ζ; has a probable role in assembly and expression of the TCR complex as well as signal transduction upon antigen triggering. TCRζ together with TCRα:β and γ: δ heterodimers and CD3-γ, -δ, and -ε, forms the TCR-CD3 complex. The ζ chain plays an important role in coupling antigen recognition to several intracellular signal-transduction pathways. Low expression of the antigen results in impaired immune response	ζ chain, CD3Z	Immunoglobulin superfamily

Compiled by Laura Herbert, Royal Free Hospital, London. Data based on CD designations made at the 7th Workshop on Human Leucocyte Differentiation Antigens, provided by Protein Reviews on the Web (www.ncbi.nlm.nih.gov/prow/).

Appendix III. Cytokines and their receptors.						
Family	**Cytokine (alternative names)**	**Size (no. of amino acids and form)**	**Receptors (c denotes common subunit)**	**Producer cells**	**Actions**	**Effect of cytokine or receptor knock-out (where known)**
Colony-stimulating factors	G-CSF	174, monomer*	G-CSFR	Fibroblasts and monocytes	Stimulates neutrophil development and differentiation	G-CSF, G-CSFR: defective neutrophil production and mobilization
	GM-CSF (granulocyte-macrophage colony-stimulating factor)	127, monomer*	CD116, βc	Macrophages, T cells	Stimulates growth and differentiation of myelomonocytic lineage cells, particularly dendritic cells	GM-CSF, GM-CSFR: pulmonary alveolar proteinosis
	M-CSF (CSF-1)	α: 224 β: 492 γ: 406 active forms are homo- or heterodimeric	CSF-1R (c-fms)	T cells, bone marrow stromal cells, osteoblasts	Stimulates growth of cells of monocytic lineage	Osteopetrosis
Interferons	IFN-α (at least 12 distinct proteins)	166, monomer	CD118, IFNAR2	Leukocytes, dendritic cells, plasmacytoid dendritic cells, conventional dendritic cells	Antiviral, increased MHC class I expression	CD118: impaired antiviral activity
	IFN-β	166, monomer	CD118, IFNAR2	Fibroblasts	Antiviral, increased MHC class I expression	IFN-β: increased susceptibility to certain viruses
	IFN-γ	143, homodimer	CD119, IFNGR2	T cells, natural killer cells	Macrophage activation, increased expression of MHC molecules and antigen processing components, Ig class switching, supresses T_H2	IFN-γ, CD119: decreased resistance to bacterial infection and tumors
Interleukins	IL-1α	159, monomer	CD121a (IL-1RI) and CD121b (IL-1RII)	Macrophages, epithelial cells	Fever, T-cell activation, macrophage activation	IL-1RI: decreased IL-6 production
	IL-1β	153, monomer	CD121a (IL-1RI) and CD121b (IL-1RII)	Macrophages, epithelial cells	Fever, T-cell activation, macrophage activation	IL-1β: impaired acute-G21 phase response
	IL-1 RA	152, monomer	CD121a	Monocytes, macrophages, neutrophils, hepatocytes	Binds to but doesn't trigger IL-1 receptor, acts as a natural antagonist of IL-1 function	IL-1RA: reduced body mass, increased sensitivity to endotoxins (septic shock)
	IL-2 (T-cell growth factor)	133, monomer	CD25α, CD122β, CD132 (γc)	T cells	T-cell proliferation	IL-2: deregulated T-cell proliferation, colitis IL-2Rα: incomplete T-cell development autoimmunity IL-2Rβ: increased T-cell autoimmunity IL-2Rγc: severe combined immunodeficiency
	IL-3 (multicolony CSF)	133, monomer	CD123, βc	T cells, thymic epithelial cells and stromal cells	Synergistic action in early hematopoiesis	IL-3: impaired eosinophil development. Bone marrow unresponsive to IL-5, GM-CSF
	IL-4 (BCGF-1, BSF-1)	129, monomer	CD124, CD132 (γc)	T cells, mast cells	B-cell activation, IgE switch, induces differentiation into T_H2 cells	IL-4: decreased IgE synthesis
	IL-5 (BCGF-2)	115, homodimer	CD125, βc	T cells, mast cells	Eosinophil growth, differentiation	IL-5: decreased IgE, IgG1 synthesis (in mice); decreased levels of IL-9, IL-10, and eosinophils
	IL-6 (IFN-B502, BSF-2, BCDF)	184, monomer	CD126, CD130	T cells, macrophages, endothelial cells	T- and B-cell growth and differentiation, acute phase protein production, fever	IL-6: decreased acute phase reaction, reduced IgA production
	IL-7	152, monomer*	CD127, CD132 (γc)	Non-T cells, stromal cells	Growth of pre-B cells and pre-T cells	IL-7: early thymic and lymphocyte expansion severely impaired
	IL-9	125, monomer*	IL-9R, CD132 (γc)	T cells	Mast-cell enhancing activity, stimulates T_H2	Defects in mast-cell expansion
	IL-10 (cytokine synthesis inhibitory factor)	160, homodimer	IL-10Rα, IL–10Rβc (CRF2-4, IL-10R2)	Monocytes	Potent suppressant of macrophage functions	IL-10 and IL20Rβc-: reduced growth, anemia, chronic enterocolitis

Family	Cytokine (alternative names)	Size (no. of amino acids and form)	Receptors (c denotes common subunit)	Producer cells	Actions	Effect of cytokine or receptor knock-out (where known)
Interleukins	IL-11	178, monomer	IL-11R, CD130	Stromal fibroblasts	Synergistic action with IL-3 and IL-4 in hematopoiesis	IL-11R: defective decidualization
	IL-12 (NK-cell stimulatory factor)	197 (p35) and 306 (p40c), heterodimer	IL-12Rβ1c + IL-12Rβ2	Macrophages, dendritic cells	Activates NK cells, induces CD4 T-cell differentiation into T_H1-like cells	IL-12: impaired IFN-γ production and T_H1 responses
	IL-13 (p600)	132, monomer	IL-13R, CD132 (γc) (may also include CD24)	T cells	B-cell growth and differentiation, inhibits macrophage inflammatory cytokine production and T_H1 cells, induces allergy/asthma	IL-13: defective regulation of isotype specific responses
	IL-15 (T-cell growth factor)	114, monomer	IL-15Rα, CD122 (IL-2Rβ) CD132 (γc)	Many non-T cells	IL-2-like, stimulates growth of intestinal epithelium, T cells, and NK cells, enhances CD8 memory T cell survival	IL-15: reduced numbers of NK cells and memory phenotype $CD8^+$ T cells IL-15Rα: lymphopenia
	IL-16	130, homotetramer	CD4	T cells, mast cells, eosinophils	Chemoattractant for CD4 T cells, monocytes, and eosinophils, anti-apoptotic for IL-2-stimulated T cells	
	IL-17A (mCTLA-8)	150, homodimer	IL-17AR (CD217)	T_H17, CD8 T cells, NK cells γ:δ T cells, neutrophils	Induces cytokine production by epithelia, endothelia, and fibroblasts, proinflamatory	IL-17R: reduced neutrophil migration into infected sites
	IL-17F (ML-1)	134, homodimer	IL-17AR (CD217)	T_H17, CD8 T cells, NK cells γ:δ T cells, neutrophils	Induces cytokine production by epithelia, endothelia, and fibroblasts, proinflamatory	
	IL-18 (IGIF, interferon-α inducing factor)	157, monomer	IL-1Rrp (IL-1R related protein)	Activated macrophages and Kupffer cells	Induces IFN-γ production by T cells and NK cells, promotes T_H1 induction	Defective NK activity and T_H1 responses
	IL-19	153, monomer	IL-20Rα + IL–10Rβc	Monocytes	Induces IL-6 and TNF-α expression by monocytes	
	IL-20	152	IL-20Rα + IL–10Rβc; IL-22Rαc + IL-10Rβc	T_H1 cells, monocytes, epithelial cells	Promotes T_H2 cells, stimulates keratinocyte proliferation and TNF-α production	
	IL-21	133	IL-21R, + CD132(γc)	T_H2 cells	Induces proliferation of B, T and NK cells	Increased IgE production
	IL-22 (IL-TIF)	146	IL-22Rαc + IL-10Rβc	NK cells, T_H17 cells, T_H22 cells	Induces liver acute-phase proteins, pro-inflammatory agents; epithelial barrier	
	IL-23	170 (p19) and 306 (p40c), heterodimer	IL-12Rβ1 + IL-23R	Dendritic cells, macrophages	Induces proliferation of T_H17 memory T cells, increased IFN-γ production	Defective inflammation
	IL-24 (MDA-7)	157	IL-22Rαc + IL-10Rβc; IL-20Rα + IL-10Rβc	Monocytes, T cells	Inhibits tumor growth, wound healing	
	IL-25 (IL-17E)	145	IL-17BR (IL-17Rh1)	T_H2 cells, mast cells, epithelial cells	Promotes T_H2 cytokine production	Defective T_H2 response
	IL-26(AK155)	150	IL-20Rα +IL-10Rβc	T cells (T_H17), NK cells	Pro-inflammatory, stimulates epithelium	
	IL-27	142 (p28) and 229 (EBI3), heterodimer	WSX-1 + CD130c	Monocytes, macrophages, dendritic cells	Induces IL-12R on T cells via T-bet induction, induces IL-10	EBI3: reduced NKT cells. WSX-1: overreaction to *Toxoplasma gondii* infection and death from inflammation
	IL-28A,B (IFN-B502,3)	175	IL-28Rαc + IL-10Rβc	Dendritic cells	Antiviral	
	IL-29 (IFN-λ1)	181	IL-28Rαc + IL-10Rβc	Dendritic cells	Antiviral	
	IL-30 (p28, IL27A, IL-27p28)	243	see IL-27			
	IL-31	164	IL31A + OSMR	T_H2	Proinflammatory, skin lesions	IL-31A: elevated OSM responsiveness
	IL-32 (NK4, TAIF)	188	Unknown	Natural killer cells, T cells, epithelial cells, monocytes	Induces TNF-α	

Family	Cytokine (alternative names)	Size (no. of amino acids and form)	Receptors (c denotes common subunit)	Producer cells	Actions	Effect of cytokine or receptor knock-out (where known)
Interleukins	IL-33 (NF-HEV)	270 heterodimer	ST2 (IL1RL1) + IL1RAP	High endothelial venules, smooth muscle	Induces T_H2 cytokines (IL-4, IL-5, IL-13)	IL-33: reduced dextran-induced colitis; reduced LPS-induced systemic inflammatory response
	IL-34 (C16orf77)	242 homodimer	CSF-1R	Many cell types	Promotes growth and developmnet of myeloid cells/osteoclasts	
	IL-35	197 (IL-12α (p35)) + 229 (EB13) heterodimer	Unknown	T_{reg}	Immunosuppressive	
	IL-36α, β, λ	(20 kDa) 155–169	IL-1Rrp2, Acp	Keratinocytes, monocytes	Pro-inflammatory stimulant of macrophages and dendritic cells	
	IL-36 Ra		IL-1Rp2, Acp		Antagonist of IL-36	
	IL-37	(17–24 kDa) homodimer	IL-18Rα?	Monocytes, dendritic cells, epithelial cells, breast tumor cells	Suppresses dendritic cell/ monocyte production of IL-1, -6, -12 etc. cytokines, synergizes with TGFs	siRNA knockdown: increases pro-inflammatory cytokines
	TSLP	140 monomer	IL-7Rα, TSLPR	Epithelial cells, especially lung and skin	Stimulates hematopoietic cells and dendritic cells to induce T_H2 responses	TSLP: resistance to induction of allergies and asthmatic reactions
	LIF (leukemia inhibitory factor)	179, monomer	LIFR, CD130	Bone marrow stroma, fibroblasts	Maintains embryonic stem cells, like IL-6, IL-11, OSM	LIFR: die at or soon after birth; decreased hematopoietic stem cells
	OSM (OM, oncostatin M)	196, monomer	OSMR or LIFR, CD130	T cells, macrophages	Stimulates Kaposi's sarcoma cells, inhibits melanoma growth	OSMR: defective liver regeneration
TNF family	TNF-α (cachectin)	157, trimers	p55 (CD120a), p75 (CD120b)	Macrophages, NK cells, T cells	Promotes inflammation, endothelial activation	p55: resistance to septic shock, susceptibility to *Listeria*, STNFαR: periodic febrile attacks
	LT-α (lymphotoxin-α)	171, trimers	p55 (CD120a), p75 (CD120b)	T cells, B cells	Killing, endothelial activation	LT-α: absent lymph nodes, decreased antibody, increased IgM
	LT-β	Transmembrane, trimerizes with LT-α	LTβR or HVEM	T cells, B cells	Lymph node development	Defective development of peripheral lymph nodes, Peyer's patches, and spleen.
	CD40 ligand (CD40L)	Trimers	CD40	T cells, mast cells	B-cell activation, class switching	CD40L: poor antibody response, no class switching, diminished T-cell priming (hyper IgM syndrome).
	Fas ligand (FasL)	Trimers	CD95 (Fas)	T cells, stroma(?)	Apoptosis, Ca^{2+}-independent cytotoxicity	Fas, FasL: mutant forms lead to lymphoproliferation, and autoimmunity
	CD27 ligand (CD27L)	Trimers (?)	CD27	T cells	Stimulates T-cell proliferation	
	CD30 ligand (CD30L)	Trimers (?)	CD30	T cells	Stimulates T- and B-cell proliferation	CD30: increased thymic size, alloreactivity
	4-1BBL	Trimers (?)	4-1BB	T cells	Co-stimulates T and B cells	
	Trail (AP0-2L)	281, trimers	DR4, DR5 DCR1, DCR2 and OPG	T cells, monocytes	Apoptosis of activated T cells and tumor cells	Tumor-prone phenotype
	OPG-L (RANK-L)	316, trimers	RANK/OPG	Osteoblasts, T cells	Stimulates osteoclasts and bone resorption	OPG-L: osteopetrotic, runted, toothless OPG: osteoporosis
	APRIL	86	TAC1 or BCMA	Activated T cells	B-cell proliferation	Impaired IgA-class switching
	LIGHT	240	HVEM, LTβR	T cells,	Dendritic cell activation	Defective CD8$^+$ T-cell expansion
	TWEAK	102	TWEAKR (Fn14)	Macrophages, EBV transformed cells	Angiogenesis	
	BAFF (CD257, BlyS)	153	TAC1 or BCMA or BR3	B cells	B-cell proliferation	BAFF: B-cell dysfunction
Unassigned	TGF-β1	112, homo- and heterotrimers	TGF-βR	Chondrocytes, monocytes, T cells	Inhibits cell growth, anti-inflammatory, induces switch to IgA production	TGF-β: lethal inflammation
	MIF	115, monomer	MIF-R	T cells, pituitary cells	Inhibits macrophage migration, stimulates macrophage activation, induces steroid resistance	MIF: resistance to septic shock, hyporesponsive to Gram-negative bacteria

* May function as dimers

Compiled by Robert Schreiber, Washington University School of Medicine, St. Louis.

Appendix IV. Chemokines and their receptors.				
Chemokine systematic name	**Common names**	**Chromosome**	**Target cell**	**Specific receptor**
CXCL (†ELR+)				
1	GROα	4	Neutrophil, fibroblast, melanoma cell	CXCR2
2	GROβ	4	Neutrophil, fibroblast, melanoma cell	CXCR2
3	GROγ	4	Neutrophil, fibroblast, melanoma cell	CXCR2
5	ENA-78	4	Neutrophil, endothelial cell	CXCR2>>1
6	GCP-2	4	Neutrophil, endothelial cell	CXCR2>1
7	NAP-2 (PBP/CTAP-III/β-B44TG)	4	Fibroblast, neutrophil, endothelial cell	CXCR2
8	IL-8	4	Neutrophil, basophil, CD8 cell subset, endothelial cell	CXCR1, E482
14	BRAK/bolekine	5	T cell, monocyte, B cell	Unknown
15	Lungkine/WECHE	5	Neutrophil, epithelial cell, endothelial cell	Unknown
(†ELR–)				
4	PF4	4	Fibroblast, endothelial cell	CXCR3B (alternative splice)
9	Mig	4	Activated T cell ($T_H1 > T_H2$), natural killer (NK) cell, B cell, endothelial cell, plasmacytoid dendritic cell	CXCR3A and B
10	IP-10	4	Activated T cell ($T_H1 > T_H2$), NK cell, B cell, endothelial cell	CXCR3A and B
11	I-TAC	4	Activated T cell ($T_H1 > T_H2$), NK cell, B cell, endothelial cell	CXCR3A and B, CXCR7
12	SDF-1α/β	10	$CD34^+$ bone marrow cell, thymocytes, monocytes/macrophages, naive activated T cell, B cell, plasma cell, neutrophil, immature dendritic cells, mature dendritic cells, plasmacytoid dendritic cells	CXCR4, CXCR7
13	BLC/BCA-1	4	Naive B cells, activated CD4 T cells, immature dendritic cells, mature dendritic cells	CXCR5>>CXCR3
16	sexckine	17	Activated T cell, natural killer T (NKT) cell, endothelial cells	CXCR6
CCL				
1	I-309	17	Neutrophil (TCA-3 only), T cell ($T_H2 > T_H1$) monocyte	CCR8
2	MCP-1	17	T cell ($T_H2 > T_H1$) monocyte, basophil, immature dendritic cells, NK cells	CCR2
3	MIP-1α/LD78	17	Monocyte/macrophage, T cell ($T_H1 > T_H2$), NK cell, basophil, immature dendritic cell, eosinophil, neutrophil, astrocyte, fibroblast, osteoclast	CCR1, 5
4	MIP-1β	17	Monocyte/macrophage, T cell ($T_H1 > T_H2$), NK cell, basophil, immature dendritic cell, eosinophil, B cell	CCR5>>1
5	RANTES	17	Monocyte/macrophage, T cell (memory T cell > T cell; $T_H1 > T_H2$), NK cell, basophil, eosinophil, immature dendritic cell	CCR1, 3, 5
6	C10/MRP-1	11 (mouse only)	Monocyte, B cell, CD4 T cell, NK cell	CCR 1
7	MCP-3	17	$T_H2 > T_H1$ T cell, monocyte, eosinophil, basophil, immature dendritic cell, NK cell	CCR1, 2, 3, 5, 10
8	MCP-2	17	$T_H2 > T_H1$ T cell, monocyte, eosinophil, basophil, immature dendritic cell, NK cell	CCR2, 3, 5>1
8	MCP-2	17	T cell, monocyte, eosinophil, basophil, immature dendritic cell, NK cell	CCR2, 3, 5>1
9	MRP-2/MIP-1γ	11 (mouse only)	T cell, monocyte, adipocyte	CCR1
11	Eotaxin	17	Eosinophil, basophil, mast cell, T_H2 cell	CCR3>>CCR5
12	MCP-5	11 (mouse only)	Eosinophil, monocyte, T cell, B cell	CCR2
13	MCP-4	17	$T_H2 > T_H1$ T cell, monocyte, eosinophil, basophil, dendritic cell	CCR1, 2, 3>5
14a	HCC-1	17	Monocyte	CCR1, 5
14b	HCC-3	17	Monocyte	Unknown

Chemokine systematic name	Common names	Chromosome	Target cell	Specific receptor
15	MIP-5/HCC-2	17	T cell, monocyte, eosinophil, dendritic cell	CCR1, 3
16	HCC-4/LEC	17	Monocyte, T cell, natural killer cell, immature dendritic cell	CCR1, 2, 5
17	TARC	16	T cell ($T_H2 > T_H1$), immature dendritic cells, thymocyte, regulatory T cell	CCR4>>8
18	DC-CK1/PARC	17	Naive T cell > activated T cell, immature dendritic cells, mantle zone B cells	Unknown
19	MIP-3β/ELC	9	Naive T cell, mature dendritic cell, B cell	CCR7
20	MIP-3α /LARC	2	T cell (memory T cell, T_H17 cells, blood mononuclear cell, immature dendritic cell, activated B cells, NKT cells, GALT development	CCR6
21	6Ckine/SLC	9	Naive T cell, B cell, thymocytes, NK cell, mature dendritic cells	CCR7
22	MDC	16	Immature dendritic cell, NK cell, T cell ($T_H2 > T_H1$), thymocyte, endothelial cells, monocyte, regulatory T cell	CCR4
23	MPIF-1/CK-β\8	17	Monocyte, T cell, resting neutrophil	CCR1, 5
24	Eotaxin-2/MPIF-2	7	Eosinophil, basophil, T cell	CCR3
25	TECK	19	Macrophage, thymocytes, dendritic cells, intraepithelial lymphocytes, IgA plasma cells, mucosal memory T cells	CCR9
26	Eotaxin-3	7	Eosinophil, basophil, fibroblast	CCR3
27	CTACK	9	Skin homing memory T cell, B cell	CCR10
28	MEC	5	T cell, eosinophil, IgA^+ B cell	CCR10>3
C and CX3C				
XCL 1	Lymphotactin	1 (1)	T cell, natural killer cell, CD8α + dendritic cell	XCR1
XCL 2	SCM-1β	1	T cell, natural killer cell, CD8α + dendritic cell	XCR1
CX3CL 1	Fractalkine	16	Activated T cell, monocyte, neutrophil, natural killer cell, immature dendritic cells, mast cells, astrocytes, neurons, microglia	CX3CR1

Atypical chemokine receptors

Chemokine ligands	Target cell	Specific receptor
Chemerin and resolvin E1	Macrophages, immature dendritic cells, mast cells, plasmacytoid dendritic cells, adipocytes, fibroblasts, endothelial cells, oral epithelial cells	CMKLR1/chem23
CCL5, CCL19 and chemerin	All hematopoietic cells, microglia, astrocytes, lung epithelial cells	CCRL2/CRAM
Inflammatory CC chemokines	Lymphatic endothelial cells	D6
Various CXC and CC chemokines	Red blood cells, Purkinje cells, blood endothelial cells, kidney epithelial cells	Duffy/DARC
CCL19, CCL21, CCL25	Thymic epithelial cells, lymph node stromal cells, keratinocytes	CCXCKR

Chromosome locations are for humans. Chemokines for which there is no human homologue are listed with the mouse chromosome.

† ELR refers to the three amino acids that precede the first cysteine residue of the CXC motif. If these amino acids are Glu-Leu-Arg (i.e. ELR+), then the chemokine is chemotactic for neutrophils; if they are not (ELR–) then the chemokine is chemotactic for lymphocytes

Compiled by Joost Oppenheim, National Cancer Institute, NIH.

BIOGRAPHIES

Emil von Behring (1854–1917) discovered antitoxin antibodies with Shibasaburo Kitasato.

Baruj Benacerraf (1920–) discovered immune response genes and collaborated in the first demonstration of MHC restriction.

Jules Bordet (1870–1961) discovered complement as a heat-labile component in normal serum that would enhance the antimicrobial potency of specific antibodies.

Frank MacFarlane Burnet (1899–1985) proposed the first generally accepted clonal selection hypothesis of adaptive immunity.

Jean Dausset (1916–2009) was an early pioneer in the study of the human major histocompatibility complex or HLA.

Peter Doherty (1940–) and **Rolf Zinkernagel** (1944–) showed that antigen recognition by T cells is MHC-restricted, thereby establishing the biological role of the proteins encoded by the major histocompatibility complex and leading to an understanding of antigen processing and its importance in the recognition of antigen by T cells.

Gerald Edelman (1929–) made crucial discoveries about the structure of immunoglobulins, including the first complete sequence of an antibody molecule.

Paul Ehrlich (1854–1915) was an early champion of humoral theories of immunity, and proposed a famous side-chain theory of antibody formation that bears a striking resemblance to current thinking about surface receptors.

James Gowans (1924–) discovered that adaptive immunity is mediated by lymphocytes, focusing the attention of immunologists on these small cells.

Michael Heidelberger (1888–1991) developed the quantitative precipitin assay, ushering in the era of quantitative immunochemistry.

Charles A. Janeway, Jr. (1945–2003) recognized the importance of co-stimulation for initiating adaptive immune responses. He predicted the existence of receptors of the innate immune system that would recognize pathogen-associated molecular patterns and would signal activation of the adaptive immune system. His laboratory discovered the first mammalian Toll-like receptor that had this function. He was also the principal original author of this textbook.

Edward Jenner (1749–1823) described the successful protection of humans against smallpox infection by vaccination with cowpox or vaccinia virus. This founded the field of immunology.

Niels Jerne (1911–1994) developed the hemolytic plaque assay and several important immunological theories, including an early version of clonal selection, a prediction that lymphocyte receptors would be inherently biased to MHC recognition, and the idiotype network.

Shibasaburo Kitasato (1852–1931) discovered antibodies in collaboration with Emil von Behring.

Robert Koch (1843–1910) defined the criteria needed to characterize an infectious disease, known as Koch's postulates.

Georges Köhler (1946–1995) pioneered monoclonal antibody production from hybrid antibody-forming cells with César Milstein.

Karl Landsteiner (1868–1943) discovered the ABO blood group antigens. He also carried out detailed studies of the specificity of antibody binding using haptens as model antigens.

Peter Medawar (1915–1987) used skin grafts to show that tolerance is an acquired characteristic of lymphoid cells, a key feature of clonal selection theory.

Elie Metchnikoff (1845–1916) was the first champion of cellular immunology, focusing his studies on the central role of phagocytes in host defense.

César Milstein (1927–2002) pioneered monoclonal antibody production with Georges Köhler.

Louis Pasteur (1822–1895) was a French microbiologist and immunologist who validated the concept of immunization first studied by Jenner. He prepared vaccines against chicken cholera and rabies.

Rodney Porter (1917–1985) worked out the polypeptide structure of the antibody molecule, laying the groundwork for its analysis by protein sequencing.

Ignác Semmelweis (1818–1865) German-Hungarian physician who first determined a connection between hospital hygiene and an infectious disease, puerperal fever, and consquently introduced antisepsis into medical practice.

George Snell (1903–1996) worked out the genetics of the murine major histocompatibility complex and generated the congenic strains needed for its biological analysis, laying the groundwork for our current understanding of the role of the MHC in T-cell biology.

Tomio Tada (1934–2010) first formulated the concept of the regulation of the immune response by 'suppressor T cells' in the 1970s, from indirect experimental evidence. The existence of such cells could not be verified at the time and the concept became discredited, but Tada was vindicated when researchers in the 1980s identified the cells now called 'regulatory T cells.'

Susumu Tonegawa (1939–) discovered the somatic recombination of immunological receptor genes that underlies the generation of diversity in human and murine antibodies and T-cell receptors.

Jürg Tschopp (1951–2011) contributed to the delineation of the complement system and T-cell cytolytic mechanisms, and made seminal contributions to the fields of apoptosis and innate immunity, in particular by discovering the inflammasome.

Don C. Wiley (1944–2001) solved the first crystal structure of an MHC I protein, providing a startling insight into how T cells recognize their antigen in the in the context of MHC molecules.

Photograph Acknowledgments

Chapter 1
Fig. 1.1 reproduced courtesy of Yale University, Harvey Cushing/John Hay Whitney Medical Library. Fig. 1.20 photographs from Mowat, A., Viney, J.: **The anatomical basis of intestinal immunity.** *Immunol. Rev*. 1997, **156**:145–166. Fig. 1.27 photographs from Kaplan, G., et al.: **Efficacy of a cell-mediated reaction to the purified protein derivative of tuberculin in the disposal of *Mycobacterium leprae* from human skin.** *PNAS* 1988, **85**:5210–5214.

Chapter 2
Fig. 2.31 photographs reproduced with permission from Bhakdi, S., et al.: **Functions and relevance of the terminal complement sequence.** *Blut* 1990, **60**:309–318. © 1990 Springer-Verlag.

Chapter 3
Fig. 3.11 structure reprinted with permission from Jin, M.S., et al.: **Crystal structure of the TLR1-TLR2 heterodimer induced by binding of a tri-acylated lipopeptide**. *Cell* 2007, **130**:1071–1082. © 2007 with permission from Elsevier. Fig. 3.12 structure reprinted with permission from Macmillan Publishers Ltd. Park, B.S., et al.: **The structural basis of lipopolysaccharide recognition by the TLR4-MD-2 complex**. *Nature* 2009, **458**:1191–1195. Fig. 3.28 model structure reprinted with permission from Macmillan Publishers Ltd. Emsley, J., et al.: **Structure of pentameric human serum amyloid P component**. *Nature* 1994, **367**:338–345.

Chapter 4
Fig. 4.4 photograph from Green, N.M.: **Electron microscopy of the immunoglobulins**. *Adv. Immunol*. 1969, **11**:1–30. © 1969 with permission from Elsevier. Fig. 4.13 and Fig. 4.22 model structures from Garcia, K.C., et al.: **An $\alpha\beta$ T cell receptor structure at 2.5 Å and its orientation in the TCR-MHC complex**. *Science* 1996, **274**:209–219. Reprinted with permission from AAAS. Fig. 4.23 from Reinherz, E.L., et al.: **The crystal structure of a T cell receptor in complex with peptide and MHC Class II**. *Science* 1999, **286**:1913–1921. Reprinted with permission from AAAS. Fig. 4.26 reprinted with permission from Macmillan Publishers Ltd. Gao, G.F., et al.: **Crystal structure of the complex between human CD8$\alpha\alpha$ and HLA-A2**. *Nature* 1997, **387**:630–634.

Chapter 6
Fig. 6.3 bottom panel from Velarde, G., et al.: **Three-dimensional structure of transporter associated with antigen processing (TAP) obtained by single particle image analysis**. *J. Biol. Chem*. 2001, **276**:46054– 46063. © 2001 ASBMB. Fig. 6.4 reprinted with permission from Macmillan Publishers Ltd. Whitby, F.G., et al.: **Structural basis for the activation of 20S proteasomes by 11S regulators**. *Nature* 2000, **408**:115–120. Fig. 6.20 structures from Mitaksov, V.E., Fremont, D.: **Structural definition of the H -2Kd peptide-binding motif**. *J. Biol. Chem*. 2006, **281**:10618–10625. © 2006 American Society of Biochemistry and Molecular Biology. Fig. 6.23 molecular model reprinted with permission from Macmillan Publishers Ltd. Fields, B.A., et al.: **Crystal structure of a T-cell receptor β-chain complexed with a superantigen**. *Nature* 1996, **384**:188–192.

Chapter 8
Fig. 8.18 photographs reprinted with permission from Macmillan Publishers Ltd. Surh, C.D., Sprent, J.: **T-cell apoptosis detected in situ during positive and negative selection in the thymus**. *Nature* 1994, **372**:100–103. Fig. 8.33 photographs from Wack, A., Kioussis, D.: **Direct visualization of thymocyte apoptosis in neglect, acute and steady-state negative selection**. *Int. Immunol*. 1996, **8**:1537–1548. © 1996 Oxford University Press.

Chapter 9
Fig. 9.9 fluorescent micrographs reprinted with permission from Macmillan Publishers Ltd. Pierre, P., Turley, S.J., et al.: **Development regulation of MHC class II transport in mouse dendritic cells**. *Nature* 1997, **388**:787–792. Fig. 9.32 panel c from Henkart, P.A., Martz, E. (eds): *Second International Workshop on Cell Mediated Cytotoxicity*. © 1985 Kluwer/Plenum Publishers. With kind permission of Springer Science and Business Media.

Chapter 10
Fig. 10.17 left panel from Szakal, A.K., et al.: **Isolated follicular dendritic cells: cytochemical antigen localization, Nomarski, SEM, and TEM morphology**. *J. Immunol*. 1985, **134**:1349–1359. © 1985 The American Association of Immunologists. Fig. 10.17 center and right panels from Szakal, A.K., et al.: **Microanantomy of lymphoid tissue during humoral immune responses: structure function relationships**. *Ann. Rev. Immunol*. 1989, **7**:91–109. © 1989 Annual Reviews www.annualreviews.org.

Chapter 12
Fig. 12.3 adapted by permission from Macmillan Publishers Ltd. Dethlefsen, L., McFall- Ngai, M., Relman, D.A.: **An ecological and evolutionary perspective on human- microbe mutualism and disease**. *Nature* 2007, **449**:811–818. © 2007. Fig. 12.7 photographs from Mowat, A., Viney, J.: **The anatomical basis of intestinal immunity.** *Immunol. Rev*. 1997, **156**:145–166. Fig. 12.9 photograph from Brandtzaeg, P., et al.: **Regional specialization in the mucosal immune system: what happens in the microcompartments?** *Immunol. Today* 1999, **20**:131–151. © 1999 with permission from Elsevier. Fig. 12.13 micrograph from Niess, J.H., et al.: **CX3CR1-mediated dendritic cell access to the intestinal lumen and bacterial clearance**. *Science* 2005, **307**:254–258. Reprinted with permission from AAAS.

Chapter 13
Fig. 13.6 top left photograph from Kaplan, G., Cohn, Z.A.: **The immunobiology of leprosy**. *Int. Rev. Exp. Pathol*. 1986, **28**:45–78. © 1986 with permission from Elsevier. Fig. 13.29 based on data from Palella, F.J., et al.: **Declining morbidity and mortality among patients with advanced human immunodeficiency virus infection. HIV Outpatient Study Investigators**. *N. Engl. J. Med*. 1998, **338**:853–860. Fig. 13.32 adapted by permission from Macmillan Publishers Ltd. Wei, X., et al.: **Viral dynamics in human immunodeficiency virus type 1 infection**. *Nature* 1995, **373**:117–122.

Chapter 14
Fig. 14.6 top photograph from Sprecher, E., et al.: **Deleterious mutations in SPINK5 in a patient with congenital ichthyosiform erythroderma: molecular testing as a helpful diagnostic tool for Netherton syndrome**. *Clin. Exp. Dermatol*. 2004, **29**:513–517. Fig. 14.7 adapted by permission from Macmillan Publishers Ltd. Cookson, W.: **The immunogenetics of asthma and eczema: a new focus on the epithelium**. *Nat. Rev. Immunol*. 2004, **4**:978–988. Fig. 14.17 photographs from Finotto, S., et al: **Development of spontaneous airway changes consistent with human asthma in mice lacking T-bet**. *Science* 2002, **295**:336–338. Reprinted with permission from AAAS. Fig. 14.27 left photograph from Mowat, A.M.,Viney, J.L.: **The anatomical basis of intestinal immunity**. *Immunol. Rev.* 1997, **156**:145–166.

Chapter 16
Fig. 16.15 photographs are reprinted from Herberman, R., Callewaert, D. (eds.): *Mechanisms of Cytotoxicity by Natural Killer Cells*, © 1985 with permission from Elsevier.

GLOSSARY

12/23 rule
The rule that gene segments of immunoglobulin or T-cell receptors can be joined only if one has a recognition signal sequence with a 12-base-pair spacer, and the other has a 23-base-pair spacer.

α
Name given to a wide variety of subunits in different proteins. In immunology it most usually refers to (1) the type of heavy chain in IgA class immunoglobulins, or (2) one of the two types of chain in the α:β T-cell receptor, or (3) the heavy chain in MHC molecules.

α:β heterodimer
The dimer of one α and one β chain that makes up the antigen-recognition portion of an α:β T-cell receptor.

α:β T cell
See **T cell**.

α:β T-cell receptor
See **T-cell receptor**.

α-defensins
A class of antimicrobial peptides. They are produced by neutrophils and epithelial cells, in particular the Paneth cells of the gut.

ABO blood group system
A set of antigens expressed on red blood cells that are used for typing human blood for transfusion. Matching is necessary because individuals who do not express A or B antigens on their red blood cells naturally form anti-A and anti-B antibodies that interact with and destroy red blood cells bearing A or B antigens if they are transfused into the bloodstream.

absorption
The removal of antibodies specific for one antigen from an antiserum to render it specific for another antigen or antigens.

accelerated rejection
The more rapid rejection of a second graft after rejection of the first graft. It was one of the pieces of evidence that showed that graft rejection was due to an adaptive immune response.

accessory effector cells
Cells that aid in an adaptive immune response but are not involved in specific antigen recognition. They include phagocytes, mast cells, and NK cells.

acquired immune deficiency syndrome (AIDS)
A disease caused by infection with the human immunodeficiency virus (HIV-1). AIDS occurs when an infected patient has lost most of his or her CD4 T cells, so that infections with opportunistic pathogens occur.

acquired immune response
See **adaptive immune response**.

activating receptor
On NK cells, a receptor whose stimulation results in activation of the cell's cytotoxic activity.

activation-induced cell death
The normal process by which all immune responses end in the death of most of the responding cells, leaving only a small number of resting memory cells.

activation-induced cytidine deaminase (AID)
Enzyme that contributes to somatic hypermutation of immunoglobulin gene variable regions by deaminating DNA directly at cytosine. Depending on how this DNA lesion is repaired, it can lead to a permanent base change at the deaminated site. The enzyme is also involved in isotype switching and gene conversion.

activation-induced cytidine deaminase deficiency
An inherited deficiency of the enzyme activation-induced cytidine deaminase (AID) that blocks both somatic hypermutation and isotype switching, leading to a type of hyper IgM immunodeficiency syndrome.

active immunization
Immunization with antigen to provoke adaptive immunity.

acute desensitization
A high-risk immunotherapeutic technique for rapidly inducing temporary tolerance to, for example, an essential drug such as insulin or penicillin in a person who is allergic to it. Also called rapid desensitization.

acute phase
The stage of HIV infection that occurs soon after a person becomes infected. It is characterized by an influenza-like illness, abundant virus in the blood, and a decrease in the number of circulating CD4 T cells.

acute-phase proteins
Proteins with innate immune function whose production is increased in the presence of an infection (the acute-phase response). They circulate in the blood and participate in early phases of host defense against infection. An example is mannose-binding lectin.

acute-phase response
A change in the proteins present in the blood that occurs during the early phases of an infection. It includes the production of acute-phase proteins, many of which are produced in the liver.

acute rejection
The rejection of a tissue or organ graft from a genetically unrelated donor that occurs within 10–13 days of transplantation unless prevented by immunosuppressant treatment.

adaptive immune response
The response of antigen-specific lymphocytes to antigen, including the development of immunological memory. Adaptive immune responses are distinct from the innate and nonadaptive phases of immunity, which are not mediated by clonal selection of antigen-specific lymphocytes.

adaptive immunity
Immunity to infection conferred by an adaptive immune response.

adaptors
Nonenzymatic proteins that form physical links between members of a signaling pathway, particularly between a receptor and other signaling proteins. They recruit members of the signaling pathway into functional protein complexes.

ADCC
See **antibody-dependent cell-mediated cytotoxicity**.

adenoids
Paired mucosal-associated lymphoid tissues located in the nasal cavity.

adenosine deaminase deficiency (ADA deficiency)
An inherited defect characterized by nonproduction of the enzyme adenosine deaminase, which leads to the accumulation of toxic purine nucleosides and nucleotides in cells, resulting in

the death of most developing lymphocytes within the thymus. It is a cause of severe combined immunodeficiency.

adhesins
Cell-surface proteins on bacteria that enable them to bind to the surfaces of host cells.

adhesion molecules
See **cell-adhesion molecules**.

adjuvant
Any substance that enhances the immune response to an antigen with which it is mixed.

adoptive immunity, adoptive immunization, adoptive transfer
Immunity conferred on a naive or irradiated recipient by transfer of lymphoid cells from an actively immunized donor.

afferent lymphatic vessels
Vessels of the lymphatic system that drain extracellular fluid from the tissues and carry antigen, macrophages, and dendritic cells from sites of infection to lymph nodes or other peripheral lymphoid organs.

affinity
The strength of binding of one molecule to another at a single site, such as the binding of a monovalent Fab fragment of antibody to a monovalent antigen. Cf. **avidity**.

affinity chromatography
The purification of a substance by means of its specific binding to another substance immobilized on a solid support. For example, an antigen can be purified by running it through a column of inert matrix to which its specific antibody molecules are covalently linked.

affinity hypothesis
Hypothesis that proposes how the choice between negative selection and positive selection of T cells in the thymus is made, according to the strength of self-peptide:MHC binding by the T-cell receptor. Low-affinity interactions rescue the cell from death by neglect, leading to positive selection; high-affinity interactions induce apoptosis and thus negative selection.

affinity maturation
The increase in affinity for their specific antigen of the antibodies produced as an adaptive immune response progresses. This phenomenon is particularly prominent in secondary and subsequent immunizations.

agammaglobulinemia
An absence of antibodies in the blood. See **X-linked agammaglobulinemia** (**XLA**).

age-related macular degeneration
A leading cause of blindness in the elderly, for which some single-nucleotide polymorphisms (SNPs) in the factor H genes confer an increased risk.

agglutination
The clumping together of particles, usually by antibody molecules binding to antigens on the surfaces of adjacent particles.

agonist selection
A process by which T cells are positively selected in the thymus by their interaction with relatively high-affinity ligands.

AID
See **activation-induced cytidine deaminase**.

AIDS
See **acquired immune deficiency syndrome**.

AIRE
Gene encoding a protein (autoimmune regulator) that is involved in the expression of numerous genes in the thymus, enabling developing T cells to be exposed to self proteins characteristic of other tissues, enabling tolerance to these proteins to develop. Deficiency of *AIRE* leads to an autoimmune disease, APECED.

airway tissue remodeling
A thickening of the airway walls due to hyperplasia and hypertrophy of the smooth muscle layer and mucus glands, with the eventual development of fibrosis, that occurs in chronic asthma.

allele
A variant form of a gene; many genes occur in several (or more) different forms within the general population. See also **polymorphism**.

allelic exclusion
In a heterozygous individual, the expression of only one of the two alternative alleles of a particular gene. In immunology, the term describes the restriction of expression of the individual chains of the antigen receptor genes, so that each individual lymphocyte produces immunoglobulin or T-cell receptors of a single antigen specificity.

allergen
Any antigen that elicits an allergic reaction.

allergen-specific immunotherapy
See **desensitization immunotherapy**.

allergic asthma
An allergic reaction to inhaled antigen, which causes constriction of the bronchi and difficulty in breathing.

allergic conjunctivitis
An allergic reaction involving the conjunctiva of the eye that occurs in sensitized individuals exposed to airborne allergens. It is usually manifested as allergic rhinoconjunctivitis or hay fever.

allergic contact dermatitis
A type IV immunological hypersensitivity reaction, seen against, for example, chemicals in the plant poison ivy, which involves T cell-mediated immune reactions, and which produces a skin rash at the points of contact with the antigen.

allergic reaction
A specific response to an innocuous environmental antigen, or allergen, that is caused by preexisting antibody or primed T cells. Allergic reactions can be caused by various mechanisms, but the most common is the binding of allergen to IgE bound to mast cells, which causes the release of histamine and other biologically active molecules from the cell that cause the symptoms of asthma, hay fever, and other common allergic reactions.

allergic rhinitis
An allergic reaction in the nasal mucosa that causes excess mucus production and sneezing.

allergy
The state in which a symptomatic immune reaction is made to a normally innocuous environmental antigen. It involves the interaction between the antigen and antibody or primed T cells produced by earlier exposure to the same antigen.

alloantibodies
Antibodies produced against antigens from a genetically nonidentical member of the same species.

alloantigens
Antigens from another genetically nonidentical member of the same species.

allogeneic
Describes two individuals or two mouse strains that differ at genes in the MHC. The term can also be used for allelic differences at other loci.

allograft
A transplant of tissue from an allogeneic (genetically nonidentical) donor of the same species. Such grafts are invariably rejected unless the recipient is immunosuppressed.

allograft rejection
The immunologically mediated rejection of grafted tissues or organs from a genetically nonidentical donor. It is due chiefly to recognition of nonself MHC molecules on the graft.

alloreaction
See **alloreactivity**.

alloreactivity (*adj.* alloreactive)
The recognition by T cells of MHC molecules other than self. Such responses are also called alloreactions or alloreactive responses.

allotypes
Allelic polymorphisms that can be detected by antibodies specific for the polymorphic gene products. In immunology, **allotypic**

differences in the constant regions of immunoglobulin molecules were important in deciphering the genetics of antibodies.

ALPS
See **autoimmune lymphoproliferative syndrome**.

altered peptide ligands
Peptides in which in which amino acid substitutions have been made in T-cell receptor contact positions that affect their binding to the receptor.

altered self
Hypothesis explaining the ability of NK cells to attack infected cells while sparing uninfected cells, which depends on changes in the MHC molecules on infected cells.

alternatively activated macrophages
Macrophages that differentiate under the influence of the T_H2-cell cytokines IL-4 and IL-13. They produce the enzyme arginase, which helps increase the contractility of intestinal smooth muscle in response to parasite infection and promotes tissue remodeling and repair. Also known as M2-type macrophages.

alternative pathway
Complement activation pathway that is triggered by the presence of a pathogen in the absence of specific antibodies, and is thus part of the innate immune system. It leads to the production of complement protein C3b and its binding to the surface of the pathogen, after which the pathway is the same as the classical and lectin pathways of complement activation.

alternative pathway C3 convertase
C3bBb, a proteolytic enzyme complex made up of C3b and the activated form (Bb) of alternative complement pathway complement factor B. It converts C3 to C3b bound to a pathogen surface and the anaphylatoxin C3a.

alum
Inorganic aluminum salts (for example aluminum phosphate and aluminum hydroxide); they act as adjuvants when mixed with antigens and are one of the few adjuvants permitted for use in humans.

amphipathic
Describes molecules that have a positively charged (or hydrophilic) region separated from a hydrophobic region.

anaphylactic shock
See **anaphylaxis**.

anaphylatoxins
Pro-inflammatory complement fragments C5a, C3a, and C4a, released by cleavage during complement activation. They are recognized by specific receptors, and recruit fluid and inflammatory cells to the site of their release.

anaphylaxis
A rapid-onset allergic reaction to antigen that occurs throughout the body, for example to insect venom injected directly into the blood stream, or to foods such as peanuts. In the most severe cases the systemic reaction leads to potentially fatal anaphylactic shock as a result of circulatory collapse and suffocation due to tracheal swelling. Anaphylaxis usually results from binding of antigen to IgE antibody on connective tissue mast cells throughout the body, leading to the disseminated release of inflammatory mediators.

anchor residues
Specific amino acid residues in peptide fragments of antigens that determine peptide binding to MHC class I molecules. Each MHC class I molecule binds different patterns of anchor residues, called anchor motifs, giving some specificity to peptide binding. Anchor residues exist but are less obvious for peptides that bind to MHC class II molecules.

anergy (*adj.* anergic)
A state of nonresponsiveness to antigen. People are said to be anergic when they cannot mount delayed-type hypersensitivity reactions to a test antigen, whereas T cells and B cells are said to be anergic when they cannot respond to their specific antigen under optimal conditions of stimulation.

antibody
A protein that binds specifically to a particular substance—called its antigen. Each antibody molecule has a unique structure that enables it to bind specifically to its corresponding antigen, but all antibodies have the same overall structure and are known collectively as immunoglobulins. Antibodies are produced by differentiated B cells (plasma cells) in response to infection or immunization, and bind to and neutralize pathogens or prepare them for uptake and destruction by phagocytes.

antibody combining site
See **antigen-binding site**.

antibody-dependent cell-mediated cytotoxicity (ADCC)
The killing of antibody-coated target cells by cells with Fc receptors that recognize the constant region of the bound antibody. Most ADCC is mediated by NK cells that have the Fc receptor FcγRIII on their surface.

antibody-directed enzyme/pro-drug therapy (ADEPT)
Treatment in which an antibody is linked to an enzyme that metabolizes a nontoxic pro-drug to the active cytotoxic drug.

antibody repertoire
The total variety of antibodies in the body of an individual.

antigen
Any molecule that can bind specifically to an antibody or generate peptide fragments that are recognized by a T-cell receptor.

antigen:antibody complexes
Noncovalently associated groups of antibody molecules bound to their specific antigen.

antigen-binding site
The site at the tip of each arm of an antibody that makes physical contact with the antigen and binds it noncovalently. The antigen specificity of the site is determined by its shape and the amino acids present.

antigen display library
A collection of cDNA clones in expression vectors or collections of bacteriophages encoding random peptide sequences that can be expressed as part of the phage coat. They are used to identify the targets of specific antibodies and, in some cases, of T cells.

antigenic determinant
That portion of an antigenic molecule that is bound by the antigen-binding site of a given antibody or antigen receptor; it is also known as an epitope.

antigenic drift
The process by which influenza virus varies genetically in minor ways from year to year. Point mutations in viral genes cause small differences in the structure of the viral surface antigens.

antigenic shift
A radical change in the surface antigens of influenza virus, caused by reassortment of their segmented genome with that of another influenza virus, often from an animal.

antigenic variation
Alterations in surface antigens that occur in some pathogens (such as African trypanosomes) from one generation to another, which renders them immune to preexisting antibodies.

antigen presentation
The display of antigen on the surface of a cell in the form of peptide fragments bound to MHC molecules. T cells recognize antigen when it is presented in this way.

antigen-presenting cells (APCs)
Highly specialized cells that can process antigens and display their peptide fragments on the cell surface together with other, co-stimulatory, proteins required for activating naive T cells. The main antigen-presenting cells for naive T cells are dendritic cells, macrophages, and B cells.

antigen processing
The intracellular degradation of foreign proteins into peptides that can bind to MHC molecules for presentation to T cells. All protein antigens must be processed into peptides before they can be presented by MHC molecules.

antigen receptor
The cell-surface receptor by which lymphocytes recognize antigen. Each individual lymphocyte bears receptors of a single antigen specificity.

anti-idiotype antibody
Antibody raised against antigenic determinants unique to the variable region of a single antibody.

anti-immunoglobulin antibody
See **anti-isotypic antibody**.

anti-isotypic antibody
Antibody specific for universal features of a given immunoglobulin constant-region isotype (such as γ or μ) of one species. They are made by immunizing a member of another species with that isotype.

anti-lymphocyte globulin
Antiserum raised in another species against human T cells. It is used in the temporary suppression of immune responses in transplantation.

antimicrobial enzymes
Enzymes that kill microorganisms by their actions. An example is lysozyme, which digests bacterial cell walls.

antimicrobial peptides
Amphipathic peptides secreted by epithelial cells and phagocytes that kill a variety of microbes nonspecifically, mainly by disrupting cell membranes. Antimicrobial peptides in humans include the defensins, the cathelicidins, and the histatins.

antiserum (plural: antisera)
The fluid component of clotted blood from an immune individual that contains antibodies against the antigen used for immunization. An antiserum contains a mixture of different antibodies that all bind the antigen, but which each have a different structure, their own epitope on the antigen, and their own set of cross-reactions. This heterogeneity makes each antiserum unique.

antivenin
Antibody raised against the venom of a poisonous snake and which can be used as an immediate treatment for snakebite to neutralize the venom.

AP-1
A transcription factor formed as one of the outcomes of intracellular signaling via the antigen receptors of lymphocytes and the TLRs of cells of innate immunity. AP-1 mainly activates the expression of genes for cytokines and chemokines.

APECED
See **autoimmune polyendocrinopathy-candidiasis-ectodermal dystrophy**.

apoptosis
A form of cell death common in the immune system, in which the cell activates an internal death program. It is characterized by nuclear DNA degradation, nuclear degeneration and condensation, and the rapid phagocytosis of cell remains. Proliferating lymphocytes experience high rates of apoptosis during their development and during immune responses.

appendix
A gut-associated lymphoid tissue located at the beginning of the colon.

apurinic/apyrimidinic endonuclease 1
A DNA repair endonuclease involved in class switch recombination.

Artemis
An endonuclease involved in the gene rearrangements that generate functional immunoglobulin and T-cell receptor genes.

Arthus reaction
A local skin reaction that can be induced to show the presence of IgG antibodies against a particular allergen. Antigen injected into the dermis reacts with IgG antibodies in the extracellular spaces, activating complement and phagocytic cells to produce a local inflammatory response.

asymptomatic phase
The phase of HIV infection in which the infection is being partly held in check and no symptoms occur; it may last for many years.

ataxia telangiectasia
A disease characterized by a staggering gait and multiple disorganized blood vessels, and often accompanied by clinical immunodeficiency. It is caused by defects in the ATM protein, which is involved in DNA repair pathways that are also used in V(D)J recombination and class-switch recombination.

atopic dermatitis, atopic eczema
An allergic skin condition seen mainly in children and characterized by patches of intensely itchy, reddened, and scaly skin.

atopy (*adj.* atopic)
A genetically based increased tendency to produce IgE-mediated allergic reactions against innocuous substances.

attenuation
The process by which human or animal pathogens are modified by growth in culture so that they can grow in their host and induce immunity without producing serious clinical disease.

atypical hemolytic uremic syndrome
A condition characterized by damage to platelets and red blood cells and inflammation of the kidneys that is caused by uncontrolled complement activation in individuals with inherited deficiencies in complement regulatory proteins.

autoantibody
An antibody specific for self antigens.

autoantigen
A self antigen to which the immune system makes a response.

autocrine
Describes a cytokine or other biologically active molecule acting on the cell that produces it.

autograft
A graft of tissue from one site to another on the same individual.

autoimmune disease
Disease in which the pathology is caused by adaptive immune responses to self antigens.

autoimmune hemolytic anemia
A pathological condition with low levels of red blood cells (anemia), which is caused by autoantibodies that bind red blood cell surface antigens and target the red blood cell for destruction.

autoimmune lymphoproliferative syndrome (ALPS)
An inherited syndrome in which a defect in the *Fas* gene leads to a failure in normal apoptosis, causing unregulated immune responses, including autoimmune responses.

autoimmune polyendocrinopathy-candidiasis-ectodermal dystrophy (APECED)
A disease characterized by a loss of tolerance to self antigens, caused by a breakdown of negative selection in the thymus. It is due to defects in the gene *AIRE*, which encodes a transcriptional regulatory protein that enables many self antigens to be expressed by thymic epithelial cells. Also called autoimmune polyglandular syndrome type I.

autoimmune thrombocytopenic purpura
An autoimmune disease in which antibodies against platelets are made. Antibody binding to platelets causes them to be taken up by cells with Fc receptors and complement receptors, resulting in a decrease in platelet count that leads to purpura (bleeding).

autoimmunity
Adaptive immunity specific for self antigens.

autoinflammatory disease
Disease due to unregulated inflammation in the absence of infection; it can have a variety of causes, including inherited genetic defects.

autophagy
The digestion and breakdown by a cell of its own organelles and proteins in lysosomes. It may be one route by which cytosolic proteins can be processed for presentation on MHC class II molecules.

autoreactivity
The generation of immune responses directed at self antigens.

avidity
The sum total of the strength of binding of two molecules or cells to one another at multiple sites. It is distinct from affinity, which is the strength of binding of one site on a molecule to its ligand.

azathioprine
A powerful cytotoxic drug that is converted to its active form *in vivo*, which then kills rapidly proliferating cells, including proliferating lymphocytes; it is used as an immunosuppressant to treat autoimmune disease and in transplantation.

4-1BB, 4-1BB ligand
A co-stimulatory receptor–ligand pair. 4-1BB is present on T cells, and 4-1BB ligand on dendritic cells.

B-1 cells
A class of atypical, self-renewing B cells (also known as CD5 B cells) found mainly in the peritoneal and pleural cavities in adults. They have a much less diverse antigen-receptor repertoire than conventional B cells.

B-2 cells
A name sometimes given to the conventional B lymphocytes with highly variable antigen receptors. Cf. **B-1 cells**.

B7 molecules, B7.1 and B7.2
Cell-surface proteins on specialized antigen-presenting cells such as dendritic cells, which are the major co-stimulatory molecules for T cells. B7.1 (CD80) and B7.2 (CD86) are closely related members of the immunoglobulin superfamily and both bind to the CD28 protein on T cells.

β
Name given to a wide variety of subunits in different proteins. In immunology it is likely to refer to (1) one of the two types of chain in the α:β T-cell receptor, or (2) the light chain of an MHC class II molecule.

β barrel
An element of protein secondary structure composed of one or more β sheets curved round to form a roughly barrel-shaped structure. Immunoglobulin domains contain this structure.

β-defensins
Antimicrobial peptides made by virtually all multicellular organisms. In mammals they are produced by the epithelia of the respiratory and urogenital tracts, skin, and tongue.

β sheet
One of the fundamental elements of secondary structure in proteins, consisting of adjacent, extended strands of amino acids (β strands) that are noncovalently bonded by interactions between backbone amide and carbonyl groups. In 'parallel' β sheets, the adjacent strands run in the same direction; in 'antiparallel' β sheets, adjacent strands run in opposite directions. Immunoglobulin domains are made up of two antiparallel β sheets arranged in the form of a β barrel.

β_2-microglobulin
The light chain of the MHC class I proteins, encoded outside the MHC. It binds noncovalently to the heavy or α chain.

B and T lymphocyte attenuator (BTLA)
An inhibitory CD28-related receptor expressed by B and T lymphocytes that interacts with the herpes virus entry molecule (HVEM), a member of the TNF receptor family.

bacteria (singular: bacterium)
A vast kingdom of unicellular prokaryotic microorganisms, some species of which cause infectious diseases in humans and animals, while others make up most of the body's commensal microbiota. Disease-causing bacteria may live in the extracellular spaces, or inside cells in vesicles or in the cytosol.

bacterial lipopolysaccharide
See **lipopolysaccharide**.

BAFF
B-cell activating factor belonging to the TNF family. It is one of the survival signals provided to B cells by follicular dendritic cells.

BALT
See **bronchus-associated lymphoid tissues**.

bare lymphocyte syndrome
See **MHC class I deficiency**, **MHC class II deficiency**.

base-excision repair
Type of DNA repair that can lead to mutation and that is involved in somatic hypermutation and class switching in B cells.

basophil
Type of white blood cell containing granules that stain with basic dyes. It is thought to have a function similar to mast cells.

Bb
See **factor B**.

B cell, B lymphocyte
One of the two types of antigen-specific lymphocytes responsible for adaptive immune responses, the other being the T cells. The function of B cells is to produce antibodies. B cells are divided into two classes. Conventional B cells have highly diverse antigen receptors and are generated in the bone marrow throughout life, emerging to populate the blood and lymphoid tissues. B-1 cells have much less diverse antigen receptors and form a population of self-renewing B cells in the peritoneal and pleural cavities.

B-cell co-receptor complex
A transmembrane signaling receptor on the B-cell surface composed of the proteins CD19, CD81, and CD21 (complement receptor 2), which binds complement fragments on bacterial antigens also bound by the B-cell receptor. Co-ligation of this complex with the B-cell receptor increases responsiveness to antigen about 100-fold.

B-cell corona
The zone of the white pulp in the spleen that is primarily made up of B cells.

B-cell mitogen
Any substance that nonspecifically causes B cells to proliferate.

B-cell receptor (BCR)
The cell-surface receptor on B cells for specific antigen. It is composed of a transmembrane immunoglobulin molecule (which recognizes antigen) associated with the invariant Igα and Igβ chains (which have a signaling function). On activation by antigen, B cells differentiate into plasma cells producing antibody molecules of the same antigen specificity as this receptor.

Bcl-2 family
Family of intracellular proteins that includes members that promote apoptosis (Bax, Bak, and Bok) and members that inhibit apoptosis (Bcl-2, Bcl-W, and Bcl-X_L).

biologics therapy
Medical treatments comprising natural proteins such as antibodies and cytokines, and antisera or whole cells.

Blau syndrome
An inherited granulomatous disease caused by gain-of-function mutations in the *NOD2* gene.

Blk
A Src-family tyrosine kinase that relays signals from the B-cell receptor.

BLNK
B-cell linker protein. A scaffold protein in B cells that recruits proteins involved in the intracellular signaling pathway from the antigen receptor.

blood group antigens
Surface molecules on red blood cells that are detectable by antibodies from other individuals. The major blood group antigen systems in humans are called ABO and Rh (Rhesus), and are used routinely to type blood. There are many other blood group antigens.

blood typing
Procedure used to determine whether donor and recipient have compatible ABO and Rh blood group antigens before blood is transfused. A cross-match, in which serum from the donor is tested on the cells of the recipient, and vice versa, is used to rule out other incompatibilities. Transfusion of incompatible blood causes a transfusion reaction, in which red blood cells are destroyed and the released hemoglobin causes toxicity.

Bloom's syndrome
A disease characterized by low T-cell numbers, reduced antibody levels, and an increased susceptibility to respiratory infections, cancer, and radiation damage. It is caused by mutations in a DNA helicase.

B lymphocyte
See **B cell**.

bone marrow
The tissue where all the cellular elements of the blood—red blood cells, white blood cells, and platelets—are initially generated from hematopoietic stem cells. The bone marrow is also the site of further B-cell development in mammals and the source of stem cells that give rise to T cells on migration to the thymus. Thus, bone marrow transplantation can restore all the cellular elements of the blood, including the cells required for adaptive immune responses.

bone marrow chimera
A mouse whose own bone marrow has been destroyed by irradiation and has been reconstituted with bone marrow from another mouse, so that all of the lymphocytes and blood cells are of donor genetic origin. Bone marrow chimeras have been crucial in elucidating the development of lymphocytes and other blood cells.

booster immunization
An additional immunization commonly given after a primary immunization, to increase the amount, or titer, of antibodies.

bradykinin
A vasoactive peptide that is produced as a result of tissue damage and acts as an inflammatory mediator.

bronchiectasis
Injury and dilation of the airways (bronchi).

bronchus-associated lymphoid tissues (BALT)
Organized lymphoid tissue found in the bronchi in some animals. Adult humans do not normally have such organized lymphoid tissue in the respiratory tract, but it may be present in some infants and children.

Bruton's tyrosine kinase (BTK)
A Tec-family kinase that is mutated in the human immunodeficiency disease X-linked agammaglobulinemia.

Bruton's X-linked agammaglobulinemia
See **X-linked agammaglobulinemia**.

BTLA
See **B and T lymphocyte attenuator**.

bursa of Fabricius
Lymphoid organ associated with the gut that is the site of B-cell development in chickens.

C1 complex
Protein complex activated as the first step in the classical pathway of complement activation. It comprises one molecule of the protein C1q bound to two molecules each of the proteases C1r and C1s. C1q initiates the classical pathway of complement activation by binding to a pathogen surface or to bound antibody. This binding activates the associated C1r, which in turn cleaves and activates C1s. The active form of C1s then cleaves the next two components in the pathway, C4 and C2.

C1 inhibitor (C1INH)
A protein that inhibits the activity of the C1 complex by binding to and inactivating C1r:C1s enzymatic activity. Deficiency in C1INH is the cause of the disease hereditary angioedema, in which the production of vasoactive peptides leads to subcutaneous and laryngeal swelling.

C1q, C1r, C1s
Complement proteins that form the C1 complex in the first step of the classical pathway of complement activation.

C2
Complement protein of the classical pathway of activation that is cleaved by the C1 complex to yield C2b and C2a. C2a is an active protease that forms part of the classical C3 convertase C4bC2a.

C3
Complement protein on which all complement activation pathways converge. C3 is cleaved to form C3b, the principal effector molecule of the complement system, which binds covalently to the surface on which it is generated. Once bound, it acts as an opsonin to promote the destruction of pathogens by phagocytes and removal of immune complexes.

C3 convertase
Enzyme complex that cleaves C3 to C3b and C3a on the surface of a pathogen. The C3 convertase of the classical and lectin pathways is formed from membrane-bound C4b complexed with the protease C2a. The alternative pathway C3 convertase is formed from membrane-bound C3b complexed with the protease Bb.

C3(H_2O)Bb
See **fluid-phase C3 convertase**.

C3b
The cleavage product of C3 that binds covalently to a pathogen surface or to soluble antigen molecules. All pathways of complement activation converge on the production of C3b.

C4
Complement protein of the classical pathway of activation; it is cleaved to C4b, which forms part of the classical C3 convertase.

C3dg
Breakdown product of iC3b that remains attached to the microbial surface, where it can bind complement receptor CR2.

C4b-binding protein (C4BP)
A complement-regulatory protein that inactivates the classical pathway C3 convertase formed on host cells by displacing C2a from the C4bC2a complex. C4BP binds C4b attached to host cells, but cannot bind C4b attached to pathogens.

C5
Complement protein that is cleaved to release the pro-inflammatory peptide C5a and a larger fragment, C5b, that initiates the formation of a membrane-attack complex from the terminal components of complement.

C5a receptor
The cell-surface receptor for the pro-inflammatory C5a fragment of complement, present on macrophages and neutrophils.

C5 convertase
Enzyme complex that cleaves C5 to C5a and C5b.

C6, C7, C8, C9
Complement proteins that form a complex with C5b to initiate the late events of complement activation that induce cell lysis. This complex inserts into the membrane and induces the polymerization of complement protein C9 to form a pore known as the membrane-attack complex.

calcineurin
A cytosolic serine/threonine phosphatase with a crucial role in signaling via the T-cell receptor. The immunosuppressive drugs cyclosporin A and tacrolimus inactivate calcineurin, suppressing T-cell responses.

calmodulin
Calcium-binding protein that is activated by binding Ca^{2+}; it is then able to bind to and regulate the activity of a wide variety of enzymes.

calnexin
A chaperone protein in the endoplasmic reticulum (ER) that binds to partly folded members of the immunoglobulin superfamily of proteins and retains them in the ER until folding is complete.

calreticulin
A chaperone protein in the endoplasmic reticulum that, together with ERp57 and tapasin, forms the peptide-loading complex that loads peptides onto newly synthesized MHC class I molecules.

cancer immunoediting
A process that occurs during the development of a cancer when it is acquiring mutations that favor its survival and escape from immune responses, such that cancer cells with these mutations are selected for survival and growth.

cancer-testis antigens
Proteins expressed by cancer cells that are normally expressed only in male germ cells in the testis.

capture ELISA
An assay for measuring antibodies or antigens. Antigens are

captured by antibodies bound to plastic (or vice versa). Antibody binding to a plate-bound antigen can be measured by using labeled antigen or anti-immunoglobulin. Antigen binding to plate-bound antibody can be measured by using an antibody that binds to a different epitope on the antigen.

CARD
See **caspase recruitment domain**.

carriers
Foreign proteins to which small nonimmunogenic antigens, or haptens, can be coupled to render the hapten immunogenic. *In vivo*, self proteins can also serve as carriers if they are correctly modified by the hapten; this is important in allergy to drugs.

caspases
A family of cysteine proteases that cleave proteins at aspartic acid residues. They have important roles in apoptosis and in the processing of cytokine pro-polypeptides.

caspase recruitment domain (CARD)
A protein domain present in some receptor tails that can dimerize with other CARD-domain-containing proteins, including caspases, thus recruiting them into signaling pathways.

cathelicidins
Family of antimicrobial peptides that in humans has one member.

CC chemokines
One of the two main classes of chemokines, distinguished by two adjacent cysteines (C) near the amino terminus. They have names CCL1, CCL2, etc. See Appendix IV for a list of individual chemokines.

CD antigens
See **clusters of differentiation**, and see Appendix II for individual CD antigens.

CD1
Small family of MHC class I-like proteins that are not encoded in the MHC and can present glycolipid antigens to CD4 T cells.

CD3 complex
The invariant proteins CD3γ, δ, and ε, and the dimeric ζ chains, which form the signaling complex of the T-cell receptor.

CD4
The co-receptor for T-cell receptors that recognize peptide antigens bound to MHC class II molecules. It binds to the lateral face of the MHC molecule.

CD4 T cells
T cells that carry the co-receptor protein CD4 and recognize peptides derived from intravesicular sources (including endocytosed extracellular antigens), which are bound to MHC class II molecules. They differentiate into a variety of different subsets of CD4 effector cells (such as T_H1 and T_H2) that secrete particular cytokines, activate macrophages, and help induce B-cell antibody production.

CD8
The co-receptor for T-cell receptors that recognize peptide antigens bound to MHC class I molecules. It binds to the lateral face of the MHC molecule.

CD8 cytotoxic T cells
T cells that carry the co-receptor CD8 and recognize antigens, for example viral antigens, that are synthesized in the cytoplasm of a cell and become bound to MHC class I molecules. CD8 T cells differentiate into cytotoxic CD8 T cells.

CD19
See **B-cell co-receptor complex**.

CD21
Another name for **complement receptor 2** (**CR2**). See also **B-cell co-receptor complex**.

CD23
A low-affinity Fc receptor for IgE.

CD27
A TNF receptor-family protein constitutively expressed on naive T cells that binds CD70 on dendritic cells and delivers a potent co-stimulatory signal to T cells early in the activation process.

CD28
An activating receptor on T cells for the B7 co-stimulatory molecules present on specialized antigen-presenting cells such as dendritic cells.

CD30, CD30 ligand
CD30 on B cells and CD30 ligand (CD30L) on helper T cells are co-stimulatory molecules involved in stimulating the proliferation of antigen-activated naive B cells.

CD31
CD31, a cell-adhesion molecule found both on lymphocytes and at endothelial cell junctions. CD31–CD31 interactions are thought to enable leukocytes to leave blood vessels and enter tissues.

CD40, CD40 ligand
CD40 on B cells and CD40 ligand (CD40L, CD154) on activated helper T cells are co-stimulatory molecules whose interaction is required for the proliferation and class switching of antigen-activated naive B cells. CD40 is also expressed by dendritic cells, and here the CD40–CD40L interaction provides co-stimulatory signals to naive T cells.

CD40 ligand deficiency
An immunodeficiency disease in which little or no IgG, IgE, or IgA antibody is produced and even IgM responses are deficient, but serum IgM levels are normal to high. It is due to a defect in the gene encoding CD40 ligand (CD154), which prevents class switching from occurring.

CD45
A transmembrane tyrosine phosphatase found on all leukocytes. It is expressed in different isoforms on different cell types, including the different subtypes of T cells. Also called leukocyte common antigen.

CD59
See **protectin**.

CD80, CD86
See **B7 molecules**.

CD81
See **B-cell co-receptor complex**.

CD94
A C-type lectin that is a subunit of the KLR-type receptors of NK cells.

CD103
Integrin α_E:β_7, a cell-surface marker on a subset of dendritic cells in the gastrointestinal tract that are involved in inducing tolerance to antigens from food and the commensal microbiota.

cDCs
See **conventional dendritic cells**.

C domain
See **constant domain**.

CDRs
See **complementarity-determining regions**.

celiac disease
A chronic condition of the upper small intestine caused by an immune response directed at gluten, a complex of proteins present in wheat, oats, and barley. The gut wall becomes chronically inflamed, the villi are destroyed, and the gut's ability to absorb nutrients is compromised.

cell-adhesion molecules
Cell-surface proteins of several different types that mediate the binding of one cell to other cells or to extracellular matrix proteins. Integrins, selectins, and members of the immunoglobulin gene superfamily (such as ICAM-1) are among the cell-adhesion molecules important in the operation of the immune system.

cell-mediated immune response
An adaptive immune response in which antigen-specific effector T cells have the main role. The immunity to infection conferred by such a response is called cell-mediated immunity. A primary cell-mediated immune response is the T-cell response that occurs the first time a particular antigen is encountered.

cell-mediated immunity
Immunity to infection in which antigen-specific T cells play the

main role. It is defined operationally as all adaptive immunity that cannot be transferred to a naive recipient by serum antibody. Cf. **humoral immunity**.

cellular immunology
The study of the cellular basis of immunity.

central lymphoid organs, central lymphoid tissues
The sites of lymphocyte development; in humans, these are the bone marrow and thymus. B lymphocytes develop in bone marrow, whereas T lymphocytes develop within the thymus from bone marrow derived progenitors. Also called the primary lymphoid organs.

central memory cells
A class of memory cells with characteristic activation properties that are thought to reside in the T-cell areas of peripheral lymphoid tissues.

central tolerance
Immunological tolerance to self antigens that is established while lymphocytes are developing in central lymphoid organs. Cf. **peripheral tolerance**.

centroblasts
Large, rapidly dividing activated B cells present in the dark zone of germinal centers in follicles of peripheral lymphoid organs.

centrocytes
Small B cells that derive from centroblasts in the germinal centers of follicles in peripheral lymphoid organs; they populate the light zone of the germinal center.

CGD
See **chronic granulomatous disease**.

checkpoint blockade
Approach to tumor therapy that attempts to interfere with the normal inhibitory signals that regulate lymphocytes.

Chediak–Higashi syndrome
A defect in phagocytic cell function caused by a defect in a protein involved in intracellular vesicle fusion. Lysosomes fail to fuse properly with phagosomes, and killing of ingested bacteria is impaired.

chemokine
Small chemoattractant protein that stimulates the migration and activation of cells, especially phagocytic cells and lymphocytes. Chemokines have a central role in inflammatory responses. Individual chemokines and their receptors are listed in Appendix IV.

chimeric
When applied to molecules, molecules with parts derived from different species, for example antibodies that have mouse variable regions and human constant and framework sequences. See also **bone marrow chimera**.

chronic allergic inflammation
The chronic inflammation of the airways present in chronic asthma, leading to permanent damage to the airways.

chronic allograft vasculopathy
Chronic damage that can lead to late failure of transplanted organs. Arteriosclerosis of graft blood vessels leads to hypoperfusion of the graft and its eventual fibrosis and atrophy.

chronic granulomatous disease (CGD)
An immunodeficiency in which multiple granulomas form as a result of defective elimination of bacteria by phagocytic cells. It is caused by defects in the NADPH oxidase system of enzymes that generate the superoxide radical involved in bacterial killing.

chronic infantile neurologic cutaneous and articular syndrome (CINCA)
An autoinflammatory disease due to defects in the gene *NLRP3*, one of the components of the inflammasome.

chronic rejection
Late failure of a transplanted organ, which can be due to nonimmunological causes.

CLA
See **cutaneous lymphocyte antigen**.

class
The class of an antibody is defined by the type of heavy chain it contains. There are five main antibody classes: IgA, IgD, IgM, IgG, and IgE, containing heavy chains α, δ, μ, γ, and ε, respectively. The IgG class has several subclasses. See also **isotype**.

class II-associated invariant chain peptide (CLIP)
A peptide of variable length cleaved from the invariant chain (Ii) by proteases. It remains associated with the MHC class II molecule in an unstable form until it is removed by the HLA-DM protein.

classical pathway
The complement-activation pathway that is initiated by C1 binding either directly to bacterial surfaces or to antibody bound to the bacteria, thus flagging the bacteria as foreign. See also **alternative pathway**, **lectin pathway**.

classical C3 convertase
The complex of activated complement components C4b2a, which cleaves C3 to C3b on pathogen surfaces in the classical pathway of complement activation.

class switching, class switch recombination
A somatic recombination process that occurs in activated B cells in which one heavy-chain constant region gene is replaced with another of a different isotype, resulting in a switch from the production of antibodies of class IgM to the production of IgG, IgA, or IgE classes. Class switching affects the effector functions of the antibodies produced but not their antigen specificity. Also known as isotype switching. Cf. **somatic hypermutation**.

CLIP
See **class II-associated invariant-chain peptide**.

clonal deletion
The elimination of immature lymphocytes when they bind to self antigens, which produces tolerance to self as required by the clonal selection theory of adaptive immunity. Clonal deletion is the main mechanism of central tolerance and can also occur in peripheral tolerance.

clonal expansion
The proliferation of antigen-specific lymphocytes in response to antigenic stimulation that precedes their differentiation into effector cells. It is an essential step in adaptive immunity, allowing rare antigen-specific cells to increase in number so that they can effectively combat the pathogen that elicited the response.

clonal selection theory
The central paradigm of adaptive immunity. It states that adaptive immune responses derive from individual antigen-specific lymphocytes that are self-tolerant. These specific lymphocytes proliferate in response to antigen and differentiate into antigen-specific effector cells that eliminate the eliciting pathogen, and into memory cells to sustain immunity. The theory was formulated by Macfarlane Burnet and in earlier forms by Niels Jerne and David Talmage.

clone
A population of cells all derived from the same progenitor cell.

clonotypic
Describes a feature unique to members of a clone. For example, the distribution of antigen receptors in the lymphocyte population is said to be clonotypic, as the cells of a given clone all have identical antigen receptors.

CLP
See **common lymphoid progenitor**.

clusters of differentiation (CD)
Groups of monoclonal antibodies that identify the same cell-surface molecule. The cell-surface molecule is then designated CD followed by a number (CD1, CD2, etc.). For a current listing of CDs see Appendix II.

coagulation system
A collection of proteases and other proteins in the blood that trigger blood clotting when blood vessels are damaged.

coding joint
The join formed in DNA by the imprecise joining of a V gene segment to a (D)J gene segment during somatic recombination of the immunoglobulin or T-cell receptor genes. It is the joint retained in the rearranged gene. Cf. **signal joint**.

codominant
Describes the situation in which the two alleles of a gene are expressed in roughly equal amounts in the heterozygote. Most genes show this property, including the highly polymorphic MHC genes.

cognate
Describes interactions between a B cell and a T cell specific for the same antigen.

co-immunoprecipitation
A technique for isolating a particular protein from a cell extract together with other proteins that bind to it, by using a labeled antibody against the first protein to precipitate the protein complex from the cell extract.

collectins
A family of calcium-dependent sugar-binding proteins (lectins) containing collagen-like sequences. An example is mannose-binding lectin (MBL).

combinatorial diversity
The diversity among antigen receptors that is generated by the combination of separate units of genetic information. Two types of combinatorial diversity operate in antigen receptor formation. First, receptor gene segments are joined in many different combinations to generate diverse receptor chains; second, two different receptor chains (heavy and light in immunoglobulins; α and β, or γ and δ, in T-cell receptors) are combined to make the antigen-recognition site.

commensal microbiota, commensal microorganisms
Microorganisms (predominantly bacteria) that normally live harmlessly in symbiosis with their host (for example the gut bacteria in humans and other animals). Many commensals confer a positive benefit on their host in some way.

common γ chain (γ_c)
A transmembrane polypeptide chain (CD132) that is common to a subgroup of cytokine receptors.

common lymphoid progenitor (CLP)
Stem cell that can give rise to all the types of lymphocytes.

common mucosal immune system
The mucosal immune system as a whole, the name reflecting the fact that lymphocytes that have been primed in one part of the mucosal system can recirculate as effector cells to other parts of the mucosal system.

common myeloid progenitor
Stem cells that can give rise to the myeloid cells of the immune system—macrophages, granulocytes, mast cells, and dendritic cells of the innate immune system. This stem cell also gives rise to megakaryocytes and red blood cells.

common variable immunodeficiency (CVID)
A relatively common deficiency in antibody production in which only one or a few isotypes are affected. It can be due to a variety of genetic defects.

competitive inhibition assay
Serological assay in which known antibodies or antigens are used as competitive inhibitors of antigen–antibody interactions, to detect and quantitate an unknown antigen or antibody, respectively.

complement, complement system
A set of plasma proteins that act together as a defense against pathogens in extracellular spaces. The pathogen becomes coated with complement proteins that facilitate its removal by phagocytes and that can also kill certain pathogens directly. Activation of the complement system can be initiated in several different ways. See **classical pathway**, **alternative pathway**, **lectin pathway**.

complement activation
The activation of the normally inactive proteins of the complement system that occurs on infection. See **classical pathway**, **alternative pathway**, **lectin pathway**.

complement proteins
See **C1**, **C2**, **C3** etc.

complement receptors (CRs)
Cell-surface proteins of various types that recognize and bind complement proteins that have become bound to an antigen such as a pathogen. Complement receptors on phagocytes enable them to identify and bind pathogens coated with complement proteins, and to ingest and destroy them. See **CR1**, **CR2**, **CR3**, **CR4**, **CRIg**, and the **C1q receptor**.

complement regulatory proteins
Proteins that control complement activity and prevent complement from being activated on the surfaces of host cells.

complementarity-determining regions (CDRs)
Parts of the V domains of immunoglobulins and T-cell receptors that determine their antigen specificity and make contact with the specific ligand. The CDRs are the most variable part of antigen receptor, and contribute to the diversity of these proteins. There are three such regions (CDR1, CDR2, and CDR3) in each V domain.

confocal fluorescent microscopy
Optical microscopy technology that produces images at very high resolution by having two sources of fluorescent light that come together at one plane in the specimen.

conformational epitope
Antigenic structure (epitope) on a protein antigen that is formed from several separate regions in the sequence of the protein brought together by protein folding. Antibodies that bind conformational epitopes bind only native folded proteins. Also called a discontinuous epitope.

conjugate vaccines
Antibacterial vaccines made from bacterial capsular polysaccharides bound to proteins of known immunogenicity, such as tetanus toxoid.

constant domain (C domain)
Type of protein domain that makes up the constant regions of each chain of an immunoglobulin molecule.

constant region (C region)
That part of an immunoglobulin or a T-cell receptor that is relatively constant in amino acid sequence between different molecules. Also known as the Fc region in antibodies. The constant region of an antibody determines its particular effector function. Cf. **variable region**.

continuous epitope
Antigenic structure (epitope) in a protein that is formed by a single small region of amino acid sequence. Antibodies that bind continuous epitopes can bind to the denatured protein. The epitopes detected by T cells are continuous. Also called a linear epitope.

conventional dendritic cells (cDCs)
The lineage of dendritic cells that mainly participates in antigen presentation to, and activation of, naive T cells. Cf. **plasmacytoid dendritic cells**.

convertase
A protease that converts a complement protein into its reactive form by cleaving it.

Coombs test
See **direct Coombs test**, **indirect Coombs test**.

cooperativity
The interaction often seen between two binding sites on the same protein, whereby the binding of ligand to one site enhances the binding of ligand to the second site.

co-receptor
Cell-surface protein that increases the sensitivity of a receptor to its ligand by binding to associated ligands and participating in signaling. The antigen receptors on T cells and B cells act in conjunction with co-receptors, which are either CD4 or CD8 on T cells, and a co-receptor complex of three proteins, one of which is the complement receptor CR2, on B cells.

cortex
The outer part of a tissue or organ; in lymph nodes it refers to the follicles, which are mainly populated by B cells.

corticosteroids
Family of drugs related to natural steroids such as cortisone. Corticosteroids can kill lymphocytes, especially developing

thymocytes, inducing apoptotic cell death. They are medically useful anti-inflammatory and immunosuppressive agents.

co-stimulation
Additional signals required for the activation of naive lymphocytes in addition to the signals generated by the binding of antigen to the antigen receptors. See **co-stimulatory signal**.

co-stimulatory molecules
Cell-surface proteins on antigen-presenting cells that deliver co-stimulatory signals to naive T cells. Examples are the B7 molecules on dendritic cells, which are ligands for CD28 on naive T cells.

co-stimulatory receptors
Cell-surface receptors on naive lymphocytes through which the cells receive signals additional to those received through the antigen receptor, and which are necessary for the full activation of the lymphocyte. Examples are CD30 and CD40 on B cells, and CD27 and CD28 on T cells.

co-stimulatory signal
Signal required in addition to the signal from the antigen receptor, to induce the activation and proliferation of naive lymphocytes when they first encounter antigen. Such signals are usually delivered to T cells by proteins on the surface of the antigen-presenting cell such as the B7 molecules. B cells can receive co-stimulatory signals from pathogen components such as lipopolysaccharide, from complement fragments, or via CD40 ligand on the surface of an activated antigen-specific helper T cell. See also **co-stimulatory molecules**.

counter-regulation hypothesis
The idea that all types of infection early in childhood might protect against the development of atopy by driving the production of cytokines such as IL-10 and transforming growth factor-β, which promote the production of regulatory T cells.

cowpox
Common name for the disease produced by vaccinia virus, which was used by Edward Jenner to vaccinate successfully against smallpox, which is caused by the related variola virus.

CR
See **complement receptor**.

CR1 (CD35)
A receptor for the complement protein C3b bound to pathogen surfaces. Present on phagocytic cells, where it stimulates phagocytosis on binding its ligand. It also has a role as a complement regulatory protein, inhibiting C3 convertase formation on host-cell surfaces.

CR2 (CD21)
Complement receptor that is part of the B-cell co-receptor complex. It binds to antigens coated with breakdown products of C3b, especially C3dg, and, by cross-linking the B-cell receptor, enhances sensitivity to antigen at least 100-fold. It is also the receptor used by the Epstein–Barr virus to infect B cells.

CR3 (CD11b:CD18)
Complement receptor 3. A β_2 integrin that acts both as an adhesion molecule and as a complement receptor. CR3 on phagocytes binds iC3b, a breakdown product of C3b on pathogen surfaces, and stimulates phagocytosis.

CR4 (CD11c:CD18)
A β_2 integrin that acts both as an adhesion molecule and as a complement receptor. CR4 on phagocytes binds iC3b, a breakdown product of C3b on pathogen surfaces, and stimulates phagocytosis.

CRAC channels
Calcium-release-activated calcium channels in the lymphocyte plasma membrane that open to let calcium flow into the cell during the response of the cell to antigen.

C region
See **constant region**.

C-reactive protein
An acute-phase protein that binds to phosphocholine, a constituent of the surface C-polysaccharide of the bacterium *Streptococcus pneumoniae* and of many other bacteria, thus opsonizing them for uptake by phagocytes.

CRIg
A complement receptor that binds to inactivated forms of C3b.

Crohn's disease
Chronic inflammatory bowel disease thought to result from an abnormal overresponsiveness to the commensal gut microbiota.

cross-linked
Describes the linking together of receptors on a cell (especially antigen receptors on lymphocytes) by a multivalent extracellular ligand.

cross-matching
A test used in blood typing and histocompatibility typing to determine whether donor and recipient have antibodies against each other's cells that might interfere with successful transfusion or grafting.

cross-presentation
The process by which extracellular proteins taken up by dendritic cells can give rise to peptides presented by MHC class I molecules. It enables antigens from extracellular sources to be presented by MHC class I molecules and activate CD8 T cells.

cross-reaction
The binding of antibody or a T-cell to an antigen not used to elicit that antibody.

cryptdins
α-Defensins (antimicrobial peptides) made by the Paneth cells of the small intestine.

cryptic epitope
Any epitope that cannot be recognized by a lymphocyte receptor until the antigen has been broken down and processed.

cryptopatches
Aggregates of lymphoid tissue in the gut wall that are thought to give rise to isolated lymphoid follicles.

c-SMAC
See **supramolecular adhesion complex**.

C-terminal Src kinase (Csk)
A kinase that phosphorylates the C-terminal tyrosine of Src-family kinases in lymphocytes, thus inactivating them.

CTLA-4
A high-affinity inhibitory receptor on T cells for B7 molecules; its binding blocks T-cell activation.

C-type lectins
Large class of carbohydrate-binding proteins that require Ca^{2+} for binding. C-type lectin domains are present in proteins of innate immunity that recognize bacterial carbohydrates, such as the collectins, many receptors, and the antimicrobial peptide RegIIIγ.

cutaneous lymphocyte antigen (CLA)
A cell-surface molecule that is involved in lymphocyte homing to the skin in humans.

CVID
See **common variable immunodeficiency**.

CXC chemokines
One of the two main classes of chemokines, distinguished by a Cys-X-Cys (CXC) motif near the amino terminus. They have names CXCL1, CXCL2, etc. See Appendix IV for a list of individual chemokines.

cyclic neutropenia
A dominantly inherited disease in which neutrophil numbers fluctuate from near normal to very low or none, with an approximate cycle time of 21 days.

cyclic reentry model
An explanation of the behavior of B cells in lymphoid follicles, proposing that activated B cells in germinal centers lose and gain expression of the chemokine receptor CXCR4 and thus move from the light zone to the dark zone and back again under the influence of the chemokine CXCL12.

cyclophosphamide
A DNA alkylating agent that is used as an immunosuppressive drug. It acts by killing rapidly dividing cells, including lymphocytes proliferating in response to antigen.

cyclosporin A
A powerful noncytotoxic immunosuppressive drug that inhibits signaling from the T-cell receptor, preventing T-cell activation and effector function. It binds to cyclophilin, and the complex formed binds to and inactivates the phosphatase calcineurin.

cytokine
In the most general sense, any small protein made by a cell that affects the behavior of other cells. Cytokines made by lymphocytes are often called interleukins (abbreviated IL). Cytokines act via specific cytokine receptors on the cells that they affect. Cytokines and their receptors are listed in Appendix III. Cf. **chemokine**.

cytokine capture
An assay for antigens and other molecules such as cytokines, in which the antigen binds to a specific antibody via one epitope, and its presence is detected by a second, labeled, antibody that is directed at a different epitope.

cytokine receptors
Cell-surface receptors for cytokines. Binding of the cytokine to the cytokine receptor induces new activities in the cell, such as growth, differentiation, or death. Cytokines and their receptors are listed in Appendix III.

cytotoxic granules
Membrane-enclosed granules containing the cytotoxic proteins perforin, granzymes, and granulysin, which are a defining characteristic of effector CD8 cytotoxic T cells and NK cells.

cytotoxic lymphocytes
See **cytotoxic T cells**, **NK cells**.

cytotoxic T cell
Type of T cell that can kill other cells. Most cytotoxic T cells are MHC class I-restricted CD8 T cells, but CD4 T cells can also kill in some cases. Cytotoxic T cells are important in host defense against intracellular pathogens that live or reproduce in the cell's cytosol.

cytotoxins
Proteins made by cytotoxic T cells and NK cells that participate in the killing of target cells. Perforins, granzymes, and granulysins are the major defined cytotoxins.

δ
(1) The type of heavy chain in IgD class immunoglobulins. (2) One of the chains (the δ chain) of the antigen receptor of a subset of T cells called $\gamma{:}\delta$ T cells.

DAG
See **diacylglycerol**.

DAP10, DAP12
Signaling chains containing ITAMS that are associated with the tails of some activating receptors on NK cells.

dark zone
See **germinal centers**.

DC-SIGN
A lectin on the dendritic-cell surface that binds ICAM-3 with high affinity.

death domain
Protein-interaction domain originally discovered in proteins involved in programmed cell death or apoptosis. As part of the intracellular domains of some adaptor proteins, death domains are involved in transmitting pro-inflammatory and/or pro-apoptotic signals.

death receptors
Cell-surface receptors whose engagement by extracellular ligands stimulates apoptosis in the receptor-bearing cell.

decay-accelerating factor (DAF, CD55)
A cell-surface protein that protects cells from lysis by complement. Its absence causes the disease paroxysmal nocturnal hemoglobinuria.

Dectin-1
A phagocytic receptor on neutrophils and macrophages that recognizes β-1,3-linked glucans, which are common components of fungal cell walls.

defective ribosomal products (DRiPs)
Peptides translated from introns in improperly spliced mRNAs, translations of frameshifts, and improperly folded proteins, which are recognized and tagged by ubiquitin for rapid degradation by the proteasome.

defensins
See **α-defensins**, **β-defensins**.

delayed-type hypersensitivity reaction
A form of cell-mediated immunity elicited by antigen in the skin that is mediated by T cells. It is called delayed-type hypersensitivity because the reaction appears hours to days after antigen is injected. Also called type IV hypersensitivity.

dendritic cells
Bone marrow derived cells found in most tissues, including lymphoid tissues. There are two main functional subsets. Conventional dendritic cells take up antigen in peripheral tissues, are activated by contact with pathogens, and travel to the peripheral lymphoid organs, where they are the most potent stimulators of T-cell responses. Plasmacytoid dendritic cells can also take up and present antigen, but their main function in an infection is to produce large amounts of the antiviral interferons as a result of pathogen recognition through receptors such as TLRs. Both these types of dendritic cells are distinct from the follicular dendritic cell that presents antigen to B cells in lymphoid follicles.

dendritic epidermal T cells (dETCs)
A specialized class of $\gamma{:}\delta$ T cells found in the skin of mice and some other species, but not humans. All dETCs have the same $\gamma{:}\delta$ T-cell receptor; their function is unknown.

depleting antibodies
Immunosuppressive monoclonal antibodies that trigger the destruction of lymphocytes *in vivo*. They are used for treating episodes of acute graft rejection.

dephosphorylation
The removal of a phosphate group from a molecule, usually a protein.

desensitization immunotherapy
A procedure in which an allergic individual is exposed to increasing doses of allergen in an attempt to inhibit their allergic reactions. Success in such treatment probably involves skewing the response away from CD4 T_H2 cells.

D gene segments
Diversity gene segments. Short DNA sequences that form a join between the V and J gene segments in rearranged immunoglobulin heavy-chain genes and in T-cell receptor β- and δ-chain genes. See **gene segments**.

diabetes
See **type 1 diabetes mellitus**.

diacylglycerol (DAG)
A lipid intracellular signaling molecule formed from membrane inositol phospholipids by the action of phospholipase C-γ after the activation of many different receptors. The diacylglycerol stays in the membrane and activates protein kinase C, which further propagates the signal.

diacyl lipoproteins
Ligands for the Toll-like receptors TLR1:TLR2 and TLR2:TLR6.

diapedesis
The movement of blood cells, particularly leukocytes, from the blood across blood vessel walls into tissues.

DIC
See **disseminated intravascular coagulation**.

DiGeorge syndrome
Recessive genetic immunodeficiency disease in which there is a failure to develop thymic epithelium. Parathyroid glands are also absent and there are anomalies in the large blood vessels.

direct allorecognition
Host recognition of a grafted tissue that involves donor antigen-presenting cells leaving the graft, migrating via the lymph to regional lymph nodes, and activating host T cells bearing the corresponding T-cell receptors.

direct Coombs test
Test used to detect the potentially harmful anti-Rh blood group antibodies produced in pregnancy as a result of Rh incompatibility between a mother and her fetus. The direct Coombs test uses anti-immunoglobulin to agglutinate red blood cells as a way of detecting whether they are coated with antibody *in vivo* as a result of autoimmunity or maternal anti-fetal immune responses (see also **indirect Coombs test**).

discontinuous epitopes
See **conformational epitopes**.

dislocation
In reference to viral defense mechanisms, the degradation of newly synthesized MHC class I molecules by viral proteins.

disseminated intravascular coagulation (DIC)
Blood clotting occurring simultaneously in small vessels throughout the body in response to disseminated TNF-α, which leads to the massive consumption of clotting proteins, so that the patient's blood cannot clot appropriately. Seen in septic shock.

diversity gene segments
See **D gene segments**.

DN1, DN2, DN3, DN4
Substages in the development of double-positive T cells in the thymus. Rearrangement of the TCRβ-chain locus starts at DN2 and is completed by DN4.

DNA-dependent protein kinase
Protein kinase in the DNA repair pathway involved in the rearrangement of immunoglobulin and T-cell receptor genes.

DNA ligase IV
Enzyme that joins together the ends of double-stranded DNA broken during the gene rearrangements that generate functional genes for immunoglobulins or T-cell receptors.

DNA vaccination
Vaccination by introduction into skin and muscle of DNA encoding the desired antigen; the expressed protein can then elicit antibody and T-cell responses.

double-negative thymocytes
Immature T cells in the thymus that lack expression of the two co-receptors CD4 and CD8. In a normal thymus, these represent about 5% of thymocytes.

double-positive thymocytes
Immature T cells in the thymus that are characterized by expression of both the CD4 and the CD8 co-receptor proteins. They represent the majority (about 80%) of thymocytes.

Down syndrome cell adhesion molecule
See **Dscam**.

draining lymph node
A lymph node downstream of a site of infection that receives antigens and microbes from the site via the lymphatic system. Draining lymph nodes often enlarge enormously during an immune response and can be palpated; they were originally called swollen glands.

Dscam
A member of the immunoglobulin superfamily that in insects is thought to opsonize invading bacteria and aid their engulfment by phagocytes. It can be made in a multiplicity of different forms as a result of alternative splicing.

ε
The heavy chain in the IgE class of immunoglobulins.

EAE
See **experimental allergic encephalomyelitis**.

early lymphoid progenitor (ELP)
A bone marrow cell that can give rise both to the common lymphoid progenitor and to T-cell precursors that migrate from the bone marrow to the thymus.

early pro-B cell
See **pro-B cells**.

eczema
A skin condition characterized by patches of itchy, scaly, reddened skin. It can be due to an allergic reaction, but also has many other causes, not all of which are known. Eczema initiated by an IgE-mediated allergic reaction is called atopic eczema or atopic dermatitis.

edema
Swelling caused by the entry of fluid and cells from the blood into the tissues; it is one of the cardinal features of inflammation.

effector caspases
Intracellular proteases that are activated as a result of an apoptotic signal and initiate the cellular changes associated with apoptosis.

effector CD4 T cells
The subset of differentiated effector T cells carrying the CD4 co-receptor molecule, which includes the T_H1, T_H2, T_H17, and regulatory T cells.

effector lymphocytes
The cells that differentiate from naive lymphocytes after initial activation by antigen and can then mediate the removal of pathogens from the body without further differentiation. They are distinct from memory lymphocytes, which must undergo further differentiation to become effector lymphocytes.

effector mechanisms
Those processes by which pathogens are destroyed and cleared from the body. Innate and adaptive immune responses use most of the same effector mechanisms to eliminate pathogens.

effector memory cells
Memory lymphocytes that are thought to be specialized for rapidly maturation into an effector T cell after restimulation with antigen.

effector T cells
The T cells that perform the functions of an immune response, such as cell killing and cell activation, that clear the infectious agent from the body. There are several different subsets, each with a specific role in an immune response.

efferent lymphatic vessel
The route by which circulating lymphocytes leave a lymph node and some other types of peripheral lymphoid organs. Cf. **afferent lymphatic vessel**.

ELISA
See **enzyme-linked immunosorbent assay**.

ELISPOT assay
An adaptation of ELISA in which cells are placed over antibodies or antigens attached to a plastic surface. The antigen or antibody traps the cells' secreted products, which can then be detected by using an enzyme-coupled antibody that cleaves a colorless substrate to make a localized colored spot.

embryonic stem cells (ES cells)
Early embryonic cells that will grow continuously in culture and that retain the ability to contribute to all cell lineages. Mouse ES cells can be genetically manipulated in tissue culture and then inserted into mouse blastocysts to generate mutant lines of mice.

endocrine
Describes the action of a biologically active molecule such as a hormone or cytokine that is secreted by one tissue into the blood and acts on a distant tissue. Cf. **autocrine**, **paracrine**.

endogenous pyrogens
Cytokines that can induce a rise in body temperature.

endoplasmic reticulum aminopeptidase associated with antigen processing (ERAAP)
Enzyme in the endoplasmic reticulum that trims polypeptides to a size at which they can bind to MHC class I molecules.

endosomes
Membrane-enclosed intracellular vesicles in which material is carried from and to the cell exterior. Antigen taken up by phagocytosis generally enters the endosomal system.

endosteum
The region in bone marrow adjacent to the inner surface of the bone; hematopoietic stem cells are initially located there.

endothelial activation
The changes that occur in the endothelial walls of small

blood vessels as a result of inflammation, such as increased permeability and the increased production of cell-adhesion molecules and cytokines.

endothelial cell
Cell type that forms the endothelium, the epithelium of a blood vessel wall.

endothelium
The epithelium that forms the walls of blood capillaries and the lining of larger blood vessels.

endotoxic shock
See **toxic shock, toxic shock syndrome**.

endotoxin
A toxin that is a structural part of a bacterial cell and is released only when the cell is damaged. The most important endotoxin medically is the lipopolysaccharide (LPS) of the outer cell membrane of Gram-negative bacteria, which is a potent inducer of cytokine synthesis. When present in large amounts in the blood it can cause a systemic shock reaction called septic shock or endotoxic shock.

enzyme-linked immunosorbent assay (ELISA)
A serological assay in which bound antigen or antibody is detected by a linked enzyme that converts a colorless substrate into a colored product. The ELISA assay is widely used in biology and medicine as well as in immunology.

eosinophil
A type of white blood cell containing granules that stain with eosin. It is thought to be important chiefly in defense against parasitic infections, but is also medically important as an effector cell in allergic reactions.

eosinophilia
An abnormally large number of eosinophils in the blood.

eotaxin-1 (CCL11), eotaxin-2 (CCL24), eotaxin-3 (CCL26)
CC chemokines that act predominantly on eosinophils.

epitope
A site on an antigen recognized by an antibody or an antigen receptor. A T-cell epitope is a short peptide derived from a protein antigen. It binds to an MHC molecule and is recognized by a particular T cell. B-cell epitopes are antigenic determinants recognized by B cells and are typically structural motifs on the surface of the antigen. Also called an antigenic determinant.

epitope spreading
Increase in diversity of responses to autoantigens as the response persists, as a result of responses being made to epitopes other than the original one.

equilibrium dialysis
Technique for determining the affinity of an antibody for its antigen.

ERAAP
See **endoplasmic reticulum aminopeptidase associated with antigen processing**.

Erk
Extracellular signal-related kinase, a protein kinase that is the MAPK in some intracellular signaling pathways leading from the T-cell receptor (and in other receptors in other cell types).

Erp57
A chaperone protein involved in loading peptide onto MHC class I molecules in the endoplasmic reticulum.

erythroblastosis fetalis
See **hemolytic disease of the newborn**.

escape mutants
Mutants of pathogens that are changed in such a way that they can evade the immune response against the original pathogen.

E-selectin
See **selectins**.

exogenous pyrogen
Any substance originating outside the body that can induce fever, such as the bacterial lipopolysaccharide LPS. Cf. **endogenous pyrogen**.

exotoxin
A protein toxin produced and secreted by a bacterium.

experimental allergic encephalomyelitis (EAE)
An inflammatory disease of the central nervous system that develops after mice are immunized with neural antigens in a strong adjuvant.

extravasation
The movement of cells or fluid from within blood vessels into the surrounding tissues.

extrinsic pathway of apoptosis
A pathway triggered by extracellular ligands binding to specific cell-surface receptors (death receptors) that signal the cell to undergo programmed cell death (apoptosis).

Fab fragment
Protein fragment comprising a single antigen-binding arm of an antibody, without the Fc region. It is composed of the complete light chain and the amino-terminal variable region and one constant region of a heavy chain, held together by an interchain disulfide bond. Produced by cleavage of IgG by the enzyme papain.

$F(ab')_2$ fragment
Protein fragment comprising two linked antigen-binding arms (Fab fragments) of an antibody molecule, without the Fc fragment. Produced by cleavage of IgG with the enzyme pepsin.

FACS®
See **fluorescence-activated cell sorter**.

factor B
Protein in the alternative pathway of complement activation, in which it is cleaved to Ba and an active protease, Bb, the latter binding to C3b to form the alternative pathway C3 convertase, C3bBb.

factor D
A serine protease in the alternative pathway of complement activation, which cleaves factor B into Ba and Bb.

factor H
Complement-regulatory protein in plasma that binds C3b and competes with factor B to displace Bb from the convertase.

factor I
Complement-regulatory protease in plasma that cleaves C3b to the inactive derivative iC3b, thus preventing the formation of a C3 convertase.

factor I deficiency
A genetically determined lack of the complement-regulatory protein factor I. This results in uncontrolled complement activation, so that complement proteins rapidly become depleted. Those with the deficiency suffer repeated bacterial infections, especially with ubiquitous pyogenic bacteria.

factor P
See **properdin**.

familial cold autoinflammatory syndrome (FCAS)
An episodic autoinflammatory disease caused by mutations in the gene *NLRP3*, encoding NLRP3, a member of the NOD-like receptor family and a component of the inflammasome. The symptoms are induced by exposure to cold.

familial hemophagocytic lymphohistiocytosis (FHL)
A progressive and potentially lethal inflammatory disease caused by an inherited deficiency of perforin. Large numbers of polyclonal CD8-positive T cells accumulate in lymphoid and other organs, and this is associated with activated macrophages that phagocytose blood cells, including erythrocytes and leukocytes.

familial Mediterranean fever (FMF)
A severe autoinflammatory disease, inherited as an autosomal recessive disorder. It is caused by mutation in the gene encoding the protein pyrin, but how this results in the disease is not known.

farmer's lung
A hypersensitivity disease caused by the interaction of IgG antibodies with large amounts of an inhaled allergen in the alveolar wall of the lung, causing alveolar wall inflammation and compromising gas exchange.

Fas
A member of the TNF receptor family that makes cells on which it is expressed susceptible to killing by cells bearing Fas ligand, a cell-surface member of the TNF family of cytokines. Binding of Fas ligand to Fas triggers apoptosis in the Fas-bearing cell.

Fas ligand (FasL)
See **Fas**.

Fc fragment, Fc region
The carboxy-terminal halves of the two heavy chains of an IgG molecule disulfide-bonded to each other by the residual hinge region. It is produced by cleavage of IgG by papain. In the complete antibody this portion is often called the Fc region.

FCAS
See **familial cold autoinflammatory syndrome**.

Fc receptor
Type of cell-surface receptor on macrophages and other cells in the immune system that binds the Fc portions of immunoglobulins. There are different Fc receptors for different isotypes: Fcγ receptors bind IgG, for example, and Fcε receptors bind IgE.

FcεRI
The receptor on the surface of mast cells and basophils that binds the Fc region of free IgE at high affinity. When antigen binds this IgE and cross-links FcεRI, it causes mast-cell activation.

FcγRI, FcγRII, FcγRIII
Cell-surface receptors that bind the Fc portion of IgG molecules. Most Fcγ receptors bind only aggregated IgG, allowing them to discriminate bound antibody from free IgG. Expressed variously on phagocytes, B lymphocytes, NK cells, and follicular dendritic cells, the Fcγ receptors have a key role in humoral immunity, linking antibody binding to effector cell functions.

FcγRIIB-1
An inhibitory receptor on B cells that recognizes the Fc portion of IgG antibodies.

FcRn
Neonatal Fc receptor, a receptor that transports IgG from mother to fetus across the placenta, and across other epithelia such as the epithelium of the gut.

FDC
See **follicular dendritic cell**.

FHL
See **familial hemophagocytic lymphohistiocytosis**.

fibrinogen-related proteins (FREPs)
Members of the immunoglobulin superfamily that are thought to have a role in innate immunity in the freshwater snail *Biomphalaria glabrata*.

ficolins
Carbohydrate-binding proteins that can initiate the lectin pathway of complement activation. They are members of the collectin family and bind to the *N*-acetylglucosamine present on the surface of some pathogens.

filaggrin
Skin protein, defects in which have been linked to an increased risk of developing atopic eczema.

fingolimod
Small-molecule immunosuppressive drug that interferes with the actions of sphingosine, leading to retention of effector T cells in lymphoid organs.

first-set rejection
The rejection of a tissue or organ graft placed in an unmatched recipient, due to an immune response by the host against foreign antigens in the graft. Cf. **second-set rejection**.

FK506
See **tacrolimus**.

fluid-phase C3 convertase
Short-lived alternative pathway C3 convertase that is continually being produced at a low level in the plasma, and which can initiate the alternative pathway of complement activation after infection.

fluorescence-activated cell sorter (FACS®)
Machine for separating and characterizing individual cells on the basis of fluorescent antibodies specific for particular surface proteins, and sorting them into different populations.

fMet-Leu-Phe receptor (fMLP receptor)
A pattern recognition receptor for the peptide fMet-Leu-Phe, which is specific to bacteria, on neutrophils and macrophages. fMet-Leu-Phe acts as a chemoattractant.

FMF
See **familial Mediterranean fever**.

fMLP receptor
See **fMet-Leu-Phe receptor**.

follicle
An area of predominantly B cells in a peripheral lymphoid organ, such as a lymph node, which also contains follicular dendritic cells.

follicle-associated epithelium
Specialized epithelium separating the lymphoid tissues of the gut wall from the intestinal lumen. As well as enterocytes it contains microfold cells, through which antigens enter the lymphoid organs from the gut.

follicular dendritic cell (FDC)
A cell type of uncertain origin found in the follicles of peripheral lymphoid organs. It is characterized by long branching processes that make intimate contact with many different B cells, and has Fc receptors that are not internalized by receptor-mediated endocytosis and thus hold antigen:antibody complexes on the surface for long periods. These cells are crucial in selecting antigen-binding B cells during antibody responses.

follicular helper T cell (T_{FH} cell)
Type of effector CD4 T cell that resides in lymphoid follicles and provides help to B cells for antibody production.

FoxP3
Transcription factor whose expression is characteristic of several types of regulatory T cells.

framework regions
Relatively invariant regions that provide a protein scaffold for the hypervariable regions in the V domains of immunoglobulins and T-cell receptors.

fungi (singular: fungus)
A kingdom of single-celled and multicellular eukaryotic organisms, including the yeasts and molds, that can cause a variety of diseases. Immunity to fungi is complex and involves both humoral and cell-mediated responses.

fusion inhibitors
Drugs that inhibit the fusion of HIV with its host cells.

Fv
See **single-chain Fv**.

Fyn
A Src-family tyrosine kinase that relays signals from the B-cell receptor.

γ
(1) The heavy chain of the IgG class of immunoglobulins. (2) One of the chains (the γ chain) of the antigen receptor of a subset of T cells called γ:δ T cells.

γ:δ T cells
Subset of T lymphocytes bearing a T-cell receptor composed of the antigen-recognition chains, γ and δ, assembled in a γ:δ heterodimer.

γ:δ T-cell receptor
Antigen receptor carried by a subset of T lymphocytes that is distinct from the α:β T-cell receptor. It is composed of a γ and a δ chain, which are produced from genes that undergo gene rearrangement.

GALT
See **gut-associated lymphoid tissues**.

γ globulins, gamma globulins
That fraction of plasma proteins (as separated by electrophoretic

mobility) that contains most antibodies. Patients who lack antibodies are said to have agammaglobulinemia.

GAP
See **GTPase-activating protein**.

GEFs
See **guanine nucleotide exchange factors**.

gene conversion
The process by which most immunoglobulin receptor diversity is generated in birds and rabbits, in which homologous inactive V gene segments exchange short sequences with an active, rearranged variable-region sequence.

gene knockout
The disabling of a specific gene in the genome, which can be achieved by various techniques. Mice carrying gene knockouts are of great value in immunological research.

gene rearrangement
The process of somatic recombination of gene segments in the immunoglobulin and T-cell receptor genetic loci to produce a functional gene. This process generates the diversity found in immunoglobulin and T-cell receptor variable regions.

gene segments
Sets of short DNA sequences at the immunoglobulin and T-cell receptor loci that encode different regions of the variable domains of antigen receptors. Gene segments of each type are joined together by somatic recombination to form a complete variable-domain exon. There are three types of gene segments: V gene segments encode the first 95 amino acids, D gene segments (in heavy-chain and TCRα chain loci only) encode about 5 amino acids, and J gene segments encode the last 10–15 amino acids of the variable domain. There are multiple copies of each type of gene segment in the germline DNA, but only one of each type is joined together to form the variable domain.

genetic locus (plural: genetic loci)
The site of a gene on a chromosome. In the case of the genes for the immunoglobulin and T-cell receptor chains, the term locus refers to the complete collection of gene segments and C-region genes for the given chain.

gene targeting
See **gene knockout**.

genome-wide association studies (GWASs)
Genetic association studies in the general population that look for a correlation between disease frequency and variant alleles by scanning the genomes of many people for the presence of informative single-nucleotide polymorphisms (SNPs).

germ-free mice
Mice that are raised in the complete absence of intestinal and other microorganisms. Such mice have very depleted immune systems, but they can respond virtually normally to any specific antigen, provided it is mixed with a strong adjuvant.

germinal centers
Sites of intense B-cell proliferation and differentiation that develop in lymphoid follicles during an adaptive immune response. Somatic hypermutation and class switching occur in germinal centers.

germline configuration
The immunoglobulin and T-cell receptor loci as they exist in the DNA of germ cells and of all somatic cells in which somatic recombination has not occurred.

germline diversity
The diversity among antigen receptors that is due to the inheritance of sets of multiple gene segments that encode the variable domains. This type of diversity is distinct from the diversity that is generated during gene rearrangement or after the expression of an antigen-receptor gene, which is somatically generated.

germline theory
A hypothesis about antibody diversity that proposed that each antibody was encoded by a separate germline gene. This is now known not to be true for humans, mice, and most other vertebrates, but cartilaginous fishes do have some rearranged V regions in the germline.

glandular fever
See **infectious mononucleosis**.

GlyCAM-1
A mucinlike molecule present on the high endothelial venules of peripheral lymphoid tissues. It is a ligand for the cell-adhesion protein L-selectin on naive lymphocytes, directing these cells to leave the blood and enter the lymphoid tissues.

gnotobiotic mice
See **germ-free mice**.

Goodpasture's syndrome
An autoimmune disease in which autoantibodies against type IV collagen (found in basement membranes) are produced, causing extensive inflammation in kidneys and lungs.

gout
Disease caused by monosodium urate crystals deposited in the cartilaginous tissues of joints, causing inflammation. Urate crystals activate the NLRP3 inflammasome, which induces inflammatory cytokines.

GPCR
See **G-protein-coupled receptor**.

G proteins
Intracellular GTPases that act as molecular switches in signaling pathways. They bind GTP and hydrolyze it to GDP, which causes a conformational change in the protein and activation of its function. There are two kinds of G proteins: the heterotrimeric (α, β, γ subunits) receptor-associated G proteins, and the small G proteins, such as Ras and Raf, which act downstream of many transmembrane signaling events.

G-protein-coupled receptor (GPCR)
Any of a large class of cell-surface receptors that associate with intracellular heterotrimeric G proteins after ligand binding, and generate an intracellular signal by activation of the G protein. They are all seven-span transmembrane proteins. Immunologically important examples are the chemokine receptors.

graft rejection
See **allograft rejection**.

graft-versus-host disease (GVHD)
An attack on the tissues of the recipient by mature T cells in a bone marrow graft from a nonidentical donor, which can cause a variety of symptoms; sometimes these are severe.

graft-versus-leukemia effect
A beneficial side-effect of bone marrow grafts given to treat leukemia, in which mature T cells in the graft recognize minor histocompatibility antigens or tumor-specific antigens on the recipient's leukemic cells and attack them.

Gram-negative binding proteins
Proteins that act as the pathogen-recognition proteins in the Toll pathway of immune defense in *Drosophila*.

granulocytes
White blood cells with multilobed nuclei and cytoplasmic granules. They comprise the neutrophils, eosinophils, and basophils. Also known as polymorphonuclear leukocytes.

granulocyte–macrophage colony-stimulating factor (GM-CSF)
A cytokine involved in the growth and differentiation of cells of the myeloid lineage, including dendritic cells, monocytes and tissue macrophages, and granulocytes.

granuloma
A site of chronic inflammation usually triggered by persistent infectious agents such as mycobacteria or by a nondegradable foreign body. Granulomas have a central area of macrophages, often fused into multinucleate giant cells, surrounded by T lymphocytes.

granulysin
A cytotoxic protein present in the cytotoxic granules of cytotoxic CD8 T cells and NK cells.

granzymes
Serine proteases that are present in cytotoxic CD8 T cells and NK cells and are involved in inducing apoptosis in the target cell.

Graves' disease
An autoimmune disease in which antibodies against the thyroid-stimulating hormone receptor cause overproduction of thyroid hormone and thus hyperthyroidism.

Griscelli syndrome
An inherited immunodeficiency disease that affects the pathway for secretion of lysosomes. It is caused by mutations in a small GTPase Rab27a, which controls the movement of vesicles within cells.

GTPase-activating proteins (GAPs)
Regulatory proteins that accelerate the intrinsic GTPase activity of G proteins and thus facilitate the conversion of G proteins from the active (GTP-bound) state to the inactive (GDP-bound) state.

guanine nucleotide exchange factors (GEFs)
Proteins that can remove the bound GDP from G proteins, thus allowing GTP to bind and activate the G protein.

gut-associated lymphoid tissues (GALT)
Organized lymphoid tissues closely associated with the gastrointestinal tract, comprising Peyer's patches, the appendix, and isolated lymphoid follicles found in the intestinal wall. They have an anatomically compartmentalized structure typical of peripheral lymphoid organs and are sites at which adaptive immune responses are initiated. The tissues are connected to the mesenteric lymph nodes by lymphatic vessels.

GVHD
See **graft-versus-host disease**.

H-2
The major histocompatibility complex of the mouse. Haplotypes are designated by a lower-case superscript, as in H-2^b.

HAART
See **highly active antiretroviral therapy**.

H antigens
See **histocompatibility**.

haploinsufficient
Describes the situation in which the presence of only one normal allele of a gene is not sufficient for normal function.

haplotype
A linked set of genes on a chromosome that are typically inherited without undergoing recombination. The term is used mainly in connection with the linked genes of the major histocompatibility complex (MHC), which is usually inherited as one haplotype from each parent.

hapten
Any small molecule that can be recognized by a specific antibody but cannot by itself elicit an immune response. A hapten must be chemically linked to a protein molecule to elicit antibody and T-cell responses.

Hashimoto's thyroiditis
An autoimmune disease characterized by persistent high levels of antibody against thyroid-specific antigens. These antibodies recruit NK cells to the thyroid, leading to damage and inflammation.

heavy chain (H chain)
One of the two types of protein chain in an immunoglobulin molecule, the other being called the light chain. There are several different classes, or isotypes, of heavy chain (α, δ, ε, γ, and μ), each of which confers a distinctive functional activity on the antibody molecule. Each immunoglobulin molecule contains two identical heavy chains.

helper CD4 T cells, helper T cells
Effector CD4 T cells that stimulate or 'help' B cells to make antibody in response to antigenic challenge. T_H2, T_H1, and the T_{FH} subsets of effector CD4 T cells can perform this function.

hemagglutination
The clumping together (agglutination) of red blood cells, which can be caused by antibodies against antigens on the surfaces of the red cells and by some other substances.

hemagglutinin
Any substance that can cause hemagglutination. The hemagglutinins in human blood are antibodies that recognize the ABO blood group antigens on red blood cells.

hematopoiesis
The generation of blood cells, which in humans occurs in the bone marrow.

hematopoietic stem cell
Type of pluripotent cell in the bone marrow that can give rise to all the different blood cell types.

hematopoietin superfamily
Large family of structurally related cytokines that includes growth factors and many interleukins with roles in both adaptive and innate immunity.

hemolytic disease of the newborn
A severe form of Rh hemolytic disease in which maternal anti-Rh antibody enters the fetus and produces a hemolytic anemia so severe that the fetus has mainly immature erythroblasts in the peripheral blood.

hemophagocytic syndrome
A dysregulated expansion of CD8-positive lymphocytes that is associated with macrophage activation. The activated macrophages phagocytose blood cells, including erythrocytes and leukocytes.

hepatobiliary route
One of the routes by which mucosally produced dimeric IgA reaches the intestine. The antibodies are taken up into the portal veins in the lamina propria, transported to the liver, and from there reach the bile duct by transcytosis. This pathway is not of great significance in humans.

heptamer
The conserved seven-nucleotide DNA sequence in the recombination signal sequences (RSSs) flanking gene segments in the immunoglobulin and T-cell receptor loci.

hereditary angioedema (HAE)
The clinical name for a genetic deficiency of the C1 inhibitor of the complement system. In the absence of C1 inhibitor, spontaneous activation of the complement system can cause diffuse fluid leakage from blood vessels, the most serious consequence of which is swelling of the larynx, leading to suffocation.

herpes virus entry molecule (HVEM)
See **B and T lymphocyte attenuator**.

heterotrimeric G proteins
See **G proteins**.

heterozygous
Describes individuals that have two different alleles of a given gene, one inherited from the mother and one from the father.

high endothelial cells, high endothelial venules (HEVs)
Specialized small venous blood vessels in lymphoid tissues. Lymphocytes migrate from the blood into lymphoid tissues by attaching to the high endothelial cells in the walls of the venules and squeezing between them.

highly active antiretroviral therapy (HAART)
A combination of drugs that is used to control HIV infection. It comprises nucleoside analogues that prevent reverse transcription, and drugs that inhibit the viral protease.

HIH/PAP (hepatocarcinoma-intestine-pancreas/pancreatitis-associated protein)
An antimicrobial C-type lectin secreted by intestinal cells in humans. Also known as RegIIIγ in mice.

hinge region
The flexible domain that joins the Fab arms to the Fc piece in an immunoglobulin. The flexibility of the hinge region in IgG and IgA molecules allows the Fab arms to adopt a wide range of angles, permitting binding to epitopes spaced variable distances apart.

histamine
A vasoactive amine stored in mast-cell granules. Histamine released by antigen binding to IgE antibodies bound to mast cells causes the dilation of local blood vessels and the contraction of smooth muscle, producing some of the symptoms of IgE-mediated allergic reactions. Antihistamines are drugs that counter histamine action.

histatins
Antimicrobial peptides constitutively produced by the parotid, sublingual, and submandibular glands in the oral cavity. Active against pathogenic fungi such as *Cryptococcus neoformans* and *Candida albicans*.

histocompatibility
The ability of tissues from one individual to be accepted (histocompatible tissues) or rejected (histoincompatible tissues) if transplanted to another individual.

Histocompatibility-2
See **H-2**.

histocompatibility antigens (H antigens)
Any tissue antigens that provoke an immune response in a genetically nonidentical member of the same species. The major histocompatibility antigens, also known as MHC molecules, are encoded in the MHC, and are also the molecules that present foreign peptides to T cells. Minor histocompatibility antigens comprise other polymorphic proteins that can be recognized as foreign by an unrelated individual. See also **MHC**, **MHC molecules**.

HIV
See **human immunodeficiency virus**.

HLA
The acronym for human leukocyte antigen. It is the genetic designation for the human MHC. Individual loci are designated by upper-case letters, as in *HLA-A*, and alleles are designated by numbers, as in *HLA-A*0201*.

HLA-DM
An invariant MHC protein in humans that is involved in loading peptides onto MHC class II molecules. It is encoded in the MHC by a set of genes resembling MHC class II genes. A homologous protein in mice is called H-2M.

HLA-DO
An atypical MHC class II molecule that acts as a negative regulator of HLA-DM, binding to it and inhibiting the release of CLIP from MHC class II molecules in intracellular vesicles.

homeostasis
The state of physiological normality. In the case of the immune system, homeostasis refers to its state (for example the numbers and proportions of lymphocytes) in an uninfected individual.

homing
The direction of a lymphocyte into a particular tissue.

homing receptors
Receptors on lymphocytes for chemokines, cytokines, and adhesion molecules specific to particular tissues, and which enable the lymphocyte to enter that tissue.

homologous recombination
The disruption of a gene's function by the gene-specific insertion of nonfunctional DNA by means of sequence-specific recombination.

host-versus-graft disease (HVGD)
Another name for the allograft rejection reaction. The term is used mainly in relation to bone marrow transplantation.

human immunodeficiency virus (HIV)
The causative agent of the acquired immune deficiency syndrome (AIDS). HIV is a retrovirus of the lentivirus family that selectively infects macrophages and CD4 T cells, leading to their slow depletion, which eventually results in immunodeficiency. There are two major strains of the virus, HIV-1 and HIV-2, of which HIV-1 causes most disease worldwide. HIV-2 is endemic to West Africa but is spreading.

humanization
The genetic engineering of mouse hypervariable loops of a desired specificity into otherwise human antibodies for use as therapeutic agents. Such antibodies are less likely to cause an immune response in people treated with them than are wholly mouse antibodies.

human leukocyte antigen
See **HLA**.

humoral immunity, humoral immune response
Immunity due to proteins circulating in the blood, such as antibodies (in adaptive immunity) or complement (in innate immunity). Adaptive humoral immunity can be transferred to unimmunized recipients by the transfer of serum containing specific antibody.

hybridomas
Hybrid cell lines formed by fusing a specific antibody-producing B lymphocyte with a myeloma cell that is selected for its ability to grow in tissue culture and for an absence of immunoglobulin chain synthesis. The antibodies produced are all of a single specificity and are called monoclonal antibodies.

hygiene hypothesis
A hypothesis first proposed in 1989 that a change in exposure to ubiquitous microorganisms was a possible cause of the increase in allergy. See also **counter-regulation hypothesis**.

hyperacute graft rejection
Immediate rejection reaction caused by preformed natural antibodies that react against antigens on the transplanted organ. The antibodies bind to endothelium and trigger the blood-clotting cascade, leading to an engorged, ischemic graft and rapid death of the organ.

hypereosinophilic syndrome
Disease caused by an overproduction of eosinophils.

hyper IgD syndrome (HIDS)
An autoinflammatory disease due to mutations that lead to a partial deficiency of mevalonate kinase.

hyper IgE syndrome (HIES)
Also called Job's syndrome. A disease characterized by recurrent skin and pulmonary infections and high serum concentrations of IgE.

hyper IgM syndromes
A group of genetic diseases in which there is overproduction of IgM antibody, among other symptoms. They are due to defects in various genes for proteins involved in class switching such as CD40 ligand and the enzyme AID. See **activation-induced cytidine deaminase deficiency**, **CD40 ligand deficiency**.

hyperimmunization
Repeated immunization to achieve a heightened state of immunity.

hyperreactivity, hyperresponsiveness
The general overreactivity of the airways to nonimmunological stimuli, such as cold or smoke, that develops in chronic asthma.

hypersensitivity
An abnormal or exaggerated reaction to the ingestion, inhalation, or contact with a substance that does not provoke such a response in most people. Not all types of hypersensitivity have an immunological basis. Immunologically based hypersensitivity to a normally innocuous antigen is known as allergy.

hypervariable regions (HV regions)
See **complementarity-determining regions**.

hypomorphic
Applied to mutations that result in reduced gene function.

iC3b
Inactive complement fragment produced by cleavage of C3b.

ICAMs
ICAM-1, **ICAM-2**, **ICAM-3**. Cell-adhesion molecules of the immunoglobulin superfamily that bind to the leukocyte integrin CD11a:CD18 (LFA-1). They are crucial in the binding of lymphocytes and other leukocytes to antigen-presenting cells and endothelial cells.

iccosomes
Small fragments of membrane coated with immune complexes that fragment off the processes of follicular dendritic cells in lymphoid follicles early in a secondary or subsequent antibody response.

ICOS
A CD28-related co-stimulatory receptor that is induced on activated T cells and can enhance T-cell responses. It binds a

co-stimulatory ligand known as ICOSL (ICOS ligand), which is distinct from the B7 molecules.

ICOSL
See **ICOS**.

i.d.
See **intradermal**.

idiotype
The set of epitopes unique to each immunoglobulin molecule.

IEL
See **intraepithelial lymphocyte**.

IFN-α, IFN-β, IFN-γ
See **interferon-α, interferon-β, interferon-γ**.

Ig
The standard abbreviation for immunoglobulin.

Igα, Igβ
See **B-cell receptor**.

IgA
The class of immunoglobulin characterized by α heavy chains. It can occur in a monomeric and a polymeric (mainly dimeric) form. Polymeric IgA is the main antibody secreted by mucosal lymphoid tissues.

IgA deficiency
An absence of antibodies of the IgA class. It is the most common inherited form of immunoglobulin deficiency in populations of European origin. In most people with this deficiency there is no obvious disease susceptibility.

IgD
The class of immunoglobulin characterized by δ heavy chains. It appears as surface immunoglobulin on mature I B cells but its function is unknown.

IgE
The class of immunoglobulin characterized by ε heavy chains. It is involved in the defense against parasite infections and in allergic reactions.

IgG
The class of immunoglobulin characterized by γ heavy chains. It is the most abundant class of immunoglobulin found in the plasma.

IgM
The class of immunoglobulin characterized by μ heavy chains. It is the first immunoglobulin to appear on the surface of B cells and the first to be secreted.

Ii
See **invariant chain**.

IκB kinase (IKK)
Protein complex in the NFκB signaling pathway triggered by TLRs and other receptors. It phosphorylates IκB, which then releases the transcription factor NFκB, enabling it to enter the nucleus.

IL
See **interleukin**.

IL-1β
Interleukin-1β, a cytokine produced by active macrophages that has many effects in the immune response, including the activation of vascular endothelium, activation of lymphocytes, and the induction of fever.

IL-2
Interleukin-2, a cytokine produced by activated naive T cells and essential for their further proliferation and differentiation. It is one of the key cytokines in the development of an adaptive immune response.

IL-4
Interleukin-4, a cytokine characteristic of CD4 T_H2 effector T cells.

IL-5
Interleukin-5, a cytokine characteristic of CD4 T_H2 effector T cells, which, among other functions, promotes the growth and differentiation of eosinophils.

IL-6
Interleukin-6, a cytokine produced by activated macrophages and which has many effects, including lymphocyte activation, the stimulation of antibody production, and the induction of fever.

IL-10
Interleukin-10, a cytokine characteristically produced by regulatory T cells and which tends to suppress lymphocyte responses.

IL-12
Interleukin-12, a cytokine produced by activated macrophages and which, among other effects, activates NK cells and induces the differentiation of CD4 T cells into T_H1 effector T cells.

IL-17
Interleukin-17, a cytokine characteristically produced by T_H17 effector T cells and which promotes inflammation.

ILLs
See **innate-like lymphocytes**.

i.m.
See **intramuscular**.

Imd pathway
A defense against Gram-negative bacteria in insects that results in the production of antimicrobial peptides such as diptericin, attacin, and cecropin.

immature B cells
B cells that have rearranged a heavy- and a light-chain V-region gene and express surface IgM, but have not yet matured sufficiently to express surface IgD as well.

immature dendritic cells
Phagocytic cells present in tissues throughout the body, which pick up antigen at a site of infection or inflammation, leave the tissues, and mature into antigen-presenting dendritic cells. See also **dendritic cells**.

immediate reaction
In an experimentally induced allergic response, the reaction that occurs within seconds of encounter with antigen. Cf. **late-phase reaction**.

immune complex
Complex formed by the binding of antibody to its antigen. Complement (C3b) is also often bound in immune complexes. Large immune complexes form when sufficient antibody is available to cross-link the antigen; these are cleared by the reticuloendothelial system of cells bearing Fc receptors and complement receptors. Small, soluble immune complexes form when antigen is in excess; these can be deposited in small blood vessels and damage them.

immune dysregulation, polyendocrinopathy, enteropathy, X-linked syndrome
See **IPEX**.

immune effector functions
All those components and functions of the immune system that restrict an infection and eliminate it, for example, complement, macrophages, neutrophils and other leukocytes, antibodies, and effector T cells.

immune modulation
The deliberate attempt to change the course of an immune response, for example by altering the bias toward T_H1 or T_H2 dominance.

immune regulation
The capacity of the immune system in normal circumstances to regulate itself so that an immune response does not go out of control and cause tissue damage, autoimmune reactions, or allergic reactions.

immune response
Any response made by an organism to defend itself against a pathogen.

immune response (Ir) gene
A term used in the past to describe a genetic polymorphism that controls the intensity of the immune response to a particular antigen. Virtually all Ir phenotypes are now known to be due to

differences between alleles of the genes for MHC molecules, especially MHC class II molecules, leading to differences in the ability of MHC molecules to bind particular peptide antigens.

immune surveillance
The recognition, and in some cases the elimination, of tumor cells by the immune system before they become clinically detectable.

immune system
The tissues, cells, and molecules involved in innate immunity and adaptive immunity.

immunity
The ability to resist infection with a particular pathogen. See also **protective immunity**.

immunization
The deliberate induction of an adaptive immune response by introducing antigen into the body. See also **active immunization**, **passive immunization**.

immunobiology
The study of the biological basis for the body's defenses against infection.

immunoblotting
A common technique in which proteins separated by gel electrophoresis are blotted onto a nitrocellulose membrane and revealed by the binding of specific labeled antibodies.

immunodeficiency disease
Any inherited or acquired disorder in which some aspect or aspects of host defense are absent or functionally defective.

immunodominant
Describes epitopes in an antigen that are preferentially recognized by T cells, such that T cells specific for those epitopes come to dominate the immune response.

immunoediting
A phase of immune surveillance thought to occur if tumor cells are not completely eliminated as a result of their initial recognition by the immune system. During this phase, further mutation of the tumor cell occurs and cells that escape elimination by the immune system are selected for survival.

immunoelectron microscopy
Electron microscopic technique for revealing ultramicroscopic structures in cells by using antibodies linked to small gold particles, which show up in the electron microscope.

immunoevasins
Viral proteins that prevent the appearance of peptide:MHC class I complexes on the infected cell, thus preventing the recognition of virus-infected cells by cytotoxic T cells.

immunofluorescence microscopy
Technique for detecting molecules by using antibodies labeled with fluorescent dyes.

immunogen (*adj.* immunogenic)
Any molecule that, on its own, is able to elicit an adaptive immune response on injection into a person or animal.

immunoglobulin (Ig)
The protein family to which antibodies and B-cell receptors belong.

immunoglobulin A
See **IgA**.

immunoglobulin D
See **IgD**.

immunoglobulin domain
Structural protein domain present in many different proteins that was first described in antibody molecules. The heavy and light chains in immunoglobulins are each composed of multiple immunoglobulin domains.

immunoglobulin E
See **IgE**.

immunoglobulin fold
The tertiary structure of an immunoglobulin domain, comprising a sandwich of two β sheets held together by a disulfide bond.

immunoglobulin G
See **IgG**.

immunoglobulin-like domain (Ig-like domain)
Protein domain structurally related to the immunoglobulin domain.

immunoglobulin-like proteins
Proteins that contain one or more immunoglobulin-like domains, which are protein domains structurally similar to those of immunoglobulins.

immunoglobulin M
See **IgM**.

immunoglobulin repertoire
The variety of antigen-specific immunoglobulins (antibodies and B-cell receptors) present in an individual. Also known as the antibody repertoire.

immunoglobulin superfamily (Ig superfamily)
Large protein family comprising many proteins involved in antigen recognition and cell–cell interaction in the immune system and other biological systems. All members of the superfamily have at least one immunoglobulin domain or immunoglobulin-like domain.

immunohistochemistry
The detection of antigens in tissues by means of visible products produced by the degradation of a colorless substrate by antibody-linked enzymes.

immunological ignorance
A form of self tolerance in which reactive lymphocytes and their target antigen are both detectable within an individual, yet no autoimmune attack occurs.

immunologically privileged sites
Certain sites in the body, such as the brain, that do not mount an immune response against tissue allografts. Immunological privilege can be due both to physical barriers to cell and antigen migration and to the presence of immunosuppressive cytokines.

immunological memory
The ability of the immune system to respond more rapidly and more effectively on a second encounter with an antigen. Immunological memory is specific for a particular antigen and is long lived.

immunological recognition
General term for the ability of the cells of the innate and adaptive immune system to recognize the presence of an infection.

immunological synapse
The highly organized interface that develops between a T cell and the target cell it is in contact with, formed by T-cell receptors binding to antigen and cell-adhesion molecules binding to their counterparts on the two cells. Also known as the supramolecular adhesion complex.

immunological tolerance
See **tolerance**.

immunology
The study of all aspects of host defense against infection and also of the adverse consequences of immune responses.

immunomodulatory therapy
Treatments that seek to modify an immune response in a beneficial way, for example to reduce or prevent an autoimmune or allergic response.

immunopathology
The damage caused to tissues as the result of an immune response.

immunophilins
Proteins in T cells that are bound by the immunosuppressive drugs cyclosporin A, tacrolimus, and rapamycin. The complexes thus formed interfere with intracellular signaling pathways and prevent the clonal expansion of lymphocytes that normally follows antigen activation.

immunoprecipitation
A technique for determining the presence of a particular protein in a cell, by using specific labeled antibodies to precipitate it from a cell extract.

immunoproteasome
A form of proteasome found in cells exposed to interferons. It contains three subunits that are different from the normal proteasome.

immunoreceptor tyrosine-based activation motifs (ITAMs)
Sequence motifs present in signaling chains of receptors, including those associated with the antigen receptors on lymphocytes, that are the site of tyrosine phosphorylation after activation of the receptor by ligand binding. The phosphorylated tyrosines then recruit other signaling proteins.

immunoreceptor tyrosine-based inhibitory motifs (ITIMs)
Sequence motifs present in the cytoplasmic tails of some inhibitory receptors. Phosphorylated tyrosines in these motifs recruit phosphatases to the signaling pathway, which remove phosphate groups added by tyrosine kinases.

immunoreceptor tyrosine-based switch motif (ITSM)
A sequence motif present in the cytoplasmic tails of some inhibitor receptors.

immunotoxins
Antibodies that are chemically coupled to toxic proteins usually derived from plants or microbes. The antibody targets the toxin moiety to the required cells. Immunotoxins are being tested as anticancer agents and as immunosuppressive drugs.

i.n.
See **intranasal**.

indirect allorecognition
Recognition of a grafted tissue that involves the uptake of allogeneic proteins by the recipient's antigen-presenting cells and their presentation to T cells by self MHC molecules.

indirect Coombs test
A variation of the direct Coombs test in which an unknown serum is tested for antibodies against normal red blood cells by first mixing the two and then washing out the serum from the red blood cells and reacting them with anti-immunoglobulin antibody. If antibody in the unknown serum binds to the red blood cells, the red blood cells will be agglutinated by the anti-immunoglobulin.

indirect immunofluorescence
See **immunofluorescence microscopy**.

indoleamine 2,3-dioxygenase
Enzyme that catabolizes tryptophan, an essential amino acid, producing the immunosuppressive metabolite kynurenine.

induced pluripotent stem cells (iPS cells)
Pluripotent stem cells that are derived from adult somatic cells by the introduction of a cocktail of transcription factors.

induced regulatory T cells (induced T_{reg} cells)
Regulatory CD4 T cells that differentiate from naive CD4 T cells in the periphery under the influence of particular environmental conditions. Several different subtypes have been distinguished. Cf. **natural regulatory T cells**.

induced responses
In the context of innate immune responses, those cellular responses that are induced—by, for example, TLR signaling—after infection. They are distinct from preexisting innate defenses such as anatomic barriers, antimicrobial enzymes, and complement, and are distinct from adaptive immunity in that they do not operate by clonal selection of rare antigen-specific lymphocytes.

inducible co-stimulatory protein
See **ICOS**.

infectious mononucleosis
The common form of infection with the Epstein–Barr virus. It consists of fever, malaise, and swollen lymph nodes. Also called glandular fever.

infectious tolerance
See **regulatory tolerance**.

inflammasome
A pro-inflammatory protein complex that is formed after stimulation of the intracellular NOD-like receptors. It contains the enzyme caspase, which is activated in the complex and can then process cytokine proproteins into active cytokines.

inflammation
General term for the local accumulation of fluid, plasma proteins, and white blood cells that is initiated by physical injury, infection, or a local immune response.

inflammatory bowel diseases
General name for a set of inflammatory conditions in the gut, such as Crohn's disease and colitis, that have an immunological component.

inflammatory cells
Cells such as macrophages, neutrophils and effector T_H1 lymphocytes that invade inflamed tissues and contribute to the inflammation.

inflammatory response
See **inflammation**.

inhibitory receptors
On NK cells, receptors whose stimulation results in suppression of the cell's cytotoxic activity.

initiator caspases
Proteases that promote apoptosis by cleaving and activating other caspases.

iNKT cells
See **invariant NK cells**.

innate immune response
That part of a response to an infection that is due to the presence of, and immediate activation of, the body's innate and relatively nonspecific defense mechanisms, in contrast to an adaptive immune response that develops later and involves antigen-specific lymphocytes.

innate immunity
The various innate resistance mechanisms that are encountered first by a pathogen, before adaptive immunity is induced, such as anatomical barriers, antimicrobial peptides, the complement system, and macrophages and neutrophils carrying nonspecific pathogen-recognition receptors. Innate immunity is present in all individuals at all times, does not increase with repeated exposure to a given pathogen, and discriminates between groups of similar pathogens, rather than responding to a particular pathogens. Cf. **adaptive immunity**.

innate-like lymphocytes (ILLs)
Various types of lymphocytes that are activated early in an infection but have immunoglobulins and T-cell receptors of very limited diversity.

innate-type allergy
The production or exacerbation of a hypersensitivity-type response to an antigen because of innate immune responses due to the activation of Toll-like receptors.

inositol 1,4,5-trisphosphate (IP_3)
A soluble second messenger produced by the cleavage of membrane inositol phospholipids by phospholipase C-γ. It acts on receptors in the endoplasmic reticulum membrane, resulting in the release of stored Ca^{2+} into the cytosol.

insulin-dependent diabetes mellitus (IDDM)
See **type 1 diabetes mellitus**.

integrase
The enzyme in the human immunodeficiency virus (HIV) and other retroviruses that mediates the integration of the DNA copy of the viral genome into the host-cell genome.

integrins
Heterodimeric cell-surface proteins involved in cell–cell and cell–matrix interactions. They are important in adhesive interactions between lymphocytes and antigen-presenting cells and in lymphocyte and leukocyte adherence to blood vessel walls and migration into tissues.

intercellular adhesion molecules
See **ICAMs**.

interdigitating dendritic cells
See **dendritic cells**.

interferon-α (IFN-α), interferon-β (IFN-β)
Antiviral cytokines produced by a wide variety of cells in response

to infection by a virus, and which also help healthy cells resist viral infection. They act through the same receptor, which signals through a Janus-family tyrosine kinase. Also known as the type I interferons.

interferon-γ (IFN-γ)
A cytokine of the interferon structural family produced by effector CD4 T_H1 cells, CD8 T cells, and NK cells. Its primary function is the activation of macrophages, and it acts through a different receptor from that of the type I interferons.

interferon-producing cells (IPCs)
A subset of dendritic cells, also called plasmacytoid dendritic cells, that are specialized to produce large amounts of interferon in response to viral infections.

interferon receptors
Class of receptors that bind the type I interferons, IFN-α and IFN-β.

interferon regulatory factors (IRF)
Transcription factors such as IRF3 and IRF7 that are activated as a result of signaling from some TLRs. IRFs promote expression of the genes for type I interferons.

interferons
Cytokines (interferon-α family, interferon-β family, and interferon-γ) that are induced in response to infection. IFN-α and IFN-β are antiviral in their effects; IFN-γ has other roles in the immune system.

interleukin (IL)
A generic name for cytokines produced by leukocytes. The more general term cytokine is used in this book, but the term interleukin is used in the naming of specific cytokines such as IL-2. Some key interleukins are listed in the glossary under their abbreviated names, for example IL-1β and IL-2. Cytokines are listed in Appendix III.

intracellular cytokine staining
Cytokines can be stained within their producing cells by permeabilizing the cell and reacting it with a labeled fluorescent anti-cytokine antibody.

intracellular signaling pathway
The set of proteins that interact with each other to carry a signal from an activated receptor to the part of the cell where the response to the signal will be made.

intradermal (i.d.)
Describes an injection that delivers antigen into the dermis of the skin.

intraepithelial lymphocytes (IELs)
Lymphocytes present in the epithelium of mucosal surfaces such as the gut. They are predominantly T cells, and in the gut are predominantly CD8 T cells.

intramuscular (i.m.)
Describes an injection that delivers antigen into muscle tissue.

intranasal (i.n.)
Describes the administration of antigen directly into the nose, usually in the form of an aerosol.

intravenous (i.v.)
Describes an injection that delivers antigen directly into a vein.

intrathymic dendritic cells
See **dendritic cells**.

intrinsic pathway of apoptosis
Signaling pathway that mediates apoptosis in response to noxious stimuli including UV irradiation, chemotherapeutic drugs, starvation, or lack of the growth factors required for survival. It is initiated by mitochondrial damage.

invariant chain (Ii)
A polypeptide that binds into the peptide-binding cleft of a newly synthesized MHC class II protein in the endoplasmic reticulum and shields the MHC molecule from binding peptides there. When the MHC molecule reaches an endosome, Ii is degraded, leaving the MHC class II molecule able to bind antigenic peptides.

invariant NKT cells (iNKT cells)
A type of innate-like lymphocyte that carries a T-cell receptor with an invariant α chain and a β chain of limited diversity that recognizes glycolipid antigens presented by CD1 MHC-like molecules. This cell type also carries the surface marker NK1.1, which is usually associated with NK cells.

IPCs
See **interferon-producing cells**.

IPEX
Immune dysregulation, polyendocrinopathy, enteropathy, X-linked syndrome. A very rare inherited condition in which CD4 CD25 regulatory T cells are lacking as a result of a mutation in the gene for the transcription factor FoxP3, leading to the development of autoimmunity.

IRAK1, IRAK4
Protein kinases that are part of the intracellular signaling pathways leading from TLRs.

IRAK4 deficiency
An immunodeficiency characterized by recurrent bacterial infections, caused by inactivating mutations in the *IRAK4* gene that result in a block in TLR signaling.

IRF
See **interferon regulatory factor**.

irradiation-sensitive SCID
A type of severe combined immunodeficiency in which there is also abnormal sensitivity to ionizing radiation. This type of SCID is due to mutations in the DNA repair pathways that normally repair radiation-damaged DNA, and which are also used in V(D)J recombination.

ISCOM
Immune stimulatory complex. A complex of antigen held within a lipid matrix that acts as an adjuvant and enables the antigen to be taken up into the cytoplasm after fusion of the lipid with the plasma membrane.

isoelectric focusing
An electrophoretic technique in which proteins migrate in a pH gradient until they reach the place in the gradient at which their net charge is neutral—their isoelectric point. Uncharged proteins no longer migrate; thus, each protein is focused at its isoelectric point.

isoforms
Different forms of the same protein, for example the different forms encoded by different alleles of the same gene.

isolated lymphoid follicles
A type of organized lymphoid tissue in the gut wall that is composed mainly of B cells.

isotype
The designation of an immunoglobulin chain in respect of the type of constant region it has. Light chains can be of either κ or λ isotype. Heavy chains can be of μ, δ, γ, α, or ε isotype. The different heavy-chain isotypes have different effector functions and determine the class and functional properties of antibodies (IgM, IgD, IgG, IgA, and IgE, respectively).

isotype switching
See **class switching**.

isotypic exclusion
Describes the use of one or other of the light-chain isotypes, κ or λ, by a given B cell or antibody.

Itk
A Tec-family kinase that activates phospholipase C-γ in T cells, involved in T-cell activation.

i.v.
See **intravenous**.

Janus-family tyrosine kinases (JAKs)
Enzymes of the JAK–STAT intracellular signaling pathways that link many cytokine receptors with gene transcription in the nucleus. The kinases phosphorylate STAT proteins in the cytosol, which then move to the nucleus and activate a variety of genes.

J chain
Small polypeptide chain synthesized in immunoglobulin-secreting cells and attached by disulfide bonds to polymeric

immunoglobulins IgM and IgA. Its presence is essential for the formation of the binding site for the polymeric immunoglobulin receptor.

J gene segments
Joining gene segments. Short DNA sequences that encode the J regions of immunoglobulin and T-cell receptor variable domains. In a rearranged light-chain locus, TCRα locus, or TCRγ locus, a J gene segment is joined to a V gene segment. In a rearranged heavy-chain locus, TCRβ locus, or TCRδ locus, a J gene segment is joined to a D gene segment.

Job's syndrome
See **hyper IgE syndrome**.

joining gene segments
See **J gene segments**.

JNK
See **Jun kinase**.

junctional diversity
The variability in sequence present in antigen-specific receptors that is created during the process of joining V, D, and J gene segments and which is due to imprecise joining and insertion of nontemplated nucleotides at the joins between gene segments.

Jun kinase (JNK)
A protein kinase that phosphorylates the transcription factor c-Jun, enabling it to bind to c-Fos to form the AP-1 transcription factor.

κ
One of the two classes or isotypes of immunoglobulin light chains.

killer cell immunoglobulin-like receptors (KIRs)
Large family of receptors present on NK cells, through which the cells' cytotoxic activity is controlled. The family contains both activating and inhibitory receptors.

killer lectin-like receptors (KLRs)
Large family of receptors present on NK cells, through which the cells' cytotoxic activity is controlled. The family contains both activating and inhibitory receptors.

kinin system
An enzymatic cascade of plasma proteins that is triggered by tissue damage to produce several inflammatory mediators, including the vasoactive peptide bradykinin.

Ku
A DNA repair protein required for immunoglobulin and T-cell receptor gene rearrangement.

Kupffer cells
Phagocytes lining the hepatic sinusoids; they remove debris and dying cells from the blood, but are not known to elicit immune responses.

λ
One of the two classes or isotypes of immunoglobulin light chains.

$\lambda 5$
See **surrogate light chain**.

LAD
See **leukocyte adhesion deficiencies**.

lamellar bodies
Lipid-rich secretory organelles in keratinocytes and lung pneumocytes that release β-defensins into the extracellular space.

lamina propria
A layer of connective tissue underlying a mucosal epithelium. It contains lymphocytes and other immune-system cells.

Langerhans cells
Phagocytic immature dendritic cells found in the epidermis, which in the presence of infection migrate to regional lymph nodes via the afferent lymphatics. In the lymph node they differentiate into mature antigen-presenting dendritic cells.

langerin
A lectin-type mannose receptor present in Langerhans cells.

large pre-B cell
Stage of B-cell development immediately after the late pre-B cell, in which the cell stops expressing the pre-B-cell receptor and stops dividing. Light-chain gene rearrangement starts.

LAT
See **linker of activation in T cells**.

latency
A state in which a virus infects a cell but does not replicate.

late-phase reaction
In an experimentally induced allergic response, the reaction that occurs some hours after initial encounter with an antigen.

late pro-B cell
Stage in B-cell development in which V_H to DJ_H joining occurs.

L chain
See **light chain**.

Lck
A Src-family tyrosine kinase that associates with the cytoplasmic tails of CD4 and CD8 and phosphorylates the cytoplasmic tails of the T-cell receptor signaling chains, thus helping to activate signaling from the T-cell receptor complex once antigen has bound.

lectin
A carbohydrate-binding protein.

lectin pathway
Complement activation pathway that is triggered by mannose-binding lectins (MBLs) or ficolins bound to bacteria.

lentiviruses
A group of retroviruses that include the human immunodeficiency virus, HIV-1. They cause disease after a long incubation period.

lepromatous leprosy
See **leprosy**.

leprosy
A disease caused by *Mycobacterium leprae* that occurs in a variety of forms. There are two polar forms: lepromatous leprosy, which is characterized by abundant replication of leprosy bacilli and abundant antibody production without cell-mediated immunity; and tuberculoid leprosy, in which few organisms are seen in the tissues, there is little or no antibody, but cell-mediated immunity is very active. The other forms of leprosy are intermediate between the polar forms.

leucine-rich repeats (LRRs)
Protein motifs that are repeated in series to form, for example, the extracellular portions of Toll-like receptors.

leukocyte
A white blood cell. Leukocytes include lymphocytes, polymorphonuclear leukocytes, and monocytes.

leukocyte adhesion deficiencies (LAD)
A class of immunodeficiency diseases in which the ability of leukocytes to enter sites infected by extracellular pathogens is affected, so that this type of infection cannot be effectively eradicated. There are several different causes, including a deficiency of the common β chain of the leukocyte integrins.

leukocyte functional antigens (LFAs)
Cell-adhesion molecules on leukocytes that were initially defined using monoclonal antibodies. LFA-1 is a β_2 integrin; LFA-2 (now usually called CD2) is a member of the immunoglobulin superfamily, as is LFA-3 (now called CD58). LFA-1 is particularly important in T-cell adhesion to endothelial cells and antigen-presenting cells.

leukocyte integrins
Those integrins typically found on leukocytes. They have a common β_2 chain with distinct α chains, and include LFA-1 and the very late activation antigens (VLAs).

leukocyte receptor complex (LRC)
A large cluster of immunoglobulin-like receptor genes that includes the killer cell immunoglobulin-like receptor (KIR) genes.

leukocytosis
The presence of increased numbers of leukocytes in the blood. It is commonly seen in acute infection.

leukotrienes
Lipid mediators of inflammation that are derived from arachidonic acid. They are produced by macrophages and other cells.

LFA-1, LFA-2, LFA-3
See **leukocyte functional antigens**.

licensing
The activation of a dendritic cell so that it is able to present antigen to naive T cells and activate them.

LICOS
The ligand for ICOS, a CD28-related protein that is induced on activated T cells and can enhance T-cell responses.

light chain (L chain)
The smaller of the two types of polypeptide chains that make up an immunoglobulin molecule. It consists of one V and one C domain, and is disulfide-bonded to the heavy chain. There are two classes, or isotypes, of light chain, known as κ and λ, which are produced from separate genetic loci.

light zone
See **germinal centers**.

limiting-dilution culture
An assay using the Poisson distribution to estimate the frequency of lymphocytes able to respond to a given antigen.

linear epitope
See **continuous epitope**.

lingual tonsils
Paired masses of organized peripheral lymphoid tissue situated at the base of the tongue, in which adaptive immune responses can be initiated. They are part of the mucosal immune system. See also **palatine tonsils**.

linked recognition
The rule that for a helper T cell to be able to activate a B cell, the epitopes recognized by the B cell and the helper T cell have to be derived from the same antigen (that is, they must originally have been physically linked).

linker of activation in T cells (LAT)
A cytoplasmic adaptor protein with several tyrosines that become phosphorylated by the tyrosine kinase ZAP-70. It helps to coordinate downstream signaling events in T-cell activation.

lipopolysaccharide (LPS)
The surface lipopolysaccharide of Gram-negative bacteria, which stimulates a Toll-like receptor on macrophages and dendritic cells.

lipoteichoic acids
Components of bacterial cell walls that are recognized by Toll-like receptors.

locus
See **genetic locus**.

LPS
See **lipopolysaccharide**.

LPS-binding protein (LBP)
Protein in blood and extracellular fluid that binds bacterial lipopolysaccharide (LPS) shed from bacteria.

LRR
See **leucine-rich repeats**.

L-selectin
Adhesion molecule of the selectin family found on lymphocytes. L-selectin binds to CD34 and GlyCAM-1 on high endothelial venules to initiate the migration of naive lymphocytes into lymphoid tissue.

LT, LT-α_2:β_1 (LT-β), LT-α_3
See **lymphotoxin**.

lymph
The extracellular fluid that accumulates in tissues and is drained by lymphatic vessels that carry it through the lymphatic system to the thoracic duct, which returns it to the blood.

lymphatic system
The system of lymph-carrying vessels and peripheral lymphoid tissues through which extracellular fluid from tissues passes before it is returned to the blood via the thoracic duct.

lymphatic vessels, lymphatics
Thin-walled vessels that carry lymph.

lymph node
A type of peripheral lymphoid organ present in many locations throughout the body where lymphatic vessels converge.

lymphoblast
A lymphocyte that has enlarged after activation and has increased its rate of RNA and protein synthesis, but is not yet fully differentiated.

lymphocyte receptor repertoire
All the highly variable antigen receptors carried by B and T lymphocytes.

lymphocytes
A class of white blood cells that bear variable cell-surface receptors for antigen and are responsible for adaptive immune responses. There are two main types—B lymphocytes (B cells) and T lymphocytes (T cells)—which mediate humoral and cell-mediated immunity, respectively. On antigen recognition, a lymphocyte enlarges to form a lymphoblast and then proliferates and differentiates into an antigen-specific effector cell.

lymphoid
Describes tissues composed mainly of lymphocytes.

lymphoid follicles
Areas of peripheral lymphoid tissues that are composed mainly of B lymphocytes and follicular dendritic cells.

lymphoid organs
Organized tissues characterized by very large numbers of lymphocytes interacting with a nonlymphoid stroma. The central, or primary, lymphoid organs, where lymphocytes are generated, are the thymus and bone marrow. The main peripheral, or secondary, lymphoid organs, in which adaptive immune responses are initiated, are the lymph nodes, spleen, and mucosa-associated lymphoid organs such as tonsils and Peyer's patches.

lymphoid tissue
Tissue composed of large numbers of lymphocytes.

lymphoid tissue inducer cells (LTi cells)
Cells of the blood lineage, which arise in the fetal liver and are carried in the blood to sites where they will form lymph nodes and other peripheral lymphoid organs.

lymphokines
Cytokines produced by lymphocytes.

lymphopoiesis
The differentiation of lymphoid cells from a common lymphoid progenitor.

lymphotoxin (LT)
A cytokine of the tumor necrosis factor (TNF) family that is directly cytotoxic for some cells. It occurs as trimers of LT-α chains (LT-α_3) and heterotrimers of LT-α and LT-β chains (LT-α_2:β_1).

Lyn
A Src-family tyrosine kinase that relays signals from the B-cell receptor.

lysosomes
Acidified organelles that contain many degradative hydrolytic enzymes. Material taken up into endosomes by phagocytosis or receptor-mediated endocytosis is eventually delivered to lysosomes.

lysozyme
Antimicrobial enzyme that degrades bacterial cell walls.

μ
The type (isotype) of heavy chain in IgM immunoglobulins.

macroautophagy
The engulfment by a cell of large quantities of its own cytoplasm, which is then delivered to the lysosomes for degradation.

macrophage
A large mononuclear phagocytic cell type important as scavenger cells, as pathogen-recognition cells, as a source

of pro-inflammatory cytokines in innate immunity, as antigen-presenting cells, and as effector phagocytic cells in humoral and cell-mediated immunity. Macrophages derive from bone marrow precursors and are found in most tissues of the body.

macrophage activation
The enhancement of the capacity of a macrophage to kill engulfed pathogens and to produce cytokines that follows its antigen-specific interaction with an effector T cell.

macropinocytosis
A process in which large amounts of extracellular fluid are taken up into an intracellular vesicle. This is one way in which dendritic cells can take up a wide variety of antigens from their surroundings.

MAdCAM-1
Mucosal cell-adhesion molecule-1. A mucosal addressin that is recognized by the lymphocyte surface proteins L-selectin and VLA-4, enabling the specific homing of lymphocytes to mucosal tissues.

MAIT
See **mucosal invariant T cells**.

major basic protein
Protein produced by eosinophils that acts on mast cells to cause their degranulation.

major histocompatibility complex (MHC)
A cluster of genes on human chromosome 6 that encodes a set of membrane glycoproteins called the MHC molecules. The MHC also encodes proteins involved in antigen processing and other aspects of host defense. The genes for the MHC molecules are the most polymorphic in the human genome, having large numbers of alleles at the various loci.

MALT
See **mucosa-associated lymphoid tissue**.

mannose-binding lectin (MBL)
Mannose-binding protein present in the blood. It can opsonize pathogens bearing mannose on their surfaces and can activate the complement system via the lectin pathway, an important part of innate immunity.

mannose receptor
A receptor on macrophages that is specific for mannose-containing carbohydrates that occur on the surfaces of pathogens but not on host cells.

mantle zone
A rim of B lymphocytes that surrounds lymphoid follicles. The precise nature and role of mantle zone lymphocytes have not yet been determined.

MAP kinases (MAPKs)
See **mitogen-activated protein kinases**.

marginal sinus
A blood-filled vascular network that branches from the central arteriole and demarcates each area of white pulp in the spleen.

marginal zone
Area of lymphoid tissue lying at the border of the white pulp in the spleen.

marginal zone B cells
A unique population of B cells found in the spleen marginal zones; they do not circulate and are distinguished from conventional B cells by a distinct set of surface proteins.

MASP-1, MASP-2
Serine proteases of the lectin pathway of complement activation that bind to mannose-binding lectin and cleave C4.

mast cell
A large granule-rich cell found in connective tissues throughout the body, most abundantly in the submucosal tissues and the dermis. The granules store bioactive molecules including the vasoactive amine histamine, which are released on mast-cell activation. Mast cells are thought to be involved in defenses against parasites and they have a crucial role in allergic reactions.

mastocytosis
The overproduction of mast cells.

mature B cell
B cell that expresses IgM and IgD on its surface and has become able to respond to antigen.

MBL
See **mannose-binding lectin**.

MBL-associated serine proteases
See **MASP-1, MASP-2**.

M cell (microfold cell)
Specialized epithelial cell type in the intestinal epithelium over Peyer's patches, through which antigens and pathogens enter from the gut.

MD-2
Accessory protein for TLR-4 activity.

MEK1
A MAPK kinase in the Raf–MEK1–Erk signaling module, which is a part of a signaling pathway in lymphocytes leading to activation of the transcription factor NFAT.

medulla
The central or collecting point of an organ. The thymic medulla is the central area of each thymic lobe, rich in bone marrow derived antigen-presenting cells and the cells of a distinctive medullary epithelium. The medulla of the lymph node is a site of macrophage and plasma cell concentration through which the lymph flows on its way to the efferent lymphatics.

membrane-attack complex
Protein complex composed of the terminal complement proteins, which assembles on pathogen surfaces to generate a membrane-spanning hydrophilic pore, damaging the membrane and causing cell lysis.

membrane cofactor of proteolysis (MCP)
A complement regulatory protein, a host-cell membrane protein that acts in conjunction with factor I to cleave C3b to its inactive derivative iC3b and thus prevent convertase formation.

membrane immunoglobulin (mIg)
Transmembrane immunoglobulin present on B cells; it is the B-cell receptor for antigen.

memory B cells
See **memory cells**.

memory cells
B and T lymphocytes that mediate immunological memory. They are more sensitive than naive lymphocytes to antigen and respond rapidly on reexposure to the antigen that originally induced them.

memory T cells
See **memory cells**.

mesenteric lymph nodes
Lymph nodes located in the connective tissue (mesentery) that tethers the intestine to the rear wall of the abdomen. They drain the GALT.

MHC
See **major histocompatibility complex**.

MHC class I
See **MHC class I molecules**.

MHC class Ib
A class of proteins encoded within the MHC that are related to the MHC class I molecules but are not highly polymorphic and present a restricted set of antigens.

MHC class I deficiency
An immunodeficiency disease in which MHC class I molecules are not present on the cell surface, usually as a result of an inherited deficiency of either TAP-1 or TAP-2.

MHC class I molecules
Polymorphic proteins encoded in the MHC and expressed on most cells in the body. They are cell-surface proteins that present antigenic peptides generated in the cytosol to CD8 T cells. They interact with the T-cell co-receptor CD8.

MHC class II
See **MHC class II molecules**.

MHC class II compartment (MIIC)
The cellular vesicles in which MHC class II molecules accumulate, encounter HLA-DM, and bind antigenic peptides, before migrating to the surface of the cell.

MHC class II deficiency
A rare immunodeficiency disease in which MHC class II molecules are not present on cells as a result of various inherited defects. Patients are severely immunodeficient and have few CD4 T cells.

MHC class II molecules
Polymorphic proteins encoded in the MHC and expressed on some cells of the immune system, primarily specialized antigen-presenting cells. They are cell-surface proteins that present antigenic peptides derived from internalized extracellular pathogens to CD4 T cells. They interact with the T-cell co-receptor CD4.

MHC class II transactivator (CIITA)
Protein that activates transcription of MHC class II genes. Defects in the *CIITA* gene are one cause of MHC class II deficiency.

MHC haplotype
A set of alleles in the MHC that is inherited unchanged (that is, without recombination) from one parent.

MHC molecules
The highly polymorphic glycoproteins encoded by MHC class I and MHC class II genes, which are involved in the presentation of peptide antigens to T cells. They are also known as histocompatibility antigens.

MHC restriction
The fact that a peptide antigen can only be recognized by a given T cell if it is bound to a particular self MHC molecule. MHC restriction is a consequence of events that occur during T-cell development.

MIC
Family of MHC class I-like molecules that are expressed in the gut under conditions of stress and are encoded within the class I region of the human MHC. They are not found in mice.

microautophagy
The continuous internalization of the cytosol into the vesicular system.

microbiota
The microorganisms normally present in any particular environment, such as the human gut.

microclusters
Assemblies of small numbers of T-cell receptors that may be involved in the initiation of T-cell receptor activation by antigen in naive T cells.

microfold cells
See **M cells**.

microorganisms
Microscopic organisms, unicellular except for some fungi, that include bacteria, yeasts and other fungi, and protozoa. All these groups contain some microorganisms that cause human disease.

mIg
See **membrane immunoglobulin**.

MIIC
See **MHC class II compartment**.

minor histocompatibility antigens (minor H antigens)
Peptides of polymorphic cellular proteins bound to MHC molecules that can lead to graft rejection when they are recognized by T cells.

mismatch repair
A type of DNA repair that causes mutations and is involved in somatic hypermutation and class switching in B cells.

mitochondrial pathway of apoptosis
See **intrinsic pathway of apoptosis**.

mitogen-activated protein kinases (MAPKs)
A series of protein kinases that become phosphorylated and activated on cellular stimulation by a variety of ligands, and lead to new gene expression by phosphorylating key transcription factors. The MAPKs are part of many signaling pathways, especially those leading to cell proliferation, and have different names in different organisms.

mixed essential cryoglobulinemia
Disease due to the production of cryoglobulins (cold-precipitable immunoglobulins), sometimes in response to chronic infections such as hepatitis C, which can lead to the deposition of immune complexes in joints and tissues.

mixed lymphocyte reaction
A test for histocompatibility in which lymphocytes from donor and recipient are cultured together. If the two people are histoincompatible, the recipient's T cells recognize the allogeneic MHC molecules on the cells of the other donor as 'foreign' and proliferate.

molecular mimicry
The similarity between some pathogen antigens and host antigens, such that antibodies and T cells produced against the former also react against host tissues. This similarity may be the cause of some autoimmunity.

monoclonal antibodies
Antibodies produced by a single clone of B lymphocytes, so that they are all identical.

monocyte
Type of white blood cell with a bean-shaped nucleus; it is a precursor of tissue macrophages.

monomorphic
Describes a gene that occurs in only one form. Cf. **polymorphic**.

mononuclear phagocytes
Macrophages and dendritic cells.

M1-type macrophages
See **alternatively activated macrophages**.

M2-type macrophages
The name sometimes given to 'conventional' pro-inflammatory macrophages. See **macrophages**.

mucins
Highly glycosylated cell-surface proteins. Mucinlike molecules are bound by L-selectin in lymphocyte homing.

Muckle–Wells syndrome
An inherited episodic autoinflammatory disease caused by mutations in the gene encoding NLRP3, a component of the inflammasome.

mucosa-associated lymphoid tissue (MALT)
Generic term for all organized lymphoid tissue found at mucosal surfaces, in which an adaptive immune response can be initiated. It comprises GALT, NALT, and BALT (when present).

mucosal epithelia
Mucus-coated epithelia lining the body's internal cavities that connect with the outside (such as the gut, airways, and vaginal tract).

mucosal immune system
The immune system that protects internal mucosal surfaces (such as the linings of the gut, respiratory tract, and urogenital tracts), which are the site of entry for virtually all pathogens and other antigens. See also **mucosa-associated lymphoid tissue**.

mucosal invariant T cells (MAIT)
Small subset of T cells found in the mucosal immune system that express antigen receptors of limited diversity and respond to antigens presented by nonclassical class I-type MHC molecules.

mucosal mast cells
Specialized mast cells present in mucosa. They produce little histamine but large amounts of prostaglandins and leukotrienes.

mucosal tolerance
The suppression of specific systemic immune responses to an antigen by the previous administration of the same antigen by a mucosal route.

mucus
Sticky solution of proteins (mucins) secreted by goblet cells of internal epithelia, forming a protective layer on the epithelial surface.

multiple sclerosis
A neurological autoimmune disease characterized by focal demyelination in the central nervous system, lymphocytic infiltration in the brain, and a chronic progressive course.

multipotent progenitor cells (MPPs)
Bone marrow cells that can give rise to both lymphoid and myeloid cells but are no longer self-renewing stem cells.

mutualism
A symbiotic relationship between two organisms in which both benefit, such as the relationship between a human and its normal resident (commensal) gut microorganisms.

Mx proteins
Interferon-inducible proteins required for cellular resistance to influenza virus replication.

myasthenia gravis
An autoimmune disease in which autoantibodies against the acetylcholine receptor on skeletal muscle cells cause a block in neuromuscular junctions, leading to progressive weakness and eventually death.

mycophenolate
An inhibitor of the synthesis of guanosine monophosphate that acts as a cytotoxic immunosuppressive drug. It acts by killing rapidly dividing cells, including lymphocytes proliferating in response to antigen.

myeloid
Refers to the lineage of blood cells that includes all leukocytes except lymphocytes.

myeloid-derived suppressor cells
Cells in tumors that can inhibit T-cell activation within the tumor.

NADPH oxidase
Multicomponent enzyme complex that is assembled and activated in the phagolysosome membrane in stimulated phagocytes. It generates superoxide in an oxygen-requiring reaction called the respiratory burst.

naive lymphocytes, naive T cells, naive B cells
Lymphocytes that have never encountered their specific antigen and thus have never responded to it, as distinct from effector and memory lymphocytes.

nasal-associated lymphoid tissue (NALT)
Organized lymphoid tissues found in the upper respiratory tract. In humans, NALT consists of Waldeyer's ring, which includes the adenoids, palatine, and lingual tonsils, plus other similarly organized lymphoid tissue located around the pharynx. It is part of the mucosal immune system.

natural antibodies
Antibodies produced by the immune system in the apparent absence of any infection. They have a broad specificity for self and microbial antigens, can react with many pathogens, and can activate complement.

natural cytotoxicity receptors (NCRs)
Activating receptors on NK cells that recognize infected cells and stimulate cell killing by the NK cell.

natural interferon-producing cells
See **plasmacytoid dendritic cells**.

natural killer cell (NK cell)
Large granular, non-T, non-B lymphocyte, which kills virus-infected cells and some tumor cells. NK cells bear a wide variety of invariant activating and inhibitory receptors, but do not rearrange immunoglobulin or T-cell receptor genes. NK cells are important in innate immunity to viruses and other intracellular pathogens, and in antibody-dependent cell-mediated cytotoxicity (ADCC).

natural regulatory T cells (natural T_{reg} cells)
Regulatory CD4 T cells that are specified in the thymus. They express FoxP3 and bear the markers CD25 and CD4 on their surface. Cf. **induced regulatory T cells**.

negative selection
The process by which self-reactive thymocytes are deleted from the repertoire during T-cell development in the thymus. Autoreactive B cells undergo a similar process in bone marrow.

NEMO
IKK-γ, a component of the IκB kinase (KK) complex that acts in the NFκB intracellular signaling pathway.

NEMO deficiency
See **X-linked hypohidrotic ectodermal dysplasia with immunodeficiency**.

neonatal Fc receptor (FcRn)
See **FcRn**.

neutralization
Inhibition of the infectivity of a virus or the toxicity of a toxin molecule by the binding of antibodies.

neutralizing antibodies
Antibodies that inhibit the infectivity of a virus or the toxicity of a toxin.

neutropenia
Abnormally low levels of neutrophils in the blood.

neutrophil
The most numerous type of white blood cell in human peripheral blood. Neutrophils are phagocytic cells with a multilobed nucleus and granules that stain with neutral stains. They enter infected tissues and engulf and kill extracellular pathogens.

neutrophil elastase
Proteolytic enzyme stored in the granules of neutrophils that is involved in the processing of antimicrobial peptides.

NFAT
See **nuclear factor of activated T cells**.

NFκB
One of the transcription factors activated by the stimulation of Toll-like receptors and also by antigen-receptor signaling.

NK cells
See **natural killer cells**.

NK receptor complex (NKC)
A cluster of genes that encode a family of receptors on NK cells.

NKG2
Family of C-type lectins that supply one of the subunits of KLR-family receptors on NK cells.

NKG2D
Activating C-type lectin receptor on NK cells, cytotoxic T cells, and γ:δ T cells that recognizes the stress-response proteins MIC-A and MIC-B.

NKT cells
See **invariant NKT cells**.

NLRP3
A member of the family of intracellular NOD-like receptor proteins that have pyrin domains. It acts as a sensor of cellular damage and is part of the inflammasome. Sometimes called NALP3.

N-nucleotides, N-regions
Nontemplated nucleotides inserted by the enzyme terminal deoxynucleotidyl transferase into the junctions between gene segments of T-cell receptor and immunoglobulin heavy-chain V regions during gene segment joining. The stretches of N-nucleotides (N-regions) are translated, and markedly increase the diversity of these receptor chains.

NOD1, NOD2
Intracellular proteins with a nucleotide-oligomerization domain (NOD) and a leucine-rich repeat domain that bind components of bacterial cell walls and activate the NFκB pathway, initiating inflammatory responses.

NOD-like receptors (NLRs)
Large family of proteins containing a nucleotide-oligomerization domain (NOD) associated with various other domains, and whose general function is the detection of microbes and of cellular stress.

nonamer
Conserved nine-nucleotide DNA sequence in the recombination signal sequences (RSSs) flanking gene segments in the immunoglobulin and T-cell receptor loci.

nondepleting antibodies
Immunosuppressive antibodies that block the function of target proteins on cells without causing the cells to be destroyed.

nonproductive rearrangements
Rearrangements of T-cell receptor or immunoglobulin gene segments that cannot encode a protein because the coding sequences are in the wrong translational reading frame.

nonreceptor kinases
Cytoplasmic protein kinases that associate with the intracellular tails of signaling receptors and help generate the signal but are not an intrinsic part of the receptor itself.

N-regions
See **N-nucleotides**.

nuclear factor of activated T cells (NFAT)
Transcription factor composed of NFATc and the Fos/Jun dimer AP-1. It is activated in response to signaling from the antigen receptor in lymphocytes.

nude
A mutation in mice that results in hairlessness and defective formation of the thymic stroma, so that mice homozygous for this mutation have no mature T cells.

occupational allergy
An allergic reaction induced to an allergen to which someone is habitually exposed in their work.

oligoadenylate synthetase
Enzyme produced in response to stimulation of cells by interferon. It synthesizes unusual nucleotide polymers, which in turn activate a ribonuclease that degrades viral RNA.

Omenn syndrome
A severe immunodeficiency disease characterized by defects in either of the *RAG* genes. Affected individuals make small amounts of functional RAG protein, allowing a small amount of V(D)J recombination.

opportunistic pathogen
A microorganism that does not normally cause disease in healthy individuals but can cause disease in individuals with compromised host defenses.

opsonization
The coating of the surface of a pathogen by antibody and/or complement that makes it more easily ingested by phagocytes.

oral tolerance
The suppression of specific systemic immune responses to an antigen by the prior administration of the same antigen by the oral (enteric) route.

original antigenic sin
The tendency of humans to make antibody responses to those epitopes shared between the first strain of a virus they encounter and subsequent related viruses, while ignoring other highly immunogenic epitopes on the second and subsequent viruses.

OX40, OX40L
Receptor–ligand pair that delivers co-stimulatory signals to T cells.

palatine tonsils
Paired masses of organized peripheral lymphoid tissues located on each side of the throat, and in which an adaptive immune response can be generated. They are part of the mucosal immune system.

PAMPs
See **pathogen-associated molecular patterns**.

Paneth cells
Specialized epithelial cells at the base of the crypts in the small intestine that secrete antimicrobial peptides.

panning
Technique by which lymphocyte subpopulations can be isolated on Petri dishes coated with monoclonal antibodies against cell-surface markers, to which the lymphocytes bind.

PAPA
See **pyogenic arthritis, pyoderma gangrenosum, and acne**.

paracortical area
The T-cell area of lymph nodes.

paracrine
Describes a cytokine or other biologically active molecule acting on cells near to those that produce it.

parasites
Organisms that obtain sustenance from a live host. In immunology, it refers to worms and protozoa, the subject matter of parasitology.

paroxysmal nocturnal hemoglobinuria (PNH)
A disease in which complement regulatory proteins are defective, so that activation of complement binding to red blood cells leads to episodes of spontaneous hemolysis.

passive immunization
The injection of antibody or immune serum into a naive recipient to provide specific immunological protection. Cf. **active immunization**.

pathogen, pathogenic microorganism
Microorganism that typically causes disease when it infects a host.

pathogen-associated molecular patterns (PAMPs)
Molecules specifically associated with groups of pathogens that are recognized by cells of the innate immune system.

pathogenesis
The origin or cause of the pathology of a disease.

pathology
(1) The scientific study of disease. (2) The detectable damage to tissues that occurs in a disease.

pattern recognition receptors (PRRs)
Receptors of the innate immune system that recognize common molecular patterns on pathogen surfaces.

PCR
See **polymerase chain reaction**.

PD-1
Programmed death-1, a receptor on T cells that when bound by its ligands **PD-L1** and **PD-L2** inhibits signaling from the antigen receptor.

pDCs
See **plasmacytoid dendritic cells**.

PECAM
See **CD31**.

pemphigus vulgaris
An autoimmune disease characterized by severe blistering of the skin and mucosal membranes.

pentraxins
A family of acute-phase proteins formed of five identical subunits, to which C-reactive protein and serum amyloid protein belong.

peptide-binding cleft
The longitudinal cleft in the top surface of an MHC molecule into which the antigenic peptide is bound. Sometimes called the peptide-binding groove.

peptide editing
In the context of antigen processing and presentation, the removal of unstably bound peptides from MHC class II molecules by HLA-DM.

peptide-loading complex
A protein complex in the endoplasmic reticulum that loads peptides onto MHC class I molecules.

peptide:MHC tetramers
Four specific peptide:MHC complexes bound to a single molecule of fluorescently labeled streptavidin, which are used to identify populations of antigen-specific T cells.

peptidoglycan
A component of bacterial cell walls that is recognized by certain receptors of the innate immune system.

perforin
A protein stored in the lytic granules of cytotoxic T cells and

NK cells, which on release polymerizes to form pores in cell membranes through which other cytotoxic proteins enter the target cell.

periarteriolar lymphoid sheath (PALS)
Part of the inner region of the white pulp of the spleen; it contains mainly T cells.

peripheral blood mononuclear cells (PBMCs)
Lymphocytes and monocytes isolated from peripheral blood, usually by Ficoll-Hypaque™ density centrifugation.

peripheral lymphoid organs, peripheral lymphoid tissues
The lymph nodes, spleen, and mucosal-associated lymphoid tissues, in which immune responses are induced, as opposed to the central lymphoid organs, in which lymphocytes develop. They are also called secondary lymphoid organs and tissues.

peripheral tolerance
Tolerance acquired by mature lymphocytes in the peripheral tissues, as opposed to central tolerance, which is acquired by immature lymphocytes during their development.

Peyer's patches
Organized peripheral lymphoid organs under the epithelium in the small intestine, especially the ileum, and in which an adaptive immune response can be initiated. They contain lymphoid follicles and T-cell areas. They are part of the gut-associated lymphoid tissues (GALT).

phage display library
Library of antibodies produced by cloning immunoglobulin V-region genes in filamentous phages, which thus express antigen-binding domains on their surfaces.

phagocyte
Cell type specialized to perform phagocytosis; examples are macrophages and neutrophils.

phagocyte oxidase
See **NADPH oxidase**.

phagocytosis
The internalization of particulate matter by cells by a process of engulfment, in which the cell membrane surrounds the material, eventually forming an intracellular vesicle (phagosome) containing the ingested material.

phagolysosome
Intracellular vesicle formed by the fusion of a phagosome (containing ingested material) and a lysosome, and in which the ingested material is broken down.

phagosome
Intracellular vesicle formed when particulate material is ingested by a phagocyte.

phosphatidylinositol kinases
Enzymes that phosphorylate the inositol headgroup on membrane lipids to produce phosphorylated derivatives that have a variety of functions in intracellular signaling.

phosphatidylinositol 3-kinase (PI 3-kinase)
Enzyme involved in intracellular signaling pathways. It phosphorylates the membrane lipid phosphatidylinositol 3,4-bisphosphate (PIP_2) to form phosphatidylinositol 3,4,5-trisphosphate (PIP_3), which can recruit signaling proteins to the membrane.

phospholipase C-γ (PLC-γ)
Key enzyme in intracellular signaling pathways leading from many different receptors. It is activated indirectly by receptor ligation, and cleaves membrane inositol phospholipids into inositol trisphosphate and diacylglycerol.

phosphorylation
Addition of a phosphate group to a molecule, usually a protein, catalyzed by enzymes called kinases.

physiological inflammation
The state of the normal healthy intestine, whose wall contains large numbers of effector lymphocytes and other cells. It is thought to be the result of continual stimulation by commensal organisms and food antigens.

pIgR
See **polymeric immunoglobulin receptor**.

PIP_2
Phosphatidylinositol 3,4-bisphosphate, a membrane-associated phospholipid that is cleaved by phospholipase C-γ to give the signaling molecules diacylglycerol and inositol trisphosphate.

PIP_3
Phosphatidylinositol 3,4,5-trisphosphate, a membrane-associated phospholipid that can recruit intracellular signaling molecules to the membrane.

PKR (protein kinase, RNA-activated)
Serine/threonine kinase activated by IFN-α and IFN-β. It phosphorylates the eukaryotic protein synthesis initiation factor eIF-2, inhibiting translation and thus contributing to the inhibition of viral replication.

plasma
The fluid component of blood, composed of water, electrolytes, and the plasma proteins.

plasmablast
A B cell in a lymph node that already shows some features of a plasma cell.

plasma cell
Terminally differentiated activated B lymphocyte. Plasma cells are the main antibody-secreting cells of the body. They are found in the medulla of the lymph nodes, in splenic red pulp, in bone marrow, and in mucosal tissues.

plasmacytoid dendritic cells (pDCs)
A distinct lineage of dendritic cells that secrete large amounts of interferon on activation by pathogens and their products via receptors such as Toll-like receptors. Cf. **conventional dendritic cells**.

platelet-activating factor (PAF)
A lipid mediator that activates the blood clotting cascade and several other components of the innate immune system.

platelets
Small cell fragments found in the blood that are crucial for blood clotting. They are formed from megakaryocytes.

PMN
See **granulocytes**.

P-nucleotides
Nucleotides found in junctions between gene segments of the rearranged V-region genes of antigen receptors. They are an inverse repeat of the sequence at the end of the adjacent gene segment, being generated from a hairpin intermediate during recombination, and hence are called palindromic or P-nucleotides.

polyclonal activation
The activation of lymphocytes by a mitogen regardless of antigen specificity, leading to the activation of clones of lymphocytes of multiple antigen specificities.

polyclonal mitogens
Agents that activate lymphocytes regardless of antigen specificity and stimulate them to proliferate, resulting in the presence of clones of activated lymphocytes of multiple antigen specificities.

polygenic
Containing several separate loci encoding proteins of identical function; applied to the MHC. Cf. **polymorphic**.

polymerase chain reaction (PCR)
A technique for replicating specific sequences in DNA *in vitro*, producing thousands of copies of the replicated sequence.

polymeric immunoglobulin receptor (pIgR)
The receptor for polymeric immunoglobulins IgA and IgM, which is present on basolateral surfaces of mucosal and glandular epithelial cells, and is responsible for the transport of IgA (or IgM) into secretions to form secretory IgA (or secretory IgM). pIgR is the membrane-bound precursor of secretory component.

polymorphic
Existing in a variety of different forms; applied to a gene, occurring in a variety of different alleles.

polymorphism
Applied to genes, variability at a gene locus in which all variants occur at a frequency greater than 1%.

polymorphonuclear leukocytes (PMNs)
See **granulocytes**.

polyspecificity
The ability of an antibody to bind to many different antigens. Also known as polyreactivity.

positive selection
A process occurring in the thymus in which only those developing T cells whose receptors can recognize antigens presented by self MHC molecules can mature.

post-transplant lymphoproliferative disorder
B-cell expansion driven by Epstein–Barr virus (EBV) in which the B cells can undergo mutations and become malignant. This can occur when patients are immunosuppressed after, for example, solid organ transplantation.

pre-B-cell receptor
Receptor produced by pre-B cells that includes an immunoglobulin heavy chain, as well as other proteins, and which induces the pre-B cell to enter the cell cycle, to turn off the *RAG* genes, to degrade the RAG proteins, and to expand by several cell divisions.

pre-B cells
Developing B lymphocytes that have rearranged their heavy-chain genes but not their light-chain genes.

precipitin reaction
The first quantitative technique for measuring antibody production. The amount of antibody in a sample was determined from the amount of precipitate obtained with a fixed amount of antigen.

prednisone
A synthetic steroid with potent anti-inflammatory and immunosuppressive activity used in treating acute graft rejection, autoimmune disease, and lymphoid tumors.

pre-T-cell receptor
Receptor protein produced by developing T lymphocytes at the pre-T-cell stage. It is composed of TCRβ chains that pair with a surrogate α chain called pTα (pre-T-cell α), and is associated with the CD3 signaling chains.

primary cell-mediated immune response
See **cell-mediated immune response**, **cell-mediated immunity**.

primary focus
Initial area of differentiating B cells that forms outside the follicles in peripheral lymphoid tissues after antigen activation.

primary granules
Granules in neutrophils that correspond to lysosomes and contain antimicrobial peptides such as defensins and other antimicrobial agents.

primary immune response
The adaptive immune response that follows the first exposure to a particular antigen.

primary immunization
See **priming**.

primary immunodeficiencies
A lack of immune function that is caused by a genetic defect.

primary interaction
The binding of antibody molecules to antigen, as distinct from secondary interactions in which binding is detected by some associated change such as the precipitation of soluble antigen or agglutination of particulate antigen.

primary lymphoid follicles
Aggregates of resting B lymphocytes in peripheral lymphoid organs. Cf. **secondary lymphoid follicle**.

primary lymphoid organs
See **central lymphoid organs**.

priming
The first encounter with a given antigen, which generates the primary adaptive immune response.

pro-B cells
A stage in B-lymphocyte development in which cells have displayed B-cell surface marker proteins but have not yet completed heavy-chain gene rearrangement.

productive rearrangements
Rearrangements of T-cell receptor or immunoglobulin gene segments that encode a functional protein.

programmed cell death
See **apoptosis**.

pro-inflammatory
Tending to induce inflammation.

propeptide
Inactive precursor form of a polypeptide or peptide, which requires proteolytic processing to produce the active peptide.

properdin
Protein secreted by activated neutrophils that binds to and stabilizes the alternative pathway C3 convertases. Also called factor P.

prostaglandins
Lipid products of the metabolism of arachidonic acid that have a variety of effects on tissues, including activities as inflammatory mediators.

proteasome
A large intracellular multisubunit protease that degrades proteins, producing peptides. Peptides that are presented by MHC class I molecules are generated by the proteasomes.

protectin (CD59)
Cell-surface protein that protects host cells from being damaged by complement. It inhibits the formation of the membrane-attack complex by preventing the binding of C8 and C9 to the C5b67 complex.

protective immunity
The resistance to a specific pathogen that results from infection or vaccination. It is due to the adaptive immune response, which sets up an immunological memory of that pathogen.

Protein A
A membrane component of *Staphylococcus aureus* that binds to the Fc region of IgG and is thought to protect the bacteria from IgG antibodies by inhibiting their interactions with complement and Fc receptors. It is useful for purifying IgG antibodies.

protein dephosphorylation
The removal of a phosphate group from a protein, performed by protein phosphatases. Like phosphorylation, this is a means of regulating protein activity.

protein-interaction domains
Protein domains, usually with no enzymatic activity themselves, that specifically interact with particular sites (such as phosphorylated tyrosines, proline-rich regions, or membrane phospholipids) on other proteins or cellular structures.

protein kinases
Enzymes that add phosphate groups to proteins at particular amino acid residues: tyrosine, threonine, or serine.

protein kinase C-θ (PKC-θ)
A serine/threonine kinase that is activated by diacylglycerol and calcium as part of the signaling pathways from the antigen receptor in lymphocytes.

protein phosphatases
Enzymes that remove phosphate groups from proteins phosphorylated on tyrosine, threonine, or serine residues by protein kinases.

protein phosphorylation
The covalent addition of a phosphate group to a specific site on a protein. Phosphorylation can alter the activity of a protein and also provides new binding sites for other proteins to interact with it.

protein tyrosine kinases
Enzymes that add phosphate groups to specific tyrosine residues on proteins.

provirus
The DNA form of a retrovirus when it is integrated into the host-cell genome, where it can remain transcriptionally inactive for long periods.

P-selectin
See **selectins**.

P-selectin glycoprotein ligand-1 (PSGL-1)
Protein expressed by activated effector T cells that is a ligand for P-selectin on endothelial cells, and may enable activated T cells to enter all tissues in small numbers.

'pseudo-dimeric' peptide:MHC complexes
Hypothetical complexes containing one antigen peptide:MHC molecule and one self-peptide:MHC molecule on the surface of the antigen-presenting cell, which have been proposed to initiate T-cell activation.

p-SMAC
See **supramolecular adhesion complex**.

pTα
See **pre-T-cell receptor**.

purine nucleotide phosphorylase (PNP) deficiency
An enzyme defect that results in severe combined immunodeficiency. The deficiency of PNP causes an intracellular accumulation of purine nucleosides, which are toxic to developing T cells.

pus
Thick yellowish-white liquid typically found at sites of infection with some types of extracellular bacteria, which is composed of the remains of dead neutrophils and other cells.

pus-forming bacteria
Capsulated bacteria that result in pus formation at the site of infection. Also called pyogenic (pus-forming) bacteria.

pyogenic arthritis, pyoderma gangrenosum, and acne (PAPA)
Autoinflammatory syndrome that is caused by mutations in a protein that interacts with the protein pyrin.

pyogenic bacteria
See **pus-forming bacteria**.

pyrin
protein encoded by the *MEFV* gene in humans, and which is mutated in the autoinflammatory disease familial Mediterranean fever.

quasi-species
The different genetic forms of certain RNA viruses that are formed by mutation during the course of an infection.

radiation bone marrow chimeras
Mice that have been heavily irradiated and then reconstituted with bone marrow cells of a different strain of mouse, so that the lymphocytes differ genetically from the environment in which they develop.

radiation-sensitive SCID
Severe combined immunodeficiency due to a defect in DNA repair pathways, which renders cells unable to perform V(D)J recombination and unable to repair radiation-induced double-strand breaks.

radioimmunoassay (RIA)
An assay for antigen or antibody in which an unlabeled antigen or antibody is attached to a solid support such as a plastic surface, and the fraction of radioactively labeled test antibody or antigen retained on the surface is determined in order to measure binding.

RAG-1, RAG-2
Recombinase proteins encoded by the recombination-activating genes *RAG1* and *RAG2*, which are essential for immunoglobulin and T-cell receptor gene rearrangement. They form a complex with other proteins that initiates V(D)J recombination.

rapamycin
An immunosuppressant drug that blocks intracellular signaling pathways involving the serine/threonine kinase mammalian target of rapamycin (mTOR) required for the inhibition of apoptosis and T-cell expansion. Also called sirolimus.

Ras
A small GTPase with important roles in intracellular signaling pathways, including those from lymphocyte antigen receptors.

reactive oxygen species (ROS)
Superoxide anion (O_2^-) and hydrogen peroxide (H_2O_2), produced by phagocytic cells such as neutrophils and macrophages after ingestion of microbes, and which help kill the ingested microbes.

receptor editing
The replacement of a light or heavy chain of a self-reactive antigen receptor on immature B cells with a chain that does not confer autoreactivity.

receptor-mediated endocytosis
The internalization into endosomes of molecules bound to cell-surface receptors.

receptor serine/threonine kinase
Receptors that have an intrinsic serine/threonine kinase activity in their cytoplasmic tails.

receptor tyrosine kinase
Receptors that have an intrinsic tyrosine kinase activity in their cytoplasmic tails.

recessive lethal gene
A gene that encodes a protein needed for the individual to develop to adulthood; when both copies are defective, the individual dies *in utero* or soon after birth.

recipient
An individual who receives transplanted cells or tissues.

recombination-activating genes
See **RAG-1, RAG-2**.

recombination signal sequences (RSSs)
Short stretches of DNA flanking the gene segments that are rearranged to generate a V-region exon in antigen-receptor gene loci and which are recognized by the RAG-1:RAG-2 recombinase. They consist of a conserved heptamer and nonamer separated by 12 or 23 base pairs.

recurrent herpes simplex encephalitis
A manifestation of a deficiency in the receptor TLR-3.

red pulp
The nonlymphoid area of the spleen in which red blood cells are broken down.

RegIIIγ
An antimicrobial protein of the C-type lectin family, produced by Paneth cells in the gut in mice.

regulatory CD4 T cells, regulatory T cells (T_{reg} cells)
Effector CD4 T cells that inhibit T-cell responses and are involved in controlling immune reactions and preventing autoimmunity. Several different subsets have been distinguished, notably the natural regulatory T-cell lineage that is produced in the thymus, and the induced regulatory T cells that differentiate from naive CD4 T cells in the periphery in certain cytokine environments.

regulatory tolerance
Tolerance due to the actions of regulatory T cells.

respiratory burst
An oxygen-requiring metabolic change in neutrophils and macrophages that have taken up opsonized particles, such as complement- or antibody-coated bacteria, by phagocytosis. It leads to the production of toxic metabolites that are involved in killing the engulfed microorganisms.

respiratory syncytial virus (RSV)
A human pathogen that is a common cause of severe chest infection in young children, often associated with wheezing, and also in immunocompromised patients.

retinoic acid
Signaling molecule derived from vitamin A with many roles in the body. It is thought to be involved in the induction of immunological tolerance in the gut.

retrograde translocation, retrotranslocation
The return of endoplasmic reticulum proteins to the cytosol.

reverse transcriptase
Viral RNA-dependent DNA polymerase that transcribes the viral genomic RNA into DNA during the life cycle of retroviruses (such as HIV).

reverse transcriptase–polymerase chain reaction (RT-PCR)
In vitro technique used to study gene expression. RNA sequences (such as total cellular mRNA) are converted to DNA by using reverse transcriptase, and the DNAs are then amplified by the polymerase chain reaction.

Rhesus (Rh) blood group antigen
Red cell membrane antigen that is also detectable on the red blood cells of rhesus monkeys. Anti-Rh antibodies do not agglutinate human red blood cells, so antibody against Rh antigen must be detected by using a Coombs test.

rheumatic fever
Disease caused by antibodies elicited by infection with some *Streptococcus* species. These antibodies cross-react with kidney, joint, and heart antigens.

rheumatoid arthritis
A common inflammatory joint disease that is probably due to an autoimmune response.

rheumatoid factor
An anti-IgG antibody of the IgM class first identified in patients with rheumatoid arthritis, but which is also found in healthy individuals.

RIA
See **radioimmunoassay**.

RIG-I-like helicases
Family of intracellular proteins (of which RIG-I is the prototype) that detect viral RNAs, initiating a signaling pathway that leads to interferon production.

R-loop
A bubble-like structure formed when transcribed RNA displaces the nontemplate strand of the DNA double helix at switch regions in the immunoglobulin constant-region gene cluster. R-loops are thought to promote class switch recombination.

ROS
See **reactive oxygen species**.

RSS
See **recombination signal sequences**.

RSV
See **respiratory syncytial viruses**.

RT-PCR
See **reverse transcriptase–polymerase chain reaction**.

sandwich ELISA
An antibody-based technique for protein antigen detection in which antibody bound to a surface binds the antigen by one of its epitopes. The trapped antigen is detected by an enzyme-linked antibody specific for a different epitope on the antigen.

s.c.
Abbreviation for subcutaneous; refers to injection into the tissues underlying the skin (epidermis and dermis).

scaffolds
Adaptor-type proteins with multiple binding sites, which bring together specific proteins into a functional signaling complex.

Scatchard analysis
A mathematical analysis of equilibrium binding that allows the affinity and valence of a receptor–ligand interaction to be determined.

scavenger receptors
Receptors on macrophages and other cells that bind to numerous ligands and remove them from the blood. The Kupffer cells in the liver are particularly rich in scavenger receptors.

SCID
See **severe combined immunodeficiency**.

scid
Mutation in mice that causes severe combined immunodeficiency. It was eventually found to be due to mutation of the DNA repair protein DNA-PK.

SDS-PAGE
The abbreviation for polyacrylamide gel electrophoresis (PAGE) of proteins dissolved in the detergent sodium dodecyl sulfate (SDS). This technique is widely used to characterize proteins, especially after labeling and immunoprecipitation.

SE
See **staphylococcal enterotoxins**.

seasonal allergic rhinoconjunctivitis
IgE-mediated allergic reaction to specific seasonally occurring antigens, for example grass pollen, which involves rhinitis and conjunctivitis. Commonly called hay fever.

secondary granules
Type of granule in neutrophils that stores certain antimicrobial peptides.

secondary immune response
The immune response that occurs in response to a second exposure to an antigen. In comparison with the primary response, it starts sooner after exposure, produces greater levels of antibody, and produces class-switched antibodies. It is generated by the reactivation of memory lymphocytes.

secondary immunization
A second or booster injection of an antigen, given some time after the initial immunization. It stimulates a secondary immune response.

secondary immunodeficiencies
Deficiencies in immune function that are a consequence of other diseases, malnutrition, etc.

secondary interactions
See **primary interaction**.

secondary lymphoid organ
See **peripheral lymphoid organ**.

secondary lymphoid follicle
A follicle containing a germinal center of proliferating activated B cells during an ongoing adaptive immune response.

second messengers
Small molecules or ions (such as Ca^{2+}) that are produced in response to a signal; they act to amplify the signal and carry it to the next stage within the cell.

second-set rejection
A rapid and vigorous rejection of grafted tissue that occurs if the recipient has already rejected a first tissue graft from the same donor.

secretory component (SC)
Fragment of the polymeric immunoglobulin receptor left attached to secreted IgA after transport across epithelial cells.

secretory IgA (SIgA)
Polymeric IgA antibody (mainly dimeric) containing bound J chain and secretory component. It is the predominant form of immunoglobulin in most human secretions.

secretory IgM (SIgM)
Polymeric IgM antibody containing bound J chain and secretory component. It is secreted through mucosal epithelia and provides antibody-mediated protection for the mucosae when secretory IgA is absent.

secretory phospholipase A_2
Antimicrobial enzyme present in tears and saliva and also secreted by the Paneth cells of the gut.

selectins
Family of cell-adhesion molecules on leukocytes and endothelial cells that bind to sugar moieties on specific glycoproteins with mucinlike features.

self antigens
The potential antigens on the tissues of an individual, against which an immune response is not usually made except in the case of autoimmunity.

self-tolerance
The failure to make an immune response against the body's own antigens.

sensitization
The acute adaptive immune response made by susceptible individuals on first exposure to an allergen. In some of these individuals, subsequent exposure to the allergen will provoke an allergic reaction.

sensitized
In allergy, describes an individual who has made an IgE response on initial encounter with an environmental antigen, which may then induce an allergic reaction on subsequent exposure.

sepsis
Bacterial infection of the bloodstream. This is a very serious and frequently fatal condition.

septic shock
Systemic shock reaction that can follow infection of the bloodstream with endotoxin-producing Gram-negative bacteria. It is caused by the systemic release of TNF-α and other cytokines. Also called endotoxic shock.

sequence motif
A pattern of nucleotides or amino acids shared by different genes or proteins that often have related functions.

serine/threonine protein kinases
Enzymes that phosphorylate proteins on either serine or threonine residues.

seroconversion
The phase of an infection when antibodies against the infecting agent are first detectable in the blood.

serology, serological assays
The use of antibodies to detect and measure antigens. So called because these assays were originally performed with serum, the fluid component of clotted blood, from immunized individuals.

serotype
Name given to a strain of bacteria, or other pathogen, that can be distinguished from other strains of the same species by specific antibodies.

serum
The fluid component of clotted blood; it contains blood proteins, such as immunoglobulins, but not cells.

serum sickness
A usually self-limiting immunological hypersensitivity reaction originally seen in response to the therapeutic injection of large amounts of foreign serum (now most usually evoked by the injection of drugs such as penicillin). It is caused by the formation of immune complexes of the antigen and the antibodies formed against it, which become deposited in the tissues, especially the kidneys.

severe combined immunodeficiency (SCID)
Type of immune deficiency (due to various causes) in which both antibody and T-cell responses are lacking; it is fatal if not treated.

severe congenital neutropenia
Inherited condition in which the neutrophil count is persistently extremely low. Also known as Kostmann disease.

SH2 domain
See **Src-family tyrosine kinases**.

shingles
Disease caused when herpes zoster virus (the virus that causes chickenpox) is reactivated later in life in a person who has had chickenpox.

SHIP
An SH2-containing inositol phosphatase that removes the phosphate from PIP_3 to give PIP_2.

shock
The potentially fatal circulatory collapse caused by the systemic actions of cytokines such as TNF-α.

SHP
An SH2-containing protein phosphatase.

Shwachman–Diamond syndrome
A rare genetic condition in which some patients have a deficiency of neutrophils.

signal joint
The noncoding joint formed in DNA by the recombination of recognition signal sequences during the assembly of a V segment with a (D)J segment in T-cell receptor and immunoglobulin gene rearrangement. The piece of chromosome containing the signal joint is excised from the chromosome as a small circle of DNA. Cf. **coding joint**.

signal transducers and activators of transcription (STATs)
See **Janus-family tyrosine kinases**.

signal transduction
The processes by which a cell transforms one type of signal, for example antigen binding to a lymphocyte antigen receptor, into the intracellular events that signal the cell to make a particular type of response.

single-chain Fv
Protein fragment comprising a V region of a heavy chain linked by a stretch of synthetic peptide to a V region of a light chain; can be made by genetic engineering.

single-nucleotide polymorphisms (SNPs)
Positions in the genome that differ by a single base between individuals.

single-positive thymocyte
A mature T cell that expresses either the CD4 or the CD8 co-receptor, but not both.

sirolimus
See **rapamycin**.

SLE
See **systemic lupus erythematosus**.

SLP-76
A scaffold protein involved in the antigen-receptor signaling pathway in lymphocytes.

SMAC
See **supramolecular adhesion complex**.

small G proteins
Single-subunit G proteins, such as Ras, that act as intracellular signaling molecules downstream of many transmembrane signaling events. Also called **small GTPases**.

small pre-B cell
Stage in B-cell development immediately after the large pre-B cell, and in which heavy-chain expression and cell proliferation ceases. Light-chain gene rearrangement then starts at this stage.

somatic diversification theory
Old hypothesis that postulated that the immunoglobulin repertoire was formed from a small number of V genes that diversified in somatic cells. This turned out to be essentially true, although at the time the mechanism of gene rearrangement was not thought of.

somatic gene therapy
The introduction of functional genes into somatic cells to treat disease.

somatic hypermutation
Extensive mutation that occurs in the V-region DNA sequence of rearranged immunoglobulin genes in activated B cells, resulting in the production of variant immunoglobulins, some of which bind antigen with a higher affinity. These mutations affect only somatic cells and are not inherited through germline transmission.

somatic recombination
DNA recombination that takes place in somatic cells (to distinguish it from the recombination that takes place during meiosis and gamete formation).

spacer
See **12/23 rule**.

specific allergen immunotherapy
See **desensitization immunotherapy**.

specificity
The ability of an antibody to distinguish a particular antigen from other antigens.

spectratyping
Technique for detecting certain types of DNA gene segments that give a repetitive spacing of three nucleotides, or one codon.

sphingosine 1-phosphate (S1P)
A lipid with chemotactic activity that controls the egress of T cells from lymph nodes.

spleen
An organ in the upper left side of the peritoneal cavity containing a red pulp, involved in removing senescent blood cells, and a white pulp of lymphoid cells that respond to antigens delivered to the spleen by the blood.

Src-family tyrosine kinases
Receptor-associated protein tyrosine kinases characterized by Src-homology protein domains (SH1, SH2, and SH3). The SH1 domain contains the kinase, the SH2 domain can bind phosphotyrosine residues, and the SH3 domain can interact with proline-rich regions in other proteins. In T cells and B cells they are involved in relaying signals from the antigen receptor.

staphylococcal enterotoxins (SEs)
Secreted toxins produced by some staphylococci, which cause food poisoning and also stimulate many T cells by binding to MHC class II molecules and the V_β domain of certain T-cell receptors, acting as superantigens.

STATs
Signal transducers and activators of transcription. See **Janus-family tyrosine kinases**.

stromal cells
The nonlymphoid cells in central and peripheral lymphoid organs that provide soluble and cell-bound signals required for lymphocyte development, survival, and migration.

subcutaneous (s.c.)
Under the skin; that is, the tissues below the epidermis and dermis.

superantigens
Molecules (including some bacterial toxins) that can stimulate large numbers of T cells by binding simultaneously MHC class II molecules and certain V_β domains of T-cell receptors.

suppressor T cells
See **regulatory CD4 T cells**.

supramolecular adhesion complex (SMAC)
Organized structure that forms at the point of contact between a T cell and its target cell, in which the ligand-bound antigen receptors are co-localized with other cell-surface signaling and adhesion molecules.

surface immunoglobulin (sIg)
The membrane-bound immunoglobulin that acts as the antigen receptor on B cells.

surface plasmon resonance (SPR)
Sensitive technique for measuring the rates of binding of molecules to each other.

surfactant proteins A and D (SP-A and SP-D)
Acute-phase proteins that help protect the epithelial surfaces of the lung against infection.

surrogate light chain
A protein in pre-B cells, made up of two subunits, VpreB and $\lambda 5$, that can pair with an in-frame heavy chain, move to the cell surface, and signal for pre-B-cell growth.

switch recombinations
See **class switching**.

switch region
Short sequence of DNA between the end of the J_H region and the heavy-chain C_μ genes, and in equivalent positions in the other C-region genes, at which class switch recombination occurs.

Syk
A cytoplasmic tyrosine kinase found in B cells that acts in the signaling pathway from the B-cell antigen receptor.

sympathetic ophthalmia
Autoimmune response that occurs in the other eye after one eye is damaged.

syngeneic graft
A graft between two genetically identical individuals. It is accepted as self.

systemic allergic contact dermatitis
A T cell-mediated allergic reaction to an ingested antigen that is manifested as a skin rash.

systemic anaphylaxis
See **anaphylaxis**.

systemic immune system
Name sometimes given to the lymph nodes and spleen to distinguish them from the mucosal immune system.

systemic lupus erythematosus (SLE)
An autoimmune disease in which autoantibodies against DNA, RNA, and proteins associated with nucleic acids form immune complexes that damage small blood vessels, especially in the kidney.

tacrolimus
An immunosuppressant polypeptide drug that inactivates T cells by inhibiting calcineurin, thus blocking activation of the transcription factor NFAT. Also called FK506.

TAP1, TAP2
Transporters associated with antigen processing. ATP-binding cassette proteins that form a heterodimeric TAP-1:TAP-2 complex in the endoplasmic reticulum membrane, through which short peptides are transported from the cytosol into the lumen of the endoplasmic reticulum, where they associate with MHC class I molecules.

tapasin
TAP-associated protein. A key molecule in the assembly of MHC class I molecules; a cell deficient in this protein has only unstable MHC class I molecules on the cell surface.

target cells
Antigen-bearing cells that are the targets of effector T cells. These cells can be B cells, which are activated to produce antibody; macrophages, which are activated to kill bacteria; or virus-infected cells that are killed by cytotoxic T cells.

T cell, T lymphocyte
One of the two types of antigen-specific lymphocytes responsible for adaptive immune responses, the other being the B cells. T cells are responsible for the cell-mediated adaptive immune reactions. They originate in the bone marrow but undergo most of their development in the thymus. The highly variable antigen receptor on T cells is called the T-cell receptor and recognizes a complex of peptide antigen bound to MHC molecules on cell surfaces. There are two main lineages of T cells: those carrying $\alpha{:}\beta$ receptors and those carrying $\gamma{:}\delta$ receptors. Effector T cells perform a variety of functions in an immune response, acting always by interacting with another cell in an antigen-specific manner. Some T cells activate macrophages, some help B cells produce antibody, and some kill cells infected with viruses and other intracellular pathogens.

T-cell antigen receptor
See **T-cell receptor**.

T-cell areas
Regions of peripheral lymphoid organs that are enriched in naive T cells and are distinct from the follicles. They are the sites at which adaptive immune responses are initiated.

T-cell clone
Cells derived from a single progenitor T cell.

T-cell hybrids
Cells formed by fusing an antigen-specific, activated T cell with a T-cell lymphoma. The hybrid cells bear the receptor of the specific T-cell parent and grow in culture like the lymphoma cells.

T-cell lines
Cultures of T cells grown by repeated cycles of stimulation, usually with antigen and antigen-presenting cells.

T-cell receptor (TCR)
The cell-surface receptor for antigen on T lymphocytes. It consists of a disulfide-linked heterodimer of the highly variable α and β chains in a complex with the invariant CD3 and ζ proteins, which have a signaling function. T cells carrying this type of receptor are often called α:β T cells. An alternative receptor made up of variable γ and δ chains is expressed with CD3 and ζ on a subset of T cells.

T-cell zones
See **T-cell areas**.

TCRα, TCRβ
The two chains of the α:β T-cell receptor.

TD antigens
See **thymus-dependent antigens**.

TdT
See **terminal deoxynucleotidyl transferase**.

Tec kinase
Src-like tyrosine kinase that links activation of the lymphocyte antigen receptors to activation of PLC-γ.

TEPs
See **thioester-containing proteins**.

terminal complement components
Complement proteins C6–C9, which can assemble to form the membrane-attack complex on cells coated with complement.

terminal deoxynucleotidyl transferase (TdT)
Enzyme that inserts nontemplated N-nucleotides into the junctions between gene segments in T-cell receptor and immunoglobulin V-region genes during their assembly.

tertiary immune response
Adaptive immune response provoked by a third injection of the same antigen. It is more rapid in onset and stronger than the primary response.

tertiary immunization
A third injection of the same antigen.

T follicular helper cell (T_{FH})
An effector T cell found in lymphoid follicles that provides help to B cells for antibody production and class switching.

TGF-β
Transforming growth factor-β, a cytokine that tends to promote the differentiation of regulatory T cells, among other effects.

T_H1
A subset of effector CD4 T cells characterized by the cytokines they produce. They are mainly involved in activating macrophages but can also help stimulate B cells to produce antibody.

T_H2
A subset of effector CD4 T cells that are characterized by the cytokines they produce. They are involved in stimulating B cells to produce antibody, and are often called helper CD4 T cells.

T_H3
A subset of regulatory CD4 T cells produced in the mucosal immune response to antigens that are presented orally. They produce TGF-β.

T_H17
A subset of CD4 T cells that are characterized by production of the cytokine IL-17. They help recruit neutrophils to sites of infection.

thioester-containing proteins (TEPs)
Homologues of complement component C3 that are found in insects and are thought to have some function in insect innate immunity.

thoracic duct
A large lymphatic vessel that runs parallel to the aorta through the thorax and drains into the left subclavian vein. The thoracic duct returns lymphatic fluid and lymphocytes to the peripheral blood circulation.

thymectomy
Surgical removal of the thymus.

thymic anlage
The tissue from which the thymic stroma develops during embryogenesis.

thymic cortex
The outer region of each thymic lobule in which thymic progenitor cells (thymocytes) proliferate, rearrange their T-cell receptor genes, and undergo thymic selection, especially positive selection on thymic cortical epithelial cells.

thymic stroma
The epithelial cells and connective tissue of the thymus that form the essential microenvironment for T-cell development.

thymocytes
Developing T cells when they are in the thymus.

thymus
A central lymphoid organ, in which T cells develop, situated in the upper part of the middle of the chest, just behind the breastbone.

thymus-dependent antigens (TD antigens)
Antigens that elicit responses only in individuals that have T cells.

thymus-dependent lymphocyte
See **T cell**.

thymus-independent antigens
Antigens that can elicit antibody production without the involvement of T cells. There are two types of TI antigens: the TI-1 antigens, which have intrinsic B-cell activating activity, and the TI-2 antigens, which activate B cells by having multiple identical epitopes that cross-link the B-cell receptor.

TI antigens, TI-1, TI-2
See **thymus-independent antigens**.

tickover
The low-level generation of C3b continually occurring in the blood in the absence of infection.

time-lapse video imaging
Imaging technique that takes snapshots of an object at fixed time intervals and then runs them together to show a version of the whole process.

tingible body macrophages
Phagocytic cells engulfing apoptotic B cells, which are produced in large numbers in germinal centers at the height of an adaptive immune response.

TIR domain
Domain in the cytoplasmic tails of the TLRs and the IL-1 receptor, which interacts with similar domains in intracellular signaling proteins.

titer
A measure of the concentration of specific antibodies in a serum, based on serial dilution to a standard end point, such as a certain level of color change in an ELISA assay. It is expressed as the dilution required to reach this point.

TLR
See **Toll-like receptors** and individual entries **TLR-1**, **TLR-2**, etc.

TLR-1
Cell-surface Toll-like receptor that acts in a heterodimer with TLR-2 to recognize lipoteichoic acid and bacterial lipoproteins.

TLR-2
Cell-surface Toll-like receptor that acts in a heterodimer with either TLR-1 or TLR-6 to recognize lipoteichoic acid and bacterial lipoproteins.

TLR-3
Endosomal Toll-like receptor that recognizes double-stranded viral RNA.

TLR-4
Cell-surface Toll-like receptor that, in conjunction with the accessory proteins MD-2 and CD14, recognizes bacterial lipopolysaccharide and lipoteichoic acid.

TLR-5
Cell-surface Toll-like receptor that recognizes the flagellin protein of bacterial flagella.

TLR-6
Cell-surface Toll-like receptor that acts in a heterodimer with TLR-2 to recognize lipoteichoic acid and bacterial lipoproteins.

TLR-7
Endosomal Toll-like receptor that recognizes single-stranded viral RNA.

TLR-8
Endosomal Toll-like receptor that recognizes single-stranded viral RNA.

TLR-9
Endosomal Toll-like receptor that recognizes DNA containing unmethylated CpG.

TLR-11
Mouse Toll-like receptor that recognizes profilin and profilin-like proteins.

T lymphocyte
See **T cell**.

TNF-α
Tumor necrosis factor-α. A cytokine produced by macrophages and T cells that has multiple functions in the immune response that help prevent the spread of an infection within the body. It is the defining member of the TNF family of cytokines and is sometimes called TNF.

TNF family
Cytokine family, the prototype of which is tumor necrosis factor-α (TNF or TNF-α). It contains both secreted (for example TNF-α and lymphotoxin) and membrane-bound (for example CD40 ligand) members.

TNF receptors (TNFRs)
Family of cytokine receptors which includes some that lead to apoptosis of the cell on which they are expressed (for example Fas and TNFR-I), whereas others lead to activation (for example CD40, 4-1BB, and TNFR-II). All of them signal as trimeric proteins.

TNF-receptor associated periodic syndrome (TRAPS)
See **familial Mediterranean fever**.

tolerance
The failure to respond to an antigen. Tolerance to self antigens is an essential feature of the immune system; when tolerance is lost, the immune system can destroy self tissues, as happens in autoimmune disease.

tolerant
Describes the state of immunological tolerance, in which the individual does not respond to a particular antigen.

tolerized
Describes the induction of a state of tolerance.

tolerogenic
Describes an antigen that induces tolerance.

Toll
Receptor protein in *Drosophila* that activates the transcription factor NFκB, leading to the production of antimicrobial peptides.

Toll-like receptors (TLRs)
Innate receptors on macrophages, dendritic cells, and some other cells, that recognize pathogens and their products, such as bacterial lipopolysaccharide. Recognition stimulates the receptor-bearing cells to produce cytokines that help initiate immune responses.

tonsils
See **lingual tonsils**, **palatine tonsils**.

toxic shock, toxic shock syndrome
A systemic toxic reaction caused by the massive production of cytokines by CD4 T cells activated by the bacterial superantigen toxic shock syndrome toxin-1 (TSST-1), which is secreted by *Staphylococcus aureus*.

toxic shock syndrome toxin-1
See **toxic shock, toxic shock syndrome**.

toxoids
Inactivated toxins that are no longer toxic but retain their immunogenicity so that they can be used for immunization.

T_R1
A type of regulatory T cell.

TRAFs
TNF receptor-associated factors. They share a domain known as a TRAF domain, and have a crucial role as signal transducers between members of the TNFR family and downstream transcription factors.

transcytosis
The active transport of molecules, such as secreted IgA, through epithelial cells from one face to the other.

transfection
The introduction of small pieces of foreign DNA into cells.

transforming growth factor-β
See **TGF-β**.

transgenesis
The introduction of novel genes into an organism's genome using recombinant DNA techniques.

transgenic mice
Mice that contain novel or mutant genes introduced into the genome by recombinant DNA techniques. They are used to study the function of the inserted gene, or transgene, and the regulation of its expression.

transitional stages
Defined stages in the further development of mature B cells in the spleen, after which the B cell expresses B-cell co-receptor component CD21.

transplantation
The grafting of organs or tissues from one individual to another. The transplanted organs or grafts can be rejected by the immune system unless the host is tolerant to the graft antigens or immunosuppressive drugs are used to prevent rejection.

transporters associated with antigen processing-1 and -2
See **TAP1, TAP2**.

T_{reg} cells
See **induced regulatory T cells**, **natural regulatory T cells**, **regulatory CD4 T cells**.

triacyl lipoproteins
Bacterial lipoproteins recognized by certain Toll-like receptors.

tropism
The characteristic of a pathogen that describes the cell types it will infect.

TSLP
Thymic stroma-derived lymphopoietin. A cytokine thought to be involved in promoting B-cell development in the embryonic liver.

TSST-1
See **toxic shock, toxic shock syndrome**.

tuberculin test
Clinical test in which a purified protein derivative (PPD) of *Mycobacterium tuberculosis*, the causative agent of tuberculosis, is injected subcutaneously. PPD elicits a delayed-type hypersensitivity reaction in individuals who have had tuberculosis or have been immunized against it.

tuberculoid leprosy
See **leprosy**.

tumor necrosis factor-α
See **TNF-α**.

tumor rejection antigens (TRAs)
Antigens on the surface of tumor cells that can be recognized by T cells, leading to attack on the tumor cells. TRAs are peptides of mutant or overexpressed cellular proteins bound to MHC class I molecules on the tumor cell surface.

TUNEL staining
TdT-dependent dUTP–biotin nick end labeling. An assay that identifies apoptotic cells *in situ* by the characteristic fragmentation of their DNA.

two-dimensional gel electrophoresis
Technique for separating different proteins in a mixture. They are

first separated by isoelectric focusing in one dimension, followed by SDS-PAGE on a slab gel at right angles to the first dimension.

two-photon scanning fluorescence microscopy
Microscopic technique allied to confocal microscopy that can provide a high-resolution image of live tissue and is less damaging to living tissue than conventional confocal microscopy.

type 1 diabetes mellitus
Disease in which the β cells of the pancreatic islets of Langerhans are destroyed so that no insulin is produced. The disease is believed to result from an autoimmune attack on the β cells. It is also known as insulin-dependent diabetes mellitus (IDDM), because the symptoms can be ameliorated by injections of insulin.

type I hypersensitivity reactions
Immunological hypersensitivity reactions involving IgE antibody triggering of mast cells.

type II hypersensitivity reactions
Immunological hypersensitivity reactions involving IgG antibodies against cell-surface or matrix antigens.

type III hypersensitivity reactions
Immunological hypersensitivity reactions involving damage caused by the formation of antigen:antibody complexes.

type IV hypersensitivity reactions
T cell-mediated immunological hypersensitivity reactions.

type I interferons
The antiviral interferons IFN-α and IFN-β.

tyrosine phosphatases
Enzymes that remove phosphate groups from phosphorylated tyrosine residues on proteins. See also **CD45**.

tyrosine protein kinases
Enzymes that specifically phosphorylate tyrosine residues in proteins. They are critical in the signaling pathways that lead to T- and B-cell activation.

ubiquitin
A small protein that can be attached to other proteins to target them for degradation in the proteasome.

ubiquitin ligase
Enzyme that attaches ubiquitin covalently to the surface of other proteins.

uracil-DNA glycosylase (UNG)
Enzyme that removes uracil bases from DNA in a DNA repair pathway that can lead to somatic hypermutation, class switch recombination or gene conversion.

urticaria
The technical term for hives, which are red, itchy skin wheals usually brought on by an allergic reaction.

vaccination
The deliberate induction of adaptive immunity to a pathogen by injecting a dead or attenuated (nonpathogenic) live form of the pathogen or its antigens (a vaccine).

vaccine
An immunogenic preparation of dead or attenuated (nonpathogenic) live pathogens, or of its antigens, together with an adjuvant, that is injected to provide immunity to that pathogen.

valence
The number of different molecules that an antigen or an antibody can combine with at one time.

variability plot
A measure of the difference between the amino acid sequences of different variants of a given protein. The most variable proteins known are antibodies and T-cell receptors.

variable domain (V domain)
The amino-terminal protein domain of the polypeptide chains of immunoglobulins and T-cell receptor, which is the most variable part of the chain.

variable gene segments
See **V gene segments**.

variable lymphocyte receptors (VLRs)
Nonimmunoglobulin LRR-containing variable receptors and secreted proteins expressed by the lymphocyte-like cells of the lamprey. They are generated by a process of somatic gene rearrangement.

variable region (V region)
The region of an immunoglobulin or T-cell receptor that is formed of the amino-terminal domains of its component polypeptide chains. These are the most variable parts of the molecule and contain the antigen-binding sites.

vascular addressins
Molecules on endothelial cells to which leukocyte adhesion molecules bind. They have a key role in selective homing of leukocytes to particular sites in the body.

VCAM-1
An adhesion molecule expressed by vascular endothelium at sites of inflammation; it binds the integrin VLA-4, which allows effector T cells to enter sites of infection.

V(D)J recombinase
The enzyme responsible for the somatic recombination that joins the gene segments of B-cell and T-cell receptor genes during gene rearrangement. It is a multiprotein complex containing the recombinases RAG-1 and RAG-2, which are lymphocyte-specific, as well as other proteins involved in cellular DNA repair.

V(D)J recombination
The process, exclusive to developing lymphocytes in vertebrates, that recombines different gene segments into sequences encoding complete protein chains of immunoglobulins and T-cell receptors.

V domain
See **variable domain**.

very late activation antigens (VLAs)
Members of the β_1 family of integrins involved in cell–cell and cell–matrix interactions. Some VLAs are important in leukocyte and lymphocyte migration.

vesicles
Small membrane-enclosed compartments (for example endosomes) within the cytosol.

V gene segments
Gene segments in immunoglobulin and T-cell receptor loci that encode the first 95 amino acids or so of the protein chain. There are multiple different V gene segments in the germline genome. To produce a complete exon encoding a V domain, one V gene segment must be rearranged to join up with a J or a rearranged DJ gene segment.

viral entry inhibitors
Drugs that inhibit the entry of HIV into its host cells.

viral integrase inhibitors
Drugs that inhibit the action of the HIV integrase, so that the virus cannot integrate into the host-cell genome.

viral protease
Enzyme encoded by the human immunodeficiency virus that cleaves the long polyprotein products of the viral genes into individual proteins.

viral set point
Level of HIV virions persisting in the blood after the acute phase of infection has passed.

viruses
Pathogens composed of a nucleic acid genome enclosed in a protein coat. They can replicate only in a living cell, because they do not possess the metabolic machinery for independent life.

VLAs
See **very late activation antigens**.

VLRs
See **variable lymphocyte receptors**.

VpreB
See **surrogate light chain**.

WAS
See **Wiskott–Aldrich syndrome**.

WASP
See **Wiskott–Aldrich syndrome**.

Weibel–Palade bodies
Granules within endothelial cells that contain P-selectin.

Western blotting
Method for detecting specific proteins in a mixture by means of electrophoretic separation followed by blotting of the protein spots onto a membrane and probing with labeled antibodies.

wheal-and-flare reaction
A skin reaction observed in an allergic individual when an allergen to which they are sensitive is injected into the dermis. It consists of a raised area of skin containing fluid, and a spreading, red, itchy inflammatory reaction around it.

white pulp
The discrete areas of lymphoid tissue in the spleen.

Wiskott–Aldrich syndrome (WAS)
An immunodeficiency disease characterized by defects in the cytoskeleton of cells due to a mutation in the protein WASP, which is involved in interactions with the actin cytoskeleton. Patients with this disease are highly susceptible to infections with pyogenic bacteria.

xenografts
Grafted organs taken from a different species than the recipient.

X-linked agammaglobulinemia (XLA)
A genetic disorder in which B-cell development is arrested at the pre-B-cell stage and no mature B cells or antibodies are formed. The disease is due to a defect in the gene encoding the protein tyrosine kinase BTK.

X-linked hyper IgM syndrome
See **CD40 ligand deficiency**.

X-linked hypohidrotic ectodermal dysplasia and immunodeficiency
A syndrome with some features resembling hyper IgM syndrome. It is caused by mutations in the protein NEMO, a component of the NFκB signaling pathway. Also called NEMO deficiency.

X-linked lymphoproliferative syndrome (XLP)
Rare immunodeficiency disease that results from mutations in the gene *SH2D1A*. Boys with this deficiency typically develop overwhelming Epstein–Barr virus infection during childhood, and sometimes lymphomas.

X-linked severe combined immunodeficiency (X-linked SCID)
An immunodeficiency disease in which T-cell development fails at an early intrathymic stage and no production of mature T cells or T-cell dependent antibody occurs. It is due to a defect in a gene that encodes the γ_c chain shared by the receptors for several different cytokines.

ζ chain
One of the signaling chains associated with the T-cell receptor.

ZAP-70
A cytoplasmic tyrosine kinase found in T cells that binds to the phosphorylated ζ chain of the T-cell receptor. The main cellular substrate of ZAP-70 is a large adaptor protein called LAT.

zoonotic
Describes a disease of animals that can be transmitted to humans.

zymogen
An inactive form of an enzyme, usually a protease, that must be modified in some way, for example by selective cleavage of the protein chain, before it can become active.

INDEX

Note: Figures in the Appendix are labeled in the form **Fig. A.1** and those in Chapters 1 to 16 are labeled in the form **Fig. 1.1**, **Fig 2.1** etc.

B

D

H

M

S

T

W

X

Y

Z